Lecture Notes in Computer Science 3648

Commenced Publication in 1973
Founding and Former Series Editors:
Gerhard Goos, Juris Hartmanis, and Jan van Leeuwen

José C. Cunha   Pedro D. Medeiros (Eds.)

Euro-Par 2005
Parallel Processing

11th International Euro-Par Conference
Lisbon, Portugal, August 30 – September 2, 2005
Proceedings

Springer

Volume Editors

José C. Cunha
Pedro D. Medeiros
Universidade Nova de Lisboa
Faculdade de Ciências e Technologia CITI Centre
Quinta da Torre, 2829-516 Caparica, Portugal
E-mail: {jcc,pm}@di.fct.unl.pt

Library of Congress Control Number: 2005931410

CR Subject Classification (1998): C.1-4, D.1-4, F.1-3, G.1-2, H.2

ISSN   0302-9743
ISBN-10   3-540-28700-0 Springer Berlin Heidelberg New York
ISBN-13   978-3-540-28700-1 Springer Berlin Heidelberg New York

Springer is a part of Springer Science+Business Media

springeronline.com

© Springer-Verlag Berlin Heidelberg 2005
Printed in Germany

Typesetting: Camera-ready by author, data conversion by Olgun Computergrafik
Printed on acid-free paper   SPIN: 11549468   06/3142   5 4 3 2 1 0

Preface

Euro-Par Conference Series

Euro-Par is an annual series of international conferences dedicated to the promotion and advancement of all aspects of parallel computing. The major themes can be divided into the broad categories of hardware, software, algorithms and applications for parallel computing. The objective of Euro-Par is to provide a forum within which to promote the development of parallel computing both as an industrial technique and an academic discipline, extending the frontier of both the state of the art and the state of the practice. This is particularly important at a time when parallel computing is undergoing strong and sustained development and experiencing real industrial take-up. The main audience for, and participants in, Euro-Par are seen as researchers in academic departments, government laboratories and industrial organizations. Euro-Par's objective is to be the primary choice of such professionals for the presentation of new results in their specific areas. Euro-Par is also interested in applications which demonstrate the effectiveness of the main Euro-Par themes. Previous Euro-Par conferences took place in Stockholm, Lyon, Passau, Southampton, Toulouse, Munich, Manchester, Paderborn, Klagenfurt, and Pisa. Next year, the conference will take place in Dresden. Euro-Par has a permanent Web site where its history and organization are described: *http://www.europar.org*. The Euro-Par conference series is traditionally organized in cooperation with the International Federation for Information Processing (IFIP), in cooperation with the Association for Computer Machinery (ACM), and in cooperation with the Institute of Electrical and Electronics Engineers (IEEE) Computer Society, Technical Committee on Parallel Processing (TCPP).

Euro-Par 2005 in Lisbon, Portugal

Euro-Par 2005 was the eleventh conference in the Euro-Par series. It was organized by the Centre for Informatics and Information Technology (CITI) and the Department of Informatics of the Faculty of Science and Technology of Universidade Nova de Lisboa, at the Campus of Monte de Caparica.

The conference included three invited tutorials: *Testing Multi-threaded and Distributed Applications* (Eitan Farchi and Shmuel Ur, IBM, Haifa); *Kerrighed, a Single System Image Cluster Operating System* (Christine Morin and Renaud Lottiaux, IRISA/INRIA); and *Creating and Managing Distributed Scientific Workflows* (presented by Omer F. Rana, joint work with Ian Taylor, Matthew Shields and David W. Walker, Cardiff University).

The conference included invited talks by José A.B. Fortes (Advanced Computing and Information Systems Lab, University of Florida), *On the Use of Virtualization and Service Technologies to Enable Grid Computing*; by José E. Moreira (IBM Systems and Technology Group, Rochester), *The Evolution of the Blue Gene/L Supercomputer*; by Omer F. Rana (Cardiff University), *Agent based Computational Grids: Research Issues and Challenges*; and by Raymond Bair (Laboratory Computing Resource Center, Mathematics and Computer Science Division, Argonne National Laboratory), *Science on a Large Scale*. A full paper of the talk given by Fortes, and the abstracts of the talks by Moreira, Rana, and Bair, are included in these proceedings.

A co-located workshop was organized by the GridCoord European initiative on grid computing, on *Really Large-Scale Grid Architecture*, gathering researchers from the CoreGrid NoE and leading researchers experienced with the actual deployment of applications at a very large scale. The workshop was based on the contribution of invited speakers, and the workshop attendance was free for Euro-Par 2005 participants. The CoreGRID EU Network of Excellence organized several working meetings during the conference, enriching the opportunities for debate of Euro-Par related topics.

Euro-Par 2005 Statistics

Euro-Par 2005 was organized according to the traditional conference format, in 16 topics covering a diversity of dimensions of parallel and distributed computing. Each topic was supervised by a committee of four persons: a global chair, a local chair, and two vice-chairs. The call for papers attracted a total of 388 submissions, representing 44 countries (based on the corresponding author's country). An average of 3.8 review reports were collected for each paper, for a grand total of 1470 review reports that involved about 700 different reviewers. A total of 121 full papers were accepted: 120 regular papers, and one distinguished paper. This year, there was no call for short papers, unlike previous conferences. Eventually, the camera-ready version of one paper was not submitted by one author, and therefore 120 papers are actually included in the proceedings. Papers were accepted from 23 different countries. The principal contributors by country were the USA (30 accepted papers), France (17 accepted papers), Spain (16 accepted papers), and Germany (13 accepted papers).

Acknowledgments

Euro-Par 2005 was made possible due to the support of many individuals and organizations. The CITI Centre, the Department of Informatics, and the Faculty of Science and Technology of Universidade Nova de Lisboa were the main conference institutional sponsors. A number of institutional and industrial sponsors gave their contributions and/or participated in organizing exhibits at the conference site. Their names and logos appear on the Euro-Par 2005 Web site:

http://europar05.di.fct.unl.pt. In particular, we gratefully acknowledge the support from IBM Portugal.

Special thanks are due to the authors of all the submitted papers, the members of the topic committees, and all reviewers in all topics, for their contributions to the success of this conference.

We are grateful to the members of the Euro-Par Steering Committee for their support. In particular, Marco Danelutto, co-organizer of Euro-Par 2004, and Harald Kosch, co-organizer of Euro-Par 2003, never failed to give us their prompt advice regarding all the organization details. We owe special thanks to Christian Lengauer, chairman of the Steering Committee, who was always available for sharing with us his experience in the organization of Euro-Par, and for giving us friendly advice, support, and encouragement. We also thank Luc Bougé, vice-chair, for his vision and contributions to improve Euro-Par conferences. We are gratefull to Springer for publishing these proceedings. In particular, to Alfred Hofmann, and also specially to Ursula Barth, for their permanent availability and willingness to solve the difficulties that appeared in the preparation of the proceedings.

Euro-Par 2005 was co-sponsored by the IFIP TC10/WG10.3, and organized in cooperation with ACM (SIGACT, SIGARCH, SIGMETRICS, SIGMM, SIG-MOBILE, SIGMOD, SIGOPS and SIGSOFT), and in technical cooperation with the IEEE Computer Society TCPP.

Euro-Par 2005 was a GridCoord and a CoreGRID event, with co-located activities from these EU initiatives, and we thank Luc Bougé and Thierry Priol for their influence in making this possible.

We gratefully acknowledge the enthusiastic support from the Rector of the University and the Dean of the Faculty. Locally, we thank the staff of the Department of Informatics and the CITI research centre, funded by the Portuguese *Ministério da Ciência, Tecnologia e Ensino Superior* and all the people from the faculty services, as well as Filipa Reis, head secretary of the department, and Madalena Almeida, from Viagens Abreu, who made possible the local organization of Euro-Par 2005. In particular, we acknowledge the excellent efforts of the local team: Jorge Custódio, Carmen Morgado, Paulo Lopes, Vítor Duarte, João Lourenço, Cecília Gomes, Rui Marques, Miguel Maurício, and the student volunteers, who were all committed to solving the numerous problems related to the conference organization.

It was our pleasure and an honor to host Euro-Par 2005 at Universidade Nova de Lisboa. We hope all the participants enjoyed the technical programme and the social events organized during the conference.

Lisbon, June 2005   José C. Cunha
                                                                 Pedro D. Medeiros

Organization

Euro-Par Steering Committee

Chair

Christian Lengauer — University of Passau, Germany

Vice-Chair

Luc Bougé — ENS Cachan, France

European Representatives

José Cunha — New University of Lisbon, Portugal
Marco Danelutto — University of Pisa, Italy
Rainer Feldmann — University of Paderborn, Germany
Christos Kaklamanis — Computer Technology Institute, Greece
Paul Kelly — Imperial College London, United Kingdom
Harald Kosch — University of Klagenfurt, Austria
Thomas Ludwig — University of Heidelberg, Germany
Emilio Luque — Universitat Autònoma of Barcelona, Spain
Luc Moreau — University of Southampton, United Kingdom
Rizos Sakellariou — University of Manchester, United Kingdom

Non-European Representatives

Jack Dongarra — University of Tennessee at Knoxville, USA
Shinji Tomita — Kyoto University, Japan

Honorary Members

Ron Perrott — Queen's University Belfast, United Kingdom
Karl Dieter Reinartz — University of Erlangen-Nuremberg, Germany

Observers

Wolfgang Nagel — Dresden University of Technology, Germany
Anne-Marie Kermarrec — IRISA Rennes, France

Euro-Par 2005 Local Organization

Euro-Par 2005 was organized by the CITI Research Centre and the Department of Informatics of the Faculty of Science and Technology of Universidade Nova de Lisboa.

Conference Chair

José C. Cunha

Conference Vice-Chair

Pedro D. Medeiros

Webmaster and Systems Management

Jorge Custódio

Technical Support

Vítor Duarte, João Lourenço, Cecília Gomes, Rui Marques, Miguel Maurício

Social Events

Carmen Morgado

Exhibits

Paulo Lopes

Secretariat and Registration

Filipa Reis
Madalena Almeida (Viagens Abreu)

Euro-Par 2005 Program Committee

Topic 1: Support Tools and Environments

Global Chair

Henryk Krawczyk — Faculty of Electronics, Telecommunications and Informatics, Technical University of Gdansk, Gdansk, Poland

Local Chair

Tomàs Margalef — Computer Architecture and Operating Systems Dept., Univ. Autònoma de Barcelona, Barcelona, Spain

Vice Chairs

Jacques Chassin de Kergommeaux — INPG-ENSIMAG, LSR-IMAG, Grenoble, France

Pierre Manneback — Faculté Polytechnique de Mons, Belgium

Topic 2: Performance Prediction and Evaluation

Global Chair

Allen D. Malony — Department of Computer and Information Science, University of Oregon, Eugene, USA

Local Chair

Luís Silva — Department of Informatics Engineering, University of Coimbra, Portugal

Vice Chairs

Thomas Fahringer — Institute for Computer Science, University of Innsbruck, Austria

Allan Snavely — San Diego Supercomputer Center, University of California, USA

Topic 3: Scheduling and Load Balancing

Global Chair

Denis Trystram — ID-IMAG, Grenoble, France

Local Chair

Luís P. Santos — Dept. of Informatics, University of Minho, Braga, Portugal

Vice Chairs

Uwe Schwiegelshohn — Computer Engineering Institute, University of Dortmund, Dortmund, Germany

Michael A. Bender — Dept. of Computer Science, State Univ. of New York at Stony Brook, USA

Topic 4: Compilers for High Performance

Global Chair

Albert Cohen — INRIA Futurs, Parc Club Orsay Université, Orsay, France

Local Chair

José Moreira — IBM Systems and Technology Group, Rochester, MN, USA

Vice Chairs

Martin Griebl — University of Passau, Germany

Michael O'Boyle — Institute for Computing Systems Architecture, University of Edinburgh, UK

Topic 5: Parallel and Distributed Databases, Data Mining and Knowledge Discovery

Global Chair

Domenico Talia — DEIS, University of Calabria, Rende CS, Italy

Local Chair

Rui Camacho — Faculty of Engineering, University of Porto, Portugal

Vice Chairs

Hillol Kargupta — Dept. of Computer Science and Electrical Engineering, University of Maryland Baltimore County, USA

Patrick Valduriez — INRIA and LINA–Université de Nantes, France

Topic 6: Grid and Cluster Computing: Models, Middleware and Architectures

Global Chair

Craig Lee — The Aerospace Corporation, El Segundo, California, USA

Local Chair

João Gabriel Silva
Dept. of Informatics Engineering, University of Coimbra, Portugal

Vice Chairs

Thilo Kielmann
Dept. of Computer Science, Vrije Universiteit, Amsterdam, The Netherlands

Laurent Lefèvre
INRIA RESO/LIP École Normale Supérieure de Lyon, France

Topic 7: Parallel Computer Architecure and Instruction-Level Parallelism

Global Chair

Theo Ungerer
Institute of Informatics, University of Augsburg, Germany

Local Chair

Pedro Trancoso
Dept. of Computer Science, University of Cyprus, Nicosia, Cyprus

Vice Chairs

Kevin Skadron
Dept. of Computer Science, University of Virginia, Charlottesville, USA

Josep-Lluis Larriba-Pey
Dept. of Computer Architecture, Universitat Politécnica de Catalunya, Barcelona, Spain

Topic 8: Distributed Systems and Algorithms

Global Chair

Marc Shapiro
Microsoft Research Cambridge, UK

Local Chair

Luís Rodrigues
Dept. of Informatics, University of Lisbon, Portugal

Vice Chairs

Felix Gaertner
Dept. for Computer Science, RWTH Aachen University, Germany

Idit Keidar
Dept. of Electrical Engineering, Technion – Israel Institute of Technology, Haifa, Israel

Topic 9: Parallel Programming: Models, Methods, and Languages

Global Chair

Marco Danelutto — Dept. of Computer Science, University of Pisa, Italy

Local Chair

Fernando Silva — Dept. of Computer Science, University of Porto, Portugal

Vice Chairs

Denis Caromel — INRIA and Institut Universitaire de France, Univ. de Nice Sophia Antipolis, France

Duane Szafron — Dept. of Computing Science, University of Alberta, Edmonton, Canada

Topic 10: Parallel Numerical Algorithms

Global Chair

Jacek Kitowski — Institute of Computer Science, AGH University of Science and Technology, Krakow, Poland

Local Chair

Filomena d'Almeida — Faculty of Engineering, University of Porto, Portugal

Vice Chairs

Boleslaw K. Szymanski — Rensselaer Polytechnic Institute, Troy, NY, USA

Andrzej M. Goscinski — School of Information Technology, Deakin University, Victoria, Australia

Topic 11: Distributed and High-Performance Multimedia

Global Chair

Laszlo Boeszoermeny — Institute for Information Technology, University of Klagenfurt, Austria

Local Chair

Nuno Correia — Dept. of Informatics, Universidade Nova de Lisboa, Portugal

Vice Chairs

Max Mühlhäuser	Technical University of Darmstadt, Germany
Geoff Coulson	Computing Department, Lancaster University, UK

Topic 12: Theory and Algorithms for Parallel Computation

Global Chair

Andrea Pietracaprina	Dipartimento di Ingegneria dell'Informazione, Università di Padova, Italy

Local Chair

Casiano Rodríguez-Leon	Universidad de La Laguna, Tenerife, Spain

Vice Chairs

Kieran Herley	Dept. of Computer Science, University College Cork, Ireland
Christos Zaroliagis	Dept. of Computer Engineering and Informatics, CTI and University of Patras, Greece

Topic 13: Routing and Communication in Interconnection Networks

Global Chair

Emilio Luque	Computer Architecture and Operating Systems Dept., Universitat Autònoma de Barcelona, Spain

Local Chair

José Legatheaux Martins	Dept. of Informatics, Universidade Nova de Lisboa, Portugal

Vice Chairs

Cruz Izu	Dept. of Computer Science, University of Adelaide, Australia
Olav Lysne	Simula Research Laboratory, Lysaker, Norway

Topic 14: Mobile and Ubiquitous Computing

Global Chair

Evaggelia Pitoura	Dept. of Computer Science, University of Ioannina, Greece

Local Chair

> Nuno Preguiça   Dept. of Informatics, Universidade Nova de Lisboa, Portugal

Vice Chairs

> Marios Dikaiakos   Dept. of Computer Science, University of Cyprus, Nicosia, Cyprus
>
> Valérie Issarny   INRIA-Rocquencourt, Domaine de Voluceau, France

Topic 15: Peer-to-Peer and Web Computing

Global Chair

> Anne-Marie Kermarrec   INRIA/IRISA, Campus Universitaire de Beaulieu, France

Local Chair

> Henrique João Domingos   Dept. of Informatics, Universidade Nova de Lisboa, Portugal

Vice Chairs

> Anthony Rowstron   Microsoft Research Cambridge, UK
>
> Márk Jelasity   Dept. of Computer Science, University of Bologna, Italy

Topic 16: Applications of High-Performance and Grid Computing

Global Chair

> Raymond Bair   Mathematics and Computer Science Division, Argonne National Laboratory, USA

Local Chair

> José Laginha Palma   Faculty of Engineering, University of Porto, Portugal

Vice Chairs

> Ed Seidel   Max-Plank Institute für Gravitationsphysik, Germany and Louisiana State University, Baton Rouge, USA
>
> Michel Daydé   IRIT-ENSEEIHT, Toulouse, France

Euro-Par 2005 Referees

Not including members of the Programme and Steering Committees.

Ahmed Abdelkhalek
Tarek Abdelrahman
Jaume Abella
Adnan Agbaria
Kunal Agrawal
Kento Aida
Reza Akbarinia
Mohammad Mursalin Akon
Marco Aldinucci
Gabrielle Allen
Francisco Almeida
JP Moitinho de Almeida
Paulo Sérgio Almeida
Martin Alt
Albano Gomes Alves
Yair Amir
Maria Andreou
Cosimo Anglano
David Angulo
Filipe Araujo
Toni Arbona
Jean-Paul Arcangeli
Esther Arkin
Álvaro F.M. Azevedo

David A. Bader
Gal Badishi
Faruk Bagci
Iris Bahar
Mark Baker
Omar Bakr
Henri Bal
Subir Bandyopadhyay
Carlos Baquero
Ranieri Baraglia
Valmir C. Barbosa
Luiz A. Barroso
Sandro Bartolini
Alessandro Bassi
Michael Bauer
Jean-François Bauwens

Olivier Beaumont
Micah Beck
Zinaida Benenson
Anne Benoit
Amit Bhaya
Ricardo Bianchini
Ganesh Bikshandi
Angelos Bilas
Vicente Blanco
François Bodin
Sabine Böhm
Lars Ailo Bongo
Edward Bortnikov
K. Boryczko
Luc Bouganim
Anu Bourgeois
Hinde Lilia Bouziane
Tim Brecht
Uwe Brinkschulte
Y. D. Bromberg
Jim Browne
Piotr Brudlo
David Bunde
Mathijs den Burger
Yann Busnel
Javier Bustos
Rajkumar Buyya

Susana Cabaço
Massimo Cafaro
Wei Cai
Yvan Calas
Lasaro Jonas Camargos
Sonia Campa
Ramon Canal
Yves Caniou
Juan Carlos Cano
Massimo Canonico
Jiannong Cao
Franck Cappello
Manuel Carro

Antonio Carzaniga
Rafael Casado
Henri Casanova
António Casimiro
Jorge Castro
Christophe Cerin
Teresa Chambel
Agnès de La Chapelle
Ricardo Chaves
Ann Chervenak
Gregory Chockler
Marcelo Cintra
Walfredo Cirne
Antonio Cisternino
Rance Cleaveland
Andrea Clematis
Raphael Clifford
Murray Cole
Raphael Collet
Michele Co
Carmela Comito
Antonio Congiusta
Massimo Coppola
Julita Corbalan
Ricardo C. Correa
Luís Correia
Paulo Correia
Alexandre di Costanzo
Vítor Santos Costa
Cedric Coulon
Patrick Crowley
Maria Cutumisu
Czarnul

Pasqua D'Ambra
Litaize Daniel
Abdelmadjid Dargham
Kei Davis
Kurt Debattista
Jérôme Décamps
Ewa Deelman
Chirag Dekate
Carole Delporte-Gallet
Camil Demetrescu
Yves Denneulin

Enrico Denti
Veerle Desmet
Frédéric Desprez
Robert Dew
Gerasimos Dimitriadis
Menno Dobber
Shlomi Dolev
Rion Dooley
Yon Doursiboure
Karel Driesen
Steven G. Dropsho
Rubing Duan
Ziyang Duan
Sérgio Duarte
Frederick Ducatelle
Jan Duennweber
Iain Duff
Franciszek A. Dul
Catalin L. Dumitrescu
Christopher Dutchyn
Pierre-Francois Dutot
Inês de Castro Dutra
Partha Dutta
Vaclav Dvorak
Sandhya Dwarkadas
W. Dzwinel

Jeff Edmonds
Lieven Eeckhout
Alexandre Eichenberger
Liane Eitan
M. Ellinas
Vincent Englebert
Dick Epema
Carsten Ernemann
Carl Esswein
Luis Angelo Estefanel

Josep Fàbrega
Carlo Fantozzi
Hugues Fauconnir
Dror Feitelson
Paolo Ferragina
Paulo Ferreira
Fabrice Le Fessant

Amos Fiat
Ludger Fiege
Irene Finocchi
Stephen Fitzpatrick
Jose Flich
Pierfrancesco Foglia
Nuno Fonseca
Victor Francisco Fonte
Philippe Fortemps
José Fortes
Dimitris Fotakis
Geoffrey Fox
Pierre Fraigniaud
Felipe M. G. França
Daniel Franco
Antonio Frangioni
Hubertus Franke
Roy Friedman
Filippo Furfaro

Estelle Gabarron
Edgar Gabriel
Efstratios Gallopoulos
Ayalvadi Ganesh
Dennis Gannon
Xiaofeng Gao
Nuno Garcia
Maria Jesus Garzaran
Thierry Gautier
Georgi Gaydadjiev
Jean-Patrick Gelas
Arpad Gellert
Giorgio Ghelli
Seth Gilbert
Roberto Giorgi
L. Giraud
Olivier Gluck
Kevin Glynn
Alfredo Goldman
Michael Goldwasser
Maria Cecília Gomes
Tom Goodale
José Gortes
Dhrubajyoti Goswami
Candelaria Hernandez Goya

Paul Grace
Maria Gradinariu
Jose Angel Gregorio
Armin Groesslinger
Roberto Grossi
Abdou Guermouche
Ronan Guivarch
Fei Guo
Jia Guo

Youssef Hamadi
Abdelkader Hameurlain
Lance Hammond
Sidath Handurukande
Audun Fosselie Hansen
Robert Harakaly
Andrew Harrison
Michael Hartle
William Hart
Akira Hatanaka
Yasushi Hayashi
Andreas Heinemann
Bruce Hendrickson
Ludovic Henrio
Andreas Herkersdorf
Porfidio Hernández
Germán Rodríguez Herrera
Elisa Heymann
Christian Hochberger
Juergen Hofer
P. Horan
Geir Horn
S.T. Huang
Kevin Huck
Daniel Hughes
Shiwen Hu
Andrei Hutanu
Zhigang Hu

Alexandru Iosup

Michael A. Jaeger
Samir Jafar
Mathieu Jan
Klaus Jansen

Detlef Jantz
Stephen Jarvis
Emmanuel Jeannot
Emmanuel Jeanvoine
Wojciech Jedruch
Jean-Pierre Jessel
Chris Jesshope
Arshad Jhumka
Gangyi Jiang
Daniel A. Jiménez
Daniel Jiménez-González
Ricardo Jimenez-Peris
Ackbar Joolia
Josep Jorba
Joaquim Jorge
Norman P. Jouppi
Alexandru Jugravu
Flavio Junqueira

Pawel Kaczmarek
Dave Kaeli
David Kaeli
Tim Kaiser
Christos Kaklamanis
Odej Kao
Helen Karatza
Wolfgang Karl
Nick Karonis
Irit Katriel
Krishna M. Kavi
Gabor Kecskemeti
Joerg Keller
Ian Kelley
Paul Kelly
Mazen Kharbutli
Artur Klauser
Gabriel Kliot
Can Emre Koksal
Georgios Koltsidas
Miriam Konkel
Spyros Kontogiannis
Alix Munier Kordon
Harald Kosch
Evangelos Kotsovinos
Andreas Krall

Dieter Kranzlmueller
Axel Krings
Mukkai Krishnamoorthy
Ajay Kshemkalyani
Archit Kulshrestha
Piyush Kumar
Pierre Kuonen
Klaus Kursawe
Shay Kutten
Georgi Kuzmanov
Amund Kvalbein
Costas Kyriacou

John Lach
Adrian Lahanas
Marco Lapegna
Gregor von Laszewski
Luciano Lavagno
Doug Lea
Ben Lee
Kevin Lee
P.A. Lee
Charles Lefurgy
Arnaud Legrand
Zhou Lei
Sebastien Leriche
Vincenzo Liberatore
Keqin Li
Alexandre A. B. Lima
Mikko Lipasti
Jinshan Liu
Xiaoming Li
Josep Llosa
Gabriel Loh
Luís Lopes
Paulo Afonso Lopes
Ricardo Lopes
Pedro Lopez
João Lourenço
Mikel Lujan
Frank Luk
Paul Lu

Jason Maassen
Steve MacDonald

Cam Macdonell
Jon MacLaren
Erik Maehle
Kaoutar El Maghraoui
Nicolas Maillard
Andrew Maloney
Marco Mamei
D. Manivannan
Rajit Manohar
Daniel Marques
Osni Marques
José F. Martínez
Xavier Martorell
Mike Marty
Carlo Mastroianni
Ivan Matosevic
Kiminori Matsuzaki
M. Matuszek
Marios Mavronicolas
Michael O. McCracken
Sally A. McKee
Pedro Medeiros
Nordine Melab
Roie Melamed
John Mellor-Crummey
Alex Mendiburu
Philippe Merle
Andre Merzky
Valentin Mesaros
Michael Messig
Norbert Meyer
Pierre Michaud
B. Scott Michel
Sam Midkiff
Jose Miguel-Alonso
Simon Miles
Mike Minkoff
Neeraj Mittal
Michael Mitzenmacher
Hashim H. Mohamed
Sonia Ben Mokhtar
Ossi Mokryn
Carlos Molina
Burkhard Monien
Sebastien Monnet

Paulo Monteiro
Alberto Montresor
Oveeyen Moonian
Anna Morajko
Luc Moreau
Jose A. Moreno
Luz Marina Moreno
Andreas Moshovos
Achour Mostefaoui
Grégory Mounié
Francisco Moura
Juan Carlos Moure
Rim Moussa
Trevor Mudge
Gero Mühl
Ioan Lucian Muntean
Amy L. Murphy
Peter Musial

Hidemoto Nakada
Jim Napolitano
Wahid Nasri
Mario Nemirovsky
Kyriacos Neocleous
Francesco Nerieri
Nuno Ferreira Neves
Tuan Anh Nguyen
Rob van Nieuwpoort
Christos Nomikos
Nils Agne Nordbotten
Mário Serafim Nunes

Rui Oliveira
Suely Oliveira
Salvatore Orlando
Pablo Montesinos Ortego
Djamila Ouelhadj
Emre Ozer
Can Ozturan

Mathias Pacher
Esther Pacitti
Gérard Padiou
Marc Pantel
Dimitris Papadias

Evangelos Papapetrou
Marina Papatriantafilou
Koulla Papavasiliou
Michael Papka
Savas Parastatidis
Nikos Parlavantzas
G. Paschos
Sarantis Paskalis
Simon Patarin
Sanjay Patel
Yale Patt
Mathias Paulin
Johnatan Pecero-Sanchez
Fernando Pedone
João Pedro
Susanna Pelagatti
Liang Peng
Lucia Draque Penso
José Pereira
Paulo Rogério Pereira
Christian Perez
Juan Carlos Pérez
Fabrizio Petrini
Jan Petzold
Gert Pfeifer
C. Pham
Chris Phillips
Guillaume Pierre
Jean-Marc Pierson
António M. S. Pina
Eduardo Pinheiro
Alexandre Pinto
Stefan Pleisch
Sabri Pllana
Stefan Podlipnig
Eleftherios Polychronopoulos
Konstantin Popov
Peter Popov
Fernando Cores Prado
Pascale Primet
Thierry Priol
Radu Prodan
Alberto Proença
Kirk Pruhs
Valentin Puente

André Puga
Diego Puppin

Jun Qin
Francesco Quaglia
Francisco J. Quiles
Martin Quinson

Bruno Raffin
Sergio Rajsbaum
Pierre Ramet
Alex Ramirez
Ruy Ramos
Omer Rana
Andrew Rau-Chaplin
Pierre-Guillaume Raverdy
Kees van Reeuwijk
Alexander Reinefeld
Sven-Arne Reinemo
Steve Reinhardt
José Renau
Carlos Ribeiro
Olivier Richard
Stefan Richter
Ana Ripoll
Etienne Riviere
Thomas Robertazzi
Antonio Robles
Ricardo Rocha
Jean-Louis Roch
Rodrigo Rodrigues
Francisco Almeida Rodriguez
Jose E. Roman
Michiel Ronsse
Brian Ropers-Huilman
Alain Roy
Peter Van Roy
Krzysztof Rzadka

Daniele Sacchetti
Francoise Sailhan
Pascal Sainrat
J. César de Sá
Rizos Sakellariou
Francisco de Sande

Philippas Tsigas
George Tsouloupas
Dean Tullsen
Georg Turban
Stefan Turek
Roland Tusch
Mayank Tyagi
Gary Tyson

Jo Ueyama
Sascha Uhrig
Augustus K. Uht
Brygg Ullmer
Gil Utard

Neil Vachharajani
Sathish S. Vadhiyar
Fernando Vallejo
Hans Vandierendonck
Paulo B. Vasconcelos
Jose Marcos Moreno Vega
Pierangelo Veltri
Kees Verstoep
Vincent Villain
Alex Villazon
Jean-Marc Vincent
Lucian Vintan
Frédéric Vivien
Berthold Vöcking
Michael Voss
Spyros Voulgaris
Jaksa Vuckovic

Jian Wang

Jinling Wang
Jon Weinberg
Zunce Wei
Michael Welzl
Matthias Werner
Matthias Westermann
Tony White
Philipp Wieder
Gerhard J. Wöginger
Nicole Wolter
Adam K. L. Wong
Patrick Worley
Joachim Worringen
Gosia Wrzesinska

Wei Xing

Ramin Yahyapour
Kun Yang
Eiko Yoneki
Eiko Yoneki
Ki Hwan Yum

Apostolos Zarras
Eberhard Zehendner
Chongjie Zhang
Hu Zhang
Mingmin Zhang
Dong Zhou
Wolfgang Ziegler
Craig Zilles
Corrado Zoccolo
Albert Zomaya

Table of Contents

Invited Talks

Topic 1 – Support Tools and Environments 17
  Henryk Krawczyk, Jacques Chassin de Kergommeaux,
  Pierre Manneback, and Tomás Margalef (Topic Chairs)

On the Use of Virtualization and Service Technologies to Enable Grid-Computing

Andréa Matsunaga, Maurício Tsugawa, Ming Zhao,
Liping Zhu, Vivekananthan Sanjeepan, Sumalatha Adabala,
Renato Figueiredo, Herman Lam, and José A.B. Fortes

Advanced Computing and Information Systems Laboratory (ACIS)
Dep. of Electrical and Computer Engineering, University of Florida, Gainesville, FL 32611
`fortes@ufl.edu`

Abstract. The In-VIGO approach to Grid-computing relies on the dynamic establishment of virtual grids on which application services are instantiated. In-VIGO was conceived to enable computational science to take place In Virtual Information Grid Organizations. Having its first version deployed on July of 2003, In-VIGO middleware is currently used by scientists from various disciplines, a noteworthy example being the computational nanoelectronics research community (http://www.nanohub.org). All components of an In-VIGO-generated virtual grid – machines, networks, applications and data – are themselves virtual and services are provided for their dynamic creation. This article reviews the In-VIGO approach to Grid-computing and overviews the associated middleware techniques and architectures for virtualizing Grid components, using services for creation of virtual grids and automatically Grid-enabling unmodified applications. The In-VIGO approach to the implementation of virtual networks and virtual application services are discussed as examples of Grid-motivated approaches to resource virtualization and Web-service creation.

1 Introduction

The future envisioned by the concept of Grid-computing is one where users will be able to securely and dependably access, use, "publish" and compose applications as services anywhere and anytime. Transparently to users, Grids will have to aggregate resources, possibly across different institutions, to provide application services. In addition, Grid middleware will have to create in the aggregated resources the execution environments where services and users can securely run or create applications of interest and access needed data. Unless properly designed, individual solutions for each of these requirements can conflict with each other, as shared resources cannot be easily reconfigured to simultaneously provide multiple execution environments securely and on-demand for different users and applications. This article argues that resource virtualization and service technologies provide ideal mechanisms to address these and other key requirements of Grid-computing, and describes components of In-VIGO, an evolving deployed system that successfully uses this approach [1], [2].

The remainder of this paper is organized as follows. The In-VIGO approach is briefly reviewed in Section 2. Virtual machines and the corresponding services for their creation and management are reviewed in Section 3. Virtual file systems and associated services are overviewed in Section 4. Virtual networking techniques are presented in Section 5. Virtual applications and virtual application services are discussed in Section 6. Section 7 describes how the different In-VIGO components are

J.C. Cunha and P.D. Medeiros (Eds.): Euro-Par 2005, LNCS 3648, pp. 1–12, 2005.

securely integrated. Conclusions and the current status of In-VIGO middleware and research are presented in Section 8.

2   The In-VIGO Approach

In-VIGO is unique in that it decouples user environments from physical resources by using technologies that virtualize all resources needed for Grid-computing, including machines, networks, applications and data (see Figure 1). Users will typically interact with In-VIGO through a portal where they can invoke applications of interest. In-VIGO delivers these applications through Web-enabled user interfaces that interact with virtual application (VA) services. VA services interact with other application services as well as other Grid-computing middleware services. VA services decouple application interfaces from application implementations thus hiding the kinds of codes and machines used to provide services. Transparently to users, VA services engage with virtualization services to create the virtual machines, file systems, networks and possibly other applications needed to generate a virtual grid with the necessary execution environment for the application delivered by the VA service. Virtualization services decouple users and execution environments needed by applications from the physical machines that provide them, thus allowing different instances of an application service to transparently run on different physical hardware.

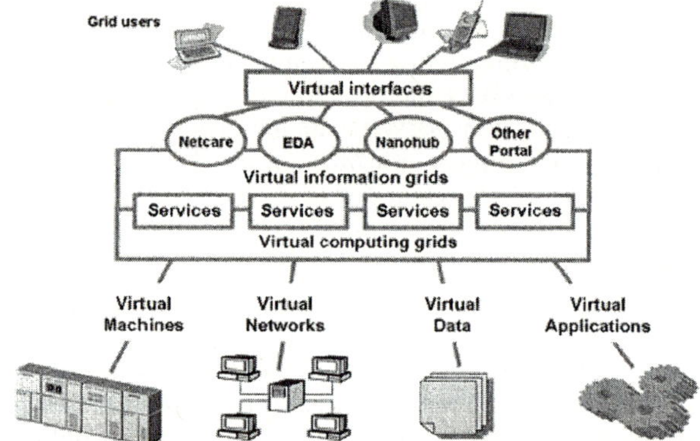

Fig. 1. High-level view of the In-VIGO approach

Ultimately, Grids will be useful only if they can provide application services for users. In-VIGO provides each user with a persistent private virtual workspace that enables him/her to both launch and develop applications, use and manage private data, and carry out conventional operating system tasks through, for example, a Unix-like shell. It is also very important that, in addition to the use of services, the process of deploying applications as services be as simple as possible. Service creation should not require application developers to know details of how Grid middleware works, and should not require the involvement of administrators. In-VIGO provides automated procedures to create application services that only require developers to provide a description of how a tool works. This description is comparable in nature and complexity to the "man pages" of an operating system command. It includes the command-line grammar and some additional information on software dependencies and other requirements of the application's execution environment.

3 Virtual Machines and Virtual Machine Services

In-VIGO supports dynamic allocation of execution environments per user and per distinct application by using virtual machine technologies (including language-based Java VMs, as well as O/S-based VMs, such as VMware, User-mode Linux) and/or "shadow" accounts. For efficiency and scalability purposes, mechanisms are provided for multiplexing virtual machines and accounts among users and applications without compromising security and customizability. Virtual machines can either be created and destroyed for every In-VIGO session or be made persistent across sessions. Virtual machines used to run applications can also be shared across several applications by using "shadow accounts" [3], which are pre-created accounts on machines that In-VIGO can use on behalf of arbitrary users.

In-VIGO manages virtual machines through a set of Grid service-based middleware components – VMShop and VMPlant [4]. The key differentiators of this approach from related work reflect the design decisions of: (1) supporting different VM technologies, such as VMware, User-Mode Linux; (2) allowing flexible, application-centric VM environment configurations using direct acyclic graph (DAG) representations; and (3) supporting dynamic "cloning" of previously-built VM images. Virtual machines managed using this middleware are highly customizable by the client, on a per-application basis. In contrast, dynamic virtual environments [5] enable the creation of VMs from a master disk (e.g. a Linux distribution with pre-installed Globus software) but do not provide mechanisms for the client to specify the desired machine's configuration.

"Classic" VMs present the image of a dedicated operating system while enabling multiple O/S configurations – completely isolated from each other – to share a single machine. This is an effective mechanism for resource consolidation, and a key reason for the renewed interest and popularity of VMs•. They also provide a flexible, powerful execution environment for Grid computing, offering isolation and security mechanisms complementary to operating systems, customization and encapsulation of entire application environments, and support for legacy applications [6], addressing a fundamental goal of Grid computing – flexible resource sharing.

VMShop provides a single logical point of contact for clients to request three core services: create a VM instance, query information about an active VM instance, and destroy (collect) an active VM instance. Requests for virtual machine creation received by VMShop contain specifications of hardware, network and software configurations. VMShop is then is responsible for selecting a VMPlant for the creation of a virtual machine. This process is implemented through a communication API and a binding protocol that allows VMShop to request and collect bids containing estimated VM creation costs from VMPlants.

The VMPlant implements the process of VM instantiation, using the VM's DAG specification provided by a client through VMShop as its input. In addition to supporting flexibility of VM configuration, the DAG aids the implementation of an efficient VM creation process by supporting partial matches of cached VM images to find a suitable match – a "golden" machine. Once a golden machine has been found, VMPlant clones the machine, and then parses the DAG to perform a series of configuration actions on the new machine. Once a machine is cloned, the configuration process returns a descriptor of the machine, which can be used by the client to make future references to the VM instance when issuing requests to VMShop.

4 Virtual File Systems and Virtual File System Services

In-VIGO uses a Grid Virtual File System (GVFS [7]) to support efficient and transparent Grid-wide data provisioning [1], [8]. GVFS presents a generic file system interface to applications by building a virtualization layer upon the de-facto NFS [9] distributed file system, and does so without changing the existing O/S clients/servers. It achieves on-demand cross-domain data transfers via the use of middleware-managed interchangeable logical user accounts [3] and file system proxy-based data access authentication, forwarding and user-identity mapping [10]. The design supports deployment of one or more proxies between a native NFS client and server. A multi-proxy setup is important to implement extensions to GVFS, provide additional functionality and improve performance.

A unique aspect of In-VIGO is how it integrates virtual machine and file system techniques to provide flexible execution environments and on-demand, transparent data access for unmodified applications. Data management has a key role in realizing the benefits of VM-based Grid computing [6] because a VM computing session typically involves data distributed across three different logical entities: the "state server", which stores VM state; the "compute server", which provides the capability of instantiating VMs; and the "data server", which stores user data. Without a virtual file system, instantiating a VM requires the explicit movement of state files to a compute server, and the explicit movement of user data to the VM once it is instantiated. In contrast, through GVFS, In-VIGO middleware creates dynamic GVFS sessions between the state and compute servers to support access of VM states for VM instantiation, and between the VM and data servers to support access to user data for application execution within the VM [7].

GVFS supports secure Grid-wide data provisioning for both VM states and user files by way of two mechanisms: private file system channels and session-key based inter-proxy authentication. Privacy and integrity are guaranteed by the SSH connection, and user authentication is independently carried out by each private file system channel. Through the use of the virtualization layer, the session key handling is completely transparent to kernel clients and servers, and it only applies to inter-proxy authentication between tunnel end-points.

Caching is especially important to exploit data locality and hide network latency in Grid environments. In each GVFS session, the client-side proxy can dynamically establish and manage a file system disk cache to complement the kernel memory buffer with much greater capacity. The cache operates at the granularity of NFS RPC calls and satisfies requests with cached file attributes and data blocks. For write requests, it can employ write-back to hide write latencies and avoid transfers of temporary data. Furthermore, GVFS caches can be customized in many aspects (including size, associativity, write policy and consistency semantics) and thus be tailored to the needs of different applications. GVFS' inherent on-demand block-based data access manner allows for partial transfer of files and can benefit many applications, especially VM monitors, which typically access only a very small part of often Gigabyte-size VM disk state. As an application, the middleware can schedule GVFS sessions with VMM-specific coherence to allow for high-performance VM instantiations. For example, a VM with non-persistent state can be read-only shared among multiple users while each user has a "clone" of the VM and independent redo logs, so that aggressive read caching for state files and write-back caching for redo logs can be employed [7].

The data management middleware mentioned above has been implemented as WSRF-compliant services to provide interoperable service interfaces and flexible state management [11]. These services include: 1) file system service, which runs on every server and controls the local file system proxies to establish and configure specific GVFS sessions; 2) data scheduler service, which provides central scheduling and customization of GVFS sessions and interacts with individual file system services to start the sessions; 3) data replication service, which creates and manages data replicas for the purpose of fault tolerance and load balancing. To initiate a VM-based computing session in In-VIGO, the VMPlant service requests the data scheduler service to prepare a GVFS session between the VM state server and the VM host to instantiate a compute VM. Afterwards, the VAS service can request the scheduling of another session between the VM and the data server, so the application can be started inside the VM and access the user files via GVFS.

5 Virtual Networking

Network connectivity is an obvious necessity in Grid-computing, as it makes remote job execution/submission possible and also allows communication between processes for parallel and/or distributed applications. However, due to firewalls and NAT devices, symmetric connectivity is often absent when resources are distributed across wide-area networks and different administrative domains.

Hosts behind firewalls or NATs can only initiate communication, i.e., they cannot receive communication initiation requests. This limits the hosts' ability to receive remote job execution requests and participate in distributed computations. Existing solutions to the asymmetric connectivity problem still face one or more of the following issues: (1) changes in firewalls or NAT configuration are required (e.g., to allow traffic in some ports or to forward ports), possibly violating security policies; (2) knowledge of network usage (e.g., transport port number) is necessary; (3) high administration overheads are implicit, since actions are required every time a new resource is added or removed from the Grid; and (4) application-transparency is not preserved. Solutions based on address/port translation require either the applications to be aware of resource discovery protocols (e.g., SOCKS [12], DPF and GCB [13]) or changes to be done in OS kernel network stack and/or in the Internet infrastructure (e.g., IPNL [14] and AVES [15]). When networking complexity is abstracted and a new API is exposed, application-transparency is lost (e.g., peer-to-peer networks and the Ibis programming environment [16]). Tunneling-based approaches have difficulties with firewalls and high administrative overhead (e.g., VPN, VNET [17], VIOLIN [18] and X-Bone [19]).

ViNe, the In-VIGO component responsible for network virtualization, has been designed to address all the above issues. It also has additional features such as support for on-demand creation, deployment and removal of isolated virtual networks that specifically connect the necessary machines for execution of a Grid application. The architecture of ViNe is based on IP-overlay on top of the Internet and resembles a site-to-site VPN setup. In each participating network, a ViNe router (VR) is placed in order to handle all ViNe traffic. VRs are responsible for intercepting IP packets destined to ViNe private address space, encapsulate them with ViNe header and forward them to the VR that can deliver the original IP packet. VRs make routing decisions (i.e., to where a packet needs to be forwarded) based on a set of routing tables, which can be updated by secure VR-to-VR communication. The secure update of the tables is the key for the on-demand definition of new virtual networks.

When a VR is placed in a network environment behind a firewall or NAT device, it is called a *limited VR*. Limited VRs cannot receive communication initiated by peer VRs, so a VR without limitations needs to be allocated as an intermediate node, which is called *queue VR*. Routing tables of all VRs are updated to forward to the queue VR the packets that are destined to the limited VR subnet. Since a limited VR can initiate communication, it is its responsibility to contact the queue VR and retrieve packets.

ViNe uses the private IP address space which is not routable in the Internet. Since ViNe nodes cannot be reached directly from the Internet, network security can be discussed with respect to external traffic and internal traffic. External traffic includes VR-to-VR communication, including encapsulated IP packets and control messages. Internal traffic includes the actual communication between hosts in ViNe space. VR-to-VR communication is secured by cryptographically authenticating all messages, and also by encrypting critical information exchange such as control messages. Internal traffic security is achieved by either implementing all security policies of an organization in the VR or delegating that function to the firewall that may be already present in the site. The latter is possible because ViNe does not modify IP packets, and the firewall can still inspect and filter ViNe internal traffic following original rules.

The first prototype of VR has been implemented in Java, with low level networking handled by C code. Hosts do not need the installation of additional software in order to join ViNe, requiring only the operating systems be able to bind additional IP addresses to a network interface and to define static routes. Those features are present in most modern operating systems, making ViNe platform independent. Experiments showed that ViNe can offer performance that is close to the physical network, both in round-trip latency and throughput.

ViNe enables machines, even if they are connected to private networks, to easily join the Grid, and also can minimize the reluctance of system administrators to share resources by not requiring changes in security policies in the existing networks (a minimal change may be necessary, i.e. allowing ViNe traffic through the shared resources; however, the ViNe traffic will undergo the same packet inspection/filtering as the regular network traffic).

6 Virtual Application Services

The In-VIGO Virtual Application (VA) framework enables developers to automatically and transparently enable unmodified legacy applications "for the Grid" and users to transparently access deployed applications using virtualized resources "on the Grid". This requires the creation of VA services capable of orchestrating the use of previously discussed virtualization and other core Grid-middleware components. GridLab's Grid Application Toolkit (GAT) [20], Application Web Services (AWS) [21] and GridPort [22] are examples of other frameworks that aggregate core Grid-middleware to facilitate execution of applications and construction of Web-portals, but that do not consider exposing each application as a Web/Grid-service.

A *virtual application* consists of a physical application (unmodified application binaries and necessary execution environment) and additional software that (1) customizes the interface of the physical application to appear as multiple different applications with different capabilities for different users, and (2) interacts with other middleware in order to enable multiple simultaneous non-conflicting application instances on Grid resources. In particular, the virtual application makes use of the resource virtualization techniques and services described in the previous sections (virtual machines, virtual file system and virtual networks) to create the execution environment required by the application.

A *virtual application service* is a virtual application whose interfaces comply with WSRF specifications. Grid-enabling is the process of turning command-line applications interfaces into services that can be integrated into Grid-portals and delivered through Web-based interfaces. Unless automated, Grid-enabling demands considerable time and programmer effort, especially for legacy applications which do not use programming technologies and practices that are well suited for Grid-computing and are not interoperable with other applications. To overcome this issue, the VA approach provides automatic Grid-enabling of legacy applications for which the following information needs to be provided: command-line syntax, description of the command line in natural language, application resource requirements, and execution environment settings.

Generated virtual application services are: (1) Consumable: the VAS can be discovered by, and made available to, other organizations in a technology-neutral manner that hides heterogeneity and allows interoperability and composition; (2) Isolated: simultaneous conflict-free execution of multiple unmodified applications is possible; (3) Customizable: VAS functionality can be customized to be the same as the original application, or it can be restricted, augmented, or composed with other applications, per user or per user-group; (4) Scalable to create and deploy: application virtualization is a one-time automated process that greatly reduces the overhead of creation and deployment of multiple application services; (5) Dynamically enabled: VAS deployment can be done in a "plug-and-play" fashion without having to bring down any part of the Grid infrastructure.

A distinct contribution of the VA approach is the VA language that allows the description of command-line applications interface with potentially complex set of parameters. The specifiable information about the command-line format includes the following: parameter types, default values, number of occurrences of a parameter, groupings of parameters, dependencies among parameters, multiple group choices, and parameter sweeping information. This language allows an application *enabler*, a special user who has knowledge of the application, but not necessarily of the underlying Grid infrastructure, to describe the application in a more comprehensive manner than solutions proposed by SoapLab [23] and Generic Application Factory Service (GAFS) [24]; thus, allowing strong parameter-type validation.

The VA architecture supports three processes: virtual application enabling, virtual application service customization and generation, and virtual application service utilization. It is divided into three tiers: the Web-portal tier which automatically generates web interfaces of the Grid-services, the virtual application tier discussed in this section and the virtual-Grid tier composed of virtualization services described in the previous sections (Fig. 2). Two solutions for the virtual application service customization and generation process were implemented in In-VIGO: (1) Generic Application Service (GAP) [25] in which a generic Grid-service dynamically configures itself according to the application information, making the interface of the specific application available to the service client using a description language developed in the In-

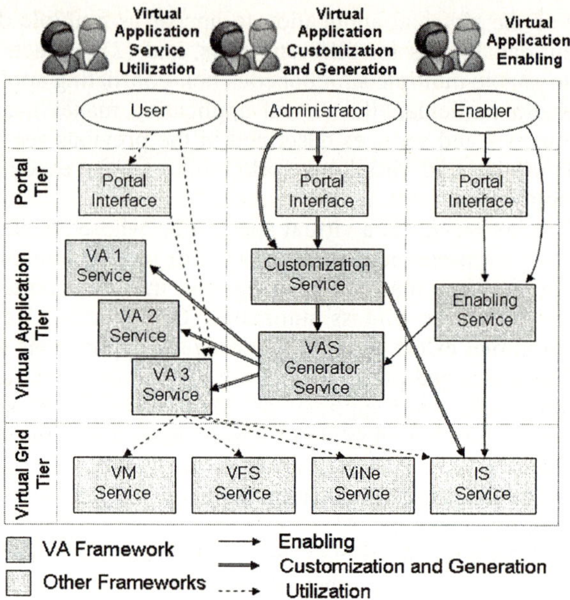

Fig. 2. VA Architecture. Components are separated into portal, virtual application and virtual Grid tiers. From right to left, the diagram depicts paths for: (1) enabling an application by an enabler, (2) customizing and generating the virtual application services VA 1, VA 2 and VA 3 by an administrator and (3) utilizing the virtual Grid (virtual machines, virtual file system, and virtual networks) to deliver the VA 3 service to a user

VIGO project, and (2) Virtual Application Service (VAS) which generates one specific Grid-service for each application so that the application interface is fully described using the standard Web Services Description Language (WSDL). The VAS framework transforms the application information into XMLSchemas fully using the expressiveness of it, including it as part of the service description (WSDL), and then it generates, compiles and deploys the service implementation. The solution makes use of third party tools like XMLBeans to generate complex binding types expressed in XMLSchemas, a modified WSDL2Java to generate the service implementation, AdminClient to deploy the service, Apache Ant to coordinate this automated process, and Apache Axis and Tomcat as containers of the generated services.

7   Building Virtual Grids: In-VIGO at Work

In order to enable sharing of geographically distributed computational and data resources with different usage policies, In-VIGO middleware shares with other Grid middleware, the requirement of interfacing with heterogeneous resource access and authentication schemes. Using resources managed by cluster or other Grid middleware, such as Globus or Condor-G, entails delegation of jobs to these middleware components using the appropriate job management syntax, and authentication and authorization scheme. This section describes the approaches used in current In-VIGO deployments to interface with multi-institutional resources for managing tasks associated with In-VIGO middleware and users.

In-VIGO users do not have direct access to, and are completely decoupled from, user accounts on Grid resources where jobs are effectively run. In-VIGO middleware has full control of all resources and is responsible for starting jobs as well as maintaining them, with complete freedom on how to dynamically map Grid users to local users, and possibly recycle local user accounts among Grid users. The approach brings advantages for both Grid users and resource providers: Grid users are freed from the need to manage several credentials; resource administrators are freed from the task of reconfiguring the access control of resources every time a user joins or leaves the Grid.

In-VIGO users authenticate themselves by presenting their username and password to the Grid portal. After login, user actions resulting in access to a Grid resource are handled by the In-VIGO middleware through the use of Role-Based Access Control (RBAC) mechanisms, offering Single Sign-On (SSO) for users. Users are grouped into roles (e.g., regular, Matlab licensed, administrator), while resources are configured by their providers with a set of permission groups which define operations (e.g., a simulator in demo, full and configuration modes). Appropriate mappings between user roles and permission groups are defined, and In-VIGO middleware enforces the mappings when accessing resources on behalf of users. For example, only users in the "Matlab licensed" role would be able to run a Matlab-based simulator in its full operation mode.

Resources, especially local user accounts, need to be isolated from each other because they are recycled among Grid users. To address this need, local accounts are either pre-created by resource providers, or created on-demand for a particular user in VMs where In-VIGO middleware has administrative privileges. In the first case, In-VIGO middleware makes sure that, at any point in time, only one user is mapped to a given local user account, and also that the account is cleaned when the job finishes. In the latter case, accounts are created and destroyed for one Grid user, without the need for recycling. Since a local account does not run processes for two different users simultaneously, user isolation at process level is guaranteed. However, local user accounts also need to have their data access privileges limited to the current assigned Grid user, as isolation is compromised if the local accounts have access to data of all Grid users. GVFS provides the necessary data isolation between Grid users. GVFS controls access at the granularity of directories so that In-VIGO middleware is able to limit the shadow account's access of data to the home directory of the Grid user allocated to it. Further data isolation, among jobs running for the same Grid user, can be achieved by limiting the access of the local user accounts running the jobs to the job working directory, which are subdirectories under the Grid user's home directory.

As the In-VIGO middleware has all the necessary credentials to access accounts (i.e., to remotely submit a job, independently of the mechanism – Condor, GSI, PBS, SSH, etc) to run jobs on behalf of the user, providing SSO access to Grid resources is trivial. More complex SSO solutions are however required when providing users access to interactive applications that require application level authentication from the user. Examples of such applications currently supported by In-VIGO include VNC sessions and a web-based file-manager. In the case of VNC, In-VIGO remotely starts its server process with a random password in a shadow account. When the user requests access to the VNC desktop, In-VIGO embeds the necessary credential into the VNC client applet and transmits it securely (through SSL) to the user. When the VNC client is run by the user, it authenticates automatically (on behalf of the user) to the server. Adding RBAC to the above process, enables In-VIGO to allow sharing of workspaces among users, i.e., it enables a group of users (belonging to a single user role) to access a given VNC session without the need for users to share credentials and/or passwords.

In-VIGO selects resources for running In-VIGO user or middleware related tasks based on the job requirements specified by the In-VIGO application enabler, and resource availability and usage policies. This resource matching is performed by In-VIGO in the case of resources directly managed by it, or may be delegated to the cluster or Grid software, such as Condor-G or PBS, managing the resources. In the latter case, In-VIGO job requirements need to be mapped to job requirements in the specification syntax of the cluster or Grid software. Allowing for direct specification of job requirements based on the specification syntax of specific cluster/Grid software requires that the application enabler be aware of the types of resources that the application can use. This problem is typically overcome by introducing a uniform specification syntax that subsumes the specification syntax of the varied cluster/Grid software. Since existing cluster/Grid specification syntax used to describe resource/ request properties are based on symmetric flat attributes [26], the uniform specification syntax inherits their shortcoming, namely the need for tight coordination between resource providers and consumers to agree upon attribute names and values. To allow for a flexible and extensible approach to resource matching in In-VIGO semantic matching of resource descriptions is used [27]. The In-VIGO job specifications and resource descriptions and usage policies are described using RDF [28] based ontologies, along with semantic entailments for matchmaking. Handlers specific to the type Cluster/Grid software are then used to map job specification and job management information to the software-specific syntax. The asymmetric description of resource and request enables VA descriptions that are decoupled from the supported resources and implementation of resource matching in In-VIGO.

8   Conclusions and In-VIGO Status

Many challenges faced in early versions of Grid middleware were due to the need to support different applications and distinct users on heterogeneous resources under separate administrative control. The use of virtualization effectively minimizes the impact of hardware and system software dependencies on Grid middleware by generating on-demand the execution environments needed for each application and user. The use of services enables customization of applications for each user while hiding implementation details, thus removing the need for multiple variants of Grid middleware. This "dual rail" decoupling greatly facilitates the management of Grid resources without interfering with other users, and the creation and provision of services without conflicts with other service implementations.

The In-VIGO research reported in this paper confirms the potential benefits of virtualization and services by devising and deploying efficient services for the creation of virtual resources and virtual grids, and providing techniques for the automatic Grid-enabling of applications as services and their on-demand instantiation. The first version of In-VIGO has been online since July of 2003; this and newer versions of In-VIGO have been the subject of research and development since August of 2001. The concepts discussed in this paper have been implemented in at least one of these versions. Extensive prototyping and experimental evaluation of these concepts have demonstrated that the overheads of using virtualization and services are either minimal or acceptable for most Grid-computing applications. In-VIGO middleware is currently being used to deliver Grid-based computational services to users in several domains of science and engineering, which include computational nanoelectronics, coastal and ocean modeling, materials science, computer architecture and parallel processing.

Acknowledgements

The In-VIGO project is supported in part by the National Science Foundation under Grants No. EIA-9975275, EIA-0224442, ACI-0219925, EEC-0228390; NSF Middleware Initiative (NMI) collaborative grants ANI-0301108/ANI-0222828, SCI-0438246; and by the Army Research Office Defense University Research Initiative in Nanotechnology. The authors also acknowledge two SUR grants from IBM and a gift from VMware Corporation. Any opinions, findings and conclusions or recommendations expressed in this material are those of the authors and do not necessarily reflect the views of the National Science Foundation, Army Research Office, IBM, or VMware.

References

1. Adabala, S., Chadha, V., Chawla, P., Figueiredo, R.J., Fortes, J.A.B., Krsul, I., Matsunaga, A., Tsugawa, M., Zhang, J., Zhao, M., Zhu, L., Zhu, X.: From Virtualized Resources to Virtual Computing Grids: The In-VIGO System. Future Generation Computing Systems, special issue on Complex Problem-Solving Environments for Grid Computing, Vol 21/6, 2005, 896–909.
2. Fortes, J.A.B., Figueiredo, R.J., Lundstrom, M.S.: Virtual Computing Infrastructures for Nanoelectronics Simulation. IEEE Proceedings: Special Issue on Blue Sky Technologies (in press), 2005.
3. Kapadia, N., Figueiredo, R.J., Fortes, J.A.B.: Enhancing the Scalability and Usability of Computational Grids via Logical User Accounts and Virtual File Systems. In Proceedings of Heterogeneous Computing Workshop at the International Parallel and Distributed Processing Symposium, April 2001.
4. Krsul, I., Ganguly, A., Zhang, J., Fortes, J., Figueiredo, R.: VMPlants: Providing and Managing Virtual Machine Execution Environments for Grid Computing. In Proceedings of Supercomputing 2004.
5. Keahey, K., Doering, K., Foster, I.: From Sandbox to Playground: Dynamic Virtual Environments in the Grid. In Proceedings of Fifth IEEE/ACM International Workshop on Grid Computing (GRID'04).
6. Figueiredo, R.J., Dinda, P.A., Fortes, J.A.B.: A Case for Grid Computing on Virtual Machines. In Proceedings of International Conference on Distributed Computing Systems, May 2003.
7. Zhao, M., Figueiredo, R.J.: Distributed File System Support for Virtual Machines in Grid Computing. In Proceedings of 13th IEEE International Symposium on High Performance Distributed Computing, June 2004.
8. Figueiredo, R.J., Kapadia, N., Fortes, J.A.B.: The PUNCH Virtual File System: Seamless Access to Decentralized Storage Services in a Computational Grid. In Proceedings of IEEE International Symposium on High Performance Distributed Computing, August 2001.
9. Pawlowski, B., Juszczak, C., Staubach, P., Smith, C., Lebel, D., Hitz, D.: NFS Version 3 Design and Implementation. In Proceedings of USENIX Summer Technical Conference, 1994.
10. Figueiredo, R.J., Kapadia, N., Fortes, J.A.B.: Seamless Access to Decentralized Storage Services in Computational Grids via a Virtual File System. In Cluster Computing, 2004.

11. Zhao, M., Chadha, V., Figueiredo, R.J.: Supporting Application-Tailored Grid File System Sessions with WSRF-Based Services. In Proceedings of the 14th IEEE International Symposium on High Performance Distributed Computing, July 2005, 202–211.
12. Leech, M., Ganis, M., Lee, Y., Kuris, R., Koblas, D., Jones, L.: SOCKS protocol version 5. RFC1928, March 1996.
13. Son, S., Livny, M.: Recovering Internet Symmetry in Distributed Computing. In Proceedings of the 3rd International Symposium on Cluster Computing and the Grid, May 2003.
14. Francis, P., Gummadi, R.: IPNL: A NAT-Extended Internet Architecture. In Proceedings of the ACM SIGCOMM 2001, August 2001.
15. Eugene Ng, T.S., Stroica, I., Zhang, H.: A Waypoint Service Approach to Connect Heterogeneous Internet Address Spaces. In Proceedings of USENIX 2001, June 2001, 319–332.
16. Denis, A., Aumage, O., Hofman, R., Verstoep, K., Kielmann, T., Bal, H.: Wide-Area Communication for Grids: An Integrated Solution to Connectivity, Performance and Security Problems. In Proceedings of 13th IEEE International Symposium on High Performance Distributed Computing, June 2004.
17. Sundararaj, A., Dinda, P.: Towards Virtual Networks for Virtual Machine Grid Computing. In Proceedings of the 3rd USENIX Virtual Machine Research and Technology Symposium, May 2004.
18. Jiang, X., Xu, D.: VIOLIN: Virtual Internetworking on Overlay Infrastructure. In Proceedings of Parallel and Distributed Processing and Applications: Second International Symposium, ISPA 2004, Hong Kong, China, December 13-15, 2004.
19. Touch, J., Hotz, S.: The X-Bone. Proc. of Global Internet Mini-Conference at Globecom, November 1998.
20. Allen, G., Davis, K., Goodale, T., Hutanu, A., Kaiser, H., Kielmann, T., Merzky, A., van Nieuwpoort, R., Reinefeld, A., Schintke, F., Schott, T., Seidel, E., Ullmer, B.: The grid application toolkit: toward generic and easy application programming interfaces for the grid. In Proceedings of the IEEE, Vol.93, Iss.3, March 2005, 534–550.
21. Pierce, M., Fox, G., Youn, C., Mock, S., Mueller, K., Balsoy, O.: Interoperable Web services for computational portals. In Proceedings of the 2002 ACM/IEEE conference on Supercomputing (Baltimore, MD, 2002), IEEE Computer Society Press, 2002, 1–12.
22. Thomas, M., Boisseau, J.: Building Grid Computing Portals: The NPACI Grid Portal Toolkit. Grid Computing: Making the Global Infrastructure a Reality, Ch 28. F. Berman, G. Fox and T. Hey, eds. John Wiley and Sons, Ltd, Chichester (2003).
23. Senger, M., Rice, P., Oinn, T.: Soaplab - a unified Sesame door to analysis tools. In Proceedings of UK e-Science All Hands Meeting September 2003, 509–513.
24. Gannon, D., Alameda, J., Chipara, O., Christie, M., Dukle, V., Fang, L., Farrellee, M., Kandaswamy, G., Kodeboyina, D., Krishnan, S., Moad, C., Pierce, M., Plale, B., Rossi, A., Simmhan, Y., Sarangi, A., Slominski, A., Shirasuna, S., Thomas, T.: Building grid portal applications from a web service component architecture. In Proceedings of the IEEE, Vol.93, Iss.3, March 2005, 551–563.
25. Sanjeepan, V., Matsunaga, A., Zhu, L., Lam, H., Fortes, J.A.B.: A Service-Oriented, Scalable Approach to Grid-Enabling of Legacy Scientific Applications. In Proceeding of International Conference on Web Services (ICWS), Industry Track, July 2005.
26. Solomon, M., Raman, R. and Livny, M.: Matchmaking distributed resource management for high throughput computing. In Proceedings of the Seventh IEEE International Symposium on High Performance Distributed Computing, Chicago, IL, July 1998.
27. Tangmunarunkit, H., Decker, S. and Kesselman, C.: Ontology-Based Resource Matching in the Grid - The Grid Meets the Semantic Web. The Semantic Web - ISWC 2003, Second International Semantic Web Conference, Sanibel Island, FL, USA, October 20-23, 2003, Proceedings. Lecture Notes in Computer Science 2870 Springer 2003, ISBN 3-540-20362-1
28. Lassila, O., and Swick, R.R.: Resource description framework (rdf) model and syntax specification. In W3C Recommendation, World Wide Web Consortium. February 1999. http://www.w3.org/TR/1999/REC-rdf-syntax-19990222.

The Evolution
of the Blue Gene/L Supercomputer

José Moreira

IBM Systems and Technology Group, Rochester, USA

Abstract. The Blue Gene project started in the final months of 1999. Five years later, during the final months of 2004, the first Blue Gene/L machines were being installed at customers. By then, Blue Gene/L had already established itself as the fastest computer in the planet, topping the TOP500 list with the breathtaking speed of over 70 Teraflops. Since the beginning of 2005, many other systems have been installed at customers, the flagship machine at Lawrence Livermore National Laboratory has greatly increased in size, and Blue Gene/L has established itself as a machine capable of breakthrough science.

We here examine how Blue Gene/L came to be. We describe how some key technical decisions were made that shaped the overall hardware and software architecture of this machine. We also describe the nature of the interactions between the teams inside and outside IBM that led to Blue Gene/L being such a successful venture. Finally, we explain why this is just the beginning, and why there is more excitement ahead of us than behind us in the Blue Gene project.

J.C. Cunha and P.D. Medeiros (Eds.): Euro-Par 2005, LNCS 3648, p. 13, 2005.

Agent Based Computational Grids: Research Issues and Challenges

Omer F. Rana

Cardiff University, UK

Abstract. As computer and computational scientists have to manage access to increasingly complex computing and data resources, this becomes a time consuming task. This is especially true for Computational Grids, which can involve the integration of resources distributed across multiple administrative domains. Deciding which systems to use, where the data resides for a particular application domain, how to migrate the data to the point of computation (or vice versa), and data rates required to maintain a particular application "behaviour" become significant. To support these, it is important to develop brokering approaches based on intelligent techniques – to support service discovery, manage performance based on data from monitoring tools, and support data selection. Although the use of broker-based techniques can be found in literature today – very few of these fully utilise the potential of an agent-based system. Intelligent agents provide a useful means to achieve the objectives outlined above. An important and emerging area within Grid computing is the role of service ontologies – especially domain specific ontologies, which may be used to capture particular application needs. Using these, scientists may be able to share and disseminate their data and software more effectively. This has been recognised as being important by both the computer and computational science community – and current efforts towards establishing "Semantic Grids" is a useful first step in this direction.

The role of agent standards and how they can be integrated with Grid computing is explored. Specialist activities that can be undertaken by agent-based computing are outlined, along with example implementation of such systems. Research challenges that still need to be addressed are highlighted, along with possible benefits that overcoming such challenges will bring.

J.C. Cunha and P.D. Medeiros (Eds.): Euro-Par 2005, LNCS 3648, p. 14, 2005.
© Springer-Verlag Berlin Heidelberg 2005

Science on a Large Scale

Raymond Bair

Mathematics and Computer Science Division, Argonne National Laboratory, USA

Abstract. The TeraGrid, the U.S. National Science Foundation's multi-year project to build a distributed national cyberinfrastructure, entered full production mode in the fall of 2004, providing a coordinated set of services for the science and engineering community. TeraGrid operates a unified user support infrastructure and software environment across its eight resource partner sites, which together provide more than 40 teraflops of computing capability and mass storage capability in the petabytes, linked by networks operating at tens of Gigabit/sec. This unified environment allows TeraGrid users to access storage and information resources as well as over a dozen major computing systems via a single allocation, either as stand-alone resources or as components of a distributed application using Grid software capabilities. Many lessons can be drawn from the dual pursuit of high performance and close integration.

The next phase will be even more exciting, with the roll out of a wide range of science gateways and additional advanced applications. Science gateway projects are aimed at supporting access to TeraGrid via web portals, desktop applications or via other grids. An initial set of 10 gateways will address new scientific opportunities in fields from bioinformatics to nanotechnology as well as interoperation between TeraGrid and other Grid infrastructures.

TeraGrid is also enabling an impressive array of large scale science applications, where researchers can perform complex simulations and manipulate enormous data sets in novel ways to gain new insights into research questions and societal problems, for example, finding the most efficient and least expensive ways to clean up groundwater pollution.

Effort in these and other related areas will allow more researchers and educators access to TeraGrid capabilities and advance compatibility between TeraGrid and other major Grid deployments such as Open Science Grid, Network for Earthquake Engineering Simulation (NEES), and major European and Asian Grid deployments.

J.C. Cunha and P.D. Medeiros (Eds.): Euro-Par 2005, LNCS 3648, p. 15, 2005.
© Springer-Verlag Berlin Heidelberg 2005

Topic 1
Support Tools and Environments

Henryk Krawczyk, Jacques Chassin de Kergommeaux,
Pierre Manneback, and Tomás Margalef

Topic Chairs

Nowadays parallel distributed programmers use different tools and environments that facilitate the design, programming, testing, debugging and performance analysis and tuning of their applications. However, they do not satisfy all user requirements, such as broad usability, high effectiveness and proper accuracy. Therefore new propositions are still being developed and their properties tested on modern environments, such as clusters and grids. Their main aim is to simplify the understanding of what-and-why happens during execution of parallel and distributed applications. An important step is to prepare semantic descriptions of system behaviour and to make progress in high quality automatic analysis of performance bottlenecks.

This year 23 papers were submitted to this topic. Overall, they address different usability aspects of parallel distributed environments and tools to improve quality of program behaviour and performance analysis in such environments. The broad scope of considerations includes efficient distributed compilation, nested loop optimisation, checkpointing, system and software configuration management. Besides, monitoring, logging and tuning procedures design for different environments as well as middleware improvements to create high quality services are presented. Among the submissions, only seven papers (30%) were finally accepted. They concentrate on improvements to the effectiveness and accuracy of the performance analysis of parallel and distributed programs. Novel approaches to these problems based on soft computing are presented. In particular, a high level query language is introduced to support performance analysis using linguistic expressions. Moreover, the performance profiling model is described to create a general algorithm for on-the-fly overhead assessment and compensation. The methods for improving performance of selected routine libraries are also discussed, and usability of analytical models and corresponding tools is also evaluated. New modelling techniques, based on occurrence and interrelationships of events, to build a data structure of a partial order of events is also given. Modification of existing middleware environments towards specification and dynamic adaptation of system services is also considered.

The qualified papers propose improvements in tools and parallel distributed environments sand are a good material for foresting a discussion during session meetings.

J.C. Cunha and P.D. Medeiros (Eds.): Euro-Par 2005, LNCS 3648, p. 17, 2005.
© Springer-Verlag Berlin Heidelberg 2005

Tolerating Message Latency Through the Early Release of Blocked Receives

Jian Ke[1], Martin Burtscher[1], and Evan Speight[2]

[1] Computer Systems Laboratory, School of Electrical & Computer Engineering,
Cornell University, Ithaca, NY 14853, USA
`{jke,burtscher}@csl.cornell.edu`
[2] Novel System Architectures, IBM Austin Research Lab, Austin, TX 78758, USA
`speight@us.ibm.com`

Abstract. Large message latencies often lead to poor performance of parallel applications. In this paper, we investigate a latency-tolerating technique that immediately releases all blocking receives, even when the message has not yet (completely) arrived, and enforces execution correctness through page protection. This approach eliminates false message data dependencies on incoming messages and allows the computation to proceed as early as possible. We implement and evaluate our early-release technique in the context of an MPI runtime library. The results show that the execution speed of MPI applications improves by up to 60% when early release is enabled. Our approach also enables faster and easier parallel programming as it frees programmers from adopting more complex nonblocking receives and from tuning message sizes to explicitly reduce false message data dependencies.

1 Introduction

Clusters of workstations can provide low-cost parallel computing platforms that achieve reasonable performance on a wide range of applications reaching from databases to scientific algorithms. To enable parallel application portability between various cluster architectures, several message-passing libraries have been designed. The Message Passing Interface (MPI) standard [10] is perhaps the most widely used of these libraries. MPI provides a rich set of interfaces for operations such as point-to-point communication, collective communication, and synchronization.

Sending and receiving messages is the basic MPI communication mechanism. The simplest receive operation has the following syntax: *MPI_Recv(buf, count, dtype, source, tag, comm, status)*. It specifies that a message of *count* elements of data type *dtype* with a tag of (*tag, comm*) should be received from the *source* process and stored in the buffer *buf*. The *status* returns a success or error code as well as the *source* and *tag* of the received message if the receiver specifies a wildcard *source/tag*. The *MPI_Recv* call blocks until the message has been completely received. The MPI standard also defines a non-blocking receive operation, which basically splits *MPI_Recv* into two calls, *MPI_Irecv* and *MPI_Wait*. The *MPI_Irecv* call returns right away whether or not the message has been received, and the *MPI_Wait* call blocks until the entire message is present. This allows application writers to insert useful computation between the *MPI_Irecv* and *MPI_Wait* calls to hide part of the message latency by overlapping the communication with necessary computation.

J.C. Cunha and P.D. Medeiros (Eds.): Euro-Par 2005, LNCS 3648, pp. 19–29, 2005.

In both cases, the computation cannot proceed past the blocking call (*MPI_Recv* or *MPI_Wait*). In our library, we immediately release (unblock) all blocked calls (*MPI_Recv* and *MPI_Wait*) even when the corresponding message has not yet been completely received, and prevent the application from reading the unfinished part of the message data through page protection. *Our early-release technique automatically delays the blocking for as long as possible, i.e., until the message data is actually used by the application, and eliminates the false message data dependency implied by the blocking calls.* As such, it provides the following benefits:

- It allows the computation to continue on the partially received message data instead of waiting for the full message to complete, thus overlapping the communication with the computation.
- All blocking receives are automatically made non-blocking. The message blocking is delayed as much as possible, benefiting even nonblocking receives with suboptimally placed *MPI_Wait* calls.
- Programmers no longer need to worry about when and how to use the nonblocking MPI calls, nor do they need to intentionally dissect a large message into multiple smaller messages. This reduces the development time of parallel applications. In addition, the resulting code is more intuitive and easier to understand and maintain, while at the same time providing or exceeding the performance of more complex code.

We implemented the early-release technique in our erMPI runtime library. Applications linked with our library instantly benefit from early release without any modification. erMPI currently supports the forty most commonly-used MPI functions, which is enough to cover the vast majority of MPI applications.

There has been much work on improving the performance of MPI runtime libraries. TMPI [12], TOMPI [1] and Tern [6] provide fast messaging between processes co-located on the same node via shared memory semantics that are hidden from the application writer. Tern [6] dynamically maps computation threads to processors according to custom thread migration policies to improve load balancing and to minimize inter-node communication for SMP clusters. Some implementations [9, 11] take advantage of user-level networks such as VIA [2] or InfiniBand [4] to drastically reduce the messaging overhead, thus reducing small-message latency. Other researchers have investigated ways to improve the performance of collective communication operations in MPI [5, 13]. Prior work by the authors has explored using message compression to increase the effective message bandwidth [7] and message prefetching to hide the communication time [8].

This paper is organized as follows. Section 2 introduces our erMPI library and describes the early-release implementation of blocked receives. Section 3 presents the experimental evaluation methodology. Section 4 discusses results obtained on two supercomputers. Section 5 presents conclusions and avenues for future work.

2 Implementation

2.1 The erMPI Library

We have implemented a commonly used subset of forty MPI functions in our erMPI library, covering most point-to-point communications, collective communications, and communicator creation APIs in the MPI specification [10]. The library is written in C and provides an interface for linking with FORTRAN applications. erMPI utilizes TCP as the underlying network protocol and creates one TCP connection between every two communicating MPI processes. Each process has one application thread as well as one message thread to handle sending to and receiving from all communication channels.

2.2 Early-Release Mechanism

The messaging thread creates an alias page block for each message receive buffer posted via an *MPI_Recv* or *MPI_Irecv* call and stores incoming message data via these alias pages. The application thread making the call to *MPI_Recv* or *MPI_Wait* never blocks when calling these routines, which is a slight departure from the spirit of these calls. However, if the message has not arrived or is only partially complete, the application thread protects the unfinished pages of the message receive buffer and immediately returns from the *MPI_Recv* or *MPI_Wait* call that would otherwise have blocked. Thus, computation can continue until the application thread touches a protected page, which causes an access exception, and the application is then blocked until the data for that page is available.

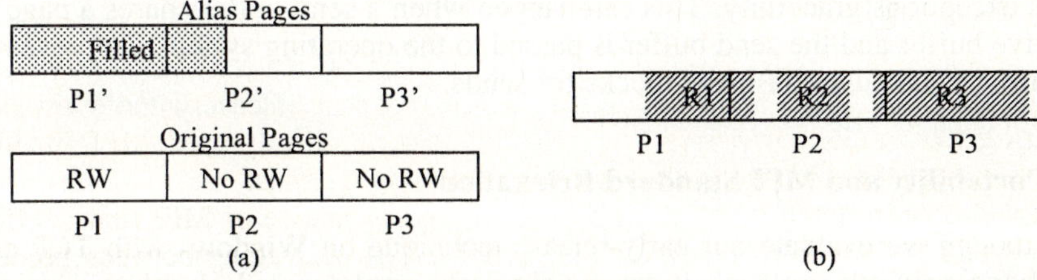

Fig. 1. Receive page examples

Figure 1 (a) shows an example of a receive buffer consisting of three pages. The virtual pages P_i and $P_i^{'}$ ($i \in \{1,2,3\}$) are mapped to the same physical page. The incoming message data is stored into the receive buffer through the alias pages $P_i^{'}$, which are created from the original buffer pages passed to *MPI_Recv* and are never protected. When the application thread calls *MPI_Recv* or *MPI_Wait* in this example, it will notice that P_1 is completely filled and therefore protects only P_2 and P_3, which will be granted *ReadWrite* access again by the messaging thread as soon as those pages are filled. The application thread returns from the *MPI_Recv* or *MPI_Wait* call without waiting for the completion of P_2 and P_3.

2.3 Implementation Issues

Shared Receive Pages. Figure 1 (b) depicts three outstanding receives, R_1, R_2 and R_3. All three receive buffers include part of page P_2. To handle such cases, we maintain a page protection count for shared receive pages to enforce the correct protection action. A page's protection count is incremented for each early-release protection. Note that a page only needs to actually be protected when the page protection count is increased from zero to one. If all three receives are released early, P_2's page protection count will be three. The count is decreased by one as soon as one of the receives completes the page. Once the count reaches zero, the page is unprotected.

There are at most two shared pages for each receive operation, one at the head of and the other at the tail of the receive buffer. For efficiency reasons, we log the page protection counts of all head and tail pages in a hash table.

Alias Page Creation. Most modern operating systems allow multiple virtual pages to be mapped to the same physical page and expose this function via system calls such as *mmap* in Unix and *MapViewOfFile* in Windows NT.

Creating an alias page is an expensive operation. To facilitate alias page reuse, we store the alias page description in a hash table. Each entry in the hash table records the starting addresses of both the original and the alias page blocks and the page block length. We hash the starting address of the original page block to index the hash table. A new alias page block is created if there is no hit or if the existing alias block is too small; otherwise a preexisting block is reused. Alias page blocks are allocated at a 16-page granularity. The page size is 4 kB in our system.

Send Operation. It is important that the protected pages be accessed only by the application thread running in user mode. If these pages are touched by a kernel or subsystem thread or in kernel mode, it may be impossible to catch and handle the access exceptions gracefully. This can happen when a send buffer shares a page with a receive buffer and the send buffer is passed to the operating system. To prevent this scenario, we also use alias page blocks for sends.

2.4 Portability and MPI Standard Relaxation

Even though we evaluate our early-release technique on Windows with TCP as the underlying network protocol, it can be similarly implemented on other systems, as long as the following requirements are met:

- The OS supports page protection calls and access violation handling.
- The network protocol can access the protected receive buffer. This is possible if the network subsystem has direct access to the physical pages or if alias pages can be used to interface with the communication protocol.
- The MPI library can be notified when a partial message arrives. This allows the protected pages to be unprotected as early as possible.

MPI_Recv returns the receive completion status in the *status* structure. It usually includes the matching send's *source* and *tag* and indicates whether the receive is a success. If a wild card *source* or *tag* is specified and the call is early released, the matching send's *source* or *tag* is typically not known. In such a case, we delay the early release until this information is available. We always return a receive success in the *status* field and force the program to terminate should an error occur.

2.5 Other Issues

Message Unpacking. In our sample applications, messages are received into the destination buffers directly, allowing the computation to proceed past the receive operation and to work on the partially received message data. For applications that first receive messages into an intermediate buffer and then unpack the message data once they have been fully received, the early-released application thread would cause an access exception and halt the execution right away due to the message unpacking step, limiting the potential of overlapping useful computation with communication.

Since unpacking adds an extra copy operation and increases the messaging latency, it should be avoided whenever possible. More advanced scatter receive operations provide better alternatives for advanced programmers and parallelizing compilers. Another possible solution is to unpack the message as needed in the computation phase instead of unpacking the whole message right after the message receive.

Correctness. To guarantee execution correctness, an early-released application thread is not allowed to affect any other application thread before all early-released receives are at least partially completed. This means that new messages are not allowed to leave an MPI process if there exists unresolved early-released receives. Otherwise, a causality loop could be formed where an early-released application thread sends a message to another MPI process, which in turn sends a message that matches the early-released receive.

3 Evaluation Methods

3.1 Systems

We performed all measurements on the Velocity + (Vplus) and the Velocity II (V2) clusters at the Cornell Theory Center [3]. Both clusters run Microsoft Windows 2000 Advanced Server. The cluster configurations are listed below.

- Vplus consists of 64 dual-processor nodes with 733 MHz Intel Pentium III processors, 256 kB L2 cache per processor and 2 GB RAM per node. The network is 100Mbps Ethernet, interconnected by 3Com 3300 24-port switches.
- V2 consists of 128 dual-processor nodes with 2.4 GHz Intel Pentium 4 processors, 512 kB L2 cache per processor and 2 GB RAM per node. The network is Force10 Gigabit Ethernet interconnected by a Force10 E1200 switch.

3.2 Applications

We evaluate the performance of early release on three representative scientific applications: PES, N-body, and M3. In general, we see small performance improvements on benchmark applications due to the message unpacking effects.

PES is an iterative 2-D Poisson solver. Each process is assigned an equal number of contiguous rows. In each iteration, every process updates its assigned rows, sends the first and last row to its top and bottom neighbors, respectively, and receives from them two ghost rows that are needed for updating the first and last row in the next iteration. We fix the two corner elements $(0,0)$, $(N-1, N-1)$ to 1.0 and the other two corner elements $(0, N-1)$, $(N-1, 0)$ to 0.0 as boundary conditions.

N-Body simulates the movement of particles under pair-wise forces between them. All particles are evenly distributed among the available processes for the force computations and the position updates. After updating the states of all assigned particles, each process sends its updated particle information to all other processes for the force computation in the next time step.

M3 is a matrix-matrix-multiplication application. In each iteration, a master process generates a random matrix A_i (emulating a data collection process), distributes slices of the matrix to slave processes for computation, and then gathers the results from all slave processes. Each slave process stores a transposed transform matrix B, which is broadcast once from the master process to all slaves when the computation starts. Each slave process first receives matrix A_{ip}, which is part of matrix A_i, then computes matrix $C_{ip} = A_{ip}*B$ and sends C_{ip} to the master. Note that this parallelization scheme is by no means the most efficient algorithm for multiplying matrices.

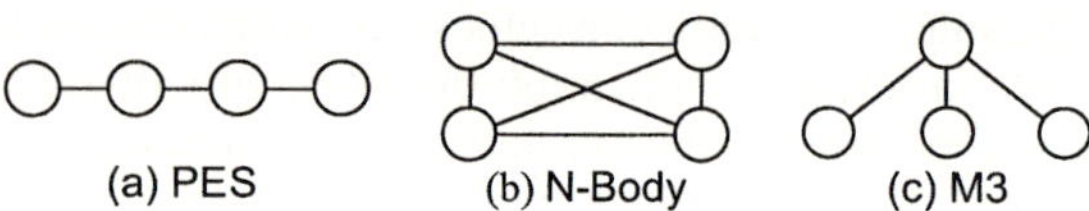

Fig. 2. Communication patterns

The communication patterns of these three applications for four-process runs are shown in Figure 2. The circles represent processes and the lines represent the communication between processes; each PES process only communicates with at most two neighboring processes; each N-Body process communicates with every other process; and each M3 slave process communicates with the master process. The messaging calls used are *MPI_Send*, *MPI_Irecv*, *MPI_Wait* and *MPI_Waitall*.

Table 1. Problem size and message size information

Program	Problem Size			Message Size (64-process run)		
	Size A	Size B	Size C	Size A	Size B	Size C
PES	5120X5120, 2000	10240X10240, 1000	20480X20480, 500	40 kB	80 kB	160 kB
N-Body	10240, 200	20480, 100	40960, 50	9 kB	18 kB	35 kB
M3	1024X1024, 400	2048X2048, 40	4096X4096, 20	128 kB	512 kB	2 MB

Table 1 lists the three problem sizes we used for each application. Size A is the smallest and Size C is the largest. In the "Problem Size" columns, the number before the comma is the matrix size for PES and M3 and the number of particles for N-Body; the number after the comma is the number of iterations or simulation time steps. We have adjusted the number of iterations so that the runtimes are reasonable. We run these applications with 16, 32, 64 and 128 processes and two processes per node. The resulting message sizes for 64-process runs are shown in the "Message Size" columns. We obtained the runtimes with three MPI libraries. *MPI-Pro* is the default MPI library on both clusters. The *erMPI-B* is the baseline version of our erMPI library, in which the early release of receives is disabled. *erMPI-ER* is the same library but with early release turned on.

4   Results

4.1   Scaling Comparison

Figure 3 (a–f) plots the scaling with problem size B of PES, N-Body and M3. Each application has two subgraphs, the left one shows Vplus and the right one V2 results. Each subgraph plots the execution speeds of the three MPI libraries against the number of processes used. The execution speeds are normalized to the 16-process run of the *erMPI* baseline library.

For a fixed problem size, the communication-to-computation ratio increases as the problem is partitioned among an increasing number of processors, which leads to worsening of parallel efficiency and scalability.

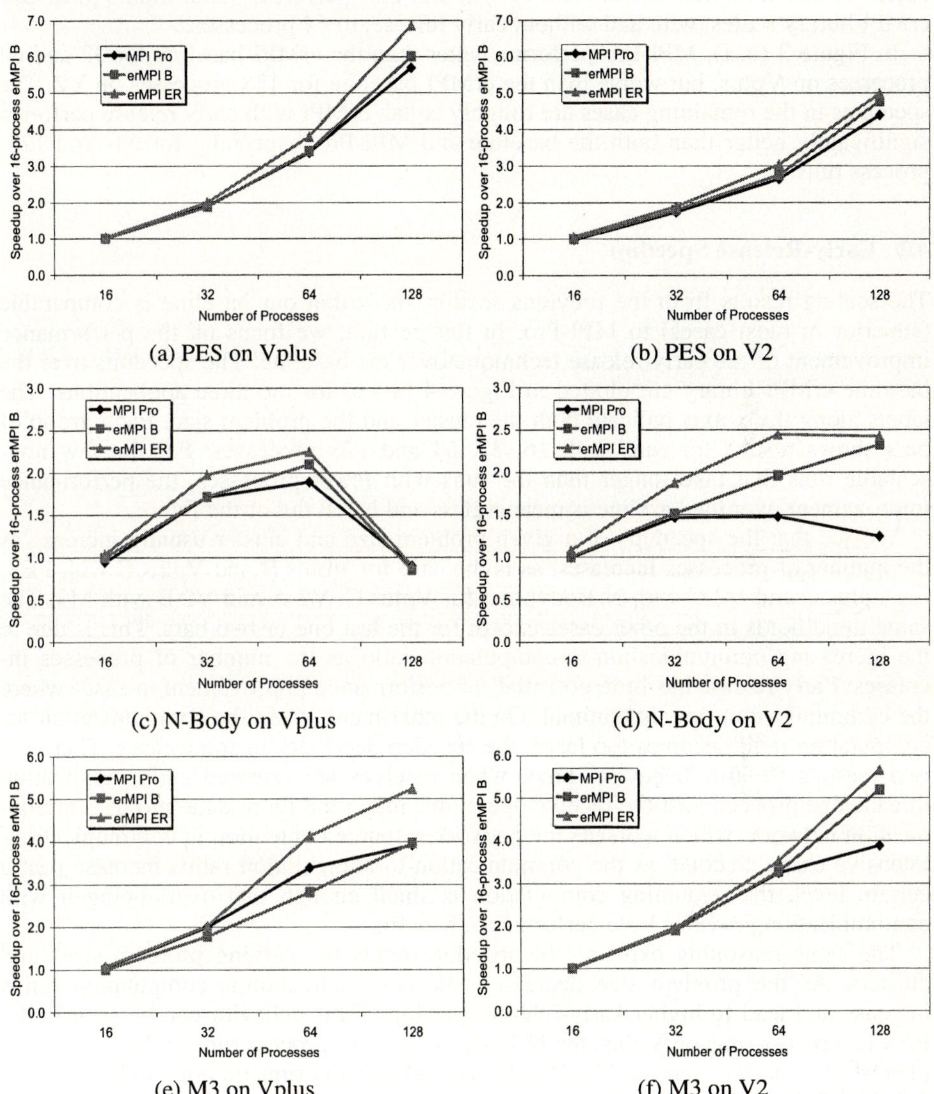

(a) PES on Vplus

(b) PES on V2

(c) N-Body on Vplus

(d) N-Body on V2

(e) M3 on Vplus

(f) M3 on V2

Fig. 3. Scaling comparisons

For PES (Figure 3 (a, b)), the erMPI early-release library scales the best among the three MPI libraries. The erMPI baseline library also scales better than MPI-Pro. PES scales better on the Vplus cluster than on V2. It appears that the higher processing power of V2 leads to a higher communication-to-computation time ratio and hence worse scalability.

N-Body is a communication intensive application. The communication dominates the computation as the number of processes increases. Figure 3 (c, d) shows that the MPI-Pro speedups start to saturate at around 32 processes and degrade at 128 processes. V2 has a higher network throughput and thus performs better than Vplus. Our erMPI library scales, with and without early release, to 64 processes.

In Figure 3 (e, f), MPI-Pro performs better than the erMPI baseline for 32 and 64 processes on Vplus, but worse than the erMPI baseline for 128 processes on V2. The speedups in the remaining cases are roughly equal. erMPI with early release performs significantly better than both the baseline and MPI-Pro, especially for 64- and 128-process runs.

4.2 Early-Release Speedup

The scaling results from the previous section show that our baseline is comparable (superior in most cases) to MPI-Pro. In this section, we focus on the performance improvement of the early-release technique over the baseline. The speedups over the baseline erMPI library are plotted in Figure 4 (a - c) for the three applications. The labels along the x-axis indicate both the cluster and the problem size. Each group of bars shows results for runs with 16, 32, 64 and 128 processes. For the few non-scalable runs that take longer than the runs with fewer processes, the performance improvement over the baseline is meaningless and is left out of the figure.

We see that the speedups of a given problem size and cluster usually increase as the number of processes increases, as is the case for Vplus.B and Vplus.C with PES; for Vplus.C and V2.C with N-Body; and for Vplus.C, V2.A and V2.B with M3. The same trend holds in the other cases except for the last one or two bars. This is due to the increasing communication-to-computation ratio as the number of processes increases. Early release has little potential for performance improvement in cases where the communication time is minimal. On the other hand, when the communication-to-computation ratio becomes too large, the speedup decreases in some cases. There are two reasons for this behavior. First, when receives are released early, application threads that proceed past the receive operations may send more data into the communication network, which worsens the network resource contention in communication-intensive cases. Second, as the communication-to-computation ratios increase past a certain level, the remaining computation is small enough that overlapping it with communication provides little performance benefit.

The same reasoning explains the speedup trends for varying problem sizes and clusters. As the problem size decreases, the communication-to-computation ratios increase and lead to higher early-release speedups. This behavior can be seen in the PES 16-process runs on Vplus, the N-Body 16- and 32-process runs on Vplus and 16-process runs on V2, and the M3 16-, 32- and 64-process runs on both Vplus and V2. The V2 cluster has a relative faster network (bandwidth) than Vplus and hence the potential for speedups due to early release is smaller. Indeed, V2 demonstrates a smaller performance improvement in most cases, except for N-Body sizes B and C, where the two negative effects start to impact the early-release speedups.

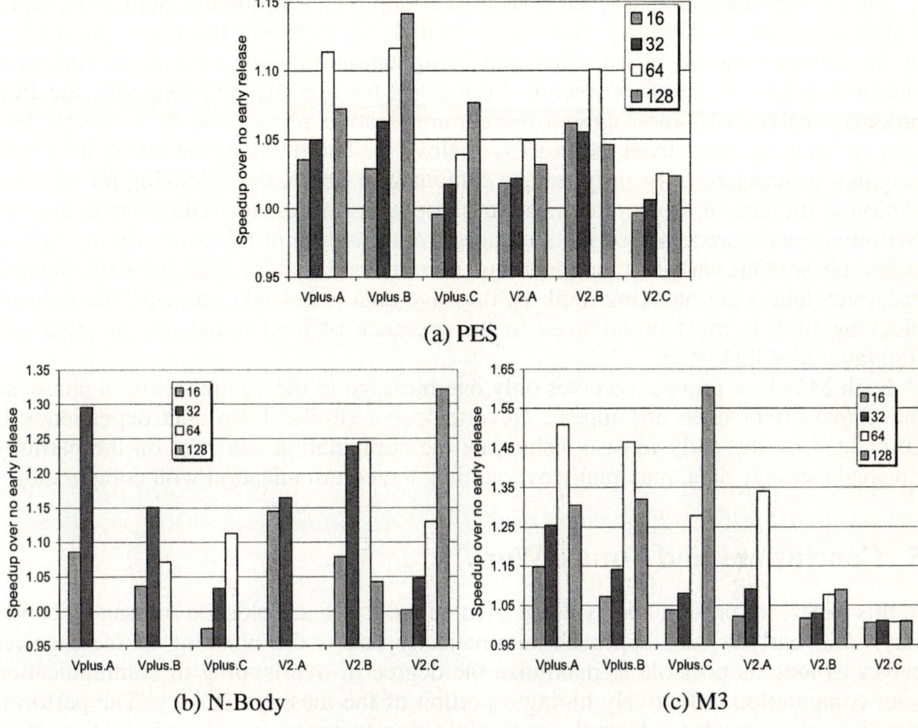

(a) PES

(b) N-Body   (c) M3

Fig. 4. Speedups due to early release

Four cases of PES show early-release speedups of over 10%. N-Body exhibits speedups of up to 32%, with four cases being over 20%. M3 reaches over 30% speedup in six cases, with a maximum speedup of 60%.

4.3 Early-Release Overhead and Benefit

The early-release overhead includes the creation of the alias page blocks and the page protection for unfinished messages. Since the alias page blocks are reused in our implementation, the overhead is amortized over multiple iterations and is negligible. Table 2 compares the page protection plus unprotection time on V2 with the raw transfer time over a 1 Gbps network. The cost of page protection is much smaller than the message communication latency. Most importantly, there is little penalty to the run time since the application thread would have been blocked waiting for the incoming message to complete anyway.

Table 2. Page protection overhead

Message Size	4 kB	16 kB	64 kB	256 kB
Protection Cost	3.0 µs	3.0 µs	3.3 µs	4.5 µs
1 Gbps Transfer Time	33 µs	131 µs	0.5 ms	2.1 ms

Parallel applications frequently consist of a loop with a communication phase and a computation phase. When a process receives multiple messages from multiple senders in the communication phase, often the computation following the communication need not access some of the received messages for a while. For example, the PES process receives two messages in the communication phase, one from the process "above" and the other from the process "below" it. The message data from the lower neighbor is accessed only at the end of computation phase, thus blocking for its completion at the end of the communication phase is not necessary. The same is true for N-Body as each process receives messages from multiple processes in the communication phase. Our early-release technique eliminates this false message data dependency and delays the blocking until the message data is indeed accessed. The reduced blocking time is most pronounced in the presence of load imbalance or processes running out of lock-step.

Each M3 slave process receives only one message in the communication phase, so the above effect does not appear. Nevertheless a similar false data dependency is eliminated by the early-release technique; the computation can start on the partially finished message data, maximally overlapping the communication with computation.

5   Conclusions and Future Work

In this paper, we present and evaluate a technique to release blocked message receives early. Our early-release approach automatically delays the blocking of message receives as long as possible to maximize the degree of overlapping of communication with computation, effectively hiding a portion of the message latency. The performance improvement depends on the communication-to-computation ratio and the extent of false message data dependencies of each application. Measurements with our erMPI library show an average early-release speedup of 11% on two supercomputing clusters for three applications with different communication patterns.

In future work, we plan to eliminate the message unpacking step for some benchmark applications and study the early-release performance on these highly tuned applications. Future research may also explore the usage of a finer early-release granularity to further improve the performance.

Acknowledgements

This work was supported in part by the National Science Foundation under Grant No. 0125987. This research was conducted using the resources of the Cornell Theory Center, which receives funding from Cornell University, New York State, federal agencies, foundations, and corporate partners.

References

1. E. D. Demaine, "A Threads-Only MPI Implementation for the Development of Parallel Programs," *Intl. Symp. on High Perf. Comp. Systems*, 7/1997, pp. 153-163.
2. D. Dunning, G. Regnier, G. McApline, D. Cameron, B. Shubert, F. Berry, A. Merritt, E. Gronke and C. Dodd, "The Virtual Interface Architecture," *IEEE Micro*, 3/1998, pp. 66-76.
3. http://www.tc.cornell.edu/

4. Infiniband Trade Association, *Infiniband Architecture Specification, Release 1.0*, Oct. 2000.
5. A. Karwande, X. Yuan and D. K. Lowenthal, "CC-MPI: A Compiled Communication Capable MPI Prototype for Ethernet Switched Clusters," *The Ninth ACM SIGPLAN Symposium on Principles and Practice of Parallel Programming*, 6/2003, pp. 95-106.
6. J. Ke, "Adapting parallel program execution in cluster computers through thread migration," *M.S. Thesis, Cornell University*, 2003.
7. J. Ke, M. Burtscher and E. Speight, "Runtime Compression of MPI Messages to Improve the Performance and Scalability of Parallel Applications," *Supercomputing,* 11/2004.
8. J. Ke, M. Burtscher and E. Speight, "Reducing Communication Time through Message Prefetching," *Intl. Conf. on Parallel and Distributed Processing Techniques and Applications,* 6/2005.
9. J. Liu, J. Wu, S. P. Kini, P. Wyckoff and D. K. Panda, "High Performance RDMA-Based MPI Implementation over InfiniBand," *Intl. Conf. on Supercomputing*, 6/2003, pp. 295-304.
10. MPI Forum, "MPI: A Message-Passing Interface Standard," *The Intl. J. of Supercomputer Applications and High Performance Computing,* 8(3/4):165-414, 1994.
11. E. Speight, H. Abdel-Shafi, and J. K. Bennett, "Realizing the Performance Potential of the Virtual Interface Architecture," *Intl. Conf. on Supercomputing*, 6/1999, pp. 184-192.
12. H. Tang and T. Yang, "Optimizing Threaded MPI Execution on SMP Clusters," *Intl. Conf. on Supercomputing*, 6/2001, pp. 381-392.
13. R. Thakur and W. Gropp, "Improving the Performance of Collective Operations in MPICH," *European PVM/MPI Users' Group Conference*, 9/2003, pp. 257-267.

Fast Convex Closure for Efficient Predicate Detection

Paul A.S. Ward and Dwight S. Bedassé*

Shoshin Distributed Systems Group
Department of Electrical and Computer Engineering
University of Waterloo
Waterloo, Ontario N2L 3G1, Canada
`{pasward,dsbedass}@shoshin.uwaterloo.ca`

Abstract. The behaviour of parallel and distributed programs can be modeled as
the occurrence of events and their interrelationship. Event data collected according to the event model is stored within a partial-order data structure, where it can
be reasoned about, enabling debugging, program steering, and autonomic feedback control of the application. Reasoning over event data, a critical requirement
for autonomic computing, is typically in the form of predicate detection, a search
mechanism able to detect and locate arbitrary predicates within the event data.
To enable hierarchical predicate detection, compound events are formed by computing the convex closure of the matching primitive events. In particular, the Xie
and Taylor convex-closure algorithm forms the basis for such an approach to
predicate detection. Unfortunately, their algorithm can be quite slow, especially
for hierarchical compound events.
In this paper, we study the cause of the problems in the Xie and Taylor algorithm. We then develop an efficient extension to their algorithm, based on a simple
caching scheme. We prove our algorithm correct. We also provide experimental
results that demonstrate that our approach reduces the execution time of the Xie
and Taylor algorithm by up to 98 percent.

Keywords: Autonomic computing, program steering, predicate detection tool.

1   Motivation

The architecture of tools for monitoring and debugging message-passing parallel programs, enabling parallel-program steering, and the autonomic observation and control of enterprise and distributed systems is broadly similar, and can be characterized
as shown in Fig.1. A variety of such tools have been built over the years, including
ATEMPT [16, 17], Object-Level Trace [13], POET [20], POTA [23], and Log and Trace
Analyzer [12]. The managed system is instrumented with monitoring code that captures
significant event data. Ideally, the information collected will include the event's process
and thread identifiers, number, and type, as well as partner-event identification, if any.
This event data is forwarded from each process to a central monitoring entity which,
using this information, incrementally builds and maintains a data structure of the partial
order of events that form the computation [21]. That data structure may be queried by
a variety of systems, the most common being visualization engines for debugging and
steering and, more recently, control entities for autonomic computing [15].

---

* The authors would like to thank IBM for supporting this work.

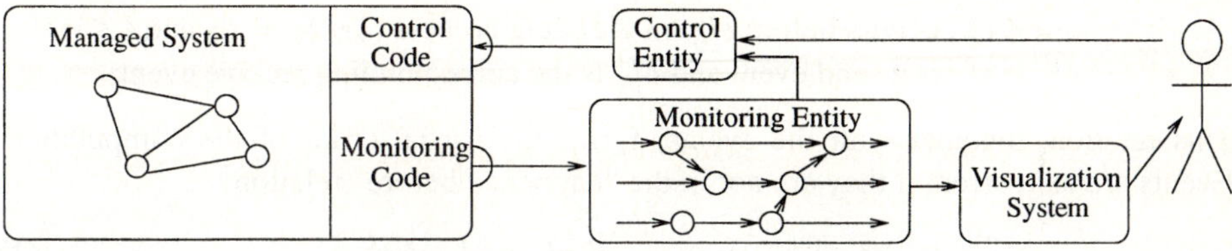

Fig. 1. Monitoring and Control Architecture

The querying of the partial-order data structure for predicate detection has the intent of either displaying predicates of interest to the user, or feeding the information directly into the controller. Rapid analysis of event data is critical for both of these uses. While there have been several approaches to predicate detection (*e.g.*, [4, 14, 22]), this paper focuses on hierarchical predicate detection based on compound events [1]. The current-best algorithm that employs this technique was developed by Xie and Taylor [24] and is implemented within the eclipse system [5].

In using the Xie and Taylor system we discovered it to be very slow in a non-trivial number of cases. Specifically, queries could take several hours to execute. In this paper, we describe a series of experiments that we performed to determine the cause of the slowness in the Xie and Taylor algorithm. As a result of our analysis, we developed a novel incremental closure algorithm that improved the performance of the predicate-detection algorithm by up to 98%.

The remainder of this paper is organized as follows. We first briefly review the operation of the Xie and Taylor algorithm, describing the basics of hierarchical predicate detection based on compound events. We discuss related work. In Sect.3 we detail in three steps the problem with the Xie and Taylor algorithm, the theoretical basis for incremental closure calculations, and finally our algorithm that solves the problem. We then provide both a theoretical and experimental analysis of our approach in Sect.3.1. We discuss related work in Sect.5, contrasting it with our approach. We conclude by observing what we have achieved and what issues remain open.

2   Fundamentals

We now describe the basics of hierarchical predicate detection based on compound events, and the Xie and Taylor approach specifically. We first briefly review the fundamentals of modeling systems as partial orders, which forms the basis of this work.

The event-based approach to modeling multi-threaded, parallel, and distributed systems abstracts computations into sequential processes[1] each of which is a sequence of four types of events: *transmit, receive, unary,* and *synchronous*. These events are considered to be atomic. Further, they form the primitive events of the computation.

The Lamport "happened before" relation [21] is then defined as the smallest transitive relation satisfying

---

[1] Throughout this paper we will use the term "process" to indicate any sequential entity. It might be a single-threaded process, a thread, a semaphore, an EJB (in the case of Object-Level Trace), a TCP stream, *etc.*

1. $e^i_{p_1} \preceq e^j_{p_2}$ if $e^i_{p_1}$ occurs before $e^j_{p_2}$ on the same process (*i.e.* $p_1 = p_2$ and $i \leq j$)
2. $e^i_{p_1} \preceq e^j_{p_2}$ if $e^i_{p_1}$ is a send event and $e^j_{p_2}$ is the corresponding receive event

This relation, together with the events, forms the partial order of the computation. Events are concurrent if they are not in the "happened before" relation.

$$e^i_{p_1} \parallel e^j_{p_2} \iff e^i_{p_1} \npreceq e^j_{p_2} \wedge e^j_{p_2} \npreceq e^i_{p_1} \tag{1}$$

Given a partial order of computation, there are two types of patterns that are typically sought. First we may seek patterns within the structure of the partial order. For example, we may wish to look for the pattern:

$$e^i_{p_1} \preceq e^j_{p_2} \wedge e^k_{p_3} \preceq e^j_{p_2} \wedge e^i_{p_1} \parallel e^k_{p_3} \; ; \; p_1 \neq p_2 \neq p_3 \tag{2}$$

This particular pattern is a crude form of race detection. We are seeking events in processes p_1 and p_3 that both precede an event in a third process p_2 but that have no synchronization between them. The events thus form a potential race condition.

This form of structural pattern searching is equivalent to directed-subgraph isomorphism. Specifically, it is equivalent to asking if the directed acyclic graph that represents the partial order of the computation contains a subgraph isomorphic to the directed graph that represents the pattern being sought. The directed graphs in this equivalence can be either the transitive reductions or the transitive closures of the respective partial orders. This problem is known to be NP-complete [8].

The second type of pattern that we may seek is a pattern within a consistent global state. There are several varieties that may be sought, such as stable predicates (once the predicate is true, it remains true), definite predicates (the predicate is true on all possible paths in the lattice), possible predicates (the predicate is true on some paths in the lattice), and so forth. From the perspective of a partial-order data structure, the primary concern is the ability to determine what is, or is not, a consistent global state. This in turn means we need the ability to determine structural patterns that are consistent global states. It is, as with the first type, NP-complete in the general case. This paper focuses solely on the problem of determining structural patterns within the partial order.

2.1 Hierarchical Predicate Detection Based on Compound Events

To alleviate the problem of NP-completeness, and to reduce the complexity of patterns, the approach taken is to seek hierarchical predicates based on compound events. In this approach, whenever a sub-pattern is matched, the events that form it are closed (according to a criteria to be described below) into a compound event. The requirements of such a compound event is that it must possess (most of) the properties of a primitive event. Specifically, given a compound event and any other event (primitive or compound), it must be possible to determine the precedence relationship between the two. The most effective way currently known of ensuring these requirements is that the compound event be convex closed [19], defined as follows:

Definition 1 (Convex Event). *An compound event c is convex if and only if*

$$\forall_{e_i, e_j \in c} \exists_{e_l} e_i \preceq e_l \wedge e_l \preceq e_j \Rightarrow e_l \in c$$

and extending the definition of precedence to compound events to:

$$c_i \preceq c_j \iff \exists_{e_i \in c_i; e_j \in c_j} e_i \preceq e_j$$

Note that the definition does not result in cyclic precedence provided the compound events are convex. Note also that a primitive event can be compared with a compound event by considering it to be a compound event with a single constituent element.

We now motivate this approach with a simple example. Consider seeking four events, e_1, e_2, e_3, and e_4 such that $e_1 \preceq e_2$, $e_3 \preceq e_4$, and yet ensuring that e_1 and e_2 are each concurrent with both e_3 and e_4. Given this requirement, a non-compound-event-based approach would require the pattern sought to be:

$$(e_1 \preceq e_2) \wedge (e_3 \preceq e_4) \wedge (e_1 \parallel e_3) \wedge (e_1 \parallel e_4) \wedge (e_2 \parallel e_3) \wedge (e_2 \parallel e_4)$$

By contrast, the compound-event-based approach seeks the pattern

$$(e_1 \preceq e_2) \parallel (e_3 \preceq e_4) .$$

Note that while the compound-event-based approach does require two convex closure operations, it requires only four precedence tests, while the alternate approach requires ten[2]. Further, observe that as predicate complexity increases, the advantage of the compound-event-based approach increases. Finally, note that in this case the matching events will be identical, regardless the method chosen. While this is not always true, we have found that it is not difficult to prune unwanted matches from the system.

2.2   The Xie and Taylor Algorithm

Given the problem of structural predicate detection, Xie and Taylor developed a straightforward naive-backtracking algorithm. A parse tree is created of the pattern sought. This tree is processed in prefix order. Whenever the parse-tree node that is matched is a precedence-relationship node, the convex closure is computed, creating a compound event at that point in the parse tree. This is treated as a matched event. This process continues until either the desired pattern is found, or there is no matching event, in which case the algorithm backtracks, matching a different event.

The key features of their algorithm are their pruning rules, necessary to limit the search space, and their convex-closure algorithm. We do not modify their pruning rules, and thus will not comment on them further other than to note that our approach is orthogonal to their pruning rules. Any revisions to the pruning rules may affect the performance of the algorithm, but will not affect the correctness of the overall system.

The critical aspect of their approach, from the perspective of this paper, is their convex-closure algorithm. This algorithm takes an input event set of primitive events, and returns as output two sets, **front** and **back**, that represent the front and back of the

---

[2] While it may seem that the precedence test cost is higher for compound events, this is not in fact the case. It is possible to assign a vector timestamp to a convex event in much the same manner as one is assigned to a primitive event, enabling precedence determination between convex events to be as efficient as it is with primitive events [18].

convex event set, respectively. For a given convex event C, $e \in \mathbf{front}(C)$ if-and-only-if $\nexists_{e'}\ e' \prec_p e$, where $e' \prec_p e$ if events e and e' are in the same process p and event e' precedes event e. Back is defined analogously:

$$e \in \mathbf{back}(C) \iff \nexists_{e'}\ e \prec_p e' \tag{3}$$

Thus, in the worse case the convex event covers all processes in the computation, and thus **front** and **back** will have size N, where N is the number of processes. In such a case, the computational complexity of their algorithm is $O(N^3)$. The full technical details of their algorithm are available in their paper [24]. From the perspective of our work, it is a black box. The primary detail specifically required in our work is that in their algorithm the input event set is composed of two (possibly compound) events. The usage of the convex-closure algorithm by their predicate-detection mechanism is such that one of the these input events is held constant, while the other is varied. The significance of this will become apparent in Sect. 3.3.

3   Incremental Predicate Detection Algorithm

As we have already observed, when using the Xie and Taylor algorithm we found it to be slow, to the point that in a non-trivial number of cases its execution time was measured in hours. We therefore set about first determining the cause of the slowness in their algorithm. Having done so, we developed a theoretically-sound solution to the problem, and then created an algorithm based on it. We now describe these three steps in detail.

3.1   Analysis of Existing Approach

To determine the cause of inefficiency in the Xie and Taylor algorithm we performed a series of experiments using a variety of predicates and data sets. In these experiments, we instrumented the Xie and Taylor code to determine how many convex closures were performed, what the input and output sets were for the given closure, and the execution time to perform the closure in question.

In analyzing the data from these experiments we discovered it was very rare for a convex closure to consist of an entirely new set of input events. Rather, in more than 90% of cases, only one of the events changed. We further discovered that in cases where one input event changes, it was typically a near successor of the input event of the prior closure. However, the Xie and Taylor algorithm made no use of this fact. Rather, it would simply recompute the closure from scratch.

This problem is best illustrated by example. Consider the set of events shown in Fig.2. The pattern $q \preceq (a \preceq d)$ is being sought, and events q and a have already been matched. All that remains is to match d, compute a convex closure between that and the matched a, and confirm that this is a successor to the matched q. If d is matched to d_1 then the convex closure $C1$ of a and d_1 is computed. Unfortunately, $C1$ is concurrent to q. As a result, the search backtracks and matches d to d_2. The compound event $C2$ is then computed as the convex closure of events a and d_2. This is found to be a successor to q, and thus the desired predicate is found. Note that $C2$ is computed without regard

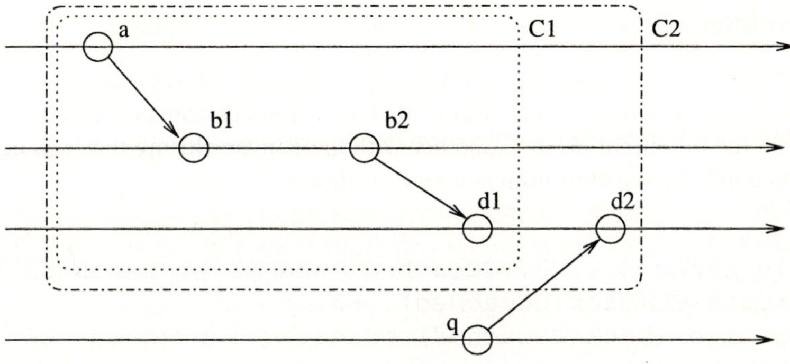

Fig. 2. Incremental Closure Computation

to the original computation of $C1$. By experimental analysis, this lack of incremental computation was found to be the major cause of inefficiency in the Xie and Taylor algorithm.

3.2   Theoretical Basis for Improvement

Having found the problem, it was necessary to determine if recomputing the closure from scratch was an inherent requirement of convex events, or if it was possible to incrementally compute such closures. Thus, considering the example of Fig.2, we wished to compute $C2$ given $C1$.

In this regard, we discovered the following theorems. To understand these theorems, we first define the following functions.

Definition 2 (Convex Closure). $CC(E)$ *is the convex closure of event set* E

Definition 3 (Location Set). $\mathrm{l}\,E$ *is the set of processes in which the various events of event set* E *occur.*

Given these definitions, we were able to prove the following theorem.

Theorem 1 (Incrementality Theorem).

$$(\mathrm{l}\,CC(E \cup \{e\}) = \mathrm{l}\,CC(E)) \wedge (CC(E) \preceq e) \implies$$
$$CC(E \cup \{e\}) = CC(E) \cup CC(\mathbf{back}(CC(E)) \cup \{e\})$$

Proof: See [2]   □

This theorem states that, as long as the location set does not change, the convex closure of an event set E together with a succeeding event e will be the union of the convex closure of the original set, together with that of the closure of e and the back of the convex closure of the original set. What this means in practice is that if the convex closure of the event set E has already been computed, then only a small addition closure needs to be computed. It is fairly trivial to show that the **front** set will remain the same as that of the closure of E, while the **back** set will be that of the closure of $\mathbf{back}(CC(E))$ together with e.

3.3   Algorithm

Given Theorem 1, we devised the following algorithm for incremental closure. First, we assume we have a small cache of closures that have already been computed. This cache will contain the two input events, together with the convex closure that was computed. Our incremental closure algorithm is then as follows:

```
CC(E1,E2) {
  if (in_cache(E1,?E3,?CCcached) and E3 precedes E2) {
    compute CC(back(CCcached), E3);
    forall e (back(CCcached) precedes e precedes E2) {
      verify e is an acceptable event;
    }
    if (no unacceptable event is found) {
      update cache as appropriate;
      return convex closure;
    }
  if (exists a non-acceptable event OR
      no matching cached closure) {
    apply Xie and Taylor;
  }
}
```

We now describe the algorithm in detail. First, we check the cache to see if there is a matching input event. In this matching, we will only check against the first of the two input events, $E1$, in the closure computation. This is because that first event is stable, while the second is varied in the backtracking search process. On finding a match in the cache, we verify that the corresponding input event $E3$ that is cached precedes the second input event, $E2$ to this convex-closure computation. If this condition is true, we compute an incremental closure between **back** of the cached closure and the second input event. This, however, is insufficient. Per the theorem, the locations sets must be identical. To satisfy this condition, we must check all events between the cached convex closure and the new input event, $E2$, to determine if any are receive or synchronous events with a partner outside of the location set of the cached convex closure. If any such event exists, and that event is a successor to the cached closure, then the event is unacceptable. Specifically, such an event means that the location set of the closure will exceed that of the location set of the cached closure. Note that only the events that are part of the incremental closure need to be checked, and not those of the cached closure. This is typically a small number of events.

If no unacceptable event is found, then the cache should be updated as appropriate, and the closure returned. We have found that a suitable cache replacement policy is to replace the closure that was just used. Specifically, this means that a small cache may be used, while still rendering most closure operations into incremental operations. The closure returned will be the **front** set of the cached closure and the **back** set of the incremental closure computation.

If no matching cache element is found, or an unacceptable event is found (that is, the location sets do not match), then we simply revert to the Xie and Taylor algorithm.

4 Analysis

We have implemented our algorithm as an eclipse plug-in, within the basic predicate-detection system implemented by Xie and Taylor. This allows us to evaluate our algorithm both experimentally and analytically.

From an analytical perspective, we can do no better than Xie and Taylor, since we degenerate to their algorithm whenever we do not have a suitable basis for an incremental-closure computation. Further, we can do worse than Xie and Taylor when we consider the worst-case scenario. In this case, we will compute an incremental closure over all but a finite number of events in the computation. We then verify this, to determine if their exist unacceptable events. In the worst case, the last event checked fails the acceptability requirement, and thus we must compute the desired closure using the Xie and Taylor algorithm. The acceptability check is thus executed $O(n)$, where n is the number of events in the computation. The cost of the acceptability check is $O(N)$, since all events in **front** must be verified for non-precedence against. In such an instance, our algorithm would be $O(nN + N^3)$, while Xie and Taylor remains at $O(N^3)$.

While analytically we are no better, and in the worst case, worse than Xie and Taylor, in practice, our algorithm is substantially superior. We have evaluated our algorithm over more than 50 different parallel and distributed computations covering a variety of different environments, including Java [10], PVM [9], DCE [6], and μC++ [3] (a language used for teaching concurrency). The PVM programs tended to be SPMD style parallel computations. As such, they frequently exhibited close neighbour communication and scatter-gather patterns. The Java programs were web-like applications, including various web-server executions. The DCE programs were sample business-application code. The μC++ were sample concurrency problems used in an educational environemnt, such as Dining Philosophers.

For each experiment we used a variety of predicates, appropriate to the computation at hand. In the experiments we computed the number of convex closures, the number of unique **front** sets, the number of successful incremental closures, and the total execution time using our algorithm and the Xie and Taylor algorithm. The cache size employed was one, while the hardware used was a Pentium III 2 GHz, with 512 MB of memory, together with eclipse version 2.1.3.

For long-running queries, defined as those whose runtime exceeded 30 minutes when using the Xie and Taylor algorithm, we have found that our algorithm reduced the runtime by more than 90%. In one instance the runtime was reduced from over four hours to less than one minute. The cause of the substantial improvement is easily comprehended when we observe that, for such queries, the cache-hit rate always exceeded 90%. Further, we observed that the number of closures per unique **fronts** averaged 15. This means that, for a given **front** set, 15 closures were computed. In the Xie and Taylor algorithm, each such closure would be recomputed from scratch. In our approach, even with a cache size of one, we effectively only incur the cost of computing the largest such closure.

While space limitations prevent the publication of the code used, it is available on request from the first author. Further details of the algorithm, it's analysis, and raw result data is available in [2] and/or from the first author.

5 Related Work

Before concluding, we first briefly review related approaches. Existing work can be broken down into two main categories, corresponding to the two main types of pattern sought, and a third, smaller, but more recent, strand. There exists a significant body of work on seeking predicates in consistent global states (*e.g.*, [4, 22]), as we have alluded to in Sect.2. While such work is clearly critical in debugging, monitoring, and controlling parallel and distributed systems, it is fundamentally different from that of seeking patterns within the partial order itself.

Pattern seeking within the partial order has historically focused on a non-compound-event-based approach. Such work includes the offline algorithm of Jaekel [14] and its online version by Fox [7]. Neither method uses the compound-event-based approach of Xie and Taylor. A variant of the pattern-seeking approach to predicate detection is Han's technique for comparing two execution histories [11]. It is unclear if our work would be of relevance to her problem. The most recent work in this area is that of Xie and Taylor, and has already been described.

A third strand of work, which is quite recent, is typified by the IBM Log and Trace Analyzer [12]. This work takes the approach of using what event data is available, rather than adding monitoring code to an application. This approach is based on the observation that most enterprise applications already possess substantial log data which represent events of significance. Further, such applications are unlikely to be instrumented according to the desires of a third-party autonomic controller. The basic approach is that the logs are gleaned for event data, which the analyzer then attempts to correlate. The value of this approach is that it requires no change to existing systems. The success of the approach is dependent on the degree to which the existing sources possess sufficient information to provide correct correlation.

6 Conclusions

In this paper we have shown how to efficiently perform hierarchical predicate detection based on compound events. Our algorithm performs incremental closure computations, effectively reusing work already done. We have both proven our algorithm correct, and have demonstrated its efficacy via experiment. While our approach applies only to structural predicate detection, we expect to study its applicability to the problem of seeking patterns in consistent global states in the near future.

References

1. A. A. Basten. Hierarchical event-based behavioural abstraction in interactive distributed debugging: A theoretical approach. Master's thesis, Eindhoven University of Technology, Eindhoven, 1993.
2. Dwight S. Bedassé. An efficient computation of convex closure on abstract events. Master's thesis, University of Waterloo, Waterloo, Ontario, 2005. Available at: http://etheses.-uwaterloo.ca/display.cfm?ethesis_id=498.

3. P. A. Buhr, Glen Ditchfield, R. A. Stroobosscher, B. M. Younger, and C. R. Zarnke. μC++: Concurrency in the Object-Oriented Language C++. *Software — Practice and Experience*, 22(2):137–172, February 1992.

4. Craig M. Chase and Vijay K. Garg. Detection of global predicates: Techniques and their limitations. *Distributed Computing*, 11:191–201, 1998.

5. Eclipse Foundation. The eclipse platform. Online documentation available at: http://www.-eclipse.org/.

6. Open Software Foundation. *Introduction to OSF/DCE*. Prentice-Hall, Englewood Cliffs, New Jersey, 1993.

7. Mark Fox. Event-predicate detection in the monitoring of distributed applications. Master's thesis, University of Waterloo, Waterloo, Ontario, 1998.

8. Michael R. Garey and David S. Johnson. *Computers and Intractability: A Guide to the Theory of NP-Completeness*. W.H. Freeman and Company, San Francisco, 1979.

9. Al Geist, Adam Begulin, Jack Dongarra, Weicheng Jiang, Robert Manchek, and Vaidy Sunderam. *PVM: Parallel Virtual Machine*. MIT Press, Cambridge, Massachusetts, 1994.

10. James Gosling, Bill Joy, and Guy Steele. *The Java Language Specification*. Addison-Wesley, 1996. Available at http://java.sun.com/docs/books/jls/.

11. Jessica Zhi Han. Automatic comparison of execution histories in the debugging of distributed applications. Master's thesis, University of Waterloo, Waterloo, Ontario, 1998.

12. IBM Corporation. Log and trace analyzer for autonomic computing. Online documentation available at: http://www.alphaworks.ibm.com/tech/logandtrace.

13. IBM Corporation. Object level trace. Online documentation available at: http://www-106.-ibm.com/developerworks/websphere/WASInfoCenter/infocenter/olt_content/olt/index.htm.

14. Christian E. Jaekl. Event-predicate detection in the debugging of distributed applications. Master's thesis, University of Waterloo, Waterloo, Ontario, 1997.

15. Jeffrey O. Kephart and David M. Chess. The vision of autonomic computing. *IEEE Computer*, 36(1):41– 50, 2003.

16. Deiter Kranzlmüller, Siegfried Grabner, R. Schall, and Jens Volkert. ATEMPT — A Tool for Event ManiPulaTion. Technical report, Institute for Computer Science, Johannes Kepler University Linz, May 1995.

17. Dieter Kranzlmüller. *Event Graph Analysis for Debugging Massively Parallel Programs*. PhD thesis, GUP Linz, Linz, Austria, 2000.

18. Thomas Kunz. *Abstract Behaviour of Distributed Executions with Applications to Visualization*. PhD thesis, Technische Hochschule Darmstadt, Darmstadt, Germany, 1994.

19. Thomas Kunz. Automatic support for understanding complex behaviour. In *Proceedings of the International Workshop on Network and Systems Management*, pages 125–132, August 1995.

20. Thomas Kunz, James P. Black, David J. Taylor, and Twan Basten. POET: Target-system independent visualisations of complex distributed-application executions. *The Computer Journal*, 40(8):499–512, 1997.

21. Leslie Lamport. Time, clocks and the ordering of events in distributed systems. *Communications of the ACM*, 21(7):558–565, 1978.

22. Alper Sen and Vijay K. Garg. On checking whether a predicate definitely holds. In *3rd International Workshop on Formal Approaches to Testing of Software (FATES 2003)*, 2003.

23. Alper Sen and Vijay K. Garg. Partial order trace analyzer (POTA) for distributed programs. In *Proc. Workshop on Runtime Verification*, 2003.

24. Ping Xie and David Taylor. Specifying and locating hierarchical patterns in event data. In *Proceedings of the 2004 CAS Conference*, pages 66–80, October 2004.

A Generic Language for Dynamic Adaptation

Assia Hachichi[1], Gaël Thomas[1], Cyril Martin[1],
Bertil Folliot[1], and Simon Patarin[2]

[1] LIP 6 - Université de Paris6
{Assia.Hachichi,Gael.Thomas,Cyril.Martin,Bertil.Folliot}@lip6.fr
[2] DSI - Università di Bologna
patarin@cs.unibo.it

Abstract. Today, component oriented middlewares are used to design,
develop and deploy distributed applications easily. They ensure the heterogeneity, interoperability, and reuse of software modules.
Several standards address this issue: CCM (CORBA Component Model),
EJB (Enterprise Java Beans) and .Net. However they offer a limited and
fixed number of system services, and their deployment and configuration
mechanisms cannot be used by any language nor API dynamically.
As a solution, we present a generic high-level language to adapt system
services dynamically in existing middlewares. This solution is based on a
highly adaptable platform which enforces adaptive behaviours, and offers
a means to specify and adapt system services dynamically. A first prototype[1] was achieved for the OpenCCM platform[2], and good performances
were obtained.

1 Introduction

Computing systems are increasingly complex and difficult to maintain. Moreover, the various elements, that constitute an environment, are often physically
distributed on heterogeneous nodes. Middlewares were introduced to solve these
difficulties, by proposing common generic system mechanisms to the distributed
applications.

The last generation of component oriented middlewares introduces the component and container concepts. A container manages system services, such as
persistence, transaction, security or naming, in a way that is transparent for
business code, which is encapsulated in components.

The adaptation of system services is often done statically, by stopping the
middleware execution, which induces a high cost for critical applications. For
this reason the dynamic adaptation is more efficient. Some platforms provide
mechanisms that can be used to adapt services dynamically. Nevertheless, these
mechanisms are specific to the targeted platforms: they are not reusable on

---

[1] This work was partially financed by the European project IST-COACH (2001-
34445).
[2] OpenCCM is an implementation of the CORBA Component Model specification [1].

J.C. Cunha and P.D. Medeiros (Eds.): Euro-Par 2005, LNCS 3648, pp. 40–49, 2005.

other middleware platforms easily. Moreover, there is no standard nor model that unifies adaptation mechanisms independently of platforms.

We propose to use a domain-specific language (DSL) [2] technique: a programming language providing high-level abstractions related to a given domain. The expertise captured in the language allows behaviours to be expressed in an intuitive and high-level manner, permits verification, and allows generation of efficient code that is automatically integrated in the target platform.

In this context, our work proposes the Container Virtual Machine (CVM) approach, which defines a generic adaptation language. The CVM includes a DSL for writing adaptation behaviours. Each adaptation need is described on CVM language then is translated to different targeted platform language on the fly, in an automatic way. This approach allows separation between the adaptation logic and its implementation, by providing a high-level language.

A first translator has been implemented on the OpenCCM platform [3], an open source implementation of the CORBA Component Model (CCM) specification defined by the Object Management Group (OMG). This prototype allows the adding and reconfiguring of new system services, and offers administrators the possibility to specify and deploy system properties dynamically even if they were not taken into consideration initially.

In the following, Section 2 presents other proposals allowing to make middlewares flexible. Then, our proposal for offering high level language for dynamic adaptation is detailed in the Section 3. Section 4 describes the CVM and examples implementation, and Section 5 presents the conclusion of our work.

2   Related Work

Several component-based models exist such as: Microsoft .Net, Sun Microsystems Enterprise Java Beans, or OMG CORBA Component Model. These models are used to design and to deploy distributed applications. However, they do not allow easy integration and adaptation of system services[3] dynamically. Moreover, no standard envisages describing the integration and adaptation of services after initial deployment of the application.

The first middlewares were not designed to be flexible. However, adaptation techniques have been proposed, such as interceptors, and Portable Object Adaptor (POA) in CORBA ([4]). The interceptors [3] allow inserting code before the reception and after sending a request. The POA allows programmers to construct object implementations that are portable between different ORB products.

Several projects aim at making CORBA more flexible. DynamicTAO [5] (based on TAO), a reflexive CORBA environment, reifies the internal elements of the ORB in the form of components called configuration components. Dynamic-TAO keeps a compatibility with CORBA applications, by offering a high degree of adaptability. One of the difficulties that this project raises is the problem of coherence when a policy is replaced by another.

---

[3] System services such as: transaction and replication service.

AspectIX [6] adopts a fragmented object model based middleware [7]. The fragments can mask the replication of a distributed object, impose real-time constraints on the communication channel, put the object information in memory cache, etc. These non-functional (system) aspects can be configured via a generic interface of the object. Each global object can be configured by a profile that specifies the aspects that the fragments must respect. Four profiles are planned, in particular a CORBA profile that allows for these AspectIX objects to interact with CORBA objects. This approach allows a clear separation between the application and the middleware over which it is deployed.

JAC (Java Component Aspect) allows to weave an aspect dynamically: the relation between the wrappers and the advice codes can be redefined on the fly. However, the number of pointcuts is not extensible dynamically: if the class is already charged in the virtual machine, there is no means to add a new pointcut.

An architecture of open containers is proposed in [8]. This architecture allows dynamic adaptation and extension of the system functions, and it allows exposing some number of container properties, using control , interception and coordination mechanisms. OpenORB [9] is a flexible architecture of component oriented middleware. OpenORB is based on the reflexion. Each Object of the system is associated to a meta-space which offers structural representation. The ORB is configured or reconfigured by using the Meta-Object protocol. Java-POD [10], is a component model which allows attaching system properties to the components. This attachment is achieved by means of open and extensible containers. Comet [11] is an events based middleware. It can be adapted by inserting pre/post hooks into the components. A language is associated to dynamic reconfiguration of the Comet middleware. However this language is not extensible dynamically, and is not generic since it is applicable only for Comet.

These various projects increase the middleware adaptation possibilities by re-coding it. Our work takes different direction, we propose a generic high-level language to adapt the system services dynamically in existing middlewares. Each adaptation behaviour is described on this high level language then is translated on all targeted platform language, in an automatic way. This abstract description allows separation between the adaptation logic and its implementation.

3   Container Virtual Machine Approach

Instead of providing adaptation behaviours that depend on the middleware platform, the Container Virtual Machine approach defines a generic language, which gives a high-level abstraction of system services adaptation behaviours, that is independent on the middleware platform. The abstraction behaviours are translated on the targeted platform, in automatic and dynamic way.

This approach allows (i) the unification of adaptation behaviours, independently on the targeted platform, (ii) the automatic generation of CVM scripts can be achieved by a design tool, and (iii) the generation of platform independent adaptation models (PIM - Platform Independent Model) and them translation on the trageted platform (PSM - Platform Specific Model).

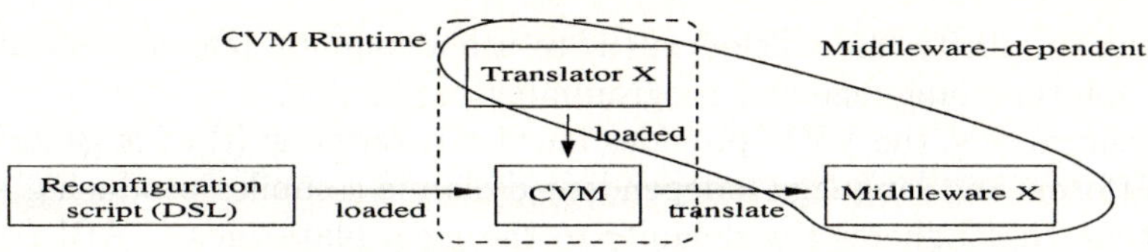

Fig. 1. CVM Concept (Container Virtual Machine)

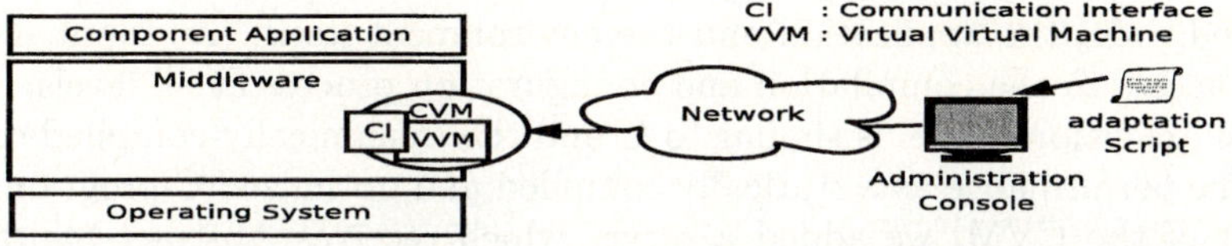

Fig. 2. CVM Processing (Container Virtual Machine)

3.1 CVM Design

The CVM approach aims to remain neutral with respect to the platform, and to separate the adaptation language from its execution. Figure 1 presents the CVM concept; its input is a configuration script, called a translator, which is dependent on middleware. The translator enables to translate an adaptation script written in CVM language for a specific middleware.

3.2 CVM Implementation

The main idea is to add an entry point to different platform middleware, at the initial deployment and in a transparent way. This entry point enables the interaction between the CVM platform and its targeted middleware platform, and is called a *Communication Interface (CI)*. It enables the translation of abstractions to the targeted middleware language. The CVM is mainly based on a *highly adaptive platform* to describe and to enforce the adaptation behaviours. This platform is generic with respect to middleware platforms, and interacts with each middleware platform through its associated *Communication Interface* (fig 2). The CVM allows to define new adaptation operations on the fly.

Virtual Virtual Machine: The CVM design requires a *highly adaptable language* to provide separation between the adaptation logic and its implementation, and to extend the access operations in order to enforce what can be adapted. The selected highly adaptable platform is the *Virtual Virtual Machine (VVM)* [12], which is a dynamic code generator that provides both a complete, reflexive language, and an execution environment. The VVM allows to modify the implemented mechanisms, to reconfigure the environment, and to extend or modify the associated language.

The main objectives of this environment are: (i) to maximize the amount of reflective accesses and intercessions, at the lowest possible software level, while preserving simplicity and efficiency; (ii) to use a common language substrate to support multiple language and programming paradigms.

To achieve this, the VVM provides four basic services: (i) code generation: a fast, platform- and language-independent dynamic compiler producing efficient native code that adheres (by default) to the local platform's C ABI (Application Binary Interface); (ii) meta-data that are kept between compilations, thus allowing higher-level software to reason about its implementation or that of the environment, and modify them dynamically; (iii) introspection on dynamically-compiled code, the application and the environment itself; (iv) input methods, giving access to the compilation and configuration process at all levels.

The execution model is similar to C and the dynamically-compiled code has the same performance as a statically compiled and optimized C program. In the context of the CVM, we added a server which receives Abstract Syntax Tree from another VVM, compiles and links them, and executes the generated code.

The use of the VVM allows the separation of the adaptation logic from its implementation. This language must be both *extensible* to new adaptive needs dynamically, and *generic* with respect to the targeted middleware in order to ensure its reusability. It allows both to reduce the possibilities of reconfiguration by limiting the language symbols, and/or to extend the language by providing introspection of the environment and the creation of new symbols.

Remote Administration of the CVM: In order to ensure the adaptation of several network nodes from a remote administration console, we built a remote adaptation environment in the VVM platform; this environment must be loaded on all VVMs. It parses/lexes scripts that reconfigure the target environment; these scripts are transformed into abstract syntax trees. These trees are sent to the VVM, which is able to receive them on a communication channel (example: a TCP socket). These trees are then compiled and executed on the second entity.

In the adaptation context, a client opens a communication channel, and parses/lexes VVM scripts that reconfigure the remote machine (server), then sends the corresponding trees to the server. When a server receives the abstract syntax trees, it compiles and executes them.

4   Qualitative Evaluation

The CVM is evaluated on a CORBA Component Model implementation written in Java: the OpenCCM [3]. This section details the prototype implementation.

4.1   OpenCCM Translator Implementation

In the case of the OpenCCM platform, the translator is achieved by using a Java native method, which launches the VVM. The communication Interface (CI) between the VVM and the standard JVM is provided by JNI (Java Native Interface [13]). JNI is an interface between the native functions and the Java virtual machine.

The VVM is executed by a Java thread in competition with those of the application. The language of the VVM is then dynamically extended: the scripts written for the VVM can then interact with the JVM directly, and the VVM is able to handle the methods and the symbols of the Java application (Fig 2).

A reconfiguration comprises two important steps:

1. The first phase consists in building methods that allow dynamic adaptation into the VVM; for example: methods that integrate or remove components.
2. The second consists in writing a CVM script that contains the adaptation needs. This script is loaded remotely by the administration console and is executed by using a CI. Scripts can either extend the reconfiguration language, or use the keywords already built in to modify the OpenCCM application.

To illustrate the use of the CVM, two examples of reconfigurations are presented in the rest of this section.

4.2   Integration of Service

We classify the system services in two classes: not-intrusive services, which do not modify the treated data, and the intrusive services, which modify the data, and requires synchronization

In this paper we present two examples for integrating services, one is a monitoring service that is not-intrusive and the other one is an encryption service that is intrusive. These integrations are based on the *Portable Interceptors* and on *System Oriented Component* respectively.

Flexible Monitoring Service: The first example illustrates the dynamic integration of a flexible monitoring service based on interceptors.

This service was designed to collect statistics on the way components interact with each other, and to make this information available to a "reconfiguration service" that will use it to adapt the platform.

The monitoring service is composed of two concurrent processes. The first one collects all available information concerning the called requests, and records them in a log file. The second process scans the log generated by the first process periodically, and calculates the statistics of the call number and the average response time for given operations. The integration of this service is based on CORBA portable interceptors.

The CORBA specification [14] defines the portable interceptor interface as a way to insert hooks directly inside the ORB. These hooks are activated for every operation performed by the broker: mainly method invocations and result returning. Hooks may be located either on the client or on the server side. We conclude that the integration of the monitoring service on the level of interceptor hooks, allows to invoke the monitoring service code at every request by extracting several metrics, such as the number of times a specific method is invoked, and sums all the invocations of methods belonging to the same component. However, no standard language or interface enable to use Portable Interceptors dynamically to achieve an integration of System Services. For this reason the CVM is used to integrate the monitoring service dynamically. This integration comprises two phases: (i) to specify, in the VVM platform, the new adaptation operations that allow adding code in the OpenCCM interceptor hooks dynamically. (ii) to write and to execute a VVM script that integrates service code in hooks through the Communication Interface.

Encryption Service: Considering an application, that contains two components "A" and "B", included in containers "CA" and "CB" respectively (see figure 4). Component "A" sends messages to "B" in a regularly way. During the execution, the administrator decides to send encrypted messages to "B". In order to achieve this, we use another mechanism to intercept requests: *System Oriented Components (SOC)*. This mechanism is used because it is generic; it can be applied for any middleware, and shows that it is possible to define other integration mechanisms on the fly.

The System Oriented Component mechanism consists in adding CCM components which containing the service code to be added, and in establishing the necessary connections with the components to which this service will apply.

In our example, the integration of an encryption service consists in integrating an encryption SOC component in container "CA", and a decryption SOC component in container "CB". Basically the integration consists in adding the necessary operations to the VVM, such as the operations which enable SOC component creation and handling the connections between any components. The second step is to write and execute the VVM script which allows us to: (i) add the encryption SOC in "CA" and another one in container "CB", (ii) disconnect "A" and "B"; (iii) establish the connections between "A", "B" and their respective SOC (see figure 4 and 3), by ensuring the synchronization. Problems that can occur are: encoded messages may be received before adding a decryption SOC, or non-encrypted messages will be sent after adding the decryption SOC.

---

1. (On_container CA
   – (Deactivate_Component A))
2. (On_container CB
   – (Deactivate_Component B))
3. (On_container CA
   – (Disconnect_components A B)
   – (Insert_SOC SocA)
   – (Connect_components A SocA))
4. (On_container CB (Insert_SOC SocB)
   – (Connect_components SocA SocB)
   – (Connect_components SocB B))
5. (On_container CA
   – (Activate_Component A))
6. (On_container CB
   – (Activate_Component B))

1. <programme>→<reconfiguration>*
2. <reconfiguration>→<SOC>*| <PI>*
3. <SOC>→ (On_container<atomeCont> <action>*)
4. <action>→ (<subaction><atome>)
5. <subaction>→Deactivate_Component| Disconnect_components <atome> | Insert_SOC | Connect_components<atome> |Activate_Component
6. <atomeCont>→ *Containerreference.*
7. <atome>→ *Compoenentreference*
8.

Fig. 3. (A) An example of the reconfiguration script that integrates encryption SOC (B) A part of the CVM grammaire.

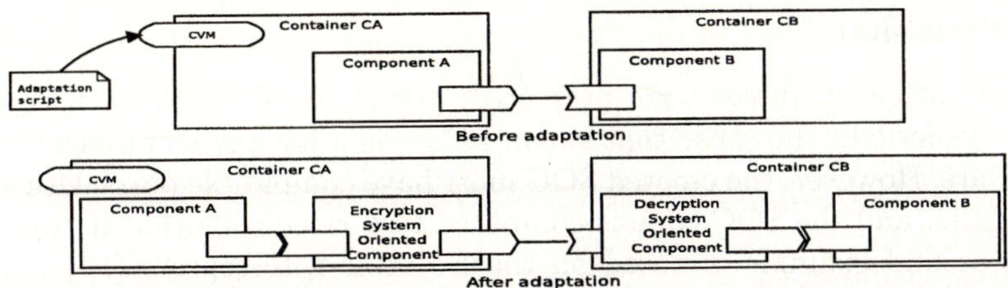

Fig. 4. Integration of the encryption service.

To avoid these problems, the synchronization must be ensured dynamically. A pseudo-algorithm that is proposed consists in deactivating "A" before "B", breaking the connections between "A" and "B", and then adding the encryption and decryption SOC in container "CA" and "CB" respectively, finally establishing the necessary connections, and activating "B" before "A".

In the case where the targeted middleware does not provide the possibility to activate or deactivate a component, we propose to use a queue. Messages from "A" will be redirected towards the queue, during integration of encryption and decryption service. Synchronization is not yet implemented in our encryption prototype.

4.3   Adaptation of the Encryption Service

To illustrate the adaptation of existing services, we adapt the encryption service of the previous example, during the execution.

Component behaviour adaptation can be achieved by replacing a component by a new one. However, it is simpler and less expensive to adapt a component by replacing some of its methods.

In the case of the encryption SOC adaptation, it is enough to adapt the Java method that contains the encryption service. The Java standard allows dynamic loading of a class and overload of the serialization methods. By coupling the Sun Java platform and CVM, we can adapt a Java method. Let us take the example of method "metA" from class "A", the adaptation of this class is done by charging a new class "A1" which inherits from "A", and which implements the new code of "metA", then redirecting all calls towards the new loaded method (Fig 5).

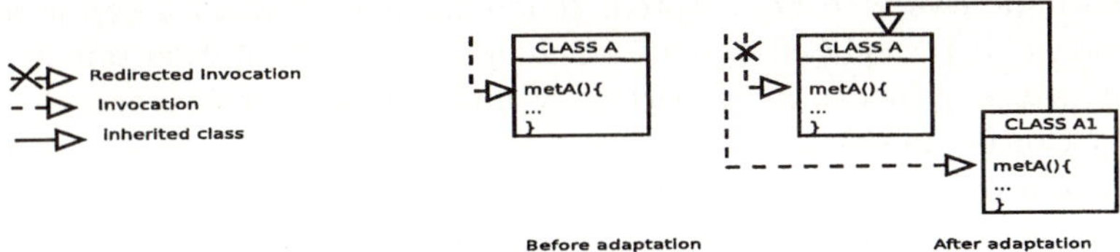

Fig. 5. Method adaptation.

4.4 Discussion

Two adaptation examples were presented; one is based on the SOC approach which is generic in the sense that it can be applied for any component-oriented middleware. However, the created SOC must have compatible ports with existing components, and the SOC code is compared to service code that it contains.

The second example is based on the Portable Interceptors (PI) approach, and is not generic, since the PI concept does not exist in all component-oriented middlewares, with EJB. But this code size is smaller than the SOC code size.

A Set of ten performance evaluation measures of the dynamic integration were performed, on a Pentium III 664MHz under Linux. These measures represent the duration between the old configuration and the new configuration after the end of service integration. (i) *The monitoring service integration average duration*, which is based on the portable interceptors, is 8.539 seconds. (ii) *The encryption SOC integration average duration* is 2.054 seconds.

Another set of ten SOC adaptation duration measures were done. This duration represents the time between the initial configuration and the end of the encryption code replacement. The average of the ten measures is $94 * 10^{-3}$ seconds.

We note that the integration based on Portable Interceptors is slower than service based on System Oriented Components. This can be explained by the cost resulting from the flowing of all requests through the interceptor layer. In [15], which studies three different ORB implementations, it is shown that the activation of portable interceptors increases latency by a factor varying between 2% and 10% and decreases request throughput by a factor ranging from 1.5% to 16%, This cost is then limited.

We note that the adaptation duration average is slower than integration service duration. However, these costs remain limited.

5   Conclusion

This paper presents the Container Virtual Machine, a platform which allows dynamic adaptation of system services and provides a generic language specific to adaptation domain (DSL) . This language offers a high-level abstraction of adaptation behaviour and is itself extensible. Adaptation CVM scripts can be translated for different target platforms during the execution automatically.

The CVM approach provides a separation between the adaptation logic and its implementation. CVM language is generic in the sense that it is independent from the middleware to be adapted. It language enables to describe any new adaptation and the related operations. It allows an adaptation remote administration which provides interoperability and synchronisation between several nodes; it can be aperated on different middleware platforms, such as EJB and CCM. The provided high-level abstractions are translated automatically for the targeted platform.

As future works, we aim to reuse the CVM language on the different platforms, such as EJB, then to refine the grammar of our DSL. To provide means that ensure the coherence, atomicity, and verification of the dynamic adaptations and of there deployment. To achieve the automatic generation of CVM scripts design tool such as Rationalrose, then to offer mechanisms that execute models automatically.

References

1. Openccm user's guide (2004) http://openccm.objectweb.org/doc/0.8.1/user_guide.html.
2. Lawall, J., Muller, G., L.P.Barreto: Caputing os expertise in an event type system: the bossa experience. In: Tenth ACM SIGOPS European Workshop (EW 2002), France, Springer-Verlag (2002) 154–61
3. OMG: Interceptors Published Draft with Corba 2.4+ Core Chapters. (2001) Document Number ptc/2001-03-04.
4. Daniel, J.: Au coeur de Corba. (2001)
5. Kon, F., Román, M., Liu, P., Mao, J., Yamane, T., Magalhães, L.C., Campbell, R.H.: Monitoring, Security, and Dynamic Configuration with the dynamicTAO Reflective ORB. In: Proceedings of the IFIP/ACM International Conference on Distributed Systems Platforms and Open Distributed Processing (Middleware'2000). Number 1795 in LNCS, New York, Springer-Verlag (2000) 121–143
6. Hauck, F.J., Becker, U., Geier, M., Meier, E., Rastofer, U., Steckermeier, M.: AspectIX: An aspect-oriented and CORBA-compliant ORB architecture. Technical Report TR-I4-98-08, Univ. of Erlangen-Nuernberg, IMMD IV (1998)
7. Makpangou, M., Gourhant, Y., Narzul, J.P.L., Shapiro, M. In: Fragmented objects for distributed abstractions. IEEE Computer Society Press (1994) 170–186
8. Vadet, M., Merle, P.: Les conteneurs ouverts dans les plates-formes à composants. Journées composants: flexibilité du système au langage (2001)
9. Blair, G.S., Costa, F.M., Coulson, G., Duran, H.A., Parlavantzas, N., Delpiano, F., Dumant, B., Horn, F., Stefani, J.B.: The Design of a Resource-Aware Reflective Middleware Architecture. In: Proceedings of the Second International Conference on Meta-Level Architectures and Reflection, France, Springer-Verlag (1999) 115–134
10. Bruneton, E., Riveill, M.: Javapod: une plate-forme à composants adaptables et extensibles. Rapport technique 3850, Inria Rhone-Alpes (2000)
11. Peschanski, F., Briot, J.P., Yonezawa, A.: Fine-grained dynamic adaptation of distributed components. Middleware 2003 (2003) 132–142
12. Ogel, F., Thomas, G., Piumarta, I., Galland, A., Folliot, B., Baillarguet, C. In: Towards Active Applications: the Virtual Virtual Machine Approach. A92 Publishing House, POLIROM Press (2003) 28–47
13. Liang, S.: The JavaTM Native Interface: Programmer's Guide and Specification. Addison Wesley Longman (1999)
14. OMG: Corba / iiop specification 3.0. formal/024206 (2002)
15. Marchetti, C., Verde, L., Baldoni, R.: Corba request portable interceptors: a performance analysis. In: Proceedings of the 3rd International Symposium on Distributed Objects and Applications, Rome, Italy (2001)

Soft Computing Approach to Performance Analysis of Parallel and Distributed Programs[*]

Hong-Linh Truong and Thomas Fahringer

Institute for Computer Science, University of Innsbruck
Technikerstrasse 21A, A-6020 Innsbruck, Austria
{truong,tf}@dps.uibk.ac.at

Abstract. This paper describes a novel approach to performance analysis for parallel and distributed systems that is based on soft computing. We introduce the concept of performance score representing the performance of code regions that is based on fuzzy logic. We propose techniques for fuzzy-based performance classification. A novel high-level query language is designed to support the search for performance problems by using linguistic expressions. We describe a fuzzy-based bottleneck search, a performance similarity measure for code regions and experiment factors, and performance similarity analysis. Our approach focuses on the support of making soft decisions on evaluation, classification, search and analysis of the performance of parallel and distributed programs.

1 Introduction

Recently, performance analysis community has focused on developing performance tools for parallel and distributed programs that are capable of supporting semi-automatic performance analysis, dealing with large performance data sets, and analyzing multiple experiments. However the development of automatic and intelligent performance analysis is still at an early stage. Current techniques in existing performance analysis tools have mainly been used to process the performance data that are in the form of precise numerical data. Firstly, these techniques always apply exact analysis methods that result in hard conclusions about performance characteristics of applications. Secondly, existing performance tools interact with the user through complex numerical values and visualizations which are not easily understood by the user. Thirdly, in the real world we largely rely on domain expertise and user-provided inputs as parameters to control the performance analysis and tuning. Such expertise and inputs may be inexact and uncertain. However, existing performance tools do not support the specification and the control of approximate and inexact parameters in data analysis techniques, in other words, these tools do not provide a mechanism to make soft decisions.

The recent emerging *soft computing* [1], however, presents another way for evaluating and analyzing data that is based on the concept of *soft, inexact, uncertainty*. Soft computing aims to support imprecision, uncertainty and approximate reasoning [1].

---

* The work described in this paper is supported in part by the Austrian Science Fund as part of the Aurora Project under contract SFBF1104 and by the European Union through the IST-2002-511385 project K-WfGrid.

J.C. Cunha and P.D. Medeiros (Eds.): Euro-Par 2005, LNCS 3648, pp. 50–60, 2005.

In this paper we present a new approach to the performance analysis that we call the *soft performance analysis*. In this approach, well known soft computing techniques such as fuzzy logic (FL), machine learning (ML) concept, and the combination of FL and ML are studied and developed for performance analysis of parallel and distributed programs. We introduce the concepts of performance score and performance similarity measure. Employing these concepts, we develop several soft techniques and methods for performance analysis such as fuzzy-based performance classification, performance search, similarity analysis, etc.

The rest of this paper is organized as follows. Section 2 outlines the so-called soft performance analysis. Section 3 presents a few preliminaries. We introduce the concept of performance score and performance similarity measure in Section 4 and Section 5, respectively. We describe soft techniques for performance analysis including fuzzy-based performance classification, query language, fuzzy-based bottleneck search and performance similarity analysis in Section 6. Section 7 discusses the related work. Section 8 gives conclusions and the future work.

2   Soft Performance Analysis

Existing performance analysis tools are based on *hard* computing model that is based on binary logic and crisp systems. For example, to classify the performance performance analysis tools normally use a *characteristic function*. That is, given a performance metric and a set of *performance characteristic term*, e.g., *poor, medium* and *good*, each term represents a performance class and is associated with a data set, the performance of a code region is classified according to characteristic terms by using a characteristic function. However, such classification is in binary form, e.g., a performance of the code region is either *good* or not, because the hard computing model does not accept imprecision and uncertainty. Since approximate search, classification and reasoning are not possible, the cycle of finding performance patterns in a large set of performance data has been lengthened because, in the real world, the boundaries between performance classes, performance search constraints, etc., are not clearly seen, thus, exact methods may not yield the expected results. Moreover, current tools focus on supporting the performance analysis through statistical graphics which are not well suited for processing large performance datasets. In practice, both performance data and expertise used in performance analysis domain can be uncertain. For example, in the case of performance classification, performance of code regions is classified into *good*, but depending on the degree of *good* the performance of code regions can be considered as *little good, fairly good* or *very good*. When we are not sure about performance data and expertise, we may accept some degrees of uncertainty and approximate in our analysis techniques.

To address the above-mentioned issues, we investigate performance analysis techniques that are based on soft computing. The soft performance analysis we propose aims to develop techniques for performance tools that can (i) extract useful performance information from large, dynamic and multi-relational performance measurement sources, (ii) support the specification and control of approximate and inexact parameters, commands and requests in existing performance analysis tools, and (iii) interact with the user through high level notions and concepts expressed in linguistic expressions.

We outline the approach as follows. Firstly, fuzzy logic (FL) can help representing and normalizing quantitative data. We can represent *performance score* of metric values by using fuzzy set (FS). By employing the concept of performance scores, we can develop several techniques that support soft, inexact and uncertainty in performance analysis. The application of FL theory also involves the concept of linguistic variables and the use of linguistic variables is particular useful for the end-user because humans employ mostly words in computing, as presented in the concept of computing with words [2]. Therefore, by using FL, performance tools can provide a way to perform the analysis and to interpret performance results with linguistic terms. Secondly, when processing large and diverse performance data, information about performance summaries, similarities and differences of data items in that data become more important as we cannot examine each data items in detail. Similarity measure techniques can be exploited to reveal the performance similarities and differences. ML techniques [3] can be utilized to discover patterns in very large performance datasets. For example, machine learning is combined with fuzzy computing to provide fuzzy clustering for performance data. Due to the space limit, this paper presents only a few points of our approach, focusing on FL and performance similarity techniques. More detail of soft performance analysis can be found in [4].

3   Preliminaries

3.1   Performance Experiment Data

A program contains a set of instrumented code regions. Performance data collected in each experiment of the program is organized into a *performance experiment data*. An experiment is associated with a set of processing units. A processing unit pu is a triple (n, p, t) where n, p and t are computational node, process identifier and thread identifier, respectively. A region summary rs is used to store performance metric records of executions of a code region cr in a processing unit pu. A performance metric record pm is represented as a tuple (m, v) where m is the metric name and v is the metric value. We denote $rs(m)$ as the value of performance metric m stored in region summary rs.

We use performance data obtained from experiments of three Fortran applications named 3DPIC (MPI program), LAPW0 (MPI program) and STOMMEL (mixed OpenMP and MPI program). All experiments are conducted on a cluster of 4CPU SMP nodes using MPICH library for Fast-Ethernet 100Mbps and Myrinet.

3.2   Representing Performance Characteristics Under Fuzzy Logic Theory

An FS is used to map metric values onto membership values in the range $[0, 1]$. An FS is expressed as a set of ordered pairs $FS = \{(v, \mu(v)) | v \in U\}$ where $\mu(v)$ is the membership function determining the degree of membership of v, and U is the universe of discourse of v. Let v be a metric value with the universal of discourse U. U is characterized by a given set of *performance characteristic terms* $T = \{t_1, t_2, \cdots, t_n\}$; performance characteristic terms are linguistic terms such as *poor, medium* and *high*. Each t_i is associated with a membership function $\mu_i(v)$ which determines the membership of v in t_i. v can be classified according to these terms. A *modifier* (e.g. *slightly*) is an operation that modifies a performance characteristic term (e.g. *bottleneck*). The modification results in a new fuzzy set represented by a new phrase (e.g. *slightly bottleneck*). In our experiments, we use the NRC-IIT FuzzyJ Toolkit [5] for fuzzy computing.

4   Performance Score

When evaluating and comparing performance of code regions most existing performance tools are normally based on quantitative measurement values and do not employ quantization or normalization techniques to evaluate multiple metrics. We present the concept of *performance score* which is used to evaluate the performance of a code region within a base, e.g. the parent code region or the whole program. The concept is based on (i) a set of selected performance metrics characterizing the performance of the code region, and (ii) a weight set representing the significance of performance metrics. Given a code region cr, let rs be the region summary of cr with a set of n performance metrics $\{m_1, m_2, \cdots, m_n\}$. Suppose the number of performance metrics measured is the same for every code regions. rs can be represented in n dimensional space. Let $v_i = rs(m_i)$ be the value of metric m_i in rs and let s_i be a score that represents the performance of rs with respect to metric m_i. We compute s_i as follows

$$s_i = \mu_i(v_i), \mu_i(v) : [0, V_{m_i}] \rightarrow [0, 1] \tag{1}$$

where $\mu_i(v)$ is the membership function determining the performance score, and V_{m_i} is the maximum observed value of m_i. V_{m_i} is dependent on the level of code region analysis. For example, if we analyze performance scores of rs with its parent rs_{parent} as the base, $V_{m_i} = rs_{parent}(m_i)$.

The value of s_i is in the range $[0, 1]$; 0 means the lowest score, 1 means the highest score. A higher performance score might be used to imply a higher performance or to indicate a lower significant impact. The exact semantics of the value of the performance score is defined by the specific implementation. As a result, performance scores can be used in various contexts such as to indicate (i) a significant impact level: the higher a performance score is, the higher impact the code region has, or (ii) a severity, the higher a performance score is, the more severe the core region is. There are several ways to select $\mu(v)$, depending on the specific analysis and approximate model used. The most simple way is to define the membership function μ as $\mu(v_i) = \frac{v_i}{V_{m_i}}$ which assumes that the score is based on linear model. We can choose trapezoid, S-function, Z-function, triangle, etc., and tool-defined function for $\mu(v)$.

Each rs is associated with a vector of performance scores $\vec{s}$. However, we may only select a subset of $\vec{s}$ as metrics to represent the performance of the code region. Like quantitative measurement values, we can compare two performance scores of two different metrics. However, because performance scores are normalized values, we can aggregate performance scores $\vec{s}$ of rs into a single score by using the overall weighted average (OWA) operator. Let $\{s_1, s_2, \cdots, s_n\}$ be performance scores of rs and $W = \{w_1, w_2, \cdots, w_n\}$ be the set of weights. w_i is a weight factor associated with metric m_i. The aggregate performance score for $\vec{s}$ may be computed as follows

$$OWA(\vec{s}) = \frac{\sum_{i=1}^{n} (|s_i w_i|)}{\sum_{i=1}^{n} w_i} \tag{2}$$

For the sake of simplicity, normally $w_i \in (0,1)$ and $\sum_{i=1}^{n} w_i = 1$. OWA score is particular useful for support of decision making in performance analysis and tuning because very often we have to decide which are the focused metrics of the code regions that should be tuned and optimized in order to achieve a better performance. Hence we use the notation (m_i, w_i) to denote m_i with its associated weight w_i.

We use performance score in ranking analysis, fuzzy C-means clustering, fuzzy rules, and similarity analysis. The former three analyses are covered in [4].

5 Performance Similarity Measure

Most existing performance tools employ numerous displays, e.g., process time-lines and histograms, to compare performance measurements and visualize that measurements. Those displays are crucial but the user has to observe the displays and perceive the similarity and the difference among these values. Moreover, it is difficult to compare multivariate data through visualization. We propose methods to compute the *performance similarity measure* which can be used as a metric to indicate the performance similarity among code regions and among experiment factors. Formally, let o_i and o_j be objects, a similarity measure is a function $sim(o_i, o_j) \rightarrow [0,1]$ that compares o_i with o_j where 0 denotes complete dissimilarity and 1 denotes complete similarity. Performance similarity measure can help uncovering similar/dissimilar performance patterns among code regions, e.g., for making decisions in dynamic performance tuning [6].

5.1 Similarity Measure for Code Regions

Let rs_i and rs_j be region summaries of cr. Let s_{il} and s_{jl} be performance score of rs_i and rs_j with respect to metric m_l, respectively. We use Equation 1 to compute s_{il} and s_{jl}. The performance similarity measure $sim_{ij}(rs_i, rs_j)$ is defined as follows

$$sim_{ij}(rs_i, rs_j) = 1 - d_{ij}, \quad d_{ij} = \sqrt{\sum_{l=1}^{n}(|s_{il} - s_{jl}|^2)} \qquad (3)$$

where d_{ij} is the distance measure between rs_i and rs_j; d_{ij} is computed based on Euclidean distance. Note that we can use other distance functions, e.g., Minkowski, Manhattan, Correlation and Chi-square, and can use weight factors associated with metrics.

To determine the performance similarity among executions of code regions across a set of experiments, we use Equation 3 to measure the performance similarity. Given a code region cr and a set of experiments $\{e_1, e_2, \cdots, e_n\}$. Let rs_i be region summary of cr in experiment e_i. We compute similarity measure $sim(rs_1, rs_i)$, $i : 2 \rightarrow n$ by using various membership functions. Given metric m_i, when determining performance score, the maximum observed value V_{m_i} is obtained from e_1 which is the base experiment.

5.2 Similarity Measure for Experiment Factors

Experiment factors which can be controllable, e.g. problem size, the number of CPUs and communication libraries, or uncontrollable such as CPU usage, have significant impact on the performance of the applications. Without considering the similarity between experiment factors, it is difficult to explain cases in which the performance of code regions is not similar because the experiment factors can be different. Therefore, initially we try to address this problem by measuring similarity between controllable factors.

Let $sim_f(e_i, e_j)$ be similarity measure for factor f between experiments e_i and e_j. Given a set of controllable factors $F = \{f_1, f_2, \cdots, f_n\}$, similarity measure is computed for each factor $f_i \in F$. There is no common way to compute sim_f as a controllable factor and its role depend on each experiment. The objective of our analysis is to find out the relationship between the performance similarity of the code regions, sim_o (e.g. $sim(rs_i, rs_j)$), and sim_{f_i}. Naturally we expect that the similarity measures of the controllable factors of two experiments and the similarity measures of the performance of these experiments behave in a similar fashion, e.g. if the controllable factors are very similar then the performance of experiments should be very similar.

6   Soft Techniques for Performance Analysis

6.1   Performance Classification

Performance classification classifies the performance of code regions according to performance characteristic terms. Formally, given a metric value v and a set of performance characteristic terms $T = \{t_1, t_2, \cdots, t_n\}$, v are classified according to that terms. In existing performance tools, the classification gives a binary result: v belongs to only one $t_i \in T$, with no degree of membership. Conversely, the fuzzy-based classification determines the degree to which v fits into t_i, for all $t_i \in T$.

To classify performance of code regions, we firstly define a set of performance metric terms for each performance metric m by partitioning the universal of discourse of metric m into segments and each segment is described by a performance metric term which is associated with a FS. Performance characteristic terms can be defined based on training data. After membership functions are determined, the membership degree of v is computed based on quantitative value v of m.

To demonstrate this analysis, we classify code regions of 3DPIC application executed on 4 processors according to performance characteristic terms $T = \{low, medium, high\}$ representing the L2 cache miss ratio. Three FSs Z-function, trapezoid and S-function are associated with *low, medium,* and *high* term, respectively, as

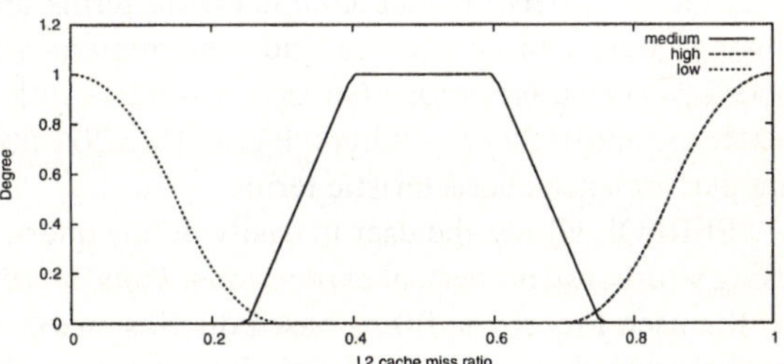

Fig. 1. Performance characteristic terms *low, medium, high* with their associated fuzzy sets.

shown in Figure 1. We then conduct the classification with a few selected code regions. Figure 2 presents the result with five selected code regions. As shown in Figure 2, the code region PARTICLE_LOAD has *high* L2 cache miss ratio. However, code region CAL_POWER is member of both *low* and *medium*.

New performance characteristic terms can also be built by combining existing ones with modifiers. For example, we can classify code regions according to *very low* L2

Performance Classification for L2_TCM/L2_TCA	
Code Region	Classes
Region 18:PARTICLE_LOAD[CR_A:191:328]	high(degree=0.937)/
Region 26:SR_E_FIELD[CR_A:653:723]	medium(degree=1)/
Region 46:IONIZE_MOVE[CR_A:1281:1788]	low(degree=0.732)/
Region 47:SET_FIELD_PAR_BACK[CR_A:1794:1928]	medium(degree=0.549)/
Region 48:CAL_POWER[CR_A:2244:2323]	low(degree=0.007)/medium(degree=0.28)/

Fig. 2. Membership in {*low, medium, high*} L2 cache miss ratio for selected code regions of 3DPIC.

cache miss ratio; the term *very* is a fuzzy modifier. The use of modifiers allows us to extend and enhance the description of performance characteristic terms.

6.2 Fuzzy Query for Performance Search

The fuzzy-based approach offers the possibility of search of performance data with words. Fuzzy-based search that uses linguistic expressions has been widely employed in database systems, information retrieval, etc., but not in existing performance tools.

We propose a fuzzy-based query language for search of performance data. Queries are constructed based on fuzzy modifiers, *AND* and *OR* operators, and performance characteristic terms.

$$\langle Statement\rangle ::= \langle Expr\rangle \mid \langle Statement\rangle \text{ OR } \langle Expr\rangle$$
$$\langle Expr\rangle \quad ::= \langle Term\rangle \mid \langle Expr\rangle \text{ AND } \langle Term\rangle$$
$$\langle Term\rangle \quad ::= (\text{METRIC is } \langle F_Expr\rangle)$$

Fig. 3. Top-level syntax of PERFQL.

Figure 3 presents the top-level syntax of our PERFQL (Performance Query Language based on fuzzy logic). METRIC is a metric name or a *metric expression*. A metric expression consists of operands and +, -, *, / arithmetic operators; operands are metric names. F_Expr describes the syntax of generic linguistic expressions (see [5] for the syntax). These expressions are constructed from performance characteristic terms and modifiers. For example, the following query can be used to find code regions which have high wallclock time and poor L2 cache miss ratio: "(wtime is HIGH_EXECUTION_TIME) AND ($\frac{L2_TCM}{L2_TCA}$ is POOR_CACHE_MISS)", where HIGH_EXECUTION_TIME and POOR_CACHE_MISS are performance characteristic terms.

PERFQL allows the user to easily define queries for search of performance data by using words, not numerical expressions. Thus, it is easy to be understood and interpreted by the user. Moreover, fuzzy-based queries enable approximate search thus interesting performance data which is slightly less or greater than the crisp condition can be easily obtained.

6.3 Fuzzy Approach to Bottleneck Search

There are several tools supporting bottleneck search, e.g., [7, 8]. These tools, however, support crisp-based searching as the search is conducted by checking crisp threshold. Given a performance metric, a threshold is pre-defined. During the search, the performance metric is evaluated against the threshold, and when the performance metric exceeds the threshold, a bottleneck is assumed to exist in the code region. There are two drawbacks of current crisp search strategy. Firstly, the search does not give the degree of severity of the bottleneck, e.g. *extremely* or *slightly* bottleneck. Secondly, there is no support to specify inexact bottleneck search statements such as *negligible bottleneck*. These statements are important as the threshold, by nature, is not an exact value.

We propose fuzzy-based bottleneck search that addresses the above-mentioned drawbacks. Figure 4 outlines the fuzzy-based bottleneck search. Given a threshold, we can use FSs to represent the severity of bottleneck and the negligible bottleneck range besides the FS representing the bottleneck threshold. For example, in Figure 4 we define a Pi-function FS used to check the negligible (close to) bottleneck points and S-function FS used to check the severity of bottleneck. When searching the bottleneck points, the value of metric used in bottleneck search is evaluated against these FSs. Not only we can locate bottleneck points as usual but also we can provide the severity of bottleneck, and are able to find negligible bottleneck points.

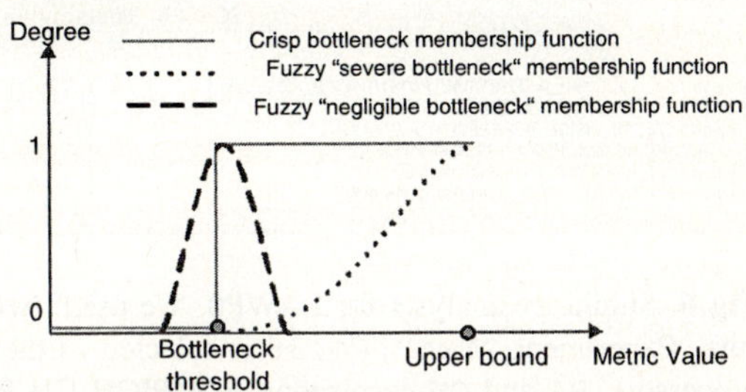

Fig. 4. Fuzzy vs crisp bottleneck search.

Fuzzy–based Search		
Code Region	L2_TCM/L2_TCA	Bottleneck
PARTICLE_LOAD	0.9497242344418835	Medium (degree=0.502)/High (degree=0.495)/

(a) Without negligible bottleneck search

Fuzzy–based Search		
Code Region	L2_TCM/L2_TCA	Bottleneck
MPI_SEND	0.6550124668722567	Negligible (degree=0.1)/
PARTICLE_LOAD	0.9497242344418835	Medium (degree=0.502)/High (degree=0.495)/

(b) With negligible bottleneck search

Fig. 5. Example of fuzzy-based bottleneck search.

Very simply, to show advantage of fuzzy-based bottleneck search, we experience with 3DPIC code to locate code regions that may have L2 cache access problems. Suppose a code region whose L2 cache miss ratio exceeds 0.7 is a bottleneck. In the first case we use a set of performance characteristic terms $T = \{low, medium, high\}$ representing the severity of the bottleneck. Three different fuzzy sets Z-function with range $[0.7, 0.8]$, Pi-function with range $[0.75, 0.95]$ and S-function with range $[0.9, 1]$ are associated with *low, medium*, and *high* term, respectively. We apply this search with 3DPIC code executed with 4 processes and we find that there is only one bottleneck as shown in Figure 5(a). The bottleneck falls into both classes *medium* and *high*, as shown in Figure 5(a). Since we are not certain about the threshold we decided to use another triangle FS with parameter $(0.65, 0.7, 0.75)$ to describe close area of the pre-defined bottleneck threshold. The result is that we find another code region as presented in Figure 5(b).

6.4 Similarity Analysis

We have implemented similarity analysis for all region summaries of a given code region in one experiment, and for region summaries of a set of selected code regions in a single or multiple experiment(s).

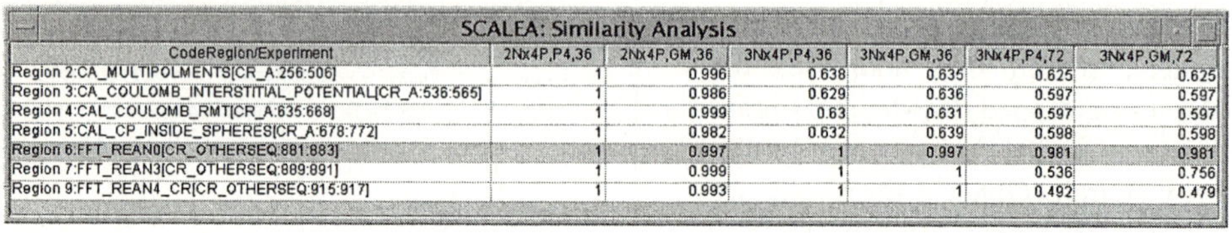

SCALEA: Similarity Analysis						
CodeRegion/Experiment	2Nx4P,P4,36	2Nx4P,GM,36	3Nx4P,P4,36	3Nx4P,GM,36	3Nx4P,P4,72	3Nx4P,GM,72
Region 2:CA_MULTIPOLMENTS[CR_A:256:506]	1	0.996	0.638	0.635	0.625	0.625
Region 3:CA_COULOMB_INTERSTITIAL_POTENTIAL[CR_A:536:565]	1	0.986	0.629	0.636	0.597	0.597
Region 4:CAL_COULOMB_RMT[CR_A:635:668]	1	0.999	0.63	0.631	0.597	0.597
Region 5:CAL_CP_INSIDE_SPHERES[CR_A:678:772]	1	0.982	0.632	0.639	0.598	0.598
Region 6:FFT_REAN0[CR_OTHERSEQ:881:883]	1	0.997	1	0.997	0.981	0.981
Region 7:FFT_REAN3[CR_OTHERSEQ:889:891]	1	0.999	1	1	0.536	0.756
Region 9:FFT_REAN4_CR[CR_OTHERSEQ:915:917]	1	0.993	1	1	0.492	0.479

Fig. 6. Similarity analysis for LAWP0. We used (`wtime`, `1.0`) to compute similarity measure. Experiment `2Nx4P,P4,36` is selected as the base. `1Nx4P` means 1 SMP node with 4 processors. `P4` and `GM` correspond to MPICH CH_P4 and Myrinet, respectively. The problem size is either 36 or 72 atoms. Distance measure is based on Euclidean function.

Figure 6 presents an example of using similarity analysis to examine selected code regions in 6 experiments. The first observation is that the performance of code region `FFT_REAN0` in the last 5 experiments is almost complete similar to the first experiment. The performance of `FFT_REAN3`, `FFT_REAN4` is almost similar in the first 4 experiments. This suggests that the performance of these code regions is not affected by changes of number of processors, communication libraries, even problem sizes (in case of `FFT_REAN0`). All code regions have similar performance in the first two experiments, suggesting the use of Myrinet does not increase much performance. This is confirmed by many cases in which communication libraries are different but the performance is very similar.

Table 1 shows an example of parameters of controllable factors. Table 2 presents the result of an example in which similarity is measured for code region `CA_MULTIPOLMENTS` in 6 experiments of LAPW0 by using parameters in Table 1. Performance score of the code region is based on S-function and distance measure is based on Euclidean function. In some cases, communication factor has very little impact on the performance, e.g., the

Table 1. Parameters for controllable factors.

Factor	Fuzzy Set	Range	Factor Category
atoms	linear	[0,72]	problem size
CPU	S-function	[0,64]	machine
network	S-function	[0,158.20]	communication

Table 2. Example of similarity analysis with experiment factors for `CA_MULTIPOLMENTS` region in 6 experiments. The first experiment is selected as the base.

Experiments	2Nx4P, P4,36	2Nx4P, GM,36	3Nx4P, P4,36	3Nx4P, GM,36	3Nx4P, P4,72	3Nx4P, GM,72
$sim_{f_{atoms}}$ ({atoms,1})	1	1	1	1	0.5	0.5
$sim_{f_{CPU}}$ ({(CPU,1)})	1	1	0.9531	0.9531	0.9531	0.9531
$sim_{f_{network}}$ ({(network,1)})	1	0.1519	1	0.1519	1	0.1519
sim_{o} ({(wtime,1)})	1	0.996	0.638	0.635	0.625	0.625

network between the first and the second experiment is quite dissimilar while other factors are very similar, but the performance is very similar. A similar result obtained if we examine the fifth and sixth experiments. The CPU factor has significant impact on some cases. E.g., factors of the third experiment are the same as those of the first experiment, except that CPU factors are slightly different. However, the performance of the code region is quite different.

7   Related Work

FL has been used in performance monitoring of parallel and distributed programs, e.g. performance contracts [9], but has not been exploited in data analysis techniques, e.g. performance classification, of existing performance tools.

APART introduces the concept of performance property [10] that characterizes a specific negative performance behavior of code regions. However, performance property is associated with a single performance metric. A performance property cannot represent a set of performance metrics. There is no concept of weight operator associated with performance properties. Also, our performance score is based on FL that allows the representation of fuzzy concepts such as *near* and *very*. Performance score can be computed based on linear and non-linear model with various membership functions.

Toward high-level scalable and intelligent analysis, classification based on machine learning has been used for classifying performance characteristics of communication in parallel programs [11]. Ahl and Vetter used multivariate statistical techniques on hardware performance metrics to characterize the system [12]. However, they do not deal with cases of multiple variables with different scales and weight factors. In [13], statistical analysis is used to study different (controllable and uncontrollable) factors that affect the mapping process of scientific computing algorithms to advanced architectures.

In [14] dispersion statistics is used to characterize the load imbalance by measuring the dissimilarity of performance metrics; metrics are normalized by measuring deviation from a mean value of a data set. Our similarity measure is based on fuzzy-based performance scores and is applied to not only code regions but also experiment factors.

In [6], historical data is used to improved automatic tuning systems. Performance score, similarity measure and fuzzy rules are fitted well for describing parameters and for improving decision making in performance tuning.

8   Conclusion and Future Work

This paper proposes a new approach to performance analysis that is based on soft computing. On the one hand, soft performance analysis techniques provide flexible, scalable and intelligent techniques for analyzing and comparing the performance of complex parallel and distributed applications. On the other hand, they interact with the user through high level notions. We complement existing work and contribute flexible and convenient methods to deal with uncertainty in the performance analysis, e.g. fuzzy-based bottleneck search, and to conduct the analysis in the form of high level notions, e.g. fuzzy-based search query. Still the soft performance analysis approach is just at an early stage, we believe it is a promising solution to provide soft, scalable and intelligent methods for automatic performance analysis.

Our future work is to study the application of soft performance analysis for dynamic performance tuning. Our proposed techniques could be applied to the performance analysis of large-scale complex dynamic Grid environments on which resources and their usage are unpredictable, performance data collected tends to be more imprecision and uncertainty. Moreover, performance similarity can be used to analyze and compare diverse Grid resources. Linguistic variables and fuzzy rules can be used in specifying and controlling service level agreements (SLAs) in the Grid.

References

1. Zadeh, L.A.: Fuzzy logic, neural networks, and soft computing. Commun. ACM **37** (1994) 77–84
2. Zadeh, L.A.: Fuzzy Logic = Computing with Words. IEEE Transactions on Fuzzy Systems **4** (1996) 103–111
3. Mitchell, T.M.: Machine Learning. McGraw Hill, New York, US (1997)
4. Truong, H.L.: Novel Techniques and Methods for Performance Measurement, Analysis and Monitoring of Cluster and Grid Applications. PhD thesis, TU WIEN, Austria (2005) http://dps.uibk.ac.at/truong/publications/linh-diss.pdf.
5. FuzzyJ Toolkit: http://ai.iit.nrc.ca/IR_public/fuzzy/fuzzyJToolkit.html (2004)
6. Chung, I.H., Hollingsworth, J.K.: Using Information from Prior Runs to Improve Automated Tuning Systems. In: ACM/IEEE SC2004, Pittsburgh, PA (2004)
7. Cain, H.W., Miller, B.P., Wylie, B.J.: A Callgraph-Based Search Strategy for Automated Performance Diagnosis. In: Euro-Par 2000 Parallel Processing. (2000) 108–122
8. Fahringer, T., Seragiotto, C.: Aksum: A performance analysis tool for parallel and distributed applications. Performance Analysis and Grid Computing (2003)
9. Vraalsen, F., Aydt, R.A., Mendes, C.L., Reed, D.A.: Performance contracts: Predicting and monitoring grid application behavior. In: Proceedings of GRID 2001. Volume LNCS 2242., Denver, Colorado, Springer-Verlag (2001) 154–165
10. Fahringer, T., Gerndt, M., Mohr, B., Wolf, F., Riley, G., Träff, J.: Knowledge Specification for Automatic Performance Analysis. Technical report, APART Working group (2001)
11. Vetter, J.: Performance analysis of distributed applications using automatic classification of communication inefficiencies. In: Conference Proceedings of the 2000 International Conference on Supercomputing, Santa Fe, New Mexico, ACM SIGARCH (2000) 245–254
12. Ahn, D.H., Vetter, J.S.: Scalable Analysis Techniques for Microprocessor Performance Counter Metrics. In: IEEE/ACM SC'2002, Baltimore, Maryland (2002)
13. Santiago, N.G., Rover, D.T., Rodriguez, D.: A Statistical Approach for the Analysis of the Relation Between Low-Level Performance Information, the Code, and the Environment. In: Proceedings of 2002 International Conference on Parallel Processing Workshops (ICPPW'02), Vancouver, B.C., Canada, IEEE Computer Society Press (2002) 282–
14. Calzarossa, M., Massari, L., Tessera, D.: A methodology towards automatic performance analysis of parallel applications. Parallel Comput. **30** (2004) 211–223

The Data Diffusion Space
for Parallel Computing in Clusters

Jorge Buenabad-Chávez and Santiago Domínguez-Domínguez

Sección de Computación
Centro de Investigación y de Estudios Avanzados del IPN
Ap. Postal 14-740, D.F. 07360, México
{jbuenabad,sdguez}@cs.cinvestav.mx

Abstract. The data diffusion space (DDS) is an all-software shared address space for parallel computing on distributed memory platforms. It is an extra address space to that of each process running a parallel application under the SPMD (Single Program Multiple Data) model. The size of DDS can be up to 2^{64} bytes, either on 32- or on 64-bit architectures. Data laid on DDS diffuses, or migrates and replicates, in the memory of each processor using the data. This data is used through an interface similar to that used to access data in files.
We have implemented DDS for PC clusters with Linux. However, being all-software, DDS should require little change to make it immediately usable in other distributed memory platforms and operating systems. We present experimental results on the performance of two applications both under DDS and under MPI (Message Passing Interface). DDS tends to perform better in larger processor counts, and is simpler to use than MPI for both in-core and out-of-core computation.

1   Introduction

Today PC clusters are widely used as platforms for parallel computing. Both message-passing and distributed shared memory environments are available for developing parallel applications on these platforms. Except for relatively simple communication patterns, message-passing programming is complicated; the programmer must specify when and which data to pass between which processing nodes. It is still more complicated for out-of-core computation, since the programmer must specify, or know, the data partitioning in disk space. However, message-passing libraries, such as MPI (Message Passing Interface) [12] and PVM (Parallel Virtual Machine) [15], are widely used because they do not require special hardware or operating system support.

A distributed shared memory (DSM) simplifies parallel programming because the location of data is not an issue. *Shared data* moves between processing nodes automatically and according to the access pattern of each application. Most DSM designs require either hardware or operating system support, which is, nonetheless, readily available in most hardware platforms and operating systems. If a DSM supports mapping files onto the *shared memory*, out-of-core

J.C. Cunha and P.D. Medeiros (Eds.): Euro-Par 2005, LNCS 3648, pp. 61–71, 2005.

computation is as simple to program as in-core computation. This will be most useful in 64-bit architectures, as in 32-bit architectures only 4 GB are available, while out-of-core applications today range in the hundreds of GB.

In this paper we present the data diffusion space (DDS), an all-software shared address space for parallel computing on clusters. It is an extra address space to the virtual address space of each process running a parallel application. DDS is for shared data only, which the programmer must explicitly specify as such through simply declaring it within a `C struct` declaration. Shared data automatically diffuses, or migrates and replicates, in the memory of each processor using the data, under a multiple-readers-single-writer protocol.

The size of DDS can be up to 2^{64} bytes, either on 32- and on 64-bit architectures. Hence shared data may not all be resident in memory. Some data will be in the disk space of processing nodes. However, the programmer uses the same interface to gain access to shared data (without specifying any location for data). This interface is similar to that used to access data in files. For a read, the programmer first calls $DDS_Read()$; for a write the programmer first calls $DDS_Write()$. The programmer then uses the data as it uses data in its local address space. After using the data the programmer must call $DDS_UnRead()$ or $DDS_UnWrite()$, respectively.

Data diffusion takes place by dynamically mapping data onto the memory of each processor using the data. Under out-of-core computation, DDS also maps shared data onto disk space in each processing node. These applications are likely to improve their performance under DDS, because DDS first tries to satisfy data requests from the memory of other nodes, instead of remote disk space.

In Section 2 we present related work. In Section 3 we present the architecture of DDS and its programming model. In Section 4 we show some empirical data on the performance of DDS compared to that of MPI for in-core and out-of-core applications. We offer some conclusions and describe future work in Section 5.

2   Related Work

A useful classification of DSM systems is that based on whether the implementation is all-hardware, mostly hardware, mostly software, or all-software [6]. All-hardware DSM moves data between processing nodes by hardware only, and at a fairly small granularity of typically 16 to 128 bytes. It includes cache-coherent non-uniform memory access (CC-NUMA) architectures, such as DASH [7] and Origin [9], and data diffusion architectures (also known as cache only memory architectures, or COMAs), such as DDM [19] and COMA-F [5]. In CC-NUMAs, data moves to the cache of each using processor, whether the data is *local* (resident in the nearest main memory node to a processor) or *remote*. In COMAs, the organisation of main memory is associative, and thus data moves to main memory nodes, and from these into processor caches, if available.

Mostly hardware DSM also moves data by hardware at a fairly small granularity, but little of its operation (e.g., gaining access to a memory region) is carried out by system software. Examples include Alewife [1] and KSR-1 [4]. Mostly software DSM is the well known virtual shared memory based on paging. Based on commodity virtual memory hardware, it has been widely investigated and improved. The first representative, IVY [8], adopted sequential consistency as its memory consistency model, incurring in general a significant communication overhead to keep data coherent. This overhead has since been reduced through the adoption of more efficient consistency models [11], such as release consistency and lazy release consistency, and optimisations relating to the implementation of the DSM [17].

All-software DSM does not rely on any hardware support other than network communication hardware. Access to shared data is controlled by software primitives (linked to the application) whose invocation is instrumented/coded either by a compiler or the application programmer. All-software, compiler assisted DSM includes Orca [2] Shasta [16], Midway [3], and CAS-DSM [10]. The C Region Library (CRL) [6] is also all-software DSM but with no compiler support. The programmer must call CRL procedures to map and gain access to shared data, and also to relinquish access to, and unmap, shared data.

DDS is similar to CRL regarding the use of shared data. However, the mapping of shared data in DDS is made only once. Another difference is that DDS manages a 2^{64} byte shared address space, both in 32- or in 64-bit architectures.

3   The Data Diffusion Space

The data diffusion space (DDS) was designed to simplify the programming of parallel applications under the SPMD (Single Program Multiple Data) model. Under this model, a process is created on each processing node to run a parallel application. With DDS, the DDSP process is also created, and runs, on each processing node. DDS is organised into a library to which a parallel application is linked.

3.1   Architecture

Figure 1 shows the DDS architecture. The data diffusion space is extra to that of each process running a parallel application. Data in the diffusion space is dynamically mapped onto the address space of whichever application process is using the data. We will use the term *shared data* to refer to data in the diffusion space from now on.

When an application process requests shared data, and this data is not resident in its local memory, the DDSP process requests the data from a remote memory node (as described in Section 3.2). When the data arrives, it is placed somewhere in the address space of the application by DDSP. The address where the data was placed is given back to the application through the DDS interface (as described in Section 3.3). DDSP processes communicate through TCP sockets, using blocks of up to 64 KB.

3.2   Protocol

Shared data diffuses under a multiple-readers-single-writer data coherency protocol. For a read request, a copy of the data is obtained; for a write request, an

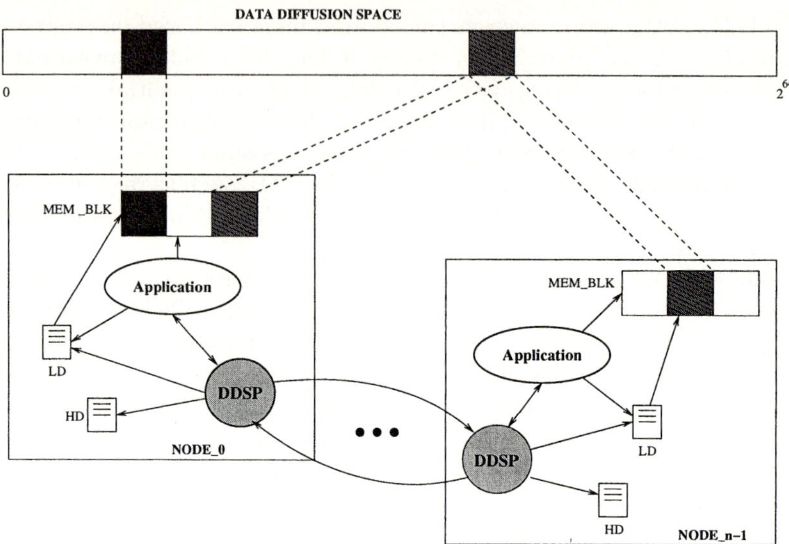

Fig. 1. DDS Architecture.

exclusive copy is obtained invalidating all other copies, thus ensuring all processors have the same view of the shared data.

The DDS protocol is similar to that of COMA-F (Cache-only Memory Architecture-Flat), an all-hardware distributed shared memory architecture [5]. It is homeless and directory-based. Data has no home location. It moves to the memory of the accessing processors and resides there, either until it is invalidated by a write by a processor or until it is evicted to give room to other data most recently used.

COMA-F uses associative main memory. Hence data has no home location therein. When a read or write misses in a memory node, a request is sent to the *home directory* of the relevant data item. This directory holds the location (node) and state information (exclusive, shared) of the item. If the home directory node is that location, it services the request; otherwise it sends the request to a node that currently has the item. A home directory is managed in each node, and some bits of each item address are used to identify a home directory.

DDS uses two directories in each node (see Figure 1). The *local directory* (LD) plays the role of an associative memory directory. A data item address is looked up there to see if the corresponding data item is in the memory. However, our local directory organisation keeps track of data not only in the memory of a node, but in both the memory and the disk space of the node. When a memory is needed to store recently used items, *exclusive* items less recently used are swapped out onto disk space. *Shared* items are just discarded.

When a read or write misses in a node, the DDS protocol sends a request to the relevant home directory, which is used and identified as described above for the COMA-F protocol.

3.3   Programming Model

Figure 2 shows the use DDS in the addition of two matrices: $C = A + B$. The programmer must define shared data within the *DDS* C structure. Before using shared data, the programmer must call *DDS_Init* as shown in that figure. In each processing node, *DDS_Init* maps the shared data to the diffusion space, initialises the local directory and the home directory, and starts the DDSP process.

In the matrix addition code, *ROWS/nprocs* rows are calculated by each processor. Before accessing data, each processor must gain access to it, through calling *DDS_Write* or *DDS_Read*. When these procedures return, the relevant data is already in the processor memory, and will remain there until the corresponding *DDS_UnWrite* or *DDS_UnRead* is issued.

```
struct DDS {                                  /* declaring shared data */
  unsigned int A[ROWS][COLUMNS];
  unsigned int B[ROWS][COLUMNS];
  unsigned int C[ROWS][COLUMNS];
};
  :
main() {
  :
  DDS_Init(sizeof(struct DDS), &myid);    /* initialising DDS */
  :
  rows = ROWS/nprocs;
  offset = myid * (ROWS/nprocs);
  for (r=0; r < rows; r++){
    i = r + offset;
    DDS_Write(DDS_C, i*COLUMNS, COLUMNS); /* gaining access */
    DDS_Read(DDS_A, i*COLUMNS, COLUMNS);  /* to shared data */
    DDS_Read(DDS_B, i*COLUMNS, COLUMNS);
    for (j=0; j<NCA; j++){                      /* using shared data */
      (dds_shmem[off_C+i])[j] = (dds_shmem[off_A+i])[j] +
                          (dds_shmem[off_B+i])[j];
    }
    DDS_UnWrite(DDS_C, i*COLUMNS, COLUMNS);
    DDS_UnRead(DDS_A, i*COLUMNS, COLUMNS);
    DDS_UnRead(DDS_B, i*COLUMNS, COLUMNS);
  }
  :
```

Fig. 2. DDS programming model example: matrix addition.

DDS_A, DDS_B and *DDS_C* are *enumeration* constants 0, 1 and 2, respectively. They refer to the order in which arrays A, B and C were declared within the *DDS* structure. They are used at run time to index the array *dds_vars*, where, for each DDS variable/array, the size of each element, the total number of elements and the initial (DDS) shared address are found. This information is used, along with the other two parameters sent to DDS_Write/DDS_Read, to calculate the DDS address of the data being accessed. The data is actually accessed through pointers held in the array dds_shmem, and the variables off_A, off_B and off_C, which are *locally* shared between the DDSP process and the application process. The variables off_A, ..., off_C (or that related with other defined shared data) are updated by DDSP according both to the address of the data requested with DDS_Read or DDS_Write, and to the actual location where that data is placed in the local memory, possibly after being requested from a remote node.

4   Performance Evaluation

To evaluate the performance of DDS we ran two applications on a 16-node PC cluster using different numbers of processors, both applications under DDS and under MPI/MPI-IO [13]. The version of MPI-IO we used is also known as ROMIO [18], and was used for our MPI version to be either in-core or out-of-core. In the out-of-core version, PVFS [14] is used and data is partitioned round robin into disk space along nodes by block (stripe in PVFS terminology). Each block is the size of n/p rows, where n is the number of rows in each array, and p is the number of processors used in each application run. Under DDS, out-of-core applications are programmed as in-core applications are. There is no need to specify a data partitioning into disk space.

It must be noted that, in programming out-of-core applications under MPI, programmers partition, or know the partition of, data into disk space. Also, programmers read data from, and write data to, disk space. That is, the programmer knows that data does not fit in memory in its entirety, and thus uses some memory only as a temporary buffer. The number of I/O requests is thus implicitly defined by the programmer. Under DDS, some reads and writes can be satisfied from copies in other memory nodes; hence the number of I/O requests can potentially be reduced.

The 16-node cluster configuration is as follows. Each node has 1 Intel Celeron 1.7 GHz processor, 512 MB RAM memory, and a hard disk drive. Hard disk drives are, however, of different make, size (1 GB, 3 GB, 4 GB and 8 GB) and speed. All nodes are interconnected by a 3COM Fast Ethernet switch with 48 ports. The operating system is Linux RedHat 9.0.

4.1   Matrix Multiplication

Our first application is a matrix multiplication (MM) algorithm: $C = A * B$. A and C are managed by rows (the C language default) and the matrix B by columns (the elements of a column are stored in consecutive localities in memory and/or in disk space). The matrices used were of size $16K \times 16K \times 8bytes$ (long type), or 2 GB each, a total of 6 GB for the three matrices. The matrices are partitioned into disk space such that each processor has $\lfloor \frac{n}{p} \rfloor$ consecutive columns

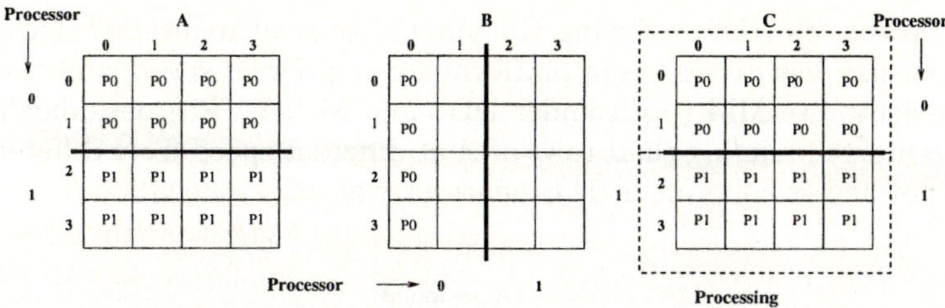

Fig. 3. Matrix multiplication: thick lines indicate data partitioning among nodes; dashed rectangles the elements in array C processed by each processor.

of B and $\lfloor \frac{n}{p} \rfloor$ rows of A and C (but C is only written). The multiplication is as follows. Each processor calculates the total value of each element in $\lfloor \frac{n}{p} \rfloor$ rows in matrix C. Each processor reads $\lfloor \frac{n}{p} \rfloor$ rows of A into memory, but only $\lfloor \frac{n}{p} \rfloor / f$ at a time, where $f = 1, 2$ and 4 for $p = 16, 8$ and 4, respectively. Then reads all columns of B, one at a time, f times, to calculate the value of all elements in n/p rows of C. Fig. 3 shows the data partitioning and processing for $n = 4$ and $p = 2$.

Table 1. MM: I/O requests under DDS and MPI-PVFS $16K * 16K$ matrices of 8-byte integers.

DDS			MPI-PVFS		
Processors	Reads	Writes	Processors	Reads	Writes
4	69632	4096	4	69632	4096
8	34816	2048	8	34816	2048
16	17408	1024	16	17408	1024

Table 1 shows the *average (total/p)* number of I/O *requests* under DDS and under MPI-PVFS. Under both, the number of I/O requests is the same on 4, 8 and 16 processors. This is somewhat surprising because it means that, under DDS, the columns of array B, which are the ones shared by all processors, were not diffused at all. The reason is as follows. In 4, 8 and 16 processors, each processor uses just above 256 MB of memory to store shared data. On the other hand, the amount of memory required by the rows of A and C that each processor holds in memory at any time is $\lfloor \frac{n}{p} \rfloor / f = (16384/4)/4 = 1024$ in all processor-count configurations (recall that for $p = 4$, $f = 4$, ... and for $p = 16$, $f = 1$). This is a total of $1024 \times 16384 \times 8$ (bytes) $= 128$ MB for each A and C. Since that many rows A and C are wired (with DDS_Read), there is very little memory for the columns of B to remain resident in main memory, and thus are evicted from main memory just after being used.

However, each processor uses all the columns of B in the same order, and thus once a column of B is resident in main memory, should it not be diffused to other processors (thus reducing the amount of read requests)? This did not happen because disk drives in our platform are of different speed, and because we did not synchronise MM (both under DDS and MPI-PVFS) periodically. Since processors started reading their rows of A at different speed from different disks, they did not access columns in B concurrently at all.

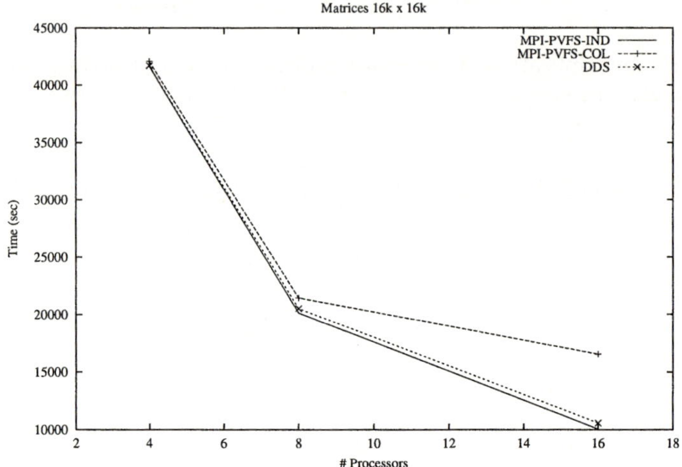

Fig. 4. MM: response time under DDS and MPI-PVFS.

Figure 4 shows the execution time of MM under DDS and MPI-PVFS, the latter both with independent I/O (MPI-PVFS-IND) and with collective I/O (MPI-PVFS-COL). DDS and MPI-PVFS-IND show almost the same performance in 4, 8 and 16 processors because they incur the same number of I/O operations and because these operations are independent in both versions. MPI-PVFS-COL also incurred the same number of I/O operations. However, synchronisation of collective operations, coupled with different speed of disk drives, increased response time.

4.2   Fast Fourier Transform

Our second application applies the Fast Fourier Transform (FFT) to restore degraded or defocused images. For an image of $N \times N$ pixels, a matrix of size $N \times N \times 8$ (float type) bytes is used. From this matrix, another matrix is created, which corresponds to an autocorrelation process of the original image that contains $M \times M$ images, where $M = 2N$ (see Figure 5). The size in bytes of this matrix is $2N \times 2N \times (N \times N) \times 8 = (N^4) \times 32$ bytes.

The image matrix is physically partitioned among processors by rows. In Figure 5, for $p = 4$, processor 0 (out of four) stores in disk space the images in the first row, processor 1 stores the images in the second row and so on.

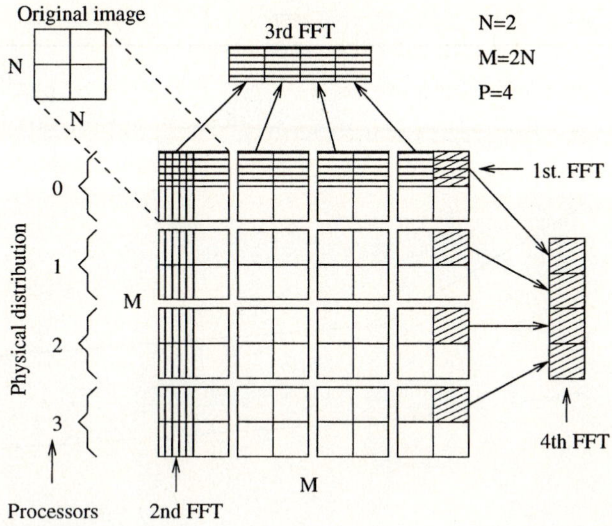

Fig. 5. FFT: data partitioning and processing of the images matrix.

Each processor applies the FFT to M/p rows and to M/p columns four times, as follows (see Figure 5): along entire rows (1st FFT), along entire columns (2nd FFT), jumping through rows (3rd FFT), and jumping through columns (4th FFT).

Table 2 shows the number of I/O requests per processor, both under MPI-PVFS and DDS on 4, 8 and 16 processors, for an original matrix of size 64×64 pixels. The *total* number of I/O requests is the same in all processor-count configurations. The images matrix is of size $((64)^4) \times 32 = 512$ MB, and could held in memory in all processor-count configurations.

Table 2. FFT: I/O requests under DDS and MPI-PVFS.

DDS			MPI-PVFS		
Processors	Reads	Writes	Processors	Reads	Writes
4	4096	4096	4	12288	12288
8	2048	2048	8	6144	6144
16	1024	1024	16	3072	3072

For each processor-count configuration, the number of I/O requests is fewer under DDS than under MPI-PVFS. Under DDS only the initial reads to load data into memory and the final writes to store results in disk space are incurred. Along the computation, other reads and writes are satisfied from copies in other memory nodes. There are more I/O requests under MPI-PVFS because each node manages only one memory buffer to hold an entire row of images at a time. As mentioned earlier, the application was programmed to manage both in-core

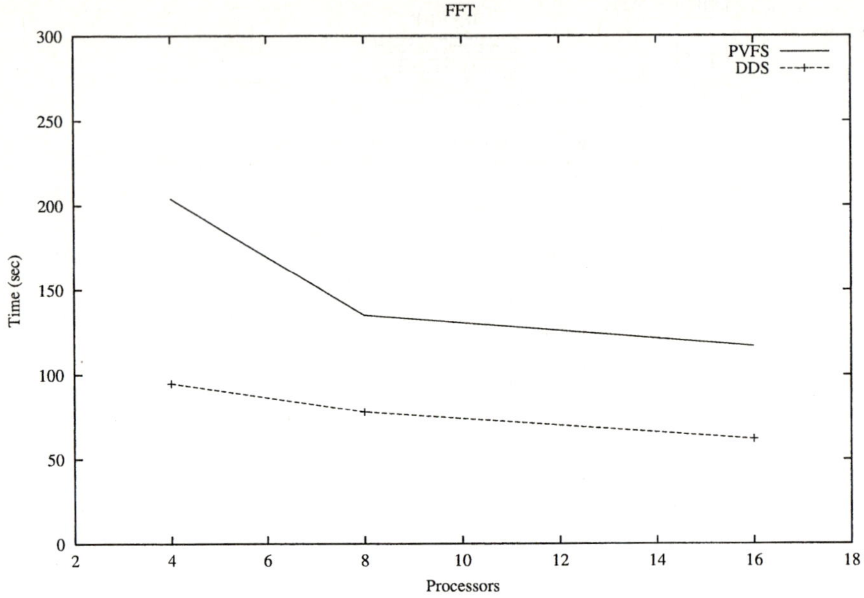

Fig. 6. FFT: response time under DDS and MPI-PVFS.

and out-of-core conditions (managing more buffers complicates programming even more).

Figure 6 shows execution of FFT under DDS and MPI-PVFS. In all processor-count configurations, DDS performs better than MPI-PVFS because it incurs fewer I/O overhead, reducing response time by half on average.

5   Conclusions and Future Work

We presented the data diffusion space (DDS), an extra shared address space for parallel computing under the SPMD model on distributed memory platforms. Compared with message passing, DDS is simpler to use and potentially offers improved performance both for in-core and out-of-core applications. On applications tested, DDS shows good performance up to 16 processors.

Programming a parallel application under DDS requires that *DDS_Read* and *DDS_UnRead*, or *DDS_Write* and *DDS_UnWrite*, functions be called to access data. DDS brings the data to the memory of the accessing processor whichever the current location of the data is, either other memory nodes or local or remote disk space.

We are currently designing a parallel file system with support to mapping files onto DDS. To support the shared memory programming model completely, we are also designing an extension to the C language and its compiler to avoid the use of the DDS interface entirely.

References

1. A. Agarwal *et al. The MIT Alewife Machine: Architecture and Performance.* In Proceedings of the 22nd ISCA (1995) 943–952.
2. H.E. Bal, M.F. Kaashoek and A.S. Tanenbaum. *ORCA: A Languaje for Parallel Programming of Distributed Systems.* IEEE Transactions on Software Engineering (March 1992) 190 – 205.
3. B.N. Bershad, M.J. Zekauskas and W. A. Sawdon. *The midway distributed shared memory system.* In Proceedings of COMPCON'93 (1993) 528–537.
4. J. Buenabad-Chávez, H.L. Muller, P.W.A. Stallard and D.H.D Warren. *Virtual memory on data diffusion architectures.* Parallel Computing 29 (2003) 1021–1052.
5. T. Joe. *COMA-F: A Non-hierarchical Cache Only Memory Architecture.* Stanford University Department of Electrical Engineering. PhD Thesis, 1995.
6. K.L. Johnson, M. F. Kaashoek, and D. A. Wallach. *CRL: High-Performance All-Software Distributed Shared Memory.* In Proceedings of the 5th SOSP (1995).
7. D. Lenoski, J. Laudon, K. Gharachorloo, A. Gupta, and J. Hennessy. *The Directory-Based Cache Coherence Protocol for the DASH Multiprocessor.* In Proceedings of the 17th ISCA (1990) 148–159.
8. K. Li. *Shared Virtual Memory Systems on Loosely Coupled Multiprocessors* (IVY). Yale University. PhD thesis,1986
9. J. Laudon and D. Lenoski. *The SGI Origin: A ccNUMA Highly Scalable Server.* In Proceedings of the 24th ISCA (1997) 241–251.
10. N. P. Manoj, K. V. Manjunath and R. Govindarajano. *CAS-DSM: A compiler assisted sofware distributed shared memory.* International Journal of Parallel Programming 32 (2004) 77–122.
11. M.D. Marino and G. Lino de Campos. *A speedup comparative study: three third generation DSM systems.* In Proceedings of the 7th International Conference on Parallel and Distributed Systems (2000) 153–158 (Workshops).
12. MPI: The Message Passing Interface. http://www-unix.mcs.anl.gov/mpi/
13. MPI-2: Extensions to the Message-Passing Interface. http://www.mpi-forum.org/docs/docs.html
14. PVFS: The Parallel Virtual File System. http://parlweb.parl.clemson.edu/pvfs/
15. PVM: Parallel Virtual Machine. http://www.epm.ornl.gov/pvm/pvm_home.html
16. D.J. Scales, K. Gharachorloo and C.A. Thekkath. *Shasta: A Low Overhead, Software-Only Approach for Supporting Fine-Grain Shared Memory.* In Proceedings of the 7th ASPLOS (1996) 174–185.
17. M. Swanson, L. Stoller and J. Carter. *Making distributed shared memory simple, yet efficient.* In proceedings of the Third International Workshop on High-Level Parallel Programming Models and Supportive Environments (1998) 2–13.
18. R. Thakur, E. Lusk, and W. Gropp. *Users Guide for ROMIO: A High-Performance, Portable MPI-IO Implementation.* Technical Report 234, Mathematics and Computer Science Division, Argonne National Laboratory, 1997.
19. D.H.D. Warren and S. Haridi. *DATA DIFFUSION MACHINE: A Scalable Shared Virtual Memory Multiprocessor.* In Proceedings of the International Conference on Fifth Generation Computer Systems (1988) 943–952.

Models for On-the-Fly Compensation
of Measurement Overhead
in Parallel Performance Profiling

Allen D. Malony and Sameer S. Shende

Performance Research Laboratory
Department of Computer and Information Science
University of Oregon, Eugene, OR, USA
{malony,sameer}@cs.uoregon.edu

Abstract. Performance profiling generates measurement overhead during parallel program execution. Measurement overhead, in turn, introduces intrusion in a program's runtime performance behavior. Intrusion can be mitigated by controlling instrumentation degree, allowing a trade-off of accuracy for detail. Alternatively, the accuracy in profile results can be improved by reducing the intrusion error due to measurement overhead. Models for compensation of measurement overhead in parallel performance profiling are described. An approach based on rational reconstruction is used to understand properties of compensation solutions for different parallel scenarios. From this analysis, a general algorithm for on-the-fly overhead assessment and compensation is derived.

Keywords: Performance measurement and analysis, parallel computing, profiling, intrusion, overhead compensation.

1 Introduction

In parallel profiling, performance measurements are made during program execution. There is an *overhead* associated with performance measurement since extra code is being executed and hardware resources (processor, memory, network) consumed. When performance overhead affects the program execution, we speak of *performance (measurement) intrusion*. Performance intrusion, no matter how small, can result in *performance perturbation* [7] where the program's measured performance behavior is "different" from its unmeasured performance. Whereas performance perturbation is difficult to assess, performance intrusion can be quantified by different metrics, the most important of which is dilation in program execution time. This type of intrusion is often reported as a percentage slowdown of total execution time, but the intrusion effects themselves will be distributed throughout the profile results.

Any performance profiling technique, be it based on *statistical profiling* methods (e.g., see [4, 14]) or *measured profiling* methods (e.g., see [2, 9]), will encounter measurement overhead and will also have limitations on what performance phenomena can and cannot be observed [7]. Until there is a systematic basis for

J.C. Cunha and P.D. Medeiros (Eds.): Euro-Par 2005, LNCS 3648, pp. 72–82, 2005.

judging the validity of differing profiling techniques, it is more productive to focus on those challenges that a profiling method faces to improve the accuracy of its measurement. In this regard, we pose the question whether it is possible to compensate for measurement overhead in performance profiling. What we mean by this is to quantify measurement overhead and remove the overhead from profile calculations. (It is important to note we are not suggesting that by doing so we are "correcting" the effects of overhead on intrusion and perturbation.) Because performance overhead occurs in both measured and statistical profiling, overhead compensation is an important topic of study.

In our Euro-Par 2004 paper [8], we presented overhead compensation techniques that were implemented in the TAU performance system [9] and demonstrated with the NAS parallel benchmarks for both flat and callpath profile analysis. While our results showed improvement in NAS profiling accuracy, as measured by the error in total execution time compared to a non-instrumented run, the compensation models were deficient for parallel execution due to their inability to account for interprocess interactions and dependencies. The contribution of this paper is the modeling of performance overhead compensation in parallel profiling and the design of on-the-fly algorithms based on these models that might be implemented in practical profiling tools.

Section §2 briefly describes the basic models from [8] and how they fail. We discuss the issues that arise with overhead interdependency in parallel execution. In Section §3, we follow a strategy to model parallel overhead compensation for message-based parallel programs based on a *rational reconstruction* of compensation solutions for specific parallel case studies. From the rationally reconstructed models, a general on-the-fly algorithm for overhead analysis and compensation is derived. Conclusions and future work are given in Section §4.

2   Basic Models for Overhead Compensation

In our earlier work [8], we developed techniques for quantifying the overhead of performance profile measurements and correcting the profiling results to compensate for the measurement error introduced. This work was done for two types of profiles: flat profiles and profiles of routine calling paths. The techniques were implemented in the TAUprofiling system [9] and demonstrated on the NAS parallel benchmarks. However, the models we developed were based on a local perspective of how measurement overhead impacted the program's execution. Profiling measurements are, typically, performed for each program thread of execution. (Here we use the term "thread" in a general sense. Shared memory threads and distributed memory processes equally apply.) By a local perspective we mean one that only regards the overhead impact on the process (thread) where the profile measurement was made and overhead incurred.

Consider a message passing parallel program composed of multiple processes. Most profiling tools would produce a separate profile for each process, showing how time was spent in its measured events. Because the profile measurements are made locally to a process, it is reasonable, as a first step, to compensate for measurement overhead in the process-local profiles only. Our original models do just that. They accounted for the measurement overhead generated during TAUprofiling for each program process (thread) and all its measured events, and then removed the overhead from the inclusive and exclusive performance results calculated during online profiling analysis. The compensation algorithm "corrected" the measurement error in the process profiles in the sense that the local overhead was not included in the local profile results.

The models we developed are necessary for compensating measurement intrusion in parallel computations, but they are not sufficient. Depending on the application's parallel execution behavior, it is possible, even likely, that intrusion effects due to measurement overhead seen on different processes will be interdependent. We use the term "intrusion" specifically here to point out that although measurement overhead occurs locally, its intrusion can have non-local effects. As a result, parallel overhead compensation is more complex. In contrast with our past research on performance perturbation analysis [10–12], here we do not want to resort to post-mortem parallel trace analysis. The problem of overhead compensation in parallel profiling using only profile measurements (not tracing) has not been addressed before. Certainly, we can learn from techniques for trace-based perturbation analysis [13], but because we must perform overhead compensation on-the-fly, the utility of these algorithms will be constrained to deterministic parallel execution, for the same reasons discussed in [7, 13].

At a minimum, algorithms for on-the-fly overhead compensation in parallel profiling must utilize a measurement infrastructure that conveys information between processes at runtime. It is important to note this is not required for trace-based perturbation analysis (since the analysis is offline) and it is what makes compensation in profiling a unique problem. Techniques similar to those used in PHOTON [15] and CCIFT [1] to embed overhead information in MPI messages may aid in the development of such measurement infrastructure. However, we first need to understand how local measurement overhead affects global performance intrusion so that we can construct compensation models and use those models to develop online algorithms.

3   Models of Parallel Overhead Compensation

To address the problem of overhead compensation in parallel execution, we must develop models that describe the effect of measurement overhead on execution intrusion. From these models we can gain insight in how the profiling overheads can then be compensated. However, unlike sequential computation, the models must identify and describe aspects of parallel interaction that may cause different intrusion behavior and, thus, lead to different methods for compensation. We know that the methods will involve the communication of information between parallel threads of execution at the time of their interaction. To be more specific, we will consider parallel compensation in message passing computation. The parallel overhead compensation models we present below allow for information about execution delay to be passed between processes during message communication. The goal is to determine exactly what information needs to be shared and how this information is to be used in compensation analysis. The modeling methodology we develop extends to shared memory parallel computing, but the case for shared memory will not be presented here.

The approach we follow below constructs an understanding of the parallel compensation problem from first principles. We first look at only two processes and then three processes. From this in-depth study, our hope is to gain modeling and analyses understanding that can extend to the general case. We will follow a strategy of *rational reconstruction* where we take scenario measurement cases and reconstruct an "actual" execution as if the measurement overhead were not present. From what we learn, we then derive a model that works for that case and look for consistent properties across the models to formulate a general algorithm for overhead compensation.

The details of overhead removal in the profile calculation are described in our earlier paper [8]. The focus below is on determining the actual overhead value to be removed for each process. These two operations together constitute overhead compensation.

3.1 Two Process Parallel Models

The simplest parallel computation involves only two processes which exchange messages during execution. Measurement-based profiling will introduce overhead and intrusion local to each process that carries between the processes as they interact. To model the intrusion and determine what information must be shared for overhead compensation, we consider the following two-process scenarios:

One send	Process P1 sends one message to process P2
Two sends	P1 sends two messages to P2
Handshake	P1 sends one message to P2, then P2 sends one message to P1
General	General message send and receive

For each scenario, we enumerate all possible cases for overhead relations between the processes (what is called the "measured execution" model) and for each case derive a representation of the execution with the overhead removed (what is called the "approximated execution" model). We determine the overhead-free approximation using a rational reconstruction of the "actual" event timings with the measurement overhead removed.

Both models are presented in diagrammatic form. In additional, we present expressions that relate the overhead, waiting, and timing parameters from the measured execution to those "corrected" parameters in the approximated execution. It is important to keep in mind that the goal is to learn from the rational reconstruction of the approximated execution how profile compensation is to be done in the other scenarios, especially the general case. For space reasons, we consider only the *One Send* and *General* scenarios in this paper.

Scenario: One Send. Consider a single message sent between two processes, P1 and P2. Figure 1 shows the two possible cases, distinguishing which process

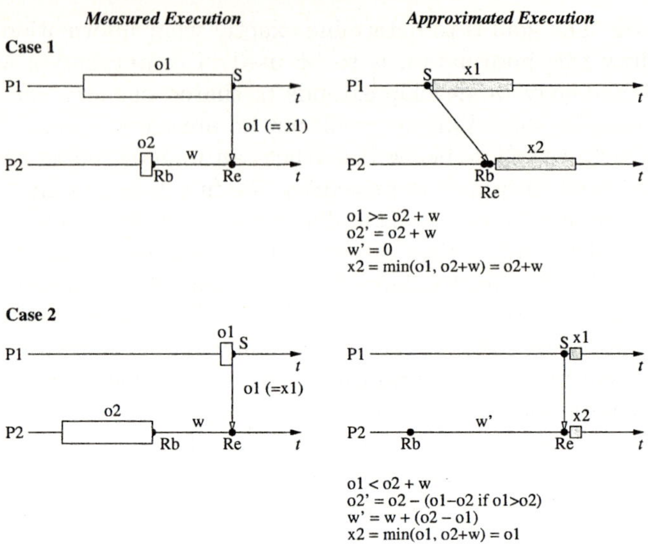

Fig. 1. Two-Process, One-Send – Models and Analysis (Case: 1, 2).

has accumulated more overhead up until the time of the message communication. Execution time advances from left to right and shown on the timelines are send events (S) and receive events (Rb, receive begin; Re, receive end). The overhead on P1 is $o1$ and the overhead on P2 is $o2$. The overhead is shown as a blocked region immediately before the S or Rb events to easily see its size in the figure, but it is actually spread out across the preceding timeline where profiled events occur. Also designated is the waiting time (w) between Rb and Re, assuming waiting time can be measured by the profiling system.

Case 1 occurs when P1's overhead is greater than or equal to P2's overhead plus the waiting time ($o1 \geq o2 + w$). A rational reconstruction of the approximated execution determines that P2 would not have waited for the message (i.e., S would occur earlier than Rb). Hence, the approximated waiting time (designated as w') should be zero, as seen in the approximated execution timeline. Of course, the problem is that P2 has already waited in the measured execution for the message to be received. In order for P2 to know P1's message would have arrived earlier, P1 must communicate this information. Clearly, the information is exactly the value $o1$, P1's overhead. This is indicated in the figure by tagging the message communication arrow with this value.

With P1's overhead information, P2 can determine what to do about the waiting time. The waiting time has already been measured and must be correctly accounted. If the approximated waiting is adjusted to zero, where should the elapsed time represented by w go? If the profiling overhead is to be correctly compensated, the measured waiting time must be attributed to P2's approximated overhead ($o2' = o2 + w$)! This is interesting because it shows how the naive overhead compensation can lead to errors without conveyance of delay information between sender and receiver. It is also important to note that Rb cannot be moved back any further in the approximated execution. This suggests that the only correction we can *ever* make in the receiver is in the waiting time.

The overhead value sent by P1 with the message conveys to P2 the information "this message was delayed being sent by $o1$ amount of time" or "this message would have been sent $o1$ time units earlier." We contend that this is exactly the information needed by P2 to correctly adjust its profiling metrics (i.e., compensate for overhead in parallel execution). We refer to the value sent by P1 as *delay* and will assign the designator x to represent its modeling and analysis that follows. For instance, P1's delay is given by $x1$. In both cases, $x1 = o1$, but it is not always true that delay will be equal to accumulated overhead, as we will see. Now an interesting question arises. How much earlier would future events on process 2 occur in the approximated execution after the message from P1 has been received? In general, each process will maintain a delay value (xi for process Pi) for it to include in its next send message to tell the receiving process how much earlier the message would have been sent. In the approximated execution, for denotational purposes, we show the $x1$ and $x2$ values for P1 and P2 as shaded regions after the last events, S and Re, respectively. We also show an expression for the calculation of $x2$ for this case.

Moving on to the second case, the overhead and waiting time in P2 is greater than what P1 reports (i.e., $o1 < o2 + w$). Rationally, this means that S happens after Rb in the approximated execution. What is the effect on w', the approximated waiting time? It is interesting to see that w' can increase or decrease, depending on the relation of $o1$ to $o2$. (Remember, $o1$ is the same as $x1$ in these cases.) However, the occurrence of Re is certainly dependent on S and, thus, $x2$ will be entirely determined by (and, in fact, equal to) $x1$.

General Scenario. The goal of the two process models is to enumerate the possible cases arising from send/receive message communication. From these cases, we can rationally reconstruct the approximated execution to determine how overhead, waiting, and delay times are to be adjusted. From this reconstruction, we can derive expressions for overhead analysis and correction. The similarity in the case results leads us to propose a general scenario for two processes. This scenario considers an arbitrary message send on one process and corresponding message receive on the other process. Thus, this is a generalization of the *One Send* scenario above. However, we now use the delay values $x1$ and $x2$ instead of the $o1$ and $o2$ overheads in the analysis. The expressions for the two cases are given below (refer to Figure 1):

Case 1	**Case 2**
x1 >= x2 + w	x1 < x2 + w
o2' = o2 + w	o2' = o2 - (x1-x2 if x1>x2)
w' = 0	w' = w + (x2-x1)
x1' = x1	x1' = x1
x2' = min(x1, x2+w) = x2 + w	x2' = min(x1, x2+w) = x1

The importance of the general scenario is the case analysis showing how the delay values are updated and what information is shared between processes during message communication. (Keep in mind that we are arbitrarily designating P1 as the sender and P2 as the receiver. The analysis also applies when P1 is the receiver and P2 the sender, with appropriate reversals of notation in the expressions.) Notice that the overhead values $o1$ (not shown) and $o2$ are accumulated overheads. The $o2$ value is updated here to account for waiting time processing, but whenever any new measurement overhead occurs on P1 or P2, the accumulated overheads $o1$ and $o2$ must be updated accordingly. Similarly, any new measurement overhead must also be added to the delay values $x1$ or $x2$.

Just to be clear, it is the overhead values that are being removed during the profiling calculations. Thus, we want these overhead to be accurately accounted. The conclusion of the two process modeling is that we can handle the parallel overhead compensation for ALL two-process scenarios by applying the general analysis described above on a message-by-message analysis, maintaining the overhead and delay values as the online analysis proceeds.

3.2   Three Process Parallel Models

The question at this point is whether that conclusion applies to three or more processes. That is, can the general two-process analysis be applied on a message-by-message basis to all send/receive messages between any two processes in a multi-process computation and, more importantly, give the desired overhead compensation result? We look at two scenarios with three processes to get a sense of the answer. These scenarios are:

Pipeline   Process P1 sends a message to P2, then P2 sends to P3
Two Receive   Process P1 and P3 sends a message each to process P2

We argue that these two scenarios are enough to elucidate all similar cases regardless of the number of processes. Again, we follow a rational reconstruction approach to determine approximated executions and then derive expressions for updating overhead, waiting time, and delay variables to match the reconstructed executions. Only the *Two Receive* scenarios is described in detail in this paper.

Scenario: Two Receive. When more than two processes are communicating, it is not hard to find a scenario that raises unpleasant issues in our ability to correct overhead intrusion under a different set of receive assumptions. These issues are brought on by the effect of intrusion on message sequencing. The *Two Receive* scenario exposes the problem. Here one process, P2, receives messages from two other processes. There are four cases to consider depending on the relatives sizes of overheads and waiting times. Figures 2 and 3 show two of the cases. For simplicity, we return to looking only at the first messages being sent and received on each process, and consider the initial overheads (not the delays values) in the analysis.

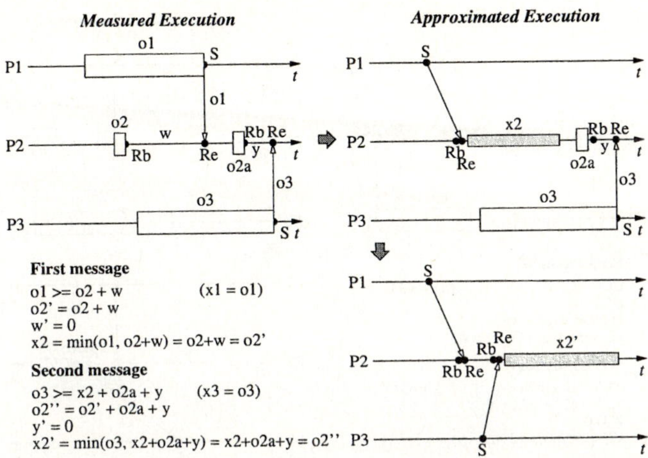

Fig. 2. Three-Process, Two Receive – Models and Analysis (Case 1).

In Figure 2, a two-part approximated execution is shown, with part one (top) giving the state after the first message is processed and part two (bottom) showing the result after the second message is processed. The analysis follows the approach we used before, with new waiting values (w' and y') being calculated and P2's delay value ($x2$) updated. In this case, no waiting time would have occurred, and no adjustment to waiting time is necessary. Otherwise, nothing particularly strange stands out in the approximated result.

What would be a surprising result? If the overhead analysis resulted in a re-ordering of send events in time, between the measured execution and the approximated execution, then there would be concerns of performance perturbation. In Figure 3, we see the send events changing order in time in the approximated execution, with P3's send taking place before P1's send. As with the other cases, our analysis reflects a message-by-message processing algorithm. In the rational reconstruction, we assume the message communication is explicit and pairs a particular sender and receiver. Under this assumption, the order of messages received by P2 must be maintained in the approximated execution. In this case, is the time reordering of send messages in Figure 3 a problem? In fact, no. It is certainly possible that a process (P2) will first receive a message from a process (P1) sent after another process (P3) sends a message to the receiving process. This just reflects the strict order of P2 receives. However, if we consider receive operations that can match any send, the send reordering exposes a problem with overhead compensation, since the message from P3 should have been received first in the "real" execution.

The application of our overhead compensation models to programs using receive operations that can match any send message results in profile analysis constrained to message orderings *as they are observed* in the measured execution. These message orderings are affected by intrusion and, thus, may not be the

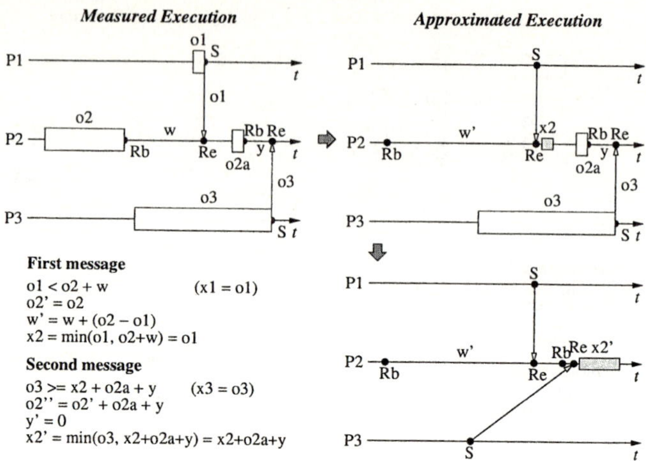

Fig. 3. Three-Process, Two Receive – Models and Analysis (Case: 2).

message orderings that occur in the absence of measurement. However, while it is actually possible to detect reordering occurrences (i.e., measured versus approximated orderings), it is not possible to correct for reordering during online overhead analysis and compensation. Why? There are two reasons. First, our analysis is unable to determine if it is correct to associate a receive event with a different send event. That is, the performance analysis does not know what type of receive is being performed, one that is for a specific sender or one that can accept any sender. Second, even if we know the type of receive operation, it is not possible to know whether changing receive order will affect future receive events. Therefore, the models must, in general, enforce message receive ordering.

3.3   Modeling Summary and General Algorithm

Our above modeling and analysis of measurement overhead in parallel message passing programs has produced three important outcomes. First, the rational reconstructions of the measurement scenarios and the analysis of the approximated executions has resulted in a robust procedure for message-by-message overhead compensation analysis in parallel profiling. It updates correctly waiting times associated with message processing and calculates per process values that capture online the amount a process has been effectively delayed due to measurement overhead and its effects. From this overhead compensation basis, the parallel profiling operations used to update inclusive and exclusive performance can be performed. Second, this analysis requires ALL send messages to be augmented with the delay value of the sender process at the time the message is sent. This information is necessary for the receiving process to apply the analysis procedures. Third, approximation models based on receive type can result in more accurate overhead handling and profile results, but the accuracy gains are anticipated to be minor compared to the processing complexity involved.

We argue that general overhead scenarios for message passing computations can all be addressed from what we learned in the two- and three-process modeling above. A general algorithm for overhead compensation effectively applies the *Two-Process, General* modeling and analysis on a message-by-message basis. The algorithm is composed of three parts:

- Updating of local overhead and delay as a result of local profile measurements.
- Updating of local overhead and delay as a result of messages received and their reported delay.
- Transmission of local delay when a process sends a message.

If the transmission of the delays values can be supported, it should be possible to incorporate this overhead compensation algorithm in a parallel profiling system such as TAU[9].

4   Conclusion and Future Work

Profiling is an important technique for the performance analysis of parallel applications. However, the measurement overhead incurred during profiling can cause intrusions in the parallel performance behavior. Generally speaking, the greater the measurement overhead, the greater the chance the measurement will result in performance intrusion. Thus, there is fundamental tradeoff in profiling methodology concerning the need for measurement detail (as determined by number of events and frequency of occurrence) versus the desired accuracy of profiling results. We argue that without an understanding of how intrusion affects performance behavior and without a way to adjust for intrusion effects in profiling calculations, the accuracy of the profiling results is uncertain. Most parallel profiling tools quantify intrusion as a percentage slowdown in the whole execution and regard this as an implicit measure of profiling goodness. This is unsatisfactory since it assumes overhead is evenly distributed across all threads of execution and all profiling results are uniformly affected.

Our early work in parallel perturbation analysis [11–13] demonstrated the ability to track performance intrusion and remove its effects in performance analysis results. However, there we had the luxury of a fully qualified event trace which included synchronization events that exposed dependent operation. This allowed us to recover execution sequences and derive performance results for an approximated "uninstrumented" execution. While the same perturbation theory applies, when profiling measurements are used, the analysis must be performed online.

This paper contributes models for measurement overhead compensation derived from a rational reconstruction of fundamental parallel profiling scenarios. Using these models we described a general on-the-fly algorithm that can be used for message passing parallel programs. The errors encountered in our earlier work on the NAS parallel benchmarks, resulting from our simpler overhead and compensation models, should now be reduced. However, implementing this algorithms requires the ability to piggyback *delay* values on send messages and to process the delay values at the receiver. We are currently developing a MPI wrapper library to support delay piggybacking that we can use to validate our approach. Our implementation is intended to be portable to all MPI implementations and will not require transmission of multiple messages. This scheme will be incorporated in the TAU performance system.

References

1. G. Bronevetsky, D. Marques, K. Pingali, and P. Stodghill, "Automated Application-level Checkpointing of MPI Programs," *Principles and Practice of Parallel Programming (PPoPP)*, 2003.
2. L. De Rose, "The Hardware Performance Monitor Toolkit," *Euro-Par Conference*, 2001.
3. A. Fagot and J. de Kergommeaux, "Systems Assessment of the Overhead of Tracing Parallel Programs," *Euromicro Workshop on Parallel and Distributed Processing*, pp. 179–186, 1996.
4. S. Graham, P. Kessler, and M. McKusick, "gprof: A Call Graph Execution Profiler," *SIGPLAN Symposium on Compiler Construction*, pp. 120–126, June 1982.
5. R. Hall, "Call Path Profiling," *International Conference on Software Engineering*, pp. 296–306, 1992.
6. D. Kranzlmüller, R. Reussner, and C. Schaubschläger, "Monitor Overhead Measurement with SKaMPI," *EuroPVM/MPI Conference*, LNCS 1697, pp. 43–50, 1999.
7. A. Malony, "Performance Observability," Ph.D. thesis, University of Illinois, Urbana-Champaign, 1991.
8. A. Malony and S. Shende, "Overhead Compensation in Performance Profiling," *Euro-Par Conference*, LNCS 3149, Springer, pp. 119–132, 2004.
9. A. Malony, et al., "Advances in the TAU Performance System," In V. Getov, M. Gerndt, A. Hoisie, A. Malony, B. Miller (eds.), *Performance Analysis and Grid Computing*, Kluwer, Norwell, MA, pp. 129–144, 2003.
10. A. Malony, D. Reed, and H. Wijshoff, "Performance Measurement Intrusion and Perturbation Analysis," *IEEE Transactions on Parallel and Distributed Systems*, 3(4):433–450, July 1992.
11. A. Malony and D. Reed, "Models for Performance Perturbation Analysis," *ACM/ONR Workshop on Parallel and Distributed Debugging*, pp. 1–12, May 1991.
12. A. Malony, "Event Based Performance Perturbation: A Case Study," *Principles and Practices of Parallel Programming (PPoPP)*, pp. 201–212, April 1991.
13. S. Sarukkai and A. Malony, "Perturbation Analysis of High-Level Instrumentation for SPMD Programs," *Principles and Practices of Parallel Programming (PPoPP)*, pp. 44–53, May 1993.
14. Unix Programmer's Manual, "prof command," Section 1, Bell Laboratories, Murray Hill, NJ, January 1979.
15. J. Vetter, "Dynamic Statistical Profiling of Communication Activity in Distributed Applications," *ACM SIGMETRICS Joint International Conference on Measurement and Modeling of Computer Systems*, ACM, 2002.

Modeling Pipeline Applications in POETRIES*

Eduardo César, Joan Sorribes, and Emilio Luque

Computer Science Department, Universitat Autònoma de Barcelona, 08193 Bellaterra, Spain
{Eduardo.Cesar,Joan.Sorribes,Emilio.Luque}@uab.es

Abstract. Parallel/Distributed application development is an extremely difficult task for non-expert programmers, and support tools are therefore needed for all phases of the development cycle of these kinds of application. This study specifically presents the development of an analytical performance model for pipelined applications. This model is intended to be used in the POETRIES distributed-program development environment, which is aimed at dynamic performance tuning based on frameworks with an associated performance model.

1 Introduction

Parallel/distributed programming constitutes a highly promising approach to the improvement of the performance of many applications. However, in comparison to sequential programming, several new problems have emerged in all phases of the development cycle of these kinds of application. One of the best ways to solve these problems would be to develop tools that support the design, coding, and analysis and/or tuning of parallel/distributed applications.

In the particular case of performance analysis and/or tuning, it is important to note that the best way for analyzing and tuning parallel/distributed applications depends on some of their behavioral characteristics. If the application being tuned behaves in a regular way, then a static analysis would be sufficient. However, if the application changes its behavior from execution to execution, or even in a single execution, then dynamic monitoring and tuning techniques should be used instead.

The key issue in dynamic monitoring and tuning is that decisions must be taken efficiently while minimizing intrusion on the application. We show that this is easier to achieve when the tuning tool uses a performance model associated to the structure of the application. Knowing the application's structure is not a problem if a programming tool, based on the use of skeletons or frameworks, is used for its development.

In this sense, we have designed a distributed-program development environment (DPSE), called POETRIES [1][2], where knowledge of the application structure is used for automatic detection and correction of its performance drawbacks at run-time. Later we will summarize the main characteristics of this tool (section 3), but at this point we can say that it consists of a programming tool based on frameworks or skeletons, plus a performance model associated to them, and a dynamic tool, called

---

* This work was supported by MCyT-Spain under contract TIN 2004 – 03388 and partially supported by the *Generalitat de Catalunya – Grup de Recerca Consolidat 2001* SGR-00218. The work developed at the University of Wisconsin has been supported by the research grant 2003/BE/00170 of the AGAUR

J.C. Cunha and P.D. Medeiros (Eds.): Euro-Par 2005, LNCS 3648, pp. 83–92, 2005.

MATE, where this model is used for dynamically improving the application's performance, in [3] we describe the implementation of this environment for M-W applications. This study represents a further step, and is focused on the development of the performance model associated to the Pipeline framework

In aiming to develop these concepts, we have organized the rest of this paper in the following way: In section 2, we present an overview of related studies. In section 3, we describe general structure of POETRIES. In section 4, we present a global analysis of the Pipeline framework. Based on it, in section 5 we present the development of its associated performance model and in section 6 some experimental results for validating it. Finally, in section 7 we set out the conclusions of this study.

2   Related Studies

It is possible to classify the related studies into two main sets; firstly those that address the problem of distributed application development in its different phases. Secondly, those related to performance issues, which in turn could be broadly divided into those that propose a static approach and those that propose a dynamic one.

In the first set, we have found some studies, like the pattern language of [4], CO_2P_3S [5], and eSkel[6], which address the whole problem, based on the fact that there is a set of design patterns which could be applied to those problems suitable to be solved in parallel. We have also found, in this set, studies, like Skil [7] or [8], that take advantage of the high abstraction degree of functional languages, as well as other important properties of these languages, such as separation of behavior and meaning, and transformation possibilities. Other, studies such as [9] suggest taking advantage of a popular modeling language (UML), adding to it extensions to model the most important constructs of parallel/distributed paradigms, plus performance annotations.

In the set of studies related to performance monitoring, analysis and tuning we have found tools that make a trace based analysis, such as Kappa-Pi [10] and EXPERT [11]. The former makes a two-step analysis based on a source of inefficiency knowledge base and generates a set of recommendations concerning inefficiencies in the application source code. The latter, a tool included in the KOJAK project, presents the complete performance behavior of the application in three dimensions: performance property, source code location and the execution phase where it occurred, and process or thread location. In this set we have also found a tool called P3T+ [12], which predicts application performance based on information gathered at compiler time, plus sequential simulation and architecture parameters.

A tool with a dynamic approach is Paradyn [13] and its Performance Consultant, which dynamically searches for performance bottlenecks using the W3 search model (Why is there a performance bottleneck? Where is it? When did it happen?).

It is worth noting that, of the programming tools, few mention the possibility of taking advantage of knowledge of the application structure to improve their performance, although the authors of eSkel [6] have recently published [14] a study of the use of this information, along with process algebras, to evaluate parallel-applications performance. Surprisingly, none of the performance analysis tools use the application structure information, even when it is available, to perform their analysis.

3   POETRIES General Structure

As mentioned in the introduction, we have designed a distributed-program development environment (DPSE) with dynamic tuning, called POETRIES, which uses knowledge of the high-level application structure to perform its task. The structure of this tool is shown in figure 1.

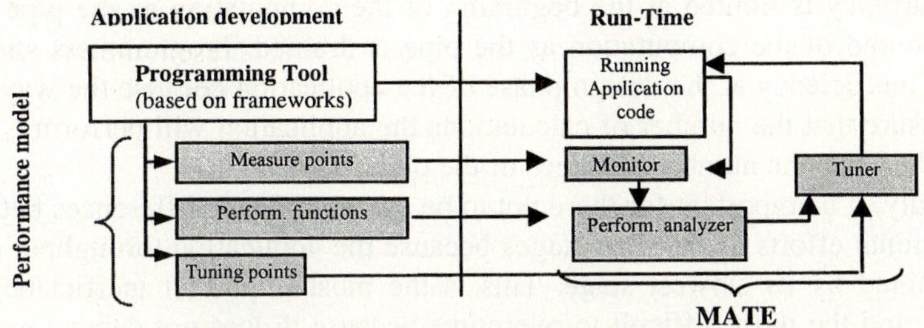

Fig. 1. Structure of the POETRIES DPSE

The main idea is that it is possible to define a *performance model* associated to the most common *application frameworks* (frameworks that are offered by many parallel development tools [5][6][7]). This model includes a set of *performance functions* (aimed at the detection of performance drawbacks), some parameters that have to be monitored (*measure points*) to evaluate these functions, and some parameters (*tuning points*) that could be modified to activate the actions that should be taken to overcome the detected performance drawbacks.

These definitions make up the static part of the environment. Then, there is a dynamic tuning environment (*Run-time*) which, at application execution time, uses this *performance model* to *monitor* the appropriate parameters to evaluate the performance functions (*performance analyzer*) and, takes the required actions to improve the application performance (*tuner*). This phase, called MATE (Monitoring, Analysis and Tuning Environment) [15], has been implemented as an independent tool.

To implement this DPSE we have created POETRIES [2] (**P**erformance **O**riented **E**nvironment for **T**ransparent **R**esource-management, **I**mplementing **E**nd-user parallel/distributed applications), which integrates a framework-based parallel/distributed programming environment with the performance model needed to perform the dynamic analysis and tuning for these kinds of applications.

4   Pipeline Framework Analysis

The Pipeline framework is a well-known parallel programming structure used as the most direct way to implement algorithms that consist of performing an orderly sequence of essentially identical calculations on a sequence of inputs. Each of these calculations can be broken down into a certain number of different stages, and these stages can be applied concurrently to different inputs.

For this study, we will assume that programmers use a linear pipeline framework, being one with every stage, but first, receiving its input from the previous stage of the pipe and sending its output to the following one, but last. This is a simplification of the more general multiple-branch pipelined structure. However, it won't significantly influence the fore coming performance analysis because analyzing a multiple-branch pipe implies an individual analysis of each branch as being an independent pipe and, in addition, being aware of performance unbalances among different branches.

The possible inefficiencies of pipelined applications are also well known. At first, the concurrency is limited at the beginning of the computation as the pipe is filled, and at the end of the computation as the pipe is drained. Programmers should deal with this inefficiency at the design phase of the application because the way to avoid it is to assure that the number of calculations the application will perform is substantially higher than the number of stages of the pipe.

Secondly, it is important for there not to be any significant differences between the computational efforts of the pipe stages because the application throughput of a pipe is determined by its slowest stage. This is the most important inefficiency of this structure, and the most difficult to overcome because it does not depend exclusively on the application design, but also on run-time conditions. Consequently, this drawback is suitable for being solved dynamically. There are different approaches for doing it depending on the target index to be optimized and the resource availability.

Therefore, we may want to improve the efficiency in the use of resources, or even try to free some underused resources to increase their availability, in this case dynamic mapping of stages could be used to group faster stages; thus improving the use of resources. On the other hand, we may want to improve the application throughput, in this case, if there are available processors, to replicate slower stages will increase its throughput, therefore decreasing the application execution time.

Furthermore, we may want to increase the application throughput but also to make an adequate use of resources. Consequently, a mixed approach could be defined, as a compromise between optimizing throughput and efficient resource management.

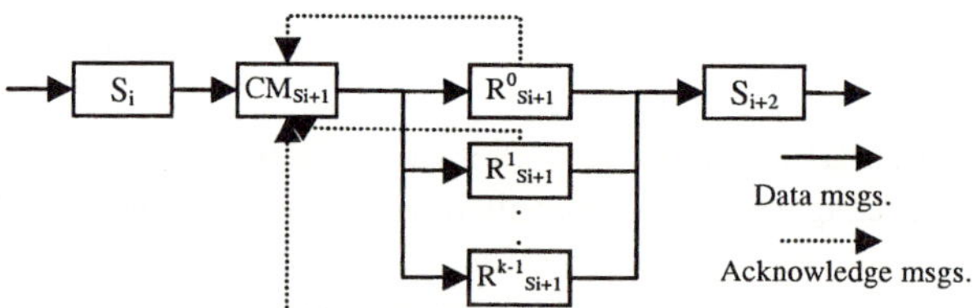

Fig. 2. Structure of a replicated stage. Stage i+1 has been replicated k times and a communication manager (CM) has been added to control the replicas' state and distribute incoming tasks

Our aim is to implement a mixed strategy, with the main objective being to optimize the application throughput but also to make reasonable use of resources. However, as a first step towards this objective we have concentrated on optimizing application throughput and, as a consequence, the model presented in this study does not include considerations about the efficiency of resource management.

Therefore, we assume that the programmer writes an application using a linear pipeline framework, and then, at run-time, our tuning tool will dynamically decide which stages should be replicated in order to improve the application's performance.

In figure 2, we show the structure of a replicated stage. It can be seen that there is a new process called *communication's manager* (CM). It is responsible for monitoring the replica's state and for distributing work, when available, to free replicas.

Finally, we should decide whether the CM should run in a separate processor or should share one with a replica. The first approach is simpler to model but could lead to a poorer use of resources. The second, in contrast, seems to lead to a better use of resources, but is more difficult to implement with some communication libraries, and is also difficult to model because the CM affects, and is affected by, the activity of the replica that shares a processor with it. We have modeled both options, but in this study we only present the model associated to the first.

5 Performance Model of the Pipeline Framework

Once the basic analysis of the framework has been performed and its structure has been defined, it is time to use the POETRIES methodology outlined in section 3 to develop the performance model associated to that framework.

Our objective is to increase the throughput of the slower stages in order to increase the global application performance. The general strategy to reach this objective will consist of calculating the best replication pattern for the current application's behaviour and available number of processors.

Consequently, if we want to increase the throughput we must minimize the time needed by each stage to process its inputs, including the time required to deliver the results to the next stage. We call this *production time*. Thus, we need expressions to find the production time each stage can reach (its *independent production time*), and also expressions that explain its observed production time due to the influence of other stages (its *dependent production time*). Moreover, we should find different expressions to make these calculations for single (5.1), and replicated stages (5.2).

In our analysis, we assume that there is just one process per processor, and we use the following terminology:

- tl, λ = fixed network overhead per message and communication cost.
- v_i = data volume sent by stage i, in bytes.
- tc_i = computation time stage i needs to process an input, in ms.
- Tr^k_i = production time of k replica of stage i.
- Tr_i = independent production time of stage i, in ms.
- rTr_i = dependent production time of stage i, in ms.

5.1 Production Time of Single Stages

A single pipe stage is one which receives messages with data, except for the first, makes its portion of calculation of this data, and sends the results to the next stage, except for the last.

The *independent production time* (Tr_i) of such a stage will depend on its position in the pipe, its computation time (tc_i), and the current communication conditions - $C(P,v_i)$- (communication protocol -P- and message size -v_i-).

This way, we can define the *independent production time* of a single stage as:

$$Tr_i = tc_i + C(P, v_i) \qquad (1)$$

Where $C(P, v_i)$ is defined as:

$$0 \qquad \text{if } (i == n-1) \ (n = \text{total number of pipe stages}).$$

The last stage will be able to process its next message just after it finishes the calculation of the previous one.

$$tl \qquad \text{if } (i < n-1) \text{ and } (P \text{ is not synchronous})$$

If the communication protocol in use does not force synchronous sends, then the stage will just have to wait to deliver the message to the library interface.

$$tl + \lambda v_i \quad \text{if } (i < n-1) \text{ and } (P \text{ is synchronous})$$

Otherwise (synchronous sends), the stage will have to wait for the whole communication to finish before going to the next receive operation.

The *dependent production* time of the current stage also depends on the dependent production times of the following and previous stages.

$$rTr_i = \begin{cases} rTr_{i-1} & \text{if } (P \text{ is sync.}) \text{ and } (((i = n-1) \text{ and } (rTr_{i-1} > Tr_i)) \text{ or} \\ & \quad ((0 < i < n-1) \text{ and } (rTr_{i-1} > Tr_i) \text{ and } (rTr_{i-1} > rTr_{i+1}))) \\ rTr_{i-1} + \lambda v & \text{if } (P \text{ is async.}) \text{ and } (((i = n-1) \text{ and } (rTr_{i-1} + \lambda v_{i-1} > Tr_i)) \text{ or} \\ & \quad ((0 < i < n-1) \text{ and } (rTr_{i-1} + \lambda v_{i-1} > Tr_i) \text{ and } (rTr_{i-1} + \lambda v_{i-1} > rTr_{i+1}))) \\ rTr_{i+1} & \text{if } ((0 < i < n-1) \text{ and } ((P \text{ is sync.}) \text{ and } (rTr_{i+1} > Tr_i) \text{ and} \\ & \quad (rTr_{i+1} > rTr_{i-1})) \text{ or } ((P \text{ is async.}) \text{ and } (rTr_{i+1} > Tr_i) \text{ and} \\ & \quad (rTr_{i+1} > rTr_{i-1} + \lambda v))) \text{ or } ((i=0) \text{ and } (rTr_{i+1} > Tr_i))) \\ Tr_i & \text{Otherwise} \end{cases} \qquad (2)$$

5.2 Production Time of Replicated Stages

A replicated pipe stage is one where data messages are received by a special process called a communication manager (CM), which is responsible for deciding which stage replica will process the data. Then the chosen replica makes the stage portion of calculation of this data and sends the results to the next stage, unless it is the last. The CM is executed on an independent processor.

To calculate the *independent production time* of such a stage, we should now consider the managing time associated with the CM (tg_i), and the waiting time for one free replica (wc_i).

The term tgi depends on the communication protocol and possibly on the message size. Basically, the CM looks at the communication channel and waits for messages that could come from the previous stage or from one of the stage replicas (indicating that the replica is free). As there could be many message sources it should look at the channel without blocking. In consequence, the managing time will be the time needed to make 1 or 2 probes of the channel with its corresponding receives plus the time needed to send the requirements to the free replica.

Therefore, if the communication protocol is synchronous then the CM should wait $2*(3tl + \lambda vi)$ to be ready to process the next requirement message. It has to spend twice the communication time because, except when filling the pipe, it has to synchronously receive the message from the previous stage ($3tl + \lambda vi$) and then synchronously send it to a free replica ($3tl + \lambda vi$). On the other hand, if the communication protocol is asynchronous then the CM will only have to wait for some network overhead before seeing if there is a new requirement message, because in this case, library buffers allow for overlapping communications.

The term wc_i depends on the processing capacity of the replicas and the managing capacity of the CM. Given m replicas, if the CM spends more time managing m input messages than the time spent by the set of replicas processing the same number of messages then there will be always free replicas ($wc_i = 0$), which is an undesirable situation because of the waste of resources.

If the CM has the capacity to feed the m replicas, then the term wc_i will less or equal to the production time of the set of replicas, plus the time needed to send the message to a replica. Except if the protocol is synchronous, because in such a case the communication time is included in tg_i. Furthermore, the production time of a given set of replicas depends on the *independent production time* of each replica Tr^k_i, which in turn is calculated in the same way as *the independent production time* of a single stage plus the time needed to send the acknowledgement message to the CM.

Summarizing, the definition of independent production time of replicated stages is:

$$Tr_i = tg_i + wc_i \tag{3}$$

$$\text{Where, } tg_i = \begin{cases} tl + c & \text{if (protocol is asynchronous)} \\ 2*(3tl + \lambda v_i) + c & \text{if not} \end{cases}$$

$$\text{and } wc_i = \begin{cases} 0 & \text{if } 1 \Big/ \sum_{k=0}^{m-1} (1/Tr_i^k) <= tg_i \\[2ex] (0, 1 \Big/ \sum_{k=0}^{m-1} (1/Tr_i^k) + \lambda v_i] & \text{if not and protocol is asynchronous} \\[2ex] (0, 1 \Big/ \sum_{k=0}^{m-1} (1/Tr_i^k)] & \text{if not and protocol is synchronous} \end{cases}$$

Finally, knowing that the *dependent production time* of a stage just defines the effect of its neighbors on the stage, we can say that the *dependent production time* of a replicated stage is defined in exactly the same way as for a single stage.

6   Experimental Validation of the Model

In this section, we first want to show the results of some relevant experiments that have been designed to validate the proposed analytical model for calculating the production time of a pipe stage. In order to get these results we have written a synthetic parametrical pipeline application. This application uses the MPI communication library and all experiments have been executed on clusters of workstations in the computer science department of the U. of Wisconsin at Madison.

Examples of figures 3 and 4 are included with the objective of showing that the expressions described in section 5 closely match the behavior of real applications, and the example of figure 5 is included with the objective of showing how the model can be used to improve the application's throughput.

Table 1. Independent, dependent, and measured production times for figure 3 pipe stages

	Stage 0	Stage 1	Stage 2	Stage 3	Stage 4
(1) Tr_i	1.002 s.	1.202 s.	1.002 s.	2.002 s.	1 s.
(2) rTr_i	2.002 s.	2.002 s.	2.002 s.	2.002 s.	2.002 s.
Measured	2.015 s.	2.015 s.	2.03 s.	2.03 s.	2.03 s

In figure 3 we can graphically see how the slower stage (3 in this case) affects all other stages. It is clear that stage 4 has to wait for the output of stage 3, but we can also see how stages 0, 1, and 2 synchronize with stage 3 due to communication synchronization.

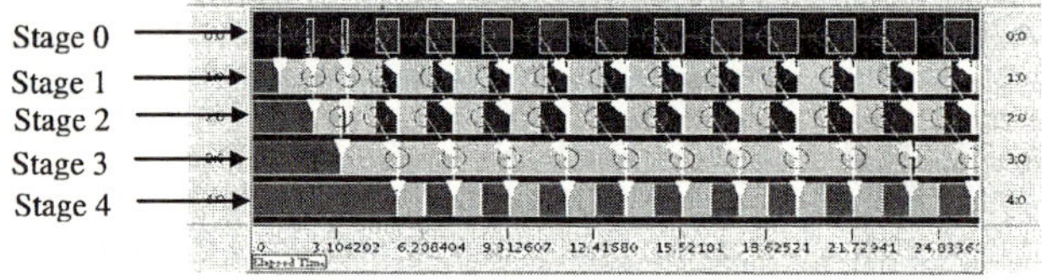

Fig. 3. Execution trace for a pipeline of five stages with message size of 200 Kb. and computing times of: 1 s. for stages 0, 2, and 4; 1.2 s. for stage 1; and 2 s. for stage 3

In table 1 we show the independent, dependent, and measured production times for all stages of the pipeline in figure 3. We should first note that expression (2) for calculating dependent production times tells us that stage 3 is the bottleneck of the application, and secondly that the difference between the dependent production times (rTr) and the measured ones is a result of the communication protocol change that MPI performs when its buffers become full.

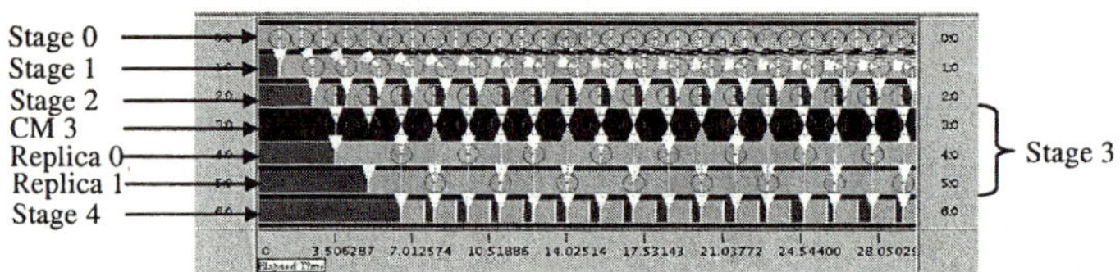

Fig. 4. Execution trace for a 5 stage pipeline, stage 3 with two replicas. Messages of 512 b. and computing times of: 1 s. (stages 0, 2, & 4); 1.5 s. (stage 1); and 3 s. for each replica of stage 3

In figure 4 we can see how replication improves the throughput of a slow stage. However, we can see that even with this replication, stage 3 does not match the independent production time of stage 4 (the last) yet.

In table 2 we show the independent, dependent, and measured production times for all stages of the pipeline in figure 4. It can be seen how replication improves the throughput of the application, but also that the model captures its behavior.

Finally, in figure 5 we can see the result of applying the performance model to an application (a) for optimizing its throughput by replication (b). In this example the computation times associated to stages 1, 2, and 3 are four, three, and two times respectively the one associated with stages 0 and 4 (the shortest). In addition, the communication protocol is asynchronous, except if forced by the communication library. In this case, if there are 10 available processors the model tells us that to optimize throughput we should replicate stage 1 four times, stage 2 three times, and stage 3 two times. However, if only 5 available processors were available the model would have advised us to introduce 2 replicas of stage 1, and 1 replica of stage 2, thus improving the throughput to that of stage 3 which is the second fastest stage.

Table 2. Independent, dependent, and measured production times for figure 4 pipe stages

	Stage 0	Stage 1	Stage 2	Stage 3	Replica 0	Replica 1	Stage 4
Tr_i	1.001 s. (1)	1.501 s. (1)	1.001 s. (1)	1.502 s.(3)	3.001 s.	3.001 s	1 s. (1)
(2) rTr_i	1.001 s.	1.501 s.	1.502 s.	1.502 s.	--	--	1.502 s.
Measured	1.0009 s.	1.5009 s.	1.5018 s.	1.5017 s.	3.0045	3.0018	1.503 s

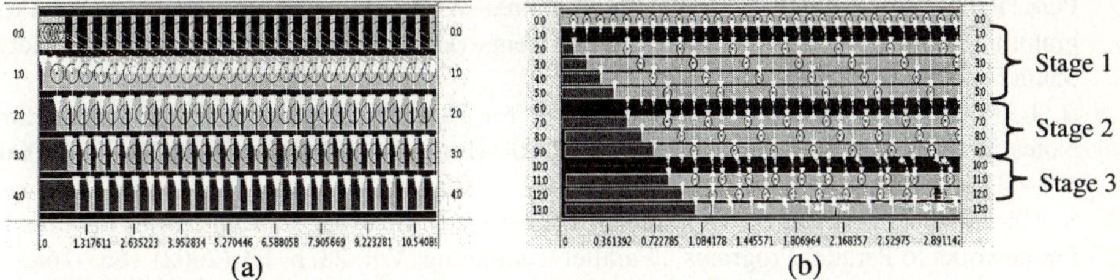

Fig. 5. Execution traces of a pipeline application of 5 stages and message size of 10 Kb. Computing time for stages 0 and 4 is 100 ms, for stage 1 is 400 ms, for stage 2 is 300 ms, and for stage 3 is 200 ms. Without replicas (a), whit the replicas indicated by the model (b)

In this example, the application throughput is improved by 3.7 times from 2.48 to 9.15. If this application were able to produce outputs at the pace determined by its fastest stage, it would have a throughput of 9.64, which is only 5% better than the one we have obtained.

7 Conclusions and Future Work

The main goal of our study was to demonstrate that advance knowledge of the structure of the application is a good way to make appropriate global decisions to dynamically improve its performance. To fulfill this goal we have designed POETRIES as a distributed-program development environment that integrates a framework based parallel/distributed programming tool with the performance model needed to perform the dynamic analysis and tuning of the applications generated using this tool.

We have defined a performance model associated to the pipeline framework with the aim being to improve the throughput of pipelined applications, and we have shown experimental results that demonstrate that it is possible to define a realistic analytical model that closely reflects the real behavior of an application developed with this framework.

Completing this model to include efficient resource management considerations is the next challenge. However, we believe that combining frameworks and model based dynamic performance tuning is a very promising approach for broadening and encouraging the use of parallel/distributed applications.

References

1. E. Cesar, A. Morajko, T. Margalef, J. Sorribes, A. Espinosa, E. Luque: Dynamic Performance Tuning Supported by Program Specification. Scientific Programming, Vol. 10. IOS Press (2002) 35-44
2. E. Cesar, J. G. Mesa, J. Sorribes, and E. Luque: POETRIES: Performance Oriented Environment for Transparent Resource-management, Implementing End-user parallel/distributed applications, Lecture Notes in Computer Science (LNCS), Vol. 2790 (Euro-Par 2003). Springer-Verlag (2003) 141-146
3. E. Cesar, J. G. Mesa, J. Sorribes, E. Luque: Modeling Master-Worker Applications in POETRIES. Proceedings of the 9th International Workshop on High-Level Parallel Programming Models and Supportive Environments (HIPS 2004). IEEE Computer Society. Santa Fe, New Mexico (April 2004) 22–30
4. B. L. Massingill et al.: A Pattern Language for Parallel Application Programs. Lecture Notes in Computer Science (LNCS), Vol. 1900 (Euro-Par 2000), Springer-Verlag (2000) 678-681
5. S. MacDonald, J. Anvik, S. Bromling, J. Schaeffer, D. Szafron, K. Tan, "From Patterns to Frameworks to Parallel Programs", Parallel Computing, Vol. 28, n. 12, (2002) 1663-1683.
6. M. Cole: Bringing Skeletons out of the Closet: A Pragmatic Manifesto for Skeletal Parallel Programming. Parallel Computing 30(3) (2004) 389-406
7. T. Richert: Skil: Programming with Algorithmic Skeletons – A Practical Point of View. Proceedings of the 12th International Workshop on Implementation of Functional Languages, Germany (2002) 15-30
8. J. Darlington, H. W. To: Building Parallel Applications without Programming. Abstract Machine Models for H. Parallel Computers, Oxford University Press (1995) 140-154
9. S. Pllana, T. Fahringer: On Customizing the UML for Modelling Performance-Oriented Applications. UML 2002 "Model Engineering, Concepts and Tools", Springer-Verlag, Dresden, Germany (September 2002)
10. A. Espinosa, T. Margalef, E. Luque: Integrating Automatic Techniques in a Performance Analysis Session. Lecture Notes in Computer Science (LNCS), Vol. 1900 (Euro-Par 2000), Springer-Verlag (2000) 173-177
11. F. Wolf, B. Morh: Automatic Performance Analysis of SPM Cluster Applications. Technical Report IB-2001-05 (2001)
12. T. Fahringer, A. Požgaj: P^3T+: A Performance Estimator for Distributed and Parallel Programs. Scientific Programming, IOS Press, Vol. 8, no. 2, the Netherlands (2000)
13. B. P. Miller et al.: The Paradyn Parallel Performance Measurement Tool. IEEE Computer 28, 11 (November 1995) 37-46
14. Anne Benoit, Murray Cole, Stephen Gilmore, Jane Hillston: Evaluating the Performance of Skeleton-Based High Level Parallel Programs. Lecture Notes in Computer Science (LNCS), Vol. 3038, Springer-Verlag (2004) 289-296
15. A. Morajko, O. Morajko, J. Jorba, T. Margalef and E. Luque: Automatic Performance Analysis and Dynamic Tuning of Distributed Applications. Parallel Processing Letters, Vol. 13 (2), World Scientific (2003) 169-187

Topic 2
Performance Prediction and Evaluation

Allen D. Malony, Thomas Fahringer, Allan Snavely, and Luís Silva

Topic Chairs

Performance is the reason for parallel computing. Achieving high performance on parallel computer systems is the product of an intimate combination of hardware architecture (processor, memory, interconnection network), system software, runtime environment, algorithms, and application design. Performance evaluation is the science of understanding these factors that contribute to the overall expression of parallel performance on real machines and on systems yet to be realized. Benchmarking and performance characterization methodologies and tools provide an empirical foundation for performance evaluation. Performance prediction techniques provide a means to model performance behaviors and properties as system, algorithm, and software features change, particularly in the context of large-scale parallelism. These two areas are closely related since most prediction requires data to be gathered from measured runs of a program, to identify application signatures or to understand the performance characteristics of current machines.

A total of twenty-nine papers were submitted to the performance prediction and evaluation topic area. The submissions covered a broad range of prediction and evaluation topics, and reflect a high level of current interest in the parallel computing community. The eleven papers accepted (38%) represent state-of-the-art results from leading parallel performance researchers in the field today. The papers cover four general themes in performance prediction and evaluation.

The first theme considers methods to explore performance properties from different evaluation contexts: data access, processor, and interconnect. The understanding gained from looking at these different performance contexts is valuable to forming a more complete performance assessment. The second theme concerns advances in measurement infrastructure for performance analysis at the application level. In particular, the three tools reported illustrate techniques for instrumenting events closely tied to parallel program operation and for capturing performance data needed to correctly interpret performance behavior. Techniques for performance prediction for large-scale parallel systems is the third theme in the topic. The contributions here on performance extrapolation from traces, performance modeling and sensitivity analysis, and performance prediction using machine learning, are especially strong and are important contributions to the field. Lastly, we consider the connection of performance evaluation in tools for performance tuning in the fourth theme. Graphical user interface support for integrated performance environments and automatic tuning for parallel program archetypes are described.

J.C. Cunha and P.D. Medeiros (Eds.): Euro-Par 2005, LNCS 3648, p. 93, 2005.
© Springer-Verlag Berlin Heidelberg 2005

Automatic Tuning of Master/Worker Applications[*]

Anna Morajko, Eduardo César, Paola Caymes-Scutari,
Tomás Margalef, Joan Sorribes, and Emilio Luque

Computer Science Department. Universitat Autònoma de Barcelona, 08193 Bellaterra, Spain
{ania,paola}@aomail.uab.es
{eduardo.cesar,tomas.margalef,joan.sorribes,emilio.luque}@uab.es

Abstract. The Master/Worker paradigm is one of the most commonly used by
parallel/distributed application developers. This paradigm is easy to understand
and is fairly close to the abstract concept of a wide range of applications. However, to obtain adequate performance indexes, such a paradigm must be managed in a very precise way. There are certain features, such as data distribution
or the number of workers, that must be tuned properly in order to obtain such
performance indexes, and in most cases they cannot be tuned statically since
they depend on the particular conditions of each execution. In this context, dynamic tuning seems to be a highly promising approach since it provides the capability to change the parameters during the execution of the application to improve performance. In this paper, we demonstrate the usage of a dynamic
tuning environment that allows for adaptation of the number of workers based
on a theoretical model of Master/Worker behavior. The results show that such
an approach significantly improves the execution time when the application
modifies its behavior during execution.

1   Introduction

The Master/Worker (M/W) paradigm is one of the most commonly used by paral-
lel/distributed application developers. In this paradigm, a master process distributes a
set of data to be processed among a set of worker processes that receives this data,
processes it and returns the results to the master. This structure fairly faithfully represents the developer abstract concept. It can be applied to a wide range of applications
and is therefore fairly easy to treat and manage. However, the actual behavior of this
structure depends on several features (target system, number of available processors,
computing capabilities, communication features, input data) that cannot be controlled
by the application developer and can only be found out during runtime. In order to
reach high performance indexes and eliminate performance bottlenecks, the behavior
of the particular application must be analyzed and problems that appear during the
execution must be determined.

One of the major performance bottlenecks in the Master/Worker paradigm is the
inadequate number of workers. When there are not enough worker processes, the
master process distributes the data and becomes idle as it waits for results. On the
other hand, if there are too many workers, the amount of data is divided into small
pieces and the communications saturate the system. Therefore, it is important to find
an optimal number of workers. This number depends on: the computing volume per

---

[*] This work has been supported by the MCyT (Spain) under contract TIC2001-2592 and has
   been partially supported by the *Generalitat de Catalunya* – GRC 2001SGR-00218

J.C. Cunha and P.D. Medeiros (Eds.): Euro-Par 2005, LNCS 3648, pp. 95–103, 2005.
© Springer-Verlag Berlin Heidelberg 2005

datum, the volume of data sent to and received from the worker processes, the computing capabilities of each of the system's processors and the latency and bandwidth of the communication network.

In many cases, these features are not completely static and change dynamically during the execution of the application (e.g., the computing requirements evolve during execution of the application or the computing capabilities of the processors change due to an additional load in the system). In these situations, the optimal number of workers is not fixed, but changes during the execution of the application and it must be tuned dynamically.

In the following sections of this paper, we present a complete performance optimization scenario that considers the problem of the number of workers in a dynamic approach. Section 2 presents example automatic analysis and tuning environments. In Section 3, we describe the performance model used to calculate the optimal number of workers. In Section 4, we analyze the tuning of the number of workers using the MATE environment that supports the dynamic tuning of parallel applications. In Section 5, we present the results of the experiments conducted in the MATE environment to dynamically tune the number of workers using the presented performance model. Finally, Section 6 shows the conclusions of this study.

2 Related Work

The optimization process requires a developer to go through the application performance analysis and the modification of critical application parameters. First, the performance measurements must be taken in order to provide information about the application. Then, the analysis of this information is carried out. It finds performance bottlenecks, deduces their causes and determines the actions to be taken to eliminate these bottlenecks. Finally, appropriate changes must be applied into the application.

To reduce developers efforts, an automatic analysis has been proposed. Tools using this type of analysis are based on the knowledge of well-known performance problems. They are able to identify critical bottlenecks and help in optimizing applications by giving suggestions to developers [1, 2, 3, 4].

Such tools require a certain degree of knowledge and experience of parallel/distributed applications. To tackle these problems, it is necessary to provide tools that automatically perform program optimizations during run time. Active Harmony [5] is a framework that allows an application for dynamic adaptation to network and resource capacities. The application must be Harmony-aware, that is, to use the API provided by the system. The project focuses on the selection of the most appropriate algorithm. Active Harmony automatically determines good values for tunable parameters by searching the parameter value space using heuristic algorithm. MATE uses a distinct approach in which performance models provide conditions and formulas that describe the application behavior and allow the system to find the optimal values. The AppLeS [6] project has developed an application-level scheduling approach. It combines dynamic system performance information with application-specific models and user specified parameters to provide better schedules. A programmer is supplied information about the computing environment and is given a library to facilitate reactions to changes in available resources. Each application then selects the resources and determines an efficient schedule, trying to improve its own performance without considering other applications. MATE is similar to AppLeS in that it tries to maximize the performance of a single application. However, MATE focuses on the efficiency of resource utilization rather than on resource scheduling.

3 Performance Model for the Number of Workers

In this section, we present the problem of determining a suitable number of workers for a M/W application. We will only consider this problem for homogeneous M/W applications, defining these as applications where all tasks (i.e. a set of data to be processed by each worker) are approximately of the same size and require the same processing time. In actual fact, these kinds of applications exhibit a similar performance to a balanced M/W application with the same total processing time and the same global communication volume, as shown in [7]. This is an important observation, because in homogeneous application it is easier to determine the appropriate number of processors to be used.

For this analysis, we have assumed that the following conditions are met:

- There is just one process (master or worker) per processing element.
- The master process distributes all available data among workers, then waits for all results and, eventually sends a new set of tasks to workers, which means that the application could be iterative.

In addition, we will use the following terminology to identify the different parameters that will form part of the performance model:

- tl = fixed network time overhead per message, in ms.
- λ = communication cost per byte (inverse bandwidth), in ms/byte.
- v_i = size of tasks sent to worker i, in bytes.
- v_m = size of results sent back to master from each worker, in bytes.
- V = total data volume $(\Sigma\ (v_i + v_m))$, in bytes.
- n = current number of workers in the application.
- tc_i = time that worker i spends processing a task, in ms.
- Tc = total computing time $(\Sigma\ tc_i\ )$
- Tt = total time spent on an application iteration (execution time). Our objective is to estimate and minimize this magnitude.
- Nopt = number of workers needed to obtain the minimum Tt (best performance).

It can be seen that the parameters that must be monitored in order to apply the performance model associated to a M/W application are:

- tl and λ which could be calculated at the beginning of the execution and should be re-evaluated periodically to make allowances for the adaptation of the system to the network load conditions.
- Task sizes (v_i) have to be captured when the master sends tasks to workers.
- Result sizes (v_m) have to be captured when the master receives results from workers.
- The time the workers spend on each task (tc_i) has to be measured in order to calculate the total computing time (Tc).

Now, we can describe the analysis performed in order to construct the performance functions associated to this kind of application. We should point out that these functions are defined to enable the optimization of the execution time of the application (Tt).

First, the master sends a set of tasks to each worker. If the communication protocol is asynchronous then the network overhead (tl) for one message overlaps with the communication time of the previous one ($\lambda * v_i$), otherwise both times should be added.

The time spent on this operation is $n*tl+\lambda*\sum_{i=0}^{n-1} v_i$ if the communication protocol is synchronous but, if the protocol is asynchronous then it depends on the relation between the network overhead (tl) and the communication time ($\lambda * v_i$).

If tl is greater than $\lambda * v_i$ (communication time of the tasks sent to one worker) then it is $n*tl$ (network overhead) $+\lambda * v_i$. This is the overhead of sending messages to all workers plus the communication time of the last message, otherwise, it is $tl+\lambda*\sum_{i=0}^{n-1} v_i$. This is the overhead of the first message plus the communication time of all messages.

Then, as every worker spends the same time processing its tasks, we just have to add the processing time of one worker (the last one to receive a task); which is tc_i.

At this point, processing has finished and we must evaluate what happens to the results sent back to the master. We only need to add the communication time for the last message, which is $tl + \lambda*v_m$ (communication time of one answer). This last statement only holds if the master has completed the data distribution before there is an answer from a worker, otherwise it will not be ready to receive messages when the last worker sends its results back.

This never happens before the optimal number of workers if $tl \geq \lambda * v_i$, but may not be true if $tl < \lambda * v_i$ or when the communication protocol is synchronous. In the latter case, the following condition must also hold: the time spent by the master to distribute the tasks ($tl+\lambda*\sum_{i=0}^{n-1} v_i$ or $n*tl+\lambda*\sum_{i=0}^{n-1} v_i$) must be greater than the response time of the first worker ($2*tl+\lambda*v_i+tc_i+\lambda*v_m$).

The expressions to calculate the total iteration time are formed by adding these quantities together, if the communication protocol is synchronous and $n*tl+\lambda*\sum_{i=0}^{n-1} v_i \geq 2*tl+\lambda*v_i+tc_i+\lambda*v_m$ then we get:

$$Tt = n*(tl+1)+\lambda*\sum_{i=0}^{i=n-1} v_i + tc_i + \lambda*v_m$$

But, if the communication protocol is asynchronous we get:

$$Tt = 2*tl+\lambda*\sum_{i=0}^{n-1} v_i + tc_i + \lambda*v_m$$

$$if\,((tl \leq \lambda*v_i)\,and\,(tl+\lambda*\sum_{i=0}^{n-1} v_i > 2*tl+\lambda*v_i+tc_i+\lambda*v_m))$$

Or

$$Tt = n*tl+\lambda*v_i + tc_i + tl + \lambda*vm \quad (if\ tl > \lambda*v_i)$$

Considering that $tc_i = Tc/n$, $v_i = p*V/n$ (a portion p of the overall data volume which is distributed among the workers), and $v_m = (1-p)*V/n$ (the remaining portion of the overall data volume which are the results that workers return to the master) we could rewrite these expressions as:

$$Tt = n*tl + \frac{Tc}{n} + tl + (n-1)*(p+1)*\lambda*\frac{V}{n} \tag{1}$$

if protocol is synchronous and $n*(tl + \lambda*V/n) > 2*tl + \lambda*V/n + Tc/n$

Or

$$Tt = \frac{(2*tl + \lambda*V*p)*n + Tc + \lambda*(1-p)*V}{n} \tag{2}$$

if protocol is asynchronous and $(tl \leq \lambda*p*V/n)$ and

$$(n \leq \lceil (\lambda*V + Tc)/(\lambda*p*V - tl) \rceil)$$

Or

$$Tt = n*tl + \frac{Tc}{n} + tl + \lambda*\frac{V}{n} \quad \text{if protocol is asynchronous and } (tl > \lambda*\frac{p*V}{n}) \tag{3}$$

If we calculate $\delta Tt/\delta n = 0$ for expression (1) then we will obtain an expression to calculate the number of workers needed to minimize Tt when the communication protocol is synchronous, which is:

$$N_{opt} = \frac{1}{2}\sqrt{(\lambda*V + 4Tc)/tl} \tag{4}$$

And, if we calculate $\delta Tt/\delta n = 0$ for expression (3) then we will obtain an expression to calculate the number of workers needed to minimize Tt when the communication protocol is asynchronous, which is:

$$N_{opt} = \sqrt{(\lambda*V + Tc)/tl} \tag{5}$$

We cannot do the same with expression (2) because it can easily be demonstrated that for this expression: $\lim_{n \to \infty} Tt = 0$. But, if the number of workers (n) grows, then the message size (v_i) decreases and, consequently: $tl > \lambda*p*V/n$ when $n > \lambda*V/(2*tl)$. This means that expression (5) can be also applied from the time this condition holds. With expressions (1), (2) and (3), we have a model of the behavior of an application, and we have expressions (4) and (5) to tune the number of workers of the application.

Figure 1 shows the expected execution time for an example M/W application considering expression (2) and compares the results of predicted values to the real execution times. This figure presents also the optimal number of workers provided by expression (4). It can be observed that the predicted behavior matches well the real behavior.

4   Tuning Number of Workers with MATE

The performance model described in the previous section provides the optimal number of workers for a particular situation. However, in many cases the developer of an M/W application cannot know all of the details needed to provide such an optimal number. Moreover, in many cases the conditions change during the execution of the application (for example, systems with shared load) and the optimal number of workers is not fixed, but evolves during the execution of the application. In these cases, number must be adjusted on the fly during the execution of the application.

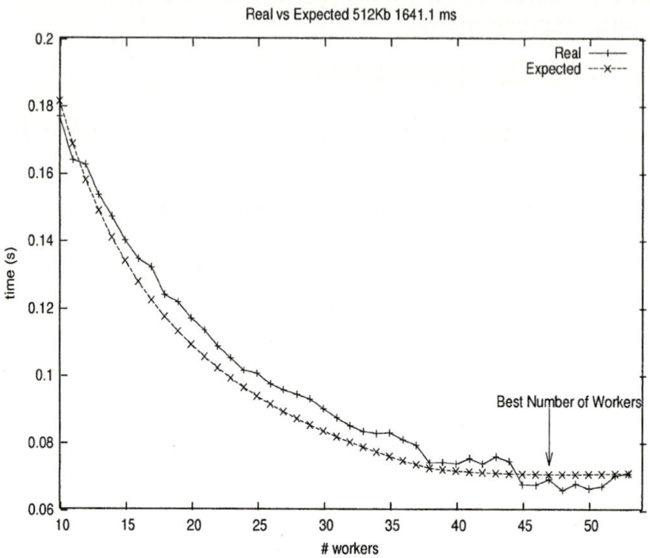

Fig. 1. Real vs. expected execution time, showing the use of expressions (2) and (4)

To provide dynamic automatic tuning of parallel/distributed applications we have developed an environment called MATE (Monitoring, Analysis and Tuning Environment) [8, 9]. MATE performs dynamic tuning in three basic and continuous phases: monitoring, performance analysis and modifications. This environment dynamically and automatically instruments a running application to gather information about the application's behavior. The technique that fulfills these requirements is called dynamic instrumentation [10]. The analysis phase receives events, searches for bottlenecks applying a performance model and determines solutions to overcome such performance bottlenecks. Finally, the application is dynamically tuned by applying the given solution. Moreover, while it is being tuned, the application does not need to be re-compiled, re-linked or restarted. The knowledge to represent the performance model of each particular performance problem is specified in a component called a "tunlet". Each tunlet includes the information about the measure points to insert instrumentation into the target application, the performance model to determine the behavior of the application and the required modifications, and finally, the tuning actions to improve the application's performance.

We have defined two main approaches to tuning: automatic and cooperative. In the automatic approach, an application is treated as a black-box, because no application-specific knowledge is provided by the programmer. This approach attempts to tune any application and does not require the developer to prepare it for tuning (the source code does not need to be adapted). The cooperative approach assumes that the application is tunable and adaptable. This means that developers must prepare the application for the possible changes.

We have conducted a variety of practical experiments on parallel/distributed applications to check whether our approach really works. We have proven that it is effective, profitable, and can be used for a real improvement in program performance. Running applications under MATE control has allowed for adaptation of their behavior to the existing conditions and improvements in their performance.

To dynamically tune the number of workers, we determined conditions that a M/W application must fulfill (as this optimization belongs to the cooperative approach) and implemented a specific tunlet. The application must be based on iterations where all processes repeatedly perform all operations. During each iteration, the master distributes tasks to a specified number of workers and then waits for the results. It must synchronize the results before the next iteration. Tasks being distributed must be independent of each other. In addition, the task processing time cannot depend on the task content, but only on the task size. Finally, worker processes cannot exchange tasks with each other in order to calculate and provide results. The condition of the iteration-based application structure implies the existence of a significant number of iterations. If there is a small number of repetitions, the tuning overhead might be high and the improvement might not be seen.

The tunlet that optimizes the number of workers requires run-time monitoring of the functions responsible for exchanging messages (send and receive), in particular: send entry/exit, receive entry/exit events in the master process, and receive entry/exit and send entry/exit in all worker processes. Instrumenting these functions we are able to perform all measurements required by the performance model presented in Section 3 (expressions (4) and (5)).

The model is evaluated after each iteration when all measurements gathered from that iteration are available. If the computed optimal number of workers differs from the current value, the associated tuning procedure is invoked. In this case, we require the application to be prepared by the developer for the potential changes. The application must contain the specific variable that represents the number of workers. MATE will change this variable automatically. During execution, the application should be aware of the current number of workers and if it is different from the previous one, the new number must be used. This can only be done between two iterations because it is difficult to change the current work distribution that is already being processed. Once the number of workers has been adjusted, the work can be distributed adequately to all running workers.

If there are any new workers to be added, the new machines (processors) are required for them. There is no sense in running a new worker on the same machine where another worker is already running. In such a situation we would not gain anything since the CPU time is divided between both workers.

5   Experimental Results

In this section, the experimental results obtained by applying the tuning environment to a real Master/Worker application are presented. To conduct the experiments, we selected an intensive computing Forest Fire Propagation application called Xfire [11]. The Xfire application is a Master/Worker PVM based implementation of the simulation of the fireline propagation. It calculates the next position of the fireline considering the current fireline position and different aspects such as weather, wind, vegetation, etc. Experiments were conducted on a cluster of homogenous Pentium 4, 1.8 Ghz, (SuSE Linux 8.0) connected by a 100Mb/sec network.

Since we need to control the load in the system to reproduce the experiments several times, we created certain load patterns, so that we can introduce and modify certain external loads to simulate the system's time-sharing. We defined load patterns and executed the application with several fixed number of workers (2, 4, 6, and successively until 26) and also under the control of the MATE tuning environment where the number of workers is adapted dynamically. In every scenario one worker was executed in the same machine as master.

We have conducted our experiments in two scenarios:

- In the first scenario, Xfire was executed on different number of workers, without any tuning.
- In the second scenario Xfire was executed under MATE applying the tuning of the number of workers. The application started with one worker and then during the execution the number is changed according to the model described in Section 3. In this scenario one machine of the cluster was dedicated to run the analyzer, so that the analysis does not introduce additional overhead in the application.

Table 1 summarizes the experimental results. These results are also presented in Figure 2.

Table 1. Execution time of Xfire (in seconds) considering different number of workers, and Xfire under MATE

#workers	1	2	4	6	8	10	12	14	16	18	20	22	24	26
Execution Time	1209	624	345	249	206	181	166	156	144	137	130	129	122	125
Xfire + MATE Execution Time	Starting with 1 worker								141					

Figure 2 shows the execution time of Xfire application considering different number of workers and in the last column the execution time of Xfire under MATE. As it is indicated before, Xfire while executed under control of MATE starts with only one worker. When MATE receives all data from the first iteration, it evaluates the performance model and immediately detects the need of adding workers to reach the optimal number related to the initial total work. Then during the execution of the application the load is changed and the number of workers is adapted to the optimal number provided by the performance model.

It can be observed that execution time of Xfire under MATE is close to the best execution times obtained by different fixed number of workers. However, the re-

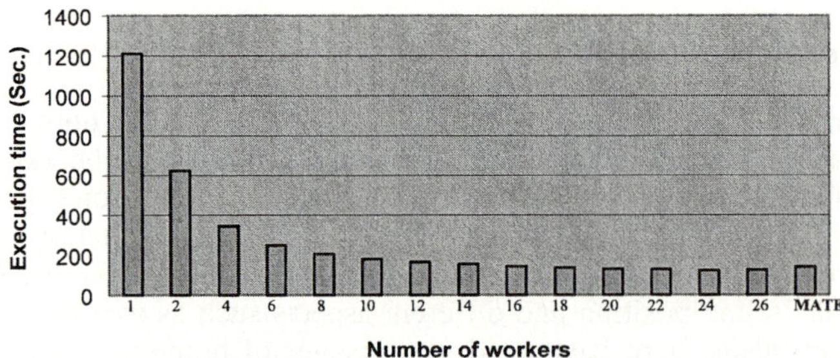

Fig. 2. Execution time of Xfire considering different number of workers and Xfire under MATE

sources devoted to the application using the MATE tuning environment are taken considering the actual requirements of the application and are used when they are really needed.

6 Conclusions

Parallel and distributed programming offer high computing capabilities to users in many scientific research fields. The performance of applications written for such environments is one of the crucial issues. Master/Worker is one of the most significant paradigms in these environments. The number of workers is a key issue in considering the performance of the application.

A performance model to evaluate the optimal number of workers has been presented. This performance model has been incorporated into the MATE automatic tuning environment by the corresponding "tunlet". The presented optimization scenario adapts the number of workers assigned to perform a specified amount of work to changing environment conditions. It requires the application to be prepared for the possible changes, i.e. adding or removing worker processes. MATE is able to estimate the application's performance by means of the analytical model, and to calculate and apply the optimal number of workers. The tuning action changes the number of workers by updating the variable value in the master process.

The experimental results show that the dynamic tuning approach significantly improves the execution times without consuming unnecessary resources when the application is executed under dynamic conditions (changes in the system load).

References

1. Espinosa, A., Margalef, T., Luque, E. "Automatic Performance Analysis of PVM applications". EuroPVM/MPI 2000, LNCS 1908, pp. 47-55. 2000.
2. Wolf, F., Mohr, B., "Automatic Performance Analysis of MPI Applications Based on Event Traces". EuroPar 2000, LNCS 1900, pp. 123-132. 2000.
3. Truong, H.L., Fahringer, T. "Scalea: A Performance Analysis Tool for Distributed and Parallel Programs". EuroPar 2002, LNCS 2400, pp. 75-85. 2002.
4. Miller, B.P., Callaghan, M.D., Cargille, J.M. Hollingswoth, J.K., Irvin, R.B., Karavanic, K.L., Kunchithapadam K., Newhall, T. "The Paradyn Parallel Performance Measurement Tool". IEEE Computer vol. 28. pp. 37-46. November 1995.
5. Tapus, C., Chung, I-H., Hollingsworth, J.K. "Active Harmony: Towards Automated Performance Tuning". SC'02. November 2002.
6. Berman, F., Wolski, R. "Scheduling From the Perspective of the Application". High Performance Distributed Computing 1996. Syracuse, NY, USA, August 1996.
7. César, E., Mesa, J.G., Sorribes, J., Luque, E. "Modeling Master-Worker Applications in POETRIES". IEEE 9th International Workshop HIPS 2004, IPDPS, pp. 22-30. April, 2004.
8. Morajko, A., Morajko, O., Jorba, J., Margalef, T., Luque, E. "Dynamic Performance Tuning of Distributed Programming Libraries". LNCS, 2660, pp. 191-200. 2003.
9. Morajko, A., Morajko, O., Margalef, T., Luque, E.. "MATE: Dynamic Performance Tuning Environment". LNCS, 3149, pp. 98-107. 2004.
10. Buck, B., Hollingsworth, J.K. "An API for Runtime Code Patching". University of Maryland, Computer Science Department, Journal of High Performance Computing Applications. 2000.
11. Jorba, J., Margalef, T., Luque, E., Andre, J, Viegas, D.X. "Application of Parallel Computing to the Simulation of Forest Fire Propagation", Proc. 3rd International Conference in Forest Fire Propagation, Vol. 1, pp. 891-900. Portugal, November 1998.

Performance Cockpit:
An Extensible GUI Platform
for Performance Tools*

Tianchao Li and Michael Gerndt

Institut für Informatik, Technische Universität München,
Boltzmannstr. 3, D-85748 Garching bei Muñchen, Germany
{lit,gerndt}@in.tum.de

Abstract. Within the EP-Cache project, the Performance Cockpit has
been developed to provide a unified GUI for a series of performance tools.
This is achieved through the establishment of a general extensible architecture and the application of standardized intermediate representations
of program structures. This paper describes the design and implementation of this platform, and discusses the future evolvement into a universal GUI platform for performance tools independent of programming
language and programming paradigms.

1   Introduction

Performance tools are commonly used in high performance computing in order
to understand and correct the performance problems of sequential and parallel
codes. Such tools monitor a program's execution and produce performance data
that can be analyzed to locate and understand areas of poor performance.

There are a number of performance tools, both research and commercial.
Many of these tools are language-dependent and can be applied to high performance programs written in one or more of FORTRAN, C, C++ etc, while some
are language-independent. There are also different programming paradigms, typically shared memory (PThread, OpenMP) and message passing (MPI, PVM).
The support for those programming paradigms also varies.

The most prevalent approach taken by these tools is to collect performance
data during program execution and then provide post-mortem analysis and display of performance information. Some tools do both steps in an integrated manner, while other tools or tool components provide just one of these functions. A
few tools also have the capability for run-time analysis, either in addition to or
instead of post-mortem analysis.

Typically, each performance tool provides a customized user interface for
showing the structure of the application, specifying the target of measurement,
controlling the measurement execution and displays the result of measurement.
The diversity of those user interfaces demands a lot of time for studying the

---

* The work presented in this paper is mainly performed in the context of the EP-
Cache Project, funded by the German Federal Ministry of Education and Research
(BMBF)

J.C. Cunha and P.D. Medeiros (Eds.): Euro-Par 2005, LNCS 3648, pp. 104–113, 2005.

usage of a new performance tool, and makes it hard to integrate and incorporate different performance tools.

While the exact sequence can be different, performance measurement typically includes the common procedures: instrumenting the monitored program (source or binary, static or dynamic), linking the instrumented program with specific runtime libraries, program execution and result retrieval (online or post-mortem), and result display and/or analysis. This provides us the possibility to set up a general infrastructure to support different performance tools. In this infrastructure, a performance tool independent platform serves as the basis that can be extended by individual modules (plug-ins) to support different performance tools.

In the German EP-Cache project, Performance Cockpit has been implemented as an extensible GUI platform for a series of performance tools. Based on the open source tooling platform Eclipse [4], this platform supports both post-mortem (i.e. CPTE) and online monitoring (i.e. EPCM) environments, and is intended to integrate other performance tools. The Performance Cockpit serves as the starting point for the development of a universal GUI that is neutral to both programming languages and programming paradigms.

The remainder of this paper is organized as follows. Section 2 introduces the performance monitoring tools developed in the EP-Cache project, and discusses the need for a common GUI platform. Section 3 discusses the major issues involved in the design and development, including the establishment of a general extensible architecture and the standardization of information representations. Section 4 presents the defined architecture, and section 5 introduces the GUI and its typical scenario of usage. Section 6 discussed activities related with our development, and Section 7 looks forward to the development of a universal platform for performance tools based on the efforts in Performance Cockpit. The paper concludes with a short summary in Section 8.

2   The Need for a Common GUI Platform

The EP-Cache (Efficient Parallel Programming of Cache Architectures) project is a three-year research project funded by the German Federal Ministry for Education and Research. The goal of this project is to develop new performance analysis tools and performance tuning techniques with which programs can be improved to efficiently utilize the underlying cache architectures, especially on SMPs. As a fundamental part of the EP-Cache project, existing performance measurement tools are evaluated and new tools either using existing hardware counters and/or based on a novel hardware monitor design [7] that reveals further details on the access behaviors for individual data structures and code regions are developed. These include the Counter-based Profiling and Tracing Environment (CPTE) [2] and the EP-Cache Monitor (EPCM) [5].

CPTE is a performance monitoring tool based on hardware performance counters. It provides profiling, tracing, and sampling for arbitrary program regions. Performance measurement with CPTE is done with the following steps -

instrumentation of the program, specification of measurements, program execution and generation of the measured values, and analysis of resulting performance data. Measurement results are produced in the form of a trace file which may contain measurements for individual instances of a region, and/or summaries of all instances of a region. The results can also be transformed for visualization with KCachegrind [6].

EPCM is a data-structure centric performance monitoring tool. It is based on a novel hardware monitor [7] designed to be integrated into cache controllers which provides counters that can be configured to measure events for certain address ranges, and record the accesses in the form of event counts and access histograms. As the hardware monitor is not available, EPCM is actually implemented on top of a simulator that provides runtime instrumentation of application binary, on-the-fly simulation of the cache access behavior and performance monitoring for multi-processor shared memory systems [14]. EPCM provides Monitoring Request Interface (MRI, ref. [8]), through which performance analysis tools can specify monitoring requests and retrieve monitoring data in online fashion. EPCM also generates trace records compatible with VAMPIR [16] that is extended with OpenMP, data structure and histogram support.

CPTE and EPCM share some similarities in that both environments are targeted to Fortran 95 OpenMP programs and extendable for other programming languages and programming paradigms provided that the specific instrumenters are available. Both require selective code-region instrumentation in user specified source files and region types. The differences between these two environments are even more evident. The post-mortem data analysis in CPTE and online monitoring and analysis in EPCM requires different procedures in the measurement. EPCM's support for code-regions involves additional code region instrumentation and different specifications of measurement targets. The measurement results are also different for CPTE and EPCM in both content and format, and are to be visualized with different visualization and analysis tools.

In order to ease the usage, graphical user interfaces (GUIs) are demanded for both CPTE and EPCM. Taken into consideration of the vast differences between those two environments, it might seem a natural choice to develop separate GUI for each of those platforms. However, this leaves many problems like duplicate work for the common features, low maintainability, inconsistent in the user interface and low inter-operability. Instead, we have chosen another approach - to implement a common GUI to support both these monitoring environments as well as other existing and future monitoring environments through the establishment of a common extensible infrastructure, namely the Performance Cockpit.

3   Key Issues in Design and Implementation

In the design and implementation of such an extensible GUI platform as the Performance Cockpit, the major issues to be considered include the establishment of a general extensible architecture and the standardization of information representations.

3.1 Define General Extensible Architecture

A general architecture should be constructed for integrating different performance tools through extension. Generality and extensibility are the major considerations of the defined architecture. While the powerful extension mechanism from Eclipse provides extensibility, the GUI elements required by different tools are to be studied and organized with respect to their nature for generality.

The generic GUI elements and the underlying supporting mechanisms form the basic platform, and the tool-specific elements are to be grouped into individual extension modules, i.e. plug-ins. Each plug-in extends the basic platform through properly defined interfaces, i.e. extension points. The interface between the basic platform and the extensions should be defined generic enough to allow possible situations of extensions.

For more details of the established architecture, please refer to Section 4.

3.2 Define Standard Representation for Relevant Information

For the interaction between Performance Cockpit and the different performance tools that are integrated, standardized representation should be defined for all relevant information. The information includes program code region structure, program instrumentation targets and/or monitoring requests, as well as the measured performance data.

For the program code region structure, we have participated in the development of Standardized Intermediate Representation (SIR) [13], a standardized abstract representation of program structure for Fortran 95, Java, C and C++ programs defined in the APART working group [1]. SIR is defined in the format of XML document; each SIR is a XML document following the DTD or XML schema definition for SIR. SIR is intended to be used by performance tools and contains only high-level information about positions and types of statements and directives (e.g. OpenMP) that represent the coarse structure of programs, as opposed to more complicated intermediate languages like WHIRL used in the Open64 compiler suite [9]. This simplicity helps keep SIR compact and applicable for both procedural and object-oriented programs of various languages.

For the information of program instrumentation and monitoring, common formats that are general enough for the performance tools of EP-Cache project are also defined.

4 The General Extensible Architecture

A general architecture for integrating different performance tools has been constructed (see Figure 1). This architecture follows a layered design and is based on the extension mechanism provided by Eclipse.

In this architecture, the generic functions including the management of monitoring projects (new project or example project), configuration of common preference and project properties, and the management of platform extensions forms the performance platform. The support for code instrumentation is provided with separate instrumentation plug-in, each for a different instrumenter. And the concrete support for different underlying monitoring platforms, either based on hardware counters (the CPTE platform), or software simulators (the EPCM platform) are implemented as separate plug-ins.

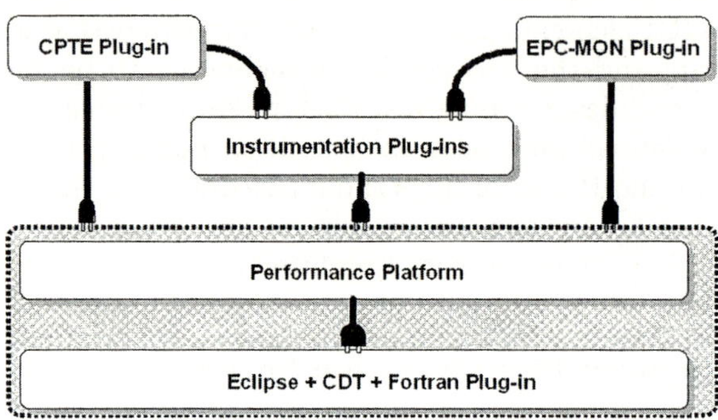

Fig. 1. The General Extensible Architecture for Performance Cockpit

4.1   Eclipse, CDT and Fortran Plug-In

Eclipse [4] is a kind of universal tooling platform - an open-source extensible IDE for the integration of various software development tools. Eclipse represents a component-based approach for software development, which promotes a view of software development in which applications are composed out of reusable, relatively large-grained, and mostly pre-existing components.

The C/C++ Development Tools (CDT) [3] provides a full functional C and C++ IDE for the Eclipse platform. It provides support for C/C++ edit, build, launch, and debug. For project building, CDT incorporates a *standard make* feature (a term used by Eclipse to represent a group of tightly related plug-ins) that support standard makefiles.

By the time of implementation, Fortran support for Eclipse is not available, and a self-developed plug-in has been developed for simple Fortran support. It implements a Fortran 95 editor with simple syntax highlighting, and supports the building of Fortran applications by reusing the incrementally build functionality provided by the makefile support of CDT. This part can be replaced once an advanced Fortran plug-in such as Photran [11] becomes mature.

4.2   The Performance Platform

The performance platform provides the basis for integration and extension for different performance tools. It is based on Eclipse, Eclipse CDT and the self-developed Fortran plug-in. The major function of this platform includes the management of the performance projects, the management of common properties and preferences, and the management of performance tools extensions. For the management of performance tools extensions, custom interfaces (i.e. extension points), are defined by the performance platform, through which the individual performance tools can be discovered and integrated.

4.3   Instrumentation Plug-Ins

Code instrumentation is a separate process from performance measurement, and is often shared among tools. For each specific code instrumenter, a separate instrumentation plug-in is to be implemented. Each of these plug-ins provides GUI elements for the instrumentation of the whole project or selected files of a certain type (e.g. Fortran programs), and directs the underlying instrumenter upon user control. The instrumenter also generates information about the source code structure, in the format of SIR, which will later be read by the GUI and the individual tools plug-ins.

4.4   Tools Plug-Ins

Each performance tool requires the development of an individual plug-in to be integrated into the performance platform. The responsibility of each plug-in includes the translation of standard-based data representation to tool-specific data formats, providing custom GUI elements, and the interaction with the underlying tools.

The plug-in for each performance tool must implement certain interfaces to be managed by the performance platform. User interactions with the common GUI elements are processed by the performance platform and translated into appropriate function calls as defined in the interfaces. User interactions with the custom GUI elements are directly handled by the tool plug-in.

5   The Performance Cockpit GUI

5.1   GUI Elements

In terms of Eclipse, the common GUI elements for performance monitoring provided by the Performance Cockpit include:

Monitoring Perspective: This perspective organizes all relevant components into a role-oriented GUI to the user of the monitoring environment.

Project Creation Wizards: These wizards help create projects that are either empty or containing example programs. Projects created with these wizards are marked with *Monitoring Nature*, which is identified later by other components of the Performance Cockpit.

Monitoring Resource Explorer: As a resource explorer customized for our measurement environment, it provides standard project and file manipulation functions; however all unnecessary details are hidden from the user, such as the

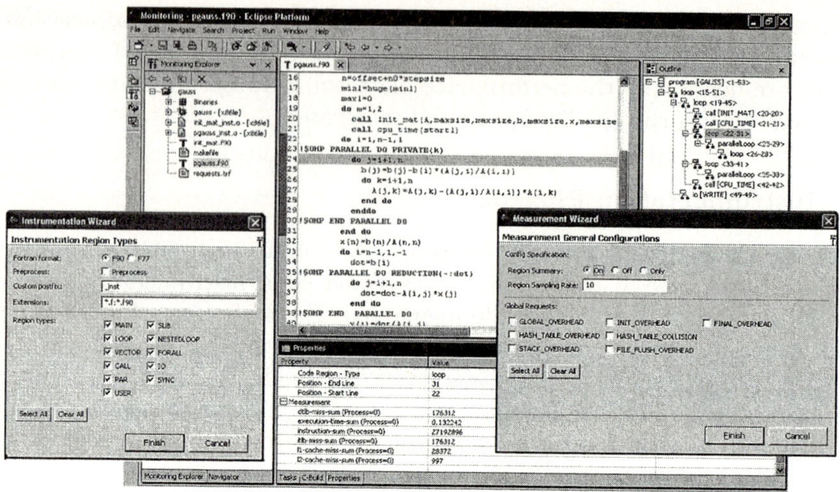

Fig. 2. Extensible GUI Platform for Performance Tools

files created during the process of instrumentation and internal configuration files.

Monitoring Environment Preferences: This enables required configurations for the monitoring environment, such as the path of instrumenter executable, the path of result format converter etc.

Instrumentation Wizards: These wizards guide the user through the process of instrumentation either for selected files or for the whole project.

Code Regions Outline View: This view provides an outline of code regions for the active editor, according to the result of instrumentation. Context menu items of this view also allow users to add/remove certain code region(s) into/from target of measurement.

Code Region Properties View: This view displays available properties and measurement results of individual code regions, in response to user selections in the Code Regions Outline View.

In the current implementation of the Performance Cockpit, the following performance tool specific GUI elements are defined by the plug-ins for CPTE and EPCM environment:

Measurement Wizards: The measurement wizards guide the user through the process of measurement. A separate wizard is defined for each of the performance tools. For example, the measurement wizard for CPTE directs the user to specify parameters and general requests for the measurement, generates configuration file, and launches the program measurement.

Measurement Result Views: These views display available properties and measurement results. For example, the EPCM displays the result of each measurement request as a single count or a histogram.

Visualization Wizards: These wizards guide the user to choose and invoke appropriate external visualizers to display the measurement results.

The above GUI elements, either provided by the common performance platform or contributed by the individual performance tools plug-ins, are seamlessly integrated with the Eclipse platform. From the user's perspective, those two types of GUI elements are not distinguishable (see Figure 2).

All components described above are grouped into a *Monitoring Feature*, which allows the whole platform to be installed and updated in a way that coexists with other Eclipse based systems.

5.2   Usage Scenario

The process of measurement using Performance Cockpit can be summarized as follows. The user creates a project with one of the Project Creation Wizards, and then the user can manipulate the content of the project with the Monitoring Resource Explorer. To do instrumentation, the user can right click on the whole project or selected files and start Project Instrumentation Wizard or Files Instrumentation Wizard from the context menu. After specifying the required parameters like region types to be instrumented, the instrumenter is invoked by the wizard. The user can then open the instrumented files in the Fortran Source Editor, examine the code regions in the Code Regions Outline View, and specify local measurement requests for specific code regions. After building the program, the user invokes the Measurement Wizard for measurement that will guide the user through out the measurement process, which varies from tool to tool. In any case, the wizard will launch the program for execution. Once the execution finishes, the user can choose appropriate Measurement Result Views to examine the individual measurement, or invoke visualization tools that are integrated in the platform through the help of Visualization Wizards.

6   Related Works

Previous attempts to construct general interfaces for instrumentation and visualization also exist in other parallel tool groups. The Pablo project [10] at the University of Illinois has implemented svPablo, a graphical interface for instrumenting source code and browsing runtime performance data. The Tool Gear project [15] at LLNL is a GUI tool and database for dynamic instrumentation and display of the instrumentation results. However, the extensibility and flexibility of such tools are not comparable to our Performance Cockpit. In fact, taken the vast differences between instrumentation and measurement tools (e.g. consider just profiling vs. tracing tools), opportunities for integration can be only guaranteed by a general extensible platform like the Performance Cockpit.

Existing Eclipse-based GUIs for performance tools include the Eclipse OProfile plug-in as a CDT contribution and the Intel VTune Performance Analyzer for Linux [17]. Both are specific to the underlying tool (OProfile and VTune), and none of them address the extensibility and coexistence with other performance tools. The tight dependence with Eclipse CDT also makes Eclipse OProfile plug-in restricted to C/C++. Intel VTune Performance Analyzer for Linux supports multiple languages, including Intel Visual Fortran, Java and languages supported by the Linux GNU Compiler Collection (GCC); however the proprietary nature of this product and its closed internal data models makes integration with other tools impractical.

The recently proposed PTP (Parallel Tools Platform) project [12] aims to extend the Eclipse framework to support a rich set of parallel programming languages and paradigms, and provide a core infrastructure for the integration of a wide variety of parallel tools. Although the PTP is still in the initial status of proposition, it casts new light on the construction of a generic platform including performance monitoring. We have expressed our interest in this project and will actively participate in the discussions to influence the design so that the work in PTP and our work can be seamlessly integrated.

7   Towards a Universal Platform for Performance Tools

The Performance Cockpit provides the basis for the future development of a universal integration platform for performance tools. Such a platform will be beneficial to users and developers of all performance tools in that it provides a consistent user experience and gentle learn curve, enables interoperability among performance tools, reduces redundant work by reusing common functions, etc. It is intended to be programming language neutral, programming paradigm neutral, and performance tool neutral.

While the extension architecture and the standardized representation of information defined in its development generally enables the step-forward towards universal platform, further efforts are required. In order to be performance tool neutral, the currently defined architecture should be refined, and standardized representation of more types of information (e.g. the trace record) should be defined. This requires of course the examination of a large amount of existing performance tools and identify the commonalities and specialties. This will also involve a lot of compromise between generality and functionality.

For Eclipse, supports to programming languages and programming paradigms are usually provided by extensions from different parties. The integration of Performance Cockpit with those diverse programming extensions constitutes another challenge, and will foreseenably result into changes to the general architecture and implementation. For example, the PTP described above that provides support for parallel programming will be integrated as part of the underlying platform.

8   Conclusions

Performance is a very important factor that drives the development of computing. Code optimization with the help of performance tools is one of the major measures to achieve better performance. However, the existing performance tools usually have different graphical user interfaces and results into difficulty in the usage and poor interoperability.

In the EP-Cache project, the Performance Cockpit, a GUI platform that provides a unified user interface for a series of performance tools, has been developed. Compared to other GUIs for performance tools, the Performance Cockpit excels in its easy learning and usage, its extensibility and interoperability. The general extensible architecture and standard representations for related information that are defined in the development of Performance Cockpit provide the basis for the future development of a universal platform for performance tools. The integration of performance tools with the Eclipse environment would also allow programmers of high-performance systems to exploit the general advantages of the integrated interactive development environment.

References

1. APART Working Group. http://www.fz-juelich.de/apart/
2. M. Gerndt, T. Li: *Automated Analysis of Memory Access Behavior*, Proceedings of HIPS-HPGC 2005, Denver Colorado, April, 2005
3. Eclipse C/C++ Development Tools. http://www.eclipse.org/cdt/
4. Eclipse. http://www.eclipse.org
5. E. Kereku, T. Li, M. Gerndt, and J. Weidendorfer: *A Selective Data Structure Monitoring Environment for Fortran OpenMP Programs*, Proceedings of Euro-Par 2004, Pisa, Italy, Aug. 31th - Sept. 3rd, 2004
6. KCachegrind. http://kcachegrind.sourceforge.net
7. M. Schulz, J. Tao, J. Jeitner, W. Karl: *A Proposal for a New Hardware Cache Monitoring Architecture*, Proceedings of MSP 2002, Berlin, Germany. June 2002
8. M. Gerndt, E. Kereku: *Monitoring Request Interface Version 1.0*, http://wwwbode.in.tum.de/~kereku/projects/epcache/pub/MRI.pdf
9. Open64 Compiler Tools. http://open64.sourceforge.net
10. Pablo Research Group. http://www.renci.unc.edu
11. Photran. http://www.photran.org
12. Eclipse Parallel Tools Platform. http://www.eclipse.org/ptp/
13. C. Seragiotto et. al.: *Standardized Interfaces for Representing, Instrumenting, and Monitoring Fortran, Java, C and C++ Programs*, Concurrency and Computation: Practice and Experience, submitted.
14. SMART: A Simulation Tool for Monitoring Cache Access Behavior on SMPs, http://wwwbode.cs.tum.edu/~lit/smart/
15. Tool Gear. http://www.llnl.gov/CASC/tool_gear/
16. VAMPIR. http://www.pallas.com/pages/vampir.htm, www.tu-dresden.de/zhr/
17. VTune Performance Analyzer for Linux. http://www.intel.com/software/products/vtune/vlin/index.htm

Apex-Map: A Synthetic Scalable Benchmark Probe to Explore Data Access Performance on Highly Parallel Systems

Erich Strohmaier and Hongzhang Shan

Future Technology Group, CRD, Lawrence Berkeley National Laboratory
One Cyclotron Road, Berkeley, CA 94720
{estrohmaier,hshan}@lbl.gov

Abstract. With the increasing gap between processor, memory, and interconnect speed, the performances of scientific applications on high performance computing systems have become dominated by the ability to move global data. However, many benchmarks in the field of high performance computing focus on measuring the achieved CPU speed in MFlop/s. In this paper, we introduced a novel benchmark, Apex-Map, which focuses on global data movement and measures how fast global data can be fed into computational units. Apex-Map is a parameterized synthetic performance probe and integrates concepts for temporal and spatial locality into its design. By measuring the Apex-Map performance for a whole range of temporal and spatial localities performance surfaces can be generated which can be used to study the characteristics of the computational platforms and which are useful for performance comparison. Results on a vector platform and two superscalar platforms clearly reflect the design differences between these two types of systems.

1 Introduction

Benchmarking of high performance computing has often focused on floating point performance. One prominent example of this is the Linpack benchmark, which is used to rank systems in the TOP500 Project [1]. However, the performance of Linpack is in general not a good performance indicator for real applications. On most platforms, Linpack can achieve over 70% of peak performance while on the same systems many real applications might only achieve substantially lower performances.

With the increasing gap between CPU speed and memory speed, the capability to load and store data locally and globally has become the dominant performance factor for many applications. System designers are spending enormous efforts to design complex memory systems and interconnect networks to increase the data transfer bandwidths and reduce latencies. However, we still lack a quantitative methodology to relate changes in computer architectures to improvements in application performances. There even still is no standard or widely accepted way to measure progress in our ability to access globally distributed data. STREAM [2] is often used to measure memory bandwidth but its use is limited to at the most a single shared memory node. Recently, the HPC Challenge benchmark [3] has included the RandomAccess benchmark, to measure the rate of integer random updates of memory. Unfortunately, this benchmark cannot easily be related to scientific applications and thus does not help much for applications performances.

In this paper, we introduced a novel synthetic memory access probe, called Apex-Map [4], to measure global data access performance. Apex-Map has three main parameters, the global memory size M used, the temporal locality α, and the spatial

J.C. Cunha and P.D. Medeiros (Eds.): Euro-Par 2005, LNCS 3648, pp. 114–123, 2005.

locality L. Our basic assumption is that an application's global memory access can be approximated by multiple data access streams, each of which can be characterized with the three parameters introduced above. The execution profile of Apex-Map can then be tuned by its set of input parameters to match the data access characteristics of a chosen scientific application. This allows us to use Apex-Map as a performance proxy for the actual codes. An advantage of our synthetic benchmark probe is that due to its simplicity it can easily be run by simulators. This allows its usage in the early stages of architecture design.

Another feature that distinguishes Apex-Map from many other benchmarks is that its input parameters can be varied independent of each other between extreme values. This allows generating continuous performance surfaces to explore the performance effects of all potential values of the characterizing parameters. By examining these surfaces, we can understand how changes in spatial or temporal locality affect the performances of applications and which factors are more important for performance. Moreover, we can compare these performance surfaces across different platforms and explore the advantages and disadvantages of each platform. Most current benchmark suits (HPCC, NAS [5], and SPEC [6]) only contain several application codes or their synthetic benchmarks have other features strongly limiting the scope of performance behaviors they can explore. The results of these application benchmarks provide very good indications how similar applications will perform on a specific platform. However, these benchmarks are not very helpful for other applications, as their performances cannot be related directly to them.

The design details of Apex-Map are described in Section 2. In Section 3, we analyze our results on our three test platforms, two superscalar platforms and one vector platform. We find that the Apex-Map performance results clearly reflect the design differences between the superscalar and the vector platforms. Finally, we analyze the scalability of these three platforms based on the Apex-Map results. Section 4 summarize our results and discusses our ongoing and future work.

2 Implementation

The parallel implementation of Apex-Map uses the same concept as the sequential version [7]. It has the same three main parameters, the global memory size M, the temporal locality α, and the spatial locality L. These parameters are related to our methodology to characterize application performances. Apex-Map assumes that the performance of a data access pattern of an application can be approximated by combining a blocked access to memory with length L with a non-uniform random address determined by α. In Apex-Map a global data-array of size M is evenly distributed across all processes as illustrated in Fig. 1. Data will be accessed in block mode, i.e., L continuous memory addresses will be accessed in succession and the block length L is used to characterize spatial locality. The starting addresses X of these data blocks are computed by using a non-uniform random address generator driven by a power function with the shape parameter α. A power function was chosen as generating function as a simple scale-invariant, one-parameter approximation for the behavior of real applications.

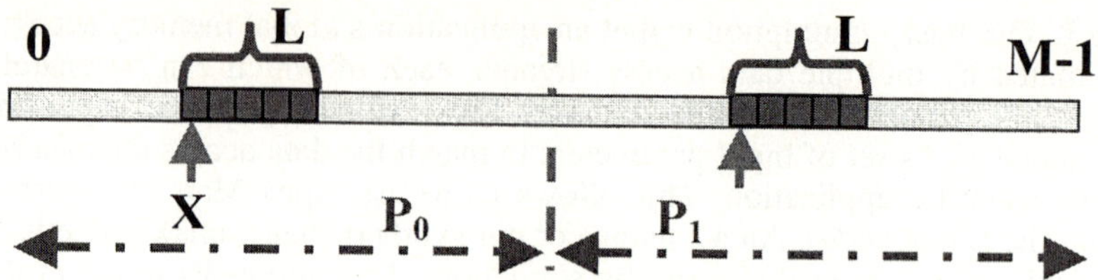

Fig. 1. Apex-Map Data Distribution and Data Access

Table 1. The flowchart of the Apex-Map implementation

Basic Parallel	MPI
Repeat N Times	Repeat N Times
Generate Index Array	Generate Index Array
CLOCK(start)	CLOCK(start)
For each Index i in the Array	For each Index i in the Array
If (not local data)	If (local data)
Get Remote Data	Compute
End If	Else
Compute	Generate Remote Request
CLOCK(end)	End If
RunningTime += end − start;	Serve Incoming Requests
End Repeat	Process Replies
	CLOCK(end)
	RunningTime += end − start
	End Repeat
	CLOCK(start)
	Wait For Finish
	CLOCK(end)
	RunningTime += end − start

The basic flowchart of the plain parallel version of Apex-Map is shown in the left side of Table 1. The indices X are generated and stored in an index array first before the measurement starts. Then, for each index it is tested, if the addressed data resides in local memory in which case the computation proceeds immediately, or if it resides in remote memory in which case it is fetched into local memory first. Apex-Map is designed to measure the rate at which global data can be fed not only into the memory or into cache but into the CPU itself. Therefore, it is essential that an actual computation is performed in the Compute module, which currently is a global sum of all accessed array elements.

The pre-computed indices X are stored in an array of size I. The indices are generated based on a power distribution based random function, which is controlled by the parameters M, L, and α. Generated addresses are shifted so that each process accesses its own memory with the highest probability. The frequency with which remote data access occurs is determined by the temporal locality parameter α. For 256 processes and $\alpha = 1$, the data accesses follow a uniform random distribution and the percentage of remote access is 255/256 (=99.6%). With the increase of temporal locality, the percentage reduces to 0.55% for $\alpha = 0.001$.

The main output of Apex-Map is the average cycles per data access for one process and the aggregate bandwidth in MB/s for the given parameters. The results are directly comparable across different platforms. By running a set of parameters, such as

$\alpha = 0.001$ to 1.0 and $L = 1$ to 16384 words, Apex-Map can generate a performance surface to explore the performance effects of temporal locality and spatial locality.

2.1 MPI Implementation

One major non-trivial issue that has not been discussed until now is how the remote access is carried out. The implementation could be highly affected by the available parallel programming paradigm and different programming styles. We assume that the operation for different indices is independent and multiple remote accesses can be executed on the fly at the same time. Our first version was developed using two-sided MPI since it is the most popular and portable parallel programming model available today.

Even if we only consider MPI, there are many implementations thinkable. One possibility is to aggregate the remote requests instead of sending them one by one. We explored several different strategies to do this in depth, but had to conclude, that we ended up only benchmarking our inventiveness for new algorithms to assemble and exchange these messages and our skills to implement them. This approach not only further complicates the code, but also conflicts with our locality concept. By extensively rearranging the order of data-accesses, the actual executed address stream will no longer show the intended features to achieve the given localities. In effect, such rearranging would substantially change the actual localities from the intended localities and would go contrary against our design principles. We therefore decided not to permit such message aggregation and to exchange messages for each remote access.

However, we permit multiple outstanding requests for data and out-of-order processing of the received data. Since in Apex-Map the process numbers for message exchanges are generated based on a non-uniform random access, non-blocking, asynchronous MPI functions are used to avoid blocking and deadlock. Given our non-deterministic random message pattern it was not clear if a scalable implementation of Apex-Map in MPI was possible. However, we succeeded with an efficient and scalable implementation, which shows increasing performance up to 1000s of processors.

Due to the unpredictable communication patterns, the flowchart becomes substantially more complex (see the right side of Table 1) and several MPI related implementation parameters have to be introduced. The first parameter is B, the number of receive buffers allocated, which are needed for each call of MPI_Irecv. It defines the maximum possible number of concurrent outstanding remote data requests per process. Another parameter is SMSG, the maximum number of outstanding send handles defined for MPI_Isend. The last parameter is NSER, with which we limit how many remote requests can be served at one time by our Serving Incoming Requests module. This parameter is especially useful when the remote request distribution is imbalanced. Without this parameter, a process may get completely stuck in serving remote requests for a long time and might not make any progress on its own local computation, which would cause a severe load-imbalance at the end of the global execution.

In summary, there are three kinds of Apex-Map parameters. The first category of parameters includes M, L and α, which are the characteristic parameters of interest. The second category includes general implementation related parameters, including the index array size I and the number of times N the experiment is repeated. The third category includes parameters related to the MPI implementation such as the number of receive buffer B, the number of send handles SMSG, and the maximum number of served requests in one iteration NSER. Fortunately, experiments on several systems indicate that our default values for all implementation parameters work reasonably well on all of them. The "Wait For Finish" module is needed for MPI because even if a process has finished its own task, it may still need to provide data for other processes and hence cannot complete its execution.

3 Results and Analysis

In this section, we first introduce the three platforms we tested, two superscalar platforms and one vector platform. Then, we analyze the relation of the results of Apex-Map and the PingPong benchmark, as a traditional measure for global communication performance. Finally, we compare the Apex-Map results between the three platforms and examine how the Apex-Map results reflect their architectural differences.

Table 2. Some characteristics of the three platforms used

	CPU	Memory Bandwidth	Network
Seaborg	IBM Power3, 375 MHz	16 GB/s /node 1 GB/s /processor	IBM Colony-II, 1 GB/s /node
Cheetah	IBM Power4, 1.3 GHz	44 GB/s /node 1.375 GB/s /processor	IBM Federation, 4 GB/s /node
Phoenix	Cray X1, 400 MHz, (800 MHz for vector units)	25.6 GB/s/ MSP	Cray SeaStar 25 GB/s /node

3.1 Three Platforms: Seaborg, Cheetah, and Phoenix

Seaborg is currently the main computing platform of NERSC, a DOE Office of Science user facility at Lawrence Berkeley National Laboratory. It is an IBM Power3 based distributed memory machine. Each node has 16 IBM Power3 processors running at the speed of 375 MHz. The peak performance of each processor is 1.5 Gflop/s. Its network switch is the IBM Colony II, which is connected to two "GX Bus Colony" network adapters per node.

Cheetah is a 27-node IBM p690 system with the IBM Federated switch, where each node has 32 Power4 processors at 1.3 GHz. The peak performance of each processor is 5.2 Gflop/s. Phoenix is a Cray X1 platform consisting of 512 multi-streaming vector processors. Each MSP has four single-stream vector processors and a 2 MB cache. Four MSPs form a node with 16 GB of shared memory. The inter-connect functions as an extension of the memory system, offering each node direct access to memories on other nodes. These two machines are currently operated by the center for Computational Sciences at Oak Ridge National Laboratory. Table 2 lists some main characteristics of these three systems.

3.2 Relationship with PingPong Performance

The PingPong benchmark performance is a well-accepted performance number of parallel systems. In this subsection, we are going to examine the relationships between Apex-Map and PingPong on the above three platforms. The inter-node Ping-Pong performance is measured with one process sending data while the other process is receiving them. The code used was obtained from the Pallas MPI benchmarks [8].

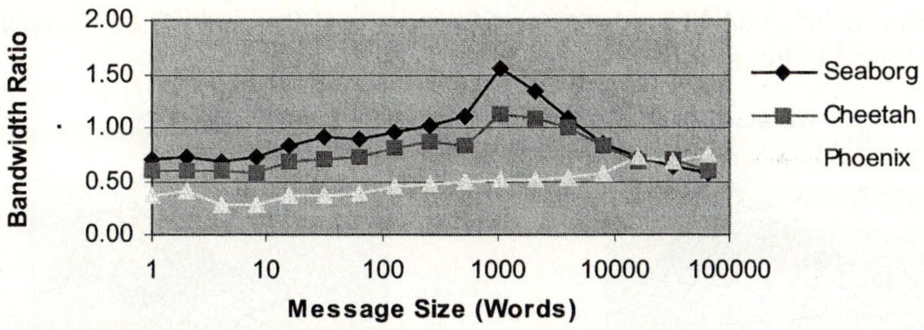

Fig. 2. The performance ratio between Apex-Map (α =1.0) and PingPong

We plot the relative performance of Apex-Map to PingPong in Fig. 2. The inter-node Apex-Map bandwidth per process is obtained with α =1.0 (uniform random data access) and M = 64 Mwords using two Apex-Map processes. Unlike PingPong, Apex-Map measures the performance of non-uniform random access. The communication pattern is unpredictable and the code overhead for it is substantially higher. These factors contribute to the lower performance of Apex-Map when the message size is small. With the increase of message size, the constant overhead becomes less and less important and the Apex-Map performance gets closer to that of PingPong. On Seaborg, Apex-Map performance becomes 60% better than PingPong when message size reaches 1024 words. If we only count the number of exchanged messages and of local memory accesses, Apex-Map should perform 200% better than PingPong since only 50% of the accesses are remote access when α = 1. However, beyond the message size of 1024 words, the performance ratio begins to drop. The main reason here is that Apex-Map measures how fast the data can be fed into the CPU. After remote data arrive in local memory, they further have to be brought into cache and registers for the global sum computation. The effect of this computation can be ignored for smaller messages but is more substantial for large messages on superscalar platforms such as Seaborg. The performance ratio on Cheetah is similar to Seaborg but the MPI-overhead seems to be more severe.

On Phoenix, the performance ratio of Apex-Map to PingPong for smaller messages is even smaller than on the IBM platforms. There also are further differences in the MPI implementations on these two different systems. On Phoenix, using multiple receive buffers in Apex-Map does not improve the performance at all while on Seaborg and Cheetah, the performances benefit substantially from using multiple buffers. Phoenix also does not exhibit the drop in the performance ratio for large messages. Experimental results indicate that the sum computation has only a minor effect on Apex-Map performance on this vector platform.

3.3 Apex-Map Performance

Different from other benchmarks, which usually provide only several performance points, Apex-Map can generate continuous performance surfaces over a whole range of temporal and spatial locality values. These surfaces can be used to study the effects of varying temporal and spatial locality and provide insight into architectural designs. Fig. 3 and 4 show the surface space for α = 0.001 to 1.0, L = 1 to 65536 words on 256 processors for M = 64 Mwords*256 on Seaborg and Phoenix. The Z-axis shows the achieved bandwidth per processes in log-scale.

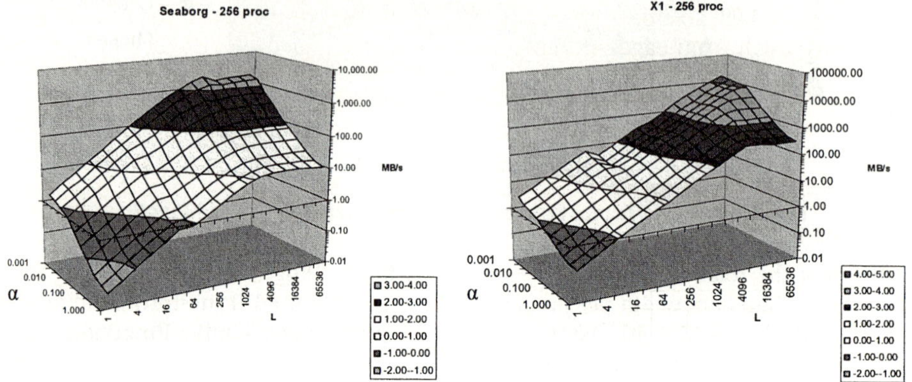

Fig. 3. The achieved bandwidth per process on Seaborg for 256 Processes

Fig. 4. The achieved bandwidth per process on Phoenix for 256 Processes

Fig. 3 shows that both temporal and spatial localities affect the bandwidth substantially. The worst performance is observed when $\alpha = 1$ and $L = 1$, which are the lowest values for temporal and spatial locality. By increasing either the temporal locality or spatial locality, the performance improves. The best performance is obtained when $\alpha = 0.001$ and $L = 4096$ Words. Further increasing L does not improve performance. This is mainly because the sum computation on this platform is less efficient for very large messages. Beyond $L = 4096$ spatial locality has only minor influence on performance while temporal locality α still has a large influence. If we look at an intermediate performance level such as 1 MB/s, we see that the temporal locality and spatial locality can be substituted by each other to some degree. To achieve 1 MB/s at high temporal locality of $\alpha = 0.005$, a very low spatial locality of $L = 1$ is sufficient. With decreasing temporal locality (increasing α), a higher spatial locality of up to $L = 85$ is needed to maintain this performance. The performance characteristics of Cheetah are very similar to Seaborg.

Fig. 5 shows the performance ratio between Cheetah and Seaborg. From Table 2 we see that the ratio of processor speeds between these two systems is 3.47, the ratio of local memory bandwidth is 1.375, and of network bandwidth is 4. For high temporal locality or high spatial locality the performance ratio of 2-4 seems to be dominated by the ratio of the respective memory bandwidth. For low localities, the performance ratio between these two systems is in the range of 6-8 and thus higher than any ratio of simple architectural parameters. In this locality range, performance is dominated by a large number of very short messages. The details of the MPI implementation as well as the cross-section bandwidth of the interconnect can be expected to have a large influence on performance in this corner of low localities where it will be notoriously difficult to achieve high absolute performance.

Fig. 4 shows the performance surface for the Cray X1 for which the effects of increasing spatial locality are significant even for values of L beyond 4096. Spatial locality affects the performance in general much stronger. For example, on Cheetah, in order to maintain the bandwidth around 10 MB/s, if we reducing the temporal locality α from 0.001 to 1, the spatial locality needs to increase 128 times. On Phoenix, it only needs to increase 16 times. We also notice that when L changes from 32 to 64, the performance drops. This is an effect of the MPI implementation on the Cray X1. When the message size becomes larger than 32 words or 256 bytes, communication in MPI will switch from eager mode to rendezvous mode and the implementation overhead increases.

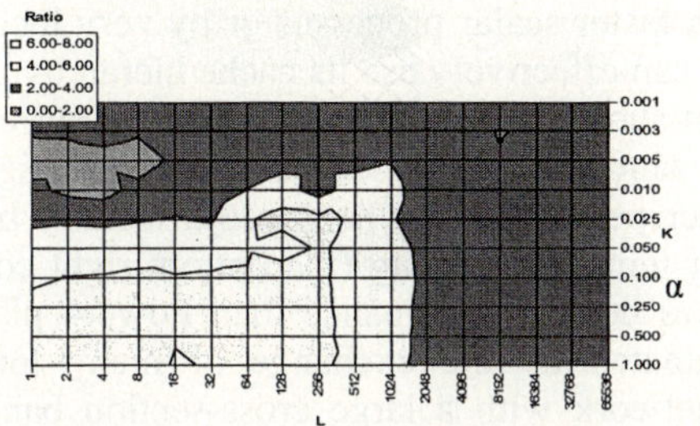

Fig. 5. The bandwidth performance ratio between Cheetah and Seaborg

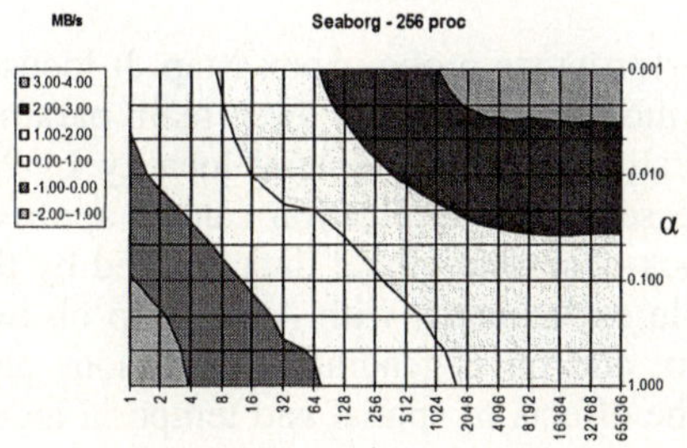

Fig. 6. The performance ratio between Phoenix and Cheetah

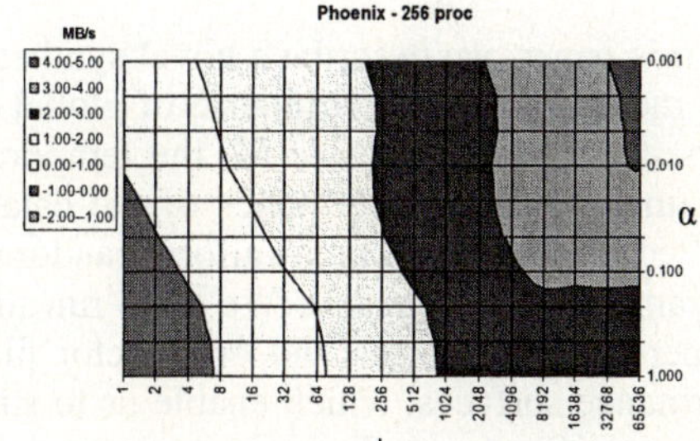

Fig. 7. Contour plots of the performance surfaces for Seaborg and Phoenix

To compare the performance surface for the superscalar IBM systems with the Cray vector system we put contour-plots of Seaborg and Phoenix next to each other in Fig 7. For the IBM systems, the area of highest performance is of rectangular shape and clearly elongated parallel to the spatial locality axis while for the Cray system it is elongated parallel to the temporal locality axis. The IBM system can tolerate a decrease in spatial locality more easily but is much more sensitive to a loss of temporal locality. This reflects the elaborate cache and memory hierarchy on the individual nodes as well as the global system hierarchy which also heavily relies on reuse of data as the interconnect bandwidth is substantially lower than the local memory bandwidth. The Cray system can tolerate a decrease in temporal locality much better but is sensitive to a loss in spatial locality. This reflects an architecture which depends very little on local caching of data and an interconnect bandwidth equal to local memory bandwidth. To see such a clear signature of the Cray architecture is even more surprising considering that we us an MPI based benchmark, which does not fully exploit the capability of this system. The lines of equal performance on the Cray system are in general more vertical than diagonal as with the IBM system, which further confirms our interpretation. These differences in our performance surfaces overall clearly reflect the different design philosophies of these two different systems and demonstrate the utility of our approach.

The performance ratio between Phoenix and Cheetah is shown in Fig. 6. Interestingly, when the spatial locality is poor or temporal locality is high, the vector processor X1 delivers less performance than the super-scalar processor Power4. In these cases, performance is dominated either by short MPI messages for which the Power 4 processor has the clear advantage of a much faster scalar processor or by very localized memory accesses for which the Power4 can effectively use its cache hierarchy. In this locality range, the Cray X1 can also not show its true potential with our current MPI based benchmark implementation. A shmem or UPC implementation might change this. The X1 shows the clearly better performance when spatial locality becomes high, especially in the area with poor temporal locality (the bottom-right corner). In the best case, it can deliver 12 times better performance than Power4 platform. Performance in this corner is dominated by the exchange of many long messages which requires an interconnect network with a large cross-section bandwidth.

4 Conclusion and Future Work

In this paper, we describe a novel synthetic performance probe, Apex-Map. It focuses on measuring the performance of global data movement and has three main parameters, the global data size M, the temporal locality α, and the spatial locality L. We assume that the performance of the data accesses of an application can be approximated by a generic, non-uniform random, block-access to global data defined by the parameters M, α, and L. We have run multiple experiments with Apex-Map on two superscalar platforms and one vector platform and have generated continuous performance surfaces, which enable us to study the effects of spatial and temporal locality on performance. The initial results on these platforms show that Apex-Map can be used to compare efficiency and scalability across different platforms and the performance surfaces generated by Apex-Map clearly reflect the design differences between these platforms.

Our first parallel implementation of Apex-Map is based on the most common parallel programming model, MPI. Currently we are implementing Apex-Map in other popular or emerging programming models, such as SHMEM and UPC, to study the effects of different programming paradigms and their relation to spatial and temporal locality. More importantly, we are also investigating methods to characterize parallel applications with the Apex-Map parameters. In our earlier work, we have successfully characterized several sequential scientific kernels [7] this way. Such a characterization allows us to use Apex-Map as a performance proxy for real scientific applications.

References

1. http://www.top500.org
2. STREAM: Sustainable Memory Bandwidth in High Performance Computers, http://www.cs.virginia.edu/stream/

3. HPC Challenge Benchmark, http://icl.cs.utk.edu/hpcc/
4. Apex-Map: Application Characterization-Memory Access Probe, http://ftg.lbl.gov
5. NAS Parallel Benchmarks, http://www.nas.nasa.gov/Software/NPB/
6. SPEC, http://www.spec.org/
7. E. Strohmaier, Hongzhang Shan, "Architecture Independent Performance Characterization and Benchmarking for Scientific Applications", International Symposium on Modeling, Analysis and Simulation of Computer and Telecommunication Systems. Volendam, The Netherlands, Oct. 2004
8. Pallas MPI Benchmarks, http://www.pallas.com/e/products/pmb/

PerfMiner: Cluster-Wide Collection, Storage and Presentation of Application Level Hardware Performance Data

Philip J. Mucci[1,2,*], Daniel Ahlin[2], Johan Danielsson[2],
Per Ekman[2], and Lars Malinowski[2]

[1] Innovative Computing Laboratory, University of Tennessee, Knoxville, TN, USA
[2] Center for Parallel Computers, Royal Institute of Technology, Stockholm, Sweden

Abstract. We present PerfMiner, a system for the transparent collection, storage and presentation of thread-level hardware performance data across an entire cluster. Every sub-process/thread spawned by the user through the batch system is measured with near zero overhead and no dilation of run-time. Performance metrics are collected at the thread level using tool built on top of the Performance Application Programming Interface (PAPI). As the hardware counters are virtualized by the OS, the resulting counts are largely unaffected by other kernel or user processes. PerfMiner correlates this performance data with metadata from the batch system and places it in a database. Through a command line and web interface, the user can make queries to the database to report information on everything from overall workload characterization and system utilization to the performance of a single thread in a specific application. This is in contrast to other monitoring systems that report aggregate system-wide metrics sampled over a period of time. In this paper, we describe our implementation of PerfMiner as well as present some results from the test deployment of PerfMiner across three different clusters at the Center for Parallel Computers at The Royal Institute of Technology in Stockholm, Sweden.

1 Introduction

Until unlimited compute power becomes pervasive, HPC systems must be carefully managed in order to maximize the users' productivity and the operating sites' return on investment. In most supercomputer installations, the cost of the machines and their maintenance is passed along to the user in terms of dollars per CPU hour. The user then either directly purchases compute time from the site or he applies for a grant from a central authority; often the same authority that funds the purchase and operation of the machine. This process is designed to balance a budget, equating an hour of CPU usage with an amortized cost of

---

* Work by this author has been partially supported by the Department of Energy Sci-DAC program (grant DE-FC02-01ER25490) and the Los Alamos Computer Science Institute (contract 86192-001-04 49).

J.C. Cunha and P.D. Medeiros (Eds.): Euro-Par 2005, LNCS 3648, pp. 124–133, 2005.

installation, operation and maintenance of a large machine. If we consider that the lifetime of a supercomputer or large cluster is about four years before it's retired, the above process appears wasteful, as it makes no attempt to optimize the use of either financial or computational resources. Compute time from user to user and group to group is treated equally; even though the amount of work that can be accomplished during each CPU hour can differ by many orders of magnitude. For example, a user with a large allocation and an inefficient code can easily 'steal' otherwise available resources from less well-funded users. The allocation is not based on computational work nor efficiency, rather it's based on a rough estimate of the number of CPU hours required to accomplish a given problem. Given the same budget, it is certainly possible that this user could solve much larger problems with an optimized code. The converse does not necessarily hold, as a user with a small budget and a large problem must strive to achieve some degree of efficiency in order to complete his work in the allotted time. If the allocation policy was biased towards actual computational resource requirements AND towards the efficient use of those resources, aggregate throughput of the system would rise and more CPU hours would be available to the community as a whole. Consider these other cases:

Purchase of a New Computing Resource. Procurements are often run in two different modes; either the customer submits a set of benchmarks to be optimized by the vendor or the vendor provides access to hardware resources on which the customer runs the benchmarks. These benchmarks run the gamut from microbenchmarks that measure particular machine parameters to full-blown applications. Benchmarks by their very nature, attempt to represent a very large code base with a very small code base. If hardware performance data could be collected for every application and correllated with data from the batch system and other sources, specific criteria that bound application performance could be used to guide the procurement process. For example, answers to questions like "Do the majority of our applications demonstrate high level 2 cache hit rates and thus are sensitive to cache size and front side bus frequency?" provide specific information about what kind of hardware should be purchased.

Improving the Software Development Cycle. While there are many excellent open source performance analysis tools available[TAU][SvPablo][Paradyn][Mucci], virtually all of them require the user to change his environment, Makefiles or source code. While simple for computer scientists, this process is fraught with potential error for scientists and engineers who are focused on their field of research. One or two failed attempts at using a performance tool is enough to permanently deter a scientist from making further efforts to characterize the behavior of his application. If the monitoring system could itself provide a completely transparent mechanism to measure important performance characteristics and the user could access that information quickly and easily, the process of application performane analysis could become an integral part of the software development process.

Performance Focused System Administration. As mentioned above, by having access to detailed performance data about all applications, system administrators could systematically address applications and their users that make inefficient use of compute resources. Centers with application specialist teams could deploy staff on the basis of low performance and high CPU hour consumption. This type of targeted optimization effort has the potential of optimizing a sites heavy users and reap continued benefits through successive generations of machines as the big users' applications receive the attention they deserve.

1.1   The Design of PerfMiner

A performance collection system must be carefully designed in order to meet the above goals. Most importantly, it must be transparent, lightweight and very efficient. Such a system can be split up into four components:

> *Integration into the User's Environment.* Changes to the user's environment should not be required by the system.
> *Collection of Hardware Performance Data.* The data must be collected at a sufficiently fine granularity to allow thread-level performance analysis.
> *Post-processing of the Data and Storage into a Scalable Database.* The database must be carefully designed to support queries that may span tables with ten's of millions of rows.
> *Presentation of the Data to the User Community.* The interface must be as simple as possible, yet should facilitate rapid "drill-down" investigation from widest granularity down to the thread level.

In order to meet the above needs, a performance collection system must be carefully designed. First and foremost, it must be focused on the simplicity of it's user interface and the speed of which it operates. As the system could be running on many clusters across a site and measure every job through the system, the amount of data could grow quite large. The system has four basic components: Integration into the user's environment and/or batch system. This must be completely transparent to the user, but yet facilitate conditional execution of monitoring for debugging and other purposes. Collection of the job and hardware performance data. This must also be completely transparent to the user with no modifications to the user's job. Post-processing of the data and insertion into a database. The database must be carefully designed to support queries that may span tables with ten's of millions of rows. Furthermore, the schema should facilitate the rapid development of reasonably complex queries in order to accommodate the demands of its user base. Presentation of the data to the users, system-administrators and managers. This interface must be as simple as possible to guarantee maximum acceptance into a daily usage cycle. Complex functionality should be hidden from the main interface yet remain accessible to those wishing to dig deeper. The interface should facilitate rapid "drill-down" investigation from widest granularity down to the thread level.

PerfMiner is an perfomance monitoring system that attempts to meet the above goals. To test our initial implementation, we deployed PerfMiner for a subset of users for three weeks across all three of PDC's clusters, Roxette, a cluster of 16 dual Pentium III nodes, Lucidor, a cluster of 90 dual Itanium 2 systems, and Beppe, a 100 processor Pentium IV cluster that is one of the six SweGrid clusters spread across Sweden. All systems have gigabit ethernet as an interconnect, with the exception of Lucidor which also contains Myrinet-D cards in every node.

In the next four sections, we describe each of the components of the PerfMiner system, working our way from the integration into the batch system to the Web interface presented to the user. Following that, we present the results discuss the relevance of a few queries made to the PerfMiner database. We then conclude with a review of related work and some comments about the future of PerfMiner.

2 Integration of PerfMiner into the Easy Batch System

One of the challenges of the implementation of PerfMiner at PDC was how to manage the integration into the batch system. PDC runs a modified version of the Easy[Easy] scheduler. At it's core, Easy is a reservation system that works by enabling the user's shell in /etc/passwd on the compute nodes. The user is free to login directly to any subset of the reserved nodes. There is no restriction on using MPI as a means to access these nodes from the front end. In this way, Easy serves the needs of PDC's data processing community who frequently submit ensembles of serial jobs, often written in Perl. Given this, we could not count on mpirun as our single point of entry to the compute nodes. This left us with only one means to guarantee the initiation of the collection process: the installation of a shell wrapper as the user's login shell, pdcsh.sh (PDC Shell). The reader may wonder why we didn't choose to use a system shell startup script. Unfortunately, the Bourne shell does not execute the system scripts in /etc when started as a non-login shell (C-Shell does). By the installation of a wrapper script, every process, whether started via ssh, kerberized telnet/rsh or MPI was guaranteed to be executed in our environment. Due to the design of the Easy scheduler, this modification was rather trivial to perform. Easy maintains two password files, password.complete and passwd. The former contains valid shells for all users. The latter contains valid shells only for that user who has reserved the node. This file is constructed on the fly by Easy when the job has come to the top of the queue.

The steps for job execution and finalization occur as follows:

First, a preamble script is initiated by Easy: (`pdcsh-pre.sh`)

1. Check if the cluster, charge group, user and host were enabled for use with PAPI Monitoring. If not, bail out.
2. Verify the existence of the output directory tree.
3. In the above directory, create two files:
   - `BUSY`, which is a zero length file that indicates that this job is running and that monitoring is taking place.
   - `METADATA`, which contains job information that is cross referenced with that from PapiEx. It contains the following fields: cluster name, job ID,

username, number of nodes reserved, charge group (CAC), start time and the finish time of the job. The finish time is filled in by the postamble script described below.

Second, Easy conditionally modifies the user's shell in the passwd files: (`adduser.py`)

1. Check if the cluster, charge group, user and host were enabled for use with PDCSH. If not, bail out.
2. Give the user PDC shell as his login shell on all reserved nodes.

When any job is started on any node, it will run under PDC shell and all subprocesses and threads will be monitored. (`pdcsh.sh`)

1. Execute a common cluster wide setup script. (for other administrative purposes)
2. Determine the following:
   - Whether or not we are a login shell.
   - The user's actual shell from passwd.complete.
3. If the cluster, charge group, user and host are enabled for PAPI Monitoring, execute the PAPI monitoring script.
4. Execute the user's actual shell appropriately. (as a login shell or not)

The PAPI monitoring script performs the following: (`papimon.sh`)

1. Check for the file that contains the prepared arguments to PapiEx.
2. Check that these arguments are correct.
3. Verify the existence of the output directory tree.
4. Set the output environment variable to PapiEx.
5. Set up the library preload environment variables.

At this point, the user's job runs to completion. The only processes not monitored are those that are either statically linked or they access PAPI or the hardware performance monitors directly. Upon completion of the job, a postamble runs on the front end. This script does the following: (`pdcsh-post.sh`)

1. Check if the cluster, charge group, user and host were enabled for use with PAPI Monitoring. If not, bail out.
2. Append the job finish time to the `METADATA` file.
3. Remove `BUSY` file .
4. Schedule the parsing and submission of collected data to the PerfMiner database and remove/backup the original files.

3 Collecting Hardware Performance Data Transparently with PapiEx

At the lowest level, PerfMiner can use any mechanism to collect application performance data. However, other methods require the user to recompile his application or use customized batch scripts. For our setup, we wanted a system that would be completely transparent to the user, requiring no modifications to user's environment, application code or run-time libraries. Existing binaries would continue to run as they did prior to the deployment of the software. To accomplish this, we decided to use PapiEx, a tool based on the PAPI[PAPI]. PapiEx can run unmodified dynamically linked binaries and monitor them with PAPI. It follows all spawned subprocesses and threads and generates output for each. In PerfMiner, the output of PapiEx is directed to a file, which is then later parsed by a perl script upon job completion.

4   Scalable Database Design

We chose to use Postgres as the database back end for PerfMiner. The primary reason for choosing Postgres was prior experience and its support for kerberized authentication. Care has been taken to avoid the use of any nonstandard SQL that could prevent the use of Mysql, Oracle or another SQL95 compliant database. Access to the database has been abstracted through the use of both Perl and PHP's DBI interface, providing further portability. Much work has been done to keep the PerfMiner database as robust as possible. In an early implementation of PerfMiner, we rather hastily built a database schema around a common set of queries we were hoping to run. We quickly realized that this was neither general nor robust enough to support queries spanning millions of rows. Thus a new database was designed, focusing on flexibility, extensibility and easy of implementation of sophisticated queries. Our goal was to have as much of the query processing be done by the database server itself instead of the client. Thus queries processing vast quantities of data can be performed on underpowered web servers.

4.1   Direct Measurements

There are only two truly static items of knowledge in the database. First, all measurements have a target (or scope) that is one of cluster, job, node, process or thread. Secondly, there is a hierarchy of these targets; a cluster contains jobs, which contain nodes, which contain processes, which contain threads. These targets can can be regarded as one to many mappings and naturally produce keys for addressing the collected data. For instance, a specific threads measurements are accessed by specifying cluster, job, node, process identifier and possibly thread identifier as the primary key. Since no assumptions of existence of any specific measurement are made, it is not possible to minimize the tables by putting all measurements of thread scope in the table that specifies which threads exist (unless you are prepared to accept null values and that the underlying database is able to insert columns in preexisting tables). Instead, each measurement resides in a separate table. The database also contains additional tables that describe the scope, type and meaning of each of the collected measurements. This ensures that no measurement is stored differently from any other. The primary advantage of this approach is that it makes it possible to combine measurements and construct reports in a uniform way. In PerfMiner, this means that any change in the data collected from PapiEx or from the batch system, results in the creation of a table and associated metainformation. Thus, no changes need be made to the database or to the query engine.

4.2   Derived Measurements

The measurement floating point operations per second (or FLOPS) is an example of a derived measurement having thread scope. It combines the direct measurement, floating point operations, with the derived measurement, duration, which in turn is derived from clockrate and total cycles. The database is designed to store information about the derived measurements in the same way that it stores the direct measurements. The query author does not have to know if a derived or direct measurement is being referenced in his query.

4.3   Problems with the Current Approach

Putting the measurements in different tables can be perceived as discarding the fact that they are collected simultaneously and belong to the same thread. When the data is harvested, the application knows that a certain value of total cycles is associated with a certain value of total floating point operations. The only way to reconstruct this information is by joining the two tables, an $O(n^2)$ operation. This can be mitigated by instructing the database to build indexes for the fields of every metric table that serve as keys. This reduces the cost of the join to $O(nlogn)$ or less depending on the method used for indexing. However, adding indexes aggravates another problem caused by the nature of the measurements. Since the target of most measurements is threads, and the key for addressing a certain thread is made up of cluster, job, node, process, thread (of which three are TEXT-fields), the key component will strongly dominate the storage demand for most tables. A solution to this is to create synthetic keys for tables where this is a problem.

5   The PerfMiner User Interface

For the current implementation of PerfMiner, the front end runs on an Apache web server with PHP and JpGraph[JpGraph] installed. JpGraph is an open source graphing library built upon PHP and the GDGD library. The user is presented with a simple interface through which he can construct queries to be visualized. The resulting graph is dynamically generated with JpGraph along with a corresponding image map, such that the user can click on a corresponding portion to "drill-down" to more interesting data. As developers, we are presented with the canonical problem of balancing functionality with interface complexity. For our initial implementation, we chose a small subset of the available data as targets of our queries. We chose to present a query interface that specified the logical-AND of any four items present in the job's METADATA file: four on which to scope the queries and one choice by which to group Cluster, Charge Group, User and Job ID. Each column is updated from the selections to it's left. Should the user choose a combination that results in the availability of a single job ID, an additional dialog is presented with the names of all the processes in that job.

6 Evaluation

PerfMiner aims to meet the needs of three different user bases (users, system administrators and managers), through a common information collection infrastructure. For the user community, we provide a simple way of providing performance information about recently submitted jobs without any changes to the user's application or environment. This information can contain the efficiency of various components, the overall processing time of each component or more details hardware performance metrics. The ultimate goal is to not only provide performance information but to provide information as to why the components of that job are performing a certain way. In Figure 1, we have used PerfMiner to plot instructions per cycle (IPC) against the executable name. This particular user has submitted a shell script to perform a run of Gamess, an ab-initio quantum chemistry package. Here we find that Gamess was the fifth most inefficient executable. This data was taken from our Xeon cluster.

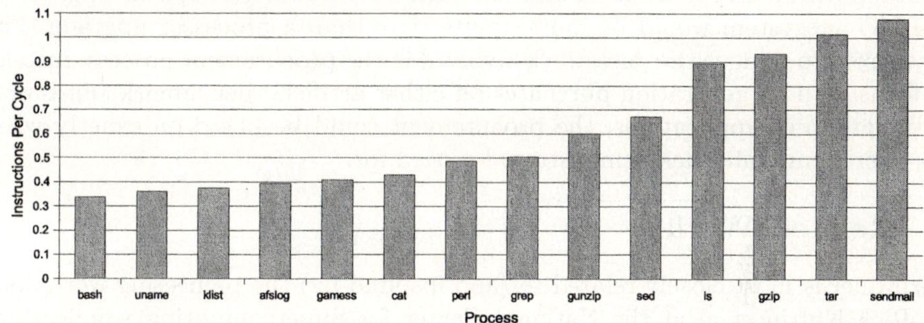

Fig. 1. PerfMiner Graph of Instructions Per Cycle of a Serial Job

For the administrator and support staff community, we may not be so interested in per-process performance, but rather the throughput of the system as a whole. In Figure 2, we have asked the system to plot the average level 1 data cache hit rate of all jobs and sort the results by user. We find that the user who has consumed the most compute cycles has the second lowest miss rate of all jobs. This kind of query is extremely powerful when aiming to maximize the throughput of a particular system. It's not hard to envision a scenario where application specialists approach a user and offer help on code optimizaton.

Lastly, PerfMiner's goal is to be able to facilitate a good understanding of exactly how the systems are being used by the various user communities. By doing so, they can plan appropriately for future procurements. The central idea here

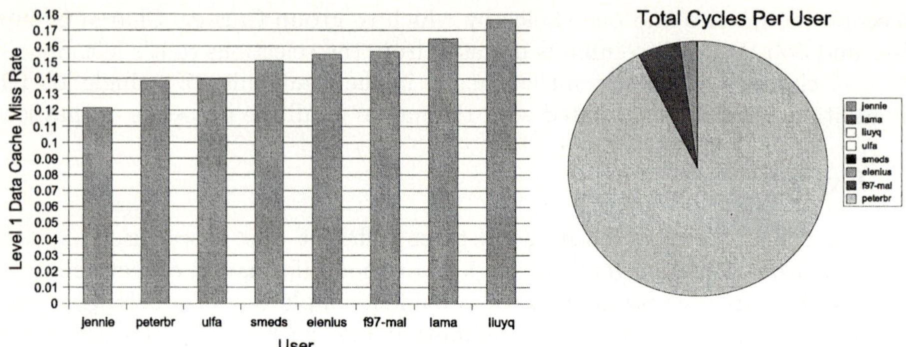

Fig. 2. PerfMiner Graphs of Level 1 Data Cache Miss Rates and Total Cycle Consumption by User

being that they can focus their procurements on having the type of hardware appropriate for the problems being solved. Should the user workload demonstrate high cache hit-rates and counts of floating point instructions, perhaps a system with a similar size cache but a higher core clock frequency and deeper floating point pipeline would be an appropriate upgrade. Should the workload demonstrate low processor utilization and low TLB-miss rates, perhaps an upgrade of the I/O subsystem would be more appropriate than a processor upgrade. The key here is to remove the guesswork involved in the procurement process. Instead of focusing next generation purchases on either artificial benchmark suites or a select group of applications, the procurement could be based on exactly what the user community has demonstrated a need for.

7   Related Work

PerfMiner is most closely related to (and inspired by) the pioneering work done by Rick Kufrin et al at the National Center for Supercomputing Applications [Kufrin1]. In that work, a locally developed PAPI based tool PerfSuite[PerfSuite] is used to collect information on jobs in the batch system. The primary differences between our work are the collection mechanism, the design of the database and the user interface. There are numerous systems in existence that do cluster-wide performance monitoring. Many of them like Ganglia[Ganglia], SuperMon[SuperMon], CluMon[CluMon], NWPerf[NWPerf] and SGI's Performance CoPilot[PCP] are extensible frameworks capable of presenting any metric. All these systems gather their metrics only on a system wide basis through a daemon process that scrapes the `/proc` filesystem.

References

[SvPablo] Reed, D. A., et al. Scalable Performance Analysis: The Pablo Performance Analysis Environment. Proc. Scalable Parallel Libraries Conf. IEEE Computer Society. (1993) 104–113

[Easy] Lifka, D., Henderson, M., Rayl, K.: Users guide to the argonne sp scheduling system. Technical Report **ANL/MCS-TM-201** (1995)

[Paradyn] Miller, B. et al. The Paradyn Parallel Performance Measurement Tool. IEEE Computer **28/11** (1995) 37–46

[TAU] Mohr, B., Malony, A., Cuny, J.: TAU Tuning and Analysis Utilities for Portable Parallel Programming. Parallel Programming using C++, M.I.T. Press. (1996)

[PAPI] Mucci, P. et al. A Scalable Cross-Platform Infrastructure for Application Performance Tuning Using Hardware Counters. Proceedings of Supercomputing 2000. (2000)

[GD] Boutell.Com, Inc. GD Graphics Library. `http://www.boutell.com/gd`

[JpGraph] Persson, J. JpGraph - OO Graph Library for PHP. `http://www.aditus.nu/jpgraph/index.php`

[PerfSuite] Kufrin, R. The PerfSuite Collection of Performance Analysis Tools. `http://perfsuite.ncsa.uiuc.edu`

[Ganglia] The Ganglia Scalable Distributed Monitoring System. `http://ganglia.sourceforge.net`

[PCP] Performance Co-Pilot `http://oss.sgi.com/projects/pcp`

[SuperMon] SuperMon High Performance Cluster Monitoring. `http://supermon.sourceforge.net`

[CluMon] Fullop, J. CluMon Cluster Monitoring System. `http://clumon.ncsa.uiuc.edu`

[NWPerf] Mooney, R. et al. NWPerf: A System Wide Performance Monitoring Tool Poster Session 31, Supercomputing 2004, Pittsburg, PA.

[Petrini] Petrini, F. et al. The Case of the Missing Supercomputer Performance:Achieving Optimal Performance on the 8,192 Processors of ASCI Q Proceedings of Supercomputing 2003. (2003)

[Mucci] Mucci, P. et al. Application Performance Analysis Tools for Linux Clusters. Linux Clusters: The HPC Revolution 2004, Austin, TX. (2004)

[Kufrin1] Kufrin, R. et al. Automating the Large-Scale Collection and Analysis of Performance Data on Linux Clusters Linux Clusters: The HPC Revolution 2004, Austin, TX. (2004)

[Kufrin2] Kufrin, R. et al. Performance Monitoring/Analysis of Overall Job Mix on Large–Scale Pentium and Itanium Linux Clusters SIAM Parallel Processing, San Francisco, CA. (2004)

[Monitor] Mucci, P., Tallent, N. Monitor - user callbacks for library, process and thread initialization/creation/destruction. `http://www.cs.utk.edu/~mucci/monitor`

Performance Evaluation of MM5 on Clusters with Modern Interconnects: Scalability and Impact*

Ranjit Noronha and Dhabaleswar K. Panda

Dept. of Computer Science and Engineering
The Ohio State University
Columbus, OH 43210

Abstract. Clusters have become a crucial technology for providing low-cost high performance computing to scientific applications like weather prediction. In addition, networks like Myrinet, InfiniBand and Quadrics have become popular as an interconnection technology for high performance clusters. The high-bandwidth, low-latency characteristics of these networks make them ideally suited to the demanding characteristics of large scale weather simulations. Additionally, these networks have features like efficient and scalable hardware broadcast, reduce and atomic operations. Some of the features have been integrated into the MPI stack for these networks, allowing the user to exploit them for improved performance. In this paper, we evaluate the communication characteristics of a popular weather simulation code MM5 using InfiniBand. We also investigate how special features of InfiniBand like scalable broadcast can benefit MM5 performance. For some workloads, we see that InfiniBand performs up to 34% better than other interconnects. It also performs better in general than other networks for all workloads.

Keywords: MM5, Myrinet, InfiniBand, Quadrics, System Area Networks, Clusters

1 Introduction

Clusters have been widely deployed for providing high-performance computing for scientific applications. The lower cost of clusters means that several thousand nodes may be deployed for running large scale applications. Achieving improved performance from applications on large scale clusters is a challenging endeavor. This is especially so, given the wide diversity of architectures and networking technologies. Understanding the characteristics and trade-offs in different cluster architectures is crucial for achieving the best performance from applications.

To exploit the benefits of parallel computers, several large scale applications have been parallelized using implementations of the popular MPI standard [1].

---

* This research is supported in part by Department of Energy's Grant #DE-FC02-01ER25506; National Science Foundation's grants #CCR-0204429 and #CCR-0311542; and equipment donations from Intel and Mellanox

J.C. Cunha and P.D. Medeiros (Eds.): Euro-Par 2005, LNCS 3648, pp. 134–145, 2005.

MPI provides an interface to the application, abstracting out details of the underlying architecture and network. Computationally demanding applications such as weather simulation, computational fluid dynamic codes and crash simulation codes have MPI parallelizations [2–4]. Implementation of MPI such as MPICH have been ported to a variety of architectures and networks. This allows the application to be run on a wide-range of platforms. The application may potentially exploit the characteristics of these architectures and networks for improved performance.

Myrinet [5], InfiniBand [6] and Quadrics [7] are some of the popular networks used in high performance computing. These networks offer low latency of a few microseconds and high-bandwidth communication. Additionally, they offer several features which may be exploited by the application or MPI layers. Networks like Myrinet and Quadrics have a programmable network interface card (NIC), which may be used to offload application or system level computation [8]. Scalable collectives may be used to enhance application performance. Exploiting these features is possible not only at the MPI level, but also by the application itself.

In this paper, we evaluate a widely used weather simulation code MM5. We attempt to study its communication characteristics. This is done by analyzing the MPI calls characteristics with increasing system size. Following that, we study the impact of varying different network parameters like latency and bandwidth on the performance of the MM5. The impact of special network features is also evaluated.

The rest of the paper is organized as follows. Section 2 gives some background on high performance networks, bus technologies and MM5. Following that in section 3, the scalability of different workloads is evaluated. In section 4, the performance of MM5 while varying various network parameters is evaluated. Some related work is discussed in section 5. Finally, in section 6, conclusions and future work is presented.

2   Background

In this section, we discuss some of the topics relating to networks and weather simulation models. In particular, in section 2.1, we first discuss the networking technology Myrinet, InfiniBand and Quadrics. Following that in section 2.2, we discuss the weather simulation model MM5.

2.1   Overview of Cluster-Networking Technologies

In the high performance computing domain, Myrinet, InfiniBand and Quadrics are three of the popular networking technologies. InfiniBand [6] uses a switched, channel-based interconnection fabric, which allows for higher bandwidth, more reliability and better QoS support. The Mellanox implementation of the InfiniBand Verbs API (VAPI) supports the basic send-receive model and the RDMA operations read and write. There is also support for atomic operations and multicast. MVAPICH [9] is an implementation of Argonne's MPICH [1] over InfiniBand. The design of MVAPICH is based on the InfiniBand RDMA primitives. MVAPICH delivers small message latency of $5.0\mu s$ and large message bandwidth of up to 900 MillionBytes/sec. MVAPICH is designed to take advantage of hardware based multicast in InfiniBand [10].

Myrinet [5] is another low latency, high-bandwidth network which uses cut-through switches. Myrinet E-cards [11] are programmable and allow up to two ports for maximum bandwidth. MPICH-GM is an implementation of MPICH over Myrinet delivering small message latency of up to 6.0 μs and large message bandwidth up to 500 MillionBytes/sec. Quadrics [7] is another high-speed network. The current generation of Quadrics is Elan 4. Quadrics has a programmable NIC which can be used to offload computation from the host. MPI/Elan4 is an implementation of MPICH over Quadrics QsNet II. MPI/Elan4 can send small messages with a latency of 2.4 μs and large messages with a bandwidth of up to 900 MillionBytes/s. The latency and bandwidth of these different networks is shown in Table 1. A basic comparison of these networks in terms of micro-benchmarks is presented in [12].

Table 1. Latency and Bandwidth for some high performance networks

Network	Latency(μs)	Bandwidth (MegaBytes/sec)
Myrinet (MPICH-GM)	6.0	500
InfiniBand(MVAPICH)	5.0	890
Quadrics(MPI/Elan4)	2.4	900

2.2   Overview of MM5

MM5 [4] is a limited area, non-hydrostatic, terrain following sigma-coordinate model designed to simulate or predict mesoscale atmospheric circulation. This regional model may be used for prediction on domains ranging from several thousand miles to a few hundred miles or less. Domains are uniform rectangular three dimensional areas of the atmosphere. The atmospheric dynamics are non-hydroscopic and use finite-difference approximations. The model is supported by several pre- and post-processing programs, which are referred to collectively as the MM5 modeling system. The MM5 modeling system software is mostly written in Fortran, and has been developed at Penn State and NCAR as a community mesoscale model with contributions from users worldwide.

The distributed-memory version of MM5 [13](MM5-MPP) has been implemented using MPI message-passing provided by the parallel Runtime System Library (RSL) [14]. RSL is a run-time system and library to support parallelization of grid-based finite-difference weather models. RSL supports *mesh refinement*. *Mesh refinement* allows the original domain to be divided into smaller areas (which may be nested). By allowing these areas to be non-uniform, computation may be focused on areas of more active interest in the domain. This usually sacrifices resolution in some areas of the domain (which may not be of interest), but reduces the computational requirements. RSL communicates results between sub-domains as shown in Figure 1. In the next section, we discuss how different workloads scale with an increase in the number of nodes.

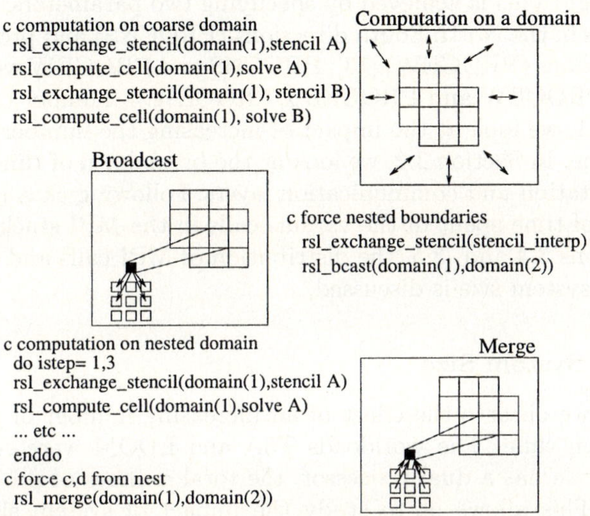

Fig. 1. Overall parallel driver for a MM5 timestep with nest interactions (courtesy J. Michalakes, et al. [13])

3 Communication Characteristics of Parallel MM5 (MM5-MPP)

In this section, we take a look at the communication patterns in the parallel version of MM5 (MM5-MPP) when using MPI over InfiniBand (MVAPICH). This was done to help us understand what parameters of a network would help us achieve better performance from this application. For example, if the application sends a lot of small messages, low network latency might help. If it sends large messages, it might be bandwidth sensitive. In addition, we would like to understand whether the application employs special operations like collectives. If it does, we would like to examine how efficient implementation of some collectives by certain networks might impact the performance of MM5-MPP. In-order to do this, we first evaluated how increasing the number of processors or system size, impacted the performance of the application. Also with increasing system size, we looked at how the distribution of MPI calls in the application changed. Finally, we also looked at how the message sizes to different MPI calls changed with increase in system size.

To evaluate these characteristics of MM5-MPP, we chose two different workloads and ran them on a 64-node, dual 2.4 GHz processor cluster with Mellanox

MT23108 InfiniBand adapters and a MVAPICH 0.9.4 installation (cluster A). The first workload is the MM5 benchmark data set [15], which specifies a 3 hour run, TIMAX = 180, with an 81 second time-step (T3A). The second is the large-domain run (LDOM) which may be obtained from [16]. MM5-MPP allows the user to divide the workload among the different processors, so as to reduce the memory usage. This is achieved by specifying two parameters; namely number of processors in the North-South directions (PROCNS) and processors in the East-West directions (PROCEW) [13]. PROCNS and PROCEW were set so that PROCNS >= PROCEW and PROCNS x PROCEW = number of processors.

In Section 3.1, we look at the impact of increasing the number of processors on execution time. In Section 3.2, we look at the breakdown of time between application computation and communication layers. Following that in Section 3.3, the breakdown of time spent in the various calls in the MPI stack is presented. Finally in Sections 3.4 and 3.5, the distribution of MPI calls and message sizes with increasing system size is discussed.

3.1 Effect of System Size

In this section, we observe the effect of an increasing number of processors on parallel execution time. The workloads T3A and LDOM were run on cluster A. Since each node has a dual processor, the total number of processors in the system is 128. This allows us to study the impact of system size up to 128 processors. The effect of increasing system size on execution time for T3A and LDOM is shown in Figure 2. It can be observed that with increasing system size for T3A, the execution time decreases up to 128 processors. For T3A, there is an approximate decrease in execution time of up to 37% when doubling the number of processors. Figure 2 also shows the scaling efficiency of the two workloads. It can be seen that for the workload T3A, the scaling efficiency starts at above 90% and then gradually decreases to a little over 65%.

For LDOM, there is a maximum decrease in execution time of 32% when doubling the number of processors. The scaling efficiency gradually decreases from 74% to approximately 49%, as shown in Figure 2. For LDOM, the benefits of an increasing system size plateaus after 64 processors. This can be attributed to the smaller problem size of LDOM compared to T3A. This leads to an increased load imbalance, which manifests itself as increased wait time. This effect is discussed in further detail in Section 3.2.

3.2 Overall Application Timing Breakdown

We will now discuss the average per-process breakdown of execution time of the workloads LDOM and T3A. For improved scalability, it is better to spend the maximum amount of time in application level computation and as little time as possible in the communication libraries or in MPI calls. How much time is spent in the communication libraries is partly dependent on the design of the application as well as the communication library. If the application uses non-blocking MPI calls, this time can be minimized. Blocking calls on the other

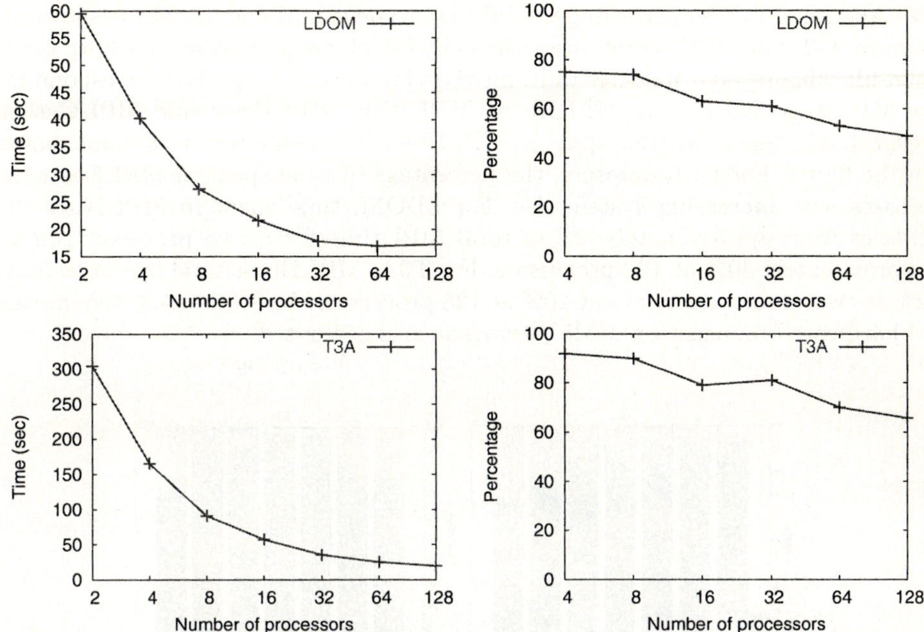

Fig. 2. MM5-MPP execution time with increasing system size of two different workloads (left) and scaling efficiency (right)

hand increase the amount of time spent in the communication libraries. The time spent in the communication library also depends partly on the nature of the progress function employed by the MPI stack.

The breakdown of timing was obtained using the lightweight profiling tool mpiP [17]. We find that for LDOM, the percentage of time spent in communication (time spent in MPI layers) increases from slightly less than 5% at two processors, to approximately 37% at 128 processors. For T3A, the percentage of communication increases from approximately 2% at two processors to 27% at 128 processors. This difference can be mainly attributed to the difference in sizes of the two workloads. LDOM is a smaller workload as compared to T3A. As a result, the computation datum assigned to each processor is smaller. This effect manifests itself as increased process skew. Overall, a large amount of time is spent in MPI layers particularly blocking MPI Receive calls. This issue will be discussed in more detail in Section 3.3.

3.3   MPI Timing Breakdown

In this section, we discuss the average per-processor distribution of time spent in different MPI calls for LDOM and T3A. Understanding the distribution of time spent in different calls, gives us insight into which network might potentially enhance the performance of MM5-MPP. This is specially true in the case of efficient implementation of collective operations in some of the stacks such as

MVAPICH [10]. The percentage of MPI time spent in different calls is shown in Figure 3. MM5-MPP largely uses the calls for blocking receive, blocking send, non-blocking receive, message wait, message broadcast and gather corresponding to MPI_Recv, MPI_Send, MPI_IRecv, MPI_Wait, MPI_Bcast and MPI_Gather respectively. Since the time spent in MPI_IRecv is not significant, it is not shown in the figure. For both datasets, the percentage of time spent in MPI_Bcast increases with increasing system size. For LDOM, time spent in MPI_Bcast increases from approximately 4% of total MPI time for a two processor run to approximately 30% at 128 processors. For T3A, MPI_Bcast time increases from 2% at two processors to about 20% at 128 processors. In Section 4.3, the impact of hardware broadcast on MM5 performance is evaluated.

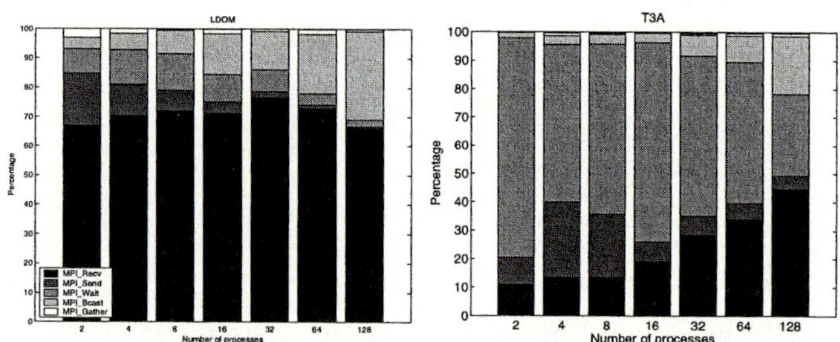

Fig. 3. Breakdown of time spent in different MPI functions for two different workloads

A large percentage of time is spent in MPI_Recv for both LDOM and T3A and increases with increasing system size. At 128 processors for LDOM and T3A, the time spent is about 30% and 44% of communication time respectively. For both cases, time spent in MPI_Wait decreases with increase in system size. This decrease is more rapid in the case of T3A. Time spent in MPI_Recv and MPI_Wait can be correlated to the amount of application wait time. This is approximately 26% for LDOM and 21% for T3A. This would suggest that MM5-MPP would benefit from dynamic load balancing, currently not implemented in this version of MM5-MPP.

3.4   MPI Call Count Distribution

In this section, we look at the average per-processor distribution of MPI calls in MM5-MPP. The distribution of MPI calls for LDOM and T3A with increasing system size is shown in Figure 4. As discussed in section 3.3, implementation of MM5-MPP makes calls to the MPI functions for blocking sends, blocking receives, non-blocking receives, broadcast and gather. These calls are MPI_Send, MPI_Recv, MPI_IRecv, MPI_Bcast and MPI_Gather respectively. Since the proportion of calls to MPI_IRecv is not significant, these calls are not shown in the graphs. For both workloads, the number of calls increases with increasing system size. For LDOM, MPI_Send has the highest count, while for T3A MPI_Bcast is the highest. For both cases, the number of calls to MPI_Send and MPI_Bcast

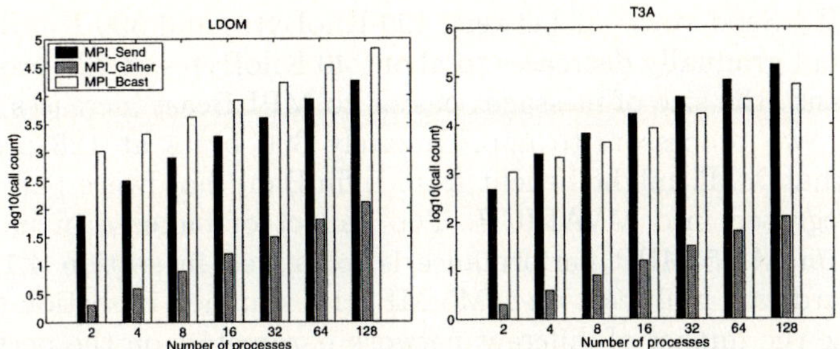

Fig. 4. Frequency of different MPI calls for the two different workloads

increases ten-fold, when the system size is increased from two processors to 128 processors.

From this we observe that MM5-MPP largely uses blocking MPI calls. MM5-MPP might benefit from a design which uses more non-blocking calls. This might be possible through the modification of the *rsl_exch_stencil* and *rsl_merge_stencil* calls in Figure 1 to use non-blocking calls. In this case, it might issue a non-blocking receive, to receive data from its adjacent neighbors. It might then continue computation on different sub-domains (assuming there is sufficient data available). Between computations, it might check if there is any additional data from its adjacent neighbors. If there is data available, it might use that to complete some computations rather than blocking. This would help us with overlap of computation and communication. This might also help reduce some of the application wait time discussed in section 3.3. We plan on investigating this in our future work.

3.5   Message Size Distribution

As discussed in Section 3.4, MM5-MPP largely makes blocking MPI calls. In this section, we look at the average sizes of messages sent from these blocking calls namely MPI_Send, MPI_Recv, MPI_Bcast, and MPI_Gather. These results are shown in Figure 5. For both workloads LDOM and T3A, the size of the message

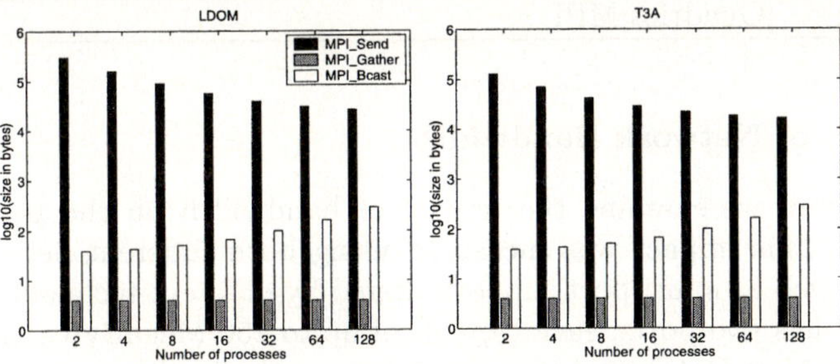

Fig. 5. Sizes of messages sent through different MPI calls in two different workloads

passed to MPI_Send starts at between 129 KiloBytes and 300 KiloBytes at two processors and gradually decreases to about 40 KiloBytes at 128 processors. On the other hand, the size of messages passed to MPI_Bcast increases from about 50 bytes at two processors to approximately 300 bytes at 128 processors. It is possible that MM5 might benefit from InfiniBand hardware based multicast support integrated into MVAPICH. The impact of increase in unidirectional bandwidth on MM5-MPP performance is examined in section 4.1, while the impact of hardware multicast on MM5-MPP is examined in section 4.3. We will now examine the impact of different network parameters on the performance of MM5-MPP.

4   Impact of Network Technology

In this section, we look at how different network parameters affect the execution time of MM5-MPP. In particular, the impact of latency, bandwidth and hardware broadcast is examined. Experiments are conducted using the workloads LDOM and T3A described in Section 3. These workloads were run on cluster B (8-node, dual 3.0 GHz processor cluster with Myrinet E-cards, Quadrics Elan-4 and Mellanox MT23108 InfiniBand adapters). All experiments were run with 16 processes on eight nodes. In Section 4.1, the effect of network bandwidth on applications is examined. Following that, we look at the impact of network latency on MM5-MPP performance in 4.2.

Table 2. Explanation of notation used in this section

Notation	Explanation
MPICHGM-1P	MPICHGM 1.2.6..14a using E-cards, with a single port activated (GM 2.0.21)
MPICHGM-2P	MPICHGM 1.2.6..14a using E-cards, with both ports activated (GM 2.1.21)
MVAPICH-1N	MVAPICH 0.9.5 with a single NIC per node
MVAPICH-HB	MVAPICH 0.9.5 with InfiniBand hardware broadcast enabled
MVAPICH-SB	MVAPICH 0.9.5 without InfiniBand hardware broadcast
MPI/Elan4	Quadrics MPI

4.1   Effect of Network Bandwidth

In this section, we examine the impact of bandwidth on the performance of MM5-MPP. This impact was measured using both different networking technologies, as well as multi-port support offered by different technologies. Myrinet E-cards [11] has two ports, each capable of up to 250 MegaBytes/sec for a total of up to 500 MegaBytes/sec. It is possible to activate either one or both ports on these cards. We use notation as explained in Table 2. For large messages,

MVAPICH-1N delivers up to 900 MegaBytes/sec. MPI/ELAN4 delivers up to 900 MegaBytes/sec [9]. The two workloads were run on cluster B, described in Section 4. The execution time across different networks for LDOM and T3A at 16 processes on 8 nodes is shown in Figure 6. For LDOM, execution time is reduced by approximately 34% when MPICHGM-2P is replaced by MVAPICH-1N. The reduction in execution time may be attributed mainly to to the reduction in time spent in MPI_Bcast (24.2%), followed by the reduction in MPI_Recv (5%), along with small reductions in MPI_Wait, MPI_Gather and MPI_Send making up the remaining 5%. Note that hardware broadcast was not enabled for MVAPICH-1N. For T3A, on replacing MPICHGM-2P with MVAPICH-1N, there is a reduction in execution time of up to 12%. Most of this reduction comes from reduced time spent in MPI_Bcast.

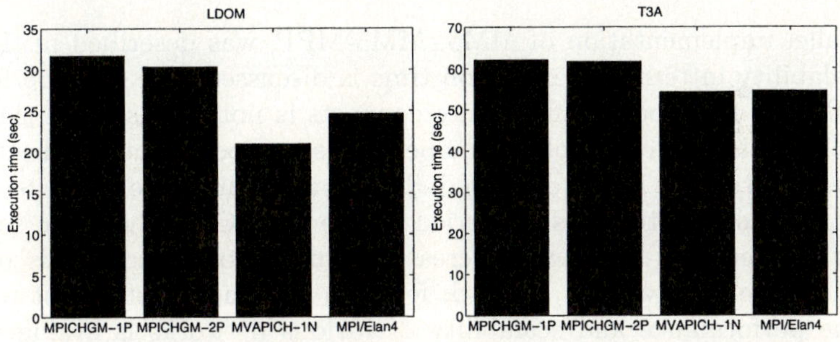

Fig. 6. MM5-MPP execution time with different networks

4.2   Effect of Network Latency

We will now examine the effect of network latency on the performance of MM5-MPP. For the different network MPI stacks, we use notation similar to that in Table 2. On cluster B, the latency of a 0-byte message for MPI/Elan4 is approximately $2\mu s$ while for MVAPICH-1N it is $5\mu s$. The bandwidth for large messages of these two networks is comparable as shown in Table 1. The execution time of the two workloads LDOM and T3A at 16 processors, on eight nodes is shown in Figure 6. At 16 processors, for LDOM MVAPICH-1N performs better than MPI/Elan4 by approximately 20%. Most of this difference may be attributed to time spent in MPI_Recv and MPI_Wait. For T3A, there is very little difference in performance between MVAPICH-1N and MPI/Elan4.

4.3   Effect of Hardware Broadcast

In this section, we evaluate the impact of hardware based broadcast in InfiniBand on the performance of MM5-MPP. As discussed in section 3.3, and shown in Figure 3, a significant amount of time spent in the blocking call MPI_Bcast. At 16 processors for LDOM, approximately 10% of time is spent in MPI_Bcast. For

T3A at 16 processors, approximately 5% of time is spent in MPI_Bcast. Also as discussed in Section 3.5, at 16 processors, the message size passed to MPI_Bcast by both T3A and LDOM is approximately 100 bytes. At this size, hardware based broadcast does better by up to 50% in terms of latency than the current software based point-to-point algorithm [10]. It seems likely that MM5-MPP could potentially benefit from InfiniBand hardware broadcast.

The workloads LDOM and T3A were evaluated with and without hardware broadcast referred to as MVAPICH-SB and MVAPICH-HB respectively on cluster B, as explained in Section 4. All runs were taken up to 16 processes on eight nodes. For LDOM there is a reduction in execution time of approximately 2.14%. For T3A, the reduction in execution time is approximately 5.1%.

5   Related Work

The parallel implementation of MM5, MM5-MPP, was described in [13]. Only basic scalability in terms of execution time is discussed here. The performance using different commodity cluster interconnects is not discussed in this paper. Also the impact of efficient collective operations in modern interconnects on application performance is not discussed. The evaluation of the MM5 benchmark T3A on various architectures is carried out in [15]. Only the basic scalability in terms of execution time with increasing number of processors is discussed here. The impact of various network features like multicast is not evaluated here. The performance and scalability of various networks is evaluated using micro-benchmarks and NAS parallel benchmarks in [18]. This study focuses on comparing Myrinet, Quadrics and InfiniBand. The relative performance of Myrinet, InfiniBand and Quadrics in terms of micro-benchmarks is evaluated in [12]. There is no application-level evaluation here.

6   Conclusions and Future Work

In this paper, we have looked at the scalability of the parallel distributed memory version of the popular weather simulation code MM5. We have also looked at the sensitivity of MM5 to network parameters like latency, bandwidth and efficient collectives like hardware broadcast in InfiniBand. MM5 uses messages sizes of the order of 100 to 300 KiloBytes for system sizes up to 16 processors. These sizes decrease with increase in system size. A considerable amount of time is spent in the collective call MPI_Bcast which increases with increasing system. We conclude that, at smaller system sizes, MM5 would benefit from increased bandwidth. Experimentation with InfiniBand shows a reduction in execution time up to 34% compared with Myrinet at 16 processors. For larger system sizes, the improved latency of hardware based broadcast might be more beneficial to the application. Experimentally on a 16 processor environment, we see an improvement of up to 5% in overall execution time when using InfiniBand based hardware broadcast. Additionally, MM5 spends substantial time waiting in the MPI calls MPI_Wait and MPI_Recv as system size increases. We would like to determine the impact of efficient communication progress functions available in stacks like Myrinet MX and Quadrics on the performance of MM5, for large scale systems. MM5 also packs and unpacks its own data structures. We would like to investigate the effect of efficient zero-copy datatypes on the performance of MM5.

References

1. Gropp, W., Lusk, E., Doss, N., Skjellum, A.: A High-Performance, Portable Implementation of the MPI, Message Passing Interface Standard. Technical report, (Argonne National Laboratory and Mississippi State University)
2. Fluent CFD. (http://www.fluent.com)
3. LSDYNA. (http://www.lstc.com)
4. G.A. Grell, J. Dudhia, and D.R. Stauffer: A Description of the Fifth-Generation Penn State/NCAR Mesoscale Model (MM5). Tech. Rep. NCAR/TN-398+STR, National Center for Atmospheric Research, Boulder, Colarado (1994)
5. Boden, N.J., Cohen, D., et al.: Myrinet: A Gigabit-per-Second Local Area Network. IEEE Micro (1995) 29–35
6. Infiniband Trade Association. (www.infinibandta.org)
7. Quadrics Ltd. (www.quadrics.com)
8. R. Noronha and N. B. Abu-Ghazaleh: Using Programmable NICs for Time Warp Optimization. IPDPS (2002)
9. MPI over InfiniBand Project. (http://nowlab.cis.ohio-state.edu/projects/mpi-iba)
10. J. Liu, A. Mamidala and D.K. Panda: Fast and Scalable MPI-Level Broadcast using InfiniBand's Hardware Multicast Support. IPDPS (2004)
11. Myrinet E-cards. (http://www.myri.com/myrinet/PCIX/m3f2-pcixe.html)
12. J. Liu, B. Chandrasekaran, W. Yu, J. Wu, D. Buntinas, S. Kini, P. Wyckoff and D. K. Panda.: Micro-Benchmark Performance Comparison of High-Speed Cluster Interconnects. IEEE Micro. (2004)
13. J. Michalakes, T. Canfield, R. Nanjundiah and S. Hammond: Parallel Implementation, Validation and Performance of MM5. Sixth Workshop on the Use of Parallel Processors in Meteorology, European Center for Medium Range Weather Forecasting, Reading, U.K. (1994)
14. Michalakes, J.: A Runtime System Library for Parallel Finite Difference Models with Nesting. Technical Report ANL/MCS-TM-197 (1997)
15. Parallel MM5 benchmarks. http://www.mmm.ucar.edu/mm5/mpp/helpdesk/20040304a.html (2004)
16. MM5 Community Model. (http://www.mmm.ucar.edu/mm5/)
17. mpiP MPI Profiling Tool. (http://www.llnl.gov/CASC/mpip)
18. R. Brightwell, D. Doerfler and K.D. Underwood: A Comparison of 4X InfiniBand and Quadrics Elan-4 Technologies. IEEE Conference on Cluster Computing. (2004)

A Performance Measurement Infrastructure for Co-array Fortran

Bernd Mohr[1], Luiz DeRose[2], and Jeffrey Vetter[3]

[1] Forschungszentrum Jülich, ZAM,
Jülich, Germany
`b.mohr@fz-juelich.de`
[2] Cray Inc.
Mendota Heights, MN, USA
`ldr@cray.com`
[3] Oak Ridge National Laboratory
Oak Ridge, TN, USA
`vetterjs@ornl.gov`

Abstract. Co-Array Fortran is a parallel programming language for scientific applications that provides a very intuitive mechanism for communication, and especially, one-sided communication. Despite the benefits of this integration of communication primitives with the language, analyzing the performance of CAF applications is not straightforward, which is due, in part, to a lack of tools for analysis of the communication behavior of Co-Array Fortran applications. In this paper, we present an extension to the KOJAK toolkit based on a source-to-source translator that supports performance instrumentation, data collection, trace generation, and performance visualization of Co-Array Fortran applications. We illustrate this approach with a performance visualization of a Co-Array Fortran version of the Halo kernel benchmark using the VAMPIR event trace visualization tool.

1 Introduction

Co-Array Fortran (CAF) [12] extends Fortran 95 providing a simple, explicit notation for data decomposition, communication, and synchronization, expressed in a natural Fortran-like syntax. These extensions provide a straightforward and powerful paradigm for parallel programming of scientific applications based on one-sided communication. One of the problems that CAF users face is the lack of tools for analysis of the communication and synchronization behavior of the application. One of the reasons for the lack of tools is because communication operations in CAF programs are not expressed through function calls, as in MPI, or via directives that are executed by a run-time library, as in OpenMP. In contrast, CAF communication operations are integrated into the language, and, on certain platforms like the Cray X1, they are implemented via remote memory access instructions provided by the hardware.

For MPI applications, performance data collection is, in general, facilitated by the existence of the MPI profiling interface (PMPI), which is used by most MPI tools [2, 7, 14]. Similarly, performance measurement of OpenMP applications can be done by instrumenting the calls to the runtime library [1, 4, 5]. However, with the challenge

J.C. Cunha and P.D. Medeiros (Eds.): Euro-Par 2005, LNCS 3648, pp. 146–155, 2005.

of CAF communication primitives being integrated into the language, and potentially implemented with special hardware instructions, the instrumentation of these communication primitives requires a different approach that is not straightforward.

In order to address this problem, we first defined PCAF, an interface specification of a set of routines intended to monitor all important aspects of CAF applications. Then, we extended the OPARI source-to-source instrumentation tool [10] to search for CAF constructs and to generate instrumented source code with the appropriate PCAF calls. Finally, we implemented the PCAF interface for the the KOJAK measurement system [13] enabling it to trace CAF communication and synchronization instructions. With this extension, the KOJAK measurement system is able to support performance instrumentation and performance data collection of CAF applications, generating trace files that can be analyzed with the VAMPIR event trace visualization tool [11]. In this paper, we describe our approach for performance measurement and analysis of CAF applications.

The remainder of this paper is organized as follows. In Section 2, we present an overview of Co-Array Fortran. In Section 3, we briefly describe the KOJAK performance measurement and analysis environment. In Section 4, we describe our approach for performance instrumentation and measurement of Co-Array Fortran applications. In Section 5, we discuss performance visualization with an example using the Halo kernel benchmark code. Finally, we present our conclusions in Section 6.

2   An Overview of Co-array Fortran

Co-array Fortran [12] is a parallel programming language extension to Fortran 95. At the highest level, CAF uses a Single Program Multiple Data (SPMD) model to allow multiple copies (*images*) of a program to execute asynchronously. Each image contains its own private set of data objects. When data objects are distributed across multiple images, the array syntax of CAF uses an additional trailing subscript in square brackets to allow explicit access to remote data (as shown in Figures 2 and 4), and it is referred to as the *co-dimension*. Data references that do not use these square brackets are strictly local accesses.The CAF compiler translates these remote data accesses into underlying communication mechanisms for each target system. CAF also includes intrinsic routines to synchronize images, to return the number of images, and to return the index of the current image. Besides functions for delimiting a critical region, CAF provides four different forms of a barrier synchronization:

SYNC_ALL(): a global barrier where every image waits for every other image.
SYNC_ALL(<wait list>): a global barrier where every image waits only for the listed
    images.
SYNC_TEAM(<team>): a barrier where a team of images wait for every other team
    member.
SYNC_TEAM(<team>, <wait list>): a barrier where a team of images wait for a
    subgroup of the team members.

CAF was originally developed on the Cray-T3D, and, as such, it is very efficient on platforms that support one-sided messaging and fast barrier operations. On systems with globally addressable memory, such as the Cray X1 or the SGI Altix 3700, these mechanisms may be as simple as load and store memory references. By contrast, on distributed memory systems that do not support efficient Remote Direct Memory Access (RDMA), these mechanisms can be implemented in MPI.

3   The KOJAK Measurement System

The KOJAK performance-analysis tool environment provides a complete tracing-based solution for automatic performance analysis of MPI, OpenMP, or hybrid applications running on parallel computers. KOJAK describes performance problems using a high level of abstraction in terms of execution patterns that result from an inefficient use of the underlying programming model(s). KOJAK's overall architecture is depicted in Figure 1. Tasks and components are represented as rectangles and their inputs and outputs are represented as boxes with rounded corners. The arrows illustrate the whole performance-analysis process from instrumentation to result presentation.

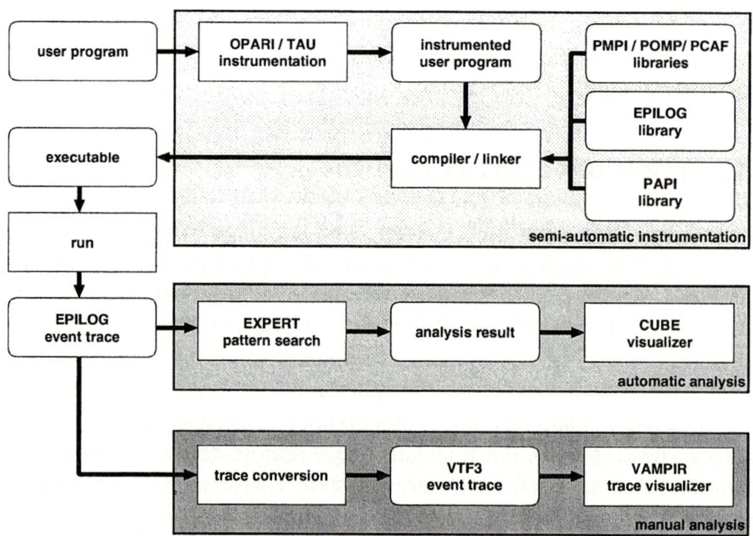

Fig. 1. KOJAK overall architecture.

The KOJAK analysis process is composed of two parts: a semi-automatic multi-level instrumentation of the user application followed by an automatic analysis of the generated performance data. The first part is considered semi-automatic because it requires the user to slightly modify the makefile.

To begin the process, the user supplies the application's source code, written in either C, C++, or Fortran, to OPARI, which is a source-to-source translation tool. OPARI performs automatic instrumentation of OpenMP constructs and redirection of OpenMP-library calls to instrumented wrapper functions on the source-code level based on the POMP OpenMP monitoring API [9]. In Section 4.2, we describe how we extended OPARI for instrumentation of CAF programs with the appropriate PCAF calls.

Instrumentation of user functions is done either during compilation by a compiler-supplied instrumentation interface or on the source-code level using TAU [2]. TAU is able to automatically instrument the source code of C, C++, and Fortran programs using a preprocessor based on the PDT toolkit [8].

Instrumentation for MPI events is accomplished with a wrapper library based on the PMPI profiling interface. All MPI, OpenMP, CAF and user-function instrumentation calls the EPILOG run-time library, which provides mechanisms for buffering and trace-file creation. The application can also be linked to the PAPI library [3] for collection of hardware counter metrics as part of the trace file. At the end of the instrumentation process, the user has a fully instrumented executable.

Running this executable generates a trace file in the EPILOG format. After program termination, the trace file is fed into the EXPERT analyzer. (See [13] for details of the automatic analysis, which is outside of the scope of this paper.) In addition, the automatic analysis can be combined with a manual analysis using VAMPIR [11], which allows the user to investigate the patterns identified by EXPERT in a time-line display via a utility that converts the EPILOG trace file into the VAMPIR VTF3 format.

4 Performance Instrumentation and Measurement Approach

In this section, we describe the event model that we use to describe the behavior of CAF applications, and the approach we take to instrument CAF programs and to collect the necessary measurement data.

4.1 An Event Model of CAF

KOJAK uses an event-based approach to analyze parallel programs. A stream or trace of events allow to describe the dynamic behavior of an application over time. If necessary, execution statistics can be calculated from that trace. The events represent all the important points in the execution of the program. Our CAF event model is based on KOJAK's basic model for one-sided communication [6]. We extended KOJAK's existing set of events, which cover describing the begin and end of user functions and MPI and OpenMP related activities, with the following events for representing the execution of CAF programs:

- Begin and end of CAF synchronization primitives
- Begin and end of remote read and write operations

For each of these events, we collect a time stamp and location. For CAF synchronization functions, we also record which function was entered or exited. For the barrier routines we also collect the group of images which participate in the barrier and the group of images waited for, if applicable. Finally, for reads and writes, we collect the amount of data which is transferred (i.e., the number of array elements) as well as the source or destination of the transfer.

The event model is also the basis for the instrumentation and measurement. The events and their attributes specify which elements of CAF programs need to be instrumented and which data has to be collected.

4.2   Performance Instrumentation

Instrumentation of CAF programs can be done on either of two levels depending on how CAF is implemented on a specific computing platform. On systems where CAF constructs and API calls are translated into calls to a run-time library, these calls could easily be instrumented by traditional techniques (e.g., linking a pre-instrumented run-time library or instrumenting the calls with a binary instrumentation tool). However, for systems like the Cray X1, where the CAF communication is executed via hardware instructions, this approach is not possible. Therefore, we extended OPARI, KOJAK's source-to-source translation tool, to also locate and instrument all CAF constructs of a program.

As Fortran is line-oriented, it is possible for OPARI to read a program line by line. Of course, it is also necessary to take continuation lines into account. Then, each line is scanned for occurrences of CAF constructs and synchronization calls (but ignoring comments and contents of strings). CAF constructs can be located by looking for pairs of brackets ([...]). The first word of the statement determines whether it is a declaration line or a statement containing a remote read or write operation. For CAF declarations, OPARI collects attributes like array dimensions, and lower and upper bounds for later use.

The handling of statements containing remote memory operations is more complex. First, all operations are located in the line. If it is an assignment statement and the operation appears before the assignment operator, it is a write operation. In all other cases it is a read. OPARI determines which CAF array is referenced by the operation, the number of elements transferred (by parsing the index specification), and the source or destination of the transfer (determined by the expression inside the brackets). Simple assignment statements containing a single remote memory operation are instrumented by inserting calls to the corresponding PCAF monitoring functions before and after the statement, which get passed in the attributes determined by OPARI. In case of more complex statements where a remote memory operation cannot be easily separated out and wrapped by the measurement calls, or when it is necessary to keep instrumentation overhead low, OPARI uses the single call version of the PCAF remote memory access monitoring functions (instead of separate begin and end calls) and inserts them either before (for reads) or after (for writes) the statement for each identified remote memory access operation.

Finally, OPARI scans the line for calls to CAF synchronization routines, and replaces them by calls to PCAF wrapper functions that will execute the original call in addition to collecting all important attributes.

Figure 2(b) shows the instrumented source code generated for the example in Figure 2(a). In this example, there is a two-dimensional array A, which is distributed on all processors. In the CAF statement `A(me,::2)[left] = me`, each processor updates the odd entries of the row corresponding to its image in the left neighbor array with its index, and then waits on a barrier. OPARI identifies the CAF statement, and adds a begin and end instrumentation event. The call to indicate the beginning of the event contains the destination of the write (normalized to the range 0 to `num_images()-1`) and the number of array elements being transferred; the end call only gets passed the destination. The barrier call (`sync_all`) is translated into a call of the corresponding *wrapper* function.

```
integer :: me, num, left        integer :: me, num, left
integer :: A(1024,1024)[*]       integer :: A(1024,1024)[*]
. . .                            . . .
me = this_image()                me = this_image()
num = num_images()               num = num_images()
left = me - 1                    left = me - 1
if ( left < 1 ) left = num       if ( left < 1 ) left = num
                                 call PCAF_rma_write_begin(-1+left, &
                                     1 * max((ubound(A,2)-         &
                                         lbound(A,2)+2)/2,0))
A(me,::2)[left] = me             A(me,::2)[left] = me
                                 call PCAF_rma_write_end(-1+left)
call sync_all()                  call PCAF_sync_all()

(a)                              (b)
```

Fig. 2. (a) Example of a CAF source code and (b) OPARI instrumented version.

4.3   Performance Measurement

Finally, the KOJAK measurement system was extended by implementing the necessary
PCAF monitoring functions and wrapper routines and adding support for the handling
of the new remote memory access event types. We chose to implement our approach
within the KOJAK framework, as KOJAK is very portable and supports all major HPC
computing platforms. Also, this way, we could re-use many of KOJAK's features like
event trace buffer management, generation, and conversion. Finally, it allows us not
only to analyze plain CAF applications but also hybrid programs using any combination
of MPI, OpenMP, and CAF. A separate, new instrumentor just for CAF would probably
be problematic in this respect, as the modifications done by two independent source-to-
source preprocessors could conflict.

The PCAF interface is shown in Figure 3. Since this monitoring API is open, and
OPARI is a stand-alone tool, other performance analysis projects could use this infrastructure to also support CAF. For example, it would be very easy to implement a version
of the PCAF monitoring library which (instead of tracing) just collects basic statistics
(number of RMA transfers, amount of data transferred) for each participating image.
Ideally, in the future, CAF compilers could support this interface directly.

5   Performance Visualization

For illustration of our performance analysis approach, we ran the Halo kernel benchmark on the Cray X1 system at the Oak Ridge National Laboratory, using 16 and 64
processors. The Halo benchmark simulates a halo border exchange with the four different synchronization methods CAF provides (see Section 2). The exchange procedure is
outlined in Figure 4. During each iteration, the following events from our event model
occur: S2 a synchronization call; S3 a remote read of n elements from the north neighbor; S4 a remote read of $2n$ elements from the south neighbor; S5 a synchronization
call; S7 another synchronization call; S8 a remote read of n elements from the west

Remote Memory Access Monitoring Routines

```
    SUBROUTINE PCAF_rma_write_begin(dest, nelem)
    SUBROUTINE PCAF_rma_write_end(dest)
    SUBROUTINE PCAF_rma_write(dest, nelem)
    SUBROUTINE PCAF_rma_read_begin(src, nelem)
    SUBROUTINE PCAF_rma_read_end(src)
    SUBROUTINE PCAF_rma_read(src, nelem)
where  INTEGER, INTENT(IN) :: dest, src, nelem
```

CAF **Synchronization Wrapper Routines**

```
    SUBROUTINE PCAF_sync_all()
    SUBROUTINE PCAF_sync_all(wait)
    SUBROUTINE PCAF_sync_team(team)
    SUBROUTINE PCAF_sync_team(team, wait)
    SUBROUTINE PCAF_sync_file(unit)
    SUBROUTINE PCAF_sync_memory()
    SUBROUTINE PCAF_start_critical()
    SUBROUTINE PCAF_end_critical()
where  INTEGER, INTENT(IN) :: unit
       INTEGER, INTENT(IN) :: wait(:), team(:)
```

Fig. 3. PCAF Measurement Function Interface Specification.

```
S1          HINS(1:3*n)     = HOEW(1:3*n)
S2          CALL synchronization method
S3          HONS(1:n)       = HINS(1:n) [MYPEN]
S4          HONS(n+1:3*n) = HINS(n+1:3*n) [MYPES]
S5          CALL synchronization method
S6          HIEW(1:3*n)     = HONS(1:3*n)
S7          CALL synchronization method
S8          HOEW(1:n)       = HIEW(1:n) [MYPEW]
S9          HOEW(n+1:3*n) = HIEW(n+1:3*n) [MYPEE]
S10         CALL synchronization method
```

Fig. 4. Pseudo-code for the halo exchange procedure.

neighbor; S9 a remote read of $2n$ elements from the east neighbor; and finally S10 a synchronization call. For each synchronization method, this procedure is repeated 5 times per iteration, with 10 iterations being executed with n varying from 2 to 1024 in powers of 2.

Figure 5 (a) shows the timeline view of the Halo benchmark running with 16 processors. The four phases of the code (marked with white lines in the figure) can easily be identified due to the different communication behavior of each of the synchronization methods. The communication pattern between processors, as well as the amount of data exchanged, can be observed with the pair-wise communication statistics view, shown in Figure 6 (left).

Figure 5 (b) and Figure 5 (c) show a section of the timeline corresponding to a full exchange (one call to the subroutine outlined in Figure 4) for sync_all and sync_team(wait) synchronization methods respectively. We observe that the re-

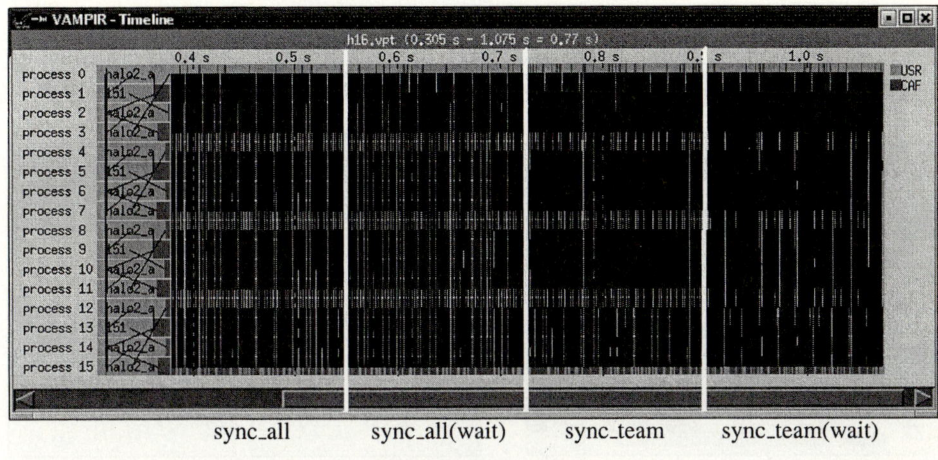

(a) Complete program

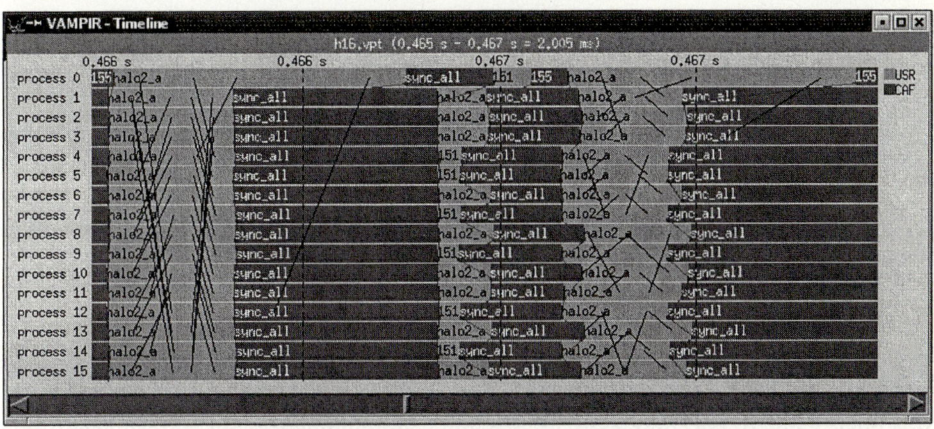

(b) One exchange using `sync_all`

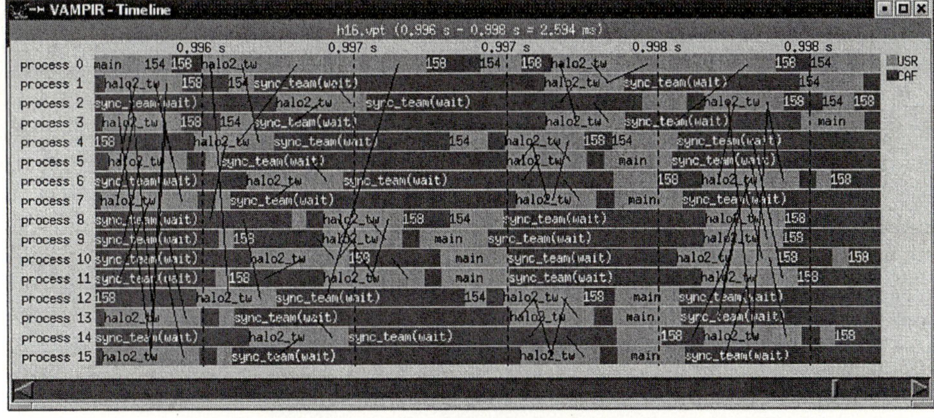

(c) One exchange using `sync_team(wait)`

Fig. 5. Timeline views of the Halo benchmark using 16 processors.

gion corresponding to the `sync_team(wait)` synchronization method is much more irregular (unsynchronized) than the one for the `sync_all`, where the waiting times are longer, due to the global synchronization.

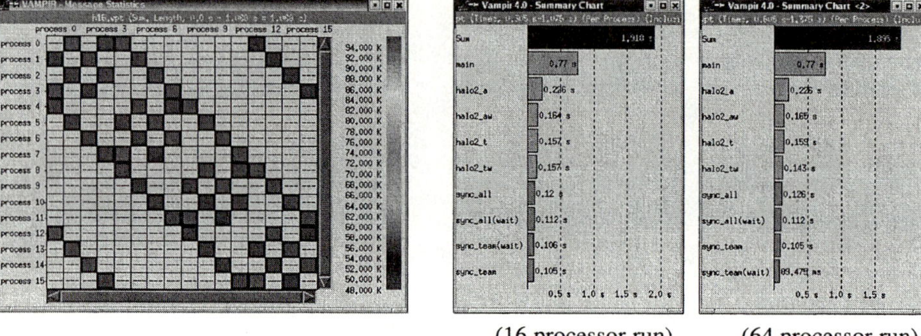

(16 processor run)   (64 processor run)

Fig. 6. Message statistics view of the Halo benchmark using 16 processors (left) and Summary Chart View of Function times running on 16 and 64 processors (right).

Finally, on Figure 6 (right), we observe the time spent on each synchronization method for the 16 and 64 processors runs respectively. We notice that with the increase of number of processors, the `sync_team(wait)` method performs significantly better than the `sync_all` method, going from about 10% faster with 16 processors to about 30% faster with 64 processors.

6   Conclusion

The CAF parallel programming language extends Fortran 95 providing a simple technique for accessing and managing distributed data objects. This language-level abstraction hides much of the complexity of managing communication, but, unfortunately, this also makes diagnosing performance problems much more difficult. In this paper, we have proposed one approach to solve this problem. Our solution uses a source-to-source translator to allow performance instrumentation, data collection, trace generation, and performance visualization of Co-Array Fortran applications implemented as an extension of the KOJAK performance analysis toolset. We illustrated this approach with performance visualization of a Co-Array Fortran version of the Halo kernel benchmark using the VAMPIR event trace visualization tool. Our initial results are promising; we can obtain statistical quantification and graphical presentation of CAF communication and synchronization characteristics. We will extend KOJAK's automated analysis to also cover CAF constructs and determine the benefits of this approach for real applications.

References

1. E. Ayguadé, M. Brorsson, H. Brunst, H.-C. Hoppe, S. Karlsson, X. Martorell, W. E. Nagel, F. Schlimbach, G. Utrera, and M. Winkler. OpenMP Performance Analysis Approach in the INTONE Project. In *Proceedings of the Third European Workshop on OpenMP - EWOMP'01*, September 2001.

2. R. Bell, A. D. Malony, and S. Shende. A Portable, Extensible, and Scalable Tool for Parallel Performance Profile Analysis. In *Proceedings of Euro-Par 2003*, pages 17–26, 2003.

3. S. Browne, J. Dongarra, N. Garner, G. Ho, and P. Mucci. A Portable Programming Interface for Performance Evaluation on Modern Processors. *The International Journal of High Performance Computing Applications*, 14(3):189–204, Fall 2000.

4. J. Caubet, J. Gimenez, J. Labarta, L. DeRose, and J. Vetter. A Dynamic Tracing Mechanism for Performance Analysis of OpenMP Applications. In *Proceedings of the Workshop on OpenMP Applications and Tools - WOMPAT 2001*, pages 53 – 67, July 2001.

5. L. DeRose, B. Mohr, and S. Seelam. Profiling and Tracing OpenMP Applications with POMP Based Monitoring Libraries. In *Proceedings of Euro-Par 2004*, pages 39–46, September 2004.

6. Marc-André Hermanns, Bernd Mohr, and Felix Wolf. Event-based Measurement and Analysis of One-sided Communication. In *Proceedings of Euro-Par 2005*, September 2005.

7. S. Kim, B. Kuhn, M. Voss, H.-C. Hoppe, and W. Nagel. VGV: Supporting Performance Analysis of Object-Oriented Mixed MPI/OpenMP Parallel Applications. In *Proceedings of the International Parallel and Distributed Processing Symposium*, April 2002.

8. K. A. Lindlan, Janice Cuny, A. D. Malony, S. Shende, B. Mohr, R. Rivenburgh, and C. Rasmussen. A Tool Framework for Static and Dynamic Analysis of Object-Oriented Software with Templates. In *Proceedings of Supercomputing 2000*, November 2000.

9. B. Mohr, A. Mallony, H.-C. Hoppe, F. Schlimbach, G. Haab, and S. Shah. A Performance Monitoring Interface for OpenMP. In *Proceedings of the fourth European Workshop on OpenMP - EWOMP'02*, September 2002.

10. Bernd Mohr, Allen Malony, Sameer Shende, and Felix Wolf. Design and Prototype of a Performance Tool Interface for OpenMP. *The Journal of Supercomputing*, 23:105–128, 2002.

11. W. Nagel, A. Arnold, M. Weber, H.-C. Hoppe, and K. Solchenbach. Vampir: Visualization and Analysis of MPI Resources. *Supercomputer*, 12:69–80, January 1996.

12. R. W. Numrich and J. K. Reid. Co-Array Fortran for Parallel Programming. *ACM Fortran Forum*, 17(2), 1998.

13. Felix Wolf and Bernd Mohr. Automatic Performance Analysis of Hybrid MPI/OpenMP Applications. *Journal of Systems Architecture, Special Issue 'Evolutions in parallel distributed and network-based processing'*, 49(10–11):421–439, November 2003.

14. C. Wu, A. Bolmarcich, M. Snir, D. Wootton, F. Parpia, A. Chan, E. Lusk, and W. Gropp. From trace generation to visualization: A performance framework for distributed parallel systems. In *Proceedings of Supercomputing 2000*, November 2000.

Event-Based Measurement and Analysis of One-Sided Communication

Marc-André Hermanns[1], Bernd Mohr[1], and Felix Wolf[2]

[1] Forschungszentrum Jülich,
Zentralinstitut für Angewandte Mathematik,
52425 Jülich, Germany
`{m.a.hermanns,b.mohr}@fz-juelich.de`
[2] University of Tennessee, ICL
1122 Volunteer Blvd Suite 413
Knoxville, TN 37996-3450, USA
`fwolf@cs.utk.edu`

Abstract. To analyze the correctness and the performance of a program, information about the dynamic behavior of all participating processes is needed. The dynamic behavior can be modeled as a stream of events required for a later analysis including appropriate attributes. Based on this idea, KOJAK, a trace-based toolkit for performance analysis, records and analyzes the activities of MPI-1 point-to-point and collective communication.

To support remote-memory access (RMA) hardware in a portable way, MPI-2 introduced a standardized interface for remote memory access. However, potential performance gains come at the expense of more complex semantics. From a programmer's point of view, an MPI-2 data transfer is only completed after a sequence of communication and associated synchronization calls.

This paper describes the integration of performance measurement and analysis methods for RMA communication into the KOJAK toolkit. Special emphasis is put on the underlying event model used to represent the dynamic behavior of MPI-2 RMA operations. We show that our model reflects the relationships between communication and synchronization more accurately than existing models. In addition, the model is general enough to also cover alternate but simpler RMA interfaces, such as SHMEM and Co-Array Fortran.

1 Introduction

Remote memory access (RMA) describes the ability of a process to directly access a part of the memory of a remote process, without explicit participation of the remote process in the data transfer. As all parameters for the data transfer are determined by one process, it is also called *one-sided* or *single-sided* communication. This distinguishes the one-sided communication from point-to-point messages, where explicit send and receive statements are required on both sides. Providing one-sided in addition to two-sided communication significantly expands the flexibility to chose a communication scheme most suitable for a given problem on a given hardware.

On platforms with special hardware providing efficient RMA support, one-sided communication is often made available to the programmer in the form of libraries,

J.C. Cunha and P.D. Medeiros (Eds.): Euro-Par 2005, LNCS 3648, pp. 156–165, 2005.

for example SHMEM (Cray), LAPI (IBM), or ELAN (Quadrics). However, these libraries are typically platform- or at least vendor-specific. The exception is SHMEM, which is offered by a group of vendors. Since this restricts portable programming, many programmers do not utilize one-sided communication.

This is one of the reasons why the MPI forum decided to define a portable one-sided communication interface as part of MPI-2. The Message Passing Interface (MPI) was defined by a group of vendors, government laboratories and universities in 1994 as a community standard [1]. This has become known as MPI-1. It is fully supported by all freely-available and commercial MPI implementations and was quickly adopted by the scientific computing community as a de-facto standard. As MPI also provides a standard monitoring interface (PMPI), there is a wide variety of tools for MPI performance analysis and visualization. In 1997, a second version of the interface (MPI-2) was defined, which added support for parallel I/O, dynamic process creation, and one-sided communication [2]. However, only now, seven years after its definition, is support for all MPI-2 features portably available for all major parallel computing platforms.

Until recently there was only rare usage of RMA features in scientific applications and, therefore, the demand for performance tools in this area was limited. As more and more programmers adopt the new features to improve the performance of their codes, this is expected to change. For example, NASA researchers report a 39% improvement in throughput after replacing MPI-1 non-blocking communication with MPI-2 one-sided communication in a global atmospheric simulation program [3].

Currently, there are only very few tools which support the measurement and analysis of one-sided communication and synchronization in a portable way on a wider range of platforms. The well-known Paradyn tool which performs an automatic on-line bottleneck search, was recently extended to support several major features of MPI-2 [4]. For RMA analysis, it collects basic, process-local, statistical data (i.e., transfer counts and execution time spent in RMA functions). It does not take inter-process relationships into account nor does it provide detailed trace data. Also, it does not support analysis of SHMEM programs. The very portable TAU performance analysis tool environment [5] supports profiling and tracing of MPI-2 and SHMEM one-sided communication. However, it only monitors the entry and exit of the RMA functions; it does not provide RMA transfer statistics nor are the transfers recorded in tracing mode. The commercial Intel Trace Collector tool (formerly known as VampirTrace) [6] records MPI execution traces. When used with MPI-2, it records enter and exits of only a subset of the RMA functions. It also traces the actual RMA transfers, but misrepresents their semantics, as defined in MPI-2. Finally, it does not record the collective nature of MPI-2 window functions. Besides these there are also some non-portable vendor tools with similar disadvantages.

KOJAK, our toolkit for automatic performance analysis [10], is jointly developed by the Central Institute for Applied Mathematics of the Research Centre Jülich and by the Innovative Computing Laboratory of the University of Tennessee. It is able to instrument and analyze OpenMP constructs and MPI-1 calls. In this paper we report on the integration of performance analysis methods for one-sided communication into the existing toolkit. We put special emphasis on the development of a new event model that realistically represents the dynamic behavior of MPI-2 RMA operations in the event stream. We show that our model reflects the relationships between communication and synchronization more accurately than existing models. In addition, the model is general enough to also cover alternate, but simpler RMA interfaces. In our new prototype implementation, we added support for measurement and analysis of parallel programs using MPI-2 and SHMEM one-sided communication and synchronization. In addition, we are also able to handle Co-Array Fortran programs [9], a small extension to Fortran 95 that provides a simple, explicit notation for one-sided communication and synchronization, expressed in a natural Fortran-like syntax. Details of this work can be found in [11].

The remainder of the paper is organized as follows: In Section 2 we give a short description of the MPI-2 RMA communication and synchronization functions. In Section 3, we present our event model, which allows the realistic representation of the dynamic behavior of vendor-specific and MPI-2 RMA operations. The extensions to KOJAK components allowing the instrumentation, measurement, analysis, and visualization of parallel programs based on one-sided communication are described in Section 4. Finally, we present conclusions and future work in Section 5.

2   MPI-2 One-Sided Communication

The interface for RMA operations defined by MPI-2 differs from the vendor-specific APIs in many respects. This is to ensure that it can be efficiently implemented on a wide variety of computing platforms even if a platform does not provide any direct hardware support for RMA. The design behind the MPI-2 RMA API specification is similar to that of weakly coherent memory systems: correct ordering of memory accesses has to be specified by the user with explicit synchronization calls; for efficiency, the implementation can delay communication operations until the synchronization calls occur.

MPI does not allow access to arbitrary memory locations with RMA operations, but only to designated parts of a process's memory, the so-called *windows*. Windows must be explicitly initialized (with a call to MPI_Win_create) and released (with MPI_Win_free) by all processes that either provide memory or want to access this memory. These calls are *collective* between all participating partners and include an internal barrier operation. MPI denotes by *origin* the process that performs an RMA read or write operation, and by *target* the process in which the memory is accessed.

There are three RMA communication calls in MPI: MPI_Put transfers data from the caller's memory to the target memory (*remote write*); MPI_Get transfers data from the target to the origin (*remote read*); and MPI_Accumulate updates locations in the target memory, for example, by replacing them with sums or products of the local and remote data values (*remote update*). These operations are *nonblocking*: the call initiates the transfer, but the transfer may continue after the call returns. The transfer is completed, both at the origin and the target, only when a subsequent synchronization call is issued by the caller on the involved window object. Only then are the transferred values (and the associated communication buffers) available to the user code. RMA communication falls in two categories: *active target* and *passive target* communication. In both modes, the parameters of the data transfer are specified only at the origin, however in active mode, both origin and target processes have to participate in the synchronization of the RMA accesses. Only in passive mode is the communication and synchronization completely one-sided.

RMA accesses to locations inside a specific window must occur only within an *access epoch* for this window. Such an access epoch starts with an RMA synchronization call, proceeds with any number of remote read, write, or update operations on this window, and finally completes with another (matching) synchronization call. Additionally, in active target communication, a target window can only be accessed within an *exposure epoch*. There is a one-to-one mapping between access epochs on origin processes and exposure epochs on target processes. Distinct epochs for a window at the same process must be disjoint. However, epochs pertaining to different windows may overlap.

MPI provides three RMA synchronization mechanisms:

Fences: The `MPI_Win_fence` collective synchronization call is used for active target communication. An access epoch on an origin process or an exposure epoch on a target process are started and completed by such a call. All processes who participated in the creation of the window synchronize, which in most cases includes a barrier. The data transfered is only accessible to user code after the fence.

General Active Target Synchronization: Here, synchronization is minimized: only pairs of communicating processes synchronize, and they do so only when needed to correctly order accesses to a window with respect to local accesses to that window. An access epoch is started at an origin process by `MPI_Win_start` and is terminated by a call to `MPI_Win_complete`. The start call specifies the group of targets for that epoch. An exposure epoch is started at a target process by `MPI_Win_post` and is completed by `MPI_Win_wait` or `MPI_Win_test`. Again, the post call specifies the group of origin processes for that epoch. Data written is only accessible after the wait call, however data can only be read after the complete operation.

Locks: Finally, shared and exclusive locks are provided through the `MPI_Lock` and `MPI_Unlock` calls. They are used for passive target communication. In addition, they also define the access epoch for this window at the origin. Data read or written is only accessible from user code after the unlock operation has completed.

It is implementation-defined whether some of the described calls are blocking or nonblocking; for example, in contrast to other shared memory programming paradigms, the lock call must not be blocking. For a complete description of MPI-2 RMA communication see [2].

3   An Event Model for One-Sided Communication

Many performance analysis tools use an *event-based* approach, that is, they instrument user applications only at specific points to collect the performance data they need for their analysis. These points, called *events*, are chosen in a way that they represent important aspects in the dynamic behavior of the application on a level of abstraction suitable for the tools' task. Trace-based tools record the occurrence of events as a stream or *trace* of event records for later analysis.

For the analysis of parallel scientific applications, events that capture the most important aspects of the parallel programming paradigm used (e.g., MPI or OpenMP) are defined. Often, to provide a context for events representing specific actions related to a parallel programming interface, the entering and leaving of surrounding user regions (e.g., functions, loops or basic blocks) are also captured.

Table 1. KOJAK's Event Types

Abstraction	Event type	Type specific Attributes
Entering / leaving a region (a function)	ENTER	region id
	EXIT	region id
Leaving a collective MPI	MPICEXIT	region id, comm id, root loc, sent, revd
or OpenMP region	OMPCEXIT	region id
Sending / receiving a message	SEND	dest loc, tag, comm id, length
	RECV	src loc, tag, comm id, length
Start / end of OpenMP parallel region	FORK	
	JOIN	
Acquiring / releasing an OpenMP lock	ALOCK	lock id
	RLOCK	lock id
Start / end / origin of RMA	PUT_1TS	window id, rma id, length, dest loc
one-sided transfers	PUT_1TE	window id, rma id, length, src loc
	GET_1TO	window id, rma id
	GET_1TS	window id, rma id, length, dest loc
	GET_1TE	window id, rma id, length, src loc
Leaving MPI GATS function	MPIWEXIT	window id, region id, group id
Leaving MPI collective RMA function	MPIWCEXIT	window id, region id, comm id
Locking / unlocking a MPI window	WLOCK	window id, lock loc, type
	WUNLOCK	window id, lock loc

Table 1 lists all event types used by the KOJAK performance analysis toolset. In the upper half, the already existing events for modeling MPI-1 and OpenMP behavior are shown. in addition to type-specific attributes for each event we also collect the *timestamp* and *location* which describe when and where the event occurred. For user regions, MPI functions, and OpenMP constructs and runtime functions, we record which region was entered or left. In the case of collective MPI functions and OpenMP constructs, instead of "normal" EXIT events, special collective events are used to capture the attributes of the collective operation. For MPI this is the communicator, the root process, and the amounts of data sent and received during this operation. MPI-1 point-to-point messages are modeled as pairs of SEND and RECV events with the source or destination of the message, the tag and communicator used, and the amount of data transferred being attributes. In OpenMP applications, FORK and JOIN events mark the start and end of parallel regions and ALOCK and RLOCK events the acquisition and release of locks. For a complete, more detailed description of KOJAK's event types and of its analysis features see [7, 10]. A similar event model is also used by most other event-based tools (e.g., by TAU).

In order to be able to also analyze RMA operations, we defined new event types to realistically model the behavior of MPI-2 as well as Co-Array Fortran and vendor-specific RMA operations. These new event types are shown in the bottom compartment of Table 1. Start and end of RMA one-sided transfers are marked with PUT_1TS and PUT_1TE (for remote writes and updates) or with GET_1TS and GET_1TE (for remote reads). For these events, we collect the source and destination and the amount of data transferred, as well as a unique RMA operation identifier which allows an easier mapping of #_1TE to the corresponding #_1TS events in the analysis stage later on. For all MPI RMA communication and synchronization operations we also collect an identification for the window on which the operation was performed. Exits of MPI-2 functions related to general active target synchronization (GATS) are marked with a MPIWEXIT event which also captures the groups of origin or target processors. For collective MPI-2 RMA functions we use a MPIWCEXIT event and record the communicator which defines the group of processes which participate in the collective operation. Finally, MPI window lock and unlock operations are marked with WLOCK and WUNLOCK events.

Based on these event types and their attributes, we now introduce two event models for describing the dynamic behavior of RMA operations. For each model, we describe its basic features and analyze its strengths and weaknesses. To illustrate the location of events and relationships between them, we use simple *time-line* diagrams. In these diagrams, time progresses from left to right. Event instances are shown as colored circles on different "time lines", one for each process involved in the execution. Invocations of functions are shown as gray boxes with the name of the function executed. Finally, relationships between events are displayed as arrows with different line styles. Following KOJAK conventions [7], relationships are always named with a suffix `ptr` (for pointer) and always point from a later event back to an earlier event related to the later one. This allows for an efficient analysis process with a single pass through the event trace.

3.1 Basic Model

In the first and simpler model, it is assumed that the RMA communication functions have a blocking behavior, that is, the data transfer is completed before the function is finished. Also, RMA synchronization functions are treated as if they were independent of the communication functions.

The invocations of RMA communication and synchronization functions are modeled with ENTER and EXIT events. To model the actual RMA transfer, the transfer-start event is associated with the source process immediately after the begin of the corresponding communication function. Accordingly, the end event is associated with the destination process shortly before the exit of the function. Finally, we define a relationship *startptr* which allows analysis tools to easily locate the matching start event from the transfer end event. Figure 1 shows the model for typical usage patterns of one-sided communication. A sequence of get and put operations is guarded by fences, barriers, or lock/unlock operations. The message line shown in the picture is not part of the model and only shown for clarity.

The advantage of this model is a straight-forward implementation because events and their attributes can be recorded at exactly the place and time where they are supposed to appear in the model. We use this model for analyzing SHMEM and Co-Array Fortran programs. However, for MPI-2 this model is not sufficient because it ignores the necessary synchronization, as described in Section 2. Since the end-of-transfer event is placed before the end of the communication function, the transfers are recorded as completed even when, for example, in the case of a nonblocking implementation, this is not true. Even if the implementation is blocking, it still does not reflect the user-visible behavior. Therefore, in case of MPI-2, we use an extended model, which is described in the next subsection.

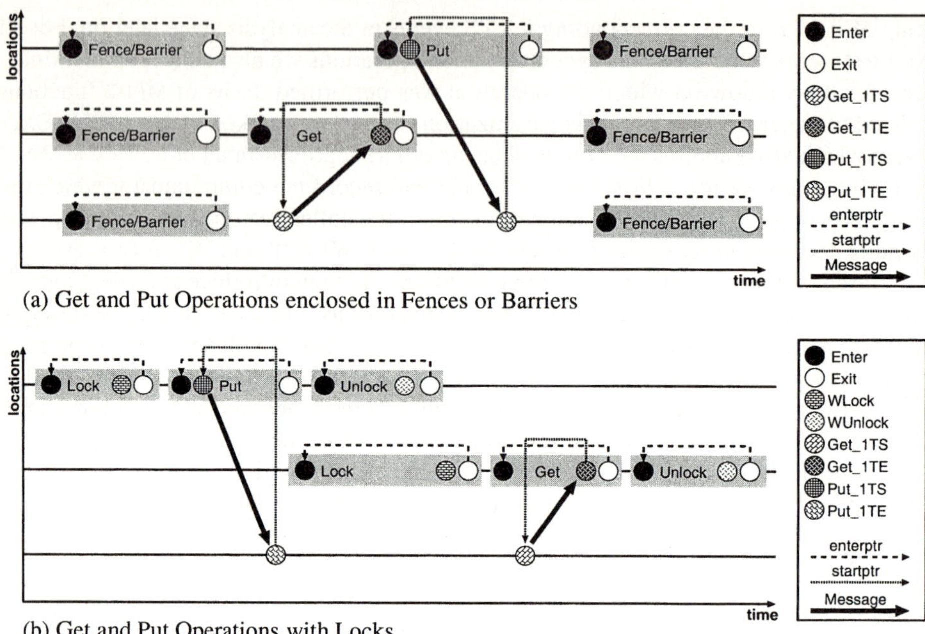

(a) Get and Put Operations enclosed in Fences or Barriers

(b) Get and Put Operations with Locks

Fig. 1. Examples for Basic Event Model

3.2   Extended Model

The *extended model* observes the MPI-2 synchronization semantics and, therefore, better reflects the user-visible behavior of MPI-2 RMA operations. Figure 2 shows the model for the three different synchronization methods defined by MPI-2. The end of fences and GATS calls is now modeled with MPIWExit or MPIWCExit respectively in order to capture their collective nature. The transfer-start event is still located at the source process immediately after the begin of the corresponding communication function (as it is in the basic model). However, the transfer-end event is now placed at the destination process shortly before the exit of the RMA synchronization function which completes the transfer according to the MPI-2 standard rules. Unfortunately, this has an undesired side effect. As one can see in the figure, this results in a separation of the data transfer for remote reads from the corresponding `MPI_Get` function. In order to rectify this situation, we introduced a new event GET_1TO, which marks the origin's location and time, as well as a new relationship *originptr* associating this new event with the start of the transfer (GET_1TS). This allows us in the analysis phase to locate all events related to RMA transfers. The extended model removes all disadvantages of the basic model, and for most MPI-2 implementations (which have a non-blocking behavior), it is even closer to reality. However, the model is more complex and the events can no longer be recorded at the location where they appear in the model. Therefore, a post-processing of the collected event trace becomes necessary.

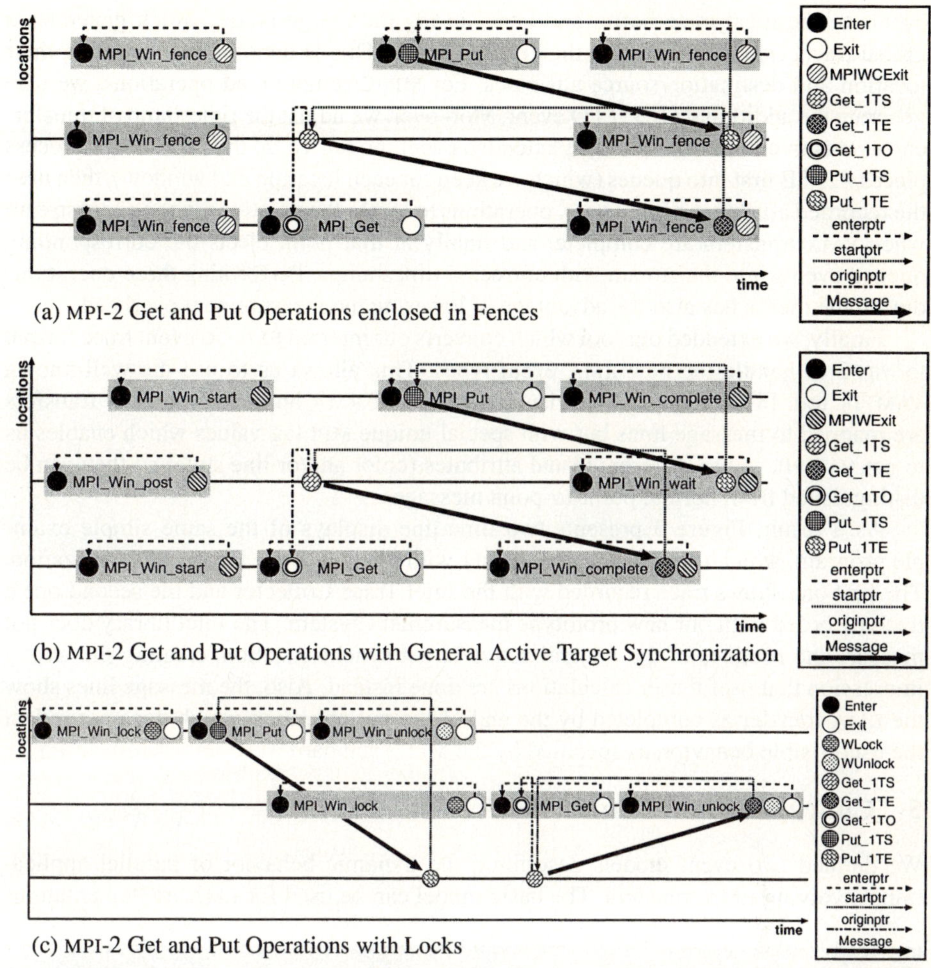

(a) MPI-2 Get and Put Operations enclosed in Fences

(b) MPI-2 Get and Put Operations with General Active Target Synchronization

(c) MPI-2 Get and Put Operations with Locks

Fig. 2. Examples for Extended Event Model

4   Analysis and Visualization

In this section, we outline the changes to KOJAK components that were necessary to implement support for the event models introduced in the last section. For a detailed description of the implementation see [13].

To record the new RMA related events, we implemented a set of wrapper functions for all SHMEM and MPI-2 communication and synchronization functions for C/C++ and Fortran. As MPI uses opaque types for representing windows and groups, we also had to add code for tracking these objects to the PMPI wrappers. Since the code for tracking the communication is only executed by the origin process, but the events for marking the start of a remote read (GET_1TS) and for the end of a remote write (PUT_1TE) are associated with the target process in our model, we cannot directly place the events in the correct trace buffer, which resides in the target process, during measurement. We solve this problem by writing temporary REMOTE_PUT_1TE and REMOTE_GET_1TS events to the local trace buffer and later, during the merge phase, which generates a global trace, replace these with the correct events. This is done by manipulating their location and destination/source attributes. For MPI-2 remote read operations, we also generate the additional GET_1TO event. Moreover, we adjust the timestamp of transfer-end events in compliance with the extended event model. To do this, the merge process places #_1TE first into queues (which we keep for each location and window), then uses the recorded attributes of MPI RMA operations to locate the positions in the event stream when RMA transfers are complete, and finally at that point ejects the corresponding queued events into the stream with corrected timestamps. Performing these operations during the merge has also the advantage of lowering the measurement overhead.

Finally, we extended our tool which converts our internal EPILOG event trace format to VTF3 to handle the new RMA event types. This allows us to use the well-known VAMPIR tool [8] to analyze and visualize traces of RMA applications. RMA transfers are mapped to message lines but with special unique MPI tag values which enables us to get VAMPIR to use different visual attributes (color and/or line style) so they can be distinguished from normal point-to-point messages.

As a result, Figure 3 presents two time-line displays of the same simple example program, which uses MPI_Put together with general active target synchronization. The first one shows trace recorded with the Intel Trace Collector and the second one a trace recorded with our new prototype measurement system. The Intel library does not measure the routines of the general active target synchronization, creating the wrong impression that useful user calculations are done instead. Also, the message lines show the RMA transfer as completed by the end of the put operation which does not reflect the user-visible behavior, as specified by the MPI-2 standard.

5   Conclusion and Future Work

We defined two event models describing the dynamic behavior of parallel applications involving RMA transfers. The basic model can be used for RMA implementations

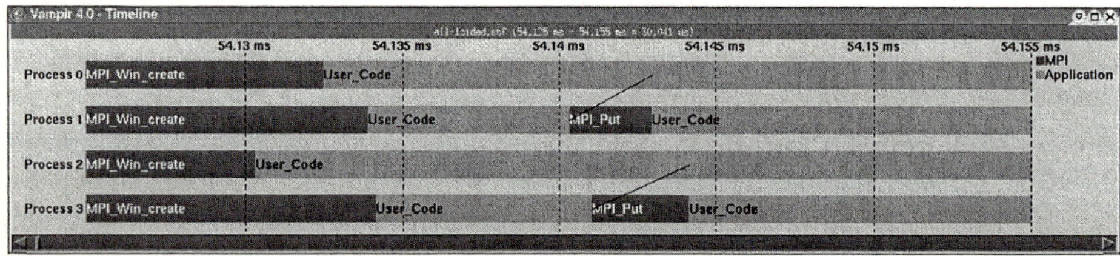

(a) Recorded with Intel Trace Collector

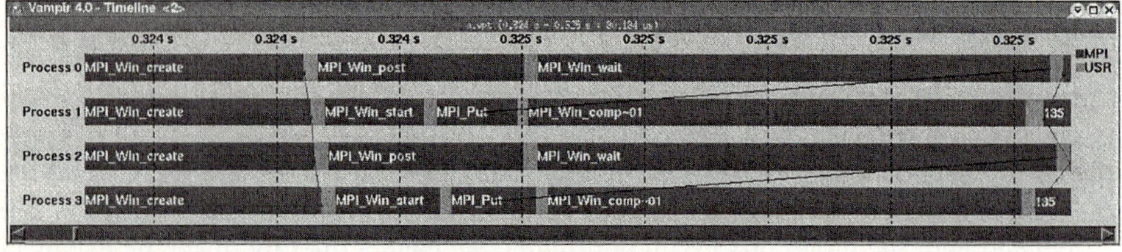

(b) Recorded with KOJAK

Fig. 3. Time-line of MPI-2 Put Operation and General Active Target Synchronization

with blocking behavior, that is, vendor-specific one-sided communication libraries like SHMEM or language extension like Co-Array Fortran and Unified Parallel C (UPC). For MPI, we defined an extended event model that reflects the user-visible behavior as specified by the MPI-2 standard. We implemented an extension to the KOJAK performance analysis toolset to instrument and trace applications based on one-sided communication and synchronization and to analyze the collected traces using the VAMPIR event trace visualizer. The next step will be to extend EXPERT [12], the automatic trace analysis component of KOJAK, to handle one-sided communication. This will include the definition of RMA-related performance properties (i.e., event patterns which represent inefficient behavior of RMA communication and synchronization).

Acknowledgments

We would like to thank Rolf Rabenseifner for helping us better understand MPI-2 one-sided communication and synchronization and for many helpful suggestions to improve our event models.

References

1. M. Snir, S. Otto, S. Huss-Lederman, D. Walker, and J. Dongarra. *MPI - the Complete Reference, Volume 1, The MPI Core.* 2nd ed., MIT Press, 1998.
2. W. Gropp, S. Huss-Lederman, A. Lumsdaine, E. Lusk, B. Nitzberg, W. Saphir, and M. Snir. *MPI - the Complete Reference, Volume 2, The MPI Extensions.* MIT Press, 1998.
3. A. Mirin and W. Sawyer. *A scalable implementation of a finite volume dynamical core in the Community Atmosphere Model.* Accepted for publication in the International Journal of High-Performance Computing Applications.
4. K. Mohror and K.L. Karavanic. *Performance Tool Support for MPI-2 on Linux.* In Proceedings of SC'04, Pittsburgh, PA, November 2004.
5. S. Shende, A. D. Malony, J. Cuny, K. Lindlan, P. Beckman, and S. Karmesin. Portable Profiling and Tracing for Parallel Scientific Applications using C++. In *Proceedings of the SIGMETRICS Symposium on Parallel and Distributed Tools*, pp. 134–145. ACM, August 1998.
6. Pallas/Intel. *The Intel Trace Collector.* 2004.
   → http://www.intel.com/software/products/cluster/tcollector/
7. F. Wolf. *Automatic Performance Analysis on Parallel Computers with SMP Nodes.* Dissertation, NIC Series, Vol. 17, Forschungszentrum Jülich, 2002.
8. W. Nagel, A. Arnold, M. Weber, H.-C. Hoppe, and K. Solchenbach. Vampir: Visualization and Analysis of MPI Resources. *Supercomputer*, 12:69–80, January 1996.
9. R. W. Numrich and J. K. Reid. Co-Array Fortran for Parallel Programming. *ACM Fortran Forum*, 17(2), 1998.
10. F. Wolf and B. Mohr. Automatic Performance Analysis of Hybrid MPI/OpenMP Applications. *Journal of Systems Architecture*, 49(10–11):421–439, November 2003.
11. B. Mohr, L. DeRose, and J. Vetter. *A Performance Measurement Infrastructure for Co-Array Fortran.* In *Proceddings of of Euro-Par 2005*, Springer, Lisboa, Portugal, September 2005.
12. F. Wolf, B. Mohr, J. Dongarra, and S. Moore. Efficient Pattern Search in Large Traces through Successive Refinement. In *Proceddings of Euro-Par 2004*, Springer, LNCS 3149, pp. 47–54, Pisa, Italy, September 2004.
13. M. -A. Hermanns. *Event-based Performance Analysis of Remote Memory Access Operations* (In German). Diploma Thesis, Forschungszentrum Jülich, 2004.

An Efficient Multi-level Trace Toolkit
for Multi-threaded Applications*

Vincent Danjean, Raymond Namyst, and Pierre-André Wacrenier

LaBRI / INRIA-Futurs, Université Bordeaux 1
351, cours de la Libération
33405 Talence Cedex, France

Abstract. Nowadays, observing and understanding the behavior and
performance of a multi-threaded application is a nontrivial task, especially within a complex multi-threaded environment such as a multi-level
thread scheduler. In this paper, we present a trace toolkit that allows
programmers to precisely analyze the behavior of a multi-threaded application. Running an application through this toolkit generates several
traces which are merged and analyzed offline. The resulting *super-trace*
contains not only classical information but also detailed informations
about thread scheduling at multiple levels.

1 Introduction

Bottleneck analysis, deadlock debugging, and performance understanding are
tasks which require a fine-grain analysis of the behavior of a parallel application. The problem becomes even more tricky when dealing with multi-level
multi-threading applications. Let us recall that there are three main families of
threads: *User-level threads* are managed by the application, offer efficient basic
operations and, most importantly, can be tailored to the particular requirements
of the application; however as the operating system knows nothing about these
threads, they have the disadvantage of not being able to use all available system resources, especially multi-processors resources. *Lightweight processes* (also
called LWPs or kernel-level threads) are managed by the kernel and have access
to kernel resources. For instance, several LWPs belonging to the same process
can be simultaneously active. The disadvantages are that they consume kernel
resources (the number of LWPs is usually limited) and tend to incur a bigger
overhead since all LWP scheduling and switching tasks require a kernel intervention. *Hybrid threads* (multi-level threads) were introduced in order to take
advantage of the two previous techniques, the key idea is to map user-level
threads onto a pool of LWPs. This leads to a two-level scheduling: the kernel
manages LWPs which themselves manage user-level threads in a distributed fashion. Although the implementation of this scheme within an operating system is
very complex [1], hybrid threads offer significant performance benefits with high
performance parallel applications involving only few I/O operations [2].

---

* This work has been supported by the *ACI Masse de données & ACI Grid*

J.C. Cunha and P.D. Medeiros (Eds.): Euro-Par 2005, LNCS 3648, pp. 166–175, 2005.
© Springer-Verlag Berlin Heidelberg 2005

Analyzing the scheduling of a multi-threaded application executed by an hybrid-thread system, and observing the behavior of a such application in its global context are difficult tasks which require support from the kernel, from the (hybrid-)thread library and from the application. For this purpose, the code must be instrumented in order to record selected events in one or several *trace buffers*. This leads to multi-level instrumentation. In this framework, we may notice the work of Shende [3] who has defined a strategy using multi-level instrumentation in order to improve the coverage of performance measurement in layered software. His approach, based on the *node/process/thread* model, was successfully implemented in the TAU portable profiling and tracing toolkit. For instance, to deal with Java's multi-threaded environment [4], each thread creation is recorded into a TAU's performance database (this requiring mutual exclusion with other threads) in order to create a per-thread performance data structure.

In [5], Xu *et al* use the dynamic environment PARADYN [6] to profile multi-threaded applications through statistics. In their approach, each thread has its own private copy of some performance counters or timers; locks are used to access the minimal set of global book-keeping data structures.

However, in the framework of the parallel environment PM^2 [2] which is based on an hybrid-thread library, our goal is to debug and to optimize low-level middlewares, such as a reactive communication library [7, 8], and tricky mechanisms like scheduler activations [9, 10]. To that effect, it is important to consider aspects such as lock mechanisms and interruption handler routines. When dealing with such low-level middlewares and parallel processes, there is no secret: the instrumentation must be as less intrusive as possible. Especially, we do not want to introduce new synchronization points within the kernel or within the thread scheduler in order to minimize interdependent intrusion effects[1]. In that respect, we have defined a lightweight multi-level instrumentation toolkit which aims to precisely trace the behavior of a multi-threaded program. In order to be efficient, this toolkit has to meet the following requirements:

To be the less intrusive as possible. The tracing overhead must be very small not only to allow an accurate performance analysis but also to minimize the intrusive effect on the global scheduling of the application (which would be the result e.g of an excessive increasing of the execution time of a critical section, some new synchronization points or some new context switch points). Therefore, system calls and high-level synchronization mechanisms must be avoided.

To deal with multi-level instrumentation. Since our goal is to study multi-level schedulers and/or high performance communication libraries, we need to record both kernel- and user-level events. For instance, we need to record all the kernel's scheduler decisions and all the thread library's scheduler decisions in order to get a complete knowledge of the scheduling of a multi-threaded application.

To deal with a huge amount of data. The toolkit may need to record a lot of events such as scheduler decisions, starting and termination points of functions

---

[1] Note that, Malony *et al* [11] have shown that while it is possible to compensate overhead due to the intrusion in a single process application, parallel overhead compensation is a more complex problem because of interdependent intrusion effects.

executed by a thread or by the kernel. This may generate several mega bytes of data per second.

In this paper, we propose a solution based on two independent buffer traces: the first one containing kernel-level events, the second one containing user-level events. We will first present the FAST KERNEL TRACE toolkit which is the basis of our work. Then we will justify our approach and give some technical details. Finally, we will analyze the introduced overhead on two applications.

2   From Kernel Tracing to Multi-level Tracing

2.1   The Fast Kernel Traces (FKT) Toolkit

Kernel instrumentation may be done at compilation time [12, 13] or dynamically at run-time like in KERNINST [14]. It is worth noting that operating systems such as LINUX 2.6.10 and SOLARIS 10 (DTRACE) already provide a dynamic instrumentation toolkit which allows to instrument the running operating system kernel. For our purpose, we chose to use the FKT toolkit [13] which is a simple and efficient SMP LINUX kernel-dedicated trace toolkit. It is based on a source-level instrumentation, which is achieved thanks to a set of macro-functions. Therefore, the modification of a tracing call requires to recompile the source code and to restart the kernel. Nevertheless, basic operations such as `tracing start`, `tracing stop` or `tracing store` can be executed from the user-level space. It is worth knowing that FKT uses a well-optimized storage mechanism [15] which allows to use TLB mechanisms to directly write buffer's pages on the disk, avoiding useless memory copy and limiting memory consumption.

```
#define FKT_PROBE2(KEYMASK,CODE,P1,P2)                    \
  do {                                                    \
      if( KEYMASK & fkt_active )                          \
        fkt_header( ((unsigned int)(CODE)),               \
            (unsigned int)(P1), (unsigned int)(P2) );\
      } while(0)
```

Fig. 1. A definition of a FKT macro for an event with two parameters.

Figure 1 shows the details of an FKT macro. The `KEYMASK` argument and the kernel variable `fkt_active` allows to enable/disable the tracing. A new system call is defined to set the variable `fkt_active` from the user-space. The `CODE` argument denotes the recorded event. `P1` and `P2` are two integer arguments left to the programmer (it is possible to record up to five integer arguments).

2.2   Meeting Hybrid Scheduling's Requirements

In order to precisely rebuild the behavior of multi-threaded programs, it is necessary to be able to determine at any time the current running user-level threads on the SMPs. Note that a kernel view is insufficient: indeed, the kernel has no knowledge about user-level threads which are scheduled by the LWP pool. On the other side, an user-space's view is also insufficient: LWPs are usually unaware of kernel's context switches, so it is difficult, from the user-space point of view, to get the identifiers of the running LWPs at a given date and to get the processor identifier on which the user code is running. To solve these problems, new system calls might be created to request the identifier of the processor which is recording an event, for instance, or to notify the kernel scheduler about the user-level scheduler's context switches. However, such a solution is too intrusive: system calls are expensive (see micro benchmarks given in Section 3.3) and, moreover, this solution would introduce a higher number of context-switch points than the uninstrumented execution would encounter. Another solution would be to define a mechanism based on *up-calls*: in order to transmit the kernel view to the user-space level, the kernel forces the application to call a given function, like the POSIX signal's mechanism does. However this solution is also expensive since the thread state must be saved at each up-call.

Our proposition is to generate a trace from both point of views. The kernel's trace will be generated by FKT and the user-level trace will be generated by FAST USER TRACE (FUT), a tool similar to FKT. The key-points of this solution are: (1) Dealing with hybrid scheduling, Kernel- and user-level traces are both necessary to get a full description of a multi-threaded application run. Both traces use the cycle counter register to stamp the events since this clock is very accurate. (2) Dealing with SMP, the cycle counter register of each processor is perfectly synchronized with each other registers at the hardware level. (3) All context switches (user's and kernel's) are recorded, so that we will be able to deduce what happens from a scheduling point of view within the system.

After the execution of the application, both traces are merged into a so-called *super-trace* which contains the following event data: the event code, time-stamp, size and parameters; the identifiers of the user-level thread, the LWP and the processor which executed the recording. By reconciling the kernel- and the user-level sides, this toolkit allows to trace multi-threaded applications and, moreover, it allows to put the application run back into its execution context, as any kernel event may be recorded. Hence it is possible to get an accurate analysis of low-level middlewares such as a multi-threaded communication library.

2.3   Description of the Tracing Toolkit

The multi-level tracing toolkit FUT has been implemented on top of the MARCEL/LINUX/X86 system which is the hybrid-thread library of the portable parallel environment PM^2 [2].

In order to instrument the kernel, users need to apply a given patch against the LINUX kernel. This patch introduces instrumented points in the kernel code allowing to record events such as context switches, starting/termination points of hardware interruption (IRQ) and software interruptions (system calls). The thread library MARCEL is instrumented in order to record user-thread scheduling decisions; for instance, events such as user-level context switches, creation and termination of LWPs are recorded. Moreover, this instrumentation allows to trace any function call of the library MARCEL. This way, one may accurately trace the performances of the library and determine the cause of the preemption of a user-level thread (elapsed time-slice, unacquired lock,...).

The API of FUT is similar to the FKT's one. Event recording is done by FUT_PROBEx() macros and some event types are already defined. A basic code instrumentation tool is also implemented in order to automatically add attributes to the starting/termination points of each function. The PROF_IN() and PROF_OUT() macros may be used to trace the call and the termination of a function. The code instrumentation may either be called directly by programmers or be inserted automatically by compilers, like GCC does.

Once both traces have been recorded, they are merged in a super-trace in which events are ordered with respect to the time-stamps. During the merge, the relationship between user-level events, user-level threads, LWPs and processors is established. However, some kernel events, such as those which are recorded during interruption routines, are not to be associated with any user-level thread.

We have developed a tool (called SIGMUND) which allows to apply filters to the super-trace in order to extract a sub-trace from it. One may filter events matching some criteria (a given kind of event, a given user-thread, a time-slice). Some basic measures may also be computed like, for instance, the (active) execution time of a given thread or the reactivity of the communication library to a given communication event (the elapsed time between the detection of a given event by the kernel and its treatment by the application). Moreover, a specific filter has been developed to translate the super-trace format into the file format of Pajé [16], a generic graphic trace viewer.

Figure 2 shows two requests getting information about the user-level thread 15 from a given super-trace. The instrumented program was executed on a SMT bi-processors machine (thus 4 logical processors, numbered from 0). For this execution, 4 LWPs were defined by the 2-level thread library to execute the user-level threads. Figure 3 shows how one may observe thread's reactivity.

3   Implementation Details and Performance Analysis

We are addressing in this section some technical issues we encountered in order to limit the intrusion of the tracing mechanisms. We will first detail the time-stamping, the trace format and the concurrent recording mechanism. Then we will discuss about the overhead introduced by the instrumentation.

3.1   About the Time-Stamping and the Trace Format

FKT and FUT use the cycle counter register as a time reference; this register stores the number of elapsed cycles since the last time the machine was started up. It is directly readable from the user-space. It is as accurate as possible and it is 64 bit wide. This leads to a 136 years period (2^{32} s) on a 4 GHz (2^{32} Hz) machine, moreover cycle counter registers of a SMP machine are synchronized.

```
$> sigmund --trace-file supertrace.log --thread 15 \
   --event CONTEXT_SWITCH --list-events
type   date_tick   pid cpu thr    code name                   param(s)
[...]
USER   97615576    7137    1    7  23014 USER_CONTEXT_SWITCH   15
USER   97757052    7137    1   15  23014 USER_CONTEXT_SWITCH   8
USER   98006248    7136    0    6  23014 USER_CONTEXT_SWITCH   15
KERN   98139183    7136    0   15  23014 KERN_CONTEXT_SWITCH   6152
KERN   98638163    2352    2    ?  23014 KERN_CONTEXT_SWITCH   7136
USER   99060185    7136    2   15  23014 USER_CONTEXT_SWITCH   7
[...]
$> sigmund --trace-file supertrace.log --thread 15 --active-time
130193845 cycles
```

type: event level – date_tick: event date – pid: LWP identifier
cpu: processor identifier – thr: user-thread level identifier
code: event code – name: event name – param(s): associated parameter values

In this example, we can see that the user-level thread 15 was firstly scheduled on LWP 7137 on CPU 1; then it was scheduled on LWP 7136 on CPU 0. Then following the preemption of LWP 7176 by the kernel (in order to schedule another application), it was scheduled on CPU 2. Then the user-level scheduler preempted the thread 15 in order to run the thread 7. Here we can see that this 2-level scheduler does not take into account the affinity of the threads.

Fig. 2. Super-trace analysis using sigmund.

Note that only 32 bits are required to stamp the kernel events. Indeed, from the first recording of the cycle register, there is enough kernel events (such as kernel scheduling decisions or clock interruptions) that are recorded during a defined period (2^{32} cycles) to infer the 32 higher bits. However, this argument does not hold for user-level threads which may not produce any event for several seconds.

In order to limit the intrusiveness, event buffers are created and initialized before the real launching of the application. An initial section containing context information (function names, running LWPs) is also recorded in both buffers. The size of the initial section is about several hundred of kilobytes.

3.2   Mutual Exclusion Mechanism

Dealing with threads and SMP machines, we have to take care of concurrent accesses to the trace buffers. Actually this problem of concurrency appears as soon as we want to record asynchronous events such as hardware interrupts or signals, even on a single processor machine. Indeed, asynchronous events may be raised at any time and we do not want to try to block them in order to avoid interferences with the scheduler. Therefore the instrumentation code must be fully reentrant. The basic idea of our approach is to atomically increment the buffer length variable. However, high-level mutual exclusion mechanisms are forbidden. We have solved this problem using the *atomic* CPU instruction

type	date_tick	pid	cpu	thr	event	param(s)
USER	5150163706	2732	2	8(work/6)	USER_CONTEXT_SWITCH	1(*daemon*)
KERN	5150169922	2732	2	1(*daemon*)	SYSTEM_CALL	142(select)
USER	5150182646	2732	2	1(*daemon*)	USER_CONTEXT_SWITCH	11(work/9)
USER	5152816866	2733	3	9(work/7)	USER_CONTEXT_SWITCH	12
KERN	5170071750	1630	0	?	IRQ	24(eth0)
USER	5176768370	2731	1	10(work/8)	USER_CONTEXT_SWITCH	5(work/3)
USER	5179394810	2732	2	11(work/9)	USER_CONTEXT_SWITCH	13(work/11)
USER	5182046038	2733	3	12(work/10)	USER_CONTEXT_SWITCH	14(work/12)
USER	5205964954	2731	1	5(work/3)	USER_CONTEXT_SWITCH	15(work/13)
USER	5208624942	2732	2	13(work/11)	USER_CONTEXT_SWITCH	17(work/15)
USER	5211315302	2733	3	14(work/12)	USER_CONTEXT_SWITCH	18(work/16)
USER	5235191514	2731	1	15(work/13)	USER_CONTEXT_SWITCH	19(work/17)
USER	5237854634	2732	2	17(work/15)	USER_CONTEXT_SWITCH	20(work/18)
USER	5240544282	2733	3	18(work/16)	USER_CONTEXT_SWITCH	21(work/19)
USER	5264421734	2731	1	19(work/17)	USER_CONTEXT_SWITCH	4(work/2)
USER	5267084698	2732	2	20(work/18)	USER_CONTEXT_SWITCH	2(work/0)
USER	5269736086	2733	3	21(work/19)	USER_CONTEXT_SWITCH	3(work/1)
USER	5293652362	2731	1	4(work/2)	USER_CONTEXT_SWITCH	6(work/4)
USER	5296353186	2732	2	2(work/0)	USER_CONTEXT_SWITCH	7(work/5)
USER	5298968062	2733	3	3(work/1)	USER_CONTEXT_SWITCH	8(work/6)
USER	5322881710	2731	1	6(work/4)	USER_CONTEXT_SWITCH	1(*daemon*)
KERN	5322893274	2731	1	1(*daemon*)	SYSTEM_CALL	142(select)
USER	5322907374	2731	1	1(*daemon*)	USER_EVENT	received_msg

Our tracing toolkit allows to emphasis the reactivity of multi-threaded applications. One can compute the elapsed time between the network message arrival (a hardware interrupt is raised by the network card) and the processing of this message by the appropriate user-level thread in the application.

This figure shows the relevant parts of a trace of a run where 20 threads are devoted to some computation (denoted **work/0** *to* **work/19***) and one special thread (denoted* **daemon***) is listening to the network in order to process the incoming messages as soon as possible. In this program, the* **daemon** *thread executes a non-blocking system call[a] select() and calls* pthread_yield() *to yield its execution in favor of another thread when no message is available.*

Here the considered algorithm leads to very bad latencies, as the **daemon** *thread has to wait for all the other threads to use their quantum before getting active even though messages could have already been received by the OS. However, most of thread libraries do not provide any mechanism to deal efficiently with this kind of problem. A description of an adequate support within thread libraries to improve thread reactivity to external asynchronous events can be found in [17].*

---

[a] blocking system call must be avoided when using user-level thread library.

Fig. 3. Using our mechanisms to observe thread's reactivity.

cmpxchgl. The idea is to store the buffer's length value in a register, then to store the new buffer's length in a second register and finally to call cmpxchgl

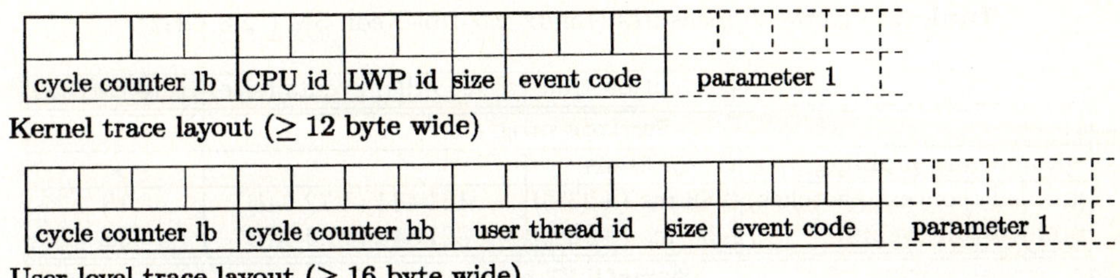

Kernel trace layout ($\geq$ 12 byte wide)

User-level trace layout ($\geq$ 16 byte wide).

Fig. 4. Kernel and user-level trace entry layout.

in order to set the new buffer's length. This subroutine is repetitively called until the `cmpxchgl` call is successful. As a result, the event trace may be not time-stamp ordered, thus the merging tool may have to reorder the trace.

3.3   Analysis of the Tracing Overhead

In table 1, we compare the cost of recording a single trace sample with the cost of a few other operations. Let us note that according to [11], the TAU measurement overhead per (flat) event is about 1400 cycles on a XEON processor.

Table 1. Micro benchmarks (Linux 2.6.4 bi-Xeon SMT 2.8 GHz).

Function/Macro		cycles
Macro	`PROF_IN`	260
System call	`getpid()`	1900
buffered io	`printf(``test'')`	672

We also measured the overhead and the size of generated traces. These two values depend on the instrumentation level and on the application. Here we have considered three instrumentation levels: no instrumentation, scheduling instrumentation and complete instrumentation (where system calls and all the functions of the application and of the hybrid-thread library MARCEL are traced). It is worth noting that there is no need to recompile the source code: the degree of instrumentation is defined through the use of the global variable `fkt_active`.

The `Sumtime` program can be seen as a torture test for the hybrid-thread library: it recursively builds a complete binary tree of threads for a given height. As a matter of fact, this program spends most of its execution time in creating, synchronizing and destructing user-level threads. Hence highly frequent scheduling events have to be recorded. This leads to a 23% overhead for the scheduler-level instrumentation and a 80% overhead for a complete instrumentation. This is the worst case, clearly this is not the best way to analyze the performances of our toolkit, however the gathered information may prove to be useful for debugging purposes. The second program is a multi-threaded direct solver for sparse

Table 2. Overhead measures (Linux 2.6.4 bi-Xeon SMT 2.8 GHz).

	execution time	# recorded events (size)	Rate (MB/s)
Sumtime program			
without any profiling	230 ms	-	-
profiled (context switches)	288 ms (+23%)	161 484 (3.72 MB)	13
profiled (all events)	430 ms (+80%)	821 844 (13.4 MB)	31
SuperLU_MT program			
without any profiling	7.17 s	-	-
profiled (context switches)	7.30 s (+1.8%)	374 (0.007 MB)	0.001
profiled (all events)	7.50 s (+4.6%)	836054 (8.39 MB)	1.1

systems of linear equations based on the library SUPERLU [18]. As there is a lot of computation within threads, the overhead of the instrumentation becomes quite reasonable.

4   Conclusion

Hybrid-thread scheduling's approach allows to efficiently exploit SMP architecture, as basic operations on threads are efficient and several user-level threads of a given application can run in a true parallel way. However, analyzing the performance of such programs is delicate, mainly because some events occur within the kernel and some others in user space. Thus, instrumentation of these programs has to be carried out at both levels. Our toolkit allows to instrument a multi-threaded program in order to conduct a precise analysis of executions of this program. It avoids the introduction of synchronization points or system calls during the execution, including basic thread operations such as creation, destruction and synchronization.

Our toolkit is available on SMP x86 / ITANIUM architectures, LINUX and the hybrid-thread library MARCEL. The required modifications of the LINUX kernel and of the library sources are localized. Therefore thread libraries such as NGPT, NPTL or LINUXTHREAD can easily be adapted to our toolkit. The implementation of our toolkit onto other CPU architectures relies on the availability of an instruction similar to the instruction `cmpxchgl` (which usually exists on modern processors) and on an accurate and CPU synchronized clock (such as cycle counter registers).

We are currently implementing our toolkit on NUMA machines where cycle counter registers are *nearly* synchronized. To deal with this problem, we have to introduce calibration steps. Some other interesting improvements include the recording of the performance of the counter registers and the translation of our trace format into other trace format such as VAMPIR.

References

1. Sun microsystems: Multithreading in the solaris operating environment. `http://www.sun.com/software/whitepapers/solaris9/multithread.pdf` (2002)
2. Namyst, R., Méhaut, J.F.: PM2: Parallel Multithreaded Machine. A computing environment for distributed architectures. In: Parallel Computing (ParCo '95), Elsevier Science Publishers (1995) 279–285
3. Shende, S.: The Role of Instrumentation and Mapping in Performance Measurement. PhD thesis, University of Oregon (2001)
4. Malony, A.D., Shende, S.: Performance Technology for Complex Parallel and Distributed Systems. In: Distributed and parallel systems: from instruction parallelism to cluster computing, Kluwer Academic Publishers (2000) 37–46
5. Xu, Z., Miller, B.P., Naim, O.: Dynamic instrumentation of threaded applications. In: PPoPP '99: Proceedings of the seventh ACM SIGPLAN symposium on Principles and practice of parallel programming, ACM Press (1999) 49–59
6. Miller, B.P., Callaghan, M.D., Cargille, J.M., Hollingsworth, J.K., Irvin, R.B., Karavanic, K.L., Kunchithapadam, K., Newhall, T.: The paradyn parallel performance measurement tool. Computer **28** (1995) 37–46
7. Aumage, O., Bougé, L., Méhaut, J.F., Namyst, R.: Madeleine II: A portable and efficient communication library for high-performance cluster computing. Parallel Computing **28** (2002) 607–626
8. Danjean, V., Namyst, R.: Controling Kernel Scheduling from User Space: an Approach to Enhancing Applications' Reactivity to I/O Events. In: HiPC '03. Volume 2913 of LNCS., Hyderabad, India, Springer-Verlag (2003) 490–499
9. Anderson, T., Bershad, B., Lazowska, E., Levy, H.: Scheduler Activations: Efficient kernel support for the user-level managment of parallelism. In: Proc. 13th ACM Symp. on Operating Systems Principles (SOSP 91). (1991) 95–105
10. Danjean, V., Namyst, R., Russell, R.: Integrating Kernel Activations in a Multithreaded Runtime System on Linux. In: (RTSPP '00. Lect. Notes in Comp. Science, Cancun, Mexico, Springer-Verlag (2000)
11. Malony, A.D., Shende, S.S.: Overhead Compensation in Performance Profiling. In: Proc. Europar 2004 Conference, LNCS (2004)
12. Yaghmour, K., Dagenais, M.R.: Measuring and Characterizing System Behavior Using Kernel-Level Event Logging. In: Proceeding of the 2000 USENIX Annual Technical Conference. (2000)
13. Russell, R.D., Chavan, M.: Fast Kernel Tracing: a Performance Evaluation Tool for Linux. In: Proc. 19th IASTED International Conference on Applied Informatics (AI 2001), IASTED (2001)
14. Tamches, A., Miller, B.P.: Using dynamic kernel instrumentation for kernel and application tuning. The International Journal of High Performance Computing Applications **13** (1999) 263–276
15. Thibault, S.: Developping a software tool for precise kernel measurements. Master's thesis, University of New Hampshire (2003)
16. de Kergommeaux, J.C., de Oliveira Stein, B.: Pajé: an extensible environment for visualizing multi-threaded programs executions, EuroPar2000 (2000)
17. Bougé, L., Danjean, V., Namyst, R.: Improving Reactivity to I/O Events in Multithreaded Environments Using a Uniform, Scheduler-Centric API. In: Euro-Par 2002. Volume 2400 of LNCS., Paderborn, Germany (2002) 605–614
18. Demmel, J.W., Gilbert, J.R., Li, X.S.: An asynchronous parallel supernodal algorithm for sparse gaussian elimination. SIAM J. Matrix Anal. Appl. **20** (1999) 915–952

Knowledge Based Automatic Scalability Analysis and Extrapolation for MPI Programs

Michael Kluge, Andreas Knüpfer, and Wolfgang E. Nagel

Technische Universität Dresden, Dresden, Germany
{kluge,knuepfer,nagel}@zhr.tu-dresden.de

Abstract. The question how well a MPI program is scaling with an increasing number of processors becomes more and more interesting, especially when these number grows to 10.000 or even 100.000 with IBM's 'Blue Gene' this year. The approach presented with this paper is able to identify locations within the source code of an application where the communication effort does not scale well with the growing number of processors. We show how traces for the same program generated with different numbers of processors can be inspected and compared automatically. An analytical approach will then identify the points within the source that do not scale as expected. At the end of this article, the benefits from this method are demonstrated on an ASCI benchmark.

1   Introduction

The number of processors available for a single application will break through the mark of 100.000 this year with IBM's supercomputer 'Blue Gene'. Parallel programs today are usually developed by using either OpenMP [1] or MPI [2] or both together. Both will generate some overhead during the program execution. Some of this additional time (compared to a serial execution) can be considered as necessary, some can be described as overhead and maybe avoidable. A necessary part for MPI communication is involved when there is no extra communication processor within the system so that the processor itself has to do execute the MPI protocol. We will show how traces obtained with Vampirtrace [3] for different numbers of processors can automatically be compared. We also demonstrate how it is possible to identify MPI scalability problems. A polynomial will be used to compare statistical data generated for each source code location (either a function name or a source file/line number combination) that calls a MPI function.

The first section gives an overview of approaches to detect MPI communication inefficiencies. The second section is dedicated to our analytical approach that is able to find those source code locations where the time needed for MPI communication does not scale as expected (with the numbers of processors). Within the third section we give an overview of the architecture of a tool that implements the ideas mentioned before. The final part of this article shows results of the new tool applied to a benchmark taken from the ASCI benchmark collection.

J.C. Cunha and P.D. Medeiros (Eds.): Euro-Par 2005, LNCS 3648, pp. 176–184, 2005.
© Springer-Verlag Berlin Heidelberg 2005

2   Detection of MPI Communication Inefficiencies

The usual way to do MPI program analysis is the post-mortem analysis. All program activities are recorded during runtime and the data generated are inspected afterwards. On one hand, this allows to keep the overhead for analysis during runtime low. On the other hand, it generates usually a large amount of data. An automatic analysis should be able to provide hints for source code optimization and should guide the user to points that are worth a manual inspection. One tool that is able to identify MPI and OpenMP performance bottlenecks is KOJAK [4]. It is based on automatic analysis of program execution patterns (implemented as C++ classes) that describe inefficient behavior of parallel programs.

The approach presented with this article will use a different way to identify MPI performance problems. By assuming a concrete hard- and software stack it can be further assumed that the execution time for a MPI function call is roughly constant when the number of processors involved and the message size is constant. Our approach derives from a repeated execution of a MPI function the 'usual' execution time for ideal circumstances for that function. The time t_{max} calculated this way is the maximum execution time we accept for this specific function call for a given number of processors and a message size. The time t_{max} is calculated for some base points, all other points are interpolated from these. The approach has been described in [5] and [6]. An other way to define this time is shown by the DIMEMAS [7] tool. It uses a latency/bandwidth and a point-to-point communication model to predict the execution time for MPI functions that transmit data.

Execution times t above t_{max} can be considered as overhead. Each source code location that calls a MPI function will be labeled with an index $l = \{0, 1, 2, \ldots\}$. Each function call recorded for this source code location will be indexed with an index $i = \{0, 1, 2, \ldots\}$. For a single program execution it is possible to calculate the total time T spent within this MPI function

$$T_l = \sum_i t \tag{1}$$

and the overhead O wasted at this source code location with

$$O_l = \sum_i \max(0, t - t_{max}). \tag{2}$$

When the O_l are sorted backwardly, the first positions in the list generated denotes the source code locations that generate most of the overhead and should give the user some clues about the MPI communication events that are worth to be optimized.

3   Comparing Multiple Trace Files

The metrics defined within the previous section help the user by automatically finding problems of MPI communication within a single program trace. Scalability investigations can only be made by comparing multiple trace files obtained with different numbers of processors.

The first thing we assume is that the MPI communication time T_P and/or the MPI communication overhead O_P is a function of the numbers of processors P used. We will use a polynomial

$$f(P) = \sum_i a_i P^i \tag{3}$$

that fits best to the communication time (or overhead). The a_i can be determined by an optimization and will be calculated for each source code location as well as for the whole application. For an application that is scaling well the time each process spent within MPI has to remain constant. From this follows that the time spent within MPI for the whole application should be at maximum a linear function of the numbers of processors. If this assumption is violated the scalability of the application is bound because the MPI communication time will grow rapidly and take most of the runtime of the application. First of all the a_i are calculated for the MPI communication time as well as the overhead for the whole application. Large values for a_i for $i > 1$ will indicate that the application may have a scalability problem. Because we are interested in if there is a scalability problem and not which degree greater than 1 of the polynomial is substantial greater than 0 we set

$$a_i = 0 \text{ for all } i > 2. \tag{4}$$

That results in a quadratic polynomial $f(P) = a_0 + a_1 P + a_2 P^2$ which describes the progression of the communication time (or the overhead). The a_i will also be ascertained for each source code location O_l. Large a_2 at the source code locations will give hints for those locations that causes the scalability problem.

The optimization process mentioned above will finish with a set of parameters $\{a_0, a_1, a_2\}$ and a set of errors e_P, one for each used numbers of processors P. The accuracy of the optimization can be described by the maximum of the ration between the error e_P and the values T_P (or O_P) as

$$e_{\text{max}} = \max_P \left[abs \left(\frac{e_P}{T_P} \right) \right] \text{ or } \max_P \left[abs \left(\frac{e_P}{O_P} \right) \right]. \tag{5}$$

4 MPI Communication Time Extrapolation

The next step to reach the target of predicting the MPI communication performance is to use the data extracted in the sections before for an extrapolation. The goal here is to replace a program execution with a large number of processors by some executions of the same program with smaller numbers of processors. The advantage gained out of this is that the communication performance of a MPI application running on large numbers of processors can be determined without actually doing the program execution.

The setup we are using is the following:

1. A MPI application is executed with five or more different numbers of processors. The used numbers of processors are labeled as $p = \{P_1, P_2, \ldots, P_N\}$
2. $N - 3$ different subsets p_S of p are built, containing $\{P_1, P_2, \ldots, P_S\}$ with $S = \{4, \ldots, N - 3\}$. To start with the first four points and try to extrapolate the fifth one is due to the following experience we made during the experiments. If a program is executed with the same number of processors multiple times all trace files generated this way will differ. Some of them will represent a minimal total runtime and have almost the same characteristics (total time spent within MPI etc.) but even they will still differ. The best trace file for this number of processors will be used for the extrapolation process. To balance the errors introduced this way we use as much trace files with different numbers of processors as possible. Due to an usual limited time and the costs associated with program executions on a parallel machine we recommend to start with at least four or five different numbers of processors.
3. The a_i for the MPI communication time and the MPI communication overhead for each subset are evaluated.

After extrapolating the fifth point from the first four points we will use the first five measured points for an extrapolation of the sixth point and so on. The sum of all this extrapolation and the stability of the parameters a_i give hints if the extrapolation itself makes sense.

The evaluation that has been done is based on the following criteria:

1. How big is the maximum error e_{max} of the optimization process?
2. Can the next point P_{S+1} be extrapolated from the subset p_S? This question can be answered by comparing the difference d between the value of $f(P)$ at the point P_{S+1} and the real value T_P (or O_P) with e_{max}.
3. Are the parameters a_i stable?

5   Proposing a Tool Architecture

To actually implement the ideas mentioned in the sections before a tool has been implemented. The first requirement that has driven the development of the tool is the adaptation of the models for the MPI communication. The parameters for the models have to be found and the maximum execution times for the MPI functions have to be evaluated (see [5]). The next point is that the tool should be independent from a specific trace library. Instead of that, Vampirtrace or PARAVER [8] or even proprietary libraries should be usable. An other point is the comparison and the analysis of multiple trace files which should be done automatically. Results from this evaluation should be available to the user for his own purposes, e.g. for graphical representations, for export (to Gnuplot) or for calculation of his own derived metrics.

An overview of the architecture of the tool is given in figure 1. Basically three different parts can be distinguished.

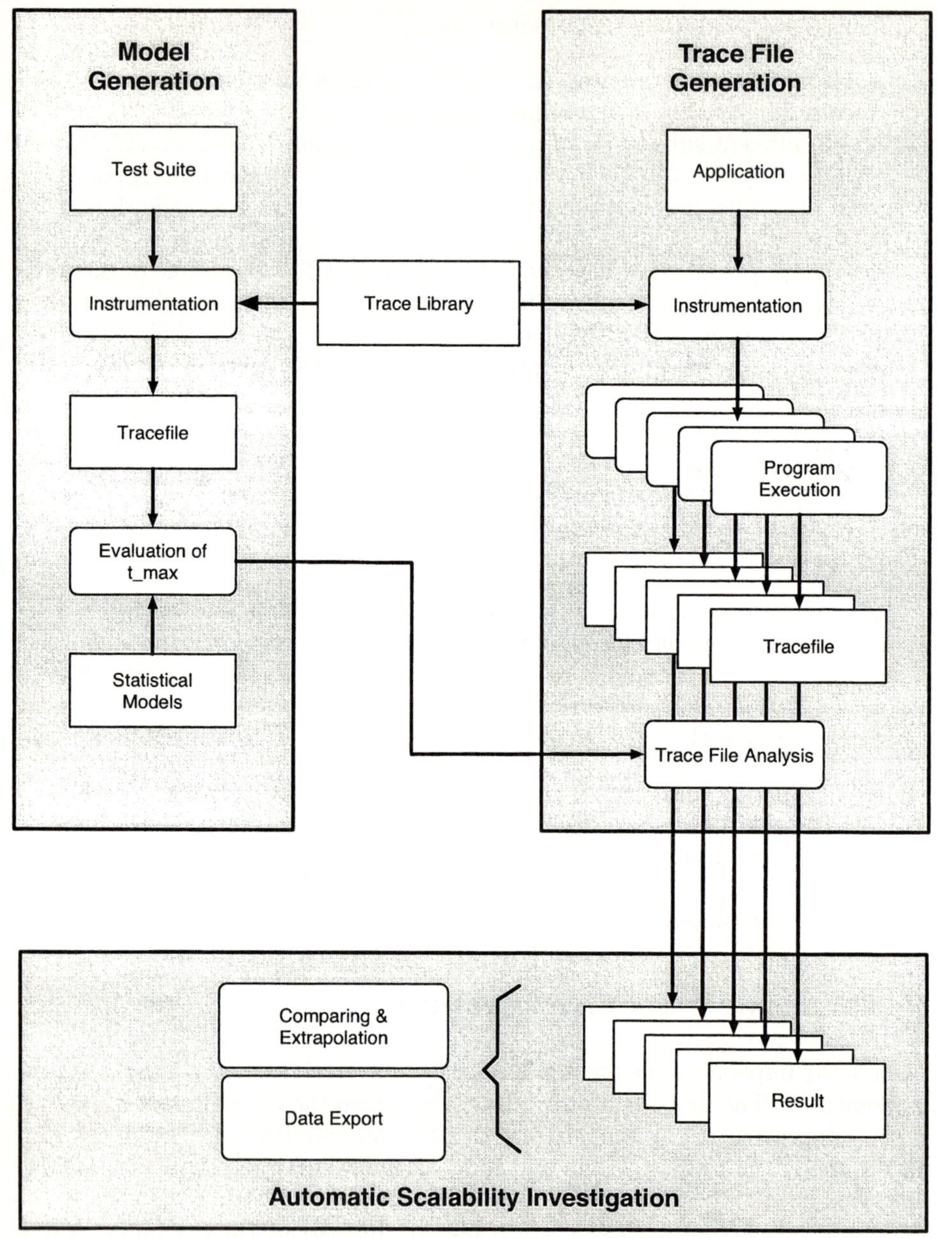

Fig. 1. The tool architecture

First of all, the paramters for the models for all MPI functions has to be
computed. This step has to be done just once for a given hardware and software stack. The model will be stored permanently and can be reused as long as

the machine configuration with respect to the communication layer(s) does not change.

The second part uses as input trace files of an application running using different numbers of processors. The models created within the first part are used to investigate each communication event, to divide the necessary communication time from the overhead and to assign all results to their associated source code locations.

The result is a list of source code locations and associated data that can now be analyzed in the third part of the program. The data collected for the same source code location but for program executions with different numbers of processors are used to compute the parameters a_i, the extrapolation and to evaluate the quality of the fitting of $f(P)$. Beside these results a report for the user will be generated.

To be able to generate data with some expressiveness it is recommended to use the same trace library to adapt the models and to trace the program execution. Usually trace libraries differ in the overhead generated. Only when using the same library for model adaptation and trace file generation this overhead keeps transparent.

6   Application to the Sweep3D Benchmark

We have chosen the Sweep3D benchmark [9] to demonstrate the usefulness of the extrapolation process. This benchmark has often been used to test new tools because its characteristics are well researched [4]. We have used a fixed global problem size of $168 \times 168 \times 168$ for all program executions. By cutting this cube along the first dimension into slices of equal size and assigning each slice to a single processor (and MPI task) we are able to execute the benchmark with numbers of processors of $4, 8, 12, 14, 21, 24$ and 28. The system we used for the benchmark is the IBM eServer pSeries 690 running at 1.7 GHz installed at the Forschungszentrum Jülich in Germany. One SMP node consists of 32 processors and delivers a heterogeneous environment which also can easily be used exclusively. Within this environment we can expect that the time the benchmark spends within MPI scales linearly with the number of processors. The use of more that one SMP node would introduce more influence parameters due to the necessary external network.

Figure 2 shows the total execution time (over all processors) including and excluding the communication time of the original program version. As is can be seen the time used for calculation remains almost constant during all used numbers of processors and the increasing part is only due to the communication.

Table 1 (and figure 3) shows the calculated parameters for the extrapolation of the overhead generated with MPI when using the first four measurements, then the first five, and so on. The remaining error e_{max} for the fitting of the quadratic polynomial is plotted as well as the errors for an extrapolation of those points that have not been used for the fitting of the polynomial. These parameters as well as e_{max} and d indicates that the benchmark does not scale linearly with the

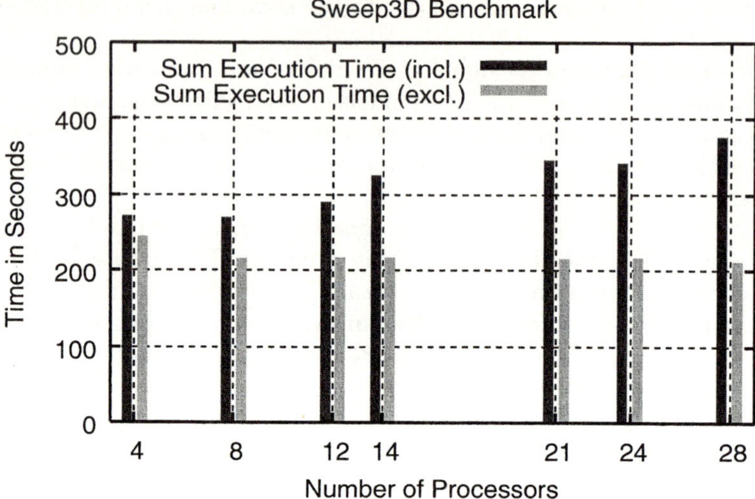

Fig. 2. Sweep3D execution time including and excluding time for MPI communication, original program version

Table 1. Sweep3D: Scalability parameters for the original source code

used executions	a_0	a_1	a_2	$e_{max}(\%)$	d
4 (4,8,12,14)	21.43	-0.00	0.41	12.58	60.93
5 (4,8,12,14,21)	-11.02	9.34	-0.13	15.14	15.61
6 (4,8,12,14,21,24)	-18.79	11.20	-0.22	17.80	23.22
7 (4,8,12,14,21,24,28)	-2.15	7.72	-0.08	15.35	-

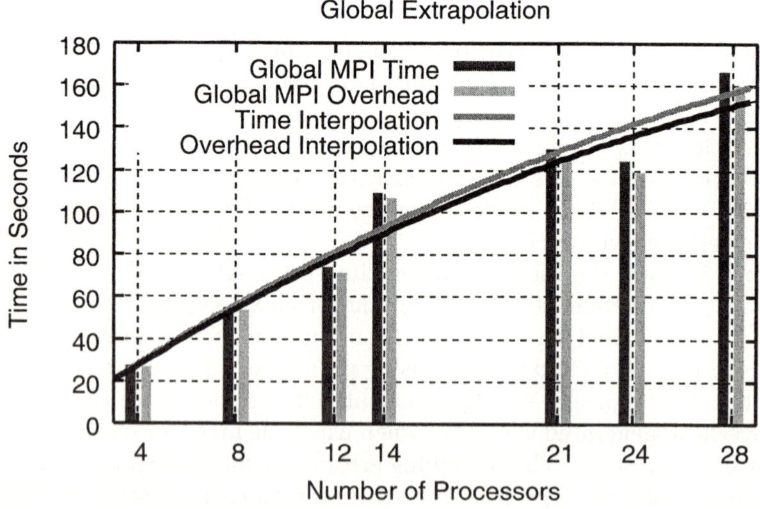

Fig. 3. Sweep3D MPI communication time, original program version

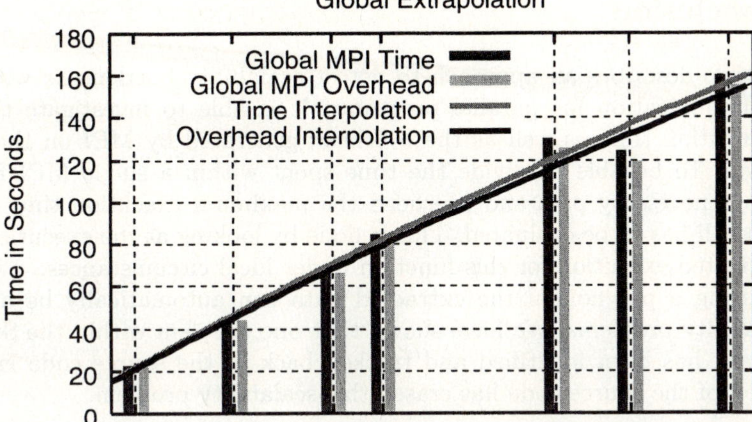

Fig. 4. Sweep3D MPI communication time, optimized program version

Table 2. Sweep3D: Scalability parameters for the optimized source code

used executions	a_0	a_1	a_2	$e_{max}(\%)$	d
4 (4,8,12,14)	3.68	4.16	0.10	2.02	6.62
5 (4,8,12,14,21)	0.15	5.17	0.04	2.32	23.35
6 (4,8,12,14,21,24)	-11.48	7.96	-0.09	12.63	10.87
7 (4,8,12,14,21,24,28)	-3.91	6.38	-0.03	10.89	-

numbers of processors. Anyhow, it is also clear that those numbers can considered to be stable (when using five different numbers of processors or more) because the approximation error e_{max} is acceptable and the communication time for the next used number of processors can be predicted using all previous (lower) numbers.

The extrapolations done at the source code level have shown that the non linear part of the growth of the MPI communication time results from two source code locations within the source code file **sweep.f**. Within the most inner loop of the algorithm two vectors are sent to/received from the (virtually) adjacent processors. To erase this non linear part we have replaced the blocking calls to MPI_Send and MPI_Recv with their non blocking equivalents and a following MPI_Waitall.

As it can be seen in figure 4 the MPI communication time as well as the MPI overhead scales now linearly with the number of used processors, a table 2 shows that the parameter a_2 is now almost 0. Even the quality of the fitting (e_{max}) has been improved as like as the quality of the prediction for the next step (d).

7   Conclusion

This article describes an approach to detect scalability bottlenecks within the MPI communication for parallel programs. It is able to investigate the MPI communication time as well as the overhead generated by MPI on the source code level. To be able to divide the time spent within a single MPI function call into a necessary part and overhead the maximum execution time for this function call has to be evaluated. This is done by looking at the execution times for a repeated execution for this function under ideal circumstances.

By using a polynomial the extracted data can automatically be analyzed for scalability problems. We have shown that one problem within the Sweep3D benchmark has been identified and tracked back to the source code level. An alteration of the source code has erased this scalability problem.

References

1. OpenMP Architecture Review Board. OpenMP Application Program Interface, Version 2.5. `http://www.openmp.org/drupal/mp-documents/draft_spec25.pdf`, November 2004.
2. Message Passing Interface Forum. MPI: A Message-Passing Interface Standard. Technical report, University of Tennessee, 1995.
3. Wolfgang E. Nagel, Alfred Arnold, Michael Weber, Hans-Christian Hoppe, and Karl Solchenbach. VAMPIR: Visualization and Analysis of MPI Resources. In *Supercomputer 63, Volume XII, Number 1*, pages 69–80, 1996.
4. S. Moore, F. Wolf, J. Dongarra, and B. Mohr. Second workshop on productivity and performance in high-end computing (p-phec) at 11th international symposium on high performance computer architecture (hpca-2005) (submitted). 2005.
5. Michael Kluge. Statistische Analyse von Programmspuren für MPI-Programme. Diploma thesis, November 2004.
6. Michael Kluge, Andreas Knüpfer and Wolfgang E. Nagel. Statistical Methods for Automatic Performance Bottleneck Detection in MPI Based Programs In *Proceedings of 5th International Conference on Computational Science*, pages 330–337, 2005.
7. European Center for Parallelism of Barcelona. Dimemas. `http://www.cepba.upc.es/dimemas/`.
8. European Center for Parallelism of Barcelona. Paraver. `http://www.cepba.upc.es/paraver/`.
9. Lawrence Livermore National Laboratory. The ASCI Sweep3D Benchmark Code. `http://www.llnl.gov/asci_benchmarks/asci/limited/sweep3d/asci_sweep3d.html`, 1995.

Performance Modeling: Understanding the Past and Predicting the Future

David H. Bailey[1,*] and Allan Snavely[2]

[1] Lawrence Berkeley National Laboratory, 1 Cyclotron Road, Berkeley, CA 94720
dhbailey@lbl.gov
[2] University of California, San Diego, 9500 Gilman Drive, La Jolla, CA
asnavely@cs.ucsd.edu

Abstract. We present an overview of current research in performance modeling, focusing on efforts underway in the Performance Evaluation Research Center (PERC). Using some new techniques, we are able to construct performance models that can be used to project the sustained performance of large-scale scientific programs on different systems, over a range of job and system sizes. Such models can be used by vendors in system designs, by computing centers in system acquisitions, and by application scientists to improve the performance of their codes.

1 Introduction

The goal of performance modeling is to gain understanding of a computer system's performance on various applications, by means of measurement and analysis, and then to encapsulate these characteristics in a compact formula. The resulting model can be used to gain greater understanding of the performance phenomena involved and to project performance to other system/application combinations.

We will focus here on large-scale scientific computation, although many of the techniques we describe below apply equally well to single-processor systems and to business-type applications. Also, this paper focuses on some work being done within the Performance Evaluation Research Center (PERC) [1], a research collaboration funded through the U.S. Department of Energy's Scientific Discovery through Advanced Computation (SciDAC) program [10]. A number of important performance modeling activities are also being done by other groups, for example at Los Alamos National Laboratory [6].

The performance profile of a given system/application combination depends on numerous factors, including: (1) system size; (2) system architecture; (3) processor speed; (4) multi-level cache latency and bandwidth; (5) interprocessor network latency and bandwidth; (6) system software efficiency; (7) type of application; (8) algorithms used; (9) programming language used; (10) problem size;

---

* This work was supported by the Director, Office of Computational and Technology Research, Division of Mathematical, Information, and Computational Sciences of the US DOE, under contract DE-AC03-76SF00098.

J.C. Cunha and P.D. Medeiros (Eds.): Euro-Par 2005, LNCS 3648, pp. 185–195, 2005.
© Springer-Verlag Berlin Heidelberg 2005

(11) amount of I/O; and others. Indeed, a comprehensive model must incorporate most if not all of the above factors. Because of the difficulty in producing a truly comprehensive model, present-day performance modeling researchers generally limit the scope of their models to a single system and application, allowing only the system size and job size to vary. Nonetheless, as we shall see below, some recent efforts appear to be effective over a broader range of system/application choices.

Performance models can be used to improve architecture design, inform procurement, and guide application tuning. Unfortunately, the process of producing performance models historically has been rather expensive, requiring large amounts of computer time and highly expert human effort. This has severely limited the number of high-end applications that can be modeled and studied. Someone has observed that, due to the difficulty of developing performance models for new applications, as well as the in-creasing complexity of new systems, our supercomputers have become better at predicting and explaining natural phenomena (such as the weather) than at predicting and explaining the performance of themselves or other computers.

2   Applications of Performance Modeling

Performance modeling can be used in numerous ways. Here is a brief summary of these usages, both present-day and future possibilities:

Runtime estimation. The most common application for a performance model is to enable a scientist to estimate the runtime of a job when the input parameters for the job are changed, or when a different number of processors is used in a parallel computer system. One can also estimate the largest size of system that can be used to run a given problem before the parallel efficiency drops to an unacceptable area.

System design. Performance models are frequently employed by computer vendors in their design of future systems. Typically engineers construct a performance model for one or two key applications, and then compare future technology options based on performance model projections. Once performance modeling techniques are better developed, it may be possible to target many more applications and technology options in the design process. As an example of such "what-if" investigations, application parameters can be used to predict how performance rates would change with a larger or more highly associative cache. In a similar way, the performance impact of various network designs can be explored. We can even imagine that vendors could provide a variety of system customizations, depending on the nature of the user's anticipated applications.

System tuning. One example of using performance modeling for system tuning is given in [4]. Here a performance model was used to diagnose and rectify a misconfigured MPI channel buffer, which yielded a doubling of network performance for programs sending short messages. Along this line, Adolfy Hoisie of LANL recalls that when a recent system was installed, its performance fell below model predictions by almost a factor of two. However, further analysis uncovered some system difficulties, which, when rectified, improved performance to almost the same level the model predicted [6]. When observed performance of a system falls short of that predicted by a performance model, it may be the system that is wrong not the model!

Application tuning. If a memory performance model is combined with application parameters, one can predict how cache hit-rates would change if a different cache-blocking factor were used in the application. Once the optimal cache blocking has been identified, then the code can be permanently changed. Simple performance models can even be incorporated into an application code, permitting on-the-fly selection of different program options.

Performance models, by providing performance expectations based on the fundamental computational characteristics of algorithms, can also enable algorithmic choice before going to the trouble to implement all the possible choices. For example, in some recent work one of the present authors employed a performance model to estimate the benefit of employing an "inspector" scheme to reorder data-structures before being accessed by a sparse-matrix solver, as part of software being developed by the SciDAC Terascale Optimal PDE Simulations (TOPS) project [13]. It turned out that the overhead of these "inspector" schemes is more than repaid provided the sparse-matrices are large and/or highly randomized.

System procurement. Arguably the most compelling application of performance modeling, but one that heretofore has not been used much, is to simplify the selection process of a new computing facility for a university or laboratory. At the present time, most large system procurements involve a comparative test of several systems, using a set of application benchmarks chosen to be typical of the expected usage. In one case that the authors are aware of, 25 separate application benchmarks were specified, and numerous other system-level benchmark tests were required as well. Preparing a set of performance benchmarks for a large laboratory acquisition is a labor-intensive process, typically involving several highly skilled staff members. Analyzing and comparing the benchmark results also requires additional effort. These steps involved are summarized in the recent HECRTF report [7].

What is often overlooked in this regard is that each of the prospective vendors must also expend a comparable (or even greater) effort to implement and tune the benchmarks on their systems. Partly due to the high personnel costs of benchmark work, computer vendors often can afford only a minimal effort to implement the bench-marks, leaving little or no resources to tune or customize the implementations for a given system, even though such tuning and/or customization would greatly benefit the customer. In any event, vendors must factor the cost of implementing and/or tuning benchmarks into the price that they must charge to the customer if successful. These costs are further multiplied because for every successful proposal, they must prepare several unsuccessful proposals.

Once a reasonably easy-to-use performance modeling facility is available, it may be possible to greatly reduce, if not eliminate, the benchmark tests that are specified in a procurement, replacing them by a measurement of certain performance model parameters for the target systems and applications. These parameters can then be used by the computer center staff to project performance rates for numerous system options. It may well be that a given center will decide not to rely completely on performance model results. But if even part of the normal application suite can be replaced, this will save considerable resources on both sides.

3   Basic Methodology

Our framework is based upon application signatures, machine profiles and convolutions. An application signature is a detailed but compact representation of the fundamental operations performed an application, independent of the target system. A machine profile is a representation of the capability of a system to carry out fundamental operations, independent of the particular application. A convolution is a means to rapidly combine application signatures with machine profiles in order to predict performance. In a nutshell, our methodology is to

1. Summarize the requirements of applications in ways that are not too expensive in terms of time/space required to gather them but still contain sufficient detail to enable modeling.
2. Obtain the application signatures automatically.
3. Generalize the signatures to represent how the application would stress arbitrary (including future) machines.
4. Extrapolate the signatures to larger problem sizes than what can be actually run at the present time.

With regards to application signatures, note that the source code of an application can be considered a high-level description, or application signature, of its computational resource requirements. However, depending on the language it may not be very compact (Matlab is compact, while Fortran is not). Also, determining the resource requirements the application from the source code may not be very easy (especially if the target machine does not exist!). Hence we need cheaper, faster, more flexible ways to obtain representations suitable for performance modeling work. A minimal goal is to combine the results of several compilation, execution, performance data analysis cycles into a signature, so these steps do not have to be repeated each time a new performance question is asked.

A dynamic instruction trace, such as a record of each memory address accessed (us-ing a tool such as Dynist [3], of the Alpha processor tool ATOM) can also be considered to be an application signature. But it is not compact-address traces alone can run to several Gbytes even for short-running applications—and it is not machine independent.

A general approach that we have developed to analyze applications, which has resulted in considerable space reduction and a measure of machine independence, is the following: (1) statically analyze, then instrument and trace an application on some set of existing machines; (2) summarize, on-the-fly, the operations performed by the application; (3) tally operations indexed to the source code structures that generated them; and (4) perform a merge operation on the summaries from each machine [4][11][12][5]. From this data, one can obtain information on memory access patterns (namely, summaries of the stride and range of memory accesses generated by individual memory operations) and communications patterns (namely, summaries of sizes and type of communications performed).

The specific scheme to acquire an application signature is as follows: (1) conduct a series of experiments tracing a program, using the techniques described above; (2) analyze the trace by pattern detection to identify recurring sequences of messages and loads/store operations; and (3) select the most important sequences of patterns. With regards to (3), infrequent paths through the program are ignored, and sequences that map to insignificant performance contributions are dropped.

As a simple example, the performance behavior of CG (the Conjugate Gradient benchmark from the NAS Parallel Benchmarks [2]), which is more 1000 lines long, can be represented from a performance standpoint by one random memory access pattern. This is because 99% of execution is spent in the following loop:

```
do k = rowstr(j), rowstr(j+1)-1
  sum = sum + a(k)*p(colidx(k))
enddo
```

This loop has two floating-point operations, two stride-1 memory access patterns, and one random memory access pattern (the indirect index of p). On almost all of today's deep memory hierarchy machines the performance cost of the random memory access pattern dominates the other patterns and the floating-point work. As a practical matter, all that is required to predict the performance of CG on a machine is the size of the problem (which level of the memory hierarchy it fits in) and the rate at which the machine can do random loads from that level of the memory. Thus a random memory access pattern succinctly represents the most important demand that CG puts on any machine.

Obviously, many full applications spend a significant amount of time in more than one loop or function, and so the several patterns must be combined and weighted. Simple addition is often not the right combining operator for these patterns, because different types of work may be involved (say memory accesses and communication). Also, our framework considers the impact of different compilers or different compiler flags in producing better code (so trace results are not machine independent). Finally, we develop models that include scaling and not just ones that work with a single problem size. For this, we use statistical methods applied to series of traces of different input sizes and/or CPU counts to derive a scaling model.

The second component of this performance modeling approach is to represent the resource capabilities of current and proposed machines, with emphasis on memory and communications capabilities, in an application-independent form suitable for parameterized modeling. In particular, we use low-level benchmarks to gather ma-chine profiles, which are high-level representations of the rates at which machines can carry out basic operations (such as memory loads and stores and message passing), including the capabilities of memory units at each level of the memory hierarchy and the ability of machines to overlap memory operations with other kinds of operations (e.g., floating-point or communications operations). We then extend machine profiles to account for reduction in capability due to sharing (for example, to express how much the memory subsystem's or communication fabric's capability is diminished by sharing these with competing processors). Finally, we extrapolate to larger systems from validated machine profiles of similar but smaller systems.

To enable time tractable modeling we employ a range of simulation techniques [4][9] to combine applications signatures with machine profiles:

1. Convolution methods for mapping application signatures to machine profiles to enable time tractable statistical simulation.
2. Techniques for modeling interactions between different memory access patterns within the same loop. For example, if a loop is 50% stride-1 and 50% random stride, we determine whether the performance is some composable function of the these two separate performance rates.
3. Techniques for modeling the effect of competition between different applications (or task parallel programs) for shared resources. For example, if program A is thrashing L3 cache with a large working set and a random memory access pattern, we determine how that impacts the performance of program B with a stride-1 access pattern and a small working set that would otherwise fits in L3.
4. Techniques for defining "performance similarity" in a meaningful way. For example, we determine whether loops that "look" the same in terms of application signatures and memory access patterns actually perform the same. If so, we define a set of loops that span the performance space.

In one sense, cycle-accurate simulation is the performance modeling baseline. Given enough time, and enough details about a machine, we can always explain and predict performance by stepping through the code instruction by instruction. However, simulation at this detail is exceedingly expensive. So we have developed fast-to-evaluate machine models for current and proposed machines, which closely approximate cycle-accurate predictions by accounting for fewer details.

Our convolution method allows for relatively rapid development of performance models (full application models take 1 or 2 months now). Performance predictions are very fast to evaluate once the models are constructed (few minutes per prediction). The results are fairly accurate. Figure 1 show qualitatively the accuracy results across a set of machines and problem sizes and CPU counts for POP, the Parallel Ocean Program.

We have carried out similar exercise for several sizes and inputs of POP problems. And we have also modeled several applications from the DOD HPCMO [8] work-load, including AVUS a CFD code, GAMESS a computational chemistry

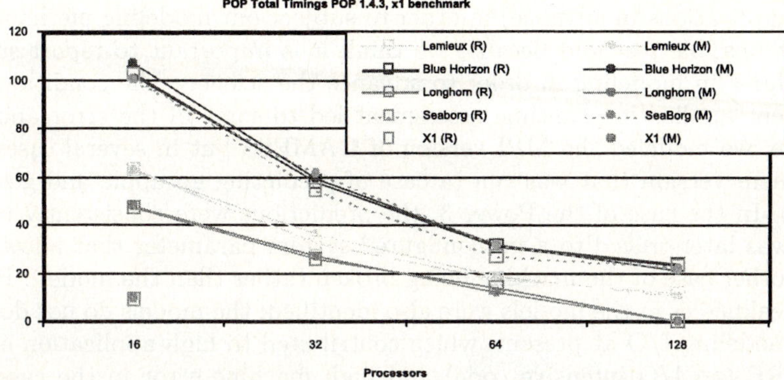

Fig. 1. Results for Parallel Ocean Program (POP). (R) is real runtime (M) is modeled (predicted) runtime

code, HYCOM a weather code, and OOCORE an out-of-core solver. In a stern test of the methods we were allowed access to DOD machines only to gather machine profiles via low-level benchmarks. We then modeled these large parallel applications at several CPU counts ranging from 16 to 384, on Power3, Power4 in two different flavors, Alpha, Xeon, and R16000 processor based supercomputers. We then predicated application performance on these machines; an d only after the predictions were issued were the application true runtimes independently ascertained by DOD personnel.

Table 1. Results of "blind" predictions of DoD HPCMO Workload Category

Category	Average Absolute Error	Standard Deviation
Overall	20.5%	18.2%
AVUS std. input	15.0%	14.2%
AVUS large input	16.5%	16.2%
GAMESS std. input	45.1%	24.2%
HYCOM std. input	21.8%	16.7%
HYCOM large input	21.4%	16.9%
OOCORE std. input	32.1%	27.5%
Power3	17.4%	17.0%
Power4 p690	12.9%	9.6%
Power4 p655	15.7%	19.9%
Alpha	29.0%	17.6%
R16000	41.0%	18.5%
Xeon	28.2%	12.3%

Table 1 above gives the overall average absolute error and standard deviation of absolute average error as well as breakdowns by application/input and architecture. We conducted this 'blind' test (without knowing the performance of the applications in advance) in order to subject our modeling methods to the sternest possible test and because we think it is important to report successes and failures in modeling in order to advance the science. The conditions of independent application runtime assessment led to some of the error above. For example, we modeled the MPI version of GAMESS but in several cases it was the shmem version that was run (a case of predicting an apple and getting an orange). In the case of the Power 3, the predictions were consistently too high which was later traced to a misconfigured system parameter that allowed paging (another case of the machine being broken rather than the model). However some weaknesses in the models were also identified; the models do not do a good job of modeling I/O at present, which contributed to high application error for OOCORE (an I/O intensive code) and high machine error in the case of the Alpha system (which has a weak I/O subsystem). Xeons were consistently over predicted for reasons that appear to have to do with weak architectural support for floating-point (few, shallow, pipelines). Augmentation of the models to address systematic errors and add additional terms for I/O and enhanced accuracy of floating-point scheduling is work in progress.

4   Performance Sensitivity Studies

Reporting the accuracy of performance models in terms of model-predicted time vs. observed time (as in the previous section) is mostly just a validating step for obtaining confidence in the model. A more interesting and useful exercise is to explain and quantify performance differences and to play "what if" using the model. For example, it is clear from Figure 1 above that Lemeiux, the Alpha-based system, is faster across-the-board on POP x1 than is Blue Horizon, the Power3 system. The question is why? Lemeuix has faster processors (1GHz vs. 375 MHz), and a lower-latency network (a measured ping-pong latency of about 5 ms vs. about 19 ms), but Blue Horizon's net-work has the higher bandwidth (a measured ping-pong bandwidth of about 350 MB/s vs. 269 MB/s). Without a model, one is left to conjecture "I guess POP performance is more sensitive to processor performance and network latency than network bandwidth," but without solid evidence.

With a model that can accurately predict application performance based on proper-ties of the code and the machine, we can carry out precise modeling experiments such as that represented in Figure 2. Here we model perturbing the Blue Horizon (BH) system (withPower3 processors and a Colony switch) into the TCS system (with Alpha ES640 processors and the Quadrics switch) by replacing components one by one. Figure 2 represents a series of cases modeling the perturbing from BH to TCS, going from left to right. The four bars for each case represent the performance of POP x1 on 16 processors, the processor and memory subsystem performance, the network band-width, and the network latency, all normalized to that of BH.

In Case 1, we model the effect of reducing the bandwidth of BH's network to that of a single rail of the Quadrics switch. There is no observable performance effect, as the POP x1 problem at this size is not sensitive to a change in peak

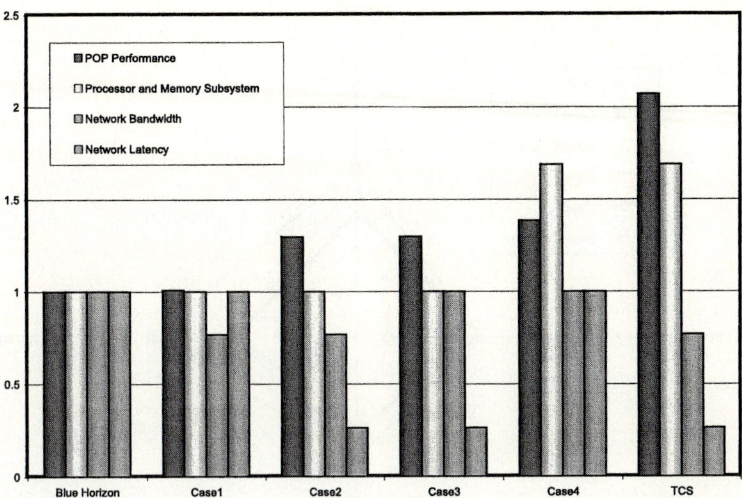

Fig. 2. Performance sensitivity study of POP applied to proposed Lemieux upgrade

network band-width from 350 MB/s to 269 MB/s. In Case 2, we model the effect of replacing the Colony switch with the Quadrics switch. Here there is a significant performance improvement, due to the 5 ms latency of the Quadrics switch versus the 20 ms latency of the Colony switch. This is evidence that the barotropic calculations in POP x1 at this size are latency sensitive. In Case 3, we use Quadrics latency but the Colony band-width just for completeness. In Case 4, we model keeping the Colony switch latency and bandwidth figures, but replacing the Power3 processors and local memory subsystem with Alpha ES640 processors and their memory subsystem. There is a substantial improvement in performance, due mainly to the faster memory subsystem of the Alpha. The Alpha can load stride-1 data from its L2 cache at about twice the rate of the Power3, and this benefits POP x1 significantly. The last set of bars show the TCS values of performance, processor and memory subsystem speed, network bandwidth and latency, as a ratio of the BH values.

The principal observation from the above exercise is that the model can quantify the performance impact of each machine hardware component.

In these studies we find that larger CPU count POP x1 problems become more network latency sensitive and remain not-very bandwidth sensitive.

We can generalize a specific architecture comparison study such as the above, by using the model to generate a machine-independent performance sensitivity study. As an example, Figure 3 indicates the performance impact on the 128-CPU POP x1 pro-gram of quadrupling the speed of the CPU-memory subsystem (lumped together we call this the processor), quadrupling the network bandwidth, reducing network latency by four, and various combinations of these four-fold hardware improvements. The data values are plotted in a logarithmic scale and normalized to one, so that the solid black quadrilateral represents the execution time, network bandwidth, network latency, CPU and memory subsys-

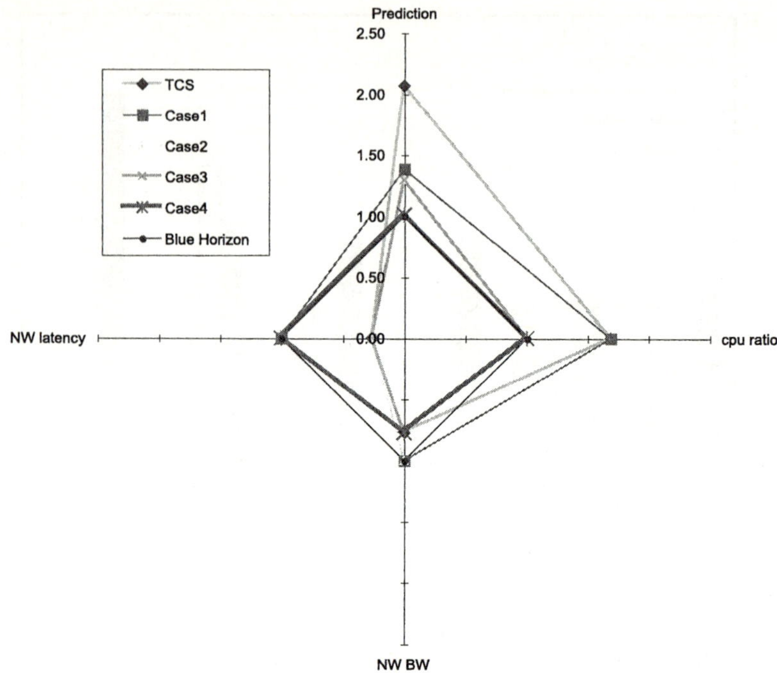

Fig. 3. A generalized performance sensitivity study

tem speed of Blue Horizon. At this size, POP x1 is quite sensitive to processor speed (a faster CPU and memory subsystem), somewhat sensitive to latency (because of the barotropic portion of the code is communications-bound, with small-messages), and fairly insensitive to bandwidth. In a similar way we can "zoom in" on the processor performance factor. In the above results for POP, the processor axis shows modeled execution time decreasing from a four-times faster CPU with respect to clock rate (implying a 4X floating-point issue rate), but also quadruple bandwidth and one-quarter latency to all levels of the memory hierarchy (unfortunately this may be hard or expensive to achieve architecturally!).

5   Conclusion

We have seen that performance models enable "what-if" analyses of the implications of improving the target machine in various dimensions. Such analyses obviously are useful to system designers, helping them optimize system architectures for the highest sustained performance on a target set of applications. They are potentially quite useful in helping computing centers select the best system in an acquisition. But these methods can also be used by application scientists to improve performance in their codes, by better understanding which tuning measures yield the most improvement in sustained performance.

With further improvements in this methodology, we can envision a future wherein these techniques are embedded in application code, or even in system software, thus enabling self-tuning applications for user codes. For example, we can conceive of an application that performs the first of many iterations using numerous cache blocking parameters, a separate combination on each processor, and then uses a simple performance model to select the most favorable combination. This combination would then be used for all remaining iterations.

Our methods have reduced the time required for performance modeling, but much work needs to be done here. Also, running an application to obtain the necessary trace information multiplies the run time by a large factor (roughly 1000). The future work in this arena will need to focus on further reducing the both the human and computer costs.

References

1. The Performance Evaluation Research Center (PERC),
   see http://www.perc.nersc.gov.
2. Bailey, D., et. al, "The NAS Parallel Benchmarks," *International Journal of Supercomputer Applications*, vol. 5 (1991), no. 3, pg. 66–73.
3. Buck, B. and J. K. Hollingsworth, "An API for Runtime Code Patching," *Journal of Supercomputing Applications*, 14(4), 2000, pg. 317–329.
4. Carrington, L., A. Snavely, X. Gao, and N. Wolter, "A Performance Prediction Framework for Scientific Applications," *ICCS Workshop on Performance Modeling and Analysis (PMA03)*, June 2003, Melbourne, Australia.
5. Carrington, L., N. Wolter, A. Snavely, and C. B. Lee, "Applying an Automated Framework to Produce Accurate Blind Performance Predictions of Full-Scale HPC Applications," *UGC 2004*, Williamsburgh, June 2004.
6. Hoisie, A., O. Lubeck, H. Wasserman, "Performance and Scalability Analysis of Teraflop-Scale Parallel Architectures Using Multidimensional Wavefront Applications," *The International Journal of High Performance Computing Applications*, vol. 14 (2000), no. 4, pg. 330–346.
7. Report of the High-End Computing Revitalization Task Force (HECRTF), see http://www.sc.doe.gov/ascr/hecrtfrpt.pdf.
8. Department of Defense High Performance Computing Modernization Office (HPCMO), see http://www.hpcmo.hpc.mil.
9. Perelman, E., G. Hamerly, M. V. Biesbrouck, T. Sherwood, and B. Calder, "Using SimPoint for accurate and efficient simulation," *ACM SIGMETRICS Performance Evaluation Review*, vol. 31 (2003), no. 1, pg. 318–319.
10. Scientific Discovery through Advanced Computing (SciDAC),
    see http://www.science.doe.gov/scidac.
11. Snavely, A., L. Carrington, N. Wolter, J. Labarta, R. Badia, and A. Purkayastha, "A Framework for Application Performance Modeling and Prediction," *Proceedings of SC2002*, Nov. 2002, Baltimore, MD.
12. Snavely, A., X. Gao, C. Lee, N. Wolter, J. Labarta, J. Gimenez, and P. Jones, "Performance Modeling of HPC Applications," *Proceedings of the Parallel Computing Conference 2003*, Oct. 2003, Dresden, Germany.
13. Terascale Optimal PDE Simulations (TOPS) project,
    see http://www-unix.mcs.anl.gov/scidac-tops.

An Approach to Performance Prediction for Parallel Applications

Engin Ipek[1], Bronis R. de Supinski[2], Martin Schulz[2], and Sally A. McKee[1]

[1] Computer Systems Lab
School of Electrical and Computer Engineering
Cornell University
{engin,sam}@csl.cornell.edu

[2] Center for Applied Scientific Computing
Lawrence Livermore National Laboratory
Livermore, CA 94551
{bronis,schulzm}@llnl.gov

Abstract. Accurately modeling and predicting performance for large-scale applications becomes increasingly difficult as system complexity scales dramatically. Analytic predictive models are useful, but are difficult to construct, usually limited in scope, and often fail to capture subtle interactions between architecture and software. In contrast, we employ multilayer neural networks trained on input data from executions on the target platform. This approach is useful for predicting many aspects of performance, and it captures full system complexity. Our models are developed automatically from the training input set, avoiding the difficult and potentially error-prone process required to develop analytic models. This study focuses on the high-performance, parallel application SMG2000, a much studied code whose variations in execution times are still not well understood. Our model predicts performance on two large-scale parallel platforms within 5%-7% error across a large, multi-dimensional parameter space.

1 Introduction

With rising architecture and software complexity, it becomes increasingly difficult to accurately model and predict performance for large-scale applications. Analytic models often fail to capture subtle interactions between architecture and software. Furthermore, they usually must be constructed manually in a long and often error-prone process. In this paper, we address these problems with the help of machine learning techniques. We gather performance samples from multiple executions of an application, and use this data to *automatically* construct performance models by training multilayer neural networks. Since we take input data from executions on the target platform, we capture full system complexity, without having to manually model architectural details. Our approach is useful for a wide range of application performance prediction problems. Our techniques are particularly well suited to mining performance databases or to extend fast, parameter-specific models.

J.C. Cunha and P.D. Medeiros (Eds.): Euro-Par 2005, LNCS 3648, pp. 196–205, 2005.
© Springer-Verlag Berlin Heidelberg 2005

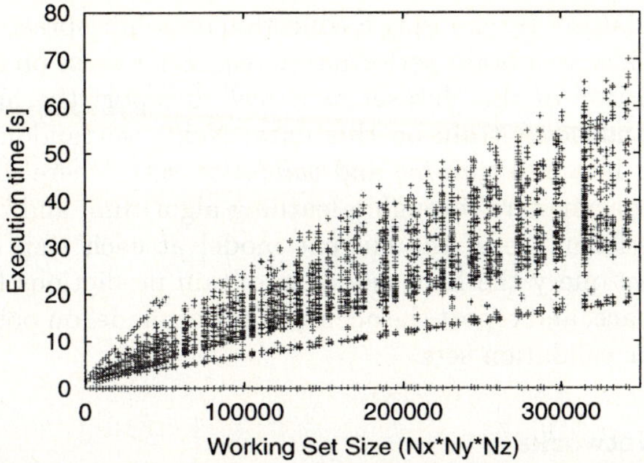

Fig. 1. Execution times for SMG2000 for varying processor workloads (Nx,Ny,Nz) and processor topologies (Px,Py,Pz) running on 512 nodes on BlueGene/L

Here we focus on SMG2000, a semicoarsening multigrid solver based on the *hypre* library [4]. We develop application-specific performance models for parallel architectures, enabling prediction of runtime or other important characteristics across a large input parameter space with high dimension. SMG's six-dimensional parameter space describes both shape of the workload per processor and logical processor topology. These parameters have substantial impact on runtime, as shown in Figure 1. For a fixed working set size—a fixed subvolume size per CPU—runtime varies by up to 5×. Although SMG has been studied extensively and an analytic model describing communication requirements exists [1], the code's variations in execution time are not well understood (partly due to SMG's complex, recursive algorithm). The analytic model is restricted to cubic workloads and only describes communication complexity; it is not designed to represent architectural details. Extending it for arbitrarily shaped workloads is possible, but would be extremely complex, and the result would likely be intractable. Worse, adding architectural features is infeasible. Our automatic, empirical modeling approach overcomes these limitations *without knowledge of the application or algorithms.*

We demonstrate how we use neural networks to construct our models, and we identify the two major challenges of this approach: avoiding noise in the dataset, and choosing an appropriate sampling technique for the training phase of the neural network. The latter is necessary to avoid a bias toward short runtimes, since those exhibit a higher relative error. To correct this skew, we develop new functions that scale error by the runtime of the training samples. The resulting model can predict SMG2000's performance on two large-scale parallel platforms within 5%-7% error across a large, multi-dimensional parameter space.

2   Approach

We use machine learning models to predict application performance across a large, multidimensional parameter space defined by program inputs. We first collect a sample dataset by choosing a collection of points spread regularly across the parameter space; we obtain performance results for these on actual hardware. We reserve a portion of this dataset as a *test* to report the final performance of our models, and never train on this data. Next, we randomly separate the remainder of the data into *training* and *validation* sets, where the former is used to adjust model parameters through a learning algorithm, and the latter is used to assess the performance of the current model at each step during training. After training, we query the final model to obtain predictions for points in the full parameter space, and report the accuracy of our model on points not included in our training or validation sets.

2.1   Neural Networks

Artificial Neural Networks (ANNs) are a class of machine learning models that map a set of input parameters to a set of target values. Figure 2 shows an example neural network architecture. The network is composed of a set of *units* that process the value at their inputs and produce a single scalar value. These values are then multiplied by a set of weights and communicated to other units within the network. Each edge in Figure 2 represents a weight, and each node represents a unit. The set of incoming edges at each unit indicates the set of values communicated to it. In this specific network architecture, the input parameters are placed at the first (lowest) layer, and information flows from bottom to top. The units that produce the final predictions are *output units*, and those that receive input parameters are *input units* (input units simply pass incoming values to all of their outgoing edges). In addition, one or more layers of *hidden units* may be part of the network architecture. Hidden units process the outputs of other units, and, in turn pass their own outputs to another set of (hidden or output) units. The representational power of a neural network (the set of functions it can represent) can be increased by adding hidden units and layers. Every unit in a given layer receives values from all units in the layer below it, and hence this type of ANN architecture is called a *multilayer fully connected feedforward* neural network. Figure 2 shows a feedforward neural network with three input units, one output unit, and a single layer of four hidden units.

At each step during training, a new example is presented at the network's input layer. At each layer, every unit forms a weighted sum of the incoming values and associated weights. This sum is then processed by an *activation function* that produces the output of that unit. In this study, we use fully connected feedforward neural networks with the sigmoid function as the activation function. Figure 3 shows the operation of the sigmoid activation function on the weighted sum of inputs (depicted immediately right of the summation in Figure 3) to form the unit's result output. After a prediction on the current example, the weights in the network are updated in proportion to their contribution to the error.

Other types of predictive models may be applied to performance (see Section 4). Here we limit our scope to ANNs for three reasons. First, ANNs permit target values and inputs/outputs to be discrete, continuous, or a mix, allowing them to perform well in both regression and classification problems and to learn

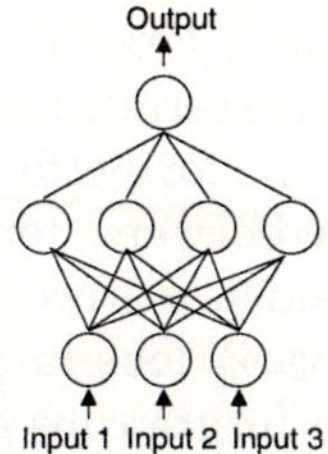

Fig. 2. A feedforward neural network with a single hidden layer

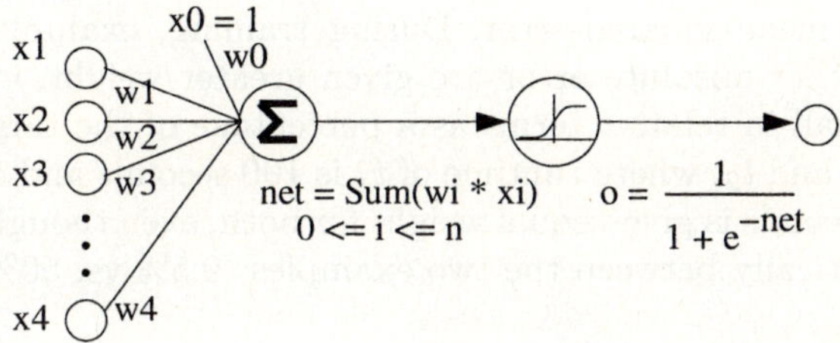

$$o = \frac{1}{1 + e^{-net}}$$

Fig. 3. A network unit with sigmoid threshold activation function (reproduced from Mitchell [8])

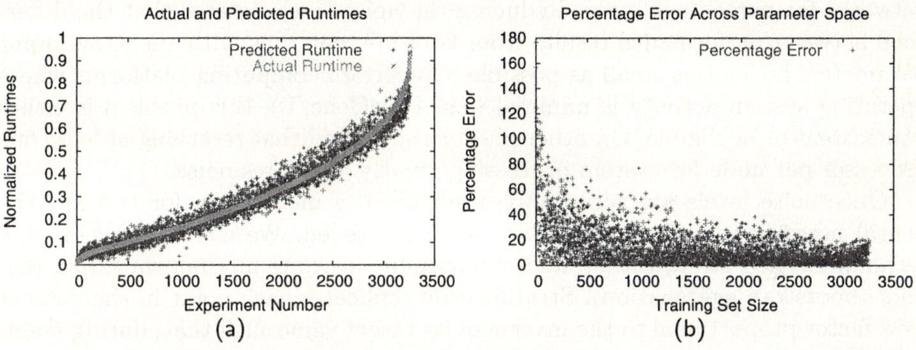

Fig. 4. Comparison of (a) predicted vs. actual performance and (b) percentage error

from various types of attributes describing performance prediction problems. Second, ANNs need not know the form of the target function in advance. Third, ANNs tend to work well with possibly noisy data, making them ideal for training on performance results collected in the presence of system noise.

2.2   Application to Performance Prediction

Figure 4 shows results of an initial performance prediction study consisting of 13.2K data points. A standard, fully connected feedforward neural network with 16 hidden units is trained on 10K points, and predictions are made on the remaining 3.2K. Despite training on a large portion of the parameter space, in most cases model accuracy is low. Average test error is 13.8% with a standard deviation of 14.8%—excessively high for performance prediction purposes.

Poor accuracy of standard feedforward neural networks on this dataset results from two factors. First, system activities sharing resources with application threads create nondeterministic variations in performance, yielding significant noise in the dataset. Accuracy on future runs can never exceed this noise level. This imposes a fundamental limit on model accuracy for future datasets. Second, the training algorithm that adjusts network weights is unsuitable for reducing percentage error. By default, the backpropagation training algorithm tries to reduce absolute mean-squared-error. During training, examples on which the model makes higher absolute error are given greater weight, even though this error may be small in relative terms as a percentage of the target value. Given two test cases t_1 and t_2, where runtime of t_1 is 100 seconds and of t_2 is 1 second, an error of 0.5 seconds is given equal weight for both, even though the percentage error varies drastically between the two examples (0.5% vs. 50%).

2.3   Required Network Refinements

Applying ANNs to application performance prediction requires both a mechanism for reducing noise during data collection and a technique to train the networks for percentage error. Reducing the noise level dictates that the difference between performance results from two different runs with the same input parameters be kept as small as possible. On certain computing platforms where operating system activity is minimal (e.g, BlueGene/L), this problem is either nonexistent or negligible. On other platforms, we find that reserving at least one processor per node for system processing greatly alleviates noise.

Once noise levels are acceptably diminished, a mechanism for training the neural network to reduce percentage error is needed. We combine a sampling technique called *stratification*, and an ensemble learning mechanism called *bagging* (bootstrap aggregation). Stratification replicates each point in the dataset by a factor proportional to the inverse of its target value such that, during training, the network sees points with small target values many more times than it sees those with large absolute values. As a result, the training algorithm puts varying amounts of emphasis on different regions of the search space, making the right tradeoffs when setting weights to minimize percentage error. We apply bagging to train an ensemble of models from the dataset, averaging predictions from the ensemble to reduce model variance.

3   Experiments

We present results of applying our technique to performance prediction of SMG on the Thunder and BlueGene/L systems at Lawrence Livermore National Laboratory. Architectural features of these systems on which data is taken are detailed in Table 1. Table 2 shows program parameters. For the BlueGene/L dataset, we

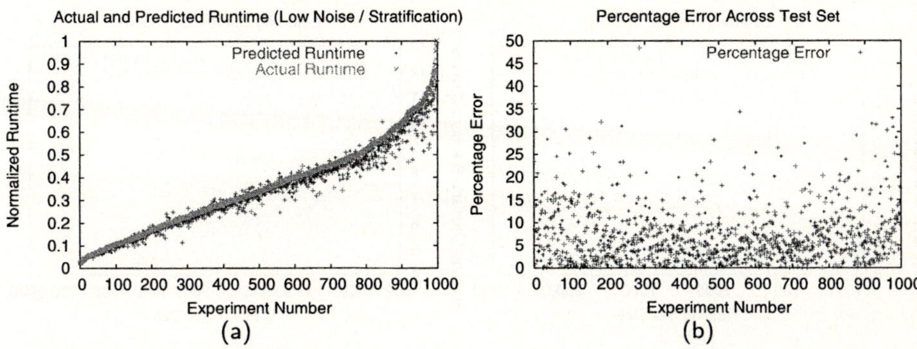

Fig. 5. Comparison of (a) predicted and actual performance and (b) percentage error

Table 1. Platform parameters

	BlueGene/L	Thunder
Processor	IBM BlueGene	Intel Itanium 2
Frequency	700MHz	1.4GHz
L1 ICache	32KB	32KB
L1 DCache	32KB	32KB
L2 Cache	2KB (Prefetch Buffer)	256KB
L3 Cache	4MB	4MB
SDRAM	512MB	8GB DDR266
Network	3D Torus + Global Combine/Broadcast Tree Network	Fat Tree (Quadrics QsNet)
Processors Used/Node	1/2	3/4
Number of Nodes Used	512	64

Table 2. Application parameters

Parameter	BlueGene/L	Thunder
Nx	10-510 in steps of 20	10-250 in steps of 30
Ny	10-510 in steps of 20	10-250 in steps of 30
Nz	10-510 in steps of 20	10-250 in steps of 30
Px	1,8,64,512	1,3,4,12,16,48,64,192
Py	1,8,64,512	1,3,4,12,16,48,64,192
Pz	1,8,64,512	1,3,4,12,16,48,64,192
Px*Py*Pz	512	192
Nx*Ny*Nz	1000>Nx*Ny*Nz>343000	216000>Nx*Ny*Nz>9261000

keep 1K random samples for final testing only (we do not train on these points) and report the accuracy of our model on this data. Similarly, we separate 1.3K data points for testing in the Thunder dataset.

Figure 6(a) shows a learning curve that indicates how the accuracy of the neural network changes as the size of the training set is increased for the Blue-Gene/L dataset. At a training set size of 250 points, the average error on the

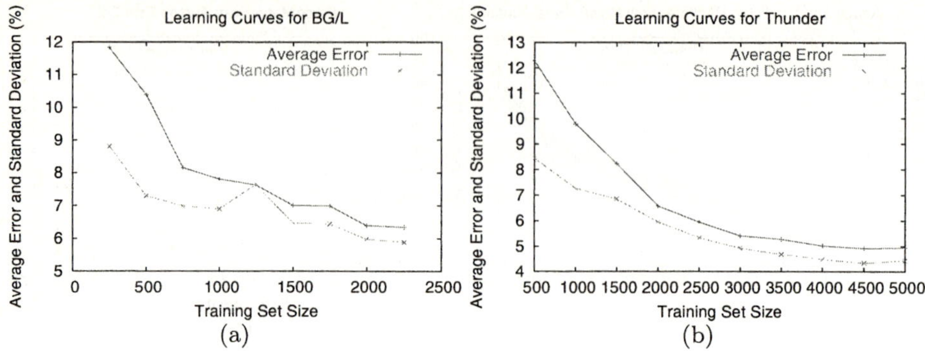

Fig. 6. Learning curves showing how average error and standard deviation improve with training set size for (a) BlueGene/L and (b) Thunder

test set is nearly 12.3%, and the standard deviation of error across the test set is 8.7%. At this point, the training set is too small and contains too little information to build a highly accurate model. As training set size increases, error decreases sharply, showing that the model benefits significantly from the additional information included in the dataset at each point. Eventually, the curves begin to flatten, as any additional data presented to the network contains only incremental new information. When 2.25K of the 3.25K total points are used for training, the error rate of the of the network falls to 6.7%. Similarly, the standard deviation of the error decreases with increasing training set size.

Thunder's learning curves (Figure 6(b)) follow the same trends. With 500 data points, the error rate on the test set is 12.28%. The error falls sharply as more data points are added, reaching 5.4% at a training set size of 3K. Further increases in training set size result in diminishing improvements, and at a training set size of 5K points, the network achieves 4.9% error. Similarly, the standard deviation ranges from 8.4%-4.4% between 500-5K points.

The results indicate that the accuracy of our approach can be quite high given enough training points. The size of the parameter space is much, much larger than the total number of points we have collected. We *sparsely* step through the SMG2000 parameters to obtain our dataset. Therefore, our approach is easily applicable to learning from performance databases that contain results for a sparse sampling of parameters. In addition,the amount of time required to train a model ranges between 1-15 minutes on a typical workstation with a 3.0GHz Pentium 4 processor and 1GB of main memory, making it easy to build parameterized performance models much more efficiently than most analytical models.

4   Related Work

Other approaches to performance prediction include analytic models. Space prevents our providing a full treatment of related work, but Karkhanis and Smith [5] give an excellent review of prior work in architectural performance prediction.

Marin and Mellor-Crummey [7] semi-automatically measure and model program characteristics, predicting application behavior based on properties of the architecture, properties of the binary, and application inputs. Their toolkit provides a set of predefined functions, and the user may add customized functions to this library if the set of existing functions is too restrictive. In contrast to our work, they vary the input size in only one dimension, and they cannot account for some important architectural parameters, such as cache associativity in their memory reuse modeling. Our six-dimensional space would make use of their approach much more difficult, significantly increasing the number of required samples as well as the search space for the best analytic function (as a weighted sum of given base functions along each parameter dimension).

Carrington et al. [2] develop a framework for predicting performance of scientific applications, demonstrating its effectiveness on LINPACK and an ocean modeling application. The approach is built on a convolution method that represents a computational mapping of an application signature onto a machine profile. Simple benchmark probes create the machine profiles, and a separate tool generates the application signatures. Extending the convolution method allows them to go from modeling kernels to whole benchmarks to full-scale HPC applications [3]. This automated approach relies on the generation of several traces, delivering predictions with accuracies of between 4.6 and 8.4%, depending on the sampling rates of those traces. Using full traces obviously gives the best performance, but such trace generation can slow application execution by almost three orders of magnitude. Some applications demonstrate better predictability than others, and for these trace reduction techniques work well: prediction accuracies range from 0.1 to 8.7% on different platforms. This work is complementary to our own, and the two approaches may work well in combination. The analytic models could provide the bootstrap data, and our models could give them full application input parameter generality.

Kerbyson et al. [6] present a highly accurate, predictive analytical model that encompasses the performance and scaling characteristics of SAGE, a multidimensional hydrodynamics code with adaptive mesh refinement. As with the model presented here, inputs to their parametric model come from machine performance information, such as latency and bandwidth, along with application characteristics, such as problem size and decomposition. They validate the prediction accuracy of the model against measurements on two large-scale ASCI systems. In addition to predicting performance, their model can yield insight into performance bottlenecks. Their application-centric modeling approach requires static analysis of the code: a detailed model must be developed for each application of interest.

Karkhanis and Smith [5] construct a first-order model of superscalar microprocessors. Their approach is intuitive, provides insight, and is reasonably accurate, finding that their performance estimates are between five and 13% accurate with respect to detailed simulations of the applications they study. The model's analytic core incorporates cache and branch predictor statistics gathered from functional-level trace driven simulation. They target uniprocessors, and while intuitive, the approach is largely ad hoc and currently limited in the architectural features it models. Their model is more appropriate for studying proposed architectures, whereas we predict performance on existing platforms.

5   Conclusions and Future Work

We have presented a machine learning approach to application performance prediction—multilayer neural networks—and have refined and adapted this approach to yield highly accurate results for SMG2000 on two different high-performance platforms. Our approach is especially attractive for its ease of use and its obliviousness to details of application internals. This makes it ideal for mining performance databases to make performance predictions. While promising, this approach still presents some challenges in making it generally useful in the absence of an existing database. The time required to gather each data point in the training set is larger than we would like, for instance. Reducing the number of points required in our training datasets is one promising direction of current research.

Acknowledgments

Part of this work was performed under the auspices of the U.S. Department of Energy by University of California Lawrence Livermore National Laboratory under contract No. W-7405-Eng-48 (LLNL Document Number UCRL-CONF-212365) and by by the National Science Foundation under award ST-HEC 0444413. Any opinions, findings, and conclusions or recommendations expressed in this publication are those of the authors and do not necessarily reflect the views of the National Science Foundation, the Lawrence Livermore National Laboratory, or the Department of Energy. The authors thank Rich Caruana and the anonymous referees for their valuable feedback on this work.

References

1. P. Brown, R.D. Falgout, and J.E. Jones. Semicoarsening multigrid on distributed memory machines. *SIAM J. Sci. Computing*, 21:1823–1834, 2000.
2. L. Carrington, A. Snavely, X.Gao, and N. Wolter. A performance prediction framework for scientific applications. In *International Conference on Computational Science Workshop on Performance Modeling and Analysis (PMA03)*, pages 926–935, June 2003.
3. L. Carrington, N. Wolter, A. Snavely, and C.B. Lee. Applying an automatic framework to produce accurate blind performance predictions of full-scale HPC applications. In *Department of Defense Users Group Conference*, June 2004.
4. R.D. Falgout and U.M. Yang. hypre: a Library of High Performance Preconditioners. In *Proceedings of the International Conference on Computational Science (ICCS), Part III, LNCS vol. 2331*, pages 632–641, April 2002.

5. T.S. Karkhanis and J.E. Smith. A first-order superscalar processor model. In *Proceedings of the 31st Annual International Symposium on Computer Architecture*, pages 338–349, June 2004.

6. D.J. Kerbyson, H.J. Alme, A. Hoisie, F. Petrini, A.J. Wasserman, and M. Gittings. Predictive performance and scalability modeling of a large-scale application. In *Proceedings of IEEE/ACM Supercomputing '01*, November 2001.

7. G. Marin and J. Mellor-Crummey. Cross-architecture performance predictions for scientific applications using parameterized models. In *Proceedings of the International Conference on Measurement and Modeling of Computer Systems (Sigmetrics '04)*, pages 2–13, June 2004.

8. T.M. Mitchell. *Machine Learning*. WCB/McGraw Hill, Boston, MA, 1997.

Topic 3
Scheduling and Load-Balancing

Denis Trystram, Michael Bender, Uwe Schwiegelshohn, and Luís Paulo Santos

Topic Chairs

More and more parallel and distributed systems (clusters, grid and global computing) are available all over the world, opening new perspectives for developers of a large range of applications including data mining, multi-media, and bio-computing. However, this very large potential of computing power remains unexploited to a large degree, mainly due to the lack of adequate and efficient software tools for managing the resources. Scheduling problems address the allocation of those resources over time to perform tasks being parts of processes and are the key components in resource management.

As processors are the source of computing power of parallelism, it is crucial to carefully managing them in order to achieve a high efficiency of parallel systems. In most new parallel architectures and distributed platforms, the processors or machines are spatially distributed and communicate via various kinds of interconnections. Therefore, the communication medium is another important resource that must be considered during scheduling. New parameters like heterogeneity, the hierarchical character of memory, versatility of the context, and large scale computing should be taken into account as well. As conventional models and techniques cannot always be used, it is necessary to propose, implement and validate new approaches.

Therefore, the classical topic of Scheduling and Load Balancing remains very active in the perspective of new parallel and distributed systems. The subjects presented in Topic 3 cover all aspects related to scheduling and load-balancing including applications, system level techniques, theoretical foundations and practical tools. Some new trends and emerging models are also presented and discussed.

There were 31 papers submitted to this topic. Each submitted paper has been reviewed by 4 reviewers, and finally 11 papers were chosen to be included into the final program. They reflect the good and necessary synergy between theoretical approaches (models, analysis of algorithms, complexity, approximability results, multi-criteria analysis) and practical realizations and tools (new methods, simulation results, actual experiments, specific tuning for an application).

Finally, we would like to express our thanks to our colleagues, experts in the fields, who helped in the reviewing process.

J.C. Cunha and P.D. Medeiros (Eds.): Euro-Par 2005, LNCS 3648, p. 207, 2005.
© Springer-Verlag Berlin Heidelberg 2005

Balancing Parallel Adaptive FEM Computations by Solving Systems of Linear Equations*

Henning Meyerhenke and Stefan Schamberger

Universität Paderborn,
Fakultät für Elektrotechnik, Informatik und Mathematik
Fürstenallee 11, D-33102 Paderborn
{henningm,schaum}@uni-paderborn.de

Abstract. Load balancing plays an important role in parallel numerical simulations. State-of-the-art libraries addressing this problem base on vertex exchange heuristics that are embedded in a multilevel scheme. However, these are hard to parallelize due to their sequential nature. Furthermore, libraries like Metis and Jostle focus on a small edge-cut and cannot obey constraints like connectivity and straight partition boundaries, which are important for some numerical solvers.

In this paper we present an alternative approach to balance the load in parallel adaptive finite element simulations. We compute a distribution that is based on solutions of linear equations. Integrated into a learning framework, we obtain a heuristic that contains a high degree of parallelism and computes well shaped connected partitions. Furthermore, our experiments indicate that we can find solutions that are comparable to those of the two state-of-the-art libraries Metis and Jostle also regarding the classic metrics like edge-cut and boundary length.

Keywords: Parallel adaptive FEM computations, load balancing, graph partitioning.

1   Introduction

Finite Element Methods (FEM) are used extensively by engineers to analyze a variety of physical processes which can be expressed via Partial Differential Equations (PDE). The domain on which the PDEs have to be solved is discretized into a mesh, and the PDEs are transformed into a set of equations defined on the mesh's elements (see e. g. [1]). These can then be solved by iterative methods such as Conjugate Gradient (CG) and Multigrid. Due to the very large amount of elements needed to obtain an accurate approximation of the original problem, this method has become a classical application for parallel computers. The parallelization of numerical simulation algorithms usually follows the Single-Program Multiple-Data (SPMD) paradigm: Each processor executes the same code on a different part of the data. This means that the mesh has to be split into P sub-domains and each sub-domain is then assigned to one of the P processors. To minimize the overall computation time, all processors should thereby roughly

* This work is supported by the German Science Foundation (DFG) project SFB-376 and by DFG Research Training Group GK-693.

J.C. Cunha and P.D. Medeiros (Eds.): Euro-Par 2005, LNCS 3648, pp. 209–219, 2005.

contain the same amount of elements. Since iterative solution algorithms perform mainly local operations, i. e. data dependencies are defined by the mesh, the parallel algorithm mainly requires communication at the partition boundaries. Hence, these should be as small as possible. Depending on the application, some areas of the simulation space require a higher resolution and therefore more elements. Since the location of these areas is not known beforehand or can even vary over time, the mesh is refined and coarsened during the computation. However, this can cause imbalance between the processors' load and therefore delay the simulation. Hence, the element distribution needs to be rebalanced. The application is interrupted and the at this point static repartitioning problem is solved. Though this interruption should be as short as possible, it is also important to find a new balanced partitioning with small boundaries that does not cause too many elements to change their processor. Migrating elements can be an extremely costly operation since large amounts of data have to be sent over communication links and stored in complex data structures.

The described problem can be expressed as a graph (re-)partitioning problem. The mesh is transformed into a graph where the vertices represent the computational work and the edges their interdependencies. Due to the complexity of the problem, the large input sizes and the given time constraints, existing libraries that address the graph (re-)partitioning problem are based on heuristics. State-of-the-art implementations like Metis [2], Jostle [3] or Party [4] follow the multilevel scheme [5]. Vertices of the graph are contracted according to a matching and a new level consisting of a smaller graph with a similar structure is generated. This is repeated, until in the lowest level only a small graph remains. The (re-)partitioning problem is then solved for this small graph and vertices in higher levels are assigned to partitions according to their representatives in the next lower level. Additionally, a local improvement heuristic is applied in every level. By exchanging vertices between partitions, it reduces the number of cut edges or the boundary size as well as balances the partition sizes. Hence, the final solution quality mainly depends on this heuristic. Implementations are usually based on the Kerninghan-Lin (KL) heuristic [6], while the local refinement in Party is derived from theoretical analysis with Helpful-Sets (HS) [7].

To address the load balancing problem during parallel computations, distributed versions of the libraries Metis and Jostle have been developed. Both of them apply about the same multilevel techniques as their single processor version, but special attention must be paid to the local improvement heuristic due to its sequential nature. As an example, a coloring of the graph's vertices assures in the parallel library ParMetis [8] that during the KL refinement no two neighboring vertices change their partition simultaneously and therefore destroy the consistency of the data structures. In contrast to Metis, where vertices stay on their partition until a new distribution has been computed, the parallel version of Jostle [9] maps each sub-domain to a single processor and vertices which migrate do so already during the computation of the repartitioning. Usually, Metis is very fast while Jostle takes longer but often computes better solutions. The HS heuristic in Party exchanges sets between partitions that sometimes contain a large number of vertices. Hence, even more overhead would be necessary to ensure data consistency in a parallel implementation.

While the global edge-cut is the classical metric that most graph partitioners optimize, it is not necessarily the best metric to follow because it does not model the real communication and runtime costs of FEM computations as described in [10]. Hence, different metrics have been implemented inside the local refinement process modeling the real objectives more closely. In [11], the costs emerging from vertex transfers is taken into consideration while Metis is also capable of minimizing the number of boundary vertices.

A completely different approach is undertaken in [12]. Since the convergence rate of the CGBI domain decomposition solver in the PadFEM environment depends on the geometric shape of a partition, the integrated load balancer iteratively decreases the aspect ratios by applying a bubble like algorithm. Although different to the multilevel-schemes, this approach also contains a strictly sequential section and suffers from some other difficulties that are described in [13]. However, the latter paper introduces an implementation that eliminates most of these problems by replacing the sequential growing mechanism of the bubble framework by a few iterations of the first order diffusion scheme (FOS) [14]. This leads to a graph partitioning algorithm that contains a high degree of parallelism and produces well shaped partitions. Unfortunately, it is unclear how many FOS iterations must be performed. This question is overcome in [15] introducing FOS/A. This diffusion scheme does not balance the load but converges to a state with a load distribution similar to the situation after a few FOS iterations. Its drawback is the long execution time, and its fine-grain parallelism is hard to exploit on today's processors.

In this paper we present the (re-)partitioning heuristic $\mathrm{MF}(\phi)$, which is based on the same framework as the implementations from [13] and [15]. However, in contrast to the latter that distribute the vertices of a graph according to their load, our approach is based on the flow over the edges. The main advantage is that the computation of a $\| \cdot \|_2$-minimal balancing flow, which is equivalent to solving a system of linear equations, has been studied very well and that a variety of methods addressing it exist. Among them are faster diffusion schemes like the second order scheme (SOS) [14] as well as algorithms that require more global knowledge like CG solvers. Thus, one can choose the most appropriate implementation according to the underlying hardware. The remaining part of the paper is organized as follows. The next section briefly recaptures the bubble framework from [12] and explains the main idea. In Sec. 3 we propose a new growing mechanism which we integrate into this framework in Sec. 4. Afterwards, we present our experiments in Sec. 5 before we give a short conclusion.

2   The Bubble Framework

The idea of the bubble framework is to start with an initial, often randomly chosen vertex (seed) per partition, and all sub-domains are then grown simultaneously in a breadth-first manner. Colliding parts form a common border and

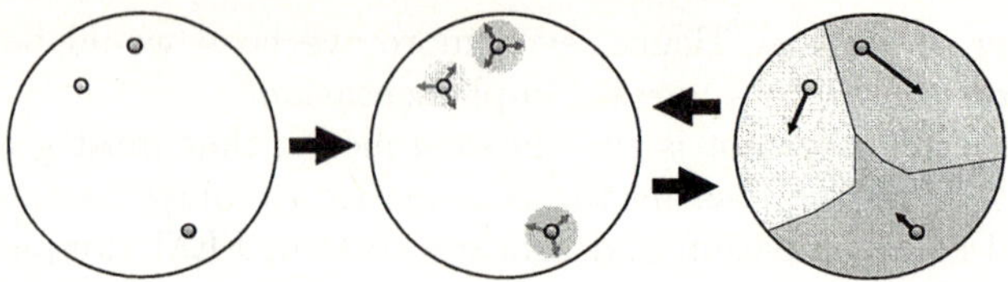

Fig. 1. The three operations of the learning bubble framework: Init: Determination of initial seeds for each partition (left). Grow: Growing around the seeds (middle). Move: Movement of the seeds to the partition centers (right).

keep on growing along this border – "just like soap bubbles". After the whole mesh has been covered and all vertices of the graph have been assigned this way, each component computes its new center that acts as the seed in the next iteration. This is usually repeated until a stable state, where the movement of all seeds is small enough, is reached. This procedure is based on the observation that within "perfect" bubbles, the center and the seed vertex coincide. Figure 1 illustrates the three main operations.

The growing mechanisms from [13] and [15] are based on diffusion. The main idea behind applying it in a graph partitioning heuristic is the fact that load primarily diffuses into densely connected regions of the graph rather than into sparsely connected ones. Following this observation one can expect to identify seeds inside such regions and therefore small partition boundaries in less dense areas. Additionally, since the load spreads around a seed vertex, the partitions should be connected and well shaped.

The remaining part of this paper is based on the following thought: If load diffuses faster into dedicated regions, then the flow over the edges directing there must be higher than the flow over edges pointing elsewhere. Hence, a $\|\cdot\|_2$-minimal flow should provide similar information as a load distribution computed by the FOS/A scheme from [15], with the advantage that a variety of faster methods are known to compute it.

3   A Growing Mechanism Based on Linear Equations

In this section we propose a new growing mechanism that is based on a $\|\cdot\|_2$-minimal flow in a network. This network G_ϕ is composed of the dual graph G corresponding to the mesh, and an extra vertex x that is connected with every other vertex of G. All edges $e \in E$ of G are assigned a weight of $w_e = 1$ while the weight of the edges incident to x are set to some constant $\phi > 0$. Now, independently for each partition p, we place a total of $|V|$ load equally on p's vertices and compute a $\|\cdot\|_2$-minimal flow f_p over the edges that transports all load to the extra vertex x. Since we minimize f_p according to the $\|\cdot\|_2$-norm, the load will not be sent directly to x, but also makes some 'detours' via other vertices in G. According to the idea mentioned in the last section, the flow thereby prefers densely connected regions of the graph. The weight constant ϕ

determines the spreading of the flow. If ϕ is large, it is cheaper to send most load directly to x, while if ϕ is small, the costs of the 'detour' into the graph are compensated by less utilized edges incident to x that can be chosen. In the extreme cases, if $\phi \to \infty$, all load is sent directly to x, while if $\phi \to 0$, the $\|\cdot\|_2$-minimal flow will converge towards the balancing flow that distributes the load equally in the original graph G.

The assignment of the vertices to the partitions is based on the amount of flow over the edges incident to x. We define a height function $h_p : V \cup \{x\} \to \mathbb{R}$ for each partition p, such that $h_p(v) = h_p(u) + f_{p_{(u,v)}} \cdot w_{(u,v)} \, \forall u \in \mathrm{adj}(v)$. Since f_p is the $\|\cdot\|_2$-minimal flow, this function is well defined and unique except for a constant, which we determine by setting the height of x to $h_p(x) = 0$. Now, we assign each vertex to that partition with the maximal height, meaning that the new partitioning π is defined by $\pi(v) = p : h_p(v) \geq h_q(v) \, \forall q \in \{1, \ldots, P\}$. If the maximum is not unique, we choose one of the eligible partitions arbitrarily.

Formally, let $G = (V, E)$ be an undirected, connected graph and $\mathbf{A} \in \{-1, 0, +1\}^{|V| \times |E|}$ its unweighted vertex-edge incidence matrix. $\mathbf{A}$ contains in each column corresponding to edge $e = (u, v)$ the entries -1 and $+1$ in the rows u and v, and 0 elsewhere. The unweighted Laplacian $\mathbf{L} \in \mathbb{Z}^{|V| \times |V|}$ is defined as $\mathbf{L} = \mathbf{A}\mathbf{A}^{\mathrm{T}}$. If we extend G by an additional vertex x and connect it to every other vertex with an edge of weight ϕ, we obtain the graph $G_\phi = (V \cup \{x\}, E \cup \{\{v, x\} : v \in V\})$ with edge weights $w_e = 1 \, \forall e \in E$ and $w_{\{v,x\}} = \phi \, \forall v \in V$. The weighted Laplacian matrix $\mathbf{L}_\phi \in \mathbb{R}^{|V|+1 \times |V|+1}$ of G_ϕ is defined as $\mathbf{L}_\phi = \mathbf{A}_\phi \mathbf{W} \mathbf{A}_\phi^{\mathrm{T}}$, where $\mathbf{A}_\phi$ denotes the unweighted vertex-edge incidence matrix of G_ϕ, and the entries of the diagonal matrix $\mathbf{W} \in \mathbb{R}^{|E|+|V| \times |E|+|V|}$ are set to $(w_{ee}) = w_e$. Hence, with $\mathbf{I}$ being the identity, $\mathbf{L}_\phi$ can be written as:

$$
\mathbf{L}_\phi = \left(\begin{array}{cccc} \left(\begin{array}{ccc} & & \\ & \mathbf{L} + \phi\mathbf{I} & \\ & & \end{array} \right) & \begin{array}{c} -\phi \\ \vdots \\ -\phi \end{array} \\ \begin{array}{ccc} -\phi & \cdots & -\phi \end{array} & |V| \cdot \phi \end{array} \right) \tag{1}
$$

Our goal is to compute a $\|\cdot\|_2$-minimal flow f_p from the vertices of the partition p to the additional vertex x. By setting the vectors $s_p, t \in \mathbb{R}^{|V|+1}$ to

$$
(s_{p_v}) = \begin{cases} |V|/|\{v : \pi_p(v) = p\}| & : \pi(v) = p \\ 0 & : \text{otherwise} \end{cases} \qquad (t_v) = \begin{cases} |V| & : v = x \\ 0 & : \text{otherwise} \end{cases}
$$

we place $|V|$ load equally on p's vertices and the corresponding 'negative' load on x. Then, we have to solve the quadratic minimization problem

$$
\min! \frac{1}{2} f_p^{\mathrm{T}} \mathbf{W}^{-1} f_p \text{ with respect to } \mathbf{A}_\phi f_p = s_p - t \; . \tag{2}
$$

Due to [16], we know that we can find the optimal f_p for (2) by first solving the linear equation

$$
\mathbf{L}_\phi \lambda_p = s_p - t \; . \tag{3}
$$

```
00   Algorithm MF(G, π, φ, l, i)
01       in each loop l
02           if π is undefined
03               π = determine-seeds(G)                    /* initial seeds */
04           else
05               parallel for each partition p             /* contraction */
06                   solve L_φ λ_p = s_p − t and compute h_p
08                   π(v) = { p : h_p(v) ≥ h_p(u) ∀u ∈ V
                             { −1 : otherwise
09               parallel for each partition p             /* consolidation */
10                   solve L_φ λ_p = s_p − t and compute h_p
12               π(v) = p : h_p(v) ≥ h_q(v) ∀q ∈ {1, . . . , P}
13           in each iteration i
14               parallel for each partition p             /* consolidation with ... */
15                   solve L_φ λ_p = s_p − t and compute h_p
17               π(v) = p : h_p(v) ≥ h_q(v) ∀q ∈ {1, . . . , P}
18               scale-balance(π)                          /* ... scale balancing */
19           greedy-balance(π)                             /* greedy balancing */
20       return smooth(π)                                 /* smoothing */
```

Fig. 2. Sketch of the MF(ϕ) heuristic.

$\mathbf{L}_\phi$ is sparse and symmetric positive semidefinite. Since $\langle s_p - t, \mathbb{1} \rangle = 0$ and the rank of $\mathbf{L}_\phi$ is $|V|$, the solution of (3) is unique except for a constant. Nevertheless, we now can determine the unique $\| \cdot \|_2$-minimal flow from the computed potential λ_p as

$$f_{P(u,v)} = w_{\{u,v\}} \cdot (\lambda_{p_u} - \lambda_{p_v}) \ . \tag{4}$$

Since we are interested in the height function $h_p(v)$, we can skip the flow computation (4) and assign $h_p(v) = \lambda_{p_v} - \lambda_{p_x}$. The new partitioning π can then be determined as described above, while the new partition seed is the vertex with the highest load according to h_p.

4   The MF(ϕ) Heuristic

In this section we describe the integration of the proposed growing mechanism into the bubble framework. The resulting algorithm is sketched in Fig. 2. It can either be invoked with or without a valid partitioning π. In the latter case, we determine initial seeds randomly (line 3). Otherwise, we contract the given partitions (lines 5-8) applying the mechanism proposed in Sec. 3. Note that in either case π only contains a single vertex for each partition when entering line 9. Following the bubble framework, we then grow the partitions from the seeds. However, if we determined single seeds right after the last contraction, these would be the same ones as before and no movement would occur. Hence, it is necessary to apply at least one consolidation (lines 9-12) between two contractions. In contrast to a contraction that determines a single vertex per partition (line 8), a consolidation results in a partitioning (lines 12/17). In the following step, the load is placed equally on the vertices of the whole partition, which causes it to move into denser regions of the graph as mentioned before.

To further enhance the solution quality, additional consolidations can be performed (lines 13-18). Furthermore, these are used for balancing by scaling the height functions h_p. If a partition is too small, h_p is multiplied by a constant $b_p > 1$, while if it is too large, a constant $b_p < 1$ is chosen. Although the choice of b is limited because no partition must become empty, this approach can find almost balanced solutions in most cases. To ensure a certain size, we perform a greedy balancing operation (line 19), where we compute a $\| \cdot \|_2$-minimal flow in the partition graph and move the vertices that cause the least error according to the height functions. The whole learning process is then repeated several times. Before returning the partitioning π, we migrate vertices if the number of their adjacent vertices in another partition is larger than the number in the current partition. This compensates numerical imprecisions that occur during the flow computation and further smoothes the partition boundaries. However, if the number of vertices in a partition is small compared to its boundary length, it might also lead to a higher imbalance.

An interesting point is the lack of an explicit objective function. Except for the balancing, the MF(ϕ) heuristic does not contain any directives what metric to minimize. This is also the case for the algorithms from [13, 15].

The run-time of MF(ϕ) greatly depends on the linear equation solver. Currently, we apply a basic CG implementation. However, due to the special structure of $\mathbf{L}_\phi$, several optimizations are possible. As indicated in lines 5, 9 and 14, all P linear systems can be solved independently. Hence, even if we apply solvers other than diffusive ones which require more global knowledge, a large amount of parallelism remains.

5   Experiments

In this section we describe our experiments with the new heuristic MF(ϕ) and compare its solutions to those of the parallel versions of the state-of-the-art graph (re-)partitioning libraries Metis and Jostle. Furthermore, we include the results of the Party/DB library from [15]. The benchmark instances are created as described in [17] and are available via [18]. Each benchmark consists of 101 frames, each containing a graph of around 15000 vertices. Though the instances are quite small, important observations can already be made. Due to space limitations we only present the data of a single benchmark here. The results of the omitted experiments are similar, however.

The libraries Metis (version 3.1) and Jostle (version 3.0) both offer a large number of options. For the presented evaluation, we chose the recommended values from their manual, respectively, and left the remaining parameters at their default. This means that Metis operates with an *itr* value of 1000.0 and Jostle uses the options *threshold = 20, matching = local, imbalance = 3*. Note that Jostle seems to ignore the imbalance setting and computes totally balanced

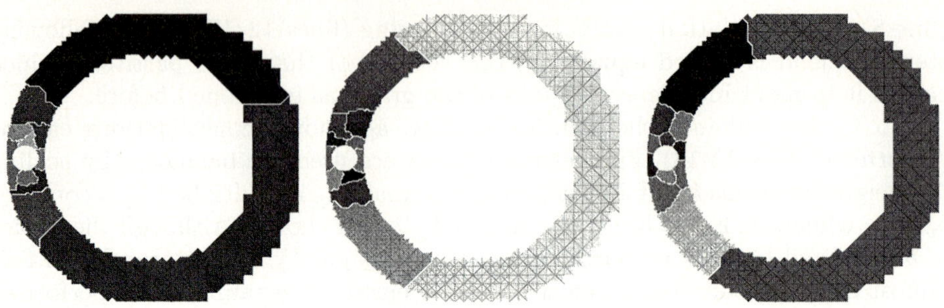

Fig. 3. Partitionings in frame 50 of the 'ring' benchmark computed by Metis (left), Jostle (middle) and the MF(ϕ) heuristic (right).

partitions, except for the initial solution where the sequential versions of the libraries are applied. The MF(ϕ) heuristic is invoked with $\phi = 0.01$ and performs 2 loops with 4 iterations, respectively.

We measure the partitioning quality according to a number of metrics, because it is known that the edge-cut does not necessarily model the real costs [10]. Depending on the application, some of the metrics described in the following might be more important than others. *External edges:* Number of edges that are incident to exactly one vertex of partition p. *Boundary vertices:* Number of vertices of partition p that are adjacent to at least one vertex from a different partition. *Send volume:* The amount of outgoing information is the sum of the adjacent partitions different to p that each vertex residing inside partition p has. *Receive volume:* The amount of incoming information is the number of vertices of partitions different to p adjacent to at least one vertex of partition p. *Diameter:* The longest shortest path between two vertices of the same partition. Infinity, if the partition is not connected. *Outgoing migration:* Number of vertices that have to be migrated to a different partition. *Incoming migration:* Number of vertices that have to be migrated from a different partition. Furthermore, the quality of a partitioning depends on its balance. A less balanced solution allows other metrics to improve further and makes comparisons less meaningful. Please note that we have omitted the run-times since our prototypic implementation is some magnitudes slower than its competitors.

In addition, for the listed metrics we consider three different norms. Given the values $x_1, \ldots, x_P$, the norms are defined as follows: $\|X\|_1 := x_1 + \ldots + x_P$, $\|X\|_2 := (x_1^2 + \cdots + x_P^2)^{1/2}$ and $\|X\|_\infty := \max_{i=1..P} x_i$. The $\|\cdot\|_1$-norm (summation norm) is a global norm. The global edge-cut belongs into this category (it equals half the external edges in this norm). In contrast to the $\|\cdot\|_1$-norm, the $\|\cdot\|_\infty$-norm (maximum norm) is a local norm only considering the worst value. This norm is favorable if synchronized processes are involved. The $\|\cdot\|_2$-norm (Euclidean norm) lays in between the $\|\cdot\|_1$ and the $\|\cdot\|_\infty$-norm and reflects the global situation as well as local peaks, but is omitted here.

Figure 3 displays a single frame from the 'ring' benchmark. In this benchmark, a circle and the refined area around it rotate through a narrow ring. One

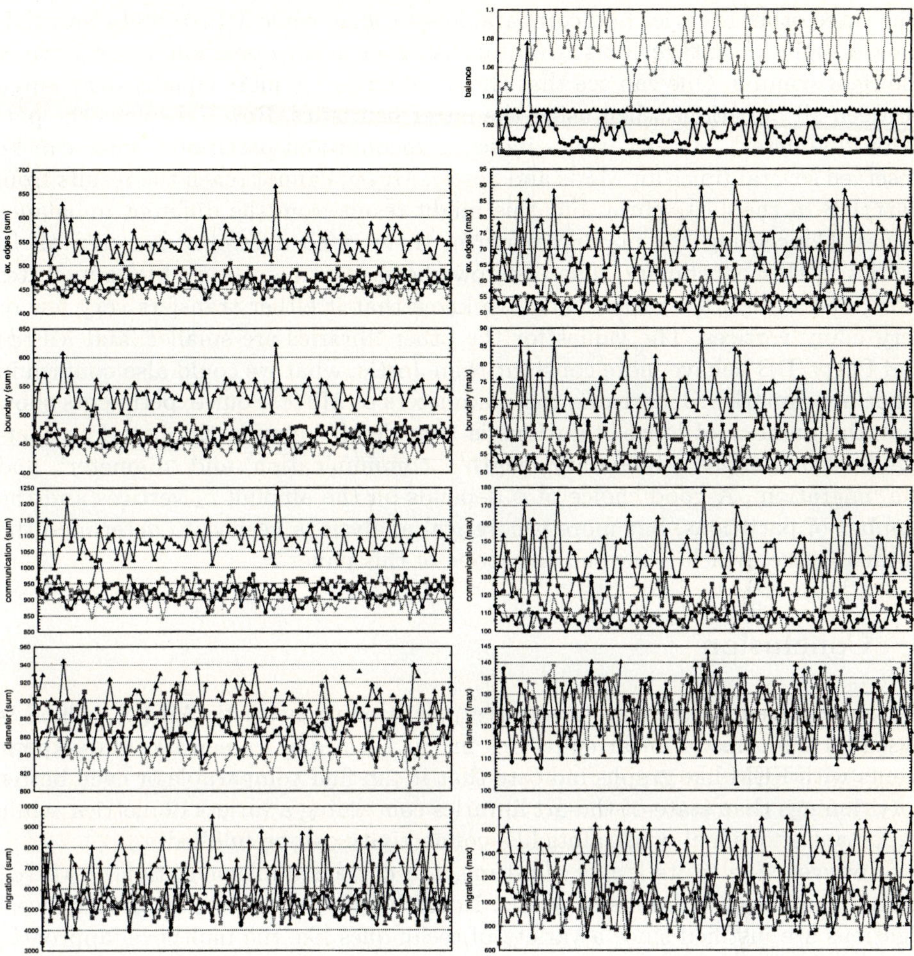

Fig. 4. Numerical results of the 'ring' benchmark for Metis (blue triangles), Jostle (red squares), Party/DB (green pentagons) and MF(ϕ) (black circles).

can see that the partitions computed by Metis have quite large fringes, while Jostle and especially MF(ϕ) find smoother partition boundaries. Though the visual display of the mesh provides a first impression of the solution quality, the numerical data of all 101 frames listed in Fig. 4 reveals many more details. Looking at the first row, we can see that Metis usually allows up to 3% imbalance, while Jostle ignores this parameter and totally equalizes the partition sizes. The solutions of the MF(ϕ) heuristic usually have an imbalance of less than 3%, while the Party/DB library has some difficulties to maintain an equal distribution. The next three rows contain the metrics 'external edges', 'boundary length' and 'communication volume'. Their values are similar. The right column, displaying the sum for all partitions, reveals that Metis computes the worst results.

The three other libraries find comparable solutions, while $MF(\phi)$ and Party/DB show a slight advantage. This advantage is larger in the maximum norm given in the right column. One can see that the boundaries are more equally distributed between all partitions when using the latter heuristics. Row 5 displays the partition 'diameter'. Missing values indicate unconnected partitions, what can be observed several times for Metis and Jostle. $MF(\phi)$ cannot reach the results from Part/DB in the $\|\cdot\|_1$-norm, but this might result from the different imbalance values. Concerning the maximum norm, there is no difference between all libraries in this benchmark. The last row shows the 'migration'. Metis migrates most, and from other experiments we know that it either transfers very few or very many vertices. The values for the other libraries are smaller, and $MF(\phi)$ and Party/DB behave more constant than Jostle, what we could also confirm in other benchmarks. Concerning the parameters of $MF(\phi)$, our experiments show that the number of loops/iterations is a trade-off between the first four metrics 'external edges', 'boundary length', 'communication' and 'diameter', and the 'migration'. A good choice of ϕ depends on the amount of vertices and the number of partitions, but more theoretical analysis is needed to determine the optimal value, which is beyond the scope of this paper.

6   Conclusion

We have presented the new graph (re-)partitioning heuristic $MF(\phi)$, which is based on solutions of linear equations inside a learning framework. Our experiments with FEM like graphs indicate that it can find comparable or even better partitionings than state-of-the-art libraries concerning a variety of metrics, while important additional constraints like connectivity can be fulfilled.

However, due to its longer run-time, the current implementation of $MF(\phi)$ cannot compete with Metis or Jostle. Nevertheless, we think that further investigations are justified since a variety of techniques like the multilevel approach, faster diffusion schemes, optimized CG preconditioners or multigrid solvers are known to speed up the computations.

References

1. G. Fox, R. Williams, and P. Messina. *Parallel Computing Works!* Morgan Kaufmann, 1994.
2. G. Karypis and V. Kumar. *MeTis: A Software Package for Partitioning Unstrctured Graphs, Partitioning Meshes, [...], Version 4.0*, 1998.
3. C. Walshaw. *The parallel JOSTLE library user guide: Version 3.0*, 2002.
4. S. Schamberger. Graph partitioning with the Party library: Helpful-sets in practice. In *Comp. Arch. and High Perf. Comp., SBAC-PAD'04*, pages 198–205, 2004.
5. B. Hendrickson and R. Leland. A multi-level algorithm for partitioning graphs. In *Supercomputing'95*, 1995.
6. B. W. Kernighan and S. Lin. An efficient heuristic for partitioning graphs. *Bell Systems Technical Journal*, 49:291–308, 1970.

7. J. Hromkovic and B. Monien. The bisection problem for graphs of degree 4. In *Math. Found. Comp. Sci. (MFCS '91)*, volume 520 of *LNCS*, pages 211–220, 1991.

8. Kirk Schloegel, George Karypis, and Vipin Kumar. Multilevel diffusion schemes for repartitioning of adaptive meshes. *J. Par. Dist. Comp.*, 47(2):109–124, 1997.

9. C. Walshaw and M. Cross. Parallel optimisation algorithms for multilevel mesh partitioning. *J. Parallel Computing*, 26(12):1635–1660, 2000.

10. B. Hendrickson. Graph partitioning and parallel solvers: Has the emperor no clothes? In *Irregular'98*, number 1457 in LNCS, pages 218–225, 1998.

11. L. Oliker and R. Biswas. PLUM: Parallel load balancing for adaptive unstructured meshes. *J. Par. Dist. Comp.*, 52(2):150–177, 1998.

12. R. Diekmann, R. Preis, F. Schlimbach, and C. Walshaw. Shape-opt. mesh part. and load bal. for par. adap. FEM. *J. Parallel Computing*, 26:1555–1581, 2000.

13. S. Schamberger. On partitioning FEM graphs using diffusion. In *HPGC, Intern. Parallel and Distributed Processing Symposium, IPDPS'04*, page 277 (CD), 2004.

14. R. Elsässer, B. Monien, and R. Preis. Diffusion schemes for load balancing on heterogeneous networks. *Theory of Computing Systems*, 35:305–320, 2002.

15. S. Schamberger. A shape optimizing load distribution heuristic for parallel adaptive FEM computations. Accepted at PACT'05.

16. Y. F. Hu and R. F. Blake. An improved diffusion algorithm for dynamic load balancing. *Parallel Computing*, 25(4):417–444, 1999.

17. O. Marquardt and S. Schamberger. Open benchmarks for load balancing heuristics in parallel adaptive finite element computations. Accepted at PDPTA'05.

18. S. Schamberger. http://www.upb.de/cs/schaum/benchmark.html.

CISNE: A New Integral Approach for Scheduling Parallel Applications on Non-dedicated Clusters[*]

Mauricio Hanzich[2], Francesc Giné[1], Porfidio Hernández[2],
Francesc Solsona[1], and Emilio Luque[2]

[1] Departamento de Informática e Ingeniería Industrial, Universitat de Lleida, Spain
{sisco,francesc}@eup.udl.es
[2] Departamento de Informática, Universitat Autònoma de Barcelona, Spain
mauricio@aows10.uab.es, {porfidio.hernandez,emilio.luque}@uab.es

Abstract. Our main interest is oriented towards keeping both local and
parallel jobs together in a non-dedicated cluster. In order to obtain some
profits from the parallel applications, it is important to consider *time*
and *space* sharing as a mean to enhance the scheduling decisions. In this
work, we introduce an integral scheduling system for non-dedicated clusters, termed CISNE. It includes both a previously developed dynamic
coscheduling system and a space-sharing job scheduler to make better
scheduling decisions than can be made separately. CISNE allows multiple parallel applications to be executed concurrently in a non dedicated
Linux cluster with a good performance, as much from the point of view
of the local user as that of the parallel application user. This is possible
without disturbing the local user and obtaining profits for the parallel
user. The good performance of CISNE has been evaluated in a Linux
cluster.

1 Introduction

There are several studies in the literature whose main aim is to determine the
interaction and effects of space-sharing (S.S.) and time-sharing (T.S.) policies.
Nevertheless, most of them are focused on dedicated environments. Furthermore,
many of these studies center on Gang Scheduling [1, 2], combined with some kind
of backfilling [1] policy for doing the job distribution.

In this work, we want to show a new scheduling approach focused on non-
dedicated cluster systems. The use of non-dedicated systems for parallel computation is based on various studies [3] that prove the effectiveness of making good
use of the idle CPU cycles by executing distributed applications.

In this article, we present a new system named CISNE. Our system combines S.S. and T.S. scheduling techniques, in order to take advantage of the idle
computer resources available across the cluster. CISNE is set up basically of a
dynamic coscheduling technique and a job scheduler.

---

[*] This work was supported by the MCyT under contract TIC 2001-2592 and partially
supported by the Generalitat de Catalunya -Grup de Recerca Consolidat 2001SGR-
00218.

J.C. Cunha and P.D. Medeiros (Eds.): Euro-Par 2005, LNCS 3648, pp. 220–230, 2005.

The dynamic coscheduling system, termed CCS, is the T.S. scheduling component. Traditional dynamic coscheduling techniques [4] rely on the communication behavior of an application, to simultaneously schedule the communicating processes of a job. Unlike those techniques, CCS takes its scheduling decisions from the occurrence of local events, such as communication, memory, Input/Output and CPU, together with foreign events received from remote nodes. This allows CCS to assure the progress of parallel jobs without disturbing local users, and even using an *MultiProgramming Level* (MPL) greater than one. In addition it is possible to re-balance the resources assigned to parallel tasks throughout the cluster. CCS was previously developed [5], and now we present the modifications that allows it to be incorporated into an integral cluster scheduling system, such as CISNE.

The job scheduler, named LoRaS, is the S.S. scheduling component of CISNE. It is responsible for distributing the parallel workload among the cluster nodes. This is performed by taking into account the state of the cluster system, the characteristics of applications already running and those of the waiting jobs. Based on those considerations and the coscheduling restrictions, different techniques for assigning jobs to processors are proposed and evaluated in this article.

CISNE was implemented in a non-dedicated Linux cluster. In this framework, we evaluated the interaction between T.S. and S.S. techniques. This experimentation shows that our proposal obtains better performance than the rest of the evaluated techniques, as much from the point of view of the local user as that of the parallel applications user and the system administrator.

The remainder of this paper is as follows: in section 2 we explain the main problems to solve and our goals for this article. In section 3 the CISNE system is presented. The efficiency measurements of CISNE are performed in Section 4. Finally, the main conclusions and future work are explained in Section 5.

2   T.S. and S.S. Interaction Problems

The choice of a dynamic coscheduler as a T.S. system is based on the fact that this kind of system is better suited to a non-dedicated environment than an explicit (or gang) T.S. coscheduling schema [6]. However, this choice has some implications for the S.S. schema, that force us to develop our own system.

The main effect could be found in the lack of an Ousterhout matrix [7], present in every explicit coscheduling system. In such a system, the parallel machine could be seen as a set of n parallel virtual machines (VM). The matrix provides information about the parallel jobs and their forming tasks, as well as the mapping onto the VMs. Every VM is synchronized to each other by means of a global context switch. Thus, there is no interaction among the VMs, which also means none between the parallel tasks.

On the other hand, in a T.S. system based on dynamic coscheduling techniques, there is no such matrix. Thus, it is not possible to apply the S.S. techniques to each row. As a consequence, and in order to improve the global performance of the system, the (now existing) interaction among the running applications has to be considered [8]. Therefore, one of the main goals of this work is to find the kind and degree of interaction between the system management components (T.S. and S.S. schemes) to achieve the maximum performance in distributed tasks without damaging the local ones.

The studies carried out by Choi et al. revealed the sensitivity of the implicit coscheduling techniques in relation to the mapping and the execution order of the parallel applications over the cluster (if the MPL is greater than one). In addition, the type of applications (CPU or communication bound) running concurrently over the cluster, and the global system state, can have a great influence on the coscheduling performance. Another factor to take into account in a non-dedicated cluster is the local user activity, which has to be monitorized periodically. The control of those factors allows the S.S. system to schedule better, while it helps the T.S. system to avoid intrusions into the local tasks.

In such a scenario, the best S.S. scheduling of a parallel workload is not obvious and hence some questions arise including how the distribution of the parallel applications over the cluster affects the coscheduler performance, how the inter-arrival time affects the turnaround time of the parallel applications, and finally, whether it is worth applying a complex scheduling policy and, if so, which. Our main goal in this work is to shed some light on those questions.

3   CISNE: Cooperative and Integrated Scheduler for Non-dedicated Environments

In order to provide a system that merges space and time sharing scheduling, we propose a new integral system called CISNE. The time sharing scheduling is done by a dynamic coscheduling technique, named CCS (Cooperating CoScheduling) [5], developed previously by our group. Concerning about the space scheduling problem, we present a system called LoRaS. This system is responsible for distributing parallel applications throughout the cluster using information about the system state, the applications to be launched and the CCS characteristics.

Fig. 1 shows the integration of CCS and LoRaS into CISNE. It shows the main components making up the virtual machine. As we can see in the figure, the interaction between the nodes follows a master-slave paradigm. There is one server node (master with the most important control and management functions), and the remaining ones interact with the server in a client (slave) mode.

In the following sections, the CCS and LoRaS systems are explained separately.

3.1   CCS (Cooperating CoScheduling) System

Our T.S. system provides an execution environment where the parallel applications could be dynamically coscheduled. Besides, the given resources are balanced and the interactive responsiveness of the local applications is totally preserved. In order to reach this situation, CCS uses the architecture shown in fig. 1, where each module goal is:

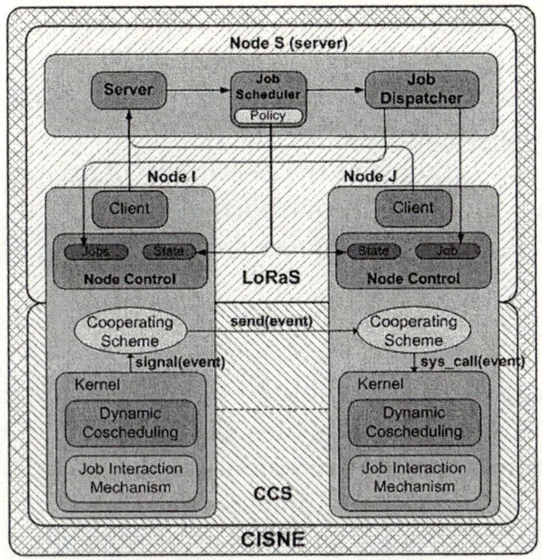

Fig. 1. LoRaS-CCS Architecture

- *Dynamic Coscheduling*: no processes should wait for a non-scheduled process for synchronization/communication. This is achieved by means of increasing the communicating task priority, even causing CPU preemption. (implemented inside the Linux kernel).
- *Job Interaction Mechanism* (JIM): preserves the local user tasks responsiveness. In order to reach its goal, this module manages the amount of resources (CPU and memory) given to the parallel tasks in the node. This is done by means of a *social contract* [9], which establish the amount of resources that could be given to the parallel and local loads, when the node is not idle (implemented inside the Linux kernel).
- *Cooperating Scheme*: this module collaborates with the JIM module in order to balance the resources (memory & CPU) given to the parallel applications throughout the cluster. It is responsible for the exchange of several events, such as the login or logout of a local user into a specific node, or the stopping (restarting) event generated by the JIM module for a specific parallel application. This happen whenever it has to preserve the local responsiveness (implemented in user space).

3.2   LoRaS (Long Range Scheduler) System

LoRaS implements a Job Scheduler in the user space, which provides a Space-Sharing scheduling mechanism. The following is the description of the LoRaS modules shown in fig.1:

- *Client*: sends a job execution request (JER) to the server module on behalf of a parallel user.
- *Server*: the admittance of new JERs to be executed in the system is performed by the *server* module. This JER is then forwarded to the *Job Scheduler* module.

- *Job Scheduler*: executes every received JER using the configured policy. It is important to mention that JER execution is conditioned by the cluster state. If there is no possibility of executing the job on its arrival, then the petition has to wait in a queue for the requested resources.
- *Policy (submodule)*: establishes the possibility of executing a JER for a given cluster state and the JER resources request. This module is designed in such a way that it is easy to change its functionality and hence the LoRaS scheduling system.
- *Job Dispatcher*: considering that every job can have its own characteristics (e.g. a PVM or MPI job), it is necessary to configure the job before launching it. Hence, this module is responsible for doing these previously required tasks.
- *Node Control*: this module has two different functions. On one hand it launches and controls the job execution. On the other hand, it gathers information from the node state and informs the *Job scheduler* (and hence, the *policy* submodule) so that it can take better scheduling decisions.

3.3   Implemented and Evaluated Policies

In this section, we propose several S.S. techniques oriented towards non-dedicated clusters. Unlike traditional techniques oriented to dedicated cluster, all our proposals are characterized by the fact of taking the cluster state into account.

The first proposed policy, named *Uniform*, is characterized by the following: (a) it merges differently oriented applications (i.e. communication or computation) in the same node and (b) it runs applications one over another in an ordered manner, whenever possible. By doing this, we expect to increase the coscheduling probability of the CCS system. By *ordering the applications* we mean to launch parallel applications in such a way that each task of a couple of parallel applications runs in the same set of nodes. This situation is depicted in fig. 2.a and we call it a *Uniform* situation.

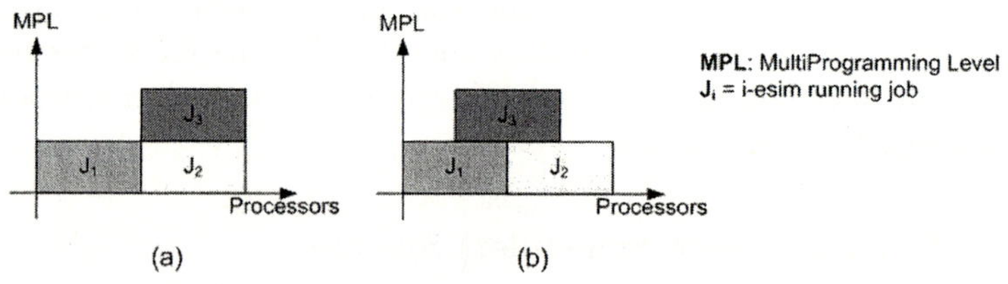

Fig. 2. Difference between a *uniform* (a) and a *normal* (b) policy

However, the *Uniform* policy executes the applications in any free place if there is no space for them in a *uniform* place. Besides, it is important to mention that in every case the policy must try to help to preserve the local user activity by not overloading nodes with some local tasks. This is done by limiting the amount of usable memory and the MPL, respecting the established social contract.

The problem of arranging different size (number of needed nodes) applications in a uniform way, was dealt with by always arranging smaller applications over bigger ones. Therefore, little applications can start sooner, while bigger applications do not notice to much effect from the coscheduling point of view.

The second proposed policy, termed *Normal*, considers the state of the system nodes, but does not consider the running job distributions as the uniform does. Thus, the resulting scheduling can reach a situation like the one depicted in fig. 2.b. where an application like J_3 shares its nodes with a couple of applications. This situation tends to diminish the coscheduling system performance and hence the application execution time is increased.

In addition, both *normal* and *uniform* policies are compared against a *Basic* policy where we execute the parallel workload with an MPL = 1, which means at most one parallel task per node. Finally, and in order to compare with a well known S.S. policy, we introduce an *EASY backfilling* [10] policy in our evaluation. The *EASY* policy executes a job not-at-the-head of the jobs queue, whenever this does not delay the start of the job at the head. By including this policy, we can show the effect of incrementing the MPL compared with the use of an *EASY* policy with an MPL = 1.

It is important to note that for every evaluated policy, we use a FCFS policy for queuing each arriving job. Doing this, we ensure the absence of starvation in the system and a fair treatment for every job.

4   Experimentation

This experimentation is divided into two sections. The first section compares our coscheduling system in relation to traditional coscheduling systems based exclusively on communication events. The second set of results shows how CISNE performs under our defined S.S. policies.

In order to simulate a non-dedicated cluster, we need two different kinds of workloads. On one hand, we need to simulate local user activity and, on the other hand, we need some parallel applications that arrive at some interval.

The local workload was carried out by running a synthetic benchmark. This allowed the CPU load, memory requirements and network traffic used by the local user to be fixed. In order to assign these values in a realistic way, we monitored the average resources used by real users. According to this monitoring, we defined two local user profiles. The first profile identifies 65% of the users with high needs on inter-activeness (called *XWindows* user: 15% CPU, 35% Mem., 0,5KB/sec LAN), while the other profile distinguishes 35% of the users with web navigation needs (called *Internet* user: 20% CPU, 60% Mem., 3KB/sec. LAN). This benchmark alternate CPU activity with interactivity by means of running several system calls and different data transfers to memory. In order to measure the level of intrusion into the local load, our benchmark provide us with the *system call latency*. Besides, and according to the values observed in the monitoring, we loaded the 25% of the nodes with local workload in our experiments.

The parallel workload was a list of 90 NAS parallel applications with a size of 2, 4 or 8 tasks that reached the system following a Poisson distribution [2]. The chosen NAS applications were: CG (mem: 55-120MB / CPU: 65-70% / time: 37-51 sec.), IS (mem: 70-260MB / CPU: 58-69% / time: 40-205 sec.), MG (mem: 60-220MB / CPU: 82-89% / time: 26-240 sec.) and BT (mem: 7-60MB / CPU: 85-93% / time: 90-180 sec.). The parallel jobs were merged so that the entire workload had a balanced requirement of computation and communication (25% of the workload composed by each application). It is important to note that the MPL reached for the workload depended on the system state at each moment, but in no case it surpassed an MPL = 4. This was established in order to respect the social contract, which was set to 50% of the resources available for each kind of load (local/parallel) [5].

Both workloads were executed in an Linux cluster composed of 16 P-IV (1,8GHz) nodes with 512MB of memory and a fast ethernet interconnection network.

4.1   Evaluating the Time-Sharing Systems

In this section we have compared the CCS policy in relation to the plain Linux scheduler and two well known communication-driven coscheduling strategies: implicit and (isolated) dynamic coscheduling. In implicit coscheduling, a process waiting for messages spins for a determined time before blocking. In contrast, dynamic coscheduling deals with all messages arrivals (like CCS, but without resource balancing). It works by increasing the receiving task priority, even causing CPU preemption of the task being executed inside.

They were evaluated by running the parallel workload for several values of MPL (1 to 4). The parallel workload was executed applying a *Normal* S.S. policy. Its performance was measured by means of the slowdown. This is the response-time ratio of a job in a non-dedicated system in relation to the time needed in a system dedicated to this job.

From fig. 3.a, we can see that the slowdown of the parallel applications is always better for our CCS coscheduling system. In fact, this difference increases with the value of the MPL. This good CCS behavior is due to the interaction of the coscheduling scheme with the adaptive and balanced resource allocation carried out by CCS. In addition, the social contract implemented by CCS maintains the response time (measured by the mean of the local benchmark system call latency in fig. 3.b) always under 400ms. This limit for the Response Time, established by [11], is an acceptable threshold before the user can notice a lack of inter-activeness.

These results encouraged us to use CCS to integrate a coscheduler into the CISNE system.

4.2   Evaluating the CISNE Integrated System

In this subsection, we want to show the performance of CISNE, by applying the described space-sharing policies to the CCS system. This interaction will be

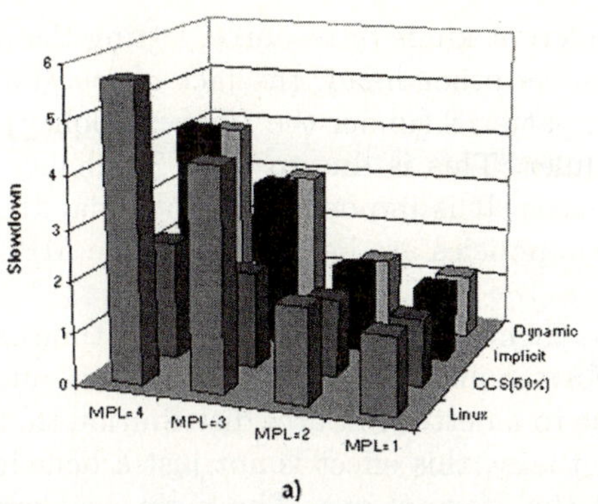
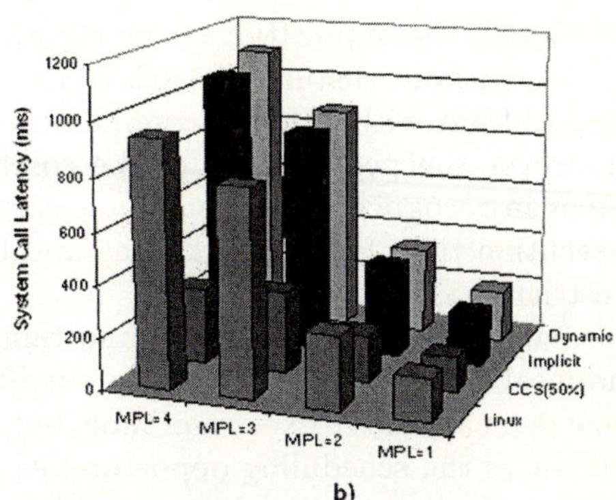

Fig. 3. Parallel applications slowdown (a) and system call latency (b) under the evaluated policies

quantified by measuring the turnaround time of the parallel applications comparing the *Uniform, Normal, Basic* and *EASY* policies. In addition, we measure the makespan of the workloads (i.e. the executing time of the whole workload). Doing this it is possible to evaluate CISNE from a system administrator's point of view.

Fig. 4.a shows the turnaround, wait and execution time for every evaluated policy. Here we can see that the *normal* and *backfilling* policies give us almost the same behavior, while the *uniform* policy performs better by reducing the execution time and hence the waiting time of the workload. From this figure, it is also clear that the turnaround time is dictated by the waiting time. On the other hand, it would be desirable to evaluate the effect of the execution time as the predominant turnaround factor. With this aim, we executed the parallel workload doubling the inter-arrival time between applications. Fig. 4.b shows the results obtained for the same policies.

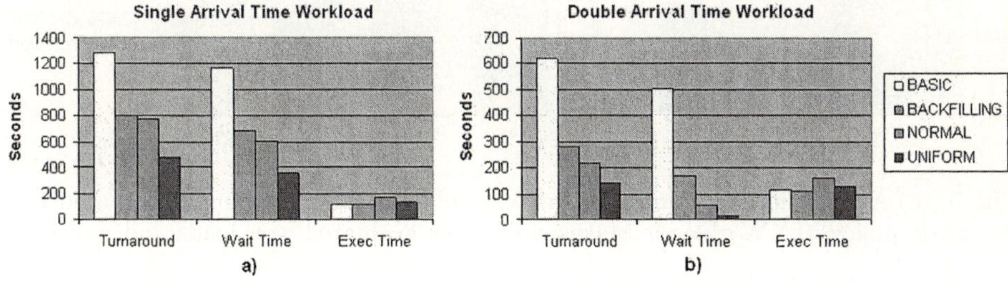

Fig. 4. Turnaround, Wait and Execution time for the exercised workloads

From those figures, it is clear that the job distribution policy has a great impact on the underlying coscheduling system performance, considering the reduction in the execution time. This effect arises for two different reasons: on one hand, the applications compete for different kinds of resource, letting them evolve without disturbing each other. On the other hand, the fact of merging applications with different communication patterns (under the *Uniform* policy), improves the performance of the coscheduler. This is due to a CCS enhancement in recognizing the communication needs. It is important to note that the execution time for the backfilling and basic policies are better due to the MPL restriction.

Another effect that it is important to mention is how the waiting time is noticeably reduced when we apply a *Uniform* policy. This effect is not only due to a decrease in the execution time, but due to a better resource distribution that enhances the scheduling opportunities. Actually, this effect is not just a benefit of the *Uniform* policy, but a problem of the *Normal* one. The main problem is that the *Normal* policy tends to distribute the resources in such a way that the total available memory throughout the cluster could be enough to execute an application, but there are not enough nodes with enough free memory for launching it. However the *Uniform* policy tends to localize the available resources and then the scheduling possibility is enhanced in the average case. This is due to the elevated percentage of small applications in the workloads tested. That fact was verified in [2] to be representative of the reality.

In order to take a closer look at the enhancement of the coscheduling performance, fig. 5.a shows how the selected policy affects the jobs slowdown. This graphic is calculated by comparing the *Normal* and *Uniform* policies with the *Basic* policy (*Slowdown* = 1), where every job is executed in isolation (except for some local activity). The figure shows how a uniform policy could reduce the slowdown from 40% (1,40) to less than 15% (1,15). This demonstrates the good performance of our coscheduling system as the close interaction with the S.S technique and, once again, that the level of resources is enough to increase the MPL with almost no detriment to the (parallel) application execution time.

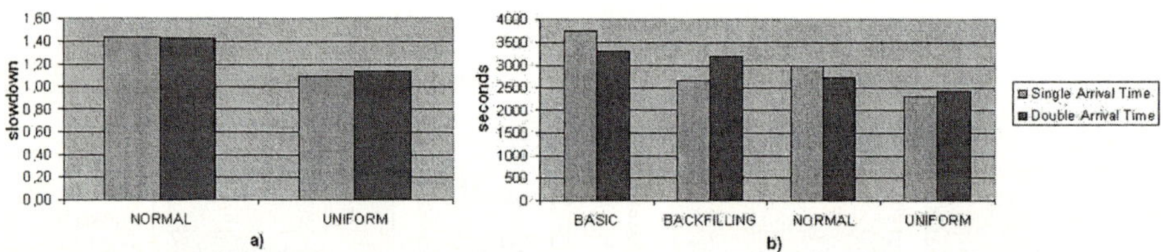

Fig. 5. (a) Applications slowdown for the *Normal* and *Uniform* policies compared with the *Basic* policy. (b) Workloads Makespan for the evaluated workloads and policies

Another aspect we want to analyze is the CISNE behavior from the system point of view (makespan). The results for both workloads (i.e. single and double arrival time), can be observed in fig. 5.b for the policies evaluated.

A couple of effects can be extracted from the figure. First of all, a *backfilling* policy behaves better with a shorter workload arrival time than with a longer one. This is due to a longer (waiting) jobs queue that enhances the backfilling opportunities. Considering the Normal and Uniform policies, it is clear that the last one has some advantages. In this case, the effect is directly related to a better resource usage and the enhancement in the application turnaround time.

5 Conclusions and Future Work

This work presents a new integral system, named CISNE, that considers both S.S and T.S. concerns, which is applied on a non-dedicated cluster. Using this framework, the paper analyzes how the performance of a dynamic coscheduling system could be affected by the distribution policy over a non-dedicated cluster. With this aim, we evaluated four policies oriented to non-dedicated clusters: *Uniform, Normal, Backfilling* and *Basic*. We found that a *Uniform* policy (i.e. a set of applications running on the same set of nodes), can dramatically diminish the turnaround time of the applications (up to 76%) compared with other approaches. In addition, the performance of a uniform distribution was evaluated considering a turnaround time limited, on one hand, by the waiting time (single arrival time workload), and on the other hand by the execution time (double arrival time workload). In both scenarios a *Uniform* policy was shown to perform well, even from the system point of view (makespan). It is important to note that those gains were obtained without disturbing the system responsiveness.

Considering our future work and taking into account that the *Uniform* and *EASY* policies attack the scheduling problem from different points of view, they could be combined in a schema where the MPL is greater than one and we also apply a backfilling policy. To do this, we have to define a prediction model to establish the execution time of a parallel application considering the cluster state and the interaction between the running applications.

References

1. Y. Zhang, H. Franke, J. Moreira and A. Sivasubramaniam. "An Integrated Approach to Parallel Scheduling Using Gang-Scheduling, Backfilling and Migration". *IEEE Transactions on Parallel and Distributed Systems, 14(3):236-247, March 2003.*
2. D. G. Feitelson. Packing schemes for gang scheduling. *In Job Scheduling Strategies for Parallel Processing, D. G. Feitelson and L. Rudolph (Eds.), Springer-Verlag, 1996, Lect. Notes Comput. Sci. vol. 1162, pp. 89-110.*
3. T. Anderson, D. Culler, D. Patterson and the NOW team. "A case for NOW (Network of Workstations)". *IEEE Micro, Vol. 15, pp. 54-64. 1995.*
4. C. Anglano. "A Comparative Evaluation of Implicit Coscheduling Strategies for Networks of Workstations". *In 9th International Symposium on High Performance Distributed Computing (HPDC'2000), pp.221-228, 2000.*
5. M. Hanzich, F. Giné, P. Hernández, F. Solsona and E. Luque. "Coscheduling and Multiprogramming Level in a Non-dedicated Cluster". *EuroPVM'2004, LNCS, vol. 3241, pp. 327-336, 2004.*

6. F. Solsona, F. Giné, P. Hernández, E. Luque. "Implementing Explicit and Implicit Coscheduling in a PVM Environment". *EuroPar 2000, LNCS, Vol 1900, pp 1164-1170. 2000.*

7. Ousterhout, J. . "Scheduling techniques for concurrent systems". *Proceedings of the Conference on Distributed Computing Systems. 1982.*

8. G. Choi, S. Agarwal, J. Kim, A. Yoo and C. Das. "Impact of Allocation Strategies on Communication-Driven Coscheduling in Clusters". *EuroPar 2003: 160*-168.

9. R.H. Arpaci, A.C. Dusseau, A.M. Vahdat, L.T. Liu, T.E. Anderson and D.A. Patterson. "The Interaction of Parallel and Sequential Workloads on a Network of Workstations". *ACM SIGMETRICS'95, pp.267-277, 1995.*

10. A. W. Mu'alem and D. G. Feitelson. "Utilization, predictability, workloads, and user runtime estimates in scheduling the IBM SP2 with backfilling ". *IEEE Trans. Parallel & Distributed Syst. 12(6), pp. 529-543, Jun 2001.*

11. Nielsen. "Advances in Human-Computer Interaction". J. Nielsen, J. (ed.), Intellect Publishers, 1995.

On Optimum Multi-installment Divisible Load Processing in Heterogeneous Distributed Systems

Maciej Drozdowski[1,*] and Marcin Lawenda[2]

[1] Institute of Computing Science, Poznań University of Technology,
ul.Piotrowo 3A, 60-965 Poznań, Poland
`Maciej.Drozdowski@cs.put.poznan.pl`
[2] Poznań Supercomputing and Networking Center,
ul.Noskowskiego 10, 61-704 Poznań, Poland
`Marcin.Lawenda@man.poznan.pl`

Abstract. In this paper we study multi-installment divisible load processing in heterogeneous distributed systems. Divisible loads are computations which can be divided into parts of arbitrary sizes, and these parts can be processed independently in parallel. In order to reduce the waiting time during the parallel computation initialization phase, load is sent to the processors in multiple small installments. In a heterogeneous system the sizes of the installments should be adjusted to the communication, and computation capabilities of the processors. We propose two algorithms that gear the load chunk sizes to different communication and computation speeds. The first one is an optimization branch and bound algorithm. The second algorithm is based on genetic search. The running times of both methods and the quality of the genetic algorithm solutions are compared. Then, we use these algorithms to analyze features of the scheduling problem solutions.

Keywords: scheduling and load balancing, divisible load, multi-installment processing, heterogeneous systems, optimization algorithms.

1   Introduction

Divisible loads are computations which can be divided into parts of arbitrary sizes, and these parts can be processed independently in parallel. This means that the grain of parallelism is small, and there are no data dependencies. The sizes of the load parts should be adjusted to the speeds of communication and computation such that processing finishes in the shortest possible time. Examples of real divisible applications include, among others, distributed searching for patterns in text, audio, graphic files, database and measurement processing, data retrieval systems, some linear algebra algorithms, and simulation. Surveys of the divisible load theory can be found in [4, 6, 11].

---

* This research has been partially supported by the Polish State Committee for Scientific Research.

J.C. Cunha and P.D. Medeiros (Eds.): Euro-Par 2005, LNCS 3648, pp. 231–240, 2005.
© Springer-Verlag Berlin Heidelberg 2005

Communication delays constitute an important part of the processing time in all distributed algorithms. To reduce the initial waiting for the data, and for initialization of the computations, load is sent in multiple small chunks rather than in a single long message. This way of divisible load distribution and execution is called *multi-installment processing* [3, 6, 8, 12]. In the earlier publications certain assumptions were usually made on the structure of the schedule. For example, messages of equal size were sent to processors in a round-robin fashion [6, 8, 12]. It has been shown [12] that this way of multi-installment processing reduces the length of the schedule in a homogeneous system at most $\frac{e-1}{e}$ times. Unequal load chunk size partitioning has been also proposed [3, 6, 13], but with a tacit assumption that the set of used processors, and their activation sequence are given and fixed. Furthermore, it was assumed that there are no idle times, neither in the communication nor in the computations [3, 4, 6, 13]. However, to our best knowledge, the problem of multi-installment divisible load processing with unequal chunk sizes adjusted to the communication and computation speeds, with selection of the set of exploited processors, and selection of their activation sequence is open. The goals of this paper are twofold: to propose algorithms for the multi-installment divisible load processing including selection of the set and sequence of processors, and to study influence of the system parameters on the quality of the scheduling problem solutions.

The rest of this paper is organized as follows. In Section 2 we formulate the multi-installment divisible load scheduling problem for heterogeneous systems. In Section 3 two algorithms are proposed: an optimization branch-and-bound algorithm, and a heuristic genetic algorithm. The results of computational experiments are presented and discussed in Section 4.

2   Problem Formulation

We will use the word *processor* to denote a processing element with CPU, memory, and communication link. In divisible load model it is classically assumed that initially some volume of load V (e.g. a file with data to be processed) resides on a processor P_0 called *originator*. The originator sends the load to its neighbors for remote processing. Each of the neighbors intercepts some part of the received load, and immediately starts computations related with the received load. The rest of the load is retransmitted to the still inactive neighbors. In this work we assume a star interconnection (a.k.a. a single level tree). In the star network the originator is located in the star center (or the root of the single level tree), and is connected to a set $P_1, \ldots, P_m$ of processors which perform computations. All communications involve the originator, and there are no direct communications between processors $P_1, \ldots, P_m$. For simplicity of presentation we assume that originator is communicating only. Otherwise, the computing ability of the originator can be represented as an additional processor. Each processor P_i is defined by the following parameters: A_i - computing rate (reciprocal of computing speed), C_i - communication rate (reciprocal of bandwidth), S_i - communication startup time. A_i, C_i can be expressed, e.g., in seconds per byte, and S_i can be expressed in seconds. Computing x units (e.g. bytes) of load on processor P_i takes xA_i units of time. Sending the same amount of load to P_i lasts $S_i + xC_i$. We assume that memory sizes of the processors are sufficiently big and do not influence the construction of the schedule. To simplify the mathematical model we assume that the results returning time is negligible. This simplification is not limiting generality of our considerations because result gathering can be included in the model (see e.g. applications [2, 5, 7, 12]). The computations start only after receiving the whole message with load. We assume that processors have independent communication hardware which allows for simultaneous communication and computations on the previously received load.

To reduce the initial waiting for the load, and for the start of the computations, load is sent to processors in multiple small chunks rather than in a single long message. Let n denote the number of chunks. If the sequence of processors receiving the load chunks is known then our problem can be reduced to a linear program. Let α_i denote size of chunk i. Let $d_i \in \{1, \ldots, m\}$ be the number of the processor receiving chunk i. We will denote by $H_i \subseteq \{i, \ldots, n\}$ the set of chunks sent to processor d_i, starting from chunk i. C_{max} denotes schedule length. Fig. 1 depicts an example schedule with multiple installments. The linear program can be formulated as follows:

minimize C_{max}

on condition that:

$$\sum_{j=1}^{i}(S_{d_j} + \alpha_j C_{d_j}) + A_{d_i} \sum_{j \in H_i} \alpha_j \leq C_{max} \quad i = 1, \ldots, n \qquad (1)$$

$$\sum_{i=1}^{n} \alpha_i = V \qquad (2)$$

In constraint (1) sum $\sum_{j=1}^{i}(S_{d_j} + \alpha_j C_{d_j})$ expresses communication time for chunks $1, \ldots, i$. $A_{d_i} \sum_{j \in H_i} \alpha_j$ is computation time on processor d_i starting from chunk i. Thus, (1) guarantees that all processors stop computations before the end of the schedule. All work is done by equation (2). Thus, it is possible to find optimum distribution of the load using formulation (1)-(2) if we know the sequence of the processor activation (i.e. values d_i for $i = 1, \ldots, n$).

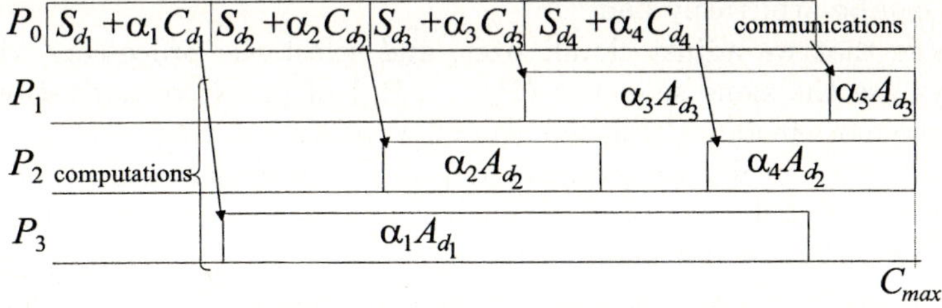

Fig. 1. Example of load distribution pattern.

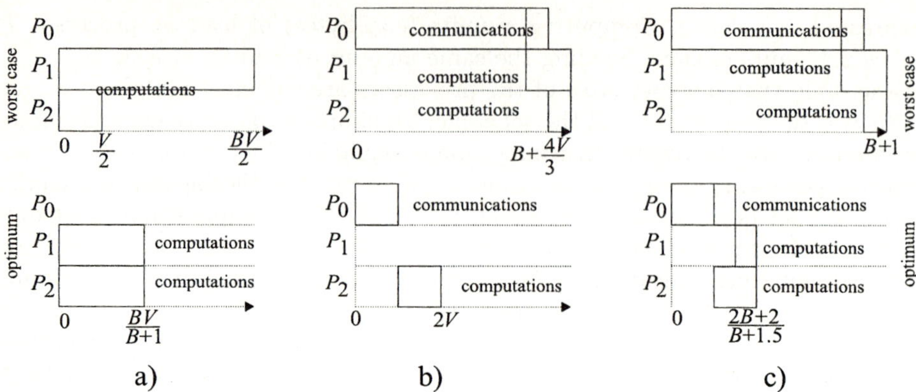

Fig. 2. The worst case examples. a) ignoring heterogeneity, b) ignoring processor set selection, c) ignoring sequencing of the processor activation.

Before proceeding to the further details let us consider *worst cases* that may appear if scheduling decisions ignore certain information. Suppose that we ignore the heterogeneity of the system, and send load parts of equal size to the processors. For instance (Fig. 2a), consider two processors P_1 with parameters $S_1 = 0, C_1 = 0, A_1 = B$, and P_2 with parameters $S_2 = 0, C_2 = 0, A_2 = 1$. We divide the load into two equal chunks of size $\frac{V}{2}$. Resulting schedule has length $\frac{BV}{2}$ but processor P_2 is idle in interval $[\frac{V}{2}, \frac{BV}{2}]$. If we use sizes $\alpha_1 = \frac{V}{B+1}, \alpha_2 = \frac{BV}{B+1}$, then both processors stop computing simultaneously, and schedule length is $\frac{BV}{B+1}$. The ratio of the two schedule lengths is $\frac{B+1}{2}$ which can be arbitrarily big. Hence, in the worst case solutions based on load equipartitioning can be arbitrarily bad in heterogeneous systems.

Suppose that we adjust chunk sizes to the parameters A_i, C_i, but all processors are always used. Let us present another example (Fig. 2b). There are two processors with parameters: $S_1 = B, A_1 = 1, C_1 = 1, S_2 = 0, A_2 = 1, C_2 = 1$. If $V < \frac{B}{2}$ then there is no point in using processor P_1 because load of this size may be processed in a shorter time than the communication activating P_1. If we use P_1 then the schedule has length at least B. If we don't, then schedule has length $V(A_2 + C_2) = 2V$. The ratio of the two lengths is at least $\frac{B}{2V}$ which can be arbitrarily big. Thus, if the set of processors is always the same, the resulting schedule can be arbitrarily bad.

Suppose that we adjust chunk sizes, and select the processors wisely, but we always use the same sequence $(P_1, \ldots, P_m)$ of processor activation. Let us analyze one more instance (Fig. 2c), $m = 2, V = 2, S_1 = 0, C_1 = B, A_1 = 1, S_2 = 0, C_2 = A_2 = 0.5$. If we use sequence (P_1, P_2) of processor activation, then the optimum load distribution is $\alpha_1 = \alpha_2 = 1$, and schedule length is $B + 1$. For sequence (P_2, P_1) the optimum distribution is $\alpha_1 = \frac{1}{B+1.5}, \alpha_2 = \frac{2B+2}{B+1.5}$, and schedule length is $\frac{2B+2}{B+1.5}$. The ratio of the two lengths is $\frac{\frac{B+1}{2-\frac{1}{B+1.5}}}$ which can be arbitrarily big.

Thus, the subset of processors $P_1, \ldots, P_m$ exploited in the computations and the targets of the communications are unknown, and must be determined. This task has combinatorial nature. In Section 3 we propose algorithms that determine destinations for the load chunks. If one ignores proper selection of the chunk destinations, the problem becomes easier to solve because only linear program (1)-(2) has to be solved for some assumed chunk destinations $d_1, d_2, \ldots, d_n$. Then, the resulting schedules can be arbitrarily bad in the worst case, as demonstrated in the preceding paragraph. How bad the solutions can be on average, if we skip the combinatorial part of the problem, is unknown. We attempt answering this question in Section 4.

3 Optimization Algorithms

3.1 Branch and Bound Algorithm

Two elements constitute a branch-and-bound algorithm. The first is *branching* procedure which divides the solution space into disjoint subsets. These subsets are either eliminated if they do not include the optimum solution, or are further divided until selecting a unique solution. Partition of the solution space can be represented as a tree. Each node is a representative of a set of solutions. Dividing such a set is equivalent to generating successors of a node. In our problem we have to select the sequence of the targets for n load chunks. In the root of the tree the sequence is empty. The first chunk may be sent to one of processors P_i, for $i = 1, \ldots, m$. Therefore, the root has m successors each representing sequences starting with a message sent to processor P_i. The second level of the tree includes two-processor sequences (P_i, P_j). Branching a node representing a leading sequence of l chunk targets consists in appending one more processor to which chunk $l + 1$ will be sent. The branching procedure is continued until constructing a sequence of the assumed length n.

The maximum number of the search tree leaves is m^n. As this number grows exponentially with n, it is necessary to prune the search tree by eliminating nodes representing solutions certainly not better than some already known solution. This procedure is the *bound* element of the algorithm. To determine if a node should be eliminated its *lower bound* of the schedule length is calculated. Suppose the node represents a sequence of l chunks. Thus values $d_1, \ldots, d_l$ are already determined. The remaining $n - l$ chunks still need to be selected. We assume that these $n - l$ chunks are sent to $n - l$ ideal target processors. The ideal target processor has parameters $A^{id} = \min_{i=1}^{m}\{A_i\}, C^{id} = \min_{i=1}^{m}\{C_i\}, S^{id} = \min_{i=1}^{m}\{S_i\}$, and processes only one load chunk. For such a sequence of l real processors, and $n - l$ ideal ones, a linear program (1)-(2) is solved for C_{max} which is the lower bound.

The best known solution used in comparisons with the lower bound is found by the algorithm itself. It is the best solution found in any leaf of the search tree. The tree is searched in the depth-first least lower bound order.

3.2   Genetic Algorithm

Genetic algorithms imitate evolution of genome. Solutions are encoded as strings of symbols analogously to the encoding of the chromosomes. Some initial population of solutions is generated randomly. Genetic operators transform populations in a direction improving quality of the solutions. Selection, crossover, and mutation are typical genetic operators. *Selection* elects better solutions for the next population. *Crossover* operation generates offspring solutions by randomly combining pieces of the parent strings. Though the offspring is constructed in a random manner, the fragments of a string encoding an optimum solution are indirectly discovered and combined due to the selection and crossover. *Mutation* changes randomly some solutions to diversify the search, and to escape local optima. Genetic search is a classic technique for solving combinatorial optimization problems, including scheduling problems. We direct interested readers to monographs [9, 10] for detailed presentation of the genetic search method.

In our implementation a chromosome is a string $(d_1, \ldots, d_n)$ of chunk destinations. The measure of a chromosome fitness is the value of schedule length C_{max} obtained from the linear program (1)-(2) formulated for the sequence of chunk targets given in the chromosome. In the crossover operation two chromosomes are randomly selected, and combined using one point crossover. For example, let $(a_1, a_2, \ldots, a_n)$, $(b_1, b_2, \ldots, b_n)$ be two parent solutions, and let k denote a randomly selected crossover point. The two offspring solutions are $(b_1, \ldots, b_{k-1}, a_k, \ldots, a_n)$ and $(a_1, \ldots, a_{k-1}, b_k, \ldots, b_n)$. The total number of new chromosomes constructed in crossover is Gp_C, where G is the size of the population, and p_C is a tunable algorithm parameter which will be called crossover probability. Mutation changes Gnp_M random genes (i.e. d_is) to different values. Gn is the total number of genes, p_M is a tunable algorithm parameter which we will call mutation probability. The selection of the chromosomes for the new population is done by a combination of elitist and roulette wheel method. The best half of the old population is always preserved. A string is passed to the second half of the new population with probability $\frac{1}{C_{max}^j} / \sum_{j=1}^{G} \frac{1}{C_{max}^j}$, where C_{max}^j is the schedule length for chromosome j. The algorithm stops after a fixed number of iterations without an improvement in the quality of the best solution ever found. There is also a limit on the total number of iterations.

4   Computational Experiments

4.1   Experiment Setting

All the experiments were performed on a PC computer with Pentium IV 1.8GHz, 512MB RAM memory, and Microsoft Windows XP. The executable code was generated by Borland C++ Builder 6.0. All LP formulations were solved by a code derived from `lp_solve` [1]. Unless stated otherwise, the test instances of the scheduling problem were generated in the following way: Processor parameters A, C, S, were generated with uniform distribution from the range [0,1]. Problem size was $V = 1E6$. The processor number was $m = 4$, and the number of chunks

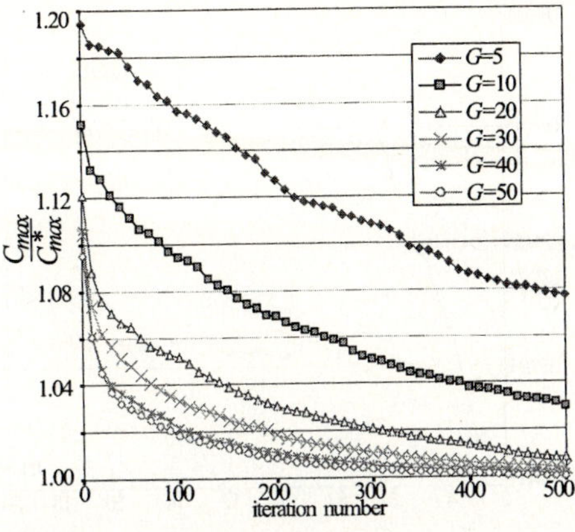

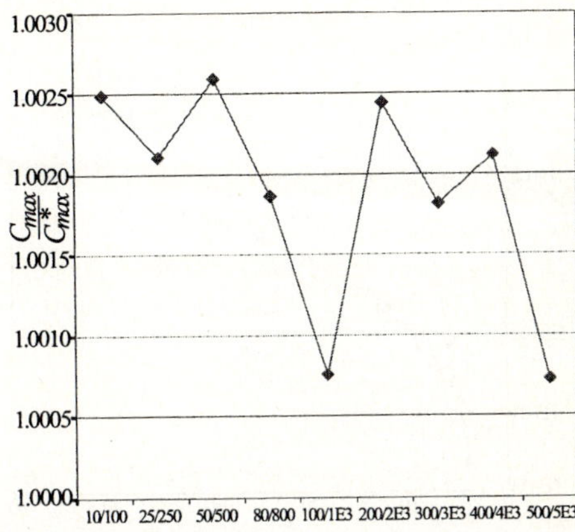

Fig. 3. Average distance from optimum vs. iteration (population) number and G.

Fig. 4. Average distance from optimum vs. iteration limits.

was $n = 8$. Each point on the following charts is an average of at least 10 instances.

In the genetic algorithm genes of the initial population were generated with uniform distribution from set $\{1, \ldots, m\}$. The following procedure has been applied to tune the genetic algorithm. A set of 100 random instances were generated as a reference benchmark. An indicator of algorithm performance was the average quality of the best solutions obtained for these benchmark instances. Population size $G = 50$ has been selected as the convergence improvement stops at this size (cf. Fig. 3). For the fixed G crossover probability $p_C = 80\%$, and then mutation probability $p_M = 3\%$ were selected. We used a limit of 10 iterations without solution improvement, and an upper limit of 100 iterations in total, which give acceptable solution quality on average (cf. Fig. 4), but still result in a shorter running time than other iteration limits combinations.

4.2   Performance of the Algorithms

Running Times. The execution times of the algorithms are collected in Fig. 5, and 6. The running time of the branch and bound is denoted by B&B, and of the genetic algorithm by GA. It can be seen that the branch and bound algorithm has exponential running time in n for fixed m (cf.Fig. 5). The execution time grows slower as a function of m for fixed n (cf.Fig. 6) because the maximum number of the search tree leaves is m^n. Nevertheless, execution time of the branch and bound algorithm allows only for solving instances with small m, and n. Execution time of the genetic algorithm grows with n (Fig. 5) because the length of the string encoding solution is n. For $m = 3, \ldots, 20$ execution time grows less than twice (Fig. 6). We also tested dependence of the execution times on size V of the problem. For small V execution time of the branch and bound was shorter than for big sizes because less processors had to be activated,

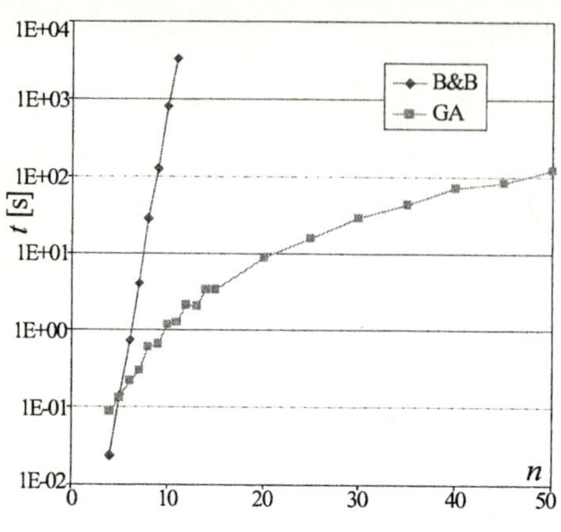

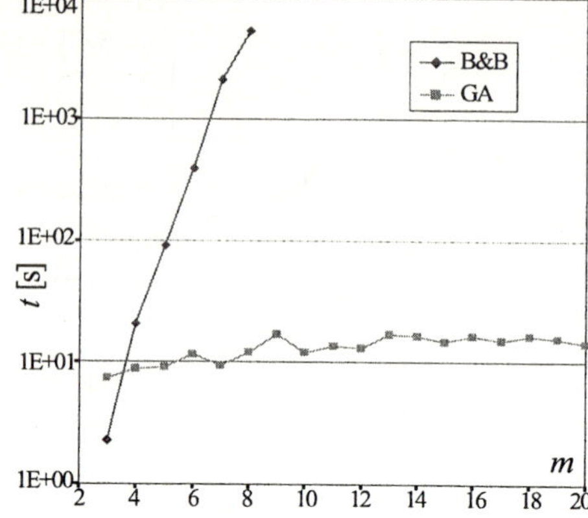

Fig. 5. Running time vs. n.   **Fig. 6.** Running time vs. m.

and therefore the search trees were smaller. The execution time of the genetic algorithm was independent of V.

Quality of the Solutions. The results of our study on the quality of solutions are collected in Fig. 7-8. The instances in Fig. 7 had A parameter equal to a given value on all processors. The remaining C, S parameters were generated as described previously. Analogously, for Fig. 8 parameter C was fixed on all processors, and A, S were randomly generated. Each figure represents quality of the solutions, i.e. the relative distance from the optimum, in three cases: the average solution of a genetic algorithm (denoted GA), the average random solution (denoted RND), and the worst selection of the chunk targets ever observed (denoted Worst). Note that the worst case (Worst) has its own 'y' axis different than RND, and GA cases. The random solutions (RND) have random chunk destinations. In all cases load chunk sizes were calculated by linear program (1)-(2).

These three cases demonstrate weaknesses and strengths of the two parts in the solution of our problem: the combinatorial part which finds targets for the chunks (d_is), and the linear programming part which calculates optimum chunk sizes (α_is) for the given destinations. It can be seen that genetic algorithm constructs solutions that are very close to the optimum. On average its solutions were not further 0.2% from the optimum. The worst solution obtained by the genetic algorithm was 1.1% away from the optimum. Thus, the genetic algorithm is a practical replacement for the optimization branch and bound algorithm which has exponential running time. The random solutions (RND) are also good on average because their distance from the optimum is not greater than approximately 30%. This is good news because solving a complex combinatorial problem of determining chunk targets (be it by a branch and bound or by a genetic algorithm) may be too time consuming and unprofitable on average. A random, or reasonable selection of processors and their activation sequence, supplemented by a linear program (1)-(2) gives solutions of acceptable quality on average. This

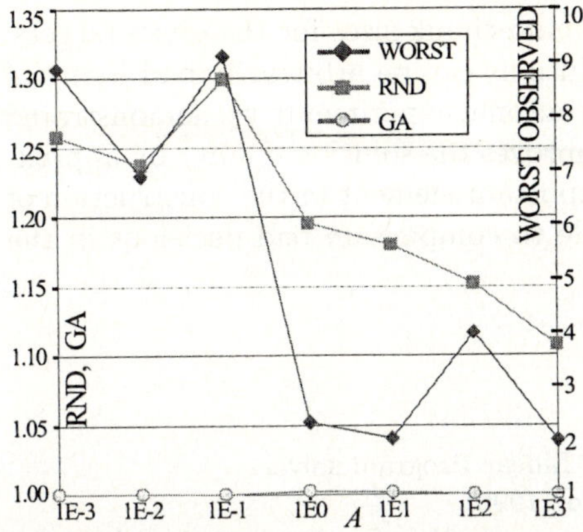
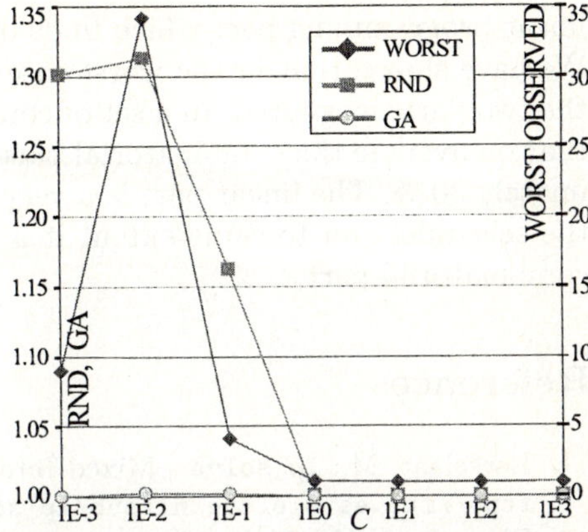

Fig. 7. Relative distance from the optimum vs. A.

Fig. 8. Relative distance from the optimum vs. C.

tells us also about the nature of the problem we are solving. Since relatively good results can be obtained only by adjusting chunk sizes (even for random chunk destinations), the chunk size selection is an important element in the solution of our problem. In other words, linear programming can compensate for some bad decisions in combinatorial part of the algorithms. It can be said that on average the combinatorial part of our problem (i.e. target selection) improves a random solution by approximately 30%. Finally, the worst case really exists. In the worst observed case of the chunk target selection a schedule 35 times worse than optimum was constructed (cf. Fig. 8).

It is possible to infer from Fig. 7-8 on the features of the solutions and performance of the algorithms. With growing A, C the quality of the random and the worst case is improving. When A is very big, the schedule length becomes dominated by the computation time. The selection of the chunk destinations is nearly meaningless because the schedule length is determined by the computation time which is approximately $\frac{AV}{m}$. Similar conclusions can be drawn for parameter C. When C is very big, chunk target selection tends to be immaterial because the schedule length is determined by the communication time which is approximately VC. We also tested dependence on S in range [1E-3,1E3]. It turned out that S constitutes at most $\approx 2\%$ of the communication time, and hence this dependence was not strong.

5   Conclusions

In this paper we studied multi-installment divisible load processing in heterogeneous distributed system. The problem we analyzed consists in determining optimum destinations for the load chunks and adjusting their sizes to the speeds of processors and communication links. Hence, we divided solution methods into two parts: combinatorial one which finds destinations for the load chunks, and linear programming part which finds optimum chunk sizes for the given targets. We have shown that in the worst case solutions can be arbitrarily bad if any of the two parts is ignored. In a set of computational experiments we demonstrated that on average the combinatorial part improves the solution quality by approximately 30 %. The linear part is a very important element in the construction of the schedule, and to some extent it is able to compensate bad decisions in the combinatorial part.

References

1. Berkelaar, M.: `lp_solve` - Mixed Integer Linear Program solver. `ftp://ftp.es.ele.tue.nl/pub/lp_solve` (1995)
2. Bharadwaj, V., Barlas, G.: Access time minimization for distributed multimedia applications. Multimedia Tools and Applications **12** (2000) 235-256
3. Bharadwaj, V., Ghose, D., Mani, V.: Multi-installment Load Distribution in Tree Networks With Delays. IEEE Transactions on Aerospace and Electronic Systems **31** (1995) 555-567
4. Bharadwaj, V., Ghose, D., Mani, V., Robertazzi, T.: Scheduling divisible loads in parallel and distributed systems. IEEE Computer Society Press, Los Alamitos CA (1996)
5. Błażewicz, J., Drozdowski, M., Markiewicz, M.: Divisible task scheduling - concept and verification. Parallel Computing **25** (1999) 87–98
6. Drozdowski, M.: Selected problems of scheduling tasks in multiprocessor computer systems. Series: Monographs, No 321, Poznań University of Technology Press, Poznań (1997). Downloadable from `http://www.cs.put.poznan.pl/mdrozdowski/txt/h.ps`
7. Drozdowski, M., Wolniewicz, P.: Experiments with Scheduling Divisible Tasks in Clusters of Workstations. In: A.Bode, T.Ludwig, W.Karl, R.Wismüller (eds.), Euro-Par 2000. Lecture Notes in Computer Science, Vol. 1900. Springer-Verlag, Berlin Heidelberg New York (2000) 311-319
8. Drozdowski, M., Wolniewicz, P.: Out-of-Core Divisible Load Processing, IEEE Trans. on Parallel and Distributed Systems **14** (2003) 1048-1056.
9. Goldberg, D.E.: Genetic Algorithms in Search, Optimization and Machine Learning, Addison-Wesley, Reading, Massachusetts (1989)
10. Michalewicz, Z.: Genetic Algorithms + Data Structures = Evolution Programs. Springer-Verlag, Berlin Heidelberg New York (1996)
11. Robertazzi, T.: Ten reasons to use divisible load theory. IEEE Computer **36** (2003) 63-68
12. Wolniewicz, P.: Divisible Job Scheduling in Systems with Limited Memory. PhD Thesis, Poznań Univ. of Technology (2003). Downloadable from `http://www.man.poznan.pl/~pawelw/phd.pdf`
13. Yang, Y., Casanova, H.: Multi-Round Algorithm for Scheduling Divisible Workload Applications: Analysis and Experimental Evaluation. Univ. of California, San Diego, Dept. of Computer Science and Engineering, Tech. Rep. CS2002-0721 (2002)

A Scalable Parallel Graph Coloring Algorithm for Distributed Memory Computers

Erik G. Boman[1], Doruk Bozdağ[2], Umit Catalyurek[2,*], Assefaw H. Gebremedhin[3,**], and Fredrik Manne[4]

[1] Sandia*** National Laboratories, USA
`egboman@sandia.gov`
[2] Ohio State University, USA
`bozdagd@ece.osu.edu, umit@bmi.osu.edu`
[3] Old Dominion University, USA
`assefaw@cs.odu.edu`
[4] University of Bergen, Norway
`Fredrik.Manne@ii.uib.no`

Abstract. In large-scale parallel applications a graph coloring is often carried out to schedule computational tasks. In this paper, we describe a new distributed-memory algorithm for doing the coloring itself in parallel. The algorithm operates in an iterative fashion; in each round vertices are speculatively colored based on limited information, and then a set of incorrectly colored vertices, to be recolored in the next round, is identified. Parallel speedup is achieved in part by reducing the frequency of communication among processors. Experimental results on a PC cluster using up to 16 processors show that the algorithm is scalable.

1 Introduction

In many parallel scientific computing applications computational dependencies are modeled using a graph, and a coloring of the vertices of the graph is used as a subroutine to identify independent tasks that can be performed concurrently. See [8] and the references therein for examples. In such cases, the computational graph is often distributed among the processors, and hence the coloring itself needs to be performed in parallel. For these applications, fast greedy coloring algorithms that work well in practice are often preferred over slower local improvement heuristics that might use fewer colors.

This paper deals with the parallelization of such fast greedy coloring algorithms and presents an efficient parallel coloring algorithmic scheme designed for distributed memory parallel computers. Several variations of the basic scheme are discussed. Our algorithms are implemented using MPI and experiments conducted on a 16-node PC cluster using several large graphs indicate that our approach is scalable.

* This research was supported in part by Sandia National Laboratories under Doc.No: 283793, Ohio Supercomputing Center #PAS0052.

** Supported by the U.S. National Science Foundation grant ACI 0203722.

*** Sandia is a multiprogram laboratory operated by Sandia Corporation, a Lockheed Martin company, for the U.S. Department of Energy's National Nuclear Security Administration under contract DE-AC04-94AL85000.

J.C. Cunha and P.D. Medeiros (Eds.): Euro-Par 2005, LNCS 3648, pp. 241–251, 2005.

The basic idea in the algorithm is to partition the graph among the available processors and let each processor be responsible for the coloring of the vertices assigned to it. Every processor colors its local vertices in steps of s vertices at a time in a sequential fashion. Between each step the processors exchange recent color information. Since a processor colors its local vertices with incomplete color information, conflicts may arise, and these are detected in a separate phase. The algorithm proceeds iteratively by recoloring vertices involved in conflicts. With an appropriate choice of a value for s, the number of ensuing conflicts can be kept low while at the same time preventing the runtime from being dominated by the sending of a large number of small messages.

2   Previous Work

A *coloring* of a graph is an assignment of positive integers (called colors) to its vertices such that no two adjacent vertices receive the same color. Finding a coloring of a general graph that minimizes the number of colors used is an NP-hard problem [6]. Moreover, the problem is difficult to approximate [4]. In practice, however, greedy sequential coloring heuristics have been found to be quite effective [3]. These greedy heuristics are inherently sequential and hence difficult to parallelize.

A number of previously suggested parallel graph coloring algorithms rely on various ways of computing an *independent set* in parallel. A characteristic feature of independent set based parallel coloring algorithms is that a vertex is assigned a color that is never changed at a later point in the algorithm. In such algorithms, while coloring a vertex v, the colors of already colored neighbors of v must be known, and none of the uncolored neighbors of v can be colored at the same time as v. The works of Jones and Plassmann [11], Gjertsen et al. [9], and Allwright et al. [1] are examples of such approaches. All of these algorithms are designed for distributed memory parallel computers and rely on partitioning a graph into the same number of components as there are processors. Each component, including information about its inter- and intra-component edges, is assigned to and colored by one processor.

To overcome the restriction that two adjacent vertices on different processors cannot be colored at the same time, Johansson [10] proposed a distributed algorithm where each processor is assigned exactly one vertex. The vertices are then colored simultaneously by randomly choosing a color from the interval $[1, \Delta + 1]$, where Δ is the maximum vertex degree in the graph. This may lead to an inconsistent coloring, and hence the process needs to be repeated recursively for the vertices that did not receive permissible colors. Finocchi et al. [5] performed extensive sequential simulations of a variant of Johansson's algorithm where the upper-bound on the range of permissible colors is initially set to be smaller than $\Delta + 1$ and then increases only when needed.

Gebremedhin and Manne [8] developed a parallel graph coloring algorithm suitable for shared memory computers. In this algorithm, each processor is assigned equally many vertices to color. A processor colors its vertices in a sequential fashion, at each step assigning a vertex the smallest color not used by any of its neighbors (both on- or off-processor). An inconsistent coloring arises only when a pair of adjacent vertices that reside on different processors is colored simultaneously. Inconsistencies are then detected in a subsequent phase and resolved in a final sequential phase.

3   A New Algorithm

Here we describe a new distributed-memory parallel graph coloring algorithm. In the spirit of the BSP model [2], the algorithm is organized as a sequence of *supersteps*. A superstep has distinct, rather than intermingled, computation and communication phases.

A partitioning of the graph among the processors classifies the vertices into *interior* and *boundary* vertices. An interior vertex is a vertex all of whose neighbors are located on the same processor as itself. A boundary vertex has at least one neighbor located on a different processor. Clearly, the subgraphs induced by interior vertices are independent of each other and hence can be colored concurrently trivially. Coloring the remainder of the graph in parallel requires communication and coordination among the processors and this is the main issue in the algorithm being described.

3.1   The Basic Scheme

At the highest level, our algorithm is iterative—it operates in *rounds*. In each round there are two phases, a *tentative coloring* and a *conflict detection* phase. The former is organized into supersteps while the latter is not, since no communication is required. In every superstep each processor colors s vertices in a sequential manner, where s is an input parameter to the algorithm, using color information available at the beginning of the superstep, and then exchanges recent color information with other processors. In particular, in the communication phase of a superstep, a processor sends the colors of its boundary vertices to other processors and receives relevant color information from other processors. In this scenario, if two adjacent vertices located on two different processors are colored during the same superstep, they may receive the same color and hence cause a conflict. The purpose of the second phase of a round is to detect such conflicts and accumulate a list of vertices on each processor to be recolored in the next round. Since it is not necessary to recolor both endpoints of a conflict edge only one of the involved processors will add a vertex to its list. The processor that will do the recoloring is determined in a random fashion in order to achieve an even distribution of the vertices to be colored in the next round.

The conflict detection phase does not require communication since every processor has acquired a complete knowledge of the colors of the neighbors of its vertices at the end of the tentative coloring phase. The algorithm terminates when there is no more processor with a nonempty list of vertices to be recolored. Algorithm 1 outlines this scheme in more detail.

The rationale for dividing the coloring phase of a round in supersteps, rather than communicating after a single vertex is colored, is to reduce communication frequency and thereby reduce communication time. However the number of supersteps used (equivalently, the number of vertices colored in a superstep) is also closely related to the likelihood of conflicts and consequently the number of rounds. The lower the number of supersteps (the higher the number of vertices colored per superstep) the higher the likelihood of conflicts and hence the higher the number of rounds required. Choosing a value for s that minimizes the overall runtime is therefore a compromise between these two contradicting requirements. An optimal value of s would depend on such factors

---

Algorithm 1 An iterative parallel graph coloring algorithm

1: **procedure** PARALLELCOLORING($G = (V, E), s$)
2:     *Initial data distribution*: V is partitioned into p subsets $V_1, \ldots, V_p$; processor P_i owns V_i, stores edges E_i incident on V_i, and stores the identity of processors hosting the other endpoints of E_i.
3:     **on each** processor $P_i, i \in P = \{1, \ldots, p\}$
4:         $U_i \leftarrow V_i$ ▷ U_i is the current set of vertices to be colored
5:         **while** $\exists j \in P, U_j \neq \emptyset$ **do**
6:             **if** $U_i \neq \emptyset$ **then**
7:                 Partition U_i into ℓ_i subsets $U_{i,1}, U_{i,2}, \ldots, U_{i,\ell_i}$, each of size s
8:                 **for** $k \leftarrow 1$ **to** ℓ_i **do**   ▷ each k corresponds to a superstep
9:                     **for each** $v \in U_{i,k}$ **do**
10:                        assign v a permissible color
11:                    Send colors of boundary vertices in $U_{i,k}$ to relevant processors
12:                    Receive color information from other processors
13:                Wait until all incoming messages are successfully received
14:                $R_i \leftarrow \emptyset$ ▷ R_i is a set of vertices to be recolored
15:                **for each** boundary vertex $v \in U_i$ **do**
16:                    **if** $\exists (v, w) \in E$ s.t. $color(v) = color(w)$ and $r(v) \leq r(w)$ **then**
17:                        $R_i \leftarrow R_i \cup \{v\}$ ▷ $r(v)$ is a random number
18:            $U_i \leftarrow R_i$
19:

---

as the size and density of the input graph, the number of processors available, and the machine architecture and network.

Note that the formulation of Algorithm 1 is general enough to encompass the algorithms of Johannsson [10], Finocchi et al. [5], and Gebremedhin and Manne [8]. Setting $p = n$ (and $s = 1$) and choosing the color of a vertex in Line 10 appropriately, gives the algorithms of Johannsson and Finocchi et al. Setting $s = 1$, restricting Algorithm 1 to *one* round, and resolving conflicts sequentially gives the algorithm of Gebremedhin and Manne.

3.2   Variations

For the sake of generality, Algorithm 1 leaves several issues unspecified. In the sequel, we discuss such issues, in each case pointing out available alternatives.

(i) Initial partitioning. In a parallel application, the graph is usually already distributed among the processors in a reasonable way. However, if this is not the case, a "good" data distribution needs to be computed. The number of conflicts in the algorithm depends on several factors including the number of boundary vertices and the number of edges between these. Thus using a graph partitioner such as Metis [12] should help reduce the number of conflicts as well as the amount of communication.

(ii) Distinguishing between interior and boundary vertices. As mentioned earlier, the subgraphs induced by interior vertices are independent of each other and can therefore

be colored concurrently without any communication. Hence, in the context of Algorithm 1, the interior vertices can be colored *before*, *after*, or *interleaved* with boundary vertices. Algorithm 1 is presented assuming the last option. Coloring the interior vertices first may produce fewer conflicts when using a regular *First-Fit* coloring scheme, since the subsequent coloring of boundary vertices is performed with a larger spectrum of available colors. Coloring boundary vertices first may be advantageous with color selection variants such as Staggered First-Fit (see the discussion later in this section).

(iii) Synchronous vs. asynchronous supersteps. In Algorithm 1, the supersteps can be made to run in a *synchronous* fashion by introducing explicit synchronization barriers at the end of each superstep. An advantage of this mode is that in the conflict detection phase, the color of a boundary vertex needs to be checked only against its neighbors colored at the same superstep. The obvious disadvantage is that the barriers, in addition to the associated overhead, cause some processors to be idle while others complete their supersteps. Alternatively, the supersteps can be made to run *asynchronously*, without explicit barriers at the end of each superstep. Each processor would then only process and use the color information that has been completely received when it is checking for incoming messages. Any color information that has not reached a processor at this stage would thus be delayed from being used until a later superstep. Due to this, in the conflict detection phase, the color of a boundary vertex needs to be checked against all of its off-processor neighbors. Also, it is possible that the asynchronous version results in more conflicts than the synchronous one since a superstep on one processor now can overlap with more than one superstep on another processor.

(iv) Choice of color. The choice of a permissible color in Line 10 of Algorithm 1 can be made in different ways. The strategy employed affects (1) the number of colors used by the algorithm, and (2) the likelihood of conflicts, and thus the number of rounds required by the algorithm. Both of these quantities are desired to be as small as possible, and a coloring strategy typically reduces one of the quantities at the expense of the other. Here, we present two strategies: *First-Fit* (FF) and *Staggered First-Fit* (SFF). In FF each processor chooses the *smallest* permissible color from the interval $[1, C]$, where C is the current largest color used. If no such color exists, the new color $C + 1$ is chosen. SFF uses an initial estimate K of the number of colors needed for the input graph. Processor P_i chooses the *smallest* permissible color from the interval $[\lceil \frac{iK}{p} \rceil, K]$. If no such color exists, then the smallest permissible color in $[1, \lfloor \frac{iK}{p} \rfloor]$ is chosen. If there is still no such color, the smallest permissible color greater than K is chosen. Unlike FF, the search for a color in SFF starts from different "base colors" for each processor. Hence the latter is likely to result in fewer conflicts than the former. Other color selection strategies that have been suggested include the *randomized* techniques of Gebremedhin et al. [7] and Finocchi et al. [5].

4   Experiments

In this section, we present results from experiments carried out on a 16-node PC cluster equipped with dual 900 MHz Intel Itanium 2 CPUs and 4 GB memory. The nodes of

246   Erik G. Boman et al.

the cluster are interconnected via switched Myrinet 2000 network. Our test set consists of 19 graphs obtained from molecular dynamics and finite element applications [8, 13]. Table 1 displays the structural properties of the test graphs, including maximum, minimum, and average degree. The table also displays the number of colors and the runtime in seconds used by a sequential FF algorithm when run on a single node of our test platform. All of the results presented in this section are average performance results over all of the graphs presented in Table 1. Each individual test is an average of 5 runs. In the timing of the parallel coloring code, we assume the graph to be initially partitioned and distributed among the nodes of the parallel machine. Hence, the times reported concern only coloring.

Table 1. Properties of the test graphs

name	$|V|$	$|E|$	Degree			Seq. First-Fit	
			max	min	avg	#colors	time
HIV-2	11,414	15,270	8	1	2.68	5	0.007
HIV-4	11,414	130,332	39	6	22.84	17	0.034
HIV-6	11,414	412,623	116	13	72.30	45	0.099
HIV-10	11,414	1,655,383	454	35	290.06	176	0.387
popc-br-2	24,916	31,449	7	1	2.52	5	0.032
popc-br-4	24,916	255,047	43	2	20.47	21	0.067
popc-br-6	24,916	850,043	125	2	68.23	49	0.206
popc-br-10	24,916	3,587,724	514	2	287.98	173	0.84
er-gre-2	36,573	53,046	8	0	2.90	5	0.022
er-gre-4	36,573	451,355	42	3	24.68	19	0.116
er-gre-6	36,573	1,482,904	116	11	81.09	47	0.357
er-gre-10	36,573	6,511,122	460	79	356.06	174	1.515
apoa1-2	92,224	139,351	8	1	3.02	5	0.057
apoa1-4	92,224	1,131,436	43	2	24.54	20	0.293
apoa1-6	92,224	3,864,429	123	13	83.81	49	0.928
apoa1-10	92,224	17,100,850	503	54	370.85	182	3.993
598a	110,971	741,934	26	5	13.37	12	0.310
144	144,649	1,074,393	26	4	14.86	11	0.219
auto	448,695	3,314,611	37	4	14.77	13	0.984

In our experiments, we considered two ways of partitioning the vertices of a graph. In the first case, the vertex set, with the vertices in their natural order (i.e. the order in which the graphs were supplied), is partitioned into p contiguous blocks of (almost) equal size. Such a *block* partitioning does not attempt to minimize cross-edges, though the structure of the natural order is exploited. In the second case, the vertex set is partitioned into p disjoint subsets of nearly equal size such that the number of cross-edges is small. For this we used the graph partitioning software Metis [12], with an option known as VMetis that also attempts to minimize the communication volume and the number of boundary vertices.

The first set of experiments, shown in Figures 1 and 2, are conducted to assess the effects of the following three issues: block partitioning using the natural order (N) vs. partitioning using VMetis (V); coloring interior vertices first (I), boundary vertices first (B), or interleaved (U); and using synchronous (S) vs. asynchronous supersteps (A). In all of these experiments we use FF for selecting the color of a vertex. A 3-letter acronym reflecting the options discussed above is used in Figures 1 and 2.

Figure 1 displays the number of conflicts (normalized with respect to the total number of vertices) for the parallel coloring algorithm for different combinations of these options while varying the superstep size and the number of processors. In Figure 1(a), we show results for the case where the number of processors is 8. Similar trends were observed for other number of processors. When varying the number of processors, the superstep size is set to 100. In the interleaved mode the superstep size gives the number of boundary vertices colored in each superstep.

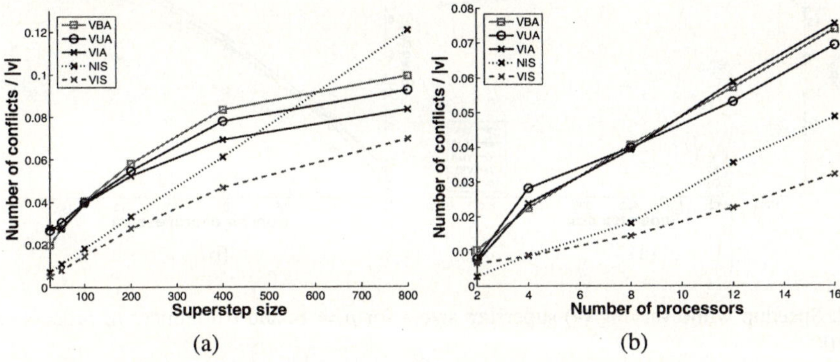

Fig. 1. Number of conflicts while varying (a) superstep size s for $p = 8$, and (b) number of processors for $s = 100$

Figure 1(a) and 1(b) shows that for all configurations the number of conflicts increases as the superstep size and the number of processors, respectively, increases. The two figures also show that asynchronous supersteps result in more conflicts than synchronous supersteps, and that graph partitioning using Metis results in fewer conflicts than block partitioning. In the case where block partitioning is used, only the combination of options (NIS) that gave the fewest conflicts is shown. When using Metis with synchronous supersteps we also only show the configuration (VIS) that gave the least number of conflicts. Using the boundary first and unordered options gave only slightly worse results than the presented ones. In terms of the number of conflicts, the results in Figures 1(a) and 1(b) suggest that the best result is obtained by partitioning the graph using Metis and using a small superstep size while running supersteps synchronously.

As can be observed from the figure in the asynchronous case, the order in which the boundary and interior vertices are colored has no major impact on the number of conflicts.

In all of our experiments, the number of rounds the algorithm has to iterate was observed to be consistently low, varying between two and five, for every configuration we tried. This is a consequence of the fact that the number of initial conflicts is small and then drops rapidly between successive rounds. As long as Metis is used the total number of conflicts is within 10% of the total number of vertices in all of the configurations considered. Thus more than 90% of the sequential work is performed in the first round. This indicates that the increase in the number of vertices that need to be colored when going from a sequential to a parallel algorithm is fairly low for the test set we use. We also note that the number of colors used stays fairly low in all of our experiments and on the average, it does not increase by more than 4% of that used by the (sequential) FF coloring scheme.

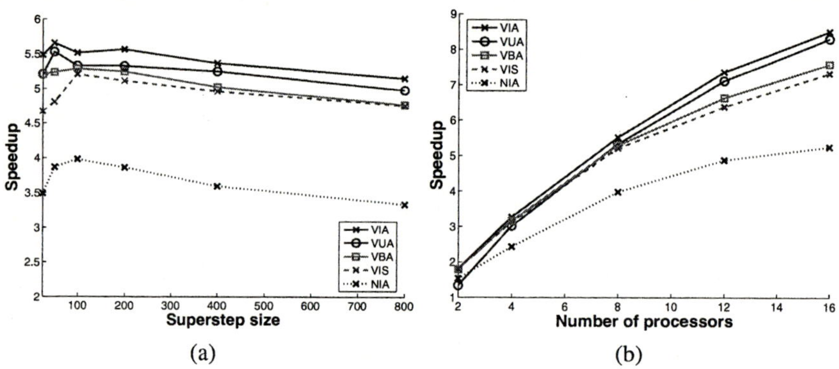

(a)   (b)

Fig. 2. Speedup while varying (a) superstep size s for $p = 8$, and (b) number of processors for $s = 100$

Figure 2(a) displays speedup values for the several variations of the parallel coloring algorithm while varying the superstep size s for a fixed number of processors $p = 8$. We show the NIA configuration (as opposed to NIS in Figure 1) as it gave the best speedup when not using Metis to partition the graph. As can be seen from the figure, the optimum value for s is close to 100 for all variants. Thus using $s = 100$ seems to be a good compromise between balancing the conflicting issues of increased message startup costs versus the number of conflicts. However, the manner in which the algorithm is configured seems to be more important than the superstep size. It is always better to use asynchronous communication than synchronous. Also, as can be seen from the figure coloring interior vertices first is slightly better than coloring the vertices interleaved which again is better than coloring the boundary vertices first.

In Figure 2(b) the speedup obtained as the number of processors is varied while using a superstep size of 100 is shown. The trends observed in Figure 2(b) are similar to those in Figure 2(a). The best *average* speedup, over all test cases, was about 8.5 while using 16 processors. However, for particular test cases, we have observed a speedup value as high as 12.5 while using 16 processors. The worst result observed was a speedup of 3.2 on 16 processors although this was a clear outlier. "Medium" dense graphs tend to give better speedup values than very sparse or very dense graphs.

Our next set of experiments concerns the different coloring schemes as discussed in Section 3. The results are shown in Figure 3(a) (conflicts), and Figure 3(b) (speedup).

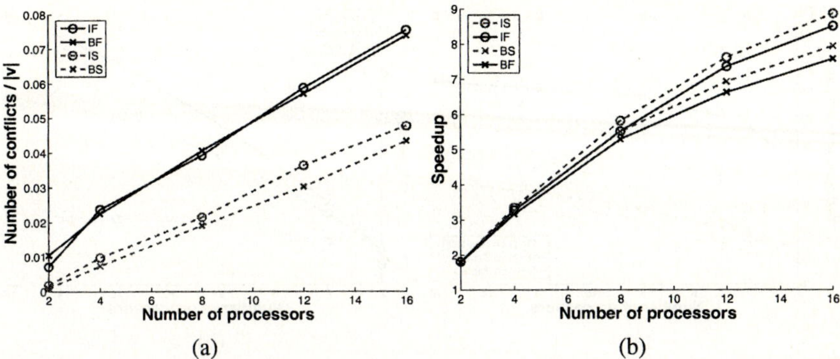

Fig. 3. Effect of the color selection algorithm on (a) the number of conflicts, and (b) speedup while using a superstep length of $s = 100$

In the figures the labels I and B show whether the interior vertices or the boundary vertices are colored first, while the second letter correspond to the FF (F) and the SFF (S) color selection scheme. In all of these experiments Metis is used for partitioning and the communication is done asynchronously. For SFF we use the number of colors found by sequential FF as our initial estimate of the number of colors. Coloring the vertices in an interleaved fashion gave similar results as those in the figures and are not shown here.

As expected, the SFF scheme gives fewer conflicts than the FF scheme. But as can be seen from Figure 3(b) in terms of speedup this is offset by the higher overhead associated with determining the correct color in the SFF scheme. Also, the SFF scheme has the disadvantage of requiring an a priori estimate on the expected number of colors.

The speedup achieved by our approach stems from two sources: partitioning and the "core" algorithm. Partitioning using Metis makes a trivial parallelization of the coloring of interior vertices possible. The "core" algorithm is a nontrivial way of coloring the boundary vertices in parallel. Figure 4(a) shows the percentage of boundary vertices for the graphs in Table 1 when using block partitioning with the natural vertex ordering, and when using Metis. As one can see the number of boundary vertices increases with the number of processors being used. Thus it is difficult to measure the particular speedup from coloring just the boundary vertices since the amount of work performed changes with the number of processors. In order to give some indication of the performance of the algorithm on the boundary vertices we present Figure 4(b). This shows the speedup when coloring three random graphs each containing 32000 vertices and with average vertex degrees 3, 20, and 70 respectively. For these experiments we used the NIA configuration with the vertices colored according to the SFF scheme. Since the vertices are ordered according to their natural order almost all the vertices become boundary vertices (see the topmost curve in Figure 4(a)). Thus this can be viewed as applying more processors while keeping the number of boundary vertices fixed. Since we are in effect traversing the graph at least twice (for coloring and verification) we cannot expect to get a speedup of more than $p/2$. Based on this the observed maximum speedup of more than 6 when using 16 processors is quite good.

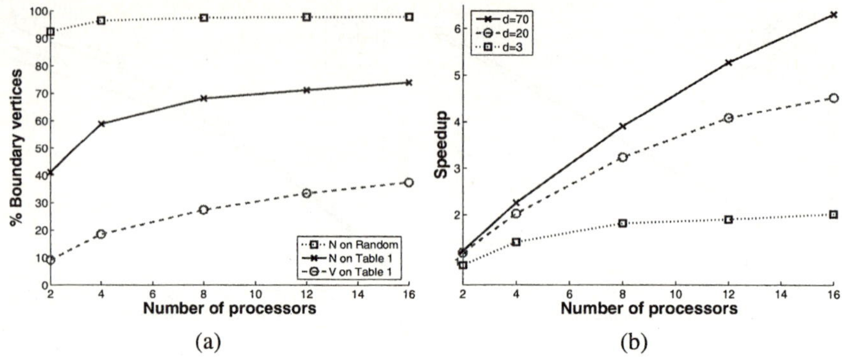

Fig. 4. (a) Percentage of boundary vertices for graphs in Table 1 (N = natural ordering, V = ordering given by Metis), and random graphs. (b) Speedup for random graphs of various average degrees

5   Conclusion

We have developed an efficient and truly scalable parallel graph coloring algorithm suitable for a distributed memory computer. The algorithm is flexible and can easily be tuned to suit the nature of the graph to be colored and the specifics of the hardware being used. The scalability of the algorithm has been experimentally demonstrated. This should be seen in light of the fact that previous distributed-memory parallel coloring algorithms, such as the algorithm of Jones and Plassmann [11], did not give any speedup when coloring the boundary vertices as more processors are applied.

Even though our main objective has been to achieve parallel speedup, being able to perform coloring in a distributed setting where the graph is already partitioned among the processors is an important functionality in itself.

In the future we plan to experiment with more sophisticated color selection schemes that may further reduce the number of conflicts. We are also considering how to generalize the algorithm to other coloring problems such as distance-2 graph coloring and hypergraph coloring, both of which have important applications in scientific computing.

References

1. J.R. Allwright, R. Bordawekar, P. D. Coddington, K. Dincer, and C.L. Martin. A comparison of parallel graph coloring algorithms. Technical Report NPAC technical report SCCS-666, Northeast Parallel Architectures Center at Syracuse University, 1994.
2. Rob H. Bisseling. *Parallel Scientific Computation: A Structured Approach Using BSP and MPI*. Oxford, 2004.
3. T. F. Coleman and J. J More. Estimation of sparse jacobian matrices and graph coloring problems. *SIAM J. Numer. Anal.*, 1(20):187–209, 1983.
4. Pierluigi Crescenzi and Viggo Kann. A compendium of NP optimization problems. http://www.nada.kth.se/~viggo/wwwcompendium/.

5. Irene Finocchi, Alessandro Panconesi, and Riccardo Silvestri. Experimental analysis of simple, distributed vertex coloring algorithms. In *Proc. 13th ACM-SIAM symposium on Discrete Algorithms (SODA 02)*, 2002.
6. M. R. Garey and D. S. Johnson. *Computers and Intractability*. Freeman, 1979.
7. Assefaw Gebremedhin, Fredrik Manne, and Alex Pothen. Parallel distance-k coloring algorithms for numerical optimization. In *proceedings of Euro-Par 2002*, volume 2400, pages 912–921. Lecture Notes in Computer Science, Springer, 2002.
8. Assefaw Hadish Gebremedhin and Fredrik Manne. Scalable parallel graph coloring algorithms. *Concurrency: Practice and Experience*, 12:1131–1146, 2000.
9. Robert K. Gjertsen Jr., Mark T. Jones, and Paul Plassmann. Parallel heuristics for improved, balanced graph colorings. *J. Par. and Dist. Comput.*, 37:171–186, 1996.
10. Öjvind Johansson. Simple distributed $\delta + 1$-coloring of graphs. *Information Processing Letters*, 70:229–232, 1999.
11. Mark T. Jones and Paul Plassmann. A parallel graph coloring heuristic. *SIAM J. Sci. Comput.*, 14(3):654–669, 1993.
12. G. Karypis and V. Kumar. A fast and high quality multilevel scheme for partitioning irregular graphs. *SIAM J. Sci. Comput.*, 20(1), 1999.
13. Michelle Mills Strout and Paul D. Hovland. Metrics and models for reordering transformations. In *Proceedings of the The Second ACM SIGPLAN Workshop on Memory System Performance (MSP)*, pages 23–34, June 8 2004.

Complexity and Approximation for the Precedence Constrained Scheduling Problem with Large Communication Delays

R. Giroudeau, J.C. König, F.K. Moulaï, and J. Palaysi

LIRMM, 161 rue Ada, 34392 Montpellier Cedex 5, France, UMR 5056

Abstract. We investigate the problem of minimizing the makespan for the multiprocessor scheduling problem. We show that there is no hope of finding a ρ-approximation with $\rho < 1 + 1/(c+4)$ (unless $\mathcal{P} = \mathcal{NP}$) for the case where all the tasks of the precedence graph have unit execution times, where the multiprocessor is composed of an unrestricted number of machines, and where c denotes the communication delay between two tasks i and j submitted to a precedence constraint and to be processed by two different machines. The problem becomes polynomial whenever the makespan is at the most $(c + 1)$. The $(c + 2)$ case is still partially opened.

1 Introduction

Scheduling theory is concerned with the *optimal allocation of scarce resources to activities over time*. The *theory* of the design of algorithms for scheduling is younger, but still has a significant history.

In this article we adopt the *classical scheduling delay model* or *homogeneous model* in which an instance of a scheduling problem is specified by a set $J = \{j_1, \ldots, j_n\}$ of n nonpreemptive tasks, a set of U of q precedence constraints (j_i, j_k) such that $G = (J, U)$ is a directed acyclic graphs (dag), the processing times $p_i, \forall j_i \in J$, and the communication times $c_{ik}, \ \forall (j_i, j_k) \in U$.

If the task j_i starts its execution at time t on processor π, and if task j_k is a successor of j_i in the dag, then either j_k starts its execution after the time $t + p_{j_i}$ on processor π, or after time $t + p_{j_k} + c_{j_i j_k}$ on some other processor. In the following we consider the case of $\forall j_k \in J, \ p_{j_k} = 1$ and $\forall (j_i, j_k) \in E, \ c_{j_i j_k} = c \geq 2$.

This model was first introduced by Rayward-Smith [13]. In this model we have a set of identical processors that are able to communicate in a uniform way. We want to use these processors in order to process a set of tasks that are subject to precedence constraints. The problem is to find a trade-off between the two extreme solutions, namely, execute all the tasks sequentially without communication, or try to use all the potential parallelism but at the cost of an increased communication overhead. This model has been extensively studies these last years both from the complexity and the (non)-approximability points of view [2].

J.C. Cunha and P.D. Medeiros (Eds.): Euro-Par 2005, LNCS 3648, pp. 252–261, 2005.

Using the *three fields* notation scheme proposed by Graham et al. [6], the problem is denoted as $\bar{P}|prec, c_{ij} = c \geq 2; p_i = 1|C_{max}$ *i.e.* we have an unbounded number of identical processors in order to schedule a dag such that each task has the same execution time and each pair of tasks have the same communication time. The aim is to minimize the length of the schedule.

1.1 Complexity Results

The problems with unitary communication delay. If we consider the problem of scheduling a precedence graph with unitary communication delays and unit execution time (UET-UCT) on an unbounded number of processors, Hoogeveen et al. [7] proved that the decision problem associated to $\bar{P}|prec; c_{ij} = 1; p_i = 1|C_{\max}$ becomes $\mathcal{NP}$-complete even for $C_{\max} \geq 6$, and that it is polynomial for $C_{\max} \leq 5$. Their proof is based on a reduction from the $\mathcal{NP}$-complete problem $3SAT$ [3]. The $\mathcal{NP}$-completeness result for $C_{\max} = 6$ implies that there is no polynomial time approximation algorithm with ratio guarantee better than 7/6, unless $\mathcal{P} = \mathcal{NP}$.

Moreover, in the presence of a bounded number of processors, Hoogeveen et al. [7] establish that whether an instance of $P|prec; c_{ij} = 1; p_i = 1|C_{max}$ has a schedule of length of at the most 4 is $\mathcal{NP}$-complete (they use a reduction from the $\mathcal{NP}$-complete problem *Clique*), whereas Picouleau [11] develops a polynomial time algorithm for the $C_{\max} = 3$. In the same way, the $\mathcal{NP}$-completeness result for $C_{\max} = 4$ implies that there is no polynomial time approximation algorithm with ratio guarantee better than 5/4, unless $\mathcal{P} = \mathcal{NP}$.

The problems with large communication delay. If we consider the problem of scheduling a precedence graph with large communication delays and unit execution time (UET-LCT), on bounded number of processors, Bampis et al. in [1] proved that the decision problem denoted by $P|prec; c_{ij} = c \geq 2; p_i = 1|C_{max}$ for $C_{\max} = c + 3$ is $\mathcal{NP}$-complete problem, and for $C_{\max} = c + 2$ (for the special case $c = 2$), they develop a polynomial time algorithm. Their proof is based on a reduction from the $\mathcal{NP}$-complete problem *Balanced Bipartite Complete Graph, BBCG* [3]. Thus, Bampis et al. [1] proved that the $P|prec; c_{ij} = c \geq 2; p_i = 1|C_{max}$ problem does not possess a polynomial time approximation algorithm with ratio guarantee better than $(1 + \frac{1}{c+3})$, unless $\mathcal{P} = \mathcal{NP}$.

Remark: Notice that in the case of an unbounded number of processors $(\bar{P}|prec; c_{ij} = c \geq 2; p_i = 1|C_{max})$, the complexity to an associated decision problem is unknown.

1.2 Approximation Results

The problems with unitary communication delay. The best known approximation algorithm for $\bar{P}|prec; c_{ij} = 1; p_i = 1|C_{max}$ is due to Munier and König [10]. They presented a (4/3)-approximation algorithm for this problem, which is based on an integer linear programming formulation. The algorithm is based on the following procedure: an integrity constraint is relaxed, and feasible schedule is produced by rounding.

Munier and Hanen [9] proposed a $(\frac{7}{3} - \frac{4}{3m})$-approximation algorithm for the problem $P|prec; c_{ij} = 1; p_i = 1|C_{max}$. They define and study a new list scheduling approximation algorithm based on the solution given on an unrestricted number of processors. They introduce the notion of *favourite successor* in order to define priorities between conflicting successors of a task. Note that, if we consider large communication delays, there is no ρ-polynomial time approximation algorithm known, except the trivial bound $(c+1)$, one whose first step consists in executing the tasks and second step in initiating communication phasis and so on . . .

Concerning the case of a restricted number of processors, an only (as known) constant 2-approximation algorithm is given by Munier [8], for the special case where the precedence graph is tree in presence of large communication delays.

The problems with large communication delay. Contrary to the complexity results, as we know, an unique approximation algorithm is given by Rapine [12]. The author gives the lower bound $O(c)$ for the list scheduling in presence of large communication delays.

1.3   Presentation of the Paper

The challenge is to determinate a threshold for approximation algorithm for the problem $\bar{P}|prec; c_{ij} = c \geq 2; p_i = 1|C_{max}$, to develop a non trivial approximation algorithm, and to improve, in the presence of a restricted number of processors, the bound given by Rapine [12].

This article is organized as follows: in the second section, we give a preliminary result. In the third section, we give the non-approximability result for the scheduling problem with the objective function of minimizing the length of the schedule. In the last section, we develop a $\frac{2(c+1)}{3}$-approximation algorithm based on the notion of expansion of the makespan of a good feasible schedule.

2   Preliminary Result

In this part, we will define a variant of SAT problem [3], denoted in the following by Π_1. The $\mathcal{NP}$-completeness of the scheduling problem $\bar{P}|prec; c_{ij} = c \geq 2; p_i = 1|C_{max}$ (see section 3), is based on a reduction from this problem.

The problem Π_1 is a variant of the well known SAT problem [3]. We will call this variant the $One\text{-}in\text{-}(2,3)SAT(2,\bar{1})$ problem. We denote by $\mathcal{V}$, the set of variables. Let n be a multiple of 3 and let $\mathcal{C}$ be a set of clauses of cardinality 2 or 3. There are n clauses of cardinality 2 and $n/3$ clauses of cardinality 3 so that:

- each clause of cardinality 2 is equal to $(x \vee \bar{y})$ for some x, $y \in \mathcal{V}$ with $x \neq y$.
- each of the n literals x (resp. of the literals $\bar{x}$) for $x \in \mathcal{V}$ belongs to one of the n clauses of cardinality 2, thus to only one of them.

- each of the n literals x belongs to one of the $n/3$ clauses of cardinality 3, thus to only one of them.
- whenever $(x \vee \bar{y})$ is a clause of cardinality 2 for some x, $y \in \mathcal{V}$, then x and y belong to different clauses of cardinality 3.

Question: Is there a truth assignment $I : \mathcal{V} \to \{0, 1\}$ such that every clause in $\mathcal{C}$ has exactly a true literal?

Example The following logic formula is a valid instance of Π_1:

$(x_0 \vee x_1 \vee x_2) \wedge (x_3 \vee x_4 \vee x_5) \wedge (\bar{x}_0 \vee x_3) \wedge (\bar{x}_3 \vee x_0) \wedge (\bar{x}_4 \vee x_2) \wedge (\bar{x}_1 \vee x_4) \wedge (\bar{x}_5 \vee x_1) \wedge (\bar{x}_2 \vee x_5)$.

The answer to Π_1 is *yes*. It suffices to choose $x_0 = 1$, $x_3 = 1$ and $x_i = 0$ for $i = \{1, 2, 4, 5\}$. This yields a truth assignment satisfying the formula, and there is exactly one true literal in every clause. For the proof of the $\mathcal{NP}$-completeness see [4].

3   Non-approximability Results

In this section, we show in the first part, that the problem denoted by $\bar{P}|prec; c_{ij} = c \geq 3; p_i = 1|C_{max}$ cannot be approximated by a polynomial time approximation algorithm with ratio guarantee better than $1 + \frac{1}{c+4}$ for the minimization of the length of the schedule.

3.1   The Minimization of Length of the Schedule

Theorem 1. *The problem of deciding whether an instance of $\bar{P}|prec; c_{ij} = c; p_i = 1|C_{max}$ has a schedule of length at most $(c + 4)$ is $\mathcal{NP}$-complete with $c \geq 3$.*

Proof. It is easy to see that $\bar{P}|prec; c_{ij} = c; p_i = 1|C_{max} = c + 4 \in \mathcal{NP}$.

Our proof is based on a reduction from Π_1. Given an instance π^* of Π_1, we construct an instance π of the problem $\bar{P}|prec; c_{ij} = c; p_i = 1|C_{max} = c + 4$, in the following way:

Remark: n designs the number of variables of π^*.

1. For all $x \in \mathcal{V}$, we introduce $(c + 6)$ variables-tasks: $\alpha_{x'\bar{x}'}$, x', $\bar{x}'$, $\hat{x}'$, β_j^x with $j \in \{1, 2, \ldots, c + 2\}$. We add the precedence constraints: $\alpha_{x'\bar{x}'} \to x'$, $\alpha_{x'\bar{x}'} \to \bar{x}'$, $\beta_1^x \to \hat{x}'$, $\beta_1^x \to \bar{x}'$, $\beta_j^x \to \beta_{j+1}^x$ with $j \in \{1, 2, \ldots, c + 1\}$.
2. For all clauses of length three denoted by $C_i = (y \vee z \vee t)$, we introduce $2 \times (2 + c)$ clauses-tasks C_j^i and A_j^i, $j \in \{1, 2, \ldots c + 2\}$, with precedence constraints: $C_j^i \to C_{j+1}^i$ and $A_j^i \to A_{j+1}^i$, $j \in \{1, 2, \ldots, c + 1\}$. We add the constraints $C_1^i \to l$ with $l \in \{y', z', t'\}$ and $l \to A_{c+2}^i$ with $l \in \{\hat{y}', \hat{z}', \hat{t}'\}$.
3. For all clauses of length two denoted by $C_i = (x \vee \bar{y})$, we introduce $2(c + 3)$ clauses-tasks D_j^i (resp. $D_j'^i$), $j \in \{1, 2, \ldots, c + 3\}$ with precedence constraints: $D_j^i \to D_{j+1}^i$ (resp. $D_j'^i \to D_{j+1}'^i$) with $j \in \{1, 2, \ldots, c + 2\}$ and $x' \to D_{c+3}^i$ (resp. $\bar{y}' \to D_{c+3}'^i$).

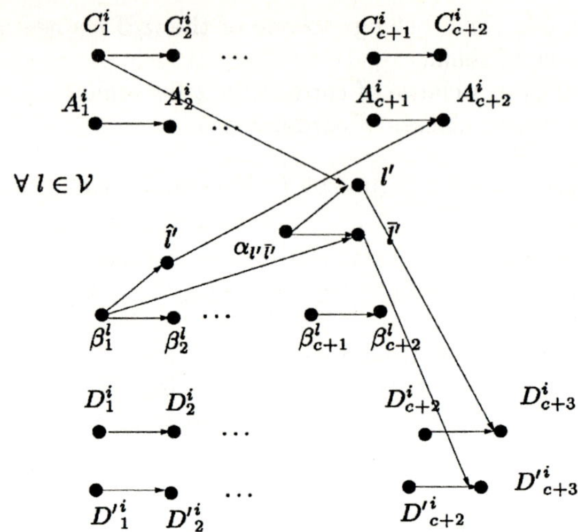

Fig. 1. A partial precedence graph for the $\mathcal{NP}$-completeness of the scheduling problem $\bar{P}|prec; c_{ij} = c \geq 3; p_i = 1|C_{max}$. **Remark:** $\bar{l}'$ is in the clause of length two associated to $D_1'^i \to D_2'^i \to \ldots D_{c+2}'^i \to D_{c+3}'^i$

The above construction is illustrated in Figure 1. This transformation can be clearly computed in polynomial time.

- Let us first assume that there is a schedule of length at most $(c+4)$. In the following, we will prove that there is a truth assignment $I : \mathcal{V} \to \{0, 1\}$ such that each clause in $\mathcal{C}$ has exactly one true literal.
  First we can remark that if $c \geq 3$ then $2c + 2 > c + 4$ and so, each path A_j^i, β_j^x, C_j^i or $D_{j'}^i$ with $j \in \{1, 2, \ldots, c+2\}$ and $j' \in \{1, 2, \ldots, c+3\}$ must be executed on the same processor. What's more, two of these paths cannot be executed on the same processor.
  Notation: In the following we denote by P_A (resp. P_C) the set of the $\frac{n}{3}$ processors which execute a path A_j^i (resp. a path C_j^i). Notice that we know by the definition of the problem Π_1, that in an instance admits $\frac{n}{3}$ clauses of length three where n denotes the number of variables. In the same way, we denote by P_β (resp. P_D) the set of the n processors which execute a path β_j^x (resp. a path D_j^i).

Lemma 1. *For $C_{max} = c + 4$: the decision to assign the true value to the variable x iff the variable-task x' is executed on a processor of the path P_C leads to a correct solution.*

Proof. In order to respect the feasible schedule of length $(c + 4)$, in the first time, we can stem from the polynomial time transformation, that the starting time of the variables-tasks l', $\bar{l}'$ and $\hat{l}'$, and that the processors on which these tasks must be executed, are given by the following remarks:
$\forall l \in \mathcal{V}$:

- Each variable-task l' is executed on a processor of P_C at slot 3 or on a processor of P_D at slot $(c+2)$ or $(c+3)$,
- Each variable-task $\bar{l}'$ is executed on a processor of P_β at slot 3 or on a processor of P_D at slot $(c+2)$ or $(c+3)$,
- Each variable-task $\hat{l}'$ is executed on a processor of P_β at slot 2 or 3 or on a processor of P_A at slot $(c+2)$ or $(c+3)$,
- The variables-tasks $\bar{l}'$ and $\hat{l}'$ cannot be executed together on a processor of P_β (they have a common predecessor).

Notation and property: For each $l \in \mathcal{V}$, we can associate the three tasks l', $\bar{l}'$, $\hat{l}'$. We denote by $X = \{l' | l \in \mathcal{V}\}$, $\bar{X} = \{\bar{l}' | l \in \mathcal{V}\}$ and $\hat{X} = \{\hat{l}' | l \in \mathcal{V}\}$ three sets of tasks. For each subset A of $\bar{X}$ (resp. $\hat{X}$), we can associate a subset B of X in the following way: $l' \in B$ *if and only if* $\bar{l}' \in A$ (resp. $\hat{l}' \in A$).

Let be the following sets: $X_1 = \{l' \backslash \pi(l') = \pi(P_C)\}$ where $\pi(l')$ (resp. $\pi(P_C)$) designs the processor on which the task l' is scheduled, $X_2 = \{l' \backslash \pi(l') = \pi(P_D)\}$, $X_3 = \{l' \backslash \pi(\bar{l}') = \pi(P_\beta)\}$, $X_4 = \{l' \backslash \pi(\bar{l}') = \pi(P_D)\}$, $X_5 = \{l' \backslash \pi(\hat{l}') = \pi(P_\beta)\}$, $X_6 = \{l' \backslash \pi(\hat{l}') = \pi(P_A)\}$.
Let be $x_i = |X_i|$ for $i \in \{1, \ldots, 6\}$.
We can stem from the construction of an instance of the scheduling problem the following table,

	P_C	P_β	P_A	P_D
x'	X_1			X_2
$\bar{x}'$		X_3		X_4
$\hat{x}'$		X_5	X_6	

From the previous table, using the variable x_i, we obtain the following in-equations system: $x_1 + x_2 = n(1), x_3 + x_4 = n(2), x_5 + x_6 = n(3), x_1 \leq \frac{n}{3}(4), x_6 \leq \frac{2n}{3}(5), x_3 + x_5 \leq n(6), x_2 + x_4 \leq n(7)$.
We will give some details about the previous system:
- For the equations(1), (2) and (3): We must execute all the tasks of the sets X, $\bar{X}$ and $\hat{X}$.
- For the equation (4), on the processor which executes the path C_j^i of the clause $C_i = (y \vee z \vee t)$, we can execute at most one of the three variables-tasks y', z', t'. Indeed, all variables-tasks l' as a successor which is executed on a processor of P_D. If it is executed on the processor which scheduled the tasks from the path P_C it cannot be executed before the slot 3 and so, the variable-task $\alpha_{l'\bar{l}'}$ must be executed on the same processor which becomes saturated. So, we have $|X_3| < |P_C|$.
- For the equation (5), each processor of the paths P_A has two free slots and $|P_A| = \frac{n}{3}$.
- For the equation (6), all the variables-tasks $\bar{l}'$ or $\hat{l}'$ which are executed on a processor of the path P_β must be finished before slot 3 (it has a successor executed on another processor). So the variable-task $\alpha_{l'\bar{l}'}$ must be executed on the same processor which becomes saturated. Therefore, at the most one task between the variables-tasks $\bar{l}'$ and $\hat{l}'$ can be executed on a processor of the path P_β and so, $|X_3| + |X_5| \leq |P_\beta|$.

- For the equation (7), it is clear that, $|P_D| = n$ and there is at the most one free slot on each processor of P_D.

On the one hand, we have $x_3 + x_5 = n$ (indeed, we have $x_3 + x_4 + x_5 + x_6 = 2n$ and $x_6 \leq \frac{2n}{3}$, $x_4 \leq \frac{n}{3}$, so $x_3 + x_5 \geq n$) and on the other hand, $\forall l'$ only one variable-task between the variables-tasks $\bar{l}'$ and $\hat{l}'$ can be executed on a processor of P_β, thus we obtain $X_3 \cap X_5 = \emptyset$. Consequently, we have $X_3 \cup X_5 = X$. As the set X_4 (resp. X_6) is the complementary of the set X_3 (resp. X_5) we have $X_4 \cup X_6 = X$. Moreover, if the variable-task l' is executed on a processor of P_C then the variable-task $\alpha_{l'\bar{l}'}$ is executed on the same processor. Thus, the variable-task $\bar{x}'$ cannot be executed before the slot $(c + 2)$, thus it is executed on a processor of P_D. We can deduce that $X_1 = X_4$ (the two sets are the same cardinality). Finally, we have $X_1 \cup X_2 = X$, $X_3 \cup X_4 = X$, $X_5 \cup X_6 = X$, $X_4 \cup X_6 = X$, $X_3 \cup X_5 = X$, $X_1 = X_4$ and therefore $X_1 = X_4 = X_5$ and $X_2 = X_3 = X_6$.

We can deduce from the previous equations that $x_1 = x_4 = x_5 = \frac{n}{3}$ and $x_2 = x_3 = x_6 = \frac{2n}{3}$.

So, if we affect the value "true" to the variable l iff the variable-task l' is executed on a processor of P_C it is trivial to see that in the clause of length 3 we have one and only one literal equal to "true".

Let be $c = (x \vee \bar{y})$, a clause of length 2.

- If $x' \in X_1 \implies y' \in X_4 \implies y' \in X_1$. The first implication (resp. the second) is due to the fact that each processor of the path P_D must be saturated ($x_2 + x_4 = n$) (resp. $X_1 = X_4$). Only the literal x is "true" between the variables x and $\bar{y}$.
- If $x' \in X_2 \implies y' \in X_3 \implies y' \in X_2$. The first (resp. the second) implication is due to the fact that there is only one free slot on each processor executing the path P_D (resp. $X_3 = X_2$). Only the literal $\bar{y}$ is "true" between the variables x and $\bar{y}$.

In conclusion, there is only one true literal per clause. *This concludes the proof of Lemma 1.*

- Conversely, we suppose that there is a truth assignment $I : \mathcal{V} \to \{0, 1\}$, such that each clause in $\mathcal{C}$ has exactly one true literal.

Suppose that the true literal in the clause $C_i = (y \vee z \vee t)$ is t. Therefore, the variable-task t' (resp. y' and z') is processed at the slot 2 (resp. at the slot $(c+2)$) on the same processor as the path P_{C_i} (resp. as the path P_D and $P_{D'}$, where D and D' indicates a clause of length two where the variables y and z occurred). The $\frac{2n}{3}$ other variables-tasks y' not yet scheduled are executed at slot 3 on processor P_β as the variable-task $\alpha_{y'\bar{y}'}$. The variable-task $\hat{t}'$ (resp. $\hat{y}'$ and $\hat{z}'$) is executed at the slot 2 (resp. $c + 2$ and $c + 3$) on a processor of the path P_β (resp. P_A). *This concludes the proof of Theorem 1.*

In the full version of this paper [5], we proved the following results:

Corollary 1. *There is no polynomial-time algorithm for the problem* $\bar{P}|prec; c_{ij} = c \geq 2; p_i = 1|C_{max}$ *with performance bound smaller than* $1 + \frac{1}{c+4}$ *unless* $\mathcal{P} \neq \mathcal{NP}$.

Theorem 2. *There is no polynomial-time algorithm for the problem* $\bar{P}|prec; c_{ij} = c \geq 2; p_i = 1| \sum_j C_j$ *with performance bound smaller than* $1 + \frac{1}{2c+5}$ *unless* $\mathcal{P} \neq \mathcal{NP}$.

Theorem 3. *The problem of deciding whether an instance of* $\bar{P}|prec; c_{ij} = c; p_i = 1|C_{max}$ *with* $c \in \{2, 3\}$ *has a schedule of length at most* $(c + 2)$ *is solvable in polynomial time.*

4 Approximation by Expansion

4.1 Introduction, Notation and Description of the Method

Notation: We denote by σ^∞, the UET-UCT schedule, and by σ_c^∞ the UET-LCT schedule. Moreover, we denote by t_i (resp. t_i^c) the starting time of the task i in the schedule σ^∞ (resp. in the schedule σ_c^∞).

Principle: We keep an assignment for the tasks given by a "good" feasible schedule on an unbounded number of processors σ^∞. We proceed to an expansion of the makespan, while preserving communication delays $(t_j^c \geq t_i^c + 1 + c)$ for two tasks, i and j with $(i, j) \in E$, processing on two different processors.

Let be a precedence graph $G = (V, E)$, we determinate a feasible schedule σ^∞, for the model UET-UCT, using an $(4/3)-$approximation algorithm proposed by Munier and König [10]. This algorithm gives a couple $\forall i \in V$, (t_i, π) on the schedule σ^∞ corresponding to: t_i the starting time of the task i for the schedule σ^∞ and π the processor on which the task i is processed at t_i.

Now, we determinate a couple $\forall i \in V$, (t_i^c, π') on the schedule σ_c^∞ in the following ways: The starting time $t_i^c = d \times t - i = \frac{(c+1)}{2} t_i$ and, $\pi = \pi'$. The justification of the expansion coefficient is given below. An illustration of the expansion is given by Figure 2.

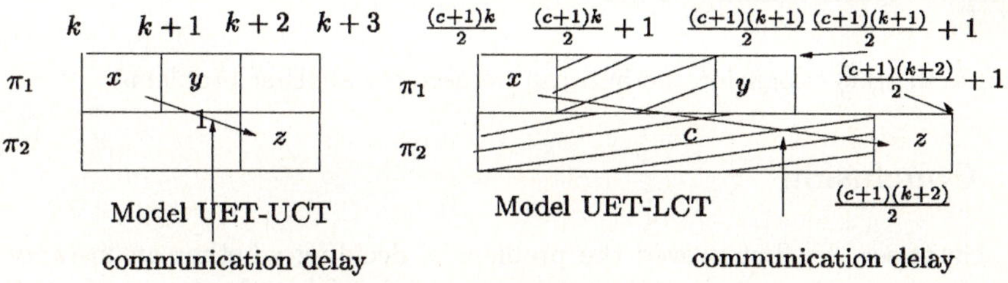

Fig. 2. Illustration of notion of an expansion

4.2 Analysis of the Method

Lemma 2. *The coefficient of an expansion is* $d = \frac{(c+1)}{2}$.

Proof. Let be two tasks i and j such that $(i, j) \in E$, which are processed on two different processors in the feasible schedule σ^∞. We are interested in having a coefficient d such that $t_i^c = d \times t_i$ and $t_j^c = d \times t_j$. After an expansion, in order to respect the precedence constraints and the communication delays we must have $t_j^c \geq t_i^c + 1 + c$, and so $d \times t_i - d \times t_j \geq c + 1$, $d \geq \frac{c+1}{t_i - t_j}$, $d \geq \frac{c+1}{2}$. It is sufficient to choose $d = \frac{(c+1)}{2}$.

Lemma 3. *An expansion algorithm gives a feasible schedule for the problem denoted by* $\bar{P}|prec; c_{ij} = c \geq 2; p_i = 1|C_{\max}$.

Proof. It sufficient to check that the solution given by an expansion algorithm produces a feasible schedule for the model UET-LCT. Let be two tasks i and j such that $(i, j) \in E$. We denote by π_i (resp. π_j) the processor on which the task i (resp. the task j) is executed in the schedule σ^∞. Moreover, we denote by π_i' (resp. π_j') the processor on which the task i (resp. the task j) is executed in the schedule σ_c^∞. Thus,

- If $\pi_i = \pi_j$ then $\pi_i' = \pi_j'$. Since the solution given by Munier and König [10] gives a feasible schedule on the model UET-UCT, then we have $t_i + 1 \leq t_j$, $\frac{2}{c+1}t_i^c + 1 \leq \frac{2}{c+1}t_j^c$; $t_i^c + 1 \leq t_i^c + \frac{c+1}{2} \leq t_j^c$.
- If $\pi_i \neq \pi_j$ then $\pi_i' \neq \pi_j'$. We have $t_i + 1 + 1 \leq t_j$, $\frac{2}{c+1}t_i^c + 2 \leq \frac{2}{c+1}t_j^c$; $t_i^c + (c+1) \leq t_j^c$.

Theorem 4. *An expansion algorithm gives a* $\frac{2(c+1)}{3}$*—approximation algorithm for the problem* $\bar{P}|prec; c_{ij} = c \geq 2; p_i = 1|C_{\max}$.

Proof. We denote by C_{max}^h (resp. C_{max}^{opt}) the makespan of the schedule computed by the Munier and König (resp. the optimal value of a schedule σ^∞). In the same way we denote by $C_{max}^{h^*}$ (resp. $C_{max}^{opt,c}$) the makespan of the schedule computed by our algorithm (resp. the optimal value of a schedule σ_c^∞).

We know that $C_{max}^h \leq \frac{4}{3}C_{max}^{opt}$. Thus, we obtain $\frac{C_{max}^{h^*}}{C_{max}^{opt,c}} = \frac{\frac{(c+1)}{2}C_{max}^h}{C_{max}^{opt,c}} \leq \frac{\frac{(c+1)}{2}C_{max}^h}{C_{max}^{opt}} \leq \frac{\frac{(c+1)}{2}\frac{4}{3}C_{max}^{opt}}{C_{max}^{opt}} \leq \frac{2(c+1)}{3}$.

Remark: this expansion method can be used for another problems.

5   Conclusion

In this paper, we first proved the problem of deciding whether an instance of $\bar{P}|prec; c_{ij} = c \geq 3; p_i = 1|C_{max}$ has a schedule of length at most $(c + 4)$ is $\mathcal{NP}$-complete. This result is to be compared with the result of [7] (resp. [1]), which states that $\bar{P}|prec; c_{ij} = 1; p_i = 1|C_{max} = 6$ (resp. $P|prec; c_{ij} = c \geq 3; p_i = 1|C_{max} = c + 3$) is $\mathcal{NP}$-complete. Our result implies that there is no ρ—approximation algorithm with $\rho < 1 + \frac{1}{c+4}$, unless $\mathcal{P} = \mathcal{NP}$. Secondly, we also propose a $\frac{2(c+1)}{3}$—approximation algorithm based on the notion of expansion. In the full version [5], we show that there is no hope of finding a ρ-approximation algorithm with ρ strictly less than $\rho < 1 + \frac{1}{2c+5}$ for the problem of the minimization of the sum of the completion time. We established that the problem of deciding whether an instance of $\bar{P}|prec; c_{ij} = c; p_i = 1|C_{max}$ with $c \in \{2, 3\}$ has a schedule of length at most $(c + 2)$ is solvable in polynomial time.

Remark: We conjecture that the problem of deciding whether an instance of $\bar{P}|prec; c_{ij} = c; p_i = 1|C_{max}$ with $c \geq 2$ has a schedule of length at most $(c + 3)$ is solvable in polynomial time.

References

1. E. Bampis, A. Giannakos, and J.C. König. On the complexity of scheduling with large communication delays. *European Journal of Operation Research*, 94:252–260, 1996.
2. B. Chen, C.N. Potts, and G.J. Woeginger. A review of machine scheduling: complexity, algorithms and approximability. Technical Report Woe-29, TU Graz, 1998.
3. M.R. Garey and D.S. Johnson. *Computers and Intractability, a Guide to the Theory of $\mathcal{NP}$-Completeness*. Freeman, 1979.
4. R. Giroudeau. *L'impact des délais de communications hiérarchiques sur la complexité et l'approximation des problèmes d'ordonnancement*. PhD thesis, Université d' Évry Val d'Essonne, 2000.
5. R. Giroudeau, J.C. König, F.K. Moulaï, and J. Palaysi. Complexity and approximation for the precedence constrained scheduling problem with large communications delays. Technical Report 11903, Laboratoire d'Informatique, de Robotique et Microélectronique de Montpellier, 2005.
6. R.L. Graham, E.L. Lawler, J.K. Lenstra, and A.H.G. Rinnooy Kan. Optimization and approximation in deterministics sequencing and scheduling theory: a survey. *Ann. Discrete Math.*, 5:287–326, 1979.
7. J.A. Hoogeveen, J.K. Lenstra, and B. Veltman. Three, four, five, six, or the complexity of scheduling with communication delays. *O. R. Lett.*, 16(3):129–137, 1994.
8. A. Munier. Approximation algorithms for scheduling trees with general communication delays. *Parallel Computing*, 25(1):41–48, January 1999.
9. A. Munier and C. Hanen. An approximation algorithm for scheduling unitary tasks on m processors with communication delays. Non publié, 1996.
10. A. Munier and J.C. König. A heuristic for a scheduling problem with communication delays. *Operations Research*, 45(1):145–148, 1997.
11. C. Picouleau. New complexity results on scheduling with small communication delays. *Discrete Applied Mathematics*, 60:331–342, 1995.
12. C. Rapine. *Algorithmes d'approximation garantie pour l'ordonnancement de tâches, Application au domaine du calcul parallèle*. PhD thesis, Institut National Polytechnique de Grenoble, 1999.
13. V.J. Rayward-Smith. UET scheduling with unit interprocessor communication delays. *Discr. App. Math.*, 18:55–71, 1987.

Batch-Scheduling Dags for Internet-Based Computing*

(Extended Abstract)

Grzegorz Malewicz[1,3] and Arnold L. Rosenberg[2]

[1] Dept. of Computer Science, Univ. of Alabama, Tuscaloosa, AL 35487, USA
[2] Dept. of Computer Science, Univ. of Massachusetts, Amherst, MA 01003, USA
[3] Div. of Mathematics and Computer Science, Argonne National Lab, Argonne, IL 60439, USA

Abstract. The process of scheduling computations for Internet-based computing presents challenges not encountered with more traditional computing platforms. The looser coupling among participating computers makes it harder to utilize remote clients well, and raises the specter of a kind of "gridlock" that ensues when a computation stalls because no new tasks are eligible for execution. This paper studies the problem of scheduling computation-dags in a manner that renders tasks eligible for execution at the maximum possible rate. Earlier work has developed a framework for such scheduling when a new task is allocated to a remote client as soon as it returns the results from an earlier task. The proof in that work that many dags cannot be scheduled optimally within this paradigm signaled the need for a companion theory that addresses the scheduling problem for all computation-dags. A new, *batched,* scheduling paradigm for Internet-based computing is developed in this work. Although optimal batched schedules always exist, computing such a schedule is NP-Hard, even for bipartite dags. In response, a polynomial-time algorithm is developed for producing optimal batched schedules for a rich family of dags obtained by "composing" tree-structured building-block dags. Finally, a fast heuristic schedule is developed for "expansive" dags.

1   Introduction

Earlier work [11, 13, 15] has developed the Internet-Computing (IC, for short) Pebble Game that abstracts the problem of scheduling computations having intertask dependencies for the several modalities of Internet-based computing, including Grid computing (cf. [1, 4, 5]), global computing (cf. [2]), and Web computing (cf. [8]). This Game was developed with the goal of formalizing the process of scheduling computations with intertask dependencies for IC. The scheduling paradigm studied in [11, 13, 15] is that a server allocates a task of the dag being computed to a remote client as soon as the task becomes eligible for allocation and the client becomes available for computation. The quality metric for schedules is to maximize the rate at which tasks are rendered eligible for allocation to remote clients, with the dual aim of maximizing the utilization of remote clients and minimizing the likelihood of the "gridlock" that can arise when a computation stalls pending completion of already-allocated tasks. These sources develop the framework for a theory of IC scheduling based on this paradigm.

---

* A portion of the research of G. Malewicz was done while visiting the Univ. of Massachusetts Amherst. The research of A. Rosenberg was supported in part by NSF Grant CCF-0342417.

J.C. Cunha and P.D. Medeiros (Eds.): Euro-Par 2005, LNCS 3648, pp. 262–271, 2005.

The present study is motivated by the demonstration in [11] that there are simple computation-dags that do not admit any optimal IC schedule. (Intuitively, any sequence of tasks that optimizes the number of eligible tasks after the first t steps of the computation is incompatible with every sequence that optimizes that number after the first t' steps.) We respond here by developing a companion scheduling theory in which every computation-dag admits an optimal schedule. This new theory is based on a *batched* scheduling paradigm, which relieves the Server from the chore of selecting a new task for allocation whenever a remote client becomes available for computation. Instead, we now assume that the Server collects requests for new tasks and then (either periodically or based on some trigger) allocates tasks for the collected requests in a batch. (This mode of operation may be inevitable if, say, tasks take extremely long to compute and enable many other tasks once completed.) The goal for the Server is to satisfy this batch of requests with a set of tasks whose execution will produce a maximal number of new eligible tasks. In contrast to the quality metric of [11, 13, 15], this new step-by-step metric can always be satisfied optimally. Moderating the news that optimality can always be achieved in the batched paradigm is our demonstration that finding such a schedule for an arbitrary computation-dag—even a bipartite one—is NP-Hard, hence likely computationally intractable (Section 3). We respond to this probable computational intractability with a polynomial-time optimal algorithm for a rich family of dags that are constructed by "composing" certain tree-structured building-block dags (Section 5). Since the preceding timing polynomial has high degree, we also develop a fast heuristic schedule for a more restricted family of "expansive" dags, whose eligible-task production rate is within a factor of 4 of optimal (Section 6).

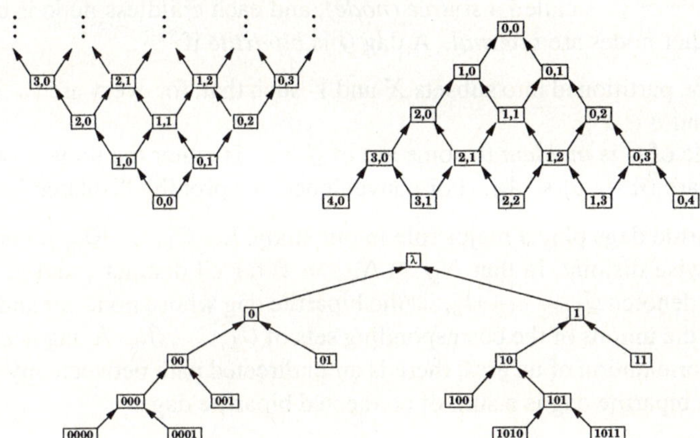

Fig. 1. Clockwise from top left: an evolving (2-dimensional) mesh, the 5-level (2-dimensional) reduction-mesh, a (binary) reduction-tree dag.

Related work. The IC Pebble Game is introduced in [13, 15], and optimal schedules are identified for the dags of Fig. 1. A framework for a theory of scheduling for IC is developed in [11], building on the principles that enable the optimal schedules of [13, 15]. Central to the framework are a formal method for composing simple dags into complex ones, together with a relation that allows one to prioritize the execution order of the constituent building-block dags of a composite dag. A probabilistic pebble game is used in [6, 9, 10] to study the problem of executing tasks on unreliable clients; our proof of the NP-hardness of batch-scheduling builds on tools from [6]. Although our goals and methodology differ significantly from those of [3, 12, 14], we owe an intellectual debt to those pioneering studies of pebbling-based scheduling models. Finally, the impetus for our study derives from the many exciting systems- and/or application-oriented studies of Internet-based computing, in sources such as [1, 2, 4, 5, 7, 8, 16].

2   A Model for Executing Dags on the Internet

2.1   Computation-Dags

Basic definitions. A *directed graph* $\mathcal{G}$ is given by a set of *nodes* $N_{\mathcal{G}}$ and a set of *arcs* (or, *directed edges*) $A_{\mathcal{G}}$, each having the form $(u \to v)$, where $u, v \in N_{\mathcal{G}}$. A *path* in $\mathcal{G}$ is a sequence of arcs that share adjacent endpoints, as in the following path from node u_1 to node u_n: $(u_1 \to u_2)$, $(u_2 \to u_3)$, $\ldots$, $(u_{n-2} \to u_{n-1})$, $(u_{n-1} \to u_n)$. A *dag* (*directed acyclic graph*) $\mathcal{G}$ is a directed graph that has no cycles; i.e., in a dag, no path of the preceding form has $u_1 = u_n$. When a dag $\mathcal{G}$ is used to model a computation, i.e., is a *computation-dag:*

- each node $v \in N_{\mathcal{G}}$ represents a task in the computation;
- an arc $(u \to v) \in A_{\mathcal{G}}$ represents the dependence of task v on task u: v cannot be executed until u is.

Given an arc $(u \to v) \in A_{\mathcal{G}}$, we call u a *parent* of v and v a *child* of u in $\mathcal{G}$. Each parentless node of $\mathcal{G}$ is called a *source (node)*, and each childless node is called a *sink (node)*; all other nodes are *internal*. A dag $\mathcal{G}$ is *bipartite* if:

1. $N_{\mathcal{G}}$ can be partitioned into subsets X and Y such that, for every arc $(u \to v) \in A_{\mathcal{G}}$, $u \in X$ and $v \in Y$;
2. each node of $\mathcal{G}$ is *incident* to some arc of $\mathcal{G}$, i.e., is either the node u or the node v of some arc $(u \to v) \in A_{\mathcal{G}}$. (For convenience, we prohibit "isolated" nodes.)

Sums of bipartite dags play a major role in our study. Let $\mathcal{G}_1, \ldots, \mathcal{G}_m$ be bipartite dags that are pairwise disjoint, in that $N_{\mathcal{G}_i} \cap N_{\mathcal{G}_j} = \emptyset$ for all distinct i and j. The *sum* of $\mathcal{G}_1, \ldots, \mathcal{G}_m$, denoted $\mathcal{G}_1 + \cdots + \mathcal{G}_m$, is the bipartite dag whose node-set and arc-set are, respectively, the unions of the corresponding sets of $\mathcal{G}_1, \ldots, \mathcal{G}_m$. A dag is *connected* if, ignoring the orientation of its arcs, there is an undirected path between any two distinct nodes. Every bipartite dag is a sum of connected bipartite dags.

Some basic building blocks. Our study focuses on dags that are built out of bipartite *building blocks* by the operation of *composition*. We present a sampler of building blocks that will illustrate the theory we begin to develop here; see Fig. 2.

A **bipartite tree-dag** $\mathcal{T}$ is a bipartite dag such that, if one ignores the orientations of $\mathcal{T}$'s arcs, then the resulting graph is a tree. The following two special classes of tree-dags generate important families of complex dags.

For each $d > 1$, the $(1, d)$-**W-dag** $\mathcal{W}_{1,d}$ has one source node and d sink nodes; its d arcs connect the source to each sink. Inductively, for positive integers a, b, the $(a+b, d)$-W-dag $\mathcal{W}_{a+b,d}$ is obtained from the (a, d)-W-dag $\mathcal{W}_{a,d}$ and the (b, d)-W-dag $\mathcal{W}_{b,d}$ by

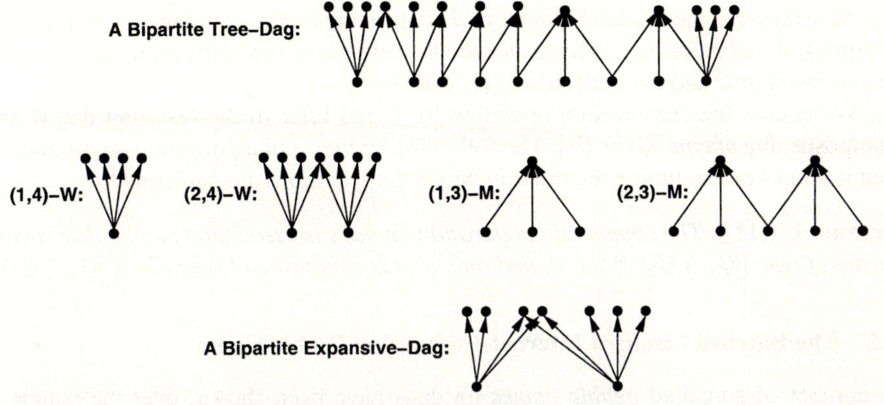

Fig. 2. Some bipartite building-block-dags.

identifying (or, merging) the rightmost sink of the former dag with the leftmost sink of the latter. W-dags epitomize "expansive" computations.

For each $d > 1$, the $(1, d)$-**M-dag** $\mathcal{M}_{1,d}$ has d source nodes and 1 sink node; its d arcs connect each source to the sink. Inductively, for positive integers a, b, the $(a+b, d)$-M-dag $\mathcal{M}_{a+b,d}$ is obtained from the (a, d)-M-dag $\mathcal{M}_{a,d}$ and the (b, d)-M-dag $\mathcal{M}_{b,d}$ by merging the rightmost source of the former dag with the leftmost source of the latter. M-dags epitomize "contractive" (or, "reductive") computations.

A large variety of significant computation-dags are "compositions" of W-dags and M-dags, including the dags in Fig. 1: The evolving mesh is constructed from its source outward by "composing" a $(1, 2)$-W-dag with a $(2, 2)$-W-dag, then a $(3, 2)$-W-dag, and so on; the reduction-mesh is similarly constructed using $(k, 2)$-M-dags for successively decreasing values of k; the reduction-tree is constructed by "composing" independent collections of $(1, 2)$-M-dags.

The following additional building blocks are highlighted in Section 6.

A **bipartite expansive-dags** $\mathcal{E}$ is a bipartite dag wherein each source v has an associated number $\varphi_v \geq 2$ such that: v has φ_v children that have no parent other than v and $\leq \varphi_v$ other children. Easily, expansive dags need not be tree-dags (cf. Fig. 2).

Compositions of bipartite dags. The following mechanism for *composing* a collection of connected bipartite dags to build complex dags is introduced in [11].

- Start with a base set $\mathcal{B}$ of connected bipartite dags.
- Given dags $\mathcal{G}_1, \mathcal{G}_2 \in \mathcal{B}$—which could be copies of the same dag with nodes renamed to achieve disjointness—one obtains a composite dag $\mathcal{G}$ as follows.
  - Let the composite dag $\mathcal{G}$ begin as the sum, $\mathcal{G}_1 + \mathcal{G}_2$, of the dags $\mathcal{G}_1, \mathcal{G}_2$. Rename nodes to ensure that $N_\mathcal{G}$ is disjoint from $N_{\mathcal{G}_1}$ and $N_{\mathcal{G}_2}$.
  - Select some set S_1 of sinks from the copy of $\mathcal{G}_1$ in the sum $\mathcal{G}_1 + \mathcal{G}_2$, and an equal-size set S_2 of sources from the copy of $\mathcal{G}_2$ in the sum. (If $S_1 = \emptyset$, then the composition operation degenerates to the operation of forming a sum dag.)
  - Pairwise identify (i.e., merge) the nodes in the sets S_1 and S_2 in some way. The resulting set of nodes is $\mathcal{G}$'s node-set; the induced set of arcs is $\mathcal{G}$'s arc-set.
- Add the dag $\mathcal{G}$ thus obtained to the base set $\mathcal{B}$.

Note the asymmetry of composition: $\mathcal{G}_1$ contributes some of its sinks, while $\mathcal{G}_2$ contributes some of its sources. The reader should note the natural correspondence between the node-set of $\mathcal{G}$ and the node-sets of $\mathcal{G}_1$ and $\mathcal{G}_2$.

We denote the composition operation by $\Uparrow$ and refer to the resulting dag $\mathcal{G}$ as a composite dag *of type* $[\mathcal{G}_1 \Uparrow \mathcal{G}_2]$. The following lemma is of algorithmic importance, in that it allows one to ignore the order in which compositions are performed.

Lemma 1 ([11]). *The composition operation on dags is associative; i.e., a dag is composite of type* $[[\mathcal{G}_1 \Uparrow \mathcal{G}_2] \Uparrow \mathcal{G}_3]$ *if, and only if, it is composite of type* $[\mathcal{G}_1 \Uparrow [\mathcal{G}_2 \Uparrow \mathcal{G}_3]]$.

2.2   The Batched *Idealized* Internet-Computing Pebble Game

A number of so-called *pebble games* on dags have been shown, over the course of several decades, to yield elegant formal analogues of a variety of problems related to scheduling dags. Such games use tokens called *pebbles* to model the progress of a computation on a dag: the placement or removal of the various available types of pebbles—which is constrained by the dependencies modeled by the dag's arcs—represents the changing (computational) status of the dag's task-nodes.

Our study is based on the Internet-Computing (IC, for short) Pebble Game of [13]. Based on studies of Internet-based computing in, for instance, [1, 7, 16], arguments are presented in [13, 15] that justify studying an idealized, simplified form of the Game. We refer the reader to these sources for both the original IC Pebble Game and for the arguments justifying its simplification. We study an idealized form of the Game here, adapted to a *batched* mode of computing.

The rules of the game. The Batched IC Pebble Game on a dag $\mathcal{G}$ involves one player S, the *Server*, who has access to unlimited supplies of two types of pebbles: ELIGIBLE pebbles, whose presence indicates a task's eligibility for execution, and EXECUTED pebbles, whose presence indicates a task's having been executed. The following rules of the Game simplify those of the original IC Pebble Game of [13, 15].

The Rules of the Batch-IC Pebble Game

- S begins by placing an ELIGIBLE pebble on each unpebbled source node of $\mathcal{G}$.
  /*Unexecuted source nodes are always eligible for execution, having no parents whose prior execution they depend on.*/
- At each step t—when there is some number, say e_t, of ELIGIBLE pebbles on $\mathcal{G}$'s nodes—S is approached by some number, say r_t, of Clients, requesting tasks. In response, S:
  - selects $\min\{e_t, r_t\}$ tasks that contain ELIGIBLE pebbles,
  - replaces those pebbles by EXECUTED pebbles,
  - places ELIGIBLE pebbles on each unpebbled node of $\mathcal{G}$ all of whose parents contain EXECUTED pebbles.
- S's goal is to allocate nodes in such a way that every node v of $\mathcal{G}$ *eventually* contains an EXECUTED pebble.
  /*This modest goal is necessitated by the possibility that $\mathcal{G}$ may be infinite.*/

For brevity, we henceforth call a node ELIGIBLE (resp., EXECUTED) when it contains an ELIGIBLE (resp., an EXECUTED) pebble. For uniformity, we henceforth talk about executing nodes rather than tasks.

The Batch-IC Scheduling (BICSO) Problem. Our goal is to play the Game in a way that maximizes the number of ELIGIBLE pebbles on $\mathcal{G}$ after every move by the Server S. In other words: for each step t of a play of the Game on a dag $\mathcal{G}$ under a schedule Σ, if there are currently e_t ELIGIBLE nodes, and if r_t Clients request tasks, then we want the Server to select a set of $\min\{e_t, r_t\}$ ELIGIBLE nodes to execute that will result in the largest possible number of ELIGIBLE nodes at step $t + 1$. We thus arrive at the following optimization problem.

Batched IC-Scheduling (Optimization version) **(BICSO)**
Instance: $\imath = \langle \mathcal{G}, X, E; r \rangle$, where:
- $\mathcal{G}$ is a computation-dag;
- X and E are disjoint subsets of $N_\mathcal{G}$ that satisfy the following;
    There is a step of some play of the Batched IC Pebble Game on $\mathcal{G}$ in which
    X is the set of EXECUTED nodes and E the set of ELIGIBLE nodes on $\mathcal{G}$.
- r is in the set[1] $[1, |E|]$.
Problem: Find a set $R \subseteq E$ of r nodes whose execution maximizes the number of ELIGIBLE nodes on $\mathcal{G}$, given that the nodes in X are already EXECUTED.

Note that solving BICSO automatically carries with it a guarantee of optimality.

The significance of BICSO—as with the IC-Scheduling Problem of [11, 13, 15]—stems from the following intuitive scenarios. (1) Schedules that produce ELIGIBLE tasks fast may reduce the chance of the "gridlock" that could occur when remote clients are slow in returning the results of their allocated tasks—so that new tasks cannot be allocated pending the return of already assigned ones. (2) If the IC Server receives a batch of requests for tasks at (roughly) the same time, then a Batched IC-optimal schedule ensures that there are maximally many tasks that are ELIGIBLE at that time, hence maximally many requests can be satisfied. This enhances the exploitation of clients' available resources. See [13, 15] for more elaborate discussions of these scheduling criteria.

3 The Intractability of BICSO Optimality

Viewed via its related decision problem, BICSO is NP-hard, even for bipartite dags. The reduction is from the problem of selecting m sets whose union has cardinality at most b from among nonempty sets $S_1, \ldots, S_n$ whose union is $[1, n]$, which is known [6] to be NP-Complete. Our reduction also uses a result that allows us to focus on a restricted class of schedules.

Lemma 2 ([11]). *Let Σ be a schedule for a dag $\mathcal{G}$. If Σ is altered to execute all of $\mathcal{G}$'s non-sinks before any of its sinks, then it produces no fewer ELIGIBLE nodes than Σ.*

Theorem 1. *BICSO is NP-hard, even when restricted to bipartite dags.*

---

[1] $[a, b] = \{a, a + 1, \ldots, b\}$.

4   Scheduling Composite Dags via Bipartite Dags

The computational intractability of BICSO (assuming that $P \neq NP$) is a mandate for seeking significant classes of dags for which one can solve BICSO efficiently. Our experience is that this goal is achievable for many classes of *bipartite* dags (such as the building blocks of Section 2). While this structural restriction is not of inherent interest, we show in this section that we can sometimes use the operation of composition to construct significant complex dags from bipartite building blocks. And, we can often solve BICSO for a composite dag $\mathcal{G}$ by solving a restricted version of BICSO for certain connected induced bipartite subdags of the bipartite dags that $\mathcal{G}$ is composed from. In the restricted version of BICSO—call it RBISCO—the bipartite subdags are connected, and all of their sources are ELIGIBLE, so the set E (of the instance of BICSO) comprises all sources of the subdag, and the set X is empty. The goal is to find an r-element subset of sources that maximizes the number of ELIGIBLE sinks—which is equivalent to solving BICSO for the restricted problem.

Theorem 2. *Let the dag $\mathcal{G}$ be a composition of bipartite dags $\mathcal{G}_1, \ldots, \mathcal{G}_m$. There is a polynomial-time algorithm that solves BICSO for $\mathcal{G}$, using as subprocedures polynomial-time algorithms for solving RBICSO for induced connected bipartite subdags of the $\mathcal{G}_i$.*

Proof Sketch. Consider instance $\imath = \langle \mathcal{G}, X, E; r \rangle$ of BICSO, where $\mathcal{G}$ is as in the theorem. We can focus on the modified goal of finding R among $\mathcal{G}$'s non-sinks. Using a result of [11], we can relate the number of ELIGIBLE nodes of $\mathcal{G}$ to the number of sinks of the $\mathcal{G}_i$ that are ELIGIBLE when the only EXECUTED nodes of $\mathcal{G}_i$ are the sources of $\mathcal{G}_i$ that correspond (in the natural manner emerging from the definition of composition) to EXECUTED nodes of $\mathcal{G}$. The latter number, however, can be calculated by focusing on a certain induced subdag of $\mathcal{G}_i$. This subdag is obtained by taking all sources of $\mathcal{G}_i$ that correspond to nodes ELIGIBLE in $\mathcal{G}$, and all sinks of $\mathcal{G}_i$ all whose parents correspond to either ELIGIBLE or EXECUTED nodes in $\mathcal{G}$ and at least one whose parent corresponds to ELIGIBLE node (These sinks are not ELIGIBLE but they may become so when we execute nodes of the $\mathcal{G}$ that we choose). The subdag is a sum of (≥ 0) isolated nodes and (≥ 0) connected bipartite dags. Let $\mathcal{S}_1, \ldots, \mathcal{S}_k$ be the connected bipartite dags obtained from the m subdags. We maximize the number of ELIGIBLE nodes by executing the r nodes of $\mathcal{G}$ that correspond to the r sources of the connected bipartite dags that maximize the number of ELIGIBLE sinks on the dags. That latter maximum can be found by first computing a maximum individually for each connected bipartite dag $\mathcal{S}_i$ and each r_i at most r, and then combining the maxima using a dynamic programming algorithm resulting from an observation that the r_i must sum up to r.

Now the goal of solving BICSO for $\mathcal{G}$ reduces to the goal of solving BICSO for the connected bipartite dags.

5   Tractable BICSO Optimality for Composite Trees

We develop a polynomial-time algorithm that solves BICSO for the family $\mathbf{T}$ of dags that are obtained from *bipartite tree-dags* via composition.

Theorem 3. *There is a polynomial-time algorithm Σ_{tree} that solves BICSO for any composite tree-dag $\mathcal{T} \in \mathbf{T}$.*

Proof. We develop a dynamic program Σ_{DP} that solves RBICSO for any bipartite tree-dag; Theorem 2 will extend Σ_{DP} to Σ_{tree}.

Lemma 3. *There is a polynomial-time algorithm Σ_{DP} that solves RBICSO for any bipartite tree-dag.*

Proof Sketch. Any bipartite tree-dag $\mathcal{T}$ arises from "folding" a (undirected, unrooted) tree T and orienting its edges. We label T's nodes "sources" and "sinks" according to their roles in $\mathcal{T}$. The key idea of Σ_{DP} is that we can find the maximum number of ELIGIBLE sinks for a "deep" tree inductively from shallow trees.

We recursively decompose T into subtrees by choosing some source w and letting it act as a root, thereby producing T_w. We traverse T_w breadth first, starting from w. Each time we descend from a sink v to a source u during the traversal, we produce a subtree, T_u, which is a copy of the subtree of T_w rooted at u. We use the natural correspondence between the node-sets of T_w and T_u to refer to corresponding nodes by the same name. We thus produce a sequence of subtrees (beginning with T_w), each including shorter ones that occur later in the sequence. Σ_{DP} processes the subtrees in the *reverse* order of this sequence, computing certain values for a subtree from analogous values for shorter ones. Σ_{DP} chooses the nodes to execute by recursively calculating the following functions. Pick any subtree T_u with, say, s sources.

- For any $r \in [1, s]$, let $E_1(T_u, r)$ be the maximum number of ELIGIBLE sinks on T_u when the root u and some other $r - 1$ of its sources are EXECUTED.

$E_1(T_u, r)$ is trivial to calculate when T_u has height 0 or 1.

- For any $r \in [0, s - 1]$, let $E_0(T_u, r)$ be the maximum number of ELIGIBLE sinks on T_u when the root u is *not* EXECUTED but some r other of its sources are.

$E_0(T_u, r) = 0$ when T_u has height 0 or 1. For $r \in [0, s]$, the maximum number of ELIGIBLE sinks in T_u when r of its sources are EXECUTED is calculated from E_0 and E_1. Σ_{DP} computes $E_0(T_w, r)$ and $E_1(T_w, r)$ for any $r \in [0, (\text{the number of sources in } T)]$, as follows. We may consider only subtrees of heights ≥ 2. We decompose trees as depicted in Fig. 3. Focus on a subtree T_u of height ≥ 2, with s sources. Consider all sinks of T_u that are linked to u. Some of these sinks—say, $v_1, \ldots, v_k$—are also linked to some other source, while some h of the sinks are not. Since T_u has height ≥ 2, we have $k \geq 1$; it is possible that $h = 0$. For any $i \in [1, k]$, sink v_i is connected to some $g_i \geq 1$ sources other than u—call them $u_{i,1}, \ldots, u_{i,g_i}$. Consider the subtrees $T_{u_{i,j}}$, for $i \in [1, k]$, $j \in [1, g_i]$; each has height strictly smaller than T_u's. Let $s_{i,j}$ be the number of sources in $T_{u_{i,j}}$, so that $s = 1 + \sum_{i=1}^{k} \sum_{j=1}^{g_i} s_{i,j}$. We can calculate E_0 and E_1 for T_u from E_0 and E_1 for each $T_{u_{i,j}}$, because we can control which of the v_i become ELIGIBLE.

We now apply Lemma 3 in Theorem 2, to complete the proof of Theorem 3.

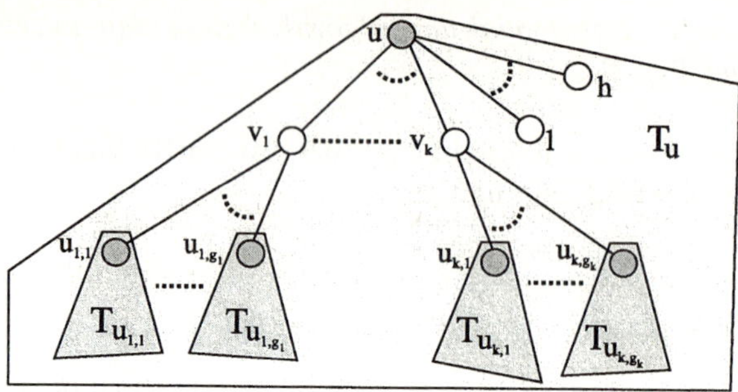

Fig. 3. Decomposing T_u: shaded nodes are sources; blank nodes are sinks.

6   Solving BICSO Efficiently for Expansive Dags

Because the timing polynomial of Σ_{tree} has high degree, we have sought nontrivial classes of dags for which we could solve BICSO *approximately* optimally, but much faster than Σ_{tree}. The initial result of our quest is Σ_{exp}, which approximates an optimal solution to BICSO for the family **E** of composite expansive dags. Σ_{exp} implements the following natural, fast heuristic. For each source v of any $\mathcal{E} \in$ **E**, say that φ_v nodes have v as their sole parent, and ψ_v nodes have other parents also. Say that $\mathcal{E}$ has $|E|$ ELIGIBLE nodes and that we must execute the best r of these. Σ_{exp} selects the r nodes that have the largest associated φ_v. This ploy solves BICSO to within a factor of 4 of optimally for the family **E**.

Theorem 4. *For any instance* $\iota = \langle \mathcal{E}, X, E; r \rangle$ *of BICSO, where* $\mathcal{E} \in$ **E**, Σ_{exp} *will, in time* $O(|E|)$, *find solution to BICSO, whose increase in the number of* ELIGIBLE *nodes is at least one-fourth the optimal increase.*

Proof Sketch. We implement Σ_{exp} by using a linear-time selection algorithm. One notes that each node v selected by an optimal algorithm adds at most $2\varphi_v$ distinct ELI-GIBLE nodes, while each node w selected by the heuristic adds at least $\frac{1}{2}\varphi_w$ such nodes.

References

1. R. Buyya, D. Abramson, J. Giddy (2001): A case for economy Grid architecture for service oriented Grid computing. *10th Heterogeneous Computing Wkshp.*
2. W. Cirne and K. Marzullo (1999): The Computational Co-Op: gathering clusters into a meta-computer. *13th Intl. Parallel Processing Symp.*, 160–166.
3. S.A. Cook (1974): An observation on time-storage tradeoff. *J. Comp. Syst. Scis. 9*, 308–316.
4. I. Foster and C. Kesselman [eds.] (2004): *The Grid: Blueprint for a New Computing Infrastructure* (2nd edition), Morgan-Kaufmann, San Francisco.
5. I. Foster, C. Kesselman, S. Tuecke (2001): The anatomy of the Grid: enabling scalable virtual organizations. *Intl. J. Supercomputer Applications.*
6. L. Gao and G. Malewicz (2004): Internet computing of tasks with dependencies using unreliable workers. *8th Intl. Conf. on Principles of Distributed Systems*, 315–325.

7. D. Kondo, H. Casanova, E. Wing, F. Berman (2002): Models and scheduling guidelines for global computing applications. *Intl. Parallel and Distr. Processing Symp.*
8. E. Korpela, D. Werthimer, D. Anderson, J. Cobb, M. Lebofsky (2000): SETI@home: massively distributed computing for SETI. In *Computing in Sci. and Engr.* (P.F. Dubois, Ed.) IEEE Computer Soc. Press, Los Alamitos, CA.
9. G. Malewicz (2005): Parallel Scheduling of Complex Dags under Uncertainty. *17th ACM Symposium on Parallelism in Algorithms and Architectures*, to appear.
10. G. Malewicz (2005): Implementation and Experiments with an Algorithm for Parallel Scheduling of Complex Dags under Uncertainty. Submitted for publication.
11. G. Malewicz, A.L. Rosenberg, M. Yurkewych (2005): On Scheduling Complex Dags for Internet-Based Computing. *IEEE Intl. Parallel and Distr. Processing Symp.*, 66.
12. M.S. Paterson, C.E. Hewitt (1970): Comparative schematology. *Project MAC Conf. on Concurrent Systems and Parallel Computation*, ACM Press, 119–127.
13. A.L. Rosenberg (2004): On scheduling mesh-structured computations for Internet-based computing. *IEEE Trans. Comput. 53*, 1176–1186.
14. A.L. Rosenberg and I.H. Sudborough (1983): Bandwidth and pebbling. *Computing 31*, 115–139.
15. A.L. Rosenberg and M. Yurkewych (2005): Guidelines for scheduling some common computation-dags for Internet-based computing. *IEEE Trans. Comput. 54*, 428–438.
16. X.-H. Sun and M. Wu (2003): GHS: A performance prediction and task scheduling system for Grid computing. *IEEE Intl. Parallel and Distributed Processing Symp.*

Scheduling Workflow Distributed Applications in JavaSymphony*

Alexandru Jugravu[1] and Thomas Fahringer[2]

[1] University of Vienna, Institute for Software Science, Liechtensteinstr. 22,
A-1090 Wien, Austria
[2] University of Innsbruck, Institute for Software Science, Technikerstr. 25/7,
A-6020 Innsbruck, Austria

Abstract. JavaSymphony is a high-level programming model for performance-oriented distributed and parallel Java applications, which allows the programmer to control parallelism, load balancing, and locality at a high level of abstraction. Recently, we have introduced new features to support the development and the deployment of workflow distributed applications for JavaSymphony. We have built a formal model of a workflow, which allows a graphical representation of the associated workflow. In this paper, we give further details about the workflow model and introduce a new theoretical framework for scheduling JavaSymphony workflow applications.

1 Introduction

Distributed heterogeneous computing has emerged as a cost-effective solution to high-performance computing on expensive parallel machines. In addition, Grid computing has been recently introduced as a worldwide generalization of distributed heterogeneous computing, which has undergone a number of significant changes in a brief time. Supporting grid middleware has expanded significantly from simple batch-processing front-ends to complex tools that provide advanced features like scheduling, reservation and information sharing.

Many complex distributed applications are today structured as workflows that consist of off-the-shelf software components, which are usually applications to be run on individual sequential or parallel machines. The specification and management of workflows is complex and currently the subject of many research projects. Typically, much of the existing work focuses on workflow languages which describe component interconnection features, on the architecture of the enactment engine which coordinates the workflow execution, or on the optimization of the execution by using complex mapping and scheduling techniques.

JavaSymphony is a programming paradigm for wide classes of heterogeneous systems that allows the programmer to control the locality, parallelism, and load balancing at a high level of abstraction without dealing with error-prone and low-level middleware

---

* This research is partially supported by the Austrian Science Fund as part of Aurora Project under contract SFBF1104.

J.C. Cunha and P.D. Medeiros (Eds.): Euro-Par 2005, LNCS 3648, pp. 272–281, 2005.

details, like creating and handling remote proxies for Java/RMI or socket communication. The JavaSymphony middleware consists of distributed objects and remote method invocations that run on distributed computing resources like workstation networks and SMP clusters. Moreover, JavaSymphony offers high level features [1] like migration, a distributed event mechanism, and distributed synchronization mechanisms, which are highly useful for developing distributed applications.

Recently, we have built new features on top of the JavaSymphony programming paradigm and runtime system, to support the widely popular workflow paradigm, which include a high-level tool that allows the graphical composition of a workflow, an expressive, yet simple workflow specification language, and an automatic scheduler and enactment engine for workflow applications. In previous work [2], we have presented a formal workflow model consisting of basic elements like activities, control and data flow links, loops, and branches. In this paper, we introduce a theoretical framework for scheduling and propose a scheduling technique for dynamic workflow applications with loops and conditional branches.

The paper is organized as follows. The next section discusses the elements of the workflow model used in JavaSymphony. Section 3 describes the framework for scheduling JavaSymphony workflow applications. Section 4 discusses related work. Finally, some concluding remarks are made and the future work is outlined in Section 5.

2   Workflow Model

A workflow consists of several interconnected computing activities. Between two computing activities there may be: (1) a control flow dependency, which means that one activity cannot start before its predecessors finished or (2) a data dependency, which means that one activity needs input data that is produced by the other. We use the terminology and specifications proposed by the Workflow Management Coalition[3] to define the workflow model and its elements. A graphical representation based on the UML Activity diagram ([4]) is associated with each workflow. In JavaSymphony workflow model, each workflow application is associated with a workflow graph defined by:
$WF = (Nodes, CEdges, DEdges, Loops, PLoops, istate, fstate)$.

$Nodes = Act \cup DAct \cup Init \cup Final \cup Branches$ comprises the vertices of the graph associated with 5 types of workflow basic elements: activities, dummy activities, initial states, final states and branches. There are 4 types of edges for the workflow graph: $CEdges, DEdges, Loops$, respectively $PLoops$ are the sets of the control links, data links, respectively loops and parallel loops of the graph. These basic workflow elements are shortly explained below.

Activities are represented as elements of the Act set. The workflow activities are placed onto computing resources and perform specific computation.

Dummy activities are represented as elements of the $DAct$ set. As a special type of activities, they evaluate complex conditional expressions that may influence the workflow schedule. On the other hand, they require only minimal computing power and therefore they run locally within the scheduler, instead of being placed onto distributed computing resources.

Control links correspond to the elements of the $CEdges$ set. A control-link between two activities means that the second activity cannot start before the first one finishes. The **control-precedence relation**, denoted by $<$, is defined over the elements of the $Nodes$ set, as the transitive closure of $CEdges$.

Data links define the **data-precedence relation** (denoted by $<_d$) over the set of the activities of a workflow. A data-link between two activities means that the second activity requires output data from the first one.

Initial and final states correspond to the elements in the sets $Init$, respectively $Final$. Each workflow has one entry and one exit point, which we call **initial state**, respectively **final state**. These are used for synchronization of activities and to mark the body of the so-called sub-workflows. They are not associated with computation.

A sub-workflow unit is delimited by a unique pair of an initial state (entry point) and a final state (exit point): $(i, f) \in Init \times Final$.

Conditional branches are represented as elements of the $Branches$ set. Due to the conditional branches, the execution plan of a workflow changes dynamically. The successors of the conditional branch correspond to the entry points (i.e. initial states) of sub-workflows. Each conditional branch exit (control link) is associated with a Boolean expression. When the execution reaches the conditional branch, the Boolean expressions are evaluated and the successors for which this expression evaluates to *false* will not be executed.

(Sequential) Loops are represented as the elements of $Loops \subset Final \times Init$ and may be attached only to entire (sub) workflow units. The body of a (sub)workflow which has a loop associated with it, is executed repeatedly for a fixed number of times (for-loops), or until an associated condition is satisfied (until-loops).

Parallel Loops are represented as the elements of $PLoops \subset Final \times Init$. They are similar with the regular loops, but model a different behaviour of the associated sub-workflow: For each parallel loop, the number of iterations n is specified, and n identical copies of the associated sub-workflow will be created and executed **in parallel**. A parallel loop can be replaced with n identical copies of the associated sub-workflow, but in this case a significantly more complex workflow graph is necessary.

3   Scheduling Workflow Applications

To build a JavaSymphony workflow application, one has to first design the workflow graph, by using the specialized graphical user interface. The developer puts together workflow activities, dummy activities, initial and final states, and connects them using control links, data links, loops and parallel loops, according to the model described in the Section 2. The result is an easy-to-understand workflow graphical representation, based on the UML Activity Diagram, which can be stored in a file by using the specific XML-based specification language. Behind the graphical representation, each element (vertices and edges of the graph) is associated with relevant workflow information. Within the same scheduling process, the workflow specification is analyzed, a resource broker determines which resources are suitable for each workflow activity, a scheduler computes the workflow execution plan, and a enactment engine manages the execution of the activities according to the execution plan. In this section, we present a theoretical framework to describe the scheduling process, and propose a scheduling technique for workflows with branches and loops.

3.1   Scheduling Workflows Without Branches and Loops

We consider first the case of scheduling workflows with no loops and branches. The graph associated with a workflow with no loops and branches becomes a static DAG. Therefore, we call such workflows DAG-based workflows. Scheduling DAGs of tasks is a problem that has been intensively studied, and consequently we can easily use one of the many already existing algorithms [5–8] for scheduling DAG-based workflows. In this section, we introduce several basic definitions and notations related to the scheduling of DAG-based workflows.

If WF is a workflow with $Loops = PLoops = \emptyset$ and $Branches = \emptyset$, then a **schedule** for WF would be a function **sched** : $Act \cup DAct \to M \times \mathbb{R}_+$, where M is the set of computing resources and $\mathbb{R}_+$ is the set of positive real numbers. $\mathbf{sched(T)} = (m_T, start_T)$ means that the activity T is started on machine m_T at the time $start_T$.

The execution time of an activity T on machine m is denoted by $\mathbf{exec(T/m)}$. We assume that the task runs exclusively on that machine. **The communication time** to send data from activity T_1 running on m_1 to activity T_2 running on m_2 is denoted by $\mathbf{comm(T_1/m_1, T_2/m_2)}$. Note that if $T \in DAct$, we may assume m_T is always a dedicated or local machine m_0 (where the scheduler is running) and we consider $exec(T/m_T)$ to be 0. We also assume that communication time for two activities running on the same machine is 0: $comm(T_1/m, T_2/m) = 0$

For a DAG-based workflow WF, a schedule $sched$ is constrained by the workflow control- and data-dependencies:

$$T_1 < T_2 \text{ implies } start_{T_1} + exec(T_1/m_{T_1}) \leq start_{T_2}$$
$$T_1 <_d T_2 \text{ implies } start_{T_1} + exec(T_1/m_{T_1}) + comm(T_1/m_{T_1}, T_2/m_{T_2}) \leq start_{T_2}$$

The goal of the scheduler is to find a schedule for each workflow application, which optimize a specific performance function, under certain constraints. Such functions are: makespan (execution time of the whole workflow application), total cost of the resources (when the resources are associated with computation/communication cost) or the throughput of the entire system.

3.2   Scheduling Workflows with Branches and Loops

The conditional branches and the loops in the workflow model enforce dynamic changes in the structure of the execution task graph associated with the application. Subsets of the activities which make up the application may be executed repeatedly several times or may not be executed at all, based on data that is available only at runtime. Consequently, scheduling techniques for static DAG-based workflows cannot be applied in this case.

Our strategy is to transform the workflow associated with the application into one with no conditional branches and loops and recursively find a schedule in the conditions of Section 3.1.

We first define two types of activities: **Unsettled activities** are the activities for which the scheduling/execution decision is taken based on data that is not (yet) available. Such activities are, for example, the activities subsequent to a conditional branch, for which the associated condition cannot be evaluated, because the parameters in the Boolean expression have not been calculated yet. Therefore, it is not sure at this point that these activities will ever be scheduled for execution. The rest of the activities are called **settled activities**. These are the activities that are planned for execution or have been executed at a specific time of the scheduling/execution process. All the activities for which it is sure that they will be scheduled for execution are considered **settled**. The two sets of activities of a workflow application are dynamically changing during execution, according to the following transformations:

Parallel loop elimination is performed before the scheduling actually starts if the number of the iterations is determined at design time. Otherwise, if the number of iterations depends on the value of workflow relevant data (e.g. variables values), the transformation is applied upon reaching the loop entry (i.e. associated initial state). The body of the parallel loop construct (i.e. the associated sub-workflow) is simply replaced with n identical copies (see Fig. 4(c)).

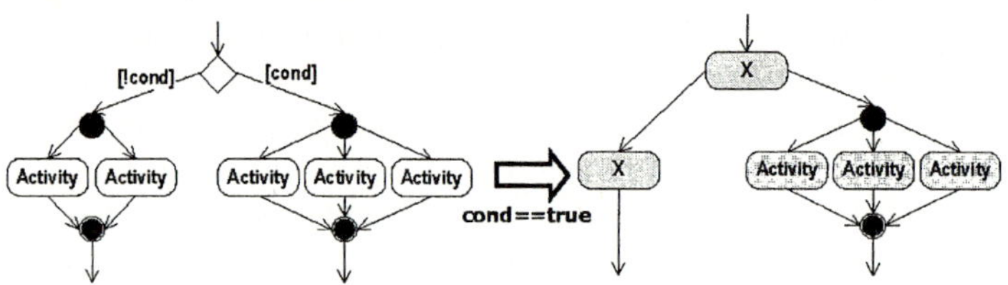

Fig. 1. Branch elimination

Branch elimination is applied when the conditions for the conditional branches are evaluated. This transformation takes place at runtime and is illustrated in Fig. 1. Note that the successors of a conditional branch are unsettled activities (uncoloured in the picture) before the evaluation of the condition, and become settled activities (coloured in the picture) after that. The branches for which the associated condition evaluates to $false$ are not executed. They are replaced by dummy activities (marked as X in the figure), which do not perform any computation.

Transformation of for-loops. The for-loops have a fixed number of iterations. This transformation may take place anytime during the scheduling process. For each iterations of the loop, clones of the activities (i.e. new activities with the same properties as the original ones) in the body of the loop and associated control/data links are added to the graph. The new activity clones preserve the settled state, if the original activities have been settled activities before the transformation.

Transformation of until-loops. The until-loops terminate when a specific condition is fulfilled. The evaluation of the condition can be performed only at runtime. This transformation is illustrated in Fig. 2. For each iteration of the loop, clones of the activities in the body of the loop and associated control/data links are added to the graph. The activities in the first iteration remain settled after the transformation if they have been settled, but the clone activities in the consequent iterations are unsettled. Any activity subsequent to an until-loop preserves its unsettled state until all the iterations of the loop are executed.

Elimination of initial and final states. The initial and final states are simply replaced by dummy activities, not associated with computation. If all their (direct) predecessors are settled activities, these become settled dummy activities.

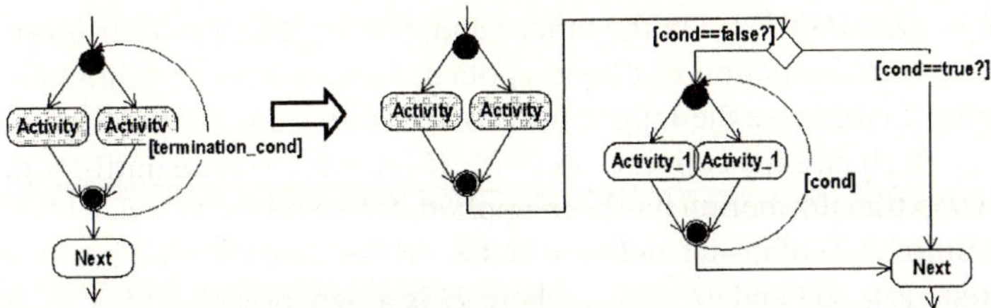

Fig. 2. Until-loops transformation

We use the notation $WF \longmapsto WF_t$ to express that WF_t is obtained from WF applying the above-mentioned transformations. We iteratively build a transformed workflow as follows: Initially (pre-scheduling), all possible transformations, except branch elimination, are applied. The workflow application is scheduled/executed until a conditional branch is reached (i.e. all predecessors of a conditional branch finished their execution). Upon this event the branch elimination is applied, followed by all the other possible transformations. The sets of settled, respectively unsettled activities are recalculated after each transformation step as following.

For $B \in Branches$ a branch node, we denote by $Next(B)$ the set of direct successors of B, which comprises all activities directly dependent via control edges on B and all the activities of the sub-workflows directly dependent via control edges on B. According to this definition, $Next(B)$ comprises all the activities that may be cancelled after reaching the conditional branch B. Note that the decision to cancel or not an activity from $Next(B)$ set can be taken only when the execution reaches B and all conditions associated with the subsequent branches are evaluated.

Consequently, the **set of unsettled activities** is $U(WF_t) = U_1 \cup U_2$, where $U_1 = \bigcup_{B \in Branches} Next(B)$ and $U_2 = \{N \in Act \cup DAct | \exists M \in U_1, M < N\}$. The **set of settled activities** is therefore $S(WF_t) = Act \cup DAct - U(WF_t)$. We denote by $DAG(WF_t) = (S(WF_t), (Edges(WF_t) \cup Loops(WF_t)) \cap S(WF_t) \times S(WF_t))$, the graph which has $S(WF_t)$ as vertices, and all the control links, and loops from WF_t that have both the targets and sources in $S(WF_t)$ as edges.

For a workflow WF, we define **a control path** as a series of activities $A_1, A_2, \ldots A_k$, where each pair (A_i, A_{i+1}) is either a control link or a sequential loop. Using the above-mentioned notations and definitions, we demonstrate the following property of $DAG(WF_t)$:

Lemma 1. $DAG(WF_t)$ *is a DAG which preserves the control paths of the initial workflow* WF.

Proof:

$DAG(WF_t)$ **has no loops.** According to the transformation of while loops, the body of a loop in WF_t has only unsettled activities. Therefore, the final state associated with a loop is not in $DAG(WF_t)$ and accordingly, the loop is not edge in $DAG(WF_t)$.

$DAG(WF_t)$ **preserves the control paths of** WF means that for each control path $A_1, A_2, \ldots A_k$ of WF, with all A_i in $S(WF_t)$, there is a corresponding control path in $DAG(WF_t)$. First, the control edges of the initial workflow are preserved by all transformations, so if $A_i, A_{i+1} \in S(WF_t)$ and $(A_i, A_{i+1}) \in CEdges$, implies (A_i, A_{i+1})

is also edge in $DAG(WF_t)$. On the other hand, if (A_i, A_{i+1}) is a for-loop, this means that a for-loop transformation has been applied, followed by an elimination of initial and final states. In this case the loop is transformed into a control link between A_i and a clone of A_{i+1}, both dummy activities in WF_t. If (A_i, A_{i+1}) is an until-loop, this means that a until-loop transformation has been applied, followed by a branch elimination and then by an elimination of initial and final states. In this case the loop is transformed into 2 control links: (A_i, B) and B, A'_{i+1}, where B is a new branch and A'_{i+1} is a clone of A_{i+1} in WF_t and all of them are (newly created) dummy activities.

---

1. Apply all possible transformations to the initial workflow $WF \longmapsto WF_t$, and compute $U(WF_t)$, $S(WF_t)$ and $DAG(WF_t)$.
2. A scheduling algorithm for DAG-based workflows (no conditional branches and loops) is applied to $DAG(WF_t)$.
3. At each scheduling event, $U(WF_t)$, $S(WF_t)$ and $DAG(WF_t)$ are recalculated. Note that termination of activities may imply adding their successors to $S(WF_t)$. Changes in $DAG(WF_t)$ automatically imply scheduling/rescheduling of unfinished activities.
4. When the execution reaches a conditional branch a branch elimination transformation is applied, followed by all the other possible transformations.
5. The result is a new WF_t, and new $U(WF_t)$, $S(WF_t)$ and $DAG(WF_t)$ are calculated. The scheduling algorithm is now applied to the new $DAG(WF_t)$.
6. The iterative scheduling/execution process finishes when all activities (in all iterations of all loops) are processed. At this point $U(WF_t) = \emptyset$, and $S(WF_t)$ comprises all the activities of WF, including the new created clones of activities (for each additional iteration of a loop) and all new created dummy activities.

Fig. 3. Strategy for scheduling workflows with loops and branches

Consequently, the dynamic scheduling strategy in Fig. 3 is adopted for workflows with conditional branches and loops.

3.3   A Sample Workflow Application

We have tested our dynamic scheduling strategy with a real-life application. WIEN2k [9] is a program package for performing structure calculations of solids using density functional theory, based on the full-potential (linearised) augmented plane-wave ((L)APW) and local orbitals (lo) method.

The components of the WIEN2k package can be organized as a workflow (Fig. 4). The *lapw1* and *lapw2_TOT* tasks can be solved in parallel by a fixed number of so-called *k-points*. This is modelled by two parallel loops in the workflow graph. Without the parallel loops, the workflow graph becomes quite complex (Fig. 4(c)). Various files are sent from one workflow activity to another, which determine complex data dependencies between the activities (Fig. 4(b)). At the end of the main sequence of the activities, a dummy activity *testconv* performs a convergence test to determine if the calculation needs to be repeated. This is modelled by the main sequential loop.

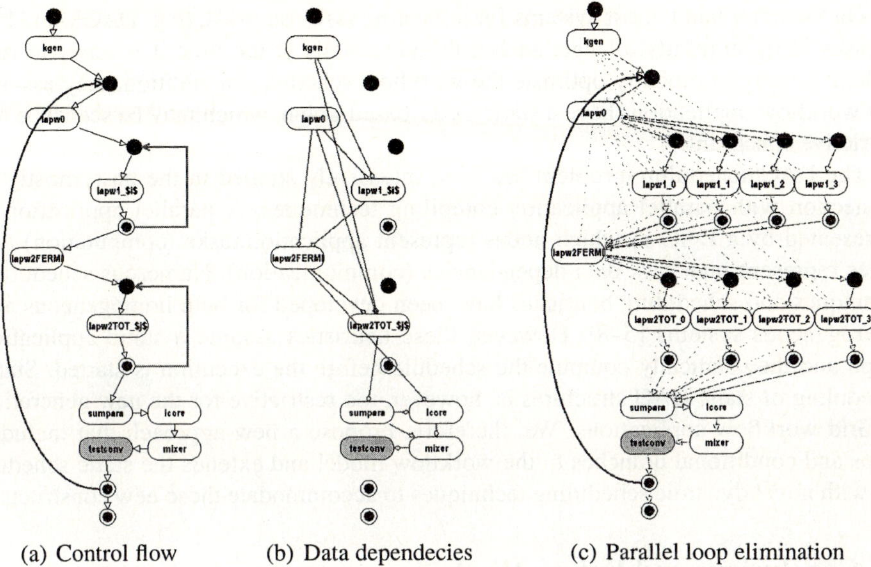

(a) Control flow   (b) Data dependecies   (c) Parallel loop elimination

Fig. 4. Wien2k workflow

We have successfully built a JavaSymphony workflow application on top of the
WIEN2k package. We have used HEFT (Heterogeneous Earliest Finish Time) [5] list
scheduling algorithm combined with the dynamic scheduling strategy described in
Fig. 3, to schedule and run this application onto a set of workstations. Due to space
limitations, in this paper we do not investigate the workflow scheduling performance.
We intend to implement several other real-life distributed applications and to investigate
several other scheduling algorithms in future work.

4   Related Work

Workflow applications have become very popular in Grid community and many research and industry groups have proposed language standards to model and develop
workflow applications [10–13]. We do not intend to compete with highly complex
workflow definition languages [10, 13]. Instead, the JavaSymphony specific XML-
based specification language for workflow applications is simple, in order to allow
an easy manipulation of the workflow structure by a scheduler. The same is valid for
the workflow graphical representation. Activity Diagrams or Petri Nets have been extensively studied as alternatives for the representation of the workflows ([14, 15]). In
[15] diverse workflow patterns are analyzed. However complex workflow specification
languages or complex workflow patters are not commonly associated with advanced
scheduling techniques for distributed workflow applications. We prefer to use a simplified graphical workflow application representation (a reduced set of workflow patterns),
in order to be able to investigate such advanced scheduling techniques.

On the other hand, most systems for allocating tasks on grids, (e.g. DAGMan [12], Pegasus [16]), currently allocate each task individually at the time it is ready to run, without aiming to globally optimise the workflow schedule. In addition, they assume that workflow applications have a static DAG-based graph, which may be seen as a too restrictive constraint.

The DAG scheduling problem has been intensively studied in the past, mostly in connection with parallel application compiling techniques. A parallel application is represented by a DAG in which nodes represent application tasks (computation) and edges represent inter-task data dependencies (communication). Numerous scheduling techniques and scheduling heuristics have been developed for both homogeneous and heterogeneous systems [5–8]. However, these heuristics assume a static application graph and they statically compute the schedule before the execution is started. Static scheduling of static DAG structures is, however, too restrictive for the new generation of Grid workflow applications. We, therefore, propose a new approach that includes loops and conditional branches to the workflow model and extends the static scheduling with novel dynamic scheduling techniques to accommodate these new constructs.

5   Conclusions and Future Work

JavaSymphony is a system designed to simplify the development of parallel and distributed Java applications on heterogeneous computing resources ranging from small-scale clusters to large scale Grid systems.

In this paper, we have presented a formal model to describe workflow applications, which allows a user-friendly graphical workflow representation based on the UML Activity Diagram, and a novel framework for scheduling workflow applications.

JavaSymphony introduces a mechanism to control loops and conditional branches in workflow applications, which is not supported by many other workflow frameworks. Furthermore, we describe a new scheduling technique for workflows which have loops and conditional branches.

We plan to evaluate this technique with several DAG-scheduling heuristics [5–8], and compare their performance with several workflow applications. We also plan to further investigate new scheduling techniques for various types of distributed applications and programming paradigms (e.g. meta-tasks, master/slave applications, etc..) and support them in JavaSymphony.

References

1. Jugravu, A., Fahringer, T.: JavaSymphony: A new programming paradigm to control and to synchronize locality,parallelism, and load balancing for parallel and distributed computing. Concurrency and Computation, Practice and Experience (2003)
2. Jugravu, A., Fahringer, T.: JavaSymphony, A Programming Model for the Grid. Future Generation Computer Systems (FGCS) **21** (2005) 239–246
3. WfMC: Workflow Management Coalition: http://www.wfmc.org/ (2003)
4. Dumas, M., Hofstede, A.: UML Activity Diagrams as a Workflow Specification Language. In: 4th International Conference on UML, LNCS 2185, Toronto, Canada, Springer Verlag (2001)

5. Topcuoglu, H., Hariri, S., Wu, M.Y.: Task scheduling algorithms for heterogeneous processors. In: Eighth Heterogeneous Computing Workshop, IEEE C.S. Press (1999) 3–14
6. Kwok, Y.K., Ahmad, I.: Benchmarking and comparison of the task graph scheduling algorithms. Journal of Parallel and Distributed Computing **59** (1999) 381–422
7. Baskiyar, S., SaiRanga, P.C.: Scheduling directed a-cyclic task graphs on heterogeneous network of workstations to minimize schedule length. In: Proc. of International Conference on Parallel Processing Workshops,Kaohsiung, Taiwan. (2003)
8. Radulescu, A., van Gemund, A.J.C.: Fast and effective task scheduling in heterogeneous systems. In: Heterogeneous Computing Workshop. (2000) 229–238
9. P.Blaha, K.Schwarz, G.Madsen, D.Kvasnicka, J.Luitz: WIEN2k: An Augmented Plane Wave plus Local Orbitals Program for Calculating Crystal Properties. Vienna University of Technology (2001)
10. Andrews, T., Curbera, F., Dholakia, H., Goland, Y., Klein, J., Leymann, F., Liu, K., Roller, D., Smith, D., Systems, S., Thatte, S., Trickovic, I., Weerawarana, S.: Business process execution language for web services (bpel4ws). Specification version 1.1, Microsoft, BEA, and IBM (2003)
11. Erwin, D.W., Snelling, D.F.: UNICORE: A Grid computing environment. Lecture Notes in Computer Science **2150** (2001) 825–??
12. The Condor Team: Dagman (directed acyclic graph manager) (2003) http://www.cs.wisc.edu/condor/dagman/.
13. Krishnan, S., Wagstrom, P., von Laszewski, G.: GSFL : A Workflow Framework for Grid Services. Technical Report, Argonne National Laboratory, 9700 S. Cass Avenue, Argonne, IL 60439, U.S.A. (2002)
14. Eshuis, R., Wieringa, R.: Comparing Petri Net and Activity Diagram Variants for Workflow Modelling - A Quest for Reactive Petri Nets. Lecture Notes in Computer Science **2472** (2003) 321–351
15. van der Aalst, W., ter Hofstede, A., Kiepuszewski, B., Barros, A.: Workflow patterns. Distributed and Parallel Databases **14(3)** (2003) 5–51
16. Deelman, E., Blythe, J., Gil, Y., Kesselman, C., Mehta, G., Vahi, K., Blackburn, K., Lazzarini, A., Arbree, A., Koranda, S.: Mapping abstract complex workflows onto grid environments. Journal of Grid Computing **1** (2003) 25–39

Tasks Mapping with Quality of Service
for Coarse Grain Parallel Applications

Patricia Pascal, Samuel Richard, Bernard Miegemolle, and Thierry Monteil

LAAS-CNRS, 7 Avenue du Colonel Roche 31077 Toulouse France
{ppascal,srichard,bmiegemo,monteil}@laas.fr

Abstract. Clusters and computational grids are opened environments on which a great number of different users can submit computational requests. Some privileged users may have strong Quality of Service requirements whereas others may be less demanding. Common mapping algorithms are not well suited to guarantee a defined quality of service, they propose at best priority systems in order to favour some applications without any guaranty. We propose a new mapping algorithm, dealing with the notion of quality of service for scheduling applications over clusters and grids over different classes of service.

This algorithm uses information on the application to map, all the unfinished applications previously mapped, the state of the execution support, and the processor access model (round robin model) to suggest a mapping which guarantees all the expressed constraints. The mapping decision is taken on-line based on the release date of all applications and the memory space used. To finish, the validation of the algorithm is performed with real log files entries simulated with Simgrid.

Keywords: scheduling, quality of service, resource manager, grid, clusters

1   Introduction

In distributed environments, resource management is very important in order to take advantage of multiple hosts and to optimize resource use. The sheduling policy commonly used on distributed systems is best effort with priorities. Different queues are created: short jobs, long jobs, high parallel jobs, etc with FIFO or more elaborated policies. This system of queues can be used to allow the differentiation of users by assigning a priority to each queue but does not guarantee any quality of service. This limitation is due to historical reasons because batch schedulers have been created for parallel computers which are generally used by few users. Clusters are more opened and also complicated environments. Due to their low cost, they can be accessed by a lot of different users that have different needs and expectations; they are also connected with network on which different policies of quality of service can be used. For this reason, batch schedulers are not well suited to ensure different qualities of service to many users using the same execution environment. In this article, a scheduling algorithm, used in distributed systems like clusters or aggregation of clusters (grid), and

J.C. Cunha and P.D. Medeiros (Eds.): Euro-Par 2005, LNCS 3648, pp. 282–291, 2005.
© Springer-Verlag Berlin Heidelberg 2005

which implements different classes of service, is presented. This algorithm is implemented in a tool called AROMA (scAlable ResOurces Manager and wAtcher) [4]. AROMA integrates a resource management system, an application launcher, a scheduler, a statistic module and an accounting system.

In the first section, a state of the art is presented; then the context of the study, the notion of quality of service and AROMA are detailed. After that, the optimization problem that has been solved is explained. To conclude, first results validating the proposed algorithm are given. The originality of this work is to mix different processes from different classes of service on the same processor at the same time.

2 Related Work

Batch and dynamic schedulings are difficult problems to solve because resources needed by an application may not be known. Resources are heterogeneous and their availability is not completely known. Moreover, the mapping algorithm must run quickly, therefore heuristics with good properties are used. Henri Casanova in [5] studies deadline scheduling on computational grid. His goal is to minimize the overall occurrences of deadline misses as well as their magnitude.

Rajkumar Buyya in [6] proposes a deadline and budget constrained cost-time optimization algorithm for scheduling on grids. The algorithm is called DBC (Deadline and Budget Constrained).

Mechanisms have been created to improve the mapping. The first one concerns resource reservation. Different types of resource reservation algorithms are studied in [1]. They evaluate the performance with or without preemption. The reservation insures that all the resources necessary to run the applications will be free. A second mechanism is the gang-scheduling. It creates time slices, that is to say parts of time that are allocated to the parallel and sequential applications [2]. With this solution, all applications progress simultaneously. Nevertheless, it could create a problem of overload and memory saturation. Finally, the backfilling [3] allows the insertion of jobs into scheduler queue. The insertion is possible if it does not perturb the other jobs. It is a way to remove the holes in resources utilization.

This article proposes a way to mix jobs requirements with different qualities of service (deadline, immediate execution, dedicated resources).

3 Mapping Algorithm

As the context of the study is ASP (Application Service Provider), the hypothesis that all the applications consuming resources are known is made : that is to say, hosts are considered dedicated to computation. All the jobs are submitted through AROMA and system tasks influence is neglected. AROMA daemons are also able to monitor running applications; this information is used to refresh estimated completion date of running jobs. The second hypothesis made is that applications are regular coarse grain parallel applications for which the time spent in communication is small and the execution time can be roughly predicted. The problem is to find the mapping of a new application knowing all the previously mapped applications that are still consumming resources in the system. The proposed mapping has to guarantee that the quality of service is respected for all applications (running and currently scheduled applications).

3.1   Quality of Service and Mapping Problem

Applications are grouped into four application classes (in order of importance):

- **Deadline applications (class 1):** this class of service guarantees that the execution will end before the deadline. Execution can be immediate or deferred.
- **High priority applications (class 2):** this class of service guarantees that the execution will be immediate.
- **Applications with dedicated resources (class 3):** this class of service guarantees that each application will be the only one to use resources during its execution. Execution can be immediate or deferred.
- **Applications without constraint (class 4):** this class of service corresponds to applications which will be executed as soon as possible with available resources. This class is also named "Best Effort".

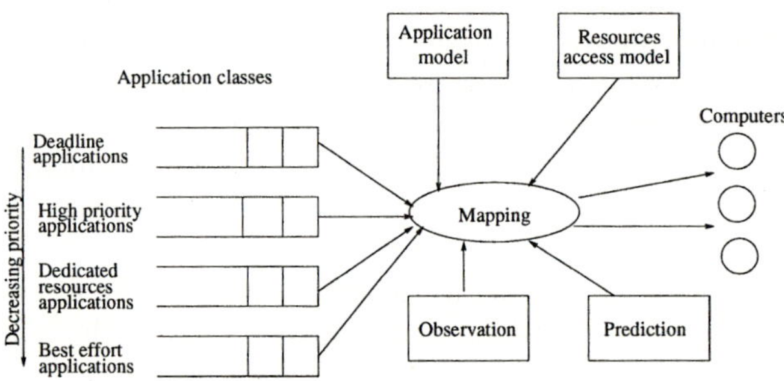

Fig. 1. The mapping problem

The mapping problem inputs are (figure 1):

- **Application model:**
  Some information describing the application needs and requirements has to be supplied to the scheduler in order to take a good mapping decision. Some elements are inputs of the algorithm while others express constraints for the mapping. Inputs are an estimation of the cpu time required by each task, the number of tasks and the size of exchanged data between the different tasks of the application. Those values can be given by the user or retrieve

from a database containing information on previous runs for the same type of application. The application model can express additional constraints like the fact that all the tasks must begin at the same date or temporal relations between the tasks (classical description in graph theory).

Software or specific hardware requirements add constraints on the execution hosts. Class 1 induces a constraint on the ending date of the application, class 2 induces a constraint on the starting date of the application and class 3 induces a constraint on the execution hosts.

All the constraints are verified by the algorithm.

– **Resources access model:**

According to the resources access policy, equations are deduced and used to predict the utilization time of the resources and the end of execution. Models developed in this article are deterministic, nevertheless models which take care of random perturbation (arrival of uncontrolled jobs, for example a direct login on the host) have been developed in [12]. The difficulty is to mix different applications from different classes of service on the same host while respecting all the constraints: several tasks may share the same host during the same period of time.

– **Observations:**

The processors and network load (percentage of processors utilization, bandwidth used), idle memory space and number of processes are monitored. They are used to refresh information used for mapping.

– **Mapping:**

Different queues exist, each corresponding to a priority level (figure 1). When several applications have to be mapped, the jobs of the highest priority queue will be treated first. If no mapping respecting the constraints is found, three cases are studied:

- the algorithm try to move a job with a weaker priority : it looks for an application in a lower priority queue which has been planned to be run in the future, then it removes it and try to map it again after the current mapping.
- if the previous case is impossible, the algorithm can stop a running application and try to find a new mapping for this application. In this case, a new constraint is created to express that this application must go on later on the same host. There is no migration.
- if the two previous cases are impossible, the mapping request is rejected.

3.2 Mathematical Expression of the Problem

The Variables

- t_0: initial date of the mapping research.
- $t_b(a, p, m)$: execution starting date of the task p of the application a on the host m.
- $t_{fwn}(a, p, m)$: end of execution of the task p of the application a on the host m when the network is neglected.

- $t_f(a, p, m)$: end of execution of the task p of the application a on the host m when an estimation of time spent on communication is done.
- $t_c(a, p, m)$: processor time requested by the task p of the application a on the host m. Coarse grain applications are considered, so "small" communication time is taken into account and the synchronization time is neglected.
- $t_c^k(a, p, m)$: remaining processor time for the task p of the application a after the event number k on the host m (events are a beginning or the end of a task).
- $D_r(a, p)$ estimation of the size of data sent and received by the task p of the application a.
- C(m) : coefficient to take care of heterogeneous processors.
- $t_c(a, p, m) = t_c(a, p) * $ C(m) : equivalence of processor time requested by the task p of the application a for the host m.
- B(i,j) : estimation of bandwidth of the network between host i and host j.
- M : number of hosts. A : number of applications to map.
- N^a : number of tasks of application a.
- $X_m(t)$: number of processes on the host m at time t.
- t_m^k : a beginning event or an ending event of a process on the host m. It is the date of the event number k.
- P(a) : set of possible mapping for application a.
- $t_f(m, s)$: the end of all the tasks mapped on the machine m after the mapping s of application a with $s \in$ P(a).
- $M(a, p)$: the machine on which the task p of the application a is executed.
- $U(m)$: number of processors available on host m.
- $M_u(m)$: total memory used on host m. $M_t(m)$: total memory on host m.
- M_f: constant to modify the weight of the memory criteria.

The Optimization Criteria: The mapping problem is an optimization problem, criterion has to be chosen. Several criteria are well known ([7][8][9][10]) : makespan (minimizing the termination date of an application), sum-flow (minimizing the quantity of resources used), max-stretch (it expresses that a task has been slowed compared to what its execution would have been on an idle server). The objective here is to optimize the use of the providers resources and to guarantee the level of quality of service required. So it has been chosen to liberate all the resources as soon as possible. Applications already mapped may be influenced by the mapping found. By consequence, all the applications (the currently mapped and the previously mapped) must finish as soon as possible.

The date of the end of resources utilization is optimized. Moreover a second criteria consists in moderating with memory space used: hosts which have the most free memory space are privileged first. The choice has been made to introduce memory in criteria because the exact amount of memory requested by an application is often unknown by users. The problem is multi-criteria by using a linear combination of different criteria (1). The first part of the addition $(t_f(m, s))$ refers to the release date of the machine. The second part of the addition $t_f(m, s) * M_u(m) * M_f/M_t(m)$ penalizes machines which have less free memory. $M_u(m)/M_t(m)$ gives an idea of memory utilization on this host. The coefficient M_f influences the weight of this part of criterion. The multiplication with $t_f(m,s)$ puts the value into the same order of value as the first part of the criterion.

$$\min_{s \in P(a)} \left(\max_m (t_f(m,s) + t_f(m,s) * M_u(m) * M_f/M_t(m)) \right) \tag{1}$$

The Mapping Algorithm: A list algorithm for applications, tasks and machines is used to reduce the combinatorial. The mapping of an already studied task of an application a is revised only if it is impossible to find a mapping for this application. Moreover, in order to reduce the time used by the algorithm, for each host, release date is saved and hosts are ordered to study which of them will give the best mapping first.

The quality of service is already respected, all the constraints induced are verified by the algorithm. If it is impossible to find a mapping corresponding to the demand, the request is refused.

The computation of $t_b(a,p,m)$ and $t_f(a,p,m)$ will now be explained. The equations are found considering that the processor access follows a round robin policy.

The Starting Date of a Task: $t_b(a,p,m)$ is computed with an iterative algorithm (at the beginning it is t_0). If, at this date, no mapping can be found, another date is searched.

$t_b^0(a,p,m) = t_0$

$t_b^{k+1}(a,p,m) = \min t_f(i,j,m)$

with $i \in [1, a-1]$, $j \in [1, N^i]$ and $t_f(i,j,m) > t_b^k(a,p,m)$ (only the tasks that can end after the last t_b studied, are considered). The idea is to search a new starting date when the system is less loaded: when a job finishes.

The End of a Task: $t_f(a,p,m)$ is computed with an iterative algorithm. Each date is studied when there is a creation or a termination of a job. t_m^0 corresponds to the arrival of the first process on the host.

for all tasks (*a goes from 1 to A and p from 1 to N^a, on m*)

if $X_m(t_m^k) > U(m)$ (*is there more processes than processors ?*)

$t_m^{k+1} = \min_{a,p} (t_b(a,p,m), t_c^k(a,p,m)*X_m(t_m^k)/U(m) + t_m^k)$ (*next event corresponds to a creation or a death of a process*)

else

$t_m^{k+1} = \min_{a,p} (t_b(a,p,m), t_c^k(a,p,m) + t_m^k)$ (*next event corresponds to a creation or a death of a process*)

if $t_b(a,p,m) > t_m^k$ and if $X_m(t_m^k) > U(m)$

$t_c^{k+1}(a,p,m) = t_c^k(a,p,m) - U(m)*(t_m^{k+1} - t_m^k)/X_m(t_m^k)$ (*estimation of the new requested time of processor for this task after this short execution on processor: it is the time sharing policy*)

$$\text{if } t_b(a,p,m) = t_m^i$$
$$\mathrm{t}_c^i(a,p,m) = \mathrm{t}_c(a,p,m)$$

(it is case of an insertion of a new process in the recurrence)

Estimation of the finished date of process without the network:

$$\text{if } \mathrm{t}_c^{k+1}(a,p,m) = 0, \ \mathrm{t}_{fwn}(a,p,m) = t_m^{k+1}$$

Time spent in communication are put inside the estimation of the end of process to advantage location of tasks on the same cluster or on the same site because the bandwidth will be better. The unmapped tasks are ignored in the estimation of the worst bandwidth used for the application a.

$$t_f(a,p,m) = t_{fwn}(a,p,m) + D_r(a,p)/\min_{i\in[1,N^a[} B(M(a,i),m)$$

The iterations continue until all the tasks of the host finish. In fact, the mapping algorithm quickly simulates the execution of jobs and can mix the different classes of service.

4   Validation

To validate the algorithm, Simgrid ([11]) simulator has been used. Real jobs submission log files have been used to estimate the behavior of the algorithm with Simgrid. Nevertheless, it is difficult to compare the algorithm to others because algorithms found do not define classes of service and do not execute processes of the different classes at the same time on the same processors.

4.1   Comparison with NQS

Feitelson logs (real logs) have been used. These logs give : the submission date of jobs, the required cpu time, the number of tasks. The log file *l_sdsc_sp2.swf* [13] is used. This Job Trace Repository is brought by the HPC Systems group of the San Diego Supercomputer Center (SDSC), which is the leading-edge site of the National Partnership for Advanced Computational Infrastructure (NPACI) [14], [15]. The real system has 128 nodes and is scheduled with NQS [16]. Jobs submissions are reproduced, the mapping research is done with the algorithms on 128 nodes and their execution is simulated with Simgrid. The mean waiting times given by the logs are compared to the mean waiting times of the algorithms with quality of service. So, 10000 and 35000 jobs are simulated. Each job was synchronized, this means that all the tasks of the same parallel application must begin at the same date and belong to the dedicated resources class. This class seems to be the nearest from NQS policy. The simulation of 10000 jobs is equivalent to an activity on the supercomputer during 112 days. The simulation of 35000 jobs is equivalent to an activity on the supercomputer during 249 days. There is no information about the communications, so they are neglected. The results show that, with the proposed algorithm, a better waiting time than NQS is obtained. In fact, with the proposed algorithm, an application can be mapped between two others because the processor time required is known and the prediction of the end of each task mapped can be done. This reduces significantly the waiting times. NQS sorts the applications into queues (based on required cpu time) and mixes the applications when a mapping is researched. This can increase the waiting time of short jobs which can be slowed by long ones. This problem is avoided with the proposed algorithm because the applications are considered in the order of submission and classes. The second reason is that the proposed algorithm uses more precise values for the requested time of cpu than NQS; by consequence, the mapping is more accurate. The comparison would be more fair if NQS had as much queues than the different times of processor requested. The results are presented in the table 1. The times are given in seconds (s).

Table 1. Comparison of the waiting time between NQS and the proposed algorithm

	10000 events	35000 events
mean waiting time with NQS(s)	10796	8979
mean waiting time with the algorithm(s)	4008	4202

4.2   Influence of Quality of Service

The same log file as previously has been used. In this log file, it is specified that there are four queues (low, normal, high, express), and for each job the queue of submission is known. So this information is used to make an arbitrary correspondence with the proposed classes of applications. The logs are used only to have an approximation of a realistic incoming rate of applications and a good sample of applications requested cpu time. So queue *low* corresponds to best effort class (class 4), queue *normal* corresponds to deadline class (class 1), queue *high* corresponds to dedicated resources class (class 3) and queue *express* corresponds to high priority class (class 2). For deadline applications (class 1), the deadline is : *submission date + 5*cpu time required*. For high priority applications (class 2), the starting date that must be respected is : *submission date + 5 seconds*. Four cases have been simulated:

- case 1 (10000 events using the four classes) : the waiting time for each class, the global waiting time, the mapping time for each class and the global mapping time are computed.
- case 2 (10000 events in the best effort class) : the global waiting time and the global mapping time are computed.
- case 3 (35000 events using the four classes) : the waiting time for each class, the global waiting time, the mapping time for each class and the global mapping time are given.
- case 4 (35000 events in the best effort class) : the global waiting time and the global mapping time are evaluated.

Doing so, case 1 can be compared with case 2 and then case 3 can be compared with case 4 to see the influence of the quality of service on the mapping

Table 2. Influence of the quality of service on the performance of the algorithm

	Case 1	Case 2	Case 3	Case 4
waiting time class 1 (s)	30.51		30.7	
waiting time class 2 (s)	0.0036		0.003	
waiting time class 3 (s)	4845		5182	
waiting time class 4 (s)	4.55		30.5	
global waiting time (s)	710.5	64.68	396	70.3
mapping time class 1 (ms)	20.22		21.1	
mapping time class 2 (ms)	3.6		3	
mapping time class 3 (ms)	269.3		334.4	
mapping time class 4 (ms)	7.2		28.6	
global mapping time (ms)	50.3	13	49.1	76.4

performances. The results are presented in the table 2. The waiting times are given in seconds (s) and the mapping times in milliseconds (ms).

The introduction of the quality of service increases the mapping times and waiting times but the algorithm still has good performances (mapping time is inferior to 80 ms). The mapping time increases because there are more constraints to verify. It takes many iterations before finding the good t_b. Globally the increase of waiting time is due to class 3 applications (dedicated resources applications) which have to wait a long time before being alone on the machines. The economical or political criteria, which define important jobs, could depreciate the global utilization of resources. In case 1, there are 21 rejects of applications belonging to class 1 (deadline). In case 3, there are 154 rejects of applications belonging to class 1 (deadline). In those cases the machines are full. Because of the constraints of deadline, the tasks can not start later like in classical batch schedulers. Applications of class 2 can be refused but a mapping has always been found. Applications of class 3 and 4 can never be refused, they will only be slowed down.

5   Conclusion

This article presents a scheduling algorithm with quality of service usable in distributed systems like clusters or grids. It is implemented in AROMA : a resource management system used in ASP model. The validation shows that the proposed algorithm is better than NQS, nevertheless, the comparison is difficult because NQS does not implement classes of service and has not access to the same information.

The algorithm always respects the quality of service required for accepted applications. When different applications are mixed, the mapping and the waiting times are more important than when there are only best effort applications. The performances of the algorithm are still very good (mapping time inferior to 80 ms). The communication weight are introduced to favour the execution of an application in the same network area.

Future work will be to improve the notion of communication model and its use into the application model. The theoretical complexity of the mapping algorithm will be studied. The main difficulty will be to explore the complexity of the estimation of the end of a task. More comparisons with other mapping algorithms will be done. Real execution of a set of jobs during many weeks will be done on a small grid over 3 different sites.

References

1. Warren Smith, Ian Foster and Valerie Taylor. *Scheduling with advanced reservations.* Proceeding of the IPDPS Conference, May 2000.
2. Dror G. Feitelson and Morris A. Jette. *Improved utilization and responsiveness with gang scheduling.* Proceeding JSSPP 1997 : 238-261. Job scheduling strategies for parallel processing, IPPS'97 workshop, Geneva, Switerlang.
3. Dmitry Zotkin and Peter J. Keleher. *Job-length estimation and performance in backfilling schedulers.* 8th Intl Symp. High Performance Distributed Comput., august 1999.
4. P.Bacquet, O.Brun, J.M.Garcia, T.Monteil, P.Pascal, S.Richard. *Telecommunication network modeling and planning tool on ASP clusters.* Proceedings of the International Conference on Computational Science (ICCS'2003) Melbourne, Australia, June 2-4, 2003.
5. Atsuko Takefusa, Satoshi Matsuoka, Henri Casanova, Francine Berman. *A study of deadline scheduling for client-server systems on the computational grid.* Proceedings of the Tenth IEEE Symposium on High Performance Distributed Computing (HPDC10) San Francisco, California, August 7-9, 2001.
6. Rajkumar Buyya. *Economic-based distributed resource management and scheduling for grid computing.* Thesis, April 2002.
7. T.L. Casavant and J.G. Kuhl. *Effects of Response and Stability on scheduling in distributed computing systems.* IEEE Transactions on software engineering, vol. 14, No 11, pp. 1578-1588, november 1988.
8. Y.C. Chow, W.H. Kohler. *Models for dynamic load balancing in a heterogeneous multiple processor system.* IEEE Transactions on computers, vol. c-28, No 5, pp. 354-361, 1979
9. C.Y. Lee. *Parallel machines scheduling with non simultaneous machine available time.* Discrete Applied Mathematic North-Holland 30, pp 53-61, 1991.
10. F. Bonomi and A. Kumar. *Adaptative optimal load balancing in a non homogeneous multiserver system with a central job scheduler.* IEEE Transactions on computers, vol. 39, No 10, pp. 1232-1250, october 1990.
11. Henri Casanova, Arnaud Legrand and Loris Marchal *Scheduling Distributed Applications: the SimGrid Simulation Framework.* Proceedings of the third IEEE International Symposium on Cluster Computing and the Grid (CCGrid'03).
12. Patricia Pascal and Thierry Monteil. *Influence of Deterministic Customers in Time Sharing Scheduler.* ACM Operating Systems Review, 37(1):34-45, January 2003.
13. http://www.cs.huji.ac.il/labs/parallel/workload/logs.htmlsdscsp2
14. http://joblog.npaci.edu/
15. http://www.cs.huji.ac.il/labs/parallel/workload/
16. B. Kingsbury. *The network queuing system.* 16 May 1998. http://pom.ucsf.edu/ srp/batch/sterling/READMEFIRST.txt.

Initiating Load Balancing Operations

Marta Beltrán, Jose L. Bosque, and Antonio Guzmán

DIET, ESCET, Rey Juan Carlos University, 28933 Móstoles, Madrid, Spain
{mbeltran,jbosque,aguzman}@escet.urjc.es

Abstract. The initiation rule of a load balancing algorithm determines when to begin a new load balancing operation. Therefore, it is critical to achieve the desired system performance. This paper proposes a generalized procedure for deriving initiation mechanisms or rules based on different objectives for the load balancing algorithm. A new metric, the initiation efficiency, is defined in order to evaluate the initiation performance and to compare the different alternatives.

1 Introduction

Load balancing is critical for achieving high performance in clusters and Grid systems because it enables an effective and efficient utilization of all the available resources ([1],[2]). Dynamic load balancing algorithms can be decomposed in different rules or policies ([3], [4], [5]). But all these decompositions have something in common: it is necessary an initiation mechanism to decide on each system node when to begin a load balancing operation. This mechanism must be efficient, scalable, low overheading, and must be capable of deciding about load balancing operations taking into consideration the available system and workload information.

Different solutions have been proposed for the initiation rule. There are sender-initiated ([5], [6], [7]), receiver-initiated ([8]), symmetric ([9]) and periodic ([10]) rules. On the other hand, some of these solutions are completely local ([5], [7], [11]), i.e, each node evaluates only its own state to determine if a load balancing operations is necessary or not, while other are global ([12]), taking into consideration the global system state.

An exhaustive analysis of all these alternatives allows to conclude that they are designed for a particular load balancing algorithm. The main contributions of this paper are a procedure for deriving initiation mechanisms from general objectives for load balancing algorithms and a performance metric for the initiation rule, the initiation efficiency (ε). It has been defined in order to evaluate the initiation mechanisms performance and to compare the different solutions. For illustration, three example objectives have been proposed to derive their correspondent initiation mechanisms and to compare their performance using the new defined metric.

The rest of this paper is organized as follows. Section 2 proposes the general methodology for obtaining initiation policies from the objectives of the load balancing algorithms. Section 3 illustrates this methodology with three different examples of load balancing objectives. Section 4 defines the initiation efficiency necessary to evaluate these initiation mechanisms performance and to establish comparisons. Section 5 presents some experimental results for the example cases and finally, Section 6 with conclusions.

J.C. Cunha and P.D. Medeiros (Eds.): Euro-Par 2005, LNCS 3648, pp. 292–301, 2005.
© Springer-Verlag Berlin Heidelberg 2005

2   Generalized Procedure for Deriving Initiation Rules

To implement the initiation mechanism for a load balancing algorithm, it is necessary to decide when a load balancing operation should be requested, considering both its possible benefits, and on the other hand, the overhead it will cause. This section proposes a general procedure for designing initiation mechanisms for load balancing algorithms taking into account their general objectives. The steps of this procedure are the following:

1. Describe quantitatively the requirements for the load balancing algorithm. Let ω be the objective that should be achieved with the algorithm to ensure that these requirements are met.
2. Identify an objective function to quantify the achievement of this objective. Let ϕ^{ω} be the objective function for ω. This objective function must depend on the available information about the local and the global state. Mathematically, it should be expressed as $\phi^{\omega}(I)$ where I is a vector composed of the system nodes load indexes (these load indexes quantify the system nodes computing capabilities and must be updated in all the system nodes with some kind of information policy).
3. Incorporate this objective function to the initiation mechanism in the load balancing algorithm. For this last step two kind of objective functions can be distinguished:
   - **Boundary functions:** In this case the objective function defines an upper or lower bound to a certain magnitude. This condition can be directly transferred to the initiation mechanism.
   - **Optimization functions:** The objective function requires the optimization of a certain magnitude. Even though this kind of functions can be sometimes easily incorporated to the initiation rule, they usually introduce too much overhead in the algorithm. To evaluate the load balancing operation it is necessary to solve an optimization problem and such computation may be very expensive for a dynamic load balancing algorithm. In such cases the optimization objective should be transformed to a boundary one, defining an upper or lower bound for the magnitude that was initially supposed to be optimized.

3   Some Initiation Rule Examples

3.1   Objective 1: Maximize System Load Balance

In this first example the aim of the algorithm is to maximize the balance among the system nodes. This is usually the main objective of any load balancing algorithm but it is not the only one as it will be seen later. Once this objective is identified (step 1 of the proposed procedure), an objective function can be defined for the step 2. Let b denote the system balance, therefore, the objective function is:

$$\phi^1 : max(b) \tag{1}$$

The larger the b value, the more balanced is the load of the system. This balance can be quantified at a given instant as the ratio of the minimum load index to the maximum.

That is, it can vary from 0 to 1. With this definition, the objective function can be denoted:

$$\phi^1(I) : max\left(\frac{I_{min}}{I_{max}}\right) \tag{2}$$

But this is an optimization function, so every time a new task arrives to a system node, its initiation mechanism must evaluate the b value for each possible allocation for this task, searching in each case the minimum and the maximum load indexes in the system. This process can suppose a great overhead for the load balancing algorithm, specially in systems with a large number of nodes. To overcome this scalability limitation, the objective function is transformed to a boundary one. This function is:

$$\phi^1(I) : \frac{I_{min}}{I_{max}} > \tau \tag{3}$$

Where τ is the algorithm tolerance value, which defines the threshold desired for the system balance. With this objective function, the step 3 of the general procedure is immediate. The initiation rule must try to allocate all the new tasks achieving this objective: the system balance must be always above the τ value after the allocation. This boundary condition can be directly transferred to the load balancing algorithm with a very simple evaluation, without solving the maximization problem.

3.2   Objective 2: Maximize System Throughput

Here the aim of the algorithm is to maximize the system throughput, that is, to minimize the individual processes elapsed time to finish as many tasks per time unit as possible. This objective is typically identified during the step 1 in systems executing independent tasks for high performance computing. It must be achieved using a load balancing algorithm, therefore, assigning each task to the system node in which its elapsed time will be the shortest. Let t_i be the elapsed time of th ith task, therefore the objective function proposed in the step 2 is:

$$\phi^2 : min(t_i) \ \ \forall i \tag{4}$$

But again ϕ^2 is an optimization function that implies evaluating the new task elapsed time for each possible allocation before assigning it to the best system node in terms of this time requirement. Therefore, this objective function is transformed to a boundary one. In this case the bound is referred to the elapsed time of the new task in its *home node*, that is, the node to which this task is initially assigned. The objective function is:

$$\phi^2 : t_i < \tau \cdot \hat{t}_i \ \ \forall i \tag{5}$$

Where $\hat{t}_i$ is the elapsed time of the ith task in its home node and τ is the algorithm tolerance to define the threshold desired for this objective. For the step 3, this objective function implies that the initiation rule tries to allocate new tasks to nodes where their elapsed time will not be more than τ times greater than in their home nodes. In this paper the DYPAP monitor ([13]) is used to provide local and global state information to the load balancing algorithm, so the objective function can be based on the load indexes values provided by this tool. The load index provided by the DYPAP monitor is based on the CPU assignment, defined as the percentage of CPU time that would be available for a new task on a system node, and on the nodes computational powers.

In the home node the objective function is achieved if the CPU assignment for the new task is:

$$\alpha = \frac{1}{\tau} \tag{6}$$

This CPU assignment is the minimum necessary to accomplish the elapsed time requirements for the new task on this node, but due to system heterogeneity, this may not be the minimum in another node (for example, on a node with more computational power). Therefore, the objective function is based on the load indexes values, which take into account the nodes computational powers. The minimum index necessary to achieve the objective is :

$$I_{min} = \alpha_{min} \cdot \frac{P_{home}}{P_{max}} = \frac{1}{\tau} \cdot \frac{P_{home}}{P_{max}} \tag{7}$$

Where P_{home} is the computational power of the home node and P_{max} is the computational power of the most powerful system node. That is:

$$\phi^2(I) : I > \frac{1}{\tau} \cdot \frac{P_{home}}{P_{max}} \tag{8}$$

The initiation mechanism of the load balancing algorithm must look for a node which fulfill this bound to allocate the new task.

3.3 Objective 3: Minimize the Application Elapsed Time

In this last example the aim of the algorithm is to obtain the best elapsed time for the application executing on the system. Once the objective is identified in the step 1, the objective function must be proposed in the step 2. Let T denote this elapsed time:

$$\phi^3 : min(T) \tag{9}$$

Again this objective must be accomplished using a load balancing algorithm, i.e., allocating new tasks in the best way for this objective function. In this case, this function can be easily implemented in the initiation mechanism despite it is an optimization function:

$$\phi^3(I) : max(I) \; or \; min(I) \tag{10}$$

That is, in the step 3, new tasks are always assigned to the system node with the lowest or greatest load index, depending on this index meaning. For example, assuming again the utilization of the DYPAP monitoring tool, the index maximization should be used because the system node with the greatest load index is the one which offers the best compromise between computational power and CPU assignment and can be easily found by the initiation mechanism without evaluating any expression or predicting the system behavior. In this example the optimization only implies the search of the system node with the greatest index value and this does not introduce too much overhead in the algorithm. Therefore, it may not be necessary to transform it into a boundary function.

4   Initiation Efficiency

In order to evaluate the different initiation rules performance and to compare the different alternatives, an initiation performance metric is needed. In this paper, the initiation efficiency (ε) is defined considering two important issues: the ratio of accepted load balancing operations to the requested operations (R) and the degree of achievement for the load balancing algorithm objective (A). Therefore, the efficiency definition is:

$$\varepsilon = R \cdot A \tag{11}$$

The first factor must be taken into account because a good initiation mechanism should begin load balancing operations only when they are going to be accepted. The rejected operations imply an unnecessary overhead to the system, specially to the network. If the mechanism is not efficient or if it is, but it does not have updated information to decide about load balancing operations, some load balancing operations might be rejected in the target node. The ratio of the accepted operations to the requested operations quantifies the efficiency of the initiation rule in this sense (the largest value being 1 in the best case):

$$R = \frac{O_{acep}}{O_{req}} \tag{12}$$

Where O_{acep} is the number of accepted load balancing operations and O_{req} the number of requested operations.

On the other hand, an efficient initiation rule should comply with the objective of the load balancing algorithm. Due to inaccuracies in the state information, to wrong initiation mechanisms or to very demanding requirements this objective might not be achieved. The degree of achievement of the load balancing algorithm objective is quantified in a different way for the boundary and optimization functions:

– **Boundary objective function:** The degree of achievement of the objective can be measured with the ratio of tasks which are assigned accomplishing the proposed objective to the total number of assigned tasks:

$$A = \frac{tasks\ accomplishing\ the\ objective}{N} \tag{13}$$

Where N denotes the total number of tasks composing the executed application.
– **Optimization objective function:** In this case, the magnitude or attribute that has to be optimized (M) gives the degree of achievement of the objective. Its value can be referred to its optimal value (M_{op}) to quantify how near is the system to the optimal situation:

$$A = \frac{M_{op}}{M}\ or\ A = \frac{M}{M_{op}} \tag{14}$$

Depending on the kind of optimization the first equation (for a minimization) or the second equation (for a maximization) must be used.

Anyway, the A value is always normalized, varying from 0 in the worst case to 1 in the best case. With these definitions, for the initiation rules proposed in the previous section, the initiation efficiency can be measured as:

- **Objective 1:**

$$\varepsilon = \frac{O_{acep}}{O_{req}} \cdot \frac{tasks\ assigned\ with\ b > \tau}{N} \tag{15}$$

- **Objective 2:**

$$\varepsilon = \frac{O_{acep}}{O_{req}} \cdot \frac{tasks\ assigned\ with\ t < \tau \cdot \hat{t}}{N} \tag{16}$$

- **Objective 3:**

$$\varepsilon = \frac{O_{acep}}{O_{req}} \cdot \frac{T_{op}}{T} \tag{17}$$

To evaluate the application elapsed time in the optimum or perfect situation, when all the system load is perfectly balanced, only some information must be known ([14]): the number of tasks that compose the executed application (N), the number of system nodes (q) and their computational powers (P_i with $i = 1, ..., q$). The optimum time value can be obtained supposing that all the workload is sequentially executed on a system with computational power equal to the total computational power of the system (P_T), therefore:

$$T_{op} = \frac{N}{\sum_{i=1}^{q} P_i} = \frac{N}{P_T} \tag{18}$$

With the given definition, a perfect initiation rule would obtain $\varepsilon = 1$. It would request load balancing operations only when they are necessary and can be performed, that is, when they are going to be accepted. And in addition, it would completely achieve the load balancing objective, assigning all the tasks to accomplish this objective or obtaining the desired optimum situation.

5   Experimental Results

This section presents some experimental results to show the influence of the initiation mechanism on the load balancing algorithm performance and to establish the utility of the initiation efficiency in selecting the best initiation rule. These experiments have been performed on a 32 nodes heterogeneous cluster called Medusa. In all the experiments an application composed by 320 tasks is executed on this system. For simplicity, these tasks are independent, i.e. there are no communications between them. In addition it is assumed that they arrive periodically to the cluster, and that they are initially assigned to system nodes between n17 and n31. These assumptions have been made only to simplify the experiments but they are not part of the general formulation presented in previous sections. The computational power (P) for the different system nodes has been computed as the inverse of the elapsed time for this application tasks on each kind of node, being $P_T=2.47$ the global system computational power.

The elapsed time for the selected application is 436 s without the load balancing algorithm, and with equation 18, the elapsed time with an optimum balance would be T_{op}=129.38 s. In this context, the implemented load balancing algorithm must dynamically balance the system workload. This load balancing is based on the DYPAP model ([13]), therefore, it includes the DYPAP monitoring tool to periodically characterize the system nodes state. An event-driven information policy has been used to exchange this state information between the system nodes. And to evaluate the three proposed initiation mechanisms, different implementations of the load balancing algorithms have been used. But the only difference between all these implementations is the initiation rule, in order to establish fair comparisons and to draw general conclusions from the obtained results.

The implemented load balancing algorithm is based on non-preemptive tasks assignment, thus, the objective functions proposed in equations 3, 8 and 10 have been directly translated to the initiation mechanism:

- **Objective 1:** When a new task arrives to a cluster node, it must be assigned to obtain a balance greater than the algorithm tolerance (τ) after its allocation.
- **Objective 2:** In this case, it is required that the new task allocation achieves an elapsed time for this task no more than τ times its elapsed time in its home node (the node to which it was initially assigned when it arrived to the cluster).
- **Objective 3:** For this last objective function, the new task is always allocated to the node with the largest I value in order to minimize the application elapsed time: it is assumed that this kind of allocation always obtains the best elapsed time for the individual tasks and, thereby, for the global application.

In these three implementations, after checking the local and remote execution of the task, if the achievement of the initial objective is not possible, this requirement is relaxed to avoid blocking a task execution, for example, if it is impossible to comply with this objective in some environment. That is why the algorithm objective achievement not always equals 1. For the objectives 1 and 2, boundary functions have been used, thus, the algorithm tolerance is in both cases an implementation parameter. Tables 1, 2 and 3 show the results obtained for the three proposed initiation mechanisms, and for the two first objectives, different tolerance values have been considered. Each table shows the number of accepted (O_{acep}), requested (O_{req}) and rejected operations (O_{rej}), the application elapsed time (T), the A and R values, and finally, the initiation efficiency (ε) for the different algorithm implementations.

In table 1, results for the first objective are shown. The larger the value of the algorithm tolerance, the more restrictive is the initiation mechanism: more load balance is required in the system. This is why for the largest τ values, more load balancing operations are requested, because they are needed to achieve these exigent load balance requirements. But it can be seen that the increase of τ leads to a decrease of both A and R, due to the difficulty in finding a proper allocation to achieve the algorithm objective. Therefore, the initiation efficiency decreases when the τ value increases. The intuitive explanation for this behavior is that the more difficult is to find a good allocation the less efficient becomes the initiation mechanism. The best elapsed time for the application is obtained with the medium tolerance values. With low τ values, the load balance

Table 1. Experimental results with the objective 1

τ	O_{acep}	O_{req}	O_{rej}	T (s)	R	A	ε
0.20	291	328	37	304	0.89	0.96	0.85
0.30	286	335	49	289	0.85	0.86	0.74
0.40	313	395	82	276	0.79	0.82	0.65
0.45	318	403	85	272	0.79	0.80	0.63
0.50	325	402	77	269	0.81	0.74	0.60
0.60	318	410	92	286	0.78	0.63	0.49
0.70	315	419	104	297	0.75	0.55	0.41

Table 2. Experimental results with the objective 2

τ	O_{acep}	O_{req}	O_{rej}	T (s)	R	A	ε
1	315	423	108	372	0.74	0.92	0.68
2	313	420	107	311	0.75	0.98	0.73
3	240	332	92	259	0.72	1	0.72
4	197	286	89	292	0.69	1	0.69
5	184	263	79	316	0.70	1	0.70
6	163	234	71	329	0.70	1	0.70
10	142	210	68	359	0.68	1	0.68
15	121	177	56	388	0.68	1	0.68

Table 3. Experimental results with the objective 3

Version	O_{acep}	O_{req}	O_{rej}	T (s)	R	A	ε
Simple	300	340	40	323	0.88	0.40	0.35
Modified, F=1.2	288	318	30	320	0.91	0.40	0.37
Modified, F=1.8	258	315	57	280	0.82	0.46	0.38

required in the system is too low to give good elapsed times, but with the largest values, the tasks assignment becomes too complicated and this has a negative influence on the elapsed times.

For the second objective (table 2), similar conclusions can be derived. But in this case low tolerance values imply more restrictive requirements, therefore, more requested load balancing operations. The main difference with the first objective is that in this case the influence of the τ value on the A, R and ε values is not so significant. Similar efficiency values can be obtained with all the considered algorithm tolerances. This is due to the specific features of the performed experiment, that is, with this system-application combination, it is easier to achieve the objective 2 than the objective 1 even with the more restrictive requirements.

Finally, in table 3 the results for the objective 3 are shown . The 'simple' implementation is based on the objective function proposed in equation 10. But this optimization function leads to a poor system performance, in terms of elapsed time and initiation efficiency. So, an easy modification is proposed (the 'modified' version), to assign new tasks to the system node with the largest load index only when this index is F times greater than in the local node. The utilization of this threshold does not affect the objective achievement and allows to avoid unnecessary load balancing operations, improving the algorithm performance specially for F values significantly different from 1.

In the proposed context, the three implementations can obtain similar elapsed time values: 269, 259 and 280 s respectively. But the third objective must be rejected due to its efficiency value, only 0.38 for this best elapsed time. For the first and second objectives, similar initiation efficiencies can be obtained, 0.60 and 0.72, therefore both objective functions could be used for this algorithm, being a little best the second objective performance for this experiment.

6   Conclusions

This paper proposes a general procedure to methodically obtain initiation mechanisms for load balancing algorithms. This methodology implies choosing a general objective for the load balancing algorithm. This objective is mathematically expressed with an objective function, which can be an optimization or a boundary one, depending on the available system state information. And finally, this function is directly translated into an initiation mechanism for the load balancing algorithm. In addition, a performance metric for this mechanism, the initiation efficiency, has been defined.

For illustration, three example objectives have been presented to derive their initiation mechanisms using the proposed methodology and to evaluate their performance with the defined metric. The presented experiments for these three objectives show the utility of the proposed procedure in implementing initiation policies and of the initiation efficiency in selecting the best alternative.

All these results highlight the fact that it is possible to find different tasks allocations with similar elapsed times values but very different values for the initiation efficiency. And in this situation, the implementation with the best initiation efficiency must be always selected because it implies a better resources utilization (less rejected load balancing operations) and a better degree of the algorithm objective achievement. And of course, it can be seen with the different examples that the best elapsed time does not necessary imply the best initiation efficiency for the algorithm, because for this performance metric the important issue is the degree of achievement for the algorithm objective and the resources utilization efficiency to obtain this degree, and not the elapsed time.

Furthermore, a general observation can be made based on all the performed experiments: the boundary objectives are the best solution for load balancing algorithms, their performance always improve the obtained with optimization objectives. The explanation for this behavior is that the optimization objectives always introduce more overhead in the initiation mechanism and are not scalable, while a good selected boundary objective can obtain better results without causing this overhead due to its simplicity. An example of this behavior is that the mechanism 3 achieves a worse elapsed time for the global application although it is its main objective, due to the utilization of an optimization function.

On the other hand, for these boundary objective functions the algorithm tolerance must be tuned, taking into account that too demanding requirements can have a negative influence on the system performance.

References

1. Rajkumar Buyya. *High Performance Cluster Computing: Architecture and Systems, Volume I*. 1999. Prentice-Hall.
2. Gregory Pfister. *In search of clusters: The Ongoing Battle in Lowly Parallel Computing*. 1998. Prentice-Hall.
3. Jerrell Watts; Mark Rieffel and Stephen Taylor. Dynamic management of heterogeneous resources. In *High Performance Computing: Grand Challenges in Computer Simulation*, pages 151–156, 1998.
4. Chengzhong Xu and Francis C. M. Lau. *Load Balancing in Parallel Computers: Theory and Practice*. 1997. Kluwer Academic Publishers.
5. Ashok Rajagolapan and Salim Hariri. An agent based dynamic load balancing system. In *Proceedings of the International Workshop on Autonomous Decentralized Systems*, pages 164–171, 2000.
6. Manish Arora; Sajal K. Das and Rupak Biswas. A de-dentralized scheduling and load balancing algorithm for heterogeneous grid environments. In *Proceedings of the International Conference on Parallel Processing Workshops*, 2002.
7. Ron Lavi and Ammon Barak. The home model and competitive algorithms for load balancing in a computing cluster. In *Proceedings of the 21st International Conference on Distributed Computing Systems*, pages 127–134, 2001.
8. L.M. Ni; C. Xu and T.B. Gendreau. A distributed drafting algorithm for load balancing. *IEEE Transactions on Software Engineering*, 11(10):1153–1161, 1985.
9. Raymond Chowkwanyun and Kai Hwang. Multicomputer load balancing for LISP execution. In *Parallel Processing for Supercomputers and Artificial Intelligence*, pages 325–366, 1989. McGraw-Hill.
10. Rami G. Melhem; Kirk R. Pruhs and Taieb F. Znati. Using spanning-trees for balancing dynamic load on a multiprocessor. In *Proceedings of the Sixth Distributed Memory Computing Conference*, pages 233–237, 1991.
11. M. Beltrán; A. Guzmán and J.L. Bosque. Dynamic tasks assignment for real heterogeneous clusters. *Parallel Processing and Applied Mathematics:5th International Conference, Lecture Notes in Computer Science*, 3019/2004:888–895. Springer Verlag.
12. Michael Mitzenmacher; Balaji Prabhakar and Devavrat Shah. Load balancing with memory. In *Proceedings of the 43rd Annual IEEE Symposium on Foundations of Computer Science*, 2002.
13. M. Beltrán and J.L. Bosque. Estimating a workstation CPU assignment with the DYPAP monitor. In *Proceedings of the 3rd IEEE International Symposium on Parallel and Distributed Computing*, 2004.
14. Luis Pastor and Jose L. Bosque. Efficiency and scalability models for heterogeneous clusters. In *Proceedings of the 3rd IEEE International Conference on Cluster Computing,*, pages 427–434. IEEE Computer Society Press, 2001.

Hierarchical Scheduling for Moldable Tasks

Pierre-François Dutot

Laboratoire ID-IMAG
38330 Montbonnot St-Martin, France
`Pierre-Francois.Dutot@imag.fr`

Abstract. The model of *moldable task* (MT) was introduced some years ago and has been proven to be an efficient way for implementing parallel applications. It considers a target application at a larger level of granularity than in other models (typically corresponding to numerical routines) where the tasks can themselves be executed in parallel on any number of processors. Clusters of SMPs (symmetric Multi-Processors) are a cost effective alternative to parallel supercomputers. Such hierarchical clusters are parallel systems made from m identical SMPs composed each by k identical processors. These architectures are more and more popular, however designing efficient software that take full advantage of such systems remains difficult. This work describes approximation algorithms for scheduling a set of tree precedence constrained moldable tasks for the minimization of the parallel execution time, with a scheme which is first used for two multi-processors and several bi-processors and then extended to the general case of any number of multi-processors. The best known approximations of competitive ratios for trees in the homogeneous case is 2.62, and although the hierarchical problem is harder our results are close as we obtain a ratio of 3.41 for two multi-processors, 3.73 for several bi-processors and 5.61 for the general case of several SMPs with a large number of processors. To our knowledge, this is the first work on precedence constrained moldable tasks on hierarchical platforms.

1   Introduction

In recent years computer hardware became increasingly affordable. This trends led to a greater number of parallel computers. However, a fast interconnection network is still very expensive. A solution to this problem is to use several processors on each motherboard connected by the network. This introduces a large difference in the time needed for on-board communications and for communications between two different motherboards.

In the case of Parallel Tasks (PT), where a task has to be processed by a fixed number of processors, the execution time of a task cannot be easily predicted on such hierarchical architectures unless some very restrictive hypothesis are made such as tasks have to be executed on one board only, or all communications are considered as long communications. We consider in this paper the related Moldable Task (MT) model, where the execution time of a task depends on the number of processors used to compute the task. However, in a hierarchical system knowing the number of processors used is not enough to predict the execution time, as communications can be local or distant. In [1], we provided a new hypothesis to deal with this problem. This placement hypothesis is

J.C. Cunha and P.D. Medeiros (Eds.): Euro-Par 2005, LNCS 3648, pp. 302–311, 2005.

recalled in Section 2. With this additional rule, the MT model is well suited to hierarchical systems.

Scheduling precedence constrained MT tasks is a NP-hard problem [2], and therefore approximation algorithms were developed to provide efficient schedules in polynomial time. The first approximation algorithm for the homogeneous case has been introduced by Lepère et al [3] with a ratio of 2.62 for tree based precedence constraints and a ratio of 5.24 for general graphs. This scheme has been recently improved by Hu Zhang in his PhD thesis [4–6] (under supervision of Pr. Jansen) achieving a 4.73 approximation ratio. In this paper, we adapted this scheduling technique of Lepère et al. in the case of tree precedence constrained moldable tasks, as a first step towards scheduling general graphs. To obtain ratios for general graphs without the improvements designed by Hu Zhang, the results presented here can be simply multiplied by a factor of 2. The recent improvements were not taken into consideration here due to the length limitation.

In the next section, we will recall the definitions of the Moldable Task model and its adaptation to hierarchical platforms. We will then briefly recall the scheduling scheme used for the homogeneous case. This scheme (and improvements by Zhang) will then be adapted for the two extremal cases of scheduling on two multi-processors and scheduling for several bi-processors. Finally a general scheme for scheduling on several multi-processors is proposed in Section 6.

2   The Moldable Tasks Model on Hierarchical Platforms

In the MT model a processor can compute only one task at a time, and the number of processors allocated to a task is constant during its whole execution. The execution time of a task depends on the number of processors allotted to it.

We consider an instance composed of n moldable tasks $\{T_1, \ldots, T_n\}$ to be scheduled on a cluster of m identical SMPs composed each of k identical processors. The tasks are linked with precedence constraints, in the form of trees (each node has at most one predecessor). The execution time of the moldable task T_i when allotted to p processors will be denoted by $t_i(p)$. Its *computational area* (or *work*) is defined as usually as the time space product $W_i(p) = pt_i(p)$. For a given allocation, we call *critical path* the maximum sum of execution times over a chain of the graph, and *work* of the graph, the sum of all the work of the tasks. The total work $W = \sum W_i(1)$ divided by mk, and the critical path L_{max} are straightforward lower bounds of the optimal makespan.

Using more than one processor to compute a task will cost some penalty for managing the communications and synchronizations. According to the usual behavior of the execution of parallel programs, we assume that the tasks are *monotonic*. This means that allocating more processors to a task will decrease its execution time and increase its computational area.

There exists a difficulty inherent to hierarchical systems due to the fact that communications inside the same SMP are faster than between processors belonging to different SMPs. In this case, the number of processors allotted to a task does not give all the informations needed to determine the execution time of a task: a task will be scheduled faster using processors inside the same SMP than using processors of different SMPs. In order to avoid this problem, we introduce below a dominant rule:

Definition (Best placement rule). *For a given number of processors, we say that a task is in its best placement if the penalty with this number of processors is the lowest possible.*

This definition is not very useful in the sense where many placements may verify the *best placement* condition, and from the definition we cannot decide where it is best to schedule the task. However, we can usually make the assumption that a task which runs on less than k processors will be in its *best placement* if all the processors allotted to the task are into the same SMP.

For tasks allotted to more than k processors, we need an additional hypothesis which is the following:

Hypothesis (Minimal penalty). *We assume in the rest of the paper that a task T_i allotted to $a_i k + b_i$ processors (with $a_i \in [0; m]$ and $b_i \in [0; k-1]$) is in its best placement if exactly a_i SMPs are dedicated to it during its execution and the remaining b_i processors are within the same SMP.*

This hypothesis is clearly verified for clusters of bi-processors, as it avoids the cases where a task is sharing more than one bi-processor with other tasks. For larger values of k, this placement minimizes the number of clusters used by a task for a given allocation, therefore it is probably not far from the optimal placement.

Remark that we do not ask the processors to be contiguous. For instance, Figure 1 represents two tasks verifying the *minimal penalty* hypothesis. The third one does not.

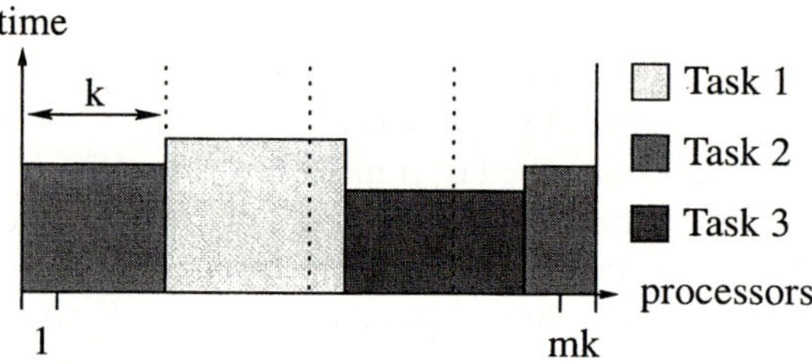

Fig. 1. Tasks 1 and 2 are in their *best placement*, whereas task 3 is not ($m = 4$).

In the rest of the paper, we will build algorithms whose output verify this best placement rule. However, the competitive ratios given are with respect to an optimal schedule which can use any kind of placement as long as the minimal penalty hypothesis holds, as the proof is based on the total workload.

3   Previous Results with Precedence Constraints

The schemes used in this paper are mainly inspired from the scheduling algorithm for the homogeneous case [3] (in this case $m = 1$). In this section, we will recall the basics of this algorithm.

In the homogeneous case, there is no placement problem ($k = 1$). The algorithm is composed of two phases. The first phase is a search for a good allocation for the moldable tasks, i.e. an allocation which realizes a trade-off between the workload and the length of the critical path in the precedence graph. This problem is related to the general class of time-cost problems where the time needed to perform a task depends on the budget allotted to it. This problem has been solved by Skutella [7] very efficiently in the case of tree precedence constraints leading to an optimal trade-off, and also has good solutions for general graphs (leading to a 2 approximation on both the work and the critical path).

Once this allocation is known, all allocations greater than a parameter μ (i.e. all tasks using more than μ processors) are reduced to μ and then the second phase is a classic list scheduling algorithm. The analysis of the algorithm is similar to the classic proof of Graham's list scheduling algorithm, and for the best possible μ the performance ratio is $(3 + \sqrt{5})/2 \simeq 2.62$ for trees and $3 + \sqrt{5} \simeq 5.24$ for general graphs [3].

4   Scheduling with Two Multi-processors ($m = 2, k > 1$)

Schedules produced by the homogeneous algorithm are usually inadequate in a multi-processor setting, because of the placement rule. For a first view of the problem, we will consider in this section the restricted case of scheduling on two multi-processors.

To keep the same construction scheme as in the homogeneous case, we have to consider how the placement rule interferes in the list scheduling. As the parameter μ is less or equal to $mk/2$ in the homogeneous case, a task in its best placement cannot use processors in both multi-processors. We now distinguish two cases depending on the value of μ.

In the first case, for $\frac{2k+1}{3} < \mu \leq k$, the schedule produced by the list algorithm can be split into two kinds of time intervals. The first kind (of total length I_1) is composed of all the time intervals during which at most $2(k - \mu) + 1$ processors are used. During these intervals, there are enough idle processors on at least one of the multi-processor to schedule a task. If those processors are idle there is no available tasks, which means that as in the original proof from Graham, a precedence constrained chain of tasks which covers all these intervals can be found. As $2(k - \mu) + 1 < \mu$, the tasks in this chain did not have their allocation reduced to μ processors. The other kind of interval (of total length I_2) is composed of all the other time intervals. We denote by ω the length of the schedule.

With these two kinds of intervals defined, we can write the following (in)equalities:

$$\omega = I_1 + I_2 \tag{1}$$

$$\omega^* \geq L^*_{max} \geq I_1 \tag{2}$$

$$2k\omega^* \geq W^* \geq I_1 + 2(k - \mu + 1)I_2 \tag{3}$$

where ω^* is the optimal makespan. The first one states that the total schedule length is the sum of all the time intervals, the second states that the critical path (and therefore the optimal schedule length) is greater than the length of the first kind of interval, and the third one is a lower bound on the workload in the optimal schedule.

A straightforward calculation proves that the ratio $\frac{\omega}{\omega^*}$ is at most equal to $\frac{4k - 2\mu + 1}{2(k - \mu + 1)}$ which takes its minimum when μ is smallest, i.e. $\mu \leq \frac{2k+4}{3}$. The ratio is therefore bounded by $4 + \frac{3}{2(k-1)}$.

In the second case, for $\mu \leq \frac{2k+1}{3}$, the schedule can be split into three different kinds of time intervals. The first kind (of total length I_1) is when less than μ processors are used, the second kind (of length I_2) when between μ and $2(k - \mu) + 1$ processors are used, and the third when at least $2(k - \mu + 1)$ processors are used.

In the first and second kind of intervals, there is enough idle processors to schedule any tasks, therefore a chain of tasks covering all these intervals is again constructible. However this time, the tasks executed during intervals of the second kind may have been reduced from their original allocation to an allocation of size μ.

The previous (in)equalities are now:

$$\omega = I_1 + I_2 + I_3 \tag{4}$$

$$\omega^* \geq L^*_{max} \geq I_1 + \frac{\mu}{2k} I_2 \tag{5}$$

$$2k\omega^* \geq W^* \geq I_1 + \mu I_2 + 2(k - \mu + 1)I_3 \tag{6}$$

To find the best upper bound for the performance ratio $\frac{\omega}{\omega^*}$, we can consider these inequalities as a set of linear programming constraints, where ω has to be maximized, and I_1, I_2 and I_3 are the variables. The dual problem is easier to solve, as there are only two variables. It is composed of the following (in)equalities:

$$z = \omega^* y_1 + 2k\omega^* y_2 \tag{7}$$

$$1 \leq y_1 + y_2 \tag{8}$$

$$1 \leq \frac{\mu}{2k} y_1 + \mu y_2 \tag{9}$$

$$1 \leq 2(k - \mu + 1)y_2 \tag{10}$$

With the new objective of minimizing z. Combining equality 7 and inequality 9 we have $\frac{z}{\omega^*} \geq \frac{2k}{\mu}$, and adding $2(k-\mu+1)$ times inequality 8 to $2k-1$ time inequality 10, we get $\frac{z}{\omega^*} \geq 1 + \frac{2k-1}{2(k-\mu+1)}$. To minimize z we have to minimize the maximum of $\frac{2k}{\mu}$ and $1 + \frac{2k-1}{2(k-\mu+1)}$. The first quantity decreases when μ increases while the second quantity has the opposite behavior. The real minimum is therefore achieved when the two are equal, and the best μ is one of the two integers closest to the solution of $\frac{2k}{\mu} = 1 + \frac{2k-1}{2(k-\mu+1)}$, which is $\frac{8k+1-\sqrt{(8k+1)^2-32k(k+1)}}{4} \simeq (2 - \sqrt{2})k + \frac{2+\sqrt{2}}{4\sqrt{2}}$. As k grows without bounds, this minimum gets close to $\frac{2}{2-\sqrt{2}} \simeq 3.41$. The value of the performance ratio for small values of k is given in Figure 2. With the exception of $k = 2$ where the ratio is 4, all the obtained performance ratio are less than $\frac{2}{2-\sqrt{2}}$, the minimum being 2.75 for k equal to four. Therefore it is always better to choose μ lower or equal to $(2k + 1)/3$ for two multiprocessors.

Remark that if $\frac{2k}{\mu} \geq 1 + \frac{2k-1}{2(k-\mu+1)}$, the ratio is reached by a schedule of a single task. Let T_1 be a highly parallel task such as $t_1(p) = \frac{t_1(1)}{p}$, its optimal execution time would be $\frac{t_1(1)}{2k}$, and the schedule produced with our algorithm has an execution time of $\frac{t_1(1)}{\mu}$, leading to the ratio $\frac{2k}{\mu}$.

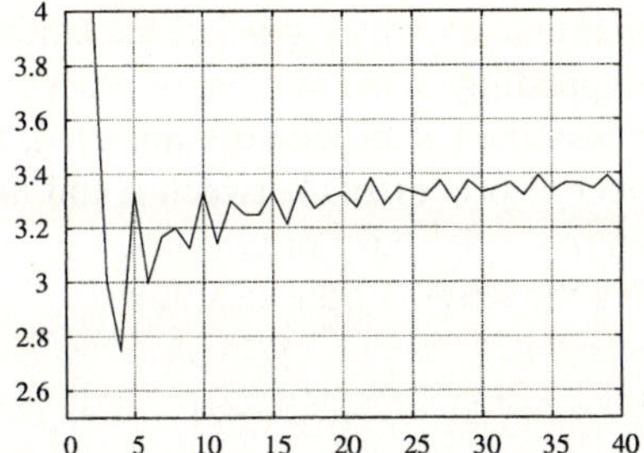

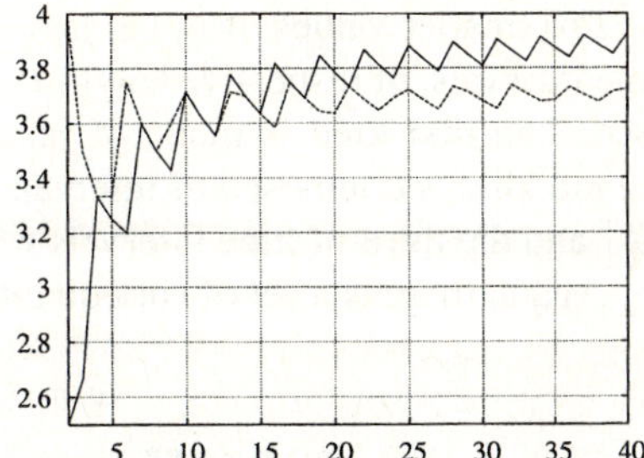

Fig. 2. Best performance ratio for two multi-processors of sizes up to 40 processors each.

Fig. 3. Best performance ratio for up to 40 bi-processors. The dotted line is for $\mu \leq \frac{2m+1}{3}$, and the solid line for $\frac{2m+1}{3} < \mu$.

5 Scheduling on Bi-processors ($m \geq 2, k = 2$)

The second restricted case which is interesting to consider before addressing the general case, is scheduling on a large number of bi-processors. In this case, restricting the allocation to a portion of a bi-processor as we did previously makes no sense. The solution we considered is to directly use the homogeneous algorithm, with a different value for μ, and try to prove that the placement constraint with bi-processors is generally satisfiable.

Let m be the number of available bi-processors. As previously, we restrict the allocations of the first phase which are greater than μ to μ. The placement rule states that to place a task of allocation a, we need to have at least $\lfloor \frac{a}{2} \rfloor$ idle bi-processors plus eventually a processor if a is odd. As we did in the previous section, we will consider two cases depending on the value of μ.

For $\frac{2m+1}{3} < \mu \leq m$, the schedule can be split into two kinds of time intervals of respective length I_1 and I_2. The first kind of time intervals is when at most $m - \lfloor \frac{\mu}{2} \rfloor$ processors are used. In these intervals, there is enough idle processors to schedule a task using μ processors. All other time intervals are counted in the other kind of time interval.

As previously, we can write some inequalities on the length ω of the schedule produced by the algorithm:

$$\omega = I_1 + I_2 \tag{11}$$

$$\omega^* \geq L_{max}^* \geq I_1 \tag{12}$$

$$2m\omega^* \geq W^* \geq I_1 + \left(m - \left\lfloor \frac{\mu}{2} \right\rfloor + 1 \right) I_2 \tag{13}$$

From these (in)equalities, it is straightforward to prove that:

$$\frac{\omega}{\omega^*} \leq \frac{3m - \left\lfloor \frac{\mu}{2} \right\rfloor}{m - \left\lfloor \frac{\mu}{2} \right\rfloor + 1} \tag{14}$$

which means that the best ratio is obtained for the smallest possible value of μ, which is $\lfloor \frac{2m+1}{3} \rfloor + 1$. This ratio is lower than 4 and tends to 4 for large values of m.

For smaller values of μ, i.e. $\mu \leq \frac{2m+1}{3}$, we again have to distinguish three kinds of time intervals, of respective length I_1, I_2 and I_3, depending on the number of processors used. The first kind is made of intervals where less then μ processors are used, the second kind is composed of intervals with a number of processors between μ and $m - \lfloor \frac{\mu}{2} \rfloor$ and the third of time intervals with more than $m - \lfloor \frac{\mu}{2} \rfloor$ busy processors.

Again, there is a set of (in)equalities describing the length of the schedule:

$$\omega = I_1 + I_2 + I_3 \tag{15}$$

$$\omega^* \geq L^*_{max} \geq I_1 + \frac{\mu}{2m} I_2 \tag{16}$$

$$2m\omega^* \geq W^* \geq I_1 + \mu I_2 + \left(m - \left\lfloor \frac{\mu}{2} \right\rfloor + 1 \right) I_3 \tag{17}$$

Which can be seen as a linear programming set of equations, and the dual is this time:

$$z = \omega^* y_1 + 2m\omega^* y_2 \tag{18}$$

$$1 \leq y_1 + y_2 \tag{19}$$

$$1 \leq \frac{\mu}{2m} y_1 + \mu y_2 \tag{20}$$

$$1 \leq \left(m - \left\lfloor \frac{\mu}{2} \right\rfloor + 1 \right) y_2 \tag{21}$$

As before, some straightforward rewriting yields to:

$$\frac{z}{\omega^*} \geq \frac{2m}{\mu} \tag{22}$$

$$\frac{z}{\omega^*} \geq 1 + \frac{2m - 1}{m - \lfloor \frac{\mu}{2} \rfloor + 1} \tag{23}$$

Again, we have to find the μ which will minimize the maximum of the two lower bounds. This time, the best μ can be bounded between two functions of m:

$$\left\lceil 4m - 1 - \sqrt{12m^2 + 4m + 1} \right\rceil - 1 \leq \mu \tag{24}$$

$$\mu \leq \left\lfloor 4m - \sqrt{12m^2 - 8m} \right\rfloor + 1 \tag{25}$$

The obtained performance ratio is presented in Figure 3, with a dotted line for small values of μ and a solid line for large values of μ. When the number of bi-processors is lower than ten, the best solution is achieved with a large μ, whereas for more bi-processors, μ has to be smaller. As m grows without bounds, $\frac{\mu}{m}$ gets close to $(4 - 2\sqrt{3})$ and the performance ratio of the algorithm tends to $\frac{1}{2-\sqrt{3}} \simeq 3.73$.

6   A General Framework ($m > 2, k > 2$)

The algorithms of the two previous sections cannot easily be extended to an arbitrary number of multi-processors with a large number of processors. The number of multi-processors m is a lower bound on the ratio of the first algorithm, as μ is always lower than k, while k is a lower bound of the ratio of the second one as m sequential tasks can prevent the execution of tasks allotted to at least k processors. A closer look shows that the first algorithm corresponds to $\mu < k$, and the second one to $\mu \geq k$.

To design efficient schedules for the general case, we have to take the best of the two previous algorithms, considering both the tasks with a large allocation and the tasks with a small allocation. The main idea is to use different values μ for small and large tasks, and then restrict the execution of the small tasks on a specific part of the platform.

For the rest of the paper, we consider m multi-processors, having k processors each. Let γ be an integer between 1 and m, γ sets the threshold between "small" and "large" tasks. Tasks allotted to less than γk processors are "small", while other tasks are "large". As we will need two different values of μ for small and large tasks, we will keep the μ notation for small tasks, and denote by δk the largest allotment allowed (hence δk plays the same role for large task as μ does for small tasks).

After the first allotment phase, the allotment of the tasks is reduced in the following way:

- Tasks allotted to a processors, with $a \leq \mu$ are kept in their original allotment.
- Tasks allotted to a processors, with $\mu < a < \gamma k$ are reduced to μ processors.
- Tasks allotted to a processors, with $\gamma k \leq a < \delta k$ are reduced to $\lfloor \frac{a}{k} \rfloor k$ processors.
- Tasks allotted to a processors, with $\delta k \leq a$ are reduced to δk processors.

Once this allotment is determined, the schedule is produced by a list scheduling algorithm, with always at most θ multi-processors[1] filled with small tasks. However, the large tasks can fill more than $(m-\theta)$ multi-processors if there is not enough small tasks. As previously, we can split the resulting schedule in several kind of time intervals, depending on occ_{small} and occ_{large} which are the number of processors used respectively by small and large tasks:

- S_1 is the set of intervals such as $1 \leq occ_{small} < \mu$ and $occ_{large} = 0$. In all the time intervals of this set, there is always a task which is part of the constructed critical path, and whose allocation has not been reduced.
- S_2 is the set of intervals such as $\mu \leq occ_{small} < \theta(k - \mu + 1)$ and $occ_{large} = 0$. In all the time intervals of this set, there is always a task which is part of the constructed critical path, and whose allocation may have been reduced to μ.
- S_3 is the set of intervals such as $\gamma k \leq occ_{large} < \delta k$ and $occ_{small} = 0$. In all the time intervals of this set, there is always a task which is part of the constructed critical path, and whose allocation has been reduced to the nearest multiple of k.
- S_4 is the set of intervals such as $\delta k \leq occ_{large} < (m - \delta + 1)k$ and $occ_{small} = 0$. In all the time intervals of this set, there is always a task which is part of the constructed critical path, and whose allocation may have been reduced to δk.
- $S_{critical}$ is the set of intervals which are not in the previous sets, and where you can still schedule a task, either small or large. Mathematically, the occupations are either $occ_{large} < (m - \theta - \delta + 1 + a)k$ and $occ_{small} \leq \theta - a$ for a between 1 and θ, or $occ_{large} < (m - \theta - \delta + 1)k$ and $occ_{small} < \theta(k - \mu + 1)$. We can redistribute all the time intervals from this set to sets S_1 to S_4, depending on the task of the interval which is considered for building the critical path.

---

[1] Please note that these θ SMPs are not fixed. If a small task is ready and less than θ SMPs are used by small tasks, any available SMP can be partially used by the small task.

- S_5 is the set of intervals such as $\theta(k - \mu + 1) \leq occ_{small}$. In these time intervals, if a task of size μ is available, it may be impossible to schedule it.
- S_6 is the set of intervals such as $(m - \delta - \theta + 1)k \leq occ_{large}$ and $m + 1 - \delta - \frac{occ_{large}}{k} \leq occ_{small}$. In these time intervals, if there is an available task of size δk, it may be impossible to schedule it.

Remark that some of these intervals may be empty, and some are overlapping. Depending on the values of θ, k and μ, S_2 can be empty. If this is the case, the upper bound on occ_{small} of S_1 is reduced to meet the upper bound of S_2. In the same way, depending on the values of m and δ, S_4 may be empty. Again, if this is the case, the upper bound of S_3 must be reduced to the upper bound of S_4. Time intervals which can be in S_5 and S_6 are put in the set S_5 if $\theta(k - \mu + 1) > (m - \delta - \theta + 1)k + \theta$ and in set S_6 otherwise.

As previously, denoting I_x the total length of the intervals in set S_x, we can bound the length of the intervals with the total workload and the critical path:

$$\omega = I_1 + I_2 + I_3 + I_4 + I_5 + I_6 \tag{26}$$

$$\omega^* \geq I_1 + \frac{\mu}{\gamma k - 1}I_2 + \frac{\gamma k}{(\gamma + 1)k - 1}I_3 + \frac{\delta}{m}I_4 \tag{27}$$

$$mk\omega^* \geq I_1 + \mu I_2 + \gamma k I_3 + \delta k I_4 + \theta(k - \mu + 1)I_5$$
$$+ ((m - \delta - \theta + 1)k + \theta)\,I_6 \tag{28}$$

And from these equations, we can write the dual problem:

$$z = \omega^* y_1 + mk\omega^* y_2 \tag{29}$$

$$1 \leq y_1 + y_2 \tag{30}$$

$$1 \leq \frac{\mu}{\gamma k - 1}y_1 + \mu y_2 \tag{31}$$

$$1 \leq \frac{\gamma k}{(\gamma + 1)k - 1}y_1 + \gamma k y_2 \tag{32}$$

$$1 \leq \frac{\delta}{m}y_1 + \delta k y_2 \tag{33}$$

$$1 \leq \theta(k - \mu + 1)y_2 \tag{34}$$

$$1 \leq ((m - \delta - \theta + 1)k + \theta)\,y_2 \tag{35}$$

Although it may seem much more complicated, this problem is still two dimensional and the extremal point of the polytope can be found. Due to the restrictions on the paper length the case analysis will not be presented here, but is instead provided in an extended version of this paper [8]. Unsurprisingly the guarantees for the general case are not as good as in the two special cases studied in the previous sections. These results are summarized in Figure 4 and Figure 5.

We can see in these figures that the performance ratio is quickly worse than 4, and does not get bigger than 5.5 for small values of k and m. For very large values of k and m, this ratio tends to 5.61.

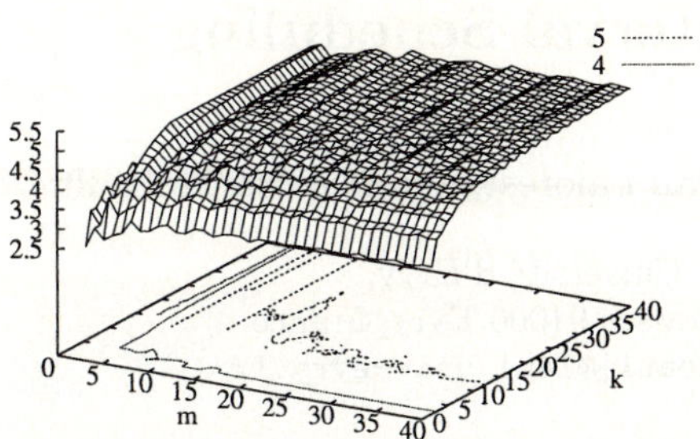

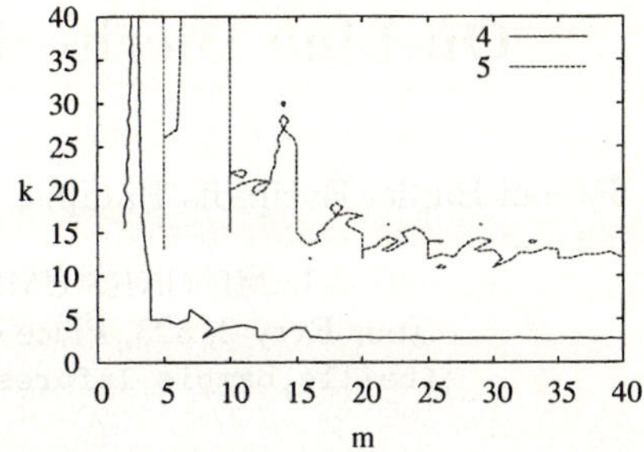

Fig. 4. Performance ratios for up to 40 SMPs having each up to 40 processors.

Fig. 5. Projections of the iso-levels 4 and 5 of Figure 4.

7   Conclusion

The algorithms presented in this paper are (to our knowledge) the first to address the problem of scheduling moldable tasks on hierarchical platforms. The next step is to add the improvements from Hu Zhang. In the longer run, we should implement the resulting algorithms in operational resource management systems. This implementation has to be preceded by a simulation phase, as the behavior of the algorithms on real workloads can be quite different from expected.

References

1. Dutot, P.F., Trystram, D.: Scheduling on hierarchical clusters using malleable tasks. In: Proceedings of the thirteenth annual ACM symposium on Parallel algorithms and architectures, ACM Press (2001) 199–208
2. Du, J., Leung, J.T.: Complexity of scheduling parallel tasks systems. SIAM Journal on Discrete Mathematics **2** (1989) 473–487
3. Lepere, R., Trystram, D., Woeginger, G.: Approximation algorithms for scheduling malleable tasks under precedence constraints. In Springer-Verlag, ed.: 9th Annual European Symposium on Algorithms - ESA 2001. Number 2161 in LNCS (2001) 146–157
4. Zhang, H.: Approximation Algorithms for Min-Max Resource Sharing and Malleable Tasks Scheduling. PhD thesis, University of Kiel, Germany (2004)
5. Jansen, K., Zhang, H.: Scheduling malleable tasks with precedence constraints. In: 17th ACM Symposium on Parallelism in Algorithms and Architectures (SPAA 2005), Las Vegas (2005)
6. Jansen, K., Zhang, H.: An approximation algorithm for scheduling malleable tasks under general precedence constraints (2005) submitted.
7. Skutella, M.: Approximation algorithms for the discrete time-cost tradeoff problem. Mathematics of Operations Research **23** (1998) 909–929
8. Dutot, P.F.: Hierarchical scheduling for moldable tasks – extended version. Technical report, Laboratory ID-IMAG (2005) www-id.imag.fr/~pfdutot/perso.html.

On-Line Bicriteria Interval Scheduling

Fabien Baille, Evripidis Bampis, Christian Laforest, and Nicolas Thibault

LaMI, CNRS UMR 8042, Université d'Evry,
Tour Evry 2, 523, Place des Terrasses 91000 Evry, France
{fbaille,bampis,laforest,nthibaul}@lami.univ-evry.fr

Abstract. We consider the problem of scheduling a sequence of *intervals* revealed *on-line* one by one in the order of their release dates on a set of k identical machines. Each interval i is associated with a processing time p_i and a pair of arbitrary weights (w_i^A, w_i^B) and may be *scheduled* on one of the k identical machines or *rejected*. The objective is to determine a valid schedule maximizing the sum of the weights of the scheduled intervals for each coordinate. We first propose a generic on-line algorithm based on the combination of two monocriteria on-line algorithms and we prove that it gives rise to a pair of competitive ratios that are function of the competitive ratios of the monocriteria algorithms in the input. We apply this technique to the special case where $w_i^A = 1$ and $w_i^B = p_i$ for every interval and as a corollary we obtain a pair of constant competitive ratios.

We consider the problem of scheduling in an on-line context a set of n intervals on k identical machines. An interval i is defined as a tuple of five positive real numbers $(r_i, p_i, d_i, w_i^A, w_i^B)$, where r_i denotes the *release date*, p_i the *processing time*, $d_i = r_i + p_i$ the *deadline* and w_i^A and w_i^B two arbitrary *weights*. We consider the following on-line context: Intervals arrive (are *revealed*) one by one in increasing order of their release dates, i.e. $r_1 \leq r_2 \leq \cdots \leq r_i \leq \cdots$, and they are not known before they are revealed. A revealed interval must either be *served* or *rejected*. An interval i is said to be *served* or *accepted* if it is alloted exclusively and without interruption (preemption is not allowed) to one of the k machines from date r_i to date d_i. Note that the acceptance of an interval may lead to the *interruption* of already scheduled intervals. A schedule O is *valid* if every served interval is scheduled at most once and if at each date every machine schedules at most one interval. There are two objective functions that we call the weight $W_A(O)$, defined as the sum of the first-coordinate-weights w_i^A of the accepted intervals, and the weight $W_B(O)$, corresponding to the sum of the second-coordinate-weights w_i^B of the accepted intervals in O. Note that if an interval is rejected or scheduled and interrupted later before its deadline, it is definitely lost and no gain is obtained from it for none of the metrics. In this model, we search for a solution/schedule that simultaneously maximizes the two objectives W_A and W_B. The particular weight function $w_i^A = 1$ (resp. $w_i^B = p_i$) corresponds to the well known SIZE (resp. PROPORTIONAL WEIGHT) problems.

Competitive ratio. In order to analyze the performance of an on-line algorithm, we use the notion of *competitive ratio* [4, 7]. Let $\sigma_1, \cdots, \sigma_n$ be any on-line sequence. For every i, $1 \leq i \leq n$, let $A(\sigma_1, \cdots, \sigma_i)$ be the schedule returned by the

J.C. Cunha and P.D. Medeiros (Eds.): Euro-Par 2005, LNCS 3648, pp. 312–322, 2005.

algorithm A at *step i*, i.e. when the first i intervals are revealed, and let O_i^* be an optimal schedule of the set $\{\sigma_1, \cdots, \sigma_i\}$ for some metric C. Then A is said to be ρ-competitive for the metric C if, for all i, $1 \le i \le n$, this inequality holds:

$$\rho C(A(\sigma_1, \cdots, \sigma_i)) \ge C(O_i^*)$$

For our bicriteria problem, an algorithm A is said to be (ρ, μ)-competitive if it is *simultaneously* ρ-competitive for W_A and μ-competitive for W_B.

Previous works. To the best of our knowledge, this is the first work considering the simultaneous maximization of two different weight functions in an on-line context. Nevertheless, the off-line version of the bicriteria problem has been treated in [2] where a $(\frac{k}{r}, \frac{k}{k-r})$-approximation algorithm $(1 \le r < k)$ has been proposed. On the contrary, the monocriteria problems have been extensively studied for both the off-line and the on-line versions. In particular, the off-line versions are polynomial (see Faigle and Nawijn [6] for the SIZE and Carlisle and Lloyd [5] or Arkin and Silverberg [1] for the WEIGHT problems). In the on-line context, the algorithm GOL of Faigle and Nawijn [6] is optimal for the SIZE problem. For the WEIGHT problem, there is a series of works going from the paper of Woeginger, in [8], who proposed a 4-competitive algorithm for the PROPORTIONAL WEIGHTS problem in a single machine system, to the paper of Bar-Noy et al. [3] who proposed the LR algorithm which is $\frac{2}{1-2\delta}$-competitive for the PROPORTIONAL WEIGHT problem in a different model than ours (instead of k machines, they consider a continuous channel where an interval requires less than a portion δ of the total channel).

Outline of the paper. In Section 1, we describe a generic on-line algorithm for the simultaneous maximization of two weight functions W_A and W_B. We prove that it is a $(\frac{k}{r}\rho, \frac{k}{k-r}\mu)$-competitive algorithm, for $1 \le r \le k$, where ρ and μ are the competitive ratios of the corresponding monocriteria algorithms. However, up to our knowledge, no on-line algorithm is available for the general WEIGHT problem. So, we focus, in Section 2, on the special case of the *size* and *proportional weights* metrics. We combine the algorithms GOL of [6] for the *size* criterion and of LR of [3] for the *proportional weights* criterion in our generic method. We thus propose a bicriteria on-line algorithm and we prove that it induces a pair of constant competitive ratios for this bicriteria case. Finally, we prove in the appendix the competitiveness of LR.

1   Our Generic Bicriteria Algorithm

In this section, we describe our generic bicriteria on-line algorithm. It uses as subroutines two on-line monocriteria algorithms having the following structure.

Structure of the monocriteria algorithms. At the release date r_i of a new interval σ_i, any on-line monocriterion algorithm can be split into two main stages. In the first one, called the *interrupting stage*, a set of already scheduled intervals are selected to be interrupted at time r_i. This set can potentially be empty meaning that no interval is interrupted when the algorithm considers σ_i. The second stage is the *scheduling stage*. Here, the algorithm can either reject the interval σ_i or schedule it on one of the available machines.

The rough idea of our generic algorithm is the following: it simulates the execution of two algorithms, say A for the maximization of the weight W_A and B for the maximization of the weight W_B on r and $k - r$ machines, respectively. By doing this, it builds its own interrupting (resp. scheduling) stage from the corresponding interrupting (resp. scheduling) stage of the input algorithms.

1.1 The Algorithm AB^k

We consider the i-th step of an arbitrary algorithm for the WEIGHT problem, i.e. the step at which interval σ_i is released. For any algorithm ALG and for every execution step i of this algorithm, let $\mathcal{O}_{i_1}(ALG)$ (resp. $\mathcal{O}_{i_2}(ALG)$) be the schedule given by ALG after the execution of its *interrupting* (resp. *scheduling*) stage of step i.

Given two algorithms A for the maximization of the weight W_A and B for the weight W_B, our generic algorithm AB^k is constructed as follows: AB^k builds the final schedule by combining the schedules returned by algorithms A and B when applied on r machines and $k - r$ machines, respectively. For the ease of presentation, we denote by A^r (resp. B^{k-r}) the algorithm A (resp. B) when applied on r (resp. $k-r$) machines. We also call *real* (resp. *virtual*) the machines involved in the algorithm AB^k (resp. A^r and B^{k-r}).

For every execution step i of AB^k, let $\mathcal{R}_{i_1}(AB^k)$ (resp. $\mathcal{R}_{i_2}(AB^k)$) be the set of scheduled intervals after the *interrupting* (resp. *scheduling*) stage of step i on the real machines associated to AB^k.

For every step i of the algorithm A^r (resp. B^{k-r}), let $\mathcal{V}_{i_1}(A^r)$ (resp. $\mathcal{V}_{i_1}(B^{k-r})$) be the set of scheduled intervals after the *interrupting stage* of step i on the r (resp. $k - r$) virtual machines associated to A^r (resp. B^{k-r}), and let $\mathcal{V}_{i_2}(A^r)$ (resp. $\mathcal{V}_{i_2}(B^{k-r})$) be the set of scheduled intervals after the *scheduling stage* of step i on the r (resp. the $k - r$) virtual machines associated to A^r (resp. B^{k-r}).

Algorithm AB^k

Input: k identical machines and an on-line sequence of intervals $\sigma_1, \ldots, \sigma_n$.

Output: After each step i $(1 \leq i \leq n)$, a valid schedule $\mathcal{O}_{i_2}(AB^k)$ involving a subset of $\sigma_1, \ldots, \sigma_i$ on k real machines.

Step 0: $\mathcal{V}_{0_2}(A^r) = \mathcal{V}_{0_2}(B^{k-r}) = \mathcal{R}_{0_2}(AB^k) = \emptyset$.

Step i (date r_i):

1. The **interrupting stage** of AB^k:
   (a) Execute the *interrupting stage* of A^r (resp. B^{k-r}) on the r (resp. $k - r$) virtual machines associated to A^r (resp. B^{k-r}) by submitting the new interval σ_i to A^r (resp. B^{k-r}). Note that the set of intervals scheduled and not interrupted by A^r (resp. B^{k-r}) is now $\mathcal{V}_{i_1}(A^r)$ (resp. $\mathcal{V}_{i_1}(B^{k-r})$).
   (b) On the k real machines associated to AB^k, interrupt the intervals of $\mathcal{R}_{(i-1)_2}(AB^k)$ such that after this interruption we get:
   $$\mathcal{R}_{i_1}(AB^k) = \mathcal{V}_{i_1}(A^r) \cup \mathcal{V}_{i_1}(B^{k-r}).$$

2. The **scheduling stage** of AB^k:

   (a) Execute the *scheduling stage* of A^r (resp. B^{k-r}) on the r (resp. $k-r$) virtual machines associated to A^r (resp. B^{k-r}) by serving or rejecting the new interval σ_i.

   (b) On the k real machines associated to AB^k, switch to the appropriate case:

      i. If A^r and B^{k-r} reject σ_i, then AB^k does not schedule (rejects) σ_i. Thus, we have:
$$\mathcal{R}_{i_2}(AB^k) = \mathcal{R}_{i_1}(AB^k).$$

      ii. If A^r or B^{k-r} serves σ_i (including the case in which both A^r and B^{k-r} serve σ_i), then AB^k schedules σ_i on any free real machine at time r_i. Thus, we have:
$$\mathcal{R}_{i_2}(AB^k) = \mathcal{R}_{i_1}(AB^k) \cup \{\sigma_i\}.$$

1.2 Competitiveness of AB^k

Here, we analyze the competitiveness of AB^k. We start with the following lemma which states that AB^k returns a valid schedule and executes the same set of intervals as the union of A^r and B^{k-r}.

Lemma 1 *For every step i of the algorithm AB^k, the schedule $\mathcal{O}_{i_2}(AB^k)$ is valid and we have:*
$$\mathcal{R}_{i_2}(AB^k) = \mathcal{V}_{i_2}(A^r) \cup \mathcal{V}_{i_2}(B^{k-r})$$

Proof. We prove this lemma by induction on the execution steps i of AB^k.
The basic case (step 0): By definition $\mathcal{V}_{0_2}(A^r) = \mathcal{V}_{0_2}(B^{k-r}) = \mathcal{R}_{0_2}(AB^k) = \emptyset$ and thus, $\mathcal{O}_{i_2}(AB^k)$ is valid and of course $\mathcal{R}_{0_2}(AB^k) = \mathcal{V}_{0_2}(A^r) \cup \mathcal{V}_{0_2}(B^{k-r})$.
The main case (step i): Let us assume that $\mathcal{O}_{(i-1)_2}(AB^k)$ is valid and that $\mathcal{R}_{(i-1)_2}(AB^k) = \mathcal{V}_{(i-1)_2}(A^r) \cup \mathcal{V}_{(i-1)_2}(B^{k-r})$ (assumption of induction).

1. The interrupting stage: We first need to prove that:
$\mathcal{R}_{i_1}(AB^k) = \mathcal{V}_{i_1}(A^r) \cup \mathcal{V}_{i_1}(B^{k-r})$ and that $\mathcal{O}_{i_1}(AB^k)$ is valid.

   (a) By definition AB^k interrupts a subset of intervals of $\mathcal{R}_{(i-1)_2}(AB^k)$ in such a way that:

$$\mathcal{R}_{i_1}(AB^k) = \mathcal{V}_{i_1}(A^r) \cup \mathcal{V}_{i_1}(B^{k-r}) \tag{1}$$

We have to show that there is always a subset of $\mathcal{R}_{(i-1)_2}(AB^k)$ that can be removed such that the above equality is possible.
Since $\mathcal{V}_{i_1}(A^r) \subseteq \mathcal{V}_{(i-1)_2}(A^r)$, $\mathcal{V}_{i_1}(B^{k-r}) \subseteq \mathcal{V}_{(i-1)_2}(B^{k-r})$ and given that $\mathcal{R}_{(i-1)_2}(AB^k) = \mathcal{V}_{(i-1)_2}(A^r) \cup \mathcal{V}_{(i-1)_2}(B^{k-r})$ (by the assumption of induction), we have $\mathcal{V}_{i_1}(A^r) \cup \mathcal{V}_{i_1}(B^{k-r}) \subseteq \mathcal{R}_{(i-1)_2}(AB^k)$.

   (b) By definition, AB^k interrupts only intervals scheduled in $\mathcal{O}_{(i-1)_2}(AB^k)$, and by the induction hypothesis, $\mathcal{O}_{(i-1)_2}(AB^k)$ is valid. Thus, $\mathcal{O}_{i_1}(AB^k)$ is clearly valid.

2. The scheduling stage: Now, we have to prove that:
   $\mathcal{R}_{i_2}(AB^k) = \mathcal{V}_{i_2}(A^r) \cup \mathcal{V}_{i_2}(B^{k-r})$ and that $\mathcal{O}_{i_2}(AB^k)$ is valid. By the definition of AB^k, several cases may occur:
   (a) If A^r and B^{k-r} reject σ_i, then AB^k does not schedule σ_i and we have:
       i. $\mathcal{R}_{i_2}(AB^k) = \mathcal{R}_{i_1}(AB^k)$ (by the definition of AB^k)
          $= \mathcal{V}_{i_1}(A^r) \cup \mathcal{V}_{i_1}(B^{k-r})$ (by (1))
          $= \mathcal{V}_{i_2}(A^r) \cup \mathcal{V}_{i_2}(B^{k-r})$
          (since A^r and B^{k-r} reject σ_i, we have:
          $\mathcal{V}_{i_1}(A^r) = \mathcal{V}_{i_2}(A^r)$ and $\mathcal{V}_{i_1}(B^{k-r}) = \mathcal{V}_{i_2}(B^{k-r})$)
       ii. $\mathcal{O}_{i_2}(AB^k) = \mathcal{O}_{i_1}(AB^k)$. Thus $\mathcal{O}_{i_2}(AB^k)$ is valid (because in item 1b of this proof, we have already seen that $\mathcal{O}_{i_1}(AB^k)$ is valid).
   (b) If A^r (resp. B^{k-r}) serves σ_i and B^{k-r} (resp. A^r) rejects σ_i, then AB^k schedules σ_i on any free real machine at time r_i. We have:
       i. $\mathcal{R}_{i_2}(AB^k) = \mathcal{R}_{i_1}(AB^k) \cup \{\sigma_i\}$ (by the definition of AB^k)
          $= \mathcal{V}_{i_1}(A^r) \cup \mathcal{V}_{i_1}(B^{k-r}) \cup \{\sigma_i\}$ (by (1))
          $= \mathcal{V}_{i_2}(A^r) \cup \mathcal{V}_{i_2}(B^{k-r})$
          (since A^r (resp. B^{k-r}) serves σ_i and B^{k-r} (resp. A^r) rejects σ_i, we have: $\mathcal{V}_{i_2}(A^r) = \mathcal{V}_{i_1}(A^r) \cup \{\sigma_i\}$ (resp. $\mathcal{V}_{i_2}(B^{k-r}) = \mathcal{V}_{i_1}(B^{k-r}) \cup \{\sigma_i\}$) and $\mathcal{V}_{i_2}(B^{k-r}) = \mathcal{V}_{i_1}(B^{k-r})$ (resp. $\mathcal{V}_{i_2}(A^r) = \mathcal{V}_{i_1}(A^r)$)).
       ii. Since $\mathcal{O}_{i_1}(AB^k)$ is a valid schedule (by the item 1b of this proof) and $\mathcal{O}_{i_2}(AB^k)$ is built by adding σ_i to $\mathcal{O}_{i_1}(AB^k)$ only once, the only reason for which $\mathcal{O}_{i_2}(AB^k)$ could not be valid would be because σ_i is scheduled by AB^k at time r_i whereas there is no free machine at time r_i, i.e. because there is at least $k+1$ intervals of $\mathcal{R}_{i_2}(AB^k)$ scheduled at time r_i by AB^k. Let us prove that this is impossible. Indeed, since A^r and B^{k-r} build at each time valid schedules, there are at most $r + k - r = k$ intervals of $\mathcal{V}_{i_2}(A^r) \cup \mathcal{V}_{i_2}(B^{k-r})$ scheduled at time r_i by A^r and B^{k-r}, and thus, there are at most k intervals of $\mathcal{R}_{i_2}(AB^k)$ scheduled at time r_i by AB^k (because we have just proved above that $\mathcal{R}_{i_2}(AB^k) = \mathcal{V}_{i_2}(A^r) \cup \mathcal{V}_{i_2}(B^{k-r})$). Thus, $\mathcal{O}_{i_2}(AB^k)$ is a valid schedule.
   (c) If A^r and B^{k-r} serve σ_i, then AB^k schedules σ_i on any idle machine at time r_i and we get:
       i. $\mathcal{R}_{i_2}(AB^k) = \mathcal{R}_{i_1}(AB^k) \cup \{\sigma_i\}$ (by the definition of AB^k)
          $= \mathcal{V}_{i_1}(A^r) \cup \mathcal{V}_{i_1}(B^{k-r}) \cup \{\sigma_i\}$ (by (1))
          $= \mathcal{V}_{i_2}(A^r) \cup \mathcal{V}_{i_2}(B^{k-r})$
          (since A^r and B^{k-r} serve σ_i, we have $\mathcal{V}_{i_2}(A^r) = \mathcal{V}_{i_1}(A^r) \cup \{\sigma_i\}$ and $\mathcal{V}_{i_2}(B^{k-r}) = \mathcal{V}_{i_1}(B^{k-r}) \cup \{\sigma_i\}$)
       ii. We prove that $\mathcal{O}_{i_2}(AB^k)$ is valid in the same way as before. $\square$

A direct consequence of Lemma 1 is that AB^k is better than A^r (resp. B^{k-r}) for the weight function that A^r (resp. B^{k-r}) maximizes.

Corollary 1 *Let W_A and W_B be two arbitrary weight functions. For every input sequence $\sigma_1, \ldots, \sigma_n$ and for each step i ($1 \leq i \leq n$) of AB^k, we have:*
$$W_A(\mathcal{V}_{i_2}(A^r)) \leq W_A(\mathcal{R}_{i_2}(AB^k)) \text{ and } W_B(\mathcal{V}_{i_2}(B^{k-r})) \leq W_B(\mathcal{R}_{i_2}(AB^k))$$

Proof. By Lemma 1, for every step i of the algorithm AB^k, we have $\mathcal{R}_{i_2}(AB^k) = \mathcal{V}_{i_2}(A^r) \cup \mathcal{V}_{i_2}(B^{k-r})$ and thus Corollary 1 is valid. $\square$

In the following lemma, we analyze, for any type of weight function W, the competitive ratio of the algorithm A applied on r ($r \leq k$) machines when compared to the optimal schedule on a system of k machines.

Lemma 2 *Let $\sigma_1, \cdots, \sigma_n$ be any on-line sequence of intervals. Let A be an on-line algorithm with competitiveness ρ on r machines ($r \leq k$) and O_k^* (resp. O_r^*) be an optimal schedule of $\sigma_1, \cdots, \sigma_n$ for the weight function W on k (resp. r) machines and O_r be the schedule returned by A_r on $\sigma_1, \cdots, \sigma_n$ on r machines. Then,*

$$W(O_k^*) \leq \tfrac{k}{r} \rho W(O_r)$$

Proof. Since A is ρ-competitive, we have by definition $W(O_r^*) \leq \rho W(O_r)$. If we multiply both sides of this inequality by $\frac{k}{r}$, we get $\frac{k}{r} W(O_r^*) \leq \frac{k}{r} \rho W(O_r)$.

Let O_1 be the schedule composed of the first r machines of O_k^* in the decreasing order of their weights. Since O_1 is an r-machine schedule, its weight is at most $W(O_r^*)$. We thus have:

$$\frac{k}{r} W(O_1) \leq \frac{k}{r} W(O_r^*) \leq \frac{k}{r} \rho W(O_r) \tag{2}$$

Since O_1 is an r-machine-schedule executing the intervals scheduled on the r machines generating the maximum weight in O_k^*, the average weight per machine in O_1 is greater than the average weight per machine in O_k^*. So, we have: $\frac{W(O_k^*)}{k} \leq \frac{W(O_1)}{r}$. Combining this result with (2), we get: $W(O_k^*) \leq \frac{k}{r} \rho W(O_r)$. $\square$

Theorem 1 *Let $\sigma_1, \cdots, \sigma_n$ be any on-line sequence of intervals. If A^r is a ρ-competitive algorithm for the weight function W_A on r machines and B^{k-r} is a μ-competitive algorithm for the weight function W_B on $k-r$ machines, then the algorithm AB^k using A^r and B^{k-r} as subroutines is $(\frac{k}{r}\rho, \frac{k}{k-r}\mu)$-competitive.*

Proof. Let $O_k^*(A)$ be an optimal schedule of $\sigma_1, \ldots, \sigma_n$ on k machines for the weight function W_A and $O_k^*(B)$ be an optimal schedule of $\sigma_1, \ldots, \sigma_n$ on k machines for the weight function W_B. By Lemma 2, we have:
$W_A(O_k^*(A)) \leq \frac{k}{r} \rho W_A(\mathcal{V}_{i_2}(A^r))$ and $W_B(O_k^*(B)) \leq \frac{k}{k-r} \mu W_B(\mathcal{V}_{i_2}(B^{k-r}))$
Moreover, using Corollary 1, we have:
$W_A(O_k^*(A)) \leq \frac{k}{r} \rho W_A(\mathcal{R}_{i_2}(AB^k))$ and $W_B(O_k^*(B)) \leq \frac{k}{k-r} \mu W_B(\mathcal{R}_{i_2}(AB^k))$
Thus AB^k is $(\frac{k}{r}\rho, \frac{k}{k-r}\mu)$-competitive. $\square$

2   Application to the SIZE and the PROPORTIONAL WEIGHT

Given that, to the best of our knowledge, we do not know on-line algorithm with constant competitive ratio for general weight functions, we focus in this section on the particular case where $w_i^A = 1$ and $w_i^B = p_i$ for every $i = 1, \ldots, n$, i.e. for the *size* and *proportional weights* metrics. We first show that the optimal on-line algorithm GOL of Faigle and Nawijn [6] can be described following the two-stages structure presented in the previous section. We also present in this form the on-line algorithm LR^k of Bar-Noy et al. [3]. Recall that GOL^k is optimal for the SIZE problem while LR deals with *proportional weights* (but for a different model than the one adopted here). Then, we use these algorithms as input of our generic method.

Here is a description of the algorithm GOL^k. It is the original algorithm GOL of [6] (using k machines) except that it is split into an interrupting stage and a scheduling stage.

Algorithm GOL^k[6]

At the arrival of interval σ_i do:

Interrupting stage: If there are k served intervals intersecting the date r_i, let σ_{max} be the one with the maximum deadline.

If σ_{max} does not exist (there is a free machine), do not interrupt any interval.

If $d_{max} \geq d_i$ then interrupt σ_{max}.

If $d_{max} < d_i$ then do not interrupt any interval.

Scheduling stage: If an interval has been interrupted (a machine became idle) or if there is a free machine, then schedule σ_i on any free machine. Else, reject σ_i.

We now adapt the algorithm LR. In [3], LR is described as an algorithm running on a *continuous* channel, where each interval requires a portion (not necessarily contiguous) of this channel. In our model we consider k machines (instead of a continuous channel), and each interval requires exactly one (discrete) machine. That is why we give the description of LR^k (the adaptation of LR on a *discrete* model of $k \geq 3$ machines). The proof of its $\frac{2}{1-\frac{2}{k}}$-competitiveness is given in the appendix because Lemma 3 and Theorem 2 are adaptations of the proof of competitiveness of LR coming from [3] to our model.

Algorithm LR^k(adaptation of [3])

We define F_t as the set of scheduled intervals containing date t.
When σ_i is revealed do:

Interrupting stage:

- If $|F_{r_i}| < k$, then do not interrupt any interval
- If $|F_{r_i}| = k$, then:
  1. Sort the $k+1$ intervals of $F_{r_i} \cup \{\sigma_i\}$ by increasing order of release dates, if several intervals have the same release date, order them in the decreasing order of their deadlines and let L be the set of the $\left\lceil \frac{k}{2} \right\rceil$ first intervals.
  2. Sort the $k+1$ intervals of $F_{r_i} \cup \{\sigma_i\}$ by decreasing order of deadlines (ties are broken arbitrarily) and let R be the set of the $\left\lfloor \frac{k}{2} \right\rfloor$ first intervals.

  If $\sigma_i \in L \cup R$ then interrupt any interval σ_j of $F_{r_i} - L \cup R$.
  Else do not interrupt any interval.

Scheduling stage:
- If $|F_{r_i}| < k$ then schedule σ_i on any free machine.
- If $|F_{r_i}| = k$, then:
  * If $\sigma_i \in L \cup R$ then schedule σ_i on the machine where σ_j was interrupted.
  * If $\sigma_i \notin L \cup R$ then reject σ_i.

Theorem 2 *For proportional weights ($w_i = p_i$) and for $k \geq 3$, LR^k is $\frac{2}{1-\frac{2}{k}}$-competitive.*

Recall that GOL^r is an optimal on-line algorithm (i.e. 1-competitive) for the SIZE and LR^{k-r} is an on-line $\frac{2}{1-\frac{2}{k-r}}$-competitive algorithm for the PROPORTIONAL WEIGHTS problem. So, applying Theorem 1, we have:

Corollary 2 *For $k \geq 4$ and for all $1 \leq r \leq k-3$, AB^k applied with $A^r = GOL^r$ and $B^{k-r} = LR^{k-r}$ is $(\frac{k}{r}, \frac{2k}{k-r-2})$-competitive for the* size *and* proportional weights *criteria.*

Note that the parameter r can be tuned in order to make AB^k more precise for one of the objectives. For example, if we set $r = \frac{k-2}{3}$, we obtain a pair of competitive ratios of $(\frac{3}{1-\frac{2}{k}}, \frac{3}{1-\frac{2}{k}}) \leq (6,6)$ and which tends to $(3,3)$ for large k. In figure 1, we show all the couples of approximation ratios that our algorithm applied with GOL and LR with $k = 20$ can reach by variations of r.

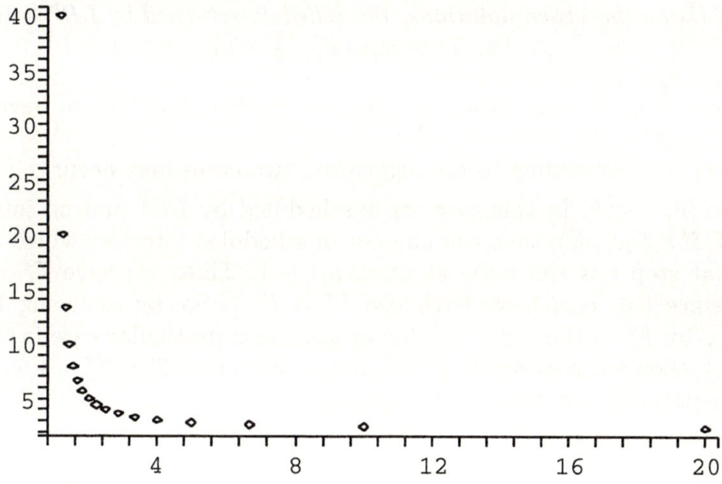

Fig. 1. Competitive ratios for the Weight (Y-axis) and the Size (X-axis) when $k = 20$.

References

1. E. ARKIN AND B. SILVERBERG, *Scheduling jobs with fixed start and end times*, Discrete Applied Mathematics, 18 (1987), pp. 1–8.
2. F. BAILLE, E. BAMPIS, AND C. LAFOREST, *A note on bicriteria schedules with optimal approximation ratios*, Parallel Processing Letters, 14 (2004), pp. 315–323.
3. A. BAR-NOY, R. CANETTI, S. KUTTEN, Y. MANSOUR, AND B. SCHIEBER, *Bandwidth allocation with preemption*, SIAM J. Comput., 28 (1999), pp. 1806–1828.
4. A. BORODIN AND R. EL-YANIV, *Online computation and competitive analysis*, Cambridge University press, 1998.
5. M. C. CARLISLE AND E. L. LLOYD, *On the k-coloring of intervals*, Discrete Applied Mathemetics, 59 (1995), pp. 225–235.
6. U. FAIGLE AND M. NAWIJN, *Note on scheduling intervals on-line*, Discrete Applied Mathematics, 58 (1995), pp. 13–17.
7. A. FIAT AND G. J. WOEGINGER, *Online algorithms: The state of the art*, LNCS no. 1442, Springer, 1998.
8. G. J. WOEGINGER, *On-line scheduling of jobs with fixed start and end times*, Theor. Comput. Sci., 130 (1994), pp. 5–16.

Appendix: Proof of Theorem 2

Let $O = LR^k(\sigma_1, \cdots, \sigma_i)$ be the schedule on k machines returned by LR^k on $\sigma_1, \cdots, \sigma_i$. Let T_i^t be the number of intervals of O containing the date t. Let F_i^t be the number of intervals of $\{\sigma_1, \cdots, \sigma_i\}$ containing the date t. For the proof of the Theorem, we need the following result:

Lemma 3 *Using the above notations, the schedule returned by LR^k satisfies:*
$$\forall i, \ \forall t, \ T_i^t \geq \min\{F_i^t, \tfrac{k}{2} - 1\}$$

Proof. We proceed by induction on i. For $i = 1$, $\forall t \in [r_1, d_1)$, we have: $T_1^t = F_1^t = 1$ and $\forall t \notin [r_1, d_1)$, $T_1^t = F_1^t = 0$.

Suppose $i > 1$. According to the algorithm, two cases may occur:

First case: $|F_{r_i}| < k$. In this case, σ_i is scheduled by LR^k and no interval is interrupted. If $t \notin [r_i, d_i)$, then the number of scheduled intervals which contain the date t at step i is the same as at step $i - 1$. Thus, we have $T_i^t = T_{i-1}^t$. Moreover, since $t \notin [r_i, d_i)$, we have also $F_i^t = F_{i-1}^t$. So, by replacing T_{i-1}^t by T_i^t and F_{i-1}^t by F_i^t in the induction hypothesis, this particular case is checked. If $t \in [r_i, d_i)$, then since σ_i has been scheduled, we have: $T_i^t = T_{i-1}^t + 1$. By the induction hypothesis, we can rewrite this equation:

$$T_i^t \geq 1 + \min\{F_{i-1}^t, \frac{k}{2} - 1\} \tag{3}$$

If $\min\{F_{i-1}^t, \tfrac{k}{2} - 1\} = \tfrac{k}{2} - 1$, then (3) becomes: $T_i^t \geq 1 + \tfrac{k}{2} - 1 = \tfrac{k}{2} > \tfrac{k}{2} - 1 \geq \min\{F_i^t, \tfrac{k}{2} - 1\}$. If $\min\{F_{i-1}^t, \tfrac{k}{2} - 1\} = F_{i-1}^t$, then (3) becomes: $T_i^t \geq 1 + F_{i-1}^t$. But since $t \in [r_i, d_i)$, we have $F_i^t = F_{i-1}^t + 1$. Thus, we have: $T_i^t \geq F_i^t - 1 + 1 = F_i^t \geq \min\{F_i^t, \tfrac{k}{2} - 1\}$.

Second case: $|F_{r_i}| = k$. In this case, three sub-cases may occur: If $\sigma_i \notin L$ and $\sigma_i \notin R$. This means that σ_i is rejected by LR^k. If $t \notin [r_i, d_i)$ then $T_i^t = T_{i-1}^t$ and $F_i^t = F_{i-1}^t$. By replacing T_{i-1}^t by T_i^t and F_{i-1}^t by F_i^t in the induction hypothesis, this particular case is checked. If $t \in [r_i, d_i)$, since $\sigma_i \notin L \cup R$, there are always at least $\lfloor \frac{k}{2} \rfloor$ intervals containing t in O. Thus, $T_i^t \geq \lfloor \frac{k}{2} \rfloor \geq \min\{F_i^t, \frac{k}{2} - 1\}$.

If $\sigma_i \in R$ (including the case where σ_i is also in L). This means that σ_i is accepted by LR^k and σ_j is rejected. Then, since σ_j is revealed before σ_i, we have $r_j \leq r_i$. Furthermore, we have $d_j \leq d_i$ otherwise, we would have $\sigma_j \in R$, contradicting the fact that σ_j is interrupted. We have then these cases: For all $t \notin [r_j, d_i)$, we have $F_i^t = F_{i-1}^t$ and $T_i^t = T_{i-1}^t$. Thus, by replacing T_{i-1}^t by T_i^t and F_{i-1}^t by F_i^t in the induction hypothesis, this particular case is checked. For all $t \in [r_j, r_i)$, since $\sigma_j \notin L$, there are at least $\lceil \frac{k}{2} \rceil$ intervals containing the date t. Thus, we have: $T_i^t \geq \lceil \frac{k}{2} \rceil > \min\{F_i^t, \frac{k}{2} - 1\}$. For all $t \in [r_i, d_j)$, we have $T_i^t = T_{i-1}^t$ because σ_j is deleted but σ_i is added. Since $\sigma_j \notin R$, there are at least $\lfloor \frac{k}{2} \rfloor$ intervals containing date t. Thus, we have $T_i^t \geq \lfloor \frac{k}{2} \rfloor \geq \min\{F_i^t, \frac{k}{2} - 1\}$. For all $t \in [d_j, d_i)$, since σ_i occupies a machine that was free at step $i - 1$ of the algorithm, we have: $T_i^t = T_{i-1}^t + 1$. By the induction hypothesis, we can rewrite this equation:

$$T_i^t \geq 1 + \min\{F_{i-1}^t, \frac{k}{2} - 1\} \tag{4}$$

If $\min\{F_{i-1}^t, \frac{k}{2} - 1\} = \frac{k}{2} - 1$, then (4) becomes: $T_i^t \geq 1 + \frac{k}{2} - 1 = \frac{k}{2} > \frac{k}{2} - 1 \geq \min\{F_i^t, \frac{k}{2} - 1\}$. If $\min\{F_{i-1}^t, \frac{k}{2} - 1\} = F_{i-1}^t$, then (4) becomes: $T_i^t \geq 1 + F_{i-1}^t$. But since $t \in [r_i, d_i)$, we have $F_i^t = F_{i-1}^t + 1$. Thus, we have: $T_i^t \geq F_i^t - 1 + 1 = F_i^t \geq \min\{F_i^t, \frac{k}{2} - 1\}$.

If $\sigma_i \in L$ and $\sigma_i \notin R$. This means that σ_i is accepted by LR^k and σ_j is rejected. By the on-line context, since the last revealed interval is σ_i, all the intervals which do not belong to L have a release date equal to r_i (otherwise they would belong to L). In particular, $\sigma_j \notin L$ because it is interrupted and thus it satisfies $r_j = r_i$. Moreover, by the manner the algorithm builds L, σ_i has also a greater deadline than σ_j (otherwise, $\sigma_j \in L$ and thus it would not be interrupted): $d_j \leq d_i$. We have 3 cases to consider: For all $t \notin [r_i, d_i)$, we have $F_i^t = F_{i-1}^t$ and $T_i^t = T_{i-1}^t$. Thus, by replacing T_{i-1}^t by T_i^t and F_{i-1}^t by F_i^t in the induction hypothesis, this particular case is checked. For all $t \in [r_i, d_j)$, we have $T_i^t = T_{i-1}^t$ because σ_j is deleted but σ_i is added. Since $\sigma_i \notin R$, there are at least $\lfloor \frac{k}{2} \rfloor$ intervals containing date t having a deadline at least d_i. Thus, we have: $T_i^t \geq \lfloor \frac{k}{2} \rfloor \geq \min\{F_i^t, \frac{k}{2} - 1\}$. For all $t \in [d_j, d_i)$, since σ_i occupies a machine that was free at step $i - 1$ of the algorithm, we have: $T_i^t = T_{i-1}^t + 1$. By the induction hypothesis, we can rewrite this equation:

$$T_i^t \geq 1 + \min\{F_{i-1}^t, \frac{k}{2} - 1\} \tag{5}$$

If $\min\{F_{i-1}^t, \frac{k}{2} - 1\} = \frac{k}{2} - 1$, then (5) becomes: $T_i^t \geq 1 + \frac{k}{2} - 1 = \frac{k}{2} > \frac{k}{2} - 1 \geq \min\{F_i^t, \frac{k}{2} - 1\}$. If $\min\{F_{i-1}^t, \frac{k}{2} - 1\} = F_{i-1}^t$, then (5) becomes: $T_i^t \geq 1 + F_{i-1}^t$. But since $t \in [r_i, d_i)$, we have $F_i^t = F_{i-1}^t + 1$. Thus, we have: $T_i^t \geq F_i^t - 1 + 1 = F_i^t \geq \min\{F_i^t, \frac{k}{2} - 1\}$. We have checked the induction step and thus the lemma. $\qquad\square$

proof of Theorem 2: Let O_i^* be the optimal (off-line) weight schedule of $\sigma_1, \ldots, \sigma_i$. Let T_i^{*t} be the number of intervals of the schedule O_i^* containing date t. Let t be a date of the schedule O returned by LR^k on the input sequence $\sigma_1, \cdots, \sigma_i$ and i be a step of the algorithm. If $\min\{F_i^t, \frac{k}{2} - 1\} = F_i^t$ then by Lemma 3, we have $T_i^t \geq F_i^t \geq T_i^{*t}$. Now, let us consider the case in which $\min\{F_i^t, \frac{k}{2} - 1\} = \frac{k}{2} - 1$. Since O_i^* is valid, we have $T_i^{*t} \leq k$. Multiplying both sides by $\frac{1 - \frac{2}{k}}{2}$, by remarking that $\frac{k}{2}\left(1 - \frac{2}{k}\right) = \frac{k}{2} - 1$ and by Lemma 3, we obtain:
$\frac{T_i^{*t}}{2}\left(1 - \frac{2}{k}\right) \leq \frac{k}{2}\left(1 - \frac{2}{k}\right) = \frac{k}{2} - 1 \leq T_i^t.$
Thus, we have for all dates t and for all steps i: $\frac{2}{1 - \frac{2}{k}} T_i^t \geq T_i^{*t}$. If we sum this inequality for all dates t, we obtain that LR^k is $\frac{2}{1 - \frac{2}{k}}$-competitive. $\qquad\square$

Topic 4
Compilers for High Performance

Albert Cohen, Michael F.P. O'Boyle, Martin Griebl, and José Moreira

Topic Chairs

This topic deals with a range of subjects concerning the compilation of programs for high performance architectures, from general-purpose machines to specific hardware designs. It includes language aspects, program analysis and transformation to optimize resource utilization or to support parallelization. Most papers study the interactions between the programming language, the compiler framework, the hardware, operating system or runtime environment.

Out of the 6 papers submitted to this topic, 2 were accepted for presentation at the conference (as regular papers). We provide a short outline of the topics addressed in these contributions.

Feedback-directed and adaptive compilation, as well as domain-specific program generation and optimization are the hot research areas for topic 4 this year.

The paper *Deciding Where to Call Performance Libraries* by Christophe Alias and Denis Barthou propose a framework to recognize library function templates from the high-level semantical analysis of scientific codes. This ambitious work demonstrates the effectiveness of the approach, replacing hot program parts in standard benchmarks by highly tuned implementations in parallel or sequential libraries.

The paper *The Periodic-Linear Model of Program Behavior Capture* by Philippe Clauss, Bénedicte Kenmei and Jean-Christophe Beyler introduces a formal model and algorithms to analyze and predict the runtime phase behavior of programs; this model can be used for adaptive and dynamic optimization.

J.C. Cunha and P.D. Medeiros (Eds.): Euro-Par 2005, LNCS 3648, p. 323, 2005.
© Springer-Verlag Berlin Heidelberg 2005

The Periodic-Linear Model
of Program Behavior Capture

Philippe Clauss, Bénédicte Kenmei, and Jean Christophe Beyler

ICPS/LSIIT, Université Louis Pasteur, Strasbourg
Pôle API, Bd Sébastien Brant
67400 Illkirch, France
`{clauss,kenmei,beyler}@icps.u-strasbg.fr`

Abstract. Understanding and controlling program behavior is a challenging objective for the design of advanced compilers and critical system development. In this paper, we propose an analysis and modeling strategy of program behavior characteristics by considering traces generated from opportune code instrumentation. The proposed models consist in periodic and linear interpolations separated into adjacent program phases. It is shown that these models exhibit apparent and useful information on program behavior. Moreover they can directly be used to guide static optimizations or to build dynamic optimization processes as it is shown for the implementation of efficient dynamic data prefetching processes for some benchmark programs.

1 Introduction

Many works have shown that software controlled policy of hardware mechanisms can significantly improve their efficiency. A compiler can be able from a static analysis of the source code to generate some instruction hints [1]. However, such an approach is only exploitable for static control and data structures as for-loops accessing multi-dimensional arrays through affine reference functions. When considering more general control structures accessing data through pointers, static optimizations generally can not be applied since essential information is not known at compile-time and can only be observed during execution. Hence dynamic analysis and optimization have become an important area of research.

In this paper, we propose an off-line analysis and modeling strategy for traces generated from opportune code instrumentation. We consider Input Independent Programs (IIPs) as programs whose execution behaviors are not influenced qualitatively by their input data. IIPs are interesting candidates for trace driven analysis and profile feedback optimizations, since information common to any of their runs and input data can be extracted. IIPs, or input independent program sub-parts, can be identified through several approaches: input-dependency analysis by abstract interpretation [2]; static code analysis of control structures and conditionals by variable propagation; comparisons of traces resulting from a sufficient number of executions and showing the same execution behavior. Notice

J.C. Cunha and P.D. Medeiros (Eds.): Euro-Par 2005, LNCS 3648, pp. 325–335, 2005.

that this last approach can never be as reliable as the previous ones, but can give useful information relevant to the most frequent case.

Traces resulting from IIPs are candidates for standard data-mining methods based on statistical and machine learning algorithms. Although some relevant information can be found from their use, programs should take advantage of more dedicated approaches. Interesting observations are necessarily related to software and/or hardware mechanisms involving some specificities: many traces can be reduced to binary data or at least integers, repetitive and/or periodic behaviors can be often expected from program executions, observations can hopefully lead to simulation models implemented as programs, ...

With this purpose of a dedicated approach, we propose a model based on periodic interpolation by intervals of the trace values. Periodic interpolation consists in interpolating a sequence of values by a periodic polynomial function. A periodic polynomial is a polynomial whose coefficients are periodic numbers, *i.e.*, a sequence of values indexed by the modulo of the variable relatively to the number of these values. For example, the periodic polynomial $[1, 2, 3]x^2 + [3, 4]x + 5$ is equal to $x^2 + 3x + 5$ if $x \; mod \; 3 = 0$ and $x \; mod \; 2 = 0$.

Periodic interpolation allows to extract periodic behavior information of the observed program as well as reduce the complexity of the interpolation function. For example, the sequence $[3, 3, 7, 13, 11, 23, 15, 33, 19, 43, 23]$ where each element is respectively indexed by $0, 1, 2, ...$ is interpolated classically by the polynomial:

$$\frac{-4}{2835}x^{10} + \frac{197}{2835}x^9 - \frac{277}{189}x^8 + \frac{16348}{945}x^7 - \frac{16912}{135}x^6 + \frac{77408}{135}x^5 - \frac{932752}{567}x^4 + \frac{7998976}{2835}x^3 - \frac{814336}{315}x^2 + \frac{59518}{63}x + 3$$

which is a quite high degree polynomial exhibiting no apparent information about periodicity. Instead, periodic interpolation would give the following interpolation function: $[2, 5]x + [3, -2]$, showing a periodic behavior of period 2 and a linear relation between 2-spaced elements.

The elements inside a large program trace generally represent several different behaviors associated to several different program phases. Hence, a unique periodic interpolation function with a low degree can rarely be found on the whole trace, but on some contiguous values in separated intervals. These successive intervals covering the whole trace are then associated to successive program phases. We target originally linear functions, i.e., polynomials of degree 1, but construct a non-linear model by recursive compositions of the linear functions.

Since each periodic coefficient of the periodic interpolation function can itself be interpreted as a trace, we recursively apply the model to the coefficients. This approach yields the definition of a multi-dimensional time space and a granularity hierarchy of the program behavior.

Our phase definition criteria are different than those of other works based either on hardware or software metrics. A phase is classically defined as intervals characterized by values staying near a given average [4, 5]. Such an approach ables the extraction of some specific hardware behavior of a program. Our approach is closer to the program semantic since it can be seen as a way trying to re-write the original program from the unique knowledge of some observation traces, but in a more "behavior-expressive" way.

Our representation model is detailed in next section where periodic-linear functions are defined, as well as periodic-linear interpolation and our notion of program phases. Model construction algorithms are presented in section 3 where important considerations related to the nature of the extracted models are discussed. Applications and experiments are presented in section 4. Conclusions and perspectives are given in section 5.

2   Formal Definition of the Periodic-Linear Model

2.1   Periodic-Linear Function

A periodic-linear function f is a function of the form $f(x) = ax + b$ where a and b are periodic numbers. A periodic number is a finite list of n numerical values $[a_1, a_2, ..., a_n]$ where the rank of the selected value at a given time to evaluate f is given by $y \bmod n$, $y \in Z$:

$$f(x) = ax + b = [a_1, a_2, ..., a_n]x + b = \begin{cases} a_1 x + b, & \text{if } y \bmod n = 0 \\ a_2 x + b, & \text{if } y \bmod n = 1 \\ ... & ... \\ a_n x + b, & \text{if } y \bmod n = n - 1 \end{cases}$$

Notice that since b is also a periodic number of m values $[b_1, b_2, ..., b_m]$, f is also defined depending on $y \bmod m$. The number of values of a periodic number is called the *period*. Two periodic numbers can be reduced to the same period equal to the lowest common multiple (*lcm*) of their respective periods.

2.2   Periodic-Linear Interpolation

A periodic-linear interpolation of a time-serie links non-overlapping successive intervals (slices of the trace) such that any element in interval i at position j, e_{ij}, is linearly dependent of $e_{i-1,j}$: $e_{ij} = e_{i-1,j} + a_j$, where a_j is constant. The number of elements in each interval is the lowest common multiple of both periods of the periodic coefficients a and b in the interpolation function f.

We distinguish 4 possible cases:

1. all intervals are adjacent and their size p is constant;
2. all intervals are adjacent and their respective sizes $p(t)$ can vary depending on the interval occurrence t;
3. intervals are not necessarily adjacent and their size is constant;
4. intervals are not necessarily adjacent and their respective sizes can vary (see figure 1).

Adjacent intervals correspond to a unique behavior model while non adjacent intervals represent several interleaved behaviors: the dots in figure 1 are other intervals interpolated by some other periodic linear functions. A model with constant size intervals considers the duration as being a criterion characterizing a behavior, saying that two behaviors are identical if their durations are equal.

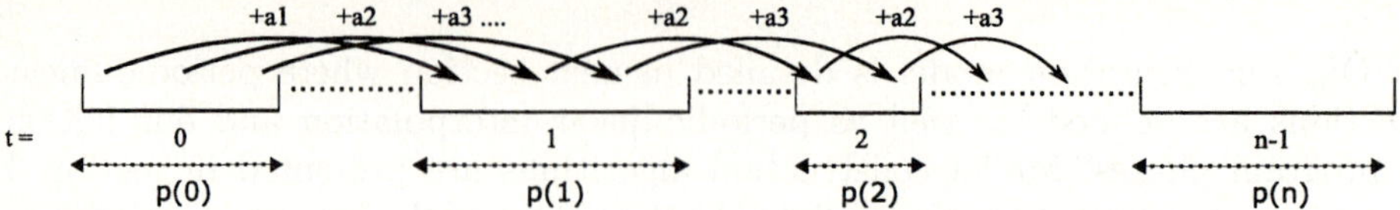

Fig. 1. an illustration of the case with non adjacent intervals of different sizes

On the other hand, with intervals of different sizes, the model does not consider the duration as being discriminant.

Consider two successive interpolated intervals i and $i - 1$. If both intervals have the same size then the definition of periodic-linear interpolation given previously holds. Otherwise, we redefine periodic-linear interpolation in the following way. Let i_{max} be the largest interval of size max over all the interpolated intervals. Then for all successive elements e_{ij} in another interval i of size s, $0 \leq j \leq s - 1$, there exist $\alpha \in Z$ and s successive elements $e_{i_{max}k}$ in interval i_{max}, $0 \leq k \leq max - 1$, such that $e_{ij} = e_{i_{max}k} + \alpha a_k$. Moreover, the value of α is uniquely associated to interval i. In other words, any interval i can be mapped onto the interval i_{max} such that the difference between each inter-mapped elements is equal to αa_k, and no other mapping onto the interval i_{max} yields the same value of α.

With each of the n intervals is associated a time instant t, $0 \leq t \leq n - 1$, defining the time space of the model. All n intervals are modeled by a periodic linear function $f(t) = at + b$ where a and b are periodic numbers. Their periods are equal to either p in cases (1) and (3), or the maximum of the $p(t)$'s in cases (2) and (4). These periodic numbers can have a large period and therefore constitute by themselves new time-serie. Hence we recursively apply our periodic-linear model to these new traces, i.e., to both periodic numbers, yielding an additional time dimension. Finally the whole application of the model yields a multi-dimensional time space $(t_1, t_2, ...)$.

Application of the model on both time-serie, or periodic numbers, a and b can result in two different periodic linear functions f_a and f_b. If their periods are different, we need to reduce them to the same period which is the lcm of their initial respective periods.

The whole recursive process yields a binary tree where each node is either a f_a or a f_b function and at each depth level is associated a time dimension. All functions of a same depth level have the same period and model simultaneously occurring traces. Finally, this multi-dimensional time model can be fully represented as a loop nest of depth d of the following general form, where the instruction of the innermost loop serves to output the element value associated to a time instant $(t_1, t_2, ..., t_d)$:

```
for t₁ = 0 to n
    for t₂ = l(t₁) to u(t₁)
        for t₃ = l(t₁, t₂) to u(t₁, t₂)
            ...
            for t_d = l(t₁, t₂, ..., t_{d−1}) to u(t₁, t₂, ..., t_{d−1})
                f(t₁, t₂, ..., t_d);
```

where $f(t_1, t_2, ..., t_d)$ is the final multi-variable function resulting from the d-depth recursive application of the model. This function is linear relatively to each variable t_i and globally non-linear. Moreover if d is maximum, i.e., the model has been applied as far as possible, then the function is no longer periodic, since any period associated to a time dimension t_i is now expressed as a loop index.

The recursive application of the model terminates as soon as the computed coefficients a and b are no more periodic numbers, i.e., are reduced to one single value. In the case where no more interpolation of any coefficients a or b is possible, with a or b consisting in at least three values, coefficients are decomposed into phases as explained in the next subsection. However the application depth can be fixed according to a chosen analysis granularity. All time dimensions can be seen as different granularity levels of the behavior model.

Functions $l(t_1, t_2, ..., t_i)$ and $u(t_1, t_2, ..., t_i)$ give the sizes of the interpolated intervals at depth $i+1$. When the model consists in constant size intervals, these functions are constants: $l = 0$ and $u+1$ is equal to the period of the interpolation function at depth $i+1$. When the model consists in non-constant sized intervals, functions $l(t_1, t_2, ..., t_i)$ and $u(t_1, t_2, ..., t_i)$ are functions interpolating the positions of the first, respectively the last, elements in all the intervals. Although these functions are generally not linear functions we limit ourselves to the cases where they are affine.

Example 1. Consider again the time-serie $[3, 3, 7, 13, 11, 23, 15, 33, 19, 43, 23]$. Following the model of adjacent intervals of constant sizes, at the first time dimension, it can be interpolated by $f_1(t_1) = [4, 10]t_1 + [3, 3], 0 \leq t_1 \leq 5$. At the second dimension, serie $[4, 10]$ and $[3, 3]$ are interpolated respectively by the functions $f_a(t_2) = 6t_2 + 4$ and $f_b(t_2) = 0t_2 + 3, 0 \leq t_2 \leq 1$. Hence the recursive process stops and the following loop nest can be generated:

for $t_1 = 0$ to 5
    for $t_2 = 0$ to 1
        $(6t_2 + 4)t_1 + 3;$

2.3   Program Phase Intervals

In our model, we define phases as the largest adjacent slices of the trace allowing periodic-linear interpolations of their elements. Hence successive phases can occur at different granularity levels yielding a hierarchy of phases. This can be represented as successive loops whose loop indices range from the first to the last element of each phase, and where each loop contains itself successive loops associated to inner level phases and so on.

The size of the generated program can be seen as a complexity measure of the modeled input trace, in the same way as it is stated in the Kolmogorov complexity theory [6].

3   Model Construction Algorithms

Our algorithms have been implemented and can obviously represent a high computation cost for large and highly irregular input traces. However modeling a very regular behavior with a few number of phases is fast. For example, each memory access information considered in [3] in several benchmarks are instantaneously modeled by our tool while giving similar results. Anyway it is generally worth the time to model critical systems behavior.

In the following and due to space limitation, only the algorithm dedicated to adjacent intervals of constant size is presented.

3.1   Quality Criteria of the Model

Since our model is hierarchically organized as a multi-dimensional time-space, the deeper we go into the hierarchy, the more accurate is the model. Interpolations involving a minimum number of phases per level is preferred since it corresponds to a minimum number of general behaviors associated to each current levels. Hence between the four model alternatives presented in subsection 2.2, a preference order related to the regular layout of the model and the number of phases is applied: (1),(2),(3) and (4).

A convenient number of phases is related to their different sizes and the size of the input trace. Each phase must include a sufficient number of elements. However since phases are related between each other through the whole interpolation model, some interesting and large phases can be coupled with some small phases. Hence an opportune quality criterion can just consider the large phases.

On the other hand, some solutions with only a few phases have to be evicted. For example, the solution consisting in modeling the whole input trace as two half traces interpolated by one periodic-linear function is obviously not interesting and does not represent any behavior specificity, since the same can be done for any sequence of numbers. Hence we constraint each phase to contain at least three interpolated intervals. Moreover, between several possible solution phases of the same size modeling the same elements, the phase containing the maximum number of interpolated intervals is selected, since it involves a lower periodicity of the interpolation function, each period corresponding to a larger number of interpolated elements.

From these observations, we can define an heuristic criterion consisting in a lower bound for the covering range of the phases. For example, we can state that at least 80% of the input trace has to be covered by all the phases whose sizes are greater than 5% of the input trace.

3.2   Phase Detection

Our model construction algorithms have the following general scheme:

1. find a periodic linear interpolation function covering the largest possible slice of the trace with at least three interpolated intervals. Define this slice as a program phase.
2. for all the remaining elements not belonging to the previously defined phases, repeat the previous step in order to define more program phases.
3. at this step, a hierarchical level has been fully modeled.

4. for each of the previously defined phases and their associated periodic-linear interpolation functions, repeat all steps with each of the periodic coefficients a and b considered themselves as traces, thus defining a deeper hierarchical level.

3.3   Adjacent Intervals of Constant Size

The frequency of linear relations between p-spaced elements in a trace can be detected by computing the autocorrelation coefficients for several values of p. The highest obtained coefficients, *i.e.*, the closest to 1, associated to given values of p, give some good indications on the best possible sizes, or periods, of the interpolated intervals. Hence our algorithm tries successively all interval sizes from their highest to their lowest associated autocorrelation coefficients in order to find the largest phase of interpolated intervals. Since at least three intervals have to take part in a phase, autocorrelation coefficients for all values of p less than $n/3$ are computed, n being the number of elements in the input trace.

The general algorithm is shown in figure 2. It is defined as a recursive function devoted to finding the largest phase of interpolated intervals from a given input trace. As it has been found, the function is recalled to find some previous or next phases necessarily smaller and covering the remaining parts of the trace. Some added comments in the figure explains some further details. When phases have been found covering the whole trace, the first time dimension has been defined and the next step consists in applying the function find_phase for each phase and their periodic-linear interpolation function to the periodic coefficients a and b. This will define the second time dimension. The same process is applied recursively until there is no remaining phase interpolated by a periodic function, i.e., the interpolation function has constant coefficients a and b.

4   Application Examples and Data Prefetching

In these experiments, we model memory addresses accessed by some time-consuming program instructions. We initially profile the execution using the *gnu gprof* tool in order to exhibit the most time consuming functions. We then instrument the code in order to store the accessed virtual memory addresses in output files. Those files are then used as input in our model construction algorithms.

4.1   Building an Hybrid Model

Hybrid models can be constructed from the observation of some input dependent events mixed with input independent events. Through abstraction of the input dependent events that cannot be modeled, a model characterizing some linear and periodic behavior of these non-deterministic occurring events can be constructed.

We consider the program `ks` from the pointer intensive benchmarks, and model memory addresses accessed through the pointer `mrB` in the most time-consuming function `FindMaxGpAndSwap`. We observe the following in the trace

function find_phase(T: input trace) $\{n$ is the trace size$\}$
$phase_{max} = NULL$
for all p such that $1 \le p \le n/3$ **do**
   compute the autocorrelation coefficient r_p of order p
for the highest to the lowest coefficient r_p and its associated period p, $r_p \ge 0.1$ **do**
   for $i=1$ to p **do**
      find the lowest integer value $\alpha \le \frac{n-3p}{p}$ such that at least the 3 elements $T[i+\alpha p]$,
      $T[i+(\alpha+1)p]$ and $T[i+(\alpha+2)p]$ are linearly dependent
      while a value α has been found **do**
         for all q such that $1 \le q \le p-1$ **do**
            check if elements $T[i+\alpha p+q]$, $T[i+(\alpha+1)p+q]$ and $T[i+(\alpha+2)p+q]$
            are also linearly dependent
         if so **then**
            extend this sequence of intervals to the right to the maximum possible size
            if size($phase_{current}$) > size($phase_{max}$) $\{$the last found phase is the largest
            at the moment$\}$
            OR (size($phase_{current}$) = size($phase_{max}$) AND $p_{current} < p_{max}$) $\{$the last
            found phase has smaller interpolated intervals$\}$ **then**
               $phase_{max} = phase_{current}; p_{max} = p_{current}$
          find the next greater value for $\alpha \le \frac{n-3p}{p}$

$\{$the following is useful to find the last phases from a few remaining elements:$\}$
if $phase_{max} = NULL$ $\{$no phase with at least 3 intervals has been found$\}$ **then**
   allow to select phases with less than 3 elements
build the periodic-linear interpolation function $f_{phase_{max}}$ of coefficients a and b and
of period p_{max}
let T_{left} be the left part of $T-phase_{max}$; let T_{right} be the right part of $T-phase_{max}$
$\{$recursive calls$\}$
find_phase(T_{left}); find_phase(T_{right})

Fig. 2. The general algorithm to find phases of adjacent interpolated intervals of constant sizes

of memory addresses: same sequences of addresses are accessed successively several times; after a sequence has been accessed, a new sequence, being the same as before but with one element less, is accessed again successively several times; after the last sequence of one element has been accessed, a completely different sequence of `numModules/2` elements is accessed successively several times, `numModules` being the second value in the input file; then the same process goes on with the same sequence having one element less; the number of times a sequence is accessed successively is equal to its number of elements; values into a sequence cannot be interpolated linearly and periodically; elements evicted from a sequence cannot be determined and depend on the considered input.

In conclusion, non-predictable sequence of known sizes are accessed in a predictable manner. Hence an hybrid model can be constructed consisting in a learning phase storing occurring address sequences, and a following prediction phase that outputs in a exact way occurring addresses. This model can be represented as the loop nest shown in figure 3.

$M = numModules/2 - 1$
for $t_1 = 0$ to N
  for $t_2 = 0$ to M
    for $t_3 = 0$ to 0 // * Learning phase *
      for $t_4 = 0$ to $M - t_2$
        $T[t_4] = accessed\ address$;// store values in an array of size M
      for $t_3 = 1$ to $M - t_2$ // * Prediction phase *
        for $t_4 = 0$ to $M - t_2$
          $T[t_4]$;

Fig. 3. Hybrid model capturing program `ks` memory behavior

4.2   Using Models for Data Prefetching

Let us consider the program `mcf` from the Spec2000 benchmarks: 31% of the whole running time is spent in function `price_out_impl`. By looking at the source code of this function, we can see that two main instructions access some data structures defined as chained lists. We then instrument the code in order to store the accessed virtual memory addresses in two output files and run the program using the `test.in` input file provided in the SPEC2000 benchmarks.

For both memory accesses, an hybrid model of adjacent intervals of different sizes, enclosed in adjacent intervals of constant size is constructed. In the first dimension t_1, the constant size intervals are identical. In the second dimension, successive intervals sizes grow from one element at each interval, and corresponding elements between successive intervals are spaced by 120 for the first instruction, and by 192 for the second instruction.

In a given interval, values are decreasing by 120, or 192, until 0. Hence, a 3-dimensional model represents entirely the whole trace. It is stated by the loop nest shown in table 1.

Table 1. Nested loop models, execution times and speedups

Program: opt. function	Model	Orig. time	Opt. time	Speedup

mcf: price_out_impl	for $t_1 = 0$ to M
	for $t_2 = 0$ to $nb_timetable_trips - 2$
	  for $t_3 = 0$ to t_2
	    $120t_2 - 120t_3 + offset$;
		512 sec.	405 sec.	20%

equake: smvp	for $t_1 = 0$ to $timesteps - 1$
	for $t_2 = 0$ to $N - 1$
	  for $t_3 = 0$ to 2
	    $128t_2 + 32t_3 + offset$;
		350 sec.	262 sec.	25%

Values M and N vary depending on the program input file. Value $M + 1$ denotes the number of constant size intervals in the first dimension and $N + 1$ denotes their size. We use the three input files provided in the SPEC2000 benchmarks, `test.in`, `train.in` and `ref.in`, and analyze the generated traces to extract the associated values of M and N, by checking the model adequacy. We notice that interval sizes are directly given by the first input parameter of the input files. In the `mcf` program documentation, this parameter is defined as being *the number of timetabled trips*. It is equal to $N + 2$. The number of intervals cannot be linked directly to an input parameter, since it rather depends qualitatively on the convergence speed of the implemented optimisation process for the considered problem. Nevertheless, a generic model can still be described, since values in dimensions t_2 and t_3 do not depend at all on t_1. Moreover, the use of the model for some dynamic optimization is not constrained at all by the ignorance of M, since the optimization process runs until the end of the whole program run.

We use both generated models to implement a dynamic prefetching mechanism for improving the program performance on an *Itanium-2* processor. This mechanism is simply built as two functions prefetching data three accesses in advance from the address computed due to our model. They are called before each memory access in function `price_out_impl`. Significant speedups are obtained for reference input runs of the whole program as it is shown in table 1. Original and optimized programs have been compiled at O3 optimization level.

In the same way, we model the program `equake` from the SPEC2000 benchmarks as shown in table 1.

Notice that other optimizations could have been constructed from these models as for example the generation of cache hints from the knowledge of data-reuse distances due to our models, as it is done from static analysis in [1].

5   Conclusion

The presented dynamic analysis and modeling approach constitutes a rich framework to formalize behavior capture of programs. The representation model facilitates program behavior understanding and analysis, and also allows the construction of efficient static or dynamic optimizations. It is for example pleasant to notice that array-like memory accesses are identified through our model, as it generates an access function of the same form as an access function resulting from a linearized multi-dimensional array accessed through affine functions indices. We argue that a lot of important behavior characteristics can be handled through our approach: as we were working on the experiments of the previous section, we observed that a lot of memory instructions can be nicely modeled. Moreover, even non-deterministic events can be considered as they are enclosed in a behavior that can be represented.

Our immediate objective is to improve performance of our algorithms in order to build a global profiling and modeling system. Such a system could then advantageously be used for applications whose performance or behavior control are critical.

References

1. K. Beyls and E. H. D'Hollander. Reuse distance-based cache hint selection. In *Euro-Par '02: Proceedings of the 8th International Euro-Par Conference on Parallel Processing*, pages 265–274. Springer-Verlag, 2002.
2. J. Gustafsson, B. Lisper, R. Kirner, and P. Puschner. Input-dependency analysis for hard real-time software. In *Proc. 9th IEEE International Workshop on Object-Oriented Real-Time Dependable Systems*, Oct. 2003.
3. I. Issenin and N. D. Dutt. Foray-gen: Automatic generation of affine functions for memory optimizations. In *DATE*, pages 808–813, 2005.
4. J. Lau, S. Schoenmackers, and B. Calder. Structures for phase classification. In *IEEE International Symposium on Performance Analysis of Systems and Software*, March 2004.
5. J. Lau, S. Schoenmackers, and B. Calder. Transition phase classification and prediction. In *11th International Symposium on High Performance Computer Architecture*, February 2005.
6. M. Li and P. Vitanyi. *An Introduction to Kolmogorov Complexity and Its Applications*. Springer-Verlag, New York, 1993.

Deciding Where to Call Performance Libraries

Christophe Alias and Denis Barthou

Laboratoire PR*i*SM, Université de Versailles, France
`{Christophe.Alias,Denis.Barthou}@prism.uvsq.fr`

Abstract. As both programs and machines are becoming more complex, writing high performance codes is an increasingly difficult task. In order to bridge the gap between the compiled-code and peak performance, resorting to domain or architecture-specific libraries has become compulsory. However, deciding when and where to use a library function must be specified by the programmer. This partition between library and user code is not questioned by the compiler although it has a great impact on performance. We propose in this paper a new method that helps the user find in its application all code fragments that can be replaced by library calls. The same technique can be used to change or fusion multiple calls into more efficient ones. The results of the alternative detection of BLAS 1 and 2 in SPEC are presented.

1   Introduction

The recent generation of microprocessors can deliver high performance thanks to a large number of mechanisms: cache hierarchies, branch prediction, specific instructions such as fused multiply add, speculative execution, predicated instructions, prefetches, etc. One way to obtain high performance code is to rely on compiler optimizations. However the complex optimizations that tap these hardware features come at the expense of performance stability. For instance, multiversioning is an optimization generating several versions for the same code fragment, these versions are selected dynamically depending on parameters such as loop iteration count or data alignment. But a bad choice for the strategy selecting the different versions can introduce important latencies. Another approach has focused on library tuning as a more reliable way to deliver performance. The assembly code is either generated by hand, using architecture specific instructions, or by adaptative code generation (*e.g.* ATLAS [14], FFTW [10] or STAPL [12]). The important compilation time is then balanced by the reusability of the libraries. In all cases, library functions can be considered as the building blocks, essential to get high performance on real codes. In general programming languages, code tuning is performed in the last stage of the development process. The selection of the library functions and the rewriting of the code falls under the responsibility of the user. The usual steps of this process are: find out code fragments and library functions that are semantically equivalent, replace these fragments by function calls with correct parameters, debug the application and finally evaluate performance. In the case of non-portable libraries, this time

J.C. Cunha and P.D. Medeiros (Eds.): Euro-Par 2005, LNCS 3648, pp. 336–345, 2005.

consuming process has to be reconducted for each target architecture. It is surprising how little the compiler helps the user in this tedious task. Compile-time optimizations neither change the partition between library and user code, nor cross library boundaries.

This paper presents an efficient method to find in a program all code fragments that match library functions. Programs under study are any C or Fortran codes, and libraries can be template libraries (with the meaning of C++ templates). In general, deciding whether two codes are semantically equivalent is undecidable. The equivalence considered in our approach does not take into account any special operator semantics, such as associativity or commutativity. Within this framework, the method presented is conservative: some of the fragments found are not semantically equivalent to the library codes, but none of the truly equivalent fragments is missed. The analysis produces "may" information: between lines 537 and 541, it may be a matrix-vector product. Combined with an exact but more expensive method [2] applied only on fragments, both analyses would produce "must" information, also providing the effective parameters for the library call and its instantiation if this is a template. Finally, we describe in this paper the conditions for which code substitution by function calls is safe. Note that as a prerequisite for the detection step, each library function has to be described by a program. We do not assume that the analysis has access to the source of the library. Instead we assume the library designer provides a public version for each function. This program must have the same semantics than the optimized, private version but the algorithm used can be completely different.

Section 2 presents some related work. Section 3 describes the new detection technique. In Section 4, we sum up, out of completion, the method used to prove the equivalence and to find the parameters of the call. We then give the conditions for a safe substitution in Section 5 and conclude in Section 6 with the results of our experiments on SPEC benchmarks.

2   Related Work

The detection of code matching library functions is related to the detection of slices, which consists of identifying all the statements contributing to a given computation. Cimetile et al. [1] propose a semi-automatic approach to extract program parts (slices) verifying pre- and post-conditions. They rely on a theorem prover which requires user interaction to assert some invariants, and has a high complexity, which makes it unrelevant to large applications.

Another approach proposed by Paul and Prakash [16] describes an extension of grep in order to find *program patterns* in source code. They use a pattern language with wild-cards on syntactic entities *e.g.* declaration, type, variable, function, expression, statement,... allowing to search for specific sequences and nested control structures. Their algorithm has a $O(n^2)$ complexity with n the code size. This detection method has the same goal as ours; one of its drawbacks is that the same pattern cannot handle variations in control (loop unroll, tiling) or in data structures (array expansion, scalar promotion) whereas this is addressed in our framework.

Finally, several approaches encode the knowledge of the functions to be identified in the form of programming plans. Top-down methods [15] use the knowledge of the goals the program is assumed to achieve and some heuristics to detect both the program slice and the library functions that can achieve these goals. Bottom-up methods [9] start from statements and try to find the corresponding plans. Wills [9] represents programs by a flow-graph, and patterns by grammar rules. The recognition is performed by parsing the program graph according to the grammar rules and has an exponential cost at worst. Metzger and Wen [13] have built a complete environment to recognize and replace algorithms. They first normalize both program and pattern abstract syntax tree by applying usual program transformations (if-conversion, loop-splitting, scalar expansion...). Then they consider all strongly connected components in the dependence graph, containing at least one `for` statement as candidate slices. Their method provides therefore a large number of candidate slices with many false detections, which is balanced by the low complexity of their equivalence test. Compared to the combination of the detection with the instantiation test we recall in this paper, they can handle fewer program variations (reuse of temporaries across loop iterations for instance is not handled) for a lower cost.

3   Detection of Library Templates

The detection of library templates consists in localizing in a code the lines that possibly correspond to a given library function or template. In the case of a template, the code detected is a possible instance of the template. We propose an efficient method based on a symbolic execution of both program and template, following the def-use chains. The method symbolically executes both program and template slices simultaneously and compares the sequence of operators along these slices, abstracting away the number of iterations of the loops.

3.1   Principle

The template and the program are assumed to be given in SSA-form, and normalized with one operator by statement. Each edge of the program SSA-graph is labeled with its operator. Loops create cycles in this graph but we abstract away the number of iterations. The sequence of operators along a path is considered as a word and the graph can be considered as a finite automaton. The idea of the algorithm is to check whether the language of operators generated by some code fragment is included in the language of operators generated by a library function. Intuitively, this ensures that the same sequence of operations can happen in the code and in the library function.

Figure 1 provides a very simple example of matching problem between a template and a program. The template and the program are assumed to be given in SSA-form, which means that the variables are assigned one time *at most* in the program text. In addition, each reference to a variable is substituted by a ϕ-function providing the set of its potential values. For example, the ϕ-function used in the assignment P_4 means that z2 = 1/z1 or z2 = 1/z3. Since statements assigning a constant such as T_1, P_1 or P_2 have no predecessors in the graph of def-use chains, they can be taken as a starting point for the inspection.

$$
\begin{array}{ll}
T_1 & \texttt{r1 = 1} \\
& \textbf{do}\ i_T\ \texttt{= 1,}n_T \\
T_2 & \quad \texttt{r2 = } X(\phi(\texttt{r1,r3})) \\
T_3 & \quad \texttt{r3 = 1+r2} \\
& \textbf{enddo} \\
T_{STOP} & \texttt{r4 = exp(}\phi(\texttt{r1,r3})\texttt{)}
\end{array}
\qquad
\begin{array}{ll}
P_1 & \texttt{z1 = 1} \\
P_2 & \texttt{t = 0} \\
P_3 & \texttt{a = tan(t)} \\
& \textbf{do}\ i_P\ \texttt{= 1,}n_P \\
P_4 & \quad \texttt{z2 = 1/(}\phi(\texttt{z1,z3})\texttt{)} \\
P_5 & \quad \texttt{z3 = 1+z2} \\
& \textbf{enddo} \\
P_6 & \texttt{r = exp(}\phi(\texttt{z1,z3})\texttt{)}
\end{array}
$$

Fig. 1. A template (left) and a program (right)

Starting from P_1, a stepping among def-use chains would follow the sequence:

$$
\xrightarrow{1} P_1 \xrightarrow{1/.} P_4 \xrightarrow{1+.} P_5 \xrightarrow{\text{exp}} P_6
$$

Likewise for the template a possible sequence of operators is:

$$
\xrightarrow{1} T_1 \xrightarrow{X(.)} T_2 \xrightarrow{1+.} T_3 \xrightarrow{\text{exp}} T_{STOP}.
$$

Walking through both program and template, with the condition that for each transition, the operator must be the same, we obtain the sequence:

$$
\xrightarrow{1} (T_1, P_1) \xrightarrow{1/.,X(.)=1/.} (T_2, P_4) \xrightarrow{1+.} (T_3, P_5) \xrightarrow{\text{exp}} (T_{STOP}, P_6).
$$

This provides the candidate slice $\{P_1, P_4, P_5, P_6\}$, that possibly corresponds to the template provided that $X(.) = 1/.$ (this condition appears on the transition). This condition is necessary for the sequence to be the same for both template and program. Note however that the method will not check the coherence between the values of template variables. Likewise, the number of iterations in loops or the branches chosen in conditionals are ignored. These important points will be checked during the exact instantiation test (see Section 4).

3.2   Detailed Algorithm

Following the idea described above, we build an automaton recognizing the sequences of operators executed by all possible instances of the template, and an automaton recognizing the sequences of operators executed by the program. The simultaneous stepping of he template and the program is then achieve by computing the Cartesian product of the template's and the program's automaton, which provides the candidate slices.

Figure 2 shows the automata built from the template and the program of the above example. The states represent the assignments, and the transitions are driven by the flow-dependences given by the ϕ-functions, and labeled by the operator used in the destination state. Template variable X can match any composition of program operators, that is why the state involving X has a loop for each program operator. Since most operators have an arity greater than 2,

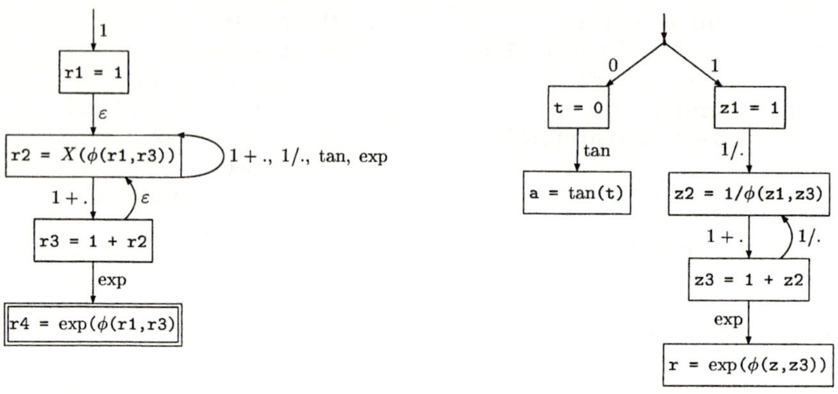

Fig. 2. Automata build associated to the template (left), and the program (right)

word automata are not expressive enough in general. Instead we build a *tree-automaton*, using the algorithm described in figure 3. There is no major differ-

---

Algorithm *Build_Automaton*

Input: *The template or the program.*
Output: *The corresponding tree automaton.*

1. Associate a new state to each assignment statement.
2. For each state:

$$q = \boxed{\texttt{r = } f(\phi(Q_1)\ldots\phi(Q_n))}$$

   Add the transitions: $f(q_1 \ldots q_n) \longrightarrow q$, for each $q_i \in Q_i$.
3. For each state:

$$q = \boxed{\texttt{r = } X(\phi(Q_1)\ldots\phi(Q_n))}$$

   Add the transitions: $q_i \longrightarrow q$, for each $q_i \in Q_i$.
   And: $f(q \ldots q) \longrightarrow q$, for each operator f used in the template and the program, including constants (0-ary operators).

Fig. 3. *Build_Automaton*

ence with the word automata: we associate a state to each assignment then we add transitions according to the dependences given by the ϕ-functions (step 2).

Remark that when $n = 1$, we obtain a word automaton since $f(q_1) \longrightarrow q$ can be interpreted as $q_1 \xrightarrow{f} q$. X is handled as a wild-card, which leads to add looping transitions with the operators used in the program (step 3).

The detection is achieved by stepping simultaneously both automata as soon as the operators are the same. Each stepping leading to the final state of the template will provide a candidate slice in the program, built of all reached program's statements. These steppings can be performed in an exhaustive manner by computing the Cartesian product $\mathcal{A}_T \times \mathcal{A}_P$ of both automata. It remains to mark the states (q_T, q_P) with a final state q_T of $\mathcal{A}_T$, and to emit the states of $\mathcal{A}_P$ on a path from the initial state as a *potential* instance.

Our method is able to detect any template variation which does not involve the semantic properties of operators such as associativity, or commutativity. Particularly we can handle any loop transformation and most control restructuring transformations. Moreover, our method is completely independent of data structure used, which allows the detection of a large amount of template variations in the program. Whether a slice detected is a real instantiation of the template is determined during the exact instantiation test.

In the worst case, the construction of the Cartesian product of the template and the program automata is computed in $O(T \times P)$ where T is the number of template statements and P is the number of program statements, i.e. the complexity is linear in the size of the program analyzed.

4   Exact Instantiation Test

Once the candidate slices are found, we have only detected a code that "may" match the library template. Either the user decides from this information to substitute or not, or another procedure decides if both program and template are indeed equivalent and finds the instantiations. We recall the main steps of this procedure eliminating false detections, described in [2].

The instantiation test follows the steps of the detection method described in Section 3. An exact instance-wise reaching definition analysis is performed. As reaching definitions may depend on the values of iteration counters, these conditions are put on the transitions of the tree-automata. Deciding if the code fragment under study is an instantiation of the template boils down to compute the loop counter values that can reach final states of the Cartesian-product automaton. Efficient heuristics [17] perform this computation.

The power of this instantiation test is assessed according to its capacity to prove the equivalence between two codes, one a variation of the other. The test handles variations coming from loop transformations (splitting, fusion, skewing, tiling, unroll,...), from data structures (scalar expansion, scalar promotion, use of temporaries), from common subexpression elimination or other factorization of computation. However, the test does not handle the semantic properties of the operators, such as commutativity or associativity.

5   Substitution

Once candidate slices are found, it remains to substitute them by a call to an optimized library. We describe thereafter an algorithm to decide whether a substitution preserves the program semantics, and to perform the substitution in case of success.

Detected slices are often interleaved with other program statements. We have first to separate them from these statements. Consider an algorithm A consisting in the set of operations $\{(A_1, I_1) \ldots (A_a, I_a)\}$, where A_i is a statement, and I_i a set of iteration vectors. Let (A_1, i_1) be its first operation, and (A_a, i_a) its last operation. Its *complementary* is the set of program operations executed between the first and the last operations of A:

$$\overline{A} = \{(S, i) \mid (A_1, i_1) \prec (S, i) \prec (A_n, i_n) \text{ and } S \text{ is not an } A_i\} \qquad (1)$$

Consider the following example (left):

```
P₁   s = 0                          P₁   s = 0
     do i = 1, 10                        do i = 1, 10
P₂ |   a(i) = a(i-1) + 1            P₂ |   a(i) = a(i-1) + 1
   |   if i >= 9 then                  |   if i ∉ { 9, 10 } then
A₁ | |   dot = dot + 2*a(i)            |     if i >= 9 then
   |   endif                        A₁ | |     dot = dot + 2*a(i)
   enddo                              | |   endif
A₂  dot = dot + b*c                   |   endif
P₃  s = s + 1                         enddo
    do i = 1, 4                    (A₂ removed)
P₄ |   s = s + b(i)                P₃  s = s + 1
A₃ |   dot = dot + a(i)*b(i)           do i = 1, 4
    enddo                          P₄ |   s = s + b(i)
                                      |   if i = 4 then call Optimized_A
                                      |   if i ∉ { 1, 2, 3, 4 } then
                                   A₃ | |   dot = dot + a(i)*b(i)
                                      |   endif
                                      enddo
```

(a). Original Program   (b). Program with substitution

where the recognized algorithm is constituted of operations:

$$A = \{(A_1, \{9, 10\}), (A_2, \{.\}), (A_3, \{1, 2, 3, 4\})\}$$

Its complementary is thus: $\overline{A} = \{(P_2, \{10\}), (P_3, \{.\}), (P_4, \{1, 2, 3, 4\})\}$. For each statement P in the program, we compute the set of corresponding operations between the first and last operations of A by giving relation (1) to a solver [5]. If it is not empty, we emit it.

Once $\overline{A}$ is computed, it remains to separate it from A in order to replace A by a call to an optimized library. A is *separable* if all dependences go exclusively from A to $\overline{A}$, or exclusively from $\overline{A}$ to A. In the first case, A can be substituted by a call to A before $\overline{A}$. In the other case, the call has to be insert after $\overline{A}$.

Otherwise, we do not perform substitution. In the example given above, A is separable and can be replaced by a call after $\overline{A}$ because of a dependence from $(P_2, 10)$ to $(A_1, 10)$. In addition, if an intermediate variable is alive outside the slice, we do not perform the substitution.

The substitution can now be performed by deleting operations of A, and placing the relevant call before, or after $\overline{A}$. Consider the above example (right). Relevant operations of statements A_1 and A_3 are disabled using a condition. Because A_2 have no nesting loops, it is just removed from the program text (step 2). As said above, the optimized call is inserted after the last operation of $\overline{A}$ $(P_4, 4)$, using a condition. A more efficient code can be produced by first reschedule operations of $\overline{A}$, and then generating efficiently the code with an appropriate method [6].

6 Experimental Results

We have implemented the detection method and the instantiation test for fortran, C and C++ applications. The C/C++ front-end is based on the LLVM compiler infrastructure [3]. We have applied our slicing algorithm to detect potential calls to the BLAS library [4] in LINPACK [7] and four programs involved in the SPEC benchmarking suite [8]. Our pattern base is constituted of direct implementations of BLAS functions from the mathematical description. After having applied our algorithm to each pair of pattern and program, we have checked by hand whether the slices are equivalent to the pattern, and if the substitution by a call to BLAS is possible. Figure 4 shows the results.

It appears that $1/2$ of candidates do not match, $1/4$ are instances of patterns for vectors of size 1, and $1/4$ of candidates are correct and can be replaced by a call to BLAS. We present different candidates involved in these categories.

Most of the incorrect detections are due to the approximation of the dependences with ϕ-functions. Neither loop iteration count, nor `if` conditions, nor complex dependences due to array index functions are handled. In addition, our method handles arrays as scalar variables, which can lead to detect a BLAS 1 xaxpy `y(i) = y(i) + a*x(i)` when there is a reduction `s = s + a(i)*a(i)`. Likewise, the method detects the same number of matrix-matrix multiplication than of matrix-vector multiplication. Note that a vector is a particular case of matrix but the code should not be substituted by a BLAS 3.

For $1/4$ of the slices, the substitution can potentially increase the program performance. Our algorithm seems to have discovered all of them, and particularly hidden candidates. Indeed, most slices found are interleaved with the source code, and deeply destructured. Our method has been able to detect a dot product in presence of a splitting and a loop unroll, which constitute important program variations that a `grep` method would not catch. The same remark applies on `equake` program. Two versions of matrix-vector product appear, one hand optimized and the other not. Both are detected whereas a method based on regular expressions fits only the second. In addition, execution times confirm that our

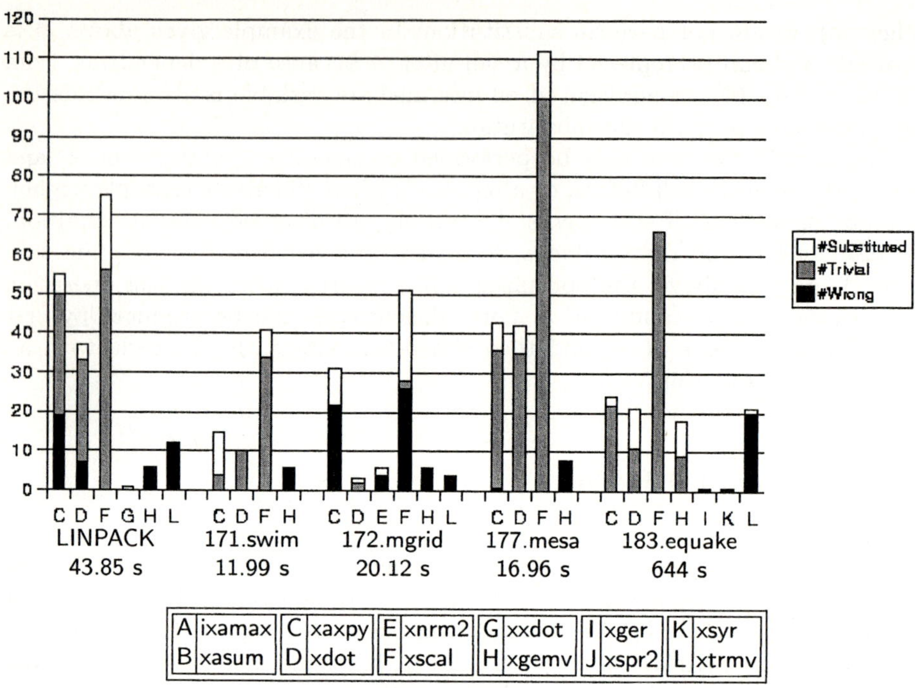

Fig. 4. For each kernel, we provide each BLAS function recognized, the number of wrong slices (# Wrong), the number of trivial detections (# Trivial), and the number of candidates interesting to replace (# Substituted). The experimentation was done on a Pentium 4 1,8 GHz with 256 MB RAM

algorithm is linear in the program size. Thus, our slicing method is scalable and can be applied to real-life applications.

7   Conclusion

The method presented shows that the compiler can help the user write or rewrite a code with high performance libraries. Combined with an instantiation test, this process can be fully automatic. The advantages are a better portability and higher productivity of the programmer. The detection only requires that each library function has a public version, in C or Fortran, semantically equivalent to the real code. The experiments on the SPEC benchmarks are encouraging: the method detects a significant number of linear algebra functions with linear complexity. The evaluation of the performance gain expected when using library calls is still however an ongoing work.

More generally, this approach can change the abstraction level of the program, replacing C code by algorithms or formulae. From this higher level of abstraction, it enables a change of algorithm [11] or simply improves code comprehension. For large scale applications, high performance cannot be at the expense of portability. The method described could be a solution to combine both and this will be the subject of future work.

References

1. A.Cimetile, A.De Lucia, and M.Munro. A Specification-driven Slicing Process for Identifying Reusable Functions. *J. of Software Maintenance: Research and Practice*, 8(3):145–178, 1996.
2. C.Alias and D.Barthou. Algorithm Recognition based on Demand-driven Data-flow Analysis. In *Working Conf. on Reverse Engineering*. IEEE, 2003.
3. C.Lattner and V.Adve. LLVM: A Compilation Framework for Lifelong Program Analysis & Transformation. In *Proceedings of CGO'2004*, Palo Alto, 2004.
4. C.Lawson, R.Hanson, D.Kincaid, and F.Krogh. Basic Linear Algebra Subprograms for Fortran usage. *Trans. on Mathematical Software*, 5(3):308–323, 1979.
5. D.Wilde. A Library for Doing Polyhedral Operations. INRIA TR 2157, 1993.
6. F.Quilleré, S.Rajopadhye, and D.Wilde. Generation of Efficient Nested Loops from Polyhedra. *Int. J. of Parallel Programming*, 28(5):469–498, 2000.
7. J.Dongarra. The LINPACK Benchmark: An Explanation. In *Supercomputing*, pages 456–474. Springer-Verlag, 1988.
8. J.Henning. SPEC CPU2000: Measuring CPU Performance in the New Millennium. *Computer*, 33(7):28–35, 2000.
9. L.Wills. *Automated Program Recognition by Graph Parsing*. PhD thesis, MIT, 1992.
10. M.Frigo and S.Johnson. FFTW: An Adaptive Software Architecture for the FFT. In *Proc. Intl. Conf. Acoustics Speech and Signal Processing*, volume 3, pages 1381–1384. IEEE, 1998.
11. M.Püschel, B.Singer, J.Xiong, J.Moura, J.Johnson, D.Padua, M.Veloso, and R.Johnson. SPIRAL: A Generator for Platform-Adapted Libraries of Signal Processing Algorithms. *J. of High Perf. Comp. and Applications*, 1(18):21–45, 2004.
12. N.Thomas, G.Tanase, O.Tkachyshyn, J.Perdue, N.Amato, and L.Rauchwerger. A Framework for Adaptive Algorithm Selection in STAPL. In *Proc. ACM PPoPP'05*, Chicago, 2005.
13. R.Metzger and Z.Wen. *Automatic Algorithm Recognition: A New Approach to Program Optimization*. MIT Press, 2000.
14. R.Whaley and J.Dongarra. Automatically Tuned Linear Algebra Software. In *SuperComputing*. Springer-Verlag, 1998.
15. S.Kim and J.Kim. An Hybrid Approach for Program Understanding based on Graph-Parsing and Expectation-driven Analysis. *J. of Applied A.I.*, 12(6):521–546, 1998.
16. S.Paul and A.Prakash. A Framework for Source Code Search using Program Patterns. *IEEE Trans. on S.E.*, 20(6):463–475, 1994.
17. W.Kelly, W.Pugh, E.Rosser, and T.Shpeisman. Transitive Closure of Infinite Graphs and its Applications. *Int. J. of Parallel Programming*, 24(6):579–598, 1996.

Topic 5
Parallel and Distributed Databases,
Data Mining and Knowledge Discovery

Domenico Talia, Hillol Kargupta, Patrick Valduriez, and Rui Camacho

Topic Chairs

To manage the very large amount of data available today, computer scientists are working on efficient systems, algorithms and applications that can handle and analyze very large databases. Intensive data consuming applications are running on very large databases (on data warehouses, on multimedia databases) with the task to extract information diamonds. Data mining is one of the key applications here. However, these intensive data consuming applications suffer from performance problems and single database sources. Introducing data distribution and parallel processing help to overcome resource bottlenecks and to achieve guaranteed throughput, quality of service, and system scalability. Distributed architectures, cluster systems and P2P systems, supported by high performance networks and intelligent middleware offer parallel and distributed databases a great opportunity to support cost-effective everyday applications.

Data processing and knowledge discovery on large data sources can benefit from parallel and distributed computing both to improve performance and quality of results. Development of data mining tools on high-performance parallel computers allows for analyzing massive databases in a reasonable time. Faster processing also means that users can experiment with more models to understand complex data. Furthermore, high performance makes it practical for users to analyze greater quantities of data. Distribution of data sources and data mining tasks is another key issue that the increasing decentralization of human activities and large availability of connection facilities are making more and more critical.

This year, 9 papers discussing some the those issues were submitted to this topic. Each paper was reviewed by at least three reviewers and, finally, we were able to select 3 regular papers. The accepted papers discuss very interesting issues such as middleware for database replication, mining global association rules on Grids, and hierarchical aggregation in networked aata management.

We would like to take the opportunity of thanking the authors who submitted a contribution, as well as the Euro-Par Organizing Committee, and the referees with there highly useful comments, whose efforts have made this conference, and Topic 5 possible.

J.C. Cunha and P.D. Medeiros (Eds.): Euro-Par 2005, LNCS 3648, p. 347, 2005.
© Springer-Verlag Berlin Heidelberg 2005

MADIS: A Slim Middleware for Database Replication*

Luis Irún-Briz[1], Hendrik Decker[1],
Rubén de Juan-Marín[1], Francisco Castro-Company[1],
Jose E. Armendáriz-Iñigo[2], and Francesc D. Muñoz-Escoí[1]

[1] Instituto Tecnológico de Informática
Universidad Politécnica de Valencia – 46071 Valencia, Spain
{lirun,hendrik,rjuan,fcastro,fmunyoz}@iti.upv.es
[2] Dpto. de Matemática e Informática
Universidad Pública de Navarra – Campus Arrosadía s/n, 31006 Pamplona, Spain
enrique.armendariz@unavarra.es

Abstract. Data replication serves to improve the availability and performance of distributed systems. The price to be paid consists of costs caused by protocols by which a sufficient degree of consistency of replicated data is maintained. Different kinds of targeted applications require different kinds of replication protocols, each one requiring a different set of metadata. We discuss the middleware architecture used in the MADIS project for maintaining the consistency of replicated databases. Instead of reinventing wheels, MADIS makes use of basic resources provided by conventional database systems (e.g. triggers, views, etc) to achieve its purpose, to a large extent. So, the underlying databases can perform more efficiently many of the routines needed to support any consistency protocol, the implementation of which thus becomes much simpler and easier. MADIS enables the databases to simultaneously maintain different metadata needed for different replication protocols, so that the latter can be chosen, plugged in and exchanged *on the fly* as online-configurable modules, in order to fit the shifting needs of given applications best, at each moment.

1   Introduction

Providing distributed access to their databases is key for banks, warehouse chains and large enterprises with geographically widespread branches. Computer applications and services for such companies must cater for shared accesses to and transactions of local and global enterprise data, which may be distributed or replicated over several sites. With databases that are fully replicated in several nodes of the network, read accesses can be local if a ROWAA [1] policy is used, so that the availability of the data and the performance of the applications is improved. Employing replication techniques also benefits the fault-tolerance of the system, improving the ability of the database to be transparent with regard to local failures and to recover seamlessly.

However, replication has some important drawbacks. The system must introduce a potential overhead for maintaining the consistency of replicated data [2]. In addition, applications making use of replication necessitate additional pieces of software in order

---

* This work has been partially supported by the Spanish grant TIC2003-09420-C02-01.

J.C. Cunha and P.D. Medeiros (Eds.): Euro-Par 2005, LNCS 3648, pp. 349–359, 2005.

to manage the access to distributed resources, thus incrementing the complexity of their development.

In this paper, we describe a new middleware architecture, called MADIS[3], for supporting the distributed replication and hence the high availability, high perfomance and high fault tolerance of databases. It is designed as a two-layered architecture, with the aim to isolate the consistency manager (CM) as a module which is independent of any underlying DBMS particularities. Additionally, MADIS takes advantadge of existing database resources for efficiently achieving its tasks, so that the implementation and execution of protocols is not overburdened by the overhead usually entailed by the consistency management in replicated databases.

The upper layer consists of the middleware providing the protocol functionalities for replication and consistency management. The lower layer is an automatically producible extension of the original schema of a given database, using exclusively standard SQL features such as triggers and stored procedures, so as to provide to the upper layer the information needed to carry out its tasks efficiently. For instance, the set of records read, written, created or deleted in a transaction is automatically stored in a particular table of the extended database schema. The consistency protocol thus is able to retrieve that information, avoiding the use of otherwise necessary additional routines, which usually tend to be complex and error-prone. As a result, the mechanisms of the middleware to manage the collection, retrieval and removal of such meta-data have become much simpler, when compared to those needed in other middleware-based systems for replicated databases, such as COPLA [4]. Of course, the performance of a middleware-based replicated database will be worse than that of a core-based one, such as Postgres-R [5], but its advantage is to be independent of, and thus much more easily portable to other DBMSs. Moreover, the upper layer can be implemented in any programming language, since the support it needs is fully in the DBMS using SQL.

MADIS supports the pluggability of protocols, such that different kinds of protocols (ranging over various paradigms, from eager [6, 7] to lazy update propagation [8], from optimistic to pessimistic concurrency control, etc) can be modularly chosen, plugged in and exchanged, according to the shifting needs of given applications. An important feature is that protocol switching is seamless and fast, since it can be performed without having to recompute the required metadata for a newly plugged-in protocol. In general, the modularity of the system and the pluggability of protocols provide an unprecedented openness of the replication middleware.

The rest of the paper is structured as follows. Section 2 describes the structure and functionality of MADIS. Section 3 describes the schema modification that MADIS proposes to aid a local consistency manager (CM). Section 4 outlines an implementation of the CM, in the form of a standard JDBC driver. In section 5, a performance analysis studies the overhead of MADIS over an unmodified PostgreSQL schema. Sections 6 and 7 compare our approach with other systems and summarize the paper.

2   The MADIS Architecture

The architecture proposed by MADIS consists of two main layers, each one providing a number of functionalities. In essence, the lower of these layers consists of a modification in the schema of the underlying database repository, in order to provide and manage additional tables. We call these tables "report-tables". The upper layer intercepts requests from the user application, and makes use of the information stored in the report-tables to perform the consistency management.

The report-tables are automatically maintained in the lower layer. They contain information accounting the execution of various transactions in the local node. The modifications done in the report-tables are managed inside the same transactional context as the transaction which these modifications refer to. As the modification of the schema only uses SQL-99 features, a high degree of portability is ensured. A set of database procedures is also provided in the schema modification, in order to hide to the upper layer the details of the schema extension.

The upper layer of the MADIS architecture is positioned between the client applications and the database. It acts as a database mediator. Common accesses to the database as well as the commit/rollback requests are intercepted, allowing the consistency protocol to take part in the process. The consistency protocol can gain access to the incremented schema of the underlying database to obtain information about executing transactions, thus performing the actions needed to provide the required consistency guarantees. Finally, the consistency protocol can also manipulate the incremented schema, making use of the provided database procedures when needed.

The implementation of the upper layer (i.e. the Consistency Manager) can be done regardless of the underlying database. In this paper, we describe a Java implementation, designed to be used by the client applications as a common JDBC driver. The functionality this driver introduces regarding consistency control over a distributed database is provided in a transparent way to the user applications. The Consistency Manager is the core of MADIS. It manages database *connections* (which may include multiple sequential transactions, working in different JDBC consistency modes) and controls a set of database replicas. Moreover, it provides the plug-in for a *consistency protocol agent*, which can be chosen according to the requirements of the given application. The supported protocols share some common characteristics. All the communication performed between the networked databases is controlled by the local consistency manager.

3   Schema Modification

The lower layer of the MADIS architecture consists of a modification in the schema of the existing database. The process for distributing an existing centralized database starts with the execution of a program that performs a schema migration at each replicated node. This migration consists of the inclusion of tables, views, triggers and database procedures designed to maintain, automatically, a number of reports about the activity performed during the lifetime of a transaction. That way, the schema modification allows the database to automatically perform the collection and maintenance of transactions writesets, as well as the metadata pertaining to the different records in the database[1]. *Optionally*, it also collects and manages transaction readsets (possibly including the information read to perform queries). If this information is not generated, a

---

[1]   As different metadata are needed by different consistency protocols, the extension caters for all of them.

consistency protocol requiring such information should perform some additional work from the upper layer.

The operations needed by the consistency protocols can be performed through a number of added database procedures, thus enabling an *ad-hoc* management (not always required) of the information automatically maintained in the database.

3.1   Modified and Added Tables

For each existing table T_i in the original schema, MADIS defines a number of modifications, relating field additions, view definitions, and others. Therefore, a new field is added for metadata purposes so as to identify a record on T_i, this new field is called *local_T_i_oid*. To this end, a field is added, defining a link to the metadata associated with each record in the table T_i

The attribute holds the local object identifier for the record. This identifier is local to a particular node in the system. Thus, it is possible for an object (identified by a *unique* global_oid) to have different *local_T_i_oid*'s within the system. A global_oid is required for the different nodes in the system, to agree in the identity of each record, regardless the local identification (sensible to local information).

In addition, MADIS creates for each table in the original schema (T_j) an extra table (named MADIS_Meta_T_j), containing the metadata needed for any protocol pluggable in the consistency manager. When a protocol is activated, MADIS executes a start-up process, to initialize each *"Meta"* table in the database. The primary key of the table consists of a unique object identifier. A typical *"Meta"* table is described as a tuple: **(local_oid *(pr.key)*, global_oid, version, transaction_id, timestamp)**.

The *MADIS_Meta_T_j* tables contain all the information needed by any replication protocol pluggable in the system. Hence, as all the fields are automatically maintained by the database manager, any of such protocols is suitable to be activated at will.

In addition to meta-tables, MADIS defines a table *MADIS_TrReport* containing a log including the activity of each transaction of the database. The table is as follows: (**trid, global_oid, field_id** *(optionally)*, **mode**). Where the primary key is composed by: (**trid, global_oid, field_id**). For each transaction, only one record per field-of-object is maintained, recording the access mode (**mode**) is recorded for each accessing transaction (**trid**), the global object identifier (**global_oid**) corresponding to the accessed record, and -maybe- the identifier for the accessed field within the record (**field_id**). In addition, once the transaction is terminated, the consistency manager eliminates from this table any record relating the concluded transaction. Note that several MVCC-based DBMSs (this is not the case of Postgress) do not use locks with record granularity, but locks that block access to entire pages or even tables. Such systems must use multiple "per transaction" temporary *TrReport* tables, including the transaction in the table name (i.e., these tables have a *<trid>_TrReport* name).

3.2   Triggers

As mentioned, MADIS introduces a set of new triggers in the database schema definition. These triggers can be classified in three main groups:

- *Writeset managers.* They are responsible for the collection of the information relating the objects written by the executing transactions.
- *Readset managers.* Collect the information related to the objects read by executing transactions. Their inclusion in the schema is optional, and when included, it is requested to be implemented by creating views.
- *Metadata automation.* These triggers are executed when the metadata stored in the MADIS extension tables must be updated. The collection and maintenance of such information is performed automatically by the triggers.

The writeset collection (WSC) is performed defining three triggers for each table T_i in the original schema. They insert in the `TrReport` table the information related to any write-access to the table performed by the executing transactions. These triggers are named `WSC_I_`T_i, `WSC_D_`T_i, and `WSC_U_`T_i, and its definition allows to intercept any write access (insert, delete or update respectively) to the T_i table, recording the event in the transaction report table (TrReport). The following example shows the definition of a basic WSC trigger, related to the insertion of a new object[2] into the table `MYTABLE`.

```
CREATE TRIGGER WSC_I_mytable
BEFORE INSERT ON mytable FOR EACH ROW EXECUTE
PROCEDURE tr_insert( mytable, getTrid(), NEW.l_mytable_oid);
```

Deletions and updates must also be intercepted by means of analogous triggers. However, as described above, the accessed fields can be optionally included in the transaction report (depending on the configuration of the MADIS middleware). To this end, a WSC trigger managing the updates should be *split* into a number of triggers, one for each field contained in the managed table (`WSC_U_mytable_field1`, `...WSC_U_mytable_field`**N**).

The second group of triggers is responsible for the transactions' readset collection. As already mentioned, this collection is optional, due to its high cost, and the fact that some consistency protocols can be accomplished without using readsets. To implement this collection, a view must be included for each table in order to compensate the lack of `TRIGGER ...BEFORE SELECT` in the SQL-99 standard. The original table must be renamed, and replaced by the new trigger. As views cannot be updated in several DBMSs, it becomes also necessary for the WSC triggers to be modified, in order to redirect the write accesses to the renamed original table. This can be done by implementing the WSC triggers as 'INSTEAD OF event' triggers, (in contrast to the basic `BEFORE event` detailed above). Finally, the `tr_insert`, `tr_update` and `tr_delete` procedures should be modified, in order to include the required redirection.

The last group of triggers added by MADIS is those responsible for the metadata management. In fact, this management can be disseminated in the WSC triggers detailed in this section. However, we describe here the metadata management implementation as independent triggers, in order to simplify the discussion. Whenever a new record is inserted, the DBMS must automatically insert the corresponding row in the metadata table. To this end, MADIS includes, for each table T_i, a trigger that inserts a row in the corresponding $MADIS_Meta_T_i$ table. As the `global_oid` is established based on the creator node identifier (i.e. the node where the object was created), and the lo-

---

[2] Note that the trigger executes the procedure `getTrid()` to obtain the transaction identifier.

cal object identifier in the creator node (managed in the $MADIS_Global$ table), all fields contained in the $MADIS_Meta_T_i$ table can be filled without intervention of any consistency protocol.

Following the life-cycle of a row, when a row is accessed in write mode, the DBMS must intercept the access, and the metadata (e.g. version, timestamp, etc) of such object must be updated. To this end, a specialized metadata maintainer (MM) trigger is included for each table. The MM trigger updates the `version`, the `transaction identifier`, and `timestamp` of the record in the given metadata table. Finally, when an object is deleted, the corresponding metadata row must be also deleted. To this end, an additional trigger is also included for each table in the original schema.

Summarizing the tasks performed by the described triggers, it is easy to see that, for each table, only three triggers must be included: BEFORE INSERT, BEFORE UP-DATE, BEFORE DELETE. Their implementation include both the transaction report management, and the metadata maintenance. If the readset management is a requirement, it is necessary to replace the definition of the triggers, implementing INSTEAD OF triggers, in contrast to BEFORE triggers. This allows the DBMS to redirect any write access to the adequate table, as well as to perform the metadata maintenance and the transaction management.

4   Consistency Manager

The architecture proposed by MADIS makes use of the database as the manager for most information related to consistency management. Moreover, the DBMS also provides the collected information to the consistency manager (CM) (situated on top of the database) with standardized structures.

Thus, the consistency management can be ported from a platform to another with a minimal effort. The rest of this section shows a Java implementation of a CM making use of the described schema modification.

Our Java implementation of the CM allows a pluggable consistency protocol to intercept any access to the underlying database, in order to coordinate both local accesses, and update propagation of committed local transactions (and, consequently, the local application of remotely committed transactions).

In our basic implementation of MADIS, we implement a JDBC driver that encapsulates an existing PostgreSQL driver, intercepting the requests performed by the user applications. The requests are transformed, and a new request is elaborated in order to obtain additional information (as metadata). The user perception of the result produced by the requests is also manipulated, in order to hide to the user applications the additionally recovered information. This mechanism allows the plugged replication protocol to be notified about any access performed by the application to the database, including query execution, row recovery, transaction termination requests (i.e. commit/rollback), etc. Thefore, the protocol has a chance to take specific actions during the transaction execution so as to accomplish its tasks.

Java user applications request a `MADIS Connection`, specifying the JDBC Driver to be used by the middleware to access the database. Query executions are also intercepted by MADIS encapsulating the `Statement` class. As response of user invocation

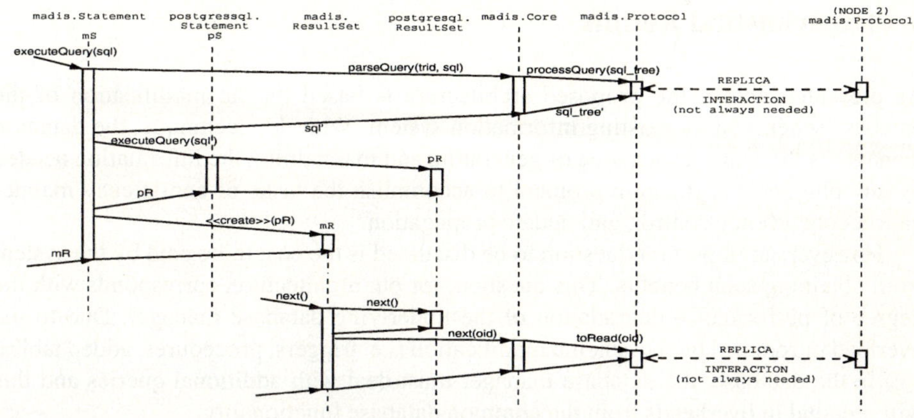

Fig. 1. Query Execution

to `createStatement` or `prepareStatement` the `MADIS Connection` generates `Statements` that manage user queries execution. When the user application requests a query execution, the request is sent to the `MADIS Core` class, which calls the `processStatement()` operation of the plugged consistency protocol.

Once this is done, the consistency protocol may modify the statement, adding to it the patches needed to retrieve some metadata, or collect additional information[3] into the transaction report. However, this statement modification is only needed by a few consistency protocols, which also have the opportunity to retrieve these metadata using additional sentences (on the "report-tables") once the original query has been completed. Optimistic consistency protocols do not need such metadata (like current object versions, or the latest update timestamps for each accessed object) until the transaction has requested its commit operation. So, they do not need these statement modifications on each query. The process for queries is depicted in figure 1.

Either if the application requests a *commit* as well as when a *rollback* is invoked, MADIS must intercept the invocation, and take additional actions. When the user application requests a commit operation, the `MADIS Connection` redirects the request to the `MADIS Core` instance. Then, the plugged protocol is notified, having then the chance to perform any action involving other nodes, access to the local database, etc. If the protocol concludes this activity with a positive result, then the transaction is suitable to commit in the local database, and the MADIS Core responds affirmatively to the `Connection` request. Finally, the `MADIS Connection` completes locally the commit, returning the completion to the user application after the notification to the MADIS Core. On the other hand, a negative result obtained from the protocol activity will be notified directly to the application, after the abortion of the local transaction. Finally, `rollback()` requests received from the user application must be also intercepted, redirected to the MADIS Core statement, and notified to the plugged protocol.

---

[3] The `ResultSet` should be also encapsulated in order to hide such included metadata.

5   Experimental Results

As presented above, the proposed architecture is based on the modification of the database schema of an existing information system. With this technique, the database manager is the main responsible of generating and maintaining the information needed by any pluggable replication protocol to accomplish the tasks of consistency maintenance, concurrency control, and update propagation.

However, an important question to be discussed is the cost to be paid by the system from obtaining such benefits. This question, for our architecture, corresponds with the degree of performance degradation of the underlying database manager. Due to the overload introduced by the schema modification (i.e. triggers, procedures, added tables, etc) in the database, the database manager must deal with additional queries and this will redound in overheads from the common database functionality.

In spatial terms, the overhead introduced by the schema modification is easy to be determined, and leads out of the scope of this paper. Regarding computational overhead, our architecture introduces a number of additional SQL sentences and calculations for each access to the database when comparing with accesses to the original schema. Summarizing, *Insertion*, *Update* and *Deletion* operations need additional insertions on the `TrReport` table, and other operations with the corresponding `MADIS_Meta_`T_j table. In contrast *Selection* overhead varies depending on the plugged protocol. The readset collection may be performed in most of the cases by the middleware, just including the $local_T_i_oid$ in the SQL sentences executed in the database. Thus, this inexpensive *oid* inclusion is often the overhead introduced in *Selection* operations. In this section, we discuss the overhead introduced in **Insertion, Update** and **Deletion** operations, due to the relevance of the overhead in these operations. We are using a dummy consistency protocol, in order to calculate just the overhead introduced by the architecture.

The experiments consisted of the execution of a Java program, performing database accesses via JDBC. The schema used by the program contains four tables (CUSTOMER, SUPPLIER, ARTICLE, and ORDER). Each article references a row in the SUPPLIER table, and each ORDER references a CUSTOMER row, as well as an ARTICLE row. Each table contains additional fields as item description (a `varchar[30]`).

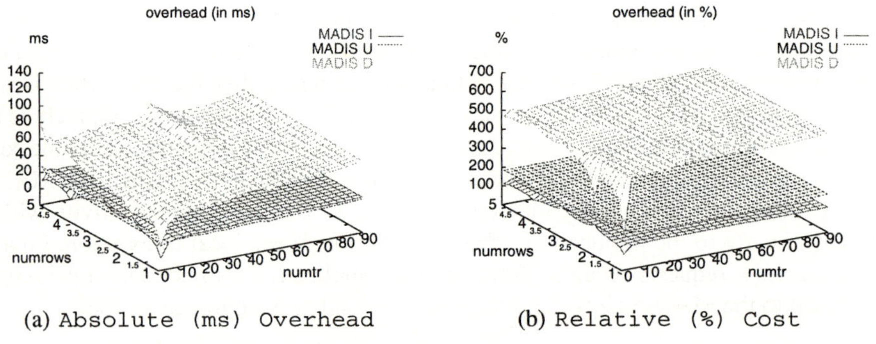

(a) Absolute (ms) Overhead   (b) Relative (%) Cost

Fig. 2. Mean Overhead

For each measurement, the experiment provides three values: the total cost of the *numtr* transactions of type I, U and D respectively, each one acting with *numrows* rows per table. We observed that deletions are the most overheaded operations in our core implementation. For a more accurate description of the overhead we calculated the time cost per transaction (figures 2 and 2(b)).

The results stabilized with a few number of transactions, which indicates that the system does not suffer appreciable performance degradation along the time. In addition, it is shown in figure 2 that the overhead per transaction is always lower than 80 ms in our experiments. Besides, figure 2(b) shows that the sensitivity for *numrows* is unappreciable (the system scales well in relation to managed rows) for any of the transaction types (I,U, and D). We concluded that our implementation of the MADIS database core introduces bounded overheads for Insertion and Update operations. However, Delete operations cause the schema modification to produce a dangerous, although bounded relative degradation of the performance (600% for 6000 rows deleted).

In GlobData, a middleware was developed to be used as a research tool in the field of replication protocols. In fact, several protocols were designed, developed, and implemented using this middleware. However, the architecture used in Globdata (COPLA) did not be conceived to provide low overheads in order to provide the required metadata to the plugged protocols. We include a comparison with COPLA. In the same conditions as the ones depicted in the previous subsections, we executed an equivalent test using COPLA. The conclusions (fig.3(a)) were that COPLA has a poor scalability for Update and Delete operations (50 and 200 times more costly than the standard schema).

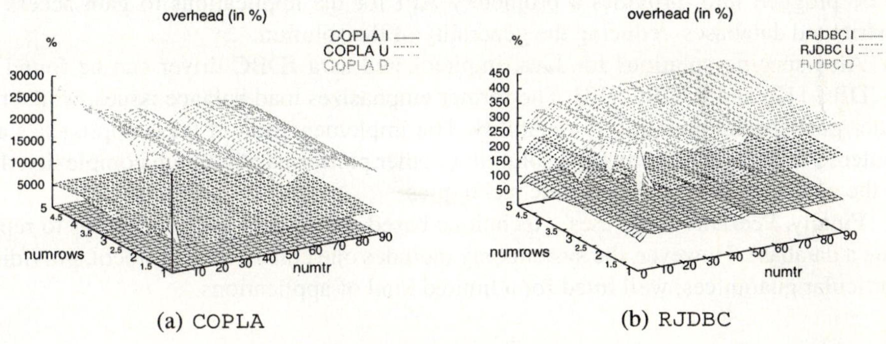

Fig. 3. Relative Overhead

Finally, the MADIS architecture was compared with RJDBC. We consider RJDBC a lower bound of the achievable results in respect of metadata collection, although such approach doesn't scale well with regard to the number of connected nodes, due to the replication technique used (eager, pessimistic, and linear interaction). In RJDBC, there is no metadata maintained in the system. In contrast, all the requests to the database are just broadcast to any node in the system. When there is a unique node (as in our experiments), the system introduces a minimal overhead, consisting in the management of the requests. The experiments showed (figure 3(b)) that the system overhead remains stable proportionally to the number of rows processed. However, it is also shown that the overhead introduced for I and U operations is comparable to the introduced by MADIS.

6   Related Work

Considering the way the metadata collection is implemented, replication approaches can be classified as *Middleware-based* (where all the work is performed by a middleware external to the database), *Trigger-based* (where the collection is performed by triggers and callbacks to external procedures), *Shadow-Table-based* (using the shadow copies in order to build the update messages needed by other replicas), and *Control-Table-based* (based on timestamping of each row of the database). Each technique has its own benefits and drawbacks, as described in [9, 10]. There have been many implementations of middleware software providing database replication services.

In Postgres-R and Dragon [7], a DBMS core is modified in order to include distributed support to the database engine. This approach has a strong dependency on the database engine for which the system is developed, and it must be reviewed each time the original DBMS software release is updated. On the other hand, its performance is generally better than the one achievable using a middleware-based architecture.

In Globdata [4, 11], a middleware providing a standard API for Java applications was presented as a general solution for distributed database access. The system also included a heavy Relational-Objectual transformation. This allows the applications to make use of an object-oriented database schema, and the system translates this schema to a relational database. The system, although allows multiple consistency protocols to be plugged into, provides a propietary API for the applications to gain access to distributed databases, reducing the generality of the solution.

Also specific solutions for Java, implemented as a JDBC driver can be found in C-JDBC [12] and RJDBC [13]. The former emphasizes load balance issues, whilst the latter puts special attention to reliability. The implementation of these approaches are centered in Java, and porting the solution to other platforms has a high complexity, due to the characteristics of the specific techniques.

Finally, PeerDirect [9] uses a technique based on triggers and procedures to replicate a database. However, the system only includes one consistency protocol, providing particular guarantees, well fitted for a limited kind of applications.

7   Conclusions

Different applications require different kinds of managing replicated information. Hence, an adequate choice of appropriate replication protocols is due. Hence, a middleware which provides flexibile support for choosing, plugging in, operating and exchanging suitable protocols, including a homogeneous access to replicated databases, is desirable for many applications.

MADIS is a platform designed to provide such functionality. It supports an ample spectrum of diferent kinds of replication protocols. It is conceived as a two layers architecture. Most of the actual work is accomplished by the lower layer, which is implemented as part of an extension of the database schema. Its implementation makes use of standard SQL-99 database resources such as tables, views, triggers, constraints and stored procedures. Being independent of the underlying DBMS, its portability is easy and smooth. The lower layer consists of the collection of all information related to the accesses performed by the database transactions of a given application.

The upper layer makes use of this automatically collected information, by notifying the transactions' accesses to the currently plugged-in replication protocol. MADIS provides and allows to choose, plug in, run and perform on-the-fly exchanges of a wide range of different protocols, each one offering a particular choice of guarantees and behaviours to the user transactions. The implementation of this upper layer is simple enough to be ported from one platform to another with a minimal cost.

In this paper, we have described the MADIS lower layer, which is implemented as a set of SQL statements that modify the original database schema. As for the upper layer, we have exemplified the outlines of an implementation providing a Java JDBC standard API. This implementation enables a transparent, standard-conform access to replicated databases, without the need to make changes to the applications' code.

References

1. Bernstein, P., Hadzilacos, V., Goodman, N.: Concurrency Control and Recovery in Database Systems. Addison-Wesley (1987)
2. Gray, J., Helland, P., O'Neil, P., Shasha, D.: The dangers of replication and a solution. In: Proc. of the 1996 ACM SIGMOD International Conference on Management of Data, Montreal, Canada (1996) 173–182
3. Instituto Tecnológico de Informática: MADIS web site. *http://www.iti.es/madis* (2004)
4. L.Irún, F.Muñoz, H.Decker, J.M.Bernabéu: COPLA: A platform for eager and lazy replication in networked databases. In: 5th Int.Conf. Enterprise Information Systems. Volume 1. (2003) 273–278
5. Kemme, B.: Database Replication for Clusters of Workstations. PhD thesis, Swiss Federal Institute of Technology, Zurich, Switzerland (2000)
6. Agrawal, D., Alonso, G., El Abbadi, A., Stanoi, I.: Exploiting atomic broadcast in replicated databases. LNCS **1300** (1997) 496–503
7. Kemme, B., Alonso, G.: A suite of database replication protocols based on group communication primitives. In: Intl.Conference on Distributed Computing Systems. (1998) 156–163
8. Ferrandina, F., Meyer, T., Zicari, R.: Implementing lazy database updates for an object database system. In: Proceedings of the Twentieth International Conference on Very Large Databases, Santiago, Chile (1994) 261–272
9. PeerDirect.: Overview & comparison of data replication architectures (white paper) (2002)
10. Sybase, Inc.: Replication strategies: Data migration, distribution and synchronization. White paper (2003) 30 pages.
11. Rodrigues, L., Miranda, H., Almeida, R., Martins, J., Vicente, P.: The GlobData fault-tolerant replicated distributed object database. In: Proceedings of the First Eurasian Conference on Advances in Information and Communication Technology, Teheran, Iran (2002)
12. ObjectWeb: C-JDBC web site. Accessible in URL: *http://c-jdbc.objectweb.org* (2004)
13. Esparza-Peidro, J., Muñoz-Escoí, F.D., Irún-Briz, L., Bernabéu-Aubán, J.M.: RJDBC: A simple database replication engine. In: 6th Int.Conf.Enterprise Information Systems. (2004)

Hierarchical Aggregation
in Networked Data Management

Pedro Furtado

DEI /CISUC, Universidade de Coimbra, Portugal
`pnf@dei.uc.pt`

Abstract. The recent trend towards peer-to-peer and networked data management raises some challenging issues regarding data placement and processing. Additionally, as data management environments change from a machine into a local area network and from there into a global inter-network, the context of application of parallel query processing changes. In this paper we analyze parallel processing of aggregation queries in different networked contexts. First we describe briefly the Node-Partitioned Data Manager architecture and the aggregation processing in that architecture. We identify a performance bottleneck in the basic typical parallel aggregation strategy and discuss the use of hierarchical aggregation to overcome the problem. We analyze and compare the strategies both analytically and experimentally by means of a model and a simulator capable of generating different networked settings. This allowed us to compare the influence of different parameters on the performance. We were able to show the increased efficiency of the strategy and also to analyze and obtain interesting results of its behavior in varied settings.

1 Introduction

Processing and performance issues in parallel and distributed databases have received lots of attention in the past. One of the interesting issues was the advantage of using clusters of lower-cost nodes to process efficiently against databases in general, typically under generic online transaction processing (OLTP) workloads and sometimes more complex analysis query workloads. The main issues in such environments include data allocation, query processing and load balancing. Not all database access patterns benefit linearly (linear speedup) from parallel architectures and some can have much less than linear speedup unless expensive massively parallel hardware is used. Extra overheads such as data exchange costs and resource access contention are reasons for this. To minimize query response time of a node-partitioned data management system in a LAN or global inter-networked environment, it is important to consider placement and processing strategies that promote simultaneous parallel processing of most of the query by component processing nodes. This is the objective of our architecture – the Node-Partitioned Data Management architecture (NPDM). In this paper we review the main features of the NPDM system and then concentrate on the processing of aggregation queries over NPDM. We identify a performance bottleneck in the basic parallel aggregation algorithm and describe an alternative solution based on hierarchical aggregation. We analyze the strategies both analytically and by means of simulation experiments to test over different networked contexts.

J.C. Cunha and P.D. Medeiros (Eds.): Euro-Par 2005, LNCS 3648, pp. 360–369, 2005.

The paper is organized as follows: section 2 discusses related work. Section 3 overviews the Node Partitioned Data Management system. Section 4 discusses the aggregation processing alternatives and a simplified model for comparative purposes. In section 5 we describe our simulation experiments and analyze experimental results for varied scenarios. Section 6 concludes the paper.

2   Related Work

A large body of work exists in applying parallel processing techniques to relational database systems. The objective is to apply either inter-query or intra-query parallelism to improve performance. In [4] the authors review parallel processing in databases, speedup and scalability issues in parallel database systems, including shared-nothing environments, as the basis for the future of high-performance database computing. Query processing in parallel and distributed databases has been the focus of much research on the database field [1, 2, 3, 11, 13]. Parallel processing of most operators, including aggregation, is typically based on parallel processing by nodes, followed by merging the results (e.g. [4]). Although standardized relational algebra operators used in databases lend themselves well to parallelization, issues are raised concerning mainly communication and merging overheads in not so fast communication mediums. These overheads can be especially relevant during the processing of distributed join and aggregation operations. Parallel Hash Join algorithms are described in [5, 8, 7, 11]. In [12] the authors discuss parallel aggregation but in the context of fast parallel machines. In contrast, we analyze aggregation in a generic networked environment, study the merging performance bottleneck, propose hierarchical aggregation and compare the alternatives in networked settings.

Data placement, although not the main focus in this paper, is another important issue in parallel and distributed database architectures [14, 15], due to the large data exchange overhead required to repartition relations for join and aggregation processing in a parallel setting (e.g. see [11]). Workload-based partitioning [9, 17, 6] aims at minimizing these overheads.

We consider that the nodes holding the schema can be in a fast local LAN, completely distributed in the global inter-network or in an intermediate layout. Using strategies such as the one in [10], it is possible to select overlay servers to hold a schema based on cost locality.

3   The Node-Partitioned Data Manager

We assume either a non-dedicated local area network (LAN) or a more generic networked environment. Given the data to be processed, a number of nodes will hold and process the schema to answer queries to submitting clients. Figure 1 provides an overview of the NPDM modules. The node manager (NM) is responsible for adding and removing nodes from the system (including data reorganization in such advents), managing addressing information at each node and overseeing replicas and node replacement in case of node failures. The Storage Manager (SM) is responsible for the administration of object persistence. This includes storage, retrieval, replication and migration of objects in response to requests. Each node holding schema data has a local database engine (DBE) which is part of the storage manager. The Data Manager (DM) offers data management typical of a DBMS on a different networked context.

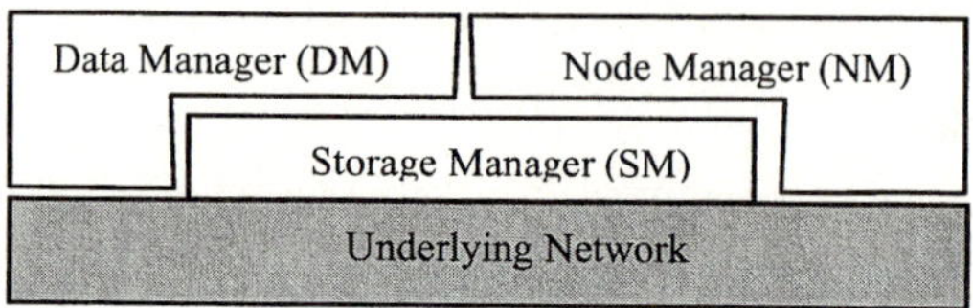

Fig. 1. NPDM Modules

Tuple Management: the NPDM implements parallel processing of tuples over the network. Nodes participating in the storage and processing of complex queries are expected to have a simple database-like engine (DBE) as part of their storage manager (SM) to handle local data management. As in relational databases, schemas comprise relations (tables) made of sets of tuples and the SM maintains the relation metadata at each node. Given a relation R with tuples of the form T(attr1,...,attrn), the relation can either be fully replicated or fully partitioned using a hash table functionality. In the last case one attribute must be chosen as the tuple identifier (tid) by the partitioning and placement algorithm [9, 17, 6]. Like in a distributed hash table (DHT) environment, nodes receive tuples (objects) determined by the hash-value and the object to be inserted is the tuple with oid=tid. This strategy enables a tuple to be looked up directly in the node holding it using its tuple identifier. Consider for instance the relation PART(partkey, name, mfgr, brand, type, size). If oid=partkey, the relation will either be fully replicated into all nodes or fully partitioned by partkey.

Partitioning and placement in this environment should be tailored to the workload and objectives of the schemas. As in typical parallel and distributed databases, a workload-based partitioning and placement approach can be used [9, 17, 6]. In distributed systems with slow interconnects (e.g. global internetworking environments) data exchange requirements should be minimized. To that end, small infrequently-modified relations can be replicated to minimize data exchange requirements for processing joins. Large data sets can instead be fully partitioned to take advantage of the parallel environment.

4   Processing of Aggregation in NPDM

Consider a generic NPDM setting with several nodes in which any node can submit a query. The typical query processing cycle of NPDM is shown in Figure 2.

The query is first rewritten in step 1. Step 2 "Send Query" forwards the query into all nodes. Step 3 computes partial results and step 4 forwards those results into a merging node. Step 5 applies the merge query. Step 6 may be necessary in some queries containing subqueries, to redistribute results into processing nodes for another processing cycle. Aggregation is a very common operation. For instance, the follow-

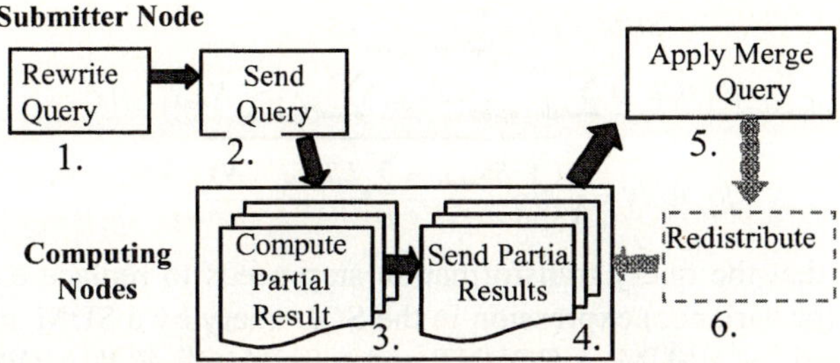

Fig. 2. Query Processing Steps in NPDM

ing SQL query in Figure 3 (from the TPC-H performance benchmark [18]) computes the sales of each brand per month:

```
SELECT       p_brand, year_month, sum(l_quantity), count(*)
FROM         JOIN lineitem LI, part P, time T, supplier S
WHERE        year_month>= '1997' AND   supplier = 'X'
GROUP BY   to_char(l_shipdate,'yyyy-mm'), p_brand, year_month;
```

Another example of aggregation is the number and duration of accesses to sites or pages grouped by day, week or month and by country or region. These queries typically contain group-by attributes that allow the aggregation to be determined for each group. Our objective is to process these queries efficiently. Aggregation produces some statistical results (sum, count, average, deviation, etc) for each group and groups are determined by grouping attributes. In a parallel or distributed setting with possibly slow interconnects, this aggregation can best be handled using the following scheme, which adheres to the diagram of Figure 2: each node needs to apply an only slightly modified query on its partial data (steps 1, 2 and 3) and the results are merged by applying the same query again at the merging node with the partial results coming from the processing nodes (steps 4 and 5). Figure 3 illustrates this process for a simple sum query:

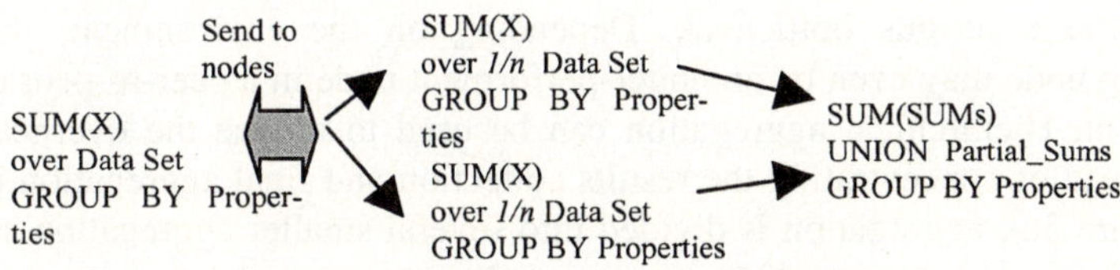

Fig. 3. Typical SUM Query over NPDM

While the sum operation was unchanged in the query rewrite step of Figure 3, other aggregation operators need slight modifications. In practice simple additive aggregation primitives are computed in each node, from which the final aggregation function is derived. The most common primitives are: (LS, SS, N, MAX, MIN - linear sum LS = sum(x); sum of squares SS = sum(x^2); number of elements N, extremes MAX and MIN).Examples of final aggregation functions are:

$$COUNT = N = \sum_{all_nodes} n_{node\ i} \tag{1}$$

$$SUM = LS = \sum\nolimits_{all_nodes} LS_{node\,i} \tag{2}$$

$$AVERAGE = \sum\nolimits_{all_nodes} LS_{node\,i} \Big/ \sum\nolimits_{all_nodes} N_{node\,i} \tag{3}$$

$$STDDEV = \sqrt{\frac{\left(\sum SS_{node_i} - \sum LS_{node_i}{}^{2} / N\right)}{N}} \tag{4}$$

This means that the query transformation step needs to replace each AVERAGE and STDDEV (or variance) expression in the SQL query by a SUM and a COUNT in the first case and by a SUM, a COUNT and a SUM_OF_SQUARES in the second case to determine the local query for each node and by the expressions (1) to (4) in the final merging query. Figure 4 shows an example of aggregation query processing steps with those steps numbered according to Figure 2.

0. Query submission:
Select sum(a), count(a), average(a), max(a), min(a),
stddev(a), group_attributes
From data set
Group by group_attributes;

4. Results sending/collecting:
Create cached table
PRqueryX(node, suma, counta, ssuma, maxa, mina,
group_attributes)
as <insert received results>;

1. Query rewriting and distribution to each node:
Select sum(a), count(a), sum(a x a), max(a), min(a), group_attributes
From data set
Group by group_attributes;

5. Results merging:
Select sum(suma), sum(counta),
sum(suma)/ sum(counta), max(maxa), min(mina)
(sum(ssuma)-sum(suma)^2)/sum(counta), ga
From UNION_ALL(PRqueryX)
Group by group_attributes;

Fig. 4. Basic Aggregation Query Steps

When aggregation results are a few set of tuples, steps 4 and 5 are not expensive. However, this is not always the case, as the number of nodes and/or groups can be very large. In such cases, while the processing was done in parallel by all nodes, the partial results collection (step 4), buffering and merging into a single node (step 5) may become a serious bottleneck. Depending on the environment, the submitter/merging node may even be an under-performant node in a peer-to-peer computing environment. Hierarchical aggregation can be used to reduce the overhead in such cases. Instead of concentrating the results collection and final aggregation in a single node (Figure 5a), aggregation is divided into several smaller aggregation steps along the way (Figure 5b). Intermediate merging nodes receive and merge the partial result sets from incoming nodes, then propagate the merged partial results into the next merging step until the final merging node.

The hierarchy can be configured on-the-fly, by the submitter node sending to each other node, together with the query, the identifier of the node to route the answer into. Next we derive a simplified cost model for comparison purposes. We consider a unit processing cost P. This processing cost accounts for buffering, reading partial results, applying the aggregation merge queries and writing the results. For simplicity, we assume that the processing cost is monotonically increasing with the size of the data set to be processed.

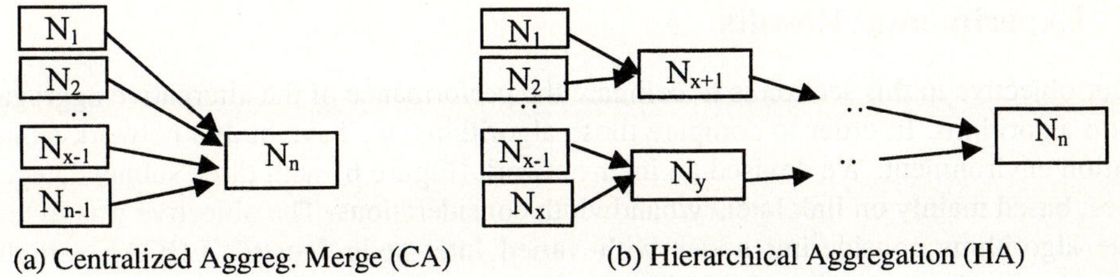

(a) Centralized Aggreg. Merge (CA)   (b) Hierarchical Aggregation (HA)

Fig. 5. Hierarchical Aggregation in NPDM

As it is hard to model precisely the cost involved in sending and queuing the data from origin nodes to a destination node, we simplify by assuming the communication cost C – the communication cost accounting for the cost of sending the data through the network medium if there were no delays - and a queueing cost resulting from many incoming fluxes going into a single node, with unitary value Q. We also assume the queuing cost to be monotonically increasing with data size. Consider also partial result sizes S and a number of nodes n.

With this simplified model the cost of the centralized aggregation scheme (CA) in Figure 6a is:

$$C + Q \times S \times (n-1) + P \times S \times (n-1) \tag{5}$$

In this expression the no-delay communication cost (C) depends only on the largest route latency. The queuing or delay cost (unit Q) is modeled as increasing linearly with the size of the data set (S) and the number of incoming streams into a single node ($n-1$). The processing cost P is also modeled as increasing linearly with S and n. Hierarchical aggregation (HA) offers the potential to reduce the overhead by essentially decreasing the queuing and processing costs, as shown in equation (6):

$$(\log_f n - 1) \times [C + Q \times f \times S + P \times f \times S] \tag{6}$$

In this expression f is the fanout of each node at each level (for simplicity we consider equal fanouts); $(\log_f n - 1)$ accounts for the number of levels in the hierarchy (minus 1). The number of hierarchical steps and both queuing and processing are assumed to be done in parallel by the decentralized merger nodes, each one with the load of $f \times S$. We assume that the worst communication cost C in each level in Figure 5b is the same as the worst cost in Figure 5a.

The comparison between expressions (5) and (6) shows that, due to increased parallelism, hierarchical aggregation decreases both queuing and processing costs each by a factor $F1 = (n-1) - \lfloor (\log_f n - 1) \times f \rfloor$. On the other hand, it increases communication costs by a factor $F2 = (\log_f n - 1) - 1$, so that (5)-(6) is:

$$\Delta(CA, HA) = F_1 \times S \times (Q + P) - F_2 C \tag{7}$$

For instance, if $f=2$ and $n=16$, $F_1=9$ (9 times faster) and $F_2=2$ (two times slower). Expression (7) shows that, with this cost model, HA is expected to improve the performance of aggregation in most cases, but if C becomes large, this effect can vanish.

5 Experimental Results

Our objective in this section is to compare the performance of the alternative aggregation algorithms. In order to compare those algorithms, we have built a network simulation environment. We devised an inter-network (Figure 6) with three subnet categories, based mainly on link latency/bandwidth considerations. The objective was to test the algorithms considering nodes with varied inter-node "costs": LOCAL - high-speed local network (LAN-like with inter-node latencies of 0.1ms); INTERMEDIATE HUBS - intermediate latency interconnects (inter-node latencies of 1ms); GLOBAL – larger latency interconnections (inter-node latencies of 4ms). This is similar to a transit-stub (TS) topology [16] but considering 3 network categories. Inter-node links were generated following a strategy similar to [16] as well.

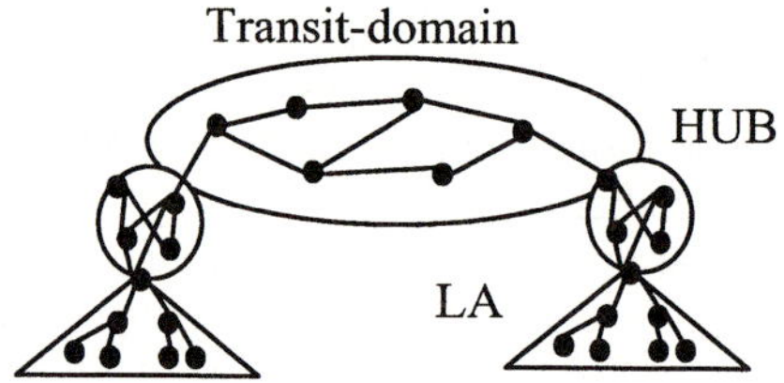

Fig. 6. Network Topology

The experimental setup was based on the generation of four 100 node hubs linked through the transit network. Each hub harbored five 200 node LANs. From these nodes we generated three node sets by picking nodes randomly: a LOCAL set based on picking nodes from a LAN; a HUB set, by picking nodes from all LANs within a single HUB; a GLOBAL set, by picking nodes randomly from all LANs, therefore going through all hubs. This setup was used to measure the time taken to exchange data (C + Q costs in the analytical expressions of (7) and (8)). The processing cost was measured by submitting the merge aggregation query against the number of partial result data sets for each case, over an Oracle database and measuring the time taken to process them and obtain the results (P). We tested "overlays" configured with 10, 25, 100 and 200 nodes for each case. The basic parallel aggregation strategy was setup by centralizing result sets into a single merge node (CA). For hierarchical aggregation (HA) we considered alternatives with different number of levels, but in this paper we discuss a single alternative with three levels (HAggreg) for lack of space. The node fanout was configured approximately equal for all nodes (with small adjustments to yield the desired number of nodes). We show and analyze only the most relevant selected results from a large pool of results we were able to obtain.

5.1 Comparing Aggreg to HAggreg

The results of Figure 7 compare the performance of Aggreg to HAggreg by measuring the response time (data exchange + merge) to aggregate a data set from 25 nodes (similar results were obtained when aggregating over 10, 100 or 200 nodes). We considered partial result set sizes of 50MB, 100MB, 200MB and 400MB, meaning

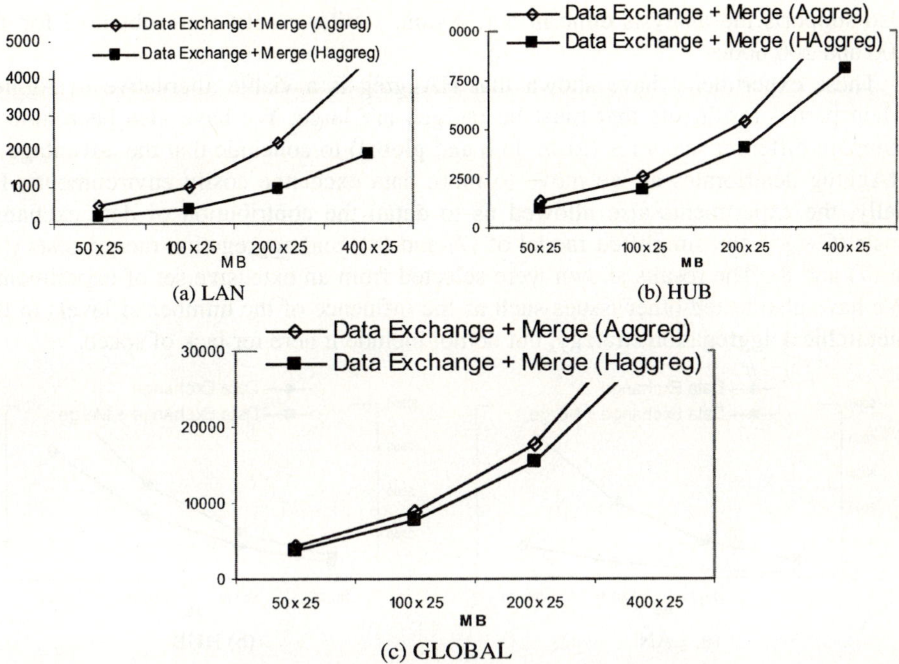

Fig. 7. Response Time Aggreg VS HAggreg (25 nodes)

that each node produces such a partial result set. The three cases shown correspond to LAN (a), HUB (b) and GLOBAL (c) environments.

As expected, these results show that HAggreg improves the response time considerably when compared with Aggreg. The comparison between the cases (a, b and c) also shows that the advantage of using HAggreg measured relative to Aggreg is larger in a LAN environment and decreases as we move from there to Hub and from Hub to Global context - HAggreg took about 40%, 75% and 85% of the time that Aggreg took, respectively (although in absolute terms the response time difference increases from LAN to GLOBAL). This effect is due to a significant increase in communication and queuing delays, as the data needs to travel the much larger HUB or GLOBAL network paths. In section 4 we also predicted this possibility using the analytical model expressed in (7) and (8). In the next subsection we detail the contribution of data exchange and merge components to the response time.

5.2   Detailing Data Exchange and Merge Costs

Figure 8 compares the data exchange cost to the total data exchange plus merge cost in the LAN, HUB and GLOBAL contexts considering Aggreg (similar comparative results were obtained for HAggreg). From these Figures it is clear that the data exchange overhead is a small component in a LAN environment (a), but gradually becomes more relevant as the environment moves into the GLOBAL case (c), which also supports the analysis done in 5.1. Again, similar results were obtained for 10, 100 and 200 nodes.

These experiments have shown that HAggreg is a viable alternative to Aggreg when partial result sets that must be merged are large. We have also been able to compare different scenarios (local, hub and global) to conclude that the advantage of HAggreg deteriorates as we move to more data exchange costly environments. Finally, the experiments also allowed us to detail the contribution of data exchange costs (C+Q in the simplified model of (7) and (8)) and aggregation merge costs ((P) in (7) and 8). The results shown were selected from an extensive set of experiments. We have also tested other issues such as the influence of the number of levels in the hierarchical aggregation strategy, but do not include it here for lack of space.

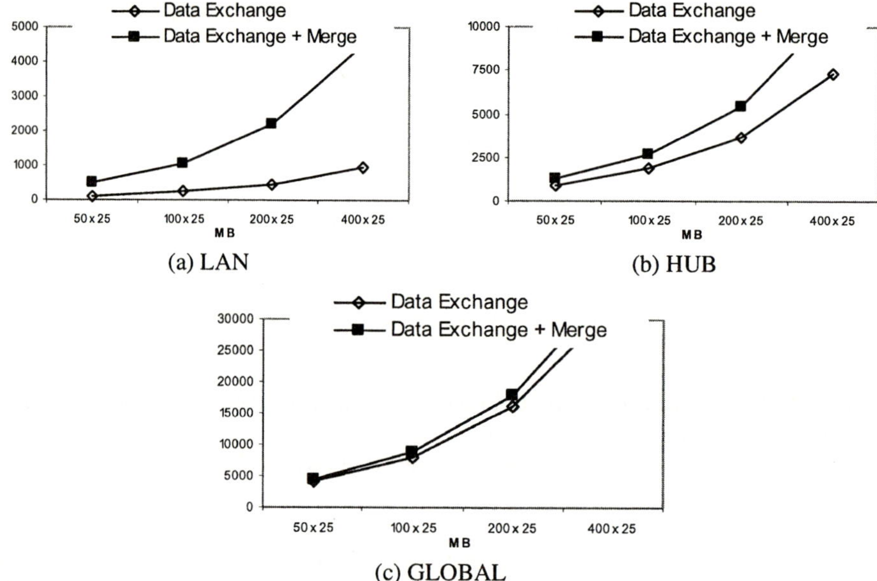

Fig. 8. Comparison Data Exchange / Merge Costs

6   Conclusions

In this paper we have studied aggregation processing in a networked data management system. We have analyzed and compared the use of two parallel aggregation alternatives: simple and hierarchical aggregation, with the objective of reducing the bottleneck related to centralized merging and therefore optimizing response time. Our experimental objectives were to compare the approaches and alternatives considering different networked environments, from a local LAN to a global inter-network. The results have allowed us to compare different settings, number of nodes and result sizes and also to analyze the contribution of data exchange and data merging operations. We conclude that the hierarchical aggregation is a useful strategy but its results are dependent on the characteristics of the networked environment. Future work on this subject includes a cost-based approach to determine the most appropriate strategy and configuration to process aggregations in such an environment.

References

1. M. Akinde, M. Bohlen et al., "Efficient OLAP Query Processing in Distributed Data Warehouses". EDBT'02, Czech Republic March 2002.
2. P.A. Bernstein et al. "Query Processing in a System for Distributed Databases (SDD-1)", ACM Trans. Database Systems, vol. 6, no. 4, pp. 602-625, Dec. 1981.
3. Chen, Hao, C. Liu: "An Efficient Algorithm for Processing Distributed Queries Using Partition Dependency", International Conference on Parallel and Distributed Systems, ICPADS 2000: 339-346.
4. David J. DeWitt, Jim Gray, "Parallel Database Systems: The Future of High Performance Database Processing", Communications of the ACM, 1992.
5. David J. DeWitt, Robert Gerber, Multiprocessor Hash-Based Join Algorithms, Procs. 11th VLDB Conference.
6. P. Furtado: "Hash-based Placement and Processing for Efficient Node Partitioned Query-Intensive Databases", International Conference on Parallel and Distributed Systems, ICPADS 2004.
7. M. Kitsuregawa, H. Tanaka, and T. Motooka. Application of hash to database machine and its architecture. New Generation Computing, 1(1):66-74, 1983.
8. Liu, Chengwen and Hao Chen, "A Hash Partition Strategy for Distributed Query Processing", EDBT 1996.
9. J. Rao et al. "Automating physical database design in a parallel database". SIGMOD Conference 2002: 558-569.
10. Ratnasamy et al., "Topologically-Aware Overlay Construction and Server Selection", INFOCOMM02.
11. Shasha, Dennis et al.: "Optimizing Equijoin Queries In Distributed Databases Where Relations Are Hash Partitioned". ACM Trans. on Database Systems, V 16, N. 2, June 1991.
12. Shatdal, A., Naughton, J. F.: "Adaptive Parallel Aggregation Algorithms". In Proceedings of the 1995 ACM SIGMOD Int'l Conference on Management of Data, San Jose, California, May 22-25, 1995.
13. Teradata Corporation. Database Computer System Concepts and Facilities. Document C02-0001-01, Teradata Corporation, Los Angeles, Oct. 1984.
14. Yu, C. T., Guh, K. C., Brill, D. and Chen, A. L. P.: Partition strategy for distributed query processing in fast local networks. IEEE Transactions on Software Engineering, Vol. 15, No. 6, pp. 780-793, June 1989.
15. Zhou S., M.H. Williams, "Data placement in parallel database systems," Parallel Database Techniques, IEEE Computer Society Press, 1997.
16. E. Zegura, K. Calvert, and S. Bhattacharjee, "How to Model an Internetwork," in Proceedings IEEE Infocom '96, CA, May 1996.
17. Zilio, Daniel C et al., "Partitioning Key Selection for a Shared-Nothing Parallel Database System". IBM Research Report RC 19820 (87739) 11/10/94.
18. TPC-H, Transaction Processing Council Benchmark TPC-H, www.tpc.org.

Mining Global Association Rules on an Oracle Grid by Scanning Once Distributed Databases

Frank Wang[1] and Na Helian[2]

[1] Centre for Grid Computing
Cambridge-Cranfield High-Performance Computing Facilities, UK
`frankwang@ieee.org`
`http://www.hpcf.cam.ac.uk/research.html`
[2] Department of Computing, Communication Technology and Mathematics
London Metropolitan University, UK
`n.helian@londonmet.ac.uk`

Abstract. Oracle 10g is a commercial infrastructure specifically designed for enterprise grid computing. On an Oracle Grid, we implement a ScanOnce algorithm to mine global association rules in support of the third generation of data mining systems on distributed and massive data. The ScanOnce algorithm does not need to ship all of local data to one site thereby not causing excessive network communication cost. The power of generating ad hoc queries in SQL ensures fast access to any desired counter.

1 Introduction

Grid computing offers an opportunity to improve the existing IT infrastructure while lowering costs. Oracle's grid infrastructure virtualizes and provisions IT resources, making them available to applications and users on demand [Shimp, 2005]. Oracle 10g ("g" stands for "grid computing") is a commercial infrastructure specifically designed for enterprise grid computing, delivering higher quality of service at a lower cost.

Knowledge discovery and data mining deal with the problem of extracting interesting associations, classifiers, clusters, and other patterns from data. Among the best known algorithms is the Apriori algorithm. An ever increasing number of organizations are installing large data warehouse using relational database technology. With the explosion of the commodity internet and the emergence of wide area high performance networks, mining distributed data is becoming recognized as a fundamental scientific challenge. The Internet, corporate intranets, sensor networks, and even scientific computing domains support this observation [Bailey, 1999].

In support of the current practice that restricts the communication cost but, if possible, without missing important rules, we proposed an efficient distributed and mobile algorithm for global association rule mining while leaving the data in place. This provides a fundamentally new technology, since, in fact, most data is distributed.

2 Data Transformation and Distributed Management of the Itemset Counter Sets

PL/SQL is Oracle's procedural extension to SQL. SQL is good in defining the structure of the database and generating ad hoc queries. However to build applications, the

J.C. Cunha and P.D. Medeiros (Eds.): Euro-Par 2005, LNCS 3648, pp. 370–378, 2005.

power of a full-fledged high-level programming language is needed. PL/SQL provides such an environment to develop application programs. It supplements SQL with several high-level programming language features such as object-oriented features [Sunderraman, 2000]. Thus, PL/SQL can be used to build sophisticated database applications, like the implementation of our proposed algorithm in this article.

2.1 Data Transformation

Since we are implementing our Distributed ScanOnce algorithm in PL/SQL, a Boolean database D needs to be transformed to relational representation. Each transaction and each item is uniquely identified by an integer (for ease of demonstration, we use alphabetic letters to identify different items in the article). The two column representation is implied by the set characteristic of the transactions. That is, the number of items typically contained in a transaction may vary largely. Moreover, a maximal transaction size may not be determined in advance.

2.2 Local Database Access Using Cursors and Records

PL/SQL provides cursors for processing a query resulting in more than one row. A PL/SQL cursor allows the program to fetch and process information from the database into the PL/SQL program. The relational transaction table should be loaded into a predefined cursor, as shown in Figure 1, one transaction (consisting of a number of rows) at a time. Cursor variables provide a pointer to the cursor work area and thereby increase the flexibility of the use of cursors. Once a cursor has been declared, it can be processed using the open, fetch, and close statements. PL/SQL provides for loop to be used with cursors. This loop is very useful in our situation where all rows of the cursor (one transaction) are to be processed.

A PL/SQL record is a composite data structure, similar to a structure in a

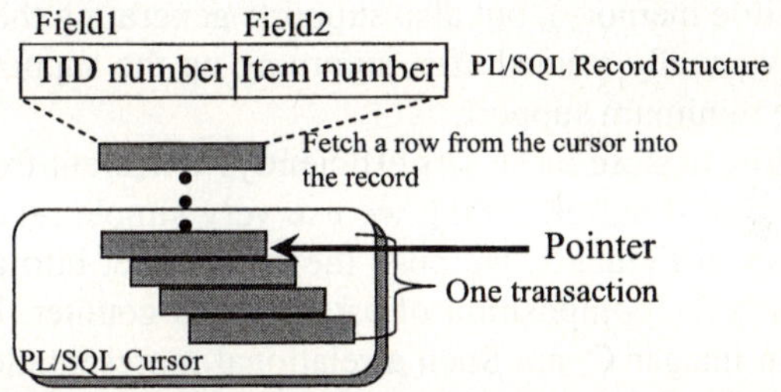

Fig. 1. A PL/SQL cursor allows the program to fetch and process transactions from the local database into the PL/SQL program. A PL/SQL record is also used, whose structure is based on the select-list of a cursor

high –level programming language. A record is used, whose structure is based on the select-list of a cursor. The record into which the cursor rows are to be fetched must be compatible with the cursor row type. The below coding list declares a cursor-based record and then processes the cursor using the record. Within the body of the loop, the individual fields of the record can be accessed. The loop terminates automatically when all rows of the cursor, i.e., the current transaction, have been scanned.

```
declare
cursor ta_line is       //cursor definition
      select item
      from transaction ta
      where tid=i;    //i is a loop variable for scanning
                    //the entire transaction database
ta_rec ta_line%rowtype;    //record definition
begin
  open ta_line;
    loop
          fetch ta_line into ta_rec;
          exit when ta_line%notfound;
              Distributed ScanOnce Algorithm in PL/SQL
      end loop;
  close ta_line;
end;
```

Declaration of a cursor-based record and then processing of the cursor using the record.

2.3 Distributed Management of the Itemset Counter Sets

After the whole local database is scanned, the counters for the all enumerated itemsets, organized in a group of itemset counter sets, is obtained, as shown in the left part of Figure 2, which is the contribution from an exemplified transaction 'abde' consisting of four singleton items 'a', 'b', 'd' and 'e'. Each entry in the set is in the form (IS, f), where IS represents an itemset and f represents the exact frequency count of IS. Organizing the counters in such sets not only allows us to store them efficiently (using little memory), but also supports generating the rules. The infrequent itemsets will be eventually pruned after summing up the support across all sites since they do not have minimum support.

How to store these sets efficiently? Different from traditional linked-list tree representation [Shaffer, 2001], we use very simple relational tables to store these sets. As shown in Figure 2, we push the counter set into a relational table "Counter", which records the composition of each itemset counter. Each counter is uniquely identified by an integer C_no. Such a relational representation of an itemset counter is flexible as the number of item identifiers in a counter may vary largely. Moreover, a maximal number of the counters may not be determined in advance. Another relational table "Support_Count" is also locally maintained for storing support count for each counter. The counter identifier (C_no) links "Counter" and "Support_Count" tables.

To locate a desired counter becomes another important issue. Not only in the first step of itemset appearance counting in each local site, but also in the second step of generating association rules from globally frequent itemsets, there are frequent accesses to those counters. The power of generating ad hoc queries in PL/SQL ensures fast access to any desired counter. The below coding list shows an example of locating a desired counter corresponding to 3-itemset 'abd'. An intermediate variable "pass", declared as support.count%type, is used to transfer the previous count number from the "Support_Count" table, which will be updated by a new count number.

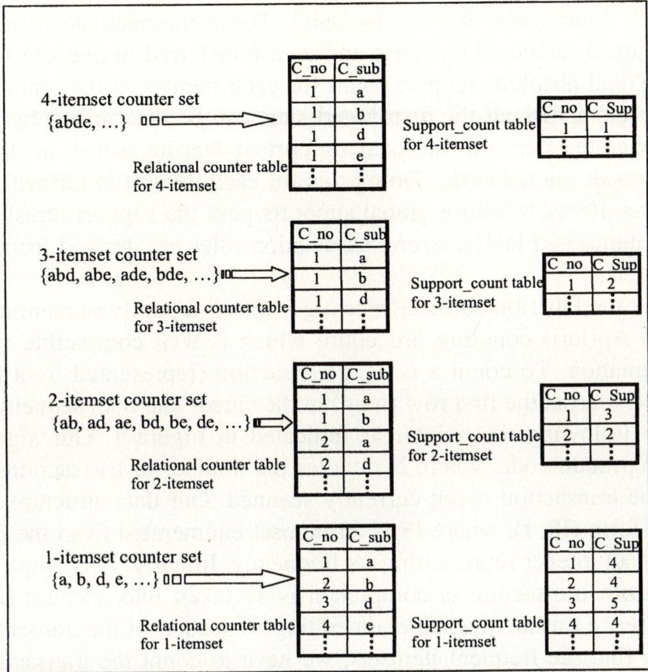

Fig. 2. The relational representation of the sets of the itemset counters

```
Declaration
...

select s.count
into pass  //pass is a predefined intermediate variable
from support s
where s.c_no=(
    select c1.c_no
    from counter c1, counter c2, counter c3
   where c1.c_sub='a' and c2.c_sub='b' and c3.c_sub='d'
     and c1.c_no=c2.c_no and c2.c_no=c3.c_no
     and c1.c_no in
         (select distinct c.c_no
         from counter c
         group by c.c_no
         having count(*)=3));   //assume 3 is minimum
         support
```

An example of locating a desired counter corresponding to 3-itemset 'abd'.

3   Distributed ScanOnce Algorithm in PL/SQL

The distributed data mining algorithms are encapsulated into SQL Server stored procedures. The algorithm is outlined as below. In each site, the local absolute support count for each enumerated itemset is found. These intermediate results are stored back into the local relational database and then transferred to one site for final processing. The global absolute support count for each itemset in the union CF of all of the local itemsets across all the distributed sites can be determined by summing up, for each enumerated itemset, the local support of that itemset in all the distributed sites (Move Result via network). Doing this for each itemset in CF will give us their global supports. Itemsets whose global supports pass the support threshold are globally frequent itemsets. Finally, strong association rules are derived from the globally frequent itemsets.

In each site, the Distributed ScanOnce algorithm is a purely sequential (rather than recursive like Apriori) counting procedure which is well compatible with the relational representation. To count a certain transaction (represented by a PL/SQL cursor), we merely start at the first row (item) in the cursor and then sequentially traverse the cursor by following the pointer as indicated in Figure 1. Our algorithm is described in the pseudo-code, where N denotes the number of transactions in the database and T the transaction being currently scanned. Our data structure A is a set of entries of the form (IS, f), where IS is an itemset enumerated from the current transaction and f is an integer representing its frequency. Initially, A is empty. The contribution from each transaction is comprehensively taken into account by growing a prefix tree for each transaction and enumerating all subsets of the transaction itemset.

In order to find the frequent itemsets, we have to count the transactions they are contained in. Our implementation is based on the idea to organize the counters for the itemsets in a special kind of prefix tree for enumeration. Each IS denotes a counter for an itemset IS. A node in the tree represents an itemset consisting of Item-IDs in that node and all its ancestors. The itemsets enclosed in a dashed rectangle share the same ancestor. For ease of illustration, symbolic letters are used here to represent the items. In practice they should be 4-byte integers. Since the common part would be a prefix if we were dealing with sequences instead of sets, such a data structure is commonly called a prefix tree. That we are dealing with sets, not sequences, is the reason, why this tree structure is unbalanced: 'abd', for instance, is the same as 'bda' and therefore only one of these counters is needed. This full prefix tree is created level by level. That is, the root node is created first. Then the second tree level is created—the children of the root and so on. Of course, in doing so, some branches of the tree can be pruned eventually after the whole database has been scanned, because by simply applying a user-specified threshold we can find out whether a branch can contain frequent itemsets or not.

Whenever a new itemset IS arrives, we first lookup A, to see whether an entry for IS already exists or not. If the lookup succeeds, we update the entry by incrementing its frequency f by one. Otherwise, we create a new entry of the form (IS, f). For an entry (IS, f), f represents the exact frequency count of IS ever since this entry was inserted into A.

The enumerated k-itemsets will be written into their corresponding k-itemset counter set, together with their support count in the form (IS, f). As mentioned in Section 4, we push the counter set into a relational table "Counter", which records the composition of each itemset counter. Another relational table "Support_Count" is also locally maintained for storing support count for each counter. The power of generating ad hoc queries in PL/SQL ensures fast access to any desired counter.

```
...
 1.A ? Ø  //A: The set of all counters
 2.T ? next transaction //T: Transaction
                   //(item-IDs)
 3.Grow subset tree for T and enumerate all subsets of the
   current transaction T
 4.IS ? each subset
 5.if (IS, f) exists in A do  //f: frequency
 6.       f ? f+1
 7.else do
 8.        insert (IS, 1) to A
 9.endif
10.Goto 2
11.
12.scan A and prune infrequent itemsets
13.if f = min_sup x N //min_sup: minimum  //support, N:
   Number of transactions
14.output (IS, f)
15.endif
16.
17.Generate rules from frequent itemsets IS satisfying
   minimum confidence c specified by the user
...
```

Pseudo-code for our Distributed ScanOnce algorithm described in this article.

In summary, we process the input data stream transaction by transaction in each local site. This is 100% sequential counting procedure and therefore there is no need at all to store and re-scan the previously-scanned transactions, which will be discarded after a single pass. In [Manku, 2002], they try to fill available main memory with as many transactions as possible, and then process recursively such a batch of transactions together. This is where our algorithm differs from that one. The amount of main memory available can be devoted to itemset counters. Their compact data structure ensures fast access to any counter in the set.

The above is so-called first step of association rule mining, in which the frequent itemsets are determined in each local site. The second step of generating global association rules from the frequent itemsets from the globally frequent itemsets after summing up the support across all sites is straightforward. Note that there is no need to re-scan the original transaction database (re-connect the network) any longer as the counters organized in relational tables have retained sufficient information for rule generating. In other words these tables are equivalent to the provided transaction database in terms of finding the frequent itemsets. In contrast, the classic Apriori algorithm requires repeated scans of the databases[Agrawal, 1993] [Agrawal, 1996] [Han, 2000] thereby resulting in unrealistically heavy network accesses particularly when considering large candidate sets.

4 Experiment Results and Further Analysis

The distributed ScanOnce association rule mining algorithm in PL/SQL described in this article is designed to economize mining efficiency and communication overhead, and we must show that this overriding concern for speed is compatible with a reasonable utilization of computer network. Our experiment with association rule mining algorithm is based on a simulation program coded in Oracle PL/SQL. The program runs on a variety of platforms.

Three desktop computers, one laptop computer and one pocket computer were used for the experiments. The desktop computers were Intel Pentium machine with Windows 2000 Advanced Server operating system and Oracle10g enterprise edition. The CPU frequency was 1.7 GHz. The desktop computers were connected by the 10 Mbps MAN Network. The laptop computer and pocket computer were used as mobile database platforms. Both of these computers were equipped with Wireless LAN cards (11 Mbps). The laptop (1.1 MHz CPU, 512MB RAM) was loaded with Oracle10g enterprise edition and Microsoft Access XP whereas the pocket computer (HP-Compaq iPAQ Pocket PC H3970, Windows CE 2.0, 400 MHz CPU, 64 MB RAM, 288MB Flash ROM) was loaded with Wireless Database 4.0 (KelBran Software) [kelbran, 2005]. Wireless Database supports a subset of the standard database SQL language. This feature allows us to access large databases (up to 1GB data) remotely and create our own query for distributed data mining with the SQL query wizard.

The experiments were conducted on a synthetic database, generated using the procedure described in [Agrawal, 1993] [Agrawal, 1996]. In this data set, the maximum transaction size and average transaction size are set to 15 and 10, respectively. The number of transactions in a single site ranges from 200,000 to 5 millions, which occupies up to 450 MB space. The experiments compared total mining time of distributed mining (Move Result) versus centralized mining (Move Data). For completeness we included the mining time on the mobile platform. The total mining time is the time it takes to transfer any data (communication), build/transfer models, and score the validation set back at the control workstation. Our preliminary results indicate that our PL/SQL implementation of our distributed ScanOnce algorithm is much faster than Apriori mining. Furthermore, distributed ScanOnce scales much better than Apriori. This is mainly because the wasteful operations of unnecessarily rescanning those previously-scanned subsets have been avoided by this new algorithm. It must be cautioned that error rates for distributed mining might be highly dependant on the organization of the data set and/or other factors. More extensive tests are currently under way.

An objective of a distributed data mining systems is to minimize both the volume of data transmitted over the network and the number of network transmissions. The time taken to send a message depends upon the length of the message and the type of network being used. It can be estimated using the formula:

Communication Time = C_0 + (no_of_bits_in_message / transmission_rate).

where C_0 is a fixed cost of initiating a message from one site to another, known as the access delay. In our instance, using an access delay of 1 second and a transmission rate of 10Mbps, we can calculate the time to send 1,000,000 transactions, each consisting of 720 bits (in average) if stored in a relational table (Table 2), as: Communication Time = $1 + (10^6 \times 720/10^7) = 73$ seconds.

Consider a simplified schema consisting of the following three transaction relations: 5,000,000 transactions stored in Site 1, 5,000,000 transactions stored in Site 2, 5,000,000 transactions stored in Site 3. Assume that the data mining computation time is negligible compared with the communication time. We give two possible strategies for association rule mining: the classic Apriori algorithm and the distributed ScanOnce algorithm proposed in this article. We calculate the response times for these two strategies as follows:

Apriori: As mentioned early, Apriori algorithm requires repeated scans of the databases and so the realistic way to handle the situation is to ship all of the data to one site. Move the transaction relation from Site 2 and Site 3 to Site 1, respectively, and then process mining there:

$$\text{Time} = 1+2\times(5\times10^6 \times 720/10^7) = 721 \text{ seconds} \tag{1}$$

Distributed ScanOnce: The local absolute support count for each itemset is found in each site respectively. These intermediate results are stored back into the support_count table (Figure 2) in local relational database and then transferred to Site 1 for final processing to obtain global frequent itemsets.

$$\text{Time} = 1 + 2\times(1\times10^6 \times 112/10^7) = 23.4 \text{ seconds} \tag{2}$$

where each counter record in support_count table occupies 112 bits.

The estimated response times vary across a wide range, in agreement with the experimental results considering the actual mining time, yet each strategy is a legitimate way to mine the data. Clearly, the communication time is significantly longer transferring the transaction databases themselves because of their big size. If the wrong strategy is chosen, then the effect can be devastating on system performance.

5 Conclusions

Most of the popular data mining algorithms are designed to work for centralized data [Agrawal, 1996] [Zheng, 2001] [Fayyad, 1996] [Kohavi, 2001][Webb, 1999]. In support of the third generation of data mining systems on distributed and massive data [Schuster, 2003][Kargupta, 2004], we proposed an efficient distributed and mobile algorithm for global association rule mining, which does not need to ship all of data to one site thereby not causing excessive network communication cost. Our preliminary results indicate that our PL/SQL implementation of our distributed ScanOnce algorithm is much faster than Apriori mining.

However it is based on a commercial product (Oracle 10g) that can limit the use of the proposed approach. A major limitation of using simple relationships is that only simple mining rules could be discovered. More information would be needed in the intermediate tables to do more complex mining. In the first step of finding frequent itemsets, we even count those itemset which are not globally frequent although they will be pruned eventually. Based on the observation that if any given set of attributes S is not adequately supported, any superset of S will also not be adequately supported and consequently any effort to calculate the support for such supersets is wasted. However, considering the advantage of performance improvement brought by avoiding shipping all of the data to one site, this new algorithm presents us with a broad range of trade-offs based on speed requirement and storage requirement, particularly in dealing with a huge distributed database of short transactions. It is worth mentioning that fewer items will be purchased at the same time in real life [Lin, 2002]. The experiments show that this Distributed ScanOnce algorithm in PL/SQL beats classic Apriori algorithm, which requires repeated scans of the databases thereby shipping all of the data to one site and consequently causing excessive network communication overhead, for large problem sizes, by factors ranging from 2 to more than 20. As the volume of transactions (preferably the depth rather than the width) grows up further, the difference between the two methods becomes larger and larger.

References

[Agrawal, 1996] Agrawal, A., Mannila, H., Srikant, R., Toivonen, H., and Verkamo, A. (1996).Fast Discovery of Association Rules. In: Fayyad et al. (1996), 307–328

[Bailey, 1999] S. Bailey, R. Grossman, H. Sivakumar, A. Turinsky, Papyrus: a system for data mining over local and wide area clusters and super-clusters, Proceedings of the 1999 ACM/IEEE conference on Supercomputing (CDROM), p.63-es, November 14-19, 1999, Portland, Oregon, United States

[Cheung, 1996] Cheung, D.W. Ng, V.T.Fu, A.W.Yongjian Fu; Efficient mining of association rules in distributed databases, IEEE Transactions on Knowledge and Data Engineering, Volume 8, Issue 6, Dec. 1996 Page(s):911 - 922

[Kargupta, 2004] H. Kargupta and H. Dutta (2004). Orthogonal Decision Trees. The Fourth IEEE International Conference on Data Mining. Brighton, UK

[Manku, 2002] Manku, G., Motwani, R., 2002. Approximate Frequency Counts over Data Streams, Proceedings of the 28th VLDB Conference, Hong Kong, China, 2002

[Schuster, 2003] Assaf Schuster, Ran Wolff, and Dan Trock, A high performance distributed algorithm for mining association rules, Proceedings of ICDM 2003, Melbourne, Florida

[Zheng, 2001] Zheng, Z., Kohavi, R., and Mason, L. (2001). Real World Performance of Association Rule Algorithms. In: Proc. 7th Int. Conf. on Knowledge Discovery and Data Mining (SIGKDD'01). New York: ACM Press

Topic 6
Grid and Cluster Computing:
Models, Middleware and Architectures

Craig A. Lee, Thilo Kielmann, Laurent Lefèvre, and João Gabriel Silva

Topic Chairs

Grid computing represents the common vision of truly general distributed computing across a ubiquitous, open-ended infrastructure supporting a wide range of different application areas. Realizing this vision will require a long-term collaboration of fundamental and applied computer science, industry, commercial infrastructure providers, and many, many application domains. Each meeting like Euro-Par is an important contribution to this long-term effort.

Grid computing has already made significant progress in the design, deployment and use of grids, but many challenges still remain before the vision can be realized. Many large scientific and engineering projects are adopting grids to support their project goals. Examples include the EU Data Grid project, NEESGrid, GriPhyN, and the NaReGI project, to name just a few. It is also possible to setup small-scale grids to accomplish flexible scheduling and workflow management. Industrial organizations, such as Sun Microsystems, IBM, Microsoft, and Oracle already have or are starting to develop grid products and solutions. Standards bodies are pushing to bring stability to the marketplace with the development of standards such as the WS-* set of specifications.

Even with this degree of progress, grids still suffer from issues of ease of deployment and use, fault tolerance and reliability, and the need for a true dominant best practice to emerge. Even though Globus is considered the de facto standard (and rightly so), there is still a wide "definition" of what constitutes a grid. Infrastructure developers, application domain experts and industry are all truly looking at an elephant. This is hampering the emergence of commonly accepted tools and best practices that can be codified in standards.

To help address these issues, we present these papers at Euro-Par that document the best work on many aspects of grid computing. This year 40 papers were submitted to Topic 6. The chairs assembled a team of 78 reviewers that produced 158 reviews. All accepted papers had at least four reviews. Many of the papers "in the middle" in effect received more than four reviews as the chairs read, reread and debated their merits. In the end, ten papers were accepted – an acceptance rate of 10%. These papers represent work in the areas of peer-to-peer, grid brokering, grid scheduling, load balancing in grids, processor farms for grids, virtual grid workspaces, grid information services, grid modeling, data replication, fault tolerance, and grid applications. We invite you now to browse these papers and closely study the ones that spark your interest and will help you in your work as much as possible.

J.C. Cunha and P.D. Medeiros (Eds.): Euro-Par 2005, LNCS 3648, p. 379, 2005.
© Springer-Verlag Berlin Heidelberg 2005

Combining Data Replication Algorithms
and Job Scheduling Heuristics in the Data Grid

Ming Tang, Bu-Sung Lee, Xueyan Tang, and Chai-Kiat Yeo

School of Computer Engineering, Nanyang Technological University, Singapore 639798
`mtang@pmail.ntu.edu.sg`, {`ebslee,asxytang,asckyeo`}`@ntu.edu.sg`

Abstract. In the Data Grid environment, the primary goal of data replication is to shorten the data access time that is experienced by the job and reduce the job turnaround time as a consequence. After introducing a Data Grid architecture that supports efficient data access for the Grid job, two dynamic data replication algorithms are put forward. Combined with different Grid scheduling heuristics, the performances of the data replication algorithms are evaluated with various simulations. The simulation results demonstrate that the dynamic replication algorithms can reduce the job turnaround time remarkably. Especially the combination of Shortest Turnaround Time (STT) scheduling heuristic and Centralized Dynamic Replication (CDR) algorithm exhibits prominent performance in diverse conditions of workload and system environment.

1 Introduction

The Grid resources, including computing facility, data storage and network bandwidth, are consumed by the jobs. For each incoming job, the Grid scheduler decides where to run the job based on the job requirements and the system status. In data-intensive computing, the locations of the data required by the job impact the Grid scheduling decision and performance greatly. Creating data replicas can reroute the data requests to certain replica servers and offer remarkably higher access speed than a single server. At the same time, the replicas provide more choices for the Grid scheduler to achieve better performance from the perspective of the job.

In this paper, a Data Grid architecture supporting efficient data replication and job scheduling is introduced. Both a centralized and a distributed dynamic data replication algorithms are put forward. The dynamic replication algorithms take into consideration the changes of the Grid environments and automatically creates new replicas for the popular data files. Three Grid scheduling heuristics are proposed and they work in an online scenario where the job submission is unknown in advance. In order to evaluate the performance of the scheduling heuristics combined with different replication algorithms, a Data Grid simulator called XDRepSim is developed. Various simulations are carried out with different system configurations and job workloads.

2 Related Work

Several recent studies have taken into account both job scheduling and data replication in the Data Grid. Ranganathan and Foster [1] modelled the External Scheduler that

J.C. Cunha and P.D. Medeiros (Eds.): Euro-Par 2005, LNCS 3648, pp. 381–390, 2005.

assigns the job to a specific computing site, and the Data Scheduler that runs at each site and dynamically creates replicas for popular data files. Various combinations of scheduling and replication strategies are evaluated with simulations. Their results show that data locality is an important factor when scheduling jobs. The simple scheduling policy of always assigning jobs to the sites that contain the required data works very well if the popular datasets are replicated dynamically. Takefusa *et al.* [2] also report similar conclusions using the Grid Datafarm architecture and the Bricks Grid simulator.

Another closely related work of [3] uses the Data Grid simulator OptorSim for studying scheduling and replication strategies. The simulated Grid architecture is similar to [1] in that a global Resource Broker schedules the jobs to the computing site and a Replica Optimiser in each site performs local replica optimization. For every data access required by the locally running job, the Replica Optimiser will determine whether the data should be replicated to local storage and which old replicas should be removed if there is not enough space. Their studied dynamic replication strategies are evolved from traditional cache replacement methods. The economic replication strategies are put forward and they attempt to improve the profits brought by the replicas and decrease the cost of data management at the same time. The simulation results show that the scheduling strategy considering both the file access cost of the job and the workload of computing resources produces the shortest mean job execution time, and the economic replication strategies can improve the Grid performance greatly.

3   System Model

The Data Grid architecture supporting data replication and job scheduling is shown in Fig. 1. The Data Grid consists of a set of domains, and each domain contains a replica server and many computing sites. The replica server provides storages for the replicas, and the computing site offers computational resources for the jobs. A computing site or a replica server is generally called a *node*. All nodes in a domain are served by a LAN, and the domains are interconnected by WAN. Some isolated replica servers might exist in the Data Grid to improve remote data access performance, and each of them constructs a domain by itself.

If a computing site and a replica server are in the same domain, then the replica server is defined as the computing site's *Primary Replica Server (PRS)* and all other replica servers are the computing site's *Secondary Replica Servers (SRS)*. In the system, there are one or more *Grid schedulers* that assign jobs to the computing sites based on particular strategies. Before the job runs in a computing site, the required input data should be loaded into the local storage in advance. If the required data is not in the computing site's data cache, the site will send the data access request to its PRS, which will search the data replicas in the system and select the one that provides the highest available bandwidth to the computing site. The selected replica will be transferred to the computing site.

In the real world, if there is no replica server in a domain, above architecture can still be applied by placing a dummy replica server in the domain. The storage size of the dummy replica server is zero so that no replicas will be created in the server. All data requests from the computing site in the domain will be forwarded to the suitable

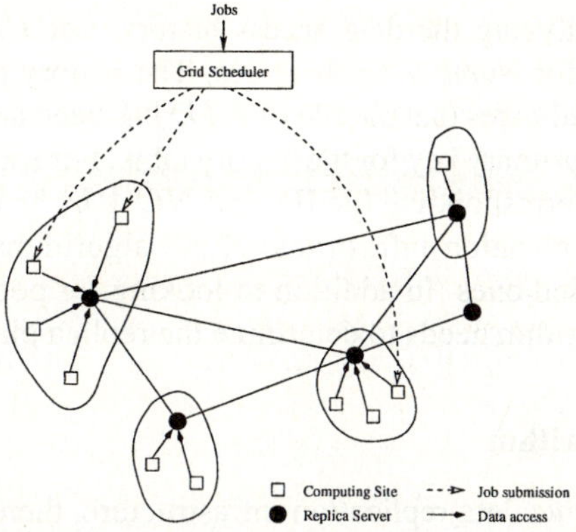

Fig. 1. System architecture.

SRS. On the other hand, if a domain contains multiple replica servers physically, we can regard them together acting as a single PRS, which aggregates the capabilities of these replica servers.

Let W be the set of domains in the Data Grid. For each domain $w \in W$, the set of computing sites located in w is denoted by $CS(w)$ and the replica server in w is denoted by $RS(w)$. For a computing site i, its computing capability is C_i.

In distributed and parallel systems, the widely used performance metrics for the job scheduling include turnaround time, throughput and utilization. Feitelson *et al.* [4] claim that turnaround time is the main metric for the open online system with endless arrival of jobs. However, system utilization and throughput are only suitable for closed systems, in which every job is re-submitted to the system when it terminates. *Turnaround time* measures how long a job takes from its submission to completion. The *geometric mean of turnaround time* is defined as $GMTT = (\prod_{k \in K} TT_k)^{\frac{1}{|K|}}$, where K is the set of jobs in the concerned scenario and $|K|$ is the number of jobs, and it is chosen by this research as the performance metric. The geometric mean equally considers the performance improvement of any job [5], so it can evaluate the scheduling performance more objectively than the arithmetic mean.

4 Dynamic Data Replication

The replication mechanism determines which file should be replicated, when to create new replicas, and where the new replicas should be placed. The dynamic replication algorithm breaks the time into *sessions*. At the beginning of each session, the replication algorithm is invoked to adjust the replica placement based on the placement in the previous session. The replica servers will be filled with replicas in long run and some replicas must be evicted to make room for new ones. In this research, LRU is applied for replica replacement with the constraint that only the replicas created in previous sessions can be evicted from the replica servers.

It is believed that the popular data in the past phase will remain popular in the short future. Thus, the dynamic data replication algorithms discussed in the paper find out the popular data by analyzing the data access history. Let FID be the abbreviation for *File ID*, and NOA for *Number Of Accesses*. The history table is in the format of $\langle FID, NOA \rangle$, which indicates that the file of FID has been accessed for NOA times. The field of FID is the primary key for the history table. For each record h in history H, let $FID(h)$ denote its corresponding file ID, and $NOA(h)$ be the number of accesses.

According to the replication infrastructure, the algorithms are classified into the centralized and distributed ones. In addition to looking for popular data files, the centralized replication algorithm needs to determine the replica placement.

4.1   Centralized Algorithm

In the centralized dynamic data replication infrastructure, there is a replication master running in the system. Every PRS collects the records of data accesses that are initiated by the computing sites in its domain. When it is time to replicate data, all PRSs send collected histories to the replication master. The master aggregates the histories and summarizes the values of NOA for the same FID. The aggregation results are stored in the history table H, which is maintained by the master. Each record h in H indicates that data file $FID(h)$ has been access for $NOA(h)$ times for the whole system since last replication session.

The centralized dynamic data replication algorithm is invoked by the replication master, which will command the replica servers to carry out the replication. The algorithm is shown as follows:

1. Compute the average number of accesses $\overline{NOA} = \frac{1}{|H|} \sum_{h \in H} NOA(h)$, where $|H|$ is the number of of records in H (also the number of data files that has been requested).
2. Remove the history records whose NOA values are less than $\overline{NOA}$. Sort the rest history records based on the field of NOA in descending order. Let l be the last record in H, and denote $MNOA = NOA(l)$.
3. While H is not empty, do:
   (a) Pop the first record h off H.
   (b) Invoke the replica placement method to create a replica for $FID(h)$.
   (c) Update record h and let $NOA(h) \leftarrow NOA(h) - MNOA$. If $NOA(h) > MNOA$, then re-insert record h into H according to the descending order of NOA field.

The average number of accesses $\overline{NOA}$ is used as the threshold to distinguish the popular data files, and only the files that have been accessed more than $\overline{NOA}$ times will be replicated. The replica placement method is introduced in following.

Replica Placement. The computing sites with higher CPU speed can finish more jobs in a given duration. Meanwhile, they access data files more frequently. The computing capability of domain w can be defined as $C_w = \sum_{i \in CS(w)} C_i$, and the computing capability of all domains is $C_W = \sum_{w \in W} C_w$. We can assume that the data request rate from domain w is proportional to its computing capability. Let θ be the factor that measures the proportional relationship between the computing capability and the data request rate for any domain, then the data request rate from domain w can be denoted by $Q(w) = \theta \cdot C_w$. Let $Prob_f$ be the request probability for data f, then the request rate from domain w for data f can be denoted by $Q(w, f) = Prob_f \cdot Q(w)$.

In the system, every computing site always accesses the replica that offers the highest bandwidth. All computing sites in the same domain should have the equivalent bandwidth to a specified replica server. Let $B_{w,j}$ be the bandwidth capacity from any node in domain w to replica server j. Let R_f be the set of replica servers that contain the replicas or the original copy of data f. The bandwidth capacity for any node in domain w to access data f can be defined as $AB(w, f) = \max_{j \in R_f} B_{w,j}$. The size of data f is denoted by $Size_f$. The average response time of all requests for data f in the system can be defined as:

$$AvgRespTime(f) = \frac{\sum_{w \in W} \frac{Q(w,f) \cdot Size_f}{AB(w,f)}}{\sum_{w \in W} Q(w, f)} = \frac{Size_f}{C_W} \cdot \sum_{w \in W} \frac{C_w}{\max_{j \in R_f} B_{w,j}}.$$

As $\frac{Size_f}{C_W}$ is constant, we only need to consider $\sum_{w \in W} \frac{C_w}{\max_{j \in R_f} B_{w,j}}$ to get the minimum average response time for data f.

Let $\mathcal{R}_f$ be the set of servers that contain the replicas created in the current replication session or the original copy of data f. To create one more replica for data file f, the replica placement method will evaluate every candidate replica server x, which has enough storages and $x \notin \mathcal{R}_f$. Attempt to let x be the choice of replica server and calculate the value of $Y(f, x) = \sum_{w \in W} \frac{C_w}{\max_{j \in \mathcal{R}_f \cup x} B_{w,j}}$. Pick replica server $\hat{x}$ that achieves the minimum of function Y to be the location of the new replica for data f. If server $\hat{x}$ has stored replica f that is created in a previous session, just stamp the replica's creation time as the current replication session, so that it will be treated as a newly created one. Otherwise, transfer data f to server $\hat{x}$ from a replica server that stores f and offers the highest available bandwidth. Let $\mathcal{R}_f \leftarrow \mathcal{R}_f \cup \hat{x}$.

4.2 Distributed Algorithm

In the distributed dynamic data replication infrastructure, for every data access request from a computing site, the PRS records the request into its history. The histories will be exchanged among all replica servers. Every replica server aggregates NOA over all domains for the same data file and creates the overall data access history of the system. At intervals, each replica server will use the replication algorithm to analyze the history and determine data replications. The distributed dynamic data replication algorithm is shown as follows:

1. Compute the average number of data accesses $\overline{NOA}$ and let $threshold = \overline{NOA} + \delta$, where $\delta \geq 0$. Remove the history records whose NOA is less than $threshold$. Sort the rest of the history records based on the field of NOA in descending order.
2. For each record h in the history according to the order, try to create a replica of $FID(h)$ to local replica server till the storages are used up.
3. Clean the history of data access.

The increment δ is used for avoiding excessive data replications that will cause heavy network contention, and it can be chosen depending on how much we are willing to compromise on the quality of replication.

Hereafter the centralized and distributed dynamic replication algorithms are referred as CDR and DDR respectively for short.

5 Grid Scheduling

For any incoming job, the Grid scheduler analyzes the system situations, communicates with the low-level local schedulers and decides which resources should be used for the job. On behalf of the end-users, the Grid scheduler submits the jobs to the local schedulers. The policy of First-Come-First-Serve (FCFS) is adopted by the local schedulers in this research. To optimize the local scheduling, jobs are allowed to move ahead provided they do not delay the first job in the queue.

We assume that the jobs are of moldable, i.e. jobs can run with multiple partition sizes. Each job is scheduled to a single computing site, and it runs on all processors in the site. The jobs are independent and every job requires a single input data file to process. For each job submission, the user needs to estimate the computational cost. The job's *computational cost* is the execution time when it runs on a benchmark computer, and it is also called the *benchmark execution time*. Let BT_k be the user estimated computational cost and $f(k)$ be the input data file for job k. The Grid scheduling heuristics studied in this paper are of online mode.

5.1 Shortest Turnaround Time

For each incoming job, the *Shortest Turnaround Time* (*STT*) heuristic estimates the job's turnaround time on every computing site, and it chooses the site that provides the shortest turnaround time. The estimated turnaround time of job k running in computing site i is denoted by $TT_{k,i} = \max\{QT(i), DT(f(k), i)\} + ET_{k,i}$, where $QT(i)$ is the queuing time, $DT(f(k), i)$ is the data transfer time, and $ET_{k,i}$ is the job execution time.

Assume that every job in the queue executes immediately when the prior one terminates, then the new job queuing time in site i can be approximated as $QT(i) = \frac{\sum_{job \in Queue(i)} BT_{job}}{C_i}$, where $Queue(i)$ is the set of jobs already in the queue of site i.

If job k's input data file $f(k)$ is in computing site i's data cache, then $DT(f(k), i) = 0$. Otherwise, let $R_{f(k)}$ be the set of replica servers that have data file $f(k)$, and $BW(i, j)$ be the available bandwidth between computing site i and replica server j. The data file downloading time can be estimated as $DT(f(k), i) = \frac{Size_{f(k)}}{\max_{j \in R_{f(k)}} BW(i,j)}$.

Let the benchmark computer's computing capability be normalized as one, then the execution time for job k running in computing site i can be estimated as $ET_{k,i} = BT_k/C_i$. Hence, $TT_{k,i}$ is figured out.

5.2 Least Relative Load

The *relative load* of computing site i is defined as $RL_i = \frac{NumOfJobs(i)+1}{C_i}$, where $NumOfJobs(i)$ is the number of jobs in computing site i at the moment, including the running and queuing jobs. The *Least Relative Load* (*LRL*) heuristic assigns the new job to the computing site that has the least relative load.

5.3 Data Present

The *Data Present* (*DP*) heuristic is an extension of JobDataPresent in [1], and it takes the data location as the major factor when assigning the job. According to different situations of the data file required by the job, DP works in following approaches:

1. **If** there are a number of computing sites having the data in their caches, then assign the job to the one with the least relative load among these sites.
2. **Else if** there are non-isolated replica servers having the data replicas, then assign the job to the computing site that is in the same domain as one of these replica servers and has the least relative load.
3. **Else**, all replicas of the data file must be in the isolated replica servers. Assign the job to the least relative load computing site in the system. In this situation, DP works the same as LRL.

6   Performance Result

In order to evaluate the performances of the dynamic replication algorithms and the Grid scheduling heuristics, a Data Grid simulator named *XDRepSim* is developed. With XDRepSim, users can easily create a Data Grid with the desired parameters of system environment and job workload. Under the same conditions of the Data Grid, various dynamic replication algorithms and Grid scheduling heuristics can be chosen and combined for performance evaluation.

There are 25 domains in the simulated Data Grid and each domain has a replica server. 200 computing sites are scattered randomly to 20 domains. Thus, there are five isolated replica servers. Let $U(x, y, \Delta)$ be a number sampled from the uniform distribution with a range from x to y, and the sampling granularity is Δ. The computing capability of each computing site is $U(1, 20, 1)$. There are 10,000 data files in the system, and each file is in the size $U(500\text{MB}, 5\text{GB}, 100\text{MB})$. The primary copy of each data file is randomly stored in a replica server. The data file popularity changes in epochs and it follows Zipf-like distribution [6] with $\alpha = 1.0$ in each epoch. Network bandwidth sharing behaviors are modelled. The bandwidth capacity between any two nodes is $U(10, 100, 10)$Mbps if they are in the same domain. Otherwise, the bandwidth capacity is $U(1, 10, 1)$Mbps. The outbound bandwidth limitation of every replica server is 200Mbps. The *relative capacity* of all replica servers is defined as $r = S/D$, where S is the sum of storage sizes of all replica servers and D is the sum of sizes of all data files in the Data Grid. In each simulation, the relative capacity of all replica servers is given. Then, each replica server is assigned with a specific size of storage in accordance with the computing capability of the domain in which the replica server is located. There are 100,000 jobs in the workload for every simulation. Job arrivals follow Poisson distribution, and a new job arrives every 10 seconds on average. The actual computational cost t_a of each job is $U(30\text{sec}, 10\text{hour}, 1\text{sec})$. The estimated computational cost t_e of the job may be different from t_a [7]. The *estimation error* of the job computational cost is defined as $e = \frac{|t_e - t_a|}{t_a}$, and the *average estimation error* for a workload is the mean value of estimation errors for all jobs. Appendix A presents the detailed simulation methods.

In order to demonstrate the advantages of the system architecture in which every domain has a replica server, the method without any data replication is studied. We refer this method as *NoR* shortly. To illustrate the performances of the dynamic replication algorithms, the static replication method is also studied. As the data file popularity changes with time, it is impossible to deduce an optimal static replication method without knowing the data access pattern in advance. Therefore, the *Random Static Replication (RSR)* policy is applied. Before each simulation is started, the data files are replicated to the servers randomly till all available storages are used up. We then evaluate the performances of RSR, CDR and DDR under the same environment. For RSR, all replicas created in the initialization step will not be changed and no new replicas will be created during the whole simulation term. On the contrary, the dynamic replication algorithms will alter the replication status with different strategies after the unified placement.

With four replication methods of NoR, RSR, CDR and DDR, and three scheduling heuristics of STT, LRL and DP, there are a total of twelve method combinations to evaluate. For each combination, we call it *policy*. The performance results of geometric mean of turnaround time (GMTT) for different policies are shown in Fig. 2(a), where the relative capacity of all replica servers is 50% and the average estimation error of the workload is 2.0.

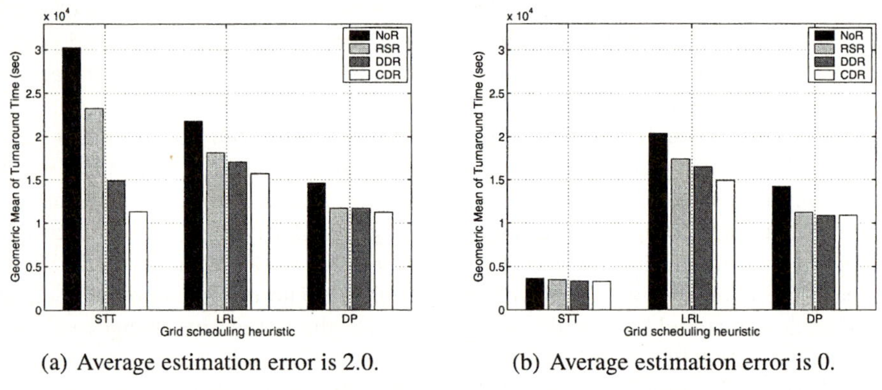

(a) Average estimation error is 2.0.   (b) Average estimation error is 0.

Fig. 2. Performance of different policies.

With the same scheduling heuristic, the performance of NoR is always the worst and its GMTT is evidently larger than any static and dynamic replication algorithm, which proves that data replication can shorten the job turnaround time. The performance of RSR is better than NoR but worse than both dynamic replication algorithms. The centralized dynamic replication algorithm CDR outperforms the distributed algorithm DDR. Generally, DP scheduling heuristic works better than STT and LRL under the simulated environment. The performance differences among the static and dynamic replication algorithms are not distinct for DP scheduling heuristic. As a whole, policy STT+CDR and DP+CDR produce the shortest GMTT.

Among the three scheduling heuristics, only STT takes the job computational cost into consideration. The inaccurate estimation of job execution time will lead to improper scheduling decision for STT. Consequently the performance of STT will be diminished. To study the properties of the policies comprehensively, we evaluate them with the workload that is same as the previous one except that the estimated job computational costs are accurate, namely the average estimation error is 0.

Figure 2(b) shows the performance results when the user estimations are accurate. It can be noted that the performance of STT with any replication method is improved prominently compared with Fig. 2(a). The GMTT values of LRL and DP also reduce when the user estimation is accurate, but the changes are slim. Overall, the performance of STT scheduling heuristics is far better than LRL and DP, and particularly STT+CDR works best among all policies.

To study the impact of the replica server storage size on the job turnaround time, the relative capacity of all replica servers r is varied from 10% to 100% in ten simulation cases. The used workload is the previous one whose average estimation error is 2.0. We only show the performance results of STT scheduling heuristic combined with each data replication method in Fig. 3.

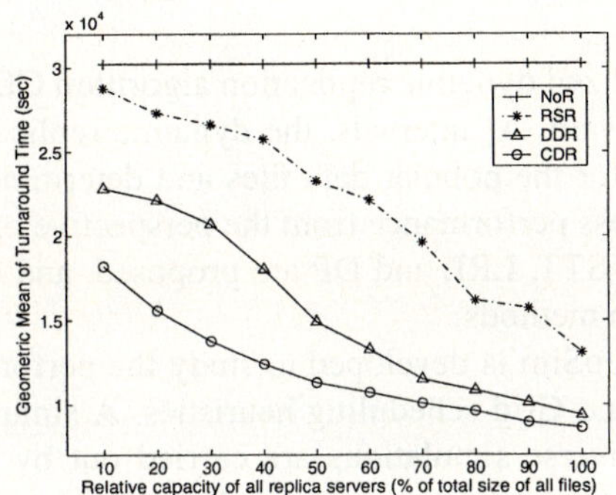

Fig. 3. Geometric mean of turnaround time vs. replica server capacity for STT.

As NoR does not create replicas, its performance remains constant irrespective of the sizes of the replica servers. The horizontal line of NoR indicates the benefit gain of data replication in reducing the job turnaround time clearly. For any storage capacity, both dynamic replication algorithms work better than RSR, and CDR performs best. By increasing the replica server capacity, the performances of the RSR and all dynamic replication algorithms are improved with different degrees.

From the simulation results we know that for the same Grid scheduling heuristic, the data replication methods of CDR, DDR and RSR can reduce the job turnaround time significantly compared with the method without any data replication (NoR). The dynamic replication algorithms perform better than the static replication algorithm because the dynamic algorithms can detect the changes of file popularity and update the data replication status in real-time.

The centralized data replication algorithm CDR outperforms the distributed data replication algorithm DDR for the same scheduling heuristic and same system configuration. The reason is that CDR makes replication decision based on the global view of the entire Data Grid environment, so the redundant replications are avoided and the storage resources are utilized efficiently. On the contrary, by applying DDR, every replica server tries to replicate the most popular data files to local storages, so the contents of every replica servers are similar. Consequently, the top hot data files are replicated in too many servers, while many medium hot data files do not have the chances to be replicated due to the storage limitation. DDR does not use the replica server capacity efficiently, and the performance differences between DDR and CDR are more distinct when the replica server capacity is small. On the other hand, the distributed replication infrastructure is more scalable than the centralized one especially when the system is in tremendous size.

When the execution time estimation is very inaccurate, the polices of DP scheduling heuristic combined with any data replication algorithm and STT+CDR outperform the others. However, when the user estimation is accurate, the performance of STT scheduling heuristic with any data replication algorithm is far better than any other policy. Overall, the policy of STT+CDR is a sound choice as it works well under various situations of the estimation accuracy and replica server capacity.

7   Conclusions

In this paper, the centralized dynamic replication algorithm CDR and distributed algorithm DDR are put forward. At intervals, the dynamic replication algorithms exploit the data access history for the popular data files and determine the replications in order to improve data access performance from the perspective of the Grid job. The Grid scheduling heuristics of STT, LRL and DP are proposed, and they are combined with different data replication methods.

The simulator XDRepSim is developed to study the performances of the dynamic replication algorithms and Grid scheduling heuristics. A simulated Data Grid is built with XDRepSim and diverse simulations are carried out by varying the settings of the job execution time estimation accuracy and the replica server capacity. The results demonstrate that the dynamic replication algorithms can shorten the job turnaround time effectively. Especially the policy of STT+CDR exhibits prominent performance under various conditions of the job workload and system environment.

References

1. Ranganathan, K., Foster, I.: Simulation studies of computation and data scheduling algorithms for data grids. Journal of Grid Computing **1** (2003) 53–62
2. Takefusa, A., Tatebe, O., Matsuoka, S., Morita, Y.: Performance analysis of scheduling and replication algorithms on grid datafarm architecture for high-energy physics applications. In: Proceedings of 12th IEEE International Symposium on High Performance Distributed Computing (HPDC'03). (2003)
3. Cameron, D.G., Millar, A.P., Nicholson, C., Carvajal-Schiaffino, R., Zini, F., Stockinger, K.: Analysis of scheduling and replica optimisation strategies for data grids using optorsim. Journal of Grid Computing (to appear)
4. Feitelson, D.G., Rudolph, L.: Metrics and benchmarking for parallel job scheduling. In: Proceedings of the Workshop on Job Scheduling Strategies for Parallel Processing. (1998) 1–24
5. Cirne, W., Berman, F.: When the herd is smart: aggregate behavior in the selection of job request. IEEE Transactions on Parallel and Distributed Systems **14** (2003) 181–192
6. Zipf, G.K.: Human Behavior and the Principles of Least Effort. Addison-Wesley (1949)
7. Lee, C.B., Schwartzman, Y., Hardy, J., Snavely, A.: Are user runtime estimates inherently inaccurate? In: Proceedings of 10th Job Scheduling Strategies for Parallel Processing. (2004)
8. Chapin, S.J., Cirne, W., Feitelson, D.G., Jones, J.P., Leutenegger, S.T., Schwiegelshohn, U., Smith, W., Talby, D.: Benchmarks and standards for the evaluation of parallel job schedulers. In: Proceedings of 5th Job Scheduling Strategies for Parallel Processing. (1999)
9. Maheswaran, M., Ali, S., Siegel, H.J., Hensgen, D., Freund, R.F.: Dynamic mapping of a class of independent tasks onto heterogeneous computing systems. Journal of Parallel and Distributed Computing **59** (1999) 107–131

Towards High-Level Grid Programming
and Load-Balancing: A Barnes-Hut Case Study

Martin Alt, Jens Müller, and Sergei Gorlatch

Westfälische Wilhelms-Universität Münster, Germany
{mnalt,jmueller,gorlatch}@uni-muenster.de

Abstract. We propose a high-level approach to grid application programming, based on generic components (*skeletons*) with prepackaged parallel and distributed implementations and integrated load-balancing mechanisms. We present an experimental Java-based programming system with skeletons and use it on a non-trivial, dynamic application – the Barnes-Hut algorithm for N-body simulation. The proposed approach hides from the application programmer many complex details of grid programming and load-balancing, and demonstrates good performance on an experimental grid testbed.

1 Introduction

Grid programmers are faced with the challenge of developing applications that can run distributed across several heterogeneous hosts. Programs must use grid resources efficiently and incorporate flexible load-balancing strategies in order to distribute tasks among hosts that are not known at compile time. The success of grid technology will depend on creating suitable programming models and middleware to liberate application programmers from the complex and low-level details that have to be taken into account during grid software development.

In this paper, we argue for a high-level approach to programming grids, which combines application program development and load-balancing strategies. We present an implementation of the approach as an experimental Java-based programming system that provides application programmers with a set of high-level, reusable components, called *skeletons*, which are customisable for particular applications by means of data and code parameters. We demonstrate the use of our system on a non-trivial case study with a complex, dynamic behaviour – the Barnes-Hut (BH) algorithm for N-body simulation. We show how the use of high-level components hides from the application programmer most of the complexity of distributing computations over the grid and how the load-balancing mechanisms incorporated in our system can evenly balance work between heterogeneous grid servers. We conclude the paper by reporting experimental results for the BH algorithm on a grid-like testbed and by discussing related work.

2 Programing and Load-Balancing with Skeletons

In our programming model, each grid server provides a set of generic algorithmic components called skeletons. When a program is executed on a client, calls to

J.C. Cunha and P.D. Medeiros (Eds.): Euro-Par 2005, LNCS 3648, pp. 391–400, 2005.

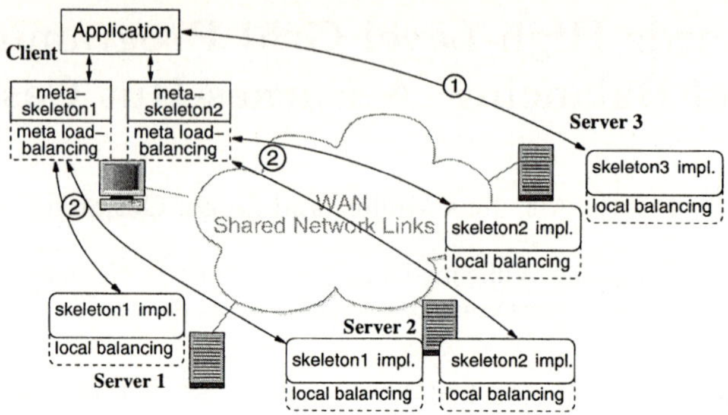

Fig. 1. Application using three skeletons executed in the Grid.

resource intensive skeletons are delegated to the servers as shown in Fig. 1. The client program either calls skeletons directly (①) or uses *meta-skeletons* which combine several skeletons of the same type running on different servers and present themselves as a single skeleton to the client (②). While each skeleton comes with a prepackaged efficient implementation on a server, meta-skeletons are implemented locally on the client; they distribute computations and perform load-balancing, coordination and monitoring of the distributed execution.

Due to space constraints, we limit our discussion to programming and load-balancing issues, and omit details about resource management in our system.

2.1   The Skeleton-Based System and Its Implementation

We present here a basic repository of skeletons, which will be used for implementing the Barnes-Hut case study, using a functional notation:

Map: Apply a unary function f to all elements of a list:
$$map(f, [x_1, \ldots, x_n]) = [f(x_1), \ldots, f(x_n)]$$
Sort: Sort all elements of an input list according to a given order $\prec$:
$$sort(\prec, [x_1, \ldots, x_n]) = [x_{i_1}, \ldots, x_{i_n}] \text{ where } x_{i_j} \prec x_{i_k} \text{ for all } j < k$$
Reduce: Compute the "sum" of a list using a binary associative operator $\oplus$:
$$reduce(\oplus, [x_1, \ldots, x_n]) = [x_1 \oplus x_2 \oplus x_3 \oplus \cdots \oplus x_n]$$
Apply: Applies a unary function f to a parameter x: $apply(f, x) = f(x)$. The *apply* skeleton is used to remotely execute a function f on a server.

In addition to these data-parallel skeletons, the system also provides personalised all-to-all communication facilities to transfer data directly between servers.

Our skeleton-based grid programming system shown in Fig. 1 is implemented on top of Java and RMI, mostly for reasons of portability (see [1] for "10 reasons to use Java in Grid computing"). Skeletons are offered as Java (remote) interfaces, implemented in an architecture-specific way on different servers. All skeletons operate on single objects or arrays which can be distributed blockwise over several servers. For each skeleton, the system provides an interface which is implemented both by client-based meta-skeletons and on the servers.

To avoid unnecessary remote communication, we developed a special mechanism, "future-based RMI" [2]: instead of sending the actual data, skeletons work as much as possible with *remote references*, which are small pointers to the actual data on the servers. Skeletons are executed asynchronously, returning such a remote reference immediately when called. This remote reference can be passed to the next server, while the first server is still executing. Upon completion of the first skeleton, the actual data is sent directly to the second server.

2.2 Integrating Load-Balancing into Skeletons

On the grid, it is critical to distribute data and computations efficiently across the potentially heterogeneous hosts. In our approach, we integrate load-balancing policies into skeleton and meta-skeleton implementations, thereby hiding the complexity of load-balancing from the application programmer. Load-balancing is realised on two levels in the system: (1) each skeleton running on a single parallel server balances the load between the processors of the server, and (2) the meta-skeleton is responsible for distributing data and balancing work between different servers. On both levels, skeleton-specific load-balancing strategies are used, which distribute data based on the skeleton structure. On the server level, implementation and architectural aspects can also be taken into account.

3 Case Study: The Barnes-Hut Algorithm

The *Barnes-Hut (BH)* algorithm [3] is a widely used approach to computing force interactions of bodies (particles) based on their mass and position in space, e.g. in astrophysical simulations. At each timestep, the pairwise interactions of all bodies have to be calculated, which implies a computational complexity of $O(n^2)$ for n bodies. The BH algorithm reduces the complexity to $O(n \cdot log n)$, by grouping distant particles: for a single particle in the BH algorithm, distant groups of particles are considered as a single object if the ratio *boxsize/distance* is smaller than a simulation-specific coefficient θ chosen by the user (see Fig. 2).

For an efficient access to the huge amount of possible groups in a simulation space with a large number of objects, the BH algorithm subdivides the 3D simulation space using a hierarchical *octree* with eight child cubes for each node

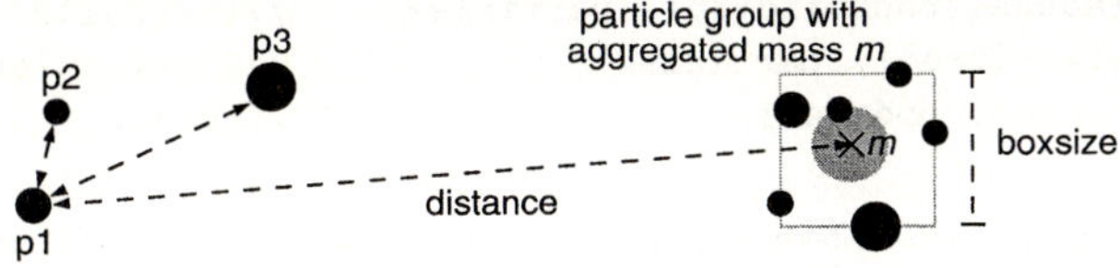

Fig. 2. When calculating forces for p1, particles p2 and p3 have to be considered individually. For a distant particle group, aggregated calculation using m is performed.

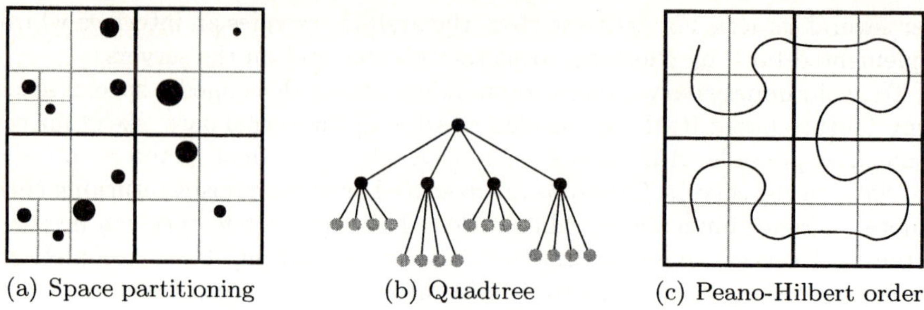

(a) Space partitioning   (b) Quadtree   (c) Peano-Hilbert order

Fig. 3. Barnes-Hut octree partition of the simulation space.

(or *quadtree* for the 2D case). The tree's leaves contain single particles, parental nodes represent the particle group of all child nodes and contain the group's centre and aggregated mass. The force calculation of a single particle then is performed by a depth-first traversal of the tree. Fig. 3(a) and 3(b) depict an example partition and the resulting quadtree for the 2D case (see [3] for further details and cost considerations).

For computing the force interactions in parallel, the particles are distributed among participating hosts. Our implementation uses the *Peano-Hilbert order* (Fig. 3(c)), providing a total linearisation of a two- or three-dimensional space [4]. In the resulting vector of particles, adjacent objects are placed close together. Thus, a blockwise distribution of the particles vector among hosts results in a contiguous particle space assigned to each host. This leads to a reduced amount of communication between hosts, because particles that are close together (and thus need to exchange information often) are likely to be on the same host.

3.1 Barnes-Hut Using Skeletons

The first step in our high-level programming approach is to express the desired application using the skeletons contained in the repository of Section 2.1. First, the particle vector is partitioned into p segments which are distributed among the p servers taking part in the computation. Then, the following five steps are performed iteratively:

```
1      bb = reduce(boundaries, particles);       //bounding box
2      map(PHIndex, particles);                   //Peano-Hilber index
3      sort(LessEqual, particles)                 //sorting
4      tree = reduce(treebuild(bb), particles);   //treebuild
5      map(interact(tree), particles);            //interaction
6      map(particles, update);                    //update
```

1. Calculation of the total boundary of the simulation space: In order to build the tree, the size of the simulation universe is calculated by the reduce operation in line 1. The function **boundaries** compares two particles or a particle and a bounding box, and returns the new bounding box.

2. Sorting: Sorting involves two steps: (1) computing an index for each particle according to the Peano-Hilbert order (using the *map* skeleton in line 2), and (2) sorting the particles vector in ascending order using this index (*sort* in line 3).

3. Building the octrees: For each remote Grid host, the local octree is built in a bottom-up fashion by combining neighbouring particles into trees of depth one. Neighbouring trees are combined into larger trees until a single tree is formed, using a merging algorithm similar to the one described in [6]. Line 4 of the code expresses this process as a reduction skeleton, using the tree merging operation as a parameter, where bb is the bounding box computed in the first step.

4. Force computation: The particle interactions are computed using the *map*-skeleton as depicted in line 5. For each particle in particles, the operation interact traverses the octree tree and adds the force effects of the current node to the velocity vector of the particle if $boxsize/distance < \theta$. If this criterion is not yet met, the eight child nodes are processed recursively.

5. Particle update: Line 6 uses a *map* invocation to update the current particle position according to the new velocity vector. For each particle, the unary function update adds the velocity, multiplied by the length of time for each iteration, to the current particle's position.

The six-line skeleton code of the BH algorithm has a clear structure, where the details of the parallelisation are hidden in the skeleton implementations. The application programmer is, therefore, liberated from low-level considerations.

3.2   Barnes-Hut on the Grid

The critical problem when bringing the BH algorithm on the grid is that it is infeasible to make the entire particle tree available on every host by replication: this would require each host to send the updated values for its own particles to all other hosts after each iteration, thus increasing communication time. Therefore, a distribution among several hosts requires a more elaborate algorithm.

We adopt a solution similar to [4]: Each host constructs a *locally essential tree* by requesting all tree nodes from other hosts that are relevant to the force computations for its own local particles. The goal is to minimise information exchange between nodes during the interaction phase, by sending all necessary particles between nodes once in advance. Each host executes three steps:

1. Send a description of the sub-space with local particles to all other hosts.
2. For each sub-space received from another host, traverse the local tree recursively and add all nodes matching the $boxsize/distance < \theta$ criterion to the result vector. Send the result vectors to all other hosts.
3. Build the locally essential tree by incorporating the particles and tree nodes received from other hosts.

The skeleton pseudocode for building the locally essential tree is as follows:

```
for each host:
  host.localtree = host.reduce(treebuild, host.particles);
  host.subspace = host.exec(constructSubspace, host.particles);
otherSubspaces = allToAll(subspaces);
for each host:
  relevantParticles = host.map(selectParticles(host.localtree),
                               host.otherSubspaces);
alltrees=personalisedAllToAll(relevantParticles);
tree = reduce(treebuild, alltrees);
```

For the amount of data communicated between hosts in the essential tree building phase, there is a trade-off between the accuracy of space approximation
sent to other servers and the remote tree-nodes received in reply. Our approach
(whose details we omit due to the lack of space) is to describe local sub-spaces by
several boxes of varying sizes, where the number of boxes used for the sub-space
approximation can be adjusted before the start of a simulation run.

4   Load-Balancing

Our approach is to aid the application programmer in the task of load-balancing
by providing generic load-balancing strategies for skeletons, which can be adapted
to the application. In our grid programming system, each single-server skeleton
implementation is responsible for distributing work equally among all processors
of the server, and meta-skeletons distribute work among all participating servers.
Load-balancing within a server depends on the server's hardware and is considered a black-box from the viewpoint of the application and the meta-skeleton.

Load-balancing at the meta-skeleton level can either be done dynamically
(e.g., using load-stealing), or statically, i.e., before the skeleton execution is
started. For our BH case study, dynamical load-balancing is not feasible because redistributing a particle would also require to redistribute parts of the
particle tree. Therefore, we will focus on statical load-balancing throughout the
remainder of the paper. Load-Balancing for meta-skeletons is done in two steps:
(1) a skeleton specific *load-balancing function* computes an optimal distribution of data according to a predefined load-balancing strategy, and (2) a generic
redistribution method is responsible for the actual communication required to
distribute the input data according to the new distribution.

In general, load-balancing for grids considers two factors: (1) the amount of
work required for an input element (application-specific), and (2) the amount
of work that a host can process per second (host-specific). Therefore, our load-
balancing functions have both application- and hardware-specific parameters.

As an example, we will discuss the load-balancing strategies implemented
for the *map* skeleton (both single-server and meta-skeleton), using a static load-
balancing approach, where the load generated by each particle is known in advance.

4.1   Example: Load-Balancing for the Map Skeleton

In order to statically balance the work among processors, it is necessary to assess the load generated by each element of the input list in advance, i.e. we need a *load-prediction function*. For our BH implementation we use the work necessary for the previous iteration as an estimate for the work of the next iteration.

Single-Server Map Skeleton. On a single server with homogeneous processors, the *map* skeleton implementation has to distribute the load equally among all processors. This is done in two steps: in the first step, for each element, the partial sum of the work required for all elements up to the current one is computed (accumulated load). The total work (i.e., the value computed for the last element) is divided by the number of processors to obtain the amount of work to be assigned to each processor (processor share). Then, the elements of the input list are assigned to processors in a blockwise fashion, assigning elements to one processor until the total work required for completing the share assigned to the processor equals the processor share computed in the first step. Thus the first data element assigned to processor p is the first element for which the partial sum of work exceeds the share that should be assigned to processor $p - 1$.

Map Meta-skeleton. For computing an efficient load distribution between several grid hosts, we take into account the heterogeneity of the hosts. We introduce a performance factor c_p for each host p, which is proportional to the number of "load-units" that host p can process per second. The amount of work for server p, w_p, is proportional to the performance factor of that server and computed as follows: $w_p = c_p / \sum_i c_i$. The load is then distributed according to the computed distribution $[w_p]$ using the same method as for the single-server case.

4.2   Load-Balancing for Barnes-Hut

The particle interaction phase which computes force interactions for each particle is the most time consuming phase of the BH algorithm: it accounts for well over 90% of the overall runtime. Compared to the interaction phase, all other phases account for only very little time, so that the overhead for balancing load for these phases can be expected to outweigh the gain in performance. Additionally, the locally essential particle-trees constructed for a particular host in the tree-build phase are specific to the particles on that host. Thus, the particles assigned to a particular host need to be the same in the tree-build and the particle interaction phase, impeding load balancing for the tree-build phase. Therefore we only do load-balancing for this phase, which is expressed as a *map* skeleton (line 5 in the code in Sect. 394). Because the distribution of the particles must not be changed between the tree building and the interaction phase, redistribution of the particles is inserted between lines 3 and 4 in the code.

We use *map* skeleton's load-balancing function presented in the previous sections, which requires two parameters: a load predictor which estimates the amount of work necessary for each particle, and a performance factor for each

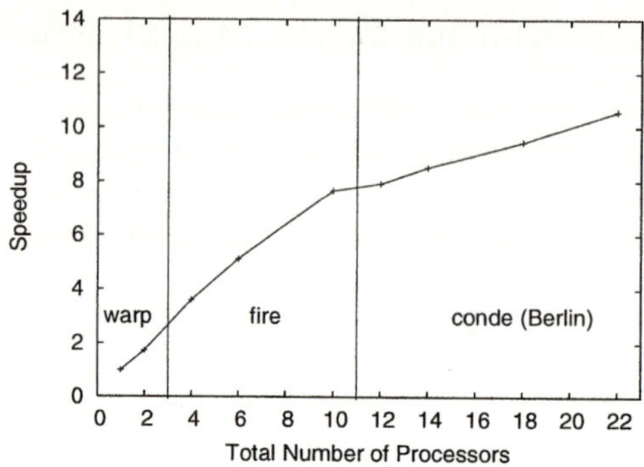

Fig. 4. Relative speedup for three shared-memory servers with a total of 22 processors and $2 \cdot 10^5$ particles ($\theta = 0.25$).

grid host (amount of work per second). For the performance factors, we use the factors obtained a-posteriori from the previous skeleton iteration. For the first iteration, the performance factor of each host is set to the number of processors available on that host. The load predictor for the BH algorithm returns for each particle the number of interactions recorded for that particle during the last interaction phase. For the first iteration, the load predictor returns '1' for each particle, assuming the same amount of work for all particles.

Note that the initial performance factors and load predictions are not very accurate and may lead to a considerable load imbalance. However, load is quickly balanced for later iterations, as demonstrated in our experiments.

5　Experimental Results

We measured the performance of our skeletal Barnes-Hut implementation using three shared-memory servers, two at the University of Muenster ("warp" and "fire"), and one at the Technical University of Berlin ("conde"). Server "warp" has two 2.8GHz Pentium4 processors and "fire" has eight UltraSparc III+ processors running at 1.2GHz, the server in Berlin has 12 900 MHz UltraSparc III+ processors. The client is a workstation with a P4, 2.6 GHz processor in Muenster. All experiments were done using SUN's JDK 1.5.0. Client and servers are connected by two LANs and the german academic internet backbone (WiN). The measured bandwidth within the LAN at Muenster was approx. 3.2MB/s, the bandwidth between Muenster and Berlin (450 km) was measured at 1.1MB/s.

Figure 4 shows the relative speedup obtained for an input size of $2 \cdot 10^5$ particles and $\theta = 0.25$, for varying numbers of processors on different hosts (compared to the performance on one processor of "warp"). The absolute runtimes ranged between 315 s on one processor of "warp" and 26 s on all 22 processors of all hosts for one iteration. Our high-level Barnes-Hut implementation shows very good speedups, also across host-boundaries. Note that the decrease in speedup at 4 and 12 processors is at least partly a result of the slower processors on hosts

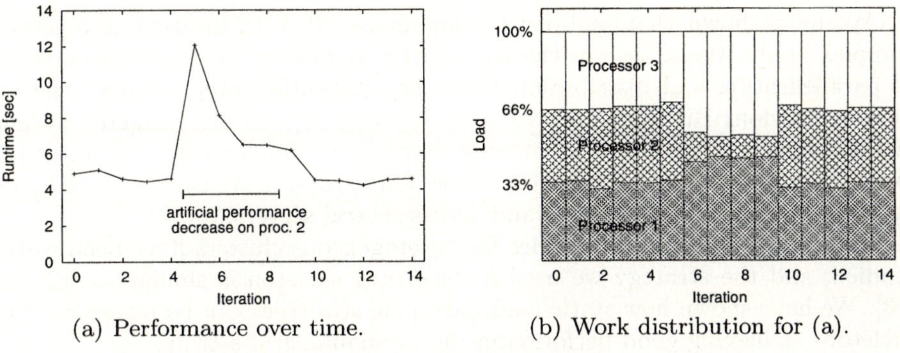

(a) Performance over time.   (b) Work distribution for (a).

Fig. 5. Load-balancing on three processors with varying load.

"fire" and "conde". Additionally, the speedup decreases between 10 and 12 processors due to the increased communication costs when adding a remote host. However, this cost is amortised for higher numbers of processors. Also note that the particle shares allocated to the two servers in Muenster are neighbouring blocks (according to the Peano-Hilbert ordering), thus reducing the amount of communication between Berlin and Muenster.

To evaluate the load-balancing strategy used for *map*, we measured the performance for 10^4 particles on three Pentium 4, 1.7 GHz PCs connected by LAN ($\theta = 0.25$). The runtimes for each of 15 iterations are shown in Fig. 5(a), and the partition of the particles between the hosts is shown in 5(b). For iterations 5 through 8, we have artificially decreased the performance of processor 2 (starting other time-consuming applications on that host). The figure shows that the work is rebalanced in iteration 6, due to the decreased performance of processor 2. The performance decrease seen in Fig. 5(a) at iteration 5 is thus compensated in the following iterations. After processor 2 is fully available again in iteration 9, work is rebalanced for iteration 10, obtaining the same performance as in the first iterations. This shows that the implemented load-balancing mechanisms are able to adapt to varying performance on different hosts.

6   Related Work and Conclusion

Our work is a step towards designing efficient applications for grids by providing high-level components (skeletons), which hide the complexity of parallelisation and distribution from the application programmer. We have demonstrated how pre-defined generic load-balancing strategies can be integrated with skeletons and then specialised for a particular application.

Our approach differs from other Java-based programming frameworks for grids, such as *ProActive* [7] and *Ibis* [8], because it provides predefined computation patterns with built-in load-balancing strategies. *Satin* [9] also provides load-balancing, but it is limited to divide-and-conquer applications. Another skeleton-based approach, *Lithium* [10], focused mainly on task-parallel skeletons rather than our data-parallel skeletons.

We have shown that high-level components allow to implement relatively complex applications, such as the Barnes-Hut algorithm, hiding the complexity of parallelisation and distribution from the application programmer. The numerous previous BH implementations (e. g. [4–6, 11]), mostly targeted parallel or homogeneous distributed architectures, while our implementation aims to be executed in heterogeneous (grid) systems. Our experiments demonstrated good performance both on one server and across several servers.

Static load-balancing strategies for heterogeneous clusters have been widely studied, and the strategy we used for the *map* skeleton is similar to [12] and [13]. We have shown how static load-balancing strategies can be integrated into skeletons, achieving good performance in a dynamic grid setting.

References

1. Getov, V., von Laszewski, G., Philippsen, M., Foster, I.: Multiparadigm communications in Java for Grid computing. Comm. of the ACM **44** (2001) 118–125
2. Alt, M., Gorlatch, S.: Future-Based RMI: Optimizing compositions of remote method calls on the Grid. Euro-Par 2003. LNCS 2790, Springer (2003) 682–693
3. Barnes, J.E., Hut, P.: A hierarchical $O(N \log N)$ force-calculation algorithm. Nature **324** (1986) 446–449
4. Grama, A.Y., Kumar, V., Sameh, A.: Scalable parallel formulations of the barnes-hut method for n-body simulations. Supercomputing '94, IEEE (1994) 439–448
5. Blelloch, G.E., Narlikar, G.: A practical comparison of N-body algorithms. In: Parallel Algorithms. American Mathmatical Society (1997)
6. Singh, J.P., Holt, C., Totsuka, T., Gupta, A., Hennessy, J.: Load balancing and data locality in adaptive hierarchical N-body methods: Barnes-Hut, fast multipole, and radiosity. Journal of Parallel and Distributed Computing **27** (1995) 118–141
7. ProActive: INRIA (1999) http://www-sop.inria.fr/oasis/ProActive.
8. van Nieuwpoort, R.V., Maassen, J., Hofman, R., Kielmann, T., Bal, H.E.: Ibis: an efficient Java-based Grid programming environment. Proc. of the 2002 joint ACM-ISCOPE conference on Java Grande, ACM Press (2002) 18–27
9. van Nieuwpoort, R.V., Kielmann, T., Bal, H.E.: Efficient load balancing for wide-area divide-and-conquer applications. Proc. Eighth ACM SIGPLAN Symp. PPoPP'01, Snowbird, UT (2001) 34–43
10. Danelutto, M., Teti, P.: Lithium: A structured parallel programming enviroment in Java. Proc. ICCS'02. LNCS 2330, Springer (2002) 844–853
11. Sun, Y., Liang, Z., Wang, C.: Distributed particle simulation method on adaptive collective system. Future Generation Computer Systems **18** (2001) 79–87
12. Crandall, P., Quinn, M.J.: Block data decomposition for data-parallel programming on a heterogeneous workstation network. HPDC. (1993) 42–49
13. Kaddoura, M., Ranka, S., Wang, A.: Array decompositions for nonuniform computational environments. Journal of Parallel and Dist. Comp. **36** (1996) 91–105

An Adaptive Skeletal Task Farm for Grids

Horacio González-Vélez*

University of Edinburgh
School of Informatics
Edinburgh, UK
`h.gv@ed.ac.uk`

Abstract. Algorithmic skeletons abstract commonly used patterns of parallel computation, communication, and interaction. By demonstrating a predictable communication and computation structure, they provide a foundation for performance modelling and estimation. Grids pose a challenge to known distributed systems techniques as a result of their dynamism. One of the most prominent research areas concerns the availability of proved programming paradigms with special emphasis on the performance side. Thus, adaptable performance improvement techniques have been the subject of intense scrutiny. Scant research has been conducted on using the skeletal predicting information to enhance performance in heterogeneous environments. We propose the use of these predicting properties to adaptively enhance the performance of skeletons, in particular of a task farm, within a computational grid.
Hence, the problem addressed in this paper is: given a skeletal task farm, find an effective way to improve its performance on a heterogeneous distributed environment by incorporating information at compile time that helps it to adapt at execution time. This work provides a grid-enabled, adaptive task farm model, using the NWS statistical predictions on bandwidth, latency and processor availability. The central case study implements an ad-hoc task farm based on C/MPI and employs PACX-MPI for inter-node communication. We present initial promising results of parallel executions of an artificially-generated numerical code in a grid.

1   Introduction

With the advent of grid computing, the availability of proved programming paradigms has become an issue in computational science. It is widely acknowledged that one of the major challenges in programming support for these environments is the prediction and improvement of performance, due to the vast aggregation of heterogeneous resources and policies. Indeed, holistic projects emphasise the need not only for reliable programming environments, but also for improved performance capabilities [1, 2].

Performance enhancement is a multidimensional activity. One of the most powerful of these dimensions relates to optimisation techniques which work adaptively on an application-specific basis. Grid adaptability is quite broad, and often

* Work partly supported by the EC-funded project HPC-Europa, contract number 506079. Special thanks are owed to Murray Cole for his suggestions and review. The comments of anonymous referees have also helped to improve this paper considerably.

J.C. Cunha and P.D. Medeiros (Eds.): Euro-Par 2005, LNCS 3648, pp. 401–410, 2005.

relates to flexibility, ability to transform and evolve, re-usability, and extensibility. It encompasses several layers in the architecture, involves a multiplicity of parameterised values, and is typically performed in a custom-built fashion either at compile-time or during execution. There have been substantial examples of the applicability of adaptable models in computational grids [3, 4]. Moreover, AppLeS [5] builds on these efforts and provides a comprehensive approach including resource discovery, selection, and scheduling . Nevertheless, one of the most fascinating open-ended questions in computational science still concerns the self-adaptation of programming structures to grids [6].

The separation of software and hardware has long been considered critical to the success of any parallel programming endeavour, as it is vital to foster the reuse of algorithms and software. Furthermore, we consider that the division of the structure from the application itself to be crucial to the goal of delivering adaptability.

Algorithmic skeletons (AS) abstract commonly used patterns of parallel computation, communication, and interaction [7]. They present a top-down structured approach where parallel programs are formed from the parametrisation of skeleton nest, also known as Structured Parallelism (SP). AS provide a clear and consistent structure across platforms by distinctly decoupling the application from the structure in a parallel program [8]. They do not rely on any specific hardware and benefit entirely from any performance improvements in the systems infrastructure.

SP is not universally applicable to the production of parallel and distributed programs, but there is an important growing number of applications to consider them an interesting research area [9]. Furthermore, skeletal methodologies inherently posses a predictable communication and computation structure, since they capture the structure of the program. They provide, by construction, a foundation for performance modelling and estimation of parallel applications.

This work is concerned with the feasibility of using this predicting capabilities of SP. In particular we propose a model to enhance the performance of skeletal task farms in grids. First, we consider related work and provide motivation, moving on to some scalability results using the farm of the Cole's eSkel [10] library. These figures are collected using a computational grid with nodes in Edinburgh and Stuttgart. This part helps us to reinforce the feasibility of the skeletal programming model for grids. Then, we have constructed our main case study using an utilitarian skeletal task farm implemented using C, MPI and the PACX-MPI [11] library for the inter-cluster communication. It also employs the Network Weather Service (NWS) [12] to monitor the grid environment, forecast processor and network availability, and adapt to the load conditions on the nodes and interconnections. While the model is embedded in the code, the NWS forecasts drive the behaviour adaptation of the task farm at execution time. Finally, future directions for this work are supplied.

2   Motivation

We would like to formulate the problem covered in this paper as: given a skeletal task farm, find an effective way to improve its performance on a heterogeneous distributed environment by incorporating information at compile time that helps it to adapt at execution time.

Compile-time optimisers formulate static decisions about the expected behaviour of an application. On the other hand, run-time optimisers do not generally possess direct knowledge of the structure (meaning) of the application. They lack specific information on its data and control flows and their operation is normally driven by load-balancing criteria. Indeed, the development of effective compilers and optimisers remains an active area in computer science.

In addition to this, there is no equivalent to a compiler or to an universal run-time optimiser for grids. Due to the complexity involved, grid optimisation techniques have usually been custom-made. They require the modification of the source code to enable its operation [5], the use of capacity characterisation methods [13], or even the creation of special-purpose languages [14] .

From a grid perspective, although different parallel solutions have traditionally exhibited skeletal constructs, their associated optimisations have not employed the structural information of the skeletons but rather modified the scheduler [15], or have not decoupled entirely the structure from the behaviour keeping the actual application interlaced [16]. On the AS side, the emerging approaches to performance optimisation in computational grids have employed process algebra methods [17] and future-based RMI mechanisms with Java [18].

AS possess a crucial property which favours performance optimisation: their structured and predictable behaviour for a given meaning (program). Nevertheless, scant research has been conducted on improving the skeletal performance by actively using this information from a systems infrastructure perspective.

Thus we would like to bring all these factors together and use the structural, forecasting capabilities of skeletons to optimise pragmatically their performance in grids. As opposed to standard optimisation at compile time, in this case the behaviour of the application is known prior to the execution. In contrast to standard run-time optimisation, the meaning is clearly defined as well. In principle, therefore, this skeletal optimiser could forecast and enhance the actual behaviour of the application by exploiting the knowledge of its structure.

The open question must be: how much can the structural forecasting information of AS help to improve the performance of parallel applications whilst executing in heterogeneous clusters and eventually in computational grids?

Hence, we argue that the AS need to evolve to include adaptive capabilities to improve its performance in the Grid. In this work, we present a pragmatic approach using an optimised ad hoc task farm skeleton, NWS, and a realistic grid environment. It is important to note that we do not intend to develop a scheduler nor to solve the general optimisation case for every structured parallel program. We concentrate on empirically finding optimisation techniques for solutions based on skeletal task farms.

This work is the first step in the development of a framework which incorporates different SP programs and its associated optimisation techniques, similar to the way a compiler incorporates optimisation techniques, to be deployed in computational grids as illustrated in Fig. 1. The ultimate objective of this ongo-

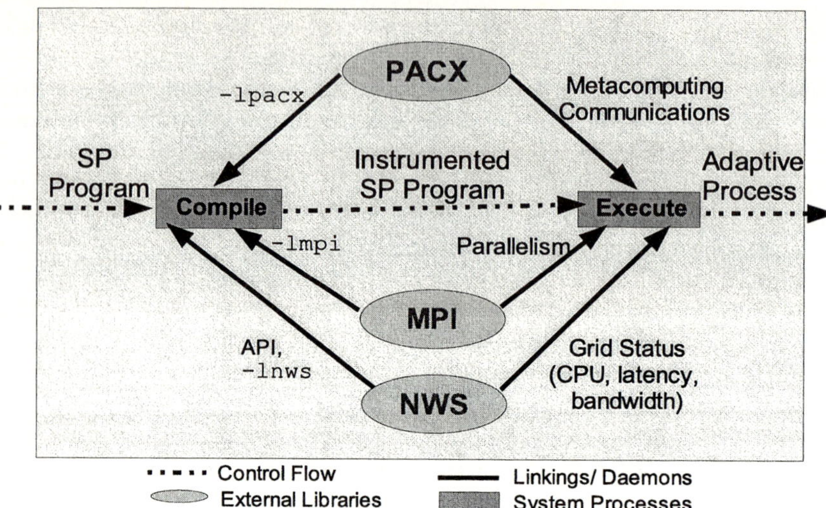

Fig. 1. Methodology to Introduce Adaptiveness into Structured Parallelism Programs

ing endeavour is to build upon theoretical performance models and design a set of adaptive techniques. The main difference to other performance approaches is that it intends to be SP-oriented, adaptable by construct, and focused on empirical, system-infrastructure methodologies.

3 The Task Farm

We have selected a task farm for this initial approach due to its applicability to the solution of a great number of problems in parallel programming and its simple structure.

A Task Farm (TF) can be roughly described as a "farmer" process which spawns N independent "worker" processes to execute a parallel workload. The TF is composed by a input I, an output O and a function F, or $TF = \langle I, O, F \rangle$. A worker executes a task by mapping F into a subset of I (task size), computing a subset of O, and then reporting back to the farmer for the next unit of work or termination. This is shown schematically in Fig. 2.

The TF construct has traditionally dealt with fine-grained data parallelism. In its canonical form, the communication times between the farmer and the workers can be adjusted to be constant and much less than the computation times [19]. All workers are allocated to dedicated processor in a parallel machine and the computation of each element O is independent and characterised by the fact that the F does not generate the same amount of work for different elements. The TF needs to keep distributing the elements in I to avoid worker starvation while minimising communication. This TF feature aims at providing the best load-balancing. Under this scenario, the number of elements of I sent to a given worker at once (task size) can be statically calculated to minimise idle times and require minimal scheduling from the farmer side [19].

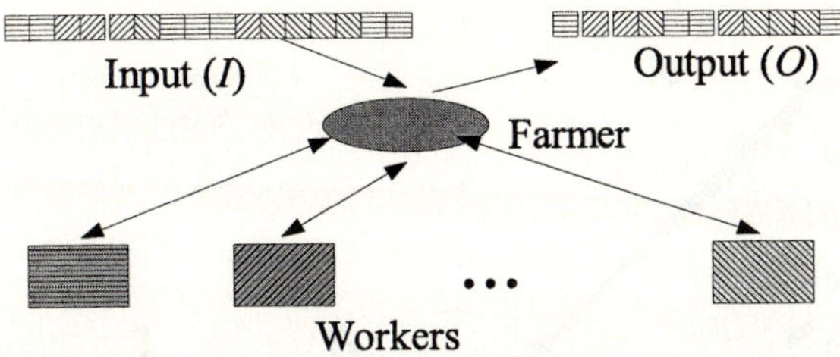

Fig. 2. A Task Farm

However, in more realistic scenarios, the farmer requires to assign different tasks sizes to workers because:

- The underlying architecture can have multiple communication links between the farmer and workers with different bandwidths and latencies
- The workers and the farmer run in non-dedicated nodes with distinct workloads in a distributed environment

Furthermore, in a computational grid, communication and computation times vary greatly, an undersized farmer, a saturated communication channel, or the sudden termination of a worker can produce unpredictable situations. Thus our particular objective is to adaptively determine the optimal task size for each worker in a task farm in order to minimise the total execution time for a given application.

4   Implementation

We have initially employed the first version of the Cole's eSkel library and an artificial integer application. The overall system is configured by evenly distributing the processes between two 16-node Beowulf clusters located at the High Performance Computing Centre (HLRS) in the University of Stuttgart and the School of Informatics in the University of Edinburgh respectively.

The farmer node has been positioned at HLRS. We have used the PACX-MPI library for interconnection with the allocation of two pairs of communication nodes to interconnect both installations. This is the standard requirement for soundly executing PACX-MPI. The MPI versions are MPICH and LAM/MPI and the nodes and communication channels are in non-dedicated mode.

Figure 3 shows the channel utilisation from the worker standpoint on the eSkel Task Farm version 1.0 and PACX-MPI 5 for a simplistic application. It presents different values of I ranging from 160B or $I = 200$ to 1.6MB or $I = 200,000$. It is important to mention that half of the workers are located in Edinburgh while the other are in Stuttgart.

Although the communication channel at 270KB/s is not saturated while working with 8 processes and 1.6 MB and 160KB vectors, equivalent to an I

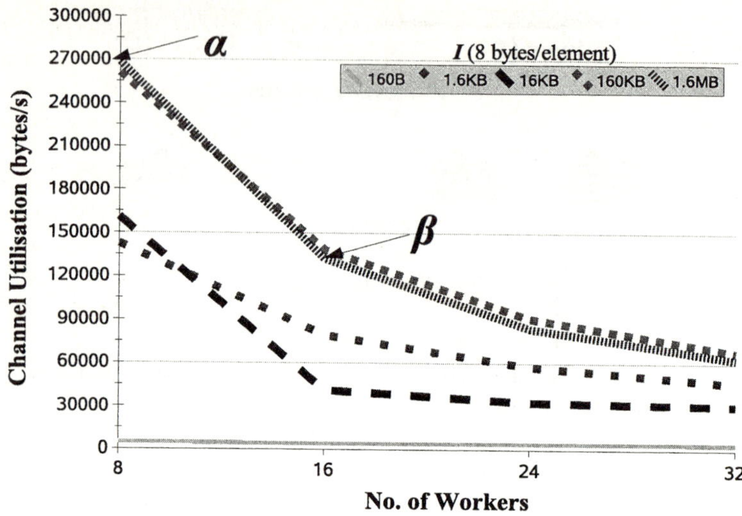

Fig. 3. Worker Channel Utilisation for different values of I, ranging from 160B or $I = 200$ to 1.6MB or $I = 200,000$, using the eSkel Task Farm version 1.0 and PACX-MPI 5. Half of the workers are located in Edinburgh and the other half in Stuttgart

of 200,000 and 20,000 8-byte data elements respectively (α in Fig. 3), the increase to 16 processes with the same amount of data implies a 50% decrement in the ability to transmit (β in Fig. 3). This is unexpected since there are more processes transmitting than the decrease in channel utilisation.

After the analysis of the communication patterns when increasing the number of processes, the performance degradation in the communications was attributed to the use of MPI collective communication operations under PACX-MPI, since the synchronisation mechanisms in PACX-MPI are not optimised for collectives across sites.

In order to address this issue, we have implemented a new skeletal TF using MPI send-receive operations only. Furthermore, this new TF incorporates the ability to adapt the task size using the NWS forecasting capabilities of the task farm.

NWS provides utilitarian forecasting figures based on the statistical time-series analysis of the processing and networking load and configuration. It presents fault-resilient capabilities to support adaptive applications, and its accuracy has been successfully tested in major grids [20].

This case study features two distinct approaches to define the task size: static and adaptive. In the former, the task size assigned to a worker is defined by the full input data size and the number of workers. Although each worker operates on the same task size, except possibly for the last one, even processing is not guaranteed due to the different node workload and network conditions.

In the adaptive strategy, the central part of this work, the task size is defined by taking into account four factors:

– Available CPU: CPU fraction allocatable to a new process
– Bandwidth: Speed to transmit data to/from the farmer and a worker
– Current CPU: CPU fraction usable by a running process
– Latency: Time (in msec) to send a TCP message from the farmer to a worker

The forecasts of the above system indicators, supplied by NWS, are composed into a fitness index FI as shown in (1). FI defines the adaptive task size assuming a certain system behaviour based on historic measurements. It is also important to mention that the heuristic to calculate the index coefficients is application dependant, and in this initial approach, based on an artificially-created application, we have employed $A = 0.4$, $B = 0.1$, $C = 0.4$, and $L = 0.1$. The intention is to have an even task processing by allocating larger tasks to the fitter nodes in terms of its processing and communications.

$$FI_i = A * avail_i + \frac{B * bandwidth_i}{max(bandwidth)} + C * current_i + \frac{L * min(latency)}{latency_i} \quad (1)$$

The algorithm provides default values which assume an unfit node, since unexpected surges in workload and latencies may affect the operation of the NWS sensor and memory processes returning no readings at execution.

5 Empirical Results

We have deployed the initial implementation employing a configuration including 32 processors distributed into two 16-node Beowulf clusters (bw240 and bw530) located at the School of Informatics in the University of Edinburgh. A summary of the hardware and software configuration of a typical node in each cluster is presented in Table 1.

Table 1. Hardware/Software Configuration of the bw240 & bw530 Beowulf Clusters located at the School of Informatics in the University of Edinburgh

CONFIGURATION	bw240	bw530
CPU	Intel Pentium 4 1.80GHz	Intel Xeon 1.70GHz
Memory	1 GB	2 GB
Linux kernel	2.4.20-24.7_1.public.1	2.4.20-31.9_v1_dice_1
gcc	gcc-2.96-112.7.1	gcc-3.2.2-5
LAM/MPI	lam-6.5.6-tcp.1	lam-6.5.8-4
PACX	PACX-5.0-beta 9/8/04	PACX-5.0-beta 9/8/04
NWS	2.10.1	2.10.1

All nodes were on non-dedicated mode during all the experiments, and their interconnection channels did not have any bandwidth reservation. We observed that the workloads, latency and bandwidth varied greatly during the experimentation periods.

There were NWS sensors running in all nodes, and there was a clique encompassing all nodes. The NWS name and memory server daemons were running in the farmer node. We allocated four nodes (two per cluster) to run the PACX-MPI communication processes. All execution time measurements were obtained using the `MPI_Wtime` function for the TF skeleton only.

The farmer, the workers and the two-pair communications hosted concurrently the series of jobs for the static and adaptive runs for each I (and therefore under similar external load and competing for the same resources). All farmer and worker had a system priority of 10.

Figure 4 graphs the execution times for $I = 512, 1024, 2048, 4096, 16384$. All entries in the graph perform $O(10^{12})$ daxpy operations, and present 8 different lines corresponding to $4, 8, 16,$ and 32 workers for the static and dynamic models. The thicker lines average the series of executions for both models.

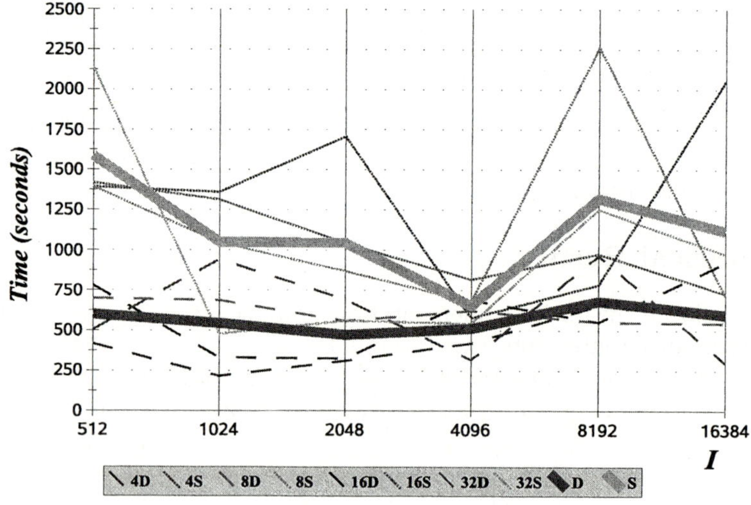

Fig. 4. Task Farm execution times (in seconds) for a series of concurrent executions. The application performs $O(10^{12})$ daxpy operations. Key: [numb][model], e.g., *4S* means 4 workers and Static model. The thick lines are the average of all executions

In an homogeneous dedicated system, one would expect a smooth line. That is to say, the peaks in the execution times in both models are chiefly defined by the non-deterministic nature of this grid, e.g. the noticeable peak in the static case with 8 workers and $I = 8192$. Hence, it is clear that the adaptive model surpasses the static one, as reflected by the fact that its averaged graph is flatter and with lower time measurements for every entry of I.

6   Future Directions

Although the nodes involved in this experiment demonstrated a certain degree of homogeneity within the same Beowulf cluster, their extremely different workload and interconnections made them distinct enough to become a representative environment for this initial case study. Even more, the uncontrolled workloads present at the execution time comprised a non-deterministic scenario. On the other hand, the model still has room for improvement on its interaction with the NWS API.

The skeletal approach has helped us to discretise and bound the parameters involved, reducing the number of combinations. The use of AS allowed us to make assumptions on the input and output sets that permit a more effective allocation of computation and communication resources.

These results may seem to be intuitive. However, we consider them significant since they provide a common ground to further tweak our model under the dynamic environment of a computational grid. We intend to keep improving it by:

- Devising a more accurate adaptiveness strategy through more comprehensive experimentation. A biomedical code is being tinkered with [21].
- Deploying a faster distribution and execution of tasks by analysing the scalability, buffering, and resource monitoring issues.
- Improving the model by incorporating new indicators such as task termination time and CPU capacity. This in turn will provide foundations to develop a methodology for the creation of the index heuristics.

References

1. Foster, I., Kesselman, C., eds.: The Grid: Blueprint for a new computing infrastructure. Second edn. Morgan Kaufmann, San Francisco, USA (2003)
2. Laforenza, D.: Grid programming: some indications where we are headed. Parallel Comput. **28** (2002) 1733–1752
3. Vetter, J.S., Reed, D.A.: Real-time performance monitoring, adaptive control, and interactive steering of computational grids. Int. J. High Perf. Comput. Appl. **14** (2000) 357–366
4. Cheng, S.W., Garlan, D., Schmerl, B., Steenkiste, P., Hu, N.: Software architecture-based adaptation for grid computing. In: HPDC-02: 11th IEEE Int Symp on High Performance Distributed Computing, Edinburgh, UK, IEEE CS (2002) 389–398
5. Berman, F., Wolski, R., Casanova, H., Cirne, W., Dail, H., Faerman, M., Figueira, S., Hayes, J., Obertelli, G., Schopf, J., Shao, G., Smallen, S., Spring, N., Su, A., Zagorodnov, D.: Adaptive computing on the grid using AppLeS. IEEE Trans. Parall. Distrib. Sys. **14** (2003) 369–382
6. Vadhiyar, S.S., Dongarra, J.J.: Self adaptivity in grid computing. Concurrency Computat. Pract. Exper. **17** (2005) 235–257
7. Cole, M.: Algorithmic Skeletons: Structured Management of Parallel Computation. MIT Press, London, UK (1989)
8. Cole, M.: Bringing skeletons out of the closet: a pragmatic manifesto for skeletal parallel programming. Parallel Comput. **30** (2004) 389–406
9. Rabhi, F.A., Gorlatch, S., eds.: Patterns and skeletons for parallel and distributed computing. Springer-Verlag, London, UK (2003)

10. Cole, M.: eSkel: The Edinburgh Skeleton library API reference manual. University of Edinburgh, UK. First edn. (2002)
    Available on-line at: http://homepages.inf.ed.ac.uk/mic/eSkel.
11. Keller, R., Gabriel, E., Krammer, B., Muller, M.S., Resch, M.M.: Towards efficient execution of MPI applications on the Grid: Porting and optimization issues. J. Grid Comput. **1** (2003) 133–149
12. Wolski, R., Spring, N., Hayes, J.: The Network Weather Service: A distributed resource performance forecasting service for metacomputing. Future Gener. Comput. Syst. **15** (1999) 757–768
13. Hey, A.J.G., Papay, J., Surridge, M.: The role of performance engineering techniques in the context of the grid. Concurrency Computat. Pract. Exper. **17** (2005) 297–316
14. Lian, C.C., Tang, F., Isaac, P., Krishnan, A.: GEL: Grid execution language. J. Parallel Distrib. Comput. **65** (2005) 857–869
15. Casanova, H., Kim, M.H., Plank, J.S., Dongarra, J.J.: Adaptive scheduling for task farming with grid middleware. Int. J. High Perf. Comput. Appl. **13** (1999) 231–240
16. Shao, G., Berman, F., Wolski, R.: Master/slave computing on the grid. In Raghavendra, C., ed.: HCW'00: 9th Heterogeneous Computing Wksp, Cancun, Mexico, IEEE CS (2000) 3–16
17. Benoit, A., Cole, M., Gilmore, S., Hillston, J.: Scheduling skeleton-based grid applications using PEPA and NWS. Comput. J. **48** (2005) 369–378
18. Aldinucci, M., Danelutto, M., Dünnweber, J., Gorlatch, S.: Optimization techniques for skeletons on grid. In Grandinetti, L., ed.: Grid Computing and New Frontiers of High Performance Processing. Elsevier Science (2005) To appear.
19. Hey, A.J.G.: Experiments in MIMD parallelism. Future Gener. Comput. Syst. **6** (1990) 185–196
20. Wolski, R.: Experiences with predicting resource performance on-line in computational grid settings. Sigmetrics Perform. Eval. Rev. **30** (2003) 41–49
21. Gonzalez-Velez, V., Gonzalez-Velez, H.: A grid-based stochastic simulation of unitary and membrane Ca^{2+} currents in spherical cells. In: 18th IEEE Int Symp on Computer-Based Medical Syst., Dublin, Ireland, IEEE CS (2005) To appear.

Developing Java Grid Applications with Ibis

Kees van Reeuwijk, Rob van Nieuwpoort, and Henri Bal

Vrije Universiteit Amsterdam
{reeuwijk,rob,bal}@cs.vu.nl

Abstract. *Ibis*[1] is a programming environment for the development of grid applications in Java. We aim to support a wide range of applications and parallel platforms, so our example programs should also go beyond small benchmarks.
In this paper we describe a number of larger applications we have developed to evaluate Ibis' suitability for writing grid applications: a cellular automata simulator, a solver for the Satisfiability problem, and grammar-based text analysis. We give an overview of the applications, we describe their implementation, and we show performance results on a number of parallel platforms, ranging from a large supercomputer cluster to a real global grid testbed.
Since all of these applications require communication between the processors during execution, it is not surprising that a supercomputer cluster proved to be the most effective platform. However, all of our applications were also efficient on a wide-area cluster system, and some of them even on a grid testbed. Since grid systems are usually only used for trivially parallel systems, we consider these results an encouraging sign that Ibis is indeed an effective environment for grid computing. In particular because for two of the three of the applications the parallelisation required very little additional program code.

1 Introduction

Traditional supercomputing offers large amounts of computational power, but requires tightly controlled homogeneous systems at a single location. For grid computing these restrictions are lifted, and it is assumed that effective computation is still possible on heterogeneous, widely distributed, and independently managed computer systems. The additional flexibility of such a configuration is attractive, but writing efficient software for grid computing is challenging: Differences in processor architecture and power, external loads on the processors, differences in network performance, geographical distance, security measures such as firewalls and proxies, and the possibility of faults in processors and networks, all complicate software development.

For grid computing it is also desirable that processors can join a running computation. This is called *open-world* computation, in contrast to traditional *closed-world* computation. However, the open-world requirement complicates the program, and restricts the way a program can be parallelised.

The complications of grid computing require a solid programming environment to hide these complexities. *Ibis* provides such an environment. It not only allows programs to be written in Java, but is also itself written in Java. Choosing Java already solves many

---

[1] Ibis is available under an open source licence, and can be downloaded from www.cs.vu.nl/ibis.

J.C. Cunha and P.D. Medeiros (Eds.): Euro-Par 2005, LNCS 3648, pp. 411–420, 2005.

software portability issues (very important in such a heterogeneous environment!), and allows the programmer to use a modern high-level language.

Standard Java has some support for distributed computing through the Remote Method Invocation (RMI) library. Ibis provides a very efficient [1] implementation, but RMI only allows client-server style parallel programming, which is not suitable for many parallel problems. Ibis therefore offers a wide range of communication models, including replicated objects, group/collective communication, and a divide-and-conquer programming model. Also, the Ibis implementation layer, primarily designed to support the higher-level models, has proved to be an effective programming model for some problems.

To provide this functionality, other systems typically compromise portability, for example by interfacing to a native MPI library. In Ibis, all parallel programming models are cleanly integrated into Java. However, behind this friendly facade a lot of work is done for the sake of efficiency: native implementations are dynamically selected for high-performance networks such as Myrinet [2], bytecode is rewritten to generate efficient communication and parallel code [3], and when possible the improved functionality of modern JVM implementations is exploited. However, portable implementations are always available as fallback.

We aim to support a wide range of programs, so it is important to go beyond small benchmarks, and try larger applications. In this paper we describe a number of larger applications we have developed to evaluate Ibis' suitability for writing grid applications: a Cellular Automata simulator (§4), a solver for the Satisfiability problem (§5), and grammar-based text analysis (§6). §2 describes our measurement setup and the way we evaluate the measurements. §3 describes the Satin divide-and-conquer framework. In §7 and §8 we show some results for wide-area systems.

2   Measurement Setup and Evaluation

As part of our application descriptions, we will show performance results on a single site of the DAS2 supercomputer cluster system [4]. Each node of this cluster is a dual 1GHz Intel Pentium III system. Unless specified otherwise, we use the IBM 1.4.1 JVM.

On homogeneous systems like the DAS2 cluster, the efficiency of a computation is easily determined. Ideally, a cluster of N processors has a *speedup* of N: it is N times faster than an individual processor. However, for computations with non-identical processors the notion of speedup is meaningless. Instead, we express the efficiency as a fraction of the ideal speed of the system. Given a system with N nodes, and execution times on individual nodes $t_1 \ldots t_N$, each processor ideally contributes to a cluster computation inversely proportional to its individual computation time. Thus, the ideal execution time is $t_{ideal} = 1/\sum_{i=1}^{N} 1/t_i$. Given a real cluster computation time t_p, the efficiency of the cluster computation is $\eta = t_{ideal}/t_p$. For a homogeneous system, the execution time t on each processor is the same, so $t_{ideal} = t/N$.

3   Satin

Satin [5, 6] is a divide-and-conquer framework similar to Cilk [7], but built on top of Ibis. In Satin, the user must annotate methods that can be executed in parallel, and provide an explicit demarcation point where the results of these methods should be available. For example, the following method recursively creates tasks, waits for them to complete (the `sync()` call), and uses the results to compute its own result:

```
// List all parallel methods in a subinterface of Spawnable
interface I extends ibis.satin.Spawnable {
    int f(int n);
}

class F extends ibis.satin.SatinObject implements I {
    int f(int n) {
        if(n<2) return n;
        int x = f(n-1); // these two methods
        int y = f(n-2); // are executed in parallel
        sync(); // Satin method: wait for the f() methods
                // to finish before using their results.
        return x+y;
    }
}
```

This implicitly parallel code is rewritten by Ibis to explicitly parallel code, but that is done on the bytecode, and is invisible to the programmer. This parallel code implements a cluster-aware work-stealing algorithm that usually is very efficient, even on a grid system.

4   A Cellular Automata Simulator

Many simulations can be described as interactions between cells, with each cell in one of a finite number of *states*. Typically the cells are arranged in a 2- or 3-dimensional rectangular matrix. The state of the system progresses in a sequence of discrete steps, where the next state of each cell is determined by the state of the cell itself and that of its immediate neighbours, according to a homogeneous, fixed set of rules. Such a problem is called a *cellular automata* (CA) problem. Such problems occur for example in biology [8] and urban planning [9].

In our example program we implement a simple ecological model where each cell represents a patch of land that can be seeded from one of the eight neighbouring patches with grass or trees, and where woodlands regularly suffer from forest fires. The structure and behaviour of the program is largely independent of the exact update rules, so our findings are applicable to a larger set of problems.

We parallelise the computation by distributing the matrix over multiple processors. Since a cell update requires the state of the neighbours, cell states must be communicated between processors. Since the computation of the new generation on a processor can only be started when its neighbours have completed the previous generation, there is a fairly close synchronisation between the processors.

In our implementation, we organise the processors in a one-dimensional array, and distribute the cells column-wise over the processors. This is simple to implement, but not optimal: a two-dimensional distribution of the cells may require less communication. However, since the gains are often small or non-existent, support is complex, and

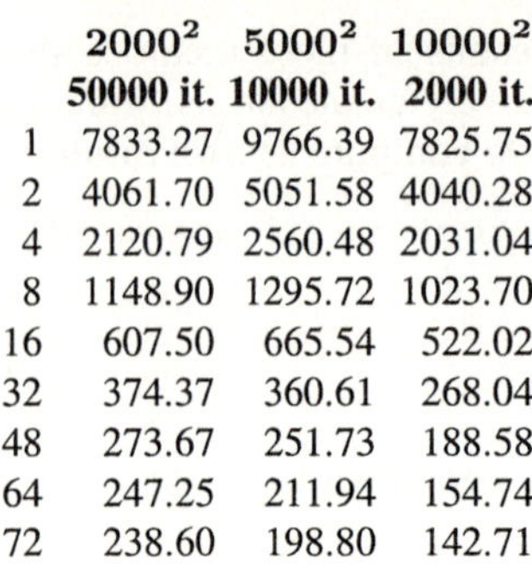

| | 2000^2 | 5000^2 | 10000^2 |
	50000 it.	10000 it.	2000 it.
1	7833.27	9766.39	7825.75
2	4061.70	5051.58	4040.28
4	2120.79	2560.48	2031.04
8	1148.90	1295.72	1023.70
16	607.50	665.54	522.02
32	374.37	360.61	268.04
48	273.67	251.73	188.58
64	247.25	211.94	154.74
72	238.60	198.80	142.71

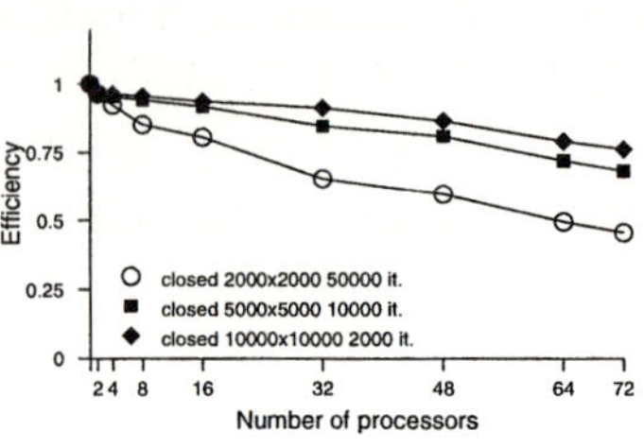

Fig. 1. Parallel execution times in seconds and efficiency of our closed-world CA simulator.

since we want to allow open-world computation, we accept the potential inefficiencies of column-wise distribution.

Our *closed-world* implementation requires a fixed number of processors that is known at the start of the program. All processors have an equal share of the matrix, and initialise their part of the matrix.

We also implemented an *open-world* version. Since the number of participating processors is not known in advance, the first processor to join the computation must initialise the entire grid; when other processors join the computation, they are sent a fair share of the matrix. This requires a significant amount of communication.

The open-world version also dynamically redistributes the computational load. When a processor has completed a generation, it sends work requests to its neighbours. A neighbour that has not completed its own computation sends some columns of the matrix to its faster neighbour. To dampen temporary disturbances, sporadic requests are only honoured by small redistributions, and repeated requests by larger redistributions.

The program was implemented using the Ibis communication layer. This layer was mainly designed to support the higher levels of Ibis, but it was also a good choice for the CA simulator. We estimate that a sequential simulator would require 200 lines of code. The closed-world version is about 400 lines of code, and the open-world version is about 1200 lines of code, mainly because of the load-balancing mechanism.

We do our simulations on a square matrix, denoted as a squaring expression, e.g. 20^2. In Fig. 1 we show execution times of the closed-world simulator for various problem sizes. As the results indicate, the program can achieve good speedups, especially for large grid sizes.

In Fig. 2 we show the results of open-world simulation on a 5000^2 matrix. For comparison we repeat the results of the closed-world simulation. As expected, the fact that one processor starts with the entire matrix, and then sends most of it to other processors, has a significant impact, particularly for larger numbers of processors. To evaluate this effect we also show the results for the last 75% of the iterations. The results indicate that after the initial redistribution, open-world simulation is as efficient as closed-world simulation.

Our measurements indicate that the dynamic redistribution system is very effective in systems with moderate imbalances. However, in extreme cases a slow processor or a high-latency communication link can determine the pace of the computation or can

	closed	open	started
1	9766.39	10628.54	7969.99
2	5051.58	5326.20	3988.71
4	2560.48	2785.53	2085.54
8	1295.72	1386.68	1028.97
16	665.54	774.53	560.16
32	360.61	460.86	313.36
48	251.73	376.60	222.29
64	211.94	344.05	178.47
72	198.80	324.44	136.17

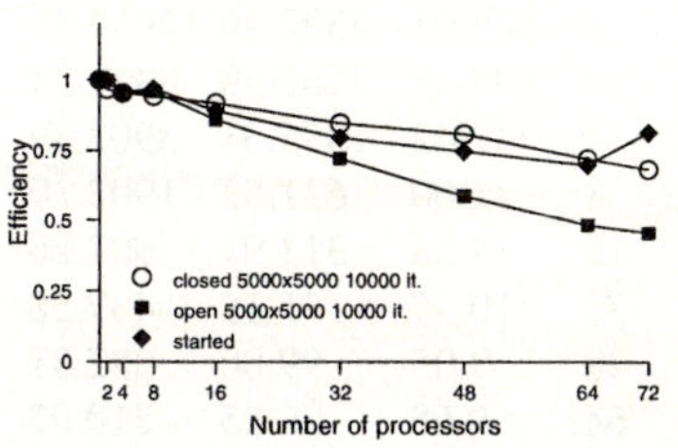

Fig. 2. Parallel execution times in seconds and efficiency for 10000 iterations of our open- and closed-world CA simulators on a 5000^2 matrix. Also, the parallel execution times and efficiency for the last 7500 iterations of our open world simulator.

render the balancing mechanism ineffective. In such cases it would be more effective to withdraw a processor from the computation, but this is currently not supported.

5   Solving the Satisfiability Problem

Given a symbolic Boolean expression, the *Satisfiability* (*SAT*) problem requires that a set of assignments to the variables of the expression is found for which the expression evaluates to true, or that it is established that no such set of assignments exists[2].

The SAT problem plays a pivotal role in theoretical computer science as a representative example of an NP-complete problem. It also occurs in a number of practical applications. Since the problem is NP-complete, no algorithm is known that is guaranteed to solve this problem in polynomial time. Nevertheless, a number of heuristics allow the development of practical SAT solvers.

A brute-force SAT solver could try all possible combinations of assignments, and see if one of them yields true. However, since an expression with n variables has 2^n combinations, this is rarely practical. A better strategy is to try to eliminate large parts of the solution space at once by evaluating partial assignments. For example, the expression $(a \vee \neg b) \wedge (b \vee c \vee \neg d)$ can never be satisfied with the assignments $a = \mathtt{false}, b = \mathtt{true}$, since the first clause cannot be satisfied for any assignment to c and d. Yet the assignments $a = \mathtt{true}, c = \mathtt{true}$ satisfy the entire Boolean expression for all assignments to b and d. Modern SAT solvers use a backtracking search that speculatively assigns values to variables until the problem is either satisfied, or until there is a conflict. Upon a conflict, the solver backtracks. The efficiency of the search process is strongly influenced by the order in which the variables are assigned [10]. A common and effective heuristic is to select variables that satisfy as many unsatisfied clauses as possible. More refined variable selection heuristics tend to require more sophisticated bookkeeping, and are often not deemed to be worth the extra trouble.

The backtracking search maps naturally to a divide-and-conquer implementation: our SAT solver, using a recursive backtracking search as described above, is implemented in 2500 lines of code. Only 25 of these are required by the Satin framework.

---

[2] For a more detailed overview of the Satisfiability problem and its solvers see for example [10].

	uuf200	FPGA12	FPGA13
1	205.01	4322.26	12635.36
2	108.18	2767.59	8349.74
4	47.42	1335.31	3903.39
8	25.00	631.83	1902.20
16	14.58	313.91	901.26
32	10.45	147.15	437.58
48	9.05	99.00	282.81
64	8.68	76.55	210.07
72	8.07	70.54	186.48

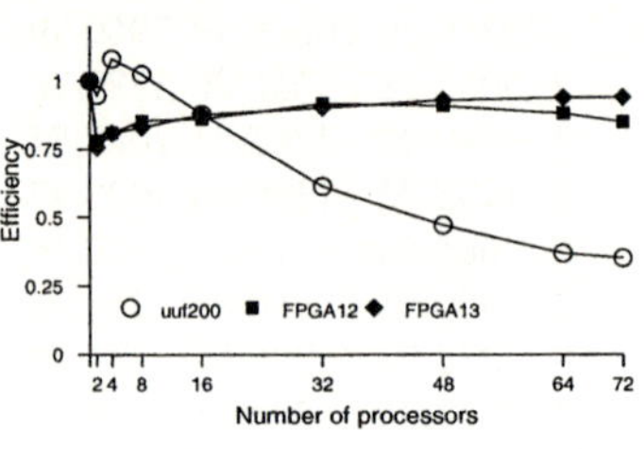

Fig. 3. Parallel execution times in seconds and efficiency for the SAT solver.

Figure 3 shows results for our solver for a number of SAT problems. The *uuf200* problem is from the SATLIB collection [11]. It is a set of 860 random clauses with 3 variables each, and 200 variables in total. Random problems are known to be difficult for many SAT solvers, since there is no structure to guide them, and problems with the chosen proportion of clauses and variables are known to be the most difficult [12]. The *FPGA12* and *FPGA13* problems are described in [13]. They represent routing problems through FPGA switchboxes. FPGA12 has 240 variables and 1344 clauses; FPGA13 has 260 and 1586. All problems are known to be unsatisfiable.

As results of Fig. 3 indicates, for the large problems parallel efficiency is very good, even for large numbers of processors. The uuf200 problem is too small to scale well.

6   Grammar-Based Text Analysis

As a final application we show *Grammy*, a program that analyses text by constructing a grammar that produces the original sentence. For example, for the sentence 'a long long time ago' we could construct the grammar

$$\textbf{start} \rightarrow \text{'a}\diamond\diamond \text{ time ago'}$$
$$\diamond \quad \rightarrow \text{' long'}$$

Constructing a compact grammar is useful for text analysis, since it infers hierarchical structure [14]. The analysis is also useful for compression. In fact, the classical compression algorithm LZ78 [15], can be viewed as constructing a grammar.

Choosing the most effective grammar rules is often difficult, since a text usually has many repeated sequences to choose from. The most obvious strategy is to repeatedly select the longest repeat, or the repeat resulting in the largest gain. However, that is not optimal, since each choice may preclude subsequent choices. We approximate optimal compression by considering a number of efficient choices at each step and looking ahead a number of steps. Since each of the possibilities can be evaluated independently, this can be implemented as a recursive parallel process. Our implementation was parallelised very effectively by using Satin: only about 20 of the 1850 lines of code of the program are required by the Satin framework.

To evaluate our program, we use the following texts: William Shakespeare, *A Midsummer Night's Dream* (96508 Bytes, *text 1*), Sir Arthur Conan Doyle, *The Adventure*

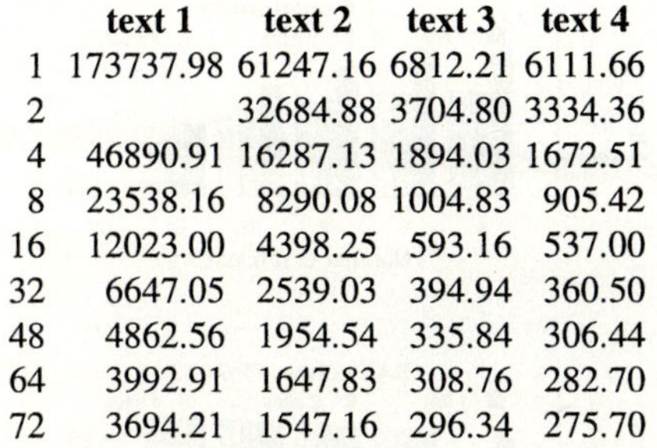

	text 1	text 2	text 3	text 4
1	173737.98	61247.16	6812.21	6111.66
2		32684.88	3704.80	3334.36
4	46890.91	16287.13	1894.03	1672.51
8	23538.16	8290.08	1004.83	905.42
16	12023.00	4398.25	593.16	537.00
32	6647.05	2539.03	394.94	360.50
48	4862.56	1954.54	335.84	306.44
64	3992.91	1647.83	308.76	282.70
72	3694.21	1547.16	296.34	275.70

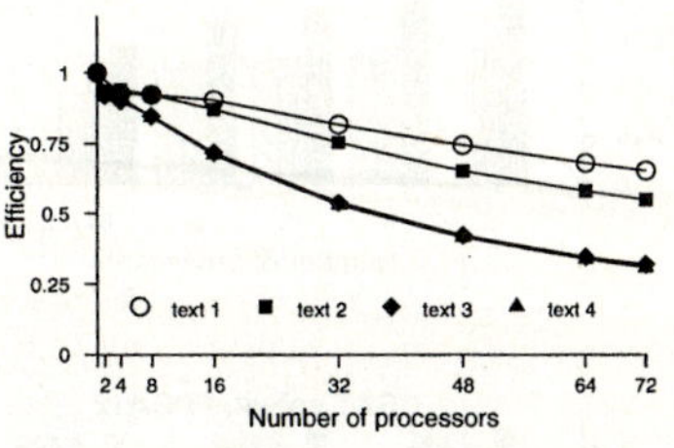

Fig. 4. Parallel execution times in seconds and efficiency for grammar construction.

of the Red Circle (53394 Bytes, *text 2*), the *grep* man page of a recent Debian system (20047 Bytes, *text 3*), and the file `SuffixArray.java` from the Grammy source code (28572 Bytes, *text 4*).

Figure 4 shows the execution times and efficiency for these texts. All runs were done with a lookahead of 5 steps and with at most 7 candidates at each step. These results show that for large texts the computation scales quite well to larger numbers of processors. This is because large texts tend to have many repeats, resulting in sufficient parallelism to keep all processors busy.

7　Results on Wide-Area Clusters

In §4, §5, and §6 we showed results for clusters of processors on a single DAS2 site. Figure 5 shows results for the same programs on clusters of processors on two and four DAS sites in the Netherlands. In all cases we use an equal number of processors on the participating sites, so if we run a program on 64 processors on four sites, each site has 16 processors. As for computation on a single site, we compute the efficiency of the parallel computation relative to computation on a single node.

Since the CA computation requires information exchange after each iteration, the processors run in tight lockstep. The larger latency of the wide-area links therefore has a noticeable influence on the efficiency of the computation. Nevertheless, the computation is efficient enough to be useful.

The SAT solver and the text analysis also require communication for work stealing, but the Satin framework was able to hide the higher latencies of the wide-area links. In fact, in a few cases wide-area execution was more efficient than execution on a single site, presumably due to reduced contention on shared resources such as local communication channels.

8　Results on a Global Grid

Finally, we have executed a number of runs on a grid testbed. Efficient computation on such a system requires a careful choice of communication structure. Often communication between grid nodes is avoided entirely, but this obviously restricts the use of grid systems to trivially parallel systems, and all of our example programs require more.

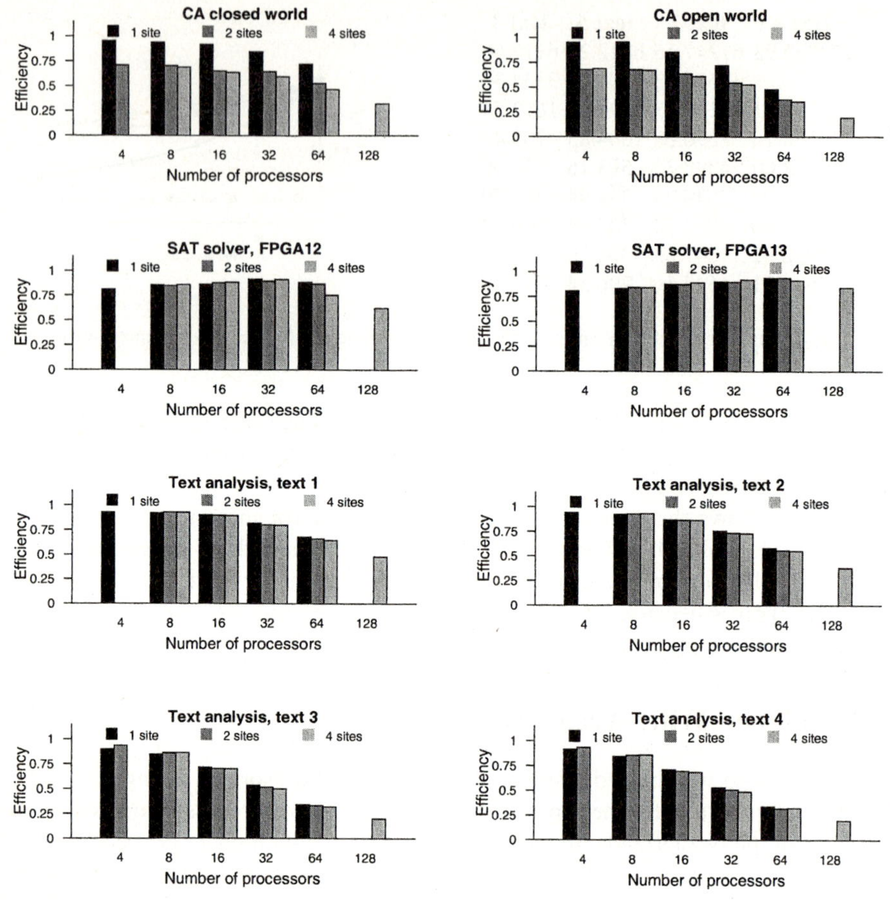

Fig. 5. Efficiency of computations on two and four DAS sites, compared to a single site.

Our closed-world CA simulator is clearly unsuitable, since it divides the computation equally over the processors. The other programs were run using the following set of systems and JVM implementations (site 1 is a DAS2 site):

site	CPUs	Architecture	JVM	Location
1	16	Intel Pentium III 1GHz	IBM 1.4.1 32 bit	the Netherlands
2	4	Intel Xeon 3GHz	SUN 1.4.2 32 bit server VM	Czech Republic
3	2	Intel IA64 (Itanium) 1.4 GHz	SUN 1.4.2 64 bit server VM	Poland
4	4	Intel Xeon 2GHz	SUN 1.4.2 32 bit server VM	Louisiana, USA
5	2	Intel Xeon CPU 2.4 GHz	SUN 1.4.2 32 bit server VM	Germany
	28	**total**		

Table 1 shows the results for these runs. The SAT solver performs very well on a world-wide grid. The other two applications performed almost as well as on the wide-area DAS-2 system after we removed two sites. We removed site 3 because the (64 bit) JVM on that site performs very poorly on the CA and text analysis applications, and did not

Table 1. Execution times in seconds on individual grid sites, on the combined set of processors (t_p), and efficiency of the various programs on a global grid.

Benchmark	system 1	2	3	4	5	t_{ideal}	t_p	η
SAT solver, FPGA12	318.88	615.41	1263.70	727.50	1294.16	129.88	146.85	0.88
SAT solver, FPGA12	318.88	615.41	-	-	1294.16	180.71	217.99	0.83
Text analysis, text 4	532.68	1052.97	-	-	2350.49	307.46	459.82	0.67
CA open 5000x5000, 2500 it	284.40	652.44	-	-	1652.83	176.87	266.39	0.66

contribute anything. Site 4 is located in the USA and has a high latency link to Europe. This interferes with the load balancing algorithm of the CA application: due to the high latency, steal requests arrive after the victim has finished the iteration. The text analysis application does not perform well if site 4 is used due to the limited amount of parallelism that is generated. Jobs are relatively small, and the transfer of jobs over the slow WAN link does not outweigh the cost. For comparison we also show the results for the SAT solver without these sites.

9 Conclusions and Future Work

In this paper we have shown the use of Ibis for a number of larger applications. Ibis proved to be very effective. Both the SAT solver and the text analyser could be developed as mainly sequential programs, with only a few additional lines of code to interface to the Satin framework. Parallelisation doesn't get much simpler than this. Although the Cellular automata simulator required more explicitly parallel code, the amount of parallel code was still limited, even for the load-balancing mechanism in the open-world version.

All programs performed well on a traditional supercomputer cluster, and a wide-area cluster system. Since all of the programs require communication between the processors, execution on a grid system was not always efficient, but even there very creditable results could be achieved, in particular for the SAT solver.

We are currently extending Ibis with support for fault tolerance, more elaborate automatic configuration, and peer-to-peer computing. Other areas of study are performance debugging and additional high-level parallel programming models.

Acknowledgements

This work was partially supported by the Dutch Organisation for Scientific research (NWO). This work was part of the Virtual Laboratory for e-Science project (www.vl-e.nl). This project is supported by a BSIK grant from the Dutch Ministry of Education, Culture and Science (OC&W) and is part of the ICT innovation program of the Ministry of Economic Affairs (EZ).

References

1. Nieuwpoort, R.V.v., Maassen, J., Wrzesinska, G., Hofman, R., Jacobs, C., Kielmann, T., Bal, H.E.: Ibis: a flexible and efficient Java-based grid programming environment. Concurrency and Computation: Practice and Experience **16** (2003) 1–29 Published online in Wiley Inter-Science (www.interscience.wiley.com). DOI 10.1002/cpe.860.
2. Aumage, O., Hofman, R., Bal, H.: Netibis: An efficient and dynamic communication system for heterogeneous grids. In: Proc. of CCGrid. (2005) (accepted for publication)
3. Nieuwpoort, R.V.v., Maassen, J., Hofman, R., Kielmann, T., Bal, H.E.: Ibis: an efficient Java-based grid programming environment. In: Joint ACM Java Grande - ISCOPE 2002 Conference, Seattle, Washington, USA (2002) 18–27
4. Bal, H., et al.: The distributed ASCI supercomputer project. ACM SIG, Operating System Review **34** (2000) 76–96
5. Nieuwpoort, R.V.v., Maassen, J., Hofman, R., Kielmann, T., Bal, H.E.: Satin: Simple and efficient Java-based grid programming. In: AGridM 2003 Workshop on Adaptive Grid Middleware, New Orleans (2003)
6. Nieuwpoort, R.V.v., Kielmann, T., Bal, H.E.: Efficient load balancing for wide-area divide-and-conquer applications. In: Proc. Eight ACM SIGPLAN Symp. on Princ. and Practice of Par. Progr. (PPoPP), Snowbird, UT, USA (2001)
7. Blumofe, R.D., Joerg, C.F., Kuszmaul, B.C., Leiserson, C.E., Randall, K.H., Zhou, Y.: Cilk: An efficient multithreaded runtime system. J. of Par. and Distr. Computing **37** (1996) 55–69
8. Ermentrout, G.B., Edelstein-Keshet, L.: Cellular automata approaches to biological modeling. J. theor. Biol. **160** (1993) 97–133
9. Colonna, A., Stefano, V.d., Lombardo, S., Papini, L., Rabino, G.A.: Learning cellular automata: Modelling urban modelling. In: Proc. 3rd Intl Conf. on GeoComputation, University of Bristol, UK (1998) 388–395
10. Zhang, L., Malik, S.: The quest for efficient boolean satisfiability solvers. In: Proc. of the 18th Intl Conf. on Automated Deduction, Copenhagen, ACM (2002) 295–313
11. Hoos, H.H., Stützle, T.: Satlib - the satisfiability library. webpage (2000) www.satlib.org.
12. Cheeseman, P., Kanefsky, B., Taylor, W.M.: Where the Really Hard Problems Are. In: Proceedings of the Twelfth International Joint Conference on Artificial Intelligence, IJCAI-91, Sidney, Australia (1991) 331–337
13. Aloul, F., Ramani, A., Markov, I., Sakallah, K.: Solving difficult SAT instances in the presence of symmetry. In: Proc. of the Design Automation Conf. (DAC), New Orleans (2002)
14. Nevill-Manning, C.G., Witten, I.H.: Compression and explanation using hierarchical grammars. The Computer Journal **40** (1997) 103–116
15. Ziv, J., Lempel, A.: Compression of individual sequences via variable-rate coding. IEEE Transactions on Information Theory **24** (1978) 530–536

Virtual Workspaces in the Grid

Katarzyna Keahey[1], Ian Foster[1,2], Timothy Freeman[2],
Xuehai Zhang[2], and Daniel Galron[3]

[1] Argonne National Laboratory
{Keahey,foster}@mcs.anl.gov
[2] The University of Chicago
{tfreeman,hai}@cs.uchicago.edu
[3] The Ohio State University
galron@cis.ohio-state.edu

Abstract. Despite significant progress in the development of Grid infrastructure, the provisioning of a customized and controllable remote execution environment remains an open issue. This paper introduces the concept of a virtual workspace, a configurable execution environment that can be created and managed as a first-class entity to reflect client requirements. Such workspaces can be dynamically deployed on a variety of resources decoupling the notion of environment and resource. We show how virtual workspaces fit into the Grid architecture, present an example implementation using virtual machines, and discuss our initial experiences using this system in practice and with applications.

1 Introduction

While significant progress has been achieved with the deployment of Grid-based applications, the preparation of a remote execution environment remains an issue. One of the reasons is that while Grids offer access to diverse software environments, an application typically requires a very specific, customized environment. As a consequence of variations in operating system, middleware versions, library environments, and file system layouts a user's application may in practice be able to use only a small fraction of the resources potentially available on the Grid. The second issue is the need to provide reliable isolation and dynamic, fine-grain control of shared resources to ensure enforcement of policies and thus provide incentive for wider sharing. The development of Grid protocols [1, 2] provides uniform ways to manage Grid entities that could represent such environments. It now remains to find ways to describe represent and implement them.

In this paper, we present the concept of a virtual workspace, which allows a Grid client to define an environment in terms of its requirements (such as resource requirements or software configuration), manage it, and then deploy the environment in the Grid. Workspaces defined in this way can be implemented in a variety of ways such as for example dynamically creating Unix accounts and using system as well as software configuration tools to enforce the required properties. Here, we focus on a particularly promising implementation of virtual workspaces: virtual machines (VMs). The use of virtual machines in Grid computing has been proposed before [3, 4]. In addition to outstanding isolation properties, VMs can provide fine-grained enforcement; and by their very nature—virtualization of the underlying hardware—

J.C. Cunha and P.D. Medeiros (Eds.): Euro-Par 2005, LNCS 3648, pp. 421–431, 2005.

they enable instantiation of a new, independently configured guest environment on a host resource. They can be rapidly suspended and their state serialized, and thus easily migrated to remote resources. Moreover, as a result of recent progress in virtual machine technology, these advantages no longer come at a performance cost to either the application or the hosting resource: systems such as Xen [5] demonstrate that they can be used with little or no performance degradation.

In the rest of the paper, we describe how virtual workspaces fit into the Grid architecture, present a prototype of this architecture based on the Globus Toolkit (GT) and experiment with two workspace implementations using Xen [5] and the VMware Workstation [6]. We describe our initial experiences with integrating VMs into the Grid infrastructure, and we present preliminary results of testing our prototype system with a bioinformatics application suite.

2 Virtual Workspaces

Interactions in present-day Grids focus on mapping jobs to resources, often with the assumption that an execution environment with suitable configuration and enforcement characteristics will be provided by means independent of the Grid infrastructure. Although such an assumption is true for closely knit groups of Grid users, it is not justified when applications or users with drastically different requirements and rights are trying to use the same resources. Recognizing this fact, we define the concept of a *virtual workspace* that can be automatically deployed on resources and provide a required execution environment. Thus, jobs can be mapped to workspaces, and workspaces can be mapped to actual resources in the Grid.

A virtual workspace (VW) is a definition of an execution environment in terms of its hardware requirements, software configuration, isolation properties, and other salient characteristics. The intent of defining a workspace is to capture the requirements for an execution environment in the Grid and then use automated tools in order to find, configure, and provide an environment best matching those requirements. We could, for example, use agreement-based tools to negotiate contracts defining workspaces and the use of actual resources and then negotiate a binding between them following the models described in [2, 7, 8]. Depending on the requirements, such contracts could be fulfilled by simply dynamically creating and configuring user accounts as in [9], by using pre-configured virtual machines or other sandboxing and virtualization technologies.

2.1 Virtual Workspace Descriptions

To describe workspaces, we use an XML Schema, which captures generic properties of every workspace, as well as properties subject to specific workspace definition or workspace implementation-specific properties.

The XML example below shows a description of a workspace. It contains the workspace category type that describes what mechanisms are used to create the workspace: we currently support implementations based on different types of virtual machines and dynamic accounts [10]. Currently, the implementation type is used to define the isolation model as well as provide a clue to services processing the workspaces to provide implementation-specific processing. Workspace state can be one of running (a workspace deployed on a resource), shutdown (for example, a "cold" VM image containing no running processes), paused (for example, a "hot" VM image), and corrupted (a workspace that cannot be deployed because of internal inconsistencies). In addition, the generic part of workspace description contains a reference that can be used to check the status of the workspace. The reference embeds information about the workspace owner's distinguished name. Other properties contained in the generic part of the schema include three time-related elements: creationTime, lifeCycle, and lastModified. CreationTime records the time when the workspace was first instantiated. LifeCycle indicates how long the workspace is available for use. Last-Modified keeps the information about when the properties of the workspace were modified last.

Further definition contains a description of different workspace aspects, such as required hardware, networking configuration, required software installations and workspace capability. Description of the "virtual resource" that represents the hardware requirements of the execution environment contains elements such as the RAM size, disk size, disk type, and accessing mode, as well as devices such as virtual CD-ROM drives. A network specification contains the description of a network connection and how to establish it (such as the method to obtain an IP address). Software descriptions contain information about the operating system (e.g., kernel version, distribution type), library signature, and programs installed. Workspace capability describes what can be done with the workspace: for example, a workspace may be configured to run a program on startup or have some programs (in particular, hosting programs) already running and be able to service requests on specific ports.

```xml
<xs:simpleType name="categoryType">
  <xs:restriction base="xs:string">
   <xs:enumeration value="dynamic account"/>
   <xs:enumeration value="vm"/>
  </xs:restriction>
</xs:simpleType>

<xs:simpleType name="stateType">
  <xs:restriction base="xs:string">
   <xs:enumeration value="shutdown"/>
   <xs:enumeration value="paused"/>
   <xs:enumeration value="running"/>
   <xs:enumeration value="corrupted"/>
  </xs:restriction>
</xs:simpleType>

<xs:element name="virtualWorkspace">
  <xs:complexType>
   <xs:sequence>
    <xs:element name="category" type="categoryType" default="vm"/>
    <xs:element name="state" type="stateType" default="shutdown"/>
    <xs:element name="EPR" type="wsa:EndpointReferenceType"/>
    <xs:element name="creationTime" type="xs:dateTime" minOccurs="0"/>
    <xs:element name="lifeCycle" type="xs:integer" minOccurs="0"/>
    <xs:element name="lastModified" type="xs:dateTime" minOccurs="0"/>

    <xs:element ref="hw:hardware" minOccurs="0"/>
    <xs:element ref="net:network" minOccurs="0"/>
    <xs:element ref="sw:software" minOccurs="0"/>
    <xs:element ref="cap:capability" minOccurs="0"/>
   </xs:sequence>
  </xs:complexType>
</xs:element>
```

Wherever applicable, workspace properties are described as a set of possible values (e.g., RAM size with min and max requirements) rather than one value. The intent is to leave open the largest possible set of mappings of workspaces to real resources. The schemas are extensible to reflect the capabilities of different implementations. For example, we use a VM-based workspace category type, extended from the generic category type, to describe the category information of a workspace implemented with virtual machines. Besides the inherited category name, it contains extra properties such as the type of virtual machine monitor (VMM), which are specific for VM-based workspaces only.

Based on the descriptions defined in the schema, workspaces can be selected, cloned, or refined. Cloning a workspace, for example, involves creating a new name (as encoded within the reference element), a new resource and copying the description metadata.

2.2 Virtual Workspaces as Virtual Machines

We have surveyed candidate technologies for workspace implementations [4] and identified two especially promising ones: configurable dynamic accounts [10, 11] and virtual machines. Although dynamic accounts present an interesting implementation option (especially when used with enforcement tools such as quota and software configuration tools such as Pacman [12]), we pursue this work elsewhere [9]. Our focus in this paper is on a virtual machine implementation of workspaces in view of their outstanding isolation and serialization properties.

A virtual machine [13] provides a virtualization of the underlying physical host machine. Software running on the host, called virtual machine monitor (VMM) or hypervisor, is responsible for supporting the perception of multiple isolated machines by intercepting and emulating privileged instructions issued by the guest machines. A VMM typically also provides an interface allowing a client to start, pause or stop multiple guests. A VM representation contains a full image of a VM's RAM, disk, and other devices, allowing its state to be fully serialized, preserved, and restored at a later date. Recent exploration of paravirtualization techniques [5] has led to substantial performance improvements in virtualization technologies, making virtual machines an attractive option for Grid computing.

The serialization properties of VMs create the potential for effortless configuration of execution environments (for example, allowing a user to configure VMs with software required by a given community and clone them). Their isolation from the host machine provides a way of running a different configuration from that of the VM host and allows guest VMs and resource owners to take advantage of enhanced security properties of VMs. Further, those abilities combined provide potential for migration. For these reasons, VMs provide an excellent implementation option for workspaces: the configuration of a VM image can reflect a workspace's software requirements while the VMM can ensure the enforcement of hardware properties.

3 Integrating Virtual Workspaces into the Grid Architecture

Virtual workspaces refine the execution environment layer in Grid architecture: rather than mapping jobs directly onto hardware resources as in [14], we map jobs to pre-configured workspaces which can then be mapped to Grid resources. Since a workspace may exist beyond its deployment, and may in fact may be deployed many times on different resources during its lifetime, we introduce two new services: VW Repository which provides a management interface to workspaces, and VW Manager which orchestrates their deployment. The figure below illustrates how workspaces work within the Grid infrastructure:

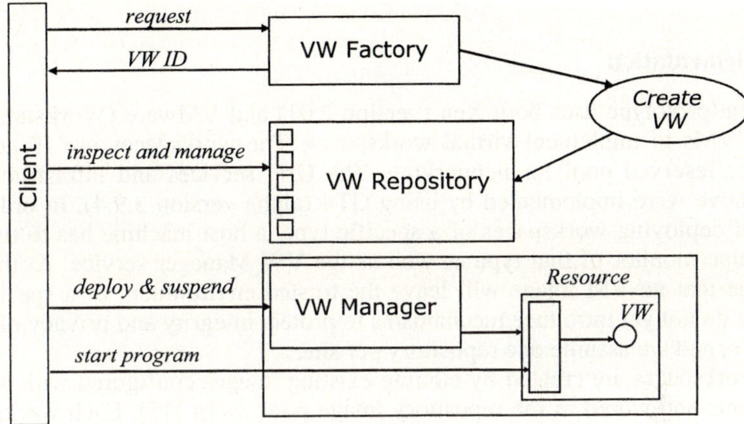

Fig. 1. Grid Interactions with Virtual Workspaces

3.1 Grid Interactions

In order to create a workspace instance, a Grid client contacts the VW Factory with a workspace description presented in Section 2.1. A negotiation process may take place to ensure that the workspace is created in a policy controlled way. The newly created workspace is registered with a VW Repository, which provides a Grid service interface allowing for inspection and management of workspaces and keeps track of resources implementing workspaces such as virtual machine images. As a result of creation the client is returned a WSRF end-point reference (EPR) to the workspace. We leverage the abstraction of a Grid service resource [1] to enable inspection and management of properties such as termination time.

To deploy a VW on a specific resource, a client contacts a VW Manager Grid service on that resource and presents it the workspace's EPR. The VW Manager allows a client to deploy/undeploy, start/stop, and (in our VM-based implementation) also pause/unpause workspaces. Deploying a workspace simply means staging all components of the workspace (such as a VM image) to the resource so that they are available to the VW Manager on that resource. Currently, "checking out" workspaces from the repository puts a lock on them, since their state might change during use. Similarly, "undeploy" releases the hold on workspace resources and the corresponding lock. After the workspaces are staged, they can be started (i.e., become available for computation). Once a VW becomes ready for execution, a program can be started by using Grid infrastructure mechanisms (e.g., Globus Resource Allocation Manager, or GRAM) or by using other methods such as preconfigured program startup or a continuation of a previous execution. The VW Manager can also stop, pause ("freeze" an ongoing computation), or undeploy a stopped or paused workspace (stage it back to the repository for example).

In a complete picture of interactions, a community broker would negotiate reservations or agreements for the use of specific resource allocations on a resource. The workspace agreements would be matched against those allocations and create binding agreements allowing the deployment of a specific workspace on a selected resource much as described in SNAP [7]. Such agreements could then be renegotiated and the workspaces migrated, as need arises.

3.2 Implementation

Our current prototype uses both Xen (version 2.01) and VMware (Workstation, version 4.5) VMs to implement virtual workspaces. The workspaces use IP addresses from a pre-reserved pool for networking. The Grid services and infrastructure described above were implemented by using GT4 (alpha version 3.9.4). In order to be capable of deploying workspaces of a specific type, a host machine has to run a virtual machine monitor of that type as well as the VW Manager service. At this point we assume that no VM image will leave the trusted environment of a specific site; that is, we do not yet introduce mechanisms to protect integrity and privacy of images themselves, and we assume one repository per site.

New workspaces are created by cloning existing images configured with the same requirements and stored in the repository image pool, as in [15]. Each workspace is configured with a certificate and a private key to authenticate itself to clients. At this stage, we do not support negotiation for workspace creation; a workspace request is either accepted or rejected based on existing policies.

The VW Manager interfaces with VMM running on a particular resource to implement to stage/unstage, start/stop, and pause/unpause operations: in Xen via an HTTP control interface and on VMware Workstation via GUI scripting. To stage a workspace, the VW Manager transfers the workspace data (including description metadata and the implementation-specific image) from the VM Repository to the host node by using GridFTP [16]. Once the workspace data transfer is complete, the VW Manager waits for the client to start the workspace, which includes creating a workspace resource, loading the VM image into memory, and booting the VM. At boot time the VM may initiate preconfigured operations such as obtaining its network address and starting programs (including the GT hosting environment). Once this step is completed, a VM can advertise the hosted Grid services such as GRAM for clients to invoke.

Although in the current prototype we do not address the issue of ensuring privacy and integrity of workspace representations (VM images), we do support standard Grid authentication and authorization mechanisms. Running workspaces and Grid clients mutually authenticate by using the Grid Security Infrastructure GSI [17]. Our infrastructure also accepts VOMS certificates [18] and is capable of extracting VOMS attributes for authorization. Authorization of workspace creation, deployment, and management is configured via access control lists based on the distinguished name and attributes of Grid entities.

4 Experiences with Virtual Workspaces

The performance impact of virtual machines on applications has been shown to be small (typically under 5% of slowdown) for different application classes [5, 19, 20].

In our preliminary evaluation, we explored the performance impact of different ways of using VMs as part of Grid infrastructure. We also conducted a preliminary evaluation of VM usage with applications.

To explore the best ways of using VMs within the Grid infrastructure, we timed the process of starting up a workspace and running a program in it on a remote node in different startup configurations. We repeated those experiments for both of our workspace implementations (Xen and VMware Workstation) and compared them with the time of job startup through a call to GT4 GRAM. In all the experiments we assume that the necessary data had already been staged to the node (i.e., the executable and input data for GRAM and a VM image in the case of workspace deployment).

All experiments were run on a dual 2.2 GHz Xeon server configured to run single-CPU guest VMs. The same VM image configuration (Debian Sarge) was used for both Xen and VMware, and the same test application was used for all experiments. Time was measured on the server side only, by using wall clock time.

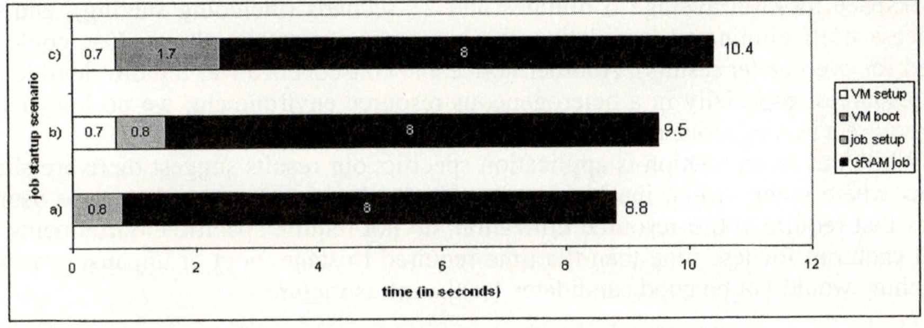

Fig. 2. Xen-based VWs versus job invocation using GRAM

Figure 2(a) shows time elapsed from the moment GRAM received a job startup request till the end of the job: as per GRAM default the job is run in a user account. Figures 2(b-c) show the combined time of deploying a workspace implemented as a Xen VM and starting a job in it under two different scenarios. In 2(b) we take advantage of the serialization property of VMs: we prepare a "hot image" (a paused image) with a hosting environment already started up. A call to GRAM is placed as soon as the environment is available. In 2(c), instead of configuring the workspace to start up a GT4 hosting environment, we start the job directly. Unless the VM is partially configured at boot time, this scenario could also be optimized by pausing a booted VM.

For comparison, we also ran a similar test on VMware Workstation, which is hard to time given its non-programmatic, opaque interface (VMware ESX/GSX tools provide more efficient and direct interfaces). The results are summarized in Figure 3; a significant amount of time is spent in a controller adapter to the Workstation version.

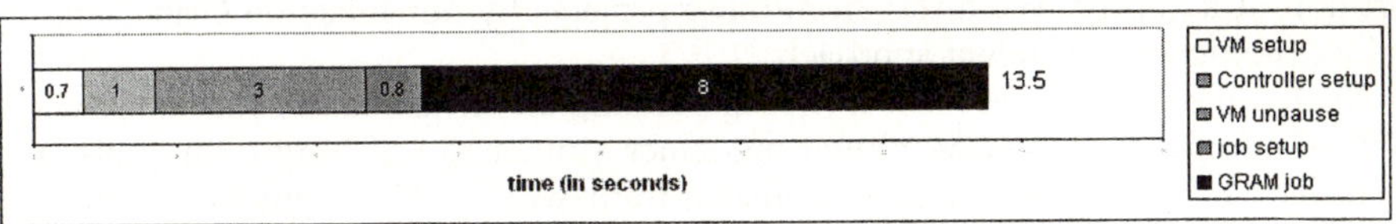

Fig. 3. VM deployment time using VMware workstation

Our preliminary experiments show that a Xen VM could be effective for a range of configurations with addition to job deployment time that can easily be absorbed by latency in the Grid environment. Startup costs of deploying a VM image are comparable to starting a job; the trade-off is that while GRAM implements a flexible job deployment strategy the VM is already pre-configured with the job to be started. On the other hand, the advantages gained by using VMs are significant. Deploying hot images ready to process requests offers the fastest solution (note that this method could be used to eliminate application initialization time as well). However, this may not always be possible; in such cases using a VM preconfigured to start a specific job is also an efficient alternative (currently no credential is delegated to such a job; the startup is based on authorization of the client that submits the VM).

To obtain a preliminary assessment of the usefulness of this infrastructure to Grid applications, we experimented with the EMBOSS [21] suite for bioinformatics applications. The most important effect was facilitating deployment: while the EMBOSS installation took roughly 45 minutes, starting a preconfigured VMware Workstation workspace took on average 6 minutes and 23 seconds (including staging), and the process itself eliminated installation errors (as per our results above, Xen could be used for even better results). Another noticeable consequence was a more flexible use of resources, especially in a heterogeneous resource environment: we no longer had to require a homogenous resource base.

While the determination is application specific, our results suggest there are situations where using virtual machine workspace implementations would not be useful. Jobs that require 100% resource utilization, do not require specific environments, or that each run for less time than the time required to stage, boot or unpause a virtual machine would not be good candidates for the infrastructure.

5 Related Work

Due to its potential, the use of virtual machines in Grid computing is attracting increasing attention [3]. The In-Vigo project [15, 22] made substantial progress in this direction while the Virtuoso [23] and VIOLIN [24] projects explored networking issues arising from use of VMs in this setting. Our approach differs in that it focuses on technology-neutral virtual workspaces, using virtual machines as only one of their potential implementations. Further, recognizing that a workspace may be deployed many times we distinguish between workspace creation and its deployment introducing a different architecture for its support.

Attempts to capture requirements of an execution environment to some extent and automate their deployment have also been made before: for example, the virtual appliance project [25] uses virtual machines configured based on descriptions in a configuration language to ease administrative burden, and the Cluster on Demand (COD) project [26] allows a user to choose from a database of configurations to configure a partition of a cluster. We differ from these projects by our focus on Grid computing and technology-independent approach.

Finally, the Xenoserver project [27] is building an infrastructure for wide-area distributed computing based on virtual machines similar to our long-term goals. Here, we differ by building within the established framework of Grid computing rather than providing new infrastructure.

6 Conclusions and Future Work

We have described the abstraction of a virtual workspace, a customizable execution environment capable of being deployed on a variety of platforms in the Grid. Workspaces are defined in terms of client requirements, such as software and hardware requirements, and are implemented in terms of technologies providing an isolated execution environment, quality of service at the granularity of a workspace (as opposed to a single process), customized software installation, and, in the case of VMs, execution serialization and migration.

We showed how workspaces can be integrated into the existing Grid infrastructure. The integration entails relatively small changes, but introduces substantial flexibility of use. The use of Grid protocols allows us to fully leverage this flexibility and create, deploy, and shut down workspaces dynamically based on policy-driven provisioning decisions. By virtue of their properties, workspaces are a promising vehicle for implementing policy-driven Grid usage.

To evaluate the feasibility of our workspace implementations, we compared the performance of GT4 GRAM, a widely used job startup service for Grid applications, to the process of starting up workspaces implemented as virtual machines in a variety of scenarios. In conjunction with the small performance impact on applications [5], our results show workspaces to be a promising abstraction for Grid computing. In addition, our experiments identified a number of job startup scenarios relevant in the workspace context, showing how they may be used in practice. Preliminary application evaluation of workspaces also proved satisfactory and fully realized our expectations for more flexible resource usage.

More work is needed in order to fully assess the usefulness of these ideas. In the short term, we will focus on the privacy and integrity of migrating workspaces. Workspace distribution (delivery to host) can require transferring images of a few gigabytes; this task can be handled by using the presence of an image on a node as a matching criterion as proposed in [15] or by transferring only partial images as in [28]. In addition to performance impact on individual applications, we are considering scalability issues that can be addressed by either using a lighter-weight workspace implementation such as [29] or mapping groups of jobs to one workspace.

Acknowledgments

This work was supported in part by the Mathematical, Information, and Computational Sciences Division subprogram of the Office of Advanced Scientific Computing Research, SciDAC Program, Office of Science, U.S. Department of Energy, under Contract W-31-109-ENG-38.

References

1. Czajkowski, K., D. Ferguson, I. Foster, J. Frey, S. Graham, I. Sedukhin, D. Snelling, S. Tuecke, and W. Vambenepe, *The WS-Resource Framework*. 2004: www.globus.org/wsrf.
2. Andrieux, A., K. Czajkowski, A. Dan, K. Keahey, H. Ludwig, J. Pruyne, J. Rofrano, S. Tuecke, and M. Xu, *Web Services Agreement Specification (WS-Agreement) Draft 20*. 2004: https://forge.gridforum.org/projects/graap-wg/.

3. Figueiredo, R., P. Dinda, and J. Fortes. *A Case for Grid Computing on Virtual Machines*. in *23rd International Conference on Distributed Computing Systems*. 2003.
4. Keahey, K., K. Doering, and I. Foster. *From Sandbox to Playground: Dynamic Virtual Environments in the Grid*. in *5th International Workshop in Grid Computing*. 2004.
5. Barham, P., B. Dragovic, K. Fraser, S. Hand, T. Harris, A. Ho, R. Neugebar, I. Pratt, and A. Warfield. *Xen and the Art of Virtualization*. in *ACM Symposium on Operating Systems Principles (SOSP)*.
6. *VMware: http://www.vmware.com*.
7. Czajkowski, K., I. Foster, V. Sander, C. Kesselman, and S. Tuecke. *SNAP: A Protocol for Negotiating Service Level Agreements and Coordinating Resource Management in Distributed Systems*. in *8th Workshop on Job Scheduling Strategies for Parallel Processing*. 2002. Edinburgh, Scotland.
8. Raman, R., M. Livny, and M. Solomon, *Matchmaking: An Extensible Framework for Distributed Resource Management*. Cluster Computing: The Journal of Networks, Software Tools and Applications, 1999. **2**: p. 129-138.
9. *Workspace Management Service*: http://www.mcs.anl.gov/workspace/.
10. Keahey, K., M. Ripeanu, and K. Doering. *Dynamic Creation and Management of Runtime Environments in the Grid*. in *Workshop on Designing and Building Web Services*. 2003. Chicago, IL.
11. McNab, A., *Grid-Based Access Control for Unix Environments, Filesystems and Web Sites*. Proceeings of the CHEP 2003 conference, 2003.
12. Youssef, S., *Pacman: A Package Manager*. 2004: http://physics.bu.edu/~youssef/pacman/.
13. Goldberg, R., *Survey of Virtual Machine Research*. IEEE Computer, 1974. **7**(6): p. 34-45.
14. Czajkowski, K., I. Foster, N. Karonis, C. Kesselman, S. Martin, W. Smith, and S. Tuecke, *A Resource Management Architecture for Metacomputing Systems*, in *4th Workshop on Job Scheduling Strategies for Parallel Processing*. 1998, Springer-Verlag. p. 62-82.
15. Krsul, I., A. Ganguly, J. Zhang, J. Fortes, and R. Figueiredo. *VMPlants: Providing and Managing Virtual Machine Execution Environments for Grid Computing*. in *SC04*. 2004. Pittsburgh, PA.
16. Allcock, W., *GridFTP: Protocol Extensions to FTP for the Grid*. 2003, Global Grid Forum.
17. Butler, R., D. Engert, I. Foster, C. Kesselman, S. Tuecke, J. Volmer, and V. Welch, *Design and Deployment of a National-Scale Authentication Infrastructure*. IEEE Computer, 2000. **33**(12): p. 60-66.
18. EU DataGrid, *VOMS Architecture v1.1*. 2003.
19. Fraser, K., S. Hand, R. Neugebar, I. Pratt, A. Warfield, and M. Williamson. *Safe Hardware Access with the Xen Virtual Machine Monitor*. in *OASIS ASPLOS 2004 workshop*. 2004.
20. Keahey, K. and K. Doering, *From Sandbox to Playground: Dynamic Virtual Environments in the Grid*. ANL/MCS-P1141-0304, 2003.
21. Rice, P., I. Longde, and A. Bleasby, *EMBOSS: The European Molecular Biology Open Software Suite Trends in Genetics*. 16, 2000. **6**: p. 276-277.
22. Adabala, S., V. Chadha, P. Chawla, R. Figueiredo, J. Fortes, I. Krsul, A. Matsunaga, M. Tsugawa, J. Zhang, M. Zhao, L. Zhu, and X. Zhu, *From Virtualized Resources to Virtual Computing Grids: The In-VIGO System*. Future Generation Computer Systems, 2004.
23. Sundararaj, A. and P. Dinda. *Towards Virtual Networks for Virtual Machine Grid Computing*. in *3rd USENIX Conference on Virtual Machine Technology*. 2004.
24. Jiang, X. and D. Xu, *VIOLIN: Virtual Internetworking on OverLay INfrastructure*. Department of Computer Sciences Technical Report CSD TR 03-027, Purdue University, 2003.
25. Sapuntzakis, C., D. Brumley, R. Chandra, N. Zeldovich, J. Chow, M.S. Lam, and M. Rosenblum. *Virtual Appliances for Deploying and Maintaining Software*. in *Proceedings of the 17th Large Installation Systems Administration Conference (LISA '03)*. 2003.
26. Chase, J., L. Grit, D. Irwin, J. Moore, and S. Sprenkle, *Dynamic Virtual Clusters in a Grid Site Manager*. accepted to the 12th International Symposium on High Performance Distributed Computing (HPDC-12), 2003.

27. Reed, D., I. Pratt, P. Menage, S. Early, and N. Stratford. *Xenoservers: Accountable Execution of Untrusted Programs.* in *7th Workshop on Hot Topics in Operating Systems.* 1999. Rio Rico, AZ: IEEE Computer Society Press.
28. Sapuntzakis, C., R. Chandra, B. Pfaff, J. Chow, M.S. Lam, and M. Rosenblum. *Optimizing the Migration of Virtual Computers.* in *5th Symposium on Operating Systems Design and Implementation.* 2002.
29. Whitaker, A., M. Shaw, and S.D. Gribble. *Denali: Lightweight Virtual Machines for Distributed and Networked Applications.* in *In Proceedings of the USENIX Annual Technical Conference.* 2002. Monterey, CA.

Modeling Machine Availability in Enterprise and Wide-Area Distributed Computing Environments

Daniel Nurmi, John Brevik, and Rich Wolski

Department of Computer Science
University of California, Santa Barbara
Santa Barbara, CA 93106

Abstract. In this paper, we consider the problem of modeling machine availability in enterprise-area and wide-area distributed computing settings. Using availability data gathered from three different environments, we detail the suitability of four potential statistical distributions for each data set: exponential, Pareto, Weibull, and hyperexponential. In each case, we use software we have developed to determine the necessary parameters automatically from each data collection.

To gauge suitability, we present both graphical and statistical evaluations of the accuracy with each distribution fits each data set. For all three data sets, we find that a hyperexponential model fits slightly more accurately than a Weibull, but that both are substantially better choices than either an exponential or Pareto.

These results indicate that either a hyperexponential or Weibull model effectively represents machine availability in enterprise and Internet computing environments.

1   Introduction

As performance-oriented distributed computing (often heralded under the moniker "Computational Grid" computing [13]) becomes more prevalent, the need to characterize accurately resource reliability emerges as a critical problem. Today's successful Grid applications uniformly rely on run-time scheduling [1, 6, 9, 10, 27, 29] to identify and acquire the fastest, least loaded resources at the time an application is executed. While these applications and systems have been able to achieve new performance heights, they all rely on the assumption that resources, once acquired, will not fail during application execution. In many resource environments such an assumption is valid, but in order to employ nationally or globally distributed resource pools (e.g. in the way SETI@Home [30] does) or enterprise-wide desktop resources (as many commercial endeavors do [3, 12, 36]), performance-oriented distributed applications must be able either to avoid or to tolerate resource failures.

Designing the next generation of Grid applications requires an accurate model of resource failure behavior. There has been a great deal of work [14, 18, 20, 21, 25] on the problem of modeling resource failure (or, equivalently, resource availability) statistically. More recently, peer-to-peer systems have used statistical

J.C. Cunha and P.D. Medeiros (Eds.): Euro-Par 2005, LNCS 3648, pp. 432–441, 2005.

distributions as the basis of their availability assumptions [32, 39]. As Plank and Elwasif point out in their landmark paper [28], however, most of these approaches assume that the underlying statistical behavior can be described by some form of exponential distribution or hyperexponential distribution [21]. In addition, they go on to note that, despite the popularity of these models, they often fail to reflect empirical observation of machine availability. In other contexts, such as process lifetime estimation [16] and network performance [22, 26, 37], researchers often advocate the use of "heavy-tailed" distributions, especially the Pareto. Other work has been done showing that a Weibull distribution is an appropriate model for various resource availability data [17, 38], but this work lacks a detailed analysis of model fitting and verification.

Our goal with this work is to develop an automatic method for modeling the availability of enterprise-wide and globally distributed resources. Automatic model determination has several important engineering applications. We plan to incorporate such models into Grid programming systems, such as the Grid Application Development Software [5] system, NetSolve [9], and APST [6], in order to enable effective resource allocation and scheduling. Commercial-enterprise computing systems such as Entropia [12], United Devices [36], and Avaki [3] will also be able to take advantage of automatically determined models as they tune themselves to the characteristics of a particular site. We believe this work will be particularly important to the development of credible and effective Grid and Autonomic Computing [19] simulations. Because Grid architectures are driven by the dynamic resource sharing of competing users, repeatable "en vivo" experiments are difficult or impossible. Several effective emulation [31] and simulation [7, 8, 33] systems have been developed for Grid environments. These systems will benefit immediately from the more accurate models our method produces.

We propose an approach to modeling machine availability based on fitting statistical distributions to observed data, which is outlined in the following manner. In Section 2 we define the statistical distributions used throughout this work and describe our method for estimating the necessary parameters from a given set of availability measurements. We also outline the three data sets used in this study in Section 2. To gauge the effectiveness of our modeling methodology, we detail and analyze the degree to which an automatically generated model fits three diverse sets of empirical observations in Section 3, in which we compare the generated models for all three data sets both visually and through the use of two Goodness of Fit (GOF) tests to complement our visual analysis. In Section 4 we discuss the conclusions we draw from this work and point to future research directions it enables.

2   Fitting a Distribution to Availability Data

In this study, the two distribution families that consistently fit the data we have gathered most accurately are the Weibull and the hyperexponential. The *Weibull distribution* is often used to model the lifetimes of objects, including physical system components [4] and also to model computer resource availability distributions [17, 38]. Hyperexponentials have been used to model machine availability previously [25] especially when observed data requires a model which can approximate a wide variety of shapes. In order to fit most of the statistical distributions used in this paper to observed data, we implemented Matlab [24] scripts which found the MLE (Maximum Likelihood Estimation) parameters. The problem of finding MLE parameters for the hyperexponential, however, tends to be numerically intractable for large data sets, so instead we use EM-pht software [2]. Following are the equations for the models we compare in this work, along with a description of how we estimate the model parameters given a sample data set.

2.1   Statistical Distributions

Throughout this paper, we will use small f for density functions and capital F for distribution functions, subscripted to differentiate among the various types of distribution. These functions, f_W and F_W respectively, for a Weibull distribution are given by

$$f_W(x) = \alpha \beta^{-\alpha} x^{\alpha-1} e^{-(x/\beta)^{\alpha}} \tag{1}$$

$$F_W(x) = 1 - e^{-(x/\beta)^{\alpha}} \tag{2}$$

The parameter α is called the *shape* parameter, and β is called the *scale* parameter[1]. Note that the Weibull distribution reduces to an exponential distribution when $\alpha = 1$.

Hyperexponentials are distributions formed as the weighted sum of exponentials, each having a different parameter. The density function is given by

$$f_H(x) = \sum_{i=1}^{k} [p_i \cdot f_{E_i}(x)] \tag{3}$$

where

$$f_{E_i}(x) = \lambda_i e^{-\lambda_i x} \tag{4}$$

defines the density function for an exponential having parameter λ_i. In the definition of $f_H(x)$, all $\lambda_i \neq \lambda_j$ for $i \neq j$, and $\sum_{i=1}^{k} p_i = 1$. The distribution function is defined as

$$F_H(x) = 1 - \sum_{i=1}^{k} p_i \cdot e^{-\lambda_i x} \tag{5}$$

for the same definition of $f_{e_i}(x)$.

The probability density and distribution functions for the exponential and Pareto distributions, respectively, are as follows:

$$f_E(x) = \lambda e^{-\lambda x} \tag{6}$$

$$F_E(x) = 1 - e^{-\lambda x} \tag{7}$$

---

[1] The general Weibull density function has a third parameter for *location*, which we can eliminate from the density simply by subtracting the minimum lifetime from all measurements. In this paper, we will work with the two-parameter formulation.

$$f_P(x) = \frac{\alpha \beta^{\alpha}}{x^{\alpha+1}} \tag{8}$$

$$F_P(x) = 1 - \left(\frac{\beta}{x}\right)^{\alpha} \tag{9}$$

2.2   Data Sets

In this work, we use three data sets which we believe exhibit availability behavior typical of hosts currently residing on the Internet. The first data set is from UCSB's CSIL computer science student lab. Each measurement records the time from when a workstation is able to run a user process to when it no longer can do so. The second data set is drawn from the Condor [34] pool running at the University of Wisconsin. Each condor availability measurement is the time from when the Condor scheduler starts one of our monitoring processes to when the allocated workstation evicts our monitor process. Finally, we have obtained the dataset from a work by Long, Muir, and Golding [23] in which they remotely measured Internet host availability. We measured 83 machines in the CSIL lab for 8 weeks. The Condor dataset comes from 210 during a 6-week period, and Long, Muir and Golding gathered data from 1170 machines over a 3-month experimental period.

Before attempting to capture the distribution behavior of our data sets, we wanted to explore the data characteristcs of independence and identical distribution. We assume data independence since we intuitively believe, for instance, that one uptime interval on some machine has no effect on the length of the next uptime interval. To inspect the identical distribution characteristics of the data, we performed a Kruskal-Wallis test for identical location between individual machines for each data set. The test strongly *rejected* the null hypothesis that the data is i.d. This is not, however, entirely surprising as the machines we are monitoring have a wide range of usage models which impact their availability. This does imply, however, that although we can use models which fit combined machine availability data, we cannot infer from these models any information about the individual machines that make up the combined data set.

3   Analysis

The goal of our study is to determine the value of using Weibull and hyperexponential distributions to model resource availability. Our method is to compare the MLE-determined Weibull and EMpht-determined hyperexponential to the MLE exponential and Pareto for each of the data sets discussed in the previous section. For reference, we have included the MLE-determined and EMpht-determined model parameters that were used for all fitted distributions discussed and shown in this work (Table 1). As we noted in the introduction, both exponential and the Pareto models have been used extensively to model resource and process lifetime. Thus the value we perceive is the degree to which the Weibull and hyperexponential models more accurately fit each data set.

In each case, we use three different techniques to evaluate model fit: graphical; the Kolmogorov-Smirnov [11] (KS) test; and the Anderson-Darling [11] (AD)

Table 1. Table of fitted model parameters

Data Set	Weibull		Hyperexponential						Exp.	Pareto	
	α	β	p_1	p_2	p_3	λ_1	λ_2	λ_3	λ	α	β
CSIL	.545	275599	.197	.389	.464	$2*10^{-4}$	$8*10^{-6}$	$1*10^{-6}$	$2*10^{6}$	.087	1
Condor	.49	2403	.592	.408	NA	$3*10^{-3}$	$7*10^{-5}$	NA	.00018	.149	1.005
Long	.61	834571	.271	.474	.282	$1*10^{-5}$	$1*10^{-6}$	$3*10^{-7}$	$7*10^{7}$	.079	1

test. Graphical evaluation is often the most compelling method [35] but it does not provide the security of a quantified result. The other two tests come under the general heading of "goodness-of-fit" tests[2].

3.1   Graphical Analysis of the Availability Measurements

To gauge the fit of a specific model distribution to a particular data set, we plot the cumulative distribution function (CDF) for the distribution and the empirical cumulative distribution for the data set. The form of the CDF for the Weibull, hyperexponential, exponential and Pareto are given by equations 2, 5, 7, and 9 respectively (*cf.* Section 2). The empirical distribution function (EDF) is the CDF of the actual data; it is calculated by ordering the observed values as $X_1 < X_2 < \cdots < X_n$ and defining

$$F_e(x) = j/n, X_j \le x < X_{(j+1)} \tag{10}$$

We start by comparing the empirical observations from the CSIL data set (as an EDF) to the CDF determined by the EMpht-estimated hyperexponential, and the MLE-estimated Weibull, exponential, and Pareto distributions (shown in Figures 1, 2, and 3). In all of the figures depicting distributions in this paper, the units associated with the x-axis are seconds of machine availability. We

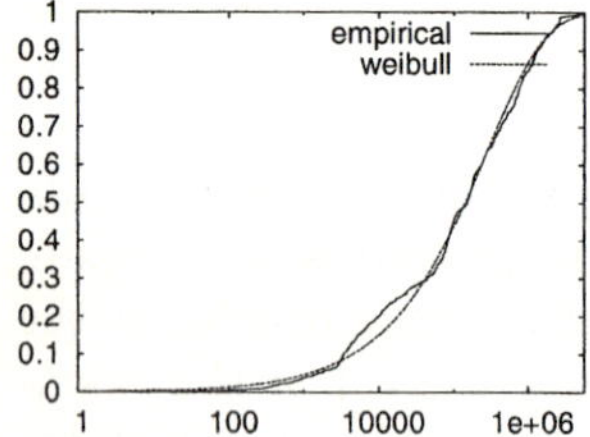

Fig. 1. CSIL data with Weibull fit

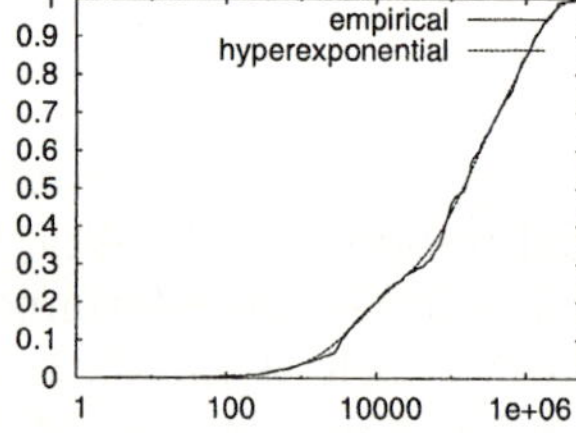

Fig. 2. CSIL data with hyperexponential fit

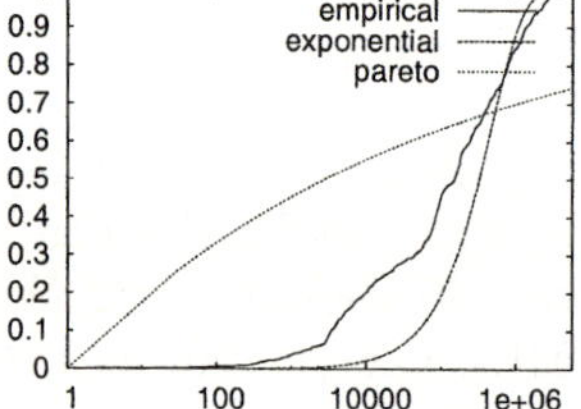

Fig. 3. CSIL data with exponential and Pareto fits

---

[2] The best known goodness-of-fit test is the Chi-squared test. Both the Kolmogorov-Smirnov and the Anderson-Darling tests are considered more appropriate for continuous distributions than the Chi-squared test, which is designed for categorical data and would thus require artifical "binning" of data. We therefore use these methods in place of the more familiar one.

use a log scale for the x-axis to better expose the nature of each fit. Both the hyperexponential and the Weibull fit the data substantially better than either an exponential or Pareto; the hyperexponential is also able to capture the slight inflection around 10,000 seconds. Since automatic selection of the number of phases to use when fitting a hyperexponential is not part of the EMpht software, we have devised our own method. To determine the number of phases, we begin with a 2-phase hyperexponential, test the resulting fit with a Kolmogorov-Smirnov test, and then repeat with an increased number of phases until the KS test result shows no improvement. In this case, for the CSIL data, the algorithm terminated using three phases.

For the Condor data set, the comparison (shown in Figures 4, 5, and 6) is more striking. Again, the hyperexponential (a 2-phase, in this case) appears to fit the shape of the curve most closely, and the Weibull appears a better choice than either exponential or Pareto. Note in particular how again the hyperexponential is able to capture the inflection points of the Condor EDF around 1000 seconds, while the Weibull is unable to do so.

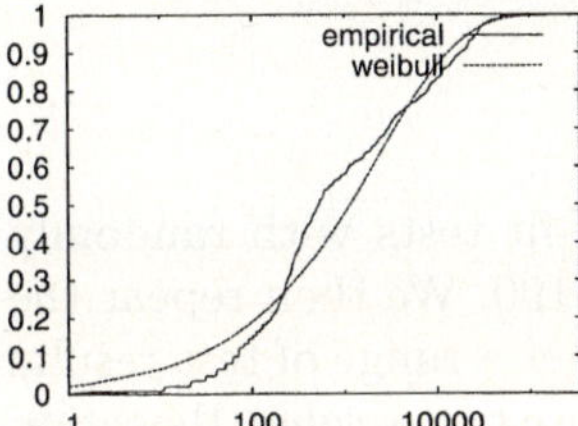

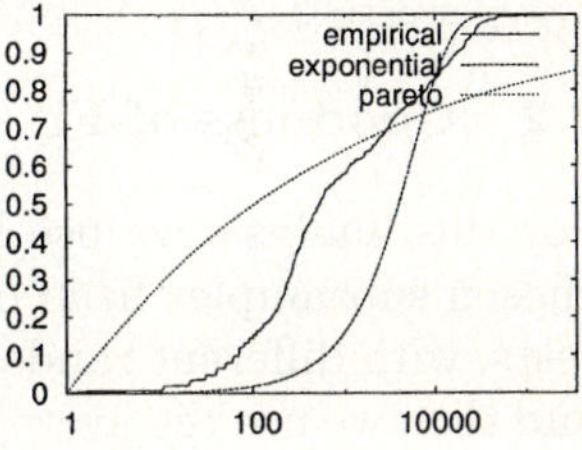

Fig. 4. Condor data with Wiebull fit

Fig. 5. Condor data with hyperexponential fit

Fig. 6. Condor data with exponential and Pareto fits

Finally, the fits (3-phase hyperexponential in this case) for the Long, Muir, and Golding data are shown in Figures 7, 8, and 9.

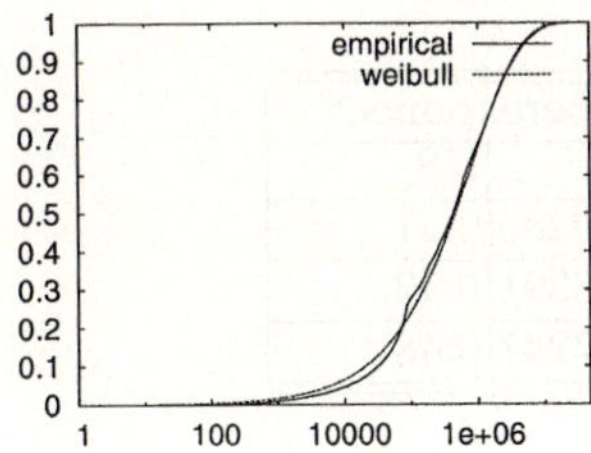

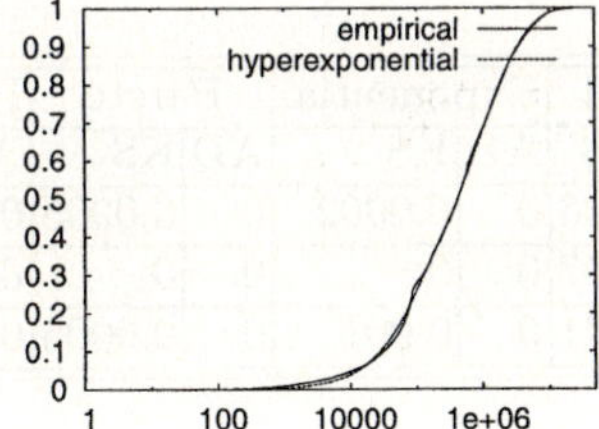

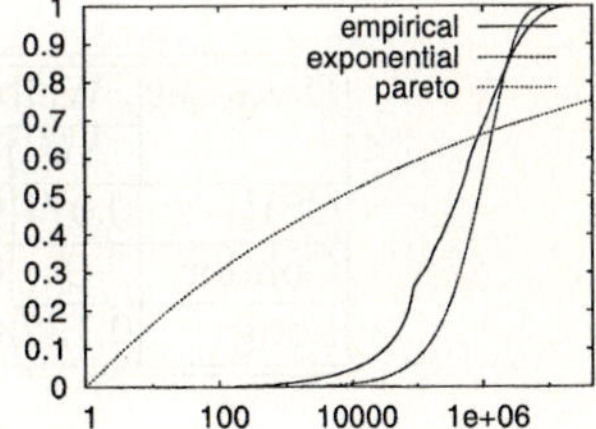

Fig. 7. Long data with Weibull fit

Fig. 8. Long data with hyperexponential fit

Fig. 9. Long data with exponential and Pareto fits

The comparison is similar to that for the CSIL data. The multi-phase hyperexponential fits slightly better than a Weibull, and both are substantially better than an exponential or Pareto.

Of particular interest are the way in which each hypthetical distribution appears to match the tail of an EDF. In many application contexts, "tail behavior" can be important, especially if the presence or absence of rare occurrences must be modeled accurately. For example, previous research [15, 16] reveals Unix process lifetimes to be "heavy-tailed" and well-modeled by a Pareto distribution. Thus schedulers and process management systems must be designed for infrequently occurring processes that have very long execution times.

According to Figures 3, 6, and 9, however, a Pareto distribution would overestimate the probability of very long-lived resources by a considerable amount. Indeed, it may be that while Unix process lifetime distributions are heavy tailed, if they are executed in distributed or global computing environments, many of them will be terminated by resource failure since the resource lifetime distributions (both EDFs and their matching Weibull and hyperexponential fits) have considerably less tail weight.

Even beyond the differences in the tails, however, we can clearly see that the general shape of the exponential and Pareto distributions do not seem to fit the sample CDFs well.

3.2 Goodness-of-Fit Analysis

For this analysis we use both KS and AD goodness-of-fit tests with randomly chosen subsamples from our data sets each having size 100. We then repeat the tests, with different random subsamples, 1000 times to get a range of test results and then we use the average test statistic value to compute the *p-value*. Rejection at size 100 indicates that with as few as 100 data points it is evident that the tested distribution is inappropriate. The addition of more data points to the test will only confirm this inappropriateness further.

Table 2 shows the GOF test results which are the average *p-values* from the 1000 iterations of the experiment.

Table 2. Table of *p-value* results from GOF tests

Data Set	Weibull		Exponential		Pareto		Hyperexponential	
	AD	KS	AD	KS	AD	KS	AD	KS
CSIL	0.071	0.36	0	0.0002	0	0.0005	0.59280	0.47
Condor	0	0.07	0	0	0	0	0.68291	0.42
Long	0.132	0.41	0	0.001	0	0.0005	0.77247	0.48

From the table, it is clear that both the exponential and Pareto perform poorly on these tests for all three data sets. This is not entirely surprising, since the visual fit was clearly inferior for all three data sets. The hyperexponential performs substantially better than all of the other models for all of the data sets; again this is not surprising since the hyperexponential model can increase its number of phases as needed. For the Weibull, we fail to reject the null hypothesis at $\alpha = 0.05$ significance level on average for subsamples of size 100 using the KS test for all three data sets. We fail to reject the null hypothesis at $\alpha = 0.05$ significance using the AD test for the CSIL and Long data sets, but reject for the Condor data set, supporting the graphical evidence that the Condor data set is less-well modeled by a Weibull than the CSIL or Long-Muir-Golding data. Although GOF tests cannot provide a positive result, note that if we were taking a random sample directly from a continous distribution, the GOF test would on average result in a p-value of 0.5. This being the case, we consider an empirical data set p-value result close to 0.5 to be essentially indistinguishable, with respect to the GOF test being used, from data actually drawn from a statistical distribution. p-values of this magnitude were clearly obtained for all three data sets and both tests when performed against a hyperexponential null hypothesis, and to a lesser degree for the CSIL and Long data sets from the KS test against the Weibull null hypothesis.

4   Conclusions

From the results presented in this paper, we feel that there is a compelling case for the superiority of Weibull or hyperexponential distributions in the modeling of resource availability data.

The need to model resource availability and to characterize groups of resources in terms of their availability is critical to desktop Grid, peer-to-peer, and global computing paradigms. Previous related work has used exponential (memoryless) or Pareto distributions, but our work shows that Weibull and hyperexponential distributions are more accurate choices. Visual evidence and GOF results (when applied repeatedly to subsamples) make a compelling case for the use of either a Weibull or hyperexponential distribution to approximate the behavior of resources in our three environments. The choice of which to use depends on the application for which the model is needed, and how complex a model can be handled under application constraints. The hyperexponential, although it generally shows a better fit for the data, is significantly more complex than the Weibull models due to its larger number of estimated parameters, the fact that the phase parameter is free and must be decided iteratively, and its resistance to the MLE methods used for the other distributions presented. The Weibull distribution, with its two MLE-computable parameters and relative mathematical simplicity, seems a better choice if speed and complexity are of interest. Regardless, the methods we use in this paper can be used to automatically decide which model is best at any given moment based on GOF analysis. Both the Weibull and hyperexponential are significantly better at capturing the distribution of availability time than the exponential or Pareto, and both can be computed automatically from availability measurement data.

From these results, we hope to generate individual resource models and to improve the quality of simulation and modeling for volatile distributed systems.

References

1. D. Abramson, J. Giddy, I. Foster, and L. Kotler. High Performance Parametric Modeling with Nimrod/G: Killer Application for Global Grid? In *The 14th International Parallel and Distributed Processing Symposium (IPDPS 2000)*, 2000.

2. S. Asmussen, O. Nerman, and M. Olsson. Fitting phase-type distributions via the em algorithm. *Scandinavian Journal of Statistics*, 23:419–441, 1996.

3. The Avaki Home Page. http://www.avaki.com, January 2001.

4. C. E. Beldica, H. H. Hilton, and R. L. Hinrichsen. Viscoelastic beam damping and piezoelectric control of deformations, probabalistic failures and survival times.

5. F. Berman, A. Chien, K. Cooper, J. Dongarra, I. Foster, L. J. Dennis Gannon, K. Kennedy, C. Kesselman, D. Reed, L. Torczon, , and R. Wolski. The GrADS project: Software support for high-level grid application development. *International Journal of High-performance Computing Applications*, 15(4):327–344, Winter 2001.

6. F. Berman, R. Wolski, H. Casanova, W. Cirne, H. Dail, M. Faerman, S. Figueira, J. Hayes, G. Obertelli, J. Schopf, G. Shao, S. Smallen, N. Spring, A. Su, and D. Zagorodnov. Adaptive computing on the grid using apples. *IEEE Transactions on Parallel and Distributed Systems*, 14(4):369–382, April 2003.

7. R. Buyya and M. Murshed. Gridsim: A toolkit for the modeling and simulation of distributed resource management and scheduling for grid computing. *Comcurrency Practice and Experience*, 14(14-15), Nov-Dec 2002.

8. H. Casanova. Simgrid: A toolkit for the simulation of application scheduling. In *Proceedings of the First IEEE/ACM International Symposium on Cluster Computing and the Grid (CCGrid 2001)*, 2001.

9. H. Casanova and J. Dongarra. NetSolve: A Network Server for Solving Computational Science Problems. *The International Journal of Supercomputer Applications and High Performance Computing*, 1997.

10. W. Chrabakh and R. Wolski. GrADSAT: A Parallel SAT Solver for the Grid. In *Proceedings of IEEE SC03*, November 2003.

11. R. B. D'Agostino and M. A. Stephens. *Goodness-Of-Fit Techniques*. Marcel Dekker Inc., 1986.

12. The Entropia Home Page. http://www.entropia.com.

13. I. Foster and C. Kesselman. *The Grid: Blueprint for a New Computing Infrastructure*. Morgan Kaufmann Publishers, Inc., 1998.

14. A. L. Goel. Software reliability models: Assumptions, limitations, and applicability. In *IEEE Trans. Software Engineering, vol SE-11, pp 1411-1423*, Dec 1985.

15. M. Harchol-Balter and A. Downey. Exploiting process lifetime distributions for dynamic load balancing. In *Proceedings of the 1996 ACM Sigmetrics Conference on Measurement and Modeling of Computer Systems*, 1996.

16. M. Harcol-Balter and A. Downey. Exploiting process lifetime distributions for dynamic load balancing. *ACM Transactions on Computer Systems*, 15(3):253–285, 1997.

17. T. Heath, P. M. Martin, and T. D. Nguyen. The shape of failure.

18. R. K. Iyer and D. J. Rossetti. Effect of system workload on operating system reliabilty: A study on ibm 3081. In *IEEE Trans. Software Engineering, vol SE-11, pp 1438-1448*, Dec 1985.

19. J. Kephart and D. Chess. The vision of autonomic computing. *IEEE Computer*, January 2003.

20. J.-C. Laprie. Dependability evaluation of software systems in operation. In *IEEE Trans. Software Engineering, vol SE-10, pp 701-714*, Nov 1984.

21. I. Lee, D. Tang, R. K. Iyer, and M. C. Hsueh. Measurement-based evaluation of operating system fault tolerance. In *IEEE Trans. on Reliability, Volume 42, Issue 2, pp 238-249*, June 1993.

22. W. Leland and T. Ott. Load-balancing heuristics and process behavior. In *Proceedings of Joint International Conference on Measurement and Modeling of Computer Systems (ACM SIGMETRICS '86)*, pages 54–69, May 1986.

23. D. Long, A. Muir, and R. Golding. A longitudinal survey of internet host reliability. In *14th Symposium on Reliable Distributed Systems*, pages 2–9, September 1995.

24. Matlab by Mathworks. http://www.matlab.com.

25. M. Mutka and M. Livny. Profiling workstations' available capacity for remote execution. In *Proceedings of Performance '87: Computer Performance Modelling, Measurement, and Evaluation, 12th IFIP WG 7.3 International Symposium*, December 1987.

26. V. Paxon and S. Floyd. Why we don't know how to simulate the internet. In *Proceedings of the Winder Communication Conference also http://citeseer.nj.nec.com/paxon97why.html*, December 1997.

27. A. Petitet, S. Blackford, J. Dongarra, B. Ellis, G. Fagg, K. Roche, and S. Vadhiyar. Numerical libraries and the grid. In *Proceedings of IEEE SC'01 Conference on High-performance Computing*, November 2001.

28. J. Plank and W. Elwasif. Experimental assessment of workstation failures and their impact on checkpointing systems. In *28th International Symposium on Fault-Tolerant Computing*, pages 48–57, June 1998.

29. M. Ripeanu, A. Iamnitchi, and I. Foster. Cactus application: Performance predictions in a grid environment. In *proceedings of European Conference on Parallel Computing (EuroPar) 2001*, August 2001.

30. SETI@home. http://setiathome.ssl.berkeley.edu, March 2001.

31. H. Song, J. Liu, D. Jakobsen, R. Bhagwan, X. Zhang, K. Taura, and A. Chien. The MicroGrid: a Scientific Tool for Modeling Computational Grids. In *Proceedings of SuperComputing 2000 (SC'00)*, Nov. 2000.

32. I. Stoica, R. Morris, D. Karger, M. F. Kaashoek, and K. Balakrishnan. Chord: A scalable peer-to-peer lookup service for internet applications. In *In Proc. SIGCOMM (2001)*, 2001.

33. A. Takefusa, O. Tatebe, S. Matsuoka, and Y. Morita. Performance analysis of scheduling and replication algorithms on grid datafarm architecture for high-energy physics applications. In *Proceedings 12th IEEE Symp. on High Performance Distributed Computing*, June 2003.

34. T. Tannenbaum and M. Litzkow. The condor distributed processing system. *Dr. Dobbs Journal*, February 1995.

35. E. Tufte. *The Visual Display of Quantitative Information, 2nd Ed.* Graphics Press, May 2001.

36. The United Devices Home Page. http://www.ud.com/home.htm, January 1999.

37. W. Willinger, M. Taqqu, R. Sherman, and D. Wilson. Self-similarity through high-variability: statistical analysis of ethernet lan traffic at the source level. In *SIGCOMM'95 Conference on Communication Architectures, Protocols, and Applications*, pages 110–113, 1995.

38. J. Xu, Z. Kalbarczyk, and R. K. Iyer. Networked windows nt system field failure data analysis.

39. B. Zhao, L. Huang, J. Stribling, S. Rhea, A. Joseph, and J. Kubiatowicz. A resilient global-scale overlay for service deployment. *(to appear) IEEE Journal on Selected Areas in Communications*.

Faults in Large Distributed Systems
and What We Can Do About Them

George Kola, Tevfik Kosar, and Miron Livny

Computer Sciences Department, University of Wisconsin-Madison
1210 West Dayton Street, Madison WI 53706
{kola,kosart,miron}@cs.wisc.edu

Abstract. Scientists are increasingly using large distributed systems built from
commodity off-the-shelf components to perform scientific computation. Grid computing has expanded the scale of such systems by spanning them across organizations. While such systems are cost-effective, the usage of large number of
commodity components causes high fault and failure rates. Some of these faults
result in silent data corruption leaving users with possibly incorrect results. In
this work, we analyzed the faults and failures that occurred in Condor pools at
UW-Madison having a few thousand CPUs and in two large distributed applications: US-CMS and BMRB BLAST, each of which used hundreds of thousands
of CPU hours. We propose 'silent-fail-stutter' fault-model to correctly model the
silent failures and detail how to handle them. Based on the model, we have designed mechanisms that automatically detect and handle silent failures and ensure
that users get correct results. Our mechanisms perform automated fault location
and can transparently adapt applications to avoid faulty machines. We also designed a data provenance mechanism that tracks the origin of the results, enabling
scientists to selectively purge results from faulty components.

1   Introduction

Scientists are increasingly using distributed systems built from commodity components
for their computing needs. Grid computing [1] has increased the scale by sharing these
computing resources across organizations. While this approach is cost-effective, hardware errors may create havoc if the system software and applications do not handle them
appropriately. For instance, most of the several thousand UW-Madison Condor pool
compute nodes have non-parity memory. A stray alpha particle may corrupt the memory leaving the scientists with incorrect results. Further, just failure of certain memory
chips may corrupt parts of the computation and this failure may go unnoticed for a long
period.

Hard-drives, RAID-controllers [2] and processors may also exhibit such faulty behavior. While detecting such faulty components is itself difficult, detecting all the corrupted results and recomputing them is even more difficult. This leaves the users with
results that may be incorrect. This is particularly troublesome for large-scale computation performed on these distributed systems. While system-administrators may detect
these faults at some point and replace the faulty components, purging the results that
touched these components is non-trivial.

J.C. Cunha and P.D. Medeiros (Eds.): Euro-Par 2005, LNCS 3648, pp. 442–453, 2005.

We analyzed the faults in large distributed systems by looking at the faults and failures that occurred in the Condor pools at UW-Madison campus, and in two large distributed applications: US-CMS and BMRB BLAST, each of which processed terabytes of data and used hundreds of thousands of CPU hours.

In this work, we present a summary of our experience and propose 'silent-fail-stutter' fault model that correctly models components that exhibit silent failures. We highlight the implications of this model and detail what a system incorporating such components should do.

Using the insights from the model, we have designed mechanisms that can automatically detect and handle silent failures taking into account user specified policies. Our mechanisms also provide fault location and can dynamically adapt applications to avoid faulty machines. They keep track of the origin of the result, including the components that interacted with the source component to generate this result. This enables users at any point to selectively purge results that interacted with faulty components. Finally, we evaluate our new model and the mechanisms on a real-life distributed workload and highlight its effectiveness.

2 Faults in Distributed Systems

In this section, we give a widely accepted definition of faults, and present different types of faults experienced by real distributed applications.

2.1 Definition of Faults

The widely accepted definition, given by Avizienis and Laprie [3] is as follows. A fault is a violation of a system's underlying assumptions. An error is an internal data state that reflects a fault. A failure is an externally visible deviation from specifications. A fault need not result in an error, nor an error in a failure. An alpha particle corrupting a memory location is a fault. If that memory location contains data, that corrupted data is an error. If a program crashes because of using that data, it is a failure.

2.2 Experienced Faults in Distributed Systems

We analyzed the faults in large distributed systems by looking at the faults and failures that occurred in two large distributed applications: US-CMS and BMRB BLAST, each of which was processing terabytes of data and using hundreds of thousands of CPU hours. We also analyzed several other small applications running in the Condor pool at UW-Madison campus having a couple of thousand compute nodes. The most common failures we have observed are:

Data Corruption. Faulty hardware in data storage, staging and compute nodes corrupted several data bits occasionally. The faults causing this problem included a bug in the raid controller firmware on the storage server, a defective PCI riser card, and a memory corruption. The main problem here was that the problem developed over a course of time, so initial hardware testing was not effecting in finding the problems. The raid controller firmware bug corrupted data only after a certain amount of data was stored on the storage server and hence was not detected immediately after installation. In almost all of these cases, the data corruption happened silently without any indication from hardware/operating system that something was wrong. Tracking down the faulty component took weeks of system administrator's time on average.

Hanging Processes. Some of the processes hang indefinitely and never return. From the submitters point of view there was no easy way of determining whether the process was making any progress or was hung for good. The most common cause of hanging data transfers was the loss of acknowledgment during third party file transfers. In BMRB BLAST, a small fraction of the processing hung and after spending a large amount of time, the operator tracked it down to an unknown problem involving the NFS server where an NFS operation would hang.

Misleading Return Values. An application returning erroneous return values is a very troublesome bug that we encountered. We found that even though an operation failed, the application returned success. This happened during some wide area transfers using a widely used data transfer protocol. We found that if the destination disk ran out of space during a transfer, a bug in the data transfer protocol caused the file transfer server to return success even though the transfer failed. This in turn resulted in failure of computational tasks dependent on these files.

Misbehaving Machines. Due to misconfigured hardware or buggy software, some machines occasionally behaved unexpectedly and acted as 'black holes'. We observed some computational nodes accepting jobs but never completing them and some completing the job but not returning the completion status. Some nodes successfully processed certain job classes but experienced failures with other classes. As a particular case, in WCER video processing pipeline [4], we found that a machine that had a corrupted FPU was failing MPEG-4 encoding whereas it was successfully completed MPEG-1 and MPEG-2 encodings.

Hardware/Software/Network Outages. Intermittent wide area network outages, outages caused by server/client machine crashes and downtime for hardware/software upgrades and bug fixes caused failure of the jobs that happened to use that feature during that time.

Over Commitment of Resources. We encountered cases where the storage server crashed because of too many concurrent write transfers. We also encountered data transfer time-outs that that were caused by storage server trashing due to too many concurrent read data transfers.

Insufficient Disk Space. Running out of disk space during data stage-in and writing output data to disk caused temporary failures of all involved computational jobs.

3   Why Do Current Fault Models Not Work Well?

Byzantine [5] and fail-stop [6] are two widely used fault models. In Byzantine model, a component can exhibit arbitrary and malicious behavior, perhaps involving collusion with other faulty components. In fail-stop model, in response to a failure, the component changes to a state that permits other components to detect a failure has occurred and then stops.

Byzantine model is too general in nature and reasoning out the different scenarios is difficult. Normal systems may not encounter such malicious behavior. However, it is useful in security where such adversarial behavior may occur. Fail-stop model is at other end of the spectrum and it is extremely simple and tractable. For this reason, most systems are built using the fail-stop model.

Unfortunately, the fail-stop model is too simple to model the behavior of many components. Arpaci-Dusseau [7] found that similar components differ widely in performance, enough to be called a performance failure. To model the behavior of such components, they introduced the fail-stutter model where a component may operate at reduced performance level in addition to failing and stopping. Fail stutter behavior is commonly seen in disks where the controller may transparently remap bad sectors and such a disk may have lower performance compared to a bad-sector free disks because of extra seeks. Fail-stutter separates failures into correctness and performance failures and correctly models components that have performance failures but are correct. Fail-stutter expects the components to behave like fail-stop when a correctness failure occurs.

Even fail-stutter does not correctly model behavior of many components. Many components on encountering a correctness failure do not immediately change to a state that allows other components to detect the failure. For instance, if a memory chip is corrupted, it does not give that information to other components. A memory tester program can detect the corruption by writing data to it and reading the written data and verifying it. Parity and error correcting does not fully address this issue. For instance, a chip with a fault that prevents data from being written to it cannot be detected by parity because the parity bits are correct for the incorrect old data.

In a distributed system, significant fractions of the applications involve a pipeline of processing [8]. If a particular component in the pipeline generates incorrect data because of a fault, other components cannot detect immediately that this component has failed.

4 Silent-Fail-Stutter: A More Accurate Fault Model

Fail-stop expects components that encounter a failure to immediately change state and let other components detect that failure. As shown in the previous section, many components do not indicate a failure immediately. To address this, we propose a new fault-model called 'silent-fail-stutter'. In this model, on failure a component may not immediately change state and convey its failure to other components. However, other components can at anytime detect if a component has failed by testing it, incurring a certain cost. Further, the stutter behavior may be an indicator of impending failure and other components can use that as a heuristic to test that component for failure. Figure 1 shows the different fault models and how they relate to each other.

We believe that silent-fail-stutter models real life behavior of components better than fail-stutter while maintaining tractability. Below, we give a few examples of real-life cases where silent-fail-stutter is more appropriate than fail-stutter.

System Memory. Most of today's system memory chips are non-parity. In non-parity memory, memory corruption cannot be detected immediately and hence fail-stop and fail-stutter are not suitable models. Silent-fail-stutter is a suitable model because a mem-

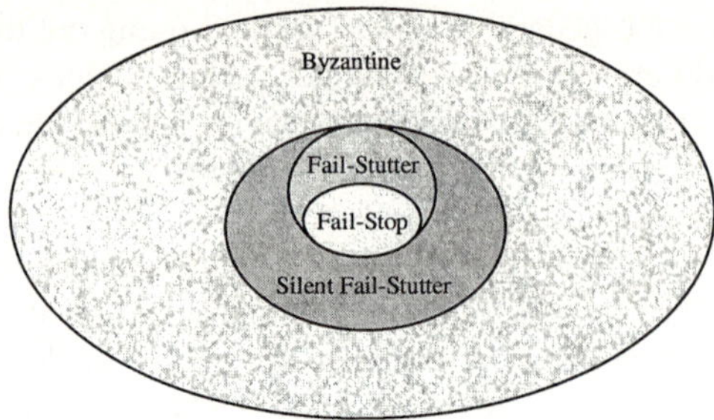

Fig. 1. Different fault models and their relation

ory tester program can detect a corrupt memory chip and doing this test incurs a cost. With parity and error-correcting memory, a chip that is faulty with respect to memory writes exhibits silent-fail-stutter. Even though most but not all read errors may be detected by parity, a corrupt chip is detected only when some earlier written data is accessed, making fail-stop inappropriate. A fail-stop component would have detected failure earlier, say during DRAM refresh cycle, and informed other components. Thus for all types of system memory, silent-fail-stutter is a suitable model.

Processor Cache. Processor caches are SRAMs and typically starting from Level 2 have error correcting code. Register [9] reports cases where Solaris operating system crashed and rebooted because the UltraSparc II had E-cache(level2 cache) parity errors. Sun's best practices guide [10] advised system administrators to log such failures and replace the processor on second such failure.

The behavior is not fail-stop, because a processor with the faulty cache does not retain information about a cache block failure across reboots even if the failure is permanent. In addition, a cache block failure is detected only when some data is written to it and fails parity test when read back.

Silent-fail-stutter is a more accurate model because even though a cache block may fail at any time, it may not be detected immediately. A program can test if any cache block is faulty by writing data to all cache blocks and reading it to verify that there is no corruption. Doing this involves a cost that silent-fail-stutter models making it a good fit.

Faults in Distributed Systems. If a system is built from silent-fail-stutter components, then silent-fail-stutter is the correct model for that system. The faults we mentioned in the section 2, were all silent-fail-stutter. The problem was that applications expected fail-stop behavior whereas the components exhibited silent-fail-stutter.

5   Implications of Silent-Fail-Stutter

Since components may fail and not convey the failure to other components, some component should periodically or on certain events determine the state of each component and if a failure is detected, report it to other components. If a single component makes the checks, designers should ensure that this component is more reliable than the ones

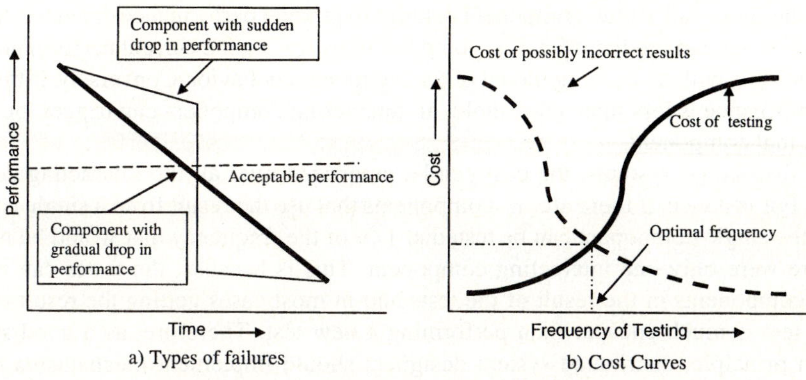

Fig. 2. a) Two types of components b) Cost Curves

it checks. All the components can co-operate to perform this check in a distributed manner.

These ways of handling silent-fail-stutter have been known informally for a while. For instance, processors during startup have the ability to test the memory for errors and in case of error report that to the user and stop. Here the processor is assumed to be more reliable than the memory and the whole system behaves like fail-stop even though the underlying component, memory, is silent-fail-stutter. However, with always-on systems, memory test during startup may not be sufficient.

Figure 2a shows how the performance changes with respect to time for two types of components. For one type, the performance drops gradually and for the other, the performance drop is minimal at first and then a sudden drastic drop. The figure also shows an acceptable performance limit. Below the acceptable performance, correctness failures may occur. Many components exhibit such gradual performance drop making it easy to predict their failure ahead of time.

Fault masking at times results in the drastic fall in performance. For example, consider a failing disk where the sectors are failing because the magnetic media is loosing its ability to retain stored data. If the hard disk employs sector remapping, it may be able to hide the bad sectors for a while by which time the media may have degraded to a point where suddenly most sectors fail causing a drastic drop in performance. A good solution to this is to expose this information about faults, so that a smarter higher-level system can use this information to take some action.

Since, components may fail silently, how often to test them is of importance. Doing the test occasionally runs the risk of not detecting the failure for a longer duration. Running the test often may result in considerable overhead.

Figure 2b shows the cost curve for component testing and using possibly incorrect results for a hypothetical system. The cost curves would depend on failure probability of the component, cost of testing and the cost of incorrect results. Here, we use the term cost loosely. In practice, we can normalize the cost to a meaningful common form like time, computation that can be performed in that time, etc. For mission critical systems, the cost of possibly incorrect result would be infinity. Similarly, for some computation that generates hints for heuristics, the cost of possibly incorrect result would be low.

If the silent-fail-stutter component belongs to gradual performance decline category, then the tester can predict when it is going to fail by testing the current performance and co-relating it with an existing model of the component behavior. Conversely, if the drop in performance is less than a threshold, an interacting component can trigger the tester to test that component.

In distributed systems, the cost can be amortized over all the interacting components. For instance, if there are 'n' components that use the result from a single component, the single component can be tested at $1/n$ of the frequency that would be needed if there were only one interacting component. This is because, the tester can inform other components of the result of the tests and in most cases getting the result of previous test is much cheaper than performing a new test. Therefore, as a good system design principle, distributed system designers should implement mechanisms to test silent-fail-stutter components and report them to interested components.

6   Failure Detection in Distributed System

Complexity of distributed systems makes failure detection difficult. There are multiple layers from the hardware to the application. Since we did not want to impose an undue burden on application developers to handle failure detection and handling, we implemented the error/failure detection on top of the application, by verifying that results generated are correct. To do this, the applications should allow multiple executions and they should produce reproducible results.

Grid applications are expected to have the ability to be run multiple times because they could be pre-empted from a resource. Further, to enable checking of outputs, they need to generate reproducible results. Most applications already do that. We need to clarify that applications produce both an output and a log of the processing. The log of the processing may include start time, information about execute machine, etc and would not be reproducible across multiple executions. However, the output, say a processed image would be the same across executions.

In addition to detecting erroneous results, we also need to detect the cause of the fault and possibly replace that faulty component. Identifying the source of the erroneous result has so far been a *'black art'* in the realm of select few system administrators and operators. This process takes considerable amount of time, usually weeks, expending considerable amount of human resources.

We classify silent failures into two types as shown in figure 3. Type I silent failures are silent failures that give incorrect results without any error status indication. Type II silent failures are silent failures in which the process or transfer just hangs. Type I gives a successful return code and shows that the process is completed but the results are incorrect. This normally happens because of interface mismatch where a component expects underlying components to be fail-stop, but they are in fact silent-fail-stutter. Type II never returns, so user cannot find out if the process will complete. This could be caused by bugs. In addition to silent failure, jobs may fail with an error status and they are easier to detect. We will first discuss about handling Type I.

We want to detect if a failure has occurred and if we need to track down the cause of that failure. A silent failure of lower level component may result in a failure higher

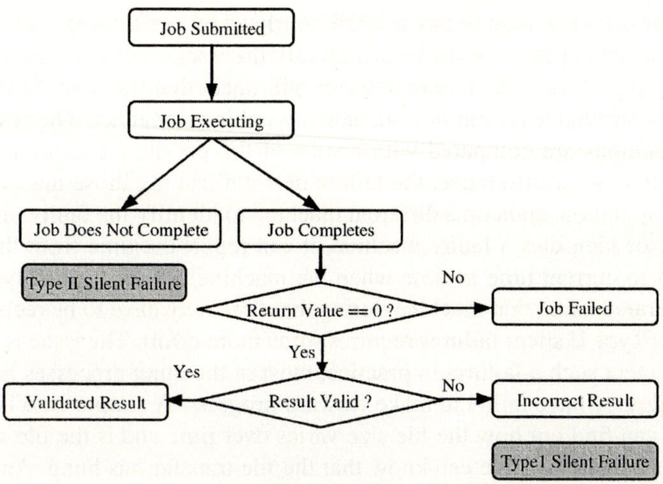

Fig. 3. Type I and Type II silent failures

up the chain and to track down the fault, we may need to go down the hierarchy. For instance, the cause of a computation failure may be because of data corruption in the intermediate storage server and this in turn may be caused by a faulty RAID controller in the storage server. We feel that automatically isolating the fault to whole system boundary is easier and this would aid the system administrator in locating the exact problem.

Consider a simple case where the user has to be 100% certain that the result is correct. A simple way of doing that is to compute the result twice and verify that they match. While doing this we need to be careful to ensure that the two computations do not overwrite the same data. Name space mapping can address this. Suppose if we find that a result is incorrect, we can pick up all the incorrect results in a given time period and all systems interacted with most of the results is the likely culprit. A simple mechanism that detects this can notify it to the system administrator who can then test that system. At the same time, the component can give feedback to higher-level planners like Pegasus [11] and/or distributed schedulers to ensure that they do not use this resource until the fault has been resolved. Verification of data transfers involves checksum generation and verifying that source and destination checksums match.

Components belonging to silent-fail-stutter allow testing to determine a failure. The methodology for testing can be inferred from "THE" multiprogramming system [12], where they had a layered structure to test that reduced the number of test cases. We believed that a conscientious distributed system designer should design such a test infrastructure. If such a test infrastructure exists, the mechanism on detecting a failure can trigger a test of the whole system to isolate the faulty component. As an alternative, to isolate machine faults at a coarse grain, a tester can periodically execute a test program that generates a known result and takes a certain deterministic amount of time on each machine. If any machine gives a wrong result or the run time deviates considerably, the system administrator can be informed of the problem.

If the user does not want to pay a 100% overhead by performing each computation twice and if testing system exists, he can specify the fraction of extra computation that he is willing to perform. The failure detector will inject that fraction of extra computation into the distributed system in a statistically unbiased manner. The results of these extra computations are compared with results of the previous execution and verified to be same. In case of difference, the failure detector can tag those machines and perform the computation again on a different machine to identify the faulty one. When the failure detector identifies a faulty machine, it can report the time from the successful machine test to current time as time when the machine was in a possibly faulty state. Results generated using that machine during that time may have to be recomputed.

Handling Type II silent failures requires some more effort. The issue is whether it is possible to detect such a failure. In practice, most of the hung processes have a way of detecting that they have failed to make forward progress. A simple case is that of data transfer, we can find out how the file size varies over time and if the file size does not change for a long period, we can know that the file transfer has hung. Another way is to come up with reasonable time-outs for operations. We can find out that a transfer or computation has hung if it does not complete in a certain period.

Most of the present day distributed workloads consist of a large number of instances of the same application. Typically, the standard deviation of execution time is of the same order of magnitude as mean if not lesser. This lends a very effective way to detecting Type II failure. Using this, mechanisms can set the threshold to be $mean + 3 \times StandardDeviation$ or some similar threshold. Users can specify policy on what fraction of the processing they are willing to re-do. If users want responsiveness, they may trade some extra processing and set a lower threshold. If they want to minimize the overhead, they would use a higher threshold.

7   Evaluation

To evaluate the effectiveness of silent-fail-stutter model and our discussion on failure detection, we implemented a prototype of the mechanism mentioned in the previous section.

We looked at how components should convey the results of test and we decided to use a database to log the results of tests and timestamp of tests. We also log the results of application execution into the database. To get this information, we parse the distributed batch scheduling system (Condor) user-job log-files and store them in a relational database. We developed the schema for doing it from our previous work [13]. Since we wanted to track down the origin of the results, we store the job description also in the database. Figure 4 shows the process.

We found that users typically submit job bundles specified as a directed acyclic graph(DAG). We categorize a job bundle as an application-class. Users normally tag application class and we can use that as well if the same application class spans across multiple job bundles.

For verification of results, we store the md5 checksum of the results in the database. To validate, we can run a simple query that checks if the checksums of results of identical jobs are the same. In case of error, we have a query that can extract out all the

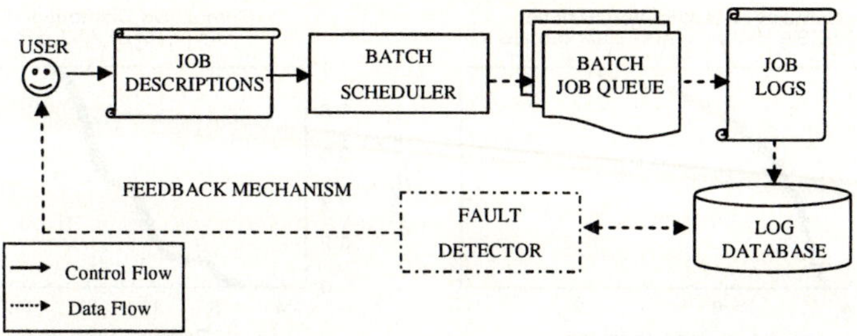

Fig. 4. Stages in performing some processing on a distributed system

Table 1. Coefficient of Variation of Execution Time

Application	Coefficient of Variation
BLAST BMRB (1MB Database)	0.19
BLAST PDB (69MB Database)	0.34
BLAST NR (2GB Database)	0.49
NCSA Sextractor Processing	2.00
NCSA Data Transfer	1.00

machines that generated suspect results and tag the machine appearing multiple times as faulty.

To evaluate our ability to identify silent Type II failures, we looked at the co-efficient of variation of executing time of some applications. Table 1 shows the co-efficient of variation of a few well-known applications. We found that the coefficient of variation of all classes we encountered were less than four.

The UW Madison condor pool consists of multiple clusters with different processor speeds, and user desktops. We did not separate the performance according to machine class as the job may be assigned to any machine depending on availability unless the jobs explicitly request certain configuration. Taking into account machine class, brought down the coefficient of variation considerably but we do not report that, as we may not be able to do so well in a general environment.

Figure 5 shows the cumulative distribution function of BLAST processing using 2 GB NR database and wide-area data transfers between NCSA and UW. Each wide-area data transfer transferred a 1.1 GB astronomy image file from NCSA to UW-Madison and that file was subsequently processed in UW condor pool.

For the blast run, a few jobs hung and took a very long time to complete, around 4 hours. The user had a hard limit of 4 hours and the jobs that exceeded 4 hours were killed and restarted. Using our mechanism, we find that we can come up with tighter bounds. In this case, $Mean + 3 \times StandardDeviation = 80$ minutes and does not require operators to magically come up with thresholds and are better than the rough guess of the user. The data transfers had a 20-minute time-out for the data transfers. There were hung transfers that succeeded second time around.

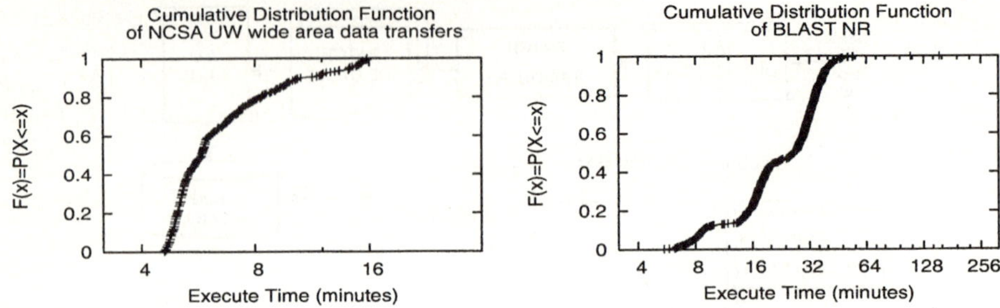

Fig. 5. Shows the cumulative distribution function of BLAST execution against the 2 GB NR database and NCSA UW wide-area data transfers

8   Future Work

We intend to develop a more rigorous theoretical analysis of our silent-fail-stutter model. We also want to deploy our mechanisms in real systems over a long period and evaluate them. The mechanism assumes that the failed fraction is significantly less than the successful fraction, which we believe would be true in practice. We would like to determine if there are limits on failure fraction that will cause the mechanisms to not work.

9   Conclusions

We have successfully analyzed the faults in large distributed systems and proposed *silent-fail-stutter* fault model to accurately model component behavior while maintaining tractability. Using insights from the model, we have developed mechanisms to automatically detect silent failures in distributed systems. We have evaluated the mechanisms and shown their effectiveness.

References

1. Foster, I., Kesselman, C., Tuecke, S.: The anatomy of the grid: Enabling scalable virtual organizations. International Journal of Supercomputing Applications (2001)
2. Patterson, D.A., Gibson, G.A., Katz, R.H.: A case for redundant arrays of inexpensive disks (raid). In Boral, H., Larson, P.Å., eds.: Proceedings of the 1988 ACM SIGMOD International Conference on Management of Data, Chicago, Illinois, June 1-3, 1988, ACM Press (1988) 109–116
3. Avizienis, A., Laprie, J.: Dependable computing: From concepts to design diversity. In: Proceeding of the IEEE. Volume 74. (1986) 629–638
4. Kola, G., Kosar, T., Livny, M.: A fully automated fault-tolerant system for distributed video processing and off-site replication. In: Proceeding of the 14th ACM International Workshop on Network and Operating Systems Support for Digital Audio and Video (Nossdav 2004), Kinsale, Ireland (2004)
5. Lamport, Shostak, Pease: The byzantine generals problem. In: Advances in Ultra-Dependable Distributed Systems, N. Suri, C. J. Walter, and M. M. Hugue (Eds.), IEEE Computer Society Press. (1995)

6. Schneider, F.B.: Implementing fault-tolerant services using the state machine approach: A tutorial. ACM Computing Surveys **22** (1990) 299–319
7. Arpaci-Dusseau, R.H., Arpaci-Dusseau, A.C.: Fail-Stutter Fault Tolerance. In: The Eighth Workshop on Hot Topics in Operating Systems (HotOS VIII), Schloss Elmau, Germany (2001) 33–38
8. Thain, D., Bent, J., Arpaci-Dusseau, A., Arpaci-Dusseau, R., Livny, M.: Pipeline and batch sharing in grid workloads. In: Proceedings of the Twelfth IEEE Symposium on High Performance Distributed Computing, Seattle, WA (2003)
9. The Register: Sun suffers UltraSparc ii cache crash headache. http://www.theregister.co.uk/2001/03/07/sun_suffers_ultra sparc_ii_cache/ (2001)
10. Sun Microsystems Inc: Best practices guide: Addressing e-cache parity errors. http://www.filibeto.org/sun/lib/hardware/enterprise_4500/ BP_Ecache_10-16-01.pdf (2001)
11. Deelman, E., Blythe, J., Gil, Y., Kesselman, C.: Pegasus: Planning for execution in grids. Technical Report 20, GriPhyN (2002)
12. Dijkstra, E.W.: The structure of the THE-multiprogramming system. Communications of the ACM **11** (1967)
13. Kola, G., Kosar, T., Livny, M.: A client-centric grid knowledgebase. In: Proceedings of Cluster 2004, San Diego, CA (2004)

A Grid Information Service
Based on Peer-to-Peer[*]

Diego Puppin, Stefano Moncelli, Ranieri Baraglia,
Nicola Tonellotto, and Fabrizio Silvestri

Institute for Information Science and Technologies
ISTI – CNR, Pisa, Italy
via Moruzzi, 56100 Pisa, Italy
`Diego.Puppin@isti.cnr.it`, `stefano7625@libero.it`,
`{Ranieri.Baraglia,Nicola.Tonellotto,Fabrizio.Silvestri}@isti.cnr.it`

Abstract. Information Services are fundamental blocks of the Grid infrastructure. They are responsible for collecting and distributing information about resource availability and status to users: the quality of these data may have a strong impact on scheduling algorithms and overall performance.

Many popular information services have a centralized structure. This clearly introduces problems related to information updating and fault tolerance. Also, in very large configurations, scalability may be an issue. In this work, we present a Grid Information Service based on the peer-to-peer technology. Our system offers a fast propagation of information and has high scalability and reliability. We implemented our system complying to the OGSA standard using the Globus Toolkit 3. Our system can run on Linux and Windows systems, with different network configurations, so to trade off between redundancy (reliability) and cost.

Keywords: Grid information service, Grid middleware, Peer-to-peer.

1 Introduction

The Grid is an emerging computing framework where resources are shared and inter-operate across the boundaries of independent organizations. In such an environment, it is very important to be able to discover efficiently which resources are available, what their status and cost are. A system where this information is outdated, approximate or difficult to access and browse may negatively affect the performance of scheduling algorithms and of final-user code.

The Grid Information Service (GIS) is the infrastructure component responsible for collecting and distributing information about the Grid. It offers some

* We thank Antonio Manglaviti, who contributed to develop the first experimental prototype. Also, we thank IIT-CNR, IMATI-CNR, University of Pisa, and University of California at San Diego, which let us use their resources for our experiments. This work has been partially supported by the MIUR GRID.it project (RBNE01KNFP) and the MIUR CNR Strategic Project L 499/97-2000.

J.C. Cunha and P.D. Medeiros (Eds.): Euro-Par 2005, LNCS 3648, pp. 454–464, 2005.
© Springer-Verlag Berlin Heidelberg 2005

tools to register resources, to query the data base, to remove lost nodes. The first implementations of a GIS used techniques based on *directories*, which are still used by Globus MDS-GT2 (LDAP). Directory-based systems suffer from a series of problems [1], including the fact that updated information does not propagate very quickly and that centralized servers may become bottle-necks or points of failure.

In this work, we introduce a Grid Information Service (GIS) based on peer-to-peer (P2P) technologies and Routing Indices (RI) [2]. There is a growing interest to the interaction of the Grid computing paradigm and the peer-to-peer technology: both work within a very dynamic and heterogeneous environment, where the role and availability of resources may quickly change; both create a virtual working environment by collecting the resources available from a series of distributed, individual entities. Even if nowadays some Grid-related tasks are performed by central servers, many authors believe that in the future many of them could be implemented as P2P services, to improve scalability, performance and fault-tolerance.

This paper, which updates the results presented in [3], is structured as follows. In Section 2, we give an overview of some existing information services, which represent the background of our work. Our infrastructure is presented in Section 3. In Section 4, we show the results of our preliminary tests. Finally, we conclude and we give an overview of future work.

2   Related Work

The importance of Information Services within the Grid infrastructure has stimulated a rich research. Due to limited space, we can cite only the works that are closer to ours. Our starting point is clearly the Information Service model of the *Globus Toolkit 3* [4]. In Globus, each entity is represented by a Grid Service, which is an extended Web Service following the new conventions introduced with OGSA. These Grid Services expose their status as a collection of Service Data (SD), composed of Service Data Elements (SDEs). Service Data replace the mechanisms offered by GRIS in MDS-GT2: they replace the GIS-enabled mechanisms present in LDAP with the OGSA mechanisms for binding. The Index Service (IS) is composed of two main parts: the *Providers* are responsible for generating SDEs; the *Aggregator* is responsible for aggregating and indexing the SDEs coming from the hosts in the VO. Typically, there is one Index Service per Virtual Organization, which is used to build a hierarchy when several institutions are connected. Every Index Service works as a cache for all the ISs below it.

In our opinion, this hierarchical structure is cause of two main limitations: (1) when a new SDE becomes available, the new information does not propagate automatically up in the hierarchy; (2) at the top levels, each IS is required to store a very large number of SDEs.

Talia et al. [5] propose a P2P-based architecture for resource discovery that extends the GT3 information service model. It is broken into two layers: the lower one is a hierarchy of information services, which publish information owned by each virtual organization; the upper one is a P2P layer, which collects and distributes this information. Queries about non-local resource are managed by the P2P nodes. The protocol used to exchange messages extends the Gnutella protocol. It uses extensive caching and merging of queries and Grid Service invocations instead of raw TCP messages. Our work differs in that we use a more advanced query forwarding strategy based on Routing Indices, and in the fact that our system never returns cached, potentially out-dated, information.

Carmen [6], developed at the HP Labs, has a structure similar to our proposal. It offers a discovery service based on P2P networks, structured in peers and super-peers. Unfortunately, we could not find in literature any results about its effectiveness. The system is apparently very complex, and no comparisons are given with other system. At the moment, we cannot know if there is an advantage, in terms of performance, with respect to centralized systems.

Another approach to data distribution on P2P environments is based on *Distributed Hash Tables (DHTs)*. This is particularly efficient for some types of resources, e.g. data files that are searched by exact name[1]. Several systems implement this solution including Tapestry [7] and Pastry [8]. They deal very well with scalability issues, but they often limit the query language to exact matches. We are verifying if our query language can be mapped to DHT or some extension thereof. Approaches based on space-filling curves, such as Squid [9], seem to offer an initial answer to this problem.

3   P2P GIS: Description of the Architecture

In this section, we present our implementation of a Grid Information Service (GIS) based on the peer-to-peer technology. Its main features are:

- peer-to-peer technologies for propagating data and elaborating queries;
- routing indices to reduce network flooding and to optimize message forwarding;
- node clustering and the use of super-peers;
- redundant configurations, when high reliability is needed.

The system is made up of two main entities (see Figure 1):

- the *Agent* is responsible for publishing information about a node to the super-peer;
- the *Aggregator* runs on the super-peer; it collects data, replies to queries and forwards them to the other super-peers; it also keeps an index about the information stored in each neighbor super-peer.

Super-peer and redundant networks are described in the next section. Then, we outline the structure of Agents and Aggregators. Routing indices and our search technique are discussed in Sections 3.3 and 3.4.

---

[1] To be more precise, DHT offers an effective way to find the hash keys obtained by manipulating the query string.

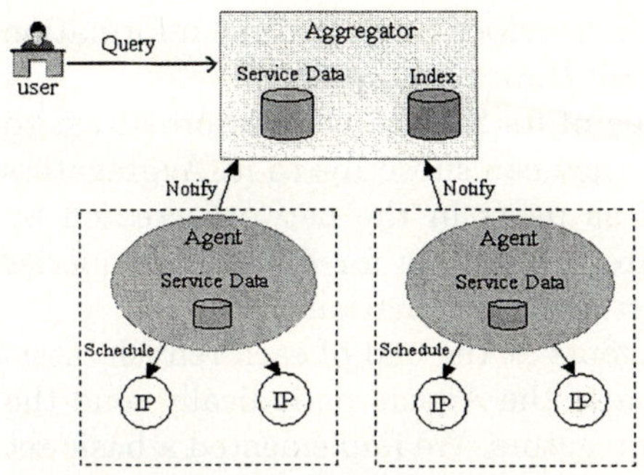

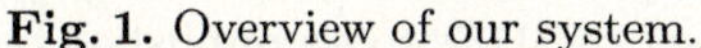

Fig. 1. Overview of our system.

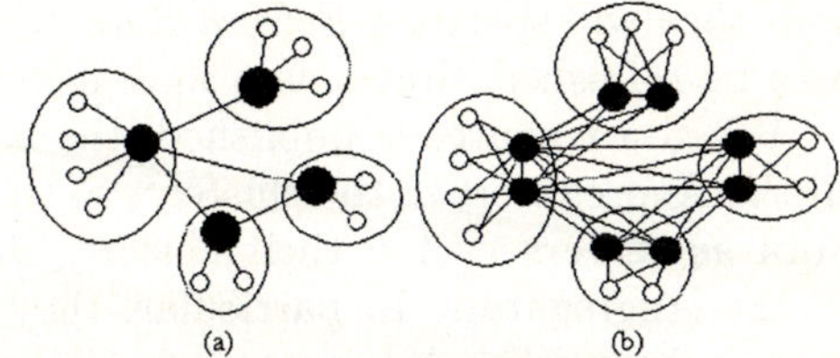

Fig. 2. Examples of super-peer networks: (a) with no redundancy, (b) with 2-redundancy. Black nodes represent super-peers. White nodes are clients. Clusters are limited by circular lines (from [10]).

3.1 Super-peer Redundant Networks

It is well known that unstructured P2P networks spend useful bandwidth in functions that can be performed by local caches [11]. This is why super-peer networks emerged as a trade-off between totally distributed systems and cache-based services [10].

Our system is set up as a super-peer network: some nodes, called *super-peers*, work as servers for a *cluster* of nodes — which usually corresponds to a virtual organization or a subset thereof — but they work as peers in a network of super-peers. Moreover, this network can be built as a redundant network, where super-peers are replicated within each cluster (see Figure 2). This solution introduces two main benefits. (1) Replicas hold a copy of the same data. In the case of failure of one replica, the system will not stop working. (2) The workload can be shared among replicas. Queries can be alternately sent (or forwarded) to each of them in turn. Also, the aggregate bandwidth for forwarded queries can be higher.

On the other side, communication costs may increase, for two reasons. First, when a new node joins a cluster or its data are updated, it has to send a message to K super-peers in a K-redundant network. Second, there are $O(K^2)$ connections between two K-replicated super-peers. The choice of K is a trade-off between reliability and cost.

3.2 Agents, Aggregators and Information Providers

The *Agent* works as a Grid Service available on each machine in the network. It publishes all relevant information, as is made available by *Information Provider* tools (IP).

The Information Providers, scheduled by the Agent, periodically query the resources and store the gathered information as Service Data Elements (SDE), according to the OGSA standard. Each SDE is tagged with a list of keywords, used for subsequent queries. In our system, there is an Information Provider for each resource. When users choose to publish information about a given resource, they will describe the type of information using our simple taxonomy. In particular, they will specify a *Refresh Rate*, which describes how often the information is to be refreshed. Static data have a Refresh Rate equal to 0.

When a resource is published, the name of its Service Data is broadcast to all the Aggregators in the cluster[2], so that they can subscribe to it. Aggregators work as servers within their cluster, and as peers in the network created by all the Aggregators. In particular, they are responsible for forwarding queries coming from other Aggregators to the most likely destination.

To prevent Aggregators from polling Agents at the end of each refresh interval, our implementation uses a *push* approach: the Agents periodically send the updated information to the subscribed Aggregators. We implemented a basic set of Information Providers (*memory, processor, processorLoad, operatingSystem, and diskSpace*). A configuration file will list a set of SDEs to be published by the Agent at launch time, but resources can be published or removed at any time by users.

A client can explicitly choose to remove its data from the super-peer database. Also, the Aggregators will scan the stored information and remove all the resources that failed to send updated information before the expiration of its validity. This way, the super-peer will always have timely information about the clients connected to it.

3.3   Routing Index

The *Routing Index* (RI) is used to improve the performance of our peer-to-peer routing, and to prevent the network from being flooded. The RI is a technique to choose the node to which a query should be forwarded: the RI represents the availability of data of a specific type in the neighbor's information base. We implemented a version of RI called *Hop-Count Routing Index* (HRI), which considers the number of *hops* needed to reach a datum. The HRI counts the number of data elements within a given number of hops. Data are then divided in classes by their keyword.

We used HRI as described in [2]: in each super-peer, the HRI is represented as a $M \times N$ table, where M is the number of neighbors and N is the horizon (maximum number of hops) of our Index: the n-th position in the m-th row is the number of data elements that can be reached going through neighbor m, within n hops.

Suppose that, from node B, we are looking for data about memory (see Figure 3). Our *goodness function* (see [2]), will give a higher value to A, because within short distance (2 hops) we can reach 6 resources. On the contrary, D could give us back information about only 3 of them.

When a new super-peer joins the network, it sends information about the data it controls to all its neighbors. They will update their table, adding the new data to those available within distance 1. Then, they will send the aggregate counts (excluding the new node) back to the new node itself. We use the techniques shown in [2] to deal with cycles in the network.

---

[2] There may be more than one Aggregator in a redundant network.

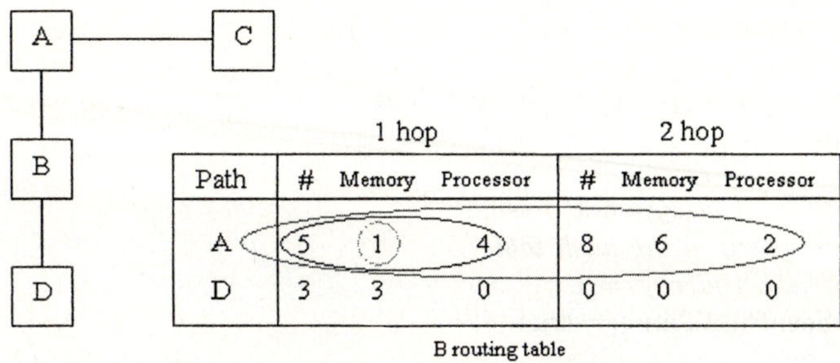

Path	#	Memory	Processor	#	Memory	Processor
A	5	1	4	8	6	2
D	3	3	0	0	0	0

Fig. 3. HRI table for node B.

3.4   Search Technique

In literature, several techniques are used for searches in P2P networks, including flooding (e.g. Gnutella), centralized servers (e.g. Napster). More effective searches are performed by systems based on distributed indices. In these configurations, each node holds a part of the index. The index optimizes the probability of finding quickly the requested information, by keeping track of the availability of data to each neighbor.

In our system, each query is submitted, by each node, only to its cluster's super-peer, which will pass it to other super-peers if needed. To this purpose, the super-peer keeps information about all the nodes in its cluster, in the form of a Hop-Count Routing Index. An outline of our algorithm is shown in Figure 4.

Each query is tagged with an expiration time. At each step, the expiration is checked. If the query is still valid, it is stored in a local hash table (*QueryStatus*), with some key information. In particular, we store what is the next neighbor to try.

The HRI is used to determine which the best neighbor aggregator is for the given query. The query is forwarded to it, while it is elaborated locally, by matching the local SDEs. This way, communication and computation are partially overlapped. The matching SDEs are sent back directly to the original requester as XML data.

If there are no available neighbors, as for C in Figure 5, the query is returned to the sender (B), which will choose the second best neighbor (D), i.e. the neighbor which has the second largest number of matching resources in the HRI. The algorithm will continue with the next best neighbor every time the query returns back (*QueryStatus*, and so *ToTry*, are increased each time).

Please note that the algorithm tries, within the given time, to find as many resources as possible. This choice is due to our goal of mimicking the behavior of the Globus Information Service. From the found resources, the user will choose those that best match his/her needs.

The current strategy suffers from two major limitations: first, under certain conditions, our algorithm may fail to find existing resources (if the query expires too early); second, it may query more Aggregators than strictly needed. Nonetheless, it offers a series of interesting features: very quick response (the

```
For each incoming query
   // check if query is still alive
   If ExpirationTime(query) < CurrentTime
     Discard

   If QueryStatus(query)=not present
     // store query in the hash table
     QueryStatus(query) := 1
     QuerySeenFirstTime := true

   ToTry := QueryStatus(query)
   // find in the Hop-Count R. Index the next
   // best neighbor of rank ToTry
   NextBestNeighbor := HRI(query, ToTry)

   If not exists NextBestNeighbor
     //the query is bounced back
     Recipient := Sender(query)
   Else
     Recipient := NextBestNeighbor
     QueryStatus(query) += 1

   Forward query to Recipient
   If QuerySeenFirstTime
     Find local matching to query
     Send local results to Requester(query)
End for
```

Fig. 4. Our search algorithm.

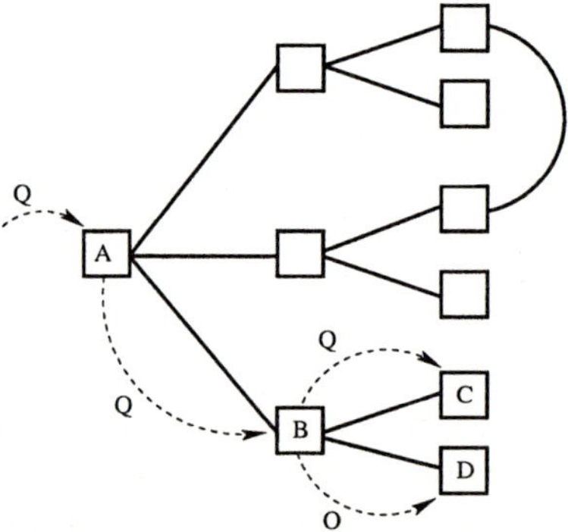

Fig. 5. A query (Q) is forwarded from A to the best neighbors (B, C, and D).

first results arrive as they are available); overlapping computation and communication; freshness of the retrieved data (which are stored very close to the resource they describe). We are investigating other algorithms, including DHT, to overcome these limitations.

4 Experimental Results

Our system was developed using Globus Toolkit 3.0.2 and Java 1.4.1. The system runs under Linux Red Hat 8 and 9, Linux Debian, and Microsoft Windows 2000. It is compliant to the OGSA standard, and uses libraries and tools from the Globus Toolkit 3. We run two tests: the first, to compare it with Globus, and the second to verify in detail its scalability.

4.1 Comparison with Globus MDS-GT3

We compared our system with Globus MDS-GT3. The results shown in Table 1 come from our preliminary tests. All the data are taken at the client side, by measuring the time passed from the beginning of the query, to the arrival of

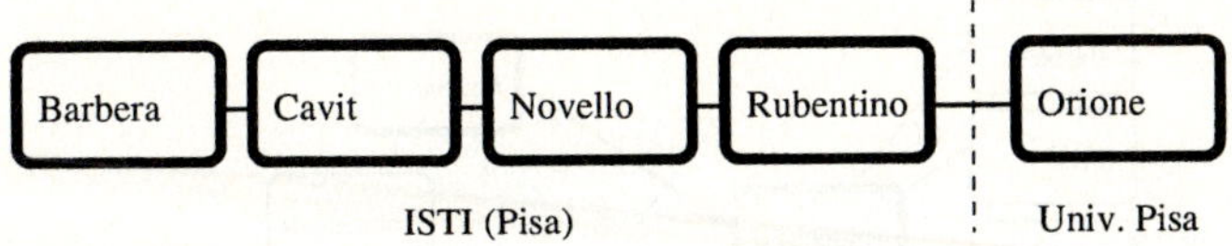

Fig. 6. Configuration for our comparison with Globus MDS. Clients are not shown.

results. Time was measured within the code, using the Java time API, for both Globus MDS and our system.

Due to problems with firewalls and Globus connection ports, we could not utilize many machines in this test. Anyway, with these limited resources, we set up a configuration that was optimal for Globus, and very hard for our system: we created a linear chain of five Aggregators (see Figure 6), and, starting from Orione (locate at the University of Pisa), we launched queries about data down the chain (located at ISTI - Pisa). Clients connected to each Aggregator are not shown. This is the worst case for our system, because clients connected to Barbera are separated by many hops from Orione.

We configured Globus Index Service (IS) with the same linear hierarchy: Cavit is subscribed to Barbera's SDEs, Novello to Cavit's and so on. In any case, all SDEs are cached by the Index Service, so the topology of ISs should not affect its performance.

We can see some interesting results. As said, our system forwards incoming queries to the best neighbors before elaborating them. This way, a query can reach the Aggregator holding the desired data very fast. Then, results are sent back directly to the requester. This is the reason of the slow growth of response time with distance in our system.

For Globus, the response time is irrespective of the distance of the resources relevant to the query, as expected (all data are cached in our experiments).

Our system, under these experimental conditions, outperforms Globus. We have to consider that, at the moment, our system is extremely light-weight, while the Globus infrastructure can support a variety of tasks. Nonetheless, we can say that our system seems to scale effectively and respond very quickly, even if data are not cached: our queries read the datum — freshly updated — available to the Aggregator closest to the resources, not a potentially stale copy.

Table 1. Comparison between our system (P2P) and Globus-MDS. Average response time (client-side) for subsequent results, about resources located at increasing distances (in milliseconds).

Hop #	P2P GIS 1st	2nd	Globus
1	743.5	801.4	3612
2	737.4	820.2	3588
3	775.5	831	3601
4	806.1	861.6	3640

4.2 System Scalability

We tested the scalability of our system by running on a Grid involving five organizations: ISTI-CNR, located in Pisa; University of Pisa; IIT-CNR, in Pisa;

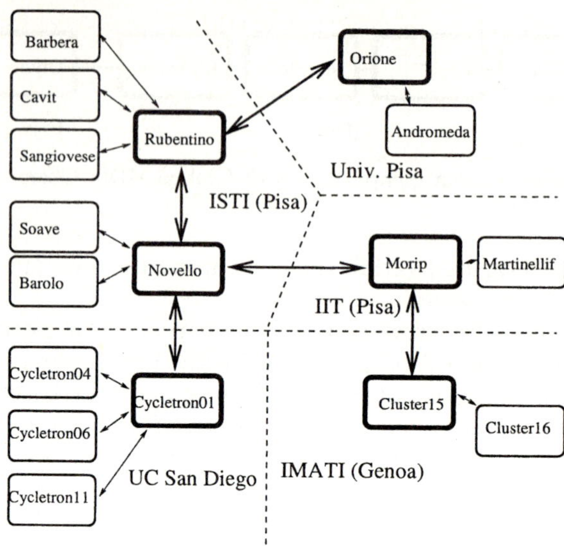

Fig. 7. Our configuration for testing scalability. An Agent is running on each machine (boxes). An Aggregator is running on thicker boxes. Arrows represent connections. The dashed line are the borders among participating institutions.

IMATI-CNR, located in Genoa; and the University of California at San Diego. The test configuration is shown in Figure 7. We artificially split ISTI-CNR into two virtual organizations by using different broadcast masks for the two subsets. This way, the Agents will connect to exactly one Aggregator.

In our tests, we verified the performance when working within the organization's borders. Queries were sent from Rubentino about the status of resources monitored by Novello. On Novello, matching SDEs are sent back to Rubentino very fast: the first result is generated within 10 ms. The results arrive regularly, within few hundred milliseconds (see Table 2(a)).

When we cross the institutions' borders, delays related to the network are more evident. We launched several queries from Orione about the status of resources within the ISTI-CNR and the IIT-CNR organizations. Queries were elaborated by Rubentino, Novello and Morip. Again, we measured that less than 10 ms are needed to generate the first matching SDE, but results take much longer to cross institutions and return to Orione. We believe that the firewall configuration, and other network effects may contribute to this large delay (see Table 2(b-c)).

For queries from distant institutions (IMATI in Genoa and UCSD), response time grows slowly with distance, and may be greater than 1 second (see Table 2(d-e)). This is a result to be expected, if we consider that the *ping* time may be 1000 times greater than among institutions in Pisa.

Our system can also be used with a *redundant configuration* for improved reliability. We run some initial tests, which showed the effectiveness of this solution: when one of the replica failed, the system continued running seamlessly.

Table 2. Average time (in milliseconds) to generate (server-side) and receive (client-side) subsequent results of a given query.

(a) Queries from Rubentino about Novello

Server side	9.1	31.9	40.0	48.3
Client side	212.2	229.2	345.4	436.4

(d) Queries from Cluster15 about Orione

Server side	10.1	40.3	52.3	64.5
Client side	890.1	905.3	958.1	1001

(b) Queries from Orione about Novello

Server side	7	34.6	49.8	65.8
Client side	767.4	826.4	935.3	981.3

(e) Queries from Cycletron01 about Orione

Server side	34.2	260.8	300.2	310.1
Client side	950.4	1104.8	1187.7	1211.7

(c) Queries from Orione about Morip

Server side	9.6	57.8	72.6	87.4
Client side	788.0	850.9	946	999

Response time did not change significantly. We expect that, in a very large configuration, redundant peers may offer a lower response time, when they are queried alternately. We are testing this hypothesis, and results will be available in the next future.

5   Conclusion

The Grid is a vast, dynamic, heterogeneous environments, where information about the status, configuration and cost of resources is extremely valuable: if users are able to find the best match to their needs, their applications will reach the best performance within the desired cost and time.

To monitor a Grid, a versatile system is needed, able to update very quickly, to satisfy a potentially very large number of users and queries, to tolerate delays and faults. Peer-to-peer systems, born out of the first file-sharing applications, evolved into very flexible frameworks, which are now gaining interest within the scientific community. The interaction between Grids and peer-to-peer systems is growing stronger, because P2P seems to be a very promising approach to some problems related to the Grid.

In this work, we presented a P2P Information System for the Grid. It is built as a network of super-peers, which aggregate the data about resources within a virtual organization. Queries performed by any client are passed among the super-peers, using optimization algorithms such as the Hop-Count Routing Index. Our system is based on Globus Toolkit 3 and complies to the OGSA standard: it can be easily integrated with any Globus-based Grid. In this first round of experiments, we used it for resource monitoring and discovery, but the same infrastructure could be used for file-sharing or other distributed applications, this way offering a P2P layer to Grid applications.

Our system was tested using a small network, split across five different institutions. In these preliminary tests, the system scaled effectively. We could not measure big delays in queries for remote resources, which are constantly monitored by their Aggregators. This way, we always have updated information available to queries. Our system also outperformed Globus MDS under our experimental conditions.

References

1. Plale, B., Jacobs, C., Jensen, S., Liu, Y., Moad, C., Parab, R., Vaidya, P.: Understanding grid resource information management through a synthetic database benchmark/workload. In: Proceedings of the 4th IEEE/ACM International Symposium on Cluster Computing and the Grid (CCGrid2004), Chicago, IL, USA (2004)
2. Crespo, A., Garcia-Molina, H.: Routing indices for peer-to-peer systems. In: Proceedings of ICDCS-02. (2002)
3. Puppin, D., Moncelli, S., Baraglia, R., Tonellotto, N., Silvestri, F.: A peer-to-peer information service for the grid. In: Proceedings of the GridNets 2004 Workshop, San José, CA (2004)
4. The Globus Alliance: Globus toolkit 3, globus information services documentation (2004) Available at http://www.globus.org/mds/.
5. Talia, D., Trunfio, P.: Web services for peer-to-peer resource discovery on the grid. In: DELOS Workshop Digital Library Architectures. (2004) 73–84
6. Marti, S., Krishnan, V.: "carmen: A dynamic service discovery architecture". Technical Report HPL-2002-257, HP Laboratories Palo Alto (2002) Available at http://www.hpl.hp.com/techreports/2002/HPL-2002-257.pdf.
7. Zhao, B.Y., Huang, L., Stribling, J., Rhea, S.C., Joseph, A.D., Kubiatowicz, J.D.: Tapestry: A resilient global-scale overlay for service deployment. IEEE Journal on Selected Areas in Communications **22** (2004)
8. Rowstron, A., Druschel, P.: Pastry: Scalable, distributed object location and routing for large-scale peer-to-peer systems. In: IFIP/ACM International Conference on Distributed Systems Platforms (Middleware), Heidelberg, Germany (2001) 329–350
9. Schmidt, C., Parashar, M.: Enabling flexible queries with guarantees in p2p systems. IEEE Internet Computing (2004) 19–26
10. Yang, B., Garcia-Molina, H.: Designing a super-peer network. In: Proceedings of the IEEE International Conference on Data Engineering. (2003)
11. Ripeanu, M.: Peer-to-peer architecture case study: Gnutella network. Technical Report TR-2001-26, University of Chicago, Department of Computer Science (2001)

GRUBER: A Grid Resource Usage SLA Broker

Catalin L. Dumitrescu[1] and Ian Foster[1,2]

[1] Computer Science Department, The University of Chicago,
5801 S. Ellis Ave., Chicago, IL, 60637
`cldumitr@cs.uchicago.edu`
[2] Mathematics and Computer Science Division, Argonne National Laboratory,
9700 S. Cass Ave., MCS/221, Argonne, IL, 60439
`foster@mcs.anl.gov`

Abstract. Resource sharing within grid collaborations usually implies specific sharing mechanisms at participating sites. Challenging policy issues can arise in such scenarios that integrate participants and resources spanning multiple physical institutions. Resource owners may wish to grant to one or more virtual organizations (VOs) the right to use certain resources subject to local usage policies and service level agreements, and each VO may then wish to use those resources subject to its usage policies. This paper describes GRUBER, an architecture and toolkit for resource usage service level agreement (SLA) specification and enforcement in a grid environment, and a series of experiments on a real grid, Grid3. The proposed mechanism allows resources at individual sites to be shared among multiple user communities.

1 Introduction

Resource sharing issues arise at multiple levels when sharing resources. Resource owners may want to grant to VOs the right to use certain amounts of their resources, and thus want to express and enforce the usage policies under which these resources are made available. We measure the impact of their introduction by means of *average resource utilization*, *average response time*, *average job completion*, *average job replanning*, and *workload completion time* [1]. We distinguish here between "resource *usage* policies" (or SLAs) and "resource *access* policies." Resource access policies typically enforce authorization rules. They specify the privileges of a specific user to *access* a specific resource or resource class, such as submitting a job to a specific site, running a particular application, or accessing a specific file. Resource access policies are typically binary: they either grant or deny access. In contrast, resource usage SLAs govern the *sharing* of specific resources among multiple groups of users. Once a user is permitted to access a resource via an access policy, then the resource usage service level agreement (SLA) steps in to govern *how much* of the resource the user is permitted to consume.

GRUBER is an architecture and toolkit for resource usage SLA specification and enforcement in a grid environment. The novelty of GRUBER consists in its capability to provide a means for automated agents to select available resources from VO level on down. It focuses on computing resources such as computers, storage, and networks; owners may be either individual scientists or sites; and VOs are collaborative groups, such as scientific collaborations. A VO [2] is a group of participants who seek

J.C. Cunha and P.D. Medeiros (Eds.): Euro-Par 2005, LNCS 3648, pp. 465–474, 2005.

to share resources for some common purpose. From the perspective of a single site in an environment such as Grid3 [3], a VO corresponds to either one or several users, depending on local access policies. However, the problem is more complex than a cluster fair-share allocation problem, because each VO has different allocations under different scheduling policies at different sites and, in parallel, each VO might have different task assignment policies. This heterogeneity makes the analogy untenable when there are many sites and VOs.

We focus in this paper on the following questions: *"How usage SLAs are handled in grid environments?"*, and *"What is the gain for taking in account such usage SLAs?"*. We build on much previous work concerning the specification and enforcement of local scheduling policies [1],[4],[5],[17], for negotiating SLAs with remote resource sites [6],[7], and for expressing and managing VO policy [8]. We describe in detail the usage SLA problem at several levels, introduce an infrastructure that allows the management of such usage SLAs for virtual and real resources, and compare various approaches for real scenarios encountered in the Grid3 context [3].

2 Motivating Scenario

In the context of grids comprising numerous participants from different administrative domains, controlled resource sharing is important because each participant wants to ensure that its goals are achieved. We have previously defined [1] three dimensions in the usage policy space: resource providers (sites, VOs, groups), resource consumers (VOs, groups, users), and time. Provider policies make resources available to consumers for specified time periods. Policy makers who participate in such collaborations define resource usage policies involving various levels in this space. We extend here the work proposed in [1],[4] with usage policies at the level of virtual organizations and beyond.

Usage SLA specification, enforcement, negotiation, and verification mechanisms are required at multiple levels within such environments. *Owners* want convenient and flexible mechanisms for expressing the policies that determine how many resources are allocated to different purposes, for enforcing those policies, and for gathering information concerning resource usage. *VOs* want to monitor SLAs under which resources are made available.

User and *group jobs* are the main interested parties in resources provided by sites and resources. They use resources in accordance with allocations specified at different level, in a hierarchic fashion and similar to LSF approach at a cluster level. Run-to-completion is a usual mode of operation, because jobs tend to be large, making swapping expensive. Also, workload preemption on several nodes is difficult in a coordinated fashion and even more difficult when a distributed file system is used for data management. Thus, we assume that jobs are preempted only when they violate certain usage rules specified by each individual provider. We note also that site resource managers such as LSF, PBS, and NQS typically only support run-to-completion policies [9].

Algorithms and *policies* capture how jobs are assigned to host machines [1],[4]. The question *"Which is the best approach for different environment models?"* is an old-age question conditioned by many parameters that vary from case to case. Usually what just appears to be an SLA "parameter" can have greater effect on the performance users get from their computing resources than various metrics reported through a monitoring system about resource availabilities.

The problem domain is expressed as follows: a grid consists of a set of resource provider *sites* and a set of *submit hosts*; each site contains a number of processors and some amount of disk space; a three-level hierarchy of *users*, *groups*, and *VOs* is defined, such that each user is a member of exactly one group, and each group is member to exactly one VO; users submit jobs for execution at submit hosts. A job is specified by four attributes: VO, Group, Required-Processor-Time, Required-Disk-space; a *site policy statement* defines site usage SLAs by specifying the number of processors and amount of disk space that sites make available to different VOs; and a *VO policy statement* defines VO usage SLAs by specifying the fraction of the VO's total processor and disk resources (i.e., the aggregate of contributions to that VO from all sites) that the VO makes available to different groups.

Within this environment, the *usage SLA-based resource sharing problem* involves deciding, at each submit host, which jobs to route to which sites, and at what time, for execution, to both (a) satisfy site and VO usage SLAs and (b) optimize metrics such as resource utilization and overall job and workload execution time. We note that this model is one of resource sub-allocation: resources are owned by sites, which apportion them to VOs. VOs in turn apportion their "virtual" resources to groups. Groups could, conceptually, apportion their sub-allocation further, among specific users. Without loss of generality, we simplify both this discussion and our implementation by sub-allocating no further than from VOs to groups.

3 GRUBER Architecture

GRUBER is composed of four principal components, as we describe. The (a) *GRUBER engine* implements various algorithms for detecting available resources and maintains a generic view of resource utilization in the grid. Our implementation is an OGSI service capable of serving multiple requests and based on all the features provided by the GT3 container (authentication, authorization, state or state-less interaction, etc). The (b) *GRUBER site monitoring component* is one of the data providers for the GRUBER engine. It is composed of a set of UNIX and Globus tools for collecting grid status elements. (c) *GRUBER site selectors* are tools that communicate with the GRUBER engine and provide answers to the question: *"which is the best site at which I can run this job?"*. Site selectors can implement various task assignment policies, such as round robin, least used, or last recently used task assignment policies. Finally, the (d) *GRUBER queue manager* is a complex GRUBER client that must reside on a submitting host. It monitors VO policies and decides how many jobs to start and when. The overall GRUBER architecture is presented in Fig. 1.

Planners, work-runners, or application infrastructures invoke GRUBER site selectors to get site recommendation, while the GRUBER queue manager is responsible for controlling job starting time. If the queue manager is not enabled, GRUBER becomes only a site recommender, without the capacity to enforce any usage SLA expressed at the VO level. The site level usage SLA is still enforced by limiting the choices a job can have and by means of removing a site for an already over-quota VO user from the list of available sites.

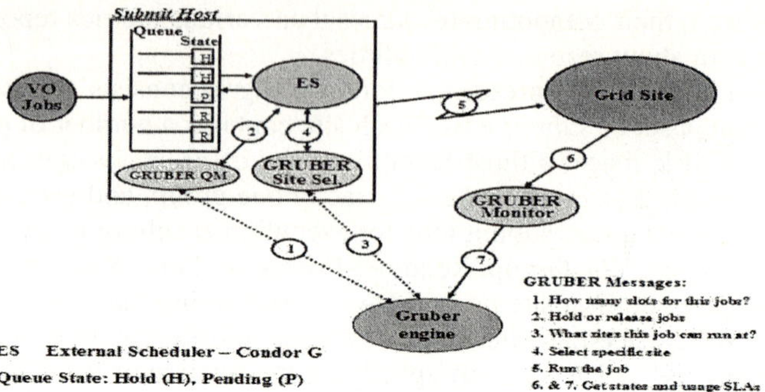

Fig. 1. GRUBER Architecture

3.1 GRUBER Engine

GRUBER decides which sites are *best* for a job by implementing the following logic:

- If there are fewer waiting jobs at a site than available CPUs, then GRUBER assumes the job will start right away if an extensible usage policy is in place [1].
- If there are more waiting jobs than available CPUs or if an extensible usage policy is not in place, then GRUBER determines the VO's allocation, the number of jobs already scheduled, and the resource manager type. Based on this information:
  - if the VO is under its allocation, GRUBER assumes that a new job can be started (in a time that depends on the local resource manager type).
  - if the VO is over its allocation, GRUBER assumes that a new job cannot be started (the running time is unknown for the jobs already running).

More precisely, for any job placement CPU-based decision a list of available sites is built and provided under the following algorithm to find those sites in the site set G that are available for use for job J (J keeps place for job characteristics as well in the following algorithm) from the VO number i.

```
fn get-avail-sites(sites G, VO i, job J)
1. for each site s in G do
2.        # Case 1: site over-used by VO_i
3.        if EA_i > EP_i for VO I at site s
4.            next
5.        # Case 2: un-allocated site
6.        else if  _k(BA_k)at s < s.TOTAL - J &&
                (BA_i + J < BP_i || extensible BP_i) then
7.            add (s, S)
8. return S
```

with the following definitions:

```
S   = Site Set ; k = index for any VO != VO_i
EP_i = Epoch Usage SLA for VO_i ; BP_i = Burst SLA for VO_i
BA_i = Burst Utilization for VO_i ; EA_i = Epoch Utilization
TOTAL = upper limit allocation on the site
```

The list of possible solutions is further provided as input to a task assignment policy algorithm that makes the actual submission decisions (e.g., round robin).

3.2   GRUBER Queue Manager / Site Selectors

GRUBER queue manager is responsible for determining how many jobs per VO or VO group can be scheduled at a certain moment in time and when to release them. Usually a VO planner is composed of a job queue, a scheduler, and job assignment and enforcement components. Here, the last two components are part of GRUBER and have multiple functionalities. The site selector component answers: *"Where is best to run next?"*, while the queue manager answers: *"How many jobs should group G_m of VO_n V be allowed to run?"* and *"When to start these jobs?"*

The queue manager is important for SLA enforcement at the VO level and beyond. This mechanism also avoids site and resource overloading due to un-controlled submissions. The GRUBER queue manager implements the following algorithm (with the assumption that all jobs are held initially at the submission host):

```
1. while (true) do
2.      if Q != empty
3.          get j from Q
4.      else
5.          next
6.      S = get-avail-sites(G, Vo(j), j)
7.      if S != empty
8.          s = schedule(j, S)
9.          run(j,s)
```

with the following definitions:

```
j = Job Id ; Q = Job Queue ; S = Site Set ;
G = All Site Set ; Vo = Mapping Function jobId -> VO
```

3.3   Disk Space Considerations

Disk space management introduces additional complexities in comparison to job management. If an entitled-to-resources job becomes available, it is usually possible to delay scheduling other jobs, or to preempt them if they are already running. In contrast, a file that has been staged to a site cannot be "delayed," it can only be deleted. Yet deleting a file that has been staged for a job can result in livelock, if a job's files are repeatedly deleted before the job runs. As a consequence, a different approach has been devised. As a concrete example, a site can become heavily loaded with a one VO jobs and because of which other jobs are either in the local queue in an idle state waiting for their turn. But this does not stop the submission of more jobs.

So far, we have considered a UNIX quota-like approach. Usually, quotas just prevent one user on a static basis from using more than his limit. There is no adaptation to make efficient use of disk in the way a site CPU resource manager adapts to make efficient use of CPU (by implementing more advanced disk space management techniques). The set of disk-available site candidates is combined with the set of CPU-available site candidates and the intersection of the two sets is used for further scheduling decisions.

3.4 GRUBER Usage SLA Language

In the experiments described in this paper we use a usage SLA representation based on Maui semantics and WS-Agreement syntax [4],[7],[11]. Allocations are made for processor time, permanent storage, or network bandwidth resources, and there are at least two-levels of resource assignments: to a VO, by a resource owner, and to a VO user or group, by a VO. We started from the Maui's semantics in providing support for fair-share rule specification [12]. Each entity has a fair share type and fair share percentage value, e.g., VO_0 15.5, VO_1 10.0+, VO_2 5.0-. The sign after the percentage indicates if the value is a target (no sign), upper limit (+), or lower limit (-).

We extend the semantics slightly by associating both a consumer and a provider with each entry; extending the specification in a recursive way to VOs, groups; and users, and allowing more complex sharing rules. In our approach, $Site_1$ makes its CPU resources available to consumer VO_0 subject to two constraints: VO_0 is entitled to 10% of the CPU power over one month; and with any burst usage up to 40% of the CPU power for intervals smaller than one day. Not all parameters are required and in a recursive fashion, a similar usage SLA is specified for VO entities.

4 Experimental Studies

We now present our experimental results. We first describe the metrics that we use to evaluate alternative strategies, afterwards introduce our experimental environment, and finally present and discuss our results.

4.1 Metrics

We use five metrics to evaluate the effectiveness of the different site selector strategies implemented in GRUBER. **Comp** is the percentage of jobs that complete successfully. **Replan** is the number of replanning operations performed. **Time** is the total execution time for the workload. **Util** is average resource utilization, the ratio of the per-job CPU resources consumed (ET_i) to the total CPU resources available, expressed as a percentage:

$$Util = \Sigma_{i=1..N} ET_i / (\#_{cpus} * \Delta t) * 100.00$$

Delay is average time per job (DT_i) that elapses from when the job arrives in a resource provider queue until it starts:

$$Delay = \Sigma_{i=1..N} DT_i / \#_{jobs}$$

4.2 Experiment Settings

We used a single job type in all our experiments, the sequence analysis program BLAST. A single BLAST job has an execution time of about an hour (the exact duration depends on the CPU), reads about 10-33 kilobytes of input, and generates about 0.7-1.5 megabytes of output: i.e., an insignificant amount of I/O. We used this BLAST job in two workload different configurations. In 1x1K, we have a single workload of 1000 independent BLAST jobs, with no inter-job dependencies. This workload is submitted once. Finally, in the 4x1K case, the 1x1K workload is run in parallel from four different hosts and under different VO groups. Also, each job can be re-planed at most four times through the submission infrastructure.

We performed all experiments on Grid3 (December 2004), which comprises around 30 sites across the U.S., of which we used 15. Each site is autonomous and managed by different local resource managers, such as Condor, PBS, and LSF. Each site enforces different usage policies which are collected by our site SLA observation point and used in scheduling workloads. We submit all jobs within the iVDGL VO, under a VO usage policy that allows a maximum of 600 CPUs. Furthermore, we submitted each individual workload under a separate iVDGL group, with the constraint than any group can not get more than 25% of iVDGL CPUs, i.e., 150.

4.3 Results

Table 1 and Table 2 give results for the 1x1K and 4x1K cases, respectively. We see considerable variation in the performance of the various site selectors.

Table 1. Performance of Four GRUBER Strategies for 1x1K

	G-RA	G-RR	G-LRU	G-LU
Comp (%)	97	96.7	85.6	99.3
Replan	1396	1679	1440	1326
Util (%)	12.85	12.28	10.63	14.56
Delay (s)	49.07	53.75	54.69	50.50
Time (hr)	8.19	10.45	22.23	9.25

In the 1x1K case, G-LU does significantly better than the others in terms of jobs completed and G-RA does significantly better than the others in execution time. G-LRU is clearly inferior to the others in both respects. Note that as a single job runs for about one hour, the minimum possible completion time is ~1000/150 ≈ 6.66 hours. Thus the best execution time achieved is ~22% worse than the minimum, which is an acceptable result. Note also the relatively high number of replanning events in each case (a mean of ~1.5 per job), another factor that requires further investigation.

Table 2. Performance of Four GRUBER Strategies for 4x1K

	G-RA	G-RR	G-LRU	G-LU
Comp (%)	98.2	98.7	87.9	91.7
Replan	1815	1789	1421	2409
Util (%)	13.51	14.02	11.05	11.52
Delay (s)	66.62	64.41	68.97	63.96
Time (hr)	11.21	10.5	13.51	13.49

In the 4x1K case, results are somewhat different. Each submitter is allocated a limit of 25% of the VO's total resource allocation of 600 CPUs. Thus, in principle, all four submitters can run concurrently, but in practice we can expect greater contention in site queues and when sites are overloaded with non-GRUBER work.

The results in Table 4 are the means across the four submitters. We see some interesting differences from Table 3. G-LU's completion rate drops precipitously, presumably for some reason relating to greater contention. Replanning events increase somewhat, as does scheduling delay. The total execution times for G-RA and G-LU increase, although more runs will be performed to determine whether these results are significant. Fig. 2 provides a more detailed picture of the job completion rates for the experiments of Table 4. We see that the inferior performance of G-LRU and G-LU is accentuated by the long time spent on a few final jobs.

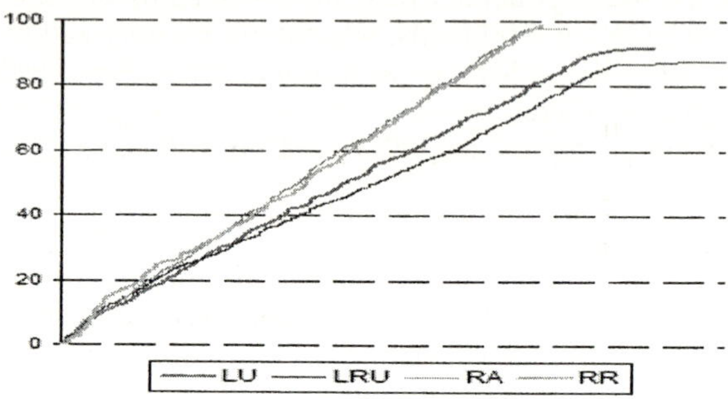

Fig. 2. Job Completion Percentage vs. Time

4.4 Results Variance

In practice, we may expect to see considerable variation in results over different runs, due to our lack of full control over the Grid3 environment. As a first step towards quantifying this variation, we ran ten identical 1x1K runs with the random assignment strategy (G-RA). Our results, in Table 3, show considerable variation for some quantities (Replan, Util, and Delay) and little variation for other quantities (Comp and Time). However, the small variance in completion time and later studies not reported here confirm our observations.

Table 3. 10 1x1K Runs with G-RA

Metric	Average	Std. Deviation	Std. Dev. As %
Comp (%)	95.11	1.23	1.3
Replan	2000	308	15.4
Util (%)	18.77	3.10	16.5
Delay (s)	78.05	15.39	19.7
Time (hr)	7.97	0.33	4.1

4.5 Site Selector Variations and Comparisons

We compare GRUBER performance with that of two other methods. In the first of the two methods, "GRUBER observant" or G-Obs, a site selector associates a further job to a site each time a job previously submitted to that site has started. In effect, the site selector fills up what a site provides by associating jobs until site's limit is reached. The second alternative, S-RA, associates each job to a site selected at random.

Table 4 shows the results obtained on Grid3 for the 1x1K workload. We find that the best standard GRUBER site selector achieves a performance comparable to that of the GRUBER observant (G-Obs) selector.

Table 4. G-LU, G-Obs and S-RA Strategies: Performance for 1x1K

	G-LU	G-Obs.	S-RA
Comp (%)	99.3	97.3	60.2
Replan	1326	284	1501
Util (%)	14.56	12.59	0.57
Delay (s)	50.50	62.01	121.0
Time (h)	9.25	11.2	22.3

The results indicate that GRUBER's automated mechanism for SLAs tuning (as described in Section 3.2) makes its site selectors comparable in terms of job scheduling performance. On the other hand, compared with GRUBER site selectors, the naive random assignment site selector policy performs two to three times worse in terms of our metrics. An important metric to observe is the number of re-planning operations. While the observant site selector had around 300 operations, GRUBER performed around 1300 re-scheduling operations and the naive site selector around 1500. The difference comes from the fact that the simple round robin algorithm selects from all the sites and not only from the candidates identified as available.

5   Related Work

Fair share scheduling strategies seek to control the distribution of resources to processes so as to allow greater predictability in process execution. These strategies were first introduced in the early 1980s in the context of mainframe batch and timeshared systems and were subsequently applied to Unix systems [13],[14].

Irwin et al. investigate the question of scheduling tasks according to a user-centric value metric – called utility [6]. Sites sell the service of executing tasks instead of raw resources. The entire framework is centered on selling and buying services, with users paying more for better services and sites paying penalties when they fail to honor the agreed commitments. The site policies are focused on finding winning bids and schedule resource accordingly. This approach is different from our work here, as a more abstract form of resources is committed under different SLAs.

In et al. describe a novel framework for policy based scheduling of grid-enabled resource allocations [15]. The framework has several features. First, the scheduling strategy can control the request assignment to grid resources by adjusting resource usage accounts or request priorities. Second, efficient resource usage management is achieved by assigning usage quotas to intended users. Third, the scheduling method supports reservation based grid resource allocation. Fourth, Quality of Service (QoS) feature allows special privileges to various classes of requests, users, groups, etc. The difference with our approach consists in the fact that we do not assume a centralized point of usage policy specification, but a more distributed approach of specification and enforcements at the site level.

6   Conclusions

We have presented and evaluated an approach to representing and managing resource allocation policies in a multi-site, multi-VO environment. We also introduced a grid resource broker, called GRUBER, and the experiments we performed with several approaches in task assignment policies. GRUBER is an architecture and toolkit for resource usage SLAs specification and enforcement in a grid-like environment. It is the prototype of a VO policy enforcement point as described by Dumitrescu et al. [1]. While the results presented here are preliminary, they are encouraging and some of the methods described here are pursued further in the Grid3 and OSG contexts.

There are still problems not fully explored in this article. For example, our analysis did not consider the case of cluster administrators that *over-subscribe* local resources, in the sense of a local policy that states that 40% of the local CPU power is available to VO_1 and 80% is available to VO_2. A second issue not discussed in this report is the hierarchic grouping and allocation of resources based on policy. Generally, VOs will group their users under different schemes. While this is an important problem for our context, we leave it as an open problem at the current stage.

References

1. Dumitrescu, C. and I. Foster. "Usage Policy-based CPU Sharing in Virtual Organizations". in *5th International Workshop in Grid Computing*. 2004.
2. Foster, I., C. Kesselman, and S. Tuecke, "The Anatomy of the Grid: Enabling Scalable Virtual Organizations", in *International Journal of Supercomputer Applications*, 2001. 15(3): p. 200-222.
3. Foster, I., et al., "The Grid2003 Production Grid: Principles and Practice", in *13th International Symposium on High Performance Distributed Computing*. 2004.
4. Dan, A., C. Dumitrescu, and M. Ripeanu. "Connecting Client Objectives with Resource Capabilities: An Essential Component for Grid Service Management Infrastructures", in *ACM International Conference on Service Oriented Computing (ICSOC'04)*. 2004. NY.
5. Altair Grid Technologies, LLC, *A Batching Queuing System,* Software Project, 2003.
6. Irwin, D., L. Grit, and J. Chase., "Balancing Risk and Reward in a Market-based Task Service", in *13th International Symposium on High Performance Distributed Computing*.
7. IBM, *WSLA Language Specification, Version 1.0*. 2003.
8. Pearlman, L., et al. "A Community Authorization Service for Group Collaboration", in *IEEE 3rd International Workshop on Policies for Distributed Systems and Networks*. '02.
9. Schroeder, B. and M. Harchol-Balter. "Evaluation of Task Assignment Policies for Super Computing Servers: The Case for Load Unbalancing and Fairness", in *Cluster Computing*.
10. Legrand, I.C., et al. "MonALISA: A Distributed Monitoring Service Architecture", in *Computing in High Energy Physics*. 2003. La Jolla, CA.
11. Ludwig, H., A. Dan, and B. Kearney. "Cremona: An Architecture and Library for Creation and Monitoring WS-Agreements", in *ACM International Conference on Service Oriented Computing (ICSOC'04)*. 2004. New York.
12. Cluster Resources, Inc., *Maui Scheduler*, Software Project, 2001-2005.
13. Henry, G.J., *A Fair Share Scheduler*. AT&T Bell Laboratory Technical Journal, 1984.
14. Kay, J. and P. Lauder, "A Fair Share Scheduler", University of Sydney, AT&T Bell Labs.
15. In, J., P. Avery, R. Cavanaugh, and S. Ranka, "Policy Based Scheduling for Simple Quality of Service in Grid Computing", in *International Parallel & Distributed Processing Symposium (IPDPS)*. April '04. Santa Fe, New Mexico.
16. Buyya, R., "GridBus: A Economy-based Grid Resource Broker", 2004, The University of Melbourne: Melbourne.
17. Dumitrescu, C. and I. Foster, "GangSim: A Simulator for Grid Scheduling Studies", in *Cluster Computing and Grid (CCGrid)*, 2005, Cardiff, UK.

An Architecture for Distributed Grid Brokering

John M. Brooke and Donal K. Fellows

University of Manchester, Manchester, UK
`j.m.brooke@manchester.ac.uk`
`http://www.unigrids.org/`

Abstract. Computational resource brokering on the Grid is the process of discovering what systems are capable of running a job, obtaining estimates for when that job may run and how much it will cost, and submitting the job to the system that best meets the users' requirements. This paper identifies how resource brokers differ from superschedulers, and describes a resource brokering architecture which is adapted to the emergent structure of the large-scale Grid. We outline the architecture of the UNICORE resource broker which is the basis of our prototype implementation, and discusses both how the existing UNICORE architecture is relevant to the wider brokering picture and what will be done in the future to bring them into closer alignment.

1 Introduction

If one examines currently accepted informal definitions of Grid computing (e.g. [1]) it is clear that the Grid approach to distributed computing does not allow us to make any assumptions about uniform policies for the management of and access to the resources provided by participation in the Grid. This potentially makes Grids very hard to use and a great deal of effort has been devoted to developing "middleware" that can hide this complexity from the users of the Grid. One important task of this middleware is to locate resources on the Grid for the purposes of performing some computational task. This is usually referred to as resource brokering and it usually also includes obtaining some form of quality-of-service (QoS) offer from those resources, so that different offers from providers may be distinguished.

This is a separate problem from the management and scheduling of tasks on Grid resources, which is referred to as "super-scheduling" since the Grid-wide scheduler may need to coordinate local scheduling systems given the potential existence of different management regimes controlling the resources of the Grid. Note that this condition makes Grid scheduling a different problem to scheduling on hierarchical clusters controlled by a uniform resource management system.

In some current resource broker implementations, e.g. in the EU DataGrid[2] and in NorduGrid[3], these two operations are merged so that the process of finding where a job can run is also the process of reserving some resources for that job. It is known that the super-scheduling problem is hard to scale especially when the local schedulers may be running different scheduling systems. The few

J.C. Cunha and P.D. Medeiros (Eds.): Euro-Par 2005, LNCS 3648, pp. 475–484, 2005.

super-schedulers that do exist (e.g. [4]) only work with certain batch or resource management systems, thus the problem is not generally solved in practice. Thus if a new site wishes to join in a larger Grid, it may have to consider changing its scheduling system and software, this breaks the autonomy of local site policy that is regarded as a key distinguishing feature of the Grid.

In this paper we argue that the resource broker problem is inherently easier to scale and present an architecture for brokering that follows natural hierarchies created when different sites or organisations which have a uniform inter-site policy join to form a Virtual Organisation (VO) by pooling their resources. We distinguish two types of scaling; scaling in magnitude allowing very large numbers of sites and resources to be organised as a Grid, and scaling in complexity allowing complex applications to federate across sites deploying different middleware and differing site policy configurations. We present preliminary results showing how our brokering architecture enables scaling in complexity and generalise from this to a proposed architecture that can support both types of scaling.

The paper is structured as follows: in Sect. 2 we examine in more detail current approaches to the scheduling and brokering problems. In Sect. 3 we present our VO-based approach to brokering and in Sect. 4 we present our reasons for choosing the UNICORE middleware as the basis for implementation. In Sect. 5 we present the detail of a prototype implementation and how this allows interoperability with other middleware systems such as Globus Toolkit versions 2 and 3 (GT2 and GT3). We also discuss how the lessons learnt have enabled a generalisation of the prototype implementation to meet the full requirements of a VO-based architecture. In Sect. 6 we present preliminary conclusions and discuss how the development of standards based around the Open Grid Services Architecture will enable the brokering architecture to be developed for a much wider range of middleware systems compliant with the emerging Open Grid Services Architecture (OGSA).

2   Existing Resource Brokers and Superschedulers

There are currently several different approaches to Grid resource brokering. Two major but similar approaches are used by the brokers developed in the DataGrid project and the NorduGrid project cited above, which are both arranged as front-ends to the job-submission system through which all jobs are required to go. In the DataGrid broker, there is a single central broker through which all submissions are made; this allows it to carry out not just brokering but also scheduling across the Grid which it controls, but it pays for this by being not scalable. The NorduGrid broker by contrast puts the brokering directly in the client toolkit; this solution is more scalable, but requires revealing large amounts of information to every client, which not something that commercial resource providers may wish to do (the precise status of a site might be commercially sensitive.) The problem with both of these approaches is that they both require collecting of large amounts of information about the state of the Grid in some centralized location, necessitating an extensive Grid monitoring architecture.

This approach does not scale well because of the amount of information that needs to be transferred and processed since a centralized broker must know the state of the system with a fair degree of accuracy or all estimates will be wrong, and a distributed broker needs to collect a large amount of information on each client request.

While both of the above two brokers are also scheduling agents, neither are as far advanced as the NaReGI[5] super-scheduler. That works through the introduction of a privileged module that can rewrite jobs into a form that can be scheduled more efficiently at individual resources. However, this approach also does not scale administratively because it requires the exposure of substantial amounts of information (much of which might be commercially sensitive) about resources on the Grid to the scheduler potentially within a different administrative domain. This approach can be extended with the use of an inter-scheduler negotiation protocol (e.g. based on ContractNet[6] or WS-Agreement[7]) but this is still difficult to scale as the system granularity changes.

The other approach in use is based on the UNICORE[8] resource brokering framework[9]). This assumes that the underlying systems used by the sites or organisations in the VO are each in control of local scheduling (e.g. through the use of a batch queue) and is oriented towards discovery of resources and the presentation of offers for a particular level of quality-of-service made by those resources. In particular, the resources may make multiple offers for a particular job and those offers do not have to precisely satisfy the requirements of the job in order to be considered; the resource making the offer may use knowledge of the application's problem-domain to create an offer based on application-specific scaling factors. The other key feature of the existing UNICORE broker is that the architecture is composable, with brokering agents being able to ask other brokering agents to work on their behalf.

3   Conceptual Basis for VO-Based Brokering

In [1] Grids are envisaged as deriving from resource sharing in Virtual Organizations. This term has no meaning unless the VO has some common policy on resource sharing, however this may be at a general level and scheduling, for example, may be a task carried out differently in different parts of the VO. If we define the VO concept recursively, i.e. VOs can be composed of sub-VOs, then we get a policy hierarchy. If we go sufficiently far down such VO trees we will eventually come to groupings of resource that can be considered to have uniform systems with respect to resource allocation and management (in the worst case this might be individual machines but economies of scale generally call for some grouping).

An actual physical computing site or organisation may have multiple sets of resources managed in substantially different ways, it can be easier to represent the site as multiple VOs, each with its own policy domain (though perhaps with a separate VO on top of them representing the federation of those resources within the overall site). Sites come together in organisations and organisations in multi-institutional collaborations but in our abstraction this is all within the recursive definition of VOs. Note that resources are not necessarily machines, but are instead a virtualization of machines. This means that a resource may also be a cluster of machines, or that a single machine may host multiple resources (much as websites may be hosted by pools of webservers, or single HTTP daemons may host multiple websites; combinations of both are also possible).

Because we have defined the structure of VOs recursively, we also define the structure of the brokering system for the Grid recursively. By arranging for the broker for a VO to operate through delegation of requests sent to it to the brokers in the sub-VOs, there is already a much substantial degree of natural scalability. Another degree of scalability can be added by loosening up the binding between a VO and its broker, so that the brokering service for a VO is actually chosen from a pool of suitable brokers.

Superschedulers are integrated into this picture by placing them at (or near) the leaves of the VO tree. This allows them to operate in highly homogeneous environment and avoid the inter-domain coupling problems found in higher-level superschedulers. It is easier to scale brokers hierarchically across administrative domain boundaries, since they do not undertake any management of resources but only make enquiries about such availability. Thus the abstract function *getResourceInformation* can be implemented without any inter-site cooperation but *scheduleTask* cannot. This is not the case, however, in the hierarchical design of the Meta Directory System v2 (MDS-2) used by Globus middleware[10] since the indexing process requires that information publishing outside the subdomains of the VO and combined at the higher levels.

4   Choice of Middleware to Build a Hierarchical Broker

To build a Grid resource broker based on VO boundaries we need support for the hierarchical structure of VOs in the middleware which provides access to the resources of the Grid. We found support for such abstractions in the UNICORE middleware. The Globus MDS-2 information provision has a hierarchical structure but this abstraction is not maintained throughout the middleware (in job submission language for example). This means that when UNICORE is installed we simultaneously gain information about resources that covers all possible task submission requests (since the middleware cannot function without this). With Globus the information provision is done separately and although the adoption of a common information schema such as the GLUE schema[11] goes some way towards providing a more information-rich Grid, it still lacks the link between resource information gathering and task submission[12].

The UNICORE[8] architecture is based around the concepts of Usites, Vsites and Abstract Job Objects (AJOs). Usites are virtualizations of resource provider sites that will normally have a shared set of policies (originally focusing on firewall and certificate authority management), Vsites are virtualizations of services providing computational resources, and AJOs are document-oriented abstractions of computational jobs that are converted by Vsites into concrete forms (through a process termed "Incarnation") before execution (see Fig. 1). The

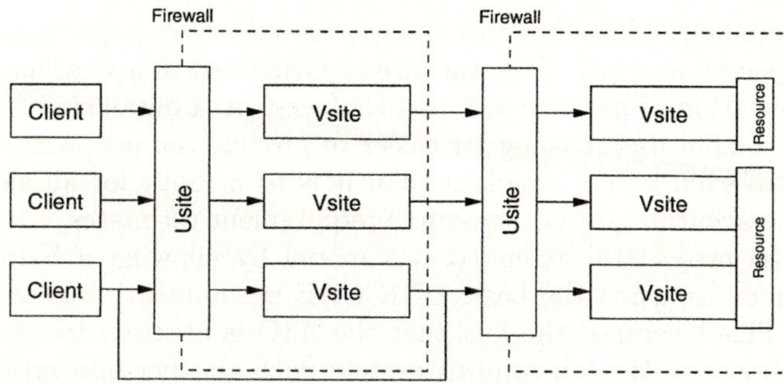

Fig. 1. The architecture of a UNICORE grid

UNICORE brokering model (developed in the EUROGRID project[9]) builds on top of the general architecture by allowing each Vsite to host a brokering service for that Vsite. This works by taking an AJO stripped of large components (like attached files) and testing to see if the resources it requests are available at the Vsite. When the resources are available, the broker then obtains an estimate for what level of quality-of-service is available for the job (obtained from the low-level job system, e.g. from a batch queue length estimator or by examining the load of the machine to get an estimate for likely processor and memory contention[1]) and then attaches it to a ticket that is handed back to the calling agent along with the QoS offer. The calling agent can then claim the offered QoS by attaching the ticket to the real job submission. Another key feature of the EUROGRID broker was support for delegation of a brokering request from one broker to another, which allowed for the deployment of a dummy Vsite which could provide brokering for a whole group of resources by delegating incoming requests to the leaf-Vsite brokers (see Fig. 2). Finally, it supports a plug-in interface which allows the broker to be enhanced with knowledge of a particular

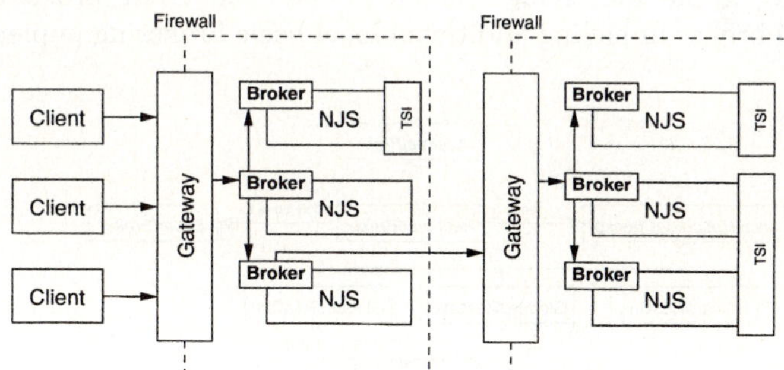

Fig. 2. The architecture of the EUROGRID broker

---

[1] Note that different kinds of systems require different kinds of QoS estimators. Batch processing systems give total control of processors to their jobs and hence the loading is irrelevant on such resources, whereas direct execution clusters will start running every job virtually immediately but will suffer from any resource contention present.

application domain. This allows for the expression of job requirements in terms of domain-specific measures (e.g. the size of Grid used in a weather simulation) and the application of performance models based on a detailed study of the actual applications in use, it being far easier to provide the user with an interface that generates such input metadata than it is to arrange for all agents on the Grid to make accurate physical resource requirement estimates.

The GRIP project[13] extended this model by allowing a Vsite to be implemented not just using the basic UNICORE mechanisms, but also on top of Globus[14]. This leveraged the fact that the AJO is abstract to allow the complete replacement of the job running system with another one with an entirely different job description language. The resource broker was also extended in GRIP to work by using Globus information services (in GT2 and GT3).

The key features of the UNICORE architecture that supported the EURO-GRID/GRIP broker were that the conceptual models of both the computational resources and the jobs running on them were abstract. By brokering jobs before they are incarnated, it is much easier to find more resources capable of running the job on a heterogeneous Grid, and the abstract resource model allows the broker to work with offers from a much wider range of resources.

5   Design and Implementation

5.1   Prototype Implementation

The EUROGRID/GRIP resource broker is implemented as a plug-in module to the UNICORE NJS[2] and consists internally of a multi-layered architecture (see Fig. 3). The outermost layer handles the communication, security and delegation model as well as providing utility and configuration services. Inside this is is the main logic module — which uses a runtime-pluggable architecture to support application-specific brokering — and the local basic brokering engine, which acts as an interface to the underlying concrete system. The GRIP broker extends the EUROGRID broker in having additional local basic brokering implementations,

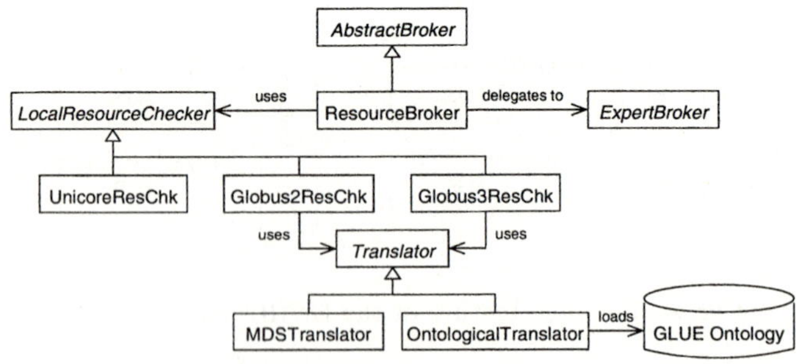

Fig. 3. The internal architecture of the EUROGRID/GRIP broker

---

[2] Network Job Supervisor, a hosting environment for UNICORE Vsites.

one that integrates with a GT2-based low-level Grid, and one that integrates with a GT3-based Grid. These local broker interfaces handle the task of brokering on a Globus-based Grid by translating from resources as requested by a UNICORE AJO to an MDS-2 or Index Service query. Each translation is performed by a pluggable translation engine; the translation to MDS-2 queries is *ad-hoc* but the translation to an Index Service query is described using an ontology developed using PCPACK[15], an ontology capture tool, which allows for much simpler maintenance of the ontology going forward in time.

To demonstrate the viability of the EUROGRID/GRIP resource broker, it has been used to broker the Deutcher Wetterdienst coupled weather simulation model across a heterogeneous grid consisting of a mixture of UNICORE-based and Globus-based nodes. The global part of the weather model was hosted on a UNICORE-based grid system, and the relocatable local weather model was transparently brokered across a Globus-based grid, with the results being reflected back to a UNICORE-based front end client for display to the user. This demonstrated both complexity of application (the primary resources over which the application was brokered were described in terms of the weather model, with translation to suitable underlying resource terms done transparently by the application-specific broker) and complexity of underlying infrastructure.

5.2   Lessons Learnt and Generalisation of the Architecture

Work on the prototype has identified three requirements to lift this infrastructure to the ideal of VO-based brokering architecture described in Sect. 3:

- The resource broker must be extended so that it can broker for more than one Vsite simultaneously without having to delegate to individual leaf brokers. This significantly reduces the degree of fan-out in a brokering request. This means that the tickets issued by the broker must be capable of inspection by services other than the issuer, though there is no need for anything outside the site that hosts such a broker to be able to carry out such an inspection.
- It must be possible to place two or more brokers in parallel and get sensible answers out of each even with simultaneous requests. This means that where a broker reserves resources for the use of a job, it must be able to make sure that the other brokers in parallel do not collide with it. This might be done by using a database to provide serialization and locking. Note that this is not necessary if no actual reservation is made for the job, such as might be the case if the brokers are just reporting estimates of how long it will take for the job to reach the head of the batch queue.
- It must be possible for an agent (whether a client or another broker) to find out an instance of a resource broker for a site or other VO. This should be done by associating a registry of some kind with the VO and placing the references to all the VO's broker instances within it. VOs that have more than one broker may wish to split the load between the brokers by arranging for different requests to the registry to return different instances (or in a different order).

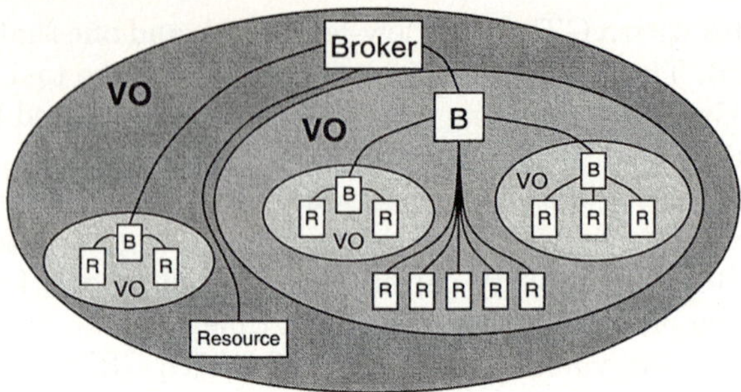

Fig. 4. An overview of the multi-VO resource broker architecture

These three architectural changes, together with the delegation model developed in the EUROGRID broker (formalized by an Explicit Trust Delegation model[16]), allow the development of a brokering infrastructure along VO lines (see Fig. 4). This work is being undertaken in the UniGrids project[17]. Note that the VO closest to the user should also be supplied with a policy description (based on Condor ClassAds[18]) that allows it to choose between the offers collected on behalf of the user. By combining that policy with any such VO-based policy, it is possible to choose a suitable offer without any further intervention from the user or instigating agent.

There are additional benefits to doing this. By moving the broker the higher level, it becomes much more efficient to use systems like R-GMA[19] or NWS[20] to estimate likely performance. The Usite level (i.e. the lowest level of VO) is a good point to introduce superschedulers like that developed by NaReGI[5] without sacrificing the simplicity of the wider brokering architecture outlined above.

The final component of the UniGrids brokering architecture is a mechanism for monitoring of submitted jobs and generation of Usage Records[21]. These usage records would then be both stored in a resource usage service (such as the one developed in the MCS project[22]) for future reference (e.g. invoicing the VO at the end of the month). The usage records are also to be fed back at periodic intervals at times when the brokers are likely to be otherwise lightly loaded.

Preliminary results indicate that the new architecture greatly increases both the throughput scalability and the management scalability of the brokering system. The key source of improvement in throughput is the reduction in the number of inter-service messages achieved through a broker being able to issue offers on behalf of an entire site at once, and the improvement in managability comes through the reduction of the number of different systems that need to be configured to bring a brokered site into service and keep it operating.

6 Conclusion and Future Work

Two major changes will be required to bring the brokering architecture within the OGSA framework. The first major change is the switch to using web-services based on WSRF[23] for the implementation and SOAP[24] for the transport protocol. This will make comparatively little difference to the brokering system because that is already highly service-oriented. We expect it to simplify the communication model between the components significantly as the resource broker does not require the transfer of large amounts of data, even for usage record reconciliation. The advantage of going to a service-oriented model will be that it will become much easier to integrate UNICORE with a more traditional web-services workflow engine like BPEL[25] and so allowing more complex workflows with multiple brokering stages (e.g. a long running job that is migrated between available platforms at regular intervals, with the broker being used to select location for job migration). It will also allow a wider range of clients written in arbitrary programming languages to make use of the brokering facilities.

The second major change will be the adoption of JSDL[26] as a job description language instead of the AJO. Resource requests expressed in AJOs or JSDL documents are largely compatible, both being abstract languages that can be rendered more complete through incarnation, and both stating what resources will be required for the job being run. JSDL is a standard language for job submission (reducing the complexity of the code to map onto non-UNICORE Grids) which supports not just simple job running but also web-services invocations and database queries. This allows the broker to be used in wider settings such as load-balancing of a pool of SOAP engines or distribution of queries across a federated database.

In the future, it will be possible to create a scheduler on top of the brokering architecture outlined in this paper. This will take advantage of the fact that the VO-based broker architecture will be able to offer both a good selection of QoS offers and, through usage record monitoring, estimates of how accurate those offers are and the likelihood of those offers being honoured. In this way, the ultimate structure of the brokered scheduled Grid will probably consist of schedulers that are very close to the top (where they take advantage of the way that the brokering architecture smooths the appearance of the Grid) and the bottom (where they can take advantage of the fine information available when deciding how to match up jobs and particular resources) of the structure[3] and a brokering network in between the schedulers.

References

1. I. Foster, C. Kesselman, and S. Tuecke. The Anatomy of the Grid: Enabl;ing Scalable Virtual Organizations. *International J. Supercomputer Applications*, 15(3), 2001.
2. The DataGrid resource broker. `http://server11.infn.it/workload-grid/`.
3. The NorduGrid resource broker. `http://www.nordugrid.org/papers.html`.
4. Maui Scheduler. `http://mauischeduler.sourceforge.net/`.

---

[3] There may also be schedulers at intermediate levels if there are organizations reselling the QoS offers, acting as a clearing house for interactions. This would parallel economic activity in many other fields within and outside the computer industry.

5. NaReGI (National Research Grid Initiative) project.
   `http://www.naregi.org/index_e.html`.
6. Contract Net specification.
   `http://www.fipa.org/specs/fipa00029/SC00029H.html`.
7. WS-Agreement specification.
   `http://www.ggf.org/Public_Comment_Docs/Documents/`
   `WS-AgreementSpecification_v2.pdf`.
8. Uniform Interface to Computing Resources. `http://www.unicore.org/`.
9. The EUROGRID project. `http://www.eurogrid.org/`.
10. K. Czajkowski, S. Fitzgerald, I. Foster, and C. Kesselman. Grid Information Services for Distributed Resource Sharing. In *Proceedings of the Tenth IEEE International Symposium on High-Performance Distributed Computing (HPDC-10)*. IEEE Press, 2001.
11. The GLUE Compute Element schema.
    `http://www.cnaf.infn.it/~sergio/datatag/glue/v11/CE/`.
12. J. Brooke, D. Fellows, K. Garwood, and C. Goble. Semantic Matching of Grid Resource Descriptions. In *Proceedings of Second European Cross-Grids Conference, Cyprus 2004*. LNCS, 2004.
13. The GRid Interoperability Project. `http://www.grid-interoperability.org/`.
14. The Globus project. `http://www.globus.org/`.
15. PCPACK. `http://www.epistemics.co.uk/Notes/55-0-0.htm`.
16. D. Snelling, S. van den Berghe, and V. Li. Explicit Trust Delegation: Security for Dynamic Grids. *Fujitsu Scientific & Technical Journal*, 40(2), 2004.
17. The UniGrids project. `http://www.unigrids.org/`.
18. Condor ClassAds. `http://www.cs.wisc.edu/condor/classad/`.
19. Relational Grid Monitoring Architecture. `http://www.r-gma.org/`.
20. Network Weather Service. `http://nws.cs.ucsb.edu/`.
21. Usage Record Working Group. `http://forge.gridforum.org/projects/ur-wg/`.
22. Markets for Computational Science project. `http://www.cs.man.ac.uk/cnc-bin/`
    `cnc_mcs.pl`.
23. WS-Resource Framework.
    `http://www.oasis-open.org/committees/tc_home.php?wg_abbrev=wsrf`.
24. Simple Object Access Protocol. `http://www.w3.org/TR/soap12/`.
25. Business Process Execution Language.
    `http://www-128.ibm.com/developerworks/library/ws-bpel/`.
26. Job Submission Description Language.
    `http://forge.gridforum.org/projects/jsdl-wg/`.

Topic 7
Parallel Computer Architecture and ILP

Theo Ungerer, Josep-Lluis Larriba-Pey, Kevin Skadron, and Pedro Trancoso

Topic Chairs

We welcome you to the Parallel Computer Architecture and Instruction Level Parallelism sessions of Euro-Par 2005 conference being held in Lisboa, Portugal.

Instruction Level Parallelism (ILP) and parallel processing techniques are present in most contemporary computing systems as they are very important and growing research fields. ILP research aims to extract fine-grained parallelism as well as thread-level parallelism not only from scientific code, but also from irregular, general code.

The scope of this topic includes parallel computer architectures, processor architecture and microarchitecture, the impact of emerging microprocessor architectures on parallel computer architectures, innovative memory designs to hide and reduce the access latency, multi-threading, and the impact of emerging applications on parallel computer architecture design.

This year 39 papers were submitted to this topic area. The majority of the papers came from the area of processor architecture and relatively few came from parallel systems. Among the submissions, 10 papers were accepted as full papers for the conference (26% acceptance rate). We are grateful to our referees for lending us their expertise and providing rigorous reviews. The accepted papers are grouped in three sessions according to the topic covered: Branch Prediction and Memory Hierarchy, Instruction Level Parallelism, and Parallel and Reconfigurable Architectures.

In the first session Monchiero and Palermo present the Combined Perceptron Branch Predictor, which consists of two concurrent perceptron-like neural networks. Moure *et al.* propose a mechanism, Target Encoding, that achieves a better ratio between the predictor accuracy and its size. Shi and Lee propose an efficient solution to scale the L1 cache based on the register-guided dynamic partition of memory reference instructions for partitioned L1 data cache. And Canal *et al.* present a scheme that compresses all values passing through a processor in order to reduce the energy consumption.

In the second session Zmily *et al.* introduce a block-aware ISA that helps accurate instruction delivery improving the energy consumption over traditional and decoupled front-ends. Sharky and Ponomarev propose a non-uniform instruction scheduler that achieves smaller scheduling delays. The same author also present an efficient wakeup-free instruction scheduler - instruction recirculation.

Finally, the third session starts with a work by Almasi *et al.* describing the early experiments on a 16384 node BlueGene/L. Vandeputte *et al.* analyze and improve the performance of state-of-the-art phase predictors, which are useful for hardware adaptation. Bardisa *et al.* present a lightweight directory architecture.

J.C. Cunha and P.D. Medeiros (Eds.): Euro-Par 2005, LNCS 3648, p. 485, 2005.
© Springer-Verlag Berlin Heidelberg 2005

The Combined Perceptron Branch Predictor

Matteo Monchiero and Gianluca Palermo

Politecnico di Milano
Dipartimento di Elettronica e Informazione, Via Ponzio 34/5, 20133 Milano, Italy
{monchier,gpalermo}@elet.polimi.it

Abstract. Previous works have shown that neural branch prediction techniques achieve far lower misprediction rate than traditional approaches. We propose a neural predictor based on two perceptron networks: the *Combined Perceptron Branch Predictor*. The predictor consists of two concurrent perceptron-like neural networks, one using as inputs branch history information, the other one using program counter bits. We carried out experiments proving that this approach provides lower misprediction rate than state-of-the-art conventional and neural predictors. In particular, when compared with an advanced path-based perceptron predictor, it features 12% improvement of the prediction accuracy.

1   Introduction

Modern high-performance microprocessors rely on sophisticated and accurate branch predictors to efficiently exploit Instruction Level Parallelism (ILP). Complex front-ends, capable of filling large instruction windows, are required to sustain high operating frequency and aggressive parallelism. Branch prediction is a key element of such a system, providing correct fetch beyond branch boundary and, therefore, large throughput instruction delivery.

Several advanced branch predictors have been suggested so far in the literature. Most of them are 2-bit counters table based predictors [1], organized in order to minimize interference which may occur in the counter tables. For example, the *2Bc-gskew* predictor [2] is composed of four 2-bit counter tables: a bimodal table (BIM), two gshare-like tables (G0 and G1) and a metapredictor table (META). Depending on the outcome of the META table, the final prediction is given either by the BIM table or by the majority vote of the predictions from the BIM, G0 and G1 tables. Other complex schemes have been recently proposed, e.g. the *Prophet/Critic* hybrid branch predictor [3], based on the combination of two components: the *prophet* and the *critic*. The critic uses both branch history and future to give a critique of the prophet prediction, which is used to make the final prediction for the current branch.

In this paper, we present an innovative branch predictor architecture, based on a neural approach, first proposed by Jimenez *et al.* in [4] (the *Perceptron* predictor). Our proposal features a novel mechanism, based on an additional *address-based* perceptron, using some PC bits as inputs, to achieve superior accuracy with respect to a single perceptron approach. Using PC bits as input

J.C. Cunha and P.D. Medeiros (Eds.): Euro-Par 2005, LNCS 3648, pp. 487–496, 2005.

of the neural network, the proposed predictor can separate branches otherwise collapsing in the same perceptron. We prove that this approach improves significantly prediction accuracy with respect to state-of-the-art conventional and neural predictors.

The paper is organized as follows: Section 2 introduces some background about neural branch prediction. In Section 3, our proposal is presented. Section 4 shows obtained experimental results. Finally, Section 5 concludes the paper.

2   Neural Branch Prediction

Branch predictors based on neural methods have been recently studied [4–7], showing that they are the most accurate predictors in the literature. In fact, neural networks can exploit much longer histories than conventional branch prediction algorithms, resulting in better performance.

The simplest neural network is the *perceptron*. For this network, the output signal, *pred*, is a non-linear function (*activation function*) of y, which is a linear combination of the network inputs, as stated by following equations:

$$pred = step(y) \tag{1}$$

$$y = \mathbf{w} \bullet [1 \ \mathbf{x}]^{\mathbf{T}} = w_0 + \sum_{i=1}^{i=N} w_i x_i \tag{2}$$

where w_i are $N + 1$ weights and x_i are N inputs; w_0 is called bias weight. The activation function can assume various shapes for a generic neural network. Perceptron uses the *step* function, which is a natural choice, when dealing with branch prediction patterns. The *step* function means taken when it is 1 and not-taken when 0. Vector $\mathbf{w} = [w_0, w_1, \cdots, w_N]$ is said *weight vector* and specifies the perceptron. Vector $\mathbf{x} = [x_1, \cdots, x_N]$ is the input vector, whose elements are the inputs of the network. Weights can be dynamically trained, so that prediction run-time adapts to the real taken/not-taken branch pattern.

The *Perceptron* predictor, presented in [4], uses perceptrons to predict branch outcome. It is a history-based predictor, since it maintains a Global Branch History Register (GBHR) and a set of local Branch History Registers (BHR), collected in a Branch History Table (BHT). A history register, obtained concatenating local and global history, is used as input of a perceptron network to perform the prediction. The perceptron to use is chosen by using the PC bits of the current branch. Weights are 8-bit integers and they are selected in a $n \times (h + 1)$ matrix, called *Weight Table* (WT). n and h are design parameters: n has the meaning of number of entries of the WT, while h is the size of the history register, which is the network input. Each row of the matrix is a $(h + 1)$-length weight vector, which determines the perceptron. When the prediction is performed, the least significant bits of the PC are used to select the row corresponding to the weight vector to use. A fast adder provides to generate the summation of the weights, according to applied inputs (see Equation 2), and a comparator makes the prediction (see Equation 1). Every time the effective

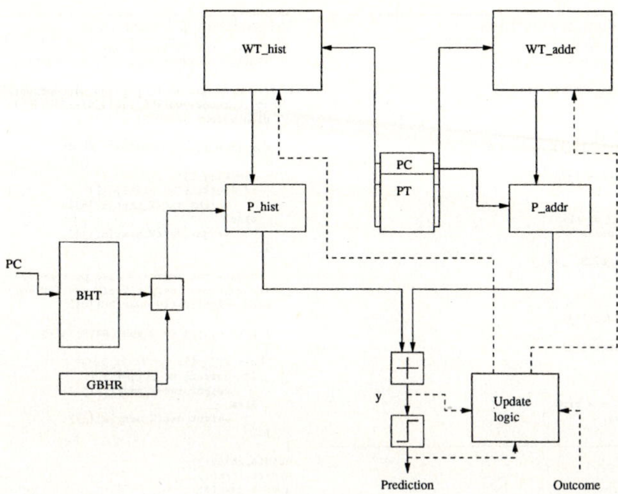

Fig. 1. Block diagram of the *Combined Perceptron Branch Predictor*

branch target is computed, the corresponding weight vector is updated, training
the weights values with the outcome pattern.

An improvement to the Perceptron predictor is the *Path-Based Perceptron*
predictor. Branch path information is used when selecting neurons to get superior
accuracy. The path of a branch is composed of the past branch PCs. In a path-
based perceptron, the i-th weight of the weight vector to use for the prediction, is
the element of the i-th column of the Weight Table, indexed by the i-th element
of the Path Table, that is $w_i = WT[PT[i]]$. This idea has been presented in [7],
where is proposed the *Fast Path-Based* Perceptron predictor. The predictor is
based on enhanced microarchitecture to minimize prediction latency. It staggers
computations in time, predicting a branch using a neuron selected dynamically
along the path to the branch, rather than selecting the neuron all at once.

In [8], Seznec proposes *pseudo-tagging*, a technique to reduce aliasing in the
perceptron table. Pseudo-tagging consists in using a few bits of the address of a
branch in the vector of weights. The author reports that this technique achieves
only a slight performance improvement with respect to the conventional perceptron predictor (2.52 % on average, with a 16KB predictor, on 10 SPEC2000
integer benchmark).

3   Proposed Predictor Architecture

The *Combined Perceptron Branch Predictor*, proposed in the paper, combines
two different kinds of Perceptron: a *history-based* one and an *address-based* one.
The address-based Perceptron has as inputs some bits of the PC. Its output is
sensitive to the branch address and, if combined with the output of the history-
based Perceptron, which is sensitive to branch history, adds a contribution which

Algorithm 1 Prediction algorithm

```
/* Calculate the output of the
   history-based perceptron */
y_hist=W_hist[PC][0];
for (j = 1; j <= HISTORY_LENGTH; j++)
{
   k = PT[j-1];
   if (history_reg[j-1])
       y_hist += W_hist[k][j];
   else
       y_hist -= W_hist[k][j];
}

/* Calculate the output of the
   address-based perceptron */
y_addr=W_addr[PC][0];
for ( j=1; j <= N_ADDR_BITS; j++)
{
   k = PT[j-1];
   if (PC[j-1])
       y_addr += W_addr[k][j];
   else
       y_addr -= W_addr[k][j];
}
y = y_hist + y_addr;

if ( y >= 0 ) prediction = true;
else prediction = false;
```

Algorithm 2 Update algorithm

```
if (last_prediction!=outcome ||
    (last_y <= THETA && last_y >= -THETA))
{

   /*update the history-based perceptron*/
   if (outcome) weight_inc(W_hist[PC][0]);
   else weight_dec(W_hist[PC][0]);

   for (j = 1; j <= HISTORY_LENGTH ; j++)
   {
      k = PT[j-1];
      if (outcome == hist[j-1])
          weight_inc(W_hist[k][j]);
      else
          weight_dec(W_hist[k][j]);
   }

   /*update the address-based perceptron*/
   if (outcome) weight_inc(W_addr[PC][0]);
   else weight_dec(W_addr[PC][0]);

   for (j = 1; j <= N_ADDR_BITS; j++)
   {
      k = PT[j-1];
      if (outcome == PC[j-1])
          weight_inc(W_addr[k][j]);
      else
          weight_dec(W_addr[k][j]);
   }
}
update_ghist();
update_lhist();
update_path();
```

Fig. 2. Algorithms for the prediction and update phase for the *Combined Perceptron Branch Predictor*

significantly improves the prediction accuracy. The basic idea of our proposal is to add to the final prediction a contribution to take into account branch address related information, dealiasing branch which collapsed in a single entry of the other component of the predictor.

Both subpredictors (history-based and address-based), which compose the whole predictor, are Perceptron predictors which exploit branch path information in the selection of the weight vector. The history-based predictor has the same structure of the Path-Based predictor described in the previous section. The address-based predictor uses a perceptron selected by the branch path, but the input of the perceptron are the least significant bits of the address of the current branch itself. We designed the update policy of this component to make that a weight is decremented/incremented if the corresponding input (that is, an address bit) has given a negative/positive contribution to the final prediction.

Furthermore, the activation function application is moved from the output of the two single subpredictors to the output of the whole predictor. In this way, each component cooperates in calculating the input value of the activation function, which is subsequently applied.

The whole predictor output is ruled by following equations, which replace Equation 1 and Equation 2.

$$pred = step(y) \tag{3}$$

$$y = \mathbf{w_{hist}} \bullet [1 \ \mathbf{h}]^{\mathbf{T}} + \mathbf{w_{addr}} \bullet [1 \ \mathbf{x}]^{\mathbf{T}} \tag{4}$$

where $\mathbf{w_{hist}}$ is the the weight vector of the history-based perceptron, while $\mathbf{w_{addr}}$ is the weight vector of the address-based perceptron. $\mathbf{h}$ and $\mathbf{x}$ are, respectively, the vectors of the input history and address bits. The activation function is the *step* function and applies the summation of the two components of the predictor.

Figure 1 shows the block diagram of the proposed predictor. We indicated as *WT_hist* and *P_hist* the Weight Table and the perceptron logic of the history-based subpredictor, while *WT_addr* and *P_addr* are related to the address-based one. Perceptron logic is substantially composed of an adder which sums selected weights depending on the inputs. The Path Table (PT) holds the branch path, that is, last branch PCs, and it is used to index into the Weight Tables. GBHR and a BHT store information related to branch outcome history and supply the history register which is the input of the history-based subpredictor. The *update logic* is the circuitry needed to update the predictor Weight Tables. Dashed lines represent data transfers needed by the update phase.

The prediction algorithm is shown, as C-like pseudocode, in Algorithm 1 (Figure 2).Weight Tables of both predictors are concurrently accessed to get weight vectors. The outputs of the perceptrons are calculated and summed together. Finally, a comparator decides the prediction whether the obtained value is greater or less than zero.

Update algorithm details are shown in Algorithm 2.The weights of the two subpredictors tables are modified on mispredictions or when the value of y is too small. A threshold is established for this purpose. Its value has been set, so that weight vectors are updated if y falls into a value range which is half of the weight range (that is, $THETA = 64$). The update is performed following the rule: $\Delta w = outcome \oplus input$, which means that a weight is incremented if the corresponding input has given positive contribution, otherwise it is decremented. The GBHR, the BHT and the PT are also updated in this phase.

3.1   Implementation Issues

Implementation of perceptron-based branch predictors has been studied in [4, 7]. The most critical component, heavily impacting on prediction latency, is the weights summation, which can be effectively implemented using a Wallace compressor. A pipelined implementation has been proposed for the Path-Based Perceptron [7], but it is feasible only considering global history. Local-global Perceptron predictors, if pipelined, would need too large hardware budget, since one pipeline per local history table (i.e. BHT) entry would be needed.

The architecture of the Combined Perceptron Branch Predictor can be implemented as shown in Figure 3. Since each column of the WTs is independently accessed by the program counter or by an element of the Path Table, WTs can be sliced and organized in banks, each of them containing as many ways as the Weight Tables (WTs) columns accessed by using the same addressing logic. The summation relative to the history-based and address-based perceptrons are implemented by a single block which generates the final value of the activation function, composed by a *Selective Inverter*, a Wallace tree and a parallel adder. The Selective Inverter, is a circuit to selectively invert fetched weights, according to the predictor inputs, i.e. the history register and a portion of the program counter bits. The Wallace tree is used to reduce the number of the operands to

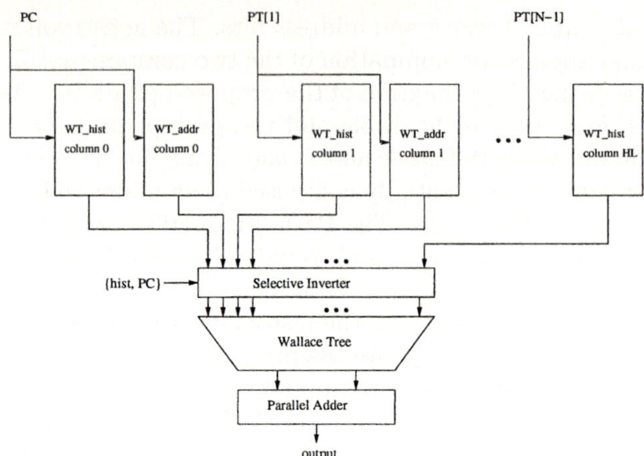

Fig. 3. Implementation of the *Combined Perceptron Branch Predictor*

2 operands, which are added by the final adder. The most significant bit of the output is the prediction.

The proposed implementation results only slightly more complex than the implementation of the Perceptron Predictor [4]. In fact, the same design can be adopted also for a Perceptron. The main difference is the width of the summation, and therefore of the Wallace tree. For example, regarding 8KB predictor as shown in Table 1, the Combined Perceptron requires 42 weights, while the Perceptron 27, resulting in 2 gate levels of the Wallace tree saving. Considering 90 nm CMOS process estimated latency for a prediction is 905 ps for the Combined Perceptron and 770 ps for the the Perceptron (-15%).

4   Experimental Results

In order to evaluate the proposed architecture, in terms of prediction accuracy, we measured misprediction rate for the proposed predictor and several different predictors. Reported results have been obtained using the Championship Branch Prediction (CBP) [9] framework, which is a trace-driven μop-based Intel IA32 simulation environment. Branch predictors were simulated on conditional branches of the given input trace. The Combined Perceptron predictor source code is available on the web [10, 11].

We used the instructions traces provided within the CBP framework. The benchmark set is composed of 20 traces, 30M instructions per trace, Each trace belongs to a specific class: INT (integer), FP (floating point), MM (multimedia), SERV (server). The INT/FP workloads are components of SPEC; the multimedia has some video/speech recognition; for the most part the server is tpcc/web server stuff.

In this paper, we compare the proposed predictor to well known state-of-the-art predictors:

Table 1. Simulated predictor configurations

		Total hardware budget		
		8KB	16KB	32KB
GShare	history length	15	16	17
2Bc-gskew	# entries (per table)	8K	16K	32K
	history length	13	14	15
Perceptron	WT # entries	304	443	630
	history length	26	36	51
Path-Based Perceptron	WT # entries	325	639	644
	global history length	17	19	40
	BHT # entries	2048	2048	2048
	local history length	4	4	7
Combined Perceptron	WT_hist # entries	137	214	493
	global history length	25	32	32
	BHT # entries	2048	2048	2048
	local history length	4	5	11
	WT_addr # entries	254	462	457
	# address bits	11	14	17

- **GShare.** It is a global two-level adaptive predictor, which uses the XOR between the global history and the branch address to minimize aliasing in the 2-bit counter table [1].
- **2Bc-gskew.** We simulated the predictor proposed by Seznec [2].
- **Perceptron.** This is the Jimenez's *Perceptron* predictor proposed in [12]. Only a global history information is used to compute the perceptron output.
- **Path-Based Perceptron.** It is an improved version of the *Perceptron* predictor. Weights are selected exploiting path information and a mixed local/global history is used.

Parameters space of the simulated predictors, including the proposed predictor, has been explored. More than 10,000 random generated configurations have been simulated and best predictors have been selected. Table 1 reports the parameters values of the optimal configurations for each predictor.

Figure 4 shows the average misprediction rate of the different branch predictors, when the size is varied. We simulated 8KB, 16KB and 32KB predictors. It can be observed that the Combined Perceptron Branch Predictor features the smallest misprediction rate for every size ranging from 3.5 to 2.8 mispredictions/K-instruction. The overall behavior of the proposed predictor is 12.5% better than the optimized configuration of the Path-Based predictor and 34% better than the GShare predictor.

In Figure 5, the plot of the misprediction rate of the simulated predictors for each benchmark, for the size of 8KB, is reported. It can be observed that the Combined Perceptron predictor achieves better performance over the other predictors for every benchmark (except for INT-4 and MM-1). A large decrease of the mispredictions is evident on integer and server benchmarks. All the simulated predictors behave homogeneously across all the benchmarks. On the other hand, the 2Bc-gskew predictor clearly results in fewer misprediction on integer programs than on server ones.

Results about exploration of different addressing modes into the Weight Tables are reported in Figure 6, for both the Perceptron Predictor and the Com-

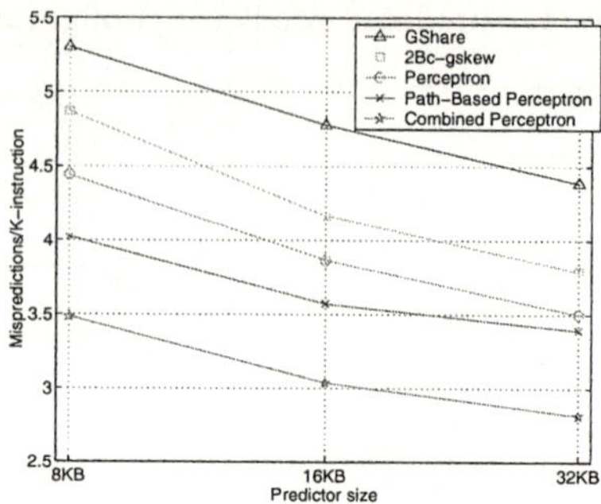

Fig. 4. Average misprediction rate of different branch predictors and sizes

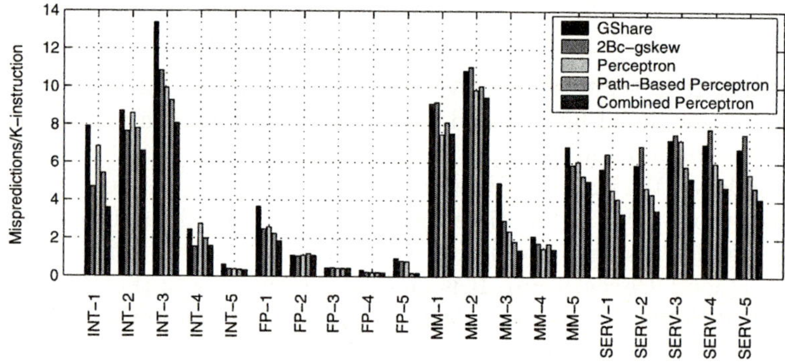

Fig. 5. Misprediction rate of the simulated branch predictors for different benchmarks and resource budget of 8KB

bined Perceptron Predictor. We consider *pure* addressing mode, which means that Weight Tables are accessed only by global history bits, and the *local-global* one, which mixes local and global history. Furthermore, the *path-based* and *local-global path-based* addressing modes, which use also path information. Observing Figure 6(a), it is evident that *path-based* solutions feature good accuracy for a relatively small history length (up to 24–26), while for longer history performance decreases rapidly. This is mainly due to path interference, which occurs in the path-based predictors, since weights ideally belonging to different path may collapse into the same weight, because some weights of the paths may physically overlap. This does not happen for the others two configurations, because a weight vector is maintained for each branch. *Local-global* history significantly impact on predictor performance. In fact, 10% accuracy improvement is achieved by using local-global addressing. Although *local-global* predictors feature more complex implementation, since pipelining is not possible, they represent a worthwhile choice.

Figure 6(b) shows addressing mode analysis for the Combined Perceptron Branch Predictor. While path interference makes that path information is not

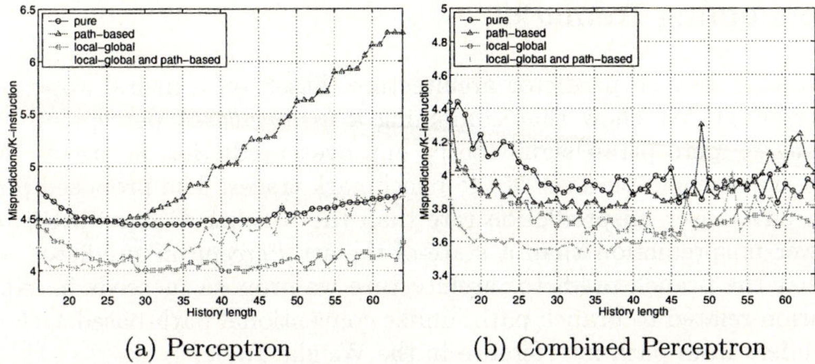

(a) Perceptron   (b) Combined Perceptron

Fig. 6. Misprediction rate versus history length, considering different addressing modes into the Weight Tables (8KB predictor size)

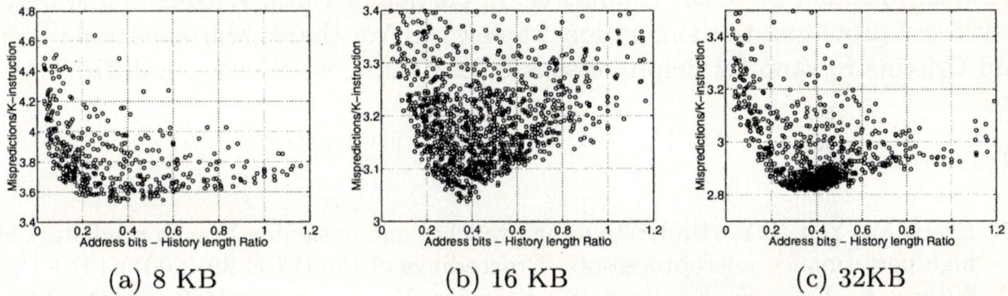

(a) 8 KB   (b) 16 KB   (c) 32KB

Fig. 7. Address bits – History length Ratio versus misprediction rate for the simulated configurations and different predictor size

effectively exploited in the Perceptron Predictor, the Combined Predictor succeeds in getting far lower misprediction rate by de-interfering paths. In fact, for history length shorter than 40 bits, up to 8% performance gain is obtained by *path-based* predictor, both for the *local-global* and the *global* one. In addition to this it is evident that the *local-global path-based* predictor reaches a misprediction rate minimum, since effectively exploit information related to both path and local-global history.

Figure 7 shows scatter plots of the misprediction rate versus the *Address bits – History length Ratio* (defined as the ratio of the number of address bits used as input of the address-based predictor and length of the history used as input of the history-based predictor) for the simulated configurations and predictor size of 8KB, 16KB, 32KB. These results show that the contribution of the *address-based* component significantly improves perceptron predictor accuracy. The phenomenon is evident for each predictor size: if the points on the Pareto curve of each plot are observed, a minimum for the misprediction rate can be found, when the value of the Address bits – History length Ratio is approximately 0.4.

5   Concluding Remarks

An innovative branch predictor architecture, based on a neural approach, has been presented. We show that combining a *history-based* perceptron with an *address-based* perceptron significantly improves prediction accuracy. We carried out experiments on a set of the benchmark traces. The proposed predictor achieves 34% lower misprediction rate than the baseline GShare predictor and 12% lower misprediction than a state-of-the-art *Perceptron* predictor. Results prove that the branch predictor architecture we propose succeeds in exploiting information related to branch path, unlike conventional path-based architecture which suffers from path interference in the Weight Table.

Acknowledgments

We wish to thank the CBP committee, in particular Chris Wilkerson and Jared Stark, for providing the simulation framework. We thank Mariagiovanna Sami and Cristina Silvano for helpful discussions. Finally, we thank μ-Lab IPECAs.

References

1. Evers, M., Yeh, T.Y.: Understanding branches and designing branch predictors for high performance microprocessors. Proceedings of the IEEE **89** (2001) 1610–1620
2. Seznec, A., Felix, S., Krishnan, V., Sazeides, Y.: Design tradeoffs for the Alpha EV8 conditional branch predictor. In: Proceedings of ISCA'02. (2002)
3. Falcon, A., Stark, J., Ramirez, A., Lai, K., Valero, M.: Prophet/critic hybrid btanch prediction. In: Proceedings of ISCA'04. (2004)
4. Jimenez, D.A., Lin, C.: Neural methods for dynamic branch prediction. ACM Transactions on Computer Systems **20** (2002) 369–397
5. Vintan, L.N., Iridon, M.: Towards a high performance neural branch predictor. In: Proceedings of the International Joint Conference on Neural Networks. (1999)
6. Egan, C., Steven, G., Quick, P., Anguera, R., Vintan, L.: Two-level branch prediction using neural networks. Journal of Systems Architecture **49** (2003) 557–570
7. Jimenez, D.: Fast path-based neural branch prediction. In: Proceedings of MICRO-36. (2003)
8. Seznec, A.: Redundant history skewed perceptron predictors: Pushing limits on global history branch predictors. Technical Report 1554, IRISA (2003)
9. CBP URL: (www.jilp.org/cbp/)
10. Monchiero, M., Palermo, G.: The combined perceptron branch predictor. Technical Report 2004.35, Politecnico di Milano (2004)
11. URL: (www.elet.polimi.it/upload/monchier/high-performance-bp.htm)
12. Jimenez, D., Lin, C.: Dynamic branch prediction with perceptrons. In: Proceedings of HPCA-7. (2001)

Target Encoding for Efficient Indirect Jump Prediction*

Juan Carlos Moure[1], Domingo Benitez[2], Dolores Isabel Rexachs[1], and Emilio Luque[1]

[1] Computer Architecture and Operating Systems Department,
Universidad Autónoma de Barcelona, 08193 Barcelona, Spain
{JuanCarlos.Moure,Dolores.Rexachs,Emilio.Luque}@uab.es
[2] University of Las Palmas G.C., 35017 Las Palmas, Spain
dbenitez@dis.ulpgc.es

Abstract. Accurate indirect jump prediction is critical for some applications. Proposed methods are not efficient in terms of chip area. Our proposal evaluates a mechanism called target encoding that provides a better ratio between prediction accuracy and the amount of bits devoted to the predictor. The idea is to encode full target addresses into shorter target identifiers, so that more entries can be stored with a fixed memory budget, and longer branch histories can be used to increase prediction accuracy. With a fixed area budget, the increase in accuracy for the proposed scheme ranges from 10% to up to 90%. On the other hand, the new scheme provides the same accuracy while reducing predictor size by between 35% and 70%.

1 Introduction

Dynamic control-flow prediction is a key task on current processors. This work proposes an efficient mechanism for predicting indirect jumps. Although they are less frequent than conditional branches, for some applications the lack of a specialized indirect jump predictor may degrade performance significantly [7], [8].

Common sources of indirect jumps are case statements and virtual function calls used in object-oriented languages. While some indirect branches jump to a unique target address during the program's execution (*monomorphic* jumps), and are easy to predict, many of them (called *polymorphic*) jump to several target addresses, depending on input data, and their prediction is complex. Accurate prediction for those jumps requires a multiple-choice predictor, rather than a mere binary (taken/not-taken) predictor, and storing several target addresses per jump.

Indirect jump predictors proposed in the literature match the scheme depicted in Fig. 1. One or more tables are indexed using the jump's address and *branch history*, which codifies the outcomes of recently executed branches (indirect or not) leading up to the jump. Tables contain full target addresses and additional data that is used to select the final predicted address. Tables are trained using the outcome of indirect jumps once they are retired from the processor pipeline.

---

* This work was supported by the MCyT-Spain under contract TIN 2004-03388, the Generalitat de Catalunya - Grup Recerca Consolidat 2001 SGR-00218, and the HiPEAC European Network of Excellence

J.C. Cunha and P.D. Medeiros (Eds.): Euro-Par 2005, LNCS 3648, pp. 497–507, 2005.

The predictor's accuracy is mostly limited by its memory size and by the length and quality of the branch history. A separate entry is allocated into the predictor's tables for each combination of jump address and branch history. As the history gets larger, the probability of containing a previous branch that correlates with the predicted branch also gets larger, increasing prediction accuracy [5]. However, more entries are required in the predictor's tables or otherwise many prediction cases will map into the same entry and create *aliasing*. The indexing and selection methods try to reduce the effect of aliasing, making efficient use of the available predictor's memory.

Although larger tables provide higher accuracy, they do not handle information efficiently, since the same long target addresses are replicated several times. We present and evaluate a method to encode full target addresses into shorter target identifiers. The proposed two-stage mechanism consists of (1) predicting a short target identifier using the scheme shown in Fig. 1, and then (2) translating it into a full target address. Since encoded targets are shorter, more entries can be stored with a fixed memory budget, and then longer histories can be used to increase prediction accuracy. The table used in the second stage holds full target addresses and still requires large entries, but since each address is stored only once, it requires considerably fewer entries.

Results obtained in simulation indicate that the design achieves a better ratio between prediction accuracy and predictor size. This increase of storage efficiency may be used to increase performance or to lower area requirements and the predictor's power consumption. The proposed two-stage scheme increases the indirect predictor's average response latency, but this increase is shown to have very little effect on performance. On a 4-way superscalar processor with a realistic memory hierarchy, with a penalty of 2 cycles for using the two-level jump predictor, the performance improvement due to target encoding ranges from 0.1% to 2.5%, depending on the benchmark.

Section 2 reviews some related work. Sections 3 and 4 describe the baseline and the proposed indirect jump predictors. Section 5 presents the experimental methodology and some preliminary results. Full results are presented and discussed in section 6. The final section outlines the conclusions and introduces future lines of research.

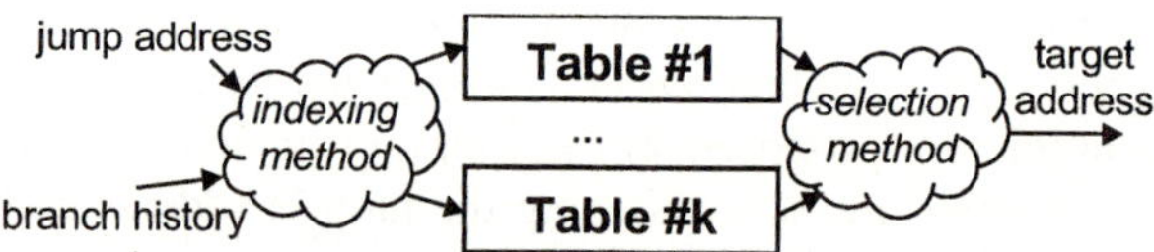

Fig. 1. General scheme of an indirect jump predictor

2 Related Work

A Branch Target Buffer (BTB) [13] provides a simple method for accurately predicting monomorphic indirect jumps or jumps whose target changes infrequently, but provides weak results for polymorphic jumps. Adding a hysteresis bit to limit the update of the target address only after two consecutive mispredictions [3] provides small gains [5, 6]. A better method for dealing with polymorphic jumps is the Target Cache (TC) [5]. It adapts the two-level prediction methods previously proposed for conditional branches [14], to indirect jumps. [5] analyzes several methods to track branch history, several indexing methods, and the use of tags on the TC.

A variation on the TC is the Cascaded Indirect Jump Predictor [6], which significantly reduces the table size needed to achieve a given accuracy. It dynamically identifies easily predicted jumps and devotes them a simple and low cost predictor, preventing insertion of these jumps into a more powerful second stage predictor. The result is that easily predicted jumps avoid most cold start misses and do not waste entries in the second stage predictor, which is better exploited by the remaining indirect jumps. Using a BTB as the first-stage filter, as in the Intel Pentium M and Pentium 4 processors [7, 8], provides good results with a simple and efficient design. We evaluate our proposal using this scheme as the baseline design (described in the next section).

Prediction by Partial Matching (PPM) [12] exploits variable-length path correlation to improve prediction accuracy. Several tables are accessed in parallel, each one addressed by an index containing a branch history of a different length. The table using the longest history and having a valid prediction provides the final target. The potential of varying the history length for each specific branch is not analyzed in this paper.

Control-flow prediction performance is improved by either increasing accuracy, width (instructions retired per prediction), or rate (predictions per cycle). We have previously proposed a two-level hierarchy [14], but aimed to increase prediction rate and not to improve the ratio between prediction accuracy and predictor size.

A method to increase prediction width is path-based next trace prediction [9]. It uses a cascaded, two-level scheme to predict instruction traces, rather than jumps or branches. Our baseline design uses the exclusive-or-fold method proposed there.

3   Baseline: A BTB-Based Cascaded Predictor

The baseline design used to evaluate our proposal is a cascaded indirect jump predictor [6, 7] using two tables (see Fig. 2). The first table is a Branch Target Buffer (BTB) [13], which identifies branches in the instruction fetch stream and provides a target address for each branch. Since the BTB always correctly predicts monomorphic jumps, a specialized Indirect Jump Predictor (IJPred) exploiting branch correlation is only required for polymorphic indirect jumps. We assume that an extra 32-entry return address stack (RAS) [11] (not shown) is used to predict indirect return jumps.

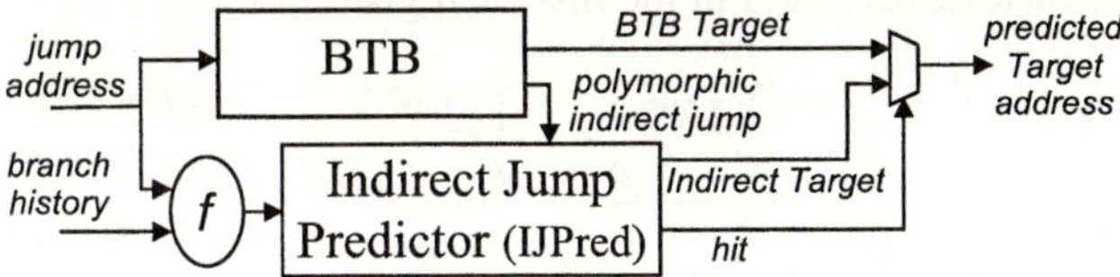

Fig. 2. Cascaded, two-level indirect jump predictor. At prediction time, a Branch Target Buffer (BTB) identifies branches and filters the use of a specialized Indirect Jump Predictor (IJPred)

Extensive simulation has been performed to obtain a realistic, highly tuned baseline design. We have simulated IJPred sizes from 256 to 16K entries, IJPred tag sizes from 0 to 16 bits, history lengths from 1 to 61, hysteresis counters from 0 to 3 bits, different ways of building and codifying branch history, and several indexing and selection algorithms. The most important results are explained in the following description.

Updating the BTB and IJPred at Retire Time (Not Speculatively)
We assumed a BTB with 4K entries and 16-bit tags, which suffers a low miss rate. At retire time, a BTB entry is allocated for each branch that misses in the BTB, and initialized with the branch type and target address. If the branch is an indirect jump (not a return), the type field is set as monomorphic. If the same indirect jump is later retired and its target address does not match the address stored in the BTB, then the type field is set as polymorphic but the target address prediction is **not** modified.

The IJPred is direct-mapped and 4-bit tagged. It is updated at retire time only for indirect jumps identified as polymorphic by the BTB, and only when their target address differs from the BTB prediction. This filtering scheme prevents prediction cases that are well-handled by the BTB from wasting IJPred memory space.

A hysteresis bit is used to avoid replacing IJPred entries that frequently provide correct predictions. The bit is cleared when the entry is first allocated, and is set on correct target predictions, and cleared on wrong predictions. On an IJPred miss, the previous entry contents are replaced by the new ones only when the hysteresis bit is found to be cleared. If the bit is found set, then it is cleared.

Using Two Types of Global Path History
Two separate 61-bit history registers are used: *cghr* is updated for each conditional branch taken, and *ighr* is updated for each indirect jump (including returns). As was noted in [12], maintaining two history registers of different types and dynamically choosing one of these for each static jump provides improved accuracy compared to using a unique history register. Our approach merges the contents of *cghr* and *ighr* using an exclusive-or operation before using history to generate the IJPred index (Fig. 4 on next page). It provides a similar improvement in accuracy (from 2% to 30%, depending on the benchmark) with a simpler implementation.

Both history registers, *cghr* and *ighr*, are speculatively updated using the outcome of the current prediction, and are corrected on branch mispredictions using a very small amount of recovery data. The update consists of shifting the contents s bits to the left and adding the exclusive-or of the s lower bits of the branch address and the branch target address, as shown in Fig. 3. The *history length*, l, is the number of branches whose histories are held in the history register, ($l = 61 / s$).

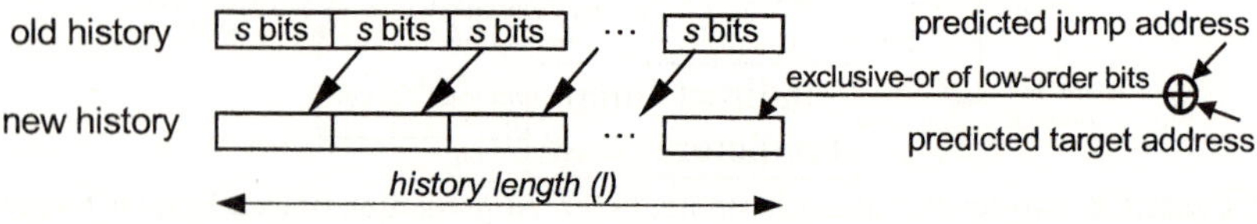

Fig. 3. Speculative update of branch history registers (corrected on branch mispredictions)

As other authors [10],[12], we have found that *l* highly influences accuracy. Also, for each IJPred table size, using the optimal *l* for each single benchmark (*BestHist*) instead of using the optimal *l* for all the benchmarks (*BestHistALL*) increases accuracy between 15% and 45%. In our experiments, we have used *BestHistALL* most of the time but has also validated that results do not significantly vary if using *BestHist*.

Indexing and Selection Algorithms at Prediction Time
The goal of an indexing scheme is to map the whole input data into *n* output bits so that the resulting index is evenly distributed, and aliasing is reduced. Fig. 4.a shows an scheme to fold a value, *v*, into an *n*-bit value using an exclusive-or-fold method [9].

The index for the BTB (Fig. 4.b) is the result of *xor-folding* the address of the branch to be predicted into chunks of different size (sizes are prime numbers) and the combination of these chunks into a large value that is again *xor-folded* into a final n-bit index. The best results are obtained with a skewed-associative scheme [1], which generates a different BTB set index for each possible BTB way.

The index for the IJPred uses the address of the jump to be predicted and the history registers (Fig. 4.c). These complex indexing methods increase accuracy slightly with respect to simpler ones but, most importantly, provide highly homogeneous results for all the configurations evaluated. The scheme is not intended to be an implementation proposal, but a reference for exploring cheaper and faster methods that merge fewer bits in this latency-critical step.

On a BTB miss, the fall-through address is predicted as the jump's target. On a hit, the BTB provides the target prediction, unless the indirect jump is identified as polymorphic or as a return. For polymorphic jumps, the IJPred is accessed and provides the prediction only if the IJPred access hits. Return jumps are handled by the RAS.

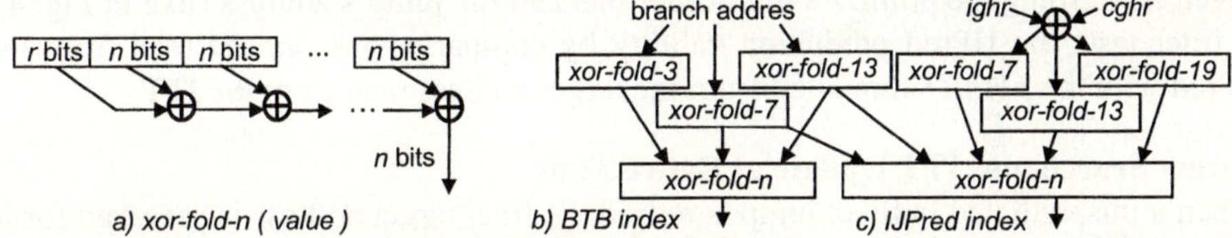

Fig. 4. Index generation scheme for BTB and IJPred using an xor-fold scheme

4 Indirect Target Encoding

The IJPred contains two types of data: (1) which of the possible paths a jump will take, and (2) at which address this path begins. Separating these two types of data on two different tables provides more efficient memory usage. If a jump can take *k* different paths, then $\log_2 k$ bits are required to codify a path identifier (*pathID*). Then, if the IJPred holds *pathID*'s instead of full addresses, it may contain more entries, use longer histories, and then increase prediction accuracy.

A second table, the Indirect Target Table (ITT), is required to provide the full target address (see Fig. 5). Using both the *pathID* and the jump's address to index the

ITT provides the best results. The ITT should ideally be able to hold all the target addresses taken by all jumps, but in practice only useful targets need to be stored, since storing those targets that rarely involve a correct prediction only marginally improves performance.

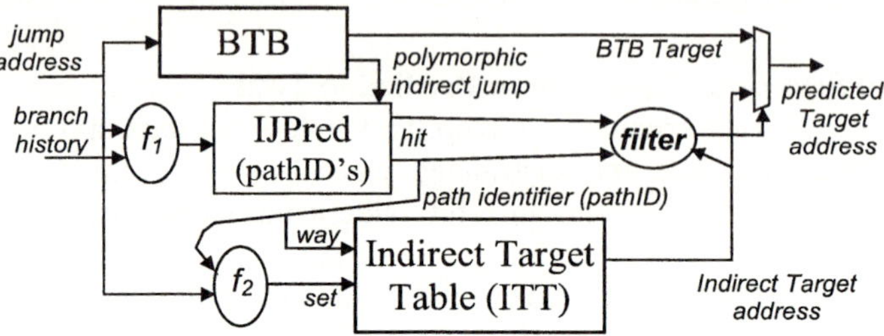

Fig. 5. Indirect Target Encoding. A short path identifier (*pathID*) replaces target addresses in the IJPred. A second-level Indirect Target Table (ITT), indexed by the jump address and the *pathID*, provides the full indirect target address

ITT Access at Prediction Time

Exploiting freedom in the target-encoding algorithm allows implementing a *k*-way set-associative ITT with the small access delay and power consumption of a direct-mapped tag-less table. The scheme is similar to the next cache line and set predictor [4]. The low order bits of the *pathID* codify the ITT way in which the target address is located. The target address is hashed (like in Fig. 4) to provide the *pathID*'s high-order bits. With this mechanism, the complexity of the associative indexing scheme is avoided at prediction time (where may affect performance), and is suffered at retire time, but only on IJPred mispredictions.

The ITT way is obtained from the *pathID* read from the IJPred. The ITT set is obtained by hashing the *pathID*'s high-order bits and the jump's address (like in Fig. 4). A filter tests the IJPred prediction validity by comparing the *pathID* read from the IJPred with the *pathID* computed from the target address read from the ITT.

Target Search and ITT Update at Retire Time

When a mispredicted indirect jump is retired, its final target address is searched for in the ITT. Since these cases are unfrequent and we will show that the ITT update latency does not affect performance, the search operation on ITT ways may be done serially to reduce H/W complexity. The *pathID*'s high-order bits are generated from the correct target address and combined with the jump's address to index each one of the ITT ways. As for the BTB, a skewed-associative scheme provided the lowest miss rate [1].

Each ITT entry contains a 4-bit saturating counter that is increased with each correct target use. When a new target address must be allocated, the counter of all the entries that are a potential placement are decremented, and only a zero counter enables the replacement. This policy prioritizes useful target addresses and reduces the number of ITT replacements, which also reduces the occurrence of ITT misses.

5 Experimental Methodology, Results, and Discussion

We use a trace-based simulation to measure prediction accuracy and tune the main design parameters. Accurate cycle-by-cycle simulation is used to measure the effect on prediction accuracy of the delayed update of prediction tables (BTB, IJPred, ITT), and the effect of prediction latency and accuracy on the processor's performance.

First, we analyze the design space of the Indirect Target Table (ITT) and select an optimal configuration. Then, we explore the best size for the IJPred tags and *pathID* field. We compare the baseline design and the proposed target encoding design with respect to prediction accuracy and predictor size. Then we verify that the effect on prediction accuracy of the delayed update of prediction tables is insignificant, and that increasing the indirect predictor's latency reduces performance slightly.

Metrics, Simulator, and Benchmarks

Prediction accuracy is measured as the average number of instructions between jump mispredictions (Kilo-instructions per misprediction). Predictor size is measured in KBytes (KB). Processor performance is measured in instructions per cycle (IPC).

We have used the Simplescalar-Alpha tool set [2] to generate the dynamic instruction trace of the first 20 billion instructions for some programs of the SPEC benchmark suites. Table 1 shows the selected benchmarks and their inputs (Alpha ISA, cc DEC 5.9, –O4). They have been selected because they have the lowest accuracy when a simple BTB is used for indirect jump prediction (6th column of Table 1), and then may benefit more from using a specialized predictor (col. 7-8 for an IJPred of 512 and 4K entries). Table 1, columns 4-5, shows the number of static polymorphic indirect jumps and targets.

Table 1. Simulation data and simulation results for selected SPEC benchmarks

Benchmark	input	SPEC	polymorphic		instructions / misprediction		
			jumps	targets	BTB	IJPred 512	IJPred 4K
crafty	reference	*int00*	15	89	877	1457	1908
eon	cook	*int00*	10	20	681	45123	> 10^5
gap	reference	*int00*	49	146	1609	46724	> 10^5
gcc	expr.i	*int00*	183	975	439	1765	3137
perl	difference	*int00*	65	659	118	763	2190
vpr	route	*int00*	4	19	4951	21291	> 10^5
m88ksim	reference	*int95*	7	26	1407	18355	56805
li	reference	*int95*	7	71	989	5381	14799
facerec	reference	*fp00*	9	47	2766	> 10^7	> 10^7
fma3d	reference	*fp00*	12	23	2977	> 10^7	> 10^7
sixtrack	reference	*fp00*	8	49	601	4248	9795

ITT Configuration

Results have shown that only a small subset of all the target addresses of polymorphic jumps needs to be held in the ITT for near-optimal performance. Although some benchmarks have more than 500 targets (see Table 1), a 64-entry ITT is enough to achieve a miss rate lower than 0.1%, except for benchmark *perl*, which requires 128 entries. For such sizes, an 8-way set-associative ITT provides the best performance.

A 4-bit *pathID* is enough to maintain the ITT miss rate below 0.1% for all benchmarks except for *gcc* and *perl*, which require a 5-bit and a 6-bit *pathID*, respectively.

IJPred Configuration

The baseline IJPred configuration, with entries containing full addresses, may afford a large tag and a large hysteresis counter to try to reduce IJPred misses. Results have shown that a tag larger than 4 bits, or a hysteresis counter larger than 1 bit improves performance very slightly on a direct-mapped IJPred organization.

When target encoding is used to improve chip area utilization, it is more effective to reduce the tag size to 3 bits and devote 5 bits to codify the *pathID*. The filter method described in section 4 detects a significant part (20-70%) of the misses not detected by the shortened tag. It also also detects (but does not correct) cases where ITT replacements have made the *pathID* in the IJPred indicate a wrong ITT way.

Accuracy Versus Storage Size

Figure 6 displays accuracy versus storage size for some representative benchmarks and for different predictor designs with an IJPred of 512-4K entries. The storage size of the baseline design accounts for the target address' size (32 bits), the tag's size (3 bits), and the hysteresis counter's size (1 bit). Target encoding replaces the target address by the *pathID* (5 bits instead of 32 bits) and must account for the ITT size (64 entries, each containing a full target address and a 4-bit replacement counter).

Results in Fig. 6 show that the ratio of accuracy versus predictor size is always better for the encoded design. Benchmarks *li* and *sixtrack* are depicted together because they have very similar results. For these benchmarks, correlation is highly effective in increasing accuracy. Given a fixed predictor size, the encoded scheme exploits correlation better than the baseline (the line depicting accuracy versus size separates for larger predictors). For example, with a 5-KB predictor, accuracy improves by 90%.

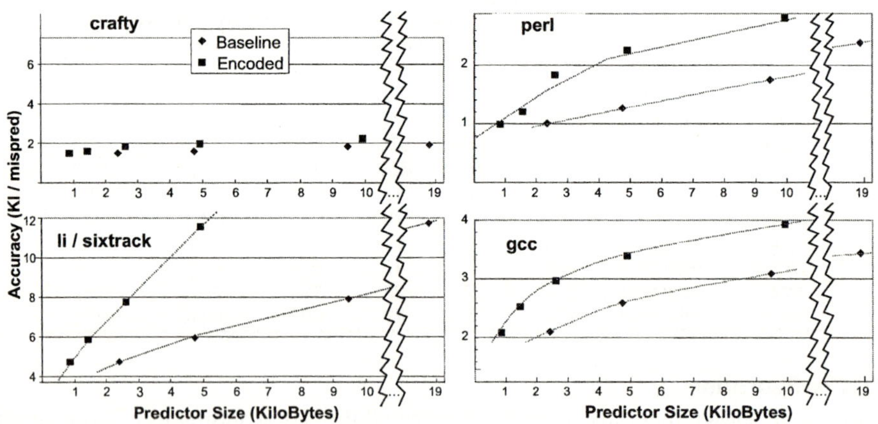

Fig. 6. Accuracy versus storage size on selected benchmarks for the baseline and encoded predictors, with IJPred sizes from 512 to 4K. 64 ITT entries, *pathID* length is 5 bits, IJPred tag length is 3 bits, history length (*l*) is *BestHistALL*, which depends on IJPred size: 512 (*l*=4), 1K (*l*=8), 2K (*l*=14), 4K (*l*=20))

The encoded scheme achieves accuracy improvements of around 50% for a fixed predictor size for benchmarks *gcc* and *perl*. With a large working set of indirect target addresses, the small ITT and short *pathID* provokes a moderate amount of ITT misses (<2%) that reduces the potential accuracy improvement by only around 10%.

For a relatively large area budget for the indirect predictor, two kind of benchmarks cannot benefit from the encoded scheme to improve processor performance. The first example is *crafty*, which benefits little from history correlation, and increases accuracy very slowly with higher storage. The other example are the benchmarks not shown in Fig. 4, which provide near-perfect prediction with a relatively small IJPred of around 1K entries. The encoded scheme, however, is still very useful in reducing storage (and power) requirements. For example, averaging for all benchmarks, a 3-KB encoded predictor provides the same accuracy as a 10-KB baseline predictor.

Table 2. Microarchitecture parameters for the cycle-accurate simulations

Front-End	Back-End (Execution Core)	Memory System
Decoupled: Fetch Target Queue (FTQ) holds up to 24 fetch blocks	Up to 4 instructions **renamed** and **dispatched** per cycle	Perfect Memory Disambiguation, Store to Load forwarding of any size
I-Cache Prefetches: two I-cache checks per cycle using FTQ contents. If not found, prefetch from L2 cache.	Up to 6 instructions **issued** and **retired** per cycle	**I-Cache**: 256 sets x 2 ways x 32B blocks **D-Cache**: 256 sets x 4 ways x 16B blocks **L2-Cache**: 1K sets x 8 ways x 64B blocks **L3-Cache**: 8K sets x 8ways x 128B blocks
Fetch Predictor: predicts one conditional branch per cycle, one return every two cycles, one indirect jump every *k* cycles.	**Fetch Queue**: 12 instructions **Issue Queue**: 24 instructions **Reorder Buffer**: 124 instructions **Load/Store Queue**: 48/24 instr.	**I-Cache / D-Cache**: 2 ports, 2-cycle latency **L2-Cache**: 7-cycle load-use latency **L3-Cache**: 40-cycle load-use latency **Memory**: 200-cycle load-use latency
Gshare: 64K entries, 2-bit counters	Operation **latencies** like Pentium IV	**Bandwidth** (L2/L3/Mem): 25,6 / 12,8 / 6,4 GB/s

Performance Measures

Cycle-level simulations have been performed by modeling a 4-way processor back-end and a realistic memory system (see Table 2). The simulated front-end is decoupled and predicts one full basic-block per cycle. Our first result was that prediction accuracy is not degraded by the delayed update of prediction tables. Accuracy varies very slightly when increasing the pipeline depth (and then the predictor update delay) from 12 to 30 cycles. As argued in other papers, a higher update delay increases the predictor's hysteresis, which does not necessarily causes a worse behavior.

Figure 7 shows the effect on performance of varying IJPred size and varying the indirect predictor's latency. On the one hand, doubling IJPred size, and therefore increasing indirect jump prediction accuracy, provides an average IPC increase of around 0.5% for the benchmarks considered in this paper (Fig. 7.a). Benchmark *perl* (Fig. 7.b) is the one that benefits most from a larger IJPred (average IPC increase of 1.2% when doubling IJPred size). The average penalty of indirect jumps has been experimentally found to be around 21 cycles, which explains why avoiding mispredictions results in a significant gain in performance.

On the other hand, a two-cycle delay penalty for using the IJPred table reduces IPC by less than 0.05% of the average. There are two main reasons for this result. First, jump mispredictions are not too frequent (less than 1 every 100 instructions), since many of the indirect jumps (30-70%) are predicted by the BTB. Second, a substantial part (more than 95%) of the delay due to using the IJPred and ITT tables instead of the BTB, is overlapped by other stalls occurring later in the pipeline. More than 60% of the overlap is due to the decoupled front-end scheme, which compensates *IJPred-use* stall cycles with cycles where a branch prediction provides several instructions (full basic blocks) to the front-end. However, as predicted by Amdhal's law, the indirect prediction latency becomes more critical for values larger than 3 cycles or if the execution width of the processor is scaled to 8 instructions per cycle.

Given that prediction latency is not critical, power consumption is afforded by delaying the IJPred access until the BTB access has been completed and a polymorphic jump has been identified. Similarly, power is saved by delaying the ITT access until a valid *pathID* from the IJPred table is read.

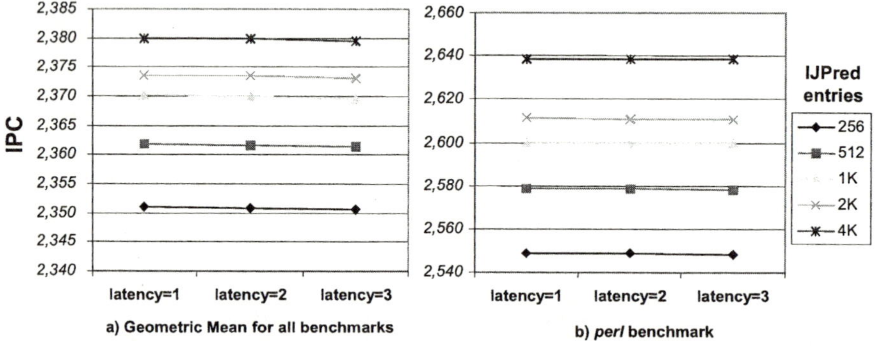

Fig. 7. IPC for varying IJPred sizes and IJPred latencies

6   Conclusions and Future Lines

We have presented and evaluated target encoding as a method for improving the indirect jump prediction accuracy to cost ratio. This improvement can be used to increase processor performance for benchmarks that execute a moderate amount of polymorphic indirect jumps. On a realistic 4-way superscalar processor with a realistic memory hierarchy, the additional latency of the proposed two-level predictor has very slight effects on performance. Assuming a two-cycle increase in latency, the performance increase ranges from 0.1% to 2.5%. For benchmarks that benefit little from larger IJPred tables, the scheme may be used to reduce chip area and power consumption. For example, a 3-KB encoded predictor (with direct-mapped access) provides the same accuracy as a 10-KB baseline predictor (with set-associative access).

The target-encoding scheme works well because indirect jumps have a small working set of target addresses, which can be effectively cached in a table with 64 entries. The careful design of the table achieves several conflictive issues: high logical associativity to reduce conflict misses, and a small latency and low power consumption due to its direct mapped access. Basically, the freedom of the target-encoding algorithm allows for the implementation of a way prediction mechanism for free. Also, the replacement policy is designed to prioritize useful targets instead of frequent targets.

The relatively small effect on performance of the enhanced indirect predictor is very related to the low frequency of indirect branches in the SPECint2000 workload. A future extension to this work is analyzing more object-oriented workloads such as SPECjvm98.

Static and profile analysis may improve indirect jump prediction in several ways. First, if the most frequent target for each indirect jump is identified, it may be used to initialize the BTB and then reduce the storage requirements of the IJPred and its usage rate. This option requires an ISA extension to allow access to the BTB. Second, embedded systems tuned at design time can use the static analysis to select the best configuration for the indirect predictor (IJPred and ITT size, *pathID*/tag length, history length, …). In particular, we have found that tuning history length for an specific benchmark may yield an accuracy improvement between 15% and 45%. Adapting history length dynamically, either for a full application, like in [10], or for each specific branch, like in [12], is another future line for improving accuracy.

References

1. Bodin, F., Seznec, A.: Skewed associativity improves program performance and enhances predictability. IEEE Trans. on Computers, vol. 46(5) (1997) 530–544
2. Burger, D., Austin, T.M.: The SimpleScalar tool set. Univ. Wisconsin-Madison Computer Science Department, Tech. Report TR-1342 (1997)
3. Calder, B., Grunwald, D.: Reducing Indirect Function Call Overhead in C+ Programs. Proc. 21th Int. Symp. on Principles of Programming Languages (1994) 397–408
4. Calder, B., Grunwald, D.: Next Cache Line and Set Prediction. Proc. 22nd Int. Symp. on Computer Architecture (1995) 287–296
5. Chang, P.-Y., Hao E., Patt, Y. N.: Target Prediction for Indirect Jumps. Proc. 24th Int. Symp. on Computer Architecture (1997) 274–283
6. Driesen, K., Hölzle, U.: The cascaded predictor: economical and adaptive branch target prediction. Proc. 31st Intl. Symp. on Microarchitecture (1998) 249–258
7. Gochman, S., et. al.: The Intel Pentium M processor: Microarchitecture and Performance. Intel Technology Journal, vol. 7(2), (2003) 21–36
8. Hinton, G., et. al.: The microarchitecture of the Pentium 4 processor. Intel Technology Journal, Q1 (2001)
9. Jacobson, Q., Rotenberg, E., Smith, J. E.: Path-based next trace prediction. Proc. 30th Int. Symp. on Microarchitecture (1997) 14–23
10. Juan, T., Sanjeevan, S., Navarro, J.J.: A third level of adaptivity for branch prediction. Proc. 25th Int. Symp. on Computer Architecture (1998) 155–166
11. Kaeli, D.R., Emma, P.G.,: Branch History Table Prediction of Moving Target Branches due Subroutine Returns. Proc. 18th Int. Symp. on Computer Architecture (1991) 34–41
12. Kalamatianos, J., Kaeli, D.R.: Predicting indirect branches via data compression. Proc. 31st Int. Symp. on Microarchitecture (1998) 272–281
13. Lee, J. K. F., Smith, A. J.: Branch Prediction Strategies and Branch Target Buffer Design. IEEE Computer Vol. 17(2) (1984) 6–22
14. Moure, J. C., Rexachs, D. I., Luque, E.: Optimizing a decoupled front-end architecture: the Indexed Fetch Target Buffer (iFTB). Lecture Notes in Computer Science, Vol. 2790. Euro-Par'03. Springer-Verlag, (2003) 566–575
15. Yeh, T.-Y., Patt, Y.: Two-Level Adaptive Branch Prediction. Proc. 24th Int. Symp. on Microarchitecture (1991) 51–61

Dynamic Partition of Memory Reference Instructions – A Register Guided Approach*

Yixin Shi and Gyungho Lee

ECE Department, University of Illinois at Chicago
`yshi7@uic.edu, ghlee@ece.uic.edu`

Abstract. A high bandwidth L-1 data cache is essential for achieving high performance in wide-issue processors. Previous studies have shown that using multiple small single-ported caches instead of a monolithic large multi-ported one for L-1 data cache can be a scalable and inexpensive way to provide higher bandwidth. Many schemes have been proposed on how to direct the memory references to these multiple caches in order to achieve a close match to the performance of an ideal multi-ported cache. However, most previous designs seldom take dynamic data access patterns into consideration and thus suffer from access conflicts within one cache and unbalanced loads between the caches. We observe that if one can group data references defined in a program into several regions (access regions) to allow parallel accesses, then providing separate small caches (access region cache) for these regions may prove to have better performance than previous multi-cache schemes. The register-guided memory reference partition approach proposed in this paper effectively identifies these semantic regions and organizes them in multiple caches in an adaptive way to maximize concurrent accesses without incurring too much overhead. In our design, the base register number, not its content, in the memory reference instruction is used as a basic guide for instruction steering. A reassignment mechanism is applied to capture the pattern when program is moving across its access regions. In addition, a distribution mechanism is introduced to further reduce residual conflicts, which adaptively enables access regions to extend or shrink among the physical caches. Our simulations of SPEC CPU2000 benchmarks have shown that the register-guided approach can reduce the conflicts effectively, distribute memory reference instructions properly, and yield considerable performance improvement in terms of IPC.

1   Introduction

Modern superscalar processors select and execute multiple independent instructions at a very high clock rate assisted by control speculation, register renaming, and dataflow execution. With ample on-chip hardware resources available, researchers have been actively proposing aggressive micro-architectures that can issue more instructions including memory reference instructions in a single clock cycle[3]. Traditional efforts were mainly focused on decreasing the cache access latency and increasing the cache capacity. However, previous studies [4][11] suggest that the capability to provide enough memory bandwidth (or cache ports) be also important to explore more instruction level parallelism[9].

---

* This work was supported in part by NSF CCR0225561

J.C. Cunha and P.D. Medeiros (Eds.): Euro-Par 2005, LNCS 3648, pp. 508–518, 2005.

Essentially the ways to achieve high memory bandwidth can be categorized into three classes. The most straightforward approach is to build an ideal multi-ported cache. This circuitry level solution often comes at the cost of complexity in memory cell and bit/word line design and possibly incurs longer cache access latency[7]. Fig.1 shows the various performance trends of a 32 KB cache modeled by CACTI 3.0 in .18um[13]. The three metrics, access time, cache area and the power consumption, increase quickly as more cache ports are introduced. Alternatively, there have been many proposals to approximate the ideal multi-ported cache including time-division multiplexing and cache replicating. These designs often suffer from either poor resource utilization or longer access latency.

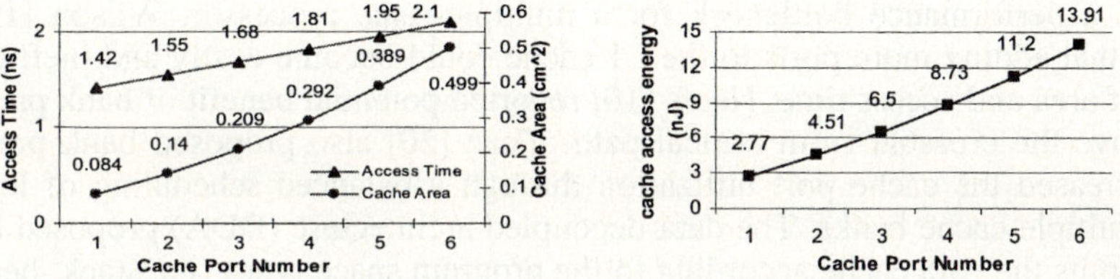

Fig. 1. Access time, area, and power consumption of a 32 KB, 32B block size, 2-way set-associative cache with different cache port number[5]

The interleaved multi-banking scheme is another way to increase the memory bandwidth with less hardware complexity. Instead of using one big ideal multi-ported cache, multiple smaller banks or caches serve as L-1 data cache. The data are simply interleaved based on word addresses or cache lines so that two or more simultaneous accesses to different banks can be supported in one clock cycle. This design typically employs an interconnection network (crossbar) to distribute memory references among the different cache banks (see Fig.3). One problem is the bank conflicts among the simultaneous accesses caused by the "random interleaving" property. Another potential problem is that the area of a crossbar in the critical path increases super-linearly when the number of banks increases. This will cause further delay when accesses are passing through the crossbar.

Other than the multi-banking solution, many schemes have been proposed to increase the bandwidth in a more scalable manner. Similar to multi-banking, multiple caches are used as L-1 data cache. However, these designs use more intelligent methods in data placement and memory reference steering rather than simply interleaving the addresses. The proposed register-guided memory partition scheme belongs to this category. It tries to exploit the semantic meaning in the program when assigning memory instructions to different caches. The key insight is that the *base register number,* not its content, can serve as the basis for instruction steering, because the register number usually reflects the data "region" on which the instruction is operating. By adaptively interpreting different registers for different regions, the data regions can be distinguished from each other and memory access parallelism can be captured from the program level. In addition, a reassignment mechanism and a distribution mechanism are applied to capture the changes in the memory reference pattern and alleviate the conflicts. Simulations show this scheme outperforms other solutions for most benchmark programs.

The remainder of this paper is organized as follows: Section 2 summarizes related works on multi-cache design; Section 3 discusses the details about the register-guided memory instructions partition scheme; Section 4 describes the scheme-specific architectural parameters, the simulation approaches, and the benchmarks used; Section 5 presents our experimental results and analysis; Section 6 provides the concluding remarks.

2   Related Work

Sohi and Franklin [5] first predicted that the L1 cache bandwidth would eventually become a performance bottleneck for a multiple-issue processor. Wilson [19] also argued that adding more ports to the L1 cache could become costly and inefficient in terms of area and access time. Neefs [10] reported potential benefit of bank prediction to remove the crossbar from critical path. Yoaz [20] also proposed bank prediction that increased the cache port utilization through a balanced scheduling of loads toward multiple cache banks. The data-decoupled architecture (DDA) proposed by Cho [4][5] splits the data cache according to the program space types (i.e. stack, heap, and data). It simply treats each area as an access region and divides the data references into two independent streams (stack and non-stack). Thakar [17] tries to further split data cache within stack cache and non-stack cache. This scheme assigns the access regions to the access region caches initially based on offline profiling and then predicted by a PC-indexed table. Redirection is used to maintain the data consistence and only one copy for a datum is allowed in the L-1 cache. The Parallel Cachelets scheme proposed by Limaye[8] also employs a PC-indexed table to determine the bank (or cachelet) number either in decode stage or execution stage. It tries to minimize contentions by reassigning the destination for memory access once a conflict occurs. To maintain consistency, a write through policy and value broadcasting are used. Racunas [11] also studied the performance impact on a partitioned L-1 data cache. They proposed a two-bit saturating instruction hysteresis counter in the prediction table to partition memory reference streams.

3   Register-Guided Memory Partition with Distribution Scheme

3.1   Motivation

The register-guided memory partition scheme is based on the concepts of Access Region and Access Region Cache first proposed in [4][17]. A key observation is that typically, there exist one or more data structures with variable sizes in a program either statically defined or generated at run-time. They can be data arrays found in FORTRAN programs or structures/unions or objects common in C/C++ programs. These data structures are called *access regions*. Our partitioning scheme tries to capture these semantically defined and logically independent access regions as atomic units in memory. Ideally, by navigating the partitioned memory reference stream, data from the different access regions are placed into physically separate caches. These multiple quasi-independent small caches working as L-1 cache are named *Access Region Caches (ARC)*[17].

We extend our previous work [4][16] by proposing a novel and more effective method to predict the destination access region cache for each memory access. Unlike some "software" solutions such as load instruction annotation [18] or static marking by compiler, our architectural level approach tries to utilize run-time information without changing the existing binaries. After investigating the prediction resources (e.g. program counter, previous branch history behavior, register number, and probably the content and offset) and their available time, we found that the base register number in the memory reference instruction can serve as a good hint.

In a typical MIPS-like architecture, the memory reference instructions, i.e. load and store, generally have the following format:

```
LOAD        destination-register,  offset(base-register)
STORE       source-register,       offset(base-register)
```

Compiler typically groups the data members belonging to one data structure by assigning a common register as their base registers. Memory reference instructions then use this register together with variable offsets to access different fields within that data structure. We also expect the memory reference instructions accessing different data regions in a short time window to have different base registers. Therefore, the base registers reflect the access regions and can be utilized to identify the data structures in the program. The partitioning based on the base register is motivated by the fact that simultaneous accesses on different data structure are usually relatively independent and can be performed concurrently. This approach ideally provides the separate spaces for the access regions that may have different access patterns. This explores opportunities to improve performance similar to separate instruction cache and data cache found in most processors today. Although some data regions might have to share one ARC due to the limited number of physical ARCs, our round-robin ARC assignment and reassignment mechanisms to be presented later can minimize this effect. Using register number to determine the ARC number in this scheme is the major difference from previously proposed schemes. Using the base register number, we can capture more program semantic meaning than just blindly using the PC or addresses. In addition, the register number is known in an early pipeline stage so that after partitioning dedicated and small hardware structures can be used to process these instructions efficiently in later pipeline stages[1].

3.2   Proposed Scheme

3.2.1   Scheme Framework

Fig.2 shows the framework of our Register Guided memory partition with Distribution scheme *(RGD)*. A register-indexed prediction table, called ARC prediction table *(ARCP)* is deployed to predict the ARC numbers in the fetch stage for memory instructions. Therefore, no crossbar is needed. The instructions are steered into multiple pipelines and Load/Store (L/S) units. Each entry in ARCP table is mapped to a random ARC cache initially and will be trained at run-time later by the prediction updating policy. The verification logic, which is activated when the effective address is known, resides in the Load/Store unit. If the ARC number is correctly predicted, the instruction goes to the cache and performs an access. Otherwise, a redirection network is used to redirect the instruction to the correct ARC with some cycles of redirection penalties. We assume a select and re-issue mechanism is employed on misprediction. Some run-time information, such as conflicts and redirection events (will be described later), is fed back from the Load/Store unit to the prediction unit to update the prediction table and adjust the steering policy.

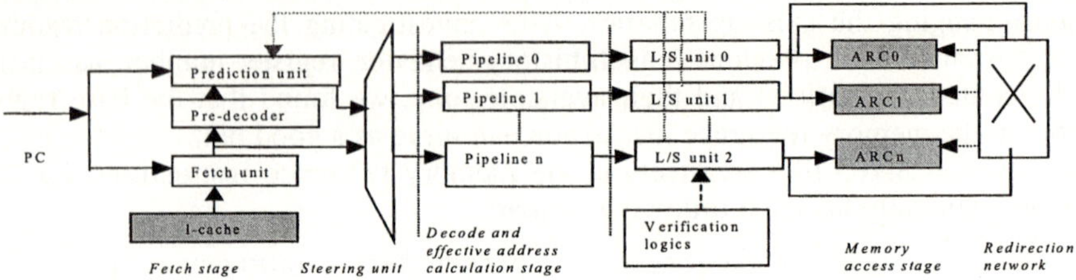

Fig. 2. The frame work of the proposed RGD scheme

We also show a typical cache-interleaving (multi-banking) scheme in Fig.3 for comparison purpose. In this scheme, the cache bank is determined *after* the effective address is calculated. Then the memory reference instruction is steered into the bank through the crossbar. Consequently the crossbar is in the critical path here while the redirection network in RGD scheme is not, provided that the ARC prediction accuracy is reasonably high.

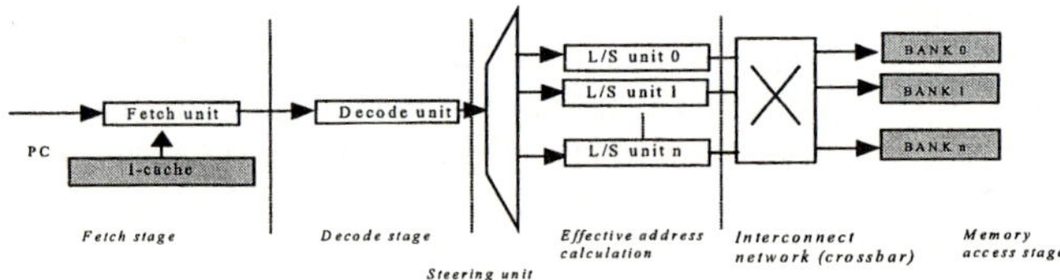

Fig. 3. Cache bank-interleaving scheme

3.2.2   Prediction Verification

In RGD scheme, every memory access must be verified against the correct access region information when the actual address is produced. The effective address is calculated during the first step of the memory-access stage. Meanwhile, access region verification is completed by comparing the tags in the cache or in a separate tag table. Unlike other schemes such as parallel cachelet[8], the RDG scheme does not allow multiple copies of a datum to exist in L-1 caches. Therefore, if the tag comparison turns out to be a mismatch, the verification unit checks other caches. This can be done by broadcasting current datum's tag to other ARCs using a bus or by maintaining a "super" tag, i.e. aggregate of all the ARC tags, in a way similar to duplicate tagging for multiple cache coherencies. If the checking results mismatch on the rest of ARCs either, a true cache miss occurs and L-2 cache access is then invoked. If the datum is found in another ARC, the instruction is redirected and reinserted into the correct memory pipeline connecting to that ARC through a redirection network as shown in Fig.2. We call such an event as ARC misprediction. In this study, as a select and re-issued approach is used, the effects of mispredictions are evaluated by imposing a penalty of a certain number of clock cycle delays for that instruction.

3.2.3   Prediction Updating

In the context of prediction on memory references, last value predictor and 2-bit saturated predictor have been studied in literature[8][11][17]. In our design, a threshold-triggered updating method is used to provide a kind of hysteresis effect to smooth the transient deviations. Rather than update the prediction table immediately when a misprediction is detected as in[8], we periodically check some interested events (including mispredictions) that are accumulated in counters during a sampling period. If any counter exceeds a pre-defined threshold, prediction updating is triggered. Following two mechanisms are implemented as updating policies.

Reassignment Mechanism: The reassignment mechanism can be used in two scenarios to improve the prediction accuracy, as shown in Fig.4. One register in a program can be utilized as the base register for different data regions at various stages of execution. This changing may cause cache misses and ARC misses (redirection events), which implies that the interested register may have been reused or spilled and it may now represent a new data region. To capture this change, a threshold, Rt (Reassignment threshold) is established for updating the ARCP table on ARC mispredictions. That is, the entry for a register in the ARCP table is reassigned to a new destination of ARC only after more than Rt redirection events have been detected in a sampling period as shown in Fig.4(a). By choosing a proper value for Rt, we can capture the moving behavior and adaptively adjust the prediction value.

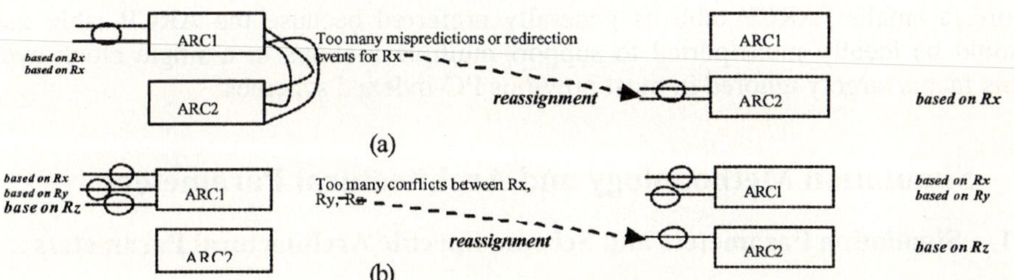

Fig. 4. Two scenarios when the reassignment mechanism is invoked

The reassignment mechanism can be also applied to reduce ARC conflicts. Similar to bank conflicts, ARC conflicts occur when two or more data regions are assigned into one physical ARC and the program happens to access these regions simultaneously as shown in Fig.4(b). In this case, one of these regions needs to migrate to another ARC to reduce the conflicts. Again, a conflict counter and a predefined threshold Ct (Conflict Threshold) are used to determine whether to update the prediction table. We direct the memory accesses of a region to the one that has the least conflicts observed. This mechanism forces one access region to leave its current ARC to avoid further conflicts.

Distribution Mechanism: We also observed that programs might reference one access region based on a *same register* intensively. For instance, a program is likely to make intensive operations on its local variables during a function call where the memory reference instructions have stack pointer or frame pointer as their base registers. In this case, the redirection mechanism will not help because all the instructions with the same base register are driven into the same ARC. To handle this, we introduce a distribution mechanism to scatter these accesses. First conflicts are classified into two types. The conflicts caused by the instructions with the same base register are named as self-conflicts and all other conflicts as interference-conflicts. The ratio of the self-conflicts over all conflicts for each base register is monitored for each register. When this ratio for one particular register reaches a pre-defined threshold, the program is identified as operating on one data region and the distribution flag is set for that entry in ARCP. The memory reference instructions based on the register are then distributed to all of the ARCs in a round-robin manner.

Two counters are employed to accumulate the number of the two types of conflicts. A parameter SIt (Self-conflicts & Interference-conflicts threshold) is used to represent the distribution threshold. Rather than calculating the ratio, the following condition, *Self-conflict number - Interference-conflict number > SIt,* is checked periodically to determine if we should distribute one data region in our simulation.

3.2.4   Hardware Cost

The hardware cost for implementing the RGD scheme is moderate. It basically consists of four counters, a small ARCP table, and some lookup and control logic. In our simulation, each entry in ARCP contains 10 fields (each of one byte). Assuming up to 32 registers can be used as base registers, the size of the ARCP table is only 32 x 10 = 320 bytes with some glue logic. In other PC-based prediction schemes, however, a modest prediction table would have 2K-4K entries totaling 10KB. Hence, the speed of accessing and updating the ARCP table in RGD scheme can be much faster. Furthermore, a smaller ARCP table is generally preferred because the ARCP table itself should be ideally multi-ported to support multiple lookups in a single clock cycle. This fact is largely ignored in most previous PC-indexed schemes.

4   Simulation Methodology and Architectural Parameters

4.1   Simulation Parameters and Scheme-Specific Architectural Parameters

In our simulation, a cycle-accurate execution driven simulator derived from the Simplescalar Tool Set 3.0[2] is modified to incorporate our design of multiple memory pipelines and ARCs. To evaluate our proposed approach as emerging trend towards aggressive ILP exploitation, an out-of-order processor model issuing up to 16 instructions per cycle is used. An ideal front-end for the processor model is assumed in order to assert a maximum data bandwidth demand on the memory system.

The L-1 Data cache are direct-mapped caches with a fixed total size of 64KB across all of the different ARC configuration and memory partitioning schemes. In order to investigate the scalability, we studied the cases of 4-ARC and 8-ARC configurations. For the 4-ARC configuration, four separate single-ported caches (ARCs) are used as the L-1 Data cache, each of 16KB; while in the 8-ARC configuration, eight ARCs are provided, each of 8KB. All caches are assumed to be lock-up free. We tested the pre-compiled Alpha binaries of both integer and floating-point benchmarks from SPECCPU2000[15] benchmark suite with *reference* inputs. To warm up the architecture, we fast-forwarded the first 500 million instructions and collected data for the next 500 million committed instructions. The parameters we assumed are summarized in Table-1.

Table 1. Architectural Parameters in our simulation model

Fetch/decode/issue/ commit width	16
Function unit size	Int ALU:16, FP ALU: 16, Int Mult: 4, FP Mult: 4
L1 I-cache	Blk size:32B; set: 512; assoc:2; access time:1 cycle;
L1 D-cache	Blk size: 32B, set: 512(4ARC), 256(8ARC); per ARC size: 16KB(4ARC), 8KB(8ARC); Total size: 64KB, access time 2 cycle;
Unified L-2 cache	Blk size:64B, set: 2048; assoc.: 4; total size: 512KB. access time: 8 cycles;
Others	Perfect branch predictor; LSQ size: 128; RUU size: 256; memory latency: 50 cycles;

Table-2 shows the scheme-specific architectural parameters in the simulation. Here, the event counters are checked when every ten memory reference instructions have been committed (SP=10). This corresponds to approximately three basic blocks. If redirection events occur roughly half the time, then reassignment is triggered (Rt = 5). Similarly, five or more conflicts also lead to migration of a data region to another ARC (Ct=5). The value for SIt is assumed to be three to determine whether to trigger distribution mechanism. These parameters, currently having fixed values, are expected to be tunable responding to different applications at run-time in the future.

Table 2. Scheme-specific Architectural Parameters in the simulation model

Parameter Name	Value	Parameter Name	Value
Sampling Period (Sp)	10	Self conflicts & Interference conflicts Threshold (SIt)	3
Redirection Threshold (Rt)	5	ARC/L-1 cache hit Time	2 cycles
Conflict Threshold (Ct)	5	Redirection Penalty	2 cycles

4.2   Schemes for Comparison

The baseline model in this study is the multi-banking schemes where data are placed in an interleave manner and the memory reference instruction is steered through a crossbar. One baseline model is the BI-2 scheme (Bank Interleaving) where 2 cycles are charged for the crossbar delay, the same as the redirection penalty in RGD scheme (see Table-2). Another one is a more aggressive multi-banking scheme, the BI-1 scheme, which charges only 1 cycle for the crossbar delay. The third scheme, the PC prediction (PCP), similar to the Parallel Cachelets[8] and Tharker's[17] design, is a general PC-based prediction scheme. It accommodates a 2KB prediction table indexed by the PC to predict the destinations for memory reference instructions. Redirection mechanism with a penalty of 2 cycles is used to maintain data consistency. A fourth scheme, called the register-guided scheme (RG), is also simulated to understand how much the distribution mechanism in RGD scheme contributes to the final performance. It is similar to the RGD scheme except that no distribution mechanism is applied. Note that the same size L-1 data caches (64KB) are used in the above four schemes as that of ARCs in RGD scheme.

5   Simulation Result and Analysis

5.1   Busy-Waiting Cycle

Fig.5 shows the busy-waiting cycles for memory reference instructions for the 4-ARC and the 8-ARC configuration. They are defined as the latencies between the time when the operands of a load or store instruction are available to the time when this instruction gets an idle port. The busy-waiting cycles include the waiting time in LSQ, redirection penalty, and the crossbar delays. It reflects the degree of bank conflicts and how well memory ports are utilized. As can been seen in Fig.5, for 4ARC-integer benchmarks, the average busy-waiting time for RGD is 0.6 to 1.5 cycles fewer than other schemes, which mainly contributes to a higher IPC. Similar results can be observed for 4ARC-INT and 8ARC-FP benchmarks. For FP programs in 8-ARC configuration, the busy-waiting cycle of RGD scheme is on average lower by about 0.5 cycle than that of BI-2, but 0.35 cycles higher than BI-1. This indicates in this case the conflict reduction by RGD scheme is not sufficient to beat the benefit obtained from a shorter crossbar delay (one cycle) we assumed in BI-1.

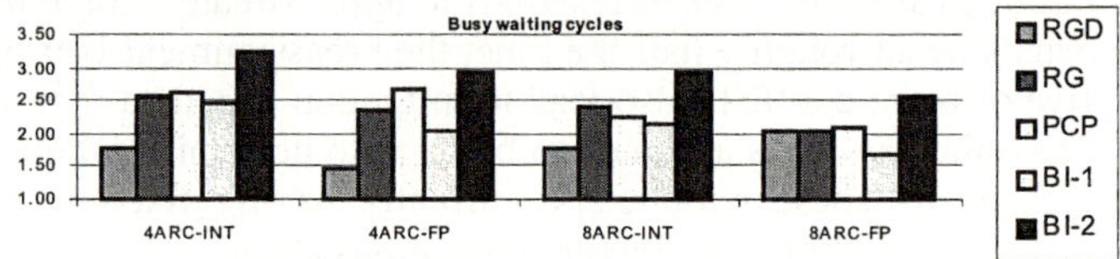

Fig. 5. Average Busy waiting cycles

5.2   ARC Prediction Accuracy and Data Cache Hit Rate

Fig.6(a) presents the ARC prediction accuracy. The RGD, RG, and PCP have similar ARC prediction accuracy of 81%, 82%, and 83.7%, respectively. Considering PCP scheme has much bigger PC-indexed prediction table, the register-guided prediction is a fair tradeoff in efficiency and accuracy. In addition, with an 81% ARC prediction accuracy on average, we can also conclude that the redirection network shown in Fig.2 is not in the critical path.

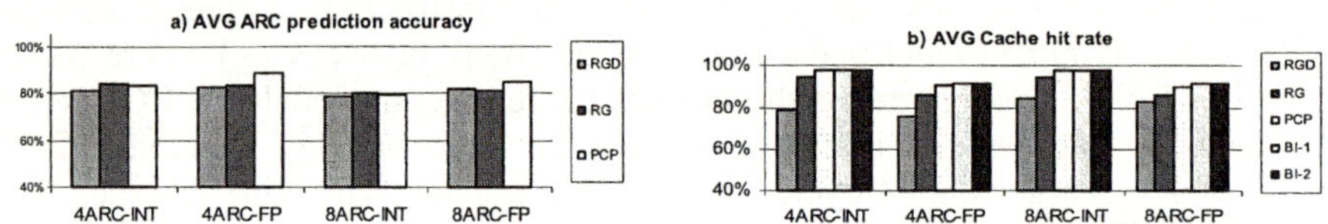

Fig. 6. Average ARC prediction accuracy and L-1 Data Cache hit rate

The overall data cache hit rate for the ARC is illustrated in Fig.6(b). The RGD scheme has about 10%-14% lower cache hit rate than that of RG, PCP, and BI scheme. This is due to the redirection and distribution mechanisms incurring considerable invalidations and thus causing extra cache misses while reducing the total number of conflicts. Note that a higher cache hit rate here does not necessarily mean higher performance, because memory reference instructions would experience redirection and conflict penalties before the final cache access occurs.

5.3   Overall IPC

Fig.7 shows the overall IPC for all of the schemes discussed so far. The simulation results indicate that with the same size of the L-1 cache and the same redirection penalty, our scheme works best for most of the benchmark programs under different ARC

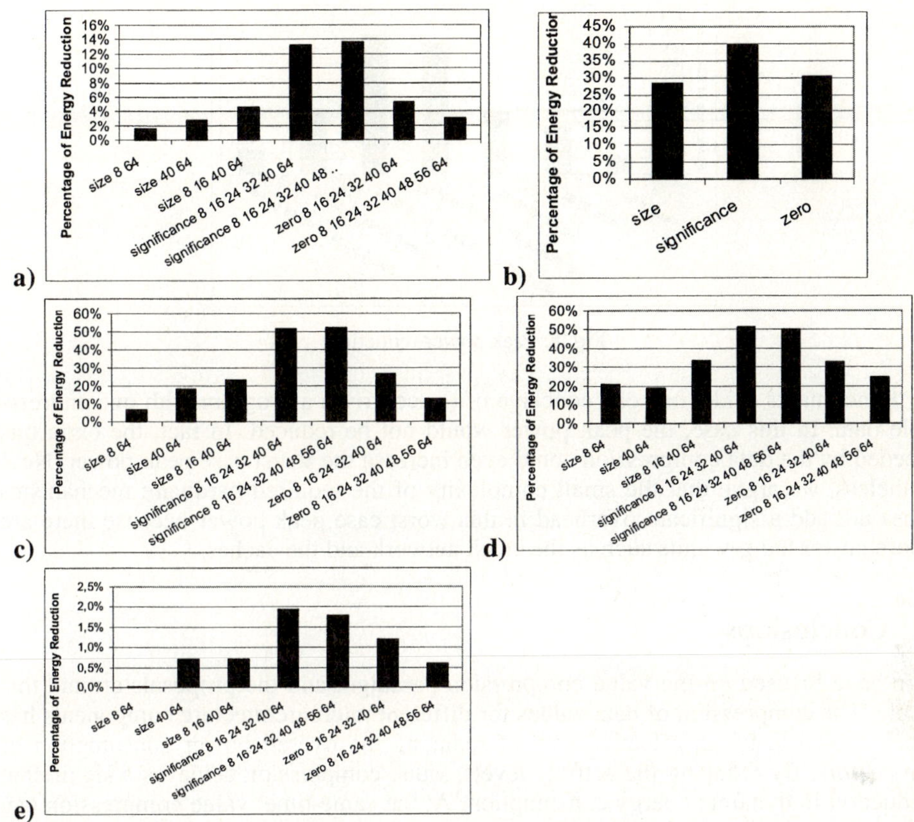

Fig. 7. Energy Savings for: (a) Data Cache (b) Instruction Cache (c) ALU (d) Register File (e) Branch Predictor

4.3 Peak Power Reduction

Peak power is an important metric because it determines the maximum possible burst of power that a processor might consume. This translates directly to hot spots and to the temperature-thermal limits of the processor. Although one may think that compressing the data may not have a direct impact on peak power because there may be cycles where every computation will need 64 bits, our experiments show that peak power is significantly reduced with the proposed compression mechanisms. The peak power shown in Fig 8 corresponds to the execution of the SpecInt2000 suite.

As in the case of the energy consumption, the significance compression mechanism achieves an 18% peak power reduction. It is interesting to see that the configuration of significance compression that achieves the highest energy reduction (see Fig 6) is not the best in terms of peak power reduction (see Fig 8) where the scheme that compresses all the bytes (significance 8,16,24,32,40,56,64) performs a little bit better. The fact that it can compress bytes within large words makes it perform better in terms of peak power. The size compression mechanism achieves, in its best configuration, an 8% peak power reduction while the zero compression mechanism stays above the 8% line.

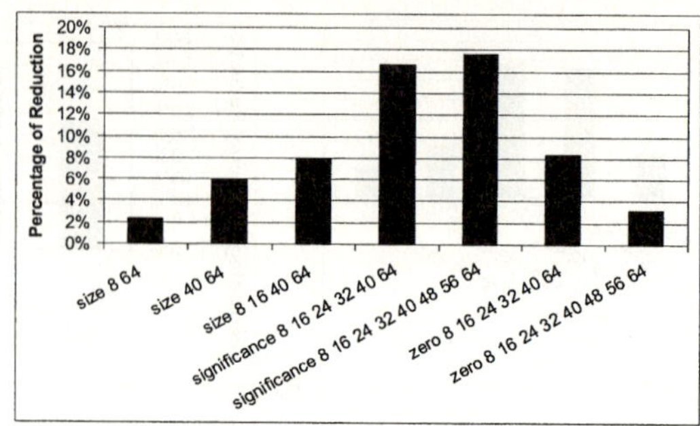

Fig. 8. Peak power reduction.

Benchmarks aside, one can conceive of (or contrive) a program with uncompressible data. In this case, the peak power would not be reduced. In fact, the extra bits needed by the data compression could even increase the worst case peak power. Nevertheless, we argue that the small complexity of the required hardware mechanisms does not add a significant overhead in this worst case peak power because there are more power hungry units such as the clock network and the caches.

5   Conclusions

We have focused on the value compression paradigm and the proposals around this topic. The compression of data values for different microarchitecture components has been shown to be an effective way of reducing the overall power consumption of processors. By reducing the activity levels, value compression achieves a significant reduction in dynamic energy consumption. At the same time, value compression can be used to make the different components of the pipeline simpler (or smaller) and thus further reducing the energy –in this case, the static energy consumption. Furthermore, we have shown that value compression can reduce the run-time peak power consumption and thus it can be a good approach for temperature-aware computing. Several studies have used different kinds of value compression mechanisms to achieve these goals. In this work, we have extended, analyzed and compared them.

References

1. D. Brooks and M. Martonosi, "Dynamically Exploiting Narrow Width Operands to Improve Processor Power and Performance", in Proc. of 5th. International Symposium on High-Performance Computer Architecture (HPCA-5), 1999.
2. D. Brooks, V. Tiwari and M. Martonosi, "Wattch: A Framework for Architectural-Level Power Analysis and Optimization", in Proc. of the 27th Annual International Symposium on Computer Architecture, June 2000.
3. G. Cai and C.H. Lim, "Architectural Level Power/Performance Optimization and Dynamic Power Estimation", Cool Chips tutorial of the 32nd Int. Symp. On Microarchitecture 1999.
4. R. Canal, A. González and J.E. Smith, "Very Low Power Pipelines using Significance Compression", in Proc. of the 33rd Int. Symposium on Microarchitecture, Dec. 2000.
5. R. Canal, A. González and J.E. Smith, "Software-Controlled Operand Gating", in Proc. of 2nd International Symposium on Code Generation and Optimization, March 2004

6. J. Choi, J. Jeon and K. Choi, "Power Minimization of Functional Units by Partially Guarded Computation", in Proc. of the 2000 International Symposium On Low Power Electronics and Design (ISLPED'00), pp. 131-136, Rapallo (Italy), August 2002.
7. R. Gonzalez and M. Horowitz, "Energy dissipation in general purpose processors", IEEE Journal of Solid State Circuits, v. 31, n. 9, pp. 1277-1284, September 1996.
8. G. Loh, "Exploiting Data-Width Locality to Increase Superscalar Execution Bandwidth", in Proc. of the 35th International Symposium on Microarchitecture (MICRO-35), pp. 395-405, Istanbul (Turkey) November 2002.
9. T. Nakra, B. Childers, and M.L.Soffa, "Width Sensitive Scheduling for Resource Contained VLIW processors", FDDO Workshop (MICRO33), Dec. 2001.
10. T. Okuma, Y. Cao, M. Muroyama and H. Yasuura, "Reducing Access Energy of On-Chip Data Memory Considering Active Data Width", in Proc. of the 2002 Int. Symp. On Low Power Electronics and Design, pp. 88-91, Monterey (CA-USA), August 2002.
11. T.Sato and I. Arita, "Table Size Reduction for Data Value Predictors by Exploiting Narrow Width Values", in Proc. of the 2000 Int. Conf. on Supercomputing, May 2000, pp.196-205.
12. J. Turley, "PowerPC Adopts Code Compression", Microprocessor Report, October 1998.
13. R. B. Tremaine, P. A. Franaszek, J. T. Robinson, C. O. Schulz, T. B. Smith, M. E. Wazlowski, and P. M. Bland, "IBM Memory Expansion Technology (MXT)", IBM Journal of Research and Development, Volume 45, Number 2, 2001, pp. 271-286.
14. L. Villa, M. Zhang, and K. Asanovic, "Dynamic Zero Compression for Cache Energy Reduction", in Proc. of the 33rd International Symposium on Microarchitecture, Dec.2000.
15. J. Yang and R. Gupta, "Energy Efficient Frequent Value Data Cache Design", in Proc. of the 35th International Symposium on Microarchitecture (MICRO-35), pp. 197-207, Istanbul (Turkey), November 2002.

Improving Instruction Delivery with a Block-Aware ISA

Ahmad Zmily, Earl Killian, and Christos Kozyrakis

Electrical Engineering Department
Stanford University
{zmily,killian,kozyraki}@stanford.edu

Abstract. Instruction delivery is a critical component for wide-issue processors since its bandwidth and accuracy place an upper limit on performance. The processor front-end accuracy and bandwidth are limited by instruction cache misses, multi-cycle instruction cache accesses, and target or direction mispredictions for control-flow operations. This paper introduces a block-aware ISA (BLISS) that helps accurate instruction delivery by defining basic block descriptors in addition to and separate from the actual instructions in a program. We show that BLISS allows for a decoupled front-end that tolerates cache latency and allows for higher speculation accuracy. This translates to a 20% IPC and 14% energy improvements over conventional front-ends. We also demonstrate that a BLISS-based front-end outperforms by 13% decoupled front-ends that detect fetched blocks dynamically in hardware, without any information from the ISA.

1   Introduction

Effective instruction delivery is vital for superscalar processors [1]. The rate and accuracy at which instructions enter the pipeline set an upper limit to sustained performance. Consequently, wide-issue designs place increased demands on the processor *front-end*, the engine responsible for control-flow prediction and instruction fetching. The front-end must handle three basic detractors: instruction cache misses that cause instruction delivery stalls; target and direction mispredictions for branches that send erroneous instructions to the execution core; and multi-cycle instruction cache accesses that cause additional uncertainty about the existence and direction of branches within the instruction stream.

To overcome these problems in high performance yet energy efficient way, we propose a block-aware instruction set architecture (BLISS). BLISS defines basic block descriptors in addition to and separately from the actual instructions in each program. A descriptor provides sufficient information for fast and accurate control-flow prediction without accessing or parsing the instruction stream. It describes the type of the control-flow operation that terminates the block, its potential target, and the number of instructions in the basic block. BLISS allows the processor front-end to access the software defined block descriptors through a small cache that replaces the block target buffer (BTB). The descriptors' cache decouples control-flow speculation from instruction cache accesses. Hence, the instruction cache latency is no longer in the critical path of accurate prediction. The fetched descriptors can be used to prefetch instructions and eliminate the impact of instruction cache misses. Furthermore, the control-flow information available in descriptors allows for judicious use of branch predictors, which

J.C. Cunha and P.D. Medeiros (Eds.): Euro-Par 2005, LNCS 3648, pp. 530–539, 2005.

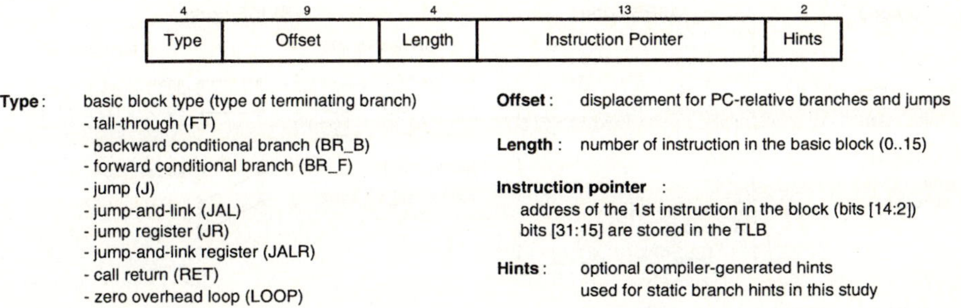

Fig. 1. The 32-bit basic block descriptor format in BLISS.

reduces interference and training time and improves overall prediction accuracy. We demonstrate that for an 8-way superscalar processor, a BLISS-based front-end allows for 20% performance improvement and 14% overall energy savings over a conventional front-end engine.

Moreover, BLISS compares favorably to advanced, hardware-based schemes for decoupled front-end engines such as the fetch-block-buffer (FTB) design [2, 3]. The FTB performs aggressive block coalescing to increase the number of instructions per control-flow prediction and increase the utilization of the BTB. The BLISS-based front-end provides higher control-flow accuracy than the FTB by removing over-speculation with block fetching and coalescing. Our experiments show that a BLISS-based 8-way processor provides 13% higher performance and 7% overall energy savings over the FTB design.

Overall, we demonstrate the potential of delegating portions of instruction delivery (accurate fetch block formation) to software using an expressive ISA.

2   Block-Aware Instruction Set Architecture

Our proposal for addressing front-end performance is based on a *block-aware instruction set (BLISS)* that explicitly describes basic blocks. A basic block (*BB*) is a sequence of instructions starting at the target or fall-through of a control-flow instruction and ending with the next control-flow instruction or before the next potential branch target.

BLISS stores the definitions for basic blocks in addition to and separately from the ordinary instructions they include. The code segment for a program is divided in two distinct sections. The first section contains descriptors that define the type and boundaries of blocks, while the second section lists the actual instructions in each block. Figure 1 presents the format of a basic block descriptor (*BBD*). Each BBD defines the type of the control-flow operation that terminates the block. The BBD also includes an offset field to be used for blocks ending with a branch or a jump with PC-relative addressing. The actual instructions in the basic block are identified by the pointer to the first instruction and the length field. The last BBD field contains optional compiler-generated hints. In this study, we make limited use of this field to convey branch prediction hints generated through profiling [4]. The overall BBD length is 32 bits.

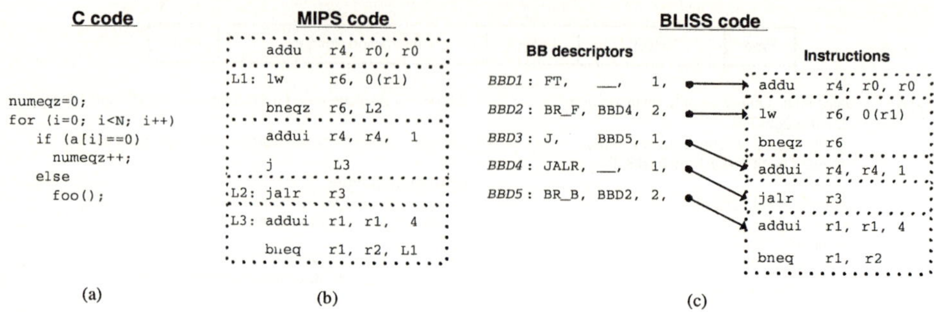

Fig. 2. Example program in (a) C source code, (b) MIPS assembly, and (c) BLISS assembly. In (b) and (c), the instructions in each basic block are identified with dotted-line boxes. Register $r3$ contains the address for the first instruction (b) or first basic block descriptor (c) of function foo. For illustration purposes, the instruction pointers in basic block descriptors are represented with arrows.

BLISS treats each basic block as an atomic unit of execution. There is a single program counter and it only points within the code segment for BBDs. The execution of all instructions associated with each descriptor updates the PC so that it points to the descriptor for the next basic block in the program order (PC+4 or PC+offset). Precise exceptions are supported similar to [5].

The BBDs provide the processor front-end with architectural information about the program control-flow in a compressed and accurate manner. Since BBDs are stored separately from instructions, their information is available for front-end tasks before instructions are fetched and decoded. The sequential block target is always at PC+4, regardless of the number of instructions in the block. The non-sequential block target (PC+offset) is also available through the offset field for all blocks terminating with a PC-relative control-flow instructions (branches – BR_B and BR_F, jumps – J and JAL, loop – LOOP). For the remaining cases (jump register – JR and JALR, return – RET), the non-sequential target is provided by the last instruction in the block through a register. BBDs provide the branch condition when it is statically determined (all jumps, return, fall-through blocks). For conditional branches, the BBD provides type information (forward, backward, loop) and hints which can assist with dynamic prediction. The actual branch condition is provided by the last instruction in the block. Finally, instruction pointer and length fields can be used for instruction (pre)fetching.

Figure 2 presents an example program that counts the number of zeros in array a and calls foo() for each non-zero element. With a RISC ISA like MIPS, the program requires 8 instructions (Figure 2.b). The 4 control-flow operations define 5 basic blocks. All branch conditions and targets are defined by the branch and jump instructions. With the BLISS equivalent of MIPS (Figure 2.c), the program requires 5 basic block descriptors and 7 instructions. All PC-relative offsets for branch and jump operations are available in BBDs. Compared to the original code, we have eliminated the j instruction. The corresponding descriptor (BBD3) defines both the control-flow type (J) and the offset, hence the jump instruction itself is redundant. However, we cannot eliminate

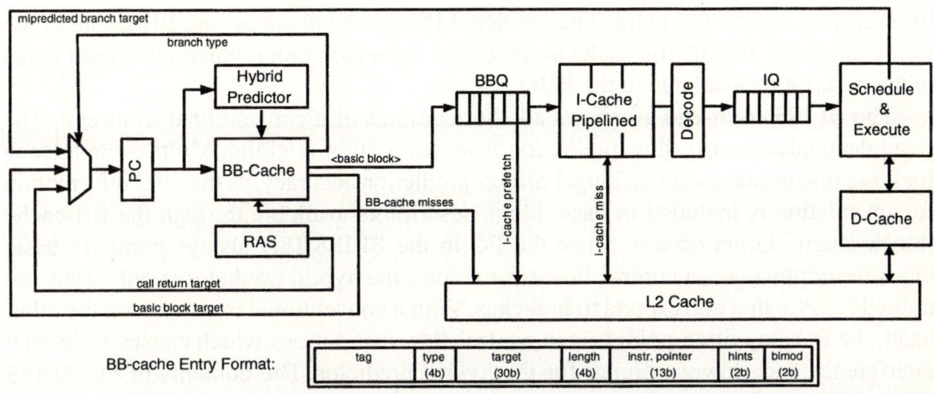

Fig. 3. A decoupled front-end for a superscalar processor based on the BLISS ISA

either of the two conditional branches (bneqz, bne). The corresponding BBDs provide the offsets but not the branch conditions, which are still specified by the regular instructions. However, the regular branch instructions no longer need an offset field, which frees a large number of instruction bits. Similarly, we have preserved the jalr instruction because it allows reading the jump target from register r3 and writing the return address in register r31.

Note that function pointers, virtual methods, jump tables, and dynamic linking are implemented in BLISS using jump-register BBDs and instructions in an identical manner to how they are implemented with conventional ISAs. For example, the target register (r3) for the jr instruction in Figure 2 could be the destination register of a previous load instruction.

3   Decoupled Front-End for the Block-Aware ISA

The BLISS ISA suggests a superscalar front-end that fetches BBDs and the associated instructions in a decoupled manner. Figure 3 presents a BLISS-based front-end that replaces branch target buffer (BTB) with a *BB-cache* that caches the block descriptors in programs. The offset field in each descriptor is stored in the BB-cache in an expanded form that identifies the full target of the terminating branch. For PC-relative branches and jumps, the expansion takes place on BB-cache refills from lower levels of the memory hierarchy, which eliminates target mispredictions even for the first time the branch is executed. For the register-based jumps, the offset field is available after the first execution of the basic block. The BB-cache stores eight sequential BBDs per cache line. Long BB-cache lines exploit spatial locality in descriptor accesses and reduce the storage overhead for tags.

The BLISS front-end operation is simple. On every cycle, the BB-cache is accessed using the PC. On a miss, the front-end stalls until the missing descriptor is retrieved from the memory hierarchy (L2 cache). On a hit, the BBD and its predicted direction/target are pushed in the *basic block queue (BBQ)*. The direction is also verified by a tag-less, hybrid predictor. The predicted PC is used to access the BB-cache in the following cycle. Instruction cache accesses use the instruction pointer and length fields in the descriptors available in the BBQ.

The BLISS front-end alleviates all shortcomings of a conventional front-end. The BBQ decouples control-flow prediction from instruction fetching. Multi-cycle latency for large instruction cache no longer affects prediction accuracy, as the vital information for speculation is included in basic-block descriptors available through the BB-cache (block length, target offset). Since the PC in the BLISS ISA always points to basic block descriptors (i.e. a control-flow instruction), the hybrid predictor is only used and trained for PCs that correspond to branches. With a conventional front-end, on the other hand, the PC may often point to non control-flow instructions which causes additional interference and slower training for the hybrid predictor. The contents of the BLISS BBQ also provide an early view into the instruction address stream and can be used for instruction prefetching and hide instruction cache misses [6].

A decoupled front-end similar to the one in Figure 3 can be implemented without the ISA support provided by BLISS. The FTB design [2, 3] describes the latest of such design. The FTB detects basic block boundaries and targets dynamically in hardware and stores them in an advanced BTB called the fetch target buffer (FTB). Block boundaries are discovered by introducing large instruction sequential blocks which are later shortened when jumps are decoded (misfetch) or branches are taken (mispredict) within the block. The FTB allows for instruction fetch decoupling and prefetching as described above. Furthermore, the FTB coalesces multiple continuous basic blocks into a single long fetch block in order to improve control-flow rate and better utilize the FTB capacity. Nevertheless, the simpler BLISS front-end outperforms the aggressive FTB design by providing a better balance between over- and under-speculation. With BLISS, block formation is statically done in software and it never introduces misfetches. In addition, the PC used to access the hybrid predictor for each block (branch) is the same. With FTB, as fetch blocks shrink dynamically when branches switch behavior, the PC used to index in the predictor and FTB for each branch changes dynamically, causing slower predictor training and additional interference.

4   Methodology

We simulate an 8-way superscalar processor in order to compare the BLISS-based front-end to conventional (base) and FTB-based front-ends. Table 1 summarizes the key architectural parameters. Note that the target prediction buffers in the three front-ends (BTB, FTB, and BB-cache) have exactly the same capacity for fairness. All other parameters are identical across the three models. We have also performed detailed experiments varying several of these parameters and the results are consistent (4-way processor, BTB size, I-cache latency, etc.). For BLISS, we fully model contention for the L2-cache bandwidth between BB-cache misses and I-cache or D-cache misses. Our graphs present two sets of results for BLISS: without (BLISS) and with (BLISS-hints) using the prediction hints in the BBDs. We do not discuss BLISS-hints in details due to space limitations.

We study 12 SPEC CPU2000 benchmarks using their reference datasets [7]. The benchmarks are compiled at the -O3 optimization level. In all cases, we skip the first

Table 1. The microarchitecture parameters for the simulations. The common parameters apply to all three models (base, FTB, BLISS).

	Base	FTB	BLISS
Fetch Width	8 instructions/cycle	1 fetch block/cycle	1 basic block/cycle
Target Predictor	BTB: 2K entries 4-way, 1-cycle access	FTB: 2K entries 4-way, 1-cycle access	BB-cache: 2K entries 4-way, 1-cycle access 8 entries per cache line
Decoupling Queue	–	FTQ: 4 entries	BBQ: 4 entries
Common Processor Parameters			
Hybrid Predictor	gshare: 4K counters PAg L1: 1K entries, PAg L2: 1K counters selector: 4K counters		
RAS	32 entries with shadow copy		
I-cache	32 KBytes, 4-way, 64B blocks, 1 port, 2-cycle access pipelined		
Issue/Commit Width	8 instructions/cycle		
IQ/RUU/LSQ Size	64/128/128 entries		
FUs	8 INT & 6 FP		
D-cache	64 KBytes, 4-way, 64B blocks, 2 ports, 2-cycle access pipelined		
L2 cache	1 MByte, 8-way, 128B blocks, 1 port, 12-cycle access, 4-cycle repeat rate		
Main memory	100-cycle access		

billion instructions and simulate another billion instructions for detailed analysis. We generate BLISS executables using a static binary translator, which can handle arbitrary programs written in any language. The generation of BLISS executable could also be done using a transparent, dynamic compilation framework [8]. Despite introducing the block descriptors, BLISS executables are actually up to 16% smaller than the original binaries, as BLISS allows aggressive code size optimizations such as branch removal and common block elimination. The evaluation of code size optimizations is omitted due to space limitations.

Our simulation framework is based on the Simplescalar/PISA 3.0 toolset [9], which we modified to add the FTB and BLISS front-end models. For energy measurements, we use the Wattch framework with the cc3 power model [10]. Energy consumption was calculated for a $0.10\mu m$ process with a 1.1V power supply. The reported *Total Energy* includes all the processor components (front-end, execution core, and all caches).

5   Evaluation

Figure 4 presents IPC and IPC improvement for the BLISS front-end over the base and FTB front-ends for the 8-way superscalar processor. BLISS outperforms the base front-end for all benchmarks with an average IPC improvement of 20%. The hardware-based FTB front-end outperforms the base for only half of the benchmarks and most of the 7% average IPC improvement is due to vortex. BLISS outperforms FTB for all benchmarks but vortex, with an average IPC advantage of 13% (up to 18% with BLISS-hints).

Figure 4 also presents total energy savings. BLISS provides a 14% total energy improvement over the base design. The advantage is mostly due to the elimination of a

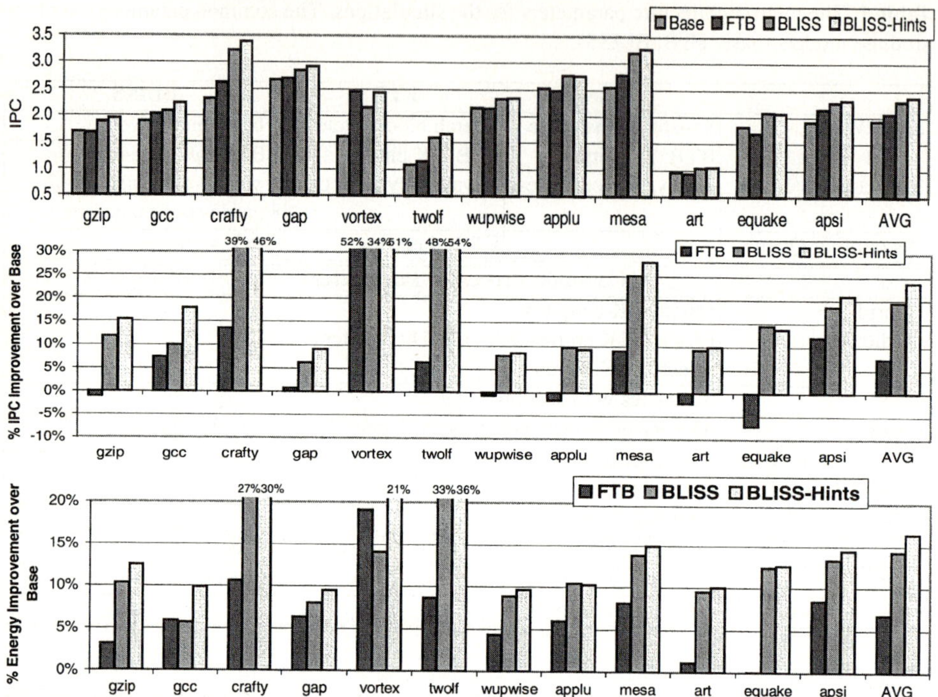

Fig. 4. IPC, percentage of IPC improvement, and percentage of total energy improvement for the FTB and BLISS front-ends over the base fornt-end design.

significant number of pipeline flushes due to control-flow misprediction. BLISS offers a 7% energy advantage over FTB which allows similar energy optimizations in the front-end but suffers from higher number of control-flow mispredictions. It is important to note from Figure 4 that BLISS provides **both** performance and energy advantages over the base and FTB.

Figure 5 explains the basic performance advantage of BLISS over the base and FTB design. Compared to the base, BLISS reduces by 36% the number of pipeline flushes due to target and direction mispredictions. These flushes have a severe performance impact as they empty the full processor pipeline. Flushes in BLISS are slightly more expensive than in the base design due to the longer pipeline, but they are less frequent. The BLISS advantage is due to the availability of control-flow information from the BB-cache regardless of I-cache latency and the accurate indexing and judicious use of the hybrid predictor. The FTB front-end has a significantly higher number of pipeline flushes compared to the BLISS front-end as block recreation affects the prediction accuracy of the hybrid predictor due to longer training and increased interference. Both BLISS and FTB allow for a decoupled front-end with instruction prefetching. BLISS enables I-cache prefetching though the BBQ which reduces the number of I-cache misses by 24% on average for the benchmarks studied. Although the BLISS L2-cache serves an additional type of misses from the BB-cache, BLISS number of

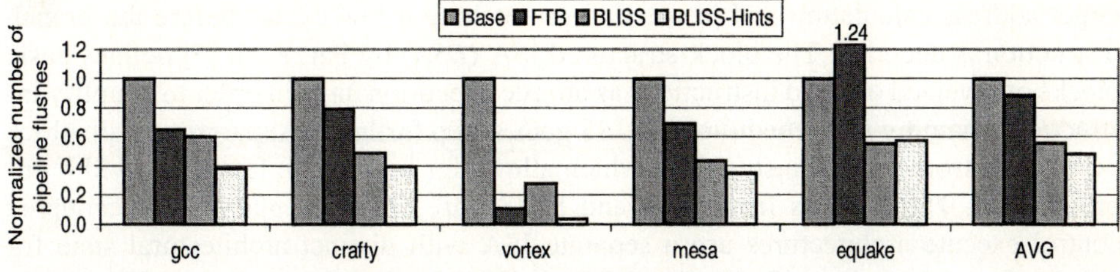

Fig. 5. Normalized number of pipeline flushes for the base, FTB, BLISS for representative benchmarks. The average is across all 12 benchmarks.

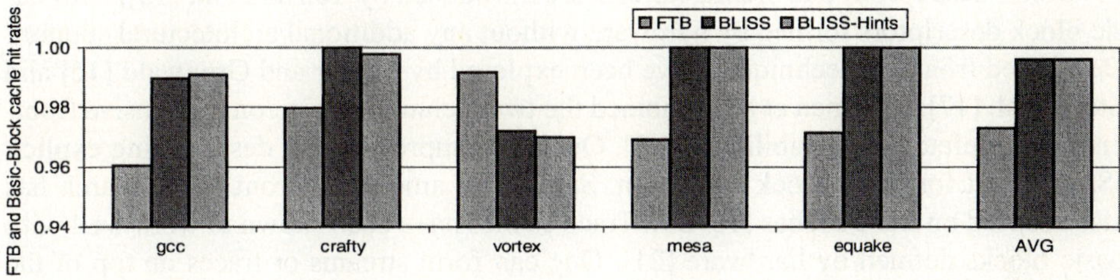

Fig. 6. Normalized FTB and BB-cache hit rates for representative benchmarks. The average is across all 12 benchmarks.

L2-cache accesses and misses are slightly better than the numbers for the FTB design. BLISS has a 10% higher number of L2-cache accesses, and 2% lower number of L2-cache misses compared to the base design for the benchmarks studied. The increased number of L2-cache accesses for BLISS and FTB designs is mainly due to instruction prefetching.

Figure 6 shows the BB-cache and FTB hit rates to evaluate the effectiveness of the FTB in forming fetch-blocks and the BB-cache in delivering BBDs. Since the FTB returns a fall-through block address even when it misses in order to avoid storing the fall-through blocks, we define its miss rate as the number of misfetches divided over the number of FTB accesses. A misfetch occurs when the decoder detects that the block fetched from the FTB is wrong and needs to be updated and a new block to be fetched. At the same storage capacity, the BLISS BB-cache achieves a 2% to 3% higher hit rate than the FTB as the BB-cache avoids block splitting and recreation that occur when branches change behavior or when the cache capacity cannot capture the working set of the benchmark. The FTB has an advantage for programs like vortex that stress the capacity of the target cache and include large fetch blocks. For vortex, the FTB packs 9.5 instructions per entry (multiple basic blocks), while the BB-cache packs 5.5 instructions per entry (single basic block).

6   Related Work

Certain ISAs allow for basic blocks descriptors, interleaved with regular operations in the instruction stream (e.g. *prepare-to-branch* instructions in [11, 12]). They allow for

target address calculation and instruction prefetching a few cycles before the branch instruction is decoded. The block-structured ISA (*BSA*) by Patt et al. [5] defines basic blocks of reversed ordered instructions as atomic execution units in order to simplify instruction renaming and scheduling. BLISS goes a step further by separating basic block descriptors from regular instructions which allows for instruction fetch bandwidth improvements. The benefits from BSA and BLISS are complimentary. The decoupled control-execute architectures use a separate ISA with distinct architectural state for control-flow calculation [13, 14]. The BBDs in BLISS are not a stand-alone ISA and do not define any state, eliminating the deadlock scenarios with decoupled control-execute ISAs.

Block-based front-end architectures were introduced by Yeh and Patt [15], with basic block descriptors formed by hardware without any additional architectural support. Decoupled front-end techniques have been explored by Calder and Grunwald [16] and Stark et al. [17]. Reinman et al. combined the two techniques in a comprehensive front-end with prefetching capabilities [2, 3]. Our work improves their design using explicit ISA support for basic block formation. Significant amount of front-end research has also focused on trace caches [18–20]. Trace caches have been shown to work well with basic blocks defined by hardware [21]. One can form streams or traces on top of the basic blocks in the BLISS ISA. BLISS provides two degrees of freedom for code layout optimizations (blocks and instructions), which could be useful for stream or trace formation. Exploring such approaches is an interesting area for future work.

7   Conclusions

We present a block-aware ISA that addresses basic challenges in the front-end of wide superscalar processors. The ISA defines basic block descriptors in addition to and separately from the actual instructions. Software-defined basic blocks allow a decoupled front-end with highly accurate control-flow speculation, which leads to 20% IPC and 14% energy advantages over conventional designs. The ISA-supported front-end also outperforms (13% IPC and 7% energy) advanced decouple front-ends that dynamically build fetch blocks in hardware. Overall, this work establishes the potential of using expressive ISAs to address difficult hardware problems in modern processors.

Acknowledgements

This work was supported by a Stanford OTL grant.

References

1. R. Ronen, A. Mendelson, et al. Coming Challenges in Microarchitecture and Architecture. *Proceedings of the IEEE*, 89(3), March 2001.
2. G. Reinman, B. Calder, and T. Austin. Fetch Directed Instruction Prefetching. In *Intl. Symposium on Microarchitecture*, Haifa, Israel, November 1999.
3. G. Reinman, C. Calder, and T. Austin. Optimizations Enabled by a Decoupled Front-End Architecture. *IEEE TC*, 50(40), April 2001.

4. A. Ramirez, J. Larriba-Pey, and M. Valero. Branch Prediction Using Profile Data. In *EuroPar Conference*, Manchester, UK, August 2001.

5. S. Melvin and Y. Patt. Enhancing Instruction Scheduling with a Block-structured ISA. *Intl. Journal on Parallel Processing*, 23(3), June 1995.

6. T. Chen and J.L. Baer. A Performance Study of Software and Hardware Data Prefetching Schemes. In *Intl. Symposium on Computer Architecture*, Chicago, IL, April 1994.

7. J. Henning. SPEC CPU2000: Measuring Performance in the New Millennium. *IEEE Computer*, 33(7), July 2000.

8. V. Bala, E. Duesterwald, and S. Banerjia. Dynamo: A Transparent Dynamic Optimization System. In *the Proceedings of the Conference on Programming Language Design and Implementation*, Vancouver, Canada, June 2000.

9. D. Burger and M. Austin. Simplescalar Tool Set, Version 2.0. Technical Report CS-TR-97-1342, University of Wisconsin, Madison, June 1997.

10. D. Brooks, V. Tiwari, , and M. Martonosi. Wattch: A Framework for Architectural-Level Power Analysis and Optimizations. In *Intl. Symposium on Computer Architecture*, Vancouver, BC, Canada, June 2000.

11. R. Wedig and M. Rose. The Reduction of Branch Instruction Execution Overhead Using Structured Control Flow. In *Intl. Symposium on Computer Architecture*, Ann Arbor, MI, June 1984.

12. V. Kathail, M. Schlansker, and B. Rau. HPL PlayDoh Architecture Specification. Technical Report HPL-93-80, HP Labs, 1994.

13. N. Topham and K. McDougall. Performance of the Decoupled ACRI-1 Architecture: the Perfect Club. In *Intl. Conference on High-Performance Computing and Networking*, Milan, Italy, May 1995.

14. R. Manohar and M. Heinrich. The Branch Processor Architecture. Technical Report CSL-TR-1999-1000, Cornell Computer Systems Laboratory, November 1999.

15. T. Yeh and Y. Patt. A Comprehensive Instruction Fetch Mechanism for a Processor Supporting Speculative Execution. In *Intl. Symposium on Microarchitecture*, Portland, OR, December 1992.

16. B. Calder and D. Grunwald. Fast and Accurate Instruction Fetch and Branch Prediction. In *Intl. Symposium on Computer Architecture*, Chicago, IL, April 1994.

17. J. Stark, P. Racunas, and Y. Patt. Reducing the Performance Impact of Instruction Cache Misses by Writing Instructions into the Reservation Stations Out-of-Order. In *Intl. Symposium on Microarchitecture*, Research Triangle Park, NC, December 1997.

18. D. Friendly, S. Patel, and Y. Patt. Alternative Fetch and Issue Techniques from the Trace Cache Mechanism. In *Intl. Symposium on Microarchitecture*, Research Triangle Park, NC, December 1997.

19. Q. Jacobson, E. Rotenberg, and J. Smith. Path-based Next Trace Prediction. In *Intl. Symposium on Microarchitecture*, Research Triangle Park, NC, December 1997.

20. S. Patel, M. Evers, and Y. Patt. Improving Trace Cache Effectiveness with Branch Promotion and Trace Packing. In *Intl. Symposium on Computer Architecture*, Barcelona, Spain, June 1998.

21. S. Jourdan et al. Extended Block Cache. In *Intl. Symposium on High-Performance Computer Architecture*, Toulouse, France, January 2000.

Non-uniform Instruction Scheduling

Joseph J. Sharkey and Dmitry V. Ponomarev

Department of Computer Science, State University of New York
Binghamton, NY 13902 USA
{jsharke,dima}@cs.binghamton.edu

Abstract. Dynamic instruction scheduling logic is one of the most critical and cycle-limiting structures in modern superscalar processors, and it is not easily pipelined without significant losses in performance. However, these perform-ance losses are incurred only due to a small fraction of instructions, which are intolerant to the non-atomic scheduling. We first perform an empirical analysis of the instruction streams to determine which instructions actually require sin-gle cycle scheduling. We then propose a *Non-Uniform Scheduler* – a design that partitions the scheduling logic into two queues, each with dedicated wakeup and selection logic: a small Fast Issue Queue (FIQ) to issue critical instructions in the back-to-back cycles and a large Slow Issue Queue (SIQ) to issue the re-maining instructions over two cycles with a one cycle bubble between depend-ent instructions. Finally, we propose and evaluate several steering mechanisms to effectively distribute instructions between the queues.

1 Introduction

It has been well documented in the recent literature that instruction wakeup and selec-tion logic form one of the most critical loops in modern superscalar processors [17,20]. Unless wakeup and selection activities are performed within a single cycle, dependent instructions can not execute in consecutive cycles, which seriously de-grades the number of instructions committed per cycle (IPC), by as much as 30% in a 4-way machine, according to our simulations. At the same time, both wakeup and selection logic have substantial delays [17], so if these activities are performed atomi-cally within a single cycle, then the designers may be forced to use lower clock fre-quency or limit the size of the instruction issue queue.

Several schemes have been recently proposed to relax the scheduling loop without seri-ously compromising the processor's performance [3,4,14,20]. Most of these designs do somewhat mitigate the problem of IPC loss due to the inability to execute dependent in-structions in consecutive cycles with pipelined schedulers. However, all of these techniques result in significant additional complexities (as we detail later) and are not easy to retrofit into existing datapaths. In this paper, we investigate a much simpler solution. The idea is based on a distributed implementation of the issue queue in the form of two separate queues: a fast, small issue queue (FIQ) to perform a 1-cycle scheduling (with atomic wakeup/select) of some instructions, and a large, slow issue queue (SIQ) to perform pipelined 2-cycle scheduling of all other instruc-tions. Each of these queues has a dedicated wakeup and selection logic, and only the dependent instructions from the FIQ are guaranteed to execute in the back-to-back cycles. In the rest of the paper, we refer to this design as the Non-Uniform Scheduler

J.C. Cunha and P.D. Medeiros (Eds.): Euro-Par 2005, LNCS 3648, pp. 540–549, 2005.

(NUS). The important feature of our design, and also the major difference from the previous proposals, is that at the time of dispatch, an instruction is steered to one of the queues and is eventually issued out of that queue. In this paper, we propose and evaluate several such steering heuristics.

2   Problem Characterization

Figure 1 presents the performance difference between the pipeline configurations with atomic (wakeup and select operations are performed within a single cycle) and pipelined (wakeup and select are pipelined over two cycles) schedulers. In general, the IPC degradation is as high as 30% for a 32-entry issue queue as seen from the graph. As the size of the issue queue is increased, this performance loss becomes smaller. For example, for a 64-entry issue queue, the performance degradation is reduced to 15% on the average. In any case, the performance impact due to the inability to execute instructions back to back is significant. Similar results were also presented by other researchers [4,20].

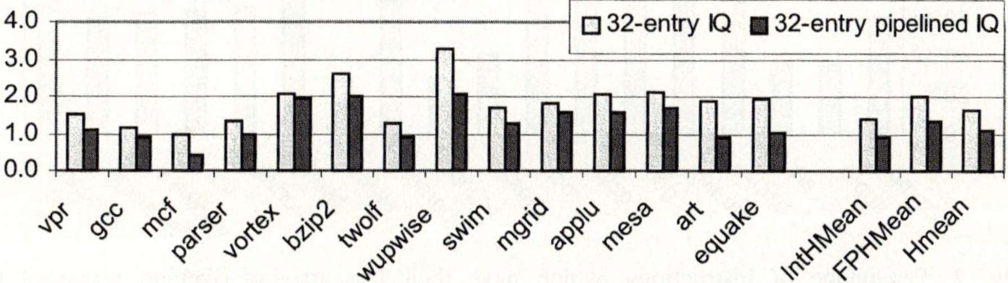

Fig. 1. IPC of 32-entry traditional queue and a 32-entry pipelined queue

Notice that it is only the dependency on a single-cycle latency operation that creates the pipeline bubble with pipelined schedulers. Moreover, it is the last arriving operand that truly awakens an instruction in the issue queue. Consequently, only the instructions with the last arriving operand produced by a single cycle latency instruction lose the ability to execute back-to-back with their parents in the presence of pipelined schedulers. In all other cases, the multi-cycle scheduling latency is completely hidden by the execution latency of the parent instructions (provided that the scheduling latency of a child does not exceed the execution latency of a parent).

To gauge the magnitude of this problem, we performed a study on the number of instructions whose last arriving operand is produced by a single-cycle latency instruction. Details of our simulation methodology are presented in Section 5. The results are shown in Figure 2. One can observe that on the average, about 60% of all instructions have a last arriving operand that is produced by a single-cycle latency instruction. These results show that there is a potential for optimizing traditional dynamic schedulers, as about half of all the instructions can tolerate 2-cycle scheduling latency.

3 Non-uniform Scheduling Logic

Non-Uniform Scheduling (NUS) logic is different from the traditional scheduling logic in that a traditional monolithic issue queue (IQ) and its associated selection logic are divided into two parts. The first is a small, fast issue queue (FIQ) and the second is a large, slow issue queue (SIQ). Instructions are steered to one of these queues at the time of dispatch according to certain heuristics. Once dispatched to a queue, the instruction waits in that queue until it is ready to execute. The steering logic is activated in parallel with the rename stage and has negligible additional overhead, as the steering heuristics that we consider are very simple. The queues have separate wakeup and selection logic, and they share functional units. During selection, priority is given to the instructions in the FIQ. The FIQ's wakeup/select loop is atomic (takes one cycle) while the SIQ's wakeup/select loop is pipelined over two cycles. The datapath incurporating the NUS scheduler is shown in Figure 3.

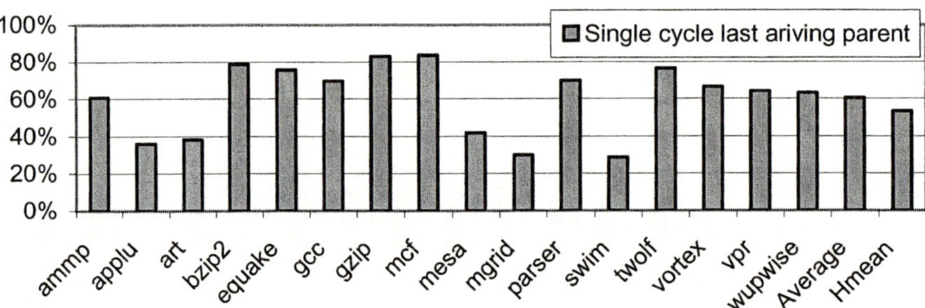

Fig. 2. Percentage of instructions, which have their last arriving operand produced by an instruction with single-cycle execution latency

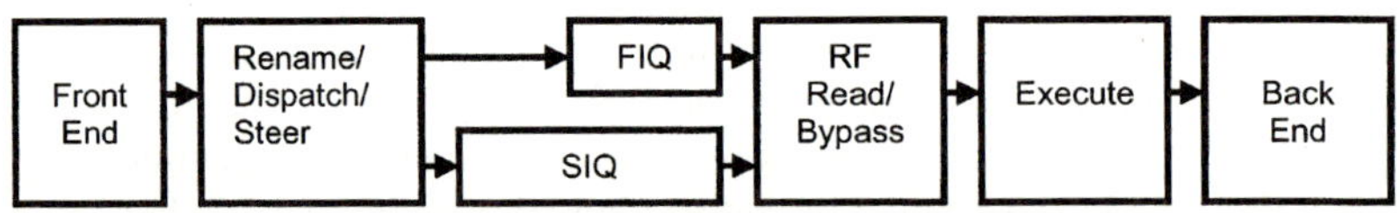

Fig. 3. Datapath Incorporating NUS Scheduling Logic

At the end of the selection cycle, combined W instructions are selected from both the FIQ and the SIQ. Then, the destination tags of all of the selected instructions are broadcast across both queues. Dependent instructions in the FIQ wakeup and get selected in the next cycle, thus allowing for the back-to-back execution. Instructions in the SIQ wakeup in the next cycle and get selected one cycle after that. Mechanisms similar to the ones described in [20] are used to ensure that an instruction does not issue prematurely, if the execution latency of its last arriving parent is higher than 2 cycles.

The selection logic in the NUS design needs to perform some arbitration between the instructions selected from both queues. For example, if N instructions are selected from the FIQ, then at most (W-N) instructions can be selected from the SIQ in the same cycle. This logic is part of the SIQ. There is enough slack in the SIQ's selection cycle (the entire cycle is still used for selection, but the SIQ's size is smaller than the size of the baseline issue queue) to perform this simple check and limit the number of instructions selected from the SIQ. The FIQ selection logic is unmodified, as all instructions selected from the FIQ are issued.

4 Steering Mechanisms for NUS

An important aspect of the NUS design is the set of heuristics used to steer the instructions between the two queues at the time of instruction dispatch. Optimally, a heuristic would be simple and easy to implement, but also be effective and schedule the key pairs of instructions back-to-back. In this paper, we examine several different steering heuristics. In all cases, if the destination queue is full, the instruction is placed into the other queue if such a possibility exists (i.e., the other queue is not full). The process of instruction dispatching blocks only when both queues saturate. The steering heuristics examined in this paper are as follows:

(1) FIQ Utilization (UTIL). Here, instructions are only steered to the SIQ if the FIQ is full. Otherwise, each instruction is steered to the FIQ. Notice that this is a greedy steering heuristic which, at first sight, may seem to be an optimal solution. However, this is not necessarily the case because the FIQ may become full with instructions that are in fact tolerant to the pipelined scheduling, forcing other instructions which are not tolerant of the pipelined scheduling to end up in the SIQ. Even in the presence of a free space in the FIQ, it could be more beneficial to steer some instruction into the SIQ so that instructions that cannot tolerate pipelined scheduling can later be placed in the FIQ. All subsequent heuristics attempt to do exactly that.

(2) Single Cycle Dependency (SCD). Here, all instructions dependent on a not-yet executed single cycle instruction are steered to the FIQ and all other instructions are steered to the SIQ.

(3) Multiple Non-ready Sources (MNR). Here, instructions with one or zero non-ready sources are steered to the SIQ and only instructions with 2 or more non-ready operands are steered to the FIQ. This heuristic is based on several observations. First, several researchers have shown that most instructions enter the scheduling window with at least one of their input operands already available [10,11,19], thus there are fewer instructions with two non-ready operands. Secondly, instructions waiting on two operands are likely to be waiting for a longer duration, and thus it could be advantageous to give these instructions scheduling priority when they do become ready to execute.

(4) Single Cycle Dependency with Multiple Non-ready Sources (SCD/MNR). This heuristic combines the previous two. An instruction is steered to the FIQ only if it is dependent on multiple not-yet-executed instructions (i.e. both source operands are not ready) and one of those instructions is a single-cycle latency operation. All other instructions are steered to the SIQ.

(5) Single Non-ready Source with a Single Cycle Dependency (SNR/SCD). Here, an instruction is steered to the FIQ only if it has exactly one non-ready source at the time of dispatch and that source will be produced by an instruction with a single-cycle execution latency. All other instructions are steered to the SIQ.

5 Simulation Methodology

Our simulation environment includes a detailed cycle accurate simulator of the microarchitecture and cache hierarchy. We used a modified version of the Simplescalar simulator [5] that implements separate structures for the issue queue, re-order buffer, load-store queue, register files, and the rename tables in order to more accurately model the operation of modern processors. All benchmarks were compiled with gcc 2.6.3 (compiler options: -O2) and linked with glibc 1.09, compiled with the same options, to generate the code in the portable ISA (PISA) format. All simulations were run on a subset of the SPEC 2000 benchmarks consisting of 7 integer and 7 floating-point benchmarks using their reference inputs. In all cases, predictors and caches were warmed up for 1 billion committed instructions and statistics were gathered for the next 500 million instructions. Table 1 presents the configuration of the baseline 4-way processor.

Table 1. Configuration of a simulated processor

Parameter	Configuration
Machine width	4-wide fetch, 4-wide issue, 4 wide commit
Window size	issue queue – as specified, 128 entry LSQ, 256–entry ROB
Function Units and Latency (total/issue)	4 Int Add (1/1), 2 Int Mult (3/1) / Div (20/19), 2 Load/Store (2/1), 2 FP Add (2), 2 FP Mult (4/1) / Div (12/12) / Sqrt (24/24)
Physical Registers	256 combined integer + floating-point physical registers
L1 I–cache	64 KB, 1–way set–associative, 128 byte line, 1 cycles hit time
L1 D–cache	64 KB, 4–way set–associative, 64 byte line, 2 cycles hit time
L2 Cache unified	2 MB, 8–way set–associative, 128 byte line, 6 cycles hit time
BTB	2048 entry, 2–way set–associative
Branch Predictor	Combined with 1K entry Gshare, 10 bit global history, 4K entry bimodal, 1K entry selector
Memory	128 bit wide, 140 cycles first chunk, 2 cycles interchunk
TLB	32 entry (I), 128 entry (D), fully associative

For estimating the delay requirements, we designed the actual VLSI layouts of the issue queue and simulated them using SPICE. The layouts were designed in a 0.18 micron 6 metal layer CMOS process (TSMC) using Cadence design tools. A Vdd of 1.8 volts was assumed for all the measurements.

6 Experimental Results

The analyses in this section are focused on comparing the appropriate NUS configurations against a 32-entry traditional atomic issue queue. We first explain how we selected the appropriate sizes of the FIQ and the SIQ in the NUS design. In order to balance the delays of both queues, we performed circuit-level simulations of *complete*, hand-crafted issue queue layouts in 0.18-micron TSMC technology. We measured the delays of both the wakeup and selection logic for the schedulers of various sizes. Results are summarized in Table 2.

Table 2. Delays of the scheduling logic

	Tag-Bus Drive (ps)	Comparator Output (ps)	Final Match Signal (ps)	Total Wakeup Delay (ps)	Selection Delay(ps)	Total Delay(ps)
64-entry	313	223	131	667	489	1156
32-entry	224	219	126	569	370	939
16-entry	131	201	114	446	252	698
8-entry	72	194	110	376	136	512

The delays of the wakeup logic comprise of three components: the delay to drive the destination tags across the issue queue, the delays in performing tag comparisons, and the delays in setting the ready bit of the entry. We assumed that traditional pull-down comparators (whose outputs are precharged every cycle and are discharged on a mismatch) are used within the issue queue to perform tag matching. The delay of the tree-structured selection logic depends on the number of levels that must be traversed in the tree, both on the way to the arbiter and on the way back plus the additional wire delays on the way [17]. According to our estimations, the delay within a single level of the selection tree is about 60ps.

To understand how we selected the sizes of the queues in the NUS scheduler, assume that the SIQ of 32-entries is used, which is equal in size to the baseline scheduler. The SIQ operates in a pipelined fashion over two cycles where the cycle time is constrained by the delay of the wakeup phase (569ps). To complement such a SIQ in the NUS design, the size of the FIQ should be chosen in such a way that the combined delays of the wakeup and selection logic in the FIQ are comparable to the wakeup delay of the SIQ. For this reason, as seen from the results presented in Table 2, an 8-entry FIQ is an appropriate match for a 32-entry SIQ. Likewise, a 16-entry FIQ is an appropriate match for a 64-entry SIQ. To summarize, the appropriate combinations of the FIQ and the SIQ sizes are such that the size of the FIQ is about one quarter of the size of the SIQ. Similar observations about the relationship of the FIQ and the SIQ sizes can be made by examining the delays presented in [17].

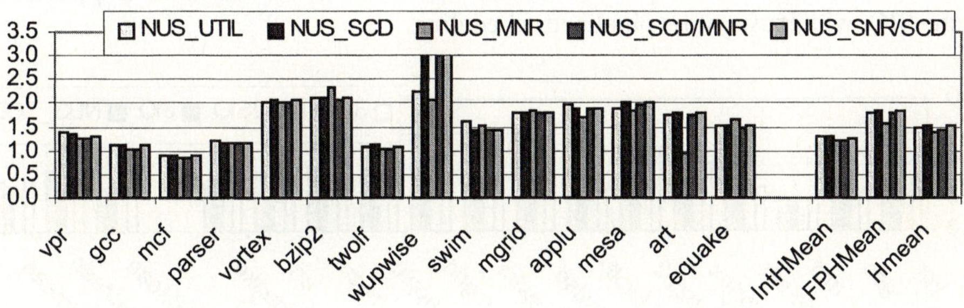

Fig. 4. Per-benchmark commit IPCs of the 8/32 NUS

We then examined the performance of various steering heuristics to be used in conjunction with the NUS scheduler, as described in Section 5. The per-benchmark results for the 8/32 NUS (8-entry FIQ and 32-entry SIQ) for each steering heuristic are presented in Figure 4. As seen from the graph, the SCD steering provides the best results - it outperforms the simple UTIL steering (which tries to fill the FIQ first) by

1% on the average. The UTIL heuristic performs reasonably well on the average, but it represents a suboptimal choice for 8 of the 14 benchmarks. For example, for *bzip2*, the MNR steering provides 10.9% better performance than simple UTIL steering. For *mcf*, SCD and SNR/SCD provide a 1.2% and 2.7% performance benefit, respectively. For *twolf*, SCD and SCD/MNR also provide a performance benefit by 1.6% and 0.5%, respectively. For *wupwise*, SCD, SCD/MNR, and SNR/SCD steering all provide better performance than the UTIL steering by 36.8%, 34.7%, and 38.2%, respectively. For *mgrid*, MNR provides 1.7% better performance than the UTIL steering. For *mesa*, SCD, SCD/MNR, and SNR/SCD perform better than the UITL steering by 6.6%, 4.6% and 6.8%, respectively. For *art*, the SCD and SNR/SCD heuristics provide better performance by 3.2% and 3.0%. Finally, the SCD/MNR steering provides 10.1% better performance for *equake*.

We now compare the performance results for the 8/32 NUS with SCD steering against traditional schedulers, both atomic and pipelined. Figure 5 compares four different scheduler architectures: the leftmost bar shows the performance of a machine with a 32-entry atomic scheduler, the next bar shows the performance of a machine with a 32-entry scheduler such that wakeup and selection are pipelined over two cycles, the third bar shows the performance of a traditional, atomic 8-entry scheduler, and finally, the rightmost bar shows the performance of a 8/32 NUS with SCD steering, as described in section 5. As expected, the combination of the two queues outperforms either queue used in isolation. On the average, the NUS scheduler outperforms the 32-entry pipelined scheduler by 37.5% and the 8-entry atomic scheduler by 10.7%. *Most importantly*, the performance of the NUS comes within 9% of the performance of a 32-entry atomic queue, but the NUS achieves significant cycle time reduction, potentially by as much as 40% according to the figures presented in Table 2 (569ps for the NUS vs. 939ps for the traditional 32-entry issue queue). Compared to the atomic scheduler with 40 entries (which in this case is the combined size of the FIQ and the SIQ), the performance of the NUS is only 9.5% lower. This is somewhat surprising because 80% of the NUS entries are implemented in the SIQ, which uses slow pipelined scheduling.

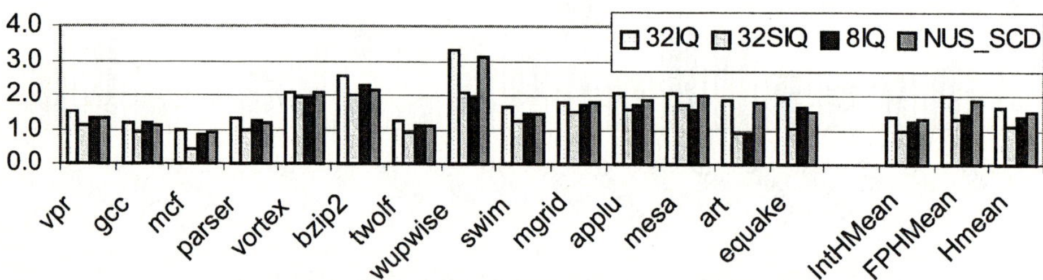

Fig. 5. Commit IPCs of the 8/32 NUS using the UTIL steering heuristic

Notice that the performance benefit of the 8/32 NUS compared to the 8-entry atomic queue is significantly higher for floating point benchmarks than it is for integer programs. This is because the floating point benchmarks significantly benefit from the presence of the larger queue and are generally more tolerant to the relaxation of the scheduling loop. Furthermore, the branch prediction accuracy is high for floating point benchmarks, which puts more pressure on the scheduler. The integer benchmarks, on the other hand, perform reasonably well even with the smaller schedulers and are significantly impacted by hindering the ability to execute instructions back to back. This is because these benchmarks are dominated by the long, serial dependency chains of mostly single-cycle latency instructions. Therefore, in many cases it may be beneficial to delay the dispatch of an instruction for a few cycles rather than dispatching it immediately to the slow queue. For example, *gcc, parser,* and *bzip2* show better performance with simply having the 8-entry issue queue compared to the NUS.

7 Related Work

Stark et.al. [20] pipelined the scheduling logic into wakeup and select stages and used the status of instruction's grandparents to wakeup the instruction earlier in a speculative manner. In [3], Brekelbaum et.al. introduced hierarchical scheduling windows (HSW) to support large number of in-flight instructions. The HSW design also relies on the use of fast and slow queues, but has a fairly complex logic for moving instructions from the slow queue (where all instructions are initially placed) to the fast queue. Our technique differs from HSW in that instructions are steered to the two queues upon dispatch. Lebeck et.al. [15] introduced a large waiting instruction buffer (WIB) to temporarily hold the load instructions that missed into the L2 cache as well as their dependents outside of the small issue queue (IQ). A somewhat similar technique, albeit implemented in a different manner, is also described in [8], where the instructions which are expected to wait for a large number of cycles before getting issued are moved to the secondary queue, called Slow Lane Instruction Queue. Brown et.al. [4] proposed to remove the selection logic from the critical path by exploiting the fact that the number of ready instructions in a given cycle is typically smaller than the processor's issue width. Kim and Lipasti [14] proposed to group two (or more) dependent single-cycle operations into so-called Macro-OP (MOP), which represents an atomic scheduling entity with multi-cycle execution latency.

Scheduling techniques based on predicting the issue cycle of an instruction [1,6,7,10,13,16] remove the wakeup delay from the critical path, but need to keep track of the cycle when each physical register will become ready. In [9], the wakeup time prediction occurs in parallel with the instruction fetching. Additional mechanisms are needed in these schemes for handling issue latency mispredictions, as the instructions executed too early need to be replayed. In [18], the use of segmented issue queues is proposed, where the broadcast and selection are limited to a smaller segment.

8 Concluding Remarks

The capability to execute the dependent instructions in the back-to-back cycles is important for sustaining high instruction throughput in modern out-of-order microprocessors. If the dependent instructions cannot execute in consecutive cycles, then the IPC impact can be very significant. We described a non-uniform scheduler design which significantly reduces the IPC penalties by using a pair of issues queues, one implementing single-cycle scheduling and the other implementing 2-cycle scheduling with pipelined wakeup and select. We evaluated several mechanisms for intelligent instruction placement between the two queues.

For a 4-way machine, the performance our design comes within 9% of an idealized atomic scheduler with potentially as much as 40% reduction in cycle time. To compare, if the idealized scheduler is simply pipelined into separate stages for wakeup and selection, then the performance loss compared to the idealized atomic situation is 30% for a 32-entry scheduler. Consequently, the NUS reduces this performance degradation by almost 70%. We evaluated several steering heuristics for the NUS and found out that, for a 32-entry queue, the single cycle dependency steering performs the best. We also found out that the simple greedy steering based on the utilization of the queues is not the optimal solution for the majority of the simulated benchmarks. Finally, we found out that most of the benefits of the NUS scheduler are achieved for floating-point benchmarks. For the majority of integer benchmarks, it is more beneficial to delay the dispatch of instructions rather than processing them through a slow queue.

Acknowledgements

We would like to thank Oguz Ergin for assistance with the VLSI layouts and Matt Yourst for help with the microarchitectural simulation environment. We would also like to thank Kanad Ghose and Deniz Balkan for useful comments on earlier drafts of this paper.

References

1. J. Abella, A.Gonzalez, "Low-Complexity Distributed Issue Queue", in Proc. of HPCA, 2004.
2. H. Akkary, R. Rajwar, S. Srinivasan, "Checkpoint Processing and Recovery: Towards Scalable Large Instruction Window Processors", in Proc. of MICRO 2003.
3. E. Brekelbaum et. al., "Hierarchical Scheduling Windows", in 35th Int'l. Symp. on Microarchitecture, 2002.
4. M. Brown, J. Stark, Y. Patt. "Select-Free Instruction Scheduling Logic", in the 34th International Symposium on Microarchitecture, 2001.
5. D. Burger and T. Austin, "The SimpleScalar tool set: Version 2.0", Tech. Report, Dept. of CS, Univ. of Wisconsin-Madison, June 1997 and documentation for all Simplescalar releases.
6. R.Canal, A. Gonzalez, "A Low-Complexity Issue Logic", in Proc. of the International Conference on Supercomputing (ICS), 2000.
7. R.Canal, A.Gonzalez, "Reducing the Complexity of the Issue Logic", in Proc, of the Int'l. Conf. on Supercomputing (ICS), 2001.

8. A. Cristal, et.al., "Out-of-Order Commit Processors", in the International Symposium on High-Perf. Comp. Arch. (HPCA), 2004.
9. T. Ehrhart, S. Patel, "Reducing the Scheduling Critical Cycle using Wakeup Prediction", in HPCA 2004.
10. D. Ernst, A. Hamel, T.Austin, "Cyclone: a Broadcast-free Dynamic Instruction Scheduler with Selective Replay", in Proc. of Int'l. Symp. On Computer Architecture (ISCA), 2003.
11. D. Ernst, T. Austin, "Efficient Dynamic Scheduling Through Tag Elimination", in the 29th Int'l. Symp. on Comp. Architecture, 2002.
12. B. Fields, R. Bodik, M. Hill. "Slack: Maximizing Performance Under Technological Constraints", in the 29th International Symposium on Computer Architecture, 2002.
13. J. Hu, N. Vijaykrishnan, M. Irwin, "Exploring Wakeup-Free Instruction Scheduling", in Proc. of the Int'l. Symp. on High Perf. Computer Architecture (HPCA), 2004.
14. I. Kim and M. Lipasti, "Macro-Op Scheduling: Relaxing Scheduling Loop Constraints", in the 36th International Symposium on Microarchitecture, 2003.
15. A. Lebeck et. al. A Large, "Fast Instruction Window for Tolerating Cache Misses", in the 29th Intl. Symp. on Comp. Arch. (ISCA), 2002.
16. P. Michaud, A. Seznec, "Data-Flow Prescheduling for Large Instruction Windows in Out-of-Order Processors", HPCA 2001.
17. S. Palacharla, N. Jouppi, J. Smith, "Complexity-Effective Superscalar Processors", in 24th Intl. Symposium on Computer Architecture, 1997.
18. S. Raasch, N.Binkert, S.Reinhardt, "A Scalable Instruction Queue Design Using Dependence Chains", in Proc. of ISCA, 2002.
19. J. Sharkey et.al., "Instruction Packing: Reducing Power and Delay of the Dynamic Scheduling Logic", in Proc. of ISLPED 2005.
20. J. Stark, M Brown, Y Patt, "On Pipelining Dynamic Instruction Scheduling Logic", in 33rd Int'l. Symp. on Microarchitecture, 2000.

Instruction Recirculation:
Eliminating Counting Logic in Wakeup-Free Schedulers

Joseph J. Sharkey and Dmitry V. Ponomarev

Department of Computer Science, State University of New York
Binghamton, NY 13902 USA
{jsharke,dima}@cs.binghamton.edu

Abstract. The dynamic instruction scheduling logic (the issue queue and the associated control logic) forms the core of an out-of-order microprocessor. Traditional scheduling mechanisms, based on tag broadcasts and associative tag matching logic within the issue queue are limited by high power consumption, large access delay and poor scalability. To address these inefficiencies, researchers have proposed various flavors of so-called *wakeup-free* scheduling logic. Such wakeup-free scheduling techniques remove the wakeup delay from the critical path, but incur other forms of complexity, essentially stemming from the need to keep track of the cycle when each physical register will become ready and when each instruction can be (speculatively) issued. We propose *instruction recirculation* – a wakeup-free instruction scheduler design that completely eliminates all counting and issue time estimation logic inherent in all previously proposed wakeup-free schedulers. This complexity reduction is also accompanied by 3.6% IPC improvement over the state-of-the-art wakeup-free scheduler.

1 Introduction

High-performance superscalar microprocessors rely on dynamic scheduling mechanisms to maximize instruction throughput across a wide variety of applications. The traditional scheduling logic operates in two phases: instruction wakeup and instruction selection. During instruction wakeup, the instructions stored within the issue queue (IQ) are associatively awakened by matching their source register addresses (called tags) against the destination tags of the instructions already selected for the execution. The selection logic, then, selects W out of N awakened instructions and issues them for execution. It has been well documented in the recent literature that the wakeup and selection logic form one of the most critical loops in modern superscalar processors [18,22]. Unless wakeup and selection activities are performed atomically (i.e. within a single cycle), dependent instructions cannot execute in consecutive cycles, which seriously degrades performance. At the same time, both wakeup and selection logic have significant delays [18], so if these activities are performed atomically, then the designers may be forced to either use the lower clock frequency or limit the size of the IQ, neither of which is desirable. In addition, traditional broadcast-oriented schedulers suffer from high power consumption, which is mainly due to broadcasting the destination tags across long, highly capacitive wakeup buses. For

J.C. Cunha and P.D. Medeiros (Eds.): Euro-Par 2005, LNCS 3648, pp. 550–559, 2005.

example, the scheduling logic of the Alpha 21264 dissipates about 18% of the total chip power [11].

To address the aforementioned deficiencies, researchers have proposed wakeup-free scheduling schemes, where the traditional tag broadcast and associative tag matching mechanisms are replaced with the capability to predict the issue cycle of an instruction based on the availability information about the source registers. Such wakeup-free scheduling techniques [5,6,8,12,17,19] remove the wakeup delay from the critical path, but need to keep track of the cycle when each physical register becomes ready so that the instructions can be issued just in time to access the value as soon as it becomes available. This is typically accomplished with the use of counters that track the availability of physical registers and also control when instructions can be issued (we describe a generic wakeup-free scheduler in more detail in Section 3). Due to the presence of a large number of these multi-bit counters and the issue time estimation logic, wakeup-free schedulers still incur substantial design complexity.

In this paper, we attempt to improve the performance/complexity trade-offs in the design of wakeup-free schedulers by exploiting the observation that most instructions that are selected for issue are typically among the few oldest in the IQ. Specifically, we introduce a technique called *Instruction Recirculation*, which uses a compacting IQ, where only N instructions at the head of the queue are considered for execution each cycle. Instead of relying on the traditional tag matching mechanisms, these N instructions determine their readiness to issue by checking the status bits associated with their source registers. Instructions at the head of the queue are recirculated back to the tail of the queue if they were not able to issue for specified number of cycles. This allows the younger instructions to be considered for scheduling in the presence of the long-latency events, such as the cache misses. Our results show that instruction recirculation achieves 3.6% better performance on the average than a state-of-the-art wakeup-free scheduler, and at the same time avoids the need to implement the counting logic for predicting the instruction issue time.

2 Simulation Methodology

Our simulation environment includes a detailed cycle-accurate simulator of the microarchitecture and cache hierarchy. We used a modified version of the Simplescalar simulator [4] that implements separate structures for the IQ, re-order buffer, load-store queue, register files, and the rename tables in order to more accurately model the operation of modern processors. All benchmarks were compiled with gcc 2.6.3 (compiler options: -O2) and linked with glibc 1.09, compiled with the same options, to generate the code in the portable ISA (PISA) format. All simulations were run on a subset of the SPEC 2000 benchmarks consisting of 8 integer and 7 floating-point benchmarks using their reference inputs. In all cases, predictors and caches were warmed up for 1 billion committed instructions and statistics were gathered for the next 500 million instructions. Table 1 presents the configuration of the baseline 4-way processor.

Table 1. Configuration of a simulated processor

Parameter	Configuration
Machine width	4-wide fetch, 4-wide issue, 4 wide commit
Window size	issue queue – as specified, 128 entry LSQ, 256–entry ROB
Function Units and Latency (total/issue)	4 Int Add (1/1), 2 Int Mult (3/1) / Div (20/19), 2 Load/Store (2/1), 2 FP Add (2), 2 FP Mult (4/1) / Div (12/12) / Sqrt (24/24)
Physical Registers	256 combined integer + floating-point physical registers
L1 I–cache	64 KB, 1–way set–associative, 128 byte line, 1 cycles hit time
L1 D–cache	64 KB, 4–way set–associative, 64 byte line, 2 cycles hit time
L2 Cache unified	2 MB, 8–way set–associative, 128 byte line, 6 cycles hit time
BTB	2048 entry, 2–way set–associative
Branch Predictor	Combined with 1K entry Gshare, 10 bit global history, 4K entry bimodal, 1K entry selector
Memory	128 bit wide, 140 cycles first chunk, 2 cycles interchunk
TLB	32 entry (I), 128 entry (D), fully associative

3 Wakeup-Free Schedulers

Wakeup-free instruction scheduling schemes have recently emerged as a viable alternative to traditional broadcast-oriented scheduling logic in complexity-aware microprocessor designs. Several variations of wakeup-free (a.k.a. broadcast-free) schedulers have been proposed in the recent literature [1,5,6,9,12,16,17]. Instead of using slow, complex and power-hungry CAM-based wakeup logic, these solutions rely on the ability to predict the cycle in which all of an instruction's input operands will become ready and issue the instruction just in time to access the values as soon as they become available. In all wakeup-free designs proposed until now, the counters are used to count down the delay between dispatch and issue for each instruction and also to keep track of the register availability information. Although the various wakeup-free scheduling schemes differ in their implementation details, they are all based on the common concept that the latencies of most instructions are deterministic and that the instruction's issue time can be fairly accurately predicted at the time of instruction dispatching.

For the analysis in this paper, we implemented a generic wakeup-free scheduling scheme, loosely based on the Cyclone scheduler [9]. At the time of instruction dispatching, a *pre-scheduler* is used to predict the number of cycles until each instruction will become ready for issue. We consider the *pre-scheduler* which is similar to that of [9], with the addition of a bimodal load hit/miss predictor and a load/store dependence predictor as proposed in [12] to improve the accuracy in scheduling load-dependent instructions. Instructions passing through the pre-scheduling stage check the availability of their source operands (by reading the availability counters) and use the maximum of these values to determine the number of cycles that will elapse before the instruction is ready for issue. This result becomes the delay counter of the instruction and is placed, along with the instruction itself, into the allocated IQ entry. We assume that the delay counter calculation can be performed in parallel with register renaming and thus it does not add an extra stage to the front end of the pipeline. If this extra stage is accounted for, the performance of the generic wakeup-free scheduler will be slightly worse than what is reported here.

Every cycle, the delay counters associated with each instruction in the IQ are decremented by one. When the delay counter falls to zero, the instruction becomes speculatively ready to execute, but must check the register ready bits of its source operands to be certain before it can be selected for execution. The hardware support for such checks is in the form of a bit-vector with one bit for each physical register. If the check succeeds, indicating that all source operands are indeed ready, the instruction is selected. We limit the number of instructions that can check their ready bits in a single cycle to only 8, requiring a register ready bit-vector with 16 read ports. Simple logic is assumed to arbitrate for these ports, using positional priority.

The IPC results for the generic wakeup-free scheduler with a 64-entry IQ are presented in Figure 1, along with the results for a 64-entry traditional atomic IQ where wakeup and select activities are implemented as an atomic operation within a single cycle. The wakeup-free scheme, as described above, exhibits 16.5% performance degradation on the average as compared to the 64-entry atomic IQ. This is consistent with the results presented in both [9] and [12], where the performance losses compared to the baseline are 17% and 14% respectively (although each of those schemes is presented for a machine configuration slightly different from ours). This performance loss can be attributed mainly to the inaccuracy in the instruction issue time estimation due to variable latency operations such as memory accesses and possible delays during instruction selection. As a result of mispredictions, non-ready instructions may deny the issue bandwidth to the ready instructions, causing a delay in the issue of those and leading to further mispredictions for the instructions dependent on the delayed ones, leading to a so-called "snowball effect". In the extreme case, all of the instructions in the queue can compete for issue in the same cycle, rendering the issue time estimation useless. In fact, we have observed such situation in our simulations on numerous occasions.

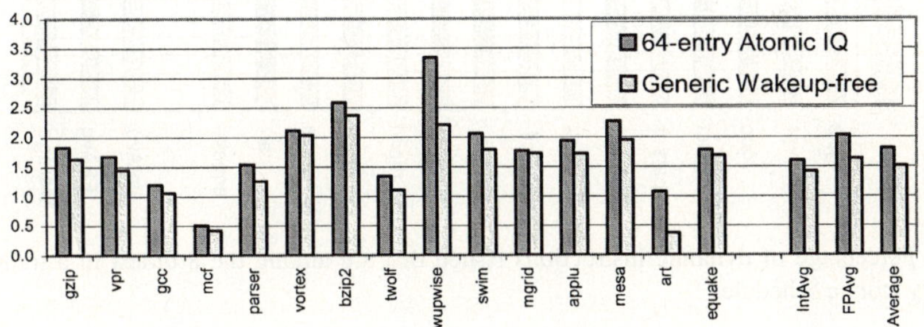

Fig. 1. IPCs of wakeup-free and traditional broadcast-based schedulers

Although the wakeup-free scheme eliminates the tag broadcast and tag matching logic in the scheduler, it does not come without a cost. Additional circuitry must be added in the front end of the pipeline to make the delay predictions. This delay prediction logic can become quite complex when multiple instructions are co-dispatched. To take dependencies into account, adders with multiple inputs (K-1 inputs in the case of a K-way machine) or an adder tree have to be used to get the initial counter values of the co-dispatched instructions in the worst case. Such adder structures may well form a critical path with the increase of the issue width.

Another source of complexity in the wakeup-free schedulers is the inherent counting logic. Two sets of counters have to be maintained– the ones that track the register availability information (one counter per physical register) and the ones that control the number of cycles that each instruction waits in the IQ before attempting to issue (one per instruction in the queue). The hardware structure that maintains the register availability counters must be multi-ported and all of these counters need to be decremented every cycle.

In the next section, we propose *Instruction Recirculation* - a wakeup-free scheduler design which achieves a better performance and completely eliminates all counting and issue time estimation logic inherent in all previously proposed wakeup-free schedulers.

4   Instruction Recirculation

The motivation for instruction recirculation stems from the observation that even in an out-or-order machine, a large percentage of issued instructions are among the few oldest instructions in the IQ. As demonstrated by Figure 2, more than 60% of all dynamic instructions that are selected for execution in a processor with a 64-entry IQ are among the 8 oldest in the scheduling window. Simply reducing the IQ size, however, is not a sufficient solution to the problem of complexity reduction because a small queue quickly saturates under a long latency event, such as a cache miss, causing performance degradation.

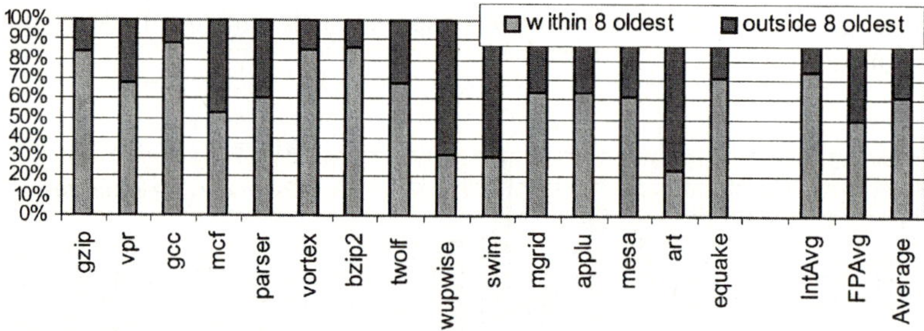

Fig. 2. Percentage of dynamic instructions issued that are among the 8 oldest in a traditional 64-entry atomic scheduler

The solution that we propose instead is to have a scheduler that examines only a small group of instructions in the window for execution, but is also capable of detecting long latency events (such as cache misses) and quickly move dependent instructions out of the front-end of the queue (if they cannot issue for a predetermined number of cycles) to allow possibly independent instructions down the stream the chance to execute. The older instructions can then be recirculated back into the tail end of the queue at a later time. We call this technique instruction recirculation. In the rest of this section we describe the details of our design.

The block diagram of instruction recirculation is shown in Figure 3. This design relies on the use of a compacting IQ, somewhat similar to the Cyclone scheduler [9], where instructions are dispatched to the tail block of the queue and work their way to the head block, from where they eventually get issued. The IQ is organized into several n-instruction blocks. Each row can compact independently of the other rows and each instruction can compact forward only one block at a time within its row. Only N instructions at the head of the queue participate in checking the register ready bit (RRB) vector and selection each cycle. Finally, an N-entry *recirculation buffer* is used. Its purpose is explained below. Conceptually, the instruction recirculation scheduler can be viewed as a wakeup-free scheduler in which the instructions present in the head block in any given cycle are predicted as "ready" and thus check the register ready bit vector.

The scheduler of Figure 3 operates in the following manner. Every cycle, N instructions within the head block of the queue check the ready bits of their corresponding source physical registers. If both sources are ready, and the instruction succeeds in acquiring the issue slot, the instruction is issued and the corresponding row within the IQ is compacted. Note that the issue rate from the head block is still limited by the issue width of the processor, which is 4 in our experiments (same as in the baseline machine). The recirculation scheduler (just as any other wakeup-free scheme), thus, still requires the selection logic to arbitrate among the instructions in the head block. However, this selection logic is much simpler than similar logic in the traditional scheduler.

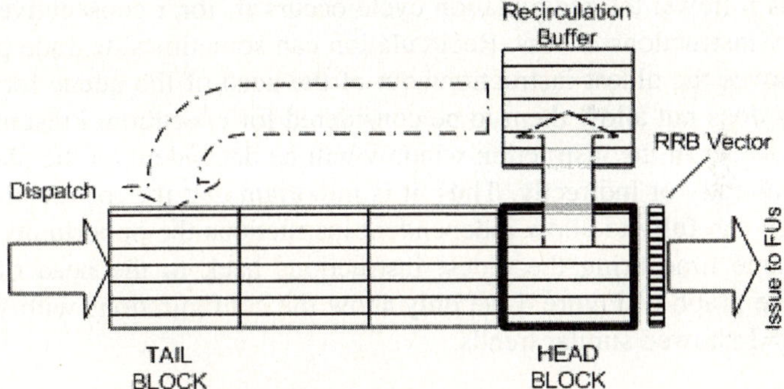

Fig. 3. Instruction Recirculation

When the instruction issue rate falls below a certain number, called the *recirculation threshold*, N instructions in the head block are moved into the recirculation buffer and the queue is compacted forward (for all rows). Such a compaction gives the next N instructions in the queue the opportunity to execute if they are ready. The scheduler remains in this state (issuing from the head block and compacting the individual rows as needed) until the recirculation threshold is reached again. At this point, the instructions in the head block are again moved, as before, into the recirculation buffer and the instructions sitting in the recirculation buffer are moved into the tail end of the queue. As instructions are circulated through the queue, they are each given a chance to execute if they are ready.

Notice that it takes a different number of recirculations for each instruction to return to its original position at the head of the queue. In the worst case, an instruction returns to the head block after K recirculations, where K is determined as the size of the IQ divided by n. Some of the original instructions from the head block can return there early if the compaction within certain rows progresses at a faster rate, i.e. the rows are compacted individually in-between block recirculations. The normal instruction dispatching continues during instruction recirculation, but the instructions that are already in the queue (i.e. recirculating) are always given higher priority over the newly dispatched instructions for the access to the tail block of the queue.

The goal of this mechanism is to quickly respond to the falling issue rates from the block of oldest instructions and consider other instructions for issue. This is useful if, for example, the instructions at the head block depend on a load that missed in the cache, but subsequent instructions are independent.

The selection of the recirculation threshold is critical to the performance of this scheme. In some cases, it may be more beneficial to wait rather than recirculate the entire block and encounter the full latency of multiple recirculations to bring these instructions back into the head block of the queue. We evaluated many configurations and present the results (Figure 4) for a few configurations that were representative of the rest. All of the presented configurations have a 56-entry IQ with an 8-entry recirculation buffer for a total of 64-entries in the scheduler to match that of the baseline case. Notice that in the interests of space we only present the averages across all benchmarks in this figure. The configurations presented are marked as x/y and can be interpreted as follows: the recirculation cycle occurs if, for x consecutive cycles, the issue rate is y instructions or less. Recirculation can sometimes degrade performance because it moves the oldest instructions out of the head of the queue for several cycles and thus does not allow them to be considered for execution. Presumably, many instructions deeper in the instruction window will be dependent on the oldest instructions, either directly or indirectly. Thus, it is important that the recirculation parameters be chosen carefully to allow independent instructions the opportunity to execute, but at the same time bring the oldest instructions back to the head of the queue quickly. In the graph of Figure 4 we only show the configurations with y=0, experiments with y=1 showed similar trends.

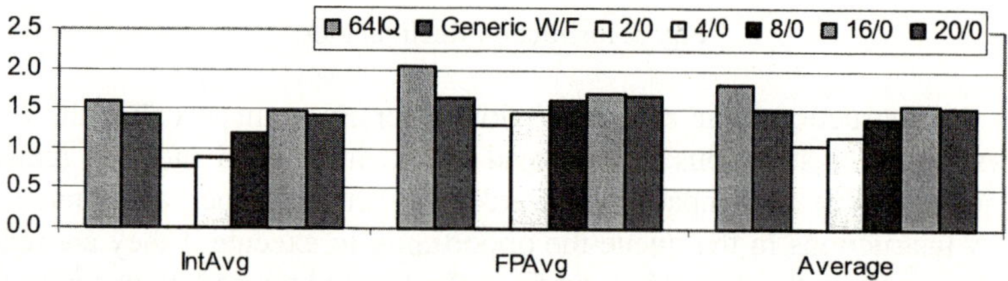

Fig. 4. IPC results of various configurations of the *instruction recirculation* scheduler. The configurations presented are marked as x/y and can be interpreted as follows: the recirculation cycle occurs if, for *x* consecutive cycles, the issue rate is *y* instructions or less

The best performance is achieved with the use of a 16/0 configuration, as shown by the graph, which degrades performance by 12.9% compared to a 64-entry traditional broadcast-based scheduler. The larger thresholds (20/0 for example) suffer because recirculations occur less often, and thus opportunities for independent instructions to execute are lost. The configurations with smaller threshold values (2/0, 4/0, 8/0) exhibit lower performance because instructions are recirculated too eagerly and may take several cycles (depending on compaction patterns, as discussed above) to return to the head block of the queue.

As a further comparison, the generic wakeup-free scheduler (as described in the previous section) degrades performance by 16.5% as compared to the same baseline, indicating that recirculation can perform 3.6% better than an aggressive state-of-the-art wakeup-free scheduler.

Figure 5 presents the per-benchmark IPC values for the 64-entry atomic queue, the generic wakeup-free scheduler from Section 3, and the best configuration for instruction recirculation. Instruction Recirculation outperforms the generic wakeup free scheduler for 10 of the examined benchmarks. The largest differences are seen on *art* and *mcf*, where recirculation shows a 149.6% and 21.3% improvement over the generic wakeup-free scheduler, respectively. This is because the poor memory behavior of these two benchmarks significantly impacts the wakeup-free scheduler's ability to predict wakeup times. Instruction Recirculation, however, does not rely on predictions and can dynamically adapt to the memory behavior of individual benchmarks. The generic wakeup-free scheduler provides higher IPC than instruction recirculation for five of the benchmarks (*vpr, vortex, wupwise, swim, mesa*), with the largest difference (8.5%) observed for the *mesa* benchmark.

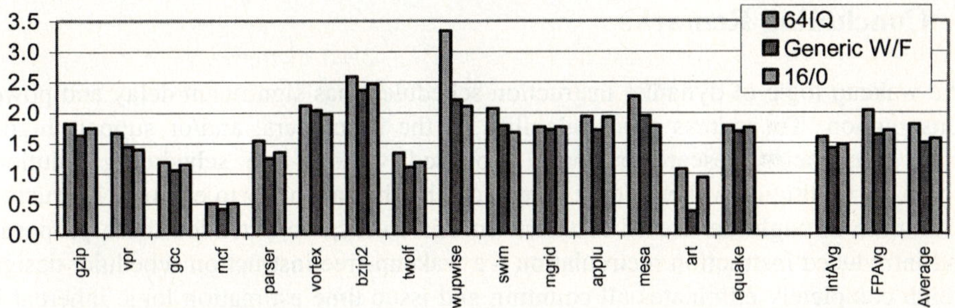

Fig. 5. Per-benchmark IPC of the best configuration of Instruction Recirculation

5   Related Work

Scheduling techniques based on predicting the issue cycle of an instruction [5,6,8,9,12,16,17] remove the wakeup delay from the critical path, but need to keep track of the cycle when each physical register will become ready. [5] proposed the "distance scheme issue logic" that reorders instructions during dispatch time based on predicted wakeup times. In [8], the wakeup time prediction occurs in parallel with the instruction fetching. [9,12] remove the counters from the IQ and instead use *pre-*

scheduling predictions to determine the placement of instructions in the IQ which, in turn, determines the number of cycles until the instruction is considered for execution. [16] also removes the counters from the IQ and uses *pre-scheduling* predictions to distribute instructions amongst the variable sized FIFOs.

Alternative mechanisms have also been proposed to reduce the complexity and access delay of the dynamic scheduling logic. To pipeline the scheduling logic without hindering the ability to execute dependent instructions back-to-back, Stark et.al. [22] proposed to use the status of an instruction's grandparents to wakeup the instruction earlier in a speculative manner. Kim and Lipasti [14] proposed grouping of two (or more) dependent single-cycle operations into so-called Macro-OP (MOP), which represents an atomic scheduling entity with multi-cycle execution latency. As a result, the scheduling logic can be pipelined with much smaller impact on the IPC. Other proposals have introduced new scheduling techniques with the goal of designing scalable dynamic schedulers [2,15,7,19,21]. Brown et.al. [3] proposed to remove the selection logic from the critical path by exploiting the fact that the number of ready instructions in a given cycle is typically smaller than the processor's issue width. Ernst et.al. [10] introduced specialized IQ entries for instructions with various numbers of non-ready operands. In [20], instruction packing was proposed to dynamically assign instructions to either a full IQ entry or a half IQ entry, depending on the number of ready sources. In [13], half of the tag comparators are offloaded from the fast wakeup bus and are connected to the slow wakeup bus, where the tags are broadcast one cycle later. In [1], instructions are issued from multiple FIFO buffers such that multiple dependency chains may be intermixed within a single FIFO.

6 Concluding Remarks

The wakeup logic of dynamic instruction schedulers has significant delay and power consumption. To address this scalability of the schedulers, and/or support higher clock frequencies, researchers have proposed wakeup-free scheduling solutions where the traditional wakeup logic is replaced by the capability to estimate instruction issue time through the use of counters. In this works, we extended these proposals and introduced instruction recirculation – a wakeup-free instruction scheduler design, which completely eliminates all counting and issue time estimation logic inherent in all previously proposed wakeup-free schedulers. This complexity reduction is accompanied by a 3.6% IPC gain over the state-of-the-art wakeup-free scheduler.

Acknowledgements

We would like to thank Matt Yourst for help with the microarchitectural simulation environment. We would also like to thank Kanad Ghose and Deniz Balkan for useful comments on earlier drafts of this paper.

References

1. J. Abella, A.Gonzalez, "Low-Complexity Distributed Issue Queue", HPCA, 2004.
2. E. Brekelbaum et. al., "Hierarchical Scheduling Windows", in Proc. of MICRO, 2002.
3. M. Brown, J. Stark, Y. Patt. "Select-Free Instruction Scheduling Logic", in the 34th International Symposium on Microarchitecture, 2001.
4. D. Burger and T. Austin, "The SimpleScalar tool set: V. 2.0", Tech. Report, Dept. of CS, Univ. of Wisconsin-Madison, June 1997 and documentation for all Simplescalar releases.
5. R.Canal, A. Gonzalez, "A Low-Complexity Issue Logic", in Proc. of the International Conference on Supercomputing (ICS), 2000.
6. R.Canal, A.Gonzalez, "Reducing the Complexity of the Issue Logic", in Proc, of the Int'l. Conf. on Supercomputing (ICS), 2001.
7. A. Cristal, et.al., "Out-of-Order Commit Processors", in Proc. of HPCA, 2004.
8. T. Ehrhart, S. Patel, "Reducing the Scheduling Critical Cycle using Wakeup Prediction", in HPCA 2004.
9. D. Ernst, A. Hamel, T.Austin, "Cyclone: a Broadcast-free Dynamic Instruction Scheduler with Selective Replay", in Proc. of Int'l. Symp. On Computer Architecture (ISCA), 2003.
10. D. Ernst, T. Austin, "Efficient Dynamic Scheduling Through Tag Elimination", in the 29th Int'l. Symp. on Comp. Architecture, 2002.
11. M.K. Gowan, L.L. Biro, D.B. Jackson, "Power considerations in the Design of the Alpha 21264 microprocessor", in the Proceedings of the 35th ACM/IEEE Design Automation Conference (DAC 98), 1998.
12. J. Hu, N. Vijaykrishnan, M. Irwin, "Exploring Wakeup-Free Instruction Scheduling", in Proc. of the Int'l. Symp. on High Perf. Computer Architecture (HPCA), 2004.
13. I.Kim, M.Lipasti, "Half-Price Architecture", in Proceedings of ISCA 2002.
14. I. Kim and M. Lipasti, "Macro-Op Scheduling: Relaxing Scheduling Loop Constraints", in the 36th International Symposium on Microarchitecture, 2003.
15. A. Lebeck et. al. "A Large, Fast Instruction Window for Tolerating Cache Misses", in the 29th Intl. Symp. on Comp. Arch. (ISCA), 2002.
16. Y.Liu, et. al.,"Scaling the Issue Window with Look-Ahead Latency Prediction" ICS 2004.
17. P. Michaud, A. Seznec, "Data-Flow Prescheduling for Large Instruction Windows in Out-of-Order Processors", HPCA 2001.
18. S. Palacharla, et.al. "Complexity-Effective Superscalar Processors", in ISCA 1997.
19. S. Raasch, N.Binkert, S.Reinhardt, "A Scalable Instruction Queue Design Using Dependence Chains", in Proc. of ISCA, 2002.
20. J. Sharkey, et. al. "Instruction Packing: Reducing Power and Delay of the Dynamic Scheduling Logic", in Proc. ISLPED 2005.
21. J. Sharkey, D. Ponomarev, "Non-Uniform Instruction Scheduling", in Proc. Euro-Par, 2005.
22. J. Stark, M Brown, Y Patt, "On Pipelining Dynamic Instruction Scheduling Logic", in 33rd Int'l. Symp. on Microarchitecture, 2000.

Early Experience with Scientific Applications on the Blue Gene/L Supercomputer

George Almasi[1], Gyan Bhanot[1], Dong Chen[1], Maria Eleftheriou[1],
Blake Fitch[1], Alan Gara[1], Robert Germain[1], John Gunnels[1], Manish Gupta[1],
Philip Heidelberg[1], Mike Pitman[1], Aleksandr Rayshubskiy[1], James Sexton[1],
Frank Suits[1], Pavlos Vranas[1], Bob Walkup[1], Chris Ward[1], Yuriy Zhestkov[1],
Alessandro Curioni[2], Wanda Andreoni[2], Charles Archer[3], José Moreira[3],
Richard Loft[4], Henry Tufo[4,5], Theron Voran[5], and Katherine Riley[6]

[1] IBM T.J. Watson Research Center, Yorktown Heights, NY, USA
`{gheorghe,gyan,chendong,mariae,bgf,alangara,rgermain,`
`gunnels,mgupta,philiph,pitman,arayshu,sextonjc,`
`suits,vranasp,walkup,tjcw,yuriyz}@us.ibm.com`
[2] IBM Zurich Research Laboratory, CH-8803 Rüschlikon, Switzerland
`{cur,and}@zurich.ibm.com`
[3] IBM Systems and Technology Group, Rochester, MN, USA
`{archerc,jmoreira}@us.ibm.com`
[4] National Center for Atmospheric Research, Boulder, CO, USA
`loft@ucar.edu`
[5] University of Colorado at Boulder, Boulder, CO, USA
`{tufo,theron.voran}@cs.colorado.edu`
[6] Argonne National Laboratory, Argonne, IL, USA
`riley@mcs.anl.gov`

Abstract. Blue Gene/L uses a large number of low power processors, together with multiple integrated interconnection networks, to build a supercomputer with low cost, space and power consumption. It uses a novel system software architecture designed with application scalability in mind. However, whether real applications will scale to tens of thousands of processors has been an open question. In this paper, we describe early experience with several applications on a 16,384 node Blue Gene/L system. This study establishes that applications from a broad variety of scientific disciplines can effectively scale to thousands of processors. The results reported in this study represent the highest performance ever demonstrated for most of these applications, and in fact, show effective scaling for the first time ever on thousands of processors.

1 Introduction

A popular approach to building supercomputers has been to build clusters of high performance nodes (based on symmetric multiprocessors or vector processors) with high performance interconnection networks. Examples of systems that have been built, or are being built, using this approach include the Earth Simulator [1], ASC Purple [2], and the Columbia [3] systems. At the highest scales, these machines consume a great deal of power and require a lot of floor space. For example, the Earth Simulator, which delivers a peak performance of 41 Teraflop/s, consumes about 7 MW of power, and occupies an area of about 70,000 square feet.

J.C. Cunha and P.D. Medeiros (Eds.): Euro-Par 2005, LNCS 3648, pp. 560–570, 2005.

The IBM Blue Gene/L (BG/L) [4] represents a different way of building supercomputers. It uses low power processors, which allows a large number of processors to be packed in a given volume (2048 processors in a rack), with aggregate heat dissipation staying within air cooling limits. Furthermore, it uses system-on-a-chip technology to integrate powerful torus and collective networks, and uses a novel software architecture [5] to support high levels of scalability. While BG/L offers the promise of making massively parallel systems accessible, and promising results have been reported on a 512 node prototype [6], how far the applications can scale has been an open question. Previous results on scaling of MPI applications on any platform have necessarily been limited (by hardware existence) to fewer than ten thousand processors. Many previous studies have shown problems with the scaling of applications to thousands of processors due to factors like computational noise [7].

This paper, together with a companion paper [8] on physics and material science applications developed by researchers at Lawrence Livermore National Laboratory, explores the scaling of applications to thousands of processors. It describes early experience with several scientific applications on a 16,384 node BG/L system. This system, which occupies less than 400 square feet of floor space, and consumes about 400 KW in power, was recently rated as the fastest supercomputer in the world (#1 on the TOP500 list), with a sustained LINPACK performance of 70.72 Teraflop/s [9]. We show that the scientific applications targeted in this study scale well on the BG/L system, thus validating the design of BG/L and establishing the ability to scale MPI applications to several thousand processors. This study also uncovers several opportunities for performance improvements through software optimizations.

2 Overview of BG/L

This section reviews some architectural features of BG/L that have a significant impact on performance.

2.1 BG/L Hardware

Each BG/L node [1] has two 32-bit embedded PowerPC (PPC) 440 processors, which have 32 KB each of L1 data and instruction caches. The BG/L nodes support prefetching in hardware, based on detection of sequential data access. The prefetch buffer for each processor holds 64 L1 cache lines (16 128byte L2/L3 cache lines) and is referred to as the L2 cache. Each chip also has a 4 MB L3 cache built from embedded DRAM, and an integrated DDR memory controller. A single BG/L node supports 512 MB memory. The PPC 440 processor does not support hardware cache coherence at the L1 level. However, there are instructions to invalidate a cache line or flush the cache, which can be used to manage coherence in software.

BG/L employs a SIMD-like extension of the PPC floating-point unit, which we refer to as the double floating point unit or DFPU [10]. The DFPU adds a secondary FPU to the primary FPU as a duplicate copy with its own register file. BG/L supports a comprehensive set of parallel instructions on double-precision floating-point data.

The BG/L ASIC supports five different networks: torus, collective, global interrupts, Ethernet, and JTAG. The main communication network for point-to-point messages is a three-dimensional torus. Each node contains six bi-directional links for direct connection with nearest neighbors. The raw hardware bandwidth for each torus link is 2 bits/cycle (175 MB/s at 700 MHz) in each direction. The torus network provides both adaptive and deterministic minimal path routing in a deadlock-free manner. The collective network implements broadcasts and reductions with a target hardware latency of 1.5 microseconds for a 64K node system. The global interrupts network supports a fast barrier operation, also with a target latency of 1.5 microseconds for a 64K node system. On BG/L, I/O is supported via special I/O nodes, which are architecturally identical to compute nodes, but are attached to the Gbit/s Ethernet network, which connects the BG/L core to external file servers and host systems. The booting, control and monitoring of the BG/L system is done over the JTAG network.

2.2 BG/L Software

The programming model supported in BG/L is *single program multiple data* (SPMD), with message passing supported via an implementation of the Message Passing Interface (MPI). A BG/L job can be submitted in one of two modes. In *co-processor mode* (CPM), which is the default mode, a single application (MPI) process runs on each compute node – one of the processors of the compute node is used for computation, and the other is used for offloading part of the communication operations. In *virtual processor mode* (VNM), two application processes are run on each compute node, one on each of the two processors.

BG/L uses a hierarchical organization of software, described in further details in [5]. User applications run exclusively on compute nodes under the supervision of a simple, minimalist *compute node kernel* (CNK). The I/O nodes run a customized version of Linux. Many system calls (such as *read* and *write*) are not directly executed in the compute node, but are function shipped through the collective network to the "parent" I/O node. The control system is implemented as a collection of processes running in an external computer, called the *service node* for the machine. All of the visible state of BG/L is maintained in a commercial database on the service node.

BG/L provides an operating environment with a very low level of "computational noise" (interference from operating system activity), about two orders of magnitude lower than traditional clusters and the ASCI Q system [11]. It also supports low latency communication (latency to nearest neighbor is about 3.3 microseconds, or 2350 processor cycles), with a low *half-bandwidth* point (half of asymptotic bandwidth is achieved at a message size smaller than 1 KB for several MPI bandwidth tests, such as point-to-point sends to all nearest neighbors and *alltoall* collective operation).

3 Applications Performance Results

This section describes the applications we used in this study and the performance of these applications on the BG/L system at IBM Rochester.

3.1 Blue Matter

The *Blue Matter* application framework has been developed at IBM Research to advance our understanding of biologically important phenomena such as protein folding through large scale simulation [12]. In addition to molecular dynamic simulations, the Blue Gene science team also aims to run using the "replica exchange" method in which a large number (32-128) of simulations are run at different temperatures and are coupled via periodic exchanges using a Metropolis Monte Carlo type criterion. This yields an application with a hierarchy of communication—tight coupling within individual trajectories and loose coupling between them. The loose coupling comes about because carrying out the exchange attempts only requires an "all gather" of a single double precision number from each replica once every few hundred to few thousand time steps. Using this technique one can use thousands of nodes with a parallel efficiency determined by that of the component molecular dynamics simulations.

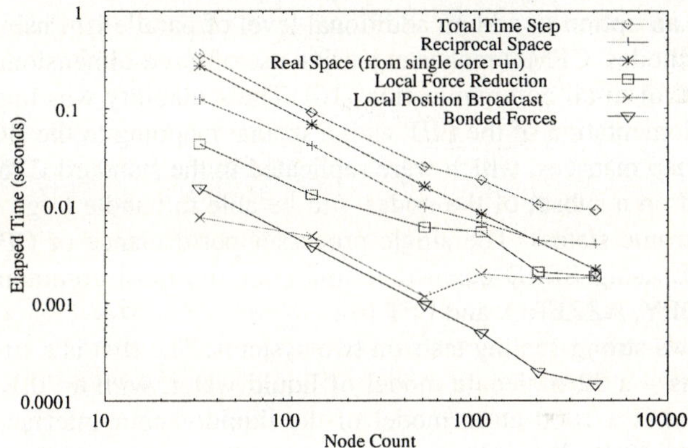

Fig. 1. Performance of Blue Matter on BG/L

Figure 1 shows the total time for a single iteration on a 43,222 atom system as a function of node count, and also the major contributions to the total time. We are now using a parallel decomposition that maps the simulation volume directly onto the 3D torus topology of BG/L and have removed our dependence on the fixed and floating point *Allreduce* collectives used in our previous work [13]. The current decomposition employs a neighborhood broadcast and reduction that is implemented using the *Alltoallv* collective. We carry out the reciprocal space operations requiring the computation of a 3D-FFT on CPU 0 while off-loading the real space non-bond computations (which require no communication) on to CPU 1. The scalability plot of the real space non-bond computation was taken from a separate run that used only one CPU because our tracing utility only functions on the first CPU.

Overall, as Figure 1 shows, we observe nearly ideal speedup up to 512 nodes and a drop in parallel efficiency to about 32% at 4096 nodes. This is scalability to a node count and value of atoms/node in a biomolecular simulation with proper treatment of electrostatics that is unique—for reference, the NAMD package scaled to 2250 nodes (on a 92K atom system) on the PSC Lemieux system before losing performance [14]. From the data plotted in Figure 1, we can extrapolate the parallel efficiency of the replica exchange technique, which uses multiple trajectories, for a 128 replica simulation of the 43K atom system using 512 nodes/replica to be over 90% on 64K nodes (normalizing the parallel efficiency to be 100% at 32 nodes for a single replica).

3.2 Car-Parrinello Molecular Dynamics (CPMD)

The Car-Parrinello Molecular Dynamics code (CPMD) originated in IBM's Zurich Research Laboratory in the early 1990s [15]. It is an implementation of density functional theory using plane waves and pseudopotentials, and is particularly designed for ab-initio molecular dynamics. The CPMD code is widely used for research in computational chemistry, materials science, and biology. It has been licensed to several thousand research groups in over 50 countries.

The application is mainly written in Fortran, parallelized for distributed-memory with MPI, with an option to add an additional level of parallelism using OpenMP for multi-processor nodes. CPMD makes extensive use of three-dimensional FFTs, which require efficient all-to-all communication [16]. The scalability was improved using a task-group implementation of the FFT with a special mapping to the BG/L torus [17]. Moreover, overlap matrices, which were replicated in the standard CPMD code, have been distributed on a subset of the nodes – to be able to handle large systems (more than 3000 electronic states). The single processor performance of CPMD was optimized for BG/L using SIMD-enabled routines for the most common calls such as DGEMM, DCOPY, AZZERO, and FFT [6].

Figure 2 shows strong-scaling tests on two systems. The first is a small test case of about 100 atoms – a 32 molecule model of liquid water, with a 70 Rydberg cutoff, while the second is a 1000 atom model of the liquid/vapour interface of methanol, with a cutoff of 140 Rydberg. In both cases a gradient corrected form (PBE) of the exchange-correlation functional was used.

For the first system (Figure 2a), a good scaling up to 512 processors is obtained (with a parallel efficiency greater than 40%). This corresponds to the limit of the data granularity of the model. Using more processors would have meant that some processors remain idle for a significant portion of the time. Low latency in the MPI layer and a total lack of system daemons contribute to very good scalability on BG/L. The execution time on 512 BG/L processors was less then 0.35 seconds per step; which is much better than the 1.5 seconds per step that we obtained on IBM p690 SMP servers (1.3 GHz) clustered via double colony switches, where scalability was limited to 128 processors (the Federation switch would enhance this performance by ~30%, but would not enhance scalability). The result on BG/L represents the highest performance obtained for ab-initio molecular dynamics simulation for a system with about 100 atoms [17]. This allows one to simulate a system of this size with fully ab-initio methods with throughput of 175 seconds per day using a fraction of a BG/L rack.

The second test case (Figure 2b) has better data granularity because it is 100 times larger in terms of the amount of data storage required. For this reason, this test case is not latency bound and good scaling up to 4096 processors was observed. The parallel efficiency is more than 90% up to 1024 processors and more than 50% up to 4096 processors. This was obtained by using a taskgroup parallelization scheme and an optimized mapping to the BG/L hardware [17].

The third test case is a prototypical case in which the linear algebra involved in the orthogonalization of the electronic states starts to dominate the computation (here we have more than 5000 electronic states) and a newly implemented distributed orthogonalization algorithm becomes essential in order to fit the problem into the available memory per node (512MB). Figure 3 shows the sustained performance (considering the number of measured floating point operations and not pseudo operations) for a path-integral ab-initio molecular dynamics simulation, where an additional parallelization layer (over the replicas) is available. We obtain a performance close to 46% of peak at 16K processors and 38% at 32K processors; this value is quite good, especially considering the application, which has memory-intensive and network-intensive algorithms like the 3D-FFT.

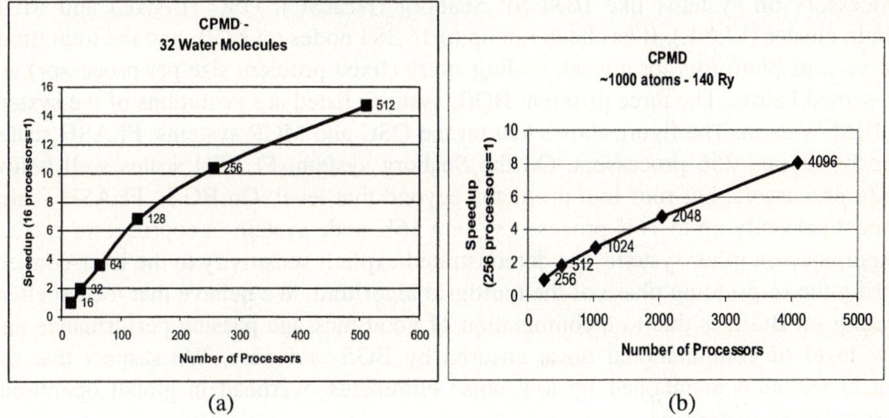

Fig. 2. Strong scaling of CPMD on BG/L: (a) ~100 atom simulation (speedup normalized to 16 procs), (b) ~1000 atom simulation (speedup normalized to 256 procs)

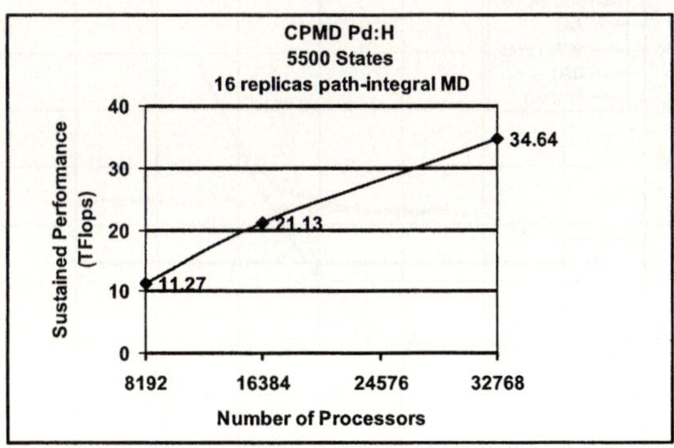

Fig. 3. Strong scaling of CPMD to 32K processors (in virtual node mode)

3.3 FLASH

FLASH is a parallel adaptive-mesh multi-physics simulation code designed to solve nuclear astrophysical problems related to exploding stars [18]. It has been developed as part of an ASC, DOE project at the University of Chicago, and won a Gordon Bell prize in 2000. The FLASH code solves the Euler equations for compressible flow and the Poisson equation for self-gravity.

The test run on BG/L was a highly resolved two-dimensional weak scaling sod (Sod, G. 1978, JCP, 27, 1) problem. The simulation exercises the core piece-wise parabolic hydrodynamics of FLASH and aggressively exercises the adaptive grid. The FLASH sod problem has been a standard benchmark for FLASH developers evaluating the scalability of a system's inter-connect. For most FLASH science runs, total memory and communication between nodes are the most important limiting factors; users need lots of memory and a scalable interconnect between it all. Therefore, weak scaling is the primary FLASH benchmark target.

Currently, simulations based on FLASH are often run on hundreds to thousands of processors on systems like IBM SP Seaborg (NERSC), QSC (LANL) and MCR Linux cluster (LLNL). It has been run up to 16,384 nodes on BG/L and the total times for various platforms for a weak scaling study (fixed problem size per processor) are presented below. The three different BG/L systems listed are evolutions of the system at IBM Watson. The figure shows that on the QSC and MCR systems, FLASH scales poorly beyond 256 processors. On the Seaborg system, FLASH scales well below 1024 processors, but runs into problems beyond that level. On BG/L, FLASH scales almost perfectly up to 16K processors (on a 16K node system in coprocessor mode). Experience on other systems has demonstrated explicit sensitivity to the inter-connect during the re-gridding phase of the multigrid algorithm. We believe that the excellent scaling on BG/L is due to a combination of good message passing performance and low level of computational noise ensured by BG/L software. We suspect that the synchronization maintained by low noise eliminates overhead in global operations normally lost to barriers.

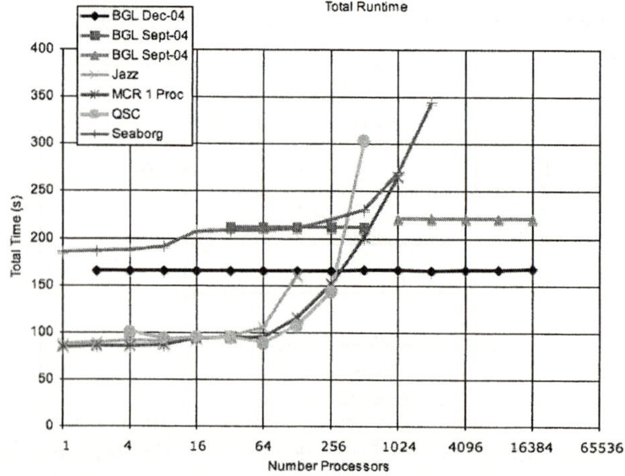

Fig. 4. Weak scaling of FLASH on different architectures

3.4 HOMME

NCAR researchers have built a scalable and efficient spectral-element-based atmospheric dynamical core using the Computational Science Section's High Order Method Modeling Environment (HOMME) [19]. Atmospheric moist processes involving phase changes of water are a fundamental component of atmospheric dynamics and are the most uncertain aspect of climate change research. Simulation of moist processes is challenging because the presence of moisture leads to a new class of fluid motions (moist convection), which is a small-scale phenomenon requiring both very high horizontal and vertical resolution (on the order of a kilometer). The moist Held-Suarez test case extends the standard (dry) Held-Suarez test of the hydrostatic primitive equations by introducing a moisture tracer and simplified physics. It is the next logical test for a dynamical core beyond dry dynamics.

HOMME is written using flexible and efficient F90 modules. The parallel implementation is hybrid MPI/OpenMP. Contiguous groups of elements are distributed to processors and computation is loosely synchronous. The spectral element kernels are similar to level 3 basic linear algebra subroutines such as matrix-matrix multiply. These operations have an O(n) flops to memory access ratio and perform well on modern cache-based microprocessor architectures where CPU-main memory bandwidth increases have significantly lagged processor speed. The physics modules rely heavily on (vector) intrinsic functions. Communication routines are built on top of the MPI message-passing library.

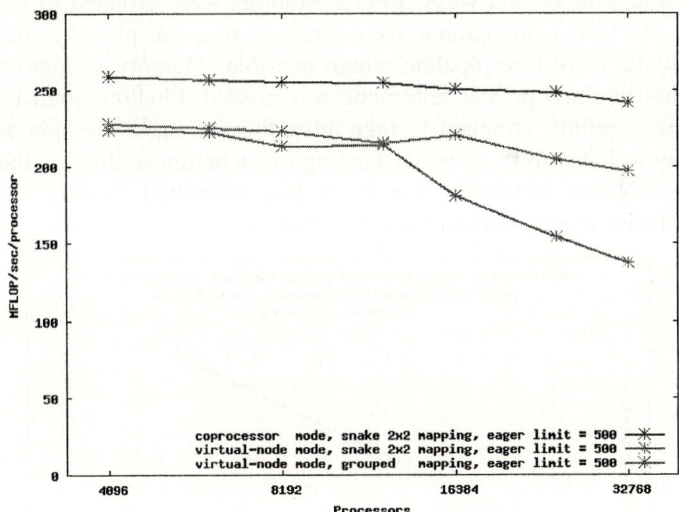

Fig. 5. Strong scaling of HOMME on BG/L: moist Held-Suarez test case, 6 x 128^2 elements, 96 vertical levels, and explicit integration with $\Delta t = 4$ seconds

To assess performance on BG/L, a moist Held-Suarez test case configured to match the resolution of the 2002 Gordon Bell AFES run which achieved 26.58 Teraflops on the Earth Simulator was used. This required a grid with 98,304 horizontal elements and 96 vertical levels. Performance results are plotted for a 32,768 node BG/L system at Rochester using both CPM and VNM mode (we were able to run HOMME on the recently installed 32K node system). The total amount of work (total flop count to solve the system) was held constant while increasing the processor count (strong scaling), until, in the 32,768 processor runs there were only three elements per processor. The data is displayed as sustained MFLOPs per processor and ideally should have a flat profile. To show the importance of properly mapping into the torus we show data for both the grouped and snake mapping strategies. In the grouped mapping elements are assigned to the torus in lexicographical order with both processors on a node filled in sequence. The snake mapping places tasks in 2 x 2 node rectangles through the torus, again filling both processors on a node in sequence. At large processor counts the snake mapping is clearly superior as the grouped mapping performance degrades sharply beyond approximately 12,000 processors.

3.5 QCD

The QCD code used in this study does a first-principles numerical simulation of Quantum Chromo Dynamics, the theory of the strong nuclear interactions, using Lattice Gauge Theory. In most full QCD calculations, more than 90% of the cycles are spent in a small kernel that is called the Wilson D-slash operator. It is therefore necessary to optimize this kernel. The kernel was optimized based on specific BG/L hardware features in several ways: FPU operations were grouped to use the DFPU instructions and FPU computations were arranged to avoid pipeline conflicts and to overlap with the load/store pipeline where possible. Memory storage ordering was chosen so that minimal pointer arithmetic was needed. Floating point load/store operations were carefully arranged to take advantage of the cache hierarchy and the CPU's ability to issue up to three outstanding loads before stalling. A thin and effective nearest-neighbor communication layer that interacted directly with the torus network hardware was used to do the data transfers.

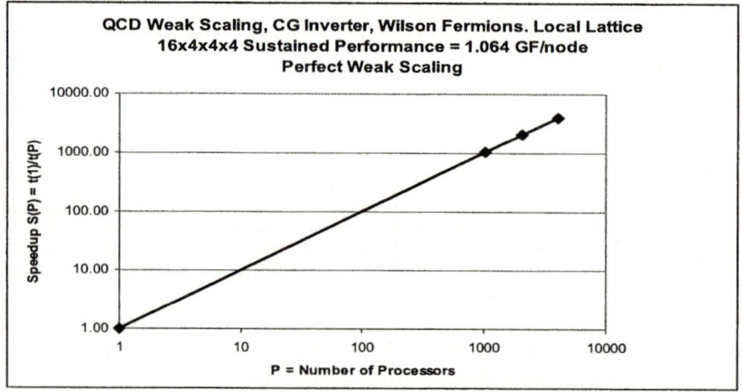

Fig. 6. Weak scaling of QCD on BG/L

Figure 6 shows weak scaling of QCD on up to 4096 processors: in June 2004, BG/L became the first system ever to sustain >1 Teraflop/s performance on QCD.

The computation was done using 1024 processors. Furthermore, the code scaled well to 4096 processors. The computational capacity of BG/L can help scientists performing QCD computations avoid approximations that make the results unreliable, and produce more reliable results to this problem.

4 Conclusions

In this paper, we have presented our experience with several applications on a 16,384 node BG/L system. Our study shows that these applications, which have been targeted to smaller systems so far, can effectively scale to the large BG/L system with a modest level of additional effort. This study represents the first time these applications have been scaled effectively to thousands of processors. These results clearly validate the scalability of the hardware and the software architecture of Blue Gene/L. We hope to improve these results further via software enhancements, and pursue scaling of these applications to levels of a hundred thousand processors in the coming year.

References

1. S. Habata, M. Yokokawa, S. Kitawaki. The Earth Simulator. In NEC Research and Development, 44(1), January 2003.
2. ASC Purple: Fifth Generation ASC Platform. http://www.llnl.gov/asci/platforms/purple/.
3. NASA Unveils its Newest, Most Powerful Supercomputer.
   http://www.nasa.gov/centers/ames/research/lifeonearth/lifeonearth-projectColumbia.html
4. N. R. Adiga et al. An overview of the BlueGene/L supercomputer. In *SC2002 – High Performance Networking and Computing*, Baltimore, MD, November 2002.
5. G. Almasi, R. Bellofatto, J. Brunheroto, C. Cascaval, J. Castaños, L. Ceze, P. Crumley, C. Erway, J. Gagliano, D. Lieber, X. Martorell, J. Moreira, A. Sanomiya, and K. Strauss. An Overview of the BlueGene/L System Software Organization (Distinguished Paper). *Proceedings of the 2003 International Conference on Parallel and Distributed Computing (Euro-Par 2003)*. August 26-29, 2003. Klagenfurt, Austria. pp. 543-555.
6. G. Almasi, S. Chatterjee, A. Gara, J. Gunnels, M. Gupta, A. Henning, J. Moreira, B. Walkup, A. Curioni, C. Archer, L. Bachega, B. Chan, B. Curtis, S. Brunett, G. Chukkapalli, R. Harkness, W. Pfeiffer. Unlocking the Performance of the BlueGene/L Supercomputer. *SC 2004: High Performance Computing, Networking and Storage Conference*, Pittsburgh, PA, November 2004.
7. F. Petrini, D. Kerbyson and S. Pakin. The Case of the Missing Supercomputer Performance: Achieving Optimal Performance on the 8,192 Processors of ASCI Q. In *IEEE/ACM SC2003*, Phoenix, AZ, November 2003.
8. G. Almasi et al. Scaling physics and material science applications on a massively parallel Blue Gene/L system. *International Conference on Supercomputing*, Cambridge, MA, June 2005.
9. TOP500 Supercomputer Sites, http://www.top500.org.
10. L. Bachega, S. Chatterjee, K. Dockser, J. Gunnels, M. Gupta, F. Gustavson, C. Lapkowski, G. Liu, M. Mendell, C. Wait, T.J.C. Ward. A High-Performance SIMD Floating Point Unit Design for BlueGene/L: Architecture, Compilation, and Algorithm Design. *Parallel Architecture and Compilation Techniques (PACT 2004)*, Antibes Juan-les-Pins, France, Sept-Oct 2004.

11. K. Davis, A. Hoisie, G. Johnson, D. Kerbyson, M. Lang, S. Pakin and F. Petrini. A Performance and Scalability Analysis of the BlueGene/L Architecture. In *IEEE/ACM SC2004*, Pittsburgh, PA, November 2004.

12. B.G. Fitch, R.S. Germain, M. Mendell, J. Pitera, M. Pitman, A. Rayshubskiy, Y. Sham, F. Suits, W. Swope, T.J.C. Ward, Y. Zhestkov, R. Zhou. Blue Matter, an application framework for molecular simulation on Blue Gene, *Journal of Parallel and Distributed Computing*, 2003, pp. 759-773.

13. R.S. Germain, et al. Early performance data on the Blue Matter molecular simulation framework. *IBM Journal of Research and Development*, 49(2/3):447–456, 2005.

14. J.C. Phillips, G. Zheng, S. Kumar, and L.V. Kale. NAMD: biomolecular simulation on thousands of processors. *Supercomputing 2002 Proceedings*, 2002.

15. CPMD home page. http://www.cpmd.org.

16. J. Hutter and A. Curioni. Dual-level parallelism for ab initio molecular dynamics: Reaching teraflop performance with the CPMD code, Parallel Computing, (31), 2005, pp. 1-17.

17. J. Hutter and A. Curioni. Car-Parrinello Molecular Dynamics on Massively Parallel Computers, ChemPhysChem, 2005, in press.

18. FLASH code. www.flash.uchicago.edu/

19. HOMME code. http://www.homme.ucar.edu/

A Detailed Study on Phase Predictors

Frederik Vandeputte, Lieven Eeckhout, and Koen De Bosschere

Ghent University, Electronics and Information Systems Department
Sint-Pietersnieuwstraat 41, B-9000 Gent, Belgium
{fgvdeput,leeckhou,kdb}@elis.UGent.be

Abstract. Most programs are repetitive, meaning that some parts of a program are executed more than once. As a result, a number of phases can be extracted in which each phase exhibits similar behavior. These phases can then be exploited for various purposes such as hardware adaptation for energy efficiency. Temporal phase classification schemes divide the execution of a program into consecutive (fixed-length) intervals. Intervals showing similar behavior are grouped into a phase. When a temporal scheme is used in an on-line system, phase predictors are necessary to predict when the next phase transition will occur and what the next phase will be. In this paper, we analyze and compare a number of existing state-of-the-art phase predictors using the SPEC CPU2000 benchmarks. The design space we explore is huge. We conclude that the 2-level burst predictor with confidence and conditional update is today's most accurate phase predictor within reasonable hardware budgets.

1   Introduction

A computer program execution typically consists of several phases of execution where each phase exhibits its own behavior. Being aware of this large-scale time-varying behavior is key to understanding the behavior of the program as a whole. Phase behavior can be exploited for various purposes, ranging from performance modeling [1], compiler optimizations [2], hardware adaptation [3][4][5][6][7], etc. For example in phase-based hardware adaptation, if we know that particular parts of the processor are unused during some program phase, we can turn off those parts during that phase resulting in a reduced energy consumption without affecting overall performance.

One way of identifying phases is to divide the complete program execution into fixed-length instruction intervals and to group instruction intervals based on the code that is being executed during those intervals—this is often referred to as the temporal approach [1]. This means that intervals that execute the same code will be grouped together in what will be called a phase. When this phase classification scheme is used in a phase-based optimization system, it is important to be able to predict when the next phase transition will occur and what the next phase will be. In other words, the phase predictor needs to predict what the phase ID of the next execution interval will be. This way, the system can proactively respond to predicted phase changes.

J.C. Cunha and P.D. Medeiros (Eds.): Euro-Par 2005, LNCS 3648, pp. 571–581, 2005.

The first contribution of this paper is to study today's state-of-the-art phase predictors in detail. The design space we explore is huge as we explore a large number of possible design parameters: the phase predictor's type, its size, its associativity, its confidence mechanism, its update mechanism, etc. We do this for two fixed-length intervals lengths, namely 1M and 8M intervals, using the complete SPEC CPU2000 benchmark suite. Our second contribution is that we improve the performance of existing phase predictor schemes by up to 14% by adding conditional update. We conclude that the 2-level burst predictor with confidence and conditional update is today's most accurate phase predictor for reasonable hardware budgets.

2   Previous Work

Duesterwald et al. [8] identify program execution phases based on hardware metrics such as CPI, miss rates, etc. They also evaluate a collection of (statistical and table-based) predictors to predict the behavior of the next phase. There is a subtle but important difference between the predictors studied by Duesterwald et al. and the phase predictors studied in this paper. Phase predictors predict the next phase ID; the predictors studied by Duesterwald et al. predict the hardware characteristics of the next phase.

Sherwood et al. [6] propose a dynamic phase classification method that is based on the code that is being executed during a fixed-length interval of execution. Per interval of execution they compute a code signature which is a hashed bitvector that keeps track of the basic blocks that are being executed. Sherwood et al. also present and evaluate several phase predictors, namely the last phase predictor, the RLE predictor and the Markov predictor.

In a follow-on study, Lau et al. [9] added confidence counters to the phase predictors to improve their accuracy. Confidence counters are n-bit saturating counters which are incremented or decremented on a correct or wrong prediction, respectively. When the confidence counter exceeds a given threshold the phase predictor is followed; if not, the default last phase prediction is taken. In their study, they also made a distinction between phase change prediction – predicting the next phase ID – and phase length prediction – predicting the length of the next phase using run length classes.

In [7], Vandeputte et al. propose an offline phase analysis methode that is capable of efficiently dealing with multi-configuration hardware where a large number of hardware units can be configured adaptively. This offline phase analysis technique determines the phases based on a fused metric that incorporates both phase predictability and phase homogeneity.

3   Methodology

We performed our analyses using the complete SPEC CPU2000 benchmark suite. The binaries were taken from the SimpleScalar website. For all our results, phase classification is done offline by profiling the program using a train

Table 1. The number of phases extracted for the SPEC2000 benchmark suite using 1M and 8M instruction intervals.

	# Phases			# Phases			# Phases	
Benchmark	1M	8M	Benchmark	1M	8M	Benchmark	1M	8M
bzip2	16	16	twolf	6	4	fma3d	2	2
crafty	2	2	vortex	2	2	galgel	9	11
eon	2	3	vpr	6	6	lucas	2	2
gap	25	10	ammp	12	13	mesa	6	13
gcc	31	27	applu	32	32	mgrid	4	18
gzip	19	13	apsi	7	2	sixtrack	6	4
mcf	10	10	art	6	3	swim	26	11
parser	10	4	equake	11	10	wupwise	9	8
perlbmk	2	2	facerec	28	16			

input—we refer to [7] how this is done. All our profiles were collected with SimpleScalar/Alpha [10]. For the phase classification, 1 million and 8 million instruction intervals are used[1]. Once the phases of the program using this training input are determined, we determine the phase sequence of each benchmark while executing the reference input; this is done by assigning a phase ID to each execution interval based on the code that is being executed. The various phase prediction schemes are then evaluated on these reference phase sequences. Table 1 shows the number of phases for the 1M and the 8M instruction intervals for all the SPEC CPU2000 benchmarks. Note that the number of unique phase IDs is fairly small here compared to [1][9]. The reason is that our offline phase analysis approach [7] balances phase predictability and phase homogeneity, whereas in [1][9], the main objective is phase homogeneity.

4 Phase Prediction

In this section, we will discuss a number of existing phase predictors. As mentioned before, the purpose of a phase predictor is to predict when the phase change will happen and to what phase the program will shift. In fact, a phase predictor predicts the phase ID of the next execution interval based on the history of phase IDs seen in recent history. The conception of these phase predictors is based on the observation that many phases tend to be stable for several consecutive execution intervals and that there exist both regular and irregular phase behavior. The predictors presented here exploit this notion. Before detailing the various phase predictors that we explore in this paper, we first want to define a *phase burst* to be a number of *consecutive* intervals belonging to the same phase, i.e. all intervals in a phase burst have the same phase ID.

4.1 Basic Phase Predictors

In this subsection we describe a number of basic phase predictors. In subsection 4.2, we will add extra features to these predictors to further increase the prediction accuracy.

[1] Actually, each interval consists of 2^{20} and 2^{23} instructions, respectively.

Last Value Predictor. The simplest predictor is the last value predictor which predicts that the phase ID of the next interval will be the same as the phase ID of the current interval. This predictor assumes that a phase burst is infinite; the predictor thus never predicts a phase change. As a result, if the average burst length is ℓ, the misprediction rate using the last value predictor is $1/\ell$. For many benchmarks this predictor performs very well. This is because these benchmarks have a rather large average burst length. For example, if the average burst length is 20, the misprediction rate for the last value predictor is only 5%.

N-Level Burst Predictor. The last value predictor gives good results in case the average burst length is large. However, if there are frequent phase changes, the misprediction rate will become very high. For example, if the average burst length is only 2, the misprediction rate of the last value predictor increases to 50%, meaning that there is a misprediction every other interval. For frequently changing phase behavior, we thus need more advanced phase predictors.

The N-level burst predictor as proposed in [6][9] uses the phase IDs of the last N phase bursts (including the current phase ID) as the history information for indexing the prediction table. This history information is hashed and the table is accessed using the lower order bits of the hash. The higher order bits are used as tag in the prediction table.

Each entry in the prediction table stores a burst length ℓ and the next phase burst ID k. This means that the current phase burst will last for ℓ execution intervals and that the following phase burst will be of phase k. In other words, the burst predictor will predict a phase change to phase k after being ℓ intervals in the current phase.

On a phase change, the entry in the prediction table is updated by writing the effective burst length and the next phase ID after the phase change. Obviously, the burst history is also updated.

N-Level RLE Predictor. Another predictor, similar to the N-level Burst Predictor is the N-level RLE Predictor [6]. The N-level RLE predictor uses the N most recent (Phase ID, burst length) pairs as history information for indexing the table. Notice the difference with the burst predictor—the N-level burst predictor only uses the phase IDs of the N most recent phase bursts, not their corresponding burst lengths. This RLE history information is hashed together of which the lower order bits are used to index the prediction table. The higher order bits are used as a tag to determine if there is a tag match. If there is a match, the phase ID stored in the table is used as phase ID for the next interval, i.e. we predict a phase change. If there is no match, the current phase ID is used as a prediction, i.e. we predict no phase change. The predictor table is updated if the actual next phase ID differs from the next phase ID stored in the phase table. A new entry is inserted if there was a phase change but no tag match. An existing entry in the phase table is removed in case it predicted a phase change when there was none.

A possible disadvantage of this scheme over the N-level burst prediction scheme might be that there is now too many history information to be hashed together for indexing the prediction table. Also, if the pattern of the burst lengths is not very regular, the learning time of all the different combinations of phase IDs with different burst lengths might increase drastically. This will be further discussed when evaluating the different phase predictors.

4.2 Phase Predictor Improvements

The N-level burst and RLE predictor are beneficial in case the phase change pattern is regular. However, if the pattern is rather irregular, predicting phase changes might be difficult using the standard burst and RLE phase predictors. Indeed, in order for a phase change prediction to be correct, both the burst length as well as the next phase ID must be predicted correctly. Mispredicting one of these can result in significant performance degradation or missed optimization opportunities. In some cases, the total number of mispredictions for the burst and RLE predictor might even end up to be larger than the last value predictor. Therefore, a number of improvements have been proposed to enhance the accuracy of these phase predictors.

Confidence. One way to reduce this amount of incorrectly predicted phase changes is to add confidence counters to each entry in the phase table [9], and to only make a phase change prediction in cases we are confident that the prediction will be correct. In all other cases, we predict no phase change. The idea is to verify a number of predictions before accepting them. In other words, only when the confidence counter exceeds a given threshold, a phase change prediction is made. When the confidence counter is lower than the given threshold, the last value prediction is taken. The rationale behind this approach is that it is better not to predict a phase change than to predict an incorrect phase change, since an incorrect phase change may initiate an expensive hardware optimization.

Conditional Update. Until now, we described phase predictors in which the information in the prediction table is updated immediately in case of a phase change misprediction. However, for phase bursts with a regular pattern, it might be useful not to change the prediction information immediately, but to wait and see if the irregularity was not just caused by noise.

This can be accomplished by implementing conditional update. This is done by adding saturating counters, which are updated in the same way as the confidence counters (i.e. plus one on a correct prediction and minus one on an incorrect prediction), and only update the phase table information when the corresponding saturating counter is zero. This way, stable phase information is not changed until we are sure that the misprediction was not caused by noise.

Confidence Combined with Conditional Update. Of course, confidence and conditional update can also be used together to take into account both the irregular patterns as well as the noise within the regular patterns. In this case, a common saturating counter is used that is incremented and decremented in the same manner as described above. With this scheme, a phase change prediction is only accepted if the counter is above some threshold, and the prediction information is only updated if the saturating counter is zero. As will be shown in the next section, this scheme gives the best results on average, both for the N-level burst predictor as well as for the N-level RLE predictor.

5   Evaluation

In the last section, we described the basic design of a number of existing phase predictors and their improvements. For each of these predictors, there remain however many parameters that can be varied and optimized. A list of these parameters and their range is shown in Table 2: the number of levels that can be used for the history information, the total number of entries in the prediction table, the associativity of the table, the number of saturating bits in case conditional update and/or confidence is used, the confidence threshold, the number of tag bits stored in each entry to identify the information that is stored in that entry, and the number of run-length bits to store the predicted burst length in case of the burst predictors[2]. This results in a very large design space that must be explored to obtain the best phase predictor for a given hardware budget. For example, in case of the N-level burst predictor, when all the parameters are varied according to Table 2, about 250,000 different configurations have to be evaluated on the complete SPEC2000 benchmark suite.

Table 2. The range of each parameter we varied for the different predictors.

Predictor Type	Levels	Entries	Assoc.	Sat. Bits	Confid. Thresh.	Tag Bits	Run-length Bits
N-Level Burst	1–4	1–4096	1–8	0–3	0–2	0–10	1–12
N-Level RLE	1–4	1–4096	1–8	0–3	0–2	0–10	0

Fig. 1 shows the average phase misprediction rates as a function of the hardware cost for the N-level burst predictor with conditional update and confidence[3] for a varying number of levels of history for 1M and 8M instruction intervals. To make a fair comparison between the predictors, we calculated the pareto-optimal predictor configurations by varying the parameters in Table 2. As can be seen in Fig. 1, adding more levels of history has an influence on the overall prediction accuracy. For small hardware budgets, using too much history information leads to higher misprediction rates because of increased aliasing as more patterns must be stored in the table. However, once the table becomes large enough, the effect of aliasing reduces and using more history becomes beneficial. Of course, using more history information also increases the learning time of the predictor. This is why the 4-level predictor does not perform much better than the 3-level predictor.

---

[2] In case the last burst length appears to be larger than the maximum burst length that can be stored, zero is used to represent ∞.

[3] We used this type of predictor because this turns out to be the best type of predictor.

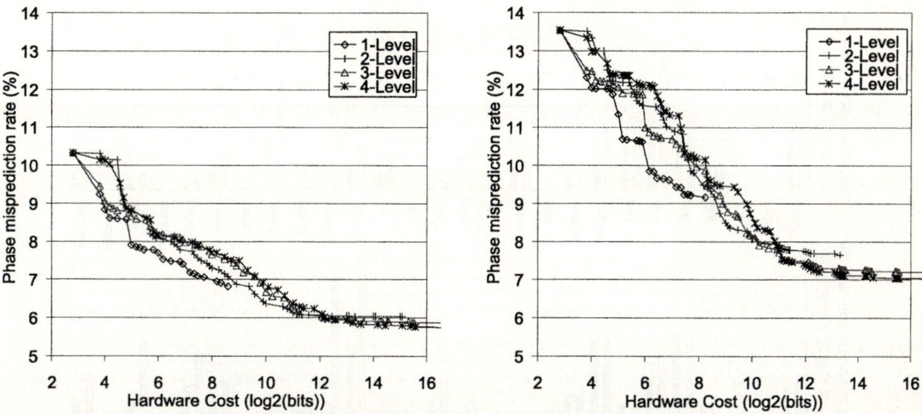

Fig. 1. The average phase misprediction rate of the N-Level burst predictor with conditional update and confidence for different values of N for different hardware budgets. The left graph shows the results for 1M instruction intervals; the right graph for 8M instruction intervals.

Taking into account the misprediction rates and the total hardware costs, one can conclude that an interesting range for the phase predictors is $2^9 \ldots 2^{12}$ bits (i.e. 64 to 512 bytes); smaller predictors result in reduced accuracy and larger predictors result in a larger hardware cost without much gain in predictability. Within this range, the 2-level predictor appears to be the best choice, so we will continue our evaluation with this type of predictor.

Fig. 2 shows the phase misprediction rates of the last value predictor, the 2-level RLE and the 2-level burst predictor (each with conditional update and confidence) for each benchmark for a reasonable hardware budget of 256 bytes. The upper and lower graph show the results for 1M and 8M instruction intervals, respectively. As can be seen, there is a large difference in misprediction rate between the benchmarks and between both interval sizes. For some benchmarks, the misprediction rate is almost zero. This is because these benchmarks have a large average burst length. Fig. 2 also shows that the burst predictor and the RLE predictor do not perform much better than the simple last value predictor for many SPECint benchmarks. For the SPECfp benchmarks however, the reduction of the misprediction rate is quite substantial. The reason for this is that the phase behavior of SPECfp programs is more regular. On average, the 2-level burst predictor and the 2-level RLE predictor can reduce the number of mispredictions by more than 40%. In other words, instead of having a phase misprediction every 7.5 intervals, we now only have a phase misprediction every 14 intervals.

Fig. 3 evaluates the impact of confidence counters and conditional update on the average phase prediction accuracy. For reasonable and large hardware budgets, adding confidence, conditional update or both improves the standard predictor by about 3%, 9% and 17% in case of the burst predictor and 2%, 7% and 13% in case of the RLE predictor. Notice that for large hardware budgets

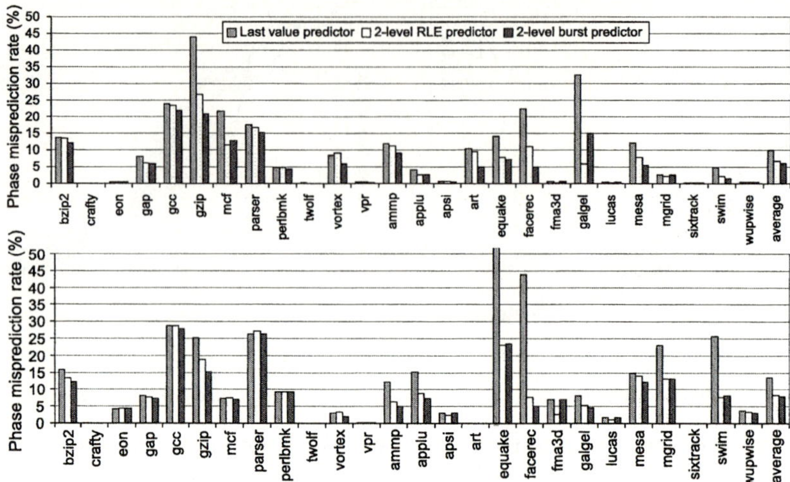

Fig. 2. The phase misprediction rate of the last value predictor, the 2-level RLE and the 2-level burst predictor (each with conditional update and confidence) for a hardware budget of 256 bytes. The upper graph shows the results for 1M instruction intervals; the lower graph for 8M instruction intervals.

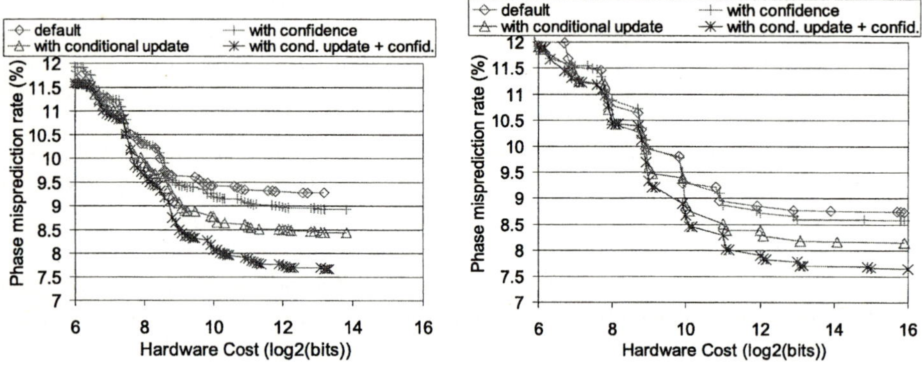

Fig. 3. The effect of applying the different versions of the 2-level burst (left graph) and the 2-level RLE predictor (right graph) on the phase misprediction rate for different hardware budgets. The results shown are for 8M instruction intervals.

the burst predictor and the RLE predictor perform equally well, whereas for smaller hardware budgets the burst predictor performs better. This is because of increased aliasing in case of the RLE predictor, as more history information is used. From these data, we conclude that adding conditional update outperforms the previously proposed predictor schemes by up to 14%.

In Fig. 4, the effect of the number of saturation bits used and the confidence threshold is shown. From this graph, some interesting conclusions can be drawn.

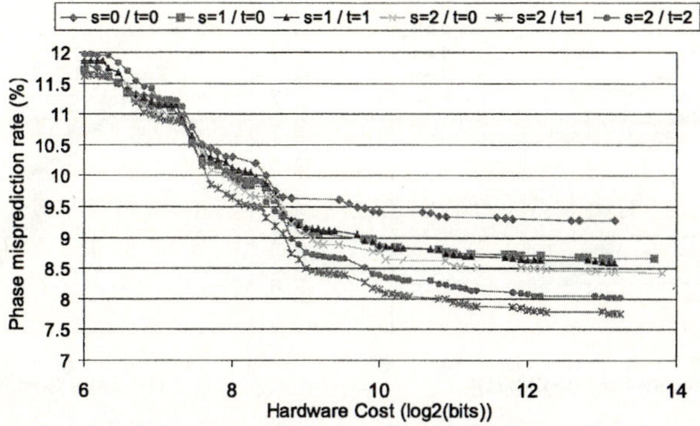

Fig. 4. The effect of the number of saturation bits (s) and the confidence threshold level (t) on the phase misprediction rate for different hardware budgets using a 2-level burst predictor with conditional update and confidence. The results shown are for 8M instruction intervals.

As can be seen, combining conditional update and confidence is only effective when using more than one saturation bit. Also, using a confidence threshold of more than 1 has a negative impact on the prediction accuracy. Using 3 saturation bits (not shown in this graph for clarity) only gives a minor increase in prediction accuracy compared to 2 saturation bits.

Another important aspect is the number of bits b used to encode the history information of the N-level burst and RLE predictor. Using more bits increases the number of bits needed to store the tag for a given table size and associativity.

In case of the N-level burst predictor, the total amount of history information is $p \times N$, where p stands for the number of bits needed for storing the phase ID and N the history depth. These $p \times N$ bits must be mapped onto the available b bits. One way to do this is by using random projection. Another way (which we used in this paper) is to partially overlap the phase IDs by shifting the i-th most recent phase ID over $\lfloor i\frac{b-p}{N-1} \rfloor$ bits ($i = 0 \ldots N - 1$), and xor-ing them together.

In case of the N-level RLE predictor, the total amount of history information is $(p+r) \times N$, where r stands for the number of bits used to represent the length of each burst. Mapping this information onto the available b bits is similar to the burst predictor.

In Fig. 5, the effect of varying the number of hashing bits b is depicted. As can be seen, the RLE predictor requires much more bits to be effective than the burst predictor, which is logical, because the former needs to encode much more information. The results shown are only for 2-level predictors; for higher level predictors, the difference is even bigger.

6   Conclusions

Most programs consist of a number of phases in which each phase exhibits similar behavior. These phases can be exploited for various purposes such as performance

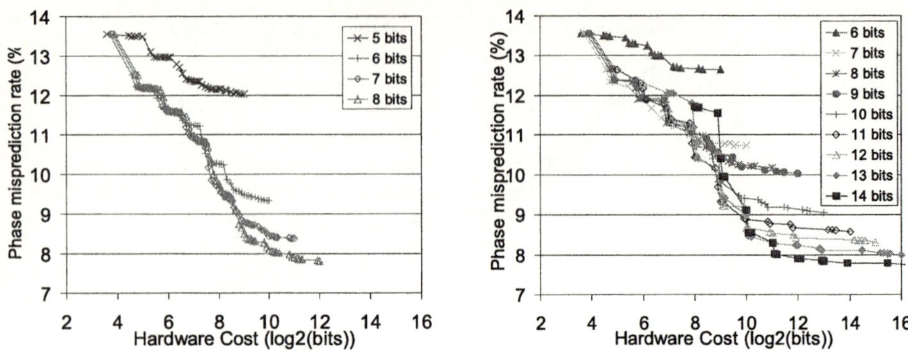

Fig. 5. The effect of the number of bits used to hash the history information used by the 2-level burst (left) and RLE predictor (right) on the phase misprediction rate for different hardware budgets. The results shown are for 8M instruction intervals.

modeling, compiler optimizations, hardware adaptation, etc. When phases are identified by dividing the program execution into fixed-length instruction intervals, and these phases are used in a phase-based optimization system, it is important to be able to predict when the next phase transition will occur and what the next phase will be.

In this paper, we studied a number of today's state-of-the-art phase predictors in detail. The design space we explore is huge as we explored a large number of possible design parameters: the phase predictor's type, its size, its associativity, its confidence mechanism, its update mechanism, etc. We did this for two fixed-length intervals lengths, namely 1M and 8M intervals, using the complete SPEC CPU2000 benchmark suite and concluded that on average the phase predictors show a consistent behavior in terms of phase misprediction reduction. Besides this, we also improved existing phase predictor schemes by 14% using conditional update. We conclude that the 2-level burst predictor with confidence and conditional update is today's most accurate phase predictor for reasonable hardware budgets, reducing the misprediction rate over the last value predictor by more than 40%.

Acknowledgements

This research was funded by Ghent University and by the Fund for Scientific Research-Flanders (FWO-Flanders).

References

1. Sherwood, T., Perelman, E., Hamerly, G., Calder, B.: Automatically characterizing large scale program behavior. In: Proc. of the Internat. Conference on Archit. support for program. languages and operating systems. (2002) 45–57

2. Barnes, R., Nystrom, E., Merten, M., Hwu, W.: Vacuum packing: Extracting hardware-detected program phases for post-link optimization. In: Proc. of the Internat. Symp. on Microarchitecture. (2002)
3. Balasubramonian, R., et al.: Memory hierarchy reconfiguration for energy and performance in general-purpose processor architectures. In: Proc. of the Internat. Symposium on Microarchitecture. (2000) 245–257
4. Dhodapkar, A.S., Smith, J.E.: Managing multi-configuration hardware via dynamic working set analysis. In: Proc. of the Internat. Symp. on Computer Arch. (2002)
5. Huang, M.C., Renau, J., Torrellas, J.: Positional adaptation of processors: application to energy reduction. In: Proc. of the Internat. Symposium on Computer Architecture. (2003) 157–168
6. Sherwood, T., Sair, S., Calder, B.: Phase tracking and prediction. In: Proc. of the Internat. Symposium on Computer architecture. (2003) 336–349
7. Vandeputte, F., Eeckhout, L., De Bosschere, K.: Offline phase analysis and optimization for multi-configuration processors. In: Proc. of the 5th SAMOS workshop. (2005)
8. Duesterwald, E., Cascaval, C., Dwarkadas, S.: Characterizing and predicting program behavior and its variability. In: Proc. of the Internat. Conf. on Parallel Architectures and Compilation Techniques. (2003)
9. Lau, J., Schoenmackers, S., Calder, B.: Transition phase classification and prediction. In: Proc. of the Internat. Symp. on High Performance Computer Arch. (2005)
10. Burger, D., Austin, T.M.: The simplescalar tool set, version 2.0. SIGARCH Comput. Archit. News 25 (1997) 13–25

A Novel Lightweight Directory Architecture for Scalable Shared-Memory Multiprocessors

Alberto Ros, Manuel E. Acacio, and José M. García

Departamento de Ingeniería y Tecnología de Computadores
Universidad de Murcia, 30071 Murcia, Spain
{a.ros,meacacio,jmgarcia}@ditec.um.es

Abstract. There are two important hurdles that restrict the scalability of directory-based shared-memory multiprocessors: the directory memory overhead and the long L2 miss latencies due to the indirection introduced by the accesses to directory information, usually stored in main memory. This work presents a lightweight directory architecture aimed at facing these two important problems. Our proposal takes advantage of the temporal locality exhibited by the accesses to the directory information and on-chip integration to design a directory protocol with the best characteristics of snoopy protocols. The lightweight directory architecture removes the directory structure from main memory and it stores directory information in the L2 cache avoiding in most cases the access to main memory. The proposed architecture is evaluated based on extensive execution-driven simulations of a 32-node cc-NUMA multiprocessor. Results demonstrate that the lightweight directory architecture achieves better performance than a non-scalable full-map directory, with a very significant reduction on directory memory overhead.

1 Introduction

Particular implementations of cache coherence protocols are quite different depending on the total number of processors of a shared-memory multiprocessor. In systems with few processors, an interconnection network with a completely ordered message delivery (such as a bus) can be used. Cache coherence in these cases is ensured by making all processors snoop the bus to obtain information regarding the blocks that are being accessed (read or written) by the other processors. This implementation of the coherence protocol is known as snooping-based protocol whereas the term Symmetric Multiprocessors (SMP) is frequently used to designate the architecture of the multiprocessor [1].

On the other hand, systems with greater number of processors are organized around a scalable point-to-point interconnection network; besides, main memory in these machines is physically distributed to ensure that memory bandwidth also scales with the number of processors. Now a directory-based cache coherence protocol is used to ensure coherence [1]. Each node of the machine (which includes the processor and a fraction of the total main memory) has a directory structure which stores coherence information for the memory blocks that are

J.C. Cunha and P.D. Medeiros (Eds.): Euro-Par 2005, LNCS 3648, pp. 582–591, 2005.
© Springer-Verlag Berlin Heidelberg 2005

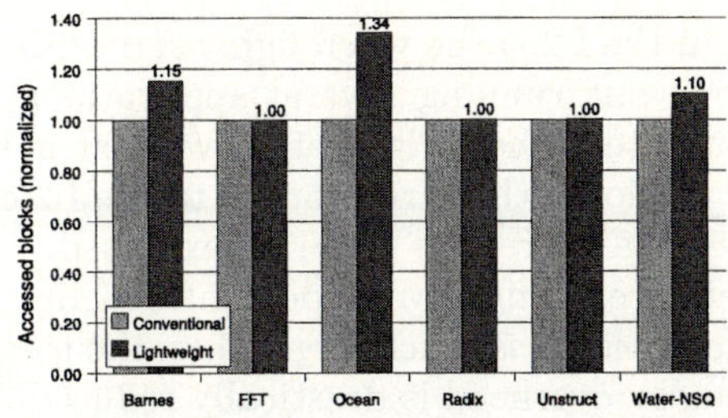

Fig. 1. Worst-case overhead introduced by the lightweight directory architecture.

allocated on it (the *home node*). In this way, L2 cache misses are sent to the corresponding home node, which acts as an ordering point and performs the actions needed to ensure coherence. Unfortunately, the accesses to the directory cause long L2 miss latencies since this structure is usually stored in main memory [2]. Additionally, the amount of extra memory required for storing directory information (directory memory overhead) could become prohibitive for a large-scale configuration of the multiprocessor if care is not taken [3]. In general, these multiprocessors have been called cc-NUMA (cache-coherent Non-Uniform Memory Access) and the best known example is the SGI Origin 2000/3000 [4].

In this paper, we propose the lightweight directory architecture, a novel architecture that takes advantage of on-chip integration to design a large scale cc-NUMA architecture with the best characteristics of SMP multiprocessors. Unlike conventional directories, which associate directory entries to memory blocks, our proposal moves directory information to the cache level where the coherence of the memory block is managed (the L2 cache in our particular case). In this way, directory information is removed from main memory. Our proposal is motivated by the observation that only a small fraction of the memory blocks are stored in the L2 caches at a particular time (temporal locality), and that in most cases, when a request for a memory block from a remote node arrives at the corresponding home node either the home node has recently accessed the block and it resides in the L2 cache, or the home node will request the block in a near future.

As in a conventional directory protocol, L2 cache misses are sent to the corresponding home node which is in charge of satisfying the miss (for example, by providing the memory block in case of a load miss). On the first reference to a memory block, however, the home node books an entry in the local L2 cache which is used to store directory information for the block and occasionally the own block. Subsequent L2 cache misses to the same block will find directory information and in some cases data in the L2 cache of the home node.

However, storing directory information in the L2 cache for each block requested by any remote node could result in a significant increase in the number of blocks being stored in the L2 cache of the corresponding home directory, and consequently, in its total number of replacements. Fortunately, the observation that motivates our proposal points out that it is not the case. We performed a preliminary study about the extra number of memory blocks in the worst case that would be stored in the L2 cache when lightweight directories are used. This study has been carried out running several applications on top of the RSIM simulator assuming infinite caches. Figure 1 shows that in the worst case the increase in the number of blocks that are brought to the L2 cache does not exceed 34% and that in general good results could be expected.

Our proposal, therefore, brings two important benefits. First of all, since the total number of memory blocks is much larger than the total number of L2 cache entries, directory memory overhead is drastically reduced by a ratio of 1024 (or more) compared to conventional directory architectures. Second, since directory entries are stored in the L2 cache of the home node, and both the L2 cache and the directory controller are integrated into the processor chip (which is common in recent processors [5] [6]), the time needed to access the directory is significantly reduced, which translates into important reductions in the latency of L2 cache misses, and therefore, improvements of up to 26% in total execution time are obtained. Moreover, we develop a coherence protocol suited to the particularities of the new directory architecture.

The rest of the paper is organized as follows. In section 2 we present a review of the related work. In sections 3 and 4 we describe the lightweight directory architecture and the coherence protocol required by it, respectively. Section 5 introduces the methodology employed in the evaluation. In section 6 we show some performance results for our proposal. And finally, section 7 concludes the paper and points out some future work.

2   Related Work

In SMP multiprocessors a shared bus is typically employed to interconnect all the processors. In this way, every processor snoops all requests to memory in the order in which they appear on the bus. Unfortunately, the bus becomes a bottleneck when the number of processors increases. Martin *et al.* proposes timestamp snooping to avoid this bottleneck [7]. Timestamp snooping allows that a snoopy protocol is implemented on top of a scalable point-to-point interconnect network by using timestamp and reordering requests at the interconnect end points.

Bandwidth Adaptive Snooping Hybrid (BASH) [8] is a hybrid coherence protocol that dynamically decides whether to act like snooping protocols (broadcast) or directory protocols (unicast) depending on the available bandwidth.

Token coherence protocols [9] avoid both the need of a totally ordered network and the indirection caused by the directory by using N tokens per memory block. In this way, a node can read a block if it has at least one token and can update the block if it has all the tokens of that block.

On the other hand, cc-NUMA multiprocessors use a scalable point-to-point interconnection network and need a directory structure to guarantee ordered memory accesses. However, directory implies memory overhead and long L2 miss latencies. Directory caches can be used to reduce the latency of L2 misses by obtaining directory information from a much faster structure than main memory [2]. Finally, several techniques have been proposed to reduce directory memory

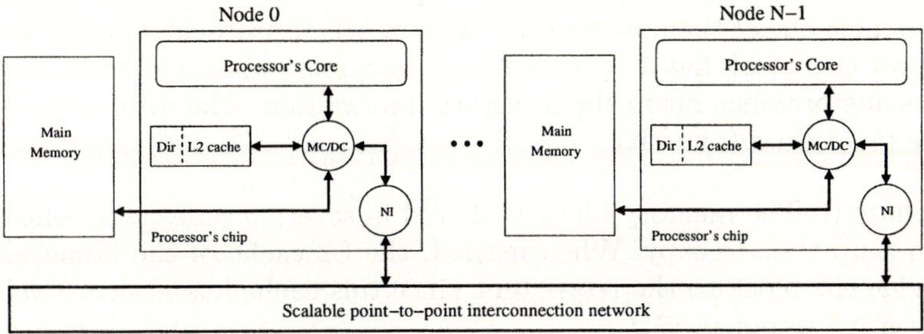

Fig. 2. The lightweight directory architecture.

overhead. Usually, they are based on compressed sharing codes, such as *coarse vector* [10], which is currently employed in the SGI Origin 2000/3000 multiprocessor, *gray-tristate* [11] or *binary tree with subtrees* [3].

3   The Lightweight Directory Architecture

The lightweight directory architecture proposed in this paper removes directory information from main memory and stores it in the L2 caches to reduce its access time. Of course, this reduction would not be so effective if the directory controller were outside the processor chip. Fortunately, current integration scale allows the inclusion of some key components of the system such as the memory controller, the coherence hardware and the network interface and router inside the processor chip (see Compaq Alpha 21364 EV7 [5] or AMD Hammer [6]). Hence, we assume in this work that the directory controller and the L2 cache with directory information are on-chip.

Figure 2 shows the proposed architecture for a N-node multiprocessor. The nodes are connected using a scalable point-to-point interconnection network through the network interface (NI). The memory and directory controller (MC/ DC) handles all inter-node memory references going into or out of the node. In this way, the L2 cache misses are sent to the memory controller of the corresponding home node, which looks for the block's directory information stored in the L2 cache tags structure speculatively in parallel with the access to the L2's structure where data is stored[1]. If the block is found in the L2 cache, directory information is obtained without going to main memory. On the other hand, if the block is not present at cache, the block is not cached by any node and it is necessary to accede to main memory.

Each cache block contains four main fields aside from the data of the block: the *tag* itself, used to identify the block, the *cache state*, the *directory state*, and the *sharing code*. The latter two fields are added by the lightweight directory structure proposed in this paper. The cache state field can take one of the four values (2 bits) used by the MESI protocol. Nevertheless, the invalid state has

---

[1] In this paper, we assume that the L2 cache is split into tags and data structures, as is commonly found in current designs.

two meanings: one of them is the same that in MESI protocol, and the other one means that this block has a valid directory information, and it takes place when there is some presence bit in the directory information. The directory state field can take two values (one bit):

- S (Shared): The memory block is shared in several caches, each one of them with a up-to-date copy. When needed, the L2 cache of the home node will provide the block to the requesters, since this cache has always a valid copy even if it has not used it.
- O (Owned): The memory block is in just one cache and could have been modified. The single valid copy of the block is held in the L2 cache of the home node, when its cache state is modified or exclusive, or alternatively, in one of the L2 cache of the remote nodes. In the latter case, the cache state for the memory block in the L2 cache of the home is invalid, and the identity of the owner is stored in the sharing code field.

Note that an additional directory state is implicit. The U state (Uncached) takes place when the memory block is not held by any cache and its only copy resides in main memory. This is the case of those memory blocks that have not been accessed by any node yet, or those that were evicted in all the caches.

The sharing code field keeps the identity of the L2 caches that hold a copy of the corresponding block. Although our lightweight directory architecture is compatible with any of the sharing codes proposed in the literature, for simplicity we have used the full-map sharing code in this paper. The election for this paper of the full-map sharing code instead of a compressed sharing code is to concentrate on the impact that lightweight directories have on performance, removing any interference caused by unnecessary coherence messages.

4   Coherence Protocol for Lightweight Directory

The proposed architecture requires a cache coherence protocol similar to MESI [1] with some minor modifications. These modifications are performed to ensure that for all memory blocks held in one or more L2 caches, directory information is present in the L2 cache of the home node. Moreover, when a memory block is evicted from the home cache, all the copies of this block must be previously invalidated. Next, we detail the modifications that are required.

We use the term local misses to refer to the L2 cache misses that take place in the home node. On the other hand, remote misses imply that the home node is not where the miss occurs. For local misses, the directory controller obtains directory information stored in cache tags, and then, the miss proceeds as usual.

On the other hand, remote misses are sent to the home node, where the directory controller checks the tags part of the L2 cache. If directory information is not in the home cache, the memory block is not cached by any node (the implicit uncached state mentioned above). Hence, the memory controller brings the block from main memory and stores an entry for it at the home cache in invalid state, just to hold directory information. Moreover, the directory state

Table 1. Where directory information and data are found when a L2 miss takes place in both conventional and lightweight directory protocols.

		Directory States		
		Uncached	Shared	Owned
Conventional	Dir. Inf.	Memory	Memory	Memory
	Data	Memory	Memory	Owner Cache
Lightweight	Dir. Inf.	-	Home Cache	Home Cache
	Data	Memory	Home Cache	Owner Cache

is set to owned because only one node will hold the copy of the block. Finally, the home node sends the block to the requester. If the directory information is in the home cache is not necessary to access to main memory. Moreover, if the directory state is shared, the home node has a valid copy and it can provide the block immediately if the request is a read one.

When a particular block in shared state is evicted from the L2 cache of its home node, the rest of the copies must first be invalidated to maintain coherence. In this way, the directory controller sends multiple invalidation requests to the sharers. Finally, the replacement proceeds once the home node has all the confirmations of the invalidations. If the evicted block has its directory state as owned, and the home node is not the owner, another node has the only valid copy of the block. Then, the directory controller requests the block to the owner. When the home node receives the block, it updates main memory.

The rest of cases are handled as in a conventional directory coherence protocol. Table 1 summarizes the advantages of our proposal. The lightweight directory avoids going to main memory when the directory state is shared, since the home node provides the block. Moreover, directory accesses in cache-to-cache transfers (owned state) are faster than in conventional architectures since the corresponding directory entry is stored in the L2 cache of the home node. Finally, we do not need directory information for uncached blocks, reducing the amount of extra memory that is required.

5 Simulation Environment

We have used a modified version of RSIM [12], a detailed execution-driven simulator, that our group has ported to the x86 architecture [13]. We have simulated a cc-NUMA system with 32 uniprocessor nodes that implements the lightweight directory protocol. Table 2 shows the parameters used to evaluate the lightweight directory architecture. We model the contention on tags and data cache accesses for the remote requests. In this way, those remote requests that try to access the tags at the same time that another request (local or remote) is in progress, will be delayed. Simulations have been performed using an optimized version of the sequential consistency model with speculative load execution following the guidelines given by Hill [14].

The benchmarks used in our simulations cover a variety of computation and communication patterns. Barnes (8192 bodies, 4 time steps), FFT (256K complex

Table 2. Base system parameters.

32-Node System - Lightweight Directory Protocol			
ILP Processor Parameters		**Memory Parameters**	
Max. fetch/retire rate	4	Memory access time	80 cycles
Instruction window	128	Memory interleaving	4-way
Branch predictor	2 bit agree, 2048 count	**Internal Bus Parameters**	
Cache Parameters		Bus width	8 bytes
Cache block size	64 bytes	Bus cycles	1 cycle
Split L1 I & D caches	16 KB, direct mapped	**Network Parameters**	
	2 hit cycles	Topology	2-dimensional mesh
Unified L2 cache	64 KB, 4-way	Flit size	8 bytes
	15 hit cycles (6 + 9)	Non-data message size	2 flits
Directory Parameters		Router speed	250 MHz
Directory controller cycle	1 cycle (on-chip)	Router's internal bus width	64 bits
Directory access time	6 cycles (L2 cache tag)	Arbitration delay	4 router cycles
Message creation time	4 cycles first, 2 next	Channel bandwidth	2 GB/s

doubles), Ocean (258x258 ocean), Radix (1M keys, 1024 radix), and Water-NSQ (512 molecules, 4 time steps) are from the SPLASH-2 benchmark suite [15] and Unstructured (Mesh.2K, 5 time steps) [16] is a computational fluid dynamics application. All experimental results reported in this work correspond to the parallel phase of these benchmarks. Input sizes have been also chosen commensurate to the total number of processors that are used, and cache sizes have been chosen so that the working set of the applications is greater than their capacity.

6   Simulation Results and Analysis

In this section, we evaluate the performance of lightweight directories in terms of total execution time as well as we analyze the effect that they have on the L2 caches, particularly whether the total number of replacements is increased or instead kept unchanged. We compare our proposal with a conventional directory based cache coherent multiprocessor, similar to the SGI Origin 2000/3000 [4], that uses full-map as the sharing code. Moreover, it is important to know the performance that can offer our proposal in ideal conditions. We called ideal case when the blocks used by a node are not affected by the allocation of remote blocks. That it is, we suppose an infinite cache size for those blocks allocated in the home node due to a request of a remote node, and a normal cache size for its local blocks[2].

Figure 3 shows the execution time for a conventional directory architecture and the ideal and realistic implementation of the lightweight directory architecture. As it can be observed, improvements in performance with the ideal implementation range from 6% to 29%. On the other hand, with the realistic 64KB L2 caches for all the blocks, reductions in terms of execution time are obtained for all the benchmarks except for Radix. Particularly, Ocean and Barnes obtain the most important reductions (20% and 19% respectively) whereas for the other benchmarks the reduction ranges from 4% to 11%. Only for Radix application execution time is increased and a degradation of 14% is observed.

---

[2] The size of the L2 cache for this paper is $64KB$

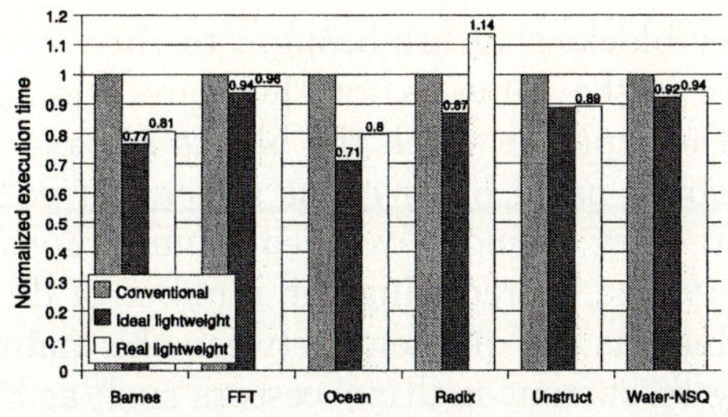

Fig. 3. Normalized execution time for conventional, ideal lightweight and real lightweight architectures.

Table 3 helps to understand the differences between the ideal and the realistic implementation. Overall, our proposal do not affect cache misses, since due to the temporal locality exhibited by the references to memory it does not cause a significant increment in the L2 cache replacements.

`Unstructured` is the nearest benchmark to the ideal case. This is because it solves almost all the cache misses without accessing to main memory. FFT has a small number of replacements in conventional case, so the ideal case only has an improvement of 6% respect to conventional case. Moreover, FFT solves half of the misses in home cache and, therefore, the real case is 2% worse that the ideal one. `Barnes` solves almost all the L2 misses in home cache and also obtains a good performance very near to the ideal case. On the other hand, the performance of `Ocean` is a 7% worse than the ideal case because most misses are solved in main memory (75%). In addition, `Ocean` maintains constant the L2 miss rate and, therefore, the number of replacements, which is translated in a considerable improvement in performance (20%) respect to the conventional case. `Water-NSQ` cannot get a very high performance improvement because this benchmark spends just a short time to solve cache misses. Finally, the benchmark `Radix` increases its L2 miss rate and most the misses must often go to main memory to obtain the directory information. This causes that the real case obtains much worse performance than the ideal case. Moreover, the accesses for the directory information to main memory in lightweight directory architecture are greater than in the conventional architecture. Hence, the performance is a 14% worse than conventional case. In the section 1, we demonstrate that the to-

Table 3. L2 cache miss rate and replacement for conventional and lightweight architecture. In a lightweight architecture miss rate is separated into misses that found directory information in home cache or in home main memory.

| Benchmark | L2 Miss rate | | | | L2 Replacements | | |
| | Conv. | Lightweight | | | Conv. | Lightweight | Ratio |
	Total	Total	Cache	Memory	Repl / Node	Repl / Node	%
Barnes	0.16	0.17	0.15	0.02	26579	29471	1.11
FFT	0.04	0.04	0.02	0.02	6956	7514	1.08
Ocean	0.16	0.16	0.04	0.12	115957	116911	1.01
Radix	0.11	0.13	0.01	0.12	38575	56190	1.46
Unstruct	0.38	0.39	0.38	0.01	42116	43851	1.04
Water-NSQ	0.20	0.20	0.08	0.12	13348	13509	1.01

tal number of memory blocks that are brought to the L2 cache for a lightweight directory architecture is the same as for the conventional one (figure 1). The only difference is in the order in which the blocks are allocated in L2 caches, so this case can be improved using other cache allocation policies.

Regarding the directory memory overhead, our proposal improves the scalability of the directory size by reducing the number of directory entries in a R ratio, where R is defined as the quotient between the main memory size and L2 cache size. According to current multiprocessors such as SGI Origin 2000/3000 [4] and AlphaServer GS320 [17], R can take a typical value of 1024, hence the memory reduction is very considerable.

7   Conclusions and Future Work

In this paper we have introduced the lightweight directory architecture, a scalable directory protocol that tries to achieve the best characteristics both of the snooping and of the directory-based protocols. Our proposal is based on current technology improvements to put the directory controller and directory information inside the processor chip. In this way, we remove the directory structure from main memory and we associate directory information to the L2 cache. Then, cache misses are satisfied by home node cache without accessing main memory, ever when some node has a valid copy of the block.

We have described the resulting architecture and a coherence protocol suited to the particularities of the architecture. In order to demonstrate the benefits derived from our proposal in terms of execution time, we have run several scientific parallel applications on top of a RSIM version that implements the lightweight directory protocol. The lightweight directory architecture presented in this paper obtains improvements of up to 20% in execution time compared to conventional architectures. Moreover, directory memory overhead is reduced by a R ratio respect to conventional directory architectures, where R is computed as the quotient between main memory size and L2 cache size. This means that our proposal drastically reduces the directory memory overhead, and in most cases improves performance.

As part of our future work, we plan to design a cache allocation algorithm, which only stores in cache some remote blocks for reducing the memory overhead caused by these blocks. Another area of interest is to study the impact of a victim cache for blocks whose replacements cause coherence actions. These blocks are those that maintain directory information in the home node cache and they have a copy in some remote node. In this way, it would not be necessary to performance coherence actions. Finally, in order to reduce even more directory memory overhead, we would like to evaluate the effect of limited pointers or compressed sharing codes.

Acknowledgments

This work has been supported by the Spanish Ministry of Ciencia y Tecnología and the European Union (Feder Funds) under grant TIC2003-08154-C06-03.

References

1. Culler, D., Singh, J., Gupta, A.: "Parallel Computer Architecture: A Hardware/Software Approach". Morgan Kaufmann Publishers, Inc. (1999)
2. Acacio, M., González, J., García, J., Duato, J.: "An Architecture for High-Performance Scalable Shared-Memory Multiprocessors Exploiting On-chip Integration". IEEE Transactions on Parallel and Distributed Systems **15** (2004) 755–768
3. Acacio, M., González, J., García, J., Duato, J.: "A Two-Level Directory Architecture for Highly Scalable cc-NUMA Multiprocessors". IEEE Transactions on Parallel and Distributed Systems **16** (2005) 67–79
4. Laudon, J., Lenosky, D.: "The SGI Origin: A cc-NUMA Highly Scalable Server". Proc. of the 24th Int'l Symposium on Computer Architecture (ISCA'97) (1997) 241–251
5. Gwennap, L.: "Alpha 21364 to Ease Memory Bottleneck". Microprocessor Report **12** (1998) 12–15
6. Ahmed, A., Conway, P., Hughes, B., Weber, F.: "AMD Opteron^{TM} Shared Memory MP Systems". Proc. 14th HotChips Symposium (2002)
7. Martin, M., Sorin, D., Ailamaki, A., Alameldeen, A., Dickson, R., Mauer, C., Moore, K., Plakal, M., Hill, M., Wood, D.: "Timestamp Snooping: An Approach for Extending SMPS". Proc. of the 9th Int'l Conference on Architectural Support for Programming Languages and Operating Systems (ASPLOS IX) (2000) 25–36
8. Martin, M., Sorin, D., Hill, M., Wood, D.: "Bandwidth Adaptive Snooping". Proc. of the 8th Int'l Symposium on High Performance Computer Architecture (HPCA-8) (2002) 251–262
9. Martin, M., Hill, M., Wood, D.: "Token Coherence: Decoupling Performance and Correctness". Proc. of the 30th Int'l Symposium on Computer Architecture (ISCA'03) (2003) 182–193
10. Gupta, A., Weber, W., Mowry, T.: "Reducing Memory Traffic Requirements for Scalable Directory-Based Cache Coherence Schemes". Proc. Int'l Conference on Parallel Processing (ICPP'90) (1990) 312–321
11. Mukherjee, S., Hill, M.: "An Evaluation of Directory Protocols for Medium-Scale Shared-Memory Multiprocessors". Proc. of the 8th Int'l Conference on Supercomputing (ICS'94) (1994) 64–74
12. Hughes, C., Pai, V., Ranganathan, P., Adve, S.: "RSIM: Simulating Shared-Memory Multiprocessors with ILP Processors". IEEE Computer **35** (2002)
13. Fernández, R., García, J.: "RSIMx86: A Cost Effective Performance Simulator". Proc. of the High Performance Computing & Simulation (HPC&S) Conference (2005)
14. Hill, M.: "Multiprocessors Should Support Simple Memory-Consistency Models". IEEE Computer **31** (1998) 28–34
15. Woo, S., Ohara, M., Torrie, E., Singh, J., Gupta, A.: "The SPLASH-2 Programs: Characterization and Methodological Considerations". Proc. of the 22nd Int'l Symposium on Computer Architecture (ISCA'95) (1995) 24–36
16. Mukherjee, S., Sharma, S., Hill, M., Larus, J., Rogers, A., Saltz, J.: "Efficient Support for Irregular Applications on Distributed-Memory Machines". Proc. of the 5th Int'l Symposium on Principles & Practice of Parallel Programming (PPOPP'95) (1995) 68–79
17. Gharachorloo, K., Sharma, M., Steely, S., Doren, S.V.: "Architecture and Design of AlphaServer GS320". Proc. of the 9th Int'l Conference on Architectural Support for Programming Languages and Operating Systems (ASPLOS IX) (2000) 13–24

Topic 8
Distributed Systems and Algorithms

Marc Shapiro, Idit Keidar, Felix Freiling geb. Gärtner, and Luís Rodrigues

Topic Chairs

Parallel computing is increasingly exposed to the challenges of distributed systems, such as asynchrony, long latencies, network partition, failures, disconnected operation, and protocol standardization. Witness the growth of peer-to-peer computing, the Grid and Web services. This topic provides a forum for research and practice, of interest to both academia and industry, about distributed computing and distributed algorithms. Submission was encouraged in all areas of distributed systems and algorithms relevant to parallel computing, with emphasis on design and practice of distributed algorithms, analysis of the behaviour of distributed systems and algorithms, distributed fault-tolerance, distributed operating systems and databases, scalability, concurrency and performance in distributed systems, resource sharing and load balancing in distributed systems, distributed algorithms in telecommunications, distributed mobile computing, resource and service discovery, security in distributed systems, and standards and middleware for the distribution of parallel computations.

Twenty-seven papers were submitted in this topic. The subjects were varied, but a common theme to many is self-organisation and fault tolerance. Other themes include mobile networks and routing, mutual exclusion and consensus algorithms, publish-subscribe networks, data replication and the dissemination of information, checkpointing, garbage collection, real time, etc. Eight papers were accepted and the paper *"Replication predicates for dependent-failure algorithms"*, by Flavio Junqueira and Keith Marzullo, was proposed as a distinguished paper.

J.C. Cunha and P.D. Medeiros (Eds.): Euro-Par 2005, LNCS 3648, p. 593, 2005.
© Springer-Verlag Berlin Heidelberg 2005

A Dynamic Distributed Algorithm
for Multicast Path Setup

Luca Gatani, Giuseppe Lo Re, and Salvatore Gaglio

Dip. di Ingegneria Informatica, Università di Palermo
Viale delle Scienze, 90128 Palermo, Italy
{gatani,lore,gaglio}@unipa.it

Abstract. In the past few years, there has been a considerable work
on multicast route selection techniques, with the aim to design scalable protocols which can guarantee an efficient use of network resources.
Steiner tree-based multicast algorithms produce optimal trees, but they
are prohibitively expensive. For this reason, heuristic methods are generally employed. Conventional centralized Steiner heuristics provide effective solutions, but they are unpractical for large networks, since they
require a complete knowledge of the network topology. In this paper, we
propose a new distributed approach that is efficient and suitable for real
network adoption. Performance evaluation indicates that it outperforms
the state-of-the-art distributed algorithms for multicast tree setup, providing good levels of competitiveness, convergence time, and communication complexity. Furthermore, we propose a novel distributed technique
for dynamically updating the multicast tree.

1   Introduction

Several multimedia networking applications, such as distance education, remote
collaboration, video-on-demand, and videoconferencing, are quickly growing in
popularity. These applications place high demands on the underlying communication network and they can become more widespread, relying on the ability of
the network to provide multicast services effectively and efficiently. Trees isolated
over the network topology are typically adopted by multicast routing protocols
for data transmission, in order to achieve a resource usage minimization by the
simultaneous sharing of links. In this context, an underlying specific multicast
routing algorithm should determine, with respect to certain optimization objectives, an efficient multicast tree, singling out the communication routes for the
participants based on the underlying network topology. Data belonging to the
source flow will reach their destinations, traversing tree edges only once and being replicated at branching points. One of the main goals of multicast routing is
to minimize the overall tree cost. Determining the optimal (i.e., minimum cost)
multicast tree connecting all the members of a group is a difficult problem: it can
be modeled as the Steiner Tree Problem in Networks (SPN) [1], which has been
proved to be NP-complete in its decisional version [2]. A number of good, inexpensive centralized heuristics for approximate Steiner trees have been proposed

J.C. Cunha and P.D. Medeiros (Eds.): Euro-Par 2005, LNCS 3648, pp. 595–605, 2005.
© Springer-Verlag Berlin Heidelberg 2005

and extensively reviewed [3–5]. Heuristic algorithms for generating the multicast tree typically balance quality of the generated tree, execution time, and storage requirements. Most of them produce solutions whose cost has been analytically demonstrated to be less than twice the cost of the optimal solution. An important distinguishing characteristic for tree generation algorithms is centralized versus distributed control. Most of the heuristics proposed are intrinsically centralized. In the centralized approach, a central node, that is aware of the state of the whole network, computes the tree. The computation is generally easy and fast. However, the overhead of maintaining, in a single node, coherent information about the state of the entire network can be prohibitive. As a consequence, centralized algorithms are not practical for large networks where complete state information is difficult to collect. In a distributed approach, on the other hand, each node of the network actively contributes to the algorithm computation. A multicast routing distributed algorithm establishes multicast connections in a decentralized manner, by exchanging messages among the nodes involved, which in turn carry out specified portions of the algorithm. These algorithms can be slower and more complex than the centralized ones, but they are more suitable for large networks, with highly dynamic multicast sessions. The development of efficient, distributed Steiner-based algorithms is of great interest at the moment, altough they have received very little attention over the last few years. Most of the versions proposed in the literature [6], [7] are based on reducing the SPN to the Minimum Spanning Tree (MST) problem, on using a distributed MST algorithm, and on pruning unnecessary leaves and branches from the resulting tree. However, these algorithms present severe shortcomings, determining trees with poor properties. Bauer and Varma [8] presented two interesting distributed algorithms for the Steiner Problem in Networks based on the centralized heuristics SPH and K-SPH. Singh and Vellanky [9] proposed a modified version of distributed K-SPH that adopts some improvements to make the fragment combination mechanism more efficient. Novak *et al.* [10] presented a distributed table-passing algorithm for SPN which is based on the centralized heuristic SPH. It is important to underline that all existing distributed algorithms suffer drawbacks such as heavy communication costs, long connection setup times, or poorer quality of the solutions produced as compared with centralized heuristics. A further classification of the Steiner Tree Problem in Networks consists of the following two categories: *a*) Static Steiner Tree Problem, in which the destination subset is fixed, and the optimal tree can be determined once only and used for the entire multicast transmission; *b*) Dynamic Steiner Tree Problem, in which the destination subset can dynamically change because of join or delete requests, and the multicast problem consists of finding a sequence of optimal trees. The dynamic Steiner Tree Problem can be solved exploiting any heuristic used for the static case, if we allow the multicast tree to be completely rebuilt after each change. This is, however, an unrealistic approach since it requires a lot of coordination among the network nodes [11], and it is very likely that a new request will come up before the new tree is ready. Several algorithms [8, 12, 13] have been proposed in order to accomplish both add, and remove requests while restricting the number of rearrangements required in order to derive a new efficient tree from the old one. In this paper, we firstly deal with the static case, presenting a distributed algorithm that is very attractive in terms of quality of computed solution, running time, and number of messages exchanged. Furthermore, we introduce a heuristic for the dynamic case, that minimizes the multicast tree cost during the overall session, while retaining light computational requirements.

The remainder of the paper is structured as follows. Section 2 presents the network model and deals with the Steiner Tree Problem. The distributed algorithm for constructing a multicast tree and a technique for its updating are presented, respectively, in Sections 3 and 4. Section 5 discusses experimental results. We conclude the paper in Section 6.

2   The Steiner Tree Problem

The Steiner Tree Problem in Network can be informally defined as the one of finding a minimum cost tree which spans the nodes belonging to a given subset of all the network nodes (the so-called destination nodes, or multicast member). The formal definition of the Static SPN can be stated as follows. Let $G = (V, E)$ be an undirected connected graph of the communication network, where V is the vertices (node) set, and E the edge (link) set, with positive weights associated with the edges. In this graph we consider a set $Z \subseteq V$ of destination nodes. The Static (minimum) Steiner Tree Problem in Networks is defined as finding a minimum cost sub-graph of G, G', such that there exists a path in G' between every pair of destination nodes. It is worth noticing that since the edge weights are positive, a solution involves isolating a subset S', disjoint of Z, of so-called Steiner nodes, which provides an optimal tree connecting all nodes in Z. We also introduce the following notations. Let $n = |V|$, $z = |Z|$, and $s = |S|$, with $S = V - Z$ the set of non-destination nodes. The minimum cost between a node i and a tree T, $c(i, T)$ is the cost of the cheapest among all paths between the node i and any node in T. We assume that each node i has a routing table that, for each destination j, provides the minimum cost $c(i, j)$ and the next hop in the path from i to j. Using this information, as provided by an underlying network layer protocol, a node is able to send messages via the minimum cost path to any destination. Since the computation of an optimal solution of the Steiner Tree Problem is NP-complete and thus not practical for real-time applications, multicast routing algorithms are based on heuristic methods, some of which have been found to perform very well [3]. However, only a subset of Steiner tree heuristics have the properties that make them suitable for distributed implementation in real networks, where nodes have limited routing information. That is, they should satisfy four criteria: a) use the existing routing information available at each node in the network, as provided by underlying unicast protocols; b) use minimal computational and network resources; c) require minimum coordination between nodes in the network; d) require a limited amount of computation by the non-member nodes. An exhaustive overview of SPN centralized heuristics can be found in [3]. Among the classical centralized path-distance heuristics, the following heuristics, in our view, seem to represent the best candidates for distributed implementation: the Shortest Path Heuristic (SPH), the Kruskal-based Shortest Path Heuristic (K-SPH), and the Average Distance Heuristic (ADH). Bauer and Varma [8] proposed a distributed implementation of SPH and K-SPH. The ADH heuristic, on the other hand, looks very attractive because of its good competitiveness values. Our distributed algorithm is based on the ADH heuristic with Full connection (ADHF).

The dynamic version of the SPN consists of finding a sequence of optimal trees, after the execution of a sequence of operations that change the multicast session membership by means of node insertions (join or add operations) or deletions (leave or remove operations). In particular, the problem of updating a Steiner tree after each insertion or deletion during the same session is known as the On-line Steiner Problem. As in the static case, the problem remains NP-complete. It can be formally defined as follows. Let $G = (V, E)$ be an undirected connected graph of the communication network, where V is the vertices set, $Z \subseteq V$ is the set of destination nodes, and E the edge set, with positive weights being associated with the edges. A vector of requests (each specifying the node and the kind of operation, join or leave) and an initial multicast tree $T(V', E')$ are also given (where $V' \subseteq V$, and $E' \subseteq E$). The On-line Steiner Tree Problem in Networks is defined as finding a sequence of trees such that each tree is obtained from nodes in T modified by all the requests so far received, and is minimum among all the possible choices.

3   A Heuristic for the Static SPN

In [14] we proposed a distributed version of the ADH (D-ADH) algorithm. Here, we describe a variant of D-ADH which exploits a full connection approach. Both algorithms are fully distributed, being designed as a set of cooperative, asynchronous, independent processes running one for each node in the network. We assume that: a) the network is connected; b) each node in the network is a router; c) each node has a unique identifier (UID); d) each node knows its minimum cost path (i.e., cost and first hop) to all other nodes in the network, via the routing table computed by an underlying unicast protocol; e) no topology changes occur during the execution of the algorithm; f) no node or link failures occur during the execution of the algorithm; g) the network delivers messages in order, in finite time, and it does not drop or corrupt messages. We note that the two last hypotheses are consistent with previous works, which have not addressed fault tolerance. A fault tolerant solution is a subject of on-going work. As in its centralized version [15], distributed ADHF (D-ADHF) starts with the forest of multicast nodes and connects them into successively larger trees until a single multicast tree has been set. We refer to the trees in the forest, which will be sub-trees of the final tree, as fragments. During algorithm execution, each node in the network is either part of a fragment, or has not yet been included in the multicast tree. It should be noticed that every Z-node is always a fragment node and every non-destination node is initially a non-fragment node. When two or more fragments merge, the nodes in these fragments, and those lying on the interconnecting paths, become the new merged fragment nodes. In order to uniquely identify fragments, each has a fragment leader and is identified by the same index (UID) as the leader. Initially each multicast member is the leader of its own one-node fragment. When two fragments merge, the node which starts the merging process assumes the leadership. Distributed ADHF processes running on the network nodes exploit the minimum cost path information, which is available on local nodes, as well as information about the multicast forest exchanged via messages. The algorithm is structured in rounds and its main steps can be detailed as follows.

1. *Initialization*
A node receives the list of multicast member UIDs from an external user. It becomes the root node for the first round and builds a data structure representing the multicast forest, initially formed only by the Z-nodes.
2. *Construction of a spanning tree*
The root node starts the construction (via a distributed algorithm) of a network spanning tree. Each node stores a reference to every node directly attached in the tree.
3. *Broadcasting along the spanning tree*
Using the spanning tree previously set, the root node sends in broadcast information about the multicast forest.
4. *Computing function f*   Using the information received from the root node and the locally available unicast routing table, each node v calculates the f function according to equations (see also [16])

$$\mu(v,r) := \frac{\sum_{j=1}^{r} c(v, T_j)}{r - 1}, \ 2 \leq r \leq k \tag{1}$$

$$f(v) := min\{\mu(v,r) \mid 2 \leq r \leq k\}, \tag{2}$$

where T_j is the ith fragment, and k is the number of current fragments. This step is achieved as follows:

   a. a node which already belongs to a fragment, determines the information about the closest external fragment;
   b. a node external to any fragment, takes into account candidate fragments in a non-descending order and determines its own f value, as well as the necessary information about the selected fragments.

The information about a fragment consists of its identifier and of the node representing the tail of the minimum cost path in the fragment.
5. *Convergecasting of the minimum value of f*
Using a converge-cast process along the spanning tree, the information about the computed minimum value of f is reported toward the root node. Ties are opportunely resolved.
6. *Election of the most central node*
When the root node receives the information from all its children on the spanning tree, it determines the best value of f, and sends a notification message to the node v^* that computed this value.

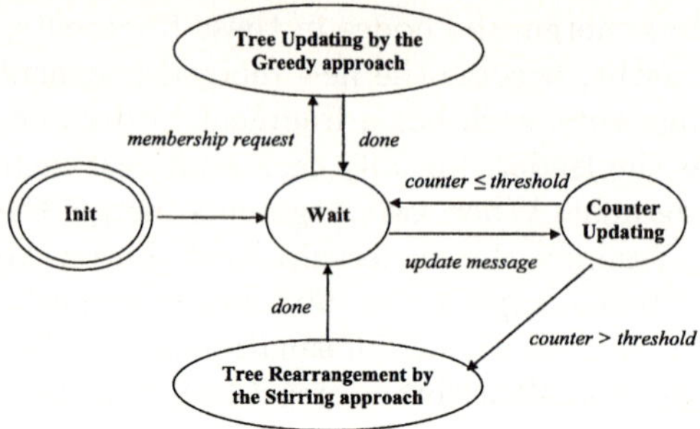

Fig. 1. The finite state automaton for the dynamic case algorithm

7. *Merging of target fragments (with full connection)*
After notification, the node v^* becomes the most central node for the current round and starts the merging process. This is carried out as follows:

  a. if v^* belongs to a fragment, it connects to itself the closest external fragment, via the minimum cost path;
  b. if v^* does not belong to any fragment, it connects the r^* closest fragments, via the minimum cost paths; r^* is the index which minimizes $\mu(v^*, r)$, according to equation 1.

During the merging process, the state of the nodes along the connecting paths and the information about the multicast forest are opportunely updated.

8. *Election of the new root node*
If all Z-nodes are in the same fragment (i.e., the forest is already connected), the algorithm terminates; if not, the node v^* becomes the new root node and starts a new round (go to step 3).

4   A Heuristic for the Dynamic SPN

In this section we present a simple heuristic for the Dynamic STP, which combines the light computational requirements of a "greedy" approach, with the ability to force a rearrangement when the competitiveness of the solution tree has degraded beyond a certain threshold. The simple, but effective dynamic heuristic Greedy [13] perturbs the existing tree as little as possible. For each add request, it connects the new member to the nearest tree node using the minimum cost path. For each remove request, Greedy deletes only leaf nodes. If this deletion creates a non-member leaf, Greedy also deletes the new leaf. This continues until no non-member leaves remain. The idea underlying our approach is that alternative and lower cost paths connecting nodes to the multicast tree can be detected after some operations executed according to Greedy. The proposed approach monitors, in a fully distributed way, the "damage" to the multicast tree. When the "damage" accumulated in a tree portion is judged too high, then a rearrangement process is started, improving the quality of the distribution tree in that area. The "damage" represents the degradation experienced by

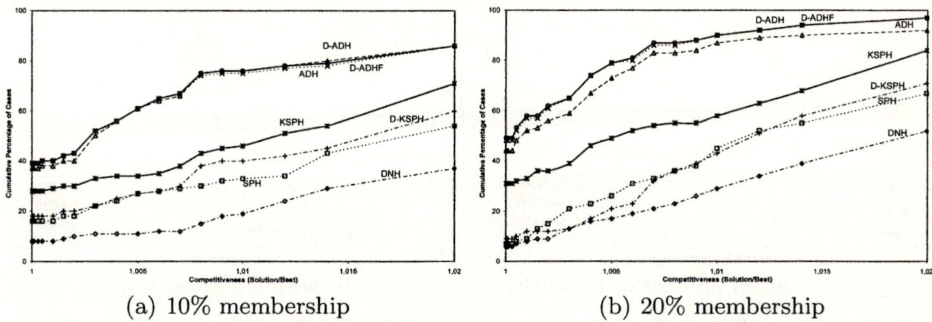

(a) 10% membership   (b) 20% membership

Fig. 2. Competitiveness distributions for different heuristics (200-node networks)

the tree when the membership dynamically changes. It can be simply measured by means of counters deployed on the nodes, which locally register the number of changes (join and leave operations) that affected the neighborhood. When a multicast node leaves, or a new node joins, the node sends a message to all its neighbors in the multicast tree. Propagations of this kind of messages is bounded in a limited region, by the adoption of a maximum hop-count field. Each node in the multicast tree, when reached by the message, increases its counter. When the counter value in a node exceeds a given threshold, the node triggers a tree re-arrangement in the local region. In this way, the effort required to maintain a low-cost tree is performed only when necessary, and only on the region that has been most affected by membership changes. The tree rearrangement process starts from the multicast tree and it iteratively improves the global cost of the tree by means of a node "stirring" technique [17]. At each step it considers the possibility of rearranging some nodes and connecting paths, according to their topological role in the tree. We point out that the distributed tree construction process of D-ADHF involves the generation of a series of partial sub-trees toward the final solution. Moreover, the Greedy approach exploited to manage dynamic membership changes applies a few rules to the current tree T_i to obtain the successive sub-tree T_{i+1}. The choices adopted in this transition are suggested by the features of T_i. When several join and leave operations occur, previous choices could reveal their limits. For this reason, we consider the possibility of recalculating the position of some nodes in the tree after some greedy changes to the multicast tree. In order to better explain the Stirring technique, we first introduce some further definitions. We define "grafting point" a node in a multicast tree which is a target node, or has a degree greater than 2 (the degree of a node is defined by the number of its links). Furthermore, we define "ancestor" of a node v in a multicast tree the node a_v which is the nearest grafting point to v, but not a descendant of v in the tree. The re-positioning process in the Stirring technique is executed for each grafting point, and it is repeated until no further improvements occur. Given the current tree, all grafting points, g_i are sequentially considered. For each node g_i, we check in a distributed way the existence of a node a_g in the tree, with $c(g_i, a_g) < c(g_i, a_c)$ where a_c is the current

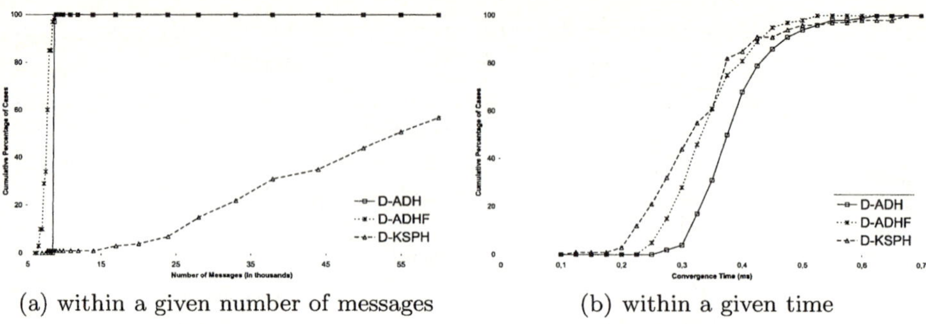

(a) within a given number of messages   (b) within a given time

Fig. 3. Cumulative percentage of networks solved

ancestor, a_g does not belong to the path $< g_i, a_c >$, and it is not a descendant of g_i. If such a node a_g is found in the tree, the path $< g_i, a_c >$ is replaced by the path $< g_i, a_g >$. It is worth noticing that path replacement may cause the insertion of new nodes into the tree. Increasing the subset of nodes, the solution space explored is extended: this can led to single out better solutions. Fig. 1 shows the finite state automaton executed by each multicast tree node, in order to trigger the stirring process execution.

5   Performance Evaluation

5.1   Evaluation Methodology

The experimental evaluation of our approach is performed both in simulation ,and on topologies deployed on a cluster of real nodes. On the one hand, simulations allow the analysis of performance on a huge set of randomly generated test networks, characterized by a large number of nodes and different topological models. On the other hand, tests on networks deployed on a cluster of real nodes allow a deeper validation of the proposed approach, by taking into account some important characteristics of real networks. In order to perform large-scale simulations, we extended the network simulator *ns-2* platform [18], implementing a new multicast agent that runs on network nodes. The micro-benchmark experiments are executed on a cluster testbed consisting of 40 nodes, under constant load, that can approximate rather well the performances in real deployment conditions. In this paper we only discuss simulation results. The measurement on topologies deployed on real nodes yields similar results which are not shown due to space limitations (some results are reported in [14]).

We compare Steiner heuristics on both randomly constructed test networks, and sub-networks extracted from a complete Internet topology. For the first group of experiments we adopt the BRITE network generator [19], and for the latter we use a simple extraction method applied to the map of Internet obtained by the project Mercator [20]. We consider several test groups, each containing 100 sparse networks, that is, networks where the number of edges is less than twice the number of nodes. On these networks multicast groups are typically constituted by 10% or 20% of the nodes. We believe that these choices describe reasonably well multicast applications in wide area networks. In [14], we investigate the scaling capability of our distributed approach. The metrics we use for comparison are cost competitiveness, convergence time, and number of messages transmitted. Competitiveness is the ratio between the heuristic tree cost and that of an optimal solution. For large networks, where explicit algorithms capable of finding optimal solutions are prohibitively expensive, we use a particular heuristic solution (or the best solution obtained by any heuristic considered), rather than an optimal solution. Convergence time is the time elapsed from the beginning of the execution to the time at which the last message reaches its destination. The number of messages is the total number of messages exchanged between nodes before convergence.

5.2   Experimental Results

Static Case. We carry out simulations on several different groups of networks, each containing 100 randomly generated or extracted topologies. On these networks, we first compare the D-ADHF cost competitiveness with that of the D-ADH algorithm we proposed in [14], of the distributed K-SPH heuristic (D-KSPH) in [8], and of some classical centralized heuristics (ADH, K-SPH, SPH, and DNH) [3]. Here we report results obtained considering 200-node networks with 20% and 40% membership sizes: simulations with different network sizes and with different membership percentages reveal similar results, which are not shown only for the sake of brevity. Fig. 2 shows that our distributed approach provides solutions that have the same level of competitiveness when compared to the centralized approach. The same charts show (according to [8]) that the distributed version of K-SPH heuristics may provide worse solutions compared to its centralized version. The second result demonstrated by Fig. 2 is that, when comparing competitiveness, both D-ADH, and D-ADHF consistently outperform both the centralized heuristics DNH, SPH, K-SPH, and the distributed algorithm proposed in [8]. Finally, we point out that, when comparing D-ADH and D-ADHF, the latter provides both lower convergence time, and fewer messages, still maintaining the competitiveness results. This result is clearly demonstrated by some charts reported in [14], and here omitted due to space limitations. Moreover, we study the communication complexity and the convergence time on several groups of large simulated networks, comparing our approach with the D-KSPH algorithm. For the sake of brevity, here we only report the results obtained in 200-node networks with 10%. Fig. 3 shows the cumulative percentage of networks with 10% membership solved within a given number of messages and within a given convergence time, respectively. The equivalent charts obtained with different network sizes and membership percentages mirror the results shown in Fig. 3. All the results indicate that both the number of messages, and the convergence time for D-ADHF fall within a more limited range as compared to the results produced by D-ADH. Moreover, we note that D-ADHF uses a number of messages that is consistently smaller when compared to the large amount of messages exchanged by the distributed K-SPH algorithm,

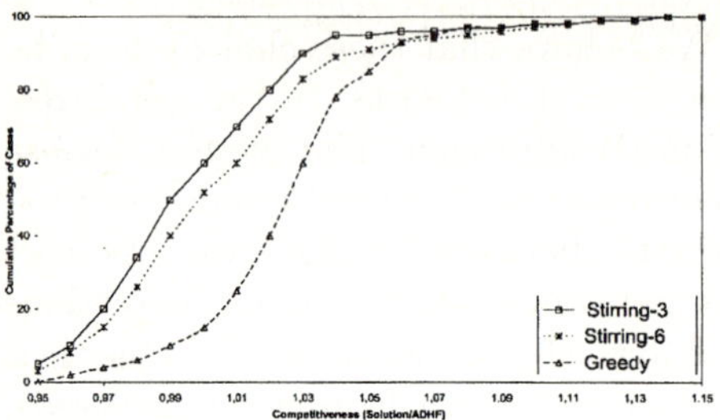

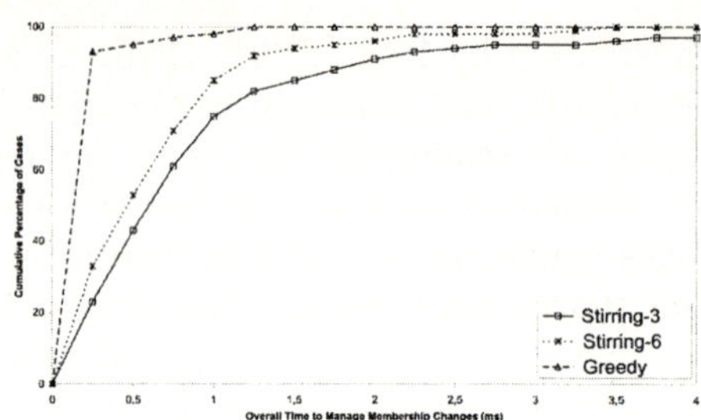

(a) Competitiveness distributions for different heuristics

(b) Cumulative percentage of networks solved within a given time

Fig. 4. Dynamic case analysis results

with a time to converge that is close to that of D-KSPH. The main reason is that the heuristic in [8] adopts a heavily parallelized approach, that can connect several fragments at the same time. Nevertheless, this approach requires a very complex algorithm, that shows severe inefficiencies in practical usage.

Dynamic Case. In order to analyze the performance of our approach for dynamically updating multicast trees, we consider a group of 200-node networks with 40 initial multicast nodes, and, for each of them, we randomly generate 60 requests to add or delete a multicast member. The kind of request is generated satisfying the following probabilistic model:

$$P_c(z) = \frac{\gamma(n - z)}{\gamma(n - z)(1 - \gamma)z} \tag{3}$$

where γ is a real number in the range $[0, 1]$ and $P_c(z)$ is the probability that the node's request is to join the multicast group. The parameter γ determines the size of the multicast group at the equilibrium. Each request is presented to the network only when the previous one is completely accomplished. We compare the algorithm Greedy with our approach, using 3 and 6 as threshold values (Stirring-3, Stirring-6). Fig. 4(a) shows the cumulative distribution for the competitiveness (defined comparing results to the heuristic ADHF), while Fig. 4(b) presents the cumulative percentage of networks whose membership changes are managed within a given execution time. Heuristic Greedy has the lightest computation requirement, but, when the number of changes increase, the tree quality heavily degrades. On the contrary, our technique strikes a good balance between frequent rearrangement and tree degradation.

6   Conclusion

In this paper, we proposed a new distributed algorithm which is capable of determining very good approximating solutions for the Steiner Tree Problem in

Networks. Our algorithm exploits the ADH tree building criterion. The experimental results showed that our algorithm consistently outperforms the state-of-the-art distributed heuristics, although it still maintains comparable performance in terms of execution time. Moreover, we introduced and evaluated a simple approach that supports dynamic multicast membership, by means of periodic improvement of locally inefficient subtrees.

References

1. Ivanov, A.O., Tuzhilin, A.A.: Minimal Networks: The Steiner Problem and Its Generalizations. CRC Press, Cleveland, OH (1994)
2. Karp, R.M.: Reducibility among combinatorial problems. In Miller, Thatcher, eds.: Complexity of Computer Computations. Plenum Prest, New York (1972) 85–103
3. Winter, P.: Steiner problem in networks: A survey. Networks **17** (1987) 129–167
4. Hwang, F., Richards, D.: Steiner tree problems. Networks **22** (1992) 55–89
5. Ramanathan, S.: Multicast tree generation in networks with asymmetric links. IEEE/ACM Transactions on Networking **4** (1996) 558–568
6. Kompella, V., Pasquale, J., Polyzos, G.: Two distributed algorithms for the constrained Steiner tree problem. In: Proc. Comput. Commun. and Netw., San Diego, CA (1993)
7. Chen, G., Houle, M., Kuo, M.: The Steiner problem in distributed computing systems. Information Sciences **74** (1993) 73–96
8. Bauer, F., Varma, A.: Distributed algorithms for multicast path setup in data networks. IEEE/ACM Transactions on Networking **4** (1996) 181–191
9. Singh, G., Vellanki, K.: A distributed protocol for constructing multicast tree. In: Proc. IEEE Int'l Conf. on Principles of Distributed Systems. (1998) 41–48
10. Novak, R., Rugelj, J., Kandus, G.: A note on distributed multicast routing in point-to-point networks. Computers & Operations Research **28** (2001) 1149–1164
11. Bauer, F., Varma, A.: Distributed algorithms for multicast path setup in data networks. In: Proc. IEEE GLOBECOM, Singapore (1995)
12. Kadirire, J., Knight, G.: Comparison of dynamic multicast routing algorithms for wide-area packet switched networks. In: Proc. IEEE INFOCOM, Boston, MA (1995)
13. Imase, M., Waxman, B.: Steiner tree problems. SIAM J. Discrete Math. **4** (1991) 369–384
14. Gatani, L., G. Lo Re, Urso, A.: Distributed algorithms for multicast tree construction. In: Proc. IEEE ISCCSP'04, Hammamet, Tunisia (2004)
15. Plesnik, J.: Worst-case relative performances of heuristics for the Steiner problem in graphs. Acta Mathematica Universitatis Comenianae **60** (1991) 269–284
16. Rayward-Smith, V.J., Clare, A.: On finding Steiner vertices. Networks **16** (1986) 283–294
17. G. Di Fatta, G. Lo Re: Efficient tree construction for the multicast problem. In: Proc. IEEE ITS '98, Sao Paolo, Brazil (1998)
18. Fall, K., Varadhan, K.: The *ns* Manual. http://www.isi.edu/nsnam/ns/doc/index.html (2003)
19. Medin, A., Lakhina, A., Matta, I., Byers, J.: BRITE Universal Topology Generation from a User's Perspective. http://www.cs.bu.edu/brite/user_manual/BritePaper.html (2001)
20. Govindan, R., Tangmunarunkit, H.: Heuristics for Internet map discovery. In: Proc. IEEE INFOCOM'00, Tel Aviv, Israel (2000) 1371–1380

Distributed Maintenance of a Spanning Tree Using Labeled Tree Encoding

Vijay K. Garg and Anurag Agarwal

University of Texas at Austin
Austin, TX 78712-1084

Abstract. Maintaining spanning trees in a distributed fashion is central to many networking applications. In this paper, we propose a self-stabilizing algorithm for maintaining a spanning tree in a distributed fashion for a completely connected topology. Our algorithm requires a node to process $O(1)$ messages of size $O(\log n)$ on average in one cycle as compared to previous algorithms which need to process messages from every neighbor, resulting in $O(n)$ work in a completely connected topology. Our algorithm also stabilizes faster than the previous approaches.

1 Introduction

Fault tolerance is a major concern in distributed systems. The self-stabilization paradigm, introduced by Dijkstra [8], is an elegant and a powerful mechanism for fault tolerance. Self-stabilizing systems tolerate transient *data faults* that can corrupt the state of the system. They ensure that a system starting from any state converges to a legal state provided the faults cease to occur.

Self-stabilizing algorithms for spanning tree construction have been extensively studied. Spanning trees have many uses in computer networks. Once a spanning tree is established in a network, it may be used in broadcast of a message, convergecast, β synchronizer, and many other algorithms. As a result, it is desirable to have an efficient self-stabilizing algorithm for spanning trees. The first algorithm in this area was given in [10, 11] which deals with building BFS tree for a graph. Other algorithms were also proposed for self-stabilizing BFS trees which dealt with different system models and assumptions [1, 2, 4, 15, 16]. Algorithms have also been proposed for other types of trees — such as DFS tree [6] and minimum spanning tree [3]. A survey of the existing self-stabilizing spanning trees can be found in [13].

In this paper, we use an extension of the well-known strategy of *detection* and *reset* [4, 5]. In this strategy, the nodes periodically test if the system is in a legal state and on detection of a fault, carry out the reset strategy. Many self-stabilizing algorithms have *local detection*, i.e., detection by each node corresponds to evaluation of a boolean predicate only on its variables and its neighbors' variables. The reset procedure may be complicated depending upon the application.

Our method is an extension of the above strategy. We view the set of *global* states as the cross-product of the *core* states and the *non-core* states. The core

J.C. Cunha and P.D. Medeiros (Eds.): Euro-Par 2005, LNCS 3648, pp. 606–616, 2005.

states satisfy the property: There exists a legal state for every *core* state. The *non-core* component of a global state is maintained only for performance reason. Given the *core* component, one could always recreate the *non-core* component. In our algorithm for maintaining a spanning tree, we use Neville's code [18] of the tree as the *core* component and the *parent* structure as the *non-core* component. Given any Neville's code, there exists a unique labeled spanning tree in a completely connected graph. Now assume that our program suffers from a data fault. The data fault could be in the core component or the non-core component. However, every value of the core component results in a valid code. Therefore, in either case, we assume that it is the non-core component that has changed. Upon detecting that the non-core component does not correspond to the core component, we simply reset the non-core component to a value corresponding to the core component. The challenge lies in identifying suitable core and non-core components and efficient detection and reset of the state when information is distributed across the network.

We assume that our system is a completely connected graph on n nodes with ids $1 \ldots n$. Such a system could be a network overlaid on a real network. Given proper routing, Internet could also be considered a fully connected topology. In such an overlaid topology, spanning trees can be used for distributing load among participants involved in the computation of a global function. For such applications, the nodes higher in the tree have to perform more computation. As a result, it is important to change the spanning tree over time so that nodes can function at different levels in the tree and every node shares the workload equally in the long run. This requirement rules out maintaining a single tree which is hardcoded in the algorithm. Our algorithm allows the application to maintain any arbitrary tree and facilitates systematically changing of the tree.

Our algorithm is designed for asynchronous message-passing systems, and does not require a central daemon [8] for scheduling decisions. Although some of our assumptions are stronger than the previous work, our algorithm has some significant advantages. In the popular shared memory model [9] for communication used by self-stabilizing spanning tree algorithms, it is assumed that a process can read/write all its shared variables including communication registers. In a completely connected topology, this means that a node can perform operations on $O(n)$ variables in $O(1)$ time which is very unreasonable especially for a message passing system. On the other hand, we assume that every communication step takes one unit of time and in this model, our algorithm stabilizes in $O(d)$ time, where d is an upper bound on the number of times a node appears in the Neville's code. It turns out that d is $O((\log n)/\log\log n)$ with high probability for a randomly chosen code. This leads to a small stabilization time and to our knowledge, it is the best stabilization time achieved by any algorithm in our model.

2   System Model

We assume that the network is a completely connected graph with n *processes* with ids from 1 to n. The processes in the system are referred to as $P_1 \ldots P_n$.

<table>
<tr><td>

```
x[1] = least node with degree 1
for i from 1 to n − 1
    y[i] = parent of x[i]
    delete edge between x[i] and y[i]
    if (degree[y[i]] = 1 ∧ y[i] ≠ n)
        x[i + 1] = y[i]
    else
        x[i + 1] = least node with degree 1
Output y as the Neville's code
```

</td><td>

```
j = least node with degree 1
for i from 1 to n − 1
    parent[j] = code[i]
    degree[j] − −
    degree[code[i]] − −
    if (degree[code[i]] = 1) then
        j = code[i]
    else
        j = least degree node with degree 1
```

</td></tr>
</table>

Fig. 1. Algorithm to compute Neville's code (y) of a labeled tree

Fig. 2. Algorithm to compute labeled tree from Neville's code

Each *process* maintains some *local variables*. The processes are connected to each other through point to point channels and communicate by passing messages to each other. The channels are assumed to be reliable and asynchronous. The *configuration* c of the system is described by the values of the *local variables* for the processes and the *messages* present in the channels. A *computation step* consists of internal computation and a single communication operation: a send or receive. From now on, we use the term *step* to refer to a computation step. A step a is said to be applicable to a configuration c iff there exists a configuration c' such that c' can be reached from c by a single step a. An execution $E = (c_1, a_1, c_2, a_2, \ldots)$ is an alternating sequence of configurations and steps such that c_i is obtained from c_{i-1} by the execution of the step a_{i-1}.

Our algorithm does not require any assumptions on the message transit time for correctness but for measuring the time complexity of our algorithm, we assume that a message can be received at the destination in the step next to the one in which it was sent. A process executes one step in one unit of time. The stabilization time of the algorithm is then given in terms of the number of time units required by the algorithm to stabilize. The reason for choosing such a model is explained later.

3 Neville's Third Encoding

To maintain a spanning tree, it is sufficient for each process to maintain a pointer to the parent but this method is not self-stabilizing as a fault in one of the parent pointers may result in an invalid structure. In this section, we present a core data structure which can be used to maintain the spanning tree in a self-stabilizing way.

For simplicity we assume that all spanning trees rooted at P_n constitute the set of legal structures. Later we explain how this assumption can be relaxed to allow any node to become the root. We represent a tree through an encoding for labeled trees called the *Neville's third encoding* [7, 18]. In this paper, we refer to Neville's third code simply as Neville's code. Each labeled spanning tree has a one-to-one correspondence with a Neville's code. This code is a sequence of $n-2$ numbers from the set $\{1 \ldots n\}$. For completeness sake, derivation of Neville's code from a labeled spanning tree is discussed. Given a labeled spanning tree with n nodes, the Neville's code can be obtained by deleting $n - 1$ edges in the

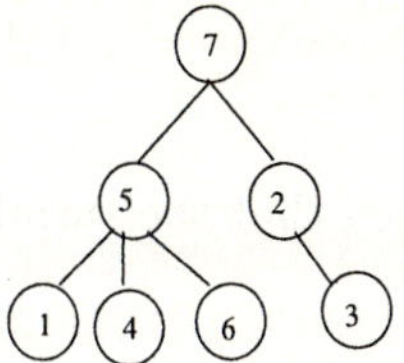

Fig. 3. A spanning tree with Neville's code (5,2,7, 5,5,7)

i	1	2	3	4	5	6	7
$parent$	2	7	5	5	7	5	0
$code$	5	2	7	5	5	7	0
f	2	3	1	4	6	5	7
z	0	2	0	0	5	0	6

Fig. 4. Structures $parent$, $code$, f and z satisfying (R1)-(R5)

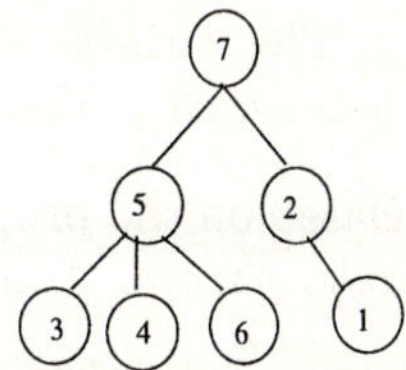

Fig. 5. Tree for $parent$ structure given in Fig. 4

tree as shown in Figure 1. The sequence $\{y[i] | 1 \le i \le n-2\}$ generated at the end of the procedure is called Neville's code.

As an example, consider the labeled tree given in Figure 3. To compute the Neville's code for the tree, we start by deleting the least leaf node, 1. Since the parent of 1 is 5, at this point the code is (5). Now 5 is still not a leaf, so we again choose the least leaf node in the remaining tree, 3. We proceed by deleting 3 and adding its parent 2 to the code. Continuing in a similar fashion, after $n - 1 = 6$ iterations of the algorithm, the code $(5, 2, 7, 5, 5, 7)$ is obtained.

Given Neville's code, the labeled spanning tree can also be computed easily. We first calculate the degree of each node v as one more than the number of times v appears in the code. For the root node n, this gives a value which is one higher than the actual degree of the root but this is intentional. Once the degree of each node is known, the procedure given in Figure 2 can be used to compute the code.

Let Neville's code of the tree be denoted by $code[i]$ for $i \in \{1 \ldots n-2\}$. We require P_i to maintain $code[i]$ as the core data structure and $parent[i]$ as the non-core data structure. If efficiency were not an issue, this would be sufficient for a self-stabilizing algorithm. Periodically, all nodes send their $code$ to P_n, P_n calculates $parent[i]$ for each node P_i and sends it back. Then P_i resets $parent[i]$ to the value received from P_n. If $parent[i]$ was corrupted, it gets reset to agree with the spanning tree given by Neville's code. Even if the variable $code[i]$ gets changed, it still results in a valid spanning tree. The $parent$ pointers are then reset to agree with the new code.

4   Non-core Data Structures for Spanning Trees

Our strategy is to introduce new data structures in the system so that by imposing a set of constraints on these data structures, we can efficiently detect and correct data faults. For this purpose, the following data structures are used:

- $parent$: The variable $parent[i]$ gives the parent of node P_i in the spanning tree.
- f: The variable $f[i]$ gives us the iteration in which the node P_i is deleted in the Neville's code generation algorithm. Therefore, $code[f[i]]$ gives us $parent[i]$. Since P_n is not deleted in first $n - 1$ iterations, we assume that $f[n] = n$.

− z: The variable $z[i]$ gives the largest value of j such that $code[j] = i$. If there is no such j, then $z[i] = 0$.

Based on the properties of Neville's code, it can be verified that the variables — $code$, $parent$, f and z — satisfy the following constraints:

(R1) $\forall i : code[f[i]] = parent[i]$
Follows from the property of function f relating it to the parent.

(R2) $(\forall i : 1 \leq i \leq n - 2 \Rightarrow 1 \leq code[i] \leq n) \wedge (code[n - 1] = n) \wedge (code[n] = 0)$
Definition of code extended to all the nodes.

(R3) **(1)** $\forall i : 1 \leq i < n \Rightarrow 1 \leq f[i] \leq n - 1$
Restricts the f values for nodes other than the root node.

(2) f is a permutation on $[1 \ldots n]$
In each iteration exactly one node is deleted and hence f values are distinct and range from $1 \ldots n$.

(R4) $\forall i : z[i] = \max\{\{j | code[j] = i\} \cup \{0\}\}$
Definition of z.

(R5) $\forall i : z[i] \neq 0 \Rightarrow (f[i] = z[i] + 1)$
If node i was not a leaf node at the starting of the algorithm, then it is deleted immediately after all its children have been deleted.

Theorems 1 and 2 show that constraints are strong enough to characterize a spanning tree, i.e., given a set of data structures $code$, $parent$, f and z which satisfy these constraints, the $parent$ structure results in a valid spanning tree regardless of the definitions of these data structures. From now on, when we consider the data structures $code$, $parent$, f and z, we just think of them as obeying a certain set of constraints and not necessarily corresponding to the original definitions that were given for them.

We deal with two sets of constraints — $\mathcal{R} = \{R1, R2, R3(1), R4, R5\}$ and $\mathcal{C} = \{R1, R2, R3, R4, R5\}$. It is evident that any algorithm which satisfies the constraint set $\mathcal{C}$ also satisfies the constraint set $\mathcal{R}$. The trees resulting from obeying these constraint sets possess different guarantees and are characterized by the following theorems.

Theorem 1. *If code, parent, f and z satisfy constraint set $\mathcal{R}$ then parent data structure forms a valid spanning tree rooted at P_n.*

Proof. Let the directed graph formed by the $parent$ relation satisfying constraints $\mathcal{R}$ be T_{parent}. The edges of T_{parent} are directed from the child to the parent. We first show that T_{parent} is acyclic.
Let $i = parent[j]$ in T_{parent} for some nodes i and j. Then,

$code[f[j]] = i$ \hspace{1em} (Using (R1))
$\Rightarrow (z[i] \neq 0) \wedge (f[j] \leq z[i])$ (Using (R4))
$\Rightarrow f[j] < f[i]$ \hspace{1em} (Using (R5) for i)

Applying this argument repeatedly shows that the ancestor of a node has a higher f value than the f value for the node itself. This implies that no node is an ancestor of itself and hence T_{parent} is acyclic.

Every node in T_{parent} has outdegree either 0 or 1 depending upon the validity of the *parent* variable. We now show that every node except P_n has a valid parent and P_n forms the root of the tree. For a node $P_i, i \neq n$,

$f[i] \neq n$ (Using (R3)(1))

$\Rightarrow 1 \leq parent[i] = code[f[i]] \leq n$ (Using (R2),(R1))

Since the graph T_{parent} is acyclic and every node except P_n has a valid parent, P_n is root of the tree.

The above theorem just ensures that the *parent* pointers form a spanning tree. It does not enforce any relationship between the structure of the tree formed by the *parent* pointers and the tree corresponding to *code*. The next theorem establishes this relationship. The proof for the theorem can be found in the technical report [14].

Theorem 2. *If code, parent, f and z satisfy constraint set C, then parent forms a rooted spanning tree isomorphic to the tree generated by code.*

The above theorem suggests that there is a possibility that the tree formed by *parent* is not same as the tree generated by *code*. For example, consider the value of the variables given in Table 4. It can be easily verified that these values satisfy the constraint set C. The tree corresponding to *code* is the one we considered earlier in Figure 3. The tree generated by *parent* is shown in Figure 5. The two trees are not the same but they are isomorphic.

5   Maintaining Constraints

Each node i maintains $parent[i]$, $code[i]$, $f[i]$ and $z[i]$ and cooperates to ensure that the required constraints are satisfied, resulting in a valid rooted spanning tree. We present a strategy for efficient detection and correction of faults for each of the constraints. We will first discuss (R3) as it turns out to be most difficult to detect and correct.

5.1   Constraint (R3)

Constraint (R3)(1) is a local constraint which can be checked easily. Violation of this constraint can be fixed by simply setting f to a random number between 1 and $n-1$. Constraint (R3)(2) requires f to be a permutation on $1 \ldots n$. This can, in turn, be modeled in terms of the following constraints:

(C1) $\forall i : 1 \leq f[i] \leq n$ (C2) $\forall i, j : f[i] \neq f[j]$

The violation of (C1) is easy to detect. Every node i checks the value $f[i]$ periodically. If it is not between 1 and n, then a fault has occurred. The constraint (C2) is more interesting. At first glance it seems counter-intuitive that we can detect violation of (C2) in $O(1)$ messages. However, by adding auxiliary variables, the above task can indeed be accomplished. We maintain $g[i]$ at each process P_i such that, in a legal global state $f[i] = j \equiv g[j] = i$. Thus, g represents the

inverse of the array f. Note that the inverse of a function exists iff it is one-one and onto which is true in this case. If each process P_i maintains $f[i]$ and $g[i]$, then it is sufficient for a node to check periodically the following constraints:
(D1) $\forall i : 1 \leq f[i] \leq n$ (D2) $\forall i : 1 \leq g[i] \leq n$ (D3) $g[f[i]] = i$

It is easy to show that (C2) is implied by (D1)-(D3). If for some distinct i and j, $f[i]$ is equal to $f[j]$, then $g[f[i]]$ and $g[f[j]]$ are also equal. This means that $(g[f[i]] = i)$ and $(g[f[j]] = j)$ cannot be true simultaneously. (D3) can be checked by P_i by sending a message to $P_{f[i]}$ periodically, prompting $P_{f[i]}$ to check whether $g[f[i]] = i$ is true. Note that by introducing additional variables we have also introduced additional sources of data faults. It may happen that requirements (C1)-(C2) are met, but due to faults in g, constraints (D1)-(D3) are not met. We believe that the advantage of local detection of a fault outweighs this disadvantage.

The above scheme has an additional attractive property: If we assume that there is a single fault in f or g, then it can also be automatically corrected. The details for this scheme are given in the technical report [14].

5.2   Other Constraints

Constraints (R1), (R2) and (R5). Constraint (R1) is trivial to check locally. Each node i inquires node $j = f[i]$ for $code[j]$. If this value does not match $parent[i]$, then the constraint $(R1)$ is violated. On violation, (R1) can be ensured by setting $parent[i]$ to $code[j]$. Constraint (R2) is also trivial to check and correct locally. Similarly, violation of (R5) can be detected easily and on a fault, $f[i]$ can be set to $z[i] + 1$.

Constraint (R4): This constraint can be modeled in terms of the following constraints:
(E1) $\forall i : (z[i] \neq 0) \Rightarrow (code[z[i]] = i)$ (E2) $\forall i, j : (code[j] = i) \Rightarrow (z[i] \geq j)$

For checking (E1), node i prompts the node $z[i]$ to verify that $code[z[i]] = i$. If the check fails, then $z[i]$ can be set to 0, which may not be the correct value for $z[i]$. If $z[i]$ is set incorrectly to 0, then constraint (E2) is also violated. As a result, while checking for (E2), $z[i]$ is set appropriately. For checking (E2), every node j sends a message to node $code[j]$ to verify that $z[code[j]] \geq j$. If (E2) is found to be violated upon receiving a message from node j, then $z[code[j]]$ is set to j.

5.3   Complete Algorithm

Depending upon the set of constraints ($\mathcal{R}$ or $\mathcal{C}$) that a process obeys, we have two versions of the algorithm. They differ in the guarantees about the resulting tree and their time complexities.

```
P_i::
  var
      code, parent, f, z: integer;

  Periodically do
      // Check (R2)
      if (i = n - 1) ∧ (code ≠ n)
          code = n
      if (i = n) ∧ (code ≠ 0)
          code = 0
      if (i ≠ n) ∧ ((code ≤ 0) ∨ (code > n))
          code = random number between 1 and n

      // Check (R3)(1)
      if (i ≠ n) ∧ ((f ≤ 0) ∨ (f ≥ n))
          f = random number between 1 and n - 1
      // First check for (R4)
      if ((z < 0) ∨ (z > n))
          z = 0
```

```
      if (z ≠ 0)
          get code from node P_z
          if P_z.code ≠ i
              z = 0
      if (code ≠ 0)
          send ("Check z", i) to node code
      // Check (R5)
      if ((z ≠ 0) ∧ (f ≠ z + 1))
          f = z + 1
      if ((z = 0) ∧ (f ≤ z))
          f = random number between 1 and n - 1

      // Check (R1)
      get code from node P_f
      if (P_f.code ≠ parent)
          parent = P_f.code

      // Second check for (R4)
      Upon receiving ("Check z", j)
          if z < j
              z = j
```

Fig. 6. Algorithm SSR for maintaining the constraint set $\mathcal{R}$

Maintaining $\mathcal{R}$. As we proved in Theorem 1, the set of constraints $\mathcal{R}$ is sufficient to maintain a spanning tree. The complete algorithm for process i to maintain the constraint set $\mathcal{R}$ is given in the Figure 6. We refer to this algorithm as SSR. In the algorithm, instead of denoting variables like $code[i]$, we have used $P_i.code$ to emphasize that the variables are local to the processes and are not shared. The algorithm checks the constraints one by one and on the violation of a constraint, it takes corrective action. For checking constraints which involve obtaining the value of another process's variable, we have used a primitive get. This involves the sender sending a request for the required variable and the receiver then replying with the appropriate value. A separate thread would be used by a process to respond to the get requests from other processes. Another point to notice in the algorithm is the asynchronous receive of the "Check z" messages. These messages would be received by a third thread which is woken up whenever a message arrives. Our system model takes this into account by assuming that a process alternates between the three threads of execution. The formal proof of correctness of the algorithm is given in the technical report [14].

At this point, we also give our reasons for choosing a different model for evaluation of our strategy. In the previous works, the *asynchronous rounds* [9, 12] model was used. The first asynchronous round in an execution E is the shortest prefix E' of E such that each process executes at least one step in E'. Let E'' be the suffix of E that follows E'. The second round of E is the first round of E'', and so on. The stabilization time of an algorithm is the maximum number of rounds it executes before the system reaches a legal state. In this model, a process waiting for a message receives the message in one round whereas if the message receive is asynchronous, it fails to provide any guarantees. In practice, running time of both the algorithms depends upon the message delivery time in a similar way and hence their time complexities should be comparable. We try to achieve this by putting a bound on the message delivery time. Our algorithm, like most other self-stabilizing algorithms, is structured as a loop that is executed periodically. We refer to this loop as a *cycle*.

The following theorems give the time and message complexity of this algorithm averaged over all the nodes.

Theorem 3. *The algorithm SSR requires $O(1)$ time per node and $O(1)$ messages per node on average in one asynchronous cycle with each message of size $O(\log n)$.*

Proof. In the algorithm *SSR*, every process sends a constant number of *get* requests and one "Check z" request. This results in a total of $O(n)$ messages. Corresponding to the *get* requests, there would be a total of $O(n)$ replies. The number of "Check z" messages received by a process i depends upon the number of times i appears in *code*. Assuming a random code, every node processes $O(1)$ messages on average. Since each node takes constant number of steps in an asynchronous cycle, every process requires $O(1)$ time on average to complete one asynchronous cycle. Moreover, since each message sends an *id* between 1 and n, each message is of size at most $O(\log n)$.

The following theorem gives the stabilization time of the algorithm in terms of our model. The proof for the theorem is given in the full version of the paper [14].

Theorem 4. *[14] The algorithm SSR stabilizes in $O(d)$ time, where d is the upper bound on the number of times a node appears in code.*

The problem of choosing the first $n - 2$ numbers of *code* at random can be considered as the problem of randomly assigning $n - 2$ balls to n bins. The following theorem is a standard result in probability theory [17][Theorem 3.1]:

Theorem 5. *If n balls are thrown randomly in n bins, then with the probability at least $1 - \frac{1}{n}$, no bin has more than $\frac{e \log n}{\log \log n}$ balls.*

For a randomly chosen code, this theorem provides an upper bound for d and hence an upper bound on the stabilization time with very high probability.

These results show that the set $\mathcal{R}$ of constraints can be maintained efficiently. The algorithm for maintaining the constraint set $\mathcal{C}$, called *SSC*, is given in the technical report [14]. The *SSC* algorithm can take upto $O(n)$ time for stabilization.

5.4   Changing the Root Node

The algorithms *SSR* and *SSC* can be easily modified to allow the root node to change dynamically i.e. any node (not necessarily n) can become the root of the tree and the root can be changed during the operation of the algorithm. This can be achieved by changing the constraints (R2) and (R3)(1) in the following way:

(R2) $(\forall i : 1 \leq i \leq n - 1 \Rightarrow 1 \leq code[i] \leq n) \wedge (code[n] = 0)$
(R3)(1) $\forall i : i \neq code[n - 1] \Rightarrow 1 \leq f[i] < n$

The modified constraints are also easy to check and maintain. In the next section we present an application which utilizes this feature.

5.5   Systematically Changing the Tree

The SSR algorithm ensures that if the code is changed, then the spanning tree stabilizes to reflect that change. This property of the algorithm could be used by an application to purposefully change the spanning tree. If we are maintaining the set of constraints $\mathcal{R}$, then changing the code value at a node may not always result in a change in the tree. To get around this problem, whenever a node i wishes to change the tree, it changes the value of $code[f[i]]$ by requesting node $f[i]$. This changes $parent[i] = code[f[i]]$ and hence the spanning tree changes. Additionally, this may result in some more changes in the spanning tree as the parent of some other nodes may also get modified. This technique could be useful for load balancing purposes.

6   Conclusion and Future Work

In this paper we presented a new technique for maintaining spanning trees using labeled tree encoding. Our method requires $O(1)$ messages per node on average in one cycle and provides fast stabilization. It also offers a method for changing the root of the tree dynamically and systematically changing the tree for load balancing purposes. This work also demonstrates the use of the concept of *core* and *non-core* states for designing self-stabilizing algorithms.

It would be interesting to extend this work for general topology. Another research direction would be to modify the algorithm so that it does not require the nodes to have labels from 1 to n.

References

1. Y. Afek, S. Kutten, and M. Yung. Memory-efficient self stabilizing protocols for general networks. In *Proc. of the 4th Int'l Workshop on Distributed Algorithms*, pages 15–28. Springer-Verlag, 1991.
2. S. Aggarwal and S. Kutten. Time optimal self-stabilizing spanning tree algorithm. In *Proc. of the 13th Conference on Foundations of Software Technology and Theoretical Computer Science*, pages 400–410, 1993.
3. G. Antonoiu and P. Srimani. Distributed self-stabilizing algorithm for minimum spanning tree construction. In *European Conference on Parallel Processing*, pages 480–487, 1997.
4. A. Arora and M. Gouda. Distributed reset. *IEEE Transactions on Computers*, 43(9):1026–1038, 1994.
5. B. Awerbuch, B. Patt-Shamir, and G. Varghese. Self-stabilization by local checking and correction (extended abstract). In *IEEE Symposium on Foundations of Computer Science*, pages 268–277, 1991.
6. Z. Collin and S. Dolev. Self-stabilizing depth-first search. *Information Processing Letters*, 49(6):297–301, 1994.
7. N. Deo and P. Micikevicius. Prufer-like codes for labeled trees. *Congressus Numerantium*, 151:65–73, 2001.
8. E. W. Dijkstra. Self-stabilizing systems in spite of distributed control. *Communications of the ACM*, 17:643–644, 1974.

9. S. Dolev. *Self-Stabilization*. MIT Press, Cambridge, MA, 2000.
10. S. Dolev, A. Israeli, and S. Moran. Self-stabilization of dynamic systems. In *MCC Workshop on Self-Stabilizing Systems*, 1989.
11. S. Dolev, A. Israeli, and S. Moran. Self-stabilization of dynamic systems assuming only read/write atomicity. In *Proc. of the ninth annual ACM symposium on Principles of Distributed Computing*, pages 103–117. ACM Press, 1990.
12. S. Dolev, A. Israeli, and S. Moran. Uniform self-stabilizing leader election. In *Proc. of the 5th Workshop on Distributed Algorithms*, pages 167–180, 1991.
13. F. C. Gaertner. A survey of self-stabilizing spanning-tree construction algorithms. Technical report, EPFL, Oct 2003.
14. V. K. Garg and A. Agarwal. Self-stabilizing spanning tree algorithm with a new design methodology. Technical report, University of Texas at Austin, 2004. Available as "http://maple.ece.utexas.edu/TechReports/2004/TR-PDS-2004-001.ps".
15. S. Huang and N. Chen. A self stabilizing algorithm for constructing breadth first trees. *Information Processing Letters*, 41:109–117, 1992.
16. C. Johnen. Memory efficient, self-stabilizing algorithm to construct bfs spanning trees. In *Proc. of the sixteenth annual ACM symposium on Principles of Distributed Computing*, page 288. ACM Press, 1997.
17. R. Motwani and P. Raghavan. *Randomized Algorithms*. Cambridge University Press, 1995.
18. E. H. Neville. The codifying of tree-structure. *Proceedings of Cambridge Philosophical Society*, 49:381–385, 1953.

Replication Predicates
for Dependent-Failure Algorithms

Flavio Junqueira and Keith Marzullo

Department of Computer Science and Engineering
University of California, San Diego
{flavio,marzullo}@cs.ucsd.edu

Abstract. To establish lower bounds on the amount of replication, there is a common partition argument used to construct indistinguishable executions such that one violates some property of interest. This violation leads to the conclusion that the lower bound on process replication is of the form $n > \lfloor kt/b \rfloor$, where t is the maximum number of process failures in any of these executions and k, b are positive integers. In this paper, we show how this argument can be extended to give bounds on replication when failures are dependent. We express these bounds in terms of our model of *cores* and *survivors sets* using set properties instead of predicates of the form $n > \lfloor kt/b \rfloor$. We give two different properties that express the same requirement for $k > 1$ and $b = 1$. One property comes directly from the argument, and the other is more useful when designing an algorithm that takes advantage of dependent failures. We also consider a somewhat unusual replication bound of $n > \lfloor 3t/2 \rfloor$ that arises in the Leader Election problem for synchronous receive-omission failures. We generalize the replication bound for dependent failures, and develop an algorithm that shows that this generalized replication bound is tight.

1   Introduction

Lower bounds for the amount of process replication are often arrived at by an argument of the following flavor:

1. Partition the n processes into k blocks, where each block has at most $\lceil t/b \rceil$ processes, $t \geq \lceil nb/k \rceil$, and k, b are positive integers such that $k > b \geq 1$.
2. Construct a set of executions. For each block A, there is at least one of the executions in which all the processes in A are faulty.
3. Given the set of executions, show that some property of interest is violated. Conclude that if the maximum number of faulty processes in an execution is never larger than t, then $t < \lceil nb/k \rceil$ and $n > \lfloor kt/b \rfloor$ [1].

Examples of such proofs include Consensus with arbitrary process failures and no digital signatures requiring $n > 3t$ ($k = 3$, $b = 1$) [16], Primary Backup with general omission failures requiring $n > 2t$ ($k = 2$, $b = 1$) [19], and Consensus with the eventually strong failure detector $\Diamond S$ requiring $n > 2t$ ($k = 2$, $b = 1$) [5, 6].

---

[1] Some authors have used different symbols, such as f, to indicate an upper bound on the number of faulty processes.

J.C. Cunha and P.D. Medeiros (Eds.): Euro-Par 2005, LNCS 3648, pp. 617–632, 2005.

We call a predicate like $n > \lfloor kt/b \rfloor$ a *replication predicate*: it gives a lower bound on the number of processes that are required given all possible sets of faulty processes. Expressing bounds in terms of t is often referred to as a *threshold model*. Using t to express the number of faulty processes is convenient, but the bounds can lead to mistaken conclusions when processes do not fail independently or do not have identical probabilities of failure. This is because one is assuming that *any* subset of t or fewer processes can be faulty, which implies that failures are independent and identically distributed (IID). To use an algorithm developed under the threshold model on a system that does not have IID failures, one can compute the maximum number of processes that can fail in any execution, and then use that number as t. On the other hand one may be able to use fewer processes if an algorithm based on non-IID failures is used instead.

In an earlier paper we introduced a method for modeling non-IID failures [13] and studied Consensus under this model. We derived replication requirements in our new model and presented protocols that showed these bounds to be tight. This paper generalizes the results of our earlier paper to protocols other than Consensus. We show how the lower bound argument given above can be easily generalized to accommodate our model of non-IID failures. This argument leads to a replication predicate that we call k–Partition, which generalizes the replication predicate $n > kt$ ($b = 1$) for when failures are not IID. The k–Partition property, however, may not prove to be very useful when designing an algorithm. An equivalent property, which we call k–Intersection, is often more useful for this purpose. It is more useful for designing algorithms because algorithms often refer to minimal sets of correct processes ($n - t$ processes when process failures are IID). These properties generalize the two properties we developed for Consensus in our earlier paper.

In this paper, after reviewing our failure model, we define the replication predicate k–Partition for $k > 1$. We then define k–Intersection and show that it is equivalent to k–Partition. We illustrate the utility of k–Intersection by showing that the *M-Availability* property [17] for Byzantine Quorum Systems is equivalent to 4–Intersection. Thus, a system that requires M-Availability has a replication predicate of 4–Partition. Finally, we examine one point in the space of replication predicates for $b > 1$. We do so by considering a weak version of the *Leader Election* problem for synchronous systems that can suffer receive-omission failures. We review a previously-given lower bound proof that argues $n > \lfloor 3t/2 \rfloor$ ($k = 3$, $b = 2$) for IID failures. This proof yields a definition that we call (3,2)–Partition. We derive an equivalent (3,2)–Intersection property and use it to develop an optimal protocol for Weak Leader Election. An immediate consequence is that the lower bound $n > \lfloor 3t/2 \rfloor$ for IID failures is tight. We believe that this is the first time this has been shown.

2   System Model

We assume a system that is amenable to the lower bound proof described in the previous section. Such systems are often comprised of processes that communicate with messages. We consider systems in which processes can be faulty (as compared, for example, to systems in which the failure of messages to be delivered are attributed to faulty links rather than omission failures of processes).

Our work is based on a model of non-independent, non-identically distributed failures. We characterize failure scenarios with *cores* and *survivor sets* [13]. A core is a minimal subset of processes such that, in every execution, there is at least one process in the core that is not faulty. A core generalizes the idea of a subset of processes of size $t + 1$ in the threshold model. A *survivor set* is a minimal set of processes such that there is an execution in which none of the processes in the set are faulty. A survivor set generalizes the idea of a subset of processes of size $n - t$ in the threshold model, where n is the total number of processes. Cores and survivor sets are duals of each other: from the set of cores one can obtain the set of survivor sets by finding all minimal subsets of processes that intersect every core.

More formally, we define cores and survivor sets as follows. Consider a system with a set $\Pi = \{p_1, p_2, \ldots, p_n\}$ of processes. Let Φ be all of the executions of a distributed algorithm *alg* run by the processes in Π, and let $Correct(\phi)$ be the set of processes that are not faulty in an execution $\phi \in \Phi$.

Definition 1. *A subset $C \subseteq \Pi$ is a core if and only if: 1) $\forall \phi \in \Phi$, $Correct(\phi) \cap C \neq \emptyset$; 2) $\forall p_i \in C$, $\exists \phi \in \Phi$ such that $C \setminus \{p_i\} \cap Correct(\phi) = \emptyset$.*

Definition 2. *A subset $S \subseteq \Pi$ is a survivor set if and only if: 1) $\exists \phi \in \Phi$, $Correct(\phi) = S$; 2) $\forall \phi \in \Phi, p_i \in S$, $Correct(\phi) \not\subseteq S \setminus \{p_i\}$.*

In [13], we defined cores and survivor sets using probabilities. In this paper, we use an alternative definition, based on executions, that is more convenient when discussing algorithms. In practice, one can use failure probabilities and a target reliability (or availability) to compute the sets of faulty processes that can be tolerated, and these sets determine the possible failures of an execution. However, one does not have to determine tolerated sets of faulty processes on probabilities. As our example below show, it can be based on a combination of quantitative and qualitative information.

We use the term *system profile* to denote a description of the tolerated failure scenarios. In the threshold model, a system profile is a pair $\langle \Pi, t \rangle$, which means that any subset of t processes in Π can be faulty. In our dependent failure model, the system profile is a triple $\langle \Pi, \mathcal{C}_\Pi, \mathcal{S}_\Pi \rangle$, where $\mathcal{C}_\Pi$ is the set of cores and $\mathcal{S}_\Pi$ is the set of survivor sets[2]. We assume that each process is a member of at least one survivor set (otherwise, that process can be faulty in each execution, and ignored by the other processes), and that no process is a member of every survivor set (otherwise, that process is never faulty). The threshold system profile $\langle \Pi, t \rangle$ is equivalent to the profile $\langle \Pi, \mathcal{C}_\Pi, \mathcal{S}_\Pi \rangle$ where $\mathcal{C}_\Pi$ is all subsets of Π of size $t + 1$ and $\mathcal{S}_\Pi$ is all subsets of Π of size $|\Pi| - t$.

We treat the *kind* of failure—crash, omission, arbitrary, etc.—as a separate part of the failure model. The kind of failure is important both in the design of algorithms and in the derivation of lower bounds. In some situations, such

---

[2] Since $\mathcal{C}_\Pi$ and $\mathcal{S}_\Pi$ can be computed from each other, in fact the system profile could contain only one of these two sets. We include both for convenience.

as with hybrid failure models (for example, [7]), separating the kind of failures from the system profile would be complex. In general, we do not assume any particular kind of failure, but we do so when discussing specific problems.

Determining the system profile requires one to consider the possible causes of process failures. For example, a process running on a particular processor fails if the processor hardware fails (crash failure). As another example, if one is concerned about software faults (bugs), then a process can fail if there is an error in one of the software packages it depends upon, and the system executes the erroneous instructions (which can result in an arbitrary failure) [9].

2.1   Determining a System Profile

We now give an example of a system profile that uses qualitative information. In the work by Castro *et al.* [4], the authors observe that independent software development ideally produces disjoint sets of software faults. This observation is the basic idea of n-version programming, whose goal is to render software failures independent. Of course, there is still a marginal probability that two or more replicas fail in the same execution, but this probability is assumed to be small enough so that it can be ignored.

Suppose we want to implement a fault-tolerant service using the State Machine approach [3, 21], and we are concerned about arbitrary failures arising from software faults. Moreover, we want to leverage the existence of multiple standalone implementations of this service we are interested in, as in BASE [4]. Thus, each replica has two components: a standalone implementation and a replica-coordination component that implements a distributed Consensus algorithm.

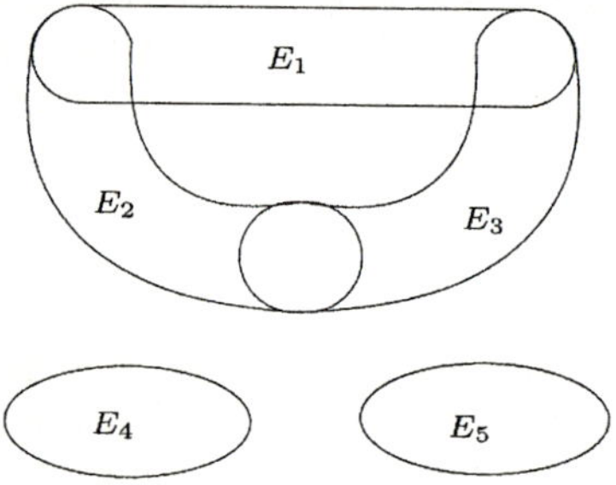

Fig. 1. E_i is the set of executions that have at least one faulty replica running version v_i

For this particular service, suppose that there are five standalone versions available: v_1 through v_5. Looking more carefully at the history of these versions, we discover that two of them reuse code from previous versions. In more detail, v_2 reuses a set X of modules from v_1, and v_3 reuses modules Y from v_1 and Z from v_2 [3]. We also assume that X, Y, and Z are disjoint sets, and that v_4 and v_5 were developed independently.

Assuming that every software module potentially has software faults, we have: 1) faults in the modules in X can affect both v_1 and v_2; 2) faults in the modules in Y can affect both v_1 and v_3; 3) faults in the modules in Z can affect both v_2 and v_3.

Consider a system in which there is at least one replica running each of the five versions. Let E_i be the set of executions in which at least one replica is faulty because of a fault in the version v_i. These sets of executions are related to each other as shown in Figure 1.

---

[3] A module is a collection of functions and data structures.

Assuming one replica for each version, and assuming that at most one software fault can be exercised in an execution, we have the profile of Example 1. This system has sufficient replication to implement Consensus in a synchronous system with arbitrarily faulty processes and no digital signatures [13]. The amount of replication is also sufficient to implement a fault-tolerant state machine for arbitrarily faulty processes using PBFT [3].

PBFT is an attractive protocol because it assumes a weak failure model. It was designed, however, assuming a threshold failure model. In the system profile given above, the smallest survivor set has three processes, which means that there are executions in which two

Example 1.

$$\Pi = \{p_1, p_2, p_3, p_4, p_5\}$$
$$\mathcal{C}_\Pi = \{\{p_1, p_2, p_3\}, \{p_4, p_5\}\}$$
$$\cup \{\{p_i, p_j\} : i \in \{1, 2, 3\} \wedge j \in \{4, 5\}\}$$
$$\mathcal{S}_\Pi = \{\{p_i, p_4, p_5\} : i \in \{1, 2, 3\}\}$$
$$\cup \{\{p_1, p_2, p_3, p_i\} : i \in \{4, 5\}\}$$

processes are faulty. Hence, there is *not* enough replication to run PBFT: seven processes are required to tolerate two faulty processes. PBFT can be used by having one process execute v_1, one process execute v_2, one process executes v_3, two processes execute v_4, and two processes execute v_5. It is easy to check that there is no more than two failures in any execution of this configuration. Alternatively, we can implement a replica coordination component with a modified version of PBFT that can be run in the five process system of the example. In this case, the PBFT implementation needs to know the system profile in the same way that an unmodified PBFT (one assuming a threshold) needs to know the maximum number of faulty processes in an execution[4].

This example illustrates an important point about dependent failures. Since IID failures can be represented as a particular system profile, lower bound proofs that hold for IID failures also hold in our model. But, if one has a system in which failures are not IID, then one should use an algorithm that explicitly uses a system profile. By using such an algorithm, it is often possible to use less replication than it requires when using an algorithm developed using the threshold model.

2.2   Survivor Sets, Fail-Prone Systems, and Adversary Structures

We are not the first to consider non-IID behaviors: quorum systems have addressed the issues of non-IID behavior for some time. In [17], the idea of *fail-prone systems* was introduced. This paper gives the following definition for a set of servers U:

A *fail-prone system* $\mathcal{B} \subseteq 2^U$ is a non-empty set of subsets of U, none of which is contained in another, such that some $B \in \mathcal{B}$ contains all the faulty servers.

This paper then observes that a fail-prone system can be used to generalize to less uniform assumptions than a typical threshold assumption. Their definition

---

[4] Although the original PBFT algorithm assumes a threshold on the number of failures, it is possible to modify it to work with cores and survivor sets. A discussion of these modifications, however, is outside of the scope of this paper.

does not give a name to the elements of $\mathcal{B}$; we call each one a *fail-prone set*. As fail-prone sets are maximal, a fail-prone set is the complement of a survivor set and $\mathcal{B} = \{\Pi \setminus S : S \in \mathcal{S}_\Pi\}$. Although both survivor sets and fail-prone sets characterize failure scenarios, survivor sets have a fundamental use: if a process is collecting messages from the other processes, it can be fruitless to wait for messages from a set larger than a survivor set. Of course, there are times when fail-prone sets are more useful. For example, if $B_{\max}$ is the largest fail-prone set, then $|B_{\max}|$ is the value of t to use if one wishes to use a threshold-based protocol.

Non-threshold protocols were also considered in the context of secure multi-party computation with adversary structures [1, 11, 15]. Adversary structures are similar to fail-prone systems. They differ in two ways. First, adversary structures can represent more than one failure mode, e.g., crash failures and arbitrary failures. Each failure mode is described with sets of possibly faulty processes (processes are referred to as *players* in this literature). Second, the sets of possibly faulty players given in an adversary structure are not necessarily maximal; all sets of possibly faulty players are given. Using all possible sets of players that can deviate from the correct protocol behavior as opposed to only maximal sets (or minimal sets of correct processes, as with survivor sets) gives one more expressiveness in modeling system failures. Using fail-prone systems or survivor sets, however, is sufficient for establishing the bounds on process replication we show in this paper. Moreover, these bounds hold even for a more expressive model such as adversary structures. This is because we use properties about the intersections of sets of correct processes. If the intersection property holds for some sets of processes A_1, A_2, ..., A_m then it holds for the sets of processes $A_1' \supset A_1$, $A_2' \supset A_2$, ..., $A_m' \supset A_m$. Hence, one only has to consider the minimal sets of correct processes in these intersection properties.

3 k Properties

In the generic lower bound proof described in Section 1, one first partitions the set of processes into k blocks, and then constructs a set of executions. For each block A, there is some execution in which all the processes in A are faulty. Being able to fail all the processes of a particular block then enables the construction of an execution in which some property is violated. For example, for Consensus, the property violated is *agreement*. For Primary-Backup algorithms, the property violated is the one that says that at any time there is at most one primary.

Having derived a contradiction, the proof concludes by stating that one cannot partition the processes in the manner that was done. With the threshold model and $b = 1$, this implies that not all processes of any subset of size $\lceil n/k \rceil$ can be faulty, and consequently $t < \lceil n/k \rceil$. In our dependent failure model, this implies that in any partition of the processes into k blocks, there is at least one block A that does not contain only faulty processes: A contains a core. More formally, let $\mathcal{P}_k(\Pi)$ be the set of partitions of Π into k blocks. We then have the following property for a system profile $\langle \Pi, \mathcal{C}_\Pi, \mathcal{S}_\Pi \rangle$:

Property 1. k–**Partition**, $k > 1$, $|\Pi| > k$: $\forall \mathcal{A} \in \mathcal{P}_k(\Pi) : \exists A_i \in \mathcal{A} : \exists C \in \mathcal{C}_\Pi : C \subseteq A_i$

Although k–Partition is useful for lower bound proofs, it is often not very useful for the design of algorithms. Survivor sets are often more convenient to refer to than cores. For example, the algorithm for Consensus by Chandra and Toueg for crash failures in asynchronous systems with failure detectors of the class $\diamond S$ assumes at least $2t + 1$ processes. For this number of processes, any pair of subsets of size $n - t$ has a non-empty intersection, and this property is crucial to avoid the violation of agreement. This is equivalent to stating that any two survivor sets intersect, or equivalently that $\mathcal{S}_\Pi$ is a coterie [8].

3.1 k–Intersection

We now state the property that we show to be equivalent to k–Partition and that references survivor sets instead of cores. We call it k–Intersection. k–Intersection states that for a system profile $\langle \Pi, \mathcal{C}_\Pi, \mathcal{S}_\Pi \rangle$, for every set $T \subset \mathcal{S}_\Pi$ of size k, there is some process that is in every element of T. Let $\mathcal{G}_x(A)$ be the set of all the subsets of A of size x; if $|A| < x$, then $\mathcal{G}_x(A) = \emptyset$. We have the following property for a system profile $\langle \Pi, \mathcal{C}_\Pi, \mathcal{S}_\Pi \rangle$:

Property 2. k–**Intersection**, $k > 1$, $|\Pi| > k$, $|\mathcal{S}_\Pi| > k$: $\forall T \in \mathcal{G}_k(\mathcal{S}_\Pi)$: $(\cap_{S \in T} S) \neq \emptyset$

As an illustration, the set $\mathcal{S}_\Pi$ in Example 1 satisfies 3–Intersection. We now show the equivalence between k–Partition and k–Intersection.

Theorem 1. k–*Partition* $\equiv k$–*Intersection*

Proof. $\Rightarrow$: Proof by contrapositive. Suppose a system profile $\langle \Pi, \mathcal{C}_\Pi, \mathcal{S}_\Pi \rangle$ such that there is a subset $\mathcal{S} = \{S_1, \ldots, S_k\} \subset \mathcal{S}_\Pi$ such that $\cap \mathcal{S} = \emptyset$. We then build a partition $\mathcal{A} = \{A_1, \ldots, A_k\}$ as in Figure 2.

Suppose without loss of generality that no A_i is empty. It is clear from the construction that no two distinct blocks A_i, A_j intersect. It remains to show that: 1) $\bigcup \mathcal{A} = \Pi$; 2) $\forall i \in \{1, \ldots, k\}$: A_i does not contain a core. To show 1), consider the derivation in Figure 3. Explaining the derivation:

$$A_1 = \Pi \setminus S_1$$
$$A_2 = \Pi \setminus (S_2 \cup A_1)$$
$$\vdots$$
$$A_i = \Pi \setminus (S_i \cup A_1 \cup A_2 \ldots \cup A_{i-1})$$
$$\vdots$$
$$A_k = \Pi \setminus (S_k \cup A_1 \ldots \cup A_{k-1})$$

Fig. 2. Partition

- Line 1 to Line 2 follows from the observation that for any subsets A, B of Π, we have that $(\Pi \setminus A) \cup (\Pi \setminus B) = \Pi \setminus (A \cap B)$;
- Line 2 to Line 3: the intersection between S_1 and A_1 has to be empty, since S_1 contains exactly the elements we removed from Π to form A_1;
- Line 3 to Line 4: by repeating inductively the process used to derive Line 3, we remove every term A_i present in the equation;
- Line 4 to Line 5: By assumption, the intersection of S_1 through S_k is empty.

To show 2), we just need to observe that any A_i is such that we removed all the elements of S_i. By the definitions of a core and of a survivor set, a subset that does not contain elements from some survivor set does not contain a core.

$\Leftarrow$: Proof also by contrapositive. Let $\langle \Pi, \mathcal{C}_\Pi, \mathcal{S}_\Pi \rangle$ be a system profile such that there is a partition $\{A_1, \ldots, A_k\}$ of Π in which no A_i contains a core. Because no block contains elements from every survivor set (no block contains a core), we have that for every A_i, there is a survivor set S_i such that $S_i \cap A_i = \emptyset$. Consequently, we have that $\cap_i S_i$ is empty, otherwise either some A_i contains an element that is in $\bigcap_i S_i$ or $\{A_1, \ldots, A_k\}$ is not a partition, either way contradicting our previous assumptions.

$$\bigcup \mathcal{A} = (\Pi \setminus S_1) \cup (\Pi \setminus (S_2 \cup A_1)) \cup \ldots$$
$$\cup (\Pi \setminus (S_k \cup A_1 \cup A_2 \ldots$$
$$\cup A_{k-1})) \tag{1}$$
$$= \Pi \setminus ((S_1 \cap (S_2 \cup A_1)) \cap \ldots$$
$$\cap (S_k \cup A_1 \cup A_2 \ldots \cup A_{k-1})) \tag{2}$$
$$= \Pi \setminus (S_1 \cap S_2 \cap \ldots$$
$$\cap (S_k \cup A_1 \cup A_2 \ldots \cup A_{k-1})) \tag{3}$$
$$\vdots$$
$$= \Pi \setminus (\cap_i S_i) \tag{4}$$
$$= \Pi \tag{5}$$

Fig. 3. Derivation

In the remainder of this section, we discuss the utility of these properties. In particular, we show the equivalence between 4–Intersection and M-Consistency [17].

3.2   4-Intersection and M-Consistency

In [17], the following *M-Consistency* property was defined. It was stated that this property was necessary for one to implement a Masking Byzantine Quorum System. This property allows a process to identify a result from a non-faulty server. The set $\mathcal{Q}$ used in this definition is the set of quorums, and $\mathcal{B}$ is the fail-prone system.

Property 3. **M-Consistency**: $\forall Q_1, Q_2 \in \mathcal{Q} : \forall B_1, B_2 \in \mathcal{B} : (Q_1 \cap Q_2) \setminus B_1 \not\subseteq B_2$

The paper then shows that if all sets in $\mathcal{B}$ have the same size t, then M-Consistency implies $n > 4t$.

We show that M-Consistency is equivalent to 4–Intersection. Since a faulty process can stop sending messages, we can use $\mathcal{S}_\Pi$ as the set of quorums: waiting to receive messages from more than a survivor set could prove fruitless. A fail-prone set is the complement of a survivor set, and for any two sets X and Y, $(X \setminus Y) \equiv (X \cap \bar{Y})$, where $\bar{Y}$ is the complement of Y. Hence, we can rewrite M-Consistency as:

$$\forall Q_1, Q_2 \in \mathcal{S}_\Pi : \forall B_1, B_2 \in \mathcal{B} : (Q_1 \cap Q_2) \cap \bar{B}_1 \not\subseteq B_2$$

Then, for any two sets X and Y, $(X \not\subseteq Y) \equiv (X \cap \bar{Y} \neq \emptyset)$, and so:

$$\forall Q_1, Q_2 \in \mathcal{S}_\Pi : \forall B_1, B_2 \in \mathcal{B} : Q_1 \cap Q_2 \cap \bar{B}_1 \cap \bar{B}_2 \neq \emptyset.$$

Since $\bar{B}_i$ is a survivor set, $B_i \in \mathcal{B}$, this can be more compactly written as:

$$\forall Q_1, Q_2, S_1, S_2 \in \mathcal{S}_\Pi : Q_1 \cap Q_2 \cap S_1 \cap S_2 \neq \emptyset.$$

which is 4–Intersection. Hence, another way to write the replication requirement stated in M-Consistency is 4–Intersection, or equivalently 4–Partition.

4   An Example of Fractional k

The results of the previous section are perhaps not surprising to those who have designed Consensus or quorum algorithms. For example, 2–Intersection states that the survivor sets are a coterie, and 3–Intersection states that the intersection of any two survivor sets contains a non-faulty process. It takes some effort to show that 4–Intersection is equivalent to M–availability, and we expect that it will not be difficult to show that the $n > 5t$ requirement of Fast Byzantine Paxos [18] can be understood from 5–Intersection. We conjecture that it is possible to define classes of algorithms that, as for Consensus [10, 12], are built on top of quorums of various strengths, and whose communication requirements are easily understood in terms of k–Intersection. Such algorithms developed for the threshold model should be easily translatable into our model of non-IID failures.

Less well understood are algorithms that have fractional replication predicates. To further motivate the utility of intersection properties, we consider a problem that we call *Weak Leader Election*. Given a synchronous system and assuming receive-omission failures (that is, a faulty process can crash or fail to receive messages), this problem requires $n > \lfloor 3t/2 \rfloor$. This lower bound is not new, but to the best of our knowledge, it has not been shown that the lower bound is tight. We show here that the bound is tight, which is a result of some theoretical value. Our primary reason for choosing this algorithm, however, is the insight we used from the intersection property to arrive at the solution.

We first specify the problem. Our specification allows for faulty (but non-crashed) processes to become elected. Such a feature is necessary because it requires more replication to detect receive-omission failures [19], and the original lower bound proof allowed such behaviors. We then discuss the lower bound on process replication for this problem using our model of dependent failures. Finally, we provide an algorithm showing that the lower bound is actually tight.

4.1   Weak Leader Election

Each process p_i has a local boolean variable $p_i.elected$ ($p_i.elected$ is false for a crashed process). We then describe Weak Leader Election with two safety and two liveness properties.

Safety: $\Box(|\{p_i \in \Pi : p_i.elected\}| < 2)$.
LE-Liveness: $\Box\Diamond(|\{p_i \in \Pi : p_i.elected\}| > 0)$.
FF-Stability: In a failure-free execution, only one process ever has *elected* set
    to true.
E-Stability: $\exists p_i \in \Pi : \Diamond\Box(\forall p_j \in \Pi : p_j.elected \Rightarrow (j = i))$.

These properties state that infinitely often some process elects itself (LE-Liveness), and no more than one process elects itself at any time (Safety). The third property states that, in a failure-free execution, a single process is ever elected. This property, however, does not rule out executions with failures in which two or more processes are elected infinitely often. We hence define E-Stability.

4.2   Lower Bound on Process Replication

In [2], the following lower bound was shown. The proof was given in the context of showing a lower bound on replication for Primary-Backup protocols.

Claim. Weak Leader Election for receive-omission failures requires $n > \lfloor 3t/2 \rfloor$.

Proof. Assume that Weak Leader Election for receive-omission failures can be solved with $n = \lfloor 3t/2 \rfloor$. Partition the processes into three blocks A, B and C, where $|A| = |B| = \lfloor t/2 \rfloor$ and $|C| = \lceil t/2 \rceil$. Consider an execution ϕ_A in which the processes in B and C initially crash. From LE-Liveness and E-Stability, eventually a process in A will be elected infinitely often. Similarly, let ϕ_B be an execution in which the processes in A and C crash. From LE-Liveness and E-Stability, eventually a process in B will be elected infinitely often.

Finally, consider an execution ϕ in which the processes in A fail to receive all messages except those sent by processes in A, and the processes in B fail to receive all messages except those sent by processes in B. This execution is indistinguishable from ϕ_A to the processes in A and is indistinguishable from ϕ_B to the processes in B. Hence, there will eventually be two processes, one in A and one in B, elected infinitely often, violating either Safety or E-Stability.

To develop the algorithm, we first generalize the replication predicate for this problem using cores and survivor sets. From the lower bound proof, we consider any partition of the processes into three blocks. Then, one constructs three executions, where in each execution all of the processes in two of the three subsets are faulty. The conclusion of the proof is the following property for $k = 3$:

Property 4. $(k, k-1)$–**Partition**, $k > 1$, $|\Pi| > 2$: $\exists k' \in \{2, \ldots, \min(k, |\Pi|)\}$: $\forall \mathcal{A} \in \mathcal{P}_{k'}(\Pi) : \exists \mathcal{A}' \in \mathcal{G}_{k'-1}(\mathcal{A}) : \exists C \in \mathcal{C}_\Pi : C \subseteq \bigcup \mathcal{A}'$

The equivalent intersection property is then:

Property 5. $(k, k-1)$–**Intersection**, $k > 1$, $|\Pi| > 2$, $|\mathcal{S}_\Pi| > 2$: $\exists k' \in \{2, \ldots, \min(k, |\Pi|)\}$: $\forall \mathcal{T} \in \mathcal{G}_{k'}(\mathcal{S}_\Pi) : \exists T \in \mathcal{G}_2(\mathcal{T}) : (\cap_{S \in T} S) \neq \emptyset$

Stated more simply, $(k, k-1)$–Intersection says that for any set of k' survivor sets, $k' \in \{2, \ldots, \min(k, |\Pi|)\}$, at least two of them have a non-empty intersection. $(k, k-1)$–Intersection and $(k, k-1)$–Partition generalize replication predicates in the threshold model of the form $n > \lfloor kt/(k-1) \rfloor$. Thus, a profile that satisfies $(k+1, k)$–Intersection must also satisfy $(k, k-1)$–Intersection. To illustrate, a system profile satisfies $(3, 2)$–Intersection if either it satisfies $(2, 1)$–Intersection or for every three survivor sets, two intersect. Also, note that $(2, 1)$–Intersection is 2–Intersection.

Consider now an example of a system that satisfies $(3, 2)$–Intersection. It is based on a simple two-cluster system. A process can fail by crashing, and there is a threshold t on the number of crash failures that can occur in a cluster. A cluster can suffer a

Example 2.

$$\Pi = \{p_{a_1}, p_{a_2}, p_{a_3}, p_{b_1}, p_{b_2}, p_{b_3}\}$$
$$C_\Pi = \{\{p_{i_1}, p_{i_2}, p_{i_3}, p_{i_4}\} : (i_1, i_2 \in \{a_1, a_2, a_3\})$$
$$\wedge (i_3, i_4 \in \{b_1, b_2, b_3\} \wedge i_1 \neq i_2 \wedge i_3 \neq i_4)\}$$
$$S_\Pi = \{\{p_{i_1}, p_{i_2}\} : ((i_1, i_2 \in \{a_1, a_2, a_3\})$$
$$\vee (i_1, i_2 \in \{b_1, b_2, b_3\})) \wedge i_1 \neq i_2\}$$

total failure, which causes all of the processes in that cluster to crash. A total failure results from the failure of a shared resource such as storage, for example. We assume that total failures are rare enough that the probability of both clusters suffering total failures is negligible. However, processes can crash in one cluster at the same time that the other cluster suffers a total failure. Assuming that each cluster has three processes and $t = 1$, we have the system profile of Example 2, where processes with identifier a_i are in one cluster and processes with identifier b_i are in the other cluster. Note that this profile satisfies $(3, 2)$–Intersection because out of any three survivor sets, at least two intersect.

The equivalence of $(k, k - 1)$–Partition and $(k, k - 1)$–Intersection can be shown with a proof similar to the one of Theorem 1, and it appears in [14]. We state the theorem here for reference purposes.

Theorem 2. $(k, k - 1)$*–Partition* $\equiv (k, k - 1)$*–Intersection*

4.3   A Weak Leader Election Algorithm

We now develop a synchronous algorithm WLE for Weak Leader Election. For this algorithm, we assume a system profile $\langle \Pi, C_\Pi, S_\Pi \rangle$ that satisfies $(3, 2)$–Intersection. WLE is round based: in each round a process receives messages sent in the previous round and then send messages to all processes. We use $p_i.M(r)$ to denote the set of messages that p_i receives in round r, and $p_i.s(r)$ to denote the set of processes from which process p_i receives messages in round r.

We developed this algorithm by first observing what $(3, 2)$–Intersection means. Given three survivor sets, at least two of them intersect. Put another way, if two survivor sets S_1 and S_2 are disjoint, then any survivor set S_3 intersects $S_1 \cup S_2$. Since a core is a minimal set that intersects every survivor set, the above implies that $S_1 \cup S_2$ contains a core. Thus, given any two disjoint survivor sets, at least one of them contains a correct process.

Our algorithm uses as a building block a weak version of Uniform Consensus that we call *RO Consensus*. We call it RO Consensus because of its resemblance to Uniform Consensus. RO Consensus, however, is tailored to suit the requirements of WLE and therefore is fundamentally different.

In RO Consensus, each process p_i has an initial value $p_i.a \in V \cup \{\bot\}$, where V is the set of initial values, and a decision value $p_i.d\,[1 \ldots n]$, where $p_i.d$ is a list and $p_i.d[j] \in V \cup \{\bot\}$. We use $v \in p_i.d$ to denote that there is some $p_\ell \in \Pi$ such that $p_i.d[\ell] = v$. If a process p_i crashes, then we assume that its decision value $p_i.d$ is $\mathcal{N}$, where $\mathcal{N}$ stands for the n element list $[\bot, \ldots, \bot]$. To avoid repetition throughout the discussion of our algorithm, we say that a process p_i decides in an execution ϕ if $p_i.d \neq \mathcal{N}$.

As we describe later, we execute our algorithm for RO Consensus, called ROC, multiple times in electing a leader. We then have that processes may crash before starting an execution ϕ of ROC. Such processes hence have initial value undefined in ϕ. We therefore use $\bot$ to denote the initial value of crashed processes. That is, if $p_i.a = \bot$, then p_i has crashed. We also use the relation $x \subseteq y$ for x and y lists of n elements to denote that: $\forall i, 1 \leq i \leq n : (x[i] \neq \bot) \Rightarrow (x[i] = y[i])$.

The specification of RO Consensus is composed of four properties as follows:

Termination: Every process that does not crash eventually decides on some value;

Agreement If $p_i.d[\ell] \neq \bot$, $p_i, p_\ell \in \Pi$, then for every non-faulty p_c, $p_i.d[\ell] = p_c.d[\ell]$;

RO Uniformity: Let *vals* be the following set: $\{d : \exists p_i \in \Pi \ s.t. \ (p_i.d = d)\} \setminus \mathcal{N}$. Then: $\bigwedge \quad 1 \leq |vals| \leq 2$
$\bigwedge \quad \forall d, d\prime \in vals : d \subseteq d\prime \vee d\prime \subseteq d$
$\bigwedge \quad \forall d_f, d_c \in vals, d_f \subseteq d_c : \exists S_f, S_c \in \mathcal{S}_\Pi:$
$\qquad \wedge \ \forall p \in S_f : (p \text{ crashes}) \ \vee \ (p.d = d_f)$
$\qquad \wedge \ \forall p \in S_c : (p.d = d_c) \ \wedge \ (p \text{ is not faulty})$

That is, there can be no more than two non-$\mathcal{N}$ decision values, and if there are two then one is a subset of the other. Furthermore, if there are two different decision values, then these are the values that processes in two disjoint survivor sets decide upon, one for the processes of each survivor set.

Validity: $\bigwedge$ If $p_j \in \Pi$ does not crash, then for all non-faulty p_i, $p_i.d[j] = p_j.a$
$\bigwedge$ If $p_j \in \Pi$ does crash, then exists $v \in \{\bot, p_j.a\}$ such that for all non-faulty p_i, $p_i.d[j] = v$
$\bigwedge$ If there are survivor sets $S_c, S_f \in \mathcal{S}_\Pi$ and values $v_c, v_f \in V$, $v_c \neq v_f$, such that: $\wedge \ \forall p \in S_f : p.a \in \{v_f, \bot\}$
$\qquad\qquad\qquad\qquad \wedge \ \forall p \in S_c : ((p.a = v_c) \ \wedge \ (p \text{ is not faulty}))$
$\qquad\qquad\qquad\qquad \wedge \ \exists p_i, p_\ell \in \Pi : p_i.d[\ell] = v_f$
$\qquad$ then for all p_j that does not crash, $v_f \in p_j.d$

That is, if a process p_i is not faulty and $p_i.d[j] \neq \bot$, then the value of $p_i.d[j]$ must be $p_j.a$. The value of $p_i.d[j]$, however, can be $\bot$ only if p_j crashes. The third case exists because we use the decision values of an execution as the initial values for another execution. From RO Uniformity, there can be two different non-$\mathcal{N}$ values d_f and d_c. If this is the case, then there is a survivor set S_c containing only correct processes such that all processes in S_c decide upon d_c, and another survivor set S_f containing only faulty processes such that all the processes in S_f either crash or decide upon d_f. Let v_f be d_f and v_c be d_c. By the third case, if some process that decides includes v_f in its decision value, then every process that does not crash also includes v_f in its decision value.

Figure 4 shows our algorithm ROC. In each round r, a process p_i collects messages and updates its list of initial values $p_i.A$. Once it updates $p_i.A$, p_i

sends a message containing $p_i.A$ to all processes. A process p_i also assigns $p_i.A$ to $p_i.A_p(r)$ once it updates $p_i.A$ at round r. This enables p_i to verify in round $r+2$ if a process p_j has received the message p_i sent in round r. As we describe below, p_i uses $p_i.A_p(r)$ to determine if it is faulty.

ROC is an adaptation of a classic round-based synchronous Consensus algorithm for crash failures. There are two main differences. First, it uses survivor sets rather than a threshold scheme. It does use a constant t to bound the number of rounds; t is the number of processes subtracted the size of the smallest survivor set. Second, it has each process verify if it has committed receive-omission failures.

There are two ways that a process can notice that it has committed an omission failure. First, processes that have not decided or crashed send messages to all processes. We then have that for all non-faulty p_i that receives messages in rounds r and $r+1$: $p_i.s(r+1) \subseteq p_i.s(r)$. If this does not hold, then p_i must have failed to receive some message. The second way uses the content of the messages that p_i receives in each round. Consider a message m that p_i receives from p_j in round $r > 1$. Unless it crashes or discovers that it is faulty, a process sends a message to all processes in each round except the last. Let m' be the message that p_i sent to p_j in round $r - 2$. If m indicates that p_j has not received m' ($p_i.A_p(r - 2) \not\subseteq m'.A$), then p_i knows that p_j is faulty. Let $p_i.sr(r)$ be the processes in $p_i.s(r)$ with all processes that p_i knows to be faulty removed. By definition, we know that there is some survivor set that contains only correct processes. If $p_i.sr(r)$ does not contain a survivor set, then there is

Algorithm ROC on input $p_i.a$, $p_i.Procs$
round 0:
    $p_i.s(0) \leftarrow p_i.Procs$; $p_i.sr(0) \leftarrow p_i.s(0)$
    $p_i.A\,[i] \leftarrow p_i.a$
    for all $p_k \in \Pi$, $p_k \neq p_i : p_i.A\,[i] \leftarrow \bot$
    $p_i.A_p(0) \leftarrow p_i.A$
    send $p_i.A$ to all
round 1:
    $p_i.sr(1) \leftarrow p_i.s(1)$
    if $\vee\ p_i.s(1) \not\subseteq p_i.s(0)$
      $\vee\ \nexists S \in \mathcal{S}_\Pi : S \subseteq p_i.sr(1)$
    then decide $\mathcal{N}$
    else for each $m \in p_i.M(1)$, $p_k \in \Pi$:
          if $(p_i.A\,[k] = \bot)$ $p_i.A\,[k] \leftarrow m.A\,[k]$
    $p_i.A_p(1) \leftarrow p_i.A$
    send $p_i.A$ to all
round r: $2 \leq r \leq t$:
    $p_i.sr(r) \leftarrow p_i.s(r) \setminus \{p_j : \exists m \in p_i.M(r) :$
        $p_i.A_p(r - 2) \not\subseteq m.A \wedge m.from = p_j\}$
    if $\vee\ p_i.s(r) \not\subseteq p_i.s(r - 1)$
      $\vee\ \nexists S \in \mathcal{S}_\Pi : S \subseteq p_i.sr(r)$
    then decide $\mathcal{N}$
    else for each $m \in p_i.M(r)$, $p_k \in \Pi$:
          if $(p_i.A\,[k] = \bot)$ $p_i.A\,[k] \leftarrow m.A\,[k]$
    $p_i.A_p(r) \leftarrow p_i.A$
    send $p_i.A$ to all
round $t + 1$:
    $p_i.sr(t + 1) \leftarrow p_i.s(t + 1) \setminus$
      $\{p_j : \exists m \in p_i.M(t + 1) :$
        $p_i.A_p(t - 1) \not\subseteq m.A \wedge m.from = p_j\}$
    if $\vee\ p_i.s(t + 1) \not\subseteq p_i.s(t)$
      $\vee\ \nexists S \in \mathcal{S}_\Pi : S \subseteq p_i.sr(t + 1)$
    then decide $\mathcal{N}$
    else for each $m \in p_i.M(t + 1)$, $p_k \in \Pi$:
          if $(p_i.A\,[k] = \bot)$ $p_i.A[k] \leftarrow m.A[k]$
    $p_i.A_p(t + 1) \leftarrow p_i.A$
    decide $p_i.A$

Fig. 4. ROC - Algorithm run by process p_i

removed. By definition, we know that there is some survivor set that contains only correct processes. If $p_i.sr(r)$ does not contain a survivor set, then there is some correct process from which p_i did not receive a message. Hence, p_i can conclude that it has failed to receive a message.

Note that RO Consensus differs from the definition of Uniform Consensus in that faulty processes may decide upon different values, although these values are not arbitrary and must be such as described by the RO Uniformity property. In the algorithm by Parvèdy and Raynal, for example, every process that decides must decide upon the same value [20].

Informally, ROC satisfies RO Uniformity because $(3,2)$–Intersection holds. To decide on a value other than $\mathcal{N}$, a process must receive in each round messages from a set of processes that contains a survivor set. $(3,2)$–Intersection implies a low enough replication that there can be a set S of non-crashed faulty processes that communicate only among themselves. But, there cannot be two such sets S and S': if S and S' do not intersect, then $(3,2)$–Intersection implies that their union contains a core, and there must be a correct process either in S or in S'.

A set S of faulty processes that communicate only among themselves will decide on a value d where $d[i] = \perp$ for $p_i \notin S$ and $d[i] = p_i.a$ for $p_i \in S$. In addition, a correct process will also decide $d[i] = p_i.a$ for $p_i \in S$. Of course, a non-crashed faulty process can read from different sets of processes in each round, but by using the two rules given above, such a process can determine that it is faulty. Hence, at worst some faulty processes will decide on a value d_f and the correct processes will decide on a value d_c such that $d_f \subseteq d_c$.

The algorithm in Figure 5 uses ROC to implement Weak Leader Election. Algorithm WLE proceeds in iterations of an infinite repeat loop, where each iteration consists of two phases. In Phase 1, processes use ROC to distribute their process identifiers. In Phase 2, they use ROC to distribute what they decided on in Phase 1.

```
Algorithm WLE
P ← Π
repeat {
p_i.elected ← FALSE
Phase 1:
    Run ROC with
        p_i.a ← i; p_i.Procs ← P
    P ← p_i.s(t + 1)
    if (p_i.d = [⊥, . . . , ⊥]) then stop
Phase 2:
    Run ROC with
        p_i.a ← p_i.d from Phase 1; p_i.Procs ← P
    P ← p_i.s(t + 1)
    if (p_i.d = N) then stop
    let x ∈ p_i.d be a value such that
        ⋀ p_i.d [x] ≠ N
        ⋀ p_i.d has the least number of non-⊥ values
    if (p_i is the first index of x such that x[i] ≠ ⊥)
        then p_i.elected ← TRUE
}
```

Fig. 5. WLE - Algorithm run by process p_i

A formal proof of WLE appears in [14]. Informally, this algorithm satisfies Safety because of the following: it is possible for a set of faulty processes S to decide on the smaller value d_f in Phase 1, but by the end of Phase 2 the correct processes will know this. By Validity and RO Uniformity, every process that finishes Phase 2 uses the same list d_f to determine whether it is the current leader or not. Having the processes decide based on the smaller list d_f forces the receive-omission faulty processes to elect the same process as the correct processes. Note though, as mentioned above, that in this case the correct processes know that the elected process is faulty (although the elected process does not know). LE-Liveness is obtained by repeatedly running the algorithm without resorting to a failure detector (which would require higher replication). If there are no faulty processes, each election will always elect the process with the lowest identifier, which implies FF-Stability. To guarantee that there is no alternating behavior in which two processes are leaders infinitely often, non-crashed processes move forward the set of processes they believe are not crashed or have not stopped. That is, the input $p_i.Procs$ in ROC takes the value $p_i.s(t+1)$ from the previous execution of ROC (Π if it is the first execution of ROC). This implies E-Stability.

5   Conclusions

In this paper we generalized a common argument used in proofs of lower bounds on process replication. The argument is based on the threshold model: it makes the assumption that, given n processes, any subset of $\lceil nb/k \rceil$ processes can be faulty. Then, after deriving a contradiction, the proof concludes that $n > \lfloor kt/b \rfloor$. In our generalization of the proof for $b = 1$, we conclude that k–Partition holds: if one partitions the processes into k subsets, then at least one of the subsets contains a core. Thus, lower bounds for many protocols can be trivially generalized for when process failures are not IID. We then gave an equivalent property, k–Intersection, that is often useful when designing a protocol that takes advantage of non-IID process failures.

We considered a problem for which the lower bound has $b = 2$. The lower bound on process replication for Weak Leader Election in a synchronous system with receive-omission failures was known to be $n > \lfloor 3t/2 \rfloor$, but this bound was not known to be tight. We showed that this bound is tight by first determining the intersection property for this replication predicate $((k, k - 1)$–Intersection, equivalent to $(k, k - 1)$–Partition, $k = 3)$ and using it to guide our development of a protocol.

As part of future work, we intend to study further replication predicates that use cores and survivor sets. In particular, we are interested in predicates that generalize $n > \lfloor kt/b \rfloor$ for other values of b.

Acknowledgements

We would like to express our gratitude to Geoff Voelker and Marcos Aguilera for useful discussions and to the anonymous reviewers for comments that improved significantly this paper. Support for this work was provided by AFOSR MURI Contract F49620-02-1-0233.

References

1. B. Altmann, M. Fitzi, and U. Maurer. Byzantine Agreement secure against general adversaries in the dual failure model. In *Proceedings of the 13th DISC*, volume 1693/1999 of *LNCS*, pages 123–139. Springer-Verlag, Sep 1999.

2. N. Budhiraja, K. Marzullo, F. Schneider, and S. Toueg. Optimal Primary-Backup protocols. In *Proceedings of the 6th WDAG*, pages 362–378, Nov 1992.
3. M. Castro and B. Liskov. Practical Byzantine fault tolerance and proactive recovery. *ACM Transactions on Computer Systems*, 20(4):398–461, Nov 2002.
4. M. Castro, R. Rodrigues, and B. Liskov. BASE: Using abstraction to improve fault tolerance. *ACM Transactions on Computer Systems*, 21(3):236–269, Aug 2003.
5. T. Chandra, V. Hadzilacos, and S. Toueg. The weakest failure detector for solving Consensus. *Journal of the ACM*, 43(4):685–722, Jul 1996.
6. T. Chandra and S. Toueg. Unreliable failure detectors for reliable distributed systems. *Journal of the ACM*, 43(2):225–267, Mar 1996.
7. F. Christian. Synchronous Atomic Broadcast for redundant broadcast channels. *Journal of Real-Time Systems*, 2:195–212, Sep 1990.
8. H. Garcia-Molina and D. Barbara. How to assign votes in a distributed system. *Journal of the ACM*, 32(4):841–860, Oct 1985.
9. J. Gray and D. Siewiorek. High-availability computer systems. *IEEE Computer*, 24(9):39–48, Sep 1991.
10. R. Guerraoui and M. Raynal. A generic framework for indulgent Consensus. In *Proceedings of 23rd IEEE ICDCS*, pages 88–95, 2003.
11. M. Hirt and U. Maurer. Complete characterization of adversaries tolerable in secure multi-party computation. In *Proceedings of the 16th ACM PODC*, pages 25–34, Aug 1997.
12. F. Junqueira and K. Marzullo. Consensus for dependent process failures. Technical Report CS2003-0737, UC San Diego, USA, Sep 2002.
13. F. Junqueira and K. Marzullo. Synchronous Consensus for dependent process failures. In *Proceedings of the 23rd IEEE ICDCS*, pages 274–283, May 2003.
14. F. Junqueira and K. Marzullo. Weak Leader Election in the receive-omission failure model. Technical Report CS2005-0829, UC San Diego, USA, Jun 2005.
15. K. Kursawe and F. Freiling. Byzantine fault tolerance on general hybrid adversary structures. Technical Report AIB-2005-09, Aachen University, Germany, Jan 2005.
16. L. Lamport, R. Shostak, and M. Pease. The Byzantine Generals Problem. *ACM Transactions on Programming Languages and Systems*, 4(3):382–401, Jul 1982.
17. D. Malkhi and M. Reiter. Byzantine Quorum Systems. In *Proceedings of the 29th ACM STOC*, pages 569–578, May 1997.
18. J.-P. Martin and L. Alvisi. Fast Byzantine Consensus. In *Proceedings of DSN*, Jun 2005.
19. S. Mullender, editor. *Distributed Systems*, chapter 8. Addison-Wesley, 2nd edition, 1995.
20. P. R. Parvèdy and M. Raynal. Optimal early stopping Uniform Consensus in synchronous systems with process omission failures. In *Proceedings of the 16th ACM SPAA*, pages 302–310, Jun 2004.
21. F. B. Schneider. Implementing fault-tolerant services using the State-Machine approach: A tutorial. *ACM Computing Surveys*, 22(4):299–319, Dec 1990.

Consistent Data Replication: Is It Feasible in WANs?

Yi Lin[1,*], Bettina Kemme[1,*],
Marta Patiño-Martínez[2,**], and Ricardo Jiménez-Peris[2,**]

[1] McGill University, School of Computer Science, Montreal, Quebec, Canada
[2] Facultad de Informatica, Universidad Politecnica de Madrid, Spain

Abstract. Recent proposals have shown that database replication providing 1-copy-serializability can have excellent performance in LAN environments by using powerful multicast primitives. In this paper, we evaluate whether a similar approach is feasible in WAN environments. We identify the most crucial bottlenecks of the existing protocols, and propose optimizations that alleviate the identified problems. Our experiments show that performance remains acceptable even for medium sized systems, and data replication guaranteeing 1-copy-serializability is a serious alternative to weaker approaches in WAN environments.

1 Introduction and Motivation

With the wide use of online transaction processing systems (e.g., online stores), comes the need for fault-tolerance, scalability, and fast response times. Replication is the most common approach to achieve these properties. For instance, if a retailer replicates its data at the different store locations, each store has fast local access, and the system can survive site crashes. For such applications, data consistency despite high update loads, and the flexibility to submit any transaction to any database replica are more important than unlimited scalability. That is, having two to ten replicas is probably a typical system configuration.

A big challenge is to keep the data replicas consistent. There is a trade-off to pay between providing full data consistency (e.g., 1-copy-serializability and atomicity) and fast response times [13]. In recent years, many replication protocols have emerged [2–4, 6, 10, 11, 14, 16, 17, 21–23] providing both data consistency and excellent performance in LANs. Many of these protocols determine the serialization order at the begin of transaction, and then serialize transactions according to this order. This is faster than traditional distributed locking and avoids an expensive 2-phase commit protocol.

However, little research has been done whether these solutions can also be applied to WANs. [1] analyzes one particular protocol. [23] uses multicast protocols that have been developed for WANs. Other solutions usually execute and commit transactions at one site, and propagate changes to other sites only some time after commit [24] allowing for faster response times. This is also referred to as lazy replication. However,

 * Research partially funded by MDER-Programme PSIIRI: Projet PSIIRI-016/ADAPT, by NSERC Rgpin 23910601, and by FQRNT 2003-NC-80398.
** Research partially funded by the European Commision under project Adapt IST-37126 and by the Spanish Research Council (MEC) under project TIN- 2004-07474-C02-01.

J.C. Cunha and P.D. Medeiros (Eds.): Euro-Par 2005, LNCS 3648, pp. 633–643, 2005.

remote sites might have stale data, and committed transactions might be lost in case of crashes. Lazy approaches also often restrict updates to be executed at a single primary site. This, however requires a client to send its update transactions over the WAN if it is not located close to the primary, again increasing response times. Other approaches allow inconsistencies between replicas [12, 24] which are difficult to resolve.

Considering these shortcomings of existing WAN solutions, this paper revisits the successful solutions developed for LANs, and evaluates how they perform in WANs. We have detected several shortcomings of these protocols when executed in WANs, and suggest improvements that help to alleviate the major bottlenecks. They show how the message overhead within the response time can be reduced through various means. Our performance results show that data consistency can be obtained with acceptable performance. In many of our experiments, response times remain below 1 second, and throughput increases over a centralized system. Furthermore, query (read-only) response times are excellent and not affected by update transactions.

2   Database Replication Strategies Developed for LANs

One common approach to provide 1-copy-serializability is to submit all SQL requests to a central middleware [3, 6] which performs concurrency control (usually on a table basis), and forwards reads to one, and updates to all replicas. This leads to communication between middleware and database replicas for each operation within a transaction.

Alternative approaches assume a replicated middleware, where a middleware instance is installed in front of each database replica. Clients contact the closest middleware instance. A client request triggers the execution of a transaction which might contain several database statements. The transaction programs either reside within the middleware or can be called from the middleware. This is an advantage for WAN replication, since client/middleware and middleware/database communication is always local – only middleware/middleware communication is across the WAN.

Most proposals following this decentralized approach ([1, 2, 10, 11, 14, 16, 21, 23]) use group communication systems [8] (GCS). All middleware replicas build a group and multicast messages which are received by all members (including the sender). Different multicast primitives provide different *ordering* and *delivery* semantics. The ordering semantics of interest for this paper are *unordered*, and *total order* (for each two members receiving m and m', both receive them in the same order). The delivery semantics are *reliable* (whenever a member receives a message m and does not fail, then all other group members will receive m unless they fail), and *uniform reliable* (whenever a member p receives a message, all other members will receive the message unless they fail –even if p fails shortly after message reception). Uniform reliable delivery provides all-or-nothing even in failure cases, while reliable delivery allows failed members to have received messages that are not received by others. Some systems (e.g., Spread [26]), call a combination of reliable and total order *agreed* delivery, and a combination of uniform reliable and total order *safe* delivery. We adopt this notation. Note that the GCS probably needs to send more than one physical message per multicast message submitted by the application, e.g., in order to determine the total order. In general, the higher the degree of ordering and/or reliability, the more internal messages will be necessary and the higher the message delay for a multicast message. In the following analysis, we only consider the number of multicast messages submitted by the database system and not the actual messages sent by the GCS since the latter depends on the particular algorithms implemented within the GCS.

We will now present a simple replication protocol using group communication. Most existing protocols are extensions of this basic protocol. The middleware distinguishes between read-only transactions (also called queries) and update transactions, e.g., by analyzing the SQL statements. It handles update transactions as follows.

    I. *Upon receiving a request for the execution of an update transaction from the client*: multicast the request to all sites with safe delivery.
    II. *Upon receiving an update request in safe delivery*: add the request to a FIFO queue.
    III. *Once a request is the first in the queue*: submit the transaction for execution.
    IV. *Upon finishing execution*: remove the request from the queue, and respond to client.

This protocol executes update transactions serially according to the total order multicast, and guarantees atomicity by relying on uniform reliable delivery. In order to allow transactions to execute concurrently, many approaches do concurrency control at the middleware. They often assume that the objects to be accessed are known in advance (e.g. by parsing the SQL statements to determine the tables to be accessed). The middleware can then, e.g., atomically request locks for all tables the transaction is going to access upon safe delivery. When all locks are granted, the transaction can start executing. In this case, non-conflicting transactions can execute concurrently. [10, 14, 16] follow this or similar approaches. We call this the *symmetric approach*. An example is shown in Fig. 1(a). Update transactions T1 and T2 are submitted concurrently to sites A and B respectively, which then multicast the request in total order. Both sites execute T1 and T2 according to the total order if they conflict otherwise in any order.

Queries only need to be executed at one replica. In a WAN this is probably the local replica to avoid messages. Since many database systems (e.g., Oracle and PostgreSQL) provide a special snapshot mode, in which queries read from a committed snapshot of the data, the middleware can immediately submit queries to the database replica without any further actions, and still provide 1-copy-serializabilty.

3   Replication Strategies in WANs

There exist many optimizations over the protocol above. They have been either analyzed only for LANs, or not at all. In this section, we look at several optimizations and their potential effect in WANs. We only look at update transactions since queries are local.

3.1   Communication Choices

Agreed vs. Safe Delivery. Safe delivery leads to long delays since it requires acknowledgements (acks) from all sites before messages can be delivered. Therefore, it is important to understand in which case agreed delivery violates transaction atomicity: a site must receive a client request, multicast it, receive it in agreed order, execute and commit the transaction, return the ok to the client and then fail while none of the surviving sites receives the request and hence, commits the transaction. Only in this case a committed transaction is "lost". This might not happen often, even in a WAN. It is up to the application to decide whether it can accept such cases or not.

Optimisitic Delivery. The idea is to deliver a message once optimistically (e.g., when it is physically received) and once final (e.g., when safe) [18]. The transaction starts executing upon the optimistic delivery but may only commit at the final delivery. This allows overlapping transaction execution and message delay. If the optimistic and final delivery orders are not the same, some transactions might have to abort. [25] delays optimistic delivery until chances are high that it has the same order as the final delivery. [19] considers uniform reliable delivery. Since out-of-order delivery is likely in a WAN, we look at a variation where the optimistic delivery is agreed (i.e., total order is established), and the final delivery is safe (i.e., uniform reliability is established).

Early Execution Variation. However, existing middleware based replication approaches based on optimistic delivery [20] do not always overlap transaction execution with the delay for final delivery. As an example, in Fig. 2(a) transaction T1 is delivered optimistically. It is executed but has to wait to commit until final delivery. T2 is delivered optimistically before T1's final delivery. If it conflicts with T1 the middleware will not start T2's execution until T1 has committed despite its optimistic delivery. Conflicts are likely if the middleware uses table level locking. However, on a tuple-basis, T2 might not conflict with T1. In LANs, this unnecessary waiting might not have a big impact because the delay between T1's optimistic and final delivery, and hence, T2's commit, is small. But in a WAN, we should take full advantage of any possible concurrency. Hence, it would be desirable to execute T2 as soon as possible We suggest as an optimization to take advantage of the concurrency control of the underlying database system if it uses tuple-based strict 2-phase locking. For example, in Figure 2(b), T2 can start execution once T1 has finished its execution. At this time, T1 has already acquired all necessary tuple locks at the database. If T2 does not conflict with T1 on any tuple, T2 can acquire all its tuple locks and execute while T1 is waiting for the final delivery of its message. If T2 and T1 conflict, T2 will block on some tuple until T1 commits and releases its locks which is the correct behavior. Hence, we will execute T1 and T2 in the correct serialization order according to the total order.

Total Order. There exist many algorithms to determine a total order (for a survey see [5, 9]). However, little has been done to evaluate them in combination with an application or in WANs. We analyze some well known algorithms. In *token* a token circulates among the sites, and only the token holder may multicast messages. Increasing sequence numbers indicate the order of these messages. In *sequencer* a sender requests a sequence number from a sequencer and then multicasts the message with this number. The *Lamport* algorithm delivers a message m at a site once this site has received messages from all other sites piggybacking acks for m. Two concurrent messages (from different senders and neither contains an ack for the other message) are delivered in predefined order, e.g., site priority. Finally, a special case of the ATOP approach [7] delivers in a *round-robin* fashion one message from each site.

3.2   Write Set Options

Alternatively to executing the entire transaction at all replicas as in the symmetric approach, execution can take place only at one replica. The resulting changes are propagated in form of a write set (i.e., set of update tuples) at the end of execution to the other sites which apply the write set. Applying writesets is usually faster than executing all

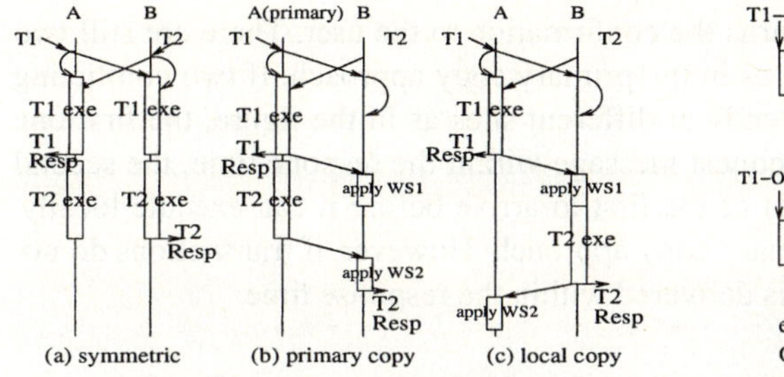
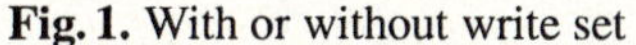

(a) symmetric   (b) primary copy   (c) local copy

Fig. 1. With or without write set

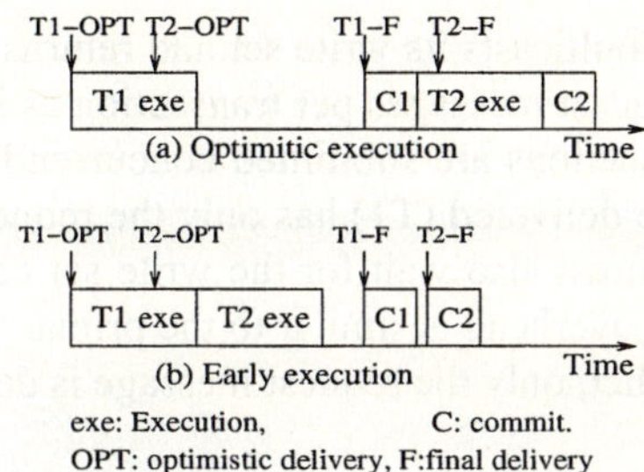

exe: Execution,   C: commit.
OPT: optimistic delivery, F:final delivery

Fig. 2. Optimitic v.s. early execution

read and update SQL statements [16]. Additionally, write sets might be the only feasible solution if there exists non-determinism (e.g., set an attribute to the current local time) which leads to data divergence if SQL statements are executed at all replicas.

Primary Copy Approach. [16] uses write sets with a primary copy approach. Each table has a primary copy, and a transaction updating this table must execute on the site holding the primary copy. If a transaction T wants to access tables that have different primaries, it is executed at the primary of one of these tables. We refer to this as the primary site of T. As before, a request for transaction T is multicast in total order to all sites and the locks requested according to its total order delivery. But only the primary site executes T and multicasts the write set using unordered reliable delivery (if the site crashes, another site can simply reexecute). The primary can commit immediately. The other sites apply the write set once the locks are granted locally. Conflicting transactions might have different primary sites, but all sites will execute or apply the write sets according to the total order since all request locks in this order. Hence, serializability is guaranteed. This approach sends two multicast messages (1 total order and 1 unordered reliable) per transaction. Fig. 1(b) depicts an example. T1 and T2 are submitted to A and B, respectively, and multicast in total order. A is primary site for both transactions. If they conflict, A executes them serially, otherwise concurrently. At the end of the execution of a transaction, A multicasts the write set to the other sites in unordered reliable order. Since T1 was submitted to A, A also returns a confirmation to the client. B, upon receiving T1 and T2, requests the locks in the correct order but does not execute the transactions. Instead it waits for the write sets from A and then applies them (serially if they conflict). It also returns the confirmation for T2 to the client. For T1 the primary is the local site, hence the response time only includes the delay of the request message. For T2 the response time also includes the delay of the write set.

Adjusting to WANs. Since two messages within transaction boundaries can increase response times significantly, we propose an optimization which provides more "locality". In the *local copy approach* depicted in Fig. 1(c), a request for an update transaction is still multicast in total order, but the transaction is executed at the local site it is submitted to. In the figure, T1 executes at A, and T2 at B. Since T1's request is received before T2's request according to the total order, A first executes T1, then multicasts the write set using unordered reliable delivery, commits and returns the confirmation to the user. If T1 and T2 conflict, B waits to receive and apply T1's write set, then executes

T2, multicasts its write set and returns the confirmation to the user. There are still two multicast messages per transaction as in the primary copy approach. If two conflicting transactions are submitted concurrently at different sites as in the figure, the first one to be delivered (T1) has only the request message within the respone time, the second one must also wait for the write set of the first to arrive before it can execute locally. This overhead is similar to the primary copy approach. However, if transactions do not conflict, only the request message is delivered within the response time.

4   Experimental Results

4.1   System Description

We have developed a modular Java based middleware that allows for easy plug-in of different replication strategies. Our middleware supports the symmetric (Sym), the primary copy (PC), the local copy (LC) approach, and the early execution variation of the symmetric approach (ESym). The middleware uses table based locking. We use PostgreSQL 7.2 as database backend. We extended PostgreSQL to provide a function to get the changes performed by a transaction (write set), and a second one that takes these changes and applies them. We used two open-source group communication systems: Spread and JavaGroups. Spread [26] uses a token protocol providing agreed and safe delivery. We integrated an optimistic delivery into Spread that basically delivers a message optimistically upon agreed delivery, and finally at safe delivery. We refer to this as optsafe delivery. JavaGroups [15] implements token based and sequencer based agreed delivery. We have further implemented the Lamport (safe delivery) and the round-robin variation of ATOP (agreed delivery) on top of JavaGroups. The last two algorithms guarantee that all sites send messages regularly by sending null messages if necessary.

4.2   Experiment Setup

The experiments were conducted across the Internet on eight machines with similiar strength (e.g. AMD 1666MHZ / 512MB memory / Red Hat Linux or Solaris), each located in a different city (four in Canada, one in Spain, one in Switzerland and two in Italy). The round trip times between machines in the same continent and between different continents are around 40 ms and 150 ms respectively.

Since all standard benchmarks have at least 50% read workload, we use our own synthetic application to stress-test the system. Our database consists of 10 tables with 10000 tuples each and most experiments use only update transactions to understand the limitations of our approach. Each update transaction modifies 10 randomly selected tuples in a randomly selected table. Since the middleware uses table-based locking, the chance that two concurrent update transactions conflict is 10%. Our scalability analysis also uses queries each of which scans a whole table performing some aggregation (*select avg(attr3), sum(attr3) from $table_i$*). Execution time of this query is three times longer than for an update transaction (as typically encountered in real applications).

In each test run, clients are equally distributed among all replicas and submit transactions concurrently and with the same rate to achieve the desired system-wide load.

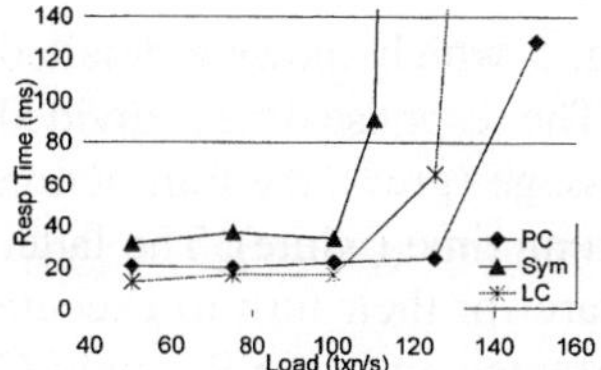

Fig. 3. Write Sets in LAN

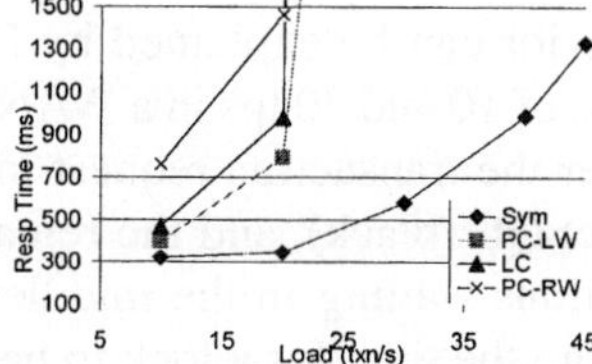

Fig. 4. Write Sets in WAN

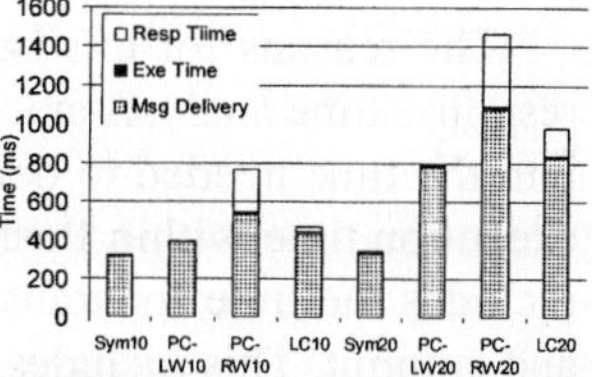

Fig. 5. Response Times WAN

Within each single machine, results were with a 95% confidence within +/- 2.5% of the results taken for our figures. However, response times for different machines varied depending on their setups, and we show the average over all machines. Most figures show the response times up to 1.5 seconds with increasing load submitted to the system. We set 1.5 seconds as an upper limit of what we consider acceptable. Often, the load could be further increased without saturation.

For baseline comparison, we conducted experiments with LAN and WAN clients that directly connect to a single database. With 100% updates, response times of LAN clients never exceed 20 ms up to the saturation point of 120 transactions per second (tps). For WAN clients, response time was over 1.5 seconds at a maximum throughput of 5 tps. This is probably due to message overhead and connection handling.

4.3 Write Set Options

We evaluated the PC, LC and Sym write set options in a LAN and a WAN with 100% updates and 5 sites. We use Spread, and safe delivery for transaction request messages. Message delivery takes a few ms in LAN but up to hundreds of ms in WAN.

Fig. 3 presents the results for the LAN. PC and LC have lower response times and achieve higher throughput than Sym. This is due to the fact that only one site executes the transactions while the others only apply the changes which requires less CPU resources. In a LAN the GCS can easily handle the high message load of LC and PC without becoming a bottleneck. LC has slightly better response time than PC at lower loads since at low loads there are few concurrent transactions, hence in LC nearly all transactions have only one message delay within their response times while for PC transactions with a remote primary always have two message delays within their response time. However, PC outperforms LC at high loads, since more concurrent transactions lead to more conflicts and hence, longer blocking times for LC.

Fig. 4 shows the results for the WAN. For PC, we have one graph showing the response time for a transaction T that was submitted to the primary of T (*local writes LW*), hence only the delay of the request message is within the response time, and one graph for a transaction T submitted by a client that is not connected to the primary of T (*remote writes RW*), hence both the delay of request and writeset message are within the response time. Contrary to the LAN environment, Sym has much better response time than the approaches using write sets. Additionally, Sym can handle up to 45 tps while PC and LC can only handle up to 20 tps with acceptable response times. PC-LW has lower response times than LC, PC-RW has the worst response time.

The reasons for this behavior can be explained by Fig. 5 which shows a detailed response time analysis at loads of 10 and 20 tps in a WAN. The response time is divided into the time needed to deliver the transaction request message (grey), the transaction execution time within PostgreSQL (black), and the remaining time (white). The latter includes the time for transactions waiting in the middleware for their turn to execute and commit. This includes, e.g., the time for a lock to be granted, or in the PC and LC approaches the time the transaction might need to wait for a write set to be delivered. The figure shows that transaction execution (black) only takes a small percentage of total response time for all approaches. That is, the PostgreSQL databases were never the bottleneck. In all cases the time for safe delivery of the request message (grey) has the biggest impact on the response time. Furthermore, PC and LC have generally longer delays for the safe delivery and resulting longer response times than Sym because Spread had to handle double as many multicast messages than with Sym, and hence, was higher loaded. This became especially significant at 20 tps. Using Sym, barely any time is spent in the middleware (white) waiting for locks or similar. In regard to PC, PC-LW transactions do not have any delays in the middleware. PC-RW Transactions have to wait in the middleware for their writesets before they can return to the user, and hence, perform worse than PC-LW transactions. In a 5-site system with evenly distributed workload, we can expect 20% of transactions to have PC-LW response times, and 80% to have PC-RW times. For LC only if there are two concurrent transactions that conflict, one has to wait for the writeset of the other to arrive, otherwise no wait is needed. Hence, the average response time is between PC-LW (do not wait for a writeset) and PC-RW (always wait for the writeset). At low load (10 tps) LC's response time is very close to PC-LW because there are barely any concurrent transactions in the system. Hence very few transactions have to wait in the middleware (very thin white stripe in the figure). At higher load (20 tps), more transactions reside concurrently in the system, hence, more conflicts occur and more transaction have to wait (larger white stripe in the figure).

As a summary, while the write set approach boosts performance in a LAN it does not in a WAN due to the additional message which leads to more contention in the GCS and additional waiting times for transactions to commit. If it is needed due to non-determinism, the local approach seems favorable over the primary copy approach.

4.4   Communication Choices

Delivery Alternatives. Fig. 6 shows the response times when using agreed, safe, or optsafe delivery in Spread. We use 5 sites, Sym, and 100% updates. For optsafe delivery, we also show the early execution variant Esym. Results for PC and LC are not shown since their relative behavior to Sym is similar as above for all delivery types.

Agreed Sym performs by far the best due to the fast delivery of reliable messages compared to uniform reliable messages. Safe Sym is worst since the full delay of safe delivery is added to the response time. Optsafe Sym will start to execute upon agreed delivery but has to wait for the safe delivery to commit, leading to better response time than Safe Sym. However, the difference is very small since transaction execution only takes small percent of response time as shown in Fig. 5. Optsafe ESym is significantly better than Optsafe Sym since it allows for more concurrent execution. Still the execution time is determined by the final safe delivery.

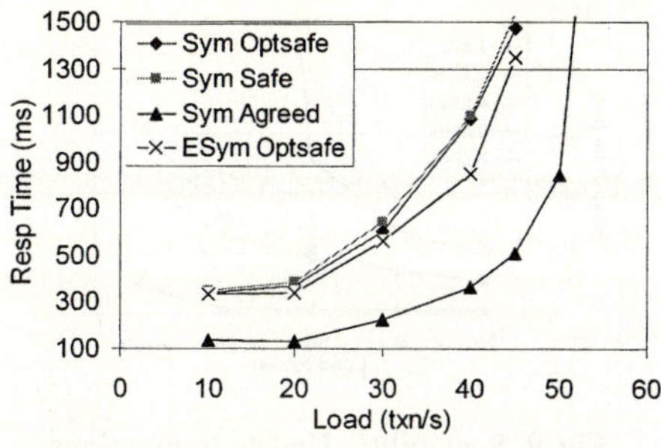

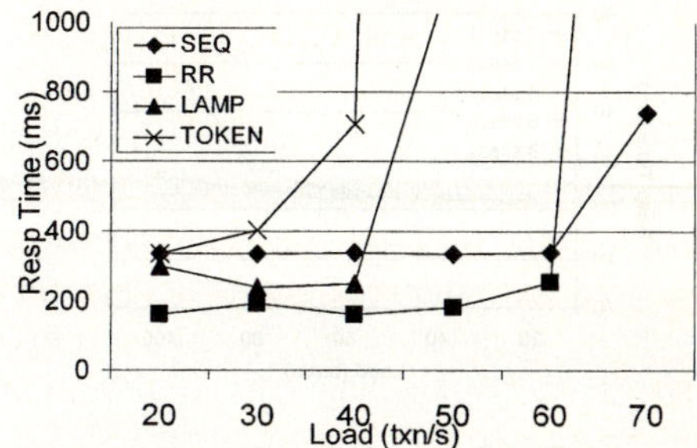

Fig. 6. Agreed, safe and optsafe delivery

Fig. 7. Total Order Implementations

These results show that best performance can be achieved if we can accept the loss of some transactions in failure cases. However, if full atomicity is needed, optimistic delivery and especially our early execution variation can increase performance.

Total Order Implementations. So far, we have used Spread which is a *token* based total order algorithm. In this section, we analyze different total order algorithms and their impact on our replicated database system using JavaGroups. Fig. 7 shows the response times for Sym, 100% updates, and 5 sites. TOKEN refers to the token based approach, SEQ to the sequencer based approach, LAMP to Lamport's algorithm, and RR to the simple round-robin algorithm. TOKEN has by far the worst performance due to the delay incurred by rotating the token. Using SEQ, despite requiring three messages per application message, and the potential bottleneck of one sequencer site, offers better performance than TOKEN. Furthermore it leads to stable response times until the sequencer becomes saturated at around 70 tps. LAMP provides faster response times than TOKEN and SEQ for low loads up to 40 tps although this protocol provides safe delivery while the others only provide agreed delivery. Response times at 30 tps are better than at 20 tps because when more messages are sent, acks arrive faster. However, LAMP saturates earlier due to CPU saturation since it has to keep track of acks. RR has the lowest response time of all since there are no additional messages and message rounds. It saturates only shortly before the sequencer due to CPU overhead. As such, RR seems the preferable choice if agreed delivery is sufficient.

In order to understand the impact of the application on the performance of the GCS we also conducted experiments with JavaGroups and a light-weight message generator client, without the replicated database. The results are similar but generally larger message throughput is achievable because there is no serious application competing with JavaGroups for resources. An interesting exception was SEQ that achieves a far higher throughput than the others without database application, but only a slightly higher throughput when run together with our replicated database system. This is because the sequencer site is saturated much earlier when resources are consumed by the database application. Looking at Spread's agreed token based algorithm from Fig. 6 and Java-Group's TOKEN, transaction response time with JavaGroups are about 2 times higher than with Spread at the same load. This is probably due to implementation choices like the programming language and internal data structures within the GCS.

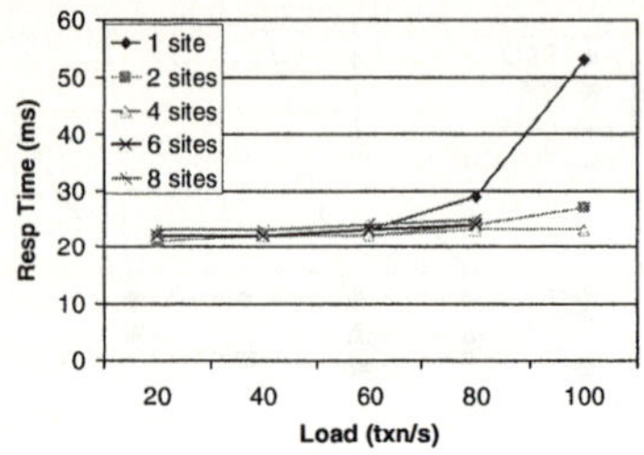

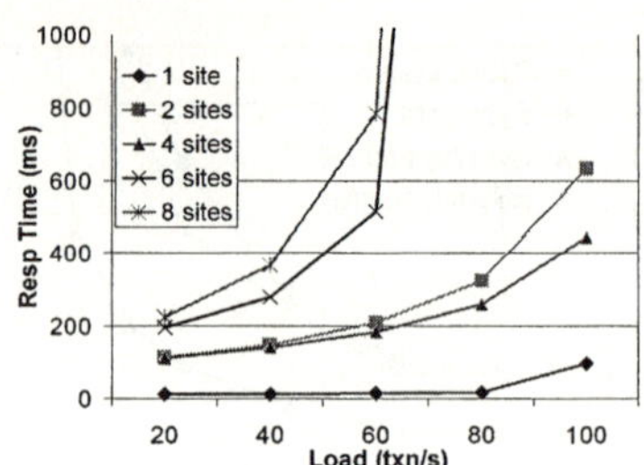

Fig. 8. Scalability: Queries   **Fig. 9.** Scalability: Update transactions

4.5   Scalability in a WAN

This experiment evaluates the performance for system configurations between one and 8 sites. This time, the workload has 50% update transactions and 50% queries. We use Spread's agreed delivery and the Sym algorithm.

Fig. 8 shows that the response times for queries are very similar for all system sizes for smaller loads. At 80 tps a 1-site system starts to saturate, and the response times for queries deteriorates while larger system sizes still provide good response times. With more sites, queries are better distributed and hence, each site is less loaded. Fig. 9 shows the response times for update transactions. A centralized system is always the best for updates since there is no communication. A 4-site system has consistently better performance than a 2-site system because the query load is distributed among more sites, and hence, response times are faster even for updates. For 6 and 8 sites, the response time increases because of the increased overhead to reach total order.

In summary, for a typical, read-intensive workload, a WAN replicated system of 8 sites can achieve an astonishingly high throughput while providing 1-copy-serial-izability. Queries are not affected by update transactions and can take advantage of distribution.

5   Conclusions

This paper presents a detailed WAN based performance analysis of group communication based data replication providing full data consistency. Since existing protocols developed for LANs have limitations in WANs due to the message overhead, we have proposed optimizations (local copy approach, early execution variant) that alleviate the problems. In general, we believe that consistent database replication is feasible.

Summarizing the alternatives, a symmetric approach is preferable over write set based approaches because of the lower message overhead. However, if non-determinism requires a writeset approach, our new local copy approach works better than a primary copy approach for the majority of transactions. Choosing agreed delivery over safe delivery can help to reduce response times. If the application requires atomicity in all situations, optsafe delivery, and especially the newly proposed early execution variation can alleviate the high message delay of safe delivery. The choice of total order algorithm and its implementation has an impact on the performance. An algorithm which simply delivers messages in round-robin appears to have good performance in WAN.

References

1. Y. Amir, C. Danilov, M. Miskin-Amir, J. Stanton, and C. Tutu. On the Performance of Consistent Wide-Area Database Replication. Technical Report CNDS-2003-3, CNDS, John Hopkins University, 2003.
2. Y. Amir and C. Tutu. From Total Order to Database Replication. In *Proc. of ICDCS*, 2002.
3. C. Amza, A. L. Cox, and W. Zwaenepoel. Conflict-Aware Scheduling for Dynamic Content Applications. In *USENIX Symp. on Internet Tech. and Sys.*, 2003.
4. T. Anderson, Y. Breitbart, H. F. Korth, and A. Wool. Replication, Consistency, and Practicality: Are These Mutually Exclusive? In *ACM SIGMOD Conf.*, 1998.
5. R. Baldoni, S. Cimmino, and C. Marchetti. Total Order Communications: A Practical Analysis. In *EDCC, Lecture Notes in Computer Science, Vol. 3463*, pages 38–54, Mar. 2005.
6. E. Cecchet, J. Marguerite, and W. Zwaenepoel. C-JDBC: Flexible database clustering middleware. In *USENIX Conference*, 2004.
7. G. V. Chockler, N. Huleihel, and D. Dolev. An Adaptive Totally Ordered Multicast Protocol that Tolerates Partitions. In *PODC*, 1998.
8. G. V. Chockler, I. Keidar, and R. Vitenberg. Group Communication Specifications: A Comprehensive Study. *ACM Computer Surveys*, 33(4), 2001.
9. X. Defago, A. Schiper, and P. Urban. Comparative Performance Analysis of Ordering Strategies in Atomic Broadcast Algorithms. *IEICE Trans. Inf. and Syst.*, E86-D(12), Dec. 2003.
10. E. Pacitti and T. Ozsu and C. Coulon. Preventive Multi-master Replication in a Cluster of Autonomous Databases. In *Euro-Par*, 2003.
11. U. Fritzke and P. Ingels. Transactions on Partially Replicated Data based on Reliable and Atomic Multicasts. In *Proc. of ICDCS*, 2001.
12. R. Goldring. A discussion of relational database replication technology. *InfoDB*, 8(1), 1994.
13. J. Gray, P. Helland, P. O'Neil, and D. Shasha. The Dangers of Replication and a Solution. In *Proc. of SIGMOD*, 1996.
14. J. Holliday, D. Agrawal, and A. E. Abbadi. The Performance of Database Replication with Group Communication. In *FTCS*, 1999.
15. Java Groups. homepage: http://www.jgroups.org/.
16. R. Jiménez-Peris, M. Patiño-Martínez, B. Kemme, and G. Alonso. Improving the Scalability of Fault-Tolerant Database Clusters. In *ICDCS*, 2002.
17. K. Böhm and T. Grabs and U. Röhm and H.J. Schek. Evaluating the Coordination Overhead of Replica Maintenance in a Cluster of Databases. In *Proc. of Euro-Par*, 2000.
18. B. Kemme, F. Pedone, G. Alonso, and A. Schiper. Processing Transactions over Optimistic Atomic Broadcast Protocols. In *ICDCS*, 1999.
19. L. Rodrigues and P. Vicente. An Indulgent Uniform Total Order Algorithm with Optimistic Delivery. In *SRDS*, 2002.
20. M. Patiño-Martínez, R. Jiménez-Peris, B. Kemme, and G. Alonso. Scalable Replication in Database Clusters. In *DISC'00*, pages 315–329, Toledo, Spain, 2000. Springer.
21. F. Pedone, R. Guerraoui, and A. Schiper. The database state machine approach. *Distributed and Parallel Databases*, 14(1), 2003.
22. C. Plattner and G. Alonso. Ganymed: Scalable replication for transactional web applications. In *Middleware*, 2004.
23. L. Rodrigues, H. Miranda, R. Almeida, J. Martins, and P. Vicente. Strong Replication in the GlobData Middleware. In *Workshop on Dependable Middleware-Based Systems*, 2002.
24. M. Shapiro and Y. Saito. Scaling optimistic replication. In *Future Directions in Distributed Computing*. Springer LNCS 2584, 2003.
25. A. Sousa, J. Pereira, F. Moura, and R. Oliveira. Optimistic Total Order in Wide Area Networks. In *SRDS*, 2002.
26. Spread. homepage: http://www.spread.org/.

A Hybrid Message Logging-CIC Protocol
for Constrained Checkpointability

Françoise Baude[1], Denis Caromel[1], Christian Delbé[1], and Ludovic Henrio[1,2]

[1] INRIA – CNRS, Univ. Nice-Sophia Antipolis
2004, Route des Lucioles, BP93, 06902 Sophia Antipolis Cedex, France
{Francoise.Baude,Denis.Caromel,Christian.Delbe,Ludovic.Henrio}@inria.fr
[2] Harrow School of Computer Science, Univ. of Westminster, Harrow HA1 3TP UK

Abstract. Communication Induced Checkpointing protocols usually make the assumption that any process can be checkpointed at any time. We propose an alternative approach which releases the constraint of always checkpointable processes, without delaying any message reception nor altering message ordering enforced by the communication layer or by the application. This protocol has been implemented within ProActive, an open source Java middleware for asynchronous and distributed objects implementing the ASP (Asynchronous Sequential Processes) model.

1 Introduction

To ensure consistency of recovery lines, Communication-Induced-Checkpointing (CIC) protocols [5, 8, 9] usually make the assumption that every process of the system can be checkpointed at any time: a reception might lead to a *forced* checkpoint. But this assumption can fail for complex or particular systems where a process is not always in a state that can be checkpointed. In particular, in the context of Java middlewares like ProActive [7], persistence can be obtained in a convenient and portable way by standard Java serialization. But, as a thread cannot be serialized, an important part of the activity[1], the threads' stacks, cannot be checkpointed without special arrangements which are discussed below.

A first solution is to use specific tools that make checkpoints possible at any time: threads persistence can be achieved by modifying the execution environment at the OS level [14] or at the virtual machine level [18], or by using a native code-based persistence library [13]. Persistence capabilities can also be added using customized compilers: they add code to capture enough informations to characterize the state of a process [2], or use compile-time reflection to provide persistence functions [12]. But those tools usually involve a loss of portability and/or efficiency. In the context of Java, it is rather unfortunate to lose portability.

A more portable and convenient solution is grounded on the possibility to identify program points at which a checkpoint is possible. In the context of a multi-threaded programming environment like in Java, it concretely means

---

[1] We prefer the term activity rather than process to identify the runtime entity.

J.C. Cunha and P.D. Medeiros (Eds.): Euro-Par 2005, LNCS 3648, pp. 644–653, 2005.

that, at those points, the state of an activity is fully characterized without any knowledge about the state of its thread(s). For instance, the existence of such states grounds the *weak* migration capability of mobile agents, as provided for instance in Voyager or Aglets [1]: a mobile agent is able to migrate only when it reaches such a state. We have identified such program points in the ASP model [6], which are called *stable states* in the following.

As already said, in CIC checkpointing, a message reception might lead to a forced checkpoint. In a fully asynchronous message-passing context, message receptions are unpredictable. So, message receptions can be artificially and simply delayed without consequence, until a checkpoint can be taken (i.e., the execution reaches a stable state). But, as soon as synchronization mechanisms through blocking message reception exists, like the *wait-by-necessity* mechanism in the ProActive middleware, or in MPI with the blocking receive routine [11], this simple solution is not applicable. Indeed, postponing a message reception could lead to a deadlock (e.g. considering that a message could be awaited by the program in order to continue the execution and finally reach a stable state, postponing this message to the next stable state would obviously yield to a deadlock).

Our work has thus consisted in reconsidering the initial simple solution of postponing message receptions, given the constraints raised by Java middlewares like ProActive and its associated computation model, the ASP calculus. Eventually, it has appeared that this new protocol applies for a wider range of contexts.

2   Context

This section describes the hypothesis for which our protocol have been designed and circumscribes the more general context in which it can be applied.

2.1   Constrained Checkpointability

ProActive being a Java middleware, it is impossible to store the state of a thread, thus to take a checkpoint at any time. However, some stable states where a checkpoint is possible can be either automatically identified at the middleware level, or explicitly defined at the application level.

Middleware Level. An activity is in a stable state when its state can be represented without any information about its thread (particularly the stack). The checkpointing can be thus performed using standard Java serialization; the persistence capability is then fully-transparent to the programmer. We have identified such states in the ASP model (see Section 2.3).

Application Level. The proposed protocol can be also used at the application level: stable states could also be specified by the programmer as in [11] ; in that case, the application would be responsible for restoring the state of the activity upon recovery. Although this second approach loses transparency for the programmer, it allows the programmer to save the *minimum* amount of data necessary to recover the activity state.

2.2 Distribution and Communication Model

ASP object calculus is based on concurrent mono-threaded activities communicating using two kinds of messages: request and reply. Each activity consists of one thread, a set of objects (which we call its *applicative state*), and a request queue. There is a master object among the applicative state that is called the *active object*. Note that the applicative state of an active object cannot be shared with another active object: there is no shared memory in our model.

In ASP, when an activity calls a method on an active object, a new request is added to the request queue of this active object. When the signature of the called method has a return value, a future is created on the sender side: this future represents the result of the request that is not known yet. Futures are generalized references that can be manipulated as classical objects. However, some operations (e.g. field access) need a real object value to be performed. Performing such operations on future objects leads to a blocked state called *wait-by-necessity*. When the receiver ends the service of a request, the associated future can be updated: the receiver sends a reply that will replace the future. Note that the impact of a message reception is different depending on the kind of the message. On one hand, a request reception modifies only the request queue of the receiver until it serves the request ; this alteration is reversible by removing this request. On the other hand, a reply reception modifies, in a *non reversible* manner, the applicative state of the receiver.

Causally ordered communications are achieved using a *rendez-vous* taking place at the beginning of each communication [4]. When an activity sends a message to another, it stops its execution until the message is in the context of the receiver. The rendez-vous implies that communications are acknowledged and has the advantage to always ensure point-to-point FIFO ordering of messages.

Our model then guarantees causal ordering of messages; the fault-tolerance protocol thus has to preserve this ordering in case of recovery of the system. More generally, as any synchronization primitive is sufficient to ensure any (partial) ordering of messages at the application level, a protocol with constrained checkpointability has to preserve during a recovery the communication order enforced by the application.

2.3 Properties and Assumptions

The ASP Calculus: In [6], we proved using the ASP calculus the two following main properties:

Property 1 *The relative order of reception of replies during a distributed execution has no consequence on the behavior of the program, assuming that no deadlock is caused by wait-by-necessity.*

Property 2 *An execution can be characterized only by the ordered lists (one for each activity) of request sender identifiers.*

The first property would not be necessary in a middleware that does not have futures. A weaker version of the second one seems to be verifiable in most service-oriented platforms: *An execution can be characterized only by the ordered lists of requests.* With such a property, one would just have to store more informations inside *promised requests* (Section 3.1) in order to use our protocol.

Assumptions on the System: We also make the following assumptions:

- activities are piecewise-deterministic [17] and fail-stop [16],
- failures are detected in an arbitrary but finite time [19],
- an available host always exists, in order to restart a failed activity,
- a stable storage, known by each activity, exists in order to save checkpoints.

Presence of Stable States: An activity is in a *stable state* when it does not serve any request. Indeed, between two request services, the thread state is not necessary to fully characterize the state of the activity. Consequently, the stable state of an activity can be recorded through standard Java serialization of the applicative state and of the request queue. Note that in practice, the presence of stable states requires that the activities never serve a request which service does not end. But this restriction is easy to tackle since an infinite service can always be imitated by a infinite sequence of self-sent requests.

2.4   Notations

Figure 1 shows two activities i and j. j calls a method on i: a request Q is sent. Eventually, this request is served on i and a reply, result of the service, is sent. A rectangle drawn using dotted lines represents the period of service of a request. Conversely, a period of stable state is represented by a simple line. Figure 2 shows a checkpoint C_i^n on an activity i, its sequence number (n) and the request queue of i ($[Q^1, Q^3, Q^4]$). An empty queue is denoted by $[\emptyset]$.

3   Principle of the Protocol

The proposed protocol is an adaptation of [5] and [8] for constrained checkpointability. A parameter TTC, the checkpointing time counter, allows each activity to periodically take checkpoints: if an activity has not taken any checkpoint

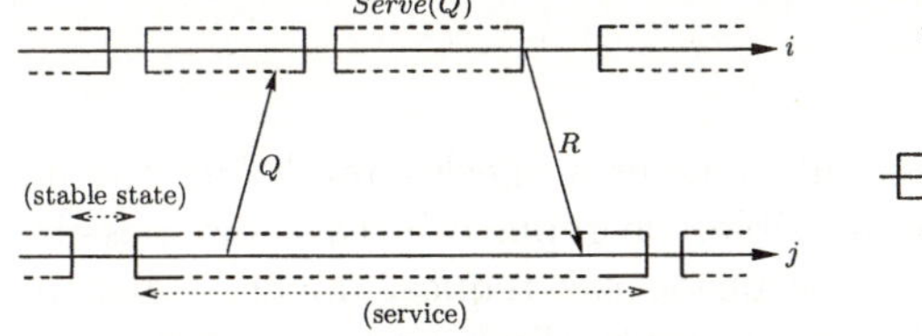

Fig. 1. Communicating activities

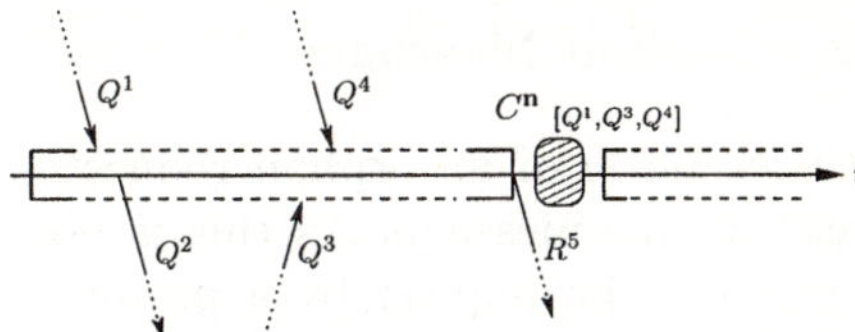

Fig. 2. Checkpoint on an activity

during TTC seconds then a checkpoint is triggered as soon as a stable state is reached. This time counter is reinitialized *each time* a checkpoint is taken. On an activity, each checkpoint is identified by an index which monotonically increases. In case of recovery, all activities have to restart from the same checkpoint index; a set of checkpoint with the same index is called a recovery line. The current checkpoint index of the sender is piggybacked on every message, and the current index of the receiver is piggybacked on every acknowledgment message. These piggybacking allow to identify potential *orphan messages* (message that have been sent after but received before the recovery line, then duplicated in case of recovery) or *in-transit messages* (messages that have been sent before but received after the recovery line, then lost in case of recovery).

In classical CIC protocols, such messages should trigger a *forced* checkpoint on the sender or on the receiver so as to ensure consistency of the currently built recovery line. In our context, it is not possible to take those forced checkpoints; the consequence of this constrained checkpointability is that recovery lines are *inconsistent*. Indeed, there might be orphan and in-transit messages. Compared to classical CIC protocols, our protocol must then handle those messages *a-posteriori*, as soon as a checkpoint is possible, to avoid lost or duplicated messages in case of recovery: we then introduce an *additional message-logging mechanism*.

Since the sending of logged messages after a recovery is obviously triggered by the protocol, the message-logging mechanism could lead to a loss of causal dependencies between messages, and then break the message ordering. Thus, as long as there might exists a message that can be logged during the first execution, the protocol has to record enough informations to be able to ensure execution equivalence in case of recovery. For that, we introduce the *request reception history*, a list of *promised requests*.

3.1   Promised Requests

A promised request is a local substitute for a request that is not yet received in the re-execution; it only contains the identity of the activity from which a request is awaited. A promised request awaited from i in the request queue of j is denoted by $Q_{i,j}^{pmd}$. The service of a promised request is subject to synchronization through a wait-by-necessity mechanism: if an activity tries to serve a promised request, it is blocked until the awaited request is received and updates the promised one.

To summarize, a promised request is a place holder for a request that will be received after a recovery and has already been received in the first execution.

3.2   Orphan Messages

The reception of an orphan request should trigger a checkpoint *before* the delivery of this message. As this is not possible, we replace in the next possible checkpoint the request by a promised one inside the request queue. When re-sent during recovery, this request will thus automatically be inserted at the right place in the request queue, like Q in case of recovery from $n+1$ in Figure 3: the use of a promised request allows to preserve the relative order of requests during the two executions, thus ensures execution equivalence.

Concerning replies, their reception order is not significant thanks to Property 1 but, as stated in Section 2.2, we cannot cancel their effect on the applicative state. However, activities being piecewise-deterministic, ensuring the equivalence of executions is sufficient for guaranteeing that the reply sent during the re-execution is the same as during the first execution, and thus can be ignored.

3.3   In-transit Messages

In-transit messages (requests and replies) can be logged in the next possible checkpoint and re-sent during the re-execution. We introduce the *re-send queue*, denoted by $\Uparrow \{Q^n, Q^m ....\}$, a queue of messages that have to be re-sent during a recovery. Figure 4 shows the logging (noted $\Uparrow \{Q\}$) of the in-transit request Q in the checkpoint n.

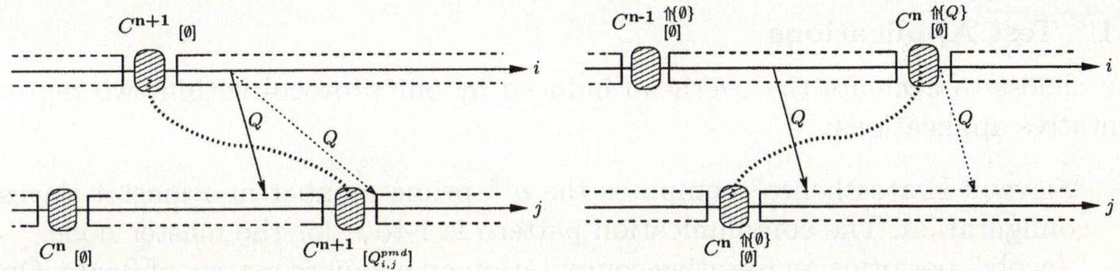

Fig. 3. The request Q is replaced by a promised request in C_j^{n+1}

Fig. 4. The request Q is logged for re-sent in the checkpoint n of i

3.4   Request Reception History

To preserve message ordering, the protocol must ensure equivalence between the first execution and the re-execution in case of recovery. By doing this, it also ensures that orphan replies are identical in the first execution and in the re-execution (Section 3.2). This equivalence must last until the completion of the currently built recovery line. Indeed, after this completion, there cannot be any in-transit nor orphan messages: there cannot be anymore causal relation loss between messages nor duplicated reply.

So as to ensure execution equivalence, we introduce the request reception history. Thanks to the Property 2, this history just needs to record the ordered list of the identity of activities that have sent requests; this information is sufficient to ensure execution equivalence. Consequently, a request reception history for an activity i and for its nth checkpoint is a list of promised requests standing for the requests received between this local checkpoint n and the *history closure*. The only constraint on this closure point is that it must occur *after* the completion of the recovery line n.

The history closure can thus be triggered by a message sent by the stable storage as soon as all the checkpoints with the same index have been received.

The consistency of the history closure line is crucial for preventing infinite wait on a promised request. It is ensured by avoiding orphan message as in [5]: the reception of a message from an activity that has already closed its history triggers on the receiver the closure *before* the delivery of this message.

Finally, when an activity recovers from a checkpoint n, it just has to append to its request queue the history of the checkpoint n: execution equivalence is then ensured as long as an inconsistency could appear.

4   Experiments

A prototype has been implemented within the ProActive Java library. These experiments are a first experimental validation of our protocol, but [3] and [10] provide also a formal presentation of the protocol and the main steps of the correctness proof. [10] proves that our protocol ensures that any re-execution from any recovery line eventually reaches a consistent global state that occurs in the first execution.

4.1   Test Applications

We choose to evaluate the overhead induced by our protocol within two representative applications:

- "Sieve of Eratosthenes" computes the nth prime number in a master-slaves configuration. The communication pattern is 1-to-n for the master node.
- "Jacobi" performs an iterative computation on a square matrix of floats. On each iteration, the value of each point is computed as a function of its value in the last iteration and the values of its neighbors. A square sub-matrix is allocated to each activity. The communication pattern is 1-to-m for all nodes; each activity communicates with its direct neighbors. Each activity is equivalent to the others.

Note that the benchmarks are performed with the same source code for standard and fault-tolerant executions, since there is no need to alter nor recompile the source of an application to make it fault-tolerant. The tests have been performed on a cluster of bi-Xeon @ 2Ghz 1 Gb RDRAM - 512 Kb L2 cache, Linux 2.4.17, interconnected with a 1 Gb/s Ethernet, on the Sun Java Virtual Machine 1.4.2.

4.2   Performance Overhead

Table 1 shows the overhead ($\frac{ExecTime_{tolerant} - ExecTime_{non-tolerant}}{ExecTime_{non-tolerant}}$) induced by the protocol for respectively the Eratosthenes and the Jacobi application running with 8 slaves (one slave per CPU) for Eratosthenes, and with 9 sub-matrix (one sub-matrix per CPU) for Jacobi. The checkpointing time counter is initialized with $TTC = 100\ sec$ for each activity. For each data size (the computed prime number or the matrix size) and *for one activity*, average checkpoint size (checkpoint of the slave for Eratosthenes), number of checkpoints performed, cumulated checkpointing time (the maximum among all activities) and average received message rate are given.

Table 1. Overhead for Jacobi (9 CPUs) and Eratosthenes (8 CPUs), $TTC = 100sec$

	Erathostenes (# Prime Number)					Jacobi (Matrix Size)			
Data Size	1000	2000	3500	5000	10000	500	1000	2000	3000
Exec. Time (s)	201	420	785	1152	2477	153	253	563	1315
Msg Rate (msg/s) (Slave)	42	42	42	42	42	78	47	22	9
Msg Rate (msg/s) (Master)	316	331	332	337	338	n/a	n/a	n/a	n/a
Ckpt Size (Mb)	0.39	0.82	1.4	1.95	4.01	0.68	1.07	7.16	15.07
# Ckpts	3	5	9	13	27	2	3	6	14
Ckpting Time (s)	0.6	1.1	3.5	7.1	33.5	0.9	1.3	3.7	3.9
Overhead (%)	4.95	6.56	6.38	7.77	7.50	2.87	2.94	4.07	4.41

The measured overhead is low: it varies from about 3% to 8% in the worst case. This overhead can be decomposed in two parts: overhead due to message treatment, and time spent for checkpointing (mainly serialization and communication with the stable storage). The higher overhead observed for Eratosthenes application is due to the higher received message rate of the master. Both overheads increase with data size because of checkpoint size. The smoother increasing of Jacobi overhead is explained by the fact that the growing of checkpoint size is counterbalanced by the decreasing of the received message rate for each activity, while the rate of the master for Eratosthenes does not lower with data size.

We also notice that the number of checkpoints performed by an activity *linearly* increases with the fault-tolerant execution time, and that each activity performs the same number of checkpoints. This stability (observed for all the applications we have experimented) is an interesting property of our protocol since a large an unpredictable number of *additional* checkpoints forced by the protocol is known to be the Achilles' heel of CIC protocols [15].

4.3   Scalability

Figure 5 presents the overhead induced by the protocol for Eratosthenes and Jacobi applications regarding the number of CPUs. Eratosthenes computes the 2000th prime number and Jacobi iterates on a 2000*2000 matrix. We observe that the overhead remains roughly constant up to 25 CPUs for both applications: this result demonstrates that the proposed protocol scales well.

4.4   Faulty Execution

Figure 6 shows the recovery time, i.e. the time spent for recovering every activities after a fault, for Eratosthenes and Jacobi regarding the number of CPUs.

The recovery time remains low, up to 25 CPUs (38 *sec* for Eratosthenes and 18 *sec* for Jacobi) and smoothly increases with the number of CPUs from 9 CPUs. The higher recovery time for Eratosthenes is linked to the higher message rate for both master and slaves. Indeed, a higher message rate leads to a longer history, then the synchronization due to promised requests lasts longer.

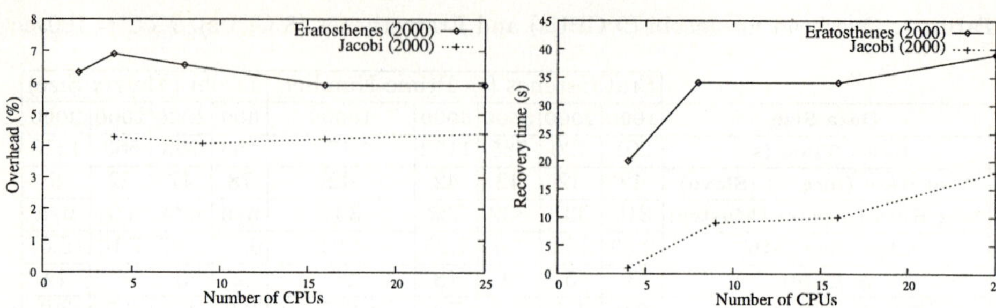

Fig. 5. Execution overhead for Eratosthenes and Jacobi

Fig. 6. Recovery time for Eratosthenes and Jacobi

5   Conclusion

In this paper, we have presented a new hybrid CIC-message logging protocol for Java middlewares that *does not assume* permanent checkpointability. It allows recovery from lines made of restrictively placed checkpoints, without delaying any message reception nor breaking message ordering. The proposed protocol:

- deals with in-transit messages thanks to message logging,
- deals with orphan messages thanks to promised requests and history,
- performs low number of checkpoints,
- is fully transparent since there is no need to alter nor recompile code application to make it fault-tolerant.

It has been implemented in a 100% Java compatible way within the middleware ProActive; as a consequence, the usage of dedicated tools for persistence can be avoided, and portability is total.

Even if the presented protocol has been designed and implemented in the context of ProActive, its main idea is the ability to recover from inconsistent recovery line without breaking any message ordering. As such, this work is applicable to other middlewares, even those using applicative-level persistence. Overall, the location of checkpoints is no more a strong constraint.

The context of this article is somehow similar to [11]; but, contrarily to Bronevetsky et al., our protocol focuses on ensuring the message ordering at recovery. Indeed, a given ordering is always ensured by ProActive but may also be enforced by some applications, even over MPI. The [11] approach may lead those applications into a state that should not exist, since *only* the receptions of orphan or in-transit messages are logged. Introducing a message reception history in [11] would allow one to also cope with this category of applications.

The practical target of our research is also large-scale distributed programming such as grids. In this context, CIC protocols are maybe not the best choice. Indeed, these protocols are more efficient for small systems with low failure rate. On the contrary, grids are large systems with a high failure rate, and a grid application is often partitioned into loosely coupled components, each component being based upon more strongly cooperating processes. In this case, we think of an adaptive approach that autonomously chooses the best combined usage of message logging and hybrid CIC-message logging. The protocol proposed here can thus be considered as a step towards such a *single parameterized* protocol.

References

1. Aglets Software Development Kit. IBM, 1999. http://www.trl.ibm.com/aglets/.
2. B.Ramkumar and V.Strumpen. Portable checkpointing for heterogenous architectures. In *Fault-Tolerant Parallel and Distributed Systems*, pages 73–92, 1998.
3. C.Delbé. Causal ordering of asynchronous request services. In *Dependable Systems and Networks - Student Forum*. IEEE, June 2004.
4. B. Charron-Bost, F. Mattern, and G. Tel. Synchronous, asynchronous, and causally ordered communications. *Distributed Computing*, 9(4):173–191, 1996.
5. D.Briatico, A.Ciuffoletti, and L.Simoncini. A distributed domino-effect free recovery algorithm. In *IEEE International Symposium on Reliability, Distributed Software, and Databases*, pages 207–215, 1984.
6. D.Caromel, L.Henrio, and B.Serpette. Asynchronous and deterministic objects. In *31st ACM Symposium on Principles of Programming Languages*. ACM Press, 2004.
7. D.Caromel, W.Klauser, and J.Vayssiere. Towards seamless computing and metacomputing in java. In Geoffrey C. Fox, editor, *Concurrency Practice and Experience*, volume 10, pages 1043–1061. Wiley & Sons, Ltd., November 1998.
8. D.Manivannan and M.Singhal. A low-overhead recovery technique using quasi-synchronous checkpointing. In *Proceedings of the 16th ICDCS*, pages 100–107, 1996.
9. D.Manivannan and M.Singhal. Quasi-synchronous checkpointing: Models, characterization, and classification. In *IEEE Transactions on Parallel and Distributed Systems*, volume 10, pages 703–713, 1999.
10. F.Baude, D.Caromel, C.Delbé, and L.Henrio. A fault tolerance protocol for asp calculus: Design and proof. Technical Report RR-5246, INRIA, 2004.
11. G.Bronevetsky, D.Marques, K.Pingali, and P.Stodghill. Automated application-level checkpointing of mpi programs. *SIGPLAN Not.*, 38(10):84–94, 2003.
12. J.C.Ruiz-Garcia, M.O.Killijian, J.C.Fabre, and S.Chiba. Optimized object state checkpointing using compile-time reflection. In *Workshop on Embedded Fault-Tolerant Systems*, pages 46–48, 1998.
13. J.Howell. Straightforward java persistence through checkpointing. In *Proceedings of the 3rd International Workshop on Persistence and Java*, pages 322–334, 1998.
14. J.S.Plank, M.Beck, G.Kingsley, and K.Li. Libckpt: Transparent checkpointing under Unix. In *Usenix Winter Technical Conference*, pages 213–223, January 1995.
15. L.Alvisi, E.N.Elnozahy, S.Rao, S.Husain, and A.De Mel. An analysis of communication induced checkpointing. In *Symposium on Fault-Tolerant Computing*, pages 242–249, 1999.
16. R.D.Schlichting and F.B.Schneider. Fail-stop processors: an approach to designing fault-tolerant computing systems. In *ACM Transactions on Computer Systems*, volume 1, pages 222–238, 1983.
17. R.E.Strom and S.Yemini. Optimistic recovery in distributed systems. In *ACM Transactions on Computer Systems*, volume 3, pages 204–226, 1985.
18. S.Bouchenak. Pickling threads state in the java system. In *Third European Research Seminar on Advances in Distributed Systems*, 1999.
19. T.D.Chandra and S.Toueg. Unreliable failure detectors for reliable distributed systems. *Journal of the ACM*, 43(2):225–267, 1996.

A Fault-Tolerant Token-Based Mutual Exclusion Algorithm Using a Dynamic Tree

Julien Sopena[1], Luciana Arantes[1], Marin Bertier[2], and Pierre Sens[1]

[1] LIP6 – Université Paris 6, INRIA, CNRS
{Julien.Sopena,Luciana.Arantes,Pierre.Sens}@lip6.fr
[2] LRI – Université Paris 11, CNRS
Marin.Bertier@lri.fr

Abstract. This article presents a fault tolerant extension for the Naimi-Trehel token-based mutual exclusion algorithm. Contrary to the extension proposed by Naimi-Trehel, our approach minimizes the use of broadcast support by exploiting the distributed queue of token requests kept by the original algorithm. It also provides good fairness since, during failure recovery, it tries to preserve the order in which token requests would have been satisfied had the failure not occurred.

1 Introduction

Mutual exclusion is a fundamental concept in distributed systems. Several algorithms have been proposed to solve the problem of mutual exclusion, serializing concurrent accesses to a shared resource. They can essentially be divided into two groups: *permission-based* (e.g. Lamport [2], Ricart-Agrawala [8], Maekawa [3]) and *token-based* (e.g. Suzuki-Kazami [9], Raymond [7], Naimi-Trehel [5]). Algorithms of the first group are based on the principle that a node may enter critical section only after having received permission from all the other nodes (or a majority of them [3]). The drawback of these algorithms is the high communication overhead. In the second group of algorithms, a system-wide unique token is shared among all nodes, and its possession gives a node the exclusive right to enter into the critical section, thus ensuring the safety property.

Some token-based algorithms, such as Raymond [7] and Naimi-Trehel [5], consider that nodes are organized in a logical tree and that a node always sends a token request to its father in the tree. Tree-based algorithms have an average lower message cost, and many of them result in a logarithmic message complexity $O(logN)$ with regard to the number of nodes. Presenting better scalabilty, they can be more easily adapted to large scale configurations like grid and peer-to-peer environments [1]. Another advantage of these algorithms is the simplicity of their local data structures. However, a tree-based algorithm is very sensitive to node failure since it cannot tolerate even a single failure of one of the nodes in a token request path.

In this paper we propose a fault tolerant extension for the Naimi-Trehel token-based mutual exclusion algorithm. Naimi-Trehel's algorithm maintains

J.C. Cunha and P.D. Medeiros (Eds.): Euro-Par 2005, LNCS 3648, pp. 654–663, 2005.
© Springer-Verlag Berlin Heidelberg 2005

two main data structures: a dynamic logical tree such that the root of the tree is always the last node that will have the token among the current requesting ones nodes, and a distributed queue that keeps token requests that have not been satisfied yet. The dynamic property of the request tree is strongly exploited in our solution. Let N be the number of nodes in the system. The new algorithm can tolerate at most $N - 1$ node failures and the message overhead for failure recovery is relatively low.

Naimi and Trehel have proposed a fault tolerant extension of their own algorithm in [6]. In the absence of failure, the original algorithm is not modified. However, recovery from failure is very expensive in terms of messages since it requires multiple broadcasts, causing a high message overhead. Furthermore, the distributed queue of token requests has to be completely rebuilt.

We have modified the original Naimi-Trehel algorithm, introducing one additional message per token request. This modification has minimal impact on the original protocol in the absence of failures. On the other hand, our algorithm presents a lower cost in terms of messages in the presence of failures since it broadcasts at most one message when compared to the multiple broadcast messages of Naimi-Trehel's algorithm [6]. In contrast to the latter, the basic idea of our algorithm, in case of failure recovery, is to reconstruct the distributed queue of token requests by assembling disconnected portions of the previous queue. In addition to the low recovery cost, this approach exhibits the fairness property since it preserves the order in which token requests were previously queued.

We should mention that Mueller also presents in [4] a fault-tolerant extension of the Naimi-Trehel algorithm without broadcast support. However in his solution a ring communication structure, which includes all nodes of the system, is used for detecting a node failure as a message circulates constantly on it. Even if his solution does not use broadcast support, it also presents a lack of scalability, since the ring exhibits a message overhead that grows linearly with the number of nodes. Furthermore, his approach tolerates only one node failure.

The organization of this paper is as follows. Section 2 presents our considered system model. Section 3 briefly describes Naimi-Trehel's algorithm and outlines the problems for making it fault tolerant. Their fault tolerant version of this algorithm is described in section 4. In Section 5, we describe our fault-tolerant extension for the original Naimi-Trehel algorithm. A performance comparison of both fault-tolerant algorithms is presented in section 6, whilst the last section concludes our work.

2   General Model

We consider a distributed system consisting of a finite set of N sites $\Pi = \{S_1, S_2, \ldots, S_N\}$ that are spread throughout a network. Sites communicate only by sending and receiving messages. Every pair of sites is assumed to be connected by means of a reliable communication channel. However, messages may be delivered in a different order than the one they were sent in. The words site and node are interchangeable.

We consider a synchronous fully-connected network where process speeds and message transmission times are bounded. T_{msg} is the maximum latency for sending a message between two sites. Contrary to Naimi-Trehel, which considers that a critical section execution takes T_{cs} in average, our algorithm makes no assumption on the time for executing critical sections.

Sites can fail by crashing only, and this crash is permanent. $N - 1$ node failures are tolerated.

3   Naimi-Trehel's Algorithm

Naimi-Trehel's algorithm [5] is a token-based algorithm. It keeps two data-structures:

1. A logical dynamic tree structure such that the root of the tree is always the last site that will get the token among the current requesting ones. Requesting sites then form a logical tree pointing by probable token owners towards the root. Initially, the root is the token holder, elected among all sites. We call this tree the *last tree*, since each site keeps a local variable called *last* that points to the probable owner of the token.

2. A distributed queue which keeps critical section (CS) requests that have not yet been satisfied. We call this queue the *next queue*, since each site S_i keeps a local variable called *next* that points to the next site to whom the token will be granted after S_i leaves the critical section.

One **invariant of Naimi-Trehel algorithm** is that the root node of the *last* tree is always the tail node of the *next* queue.

When a site S_i wants to enter the critical section, it sends a request to its *last*. S_i then sets its *last* variable to itself and waits for the token. Site S_i becomes the new root of the tree.

Receiving S_i's token request message, site S_j can take one of the following actions: (1) S_j is not the root of the tree. It forwards the request to its *last* and then updates its *last* variable to S_i. Notice that the *last tree* is modified dynamically; (2) S_j is the root of the tree. If S_j is holding an idle token, it sends it back to S_i directly. On the other hand, if S_j holds the token but is in the critical section or is waiting for the token, S_j sets its *next* variable to S_i. At the end of the execution of critical section, S_j sends the token to its *next*.

An example of Naimi-Trehel's algorithm execution with four nodes is shown in Figure 1. Initially (a), site A is the root which holds the token. The local *last* variable of all nodes points to A. In (b), node B asks for the token by sending a request to its *last* ($last_B = A$). B becomes the new root ($last_B = B$). Then, A updates its *next* and *last* variables to point to B. In (c), C asks A for the token. The request is forwarded to B which updates its *next* to C ($next_B = C$). Both A and B update their *last* to C, since the latter is the last requester of the token (C becomes the new root of the tree). When A releases the critical section, the token will be sent to B as $next_A = B$.

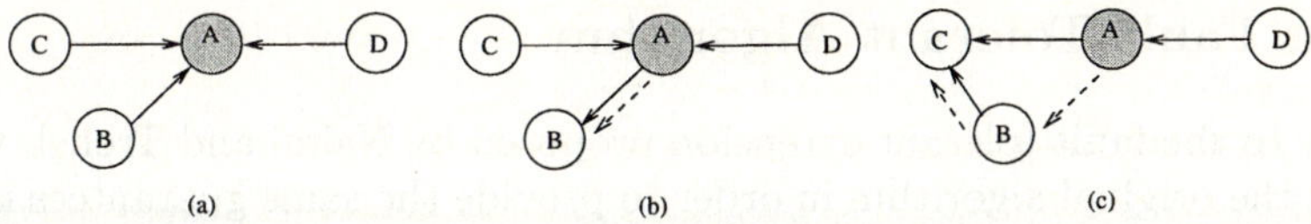

Fig. 1. Example of Naimi-Trehel's algorithm execution

The major challenges for making Naimi-Trehel algorithm fault-tolerant are:

1. *The faulty node is an intermediate node of the last tree.* In this case, if the faulty node were used for forwarding the token request before the failure, the request should be resent. Furthermore, we must be sure that the state of *last tree* is consistent before re-sending the request. However, in the Naimi-Trehel algorithm, while a request message is in transit, the tree is temporarily broken into several smaller rooted trees. Thus, finding a right path to the root may be impossible if the failure has occurred when the tree was in an unstable state.

2. *The faulty node belongs to the next queue.* In this case, it is not possible to known the path for token transmission anymore. Therefore, *next queue* must either be rebuilt from the beginning or by gathering disconnected portions of the queue which existed before the failure.

3. *The faulty node had the token.* In this case, the token must be regenerated and the uniqueness of the token must be guaranteed.

4 Naimi-Trehel Fault Tolerant Extension

In [6], Naimi and Trehel propose a fault-tolerant version of their algorithm. The original algorithm is not modified but some extensions are included in the algorithm to detect site failures, recover from failures, and regenerate the token.

To detect a site failure, site S_i, which requests the critical section, arms a timer T_{wait}. This timer depends on latency communication time (T_{msg}) and the average time (T_{cs}) for executing the critical section. If S_i does not receive the token after the expiration of T_{wait}, it suspects that a failure has occurred. Therefore, S_i broadcasts a *CONSULT* message to ask for the state of the other sites and arms a new timer, T_{elec}. When a site S_j receives this message, it answers to S_i only if the latter is its *next*. At the expiration of T_{elec}, if S_i does not receive any response, it is sure that a failure has occurred. S_i then broadcasts a *FAILURE* message to detect the presence of the token in one of the sites. A site replies to S_i if it owns the token.

If after a new T_{elec} delay, S_i has not received any answer to its failure message, it considers that the token is lost and it becomes a candidate to regenerate the token. It then broadcasts an *ELECTION* message. In case of concurrent election messages, the site with the smallest identifier is chosen. At the end, an *ELECTED* message is broadcasted to inform all sites of the new token owner. Finally, the identification of the *last* of each site is set to the new token owner. Notice that each node having requested the token before the failure has to re-send its token request.

5   Our Fault-Tolerant Algorithm

Contrary to the fault tolerant extension proposed by Naimi and Trehel, we have modified the original algorithm in order to provide the same guarantees in terms of fault tolerance and to optimise efficiency and complexity in the occurence of failures.

5.1   Principle of the Algorithm

The guiding principle of our algorithm is to reconstruct the *next queue* by gathering intact portions of the previous *next queue* which existed just before the failure. The aim of this reconstruction is to preserve the initial order of token requests as much as possible and to avoid request retransmitions, as is the case in Naimi-Trehel's solution. On the other hand, if the reconstruction is not possible, a new *next queue* will be created as well as a new *last tree*. The latter needs to be consistent with the former guaranteeing the invariant mentioned in section 3.

Considering the original algorithm, a site always knows, through its *next* variable, which site will receive the token after it, i.e. its successor in *next queue*. However, it is not aware of which site will grant the token to it, neither which sites will get the token before it. In other words, it is not aware of its predecessors in *next queue*. Thus, in order to inform a node of its predecessor in *next queue*, we have added a confirmation mechanism to the original algorithm for each token request. Whenever a site S_j updates its *next* variable, i.e. S_j is in the last element of the *next* queue and received a token request, it sends a *COMMIT* message to the requester in order to confirm the reception of the request and to communicate the identification of its predecessors. The *next queue* then keeps a ordering where the smallest position corresponds to the site which has the token. A site loses its position when it leaves the queue. Initially the token holder has position zero. A *COMMIT* message sent to the requester S_i, by site S_j, contains the two following informations:
– *The k predecessors of S_i*: k is a configurable parameter, indicating how many failures the algorithm can recover by using mechanism **M1**, described below.
– *S_i's position in the queue*: equals to S_i's closest predecessor's position + 1.

The cost of having a predecessor information mechanism is low in terms of messages. We have added just one message per token request. Thus, the message complexity of the algorithm only grows from $log(N)$ to $log(N) + 1$ and thus remains **$\mathcal{O}(log(N))$**. However, this mechanism enables the detection of failures more effectively than Naimi-Trehel's fault tolerant extension. In their approach, the reception of the token is controlled by a timer T_{wait}, which depends both on latency (T_{msg}) and the time (T_{cs}) for executing the critical section. In our approach, the same timer depends only on latency (T_{msg}). After receiving a *COMMIT* message, S_i periodically checks the liveness of its closest predecessor.

After detecting a failure, site S_i will start a failure recovery by executing a different mecanism for each of the three following cases:

– **Mechanism 1 (M1)**. Site S_i has received a *COMMIT* message and there are less than k consecutive faulty sites in *next queue*.

– **Mechanism 2 (M2).** Site S_i received a *COMMIT* message, but there are more than k consecutive faulty sites in the *next queue*.

– **Mechanism 3 (M3).** The site did not receive any *COMMIT* message.

We now detail how to recover from failures in the three cases. **M1.** When site S_i detects a failure of its closest predecessor, it sends an *ARE_YOU_ALIVE* message to each of its predecessors from the closest to the farthest, so as to check if they are still alive. It stops querying when it obtains an I_AM_ALIVE message from one of its predecessors. The latter then takes S_i as its new successor, i.e. it sets its *next* variable to S_i. The *next queue* is then reconstructed and the order is preserved. Furthermore, the *last tree* remains consistent with the *next queue* and the invariant mentioned in section 3 is asserted.

M2. If no predecessor responded to the *ARE_YOU_ALIVE* message, S_i will try to reconnect itself to *next queue* by diffusing a *SEARCH_PREV* message which contains S_i's position. S_i then arms a timer ($2 * T_{msg}$), waiting for the answer messages. All sites having a smaller position then S_i's will answer to it. After waiting $2 * T_{msg}$, S_i will choose among these sites, the one which has the greatest position to become its closest predecessor. Then, S_i reconnects itself to this chosen site by sending a *CONNECTION* message to it. If S_i does not receive any answer at all after $2 * T_{msg}$, it concludes that it has no predecessors and consequently the token has been lost. S_i should then regenerate the token, initializing its position to zero.

Observe that in both mechanisms **M1** and **M2**, due to our predecessor information approach, the order of *next queue* is preserved.

M3. We must consider now the case where the site which detects the failure has not received the *COMMIT* message yet, and therefore has no position in *next queue*. Moreover, in the absence of such information, several sites can detect the same failure simultaneously.

M3.a We initially consider the situation when just one site S_i detects the failure. In order to reconnect itself to *next queue*, S_i will search for the site which has the greatest position. This search is initiated by the diffusion of a *SEARCH_QUEUE* message. S_i then arms a timer ($2 * T_{msg}$), waiting for the answer messages. A site that has a position in *next queue* answers to S_i with an *ACK_SEARCH_QUEUE* message which contains its position in the *next queue*, as well as whether or not it has a *next*. Among all the received answers within $2 * T_{msg}$, S_i will select the site S_j with the greatest position. S_i then considers three possibilities:

(i): S_j *has informed that it has no next.* S_i then resends a token request to S_j. Notice that, since this request is sent directly to a node at the tail of *next queue*, S_i does not use *last tree* to send a token request. Thus, we avoid the problem mentioned in section 3 concerning the instability of *last tree* when token requests are in transit.

(ii): S_j *has informed that it has a next.* S_i can conclude that S_j's *next* has failed. S_i then sends a *CONNECTION* message to S_j in order to force S_j to reconnect itself to S_i; i.e. S_j will set its *next* to S_i.

(iii): If site S_i has not received any answer, it concludes that it has no more predecessors and that the token has been lost. S_i can then regenerate the token, initializing its position to 0. It is sure to be the only site to regenerate the token.

M3.b We now discuss the situation when several sites detect the node failure concurrently. They will start tracking the *next queue*, and will even generate a new token, which may bring *next queue* to an inconsistent state or the loss of the token uniqueness property. An election mechanism then is necessary. We consider that a site is elected if it is always candidate after a time of $2 * T_{msg}$.

Having sent a *SEARCH_QUEUE* message to the other sites as described above, site S_i is a candidate to reconnect to *next queue*. However, if S_i receives a *SEARCH_QUEUE* message from node S_j, it knows that another site S_j is also a candidate for reconnection. Thus, if S_j has made fewer accesses to the critical section than S_i (this information is included in the *SEARCH_QUEUE* message) or S_i's access number is equal to S_j's but S_j has a greater identifier than S_i, the latter loses the election, sending a token request to S_j. In turn, S_j will be responsible for reconnecting itself to *next queue*. If S_j later loses the election, it will behave like S_i. However, if it wins the election, it finds itself in the situation of mechanism **M3.a**. *Next queue* is thus repaired[1].

Contrary to the first two mechanisms, the order of previous token requests is not preserved in mechanism **M3**. Thus, *last tree* must be reconstructed to be consistent with the new *next queue*. However, this reconstruction is done dynamically, without any additional overhead in terms of message and latency, since all the information a site needs has been transmitted to it in the *SEARCH_QUEUE* message. Considering that S_i is the single site that suspected the fault (**M3.a**), or the one that wins the election (**M3.b**), *last tree* is reconstructed as follows:

I: all sites which do not wait for the token set their *last* variable to S_i.

II: all sites that have a position in *next queue* set their *last* variable to S_i.

III: all sites without a position, but in wait for the token, set their *next* variable to the same value as their *last* variable.

An example of failure recovery based on mechanism **M3** is shown in figure 2. We consider that there are two faulty sites, as shown in figure 2.a. The *next queue* is broken into two portions (G,H and A,B,C). Sites G and H have already obtained a position in *next queue*, but sites A, B and C have not. The token is held by site G, first site of the *next queue*. We also consider that site D had sent a token request to one of the two faulty sites and while waiting for the *COMMIT* message, it accepted a token request from E. Thus, there is a second queue of sites waiting for the token, but it is not connected to *next queue* yet. Notice that in such configuration *last tree* is also broken (the *last* variable values of sites F, G, H and D have become useless).

Suppose that sites A and D detect a node failure concurrently. Both of them broadcast a *SEARCH_QUEUE* message. We then say that A wins the election, i.e. A is the elected node. In figure 2.b, we can see how some of the *last* variables are updated. Having a position, sites G and H update their *last* to A (see II), as

---

[1] To ensure that two sites do not get the same position, message ordering is controlled by using Lamport's timestamp.

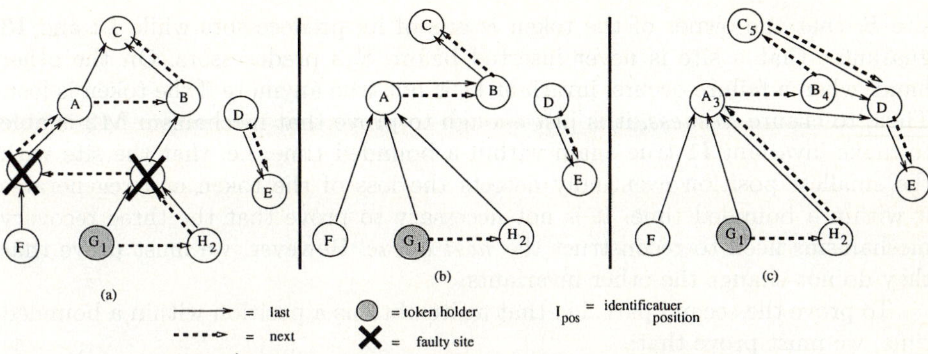

Fig. 2. Example of our fault-tolerant algorithm execution

well as site F, which was not waiting for the token (see I). However, A, B and E, which are waiting for the token but do not have a position, update their *last* variables to the same value as their respective *next* variables (see III).

When receiving the *ACK_SEARCH_QUEUE* message from H, site A can conclude that the *next* of H is a faulty site (see *ii*). A then sends a *CONNEC-TION* message to H. When the latter receives such a message, it sets its *next* variable to A. On the other hand, since site D lost the election, it sends a token request to the elected node (A). When receiving D's request, A forwards this request to its *last* ($last_A = B$). The request travels along *last tree*, arriving at C, the root of the tree. All sites belonging to *last path* which received the request update their *last* variable to D. Site C sets its *next* variable to D, sending a *COMMIT* message to it. Figure 2.c shows the final configuration, considering that sites D et E have not received a *COMMIT* message yet.

5.2   Sketch of Proof

We are just going to give the outline of the correctness proof of our algorithm. For this purpose, we should prove its *safety* and *liveness* properties:
– **Safety:** there is always at most one token in the system, which guarantees that at most one site can execute the critical section at any time.
– **Liveness:** A site requesting entry to the critical section will eventually succeed within a bounded time.

Proof of liveness: This proof comprises two parts. Firstly, we should prove the *liveness* for a site which has a position, and then that a site eventually obtains a position within a bounded time.

In the absence of failures, we can identify the following four invariants, which are easily proven by induction: **I1** - the site with the smallest position has the token; **I2** - the position ordering respects the order of *next queue*; **I3** - after S_i got its position, no site can get a new position which is smaller than S_i's; **I4** - two sites cannot have the same position.

In the absence of failures, these invariants ensure that a site S_i holding a position will receive the token within a finite time. Indeed, **I1** and **I2** ensure site S_i that the owner of the token is one of its predecessors while **I2** and **I3** guarantee that a site is never inserted before S_i's predecessors. On the other hand, when a failure occurs, invariant **I1** is not true anymore if the token is lost. Thus, to ensure *liveness*, it is just enough to prove that mechanism **M2** is able to make invariant **I1** true again within a bounded time, i.e. that the site with the smallest position eventually detects the loss of the token and regenerates it within a bounded time. It is not necessary to prove that the three recovery mechanisms need to reconstruct the *next queue*. However, we must prove that they do not change the other invariants.

To prove the second part, i.e. that a site obtains a position within a bounded time, we must prove that:

– *A site whose token request is lost retransmits it within a bounded time using a set of data structures which is consistent with the original algorithm.* The failure detection of mechanism **M3** takes place within at most $N * Tmsg$. Moreover, it is also possible to show a scenario in the absence of failure which is similar to the one resulting from the execution **M3** that rebuilt the *last tree* and the *next queue*.

– *A request can be lost a finite number of times.* This is ensured by our model, i.e. there can be at most $N-1$ permanent crashes. However, if there is an infinite number of failures, this property keeps true if and only if the system has periods of stability of at least $N * T_{msg}$.

Proof of safety: in the original Naimi-Trehel algorithm there is always only one token. However, in our approach, **M2** and **M3** may regenerate a token. Thus we need to prove that in these mechanisms:

– *A site regenerates a token only when the latter has been lost:* in mechanism **M2** (resp. **M3**), a site regenerates a token, if and only if, it did not receive an answer ACK_SEARCH_PREV (resp. ACK_SEARCH_QUEUE). To prove that "no answer implies no more token", its contrapositive can easily be proven. This can be done by using invariant **I1**, which implies that if there is at least one token, then the site with the smallest position has one of these tokens.

– *Only one site regenerates the token:* we can prove by contradiction that **M2** and **M3** are not compatible. In the same way, we can prove that two sites cannot regenerate the token by using both mechanism **M2** (resp. **M3**).

6   Performance Issues

Considering failure detection and recovery, we can compare our algorithm to Naimi-Trehel's fault-tolerant one (see section 4).

– **Message complexity:** in the worst case of failure recovery, Naimi-Trehel's fault tolerant algorithm broadcasts four messages. Our solution sends one *COM-MIT* message per token request in order to keep control of *next queue* and a broadcast *SEARCH_PREV* message, if necessary. It is also worth mentioning that in the former all the successors of the faulty node must resend their token request, while in our algorithm only lost requests are resent.

– **Time:** so as to detect faulty sites, Naimi-Trehel's algorithm controls the reception of the token. In the worst case, it waits $(N-1) * (T_{msg} + T_{cs})$. Our algorithm controls the arrival of the token request at the tail node of *next queue* as well as the reception of the *COMMIT* message by the requester. It then waits at most $((N-1)+1)*T_{msg}$ for suspecting a faulty node, which does not depend on T_{cs}. Another important point is that our algorithm has fewer phases than Naimi-Trehel's during failure recovery, reducing recovery time. Moreover, it may happen that such a recovery is done while the algorithm goes on executing normally as if no failure had occurred; i.e it does not need to wait for a stable *last tree* as in the Naimi-Trehel approach. The failure recovery time is then covered up by the time that a site waits for a token.

– **Fairness:** in Naimi-Trehel's approach, *next queue* is rebuilt from the beginning at each failure recovery and the original ordering is not preserved. In our approach, after receiving the *COMMIT* message, a site has its position p in *next queue* which ensures that it will access the critical section after at most $p-1$ other critical section accesses.

7   Conclusion

We presented in this paper a new fault-tolerant algorithm for mutual exclusion. This algorithm is an extension of the Naimi-Trehel token-based algorithm. Compared to the solution proposed in [6], our algorithm has two main properties: a short recovery delay and a resilient fairness of requests. In case of failure, we reconstruct the distributed request queue by assembling portions of the previous queue. Our algorithm requires at most one broadcast and the order of critical section requests is preserved as much as possible despite failures.

References

1. M. Bertier, L. Arantes, and P. Sens. Hierarchical token based mutual exclusion algorithms. In *4th IEEE/ACM CCGrid04*, 10 April 2004.
2. L. Lamport. Time, clocks, and the ordering of events in a distributed system. *Communications of the ACM*, 21(7):558–564, July 1978.
3. M. Maekawa. A $\sqrt{N}$ algorithm for mutual exclusion in decentralized systems. *ACM Transactions on Computer Systems*, 3(2):145–159, May 1985.
4. Frank Mueller. Fault tolerance for token-based synchronization protocols. *Workshop on Fault-Tolerant Parallel and Distributed Systems, IEEE*, april 2001.
5. M. Naimi, M. Trehel, and A. Arnold. A log (N) distributed mutual exclusion algorithm based on path reversal. *Journal of Parallel and Distributed Computing*, 34(1):1–13, 10 April 1996.
6. Mohamed Naimi and Michel Trehel. How to detect a failure and regenerate the token in the log(n) distributed algorithm for mutual exclusion. *Lecture Notes In Computer Science LNCS*, 312:155–166, 1987.
7. K. Raymond. A tree-based algorithm for distributed mutual exclusion. *ACM Transactions on Computer Systems (TOCS)*, 7(1):61–77, 1989.
8. G. Ricart and A. Agrawala. An optimal algorithm for mutual exclusion in computer networks. *CACM: Communications of the ACM*, 24, 1981.
9. I. Suzuki and T. Kasami. A distributed mutual exclusion algorithm. *ACM Transactions on Computer Systems (TOCS)*, 3(4):344–349, 1985.

Self-stabilizing Publish/Subscribe Systems:
Algorithms and Evaluation

Gero Mühl[1,*], Michael A. Jaeger[1,**], Klaus Herrmann[1,*],
Torben Weis[1,**], Andreas Ulbrich[1,*], and Ludger Fiege[2]

[1] TU Berlin, EN6, Einsteinufer 17, 10587 Berlin, Germany
{g_muehl,michael.jaeger,klaus.herrmann}@acm.org,
{weis,ulbi}@ivs.tu-berlin.de
[2] TU Darmstadt, Wilhelminenstraße 7, 64283 Darmstadt, Germany
fiege@acm.org

Abstract. Most research in the area of publish/subscribe systems has not considered fault-tolerance as a central design issues. However, faults do obviously occur and masking all faults is at least expensive if not impossible. A potential alternative (or sensible supplementation) to fault masking is self-stabilization which allows a system to recover from arbitrary transient faults such as memory perturbations, communication errors, and process crashes with subsequent recoveries.

In this paper we discuss how publish/subscribe systems can be made self-stabilizing by using self-stabilizing content-based routing. When the time between consecutive faults is long enough, corrupted parts of the routing tables are removed, while correct parts are refreshed in time, and missing parts are inserted. To judge the efficiency of self-stabilizing content-based routing, we compare it to flooding, which is the naïve implementation of a self-stabilizing publish/subscribe system. We show that our approach is superior to flooding for a large range of practical settings.

1 Introduction

In many applications, independently created components have to be integrated into complex information systems. Especially in large-scale distributed applications, a loosely-coupled event-based style of communication has many advantages. It allows the clear separation of communication from computation and eases the integration of autonomous, heterogeneous components.

In publish/subscribe systems individual processing entities, which we call *clients*, can *publish* information without specifying a particular destination. Similarly, clients can express their interest in receiving certain types of information by *subscribing*. Clients can be *producers* and *consumers* at the same time. Information is encapsulated in *notifications* and the *notification service* is responsible for notifying each consumer about all occurrences of notifications which match one of its active subscriptions.

---

* Funded by Deutsche Telekom.
** Funded by Deutsche Telekom Stiftung.

J.C. Cunha and P.D. Medeiros (Eds.): Euro-Par 2005, LNCS 3648, pp. 664–674, 2005.

Many research prototypes of notifications services exist including SIENA [2], Gryphon [9], Hermes [10], and REBECA [7]. The Java Message Service (JMS) [12] and the CORBA Notification Service [8] are two prominent examples of industrial specifications of notification services. However, in most research prototypes and industrial specifications fault-tolerance has not been a central design issue as the focus was mostly put on the efficiency of routing. Obviously, faults do occur and considering all kinds of faults when implementing fault masking is at least expensive if not impossible.

A potential alternative (or sensible supplementation) to fault masking is *self-stabilization*, a concept introduced by Dijkstra [3] in 1974. He defined a system as being self-stabilizing if "regardless of its initial state, it is guaranteed to arrive at a legitimate state in a finite number of steps". In contrast to that, a system which is not self-stabilizing may stay in illegitimate states forever leading to a permanent failure of the system. Self-stabilization models the ability of a system to recover from arbitrary transient faults within a finite time without any intervention from the outside. If the time between consecutive faults is long enough, the system will start to work correctly again. Transient faults include temporary network link failures resulting in message duplication, loss, corruption, or insertion, arbitrary sequences of process crashes and subsequent recoveries, and arbitrary perturbations of the data structures of any fraction of the processes. The program code running at the nodes and inputs from the outside, however, cannot be corrupted. Dolev [4] gives a comprehensive discussion of self-stabilization.

The remainder of this paper is structured as follows: In Sect. 2 we introduce the notion of self-stabilizing publish/subscribe systems. In Sect. 3, we show how specific routing algorithms can be made self-stabilizing. Sect. 4 presents our comparison of self-stabilizing identity-based routing with flooding. Sect. 5 presents some related work. We close with conclusions and give an outlook in Sec. 6.

2   Self-stabilizing Publish/Subscribe Systems

In previous work, Mühl, Fiege, and Gärtner presented a formalization of publish/subscribe systems as a requirement specification [5, 7] consisting of safety and liveness properties. Due to spatial restrictions, we only give an informal definition of our specification here:

Definition 1. *A* publish/subscribe system *is a system satisfying the following requirements:*

1. *Safety Property*
   (a) A notification is only delivered to a client at most once.
   (b) A client only receives notifications that have previously been published.
   (c) A client only receives notifications it is subscribed for.
2. *Liveness Property: When a client subscribes to a filter and does not issue an unsubscription for this filter, then, from some time on, every notification that is published thereafter and matches the filter will be delivered to the subscribing client.*

Content-based routing is one possibility to implement a distributed notification service. In this case, the notification service is realized by a set of *brokers* forming an overlay network. Here, we restrict ourselves to acyclic connected topologies. This restriction can be circumvented, e.g. by running a spanning tree algorithm on the original (potentially cyclic) topology. Each broker B communicates with its neighbor brokers N_B using *asynchronous message passing* and with its mutually exclusive set of local clients L_B using *local synchronous procedure calls*. The private *routing table* T_B of a broker B determines to which neighbors and local clients broker B forwards a notification that it processes. Each *routing entry* is a pair (F, D) consisting of a filter F having a unique id $id(F)$ and a *destination* $D \in N_B \cup L_B$. A broker sends a notification that it processes to all destinations for which a matching filter exists. However, if a notification is received from a neighbor broker, it is not sent back to this broker.

The routing table determines the current *routing configuration* of a publish/subscribe system. A *routing algorithm* starts from an eligible initial routing configuration and subsequently adapts it. To achieve this, *update messages* are propagated through the broker network when clients issue new or cancel existing subscriptions. Intuitively, a routing algorithm is *valid* if it adapts the routing configuration such that the resulting system satisfies the safety and the liveness property of Def. 1. Several content-based routing algorithms are known, including simple, identity-based, covering-based, and merging-based routing [7]. These algorithms exist in a peer-to-peer and in a hierarchical variant [1].

Definition 1 requires that the system is correct, i.e. exhibits the desired functionality at its interface, under all circumstances. Thus, all occurring faults would have to be masked. Provided that a temporary failure of the system can be accepted, making a system self-stabilizing is an attractive alternative to fault masking. However, it is in general impossible under the fault assumption of self-stabilization to require *any* property that prohibits certain states, i.e. safety properties. For example, the system could deliver a notification n to a client X although X has no active subscription matching n because a fault corrupted the state of the system such that that it "thinks" that X subscribed to n. Therefore, we require that a self-stabilizing publish/subscribe system satisfies the safety property of Def. 1 only eventually. This ensures that the system starting from any state will eventually satisfy the actual safety property and continue to do so if no faults occur. The liveness property of Def. 1 can be left unchanged. This leads to the following definition:

Definition 2. *A* self-stabilizing publish/subscribe system *is a system satisfying the following requirements:*

1. *Eventual Safety Property: Starting from any state, it eventually satisfies the safety property of Def. 1.*
2. *Liveness Property: Starting from any state, it satisfies the liveness property of Def. 1.*

In the following section, we discuss how self-stabilizing publish/subscribe systems can be realized using self-stabilizing content-based routing algorithms.

3   Self-stabilizing Content-Based Routing

Under the fault assumption of self-stabilization, the routing configuration can arbitrarily be corrupted by transient faults. Therefore, the routing algorithm must ensure that corrupted routing entries are corrected or deleted from the routing table and that missing routing entries are inserted into the routing table.

For spatial reasons we assume in this paper that each broker stores the information about its neighbors in its ROM. This ensures that this information cannot be corrupted. If it would be stored in RAM or on harddisk, it could also be corrupted by a fault. In this case, we would have to layer self-stabilizing content-based routing on top of a self-stabilizing spanning tree algorithm. Layered composition of self-stabilizing algorithms is a standard technique which is easy to realize when the individual layers have no cyclic state dependencies [4]. In this case, the stabilization time would be bounded by the sum of the stabilization times of the individual layers.

3.1   Basic Idea

The basic idea for making content-based routing self-stabilizing is that routing entries are only *leased*. To keep a routing entry, it must be renewed before the *leasing period* π has expired. If a routing entry is not renewed in time, it is removed from the routing table. Interestingly, this approach does not only allow the publish/subscribe system to recover from internal faults but also from certain external faults. For example, if a client crashes, its subscriptions are automatically removed after their leases have expired.

To support leasing of routing table entries, we use a *second chance* algorithm. Routing entries are extended by a *flag* that can only take the two values 1 and 0. Before a routing entry is (re)inserted into the routing table, all existing routing entries whose filter has the same *id* (as the *id* of the filter of the routing entry to be inserted) are removed from the routing table. This is necessary as the *id*s of the routing entries can be corrupted, too. We assume that the clock of a broker can only take values between 0 and $\pi - 1$ to ensures that if the clock is corrupted, it can diverge from the correct clock value by at most π. When its clock overruns, a broker deletes all routing entries whose flag has the value 0 from the routing table and sets the flag of all remaining routing entries to 0 thereafter (new subscriptions have the flag set to 1 initially). Hence, it must be ensured that an entry is renewed once in π to prevent its expiry. On the other hand, it is guaranteed that an entry which is not renewed will be removed from the routing table after at most 2π.

The renewal of routing entries originates at the clients. To maintain its subscriptions without interruption, a client must renew the lease for each of its subscriptions by "resubscribing" to the respective filter once in a *refresh period* ρ. Resubscribing to a filter is done in the same way as subscribing. In general, π must be chosen to be greater than ρ due to varying link delays. The *link delay* δ is the amount of time needed to forward a message over a communication link and to process this message at the receiving broker. In our model, it is considered a fault when δ is not in the range between $\delta_{\min}$ and $\delta_{\max}$. It is important to note, that assuming an upper bound for the link delay is a necessary precondition for realizing self-stabilization.

3.2 Flooding

The naïve implementation of a self-stabilizing publish/subscribe system is *flooding*: When a broker receives a notification from a local client, the broker forwards the notification to all neighbor brokers. When it receives a notification from a neighbor broker, the notification is forwarded to all other neighbor brokers. Additionally, each processed notification is delivered to all local clients with a matching subscription. Flooding only requires a broker to keep state about the subscriptions of its local clients. Therefore, errors in this state can be corrected locally by forcing clients to renew their subscriptions once in a leasing period. This means that $\rho = \pi$. The main advantage of this scheme is that a coordination among neighboring brokers is not necessary. Hence, no additional network traffic is generated. Additionally, new subscriptions become active immediately. While a corrupted or erroneously inserted subscription survives at most 2π in a routing table and a missing subscription is reinserted after at most π, an erroneously inserted or corrupted notification disappears from the network after at most $d \cdot \delta_{\max}$ where d is the *network diameter*, i.e. the length of the longest path a message can take in the broker network. Hence, for flooding, the *stabilization time* Δ, i.e. the time it takes for the system to reach a legitimate state starting from an arbitrary state, equals $\max\{2\pi, d \cdot \delta_{\max}\}$.

3.3 Simple Routing

The solution for flooding can be extended to simple routing. *Simple routing* treats each subscription independently of other subscriptions. A (un)subscription is inserted into (removed from) the routing table and flooded into the broker network. If a broker receives a (un)subscription from a local client, it is forwarded to all neighbor brokers. If it was received from a neighbor broker, it is forwarded to all other neighbor brokers. Thus, simple routing is idempotent to resubscriptions and a subscription is redistributed through the broker network when it is renewed by the client. Note that here subscriptions become active only gradually.

A critical issue is that the timing assumptions must allow the clients to renew their leases everywhere in the network before they expire. How large must π be with respect to ρ in this case? To answer this question, consider two brokers B and B' connected by the longest path a message can take in the broker network. This situation is illustrated in Fig. 1. Assume a local client X of B leases a routing table entry of B at time t_0 and renews this lease at time $t_1 = t_0 + \rho$. X's lease causes other leases to be granted all along the path to broker B'. Considering the best and worst cases of the link delay, the first lease reaches B' at time $a_0 = t_0 + d \cdot \delta_{\min}$ in the best case and the lease renewal reaches B' at time $a_1 = t_1 + d \cdot \delta_{\max}$ in the worst case. If X refreshes its leases after ρ time and if network delays are unfavorable, two lease renewals will arrive at B' within at

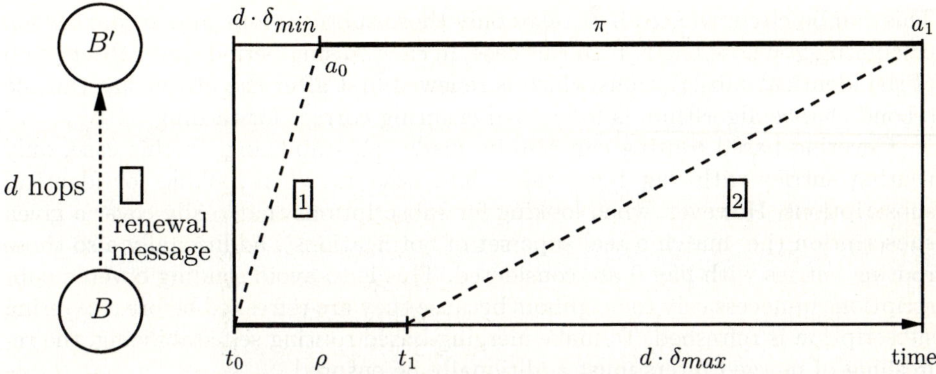

Fig. 1. Deriving the Minimum Leasing Time.

most $a_1 - a_0$. Hence, $\pi > a_1 - a_0$ must hold to ensure that the entry is renewed in time. Thus, we get $\pi > \rho + d \cdot (\delta_{\max} - \delta_{\min})$.

The stabilization time Δ depends on the value of π. Since corrupted or erroneously inserted messages can contaminate the network, a delay of $d \cdot \delta_{\max}$ must be assumed before their processing is finished. After at most 2π, their effects will be removed everywhere. Overall, the stabilization time sums up to $\Delta = d \cdot (\delta_{\max} - \delta_{\min}) + 2\pi$. For example, assume that $d = 10$, $\delta_{\max} = 25\,ms$, and $\delta_{\min} = 5\,ms$. To guarantee a stabilization time of $\Delta = 30\,s$, $\pi = 14.9\,s$ and thus $\rho = 14.7\,s$ follows. There is a tradeoff between π and ρ. To have low message overhead, ρ should be as large as possible. However, this implies a large value of π, but π should be as small as possible to facilitate fast recovery.

3.4   Advanced Routing Algorithms

The situation is more complicated if advanced content-based routing algorithms such as identity-based, covering-based, or merging-based routing are applied. Contrary to flooding and simple routing these algorithms are – at least the versions presented so far – not idempotent with respect to resubscriptions. However, they can be made idempotent with some minor changes. Note that the maximum stabilization time Δ is not affected by whether an advanced routing algorithm or simple routing is applied because in the worst case a filter will nevertheless travel all along the longest path in the network.

Consider identity-based routing (for more details we refer to [7]). When a broker B processes a new or canceled subscription F from destination D, it counts the number d of destinations $D' \neq D$ for which a subscription matching the same set of notifications exists in T_B. Depending on the value of d, F is forwarded differently. If $d = 0$, F is forwarded to all neighbors if $D \in L_B$ and to all neighbors except D if $D \in N_B$. If $d = 1$ and $D' \in N_B$, F is forwarded only to D'. If $d = 1$ and $D' \in L_B$ or if $d \geq 2$, F is not forwarded at all. This scheme is not idempotent to resubscriptions because if $d \geq 2$ and one of the identical subscriptions is renewed at B, none of those subscriptions will be forwarded.

This can be circumvented if B takes only those subscriptions into account when calculating d whose flag is 1. In this case, in each leasing period that subscription of the identical subscriptions which is renewed first after the broker has run the second chance algorithm, is forwarded ensuring correct forwarding.

Covering-based routing can also be made self-stabilizing. In this case, only routing entries with flag 1 are taken into account when looking for identical subscriptions. However, when looking for subscriptions that really cover a given subscription (i.e. match a real superset of notifications), additionally also those routing entries with flag 0 are considered. This is to avoid sending covered subscriptions unnecessarily to neighbors because they are refreshed before a covering subscription is refreshed. To make merging-based routing self-stabilizing, the refreshing of merged filters must additionally be ensured.

3.5   Discussion

The values of π and ρ depend on the delay of the links in the network. So far, we assumed that these values are fixed and equal for every broker in the system. In many scenarios, link delays vary a lot such that it could be advantageous to incorporate this property into the algorithm. We assume that the value of link delay stored at every adjacent broker can not be corrupted (i.e. it is stored in ROM). The values of π and ρ has then to be calculated individually for every subscription, depending on where the publishers are. Additionally, π and ρ have to be refreshed the same way as described previously for subscriptions. Advertisements that are sent periodically by the publishers could be used for this purpose. Taking this approach, the broker algorithm can take advantage of faster links and stabilizes subtrees of the broker topology faster if the links allow for this. The application of leasing is a common way to keep soft states. This technique is used in many protocols and algorithms such as the Routing Information Protocol (RIP, RFC2453) and Directed Diffusion [6].

4   Simulation

We carried out a discrete event simulation to compare self-stabilizing content-based routing to flooding with respect to their message complexity. Before we discuss the results, we describe the setup of the experiments.

4.1   Setup

We consider a broker hierarchy being a completely filled 3-ary tree with 5 levels. Hence, the hierarchy consists of 121 brokers of which 81 are leaf brokers. Since we use a tree for routing, this implies a total number of 120 communication links. We use hierarchical routing but similar results can be obtained for peer-to-peer routing, too. With hierarchical routing, subscriptions are only propagated from the broker to which the subscribing client is connected towards the root broker. This suffices because every notification is routed through the root broker. Hence, control messages travel over at most 4 links. We use identity-based routing and consider 1000 different filter classes (e.g. stocks) to which clients can subscribe.

Subscribers only attach to leaf brokers. Results for scenarios where clients can attach to every broker in the hierarchy can be derived similarly. Instead of dealing with clients directly, we assume independent arrivals of new subscriptions with exponentially distributed interarrival times and an expected time of λ^{-1} between consecutive arrivals. When a new subscription arrives, it is assigned randomly to one of the leaf brokers and one of the filter classes is randomly chosen. The lifetime of individual subscriptions is exponentially distributed with an expected lifetime of μ^{-1}. Each notification is published at a randomly chosen leaf broker. Hence, notifications travel over at most 8 links. The corresponding filter class is also chosen randomly. The interarrival times between consecutive publications are exponentially distributed with an expected delay of ω^{-1}. We assume a constant delay in the overlay network of $\delta = 25\,ms$ including the communication and the processing delay caused by the receiving broker.

To illustrate the effects of changing the parameters, we considered two possible values for some of the system parameters: For each of the 1000 filter classes, a publication is expected every $1\,s$ ($10\,s$), i.e. $\omega_1 = 1000\,s^{-1}$ ($\omega_2 = 100\,s^{-1}$). The expected subscription lifetime is $600\,s$ ($60\,s$), i.e. $\mu_1 = (600\,s)^{-1}$ ($\mu_2 = (60\,s)^{-1}$). Each client refreshes its subscriptions once in $60\,s$ ($600\,s$), i.e. a refresh period of $\rho_1 = 60\,s$ ($\rho_2 = 600\,s$). Since $d = 8$ in our scenario, the leasing period is $\pi_1 = 60.2\,s$ ($\pi_2 = 600.2\,s$) for ρ_1 (ρ_2). Hence, a subscription will on average be refreshed 10 (100) times before it is canceled by the subscribing client if $\mu = (600\,s)^{-1}$. The resulting stabilization time is $\Delta_1 = 120.6\,s$ ($\Delta_2 = 1200.6\,s$).

We are interested in how the system behaves in equilibrium for different numbers of active subscriptions N. In equilibrium, $dN/dt = 0$ where $dN/dt = \lambda - \mu \cdot N(t)$, implying $N = \lambda/\mu$. Thus, if N and μ is given, λ can be determined. If the system was started with no active subscriptions, we would have to wait until the system approximately reached equilibrium before we begin the measurements. However, in our scenario it is possible to start the system right in the equilibrium. At time 0, we create N subscriptions. For each of these subscriptions, we determine how long it will live, for which filter class it is, and at which leaf broker it is allocated. Since we use an exponential distribution for the lifetime, this approach is feasible because the exponential distribution is memoryless.

4.2   Results

The results of our simulation are depicted in Fig. 2. Note that the right plot is a magnification of the most interesting part of the left plot. $b_{s1/2}$ is the notification bandwidth saved if filtering is applied instead of flooding. The figure shows b_{s1} and b_{s2} which correspond to the publication rate ω_1 and ω_2, respectively. Because b_s linearly depends on ω, a decrease of ω by a factor of 10 leads to 10 times less saving of notification bandwidth. If there are no subscriptions in the system, $b_{s1} = 116,000\,s^{-1}$ and $b_{s2} = 11,600\,s^{-1}$, respectively. These numbers are $4,000\,s^{-1}$ and $400\,s^{-1}$ less than the overall number of notifications

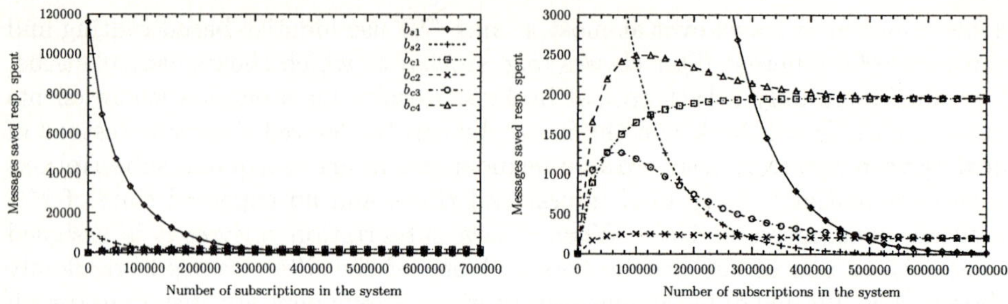

Fig. 2. Notification bandwidth saved by doing filtering instead of flooding ($b_{s1} : \omega_1 = 1000\ s^{-1}, b_{s2} : \omega_2 = 100\ s^{-1}$) and control traffic caused by filtering and leasing ($b_{c1}, b_{c4},: \rho_1 = 60\ s, b_{c2}, b_{c3},: \rho_2 = 600\ s, b_{c1}, b_{c2} : \mu_1 = (600\ s)^{-1}, b_{c3}, b_{c4} : \mu_2 = (60\ s)^{-1}$).

published per second. This is because with hierarchical routing, a notification is always propagated to the root broker. The control traffic b_c is caused by subscribing, refreshing, and unsubscribing clients. It only arises if filtering is used. The figure shows b_{c1}, b_{c2}, b_{c3}, and b_{c4} which result from the different combinations of μ and ρ. The value to which b_c converges for large numbers of subscriptions, mainly depends on the refresh period ρ. Thus, b_{c1} and b_{c3} converge to $120,000/\rho_1 = 2,000s^{-1}$, while b_{c2} and b_{c4} converge to $120,000/\rho_2 = 200s^{-1}$. The evolution of b_c for numbers of subscriptions in the range between 0 and 200,000 is largely influenced by the value of μ. A small μ such as μ_2 leads to a hump (cf. b_{c3} and b_{c4} in Fig. 2). Filtering saves bandwidth compared to flooding if b_s exceeds b_c. The points where the curve of the respective variants of b_s and b_c intersects are important: If the number of subscriptions is smaller than at the intersection point, filtering is superior, while for larger numbers flooding is better. For example, the curves of b_{s1} and b_{c1} intersect for about 300,000 subscriptions. Thus, filtering is superior for less than 300,000 subscriptions, while flooding is superior for more than 300,000 subscriptions. Since we consider 8 scenarios, we have 8 intersection points in Fig. 2.

The results gained through the simulation show, that applying self-stabilizing filtering makes sense if the average number of subscriptions in the system does not grow beyond a certain point. However, it is important to note, that all assumptions taken for the simulation depict worst-case scenarios. For example, the equal distribution of subscriptions to leaf brokers is disadvantageous for filtering. If there was locality in the interests of the clients, filtering would always save a portion of the notification traffic regardless how large the number of subscriptions grows [7] and the control traffic would also be smaller. In such scenarios, filtering can be superior to flooding for all numbers of subscriptions.

5   Related Work

Many self-stabilizing algorithms have been proposed for various kinds of scenarios whilst there are only a few contributions that cover publish/subscribe systems. In this area self-stabilization was first considered by Mühl [7]. This work was used as the basis for this paper. Recently, Shen and Tirthapura [11] presented an alternative approach for self-stabilizing content-based routing. In their approach, all pairs of neighboring brokers periodically exchange sketches of those parts of their routing tables concerning their other neighbors to detect corruption. The sketches that are exchanged are lossy because they are based on bloom filters (which are a generalization of hash functions). However, due to the information loss, it is not guaranteed that an existing corruption is detected deterministically. Hence, the algorithm is not self-stabilizing in the usual sense. Moreover, although generally all data structures can be corrupted arbitrarily, the authors' algorithm computes the bloom filters incrementally. Thus, once a bloom filter is corrupted, it may never be corrected. Furthermore, in their algorithm, clients do not renew their subscriptions. Without this, corrupted routing entries regarding local clients are never corrected. Finally, their algorithm is restricted to simple routing in its current form.

6   Conclusions and Outlook

To make publish/subscribe systems self-stabilizing, we applied a leasing mechanism ensuring that the routing tables are always refreshed in time provided that no faults occur. When faults do occur, the leasing mechanism ensures (a) that corrupted parts of routing tables are either corrected or removed and (b) that missing part are inserted. This way, routing tables recover. We described how flooding and simple routing can be made self-stabilizing. In both cases, we calculated the maximum stabilization time, i.e. the time the system needs to recover from an error. We also described how the stabilization time depends on the leasing period and how the refresh period must be chosen to ensure that in a correct system no routing entries expire. Furthermore, we sketched how advanced routing algorithms can be made self-stabilizing. Our contributions in this paper enable the designers of publish/subscribe systems to render their system self-stabilizing. Therefore, designers and implementers need not consider explicit fault management mechanisms if fault masking is not an issue.

Using a simulation we tested the effectiveness of our approach in an example scenario and showed, that it depends on the number of subscriptions in the system. In future work, it would be interesting to take an analytical approach to judge the proposed algorithms without employing simulations.

In this paper, we assumed for spatial reasons that the broker topology is statically stored in ROM. Currently, we work on an algorithm for a self-stabilizing broker topology which ensures the correct behavior of the system even if nodes or links are added or removed from the broker topology. Besides this, we are investigating self-organizing and self-optimizing algorithms for managing the broker topology. These management algorithms decide on which hosts brokers are started and to which neighbor brokers a broker connects.

References

1. A. Carzaniga. *Architectures for an Event Notification Service Scalable to Wide-area Networks*. PhD thesis, Politecnico di Milano, Milano, Italy, Dec. 1998.
2. A. Carzaniga, D. S. Rosenblum, and A. L. Wolf. Design and evaluation of a wide-area event notification service. *ACM Transactions on Computer Systems*, 19(3):332–383, 2001.
3. E. W. Dijkstra. Self-stabilizing systems in spite of distributed control. *Communications of the ACM*, 17(11):643–644, 1974.
4. S. Dolev. *Self-Stabilization*. MIT Press, 2000.
5. L. Fiege, G. Mühl, and F. C. Gärtner. Modular event-based systems. *The Knowledge Engineering Review*, 17(4):359–388, 2003.
6. C. Intanagonwiwat, R. Govindan, D. Estrin, J. Heidemann, and F. Silva. Directed diffusion for wireless sensor networking. *IEEE/ACM Transactions on Networking (TON)*, 11(1):2–16, 2003.
7. G. Mühl. *Large-Scale Content-Based Publish/Subscribe Systems*. PhD thesis, Darmstadt University of Technology, 2002. http://elib.tu-darmstadt.de/diss/000274/.
8. OMG. CORBA notification service, version 1.0.1. OMG Document formal/2002-08-04, 2002.
9. L. Opyrchal, M. Astley, J. Auerbach, G. Banavar, R. Strom, and D. Sturman. Exploiting IP multicast in content-based publish-subscribe systems. In *IFIP/ACM International Conference on Distributed Systems Platforms (Middleware 2000)*, volume 1795 of *LNCS*, pages 185–207. Springer-Verlag, 2000.
10. P. Pietzuch and J. Bacon. Hermes: A distributed event-based middleware architecture. In *In Proceedings of the 1st International Workshop on Distributed Event-Based Systems (DEBS'02)*, July 2002.
11. Z. Shen and S. Tirthapura. Self-stabilizing routing in publish-subscribe systems. In *3rd International Workshop on Distributed Event-Based Systems (DEBS 2004)*, Edinburgh, Scotland, UK, May 2004.
12. Sun Microsystems, Inc. Java Message Service (JMS) Specification 1.1, 2002.

A Checkpoint/Recovery Model for Heterogeneous Dataflow Computations Using Work-Stealing[*]

Samir Jafar[1,**], Thierry Gautier[1], Axel Krings[2], and Jean-Louis Roch[1]

[1] Laboratoire ID – IMAG, Pre-project MOAIS (CNRS-INRIA, INPG-UJF)
51, Avenue Jean Kuntzmann, 38330 Montbonnot St. Martin, France
{Samir.Jafar,Thierry.Gautier,Jean-Louis.Roch}@imag.fr
[2] Computer Science Dept., University of Idaho, Moscow, ID 83844-1010, USA
krings@cs.uidaho.edu

Abstract. This paper presents a new checkpoint/recovery method for dataflow computations using work-stealing in heterogeneous environments as found in grid or cluster computing. Basing the state of the computation on a dynamic macro dataflow graph, it is shown that the mechanisms provide effective checkpointing for multithreaded applications in heterogeneous environments. Two methods, *Systematic Event Logging* and *Theft-Induced Checkpointing*, are presented that are efficient and extremely flexible under the system-state model, allowing for recovery on different platforms under different number of processors. A formal analysis of the overhead induced by both methods is presented, followed by an experimental evaluation in a large cluster. It is shown that both methods have very small overhead and that trade-offs between checkpointing and recovery cost can be controlled.

1   Introduction and Background

Grid and cluster architectures are gaining in popularity for scientific computing applications. The distributed computations, as well as their underlying infrastructure consisting of a large number of computers, storage and networking devices, pose challenges in overcoming the effects of node and communication link failures. Since the computation times are often significant, effective fault-tolerance mechanisms are required to recover from faults in a fashion that avoids costly restarts.

Fault-tolerance is an effective method to address the possibility of faults in large systems. This is especially important in the case of grids and clusters since in the absence of fault-tolerance the probability of failure, and thus the unreliability of such architecture, increase with the number of components that can fail [21]. Recovery from faults imply the existence of redundancy, e.g. time, information, or spatial redundancy. In the case of large heterogeneous environments, redundancy mechanisms must address the specific requirements associated with recovery mechanisms, taking into account a dynamic number of possibly dissimilar computational nodes.

---

[*] This work is supported by the CNRS, ACI-GRID DOC-G and Région Rhône-Alpes project RAGTIME.
[**] The author is supported by Damascus University.

J.C. Cunha and P.D. Medeiros (Eds.): Euro-Par 2005, LNCS 3648, pp. 675–684, 2005.
© Springer-Verlag Berlin Heidelberg 2005

Many possible solutions based on fault-tolerance have been studied in the literature. Approaches based on duplication [15] can only tolerate a fixed number of faults. All other protocols are based on saving the state of the processes and on constructing a consistent global state [11], i.e. log-based and checkpoint-based protocols [5]. The various protocols can be compared based on three fundamental criteria. The first criterion is *coordination*, where processes coordinate each other in order to build a consistent global state at the time of checkpointing or recovery. The second is *heterogeneity*, which implies that the checkpoint state can be restored on a variety of platforms, e.g. node architecture or operating system. In the contrary, one speaks of homogeneity. The last criterion addresses the *scope of the recovery*, i.e. global or local recovery. If a single fault causes the roll-back of all processes in the application, one speaks of global recovery. Local recovery implies that only the roll-back of the crashed process is necessary.

We focus on roll-back strategies under consideration of crash faults and present two major mechanisms: *log-based* and *checkpoint-based* rollback-recovery.

1.1 Log-Based Protocols

Message logging [12] is based on the fact that a process can be modelled by a sequence of interval states, each one representing a non-deterministic event [16]. Under the hypothesis that each non-deterministic event can be identified, their logging allows a crashed process to be recovered by (1) restoring it to the initial state and (2) replaying messages to it in the same order they were delivered before the crash. To avoid a roll-back to the initial state of a process and to limit the amount of messages that need to be replayed, each process periodically saves its local state. Examples of systems based on this method include MPICH-V2 [4], and FTL-Charm++ [18]. For applications with extensive inter-process communication, log-based protocols are burdened with the possibly large overhead, with respect to space and time, induced by the logging of messages.

1.2 Checkpoint-Based Protocols

Checkpointing methods are based on periodically saving a global state [11] of the computation to stable storage. In case of a fault, the computation is restarted from one of these previously saved states. Checkpointing-based methods differ in the way processes are coordinated and on the interpretation of a consistent global state.

Coordinated checkpointing requires the coordination of all processes for building a consistent global state before writing the checkpoints to stable storage. The disadvantage of coordinated checkpointing is the large latency due to coordination in order to achieve a consistent checkpoint. Its advantage is the simplified recovery without roll-back propagation and minimal storage overhead, since there is only one checkpoint per process. This protocol is included in [6, 19].

Uncoordinated checkpointing assumes that each process independently saves its state and a consistent global state is achieved in the recovery phase [5]. The advantage of this method is that each process can make a checkpoint when its state is small. However, there are two main disadvantages. First, there is a possibility of rollback propagation which can result in a domino effect, i.e. rollback to the beginning of the computation. Second, the possibility of rollback propagation requires the storage of multiple checkpoints for each process.

Communication-induced checkpointing is a compromise between coordinated and uncoordinated checkpointing. To avoid a domino effect that can result from independent checkpoints of different processes, a consistent global state is achieved by forcing each process to take additional checkpoints based on some information piggybacked on the application messages [2]. The disadvantage of this approach is the possibly large number of forced checkpoints and the overhead associated with storing them.

There are only few approached supporting portability, multi-threading, local recovery and cost models [4, 13, 20]. However, portability of existing checkpointing tools is achieved by using portable languages like Java or by re-compilation to support heterogeneity [20], but not by the checkpointing mechanism itself.

2 Dataflow Work-Stealing for Grid Computations

Dataflow graphs [9] allow for a natural representation of a parallel execution, and they can be exploited to achieve fault-tolerance [1]. At runtime, ready-to-execute instructions are executed depending on the availability of data. Formally, a dataflow graph is a directed graph $G = (\mathcal{V}, \mathcal{E})$, where $\mathcal{V}$ is a finite set of vertices and $\mathcal{E}$ is a set of edges representing precedence relations between vertices. The vertex set consists of computational tasks, as seen in the traditional context of task scheduling, and the edge set represents the data dependencies between the tasks. Within the context of this research G is a dynamic dataflow graph, generated at runtime, as described in [7].

We adopt an efficient online scheduling algorithm called work-stealing [14]. The principle is simple, when a processor becomes idle it tries to *steal* work from other processors. In Cilk [14], a theoretical upper bound on the makespan is given for the case of *multithreaded computation*. This result was extended in [7] to our dynamic dataflow graph, and in [3] to consider heterogeneous systems.

2.1 Dataflow and Work-Stealing in KAAPI

The Kernel for Adaptive, Asynchronous Parallel Interface (KAAPI) used in this research is a C++ library that allows to program and execute multithreaded computations with dataflow synchronization between threads. The library is able to schedule programs at fine or medium granularity in a distributed environment.

In the KAAPI execution model a multi-processor system is viewed as a collection of so-called K-processors, which can be thought of as kernel threads. A process may consist of several K-processors. A K-processor in turn executes so-called K-threads, which can be thought of as application-level user threads. On a K-processor only one K-thread is active at a given time. The thread of control is a sequence of non-interruptible tasks. A K-processor becomes idle if there are no ready-tasks, i.e. either all tasks have finished execution or they are waiting for data as the result of synchronization. Under the work-stealing strategy, an idle K-processor tries to steal a task of a K-thread from a randomly selected K-processor called *victim*.

2.2 KAAPI Model Analysis

The KAAPI cost model will be the frame of reference for the analysis in Section 4. The time of a sequential execution of a program is denoted by T_1. It is the total time to execute all the operations in the computation on a single processor, with no scheduling overhead. Furthermore, let T_∞ be the execution time of the application as executed on an unbounded number of processors. Thus T_∞ represents the execution time associated with the critical-path.

For the execution of a KAAPI program on p identical physical processors, the corresponding execution time T_p is affected by T_1, T_∞ as well as the overhead associated with loading and managing the data-structures and scheduling using work stealing. We will adopt the simplified model of Cilk-5 [14], which utilizes Graham's bound [8], and is also valid for KAAPI. Then, T_p is bounded by (see Equation 2 in [14]):

$$\frac{T_1}{p} \leq T_p \leq \frac{T_1}{p} + c_\infty T_\infty \tag{1}$$

The constant c_∞ defines a bound on the overhead associated with the critical-path. In the remainder of the text, we assume that each physical processor executes only one K-processor.

3   Checkpoint/Recovery Model

Before describing the two fault-tolerance mechanisms we have integrated into KAAPI we need to define the *state of an execution*. This definition is perhaps the most important difference between this work and the related works (see Section 1) and is the basis for allowing checkpointing in a heterogeneous environment with the flexibility of recovery on any type or number of processors.

3.1   Definition of Execution State

We use a macro dataflow graph to define the state of the application's execution. The graph is a representation of the computational tasks to be carried out along with the associated data, which constitute the inputs and outputs. The dataflow is dynamic, changing during execution of the program, e.g. at the invocation of a task, and it is platform-independent. As a result the graph or portions of it can be moved across platforms during execution. Formally, at any instance of time t, the macro dataflow graph G describes a platform-independent, and thus portable, consistent global state of the execution of an application.

Whereas graph G is viewed as a single virtual dataflow graph, its implementation is in fact distributed. Specifically, each process i contains and executes a subgraph G_i of G. Within this representation lies the flexibility of restarting individual processes: in the case of a single fault, one does not have to perform a global roll-back. This is due to the fact that G_i, by definition of the principle of macro dataflow, contains all information necessary to identify exactly which data is missing. Note that this also includes the information associated with dependencies between G_i and G_j, $i \neq j$.

The instant of time at which a checkpoint can be taken is either before or after the execution of an application task. The checkpoint itself is a snapshot of G, which consists of tasks, specifically their function IDs, and their associated inputs. It does not consist of the task execution context itself. Understanding this difference between the two concepts is crucial. *Checkpointing a task* and its inputs simply requires to store the task's function ID and its input data. *Checkpointing the execution of a task* usually consists of storing the context of the processor, i.e. processor registers (such as program counters and stack pointers) and data. In the first case, it is possible to move a task and its inputs, assuming that both are represented in a platform-independent fashion. In the latter case the fact that the process context is platform-dependent requires a homogeneous system in order to perform a restore operation.

The checkpointed macro dataflow graph contains only the future of the execution, i.e. the tasks to be carried out and the necessary data. Certain temporary data associated with the execution of the task are not necessary to the future of the execution and are not checkpointed. The result is a reduction of the checkpoint size.

3.2 Systematic Event Logging

Systematic Event Logging (SEL) is derived from a log-based method [12]. Only the state-change events, i.e. additions and deletions of nodes in the macro dataflow graph, are logged. A recovery consists of simply loading and rebuilding subgraph G_i associated with the failed process i from the respective checkpoint file. The advantage of this approach is that during recovery it allows the re-execution of single tasks, which is interesting for applications requiring the certification of computations and results [10].

In the implementation of SEL, the events that trigger the change of the state of the macro dataflow graph are either the creation or deletion of tasks or the data dependencies they produce. Recall that tasks and data dependencies constitute the two principal components of the graph. These events, together with a uniquely assigned identifier allowing their association with the node in the graph, are stored in stable storage.

3.3 Theft-Induced Checkpointing

Theft-induced checkpointing (TIC) is based on the method presented in [2]. The creation of checkpoints can be initiated (1) at specific checkpointing periods or (2) by the theft of a task. In the first case, checkpoints of the macro dataflow graph G, i.e. G_i on process i, are stored periodically[1] at pre-defined periods τ. In the second case, the state of the macro dataflow graph is checkpointed as the result of communication between processes. In the presence of work stealing, each theft will cause such communication, thus resulting in a so-called *forced* checkpoint. The communication due to work stealing accounts for the only communication of the application. The checkpoint is generated at the time of a theft operation. A recovery consists of loading G_i from the checkpoint file related to the crashed process.

Recall that in KAAPI a process executes on a collection of K-processors, which in turn execute a certain number of K-threads. In the implementation of TIC in KAAPI the checkpoint of a process is implemented by checkpointing its associated K-processor. Each K-processor generates incremental checkpoints for each associated K-thread. At the expiration of period τ, each process checkpoints its state represented by G_i. In case of a task theft, only the K-processor from which the task was stolen forces a checkpoint.

---

[1] Recall that checkpointing is performed at the task level. This should not be confused with preemptive periodic scheduling, where the context of the preempted tasks are stored.

4 Model Analysis

In the analysis of the overhead associated with SEL and TIC we differentiate between executions without and with faults. Furthermore, we assume that $\frac{T_1}{p} \gg c_\infty T_\infty$, which will be referred to as the *parallel slackness assumption* [14]. In the presence of work-stealing this leads to a linear speedup of $T_p \approx \frac{T_1}{p}$.

4.1 Analysis of Fault-Free Execution

If we add a checkpointing mechanism, it is of special interest to analyze its overhead associated with fault-free execution, since the occurrence of faults is considered to be the rare exception rather than the norm.

Analysis of SEL: In SEL a log is initiated for each node created. Thus the overhead associated with logging depends on dataflow graph G. Let T_P^{SEL} denote the execution of a KAAPI program on p processors under consideration of logging overhead. Then,

$$T_P^{SEL} \leq \frac{T_1^{SEL}}{p} + c_\infty T_\infty^{SEL}. \tag{2}$$

T_∞^{SEL} denotes the critical-path under SEL, where $T_\infty \approx T_\infty^{SEL}$. Furthermore, T_1^{SEL} denotes the time of a sequential execution of a program under consideration of the overhead induced by logging, i.e. $T_1^{SEL} = T_1 +$ *logging overhead*. This overhead is a function of two parameters. First, it depends on the size of G. Specifically, it depends on the number of tasks and data dependencies, as well as the size of the latter. Second, it depends on the time of an elementary access to stable storage, denoted by t_s. Therefore,

$$T_1^{SEL} = T_1 + f_{overhead}^{SEL}(|G|, t_s). \tag{3}$$

The real measure of SEL overhead is thus $T_1^{SEL} - T_1$, which in turn allows the derivation of the overhead in the parallel execution, i.e. $T_P^{SEL} - T_p$.

Analysis of TIC: In theft-induced checkpointing, a checkpoint is performed periodically for each process, as dictated by period τ, and as the result of task stealing. Let T_P^{TIC} denote the execution of a KAAPI program on p processors under TIC. Thus,

$$T_P^{TIC} \leq T_p + \max_{i=1,\ldots,p} \{CheckpointOverhead_i\}, \tag{4}$$

where $CheckpointOverhead_i$ denotes the total checkpointing overhead on processor i. This overhead depends on the total number of checkpoints taken on processor i and the overhead of a single checkpoint. The maximal number of checkpoints performed by a processor is $[T_P^{TIC}/\tau + N_{theft}]$, where T_P^{TIC}/τ indicates the number of checkpoints due to period τ and N_{theft} is the maximal number of thefts performed by any processor.

The overhead of a single checkpoint in TIC is different from that in SEL, since now the checkpoint constitutes the collection of tasks in G_i, rather than a single task.

The number of tasks in G_i has an upper bound of N_∞, which denotes the maximum number of tasks in a path of G [14]. The checkpoint overhead is thus bound by $\max_{i=1,...,p}\{CheckpointOverhead_i\} \leq [T_P^{TIC}/\tau + N_{theft}]\, f_{overhead}^{TIC}(N_\infty, t_s)$. Note that function $f_{overhead}^{TIC}()$, which indicates the overhead associated with a single checkpoint, depends only on G, or more preceisely N_∞ and t_s. Thus, the checkpointing overhead is

$$T_P^{TIC} \leq T_p + [T_P^{TIC}/\tau + N_{theft}]\, f_{overhead}^{TIC}(N_\infty, t_s). \tag{5}$$

Under the parallel slackness assumption, it is important to note that the number of thefts, N_{theft}, resulting in forced checkpoints is bound and small for many applications [14, 17]. Then, by selecting an appropriate τ the number of local checkpoints can be adjusted in order to obtain $T_P^{TIC} \approx T_p$.

4.2   Analysis of Executions Containing Faults

The overhead associated with fault-free execution is the penalty one pays for having a recovery mechanism. It remains to be shown how much overhead is associated with recovery as the result of a fault and how much execution time can be lost under different strategies.

The overhead associated with recovery is due to loading and rebuilding G. This can be effectively achieved by loading G_i of the affected processes. The time depends on the size of G_i and is dominated by the size of the data representing the task inputs. Thus, the time of recovery of a single process i, denoted by $t_{recovery}^i$, depends only on the size of its associated subgraph G_i. Therefore, $t_{recovery}^i$ is of the order of the size of the subgraph, i.e. $t_{recovery}^i = O(|G_i|)$. Note that for a global recovery, as the result of the failure of the entire application, this translates to $max(t_{recovery}^i)$ and not to $\sum t_{recovery}^i$.

The advantage of SEL is that, due to its fine granularity, the maximum amount of execution time lost is that of a single task. Furthermore, the rollback only requires the recovery overhead of a single task. However, this comes at the cost of higher logging overhead, as was addressed in Equation 3.

For TIC the maximum amount of lost execution time is generally higher than for SEL and is bound by period τ. The recovery overhead depends on the size of the graph that need to be loaded and rebuilt. However, note that by appropriately selecting τ one can exercise control over the recovery overhead. The trade-off between lower checkpointing overhead and slower recovery will need to be determined by the application.

It should be noted that we do not consider the time lost due to fault-detection. Whereas the fault-detection time is an important issue, its impact is not the subject of this research; any detection mechanism may be used.

5   Experimental Results

The performance and overhead of the SEL and TIC mechanisms were experimentally determined for the *Quadratic Assignment Problem* (instance[2] NUGENT 22) which was

---

[2] see http://www.opt.math.tu-graz.ac.at/qaplib/

parallelized in KAAPI. The experiments were conducted on the iCluster2[3]. The cluster consists of 104 nodes interconnected by a 100Mbps Ethernet network. Each node features two Itanium-2 processors (900 MHz) and 3 GB of local memory.

In order to take advantage of the distributed fashion of the checkpoint, i.e. G_i, each processor keeps a local copy of its checkpoint. To eliminate this single source of failure, it is assumed that the checkpoint of each G_i is replicated on other nodes [6]. This configuration has two advantages. First, it reflects the theoretical assumptions of the previous section and second, the actual overhead of the checkpointing mechanism is measured, rather than the overhead associated with a centralized checkpoint server.

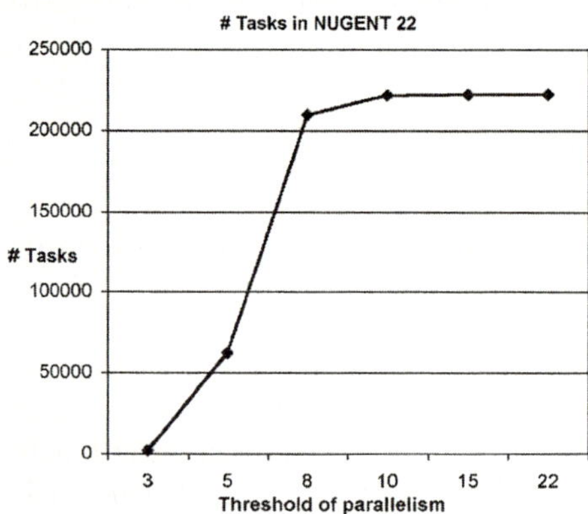

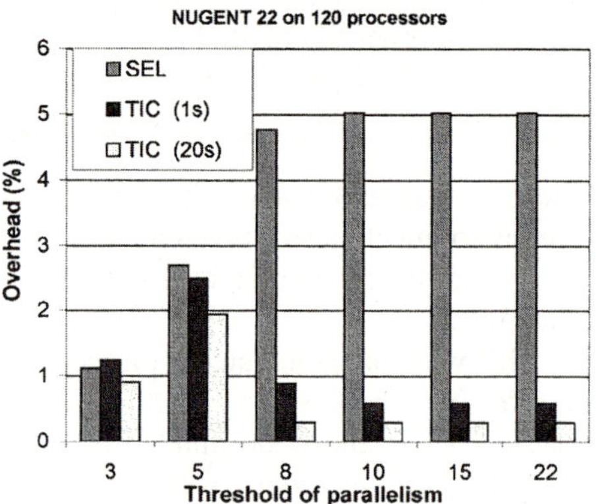

Fig. 1. Impact of threshold of parallelism   **Fig. 2.** Checkpoint overhead

The application recursively generates tasks and the degree of parallelism can be adjusted. After a given depth of recursion no more tasks are generated. The sequential execution time without KAAPI was 34,695 seconds. With KAAPI, at fine grain (threshold ≥ 10), the execution on a single processor generated 225,195 tasks and ran in 34,845 seconds. The impact of the degree of parallelism can be seen in Figure 1 and 2. The number of parallel tasks generated for different thresholds of parallelism is shown in Figure 1. The degree of parallelism increases drastically for threshold 5 and approaches its maximum at threshold 10.

The number of tasks has direct implications on the cost of the checkpointing mechanism. Figure 2 shows that the cost of SEL is very susceptible to the total number of tasks, as predicted by Equation (2) and (3), which showed the overhead as a function of the number of tasks.

Figure 2 also shows the impact of parallelism on the overhead of TIC for period τ equal to 1 and 20 seconds. As shown in Equation 5, the overhead is dependent on the critical-path, i.e. T_∞, and N_∞. As parallelism increases, and thus both T_∞ and N_∞ drastically decrease, the checkpointing overhead is reduced. As predicted, the longer period results in lower cost.

---

[3] http://www.inrialpes.fr/sed/i-cluster2

Figure 3 demonstrates that the checkpoint mechanisms as well as the application are scalable. As the number of processors increase, the different protocols show little change in cost. The impact of the number of faults on the cost of recovery for SEL can

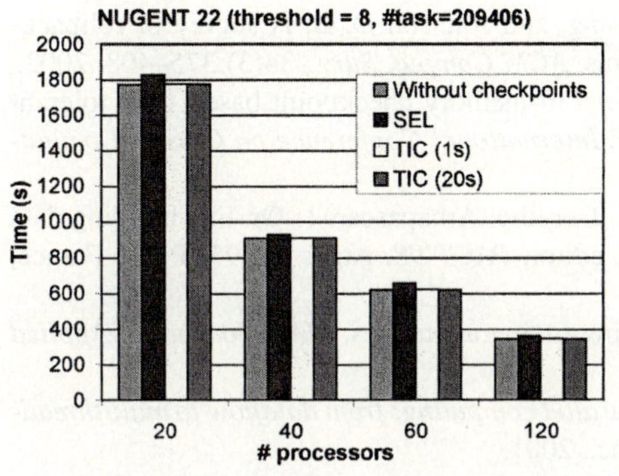

Fig. 3. Scalability of the checkpointing   **Fig. 4.** Recovery cost for SEL

be seen in Figure 4. In fact, the overhead due to restart increases linearly. The recovery times are derived from two measures. The first is the computation time averaging 0.25s per computational task. The second is the overhead due to loading the checkpoint file, averaging 7 MBytes, for rebuilding each G_i. Since single process roll-back was hardly measurable, the experiment shows faults and restarts of all processors.

6   Conclusions

Two portable fault-tolerance mechanisms, systematic event logging and theft-induced checkpointing, have been introduced for heterogeneous multithreaded applications. The flexibility of macro dataflow graphs has been exploited to allow for a platform-independent description of the application state. This description resulted in flexible, portable, recovery strategies. Systematic event logging allowed for rollback at lowest level of granularity, with a maximal computation loss of one task. However, its overhead was sensitive to the size of the application graph, i.e. the number of tasks. Theft-induced checkpointing has lower overhead, related to work-stealing, which was shown bound to the critical-path. The experimental results demonstrated low overhead of both approaches and confirmed the theoretical analysis.

References

1. M. Hyett A. Nguyen-Tuong, A. S. Grimshaw. Exploiting data-flow for fault-tolerance in a wide-area parallel system. In *Proceedings 15 th Symposium on Reliable Distributed Systesm*, pages 2–11, 1996.
2. R. Baldoni. A communication-induced checkpointing protocol that ensures rollback-dependency trackability. In *Proceedings of the 27th International Symposium on Fault-Tolerant Computing (FTCS '97)*, page 68. IEEE Computer Society, 1997.

3. M. Bender and M. Rabin. Online scheduling of parallel programs on heterogeneous systems with applications to cilk, 2002.

4. A. Bouteiller, F. Cappello, T. Hérault, P. Lemarinier, G. Krawezik, and F. Magniette. Mpich-v2: a fault tolerant mpi for volatile nodes based on the pessimistic sender based message logging. In *SuperComputing*, Phoenix, USA, 2003.

5. E. N. Mootaz Elnozahy, L. Alvisi, Y.-M. Wang, and Johnson D. B. A survey of rollback-recovery protocols in message-passing systems. *ACM Comput. Surv.*, 34(3):375–408, 2002.

6. L. V. Kalé G. Zheng, L. Shi. Ftc-charm++: An in-memory checkpoint-based fault tolerant runtime for charm++ and mpi. In *2004 IEEE International Conference on Cluster Computing*, San Dieago, CA, September 2004.

7. F. Galilée, J.-L. Roch, G. Cavalheiro, and M. Doreille. Athapascan-1: On-line building data flow graph in a parallel language. In IEEE, editor, *PACT'98*, pages 88–95, Paris, France, October 1998.

8. Ronald L. Graham. Bounds on multiprocessing timing anomalies. *SIAM Journal of Applied Mathematics*, 17(2):416–429, 1969.

9. T. Ungerer J. Silc, B. Robic. *Asynchrony in parallel computing: from dataflow to multithreading*, pages 1–33. Nova Science Publishers, Inc., 2001.

10. S. Jafar, S. Varrette, and J.-L. Roch. Using data-flow analysis for resilience and result checking in peer-to-peer computations. In *IEEE DEXA'2004*, Zaragoza, Spain, August 2004.

11. L. Lamport K. M. Chandy. Distributed snapshots: determining global states of distributed systems. *ACM Trans. Comput. Syst.*, 3(1):63–75, 1985.

12. K. Marzullo L. Alvisi. Message logging: Pessimistic, optimistic, causal and optimal. *TSE*, 24(2):149–159, 1998. Transactions on Software Engineering.

13. M. Litzkow, T. Tannenbaum, J. Basney, and M. Livny. Checkpoint and migration of unix processes in the condor distributed processing system. Technical Report CS-TR-97-1346, Univ. Wisconsin, Madison, 1997.

14. K. H. Randall M. Frigo, C. E. Leiserson. The implementation of the cilk-5 multithreaded language. In *Proceedings of the ACM SIGPLAN 1998 conference on Programming language design and implementation*, pages 212–223. ACM Press, 1998.

15. A .Schipper M.Wiesmann, F. Pedonne. A systematic classification of replited database protocols based on atomic broadcast. *In Proceedings of the 3th European Research Seminar on Advances in Distributed Systems(ERSADS99)*, pages 351–360, 1999.

16. S. Yemini R. Strom. Optimistic recovery in distributed systems. *ACM Trans. Comput. Syst.*, 3(3):204–226, 1985.

17. R. Revire. *Ordonnancement de graphe dynamique de tâches sur architecture de grande taille. Régulation par dégénération séquentielle et distribuée*. Thèse de doctorat en informatique, INPG, septembre 2004.

18. L. V. Kale S. Chakravorty. A fault tolerant protocol for massively parallel machines. In *FT-PDS Workshop for IPDPS 2004*. IEEE Press, 2004.

19. G. Stellner. CoCheck: Checkpointing and Process Migration for MPI. In *Proceedings of the 10th International Parallel Processing Symposium (IPPS '96)*, Honolulu, Hawaii, 1996.

20. V. Strumpen. Compiler technology for portable checkpoints. Technical Report MA-02139, MIT Laboratory for Computer Science, Cambridge, 1998.

21. K. S. Trivedi. *Probability and Statistics with Reliability, Queuing, and Computer Science Applications*. John Wiley and Sons, New York, 2001.

Topic 9
Parallel Programming:
Models, Methods and Languages

Marco Danelutto, Denis Caromel, Duane Szafron, and Fernando Silva

Topic Chairs

This topic covers innovative aspects as well as improvements in already known techniques in algorithms, programming models, design methods and languages that relate to the development of parallel programs. In the call-for-papers, we stressed several innovative aspects including novel techniques to assemble parallel software from reusable parallel components or from existing sequential code without compromising efficiency, and techniques to adapt parallel software to available resources as well as to the features of the problem being solved. A total of 36 papers were submitted for this topic and, after reviewing, 10 full papers were accepted (28% rate). We recognized really promising work in many of the papers that could not be accepted and fully encourage the authors to use the referees' suggestions to improve and resubmit their work. The accepted papers will be presented at the conference grouped into four sessions.

The paper by Wise et. al. discusses an innovative approach to the implementation of parallel matrix algorithms and will be presented in the session hosting papers from Topic 4.

A second session hosts papers dealing with parallelism in the context of shared memory machines. Chan et. al. focus on the implementation of asynchronous handlers in co-begin statements in the context of a Java implementation of SR named JR. Wang et. al. describe the design and use of source-level streaming pre-computation techniques to improve the performance of memory-bound scientific applications on SMT processors with limited resources. Nieplocha et. al. deal with the implementation of Cray symmetric objects in Fortan95.

A third session hosts papers discussing the usage of aspect-oriented techniques and of search parallelism. Carvalho Junior and Dueire Lins explore the possibilities offered by aspect-oriented programming to incorporate procedural language computations in the $Haskell_{\#}$ implementation of the $\#$ programming model. Copy and Ur exploit aspect-oriented programming techniques in the implementation of testing tools. Last, Viet Le and Pontelli describe the implementation of a parallel Answer Set solver using search parallelism.

The last session hosts papers related to structured parallelism. Benoit et. al. discuss how skeletons can be exploited using the eSkel. Aldinucci et. al. describe experiments in automatic adaptation of structured parallel code to changes in the target architecture features. González-Escribano et. al. describe an XML based intermediate representation for nested-parallel programming languages.

J.C. Cunha and P.D. Medeiros (Eds.): Euro-Par 2005, LNCS 3648, p. 685, 2005.
© Springer-Verlag Berlin Heidelberg 2005

A Paradigm for Parallel Matrix Algorithms:[*]
Scalable Cholesky

David S. Wise[**], Craig Citro[***], Joshua Hursey[†],
Fang Liu[†], and Michael Rainey[‡]

Indiana University, Bloomington

Abstract. A style for programming problems from matrix algebra is developed with a familiar example and new tools, yielding high performance with a couple of surprising exceptions. The underlying philosophy is to use block recursion as the exclusive control structure, down to a $2^p \times 2^p$ base case anyway, where hardware favors iterative style to fill its pipe. Use of Morton-ordered matrices yields excellent locality within the memory hierarchy—including block sharing among distributed computers. The recursion generalizes nicely to an SPMD program where such sharing is the only communication.

Cholesky factorization of an $n \times n$ SPD matrix is used as a simple non-trivial example to expose the paradigm. The program amounts to four functions, two of which are finalizers for the other two. This insight allows final blocks to be shared with inter-node communication $\in \Theta(n^2)$ for this algorithm $\in \Theta(n^3)$ FLOPs.

CCS Categories and subject descriptors: C.1.2 [Processor Architectures]:Multiprocessors—Single-program, multiple-data-stream processors (SPMD); D.1.m[Programming Techniques]: Miscellaneous; E.1 [Data Structures]: Arrays; E.2 [Data Storage Representations]: contiguous representations; F.2.1 [Analysis of Algorithms and Problem Complexity]: Numerical algorithms and problems–computations on matrices.
General Term: Design, Languages, Performance
Additional Key Words and Phrases: quadtrees, Cholesky factorization, Morton order, finalizer.

1   Introduction

A methodology for portable, scalable algorithms for linear algebra is developed on divide-and-conquer paradigm. Performance for Cholesky factorization on a

---

[*] Supported, in part, by the National Science Foundation under grants numbered CCR-0073491, ACI–0219884, and EIA–0202048. Copyright on twelve pages intact transferred, with rights reserved for anyone to make digital or hard copies of part or all of this work for personal or classroom use, provided that copies are not made or distributed for profit or commercial advantage and that copies bear this notice and the full Springer citation on the first page. Rights are similarly reserved for any library to share a hard copy through interlibrary loan.
[**] Supported, in part, by NSF grants CCR–0073491 and ACI–0219884.
[***] Supported, in part, by NSF grant CCR-0107395.
[†] Supported, in part, by NSF grant number ACI–0219884.
[‡] Supported, in part, by NSF grant CCR03-34593.

J.C. Cunha and P.D. Medeiros (Eds.): Euro-Par 2005, LNCS 3648, pp. 687–698, 2005.
© Springer-Verlag Berlin Heidelberg 2005

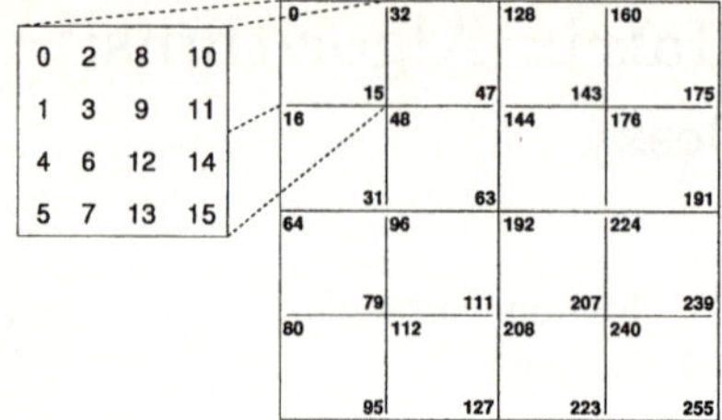

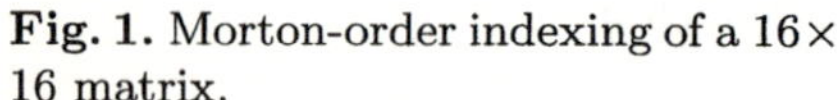

Fig. 1. Morton-order indexing of a 16×16 matrix.

Fig. 2. Ahnentafel indexing of an order-4 matrix.

cluster of eight Xeon-powered nodes is presented; it is remarkable for a couple of reasons beyond the clean coding. First, the performance answers direct challenges to the performance of Morton-order representation [1, 2]. This paradigm brings C-coded performance on those matrices very close to that of Intel's hand-coded BLAS library [3]. Second, the recursive style does so well with locality of reference in MPI multiprocessing that the burden of interprocessor communication disappears—with cheap ol' Ethernet, that is. This eight-node multiprocessor also has Infiniband, Myrinet, and Quadrics interconnects, where the tests perform surprisingly irregularly. Only Ethernet and Quadrics appear here.

Recursion is the essential tool for divide-and-conquer. The paradigm uses recursive data structures for locality, recursive programming to develop a partitioned algorithm, and SPMD recursion to balance runtime parallelism without any extra communication. The only communication for this $\Theta(n^3)$ algorithm is the $\Theta(n^2)$ sharing of blocks as they are finalized.

The underlying motivation for this paper is this straightforward paradigm of programming, yielding an understandable, portable code. We show how excellent performance arises from good locality on a uniprocessor, and demonstrate its support for distributed multiprocessing.

Three tools are developed as part of this paradigm. The first is Morton ordering for matrix representation [4, 2, 5, 1], with Ahnentafel indexing for the control of block-recursive programs [6, 7]. Figure 1 illustrates how it can represent a dense, two-dimensional array of any size in a contiguous block of address space. The properties of Morton ordering provide simple bounds checking, allow for several indexing schemes—including Cartesian indices via dilated integers [8, 9]—and guarantee that (for all p) any block of order 2^p at an address which is a multiple of 4^p resides in a block of contiguous addresses. Rectangular matrices are handled easily because Morton indices are monotonic across a row and down a column. Importantly, the matrix may occupy **address** space that is much bigger than the minimum to hold dense data. Just as importantly, however, blocks of unused addresses reside permanently in the most remote levels of the memory hierarchy—for the largest matrices as unallocated sectors on swapping disk.

A close relative of Morton-ordered arrays is Ahnentafel indexing, illustrated in Figure 2. Their difference amounts to two high-order bits that make all block indices unique, and still allow masking off the even or odd bits into dilated integers that are cartesian indices. Thereby, bounds checking at all levels of the tree becomes straightforward. Instead of one pair of bounds, a vector of precomputed bounds is indexed by the level of the quadtree.

The second tool is recursive divide-and-conquer for the matrix operations [10, 1, 7, 11]. Section 2 illustrates the development of one for Cholesky factorization. An early lesson from such algorithms is to attenuate recursion above the 1×1 base cases because the last few levels are exponentially expensive, as Spieß observed long ago for Straßen's algorithm [12]. The base cases are expanded in Section 3.

The third tool, developed directly from the recursive algorithm, is the single-program–multiple-data (SPMD) recursion for multiprocessing. All processors execute exactly the same program—in this case, four recursive functions—that seem to synchronize their recursions even though most remain idle relative to work high in the process tree. The processors, themselves, are arranged as a binary tree (Figure 7) that restricts communication to parent-child links, and the control amounts to a parallel descent into this tree by each of the 2^p processors. One can envision each descending a different path p levels into the computation tree, to the level where each node has a singleton processor executing uniprocessor code.

Significantly, the algorithm is arranged so that the processor tree is an overlay of the quadtree that is the data structure, so that each of those singletons finds its work in a local chunk of memory. Interestingly, the bulk of the work (rank-k updates) can be performed in local memories without any communication at all. The aggregate $\Theta(n^3)$ processing on an order-n matrix can be carried out without any communication. Only after these results are finalized need they be shared, and so that sharing is deferred—but then that finalization and communication is only $\in \Theta(n^2)$. The sharing is explained in Section 4.

A few definitions are necessary for context, but lots of the details are available in citations.

Definition 1. [6] *The base of a matrix has* Morton-order *index 0. A submatrix (block) at Morton-order index i is either an element (scalar), or it is composed of 4 submatrices, with indices $4i + 0, 4i + 1, 4i + 2, 4i + 3$, labeled* northwest, southwest, northeast, *and* southeast.

Definition 2. [6] *An entire matrix has* Ahnentafel *index 3. A submatrix (block) at Ahnentafel index i is either an element (scalar), or it is composed of 4 submatrices, with Ahnentafel indices $4i + 0, 4i + 1, 4i + 2, 4i + 3$.*

Morton order is used to represent an entire matrix, whether it is rectangular or square, and regardless of its order. Ahnentafel indices are used for control; conversion to or from Morton order is easy, and simple bounds checking is available with either one [6]. In rectangular graphics that are wider than tall Morton order is often rendered in Z order. Because matrices tend to be taller than they are wide, we use И order. They are equivalent; both allow dilated integers to be used for cartesian indices and bounds checking.

```
#define evenBits (((unsigned int)-1) /3)          #define e2w(a)    ((a) -2)
#define oddBits     (evenBits<<1)                  #define prnt(a)   ((a)>>2)
#define diag(a) (3*((a) & evenBits))
                                                   #define quadBd(soloBound)     (soloBound)
#define nw(a)    ( (a)*4    )                       #define rectBd(soloBound) (4*(soloBound))
#define sw(a)    ( (a)*4  +1)                       #define square    1
#define ne(a)    ( (a)*4  +2)                       #define rectangle 2
#define se(a)    ( (a)*4  +3)
                                                   #define amLeftChild(me,lgProcs)  \
#define we(a)    (((a)<<2)+0)                          (((me)&(1<<(lgProcResource-(lgProcs)-1))) ==0)
#define ea(a)    (((a)<<2)+2)                           /*  Is "me" a left child at this level of the tree?
#define no(a)    ( (a)    +0)                               Root 0 is understood leftChild by default. */
#define so(a)    ( (a)    +1)                       #define child(procRank,lgProcs)  \
                                                      ( (procRank)+(1<<(lgProcResource-(lgProcs)  )) )
                                                   #define parent(procRank,lgProcs) \
                                                      ( (procRank)-(1<<(lgProcResource-(lgProcs)-1)) )
```

Fig. 3. Helpful macros for Morton indexing.

Theorem 1. [4, 13] *The Morton index into a matrix (2-dimensional array) is* $\sum_{\ell=0}^{w-1} q_\ell 4^\ell = 2 \sum_{\ell=0}^{w-1} i_\ell 4^\ell + \sum_{\ell=0}^{w-1} j_\ell 4^\ell$ *corresponds to the cartesian index for* *row* $i = \sum_{\ell=0}^{w-1} i_\ell 2^\ell$ *and column* $j = \sum_{\ell=0}^{w-1} j_k 2^\ell$.

The three strengths of such indexing, as in real estate, is their inherent locality, locality, locality. Base cases, caches, RAM load, disk pages, and interprocessor communication all take advantage of the sequential storage of blocks of all sizes. And any block can be sent as an unbuffered stream.

The remainder of this paper is in five parts. The next section develops the outline of a recursive Cholesky factorization. Section 3 visits fast base cases for the recurrence. Then Section 4 expands that code toward the parallel implementation. Section 5 describes the times for the resulting algorithm with different MPI and interconnects. Finally, Section 6 offers conclusions.

2   Block-Recursive Cholesky

2.1   Some Macros

Macros that are used to orient the quadtrees and their processes appear in Figure 3. The reader is referred to the literature for basic operations on Morton indices and the important role of dilated integers [8, 9, 6]. Masking the even or odd bits from a Morton or Ahnentafel index yields an (even or odd) dilated integer to the row or column, respectively. Higher in the quadtree, those identify stripes of contiguous rows for bounds checking [6].

After cleaving a square block into quadrants, they are labeled by points of the compass (nw, sw, ne, se). The binary tree of processes follows this cleaving, as well. Results of processes are first split west/east into rectangles and then north/south into quadrants again.

The last few macros answer questions controlling parallelism in the processor tree:

- Is this process (rank) a left child at this level?
- What is the right child of a process at this level?
- Which is the parent of a process at this level?

```
static void doCholeskyBlk(int quad)
{
  if ( quad >= soloBound ) doCholeskyUniproc(quad);
  else
  {
    doCholeskyBlk(                          nw(quad));
    triSolveBlk  (              sw(quad), nw(quad));
    schurTri     (/*se(quad),*/ sw(quad)         );
    doCholeskyBlk(  se(quad)                      );
  }
  return;
}


static void triSolveBlk(int sQuad, int nQuad)
{
  if ( sQuad>=soloBound ) triSolveUniproc(sQuad, nQuad);
  else
  {
    triSolveBlk(so(  we(sQuad)), no(we(nQuad)) );     /*D*/
    triSolveBlk( no(we(sQuad)), no(we(nQuad)) );      /*A*/

    schurBlk   (so(  ea(sQuad)),     so(we(sQuad)) ); /*E*/
    triSolveBlk(so(  ea(sQuad)), so(ea(nQuad)) );     /*F*/
    schurBlk   ( no(ea(sQuad)),   no  (we(sQuad)) ); /*B*/
    triSolveBlk( no(ea(sQuad)), so(ea(nQuad)) );      /*C*/
  }
  return;
}


static void schurBlk(int eQuad, int wQuad)
{
  if ( eQuad >= soloBound ) schurBlkUniproc(eQuad, wQuad);
  else
  {
    schurBlk( so(we(eQuad)), e2w(so(ea(wQuad))) );
    schurBlk( so(we(eQuad)),     so(ea(wQuad))  );

    schurBlk( no(we(eQuad)), e2w(no(ea(wQuad))) );
    schurBlk( no(we(eQuad)),     no(ea(wQuad))  );

    schurBlk( so(ea(eQuad)), e2w(so(ea(wQuad))) );
    schurBlk( so(ea(eQuad)),     so(ea(wQuad))  );

    schurBlk( no(ea(eQuad)), e2w(no(ea(wQuad))) );
    schurBlk( no(ea(eQuad)),     no(ea(wQuad))  );
  }
  return;
}
static void schurTri(              int wQuad)
{
  if ( wQuad >= soloBound ) schurTriUniproc(wQuad);
  else
  {
    schurBlk( sw(diag(prnt(ea(wQuad)))), e2w(so(ea(wQuad))) );
    schurBlk( sw(diag(prnt(ea(wQuad)))),     so(ea(wQuad))  );
    schurTri(                            e2w(no(ea(wQuad))) );
    schurTri(                                no(ea(wQuad))  );

    schurTri(                            e2w(so(ea(wQuad))) );
    schurTri(                                so(ea(wQuad))  );
  }
  return;
}
```

Fig. 4. C code for the rudimentary four functions.

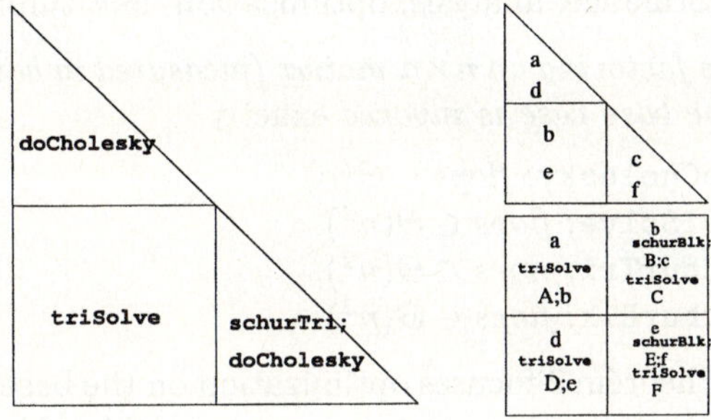

Fig. 5. Recursive decomposition of `doCholesky` and `triSolve`. In the latter, lower-case letters identify argument and upper-case indicate results of the six recursive calls.

2.2   The Algorithm

One finds iterative codes in textbooks covering Cholesky factorization. The simplicity of three nested loops is intoxicating, but it doesn't help much in balancing and scheduling processes. Good recursions can be found by abstracting the problem to be of size $2^p \times 2^p$. With that approach one arrives at the recursive algorithm in Figure 4.

Figure 5 sketches the recursion for two of the four functions presented in Figure 4. The other two are rank-k updates called `schurBlk` for a square result, and `schurTri` when the result lands on the diagonal of the symmetric matrix; it's half the work of `schurBlk`.

In most cases, the first parameter identifies the quadrant receiving an update (or side-effect) by the named code. Similarly, the left of Figure 5 identifies the blocks of the top-level matrix by the functions that update them. The computationally intensive `schurBlk`, however, does not yet appear.

The right of Figure 5 illustrates the source of all its $\Theta(n^3)$ calls: `triSolve`. There, lowercase letters indicate arguments to the six calls identified by right-margin comments in Figure 4; upper case indicates their in-place effects. Not all three arguments to, say, `schurBlk` are required as its parameters—because two Ahnentafel indices suffice to compute the third. That is also true of `triSolve`, itself; the Ahnentafel index, `a`, of a block labeled A, D, C, or F is sufficient to determine the index of the diagonal block immediately above it (*e.g.* `(a&oddBits)/2*3`) —so the latter does not appear later in Figure 12. Similarly, `schurTri` in Figure 4 only has one parameter that does *not* index the side-effected block, whose index is calculated from that of the block A, D, C, or F in Figure 5 using `diag(a)` from Figure 3. We have found that deriving such dependent indices with dilated integers avoids a lot of incompatibility errors.

3   Base Cases

Excellence performance at the base cases is essential for high performance of recursive programs; anything less has an explosive cost. Three steps are necessary to achieve that performance: analysis, optimization, and tuning.

Theorem 2. *While factoring an $n \times n$ matrix (measured in base blocks—whether 1×1 or 32×32) the base case is invoked* exactly

- $\binom{n}{1}$ *times for* `doCholesky`; *flops* $\in \Theta(n)$.
- $\binom{n}{2}$ *times for* `triSolve`; *flops* $\in \Theta(n^2)$.
- $\binom{n}{2}$ *times for* `schurTri`; *flops* $\in \Theta(n^2)$.
- $\binom{n}{3}$ *times for* `schurBlk`; *flops* $\in \Theta(n^3)$.

The analysis of Theorem 2 focuses optimization on the base case of `schurBlk`. C and FORTRAN optimizers, however, are not set up for Morton order, so we wrote a macro for that Schur complement to generate code for RISC technology. With f and p as integer tuning parameters, it generates iterative C code for a $2^p \times 2^p$ base block with 2^f in-line flops; choice of p depends on data cache, and of f on instruction cache, decoding, and the pipeline. Experiments a range of values for a large problem select a 32×32 base block and 256 inline flops. All source code is in C; the only assembly code was used to gain SSE2 performance over the compiler's scalar code.

Figure 6 presents the uniprocessor timings that result from this tuning on the coding techniques described elsewhere [6, 7]. Consistently with presentations of other cache-oblivious algorithms [10], it plots time to perform an order-n Cholesky factorization divided by the $\frac{1}{3}n^3$ FLOPs necessary for the algorithm. In other words, it normalizes to a hypothetical leading coefficient for the cubic equation that would express the time as a function of n [14], which ought to be constant with good scaling.[1] Read from top-to-bottom, Figure 6 plots the quadtree divide-and-conquer C-codes extended from Figure 5 running on the usual

---

[1] This plot is valuable also because it exposes relative performance on small tests.

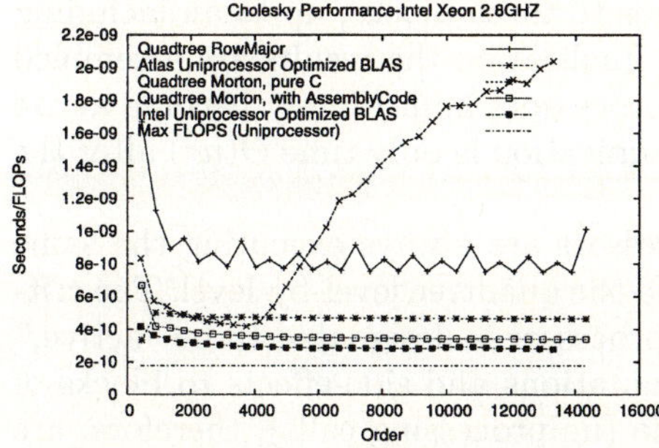

Fig. 6. Tuned uniprocessor performance.

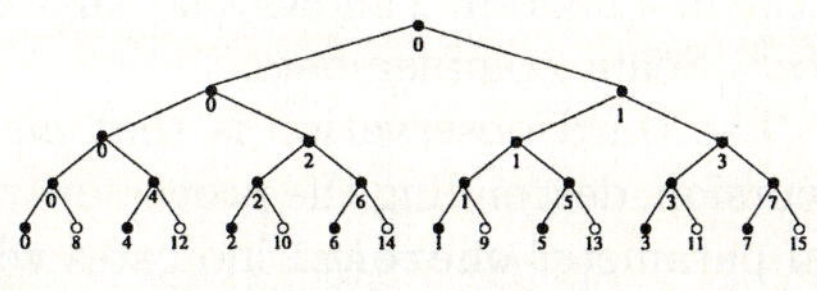

Fig. 7. A tree labeled for 16 processors, each communicating only with its parent and right child.

row-major implementation (almost flat), and ATLAS-optimized BLAS3 `dpotrf` [15] (not flat) whose performance degrades as it bleeds into outer caches. Below order 4000 it improves slightly on Morton-ordered recursions[2] (very flat) even though they do not compile to SSE2 instructions. Substituting eight SSE2/MMX packed-`double` instructions (instead of `double`) into `schurBlkBase`'s symbolic code gives the next plot. Only slightly better is Intel's hand-coded `dpotrf` [16, 3] (flat) and the idealized maximum FLOPs for our Xeon processor.[2]

4   Developing the Code

This section offers a terse description of the expansion of the recursive algorithm in Figure 5 to the SPMD version that offers perfectly balanced and scalable parallelism. This description is again written as if matrices were of order 2^p, which is an unnecessary constraint. Without it, however, another derivative of the code in Figure 5 with bounds checking is necessary to peel the unbalanced southeast perimeter from the matrix.

Three observations take us to the parallel code. First is the processor tree of Figure 7. The root is identified as the processor with MPI rank 0. Beneath it is a tree of processors indexed by a bit-reversal of level ordering: every processor is its own left child and its right child is easily computed from the level. This indexing expands to 2^p processors (for any p), and localizes communication within subtrees. Each processor communicates *only* with its parent and right child. With MPI ranks assigned cyclically, processors at the leaves (the open circles) can even share RAM on a single node.

Second is the recognition that the four functions are paired with respect to a locality operation. That is, the blocks that result from `schurTri` are not needed by any processes until they are finalized by an application of `doCholesky` there. These functions, however, are not applied very often (Theorem 2) and so are computed by the root (Rank 0) processor. Far more interesting are the results of `schurBlk` that are not shared by other processes until finalized by an application of `triSolve`. They are distributed out to RAM on remote processors

---

[2] These two plots appear as referents in later graphs.

in a pattern that will receive all the updates to these blocks, up to an including the finalization. As part of that `triSolve` finalization the results are assembled up and down the tree—so that all other processors immediately become aware of them. Theorem 2 shows that that communication is only time $\Theta(n^2)$ after the $\Theta(n^3)$ Schur complements.

The third observation is that *all* processors are always executing the same recursion, descending the processor tree and the quadtree level-by-level. The critical parameter `whereAmI` indicates which processor is designated to be "active," in the sense of actually carrying out computations and side-effects to blocks of the matrix. Base steps in the parallel code (uniprocessing calls), therefore, are always protected by a conditional comparison with it and the local processor's rank. Even non-base steps use it as if to "fork" work to two children, effected by locally computing that argument.

That is, in the SPMD recursion all processors are always executing the same code at the same level, with `whereAmI` identifying an ancestor and the local processor idle—aside from the synchronized recursion. Alternatively, it identifies the rank of the local processor, and it is actively computing on the blocks identified by the Ahnentafel indices. There is no need for interprocess communication to fork and join since all control is implicit in the recursion.

The necessary `MPI_sends` and `MPI_receives` of data are embedded in the same recursion and similarly synchronized. Two functions are used locally to share information:

- `sendDownAll` sends a square or rectangle block down to all descendants of the local processor. It is invoked after receiving a finalized block from one's parent.
- `assembleBottomUpDown` is invoked immediately before a finalizing call to assemble distributed data onto all the local RAMs in the subtree. Typically, it happens just before `triSolve`.

This provides the only communication necessary to the SPMD parallelism.

Figure 4 then expands first into Figure 8 with provision to assemble distributed Schur complements. The rest of the parallel code appears as Figures 11 and 12, which are formatted to be read side-by-side. That way the control is seen to be identical, and the absence of any communication in Figure 11 is apparent. The rank-k updates are accumulated in distributed memory in sequential blocks, each of which exists on a unique machine in the cluster; the programmer can be oblivious as to just which one—recursion determines it. No sharing occurs until just before such blocks are finalized by the code in Figure 12, and after finalization the block is implicitly shared by all processors in the subtree of the finalizing one.

Figure 11 presents the rank-k updates without any interprocess communication either for control (the SPMD recursion suffices) or for sharing. There again appear the guards on the base cases, performed only by the process selected by the parameter `whereAmI`, selecting the active processor. Here also is seen how the recursion forks within `schurBlk`, based on a bit masked from *every* processor's rank. Half the processors recur to the left subtree and half to the right,

```
static void doCholeskyBlk(int quad)
{
  if ( lgProcResource==0 || quad >= quadBd(soloBound) )
  {
    if (me != 0  /*whereAmI*/) {} else
      doCholeskyUniproc(quad);
    sendDownAll(square, quad,  0, lgProcResource);
      /* assuring that all processors have the factorization */
  }
  else
  {
      doCholeskyBlk(nw(quad));
      /* All  processors have the partial result. */

      /* Upon arriving here (except at topmost call) we must assume
         that rank-k updates (shurBlk calls) have scattered the current
         information in sw(quad) across remote memories, quad-by-quad.
         The first task is to assemble current information in all memories.
      */
      assembleBottomUpDown( sw(quad), 0, lgProcResource);
      triSolveBlk         ( sw(quad), 0, lgProcResource);
      /* All  processors have the partial result. */

      schurTri   (          sw(quad)                     );
      /* Not all processors have the rank-k updates.  They are local to... */
      doCholeskyBlk(se(quad)                             );
      /* All  processors have the partial result. */
  }
  return;
}
```

Fig. 8. The top level of parallelism.

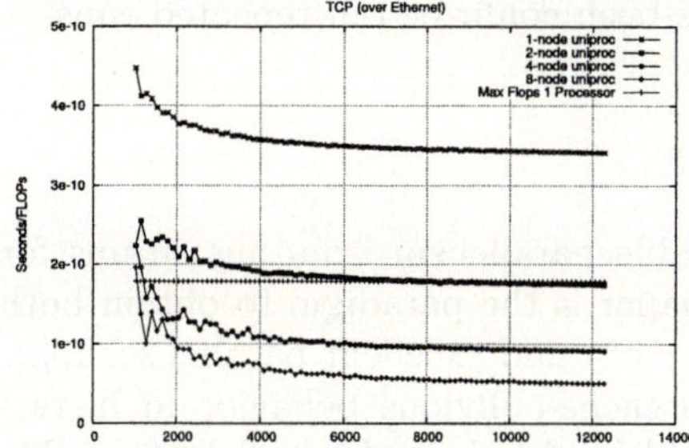

Fig. 9. Normalized time for Ethernet multiprocessing. Hardware-cost rating: **$**

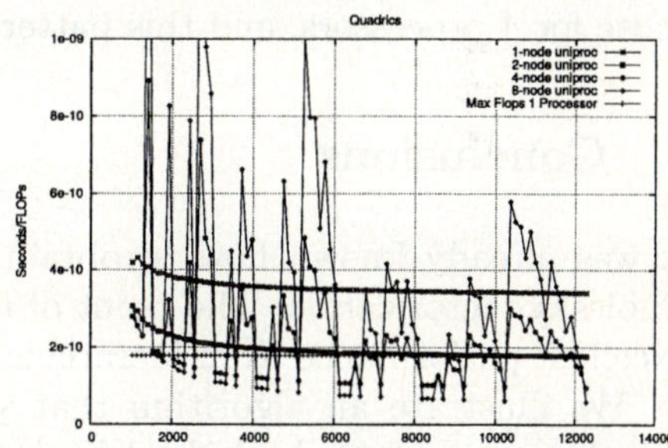

Fig. 10. Normalized time for Quadrics multiprocessing. Hardware cost: **$$$$**

one of which is newly invigorated as second argument for `whereAmI`; it allows computation when it arrives at the processor identified locally as `me`.

5   Experimental Results

The codes were tested on an Aspen Systems cluster of eight, dual-processor 2.8GHz Intel Xeons, each with 2GB of memory and 8KB L1, 512KB L2 cache. All the uniprocessing code was compiled using the native ICC compiler with -O3 optimization for the Xeon. Section 3's hand conversion of scalar instructions introduced SSE2 packed-`doubles` into the symbolic code for `schurBlkBase`.

The multiprocessing code was written using MPI [17] to distribute computation via four different interconnects. (The codes are identical, but there is room for only two plots here.) Since this work is motivated by programming style, we only sought confirmation of performance. The differences in performance among the interconnects were completely unexpected.

- TCP via on-board GIGABIT ETHERNET.
- Infiniband [18] through InfiniCon INFINIIO 2000 [19].
- Myrinet M3-E64 [20].
- Quadrics ASNET II [21].

Multiprocessing code for the last was compiled with Quadrics's release of MPI [22]. The other three were compiled with LAM/MPI, 7.0.1 here [17]. All tests were run twice, most of them thrice. There were minor variances, therefore, that are not shown here; the shapes of these plots are all faithful and reproducible.

Times are presented here in units of seconds-per-FLOP as above [14].

Figure 9 indicates excellent performance. Although Ethernet is the cheapest interconnect and presumably the slowest, its plots land at $\frac{1}{2}, \frac{1}{4}$, and $\frac{1}{8}$ of Figure 6's baseline: perfect scaling for $2, 4, 8$ processors. With the timings obtained, this figure alone demonstrates the success of the paradigm and its resulting code.

Quadrics, the most expensive, scales as perfectly for 2 uniprocessor nodes in Figure 10. Normalized times for 4 and 8 nodes tend toward a nicely scaled asymptote, but leap raggedly above it. The 8-processor leaps consistently double those for 4 processors, and this pattern has been confirmed in repeated runs.

6   Conclusions

It was already known how to obtain scalable parallelism from algorithms for Cholesky factorization. The point of this paper is the paradigm to obtain both excellent performance from hierarchical memory and excellent parallel scaling.

We illustrate an algorithm that yields cache-oblivious behavior in hierarchical memory, and similarly delivers scalable and balanced parallelism on distributed multiprocessors. We obtain both locality in hierarchical memory and excellent parallel scaling from the same style, using quadtree recursion on Morton-ordered matrices. We delivered excellent performance in moving from shared-memory parallel `dpotrf` to a SPMD a distributed-memory parallel implementation. Along the way we expose some strange behavior from current Linux/MPI implementations. The best behavior is achieved on the cheapest interconnection.

The perfect scaling of Figure 9 demonstrates success for our recursive paradigm for high-performance code. Single-recursion multiple-data (here dubbed SRMD), Morton-ordered matrices, and good style are winners.

Acknowledgement

The first author thanks his hosts at the University of Washington where Section 3 was thrashed out.

References

1. Chatterjee, S., Lebeck, A.R., Patnala, P.K., Thottenthodi, M.: Recursive array layouts and fast parallel matrix multiplication. IEEE Trans. Parallel Distrib. Syst. **13** (2002) 1105–1123 http://dx.doi.org/10.1109/TPDS.2002.1058095

2. Thiyagalingam, J., Beckmann, O., Kelly, P.H.J.: Is Morton layout competitive for large two-dimensional arrays, yet? Concur. Comput. Prac. Exper. (2004) To appear in special issue on Compilers for Parallel Computing. http://www.doc.ic.ac.uk/~phjk/Publications/IsMortonYetCCPandE2004.pdf

3. Goto, K., van de Geijn, R.: On reducing TLB misses in matrix multiplication. FLAME Working Note 9, Univ. of Texas, Austin (2002) http://www.cs.utexas.edu/users/flame/pubs/GOTO.ps.gz

4. Morton, G.M.: A computer oriented geodetic data base and a new technique in file sequencing. Technical report, IBM Ltd., Ottawa, Ontario (1966)

5. Drakenberg, P., Lundevall, F., Lisper, B.: An efficient semi-hierarchical array layout. In Lee, G., Yew, P.C., eds.: Interaction between Compilers and Computer Architectures. Volume 613 of Kluwer Intl. Series in Engineering and Computer Science. Kluwer, Deventer, Netherlands (2001) http://www.mrtc.mdh.se/publications/0313.pdf

6. Wise, D.S.: Ahnentafel indexing into Morton-ordered arrays, or matrix locality for free. In Bode, A., Ludwig, T., Karl, W., Wismüller, R., eds.: Euro-Par 2000 – Parallel Processing. Volume 1900 of Lecture Notes in Comput. Sci. Springer, Heidelberg (2000) 774–883 http://www.springerlink.com/link.asp?id=0pc0e9gfk4x9j5fa

7. Wise, D.S., Frens, J.D., Gu, Y., Alexander, G.A.: Language support for Morton-order matrices. Proc. 8th ACM SIGPLAN Symp. on Principles and Practice of Parallel Program., SIGPLAN Not. **36** (2001) 24–33 http://doi.acm.org/10.1145/379539.379559

8. Schrack, G.: Finding neighbors of equal size in linear quadtrees and octrees in constant time. CVGIP: Image Underst. **55** (1992) 221–230

9. Raman, R., Wise, D.S.: Converting to and from dilated integers. Submitted for publication (2004) http://www.cs.indiana.edu/~dswise/Arcee/castingDilated-comb.pdf

10. Frigo, M., Leiserson, C.E., Prokop, H., Ramachandran, S.: Cache–oblivious algorithms. In: Proc. 40th Ann. Symp. Foundations of Computer Science. IEEE Computer Soc. Press, Washington, DC (1999) 285–298 http://dx.doi.org/10.1109/SFFCS.1999.814600

11. Frens, J.D.: Matrix Factorization Using a Block-Recursive Structure and Block-Recursive Algorithms. PhD thesis, Indiana Univ., Bloomington (2002) http://www.cs.indiana.edu/cgi-bin/techreports/TRNNN.cgi?trnum=TR568

12. Spieß, J.: Untersuchungen des Zeitgewinns durch neue Algorithmen zur Matrix-Multiplication. Computing **17** (1976) 23–36

13. Tocher, K.D.: The application of automatic computers to sampling experiments. J. Roy. Statist. Soc. Ser. B **16** (1954) 39–61 See pp. 53–55.

14. Johnson, D.S.: A theoretician's guide to the experimental analysis of algorithms. In Goldwasser, M.H., Johnson, D.S., McGeoch, C.C., eds.: Data Structures, Near Neighbor Searches, and Methodology: 5th & 6th DIMACS Implementation Challenges. Volume 59 of DIMACS Ser. Discrete Math. Theoret. Comput. Sci. Amer. Math. Soc., Providence (2002) 215–250 http://www.research.att.com/~dsj/papers.html

15. Whaley, R.C., Dongarra, J.J.: Automatically tuned linear algebra software. In: Proc. Supercomputing '98. IEEE Computer Soc. Press, Washington, DC (1998) 38 http://dx.doi.org/10.1109/SC.1998.10004

16. Intel Corp. Santa Clara, CA: Intel Math Kernel Library. (2003) http://www.intel.com/software/products/mkl/

17. LAM/MPI Bloomington, IN: www.lam-mpi.org. (2004)

18. InfiniBand Trade Assn. Portland, OR: www.infinibandta.org. (2004)

19. InfiniCon Systems King of Prussia, PA: www.infinicon.com. (2004)

20. Myricom Inc. Arcadia, CA: www.myri.com. (2004)

21. Quadrics Ltd. Bristol, UK: www.quadrics.com. (2004)

22. Quadrics Ltd. Bristol, UK: Quadrics Release of MPICH 1.24. (2004) www.quadrics.com

```
// It is a fact of life that the locally active process has      whereAmI==0 .
// All processors are running identical code!
// Passing schurTri down to be done locally at every child.

static void schurBlk(int eQuad, int wQuad, int whereAmI, int lgProcs)
{
  if ( lgProcs==0 || eQuad>=quadBd(soloBound) )
    if (me != whereAmI) {} else
      schurBlkUniproc(eQuad, wQuad);
  else
  {
    if ( amLeftChild(me,lgProcs-1) )                /* West on self. */
      schurBlkWE(we(eQuad),    ea(wQuad),        whereAmI,              lgProcs-1);
    else                                            /* East on child. */
      schurBlkWE(  ea(eQuad), ea(wQuad), child(whereAmI, lgProcs), lgProcs-1);
  }
  return;
}

static void schurBlkWE(int eRect, int wRect, int whereAmI, int lgProcs)
{
  if ( lgProcs==0 || eRect>=rectBd(soloBound) )
    if (me != whereAmI) {} else
    {
      schurBlkUniproc(so(eRect),      so(wRect) );
      schurBlkUniproc(so(eRect), e2w(so(wRect)));
      schurBlkUniproc(no(eRect), e2w(no(wRect)));
      schurBlkUniproc(no(eRect),      no(wRect) );
    }
  else
    if ( amLeftChild(me,lgProcs-1) )                /* South on self. */
      schurBlkSN(so(eRect), so(wRect),       whereAmI,            lgProcs-1);
    else                                            /* North on child.*/
      schurBlkSN(  no(eRect), no(wRect), child(whereAmI, lgProcs), lgProcs-1);
  return;
}

static void schurBlkSN(int eQuad, int wQuad, int whereAmI, int lgProcs)
{
  schurBlk(eQuad, e2w(wQuad), whereAmI, lgProcs);
  schurBlk(eQuad,     wQuad , whereAmI, lgProcs);
  return;
}
```

```
static void schurTri(int wQuad)
{
  if ( lgProcResource==0 || wQuad >= quadBd(soloBound) )
    if (me != 0 /*whereAmI*/ ) {} else
      schurTriUniproc(wQuad);
  else
  { /*Parallelism possible here, but declined to set up for doCholesky finalizer. */

      schurTriW( ea(wQuad) );

      schurTriE( ea(wQuad) );
  }
  return;
}

static void schurTriW  (int wRect)
{
  if ( lgProcResource==0 || wRect >= rectBd(soloBound) )
    if (me != 0 /*whereAmI*/ ) {} else
    {
      schurBlkUniproc( sw(diag(prnt(wRect))),      so(wRect) );
      schurBlkUniproc( sw(diag(prnt(wRect))), e2w(so(wRect)));
      schurTriUniproc(                        e2w(no(wRect)));
      schurTriUniproc(                            no(wRect) );
    }
  else
  {/*Parallelism possible here, but declined to set up for doCholesky finalizer. */
    schurBlkSN(sw(diag(prnt(wRect))), so(wRect), 0, lgProcResource);
    schurTriSN(                       no(wRect)                    );
  }
  return;
}

static void schurTriE  (int wRect)
{
  if ( lgProcResource==0 || wRect >= rectBd(soloBound) )
    if (me != 0 /*whereAmI*/ ) {} else
    {
      schurTriUniproc ( e2w(so(wRect)));
      schurTriUniproc (     so(wRect) );
    }
  else
  {
    schurTriSN( so(wRect) );
  }
  return;
}

static void schurTriSN (int wQuad)
{
  schurTri( e2w(wQuad) );
  schurTri(     wQuad  );
  return;
}
```

Fig. 11. Aligned C code for `schurBlk` and `schurTri`. No communication at all occurs among the highly parallel `schurBlk` recursions.

```
static void triSolveBlk(int quad, int whereAmI, int lgProcs)
{
  if ( lgProcs==0 || quad>=quadBd(soloBound) )
    if (me != whereAmI) {} else   triSolveUniproc(quad);
  else
  {
    triSolveW        (we (quad),      whereAmI,      lgProcs);
    triSolveE        ( ea(quad),      whereAmI,      lgProcs);
  }
  return;
}

static void triSolveW(int rect, int whereAmI, int lgProcs)
{
  if ( lgProcs==0 || rect>=rectBd(soloBound) )
    if (me != whereAmI) {} else
    {

      triSolveUniproc( no(rect) );                                    /*A*/
      triSolveUniproc( so(rect) );                                    /*D*/
    }
  else
    if ( amLeftChild(me,lgProcs-1) )           /* South on self. */
    {

      triSolveBlk(so(rect),        whereAmI    , lgProcs-1 );         /*D*/
      if (me != whereAmI) {} else
        mpi_send_blocked_receive(square, so(rect),no(rect), child(whereAmI,lgProcs));
      sendDownAll(square,      no(rect),     whereAmI, lgProcs-1);   /* ^ */
    }                                                               /* | */
    else                                                            /* | */
    {                                                               /* | */
                                                                    /* | */
                                                                    /* | */
                                                                    /* | */
                                                                    /* | */
                                                                    /* | */
                                                                    /* | */
                                                                    /* | */
      triSolveBlk(no(rect), child(whereAmI,lgProcs), lgProcs-1);  /*A*/ /* | */
      if (me != child(whereAmI,lgProcs)) {} else                    /* | */
        mpi_send_blocked_receive(square, no(rect),so(rect),  whereAmI ); /* <-+ */
      sendDownAll(square,     so(rect), child(whereAmI,lgProcs), lgProcs-1);
    }
  return;
}
```

```
static void triSolveE(int rect, int whereAmI, int lgProcs)
{
  if ( lgProcs==0 || rect>=rectBd(soloBound) )
    if (me != whereAmI) {} else
    {
      schurBlkUniproc( so(rect), sw(prnt(rect)) );                    /*E*/
      schurBlkUniproc ( no(rect), nw(prnt(rect)) );                   /*B*/
      triSolveUniproc( no(rect) );                                    /*C*/
      triSolveUniproc( so(rect) );                                    /*F*/
    }
  else
    if ( amLeftChild(me,lgProcs-1) )           /* South on self. */
    {
      schurBlk    (so(rect), sw(prnt(rect)),     whereAmI, lgProcs-1);  /*E*/

      /* Upon arriving here, we must assume
         that rank-k updates (schurBlk calls) have scattered the current
         information in sw(quad) across remote memories, quad-by-quad.
         The first task is to assemble current information in all memories.
      */
      assembleBottomUpDown(so(rect),          whereAmI, lgProcs-1);
                                       /* South on self. */            /*F*/
      triSolveBlk         (so(rect),          whereAmI, lgProcs-1);
      if (me != whereAmI) {} else
        mpi_send_blocked_receive(square, so(rect),no(rect), child(whereAmI,lgProcs));
      sendDownAll(square,      no(rect),      whereAmI, lgProcs-1);   /* ^ */
    }                                                                /* | */
    else                                                  /*B*/      /* | */
    {                                                                /* | */
      schurBlk(no(rect), nw(prnt(rect)), child(whereAmI,lgProcs), lgProcs-1);/* | */
                                                                     /* | */
      /* Upon arriving here, we must assume
         that rank-k updates (schurBlk calls) have scattered the current
         information in sw(quad) across remote memories, quad-by-quad.
         The first task is to assemble current information in all memories.
      */
      assembleBottomUpDown(no(rect),     child(whereAmI,lgProcs), lgProcs-1);/* | */
                                      /* North on child. */        /*C*/  /* | */
      triSolveBlk         (no(rect),     child(whereAmI,lgProcs), lgProcs-1); /* | */
      if (me != child(whereAmI,lgProcs)) {} else                   /* | */
        mpi_send_blocked_receive(square, no(rect),so(rect), whereAmI ); /* <-+ */
      sendDownAll(square,     so(rect), child(whereAmI,lgProcs), lgProcs-1);
    }
  return;
}
```

Fig. 12. Aligned C code for `triSolve`. The subtree of memories must be synched for finalization after distributed `schurBlks`.

An Exception Handling Mechanism
for the Concurrent Invocation Statement

Hiu Ning (Angela) Chan[1], Esteban Pauli[1], Billy Yan-Kit Man[1],
Aaron W. Keen[2], and Ronald A. Olsson[1]

[1] Department of Computer Science, University of California, Davis
Davis, CA 95616 USA
`{chanhn,pauli,many,olsson}@cs.ucdavis.edu`
[2] Computer Science Department, California Polytechnic State University
San Luis Obispo, CA 93407 USA
`akeen@csc.calpoly.edu`

Abstract. Several concurrent programming languages and systems –
e.g., MPI, .NET, and SR – provide mechanisms to facilitate communication between one process and a group of others. One such mechanism is
SR's concurrent invocation statement (co statement). It specifies a group
of operation invocations and normally terminates when all of its invocations have completed. To make the co statement more flexible, it can
specify code in the invoker to execute as each invocation completes or to
terminate the entire co statement before all of its invocations have completed. We have added an SR-like co statement to JR. Unlike SR, JR provides exception handling mechanisms, which are integrated with Java's
exception handling mechanism. However, JR needs additional mechanisms to deal with sources of asynchrony. The co statement introduces
additional such sources of asynchrony for the invocations it initiates. This
paper describes the design and implementation of an exception handling
mechanism for JR's co statement.

1   Introduction

Communication between one process and a group of others is important in many
concurrent programs. Several concurrent programming languages and systems
provide mechanisms to facilitate such communication, e.g., MPI's [9] collective
communication; .NET's [10] delegates and asynchronous invocations, which can
be used to simulate collective communication; and SR's [1, 2] concurrent invocation statement (*co statement*, for short), which can be viewed as a form of
collective communication. These mechanisms can be used to broadcast data to
a group of processes and gather back results from each in the group, for instance, in applications such as initiating subparts of numerical computations,
reading/writing a distributed database, and distributed voting schemes.

This paper focuses on the co statement. It specifies a group of operation
invocations and normally terminates when all invocations have completed. The
co statement also allows quantifiers to deal with groups of related operations and

J.C. Cunha and P.D. Medeiros (Eds.): Euro-Par 2005, LNCS 3648, pp. 699–709, 2005.
© Springer-Verlag Berlin Heidelberg 2005

post-processing code (PPC) that is executed as each invocation completes. The co statement is allowed to terminate before all its invocations have completed. All these features are useful in practice, as seen in the examples in SR [1, 2] and later in this paper.

We have added a co statement, similar to SR's, to the JR concurrent programming language [7, 11]. Unlike SR, JR provides exception handling mechanisms, which are integrated with Java's exception handling mechanism. However, JR needs additional mechanisms to deal with sources of asynchrony, as described in Reference [8]. The concurrent invocation statement introduces additional such sources of asynchrony for the invocations it initiates. Thus, we have also designed and implemented an exception handling mechanism for the concurrent invocation statement and added it to JR (available on the web [5]). Our work should benefit others considering adding exception handing mechanisms to other concurrent programming languages and systems.

The rest of this paper is organized as follows. Section 2 presents the concurrent invocation statement without exceptions, introduces our running example, and summarizes our previous work with handling exceptions during asynchronous method invocation. Section 3 gives an overview of our approach, illustrates it by extending our running example to use exceptions, and discusses and justifies our design decisions. Section 4 presents an overview of our implementation and discusses its reasonable performance. Finally, Section 5 concludes.

2   Background

2.1   Concurrent Invocation Statement (Without Exceptions)

Figure 1 shows a small JR program with a simple co statement. It simulates a two-person election, where each person announces his or her vote. This co statement contains two arms. The process executing the co statement initiates the two *co invocations* (one of `aliceVote` and one of `bobVote`); it then waits until both invocations have finished. Thus, the individual "votes" outputs are guaranteed to occur before the "over" output.

```
public class main {
  public static void main(String [] args){
    co aliceVote();
    [] bobVote();
    System.out.println("Election is over.");
  }
  public static op void aliceVote() {
    ...
    System.out.println("Alice votes yes.");
    ...
  }
  public static op void bobVote() {
    ...
    System.out.println("Bob votes no.");
    ...
  }
}
```

Fig. 1. A simple election program.

```
int yesCount = 0, noCount = 0;  boolean vote[] = new boolean [voters];

co ((int i = 0; i < voters; i++)) vote[i] = getVote[i]();

for (int i = 0; i < voters; i++) {
  if(vote[i]) ++yesCount;  else ++noCount;
}
// announce decision
System.out.println("votes  For: " + yesCount + "  Against:  " + noCount);
if (yesCount > noCount)
  System.out.println("Victory!");
else
  System.out.println("Defeat ;-(");
```

Fig. 2. Election with tallying of votes after voting.

```
boolean vote;

co ((int i = 0; i < voters; i++)) vote = getVote[i](){
  System.out.println("Voter "+i+ " voted "+vote);
  if (vote)
    ++yesCount;
  else
      ++noCount;
}
// announce decision -- same code as in earlier figure
```

Fig. 3. Election with tallying of votes during voting.

```
int yesCount = 0, noCount  = 0;  boolean vote = false;

co ((int i = 0; i < voters; i++)) vote = getVote[i](){
  if (vote) {
    if (++yesCount > voters/2) break;
  }
  else
      if (++noCount >= (voters+1)/2) break; // tie -> No
}
// announce decision -- same code as in earlier figure
```

Fig. 4. Election with decision announced as soon as majority has decided.

Figure 2 shows a more interesting voting program fragment. Here, votes are received from an array of voters' operations, `getVote`. This co statement uses quantifier notation to initiate all invocations. As each invocation completes, the result is recorded in the `vote` array. After all votes are received, the votes are tallied and the decision is announced.

Figure 3 shows how votes can be tallied *as they are received*. This co statement specifies *post-processing code (PPC)*; the scope of the quantifier variable for the arm includes the arm's PPC. As each invocation completes, the corresponding PPC is executed by the *same* process that initiates the co statement. Thus, execution of PPCs is serial and variables local to this process used within the PPC (e.g., `vote`) are not subject to race conditions. (The assignment to `vote` is considered a part of the PPC; it is executed before the rest of the PPC.)

The co statement has an iterative nature with respect to it executing its PPCs, so a break statement makes sense within a PPC. Figure 4 shows a co statement that announces the election decision as soon as a majority of voters has decided the election. A co statement now terminates when: all its invocations have completed and their corresponding PPCs have terminated; or execution of a PPC has executed a transfer of control out of the co statement. Note that, in this example, invocations whose results are not tallied do continue to execute, even if the co statement has terminated. However, their subsequent completion has no effect on the invoking process; indeed, the invoking process may have completed and no longer exist.

2.2   Simulation of the co Statement Using Existing JR Mechanisms

The co statement can be simulated with other JR language mechanisms, but doing so is cumbersome for the programmer. For example, Figure 5 shows how to rewrite Figure 1. It uses send invocations (which are non-blocking) to initiate the voting and the receive statement to wait until both voters have voted. Here, and in general, the code requires changing the interface to the vote operations to have an extra parameter, so that the voter can notify "election central" when it has finished voting. This extra parameter is a capability (a special kind of reference) to an operation.

```
public class main {
  public static void main(String [] args){
    op void aliceVoted();
    op void bobVoted();
    send aliceVote(aliceVoted);
    send bobVote(bobVoted);
    receive aliceVoted();
    receive bobVoted();
    System.out.println("Election is over.");
  }
  public static op void aliceVote(cap void() voted) {
    ...
    System.out.println("Alice votes yes.");
    ...
    send voted();
  }
  public static op void bobVote(cap void() voted) {
    ...
    System.out.println("Bob votes no.");
    ...
    send voted();
  }
}
```

Fig. 5. Hand-coded simulation of Fig. 1.

The general simulation of a co statement with a PPC is more complicated. To illustrate, Figure 6 shows how to rewrite Figure 4. Figure 6's code properly simulates Figure 4's co statement because the PPC code is executed sequentially, but it again requires the interface to be changed, e.g., the signatures of `getVote` and the body of that operation (not shown). Although this simulation works fine for the program in Figure 4, the general simulation is more complicated. For example, suppose the co statement uses the quantifier variable in its PPC, as in Figure 3. At first look, the code in Figure 6 might seem to work, but the index variable in the second for loop (where the print statement would be

```
op void voted(boolean);
for (int i = 0; i < voters; i++) {
  send getVote[i](voted);
}
for (int i = 0; i < voters; i++) {
  boolean vote;
  receive voted(vote);
  if (vote) {
    if (++yesCount > voters/2) break;
  }
  else
    if (++noCount >= (voters+1)/2) break; // tie -> No
}
// announce decision -- same code as in earlier figure
```

Fig. 6. Hand-coded simulation of Fig. 4.

placed) has no connection to the index variable in the first for loop. Also, if the co statement has multiple arms, a simple receive statement no longer suffices. Section 4.1 discusses how to deal with these problems in general.

2.3   Handling Exceptions During Asynchronous Method Invocation

This section summarizes our earlier work that shows how to handle exceptions during asynchronous method invocation [8, 11]. Our approach bears some resemblance to that provided in both ABCL/1 [3] and Arche [4].

JR provides asynchronous method invocation via the send statement. To facilitate the handling of exceptions thrown from an asynchronously invoked method, JR requires the specification of a *handler* object as part of a **send**. Any exceptions propagated out of the invoked method are directed to the handler object. To be used as a handler, an object must implement JR's **Handler** interface and define a number of handler methods. A method is defined as a handler through the use of the **handler** modifier (much like the **public** modifier). A handler method takes only a single argument: a reference to an exception object. Each handler method specifies the exact exception type that it can handle. When an exception is delivered to a handler object, it is handled by the handler method of the appropriate type.

```
public class IOHandler implements edu.ucdavis.jr.Handler {
    public handler void handleEOF(java.io.EOFException e)
    { /* handle exception */ }
    public handler void handleNotFound(java.io.FileNotFoundException e)
    { /* handle exception */ }
}

IOHandler iohandler = new IOHandler();
...
send readFile("/etc/passwd") handler iohandler;
...
```

Fig. 7. Class definition for and use of a simple handler object.

Figure 7 shows an example definition of a handler object's class and how it is used. In this example, handler objects of type **IOHandler** can handle end-of-file and file-not-found exceptions. An exception of type **java.io.EOFException** directed to such a handler object will be handled by the `handleEOF` method. As seen in Figure 7, a send statement must specify, using a handler clause, its handler object. The JR compiler statically checks that the specified handler object can handle each of the potentially thrown exceptions.

3   Design

3.1   Overview of Exceptions in the Concurrent Invocation Statement

A key observation in integrating exception handling with the concurrent invocation statement is that the co statement adds another source of asynchrony in the same sense as for the send statement (Section 2.3). Section 2.1 described how if the co statement's PPC contains a break statement, then the invoking process may not even exist when one of its invocations completes; the same now also pertains to an exception that occurs for one of its co invocations. Therefore, we add a handler to co invocations that can throw exceptions.

Section 2.1 described when a co statement terminates. A co invocation that throws an exception does *not* cause its associated PPC to be executed (discussed further in Section 3.2). But, that invocation is now considered to have completed and contributes toward the co statement's overall termination.

```
boolean decided = false;

MyHandler mh = new MyHandler();

co ((int i = 0; i < voters; i++)) vote = getVote[i]() handler mh : {
  if (vote){
    if (++yesCount > voters/2) {decided = true; break;}
  }
  else
    if (++noCount >= (voters+1)/2) {decided = true; break;} // tie -> No
}

System.out.println("votes  For: " + yesCount + "  Against:   " + noCount);
if (!decided)
  System.out.println("Too many non-participating voters to decide election");
else  {
  if (yesCount > noCount)
    System.out.println("Victory!");
  else
    System.out.println("Defeat ;-(");
}

public class MyHandler implements edu.ucdavis.jr.Handler {
  public handler void handleNonParticipVoter(NonParticipVoterException e) {
    System.out.println("Non-participating voter");
  }
}

public class NonParticipVoterException extends java.lang.Exception {
}
```

Fig. 8. Fig. 4 extended to handle exceptions.

Figure 8 shows how to extend the program in Figure 4 for when the `getVote` operation can throw exceptions. The co invocation now specifies a handler, which

just outputs an error message. The code that outputs the results now makes sure
that enough voters actually voted yes or no.

3.2   Design Decisions

A simpler approach than using handler objects (Section 3.1) is to just enclose
the co statement, such as the one in Figure 4, within a try/catch statement.
However, if an exception occurs for an invocation, then control transfers to the
catch block and the entire co statement terminates. Thus, the results of those
invocations that complete normally after the exception occurs would be lost.
That would make dealing with code that can throw exceptions, such as that in
Figure 8, much more difficult.

Having invocation-specific handlers allows the co statement to continue in
such cases and terminate cleanly. Moreover, because the PPC can contain statements such as break, exceptions on invocations must be handled somewhere, as
noted in Section 3.1. Thus, if an op can throw an exception, its invocation within
an arm of a co statement must have a handler.

Figure 8 illustrated the use of a handler that is specified for each invocation.
The co statement also allows a default handler for the entire statement so that
exceptions from all invocations are handled by the same handler object. The
default handler is used for any invocation that requires a handler but does not
itself specify a handler. Allowing a default handler is convenient so that users
do not need to specify a handler object for each arm while they can occasionally
provide a special handler for some arms to handle their exceptions. The default
handler is used only if an invocation-specific handler is not specified for a particular invocation. An alternative is to allow the default handler to be used in
addition to the invocation-specific handler, so that it can handle some types of
exception that a handler object for a specific invocation cannot handle. However,
we have not yet seen a real need for that functionality.

If an exception occurs during execution of a PPC, then execution of the
current block will terminate and control transfers out of the co statement, thus
terminating the co statement. Consider, for example, the following co statement

```
co f() {
    ... // PPCf -- throws exception (but contains no try/catch)
}
[] g() {
    ... // PPCg
}
```

If the invocation of f finishes before the invocation of g and PPCf throws an
exception that is not caught within PPCf, then PPCg will not be executed when
the invocation of g finishes. This behavior is consistent with exceptions in Java;
e.g., if an exception occurs within a loop in Java code and is not caught within
the loop, the rest of the loop is not executed.

If an exception occurs for a co invocation, the associated PPC is not executed.
This behavior was seen in Figure 8. In addition, if the co invocation assigns to a
variable (vote in Figure 8), that assignment is considered part of the PPC and
is not executed if an exception occurs. An alternative would, of course, be to

allow the PPC to execute, but it would need some way to distinguish between success and exception (e.g., `if "exceptionOccurred" ...`), so that, for example, it would know whether the variable was assigned.

4   Implementation

4.1   Internal Transformations

Section 2.2 showed how simple co statements can be simulated using other JR language mechanisms. Internally, the JR translator transforms a co statement in a way similar to those examples; however, the transformation handles the necessary change of interface and deals with multiple arms and quantifier variables.

```
co f(5) {PPCf} [] g() {break;} [] ((int i = 0; i < N; i++)) x[i] = h(i) {PPCh}
```

Fig. 9. Example co statement.

```
cap void (void) f_retOp = f.cocall(...);
cap void (void) g_retOp = g.cocall(...);
cap void (void) h_retOp [] = new cap void (void) [N];
for (int i = 0; i < N; i++) { h_retOp[i] = h.cocall(...); }
for (int JR_CO_COUNTER = 0; JR_CO_COUNTER < 2+N; JR_CO_COUNTER++) {
  inni void f_retOp() {PPCf}
  [] void g_retOp() {break;}
  [] ((int i = 0; i < N; i++)) void h_retOp[i](int retVal) {x[i] = retVal; PPCh }
}
```

Fig. 10. Transformed version of Figure 9.

For example, the co statement in Figure 9 is translated internally to roughly the code in Figure 10. This transformed code first initiates the invocations of all 2+N operations[1]. Internally, JR operations are objects with methods for various ways of invoking them [7]. The new `cocall` method initiates an invocation of its operation, but it does not block. It returns a capability for an operation that will be invoked when the initiating invocation completes. The code in Figure 10 then uses a loop to wait for all of the co's invocations to complete (i.e., the `_retOp` operations to be invoked), but the loop can be exited early. Its body contains an inni statement, which is JR's multi-way receive statement: each execution of `inni` services an invocation for one of its arms. A multi-way receive is needed here because the order in which the invocations complete is unknown; indeed, not all invocations need to complete for a co statement to terminate, as illustrated in Figure 4. Note how the PPCs in the original program in Figure 9 simply become blocks of code in the `inni` in Figure 10; in particular, the break statement in the co's PPC simply becomes a break statement that now applies to the for loop. Also note how the third arm of the `inni` has a quantifier that is identical to the quantifier in the original program; thus, the quantifier variable's value is the same in the original invocation and in the PPC executed for the corresponding `_retOp` invocation.

---

[1] The code records for its later use the actual number of co invocations in case any of the expressions in the quantifiers or co invocations have side effects.

```
MyHandler mh = new MyHandler();
co f(5) handler mh : {PPCf} [] g() {break;} [] ((int i = 0; i < N; i++)) x[i] = h(i) {PPCh}
```

Fig. 11. Example co statement with exception handler.

```
MyHandler mh = new MyHandler();
cap void (void) co_fail_retOp = new op void(void); // new op
cap void (void) f_retOp = f.cocall(mh, co_fail_retOp, ...); // extra parameters
cap void (void) g_retOp = g.cocall(...);
cap void (void) h_retOp [] = new cap void (void) [N];
for (int i = 0; i < N; i++) { h_retOp[i] = h.cocall(...); }
for (int JR_CO_COUNTER = 0; JR_CO_COUNTER < 2+N; JR_CO_COUNTER++) {
  inni void f_retOp() {PPCf}
  [] void g_retOp() {break;}
  [] ((int i = 0; i < N; i++)) void h_retOp[i](int retVal) {x[i] = retVal; PPCh }
  [] void co_fail_retOp() {} // new arm -- no body needed
}
```

Fig. 12. Transformed version of Figure 11.

The implementation supports exception handlers in co statements by extending the above scheme. For example, consider adding a handler to the invocation of f in Figure 9, as shown in Figure 11. If an exception occurs for the invocation of f, then, without any change in the above scheme, the inni statement in Figure 10 would block forever waiting for `f_retOp` to be invoked. The extension, then, prevents that by defining an additional operation that is invoked when an exception occurs. Thus, the co statement in Figure 11 is translated internally to roughly the code in Figure 12. Note how operation `co_fail_retOp` is created, passed with the handler `mh` as extra parameters to the cocall method (which is now overloaded to allow such), and appears in a new arm in the `inni`. If execution of the invocation of f throws an exception, then both the handler `mh` is executed and the `co_fail_retOp` operation is invoked; otherwise, only the `f_retOp` operation is invoked.

4.2   Performance

We ran several micro- and macrobenchmarks to assess the performance of our co statement implementation. We ran these benchmarks on various PCs (1.4GHz and 2.0GHz uniprocessors; 2.4GHz and 2.8GHz dual-processors) running Linux; specific results, of course, varied according to platform, but the overall trends were the same.

The benchmarks confirmed that our implementation of the co statement had no noticeable impact on regular invocations; and that the cost of executing a co statement with no exceptions is nearly the same as executing a co statement with exceptions but with no exceptions actually thrown during its execution.

The benchmarks also showed that the execution costs for our implementation of the co statement were nearly identical to the hand-coded simulations of the co statement. For example, the execution costs of the programs in Figures 4 and 6 were about the same for various numbers of voters (10, 100, 500, 1000, 1200, and 1500). (Actual code and execution times are available on the web [6].)

Our initial implementation did not perform as well as our current implementation for larger numbers of voters. It used the more straightforward approach of creating a `_fail_retOp` operation for each operation in the co statement that might throw an exception, including an array of such failure operations for each quantified operation. The cost of allocation of these additional operations was high (e.g., for large numbers of voters or when done repeatedly within a loop), so our current implementation eliminates them.

We considered measuring the performance of our co implementation against those for other language implementations, but as noted in Section 1, only SR provides a co statement and it does not provide exception handling. We could measure the costs of SR's co statement versus JR's co statement without exception handling, but that would really measure more the differences of C (in which SR is implemented) versus Java (in which JR is implemented).

5   Conclusion

We have extended the JR programming language with a concurrent invocation statement that includes support for exception handling. This paper described the design tradeoffs, the implementation and performance of the new exception handling mechanism, and some examples illustrating how to use this mechanism. This new feature has been incorporated into the standard JR language release, which is available on the web [5]. We hope to extend our work by considering more formal semantics for the concurrent invocation statement and exception handling for it, and with more comprehensive performance evaluation.

Acknowledgements

Others in the JR group – Erik Staab, Ingwar Wirjawan, Steven Chau, Andre Nash, Yi Lin (William) Au Yeung, Zhi-Wen Ouyang, Alex Wen, and Edson Wong – assisted with this work. The referees provided thoughtful comments.

References

1. G. R. Andrews and R. A. Olsson. *The SR Programming Language: Concurrency in Practice*. The Benjamin/Cummings Publishing Co., Redwood City, CA, 1993.
2. G. R. Andrews, R. A. Olsson, M. Coffin, I. Elshoff, K. Nilsen, T. Purdin, and G. Townsend. An overview of the SR language and implementation. *ACM Transactions on Programming Languages and Systems*, 10(1):51–86, January 1988.
3. Y. Ichisugi and A. Yonezawa. Exception handling and real time features in an object-oriented concurrent language. In *Proceedings of the UK/Japan Workshop on Concurrency: Theory, Language, and Architecture*, pages 604–615, 1990.
4. V. Issarny. An exception handling mechanism for parallel object-oriented programming: toward reusable, robust distributed software. *Journal of Object-Oriented Programming*, 6(6):29–40, 1993.
5. JR distribution. `http://www.cs.ucdavis.edu/~olsson/research/jr/`.

6. Code and data for benchmarks. http://www.cs.ucdavis.edu/~olsson/research/jr/papers/jrcoexceptsAppendix.

7. A. W. Keen, T. Ge, J. T. Maris, and R. A. Olsson. JR: Flexible distributed programming in an extended Java. *ACM Transactions on Programming Languages and Systems*, pages 578–608, May 2004.

8. A. W. Keen and R. A. Olsson. Exception handling during asynchronous method invocation. In B. Monien and R. Feldmann, editors, *Euro-Par 2002 Parallel Processing*, number 2400 in Lecture Notes in Computer Science, pages 656–660, Paderborn, Germany, August 2002. Springer–Verlag.

9. Message Passing Interface Forum. http://www.mpi-forum.org/.

10. .NET framework developer's guide. http://msdn.microsoft.com/library/default.asp?url=/library/en-us/cpguide/html/cpconusingdelegates.asp.

11. R. A. Olsson and A. W. Keen. *The JR Programming Language: Concurrent Programming in an Extended Java*. Kluwer Academic Publishers, Inc., 2004. ISBN 1-4020-8085-9.

smt-*SPRINTS*: Software *Pre*computation with *Int*elligent Streaming for Resource-Constrained SMTs

Tanping Wang, Christos D. Antonopoulos, and Dimitrios S. Nikolopoulos

Department of Computer Science
The College of William and Mary
McGlothlin–Street Hall. Williamsburg, VA 23187–8795
{twang,cda,dsn}@cs.wm.edu

Abstract. We present *SPRINTS*, a source-level speculative precomputation framework for scientific applications running on SMTs with two execution contexts. Our framework targets memory-bound applications and reduces memory latency by prefetching long streams of delinquent data accesses. A unique aspect of *SPRINTS* is that it requires neither hardware nor compiler support. It is based on partial cache simulation and a compression algorithm which can accurately summarize very long streams of cache misses. *SPRINTS* extracts patterns from the streams, which are in turn used to generate source-level, highly optimized precomputation code. *SPRINTS* achieves significant performance improvements over plain thread-level parallelization and indiscriminate precomputation based on code cloning. We demonstrate these improvements using two realistic scientific applications.

1   Introduction

Simultaneous multithreading (SMT) allows multiple threads to concurrently issue instructions to different execution units of the same physical processor. SMT has been recently used as a core architecture by several processor manufacturers [7, 11, 14], since it has the potential of achieving better performance than conventional superscalar processors, at a minimal additional cost. The main reason for the cost effectiveness of SMT processors is that threads share a common set of execution resources. The major shortcoming of resource sharing is that it may result to performance loss, should threads on the processor end up competing for resources such as execution units, instruction buffers and cache space. This performance loss is most noticeable in parallel scientific computations, in which threads tend to be memory-bound and to have identical resource requirements.

Speculative precomputation (SPR) [4] is a technique which uses thread contexts in an SMT to eliminate L2 cache misses from the main computation threads, by pre-executing future memory accesses. SPR has demonstrated the potential of speeding up pointer-based, single-threaded code on multithreaded processors and several hardware and software implementations have been investigated in the related literature [3, 8–10, 12]. This paper makes a case for using SPR as an alternative to thread-level parallel execution on SMTs with two hardware contexts and limited execution resources. The motivation for using SPR in scientific codes stems from two observations. First, the

J.C. Cunha and P.D. Medeiros (Eds.): Euro-Par 2005, LNCS 3648, pp. 710–719, 2005.

hardware of existing SMTs can not handle the resource pressure from multiple memory- and execution unit-bound threads. A carefully designed SPR scheme can reduce this pressure to a minimum, while still reducing memory latency suffered by co-executing application code. Second, most memory-bound scientific codes suffer from memory latency caused by long, but quite predictable streams of memory accesses. SPR is a mechanism which can effectively prefetch such streams.

The contribution of this paper is a user-level software SPR framework, which supports stream-based SPR for scientific applications, with no hardware or compiler support. We named our framework *SPRINTS* (*S*oftware *PR*ecomputation with *INT*elligent *S*treaming). In the heart of this framework lies a compression and pattern extraction algorithm, the purpose of which is to identify streams of delinquent loads which can be directly mapped to streams of data accesses in the source code of the program. Although multiple forms of streams exist in a program (such as dynamic instruction streams, streams of data addresses and so on), our framework opts for identifying a form of streams which can be directly mapped back to computation and data structures in the source code. *SPRINTS* represents streams of L2 cache misses as strings of inter-miss iteration distances, using feedback from a cache simulator. It uses the compression grammar to identify strong patterns in the loop iterations that incur L2 cache misses and feeds these patterns back to a source code generator. The source code generator translates streams into precomputation loops which have small instruction working sets and are amenable to optimization by the back-end compiler.

SPRINTS has a number of advantages, both as a self-contained tool and compared to other precomputation strategies. The first is simplicity, as there is no requirement for compile-time analysis, or additional hardware to trace the code. The second is automation and transparency to the programmer. The third is portability across SMT architectures. The stream identification and compression/decompression engine is independent of the target architecture and works end-to-end using only source code. Finally, *SPRINTS* is engineered for high performance, using optimizations such as store removal, prefetch distance control, and prefetch target selection. We have evaluated *SPRINTS* on a Hyperthreaded Intel Xeon. Our results show that *SPRINTS* speeds up scientific applications for which thread-level parallelization performs poorly.

The rest of the paper is organized as follows: Section 2 introduces *SPRINTS* and provides implementation details. The experimental evaluation of *SPRINTS* is presented in section 3. In section 4 we discuss related work. Finally, section 5 concludes the paper.

2   Precomputation with Intelligent Streaming

SPR implementations [4, 15] adopt a "top-10" approach for identifying delinquent loads and emitting precomputation code. More specifically, architectural simulation is used to identify a few loads which are responsible for most cache misses. Once these loads are identified, code cloning and slicing are performed to issue the instruction paths that lead to the delinquent loads and the loads themselves in the precomputation thread. The "top-10" approach works well in practice because in most codes a few static loads are responsible for large fractions (e.g. more than 90%) of cache misses. Another reason for using this approach is that precomputation threads interfere with their sibling computation threads, sharing execution units and other resources in the processor. Indiscriminate code cloning in precomputation threads would cause excessive interference, whereas highly selective code cloning and further optimization of the precomputation code will reduce the interference. For these reasons we adopt the "top-10" approach for precomputation in *SPRINTS*.

SPRINTS uses cache simulation to identify delinquent loads. The practical impact of using simulation is that *SPRINTS* can accurately identify critical memory accesses that need to be prefetched. In contrast, mere code profiling would only discover dominant reference streams without necessarily revealing information on the cache behavior of these streams. Cache simulation is a generic method which makes *SPRINTS* portable with reasonable effort between SMT architectures. Porting *SPRINTS* requires porting of the cache simulator to accept memory reference traces from a different ISA and adaptation of the architectural parameters of the targeted processor's cache. The porting process can be facilitated by several existing simulation tools.

The profiling mechanism of *SPRINTS* has two distinctive features. Besides detecting delinquent loads, it also recognizes repetitive patterns in these loads. Moreover, it maps delinquent loads identified via profiling back to source code and actually emits source code (at the language level) in precomputation threads. This code prefetches directly elements of application-level data structures. Currently we map misses back to array elements, but the same tool can be used to map back to elements of other data structures as well. *SPRINTS* uses partial simulation, taking advantage of the iterative structure of scientific codes to simulate only a few outermost iterations of the dominant loops and save significantly on simulation time.

SPRINTS targets loop-based scientific applications, which exhibit strong patterns in several aspects of their control and data flow. In particular, the loop-intensive structure of scientific applications, and the fact that delinquent loads tend to occur in heavily traversed loops [8] motivate the use of a loop-based approach to precomputation, in which the speculative thread prefetches *streams* of data that would otherwise be streams of cache misses. *SPRINTS* uses a grammar which detects such streams, by tracking the loop iterations in which delinquent loads occur and identifying patterns of distances between delinquent loads, measured in loop iterations. The rationale for this technique is that long streams of loop iterations with delinquent loads, when mapped back to source code, can be directly translated to highly optimizable source code loops. Furthermore, using loops for precomputation allows *SPRINTS* to trigger precomputation in synch with the sibling computation threads, using loop levels as natural synchronization boundaries and specific loop iterations as natural trigger points. This property is desirable because it allows for accurate and effective control of the *runahead distance* between precomputation and sibling computation, which is in turn critical for timely prefetching [8, 15]. The following sections outline the main components of *SPRINTS*.

2.1   Cache Simulation and Trace Collection

SPRINTS uses a cache simulator based on Cachegrind, the cache simulation component of Valgrind[1], to obtain complete traces of cache misses. We have modified Cachegrind

---

[1] `http://developer.kde.org/~sewardj/docs-2.2.0/manual.html`

to analyze the instruction address stream and detect backward arcs in the dynamic control flow graph of the program. Backward arcs may correspond to loops, however they may also correspond to other control structures. *SPRINTS* uses *objdump*, a GNU development tool, to uniquely identify loops in the program. The tool is used to disassemble the object file and extract the first instruction address of the body of each loop. The simulator uses these instructions as anchors in order to both correctly identify loops and keep track of the loop iteration count. We have introduced a new module in Cachegrind to map uniquely memory references that miss in the cache to array elements, using the current loop iteration count as input.

2.2 Delinquent Load Identification and SPR Code Generation

Following profiling, *SPRINTS* reorganizes the trace of cache misses, and groups misses by associating them with the addresses of the instructions that trigger them. Typically, few static instructions are responsible for most misses and misses from the same instruction tend to exhibit strong repetitive patterns.

SPRINTS uses the Sequitur [1] grammar to compress the trace of misses into a compact representation. The representation stores the misses as strings of symbols, with each string corresponding to one array reference accessed in one loop nest. Each symbol in a string represents the distance (in loop iterations, measured after loop linearization) from the previous miss for the same data object in the same nest. The grammar is composed of rules in which terminals (symbols) represent unique distances between consecutive misses and non-terminals represent concatenations of terminals which uniquely identify the stream of cache misses for each array reference. Sequitur constructs a context-free grammar with exactly one word for each reference. The grammar can be represented as a set of DAGs and the whole stream of misses can be reproduced (uncompressed) from the grammar with one preorder pass, in time linear to the length of the grammar. The length of each string is equal to the number of misses incurred by the corresponding reference, which can then be quantified as a fraction of the total number of cache misses incurred in the whole program and used to classify the reference as delinquent or not. In order to identify strong patterns with Sequitur, it suffices to find non-terminals (sub-strings of the grammar) with multiple occurrences. Each such substring can be translated to a loop which prefetches the stream. These loops are highly optimizable[2] and even parallelizable from a standard back-end compiler.

Consider Figure 1, which shows a loop from the NAS BT benchmark. The loop belongs to the `x_backsubstitute` function of the `x_solve` module of BT. The right part of the figure shows the rules of the Sequitur grammar which describes all the cache misses incurred in accesses to elements of `lhs`, in a single string of inter-miss loop iteration distances. For further details on how this grammar is constructed the reader is referred to [1]. In the illustrated example, all integers prefixed with an ampersand are terminal symbols and all other symbols are non-terminals. One can easily observe that after the first miss, cache misses exhibit a very strong pattern with inter-miss distances

---

[2] Dead-code elimination is the only optimization that needs to be precluded in precomputation loops. Furthermore, *SPRINTS* replaces delinquent stores with loads to preserve correctness in the architectural state.

```
for (i = grid_points[0]-2; i >= 0; i--)
  for (j = 1; j < grid_points[1]-1; j++)
    for (k = 1; k < grid_points[2]-1; k++)
      for (m = 0; m < BLOCK_SIZE; m++)
        for (n = 0; n < BLOCK_SIZE; n)
          rhs[i][j][k][m] -=
                lhs[i][j][k][CC][m][n]
              *rhs[i+1][j][k][n];
```

```
&2  1  1  2  3  3  3  4  5  6  7  8  &5  1
1  ->  &8  &8
2  ->  1   &8
3  ->  4   4
4  ->  9   9
5  ->  6   6
6  ->  10  10
7  ->  8   8
8  ->  11  11
9  ->  5   5
10 ->  12  12
11 ->  13  13
12 ->  7   7
13 ->  &5  2  2  &8
```

Fig. 1. Sample loop of x_backsubstitute in NAS BT and compression grammar for the cache misses incurred by elements of lhs, during execution with the Class A problem size.

predominantly equal to 8 iterations, and sporadically equal to 5 iterations. The grammar given in this example describes a total of 6 million cache misses on element lhs (spread over 200 iterations executed by BT, with approximately 30 thousand misses each) with only 13 rules and a couple of hundreds bytes of storage. A back of the envelope calculation will show that the entire cache miss sequence of the specific data access is represented uniquely with the string: $28^7 A^{3072} A^{512} A^{128} A^{64} A^8 A^4 58^2$, where $A = 58^7$. The grammar is easily translated into tight loops for prefetching the cache-missing elements of lhs using a recursive algorithm which visits each rule of the grammar in order.

The precomputation code generation phase of *SPRINTS* uses the loop iterations as natural units for controlling the distance between the precomputation and sibling computation threads. Furthermore, it uses loop iterations to throttle the precomputation thread, so that the data fetched in a stream do not overflow the L2 cache. Both techniques (runahead distance control and throttling) derive from our earlier work [15]. Another optimization applied by *SPRINTS* is the release of processor resources held by a precomputation thread when the latter is idling and not fetching streams.

3   Experimental Evaluation

We present experiments obtained with the OpenMP, C versions of BT and FT, two realistic application codes from the NAS benchmarks suite [6], both using the class A problem size. BT is a simulated CFD application which uses an implicit finite-difference algorithm based on the alternate direction implicit method, to solve 3-dimensional compressible Navier-Stokes equations. FT implements a solver for a class of PDEs using a 3-dimensional bidirectional (forward and inverse) complex FFT. BT and in are good candidates for speculative precomputation techniques, because their parallelized versions exhibit performance degradation (in the case of BT), or very modest performance gains (in the case of FT), when executed on SMTs with two execution contexts. The performance bottlenecks of parallelization stems from contention for execution units and cache space. A speculative precomputation thread can alleviate these problems and provide speedup by reducing memory latency. The applications have been compiled with the Intel C/C++ OpenMP compiler, using the highest level of optimization. Our hardware platform is a four-way SMP with Intel's Hyperthreaded Xeon processors, clocked at 1.4 GHz. Each processor offers two execution contexts and is equipped with 8KB L1 data cache, 12KB L1 instruction trace cache and 256 KB unified L2 cache.

The Hyperthreaded processors include a hardwired hardware prefetching engine, which was active throughout all the experiments. The hardware prefetching engine may interfere with software prefetcing engines, such as *SPRINTS*, by detecting and prefetching some of the references prefetched also by the software prefetching engine. This effect can not be quantified with the tools available on the specific processor. It must be noted that the automatic software prefetching engine of the Intel compiler was activated in the baseline sequential execution of the benchmarks, as well as in parallel executions of the benchmarks with two execution contexts per processor. However, the Intel's prefetching engine was deactivated while generating code with *SPRINTS*. We have also experimented with manual, non-speculative software prefetching via directives to the Intel compiler in both single-threaded and multithreaded versions of the codes, but we have not seen appreciable performance improvements. In the experiments with *SPRINTS*, we have used a runahead distance of one iteration for each loop targeted by the software precomputation engine. The distance was controlled without synchronization, by having the precomputation thread prefetch references from the second iteration onwards.

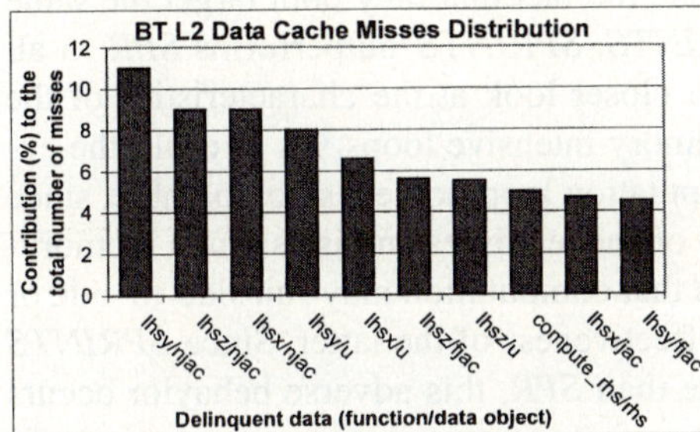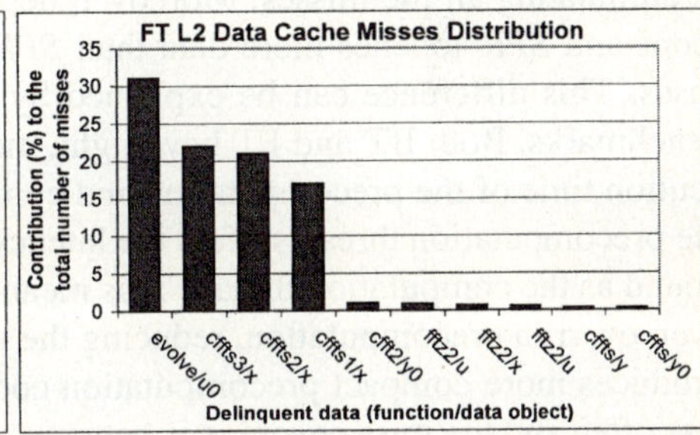

Fig. 2. The top delinquent data objects and their contribution to the total number of L2 data cache misses for BT (left diagram) and FT (right diagram).

Figure 2 depicts the contribution of the top 10 delinquent data accesses of BT and FT, to the total number of L2 data cache misses for the two applications. Those objects are responsible for 85% and 91% of the total cache misses in BT and FT respectively. In fact in FT, 4 objects generate 85.5% of the cache misses. It is thus reasonable to generate precomputation code targeting just the top few delinquent objects.

Following, we evaluate the impact of 4 different execution strategies to the number of L2 data cache misses suffered by the applications. The results are depicted in figure 3. *ST* stands for the single-threaded execution with one execution context in the processor. In *TLP* (*T*hread *L*evel *P*arallelism) mode, applications are executed in parallel by two threads, each one on a different execution context of the processor. The *SPR* (*S*peculative *PR*ecomputation) scheme exploits one of the contexts to execute a precomputation thread, which indiscriminately preexecutes all the memory references of the computation thread in each loop nest where *SPR* is applied. Finally, *SPRINTS* stands for the execution of the application using our precomputation framework. The precomputation thread of *SPR* executes exactly the same loops as *SPRINTS*.

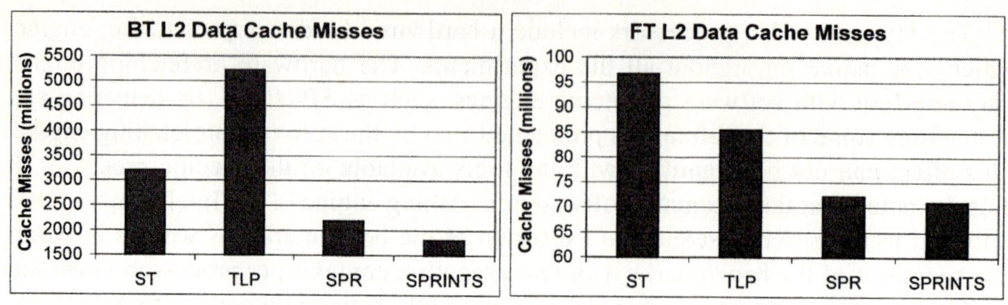

Fig. 3. L2 data cache misses under the four different execution strategies for BT and FT.

As expected, both *SPR* and *SPRINTS* significantly reduce the number of L2 data cache misses. *SPR* results to 31.6% and 25.3% less misses for BT and FT respectively. The corresponding percentages for *SPRINTS* are 42.7% and 25.6%. Although the data accesses targeted by the precomputation strategies are responsible for 85% and 91% of the misses triggered by BT and FT respectively, none of the strategies is successful in eliminating all the misses. Moreover, despite the fact that they both target the same loops and *SPR* touches more data than *SPRINTS*, *SPRINTS* outperforms *SPR* in all cases. This difference can be explained by a closer look at the characteristics of the benchmarks. Both BT and FT have tight, memory intensive loops. As a result, the execution time of the precomputation and computation loop bodies is comparable, since the precomputation thread suffers the latency of cache misses and is as much memory-bound as the computation thread. This means that computation may run side-to-side or even overrun precomputation, reducing the effectiveness of the latter. Since *SPRINTS* produces more compact precomputation code than *SPR*, this adverse behavior occurs less often and the miss coverage is better.

The effect of *TLP* on cache performance is also highly dependent on the characteristics of applications. The two threads of BT contend for L2 cache space, since their working sets do not fit in the cache. This results to a dramatic increase of 63% in cache misses. Contrary to BT, the threads of FT have smaller working sets that fit in the L2 cache. Moreover, they share data and each thread benefits from data prefetched to the cache by the other thread. As a consequence, the multithreaded execution suffers less L2 data cache misses than the sequential execution.

Table 1. Speedups over the single threaded execution using the alternative execution strategies.

	TLP	*SPR*	*SPRINTS*
BT	0.76	1.02	1.08
FT	1.03	1.03	1.05

Table 1 shows the speedups achieved by the three execution strategies which exploit both execution contexts of the processor over the single-threaded execution. The performance of the *TLP* version of BT is poor because of severe cache thrashing, as shown in figure 3. The outcome is a slowdown of 1.32 over the single-threaded execution. In the case of FT, multithreading is beneficial for cache performance, however it yields a marginal speedup of 1.03. The extensive resource sharing in Intel Hyperthreaded processors clearly does not allow effective exploitation of loop-level parallelism. For both benchmarks, the latency overlap achieved with multithreaded execution and the additional instruction-level parallelism do not measure up to the memory latency reduction achieved by precomputation. The overall performance of *SPR* is slightly better. *SPRINTS*, outperforms both *TLP* and *SPR*. Beyond the higher impact of *SPRINTS* on cache performance, the generation of efficient source code for precomputation results to smaller instruction streams and instruction working sets for the precomputation thread. This reduces the pressure on shared execution units, to the benefit of the computation thread. It must be noted that the magnitude of these speedups should be placed in the context of the capabilities of Intel's Hyperthreaded processors. The speedups attained with *SPRINTS* are comparable or higher than the speedups reported so far from physical experimentation with these processors [8].

4   Related Work

Research on SPR can be broadly classified into two classes: hardware-based SPR and software-based SPR. Hardware schemes identify accesses to precompute dynamically, by recording loads and their latencies at either the instruction fetch or the instruction retirement stage. Hardware schemes compose SPR code from the recorded delinquent loads and issue this code dynamically to hardware-triggered threads [3, 13]. The most aggressive hardware designs provide also a register communication mechanism to trigger SPR threads efficiently [12] without involving the operating system, and use manual or semi-automated construction of SPR instruction sequences. *SPRINTS* shares similarities with p-slices of Roth and Sohi [12] in that conceptually, both techniques try to derive highly optimized sequences of precomputation instructions and they both use results from simulation to drive the hadware/software precomputation engine. However, *SPRINTS* is a software technique which requires no hardware or compiler support.

Software SPR schemes can be based on programmer hints [10], compiler techniques [8] or binary modification techniques at load time [9]. Compiler and programmer-assisted techniques are more portable than binary modification techniques. Compiler techniques are preferable to programmer-assisted techniques because they are easy to use. *SPRINTS* shares this advantage with compiler techniques, but at the same time it differs in some important aspects: *SPRINTS* does not apply program analysis or runtime code profiling to detect delinquent loads, or perform any other SPR-specific optimization. It uses off-line cache simulation to identify all memory accesses that incur L2 misses and a compression grammar coupled with simple heuristics to pick those accesses that are responsible for dominant streams of L2 misses. The speculative streaming code is generated in the same high-level language as the sequential code, and can be optimized and executed efficiently from an unmodified compiler back-end and a standard multithreading runtime system. *SPRINTS* does not require program slicing, array access analysis, or other advanced compiler support to identify potential cache misses. Finally, *SPRINTS* targets specifically memory-bound scientific applications, which have not been targeted earlier compiler-based SPR schemes.

SPRINTS borrows the algorithm and the Sequitur grammar for compressing streams of delinquent memory references from earlier work on dynamic hot data stream prefetching [2]. *SPRINTS* differentiates from dynamic hot data stream prefetching in the following aspects: First, *SPRINTS* uses offline analysis of traces of memory references that miss in the L2 cache, rather than online analysis of complete traces of memory references as they appear in the program. In other words, *SPRINTS* compresses traces of misses rather than traces of accesses. This decision is mandated by the tight time constraints of prefetching, which in turn calls for high prefetching accuracy and timeliness. Second, *SPRINTS* uses offline, rather than online analysis of traces. This is dictated by the use of simulation, which is an inherently slow technique for detecting streams of misses, but detects accurately such streams. An online application of *SPRINTS* would be possible with additional hardware support for buffering streams of cache misses and the associated target memory addresses. Intel Itanium processors provide such functionality [5]. Third, *SPRINTS* exploits simultaneous multithreading, while dynamic hot data stream prefetching uses a single-threaded prefetching mechanism. Finally, in contrast to dynamic hot data stream prefetching which targets sequential codes dominated by pointer-chasing, *SPRINTS* targets memory-intensive scientific codes, which are dominated by streams of memory references with predictable patterns.

5 Conclusions

This paper presented *SPRINTS*, a source-level streaming precomputation technique designed to improve the performance of memory-bound scientific applications on SMT processors with limited resources. Resource sharing often renders the execution engine incapable of achieving high-performance from regular, thread-level parallelization on these processors. *SPRINTS* requires no compiler or hardware support. It uses a compact representation of traces of cache misses and exploits this representation to associate delinquent memory accesses to data elements in the source code and produce highly efficient, source-level precomputation code. Experiments with realistic scientific applications show that *SPRINTS* clearly outperforms both TLP and indiscriminate speculative precomputation on Intel's Hyperthreaded processors. In the near future we plan to address a number of design and implementation issues of *SPRINTS*, including the use of lossy compression to improve the quality of streams by filtering out noisy irregular references, the use of mechanisms that can project the miss streams for multiple data inputs from one cache simulation with a single representative input, and the deployment of *SPRINTS* in multi-SMT systems.

Acknowledgements

This work is supported by an NSF CAREER Award (NSF CCF–0346867), an NSF ITR grant (NSF ACI–0312980) and the College of William and Mary.

References

1. T. Chilimbi. Efficient Representations and Abstractions for Quantifying and Exploitiing Data Reference Locality. In *Proc. of the 2001 ACM SIGPLAN Conference on Programming Languages Design and Implementation (PLDI)*, pages 191–202, Snowbird, UT, June 2001.

2. T. Chilimbi and M. Hirzel. Dynamic Hot Data Stream Prefetching for General Purpose Programs. In *Proc. of the 2002 ACM SIGPLAN Conference on Programming Languages Design and Implementation (PLDI'2002)*, pages 199–209, Berlin, Germany, June 2002.

3. J. Collins, D. Tullsen, H. Wang, and J. Shen. Dynamic Speculative Precomputation. In *Proc. of the 34th Annual ACM/IEEE International Symposium on Microarchitecture (MICRO-34)*, pages 306–317, Austin, TX, December 2001.

4. J. Collins, H. Wang, D. Tullsen, C. Hughes, Y. Lee, D. Lavery, and J. Shen. Speculative Precomputation: Long-Range Prefetching of Delinquent Loads. In *Proc. of the 28th Annual International Symposium on Computer Architecture (ISCA–28)*, pages 14–25, Göteborg, Sweden, July 2001.

5. S. Eranian. The Perfmon2 Interface Specification. Technical Report HPL-2004-200R1, HP Labs, February 2005.

6. H. Jin, M. Frumkin, and J. Yan. The OpenMP Implementation of the NAS Parallel Benchmarks and its Performance. Technical Report NAS-99-011, NASA Ames Research Center, October 1999.

7. Ron Kalla, Balaram Sinharoy, and Joel M. Tendler. IBM Power5 Chip: A Dual-Core Multithreaded Processor. *IEEE Micro*, 24(2):40–47, March/April 2004.

8. D. Kim and D. Yeung. A Study of Source-Level Compiler Algorithms for Automatic Construction of Pre-Execution Code. *ACM Transactions on Computer Systems*, 22(2):326–379, 2004.

9. S. Liao, P. Wang, H. Wang, G. Hoflehner, D. Lavery, and J. Shen. Post-Bass Binary Adaptation for Software-Based Speculative Precomputation. In *Proc. of the 2002 ACM SIGPLAN Conference on Programming Languages Design and Implementation (PLDI'2002)*, Berlin, Germany, June 2002.

10. C. Luk. Tolerating Memory Latency through Software Controlled Preexecution on Simultaneous Multithreading Processors. In *Proc. of the 28th Annual International Symposium on Computer Architecture (ISCA'01)*, pages 40–51, Göteborg, Sweden, July 2001.

11. Deborah T. Marr, Frank Binns, David L. Hill, Glenn Hinton, David A. Koufaty, J. Alan Miller, and Michael Upton. Hyper-Threading Technology Architecture and Microarchitecture. *Intel Technology Journal*, 6(1), February 2002.

12. A. Roth and G. Sohi. A Quantitative Framework for Quantitative Pre-Execution Thread Selection. In *Proc. of the 35th IEEE/ACM Annual International Symposium on Microarchitecture (MICRO–35)*, Istanbul, Turkey, November 2002.

13. K. Sundaramoorthy, Z. Purser, and E. Rotenberg. Slipstream Processors: Improving both Performance and Fault Tolerance. In *Proc. of the 9th International Conference on Architectural Support for Programming Languages and Operating Systems (ASPLOS-IX)*, pages 191–202, Cambridge, MA, November 2000.

14. UltraSPARC©IV Processor Architecture Overview. Technical report, Sun Microsystems, February 2004.

15. T. Wang, F. Blagojevic, and D. Nikolopoulos. Runtime Support for Integrating Precomputation and Thread–Level Parallelism on Simultaneous Multithreaded Processors. In *Proc. of the 7th ACM SIGPLAN Workshop on Languages, Compilers and Runtime Support for Scalable Systems (LCR'2004)*, Houston, TX, October 2004.

Symmetric Data Objects and Remote Memory Access Communication for Fortran-95 Applications

Jarek Nieplocha[1], Doug Baxter[1], Vinod Tipparaju[1],
Craig Rasmunssen[2], and Robert W. Numrich[3]

[1] Pacific Northwest National Laboratory
Richland, WA, USA
[2] Los Alamos National Laboratory
Los Alamos, NM, USA
[3] Minnesota Supercomputing Institute, University of Minnesota
Minneapolis, MN, USA

Abstract. Symmetric data objects have been introduced by Cray Inc. in context of SHMEM remote memory access communication on Cray T3D/E systems and later adopted by SGI for their Origin servers. Symmetric data objects greatly simplify parallel programming by allowing programmers to reference remote instance of a data structure by specifying address of the local counterpart. The current paper describes how symmetric data objects and remote memory access communication could be implemented in Fortran-95 without requiring specialized hardware or compiler support. NAS Multi-Grid parallel benchmark was used as an application example and demonstrated competitive performance to the standard MPI implementation.

1 Introduction

Fortran is an integral part of the computing environment at major scientific institutions. It is often the language of choice for developing applications that model complex physical, chemical, and biological systems. In addition, Fortran is an evolving language [1]. The Fortran 90/95 standard introduced many new constructs, including derived-data types, new array features and operations, pointers, increased support for code modularization, and enhanced type safety. These features are advantageous to scientific applications and improve the programmer productivity.

Remote memory access (RMA) operations facilitate an intermediate programming model between message passing and shared memory. This model combines some advantages of shared memory, such as direct access to shared/global data, and the message-passing model, namely the control over locality and data distribution. Certain types of shared memory applications can be implemented using this approach. In some other cases, remote memory operations can be used as a high-performance alternative to message passing. On many modern platforms, RMA is directly supported by hardware and is the lowest-level and often most efficient communication paradigm available. The RMA has been offered in numerous portable interfaces ranging from SHMEM [10, 11, 16], ARMCI [12], and MPI-2 [13]. Among these, Cray SHMEM has been the most widely used interface and offered by hardware vendors such as Cray, IBM, HP for their architectures. Some important characteristics of SHMEM are the ease of use, simplicity, demonstrated potential for achieving high performance.

J.C. Cunha and P.D. Medeiros (Eds.): Euro-Par 2005, LNCS 3648, pp. 720–729, 2005.

One of the important characteristics of SHMEM is support for *symmetric data objects*. This concept allows the programmer to access remote instances of data structures through references to the local instance. In particular, the programmer is not required to keep track of addresses on remote processors as mandated by other RMA models such as LAPI[15] on the IBM SP where addresses for remote instances of the same data object can be different and thus need to be exchanged and stored on all processors. Implementation of symmetric data objects is difficult without hardware and/or OS assistance on clustered systems. This is because the virtual memory addresses allocated by the operating system for storing instances of the same data structure in a cluster can be different across the machine. Without symmetric data objects, the programmer would be required to store $O(P^2)$ addresses on the machine. In addition, the overall programming model is harder to use and the application codes become more error prone.

In this paper, we take advantage of the new Fortran-95 features to provide high-level interfaces to one-sided operations on multidimensional arrays consistent with symmetric data-object model of SHMEM. This work is inspired by Co-Array Fortran (CAF) [14] with its ability to reference arbitrary sections of so called *co-arrays* using high-level array assignments. *Co-arrays* represent a special type of Fortran-95 arrays defined on all tasks in the SPMD program. The main contributions of this paper are: 1) definition of an interface that support important features of SHMEM and CAF using a library- rather than compiler-based approach: symmetric data objects of SHMEM and one-sided high-level access to multidimensional arrays that CAF offers (SHMEM does not support it), 2) a description of a portable implementation of these features without relying on hardware or OS support, and 3) demonstration that the proposed approach can deliver high performance, both in context of microbenchmarks as well as the NAS NPB Multigrid (MG) benchmark [4].

The remainder of the paper is organized as follows. Section 2 describes the proposed interface and discusses its characteristics. Section 3 describes the implementation based on Chasm and the ARMCI one-sided communication library. Section 4 reports experimental results on the Linux cluster with Myrinet that demonstrate that our implementation outperforms the NAS NPB version of MG.

2 Proposed Approach

We propose to support symmetric data objects and RMA for Fortran-95 applications based on Fortran-95 array pointers with special memory allocation interface and a set of remote memory access communication interfaces handling slices (sections) of Fortran-95 arrays. These interfaces allow users to allocate/free multidimensional Fortran-95 arrays and to communicate data held in this memory using simple get/put semantics. In addition, the reference to the remote instance of arrays does not require users to keep track of addresses on remote note. A single Fortran-95 pointer is used to represent local and remote instances of a multidimensional array. A unique feature of these interfaces is that they allow users to take full advantage of Fortran-95 array mechanisms (like array-valued expressions).

The Fortran interfaces are as follows. Memory allocation is done with calls to the generic interfaces `Malloc_fa` and `Free_fa` (shown below for real, two-dimensional arrays only),

```fortran
module Mem_F95

   interface Malloc_fa
     subroutine Malloc_2DR(a, lb, ub, rc)
       real, pointer :: a(:,:)
       integer, intent(in) :: lb(2), ub(2)
       integer, intent(out) :: rc
     end subroutine Malloc_2DR
   end interface

   interface Free_fa
     subroutine Free_2DR(a, rc)
       real, pointer :: a(:,:)
       integer, intent(out) :: rc
     end subroutine Free_2DR
   end interface

end module Mem_F95
```

In the above, the arrays `lb` and `ub` contain the lower and upper bounds of the array to
be allocated and the parameter `rc` is an error code. Similarly, the generic interfaces
for RMA communication are `Put_fa` and `Get_fa`. To save space, here we only
present the interface to the first one for the double precision two dimensional arrays
(the get interface is similar).

```fortran
module Types_fa
type Slice_fa
     integer :: lo(7)
     integer :: hi(7)
     integer :: stride(7)
   end type Slice
end module Types_fa

module Mov_F95
   interface Put_fa
     subroutine Put_2DR(src, src_slc, dst, dst_slc,
                                        proc, rc)
       use Types_fa
       real, pointer :: src(:,:), dst(:,:)
       type(Slice_fa), intent(in) :: src_slc, dst_slc
       integer, intent(in) :: proc, rank
       integer, intent(out) :: rc
     end subroutine Put_2DR
   end interface
```

In the communication interfaces, `src` and `dst` are the source and destination arrays respectively, `src_slc` and `dst_slc` contain information about the memory
portion (array section) of the source and destination arrays to be used, `proc` is the
processor number of the destination array, and `rc` is an error return code. In addition
to being able access sections of multidimensional arrays, to be consistent with the
Fortran-95 capabilities for arrays, the user can also specify stride information.

The current implementation supports integer, floating, and complex data types of the 8- and 4-byte kinds but it can also be extended to other Fortran data types (character and logical). Array dimensions ranging from one to seven (Fortran limit) are handled. By exploiting Fortran-95 function name overloading, we can use a single name for the put operation `Put_fa` to handles all data types and array dimensions using a single interface. The defined memory allocation interfaces defined above are mandatory for allocating memory to be accessed remotely from other processors. However, they are not required for local arrays: source of data in *put*, and destination in *get*.

The semantics of the RMA operations (progress, ordering) follow closely that of the Cray SHMEM. In order to provide the application programmer with abilities to hide latency, we introduced nonblocking interfaces to put/get calls. A nonblocking call returns before the user buffer can be accessed and requires a special *wait* function to complete. This feature is not available in SHMEM. (Although the CAF standard does not offer this capability, the Rice CAF compiler adds directives that change semantics of the array assignments to nonblocking in so called non-blocking regions.)

3   Implementation

Unfortunately, the Fortran-95 standard *alone* does not provide sufficient capabilities to implement the memory management required to support symmetric data objects. However, this is made possible by the use of the Chasm array-descriptor library [9]. In addition, we use the ARMCI portable RMA library to handle communication. Our approach also relies on MPI for job startup and control. In fact, the user can use the interfaces described in the previous section in the MPI programs and take advantage of the full capabilities of MPI e.g., collective operations.

Chasm

Chasm [12, 3] is language transformation system providing language interoperability between Fortran and C/C++. Language interoperability is provided by stub and skeleton interfaces. This code is generated by language transformation programs taking as input existing user C, C++ or Fortran source code and generating the stub and skeleton interfaces to the input code as output [3].

One of the challenges of language interoperability with Fortran is that Fortran assumed-shape array arguments are passed by an array descriptor, rather than as a simple memory address. Array descriptors contain meta data about the array, including the base address of the array, the lower and upper bounds for each dimension of the array, and *sometimes*, the rank of the array and the type of an array element. The key point is that the format of the array descriptor is not specified by the language standard, but is left to be specified by the vendor of the Fortran compiler. Chasm provides generic C interfaces to the Fortran, vendor-specific array descriptors. Without the Chasm array-descriptor library, there would be no way to call the ARMCI library from Fortran and allocate Fortran-95 arrays using the special ARMCI memory necessary for remote communication.

It should be noted that the need for the Chasm array-descriptor library will be reduced somewhat once compiler vendors have implemented the Fortran 2003 standard [8]. Fortran-2003 contains standard mechanisms for interoperating with C that allow

Fortran array pointers to be associated with memory allocated from C. In addition, a modified version of the Chasm, array-descriptor interface has been accepted by the Fortran J3 committee [1] for possible inclusion in the next Fortran standard. This would then allow the Fortran interfaces, introduced in the previous section, to be used in a language standard way, with no additional stub or skeleton code needed. Until this time, either Chasm or Fortran-2003 compilers (with slightly modified Fortran stub code) will be needed.

ARMCI

The Aggregate Remote Memory Copy Interface (ARMCI) [6] is a portable RMA communication library. It has been used for implementing distributed array libraries such as Global Arrays, other communication libraries such as Generalized Portable SHMEM [10], and compiler run-time systems such as the Adlib [17] or the portable Co-Array Fortran compiler at Rice University [5]. ARMCI offers an extensive set of functionality in the area of RMA communication: 1) data transfer operations; 2) atomic operations; 3) memory management and synchronization operations; and 4) locks. In scientific computing, applications often require transfers of noncontiguous data that corresponds to fragments of multidimensional arrays, sparse matrices, or other more complex data structures. With remote memory communication APIs that support only contiguous data transfers, it is necessary to transfer noncontiguous data using multiple communication operations. This often leads to inefficient network utilization and involves increased overhead. ARMCI, however, offers explicit non-contiguous data interfaces: strided and generalized I/O vector that allow description of the data layout so that it could, in principle, be transferred in a single message. Of course, the effectiveness of actual transfers depends on the ability of underlying net-works to deal with noncontiguous data (e.g., scatter/gather operations). However, even when scatter/gather operations are not supported by the network, the ARMCI strided and vector operations take advantage of the information – for example, at the level of data packing/unpacking – so that the overall number of messages and network packets is reduced. The strided interfaces are important for Fortran-95 applications that use multidimensional arrays.

Fortran-95 Interfaces

The C side of the implementation is composed of 10 functions. Four of the functions are administrative functions for initializing Fortran array descriptor information, cleaning up, terminating, and synchronizing that take no arguments. The other six are functions for allocating Fortran-95 arrays that the ARMCI data movement routines can handle, blocking *put* and *get* operations for the data movement, their nonblocking analogs, and a function to free the allocated deferred shape arrays.

These functions assume the following Fortran calling convention. When calling routines with the deferred shape arrays (allocatable) as arguments, each deferred shape array argument contributes two addresses to the actual argument list. The first is the data address of the first element of the array and is in order specified in the arguments of the Fortran call/function reference. The second is the address of the dope vector describing the deferred shape array and is placed after the end of the arguments listed in the Fortran call or function reference. Routines with more than one deferred shape array have all of the addresses of the dope vectors concatenated at the end of the argument list appearing in the same relative order as the corresponding deferred shape array in the Fortran argument list. To support symmetric data objects even on clusters with virtual memory nodes, we allocate extra array memory (in addition to the user specific portion) to store array pointers on the remote nodes. When user specifies pointer to the local instance of the Fortran-95 array, we access the appropriate pointer for the specified processor and pass the required information to ARMCI put/get calls.

On the Fortran-95 side of the interface, there are corresponding routines to allocate, put, get (blocking/nonblocking) and free array memory. Fortran-95 does not have the notion of a generic pointer type, the equivalent of a (void *) in C. Each pointer in Fortran must point to an array of specified type and dimension (number of indices used to reference elements in the array). Module procedures are used to overload the C functionality of void *, giving a similar interface on the Fortran side where the user does not have to use a different function name for using ARMCI calls on different data types. Six types of elementary data are supported for one to seven dimensions yielding 42 Fortran routines for each corresponding C function (*allocate, free, put, get* and nonblocking *put* and *get*). The six Fortran data types supported are four (I4) and eight- (I8) byte integers, four- (R4) and eight- (R8) byte floating point numbers and eight (C4) and sixteen (C8) byte complex numbers. The following parameters provide a portable shorthand for defining these types and are found in the definekind Fortran-95 file:

```
module definekind
    integer, parameter :: I4 = SELECTED_INT_KIND(9)
    integer, parameter :: I8 = SELECTED_INT_KIND(16)
    integer, parameter :: R4 = SELECTED_REAL_KIND(5)
    integer, parameter :: R8 = SELECTED_REAL_KIND(12)
    integer, parameter :: C4 = SELECTED_REAL_KIND(5)
    integer, parameter :: C8 = SELECTED_REAL_KIND(12)
end module definekind
```

For each operation in each of the 42 flavors, the `definekind` module is included and the appropriate type and dimension arguments are declared in an interface block to the generic C routine.

Sample RMA Code Using Fortran-95 Interfaces

Below is a sample code snippet that allocates a couple of 50X50 arrays of integers, `src_arr` and `dst_arr`, and does a put operation from one to another. These arrays are first allocated with Malloc_fa interface and then `src` and `dst` slice information is filled up before doing the put communication.

```
integer(kind=4),pointer::src_arr(:,:),dst_arr(:,:)
type(Slice_fa) :: src_sl,dst_sl
integer :: lb(2), ub(2), ierr
lb(:) = 1
ub(:) = 50
call Malloc_fa(src_arr,lb,ub,ierr)
if (ierr .ne. 0) call myerror()
```

```
call Malloc_fa(dst_arr,lb,ub,ierr)
if (ierr .ne. 0) call myerror()
src_sl%lo(:) = 1
src_sl%hi(:) = 25
src_sl%stride(:) = 2
dst_sl%lo(:) = 25
dst_sl%hi(:) = 50
dst_sl%stride(:) = 2
Put_fa(src_arr,drc_sl,dst_arr,dst_sl,dst_proc,ierr)
```

4 Experimental Evaluation

We measured the latency and bandwidth of the Fortran 95 RMA calls with micro-benchmarks. We also ported the NAS MG benchmark to use Fortran-95. The experimental evaluation was carried out on a 24-node dual processor Intel Itanium2 1GHz cluster interconnected with Myrinet network [7]. The cluster was running Linux version 2.4.20 operating system. We used the GM dual port E cards, GM 2.1.4 and MPICH 1.2.5..12. For this test, we used Intel IFC Fortran 7.0 compilers and the 2.96 version of the GNU C compiler. We also used ARMCI 1.1 and Chasm 1.1.0 for the implementation.

Microbenchmarks

We measured the latency and bandwidth of Fortran-95 RMA interfaces with a micro-benchmark that does consecutive put and get operations from different memory locations and averages the time taken for each operation. This is a simple microbench-mark that shows the bandwidth and latency of the Fortran-95 RMA interfaces. In addition we also used a similar microbenchmark to measure the bandwidth and latency of ARMCI put and get operations in order to measure the overhead from using Fortran-95 interfaces that involved interface mapping and all the dope vector manipulations. The bandwidth of the put and get operations for the Fortran-95 RMA interfaces is shown in Figure 1. The figure also includes the bandwidth of the corresponding ARMCI put and get calls. The overhead is independent of the message size and it is related to the cost of duplicating dope vector through Chasm that includes *calloc* system call. Based on these findings, the next version of Chasm will include an alternative mechanism for accessing some of the information stored in the dope vector that will be based on portable macros rather than duplication of the dope vector. The asymptotic bandwidth in the above microbenchmarks is consistent with the bandwidth results of the Myricom GM [7].

NAS MG Benchmark

The Numerical Aerodynamic Simulation (NAS) parallel benchmarks (NPB) are a set of programs designed at NASA. Our starting point was NPB 2.4 [4] implementation written in MPI and distributed by NASA, we modified it to be compiled as a Fortran-95 file. We replaced MPI calls with the Fortran-95 non-blocking RMA interfaces. In addition to the mere replacement of the point-to-point message passing communications part of the current message-passing version of MG NAS kernels, an additional

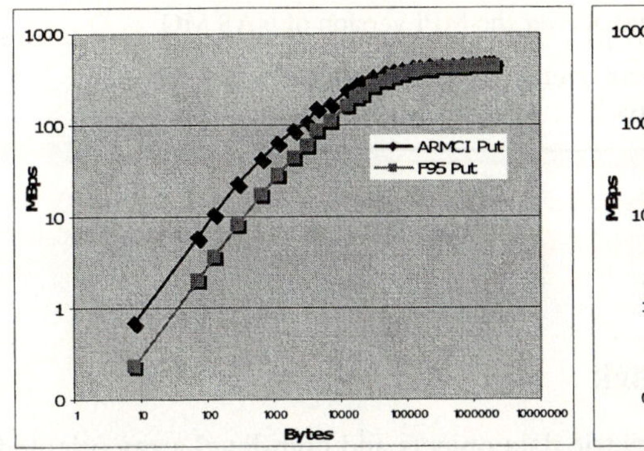
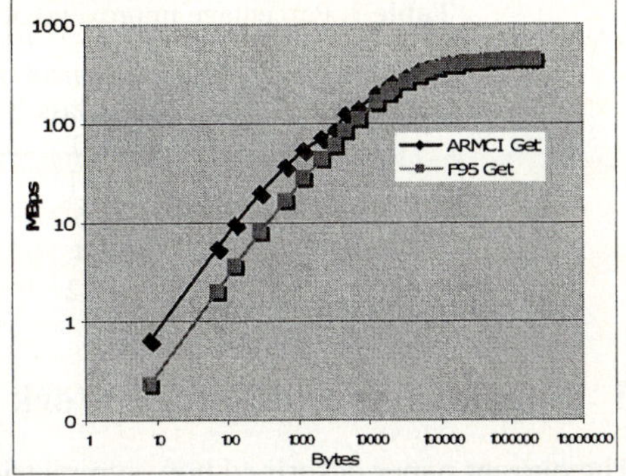

Fig. 1. Left: Bandwidth of a contiguous Fortran-95 (labled F95 in the figure) Put_fa compared to ARMCI put. Right: Bandwidth of the Fortran-95 Get_fa compared to ARMCI Get operation

set of communication buffers were used to better utilize the one-sided nature of the RMA interfaces. Figure 2 shows the performance of NAS Fortran-95 MG version written with Fortran-95 RMA interface and is compared to the original MPI implementation of NAS which has been compiled with Intel Fortran-95 compiler as an Fortran-95 file. Despite the overhead Fortran-95 interfaces involve, the RMA Fortran-95 RMA interface version of the MG benchmark outperforms the MPI version of NAS MG benchmark for Class B and Class C and performs in par with the MPI version for the Class A version of the benchmarks. The performance gains are contributed to the increased asynchronicity of the RMA model as compared to the two-sided message passing implementation of the NAS NPB MG benchmark. Table 1 shows the percentage improvement shown by the Fortran-95 RMA interface implementation of NAS MG over the standard MPI implementation of NAS MG. Up to 30% improvement was observed.

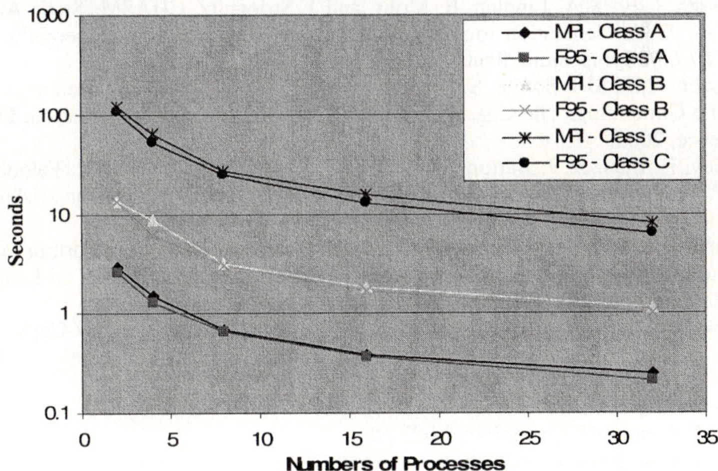

Fig. 2. Fortran-95 (labled as F95) RMA interfaces vs. MPI implementation of NAS MG benchmark for Class A, B and C

Table 1. Percentage improvement over the MPI version of NAS MG

NPROC	%improvement over MPI-Class B	%improvement over MPI-Class C
2	2.0	10.8
4	30.1	19.9
8	5.6	8.3
16	4.4	18.1
32	12.2	21.4

5 Conclusions and Future Work

The current paper described how symmetric data objects and high-level array oriented RMA interfaces can be implemented for Fortran-95 applications. The proposed approach leads to simple yet efficient code, as demonstrated in the context of the NAS NPB Multi-Grid benchmark. In the process of developing the interface we identified sources of overhead involved in accessing elements of the dope vector through Chasm. The next version of Chasm will address them by providing macros for direct access to the information stored in the dope vector required by these interfaces. Our future work in addition to these performance optimizations will include performance comparisons with the Co-Array Fortran code on the Cray X1 where the native Co-Array Fortran compiler is available as well as to the Rice compiler on Linux clusters. The implementation of NAS MG using these Fortran-95 RMA interfaces outperforms the MPI version of the benchmark demonstrating that advantages of the asynchronous RMA communication outweighs the overhead involved in pointer calculations and dope vector manipulations.

References

1. http://www.j3-fortran.org
2. Rasmussen, C.E., K.A. Lindlan, B. Mohr, and J. Striegnitz, CHASM: Static Analysis and Automatic Code Generation for Improved Fortran 90 and C++ Interoperability, Proceedings of LACSI Symposium, Santa Fe, NM, 2004.
3. Rasmussen, C.E., M.J. Sottile, S. Shende, and A.D. Malony, Bridging the Language Gap in Scientific Computing: The Chasm Approach, Concurrency and Computation: Practice and Experience, 2005.
4. D. Bailey, E. Barszcz, J. Barton, D. Browning, R. Carter, L. Dagum, R. Fatoohi, S. Fineberg, P. Frederickson, T. Lasinski, R. Schreiber, H. Simon, V. Venkatakrishnan, and S. Weeratunga, The NAS parallel benchmarks, RNR-94-007, NASA 1994.
5. C. Coarfa, Y. Dotsenko, J. Eckhardt, J. Mellor-Crummey, Co-Array Fortran Performance and Potential: An NPB Experimental Study,16th International Workshop on Languages and Compilers for Parallel Computing. 2003.
6. J. Nieplocha and B. Carpenter, "ARMCI: A Portable Remote Memory Copy Library for Distributed Array Libraries and Compiler Run-time Systems", Parallel and Distributed Processing, Jose Rolim (Eds.), Springer LNCS-1586, 1999.
7. http://www.myri.com
8. http://j3-fortran.org/doc/year/04/04-007.pdf
9. http://sourceforge.net/projects/chasm-interop/
10. K. Parzyszek, J. Nieplocha, and R.A. Kendall, "A generalized portable SHMEM library for high performance computing", in Proc. PDCS'2000, 2000

11. F. Petrini, S. Coll, E. Frachtenberg, and A. Hoisie "Performance Evaluation of the Quadrics Interconnection Network", *Journal of Cluster Computing*, 6(2): 125-142, 2003
12. Rasmussen, C.E.; Lindlan, K.A.; Mohr, B.; Striegnitz, J., CHASM: Static Analysis and Automatic Code Generation for Improved Fortran-90 and C++ Interoperability, Proceedings of the 2nd LACSI Symposium, 2001.
13. http://www.mpi-forum.org/
14. Robert W. Numrich and John K. Reid, Co-Array Fortran for parallel programming. ACM Fortran Forum, vol. 17, 2, 1998.
15. G. Shah, J. Nieplocha, J. Mirza, C. Kim, R. Harrison, R.K. Govindaraju, K. Gildea, P. DiNicola, and C. Bender, "Performance and experience with LAPI: a new high-performance communication library for the IBM RS/6000 SP", Proc., IPPS '98, *1998*.
16. R. Bariuso and A. Knies, SHMEM User's Guide, Cray Research, Inc., SN-2516, 1994.
17. D. B. Carpenter. Adlib: A distributed array library to support HPF translation. Proc. 5th International Workshop on Compilers for Parallel Computers. 1995.

Using Aspects for Supporting Procedural Modules in # Programming

Francisco Heron de Carvalho Junior[1] and Rafael Dueire Lins[2]

[1] Departamento de Computação, Universidade Federal do Ceará
Campus do Pici, Bloco 910, Fortaleza, Brazil
`heron@lia.ufc.br`
[2] Depart. de Eletrônica e Sistemas, Universidade Federal de Pernambuco
Av. Acadêmico Hélio Ramos s/n, Recife, Brazil
`rdl@ufpe.br`

Abstract. Parallel programming still demands for higher-level languages, models, and tools that do not incur in performance penalties. The # programming model aims to meet those claims in large-scale programs. This paper describes how the # programming model works with procedural languages by using techniques from AOP (Aspect Oriented Programming). Performance comparisons with MPI are presented.

1 Introduction

High performance computing (HPC) architectures of today may be split into three classes: *capability computing* (MPP's[1]), *cluster computing* [6] and *grid computing* [14] architectures. Deep memory and source hierarchies can be supported in all classes. Grids, for example, may have clusters and MPP's as processing nodes, which may be formed by multiprocessors. Individual processors may implement vector and super-scalar processing. The consolidation of distributed architectures for HPC have brought new challenges. Efficient parallel programming on these architectures is not a trivial task using the tools available today. Despite having to specify computations, like in sequential programming, programmers must partition the application functionality and/or data, according to the features of the target architecture, and implement process synchronization. There are no consensual models for programming parallel architectures.

The evolution of parallel programming technology may be divided into three phases. The first phase was marked by the use of low level architecture-specific message passing interfaces. The start of the second phase is marked by the creation of CRPC (Center for Research on Parallel Computation), in 1989. From that milestone on, research efforts started to be coordinated, culminating with the development of several *efficient* and *portable* tools, including libraries for message passing (MPI [17] and PVM [15]), parallel extensions of Fortran (HPF [12] and Fortran M [13]) and specific-purpose scientific computing libraries (PETSc [3], ScaLAPACK [5], and many others [11]). The third phase searches

---

[1] Massively parallel processors, the supercomputers.

J.C. Cunha and P.D. Medeiros (Eds.): Euro-Par 2005, LNCS 3648, pp. 730–739, 2005.

for models and languages for programming distributed high performance architectures, reconciling requirements of *generality* (**G**), high level of *abstraction* (**A**), *portability* (**P**) and *efficiency* (**E**), allowing to apply advanced software engineering concepts into the development of HPC software. Despite the efforts promoted in the second phase, and also due to the expansion in scale of HPC applications caused by *cluster* and *grid* computing, reaching the aims of the latter phase is still one of the most important challenges in parallel computing [11, 18].

The # parallel programming model provides a structured way to work with explicit message passing programming. The # parallel programming environment supports the analysis of large scale parallel programs by using Petri nets [9], including "debugging" and simulation facilities, proof of formal properties, and performance evaluation. The idea behind the # environment is to offer a "glue" for integrating existing high performance computing programming technologies in a common **component-based** framework, where advanced software engineering techniques may be successfully applied. The current prototype implementation of the # model is Haskell$_\#$ [8]. Haskell$_\#$ is a coordination language for distributing functional computations in clusters. Computations are described in Haskell, a pure lazy functional language. Haskell was initially adopted because it provides a clean orthogonal interface between coordination and computation media through lazy streams, besides allowing the analysis of formal properties of programs at computation level.

This work presents an approach based on AOP (Aspect Oriented Programming) [16] for incorporating computations written in *procedural languages* into the # programming environment. Procedural languages may either be *imperative* (such as C and Fortran), or *object oriented* (such as C++, Java, and C#). They are widely used for high performance programming, as they provide good time and space performance for scientific computations. Thus, it is possible to think about multi-lingual implementations of the # programming environment. This feature is highly desirable in large scale programming for grids.

This paper comprises three more sections. Section 2 presents an overview of the # programming model. Section 3 shows how procedural modules were introduced to the # programming environment. Section 4 benchmarks the proposed approach. Conclusions and lines for further works are presented in Section 5.

2   The # Component Model

The # programming model moves parallel programming from a *process-based perspective* to an orthogonal *concern-oriented perspective*. From the process-based perspective, a parallel program is a collection of processes synchronizing by means of communication primitives. For improving practice of parallel programming, it had been tried to lift level of abstraction for dealing with these primitives, resulting in efficiency losses. *Concerns* are scattered along implementation of processes, since they are orthogonal to processes. In fact, a process may be viewed as a set of *slices*, each one describing the role of the process with

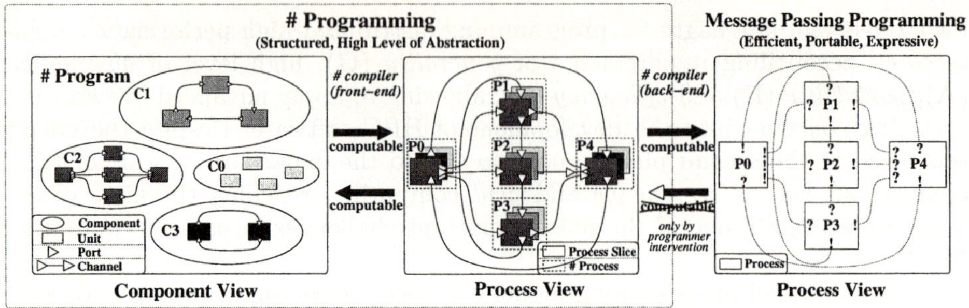

Fig. 1. Component Perspective versus Process Perspective

respect to a given concern. In this context, concerns are *decomposition criteria* for *slicing* processes [19]. Thus, they may be viewed as sets of related slices, probably from distinct processes. From the concern-oriented perspective of parallel programming, proposed by the # model, ***components*** are programming abstractions that address functional and non-functional concerns. We believe that a concern-oriented perspective of parallel programming fits contemporary advanced software engineering artifacts better than a process-based perspective.

In # programming, the slices that comprise a component are called *units*. They are connected in a communication topology, formed by one-direction, point-to-point, and typed channels. For that, a unit has a set of input and output ports, whose activation order is dictated by a protocol, specified using a formalism with expressiveness of labelled Petri nets. In # programming, concerns about parallelism and computations are separated in composed and simple components, respectively. Composed components comprise the ***coordination medium*** of # programs. They are specified in terms of units and channels, possibly by composition of existing components, by using some language that supports the coordination level abstractions of the # model. Today, there are a textual notation, called HCL (# configuration language), and a visual notation, called HVL (# visual language). Simple components are specified using Turing-computable languages, comprising the ***computation medium*** of # programs. They are the atoms of functionality in # programs. Simple components may be assigned to units of composed components in order to configure computations

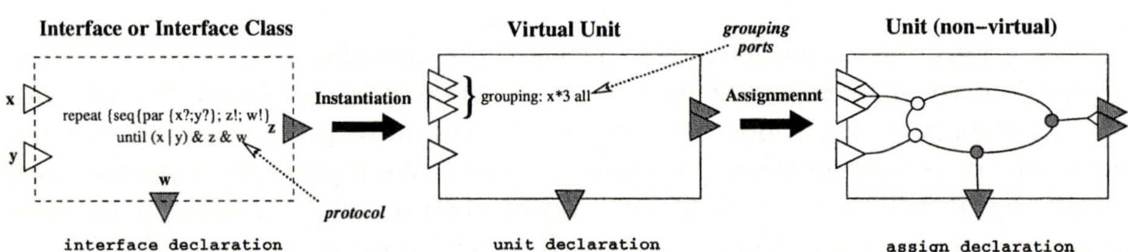

Fig. 2. Configuring a Unit

```
component CPipeLine <N> with

iterator i range [1,N]

interface ICPipe where
        ports: i* → o*
        protocol: repeat seq{o!; i?} until <o & i>

[/ unit pipe[i] where ports: ICPipe /]

connect pipe[i]→o to pipe[i+1]←i, buffered
────────────────────────────────────────────────
component Torus <N> with

use Skeletons.Common.CPipeLine

iterator i, j range [1,N]

interface ITorus where
        ports: ICPipe @ n → s # ICPipe @ e → w
        protocol: repeat seq {par {s!; w!}; par {n?; e?}}
                until <n & e & s & w>

[/ unit vpipe[i]; assign CPipeLine<N> to vpipe[i] /]
[/ unit hpipe[j]; assign CPipeLine<N> to hpipe[j] /]

[/ unify vpipe[i].pipe[j], hpile[j].pipe[i]
        to node[i][j] where ports: ITorus /]
────────────────────────────────────────────────
component Farm<N> with

unit distributor where ports: () → job
unit worker where ports: job → result
                    protocol: seq {job?; result!}
unit collector where ports: result → ()

connect distributor.job to worker.job, synchronous
connect worker.result to collector.result, synchronous

replicate N: worker
────────────────────────────────────────────────
```

```
component SqMatMult<N> where

iterator i, j range [1,N]

use Skeletons.Common.{Torus, Farm}
use MMShift, SPMD

interface ISqMatMult where
    ports: j → r # ITorus
    protocol: seq { j?; repeat seq {par {s!;e!};
                                    par {n?;w?}}
                    counter N;
                r! }

unit mm_torus; assign Torus<N> to mm_torus
unit mm_farm; assign Farm<N> to mm_farm

[/ unify farm.worker[i + j × N], torus.node[i][j]
        to sqmm[i][j] where ports: ISqMatMult /]

unify farm.distributor, farm.collector, sqmm[0][0]
    to sqmm_root where
        ports: () → ab # c → () #
            ISqMatMult @ mm
        protocol: seq {ab!; do mm; c? }

unit spmd; assign SPMD<N × N> to spmd
supersede sqmm to spmd.peer

[/ assign MMShift to sqmm[i][j] /]
────────────────────────────────────────────────
module MMShift(main) where

main :: Num t ⇒ t → t → [t] → [t] → ([t],[t],t)
main a b as_i bs_i = (as_o,bs_o,c)
    where
            c = matmult as_o bs_o
            (as_o, bs_o) = (a:as_i, b:bs_i)

matmult :: Num t ⇒ t → [t] → [t] → t
matmult [] [] = 0
matmult (a:as) (b:bs) = a*b + matmult as bs
────────────────────────────────────────────────
```

Fig. 3. Configuration Code of Matrix Multiplication on a Torus

performed by slices. Skeletons [10] are supported by allowing units with no component assigned, called *virtual units*, giving support for high level of abstraction without loss in efficiency and portability. Nested composition of components is possible by allowing to assign composed components to units of other composed components. Besides to give support for non-functional concerns and skeletons, another important distinguishing feature of the # component model in relation to other component models [1, 4] is its ability to combine components by overlapping them. For that, it is possible to unify units from different composed components. Component models of today allow only nesting composition. Components are black-boxes addressing functional concerns. Whenever supported, non-functional concerns are introduced by means of orthogonal language extensions or by using tangling code cross-cutting component modules, like in sequential programming. However, cross-cutting concerns are not exceptions in parallel programming. The ability to overlap components makes possible to treat cross-cutting concerns as first-class citizens when parallelizing of applications.

Figure 3 presents a simple, yet illustrative, process topology of a composed component, named SqMatMult, that implements a parallel matrix multiplication strategy based on a systolic interaction pattern amongst processes organized in a torus. The code is written in HCL, the textual realization of the # coordination level abstraction. The component SqMatMult is composed by overlapping skeletons Torus and Farm. A $N \times N$ Torus is defined by overlapping $N + N$ instances of CPipeLine. The configuration code of components Torus, Farm, CPipeLine and SqMatMult, in HCL, are also presented in Figure 3.

3   Procedural Modules as Simple Components

In Haskell# [8], simple components are *functional modules* written in the pure lazy functional language Haskell. Haskell provides the simplest technique for linking computation to coordination media without neither intermediate constructors nor extensions to the language Haskell. Functional modules neither make any reference to HCL constructors nor need to import libraries. They are standard Haskell modules, exporting the function *main*, whose arguments and elements of the returned tuple correspond to arguments and return points of the simple component. This is possible due to the Haskell support for *lazy lists*, which are associated to *streams* at coordination level [7]. However, using a language without lazy semantics, other approaches may be applied for keeping orthogonal the separation between coordination and computation media.

Procedural modules are simple components written in procedural languages, encompassing imperative and object oriented (OO) paradigms. They are implemented as *abstract data types* (imperative languages), or *objects* (OO languages). The routines, or methods, declared in procedural modules change the data structure state in the progress of computation. It is needed to define how procedural module routines (or methods) are invoked in response to events at coordination level and to define their arguments and return points. Techniques from Aspect Oriented Programming (AOP) [16] are used for the first purpose. For instance, a procedural module may be associated to aspect configurations, written in the # Aspect Language (HAL). In AOP, programmers may define *pointcut designators* that "identify particular *join points* by filtering out a subset of all the join points within the <u>program</u> flow". In the # terminology, the term <u>program</u> corresponds to the *protocol* of the *unit* for which the procedural module is *assigned*. *Join points* correspond to the actions in the protocol. Thus, pointcut designators stand for sub-sets of these actions. For defining them, *labels* and *pattern matching operators* may identify and filtering actions (joint points) in protocols. Labels extend HCL syntax for allowing to associate identifiers to actions. Pattern matching operators may be used for filtering sets of actions according to a given pattern. For example, the operator "$_? \mid _!$" stands for every communication action in a protocol, while the operator "**seq** $\{p!;_;_;\ldots\}$" stands for any sequential action, encompassing at least three actions, that begins with the activation of output port p. A pointcut is enabled whenever one of its join points (actions) is reached when executing the protocol. Routines in the procedural module are associated to *pointcut designators*. They may execute before or after to enable the pointcut.

Figure 4(a) presents a C version for MMShift. The HAL code presented in (b) defines three *pointcut* designators: Initial, Computation, and Tracing. For instance, the pointcut Computation is enabled whenever the actions

/* MMShift.c */	{- MMShift.hal -}	*Wire functions* (in MMShift.c)

```int a,b sum;```
 	
 ```void initial (void) {```
 ``` sum = 0;```
 ```}```
 ```void accumulate (void) {```
 ``` sum = sum + a*b;```
 ```}```
 ```void show_progress (void) {```
 ``` printf("sum = %d\n", sum);```
 ```}```
	point cut INITIAL **for** <u>A</u>
	 point cut COMPUTATION **for** <u>B</u> || <u>C</u>
	 point cut TRACING **for** _ ! || _ ?
	
	 before INITIAL , **call** *"initial()"*
	 after COMPUTATION, **call** *"accumulate()"*
	 after TRACING , **call** *"show_progress()"*
	 before TRACING , **call** *"show_progress()"*
		```void j(int x, int y) { a = x; b = y; }```
		 ```void s(int x) { a = x; }```
		 ```void e(int x) { b = x; }```
		 ```int n(void) { return a; }```
		 ```int w(void) { return b; }```
		 ```int r(void) { return sum; }```
		
		 /* NOTE: Wire functions are exposed to the # compiler using a header file, named *MMShift.wf.h*, where their function prototypes are provided. */

(a)	(b)	(c)

Fig. 4. C Version of the Functional Module MMSHIFT

(join points) labelled by B and C are reached in the protocol of the unit *sqmm* of SQMATMULT. A call to the subroutine *accumulate* is performed after COMPUTATION is enabled. The pointcut designator INITIAL has an analogous description. The pointcut designator TRACING is enabled in response to port activation. Before and after these events, the routine *show_progress* is invoked. No dynamic binding of routines to coordination events are needed, minimizing overheads. The # compiler is a *static weaver*, using the aspect configuration for generating code that calls specified routines at appropriate join points.

Arguments and return points of procedural modules are defined by means of *wire functions*. Essentially, wire functions compute the values to be transmitted through ports from the encapsulated state of the procedural module. Wire functions are declared in the procedural module and exposed by a header file listing their prototypes. Figure 4(a) exemplifies wire functions for unit *sqmm*.

4   Performance Evaluation Using NPB Kernels

A sub-set of the NPB kernels (*NAS Parallel Bechmarks*) [2] was implemented in # programming[2] by using AOP for linking imperative computations to # coordination medium: EP (Embarrassingly parallel), IS (Integer Sorting) and CG (Conjugate Gradient). They are used to compare the performance of # programs to their C/MPI (IS) and Fortran/MPI (EP and CG) counterparts. This experiment exemplifies how to design SPMD programs, a class where most of HPC programs fit, using the # approach. It also demonstrates how to translate MPI programs to the # model with minor performance penalties, despite gains in modularity and abstraction. The NPB kernels allow evaluating the use of collective communication skeletons for composing topologies and for automatically generating efficient code using lower level collective MPI primitives.

The composed components EP, IS, and CG address the functionality of the respective kernels, implementing the same strategies of parallelism adopted in the original versions. The differences lay on the separation of concerns between parallelism and computation in composed and simple components. The coor-

---

[2] Implementation codes of NPB kernels are available at
    `http://www.lia.ufc.br/ heron/npb_hash_code.html`.

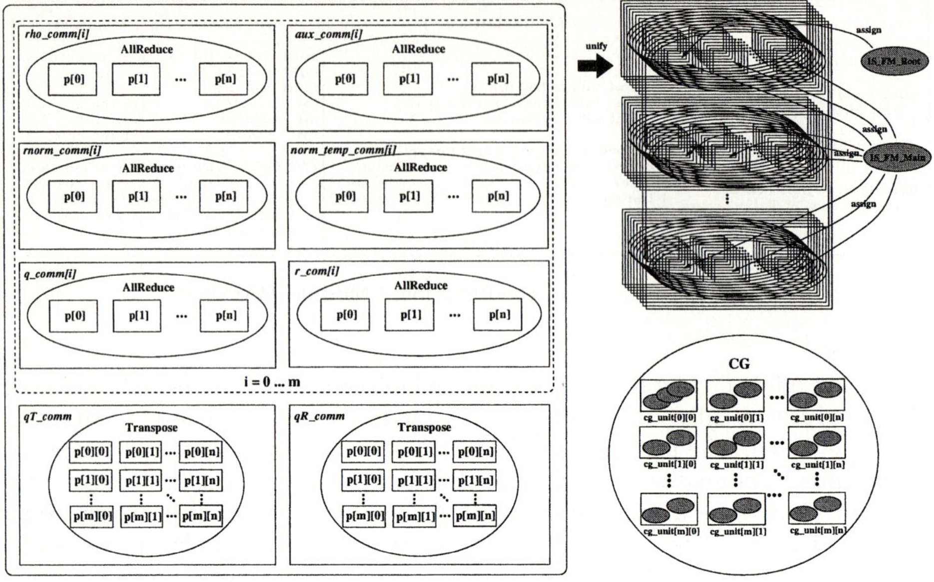

Fig. 5. The Topology of Component CG

dination medium specified is composed by overlapping composed components that implement collective communication skeletons (Figure 5). The resulting unit supersedes *peer* unit of a cluster for which component SPMD is assigned, informing the compiler about the "Single Program, Multiple Data" nature of the kernels. The procedural modules FM_EP, FM_IS, and FM_CG implement computations. Their routines are invoked according to events at coordination level, associated by means of aspect configurations (Section 3).

In the original versions of the NPB kernels, timing concern is implemented as calls to low-level timing routines intertwined with the code of computations. Using the # approach, a reusable component, called TIMER, was designed for addressing the concern of *execution timing*. It was designed for synchronizing processes before timing begins, measuring duration of computation and communication/synchronization phases in a SPMD parallel program, and finally, providing timing summaries at the end of the execution. The component TIMER is overlapped to the application components EP, IS and CG, yielding timed versions of them, called TIMED_EP, TIMED_IS and TIMED_CG, by using unification. Using the same approach, it might be possible to design other reusable components to address cross-cutting concerns, such as *debugging*, placement and *load balancing* strategies, *security policies*, etc.

4.1 Performance Measures and Discussion

Figure 6 presents the performance figures for the NPB kernels EP, IS and CG. Standard problem sizes A, B, and C, defined in kernel documentation, is con-

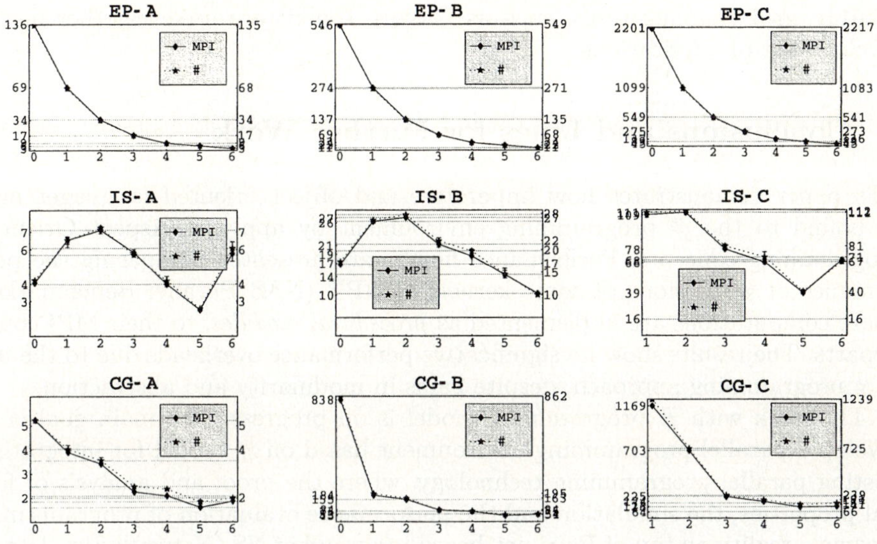

Fig. 6. Performance Figures for NPB (# vs. MPI)

sidered. For one process, IS and CG exhaust physical memory of cluster nodes ($> 1GB$). The architecture used was an Itautec cluster comprising 28 Intel Xeon nodes, each one with four processors, connected through a Gigabit Ethernet. It is installed at Computation Department of Federal University at Ceará, Brazil. MPICH 1.2.4 was used on top of TCP/IP. The data presented show no significative overheads for # versions in comparison to original ones, despite the gains in modularity advocated in the previous section. The minor differences are due to the partitioning of the monolithic original code in several routines scattered over distinct source files, affecting cache performance, causing larger number of function calls, and reducing some opportunities for compiler optimizations.

The presented empirical study may not be extended to all # programs and problem sizes. Indeed, it is not possible to define an exhaustive set of benchmarks that prove it, for any programming technology or architecture. However, it is possible to enumerate some reasons that may strengthen the reliability of the presented results to predict # performance in other situations: (**1**) virtually, any parallel programming technology on top of message-passing may be encapsulated as components in the # programming; (**2**) # programmers may implement the same parallelization strategies that they would implement by using the underlying parallel programming technology. In NPB kernels, # versions were produced by refactoring the original MPI versions, reusing all computation code without either modifications or reimplementation; (**3**) The # compiler does not add any kind of run-time support to the one provided by the underlying parallel programming technology; (**4**) The # compiler may allow the use of several parallel programming technologies in the same application, on top of the same component abstraction. In fact, this is a realistic assumption in current parallel programming practice, where non-modular combinations of MPI, openMP, and possibly, grid enabling tools, such as Globus Toolkit are used together in the development of applications.

5   Conclusions and Lines for Further Work

This paper demonstrates how imperative and object oriented languages may be bound to the # programming environment by applying Aspect Oriented Programming concepts. Performance figures are presented comparing the performance of # versions of some kernels of NPB (NAS Parallel Benchmarks), where computations are implemented as *procedural modules*, to their MPI counterparts. The results show no significative performance overheads due to the use of # programming approach, despite gains in modularity and abstraction.

The work with # programming model is on progress. The main goal is to develop a parallel programming environment based on # model for integrating existing parallel programming technology, where the proof and analysis of formal properties, the simulation and the performance evaluation of programs may become a reality on top of Petri net-based tools and of NS (Network Simulator).

References

1. R. Armstrong, D. Gannon, A. Geist, K. Keahey, S. Kohn, L. McInnes, S. Parker, and B. Smolinski. Towards a Common Component Architecture for High-Performance Scientific Computing. In *The Eighth IEEE International Symposium on High Performance Distributed Computing*. IEEE Computer Society, 1999.

2. D. H. Bailey, T. Harris, W. Shapir, R. van der Wijngaart, A. Woo, and M. Yarrow. The NAS Parallel Benchmarks 2.0. Technical Report NAS-95-020, NASA Ames Research Center, December 1995. *http://www.nas.nasa.org/NAS/NPB*.

3. S. Balay, K. Buschelman, W. Gropp, D. Kaushik, M. Knepley, L. C. McInnes, B. Smith, and H. Zhang. PETSc Users Manual. Technical Report ANL-95/11 Revision 2.1.3, Argonne National Laboratory, Argonne, Illinois, 1996. http://www.mcs.anl.gov/petsc.

4. F. Baude, D. Caromel, and M. Morel. From Distributed Objects to Hierarchical Grid Components. In *International Symposium on Distributed Objects and Applications*. Springer-Verlag, 2003.

5. L. S. Blackford, J. Choi, A. Cleary, E. D'Azevedo, J. Demmel, I. Dhillon, J. Dongarra, S. Hammarling, G. Henry, A. Petitet, K. Stanley, D. Walker, and R. C. Whaley. *ScaLAPACK User's Guide*. Society for Industrial and Applied Mathematics (SIAM), 1997.

6. R. Buyya (ed.). *High Performance Cluster Computing: Architectures and Systems*. Prentice Hall, 1999.

7. F. H. Carvalho Junior, R. M. F. Lima, and R. D. Lins. Coordinating Functional Processes with Haskell#. In ACM Press, editor, *ACM Symposium on Applied Computing, Track on Coordination Languages, Models and Applications*, pages 393–400, March 2002.

8. F. H. Carvalho Junior and R. D. Lins. Haskell#: Parallel Programming Made Simple and Efficient. *Journal of Universal Computer Science*, 9(8):776–794, August 2003.

9. F. H. Carvalho Junior, R. D. Lins, and R. M. F. Lima. Translating Haskell$_\#$ Programs into Petri Nets. *Lecture Notes in Computer Science (VECPAR'2002)*, 2565:635–649, 2002.

10. M. Cole. *Algorithm Skeletons: Structured Management of Paralell Computation.* Pitman, 1989.

11. J. Dongarra, I. Foster, G. Fox, W. Gropp, K. Kennedy, L. Torczon, and A. White. *Sourcebook of Parallel Computing.* Morgan Kauffman Publishers, 2003.

12. High Performance Fortran Forum. *High Performance Fortran, Language Specification, Version 2.0,* January 1997.

13. I. Foster and K. M. Chandy. Fortran M: A Language for Modular Parallel Programming. Technical Report MCS-P327-0992, Argonne National Laboratory, June 1992.

14. I. Foster and C. Kesselman. *The Grid 2: Blueprint for a New Computing Infrastructure.* M. Kauffman, 2004.

15. G.A. Geist, A. Beguelin, J. Dongarra, W. Jiang, R. Manchek, and V. S. Sunderam. PVM: Parallel Virtual Machine - A User's Guide and Tutorial for Networked Parallel Computing. *MIT Press, Cambridge,* 1994.

16. G. Kiczales, J. Lamping, Menhdhekar A., Maeda C., C. Lopes, J. Loingtier, and J. Irwin. Aspect-Oriented Programming. In *Lecture Notes in Computer Science (Object-Oriented Programming 11th European Conference – ECOOP '97),* pages 220–242. Springer-Verlag, November 1997.

17. Message Passing Interface Forum. MPI: A Message-Passing Interface Standard. *International Journal of Supercomputer Applications and High Performance Computing,* 8(3-4):169–416, 1994.

18. A. Skjellum, P. Bangalore, J. Gray, and Bryant B. Reinventing Explicit Parallel Programming for Improved Engineering of High Performance Computing Software. In *International Workshop on Software Engineering for High Performance Computing System Applications,* pages 59–63. ACM, May 2004. Edinburgh.

19. F. Tip. A Survey of Program Slicing Techniques. *Journal of Programming Languages,* 3:121–189, 1995.

Multi-threaded Testing with AOP Is Easy,
and It Finds Bugs!

Shady Copty and Shmuel Ur

IBM Haifa Research Lab, Haifs University Campus, Haifa 31905, Israel
{shady,ur}@il.ibm.com

Abstract. We investigate the suitability of AOP (Aspect Oriented Programming) for testing tools by trying to implement the ConTest testing tool using AspectJ, a tool that implements AOP for the Java programming language. We examine whether the entire set of features can be implemented this way, in the context of the larger problem where moving to a higher level of abstraction means that some details cannot be implemented.

Our conclusion from this exercise is that AOP is very suitable for the implementation of a number of classes of test tools. These include multithreaded noise makers such as ConTest, in addition to coverage analyzers, data-race detectors, network traffic simulators, runtime bug pattern detectors, and others. The main advantage is that the instrumentation part of the tool creating method, which usually contains little scientific contribution but consumes most of the work, becomes much easier to perform and requires less expertise. In our specific exercise, a task that took more than half a year and required specialized knowledge, was reduced to two weeks work by a relative novice.

1   Introduction

The increasing popularity of concurrent Java programming – on the Internet as well as on the server side – has brought the issue of concurrent defect analysis to the forefront. Concurrent defects, such as unintentional race conditions or deadlocks are difficult and expensive to uncover and analyze, and such faults often escape to the field.

One reason for this difficulty is that the set of possible interleavings is huge, and it is not practical to try all of them. Only a few of the interleavings actually produce concurrent faults; thus, the probability of producing one is very low. Since the scheduler is deterministic, executing the same tests many times will not help, because the same interleaving is usually created. The problem of testing multi-threaded programs is compounded by the fact that tests that reveal a concurrent fault in the field or in stress test are usually long and run under different environmental conditions. As a result, such tests are not necessarily repeatable, and when a fault is detected, much effort must be invested in recreating the conditions under which it occurred.

Much research has been done on testing multi-threaded programs. Research has examined detecting data races[13], [14], [8], replaying in several distributed

J.C. Cunha and P.D. Medeiros (Eds.): Euro-Par 2005, LNCS 3648, pp. 740–749, 2005.

and concurrent contexts[2], static analysis [16] [7] [3], and the problem of generating different interleavings for the purpose of revealing concurrent faults [4] [17]. Model checking [15], and coverage analysis [11] [5], and cloning [6] are used to improve testing in this domain.

AOP is a relatively new technology that allows the creation of generic instrumentation [10]. Rather than a complicated program that requires lots of expertise, instrumentation with AOP is something everyone can use. The central idea of AOP is that while the hierarchical modularity mechanisms of objectoriented languages are useful, they are unable to modularize all concerns of interest in complex systems. In the implementation of any complex system, there will be concerns that inherently crosscut the natural modularity of the rest of the implementation. AOP provides language mechanisms that explicitly capture crosscutting structures. This makes it possible to program crosscutting concerns in a modular way, and achieve the usual benefits of improved modularity [9]. From the perspective of our work, AOP is a simple declarative way to write the instrumentation engine.

In this paper, we demonstrate how to implement a ConTest-like tool using AOP, specifically AspectJ. We list the features of ConTest and for each feature, we either show how to implement it or explain why it is not related to instrumentation. For the few ConTest features that cannot be implemented due to limitations in AspectJ, we explain the shortcomings of AspectJ and how it can be extended. We go through the entire Contest feature list to show that we use AspectJ to solve a real problem and not an approximation. This paper shows how to use AOP to create testing tools, and examines the appropriateness of AOP for the testing domain in general. This is the first time, to our knowledge, that AOP is used to create a testing tool not specific to logging. This probably explains the missing features of AspectJ. The analysis contained will be used by the AspectJ people to make the tool more useful to the testing community.

2   ConTest

ConTest is a tool for finding bugs caused by concurrency. It alleviates the need to create a complex testing environment with many processors and applications, and works by instrumenting the bytecode of the application with heuristically controlled conditional sleep and yield instructions. ConTest is used by more than fifty testing and developer teams in IBM. In this section, we describe the features of ConTest.

2.1   Instrumentation Scheme

ConTest instruments Java bytecode using cfparse [12]. It adds instrumentation points before and after every potential concurrent event, and other nonconcurrent points due to coverage consideration. An event is called 'concurrent' if the order of its execution may impact the outcome of the program. For example, if two threads are executing but they do not share variables, and each thread outputs to a different output stream, this program will have no concurrent events. However, if they work with different variables but print to the same output stream, the printing statements are concurrent events, as the ordering between the events impacts the output by deciding which threads output is first. Concurrent events can potentially occur at the following locations:

1. Shared variable accesses.
2. Synchronization functions, such as join(), start(), wait() and notify() primitives.
3. Synchronization block, a block or method protected with the keyword *synchronize*.

In addition, ConTest has coverage related instrumentations.

1. At the entry to every method - used by the method coverage model.
2. At the entry to every basic block - used by the branch coverage model.

It is possible to perform selective instrumentation based on user directives given as flags to the instrumentation applications.

2.2   Runtime Features

Almost all the ConTest runtime features are based on instrumentation, demonstrating why all the instrumentation features are necessary.

1. Coverage information is collected and logged. There are a number of types of coverage models, each of which has it own trace. Sometimes it is necessary to create more than one coverage trace (for each coverage type) in a run. For example, a server application is tested by repeatedly running a client program. All those runs are handled by one JVM running the (instrumented) server application. Therefore, by default, they will write only one coverage file. If the user considers each run of the client program a different test, and a different coverage file is required for each run. It is possible to create many coverage files with one execution, using callback threads (see below). All coverage files thus created will have the same timestamp in their name, but they will have different serial numbers.
The different coverage models are:
   (a) **Basic coverage** models. Logs of the program executing instrumentation coverage statements, including method, branch, and concurrent point coverage.
   (b) **Synchronization coverage.** Collects temporal data, which depends on the order of the specific execution. For example, a synchronization is considered used if the point was reached while another thread was inside the synchronization block on the same object.
   (c) **Interfered location pairs coverage.** Each line in the trace file of this coverage type contains a pair of program locations that were encountered consecutively in the run, and a third field that is "t" or "f". The field is "f" if the two locations were run by the same thread, and "t" otherwise. That is, "t" means a context switch occurred.
   (d) **Shared variables coverage.** Collects the names of variables detected as shared in a given run.(i.e., accessed by more than one thread).

2. ConTest uses many heuristics to try to increase the likelihood of finding bugs. These heuristics differ in the type of delay added, the frequency, and the activation rules. The general details are as follows:

   (a) ConTest currently supports three different types of noise: yields, sleeps, and synchYields.

   (b) The amount of noise can be controlled by two properties: noiseFrequency which determines the probability that noise will be generated and strenght, which determines the amount.

   (c) ConTest can attempt to identify which variables are accessed by more than one thread, and do the heuristic noise only on accesses to those variables. A variable is determined to be shared when two different threads accessed it (read or write). With this option on, the heuristic noise can be any one of those described above (sleep, yield, etc.), with any strength.

   (d) Halt-one-thread heuristic. This heuristic occasionally causes one thread to stop executing for a long time, until no other thread can advance. It can be powerful in revealing certain types of concurrent bugs.

   (e) Tampering with time-out. The method java.lang.Thread.sleep(long millis) causes the current thread to stop executing for the specified duration. This time duration alone should not be counted upon, It should not be assumed that other threads have completed some tasks by the time the sleep returns. ConTest helps test that such wrong assumptions are not made, by randomly reducing the time-out used by these methods. This simulates a condition in which other threads work more slowly.

   (f) Deferring noise generation to a late point in the execution. It may be desirable for ConTest not to begin its perturbations just as the tested program begins. For example, maybe a bug was found using ConTest in the initialization of the program, but the testing the rest of the program should proceed without ConTest causing the bug to appear again and again. ConTest can receive the class name or the method name as a string. Until this class or method is seen, ConTest makes no noise.

3. ConTest provides a number of debugging aids that can be used to pinpoint a bug once it has been found. These include the following:

   (a) Deadlock support. To help debug a deadlock, a report can be generated containing the following information: a list of threads waiting on a given lock, a list of which thread is holding each lock, the current line number of each thread.

   (b) Orange box. When the program fails, it is often useful to know something about the behavior of the program in the last segments of its execution, similar to the black box on an airplane. The "orange box" keeps a record of the last n accesses (read or write) to each non-local variable. For example, if the program execution was terminated due to an exception caused by variable foo, having the value null, you could use the orange box feature to check where this value was set.

   (c) Partial replay. Replay for multi-threaded applications is very important for debugging, increasing the likelihood that a bug will reoccur.

(d) Callback thread. By opening a "callback thread", several kinds of requests from ConTest can be made from the outside, while the tested program is running: debug reports, closing and restarting coverage sessions, and controlling fault injection. There are two ways to pass requests to the callback thread: through standard input (normally, the keyboard), or through a network port (i.e., from another process).

3   AspectJ

AspectJ is an aspect-oriented extension to Java. With just a few new constructs, AspectJ extend Java to provide support for the modular implementation of a range of crosscutting concerns. Dynamic crosscutting makes it possible to define additional implementation to run at certain well-defined points in the execution of the program. Static crosscutting makes it possible to define new operations on existing types; it's called static because it affects the static type signature of the program. Dynamic crosscutting in AspectJ is based on a small but powerful set of constructs: join points are well-defined points in the execution of the program; pointcuts are a means of referring to collections of join points and certain values at those join points; advice are method-like constructs used to define additional behavior at join points; and aspects are units of modular crosscutting implementation, composed of pointcuts, advice, and ordinary Java member declarations. We use dynamic crosscutting to implement the features of ConTest using AspectJ, in a manner similar to that used by ConTest's instrumentor. [9]

In AspectJ, pointcuts pick out certain join points in the program flow. For example, the pointcut call(void Point.setX(int)) picks out each join point that is a call to a method with the signature void Point.setX(int) (i.e., Point's void setX method with a single int parameter). A pointcut can be built out of other pointcuts with: *and, or,* and *not.* [1] AspectJ also lets you define pointcuts using wildcards. For example, set(* *) defines all the assignments to all the variables in the program. Pointcuts pick out join points, but they don't do anything else. To implement crosscutting behavior, we use advice. Advice brings together a pointcut (to pick out join points) and a body of code (to run at each of those join points). AspectJ has several different kinds of advice. 'Before advice' runs as a join point is reached, before the program proceeds with the join point. 'After advice' runs after the program proceeds with that join point. "Around advice" on a join point runs as the join point is reached [1]. The pointcut and the advice type define *where* the instrumentation is done, and the advice body defines *what* will actually be instrumented.

3.1   ConTest Instrumentation Scheme Using AspectJ

In Section 2.1 we described the instrumentation that ConTest performs on the program. We now explain how to implement these in AspectJ. The implementation of the runtime features of ConTest is outside the scope of this paper, and therefore is not discussed. It is easy to see that the runtime features based on the instrumentation described could be implemented.

1. Shared variable accesses. The join point *set(* *) or get(* *)* instruments all accesses to variables.
2. Synchronization functions, such as join(), start(), wait() and notify() primitives. These can be instrumented using the point cut *call(signature)* for each method.
3. Synchronization block. This kind of point cut is not supported by AspectJ.
4. Additional coverage related instrumentation. These are added due to specific coverage related user requests.
   (a) At the entry to every method - can be easily implemented using the *call* point cut.
   (b) At the entry to every basic block - not supported by AspectJ.

It is possible to do selective instrumentation in AspectJ using the point cuts that relate to the static nature of the code.

4   Implementing the Tool

We implemented a few aspects to demonstrate the capabilities of an AOP language such as AspectJ. These aspects alter the class files to increase the likelihood of catching concurrent bugs, using ideas already implemented in ConTest. A special emphasis is put on the instrumentation capabilities. The following is the source code for the aspect SleepNoise:

```
public aspect SleepNoise extends Thread{
    private static Random rand = new Random();
    pointcut noiseVictem():
        ((get(* *) || set (* *))&& within(!SleepNoise));
    after(): noiseVictem() {
        try{// noise
        if (rand.nextInt(100) == 1){ // activation
            sleep(rand.nextInt(50)); // type
          }
        } catch (Exception e) {};
    }
}
```

SleepNoise is a simple aspect based on a single pointcut and a single advice. The pointcut defines where the instrumentation is being done. In our example, we are weaving the after() advice on all the gets and sets for variables in the instrumented program. The advice is a call to sleep() with a random parameter in the range [0,50] with a probability of 1% for invoking the sleep method. This adds noise to the instrumented application as done by ConTest's instrumentor. The difference, however, is that this aspect inlines the noise, whereas ConTest instruments call back methods, which add some runtime overhead. This example could easily be expanded to instrument special concurrent related methods, such as sleep, yield, notify, notifyAll, and so on. In addition, the type of noise could be altered to other types of noise that affect the interleaving of the program, all creating different kinds of heuristics.

SleepMutator is an aspect that implements the "Tampering with time-out" heuristic discussed earlier. The following is its source code:

```
public aspect SleepMutator extends Thread{
    pointcut noiseVictem(long i):
      call(void sleep(long)) && args(i) && within(!SleepMutator);
    private static Random rand = new Random();
    void around(long i): noiseVictem(i) {
    try{
      long newSleepTime = rand.nextInt((int)i*3); // [0,3*sleep]
      if (rand.nextInt(5) == 1)
        proceed(newSleepTime);
      } catch (Exception e) {}
    }
};
```

The pointcut we defined for this heuristic is a sleep call, taking into account its parameter and using the *proceed* keyword. The advice calls the sleep method with a new sleep parameter, chosen randomly in the range of [0, 3*oldParam].

ConTest provides users with coverage information. Using AspectJ, we can only implement ConTest's method coverage because AspectJ doesn't provide a pointcut for synchronization blocks or for simple blocks.

The following aspect prints the methods called to System.out. Of course, we could have created a new class that remembers which strings were already printed and not print them more than once. This aspect could be used for method coverage. As we started working on this paper, there was no way to retrieve all the points that were instrumented in AspectJ. This meant that we had no way of knowing when we reached 100% coverage. In AspectJ 1.2.1c, there is a new feature added to AspectJ's compiler (-showWeaveInfo) that accomplishes this.

```
public aspect Coverage extends Thread{
    pointcut methodExecution():
      ((execution(* *(..)))&& within(!Coverage));
    after(): methodExecution(){
        System.out.println("called: " +
          thisJoinPointStaticPart.getSignature());
    }
}
```

5   Experimental Results

We tested the aspect SleepNoise against several programs with documented bugs. These programs received a single parameter: the number of threads running simultaneously, and are categorized as low, medium, or high. We ran the tests

10 times for each category and for each configuration: once as the uninstrumented program, called original in the figure, then using ConTest with simple noise, and finally with the SleepNoise aspect.

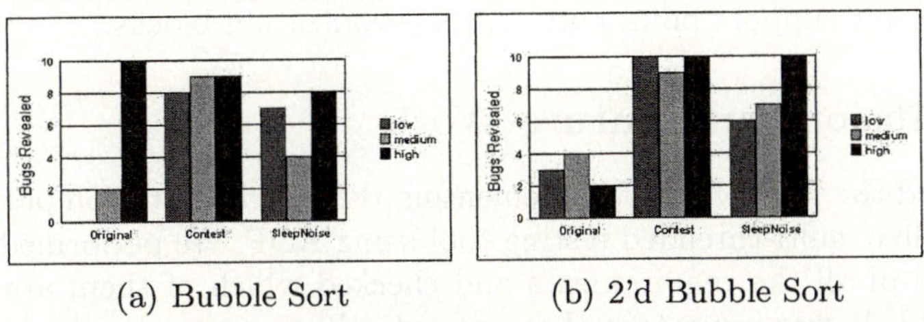

(a) Bubble Sort   (b) 2'd Bubble Sort

Fig. 1. Bubble sorts

Our first program is a concurrent version of the bubble sort algorithm. The bug in this program is that the programmer assumed the threads would finish their work without interruption and accessed the shared resources without synchronization. Figure 1a shows the results for this program. We can clearly see that using SleepNoise increases the chance of concurrent bugs being manifested.

Our second program is also a bubble sort program, with a different bug. The programmer used a sleep statement to initialize all threads before they started working. The results (Figure 1b) show the benefit in using this type of testing. We see that SleepNoise increased the chance of finding bugs. Note that the heuristic is very simple and was not modified to suit the specific program. By tuning the frequency of adding noise and the type of noise, we can achieve better results.

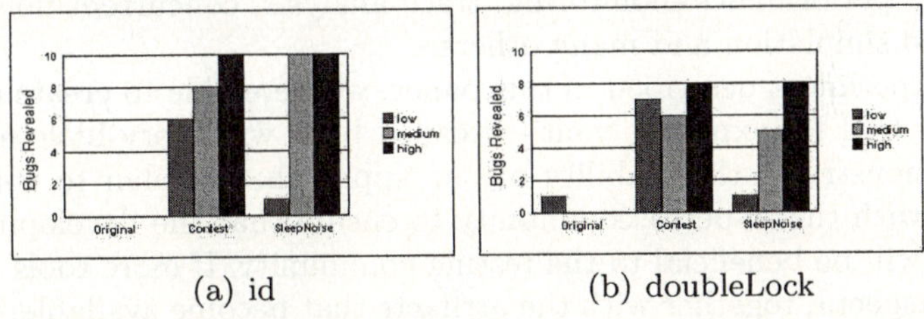

(a) id   (b) doubleLock

Fig. 2. ID Manager and Double Lock

The third program is one that issues IDs for users. Each user requests and receives a unique ID. The bug is that the programmer assumed that incrementing the ID counter was an atomic operation and didn't protect it. We see in Figure 2a that SleepNoise surpassed ConTest's performance for medium concurrency, which could be the result of a low activation parameter used for ConTest.

In the last program, the programmer wanted to obtain the locks for two files in different places, but didn't maintain a global partial order on the way she obtained these locks. This introduced a deadlock to the program. SleepNoise increased the chance of the bug being manifested. Had we been able to instrument synchronization blocks, we would have been able to find this bug much more easily, by adding a sleep exactly after the first lock was obtained in one of the two accesses. This would cause the bug to manifest instantly. Unfortunately, AspectJ doesn't support point cuts on synchronization blocks.

6   Conclusions and Future Work

We examined the possibility of implementing the instrumentation part of a commercial quality multi-threaded testing tool using AOP. We performed a detailed examination of all the requirements and checked which of them are satisfiable with AspectJ. It was important that not only the concepts be checked, but the fine details. Each application had some features that could not be implemented, and implementing them in the traditional way reduced much of the benefits of going to a higher abstraction level.

We found AOP in general, and AspectJ in particular, to be very well suited for this work. We did find two missing features in AspectJ that are required for complete implementation of the ConTest tool's instrumentation needs. One of those was added while this paper was being written and is now available. The second one, identifying synchronization blocks as places for instrumentation, is still unavailable. Had we decided to move to AspectJ, this point could have been partially mitigated due to the fact that AspectJ is open source and we could have added this capacity ourselves.

We found that with AspectJ, we could implement a useful testing tool in two weeks, which would otherwise have taken more than half a year. We consider AOP and AspectJ to be very important for implementing high quality open source and academic testing tools, in the domains of data race detection, coverage analysis, performance monitoring, trace analysis, concurrent noise making, network load simulation and many others.

In the experiment described in this paper, we were able to create a commercially useful tool for exposing multi-threaded bugs with very little effort. This exercise demonstrates the viability of our approach. We plan to continue our discussions with the AspectJ community to ensure that the development directions taken will be beneficial to the testing community. If more tools follow this approach, AspectJ, together with the artifacts that become available, will create an incentive for additional research in the area.

We add the work presented in this paper to the testing benchmark[1], in the hope that this resource will be utilized by researchers to expand on our work.

References

1. Aspectj getting started, http://www.elipse.org/aspectj.
2. J.-D. Choi and H. Srinivasan. Deterministic replay of java multithreaded applications. In *Proceedings of the SIGMETRICS Symposium on Parallel and Distributed Tools*, August 1998.

---

[1] https://qp.research.ibm.com/QuickPlace/concurrency_testing/Main.nsf

3. J. C. Corbett, M. Dwyer, J. Hatcliff, C. Pasareanu, Robby, S. Laubach, and H. Zheng. Bandera: Extracting finite-state models from Java source code. In *Proc. 22nd International Conference on Software Engineering (ICSE)*. ACM Press, June 2000.

4. O. Edelstein, E. Farchi, E. Goldin, Y. Nir, G. Ratsaby, and S. Ur. Testing multi-threaded java programs. *submitted to the IBM System Journal Special Issue on Software Testing*, February 2002.

5. O. Edelstein, E. Farchi, Y. Nir, G. Ratsaby, and S. Ur. Multithreaded Java program test generation. *IBM Systems Journal*, 41(1):111–125, 2002. Also available as `http://www.research.ibm.com/journal/sj/411/edelstein.html`.

6. A. Hartman, A. Kirshin, and K. Nagin. A test execution environment running abstract tests for distributed software. In *Proceedings of Software Engineering and Applications, SEA 2002*, 2002.

7. K. Havelund and T. Pressburger. Model checking java programs using java pathfinder. *International Journal on Software Tools for Technology Transfer, STTT*, 2(4), April 2000.

8. E. Itzkovitz, A. Schuster, and O. Zeev-Ben-Mordehai. Towards integration of data-race detection in dsm systems. *Journal of Parallel and Distributed Computing. Special Issue on Software Support for Distributed Computing*, 59(2):180–203, Nov 1999.

9. G. Kiczales, E. Hilsdale, J. Hugunin, M. Kersten, J. Palm, and W. G. Griswold. An overview of AspectJ. *Lecture Notes in Computer Science*, 2072:327–355, 2001.

10. G. Kiczales, J. Lamping, A. Menhdhekar, C. Maeda, C. Lopes, J.-M. Loingtier, and J. Irwin. Aspect-oriented programming. In M. Akşit and S. Matsuoka, editors, *Proceedings European Conference on Object-Oriented Programming*, volume 1241, pages 220–242. Springer-Verlag, Berlin, Heidelberg, and New York, 1997.

11. Y. Malaiya, N. Li, J. Bieman, R. Karcich, and B. Skibbe. Software test coverage and reliability. Technical report, Colorado State University, 1996.

12. S. Porat, M. Biberstein, L. Koved, and B. Mendelson. Automatic detection of immutable fields in java. In *Proceedings of the 2000 conference of the Centre for Advanced Studies on Collaborative research*, page 10. IBM Press, 2000.

13. B. Richards and J. R. Larus. Protocol-based data-race detection. In *Proceedings of the 2nd SIGMETRICS Symposium on Parallel and Distributed Tools*, August 1998.

14. S. Savage. Eraser: A dynamic race detector for multithreaded programs. *ACM Transactions on Computer Systems*, 15(4):391–411, November 1997.

15. S. D. Stoller. Model-checking multi-threaded distributed java programs. In *Proceedings of the 7th International SPIN Workshop on Model Checking*, 2000.

16. S. D. Stoller. Model-checking multi-threaded distributed Java programs. *International Journal on Software Tools for Technology Transfer*, 4(1):71–91, Oct. 2002.

17. S. D. Stoller. Testing concurrent java programs using randomized scheduling. In *In Proceedings of the Second Workshop on Runtime Verification (RV)*, volume 70(4) of Electronic Notes in Theoretical Computer Science. Elsevier, 2002.

An Investigation of Sharing Strategies
for Answer Set Solvers and SAT Solvers

Hung Viet Le and Enrico Pontelli

Department of Computer Science
New Mexico State University
{hle,epontell}@cs.nmsu.edu

Abstract. This paper describes a parallel engine for Answer Set solving, based on exploitation of search parallelism. The work explores a range of alternative strategies for work sharing, describing their implementations and comparing their efficiency. These results indicate methodologies to combine sharing strategies and select the most effective one depending on properties of the problem.

1 Introduction

In recent years, there has been a significant increase of interest towards the application of logic-based technology—in particular, technology based on propositional and SAT solving—in a variety of application domains. This renewed interest has also been guided by the development of formal modeling and programming paradigms based on these concepts, such as the widely used *Answer Set Programming (ASP)* [9]. ASP builds on logic programming and answer set semantics [5], to provide a set-oriented programming paradigm. In ASP and SAT, the problem is modeled using clauses of a propositional theory, and solutions are represented by minimal or *stable* [5] models of the theory. A significant push to these efforts comes from the development of efficient implementations of ASP and SAT solvers. These implementations are practical and scalable to real-life domains. The execution mechanisms employed by ASP are analogous to those used in general SAT solving, and are based on highly optimized and/or specialized versions of the Davis-Putnam procedure. In this work, we will focus on a state-of-the-art ASP engine—i.e., an implementation of the SMODELS algorithms [13]—though the proposed ideas are applicable to related systems.

In spite of the efficiency provided by existing systems, there are areas where ASP and SAT solving provide elegant, compact, and highly declarative solutions, but whose execution requirements are beyond the capabilities of existing systems. E.g., ASP is widely used in planning in complex domains [7], but the high computational requirements limit the domains and goals that can be effectively addressed. *Parallelism* has been identified as a natural avenue to further improve applicability of ASP and SAT solving to real-world problems. Preliminary steps have been taken in the design of parallel ASP [4, 11] and SAT solvers [1, 2, 15].

The literature is rich of studies related to the design of parallel engines for traditional logic programming (mostly Prolog) [6], theorem provers [2], and constraint solvers [10]. Nevertheless, recent investigations [11] have highlighted that results from these

J.C. Cunha and P.D. Medeiros (Eds.): Euro-Par 2005, LNCS 3648, pp. 750–760, 2005.

related areas are not directly transferable to ASP—e.g., task sharing techniques proved optimal in the context of Prolog have provided suboptimal performances in ASP.

In this paper, we are interested in the development of techniques to exploit parallelism from ASP at the *search level* (a.k.a. *or-parallelism*) and in the context of distributed search strategies [2]. Parallelism at the search level implies the presence of multiple processes (*search agents*) that search in parallel the solution space of the problem; in ASP and SAT, this corresponds to the concurrent construction of *distinct models* of the propositional theory, where each agent explores different truth assignments to the logical variables of the theory. Theoretical research in the area of search parallelism [6, 12] underlined that dynamic distribution of work is vital to achieve adequate parallel performance, and two components of the execution model have the greatest impact: the *sharing* strategy and the *scheduling* strategy. Scheduling determines the policy to be used to select tasks to be exchanged during execution, while the sharing strategy determines how the exchange of tasks takes place. Our focus is on the latter.

The contribution of this paper is the design of the first complete parallel ASP system (supporting the *complete* SMODELS language) on *Beowulfs*. Our focus is on investigating the impact of different sharing strategies. We explore a variety of alternatives, some adaptations to ASP of known methodologies and some novel, and study their behavior on a representative set of benchmarks. The results suggest that flexible dynamic selection of sharing strategies is vital to guarantee high parallel performance.

2   A Parallel Engine for ASP

Sequential Execution Model: The objective of a computation is, given a propositional theory (extended Horn clauses in the case of ASP), to determine one or more minimal models of the theory. For ASP, we are seeking a special set of minimal models, called *stable models* [5]. The execution is a fixpoint computation which alternates two phases: *boolean constraint propagation* and *atom splitting* [2]. During constraint propagation, clauses in the theory are used to extend a partial model, adding to the model those atoms whose truth value is uniquely determined by the theory (w.r.t. the partial model). Whenever constraint propagation is not possible, the system performs *atom splitting*, by selecting an unknown atom and "guessing" its truth value; this corresponds to the creation of a choice-point, since backtracking needs to explore both alternative truth values for such atom. Heuristic strategies (based on estimating the size of the subtrees) are employed in the selection of the atom during splitting, to guarantee effective propagation. Because of the non-determinism of atom splitting, the computation can be visualized as a search tree, where the nodes correspond to the choice-points created by splitting.

Organization of the Parallel Computation: Forcing an atom into the partial model, via atom splitting, will create two branches of the search tree, corresponding to the two roles of the atom (true and false). The system needs to completely traverse both sides of this computation subtree separately, and this is the core of the parallel search process—i.e., assign distinct branches to different search agents. The left (right) branch of each node corresponds to setting up the atom's value to true (false). In search parallelism, the two branches may be explored concurrently by separate agents.

The initial steps of the parallel computations are performed following a static partition of work. A divide and conquer scheme is applied to the first $\approx \lg_2 n$ levels of the search tree (n is the number of agents). Initially all agents independently perform the initial boolean propagation; at the first splitting, processors are partitioned along the two branches—distribution of processors between the two branches is based on the estimated size of the two subtrees (using the same heuristic strategy of [13]). The process continues until individual agents are assigned to distinct subtrees. This process does not require inter-process communication and resembles the *guiding path* generation process of PSATO [15]. After this initial setting, the system switches to a fully dynamic distributed scheduling strategy, where an idle agent secures new tasks by performing a *sharing* operation with another agent. We consider scheduling strategies that are *receiver-initiated* (i.e., idle agents initiate scheduling). The dynamic scheduler (briefly described in Sect. 3.3) is expected to determine a pair $\langle \mathcal{P}, \mathcal{N} \rangle$, where $\mathcal{P}$ is the active agent from where tasks should be acquired (*work-sender*), and $\mathcal{N}$ is the node (*target node*) in the tree containing the task to be exchanged. Once the pair $\langle \mathcal{P}, \mathcal{N} \rangle$ has been determined, a sharing operations has to be performed between the idle agent and $\mathcal{P}$. We will refer to the idle agent as the *work-receiver*. An idle agent which is searching for new tasks is called *work-checker*. The location of the idle agent in the search tree is called *start node*. A node with an unexplored alternative is said to be *open*.

Essential Data Structures: The basic data structures employed extend the traditional design used for linear time computation of declarative closures. Inter-connected objects are used to represent each rule and each atom. The computation is based on the use of a *stack*, which maintains a representation of the partially constructed stable model (as references to atoms). If an atom is determined to be true/false, it is pushed on the stack along with its truth value. An array (*history record*) has been introduced to record the choice-points created during the construction of the current branch of the search tree—i.e., it provides a compact representation of the branch in the search tree built by the agent. During the execution, search agents operate on the leaves of the search tree; to assist the sharing operations, each search agent maintains an array (*relative positions*) recording the *nearest common ancestor (nca)* nodes between the agent's leaf and the leaf of any other agent. In a distributed setting, determination of the *relative positions* require inter-process communication. In this work, the updates are lazily done via broadcast messages during the sharing operations.

3   Parallel Work Sharing Strategies

The objective of the sharing operation is to transfer tasks between two agents. In order to allow an agent to restart its computation from a different node, it is necessary to reconstruct the correct execution state; i.e., if the agent moves to node $\mathcal{N}$, then it will have to instantiate its data structures to reflect the partial model associated to node $\mathcal{N}$. [6] surveys over 20 different schemes to address the sharing problem for Prolog.

The formal analysis of search parallelism [12] recognized methods based on constant-time access to the partial model and constant-time alternative selection to be optimal. This restricts our choice to sharing methods that maintain the same model and program representation as in sequential models, and that reconstruct the necessary segments of the computation *only* when sharing takes place. We can recognize two key approaches: *copying*—where the necessary segments of computation are copied from the work-sender to the work-receiver—and *recomputing*—where the work-receiver rebuilds the segments of computation with minimal information from the work-sender. Copying requires significant communication but it minimizes the amount of work done by the receiver, while recomputing minimizes communication but requires high computation effort by the receiver. Copying has been widely adopted in parallel logic programming and considered as the most effective scheme [6]. We explore a solution where copying and recomputation can alternate depending on the application.

3.1   Recomputation-Based Methods

In the recomputation-based methods, the idle agent acquires work by recreating the state of computation existing at the time the node where a new alternative is available was created. In its most direct version, the content of the data structures representing the clauses and atoms and the content of the stack (representing the partial model) are recomputed, starting either from the root of the search tree or from the nca node of the target node and the start node, and ending in the target node [6]. The *work-sender* only needs to send a part of its history record, used to guide the *work-receiver* in making the correct choices during recomputation (i.e., what alternative to take in each choice-point created during recomputation). We introduce 4 recomputation methods, which differ in the amount of recomputation performed and in the way agents move in the search tree.

Method 1: Recomputation with Backtracking (*ReBack*). The key requirement in this method is that each agent maintains its *relative positions* in the search tree w.r.t. the other agents. The *work-receiver* acquires work by first backtracking to the nca node of the start node and the target node (a.k.a. the *relative position*), and then initiating the recomputation operation from there. The stack and the other data structures, including the rule and atom objects, are reconstructed from the *relative positions*. The path that runs from the nca node to the target node in the search tree is called the *connection*. The description of the connection is present in the work-sender—it is a segment of its *history record* (Sect. 2). The connection is exchanged between the two agents, and used by the work-receiver to perform recomputation. Both the *work-receiver* and the *work-sender* will also mark the target node as explored (to avoid duplication of work). The *relative positions* between agents also need to be updated during the sharing operation. When the receiver has completed the recomputation process, it will replace its *relative positions* array with a copy of the *relative positions* of the work-sender. Meanwhile, the *work-sender* broadcasts the new position of the work-receiver to all agents, allowing them to update their own *relative positions* arrays (w.r.t. the work-receiver).

The lack of guarantee regarding the order of arrival of messages sent by different agents might create situations where agents incorrectly update their *relative positions*. The overall effect is that the system may be unable to schedule work to idle agents. We refer to this problem as the *mismatch situation*.

Method 2: Recomputation by Backtracking-Compare History (*ReBackHis*). In this method, instead of maintaining the *relative positions* between each pair of agents, the agents explicitly exchange their history records at the beginning of the sharing operation, determining at such moment the nca node in the search tree w.r.t. their positions. Whenever two agents exchange work, the *work-receiver* sends its own history records to the *work-sender*. Having compared to the received list, the *work-sender* figures out and sends back to the work-receiver the nca node and the *connection*. The main advantage of this alternative is that the work exchange is done locally, between two agents *without notifying anyone else*. On the other hand, the agents do not know their relative positions, and they may attempt to seek work from agents that have a nca node close to the root (i.e., agents that are "far" from the receiver). This may cause longer backtrack/recomputation phases, but we avoid the messages to maintain the relative positions between agents, reducing traffic and avoiding the mismatch situation.

Method 3: Recomputation Reset (*ReReset*). In the recomputation reset scheme, the initial backtracking to the nca node between the positions of the two agents is avoided. The backtracking step is replaced by a recomputation that starts *always* from the root of the search tree. Such scheme requires each agent to store its state at the root of the tree—i.e., the result of the first constraint propagation—and the ability to make an agent efficiently switch back to such initial state. This operation is called *Reset*, and it can be accomplished by *(a)* emptying the stack; *(b)* over-writing the rule and atom objects with a permanently saved copy of their state at the root node; *(c)* removing all the atoms present in the various temporary queues used by the boolean constraint propagation procedures. The relative positions between the agents are no longer necessary, and the only communication required is the connection from the root to the target node (which might be significantly bigger than the connection in *ReBackHis*).

Method 4: Recomputation Reset Split (*ReResetSplit*). In this method, instead of sharing the highest node which contains alternatives, the sender sends to the work-receiver the complete path from the root to the lowest node in its branch. All the nodes with unexplored alternatives along the branch are distributed between the two agents according to an interleaved scheme—i.e., the first open node is kept by the *work-receiver*, the *sender* keeps the second one, the *work-receiver* keeps the third, etc. This scheme bears some similarities to the *stack splitting* scheme used in some parallel Prolog systems [14], where the partitioning is made in contiguous blocks of choice-points; our scheme makes use of an interleaved distribution—impractical in Prolog (due to the need of handling side-effects) but effective in ASP, and expected to give rise to more balanced distributions—since it is unpredictable whether the "richer" nodes are in the upper or lower levels of the search tree. Compared to the other recomputation methods, *ReResetSplit* requires larger amount of data exchanged between agents (the complete history record of the work-sender), and the sender is also required to travel more to get to the new computation state. However, the agents can share in a single interaction a large number of open nodes, quickly accessible via local backtracking.

3.2   Copy-Based Methods

In copying, instead of reconstructing the computation state, the idle agent acquires work by copying the data structures stored in the work-sender agent. In most of the cases, the data to be transfered include the components of the rule and atom objects that are part of the computation state (e.g., the truth value of the atoms, the counters in the objects used to keep track of the state of the clause), the stack, the history record, and the *relative positions* of the agent. In order to facilitate the copy process, we have separated the rule objects in two parts—a static part (set during the initialization phase and never copied) and a dynamic part (modified during the remainder of the execution). The dynamic parts are collected in arrays to facilitate copying. Also, changes to atom objects are *trailed*, and only the modifications are copied. Observe that, while in recomputation the execution develops from the root of the tree towards the open node, in copying the computation restarts from the open node and moves up via backtracking.

Method 5: Incremental Copying Split-Maintain the Relative Positions (*IncCopySplit*). In this method, we exploit the basic copying mechanism, enriched with the idea of *incrementality*—incrementality derives from the fact that the idle worker has already traversed the part of the branch in the search tree from the root node to the nca node, thus, there is no need to copy it. Like in the case of recomputation by backtracking, each agent maintains its *relative positions* with the other agents. The *work-receiver* backtracks to the nca node, while the *work-sender* transfers the part of the stack from the nca node to the selected open node. The set of clauses and atoms with up-to-date parameters and links are included in this copying operation. The *work-receiver* unpacks the data set and updates its data structures. The update process is performed by replacing the dynamic parts of rules and atoms with the received ones and adding the content of the received stack to the local stack. Similarly to the *ReBack* case, all other agents have to be notified (via broadcast) of the new position of work-receiver in the search tree. Concurrently to the exchange of data, the agents perform an interleaved partition of the open choice-points as described in the method *ReResetSplit*.

Method 6: Incremental Copying Split-Compare History (*IncCopySplitHis*). This method is simpler than *IncCopySplit*. The *relative positions* between the agents, employed to support the incremental copying behavior, are no longer required. Analogously to the *ReBackHis* scheme, the agents exchange their history records to find out the nca node. As soon as such node has been determined, the sharing process proceeds exactly as in the *IncCopySplit* scheme. The only additional difference is that the broadcast messages to notify the change of position of the work-receive agents are not required. As for the *ReBackHis* scheme, scheduling has to be performed blindly, without knowledge of the relative positions of the agents, but the message traffic is dramatically reduced and the mismatch situations are avoided.

Method 7: Copying All Split (*CopyAll*). This method is the simplest between our copying approaches. Whenever a sharing operation is invoked, the sender transfers the complete database of up-to-date rules and atoms (dynamic parts only), along with the complete stack. The receiver empties its stack and it installs the received data. Compared to the *IncCopySplitHis* scheme, the *CopyAll* allows the same type of blind scheduling (with the same advantage in terms of reduced traffic), but it does not require the initial exchange of history records (since knowledge of the nca node is not required). On the other hand, the method requires the copying of the complete stack, which might contain a substantially larger amount of data than in the incremental case.

3.3   Further Implementation Details – Scheduling and Termination

The irregular structure of ASP search trees requires the use of a dynamic distributed scheduling scheme [6, 14]. Each agent alternates between execution phases and scheduling phases. The lack of a central scheduler leads to a situation in which agents do not have knowledge about the location and status of other agents in the tree. Consequently, they need to exchange information (e.g., the *relative positions* broadcasts) to ensure a good load balancing. The scheduler addresses two aspects: *(i)* establishes policies for exchange of scheduling information; *(ii)* determines global termination (using a token-ring termination detection). All communications rely on a time-out mechanism, where unanswered task requests are discarded by the idle agent—this has shown large improvements in performance w.r.t. schemes requiring acknowledgments.

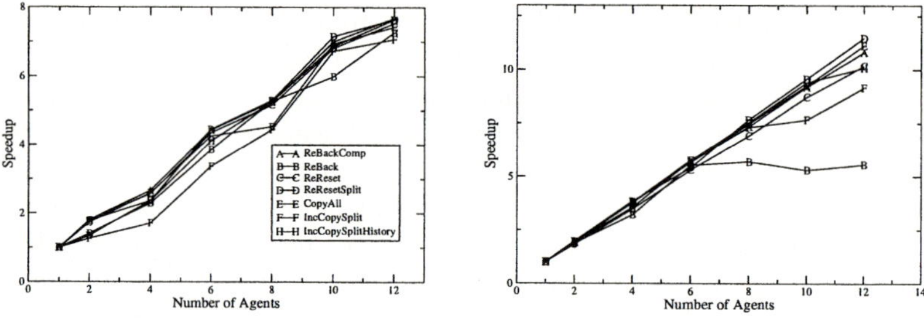

Fig. 1. Puzzle (left) and Queens Benchmarks

The Searching Agent: Whenever an agent has completely exploited all the alternatives locally available in its branch, it starts requesting work. The work requests are arranged according to a dynamic ordering of agents (stored in a *priority array*). By using a good ordering, a searching agent can reduce the volume of data in copying or recomputation. Depending on the individual approach, there are alternative strategies to compute the priority array. E.g., the methods that keep the *relative positions* of agents, sort the priority array based on the level of the nca nodes. The other methods start with a random order and re-sort the array based on the observed communications (e.g., a late message indicating tasks availability will increase the priority of that agent). Agents are contacted in the order they appear in the priority array. Time-outs lead to generation of a new request sent to the next agent in the array. The agent enters the idle state when all its attempts to get work failed. If the idle agent receives the white token, it will forward the token to the next agent in the ring. If it receives a black token, then it will know that agents are still working, and it will revert to work-search and scheduling.

4   Experimental Results

The system has been implemented on a Beowulf cluster (Xeon 1.7GHz, 1GB RAM, Linux, and Myrinet), using Java+MPIJava. The parallel system has an average overhead of less than 10% over the sequential solver (the SMODELS [13]). We tested the

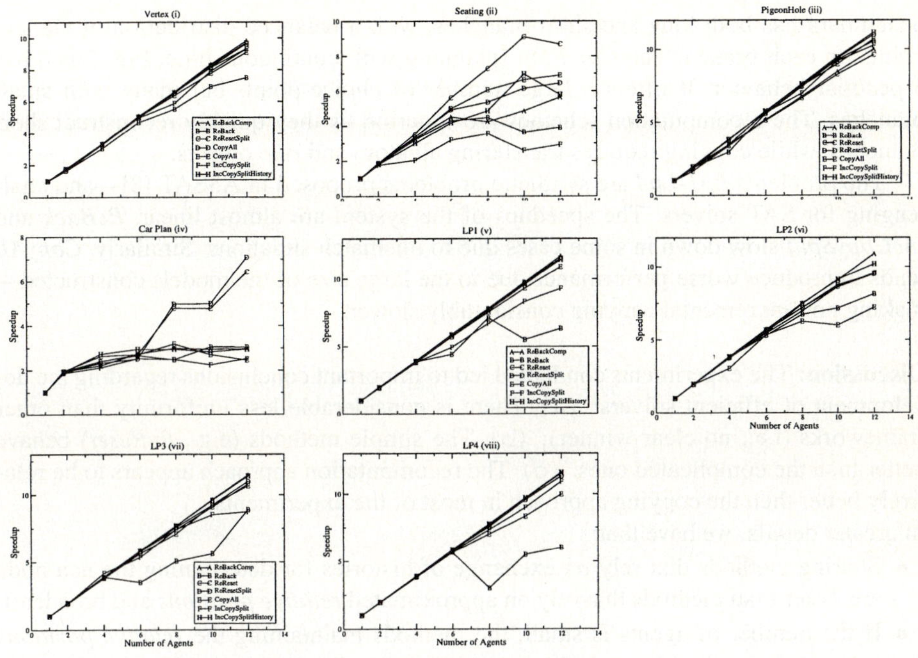

Fig. 2. Speedups for Benchmarks

implementation on a set of benchmarks, focusing on evaluating the different sharing strategies. All the timings presented are averages over 5 runs.

The *Puzzle* benchmark (4×4 numeric puzzle) is characterized by a balanced search tree with fairly short branches and a balanced distribution of choice-points in the different branches; as a result (Fig. 1) the performance and the scalability is good, and all methods provide comparable results. On the other hand, in the *Queens* benchmark (14-queen problem, Fig. 1), we can observe that the speedup of *ReBack* is significant lower due to (1) the mismatch problem and (2) the high cost of recomputation. In this problem, the branches are long and the cost for generating them is comparatively higher than just copying the corresponding data structures. Furthermore, splitting methods tend to perform worse; when the system works near the end of the branches, the agents exchange work more frequently and the work has "poorer" quality; in this case splitting schemes tend to generate larger messages (to secure larger chunks of work) without being balanced by advantages in terms of work obtained. The analogous behavior occurs in the *Vertex* problem (30-node vertex covering, Fig. 2(i)). The speedups are higher thanks to the large number of choice-points generated (towards the root part of the tree).

In the *Seating* party benchmark (Fig. 2(ii)) we see marked differences in performance. All speedups tend to flatten for many processors, due to the small number of choice-points, located in the top part of the tree. In this experiment, the winners are the simple methods: *ReReset* and *CopyAll* where the idle agents start from the root of computation tree. In the *Pigeon Hole* experiment (pigeonhole problem with 9 holes and 10 pigeons, Fig. 2(iii)), all methods reach excellent speedups (11 using 12 agents). The benchmark has both long and short branches, with a balanced distribution of choice-points on each branch. The *Car Plan* (planning with continuous time, Fig. 2(iv)) has a peculiar behavior. It offers a large number of choice-points but many with small branches. The recomputation schemes are superior, as they quickly reconstruct short branches, while copying requires transfering all atom and rule objects.

The problems *Lp1-Lp4* are synthetic problems proposed in ASSAT [8]—and challenging for SAT solvers. The speedups of the system are almost linear. *ReBack* and *IncCopySplit* slow down in some cases due to mismatch situations. Similarly, *CopyAll* tends to produce worse performance due to the large size of the models constructed—making non-incremental copying considerably slower.

Discussion: The experiments conducted led to important conclusions regarding the development of efficient solvers: (a) There is considerable less uniformity than other frameworks (i.e., no clear winner); (b) The simple methods (e.g., *ReReset*) behave better than the complicated ones. (c) The recomputation approach appears to be relatively better then the copying approach in most of the experiments.
In greater details, we have that:

- Sharing methods that rely on exchange of histories for determining the nca node are faster than methods that rely on approximated *relative positions* and broadcast.
- If the number of agents is small, the methods maintaining the *relative positions* behave very well. Although the system must handle more messages, the speedups show the benefits of choosing the "right" agents to share work with.
- Short executions are not suitable to copying (due to messages overhead).
- In recomputation, stack splitting does not greatly improve performance. In *ReReset*, the *receiver* stops recomputing at the first open node, while in *ReResetSplit* it has to recompute until the last node. The gain from the extra open nodes has to balance the extra cost; this is not the case in search trees with work near the root.

This indicates the need to support different sharing strategies in the same solver; different strategies might lead over 60% variations in speedups. The most suitable strategy can be selected based on various factors. At the application level, relevant factors include the *number of agents* (if small, methods based on *relative positions* are better), the *hardware platform* (if the ratio of cpu speed vs. interconnection speed is high, then recomputation is preferred), and the *size of the application* (recomputation seems better for small applications). Factors arising *during* the computation can also be employed to dynamically select the most effective strategy; let *expand_rate* denote the ratio $\frac{numberOfAtoms}{numberGuessedAtoms}$, where *numberGuessedAtoms* is the number atom splits and *numberOfAtoms* is the total number of atoms in the stack (size of the partial model).

0. If *expand_rate* is high then copying is better than recomputation. If the test is performed dynamically, and the focus is on incremental/relative positions methods, then *expand_rate* is limited to the part of the search tree branch below the nca node.
1. If recomputation is chosen, then backtracking can be selected if the path from the root to the nca node is significantly longer then the path from the start node to the nca node, otherwise reset can be employed. In copying, the copy-all schemes are more effective than incremental-copy if the nca node is closer to the root.

2. Based on the number of atoms between contiguous choice-points, an agent can decide whether to use splitting. E.g., in recomputation, if the sender estimates that the number of atoms between the consecutive open nodes is large, or the *expand-rate* is large, and there are only a few open nodes, then it transfers work without splitting.
3. If the system uses the *relative positions* and a mismatch situation occurs, then it can temporarily, or permanently, switch to a reset or an history comparison scheme.

Work is in progress to build a self-adapting sharing procedure based on these ideas.

Communication Frequency: All the communications between agents are asynchronous. It is very important to determine the frequency used by the agents to check for incoming messages, and the interval used to wait for a reply before giving up. Our experiments indicate that the frequency and delay intervals should be tied to the estimated size of the tasks available in the open-nodes—to avoid giving away small tasks or depriving the work-sender of work. Another significant factor is the sharing strategy adopted, in particular: *(1)* the presence of splitting requires a lower frequency; *(2)* approaches having higher communication requirements (e.g., copying or methods using *relative positions*) impose lower frequencies and higher delays; *(3)* approaches requiring more time to reach the alternative (e.g., recomputation) need longer delays.

5   Conclusions and Future Work

In this paper, we presented preliminary results from an investigation of efficient methodologies for the execution of ASP and SAT solvers on Beowulf clusters. In particular, we focused on the analysis of distinct task sharing strategies, obtained as variations of the *copying* and the *recomputation* schemes—which have been theoretically proved [12] to be optimal in the context of search parallelism. The different schemes have been developed to cover a significant spectrum of alternatives, balancing computation and communication. The ideas have been developed in a complete parallel ASP solver, and evaluated on a variety of benchmarks.

Relatively little work has appeared in the literature regarding parallelization of SAT and ASP solvers. A preliminary proposal in this area has been presented in [11], based on a simple solver; [4] presented a master-slave ASP search parallel engine, based on PVM (with limited evaluation). [15] describes a distributed implementation of the SATO SAT solver, based on a master-slave structure; the model relies on a fairly standard copying scheme. PaSAT [1] is a parallel SAT solver based on shared memory and dynamic scheduling. [3] presents a large scale copy-based SAT solver.

Current work is focused on analyzing dynamic scheduling strategies and on the investigation of how scheduling interacts with the sharing strategies presented here.

References

1. W. Blochinger et al. Parallel SAT-Checking with Lemma Exchange: Implementation and Applications. *Theory & Apps. of Satisfiability Testing*, ENDM, 2001.
2. M. Bonacina. Taxonomy of Parallel Strategies for Deduction. *Annals Math & AI*, 29, 2000.
3. W. Chrabakh, R. Wolski. A Parallel SAT solver for the Grid. UCSB TR 2003-05, 2003.

4. R. Finkel et al. Computing Stable Models in Parallel. *AAAI Spring Symposium*, 2001.
5. M. Gelfond and V. Lifschitz. The Stable Model Semantics for Logic Programs. *ILPS*, 1988.
6. G. Gupta et al. Parallel Execution of Prolog Programs. *ACM TOPLAS*, 23(4):472–602, 2001.
7. V. Lifschitz. Answer Set Planning. *LPNMR*, 373–374. Springer, 1999.
8. F. Lin and Y. Zhao. Computing Answer Sets By SAT Solvers. *AAAI*, 2002.
9. V.W. Marek and M. Truszczyński. Stable Models and an Alternative Logic Programming Paradigm. *The Logic Programming Paradigm*. Springer Verlag, 1999.
10. L. Perron. Search and Parallelism in Constraint Programming. *CP*, Springer, 1999.
11. E. Pontelli and O. El-Kathib. Construction of a Parallel Engine for ASP. *PADL*, 2001.
12. D. Ranjan et al. On the Complexity of Or-Parallelism. *NGC*, 17(3):285–308, 1999.
13. P. Simons. *Extending and Implementing the Stable Model Semantics*. PhD, HUT, 2000.
14. K. Villaverde et al. A methodology for order-sensitive execution of non-deterministic languages on beowulf platforms. *Euro-Par*, pages 694–703, 2003.
15. H. Zhang et al. PSATO: a Distributed Propositional Prover and its Application to Quasigroup Problems. *J. Symbolic Computation*, 11:1–18, 1996.

Flexible Skeletal Programming with eSkel

Anne Benoit, Murray Cole, Stephen Gilmore, and Jane Hillston

School of Informatics, The University of Edinburgh, James Clerk Maxwell Building
The King's Buildings, Mayfield Road, Edinburgh EH9 3JZ, UK
enhancers@inf.ed.ac.uk
http://homepages.inf.ed.ac.uk/mic/eSkel

Abstract. We present an overview of *eSkel*, a library for skeletal parallel programming. *eSkel* aims to maximise the conceptual flexibility afforded by its component skeletons and to facilitate dynamic selection of skeleton compositions. We present simple examples which illustrate these properties, and discuss the implementation challenges which the model poses.

1   Introduction

The skeletal approach to parallel programming is well documented in the research literature (see [1–4] for surveys). It observes that many parallel algorithms can be characterised and classified by their adherence to one or more of a number of generic patterns of computation and interaction. For instance, a variety of applications in image and signal processing are naturally expressed as process pipelines, with parallelism both between pipeline stages, and within each stage by replication and/or more traditional data parallelism [5].

Skeletal programming proposes that such patterns be abstracted and provided as a programmer's toolkit, with specifications which transcend architectural variations but implementations which recognise these to enhance performance. This level of abstraction makes it easier for the disciplined programmer to experiment with a variety of parallel structurings for a given application, enabling a clean separation between structural aspects and the application specific details. Meanwhile, the explicit structural information provided by skeletons enables static and dynamic optimisations of the underlying implementation.

In [2] we presented a number of observations and proposals designed to facilitate the popularisation of the skeletal approach, and outlined the first prototype of *eSkel*, a library of skeleton operations for C and MPI. In the light of subsequent experiences we isolated and described two key concepts, *nesting mode* and *interaction mode*, which capture important choices in the design of any skeletal programming framework [6]. This caused us to substantially refine and partly redesign *eSkel*. While these and other concepts embedded in *eSkel* may be found in various combinations in previous skeletal systems, we believe that this paper discusses the first attempt to enumerate and compose them systematically and explicitly within a single framework.

This paper describes current work on the latest instantiation of *eSkel*. We describe the ways in which key concepts have been incorporated into the programming model and API in order to maximise the flexibility with which the

J.C. Cunha and P.D. Medeiros (Eds.): Euro-Par 2005, LNCS 3648, pp. 761–770, 2005.

programmer can shape and reshape skeletal programs, and on the substantial implementation challenges which this flexibility creates.

2   Dynamic Selection of Skeletons and Modes

eSkel is designed to provide the skeletal parallel programmer with flexibility in a number of ways. As in the initial prototype [2], a skeleton abstracts the pattern of interactions between a number of component *activities*, each of which may be internally parallel, and which may choose to invoke further skeletons. The skeletons provided each encompass the ability to specify the use of either localised or distributed data (the data *spread*), and the choice of either explicit or implicit *interaction mode* [6] to trigger interactions between the various activities. The aspects of the interface which capture these choices have been rationalised. Guided by the concept of *nesting mode* [6], the programmer can choose to incorporate both *transient* and *persistent* skeleton calls within the same skeleton nest. Crucially, all of these decisions, as well as the choice of overall skeleton structure for each phase of a program, are made dynamically, and can therefore be based, where appropriate, on information gathered during execution (for example, data values or sizes). In summary the significant improvements over *eSkel*'s first version [2] are as follows.

- The definition of 'families' of related skeletons, capturing distinctions in interaction style, have been replaced by the explicit selection of interaction mode. For example, there is now only one pipeline skeleton, which can be parameterized to behave as either of its predecessors.
- All skeletons in the original *eSkel* were (implicitly) in transient mode. There were no persistent nestings. Both transient and persistent modes are now available (by setting the corresponding parameter).
- The data model has been revised and extended to incorporate the concept of *molecules*, enabling the expression of skeletons (such as `Haloswap`) in which several distinct pieces of data are exchanged at each interaction.

2.1   *eSkel* Interface

The current specification of *eSkel* defines five skeletons, each with flexibility in the dimensions discussed above. We will focus on the two of these, `Pipeline` and `Deal`, which have so far been implemented in the current prototype, since in combination these can illustrate all the points we wish to make. The reader is referred to the *eSkel* homepage [7] for discussion of the `Farm`, `HaloSwap` and `Butterfly` skeletons.

The `Pipeline` skeleton abstracts classical pipeline parallelism. It allows an indexed set of activities (called "stages") to be chained together and applied to a sequence of inputs. Data is passed from a stage to its successor following the natural ordering on stage indices. The user can choose between an implicit pipeline, in which the transfer of the data is done automatically, constraining the stages to produce one result for each input, or an explicit mode, in which a stage can produce arbitrarily many results without necessarily consuming data (for example, a generator or filter stage). In the latter case, the data transfer is controlled by the programmer, by calls to the generic interaction functions `Give` and `Take`.

The `Deal` skeleton is similar to a traditional farm, but with tasks distributed strictly cyclically to workers. Thus it is most appropriately used when the workload associated with individual tasks is expected to be homogeneous. This skeleton is useful nested in a pipeline, in order to internally replicate a stage. `Deal` semantics require the ordering of outputs from the skeleton to match that of the corresponding inputs, irrespective of the internal speed of the workers.

The price for this flexibility is paid in the complexity of the API, where each skeleton call must specify its choices. In addition to the issues already discussed, the pragmatic decision to borrow and build upon MPI's data model with its already substantial parameter set, leads to rather long parameter lists. However, when understood as a set of groups, each addressing orthogonal issues, the parameters become more conceptually manageable. We now present these groupings informally. Their formal use will be illustrated in Section 2.2.

One set of parameters deals with the called skeleton's *data buffers*. As with any MPI collective operation, there are distributed input and output buffers, each requiring the standard MPI pointer, type, length definition. These, together with the enclosing MPI communicator, capture the call's interface to the rest of the program. A second set of parameters deals with the call's *internal structure and interfaces*. For example, in a pipeline, we must describe the number of stages and the types and data modes on their interfaces, and the allocation of processes from the calling group to the distinct stages (captured by borrowing MPI's *colouring* parameter mechanism to describe the required sub-groups). Finally, a group of parameters describe the details of the skeleton's various *activities* (stages in a `Pipeline`, workers in a `Deal`) with a choice of interaction mode and a pointer to a function for each.

In the interests of regularity, we have chosen a single generic interface for all functions which contain code for skeleton activities. The essence of activities is that they *interact* in predefined patterns, via the skeleton infrastructure, with each other and with skeleton calling programs. Each such interaction may involve different numbers of data atoms, according to the semantics of the skeleton. For example, "getting the next task" in a `Deal` skeleton involves a single atom, whereas "updating the halo" in a `HaloSwap` skeleton will involve two atoms for each neighbour (one arriving, one leaving). The `eSkel_molecule_t` type collects atoms involved in an interaction into a single sequence. Activity functions both accept and return a single item, of type `eSkel_molecule_t *`. The specification of each skeleton defines its detailed molecule usage.

2.2   Examples

We present two variants of a toy program in order to concisely illustrate the ease with which the *eSkel* programmer can describe and revise the parallel structure of an application. The first version illustrates the use of persistent nesting and implicit interactions. The second uses transient nesting and explicit interactions.

```c
1   #define STAGES  4          // Number of pipeline stages
2   #define DEALWORKERS 2       // Number of workers in the deal
3   #define INPUTNB 4           // Input multiplicity
4   #define INPUTSZ 5           // Size of each input datum, per process
5
6   eSkel_molecule_t * SimpleStage (eSkel_molecule_t *thingy) {
7     int i;
8     for (i=0;i<thingy->len[0];i++) {
9         ((int *) thingy->data[0])[i] += 1000;
10    }
11    return thingy;
12  }
13
14  eSkel_molecule_t * DealWorker (eSkel_molecule_t *thingy) {
15    int i;
16    for (i=0;i<thingy->len[0];i++) {
17        ((int *) thingy->data[0])[i] += 1000;
18    }
19    return thingy;
20  }
21
22  void DealStage (void) {
23    int workercol, outmul;
24    if ((myrank() < 1))  workercol = 0; else workercol = 1;
25
26    Deal (DEALWORKERS, IMPL, DealWorker, workercol, STRM,
27          NULL, 0, 0, SPGLOBAL, MPI_INT,
28          NULL, 0, &outmul, SPGLOBAL, MPI_INT, 0, mycomm());
29  }
30
31  int main (int argc, char *argv[])
32  {
33    int i, j, p, next, outmul, mystagenum, mymult, *inputs, *results;
34    spread_t spreads[STAGES+1]   = {SPGLOBAL, SPGLOBAL, SPGLOBAL, SPGLOBAL, SPGLOBAL};
35    MPI_Datatype types[STAGES+1] = {MPI_INT, MPI_INT, MPI_INT, MPI_INT, MPI_INT};
36    Imode_t imodes[STAGES]       = {IMPL, DEV, IMPL, IMPL};
37
38    eSkel_molecule_t *(*stages[STAGES])(eSkel_molecule_t *) =
39      {SimpleStage, (eSkel_molecule_t * (*)(eSkel_molecule_t *))DealStage, SimpleStage, SimpleStage};
40
41    MPI_Init(&argc, &argv); SkelLibInit();
42    inputs = (int *) malloc (sizeof(int)*INPUTNB*INPUTSZ);
43    for (i=0; i<INPUTNB*INPUTSZ; i++) inputs[i] = myrank()*100 + i + 1;
44
45    if      (myrank()<2)  mystagenum = 0;
46    else if (myrank()<5)  mystagenum = 1;
47    else if (myrank()==6) mystagenum = 2;
48    else                  mystagenum = 3;
49
50    results = (int *) malloc (sizeof(int)*INPUTNB*INPUTSZ);  // Create output buffer
51
52    Pipeline (STAGES,  imodes, stages, mystagenum, BUF, spreads, types,
53             (void *) inputs, INPUTSZ, INPUTNB, (void *) results, INPUTSZ, &outmul,
54             INPUTNB*INPUTSZ, mycomm());
55
56    MPI_Finalize();  return 0;
57  }
```

Fig. 1. Example 1

Example 1: Persistent Nesting and Implicit Interactions

This program calls a `Pipeline` skeleton, in which the second stage contains a
persistently nested `Deal`. Some of the activities (pipeline stages and deal workers)
are assigned several processes, and process data with "global spread" (i.e. the
data is distributed across such processes and accessed in SPMD style).

The constant declarations define the number of pipeline stages (`STAGES`),
the number of workers (`DEALWORKERS`) in the deal, and the number and size of

input items (INPUTNB, INPUTSZ). When run with 8 processes, the input to this pipeline consists of a total of (8*4*5) integers, which are treated as 4 (INPUTNB) inputs to the pipeline, each of which is constructed from a 5 (INPUTSZ) integer contribution per process. The first stage in the pipeline has been assigned two processes. Thus, since we are in "global spread" mode, each pipeline input will be fed to the two first stage processes as a block of (8*5)/2 integers each.

The main function (lines 31-57) makes the pipeline call. It defines all the pipeline parameters and assigns processes to the stages. Lines 34-39 describe the structure of the pipeline: all inter-stage data transfers use global spread (SPGLOBAL) and involve sequence of integers. All activities except the second stage use "implicit" mode (IMPL) transfers (in other words, handled implicitly by the skeleton). The second stage has "devolved" mode (DEV), indicating that this will be inherited from a persistently nested skeleton (the deal in this case). All stages will execute the SimpleStage activity, except the second, which executes DealStage. Lines 42-43 create some (distributed) input data. Lines 45-48 compute an assignment of processes to stages. Line 50 creates the (distributed) output buffer. The pipeline call itself is on lines 52-54, and is passed the various parameters as created, together with details of the input and output buffers in the style of a conventional MPI collective operation.

Function SimpleStage, on lines 6-12, describes the actions of the first, third and fourth stages. Since implicit interaction mode is used, the function takes an *eSkel* molecule as a parameter and returns a molecule. Here we simply add 1000 to each integer in the block and return the modified molecule. In contrast, the second stage is described by DealStage. It computes an assignment of processes (from the stage) to workers in the internal deal, then calls the Deal skeleton on lines 26-28. The data mode is stream (STRM) indicating that the data are streamed into the deal from the *parent* skeleton. The input and output buffer related parameters are therefore redundant (all the NULLs and 0s). Deal workers run the function DealWorker. In our simple example this is identical to SimpleStage. We leave the duplication to emphasise that these functions can be programmed independently, if appropriate.

Figure 2 illustrates the requested assignment of processes to stages and workers, and the different communicators involved. This will help to clarify the discussion of *eSkel* implementation in Section 3. We have a 4-stage pipeline where the second stage is a deal with two workers, of which one has two processes. **Px** represents process x $(0 \leq x \leq 7)$, as indexed in the original MPI_COMM_WORLD. **Si** represents stage i $(0 \leq i \leq 3)$, and **Woj** represents worker j $(0 \leq j \leq 1)$.

Example 2: Transient Nesting and Explicit Interactions
Our second example demonstrates the use transient nesting, explicit interactions and local data spread, and also illustrates the ease with which the programmer can experiment with the high level structure of an application. Working from the program of example 1, the programmer decides to

— turn the second stage (formerly the Deal) into a simple single process stage;
— reallocate the two spare processes from the second stage to the first stage;

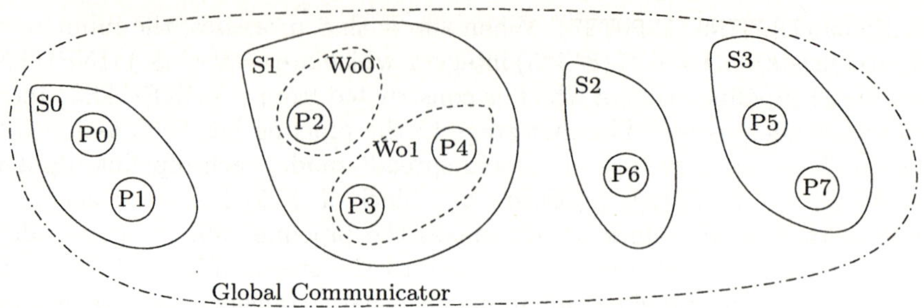

Fig. 2. Example 1 - Mapping of the processes

- adapt the first stage (which now has four processes) so that every second input is processed by a dynamically created four stage pipeline, with one process per stage. These are transiently nested pipelines, in contrast to the persistently nested `Deal` of example 1. Other inputs are treated as in the original simple stages.

The structural changes at the outer level are captured in lines of 3, 61, 63, 64, 70 and 71 of Figure 3. Line numbers in parentheses are those of the corresponding lines in Figure 2. The rest of `main` is unchanged, and is omitted here. The behaviour of the new first stage, and its sub-stages are described by the new functions `StrangeStage` and `SubStage`. `StrangeStage` creates and calls the nested pipeline for every second input (controlled by line 44). The nested sub-stages, called with explicit (**EXPL**) interaction mode (line 35), use *eSkel* functions `Give` and `Take` to control interactions with neighbouring stages (lines 21 and 26).

3   Implementation Challenges

We now consider the challenges raised by *eSkel*'s implementation, focusing particularly on the call tree, which is built dynamically, and allows processes to find communication partners. We also discuss the management of the input and output buffers.

3.1   Building the Call Tree Dynamically

In order to correctly implement the communications which have been abstracted by the skeleton calls, it is necessary for the processes assigned to the various activities to understand their position within the overall skeleton nest. Since the structure of the nest emerges dynamically, and independently within distinct branches, this investigation can only be performed dynamically, at the point at which no further changes to the structure are possible. In *eSkel* this occurs when each activity calls its first explicit (or implicit) interaction. Building the call tree requires collaborative communications between all processes in the nest.

```
  3  #define SUBSTAGES 4      // Number of stages in the nested pipeline

 15  void SubStage (void) {
 16    int i, **len;
 17    eSkel_molecule_t *tempitem;
 18
 19    len = (int **) malloc (sizeof (int *));
 20    tempitem = (eSkel_molecule_t *) malloc (sizeof (eSkel_molecule_t));
 21    while (tempitem->data = Take(len)) {
 22      tempitem->len = *len;
 23      for (i=0;i<tempitem->len[0];i++) {
 24        ((int *) tempitem->data[0])[i] += 250;
 25      }
 26      Give(tempitem->data, tempitem->len);
 27    }
 28  }
 29
 30  eSkel_molecule_t * StrangeStage (eSkel_molecule_t *thingy) {
 31    static int count=0;
 32    int i, outmul;
 33    spread_t spreads[SUBSTAGES+1]    = {SPLOCAL, SPLOCAL, SPLOCAL, SPLOCAL, SPLOCAL};
 34    MPI_Datatype types[SUBSTAGES+1] = {MPI_INT, MPI_INT, MPI_INT, MPI_INT, MPI_INT};
 35    Imode_t imodes[SUBSTAGES]       = {EXPL, EXPL, EXPL, EXPL};
 36
 37    eSkel_molecule_t *(*stages[SUBSTAGES])(eSkel_molecule_t *) =
 38      {(eSkel_molecule_t * (*)(eSkel_molecule_t *))SubStage,
 39       (eSkel_molecule_t * (*)(eSkel_molecule_t *))SubStage,
 40       (eSkel_molecule_t * (*)(eSkel_molecule_t *))SubStage,
 41       (eSkel_molecule_t * (*)(eSkel_molecule_t *))SubStage};
 42
 43
 44    if (count++ % 2) {  // handle it directly
 45      for (i=0;i<thingy->len[0];i++) {
 46        ((int *) thingy->data[0])[i] += 1000;
 47      }
 48    } else {            // handle with a transient pipeline!
 49      Pipeline (SUBSTAGES,  imodes, stages, myrank(), BUF, spreads, types,
 50                thingy->data[0], 1, thingy->len[0],  thingy->data[0], 1 , &outmul,
 51                thingy->len[0], mycomm());
 52    }
 53    return thingy;
 54  }

(36)61    Imode_t imodes[STAGES]      = {IMPL, IMPL, IMPL, IMPL};

(38)63    eSkel_molecule_t *(*stages[STAGES])(eSkel_molecule_t *)
(39)64      = {StrangeStage, SimpleStage, SimpleStage, SimpleStage};

(45)70    if (myrank()<4)       mystagenum = 0;
(46)71    else if (myrank()<5)  mystagenum = 1;
```

Fig. 3. Example 2 - changes from example 1

The root of the tree corresponds to the main skeleton call. The children describe the different activities, and if there are some persistently nested skeletons, they appear in the tree. The transiently nested structure is not built in the main tree. The sub-tree will be built dynamically when the transient skeleton call is performed. Fig. 4 represents the call tree for the examples introduced in the previous section. The tree of the nested pipeline of Fig. 4b is created at each call of this pipeline.

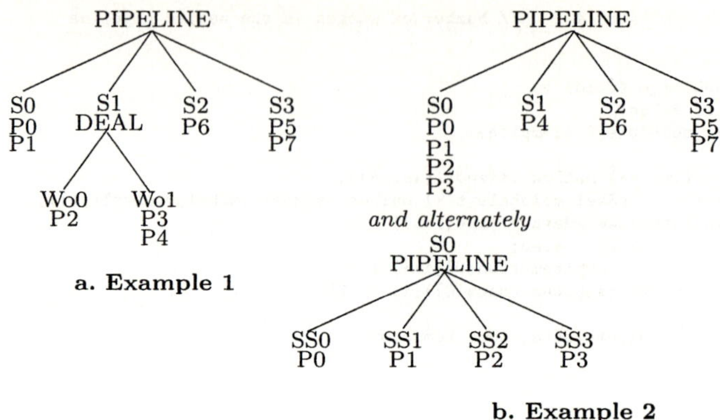

Fig. 4. Call trees for the two examples

3.2   Finding Communication Partners

Communication partners are found by traversing the call tree. For example, in Fig. 4a, we discover that Wo1 takes data from S0 and gives data to S2. Interactions with a neighbouring deal require different partners at each interaction (because of the cyclic nature of interactions in that skeleton). In our example, when S0 performs a Give, it needs to give data alternately to Wo0 and Wo1. To ensure this cyclic distribution we keep track of the number of Give and Take calls performed. This allows us to find the appropriate partner.

Particular attention must be paid to boundary cases. When S0 performs a Take, it cannot find a communication partner in the tree. It needs instead to take the data from the input buffer to the pipeline call. Similarly, S3's Give must store data in the output buffer. We detail the management of these buffers below.

3.3   Managing the Input and Output Buffers

The *input buffer* must be created collaboratively by all the processes, since each of them potentially has some of the data at the point of the skeleton call. It is created just after the call tree has been built, during the first interaction of each activity. At that time, the processes which have to take data from the input buffer (processes of S0 in our pipeline examples) collect all the data through group communication, and build the input buffer.

The *output buffer* is created by the processes which write results into the buffer (processes of S3 in the first example). After completion of the skeleton nest, the results are distributed among all the processes.

More care must be taken when the group of processes dealing with the input or output buffer are part of a deal. For example, if the nested deal was in S0 instead of S1, both Wo0 and Wo1 need to take data from the input buffer while maintaining the cyclic order. To address this issue, we dynamically create a thread which distributes the data to the workers. A symmetrical approach is taken for the output buffer.

3.4   Context of Use

The *eSkel* library operates within the context of a running MPI program which has already created its fixed number of processes. Moreover, a fully thread safe version of MPI (in the sense of `MPI_THREAD_MULTIPLE`) must be used. We are developing *eSkel* using the Los Alamos Message Passing Interface LA-MPI [8] and the Sun HPC Cluster Tools [9].

4   Future Work

A full implementation of *eSkel* is in progress, with the current status reported on the project web pages [7]. After completion of the initial prototype, work will focus on development of a set of demonstrator applications, internal optimisations and expansion of the skeleton set. We will also consider ways in which the API can be simplified, for example by moving to a more powerfully descriptive host language such as Java, as and when the Java-MPI combination gains more widespread practical acceptance. Current work on *eSkel* is conducted under the umbrella of the *Enhance* project [10], which exploits the modelling power of the Performance Evaluation Process Algebra (PEPA) [11] to predict the performance of skeletally structured Grid applications, with a view to assisting in their scheduling and re-scheduling in the presence of the dynamically heterogeneous performance and availability characteristics of Grid computing platforms.

Acknowledgements

The authors are supported by the Enhance project ("Enhancing the Performance Predictability of Grid Applications with Patterns and Process Algebras") funded by the Engineering and Physical Sciences Research Council under grant number GR/S21717/01.

References

1. Cole, M.: Algorithmic Skeletons: Structured Management of Parallel Computation. MIT Press & Pitman, ISBN 0-262-53086-4 (1989)
2. Cole, M.: Bringing Skeletons out of the Closet: A Pragmatic Manifesto for Skeletal Parallel Programming. Parallel Computing **30** (2004) 389–406
3. Pelagatti, S.: Structured Development of Parallel Programs. Taylor & Francis, London (1998)
4. Rabhi, F., Gorlatch, S., eds.: Patterns and Skeletons for Parallel and Distributed Computing. Springer Verlag, ISBN 1-85233-506-8 (2003)
5. Subhlok, J., O'Hallaron, D., Gross, T., Dinda, P., Webb, J.: Communication and memory requirements as the basis for mapping task and data parallel programs. In: Proceedings of Supercomputing '94, Washington, DC (1994) 330–339

6. Benoit, A., Cole, M.: Two Fundamental Concepts in Skeletal Parallel Programming. In: Computational Science - ICCS 2005. Number 3515 in LNCS, Atlanta, USA, Springer (2005) 764–771
7. Benoit, A., Cole, M.: `http://homepages.inf.ed.ac.uk/mic/eSkel` (2005)
8. Aulwes, R.T., Daniel, D.J., Desai, N.N., Graham, R.L., Taylor, L.D.R.M.A., Woodall, T.S., Sukalski, M.W.: Architecture of LA-MPI, A Network-Fault-Tolerant MPI. In: Proceedings of the 18th International Parallel and Distributed Processing Symposium (IPDPS'04), Santa Fe, New Mexico, IEEE Computer Society (2004)
9. Sun Microsystems, 901 San Antonio Road, Palo Alto, CA 94303-4900, USA: Sun HPC ClusterTools 3.1 User's Guide. (2000)
10. Benoit, A., Cole, M., Gilmore, S., Hillston, J.: `http://groups.inf.ed.ac.uk/enhance` (2004)
11. Gilmore, S., Hillston, J.: The PEPA Workbench: A Tool to Support a Process Algebra-based Approach to Performance Modelling. In: Proc. of the 7th Int. Conf. on Modelling Techniques and Tools for Computer Performance Evaluation. Number 794 in LNCS, Vienna, Springer-Verlag (1994) 353–368

Dynamic Reconfiguration
of Grid-Aware Applications in ASSIST*

Marco Aldinucci[1], Alessandro Petrocelli[2], Edoardo Pistoletti[2],
Massimo Torquati[2], Marco Vanneschi[2], Luca Veraldi[2], and Corrado Zoccolo[2]

[1] Inst. of Information Science and Technologies – CNR, Via Moruzzi 1, Pisa, Italy
[2] Dept. of Computer Science – University of Pisa – Largo B. Pontecorvo 3, Pisa, Italy

Abstract. Current grid-aware applications are implemented on top of low-level libraries by developers who are experts on grid middleware architecture. This approach can hardly support the additional complexity of QoS control in real applications. We discuss a novel approach used in the ASSIST programming environment to implement/guarantee user provided QoS contracts in a transparent and effective way. Our approach is based on the implementation of automatic run-time reconfiguration of ASSIST application executions triggered by mismatch between the user provided QoS contract and the actual performance values achieved.

Keywords: Structured Parallel Programming, grid, QoS contract, Adaptive Applications

1 Introduction

A grid system is a geographically distributed collection of possibly parallel, interconnected processing elements that all run some kind of common grid middleware (e.g. Globus services). Such platforms are characterized by heterogeneity of nodes, and by dynamicity in resource management and allocation [1].

One popular approach to grid programming consists in directly exploiting middleware services within a standard programming language. This approach rapidly leads to an intolerable complexity as soon as the application is both complex and requested to exploit an user-defined QoS.

Large scale grid-aware application will require developing toolkits which support reconfigurable code and the binding of legacy code. They should enforce the minimization of developing cost by enabling the (static and dynamic) reconfiguration of the application to target different customer scenarios, while exploiting a good performance/cost ratio over a broad class of hardware platforms. They should also cope with code reuse providing the programmer with suitable bridges to interoperate with legacy code (Corba, CCM, Java Beans, DCOM, etc.). In this context, an advanced run-time support should seamlessly adapt the application structure to the current grid status, and transparently manage faulty

---

* This work has been supported by the Italian MIUR FIRB *Grid.it* project No. RBNE01KNFP, and Italian Project "legge 449/97" No. 02.00640.ST97.

J.C. Cunha and P.D. Medeiros (Eds.): Euro-Par 2005, LNCS 3648, pp. 771–781, 2005.
© Springer-Verlag Berlin Heidelberg 2005

events and performance degradations of the underlying platform, which should be considered the standard behavior of a large-scale distributed platform.

High-level programming environments for grid aim at moving most of the grid specific efforts needed while developing high-performance grid applications from programmers to grid tools and run time systems. Among them, we mention ASSIST, GrADS, ProActive, Condor, Ibis [2–6]. In particular, ASSIST [2] has been designed along these guidelines. It supports the development of interoperable applications [7] onto heterogeneous platforms [8].

We present here a novel extension of the ASSIST environment exploiting a self-optimizing run-time targeted to fulfill QoS requirements, which are expressed at the language level by means of a *QoS contract*. QoS contract is transparently managed by the ASSIST compiler that preprocesses it and generates all the support code needed to enforce it at run-time by controlling parallelism degree, processes remapping, and algorithm selection. Programmer is just asked to express a composition of ASSIST modules, and declare a QoS contract either on each module or on the *whole* application. We experimentally show that both stateless and stateful computations may be suitably reconfigured to fulfill several kinds of contracts, and that these reconfigurations might have negligible cost enabling fine grain control on the application dynamic evolution.

ASSIST is briefly introduced in Sect. 2. QoS contracts and their managing strategy are presented in Sect. 3. ASSIST QoS-enabled run-time support is build on a set of *mechanisms* aiming to enable the run-time reconfiguration of applications. We believe that these mechanisms are an enabling technology for QoS control of grid-aware applications. Different *policies* may drive these mechanisms to target different QoS goals, such as performance and fault-tolerance. In this paper we mainly focus on these mechanisms, which are presented and evaluated in Sect. 4. Some examples of performance policies are sketched in Sect. 5, a full discussion on performance policies is outside the scope of this paper (due to lack of space). The presented policies are validated through experiments in Sect. 6.

2   The ASSIST Environment and Its Run-Time

ASSIST applications are described by means of a coordination language, which can express arbitrary graphs of modules, interconnected by typed streams of data. Modules can be either sequential or parallel. A sequential module wraps a sequential function. A parallel module *(parmod)* can be used to describe the parallel execution of a number of sequential functions, that are activated and run as *Virtual Processes* on items arrival onto input streams. The sequential functions can be programmed by using a standard sequential language (C, C++, Fortran). Virtual Processes may synchronize with one other by barriers. Overall, a *parmod* may behave in a data-parallel (e.g. SPMD/for-all/apply-to-all) or task-parallel (e.g. farm) way and may exploit a distributed shared state which survive to Virtual Processes lifespan. A module can non deterministically accept from one or more input streams a number of input items, which may be decomposed in parts and used as function parameters to instantiate Virtual Processes, according to

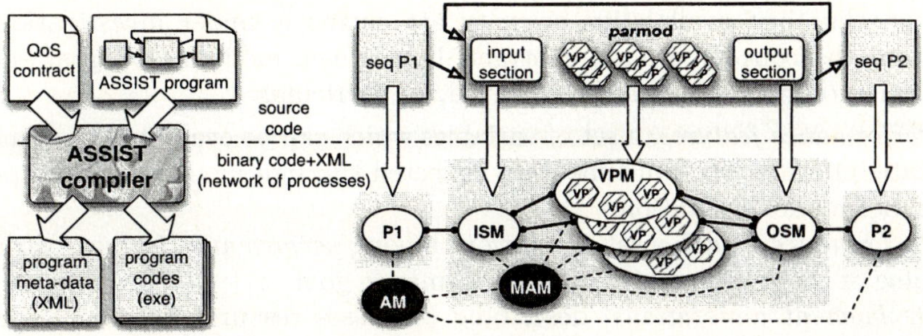

Fig. 1. An ASSIST application and a QoS contract is compiled in a set of executable codes and its meta-data [9]. These informations are used to set up a processes network at launch time: hexagons represents Virtual Processes, ovals represents processes, solid edges represent data channels, dashed edges managements channels.

the input and distribution rules specified in the parmod. Virtual Processes may send onto output streams items or parts of them which are gathered according to the output rules. More details on ASSIST environment can be found in [2, 9].

The ASSIST compiler translates a graph of modules into a network of processes. As sketched in Fig. 1, sequential modules are translated into sequential processes, while parallel modules are translated into a parametric (w.r.t. the parallelism degree) network of processes: one *Input Section Manager* (ISM), one *Output Section Manager* (OSM), and a set of *Virtual Processes Managers* (VPMs, each of them running a set of Virtual Processes). The actual parallelism degree of a parmod instance is given by the number of VPMs. ASSIST run-time support also include a *Module Adaptation Manager* per parmod (MAM), which monitors the performances of the parmod, and implements reconfiguration policies; and an *Application Manager* (AM), which coordinates the QoS at the level of the whole application by coordinating MAMs. In this work we focus on MAM design. AM design is subject of current research, we refer back to Sect. 8 and [9] for a general description.

3   ASSIST Autonomic Run-Time and QoS Contracts

The initial configuration of an ASSIST program is specified by the set of processes that are co-allocated at launch time. The configuration of a parmod is managed by its MAM, which dynamically decides the number of VPMs, and their mapping onto grid Processing Elements (PEs) acquired through grid middleware. The ASSIST compiler prepares a *QoS contract* for each parmod and bind them to MAMs. Moreover, a MAM can asynchronously receive a different QoS contract from the AM in any moment along the application run.

Among all possible QoS goals, in this work we mainly focus on performance related ones that are achievable through adaptation within each parallel module. All aspects regarding modules coordination, as well as other QoS measures such as reliability, availability, security are currently under investigation. We introduce the concept of QoS contract. It carries a module QoS goal and the description on how it should be achieved. In particular:

- *Performance features:* a set of variables which can be evaluated from module static information, run-time data collected through monitoring, and performance model evaluation.
- *Performance model:* a set of relations among *performance features* variables, some of them representing the performance goal.
- *Deployment annotations]* describing processes resource needs, such as required hardware (platform kind, memory and disk size, network configuration, etc.), required software (O.S., libraries, local services, etc.), and other all strictly required constraints to enforce code correctness.
- *Adaptation policy:* a reference to the desired adaptation policy chosen among the ones available for the module. Standard adaptation policies are represented as algorithms and embedded within MAM code at compile time.

The following is the QoS contract used in experiment Fig. 3 ❷:

Perf. features	QL_i (input queue level), QL_o (input queue level), T_{ISM} (ISM service time), T_{OSM} (OSM service time), N_w (number of VPMs), $T_w[i]$ (VPM$_i$ avg. service time), T_p (parmod avg. service time)
Perf. model	$T_p = \max\{T_{ISM}, \sum_{i=1}^{n} T_w[i]/n, T_{OSM}\}$, $T_p < K$ (goal)
Deployment	arch = (i686-pc-linux-gnu $\vee$ powerpc-apple-darwin*)
Adapt. policy	goal_based

MAM run-time behavior may be conveniently sketched in terms of autonomic control loops [10]. In order to handle situations in which resource availability affects the performance of the applications, the run-time system of the running application has to:

1. **monitor** application actual status by collecting raw sensible performance data, and synthesize *performance features* variables;
2. evaluate *performance model*, and if it is unsatisfied, **analyze** it to discover possible causes;
3. if needed, **plan** a reconfiguration strategy according to the *adaptation policy*, with the goal of re-conveying the application in a legal status;
4. **execute** the reconfiguration, possibly allocating new resources/rebalancing the computation, possibly migrating entire modules.

4   Reconfiguration Key Concepts and Mechanisms

The modular nature of ASSIST applications and their management enable the reconfiguration of a subset of modules while neither affecting nor stopping the ones not involved in the reconfiguration, which can be distinguished in two categories: (a) involving the alteration of mapping between application activities and PEs[1]; (b) involving the variation of process graph structure, including modules

---

[1] ASSIST parmod supports the migration of Virtual Processes between VPMs, and VPMs between PEs, possibly migrating or remapping associated data.

parallelism degree[2]. Observe that, load balancing within a parmod can be managed by reconfigurations of kind (a): the load of a VPM (and the PE hosting it) may be decreased by moving some of its Virtual Processes to another VPM.

Independently of when the MAM decides to trigger a reconfiguration, the module is actually reconfigured on the next *reconf-safe* point. These are the time windows during a given parmod run in which its internal attributes are completely defined by the set of local attributes. Notably, the runtime does not introduce any additional synchronization w.r.t. the ones required by program semantics. It rather delays reconfiguration execution just after next natural reconf-safe point is reached. We distinguish between two kinds of reconf-safe points:

- *on-stream-item*: A new item is available in any input streams. A complete systolic synchronization is induced by the ISM process within the parmod. If needed, shared state is consolidated along the synchronization process.
- *on-barrier*: A complete synchronization has happened within a parmod due to either an explicit or implicit barrier. Barriers are issued by the programmer or the compiler to enforce the consolidation of shared state (e.g. at each step of a data-parallel iterative program).

No other points can be considered reconf-safe. Since the reconfiguration process is designed to be transparent to the programmer, we exclude the possibility of reconfiguring the parmod during the execution of an user defined function. In this way, we avoid the instrumentation of legacy code and the adoption of process dumping techniques that are hardly effective on heterogeneous platforms.

4.1   Reconfiguration Protocol

The MAM triggers a parmod reconfiguration raising a command toward all interested processes which participate, with the MAM, to a distributed reconfiguration protocol. All data exchanges (data or computation migrations) happen among VPMs following a communication schema encoded and optimized for the particular parmod semantics at compile time. These regard the static instrumentation of reconf-safe points with the minimum needed reconfiguration actions, e.g. a farm stateless parmod is not instrumented with data migration code.

The MAM participates to the protocol in order to mediate and orchestrate the interactions between AM and the Grid Abstract Machine, and to enforce all processes involved are aware and ready to start a reconfiguration at the next reconf-safe point. This should be enforced also when a complete barrier is not needed to ensure data integrity (e.g. master-slave). The latter property is guaranteed by the MAM accordingly to the following behavioral schema:

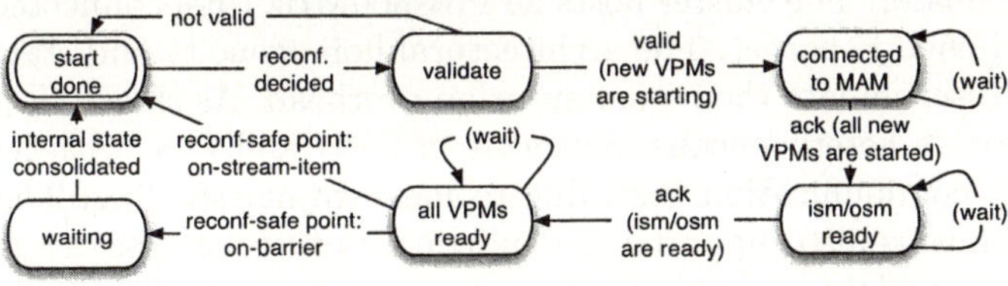

---

[2] ASSIST support an increment or decrement of the number of VPMs in parmods.

A parmod reconfiguration is initiated by its MAM. The reconfiguration plan is build accordingly to the proper strategy (see Sect. 5), e.g. increase the number of PEs, move to it a given number of Virtual Processes. Possibly some resources are asked to grid middleware through the Grid Abstract Machine. A reconfiguration command is synthesized, then validated (e.g. do not remove the last VPM, etc.).

The MAM waits in sequence that new started processes are connected to it; and the ISM, OSM, and all involved VPMs acknowledge the reconfiguration command. Eventually the MAM waits a reconf-safe point is reached and enforces all data and computation redistribution is completed.

4.2 Evaluation of Reconfiguration Overhead

We evaluate the cost of reconfiguration mechanisms against the following metrics (the former two are illustrated in Fig. 2):

- *Reconfiguration time* (R_t): refers to the total time spent to reconfigure the application, from the time a MAM decides the reconfiguration to the time it is completed.
- *Reconfiguration latency* (R_l): the time elapsed from the point a parmod is stopped for a reconfiguration to the time it is resumed. This is the foremost measure from users' viewpoint.
- *Reconfigurable code overhead* (R_o): the slowdown of an application when instrumented with the additional code needed to make it reconfigurable.

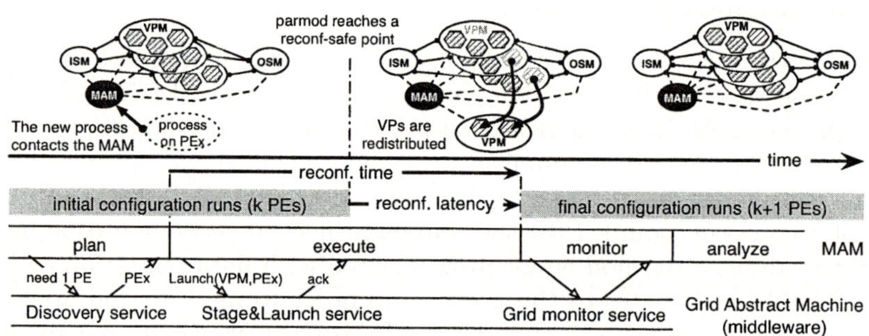

Fig. 2. Reconfiguration dynamics and metrics.

These metrics are evaluated on two different ASSIST applications on a dedicated Linux cluster. The cluster hosts 24 P3@800MHz PEs, connected through a 100MBit switched Ethernet. The architectural homogeneity and stability enable to precisely discriminate the reconfiguration overhead. As shown in [8], ASSIST already supports heterogeneous platforms in CPU and O.S. with less than 7% of additional communication cost, due to message marshalling. The reconfiguration mechanisms also support the deployment on to heterogeneous platforms with TCP/IP or Globus provided communication channels. The two applications are composed by one parmod and two sequential modules.

The first is a data-parallel application receiving a stream of integer arrays and computing a forall of simple function for each stream item; the matrix is stored in the parmod shared state. In this case the overall shared state has a negligible size since the experiment is designed to evaluate mechanism overhead. Preliminary experiments with realistic state size show a linear dependency between R_t and the global size of parmod shared state.

The second is a farm application computing a simple function on different stream items. Since R_t also depends on sequential function cost, in both cases we choose sequential functions with a close to zero computational cost in order to evaluate mechanism on the finest possible grain.

The reconfiguration overhead (R_o) measured during our experiments, without any reconfiguration change actually performed, is practically negligible, remaining under the limit of 0,004%, the measurement of the other two metrics are reported in Table 1.

Table 1. Evaluation of reconfiguration overheads (ms). On this cluster, 50 ms are needed to ping 200KB between two PEs, or to compute a 1M integer additions.

parmod kind	Data-parallel (with shared state)						Farm (without shared state)					
reconf. kind	add_PEs			remove_PEs			add_PEs			remove_PEs		
# of PEs involved	1→2	2→4	4→8	2→1	4→2	8→4	1→2	2→4	4→8	2→1	4→2	8→4
R_l on-barrier	1.2	1.6	2.3	0.8	1.4	3.7	–	–	–	–	–	–
R_l on-stream-item	4.7	12.0	33.9	3.9	6.5	19.1	~0	~0	~0	~0	~0	~0
R_t	24.4	30.5	36.6	21.2	35.3	43.5	24.0	32.7	48.6	17.1	21.6	31.9

Consider the reconfiguration $x \rightarrow y$, where x and y are the number of PEs before and after the reconfiguration respectively. For the data-parallel parmod, R_l grows linearly with $(x + y)$ for both kinds of reconf-safe points, and depends on shared state size and mapping. Shared state is kept distributed during all the reconfiguration process. Farm parmod cannot be reconfigured on-barrier since it has no barriers, and achieves a negligible R_l (below 10^{-3} ms). This is due to the fact that no processes are stopped in the transition from one configuration to the next. R_t, which includes both the protocol cost and time to reach next reconf-safe point, grows linearly with $(x + y)$ for the former cost and depends on user-function for the latter.

5   Adaptation Policies

Adaptation policies are implemented as algorithms, actually methods of the MAM automatically generated by the compiler. ASSIST provides the programmer with hooks to add user-defined policies. The definition of a set of suitable policies and models to drive MAM and AM **analyze** and **plan** phases is subject of current research. For the sake of brevity, we present here a policy to automatically drive MAM of a farm-like parmod, adaptation policies for data-parallel stateful parmod have been presented elsewhere [11]. Farm parmod exploits on-demand task scheduling that guarantees load-balancing also in case of heterogeneous platforms, thus the MAM does not need to care about it. As discussed in Sect. 3, it is worth distinguishing two kinds of goals, and their adaptation policies. Ensure: (i) a *desired* service time; (ii) the *best effort* in the performance/resource trade-off.

In general, a policy should first analyze causes of module misbehavior reasoning on performance features values, then use the performance model to forecast if an adaptation may lead to contract satisfation. Eventually use mechanisms API (e.g. add_PEs) to reconfigure the module. Different policies can lead to different decisions in the same configuration: when the QoS contract is fulfilled, a policy of kind (i) would not increase the resource assigned to the module, even if it could exploit them. A best-effort policy in this case would pursue the maximum performance. As well, when incoming data rate decreases, so that some resources could be released because the module is over-dimensioned w.r.t. the input rate, a best-effort strategy will promptly release the resources, in order to optimize their usage, while a goal based policy would not, in the eventuality that the input rate will raise again.

When operating in best effort mode, the parmod acquires a new resource if the input queue is filling (its utilization is close to 1) and the output queue is emptying, i.e. the slower stage in the parmod is the processing one. It releases a certain amount of resources if exists a proper subset R of the set of VPMs that provides enough computing power:

$$B_{ISM} < \sum_{i \in R} B_w[i], \text{where} \quad B_{ISM} = \frac{1}{T_{ISM}}, B_w[i] = \frac{1}{T_w[i]}$$

in that case, the resources not in R can be released with no loss in performance.

When pursuing target (i), in the condition to release the resource, the actual bandwidth B_{ISM} is substituted by the contractually specified one: T_p. In this setting, the parmod acquires a new resource if the contract is not satisfied and the slower stage in the parmod is the processing one (as in the best-effort case); the conjunction of the two conditions prevents from adding new resources if the contract is not satisfied but due to other modules low performances.

These policies can clearly distinguish when a contract is violated due to another module misbehavior: in such case, the module manager is aware that no actions could be performed to solve the problem: we are investigating cooperation strategies among managers, to address these issues.

6   Experiments

To evaluate the effectiveness of proposed reconfiguration mechanisms and policies we tested a farm and a data-parallel parmods on several scenarios. The former parmod farms out a dummy sequential function with 2s average service time (experiments in Fig. 3 ❶, ❷, ❸). The latter computes a shortest-path like algorithm exploiting 640KB of shared state (Fig. 3 ❹). Tests are performed on the cluster described in Sect. 4.2, results are shown in Fig. 3:

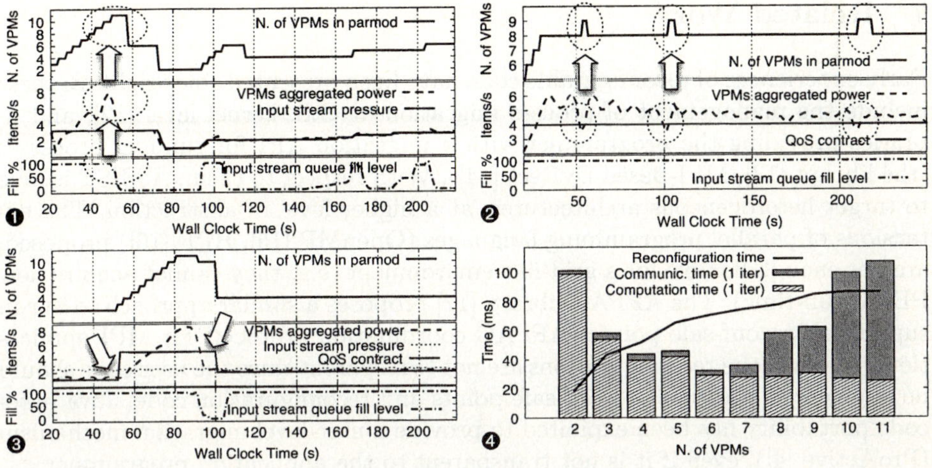

Fig. 3. Experiments on parmod reconfiguration (see Sect. 6).

❶ Farm: best effort mode: the Input stream pressure, i.e. the frequency at which parmod receive stream items, is changed along the program run. The parmod input queue tends to fill when the VPMs consume stream items slower than they are received, and vice-versa. The MAM tries to match the service time of VPMs and items arrival time by increasing or decreasing the number of VPMs. Transient VPMs may be exploited to bring back queue to a safe level.

❷ Farm: a fixed service time specified in the contract and with a fixed input pressure. Three times, along the program run, a PE is externally overloaded causing a contract violation. The MAM reacts by adding as many VPMs (one in the figure) mapped onto fresh PEs until the contract is satisfied. The MAM also knows (see Sect. 5) that the contract continues to be satisfied if the overloaded PE is removed, and after a while removes it. On the whole a VPM migrates from one PE to another without stopping the parmod.

❸ Farm: a fixed service time specified in the contract and with a fixed input pressure, but the contract is changed by the AM three times along the program run. Each time, the MAM reacts by adapting the number of VPMs in order to satisfy the new contract.

❹ Data-parallel: on-barrier reconfiguration during the execution of a single forall. The MAM receives, during the program run, different contracts with fixed number of PEs (ranging from 2 to 11); it reacts by asking each time a fresh PE, mapping on it a VPM, and triggering the suitable computation and shared data redistribution. Observe that in this case the optimal number of PEs may heuristically be decided since the iteration time (computation time + communication time) exhibit a quite regular behavior. The figure also show that reconfiguration is quite efficient since its overhead is comparable to a single iteration time.

The experiments show that the approach is feasible for data-parallel computations, and that good results are obtained for the task-parallel ones.

7   Related Work

Early experiences of reconfigurable code have been presented since eighties; these include the management of process migration at O.S. kernel level [12], and libraries providing the programmer with a migration API for running processes (the libckpt [13], MPI-based DyRect [14]). With respect to them, ASSIST is able to target heterogeneous architectures, at a higher level of abstraction. The extensions of parallel programming languages (OpenMP [15], HPF [16]) proposed, are not enough flexible for a grid-like environment (e.g. they cannot acquire new PEs at run-time). The AFPAC library [17] proposes a similar approach to our in supporting reconf-safe points (AFPAC coordination protocol) for MPI applications, however the reconfigurations are not transparent since the user code should be augmented with both reconf-safe points and reconfiguration code. Java byte-code portability has been exploited to provide a user-level migration mechanism (ProActive [4]), even if it is not transparent to the application programmer.

We followed a similar approach to the GrADS project, which exploits a complete environment, including a monitoring architecture, contract negotiators and configuration optimizer. Differently from GrADS we can reconfigure applications in transparent manner , and with a sensibly better performance (we can join additional resources without completely stopping the application). In particular [3], reports cost of minutes for reconfiguring a data-parallel application while ASSIST overheads ranges in milliseconds–seconds span. The lower reconfiguration cost diminishes the criticality of deciding a reconfiguration, and enables the use of heuristic "try-and-see" approach whether analytic modeling fails.

8   Conclusions and Future Work

We presented a novel extension of the ASSIST environment that seamlessly support application reconfiguration at run-time. Application reconfiguration is achieved efficiently and transparently to the application programmers through parmod reconfiguration. ASSIST parmod are self-optimizing parallel entities that can be able to respect a dynamically received QoS contract. A parmod reconfiguration does not have any direct impact on other parmods in the same application. Also, we shown a set of policies to deal with QoS contracts enabling parmods to self-adapt to a running environment that is heterogeneous and unreliable in provided performance.

On this ground we are extending ASSIST to full grid support. In particular, all MAMs can be organized in a hierarchy of managers, the root being the Application Manager (AM), that enforces a QoS contract for the whole application [9]. The AM works – in the large – similarly to MAM (see Sect. 3), and leverages on MAMs to detect parmods behaving as bottlenecks for the application: it reacts sending their MAMs a suitable new contract.

References

1. Foster, I., Kesselmann, C., eds.: The Grid 2: Blueprint for a New Computing Infrastructure. Morgan Kaufmann (2003)
2. Vanneschi, M.: The programming model of ASSIST, an environment for parallel and distributed portable applications. Parallel Computing **28** (2002) 1709–1732
3. Vadhiyar, S., Dongarra, J.: Self adaptability in grid computing. International Journal Computation and Currency: Practice and Experience (2005) To appear.
4. Baude, F., Caromel, D., Morel, M.: On hierarchical, parallel and distributed components for Grid programming. In: Workshop on component Models and Systems for Grid Applications. (2005)
5. Thain, D., Tannenbaum, T., Livny, M.: Condor and the grid. In: Grid Computing: Making the Global Infrastructure a Reality. John Wiley & Sons Inc. (2002)
6. van Nieuwpoort, R.V., Maassen, J., Wrzesinska, G., Hofman, R., Jacobs, C., Kielmann, T., Bal, H.E.: Ibis: a flexible and efficient Java-based grid programming environment. Concurrency & Computation: Practice & Experience (2005)
7. Magini, S., Pesciullesi, P., Zoccolo, C.: Parallel software interoperability by means of CORBA in the ASSIST programming environment. In: Proc. of Euro-Par 2004. Volume 3149 of LNCS., Springer (2004) 679–688
8. Aldinucci, M., Campa, S., Coppola, M., Magini, S., Pesciullesi, P., Potiti, L., Ravazzolo, R., Torquati, M., Zoccolo, C.: Targeting heterogeneous architectures in ASSIST: experimental results. In: Proc. of Euro-Par 2004. Volume 3149 of LNCS., Springer (2004) 638–643
9. Aldinucci, M., Coppola, M., Danelutto, M., Vanneschi, M., Zoccolo, C.: ASSIST as a research framework for high-performance Grid programming environments. In: Grid Computing: Software environments and Tools. Springer (2005)
10. Kephart, J.O., Chess, D.M.: The vision of autonomic computing. IEEE Computer **36** (2003) 41–50
11. Aldinucci, M., Campa, S., Coppola, M., Danelutto, M., Laforenza, D., Puppin, D., Scarponi, L., Vanneschi, M., Zoccolo, C.: Components for high performance Grid programming in Grid.it. In: Proc. of the Workshop on Component Models and Systems for Grid Applications. CoreGRID series. Springer (2005)
12. Zayas, E.R.: Attacking the process migration bottleneck. In: Proc. of the 11th ACM Symposium on Operating System Principles. (1987)
13. Litzkow, M.: Supporting checkpointing and process migration outside the unix kernel. In: Usenix Winter Conference. (1992)
14. E. Godard, S. Setia, E.W.: Dyrect: Software support for adaptive parallelism on nows. In: Proc. of IPDPS Workshop on Runtime Systems for Parallel Programming. (2000)
15. Scherer, A., Lui, H., Gross, T., Zwaenepoel, W.: Transparent adaptive parallelism on nows using OpenMP. In: Proc. of Principles and Practice of Parallel Programming. (1999)
16. Edjlali, G., Agrawal, G., Sussman, A., Humphries, J., Saltz, J.: Compiler and runtime support for programming in adaptive parallel environments scientific programming. Scientific Programming **6** (1997)
17. André, F., Buisson, J., Pazat, J.L.: Dynamic adaptation of parallel codes: toward self-adaptable components for the Grid. In: Workshop on component Models and Systems for Grid Applications. (2005)

SPC-XML: A Structured Representation for Nested-Parallel Programming Languages*

Arturo González-Escribano[1],
Arjan J.C. van Gemund[2], and Valentín Cardeñoso-Payo[1]

[1] Dept. de Informática, Universidad de Valladolid
E.T.I.T. Campus Miguel Delibes, 47011 - Valladolid, Spain
Phone: +34 983 423270
{arturo,valen}@infor.uva.es
[2] Embedded Software Lab, Software Technology Department
Faculty of Electrical Engineering, Mathematics and Computer Science
P.O.Box 5031, NL-2600 GA Delft, The Netherlands
Phone: +31 15 2786144
a.j.c.vangemund@ewi.tudelft.nl

Abstract. Nested-parallelism programming models, where the task graph associated to a computation is series-parallel, present good analysis properties that can be exploited for scheduling, cost estimation or automatic mapping to different architectures.

In this paper we present an XML intermediate representation for nested-parallel programming languages from which the application task-graph can be easily derived. We introduce some design principles oriented to allow the compiler to exploit information about the task synchronization structure, automatically determine implicit communication structures, apply different scheduling policies, and generate lower-level code using different models or communication tools. Results obtained for simple applications, using an extensible prototype compiler framework, show how this flexible approach can lead to portable and efficient implementations.

1   Introduction

A common practice in high-performance computing is to program applications in terms of the low-level concurrent programming model provided by the target machine, trying to exploit the maximum possible performance. Portable APIs, such as message-passing interfaces (e.g. MPI, PVM) propose an abstraction of the machine architecture, still obtaining good performance. However, programming in terms of these unrestricted coordination models can be extremely error-prone and inefficient, as the synchronization dependencies that a program can generate are complex and difficult to analyze by humans or compilers [14]. Important

---

* This work has been partially supported by: JCyL under contract number VA-053A05, the EC (FEDER) and the Spanish MCyT (Plan Nacional de I+D+I, TIC2002-04498-C05-05 and TIC2002-04400-C03). A preliminary version of this paper was presented at the 11th Int. Workshop on Compilers for Parallel Computers.

J.C. Cunha and P.D. Medeiros (Eds.): Euro-Par 2005, LNCS 3648, pp. 782–792, 2005.

decisions in the implementation trajectory, such as scheduling or data-layout become extremely difficult to optimize. Considering these problems, more abstract programming models, which restrict the possible synchronization and communication structures available to the programmer, are proposed and studied (see e.g. [21]). These models, due to their restrictions, are easier to understand and program, and can provide tools and techniques that help in mapping decisions.

Nested-parallelism models present a middle point between expressiveness, complexity and easily programming [21]. They restrict the coordination structures and dependencies to those that can be represented by series-parallel (SP) task-graphs (DAGs). Thus, nested-parallelism is also called SP programming. Due to the inherent properties of SP structures [22], they provide clear semantics and analyzability characteristics [16], a simple compositional cost model [9, 19, 20] and efficient scheduling [1, 7]. These properties can lead to automatic compilation techniques that increase portability and performance. Examples of parallel programming models based on nested-parallelism include BSP [23], nested BSP (e.g. NestStep [15], PUB [2]), BMF [20], skeleton based (e.g. SCL [5], Frame [4], OTOSP/LLC [6]), SPC [9], and Cilk [1].

In previous work (see e.g. [10]) we have shown that many applications classes, including some typical irregular scientific applications, may be efficiently mapped to nested-parallelism due to some inherent load or synchronization regularities, or using simple balancing techniques. In [12] we also discussed how using an structured XML representation to specify the coordination of a parallel program may lead to a flexible an extensible compiler framework, where scheduling or mapping plug-ins may be automatically selected by the application structure properties (for a further discussion about extensible programming see also [24]).

SPC-XML is an evolving highly-abstract XML intermediate-representation of nested-parallel programs. Although it is not intended to be used directly by the programmer, but through front-end translators, it is a complete parallel coordination language. It is designed as a portable, easy to parse, and extensible language to simplify experimentation with nested-parallel compiler technology.

SPC-XML uses a common specification syntax for coarse-grain computations (like in BSP [23] models) or fine-grain computations (like in data-parallel models such as HPF [3]). By means of simple recursive specifications it delegates to the compiler the selection of the appropriate granularity for a given application and target machine. Since no specific memory-model is forced in the language computations may be adapted to a particular target architecture, supporting both, distributed and shared-memory models. Due to its extensibility features it may support, but it is not limited to, any specific set of compile-time scheduling and data-layout techniques (such in OpenMP [17]), or a given generic run-time scheduling system for dynamically spawned processes (such in Cilk [1]). New scheduling or mapping techniques may be applied, and the compilation strategy may be guided by simple and accurate cost-models.

This XML specification is designed to simplify the reconstruction of the synchronization task-graph of the application, and to obtain dependency and data-flow information which help the compiler in determining the implicit communication structure. It is possible to: (a) express the different types of collective or structured communication operations found in other languages (e.g. message-passing interfaces [14] or OTOSP/LLC [6]), and (b) inherently support structured programming or parallel skeletons (e.g. [4, 18]). However, it achieves it using only one implicit synchronization mechanism (the end of a parallel section) in combination with the explicit data-flow information naturally exposed in the standard parameter-substitution of processes invocations.

In this paper we also show how communication structures may also be restructured and optimized at the low-level, obtaining solutions similar to manually coded message-passing programs. We present preliminary results for simple applications, generated with a source to source translator prototype which exploits these techniques. Performance is comparable to that obtained with manually developed MPI codes.

The paper is organized as follows: Section 2 introduces our new proposal for a tag-based coordination language. In section 3 we discuss how the features of this language help at the different stages of a completely-automatic compilation-path. Experiments and results with an example application are presented in section 4. Finally, section 5 draws our conclusion.

2   SPC-XML: A Tag-Based Coordination Language

In this section we introduce an intermediate highly-structured coordination language based on XML. It is named SPC-XML, due to the SPC nested-parallel model [9] (Series-Parallel and Contention model). The full description of the language including its DTD can be found in [13]. This full parallel synchronization language is designed to support any feature to be found in a nested-parallel environment such as recursion, critical sections, distributed or shared-memory models, and manual data-layout specifications. In Fig. 1 we show an excerpt of a simple cellular-automata program representation. The language is highly verbose and it is not designed to be written directly by a programmer, but as a convenient intermediate representation. Front-ends to translate legacy code written in any nested-parallel language to this representation would be a straightforward development effort. Nevertheless, standard XML tools may be used to edit, visualize or check consistency of SPC-XML representations. Although its main functionalities and semantics are clearly defined, SPC-XML is still syntactically evolving to find a more mature level. The design principles of SPC-XML are:

1. SPC-XML model implements the same semantics as the SPC model and its underlying process algebra (see e.g. [9]). Processes are composed to form a program with only two possible operators – sequential and parallel –, which may be freely nested.
2. It uses XML tags to directly represent the parallelism structure. Nested-Parallelism is a hierarchical structured form of expressing parallelism. Thus, a representation using an XML structured document is natural.
3. It is a coordination language [8]. Tags represent only the synchronization structure. Any classical sequential language may be used to code sequential sections. This separation simplifies the recognition of the parallel structure.

4. The program must specify explicit input/output interfaces for sequential sections and logical processes. The behavior of a data item in an interface (input and/or output) will be compulsory specified. This will help the compiler to compute data-flow and automatically derive communications for a given task decomposition and mapping.

5. Tag and attribute names are fully readable to help humans recognize the main program structures easily.

2.1 Document Structure

An SPC-XML document contains a HEADER tag and a BODY tag. The first one is used mainly for documentation and to specify low-level sequential code blocks to be included at the beginning of the target program. The body of the document

```
<SPC-XML version="0.4"> <!-- *** Cellular-Automata example *** -->
  <HEADER> <TITLE>cellAutom</TITLE> </HEADER>
  <UNIT-BODY>
    <FUNCTION name="seqCellComp" workload="myData#size">
      <INOUT name="myData" baseType="double" />
      <IN name="myUpBorder" baseType="double" />
      <IN name="myDownBorder" baseType="double" />
      <CODE> <!-- C code to compute a stage of the cellular automata -->
        ...
      </CODE>
    </FUNCTION>

    <PROCESS name="main">
      <LOCAL> <VAR name="data" baseType="double" dim="[16384][16384]" /> </LOCAL>
      <BODY>
        <CALL name="initData"> <PARAMETER value="data" /> </CALL>
        <CALL name="parallelCell"> <PARAMETER value="data" /> </CALL>
        <CALL name="printData"> <PARAMETER value="data" /> </CALL>
      </BODY>
    </PROCESS>

    <PROCESS name="parallelCell">    <!-- Row mapped parallel cellular-automata -->
      <INOUT name="data" baseType="double" />
      <LOCAL>
        <VAR-OVERLAP srcVar="data" name="mainPart[#P]" layout="[#blocks][:]" />
        <VAR-OVERLAP srcVar="data" name="upBorders[#P]" layout="mainPart[#i][0][:]" />
        <VAR-OVERLAP srcVar="data" name="downBorders[#P]" layout="mainPart[#i][#$][:]" />
      </LOCAL>
      <BODY>
        <LOOP n="1000">
          <PARALLEL p="#P"> <!-- As many blocks as processors available -->
            <PARBLOCK p="*"> <!-- Default parallel block -->
              <CALL name="seqCellcomp">
                <PARAMETER value="mainPart[p]" />
                <PARAMETER value="downBorders[p-1]" />
                <PARAMETER value="upBorders[p+1]" />
              </CALL>
            </PARBLOCK>
          </PARALLEL>
        </LOOP>
      </BODY>
    </PROCESS>
  </UNIT-BODY>
</SPC-XML>
```

Fig. 1. SPC-XML example: Excerpt of a cellular-automata program representation

contains a collection of elemental units (functions and processes). If a process called main is found inside a document, this document represents a program which execution begins at that main process. If the document does not include a main process, it is a library which may not be compiled alone. All elemental units may have a DOC tag containing extra documentation tags for automatic documenting tools. Programming comments are included with the usual XML tags <!- - - ->.

2.2   Logical Processes and Functions

SPC-XML has a PROCESS tag to define the content of a logical process. Each process has a well-defined interface, whose formal parameters are defined with the IN, OUT, INOUT tags, declaring the input/output intent of the data item in the interface. Processes may contain also a LOCAL tag to declare variables, with process visibility and scope. The BODY tag of a process contains the tags which define its behavior. The serial composition is implied in the tags declaration order as in many common procedural languages. For programmer convenience, other common control-flow statements are supported by the IF/THEN/ELSE, WHILE, REPEAT, and LOOP tags. The parallel composition operator is implemented using the PARALLEL tag. This tag may only contain one or more PARBLOCK tags, whose contents are composed in parallel. The PARALLEL tag, have an optional attribute named p="number", to specify a fixed number of parallel-blocks to spawn. The PARBLOCK tags may also have a p="number" attribute, to associate the tag content with a logical identifier inside the local subgroup of blocks. A default parallel-block (specify with p="*") may be used to fill-up with its content all the non-associated blocks in a parallel region. The closing parallel tag implies a synchronization of all the blocks to proceed. At this logical synchronization point, modifications done to the same variable in different blocks are made consistent. By default, the modifications done by only one block will persist. Nevertheless, optional attributes may be used in the PARALLEL tag to use typical reduction operations for given data elements. As in any nested-parallel model, other PARALLEL tags may appear inside a block, or in other processes called inside a block. Sequential codes are enclosed inside a FUNCTION tag, to distinguish them from the logical processes. However, the input/output interface is defined exactly in the same way. For symmetry, processes and functions are invoked with the same CALL tags, containing PARAMETER tags to specify the formal to real parameter substitutions. The language includes some more features to help in the mapping trajectory. For instance, the programmer, or a profiling tool, may provide to the scheduling modules with a hint, adding a performance estimation to any sequential functions using the optional workload="number" attribute.

2.3   Data Representation and Memory Model

As a coordination model, SPC-XML tags do not imply data manipulation. The only purpose to define data containers is to describe data-flow between logical processes and through sequential codes. Thus, SPC-XML variables are generic multi-dimensional arrays of a given data type defined in the sequential language used. Variables are defined by VAR tags, always inside the LOCAL tag of a process. There are no expression or data manipulation tags. The only tags which imply code execution are CALL tags referring to function names (sequential tasks). Special variables called *overlaps* are supported to define logical data partitions over other SPC-XML array variables. Thus, simple data-layouts may be devised. Each overlap element becomes an alias for a subarray of an associated SPC-XML variable. They are defined in the local section with the VAR-OVERLAP tag, which attributes specify the name of the associated variable and the layout of the overlap pieces. Subarray specifications may be expressed with a *colon-notation* similar to Fortran90 and some macro-definitions. Moreover, the language provides with some special terms which represent typical *block* or *stride* layouts.

SPC-XML provides a generic distributed-memory model. A process or function works with local copies of the parameters obtained from the caller. However, when the programmer knows it is safe to work with shared-memory instead of creating local copies of a variable, she may give a hint to the compiler, using the special shared="yes" attribute on that variable. Nevertheless, the programmer should not rely on shared-memory for communication as the underlying architecture or back-end may not support it. The only communications between parallel tasks should be driven through processes/functions interfaces.

3 Task-Graph Reconstruction and Mapping

Task decomposition is directly derived from the tag structure of an SPC-XML specification. In the case of applications which synchronization structure may be determined in compile-time, the whole task graph may be expanded and classical graph-scheduling techniques may be applied. For more data-dependent or dynamic programs, the graph represents only the potential structure, and suitable run-time scheduling tools should be included and activated in the target code to obtain the appropriate program behavior. In this paper we focus on exploiting the language characteristics for applications in the first case.

3.1 Task Decomposition and Synchronization Structure Detection

The lexical/syntactical parsing of an SPC-XML document may be completely done by generic and portable XML parsers. Due to the clear semantics and highly structured form of the documents, an application graph is easily reconstructed from its high-level specification. When considering static programs, the number of tasks and the the exact shape of the graph is completely deterministic. In this task-graph we only consider one type of *nodes* (tasks which may contain computation functions) and *edges* (precedence dependencies which may derive in data-flow at lower implementation levels). The structure of a static application is then reconstructed as a *DAG (Direct Acyclic Graph)* along the following guidelines: (1) A process invocation is always expanded, inlining its content where the CALL tag is found, also doing parameter substitution; (2) a *task-node* is delimited by consecutive PARALLEL ending and opening tags, and it contains the function calls found inside the code it represents; (3) the content

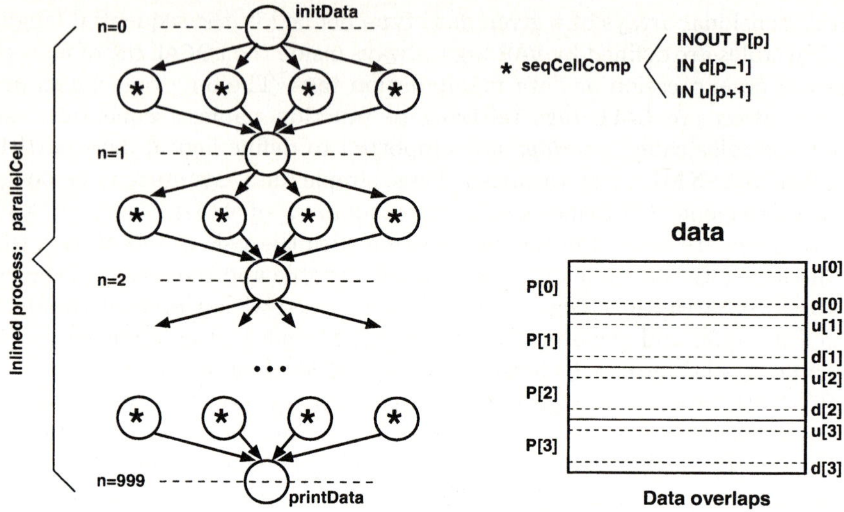

Fig. 2. Example of graph reconstruction for a cellular-automata (4 processors)

of a PARBLOCK tag is processed as a subgraph; (4) loops with a deterministic number of iterations are expanded. The closing tag of the loop is in the same task as its opening tag in the next iteration. An example of their application to of our cellular-automata program is shown in Fig. 2.

3.2   Mapping and Scheduling Strategies

The mapping module of an SPC-XML compiler should use a *machine-model* (which will be also specified in another XML application). The simplest machine-model may contain only one data-item: the number of processors; while more complex models may contain other resources information or allocating costs (e.g. communication parameters, different performances on heterogeneous clusters). The module may check graph-structure characteristics to select the best suitable scheduling or mapping technique. The scheduling algorithm selected will supply the graph with annotations about task-to-processor bindings. Workload annotations, introduced at the specification level through appropriate attributes, will be a key-point to obtain efficient schedules and accurate cost models.

Mapping decisions are also guided by communication costs. Thus, mapping modules should compute the communication-volume generated by a given mapping, using the information about the input/output interfaces of the functions contained inside the nodes. At this level, several implementation variants may be compared to decide the best one for the given target machine. (see e.g. [11]).

The top diagram in Fig. 3 represents the information obtained from the SPC-XML function interface tags, for one iteration of the cellular-automata program. This information clearly determines the data-flow between tasks. However, many low-level parallel-tools or communication-layers are non-restricted to

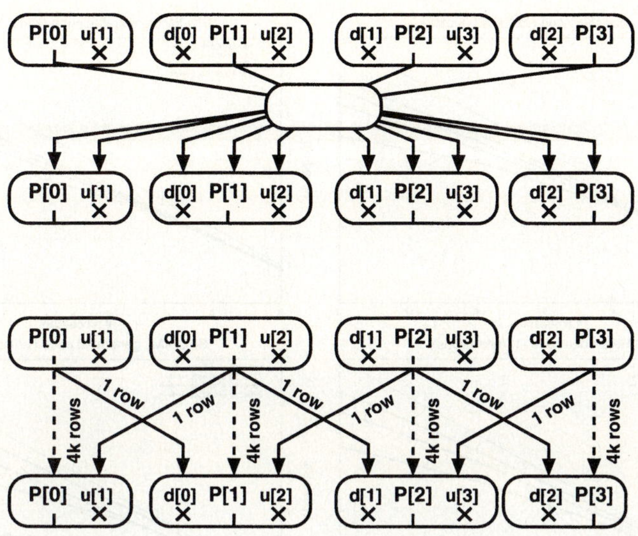

Fig. 3. Communication information and a non-nested-parallel mapping solution

nested-parallelism. Compilers based on highly-abstract nested-parallel models, may optimize communication interchange, specially at synchronization points, if global information is available (see e.g. [2]). SPC-XML may provide enough data about communication to allow very simple but efficient optimization techniques. For instance, the mapping module may detect task nodes where some received data is not used, but only redirected to other tasks. In this case, data may be sent directly to the task where it is used. In the bottom diagram in Fig. 3 we represent such a mapping solution. The synchronization node completely disappears, because every communication is redirected to further target nodes. Moreover, the mapping module may use information about data-sizes to reduce the total communication-volume. In the example, the dashed lines have 4096 more times data-volume than the non-dashed lines. Thus, the tasks vertically aligned in the figure may be scheduled to the same processor to improve locality.

This final solution, automatically obtained from the high-level nested-parallel specification, is the typical best-solution designed when programming a cellular-automata program with a low-level message-passing library. As applications become more complex, this approach may reduce development and debugging effort. Moreover, porting an optimizing an application to a different target machine (for instance a heterogeneous cluster with different resource-power per node) may not imply changes in the high-level code as data partition sizes and communications are adapted if needed.

4 Experiments and Results

We have built a source to source translator prototype implementing some of the techniques previously discussed. Using only the number of processors of the

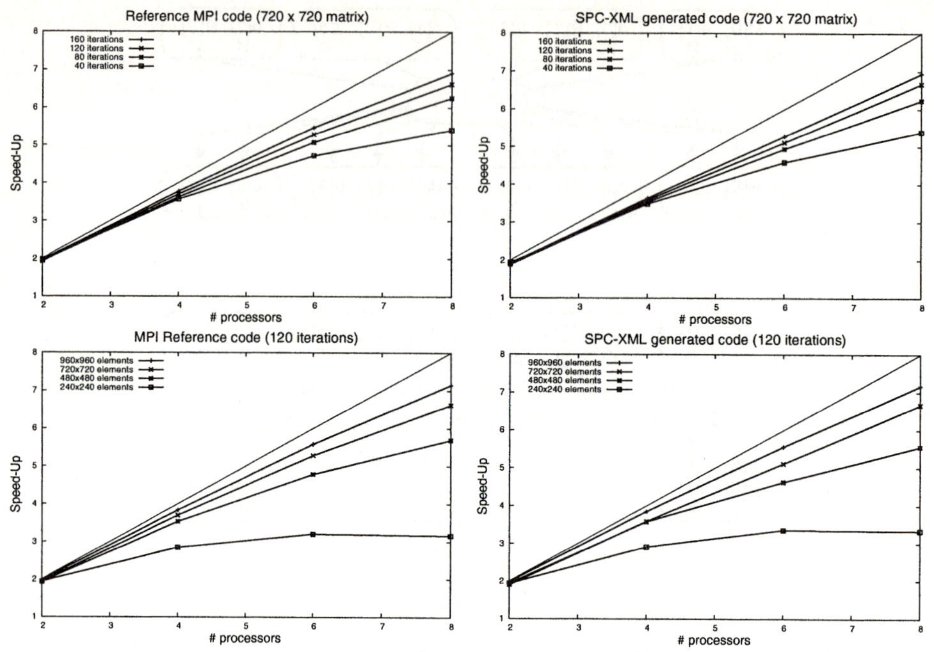

Fig. 4. Speed-Up of the MPI reference codes and the SPC-XML generated codes

target-machine a simple scheduling is applied and communication redirections computed. The mapped graph is then translated by a straightforward back-end to an MPI implementation. An approach to a flexible and extensible compiler framework exploiting more SPC-XML characteristics is also presented in [12].

We compare results obtained when executing a manually-developed and optimized cellular-automata MPI program, and the solutions generated by our prototype for different matrix sizes. The target-machine is a heterogeneous Beowulf cluster connected by a 100Mbit/sec. Ethernet network, and composed by: 3 workstations with an AMD-800Mhz processor; 1 with an AMD-750 Mhz; and 4 with an AMD-500 MHz. We present results of a configuration where processors with different speeds are evenly interleaved. This configuration shows the smoothest scalability effects. We have used input matrices up to 960×960 elements, and a number of iterations in the range $[40, 160]$.

In Fig. 4 we show the speed-ups obtained when executing the MPI reference codes and the corresponding codes automatically-generated from SPC-XML specifications. The plots show how the scaling effects obtained with different matrix sizes and different number of iterations are similar. We remark that we have detected some inefficiencies in the data-buffering and sequential treatment introduced by our simple back-end in the automatically-generated codes. They produce small performance-losses, specifically for 4 or 6 processors due to the heterogeneous nature of the cluster. However, the performance-loss is always less than 3.5% comparing with the corresponding reference code. This shows that sequential code optimization may have more impact on the performance of a parallel application than choosing a high-level structured programming model.

The results presented in this paper are extensible to most regular applications programmed with a coarse-grain style, which derive in static synchronization structures. In [12] we present similar experiments with a more complicated unstructured application, using a graph-partitioning technique to balance an irregular sparse-matrix computation.

5 Conclusion

In this paper we have discussed an XML intermediate representation for nested-parallel programming languages, named SPC-XML. The design principles of this representation allow the compiler to exploit information about the synchronization structure of an application, automatically reconstructing its task-graph. Different mapping or scheduling techniques may be automatically selected as a function of the structural details of the graph. Moreover, information about data-flow and implicit communication structures are also exposed and may be easily optimized, to generate low-level codes adapted to a specific target-machine. Results obtained for some applications, using a prototype compiler, show how this flexible approach may reduce the development effort, leading to efficient implementations from portable and high-level nested-parallel specifications. SPC-XML is the base for a much more generic framework. Future work will include a further development of mapping techniques, focusing on more powerful symbolic mapping or scheduling strategies, available for SP programming models.

References

1. R.D. Blumofe and C.E. Leiserson. Scheduling multithreaded computations by work stealing. In *Proc. Annual Symp. on FoCS*, pages 356–368, Nov 1994.
2. O. Bonorden, B. Juurlink, I. von Otte, and I. Rieping. The Paderborn University BSP (PUB) library - design, implementation, and performance. In *Proc. IPPS/SPDP'99*, San Juan, Puerto Rico, Apr 1999. Computer Society, IEEE.
3. P. Brinch Hansen. An evaluation of high performance Fortran. *ACM SigPlan*, 33(3):57–64, Mar 1998.
4. M. Cole. Frame: An imperative coordination language for parallel programming. Technical Report EDI-INF-RR-0026, Div. Informatics, Univ. of Edinburgh, Sep 2000.
5. J. Darlington, Y. Guo, H.W. To, and J. Yang. Functional skeletons for parallel coordination. In *Europar'95*, LNCS, pages 55–69, 1995.
6. A.J. Dorta, J.A. González, C. Rodríguez, and F. de Sande. LLC: a parallel skeletal language. *Parallel Processing Letters*, 13(3):437–448, Sept 2003.
7. L. Finta, Z. Liu, I. Milis, and E. Bampis. Scheduling UET–UCT series–parallel graphs on two processors. *Theoretical Computer Science*, 162:323–340, Aug 1996.
8. D. Gelernter and N. Carriero. Coordination languages and their significance. *Communications of the ACM*, 35(2):97–107, Feb 1992.

9. A.J.C. van Gemund. The importance of synchronization structure in parallel program optimization. In *Proc. 11th ACM ICS*, pages 164–171, Vienna, Jul 1997.

10. A. González-Escribano. *Synchronization Architecture in Parallel Programming Models*. Phd thesis, Dpto. Informática, University of Valladolid, Jul 2003.

11. A. González-Escribano, A.J.C. van Gemund, and V. Cardeñoso. Predicting the impact of implementation level aspects on parallel application performance. In *Proc. CPC'2001 Ninth Int. Workshop on Compilers for Parallel Computing*, pages 367–374, Edinburgh, Scotland UK, Jun 2001.

12. A. González-Escribano, A.J.C. van Gemund, V. Cardeñoso-Payo, R. Portales-Fernández, and J.A. Caminero-Granja. A preliminary nested-parallel framework to efficiently implement scientific applications. In M. Daydé et al., editor, *VEC-PAR 2004*, number 3402 in LNCS, pages 541–555. Springer, Apr 2005.

13. A. González-Escribano, A.J.C. van Gemund, V. Cardeñoso-Payo, and R. Portales-Fernández. SPC-XML(v0.4): An intermediate structured language for nested-parallel programming environments. Technical Report IT-DI-2005-0001, Dept. Computer Science, Univ. of Valladolid, Jan 2005.

14. S. Gorlatch. Send-Recv considered harmful? myths and truths about parallel programming. In V. Malyshkin, editor, *PaCT'2001*, volume 2127 of *LNCS*, pages 243–257. Springer-Verlag, 2001.

15. C.W. Kessler. NestStep: nested parallelism and virtual shared memory for the BSP model. In *Int. Conf. on Parallel and Distributed Processing Techniques and Applications (PDPTA'99)*, Las Vegas (USA), Jun-Jul 1999.

16. K. Lodaya and P. Weil. Series-parallel posets: Algebra, automata, and languages. In *Proc. STACS'98*, volume 1373 of *LNCS*, pages 555–565, Paris, France, 1998. Springer.

17. OpenMP ARB. OpenMP version 2.5 specification. WWW. On http://www.openmp.org/ (last access May 2005).

18. S. Pelagatti. *Structured Development of Parallel Programs*. Taylor & Francis, 1998.

19. R.A. Sahner and K.S. Trivedi. Performance and reliability analysis using directed acyclic graphs. *IEEE Trans. on Software Eng.*, 13(10):1105–1114, Oct 1987.

20. D.B. Skillicorn. A cost calculus for parallel functional programming. *Journal of Parallel and Distributed Computing*, 28:65–83, 1995.

21. D.B. Skillicorn and D. Talia. Models and languages for parallel computation. *ACM Computing Surveys*, 30(2):123–169, Jun 1998.

22. J. Valdés, R.E. Tarjan, and E.L. Lawler. The recognition of series parallel digraphs. *SIAM Journal of Computing*, 11(2):298–313, May 1982.

23. L.G. Valiant. A bridging model for parallel computation. *Comm.ACM*, 33(8):103–111, Aug 1990.

24. G. Wilson. Extensible programming fot the 21st century. *ACM Queue*, 2(9):48–57, December–January 2004-2005.

Topic 10
Parallel Numerical Algorithms

Jacek Kitowski, Andrzej M. Goscinski,
Boleslaw K. Szymanski, and Filomena d'Almeida

Topic Chairs

Fast and reliable parallel algorithms for the basic problems of numerical mathematics and their effective implementation in easy-to-use portable software components are crucial for computational solution of scientific and engineering problems. Since in all domains of high performance computing, parallel execution routinely is considered as one of the major sources of performance, the sessions under this topic are a forum for presentation and discussion of new developments in the field of parallel numerical algorithms, covering all aspects from basic algorithms, efficient implementation on modern parallel, distributed and network-based architectures.

This year, 16 papers were submitted, 9 of which have been accepted (acceptance rate: 56.3%) but one of the authors didn't submit the camera ready version. The topic is subdivided into three sessions.

For the first session three papers have been selected: *"Performance Measurements of the 3D FFT on the Blue Gene/L Supercomputer"*, which presents performance characteristics of a communication-intensive 3D FFT kernel on 2048-node machine with two different implementations, *"Parallel Solution of Sparse Linear Systems Arising in Advection-Diffusion Problems"*, that deals with prefetching and preconditioning techniques using a SP4 machine and a Linux cluster, and *"Parallelization of Implicit-Explicit Runge-Kutta Methods for Clusters of PCs"* with parallel efficiency about 90% obtained with a sufficiency large number of grid points.

The second session consists of *"Comparison of different parallel modified Gram-Schmidt algorithms"*, that presents computation/communication overlapping technique in two different implementations, showing substantial increase in performance, and *"Automatic Tuning of PDGEMM towards Optimal Performance"*, that shows possibility of relieving users of arduous parameter study.

The third session holds *"Parallelization of Divide-and-Conquer Eigenvector Accumulation"*, that discusses different strategies for parallelization (with superlinear speedup obtained), *"Parallel Order Reduction via Balanced Truncation for Optimal Cooling of Steel Profiles"* with multilayered architecture of libraries for model reduction proposed and *"Broadcast-Based Parallel LU Factorization"* with overlapped communication.

Acknowledgements. The topic chairs would like to thank the external reviewers for their help. Special thanks go to José C. Cunha for his frequent, ambitious and always friendly support and advice.

J.C. Cunha and P.D. Medeiros (Eds.): Euro-Par 2005, LNCS 3648, p. 793, 2005.

Performance Measurements of the 3D FFT on the Blue Gene/L Supercomputer

Maria Eleftheriou, Blake Fitch, Aleksandr Rayshubskiy,
T.J. Christopher Ward, and Robert Germain

IBM Thomas J. Watson Research Center
Yorktown Heights, NY 10598-0218
{mariae,bgf,arayshu,tjcw,rgermain}@us.ibm.com

Abstract. This paper presents performance characteristics of a communications-intensive kernel, the complex data 3D FFT, running on the Blue Gene/L architecture. Two implementations of the volumetric FFT algorithm were characterized, one built on the MPI library using an optimized collective `all-to-all` operation [2] and another built on a low-level System Programming Interface (SPI) of the Blue Gene/L Advanced Diagnostics Environment (BG/L ADE) [17]. We compare the current results to those obtained using a reference MPI implementation (MPICH2 ported to BG/L with unoptimized collectives) and to a port of version 2.1.5 the FFTW library [14]. Performance experiments on the Blue Gene/L prototype indicate that both of our implementations scale well and the current MPI-based implementation shows a speedup of 730 on 2048 nodes for 3D FFTs of size $128 \times 128 \times 128$. Moreover, the volumetric FFT outperforms FFTW port by a factor 8 for a $128 \times 128 \times 128$ complex FFT on 2048 nodes.

1   Introduction

The primary goal of the computational science portion of the Blue Gene project is to advance our understanding of biological phenomena such as protein folding via large scale computer simulations[10] and as part of this effort the Blue Gene application group has developed a Molecular Dynamics framework, *Blue Matter* [12]. Molecular dynamics (MD) is a well-established computational method used to study complex bimolecular systems. MD permits the computation of thermodynamic and kinetic properties of biomolecular systems[19]. Such studies can enhance our understanding of biological functions and provide insights into processes related to the action of potential new medications.

Blue Gene/L[15] is a massively parallel supercomputer developed at the IBM T.J. Watson Research Center in collaboration with Lawrence Livermore National Laboratory. BG/L has five networks, two of which are of particular interest to the application developers: the torus and the collective network. The three-dimensional torus has links between each node and its six neighbors while the collective network enables low latency broadcast and reduction operations as well as providing the path for i/o to external devices. Each node operates at a relatively low clock frequency of 700 MHz and has two PowerPC 440 CPUs on the same chip with associated dual floating point units. Two modes of operation are supported by the system software: (1) "coprocessor mode"

J.C. Cunha and P.D. Medeiros (Eds.): Euro-Par 2005, LNCS 3648, pp. 795–803, 2005.

which runs a single MPI task, and "virtual node mode" which runs two MPI-tasks on each node.

The operating assumption of the Blue Gene/L architecture is that performance gains will come from massive parallelism rather than increases in processor clock speed. This implies that algorithms must achieve good scalability across many thousands of processors in order to fully exploit the power of this massively parallel supercomputer. One of the major factors limiting the scalability of parallel molecular dynamics algorithms is the calculation of full electrostatic forces using Particle-Particle-Particle-Mesh-Ewald (P3ME)[5] techniques. P3ME methods involve the calculation of a convolution using two 3D-FFTs for each time step. Typical simulations require 10^6 to 10^9 time steps for 10,000-100,000 atoms system. For this reason, efficient computation of the 3D-FFT algorithm is neccessary. However, efficient implementation of the 3D-FFT presents a considerable challenge for the communications infrastructure of a parallel machine because of the all-to-all nature of the distributed transposes required.

The parallel computation of three-dimensional FFTs has been carried out using two approaches. In the first approach one-dimensional FFTs are computed in distributed fashion [7, 23, 24] and in the second approach, the transpose method, uses successive evaluations of independent one-dimensional local FFTs along each direction to evaluate the 3D-FFT [4, 6, 13, 14, 18]. A detailed description of our implementation of the volumetric FFT along with early performance data on the BG/L architecture has recently appeared[8]. A description of an earlier version of the volumetric FFT implementation and its performance on a more conventional (Power4/full bisectional bandwidth switch) machine were reported previously[9].

In this paper we present the performance characteristics of the volumetric FFT on BG/L as implemented on two different communication layers: (1) MPI with unoptimized and optimized all-to-all collective implementations and (2) a low-level System Programming Interface (SPI) of the Blue Gene/L Advanced Diagnostics Environment (BG/L ADE) [17]. The measured performance is compared with limits inherent to the hardware capabilities of the machine such as the bisectional bandwidth of the BG/L torus communications network. On BG/L the 3D-FFT of size of 128^3 continues to speed up through 16,384 nodes for the SPI-based version and up to 8,192 nodes for the optimized MPI-based vesion. On a single node, the volumetric FFT implementations are within 50% of the performance of the ported FFTW version 2.1.5, while outperforming FFTW at larger node counts. It is important to note that the MPI implementation of the volumetric FFT uses BGL/FFTW-GEL [21] 1D serial FFT as a building block. In the sections that follow, we describe the hardware limits to ultimate performance of the 3D FFT on a Blue Gene/L machine based on the hardware "speeds and feeds"[11], present the experimental measurements and discuss the corresponding results.

2   Hardware Limits on 3D FFT Performance

Lower bounds can be placed on the communication time for a 3D FFT implemented using the transpose technique by assuming that three all-to-all communications (along a row or within a plane of the processor mesh) are required. Detailed simulations of the BG/L torus network indicate that the time required for an all-to-all communication involving a set of nodes in a line, plane, or volume can be estimated using the expression[16]:

$$T_{all-to-all} = \frac{V_{received}\,\overline{N}_{hops}}{N_{links}\,BW\,f}$$

where $V_{received}$ is the volume of data received by each node, $\overline{N}_{hops}$ is the average number of hops required (for a three dimensional torus where each dimension is p, $\overline{N}_{hops} = p/4$ for all-to-all in a line, $\overline{N}_{hops} = p/2$ for all-to-all in a plane, and $\overline{N}_{hops} = 3p/4$ for all-to-all in a volume), N_{links} is the number of links available to each node (2 for linear communication, 4 for planar communication, and 6 for volumetric communication), BW is the raw bandwidth of the torus per link (2 bits per processor clock cycle), and f is the link utilization (simulations indicate that this should be about 80%). This expression indicates that the time required for all-to-all communication is independent of the dimensionality of the communication because of the increase of the average hop count with dimensionality is balanced by the increase in the number of links available. At the limits of scalability, where messages consist of a single complex number, the above expression based on bandwidth considerations will become inadequate because the hardware and software latencies associated with sending a packet will become significant.

For an idealized bound on the computation time required to compute a single 1D FFT of length N, we assume $8N\log_2 N$ cycles for a fused multiply-add machine (although the floating point operation count for a 1D FFT is $5N\log_2 N$, data dependencies force a fused multiply-add machine to use 8 cycles). Highly optimized FFT implementations, such as the library developed by the team at the Technical University of Vienna[20] can approach this idealized value. However, the 3D-FFT is dominated by the communication cost of the transposes for even small node counts and the performance of the 1D-FFT building block is not critical.

3   Parallel Performance Analysis

In this section we present performance measurements of the 3D-FFT kernel. Most of the benchmarks reported in this paper were performed on the 20480 node Blue Gene/L system at the IBM T.J. Watson Research Center. Each rack consists of two 512-node mid-planes and each mid-plane comprises 16 node cards. The Blue Gene/L system can be configured to complete a torus in all three dimension only when it is partitioned as 512 nodes and higher, where the number of nodes is power of 2. All of the benchmarks executed in 512-node and larger partitions used the torus network, while benchmarks executed on 32 and 128-node partitions had to use the mesh topology. The performance metric used here is the "total time to solution" for the problem. Our use cases for the 3D FFT from molecular simulations require strong scaling for relatively small data sizes such as 32^3, 64^3 and 128^3.

We compare the performance of two MPI implementations. The first is a port of MPICH2[22],[1] on BG/L, while the second is the optimized version for the BG/L architecture [3] and [2]. The MPI collective of interest to us is MPI_Alltoallv. The default MPICH2 MPI_Alltoallv implementation is based on non-blocking communication calls MPI_irecv and MPI_isend and thus requires message matching. However, the optimized implementation runs on a lower level messaging protocol that avoids the overheads of MPI point-to-point messages. Additional optimizations include avoiding sending zero size messages across the network and minimizing the number of cache misses. All the performance numbers reported here were obtained using co-processor mode, where both Power PC 440 processors on each node execute a single MPI task. The MPI tasks were organized in a three-dimensional grid using the MPI Cartesian topology constructs. By mapping the MPI tasks in our domain decomposition "naturally" to the physical machine topology, we can achieve substantial performance improvements since locality plays an important factor in the communication performance.

Figure 1 shows the measured execution time for the volumetric 3D-FFT as implemented using MPI collective communications (both unoptimized and optimized) on the 20480-node BG/L system. Speedup in excess of 700 is observed for 128^3 FFT at 2048 nodes and reaches over 900 at 8192 nodes for the MPI-based implementation. The optimized MPI outperforms the ported MPICH2 version when the number of tasks is 256 or more.

Although optimizations of the MPI collectives for the Blue Gene/L architecture give improved performance, it may not yet reflect the full capabilities of the hardware. To investigate whether even better performance is possible, we have implemented the `all-to-allv` collective required by the FFT via the low-level System Programming Interface (SPI) of the Blue Gene/L (ADE) [17]. The BG/L ADE was developed and is utilized by the BG/L hardware group for running diagnostics and manufacturing tests. The SPI packet-level interface provides direct access to the BG/L network-hardware. In our `all-to-allv` collective the packets were directly injected into the send FIFOs and then directly read by the receive FIFOs by using all six links concurrently. Moreover, we prepare the destination lists for the packet-headers in the plan phase, while in the MPI-based implementation the packet-headers have to be evaluated at each MPI call.

In an attempt to confine the differences in the implementations to the communications layer, the FFT code is written using C++ templates and takes a communications class that encapsulates the details of the communications implementation as a template parameter.

In Figure 2, we characterize the scalability behavior of the 3D-FFT algorithm on both MPI and BlueGene/L Advanced Diagnostics Environment SPI. The FFT scales well on both communications layers. However, as the limits of scalability are approached (where the node count $P = N^2$ for a $N \times N \times N$ FFT and only a single 1D-FFT is performed on each node between transposes) the performance of the SPI implementation still exceeds that of the optimized MPI version. The difference in the performace is probably due to the MPI overhead, related to cash misses on the user's data and dynamically creating destination lists during every MPI collective calls. Whether further improvements are possible in the MPI and/or SPI implementations is a subject for further investigation.

Comparisons of the volumetric 3D FFT on two different architectures, SP and early BG/L prototype, to the FFTW[14] library have been reported previously[8]. In this paper the authors repeat the benchmark on BG/L using the optimized MPI-version and the new SPI implementation of the `all-to-all` collective operation.

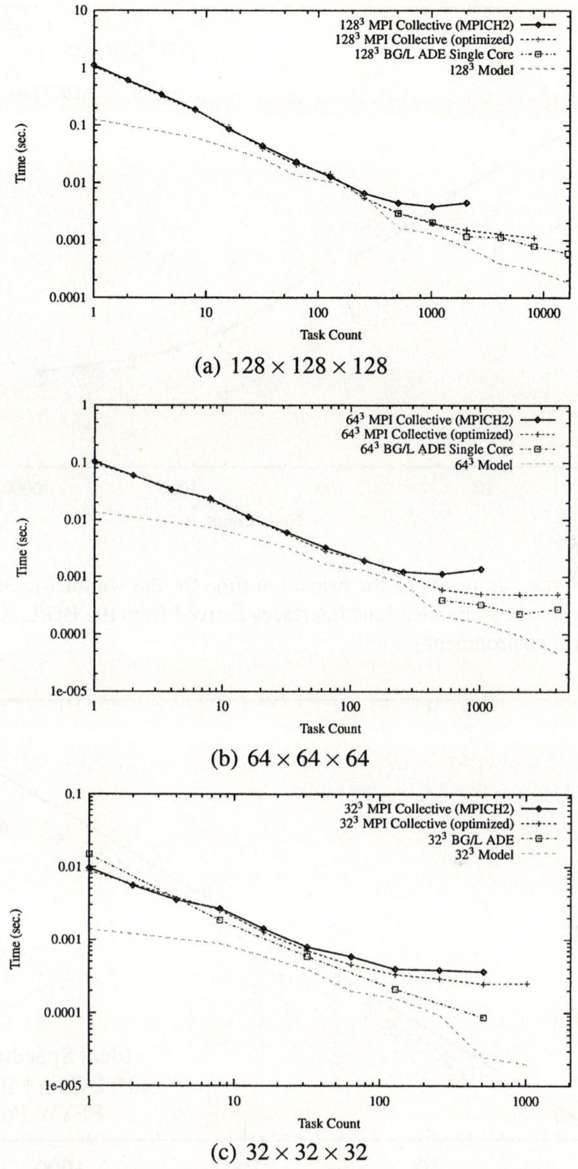

Fig. 1. Measured execution times for the volumetric FFT, for a series of problem sizes (32^3, 64^3, 128^3) as a function of number nodes. Two sets of MPI data are shown. The default MPICH2 implementation data were taken in July 2004 while the optimized MPI data were taken in May 2005 after significant optimization of the all-to-all collectives used by the FFT as were the measurements of the implementation using the BG/L Advanced Diagnostic Environment (BG/L ADE). The limits to execution time using simple estimates for computation and communication costs are also shown (the limits shown assume mesh bandwidths which are half of those available on a torus for node count below 512 and torus bandwidths for larger node counts).

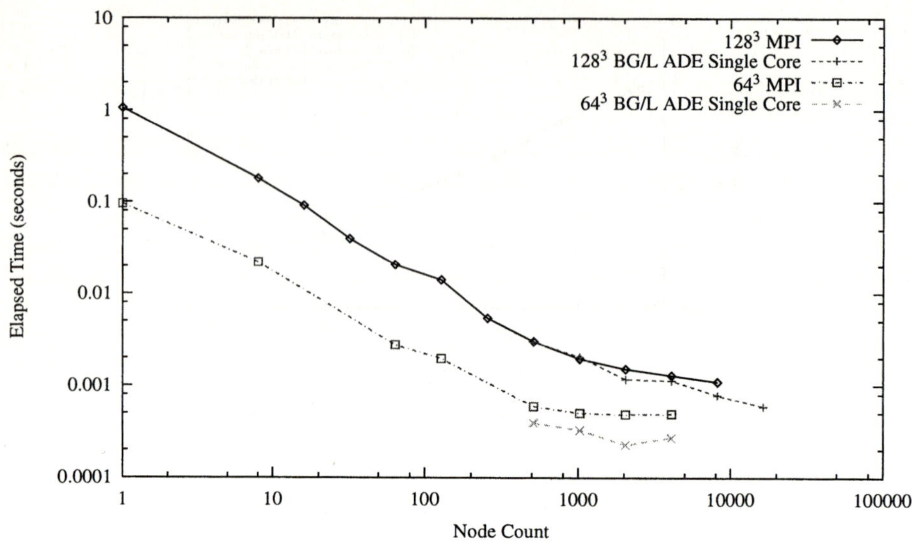

Fig. 2. Performance measurements of the execution time for the volumetric 3D-FFT[8] running on MPI and on low level communications interfaces derived from the BG/L Advanced Diagnostics Environment[17] environment.

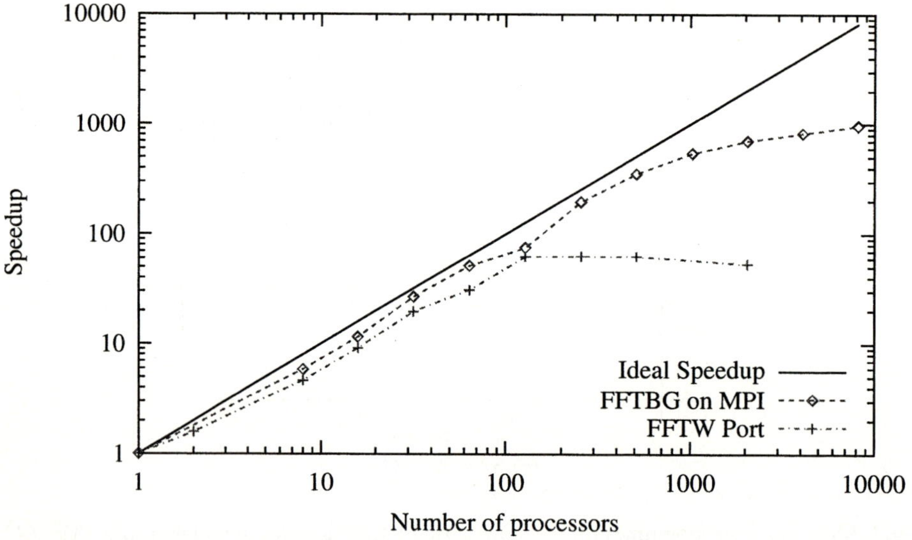

Fig. 3. This plot shows a comparison of the speedups of the two different FFT libraries, volumetric FFT (FFTBG) using the optimized MPI implementation and a straightforward port of version 2.1.5 of FFTW library on Blue Gene/L, as a function of number of processors. The 3D-FFT has a grid size of ($128 \times 128 \times 128$).

Figure 3 compares the speedup of the volumetric FFT on both SPI and MPI, and a port of the FFTW library. The two versions of the volumetric code and the FFTW library exhibit similar scalability up to 128 nodes. Since the FFTW implementation is based on the slab decomposition, the speedup flattens after 128 nodes. However, the volumetric FFT continues to exhibit good speedups through 2048 nodes for the $128 \times 128 \times 128$ size FFT. Both volumetric FFT versions scale up to 8,192 nodes with the SPI-based version performing about 20% better on 8,192 nodes.

The comparison of the bounds on execution time based on hardware bandwidths and idealized 1D serial FFT execution times with the measured total times to solution in Figure 1 shows significant divergences at small node counts. We conjecture that this divergence is caused by memory hierarchy effects since floating point efficiencies of the 1D-FFTs used as building blocks for the 3D-FFT implementation are fairly high, e.g. for a 64-point FFT, the efficiency of the FFT from the Blue Gene/L FFT library supplied by the Technical University of Vienna is over 60%.

The memory access pattern of the in-memory transpose required for the 3D-FFT is presumably very unfavorable for pre-fetching and at low node counts the memory footprint per node of the larger size 3D-FFTs will certainly spill out of the 4MB L3 cache. No effort has been made to tile the implementation of the memory transposes in the volumetric FFT. Of course, at the high node counts that represent the limits to scalability for the FFT, the data will sit in cache and the measured performance more closely approaches the bandwidth limited constraints on performance. Note that even without any efforts at tiling, the uniprocessor performance of the volumetric FFT implementation is within 50% of that of the FFTW library implementation. We intend to use instrumentation that can access the hardware performance counters available on the Blue Gene/L chip to eventually measure the memory hierarchy effects

4   Summary

We have compared the performance measurements of the volumetric 3D-FFT algorithm, on two communications communication layers, MPI and SPI. Our measurements shows that the volumetric algorithm performs impressively well on both the optimized MPI and SPI communication layer. Moreover, we found that the volumetric FFT outperforms a port of the widely used FFTW library (based on a slab decomposition) by a significant margin on large numbers of nodes.

At the limits of scalability, approached by the 128^3 FFT on 16,384 nodes, where each node sends packets of the size of single complex number the code still scales, with the BG/L ADE SPI implementation still being faster than the MPI-based FFT using the optimized MPI collectives. Future work will involve instrumenting the code to understand the role of memory access patterns in the performance at small node counts and continuing optimization of the implementations on both communications layers. Finally, we plan to reproduce and report the MPI based FFT benchmarks in the virtual node mode in which each PPC440 core on the chip executes a separate MPI task.

Acknowledgments

We would like to acknowledge the contributions of Georghe Almasi, Charles Archer, and Mark Giampappa for their work on the high performance communications layers implemented on Blue Gene/L. We would like to acknowledge the proof-reading of this paper by Frank Suits and Alan Grossfield from the Blue Gene/L application and science team.

References

1. G. Almaśi, C. Archer, J. Castanos, M. Gupta, X. Martorell, J. E. Moreira, Gropp W, S. Rus, and B. Toonen. MPI on Blue Gene/L:Designing an Efficient General Purpose Messaging Solution for a Large Cellular System. In *Proceedings of the 10th EuroPVM/MPI conference, Lecture Notes in Computer Science, Klagenfurt, Austria*, 2003.

2. G. Almaśi, C. Archer, C. Chris Eway, Philip Heidelberger, X. Martorell, J. E. Moreira, B. D. Steinmacher-Burow, and Yili Zheng. Optimization of MPI collective operations on Blue-Geneĺsystems. 2005. To appear at ICS05.

3. G. Almasi et al. Design and implementation of message-passing services for the Blue Gene/L supercomputer. *IBM Journal of Research and Development*, 49(2/3):393–406, 2005.

4. C. E. Cramer and J. A. Board. The Development and Integration of a Distributed 3D FFT for a cluster of workstations. In *4th Annual Linux Showcase and Conference*, pages 121–128, Atlanta, GA, October 2000.

5. Markus Deserno and Christian Holm. How to mesh up ewald sums. i. a theoretical and numerical comparison of various particle mesh routines. *J. Chem. Phys.*, 109(18):7678–7693, 1998.

6. H. Q. Ding, R. D. Ferraro, and D. B. Gennery. A portable 3D FFT Package for Distributed-Memory Parallel Architecture. In *SIAM Conference on Parallel Processing for Scientific Computing*, 1995.

7. A. Edelman, P. McCorquodale, and S. Toledo. The future fast Fourier transform? In *SIAM J. Sci. Comput.*, volume 20, pages 1094–1114, 1999.

8. M. Eleftheriou, B.G Fitch, A. Rayshubskiy, T.J.C. Ward, and R.S. Germain. Scalable framework for 3d FFTs on the Blue Gene/L supercomputer: Implementation and early performance measurements. *IBM Journal of Research and Development*, 49(2/3):457–464, 2005.

9. Maria Eleftheriou, José E. Moreira, Blake G. Fitch, and Robert S. Germain. A Volumetric FFT for BlueGene/L. In *High Performance Computing – HiPC 2003*, pages 194–203, 2003.

10. F. Allen *et al.* Blue Gene: a vision for protein science using a petaflop supercomputer. *IBM Systems Journal*, 40(2):310–327, 2001.

11. N.R. Adiga *et al.* An overview of the Blue Gene/L supercomputer. In *Supercomputing 2002 Proceedings*, November 2002.
    `http://www.sc-2002.org/paperpdfs/pap.pap207.pdf`.

12. B.G. Fitch, R.S. Germain, M. Mendell, J. Pitera, M. Pitman, A. Rayshubskiy, Y. Sham, F. Suits, W. Swope, T.J.C. Ward, Y. Zhestkov, and R. Zhou. Blue Matter, an application framework for molecular simulation on Blue Gene. *Journal of Parallel and Distributed Computing*, 63:759–773, 2003.

13. M. Frigo and S. G. Johnson. The Fastest Fourier Transform in the West. Technical Report MIT-LCS-TR-728, Laboratory for Computing Sciences,MIT, Cambridge, MA, 1997.

14. M. Frigo and S. G. Johnson. FFTW: An Adaptive Software Architecture for the FFT. In *Proceedings of the IEEE International Conference on Acoustics, Speech and Signal Processing (ICASSP)*, volume 3, pages 1381–1384, 1998.

15. A. Gara et al. Overview of the Blue Gene/L system architecture. *IBM Journal of Research and Development*, 49(2/3):195–212, 2005.

16. A. Gara, P. Heidelberger, and B. Steinmacher-burow. private communication.

17. M.E. Giampapa et al. Blue Gene/L advanced diagnostics environment. *IBM Journal of Research and Development*, 49(2/3):319–332, 2005.

18. P. D. Haynes and M. Cote. Parallel Fast Fourier Transforms for electronic structure calculations. *Comp. Phys. Comm.*, 130:121, 2000.

19. M. Karplus and J.A. McCammon. Molecular dynamics simulations of biomolecules. *Nature Structural Biology*, 9(9):646–652, September 2002.

20. Stefan Kral, Franz Franchetti, Juergen Lorenz, Christoph W. Ueberhuber, and Peter Wurzinger. FFT Compiler Techniques. In *Compiler Construction: 13th International Conference, CC 2004, Held as Part of the Joint European Conferences on Theory and Practice of Software, ETAPS 2004, Barcelona, Spain, March 29 - April 2, 2004. Proceedings*, pages 217–231, 2004.

21. J. Lorenz, S. Kral, F. Franchett, and C. W. Ueberhuber. Vectorization techniques for the Blue Gene/L double FPU. *IBM Journal of Research and Development*, 49(2/3), 2005.

22. The MPICH and MPICH2 homepage. {http://www-unix.mcs.anl.gov/mpi/mpich}, January 2004.

23. Mohammad Zubair Ramesh C. Agarwal, Fred G. Gustavson. A high performance parallel algorithm for 1D-FFT. 1994.

24. E. L. Zapata, F. F. Rivera, J. Benavides, J. M. Garazo, and R. Peskin. Multidimensional Fast Fourier Transform into fixed size hypercubes. *IEE Proceedings*, 137(4):253–260, July 1990.

Parallel Solution of Sparse Linear Systems Arising in Advection–Diffusion Problems

Luca Bergamaschi[1], Giorgio Pini[1], and Flavio Sartoretto[2]

[1] Dipartimento di Metodi e Modelli Matematici per le Scienze Applicate
Universita' degli Studi, Via Belzoni 7, 35131 Padova, Italy
`{berga,pini}@dmsa.unipd.it`
[2] Dipartimento di Informatica, Universitá di Venezia
Via Torino 155, 30173 Mestre VE
`sartoret@dsi.unive.it`

Abstract. Flow problems permeate hydraulic engineering. In order to solve real–life problems, parallel solutions must be engaged, for attaining large storage amounts and small wall–clock time. In this communication, we discuss valuable key points which allow for the efficient, parallel solution of our large, sparse linear systems, arising from the discretization of advection–diffusion problems. We show that data pre-fetching is an effective technique to improve the efficiency of the sparse matrix–vector product, a time consuming kernel of iterative solvers, which are the best choice for our problems. Preconditioning is another key topic for the efficient solution of large, sparse, ill–conditioned systems. Up to now, no extensive theory for choosing the best preconditioner is available, thus ad–hoc recipes and sound based experience is mandatory. We compare many preconditioners in order to show their efficiency and allowing a good choice when attacking problems like ours.

1 Introduction

The advection–diffusion equations are [1]

$$\frac{\partial u}{\partial t} = \nabla \cdot (\mathbf{K}\nabla u - \mathbf{v}u) + \mathbf{f}, \tag{1}$$

where u is the unknown function, $\mathbf{K}$ is the diffusion tensor, $\mathbf{v}$ is a given velocity, and $\mathbf{f}$ is a source or sink term. Dirichlet and Neumann boundary conditions must be given to identify a well posed mathematical formulation of the flow problem. Finite Element (FE) integration in space over a 3D FE N–node grid is performed. Further integration in time by finite difference methods is performed by Crank-Nicolson scheme when $\mathbf{v} = 0$ [1], implicit Euler otherwise. One obtains a sequence of $N \times N$ linear algebraic systems, $Ax = b$. Classical FE methods yield large, sparse linear systems. When the flow velocity is to be accurately computed, the Mixed Hybrid Finite Element (MHFE) method are exploited. MHFE provides simultaneous solution of fluid pressure and velocity. In our framework, piecewise-constant pressure is considered, while velocities are approximated using the lowest order Raviart-Thomas elements [2]. MHFE requires the solution

J.C. Cunha and P.D. Medeiros (Eds.): Euro-Par 2005, LNCS 3648, pp. 804–814, 2005.
© Springer-Verlag Berlin Heidelberg 2005

of sparse linear systems, which for a given mesh are 7~8 times larger than FE ones.

Iterative methods are the best choice for solving our test problems, provided efficient preconditioners are available. When $\mathbf{v} = 0$, Symmetric Positive Definite (SPD) matrices are obtained; unsymmetric ones otherwise. For SPD matrices, the best available iterative algorithm is Preconditioned Conjugate Gradient (PCG), while for general, unsymmetric matrices no *best* iterative algorithm is available. On the ground of our experience, we selected the BiCGSTAB algorithm [3], which displays robustness and efficiency when attacking our problems.

A core, time consuming, sub–task inside all iterative methods is the matrix–vector product. We tested Algorithm 2 after Geus and Rollin [4], which attempts to enhance cache usage by data pre-fetching (DP) techniques. Table 1 shows our implementation of the algorithm.

```
      subroutine matvec(n, ia, ja, a, x, y)
c Matrix-vector product y = A x, with data pre-fetching.
c The matrix A is stored in CSR format.
c
      implicit none
      integer n, i, j, j1, k, k1, l
      integer ia(*), ja(*)
      real*8 a(*), x(*), y(*), s, v, v1
c
      k = 1
      do i = 1, n
        s = 0.
        k1 = ia(i+1)
        if (k .lt. k1) then
          j = ja(k) ! pre-fetch
          v = a(k) ! pre-fetch
          k = k+1
          do  while (k .lt. k1)
              j1 = ja(k) ! pre-fetch
              v1 = a(k) ! pre-fetch
              s  =  s + v * x(j)
              j  =  j1 ! pre-fetch
              v  =  v1 ! pre-fetch
              k  =  k + 1
          end do
          s = s + v * x(j)
        endif
        y(i) = s
      end do
      return
      end
```

Fig. 1. Our implementation of Algorithm 2 after Geus and Röllin.

Preconditioning is another key issue for the efficient solution of large linear systems. We exploited classical Jacobi, which does not improve convergence very much, but is both not storage consuming and easily efficiently parallelizable; we also tested the more powerful FSAI preconditioners [5], which we computed by our efficient, parallel implementation. For a typical range of Peclet number values, the quality of ILU(0) and ILUT [6] preconditioners is analyzed in [7]. The results can be easily extended to FSAI and pARMS type preconditioners.

At present, no general rules to identify the best solver for either diffusion or advection dominated problems are available. This is another motivation to our study, which is aimed to suggest good solution strategies for several situations.

2 The FSAI Preconditioner

Given a SPD matrix A, let $A = L_A L_A^T$ be its Cholesky factorization. The FSAI method computes an approximate inverse of A in the factorized form $H = G_L^T G_L$, where G_L is a sparse nonsingular lower triangular matrix approximating L_A^{-1}. To attain G_L, one must first prescribe a sparsity pattern $S_L \subseteq \{(i,j) : 1 \le i \ne j \le N\}$, such that $\{(i,j) : i < j\} \subseteq S_L$. A lower triangular matrix $\hat{G}_L$ is computed by solving the equations

$$(\hat{G}_L A)_{ij} = \delta_{ij}, \qquad (i,j) \notin S_L. \tag{2}$$

The diagonal entries of $\hat{G}_L$ are all positive. Defining $D = [\mathrm{diag}(\hat{G}_L)]^{-1/2}$ and setting $G_L = D\hat{G}_L$, the preconditioned matrix $G_L A G_L^T$ is SPD and has diagonal entries all equal to 1. A common choice for the sparsity pattern is to allow non zeros in G_L only in positions corresponding to non zeros in the lower triangular part of A^k, where k is a small positive integer, e.g., $k = 1, 2, 3$; see [8]. The extension of FSAI to the non symmetric case is straightforward; however the re-solvability of the local linear systems and the non singularity of the approximate inverse is only guaranteed if all the principal sub-matrix of A are non singular (which holds true, for instance, when $A + A^T$ is SPD).

While the approximate inverses corresponding to A^k, $k > 1$, are often better than the one corresponding to $k = 1$, they may be too expensive to compute and apply. In [9] a simple approach, called *post-filtration*, was proposed to improve the quality of FSAI preconditioners in the SPD case. The method is based on a posteriori sparsification, by using a drop–tolerance parameter. We found that the quality of the preconditioner does not heavily depend upon its value, which ranges in the interval $[0, 1]$. The aim is to reduce the number of nonzero elements of the preconditioning factors, in order to decrease the arithmetic complexity of the iteration phase. In a parallel environment, a substantial reduction of the communication complexity of the preconditioner-by-vector multiplication can be achieved.

In the non symmetric case both preconditioner factors, G_L and G_U, must be sparsified. Non symmetric matrices with a symmetric nonzero pattern are considered, i.e. $S_L = S_U^T$ is assumed, and a symmetric filtration of factors G_L and G_U is performed.

3   Parallel Implementation

Our parallel implementation of the algorithms rely upon a data splitting approach, designed for sparse FE/MHFE matrix–vector (MV) products. The code is written in FORTRAN 90 and exploits MPI 1.0 calls for exchanging data among the processors. All our matrices are statically stored into CSR formatted data structures.

BiCGSTAB and PCG algorithms can be decomposed into a number of scalar products, `daxpy`-like linear combinations of vectors, $\alpha\mathbf{v} + \beta\mathbf{w}$, and MV products.

Scalar products, $\mathbf{v} \cdot \mathbf{w}$, were distributed among the P processors.

Concerning matrix splitting, note that uniform block mappings, like those exploited in High Performance Fortran `cyclic` directive, are not suitable for our sparse matrices. We splitted our matrices by a uniform, row–wise block mapping. Such distribution is ideal for our problems, since it allows for performing a piece of MV product on each processor. Moreover, our sparse matrices have quite the same number of non–zero entries per row, hence blocks consisting of the same number of rows consist of quite the same amount of bytes. We exploited blocks of *contiguous* rows. Non–contiguous row distributions yield more complex algorithms, which moreover do not perform well on (old) machines where the communication time changes with the relative position of processors in the communication net. We improved MV evaluation by using a technique for minimizing data communication between processors [10]. In the greedy matrix-vector algorithm, each processor communicates with each other. Using our approach with our sparse matrices, usually each processor sends/receives data to/from at most 2 other processors. Moreover, when running on P processors, the amount of data exchanged, when dealing with a matrix featuring M non–zero entries, is far smaller than $[M/P]$.

3.1   Parallel Implementation of FSAI

We implemented the FSAI preconditioner computation, both for SPD, and non symmetric matrices. Our code allows for the specification of either A or A^2 sparsity patterns. We used a block row distribution of matrices A, G_L (and also G_U in the non symmetric case). Complete rows are assigned to different processors.

Let n_i be the number of non zeros allowed in the i-th row of G_L. In the SPD case, any row i of the G_L matrix can be computed independently of each other, by solving a small SPD dense linear system of size n_i. To attain parallelism, the processor that computes row i must access n_i rows of A. Since the number of non local rows needed by each processor is relatively small, we temporarily replicate the non local rows on auxiliary data structures. The dense factorizations needed are carried out using BLAS3 routines from LAPACK. Once G_L is obtained, a parallel transposition routine provides to every processor the eligible part of G_L^T.

In the non symmetric case, recall that we assume a symmetric non zero pattern for matrix A, i.e. we ideally set $S_L = S_U^T$. The preconditioner factor G_L is computed as described before, while G_U is computed by columns. Hence, no additional row exchange is needed with respect to the SPD case. Every processor performs a fully parallel computation both of a set of rows into G_L, and of a set of columns into G_U.

4   pARMS

The parallel Algebraic Recursive Multilevel Solvers (pARMS) package [11, 12] is an interesting effort in devising distributed preconditioners for iterative solvers. It works in the framework of distributed linear systems, which provides an algebraic representation for the parallel solution of linear systems, $Ax = b$, arising in Domain Decomposition Methods. The coefficient matrix A is split among the available processors. A local $N_i \times N_i$ matrix, A_i, and an interface matrix, X_i are assigned to the i-th processor. Each local vector of unknowns, x_i, is split into a sub–vector u_i of interior variable contributions, and a sub-vector y_i of inter–domain interface variables. Analogously, each local right–hand side vector, b_i, is chopped into f_i and g_i contributions. The equations assigned to processor i can be written

$$\begin{pmatrix} B_i & F_i \\ E_i & C_i \end{pmatrix} \begin{pmatrix} u_i \\ y_i \end{pmatrix} + \begin{pmatrix} 0 \\ \sum_{j \in N_i} E_{ij} y_j \end{pmatrix} = \begin{pmatrix} f_i \\ g_i \end{pmatrix}. \tag{3}$$

where the matrices B_i, F_i, E_i, C_i compose a block–splitting of A_i. Additive Schwarz techniques (with or without overlapping), can be exploited, as well as Schur complement–type ones. It is well known that scalability and robustness of Additive Schwarz can be very poor [12]. We found that Schur techniques are better suited to our problems. These latter techniques rely upon Schur complement systems. They are derived by eliminating the variables u_i in equation (3), using $u_i = B_i^{-1}(f_i - F_i y_i)$. By substitution in the second equation, one gets

$$S_i y_i = \sum_{j \in N_i} E_{ij} y_j = g_i - E_i B_i^{-1} f_i = g_i', \tag{4}$$

where S_i is the *local* Schur complement

$$S_i = C_i - E_i B_i^{-1} F_i.$$

Assembling equations (4) over all processors, the *global Schur complement system*

$$Sy = \begin{pmatrix} S_1 & E_{1,2} & \dots & E_{1,p} \\ E_{2,1} & S_2 & \dots & E_{2,p} \\ \vdots & \vdots & \dots & \vdots \\ E_{p,1} & E_{p-1,2} & \dots & S_p \end{pmatrix} \begin{pmatrix} y_1 \\ y_2 \\ \vdots \\ y_p \end{pmatrix} = \begin{pmatrix} g_1' \\ g_2' \\ \vdots \\ g_p' \end{pmatrix}$$

is obtained. The matrix S is the *global* Schur complement. Once the system is approximately solved, each processor works out the system $B_i u_i = f_i - E_i y_i$, hence attaining an approximated solution x of the original problem. By extensive testing, we found that for our problems the `lsch_ilut` (left–Schur complement) algorithm performs better w.r.t. the Additive Schwarz preconditioners, right Schur complement preconditioners, and Gauss–Seidel preconditioners, which are described in [12]. The `lsch_ilut` approach solves the global Schur complement system via Block–Jacobi preconditioning. Note that pARMS package suffers from the need of assessing a large number of parameters (about 15), corresponding to the high number of possibilities that can be exploited.

pARMS algorithms are intrinsically parallel. Note that the number of iterations performed heavily depend upon the number of engaged processors, hence pARMS parallel performance analysis cannot naively rely upon classical parameters, like speedup.

5   Numerical Results

Table 1 shows the main characteristics of our test problems.

Table 1. Main characteristics of our test matrices. N=size, n_z=number of non–zero elements, dd= "is it a diagonal dominant matrix?"; HB=half–bandwidth; type = Matrix type: spd= symmetric positive definite, problem = problem type, algorithm = discretization algorithm, uns= unsymmetrical matrix. Solvers: P=PCG, B=BiCGSTAB.

#	N	n_z	dd	HB	type	problem	algorithm	solver
1	268,515	3,926,823	Y	5265	spd	diffusion	FE	P
2	390,160	2,694,160	N	18062	spd	diffusion	MHFE	P
3	531,765	7,817,373	Y	5265	spd	diffusion	FE	P
4	1,059,219	15,605,175	N	20769	spd	diffusion	FE	P
5	1,317,141	19,458,621	N	13041	uns	adv-diff	FE	B
6	2,097,669	31,066,125	N	20769	spd	diffusion	FE	P
7	2,635,731	38,927,991	N	51681	uns	adv-diff	FE	B
8	3,096,640	21,528,640	N	71966	spd	diffusion	MHFE	P

We stopped the iterations when the euclidean norm of the residual, $r_k = b - Ax_k$, satisfies $\|r_k\| \simeq 10^{-12}$.

We performed our runs on the IBM SP4 system and the IBM Linux Cluster 1350 machine, both located at CINECA supercomputing center, Italy.

The SP4 machine features 16 nodes, each one including 32 POWER 4, 1300 MHz processors. Each node is equipped with a 64 GB memory, one only node featuring a 128 GB core memory. The nodes are connected with 2 interfaces to a dual plane set of Colony switches.

The IBM Linux Cluster 1350, CLX for short, is a 256 node machine. Each node encloses a 2GB DRAM (32 nodes have a 4GB DRAM), and two Intel Xeon Pentium IV 3.055 GHz processors (10 I/O nodes feature 2.8 GHz processors). Each processor is equipped with a 512Kb L2 Cache. Disk space is 5.5 TB. The internal network is a *Myrinet IPC* one.

Table 2. Wall–clock seconds spent on the SP4, to solve our test problems up to $\|r_k\| \simeq 10^{-12}$ accuracy. D = PCG + diagonal preconditioning; F = PCG + FSAI(A) preconditioning; F1 = PCG + FSAI(A^2) preconditioning; D*, F*, F1* are D, F, F1, respectively, where *no data prefetching* inside matrix–vector products was exploited.

N	alg	T_1	T_2	T_4	T_8	T_{16}	T_{32}	S_2	S_4	S_8	S_{16}	S_{32}
2,097,669	D*	2611.0	1391.2	834.0	534.9	287.4	151.8	1.88	3.13	4.88	9.08	17.20
	F*	1976.0	1111.3	622.4	408.2	205.2	110.2	1.78	3.17	4.84	9.63	17.93
	F1*	1548.4	858.6	470.6	347.2	176.3	106.8	1.80	3.29	4.46	8.78	14.50
	averages	2045.1	1120.4	642.3	430.1	223.0	122.9	1.82	3.20	4.73	9.17	16.54
2,097,669	D	1822.7	985.5	555.1	375.1	201.6	102.4	1.85	3.28	4.86	9.04	17.80
	F	1236.6	655.6	369.6	243.3	144.2	91.2	1.89	3.35	5.08	8.58	13.56
	F1	1067.7	570.5	315.0	211.2	118.6	79.6	1.87	3.39	5.06	9.00	13.41
	averages	1375.7	737.2	413.2	276.5	154.8	91.1	1.87	3.34	5.00	8.87	14.92

Table 2 compares, on an appropriate test matrix, the performance on the SP4 of our PCG code either with or without, data pre-fetching (DP) by Geus & Rollin. The value T_p is the wall–clock seconds spent to solve a problem; $S_p = T_1/T_p$ is the classical speedup value. One can see that appreciably less wall–clock seconds are spent to solve our test problems when the DP technique is exploited. The average time over all tests goes down from 764.0 seconds when no DP is used, to as less as 508.1, which is only 67% of the former time, when DP is exploited. On the other hand, the speedup values are quite similar; their average values over all tests are 7.09 for no DP, vs 6.80 with DP. Concerning the assessment of the parameters in pARMS, we extensively engaged the package on our problems. We tested $s = 30, 50, 80, 100$ Krylov subspace sizes. We found that a good choice is using flexible GMRES (FGMRES) [13], together with $s = 100$. The `lsch_ilut` preconditioner with overlapping was enrolled. The fill-in parameter was set to `lfill=60` for all the recursion levels, and the dropping tolerance was `tol`$=10^{-4}$. The group independent set size was set to 5000, while the maximum number of internal iterations was 5. These values are also suggested in [11].

Table 3 shows the wall-clock times and *relative* speedup values, $S_p^{(r)} = T_{p/2}/T_p$, $p = 2, 4, 8, 16, 32$, recorded on the SP4 when solving our test problems (all obtained by exploiting DP technique). Note that the I/O time needed for data input is not considered. The time for printing output results is negligible.

The smallest matrices (problems 1 and 2) were solved on up to 16 processors. A larger number of processors would assign a too small data set to each one. To solve the larger problems, up to 32 processors were engaged. Inspecting Table 3 one can see that FSAI(A) allows for a slight decrease of the computing time, over Jacobi, while FSAI(A^2) with drop–tolerance value 0.1 provides appreciable enhancements over FSAI(A) and Jacobi.

One can see that pARMS in the smaller problems (1–6) is usually more time consuming than the other methods, while it is comparably expensive in the larger problem 7. In spite of the fact that we made extensive parameter space analysis, we could not attain pARMS convergence on problem 8. It is well known that pARMS suffer from high changes in the iteration number, depend-

Table 3. Analogous to the previous Table. Du = BiCGSTAB + Jacobi; Fu = BiCGSTAB + FSAI(A); F1u = BiCGSTAB + FSAI(A^2); pA = pARMS. Legend for the symbols: "*" = no convergence attained; "-" = value not computed, "/" = value not computable.

#	N	alg	T_1	T_2	T_4	T_8	T_{16}	T_{32}	$S_2^{(r)}$	$S_4^{(r)}$	$S_8^{(r)}$	$S_{16}^{(r)}$	$S_{32}^{(r)}$
1	268,515	D	448.0	230.6	136.5	84.5	52.2	-	1.94	1.69	1.62	1.62	/
		F	382.9	198.4	116.7	79.8	49.1	-	1.93	1.70	1.46	1.63	/
		F1	86.2	45.3	25.6	16.1	9.3	-	1.90	1.77	1.59	1.73	/
		pA	131.3	91.9	86.8	62.2	50.7	-	1.43	1.06	1.40	1.23	/
2	390,160	D	201.2	105.8	62.5	42.9	24.3	-	1.90	1.69	1.46	1.77	/
		F	183.2	98.1	54.8	37.6	21.3	-	1.87	1.79	1.46	1.77	/
		F1	140.1	73.4	42.3	28.4	15.0	-	1.91	1.74	1.49	1.89	/
		pA	*	235.8	189.8	217.9	*	-	/	1.24	0.87	/	/
3	531,765	D	1519.3	824.9	459.1	299.7	138.8	88.3	1.84	1.80	1.53	2.16	1.57
		F	1542.6	867.9	467.5	323.1	150.4	98.7	1.78	1.86	1.45	2.15	1.52
		F1	236.0	126.7	69.9	45.3	21.6	13.5	1.86	1.81	1.54	2.10	1.60
		pA	554.3	169.2	140.3	183.9	136.2	131.7	3.28	1.21	0.76	1.35	1.03
4	1,059,219	D	2522.7	1344.6	778.2	549.2	270.4	140.3	1.88	1.73	1.42	2.03	1.93
		F	2044.1	1050.5	598.9	479.4	253.7	136.1	1.95	1.75	1.25	1.89	1.86
		F1	770.9	412.3	229.1	174.4	93.6	52.3	1.87	1.80	1.31	1.86	1.79
		pA	*	351.8	245.0	216.8	176.3	92.3	/	1.44	1.13	1.23	1.91
5	1,317,141	Du	666.6	369.1	192.0	132.7	70.8	46.5	1.81	1.92	1.45	1.87	1.52
		Fu	527.6	280.3	163.4	113.2	64.9	45.3	1.88	1.72	1.44	1.74	1.43
		F1u	*	*	170.1	102.6	62.0	41.3	/	/	1.66	1.65	1.50
		pA	*	*	*	202.8	133.6	79.7	/	/	/	1.52	1.68
6	2,097,669	D	1822.7	985.5	555.1	375.1	201.6	102.4	1.85	1.78	1.48	1.86	1.97
		F	1236.6	655.6	369.6	243.3	144.2	91.2	1.89	1.77	1.52	1.69	1.58
		F1	1067.7	570.5	315.0	211.2	118.6	79.6	1.87	1.81	1.49	1.78	1.49
		pA	*	*	*	652.7	376.2	226.5	/	/	/	1.73	1.66
7	2,635,731	Du	*	*	768.9	546.6	380.2	295.2	/	/	1.41	1.44	1.29
		Fu	*	*	*	536.0	321.6	271.4	/	/	/	1.67	1.18
		F1u	*	*	*	419.3	254.1	212.4	/	/	/	1.65	1.20
		pA	*	*	*	419.7	295.3	191.2	/	/	/	1.42	1.54
8	3,096,640	D	6202.0	3423.4	1949.1	1612.2	714.7	350.7	1.81	1.76	1.21	2.26	2.04
		F	4950.2	2656.0	1506.4	1132.2	574.6	307.2	1.86	1.76	1.33	1.97	1.87
		F1	3801.6	2008.9	1176.0	924.3	428.7	242.8	1.89	1.71	1.27	2.16	1.77
		pA	*	*	*	*	*	*	/	/	/	/	/
		averages	1410.8	715.7	418.0	337.6	186.8	145.1	1.92	1.68	1.38	1.76	1.61

ing upon the number of processors. Note that in many problems pARMS could not be run on 1 or 2 processors, due to lack of core memory. When the relative speedup can be measured, it displays large oscillations, and questionable values (e.g. $S_2^{(r)} = 3.28$ for N=531,765). From this point of view, PCG and BiCGSTAB are more robust on this kind of problems. The average standard deviation in the number of iterations, counting all our PCG and BiCGSTAB tests, is 12.9, while for pARMS is 90.3. We feel that solving even larger problems on

Table 4. Analogous to Table 3. Problem 3, N=531,765, results obtained on the CLX system.

N	alg	T_1	T_2	T_4	T_8	T_{16}	T_{32}	$S_2^{(r)}$	$S_4^{(r)}$	$S_8^{(r)}$	$S_{16}^{(r)}$	$S_{32}^{(r)}$
531,765	D	1726.7	908.7	473.3	250.7	133.2	74.2	1.90	1.92	1.89	1.88	1.80
	F	1560.3	821.2	437.4	237.7	139.6	83.3	1.90	1.88	1.84	1.70	1.68
	F1	287.9	163.3	90.9	51.1	28.7	16.4	1.76	1.80	1.78	1.78	1.75
	pA	521.3	141.6	114.6	166.4	92.3	80.7	3.68	1.24	0.69	1.80	1.14

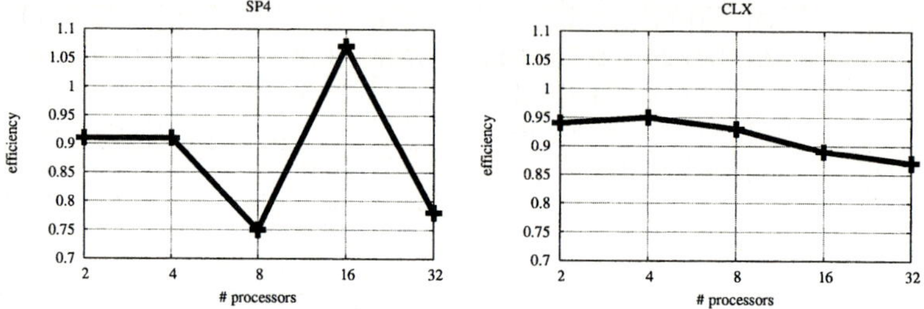

Fig. 2. Problem 3, N=531,765. Average relative efficiency on D, F, and F1 methods, obtained when running on the SP4 (left frame), and CLX (right frame).

a larger number of processors, pARMS could perform better. Figure 2 shows the average *relative* efficiency, $E_p^{(r)} = S_p^{(r)}/2$, on problem 3, recorded either on the SP4 (left frame) or on the CLX (right frame). The average was performed on D, F, and F1 algorithms; the pA technique was not considered: recall that the cost of the algorithm heavily depends upon the number of running processors, hence plain efficiency is meaningless. Usually, the relative efficiency on the SP4 is quasi optimal for $p = 2$ processors, good for 4, worsens when doubling to 8 processors, acceptable on more than 8. Such a behavior is typical for our problems, when running on CINECA's machine. The interconnecting network is not so fast as to allow high speedup values on a large number of processors (see also [4, 14]). The performance degradation when going from 4 to 8 processors occurred in all our parallel experiences on this machine, due to hard/soft processor aggregation into virtual/physical nodes. Since 8 processors share the same node core memory, when all 8 are engaged on unstructured matrix computations, many memory conflicts are raised. For comparison, Table 4 shows the time and speedup recorded on the CLX system, for problem 3 (N=531,765). Comparing with the results on the SP4 after Table 3, one can see that the performance is usually better on the CLX. A slight performance decrease is recorded when going from 1 to 2 processors. Recall that a CLX node encompasses two processors, sharing the node core memory. One can see that on the CLX the efficiency behavior matches the parallel expert feeling. This result confirms that the disturbing low performance on the SP4, when running on 8 processors, is due to the machine architecture, rather than to our algorithm, which performs well on other architectures, like CLX.

Summarizing, the parallel performance on the SP4 is more erratic than on the CLX, but note that the largest problems cannot run on a 2GB CLX node, unless a suitably large number of processors is engaged.

The parallel degrees obtained on the SP4 are compatible with the exploited machine, in accordance with the degrees shown e.g. in [4].

6 Conclusions

Summarizing, the PCG algorithm for SPD problems and BiCGSTAB for unsymmetric ones, equipped with $FSAI(A^2)$ preconditioning, prove to be the best parallel solvers for our problems, on our tests.

Data pre-fetching allows for appreciably improving the efficiency of our sparse matrix–vector products.

The parallel efficiency of our code on the SP4 can be rated satisfactory. Our results provide a guideline for the parallel performance that one can expect when running FE codes. Parallel performance losses can be recorded running on 8 processors, due to the complex, highly non uniform, SP4 architecture. This problem does not occur on the CLX, where typical parallel performance results are achieved.

Acknowledgments

This work has been supported by the italian MIUR project *Numerical models for multi-phase flow and deformation in porous media*.

References

1. Gambolati, G., Pini, G., Tucciarelli, T.: A 3-D finite element conjugate gradient model of subsurface flow with automatic mesh generation. Adv. Water Resources **3** (1986) 34–41
2. Brezzi, F., Fortin, M.: Mixed and Hybrid Finite Element Methods. Springer-Verlag, Berlin (1991)
3. van der Vorst, H.A.: Bi-CGSTAB: A fast and smoothly converging variant of BI-CG for the solution of nonsymmetric linear systems. SIAM J. Sci. Stat. Comput. **13** (1992) 631–644
4. Geus, R., Röllin, S.: Towards a fast parallel sparse symmetric matrix-vector multiplication. Parallel Computing **27** (2001) 883–896
5. Yu. Kolotilina, L., Yu. Yeremin, A.: Factorized sparse approximate inverse preconditionings I. Theory. SIAM J. Matrix Anal. Appl. **14** (1993) 45–58
6. Saad, Y.: ILUT: A dual threshold incomplete lu factorization. Numer. Linear Alg. Appl. **1** (1994) 387–402
7. Pini, G., Putti, M.: Krylov methods in the finite element solution of groundwater transport problems. In Peters, A., Wittum, G., Herrling, B., Meissner, U., Brebbia, C.A., Gray, W.G., Pinder, G.F., eds.: Computational Methods in Water Resources X, Volume 1, Dordrecht, Holland, Kluwer Academic (1994) 1431–1438

8. Kaporin, I.E.: New convergence results and preconditioning strategies for the conjugate gradient method. Numer. Linear Alg. Appl. **1** (1994) 179–210
9. Yu. Kolotilina, L., Nikishin, A.A., Yu. Yeremin, A.: Factorized sparse approximate inverse preconditionings IV. Simple approaches to rising efficiency. Numer. Linear Alg. Appl. **6** (1999) 515–531
10. Bergamaschi, L., Putti, M.: Efficient parallelization of preconditioned conjugate gradient schemes for matrices arising from discretizations of diffusion equations. In: Proceedings of the Ninth SIAM Conference on Parallel Processing for Scientific Computing. (March, 1999) (CD–ROM).
11. Li, Z., Saad, Y., Sosonkina, M.: pARMS: a parallel version of the algebraic recursive multilevel solver. Numer. Linear Alg. Appl. **10** (2003) 485–509
12. Saad, Y., Sosonkina, M.: pARMS: a package for solving general sparse linear systems of equations. In Wyrzykowski, R., Dongarra, J., Paprzycki, M., Wasniewski, J., eds.: Parallel Processing and Applied Mathematics. Volume 2328 of Lecture Notes in Computer Science., Berlin, Springer-Verlag (2002) 446–457
13. Saad, Y.: Iterative Methods for Sparse Linear Systems. Second edition. SIAM, Philadelphia, PA (2003)
14. Bergamaschi, L., Pini, G., Sartoretto, F.: Computational experience with sequential and parallel preconditioned Jacobi Davidson for large sparse symmetric matrices. J. Comput. Phys. **188** (2003) 318–331

Parallelization of Implicit-Explicit Runge-Kutta Methods for Cluster of PCs

José Miguel Mantas[1], Pedro González[2], and José A. Carrillo[3]

[1] Software Engineering Department. University of Granada
C/ P. Daniel de Saucedo s/n. E-18071 Granada, Spain
`jmmantas@ugr.es`
[2] Department of Applied Mathematics. University of Granada
Avda. Fuentenueva s/n. E-18071 Granada, Spain
`prodelas@ugr.es`
[3] Departament de Matemàtiques - ICREA. Universitat Autònoma de Barcelona
Bellaterra E-08193
`carrillo@mat.uab.es`

Abstract. Several physical phenomena of great importance in science and engineering are described by large partly stiff differential systems where the stiff terms can be easily separated from the remaining terms. Implicit-Explicit Runge-Kutta (IMEXRK) methods have proven to be useful solving these systems efficiently. However, the application of these methods still requires a large computational effort and their parallel implementation constitutes a suitable way to achieve acceptable response times. In this paper, a technique to parallelize and implement efficiently IMEXRK methods on PC clusters is proposed. This technique has been used to parallelize a particular IMEXRK method and an efficient parallel implementation of the resultant scheme has been derived in a structured manner by following a component-based approach. Several numerical experiments which have been performed on a cluster of dual PCs reveal the good speedup and the satisfactory scalability of the parallel solver obtained.

1 Introduction

The spatial discretization of a great variety of time-dependent partial differential equations (PDEs) by the method of lines leads to large systems of ordinary differential equations (ODEs) with this form:

$$\frac{dy}{dt} = \mathbf{f}(y) + \mathbf{g}(y), \qquad y(0) = y_0 \in \mathrm{I\!R}^d, \qquad t > 0 \tag{1}$$

where $y = y(t) \in \mathrm{I\!R}^d$ is the unknown function of a d-dimensional ODE system which is defined by the component functions $\mathbf{f}, \mathbf{g} : \mathrm{I\!R}^d \longrightarrow \mathrm{I\!R}^d$. The function $\mathbf{g}(y)$ results from the discretization of the stiff terms and $\mathbf{f}(y)$ results from the discretization of the remaining terms. The function $\mathbf{g}$ is usually written as $(1/\epsilon)\tilde{\mathbf{g}}$ ($\tilde{\mathbf{g}} : \mathrm{I\!R}^d \longrightarrow \mathrm{I\!R}^d$), where $\epsilon > 0$ is the stiffness parameter [5].

J.C. Cunha and P.D. Medeiros (Eds.): Euro-Par 2005, LNCS 3648, pp. 815–825, 2005.
© Springer-Verlag Berlin Heidelberg 2005

There are many practical problems where it can be advantageous to integrate $\mathbf{f}$ explicitly, to reduce computational costs, and $\mathbf{g}$ implicitly, to avoid excessively small time steps. In fact, the Jacobian of $\mathbf{g}$ in problems of form (1) frequently exhibits a particular structure (positive definite, symmetric, and sparse) whose exploitation would make it possible a considerable saving of computational effort if $\mathbf{g}$ is implicitly integrated. This special structure can be lost if global implicit methods are used to integrate both terms $(\mathbf{f}(y)+\mathbf{g}(y))$. Therefore, it often makes sense to integrate $\mathbf{g}(y)$ implicitly and $\mathbf{f}(y)$ explicitly in these problems when they exhibit this structure.

A clear example of this type of system appears in reaction-diffusion and convection-diffusion problems [1, 7, 12, 13] arising in multiple areas of science and engineering. In these problems, an explicit scheme would be used for the reaction (resp. convection) term and an implicit scheme for the diffusion term.

There exists Runge-Kutta methods which are specially suitable for systems of form (1). These schemes, known as *Implicit-Explicit Runge-Kutta Methods (IMEXRK)*, apply an implicit discretization for $\mathbf{g}$ and an explicit one for $\mathbf{f}$, simultaneously, in the same time step and using identical time step size. A Diagonally Implicit Runge-Kutta method (DIRK) [2] is usually considered to integrate $\mathbf{g}$, given the importance of the efficiency in the solution of the stiff part of the equation [1].

However, the application of an IMEXRK method, together with the complexity of these systems, demands a great deal of computing power which can be easily achieved by using efficient parallel implementations running on cluster of Personal Computers (PCs). In this paper, the development of parallel software based on IMEXRK methods is tackled.

In section 2, the structure of the most relevant IMEXRK methods is briefly presented. A general technique to parallelize these numerical methods will be described in section 3. This technique will be applied to a particular IMEXRK method in section 4, to obtain a new parallel scheme. A component based methodological approach for deriving group parallel ODE solvers [9, 10] is used in section 5 to develop an efficient parallel implementation of this numerical scheme. This approach enables the exploitation of the multilevel parallelism which exhibits the numerical scheme and the ODE system in an structured manner. This implementation is adapted to solve a 1D rarefied gas shock profile on a cluster of dual PCs. Section 6 presents the experimental results obtained with the parallel solver. Finally, section 7 gives the main conclusions of the work.

2   Implicit-Explicit Runge-Kutta Methods

We have considered IMEXRK methods where the implicit solver (which is used to integrate $\mathbf{g}$) is a DIRK method. With this requirement, we have identified two relevant types of IMEXRK methods: pure IMEXRK methods [1, 8] and Additive Semi-Implicit Runge-Kutta methods of type A (ASIRK-A) [13].

A pure s-stage IMEXRK scheme is characterized by two matrices $\tilde{A}, A \in \mathbb{R}^{s \times s}$ $(\tilde{A} = (\tilde{a}_{ij}),\ A = (a_{ij}))$ and two coefficient vectors $\tilde{b}, b \in \mathbb{R}^s$ $(\tilde{b} =$

$(\tilde{b}_1, \ldots, \tilde{b}_s)$, $b = (b_1, \ldots, b_s)$). The matrix $\tilde{A}$ is strictly lower triangular ($\tilde{a}_{ij} = 0$, for $j \geq i$) and A is lower triangular ($\tilde{a}_{ij} = 0$, for $j > i$). When these parameters are applied to a system of form (1), we obtain the numerical scheme which appears below. This scheme describes how to obtain the vector $y_n \in \mathbb{R}^d$, which approximates $y(t_n)$, from the approximation given by the previous integration step $y_{n-1} \approx y(t_{n-1})$:

$$Y_{n,i} = y_{n-1} + h_n \left(\sum_{j=1}^{i-1} \tilde{a}_{ij} \mathbf{f}(Y_{n,j}) + \sum_{j=1}^{i} a_{ij} \mathbf{g}(Y_{n,j}) \right), \qquad i = 1, \ldots, s \qquad (2)$$

$$y_n = y_{n-1} + h_n \left(\sum_{i=1}^{s} \tilde{b}_i \mathbf{f}(Y_{n,i}) + \sum_{i=1}^{s} b_i \mathbf{g}(Y_{n,i}) \right), \qquad n = 1, \ldots, N_{steps}$$

where $Y_{n,i} \in \mathbb{R}^d$, $i = 1, \ldots, s$ and h_n is the *size* of the n-th time step ($h_n = t_n - t_{n-1}$). N_{steps} is the total number of time steps.

An s-stage ASIRK-A scheme (ASIRK-sA) is represented with the same kind of parameters as an s-stage IMEXRK method. The only difference is that in ASIRK-A methods, only a coefficient vector $b \in \mathbb{R}^s$ is necessary, in addition to the matrices $\tilde{A}, A \in \mathbb{R}^{s \times s}$ which maintain the same previously mentioned structure. When these parameters are applied to a system of form (1), we obtain:

$$Y_{n,i} = h_n \left[\mathbf{f} \left(y_{n-1} + \sum_{j=1}^{i-1} \tilde{a}_{ij} Y_{n,j} \right) + \mathbf{g} \left(y_{n-1} + \sum_{j=1}^{i} a_{ij} Y_{n,j} \right) \right], i = 1, .., s \quad (3)$$

$$y_n = y_{n-1} + \sum_{i=1}^{s} b_i Y_{n,i}, \qquad n = 1, \ldots, N_{steps}$$

3   An Approach to Parallelize IMEXRK Methods

When an IMEXRK method is applied to a system of form (1), s connected d-dimensional nonlinear systems must be solved sequentially (see (2) and (3)).

If the modified Newton method [2] is applied to solve each of these systems, we obtain s Newton iterations where the i-th iteration ($i = 1, \ldots, s$) computes an approximation to the vector $Y_{n,i}$. Each iteration must be solved before the next one, because the i-th iteration depends on vectors $Y_{n,j}$, $j = 1, \ldots, i-1$, which must be computed in previous iterations. As a result, the following numerical scheme is obtained:

$$\boxed{\begin{array}{l}
\textbf{for } n = 1, \ldots, N_{steps} \\
\quad \textbf{for } i = 1, \ldots, s \\
\qquad Y_{n,i}^{(0)} \text{ is computed by using a predictor formula} \\
\qquad \textbf{for } v = 1, \ldots, m_i \\
\qquad\quad R_i = Q_i - Y_{n,i}^{(v-1)} \\
\qquad\quad Y_{n,i}^{(v)} = Y_{n,i}^{(v-1)} + \left(I_d - a_{ii} h_n \mathbf{J_g}(k_i) \right)^{-1} R_i
\end{array}}$$

The values m_i of this method are dynamically determined to ensure that $Y_{n,i}^{(m_i)}$ is a good approximation of $Y_{n,i}$. $\mathbf{J_g}(k_i)$ denotes an approximation to the Jacobian of $\mathbf{g}$ evaluated at k_i and I_d denotes the d-dimensional identity matrix.

The values of Q_i and k_i vary depending on the type of IMEXRK scheme which is considered. So, for pure IMEXRK schemes, $k_i = y_{n-1}$ and

$$Q_i = y_{n-1} + h_n \left(\sum_{j=1}^{i-1} \tilde{a}_{ij} \mathbf{f}(Y_{n,j}) + \sum_{j=1}^{i-1} a_{ij} \mathbf{g}(Y_{n,j}) + a_{ii} \mathbf{g}(Y_{n,i}^{(v-1)}) \right),$$

and for ASIRK-A schemes, $k_i = y_{n-1} + \sum_{j=1}^{i-1} a_{ij} Y_{n,j}$ and

$$Q_i = h_n \left[\mathbf{f}\left(y_{n-1} + \sum_{j=1}^{i-1} \tilde{a}_{ij} Y_{n,j} \right) + \mathbf{g}\left(y_{n-1} + \sum_{j=1}^{i-1} a_{ij} Y_{n,j} + a_{ii} Y_{n,i}^{(v-1)} \right) \right].$$

Since the iterations must be solved sequentially, this scheme does not exhibit a lot of task parallelism exploitable across the method [2]. In order to decouple the calculations associated to each stage, we propose to approximate the solution of this scheme by using a similar scheme in which the calculations of each stage can be performed in parallel. The new scheme introduces some of redundant computation, but this additional cost is relatively small. This scheme is based on the following reasonable assumptions:

- For ASIRK-A schemes, the Jacobian of $\mathbf{g}$ in $y_{n-1} + \sum_{j=1}^{i-1} a_{ij} Y_{n,j}$ is approximated by the Jacobian evaluated in y_{n-1} ($k_i = y_{n-1}$). This approximation does not involve a considerable loss of accuracy and makes it possible to use the same Jacobian matrix in the calculations of all the stages.
- The calculations of each stage can be performed in a synchronous and concurrent manner if, in the v-th iteration of the original scheme, we approximate $Y_{n,j}$ by $Y_{n,j}^{(v-1)}$ and we consider the same number of Newton iterations for all the stages ($v = 1, \ldots, m$).

Now, the parallel numerical schemes which result from these assumptions are described. In these descriptions, the term $Y_{n,j}^{(v-1)}$ appears highlighted to emphasize the modification of the original schemes.

$$
\boxed{
\begin{aligned}
&\textbf{for } n = 1, \ldots, N_{steps} \\
&\quad Y_n^{(0)} = Pred(Y_{n-1}) \qquad (Pred(\cdot) \text{ denotes a predictor formula}) \\
&\quad \textbf{for } v = 1, \ldots, m \\
&\qquad \textbf{parfor } i = 1, \ldots, s \\
&\qquad\quad R_i = Q_i - Y_{n,i}^{(v-1)} \\
&\qquad\quad Y_{n,i}^{(v)} = Y_{n,i}^{(v-1)} + \left(I_d - a_{ii} h_n \mathbf{J_g}(y_{n-1}) \right)^{-1} R_i
\end{aligned}
}
$$

where $Y_n \in \mathbb{R}^{sd}$ is the so-called *stage vector*, which contains s d-dimensional components $Y_{n,i}$, $\quad i = 1, \ldots, s$ $(Y_n = (Y_{n,1}, \ldots, Y_{n,s})^T)$. The value of m must be dynamically determined to ensure that $(Y_{n,1}^{(m)}, \ldots, Y_{n,s}^{(m)})^T$ is a good approximation of Y_n. Here the term Q_i is different. For IMEXRK schemes:

$$Q_i = y_{n-1} + h_n \left(\sum_{j=1}^{i-1} \tilde{a}_{ij} \mathbf{f}\left(\boxed{Y_{n,j}^{(v-1)}}\right) + \sum_{j=1}^{i-1} a_{ij} \mathbf{g}\left(\boxed{Y_{n,j}^{(v-1)}}\right) + a_{ii} \mathbf{g}(Y_{n,i}^{(v-1)}) \right)$$

and for ASIRK-A schemes:

$$Q_i = h_n \left[\mathbf{f}\left(y_{n-1} + \sum_{j=1}^{i-1} \tilde{a}_{ij} \boxed{Y_{n,j}^{(v-1)}} \right) + \mathbf{g}\left(y_{n-1} + \sum_{j=1}^{i-1} a_{ij} \boxed{Y_{n,j}^{(v-1)}} + a_{ii} Y_{n,i}^{(v-1)} \right) \right]$$

With this scheme, the total number of Newton iterations would be no less than in the original scheme $(m \geq \max_{1 \leq i \leq s}(m_i))$. However, this additional number of iterations does not involve an excessive loss of efficiency.

4   Application to a Particular IMEXRK Method

Now the derivation of a particular parallel IMEXRK method of second order is performed. The selected method is termed LRR(3,2,2) [8] (3 stages are used in the implicit scheme, 2 stages in the explicit scheme and the convergence order is 2). This method is very suitable in applications where the stiff terms can be easily separated from the rest of the equations and a high accuracy is not required in the time discretization [1]. The parameters which characterize this method [8] lead to the following numerical scheme:

$$Y_{n,1} = y_{n-1} + \frac{1}{2} h_n [\mathbf{f}(y_{n-1}) + \mathbf{g}(Y_{n,1})], \qquad Y_{n,2} = y_{n-1} + \frac{1}{3} h_n [\mathbf{f}(y_{n-1}) + \mathbf{g}(Y_{n,2})]$$

$$y_n = Y_{n,3} = y_{n-1} + h_n [\mathbf{f}(Y_{n,1}) + \frac{3}{4} \mathbf{g}(Y_{n,2}) + \frac{1}{4} \mathbf{g}(Y_{n,3})] \qquad (4)$$

This scheme exhibits exploitable task parallelism itself, because the computation of vectors $Y_{n,1}$ and $Y_{n,2}$ can be performed simultaneously. However, the computation of $y_n = Y_{n,3}$ requires the previous computation of those two vectors. We have applied the previously described general technique to enable a higher degree of concurrency, obtaining the following parallel scheme, which has been termed *PIMEXRK3* (Parallel IMEXRK method with 3 stages).

$$
\boxed{
\begin{array}{l}
\textbf{for } n = 1, \ldots, N_{steps} \; \{ \quad Y_n^{(0)} = Pred(Y_{n-1}) \qquad // \; a_{11} = \frac{1}{2}, \; a_{22} = \frac{1}{3}, \; a_{33} = \frac{1}{4} \\
\quad \textbf{for } v = 1, \ldots, m \; \{ \\[4pt]
\qquad \textbf{par } \{ \; \textbf{parfor } i = 1, 2 \; \{ R_i = y_{n-1} + a_{ii} h_n \left[\mathbf{f}(y_{n-1}) + \mathbf{g}(Y_{n,i}^{(v-1)}) \right] - Y_{n,i}^{(v-1)} \} \\[8pt]
\qquad\quad R_3 = y_{n-1} + h_n \left[\mathbf{f}(Y_{n,1}^{(v-1)}) + \frac{3}{4}\mathbf{g}(Y_{n,2}^{(v-1)}) + a_{33}\mathbf{g}(Y_{n,3}^{(v-1)}) \right] - Y_{n,3}^{(v-1)} \} \\[8pt]
\qquad \textbf{parfor } i = 1, 2, 3 \; \{ \; Y_{n,i}^{(v)} = Y_{n,i}^{(v-1)} + (I_d - a_{ii} h_n \mathbf{J_g}(y_{n-1}))^{-1} R_i \; \} \; \} \\[4pt]
\}
\end{array}
}
$$

5 Derivation of Parallel Implementations of the PIMEXRK3 Scheme

Following a component-based approach to derive parallel ODE solvers [9, 10], termed COMPODES, a distributed implementation of the PIMEXRK3 scheme has been obtained to solve a particular problem on a cluster of dual PCs, by appplying 3 phases in sequence.

1. Component-Based Generic Description of the Numerical Scheme

From the mathematical description of the numerical scheme, the first phase of COMPODES is applied. For that purpose, several abstract operations are selected and combined suitably to describe the algorithm and to express the maximum degree of task parallelism. A summarized generic description of the PIMEXRK3 method, based on abstract operations, is shown in Figure 1a), where the edges denote data dependencies and the main sources of task parallelism are represented with concurrent loops ($PAR\ i = 1, 3$). We have selected a direct method based on LU decomposition to solve the linear systems, because this choice enables the reuse of the same LU decomposition for the solution of all the linear systems which arise in one time step. In fact, The operation $\texttt{LUdecomp}(.., A, ..)$ denotes the LU factorization of A, $\texttt{SolveSystem}(.., A, ..., X)$ denotes the computation of $X \longleftarrow A^{-1}X$ (assuming $\texttt{LUdecomp}(.., A, ..)$) and $\texttt{Feval}(.., t, f, y, dy)$ denotes the evaluation of a function f.

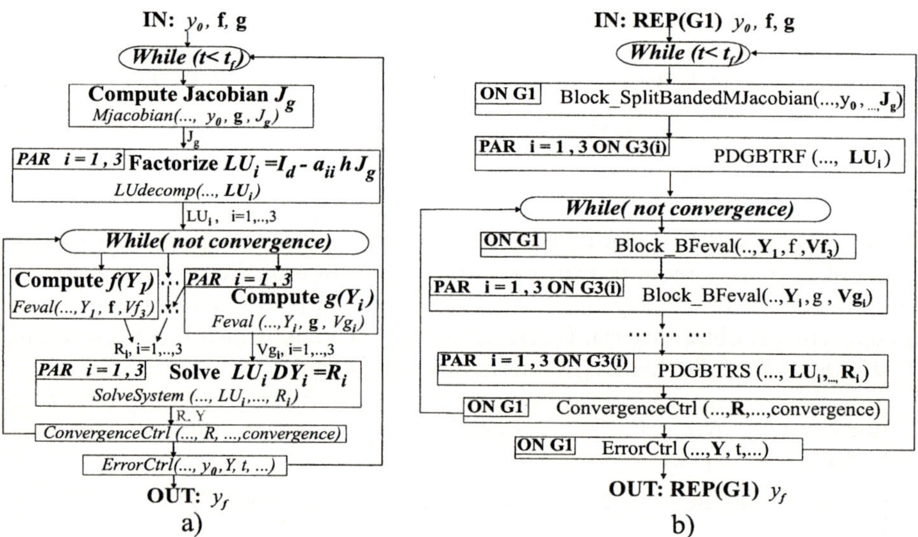

Fig. 1. a) Generic description of PIMEXRK3, b) Description of task scheduling

2. Adaptation to a Particular ODE System

The generic description of the PIMEXRK3 method has been adapted, following the second COMPODES phase, to perform the time integration of a hydrodynamical model of the Boltzmann equation for rarefied gases in 1D [11]. This model is based on a system of 5 PDEs, termed the relaxed Burnett system [6], which can be written in the following form:

$$\begin{pmatrix} U_t \\ V_t \end{pmatrix} = \begin{pmatrix} -F(U,V)_x \\ -G(U,V,U_x,V_x) + D(U,V,U_x,V_x)_x \end{pmatrix}, \text{ where} \qquad (5)$$

$$U = \begin{pmatrix} \rho \\ m = \rho u \\ z = \frac{1}{2}\rho u^2 + \frac{3}{2}p \end{pmatrix}, \quad V = \begin{pmatrix} \sigma \\ q \end{pmatrix}, \quad F(U,V) = \begin{pmatrix} \rho u \\ \rho u^2 + p + \sigma \\ \frac{1}{2}\rho u^3 + \frac{5}{2}up + \sigma u + q \end{pmatrix}.$$

The terms $G(U,V,U_x,V_x)$ and $D(U,V,U_x,V_x)$ are defined in [6]. In this system, we have five independent variables ρ (mass density), $m = \rho u$ (momentum, where u is the macroscopic velocity), $z = \frac{1}{2}\rho u^2 + \frac{3}{2}p$ (total energy, where p is the normal pressure), σ (pressure deviation tensor) and q (heat flux vector).

A spatial discretization of this system is proposed in [11]. This discretization is based on combining relaxation schemes for the conservative part and standard second order central differences for the non conservative part. The resulting system has 5N ODEs when the 1D space is discretized by using N grid points. A suitable arrangement of the equations leads to a a narrow banded ODE system whose Jacobian matrix has 9 subdiagonals and 7 superdiagonals.

The stiff and nonstiff terms in (5) have the following form:

$$\mathbf{f}(U,V) = (-F(U,V)_x, 0)^T, \quad \mathbf{g}(U,V) = (0, D(U,V,U_x,V_x)_x - G(U,V,U_x,V_x))^T.$$

If we maintain the same spatial discretization and arrangement which it is proposed in [11], the subsystem associated to $\mathbf{g}$ has a banded structure and the Jacobian of $\mathbf{g}$ has 9 subdiagonals and 5 superdiagonals. Therefore, the banded structure of the function $\mathbf{g}$ and its Jacobian is narrower than in the original system ($f = \mathbf{f} + \mathbf{g}$). Since the structure of the Jacobian matrix for $\mathbf{g}$ determines the complexity of the more costly calculations in the implicit time integration, the use of an IMEXRK method involves an important reduction of computational costs.

The initial vector y_0 of the ODE system captures the state before a one-dimensional shock profile with Mach number 10 [11].

In order to enable the exploitation of the particular structure of the subsystems $\mathbf{f}$ and $\mathbf{g}$, the generic specification of the PIMEXRK3 scheme has been adapted by replacing several operations (for instance, `LUdecomp`, `MJacobian` and `Feval`) by specializations which assume a banded structure [10].

3. Parallel Design Decisions

Following the third COMPODES phase, several parallel design decisions have been made in order to compute efficiently the specialized PIMEXRK3 scheme on a processor number, P, which is multiple of 3. These decisions include the scheduling of the tasks and the selection of the best data parallel implementation and data distribution to realize each operation. An approximation method has been proposed to make these decisions systematically [10].

A summarized graphical description of some of these decisions is shown in Figure 1b). Several groups of processors have been considered to schedule the tasks: a global group with P processors (**G1**), and 3 disjointed subgroup with $P/3$ processors (**G3(i)**, i=1,2,3). To compute an approximation of the Jacobian of **g**, we chose an optimal implementation for banded Jacobians `Block_SplitBandedMJacobian` on the global group **G1**. This implementation generates a block column distribution of a compact representation of the Jacobian. The evaluations of the function **f** are performed on the group **G1** while the evaluations of **g** for the i-th stage ($i = 1, 2, 3$) are performed on **G3(i)**. These evaluations are implemented by using a parallel block routine (`Block_BFeval`) which takes into account the banded structure of the ODE system terms in order to reduce the remote communication. The LU decompositions and system solutions for the i-th stage are computed on **G3(i)** by using the routines of the ScaLAPACK library [3] `PDGBTRF` (banded LU Factorization) and `PDGBTRS` (banded system solution). These routines takes advantage of the banded structure of the system and follows a block column distribution.

These decisions have been translated into a parallel program which is expressed in Fortran augmented with routines of ScaLAPACK and *MPI* [4].

6   Numerical Experiments

We have performed several numerical experiments on a cluster of 8 dual AMD processors 2.5Ghz, running Linux, connected via a Gigabit ethernet switch.

Table 1. Comparison among LRR(3,2,2) and PIMEXRK3 numerical solutions

$h_n=$	0.5	0.25	0.125	0.0625
$\|\|y_{LRR} - y_{P3}\|\|_2$	$5.599 \cdot 10^{-2}$	$7.956 \cdot 10^{-5}$	$2.223 \cdot 10^{-5}$	$5.970 \cdot 10^{-6}$

In order to show the accuracy of the numerical results obtained with the parallel solver, we compare the numerical solutions obtained with a sequential implementation of the LRR(3,2,2) method (y_{LRR}) and with an implementation of the PIMEXRK3 scheme running on 6 processors (y_{P3}). Table 1 shows the L^2-norm of the difference between the numerical solutions obtained with both solvers ($\|\|y_{LRR} - y_{P3}\|\|_2$). These results have been obtained for $t = 5.0$ and $N = 200$. Several experiments has been performed with a fixed step size although a different step size has been used in each experiment. The results prove that there is a great agreement between the solutions obtained by both methods.

The time results (in seconds) obtained for different values of N on several processor numbers P are shown in Table 2. The speedup results are graphically shown in Figure 2. These speedup results have been obtained by comparing the parallel execution time for several values of P, with the execution time of a sequential implementation of the $LRR(3, 2, 2)$ scheme running on a single

Table 2. Time results for the PIMEXRK3 scheme applied to the test problem

N =	72	396	900	1296	1800	2520	3600	3996	7200	9000
LRR($P = 1$)	0.258	1.417	3.222	4.644	6.451	9.026	12.94	14.37	25.89	32.37
$P = 3$	0.0899	0.476	1.093	1.585	2.1932	3.074	4.383	4.867	8.76	11.76
$P = 6$	0.0559	0.259	0.572	0.824	1.1439	1.591	2.292	2.527	4.55	5.71
$P = 9$	0.063	0.227	0.407	0.569	0.791	1.108	1.562	1.756	3.09	3.78
$P = 12$	0.0424	0.153	0.311	0.435	0.6013	0.822	1.173	1.286	2.32	2.87
$P = 15$	0.1906	0.273	0.298	0.399	0.5167	0.707	0.970	1.067	1.94	2.36

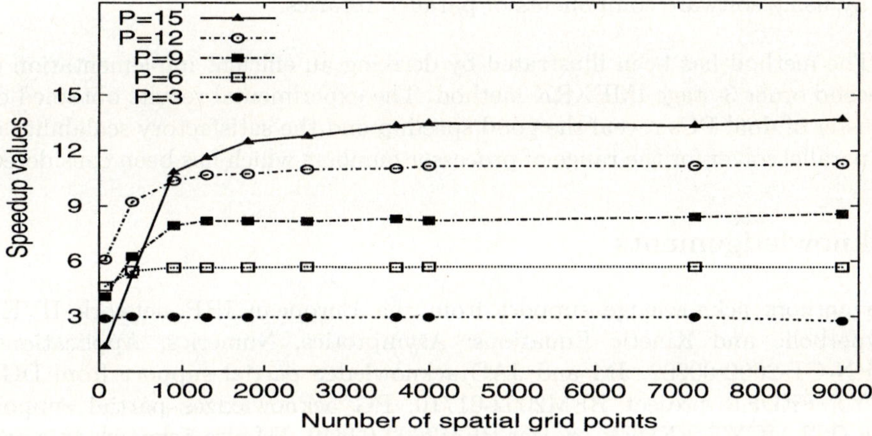

Fig. 2. Speedup results for several spatial grid sizes

processor. This implementation of the $LRR(3, 2, 2)$ scheme also takes advantage of the banded structure of the subsystems. The time results were taken for one time step and meshes varying from $N = 80$ to $N = 4000$ grid points.

The experiments show that a speedup close to the linear speedup can be achieved when N is greater than 500. In general, efficiencies higher than 90% are achieved with a sufficiently large number of grid points, except on $P = 15$ processors where a larger problem size would be necessary to achieve this efficiency. The results reveal the parallelism which exhibits the new parallel numerical scheme and the satisfactory scalability which offers the implementation derived by the COMPODES approach.

7 Conclusions

A method to obtain efficient implementations of IMEXRK methods for PC clusters has been described. This method incorporates two procedures:

1. A technique to decouple the computation associated with each stage, when an s-stage IMEXRK method is applied to a partly stiff system with d ODEs,

has been introduced. The technique is based on considering the s coupled d-dimensional nonlinear systems, which arise when an IMEXRK method is applied, as only one sd-dimensional nonlinear system, and imposing several reasonable approximation assumptions when the modified Newton method is used to solve the nonlinear system. As a result, the calculations associated with each stage of the method can be performed in parallel for each round of the Newton iteration.

2. A component based approach [10] can be applied to easily derive efficient implementations of the previously defined parallel schemes. This approach allows us to exploit the task and data parallelism which exhibits the scheme by using software components of parallel libraries.

The method has been illustrated by deriving an efficient implementation of a second order 3-stage IMEXRK method. The experimental results obtained on a cluster of dual PCs reveal the good speedup and the satisfactory scalability of the parallel solver for the range of processor numbers which has been considered.

Acknowledgements

The authors acknowledge support from the European IHP network HYKE "Hyperbolic and Kinetic Equations: Asymptotics, Numerics, Applications" HPRN-CT-2002-00282. JM and JAC acknowledge partial support from DGI-MCYT/FEDER project BFM2002-01710. PG acknowledges partial support from DGI-MCYT/FEDER project BFM2002-02649. JM also acknowledges partial support from MEC/FEDER project TIN2004-07672-c03-02.

References

1. Ascher, U. M., Ruuth, S. J., Spiteri, R. J.: Implicit-Explicit Runge-Kutta Methods for time-dependent Partial Differential Equations. Applied Numerical Mathematics. **25** (1997) 151-167
2. Burrage, K.: Parallel and Sequential Methods for Ordinary Differential Equations. Oxford Science Publications. (1995)
3. Dongarra, J., Walker, D. W.: Software libraries for linear Algebra Computations on High Performance Computers. SIAM Review. **37** (1995) 151-180
4. Message Passing Interface Forum. MPI: A Message Passing Interface Standard. Univ. of Tennessee, Knoxville, Tennessee, (1995)
5. Hairer, E., Wanner, G.: Solving Ordinary Differential Equations II: Stiff and Differential Algebraic Problems. Springer-Verlag. (1996).
6. Jin, S., Pareschi, L., Slemrod, M.: A Relaxation Scheme for Solving the Boltzmann Equation Based on the Chapman-Enskog Expansion. Acta Mathematicas Applicatae Sinica (English Series). **18** (2002) 37-62
7. Kennedy, C. A., Carpenter, M. H.: Additive Runge-Kutta schemes for convection-diffusion-reaction equations. Applied Numerical Mathematics. **1** (2003) 139-181
8. Pareschi, L., Russo, G.: Implicit-Explicit Runge-Kutta schemes for stiff systems of differential equations. In Recent Trends in Numerical Analysis. **3** (2000) 269-289

9. Mantas, J. M., Ortega, J., Carrillo, J. A.: Component-Based Derivation of a Stiff ODE Solver implemented on a PC Cluster. International Journal of Parallel Programming. **30** (2002) 99-148
10. Mantas, J. M., Ortega, J., Carrillo, J. A.: Integrating Multiple Implementations and Structure Exploitation in the Component-based Design of Parallel ODE Solvers. Recent Advances in Parallel Virtual Machine and Message-Passing Interface. Lecture Notes in Computer Science. **2840** (2003) 438-446
11. Mantas, J. M. , Pareschi, L., Carrillo, J. A., Ortega, J.: Parallel Integration of Hydrodynamical Approximations of the Boltzmann Equation for rarefied gases on a Cluster of Computers. J. Comp. Methods in Science and Engineering. **4** (2004) 33-41
12. Verwer, J.G., Sommeijer, B. : An implicit-explicit Runge-Kutta-Chebyshev scheme for diffusion-reaction equations. SIAM J. of Sci. Comp. **25** (2004) 1824-1835
13. Zhong, X.: Additive Semi-Implicit Runge-Kutta Methods for Computing High-Speed Nonequilibrium Reactive Flows. Journal of Comp. Physics. **128** (1996) 19-31

Comparison of Different Parallel Modified Gram-Schmidt Algorithms

Gudula Rünger and Michael Schwind

Department of Computer Science, Technical University Chemnitz
09107 Chemnitz, Germany
{ruenger,schwi}@informatik.tu-chemnitz.de

Abstract. The modified Gram-Schmidt algorithm (MGS) is used in many fields of computational science as a basic part for problems which relate to Numerical Linear Algebra. In this paper we describe different parallel implementations (blocked and unblocked) of the MGS-algorithm and show how computation and calculation overlap can increase the performance up to 38 percent on the two different Clusters platforms which where used for performance evaluation.

1 Introduction

The modified Gram-Schmidt (MGS) algorithm solves the problem of decomposing a matrix $A \in \mathbf{R}^{m \times n}$ (here we consider the case $n \leq m$) into matrices $Q \in \mathbf{R}^{m \times n}$ and $R \in \mathbf{R}^{n \times n}$, so that $A = QR$. The matrix Q computed by the algorithm is an orthogonal matrix composed of orthogonal vectors $Q = \{q_1, \ldots, q_n\}$, the matrix R is upper triangular.

The algorithms presented in this article were mainly developed for the full-rank QR-problem (m=n), which arises in derivation of Lyaponov -vectors and -exponents in computational physics [1], where the MGS-algorithm is part of a repeated orthogonalization process inside an integration process. But there are many other areas of application for the MGS-algorithm, for example the solution of linear least squares problems [2] or as part for the iterative solution of linear systems.

Besides the MGS-algorithm there exist several other algorithms for QR-decomposition, for example the QR-algorithm with Householder reflectors or the classical Gram-Schmidt algorithm (CGS). The Householder algorithm has a higher accuracy in the orthogonal vectors q_i, so that the norm $\|I - Q^T Q\|$ is in the range of machine-accuracy (I is the identity matrix), but it needs more floating-point-operations $\frac{8}{3}m^3$ then MGS $2m^3$ ($m = n$). The CGS- and the MGS-algorithm are mathematically equivalent but behave differently in the orthogonality of the computed matrix Q. In the MGS-algorithm the norm $\|I - Q^T Q\|$ can be predicted by an upper bound [2] but there exists no such bound for the CGS-algorithm.

The algorithm used in this article is called the row-wise MGS-algorithm because it constructs the R-matrix of the QR-decomposition row by row. There exists another version of the MGS-algorithm called the column-wise version, but it is difficult to derive a efficient parallel implementation for the column-wise algorithm.

The parallel algorithms are realized in the SPMD programming style with MPI as communication library for message passing. The parallel realization of the algorithms uses the block-cyclic-mapping over a two dimensional processor grid of dimension $(P_M \times P_N)$, with P_M processor rows and P_N processor columns. We have chosen this layout because experiments have shown that this leads to the best performance and speedup values for the underlying cluster platforms.

For using todays computers with deep memory hierarchies most efficiently, we have developed two variants of Level-3 implementations of the algorithm based on the sequential algorithm described in [2, 3]. The distinction between Level-2 and Level-3 algorithms is based on the operations which the algorithms use. The Level-2 algorithms use matrix-vector operations and the Level-3 algorithms use in addition matrix-matrix operations [4].

Early work on parallel Gram-Schmidt orthogonalization is described in [5–7]. Actual work on the topic of the parallel modified Gram-Schmidt algorithm often concentrates on the case of having a matrix to orthogonalize with $m >> n$. In this situation often row-wise and block or cyclic column-wise distributions are used. Our investigations have shown that for the case of having square matrices a block-cyclic distribution over a two dimensional processor grid gives the best performance for the Level-3 algorithms. In [8, 9] different parallel block Gram-Schmidt algorithms for a row-wise distribution have been presented, where the vectors have been grouped to a non constant block-size to have similar accuracy as MGS. Our algorithm in this work uses the iterated classical Gram-Schmidt algorithm [10] to increase accuracy. In [11] different partitioning schemes for MGS including row-wise, block and cyclic column-wise partitionings were analyzed and they use ring-communication (pipelined algorithm) which we adapt for our parallel Level-3 algorithm (Algorithm 4) for the double-cyclic distribution.

2   Sequential-Algorithms

2.1   Level-2 MGS

The MGS algorithm orthogonalizes a matrix A through a series of transformations of the column-vectors a_i, $i = 1, \ldots, n$ with previously computed orthogonal vectors q_j $(1 \leq j < i)$ in the following way:

$$q_i^j = q_i^{j-1} - \underbrace{(q_i^{j-1}, q_{j-1})}_{r_{ij}} q_{j-1} = \underbrace{(I - q_{j-1} q_{j-1}^T)}_{P_{j-1}} q_i^{j-1}, \qquad (1)$$

$$\text{with} \quad q_i = q_i^i / \underbrace{\|q_i^i\|}_{r_{ii}} \quad \text{and} \quad q_i^0 = a_i$$

The superscript j indicates that a vector has been transformed with the vectors $q_1, \ldots, q_j$. When a vector $q_i^0 = a_i$ is transformed by $i - 1$ orthogonal vectors q_j,

$1 \leq j < i$ it is normalized and stored as the new vector q_i. The norm $\|q_i^i\|$ is the element r_{ii} and the inner-product $(\cdot, \cdot)$ in Formula (1) is the element r_{ij} of R. The orthogonalization can be seen as a series of multiplication with transformation matrices P_j (see Formula (1)). Algorithm 1 shows the pseudo-code of the row-wise-algorithm which uses a matlab like notation to represent sub-matrices. In a first step, the matrix A is copied into Q so that the algorithm applies all the transformations to Q. The vector q_i is normalized in the i-th step of the main loop and then used to transform the vectors $q_{i+1}, \ldots, q_n$ with matrix-vector operations. For the transformation a vector r of size $n - i + 1$ is calculated. This vector is then used for a rank-1-update of the column-vectors $q_{i+1}, \ldots, q_n$. The symbol "$- =$" in Algorithm 1 describes that the matrix on the left is replaced by itself minus the matrix on the right. The norm $\|q_i\|$ and the vector r build the elements i to n of the i-th row of matrix R and are stored in line 4 and 10. Algorithm 1 requires $2mn^2$ floating-point-operations.

Algorithm 1: MGS	**Algorithm 2: BMGS**
Procedure MGS(A,Q,R)	**Procedure** BMGS(A,Q,R)
begin	**begin**
1 $Q = A$	1 $Q = A$
2 **for** $i = 1$ **to** n **do**	2 **for** $i = 1$ **to** n **Step** b **do**
3 $q = Q[:, i]$	3 $\bar{Q} = Q[:, i : i + b - 1]$
4 $R[i, i] = \|q\|$	4 ICGS $(\bar{Q}, R[i : i + b - 1, i : i + b - 1])$
5 $q = q/\|q\|$	5 $Q[:, i : i + b - 1] = \bar{Q}$
6 $Q[:, i] = q$	6 **if** $i < n - b$ **then**
7 **if** $i < n$ **then**	7 $\bar{R} = \bar{Q}^T \cdot Q[:, i + b : n]$
8 $r = Q[:, i + 1 : n]^T \cdot q$	8 $R[i : i + b - 1, i + b : n] = \bar{R}$
9 $Q[:, i + 1 : n] - = q \cdot r^T$	9 $Q[:, i + b : n] - = \bar{Q} \cdot \bar{R}$
10 $R[i, i + 1 : n] = r^T$	
end	**end**

2.2   Level-3 MGS

A Level-3 formulation of the MGS algorithm can be derived by replacing many transformations of Q through P_j (see Formula (1)) with transformations by matrices L_k which are defined as follows:

$$L_k = I - \bar{Q}_k \bar{Q}_k^T \,, \qquad 1 \leq k < (n/b) \qquad (2)$$

The block-matrix $\bar{Q}_k$ consists of b orthogonal vectors $\bar{Q}_k = (q_{i_1}, \ldots, q_{i_2})$, $i_1 = (k - 1) \cdot b + 1$, $i_2 = kb$, which can be calculated with a Level-2 algorithm. Without loss of generality n can be assumed to be a multiple of b.

To get numerical properties comparable to the Level-2 MGS-algorithm it is important to use a Level-2-algorithm which minimizes the norm $\|\bar{Q}_k^T \bar{Q}_k - I\|$. [2] suggests to use the iterated classical Gram-Schmidt algorithm (ICGS) for the Level-2 transformation, which orthogonalizes the columns of $\bar{Q}_k$ a second

---

Algorithm 3: applyTransform

Procedure applyTransform(Q^l,R^l,$\bar{Q}^l$,s,e)

1 if MyRank owns parts of columns of Q between s and e **then**
2   -generate a local block of inner products with a local matrix-multiplication with the parts of $\bar{Q}^l$ transposed and parts of Q^l in range from s to e and store it in $\bar{R}^l$
3   -sum $\bar{R}^l$ in process column and broadcast the result in process column
4   -local Rank-k-update of Q^l with $\bar{Q}^l$ and $\bar{R}^l$
5   -local store parts from $\bar{R}^l$ into R^l

Algorithm 4: PBMGS

Procedure PBMGS(A^l,Q^l,R^l)
  begin

1   - copy local parts of A^l into Q^l
2   for $i = 1$ *to* $n - b$ *Step* b **do**
3        $k = (i - 1)/b \bmod P_N$
4   if *MyRank is in process column* k **then**
5   -Level-2 factorization of Q in range of vectors i to $i+b-1$ store the result in $\bar{Q}^l$ and the generated elements of R
6   if $i < n - b$ **then**
7   -broadcast $\bar{Q}^l$ within process row; root of broadcast are processes of process column k
8   -applyTransform(Q^l,R^l,$\bar{Q}^l$,$i+b$,n)

  end

Algorithm 5: PBMGS2

Procedure PBMGS2(A^l,Q^l,R^l)
  begin

1   - copy local parts of A^l into Q^l
2   for $i = 1$ *to* $n - b$ *Step* b **do**
      $k = (i - 1)/b \bmod P_N$
3   if *MyRank is in process column* k **then**
4   if $i > b$ **then**
5   applyTransform(Q^l,R^l,$\bar{Q}^l_{old}$,i,$i+b-1$)
6   -Level-2 factorisation of Q in range of vectors i to $i+b-1$ store the result in $\bar{Q}^l_{new}$
7   -start Broadcast of $\bar{Q}^l_{new}$ asynchronously in process row; root of broadcast are processes of process column k
8   if $i > b$ *AND* $i < n - b$ **then**
9   -applyTransform(Q^l,R^l,$\bar{Q}^l_{old}$,$i + b$,n)
10   -end Broadcast of $\bar{Q}^l_{new}$
11       $\bar{Q}^l_{old} = \bar{Q}^l_{new}$

  end

---

time with the classical Gram-Schmidt algorithm (CGS) depending on an easy-to-compute re-orthogonalization criterion [10].

A pseudo-code-implementation of a Level-3-version of the MGS algorithm is presented in Algorithm 2. In line 3 the ICGS-algorithm is called to form $\bar{Q}$ from columns i to $i + b - 1$ of Q. While calculating $\bar{Q}$ the upper diagonal elements of an $b \times b$ sub-matrix of R are created and are stored in $R[i : i+b-1, i : i+b-1]$. The block-matrix $\bar{Q}$ is copied back into Q after the orthogonalization in line 5. After the ICGS-phase, the resulting matrix $Q[:, i + b : n]$ is transformed by calculating a block of inner-products through a matrix-matrix-multiplication in line 7 of Algorithm 2. The resulting matrix $\bar{R}$ has dimension $b \times (n - i - b + 1)$ and stores the elements of the sub-matrix $R[i : i+b-1, i+b : n]$ of R in line 8. In line 9 a rank-k-update of the columns $i + b$ to n of Q is made with $\bar{Q}$ and $\bar{R}$ to complete the block-transformation.

The number of floating-point-operations of the Level-3-MGS-algorithm depends on the number of orthogonalization steps in the ICGS-step. If only one orthogonalization is required the number of floating-point-operation is nearly the same as for MGS; if every vector needs a second orthogonalization step the number of floating-point-operations is about $2mn(n + b)$.

Figure 2 D shows the ratio between the number of re-orthogonalizations and the number of columns for different 300×300 random-matrices with different condition numbers. The matrices have been generated with the routine DLATMS of the Lapack [12] testing library. A value of 1 means that every vector has to be orthogonalized twice in the ICGS step, whereas a value of 0 means that every vector has to be orthogonalized only once. There is a strong decency of the ratio from the block-size and condition number.

3   Parallel-Algorithms

We describe only the parallel implementation of two Level-3 algorithms, but show in the next section the performance of the Level-2 algorithm too. The parallel Level-2-algorithm can be derived when setting the block-length of the Level-3 algorithm to one. For the parallel Level-3 algorithm we have chosen to use the column-block length of the underlying block-cyclic-mapping as block length of the Level-3-algorithm. This reduces the number of messages to send in the Level-2-phase and is typical for many other parallel software.

A pseudo-code of the parallel Level-3 algorithm is described in Algorithm 4; it is a "straight forward" parallelization of Algorithm 2. A superscript l denotes the local parts of a distributed matrix or vector. In the following we describe the main steps from Algorithm 4:

1. **ICGS-step:** This step performs the Level-2-Transformations in line 4. Since the block-length of the algorithm is equal to the block length of the distribution, this process is only done in one process column $k = i/b \bmod P_N$, which changes in every step cyclically. For the ICGS-step communication with combined reduction/broadcast operations are required for building vector norms and inner-products of column-vectors through matrix-vector-products.

2. **Broadcast:** The matrix $\bar{Q}$ generated in the previous step is needed to form the product in line 7 of the sequential algorithm. Since $\bar{Q}$ is located only in one column, every process in the process column k which holds $\bar{Q}$ must distribute its local parts of $\bar{Q}$ to all other processes in his process row.

3. **Transformation:** The transformation of the columns of Q in range $i + b$ to n through $\bar{R}$ and $\bar{Q}$ in lines 6 to 9 of the sequential Algorithm 2 is described in Algorithm 3.

The parameters of the algorithm for the transformation of the matrix Q (Algorithm 3) are the local parts of $Q, \bar{R}, \bar{Q}$, and a range (global indices) of column-vectors of Q to which the transformation will be applied. Line 1 of Algorithm 3 checks whether a processor column runs out of data. The following steps of Algorithm 3 are as follows:

1. **Matrix-Multiplication:** The matrix multiplication of line 7 (Alg. 2) can be parallelized by the decomposition of the inner-products which build the matrix-product globally. It consists of two steps:

   (a) **Local Matrix-Multiplication:** Build local inner-products through a matrix-multiplication of the local parts from $\bar{Q}$ (transposed) with local parts of Q.

   (b) **Reduction/Broadcast:** A summation of the local-inner products by a global communication operation in the process columns yields the global inner products. Since the result is needed in the next step a broadcast of the results is done within the process columns.

2. **Local Rank-k-Update:** The operand matrices $\bar{Q}$ and $\bar{R}$ are used to make a global Level-3-update of the matrix Q by local updates.

3. **Building R:** When the parameter for the distribution (grid-dimension, block length) is the same for the matrices Q and R then this step needs no redistribution of the matrix $\bar{R}$ which holds parts of R. Some attention must be taken, when copying local parts from $\bar{R}$ into local parts of R. Since the row-block-length may be different from b, there is the possibility that rows from $\bar{R}$ must be stored on different process rows.

3.1 Communication Patterns

Efficient communication is essential for reaching high efficiency on today's hardware architectures. A property of the parallel algorithms presented above is that communication takes place only in the process rows or in the process columns, which can be potentially independent. In [11] the runtime of the algorithms is analyzed for the strict column-block-cyclic and row-block-wise case. For the column-cyclic distribution they use a ring-broadcast for communication and for the row-wise case a tree-implementation for reduce/broadcast. Since our algorithms use the block-cyclic distribution, we use the combination of both communication-patters. For the broadcast within process rows we use a ring-broadcast and for the reduce/broadcast within the process columns we use a tree-implementation.

3.2   Communication/Calculation Overlap

In the parallel algorithm PBMGS there is a strict order between computation and communication. But it is possible to overlap these two. [13] presents a scheme for block matrix factorizations and presents results for QR-/LU- and Cholesky-factorizations. This scheme was specified for pure column-block-cyclic distributions but can be extended to block-cyclic-distributions too. The idea in [13], when transferred to Level-3-MGS, is to not transform the column-vectors of Q with $\bar{Q}$ on all process columns fully in one step. The process column which is responsible for the next Level-2-transform updates only a small stripe with length b with $\bar{Q}$ and then does the Level-2-transform on that stripe. The result of this transform $\bar{Q}_{new}$ is send via a broadcast asynchronously, while the other process columns do the update with the old $\bar{Q}$. For Algorithm 5 we adopt the idea of [13] to the Level-3 modified Gram-Schmidt orthogonalization but implement it over a two dimensional processor grid using block-cyclic distribution. Algorithm 5 uses block-matrices $\bar{Q}_{old}$ and $\bar{Q}_{new}$. The matrix $\bar{Q}_{old}$ is used for the Level-3-transformations in the i-th step of the main-loop, while $\bar{Q}_{new}$ is calculated and distributed at that step. The asynchronous broadcast is implemented through a sequential broadcast using standard MPI_Isend-/MPI_Irecv-routines for asynchronous communication.

4   Performance Evaluation

The performance evaluation has been done on two different machines, a Beowulf-Cluster (CLiC) with 528 Pentium III processors running at 800 MHz connected with Fast Ethernet and a SMP-Cluster of 16 dual XEON nodes running at 2GHz connected with SCI in a 4×4 torus. The CLiC-Cluster uses LAM Version 6.5.2 and the XEON-Cluster uses ScaMPI for message passing. Both clusters use libgoto [14] for local operations.

The performance in GFlop/s for 16 processors for different sizes of the input-matrix is shown in Figure 1. For this measurements only one processor per node is used on the XEON-Cluster. A label with It=1 means that the curve was measured with 1 orthogonalization in the ICGS-step per vector and It=2 means there have been two orthogonalizations per vector. The runtime for a typical run of the algorithms lies between these 2 curves, depending on the input-matrix.

In all figures it can be seen that the flop-rate of the Level-3 algorithm increases with higher matrix-sizes because the ratio of communication-time to computation-time shifts towards computation. The maximum flop rate measured with 16 processors on the XEON-Cluster is 38.4 GFlop/s and on the CLiC 4.73 GFlop/s for the PBMGS2 algorithm (see Figure 1 A2,B2). This is about 60% on the XEON- and about 37% on the CLiC-Cluster of the total peak performance, which is 64 GFlop/s on the XEON and 12.8 GFlop/s on the CLiC for 16 processors. The performance of the algorithms will be higher if we measure the performance for higher matrix-dimensions, but the highest matrix-dimensions (6000×6000) in Figure 1 is the maximum which is needed for the application of the algorithms described in [1].

For both Level-3 algorithms the optimal processor grid configuration (diagrams A2,B2) is the square configuration for 16 processors. As it can be seen the row-wise or the column-block-cyclic distribution are not optimal for the Level-3 algorithms. In the square configuration the algorithm with the overlap of communication and calculation (Alg. 5) reaches an up to 38% better performance on the XEON-Cluster ($m = n = 1500$, $It = 2$) and an up to 30% better performance on the CLiC-Cluster ($m = n = 1500, It = 2$) than the Algorithm 4, in Figure 1 A2,B2. It can be observed that algorithm PBMGS2 can increase the percentage of peak floating-point performance up to 10% on the XEON and up to 4.7% on the CLiC-Cluster compared to algorithm PBMGS. While the performance of PBMGS2 with the optimum grid-configuration is better than algorithm PBMGS (Alg. 4), there is a performance loss in the implementation of algorithm PBMGS2 when using the column-block-cyclic distribution, since the sequential broadcast in the process row which should be overlapped with the calculation in the algorithm PBMGS2 takes too long in this configuration to have an advantage in overlapping communication with the calculation.

The curves with two iterations per vector (It=2) in the ICGS-step show a slight performance decrease for both Level-3 implementations (PBMGS, PBMGS2). This is expected since there are more floating-point-operations and more communication. For the row-wise distribution the decreased performance can be explained by the extra communication introduced for the additional inner-products. For the column-block-cyclic distribution (grid: 1×16) idle-waiting occurs since all process columns wait for the result of the ICGS-step which is calculated in a single process column. For the square process grid both effects, the higher communication time and the higher idle-waiting time, are the reason for the difference in performance for the curves with one orthogonalization ($It = 1$) and with two orthogonalizations per vector ($It = 2$) in the ICGS-step on both machines.

Figure 2 C1 shows the performance of the algorithms for different processor numbers of the clusters in percent of the peak floating-point performance normalized to the processor number. The performance has been determined for different block-sizes and grid-configurations and the highest performance is shown. It is shown that the CLiC-Cluster reaches a smaller utilization of the processors as the XEON-Custer for both Level-3 algorithms when more than one processor is used. The reason for this is the worse ratio of performance of local operations to network bandwidth on the CLiC.

Figure 2 C2 shows the parallel speedup for the measurements of Figure 2 C1. The highest speedup for the algorithms PBMGS and PBMGS2 on the XEON-Cluster is 19.8 and 22.6 which is a parallel efficiency of about 61.9% and 70.6%, respectively. On the CLiC-Cluster with 64 processors the highest speedup is 19.6 for PBMGS and 21 for PBMGS2. The parallel efficiency on the CLiC is much lower and is 30.7% (PBMGS) and 32.8% (PBMGS2) for 64 processor. On the CLiC-Cluster with a processor number higher than 64 processor the performance of algorithms PBMGS and PBMGS2 is equal. Local calculations seem to be too short to overlap with the dominating communication times of the sequential column-broadcast.

The performance of the Level-2 algorithm is not higher than 10% of the peak floating-point performance on both Clusters (see Figure 1 and 2 C1). The reason for this is the poor cache utilization of the Level-2 operations which results in a stall of the processors because of memory operations. By increasing the number of processors the memory bandwidth increases and the Level-2 algorithm has a high speedup on the CLiC as shown in Figure 2 C2. On XEON-Cluster the PMGS algorithm has a speedup drop for 24 and 32 processors because in this range two processors per node are used. Since matrix-vector operations on the XEON cannot utilize the cache efficiently and the two processor per node share the memory bandwidth the speedup drop appears.

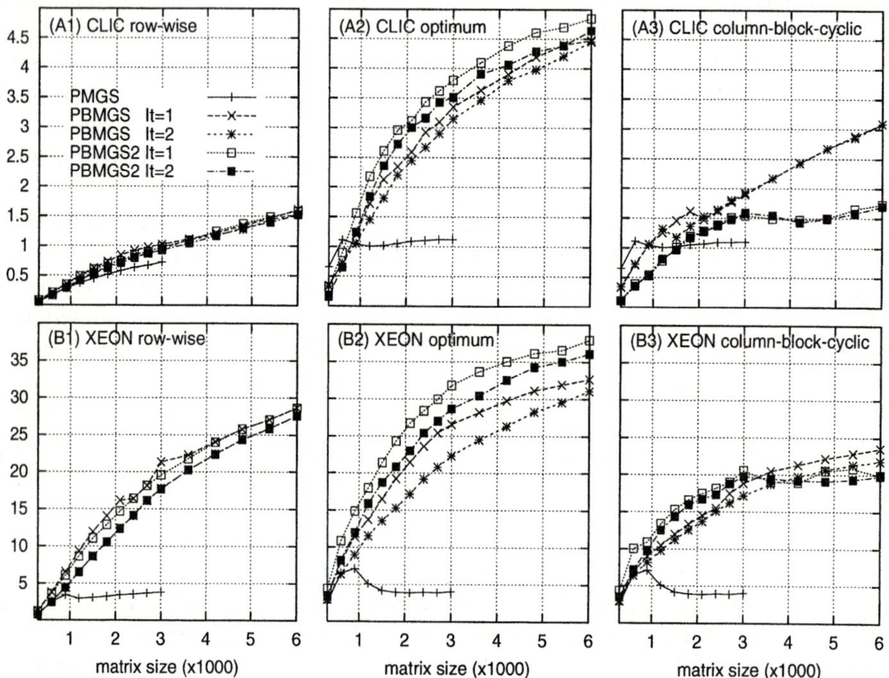

Fig. 1. Performance in GFlop/s for 16 processors for different matrix-sizes on the CLiC (top) and on the XEON (bottom) for the row-wise- (grid: 16×1), optimal- and column-block-cyclic-distribution (grid: 1×16) (from left to right). The optimal grid is the 4×4 grid for PBMGS and PBMGS2 and the 1×16 grid for PMGS

5   Conclusion

We have developed a Level-3 modified Gram-Schmidt algorithm (PBMGS2) which has the advantage of overlapping communication and calculation. The performance of PBMGS2 has been compared with a Level-2 (PMGS) and a Level-3 implementation (PBMGS) which where derived from a "straight forward" parallelization. A comparison of the two Level-3 algorithms on the two different

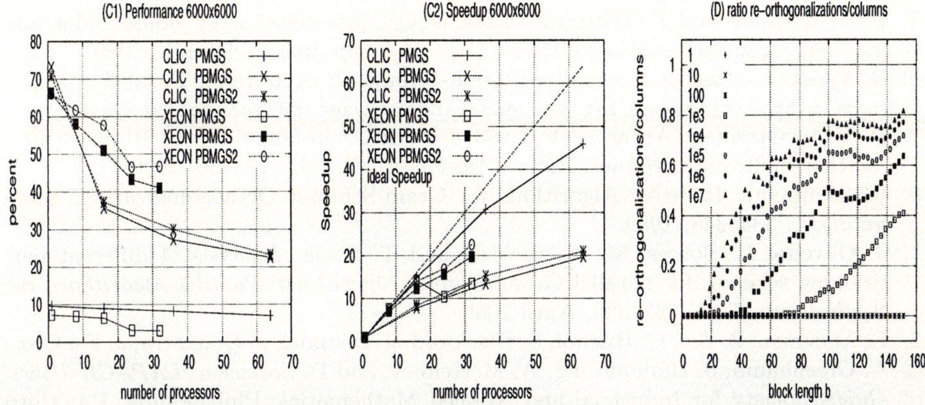

Fig. 2. Performance in percent of the total peak floating-point for the specific processor number (left) and Speedup (middle) for different number of processors for an $m = n = 6000$ matrix. right: Ratio between orthogonalizations in the ICGS-step and number of columns of the input-matrix for the Level-3-algorithms for different block lengths and different condition numbers of a m=n=300 random input-matrix

Clusters shows that the algorithm with the overlapping of communication and calculation can increase the performance up to 38% compared to the parallel Level-3 implementation without communication calculation overlap. We also show that the often described row-wise- and column-block-cyclic distributions for parallel modified Gram-Schmidt are not optimal for the Level-3 algorithms (PBMGS, PBMGS2) for the full rank QR-problem on Cluster-platforms and we have realized a more efficient two dimensional block-cyclic distribution for the Level-3 algorithm.

References

1. G. Radons, G. Rünger, M. Schwind, and H. Yang. Parallel Algorithms for the Determination of Lyapunov Characteristics of Large Nonlinear Dynamical Systems. In *Extended Abstracts: PARA'04 Workshop on State-of-the-Art in Scientific Computing*, Copenhagen, Denmark, 2004. CDROM.
2. Å. Björck. Numerics of Gram-Schmidt Orthogonalization. *Linear Algebra Appl.*, 197–198:297–316, 1994.
3. W. Jalby and B. Philippe. Stability Analysis and Improvment of the Block Gram-schmidt Algorithm. *SIAM Journal on Scientific and Statistical Computing*, 12:1058–1073, 1991.
4. J. Dongarra, J. Du Croz, I. Duff, and S. Hammarling. A set of Level 3 Basic Linear Algebra Subprograms. *ACM Trans. Math. Soft.*, 16(1):1–17, March 1990.
5. E. L. Zapata, J. A. Lamas, F. F. Rivera, and O. G. Plata. Modified Gram-Schmidt QR factorization on Hypercube SIMD Computers. *Journal of Parallel and Distributed Computing*, 12:60–69, 1991.
6. H. De Meyer, C. Niyokindi, and G. Vanden Berghe. The implementation of parallel gram-schmidt orthogonalisation algorithms on a ring of transputers. *Computers and Mathematics with Applications*, 25:65–72, 1993.

7. D. P. O'Leary and P. Whitman. Parallel QR factorization by householder and modified Gram-Schmidt algorithms. *Parallel Computing*, 16:99–112, 1990.
8. D. Vanderstraeten. A Generalized Gram-Schmidt Procedure for Parallel Applications. http://citeseer.ist.psu.edu/vanderstraeten97generalized.html.
9. D. Vanderstraeten. An accurate parallel block Gram-Schmidt algorithm without reorthogonalization. *Numer. Linear Algebra Appl.*, 7(4):219–236, 2000.
10. W. Hoffmann. Iterative Algorithms for Gram-Schmidt Orthogonalization. *Computing*, 41:334–348, 1989.
11. S. Oliveira, L. Borges, M. Holzrichter, and T. Soma. Analysis of different partitioning schemes for parallel Gram-Schmidt algorithms. *Parallel Algorithms and Applications*, 14(4):293–320, April 2000.
12. E. Anderson, Z. Bai, C. Bischof, S. Blackford, J. Demmel, J. Dongarra, J. Du Croz, A. Greenbaum, S. Hammarling, A. McKenney, and D. Sorensen. *LAPACK Users' Guide*. Society for Industrial and Applied Mathematics, Philadelphia, PA, third edition, 1999.
13. K.Dackland, E. Elmroth, and B. Kågström. A Ring-Oriented Approach for Block Matrix Factorizations on Shared and Distributed Memory Architectures. In *PPSC*, pages 330–338, 1993.
14. K. Goto and R. van de Geijn. On reducing tlb misses in matrix multiplication. Technical Report TR-2002-55, 2002.

Automatic Tuning of PDGEMM
Towards Optimal Performance

Sascha Hunold and Thomas Rauber

Department of Mathematics and Physics
University of Bayreuth, Germany
{hunold,rauber}@uni-bayreuth.de

Abstract. Sophisticated parallel matrix multiplication algorithms like PDGEMM exhibit a complex structure and can be controlled by a large set of parameters including blocking factors and block sizes used for the serial execution on one of the participating processors. But it requires a deep understanding of both the parallel algorithm and the execution platform to select the parameters such that a minimum execution time results. In this article, we describe a simple mechanism that automatically selects a suitable set of parameters for PDGEMM which leads to a minimum execution time in most cases.

1   Introduction

There is usually a complex dependency between the computations and the memory accesses performed by a computation-intensive program, the required data exchanges between neighboring processors, and the computation and communication characteristics of the execution platform. Parallel numerical libraries like ScaLAPACK (Scalable LAPACK) [1, 2] or PETSc [3] try to cope with these dependencies by providing a set of parameters which allow the user to adjust the execution behavior of the library routines to the characteristics of the execution platform such that the parallel execution time is reduced as far as possible. By selecting appropriate parameter values, the library routines can run very efficiently on most parallel execution platforms. But it is often quite difficult for the user to select suitable parameter values, since this requires a deep understanding of the algorithmic behavior of the library routines. In many situations, the user wants to use the library as black-box and does not have time to learn more about the internals of the algorithm. Moreover, even knowing the algorithmic details of a library routine does not necessarily yield a suitable set of parameters to use. The complex dependency between the algorithm and the characteristics of the execution platform is still present and it is usually necessary to perform runtime experiments with different parameter settings before a suitable set of parameters can be identified. It even might be the case that for different numbers of processors different parameter values lead to the best performance.

In this situation, it would be useful to have a tool that automatically selects a suitable set of parameters, thus relieving the user from this time-consuming

J.C. Cunha and P.D. Medeiros (Eds.): Euro-Par 2005, LNCS 3648, pp. 837–846, 2005.
© Springer-Verlag Berlin Heidelberg 2005

work. In this article, we address this issue. In particular, we consider the matrix-matrix multiplication routine PDGEMM of ScaLAPACK. PDGEMM is part of PBLAS which is the parallel implementation of BLAS (Basic Linear Algebra Subprograms) for distributed memory machines. It can be used as a building block in a parallel version of Strassen's algorithm [4] as well as in many advanced algorithms from scientific computing. We investigate which parameters have a major impact on the overall performance and should therefore be considered in more detail. In particular, we present an approach to tune the PDGEMM routine by adjusting the parameters gradually to the given execution platform. The Automatically Tuned Linear Algebra Software (ATLAS) [5] provides an approach for the sequential case, but there is no solution for a parallel execution yet. An experimental evaluation shows that the proposed method selects a parameter setting that leads to significant performance gains compared to the default setting for different test platforms.

The rest of the paper is organized as follows. Section 2 gives an overview of the algorithmic details of the PDGEMM routine and shows examples of how to use PDGEMM routines in a C environment. Section 3 analyzes the impact of different parameters for PDGEMM. In Section 4 we describe an heuristic method for an automatic selection of suitable PDGEMM parameters to optimize PDGEMM. Section 5 evaluates the experimental results and Section 6 concludes.

2   Algorithmic Details

To improve the performance of the PDGEMM routine efficiently we must first consider some algorithmic details. The PDGEMM routine which is part of ScaLAPACK is derived from the DIMMA algorithm (Distribution-Independent Matrix Multiplication). DIMMA is an enhanced version of SUMMA (Scalable Universal Matrix Multiplication Algorithm), see [6] for a detailed description of SUMMA and [7] for an introduction of DIMMA. We also refer to [8] for an overview of basic parallel matrix-matrix multiplication algorithms such as the algorithms of Cannon or Fox.

In the following, we summarize the basic ideas which make PDGEMM (using DIMMA) a well-performing algorithmic option in many cases. An example of the starting configuration of matrix A for DIMMA and SUMMA is shown in Figure 1. In SUMMA, the processors P_0 and P_3 broadcast the first column of A along their row, i.e., P_0 sends its first column to processors P_1 and P_2. At the same time, the first row of matrix B which is distributed likewise is broadcasted along the processor columns. When the broadcasts are performed on a logical ring, SUMMA takes advantage of a pipelined communication scheme. The authors of DIMMA state that SUMMA contains extra waiting times between two communication procedures [7]. Hence, DIMMA improves the communication scheme and eliminates the extra waiting time by proceeding to send blocks of columns (rows) from the current column (row) of the processor grid. That means, in SUMMA the processors P_1 and P_4 broadcast column 1 directly after receiving column 0 from P_0 and P_3, respectively. In case of DIMMA, P_0 and P_3 continue with broadcasting another column whose distance is LCM blocks where LCM

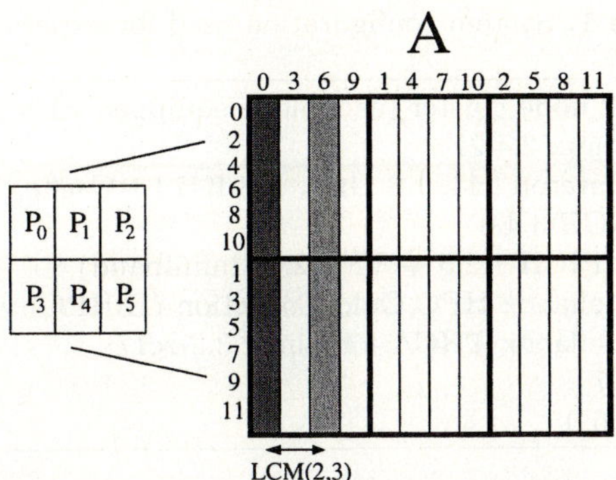

Fig. 1. DIMMA snapshot for a 2×3 processor grid. DIMMA uses a block cyclic distribution of matrix A onto the processors.

is the least common multiple of the grid dimensions p and q. For a further performance improvement, SUMMA as well as DIMMA use blocks of columns (rows) rather than single columns (rows).

Since it is not very straightforward to use ScaLAPACK routines from C we would like to demonstrate the calling conventions from within C. We indicate that this method is compiler specific. Because Fortran 77 uses the call-by-reference paradigm we need to pass the address of each parameter to Fortran functions. The basic problem of calling Fortran routines from C is the conversion of strings. Figure 2 shows a sample call of PDGEMM from C for our compiler collection, see Table 1. The function `c2f_char` copies a character into a character buffer. Since there is no header file for C, an underscore is required for the linker to resolve the PDGEMM function. But primarily we want to emphasize the use of hidden parameters. The last two arguments of the call to PDGEMM are hidden parameters because they need to be passed in order to make the Fortran function work but the arguments are not part of PDGEMM's Fortran interface. The hidden parameters 1L denote the length of the passed strings, i.e. the length of `cha` and `chb`. If we had a function with three string parameters we would need to pass three hidden parameters as well.

```
c2f_char ( cha , 'N' );
c2f_char ( chb , 'N' );
pdgemm_( cha , chb , &m, &k, &n, &scalar , a[0] ,
        &one , &one , &desca , b[0] , &one , &one , &descb ,
        &scalar , c[0] , &one , &one , &descc , 1L, 1L );
```

Fig. 2. Call to PDGEMM from C.

3   Parameter Analysis and Optimization Strategies

In this section we examine the performance dependencies of PDGEMM from different parameters. The experiments were performed on a cluster with 32 Dual Opteron nodes. An overview of the configuration is given in Table 1.

Table 1. System configuration used for experiments.

System	32 node cluster (each node equipped with 2 AMD Opterons) Linux 2.4.21
Processor	Opteron 244, 1.8 GHz, 128 KB L1-Cache, 1024 KB L2-Cache
C/F77 Compiler	GCC 3.4.0
MPI version	MPICH 1.2.5 + VMI 2.0 (Infiniband)
Infiniband driver	Mellanox HPC Gold Collection (IBHPC) v0.5.0 for Linux Mellanox THCA for Linux 3.2-rc17
ScaLAPACK	1.7
ATLAS	3.6.0

As the sequential computation of PDGEMM is based on BLAS [9], the right choice of the BLAS implementation is crucial for the overall parallel performance of PDGEMM. There is a tremendous performance difference between hardware-optimized BLAS routines and the standard routines. When the user has no access to a vendor-provided BLAS library like ESSL, we recommend using the ATLAS library [5]. All local computations used in the evaluation of the parallel algorithms are performed by ATLAS. Moreover, each experiment reported herein was repeated at least three times.

Since all variables in the parameter list of PDGEMM and also the logical block size defined inside ScaLAPACK may influence the runtime, the search space for optimization is extremely large. To obtain a satisfactory parallel performance, it is necessary to use a local computation kernel which almost achieves the peak performance of the processor.

The parameters with the biggest influence on the performance of the algorithm are the dimensions of the input matrices, the number of processors and their arrangement within the processor grid.

There are also other parameters that strongly influence the MFLOPS rate of the algorithm but are not obvious. These are

1. the three blocking factors mb, nb and kb of the block-cyclic distribution of matrices A, B and C which are of size $m \times k$, $k \times n$ and $m \times n$,
2. and the logical block size lb.

Blocking factor The blocking factor is used to distribute the rows and columns of the matrices onto the processor grid. A blocking factor of b means that blocks of matrix $\mathcal{M}$ of size $b \times b$ are distributed block-cyclicly. It is also possible to have distinct blocking factors for each matrix dimension.

Logical block size The logical block size denotes the size of the sub-matrix of C which is computed by each processor per parallel step of PDGEMM. Let the logical block size be lb. In each parallel step, a processor P_i gathers lb rows of A and lb columns of B and computes a part of the result matrix C of size $lb \times lb$.

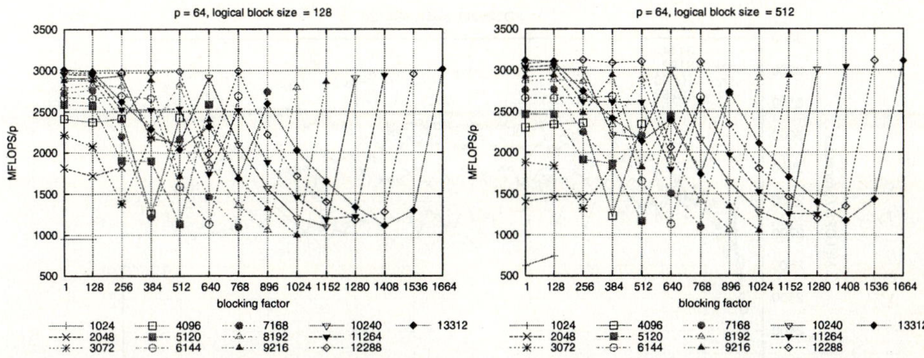

Fig. 3. Performance of PDGEMM, number of processors $p = 64$ for matrix sizes from 1024 to 13312 for different blocking factors.

3.1 Impact of the Blocking Factor

In order to analyze the performance dependency on the blocking factor, we have performed several tests of PDGEMM with varying block sizes. Due to the huge number of degrees of freedom we limited the test cases to square input matrices and square matrix blocks where $mb = nb = kb$. We performed numerous tests with blocking factors ranging from 1 to a logical maximum which is defined by the matrix dimensions and the processor grid. Furthermore, we also examined how the logical block size is reflected in the runtime of PDGEMM for each range. Figure 3 shows the performance of PDGEMM for 64 processors and for different blocking factors using logical block sizes of 128 and 512. The experiment was repeated for 8, 16 and 32 processors as well, using logical block sizes of 32, 64, and 256. We observed that the coarse characteristics of the resulting MFLOPS rate does not depend on the logical block size, but the MFLOPS rate is only slightly increased (decreased) for a smaller (larger) value of the logical block size. Hence, big differences in the MFLOPS like at 1536 and 1664 in Figure 3 can not be compensated by adjusting only the logical block size. But let us have a closer look at Figure 3. The MFLOPS rates for matrix dimension 13312 show peaks for blocking factors 1, 128 and 1664. This behavior can be explained as follows: The 64 processors are arranged in a grid of 8×8 elements. Thus, in each dimension the matrix is distributed evenly among the processors if $\frac{13312}{8} = 1664$ is a multiple of the blocking factor. And indeed, 1664 is a multiple of 1, 128 and 1664. In these cases, the data is uniformly distributed over all processors in the grid which leads to a balanced workload on homogeneous systems. So, when in doubt which blocking factor to use, it is a good choice to use the possible maximum $\mathcal{M}$ which is $\mathcal{M} = \dfrac{\text{matrix dimension}}{\#\text{ processors in row (colum)}}$.

It is not surprising that high performance is achieved when the matrix dimensions are multiples of the blocking factor. But choosing the blocking factor appropriately does not necessarily lead to best performance. The central parameter to optimize is therefore the logical block size.

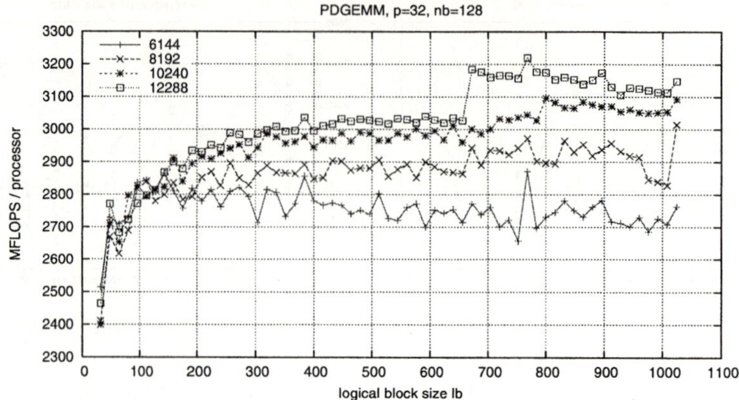

Fig. 4. MFLOPS achieved by PDGEMM with different logical block sizes, number of processors $p = 32$, blocking factor $nb = 128$ (Infiniband).

3.2 Impact of the Logical Block Size

The logical block size directly influences the overall performance of PDGEMM. A small value of the logical block size (lb) will not only cause more communication but worse, the local matrix updates (multiplications) will not reach the processor's peak performance. On the other hand, choosing a very large value may hamper the pipelined communication scheme and so the overlapping of communication and computation as well.

Finding the best value for the logical block size is highly machine-dependent and the impact on the resulting execution time can only be determined experimentally. We ran a series of tests with PDGEMM on the cluster system for varying values of lb. Since the value lb is hard-coded in file `pilaenv.f` of the ScaLAPACK distribution, the value lb in `pilaenv.f` needs to be changed and ScaLAPACK must be recompiled for each test. The results of this experiment is shown in Figure 4 and Figure 5. Figure 4 contains the MFLOPS rate achieved by PDGEMM using the Infiniband network. Figure 5 shows the results for the Gigabit Ethernet. We can observe that the plots in both figures have similar characteristics, i.e., the bandwidth and latency of the interconnection network plays a minor role. As example, we consider the steep increase of the MFLOPS rate at block size 672 for a matrix dimension of 12288. The 32 processors are arranged in a 4×8 grid and each processor stores $\frac{12288}{4} = 3072$ rows and $\frac{12288}{8} = 1536$ columns of the matrices, if a suitable blocking factor nb has been chosen. Let us examine the case where the biggest performance enhancement has been observed. For $lb = 672$, PDGEMM performs a series of local matrix-multiplications using DGEMM where the matrix A is of size 3072×672 and Matrix B has 672×1536 elements. One local matrix update with these parameters achieves about 3508 MFLOPS on a single Opteron processor. In comparison, with a logical block size of 656 the routine DGEMM achieves 3380 MFLOPS only, which is about 4%

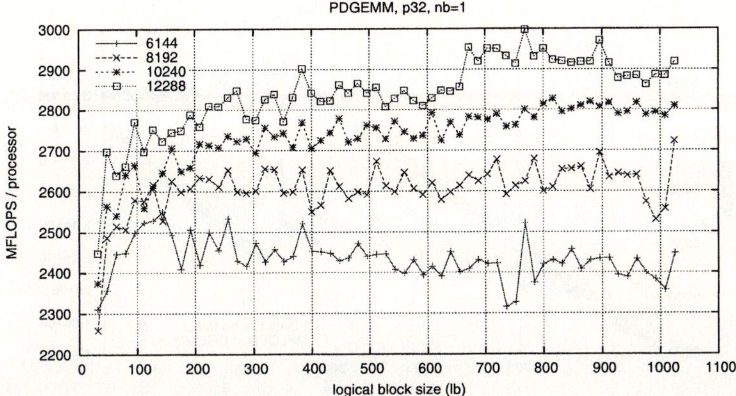

Fig. 5. MFLOPS achieved by PDGEMM with different logical block sizes, number of processors $p = 32$, blocking factor $nb = 1$ (Gigabit Ethernet).

slower. We measured the cache misses generated by DGEMM using PAPI [10]. Figure 6 clearly shows that the weak performance is the result of producing more L2 cache misses which is caused by the fact that DGEMM (ATLAS) generates a different call tree for $lb = 656$ and $lb = 672$.

Instead of ignoring possible drops in the MFLOPS rate, we present an approach to avoid choosing an unfavorable logical block size which is discussed in Section 4.

4   Automatic Parameter Tuning

In this section, we present an approach for selection a favorable logical block size. The basic problem of optimizing the logical block size is the huge search space. The logical block size depends on the network parameters, the dimension of the matrices, the matrix ordering, the processor grid, the number of processors and the BLAS implementation. Additional informations about the hardware, e.g. cache size, will surely decrease the search space but it remains too large to evaluate all possible combinations.

We present an heuristic method based on an evaluation on a single processor of the parallel execution platform. The approach is simple, but it turns out to be fast and provides a suitable logical block size for the parallel case. The algorithm keeps two matrix dimensions fixed and varies the third one. The size of the matrix dimensions which are kept fixed is computed by the information provided by the user, e.g. the number of processors, the typical size of matrices and the preferred processor grid. The third and varying dimension represents the logical block size. After the series of tests on a single processor has been completed, the optimization algorithm selects the smallest logical block size which is only less than a fixed percentage f slower than the best logical block size (which is in general the largest, but, as we said before, for the parallel execution the largest

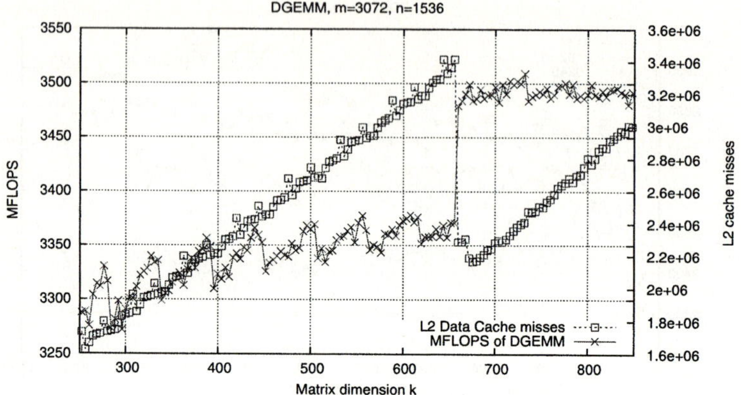

Fig. 6. MFLOPS vs. L2 cache misses by DGEMM (ATLAS).

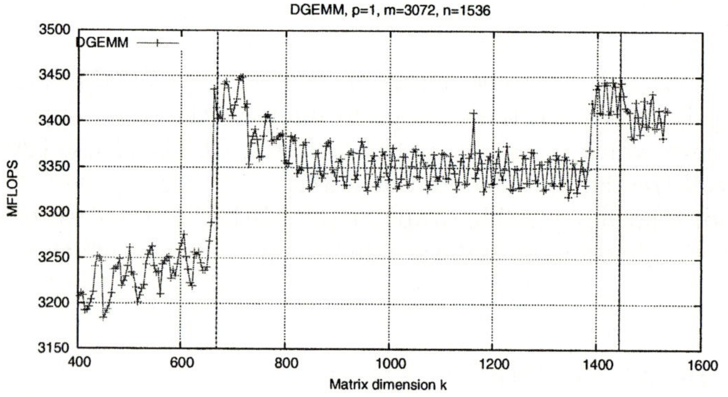

Fig. 7. MFLOPS by DGEMM (ATLAS). Vertical lines at 668 and 1444.

logical block size is not always the fastest). For a specific parallel platform, the algorithm only needs to run once on one processor in a pre-computation phase to determine a suitable value of the logical block size. Then, this block size can be used for all parallel executions. In our experiments, we obtained the best performance for $f = 2\%$.

5   Experimental Evaluation

We now consider the experimental evaluation of the proposed method on the cluster system from Table 1. The cluster consists of 32 processors and we want to use a rectangular grid, e.g. 4 rows and 8 columns. The typical size of matrices is set to 12288 in each dimension. Hence, PDGEMM will deal with submatrices of size $3072 \times lb$ and $lb \times 1536$. The tuning algorithm will test all possible logical block sizes in the range $2 \dots 1536$ (odd numbers are not considered). The test results are shown in Figure 7. The function line has a maximum at 1444. The

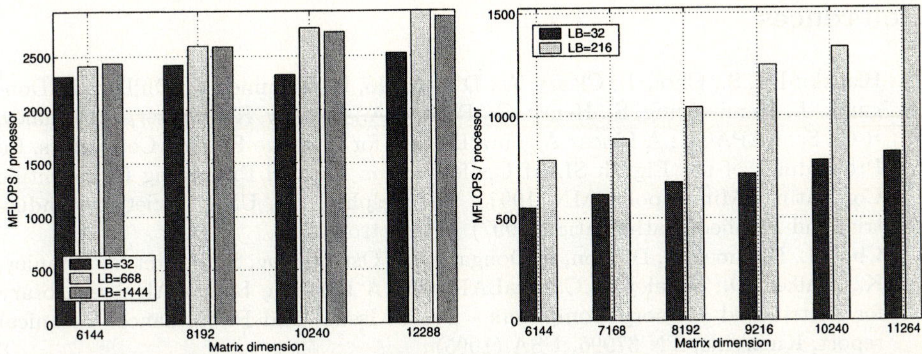

Fig. 8. MFLOPS achieved by PDGEMM on 32 processors. Left: results for Opteron cluster (Infiniband) for logical block size $lb = 32, 668, 1444$; right: performance comparison of PDGEMM for a logical block size of 32 and 216 on Xeon cluster (SCI).

smallest matrix dimension which is less than 2% slower than 1444 is 668. We marked both values in the plot. The value of 668 is the optimized logical block size. Surprisingly, the value that has been found is very close to the crucial value of 656. For a final evaluation of the logical block size, Figure 8 (left) compares the performance of PDGEMM achieved with the default value of the logical block size ($lb = 32$) and with the automatically selected value of 668. The plot also includes the MFLOPS rates for a logical block size of 1444 for comparison reasons. It can be observed that the block size which achieves best results on a single processor is not necessarily gaining the maximum performance in parallel. For matrix dimensions 10240 and 12288 in Figure 8 on the left, the performance gain for the automatically selected logical block size is 18% and 15%, respectively. Additional tests have been performed on a cluster consisting of 16 nodes (Dual Xeon 2 GHz). The nodes are running Linux and are connected via an SCI network. The resulting MFLOPS achieved by PDGEMM for this cluster are shown on the right-hand side of Figure 8. The tuning algorithm selects a logical block size of 216. This block size clearly outperforms the default settings of PDGEMM. In this experiment, the automatically selected value of $lb = 216$ reduces the runtime of PDGEMM by up to 47% for a matrix dimension of 11264.

6   Conclusions

The performance of ScaLAPACK routines strongly depends on the logical block size. In this article we have shown how to use the function PDGEMM and how to improve its performance by selecting a well-suited blocking factor and a logical block size automatically. The experimental results confirm that the heuristic method of selecting a logical block size leads to a significant performance enhancement of PDGEMM.

References

1. Blackford, L.S., Choi, J., Cleary, A., D'Azevedo, E., Demmel, J., Dhillon, I., Dongarra, J., Hammarling, S., Henry, G., Petitet, A., Stanley, K., Walker, D., Whaley, R.C.: ScaLAPACK: A Linear Algebra Library for Message-Passing Computers. In: Proceedings of the Eighth SIAM Conference on Parallel Processing for Scientific Computing (Minneapolis, MN, 1997), Philadelphia, PA, USA, Society for Industrial and Applied Mathematics (1997) 15 (electronic)
2. Choi, J., Demmel, J., Dhillon, I., Dongarra, J., Ostrouchov, S., Petitet, A., Stanley, K., Walker, D., Whaley, R.C.: ScaLAPACK: A Portable Linear Algebra Library for Distributed Memory Computers - Design Issues and Performance. Technical report, Knoxville, TN 37996, USA (1995)
3. Balay, S., Gropp, W.D., McInnes, L.C., Smith, B.F.: PETSc 2.0 User Manual. Argonne National Laboratory, http://www.mcs.anl.gov/petsc/. (1997)
4. Hunold, S., Rauber, T., Rünger, G.: Multilevel Hierarchical Matrix Multiplication on Clusters. In: Proceedings of the 18th Annual ACM International Conference on Supercomputing, ICS'04. (2004) 136–145
5. Whaley, R.C., Dongarra, J.J.: Automatically Tuned Linear Algebra Software. Technical Report UT-CS-97-366, University of Tennessee (1997)
6. Geijn, R.A.V.D., Watts, J.: SUMMA: Scalable Universal Matrix Multiplication Algorithm. Concurrency: Practice and Experience 9 (1997) 255–274
7. Choi, J.: A New Parallel Matrix Multiplication Algorithm on Distributed-Memory Concurrent Computers. Concurrency: Practice and Experience 10 (1998) 655–670
8. Golub, G.H., Van Loan, C.F.: Matrix Computations, Third Edition. John Hopkins University Press (1998)
9. Dongarra, J., Croz, J.D., Hammarling, S., Duff, I.: A Set of Level 3 Basis Linear Algebra Subprograms. ACM Transactions on Mathematical Software 16 (1990) 1–17
10. Browne, S., Dongarra, J., Garner, N., London, K., Mucci, P.: A Scalable Cross-Platform Infrastructure for Application Performance Tuning Using Hardware Counters. In: Supercomputing '00: Proceedings of the 2000 ACM/IEEE conference on Supercomputing (CDROM), IEEE Computer Society (2000) 42

Parallelization of Divide-and-Conquer Eigenvector Accumulation

Wilfried N. Gansterer and Joachim Zottl*

Institute for Distributed and Multimedia Systems
University of Vienna
Lenaugasse 2/8, A-1080 Vienna, Austria
{wilfried.gansterer,joachim.zottl}@univie.ac.at

Abstract. This paper discusses and compares several parallelization strategies for tree-structured computations. In particular, we focus on the parallelization of the eigenvector accumulation process in divide-and-conquer eigensolvers, such as the recently developed block divide-and-conquer (BD&C) eigensolver. We describe a model algorithm for evaluating the performance of several parallel variants of this accumulation process, and we develop a block parallel approach which is shown to achieve good speedup in experiments on PC clusters.

1 Introduction

In this paper, we summarize a comparison of several strategies for efficiently parallelizing *tree-structured* algorithms, that is, algorithms, where the dependencies between individual tasks have the structure of a tree. In particular, we focus on the process of accumulating eigenvectors in divide-and-conquer based eigensolvers [1, 2] as an important application scenario.

Approach. The eigenvector accumulation phase of the *block divide-and-conquer* (*BD&C*) eigensolver [2] is a special case of a tree-structured algorithm. In this case, the input to each task in a node of the tree computation consists of the two (eigenvector) matrices V_{2i-1} and V_{2i}, where V_{2i-1} has special (block diagonal) structure. The task itself is to multiply these two matrices, and the output is a combined eigenvector matrix which is input at the next level of the computation tree (see Fig. 1). The size of the matrices to be multiplied in each node grows as the computation proceeds towards the root, and thus the computational complexity grows cubically with the matrix size.

Obviously, this process could be parallelized over the nodes of the tree by assigning different tasks to different processors. However, with the decreasing number of tasks towards the root of the tree, fewer and fewer processors are active and thus this approach becomes less and less efficient for increasing numbers of processors. Another possibility is to parallelize each task in a data parallel approach over all processors. The efficiency of this approach depends on the size

---

* Partly supported by the EC-funded project HPC-Europa, contract number 5060079.

J.C. Cunha and P.D. Medeiros (Eds.): Euro-Par 2005, LNCS 3648, pp. 847–856, 2005.

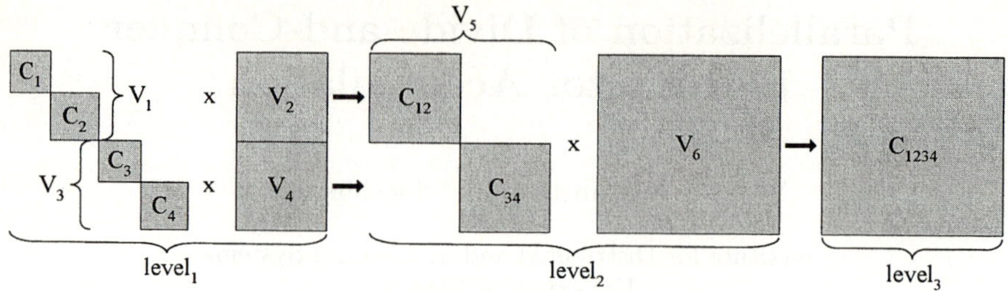

Fig. 1. Tree structured eigenvector accumulation problem

of the data left for each node and tends to be efficient only for large problems
and/or for a small or medium number of processors. Thus, an efficient approach
has to focus on task parallelism initially and then gradually put more emphasis
on data parallelism as the computation approaches the root of the tree.

Our main objective in this paper is to compare and evaluate different strategies for efficiently parallelizing the divide-and-conquer eigenvector accumulation
process. For this purpose, we have developed and implemented a prototypical
algorithmic model of this process which accurately reflects its dominant performance and parallelization properties and thus allows us to evaluate various
parallelization strategies. The evaluation of this prototypical algorithmic model
is also relevant for the parallelization of the BD&C eigensolver as a whole, since
its computational effort tends to be dominated by the eigenvector accumulation process. For simplicity, we restrict ourselves to situations where the number
of processors p available is $p = 2^h$ with $h \in \mathbb{N}$ (balanced binary tree). Direct
generalization to an arbitrary number of processors is possible.

Motivation. Potentially very large eigenvalue problems with banded structure
arise in numerous important applications. Parallel processing is often essential
in order to solve realistic cases, causing high demand for efficient and scalable
parallel eigensolvers. This motivates research in parallelizing the recently developed BD&C eigensolver [2], which does not require tridiagonalization of the
given block tridiagonal or banded matrix. Its eigenvector accumulation phase
investigated here is the most difficult part to parallelize.

The concepts investigated in this paper are also relevant for all classes of
algorithms, where the results of two neighboring tasks in the tree are passed
on to the next higher level in the tree, where they form the inputs of a task at
this higher level (divide-and-conquer-type algorithms, tree-structured algorithms
based on binary trees, etc.).

Related Work. A parallel version of the *tridiagonal* divide-and-conquer eigensolver was presented in [1]. In this paper, we are considering the more general
block tridiagonal situation, we are pursuing a different approach in the "start-up
phase" (when there are more blocks than processors), and we are also investigating modifications of the block cyclic data distribution used in [1].

The core of the eigenvector accumulation process is a special matrix multiplication. Many parallel algorithms for performing this operation on distributed memory computers have been developed. We mention Fox's algorithm [3] and one of its implementations, PUMMA [4], for block cyclic data distribution, both operating on square processor grids. Important extensions of Fox's algorithm for more general rectangular processor grids are SUMMA [5] and DIMMA [6].

Synopsis. The rest of the paper is structured as follows. In Section 2, the basic algorithmic structure of the (sequential) BD&C algorithm and of its eigenvector accumulation phase are described. In Section 3, our prototypical model for the eigenvector accumulation, various parallelization strategies, and their implementation are discussed. In Section 4 we summarize experimental results. Finally, Section 5 contains conclusions and future work.

2 Divide-and-Conquer Type Eigensolvers

Divide-and-conquer eigensolvers have evolved into highly efficient and competitive methods for computing spectral information of symmetric matrices. Their basic idea is to split the given matrix into parts, to compute the spectral information of each of these parts "conventionally", and then to synthesize the spectral information of the original problem from the spectral information of the subproblems.

This synthesis is based on special algebraic matrix properties [7]. Divide-and-conquer eigensolvers for matrices with special structure have been developed: tridiagonal divide-and-conquer (TD&C) for tridiagonal matrices [8], and, more recently, the BD&C eigensolver for block tridiagonal matrices [2, 9]. A parallel version of TD&C is already available [1], whereas a parallel version of BD&C is currently under development.

2.1 Sequential BD&C

As mentioned before, BD&C efficiently computes approximate eigenpairs of a *block tridiagonal* matrix. A block tridiagonal matrix is a natural generalization of a (scalar) tridiagonal matrix, where every scalar is "expanded" into a block–thus, any symmetric banded matrix is a block tridiagonal matrix. BD&C is especially competitive in situations where it suffices to *approximate* spectral information to low or medium accuracy, i. e., where full accuracy is not required. It involves the following three phases:

1. Subdivision of the problem into p subproblems, which mainly involves independent operations on each of the off-diagonal blocks and correction operations involving neighboring blocks;
2. eigendecomposition of each of the subproblems, which can be performed completely independently; and
3. synthesis of the solutions of the p subproblems.

The synthesis phase tends to dominate the execution time for the entire BD&C eigensolver, and its main part is the accumulation of the eigenvector matrices of the subproblems into the final eigenvector matrix of the original problem.

2.2   Eigenvector Accumulation

The eigenvector accumulation process in the current BD&C algorithm consists of a sequence of matrix multiplications with the structure of a binary tree.

For illustration purposes, we consider a special case where the original problem has been subdivided into four subproblems, numbered 1 through 4 (cf. Fig. 1). Given the eigenvector matrices C_1 and C_2 of subproblems 1 and 2 as well as the eigenvector matrix V_2 of the dependencies between these subproblems, the eigenvectors C_{12} of the combination of subproblems 1 and 2 can be computed as the matrix product

$$C_{12} = \begin{pmatrix} C_1 & \\ & C_2 \end{pmatrix} V_2. \tag{1}$$

Analogously, the eigenvectors C_{34} of the combination of subproblems 3 and 4 can be computed. With the eigenvector matrix V_6 of the dependencies between subproblems 2 and 3, the eigenvectors of the combination of all four subproblems can be computed as

$$C_{1234} = \begin{pmatrix} C_{12} & \\ & C_{34} \end{pmatrix} V_6.$$

Obviously, this process has the structure of a binary tree. If the number of subproblems is $p = 2^h$ with $h \in \mathbb{N}$ then the tree is balanced. Important properties of this process, such as computational complexity, optimal orderings of the subproblems, etc., have been analyzed in [2] and [9].

The focus in this paper is on parallelization. The operations in the subdivision phase as well as the eigendecompositions of the diagonal blocks are in general easy to parallelize over the blocks involved, exploiting the natural parallelization potential inherent in a divide-and-conquer approach. In the synthesis phase all the dependencies are collected, and thus this part, in particular the eigenvector accumulation process, is most difficult to parallelize efficiently. In the next section we will discuss and compare several strategies for achieving this goal.

3   Parallel Eigenvector Accumulation

The main parallelization challenge of the BD&C eigensolver arises in the synthesis phase. By far the most important and dominating in terms of complexity and computation time is the accumulation process of the eigenvectors as described in Section 2.2. This process exhibits complicated data dependencies with strongly varying computation and communication costs since the size of the matrices involved increases while the number of the matrices involved decreases.

It is difficult to parallelize the accumulation process efficiently and to achieve scalability to high numbers of processors. Thus, this process is also an interesting parallel benchmarking problem. Important aspects, such as the workload and the communication/computation ratio can be adjusted via the number of blocks and the sizes of the blocks in the block tridiagonal matrix (cf. [10]).

3.1   A Prototypical Performance Model

In order to be able to focus on the central parallelization aspects we have designed a prototypical model of the basic structure of the eigenvector accumulation process which reflects all essential properties relevant for the performance and scalability of a parallel eigenvector accumulation process. This model is described in the following.

Based on the input parameters k (number of diagonal blocks) and b_i, $i = 1, 2, \ldots, k$ (sizes of the diagonal blocks, $\sum_{i=1}^{k} b_i = n$) we generate a symmetric block tridiagonal matrix with random entries. Assuming that $p < k$ processors are available, in the first phase we split the sequence of diagonal blocks into p consecutive portions such that the sum of the block sizes in each portion is as close to n/p as possible (thus balancing the workload over the available processors). Each processor then combines all the blocks assigned to him locally and sequentially according to repeated application of Equation (1). All the matrices representing the dependencies between subproblems are created randomly. At the end of this phase, each processor has computed a full matrix with size equal to the sum of the sizes of his local blocks. Since all this is done locally and independently, the values of k and b_i do not have an impact on parallel performance. In the second phase, the process continues analogously in parallel, according to Equation (1). Again, matrices representing dependencies between subproblems are created randomly when required. Finally, the $n \times n$ combination matrix of all k original subproblems is computed.

This model algorithm reflects the basic operations of the eigenvector accumulation process in the BD&C eigensolver. It never computes actual eigenvector data but uses random numbers instead. However, since in terms of computational effort the creation of the actual eigenvector data in the BD&C is asymptotically less important than the accumulation of this data, we can expect the model algorithm to properly reflect the asymptotic runtime behavior and thus to be a sound basis for evaluating parallelization strategies.

3.2   Parallelization Strategies

We have implemented and investigated several strategies for parallelizing this model algorithm: (i) a purely *task parallel* variant, (ii) a purely *data parallel* variant, (iii) a *mixed* variant which switches from task parallel to data parallel at a certain stage of the process, and (iv) a pipelined combination of task and data parallel aspects which we call *block parallel* variant.

Task Parallel. Each new task (multiplying two eigenvector matrices according to Equation (1)) to be completed is assigned to an available processor. Once the task is completed, the processor becomes idle (cf. Fig. 2(a)).

This variant has potential advantages in the early stages of the accumulation process when many small tasks are to be completed, but it has disadvantages in the later stages, since gradually the number of tasks decreases and consequently more and more available processors become idle.

Data Parallel. In this variant, several processors work on each task simultaneously in a data parallel fashion. This can be implemented by using a parallel matrix multiplication method for each combination operation (1). We are using the SUMMA algorithm [5] which is very competitive in terms of communication costs, memory efficiency, and flexible processor grid usage. In our current implementation, we always group four processors together to work in a data parallel fashion in order not to occupy too many processors in a single branch of the tree. As the computation proceeds, the number of tasks decreases, the size of tasks increases, and thus it would be possible to group more processors together.

In contrast to the task parallel variant, the data parallel approach keeps more processors busy all the time and it has potential performance advantages in the late stages of the accumulation process when the tasks are big. However, it has disadvantages in the early stages of the accumulation process when the tasks are small, especially for larger processor numbers.

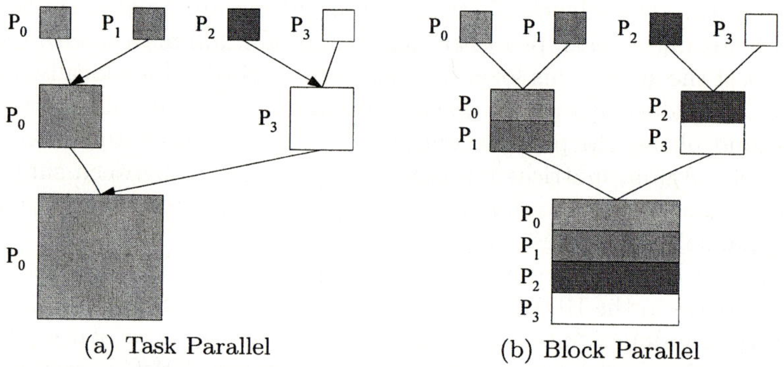

Fig. 2. Two variants for parallelizing the binary tree structure problem arising in divide-and-conquer eigenvector accumulation on $p = 4$ processors

Mixed. The previous considerations suggest a variant which explicitly switches from a task parallel to a data parallel approach when—in relation to the number of processors available—the number of tasks is considered too small and/or when a task to be completed is considered big enough. Obviously, the question arises how to determine the appropriate "switching point". In our implementation, this switching is based on the size of the block resulting from the combination operation (1) and we have investigated the sensitivity of the parallel performance in terms of this switching parameter (see Section 4).

Block Parallel. The central ideas of the block parallel variant, similar to the SUMMA algorithm for matrix multiplication, are illustrated using the notation from Equation (1): (i) Distribute the matrices C_1 and C_2 block row wise, V_2

column block wise. This distribution does not imply any additional communication overhead, since the eigenvectors contained in V_2 can be computed such that they are located properly. (*ii*) Accumulate the local block portions of the matrix product (1) on all processors simultaneously. The resulting matrix C_{12} is distributed block row wise over all processors (see Fig. 2(b)). (*iii*) Shift the column blocks of V_2 cyclically over all processors in order to accumulate the next local blocks (see Fig. 3). The communication requirements of this strategy are low and mostly localized, and *all* processors are busy during the *entire* accumulation process. Our experiments confirm that this variant is suited best among the variants considered for parallelizing the eigenvector accumulation process.

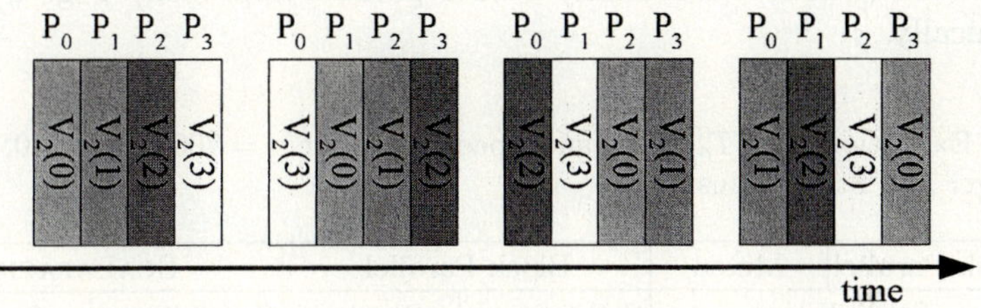

Fig. 3. Cyclic shifting of subblocks of V_{2i} in the block parallel variant

4   Experiments

In this section, we summarize and discuss our experimental results. First, we give a brief overview of the hard- and software used. Then we summarize our results, comparing different parallelization strategies among each other and with related SCALAPACK routines [11].

4.1   Hard- and Software Environment

Hardware. Experimental results are shown for the *Schrödinger II* cluster at the University of Vienna and for the *Strider* cluster at the University of Stuttgart. Schrödinger consists of 192 compute nodes in total. For a batch job the number of nodes available to us was limited to 32. Each node is equipped with an Intel Pentium 4 (2.53 GHz) processor and 1 GB main memory. The nodes are connected via a Gigabit Ethernet. Strider consists of 125 compute nodes. Each node has two 64 Bit Dual Opteron (2 GHz) and 4 GB main memory. The nodes are connected via Myrinet 2000.

Software. We have written our programs in C, using MPI (LAM 7.1.1 or MPICH 1.2.6) for the parallelization. For the matrix-matrix multiplications we used the `xgemm` routines from ATLAS CBLAS.

We also compared the speedup achieved with our routines with the speedup of related SCALAPACK routines, in particular with `pxsytrd` and `pxstedc` for computing eigenvalues and eigenvectors of a symmetric/symmetric tridiagonal matrix. It has to be emphasized that the SCALAPACK routines provide a different functionality than our model algorithm and thus it is not suitable to compare runtimes achieved or to strictly relate speedup values. The purpose of including these routines is to give an idea about the big picture and about the general parallelization behavior of our model algorithm in relation to the well established SCALAPACK standard.

4.2   Results

Table 1 summarizes execution time T_p and speedup S_p of three of the strategies discussed in Section 3. In addition, we also give corresponding data for SCALAPACK routines (which is essentially a data parallel approach). Fig. 4 illustrates S_p graphically.

Table 1. Execution time T_p in [s], and Speedup S_p for $n = 7000$ ($k = 140$, $b = 50$) at Schrödinger and Strider cluster

| | Task Parallel | | Mixed | | Block Parallel | | | | SCALAPACK | | | |
| | Schrödinger | | | | Schrödinger | | Strider | | Schrödinger | | Strider | |
p	T_p	S_p	T_p	S_p	T_p	S_p	T_p	S_p	T_p	S_p	T_p	S_p
1	140.30	1.00	138.28	1.00	138.57	1.00	71.04	1.00	267.26	1.00	245.95	1.00
2	74.95	1.87	75.61	1.83	45.42	3.05	36.10	1.97	128.08	2.08	160.34	1.53
4	66.52	2.11	66.51	2.08	23.06	6.01	19.52	3.64	60.48	4.42	75.08	3.28
8	65.42	2.15	52.69	2.62	18.44	7.52	10.75	6.61	31.38	8.52	40.15	6.13
16	63.39	2.21	38.59	3.58	11.37	12.19	6.55	10.85	15.68	17.05	24.25	10.14
32	65.89	2.13	39.98	3.46	7.92	17.50	4.80	14.56	8.34	32.05	17.72	13.88

We notice that for eight and more processors the mixed variant is a little bit faster than the task parallel one. This is explained by the switching strategy which checks for three conditions: (i) the number of total active processors must be at least eight, (ii) at least four processors have to be available for data parallel processing, and (iii) the size of the task must be greater than 1000. These settings are the result of experimentally investigating the behavior of the mixed variant for varying the switching point s. Those experiments showed that $s \approx 1000$ is a good choice for most parameter settings considered. For example, for $n = 5000$ and one to 32 processors the execution time is increasing linearly for $s > 1000$ with a break-even point at $s = 1500$.

The SCALAPACK routines have significantly longer sequential runtimes which leads to higher speedup values. On the Strider cluster, the speedup values of the block parallel variant are slightly better than those of the SCALAPACK routines. However, as indicated before, the functionality of these routines differs significantly from our routines, and thus those data cannot be used for direct performance comparisons. Generally speaking, Fig. 4 indicates that the block parallel variant has by far the best parallel performance among the variants investigated in this paper and that it achieves a scaling behavior similar to related

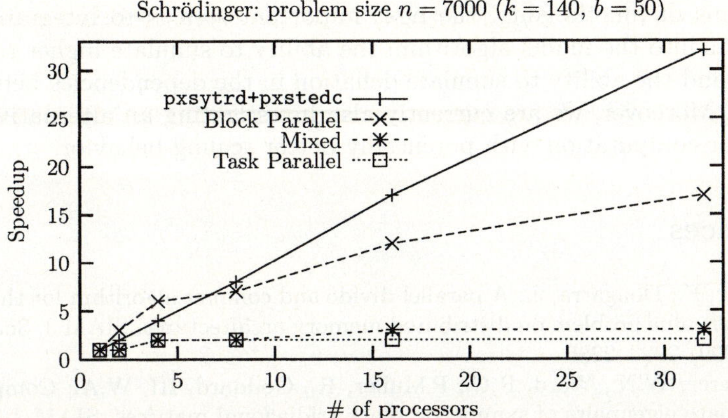

Fig. 4. Parallelization strategies for eigenvector accumulation and related SCALAPACK routines at Schrödinger cluster for $n = 7000$ ($k = 140$, $b = 50$). In the block parallel variant and in the SCALAPACK routines all processors are busy all the time

SCALAPACK routines which are a well established standard for many problems of this type.

5   Conclusions and Future Work

We have developed and compared several strategies for parallelizing the eigenvector accumulation process in the BD&C eigensolver based on a model algorithm developed for this tree structured problem. Experimental evaluation showed that a block parallel approach achieves by far the best parallel performance. It is much faster and scales much better than a task parallel and a mixed variant (a combination of task and data parallel variants).

This functionality is currently not available. The standard alternative provided by SCALAPACK is tridiagonalization-based. Although absolute runtimes are not comparable since the actual functionality provided is very different, we have shown that the scaling behavior of our model algorithm for block tridiagonal matrices is comparable to that of the tridiagonalization-based SCALAPACK approach. Speedup values are high for a small or medium number of processors. Due to cache-effects, SCALAPACK routines sometimes even exhibit superlinear speedup.

For large processor numbers (and for a constant problem size) we have to expect a decreasing tendency in the speedup values, both for our model algorithm as well as for the related SCALAPACK routines. Benchmarking on more and diverse parallel architectures will be required to investigate the performance behavior in more detail.

Future Work. Our final goal is to develop a fully parallel BD&C eigensolver. Thus, we will also work on the parallelization of the other phases of BD&C.

In terms of functionality, the next important step is to integrate two improvements into the model algorithm: the ability to simulate higher rank modifications, and the ability to simulate deflation in the dependencies between subproblems. Moreover, we are currently also investigating an alternative parallel eigenvector computation with potentially better scaling behavior.

References

1. Tisseur, F., Dongarra, J.: A parallel divide and conquer algorithm for the symmetric eigenvalue problem on distributed memory architectures. SIAM J. Sci. Comput. **20** (1999) 2223–2236
2. Gansterer, W.N., Ward, R.C., P.Muller, R., Goddard, III, W.A.: Computing approximate eigenpairs of symmetric block tridiagonal matrices. SIAM J. Sci. Comput. **25** (2003) 65–85
3. Fox, G., Otto, S., Hey, A.: Matrix algorithms on a hypercube i: Matrix multiplication. Parallel Computing **4** (1987) 17–31
4. Choi, J., Dongarra, J.J., Walker, D.: Pumma: Parallel universal matrix multiplication algorithms on distributed memory concurrent computers. Concurrency: Practice and Experience **6** (1994) 543–570
5. van de Geijn, R.A., Watts, J.: Summa: Scalable universal matrix multiplication algorithm. Concurrency: Practice and Experience **9** (1997) 255–274
6. Choi, J.: A fast scalable universal matrix multiplication algorithm on distributed-memory concurrent computers. In: Proceedings of the 11th International Parallel Processing Symposium, IEEE Press (1997) 310–314
7. Gu, M., Eisenstat, S.C.: A stable and efficient algorithm for the rank-one modification of the symmetric eigenproblem. SIAM J. Matrix Anal. Appl. **15** (1994) 1266–1276
8. Gu, M., Eisenstat, S.C.: A divide-and-conquer algorithm for the symmetric tridiagonal eigenproblem. SIAM J. Matrix Anal. Appl. **16** (1995) 172–191
9. Gansterer, W.N., Ward, R.C., Muller, R.P.: An extension of the divide-and-conquer method for a class of symmetric block-tridiagonal eigenproblems. ACM Trans. Math. Softw. **28** (2002) 45–58
10. Gansterer, W.N., Zottl, J.: Message passing vs. virtual shared memory - a performance comparison. In Juhász, Z., Kacsuk, P., Kranzlmüller, D., eds.: Distributed and Parallel Systems: Cluster and Grid Computing. Volume 777 of The Kluwer International Series in Engineering and Computer Science. Springer-Verlag (2005) 39–46
11. Blackford, L.S., Choi, J., Cleary, A., D'Azevedo, E., Demmel, J., Dhillon, I., Dongarra, J., Hammarling, S., Henry, G., Petitet, A., Stanley, K., Walker, D., Whaley, R.C.: ScaLAPACK Users' Guide. Society for Industrial and Applied Mathematics (1997)

Parallel Order Reduction
via Balanced Truncation
for Optimal Cooling of Steel Profiles*

José M. Badía[1], Peter Benner[2], Rafael Mayo[1], Enrique S. Quintana-Ortí[1],
Gregorio Quintana-Ortí[1], and Jens Saak[2]

[1] Depto. de Ingeniería y Ciencia de Computadores, Universidad Jaume I
12.071–Castellón, Spain
{badia,mayo,quintana,gquintan}@icc.uji.es
[2] Fakultät für Mathematik, Technische Universität Chemnitz
D-09107 Chemnitz, Germany
{benner,jens.saak}@mathematik.tu-chemnitz.de

Abstract. We employ two efficient parallel approaches to reduce a
model arising from a semi-discretization of a controlled heat transfer
process for optimal cooling of a steel profile. Both algorithms are based
on balanced truncation but differ in the numerical method that is used
to solve two dual generalized Lyapunov equations, which is the major
computational task. Experimental results on a cluster of Intel Xeon processors compare the efficacy of the parallel model reduction algorithms.

Keywords: Linear dynamical systems, 2D heat equation, model reduction, balanced truncation, generalized Lyapunov equations

1   Introduction

We consider the problem of optimal cooling of steel profiles. This problem arises
in a rolling mill where different phases in the production process require different temperatures of the raw material. To achieve a high production rate, the
temperature has to be reduced rapidly to the level required by the next phase.
The cooling process, accelerated by spraying cooling fluids on the surface of the
profile, has to be controlled since large gradients in the temperature lead to
unwanted deformations, brittleness, loss of rigidity, and other undesirable properties.

The heat distribution in the profile is modeled by an instationary linear
heat equation. The standard Galerkin approach for discretizing the heat transfer

---

* J.M. Badía, R. Mayo, E.S. Quintana-Ortí, and G. Quintana-Ortí were supported
by the CICYT project No. TIC2002-004400-C03-01 and FEDER, and project No.
P1B-2004-6 of the *Fundación Caixa-Castellón/Bancaixa and UJI*. P. Benner and J.
Saak were supported by the DFG Sonderforschungsbereich 393 *Parallele Numerische
Simulation für Physik und Kontinuumsmechanik* at TU Chemnitz.

J.C. Cunha and P.D. Medeiros (Eds.): Euro-Par 2005, LNCS 3648, pp. 857–866, 2005.
© Springer-Verlag Berlin Heidelberg 2005

model in space, as described in Section 1.1, results in a first-order ordinary differential equation of the form:

$$\begin{aligned}
E\dot{x}(t) &= Ax(t) + Bu(t), \quad t > 0, \quad x(0) = x^0, \\
y(t) &= Cx(t) + Du(t), \quad t \geq 0.
\end{aligned} \tag{1}$$

Here, $x^0 \in \mathbb{R}^n$ contains the initial temperature distribution in the profile, $u(t)$ and $y(t)$ are vectors for the inputs (i.e., temperatures of the cooling fluid) and outputs (i.e., approximate temperature gradients) of the system, respectively, and $E, A \in \mathbb{R}^{n \times n}$, $B \in \mathbb{R}^{n \times m}$, $C \in \mathbb{R}^{p \times n}$, $D \in \mathbb{R}^{p \times m}$. The system in (1) can also be modeled by the transfer function matrix (TFM) $G(s) = C(sE - A)^{-1}B + D$. The number of states, n, is known as the state-space dimension (or the order) of the system.

The goal of model reduction is to find a reduced-order LTI system,

$$\begin{aligned}
\hat{E}\dot{\hat{x}}(t) &= \hat{A}\hat{x}(t) + \hat{B}u(t), \quad t > 0, \quad \hat{x}(0) = \hat{x}^0, \\
\hat{y}(t) &= \hat{C}\hat{x}(t) + \hat{D}u(t), \quad t \geq 0,
\end{aligned} \tag{2}$$

of order r, with $r \ll n$, and associated TFM $\hat{G}(s) = \hat{C}(sE - \hat{A})^{-1}\hat{B} + \hat{D}$ which approximates $G(s)$. The reduced-order realization can then replace the original high-order system in the design of the optimal controller, thus simplifying by much this phase of the problem.

It is the goal of this paper to demonstrate that model reduction algorithms based on balanced truncation (BT) can effectively be applied to problems of size up to $n = 100,000$ (and even higher) when appropriate numerical algorithms are used.

Traditional algorithms for BT model reduction are available, e.g., in libraries such as SLICOT[1] or the MATLAB control-related toolboxes, and can be employed to reduce models with a few hundreds of state-space variables on current desktop computers. Here we employ two different parallel BT algorithms that allow the reduction of much larger systems, as those arising in the optimal cooling problem. These algorithms are integrated into the parallel libraries for model reduction of large-scale dense and sparse linear systems, PLiCMR [3] and SpaRed [1], respectively.

The paper is structured as follows. We conclude this section by describing with some detail the discretization procedure of the heat transfer model for the steel profile. In Section 2 we review our algorithms for BT model reduction of linear systems. In Section 3 we briefly describe the multilayered architecture of libraries employed by our model reduction algorithms. Finally, the efficacy of the algorithms is reported in Section 4, and some concluding remarks follow in Section 5.

1.1   Discretization of the Optimal Cooling Problem

The optimal cooling problem was discretized using the ALBERTA-1.2 fem-toolbox [11]. We applied linear Lagrange elements and used a projection method for the

---

[1] Available from `http://www.win.tue.nl/niconet/NIC2/slicot.html`.

curved boundaries. An initial mesh (see Fig. 1) was produced using MATLAB `pdetool` function, which implements a Delaunay triangulation algorithm. Finer discretizations were then obtained by global mesh refinement using a bisection refinement model. As a result, we obtained four different mesh resolutions and associated systems of order $n = 1{,}357$, $5{,}177$, $20{,}209$, and $79{,}841$, corresponding to maximum mesh widths (or edge sizes) $h = 5.5280 \times 10^{-2}$, 2.7640×10^{-2}, 1.3820×10^{-2}, and 6.9100×10^{-3}, respectively. The number of nonzero elements in matrices A and E range from about 4% for the smallest model to 0.009% for the largest one. A more detailed description of the model is given in [5].

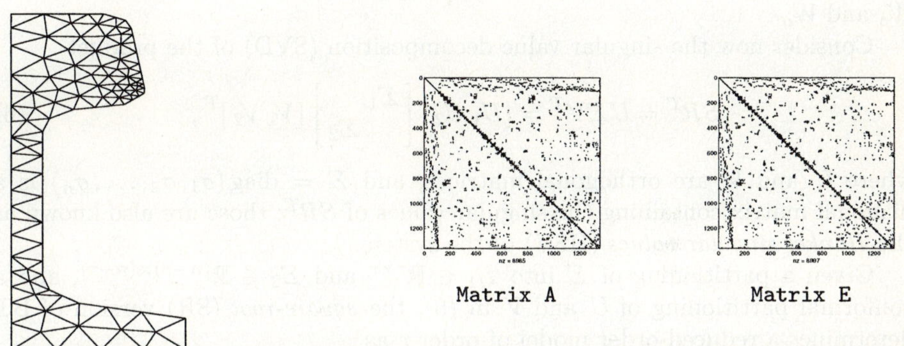

Fig. 1. Finite element discretization of the profile (left) and sparsity pattern (right) for the model of order $n=1{,}357$.

In our case, the best approximation error of the finite element discretization that one can expect is of order $\mathcal{O}(h^s)$ for an $s \in [1,2)$. Here s is very close to 2, for the boundary is almost of class $\mathcal{C}^\infty$, i.e., if the two rightmost edges were smoothly topped off we would have sufficient regularity to obtain $s = 2$ at best. Thus, reducing the model with an error bound smaller than h^2 should not contribute any significant additional error. The reduced-order models presented in Section 4 meet this requirement.

2   Model Reduction of Large Linear Systems

2.1   The Square-Root BT Method

BT model reduction [7, 10, 12, 13] belongs to the family of absolute error methods, which aim at minimizing $\|\Delta_a\|_\infty = \|G - \hat{G}\|_\infty$. Here $\|G\|_\infty$ denotes the $\mathcal{L}_\infty$- or $\mathcal{H}_\infty$-norm of a stable, rational matrix function defined as

$$\|G\|_\infty = \sup_{\omega \in \mathbb{R}} \sigma_{\max}(G(\jmath\omega)), \tag{3}$$

where $\jmath := \sqrt{-1}$ and $\sigma_{\max}(M)$ stands for the largest singular value of M.

BT methods are strongly related to the controllability Gramian W_c and the observability Gramian W_o of the system (1), given by the solutions of the two dual generalized Lyapunov matrix equations

$$AW_cE^T + EW_cA^T + BB^T = 0, \tag{4}$$
$$A^T\tilde{W}_oE + E^T\tilde{W}_oA + C^TC = 0. \tag{5}$$

In the optimal cooling problem, the matrix pair (A, E) is stable (all its generalized eigenvalues are in the open left complex plane), so that W_c and $W_o = E^T\tilde{W}_oE$ are positive semidefinite, and therefore can be factored as $W_c = S^TS$ and $W_o = R^TR$. Here, S and R are usually refered to as the *Cholesky factors* of W_c and W_o.

Consider now the singular value decomposition (SVD) of the product

$$SR^T = U\Sigma V^T = [\,U_1\ U_2\,]\begin{bmatrix}\Sigma_1 & \\ & \Sigma_2\end{bmatrix}[\,V_1\ V_2\,]^T, \tag{6}$$

where U and V are orthogonal matrices, and $\Sigma = \mathrm{diag}\,(\sigma_1, \sigma_2, \ldots, \sigma_n)$ is a diagonal matrix containing the singular values of SR^T; those are also known as the *Hankel singular values* (HSV) of the system.

Given a partitioning of Σ into $\Sigma_1 \in \mathbb{R}^{r\times r}$ and $\Sigma_2 \in \mathbb{R}^{(n-r)\times(n-r)}$, and a conformal partitioning of U and V in (6), the *square-root* (SR) version of BT determines a reduced-order model of order r as

$$\hat{E} = T_lET_r, \quad \hat{A} = T_lAT_r, \quad \hat{B} = T_lB, \quad \hat{C} = CT_r, \quad \hat{D} = D, \tag{7}$$

with the projection matrices T_l and T_r given by

$$T_l = \Sigma_1^{-1/2}V_1^TRE^{-1} \quad \text{and} \quad T_r = S^TU_1\Sigma_1^{-1/2}. \tag{8}$$

The state-space dimension r of the reduced-order model can be chosen adaptively as this method provides a realization $\hat{G}$ satisfying

$$\|\Delta_a\|_\infty = \|G - \hat{G}\|_\infty \leq 2 \sum_{j=r+1}^{n} \sigma_j. \tag{9}$$

In the following subsections we revisit two generalized Lyapunov solvers introduced in [2, 6, 8] which provide low-rank approximations to a Cholesky or full-rank factor of the solution matrix. These approximations can reliably substitute S and R in the computation of (6) and (8). For simplicity, we only describe those solvers that obtain approximations of the Cholesky factor of W_c in (4). Analogous iterations provide approximations for the Cholesky factor of W_o in (5).

2.2 Solution of Generalized Lyapunov Equations via the Matrix Sign Function

Since its introduction in [9], the sign function has proved useful in a variety of numerical linear algebra problems. In particular, the following variant of the

Newton iteration for the matrix sign function can be used for the solution of the generalized Lyapunov equation (4):

$$A_0 \leftarrow A$$
$$\tilde{S}_0 \leftarrow B$$
$$k \leftarrow 0$$
repeat
$$A_{k+1} \leftarrow \frac{1}{\sqrt{2}}\left(A_k + EA_k^{-1}E\right)$$

Compute the rank-revealing QR (RRQR) decomposition
$$\frac{1}{\sqrt{2}}\left[\tilde{S}_k \ \ EA_k^{-1}\tilde{S}_k\right]^T = Q_s \begin{bmatrix} R_s \\ 0 \end{bmatrix} \Pi_s$$
$$\tilde{S}_{k+1} \leftarrow R_s\Pi_s$$
$$k \leftarrow k+1$$
until $\|A_k - E\|_1 < \tau\|A_k\|_1$

Here τ is a tolerance threshold for the iteration that is usually set as a function of the problem dimension and the machine precision. Convergence can be improved by using several acceleration techniques. In our case, we employ an approximation of the norm scaling [4]. The RRQR decomposition can be obtained by means of the traditional QR factorization with column pivoting.

Sign function iterations usually present a fast convergence rate, which is ultimately quadratic. However, the inverse of a sparse matrix is in general dense and therefore the previous iterative scheme cannot exploit the sparsity of the matrix A. Note that, on the other hand, we can easily take advantage of the sparsity of E as this matrix is involved in matrix products and it is not modified during the iteration.

On convergence, after j iterations, $\tilde{S} = \frac{1}{\sqrt{2}}\tilde{S}_j E^{-T}$, of dimension $\tilde{l} \times n$, is a full (row-)rank approximation of S^T so that $W_c = S^T S \approx \tilde{S}\tilde{S}^T$.

2.3 Low Rank Solution of Generalized Lyapunov Equations

The cyclic *low-rank alternating direction implicit* (LR-ADI) iteration proposed in [8] can be reformulated for the generalized Lyapunov equation (4) as follows:

$$V_0 \leftarrow (A + p_1 E)^{-1}B$$
$$\hat{S}_0 \leftarrow \sqrt{-2\,\alpha_1}\,V_0$$
$$k \leftarrow 0$$
repeat
$$V_{k+1} \leftarrow V_k - \delta_k(A + p_{k+1}E)^{-1}EV_k$$
$$\hat{S}_{k+1} \leftarrow \left[\hat{S}_k\ ,\ \gamma_k V_{k+1}\right]$$
$$k \leftarrow k+1$$
until $\|\gamma_k V_k\|_1 < \tau\|\hat{S}_k\|_1$

In the iteration, $\{p_1, p_2, \ldots\}$, $p_k = \alpha_k + \beta_k\jmath$, is a cyclic set of (complex) shift parameters (that is, $p_k = p_{k+t}$ for a given period t), $\gamma_k = \sqrt{\alpha_{k+1}/\alpha_k}$, and $\delta_k = p_{k+1}+\overline{p_k}$, $\overline{p_k}$ being the conjugate of p_k. This iteration may suffer from a slow convergence rate, which is super-linear at best. Nevertheless, the iteration only requires the solution of linear systems with sparse coefficient matrices and matrix products. The use of sparse direct solvers is recommended here as iterations k and $k + t$ share the same coefficient matrices for the linear systems.

The performance of the LR-ADI iteration strongly depends on the selection of the shift parameters. For further details, see [6, 8, 15].

On convergence, after j iterations, a low-rank matrix $\hat{S} = \hat{S}_j$ of order $n \times \hat{l} = n \times (jm)$, is computed such that $\hat{S}\hat{S}^T$ approximates $W_c = S^T S$.

It should be emphasized that the iterative methods described in the previous two subsections for solving (4)–(5) significantly differ from standard methods used in the MATLAB toolboxes or SLICOT [14]. As the iterative solvers produce low-rank approximations to the full-rank or Cholesky factors, the computation of the SVD in (6) is usually much more efficient: instead of a computational cost of $\mathcal{O}(n^3)$ flops (floating-point arithmetic operations) when using the Cholesky factors, this approach leads to an $\mathcal{O}(m \cdot p \cdot n)$ cost where, in model reduction, often $m, p \ll n$; see [3].

3 Parallel Implementation

The matrix sign function-based iteration basically requires dense linear algebra operations such as matrix products and the solution of linear systems (via matrix factorizations). On the other hand, the LR-ADI iteration is composed of sparse linear algebra operations as matrix-vector products and the solution of sparse linear systems (via direct methods). Once the generalized Lyapunov equations are solved, the final stages of the SR BT method require the computation of an SVD and a few matrix products.

Our approach for dealing with these matrix operations is based on the use of existing parallel linear algebra and communication libraries. (For an extensive list, see http://www.netlib.org/utk/people/JackDongarra/la-sw.html.) In Fig. 2 we display the multilayered architecture of libraries employed by our parallel model reduction codes for large-scale dense and sparse systems in PLiCMR and SpaRed, respectively. All model reduction codes employ the parallel dense linear algebra libraries ScaLAPACK and PBLAS. Depending on the structure of the state-space matrix pair (A, E) the kernels in SpaRed also use the banded linear system solvers in ScaLAPACK or the sparse linear system solvers in the packages MUMPS or SuperLU. PARPACK is our key to compute eigenvalue information of large sparse matrix pairs.

4 Experimental Results

All the experiments presented in this section were performed on a cluster of $n_p = 16$ nodes using IEEE double-precision floating-point arithmetic ($\varepsilon \approx 2.2204 \times 10^{-16}$). Each node consists of an Intel Xeon processor@2.4 GHz with 1 GByte of RAM. We employ a BLAS library specially tuned for this processor that

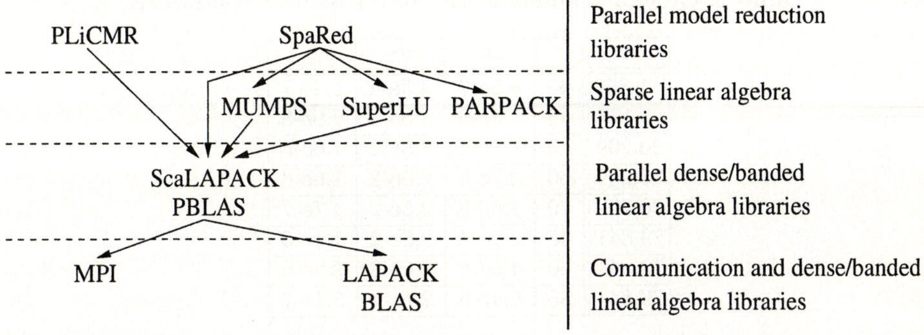

Fig. 2. Multilayered architecture of libraries for model reduction.

achieves around 3800 Mflops (millions of flops per second) for the matrix product (routine DGEMM from http://www.cs.utexas.edu/users/kgoto).The nodes are connected via a *Myrinet* multistage network and the MPI communication library is specially developed and tuned for this network. The performance of the interconnection network was measured by a simple loop-back message transfer resulting in a latency of 18 μsec. and a bandwidth of 1.4 Gbit/sec.

In this section, we compare the BT parallel routines in libraries PLiCMR and SpaRed (hereafter, PLiCMR-BT and SpaRed-BT). Given the amount of computational resources available and the memory requirements of the sign function-based iteration, we could only apply the PLiCMR-BT routine to the two smaller cases, $n = 1,357$ and $5,177$.

In order to reduce the models, we first compute the HSVs of the system and, from there, we select the order r of the reduced realization. Obviously, a larger order provides a more accurate model, but also increases the cost of those stages involving the reduced system. Table 1 reports the order r of the reduced realization, a bound for the absolute error of the reduced-order realization (computed as in (9)), and the first and $(r + 1)$-th HSV of the system. All these data were obtained by applying the SpaRed-BT routine to the system. No difference was found when the PLiCMR-BT routine was applied to compute these parameters for the two smaller problems. Our results hereafter refer to the reduced realizations of order $r = 45, 45, 70$, and 80 for the $n = 1,357, 5,177, 20,209$, and $79,841$ cases, respectively.

In order to measure the numerical accuracy of the reduced realizations, in our next experiment we compare the frequency response of the original system, G, with that of the reduced-order realization computed with the PLiCMR-BT and SpaRed-BT routines, $\hat{G}$. Figure 3 reports the absolute error $\|G - \hat{G}\|_\infty$, where the norm is defined as in (3). All four figures show that the absolute error is well below the theoretical bound. For the two smallest problems, there is no notable difference between the PLiCMR-BT and Spared-BT routines.

We next report in Tables 2 and 3 the specific results for each one of the model reduction methods. In particular, we report the number of iterations required

Table 1. Order and absolute error of the reduced realizations.

n	r	Δ_a	σ_1	σ_{r+1}
1,357	45	6.5e-7	3.5e-4	5.2e-7
5,177	45	1.3e-6	3.5e-4	9.0e-8
20,209	45	1.4e-4	2.5e-2	7.0e-6
20,209	60	2.7e-5	2.5e-2	1.6e-6
20,209	70	8.6e-6	2.5e-2	4.7e-7
79,841	45	2.2e-4	2.6e-2	1.1e-5
79,841	60	4.6e-5	2.6e-2	2.4e-6
79,841	80	5.4e-6	2.6e-2	3.1e-7

Table 2. Results for the PLiCMR-BT routine.

n	#iter.	l	k	$\mathcal{R}_{W_c}(S)$	$\mathcal{R}_{W_o}(R)$	n_p	Time (sec.)
1,357	8	310	181	1.04e-22	7.20e-14	4	21.1
5,177	8	351	209	1.48e-21	9.95e-14	16	142.6

for convergence (#iter.), the dimensions of the low-rank approximations of S and R (labeled as l and k, respectively), and the absolute residuals of these approximations, computed as

$$\mathcal{R}_{W_c}(S) := \|A(S^T S)E^T + E(S^T S)A^T + BB^T\|_F, \quad \text{and}$$

$$\mathcal{R}_{W_o}(R) := \|A^T(E^{-T}R^T RE^{-1})E + E^T(E^{-T}R^T RE^{-1})A + C^T C\|_F.$$

Finally, we also provide the number of nodes involved in the reduction (n_p), and the execution time required by the corresponding algorithm. A comparison of the results in both tables show a much slower convergence rate for the SpaRed-BT algorithm which then provides approximations to the Cholesky factors of much larger order than the PLiCMR-BT algorithm. On the other hand, the use of an approximation of larger order also explains, in part, the slightly better absolute residuals for the SpaRed-BT algorithm. However, the most important difference between the two methods lies in the execution times. For the two smallest problems, using a smaller number of computational resources (processors), the SpaRed-BT algorithm provides the reduced realization faster than the PLiCMR-BT algorithm. Furthermore, only the kernel from the SpaRed library can reduce the largest two cases, while doing so with the PLiCMR kernel would require a number of resources much larger than available in our cluster.

5   Concluding Remarks

We have compared two parallel BT model reduction algorithms using a problem arising from a semi-discretization of a controlled heat transfer process for optimal cooling of a steel profile. The algorithms are included in the parallel libraries for model reduction PLiCMR and SpaRed and basically differ in the generalized Lyapunov equation solver that is employed in each case: the PLiCMR-BT

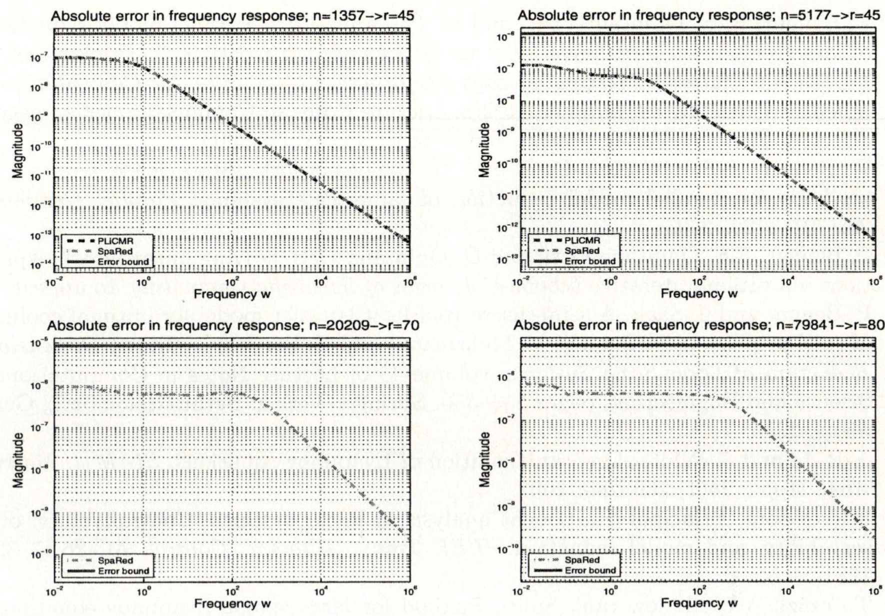

Fig. 3. Absolute error in the frequency response for the reduced-order realizations.

Table 3. Results for the SpaRed-BT routine.

n	#iter.	l	k	$\mathcal{R}_{W_c}(\hat{S})$	$\mathcal{R}_{W_o}(\hat{R})$	n_p	Time (sec.)
1,357	82	574	492	4.9e-24	1.0e-14	1	14.2
5,177	98	686	588	8.5e-23	1.5e-14	1	35.4
20,209	76	532	456	1.5e-14	2.0e-13	4	151.4
79,841	78	546	468	3.9e-13	9.6e-14	16	484.7

method is based on the sign function iteration while the SpaRed BT method employs a generalization of the LR-ADI iteration.

The experimental results show that, for the optimal cooling of steel profiles problem, the SpaRed-BT algorithm is clearly the best option. By exploiting the sparsity of the problem this algorithm requires less computational resources, provides the answer faster, and allows the reduction of problems which could not be dealt with using the PLiCMR-BT algorithm. On the other hand, for the small-dimension problems, the models computed by the PLiCMR-BT algorithm are presumably more accurate and can serve as a reference for those computed with SpaRed.

References

1. R.M. Badía, P. Benner, R. Mayo, and E.S. Quintana-Ortí. Parallel algorithms for balanced truncation model reduction of sparse systems. In *Proc. of PARA'04 – Workshop on State-of-the-Art in Scientific Computing*, Lecture Notes in Computer Science. Springer-Verlag, Berlin, Heidelberg, New York, to appear.

2. P. Benner, E. S. Quintana-Ortí, and G. Quintana-Ortí. Parallel model reduction of large-scale descriptor linear systems via balanced truncation. In *Proceedings of the 6th International Meeting on High Performance Computing for Computational Science. VECPAR'04*, number 3402 in Lecture Notes in Computer Science, pages 340–353. Springer-Verlag, Berlin, Heidelberg, New York, 2005.
3. P. Benner, E.S. Quintana-Ortí, and G. Quintana-Ortí. State-space truncation methods for parallel model reduction of large-scale systems. *Parallel Comput.*, 29:1701–1722, 2003.
4. P. Benner, E.S. Quintana-Ortí, and G. Quintana-Ortí. Solving linear matrix equations via rational iterative schemes. *Journal of Scientific Computing*, to appear.
5. P. Benner and J. Saak. A semi-discretized heat transfer model for optimal cooling of steel profiles. In P. Benner, V. Mehrmann, and D. Sorensen, editors, *Dimension Reduction of Large-Scale Systems*, volume 45 of Lecture Notes in Computational Science and Engineering, pages 353–356. Springer-Verlag, Berlin/Heidelberg, Germany, 2005.
6. J.-R. Li and J. White. Low rank solution of Lyapunov equations. *SIAM J. Matrix Anal. Appl.*, 24(1):260–280, 2002.
7. B.C. Moore. Principal component analysis in linear systems: Controllability, observability, and model reduction. *IEEE Trans. Automat. Control*, AC-26:17–32, 1981.
8. T. Penzl. A cyclic low rank Smith method for large sparse Lyapunov equations. *SIAM J. Sci. Comput.*, 21(4):1401–1418, 2000.
9. J.D. Roberts. Linear model reduction and solution of the algebraic Riccati equation by use of the sign function. *Internat. J. Control*, 32:677–687, 1980.
10. M.G. Safonov and R.Y. Chiang. A Schur method for balanced-truncation model reduction. *IEEE Trans. Automat. Control*, AC–34:729–733, 1989.
11. A. Schmidt and K. Siebert. *Design of Adaptive Finite Element Software; The Finite Element Toolbox ALBERTA*, volume 42 of *Lecture Notes in Computational Science and Engineering*. Springer-Verlag, Berlin/Heidelberg, 2005.
12. M.S. Tombs and I. Postlethwaite. Truncated balanced realization of a stable nonminimal state-space system. *Internat. J. Control*, 46(4):1319–1330, 1987.
13. A. Varga. Efficient minimal realization procedure based on balancing. In *Prepr. of the IMACS Symp. on Modelling and Control of Technological Systems*, volume 2, pages 42–47, 1991.
14. A. Varga. Model reduction software in the SLICOT library. In B.N. Datta, editor, *Applied and Computational Control, Signals, and Circuits*, volume 629 of *The Kluwer International Series in Engineering and Computer Science*, pages 239–282. Kluwer Academic Publishers, Boston, MA, 2001.
15. E.L. Wachspress. The ADI model problem, 1995. Available from the author.

Broadcast-Based Parallel LU Factorization

Fernando G. Tinetti* and Armando E. De Giusti

III-LIDI, Facultad de informática
Universidad Nacional de La Plata
50 y 115, 1er. Piso
1900 La Plata, Argentina

Abstract. This paper presents a parallel LU factorization algorithm designed to take advantage of physical broadcast communication facilities as well as overlapping of communication and computing. Physical broadcast is directly available on Ethernet networks hardware, one of the most used interconnection networks in current clusters installed for parallel computing. Overlapped communication is a well-known strategy for hiding communication latency, which is one of the most common source of parallel performance penalization. Performance analysis and experimentation of the proposed parallel LU factorization algorithm are presented. Also, the performance of the proposed algorithm is compared with that of the algorithm used in ScaLAPACK (Scalable LAPACK), which is commonly accepted as having optimized performance.

1   Introduction

Parallel computing on low-cost clusters is now a common approach in many scientific areas [3] [4]. Problems of linear algebra in general, and systems of equations in particular, have usually taken advantage of such parallel computing platforms to reduce the time needed to obtain a solution. In this context of linear algebra, there are some libraries avilable for parallel computing, such as ScaLAPACK [6] and PLAPACK [22]. Libraries specifically designed for linear algebra computing are used from many years ago. LAPACK (Linear Algebra PACKage) [2] is considered the *de facto* standard for the whole area of linear algebra applications, and the BLAS (Basic Linear Algebra Subroutines) library is well-suited for performance optimization, including parallelization [9].

Solving dense systems of equations is one of the most important tasks in the field of linear algebra. Furthermore, this problem is used as a benchmark for supercomputers, and the list of the 500 fastest computers is basically made by measuring the time to solve large systems of equations [12]. From the point of view of numerical processing, LU factorization is made following the well-known block processing approach, which is adopted for most (if not all) of the computing subroutines/algorithms in the field of linear algebra [16] [13]. For the parallel approach, bidimensional block cyclic decomposition is used in most of the approaches, included that of ScaLAPACK [10] [6]. Currently, the ScaLAPACK

---

* Investigador Asistente CICPBA

J.C. Cunha and P.D. Medeiros (Eds.): Euro-Par 2005, LNCS 3648, pp. 867–876, 2005.
© Springer-Verlag Berlin Heidelberg 2005

approach is considered good enough to be well established and used, and research is being conducted for workload balance in heterogeneous and/or dynamically changing environments [5] [7] [8]. A parallel LU factorization algorithm based on broadcast messages is proposed taking into account that:

1) Many of the current parallel computing platforms are low-cost clusters of computers interconnected by Ethernet [15]. On these clusters, physical broadcast is available even when switching (with Ethernet switches) is used to allow multiple point-to-point simultaneous communications. Furthermore, the standard Ethernet is being upgraded for better performance [20] [1] maintaining backward compatibility in general, and the broadcast address in particular. From the theoretical point of view, physical broadcasts are very interesting because of their natural scalability. In practical terms, however, synchronizations and/or acknoledgements are necessary and the time taken for these tasks is usually dependent on the number of computers involved in broadcasts. Also, some other interconnection networks support physical broadcasts or multicasts, in top of which a broadcast routine can be designed [17] [14].

2) The block processing LU factorization defines a very clear pattern of data dependence, since it is defined in terms of iterations with [2] [16]: a) single *current* block processing, on a small submatrix of the matrix being factorized, and b) *trailing* matrix update with the processed *current* block. This implies that a *current* block is needed for most of the processing in all the iterations and, if the matrix is distributed amongst two or more computers, the *current* block should be broadcasted to the computers with elements to be updated.

3) The LU factorization is similar enough to other factorizations such as QR, Cholesky, etc. to expect that the same approach could be applied at least in those factorizations with a similar processing pattern. Taking the QR factorization as an example, the processing pattern is almost the same as that of the LU factorization, even when the processing on a single block and the trailing matrix update are not the same for the two factorizations [10].

2   Broadcast-Based Parallel LU Factorization Algorithm

Some issues should be addressed for a broadcast-based parallel LU factorization:

1) High message latency (or startup) times penalizing parallel performance. Message latency on clusters are mainly due to a combination of the message passing interface/library/operating system overhead plus the latency of the interconnection hardware involving interface cards and, usually, switches or hubs.

2) The broadcast message routine implemented in message passing libraries does not usually take advantage of the physical broadcast facilities provided by interconnection networks (e.g. Ethernet). Message passing libraries such as the freely available implementations of MPI [19] are not focused on optimizing the broadcast message routine by using physical broadcast or multicast facilities. Instead, message passing libraries implementations usually optimize point-to-point routine/s and implement broadcast in terms of spanning trees [17] or just multiple messages from the broadcast sender (the broadcast *root*).

3) The parallel algorithm itself, since most (if not all) current parallel algorithms are based on bidimensional data distributions and simultaneous broadcasts on the two dimensions or broadcasts and point-to-point communications expected to be carried out simultaneously, such as in ScaLAPACK [6]. Given that physical broadcasts are going to be used to optimize broadcast messages, more than one message at a given time should not be scheduled, since it would imply performance penalization by sequential communications. It is clear that a single computer cannot receive a physical broadcast and another frame/packet at the same time.

Each one of the above three issues have been addressed fairly straightforward, and decisions can be evaluated either analytically or by experimentation. The strategy to avoid the *high message latency* penalization has been overlapping local computing with communication, which is very useful on parallel computing. This overlapping should be introduced in the algorithm, thus becoming a constraint/guideline for designing the algorithm. A *broadcast message routine* has been designed and implemented on top of the User Datagram Protocol (UDP) [18], whose broadcast facilities are implemented by most of the operating systems (e.g. Linux, Solaris, AIX) with physical broadcast on Ethernet networks. The parallel algorithm is designed with one-dimensional data distribution, which in the specicic case of LU factorization is useful also to avoid communications for selecting the pivot/s on each iteration. More specifically, a *row block cyclic partitioning* is made, with the block size (or *bs*, number of consecutive rows in a block) defined for performance tuning, as in ScaLAPACK. In either case, the block size is small enough to have many more blocks than computers, and blocks are cyclically distributed. The *row block cyclic partitioning* used for LU matrix factorization amongst computers P_0, P_1, and P_2 is shown in Fig. 1-a). A simplified pseudocode of the process running on computer P_i is shown in Fig. 1-b), where `nblocks` $= n/bs$, `Factorize` implies LU factorization with pivoting on a single block, `send_b` and `recv_b` are the routines to send and receive a broadcast message respectively, and `Update` -on a single block or on local blocks- implies several tasks: a) applying pivots, b) a triangular system solve, and c) a matrix multiplication. In fact, numerical processing (i.e. factorization and update) is the same as in the sequential case [2] except that every computer modifies only local data. More details on the sequential as well as parallel LU factorization algorithm can ba found in [21].

The idealized case in which computing and communication are overlapped and there is no overhead due to broadcast messages (except for the communication of the first block) is shown in Fig. 2. Most of the local computing time in each computer (shown as Proc$_i$ on Fig. 2) is mainly due the trailing matrix update which includes a matrix multiplication.

Time Required by Floating Point Operations. Processing requirements of the trailing matrix update depend on the iteration. Given a matrix of $n \times n$ elements, the block size bs, and starting iterations with $i = 1$, the matrix update in the i_th iteration is made on a submatrix of $(n - i * bs) \times (n - i * bs)$ elements. Furthermore, the trailing matrix update is defined as

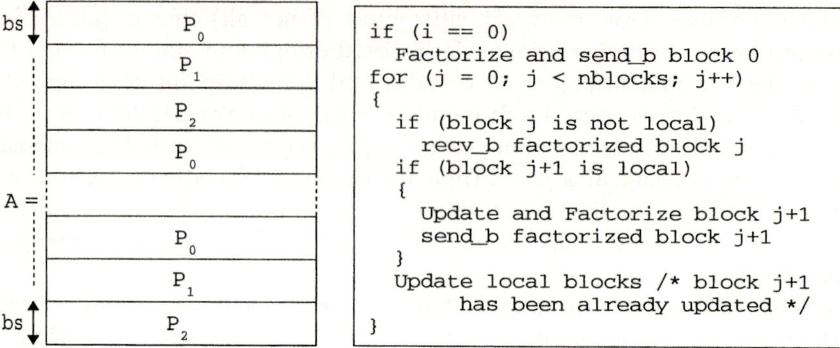

a) Row Block Cyclic Distribution.   b) Parallel LU Factorization Pseudocode.

Fig. 1. Row Block Cyclic Partitioning.

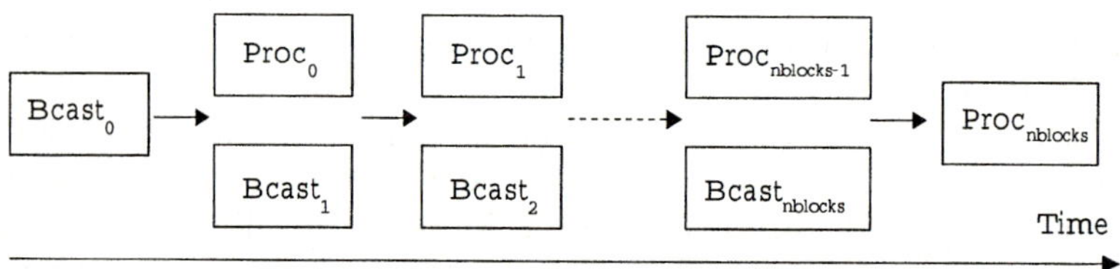

Fig. 2. Overlapped Computing and Communication.

$$tA - tL \times tU; \quad tA \in I\!R^{(n-i*bs)\times(n-i*bs)}, \ tL \in I\!R^{(n-i*bs)\times bs}, \ tU \in I\!R^{bs\times(n-i*bs)}$$

Thus, the number of floating point operations required for the trailing matrix update in iteration i, $FlopUpd(i)$, is given by

$$FlopUpd(i) = 2 * bs * (n - i * bs)^2 \tag{1}$$

because the matrix multiplication $tL \times tU$ requires $(n - i * bs)^2 * (2 * bs - 1)$ floating point operations and the matrix subtraction requires $(n - i * bs)^2$ floating point operations. Taking into account that the processing workload is evenly distributed amongst p computers, the time required for floating point operations on each computer in iteration i is given by

$$t(FlopUpd(i), p) = \frac{tf * FlopUpd(i)}{p} \tag{2}$$

where tf is the time required for a single floating point operation.

Time Required by Broadcast Communications. The timing model of point-to-point communications can be taken as a starting point:

$$t(m) = \alpha + \beta * m \tag{3}$$

where m is the amount of data to transfer, α is the communication latency or startup cost, and $1/\beta$ is the network communication bandwidth. Timing models for broadcast communication usually depend on the implementation selected for the broadcast routine in a specific implementation. In the specific case of the research presented in this paper, the broadcast message implementation is such that: a) data is physically broadcasted using UDP, and b) acknowledgements are received at the broadcast root from all the receivers to provide a reliable broadcast message. These details are hidden to the user (pertain to the broadcast implementation). Given that data is sent as in a point-to-point operation and there is a very low rate of message loss, the time required for data transmission through the network can be modeled as in the poit-to-point messages, i.e. $(\beta * m)$ in Eq (3), with $m = bs * (n - i * bs)$ on iteration i. However, ackowledgements sent from receivers to the root attempt against scalability, because these messages cannot be received simultaneously at the broadcast root. Even when there are multiple ways of avoiding such a performance drawback, it is still possible to analyze the time required by broadcast messages. Summarizing, the timing model for the broadcast operation in the i_th iteration is

$$t(bcast(i,p)) = \alpha_b + lpp * (p-1) + \beta * bs * (n - i * bs) \tag{4}$$

where α_b is the latency of the broadcast implementation independently of the number of computers involved, lpp is the latency per processor of the broadcast implementation, and $p - 1$ is the number of receivers in a broadcast operation. Summarizing, Eq. (4) is the timing model of the broadcast implementation made in the context of the research related with this paper.

Time Required by the Algorithm. Taking into account the pseudocode of Fig. 1-b) and the algorithm behavior described in Fig. 2, the time required to complete the parallel algorithm on p processors is given by

$$t(parLU,p) = \sum_{i=1}^{n/bs} max(t(FlopUpd(i),p), t(bcast(i,p))) \tag{5}$$

It is expected that the numerical computing time in the first iterations is greater than the time required by broadcast communications. Also, given that a) the trailing matrix is made smaller as more iterations are completed, and b) broadcast message latency is constant from the point of view of trailing matrix size,

$$t(FlopUpd(i),p) \geq t(bcast(i,p)); \quad i \leq k$$
$$t(FlopUpd(i),p) < t(bcast(i,p)); \quad i > k$$

and Eq. (5) *becomes*

$$t(parLU,p) = \sum_{i=1}^{k} t(FlopUpd(i),p) + \sum_{i=k+1}^{n/bs} t(bcast(i,p)) \tag{6}$$

3   Comparison with ScaLAPACK: Expected Time and Experimentation

The expected time for the Scalapack LU factorization algorithm is well known:

$$t(ScaLU, p) = \frac{2 * n^3}{3 * p} tf + \frac{(3 + log_2(p)/4) * n^2}{\sqrt{p}} \beta + (6 + log_2(p)) * n * \alpha_{ptp} \quad (7)$$

where α_{ptp} is the message latency for a point-to-point message [7] [8] [6], and the rest of parameters/coefficients have already been explained and used.

Some different points of view prevent a direct comparison among Eq. (7) and Eq. (6) above. The first term of Eq. (7) reflects the number of floating point operations in ScaLAPACK's timing model: $2/3 * n^3$. This is the *traditional* number of operations for the sequential LU factorization as given in the literature [16]. The timing model given for the proposed parallel algorithm takes into account that most of the computing time is needed for the trailing matrix update whose number of operations is given in Eq. (1) for the i_th iteration. However, both algorithms are directly based on the blocked LU factorization, so the number of floating point operations should be the same and it is not necessary a deeper comparison analysis to determine which one -Eq. (7) or Eq.(6)- is more accurate.

The ScaLAPACK timing model for communication is reflected in the second and third terms of Eq. (7). ScaLAPACK's communication costs are taken into account for every block/element of the matrix. On the other hand, for the approach proposed in this paper, Eq. (5) and Eq. (6) directly reflect that a broadcast communication adds time to the total expected algorithm time only when it is greater than the corresponding trailing matrix update time. Even when the numerical computing time is greater than the broadcast time in only a few iterations -e.g. $k < 20$ or $k < 30$ in Eq. (6)- the communication time (in those iterations) does not add time to the total processing time. However, the broadcast timing model of Eq. (4) is far from optimal and implies at least that the latency grows linearly with the number of processors. On the other hand, ScaLAPACK relies on spanning trees and, thus, the timing model implies a logarithmic growth depending on the number of processors.

Table 1. Cluster Characteristics.

Clock	Mem	Mflop/s (DGETRF)
2.4 GHz	1 GB	$\cong 2500$

Some simple experimentation will clarify the comparison on a real environment. Computers (PCs) used for experimentation have the characteristics summarized in Table 1, and the interconnection network is 100 Mb/s Ethernet with complete switching. Performance in Table 1 is given in Mflop/s using DGETRF, the sequential LU matrix factorization with double precision floating point number representation. The total number of available computers is 20, and experiments were made with 2, 4, 8, 16, and 20 computers. Matrix sizes are scaled up according to the number of computers and memory available. Local/sequential computing is made by using fully optimized ATLAS BLAS (Automatically Tuned Linear Algebra Software BLAS) [23]. ScaLAPACK communication is made as usual: BLACS (Basic Linear Algebra Communication Subroutines) implemented on top of MPICH implementation of MPI. Every possible bidimensional processors grid $P \times Q$ was considered for ScaLAPACK routines, e.g. for 16 processors, the experimental grids were: 1×16, 16×1, 2×8, 8×2, and 4×4. Also, square block sizes were used for ScaLAPACK routines: 16, 32, 64, 100 and 128. The proposed algorithm does not need to define a bidimensional processors grid, and the block values used for experimentation are the same as those used for ScaLAPACK routines.

Figure 3 shows the parallel perfomance measured as efficiency for LU matrix factorization on different number of computers from 2 to 20. The matrix order (size) for each number of computers is shown in parenthesis on the x axis. Bars show the best efficiency value obtained by the algorithms for each number of computers. Light gray bars labeled as "Sca" correspond to values obtained by ScaLAPACK's PDGETRF. Dark gray bars labeled as "Prop" correspond to values obtained by the proposed parallel LU matrix factorization algorithm. The proposed algorithm performance is better than that implemented in ScaLAPACK from the point of view of "raw" efficiency and performance degradation from 2 to 20 computers. It is worth to mention the similarity among the ScaLAPACK's results shown in Fig. 3 with those in [8], where ScaLAPACK is used for LU matrix decomposition and linear equation system solving. In Fig. 3 as well as in [8] the efficiency is about 0.5 (or 50% of the total available computing power). It is possible now to analyze the specific results regarding, for example, the advantage of broadcast overlapping of the proposed algorithm. For 20 computers, for example, the specific experimentation values are: $n = 45000$, $bs = 64$, the total number of blocks and iterations is 704 and

$$t(FlopUpd(i), p) \geq t(bcast(i, p)); \quad i = 1, \ldots, 363$$
$$t(FlopUpd(i), p) < t(bcast(i, p)); \quad i = 364, \ldots, 704$$

i.e. the first 363 broadcast messages do not add any time to the total elapsed time of the parallel algorithm on 20 computers. Thus, more than 51% of the broadcasts are completely made in background and this explains the very good parallel performance values shown in Fig. 3.

LU matrix factorization is specially penalized in ScaLAPACK's two dimensional matrix distribution due to the partial pivoting needed for numerical stability. Partial pivoting implies a collective communication in a row or a column of processors (for pivot selection) which implies a group communication penalization in an algorithm defined mainly for point-to-point communications. Given that the proposed parallel LU matrix factorization distributes data by column block or row block, this penalization is not found. Finally, the proposed parallel LU matrix factorization algorithm efficiency for 20 computers is about 7% worse than the efficiency for 2 computers, while SaLAPACK efficiency for 20 computers is about 23% worse than the efficiency for 2 computers.

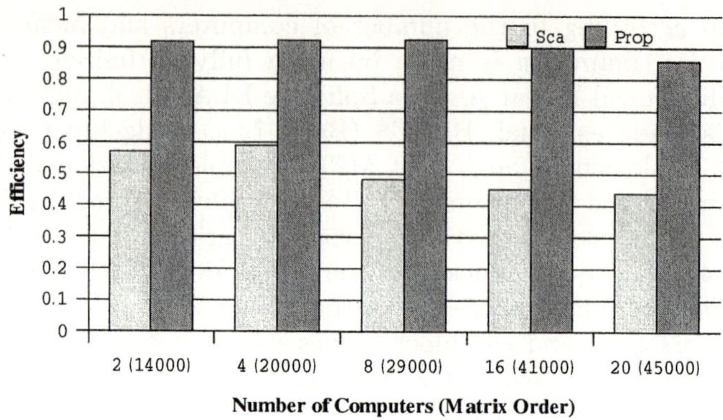

Fig. 3. LU Matrix Factorization Efficiency.

4   Conclusions and Further Work

A parallel LU factorization algorithm based on broadcast messages seems to be a good idea to optimize performance at least on clusters interconnected by networks with broadcast/multicast facilities, such as Ethernet and InfiniBand. Furthermore, the parallel LU matrix factorization algorithm presented in this paper is very simple and, thus, easy to understand and implement on clusters. A specific broadcast message routine has been implemented focusing Ethernet networks, and this routine has been successfully used in the parallel algorithm.

The performance analysis has been presented regarding the computing time of the proposed algorithm, which is dependent on the broadcast message implementation. Very good performance values are obtained in experiments by taking advantage of physical broadcast as well as overlapping communication with computing in the proposed parallel algorithm. These performance values are shown to be better than those obtained by the ScaLAPACK LU factorization algorithm, which is currently assumed to be optimized for distributed memory parallel computers. Furthermore, the proposed algorithm not only obtains better performance than that implemented in ScaLAPACK but also have better performance scalability as the number of computers is increased.

Other factorization algorithms from the field of linear algebra seem to be well suited for parallelization beased on broadcast messages. Factorization methods such as QR and Cholesky are among the inmediate candidates given their similarity in the numerical computing pattern, even when the individual operations are different from those in the LU factorization.

From the point of view of parallel hardware and clusters, it would be highly beneficial to experiment on more powerful clusters. The cluster computing power could be increased by having more computers as well as with computers with more processing power. Also, interconnection networks with performance better than Ethernet 100 Mb/s should be used. Immediate candidates in this sense are Ethernet 1 and 10 Gb/s.

Acknowledgements

Experimentation presented in this paper has been carried out at the Universidad Nacional de La Plata's JavaLab, with 20 IBM NetVista 8305-HRY, donated by IBM, as part of the Academic Cooperation Agreement signed between the University and IBM Argentina.

References

1. 10 Gigabit Ethernet Alliance (10GEA), 10 Gigabit Ethernet Technology Overview White Paper, May 2002. Available at http://www.10gea.org/10GEA_White_Paper_ Rev1_May2001.pdf.
2. Anderson E., Z. Bai, C. Bischof, J. Demmel, J. Dongarra, J. Du Croz, A. Greenbaum, S. Hammarling, A. McKenney, S. Ostrouchov, D. Sorensen, LAPACK Users' Guide (Second Edition), SIAM Philadelphia, 1995.
3. Anderson T., D. Culler, D. Patterson, and the NOW Team, "A Case for Networks of Workstations: NOW", IEEE Micro, Feb. 1995.
4. Baker M., R. Buyya, "Cluster Computing at a Glance", in R. Buyya Ed., High Performance Cluster Computing: Architectures and Systems, Vol. 1, Prentice-Hall, Upper Saddle River, NJ, USA, pp. 3-47, 1999.
5. Beaumont O., V. Boudet, A. Petitet, F. Rastello, Y. Robert, "A Proposal for a Heterogeneous Cluster ScaLAPACK (Dense Linear Solvers). IEEE Transactions on Computers, 50(10):1052-1070, 2001.
6. Blackford L., J. Choi, A. Cleary, E. D'Azevedo, J. Demmel, I. Dhillon, J. Dongarra, S. Hammarling, G. Henry, A. Petitet, K. Stanley, D. Walker, R. Whaley, ScaLAPACK Users' Guide, SIAM, Philadelphia, 1997.
7. Chen Z., J. Dongarra, P. Luszczek, K. Roche, "Self adapting software for numerical linear algebra and LAPACK for clusters", Parallel Computing 29, pp. 1723-1743, Elsevier B.V., 2003.
8. Chen Z., J. Dongarra, P. Luszczek, K. Roche, "The LAPACK for Clusters Project: an Example of Self Adapting Numerical Software", Proceedings of the 37th Hawaii International Conference on System Sciences, pp. 1-10, 0-7695-2056-1/04, IEEE, 2004.
9. Choi J., J. Dongarra, S. Ostrouchov, A. Petitet, D. Walker, R. Whaley, "A proposal for a set of parallel basic linear algebra subprograms", Technical Report CS-95-292, University of Tennessee Knoxville, LAPACK Working Note 100, May 1995.
10. Choi J., J. Dongarra, L. Ostrouchov, A. Petitet, D. Walker, R. Whaley, "The Design and Implementation of the ScaLAPACK LU, QR, and Cholesky Factorization Routines", Report ORNL/TM-12470, Sep. 1994.
11. K. Dackland and E. Elmroth. "Design and Performance Modeling of Parallel Block Matrix Factorizations for Distributed Memory Multicomputers", Proceedings of the Industrial Mathematics Week, pp 102-116, Trondheim, 1992.
12. Dongarra J., "Performance of Various Computers Using Standard Linear Equations Software", University of Tennessee, Knoxville TN, 37996, Computer Science Technical Report Number CS - 89 - 85, January 2005, http://www.netlib.org/benchmark/performance.ps.
13. Dongarra J., D. Walker, "Libraries for Linear Algebra", in Sabot G. W. (Ed.), High Performance Computing: Problem Solving with Parallel and Vector Architectures, Addison-Wesley Publishing Company, Inc., pp. 93-134, 1995.

14. InfiniBand Trade Association, InfiniBand Architecture Specification, Release 1.0, October 24 2000.
15. Institute of Electrical and Electronics Engineers, Local Area Network - CSMA/CD Access Method and Physical Layer Specifications ANSI/IEEE 802.3 - IEEE Computer Society, 1985.
16. Golub G., C. Van Loan, Matrix Computations, 2nd Edition, The John Hopkins University Press, 1989.
17. Liu J., A. R, Mamidala, D. K. Panda, "Fast and Scalable MPI-Level Broadcast Using InfiniBand's Hardware Multicast Support", Proc. 18th International Parallel and Distributed Processing Symposium (IPDPS'04), p. 10b, Santa Fe, New Mexico, USA, April 2004.
18. Postel J., "User Datagram Protocol", RFC 768, USC/Information Sciences Institute, Aug. 1980.
19. Snir M., S. Otto, S. Huss-Lederman, D. Walker, and J. Dongarra., MPI: The Complete Reference, Volume 1 - The MPI-1 Core, 2nd edition. The MIT Press, 1998.
20. Spurgeon C. E., Ethernet: The Definitive Guide, O'Reilly and Associates, ISBN 1565926609, 2000.
21. Tinetti F. G., "LU Factorization: Number of Floating Point Operations and Parallel Processing in Clusters", Technical Report LIDI PLA-001-2003, May 2003, available at https://lidi.info.unlp.edu.ar/~fernando/publis/LUops.pdf
22. van de Geijn R., Using PLAPACK: Parallel Linear Algebra Package, The MIT Press, 1997.
23. Whaley R. C., A. Petitet, J. J. Dongarra, Automated Empirical Optimization of Software and the ATLAS Project. Available at http://www.netlib.org/lapack/lawns/lawn147.ps

Topic 11
Distributed and High-Performance Multimedia

Laszlo Böszörmenyi, Max Mühlhäuser, Geoff Coulson, and Nuno Correia

Topic Chairs

Efficient resource management to handle multimedia data is one of the most important challenges of the next decade. The needs implied by multimedia sources may easily lead to data and processing explosion. The requirement to store, process, and manage large data sets naturally pose the question of programmable parallel processing systems in supporting and enabling multimedia technology. Furthermore, the indexing and retrieval of multimedia data includes time-consuming algorithms, thus high-performance architectures and algorithms are necessary in order to allow the use of multimedia databases and archives in real-world scenarios.

A number of novel and hard questions arise in this context, which can be answered only by applying techniques of parallel and/or distributed computing. The scope of this topic embraces issues from high-performance coding and retrieving over parallel architectures for multimedia servers, databases and information systems (including grids), up to highly distributed architectures in heterogeneous, wired and wireless networks. In short, the two main areas that were considered for this topic are (1) High-Performance Multimedia, including parallel and distributed algorithms for fast coding and retrieval of multimedia data or metadata and architectures and algorithms for multimedia servers, databases and information systems (including grids); (2) Distributed Multimedia, in particular, architectures and algorithms for QoS- and context-awareness in heterogeneous (wired and wireless) networks and distributed architectures related to MPEG, including novel ideas regarding "Universal Media Access" as defined in MPEG-21.

This year 13 papers of high value were submitted to this topic area. We thank all authors for their submissions. All papers were reviewed by 4 reviewers. 5 papers were selected to be presented at EuroPar 2005, in two sessions. One session is devoted to *Architectures for distributed multimedia services* and contains the papers *"Dynamic distributed collaborative merging policy to optimize the multicasting delivery scheme"*, *"Dynamic proxy-cache multiplication inside LANs"* and the *"Perspective for Lecture Videos"*. The second session, on *Coding based enhancements of video services*, contains the papers *"A Scene-based Bandwidth Allocation Scheme for Transferring VBR Streams on Clustered Video Servers"* and *"DCT Block Conversion for H.264/AVC video transcoding"*.

J.C. Cunha and P.D. Medeiros (Eds.): Euro-Par 2005, LNCS 3648, p. 877, 2005.
© Springer-Verlag Berlin Heidelberg 2005

Dynamic Distributed Collaborative Merging Policy to Optimize the Multicasting Delivery Scheme[*]

X.Y. Yang[1], Porfidio Hernández[1], F. Cores[2], A. Ripoll[1],
R. Suppi[1], and Emilio Luque[1]

[1] Computer Science Department, ETSE, Universitat Autònoma de Barcelona
08193-Bellaterra, Barcelona, Spain
[2] Computer Science & Industrial Engineering Department, EPS
Universitat de Lleida, 25001, Lleida, Spain

Abstract. The advance of Internet 2 and the proliferation of switches and routers with level three functionalities made the multicast one of the most feasible video streaming delivering techniques for the near future. Assuming this to be true, this study addressed the over-load situation that a streaming server could suffer due to client requests. As a solution, we proposed new multicast delivery scheme that allows every active client to collaborate with the server regardless of the video that they are watching, alleviating server loads, and therefore server resource requirements. The solution combined the multicast delivery scheme and client-side buffer collaboration in order to decentralize the delivery process. The new video delivering scheme was designed as two separate policies: the first policy used client collaboration to deliver first part of videos and the second policy could merge two or more multicast channels using distributed collaboration between a group of clients. Experimental results show that this scheme is better than previous schemes in terms of resource requirements and scalability.

1   Introduction

The high increase in the commercial use of the Internet (distance learning, Video on Demand (VoD) and digital video libraries) has generated a substantial growth in the demand for video streaming systems. In video streaming environments, users request the videos they desire and a server delivers the requested video information; allocating, using the most simple delivery technique, a dedicated server unicast channel for each video request. Even though the unicast delivery scheme is easy to implement, it is excessively expensive and there is a lack of scalability.

In order to reduce the cost of video-delivery and attain high server scalability, three complementary research approaches have been investigated: (1) server

---

[*] This work was supported by the MCyT-Spain under contract TIC 2001-2592 and partially supported by the Generalitat de Catalunya- Grup de Recerca Consolidat 2001SGR-00218.

J.C. Cunha and P.D. Medeiros (Eds.): Euro-Par 2005, LNCS 3648, pp. 879–889, 2005.

transmission schemes using multicast, this strategy allows users to share server and network bandwidth to reduce the individual service cost; (2) video streaming technique with application layer multicast enables multicast transmission schemes beyond a local area network, assuming only IP unicast at the network layer; and (3) proxy caching [6], enabling high scalability for clients dispersed across a wide-area. The main focus of this study is the design of multicast delivery in order to reduce the individual service cost, specially, we proposed a delivery scheme that is able to offer true VoD services[10].

Sophisticated video delivery techniques based on multicast have appeared such as Batching [4], Patching [1][2], Adaptive Piggybacking [3], Merging[5], Chaining [7] and Cooperative Video Cache(CVC)[8].

With a Batching technique, video requests for the same video that are submitted in the same short interval time are served by a single multicast channel. Clients suffer a certain period of waiting time and the average length of waiting time depends on the policy of selecting the clients to serve with the first available channel. Due to this waiting time, a Batching approach only provides near-VoD service. A Batching approach is also called static multicast since late coming requests are not allowed to join any already on-going multicast channel. With Patching, however, clients are dynamically assigned to join multicast channels. Since late coming clients miss part of the video information, a separate unicast channel, called a patch stream, is needed to deliver the first part of the video. The Patching approach assumes that clients can simultaneously download two streams and has a local buffer, capable of saving t minutes of video. While a client is watching a video from the patch stream, the video information arriving from multicast channels is buffered. Even though the Patching policy provides true-VoD service, the server resource requirement increases depending on the request arrival frequency due to unicast channels. Furthermore, a request is only able to join a multicast channel if the difference between the request arrival time and the multicast channel start-time is lower than t.

Like Patching, Adaptive Piggybacking and Merging are also dynamic multicast approaches. In the Piggybacking policy, the server slows down and speeds up the delivery rate of two consecutive multicast channels in order to merge two multicast channels into one. The number of channels that Piggybacking can merge is limited by the fact that less than 5% adjustment of the delivery rate is allowed, in order to preserve the display quality that clients receive. The Merging policy, however, does not change the display quality. Two multicast channels are merged using the client buffer. In the Merging policy, while clients are playing the video, they try to buffer video information from a previous multicast channel. This policy can only merge channels that are started in a period of time no longer than the length of video information that each client is able to save in their buffer.

The main ideas behind Chaining and CVC are fairly similar. Both policies are based on the creation of a delivery chain in which video information is forwarded from one client to another. With these policies, a new client receives the video from an existing chain and does not consume any server bandwidth. However, delivery chains could only be formed if the interarrival times of client requests are short, limited by the size of each individual client's buffer. Furthermore, only clients that are watching the same video can take part in the formation of the chain.

In this paper, we propose a new delivery technique called Dynamic Distributed Collaborative Merging (DDCM). The DDCM technique is based on the peer-to-peer paradigm and allows every active client to collaborate with the server regardless of the video that they are watching. The client collaborations are performed by two complementary delivery policies. Under the first policy, while successive incoming requests are allowed to join an existing multicast channel, the missed video information (patch stream) is delivered by another client who is playing the same video. Unlike the Patching policy, the patch stream does not consume the server's resources. The aim of the second policy is to dynamically merge multicast channels using distributed buffers. More than one client of different videos could be used in the merging process of two channels. The merge policy is able to merge multicast channels regardless of the time between their start-times. The merge policy enables clients of unpopular videos to help the server to merge the channels of more popular videos, and vice versa.

The rest of the paper is organized as follows: in section 2, we show the key ideas behind DDCM. Performance evaluation is shown in section 3. In section 4, we indicate the main conclusions of our results and future work is explained in the final section.

2 Dynamic Distributed Collaborative Merging Scheme

In the delivery scheme design, we assume that clients are able to hold two symmetric channels. We assume that video information is encoded with Constant Bitrate (CBR) and that each client channel is able to receive/send one video stream. We refer to network unicast channel that delivers the first part of a video as patch stream and the multicast channel that delivers the information for the complete video as complete stream.

The DDCM delivery scheme is designed as two separate policies: 1) Patch Stream Manager (PSM) whose main role is to deliver patch streams using client collaborations. 2) Complete Stream Manager (CSM). The main function of this second policy is to try to merge two or more complete streams into one.

2.1 Patch Stream Manager Design

When the first request from client C_i arrives time t_i, the server opens a new complete stream($M1$) for the client. When a second request from client C_{i+1} arrives in time t_{i+1}, the server decides whether or not the client can be served by using a previous complete stream ($M1$). In order to serve by using a previous complete stream, client(C_{i+1}) must have enough buffer to save more than $(t_{i+1} - t_i)$ seconds of video information from the complete stream. If not, the server will open a new complete channel. In the other case, a patch stream is needed to send video information from 0 to $(t_{i+1} - t_i)$. The remaining video information $((t_{i+1} - t_i)$ to the end) will be sent by the previous complete stream.

For each new patch stream, the PSM policy searches for an active client that has the first part of the video in their local buffer. In such a case, a collaborative client will open a patch stream and send the first part of the video. Client C_{i+1} will also join the previous multicast complete stream. Should there be no such client, the server starts a new patch stream using server bandwidth.

Fig 1 shows the delivery process following PSM for 6 clients. Each client has 3 minutes of buffer and client requests arrive at minutes 1, 2, 3, 5, 6, and 7. Under the clients name, we indicate the length of buffer that each client is dedicated for the collaboration. For example, C3's collaboration buffer is 1 minute, while C2 collaborates with a buffer of 2 minutes. These values depend on the length of the patch stream that the client needs for the delivery process. In the case of C1, no patch stream is needed, so the full buffer (3 minutes) is used for collaboration. In order to know the buffer size that each client dedicates to collaboration, each client sends a control message to the server when the client has filled the buffer with the first part of video.

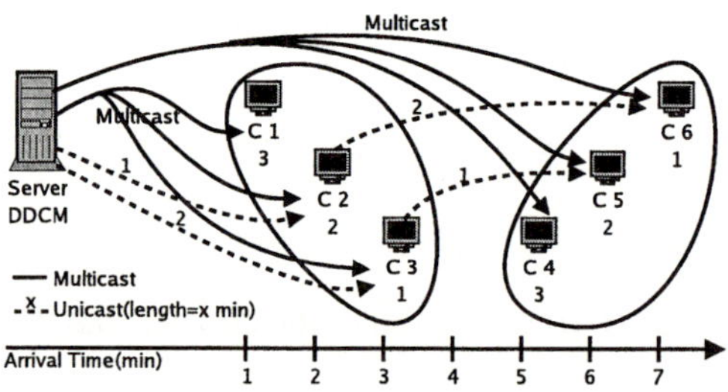

Fig. 1. Patch Stream Manager

Fig 1 shows that C2 and C3 are served using one multicast channel and 2 patch streams using server bandwidth. In the case of C5 and C6, the patch streams are delivered by C3 and C2 respectively. As we can see in Fig 1, the PSM policy is capable to deliver patch streams without consuming the server's bandwidth after minute 5.

The more clients are accepted by the server, the more client collaborations will be produced with PSM policy. This characteristic makes the PSM especially suitable as a delivery scheme for highly demanded video where a lot of patch streams are needed. However, after several minutes, the server has more than one client that is able to serve the same patch stream. This redundancy implies poor client resource utilization since many of clients will not be involved in the collaboration mechanism of PSM.

2.2   Complete Stream Manager Design

The CSM's aim is to merge the existing complete streams. Once a complete stream is merged into another, the complete stream will not consume any server resources. Since complete streams are usually long, and therefore demand most of

the server's resources, the CSM efficiently replaces the server resource demand
with client collaborations. The CSM scheme achieves a high degree of client
resource utilization since almost every client is involved in the collaboration
mechanism regardless of the video that they are watching.

Given two multicast channels (M1 and M2), the key idea of CSM is that
a group of clients form a collaborative buffer to merge M2 with M1. Then,
the multicasting channel ($M2$) from the server is replaced by a channel ($M2^1$)
from the collaborative clients. Since more than one client could be used in the
merging process, the CSM is able to merge multicast channels regardless of the
time between their start-times.

Fig 2 shows the collaborative buffer created between clients C1, C2, C3 and
C6 that collaborate by providing 3, 2, 1 and 1 minutes of buffer respectively.
A total of 7 minutes of video information could be saved in this collaborative
buffer.

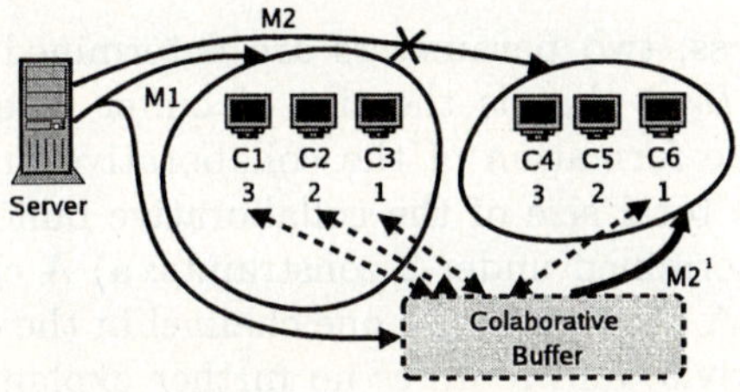

Fig. 2. Complete Stream Manager

Each client of the collaborative group successively saves video information
from M1 and then delivers the information to M2 when clients of this channel
need it. Fig 3 shows the delivery process of two multicasting channels($M1$ and
$M2$). The $M2$ was started 4 minutes(S) later than $M1$. The merging process
starts when block C and 8 are being delivered to $M1$ and $M2$ respectively.
Channel $M2$ is closed in time 4, and a new channel $M2^1$ is opened.

In order to know what information each client has to save, the video information of M2 is divided into blocks and enumerated. The CSM decides the list
of blocks that a client has to save. For example, $C1$ saves blocks [C,D,E], [J,K,L]
and so on. After saving the blocks [C,D,E] in time 0-3, $C1$ waits 1 minute before
starting to send the video information to $M2^1$. Block E is sent to $M2^1$ in time
6 and, after that, the $C1$ starts to save blocks [J,K,L].

Once the channel $M2$ is merged with $M1$, the CSM has to guarantee that
while a client is delivering video blocks, there are enough other clients that are
saving other video blocks being delivered by the server with $M1$. Since, each
client can only use one stream in the collaboration, either to deliver or to save
video information, the two processes (delivery and saving) have to be performed
separately. In the case of Fig 3, while C4 is delivering block I, it should not
receive any video information except the video that C4 is playing.

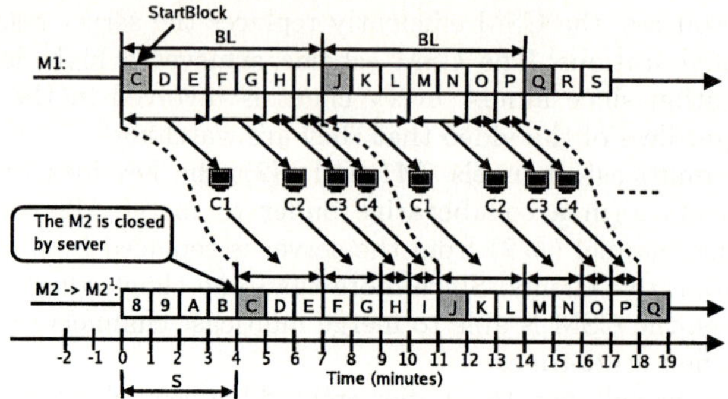

Fig. 3. CSM delivery process

2.3   Client Collaboration Group Construction Process

In the merging process, two parameters are determined by the CSM: 1)The client collaboration (B_{C_i}) that is the size of buffer of each client C_i that is to be dedicated to the formation of the collaborative buffer. 2) Accumulated buffer size that is the total size of the collaborative buffer. The value of these two parameters is determined under 2 constraints: a) A client cannot use more buffer than it has. b) A client only use one channel in the collaboration process.

Constraint a) is trivial and requires no further explanation. We established two conditions for the CSM group construction process in order to satisfy constraint b). Supposing that the CSM is interested in merging two channels that are separated into S units of time. We can formulate these 2 conditions as follows: Given a collaboration group CG of clients $\{C_1, C_2, ..., C_n\}$ [1] in which the buffer collaborations for each client are $\{B_{C_1}, B_{C_2}, ..., B_{C_n}\}$, the CSM has to satisfy:

1. Maximum collaboration: the collaboration (B_{C_i}) of a client C_i can not be greater than the value of S.

$$B_{C_i} \leq S \qquad \text{for all } C_i \in CG \tag{1}$$

2. Minimum accumulated buffer size: The total accumulated buffer size(BL) has to be bigger or equal ($S + \max\{B_{C_i}\}$).

$$\left(\sum_{C_i \in CG} B_{C_i} = BL\right) \geq S + \max\{B_{C_i}\} \text{ for all } C_i \in CG \tag{2}$$

Satisfying conditions (1) and (2), unconditioned by S, we get:

$$BL - B_{C_i} \geq S \geq B_{C_i} \text{ for all } C_i \in CG \tag{3}$$

---

[1] In the selection of clients, local network connection distance is considered in order to minimize the local network overhead.

The condition (3) indicates that the accumulated buffer size of all groups except for a client C_i is always bigger than C_i's collaboration(B_{C_i}) and is bigger than S. This means that the CSM guarantees that while a client C_i is sending video information there are enough other clients saving video information from the earlier multicast channel. Furthermore, while a client C_i is saving video information, there are enough other clients sending information. Since a client does not to have save information while it is sending, or vice versa, the CSM guarantees that in the collaboration process, each client will not use up more than one channel, leaving another one for playback. In this way, the CSM constructs the collaboration group in accordance with the following steps:

Step 1: The CSM calculates S of every pair of channels which could be merged and chooses the pair with the smallest S as channels to be merged.

Step 2: Satisfying conditions (2), the DDCM forms a list of clients $\{C_1, C_2, ...,$ $C_n\}$. In this step, the maximum collaboration of each client C_i is limited by the condition (1).

Step 3: Blocks of video (Vb_j) that a client C_i has to save and deliver are determined by: $(Vb_j - StartBlock) \bmod BL \geq \sum_{m=1}^{i-1} B_{C_m}$ and $(Vb_j - StartBlock) \bmod BL < \sum_{m=1}^{i} B_{C_m}$ where $BL = \sum_{i=1}^{n} B_{C_i}$ (the total accumulated size of collaborative buffer), B_{C_i} is the collaboration of client C_i and $StartBlock$ is the block number which indicates the starting point of the merging process.

3   Performance Evaluation

We have used our prototype to evaluate the performance of the DDCM. There are three key questions that we are interested in addressing: 1) how much reduction in server bandwidth could be achieved using DDCM in accordance with the video's popularity? 2) How much server bandwidth is required using DDCM when the system is offering more than one video? 3) How could the client collaboration following the DDCM scheme help in a high-demand situation?

The DDCM is implemented in our prototype using C++ language under Linux system. We have implemented the entire necessary client feature in a Xine player plug-in[9]. In the experimentation, clients are emulated using a cluster of PCs and client requests are generated following a Poisson$(P_k = \frac{\lambda^k}{k!} \cdot e^{-\lambda})$process. The Zips-like$(P_x = \frac{1}{x^z \cdot \sum_{i=1}^{Sv} \frac{1}{i^z}})$ distribution is used in order to assign the popularity of videos. We assume that the video length is 90 minutes and clients is able to save up to 5 minutes of video information.

3.1   Server Bandwidth Requirement According to Video Popularity

Fig 4 shows the server bandwidth requirement, in number of streams, using Patching, merging, PSM and DDCM(PSM+CSM). We perform this experiment under various request rates, which are normalized as the number of requests arriving during 90 minutes (video length). The resource requirement of a Merging

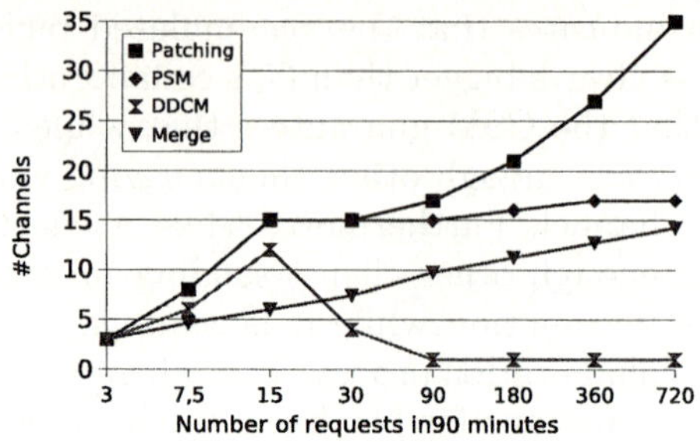

Fig. 4. Server Bandwidth Requirement for one video

policy is determined by results from [5]. We should point out that no buffer constraint is considered in Merging policy and, in a real case scenario, the Merging policy could only merge two streams separated by no more than client buffer length, so the performance will not be as good. The key observations from Fig 4 are:

1) Using Patching policy, the bandwidth requirement increases with more requests. This makes the Patching policy unsuitable for a high demand video service.

2) Under PSM, clients can collaborate with the server to deliver patch streams. Regardless of the interarrival time, the server does not need any more than 18 streams to serve a video. This makes the PSM more suitable than a Patching policy for serving popular videos.

3) The main virtue of the DDCM(PSM+CSM) could be summarised as, 'more request less server bandwidth'. As we can see in Fig 4, the service bandwidth requirement of the DDCM increases up to 12 streams. Up to this point, there are not so many client resources that can be used to merge complete streams. As soon as a critical mass of client resources are collected, the CSM tries to merge consecutive streams and the bandwidth requirement drastically drops to 1-2 streams per video. Compared with a Patching policy, the DDCM(PSM+CSM) is able to achieve a resource reduction of 73% (4 vs. 15 streams) if there are 30 requests during 90 minutes (one request per 3 minutes). Reduction of 92.5% (2 vs. 27 streams) is achieved when there are 360 requests. Comparing with Merging policy, the DDCM(PSM+CSM) does not reduce the required resource until 30 requests. However, with 180-720 requests, the Merging policy gets closer to PSM (10-14 streams) and the DDCM(PSM+CSM) reduces the bandwidth consumption up to 85.71% (2 vs. 14 streams).

3.2   Service Bandwidth Requirement for Multiple Videos

In order to measure the bandwidth requirement of a server that is offering more than one video, we suppose that the catalog is 30 to 550 videos. We consider that the number of requests that arrive during the video length(90Minutes) is from 90 (low client activity) to 4500 (high client activity).

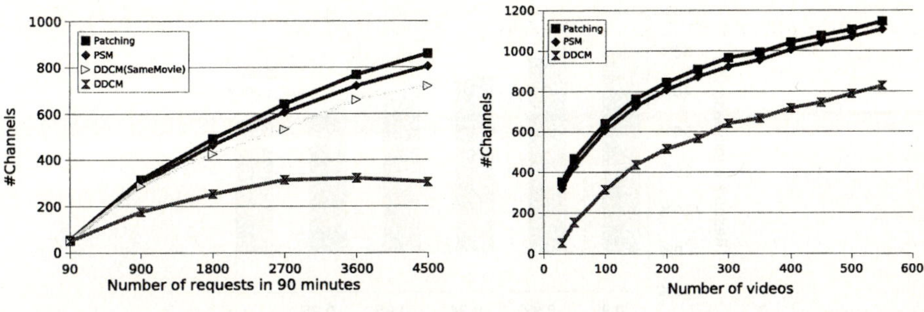

a)100 videos, 90-4500 requests in 90min   b)30-550 videos, 2700 requests in 90min

Fig. 5. Server bandwidth requirement for multiple videos

Fig 5 a) shows the server bandwidth needed to serve 100 videos. The Patching policy shows an increase in bandwidth requirement where there is high client activity (768 and 859 channels in order to serve 3600 and 4500 requests in 90 minutes). The PSM reduces the required stream up to 6.4%(804 vs. 859), while DDCM(PSM+CSM) reduce up to 64.61% (304 vs. 859).

We also obtained the bandwidth requirement if a client is not able to collaborate with the server to merge the channels that are not delivering the same video as the one that the client is playing. As we can see in Fig 5 a) (DDCM(SameMovie)) the requirement is clearly higher than the DDCM without this restriction. These results justify our delivery policy design.

Fig 5 b) shows the server bandwidth requirement according to the size of the catalog. We have supposed that 2700 requests arrive in 90 Minutes. Regardless of the delivery policy, the bandwidth requirement increases in accordance with the number of videos. The DDCM shows a requirement reduction of between 85.23%(30 videos) to 27.77%(550 videos). The DDCM is able to reduce the number of required channels by 300-344.

3.3   Circumstantial Workload Variations

In this section we are interested in measuring the server's capacity to face circumstantial workload variations. Suppose the following situation: we are designing a VoD system for 3600 clients and, for most of the time, only 50% of the clients are active. Taking the equipment cost into consideration, the VoD server could be designed for a particular, acceptable blocking probability. Most of times, the server is able to attend to all the client requests(20 requests/minute). However, in special situations, such as the Olympic Games, all 3600 clients may decide to request videos at same time (40 requests/minute). Furthermore, since the most of population is interested in this event, the video's popularity distribution could change, increasing the skew parameter of Zipf-like distribution.

Fig 6 shows the requirement variation when twice the number of client requests reach the server. With a Patching policy, the resource requirement increases by 51.58% to 34.74% depending on the skew parameters variation. As

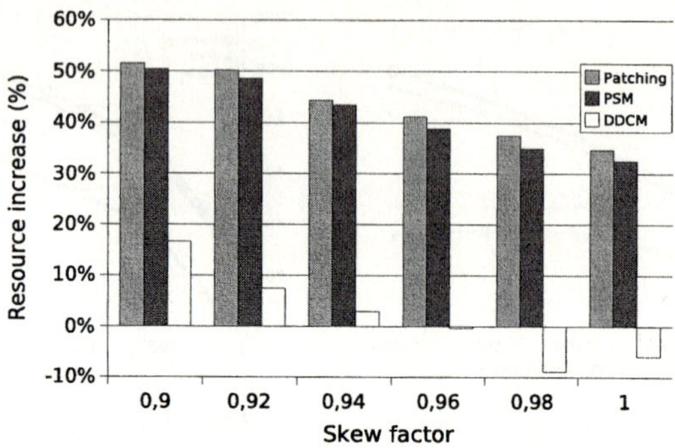

Fig. 6. Requirement increase in circumstantial workload variation

the skew fact increases, the resource requirement variation gets lower. With PSM, the resource increases 32.60% when the skew fact increases to 1 from 0.9. In this case, the PSM is 6.1% better than a Patching policy that produces an increase of 34.74%. DDCM policy produces a maximum increase of 16.67% if there is no variation in popularity distribution. In the worst case (skew fact 0.9), the DDCM is 67.68% better than a Patching policy in terms of increase in resource requirement.

4   Conclusions

We have proposed and evaluated a new video delivery technique called Dynamic Distributed Collaborative Merging that enables clients to efficiently collaborate with VoD servers. With DDCM policy, every client is able to collaborate with server, regardless of the video that they are watching. Instead of independent collaborations between the server and a client, the DDCM synchronizes a group of clients in order to merge multicast channels to achieve a better network efficiency.

Our experimental results show that DDCM has lower resource requirements than Patching policy, achieving reduction up to 92.5%. Offering multiple videos with high client activity, the DDCM is able to reduce the resource requirement up to 64.61%. These results corroborate the high scalability of DDCM when the number of requests is high. The DDCM achieves a more suitable investment in VoD server resources, since the client's punctual variation in the demand is covered by client contributions. Experimental results show that the DDCM is 67.68% better than Patching policy in terms of increase in resource requirement, suggesting that DDCM is more suitable delivery policy for VoD, in which the number of active clients changes over time.

5   Future Work

In this study, we have not considered the client network load that will suffer due to our policies and more research will be needed. However, we would like to point out that multicast schemes are usually effective in local networks. Fault tolerance is another pending question that should be carefully analyzed.

References

1. Y. C. K. A. Hua and S. Sheu. Patching: A multicast technique for true video-on-demand services. In *ACM Multimedia Conf., Bristol, U.K.*, 1998.
2. L. Gao and D. Towsley. Threshold-based multicast for continuous media delivery, 2001.
3. L. Golubchik, J. C. S. Lui and R. R. Muntz, Adaptive piggybacking: a novel technique for data sharing in video-on-demand storage servers, in *Proc. of ACM Multimedia Systems,* 1996, pp. 140-155
4. C. C. Aggarwal, J. L. Wolf and P. S. Yu, "On Optimal Batching Policies for Video-on-Demand Storage Servers", *In Proc. of the IEEE International Conference on Multimedia Computing and Systems, Hiroshima, Japan,* June, 1996.
5. M. K. V. D. L. Eager and J. Zahorjan. Bandwidth skimming: A technique for cost-effective video-on-demand. *Multimedia Computing and Networking 2000, San Jose, CA,* 2000.
6. F. Cores, A. Ripoll, E. Luque, Double P-Tree: A Distributed Architecture for Large-Scale Video-on-Demand, *Euro-Par 2002, LNCS 2400,* pp. 816-825, Aug. 2002.
7. K. A. Hua, S. Sheu, and J. Z. Wang, *Earthworm: A Network Memory Management Technique for Large-Scale Distributed Multimedia Applications,* Proc. IEEE INFOCOM'97, Kobe, Japan, Abril 7-11, 1997, pp. 58-66
8. de Pinho, Leonardo Bidese, Ishikawa, Edison, de Amorim, Claudio Luis, *Glove: A Distributed Environment for Scalable Video-on-Demand Systems,* International Journal of High Performance Computing Applications 2003 17: 147-161
9. http://xinehq.de/, January, 2005
10. H. Ma and K. G. Shin, *Multicast Video-on-Demand Services,* ACM Communication Review, pp. 31-42, January, 2002

Dynamic Proxy-Cache Multiplication Inside LANs

Claudiu Cobârzan*

"Babeş-Bolyai" University, Computer Science Department,
Mihail Kogălniceanu 1, 400084 Cluj-Napoca, Romania
`claudiu@cs.ubbcluj.ro`

Abstract. Proxy-cache deployment in LANs has become a current practice with well known benefits. For situations when a proxy-cache comes under constraints, due to increased load, and has to drop requests or perform cache replacement, we propose the alternative solution of cache-splitting. This means to dynamically deploy additional proxy-caches inside the LAN, and divert towards them some of the requests addressed to the original proxy-cache(s). By doing this, better response time, load balancing, higher availability and robustness of the service can be achieved than when using a single proxy-cache.

1 Introduction

The constant increase in both volume and demand of multimedia data in the Internet tends to stress the existing infrastructure. The main factors are the characteristics of multimedia data (e.g. size, bandwidth requirements) which highly differ from those of typical web data. The traditional way to cope with such situations is to deploy proxy-caches at LAN edges. Under certain conditions a single proxy-cache does not suffice, so multiple proxy-caches have to be used. Cooperative caching has been introduced for web caches, e.g. Harvest [4] and Squid [18], and for video caches as well, e.g. by Brubeck and Rowe [3] and MiddleMan [1].

Our paper proposes a novel proxy-cache system that is able to "spawn" new proxies via split operations whenever the actual situation demands it. Examples of such situations include extremely high load and severe storage constraints on the proxy. In those cases, one additional proxy-cache in the LAN would help lower the load on already running proxy-caches as well as increase the capacity of the "federate" cache.

The system we propose dynamically adjusts the number of running proxies in the LAN, depending on the load and on client request patterns, by either spawning new proxies on periods with high activity or putting them in a "hibernate" state or even stopping them on periods with low activity.

---

* The work was done during the stay at Klagenfurt University, Austria, within the CEEPUS (Central European Exchange Program for University Studies) framework, in winter 2004-2005.

J.C. Cunha and P.D. Medeiros (Eds.): Euro-Par 2005, LNCS 3648, pp. 890–900, 2005.

2   Proxy-Cache Splitting

There are situations when having a single proxy-cache in the LAN does not suffice, for example when servicing large, popular content to many clients, or when the volume of requested data puts the proxy-cache under constraints (cpu, mem, storage, etc.). In those cases, requests have to be rejected in order to lower the load on the machine and cache replacement has to be performed in order to free disk space. We state that in some circumstances it would be more beneficial to just deploy an additional proxy-cache inside the LAN. This new proxy-cache could take over some of the load on the existing proxy-cache(s) and by doing so, avoid both request dropping and cache replacement. On the other hand, if current and maybe predicted future load could be handled by a smaller number of proxy-caches than those currently active, then some of them could enter a "hibernating" state or could be shut down (stopped).

The distributed architecture we propose, assumes the deployment of two types of entities: the `dispatchers` and the `daemons`. The `dispatchers` are processes/threads that run on the same node as the proxy-cache and can be seen as front-ends of the proxy-caches which:

- *handle incoming requests* - serve them either from the local cache, or from the origin server; if this is not possible, the requests are forwarded to other active `dispatchers`/proxies in the LAN or they are discarded; a request is forwarded to the best candidate (the `dispatcher` that has a cached copy of the requested object or the one with the smallest load);
- *manage the proxy code* - archive the proxy code and send it to the location on which a new proxy-cache is to be spawned (using the `daemon` running on the selected target)
- *manage the "child" proxy-cache processes* - the `dispatchers` are responsible with stopping/ pausing/ restarting a "child" proxy-cache depending on various conditions (global load, volume of the clients' requests, volume of streamed/stored data, etc.)

The `daemon` processes/threads run in the ideal case on every node of the LAN and are responsible for:

- *managing clients' requests* - the `daemon` either directly receives, or it intercepts the client requests and then decides to forward them to the appropriate proxy-cache, depending on the local available knowledge about the global state of the proxy-cache "federation"
- *managing the proxy code* - the `daemon` receives/compiles the code sent by a `dispatcher` that initiates a proxy split operation;
- *managing the local proxy-cache process* - stops/ pauses/ restarts it, either as a result of incoming requests from its "parent" proxy, or depending on specific local conditions (load, storage capacity).

In the case the `daemon` thread/process hosted on a certain node crashes, the clients from that node still have access to the federate cache as long as the proxy-cache(s) set as default in the client's browser is/are still running. This is also true for the clients from nodes with no running **daemons** at all. The **daemon** is essential in the proxy-cache splitting process, as it is used by the "parent" to transfer its code if it is not already available at the selected node.

2.1 The Model of the Proposed Distributed Proxy-Cache Architecture

We consider that the number of nodes in the LAN is **n**. Let **P** be the set of available proxy-caches (there is at least one running proxy cache in the LAN):

$$P = \bigcup_{i=1}^{k} P_i, k = |P|, 1 \leq k \leq n$$

A proxy-cache $\mathbf{P_i}$ is defined as follows:
$$P_i = (maxResources_i, minResources_i, LC_i), i = 1..k$$
where:

- $maxResources_i$ - represents the maximum amount of resources that can be used by the proxy-cache:
  $$maxResource_i = (maxCpu_i, maxMem_i, maxCapacity_i, maxLan_i)$$
  namely the maximum amount of CPU power, memory, storage space and external bandwidth;
- $minResources_i$ - represents the minimum amount of resources that have to be used in order to serve any client's request. It is defined in a similar mode with the $maxResources_i$;
- LC_i - the content of the local cache
  $$LC_i = \{c_{ij}, j = 1..q\}, q = \text{the number of cached objects}$$

An object c_{ij} is defined as:
$$c_{ij} = (size(c_{ij}), timeLastAccess(c_{ij}), hitCount(c_{ij}), qualityValue(c_{ij}))$$
where $size(c_{ij})$ is the size of the object, $timeLastAccess(c_{ij})$ indicates the last time the object has been requested, $hitCount(c_{ij})$ shows the number of times the object has been served from the cache and $qualityValue(c_{ij}) \in [0..1]$ is the measure of the object's quality (based on the actual characteristics of the video object, such as resolution, color information, etc.)

The $qualityValue$ is a relative value that shows the degree in which the cached object matches the desired quality of a certain class of users. A value equal or close to 1 corresponds to the objects that have exactly or almost the desired quality, while values close to 0 are assigned to objects that show the most drastic difference between actual and desired quality. High absolute quality does not necessarily mean that the $qualityValue$ is close to 1. For example, if the vast majority of the users have only limited display size, say 800x600, a video object encoded at 1280x1024 will have a qualityValue closer to 0 than to 1, because further operations (e.g. transcoding) have to be performed in order to deliver the object to the requesting clients.

For each object c_{ij}, a *utility* value can be computed using a function $u : LC_i \rightarrow \mathbf{R}$:

$$u(c_{ij}) = const_1 * size(c_{ij}) + const_2 * \frac{1}{timeLastAccess(c_{ij})} +$$
$$+const_3 * hitCount(c_{ij}) + const_4 * qualityValue(c_{ij})$$

where $const_1, const_2, const_3, const_4 \in [0,1]$ and $const_1 + const_2 + const_3 + const_4 = 1$ ($u(c_{ij})$ is computed as a weighted average of the different characteristics of the cached video object).

Those constants can be fixed when the proxy-cache is started and remain the same during the run period of the proxy-cache. Another possibility that needs further investigation would be to dynamically modify those values when traffic conditions, load level, request rate, etc. reaches certain values, in order to maximize the byte hit ratio. The utility value of the cached objects is used to decide which objects get discarded when performing cache replacement.

We use **D** to denote the set of `dispatchers`:

$$D = \bigcup_{i=1}^{k} D_i, k = |P|, 1 \le k \le n.$$

As each `dispatcher` corresponds to a certain proxy-cache, there is a function f (bijection), $f : D \to P, f(D_i) = P_i, \forall i \in \{1,..,k\}$ (a proxy P has exactly one `dispatcher` D). One `dispatcher` **D**ᵢ is defined as follows:

$$D_i = (P_i, GC, GU, siblings_i), \forall i \in \{1, ..., k\}, k = |P|, 1 \le k \le n$$

where:

- P_i - the corresponding proxy
- GC - the content of the global cache (viewed as the union of all local caches)

$$GC = \bigcup_{i=1}^{k} LC_i, \forall i \in \{1, ..., k\}, k = |P|, 1 \le k \le n$$

- GU - the *utility* values for the objects in GC

$$GU = \bigcup_{i=1}^{k} LU_i, \forall i \in \{1, ..., k\}, k = |P|, 1 \le k \le n$$

where $LU_i =$ the set of utility values for the objects in LC_i
$$LU_i = \{u(c_{ij}) | c_{ij} \in LC_i, j \in \{1, ..., q\}, q = |LC_i|)\}, i \in \{1, ..., k\}$$
- $siblings_i$ - the rest of the running proxies/dispatchers ($siblings_i = P \setminus \{P_i\}$).

We denote by **A**, the set of **daemons**, ideally running on each node of the LAN.

$$A = \bigcup_{i=1}^{n} DA_i$$

There is a function g (bijection), $g : [1..n] \to A, g(i) = DA_i, \forall i \in \{1,..,n\}$ which assigns each node in the LAN a running **daemon**. One **daemon** **DA**ᵢ is defined as follows:

$$DA_i = (P\prime_i, LOAD(P\prime_i), \bigcup_{p \in P\prime} MLC_p(m)), \forall i \in \{1, ..., n\}$$

where:

- Pl_i - a subset of the proxy-cache set P ($Pl_i \subseteq P$)
- $LOAD(Pl_i)$ - the load of the proxy-caches in the subset Pl

$$LOAD(Pl_i) = \bigcup_{p \in Pl} LOAD(p)$$

where $LOAD(p)$ represents the current load of the proxy $p \in Pl$
- $MLC_p(m)$ - the most "useful" m objects stored in the cache $p \in Pl$

$$MLC_p(m) = \bigcup_{i=1}^{m} c_{ij}, u(c_{ij}) \geq u(c_{ij+1}), \forall j \in \{1, ..., q-1\}$$

where c_{ij} represents the cached object and $u(c_{ij})$ the value returned by the utility function defined above for the object.

2.2   Proxy Splitting Scenarios

As mentioned before we intend to perform a splitting operation under two conditions: when the proxy-cache is under storage constraints or under load constraints. The question is how to decide that a splitting operation is more appropriate than performing cache replacement or reject the incoming requests? We propose the following two conditions:

A. In the Case of Storage Constraints
If $\forall i \in \{1, .., k\}, \forall m, s \in \{1, .., q\}(m \neq s), k = |P|, q = |LC_i|$

$$|u(c_{im}) - u(c_{is})| < \delta \tag{1}$$

then perform splitting, else perform cache replacement.

In other words, splitting is performed when all cached objects are essentially "equally" useful - the difference between the utility of all objects in the cache is smaller than a certain fixed limit δ. The condition could be relaxed, if considering that not for all, but for a certain fraction of the cached object set, the above mentioned condition holds.

If the condition does not hold, than cache replacement should be performed with regard to the utility of the objects. As an observation, if $const_1 = const_3 = const_4 = 0$ then the cache replacement strategy is basically LRU (Least Recently Used), and if $const_1 = const_2 = const_4 = 0$, the cache replacement strategy is LFU (Least Frequently Used).

We present a short example of how the values of those constants could influence the decision of making either a split operation or perform cache replacement. Consider that the cache contains only 5 objects with the characteristics described in Table 1. Consider now Table 2 with four values configurations for the constants that appear in the definition of the utility function (see Subsection 2.1).

Table 1. Characteristics of the cached objects

Cached objects	Size (MB)	Time of last access	Hit count	Quality value
c_1	100	1	60	1
c_2	100	5	70	1
c_3	100	10	80	1
c_4	100	20	90	1
c_5	100	30	100	1

Table 2. Values for the coefficients used by the *utility* function u

Configuration	$const_1$	$const_2$	$const_3$	$const_4$
$conf_1$	0.25	0.25	0.25	0.25
$conf_2$	0.10	0.40	0.40	0.10
$conf_3(LRU)$	0	1	0	0
$conf_4(LFU)$	0	0	1	0

The graphical representation of the utility values corresponding to the data in Table 1 and Table 2 can be seen in Figure 1.

It can be seen that the decision to perform either cache replacement or a split operation highly depends on the value configuration of the coefficients. For example, if $\delta = 15$ and the proxy is under storage constraints, cache replacement will be performed if configuration 3 or 4 are used, but a split operation will be initiated if configuration 1 or 2 are considered.

B. In the Case of Load Constraints

When servicing a request for an object c_{ij} a certain amount of resources must be available. If $\forall P_i \in P$ the available resources are not enough to service a request r_i, than r_i is discarded and the particular time t_i is marked.

If $\forall i \in \{1, .., p-1\}$ (p fixed) we have

$$t_{i+1} - t_i < \xi \tag{2}$$

(the time interval between p consecutive discarded requests is smaller than a fixed threshold ξ), than we make a split operation.

It is to investigate in a real time environment how different values for δ and ξ influence the dynamics of the system.

2.3   Additional Costs Induced by the Proposed Architecture: Best-Case/Worst-Case Scenarios

Inside the system, the message exchange cost can be viewed with regard to the required time to transmit a message, with regard to the amount of data that is transferred, or as a combination of the two (both time and data volume).

In the following, we give a short analysis of the best/worst case scenarios from the point of view of the latency perceived by the client. For a similar analysis on the amount of data transferred within the system, please see [5].

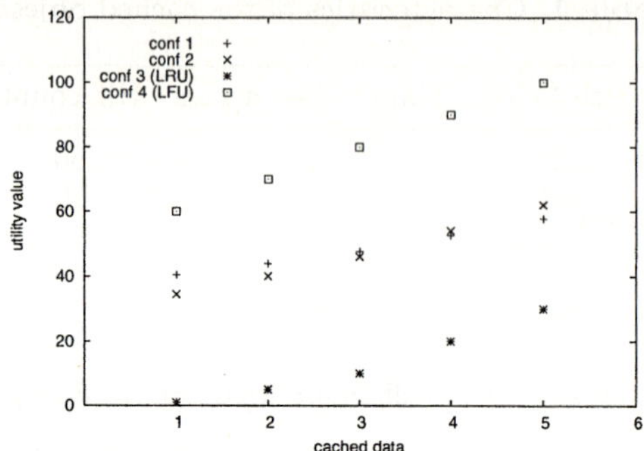

Fig. 1. The utility values for the cache configuration from Table 1 computed using the constants values from Table 2

The delay perceived by the client depends on the delay introduced by the LAN communication, the one introduced by the WAN communication, as well as on the delay introduced by searching the local caches and the server repository. We make the following notations:

- *Delay* - the total delay as perceived by the client
- d_{lan} - the delay introduced when transmitting a message in the LAN;
- d_{wan} - the delay introduced when transmitting a message in the WAN;
- d_{cache} - the delay introduced when searching the local cache;
- d_{server} - the delay introduced when searching the server repository/performing admission control;
- $d_{timeOut}$ - the time out interval fixed for the proxy-server communication

We propose the following forwarding algorithm for requests passed from one proxy to another inside the LAN: when a proxy receives a request, it first checks the local cache and returns the appropriate object in case of a hit. Otherwise (local miss) it checks the list with cached objects at siblings' sites in order to see if the requested object is cached in the federate cache. If it does, it marks the request and sends it to the appropriate sibling. The decision to forward a request to a certain proxy is made based on the locally available information on the global state of the federate cache. It may happen that this information is outdated and that by the time a forwarded request reaches the sibling, the requested object does not exist anymore on the sibling site. The *worst case* would be when a request received by a `dispatcher` D_i is forwarded from one dispatcher to the other until it returns to D_i. In this case, supposing that the client didn't cancelled the request, it is forwarded by D_i to the origin server S.

Suppose there are k active proxy-caches, and using the above mentioned notations, we distinguish the following two *worst cases*, when it comes to the user perceived latency:

- bouncing request and server down

$$Delay = (k+1)d_{lan} + kd_{cache} + d_{timeOut}$$

– bouncing request and server can't serve incoming requests

$$Delay = k(d_{lan} + d_{cache}) + 2(d_{lan} + d_{wan}) + d_{server}$$

The *best case* is of course when the first proxy receiving the client request, can serve it from the local cache. In this case we have:

$$Delay = 2d_{lan} + d_{cache}$$

Assuming the following two configurations, *conf1* with $d_{lan} = 0.1, d_{cache} = 0.001, d_{wan} = 0.5, d_{server} = 0.005$, and *conf2* with $d_{lan} = 0.01, d_{cache} = 0.001, d_{wan} = 0.05, d_{server} = 0.005$ (measurements in seconds) the maximum introduced delay in the case up to 11 proxy-caches are active inside a LAN is showed in Figure 2.

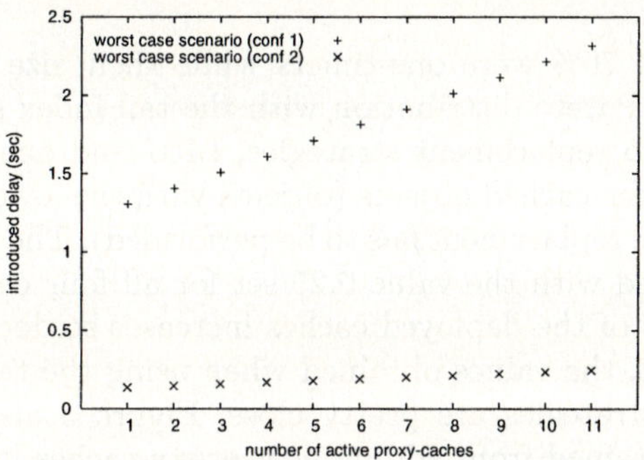

Fig. 2. Maximum introduced delay

It can be seen from the above example that, if the load on both proxy-caches and server(s) is more or less constant, then only variations in network conditions (local and external) makes the delay vary. In a well connected high speed LAN/WAN, the more realistic configuration would be one similar to *conf2*, but when no control over internal/external network can be assumed, a configuration like *conf1* is very probable. The values for *conf1* and *conf2* were measured at Klagenfurt University during normal working hours.

This means that even without constraints regarding the available external bandwidth, it is highly probable for the maximum number of active proxy-caches to be limited by the additional latency that would be induced in the worst case scenario. This holds especially if the network conditions are not very good (we have high induced latencies for both LAN and WAN) as the delay is highly dependable on those conditions.

We have performed a series of simulation experiments [6] using synthetic log traces generated with WebTraff [11]. Figure 3 shows the variation of byte-hit ratio with the number of active proxy-caches inside the LAN (after up to four split operations). The log we used for this particular simulation consisted of 1000 requests following a Zipf distribution with $\alpha = 0.3$ for a number of 300 video

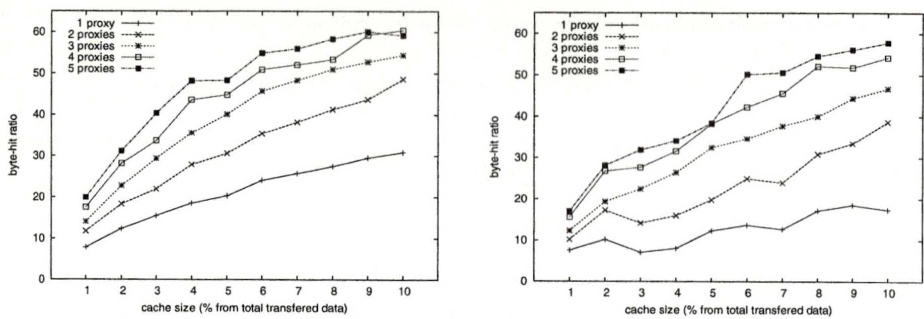

Fig. 3. The variation of the byte-hit ratio with the number of active proxy-caches and cache size. **(Left)** Cache replacement strategy set to LRU. **(Right)** Cache replacement strategy uses the utility values of the objects

objects. From those, 70% were one-timers while their size was approximately 3GB and followed a Pareto distribution with the tail index set to 1.2.

We simulated two replacement strategies, LRU and one strategy that used the utility values of the cached objects (objects with the lowest utility values are discarded when cache replacement has to be performed). The utility of the cached objects was computed with the value 0.25 set for all four coefficients. It can be seen that as the size of the deployed caches increases so does the byte-hit ratio but, more important, the values obtained when using the two above-mentioned cache replacement strategies are pretty close. Figure 3 also seems to suggest that the benefits obtained from adding new proxy-caches inside the LAN tend to diminish as the number of active proxies increases (the increase in byte-hit ratio is greater when moving from 1 to 2, or even from 2 to 3 active proxies than it is when moving from 3 to 4 or from 4 to 5 active proxies).

There is a trade-off between costs and benefits, the best *cost/benefit* ratio seems to be achieved at a moderate number of proxies, as Figure 2 and Figure 3 suggest. We intend to validate this assumption in a real time environment, once our implementation of the system (which is based on the existing implementation of the QBIX proxy-cache [16]) is completed.

3   Related Work

The last few years have brought an increasing interest in video caching as a result of the rising popularity and availability of multimedia content on the Web. The vast majority of the research concentrates on partial video caching, approach that considers specific parts of videos or is done with respect to the quality of the videos. Examples of proposals for partial video caching include caching of a prefix [17], caching of a prefix and of selected frames [10], caching of a prefix combined with periodic broadcast [8], caching of hotspot segments [7]. Other approaches consider the caching of a prefix based on popularity [12], segment-based prefix caching [19] or variable sized chunk caching [2].

Quality based video caching proposals include periodic caching of layered coded videos [9], adaptive caching of layered coded videos in combination with congestion control [14], quality adjusted caching of GoPs (group of pictures) [15] or simple replacement strategies (patterns) for videos consisting of different quality steps [13].

Regarding distributed video caching we have, among others, the work of Brubeck and Rowe [3] proposing multiple video servers accessible via the web and which manage tertiary storage systems as well as the MiddleMan [1] system which proposes a cooperative caching video server.

Our proposal, though having similarities with that in [1], differs from previous work by the fact that our system is dynamic and able to adjust the number of running proxy-caches in the LAN in a fully distributed fashion depending on a number of factors including current load, storage constraints, request patterns.

4   Conclusion and Future Work

We have presented a distributed proxy-cache architecture which aims at providing better service to LAN clients. The feature that distinguishes our proposal from those made in the past is the dynamic characteristic of our system which is able to adapt itself to changes in access, request and response patterns as well as to changes in network conditions.

Future work will focus on finishing the implementation of the system, evaluating its performance in real-life situations and compare the performance with the case in which a single proxy-cache is used. Other points of interest are represented by the conditions triggering the split, hibernate and shut down operations. Another interesting problem is what happens in a system like the one we described, when multiple outgoing links with different capacities are available.

Acknowledgements

I would like to thank my supervisor from Klagenfurt University, prof. dr. Laszlo Böszörmenyi for his constant help, support and guidance during my research work.

References

1. Acharya, S., Smith, B.: Middleman: A Video Caching Proxy Server. In: Proceedings of the 10th International Workshop on Network and Operating System Support for Digital Audio and Video (2002)
2. Balafoutis, E., Panagakis, A., Laoutaris, N., and Stavrakakis, I.: The impact of replacement granularity on video caching. In: IFIP Networking 2002. Lecture Notes in Computer Science, vol. 2345. Springer-Verlag, Berlin, Germany, (2002) 214-225
3. Brubeck, D.W., Rowe, L.A.: Hierarchical Storage Management in a Distributed VOD System, In: IEEE MultiMedia, Fall 1996, Vol. 3, No. 3

4. Chankhunthod, A., Danzig, P, Neerdaels, C., Schwartz, M., Worrell, K.: A Hierarchical Internet Object Cache. In: Proceedings of the 1996 USENIX Technical Conference (1996)
5. Cobârzan, C., Böszörmenyi, L.: Dynamic proxy-cache multiplication inside LANs. Technical Reports of the Institute of Information Technology, University Klagenfurt, TR/ITEC/05/2.02
6. Cobârzan, C., Böszörmenyi, L.: Measurements on byte-hit ratio variation in LANs deploying multiple proxy-caches. Technical Reports of the Institute of Information Technology, University Klagenfurt, TR/ITEC/05/2.05
7. Fahmi, H., Latif, M., Sedigh-Ali, S., Ghafoor, A., Liu, P., Hsu, L.H.: Proxy Servers for Scalable Interactive Video Support. In: IEEE Computer, 43(9): (2001) 54-60
8. Guo, Y., Sen, S., Towsley, D.: Prefix Caching Assisted Periodic Broadcast for Streaming Popular Videos. In: Proceedings of ICC (International Conference on Communications) (2002)
9. Kangasharju, J., Hartanto, F., Reisslein, M., Ross, K.W.: Distributing Layered Encoded Video through Caches. In: Proceedings of IEEE INFOCOM (2001)
10. Ma, W.-H., Du, D.H.-C.: Reducing Bandwidth Requirement for Delivering Video over Wide Area Networks with Proxy Server. In: IEEE International Conference on Multimedia and Expo, (2000) 991-994
11. Markatchev, N., Williamson, C.: WebTraff: A GUI for Web Proxy Cache Workload Modeling and Analysis. In: IEEE International Symposium on Modeling, Analysis, and Simulation of Computer and Telecommunications Systems, Vol. 10, (2002) 356-363
12. Park, H. S., Chung, K.D., Lim, E.J.: Popularity-based Partial Caching for VOD Systems using a Proxy Server. In: Workshop on Parallel and Distributed Computing in Image Processing, Video and Multimedia (2001)
13. Podlipnig, S., Böszörmenyi, L.: Replacement strategies for quality based video caching. In: Proceedings of the IEEE International Conference on Multimedia and Expo (ICME). Vol. 2. IEEE Computer Society, Piscataway, NJ, (2002) 49-53
14. Rejaie, R., Kangasharju, J.: A Quality Adaptive Multimedia Proxy Cache for Internet Streaming. In: Proceedings of the International Workshop on Network and Operating Systems Support for Digital Audio and Video (2001)
15. Sasabe, M., Wakamiya, N., Murata, M., Miyahara, H.: Proxy Caching Mechanisms With Video Quality Adjustment. In: Proceedings of the SPIE Conference on Internet Multimedia Management Systems (2001) 276-284
16. Schojer, P., Böszörmenyi, L., Hellwagner, H, Penz, B., Podlipnig, S.: Architecture of a quality Based Intelligent Proxy (QBIX) for MPEG-4 Videos, World Wide Web Conference, (2003) 394-402
17. Sen, S., Rexford, J., Towsley, D.: Proxy Prefix Caching for Multimedia Streams. In: Proceedings of the IEEE INFOCOM99. (1999) 1310-1319
18. Wessels, D.: Web Caching. O'Reilly, 2001
19. Wu, K.-L., Yu, P.S., Wolf, J.L.: Segment-Based Proxy Caching of Multimedia Streams. In: Proceedings of the Tenth International World Wide Web Conference (2001)

Perspectives for Lecture Videos[*]

Michael Hartle, Henning Bär, Christoph Trompler, and Guido Rößling

Darmstadt University of Technology, Department of Computer Science,
Hochschulstr. 10, 64289 Darmstadt, Germany
{mhartle,hcbaer,trompler,roessling}@informatik.tu-darmstadt.de
http://www.dlh.informatik.tu-darmstadt.de

Abstract. This article presents an architecture for automated multi-perspective lecture recordings. The implementation switches between several video streams showing different perspectives in order to make the recording more vivid and to reduce the tunnel vision effect of single-perspective recordings. Automatic switching can be based on time intervals or, at a later stage of development, via simple recording rules.

1 Introduction

Recently, lecture recordings are increasingly put online in order to support the students with their examination preparations. This is also the goal of the OpenCourse-Ware Project of the MIT [1].

A survey conducted at the Oxford Brooks University [2] has shown the strong interest of students in such an offering. Our own surveys among more than 500 students yielded the result that more than half of them considered lecture recordings as helpful. Yet, our own experience demonstrates that the acceptance is based on the quality of the material presented. Reductions in quality, e.g. if slides are not readable or the lecturer leaves the field of view being recorded, result in a notable drop of acceptance.

Lecture recordings containing an occasional change of perspective are more vivid for the viewer than a video stream fixed on the lecturer. Students also obtain an impression on the surroundings of the lecture if the recording also includes scenes of the audience present.

For the multi-perspective recording system presented in this article, a lecture hall was equipped with several cameras. These cameras can be operated remotely and can be turned and pitched. Furthermore, the lecture hall was also provided with a recording server for the video streams.

After a brief overview of related systems, we describe the processing steps, our recording approach and post-processing operations.

2 Related Work

At the University of Wisconsin - Madison [3], the lecture recordings are costly post-processed, based on knowledge about following scenes and digital zooming.

---

[*] The underlying project for this report was supported by the German Federal Ministry of Education and Research under code 08NM208. The responsibility for the content of this publication lies with the respective authors

J.C. Cunha and P.D. Medeiros (Eds.): Euro-Par 2005, LNCS 3648, pp. 901–908, 2005.

Microsoft [4] presents a system which localizes both the lecturer during his presentation and even students posing questions in order to direct the cameras automatically. The publication uses the advantages of a non-intrusive localization method which is also a research topic pursued in our research group. In [5], a localization method is described using an additional camera.

In [6], a system is presented which synchronizes video streams with provided slides. Another goal is the improvement of the quality of slides recorded via the video stream. Therefore, the actual lecture recording is generated in a post-processing step.

At the University of Queensland [7], several multimedia streams are synchronized and offered as a single document. This system is not limited to video streams and slides, but can also incorporate other media such as texts, pictures or images.

Several additional recording systems are being presented in [8]. Among other systems, Lecturnity (http://www.lecturnity.de) and the MS Producer (http://www.microsoft.com/office/powerpoint/producer) are installed on a computer system also used for presentation which simplifies a mobile deployment. However, a fixed installation with support for changes of perspective and tracking of lecturers is complicated by using both the recording and presentation system on the same computer.

The FAME Room [9] is a meeting room at Karlsruhe, Germany, where the speaker is located via audio-tracking. A digital "lecture assistant" is employed which handles other supporting tasks besides the audio-tracking such as controlling light levels and audio volume.

Typical recording systems are limited to a proprietary format such as Real Media or Microsoft PowerPoint. This severely limits new innovative methods of presentations offered by using open media format alternatives. Furthermore, some systems require additional hardware such as separate cameras needed for localization. Finally, necessary manual post-processing does not scale adequately for practical applications. Many publications do not consider the automation of this process which is essential for keeping the required manual amount of work to a minimum for a large number of recordings.

3   Management of the Recording Process

The recording process of our system can be described as a composition of several steps. This section regards the control of the recording process. Section 4 describes the actual recording, followed by the post-processing steps in Section 5. As the emphasis of a slide-based presentation is placed on the readability of the slides, these are captured separately and combined with the video stream. Further details regarding the required technology are presented in Section 6.

The recording is started and stopped manually by an operator. The functionality can also be available on the presentation system, so the lecturer is also able to control the process. The editing process creates the resulting video file which covers the whole lecture.

For starting the recording process, a management application is provided which controls the recording server via local-area network. The operator, either the lecturer himself or an assistant, clicks on the *Start Preview* button. This results in a signal for the recording server to send a video preview as a live stream. Using the preview, the setup can be tested, e.g. whether the audio signal is being properly transmitted from the lecturer's wireless microphone, or whether the correct field of view or the correct perspective is being recorded.

When the lecture begins, the *Start Section* button of the management application has to be triggered, which again sends a signal to the recording server. The server stores the current time into a configuration file necessary for the later post-processing. The lecture recording can be paused by using the respective buttons of the management application.

Until the end of the lecture, the name of the lecture to be associated with the recording can be chosen from a list. At the end of the lecture, the *Stop Lecture* button has to be triggered and the recording server is ordered to finalize the configuration file for the post-processing step and to copy the recorded video stream to a central server. On this server, incoming recordings are accepted via a web service and automatically sorted based on date, lecture name and a unique ID of the recording. Stored on the central server, the recordings are held available for post-production via the video editing application.

4   Real-Time Video Recording

The recording system was developed based on the architecture shown in Figure 1. The *Capturing Controller* is responsible for the generation of the multi-perspective video. It controls the other modules depending on the necessary steps to be performed. The *Camera Controller* handles commands for turning and pitching the cameras ordered by the Capturing Controller. The *Perspective Selector* changes the perspective when a corresponding command was issued by the Capturing Controller. The *Locator* starts the search for the lecturer only on request based on the camera that was marked as available by the Camera Controller. The *Recorder* is a standard video recording application.

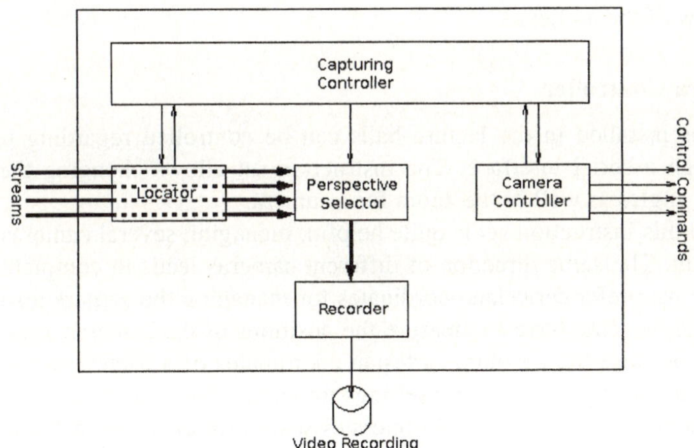

Fig. 1. Recording architecture

Achieving a lecture recording of acceptable quality requires a good integration and cooperation of these modules. They can be installed as a single software package on one machine or distributed over several machines, depending on the requirements of the available infrastructure. The communication between the modules is based on web services as an open protocol in order to facilitate the modularity and exchange-ability of the implementation of entire modules. Some of these modules are written in Java, which allows modules to be utilized on multiple platforms. For example, the Java-based Camera Controller in our infrastructure is running on a Linux system, but can also easily be deployed on a Microsoft Windows system.

The following sub-sections explain these modules in detail.

4.1 Capturing Controller

The Capture Controller is responsible for all scenes and for the choice of an appropriate perspective for each scene, where a scene is defined as the sequence between two changes of perspective. The length of a scene varies in accordance with the propositions of [4] between several minutes for perspectives showing the lecturer and several seconds for scenes showing the audience. Typically, neither panning nor zooming is performed while a scene is recorded, although there are constellation-specific exceptions. The Capturing Controller manages all other modules. It can order the Locator to localize the lecturer using a separate camera. Using the result of the Locator, the Capturing Controller requests the Camera Controller to point to and zoom into the close-up range of the lecturer with an available camera. Finally, it uses the Perspective Selector to switch the perspective to the new prepared camera.

This example is a typical constellation generating one of many imaginable recordings. The following constellations are examples which can be of use for a lecture recording:

- Close-up recordings of the lecturer
- Overview of the lecture hall centered on the lecturer
- Slow panning over the audience
- Overview of the audience

4.2 Camera Controller

The cameras installed in the lecture halls can be controlled regarding pan, tilt and zoom through a serial interface. The instruction set allows changing the horizontal and vertical angles as well as the zoom for a camera.

Although this instruction set is quite helpful, managing several cameras on this basis is difficult. The same direction of different cameras leads to completely different views. Thus we prefer cartesian coordinates for managing the camera positioning. As a prerequisite, we first have to measure the positions of the cameras after the camera installation, because the absolute cartesian coordinates of a camera target have to be computed into corresponding angle settings for each camera involved.

The Camera Controller internally operates with different spherical coordinate systems, where each system`s center is defined by a camera. Using cartesian coordinates is easier to work with especially when describing planar surfaces like the black board or the walls. So an easy command offered by the interface is *lookAt*, which directs the camera to a desired target.

Close-up views, overviews and localization are examples for different standard settings, which are necessary for a recording. The Camera Controller knows several of such settings, like setCloseRec, which offers a close-up view of the lecturer.

Currently this control is performed manually. The network interfaces allow an easy replacement of the calls from a graphical user interfaces by those of an automated process. A diploma thesis is being written about this automatic Camera Controller.

4.3 Locator and Perspective Selector

The Locator is needed when the Capturing Controller decides to record a close-up view of the lecturer. The Capturing Controller first chooses a camera for the localization and one for the recording. The current status of development requires setting the localization camera to a predefined localization position. The Locator is then ordered to find the lecturer with the available camera. In a lecture hall, the lecturer usually moves in an area close to the black board or the presentation surface. The Locator only takes this area into account. Other areas like the auditorium which are most likely irrelevant for the localization of the lecturer can be blanked out. For the remaining area, a simple comparison of adjacent frames is sufficient to figure out those areas where movement takes place. The Locator assumes that the lecturer most likely caused those movements and thus can be found at that position. The quality of the Locator has been proven in first tests. In the next term, the Locator will be tested in first real lectures.

During recording, the Perspective Selector chooses the digital video streams which shall be made visible. It offers the currently selected stream as a virtual capture device to the operating system, which can then be used by further modules.

4.6 Recorder

Basically any recording software can be chosen for the encoding and storage of the video and audio signals. To eventually synchronize the slides with the video more easily, we customized the recording software so it writes the time at which the recording was started to a file beneath the actual video. This is not strictly required, but makes the manual synchronization with the slides much easier. The procedure of cutting the video depends very much on the recorded format. In our installation, we use *mpeg4ip* [10] to record our video, because

- mepg4ip is available as open-source project, which allowed us to modify it so that the time can be stored at which the recording was started;
- mpeg4ip provides us with a suited video codec and file format on the basis of an international normalized standard, which can easily be post-processed with existing software products;
- mpeg4ip can generate both local files and video streams which can be transmitted to double-check the recording while it is in progress.

5 Post-processing the Videos

All presented slides have to be included in these recordings to enable students to properly prepare for examinations. They are visible as fixed images. We use a presentation environment [11] to create those fixed images, with which images are created out of the slides no matter which presentation software is being used. The file names of the images are based on the time when the according slide was shown.

The creation of the lecture recordings is done in two steps. First, the video is being cut, then both the cut video and the slide images are being referenced by a SMIL [13] file in accordance with their interplay during the lecture.

The editing of the videos is done via Apple's QuickTime SDK [14]. The video is cut to properly match the beginning and ending of the recorded lecture and to remove all sections of the lecture where the cameras provided a video stream, yet the recording was paused via the management application. Finally the edited video stream is put back together without any re-encoding taking place and thus without additional quality loss or computational expenses. The information at which time the cuts are planned can be taken out of the configuration file which is generated by serializing events like starting, stopping, and pausing the recording.

The changing slide images are created in an independent fashion, as mentioned at the beginning of this chapter, and thus are available separately. The video can be referenced directly in the SMIL file, which connects the edited video stream and the presented slides. The synchronization of both media is gathered on the basis of the slide presentation time encoded in the file name of each slide and the starting time of the video. The beginning and ending points of time of the video stream are obtained from the configuration file.

The SMIL file can be downloaded from a dynamically generated website. So we realized an interface to the interaction software TVremote [12], which allows students to use their mobile devices during the lecture to tag a section for later video review. They can access a list of their personal tags and start the lecture video by clicking on one of those tags.

An example of a SMIL-based lecture recording combining and synchronizing separate fixed images and video stream can be seen in Figure 2, shown with Apple's QuickTime Player.

6 Practical Operation of the System

We have been offering lecture recordings to our students for more than two years. In the beginning, the cameras had to be controlled manually. Since the winter term 2003/2004, our multi-perspective system is used practically and the cameras can be controlled from remote. That is a great assistance, as controlling the cameras can even be delegated to trained student helpers without in-depth technical knowledge. In the winter term 2002/2003, a poll of the didactical team of our university found out that more than 50% of the students experienced the recordings as helpful. At that time, the recordings were made using only a single perspective. Members of the didactical team recommended including changes of the perspective in our systems to offer an overall visual impression of the lecture hall. In winter term 2003/2004, lecture videos were made using changes of the perspective. The result of a poll thereafter showed that more than 70% of the students found the recordings partially or very helpful for the test.

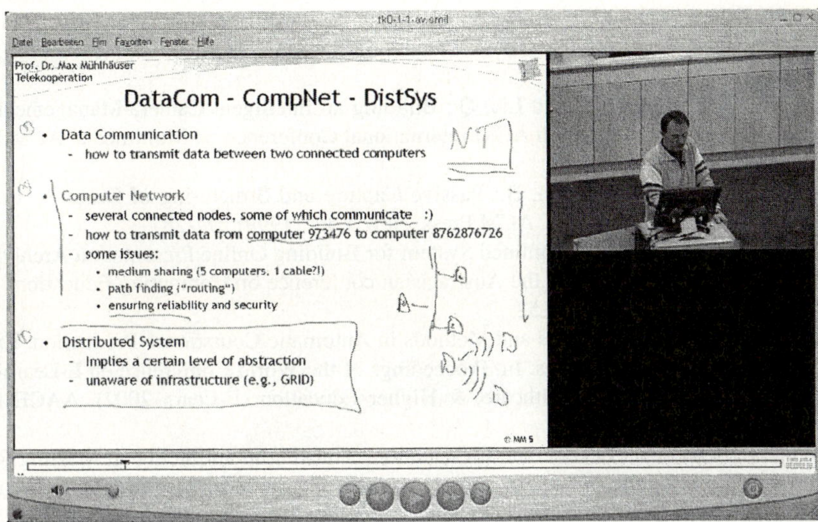

Fig. 2. A lecture recording ready for distribution among students. The hand-written annotations on the slide were added during the course of the lecture

7 Outlook

The work of recent years focused on the automation of the described recording process in order to increase the amount of lecture recordings without increasing the workload for the staff involved. Further works are intended to increase the quality and the cooperation of the modules presented in section 4.

Several features and ideas have led to an investigation of solutions based on Apple's QuickTime file format [15] as an alternative to the currently used SMIL-based system. Solutions for goals such as interactive content-based navigation or the combination of both fixed images and edited video stream into a single file are currently investigated. Concerning these features, the QuickTime specification seems to provide promising approaches.

Although we are investigating these solutions, no stable and satisfactory results regarding the creation and playback of such content are currently at hand. With their availability, these results are expected to be the subject of a future publication.

References

1. MIT OpenCourseWare – Fact Sheet. Online at
   http://web.mit.edu/newsoffice/2001/ocw-facts.html. (2001) (seen on July 12, 2004)

2. Phillimore, R.: Face to Face or eContent: Student and Staff Perspective. In: Proceedings of the International Conference on Computers in Education. IEEE Press (2002) 211-212

3. Gleicher, M., Heck, R., und Wallick, M.: A Framework for Virtual Videography. In: Proceedings of the 2nd International Symposium on Smart Graphics. ACM Press, New York (2002) 9-16

4. Rui, Y., Gupta, A. und Grudin, J.: Videography for Telepresentations. In: Proceedings of the conference on Human Factors in Computing Systems. ACM Press, New York (2003) 457-464

5. Rui, Y., He, L., Gupta, A. und Liu, Q.: Building an Intelligent Camera Management System. In: Proceedings of the 9th ACM International Conference on Multimedia. ACM Press, New York (2001) 2-11

6. Mukhopadhyay, S. and Smith, B.: Passive Capture and Structuring of Lectures. In: Proceedings of ACM Multimedia. ACM Press, New York (1999) 277-287

7. James, D., Hunter, J.: A Streamlined System for Building Online Presentation Archives using SMIL. In: Proceedings of the Australasian conference on Computing Education. ACM Press, New York (2000) 145-152

8. Lauer, T., Ottmann, T.: Means and Methods in Automatic Courseware Production: Experience and Technical Challenges. In: Proceedings of the World Conference on E-Learning in Corporate, Government, Healthcare, & Higher Education (E-Learn 2002). AACE Press, Charlottesville, VA (2002)

9. Rogina, I., Schaaf, T.: Lecture and Presentations Tracking in an Intelligent Meeting Room. In: Proceedings of the Fourth IEEE International Conference on Multimodal Interfaces (ICMI'02). IEEE Press (2002) 47-52

10. MPEG4IP – Open Streaming Video and Audio. Online at http://mpeg4ip.net. (seen on February 8, 2004)

11. Rößling, G., Trompler, C., Mühlhäuser, M., Köbler, S. und Wolf, S.: Enhancing Classroom Lectures with Digital Sliding Blackboards. In: Proceedings of the Annual Conference on Innovation and Technology in Computer Science Education (ITiCSE 2004), Leeds, UK. ACM Press, New York (2004) 218-222

12. Bär, H., Rößling, G., Mühlhäuser, M.: Improving Interaction During Lectures: A Minimal-Distraction Approach. In: Proceedings of the World Conference on Educational Multimedia, Hypermedia & Telecommunication (ED-MEDIA). AACE Press, Charlottesville, VA (2004) 1250-1255

13. W3C Synchronized Multimedia Homepage. Online at http://www.w3.org/AudioVideo/ (seen on May 30, 2005)

14. Apple QuickTime Developer Connection. Online at http://developer.apple.com/quicktime/ (seen on May 30, 2005)

15. Apple QuickTime Developer Connection, QuickTime File Format. Online at http://developer.apple.com/documentation/QuickTime/QTFF/qtff.pdf (seen on May 30, 2005)

A Scene-Based Bandwidth Allocation Scheme for Transferring VBR-Encoded Videos[*]

Dafu Deng and Hai Jin

Cluster and Grid Computing Lab
Huazhong University of Science and Technology, Wuhan, 430074, China
{dfdeng,hjin}@hust.edu.cn

Abstract. This paper classifies the *variable-bit-rate* traffic characteristics into two parts—the *intra-scene* VBR and the *inter-scene* VBR, and quantitatively analyzes the effects of these two kinds of VBR issues on the network traffic characteristics. Based on the analysis results, we propose a scene-based bandwidth allocation scheme to handle the bandwidth burstness problem. Performance evaluation based on real-life MPEG-IV video traces shows that the proposed scheme can achieve good tradeoff among the different objectives such as the performance measurements of network traffic burstness, initial delay, and client buffer requirement.

1   Introduction

Due to the inherent bit rate variety, *variable-bit-rate* (VBR) encoded streams are very difficult to be transmitted through the current best-effort Internet while guaranteeing *quality-of-service* (QoS)[1]. To reduce the network traffic burstness, previous works have proposed two categories of schemes: the renegotiation schemes, such as RCBR [3] and RED-VBR [11], as well as the *constant bit rate* (CBR) transmission with client prefetching schemes. The renegotiation schemes are often used in live streaming services.

Capitalizing on a priori knowledge of the size of frames contained in pre-recorded videos, the second schemes, such as MVBA [8], MCBA [10], CRTT[5], e-PCRTT[4], and PCRTT-DP[6], reduce the network traffic by smoothing the peak bit rate. It works by calculating the minimum bandwidth to prevent client buffer underflow. Further, considering both the bandwidth and the client buffer size, it uses dynamic methods to determine the amount of data to be transmitted in advance and allocate data transferring bandwidth. Paper [2] has analyzed the performance of those schemes in detail and shown that: MCBA and MVBA algorithms exhibit similar performance for the peak rate smoothing and the variability of bandwidth allocations, while MCBA algorithm is much more effective at reducing the total number of rate changes; CRTT, e-PCRTT, and PCRTT-DP schemes need small client buffer requirement and introduce large bandwidth variability and large peak rate requirement.

---

[*] This paper is supported by National 863 Hi-Tech R&D Program under grant 2002AA1Z2102.

J.C. Cunha and P.D. Medeiros (Eds.): Euro-Par 2005, LNCS 3648, pp. 909–918, 2005.

In this paper, we propose a scene-based bandwidth allocation scheme called SBA for pre-recorded VBR video objects. SBA scheme can significantly reduce the network traffic burstness caused by bit-rate variety while achieving small client buffer requirement and short initial delay. The subsequent sections are organized as follows. In section 2, we analyze the traffic characteristic of MPEG videos. Section 3 describes the SBA scheme in detail. In section 4, we evaluate the performance of the proposed scheme via real-life MPEG-IV traces. Finally, section 5 ends with conclusions and future works.

2 Traffic Characteristic of MPEG Videos

To achieve a high compression rate and a high playback quality, MPEG coding technique encodes the video data into three kinds of frames: Intra-coded (I) frames, Predictive (P) frames, and Bi-directionally predictive (B) frames. The I-frames are coded independently by removing spatial redundancy and providing access point to the coded sequence where decoding begins. The P-frames are coded by referencing the preceding I- or P- frame. The temporal and spatial redundancies with respect to the preceding I (or P) frame are removed. The B-frames obtain prediction from both a previous and a future I (or P) frame. The temporal and spatial redundancies with respect to both the preceding and future I (or P) frames are removed. These three kinds of frames are organized into groups, called *group of pictures* (GOP). Usually, each GOP contains one I-frame, and optionally one or more P and B frames. The playback time interval for each frame is a constant value, which usually equals to 40 ms (i.e. frame rate: 25 fps) or 33ms (i.e. frame rate: 30 fps). The fluctuation of transmission rate is caused by the frame size variety.

There are two issues causing frame size variety. One is the *intra-scene* issue, namely the *intra-scene* VBR (Ia-VBR). Usually, one visual scene (with similar picture content) is comprised of one or several GOPs. In the same visual scene, I-frames have very similar frame sizes because they are independently encoded on the basic of picture content. The size of P-frames is smaller than that of I-frames because the temporal redundancy referring to the preceding I-, or P- frame is removed. B-frames have the smallest frame size because the temporal redundancy referring to both the previous and the future I (or P) are removed. The other is the *inter-scene* issue, namely the *inter-scene* VBR (Ie-VBR). When there is a major change in picture content, the temporal correlation may be poor. It requires that a new I, P, and B frames sequence be encoded. In the same video object, the compression rate for removing spatial redundancy (by DCT algorithm) is approximately a constant value. Hence, the more complexity the picture content has, the larger the size of corresponding I (P, B)-frames will be.

The following statistical coefficients can be used to quantitatively describe the effect of the Ia-VBR and the Ie-VBR on the network traffic characteristic.

$v(Ie)_i$ is the ratio of the peak bandwidth for transferring the first I-frame of the scene with the most complex picture content to the mean bandwidth for transferring first I-frames of all scenes contained in the i-th video object, i.e.

$$v(Ie)_i = \frac{\max\{S_I^k\}}{E[S_I^k]} \quad k \in \{1, \ldots, \eta_i\} \tag{1}$$

where η_i is the number of scenes contained in the i-th video object and S_I^k is the size of first I-frame contained in the k-th scene. Since I-frames are independently encoded on the basic picture content of scenes, $v(Ie)_i$ indicates the effect of the Ie-VBR on the network traffic burstness.

$v(Ia)_i$ is the mean *intra-scene* bandwidth burstness. For each scene, the bandwidth burstness is a ratio of the peak bandwidth to the mean bandwidth of that scene. Hence, $v(Ia)_i$ can be described as

$$v(Ia)_i = E[\frac{\max\{S_k^m\}}{E[S_k^m]}] \quad k \in \{1, \ldots, \eta_i\}, m \in \{1, \ldots, \varepsilon_k\} \tag{2}$$

where ε_k is the number of frames contained in the k-th scene and S_k^m is the size of the m-th frame in the k-th scene. It indicates the effect of the Ia-VBR issue on the network traffic burstness.

$\xi(Ie)_i$ is the standard deviation of bandwidth fluctuation among different scenes, i.e.

$$\xi(Ie)_i = \sqrt{E[(S_I^k - E[S_I^k])^2]} \quad k \in \{1, \ldots, \eta_i\} \tag{3}$$

It indicates the bandwidth variation resulting from the Ie-VBR of the i-th video object.

$\xi(Ia)_i$ is the mean standard deviation of bandwidth fluctuation in the internal scene, i.e.

$$\xi(Ia)_i = E[\sqrt{E[(S_k^m - E[S_k^m])^2]}] \tag{4}$$

where $k \in \{1, \ldots, \eta_i\}$ and $m \in \{1, \ldots, \varepsilon_k\}$. It represents the bandwidth variation caused by the Ia-VBR of the i-th video object.

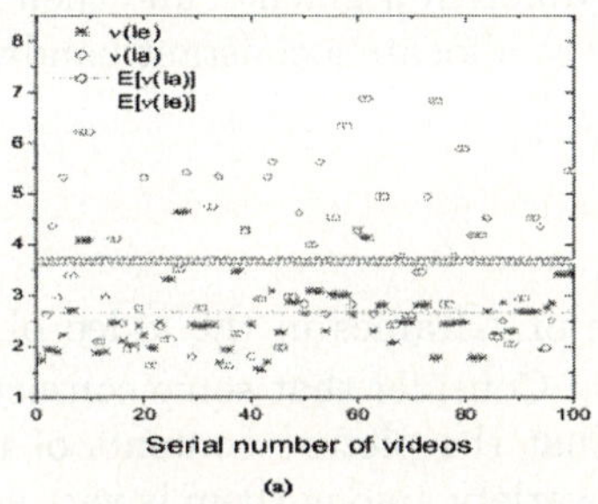
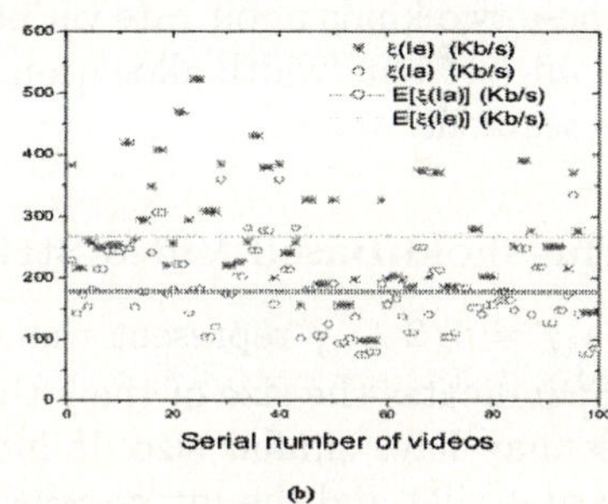

Fig. 1. The distribution of statistical coefficients (a) $v(Ie), v(Ia)$ and (b) $\xi(Ie), \xi(Ia)$ among real-life MPEG-IV video traces.

Fig.1 plots the distribution of these statistical coefficients among 100 real-life MPEG-IV video traces[1] with different contents including *Movies, Sports,*

---

[1] Some traces are obtained from the web site:
http://www-tkn.ee.tu-berlin.de/~fitzek/TRACE/pics. Others are extracted from real-life DivX movies.

News, Talk show, several Episodes, music and *Cartoons*. Here, one visual scene is assumed to be one GOP. Referring to part (a) of Fig.1, it can be easily found that the value of $v(Ie)_i$ is smaller than that of $v(Ia)_i$ with high probability. The mean value of $v(Ia)_i$ (≈ 3.67) is near 1.38 times as the mean value of $v(Ie)_i$ (≈ 2.66). It indicates that the effect of the Ie-VBR on network traffic burstness is smaller than that of the Ia-VBR. However, as shown in part (b) of Fig.1, $\xi(Ie)_i$ is far larger than $\xi(Ia)_i$ for all video traces and the expected value of $\xi(Ie)_i$ ($\approx 268.2Kb/s$) among all video traces is 1.51 times as the expected value of $\xi(Ia)_i$ ($\approx 176.8Kb/s$). This represents that the mean bandwidth variation caused by the Ie-VBR is far larger than that resulting from the Ia-VBR.

Intuitively, the bandwidth burstness indicates the level of difficulty to guarantee lossless transmission when transferring video data through Internet, whereas the bandwidth variation indicates the required "cost" (i.e. client buffer, initial buffer delay, etc.) for smoothing the bandwidth burstness. Hence, smoothing the bandwidth burstness that results from the Ie-VBR may consume large cost and just achieve small smoothing result. However, if we divide these two kinds of VBR characteristics and smoothing the bandwidth burstness resulting from the Ia-VBR first, large bandwidth smoothing result may be obtained with low cost consumption.

3   Scene-Based Bandwidth Allocation Scheme

The basic idea of the SBA scheme is that first stripping the bit rate variety caused by Ie-VBR and Ia-VBR into different video segments; and then, smoothing the Ia-VBR contained in each video segment such that the impact of bandwidth fluctuation on the network traffic can be significantly reduced with low smoothing cost. In this section, we first give a threshold-based video striping scheme to divide these two kinds of bit rate varieties into different segments, and then illustrate an *off-line* bandwidth allocation strategy to allocate a constant bandwidth for each segment.

3.1   Threshold-Based Video Striping

Let$\{S_I^j : j = 1, 2, \dots\}$ represent the sequence of I-frames in the video object, where S_I^j indicates the size of the j-th I-frame. Consider that some consecutive I-frames may have similar size. It indicates that the picture contents of those frames are similar and the *inter-scene* bit rate variety among them is very small. Thus, it is not necessary to divide them into different segment. We define the *logical visual scene (LVS)* to be the consecutive frame sequence in which the maximum size warp among I-frames is less than $2 \times p \times S_I^k$, where S_I^k represents the size of the first I-frame included in that LVS and p is a threshold ($p \geq 0$).

The threshold-based video striping scheme is to divide different LVSs into different segments. Obviously, the first segment starts at the first I-frame. Suppose that the current segment is the i-th segment that starts with the k-th I-frame. The $(n+k+1)$-th I-frame of the sequence $\{S_I^j : j = 1, 2, \dots\}$ indicates the start point of the $(i+1)$-th segment if

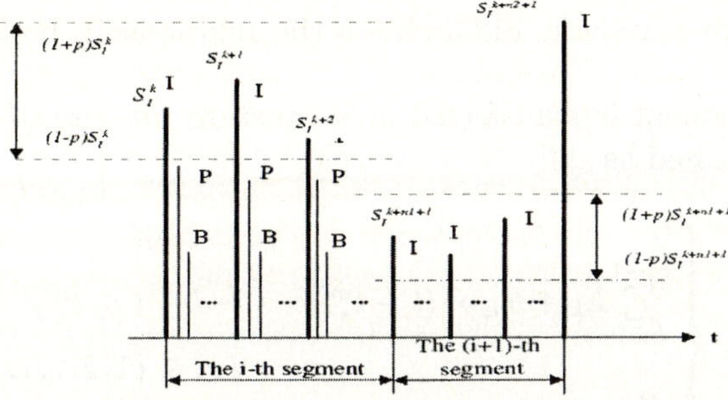

Fig. 2. The principle of the threshold-based video striping scheme.

$$|S_I^{n+k+1} - S_I^k| \geq p \times S_I^k \tag{5}$$

Fig.2 shows the principle of the threshold-based video striping scheme. In this figure, one unit of time on the x-axis corresponds to a frame playback time interval. The solid vertical lines represent the corresponding frame sizes. The dotted horizontal lines indicate the size boundaries of I-frames for different segments. In this video splitting scheme, the size of all I-frames within the i-th segment is located between $(1 - p) \times S_I^k$ and $(1 + p) \times S_I^k$.

3.2   Bandwidth Allocation

Suppose that the i-th video object is the requested one. The playback starting time of the i-th video object is 0 and the transmission starting time of the i-th video object is $-d$, where d represents the initial buffer delay used for prefetching video data into client buffer. On the server side, assume that M_i is the total number of striped video segments. $L(m)$ and $l(m)$ represent the amount of video data and the serial number of the last frame contained in the m-th segment ($m \in \{1, 2, \ldots, M_i\}$), respectively. Let χ_j and T_j represent the amount of video data contained in the j-th frame and the playback time for the j-th frame, respectively. b_m, t_s^m, and t_e^m is the allocated constant bit rate, the transmission starting time and the transmission ending time of the m-th segment, respectively. Obviously, to guarantee consecutive transmission, the value of t_s^m must be equal to the value of t_e^{m-1}, i.e.

$$t_s^m = \begin{cases} -d & m=1 \\ t_s^{m-1} + \dfrac{\sum\limits_{k=1}^{m-1} L_k}{b_{m-1}} & m = 2, 3, \ldots, M_i \end{cases} \tag{6}$$

On the client side, a playback buffer with a capacity of B bits is supplied to receive the requested video data. The received video data is played back at a fixed rate of F frames per second. Let $B(t)$ represent the buffer occupancy at time t. $A(t)$ and $C(t)$ represent the cumulated video data arrived at the playback buffer at time t and the cumulated video data that have been consumed at time t, respectively. B_{max} is used to represent the maximum client buffer occupancy during the stream playback. Fig.3 shows the relationship between $A(t)$, $C(t)$, and $B(t)$.

Since each segment is transferred at a constant bit rate, the arrival stream $A(t)$ can be expressed as

$$A(t) = \begin{cases} 0 & t < -d \\ \sum_{k=1}^{m-1} L_k + b_m \times (t - t_s^m) & -d \leq t \leq t_e^m, \\ & (m \in \{1, 2, \ldots, M_i\}) \\ \sum_{k=1}^{M_i} L_k & t > t_e^{M_i} \end{cases} \quad (7)$$

where term $\sum_{k=1}^{m-1} L_k$ represents the cumulated video data that have been transferred to the playback buffer during time period $[t_s^1, t_e^{m-1}]$, term $b_m \times (t - t_e^{m-1})$ represents the video data contained in the m-th segment that have been transferred to the playback buffer during time period $[t_e^{m-1}, t]$.

As the video data is played back at the fixed rate F frame/s, the playback stream $C(t)$ can be described as

$$C(t) = \begin{cases} 0 & t < 0 \\ \sum_{k=1}^{tF} \chi_k & 0 \leq t \leq \frac{N_i}{F} \\ \sum_{k=1}^{M_i} L_k & t > \frac{N_i}{F} \end{cases} \quad (8)$$

Then, the playback buffer occupancy $B(t)$ is given by

$$\begin{aligned} B(t) &= A(t) - C(t) \\ &= \begin{cases} 0 & t < -d \\ b_m \times t & -d \leq t < 0 \\ \sum_{k=1}^{M_i} L_k + b_m \times (t - t_s^m) - \sum_{j=1}^{tF} \chi_j & 0 \leq t \leq t_e^m, \\ & (m \in \{1, 2, \ldots, M_i\}) \\ \sum_{k=1}^{M_i} L_k - \sum_{j=1}^{tF} \chi_j & t_e^{M_i} < t \leq \frac{N_i}{F} \\ 0 & t > \frac{N_i}{F} \end{cases} \end{aligned} \quad (9)$$

We use the following *off-line* method to allocate bandwidth. First, we set the mean bit rate of the first segment to be the allocated constant bit rate b_1. i.e.

$$b_1 = \frac{L_1 \times F}{l(1)} \quad (10)$$

where L_1 represents the total number of video data contained in the first segment and $\frac{l(1)}{F}$ represent the cumulated time for playing back the first segment. In order to guarantee no buffer underflow during the time interval when the first segment

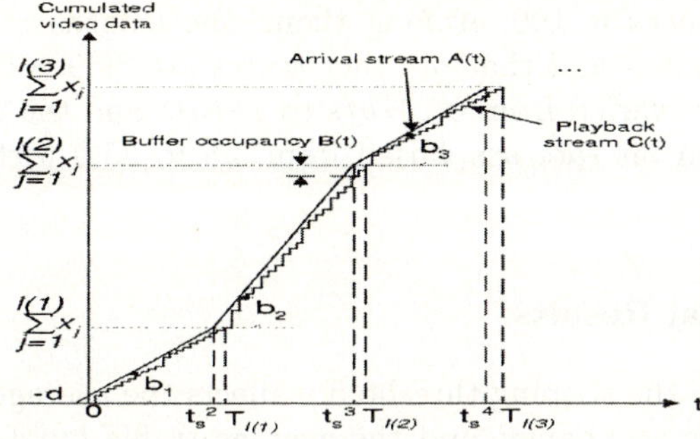

Fig. 3. The relationship between arrival stream $A(t)$, playback stream $C(t)$, and buffer occupancy $B(t)$.

is in playback, the condition $B(t) \geq 0$ must be met. Combining equation (6), (9), and (10), we derive that the minimum required initial buffer delay d as

$$d = \frac{l(1) \times \sum_{j=1}^{tF} \chi_j}{F \times L_1} - t \quad 0 \leq t \leq T_{l(1)} \tag{11}$$

Secondly, the arrival stream $A(t)$ for the m-th segment can be described as a line from the start point $(t_s^m, \sum_{k=1}^{m-1} L_k)$ to the end point $(t_e^m, \sum_{k=1}^{m} L_k)$. The slope of this line is exactly the allocated transmitting bit rate for the m-th segment (as showed in Fig.3). Obviously, larger transmission bit rate results in larger play buffer occupancy. Thus, we allocate the transmitting bandwidth for the m-th segment b_m $(m \in \{2, 3, \ldots, M_i\})$ to be the minimum required bit rate to guarantee no buffer underflow during the requested video object being playback. According to the equation (9), we obtain

$$b_m = \max\{\frac{\sum_{j=1}^{tF} \chi_j - \sum_{k=1}^{m-1} L_k}{t - t_s^m}\} \tag{12}$$

where $t \in \{T_{l(m-1)+1}, T_{l(m-1)+2}, \ldots, T_{l(m)}\}$.

4   Performance Evaluation

For illustration, we evaluate the performance of SBA scheme via experiments and compare it with the MCBA scheme, the e-PCRTT scheme, and the CRTT scheme.

4.1   Experimental Environment

Experiment is based on real-life MPEG-IV video traces with different contents including *Movies, Sports, News, Talk show, several Episodes, Music* and *Cartoons*. All traces are played back at a rate of $F = 25$ frames/s and the total number of video traces is 100. Among them, the lengths of *News* and some *Sports* are 30334 frames and that of other traces are 89,998 frames. The mean bit rates of them are varied from $0.1Mb/s$ to $1Mb/s$ and the ratios of the peak bit rate to the mean bit rate are varied from 3.5 to 35.0 with the mean value 9.99.

4.2   Experimental Results

Fig.4 illustrates how the striping threshold p affects the average initial delay, the average client buffer occupancy, and the average traffic burstness of all traces, respectively. From part (a), we find that, by the increment of p, the average buffer delay $\overline{D}$ is increased immediately. On the one hand, the larger p results in that the first segment of requested video object contains more bit rate variation of *inter-scene* VBR characteristic and requires more smoothing cost–$\overline{D}$. For the same reason, as shown in part (b), the average client buffer requirement $\overline{B_{max}}$ is also dramatically increased by the increment of p. However, referring to part (c), it can be found that the average traffic burstness is decreased by the increment of p due to that more bit rate variation of inter-scene VBR is smoothed. Overall, $p \in [0.4, 0.6]$ can achieve a good tradeoff among the performance of the initial delay, the client buffer requirement, and the network traffic burstness.

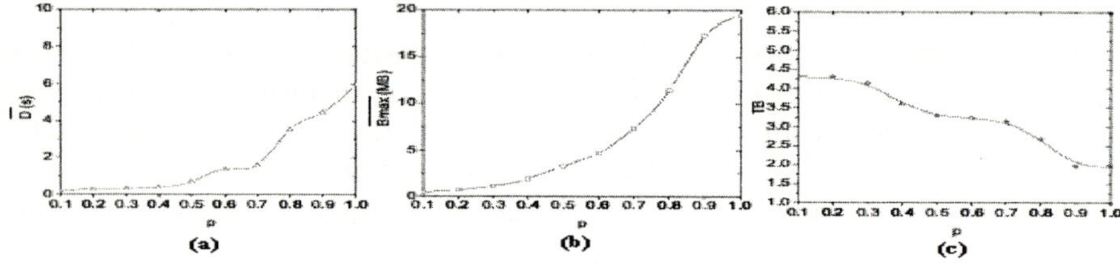

Fig. 4. The effect of striping threshold p on (a) the average initial delay $\overline{D}$, (b) the average client buffer occupancy $\overline{B_{max}}$, (c) the average traffic burstness TB.

Parts (a), (b), and (c) of Fig.5 plot the comparison of distributions of the initial delay, the network traffic burstness, and the maximum buffer occupancy on 100 MPEG-IV video traces among the CRTT scheme, the e-PCRTT scheme, the MCBA scheme, and the SBA scheme with striping threshold $p = 0.4$, respectively, where the internal figure of each part illustrates the comparison of the corresponding mean value. For the network traffic burstness, referring to part (b), it can be easily found that the CRTT scheme can achieve optimal performance since it uses the constant mean bandwidth to transfer video data. With high probability, the MCBA scheme and the SBA scheme with stripping threshold $p = 0.4$ have very similar performance and the e-PCRTT scheme with given parameters $d = 60$ and $B = 8M$ has the worst smoothing effect. For example, as shown in the internal figure of part (b), among 100 MPEG-IV video traces, the mean traffic burstness of the e-PCRTT scheme is 5.1, while those of the

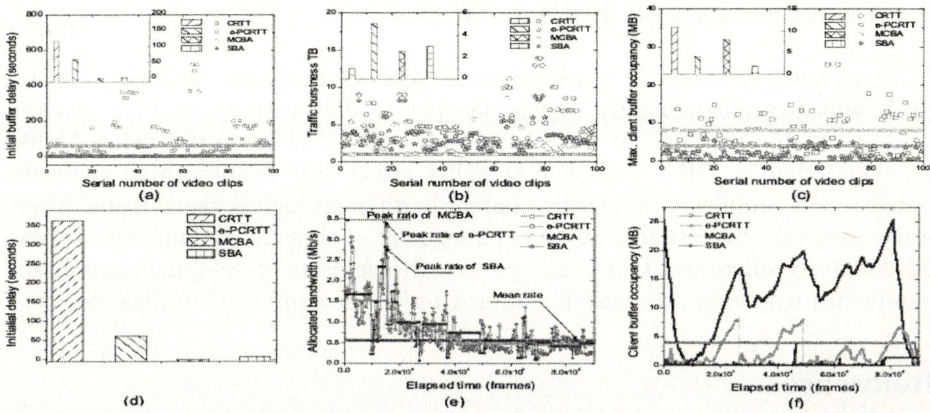

Fig. 5. The comparison of distributions of (a) the initial buffer delay, (b) the network traffic burstness, as well as (c) the maximum client buffer occupancy on 100 MPEG-IV video traces; and the individual comparison of (d) the initial delay, (e) the allocated bandwidth, and (f) the client buffer occupancy for the movie clip *Silence of Lambs* among the CRTT scheme, the e-PCRTT scheme, the MCBA scheme, and the SBA scheme with striping threshold $p = 0.4$.

MCBA scheme and the SBA scheme are just 2.6 and 3.0, respectively. Referring to part (a) and part (c), however, it can be seen that the smoothing costs of the CRTT scheme, the MCBA scheme, and the e-PCRTT scheme are very huge. As shown in the internal figures of parts (a) and (c), among 100 MPEG-IV traces, the mean initial buffer delay and the mean maximum client buffer occupancy required by the CRTT scheme are 118 seconds and 10MB. For each video trace, the maximum client buffer occupancy resulting from the MCBA scheme is $8MB$, whereas the e-PCRTT needs nearly $4.1MB$ maximum buffer consumption and $60s$ initial buffer delay. In contrast, the mean initial buffer delay and the mean maximum buffer occupancy needed by the SBA scheme are just 7.7 seconds and $1.8MB$.

Parts (d), (e), and (f) of Fig.5 illustrate the comparison of the initial buffer delay, the allocated bandwidth, and the client buffer occupancy for the movie clip *Silence of the Lambs* among those schemes, respectively. In this case, as shown in the part (e), the network traffic burstness of the SBA scheme is somewhat less than those of the MCBA scheme and the e-PCRTT scheme. As shown in part(d) and part(f), the required initial buffer delay and the maximum client buffer occupancy are just 7 seconds and 3.5 MB, respectively, while the initial buffer delay and the maximum client buffer occupancy resulting from the e-PCRTT scheme and the MCBA scheme are (60s,$4MB$), and (0s, $8MB$), respectively. The CRTT scheme must use 360 seconds initial buffer delay and consume 25MB client buffer. The smoothing cost is too big to be endured by the true VoD systems.

5 Conclusions and Future Works

In this paper, we propose a scene-based bandwidth allocation scheme, called
SBA, which can significantly reduce network traffic burstness of VBR streams
while achieving small client buffer requirement, and short initial delay. Unlike
previous schemes, SBA scheme first splits video objects into small segments
based on the complexity of picture content in different logical visual scene. Then,
it uses constant bit rate to transfer each segment so that the variable bit rate can
be effectively smoothed. Our ongoing researches focus on developing scene-based
bandwidth reserving protocols for improving server bandwidth utilization.

References

1. S. V. Anastasiadis, K. C. Sevcik, and M. Stumm, Server-Based Smoothing of Variable Bit-Rate Streams, *Proc. of the ACM MM'01*, pp. 147-158, 2001.
2. W. Feng and J. Rexford, Performance Evaluation of Smoothing Algorithms for Transmitting Prerecorded Variable-Bit-Rate Video. *IEEE Trans. on Multimedia*, 1(3):302-312, Sept. 1999.
3. M. Grossglauser, S. Keshav, and D. N. C. Tse, RCBR: A Simple and Efficient Service for Multiple Time-Scale Traffic, *IEEE/ACM Transaction on Networking*, 5(6):741-751, Dec. 1997.
4. O. Hadar and R. Cohen, PCRTT Enhancement for off-line video smoothing, *The Journal of Real-Time Imaging*, 7(3): 301-314, Jun. 2001.
5. S. Kang and H. Y. Yeom, Transmission of video streams with constant bandwidth allocation, *Computer Communications*, 22(2):173-180, Feb. 1999.
6. J. M. McManus and K. W. Ross, A dynamic programming methodology for managing prerecorded VBR sources in packet-switched networks, *Telecommun. Syst.*, Vol.9, 1998.
7. W. C. Poon and K. T. Lo, The transportation of VBR-encoded video using deterministic characterization and bandwidth renegotiation, *Computer Communications*, 25(8):731-740, May 2002.
8. A. R. Reibman and A. W. Berger, Traffic descriptors for VBR video teleconferencing over ATM networks, *IEEE/ACM Transactions on Networking*, 3(3):329-339, Jun. 1995.
9. A. R. Reibman and B. G. Haskell, Constraints on Variable Bit-rate Video for ATM Network, *IEEE Trans. on Cricuits and Systems for Video Technology*, 2(4):361-372, Dec. 1992.
10. J. Rexford and D. Towsley, Smoothing Variable-Bit-Rate Video in an Internetwork. *IEEE Trans. on Networking*, 7(2):202-215, Apr. 1999.
11. H. Zhang and E. W. Knightly, Red-vbr: a renegotiation-based approach to support delay-sensitive vbr video, *ACM Multimedia Systems Journal*, Vol. 5, pp. 164-176, 1997.

DCT Block Conversion
for H.264/AVC Video Transcoding

Joo-Kyong Lee and Ki-Dong Chung

Dept. of Computer Engneering, Pusan National Univ.,
Jangjeon-dong, Geumjeong-gu, Busan, 609-735, Korea
`jklee@melon.cs.pusan.ac.kr, kdchung@pusan.ac.kr`

Abstract. In H.264/AVC [1], integer transforms are applied instead of the 8×8 discrete cosine transform (DCT) of previous standards to avoid inverse transform mismatch problems. However, these transforms make it difficult to transcode the precoded video contents with the earlier video coding standards into H.264/AVC in DCT-domain, thus causing cascaded re-encoding in the pixel-domain. There are strong requests for an efficient way to solve this problem. In this paper, we propose a new conversion scheme from an 8×8 DCT block into four 4×4 DCT blocks in H.264/AVC. Experimental results show that the proposed scheme improves PSNR up to 0.1dB and a little computational complexity compared with the cascaded pixel-domain IDCT/DCT. Moreover, other DCT-domain transcoding methods are applicable based on our scheme.

1 Introduction

As the number of networks, types of devices, and video representation formats increase, interoperability between different systems and different networks is becoming more important [2]. To provide a seamless interaction between senders and receivers, diverse research on video transcoding such as bit-rate reduction [3], spatial resolution reduction [4], frame skipping [5][6], and simple video format conversion [7] has been studied. Recently, many DCT-domain approaches have been proposed to reduce the computation complexity [8][9][10][11].

H.264/AVC is the latest international video coding standard approved by ISO/IEC as International Standard 14496-10 (MPEG-4 part 10) Advanced Video Coding (AVC) and by ITU-T as Recommendation H.264. Its original aim was to provide enhanced compression efficiency and improved support for reliable transmission compared with earlier standards such as H.263+ and MPEG-4 visual and its target applications include two-way video communication (video telephony) and non-interactive applications (broadcast, streaming) [12]. With respect to coding efficiency, H.264/AVC outperforms MPEG-2 and MPEG-4 by 63% and 37%, respectively [13]. With this reason, H.264/AVC is expected to replace soon other video coding standards such as MPEG-2, MPEG-4 for video telephony, broadcast and streaming of standard definition TV. However, the various coding tools like 1/4 pel motion compensation, deblocking filter or integer transform what H.264/AVC adopts to improve coding efficiency, brought H.264/AVC in the most algorithmic discontinuities in the evolution of standardized video coding [13]. That is, these state-of-the-art tools are obstacles to transcode video contents compressed by earlier standards into H.264/AVC in the common-domain. For instance, the 4×4 integer transform makes it difficult to transcode the prior compressed video contents in DCT-domain. This is

J.C. Cunha and P.D. Medeiros (Eds.): Euro-Par 2005, LNCS 3648, pp. 919–927, 2005.

because the transform matrix of 4×4 DCT was modified into integer elemental matrix and the rest floating point values are absorbed by quantization step, thus avoiding inverse transform mismatch problems, in addition to reduction of computation complexity by just using add operations and shift operations.

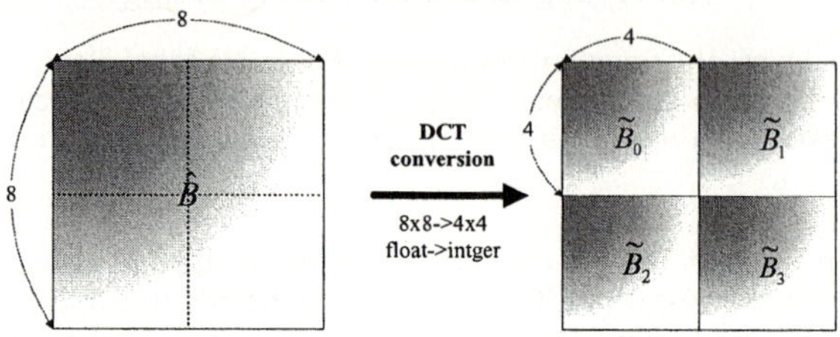

Fig. 1. The 8×8 DCT block conversion into four 4x4 approximate DCT blocks

In this paper, we propose a new algorithm that converts an 8×8 DCT block into four 4×4 approximate DCT blocks adopted in H.264/AVC in order to support DCT-domain transcoding as shown in Fig. 1, where $\hat{B}$ is 8×8 DCT block and $\tilde{B}_i$, $(i=0,..,3)$ is a 4×4 approximate DCT block. This scheme has advantages in computation complexity and video quality compared with cascaded pixel-domain transcoding.

This paper is organized as follows. In Section 2 we describe the differences between DCT of earlier video standards and 4×4 approximate DCT of H.264/AVC. In Section 3 we show our DCT conversion algorithm. Experimental results will be presented in Section 4, and the conclusion is shown in Section 5.

2   Transform in Video Coding Standards

DCT separates the image into parts of differing importance with respect to the image's visual quality. Almost video coding standards adopts 8×8 block DCT, except the H.264/AVC standard which adopts 4×4 integer transform. In this section we examine two different transforms, respectively.

2.1   DCT for Previous Standards

DCT represents an image as a sum of sinusoids of varying magnitudes and frequencies. DCT maps X, a block of N×N samples into a new block Y, N×N block of coefficients. DCT can be described in terms of a transform matrix H, that is, the transformed block $Y=HXH^T$, where the pth row and q^{th} column of H is defined by Eq. 1 and superscript T denotes transposition of matrix [14]. Since the transform matrix is orthogonal the inverse DCT is given by $X=H^T YH$.

$$H_{pq} = H(p,q) = \alpha_p \cos\left[\left(q+\frac{1}{2}\right)\frac{p\pi}{N}\right], \quad \begin{array}{l} 0 \le p \le N-1 \\ 0 \le q \le N-1 \end{array}$$

$$\alpha_p = \begin{cases} \sqrt{\dfrac{1}{N}}, \, p=0 \\ \sqrt{\dfrac{2}{N}}, \, 1 \le p \le N-1 \end{cases} \tag{1}$$

2.2   Transforms in H.264/AVC [12][14]

H.264/AVC uses three different types of transforms depending on the residual data type to be coded. If a macroblock is predicted using the type INTRA_16×16, the 16 DC coefficients of the luminance component are transformed by 4×4 Hardamard transform and 4 DC coefficients of each chrominance component are transformed by 2×2 Hardamard transform. For all other 4×4 blocks, 4×4 approximate DCT is applied. In this paper, we do not consider the Hardamard transform because it is only applied to the DC coefficients in INTRA_16×16 macroblock, and it cannot be target to convert because it was newly adopted in H.264/AVC.

The 4×4 approximate DCT in H.264/AVC is modified from the original DCT. The 4×4 transform matrix H is illustrated in Eq. 2 by setting N by 4 of H in Eq. 1. This matrix multiplication in Eq. 2 can be described in the form of Eq. 3, where the operator $\otimes$ indicates that each elements of CXC^T is multiplied by the element in the same position in matrix R.

$$Y = HXH^T, H = \begin{bmatrix} a & a & a & a \\ b & c & -c & -b \\ a & -a & -a & a \\ c & -b & b & -c \end{bmatrix}$$

(2)

$$where: \quad a = \frac{1}{2}, \quad b = \sqrt{\frac{1}{2}}\cos(\frac{\pi}{8}), \quad c = \sqrt{\frac{1}{2}}\cos(\frac{3\pi}{8})$$

It is desirable to replace the transform matrix C of Eq. 3 by an orthogonal matrix with integer entries to avoid inverse transform mismatch problems and to reduce the computation complexity. In Eq. 4, to simplify the implementation of the transform, b and d are modified as $\sqrt{2/5}$ and $\frac{1}{2}$, respectively, and the 2nd and 4th rows of matrix C are scaled by a factor of two and the matrix R is scaled down to compensate. This transform is an approximation to the 4×4 DCT. However, the result of the transform is not identical to the 4×4 DCT because the factors b and d are changed. Consequently, when we convert the 8×8 DCT block into the 4×4 approximate DCT block, we should compensate the differences.

$$Y = HXH^T = (CXC^T) \otimes R$$

$$= \begin{bmatrix} 1 & 1 & 1 & 1 \\ 1 & d & -d & -1 \\ 1 & -1 & -1 & 1 \\ d & -1 & 1 & -d \end{bmatrix} X \begin{bmatrix} 1 & 1 & 1 & d \\ 1 & d & -1 & -1 \\ 1 & -d & -1 & 1 \\ 1 & -1 & 1 & -d \end{bmatrix} \otimes \begin{bmatrix} a^2 & ab & a^2 & ab \\ ab & b^2 & ab & b^2 \\ a^2 & ab & a^2 & ab \\ ab & b^2 & ab & b^2 \end{bmatrix}$$

(3)

$$where: \quad a = \frac{1}{2}, \quad b = \sqrt{\frac{1}{2}}\cos(\frac{\pi}{8}) \approx 0.6533, \quad d = \frac{c}{b} \approx 0.414$$

$$Y = (C_f X C_f^{\ T}) \otimes R_f$$

$$= \begin{bmatrix} 1 & 1 & 1 & 1 \\ 2 & 1 & -1 & -2 \\ 1 & -1 & -1 & 1 \\ 1 & -2 & 2 & -1 \end{bmatrix} X \begin{bmatrix} 1 & 2 & 1 & 1 \\ 1 & 1 & -1 & -2 \\ 1 & -1 & -1 & 2 \\ 1 & -2 & 1 & -1 \end{bmatrix} \otimes \begin{bmatrix} a^2 & \dfrac{ab}{2} & a^2 & \dfrac{ab}{2} \\ \dfrac{ab}{2} & \dfrac{b^2}{4} & \dfrac{ab}{2} & \dfrac{b^2}{4} \\ a^2 & ab & a^2 & ab \\ \dfrac{ab}{2} & \dfrac{b^2}{4} & \dfrac{ab}{2} & \dfrac{b^2}{4} \end{bmatrix} \qquad (4)$$

$where:$

$$a = \frac{1}{2}, \quad b = \sqrt{\frac{2}{5}} \approx 0.6325, \quad d = \frac{1}{2}$$

3   The 8x8 DCT Block Conversion to the 4x4 Approximate DCT

The goal of this paper is to convert exactly an 8×8 DCT block into four 4×4 approximate DCT blocks in H.264/AVC without using a cascaded pixel-domain approach that performs the inverse DCT(IDCT)/DCT. Specifically, one 8×8 DCT block is converted to four 4×4 approximate DCT blocks described as transformed matrix Y in Eq. 4. Fig. 1 describes (a) the simple cascaded pixel-domain transcoding architecture and (b) the proposed DCT conversion architecture. Other DCT-domain transcoding schemes can be applied to H.264/AVC transcoding based on this DCT conversion scheme.

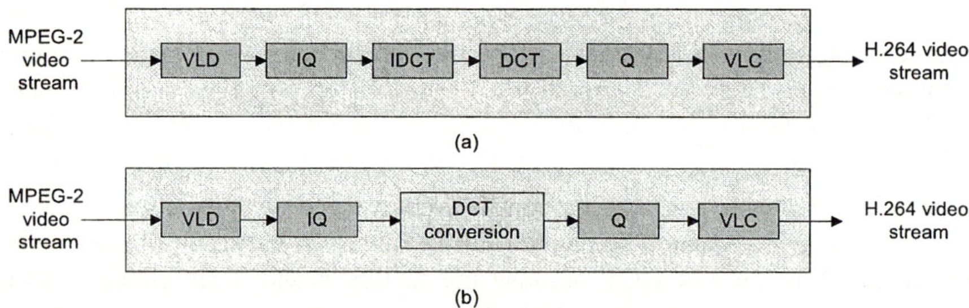

Fig. 2. Illustration of simple transcoding architecture between MPEG-2 and H.264/AVC: (a) cascaded pixel-domain transcoding architecture (b) proposed DCT conversion architecture

3.1   Extracting 4×4 Block in the Pixel-Domain

The 8×8 DCT block conversion into four 4×4 approximate DCT blocks is extended from the pixel-domain extracting. In the pixel-domain, extracting of the 4×4 sub-block B_i from B is defined by Eq. 5, where B is an 8×8 motion compensated (MC) block or a non-MC block in the pixel-domain with displacement matrices L_i, R_i, $i = 0,..,3$ that perform vertical and horizontal filtering, respectively. The location of sub-block B_i in B can be seen in Fig. 1. The matrices L_i of size 4×8 and R_i of size 8×4 are defined in Eq. 6, where $I_{4 \times 4}$ is identity matrix of size 4. The pre-multiplication by L_i extracts a sub-block size of 4×8 from B vertically whereas, the post-multiplication by R_i extracts 4×4 block from the resultant block horizontally.

$$B_i = L_i \cdot B \cdot R_i \,(0 \le i \le 3) \tag{5}$$

$$L_0 = L_1 = \left(I_{4\times4} \quad 0\right)_{4\times8}, \quad L_2 = L_3 = \left(0 \quad I_{4\times4}\right)_{4\times8}$$

$$R_0 = R_2 = \begin{pmatrix} I_{4\times4} \\ 0 \end{pmatrix}_{8\times4}, \quad R_1 = R_3 = \begin{pmatrix} 0 \\ I_{4\times4} \end{pmatrix}_{8\times4} \tag{6}$$

3.2   Extracting a 4×4 Block in the DCT-Domain

The 4×4 DCT block is calculated by performing DCT on the extracted sub-block, B_i of Eq. 5. Due to the distributive property of matrix multiplication with respect to the DCT, transformed matrix of B_i is given by $\hat{B}_i = \hat{L}_i \cdot \hat{B} \cdot \hat{R}_i$ as represented in Eq. 7.

$$DCT(B_i) = DCT(L_i \cdot B \cdot R_i), \quad (0 \le i \le 3)$$
$$= DCT(L_i) \cdot DCT(B) \cdot DCT(R_i) \tag{7}$$
$$= \hat{B}_i = \hat{L}_i \cdot \hat{B} \cdot \hat{R}_i$$

where $\hat{B}_i$, $\hat{L}_i$, $\hat{B}$, and $\hat{R}_i$ are the DCT representations of B_i, L_i, B and R_i respectively. Especially, $\hat{L}_i$ and $\hat{R}_i$ can be written in the form of $H_{4\times4} \cdot L_i \cdot H_{8\times8}^T$ and $H_{8\times8} \cdot R_i \cdot H_{4\times4}^T$ where $H_{4\times4}$ and $H_{8\times8}$ are the transform matrices produced from Eq. 1 for 4×4 and 8×8 DCT, respectively. $H_{4\times4}^T$ indicates the transpose of $H_{4\times4}$.

Consequently, $\hat{B}_i$ can be rewritten in the form of Eq. 8. In the Equation, the product of matrices, $H_{4\times4} \cdot L_i \cdot H_{8\times8}^T$ and $H_{8\times8} \cdot R_i \cdot H_{4\times4}^T$ can be pre-computed and stored in the memory as look-up table. According to Eq. 8 we can convert the 8×8 DCT block, $\hat{B}$ to four 4×4 DCT blocks, $\hat{B}_i\,(0 \le i \le 3)$. However, we should convert $\hat{B}$ to the 4×4 approximate DCT blocks of H.264/AVC, because there exists the difference between 4×4 DCT and 4×4 approximate DCT. This difference is caused by the differences of b and d in Eq. 3 and Eq. 4.

$$\hat{B}_i = (H_{4\times4} \cdot L_i \cdot H_{8\times8}^T) \cdot (\hat{B}) \cdot (H_{8\times8} \cdot R_i \cdot H_{4\times4}^T) \,(0 \le i \le 3) \tag{8}$$

At last, we can obtain a 4×4 transformed block, $\tilde{B}_i$, in H.264/AVC by replacing $H_{4\times4}$ with the modified transform matrix, $H'_{4\times4}$ by letting $b = \sqrt{2/5}$ and $d = 1/2$ instead of $b = \sqrt{1/2}\cos(\frac{\pi}{8}) \approx 0.6533$ and $d = c/b \approx 0.414$ (see Eq. 9). With this modified transform matrix we can exactly converts an 8×8 DCT block into a 4×4 approximate DCT block of H.264/AVC.

Let us show more specific derivation processes. In Eq. 10, the sub-block B_i of the raw block B (before any transform) is transformed to the 4×4 approximate DCT block, $\tilde{B}_i$, in H.264/AVC. This $\tilde{B}_i$ is what we want to get the result of DCT conversion. In Eq. 11, B_i is described as the product of three matrices, $L_i \cdot B \cdot R_i$. The detail explanation about this form is described in Eq. 5 and Eq. 6. With associative property of matrix multiplication, Eq. 11 can be described as Eq. 12 which equals to Eq. 13 by the orthogonal property of the DCT transform matrix such as $H^T H = HH^T = I_n$ where I_n is the identity matrix of order n. Again, by applying the associative property of matrix multiplication to Eq. 13, we can get the form of Eq. 14. This Equation is equal to Eq. 8, because the matrix multiplication $H_{8\times8} \cdot B \cdot H_{8\times8}^T$ is identical with $\hat{B}$.

$$\tilde{B}_i = (H'_{4\times4} \cdot L_i \cdot H_{8\times8}^T) \cdot (\hat{B}) \cdot (H_{8\times8} \cdot R_i \cdot H_{4\times4}'^T) \ (0 \le i \le 3) \tag{9}$$

$$\tilde{B}_i = H'_{4\times4} \cdot B_i \cdot H_{4\times4}'^T \tag{10}$$

$$= H'_{4\times4} \cdot (L_i \cdot B \cdot R_i) \cdot H_{4\times4}'^T \tag{11}$$

$$= (H'_{4\times4} \cdot L_i) \cdot B \cdot (R_i \cdot H_{4\times4}'^T) \tag{12}$$

$$= (H'_{4\times4} \cdot L_i) \cdot (H_{8\times8}^T \cdot H_{8\times8}) \cdot B \cdot (H_{8\times8}^T \cdot H_{8\times8}) \cdot (R_i \cdot H_{4\times4}'^T) \tag{13}$$

$$= (H'_{4\times4} \cdot L_i \cdot H_{8\times8}^T) \cdot (H_{8\times8} \cdot B \cdot H_{8\times8}^T) \cdot (H_{8\times8} \cdot R_i \cdot H_{4\times4}'^T) \tag{14}$$

To recapitulate, we proposed a new scheme to convert an 8×8 DCT block into four 4×4 approximate DCT blocks in H.264.AVC. Using the proposed scheme, we can exactly convert with lower complexity than cascaded re-encoding method in pixel-domain

4   Experimental Results

In this section, we compare the performance of our proposed conversion algorithm with the cascaded re-encoding method of pixel-domain (we call this method as "baseline" method). We assume that input bitstreams for the transcoders are compressed in the MPEG-2 encoder. To simulate the transcoders that adopt the different methods respectively, we modified MPEG-2 TM5 encoder and decoder [15] and implemented the architectures of Fig. 2. In the MPEG-2 encoder, we only use integer pixel motion vector to avoid different half pixel calculation between MPEG-2 and H.264/AVC. Even though it is out of our focus in this paper, it is an important research subject to transcode MPEG-2 video sequence to H.264/AVC.

In the experiments, we tested the objective performance and the computation complexity with different motion characteristic sequences: FOOTBALL, CARPHONE, and CLAIRE in Quarter Common Intermediate Format (QCIF). We compressed the video sequences in MPEG-2 encoder with GOP size 6, I/P distance 3 as varying the quantization step size or Qstep from 1 to 30. In the transcoder, we simulated transcoding process by changing the re-quantization step size or re-Qstep from 1 to 40.

Fig. 3 shows the peak signal-to-noise ratio (PSNR) by changing the re-Qstep from 1 to 40 in the transcoder at fixed Qstep (a) 4 and (b) 20 of the MPEG-2 encoder. It is difficult to distinguish the performances of the proposed scheme from those of baseline method in the figure. However, our proposed scheme slightly outperforms the base method by 0.05~0.1dB numerically. Fig. 4 shows the average PSNR of our scheme for CLAIRE sequence is higher than that of baseline method where Qstep 4 in the MPGE-2 encoder and re-Qstep 6. In Fig. 5, we show the real images reconstructed from baseline method and our proposed scheme when the Qstep=4 in the MPEG-2 encoder and the re-Qstep is 6 in the transcoder. The sample images show almost same visual quality.

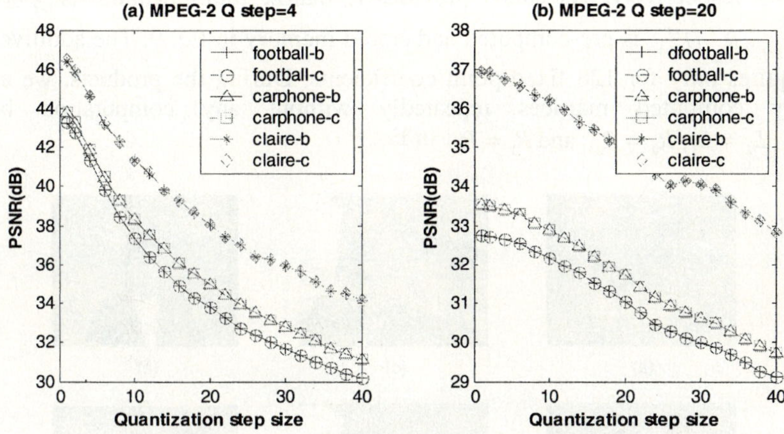

Fig. 3. Change of re-Qstep from 1 to 40 in the cascaded pixel domain transcoder and in the proposed transcoder at fixed Qstep (a) 4 and (b) 20 in the MPEG-2 encoder

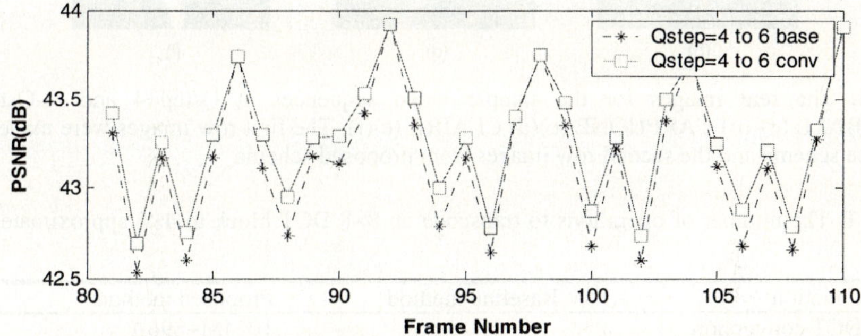

Fig. 4. PSNR comparison for CLAIRE sequence ranging from 81st to 110th frame where the first quantization value is set to 4 and re-quantization value is set to 6

Lastly, let us show you the computation complexity comparison between the baseline scheme and our proposed scheme. For computation complexity, we do no consider fast DCT/IDCT methods but just consider the reference DCT/IDCT methods. However our proposed scheme also can be applied by the same kind of fast method to speed up the conversion.

Table. 1 shows the number of operations to transcode the 8×8 DCT block to four 4×4 approximate DCT blocks. "M" stands for the number of multiplication operations and "A" for adds. Baseline method needs 64×(8M+7A)×2=1024M+896A for 8×8 IDCT because 8×8 IDCT is represented by products of three 8×8 blocks In H.264/AVC the transform can be implemented by only shift and addition operations can be reduced by the symmetric property. In JM 8.2 the 4×4 block requires 64 additions and 16 shifts. Therefore, 256 additions and 64 shifts are required for an 8×8 block. So, this method requires 1024 multiplications, 1152 adds and 64 shifts. The proposed methods using Eq. 9 requires 1024 multiplication operations and 896 add operations. This is calculated from fourfold of products of three matrices size of 4×8, 8×8, 8×4 respectively. As stated previously, matrix multiplication $H'_{4\times4} \cdot L_i \cdot H^T_{8\times8}$ and $H_{8\times8} \cdot R_i \cdot H'^T_{4\times4}$ is pre-computed and stored memory in Eq. 9. The additive memory requires only for 128 fixed-point coefficients. During the products, we can use partially computed matrices repeatedly without any computation because $L_0 = L_1, L_2 = L_3, R_0 = R_2$, and $R_1 = R_3$ in Eq. 9.

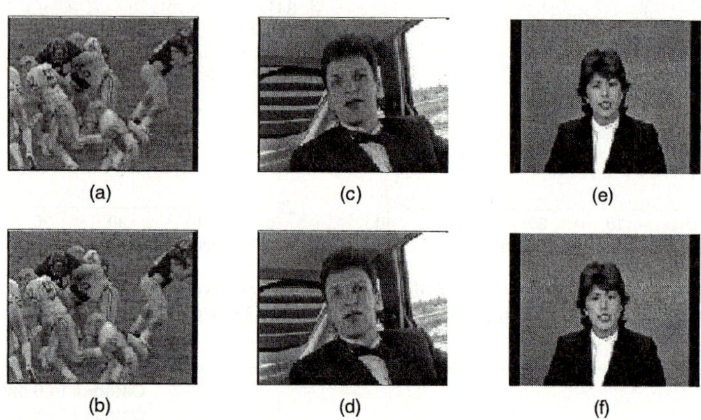

Fig. 5. The real images for the sample video sequences at Qstep=4 and re-Qstep=6: FOOTBALL (a)-(b) CARPHONE (c)(d) CLAIRE (e)(f). The first row images were made from baseline scheme and the second row images from proposed scheme

Table 1. The number of operations to transcode an 8×8 DCT block to 4×4 approximate DCT block

Function	Baseline method	Proposed method
DCT conversion	-	1024M+896A
8×8 IDCT	1024M+896A	-
4×4 approximate DCT	256A+64S	-
Total operations	1024M+1152A+64S	1024M+896A

5   Conclusion

In this paper, we proposed an efficient and exact conversion algorithm for the 8×8 DCT block into four 4×4 approximate DCT blocks to transcode the precoded video contents with the earlier video coding standards such as MPEG-2, H.263 and MPEG-4 into H.264/AVC. Based on the proposed scheme, other DCT-domain transcoding methods can be applied to H.264/AVC transcoding.

The simulation results show that PSNR of the proposed method outperforms the cascaded re-recoding method in pixel-domain by maximum 0.1 dB and that computation complexity is saved by 256A+64S per 8×8 block.

Even though we only considered the DCT conversion in this paper, in the near future we will propose a new transcoder architecture that includes both quantization conversion and DCT conversion to improve video quality and computation complexity.

References

1. ISO/IEC 14496-10:2003: Coding of Audiovisual Objects –Part 10: Advanced Video coding. 2003 and ITU-T Recommendation H.264: Advanced video coding for generic audiovisual services.
2. Vetro, A.; Christopoulos, C.; Sun, H.: Video Transcoding Architectures and Techniques: An Overview. IEEE Signal Processing Magazine, Vol. 20. Issue 2. 18-29, March 2003
3. H. Sun, W. Kwok, and J. Zdepski: architectures for MPEG compressed bitstream scaling, IEEE Trans. Circuits Syst. Video Technol., Vol. 5, 191-199, Apr. 1996
4. N. Bjork and C. Christopoulos: Trascoder architectures for video coding, IEEE Trans. Consumer Electron., vol.44, 88-98, Feb. 1998.
5. J. Youn, M.T. Sun, and C.W. Lin: Motion vector refinement for high performance transcoding. IEEE Trans. Multimedia, vol. 1, Mar. 1999, 30-40
6. J.N. Hwang, T.D. Wu, and C.W. Lin: Dynamic frame-skipping in video transcoding. In Proc. IEEE Workshop Multimedia signal processing, Redondo Beach, CA, DEC. 1998, 616-621
7. Kalva, H., Vetro, A.. Sun, H., Performance Optimization of the MPEG-2 to MPEG-4 Video Transcoder, SPIE Conference on VLSI Circuits and Systems, Vol. 5117, May 2003, 341-350
8. S.F. Chang and D.G. Messerschmidt: Manipulation and compsiting of MC-DCT compressed video, IEEE. J. Select. Areas Commun., Vol. 13, Jan. 1995.
9. C.W. Lin and Y.R. Lee: Fast algorithms for DCT-domain video transcoding, in Proc. IEEE Int. Conf. Image Processing, Thessaloniki, Greece, Vol. 1, Sept. 2001, 421-424
10. Haiyan Shu: An Efficient Arbitrary Downsizing Algorithm for Video Transcoding, IEEE Trans. Circuits Syst. Video Technol., Vol.14, No.6, Jun. 2004
11. Kwang-deok Seo, Jae-Kyoon Kim: Fast motion vector re-estimation for transcoding MPEG-1 into MPEG-4 with lower spatial resolution in DCT-domain, Signal Processing: Image Communication 19(2004) 299-312
12. Iain E.G. Richardson: H.264 and MPEG-4 video compression. Willey (2003)
13. J. Ostermann, J. Bormans, P. List, D. Marpe, M. Narroschke, F. Pereira, T. Stockhammer, and T. Wedi: Video Coding with H.264 / AVC: Tools, Performance, and Complexity, IEEE Circuits and Systems Magazine, Vol. 4, No. 1, Apr. 2004, 7-28
14. Henrique S. Malvar:Low-Complexity Transform and Quantization in H.264/AVC, IEEE Trans. Circuits.Syst. Video Technol. Vol. 13, No.7, Jul. 2003, 598-602
15. http://www.mpeg.org/MPEG/MSSG/tm5/

Topic 12
Theory and Algorithms for Parallel Computation

Andrea Pietracaprina, Kieran Herley,
Christos Zaroliagis, and Casiano Rodriguez-Leon

Topic Chairs

The study of theoretical aspects related to the design, analysis and experimentation of efficient algorithms, and to the identification of effective models of computation, represents a fundamental research area in parallel computing, which has been alive and productive for over two decades and well represented in the Euro-Par community. A distinctive characteristic of this Topic 12 is the variety of contributions addressing classical problems as well as the new challenges posed by recent technological advances and emerging computing paradigms.

This year 13 papers were submitted to the topic, investigating a variety of algorithmic and modeling problems for parallel computation and communication. Among all submissions, 4 papers were accepted as full papers for the conference, resulting in a 31% acceptance rate.

Accepted papers contain the following contributions: new centralized and distributed algorithms for bufferless routing in leveled networks, which attain optimal performance within logarithmic factors; results concerning the existence and the design of truthful mechanisms for the computation of shortest path trees in communication networks where edges are owned by selfish agents, under both utilitarian and non-utilitarian scenarios; embeddings of the hypercube in the partitioned optical passive starts network consisting of g groups of d processors each, which are optimal for all values of g and d; on-line algorithms to serve sequences of adversarial access requests to a shared memory page issued by n processors moving in a certain metric space, which attain good competitive ratios with respect to communication costs.

J.C. Cunha and P.D. Medeiros (Eds.): Euro-Par 2005, LNCS 3648, p. 929, 2005.
© Springer-Verlag Berlin Heidelberg 2005

Efficient Bufferless Routing on Leveled Networks

Costas Busch, Shailesh Kelkar, and Malik Magdon-Ismail

Department of Computer Science, Rensselaer Polytechnic Institute, Troy, NY, USA
{`buschc,kelkas,magdon`}`@cs.rpi.edu`

Abstract. We give near optimal bufferless routing algorithms for *leveled networks*. N packets with preselected paths are given, and once injected, the packets may not be buffered while in transit to their destination. For the preselected paths, the *dilation* D is the maximum path length, and the *congestion* C is the maximum number of times an edge is used. We give two bufferless routing algorithms for leveled networks:
(i) a *centralized* algorithm with routing time $O((C + D) \log(DN))$;
(ii) a *distributed* algorithm with routing time $O((C + D) \log^2(DN))$.
The distributed algorithm uses a new technique, *reverse-simulation*, which is used to obtain a distributed emulation of the centralized algorithm. Since a well known lower bound on the routing time is $\Omega(C + D)$, our results are at most one or two logarithmic factors from optimal.

1 Introduction

We study bufferless routing on leveled networks, where packets cannot be stored at nodes while in transit to their destination. In particular we consider *hot-potato* (or *deflection*) routing [2], in which packets get "deflected" (like a "hot-potato") if they cannot make progress toward their destination. Buffereless routing is appropriate when buffering is costly or impossible, for example in optical networks.

A leveled network with *depth L* has $L + 1$ *levels* of nodes, numbered 0 to L. Every node belongs to exactly one level, and the only edges are between nodes at consecutive levels (Figure 1). Many routing problems on multiprocessor networks can be represented as routing problems on leveled networks, for example routing problems on the Butterfly, the *Mesh* (Figure 1), shuffle-exchange networks, multidimensional arrays, the hypercube, fat-trees, de Bruijn networks, etc. (see [7, 14] for more details).

We assume a *synchronous* routing model in which at each discrete time step, a node forwards at most one packet down any link (two packets may use a link, one in each direction). We study *many-to-one batch routing problems*: we are given N packets with *preselected* paths; each node is the source of at most one packet, but may be the destination of many packets. Every preselected path is *monotonic* in the sense that every edge in a path connects a lower level node with a node in the next higher level, i.e., a path moves from left to right on the general leveled network depicted in Figure 1. Here we are only concerned with scheduling the packets given the paths, and not how to obtain the paths.

The *routing time* is the time at which the last packet reaches its destination. For the preselected paths, the *congestion C* is the maximum number of packets

J.C. Cunha and P.D. Medeiros (Eds.): Euro-Par 2005, LNCS 3648, pp. 931–940, 2005.

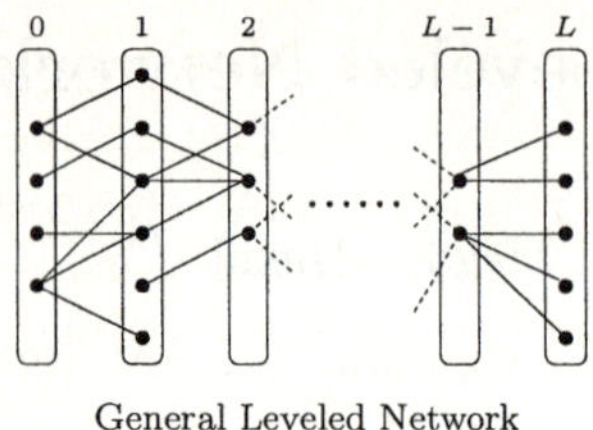

General Leveled Network

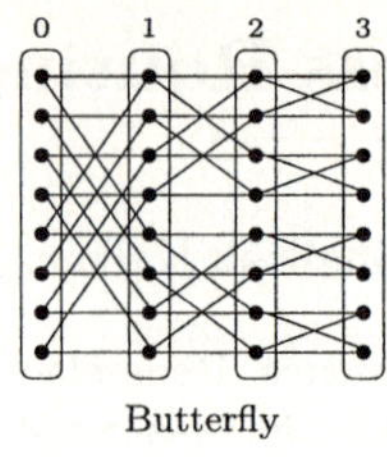

Butterfly

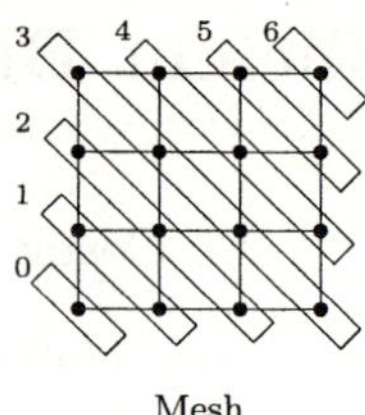

Mesh

Fig. 1. Leveled networks

that traverse any edge, and the *dilation D* is the maximum path length. At most one packet can traverse any edge per time step, a lower bound on the routing time is $\Omega(C + D)$, and routing times close to this are optimal.

If two or more packets wish to follow the same link at the same time step, then a *conflict* occurs. Only one packet can follow this link, and the others must be *deflected* along alternative links (since there are no buffers). Deflections may change the preselected paths, however, we only consider bufferless algorithms in which the final path followed by the packets contains every edge in its preselected path. A routing time close to $(C + D)$ is still optimal with respect to any routing algorithm, buffered or not, even when we allow path deformation, provided that the final paths contain the preselected paths.

Our Contributions. We present two new bufferless routing algorithms for many-to-one routing problems in leveled networks. The first algorithm is centralized, and has routing time $O(C \log(DN) + D)$, which is at most a logarithmic factor from optimal. The second algorithm is distributed and has routing time $O(C \log^2(DN) + D \log(DN))$, a logarithmic factor worse than the centralized algorithm. In the distributed algorithm, all routing decisions are made locally. Both results hold with high probability (w.h.p.), i.e., with probability at least $1 - O(1/DN)$. The distributed algorithm relies on a new technique, *reverse-simulation*, which efficiently emulates the centralized algorithm. The final paths used by the packets contain the preselected paths, which is useful to provide guarantees for the delivery time of the packets. Further, for the centralized algorithm, a packet never strays away from its preselected path.

The fundamental idea behind the centralized algorithm is to partition the network into *frames* each containing $O(\log(DN))$ levels. Packets are divided into $O(C)$ sets that move from frame to frame. The packets of a particular set are routed from frame to frame by coloring their dependency graph, which is a graph representing the conflicts between packet paths. It takes $O(\log(DN))$ time steps to move all the packets from one frame to the next, and since a packet traverses at most $O(D/\log(DN))$ frames to its destination, once a packet is injected, it is delivered in $O(D)$ time steps. The packet sets are injected sequentially, spaced by the time it takes to move packets from one frame to the next, so the last packet is injected at time $O(C \log(DN))$, resulting in a routing time $O(C \log(DN) + D)$. Note that packets are not injected all simultaneously, but rather each packet is injected at an appropriate time after which it moves from frame to frame.

The main idea behind the distributed algorithm is to color the dependency graph in a distributed way: packets randomly select colors and if the coloring is valid, they will make it to their destination. If not, they use *reverse simulation* to trace their paths backwards, recompute colors, and the process repeats. The distributed coloring imposes an extra logarithmic factor in the routing time.

Related Work. Bufferless routing algorithms have been studied for various specific network topologies, [1, 3–8, 10–12, 15]. Most related to our work is [7], which gives a distributed algorithm for leveled networks with routing time $O(C+L)\log^9(LN)$. By using refined techniques, we improve this result by *seven* logarithmic factors, and obtain a result in terms of D, rather than L. A recent result in [9] gives a general bufferless routing algorithm for arbitrary networks with routing time $O((C + D)\log^3(n + N))$, where n is the size of the network. By taking advantage of the special structure of leveled networks, we can obtain a better routing time. Further, our centralized algorithm is one of the few hot-potato routing algorithms that keep the packets on their original preselected paths. Store-and-forward (buffered) routing algorithms exist with near optimal routing time for leveled as well as arbitrary networks (see for example [13]).

Paper Outline. We begin with some preliminaries (Section 2), followed by the centralized (Section 3) and the distributed algorithm (Section 4).

2   Preliminaries

We begin with some preliminaries regarding packet paths, oscillations, frames, and the dependencies between packets.

Paths. Every packet π has a preselected path p with path length $|p|$. Its *current path* at time step t, denoted $p(t)$, is defined as follows; we assume that the current path is maintained in the header of the packet, and is used for deciding where to send the packet at each time step. At time 0, $p(0) = p$, its preselected path. If at time t, packet π is in node v_i, with current path $p(t) = (v_i, v_{i+1}, \ldots, v_k)$, and packet π successfully follows the first edge (v_i, v_{i+1}) (the packet moves forward), then, at time $t + 1$, packet π appears in node v_{i+1} with current path $p(t + 1) = (v_{i+1}, \ldots, v_k)$. If, however, at time $t + 1$ packet π is deflected toward a node v_j, then at time $t + 1$ it appears in node v_j with current path $p(t + 1) = (v_j, v_i, v_{i+1}, \ldots, v_k)$. If the packet moves forward, $|p(t + 1)| = |p(t)| - 1$ and if it is deflected, then $|p(t + 1)| = |p(t)| + 1$.

Oscillations. Suppose packet π has current path $(v_i, v_{i+1}, \ldots, v_k)$. π *oscillates* on edge $e = (v_i, v_{i+1})$ if it moves back and forth on e: if at time t, π appears in v_i, then at time $t + 1$, π appears in v_{i+1}, and at time $t + 2$ it is back in v_i, and so on. When a packet oscillates, the length of its current path increases and decreases by one each time. Oscillations are useful because they provide a way to "buffer" packets on edges instead of at nodes.

Frames. We partition the levels of the network into γ non overlapping *frames* $F_1, F_2, \ldots, F_\gamma$, each containing λ levels (except for the last frame, which may contain fewer). Frame F_i, $1 \leq i < \gamma$, consists of the λ levels $(i-1)\lambda, \ldots, i\lambda - 1$. Frame F_γ consists of the levels $(\gamma - 1)\lambda, \ldots, L$. Note that $\gamma = \lceil (L+1)/\lambda \rceil$. We will pick $\lambda = 4\alpha \log(DN)$, where α is an integer constant that will be defined later; thus, the frames have logarithmic size. (If $\log(DN)$ is not an integer, we use $\lceil \log(DN) \rceil$.)

We call the levels in frame F_i the *inner-levels* of F_i, and we number them from 1 to λ. Thus, inner-level k of frame F_i corresponds to real level $(i-1)\lambda + (k-1)$, where $1 \leq k \leq \lambda$. The *odd inner-levels* are numbered $1, 3, \ldots, \lambda - 1$ (recall that λ is even). The inner level of an edge is the smaller of the inner-levels of the nodes it is incident with. Thus, corresponding to odd inner-levels are *odd inner-edges*, and similarly even inner-levels and even inner-edges.

Packet Sets and Dependency Graphs. We partition the set of packets Π into $s = 8\alpha eC$ sets, $\Pi_1, \Pi_2, \ldots, \Pi_s$. Each packet is placed into one of these sets uniformly at random. Thus, $\Pi = \bigcup_{i=1}^{s} \Pi_i$, and $\Pi_i \cap \Pi_j = \emptyset$ for $i \neq j$, so $|\Pi| = \sum_{i=1}^{s} |\Pi_i| = N$.

Consider the packets in Π_i, and two consecutive frames F_j and F_{j+1}. For each packet $\pi \in \Pi_i$ denote by q_π the sub-path of its preselected path that consists only of edges in F_j and F_{j+1}. We define the packet dependency graph $G_{(i,j)} = (V_{(i,j)}, E_{(i,j)})$ as follows. The nodes of $V_{(i,j)}$ correspond to the packets in Π_i, so $|V_{(i,j)}| = |\Pi_i|$. Let $\pi, \sigma \in \Pi_i$, then $(\pi, \sigma) \in E_{(i,j)}$ if and only if the paths q_π and q_σ share some edge in (F_j, F_{j+1}), i.e., if the paths collide.

The *degree* of a packet π in $G_{(i,j)}$, denoted $d_{(i,j)}(\pi)$, is the number of edges incident with π. The degree of $G_{(i,j)}$, denoted $d_{(i,j)}$, is the maximum degree of any packet in $V_{(i,j)}$. Let $d = \max_{\{i,j\}} d_{(i,j)}$, i.e., d is the maximum degree of any of the graphs $G_{(i,j)}$, for any i and j.

We show that d cannot be too big. In fact, a packet path collides with at most $2\lambda C$ other paths over two consecutive frames. Only approximately $2\lambda C/s = O(\lambda/\alpha)$ of these packets are in the same set, so we expect that $d = O(\lambda/\alpha)$:

Lemma 1. $d \leq \lambda/\alpha = 4\log(DN)$, *with probability at least* $1 - 1/DN$.

Groups. We partition the network into *groups*, such that each group is a collection of γ' consecutive frames, where $\gamma' = 2\lceil D/\lambda \rceil$ (namely, the group consists of at most $2D + 2\lambda$ levels). We define two sets of groups. The first set of groups is $S_1 = \{g_1, g_2, \ldots, g_{k_1}\}$, where group g_i consists of frames $F_{(i-1)\gamma'+1}, \ldots, F_{i\gamma'}$. The group g_{k_1} consists of the rightmost frames in the network and may contain fewer than γ' frames. Note that the groups in S_1 do not share any levels. The second set of groups is $S_2 = \{h_1, h_2, \ldots, h_{k_2}\}$, where group h_i consists of frames $F_{(i-1+1/2)\gamma'+1}, \ldots, F_{(i+1/2)\gamma'}$. The group h_{k_2}, consists of the rightmost frames in the network and may contain fewer than γ' frames. Note that the groups in S_2 are shifted by $\gamma'/2$ frames with respect to the groups in S_1.

A packet belongs to a group if its path lies entirely within the group. A packet belongs to at least one group (since its preselected path length is at most D and the groups have size $\geq 2D$). If the packet belongs to a group in S_1, we assign it to S_1, otherwise it belongs to a group in S_2, and we assign it to S_2. Note that a packet may belong to a group in S_1 and to a group in S_2 if its path is in the intersection of the two groups. In this case, it is assigned to S_1. We denote by $\Pi(S_i)$ the packets that belong to group S_i, and by $\Pi(x, S_j)$ the set of packets that belong to group x of S_j.

3 Centralized Algorithm

In the centralized algorithm, we route the packets in two consecutive *sessions*. First, we route the packets $\Pi(S_1)$ (belonging to groups in S_1), and then the packets $\Pi(S_2)$. Since the packets in $\Pi(S_2)$ are routed after the packets in $\Pi(S_1)$ have reached their destinations, they cannot possibly interfere with each other.

A particular session contains packets in various groups. Since a packet's path is contained in a single group, and since the groups are level-wise disjoint, the packets in one group can be routed simultaneously with all the packets in another group without any possibility of interfering. Thus, it suffices to describe the algorithm to route the packets in any one group. We will focus on the particular group $x = g_1$ of S_1. The algorithm for other groups is identical. We will simplify the notation by dropping the x and $\mathsf{S_j}$ dependence. Hence, from now on, Π will denote $\Pi(g_1, S_1)$, and Π_i will denote $\Pi_i(g_1, S_1)$.

The session consists of m phases, each of duration τ time steps. Packets move on waves, from left to right, one frame per phase. Each packet set Π_i is associated with a particular wave, and each packet in Π_i uses this wave until it reaches its destination. Packets are assigned colors with respect to the dependency graph. Packets of the same color are routed together on a *boat* (level) in the wave. Different colors use different boats.

3.1 Waves

A *wave* ω is a pointer to a frame. Initially the wave is NULL. The wave enters the network (points to frame F_1) at some phase ϕ_i, and points to the next higher frame at each subsequent phase, so in phase ϕ_{i+k}, it points to frame F_{k+1}. Eventually, ω points to the last frame $F_{\gamma'}$, and then leaves the network (becomes NULL). There are s waves $\omega_1, \ldots, \omega_s$ (equal to the number of packet sets). Wave ω_i enters the network at phase ϕ_{2i-1}. The last wave ω_s enters in phase ϕ_{2s-1} and after γ' phases, it has left the network, so the number of phases is $m = 2s + \gamma' - 1$. We use the wave to also denote the frame it points to.

The purpose of wave ω_i is to route the packets in set Π_i along with it, as it moves from lower to higher levels. Packet $\pi \in \Pi_i$ is injected when wave ω_i contains π's source. The packet then moves along its wave and is absorbed either when the wave contains its destination or its destination is one frame ahead of the wave. Note that waves are spaced 2 frames apart in order to avoid interference of packets in different waves while the waves move from frame to frame.

At the beginning of each phase, packets appear inside their respective waves, and frames between waves are empty of packets; this property is essential for moving packets along their waves. Consider a phase ϕ during which wave ω_i points to frame F_j. At the beginning of ϕ, F_j contains only packets from Π_i, and F_{j+1} is empty of packets. By the end of phase ϕ, the packets in F_j will move from frame F_j to frame F_{j+1}. Thus, at the beginning of the next phase, all these packets are still in the wave ω_i, and frame F_j is empty (which allows packets of Π_{i+1} to move along wave ω_{i+1}). We continue by describing in detail how the packets of Π_i move from F_j to F_{j+1} during phase ϕ.

3.2　Initial and Target Levels

Suppose that phase ϕ consists of time steps $t_1, t_2, \ldots, t_\tau$. At the beginning of phase ϕ, the packets of Π_i that are already in wave ω_i are oscillating on odd inner-edges of F_j. Suppose $\pi \in \Pi_i$ is oscillating on odd inner-edge $e = (v_\ell, v_{\ell+1})$ of F_j, where the inner level of v_ℓ is ℓ (which is odd). The packet oscillates on e so that at odd time steps $t_1, t_3, \ldots$, packet π appears in v_ℓ. We say that π *oscillates* at inner-level ℓ, which is the *initial* inner-level of π in phase ϕ.

　　Now suppose that the current path of π at its initial inner-level ℓ is a subpath of its preselected path. During phase ϕ, packet π will follow its current path until it reaches a *target* inner-level ℓ' in F_{j+1}, where it will oscillate for the remainder of the phase. At its target level, π's current path will remain a sub-path of its preselected path. The target level will become the new initial level at the next phase, when the wave ω_i points to F_{j+1}.

　　We define $\chi_{(i,j)}$ different target inner-levels $\ell_1, \ell_2, \ldots, \ell_{\chi_{(i,j)}}$ in F_{j+1}, where ℓ_k is inner-level $\lambda - (2k-1)$ in F_{j+1}. (Note that target inner-levels are odd, because λ is even.) The parameter $\chi_{(i,j)}$ is the chromatic number of the dependency graph $G_{(i,j)}$. Since $d_{(i,j)} \leq d$, a trivial polynomial time coloring algorithm using $d+1$ colors shows that $\chi_{(i,j)} \leq \chi = d+1$. Each packet in Π_i is thus assigned a color between 1 and $\chi_{(i,j)}$. Denote by $\Pi_i(k)$ the respective subset of Π_i with color k. Packets in $\Pi_i(k)$ have target level ℓ_k. Note that in the above discussion we assume that $j < \gamma'$. If $j = \gamma'$ then all the target inner-levels are set to real level $2D - 1$, which are still in F_j. By construction, the paths of packets of same color are *conflict-free*, i.e. do not share any edge, and thus can be routed together in "boats" (see below). Further, the fact that the last frame extends beyond level $2D$ does not cause a problem because no packet will ever need to move into that region, as it will be absorbed before that.

3.3　Boats

A *boat* b is a pointer to a level. We have $\chi_{(i,j)}$ boats $b_1, \ldots, b_{\chi_{(i,j)}}$. Initially, b_k is NULL. At time step t_{4k-3}, boat b_k points to the first inner-level of F_j (the boat enters the wave). At each subsequent step, the boat points to the next higher inner-level, so that at time step t_{4k-3+l} it points to inner-level $l + 1$. After the boat reaches the last inner-level of F_j it continues to the inner-levels of F_{j+1} until the boat reaches the target level ℓ_k of F_{j+1}, after which b_k becomes NULL again. Note that boats are spaced 4 levels away from each other, which will be important when an oscillating packet needs to be deflected (see below). When the context is clear, we use the term boat to refer to the inner-level it points to. Note that the last boat enters at time $t_{4\chi(i,j)-3}$, and takes $2\lambda - 2\chi_{(i,j)}+1$ steps to leave the wave, so the number of time steps per phase is $\tau = 2(\lambda + \max_{i,j}\chi_{(i,j)} - 1) \leq 2(\lambda + \chi - 1)$.

The packets of $\Pi_i(k)$ will use boat b_k to move to their target level ℓ_k in F_{j+1}. Suppose $\pi \in \Pi_i(k)$ is oscillating with initial level ℓ at the beginning of phase ϕ. Packet π will continue to oscillate until its boat b_k is at inner-level ℓ, at which time packet π will "catch its boat" and move along with it. While on its boat b_k, π follows its current path until it reaches its target inner-level ℓ_k in F_{j+1}. If, during this trip, π passes through its target node it is absorbed; otherwise π reaches its target inner-level ℓ_k at which it will oscillate for the remainder of the phase. Note that b_k passes through odd inner-levels (in particular π's initial level) at odd time steps, so π is at its initial level when b_k passes through it.

Packet Injection. A packet $\pi \in \Pi_i(k)$ with source node in frame F_j, is injected into the network when its boat b_k passes through its source node. π then moves along with b_k, following its current path, until it reaches its target level ℓ_k. While packets move along their boats they may conflict with other packets; we now describe how to handle such conflicts.

3.4 Packet Conflicts

Suppose $\pi \in \Pi_i(k)$ is on its boat b_k, progressing along its current path to its target level ℓ_k. π cannot conflict with another packet of $\Pi_i(k)$ because their current paths are conflict-free ($\Pi_i(k)$ is an independent set in $G_{(i,j)}$). Earlier boats $b_{k'}$ with $k' < k$ are ahead of b_k, so π cannot conflict with packets in $\Pi_i(k')$. π can only conflict with packets in $\Pi_i(k'')$ for $k'' > k$, which are oscillating in F_j. In such a conflict, the oscillating packet is deflected (i.e., oscillating packets have lower priority than packets on boats). We show below that this does not disrupt the algorithm.

Suppose π deflects packet $\sigma \in \Pi_i(k'')$ which oscillates on edge $e = (v_\ell, v_{\ell+1})$ (ℓ is σ's inner-level in F_j). Packet π deflects σ at the (odd) time step t_k at which π passes through ℓ. Assume that σ followed edge $e' = (v_{l-1}, v_l)$ to reach v_ℓ. We deflect σ along edge e' to inner-level $\ell - 1$, (so that at time step t_{k+1}, σ appears in v_l). Note that this is always possible because no other packet oscillating at v_ℓ arrived there using edge e', because the packets that are oscillating at v_ℓ all followed the same boat, and hence had edge disjoint paths. Note also that a packet oscillating on the first inner-level may be deflected into the previous frame F_{j-1} by an injected packet, but this causes no problem. Packet σ now follows edge e' to appear back in v_l at the (odd) time step t_{k+2}. This is possible because at time step t_{k+1} there is no boat passing through inner-level $\ell - 1$ (boat b_{k+1} is two levels away), and thus σ cannot be deflected further. When packet σ is back at inner-level ℓ, it continues to oscillate in ℓ. Therefore, σ is always at level ℓ at odd time steps, and thus it can move with boat $b_{k'}$, when it passes through ℓ. Clearly, deflected packets remain on their path.

3.5 Routing Time

Since λ must be large enough to accommodate 2χ levels in F_{j+1} (at least χ odd target inner-levels), and $\chi \leq d + 1$, $\lambda = 4\alpha \log(DN) \geq 2(d+1)$. From Lemma 1,

$d \leq 4\log(DN)$ w.h.p., so we can choose $\alpha = 3$. The routing time is $O(m \cdot \tau)$ (m phases, each of duration τ). Using $m = O(s + \gamma') = O(C + D/\log(DN))$, and $\tau = O(\lambda + \chi) = O(\log(DN))$ w.h.p. (Lemma 1), we get

Theorem 1. *The routing time of the centralized algorithm is $O(C\log(DN) + D)$, with probability at least $1 - 1/DN$.*

4   Distributed Algorithm

We show how to make the centralized algorithm (Section 3) distributed when all nodes know C, D, and N (a commonnly made assumption [7, 13]). Given C, D, N, nodes can compute $\lambda, \gamma', s, m, \tau$. (Nodes do not need information about the paths of packets other than the one they inject.)

The setup is similar to the centralized algorithm: packets follow boats on waves to their destinations. The major difference with the centralized algorithm is that the new algorithm provides a distributed coloring of the dependency graphs $G_{(i,j)}$. The distributed coloring is accomplished with the method of reverse simulation that is described below.

4.1   Reverse Simulation

In the distributed algorithm, we define $\chi = 2\lambda/\alpha$ (with $\alpha = 12$) which will be an upper bound on the number of colors assigned to the packets. Packets of set Π_i follow wave ω_i. Suppose ω_i points to frame F_j. We define the initial and target levels in F_j, F_{j+1} as in the centralized algorithm. The set of packets $A \subseteq \Pi_i$ which are oscillating at their initial inner levels in frame F_j at the beginning of the phase will move to F_{j+1}, where they will oscillate at their target levels.

A phase is divided into ξ rounds $r_1, \ldots, r_\xi$, each of length 2τ time steps, twice as long as a phase in the centralized algorithm. Each round has χ boats and target levels as in the centralized algorithm. At the beginning of round r_1, each packet in A chooses a color randomly among χ colors. Let A_1 be the set of packets with a valid color, and A_1' the packets with an invalid color. ($A = A_1 \cup A_1'$.)

During round r_1, *all* packets in A will follow their respective boats. The packets in A_1 will not be deflected, and they follow their respective boats to successfully reach their target levels where they will oscillate for the rest of the round. Some packets, $A_1'' \subseteq A_1'$, will collide with non-oscillating packets as they follow their boats. Such packets can mark themselves as members of A_1'. These packets need to choose new colors and try again. At the end of round r_1, *all* packets in A return to their initial level (see below). In round r_2, packets in set A_1'' choose a new color, and a subset $A_2' \subseteq A_1''$ will still have an invalid color. A subset $A_2'' \subseteq A_2'$ will collide with non-oscillating packets, and will need to choose new colors in the next round. Continuing in this way, in round k, the packets in A_{k-1}'' choose new colors, and those in $A_k' \subseteq A_{k-1}''$ still do not have a valid color. Of these packets, A_k'' will collide with non-oscillating packets. We will show A_ξ' is empty w.h.p, i.e., all packets have a valid color by the last round. Thus, in the last round, all the packets reach their target inner-levels, where they will oscillate till the next phase. We give the details below.

We define 4 levels of priority, $0, 1, 2, 3$. When two or more packets collide, the packet with highest priority always wins, and ties are broken randomly. A packet which successfully reachs its target level in round k (without being deflected by non-oscillating packets) keeps its color in all subsequent rounds and attains priority 3 for the remainder of the phase, whenever it is not oscillating. An oscillating packet has priority 1. A packet that chooses a new color in a round attains priority 2 for the round. If, during the round, it collides with any priority 2 or 3 packet, it immediately attains priority 0 for the remainder of the round, and will select a new color in the next round. Such priority 0 packets do not "distract" other forward going packets, and they follow arbitrary paths, due to deflections, for the remainder of the round.

At the end of a round, all packets in A (with valid or invalid coloring) need to appear back at their initial levels. Let t be the time step that the last boat in the round leaves the network. After time t, all packets follow, in reverse, the path that they followed from the beginning of the round. Thus, by the end of the round, they appear at their initial level where they oscillate until the next round. The path reversal is accommodated by having the nodes store all their computations from the beginning of the round up to time t. After time t, the nodes simply do the reverse computations, since routing is a reversible operation. (This is why we need the round to be twice as long as τ.)

4.2   Packet Injections

So far we considered only the oscillating packets in Π_i, that already appear in F_j at the beginning of phase ϕ. We also need to consider the set of packets $B \subseteq \Pi_i$ that will be injected in F_j during ϕ. Packets of B can be further partitioned into two sets: B_1, which are the packets of B whose source are at odd inner-levels of F_j, and B_2, which have sources at even inner-levels of F_j. Packets of B_1 and B_2 are treated separately so that they can not interfere with each other.

We divide phase ϕ into three sub-phases ϕ_A, ϕ_{B_1}, and ϕ_{B_2} in which we send the packets of the respective sets A, B_1 and B_2 to F_{j+1}. Each sub-phase consists of ξ rounds. We also divide the frame F_{j+1} into three disjoint regions F_A, F_{B_1}, and F_{B_2}, each consisting of 2χ inner-levels and containing χ target levels. Region F_A occupies the upper one-third (right) inner-levels of F_{j+1}, F_{B_1} the middle one-third inner-levels, and F_{B_2} the lower (left) one-third inner-levels. Packets of set A, B_1 and B_2, have their target levels in F_A, F_{B_1} and F_{B_2}, respectively.

During phase ϕ_A the packets of set A will move to region F_A, using the algorithm we described in Section 4.1. During ϕ_{B_1}, the packets of B_1 are injected into the network, and then they move to their target levels in region F_{B_1} using the reverse simulation technique that was used for packets in set A. The initial levels of the packets in B_1 are the inner-levels of their sources, and the packets are injected at the beginning of phase ϕ_{B_1}. Since a node injects at most 1 packet, the packets are guaranteed to be able to oscillate on their initial inner-levels during the reverse simulation. At the beginning of phase ϕ_{B_2}, the packets of set B_2 are injected into the network. Those packets will move to their target levels in region F_{B_2} during phase ϕ_{B_2} using the reverse simulation technique that was used for packets in set A. Those packets will also oscillate on their initial inner-levels, which are even (as opposed to packets in A and B_1 which have odd initial inner-levels). In order to handle the even levels, during this phase the boats enter the frame F_j from inner-level 2.

4.3   Routing Time

Since $\chi \geq 2d$ w.h.p, a packet picks a valid color with probability $\geq \frac{1}{2}$, thus only $O(\log(DN))$ rounds are needed for every packet to pick a valid color, adding an extra factor of $\log(DN)$ to the centralized routing time. (We omit the details.)

Theorem 2. *The routing time of the distributed algorithm is* $O(C \log^2(DN) + D \log(DN))$, *with probability* $1 - O(1/DN)$.

References

1. S. Alstrup, J. Holm, K. de Lichtenberg, and M. Thorup. Direct routing on trees. In *Proc. SODA*, pages 342–349, 1999.
2. P. Baran. On distributed communications networks. *IEEE Transactions on Communications*, pages 1–9, 1964.
3. A. Ben-Dor, S. Halevi, and A. Schuster. Potential function analysis of greedy hot-potato routing. *Theory of Computing Systems*, 31(1):41–61, Jan. / Feb 1998.
4. S. N. Bhatt, G. Bilardi, G. Pucci, A. G. Ranade, A. L. Rosenberg, and E. J. Schwabe. On bufferless routing of variable-length message in leveled networks. *IEEE Trans. Comput.*, 45:714–729, 1996.
5. A. Borodin, Y. Rabani, and B. Schieber. Deterministic many-to-many hot potato routing. *IEEE Tran. on Parallel and Dist. Sys.*, 8(6):587–596, June 1997.
6. A. Broder and E. Upfal. Dynamic deflection routing on arrays. In *Proc. STOC*, pages 348–358, May 1996.
7. C. Busch. $\tilde{O}$(congestion + dilation) hot-potato routing on leveled networks. *Theory Comput. Syst.*, 37(3):371–396, 2004.
8. C. Busch, M. Herlihy, and R. Wattenhofer. Hard-potato routing. In *Proc. STOC*, pages 278–285, May 2000.
9. C. Busch, M. Magdon-Ismail, and M. Mavronicolas. Universal bufferless routing. In *Proc. 2nd Workshop on Approximation and Online Algorithms (WAOA)*, pages 239–252, 2004.
10. C. Busch, M. Magdon-Ismail, M. Mavronicolas, and R. Wattenhofer. Near-optimal hot-potato routing on trees. In *Proc. Euro-Par*, pages 820–827, 2004.
11. U. Feige and P. Raghavan. Exact analysis of hot-potato routing. In *Proc. FOCS*, pages 553–562, 1992.
12. C. Kaklamanis, D. Krizanc, and S. Rao. Hot-potato routing on processor arrays. In *Proc. SPAA*, pages 273–282, 1993.
13. F. T. Leighton, B. M. Maggs, A. G. Ranade, and S. B. Rao. Randomized routing and sorting on fixed-connection networks. *J. Algorithms*, 17(1):157–205, 1994.
14. F. Thomson Leighton. *Introduction to Parallel Algorithms and Architectures: Arrays - Trees - Hypercubes*. Morgan Kaufmann, San Mateo, 1992.
15. F. Meyer auf der Heide and C. Scheideler. Routing with bounded buffers and hot-potato routing in vertex-symmetric networks. In *Proc. ESA*, 1995.

Efficient Truthful Mechanisms for the Single-Source Shortest Paths Tree Problem[*]

Luciano Gualà[1] and Guido Proietti[1,2]

[1] Dipartimento di Informatica, Università di L'Aquila, Italy
{guala,proietti}@di.univaq.it
[2] Istituto di Analisi dei Sistemi ed Informatica, CNR, Roma, Italy

Abstract. Let a communication network be modelled by an undirected graph $G = (V, E)$ of n nodes and m edges, and assume that each edge is controlled by a selfish agent. In this paper we analyze the problem of designing a truthful mechanism for computing one of the most used structures in communication networks, i.e., the *single-source shortest paths tree*. More precisely, we will show that under various realistic agents' behavior scenarios, it can be guaranteed not only the existence, but also the efficiency (in terms of running time complexity) of such mechanisms. In particular, for the fundamental case in which the problem is *utilitarian*, we will show that a truthful mechanism can be computed in $O(mn \log \alpha(m, n))$ time, where $\alpha(m, n)$ is the classic inverse of the Ackermann's function.

Keywords: Equilibria in Distributed Systems, Single-Source Shortest Paths Tree, Selfish Agents, Algorithmic Mechanism Design, Truthful Mechanisms.

1 Introduction

Mechanisms are a classical concept of the theory of non-cooperative games [16]. In these games there are several independent agents that have to work together in order to optimize a global objective function. However, each agent has her own valuation function and may lie in hope of getting a higher profit. This leads to economically suboptimal resource allocation and is therefore undesirable. The main objective of *mechanism design* theory is to study how to incentive the agents in order to cooperate with the solving algorithm. A *mechanism* is a pair $\mathcal{M} = \langle g(\cdot), p(\cdot) \rangle$, where $g(\cdot)$ is an algorithm computing a solution, and $p(\cdot)$ specifies the payments provided to the agents. Informally, a mechanism is *truthful* if its payments guarantee that agents are not stimulated to lie. Then, the problem of combining the game theoretic concept of designing a truthful mechanism, with the computational complexity requirement of designing an efficient algorithm, is exactly the topic of the *algorithmic mechanism design* (AMD) for selfish agents.

---

[*] Work partially supported by the Research Project GRID.IT, funded by the Italian Ministry of Education, University and Research.

J.C. Cunha and P.D. Medeiros (Eds.): Euro-Par 2005, LNCS 3648, pp. 941–951, 2005.

In their seminal paper concerned with AMD [15], Nisan and Ronen addressed the classic *shortest path* problem. This problem enjoys the property of being *utilitarian*. For utilitarian problems, there exists a well-known class of truthful mechanisms, i.e., the *Vickrey-Clarke-Groves (VCG) mechanisms* [4, 6, 21], and therefore the shortest path problem can be solved optimally. Afterwards, in a sequel of papers, efficient truthful mechanism have been designed for solving several other network design problems [8, 9, 12, 14, 15, 19].

In this paper, we focus on one of the most popular network topologies, that is the *single-source shortest paths tree* (SPT). What is interesting here is that an SPT naturally admits both utilitarian and non-utilitarian formulations. Indeed, as we will discuss later in the paper, it can well happen that an agent gives an evaluation of her contribution which is simply proportional to her private type, and this unavoidably makes the problem non-utilitarian. Therefore, we analyze both the scenarios, and we provide the following main results:

- In the utilitarian case, we provide a VCG-mechanism which can be implemented in $O(mn \log \alpha(m, n))$ time on a RAM, and in $O(mn \, \alpha(m, n))$ time on a pointer machine, where $\alpha(m, n)$ is the inverse of the Ackermann's function defined in [20];
- In the non-utilitarian case, we provide: (i) an n-approximate VCG-mechanism which can be implemented in almost optimal $O(m \, \alpha(m, n))$ time, and (ii) a mechanism guaranteeing both truthfulness and positive utilities for the agents which can be implemented in $O(m + n \log n)$ time.

The paper is organized as follows: In Section 2 we recall some basic definitions from both graph theory and algorithmic mechanism design; in Section 3 we deal with the utilitarian version of our problem, while in Section 4 we analyze the different solutions for the non-utilitarian case.

2   Basic Definitions

Let $G = (V, E)$ be an undirected graph, with $|V| = n$ nodes and $|E| = m$ edges, and with a positive real weight associated with each edge $e \in E$. Given a source node s and a destination node z in G, a path in G between s and z is a *shortest path*, say $P_G(s, z)$, if the sum of its edge weights (called *distance* in G between s and z, and denoted by $d_G(s, z)$) is minimum. Given a source node $s \in V$, we denote by $S_G(s)$ a *single-source shortest paths tree* (SPT) of G rooted in s, i.e, the union of all the shortest paths from s to every $v \in V \setminus \{s\}$. Given $u, v \in V$, we denote by $\text{LCA}(u, v)$ the *least common ancestor* of u and v in $S_G(s)$, i.e, the ancestor of both u and v in $S_G(s)$ which is farthest from s.

Let $e = (u, v) \in S_G(s)$ be a tree edge, with u closer to s than v. Let $M(e)$ denote the set of nodes in $S_G(s)$ reachable from s without passing through edge e, and let $N(e) = V \setminus M(e)$ be the remaining nodes. Sets $M(e)$ and $N(e)$ define a cut in G, and $C(e) = \{(x, y) \in E \setminus \{e\} \mid (x \in M(e)) \wedge (y \in N(e))\}$ is the set of the edges crossing the cut. Moreover, we denote by $\|e\|$ the cardinality of $N(e)$.

Let a communication network be modelled by a 2-edge-connected graph G, and assume that each edge is owned by a selfish agent A_e, which holds a private information t_e. We call this value *input type* of the agent A_e. This value depends on various factors (e.g., bandwidth, reliability, etc.). Only agent A_e knows t_e, while everything else is public knowledge. Each agent has to declare a public *reported type* r_e to the mechanism. We will denote by t the vector of input types, and by r the vector of bids.

For a given optimization problem defined on G, there exists some set of feasible solutions F that the mechanism is allowed to choose. For each feasible solution $x \in F$, some measure function $\mu(x,t)$ is defined, which depends on the true types. The mechanism tries to optimize $\mu(x,t)$, but of course it does not know t directly.

For every agent A_e, a function $v_e(t_e, x)$ expresses A_e's *valuation* with respect to an output $x \in F$: this is a quantification of the service carried on by A_e into x. While t_e is known only by the agent A_e, the valuation function is public. In order to offset the costs deriving from these services, the mechanism provides some reward to agents participating to the computed solution, i.e., the mechanism makes a *payment* $p_e(r)$ to the agent A_e for the service provided in a solution which is computed as a function of the reported vector r.

A mechanism is a pair $\mathcal{M} = \langle g(r), p(r) \rangle$, where $g(r)$ is an algorithm that, given agents' bids, computes a feasible solution in F, and $p(b)$ is a scheme which describes the payments provided to the agents.

For each agent A_e and for each solution $g(r)$ computed by the mechanism, the *utility* function of A_e is defined as $u_e(t_e, r) = p_e(r) - v_e(t_e, g(r))$. We assume that each agent is selfish, i.e., she always attempts to maximize her utility. Let r_{-e} denote the vector of all bids besides r_e; the pair (r_{-e}, r_e) will denote the vector r. We say that *truth-telling* is a *dominant strategy* for agent A_e if bidding t_e always maximizes her utility, regardless of what the other agents bid, i.e., $u_e(t_e, (r_{-e}, t_e)) \geq u_e(t_e, (r_{-e}, r_e))$, for all r_{-e} and r_e. A mechanism is said *truthful* if, for every agent, truth-telling is a dominant strategy. Moreover, let $\varepsilon(\sigma)$ denote a positive real function of the input size σ. Then, an $\varepsilon(\sigma)$-*approximation mechanism* is a mechanism which returns a solution $g(r)$ which comes within a factor $\varepsilon(\sigma)$ from the optimum, i.e., $\mu(g(r), t) \leq \varepsilon(\sigma) \cdot \mu(x^*, t)$, where x^* is an optimal solution with respect to the vector t. We say that a mechanism is *poly-time computable* if $g(\cdot)$ and $p(\cdot)$ are computable in polynomial time and that satisfies the *voluntary participation* condition if agents never incur in a net loss.

One of the most important results in mechanism design theory are the well-know *Vickrey-Clarke-Groves* (VCG) mechanisms. A VCG-mechanism applies to mechanism design problems called *utilitarians* and enjoys the fundamental property of being truthful. A mechanism design problem is called *utilitarian* if its measure function satisfies $\mu(x,t) = \sum_{e \in E} v_e(t_e, x)$.

Definition 1 (VCG-mechanisms). *A mechanism is of the VCG-family if:*

1. $g(r) \in \arg\min_{x \in F} \left\{ \sum_{e \in E} v_e(r_e, x) \right\}$.
2. $p_e(r) = \sum_{e' \neq e} v_{e'}(r_{e'}, g(r)) + h_e(r_{-e})$, *where* $h_e(r_{-e})$ *is an arbitrary function independent of* r_e.

3   The Utilitarian Case

Let be given a communication network modelled by an undirected graph $G = (V, E)$ in which each edge $e \in E$ is owned by a selfish agent. In the following, we will denote by G and $\widetilde{G}$ the input graph as weighted with respect to the reported values and the input types, respectively.

Suppose that A_e holds, as the private type t_e for the owned edge e, the length of the communication link, and thus the time needed to cross it, and assume that the system-wide goal is to minimize the completion time for delivering a message from a distinguished node $s \in V$ to every node $v \in V \setminus \{s\}$. This means that the system looks for an SPT rooted in s of $\widetilde{G}$.

3.1   A (Truthful) VCG-Mechanism

By using the notation introduced in the previous section, the problem can be formalized as follows. The set of feasible solutions F is the set of all the spanning trees (considered in the following as rooted in s) of $\widetilde{G}$, and a measure of a solution $T \in F$ is

$$\mu(T, t) = \sum_{v \in V} d_T(s, v). \tag{1}$$

To complete the description of the problem, we have to define the agents' valuation. It is clear that, if an agent A_e participates to the output with her edge e, she will incur in a transmission cost (i.e., the cost for forwarding a message through that edge). In our scenario it is reasonable to assume that the transmission cost is proportional to the length of the edge, i.e., proportional to the value t_e. Notice that the TCP/IP protocol used in Internet for broadcasting a message is the so-called *unicast*. In this protocol, if a source wants to send a message to a set of recipients, it must send a copy of the message for each destination. Therefore, if any solution $T \in F$ is used for broadcasting a message from s to all the other nodes, then the cost for the agent A_e can be expressed as follows:

$$v_e(t_e, T) = \begin{cases} t_e \|e\| & \text{if } e \in E(T); \\ 0 & \text{otherwise.} \end{cases}$$

Indeed, if an agent A_e participates to the output T, she will incur a transmission cost of t_e for each message which passes through e, and the number of these messages is exactly $\|e\|$.

From the above assumptions, it immediately follows that the problem is utilitarian. Indeed, the measure function (1) can be rewritten as

$$\mu(T, t) = \sum_{v \in V} d_T(s, v) = \sum_{e \in E(T)} t_e \|e\| = \sum_{e \in E} v_e(t_e, T).$$

This means that we can use a VCG-mechanism to solve the problem. Therefore, let $\mathcal{M}_1$ be a mechanism defined as follows:

1. The algorithmic output specification selects an SPT $S_G(s)$ of G;
2. Let $G - e = (V, E\backslash\{e\})$. Then, the payment function for A_e is defined as

$$p_e(r) = \begin{cases} \sum_{v \in V} d_{G-e}(s, v) - (\mu(S_G(s), r) - r_e\|e\|) & \text{if } e \in E(S_G(s)); \\ 0 & \text{otherwise.} \end{cases}$$

It is clear that the above payments obey to Definition 1 and then the mechanism is a (truthful) VCG-mechanism. Furthermore, this is a payment scheme inducing a so-called *pivotal mechanism*, which can be shown to satisfy the voluntary participation [4].

3.2   Mechanism Time Complexity

The algorithmic question is now the following: how fast can the above mechanism be computed? We start analyzing the cost for computing the payment scheme.

To compute $p_e(r)$ for each $e \in E(S_G(s))$, the bottle-neck is to find all the distances $d_{G-e}(s, v)$, for every $v \in V$. Indeed, it is not hard to see that the term $\mu(r, S_G(s)) - r_e\|e\|$ can be found, for all the edges $e \in E(S_G(s))$, in $O(n)$ time. A trivial solution consists in computing a new SPT of the graph $G - e$ from scratch, once for each edge $e \in E(S_G(s))$. This solution clearly takes $O(mn + n^2 \log n)$ time. We now show how to improve (on a RAM) the above bound to $O(mn \log \alpha(m, n))$ time.

We start by computing, for all the pairs $u, v \in V$, the distance $d_G(u, v)$. This can be done in $O(mn \log \alpha(m, n))$ time [17]. Then, we solve $n - 1$ subproblems. Each subproblem is identified by a distinct destination node $z \in V$, and asks for computing the distance $d_{G-e}(s, z)$ for each edge e of the path in $S_G(s)$ between s and z. We have to solve exactly $n-1$ subproblems, one for each $z \in V\backslash\{s\}$, since the distance from s to z may increase – as a consequence of deleting the edge e – only if e belongs to $P_G(s, z)$. We will solve each subproblem in $O(m \log \alpha(m, n))$ time, by achieving a bound of $O(mn \log \alpha(m, n))$ time for the original problem.

Let $P_{G-e}(s, z)$ be a replacement shortest path for the edge e, i.e., a path from s to z in $G - e$ of (minimum) length $d_{G-e}(s, z)$. The problem of finding all the replacement shortest paths, one for each edge of $P_G(s, z)$, has been efficiently solved in $O(m + n \log n)$ time on a pointer machine [10], and $O(m \alpha(m, n))$ time on a word RAM [11], respectively. Both algorithms are based on a pre-computation of the SPTs $S_G(s)$ and $S_G(z)$. We now show how to improve the above results to $O(m \log \alpha(m, n))$ time, by using a powerful structure called *Split-Findmin* [17].

Let $e = (u, v)$ be an edge on $P_G(s, z)$, with u closer to s than v. Since a replacement shortest path $P_{G-e}(s, z)$ joining s and z must contain an edge in $C(e)$, it follows that it corresponds to a path of length

$$d_{G-e}(s, z) = \min_{f=(x,y) \in C(e)} \{k(f) := d_{G-e}(s, x) + r_f + d_{G-e}(y, z)\},$$

which can be shown [10] to be equivalent to

$$d_{G-e}(s, z) = \min_{f \in C(e)}\{d_{S_G(s)}(s, x) + r_f + d_{S_G(z)}(y, z)\} = \min_{f \in C(e)}\{d_G(s, x) + r_f + d_G(y, z)\}. \quad (2)$$

Hence, since we have pre-computed the all-pairs distances in G, $k(f)$ is available in $O(1)$ time for fixed f. It then remains to select the minimum over $C(e)$. To do this efficiently, we use a Split-Findmin structure. This is a structure operating on a collection of disjoint sequences of n elements. Initially, there is only one sequence containing all the elements, and each element u has a key $k(u) := +\infty$. Then, the structure supports the following operations:

$\texttt{split}(u)$: Split the sequence containing u into two sequences of elements: one up to and including u, the other sequence taking the rest;

$\texttt{findmin}(u)$: Return the element (and the associated key) in u's sequence with minimum key;

$\texttt{decrease-key}(u, k')$: Set $k(u) := \min\{k(u), k'\}$.

We find all the distance $d_{G-e}(s, z)$ as follows. First, we label each non-tree edge f with the value (2). Then, we initialize a Split-Findmin structure, where the initial n-elements sequence consists of the vertices of $S_G(s)$ as sorted in any arbitrary post-order. We maintain two invariants: (1) every sequence in the Split-Findmin structure corresponds to some rooted subtree of $S_G(s)$, and (2) $k(u)$ corresponds to the label of a min-label edge connecting u to a vertex outside u's sequence (i.e., outside the subtree of $S_G(s)$ currently containing u).

Let now $e = (u, v) \in P_G(s, z)$. By invariants (1) and (2), if S is a sequence in the Split-Findmin structure and v is the root of the subtree corresponding to S, then $\texttt{findmin}(v)$ will return a key $k(f_e)$, where f_e is a non-tree edge belonging to $P_{G-e}(s, z)$ and $k(f_e)$ is exactly the distance $d_{G-e}(s, z)$. Once f_e is determined, we proceed to solve the problem for the children of v along the path $P_G(s, z)$. Because of the post-order arrangement of the nodes, v is the rightmost element in its sequence. Then, we perform one split centered at the element preceding v in the sequence (this will sever v), and one additional split (in any arbitrary order) for each of the children of v in $S_G(s)$, to reestablish invariant (1). After, we focus on the sequence associated with the children of v in $P_G(s, z)$, say w, and we restore invariant (2) by performing a number of decrease-key operations. More precisely, for each edge $f = (w', y)$ such that $\mathrm{LCA}(w', y) = v$ and w' is a descendant of w in $S_G(s)$, we issue the operation $\texttt{decrease-key}(w', k(f))$ (see Figure 1). Concerning the time complexity, the following lemma holds:

Lemma 1. *Let $P_G(s, z)$ be a shortest path joining s and z. Given all the distances $d_G(s, x)$ and $d_G(z, x)$ for each $x \in V$, all the distances $d_{G-e}(s, z)$ for each $e \in P_G(s, z)$ can be determined in $O(m \log \alpha(m, n))$ time.*

Proof. Since $k(f)$ is available in $O(1)$ time for a fixed non-tree edge f, labelling all the non-tree edges takes $O(m)$ time. Concerning the Split-Findmin operations, in total there are $O(m)$ operations: $O(n)$ splits (one for each subtree whose root is adjacent to some node of $P_G(s, z)$), $O(n)$ findmins (one for each node of $P_G(s, z)$), and $O(m)$ decrease-keys (at most one for each non-tree edge). This takes $O(m \log \alpha(m, n))$ time [17]. Other costs, such as the post-order traversal and finding least common ancestors, are linear [2]. $\qquad\square$

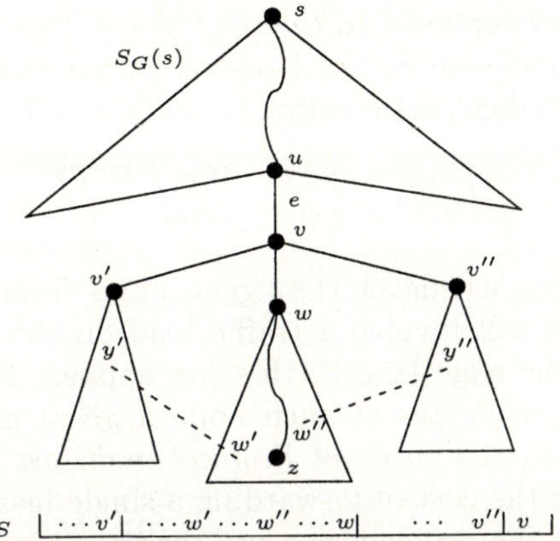

Fig. 1. The sequence S corresponding to the subtree of $S_G(s)$ rooted at v is split after the `findmin`(v) operation. Dashed edges are those for which a decrease-key operation is performed.

We are now ready to prove the main result:

Theorem 1. *The mechanism $\mathcal{M}_1$ is a truthful mechanism for the utilitarian SPT problem, and can be computed on a RAM in $O(mn \log \alpha(m, n))$ time.*

Proof. The mechanism belongs to the VCG-family, and therefore it is truthful. Concerning the output specification, the fastest solution for computing an SPT is the classic Dijkstra's algorithm implemented with Fibonacci heaps, which runs in $O(m + n \log n)$ time [5]. On the other hand, as far as the payment scheme is concerned, we proceed as follows. First, we find the all-pairs distances in G in $O(mn \log \alpha(m, n))$ time [17], and we solve each of the above described subproblems in $O(m \log \alpha(m, n))$ time. Then, for each edge $e = (u, v) \in E(S_G(s))$, we extract from the solutions of the subproblems all the distances $d_{G-e}(s, x)$, for every x in the subtree of $S_G(s)$ rooted at v (all the other nodes clearly maintain their distance from s in $G - e$). Thus, we can easily compute $p_e(r)$ in $O(n)$ time, since $\mu(r, S_G(s))$ and $r_e\|e\|$ can be obtained in $O(n)$ time by a trivial modified post-order visit of $S_G(s)$. Since we have to compute exactly $n - 1$ payment functions, one for each tree edge, the claim follows. $\square$

Notice that on a pure pointer machine model, the mechanism $\mathcal{M}_1$ can be computed in $O(mn \, \alpha(m, n))$ time, since in this case the Split-Findmin data structure requires $O(m \, \alpha(m, n))$ time for solving any given subproblem [17].

4   The Non-utilitarian Case

The utilitarian scenario assumes that each agent, in doing her valuation, starts from the assumption that each atomic operation will involve a traffic load on the owned edge which is proportional to the edge length times the size of the corresponding appended subtree of $S_G(s)$. However, in another reasonable scenario, an agent might evaluate her participation to an output $T \in F$ simply as follows:

$$v_e(t_e, T) = \begin{cases} t_e & \text{if } e \in E(T); \\ 0 & \text{otherwise.} \end{cases} \tag{3}$$

This scenario is realistic whenever the agent starts from the assumption that each atomic operation will involve a traffic load on the owned edge which is proportional only to the edge length (this can happen, for instance, when the transmission protocol replicates at each node a given message once for each descending node, like in the *Internet Protocol multicast* [3], so that each tree edge will simply afford the cost of forwarding a single message).

This setting makes the problem non-utilitarian, since the measure function associated with the SPT problem does not equal the sum of the agents' valuations. In the following, we show how to approach the problem from two different perspectives. In both cases, we make use of the pointer machine computational model, since we cannot take advantage of the addressing capabilities of a RAM, as we did for the utilitarian case.

4.1   An Approximate VCG-Mechanism

A brute-force solution consists in designing a mechanism from the VCG-family. Since the algorithmic output specification has to minimize the sum of the agents' valuations, this will clearly return an MST of $\widetilde{G}$. More formally, let $\mathcal{M}_2$ be the mechanism defined as follows:

1. The algorithmic output specification computes an MST of G;
2. Let $w(T_G) = \sum_{e \in E(T_G)} r_e$ denote the total weight of the solution T_G, as computed in G. Then, the payment function for A_e is defined as

$$p_e(r) = \begin{cases} w(T_{G-e}) - \big(w(T_G) - r_e\big) & \text{if } e \in E(T_G); \\ 0 & \text{otherwise.} \end{cases}$$

Theorem 2. *The mechanism $\mathcal{M}_2$ is a truthful n-approximation mechanism for the non-utilitarian SPT problem, and is computable on a pointer machine in $O(t^*(T_G)) = O(m\,\alpha(m,n))$ time, where $t^*(T_G)$ is the time needed to solve optimally the MST problem.*

Proof. It is easy to see that $g(\cdot)$ minimizes the sum of the agents' valuations and that the mechanism belongs to the VCG-family. Indeed, the above payments obey to Definition 1 and then the mechanism is truthful (and consequently T_G is an MST of $\widetilde{G}$). Concerning the approximation ratio, let $S_{\widetilde{G}}(s)$ be an optimal solution for the SPT problem. We now show that the solution returned by the VCG-mechanism is a factor-n approximation. Indeed, if we consider $T_{\widetilde{G}}$ as rooted in s, we have

$$\mu(t, T_{\widetilde{G}}) = \sum_{e \in E(T_{\widetilde{G}})} t_e \|e\| \le n \sum_{e \in E(T_{\widetilde{G}})} t_e = n\,w\big(T_{\widetilde{G}}\big) \le n\,w\big(S_{\widetilde{G}}(s)\big) \le n\,\mu\big(t, S_{\widetilde{G}}(s)\big).$$

Concerning the time complexity, observe that the mechanism essentially requires the computation on a pointer machine of an MST of G (which can be done optimally through the algorithm presented in [18], which has an $O(m\,\alpha(m,n))$ runtime, but for which a tighter analysis is not known), and the solution of a sensitivity analysis problem on T_G. As showed in [7], such a problem can be solved in optimal time as well, but still a tight analysis cannot be provided, thought it is known that such a problem is not harder than the MST one. Summarizing, the time complexity is $O(t^*(T_G)) = O(m\,\alpha(m,n))$. $\square$

Notice that the above approximation ratio is tight, since it is easy to exhibit an example in which an MST is an n-approximation of an SPT. This means that we cannot hope to get a better approximate result by means of VCG-mechanisms.

4.2 An Exact Truthful Mechanism Satisfying the Voluntary Participation

The alternative solution we propose is inspired to the results in [1], where the authors show how to design truthful mechanisms for those problems in which each agent's valuation has the form $v_e(t_e, x) = t_e\, w_e(r)$, where $w_e(r)$ is called *work curve* for agent A_e and it is some amount of work that depends on the algorithmic output specification, which in its turn is a function of the reported types vector r. We say the output algorithm $g(r)$ is *decreasing* if each of the associated work curves is decreasing (i.e., $w_e(r_{-e}, r_e)$ is a decreasing function of r_e, for all A_e and fixed r_{-e}).

In [1] it is shown that a mechanism is truthful for a problem where each agent's valuation has the above form if and only if the output algorithm $g(r)$ is decreasing, and the payments are given by an explicit formula involving an integral of the the work curve. In particular, if $\int_0^{+\infty} w_e(r_{-e}, z)\, dz$ is bounded for all A_e and r_{-e}, the mechanism satisfies also the voluntary participation and the payment function is of the form

$$p_e(r_{-e}, r_e) = r_e\, w_e(r_{-e}, r_e) + \int_{r_e}^{+\infty} w_e(r_{-e}, z)\, dz. \qquad (4)$$

As far as the SPT problem is concerned, let now $g(r)$ denote the output specification of an algorithm computing an SPT of G. Then, for each agent A_e, we can rewrite the valuation (3) as $v_e(t_e, g(r)) = t_e\, w_e(r)$, where

$$w_e(r) = \begin{cases} 1 & \text{if } e \in E(g(r)); \\ 0 & \text{otherwise.} \end{cases} \qquad (5)$$

Then, we can formally define the following mechanism $\mathcal{M}_3$:

1. The algorithmic output specification selects an SPT $S_G(s)$ of G;
2. The payment function for A_e is defined as $(4)^1$.

---

[1] From the assumption that the graph G is 2-edge-connected, it follows that (5) satisfies $\int_0^{+\infty} w_e(r_{-e}, z)\, dz < +\infty$, thus implying that we can apply (4), which satisfies voluntary participation.

We can now prove the following result:

Theorem 3. *The mechanism $\mathcal{M}_3$ is a truthful mechanism for the non-utilitarian SPT problem, and is computable on a pointer machine in $O(m + n \log n)$ time.*

Proof. The truthfulness follows from the fact that the output function is decreasing. Indeed, for each agent A_e, if we denote by $\hat{r}_e$ the maximum reported type for e such that e belongs to the computed solution, then the function $w_e(r_{-e}, r_e)$ is equal to 1 for $0 \leq r_e \leq \hat{r}_e$, and is equal to 0 for any $r_e > \hat{r}_e$. Thus, this is a mechanism in the class defined in [1].

From the time complexity point of view, once again the output specification can be computed in $O(m + n \log n)$ time. Concerning the payments, we have to compute all the integrals $\int_{r_e}^{+\infty} w_e(r_{-e}, z)\, dz$, one for each $e \in E(S_G(s))$ (for all other edges, the payment (4) is obviously equal to 0). By definition, we have

$$\int_{r_e}^{+\infty} w_e(r_{-e}, z)\, dz = \hat{r}_e - r_e,$$

from which it follows that for a tree edge, we have that $p_e = \hat{r}_e$. Let now $e = (u, v) \in E(S_G(s))$, with u closer to s that v. Then, e remains in $S_G(s)$ as long as $d_G(s, u) + r_e \leq d_{G-e}(s, v)$, from which it follows that $\hat{r}_e = d_{G-e}(s, v) - d_G(s, u)$. As shown in [12], computing $d_{G-e}(s, v)$ is equivalent to select a non-tree edge such that

$$d_{G-e}(s, v) = \min_{f=(x,y)\in C(e)} \{d_{G-e}(s, x) + r_f + d_{G-e}(y, v)\}. \tag{6}$$

The selection of all the non-tree edges (one for each tree edge) satisfying (6) costs $O(m\,\alpha(m, n))$ time [12]. This means that we can compute all the payments in $O(m\,\alpha(m, n)) = O(m + n \log n)$ time, from which the claim follows. $\square$

References

1. A. Archer and É. Tardos, Truthful mechanisms for one-parameter agents, *Proc. 42nd IEEE Symp. on Foundations of Computer Science (FOCS'01)*, 482–491, 2001.
2. A.L. Buchsbaum, H. Kaplan, A. Rogers, and J. Westbrook, Linear-time pointer-machine algorithms for least common ancestors, MST verification, and dominators, *Proc. 30th ACM Symp. on Theory of Computing (STOC'98)*, 279–288, 1998.
3. Cisco Systems Inc.©, *Internetworking Technologies Handbook*, Cisco Press, 2004.
4. E. Clarke, Multipart pricing of public goods, *Public Choice*, 8:17–33, 1971.
5. M.L. Fredman and R.E. Tarjan, Fibonacci heaps and their uses in improved network optimization algorithms, *J. of the ACM*, 34(3):596–615, 1987.
6. T. Groves, Incentives in teams, *Econometrica*, 41(4):617–631, 1973.
7. L. Gualà and G. Proietti, Optimal MST sensitivity analysis on a pointer machine, *manuscript submitted for publication*, 2005.

8. L. Gualà and G. Proietti, A truthful $(2-2/k)$-approximation mechanism for the Steiner tree problem with k terminals, *11th Int. Computing and Combinatorics Conference (COCOON'05)*, to appear.

9. J. Hershberger and S. Suri, Vickrey prices and shortest paths: what is an edge worth?, *Proc. 42nd IEEE Symp. on Foundations of Computer Science (FOCS 2001)*, 252–259.

10. K. Malik, A.K. Mittal, and S.K. Gupta, The k most vital arcs in the shortest path problem, *Oper. Res. Letters*, 8:223–227, 1989.

11. E. Nardelli, G. Proietti, and P. Widmayer, A faster computation of the most vital edge of a shortest path, *Info. Proc. Letters*, 79(2):81–85, 2001.

12. E. Nardelli, G. Proietti, and P. Widmayer, Swapping a failing edge of a single source shortest paths tree is good and fast, *Algorithmica*, 36(4):361–374, 2003.

13. E. Nardelli, G. Proietti, and P. Widmayer, Finding the most vital node of a shortest path, *Theoretical Computer Science*, 296(1) (2003) 167–177.

14. E. Nardelli, G. Proietti, and P. Widmayer, Nearly linear time minimum spanning tree maintenance for transient node failures, *Algorithmica*, 40(2):119–132, 2004.

15. N. Nisan and A. Ronen, Algorithmic mechanism design, *Games and Economic Behaviour*, 35:166–196, 2001.

16. M.J. Osborne and A. Rubinstein, A course in Game Theory, *MIT Press*, 1994.

17. S. Pettie and V. Ramachandran, Computing shortest paths with comparisons and additions, *Proc. 13th ACM Symp. on Discrete Algorithms (SODA'02)*, 267–276, 2002.

18. S. Pettie and V. Ramachandran, An optimal minimum spanning tree algorithm, *J. of the ACM*, 49(1):16–34, 2002.

19. G. Proietti and P. Widmayer, A truthful mechanism for the non-utilitarian minimum radius spanning tree problem, *17th ACM Symp. on Parallelism in Algorithms and Architectures (SPAA'05)*, to appear.

20. R.E. Tarjan, Efficiency of a good but not linear set union algorithm, *J. of the ACM*, 22(2):215–225, 1975.

21. W. Vickrey, Counterspeculation, auctions and competitive sealed tenders, *J. of Finance*, 16:8–37, 1961.

Optimal Embedding of the Hypercube on Partitioned Optical Passive Stars Networks*

Christos Kaklamanis and Charalampos Konstantopoulos

Research Academic Computer Technology Institute and
Computer Engineering and Informatics Department,
University of Patras, 26500 Rio, Greece
{konstant,kakl}@cti.gr

Abstract. Partitioned Optimal Passive Stars network, POPS(d,g), is an optical interconnection network of N processors ($N = dg$) which uses g^2 optical passive star couplers. The processors of this network are partitioned into g groups of d processors each and the g^2 couplers are used for connecting each group with each of the groups, including itself. In this paper, we present an optimal embedding of the hypercube on this network for all combinations of values of d and g. Specifically, we show how to optimally simulate the most common hypercube communication pattern where each hypercube node sends a packet along the same dimension. Optimal simulation of this communication on the POPS(d,g) network has already been presented for $d \leq g$ in the literature, but for the case $d > g$, the optimality remained an open problem. Now, we show that an optimal simulation is feasible in this case too.

1   Introduction

The partitioned optical passive stars (POPS) network is an optical interconnection network used in multiprocessor systems that has recently attracted the interest of many researchers [1–10]. A POPS(d,g) network comprises g^2 optical passive stars (OPS) couplers that connect N processors where $N = dg$. Processors are split into g groups of d processors each. Processors of group j can send messages to all processors of group i through the OPS coupler OPS(i,j) $(i,j = 0,\ldots,g-1)$. However, at each communication step (termed also slot), only one of the processors of group j can send a message through the coupler OPS(i,j). Then, this message is broadcast to all the processors of group i. In Fig. 1, we see that each processor can send to or receive from more than one coupler during the same slot. However, this increases the hardware complexity in each processor. So, in this paper we make the practical assumption that each processor can send or receive at most one message at each slot.

In this paper, we present a simulation of hypercube communications on the POPS(d,g) network. Specifically, we show an optimal technique for one-hop

* This work was supported in part by the European Union under the FET IST project CRESCCO.

J.C. Cunha and P.D. Medeiros (Eds.): Euro-Par 2005, LNCS 3648, pp. 952–961, 2005.

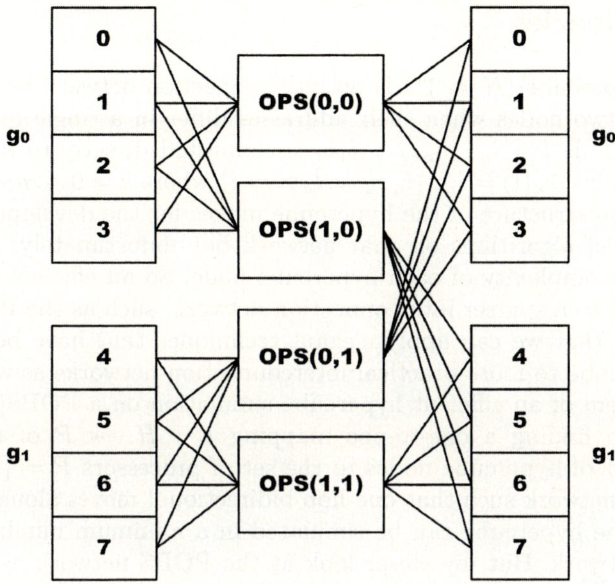

Fig. 1. A POPS$(4, 2)$ network of 8 processors

movements where each hypercube node sends a packet along the same dimension. This communication pattern is the most commonly met in parallel hypercube algorithms. Hypercube is a powerful network that has been widely used for development of many elegant parallel algorithms [11]. So an efficient hypercube simulation on a POPS(d, g) network enables the use of the these algorithms on the POPS(d, g) network too.

When performing a permutation, $\lceil \frac{d}{g} \rceil$ slots is a trivial lower bound on the POPS(d, g) network. So, the simulation of movements between neighboring nodes on the hypercube takes at least that much time. Actually, for $d = g$, it has been proved in [9] that hypercube simulation demands $2 \cdot \lceil \frac{d}{g} \rceil = 2$ slots at least.

Sahni [4] was the first that dealt with the problem of hypercube simulation on the POPS(d, g) network. Assuming that processor i simulates the node i of the hypercube, he proved that a one-hop movement on the hypercube along the same dimension can be simulated in $2 \cdot \lceil \frac{d}{g} \rceil$ slots. Mei et al. [9] used a different mapping and reduced the complexity to the lower bound for $d \leq g$. However, for the case $d > g$, their simulation achieves the lower bound only for the mono-directional moves, that is when each node of a particular $\frac{N}{2}$-node subhypercube sends messages to the corresponding node of the complementary subhypercube. However, the simulation of bidirectional moves still requires $2 \cdot \lceil \frac{d}{g} \rceil$ slots, two times higher than the lower bound for $d > g$.

Our simulation closes this gap and is optimal for all values of d and g. Specifically, for $d < g$ and $d > g$, $\lceil \frac{d}{g} \rceil$ slots are needed, whereas for $d = g$, $2 \cdot \lceil \frac{d}{g} \rceil$ slots are enough.

2 Preliminaries

A N-node hypercube ($N = 2^n$) is an interconnection network where there is a link between two nodes when their addresses differ in a single bit position. In other words, node $i = i_{n-1}i_{n-2}\ldots i_1 i_0$ is connected directly to n other nodes, namely the nodes $\oplus_k(i) = i_{n-1}i_{n-2}\ldots \bar{i}_k \ldots i_1 i_0$ where $k = 0 \ldots n-1$. The dense interconnection structure of the hypercube makes for the development of many efficient parallel algorithms on that network but unfortunately, this increases the hardware complexity of each hypercube node. So an efficient simulation of the hypercube to a sparser interconnection network, such as the POPS network simply means that we can apply elegant techniques that have been successful for the hypercube to more practical interconnection networks as well.

The problem of an efficient hypercube simulation on a POPS(d, g) network boils down to finding a one-to-one mapping $\mu : H \rightarrow P$ of the set $H = \{0, \ldots, N-1\}$ of hypercube nodes to the set of processors $P = \{0, \ldots, N-1\}$ of the POPS network such that one-hop bidirectional moves along the same dimension on the hypercube can be simulated in a minimum number of slots on the POPS network. But, by closer look at the POPS network, we realize that what really matters is not the specific POPS processor which simulates a hypercube node, but the group containing that processor, since all processors of a group are indistinguishable as far as the routing function of the POPS network is concerned. So, it is enough to find a many-to-one mapping $\mu : H \rightarrow G$ where $G = \{0, 1, \ldots, g-1\}$ is the set of groups of the POPS(d, g) network. Notice also that d and g are powers of two in this paper, since their product, N, is a power of two as well.

Before presenting our simulation technique, we should introduce some useful notations and definitions. So first we define two functions c_{01} and c_{10} as follows:

$$c_{01}(x) = \begin{cases} 01 & \text{if } x \text{ has an even number of 1s} \\ 10 & \text{if } x \text{ has an odd number of 1s} \end{cases}$$

and

$$c_{10}(x) = \begin{cases} 10 & \text{if } x \text{ has an even number of 1s} \\ 01 & \text{if } x \text{ has an odd number of 1s} \end{cases}$$

For convenience, we also define the constant functions $c_{00}(x) = 00$ and $c_{11}(x) = 11$.

We can also define the complements of these functions as $\overline{c_{01}}(x) = c_{10}(x)$, $\overline{c_{10}}(x) = c_{01}(x)$, $\overline{c_{00}}(x) = c_{11}(x)$ and $\overline{c_{11}}(x) = c_{00}(x)$. When x is clear from the context, we will use the shorter notations c_{00}, c_{01}, c_{10} and c_{11}. So, for example if x is a k-bit number, the $4+k$-bit number $c_{01}(x)c_{10}(x)x$ will be simply written as $c_{01}c_{10}x$.

Finally, by using the functions above, we define the following function

$$inv(c_i) = \begin{cases} \overline{c_i} & \text{if } i = 01 \text{ or } 10 \\ c_i & \text{if } i = 00 \text{ or } 11 \end{cases}$$

3   The Simulation Technique

Next, we present our simulation of the hypercube network on a POPS(d, g) network. First we describe the technique for the case $d > g$ and then we show how we can easily generalize for the other two cases.

3.1   Hypercube Simulation for $d > g$

The processors of a POPS(d, g) network are separated into g groups of d processors each. As mentioned before, a permutation routing on a POPS(d, g) network takes at least $\lceil \frac{d}{g} \rceil = \frac{d}{g}$ slots, since the best we can do is to partition the d processors of a group into g subgroups of $\frac{d}{g}$ processors each and then send messages by using a different coupler for each subgroup. A bidirectional move between neighboring nodes along the same dimension on the hypercube, is actually a permutation, and so a lower bound for this simulation is $\frac{d}{g}$ slots again. For reaching this lower bound, our mapping of the hypercube on the POPS(d, g) network should be such that the neighbors of all hypercube nodes mapped on the same group of the POPS(d, g) network are equally distributed among the g groups, i.e. $\frac{d}{g}$ neighbors per group. Furthermore, this property should hold for the simulation of movements along any hypercube dimension.

Now, we define a mapping μ that has the desired properties. The mapping is recursive, however, due to space limitation we give a closed form directly.

Definition 1. *For $N = 2^n$, $N = dg$, $g = 2^b$, $d = 2^a$, $d > g$, let $c_{j_{i,l_{b-1}}} \ldots c_{j_{i,l_1}}$ $c_{j_{i,l_0}} x$ ($l = 0 \ldots g - 1$, $l = l_{b-1}, l_{b-2} \ldots l_0$, $x = 0 \cdots 2^{n-2b} - 1$) be the addresses of the d hypercube nodes mapped onto the processors of group g_i ($i = 0 \cdots g - 1$, $i = i_{b-1} i_{b-2} \ldots i_0$). If r is a bit position in i ($r = 0 \ldots b - 1$), we have the following cases:*

- *$i_r = 0$. In this case, the pair of bits $c_{j_{i,l_r}}$ is given as follows:*

$$
c_{j_{i,l_r}} = \begin{cases} c_{00} & \text{if } l_r = 0 \\ c_{01} & \text{if } l_r = 1 \text{ and } l \text{ has an odd number of 1s} \\ c_{10} & \text{if } l_r = 1 \text{ and } l \text{ has an even number of 1s} \end{cases} \tag{1}
$$

- *$i_r = 1$. For this case, the pair of bits $c_{j_{i,l_r}}$ is given as follows:*

$$
c_{j_{i,l_r}} = \begin{cases} c_{11} & \text{if } l_r = 0 \\ c_{10} & \text{if } l_r = 1 \text{ and } l \text{ has an odd number of 1s} \\ c_{01} & \text{if } l_r = 1 \text{ and } l \text{ has an even number of 1s} \end{cases} \tag{2}
$$

For example, for $N = 2^n$, $g = 4$, $d = 2^{n-2}$ and for all possible values of x in the range $[0 \cdots 2^{n-4} - 1]$ the mapping is as follows:

$$
\begin{array}{cccc}
c_{00}c_{00}x & c_{00}c_{11}x & c_{11}c_{00}x & c_{11}c_{11}x \\
c_{00}c_{01}x & c_{00}c_{10}x & c_{11}c_{01}x & c_{11}c_{10}x \\
c_{01}c_{00}x & c_{01}c_{11}x & c_{10}c_{00}x & c_{10}c_{11}x \\
\underbrace{c_{10}c_{10}x}_{g_0} & \underbrace{c_{10}c_{01}x}_{g_1} & \underbrace{c_{01}c_{10}x}_{g_2} & \underbrace{c_{01}c_{01}x}_{g_3}
\end{array}
$$

Notice that each expression above represents 2^{n-4} hypercube nodes, i.e. as much as the number of possible values of x. For instance, if $N = 32$, x is 0 or 1, and so we have the following mapping

$$
\begin{array}{cccc}
00000 & 00110 & 11000 & 11110 \\
00001 & 00111 & 11001 & 11111 \\
00010 & 00100 & 11010 & 11100 \\
00101 & 00011 & 11101 & 11011 \\
01000 & 01110 & 10000 & 10110 \\
10001 & 10111 & 01001 & 01111 \\
10100 & 10010 & 01100 & 01010 \\
01011 & 01101 & 10011 & 10101 \\
\underbrace{}_{g_0} & \underbrace{}_{g_1} & \underbrace{}_{g_2} & \underbrace{}_{g_3}
\end{array}
$$

Also, note that the concept of odd and even parity has been used in [9] as well, but now we use that concept differently for defining our mapping.

Next, we prove that the mapping above is optimal.

Theorem 1. *One-hop movements along any dimension k ($k = 0 \ldots n-1$) on the hypercube are optimally simulated on a POPS(d, g) network in $\frac{d}{g}$ slots for any $d = 2^a$, $d > g$ and $g = 2^b$.*

Proof. We have to prove that for any hypercube dimension k ($k = 0 \ldots n-1$), the neighbors of nodes belonging to a group are evenly distributed among all groups including itself. Specifically, we will prove that each group possesses $\frac{d}{g}$ neighbors along dimension k of nodes belonging to each of the g groups. For convenience, we focus on group g_0 and where the neighbors of the nodes of that particular group can be found across the g groups of the POPS(d, g) network. Similar statements can be made for all other groups as well.

Let $c_{j_0, l_{b-1}} c_{j_0, l_{b-2}} \cdots c_{j_0, l_1} c_{j_0, l_0} x$ ($l = 0 \ldots g-1$) be the nodes of g_0. Let also g_i ($i \in [0 \ldots g-1]$) be some other group and $c_{j_i, l_{b-1}} c_{j_i, l_{b-2}} \cdots c_{j_i, l_1} c_{j_i, l_0} x$ be its nodes. We handle two cases: a) $k \in [0 \ldots n-2b-1]$ and b) $k \in [n-2b \ldots n-1]$.

In the first case, k is a bit position belonging to x, so two hypercube nodes $c_{j_0, p_{b-1}} c_{j_0, p_{b-2}} \cdots c_{j_0, p_1} c_{j_0, p_0} x$ and $c_{j_i, q_{b-1}} c_{j_i, q_{b-2}} \cdots c_{j_i, q_1} c_{j_i, q_0} x$ of group g_0 and g_i respectively are neighbors along dimension k if and only if $c_{j_i, q_m} = inv(c_{j_0, p_m})$, $m = 0 \ldots b-1$.

Consider now a bit position r in the binary representation of i and assume first that i has a 1 at this position. From Definition 1, we can easily see that c_{j_i, q_r} can be c_{11}, c_{01} or c_{10} and so c_{j_0, p_r} can be c_{11}, c_{10} or c_{01} respectively. However, c_{11} is not possible for c_{j_0, p_r} and the only actual possibilities are c_{10} and c_{01}. If i has 0 at position r, c_{j_i, q_r} might be c_{00}, c_{01} or c_{10} and and hence c_{j_0, p_r} will be c_{00}, c_{10} or c_{01} respectively. However, the last two combinations (c_{01}, c_{10}), (c_{10}, c_{01}) are not possible, because it can be easily checked that p and q have always the same number of 1s. So eventually, only c_{00} is possible for c_{j_0, p_r} and c_{j_i, q_r}.

So, we have determined that $c_{j_0, p_{b-1}} c_{j_0, p_{b-2}} \cdots c_{j_0, p_1} c_{j_0, p_0} x$ has c_{01} or c_{10} at the same positions where i has 1 and c_{00} at all other positions. With this information and by Definition 1, we can easily find that parameter p has 1s and 0s at the same position as i, which simply means that $i = p$. Note also that $q = p$. So we have actually determined the nodes of g_0 with neighbors in g_i. By using Definition 1 again, we can also easily prove that $c_{j0,p_{b-1}} c_{j0,p_{b-2}} \cdots c_{j0,p_1} c_{j0,p_0} x$ and $c_{ji,q_{b-1}} c_{ji,q_{b-2}} \cdots c_{ji,q_1} c_{ji,q_0} x$ are actually nodes of g_0 and g_i respectively. Since x can take $2^{n-2b} = \frac{d}{g}$ possible values, the number of messages passing through each of the couplers $\mathrm{OPS}(i,0)$, $\mathrm{OPS}(0,i)$ during the simulation of one-hop movements along dimension k is $\frac{d}{g}$ too.

Notice also that the analysis above still holds when i does not have 1s in its binary representation, i.e. $i = 0$. In this case, all $\frac{d}{g}$ messages among nodes of group g_0 are passing through the coupler $\mathrm{OPS}(0,0)$.

Now, we consider the case $k \in [n-2b \ldots n-1]$. k is one of the $2b$ most significant bits of node addresses determined by the c functions. Let c_{j0,p_r} and c_{ji,q_r} be the pair of bits of nodes $c_{j0,p_{b-1}} c_{j0,p_{b-2}} \cdots c_{j0,p_1} c_{j0,p_0} x$ and $c_{ji,q_{b-1}} c_{ji,q_{b-2}} \cdots c_{ji,q_1} c_{ji,q_0} x$ inside which bit complement is carried out. Clearly, we have that $r = \lfloor \frac{(k-n+2b)}{2} \rfloor$. Let also $\{s_0, s_1, \ldots, s_t\}$ $(0 \le t \le b-1)$ be the (possibly empty) set of bit-positions where i has 1s. We differentiate between two cases:

$r \notin \{s_0, s_1, \ldots, s_t\}$. For the pairs c_{j0,p_r}, c_{ji,q_r} we have four possible combinations $(c_{j0,p_r}, c_{ji,q_r}) = (c_{00}, c_{01}), (c_{00}, c_{10}), (c_{01}, c_{00}), (c_{10}, c_{00})$. Due to Definition 1, this implies that p and q differ in bit r.

For $z \ne r$, it should be true that $c_{j0,p_z} = c_{ji,q_z}$. If $z \in \{s_0, s_1, \ldots, s_t\}$, c_{ji,q_z} could be c_{11}, c_{01} or c_{10}. c_{11} is not possible for g_0 and thus we have two possibilities only, namely c_{01} and c_{10}, which means that p and q both have 1 at position z.

If now $z \notin \{s_0, s_1, \ldots, s_t\}$, c_{ji,q_z} is the same as c_{j0,p_z} and could be c_{00}, c_{01} or c_{10}. Regardless of the particular c_{00}, c_{01} or c_{10} at position z, we can easily see that p and q do not have the same parity, that is either p or q has an even number of 1s but not both. This in turn implies that $c_{ji,q_z} = inv(c_{j0,p_z})$ and so the only possible case is the c_{00}.

So, we have determined the values of bits of p and q in all positions but r. Firstly, if $c_{j0,p_r} = c_{00}$, we can easily determine which pair, c_{01} or c_{10}, c_{ji,q_r} is. Now the rth bit in p is 0 and the corresponding bit in q is 1. From Definition 1 also, c_{ji,q_r} is c_{01} if the number of 1s in q is odd, c_{10} otherwise. By the same token, we can specify c_{j0,p_r} when c_{ji,q_r} is c_{00}.

To sum up, we finally found two values of parameter p, say p' and p'', which satisfy our conditions. They differ only at position r: p' has a 0 at bit r whereas p'' has a 1 at the same position. Correspondingly, we have two values for parameter q, q' with 0 at bit r and q'' with 1 at the same bit. Apparently, $p' = q'$ and $p'' = q''$. Furthermore, we can easily see that $p' = i$ or $\oplus_r i$. The nodes $c_{j0,p'_{b-1}} c_{j0,p'_{b-2}} \cdots c_{j0,p'_1} c_{j0,p'_0} x$, $c_{j0,p''_{b-1}} c_{j0,p''_{b-2}} \cdots c_{j0,p''_1} c_{j0,p''_0} x$ are neighbors of the nodes $c_{ji,q''_{b-1}} c_{ji,q''_{b-2}} \cdots c_{ji,q''_1} c_{ji,q''_0} x$, $c_{ji,q'_{b-1}} c_{ji,q'_{b-2}} \cdots c_{ji,q'_1} c_{ji,q'_0} x$ respectively. During the simulation of one-hop movements along dimension k of the hypercube, $\frac{d}{2g}$ messages are starting from nodes $c_{j0,p'_{b-1}} c_{j0,p'_{b-2}} \cdots c_{j0,p'_1} c_{j0,p'_0} x$ and another $\frac{d}{2g}$ messages from nodes $c_{j0,p''_{b-1}} c_{j0,p''_{b-2}} \cdots c_{j0,p''_1} c_{j0,p''_0} x$. All these messages are passing through the OPS$(i,0)$ coupler and so $\frac{d}{g}$ slots are needed in total for this transfer. Apparently, the number of messages through the OPS$(0,i)$ coupler is $\frac{d}{g}$ too.

When the set $\{s_0, s_1, \ldots, s_t\}$ is empty, i.e. $i = 0$, the analysis above still holds. In this case, all $\frac{d}{g}$ messages are among nodes of g_0 and are passing through the coupler OPS$(0,0)$ in $\frac{d}{g}$ slots.

$\mathbf{r} \in \{\mathbf{s_0}, \mathbf{s_1}, \ldots, \mathbf{s_t}\}$. The details for this case are basically the same as previously. The only difference is the possibilities for $c_{j_0,pr}$ and $c_{j_i,qr}$. Now, we have that $(c_{j_0,pr}, c_{j_i,qr}) = (c_{00}, c_{01}), (c_{00}, c_{10}), (c_{01}, c_{11}), (c_{10}, c_{11})$. In a similar way, we can find again two p and q values satisfying our conditions. The number of messages passing through the couplers OPS$(i,0)$, OPS$(0,i)$ is $\frac{d}{g}$ once more and that many slots are needed for this routing. Hence, the theorem follows. $\square$

So far, we have seen that our mapping technique is optimal for $d > g$. In the following, we consider the case $d \leq g$ and show the optimality of our mapping in this case too.

3.2 Hypercube Simulation for $d \leq g$

Due to [9], we know that for $d = g$, we need at least two slots for the simulation of one-hop hypercube movements. For $d < g$, the lower bound is just the trivial lower bound of 1 slot. Now, we will show how to achieve these lower bounds by extending our technique. We first deal with the case $d < g$. The case $d = g$ easily follows then.

The idea is to reverse the roles of d and g in the mapping presented above. Specifically, if $d = 2^a$ and $g = 2^b$ with $a < b$, we can write the nodes of the N-node hypercube ($N = dg$, $N = 2^n$) as $c_{j_{a-1}} c_{j_{a-2}} \ldots c_{j_0} x$ where $x = 0 \ldots \frac{g}{d} - 1$. By reversing the roles of d and g in our mapping technique, we divide the nodes of the hypercube into d sets of g nodes each. The ith set[1] s_i ($i = 0 \ldots d - 1$) contains the nodes $c_{j_{i,l_{a-1}}} c_{j_{i,l_{a-2}}} \ldots c_{j_{i,l_0}} x$ where $l = 0 \ldots d - 1$. For any two sets s_u and s_v and any hypercube dimension k, there are exactly $\frac{g}{d}$ nodes in s_u whose neighbors along dimension k are inside s_v and vice versa. We also have $\frac{g}{2d}$ pairs of neighboring nodes along the same dimension inside each set.

In order to get g groups of d nodes each, we should divide each set i in $\frac{g}{d}$ subsets. Each of these subsets should not contain neighboring nodes along any dimension and also for any two subsets, the first subset should contain at most one node with neighbor in the other subset.

Since d and g are both powers of 2, we can get $\frac{g}{d}$ subsets by halving the set i $log_2 \frac{g}{d}$ times as follows:

- In the first halving step, we divide each set s_i into two subsets s_{i0} and s_{i1}. Subset s_{it} ($t = 0, 1$) is defined as:

$$s_{it} = \{c_{j_{i,l_{a-1}}} c_{j_{i,l_{a-2}}} \ldots c_{j_{i,l_0}} x_l^t | l = 0 \ldots d - 1\} \tag{3}$$

---

[1] We use the term set instead of group in order to avoid confusion.

where

$$x_l^t = \begin{cases} \{x|x \text{ a } (n-2a)\text{-bit number with odd number of 1s}\} & \text{if } l \text{ has an odd} \\ & \text{number of 1s} \\ & \text{and } t = 0 \\ \{x|x \text{ a } (n-2a)\text{-bit number with even number of 1s}\} & \text{if } l \text{ has an even} \\ & \text{number of 1s} \\ & \text{and } t = 0 \\ \{x|x \text{ a } (n-2a)\text{-bit number with even number of 1s}\} & \text{if } l \text{ has an odd} \\ & \text{number of 1s} \\ & \text{and } t = 1 \\ \{x|x \text{ a } (n-2a)\text{-bit number with odd number of 1s}\} & \text{if } l \text{ has an even} \\ & \text{number of 1s} \\ & \text{and } t = 1 \end{cases} \quad (4)$$

– At each of the remaining halving steps, let $c_{j_{i,l_{a-1}}} c_{j_{i,l_{a-2}}} \cdots c_{j_{i,l_0}} x_{i,l}$ ($l = 0 \cdots d - 1$) be the nodes of set s_i that has resulted from the previous step, where $x_{i,l}$ is the set of x's corresponding to the particular values of i and l. For example, after the first halving step, $x_{i,l}$ is either equal to x_l^0 or x_l^1. If $x_{i,l}^0$ and $x_{i,l}^1$ is any equipartition of $x_{i,l}$ into two sets, then the two new subsets s_{i0}, s_{i1} of set s_i are given as follows:

$$s_{it} = \{c_{j_{i,l_{a-1}}} c_{j_{i,l_{a-2}}} \cdots c_{j_{i,l_0}} x_{i,l}^t | l = 0 \ldots d - 1, t = 0, 1\} \ .$$

It should be noted that now it does not matter how we partition $x_{i,l}$ in two equal parts. For example $x_{i,l}^0$ could contain the smallest numbers in $x_{i,l}$ and $x_{i,l}^1$ the largest ones. So, in comparison with the mapping in [9], our mapping is more general, since it possesses more "degrees of freedom".

Now, we prove the following theorem.

Theorem 2. *After the first halving step:*

1. *Each set does not contain pairs of neighboring nodes and*
2. *between any two sets, the number of nodes that are neighbors along any hypercube dimension is reduced at least by half.*

Proof. For the first statement, we can easily verify that the addresses of the nodes of each newly created subset differ in at least two bits, and hence nodes belonging to the same subset cannot be neighbors on the hypercube.

For the second statement, we should check how the neighbors of nodes of the new subsets are distributed among the subsets. Without loss of generality, we focus on the set s_0 and its two subsets s_{00} and s_{01}. From the discussion in the proof of Theorem 1, we know that in order to determine the neighboring nodes along dimension k between set s_0 and another set s_i, we should differentiate between two ranges of values of k. Specifically if:

– $k \in [0 \ldots n - 2a - 1]$, nodes $c_{j_{0,p_{a-1}}} c_{j_{0,p_{a-2}}} \cdots c_{j_{0,p_1}} c_{j_{0,p_0}} x$ of s_0 are neighbors with nodes $c_{j_{i,p_{a-1}}} c_{j_{i,p_{a-2}}} \cdots c_{j_{i,p_1}} c_{j_{i,p_0}} x$ of s_i where $p = i$,

960   Christos Kaklamanis and Charalampos Konstantopoulos

$-$ $k \in [n - 2a \ldots n - 1]$, nodes $c_{j0,p_{a-1}} c_{j0,p_{a-2}} \cdots c_{j0,p_1} c_{j0,p_0} x_p^t$ and $c_{j0,\oplus_r p_{a-1}}$ $c_{j0,\oplus_r p_{a-2}} \cdots c_{j0,\oplus_r p_1} c_{j0,\oplus_r p_0} x_p^t$ of set s_0 are neighbors along dimension k of nodes $c_{ji,\oplus_r p_{a-1}} c_{ji,\oplus_r p_{a-2}} \cdots c_{ji,\oplus_r p_1} c_{ji,\oplus_r p_0} x_p^t$ and $c_{ji,p_{a-1}} c_{ji,p_{a-2}} \cdots c_{ji,p_1} c_{ji,p_0}$ x_p^t respectively of set s_i where $p = i$, $t = (n - k) \bmod 2$, x_p^t is given by (4) and $r = \lfloor \frac{k-n+2a}{2} \rfloor$.

For the first case, after the first halving step, nodes $c_{j0,p_{a-1}} c_{j0,p_{a-2}} \cdots c_{j0,p_1}$ $c_{j0,p_0} x_p^t$ will end up in the subset s_{0t} $(t = 0,1)$ of set s_0 and similarly nodes $c_{ji,p_{a-1}} c_{ji,p_{a-2}} \cdots c_{ji,p_1} c_{ji,p_0} x_p^t$ will be in the subset s_{it} $(t = 0,1)$ of set s_i. Now, the neighbors of nodes of s_{00} along dimension k are inside s_{i1} and the neighbors of s_{01} can be found in s_{i0}, which in turn implies that the number of neighbors between the subsets has really decreased at least by half.

For the second case, since p and $\oplus_r p$ always have a different parity, we can be sure that nodes $c_{j0,p_{a-1}} c_{j0,p_{a-2}} \cdots c_{j0,p_1} c_{j0,p_0} x_p^t$ and $c_{j0,\oplus_r p_{a-1}} c_{j0,\oplus_r p_{a-2}} \cdots$ $c_{j0,\oplus_r p_1} c_{j0,\oplus_r p_0} x_p^t$ will end up in different subsets of set s_0. The same is true for the set s_i. So, again the number of neighbors between any two subsets is at least halved after the first step.

With regard to the neighboring nodes between the two newly created subsets s_{00} and s_{01} of s_0, we can easily see that the pairs of neighboring nodes of the set s_0 have all been split between the two subsets and so again we have reduced the number of neighboring nodes by half. $\square$

At each of the remaining halving steps, we are free to take any equipartition of sets $x_{i,l}$ in two parts. The reason is the following. After the first halving step, the neighboring nodes along a hypercube dimension between any two sets, s_u and s_v, are the nodes $c_{j_u,p_{a-1}} c_{j_u,p_{a-2}} \cdots c_{j_u,p_1} c_{j_u,p_0} x_{u,p}$, $c_{j_v,q_{a-1}} c_{j_v,q_{a-2}} \cdots c_{j_v,q_1} c_{j_v,q_0} x_{v,q}$ for proper p, q, $x_{u,p}$ and $x_{v,q}$. Now, the second halving step partitions $x_{u,p}$ $(x_{v,q})$ into two equal parts, allocating half of the nodes $c_{j_u,p_{a-1}} c_{j_u,p_{a-2}} \cdots c_{j_u,p_1} c_{j_u,p_0} x_{u,p}$ $(c_{j_v,q_{a-1}} c_{j_v,q_{a-2}} \cdots c_{j_v,q_1} c_{j_v,q_0} x_{v,q})$ to s_{u0} (s_{v0}) and the other to s_{u1} (s_{v1}). We can easily see that no matter how we halve $x_{u,p}$ and $x_{v,q}$, the new subsets s_{u0}, s_{u1}, s_{v0}, s_{v1} have at most half neighboring nodes in comparison to sets s_u and s_v.

So, after applying $\log_2 \frac{g}{d}$ halving steps, we get g sets of d hypercube nodes each. These sets have the desired properties with regard to the optimal simulation of the hypercube on a POPS(d, g) network $(d < g)$. Clearly also, the mapping of these sets to the groups of the POPS(d, g) network is direct.

We conclude this section, by showing how to achieve optimal simulation of the hypercube on the POPS(d, g) network when $d = g$. We start with the mapping of the hypercube to the POPS(d, g) network for $d = 4g$ where $g \geq 1$. Clearly, for these values of d and g, we need 4 slots for simulating the hypercube on the POPS(d, g) network. Now, if we halve each group in two subgroups by using the technique of the first halving step presented above, we get $2g$ groups of $2g$ nodes each. From the discussion above also, it is clear that the simulation of the hypercube on the POPS$(2g, 2g)$ network now takes 2 slots, i.e. as much as the lower bound for $d = g$. So we can simulate the hypercube on the POPS(d, g) network with $d = g$ in a optimal number of slots.

4   Conclusions

We have presented an optimal simulation for the most common hypercube communication pattern on the POPS(d, g) network for the whole range of values of d, g. We first showed how to solve the existing open problem for the case $d > g$ and then we extended our technique for the case $d \leq g$. The optimal simulation of the hypercube network on more practical interconnection networks implies that now we are free to apply a number of elegant parallel algorithmic techniques developed on the hypercube to more "real" multiprocessor systems as well.

References

1. G. Gravenstreter and R. G. Melhem and D. Chiarulli and S. Levitan and J. Teza, The Partitioned Optical Passive Stars (POPS) Topology, Proceedings of the Ninth International Parallel Processing Symposium,1995,pp. 4-10
2. G. Gravenstreter and R. G. Melhem, Realizing Common Communication Patterns in Partitioned Optical Passive Stars (POPS) Networks, IEEE Transactions on Computers, vol. 47, no. 9, September, 1998, pp. 998-1013
3. R. G. Melhem and G. Gravenstreter and D. Chiarulli and S. Levitan, The Communication Capabilities of Partitioned Optical Passive Star Networks, Kluwer Academics Publishers, pp. 77-98, 1998, Parallel Computing Using Optical Interconnections, ed. K. Li and Y. Pan and S. Zheng
4. S. Sahni, The Partitioned Optical Passive Stars Network: Simulations and Fundamental Operations, IEEE Transactions on Parallel and Distributed Systems, vol. 11, no. 7, July, 2000, pp. 739-748
5. S. Sahni, Matrix Multiplication and Data Routing Using a Partitioned Optical Passive Stars Network, IEEE Transactions on Parallel and Distributed Systems, vol. 11, no. 7, July, 2000, pp. 720-728
6. A. Datta and S. Soundaralakshmi, Summation and Routing on a Partitioned Optical Passive Stars Network with Large Group Size, IEEE Transactions on Parallel and Distributed Systems, vol. 14, no. 12, December, 2003, pp. 1275-1285
7. A. Mei and R. Rizzi, Routing Permutations in Partitioned Optical Passive Stars Networks, Proceedings of the IEEE International Parallel and Distributed Processing Symposium (IPDPS 2002), 2002, April
8. P. Berthome and A. Ferreira, Improved Embeddings in POPS Networks through Stack-Graph Models, Proceedings of the Third International Workshop on Massively Parallel Processing Using Optical Interconnections, 1996, pp. 130-136
9. A. Mei and R. Rizzi, Mapping Hypercube Computations onto Partitioned Optical Passive Star Networks, Proceedings of High Performance Computing (HiPC 2003), 2003, pp. 95-104
10. A. Mei and R. Rizzi, Routing Permutations in Partitioned Optical Passive Stars Networks, Journal of Parallel Distributed Computing, vol. 63, no. 9, 2003, pp. 847-852
11. F.T. Leighton, Introduction to Parallel Algorithms and Architectures: Arrays-Trees-Hypercubes, Morgan Kauffman Publishers, 1992, San Mateo, California

Dynamic Page Migration
Under Brownian Motion*

Marcin Bienkowski and Miroslaw Korzeniowski

International Graduate School of Dynamic Intelligent Systems,
University of Paderborn, Germany
{young,rudy}@upb.de

Abstract. We consider *Dynamic Page Migration* (DPM) problem, one
of the fundamental subproblems of data management in dynamically
changing networks. We investigate a hybrid scenario, where access patterns to the shared object are dictated by an adversary, and each processor performs a random walk in $\mathcal{X}$. We extend the previous results of [4]:
we develop algorithms for the case where $\mathcal{X}$ is a ring, and prove that with
high probability they achieve a competitive ratio of $\tilde{\mathcal{O}}(\min\{\sqrt[4]{D}, n\})$,
where D is the size of the shared object and n is the number of nodes in
the network. These results hold also for any d-dimensional torus or mesh
with diameter at least $\tilde{\Omega}(\sqrt{D})$.

1 Introduction

The *Dynamic Page Migration* problem [3, 4] arises in a network of processors
which share some global data. Shared variables or memory pages are stored
in the local memory of these processors. If a processor wants to access (read
or write) a single unit of data from a page, and the page is not stored in its
local memory, it has to send a request to the processor holding the page, and
appropriate data is sent back. Such transactions incur a cost which is defined
to be the distance between these two processors plus a constant overhead for
communication. To avoid the problem of maintaining consistency among multiple
copies of the page, the model allows only one copy of the page to be stored within
the network. Additionally, nodes can move with a bounded speed, thus changing
the communication costs. This is typical in mobile wireless networks but also
attempts to capture the dynamics of wired ones.

To reduce the communication cost, the system can migrate the page between
processors. The migration cost is proportional to the cost of sending one unit of
data times the size of the memory page. The problem is to decide, online, when
and where to move the page in order to minimize the total cost of communication
for all possible sequences of requests and network changes. The performance of

* Partially supported by DFG-Sonderforschungsbereich 376 "Massive Parallelität: Algorithmen Entwurfsmethoden Anwendungen" and by the Future and Emerging
Technologies programme of the EU under EU Contract 001907 DELIS "Dynamically Evolving, Large Scale Information Systems".

J.C. Cunha and P.D. Medeiros (Eds.): Euro-Par 2005, LNCS 3648, pp. 962–971, 2005.

an online algorithm is measured by competitive analysis [6, 11], i.e. by comparing its total cost to the total cost of the optimal offline algorithm on the same input sequence.

Since the input consists of two independent sequences, one describing access patterns and one for mobility of the network, it is reasonable to assume that they are created by two adversarial entities; by a request and a network adversary, respectively. The thoroughly studied problem of *page migration* (PM) [1, 2, 5, 7, 8, 12] is a special case of DPM, in which the network is static, and only the request adversary is present[1].

Whereas there exist $\mathcal{O}(1)$-competitive algorithms for PM problem [1, 2, 12] in general networks, in [3, 4] much larger lower bounds for the DPM are given. They are $\Omega(\min\{\sqrt{D} \cdot n, D\})$ if both adversaries are adaptive, and $\Omega(\sqrt{D \cdot \log n}, D^{2/3})$ if they are oblivious. D denotes the size of the page and n is the number of processors in the network. The size of this ratio motivates us to search for a reasonable restriction on the adversary. Following the approach of [4] we consider the *Brownian motion* scenario. In this scenario the network adversary is replaced by a random process – each node performs a random walk on a 1-dimensional torus, i.e., on a ring.

The Model. Following [3, 4], we describe the DPM problem formally below. The network is modelled as a set V of n nodes (processors) labelled $v_1, v_2, \ldots, v_n$ placed in a metric space $(\mathcal{X}, d)$, where d is the distance function. In our case $\mathcal{X}$ is one-dimensional discrete torus (or alternatively speaking, discrete ring) of diameter (size) B. We assume discrete time steps $t = 1, 2, \ldots$. We denote the distance between two nodes v_x and v_y in time step t as $d_t(v_x, v_y)$. Since the nodes can move, this distance can change with time. A tuple C_t describing the positions of all nodes in a time step t is called a *configuration at time t.*

An input consists of a *configuration sequence* (C_t) and a *request sequence* (σ_t), where σ_t denotes the node issuing the request at time t. These sequences are chosen as follows. First, the adversary picks the request sequence (σ_t) and the initial configuration C_0 of n points on the ring. The rest of the configuration sequence (C_t) is generated randomly, i.e. for each t, C_{t+1} is generated from C_t in the following way. For each node v its new position is chosen. Let $x_v(t)$ denote the position (coordinate) of v at step t. For each node v we define a random variable $Z_v(t)$.

$$Z_v(t) = \begin{cases} -1 \text{ with probability } 1/3 \\ 0 \text{ with probability } 1/3 \\ 1 \text{ with probability } 1/3 \end{cases} \qquad (1)$$

The position of v in step $t + 1$ is defined as $x_v(t + 1) = x_v(t) + Z_v(t)$. This is further referred to as the *movement rule.*

Any two nodes are able to communicate directly with each other. The cost of sending a unit of data from node v_x to node v_y at time step t is defined in the following way by a cost function $c_t(v_x, v_y)$. If v_x and v_y are the same node (which

---

[1] In fact, in the PM model we do not have a constant overhead, but this may change the competitive ratio in this model only by a constant factor.

we denote by $v_x \equiv v_y$), then $c_t(v_x, v_y) = 0$. Otherwise, $c_t(v_x, v_y) = d_t(v_x, v_y) + 1$. We have one shared, indivisible memory page of size D, initially stored at the node v_1. The cost of moving the whole page from v_x to v_y in time step t is equal to $D \cdot c_t(v_x, v_y)$.

In time step $t \geq 1$, the positions of the nodes are set according to C_t, and then a request is issued at the node σ_t. Let $P_{\mathrm{ALG}}(t)$ denote a node keeping the algorithm ALG's page. First, ALG has to pay $c_t(P_{\mathrm{ALG}}(t), \sigma_t)$ for serving the request. Then it can optionally move the page to a new position $P'_{\mathrm{ALG}}(t)$ paying the cost $D \cdot c_t(P_{\mathrm{ALG}}(t), P'_{\mathrm{ALG}}(t))$. Sometimes, we will abuse the notation by writing that an algorithm is at v_i or moves to v_j, meaning that the algorithm's page is at v_i or the algorithm moves its page to v_j.

We consider only online algorithms, i.e. the ones which make decision in step t solely on the basis of the initial part of the input up to step t, i.e. on the sequence $C_1, \sigma_1, C_2, \sigma_2, \ldots, C_t, \sigma_t$.

In order to analyze the performance in Brownian motion scenario, we follow [4] and adapt classical competitive analysis [6, 11] for the model, where the input sequence is created both by the adversary and the stochastic process. We say that an algorithm ALG achieves competitive ratio $\mathcal{R}$ (or is $\mathcal{R}$-competitive) with probability p, if there exists a constant A, s.t. for all request sequences (σ_t) holds

$$\mathbf{Pr}_{(C_t)} \left[C_{\mathrm{ALG}}(C_t, \sigma_t) \leq \mathcal{R} \cdot C_{\mathrm{OPT}}(C_t, \sigma_t) + A \right] \geq p \ ,$$

where $C_{\mathrm{ALG}}(C_t, \sigma_t)$ and $C_{\mathrm{OPT}}(C_t, \sigma_t)$ are costs of ALG and the optimal algorithm, respectively. The probability is taken over all possible configuration sequences generated by the random movement (1).

Contribution of the Paper. We present three deterministic online algorithms: MAJL, MAJM, and MAJS, for long, middle and short diameters, respectively. Let $\gamma = \sqrt{2 \cdot \ln(n \cdot B^4)}$ and $Q = \min\{\sqrt{B}, \sqrt{D/B}, n\}$. We prove that these algorithms, with high probability, attain the competitive ratio $\mathcal{O}(\mathcal{R}) = \tilde{\mathcal{O}}(\min\{\sqrt[4]{D}, n\})$.

Algorithm	diameter	$\mathcal{R}$
MAJL	long: $B \geq 64 \cdot \gamma\sqrt{D}$	$\gamma^2 \cdot \max\{1, Q\}$
MAJM	middle: $\sqrt[3]{D} \leq B \leq 64 \cdot \gamma\sqrt{D}$	$\gamma^2 \cdot \max\{1, Q\} \cdot \log B$
MAJS	short: $B \leq \sqrt[3]{D}$	γ^2

This extends the result of [4], where an $\mathcal{O}(\log^2 D)$-competitive algorithm, working only for $B = \Theta(\sqrt{D})$ and for a constant n was presented. Furthermore, similarly to [4], it is possible to extend the result for long diameters to any d-dimensional torus or mesh of diameter $B = \Omega(\gamma \cdot \sqrt{D})$, losing only a factor of d in the competitive ratio[2].

---

[2] In this case, each node performs a random walk which is a superposition of independent random walks (1) in each dimension.

2   The Algorithms

In this section we present MAJL, MAJM and MAJS. Although they differ in details, their framework is essentially the same. For the sake of this presentation, we denote the algorithm by MAJ; the three algorithms will be just refinements of the MAJ framework.

MAJ works in phases of fixed length K. $K = D$ for MAJL, $K = B^2 \cdot \log B$ for MAJM, and $K = \Theta(\frac{D}{B} \cdot \log(B \cdot n))$ for MAJS. In a phase P, MAJ remains in one node denoted $P_{\mathrm{MAJ}}(P)$. For any time interval I and a node v, *weight* of v in I, denoted by $w_I(v)$, is defined as the number of requests issued by v during I. The name MAJ is an abbreviation of *Majority*. Namely, if there exists a node $v^* \neq P_{\mathrm{MAJ}}(P)$ s.t. $w_P(v^*) \geq K/2$, then MAJ decides to move to v^*. For long diameters it moves immediately in the last step of P; for middle and short ones it waits for a good opportunity for the next $6 \cdot B^2 \cdot \log B$ steps. These steps are called *migration sequence*. Good opportunity means that $P_{\mathrm{MAJ}}(P)$ and v^* come to each other at the distance of at most 1. If this occurs, MAJ moves in this case to v^*, otherwise it moves at the end of the migration sequence. The next phase begins right after the migration sequence.

Theorem 1. MAJ *achieves competitive ratio* $\mathcal{O}(\mathcal{R})$ *in the Brownian motion scenario of the DPM, with high probability (w.h.p.).*

We show that there exists a constant $c_{B,D,n}$ (depending on B, D, and n), s.t. for any α, any input sequence (σ_t) and any starting configuration C_0, if (C_t) is generated according to (1), then for $\mathcal{S} = ((C_t), (\sigma_t))$

$$\mathbf{Pr}[C_{\mathrm{MAJ}}(\mathcal{S}) \leq \mathcal{O}(\mathcal{R}) \cdot C_{\mathrm{OPT}}(\mathcal{S}) + \alpha \cdot \mathcal{O}(c_{B,D,n})] \geq 1 - 2 \cdot D^{-\alpha} \ . \qquad (2)$$

Let $c_{B,D,n} = D \cdot B + \mathcal{R} \cdot B^3 \cdot \log(B \cdot D \cdot n)$. We group phases (and corresponding migration sequences) in epochs. For MAJL an epoch consists of $\lceil B^2/D \rceil$ phases; for MAJM and MAJS an epoch consists of just one phase, optionally with its migration sequence. This guarantees that each epoch's length is at least B^2 and at most $L_p = \mathcal{O}((D/B + B^2) \cdot \log(Bn))$. An important property of such division into phases and epochs is the independence of the configuration sequence or the algorithm, i.e. the division can be determined entirely on the basis of the request sequence (σ_t).

We fix any input sequence $\mathcal{S}$ and divide it into epochs $M_1, M_2, \ldots$. We note that the cost of communication on the ring is bounded by $\lceil B/2 \rceil + 1 \leq B$ (we may assume that $B \geq 2$), and thus the cost of moving the page is at most $D \cdot B$. For any epoch, MAJL moves the page at most $\lceil B^2/D \rceil$ times, whereas MAJM and MAJS move the page at most once. Since the total cost of serving requests is at most $L_p \cdot B$, the total cost of MAJ in the first two epochs is bounded by $\mathcal{O}(D \cdot B + B^3 \cdot \log(Bn)) = \mathcal{O}(c_{B,D,n})$. We may also safely assume that the input sequence consists of finished epochs only, because we can hide the MAJ's cost in the last (unfinished) epoch in the additive term of (2), too.

Thus, it is sufficient to relate the cost of MAJ to the cost of optimal offline algorithm OPT in any epoch, but the first and the second one.

Lemma 1 (Crucial Lemma). *For any $j \geq 3$ holds $\mathbf{E}[C_{\mathrm{MAJ}}(M_j)] \leq \mathcal{O}(R) \cdot \mathbf{E}[C_{\mathrm{OPT}}(M_{j-1} \uplus M_j)]$. The expected value is taken over all random movements in M_{j-2}, M_{j-1}, and M_j.*

We defer the proof of the Crucial Lemma to the next subsection, and we sketch the proof how Theorem 1 follows from this lemma. The complete proof will be presented in the full version of the paper.

Proof (of Theorem 1). Fix any constant α. We already bound the cost in the two first epochs. We divide the remaining ones into three disjoint sets $\mathcal{M}_0$, $\mathcal{M}_1$, and $\mathcal{M}_2$. $\mathcal{M}_k := \{M_j : j \equiv k \mod 3\}$. From the average argument, there exists $\chi \in \{0, 1, 2\}$, s.t. $C_{\mathrm{MAJ}}(\mathcal{M}_\chi) \geq \frac{1}{3} \cdot C_{\mathrm{MAJ}}(\mathcal{M}_0 \uplus \mathcal{M}_1 \uplus \mathcal{M}_2)$. By Lemma 1 we have $\mathbf{E}[C_{\mathrm{MAJ}}(\mathcal{M}_\chi)] \leq \mathcal{O}(\mathcal{R}) \cdot \mathbf{E}[C_{\mathrm{OPT}}(\mathcal{S})]$.

We consider the value of $C_{\mathrm{MAJ}}(\mathcal{M}_\chi)$. If it is smaller than $\alpha \cdot \mathcal{O}(\mathcal{R} \cdot B^3 \cdot \ln D)$, then the whole cost of MAJ might be hidden in the additive constant of (2). Otherwise, for all each epoch $M \in \mathcal{M}_\chi$, the random variables $C_{\mathrm{MAJ}}(M_j)$ and $C_{\mathrm{OPT}}(M_j)$ are independent. Formally speaking, they are not independent, but the bounds on the costs of OPT and MAJ, which we use there, depend only on the randomness of the random walks in M_j and two preceding epochs (i.e. depend on disjoint events). Thus, we may apply Hoeffding bound [9] to get that both costs of $C_{\mathrm{OPT}}(\mathcal{S})$ and $C_{\mathrm{MAJ}}(\mathcal{M}_\chi)$ deviate by more than a constant factor from their expected values with probability at most $D^{-\alpha}$. The calculations are similar to the calculations presented in [4]. Thus, MAJ is $\mathcal{O}(\mathcal{R})$-competitive with probability $1 - 2 \cdot D^{-\alpha}$. $\qquad\qquad\square$

3   Proof of the Crucial Lemma

Fix any time interval I. We introduce a simple but useful notion of *auxiliary weight*. As we see later, we can use this notion to obtain a lower bound for $C_{\mathrm{OPT}}(I)$, and an upper bound for $C_{\mathrm{MAJ}}(I)$. Let $v_{\max}$ be the node which has the maximal weight in interval I, with ties broken arbitrarily. We define *auxiliary weight* of I as $W_{\mathrm{A}}(I) = |I| - w_I(v_{\max})$. We note that $W_{\mathrm{A}}(I)$ is a measure of requests' discrepancy in I. If it is low, then there exists a node $v_{\max}$, s.t. the algorithm, which remains in this node within I, pays relatively few. On the other hand, if it is high, there is no good single position for the page. Naturally, it can happen, that even if $W_{\mathrm{A}}(I)$ is high, all nodes are very close to each other, which means that the cost of the algorithm in I could be very low. However, such a configuration sequence is very unlikely to occur.

We keep this rationale in mind, while describing a rough idea of the Crucial Lemma's proof. For a set of intervals $\mathcal{I}$, let $\mathsf{span}(\mathcal{I})$ denote the shortest time interval containing all $I \in \mathcal{I}$. For any epoch M_j (for $j \geq 3$) we prove the existence of a so-called *critical set* of disjoint intervals $\mathcal{I}(M_j)$, s.t.

1. $\mathsf{span}(\mathcal{I}(M_j)) \subseteq M_{j-1} \uplus M_j$,
2. $\mathbf{E}[C_{\mathrm{MAJ}}(M_j)] = \mathcal{O}(\mathcal{R}_1) \cdot B \cdot \sum_{I_i \in \mathcal{I}(M_j)} W_{\mathrm{A}}(I_i)$,
3. $\mathbf{E}[C_{\mathrm{OPT}}(M_j)] = \Omega(1/\mathcal{R}_2) \cdot B \cdot \sum_{I_i \in \mathcal{I}(M_j)} W_{\mathrm{A}}(I_i)$,

where the expected values are taken over the random walks in M_{j-2}, M_{j-1}, and M_j. $\mathcal{R}_1$ and $\mathcal{R}_2$ are defined as follows.

Algorithm	$\mathcal{R}_1$	$\mathcal{R}_2$
MAJL	$\gamma^2 \cdot \max\{1, Q\}$	1
MAJM	$\gamma^2 \cdot \max\{1, Q\} \cdot \log B$	1
MAJS	1	γ^2

Clearly, if the conditions above are fulfilled, then the Crucial Lemma holds, since $\mathcal{R} = \mathcal{R}_1 \cdot \mathcal{R}_2$.

In this paper, we focus on proving the Crucial Lemma for long and middle diameters. The proof for short ones can be found in the full version of the paper. Due to space limitations we also moved the proofs of technical lemmas to the full version of the paper.

For any fixed epoch M_j $(j \geq 3)$, we construct the critical set $\mathcal{I}(M_j)$, with the properties described above. Additionally, we will have

1. $|\text{span}(\mathcal{I}(M_j))| \leq \frac{B^2}{(32\gamma)^2}$, and
2. for all $I_i \in \mathcal{I}(M_j)$, holds $|I_i| \leq \frac{B^2}{(16\gamma)^2 \cdot Q}$ and $|I_i| = \mathcal{O}(D)$.

3.1 Relating C_{MAJ} to Auxiliary Weight

First, we present a useful characterization of the MAJ's phases. Fix any phase P of length K. By P_{migr} we denote the corresponding migration sequence (if there is none, then $P_{\text{migr}} = \emptyset$). We distinguish between three cases.

1. *Wait phase* occurs, if $w_P(P_{\text{MAJ}}(P)) > K/2$. MAJ does not move and pays only for requests issued not at $P_{\text{MAJ}}(P)$. Since each request incurs a cost of at most B, $C_{\text{MAJ}}(P) \leq (K - w_P(P_{\text{MAJ}}(P)) \cdot B$.
2. *Mixed phase* occurs, if for all nodes v, $w_P(v) \leq K/2$. We have a trivial upper bound, $C_{\text{MAJ}}(P) \leq B \cdot K$.
3. *Change phase* occurs, if there exists a node $v^* \neq P_{\text{MAJ}}(P)$ s.t. $w_P(v^*) > K/2$. $C_{\text{MAJ}}(P) \leq B \cdot K$. For long diameters MAJ pays at most $B \cdot D = B \cdot K$ for moving the page. For middle ones it pays $B \cdot 6 \cdot B^2 \cdot \log B = \mathcal{O}(B \cdot K)$ for requests in P_{migr} and the expected cost for moving the page is $\frac{1}{B} \cdot B \cdot D + (1 - \frac{1}{B}) \cdot D = \mathcal{O}(D) = \mathcal{O}(B \cdot K/\log B)$ as follows from the technical lemma below.

Lemma 2. *Consider any two nodes v_a and v_b. If both move according to the movement rule, then with probability at least $1 - 1/B$ there exists a time step t within next $6 \cdot B^2 \cdot \log B$ steps s.t. $d_t(v_a, v_b) \leq 1$.*

For any phase P, we denote the phase preceding it by P_{prev}. As a conclusion from the phase characterization we get the following lemma.

Lemma 3. *For any phase P, $\mathbf{E}[C_{\text{MAJ}}(P \uplus P_{\text{migr}})] = \mathcal{O}(1) \cdot B \cdot W_A(P_{\text{prev}} \uplus P)$.*

Proof. If P is a wait phase, then $C_{\mathrm{MAJ}}(P) \leq B \cdot W_A(P)$. If P is a mixed phase, then $w_P(v) \leq K/2$ for all v, and hence $W_A(P) \geq K/2$. It follows that $\mathbf{E}[C_{\mathrm{MAJ}}(P)] \leq 2 \cdot B \cdot W_A(P)$. Since $P_{\mathrm{migr}} = \emptyset$ and $W_A(\cdot)$ is monotonic, the lemma holds in these cases.

If P is a change phase, then there exists a node $v^* \neq P_{\mathrm{MAJ}}(P)$, to which MAJ moved at the end of P. However, $w_{P_{\mathrm{prev}}}(v^*) \leq K/2$, because otherwise MAJ would have moved to v^* after phase P_{prev}, and would have been in v^* in the whole phase P. Therefore, $w_{P_{\mathrm{prev}} \uplus P}(v^*) \leq \frac{3}{2} \cdot K$. This inequality holds also for any v_i. Indeed, since $w_P(v^*) > K/2$, for any node $v_i \neq v^*$ holds $w_P(v_i) < K/2$, and hence $w_{P_{\mathrm{prev}} \uplus P}(v_i) < \frac{3}{2} \cdot K$. Therefore, $W_A(P_{\mathrm{prev}} \uplus P) \geq K/2$, and thus the lemma holds. $\qquad\square$

Constructing $\mathcal{I}(M_j)$ for middle diameters. For middle diameters, an epoch M_j consists of only one phase. Therefore, $\mathbf{E}[C_{\mathrm{MAJ}}(M_j)] = \mathcal{O}(1) \cdot B \cdot W_A(M_{j-1} \uplus M_j)$. The set $\{M_{j-1} \uplus M_j\}$ could be our critical set, consisting of one interval, but this interval is too long, i.e. has length $\Theta(B^2 \cdot \log B)$. We may shorten it to the desired length $\min\{\frac{B^2}{(32\gamma)^2}, \frac{B^2}{(16\gamma)^2 \cdot Q}\}$ losing at most a factor of $\mathcal{O}(\gamma^2 \cdot \max\{1, Q\} \cdot \log B) = \mathcal{O}(R_1)$ in auxiliary weight, using the following technical lemma.

Lemma 4. *For any interval I and any length $3 \leq \ell \leq |I|$, there exists an interval $J \subseteq I$ of length ℓ, s.t. $W_A(J) \geq \Omega(1) \cdot \frac{\ell}{|I|} \cdot W_A(I)$.*

Constructing $\mathcal{I}(M_j)$ for long diameters. In case of long diameters finding critical set is more complicated, because each epoch consists of multiple phases, and we cannot apply Lemma 3 directly. M_j consists of $\kappa := \lceil \frac{B^2}{D} \rceil$ phases $P_1, P_2, \ldots, P_\kappa$, each of length D. Let P_0 be the last phase of M_{j-1}. Let $L = B^2/(32\gamma)^2 \geq 4 \cdot D$ be the desired $\mathsf{span}(\mathcal{I}(M_j))$ length. First, we find a contiguous sequence A of phases, such that the cost of MAJ in A is large. Precisely, there is a sequence A of $\lfloor L/D \rfloor - 1$ phases from M_j such that $C_{\mathrm{MAJ}}(A) \geq \Omega(\frac{\lfloor L/D \rfloor - 1}{\kappa}) \cdot C_{\mathrm{MAJ}}(M_j) = \Omega(\frac{1}{\gamma^2}) \cdot C_{\mathrm{MAJ}}(M_j)$.

Let A' be a subset of A created by taking each second phase from A, in such way that $C_{\mathrm{MAJ}}(A') \geq \frac{1}{2} \cdot C_{\mathrm{MAJ}}(A)$. As a consequence, each two change phases from A are separated by at least one phase not belonging to A'. Let $\mathcal{I}_0 := \{(P_{\mathrm{prev}} \uplus P) : P \in A'\}$. All intervals from $\mathcal{I}_0$ are disjoint and their union contains whole A. Moreover, $\mathcal{I}_0$ is contained in $\lfloor L/D \rfloor$ consecutive phases, and hence $|\mathsf{span}(\mathcal{I}_0)| \leq L$. By Lemma 3 we get $\sum_{I_i \in \mathcal{I}_0} W_A(I_i) \geq \sum_{I_i \in \mathcal{I}_0} \Omega(\frac{1}{B}) \cdot C_{\mathrm{MAJ}}(I_i) = \Omega(\frac{1}{B}) \cdot C_{\mathrm{MAJ}}(A) = \Omega(\frac{1}{\gamma^2}) \cdot \frac{1}{B} \cdot C_{\mathrm{MAJ}}(M_j)$.

Since each interval from $\mathcal{I}_0$ has length at most $\frac{B^2}{(32\gamma)^2}$, we can use Lemma 4 to shorten each $I_i \in \mathcal{I}_0$ to length $\frac{B^2}{(16\gamma)^2 \cdot Q}$ losing additional factor of $\max\{1, Q\}$. Let $\mathcal{I}(M_j)$ be the set of shortened intervals from $\mathcal{I}_0$. Then, $C_{\mathrm{MAJ}}(M_j) = B \cdot \mathcal{O}(\gamma^2 \cdot \max\{1, Q\}) \cdot \sum_{I_i \in \mathcal{I}(M_j)} W_A(I_i)$.

3.2   Relating C_{OPT} to Auxiliary Weight

Let $\mathcal{I}(M_i)$ be the critical set chosen as described in the previous subsection. Consider any single interval $I_i \in \mathcal{I}(M_j)$. Informally, a condition for incurring a

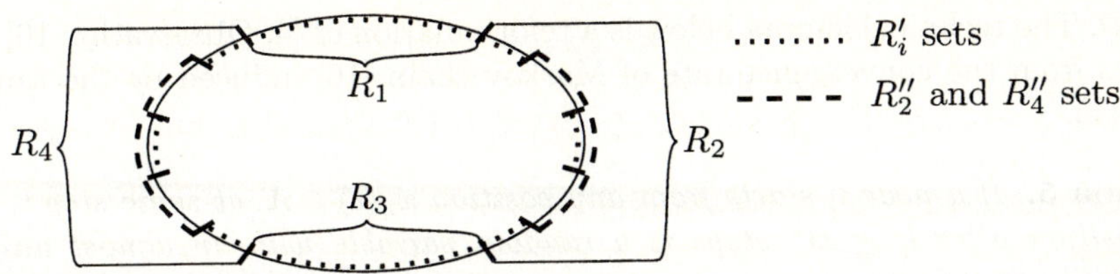

Fig. 1. Ring partitioning

high cost on any algorithm (in particular, on the optimal offline) in I_i, is that the nodes have to be distributed on the ring, so that the requests are issued from the different parts of the ring. This would assure, that an algorithm which remains at any node v in I_i pays $\Omega(B)$ for any request not at v, which amounts to at least $\Omega(B) \cdot W_A(I_i)$.

We show that it is sufficient that nodes are well distributed at the beginning of $\mathsf{span}(\mathcal{I}(M_j))$, and that they behave nicely, i.e. they never run away quickly from their starting positions. First, we formally define these two properties. Then we prove that for a single fixed interval $I_i \in \mathcal{I}(M_j)$, these properties are fulfilled with a constant probability. Finally, we show that if they hold, then $C_{\mathrm{OPT}}(I_i) = \Omega(1) \cdot B \cdot W_A(I_i)$. From this immediately follows that $\mathbf{E}[C_{\mathrm{OPT}}(I_i)] = \Omega(1) \cdot B \cdot W_A(I_i)$, and by linearity of expectation we get $\mathbf{E}[C_{\mathrm{OPT}}(M_{j-1} \uplus M_j)] = \Omega(1) \cdot B \cdot \sum_{I_i \in \mathcal{I}(M_j)} W_A(I_i)$. This would finish the proof of the Crucial Lemma.

Definition 1. *Fix any nodes configuration C and an interval I. We say that the nodes are I-distributed, if it is possible to partition the ring into 4 disjoint contiguous parts R_1, R_2, R_3, R_4, each containing $B/4$ points from $\mathcal{X}$, s.t. both $w_I(R_1)$ and $w_I(R_3)$ are at least $\frac{1}{16} \cdot W_A(I)$. $w_I(R_i)$ denotes the total weight accumulated in the part R_i, i.e., $w_I(R_i) = \sum_{v \in R_i} w_I(v)$* [3].

Definition 2. *We call a configuration sequence of length ℓ convergent, if for any $1 \le i < j \le \ell$ and any node v, positions of node v in time step i and j differ by at most $\gamma \cdot \sqrt{j - i}$. For any time interval I, we denote the event that a configuration sequence is convergent in I by* $\mathsf{conv}(I)$.

Clearly, between the beginning of M_{j-2} (called a *starting point*) and the beginning of $\mathsf{span}(\mathcal{I}(M_j)) \subseteq M_{j-1} \uplus M_j$, there are at least B^2 steps. We call them a *mixing sequence*. We observe, that for any configuration at the starting point, at the beginning of $\mathsf{span}(\mathcal{I}(M_i))$ the position of any node is a random variable with an almost uniform distribution. Formally, let $\mathcal{D}$ be the set of all probability distributions over our space $\mathcal{X}$, whose variation distance[4] to the uniform distribution on $\mathcal{X}$ is at most $1/64$. Any $\nu \in \mathcal{D}$ we call an *almost uniform distribution*. If for each node $v \in V$, its position is a random variable with distribution $\nu \in \mathcal{D}$, and all these variables are independent, then we denote it by

---

[3] By $v \in A$ we mean that the position of v is in the set A.

[4] In our discrete case, the variation distance between two probability distributions ν_1 and ν_2 is defined as $\|\nu_1 - \nu_2\| := \max_{A \subseteq \mathcal{X}} |\nu_1(A) - \nu_2(A)|$.

$V \sim \mathcal{D}$. The technical lemma below is a reformulation of [4, Observation 10], and follows from the convergence rate of Markov chain [10] induced by the random walk (1).

Lemma 5. *If a node v starts from any position $x_v(t) \in \mathcal{X}$ at some step t, then its position after $k \geq B^2$ steps is a random variable with an almost uniform probability distribution.*

Thus, we may safely assume that $V \sim \mathcal{D}$ at the beginning of $\mathsf{span}(\mathcal{I}(M_j))$. The next two lemmas are proven in the full version of the paper.

Lemma 6. *Fix any interval I. If $V \sim \mathcal{D}$, then the probability that nodes are I-distributed is at least $1/7$.*

Lemma 7. *For any time interval I starting with any configuration C, if $|I| \leq B^2$, then $\mathbf{Pr}[\mathsf{conv}(I)] \geq 1/2$.*

We are interested in the events $\mathsf{conv}(\mathcal{I}(M_j))$ and that nodes are I_i-distributed at the beginning of $\mathsf{span}(\mathcal{I}(M_j))$. These events are independent, as they rely on disjoint random experiments (random walk inside and before $\mathsf{span}(\mathcal{I}(M_j))$, respectively). Thus, their intersection occurs with probability $1/14$. It remains to show that, if they both occur, then $C_{\mathrm{OPT}}(I_i) = \Omega(1) \cdot B \cdot W_{\mathrm{A}}(I_i)$

We observe, that if $\mathsf{conv}(\mathcal{I}(M_j))$, then the speed restriction imposed on the nodes' movement creates a tradeoff: an algorithm either moves its page from one point of $\mathcal{X}$ to another slowly, or it has to pay much. To formalize this observation we need the following definition.

Definition 3 (Trails). *Fix any interval I. By a trail $\mathcal{T}(I)$ we denote the sequence of points of $\mathcal{X}$, in which OPT had its page in interval I. The trail in one step t, $\mathcal{T}(t)$ is defined as $(P_{\mathrm{OPT}}(t))$ if OPT does not move, and as the sequence of points on the shortest path between $P_{\mathrm{OPT}}(t)$ and $P'_{\mathrm{OPT}}(t)$ if OPT moves.*

Lemma 8. *Fix any time interval I of length $\ell \leq \frac{B^2}{(16\gamma)^2 \cdot Q}$. If $\mathsf{conv}(I)$ and OPT's trail $\mathcal{T}(I)$ contains two points from $\mathcal{X}$ lying at the distance of at least $B/8$, then $C_{\mathrm{OPT}}(I) = \Omega(\min\{1, 1/Q\}) \cdot D \cdot B$.*

At the beginning of $\mathsf{span}(\mathcal{I}(M_j))$, it is possible to partition the ring into four parts R_1, R_2, R_3, R_4 (see Fig. 1), s.t. $w_{I_i}(R_1), w_{I_i}(R_3) \geq W_A(I_i)/16$. Intuitively, since the configuration sequence in $\mathsf{span}(\mathcal{I}(M_j))$ is convergent, this partition is approximately preserved within whole $\mathsf{span}(\mathcal{I}(M_j))$, and thus in I_i. Formally, we define sets R'_1, R'_2, R'_3 and R'_4 as shown in Fig. 1. R'_1 (or respectively R'_3) has R_1 (R_3) in its center and contains $B/4 + 2 \cdot B/32$ points. R'_2 (or R'_4) is located in the center of R_2 (or R_4) and contains $B/8$ points. It follows that each pair of points from different R'_i sets is separated by a distance at least $B/32$. We define R''_1 (or respectively R''_3) as the part of length $3/8 \cdot B$ having R'_1 (R'_3) in its center. Thus, R''_1, R'_2, R''_3 and R'_4 create a partition of the whole ring. We make two key observations.

First, each node initially placed in R_1 (or respectively in R_3) can move by at most $\gamma \cdot \sqrt{|\mathsf{span}(\mathcal{I}(M_j))|} \leq B/32$, and thus remains within the set R_1' (or R_3') during the whole $\mathsf{span}(\mathcal{I}(M_j))$. This means that the number of requests issued in I_i at points from R_1' (or R_3') is at least $w_{I_i}(R_1) \geq W_A(I_i)/16$.

Second, OPT can either remain in $R_1'' \uplus R_2' \uplus R_4'$, remain in $R_3'' \uplus R_2' \uplus R_4'$ or its trail has to contain either all the points from R_2' or all the points from R_4'. We consider two cases, the other two are symmetric.

1. OPT remains in $R_1'' \uplus R_2' \uplus R_4'$ for the whole I_i. Then it has to pay at least $B/32$ for each of the requests issued at the points from R_3', i.e. for at least $W_A(I_i)/16$ requests. Thus, $C_{\text{OPT}}(I_i) = \Omega(B \cdot W_A(I_i))$.
2. The trail of OPT's page contain whole C_2'. Since $|I_i| \leq \frac{1}{(16\gamma)^2 \cdot Q} \cdot B^2$, we can apply Lemma 8 to get $C_{\text{OPT}}(I_i) = \Omega(\min\{1, 1/Q\}) \cdot D \cdot B$.

Thus, $C_{\text{OPT}}(I_i) \geq \Omega(B \cdot \min\{D, D/Q, W_A(I_i)\}) = \Omega(B \cdot W_A(I_i))$, which finishes the proof.

References

1. B. Awerbuch, Y. Bartal, and A. Fiat. Competitive distributed file allocation. In *Proc. of the 25th ACM Symp. on Theory of Computing (STOC)*, pages 164–173, 1993.
2. Y. Bartal, M. Charikar, and P. Indyk. On page migration and other relaxed task systems. *Theoretical Computer Science*, 268(1):43–66, 2001.
3. M. Bienkowski, M. Dynia, and M. Korzeniowski. Improved algorithms for dynamic page migration. In *Proc. of the 22nd Symp. on Theoretical Aspects of Computer Science (STACS)*, pages 365–376, 2005.
4. M. Bienkowski, M. Korzeniowski, and F. Meyer auf der Heide. Fighting against two adversaries: Page migration in dynamic networks. In *Proc. of the 16th ACM Symp. on Parallelism in Algorithms and Architectures (SPAA)*, pages 64–73, 2004.
5. D. L. Black and D. D. Sleator. Competitive algorithms for replication and migration problems. Technical Report CMU-CS-89-201, Department of Computer Science, Carnegie-Mellon University, 1989.
6. A. Borodin and R. El-Yaniv. *Online Computation and Competitive Analysis*. Cambridge University Press, 1998.
7. M. Chrobak, L. L. Larmore, N. Reingold, and J. Westbrook. Page migration algorithms using work functions. In *Proc. of the 4th Int. Symp. on Algorithms and Computation (ISAAC)*, pages 406–415, 1993.
8. C. Lund, N. Reingold, J. Westbrook, and D. C. K. Yan. Competitive on-line algorithms for distributed data management. *SIAM Journal on Computing*, 28(3):1086–1111, 1999.
9. S. Rajesekaran, P. M. Pardalos, J. H. Reif, and J. Rolim. *Handbook of Randomized Computing*, volume II. Kluwer Academic Publishers, 2001.
10. J. S. Rosenthal. Convergence rates for Markov chains. *SIAM Review*, 37(3):387–405, 1995.
11. D. D. Sleator and R. E. Tarjan. Amortized efficiency of list update and paging rules. *Communications of the ACM*, 28(2):202–208, 1985.
12. J. Westbrook. Randomized algorithms for multiprocessor page migration. *DIMACS Series in Discrete Mathematics and Theoretical Computer Science*, 7:135–150, 1992.

Topic 13
Routing and Communication
in Interconnection Networks

Emilio Luque, Cruz Izu, Olav Lysne, and José Legatheaux

Topic Chairs

Parallel and distributed computing has expanded its scope and application areas during the last decade. In addition to the traditional parallel computers, the availability of new set of services through Internet together with continuous improvement in communication bandwidth has led to new programming paradigms, deployment of high-speed communication infrastructure, and installation of very large server, usually based on cluster of processors interconnected by means of some interconnection network.

Communication networks, protocols, routing and communication policies are crucial factors for the performance of parallel and distributed computations.

This topic of Euro-Par 2005 is devoted to all aspect of communication in on-chip interconnects, parallel computers, networks of workstations and more widely distributed systems such as grids. Power-efficient interconnects, I/O architectures and storage area networks, switch architectures, as well as multimedia and QoS aware communication are new topics introduced in the last years and now included in this topic.

Papers were solicited that examine the design and implementation of interconnection networks and communication protocols, advances in system area and storage area network, routing and communication algorithms, and the cost of parallel and distributed algorithms.

The Call for Papers attracted 32 submissions for this topic, of which 8 (25%) were selected for publication and presentation at the conference. The selected papers cover a wide scope, ranging from performance evaluation and simulation of interconnection networks to optical multistage interconnection networks also covering topics like routing and scheduling policies, QoS on clusters and congestion management strategies.

We would like to thank all the authors for their submissions to this Euro-Par 2005 topic. We owe special thanks to the 76 external referees who provided competent and timely review reports. Their effort ensured the high quality of this track of Euro-Par 2005. We expect you will find this topic to be highly stimulating and quite informative.

J.C. Cunha and P.D. Medeiros (Eds.): Euro-Par 2005, LNCS 3648, p. 973, 2005.
© Springer-Verlag Berlin Heidelberg 2005

Transport Time Distribution
for Deflection Routing on an Odd Torus[*]

J.M. Fourneau[1] and T. Czachórski[2]

[1] PRiSM, Université de Versailles Saint-Quentin,
45 Av. des Etats Unis, 78000 Versailles, France
[2] IITIS-Polska Akademia Nauk,
Ul. Balticka 5, 44-100 Gliwice, Poland

Abstract. We analyze the performance of all optical packet networks. As optical storage of packets is not available, we assume that the routing protocol is based on deflection. This routing strategy does not allow packets loss. However it keeps the packets inside the network, increases the delay and reduces the bandwidth. Thus the transport delay distribution is the key performance issue for these networks. Here, we consider a 2D torus the size of which is odd. The method is based on a fixed point system between two sub-models. The first subsystem describes the global network performance while the other one models the stochastic behavior of two types of packets.

1 Introduction

All optical packets networks have received considerable attention during the last years. However with actual technology, all-optical networks do not allow the buffering of packets inside the network. Therefore packets have to be sent immediately to the next switch along the path. Old algorithm like Deflection Routing [2] have recently received attention to overcome this weakness [8, 9]. This routing strategy does not allow packets loss. However it keeps the packets inside the network, increase the delay and reduce the bandwidth. In Shortest-Path Deflection Routing, switches attempt to forward packets along a shortest hop path to their destination. Each link can send a finite number of packets per time-slot (the link capacity). If the number of packets which require a link is larger than the capacity, only some of them will use the link they ask for and the other ones have to be misdirected or deflected and they will travel through longer paths. This is the major drawback of this technique.

The tail of the transportation delay and the average usable bandwidth are therefore two major measures of interest. The mean number of deflections is not that large but a significant fraction of the number of packets is heavily deflected. We have observed several packets with more than 1000 deflections during a simulation of a 10×10 2D-mesh with unbalanced traffic [3]. As acknowledgments in

[*] This work has been partially supported by the French RNRT project *ROMEO*, by a research project between CNRS and Polish Akademia Nauk and by the EURO NGI European Network of Excellence.

J.C. Cunha and P.D. Medeiros (Eds.): Euro-Par 2005, LNCS 3648, pp. 975–983, 2005.
© Springer-Verlag Berlin Heidelberg 2005

networking protocols must arrive before some timer expiration, heavily deflected packets will be considered as lost because they experience delays larger than the transport time-out. Packets are never physically lost due to physical errors or buffer congestion, but they can be logically lost because the transport delay is too large.

Previous analytical studies of deflection [1, 4] have proposed models for networks based on 2×2 switching blocks without the queueing of new packets. Recently, Fabrega and Muñoz [7] have modeled a network with deflection routing using an approximate model based on Markov chains. However, they have only considered 2×2 switches and a topology such that only one shortest path exists between the source and the destination. Yao et al. have presented in [10] an approximate model for more general topologies which do not contain any directed cycle (again a quite restrictive topology). Clearly, all the models proposed so far have used some unrealistic assumptions about the network topology and switches. Furthermore, all these methods only estimate the mean delay while the important measure is the tail of the delay distribution. Therefore, new methods to obtain the distribution of the delay are still necessary.

In this paper, we consider 4×4 switching elements and a 2D torus topology which was considered as a reasonable topology by the ROM project [8]. We also assume that the size of the torus is odd $(2Z + 1)$ and that there are no optical converters. Following the ROM conclusions, we consider fixed size packets and the network is logically synchronous. We model explicitly the routing algorithm with minimal number of deflections per time slot which has been introduced recently by Alcatel [5, 9]. We explain at the end of the paper how to model even size torus.

We model the network by an aggregate representation of the optical packets. First, we represent the vector distance to destination and we gather the packets into two sets according to the number of favorable directions for the next hop. We assume that the packets try to follow a shortest path. Thus only some directions (among the four existing in the torus) are consistent. In an odd torus, we may have packets which have only one possible direction and packets which have two possible directions. In this paper, they are respectively denoted as type 1 and type 2 packets (see Fig. 1). Our analysis is based on the construction of a Markov chain which represents the evolution of a typical packet. The state space takes into account the packet type and the distance vector to the destination.

First, we model the path of a tagged packet inside the net using a Markov chain. The distribution of the transport time can be computed numerically once the deflection probabilities for both types of packets are obtained. The probabilities are the solution of a fixed point system based on the flows of type 1 and type 2 packets. We present an algorithm, numerical results and some simulations to check the accuracy.

This paper is organized as follows: in Section 2, we present the model of a tagged packet based on the topological properties of the torus and the traffic assumptions. Section 3 is devoted to the model of the packet flows. The two sets of equations provide a fixed point system. In Section 4 we present an algorithm and we compare numerical results with simulations.

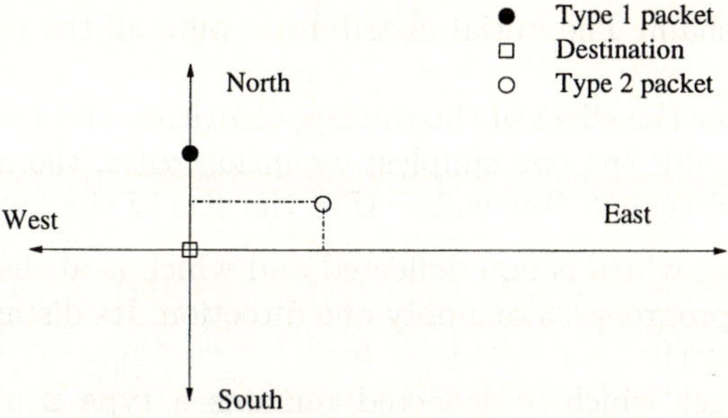

Fig. 1. Type of Packets and Routing

2   Model of a Tagged Packet

Remember that we have divided the packets into two sets: the packets which have only one possible direction (type 1) and the packets which have two possible directions (type 2). A type 1 packet has reached one coordinate of its destination while a type 2 packets must progress in two directions to reach its exit. Of course, at each step, packets may change their types according to their distance to destination and the issue of the deflection algorithm. Borrero and Quessette have proposed the following heuristic which has been proved to be optimal in [5]. The optimality criteria is the number of deflections at each time slot.

Lemma 1 (Degree 1 node of V_1). *[5] If a packet has only one possible direction, we must consider its request before looking at packets of type 2.*

Now, let us turn to the model for routing. We assume that the selection of packet during the routing algorithm is only based on the type of packets. Let $p1$ and $p2$ be respectively the probability that a type 1 and type 2 packet will be deflected at one step. These quantities will be computed in the next section.

Let us now model the evolution of a tagged packet inside the net. First, we build the transition matrix R of a typical packet and we show it for a torus with 7 rows and columns). The model is based on the following set of states: the initial state (state 0) before the packet enters the input node and the completion state (state 1) where the packet leaves the network. State 1 is an absorbing state. We represent explicitly the packet type (denoted as $t1$ and $t2$ in the state description) and the distance vector to destination. Of course, at each step, packets may change their type according to their vector of distances. Due to traffic assumptions, we aggregate states with equivalent vector of distances: for instance $(1, 2)$ and $(2, 1)$. Thus the chain has less states to represent the evolution of the packet inside the net. For instance, for the 7x7 torus, these states are (t1,1,0), (t1,2,0), (t1,3,0), (t2,1,1), (t2,1,2), (t2,1,3), (t2,2,2), (t2,2,3) and (t2,3,3). The chain has 11 states. The transport time is the time of a sample-path beginning at state 0 and finishing at state 1. The PDF of the transport time is obtained by successive multiplications of the distribution probability by transition matrix

R of the Markov chain. The initial distribution puts all the probability mass in state 0.

Now, let us show the effect of the routing algorithm and the deflection on the states. Let us consider the two simplest evolution rules: the non deflected type 1 and the deflected type 2. Remember that the size of the torus is $2Z + 1$.

- A type 1 packet which is non deflected and which is at distance k is kept as a type 1 as it progresses along only one direction. Its distance to destination is therefore $k - 1$.
- A type 2 packet which is deflected remains a type 2 packet. In general, the deflection increases by one the distance to destination, except on the boundary of the torus (see figure 2). Each of the components in the distance vector may be increased with probability $1/2$.

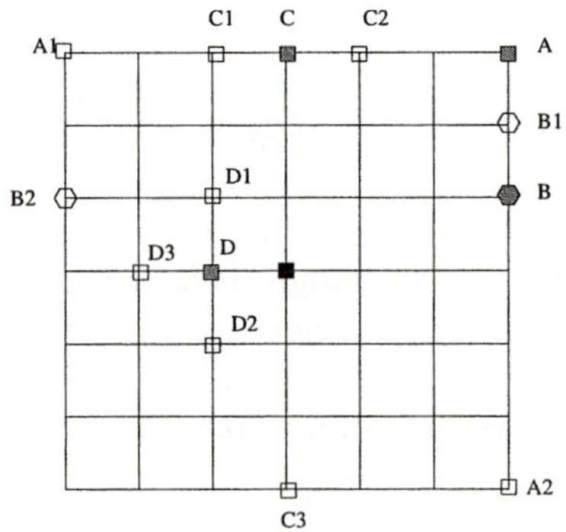

Fig. 2. Deflection on the torus: the destination is the black square. If a packet in A on the boundary is deflected, it will reach nodes $A1$ or $A2$ at next step. But A, $A1$ and $A2$ are at the same distance to destination (i.e. $2Z$). For node B, the situation is even more complex, a deflection in B implies that the packet joins node $B1$ or $B2$. At $B2$, the distance is the same, while from node $B1$ it has increased. A deflected type 1 packet in node D may become type 2 (2 possible directions) or stay type 1 (one possible direction). The distance to destination is now increased by one. But if the deflected type 1 comes from the boundary (node C) and it moves to $C3$, then its distance to destination is kept unchanged

Now consider the two other cases: a deflected type 1 packet and a non deflected type 2 packet.

- A type 1 packet at distance k which is deflected has three possible directions. Two directions lead to a type 2 packet (see figure 2) and one direction to a type 1 packet. If we assume equiprobable choices for the directions, the transition rule keeps the packet as type 1 with probability $1/3$ and changes it to type 2 with probability $2/3$. Its distance to destination is therefore $(1, k)$ except in the following case: when the packet is on the boundary of the torus and it is kept as type 1, the packet is still at distance k along the opposite direction because the torus size is odd.

– A type 2 packet at distance (m, k) which is not deflected decreases its distance to $(m, k - 1)$ or $(m - 1, k)$. And according to its position and the direction selected it may become a type 1 (if $m - 1 = 0$ or $k - 1 = 0$) or stay a type 2 packet otherwise.

Thus the transitions in R can be easily obtained from the rules formerly shown and the deflection probabilities $p1$ and $p2$. The last part of the matrix still missing gathers the transitions out of states 0 (the arrivals inside the net). Each destination node in the net (except the source of the packet) has the same probability (i.e. $\frac{1}{N^2-1}$) to be the destination. So, we must count the number of nodes of type 1 and 2 at distance $(k, 0)$ or (m, k).

$$R(0, (t1, k)) = \frac{4}{N^2 - 1} \quad and \quad R(0, (t2, m, k)) = \frac{x2(k + m)}{N^2 - 1} \tag{1}$$

where $x2(k + m)$ is the number of nodes in the torus at distance k of their destination which may contain packets of type 2. Clearly, $x2(m+k) = 4k+4m-4$ if $k + m \leq Z$ and $x2(m + k) = 8Z + 4 - 4k - 4m$ when $Z \leq m + k \leq 2Z$. Finally for the 7×7 torus we get (with $q1 = 1 - p1$ and $q2 = 1 - p2$):

$$R = \left[\begin{array}{c|ccc|cccccc}
 & 1/12 & 1/12 & 1/12 & 1/12 & 1/6 & 1/6 & 1/12 & 1/6 & 1/12 \\
\hline
1 & & & & & & & & & \\
\hline
q1 & p1/3 & & & 2p1/3 & & & & & \\
 & q1 & p1/3 & & & 2p1/3 & & & & \\
 & & q1 & p1/3 & & & 2p1/3 & & & \\
\hline
 & q2 & & & & p2 & & & & \\
 & & q2/2 & & q2/2 & & p2/2 & p2/2 & & \\
 & & & q2/2 & & q2/2 & p2/2 & & p2/2 & \\
 & & & & & & q2 & & & p2 \\
 & & & & & & q2/2 & q2/2 & p2/2 & p2/2 \\
 & & & & & & & & q22 & p2
\end{array}\right]$$

3 Macroscopic Model of the Flows

Clearly, the first model does not take into account the arrival process because we assume that the packet is in the network. We now study the flow of packets. Note that due to the topology and the traffic assumptions all the switches are statistically equivalent. The probabilities of deflection are computed by conditioning on the arrivals. Then these probabilities are shown to be related to the load of the link using an independence assumption. These last relations provide a fixed point system for $p1$ and $p2$.

3.1 Deflection Probability

The best routing algorithm must route type 1 packets with a higher priority [9]. Therefore it is sufficient to compute the deflection probability of a tagged type 1 packet knowing the exact number of type 1 packets in the switch. Note that the upper bound of the index is 3 because the tagged customer uses one input link of the switch.

$$p1 = \sum_{i=0}^{3} Pr(i \ type \ 1 \ arrivals)d1(i)$$

where $d1(i)$ is the probability that the tagged packet of type 1 will be deflected if another type 1 packet arrives. The probabilities of arrivals are obtained by an independence assumption. Let us denote by u_1 the utilization of an arbitrary link by type 1 packets.

$$Pr(i \ type \ 1 \ arrivals) = C(3,i)(u_1)^i(1 - u_1)^{3-i} \tag{2}$$

As type 2 customers have a lower priority in the routing algorithm, their deflection probability requires a conditioning on a more complex set of arrivals.

$$p2 = \sum_{i=0,j=0}^{i+j=3} Pr(i \ type \ 1 \ and \ j \ type \ 2 \ arrivals)d2(i,j) \tag{3}$$

where $d2(i,j)$ denotes the probability that the tagged type 2 packets will be deflected due to the arrivals of i type 1 and j type 2 packets. Similarly the probability of arrivals follows a multinomial distribution because of the independence assumption ($B(3,i,j)$ is the multinomial coefficient):

$$Pr(i \ type \ 1 \ and \ j \ type \ 2 \ arrivals) = B(3,i,j)(u_1)^i(u_2)^j(1 - u_1 - u_2)^{3-i-j}$$

We now have to obtain the elementary probabilities $d1(i)$ and $d2(i,j)$. First we consider an arbitrary tagged type 1 packet entering into an arbitrary switch. Clearly, $d1(0) = 0$ and $d2(0,0) = 0$ because there is no competition and $d2(1,0) = 0$, and $d2(0,1) = 0$ as a type 1 packet or a type 2 packet is not sufficient to deflect another type 2 packet. To compute the other values, we assume equiprobable choices when several packets of the same type request the same output. For the sake of conciseness, we omit the computation of $d1$ and $d2$ (see [6] for more details) and we give the results for positive values in Table 1. During this computation, we take care of some properties of routing on an odd torus. For instance, we have:

Lemma 2. *All configurations of requests are not possible due to the routing algorithm and the topology. For shortest path deflection routing in an odd torus, a type 2 packet can not ask for two opposite directions (for instance North and South).*

3.2   Average Distance and Deflection Probabilities

Let us now establish new relations between the link utilization u and the deflection probabilities. We must consider now the number of packets $\vec{n1}(k)$ and

Table 1. Table for $d1$ and $d2(i,j)$

d1(1)	d1(2)	d1(3)	d2(0,2)	d2(0,3)	d2(1,1)	d2(1,2)	d2(2,0)	d2(2,1)	d2(3,0)
$\frac{1}{8}$	$\frac{11}{48}$	$\frac{81}{256}$	$\frac{1}{8}$	$\frac{9}{32}$	$\frac{1}{16}$	$\frac{15}{64}$	$\frac{1}{48}$	$\frac{23}{192}$	$\frac{13}{128}$

$\vec{n2}(m,k)$ rather than the state of a single tagged packet. However, the evolution is modeled by a stochastic matrix M that we can deduce from R. For transitions inside the network, $M(i,j)$ is equal to $R(i,j)$. Indeed, the average numbers of customers obey the same evolution rules than a single packet. The only differences are in the transition between the network and the outside which reflects the arrival rate. We remove the first two states from R and modify the first row to take into account the flow entering the network. We assume Poisson arrivals with rate λ. We need to compute the number of fresh packets of type 1 or 2 entering the network at distance $(k,0)$ or (m,k). Let us denote $a1(k)$ and $a2(m,k)$ these numbers. The average number of packets entering the network is also the average number of packets entering the electronic buffers, if the system is stable. Therefore it is equal to λN^2.

$$a1(k) = \lambda N^2 R(0,(t1,k)) \quad and \quad a2(m,k) = \lambda N^2 R(0,(t2,m,k)) \qquad (4)$$

But, the flow entering the network must be equal to the flow leaving the switches with a successful transition from a node at distance 1. Therefore: $\lambda N^2 = \vec{n1}(1) * (1 - p1)$. Finally, the average population vector is the solution of the linear system:

$$(\vec{n1}, \vec{n2}) = (\vec{n1}, \vec{n2})M \quad and \quad \vec{n1}(1) = \frac{\lambda N^2}{(1 - p1)} \qquad (5)$$

Let us now turn back to the link utilization $u1$ and $u2$. As all the links are equivalent due to the topology and the traffic assumptions, we get:

$$u1 = \frac{\sum_{k=1}^{Z} \vec{n1}(k)}{4N^2} \quad and \quad u2 = \frac{\sum_{k=2}^{2Z} \vec{n2}(k)}{4N^2}$$

Thus we obtain a fixed point. We have proved the existence of a solution to this system using Brouwer's fixed point theory, the continuity of steady-state distribution proved by Malyšev and the convexity of $p1$ and $p2$ (see equations 2 and 3). For the sake of conciseness, the proof is omitted (see [6] for more details).

4   Experimental Results

Let us now turn to the numerical algorithm. The computation is iterative: at each step we compute the new transition probability matrix M and a new set of values for $\vec{n1}(i)$ and $\vec{n2}(i,j)$. Then we get $p1$ and $p2$ and compute the difference with their former values. If this difference is smaller than 10^{-9}, we stop the iterations. Initially, $p1$ and $p2$ are equal to 0. The computation of vectors $\vec{n1}$ and $\vec{n2}$

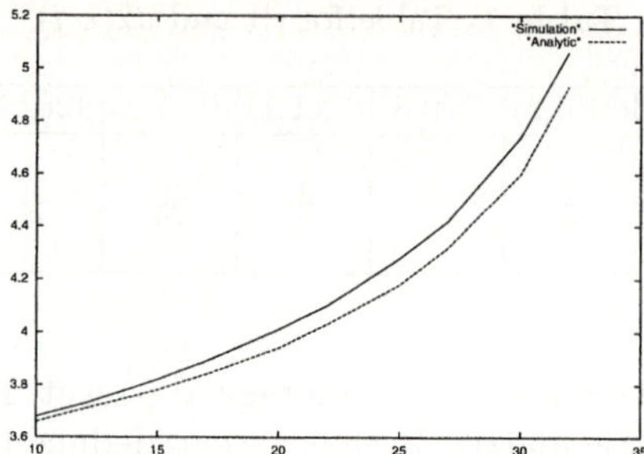

Fig. 3. Comparison of simulation and analytical results: average transport time (in hops) versus load (in packets arriving in the global networks per time slot)

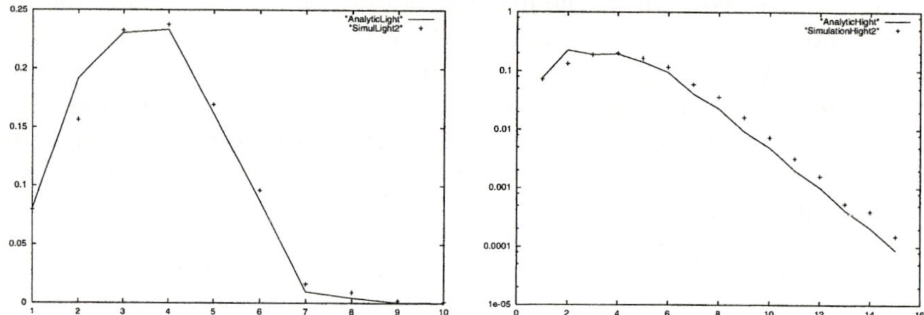

Fig. 4. Comparison of simulation and analytical results: distribution of the transport time for arrival rate=10 (left) and 25 (right)

is performed by an usual direct elimination solver. The number of iterations is usually very low and the matrix is very small: we observe a convergence before ten iterations. Once the fixed point is found, we obtain the transport time distribution using the probabilities $p1$ and $p2$ and few vector matrix multiplications.

We compare the numerical results obtained by our approach to simulation results for a 7×7 torus. In Fig. 3, we present the evolution of the average transport time (in hops) versus the global arrival rate (in number of packets for the whole network). Clearly, the approximations are quite accurate, even if we have used a large scale to emphasis the difference. The analytical results look optimistic. We now present the distribution of the transport time at light and moderate load (Fig. 4). Again we depict the simulation and the analytical results. And the figures show the accuracy of our method, even for the distribution of the transport delay.

5   Conclusions

To the best of our knowledge, it is the first approach to analyze the transport time distribution for more general switches and torus. It is possible to model even torus instead of odd ones. We must slightly change the first model. In an even torus, packets may have more than 2 good directions and the probabilities used to define matrix R are slightly different. To complete the approach, one must also study the distribution of the waiting time before entering the network. Diffusion models of these queues are currently under development.

References

1. J. Bannister, F. Borgonovo, L. Fratta, M. Gerla, A versatile model for predicting the performance of deflection routing networks, Performance Evaluation, V16, pp 201-222, 1992.
2. P. Baran. On distributed communication networks. *IEEE Transactions on Communication Systems*, CS-12, 1964.
3. D. Barth, P. Berthomé, A. Borrero, J.M. Fourneau, C. Laforest, F. Quessette, and S. Vial. Performance comparisons of Eulerian routing and deflection routing in a 2d-mesh all optical network. In *ESM'2001*, 2001.
4. A. Bonnoni, P.P. Prucnal, Analytical evaluation of improved access techniques in deflection routing networks, IEEE/ACM Trans. on Networking, V4, N5, 1996.
5. A. Borrero, J.M. Fourneau and F. Quessette Packet Selection in a deflection routing algorithm. *ISCIS 2002*, Orlando, USA, CRC Press.
6. T. Czachórski and J.M. Fourneau, Performance evaluation of an optimal deflection routing algorithm on an odd torus, Archiwum informatyki teoretycznej i stosowanej, V 16, N 4, pp 257-277, 2004.
7. J. Fabrega and X. Muñoz, A study of Network Capacity under Deflection routing schemes, Europar 03, LNCS 2790, pp 989–994.
8. P. Gravey, S. Gosselin, C. Guillemot, D. Chiaroni, N. Le Sauze, A. Jourdan, E. Dotaro, D. Barth, P. Berthomé, C. Laforest, S. Vial, T. Atmaca, G. Hébuterne, H. El Biaze, R. Laalaoua, E. Gangloff, and I. Kotuliak. Multiservice optical network: Main concepts and first achievements of the ROM program. *Journal of Ligthwave Technology*, 19:23–31, January 2001.
9. S. Mneimeh and F. Quessette. *Minimum Deflection Routing Algorithm*, Alcatel Patent Application #135945, 2002
10. S. Yao, B. Mukherjee, and S. Dixit. *Plato: a generic modeling technique for optical packet switched networks* International Journal on Wireless and Optical Communications, V1, N1, 2003, pp 91-101. World Scientific Publishing.

Routing and Scheduling for a Novel Optical Multistage Interconnection Network

Siu-Cheung Chau[1], Tiehong Xiao[2], and Ada Wai-Chee Fu[3]

[1] Dept. of Physics and Computer Science, Wilfrid Laurier University,
Waterloo, Ontario, Canada, N2L 3C5
schau@wlu.ca
[2] Dept. of Computing and Information Science
University of Guelph, Guelph, Ontario, Canada, N1G 2W1
txiao@uoguelph.ca
[3] Dept. of Computer Science and Engineering,
Chinese University of Hong Kong, Shatin, Hong Kong
adafu@cse.cuhk.edu.hk

Abstract. Multistage Interconnection Networks (MINs) are popular in computing and communication applications. Recently, there have also been significant advances in electro-optic switches that have made Optical MINs (OMINs) a good choice for the high channel bandwidth and low communication latency of high performance computing and communication applications. However, OMINs introduce crosstalk which results from coupling two signals within one switching element. Under the constraint of avoiding crosstalk, what we are interested in is how to realize a permutation that requires the minimum number of passes. This routing problem is an NP-hard one, and many heuristic algorithms have been devised to find a solution. In [9], Chau and Xiao have proposed an algorithm, called the Remove Last Pass (RLP) algorithm, to avoid crosstalk and route the traffic in an OMIN more efficiently. In this paper, we focus on the routing and scheduling of a novel OMIN, the base-2 MIN, propounded by Chau and Fu in [8]. Our experiments prove that any permutation can be realized in no more than three passes in the base-2 OMIN by using the RLP algorithm, when the network has no more than 512 nodes. The base-2 OMIN requires only $n(\log n+1)$ switching elements (SEs) for an $n{\times}n$ network, compared to the crossbar, which requires $O(n^2)$ SEs for an $n{\times}n$ network. Therefore, the base-2 MIN should be a good candidate for communication subsystems in a parallel computing environment.

1 Introduction

The research literature on interconnection networks embodies studies of a large number of direct networks such as the hypercube, mesh and torus. Although the algorithm for the hypercube network can attain an optimal time complexity, the network suffers from unbounded node degrees, and thus, has a poor scalability. Mesh and torus networks have a constant node degree and better scalability, but more time is needed to pass a message from one node to another node.

J.C. Cunha and P.D. Medeiros (Eds.): Euro-Par 2005, LNCS 3648, pp. 984–993, 2005.

Also, researchers have focused on indirect networks such as crossbar networks and MINs. Crossbar networks can realize all the possible permutations between the inputs and outputs with a constant communication latency, but the network costs of an $n{\times}n$ crossbar is O(n^2), which is prohibitively high for a large system. On the other hand, a MIN usually has fewer hardware costs and can also realize all permutations, rendering it a good candidate for a communication subsystem.

In addition, electronic MINs have been studied extensively as important interconnecting schemes for communication and parallel computing systems. MINs with electro-optic switching elements (SEs) can offer even a higher channel bandwidth and lower communication latency than MINs with electronic SEs. A large body of work has resulted from examining the structure, operation, and performance of these optical MINs (OMINs) [1][2][3][4][5][6][8][9][10]. The topology of an OMIN is similar to its electronic peer except that electro-optic switches are used instead of electronic switches. In this paper, we consider only electronically controlled optical switches such as Lithium Niobate directional couplers which are illustrated in Figure 1. Depending on the amount of voltage at the junction of the two waveguides that carry the two input signals, either of the two inputs can be coupled to either of the two outputs.

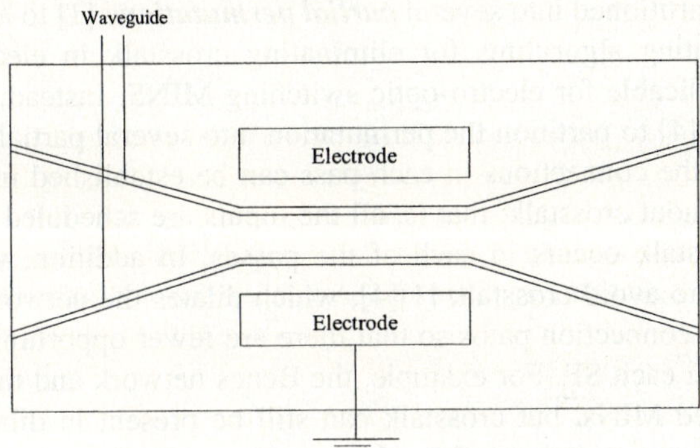

Fig. 1. A Lithium Niobate 2×2 switch.

Unlike their counterpart, electronic MINs, OMINs introduce crosstalk, the result of coupling two signals within one SE. When crosstalk occurs, a minute amount of the input signal power can be detected at another output, although the main signal is injected at the appropriate output. For this reason, the input signal is distorted at the output due to the loss and crosstalk introduced to the path. Studies also indicate that crosstalk is the most significant factor that reduces the signal-to-noise ratio and limits the size of a network [1].

We can eliminate first order crosstalk by ensuring that a switch is not used by two input signals simultaneously [1][6]. Thus, only one input can be sent to an electro-optic switch at any given time.

For an $n{\times}n$ MIN, a *permutation* is a pairing of its inputs and outputs such that each input is connected to a different output. In other words, a permutation is a one-to-one mapping between the network inputs and outputs [1]. A permutation is exemplified in Figure 2(a).

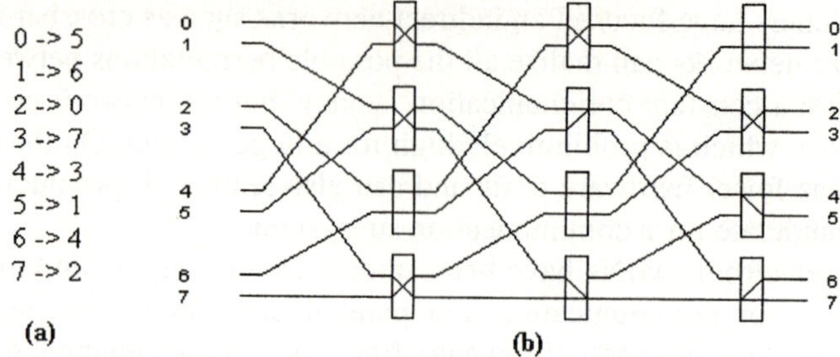

Fig. 2. (a) A permutation, and (b) routing the permutation in an 8 × 8 Omega network.

If we route all the messages for the permutation in Figure 2(a) simultaneously in an 8×8 Omega network, crosstalk will occur in each SE as reflected in Figure 2(b). Apparently, a permutation cannot be realized in one pass in an Omega network, since the two input links or the two output links in an SE cannot be active at the same time. It is possible that a permutation can be realized in one pass in the base-2 MIN [8], but typically, more passes are required to realize a permutation. Consequently, a permutation needs to be partitioned into several *partial permutations* [1] to avoid crosstalk.

Traditional routing algorithms for eliminating crosstalk in electronic switching MINs are not applicable for electro-optic switching MINS. Instead, we adopt a time domain approach [4] to partition the permutation into several partial permutations, or *passes,* such that the connections in each pass can be established in the network, simultaneously, without crosstalk; that is, all the inputs are scheduled in several passes such that no crosstalk occurs in each of the passes. In addition, we employ space domain approach to avoid crosstalk [1][4], which dilates the network by using more SEs to form more connection paths so that there are fewer opportunities for crosstalk to be introduced in each SE. For example, the Benes network and the dilated Banyan network are dilated MINs, but crosstalk can still be present in dilated networks. In order to realize any permutation efficiently, we can use the time domain approach and space domain approach simultaneously. For example, a base-2 MIN requires more SEs and more than one pass to realize a permutation

2   Routing and Scheduling in Previously Proposed MINs

Since baseline networks, Omega networks, butterfly networks and cube networks are topologically equivalent, we have selected the Omega network for our comparison. We use the Window Method [10] to find the conflicts among the messages to be sent in the Omega network. Then, we can draw a conflict graph [10], where we connect two nodes by using an edge if there is a conflict between them. We can compute the degree of conflict for each node in the graph to determine the scheduling order of the degree ascending algorithm, as well as the degree-descending algorithm.

There are many heuristic algorithms that have been designed to perform the scheduling [6]: the sequence increasing algorithm, the sequence decreasing algorithm, the degree ascending algorithm, the degree descending algorithm, the Genetic Algorithm

(GA), the Simulated Annealing (SA) algorithm, and the Remove Last Pass (RLP) algorithm. The RLP was propounded by Chau and Xiao [9]. For our investigation, we have tested all the algorithms, except the GA, on the novel base-2 optical MIN, proposed by Chau and Fu [8].

The concepts of these algorithms are the same: select one pair, an input and an output, from the permutation for the current pass based on some order, and if no crosstalk occurs, the pair is scheduled in current pass; otherwise, schedule it in the next pass. There is, though, a difference in the order by which the pair is selected. The sequence increasing algorithm selects the next input sequentially, from low to high, whereas the sequence decreasing algorithm schedules the next input sequentially in reverse order, from high to low. Although the degree ascending algorithm chooses the next input with the lowest number of degrees first, the degree descending algorithm prefers the next input with the highest number of degrees first. The Genetic Algorithm (GA) schedules the messages according to the results from the GA, and the Simulated Annealing (SA) algorithm schedules the messages based on the results from the Simulated Annealing algorithm.

2.1 Maximal Conflict Number

The Maximal Conflict Number (MCN) refers to the SE with the highest number of connections, when all the inputs of a permutation are routed simultaneously in a given OMIN [9]. For example, Figure 3(a) exhibits a permutation. If we route all the messages simultaneously in an Omega network, we will obtain the connection setting in Figure 3(b). The MCN in this case is four, and it exists in the circled SE. As it was mentioned in Section 1, to avoid crosstalk, no two inputs are allowed to inject to the same SE. Consequently, the number of passes to realize a permutation cannot be fewer than the MCN of the permutation for each of the known algorithms except the RLP algorithm [9].

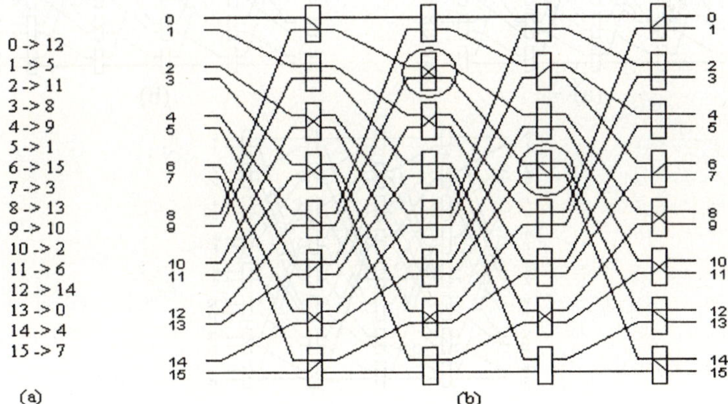

Fig. 3. (a) A permutation, and (b) routing the permutation in a 16×16 Omega network, the MCN is 4, which occurs in the circled switch element.

2.2 Remove Last Pass (RLP) Algorithm

In [9], Chau and Xiao have proposed the Remove Last Pass (RLP) algorithm that can route a permutation in fewer passes than the MCN. The RLP algorithm adopts the unused paths in the network to relay messages in the permutation. After the initial solution is derived by the SA algorithm [9], the RLP algorithm is used to remove the last pass by relaying its messages to the unused paths of the previous passes. The process is repeated until no more passes can be eliminated. For example, for the permutation in Figure 3(a), a possible solution is to use the SA algorithm (see Figure 4). We can observe that there is only one message in the last pass. If we use the empty paths in pass 1, pass 2, and pass 3, the last pass can be eliminated. For instance, node 1 sends a message in pass 1, but node 1 does not receive any message. Therefore, we can use node 1 as a receiver in pass 1 and as a sender in a later pass such that the message can be relayed from node 12 node 14 in the last pass. Figure indicates that no crosstalk is created, and we can route the permutation in fewer than the MCN passes. Other heuristic algorithms can be used to attain the initial solution for the RLP algorithm, but the SA algorithm yields better results than the other algorithms do.

Pass 1: 0 -> 12, 1 -> 5, 2 -> 11, 3 -> 8, 7 -> 3
Pass 2: 4 -> 9, 5 -> 1, 6 -> 15, 9 -> 10, 10 -> 2, 11 -> 6
Pass 3: 8 -> 13, 13 -> 0, 14 -> 4, 15 -> 7
Pass 4: 12 -> 14

Fig. 4. A possible solution by the SA.

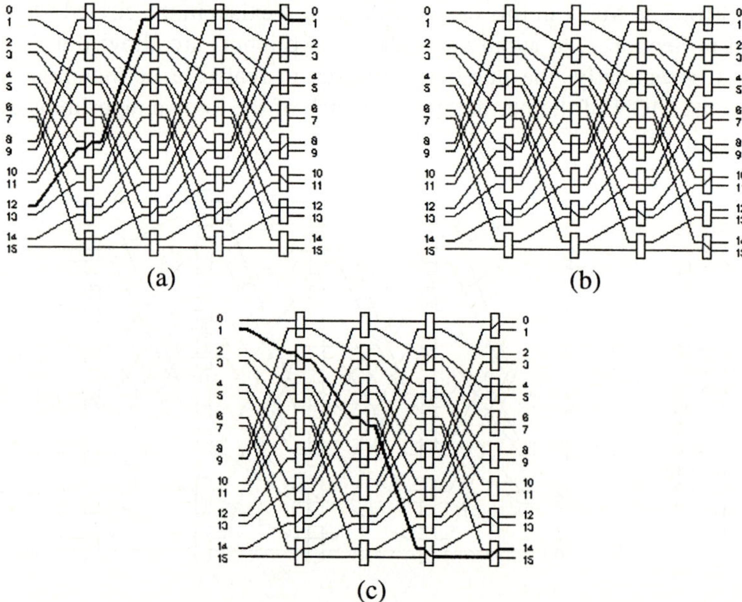

(a)   (b)

(c)

Fig. 5. (a) In pass 1, one message is added which is from node 12 to node 1. (b) In pass 2, it is the same as pass 2 in Figure 4. In pass 3, node 1 relays the message to node 14 without creating crosstalk and pass 4 is eliminated.

2.3 Experimental Results for the Omega Network

We have tested all the previous algorithms and the RLP algorithm [9] for the Omega network and have run each of them 1000 times. A random permutation is generated for each run. The average number of passes and the average computing time is listed in Table 1 and Table 2, respectively.

Table 1. The average number of passes for a 1000-round runs in an Omega network.

Network Size	Seq Inc	Seq Dec	Degree Asc	Degree Des	SA	RLP
4	2.0	2.0	2.0	2.0	2.0	2.0
8	2.78	2.78	2.91	2.63	2.61	2.58
16	3.69	3.74	3.8	3.54	3.38	3.36
32	4.52	4.52	4.7	4.26	4.1	4.0
64	5.33	5.34	5.66	5.04	4.83	4.64
128	6.19	6.21	6.6	5.79	5.58	5.17
256	6.99	6.99	7.56	6.45	6.44	5.67
512	7.75	7.77	8.53	7.15	7.29	6.02

Table 2. The average time (in milliseconds) for a 1000-round runs in an Omega network.

Network Size	Seq Inc	Seq Dec	Degree Asc	Degree Des	SA	RLP
4	0.08	0.05	0.08	0.05	0.08	0.06
8	0.06	0.02	0.05	0.13	0.42	0.39
16	0.03	0.03	0.49	0.47	0.28	0.13
32	0.13	0.16	2.89	3.19	0.95	0.89
64	0.2	0.12	22.73	23.28	2.7	2.7
128	0.37	0.5	178.37	183.92	5.14	4.58
256	0.51	0.59	1485.9	1523.1	20.11	19.84
512	1.34	0.86	12288	12679	193.67	211.66

3 The Novel Base-2 MIN

In this section, we will review the topology and characteristics of a novel base-2 MIN which has been proposed by Chau and Fu [8]. The base-2 MIN has more than twice the number of switches than multistage networks such as baseline, Omega, butterfly and cube networks do. Therefore, there are more connection paths in the base-2 MIN than in the other MINs, but the base-2 MIN has a better permutation capability than other MINs have. The base-2 MIN can realize an all-to-all personalized exchange in n-1 passes [8], whereas other MINs require a minimum of $2n$ passes.

The topology of the base-2 MIN is reflected in Figure ; it is a unidirectional blocking MIN. The base-2 MIN has n inputs, n outputs, and $(\log n + 1)$ stages. The stages are numbered from 0 to $(\log n)$, and each has n 2×2 SEs. In total, there are $n \times (\log n + 1)$ SEs in the base-2 MIN. In stage 0, only one input is injected into each SE. In the kth stage, one output of the ith switch is connected to the input of the ith switch in the $(k+1)^{th}$ stage. The other output is connected to the input of the $((i+2^k) \bmod n)$ switch in the $(k+1)^{th}$ stage. In the last stage, only one output is used, and the other output is void.

3.1 Routing for the Novel Base-2 MIN

Since we only need to find the difference c between the source s and the destination d, the routing for the base-2 MIN is very simple. If $d > s$, take the value of c and convert it to $a_0 \times 2^0 + a_1 \times 2^1 + ... + a_{n-1} \times 2^{n-1}$. If $a_i = 0$, where i is in $\{0...(n-1)\}$, connect the $((s + a_0 \times 2^0 + a_1 \times 2^1 + ... + a_{i-1} \times 2^{i-1}) \bmod n)^{th}$ switch in stage i to the first output. If $a_i = 1$, connect the $((s + a_0 \times 2^0 + a_1 \times 2^1 + ... + a_{i-1} \times 2^{i-1}) \bmod n)^{th}$ switch in stage i to the second output. If $s > d$, add n to d and find the value of c and convert it to $a_0 \times 2^0 + a_1 \times 2^1 + ... + a_{n-1} \times 2^{n-1}$. If $a_i = 0$, where i is in $\{0...(n-1)\}$, connect the $((s + a_0 \times 2^0 + a_1 \times 2^1 + ... + a_{i-1} \times 2^{i-1}) \bmod n)^{th}$ switch in stage i to the first output. If $a_i = 1$, connect the $((s + a_0 \times 2^0 + a_1 \times 2^1 + ... + a_{i-1} \times 2^{i-1}) \bmod n)^{th}$ switch in stage i to the second output.

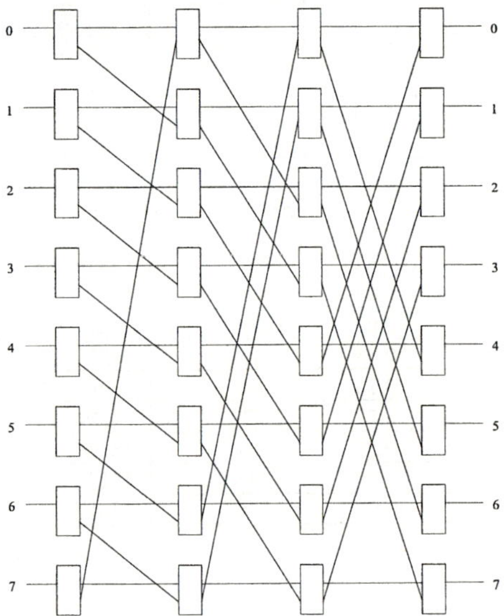

Fig. 6. A novel 8×8 unidirectional blocking OMIN: base-2 MIN.

The base-2 MIN does not fit the definition of self-routable because the routing tag has to be determined by both the input and output. In a self-routable network, only the output is necessary. However, the routing tag can be found automatically, instead of manually, according to the paired input and output. As a result, we still can say that the base-2 MIN is a type of self-routable MIN.

For example, in an 8-node base-2 MIN network, there are messages from 1 to 6 and messages from 5 to 3. For the first pair, $d > s$, we attain the value $c = 5$ and convert it into $a_0 a_{1...} a_{n-1}$, which is "101". Thus, the routing tag for the first pair is 1010, because we use the upper output in the last stage. This tag indicates that the packet must take the upper switch output at stages 1, and 3 and the lower switch output at stage 1 and 1. For the second pair $d < s$, we calculate that $c = 6$ and convert it into $a_0 a_{1...} a_{n-1}$ which is "011". Thus, the routing tag for the second pair is 0110. This tag indicates that the packet must take the upper switch output at stages 0, and 3 and the lower switch output at stage1 and 3.

4   Routing and Scheduling in the Novel Base-2 MIN

We need to test all six algorithms for the base-2 MIN network and run each of them 1000 times as we did in the Omega network. The average number of passes and the average computing time is summarized in Table 3 and Table 4, respectively. From Table 3, we can see that the SA algorithm gives far better results than the first five algorithms. Also, it is obvious that the RLP algorithm is superior to the other five algorithms in terms of the number of passes required. The larger the network size is, the better the results from the RLP algorithm.

Table 3. The average number of passes for a 1000-round runs in a base-2 MIN.

Network size	Seq Inc	Seq Dec	Degree Asc	Degree Des	SA	RLP
4	1.68	1.67	1.68	1.66	1.65	1.65
8	2.05	2.03	2.18	2.0	2.0	2.0
16	2.55	2.58	2.8	2.43	2.4	2.29
32	3.15	3.17	3.37	3.07	3.03	2.87
64	3.77	3.8	4.1	3.59	3.53	3.02
128	4.36	4.33	4.77	4.17	4.07	3.04
256	5.01	4.95	5.47	4.7	4.62	3.1
512	5.51	5.54	6.15	5.24	5.22	3.4

Table 4. The average time (in milliseconds) for a 1000-round runs in a base-2 MIN.

Network size	Seq Inc	Seq Dec	Degree Asc	Degree Des	SA	RLP
4	0.14	0.09	0.17	0.14	0.09	0.27
8	0.06	0.06	0.28	0.22	0.08	0.08
16	0.06	0.17	0.87	0.86	1.13	0.96
32	0.24	0.23	6.25	6.24	1.91	1.19
64	0.3	0.45	46.19	49.52	0.75	1.26
128	0.41	0.55	374.4	397.58	7.95	9.06
256	1.24	1.33	3041.3	3180.0	10.24	11.13
512	2.48	2.76	24323	25247	227.72	243.3

Table 4 indicates that the Sequence Increasing algorithm and Sequence Decreasing algorithm have close to the same time complexity, which is the best among all the six algorithms. Degree Ascending algorithm and Degree Descending algorithm always take more time than all the other algorithms. The SA algorithm and the RLP algorithm need about the same time for computing.

Since the solutions of the SA algorithm are randomly generated, a different solution can result each time the algorithm is run on the same permutation. Moreover, the empty paths in each solution can also be different, affecting the final solution that is generated by the RLP algorithm. In order to achieve the best solution, we can run the RLP algorithm more than once and select the best solution from each run. The results are exhibited in Table 5.

From Table 5, we can see that running the RLP algorithm many times does yield a better solution than running it only once, when the network size is greater than eight. For the Omega network, there is no difference between running the RLP algorithm three, five, or ten times. However, better results can be obtained for the base-2 MIN.

We also find that the RLP algorithm can realize any permutation in no more than three passes for the base-2 MIN, and no more than six passes for the Omega network. In fact, no permutation needs more than three passes for a base-2 MIN of a size that is less than or equal to 512.

Table 5. The average passes for 1000-round cases when running the RLP algorithm many times to get a final solution.

Size	Base-2 MIN				Omega Network			
	Once	3 times	5 times	10 times	once	3 times	5 times	10 times
4	1.65	1.67	1.69	1.64	2.0	2.0	2.0	2.0
8	2.0	2.0	1.99	2.0	2.6	2.57	2.58	2.6
16	2.29	2.2	2.17	2.14	3.33	3.23	3.21	3.22
32	2.87	2.78	2.74	2.66	3.97	3.9	3.91	3.9
64	3.02	3.0	3.0	3.0	4.5	4.27	4.27	4.26
128	3.04	3.0	3.0	3.0	5.04	4.97	4.96	4.97
256	3.1	3.01	3.0	3.0	5.29	5.01	5.01	5.01
512	3.4	3.09	3.03	3.0	5.76	5.32	5.32	5.32

If we group two MINs, such as Omega networks, together so that each handles a partial permutation, no more than three passes are needed to realize a permutation. However, the combination of two Omega networks requires $n \times \log n$ 2×2 switches. Moreover, we have to add an additional column of 2×2 switches before the first stage and another column after the last stage as indicated in Figure 7. Therefore, $n \times (\log n + 2)$ 2×2 switches are necessary for the dilated Omega network. The number of switches is n more than those used in the novel $n \times n$ base-2 MIN. The use of two Omega networks will have $(\log n + 2)$ stages instead of $\log n$ stages for one network. The novel $n \times n$ base-2 MIN has one fewer stage than the dilated Omega network does. As a result, the base-2 MIN requires less hardware than the dilated Omega network, and the switching control of the base-2 MIN is much simpler than that of the dilated Omega network.

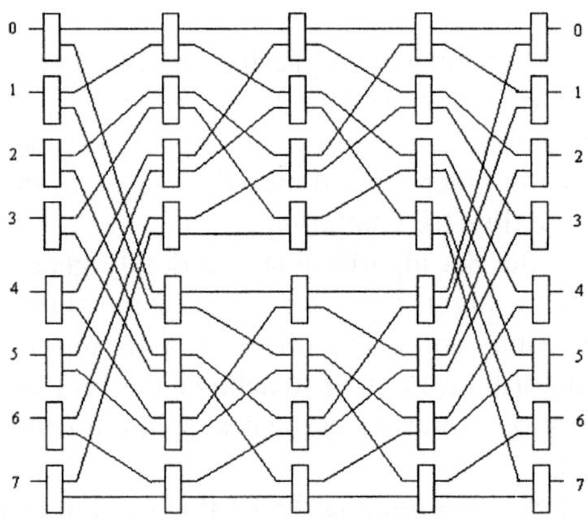

Fig. 7. An 8×8 Dilated Omega network.

6 Conclusion

In this paper, we have investigated the permutation capability of OMINs such as the Omega network and the base-2 MIN. Also the RLP algorithm has been tested in these networks to see how it can improve the performance of routing and scheduling. The RLP algorithm utilizes unused paths in the network to reduce the total number of passes required for a permutation. From our experiments, we know that any permutation can be realized in no more than three passes in the base-2 MIN, when the network size is not greater than 512. The base-2 MIN is a better choice for communication subsystems than the crossbar network in terms of hardware costs and transmission time. Finally, the base-2 MIN also outperforms the combination of two MINs such as the Omega, baseline, cubes and butterfly networks. We can conclude that the base-2 MIN can be a good choice for the high channel bandwidth and low communication latency of high performance computing and communication applications.

References

1. Y. Yang, J. Wang, and Y. Pan, Permutation Capability of Optical Multistage Interconnection Networks, *Journal of Parallel and Distributed Computing,* vol. 60, no. 1, Jan. 2000, 72-91.
2. Y. Yang and J.Wang, Optimal All-to-All Personalized Exchange in a Class of Optical Multistage Networks, *IEEE Transactions on Parallel and Distributed Systems,* Vol. 12, No. 6, June 2001, 567-582.
3. C. Qiao, A two-level process for diagnosing crosstalk in photonic dilated Benes network, *Journal of Parallel and Distributed Computing,* vol. 41, no. 1, 1997, 53-66.
4. C. Qiao and R. Melhem, A time domain approach for avoiding crosstalk in optical blocking multistage interconnection networks, *J. Lightwave Technology,* vol. 12. no. 10, Oct. 1994, 1854-1862.
5. Y. Pan, C. Qiao, and Y. Yang, Optical Multistage Interconnection Networks: New Challenges and Approaches, *IEEE Communications Magazine, Feature Topic on Optical Networks, Communication Systems and Devices,* Vol. 37, No. 2, Feb. 1999, 50-56.
6. A.K. Katangur, Y. Pan, and M.D. Fraser, Message Routing and Scheduling in Optical Multistage Networks Using Simulated Annealing, *IPDPS,* April 2002.
7. H Nakajima, Development on Guided-Wave Switch Arrays, *IEICE TRANS. COMMUN.,* vol. E82-B, no. 2, February 1999, 349-356.
8. Siu-Cheung Chau and Ada Wai-Chee Fu, Optimal All-to-All Personalized Exchange in a Novel Optical Multistage Interconnection Network, *International Journal of High Performance Computing and Networking, pending publication.*
9. Siu-Cheung Chau and Tiehong Xiao, A New Algorithm for Message Routing and Scheduling in Optical Multistage Interconnection Network, *Proceedings of the IASTED International Conference on Optical Communications Systems and Networks (OCSN 2004),* July 2004, 749-755.
10. X. Shen, F. Yang, and Y. Pan, Equivalent Permutation Capabilities between Time Division Optical Omega Networks and Non-optical Extra-Stage Omega Networks, *Proceedings of 1999 IEEE International Performance, Computing, and Communications Conference,* February 1999, 356-361.

Topology-Based Hypercube Structures
for Global Communication in Heterogeneous Networks*

Silvia M. Figueira[1] and Vijay Janapa Reddi[2]

[1] Department of Computer Engineering, Santa Clara University
Santa Clara, CA 95053-0566, USA
`sfigueira@scu.edu`
[2] Department of Computer Engineering, University of Colorado, Boulder
Boulder, CO 80309, USA
`janapare@colorado.edu`

Abstract. Hypercube structures are heavily used by parallel algorithms that require all-to-all communication. When communicating over a heterogeneous and irregular network, the performance obtained by the hypercube structure will depend on the matching of the hypercube structure to the topology of the underlying network. In this paper, we present strategies to build topology-based hypercubes structures. These strategies do not assume any kind of topology. They take into account the communication cost between pair of nodes to provide a performance-efficient hypercube structure. These enhanced hypercube structures help improve the performance of parallel applications that require all-to-all communication in heterogeneous networks by up to ~30%.

1   Introduction

Hypercube structures allow a computation that requires all-to-all communication among P tasks to be performed in $\log P$ steps. All-to-all communication is used by a variety of parallel algorithms, such as barrier synchronization, vector reduction, matrix multiplication, and sorting, making the hypercube one of the most useful structures in parallel computation [7]. In fact, many popular parallel algorithms use a hypercube communication structure, as shown by Leighton [17]. Some examples are the fast Fourier transform [19, 25], parallel prefix [24], and various computer vision [24] and linear algebra computations [13]. In addition, the MPICH's implementation [22] of the barrier synchronization operation is based on hypercubes.

To use a hypercube structure in an all-to-all operation, the processes are organized in a hypercube, i.e., the processes are assigned to positions in the hypercube structure and communicate with the nodes assigned to neighbor positions only. In homogeneous clusters, this assignment is generally done according to their nodes' identifiers, which are usually assigned *blindly* (independently of any performance measure) to the nodes [7, 11, 22]. For example, consider a cluster formed by 4 nodes located in the same LAN. These nodes are assigned identifiers 0, 1, 2, and 3. According to these identifiers, a 2D hypercube can be created, as shown in Figure 1, where node 0 communicates with nodes 1 and 2, but not to node 3, which also communicates with nodes 1 and 2.

---

* This research was supported in part by NSF cooperative agreement ACI-9619020 through computing resources provided by the National Partnership for Advanced Computational Infrastructure at the San Diego Supercomputer Center

J.C. Cunha and P.D. Medeiros (Eds.): Euro-Par 2005, LNCS 3648, pp. 994–1004, 2005.

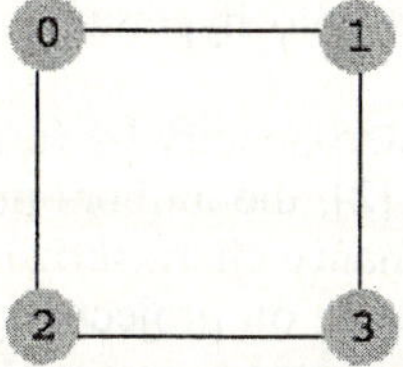

Fig. 1. 2D hypercube formed according to the nodes' identifiers

When using a hypercube structure, communication takes place between nodes that are connected in the hypercube. For this reason, the organization of the hypercube is a key process in the all-to-all communication. In regular-topology platforms, organizing the hypercube blindly may lead to good performance. However, in heterogeneous, irregular networks, we need a more sophisticated strategy to build the hypercube. Intuitively, the algorithm would execute more efficiently if we could form the hypercube by having communication take place only between nodes that are separated by a low-communication cost path, as shown in [6] for binomial trees.

This paper presents possible solutions to embed a given network topology into a hypercube template to be used in all-to-all communication. We evaluate the efficiency of the strategies proposed by comparing both the cost of the hypercubes generated and the time to execute a barrier-synchronization operation using the different hypercubes. The strategies discussed do not assume any kind of topology, i.e., the machines may be organized in any topology. It takes into account the communication cost between the nodes to come up with a performance-efficient hypercube to be used by the barrier-synchronization algorithm.

The strategies proposed are independent of the measure used as communication cost. In our experiments, latency is used but, in fact, the measure to be used should reflect the application characteristics. For example, for an application that sends a large number of short messages, the communication cost should reflect the latency. For an application that sends a small number of large messages, the communication should reflect the bandwidth. For an application that sends a large number of large messages, the communication cost should reflect both the latency and the bandwidth which, combined with the message size used by the application, provide an accurate model for communication cost. Note that these costs can be obtained with a performance prediction tool, such as the Network Weather Service [27].

This paper is organized as follows. Section 2 discusses related work. Section 3 presents the hypercube-based all-to-all communication algorithm used in the paper. Section 4 presents algorithms for creating topology-based hypercubes to be used in all-to-all communication operations. Section 5 shows experiments performed and results obtained. Section 6 concludes.

2   Related Work

Hypercube structures play an important role in global communication operations, and have been the subject of several research papers. Bertsekas et al have described optimal communication algorithms for hypercubes [4]. In [3], the authors presents a study on algorithms for collective operations for homogeneous parallel environments, and embeddings of different structures into hypercubes have been the subject of [5, 12, 18, 26].

In [1], the authors deal with heterogeneity by forming broadcast trees according to the capacity of each machine. In [2], the authors present a communication model of heterogeneous clusters for performance characterization of collective operations.

Several groups have been working on projects related to global communication in hierarchical topologies. In [20], the authors present ECO, a packet containing efficient collective operations for interconnected clusters. ECO groups hosts according to the network topology, i.e., each group contains nodes that belong to the same (homogeneous) cluster. Based on these groups, ECO implements the collective operation using a specific algorithm for each LAN. In [16], the authors also present a solution for more efficient global communication in hierarchical networks of workstations. They also group the hosts according to the network topology, but they use a binomial tree for each LAN. In [15], the authors describe MagPIe, a communication library that uses performance-efficient structures for global-communication operations in GRID environments. Global communication in WAN environments, such as the GRID, have been the subject of many other papers [8, 14], which employ a hierarchical structure to reflect the hierarchy existent in these environments.

PVM (Private Virtual Machine) [11] and MPICH [22], which is an implementation of MPI (Message Passing Interface) [21], are libraries used in practice by scientific applications using heterogeneous networks. Both of them provide various collective communication operations, such as one-to-all and all-to-all. Originally, they did not take the network topology into account. However, the Globus group have proposed the enhancement of MPICH to accommodate for the hierarchical structure of GRID environments. They use a hierarchical structure, as discussed in [9].

3 Hypercube-Based Communication Algorithm

In [7], Foster proposes a hypercube-based algorithm for all-to-all communication. It uses a hypercube communication template. The algorithm is executed by each task in a hypercube communication structure (obtained by the processes' identifiers). This algorithm allows an operation that requires all-to-all communication among P processes to be performed in $logP$ steps. The algorithm is presented below:

```
procedure hypercube (myid, input, logp, output)
begin
    state = input
    for i = 0 to logp - 1
        dest = myid XOR 2^i
        send state to dest
        receive message from dest
        state = OP (state, message)
    endfor
    output = state
end
```

The value $logp$ represents the size of the hypercube, and $myid$ represents the node's identifier. XOR denotes an exclusive OR operation, and OP is the user-supplied operator, used to combine local data with data arriving from the ith neighbor in the hypercube. In each step of the algorithm, each process exchanges its local $state$ (which embeds its local $input$ with the information received so far from its neighbors) with one of its neighbors in the hypercube and, then, combines the $message$ received from that neighbor with *state* to generate a new *state*. Note that, at each step, each node communicates with the neighbor indicated by *dest* = *myid* XOR 2^i, which does not depend on the network topology.

As shown in [7], this algorithm can be used efficiently, in regular-topology platforms, for vector reductions, matrix transpositions, merge sorts, and so on. However, as shown in Section 5, in heterogeneous networks, the algorithm's performance depends on the organization of the hypercube and, at each step, *dest* should be a node selected according to the topology of the network.

4 Enhancing Hypercube Structures

Our strategies to reorganize the nodes to form a more performance-efficient hypercube are based on the communication cost between the nodes. The hypercube-based algorithm works synchronously and, at each step, each node communicates with a specific node in the same subcube. In this case, placing nodes that are connected by a low-cost path in communicating positions, so that communication in each step of the algorithm uses a low communication-cost path, will decrease communication costs and lower the total cost of the all-to-all operation.

The following subsections present our three algorithms developed to provide topology- based hypercubes. The algorithms take the communication cost between pairs of nodes into account to form a performance-efficient hypercube.

4.1 Dim2_Cube

This algorithm tries to optimize the first dimension of the hypercube to decrease the cost in the first step of the all-to-all operation. The algorithm, which is shown below, is based on the following procedure: For every even position i, assign the first unmarked node n and select n's closest unmarked node to be placed at i's neighbor position in the first dimension.

```
Dim2_Cubes ( )
begin
    unmark all nodes
    for i = 0, 2, 4, ..., N-2
        n = 0
        while n is marked
            n = n + 1
        end while
        assign node n to position i
        mark node n
        j = non-used closest node to n
        make node j the first neighbor of node i
        mark node j
    end for
end
```

Figure 2 shows an example of a hypercube obtained with the algorithm above. Figure 2 (left) shows a network in which every link has the same cost. Figure 2 (left) illustrates the first three steps of the algorithm: node 0 is placed in position 0 with node 6 as first neighbor, node 1 is placed in position 2 with node 7 as first neighbor, and node 2 is placed in position 4 with node 3 as first neighbor. Figure 2 (middle) shows the last step of the algorithm, in which node 4 is placed in position 6 with node 5 (the last unmarked node) as first neighbor. Figure 2 (right) shows the hypercube obtained. Note that three out of the four edges in the first dimension of the hypercube shown in Figure 2 (middle) are optimum. However, the last edge obtained compromises the efficiency of the hypercube.

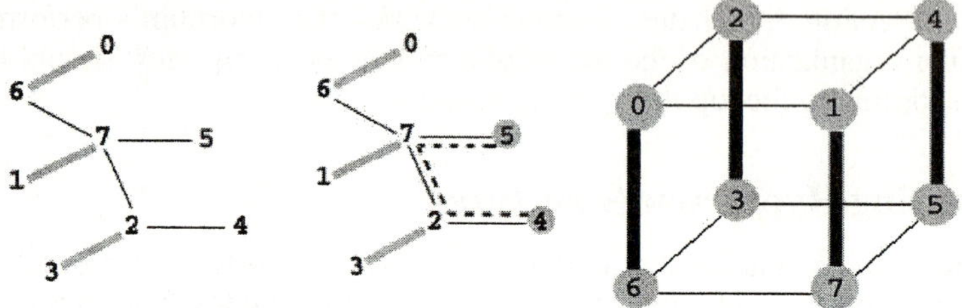

Fig. 2. Following the Dim2_Cube algorithm

This algorithm executes in time $O(n^2)$, where n is the number of nodes and, because it focuses on the first dimension only, the gains obtained by the hypercubes generated are limited (see Section 5).

4.2 TSTS_Cube

This algorithm uses a Traveling Salesman Tour with Shortcuts (TSTS), produced from a Minimal Spanning Tree (MST), to generate a *line* of nodes, from which each node is assigned to the corresponding position of the hypercube, according to the gray code [23].

The Minimal Spanning Tree (MST) is a tree that contains all the nodes in a network, and it is formed in such a way that the cost from the root to each of the nodes is minimum. The TSTS can be obtained from the MST by traversing it, visiting every node just once. A detailed explanation of how to create a TSTS from an MST is found in [10], which proves that the TSTS obtained has a length that is at most twice the length of the optimum tour.

The algorithm presented below traverses the hypercube positions using the gray code [23] and assigns nodes to each position according to the TSTS, generated from the MST corresponding to the network topology. The MST provides nodes that are close to each other, using locality to benefit the hypercube.

```
TSTS_Cube ( )
begin
      generate the MST and the TSTS
      h = 0, is the first position in the hypercube
      n = 0, is the first node in the TSTS
      for i = 0 to N-1
            assign node n to position h
            h = next position in the hypercube
            n = next node in the TSTS
      end for
end
```

The figures below illustrate the TSTS_Cube algorithm. The algorithm generates the MST from a given graph, as the one shown in Figure 3 (left). Starting with position 0, follow the gray code indexes and the nodes in the TSTS, assigning each node to the corresponding position. In Figure 3 (middle), the numbers on the edges indicate the sequence in which the nodes are visited, according to the TSTS order. Each node visited is placed in the next consecutive position, according to the gray code. Nodes are selected until the hypercube is complete, as shown in Figure 3 (right).

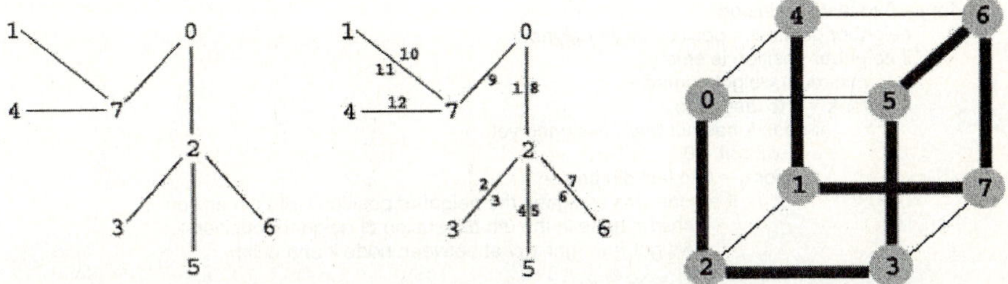

Fig. 3. Following the TSTS_Cube algorithm

This algorithm executes in time $O(n^2)$, where n is the number of nodes and, even though the MST provides the algorithm with locality, this algorithm does not scale well. The results show that the hypercubes generated achieves an average gain of about 10% over the hypercubes generated blindly for up to 16 nodes, but the gain decreases as the number of nodes increase (see Section 5). This happens because nodes that are part of a branch of the MST (which means they are connected by a low communication-cost path), will be spread in a line and the locality provided by the MST will be lost. Also the TSTS does not reflect the MST perfectly because of the detours.

4.3 Eff_Cube

The previous algorithms are based on optimizing each neighbor only. However, due to the strong coupling in hypercubes, this may not be the best approach, particularly for large networks. For example, if a node C is a neighbor of both nodes A and B's, it may be a good neighbor for node A, but may not necessarily be a good neighbor for node B. This strong coupling between neighboring nodes and their edges leads us to think in terms of selecting neighbor nodes based on the overall efficiency of the selected node.

The Eff_Cube algorithm takes into account the *local cost* before assigning a node to a position. Thus, for a given location, the node that generates the least local cost (sum of the weights on the edge in each dimension) will be chosen for a given position.

The algorithm works as follows. Neighbors of position 0 are consecutively assigned nodes 0, 1, ..., d, where d is the maximum dimension of the hypercube, since neighbors for these positions have not been assigned yet. Then, starting with the next consecutive position, the algorithm traverses all its dimensions in search of empty neighbors. Let Y be the current position, and Y.n be its nth neighbor. The algorithm searches for a node X from the list of unused nodes that gives the least local cost. If the cost of the edges between pairs of nodes are (X $\leftrightarrow$ Y.0) = C_0, (X $\leftrightarrow$ Y.1) = C_1, (X $\leftrightarrow$ Y.2) = C_2, ..., and (X $\leftrightarrow$ Y.n) = C_n, then the local cost to be minimized is

$C_0+C_1+C_2+...+C_n$. The algorithm assigns neighbors to the current node until all adjacent neighbors of the current index are filled with optimum nodes.

```
Eff_Cube ( )
begin
      for i = 0 to last position
            for j = 0 to last dimension
                  neighbor position = position i's jth neighbor
                  if neighbor position is empty
                        minimum weight = infinite
                        for k = 0 to last node
                              if node k has not been assigned yet
                                    weight = 0
                                    for d = 0 to last dimension
                                          if a node was assigned the neighbor position's dth dimension
                                                child = node in the dth dimension of neighbor position
                                                weight = weight + cost between node k and child
                                          end if
                                    end for
                                    if weight < minimum weight
                                          minimum weight = weight
                                          temp neighbor node = k
                                    end if
                              end if
                        end for
                        position i's jth neighbor = temp neighbor node (assign temp neighbor node to neighbor position)
                  end if
            end for
      end for
end
```

The figures below show an example of the Eff_Cube algorithm executing on a 3D hypercube. All neighbors of the starting position 0 are assigned nodes 0, 1 and 2. (Figure 4, left). The algorithm proceeds to position 1. For dimension 0, the algorithm starts traversing through the list of unused nodes to find the best suitable node. All nodes besides 0, 1 and 2 are checked as they are unused. Selection of the node is shown in Figure 4 (middle), Figure 4 (right), and Figure 5 (left), using as an example nodes 3, 5 and 7. For node 3 (Figure 4, middle), $(3 \leftrightarrow 0) = 13$, $(3 \leftrightarrow 1) = 12$, $(3 \leftrightarrow 2) = 0$, and the local cost = 25. For node 5 (Figure 4, right), $(5 \leftrightarrow 0) = 11$, $(5 \leftrightarrow 1) = 9$, $(5 \leftrightarrow 2) = 3$, and the local cost = 23. For node 7 (Figure 5, left), $(7 \leftrightarrow 0) = 10$, $(7 \leftrightarrow 1) = 4$, $(7 \leftrightarrow 2) = 3$, and the local cost = 17. Of all these nodes, node 7 yields the lowest local cost. Therefore, node 7 is chosen to be placed in the current empty index. All adjacent neighbors (Figure 5, right), dimension after dimension, are inserted in the same way, with the node that provides the least local cost.

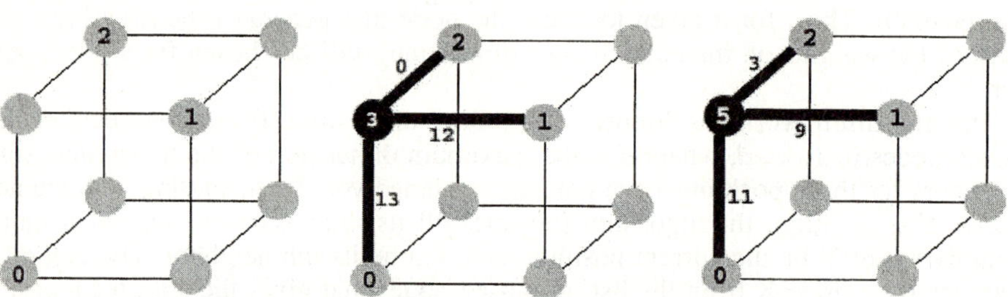

Fig. 4. Following the Eff_Cube algorithm

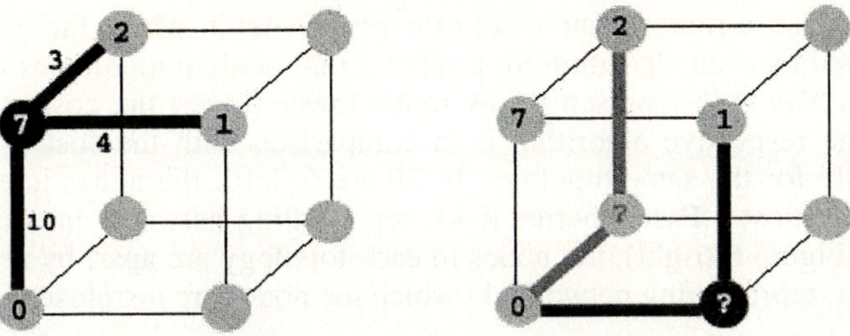

Fig. 5. Following the Eff_Cube algorithm

This algorithm executes in time $O(n^2(logn)^2)$, where n is the number of nodes, and the gains obtained by the improved hypercubes is on average around 30% for large clusters (with 1024 nodes). This algorithm scales well. In fact, the average gain starts at about 10% for 8 nodes and increases consistently until 1024 nodes. Section 5 shows a representative set of the experiments performed.

5   Results

To show the effectiveness of the algorithms proposed, we have executed two kinds of experiments. First, we compare the cost of topology-based hypercubes with the cost of hypercubes obtained blindly by the system. This comparison is made for a series of topologies generated randomly. Then, we compare the time to execute a barrier synchronization operation using topology-based hypercubes with the time to execute the same operation using blindly generated hypercubes. This comparison is also made for a series of topologies generated randomly.

5.1   Comparing Costs

The cost of a hypercube, which is the measure used in the comparison, is given by the cost of the node with the maximum cost, where the cost of each node is calculated as the sum of the weights between the node and each neighbor. Note that the cost for each node is calculated dimension-by-dimension, and that delays in previous dimensions, in which neighbors may have higher costs to their own neighbors, must be incorporated in the calculation of the cost for each dimension. The algorithm is shown below:

```
calc_cost ( )
begin
    for each node
        cost of node = weight between the node and the neighbor in the 1st dimension
    for i = 2nd to last dimension
        for each node n
            cost of n = max (cost of n, cost of n's neighbor in the ith dimension)
        for each node n
            cost of n = cost of n + weight between n and its neighbor in the ith dimension
        end for
    end for
    cost of hypercube = cost of the node with the maximum cost
end
```

Figure 6 shows a representative set of experiments, in which the average gain in cost is obtained by each algorithm for a set of 1,000 random topologies for each number of nodes. The gain represents how much lower (in %) the cost of a hypercube created by the respective algorithm is in comparison with the cost of a hypercube created blindly for the same topology. In Figure 6 (left), the nodes in each topology are apart by at most 5 Fast Ethernet links, representing networks in which the nodes are close. In Figure 6 (right), the nodes in each topology are apart by at most 20 Fast Ethernet links, representing networks in which the nodes are not close together.

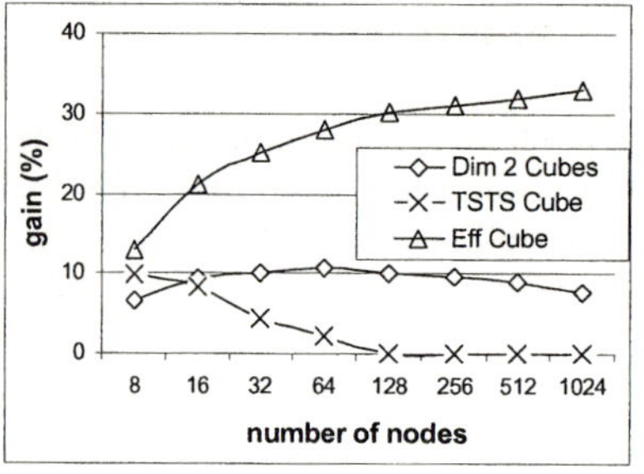

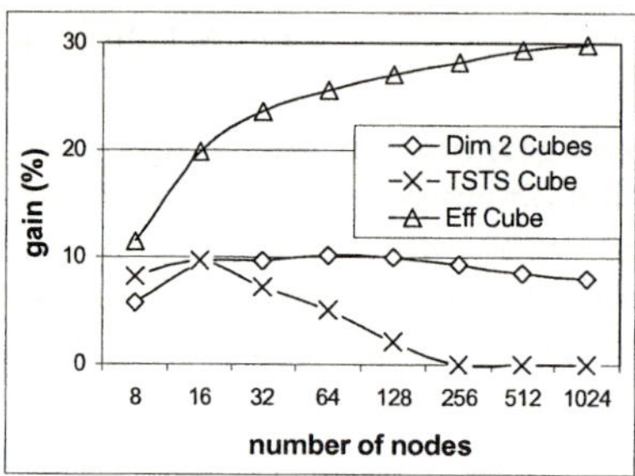

Fig. 6. Maximum cost between each pair of nodes is 5 links (left) and 20 links (right)

The graphs consistently show that the Dim2_Cube algorithm achieves an average of about 10% independently of the number of nodes, while the TSTS_Cube achieves a gain of about 10% for small networks, but does not scale well. The Eff_Cube algorithm achieves the highest gains and scales well, achieving an average gain of around 30% for 1024 nodes. These experiments show that the Eff_Cube algorithm implements the most promising approach, and that considering local costs leads to more efficient hypercubes. These graphs also show that the algorithms' behavior is independent of the maximum communication cost between the nodes.

5.2   Comparing Execution Times

Besides comparing costs of hypercubes, we have also executed experiments to compare the time to execute a barrier synchronization operation with the different hypercubes. The experiments were executed on the Blue Horizon IBM SP at the San Diego Supercomputer Center. The IBM-SP is a LAN-based, regular-topology cluster, in which we have emulated diverse heterogeneous networks by enforcing different latencies between the nodes. Our emulator was implemented using MPI [21]. Given the number of nodes and a maximum communication cost between any pair of nodes, the emulator generates random heterogeneous networks, which have various topologies. The emulator generates the topologies by generating the communication cost between every pair of nodes. Communication cost are in number of links. The latencies to be enforced are obtained by multiplying each random cost and a *base latency*, which is equivalent to the latency between two workstations connected by fast Ethernet (100Mbps), i.e., ~35ms. Each emulation executes 10,000 barrier synchronization operations.

Each graph below shows, for each algorithm, the gain obtained when comparing the improved hypercube with the one obtained blindly. For each algorithm, the topology used was the one that provided the best result among the experiments presented in Subsection 5.1. The graphs show the cost and the emulation time gains. Note that they match, showing that our cost calculation is accurate.

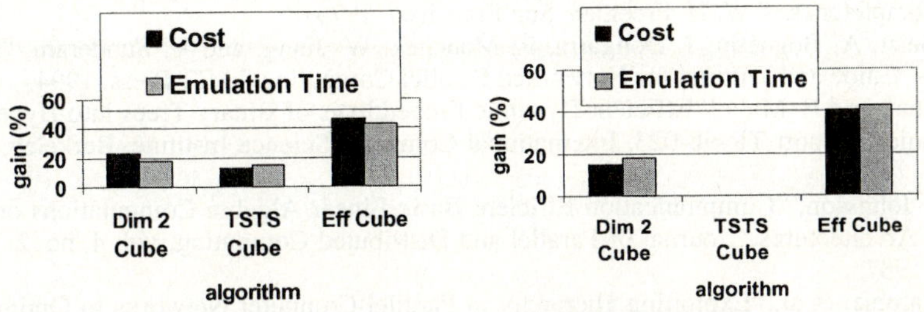

Fig. 7. 128 nodes (left) and 1024 nodes (right)

The Eff_Cube algorithm is the best option and achieves gains up to ~40% (Figure 7, left and right), corroborating the results shown in Subsection 5.1. The Dim2_Cube algorithm provides gains of up to ~20% (Figure 7, left and right). The TSTS_Cube algorithm achieves a gain of up to ~15% for small networks (Figure 7, left), but it does not provide any gain for large networks (Figure 7, right).

6 Conclusion

All-to-all communication is extensively used by parallel algorithms, and adapting these algorithms to execute efficiently in heterogeneous networks is crucial to improve their performance in this kind of environment.

In this paper, we have presented strategies to organize the nodes in a heterogeneous network into a hypercube. The strategies are based on the communication cost between the nodes in the network. The experiments performed have shown that the Eff_Cube algorithm presented helps to lower communication costs and, consequently, to improve the performance of algorithms based on all-to-all communication.

References

1. M. Banikazemi, V. Moorthy, and D. K. Panda, "Efficient Collective Communication on Heterogeneous Networks of Workstations," in Proceedings of the ICPP'98, August 1998.
2. M. Banikazemi, et al, "Communication Modeling of Heterogeneous Networks of Workstations for Performance Characterization of Collective Operations," in Proceedings of the Heterogeneous Computing Workshop, April 1999.
3. M. Bernaschi and G. Iannello, "Collective Communication Operations: Experimental Results vs. Theory," Concurrency: Practice and Experience, vol. 10, no. 5, pp. 359-386, 1998.
4. D. P. Bertsekas, et al, "Optimal Communication Algorithms for Hypercubes," Journal of Parallel and Distributed Computing, vol. 11, pp. 263-275, 1991.
5. C. Chen and R. Chen., "Compact Embeddings of Binary Trees into Hypercubes," Information Processing Letters, vol. 54, no. 2, pp. 69-72, April 1995.
6. S. M. Figueira and C. Mendes, "Dynamically Adaptive Binomial Trees for Broadcasting in Heterogeneous Networks of Workstations," in Proceedings of the VECPAR, June 2004.
7. I. Foster, "Designing and Building Parallel Programs - Concepts and Tools for Parallel Software Engineering," Addison Wesley Publishing Company, 1995.
8. I. Foster, et al, "Wide-Area Implementation of the Message Passing Interface," Parallel Computing, vol. 24, no. 12, pp. 1735-1749, 1998.

9. I. Foster and N. Karonis, "A Grid-Enabled MPI: Message Passing in Heterogeneous Distributed Computing Systems," Proceedings of the Supercomputing'98, November 1998.

10. M. R. Garey and D. S. Johnson, "Computers and Intractability: A Guide to the Theory of NPCompleteness," W. H. Freeman, San Francisco, 1979.

11. A. Geist, A. Beguelin, J. Dongarra, R. Manchek, W. Jiang, and V. Sunderam, "PVM: A Users' Guide and Tutorial for Networked Parallel Computing," MIT Press, 1994.

12. V. Heun and E. Mayr, "Efficient Dynamic Embeddings of Binary Trees into Hypercubes," Technical Report TR-98-023, International Computer Science Institute, Berkeley, California.

13. S. L. Johnsson, "Communication Efficient Basic Linear Algebra Computations on Hypercube Architectures," Journal of Parallel and Distributed Computing, vol. 4, no. 2, pp. 133-172, 1987.

14. N. Karonis, et al, "Exploiting Hierarchy in Parallel Computer Networks to Optimize Collective Operation Performance," Proceedings of the 14th IPDPS, pp. 377-84, May 2000.

15. T. Kielmann, et al, "MagPIe: MPI's Collective Communication Operations for Clustered Wide Area Systems," in Proceedings of the PPoPP'99, May 1999.

16. T. Kielmann, H. E. Bal, and S. Gorlatch, "Bandwidth-Efficient Collective Communication for Clustered Wide Area Systems," in Proceedings of the IPDPS'00, May 2000.

17. F. T. Leighton, "Introduction to Parallel Algorithms and Architectures," Morgan Kaufmann, 1992.

18. M. Livingston and Q. Stout, "Embeddings in Hypercubes," Mathematical and Computational Modeling, vol. 11, pp. 222-227, 1988.

19. C. Loan, "Computational Frameworks for the Fast Fourier Transform," SIAM, 1992.

20. B. Lowekamp and A. Beguelin, "ECO: Efficient Collective Operations for Communication on Heterogeneous Networks," in Proceedings of the 10th International Parallel Processing Symposium, April 1996.

21. Message-Passing Interface Forum, "MPI: A Message-Passing Interface Standard," International Journal of Supercomputing Applications, 8(3/4), 1994.

22. MPICH-A Portable Implementation of MPI, http://www-unix.mcs.anl.gov/mpi/mpich/.

23. M. Quinn, "Parallel Computing - Theory and Practice," McGraw-Hill, 1994.

24. S. Ranka and S. Sahni, "Hypercube Algorithms for Image Processing and Pattern Recognition," Springer-Verlag, 1990.

25. P. Swarztrauber, "Multiprocessor FFTs," Parallel Computing, vol. 5, pp. 197-210, 1987.

26. Y. Tseng, et al, "Low-Congestion Embedding of Multiple Graphs in a Hypercube," International Conference on Parallel and Distributed Systems, pp. 378-385, 1992.

27. R. Wolski, N. Spring,. and J. Hayes, "The Network Weather Service: A Distributed Resource Performance Forecasting Service for Metacomputing," in *Journal of Future Generation Computer Systems*, 1999.

Performance Effects of Node Mappings on the IBM BlueGene/L Machine

Brian E. Smith[1] and Brett Bode[2]

[1] IBM Rochester, 3605 Highway 52 North, Rochester MN 55901
smithbr@us.ibm.com
[2] Ames Laboratory, 329 Wilhelm Hall, Ames IA 50011
brett@scl.ameslab.gov

Abstract. The IBM BlueGene/L supercomputer consists of up to 65536 compute nodes connected by several networks including a three-dimensional torus. The BlueGene/L control system allows a user to re-map MPI ranks to different physical torus coordinates at run-time. Effects of node mapping on application performance are investigated for Gray-code mappings with differing aspect ratios, permutations of the X, Y, and Z coordinates, random mappings and four new mapping types. Results are presented for three NAS parallel benchmarks - BT, CG, and MG - on 128-way partitions in co-processor mode and virtual node mode on the prototype BlueGene/L hardware.

1   Introduction

Scientific application developers are constantly looking for ways to get better performance from their code to allow them to solve larger problems. One relatively easy method of increasing performance is to ensure that the system running the application is the best system available. For example, if the application code assumes a certain network topology, running on a different network topology would not provide the best performance benefits.

The IBM BlueGene/L supercomputer allows a user assign physical torus coordinates to logical MPI ranks. While not changing the physical topology, this logical re-arranging allows a user to make more assumptions in the code or port code from other platforms with less work. And, performance of existing codes might be improved by providing a more optimal node map for a given application. This paper looks at the performance benefits of several new mapping strategies and several existing mapping strategies on three of the NAS parallel benchmarks.

2   BlueGene/L Overview

2.1   The IBM BlueGene/L Hardware

The BlueGene/L supercomputer is a new parallel system from IBM [1],[2] . The system consists of up to 65,536 relatively modest compute nodes. Each node

J.C. Cunha and P.D. Medeiros (Eds.): Euro-Par 2005, LNCS 3648, pp. 1005–1013, 2005.

consists of two PowerPC 440 cores - each with two 64-bit floating point units, five network controllers. Each node has a total of 512 MB of memory. The three user-level networks are a three dimensional torus for high-performance point-to-point message passing, a global combining/broadcast tree for high-speed collective operations, and a global interrupt network for extremely fast barriers. The other two networks for low-level machine control and access to external file servers.

2.2   IBM BlueGene/L Software

The compute nodes run a custom, lightweight kernel. The kernel provides a single address space to one running application. Therefore, there is no context switching overhead and applications can use the majority of the 512 MB of memory. The kernel function-ships system calls to the I/O nodes and handles application generated signals. For more information on the BG/L software, see [3].

The standard application programming interface (API) for BG/L application developers is MPI. Specifically, BG/L uses an optimized MPICH [4] implementation where many collective operations are optimized to make use of the underlying BG/L hardware.[5] Code is compiled with the IBM XL compilers and the system is rather "Linux-like" in supported system calls.

2.3   IBM BlueGene/L Control System

The BG/L control system is responsible for booting the system and running jobs. There are two main operating modes. The standard mode is called co-processor mode. In this mode, one core is responsible for most computations while the second core is used for message passing. The other operating mode is called "virtual node mode" (VNM). In this mode, all of the physical resources of a node are split in half between the two cores. Since there are half as many packet send/receive queues per core, there can be performance degradation on communications-intensive applications. VNM can be very beneficial to applications that are not memory intensive, nor communications-bound.

3   Mappings

The BG/L control system pieces work together to provide compute processes their MPI rank based on their physical coordinates in the torus. For example, a node at physical coordinates $(0,0,0)$ would become MPI rank 0. A node at $(1,0,0)$ would then become MPI rank 1. In virtual node mode, a fourth coordinate is required (the processor or core ID, usually referred to as 'T'). A user may specify an arbitrary mapping file (as long as all nodes in a partition are uniquely accounted for in the file, even if they are not used). This allows for some interesting possibilities for improving application performance. For example, many applications might assume a two-dimensional mesh of processors. Given a proper mapping file, the three-dimensional BG/L torus can be made to look like a two-dimensional mesh with no link contention. This paper looks at the performance effects of the standard permutations of X, Y, and Z coordinates (XYZ, XZY, YXZ, YZX, ZXY, ZXY. In virtual node mode, the core ID can come first or last), along with Gray-code mesh mappings, and two other unique mapping strategies with several variations. A random mapping was also used to compare results.

3.1   Gray-Code Mappings

When hyper-cubic machine topologies were more common in the early 1990s, work was done on embedding different order hyper-cubic graphs in whatever dimension hypercube was available on the physical hardware of a given machine. See [6] or [7]. This was done using Gray codes [8]. Gray code sequences are usually constructed using a binary reflected Gray code algorithm. For example, the sequence for 8 values (3 bits) would be: 000, 001, 011, 010, 110, 111, 101, 100. For a simple discussion on embedding a lower-order graph in a higher order topology, see [6]. The general strategy is to realize that a k-way partition can be written as $k = 2^n$ for some n. Each node in the partition then has an n-bit "address". If each node exists in a three-dimensional torus, then the n-bit address can be broken down into three sub-addresses (x-bits, y-bits, z-bits) for the physical coordinates. To embed a two dimensional mesh in the three-dimensional torus, the three sub-addresses are split into two sub-addresses with m-bits and n-bits for location information. This is shown with an example. The 128-way partitions on the BG/L development machine are 8x4x4 nodes. Each node can then be represented as $n = x_1x_2x_3y_1y_2z_1z_2$. To embed a 16x8 mesh we convert each node "address" n to look like $n = a_1a_2a_3a_4b_1b_2b_3$. For example, a Z-first Gray-code mesh would have $n = z_1z_2x_1x_2x_3y_1y_2$ to then convert to a mesh "address". The T Gray-code meshes have the core ID first.

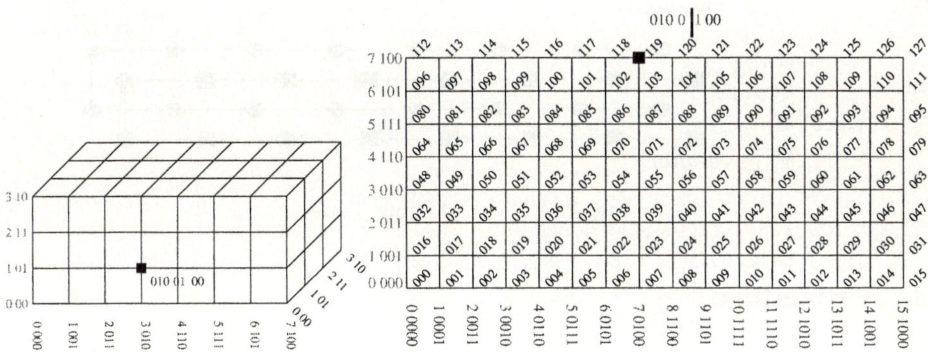

Fig. 1. Embedding 16 by 8 mesh in 128-Way BG/L Partition

The utility developed to generate all of the maps also allows the user to specify the dimensions of the mesh to control the aspect ratio of the 2D mesh.

3.2   Other Mapping Strategies

Because the Gray-code mesh mappings move some nodes away from each other that would normally be neighbors, another map type called "unfolding" was developed to try to keep more neighbor nodes next to each other while creating a mesh-like mapping. Ranks increase along the orientation axis, then increase in a perpendicular axis and back-track along the orientation axis. This continues until all nodes are mapped.

In virtual node mode, three different ways of counting the core ID - corefirst (CF), corelast (CL), and coreplanes (CP) were investigated. In corefirst unfolds, MPI rank 0 corresponds to physical coordinates $(0, 0, 0, 0)$ while MPI rank 1 corresponds to physical coordinates $(0, 0, 0, 1)$.

In corelast unfolds, the ranks progress along the orientation axis, then move to the second core and progress back along the orientation axis before increasing the next orthogonal coordinate.

In coreplane unfolds, the ranks increment along the orientation axis, then increase on the orthogonal axis, and continue increasing as the physical coordinate decreases. When the physical coordinate is back at 0, the processor core is increased and the process repeats. These three mappings are shown in 2 for two rows from an 8x8x8 torus.

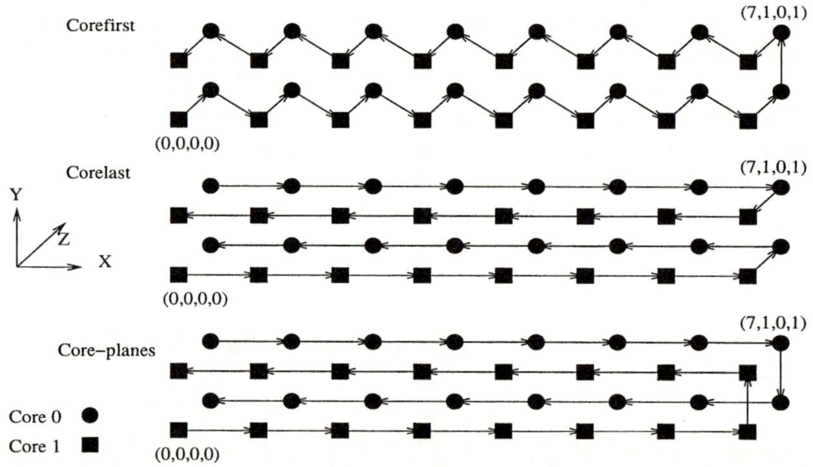

Fig. 2. VNM Unfolding Types

4   Procedure

In Sect. 5, results from three NAS benchmarks are presented. The three benchmarks included show the most variability between mappings. They are BT (block tri-diagonal), CG (conjugate gradient), and MG (multi-grid). For more information on the NAS benchmarks see [9], [10], and [11]. For the results of other NAS benchmarks (and other benchmarks in general), see [12].

The NAS benchmarks come in multiple problem sizes called classes. Results are shown for classes A, B, and C. Class A is the smallest class available for parallel machines and tends to be communications-bound on larger partitions such as 128-way. Class C was the largest class that still showed variation between mappings on 128-way partitions (in co-processor or virtual node mode), though there is a class D version of most of the NAS benchmarks. Results are presented for 128-way BG/L partitions in co-processor mode and virtual node mode (therefore utilizing 128 or 256 processors).

The result reported by the NAS benchmarks is "MOp/s" (millions of operations per second). The graphs in Sect. 5 show the MOp/s divided by number of processors used. Larger values of MOP/s indicate better performance. The NAS BT benchmark requires a square number of processors so only 121 processors are used for the co-processor mode results. However, the results get reported as if 128 processors were actively used. Therefore the MOp/s per processor values are recalculated for only 121 nodes.

The benchmarks were all run on first generation prototype hardware and were compiled with the significantly less efficient (compared to IBM XL) GNU compilers. Because of this, the numbers are not meant to represent absolute BG/L performance. Instead, they are meant to show the relative effects of mappings on performance. Re-running the benchmarks with the XL compilers should lessen computational time without affecting communications so the results should show differences between mappings more clearly.

5   Results

Typically, a smaller problem class performs fewer MOP/s per processor so it is possible to plot all three classes on the same graph to conserve space. However, a few smaller problem classes out-performed or performed similarly to larger class problems for a given mapping. This obscures the larger problem results. This is especially true in the CG benchmark.

Figure 3 shows the results for the NAS BT benchmark in co-processor mode for classes A, B, and C.

The most variation between mappings is seen in the class A runs. This is probably because less time is required for computation in the small problem sizes. Because the 128-way partition is physically arranged as 8 by 4 by 4 nodes, the mappings that hide the latency from the fact that there are an average of twice as many hops in the X direction as Y or Z show reasonably good performance increases. For example, the YZX and ZYX stock mappings show approximately a fifteen percent improvement over the default XYZ mapping. Similarly the X-first Gray-codes have the X coordinates as the most significant bit so the physical X location changes slowly through the MPI ranks. The unfold Y map counts along the Y axis, then Z before incrementing X. Finally, the Blocks - X mapping also increases the X coordinate most slowly. Figure 4 shows the results of running the CG benchmark with the mappings and Fig. 5 shows the results from MG with 128 processors in co-processor mode. Both results are somewhat

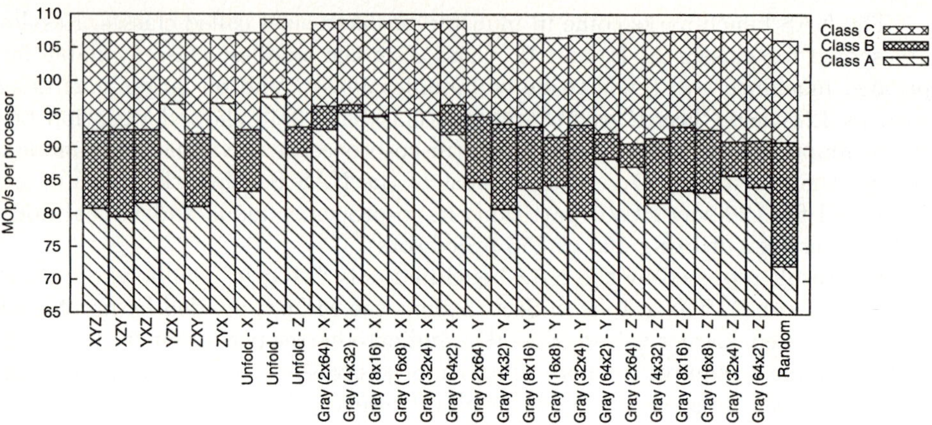

Fig. 3. BT 128 Processors

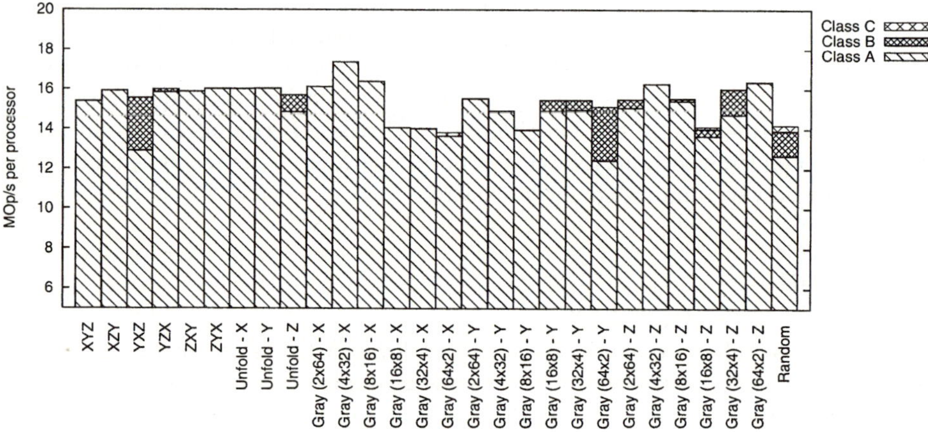

Fig. 4. CG 128 Processors

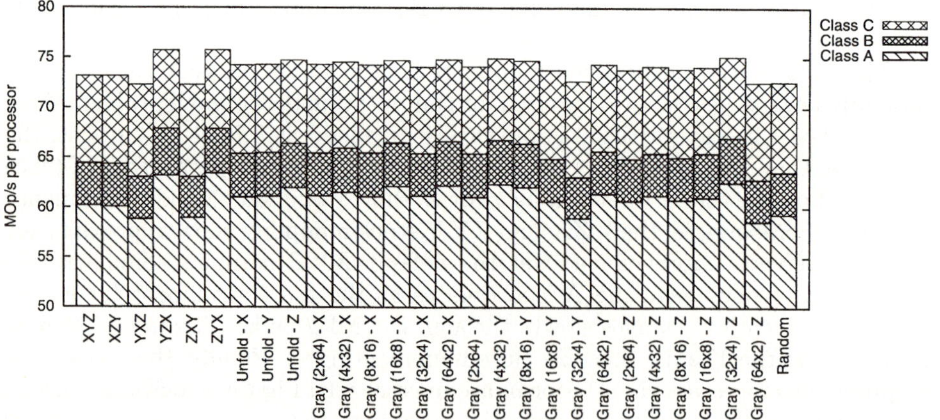

Fig. 5. MG 128 Processors

similar in that there was not much variation between the mappings, especially MG. The MG results are similar to the BT results - the mappings that have the X coordinate varying the slowest had the best performance.

Figure 6 shows the BT results with a 128-way partition booted in virtual node mode. Because BT assumes a square mesh of processes, it makes sense that the square meshes are the best Gray-code performers. As in the co-processor case, the maps that hide the latency of twice as many hops in X showed the best performance in general. Figure 8 shows the MG results. As in the 128-way case, there is less variation than with BT. There was very little variation between the Gray-code mappings as a function of aspect ratio. The virtual node mode unfold maps oriented along Y performed well since those mappings cause the physical X coordinate to vary slowly. Figure 7 shows the results for CG. The results were somewhat unexpected. Gray-code meshes that were more rectangular tended to perform better than the square meshes. However, the unfold and stock results were similar to the BT and MG results. The Gray-code mapping results will require more investigation of the CG code.

6   Conclusions

Changing the MPI rank mapping on BlueGene/L can show very good performance increases (as much as fifteen percent over the default XYZ mapping). This is especially true if the mapping hides the latency from extra network hops in asymmetric partition sizes. The majority of BlueGene/L partition sizes are not perfect cubes so there is potential for performance increases with minimal effort for most configurations.

Changing the logical node mapping is also an easy thing for an application developer to try. At the very minimum, the BG/L control system provides the stock permutations to the user, and developing mapping files is not difficult.

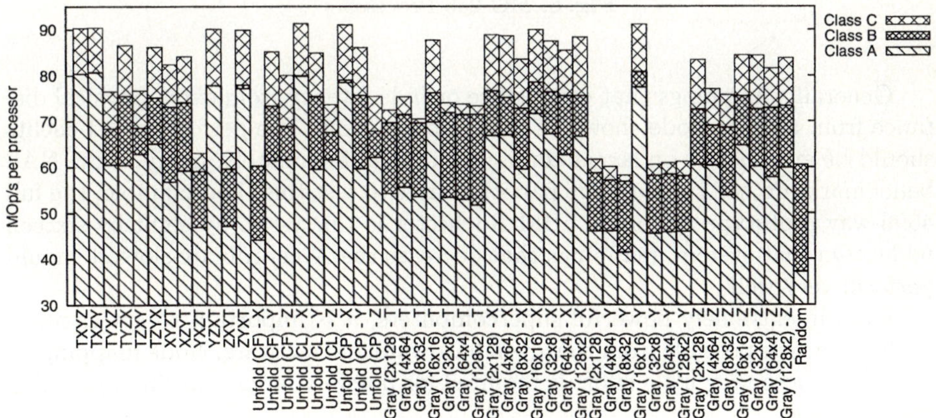

Fig. 6. BT 256 Processors

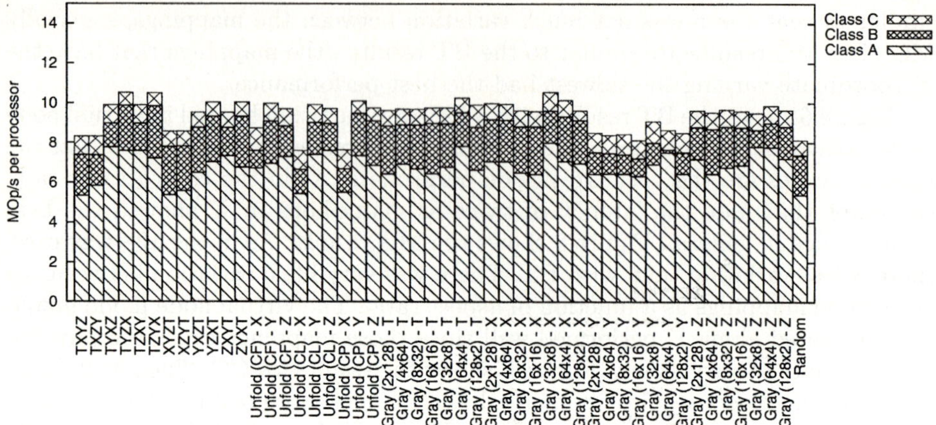

Fig. 7. CG 256 Processors

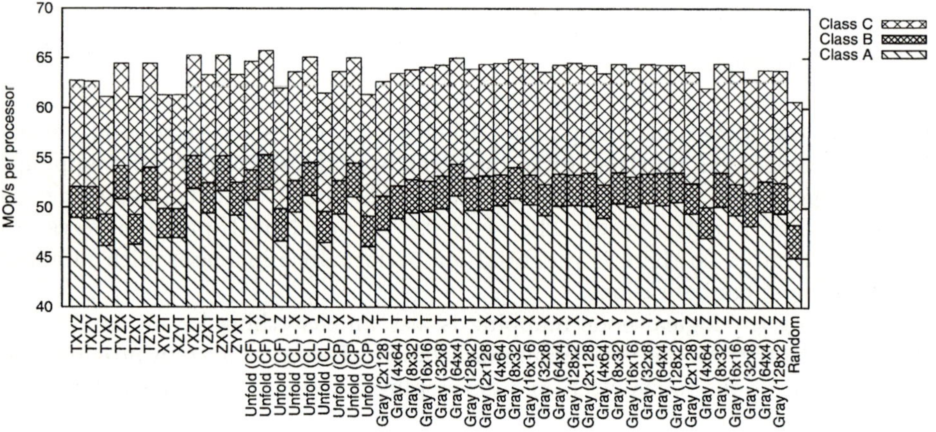

Fig. 8. MG 256 Processors

Generally, mappings that can reduce or hide latency from extra physical distance from node to node show the best improvement. The performance benefits should be even better on larger configurations (unfortunately, many of the NAS benchmarks do not run on large partition sizes to test this) For example, the full 4096-way prototype system at IBM Rochester is thirty-two by eight by sixteen nodes. On that system, mappings that decrease the X coordinate slowest should perform very well.

Even in extremely powerful supercomputers, it is important to get as much performance as possible from the given hardware and software. Node mappings is one area where very little effort is required for possibly large gains in application performance.

Acknowledgments

This work was supported in part by U.S. Department of Energy. This manuscript has been authored by Iowa State University of Science and Technology under Contract No. W-7405-ENG-82 with the U.S. Department of Energy. The authors would also like to thank IBM for access to BlueGene/L hardware, especially Sam Ellis, Charles Archer, and Dr. José Moreira.

References

1. The IBM BlueGene/L Team. An overview of the BlueGene/L supercomputer. In *Supercomputing 2002*, 2002.
2. F. Allen et al. Blue gene: A vision for protein science using a petaflop supercomputer. *IBM Systems Journal*, 40(2), 2001.
3. Gheorghe Almási et al. An overview of the BlueGene/L system software organization. *Parallel Processing Letters*, 13(4):561–574, 2003.
4. William Gropp, Ewing Lusk, Nathan Doss, and Anthony Skjellum. A high-performance, portable implementation of the MPI message passing interface standard. *Parallel Computing*, 22(6):789–828, 1996.
5. Gheorghe Almási et al. Implementing MPI on the BlueGene/L supercomputer. In *Proceedings of Euro-par 2004*, 2004.
6. Youcef Saad and Martin H. Schultz. Topological properties of hypercubes. *IEEE Transactions on Computers*, 37(7):867–872, 1988.
7. M. Y. Chan and F. Y. L. Chin. On embedding rectangular grids in hypercubes. *IEEE Transactions on Computers*, 37(10):1285–1288, 1988.
8. James R. Bitner, Gideon Ehrlich, and Edward M. Reingold. Efficient generation of the binary reflected Gray code and its applications. *Communications of the ACM*, 19(9):517–521, 1976.
9. Rob F. Van der Wijingaart. NAS parallel benchmarks version 2.4. Technical report, NASA Advanced Supercomputing Division, Ames Research Center, Moffett Field, CA 94035-1000, 2002. NAS-02-007.
10. D. Bailey et al. The NAS parallel benchmarks 2.0. Technical report, NASA Advanced Supercomputing Division, Ames Research Center, Moffett Field, CA 94035-1000, 1995. NAS-95-020.
11. D. Bailey et al. The NAS parallel benchmarks. Technical report, NASA Advanced Supercomputing Division, Ames Research Center, Moffett Field, CA 94035-1000, 1994. NAS-94-007.
12. Brian E Smith. Performance effects of node mapping on the IBM BlueGene/L machine, June 2005.

INSEE: An Interconnection Network Simulation and Evaluation Environment

Fco. Javier Ridruejo Perez and José Miguel-Alonso

The University of the Basque Country
Department of Computer Architecture and Technology
P.O. Box 649, 20080 San Sebastian, Spain
{acbripef,miguel}@si.ehu.es

Abstract. In this paper we introduce INSEE, an environment to help in the design of interconnection networks for parallel computing systems. It contains two main modules: a system to generate traffic (TrGen) and a lightweight functional simulator (FSIN). Additionally, external tools can be integrated into the environment. Examples are SICOSYS (a sophisticated network simulator that provides accurate timing information) and SIMICS (a complete, detailed computer simulator). This environment has been used to conduct some studies of interest in the design of interconnection networks, such as the effect of head-of-line blocking in the injection queues of the network routers, the effects of the injection interface in network performance, and the characteristics of topologies with skewed wraparound links.

1 Introduction

When designing and building a parallel computer, the decision of which interconnection network will be used to link all the processing elements is of crucial importance, because it greatly affects the performance of the system. In the initial design stages, most architectural proposals (routing algorithms, deadlock-avoidance strategies, topologies, etc.) are commonly tested using fast, functional simulators that do not incorporate all the details of the hardware—while considering the most relevant ones, in such a way that a selection of promising approaches can be made. Then we can proceed to use more detailed simulators, or even develop hardware prototypes.

As important as a good model of the interconnection network is a good characterization of the workload it will have to support, in the same way that processors' design are made taking into account the programs that will run on them. The network has to be designed bearing in mind the way it will be used by parallel applications. The networks for these systems will probably be very different: **(1)** a small-sized SMP; **(2)** a larger CC-NUMA multiprocessor; **(3)** a massively parallel computer (MPP), and **(4)** a distributed system based on web services. In fact, market provides different solutions for each of these needs. We can even go further: it is not the same what we expect from the network when running master-slave applications with infrequent interchange of long messages, that what we demand when running a fine-grained scientific application where messages are short but interchanged very often. In the first case we need high throughput, while in the second latency is the main constraint.

For these reasons, a set of tools to simulate-evaluate proposals for interconnection networks require a choice of simulators as well as a choice of traffic-generation methods to feed them. This is precisely what we introduce in this paper: INSEE, Intercon-

J.C. Cunha and P.D. Medeiros (Eds.): Euro-Par 2005, LNCS 3648, pp. 1014–1023, 2005.

nection Network Simulation and Evaluation Environment. It includes a fast, functional simulator (FSIN) and a flexible mechanism to generate traffic (TrGen). INSEE is modular, and can be augmented in many ways: with new modules, or adding capabilities to existing modules.

INSEE tools have been successfully used in our research group to carry out different studies of interest in the design of interconnection networks: the effect of head-of-line blocking in the injection queues of the network routers, the effects of the injection interface in network performance, and the performance of topologies with skewed wraparound links. Our aim is to continue improving this tool, and to make it available to other researchers.

The rest of this paper is organized as follows. Section 2 describes the overall design of INSEE and its basic tools: FSIN and TrGen. Section 3 enumerates several lines of research that are being developed with these tools. Section 4 describes tools with the same purpose than INSEE and compares them with our tools. Section 5 summarizes this paper, and indicates work in progress.

2   Structure and Elements of INSEE

INSEE is organized as a collection of modules, many of which can be used as stand-alone applications. Fig. 1 represents the interactions between all the modules. FSIN (Functional Simulator of Interconnection Networks) and TrGen (Traffic Generator) are the core modules. SICOSYS [13], another network simulator, was built before coining the INSEE concept, but can be considered a part of it. The rest of the modules have been developed externally, but can (with some glue code) enhance the capabilities of INSEE, or use it as a source of traffic.

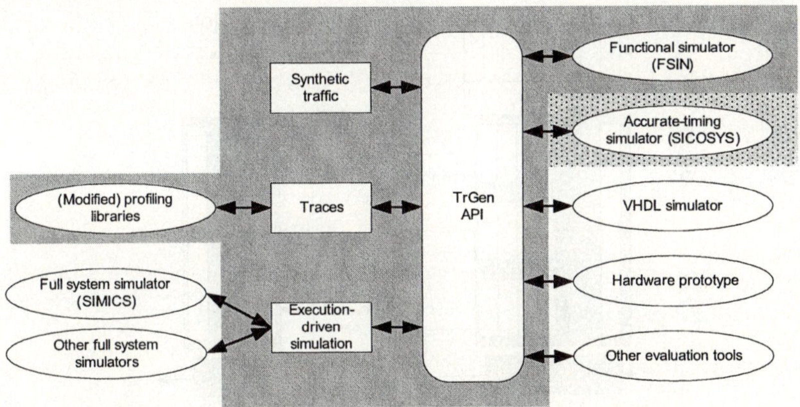

Fig. 1. Overall design of INSEE. Elements with grey background are discussed in this paper. SICOSYS, although developed independently, is considered part of this environment. The remaining parts are external modules

The current version of INSEE uses SIMICS [7] as a full system simulator, and a modified version of MPICH [5] as the source of traces. This will be explained with more detail in the following sections.

2.1 FSIN

FSIN (Functional Simulator of Interconnection Networks) is a flexible tool to help in the design of communication subsystems for parallel computers. Our research group already has a tool like this, SICOSYS [13]. In many aspects, FSIN could be considered a scaled-down version of SICOSYS, because their goals are the same: the general design of what an interconnection network is, and how routers for these networks are. However, FSIN is much faster, because it does not simulate the details of the hardware (that is the reason we call it a "functional" simulator) and consumes much less memory, thus allowing the simulation of large networks: with FSIN we are routinely simulating networks of 32K nodes using off-the-self equipment, while with SICOSYS the reasonable maximum is 1024.

The downside of FSIN is the accuracy when providing timing information. While SICOSYS includes a very detailed timing model, FSIN only has the notion of "cycle": all the relevant events in the network take exactly one cycle. Related to this, SICOSYS allows for the simulation of pipelined routers, while FSIN does not.

Fig. 2 depicts the basic elements of the simulated routers. Most of these elements can be parameterized: sizes of buffers and queues, number of dimensions (1, 2 or 3), number of virtual channels (VC) per physical link, uni- or bi-directional links, etc. This is, though, just a subset of the parameters the designer can modify. A non-exhaustive list follows: (1) packet size (measured in phits); (2) topology of the network: torus, mesh, midimew [2], twisted_torus; (3) network size; (4) VC management strategy: Dally [4], bubble [12]; (5) VC request policy (routing): several options, depending on the previous parameter; (6) VC assignment policy: round-robin, oldest, longest, random.

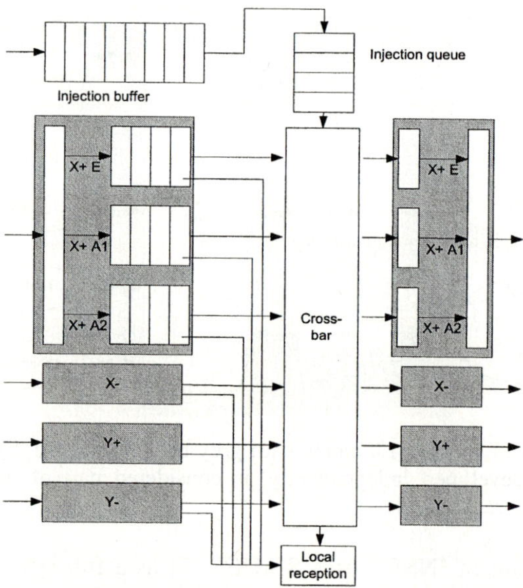

Fig. 2. Router model simulated by FSIN. This particular one is a 2-D router with 3 VCs per physical channel, and 4-phits buffers per VC

FSIN includes a built-in traffic generator that injects packet with fixed length, a Bernoulli temporal distribution, and a small choice of spatial distributions: uniform, hot-region, transpose, distribute. For more choices, it can be connected to TrGen.

This tool has been developed in ANSI C, so is fully portable to any computing environment. The verbosity of the output report can also be configured: from a short, final summary to a detailed list of all the relevant simulation events.

The most common way of using FSIN is via a batch system (it does not have a graphical interface) to run a collection of experiments in which one or several of the input parameters are changed, and then observe the impact of those changes in the performance of the simulated network. We often focus our attention on the load accepted by the network ("Load Acc.") when the offered load ("Load Prov.") is beyond saturation. This way we obtain an indicator of the maximum traffic the network can manage for a given selection of topology, VC management policies, traffic pattern, etc.

In Fig. 3 (left) we show some results obtained by FSIN, compared to those of SICOSYS. Input parameters are: 2-D torus of 4x4, 8x8 and 16x16 nodes; 3 virtual channels per physical channel (one is the bubble-managed escape channel, using oblivious DOR routing, while the other two are adaptive); packets of 16 phits; queues of 8 packets; applied load between 0.01 and 1.0; 100.000 simulation cycles. The curves show the load accepted by the network. Notice that results given by the two tools show exactly the same trends and very similar values. Fig. 3 (right) shows the resources used by FSIN compared to those used by SICOSYS, both in memory used and execution time for the longest experiments. Here is where FSIN shows its advantages: experiments consume much less resources.

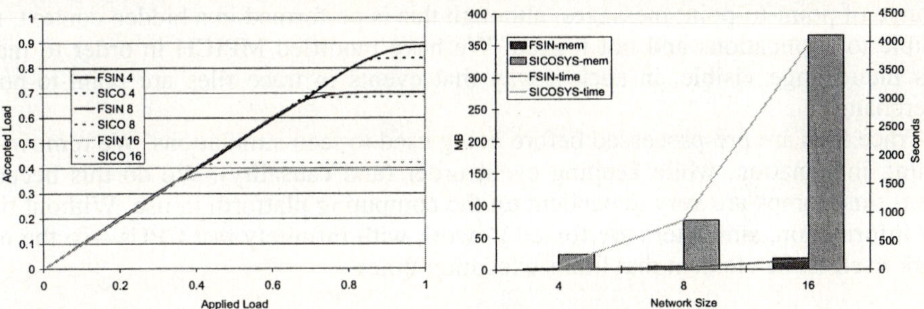

Fig. 3. Left: comparison of results provided by FSIN and SICOSYS for the same input parameters. Right: comparison of resources (memory, execution time) used by these tools

2.2 TrGen

An actual interconnection network will not be used to randomly move packets from one node to another: it will be used to deliver actual traffic generated by actual applications. The characteristics of this traffic have an enormous impact in network performance and, for this reason, a realistic evaluation of an interconnection network must be performed with actual traffic. There are many reasons, though, to work with approximations: **(1)** actual traffic may be unknown or unavailable; **(2)** agile advance in the initial stages of network development; **(3)** specific testing of particular network characteristics, etc.

TrGen is a traffic generation tool that, via a unified API, allows us to tests our designs with a large variety of traffic sources, as depicted in Fig. 1: synthetic sources, actual traffic taken from traces, and actual traffic taken from an execution-driven simulation.

Synthetic Traffic

Synthetic traffic is characterized by three distributions: temporal (that determines the packet inter-arrival times), spatial (destination of packets) and packet-size. The choice of distributions available is extensible, and they can be parameterized. Currently available options are:

- Temporal: Bernoulli, constant bursts, Markov (with several variants).
- Spatial: uniform, distribute, zipf, hot-region, constant (transpose, butterfly, perfect shuffle, inverse, etc.)
- Size: constant, uniform, polynomial.

Traces

Once we have traces taken from the actual execution of a parallel application, we can use them through the TrGen interface to feed our simulations. In order to do this, we have used the profiling capabilities of MPICH [5], a widely used MPI [8] implementation.

One of the limitations of most profiling mechanisms for MPI applications is that they log collective operations as a single event, without showing the actual packets that traversed the network. MPICH implements collective operations via the interchange of point-to-point messages, although this is performed in a hidden context, not visible to applications and not logged. We have modified MPICH in order to make this interchange visible, in such a way that events in trace files are point-to-point interchanges.

Trace files are pre-processed before being used to feed simulations. We remove all timing information, while keeping event order (and causality). We do this because event timestamps are very dependent on the computing platform in use. Without timing information, simulators are forced to work with infinitely fast CPUs—so the network itself is the element that limits execution times.

Execution-Driven Simulation

The third source of traffic available with TrGen is actual traffic generated and consumed by running applications. To close the simulation circle, we use full system simulators that host simulated computers (guests), which act as the computing nodes running parallel applications. In particular, we use SIMICS [7] because of its flexibility simulating different architectures. Currently we used Pentium 4 (host) machines to simulate generic x86 (guest) machines running Linux.

Fig. 4 depicts all the elements involved in an execution-driven simulation. The grey area (left) runs in a SIMICS environment. At the right side we have a network simulator (such as FSIN or SICOSYS) and, in the middle, TrGen acting as the interface between the two worlds. To make this setup work, it has been necessary to develop two pieces of software:

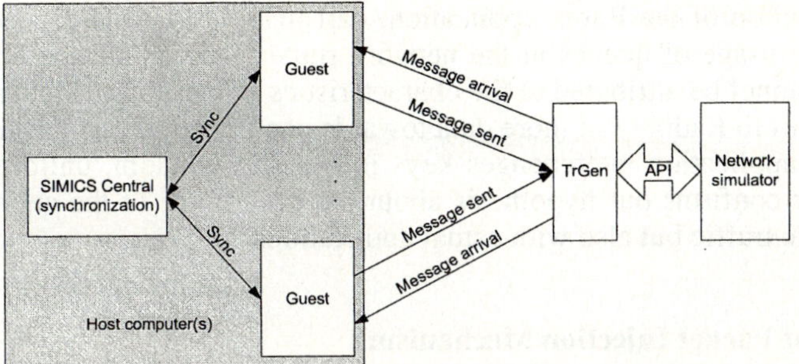

Fig. 4. Execution-driven simulation. A full system simulator (SIMICS) is connected, through TrGen, to a network simulator

- A SIMICS module that implements a PCI network adapter. A guest sees this adapter as an additional Ethernet card. All MPI communication goes through this adapter. SIMICS is instructed to communicate all activity in this adapter to TrGen.
- A Linux kernel driver to allow the guest machines to use the (simulated) PCI network adapter to interchange messages.

SIMICS Central, a part of the SIMICS environment, also plays an important role, keeping all guest machines synchronized.

3 Research Performed with INSEE

We plan to make INSEE tools available to the research community under a liberal licensing (such as GPL). However, some aspects of the code are not yet ready for distribution: user interface is not intuitive, and portions of the code are not yet fully tested. None of those aspects affect the functionality claimed in this paper.

Interested readers can obtain a copy of the software by requesting it directly to the authors. We don't have the (human) resources to provide technical support to potential users; however, comments on possible improvements, contributed code, and bug reports will be welcomed.

The next subsections introduce work already performed and current, in the context of our research group, where INSEE tools play a crucial role.

3.1 Load Unbalance in Queue Usage

In [10] we report a study of the effect that HOLB (Head-of-line blocking) in the packet injection queue has on the performance of bidirectional k-ary n-cubes, for values of k over a certain threshold (around 20). The HOLB causes an unbalanced use of the channels corresponding to the two directions of bidirectional links, which is responsible for a drop in the network throughput and a rise in the network delay. Simulation results obtained with FSIN (using its built-in uniform traffic generator) show that this anomaly only appears in those rings where most injections are performed (normally, those in the X axis), and that the elimination of the HOLB in the injection queue enables the network to sustain peak throughput after saturation.

This unbalance is also present under actual workloads. Using RSIM (a simulator of multiprocessor systems [11]) integrated with SICOSYS we performed an execution-driven simulation of the Radix application, part of the SPLASH-2 benchmark suite. Checking the usage of queues in the network routers, the unbalance was clearly noticeable. It cannot be attributed to the characteristics of the Radix application (it is not true that nodes in Radix send more data towards one direction that towards the other), because the application interchanges keys in a highly random, uniform way. Thus, this behavior confirms our hypothesis about the occurrence of the anomaly not only with synthetic traffic but also with actual applications.

3.2 Study of Packet Injection Mechanisms

In [6] we analyze the impact that the injection interface has on maximum sustained throughput in an adaptive cut-through torus network. The work described in the previous section pointed out that HOLB at the network interface, due to the use of single FIFO injection queues, may prevent a network from sustaining its peak throughput at heavy loads. Meanwhile, we observed that most recent commercial parallel systems use multiple injection queues [1], but little is known about the rationale behind these design decisions, or their implementation details.

Using FSIN-TrGen we modeled and thoroughly analyzed the effect on performance of the following factors: the number of injection queues (from 1 to 4), the allocation of packets to queues (testing different selection policies, with or without pre-routing at the interface) and the mapping of queues to the available number of injection channels (virtual injection channels vs. physical injection channels). Network evaluations for medium to large size 2D tori showed that designs with multiple FIFO injection queues do not improve performance under uniform traffic, when compared with the simple, single-FIFO interface. On the contrary, for some injection policies, throughput loss increases for loads beyond the saturation point. At the heart of this behavior was network congestion: more injection ports results in more pressure from the injection interface to acquire the scarce network resources of an already clogged system.

We concluded that new, restrictive injection policies are required that prevent processing nodes from overflowing routers with new packets for loads beyond the network's saturation point. Interestingly, for small networks, a single injection FIFO queue, with the HOLB it entails, may actually be a good design choice as it indirectly provides the much-needed injection control. For networks with thousands of nodes, as those being implemented in current massively parallel processors, this basic form of congestion control is not enough. Regardless of the number of injectors, an injection-throttling mechanism is essential to reduce throughput losses and maintain, or even increase, maximum sustained throughput.

3.3 Study of Torus-Like Topologies

The interconnection network literature includes many studies devoted to select the best topology for a given parallel computer, application, or combination of those. Most current systems use indirect networks (fat-trees, omega) or direct networks of the k-ary n-cube family. Examples of the later are the 3D-cube for above-mentioned BG/L, or the 3D-mesh planned for Cray's Red Storm.

Still, the work in topologies is not complete: it is still possible to obtain performance gains using the right choices of wiring. Our research group has been for a long time studying the characteristics of networks such as midimews and twisted tori (see Fig. 5), which are similar to tori, but with skewed wrap-around links [2,9].

To demonstrate the properties of this family of networks we are using mathematical analysis as well as simulation—using INSEE.

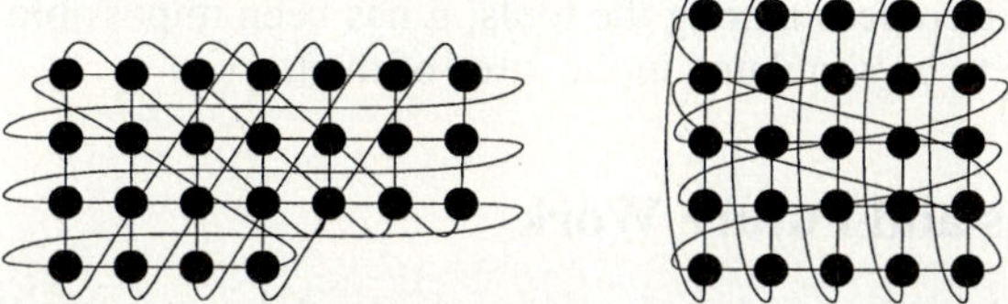

Fig. 5. Left: midimew network. Right: twisted torus with skews 3 (X-axis) and 0 (Y-axis)

4 Related Work

There are several tools available from research groups with the same purpose of INSEE. Two of them FlexSim [14] and the Chaos Router Simulator (ChaosSim) [3] are very close to FSIN.

FlexSim is a powerful tool to simulate wormhole networks. Some of its most salient advantages over FSIN are: **(1)** supports full or half-duplex links; **(2)** models faults in nodes / links; **(3)** allows the specification of several delays: routing delay, cut-through delay, switch-to-switch delay, hand-shaking delay and link arbitration delay for half-duplex channels; **(4)** includes a large set of routing and selection functions. However, it has some downsides. In particular, it neither supports virtual cut-through networks, nor bubble routing. It cannot take traffic from full system simulators, and output information is very limited.

ChaosSim is in many aspects close to the initial versions of FSIN. Its development stopped around 1993, so it does not incorporate most recent advances in interconnection network design. Positive aspects of ChaosSim are its ability of supporting full and half-duplex links, and to simulate wormhole as well as VCT networks. It is also possible to animate simulations using a graphical interface. On the downside, traffic sources are limited to the usual set (uniform, hot-spot and some permutations), and all characteristics of the simulated router have to be included in source files, so a change in a design parameter requires recompilation.

We could have decided to take one these simulators as the foundation for FSIN, adapting it to our needs. However, that was not an easy task: FlexSim is too circumscribed to wormhole routing, so it is not trivial to adjust it to work with virtual cut-through networks (the ones we are working with). Something similar happens with ChaosSim: it was developed for a particular design of router, so we would need to change too many things to simulate our routers.

In addition to this, FSIN has evolved very rapidly from its origin to its present state; these modifications have been relatively easy because we have a thorough knowledge of its internals. It would be necessary to fully understand even the most minor detail of ChaosSim or FlexSim to adapt them to our needs—and that is not an easy task to do with external code. The design and source code of FlexSim is not documented, so it is very hard to understand; in contrast, ChaosSim is better organized and documented.

As we stated before, it is our intention to release the source code of FSIN (and all the INSEE modules) under a liberal license, so we are taking special care providing a well organized and documented product.

We would like to provide a comparison of FSIN, FlexSim and ChaosSim in terms of metrics such as execution time and memory usage for a given configuration. However, due to the differences among the tools, it has been impossible to find a network definition suitable to be simulated in the three of them.

5   Conclusions and Future Work

Although INSEE is still a work in progress, the modules that constitute this environment have already proven their usefulness when researching interconnection network topics. We are pleased with the performance of the standalone FSIN and integration tests with TrGen are also satisfactory.

Clearly, future work includes the improvement of each tool, as well as the integration capabilities with external tools. Some ideas already stated in this paper are summarized here. We want to improve TrGen in terms of additional sources of synthetic traffic, more accurate time modeling in trace-driven simulation, and full integration with SICOSYS. Immediate plans for FSIN include the ability of modeling pipelined routers. All these improvements will have a common goal: maximization of our capabilities to model, simulate and evaluate interconnection networks.

Acknowledgements

This work has been done with the support of the Ministerio de Educación y Ciencia, Spain, under grant TIN2004-07440-C02-02, and also by the Diputación Foral de Gipuzkoa under grant OF-846/2004.

SICOSYS has been designed by the ATC Group at the University of Cantabria (Spain). This group provided the data for Fig. 3, as well as invaluable help for the design and implementation of the tools described in this paper.

Work described in Section 4 has been carried out in collaboration with the ATC Group and with Dr. Cruz Izu (U. of Adelaide).

References

1. N.R. Adiga et al. (2002). An overview of the BlueGene/L Supercomputer. Supercomputing 2002 Technical Papers, Available at http://sc-2002.org/paperpdfs/pap.pap207.pdf.
2. R. Beivide E. Herrada, J.L. Balcazar, A. Arruabarrena (1991). Optimal distance networks of low degree for parallel computers. IEEE Transactions on Computers. Vol. 40, No. 10.
3. The Chaotic Routing Project at the U. of Washington. Chaos Router Simulator. Available at http://www.cs.washington.edu/research/projects/lis/chaos/www/chaos.html
4. W. J. Dally and C. L. Seitz (1987). Deadlock-free message routing in multiprocessor interconnection networks, IEEE Transactions on Computers, vol. 36, no.5.
5. W. Gropp, E. Lusk, N. Doss, A. Skjellum (1996). A high-performance, portable implementation of the MPI message passing interface standard, in Parallel Computing, vol. 22, no. 6.

6. C. Izu, J. Miguel, J.A. Gregorio, R. Beivide (2005). Packet Injection Mechanisms and their Impact on Network Throughput. Technical report EHU-KAT-IK-01-05. Department of Computer Architecture and Technology, The University of the Basque Country. Available at http://www.sc.ehu.es/acwmialj/papers/ehu_kat_ik_01_05.pdf
7. Peter S. Magnusson, Magnus Christensson, Jesper Eskilson, Daniel Forsgren, Gustav Hållberg, Johan Högberg, Fredrik Larsson, Andreas Moestedt, Bengt Werner, (2002). Simics: A Full System Simulation Platform, IEEE Computer, February.
8. Message Passing Interface Forum. MPI: A Message-Passing Interface Standard. Available at http://www-unix.mcs.anl.gov/mpi/standard.html.
9. J. Miguel, C. Izu, A. Arruabarrena, J. García-Abajo and R. Beivide (1991). Toroidal networks for multicomputer systems. Proc. of the ISMM International Workshop on Parallel Computing. Trani (Italy).
10. J. Miguel-Alonso, J.A. Gregorio, V. Puente, F. Vallejo and R. Beivide (2004) Load Unbalance in k-ary n-cube Networks. Lecture Notes in Computer Science 3149.
11. V.S. Pai, P. Ranganathan, and S.V.Adve (1997). RSIM: An Execution-Driven Simulator for ILP-Based Shared-Memory Multiprocessors and Uniprocessors. IEEE TCCA New., Oct.
12. V. Puente, C. Izu, R. Beivide, J.A. Gregorio, F. Vallejo and J.M. Prellezo (2001). The Adaptative Bubble Router. Journal of Parallel and Distributed Computing. Vol 61 - n. 9.
13. V. Puente, J.A. Gregorio, R.Beivide (2002). SICOSYS: An Integrated Framework for studying Interconnection Network in Multiprocessor Systems, Proceedings of the IEEE 10th Euromicro Workshop on Parallel and Distributed Processing. Gran Canaria, Spain.
14. SMART group at the U. of Southern California. FlexSim 1.2. Available at http://ceng.usc.edu/smart/FlexSim/flexsim.html

Cost / Performance Trade-Offs and Fairness Evaluation of Queue Mapping Policies*

Teresa Nachiondo[1], José Flich[1], José Duato[1], and Mitchell Gusat[2]

[1] Dept. of Computer Engineering, Universidad Politécnica de Valencia, 46071–Valencia, Spain
{tnachion,jflich,jduato}@gap.upv.es
[2] IBM, Research, Zurich Research Lab. GmbH, Saeumerstr. 4,
CH-8803, Rueschlikon, Switzerland
mig@zurich.ibm.com

Abstract. Whereas the established interconnection networks (ICTN) achieve low latency by operating in the linear region, i.e. oversizing the fabric, the recent strict cost and power constrains demand more efficient utilization of future networks. Increasing the utilization of lossless ICTNs may, however, lead to saturation and performance degradation owing to HOL-blocking. The current solution to HOL-blocking consists of using Virtual Output Queueing (VOQ), whose quadratical scalability is expensive in large networks. To improve VOQ's scalability we have proposed the Destination-Based Buffer Management (DBBM), a scheme that compares well with VOQ. Whereas previously we have analyzed DBBM's basic operation and performance, in this paper we have set two different goals. First we focus on how the different DBBM mappings can impact the cost/performance of multistage ICTNs. Next, because DBBM can introduce unfairness, this constitutes the second theme of our paper. The new results show that DBBM with modulo-4/8 mapping performs very well for only a fraction of the VOQ cost. Also in terms of fairness DBBM shows promise, because it (i) keeps the unfairness degree independent of both topology and routing, while (ii) minimizing the number of flows affected by unfairness.

1 Introduction

Proprietary lossless ICTNs are frequently used to build large supercomputers such as BlueGene/L [7] and the Earth Simulator [6]. Alternatively, commercial ICTNs like InfiniBand [8], Myrinet [11], and Quadrics [14] are used to build large clusters such as the Myrinet-based Mare Nostrum IBM cluster [1] recently ranked 4th in the Top500 supercomputers [16]. However, these interconnect technologies are not following the cost/performance curve of other components, and therefore they have remained expensive relative to processor, memory and storage.

Hence the need to reduce the cost of the interconnect. One can drastically reduce the ICTN costs by decreasing the number of components – adapters, links and switches – and proportionally increasing the utilization of the remaining parts. The increased load, however, may require to operate the network as close to saturation as possible, which raises the probability of creating saturation trees and *congestion collapse* [13].

---

* This work was supported by CICYT under Grant TIC2003-08154-C06.

J.C. Cunha and P.D. Medeiros (Eds.): Euro-Par 2005, LNCS 3648, pp. 1024–1034, 2005.

The problem is derived from the initially blocked packets (addressed to the congested "hot" destination) that will also block packets addressed to other cold destinations. Known as Head-of-line (HOL) blocking, this is a key issue in packet switching. A blocked packet at the head of a queue prevents packets behind it from reaching idle outputs, thus leading to potentially severe throughput degradation.

2 Motivation

In addition to cost, more recent power constrains [15] also call for a higher utilization of the ICTN components (i.e. switches and links). With bursty traffic, however, increasing the link utilization can lead to saturation and performance collapse due to interference between flows and packet HOL-blocking. Factually any HOL blocking will reduce the ICTN throughput-delay performance.

First order HOL blocking results from using the FIFO queuing discipline into a switch element. Ethernet is an example of a switching standard that is widely deployed with FIFO queuing. Because Ethernet is not typically required to be lossless, as soon as HOL would occur, packets can be dropped. However, dropping packets is not an option in our study, i.e. lossless ICTNs

VOQ at switch level –referred in this paper to as VOQ_SW– solves HOL blocking at a reasonable cost [10] for single stage switches. Normally VOQ_SW strictly removes the first –but no higher– order HOL blocking [9]. The best performance would be reached by applying a global VOQ: at each queuing point there are as many queues as there are endnode destination ports. This resolves the higher order HOL blocking. Whereas attractive, the latter solution –called VOQ_Net in this paper– is not practically implementable for a large number of ports.

A new queuing discipline was described in [4, 12]. DBBM uses approximately the same number of queues as VOQ_SW or Virtual Channels (VC) [2], but has no direct association of these queues to the next stage output ports as VOQ_SW does or to bandwidth in case of VC. In this paper we investigate how the additional degree of freedom of DBBM –the mapping of queues to packet flows sorted per destination– can be exploited to address the higher order HOL blocking [9]. We do this by studying various mapping options across a variety of multistage topologies, also taking fairness into account. For reference we compare our results not only to the FIFO and the ideal global VOQ queuing mechanisms, but also to a few other relevant schemes.

The rest of the paper is organized as follows. In Sect. 3 we present the two typical queuing options in use today. In Sect. 4 DBBM and its main features are described. In Sect. 5 different VOQ and DBBM schemes are evaluated in terms of performance, scalability and fairness. In Sect. 6 conclusions are extracted and future work is outlined.

3 Traditional Queuing Options in Lossless ICTNs

As mentioned above, every switch of a modern ICTN will have a limited set of queues associated with every input and/or output port; normally the set cardinality is lower than the number of network endpoints. In some cases switches will have only one queue per input port, e.g. Myrinet. Otherwise, whenever using multiple queues per switch port, a queuing architecture and a suitable mapping policy must be selected. Here we consider three alternatives.

The first one is to use VOQ at the switch/link level, i.e. hop-by-hop. Every input port will have as many queues as output ports, and an incoming packet will be mapped to the queue associated with the requested output port. Thus HOL-blocking at the switch and link level is eliminated. HOL -blocking can still occur, however, between flows sharing a subset of consecutive links along their paths. With switch-level VOQ and no special FC means [5], packet switching ICTNs are exposed to a form of flow interference known as high-order HOL -blocking [9].

As a second option we can use virtual channels (VC) -i.e. different queues with dedicated FC– as introduced in [2]. These channels can be load-balanced by allocating each outgoing packet to the (currently) emptiest VC. However, mis-order among the packets belonging to the same flow can be introduced, thus requiring a resequencing solution.

4   Destination-Based Buffer Management (DBBM)

As a third option, in [4] we have introduced DBBM as a scheme to reduce HOL-blocking in ICTNs. Temporal and spatial locality in the packet destination distribution suggest that a small number of queues could be sufficient for storing all the incoming packets at each switch - while still classifying and demultiplexing them according to their destination. This allows DBBM, when used in conjunction with a suitable *mapping* strategy, to practically eliminate most –if not all– of the HOL-blocking. E.g., a simple mapping method will allow multiple flows to share –cyclically or in linear blocks– a single DBBM queue. Although some HOL-blocking will be introduced among the flows having to share the same queue, this approach can radically decrease the set of queues –and the cost– to be built and maintained in hardware.

Albeit more complex than the direct 1:1 mapping of flows to queues inherent to the VOQ disciplines, a simple DBBM mapping method could be based on the destination encoding. E.g. some bits of the destination address field in the packet header will select the DBBM queue where the packet is to be stored. As a subset of the lowest order bits are used, here this method is referred to as 'modulo' mapping; e.g, for a network with 256 destinations and 16-queue DBBM, the four least significant bits of the destination ID (8-bit field) will point to the queue to map the flow into.

With DBBM, different mappings trade-off between performance and implementation cost - expressed in the number of hardware queues. In [12] we have performed an evaluation of the 'modulo' mapping in some multistage networks. We have shown that, to practically reach the maximum throughput, the number of DBBM queues required per switch was 8 times lower than for a full VOQ. However, DBBM's mapping policy now primarily determines the system performance.

Since the number of DBBM queues is lower than the number of ICTN endpoints, irrespective of the mapping policy in use any DBBM can –and eventually will– map flows addressed to different destinations to the same queue (i.e. intrinsic HOL-blocking).

DBBM's principle of operation is depicted in Fig. 1.(a); more details in [4, 12]. Its main functions are:

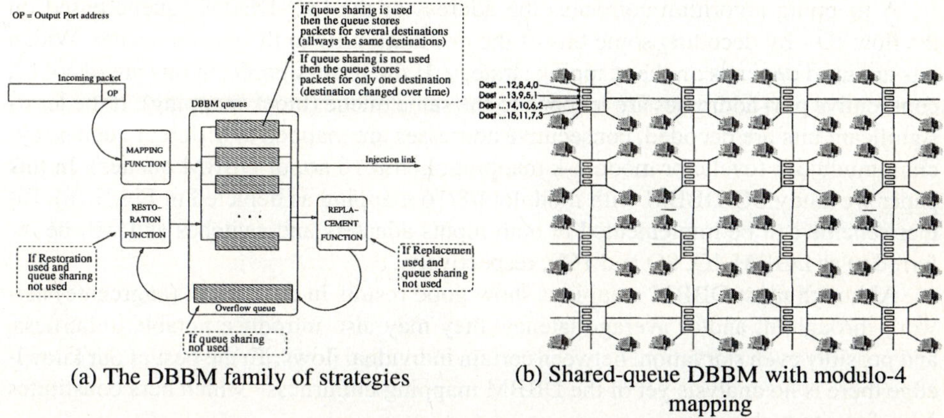

(a) The DBBM family of strategies   (b) Shared-queue DBBM with modulo-4 mapping

Fig. 1. DBBM description.

- **Queue sharing:** indicates whether packets with different destinations can be concurrently stored in the same queue or not. Both cases require a careful mapping of packet destinations to queues. In the first case to minimize the HOL-blocking. In the second one, when regular queues are no longer available, an auxiliary "overflow" queue will store either the newly incoming packets or the packets relocated from a regular queue. The overflow queue operation, however, introduces a few practical issues of implementation and re-ordering.
- **Mapping method:** determins the queue where an incoming packet will be stored. E.g., a mapping method may indicate a set of queues from which a free one will be selected (set-associative).
- **Replacement:** a binary value that indicates whether already stored packets can be relocated from a regular queue when an incoming packet requests that queue. Used only if queue sharing is not enabled.
- **Replacement function:** selects one of the queues indicated by the mapping method to (i) relocate a previously stored packet in order to (ii) store the incoming packet.
- **Restoration:** a binary value that indicates whether packets in the overflow buffers are allowed to move back to a regular queue when this has room. Used only if queue sharing is not enabled.
- **Restoration function:** selects the packets (with the same destination) to be relocated back into their initial regular queue, if restoration is enabled.

The simplest DBBM strategy allows queue sharing. With shared-queue DBBM (SQ-DBBM) each queue can only store packets for a subset of the destination ports. I.e. as if (i) the physical output ports of the network were virtually grouped into a smaller set of logical output ports and (ii) each SQ-DBBM queue stores packets destined *only* to a particular logical output port. Thus SQ-DBBM implements a 'set-VOQ' architecture organized on logical, instead of the physical, fabric ports. While this strategy does not directly avoid HOL-blocking, it may reduce it down to negligible values when a suitable mapping method is used. In-order delivery is simplified as all the packets of a flow will be mapped to the same queue, where they are stored in arrival order.

A mapping algorithm computes the address of the [SQ-DBBM] queue based on the flow ID - by decoding some bits of the destination ID in the packet header. Which bits are used depends on the mapping strategy. If the most-significant bits are decoded, consecutive port addresses are mapped to the same queue (block mapping); if the least-significant bits are decoded, consecutive addresses are mapped to different queues, cycling modulo-k (cyclic or modulo-k mapping; k=4,8,16 no. of DBBM queues). In this paper we study SQ-DBBM with modulo-4/8/16 mapping as depicted in Fig. 1.(b). Either scheme can be implemented in both inputs adapters and switches and will be referred to as DBBM 4Q, 8Q and 16Q, respectively.

Although most DBBM mappings show good results in the overall (aggregate) network throughput and/or average latency, they may also introduce notable unfairness, and possibly even starvation, between certain individual flows. To the best of our knowledge there is no analysis yet of the DBBM mapping unfairness - which here constitutes one of our two objectives.

Also in this paper we will analyze how different mappings impact the ICTN cost. We will evaluate the performance of each mapping method while varying number of endpoints attached per switch. Our goal is to maximize the system performance when minimizing the ICTN hardware resources available for a constant number of endpoints connected in 2D and 3D mesh topologies.

5   Performance Evaluation

To achieve our above stated goals we will compare the performance of 6 queuing and mapping schemes. Ordered per increasing cost, they are: (a) single FIFO (1Q), (b) DBBM 4Q, (c) DBBM 8Q, (d) load-balanced (EMPTIEST 8Q), (e) VOQ_SW and (f) VOQ_Net.

1Q (a) sets the lower bound of performance, that of a single queue with FIFO service. As the simplest queuing structure, even in single-stage fabrics with uniform traffic the 1Q scheme is theoretically limited at 58% throughput. DBBM is represented by (b,c), id est SQ-DBBM with modulo-4/8 mapping, respectively. This scheme was briefly described above. EMPTIEST 8Q (d) is a load-balancing scheduling strategy with 8 queues per input port. I.e., packets will be always mapped to the queue with the lowest current occupancy. Whereas such load-balancing is based on the queue status of the next/downstream switch, this mapping is destination/port-oblivious, and thus it represents the opposite of VOQ (DST-based, load-independent). VOQ_SW (e) is the typical VOQ scheme implemented in some modern ICTNs. It applies a link-level VOQ at every hop; i.e. switch will have at every input port as many queues as output ports. VOQ_Net (f) sets the upper bound of performance, that of an end-to-end VOQ scheme globally applied across the entire ICTN. VOQ_Net requires in every switch and IA as many queues as destinations in the network. Its main use here is as a reference for other, more practical, schemes.

5.1   Simulation Model

We have developed a detailed event-driven simulator that allows us to model the network at a level adequate for our study. The simulator models an ICTN with switches, nodes, and links. Buffers up to 4KB are modeled for both the input and the output ports of every switch. The buffer capacity is statically divided by the number of queues defined by each of the six schemes above, resulting in a fixed size per queue.

At every switch packets are forwarded from any input queue to any output queue through a multiplexed crossbar. We have considered a crossbar bandwidth of 1.5 GB/s with a speedup of 1.5. The crossbar is controlled by a scheduler that receives requests from the packets at the head of any input queue. A requesting packet is forwarded only if the corresponding crossbar input and crossbar output are free. At each output port a weighted round-robin arbiter selects the output queue to be served.

For links we assume serial full-duplex pipelined transmissions with 1 GB/s effective bandwidth. The link-level flow control (LL-FC) protocol is credit-based; a packet can be transmitted downstream only if a credit is available. Whenever a packet frees an input buffer location a new credit is sent to the output port upstream. A similar flow control scheme has been implemented for the internal switch traversal (input-output packet forwarding). The maximum number of credits per output (input) port depends on the buffer size at the next input (output) port and the total number of queues. The LL-FC packets share the link bandwidth with data traffic.

The endpoints are connected to switches using Input Adapters (IAs). Every IA is modeled by (i) a *fixed* number N of message *admittance* queues organized in VOQ; (ii) and a *variable* number of *injection* queues organized similarly to the output ports of a switch. When a new message is generated, first it is stored completely in the admittance queue assigned to its destination; then it is segmented into 64B packets before being transferred to an injection queue. The transfer from admittance queues to injection queues are controlled by a round-robin arbiter. The transmission of packets from injection queues into the network is controlled by a weighted round-robin arbiter.

5.2   Topologies and Traffic Patterns

In [12] performance of DBBM with modulo mapping was evaluated in different multistage ICTNs. Now, 2D/3D meshes and a bidirectional multistage network (BMIN) will be evaluated for performance and fairness. In all the cases deterministic routing is used; for the 2D and 3D meshes we use the Dimension Order Routing (DOR). The BMIN is built from 8-port switches interconnected in a perfect shuffle topology.

We have defined 8 different scenarios based on synthetic traffic patterns as (partially) shown in Table 1. All the cases cause a congestion tree by oversubscribing the hotspotted endpoint; for background traffic 70% of the sources inject at 20% of link bandwidth to randomly selected destinations, while the remaining 30% of sources inject full rate to a randomly selected hotspot destination. As the background traffic shares links and queues with the flows belonging to the congestion tree, substantial HOL-blocking is introduced in multiple switches.

5.3   Evaluation Results

First we analyze the overall performance achieved by each of the 6 schemes. Then the network (cost) is reduced by removing some switches and links. Finally we focus on fairness by analyzing the goodput patterns, i.e. the traffic arrived at each destination.

Table 1. Topologies and synthetic traffic patterns evaluated.

Case	Network evaluated	# Total endpoints	# Endpoints attached per switch	Traffic Injected			
				to random destinations		to hotspot	
				% of injecting Sources	Injection rate (% of link BW)	% of injecting Sources	Injection rate (% of link BW)
#1	8×8	64	1	70%	20%	30%	100%
#2	$8 \times 8 \times 4$	256	1	70%	20%	30%	100%
#3	BMIN (64×64)	64	4	70%	60%	30%	100%
#4	4×4	64	4	70%	20%	30%	100%
#5	16×16	256	1	70%	20%	30%	100%
#6	8×8	256	4	70%	20%	30%	100%
#7	4×4	256	16	70%	5%	30%	100%
#8	$4 \times 4 \times 4$	256	4	70%	20%	30%	100%

Overall Performance. Hotspot traffic (cases #1 and #2, Fig.s 2.(a) and 2.(b)) in 2D and 3D meshes show that VOQ_Net achieves the maximum throughput whereas 1Q and EMPTIEST perform the worst. Reason for EMPTIEST's poor performance: eventually most of its queues will be backlogged with packets belonging to the congestion tree. Similar results have been observed in all the studied cases that –for space reasons– can not be shown here; henceforth results for 1Q and EMPTIEST will be only plotted for case #1.

VOQ_SW achieves 77% of the VOQ_Net throughput, whereas DBBM-4Q performs better. Overall DBBM matches VOQ_Net; e.g., for #1, DBBM-4/8Q achieves 90/95% of the VOQ_Net performance. Similar for #2, despite the increase in the number of endpoints. DBBM-4/8Q achieves 86/91% of the VOQ_Net throughput. However, with 16 queues (a reduction factor of 16 of VOQ_Net queues) DBBM achieves 97% of the VOQ_Net throughput.

Whereas in Fig. 2.(c) (case #3) VOQ_SW achieves 71% of the VOQ_Net, DBBM roughly matches the VOQ_Net performance - 92% and 96% with 4, resp. 8 queues. Confirming our results from previous work, regardless of the topology, DBBM can match VOQ_Net in performance - while using a reduced set of queues.

Performance on Reduced Networks. One way to reduce the ICTN cost is by sharing: connect more endpoints to each switch, thus also increasing the HOL-blocking probability. This is confirmed in cases #5 (Fig. 2.(d)), #6 (Fig. 2.(e)), and #7 (not shown); in each, 256 endpoints attached to 2D meshes with different sizes. In all them VOQ_SW shows worse performance than DBBM. DBBM-8Q reaches 91% of VOQ_Net for cases #5 and #7, and 83% for case #6.

For a $4 \times 4 \times 4$ mesh with 256 endpoints (traffic case #8, Fig. 2.(f)) DBBM-16Q achieves 97% of the VOQ_Net. VOQ_SW reaches 61% of the VOQ_Net, whereas with a larger network ($8 \times 8 \times 4$) and with the same number of endpoints (256) it achieves 77% of the VOQ_Net throughput. On the other hand, DBBM has constant performance, independent of the network size. Again, also in reduced networks DBBM achieves the VOQ_Net performance.

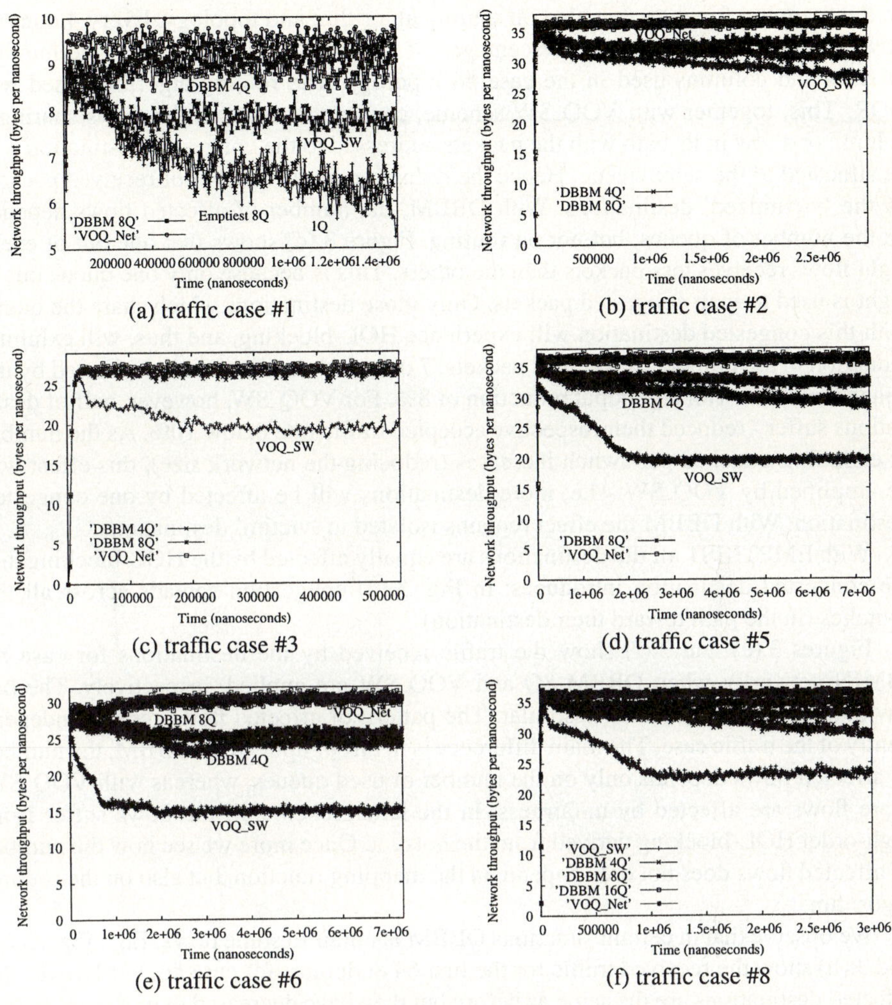

Fig. 2. Accepted traffic vs. simulated time.

Fairness. Thus far DBBM and VOQ_SW exhibit good performance for hotspot traffic, DBBM being more efficient. Also they perform well when the network size is reduced. However, as they map different flows to the same queue, they introduce a degree of unfairness. We analyze this effect by plotting for each scheme the accepted traffic per endpoint.

Figures 3.(a), 3.(b), 3.(c), and 3.(d) show the traffic received by each endpoint for traffic case #4 (4 × 4 mesh with 4 nodes/switch), when VOQ_Net, VOQ_SW, DBBM-8Q and EMPTIEST scheme is used, respectively. The *highest bar represents the hotspot* (endpoint 30), which reaches 90% of received traffic (axes are truncated at 50%). With VOQ_Net every destination, except the hotspot, receives roughly the same goodput. With VOQ_SW, all the flows that share two consecutive links with the congested flow suffer from HOL-blocking. Thus the number of affected flows does not only depend on the mapping function, but also on routing algorithm and topology. Every 4 consecutive endpoints exhibit similar percentages of accepted traffic, since 4 is the number of rows and columns used in the case #4 topology. The routing algorithm used was DOR. This, together with VOQ_SW scheme, causes that most of the flows sharing a column or a row in its path with the packets addressed to the congested destination will be allocated to the same queue. Hence the reduction in the number of received packets by the 'victimized' destinations. With DBBM, the number of affected flows depends on the number of queues, but not on routing. Figure 3.(c) shows that one out of every eight flows receives less packets than the others. This is because only one queue out of eight is used to map congested packets. Only those destinations which share the queue with this congested destination will experience HOL-blocking, and thus, will exhibit a reduction in the number of received packets. 7 out of 64 destinations are affected by the congestion tree, with a goodput reduction of 8%. For VOQ_SW, however, half of destinations suffer - reduced their respective accepted traffic rate below 10%. As the number of endpoints attached per switch increases (reducing the network size), this effect will be amplified by VOQ_SW - i.e. more destinations will be affected by one congested destination. With DBBM the effect remains isolated to 'victim' destinations.

With EMPTIEST all the destinations are equally affected by the HOL-blocking that the congested destination introduces; in Fig. 3.(d) congestion spreads across all the switches (in the path toward their destination).

Figures 3.(e) and 3.(f) show the traffic received by the destinations for case #3 (BMIN network), when DBBM-8Q and VOQ_SW are applied, respectively. The behaviour for cases #3 and #4 is similar. The pattern of affected flows repeats independently of the traffic case. The main difference is in throughput. With DBBM, the number of affected flows depends only on the number of used queues, whereas with VOQ_SW more flows are affected by unfairness. In the latter scheme all the flows suffer from high-order HOL-blocking derived from the hotspot. Once more we see how the number of affected flows does not only depend on the mapping function, but also on the routing algorithm.

We observe that in certain situations DBBM is unfair to some flows. E.g., Fig.s 3.(g) and 3.(h) show the received traffic for the first 64 endpoints for case #6. For DBBM the affected destinations are the same as before but they have decreased their reception rate below 5%. With the same traffic, VOQ_SW behaves worse: half of the endpoints have a traffic percentage lower than 5%.

To conclude, excepting VOQ_Net, all the other schemes introduce some degree of unfairness under high load and congestion. However, DBBM is the only one that keeps the unfairness degree independent of the topology and routing used.

6   Conclusions

In order to reduce HOL-blocking a number of queuing schemes and mapping methods have been proposed. Theoretically only a full end to end VOQ, or at least a subset of non-interfering flows [3] is able to eliminate completely HOL-blocking. However, this solution is not scalable to large ICTNs. In order to overcome these problems, other mapping strategies have been proposed and evaluated. In these evaluations we have studied the trade-offs between performance and the number of required queues. We

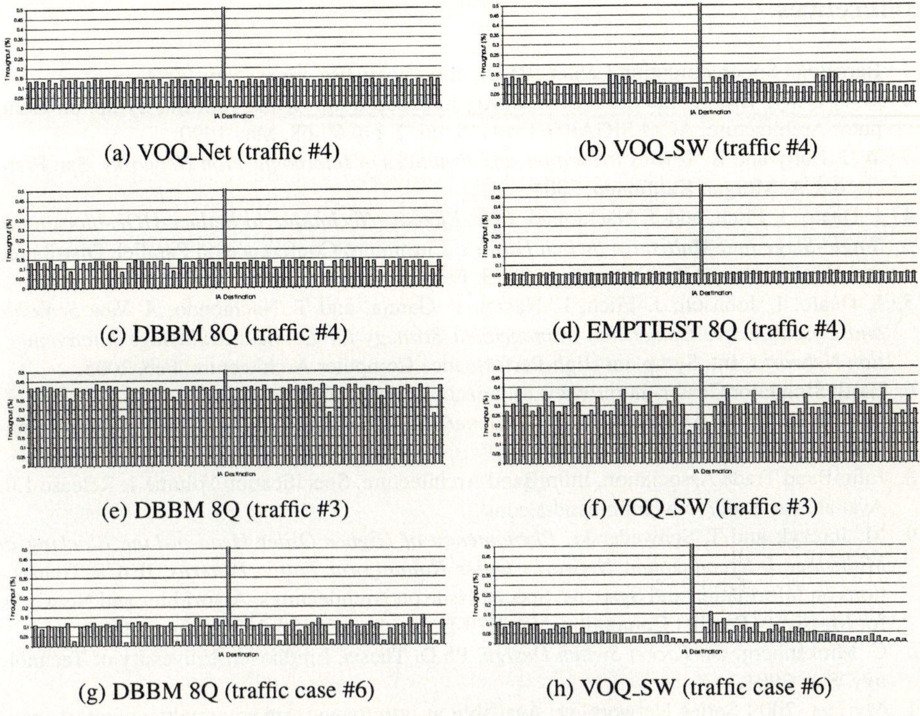

(a) VOQ_Net (traffic #4)   (b) VOQ_SW (traffic #4)

(c) DBBM 8Q (traffic #4)   (d) EMPTIEST 8Q (traffic #4)

(e) DBBM 8Q (traffic #3)   (f) VOQ_SW (traffic #3)

(g) DBBM 8Q (traffic case #6)   (h) VOQ_SW (traffic case #6)

Fig. 3. Accepted traffic per destination.

have analyzed the unfairness that a mapping strategy can introduce in lossless ICTNs. Also, we have analyzed how the different mapping methods can help in reducing the cost of the ICTNs.

Simulation results have confirmed that both link/switch-level VOQ (VOQ_SW) and destination-oblivious load-balancing (EMPTIEST) schemes suffer from high-order HOL-blocking. On the other hand, for a moderate increase in complexity DBBM shows clear improvements, linearly proportional to the number of operating queues. Independent of the network size, DBBM with 8 queues has achieved roughly the same throughput as the 'ideal' VOQ, while using only a small fraction of the queues.

Excepting VOQ_Net, all the mapping strategies introduce some degree of unfairness. However, DBBM kept the unfairness independent of the topology and the routing in use. For DBBM, the number of affected flows by the congestion tree depends on the number of used queues, whereas for VOQ_SW the affected flows not only depend on the mapping function but also on the routing algorithm. As future work we are currently exploring other DBBM schemes, such as combinations of block and cyclical mapping.

Acknowledgments

We are indebted to Ton Engbersen and Ronald Luijten for significant contributions and careful review.

References

1. Barcelona Supercomputing Center (BSC), http://www.bsc.org.es, Nov. 2004.
2. W. J. Dally, *Virtual-channel Flow Control,* in Proceedings of the 17th Int. Symp. on Computer Architecture, ACM SIGARCH vol. 18, no. 2, pp. 60-68, May 1990.
3. W. J. Dally and B. Towles *Principles and Practices of Interconnection Networks,* San Francisco, CA, Morgan Kaufmann, 2004.
4. J. Duato, J. Flich, and T. Nachiondo, *Cost-Effective Technique to Reduce HOL-blocking in Single-Stage and Multistage Switch Fabrics,* Euromicro Conference on Parallel, Distributed and Network-based Processing, pp. 48-53, Feb. 2004.
5. J. Duato, I. Johnson, J. Flich, F. Naven, P. García, and T. Nachiondo, *A New Scalable and Cost-Effective Congestion Management Strategy for Lossless Multistage Interconnection Networks,* Int. Symp. on High-Performance Computer Architecture, Feb. 2005.
6. Earth Simulator Center. http://www.es.jamstec.go.jp/esc/eng/index.html.
7. IBM BG/L Team, *An Overview of BlueGene/L Supercomputer,* ACM Supercomputing Conference, 2002.
8. InfiniBand Trade Association, InfiniBand Architecture. Specification Volume 1. Release 1.0. Available at http://www.infinibandta.com/.
9. M. Jurczyk and T. Schwederski, *Phenomenon of Higher Order Head-of-Line Blocking in Multistage Interconnection Networks under Nonuniform Traffic Patterns,* IEICE Transactions on Information and Systems, Special Issue on Architectures, Algorithms and Networks for Massively Parallel Computing, Vol. E79-D, No. 8, pp. 1124-1129, August 1996.
10. C. Minkenberg, *On Packet Switch Design,* Ph.D. Thesis, Eindhoven University of Technology, Sep. 2001.
11. Myrinet, 2000 Series Networking. Available at http://www.cspi.com/multicomputer/ products/2000 _series_networking/ 2000_networking.htm.
12. T. Nachiondo, J. Flich, and J. Duato, *Efficient Reduction of HOL blocking in Multistage Networks,* Workshop on Communication Architecture for Clusters (CAC 2005), April 2005.
13. G. F. Pfister and V. A. Norton,*Hot Spot Contention and Combining in Multistage Interconnection Networks*, IEEE Transactions on Computers, vol. C-34:10,pp. 943-948, Oct.1985.
14. Quadrics QsNet. Available at http://doc.quadrics.com.
15. L. Shang, L. S. Peh, and N. K. Jha, *Dynamic Voltage Scaling with Links for Power Optimization of Interconnection Networks*, Proc. Int. Symp. on High-Performance Computer Architecture, pp. 91–102, Feb. 2003.
16. http://www.top500.org

On the Correct Sizing on Meshes Through an Effective Congestion Management Strategy*

Pedro Javier García[1], José Flich[2], José Duato[2], Francisco José Quiles[1], Ian Johnson[3], and F. Naven[3]

[1] Dept. de Informática. Escuela Politécnica Superior
Universidad de Castilla-La Mancha 02071-Albacete, Spain
`{pgarcia,paco}@info-ab.uclm.es`
[2] Dept. of Computer Science, Univ. Politécnica de Valencia 46071-Valencia, Spain
[3] Xyratex, Haven, UK

Abstract. Interconnection networks used in clusters of PCs are often dimensioned with certain restrictions. One restriction could be the reduction of power consumption and overall cost. In this sense, the network size must be reduced. Another restriction is to guarantee that the system offers a minimum bandwidth. In this case, the network size must be increased. In both cases, the head-of-line (HOL) blocking effect (related to network congestion) may appear, degrading network performance and thus, preventing the correct sizing of the network. Therefore, some mechanisms should be implemented for reducing or eliminating this problem, in order to dimension the network as desired while keeping network performance at maximum. In this paper we analyze the impact on network performance when using different mechanisms for handling HOL blocking when interconnection networks with mesh topology are dimensioned in several ways. We show that the previously proposed RECN congestion control mechanism is key in order to efficiently eliminate HOL blocking in meshes and, therefore, it allows the correct network sizing.

1 Introduction

In the last years, clusters of PCs are becoming a challenging alternative to massive parallel computers dedicated to high performance computing (HPC). Also,cluster of PCs are becoming an alternative to build large Internet servers. The attractive performance/cost ratio of PCs makes building large cluster-based systems an interesting solution. Examples of clusters for HPC can be obtained from the top500 list [1] where 294 systems out of 500 are clusters of PCs (three in the top five list). Also, commercial Internet portal servers using clusters are being used at AOL, Google, Amazon or Yahoo.

In such systems, the interconnection network plays a key role in the performance achieved. For this, it is common to use high-speed interconnect networks like Myrinet [2], InfiniBand [3], and Quadrics [12]. Such networks provide high

---

* This work was supported by CICYT under Grant TIC2003-08154-C06 and by UPV under Grant 20040937.

J.C. Cunha and P.D. Medeiros (Eds.): Euro-Par 2005, LNCS 3648, pp. 1035–1045, 2005.

bandwidth and low latencies. However, these networks offer additional features. One of the most interesting feature is that network topology can be as flexible as needed. Indeed, nothing prevents a network designer to attach several endnodes to the same switch or to build a complete irregular network. This capability makes system scalability a reality in cluster-based systems.

However, new problems arise in these networks that may affect scalability. One of the main problems is congestion. As these networks usually do not drop packets (lossless networks), whenever a packet blocks (as it requests a resource not available) it may block other packets stored behind it, even in the case they could make forward progress (they would request available resources). This effect is referred to as head-of-line (HOL) blocking. HOL blocking will be propagated quickly because flow control spreads the congestion, thus collapsing the network.

The de facto solution to avoid congestion has traditionally been to overdimension the interconnection network. The introduction of wormhole switching [4] made feasible to integrate a switch into a single chip. In turn, this allowed such a dramatic increase in link bandwidth that interconnection networks could be overdimensioned at a low cost. As a consequence, reported network utilization in parallel machines and clusters has been quite low for almost two decades.

On the other hand, power consumption is becoming increasingly important. As VLSI technology advances and link speed increases, interconnects are consuming an increasing fraction of the total system power [5]. Taking this into account, there are only two ways of reducing network power consumption: a) reducing the number of links in the network (and using the remaining links more efficiently), and b) using some frequency/voltage scaling technique to reduce link power consumption [5]. Unfortunately, dynamic voltage scaling (DVS) techniques are quite inefficient due to their extremely slow response in the presence of traffic variations and the suboptimal frequency/voltage settings during transitions [6]. In fact, a recent paper shows that static voltage scaling (SVS) combined with adaptive routing achieves higher performance and lower power consumption than DVS techniques, as far as the network does not saturate [6].

Thus, the simplest way to reduce cost and power consumption is reducing the network size (less switches and links). Obviously, as the network size is reduced, the offered network bandwidth will be also lower. Therefore, the network must be dimensioned accordingly to the estimated bandwidth required by the endnodes of the system. For instance, the network designer can rearrange the endnodes and attach a higher number of endnodes at each switch. Imagine a system with 256 endnodes attached through a 16×16 mesh network using links with 1Gbps capacity. This network can be reduced to a 4×4 mesh network (attaching groups of 16 endnodes to the same switch) only if the available network bandwidth (16 Gbps[1]) is enough for communicating the endnodes. Notice that this can be the case for a server system where the traffic is highly local (applications are exclusively run in groups of 16 endnodes attached to the same switch).

---

[1] Theoretically, for uniform traffic, the offered bandwidth of a mesh network is $2 \times BW_{bisection}$.

However, as the network size is reduced, the link utilization will be higher and, thus, the network will work closely to the saturation point. Additionally, notice that traffic is usually bursty (temporal congestion trees will be common). In this scenario, it will be usual that the network will be working beyond its saturation point. Therefore, it will be required an effective congestion control mechanism in order to allow an effective reduction in the network size.

Another restriction for network dimensioning is when more bandwidth is required by the system. This is the case of HPC systems where it is expected to have an intense traffic among all endnodes (working on the same application). Therefore, the way to dimension these systems is to put as much switches and links as necessary to meet the required traffic conditions. As an example, a system with 16 endnodes attached to a 4×4 mesh network can be scaled up by building a larger system with 256 endnodes attached to a 16×16 mesh network. At first sight it seems that there will be no problems when using more network components. This can be deduced from the fact that the utilization of links will be lower (as the network is overdimensioned). However, notice that as network size is increased, the average path length will also increase. Therefore, packets will travel longer distances and, in the presence of congestion trees, they will have more chances of being affected by the HOL blocking effect. Thus, again it will be necessary an effective congestion management technique that eliminates those effects in order to allow an effective increase in the network size without degrading performance.

HOL blocking has been studied for long, and very efficient techniques exist for avoiding it within a single switch (e.g. virtual output queues (VOQs) [7], dynamically allocated multiqueues (DAMQs) [8], congestion buffers [9], etc.). These techniques work by allocating separate buffers for packets destined to different output ports or by providing a way for non-blocked packets to pass blocked packets. However, these solutions either do not work efficiently for multihop networks (e.g. DAMQs) or are not scalable at all because the number of buffers required at every switch port increases linearly with the number of endpoints attached to the network. Thus, overall buffer capacity increases at least quadratically with the number of network endpoints. Although some implementations of network-level VOQs exist [10], they are very expensive and do not scale, and may even become infeasible beyond certain network size.

An intermediate solution is to use VOQ at the switch level. With this solution, every switch port has as many queues as output ports of the switch, and whenever a packet arrives to the port it is stored in the queue assigned to its requested output port. Although this solution does not eliminate completely HOL blocking it can minimize its impact. This solution will be referred to as VOQ_{sw}.

In [13] we proposed a new congestion management technique, referred to as RECN (Regional Explicit Congestion Management), focused in eliminating the HOL blocking effect produced by congestion trees rather than eliminating congestion. In particular, once incipient congestion is detected within the network, RECN assigns new queues to the congested points and thus, the congested traffic is isolated and the HOL blocking is avoided.

A recent technique has also been proposed in [11], referred to as DBBM (Destination-Based Buffer Management). In this approach, the whole set of network endpoints are divided into several sets, and all the packets addressed to a set of destinations are stored in the same queue. Thus, HOL blocking is avoided among destinations grouped in different sets. Notice that RECN differs from DBBM in the sense that dynamic queues are allocated for congestion trees whereas in DBBM queues are statically allocated to groups of destinations. Although DBBM is very efficient in the general case, there may be some special traffic situations that may introduce HOL blocking.

In this paper we take on different challenges. Firstly, we apply the RECN mechanism to mesh networks. By doing this, we analyze the benefits that RECN will give to applications run on such networks. Secondly, we will analyze up to what extent the traditional VOQ_{sw} solution is able to efficiently handle the HOL blocking introduced when the network is dimensioned in different ways (downsizing the network to reduce cost and power consumption and upsizing the network to achieve a certain bandwidth). As a third challenge we will evaluate RECN as a way to allow an efficient system sizing. We will show that, contrary to the VOQ_{sw} solution, RECN allows to achieve ideal network sizing.

The rest of the paper is organized as follows. In Section 2, RECN is described. In Section 3, scalability issues by using RECN and VOQ_{sw} are analyzed in detail by means of simulation results of network performance under different situations of traffic, network size and congestion control mechanisms used. Finally, in Section 4 some conclusions are drawn.

2   RECN Description

RECN (Regional Explicit Congestion Notification)[13] is a new congestion management strategy that focuses on eliminating the main negative effect of congestion: the HOL blocking. In order to achieve it, RECN detects congestion and dynamically allocates separate buffers for each congested flow, assuming that packets from non-congested flows can be mixed in the same buffer without producing significant HOL blocking.

RECN requires the use of a kind of deterministic routing that makes possible to address a particular network point from any other point in the network. In fact, RECN has been designed for PCI Express Advanced Switching (AS) [14, 15], a technology that uses source routing. AS packet headers include a turnpool made up of 31 bits, that contains all the turns (offset from the incoming port to the outgoing port) for every switch in a route. Thus, a switch, by inspecting the appropriate turnpool bits, can know in advance if a packet that is coming through one of its incoming ports will pass through a particular network point.

In order to separate congested and non-congested flows, RECN adds a set of additional queues at every input (ingress) and output (egress) port of a switch. These queues (referred to as Set Aside Queues or SAQs) are dynamically allocated and used to store packets passing through a congested point. To do this, RECN associates a CAM memory to each set of queues. The CAM contains all the control info required to identify the congested point and to manage the corresponding SAQ. In the aim of guaranteeing in order delivery, whenever a new SAQ is allocated, forwarding packets from that queue is disabled until the last packet of the standard queue (at the moment of the SAQ allocation) is forwarded. This is implemented by a simple pointer associated to the last packet in the standard queue and pointing to the blocked SAQ.

Whenever an ingress or egress queue receives a packet and fills over a given threshold, a RECN notification is sent to the sender port indicating that an output port is congested. When congestion is detected at the egress side, the congested point is this egress port. In order to detect congestion at the ingress side, the standard queue is replaced by a set of detection queues. The detection queues are structured at the switch level: there are as many detection queues as output ports in the switch, and packets heading to a particular output port are directed to the associated detection queue. So, when a detection queue reaches a threshold, it means that the associated output port is congested.

RECN notifications also include the routing information (a turnpool) to reach the congested output port from the notified port. Upon reception of a notification, each port maps a new SAQ and fills the corresponding CAM line with the received turnpool. From that moment, every incoming packet that will pass through the congested point (easily detected from the turnpool of the packet) will be stored in the newly allocated SAQ, thus eliminating the HOL blocking it may cause. If a SAQ becomes subsequently congested, a new notification will be sent upstream to some port that will react in the same way, allocating a new SAQ, and so on. As the notifications go upstream, the included information indicating the route to the congestion point is updated accordingly, in such a way that growing sequences of turns (turnpools) are stored in the corresponding CAM lines. So, the congestion detection is quickly propagated through all the branches of a congestion tree. Apart from the SAQs allocated due to notifications, when congestion is detected at the ingress side, a SAQ is also allocated at this port, and the detection queue and the new allocated SAQ are swapped.

RECN keeps track (with a control bit on each CAM line) of the network points that are a leaf of a congestion tree. Whenever a SAQ with the leaf bit set empties, the queue is deallocated and a notification packet is sent downstream, repeating the process until the root of the congestion tree is reached.

Regarding flow control, RECN uses for each individual SAQ a level-based flow control (Xon/Xoff). This mechanism is different from the credit-based flow control used for standard queues, that considers all the unused space of the port data memory available for each individual queue. Xon/Xoff scheme guarantees that the number of packets in a SAQ will be always below a certain threshold. Further details about RECN can be found in [13].

3 Performance Evaluation

In this section we will evaluate the performance of the network, in several scenarios of network size, traffic load, and different mechanisms focused in reducing the HOL-blocking: VOQ at the network level (VOQnet), VOQ at the switch level (VOQsw) and RECN. For this purpose we have developed a detailed event-driven simulator that allows us to model the network at the register transfer level. Firstly, we will describe the main simulation parameters and the modeling considerations we have used in all the evaluations. Secondly, we will present the evaluation results and analyze them.

3.1 Simulation Model

The simulator models square meshes consisting of a variable number of switches and bidirectional links that connect a variable number of endnodes. Specifically, we have used five network configurations, shown in Table 1. In all the cases X-Y deterministic routing is used.

Table 1. Network configurations and traffic cases evaluated.

Network	Top	#sw	#endnodes/sw
#1	16 × 16	256	1
#2	8 × 8	64	4
#3	4 × 4	16	16
#4	8 × 8	64	1
#5	4 × 4	16	1

	normal traffic		congestion tree	
Traffic	#sources	dst	#sources	dst
#1	100%	random	-	-
#2	87.5%	random	12.5%	hot-spot
#3	75%	random	25%	hot-spot

Due to the different number of endnodes per switch, the number of bidirectional ports of the switches varies depending on network configuration. At these ports, the simulator models a 128 KB memory for both input and output ports. When VOQ is used, the total memory size per port is equally divided into as many queues as endnodes (VOQnet) or into as many queues as ports in the switch (VOQsw).

RECN has been modeled in detail. The memory is shared by all the queues (detection or standard queues and SAQs) defined at this port at a given time, in such a way that memory cells are dynamically allocated (or deallocated) for any queue when it is required. In order to support the RECN detection at ingress ports, several detection queues are defined at ingress ports, and one standard queue at egress ports.

To model the links, we have assumed serial full-duplex pipelined links with 8 Gbps bandwidth. Inside every switch, packets are forwarded from any input queue to the corresponding output queue through a multiplexed crossbar. The crossbar access is controlled by an arbiter that receives requests from packets at the head of any input queue. A requesting packet is forwarded only when the corresponding crossbar input and crossbar output are free. Requests from packets in detection queues have preference over requests from packets in SAQs.

Regarding flow control, we have modeled several mechanisms. RECN uses credit-based flow control at the port level. So, whenever a new packet is transmitted from an output port to the corresponding input port of the next switch, a credit is consumed. When a packet leaves an input port, a new credit is granted to the previous output port at the upstream switch or endnode. Output port credits can be consumed for transmitting packets from the standard queue or SAQs at this port. A similar flow control scheme has been implemented for the internal (input-output) switch packet forwarding. So, the maximum number of credits per output (or input) port depends on the total memory size at the next input (or output) port. In addition, Xon/Xoff flow control has been modeled for limiting the injection of packets between SAQs. When the occupancy of a SAQ grows up to a given threshold[2], an Xoff packet is sent to the corresponding upstream SAQ. Any SAQ that receives an Xoff packet stops the injection of packets until the reception of an Xon packet. Any SAQ that previously sent an Xoff packet sends an Xon packet when its occupancy goes below a given threshold. On the other hand, a credit-based flow control at the queue level has been implemented for the VOQs mechanisms. In these cases, the maximum number of credits per queue depends on the total memory size at the next input (or output) port and the number of queues at this port. Flow control packets have been modeled and they share the link bandwidth with data packets.

Endnodes are connected to switches using Input Adapters (IAs). Every IA is modeled with a fixed number of message admittance queues following a VOQnet scheme, and a variable number of injection queues, that follow a scheme similar to that of the output ports of a switch. So, SAQs can be allocated dynamically at the output side of input adapters when the RECN mechanism is used. When a message is generated, it is stored completely in the admittance queue assigned to its destination, and is packetized before being transferred to an injection queue. We have used 64-byte packets. The transfer from admittance queues to injection queues is controlled by an arbiter that follows a round-robin scheme. The injection of packets from injection queues to the network is also controlled by an arbiter that selects the next packet to be transmitted, using a round robin scheme among all the queues.

3.2   Traffic Load

For all the network configurations we have made experiments under several traffic scenarios. We have used synthetic traffic patterns modeling simple but significant traffic situations in order to check how the analyzed mechanisms react to different traffic loads. Table 1 shows the traffic parameters of each traffic case.

For each traffic case, there is a variable percentage of sources injecting traffic to random destinations. This percentage is 100% in traffic case #1, but it is lower in traffic cases #2 and #3. In these cases, the rest of sources inject traffic to the same destination (endnode 32 for network configurations #1, #2, #3 and #4; endnode 10 for network configuration #5). Thus, in these cases, congestion trees will be formed in the network. All the endnodes inject traffic at the same rate during all the simulation period. This rate has been varied in an incremental way for obtaining a metric of the network performance under different loads of normal and congested traffic.

---

[2] Although several thresholds have been tested, all of them gave us similar performance results. Therefore, we fixed threshold to 1% of total port memory.

For all the cases evaluated, the network relative throughput[3] as a percentage will be shown (for different injection rates). This will allow direct comparisons among different network configurations.

3.3   Performance Evaluation

In the following subsections we will show simulation results that allow us to analyze the impact of RECN and VOQsw when used as a mechanism to reach the maximum performance when sizing the network in different ways. Moreover, results for VOQnet will be also shown as a reference for maximum performance (no possible HOL blocking). Specifically, we will analyze first the impact of such mechanisms when the network is downsized while keeping constant the number of endnodes. Next, we will analyze their impact when the network size is increased in order to achieve higher bandwidth.

Reducing Network Cost and Consumption. Figure 1 shows the performance results for network configurations #1, #2 and #3 for different traffic patterns. For traffic case #1 (Figures 1.a, Figures 1.b, and Figures 1.c), all the mechanisms evaluated achieve roughly the maximum performance, although the performance slightly decreases when VOQsw is used for high traffic loads. This is because VOQsw does not correctly handle all the traffic, and some HOL blocking appears. Additionally, for higher traffic loads (beyond saturation point; not shown), VOQsw even significantly degrades performance. On the other hand, in these situations, RECN keeps relative throughput above 90%.

From the previous results, it could be deduced that VOQsw is an effective mechanism that allows to achieve maximum performance for low or medium traffic loads. However, real traffic is usually bursty, and a different behavior could be expected. Indeed, Figures 1.d, 1.e, and 1.f show the results for network configurations #1, #2 and #3 when a light hot-spot traffic pattern (traffic case #2) is present in the network. For all the network configurations, VOQsw is not able to obtain maximum performance, regardless of the injection rate. Indeed, for network configuration #1, it achieves only 50% of relative throughput. It can be seen that, as network size decreases, VOQsw tends to achieve higher relative throughput. This is due to the shorter average routes on the network, that reduce the HOL blocking effect. On the opposite side, RECN achieves roughly maximum throughput (90% of relative network throughput in the worst case). So, RECN eliminates the HOL blocking introduced by the congestion tree and, as a consequence, it uses efficiently all the bandwidth offered by the network.

For a more intense hot-spot traffic pattern (traffic case #3), the behavior is similar but more dramatic for VOQsw. Results for this traffic case are shown in Figures 1.g, 1.h, and 1.i for network configurations #1, #2 and #3, respectively.

---

[3] Network relative throughput is computed as the network absolute throughput divided by the maximum theoretical throughput ($2 \times BW_{bisection}$). The maximum theoretical throughput for the $N \times N$ mesh is $4 \times N$ bytes/ns.

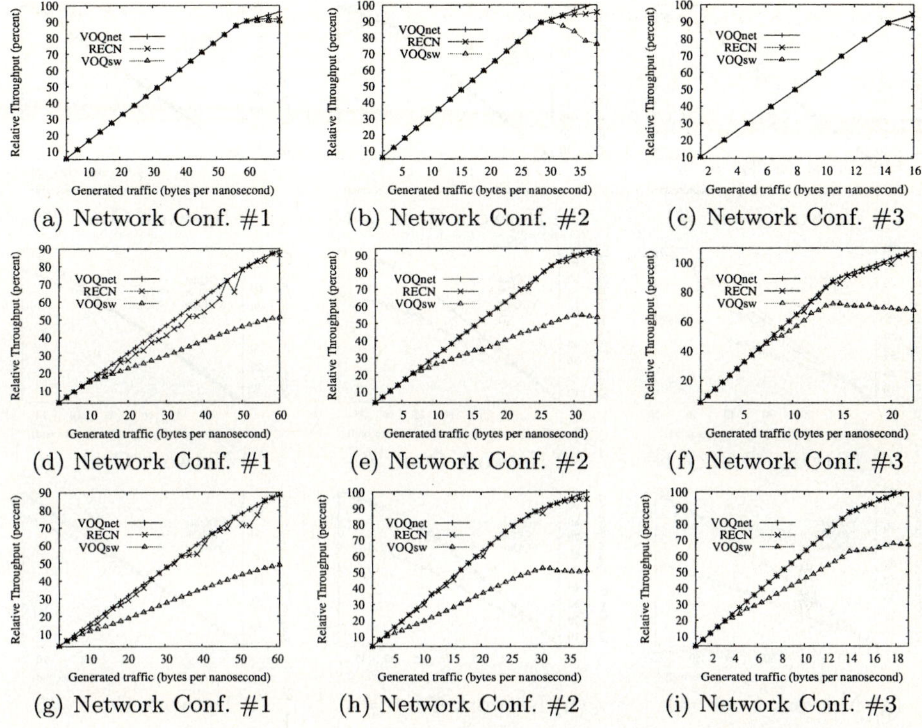

Fig. 1. Relative throughput for network configurations #1, #2 and #3, traffic case (a,b,c) #1 (uniform), (d,e,f) #2 (light hot-spot), and (g,h,i) #3 (heavy hot-spot).

To sum up, RECN allows to achieve the maximum bandwidth offered by the network by virtually eliminating the HOL blocking introduced by the higher use of links when network is downsized. VOQsw is far from achieving the maximum offered network bandwidth as it does not handle properly HOL blocking.

Increasing Network Size and Bandwidth. Now, we will evaluate how VOQsw and RECN behave when they are used as a technique to achieve maximum throughput when overall network bandwidth is increased by upsizing the network. For all the network configurations evaluated in this section, one endnode is attached to each switch. Thus, as the network size increases, the number of endnodes also increases, and so does the average length of routes (potentially increasing HOL blocking).

Figure 2 shows the performance for different network configurations (#1, #4, and #5) and different traffic patterns. For uniform traffic pattern (traffic case #1, Figures 2.a, 2.b, and 2.c) it can be deduced that the VOQsw solution behaves roughly as well as RECN and VOQnet. From this fact, it could be deduced also that VOQsw is a good solution in order to efficiently upsize the network. However, again, this deduction is not valid when a congestion spot is present in the network (modeling bursty traffic). For a light congestion tree (traffic case

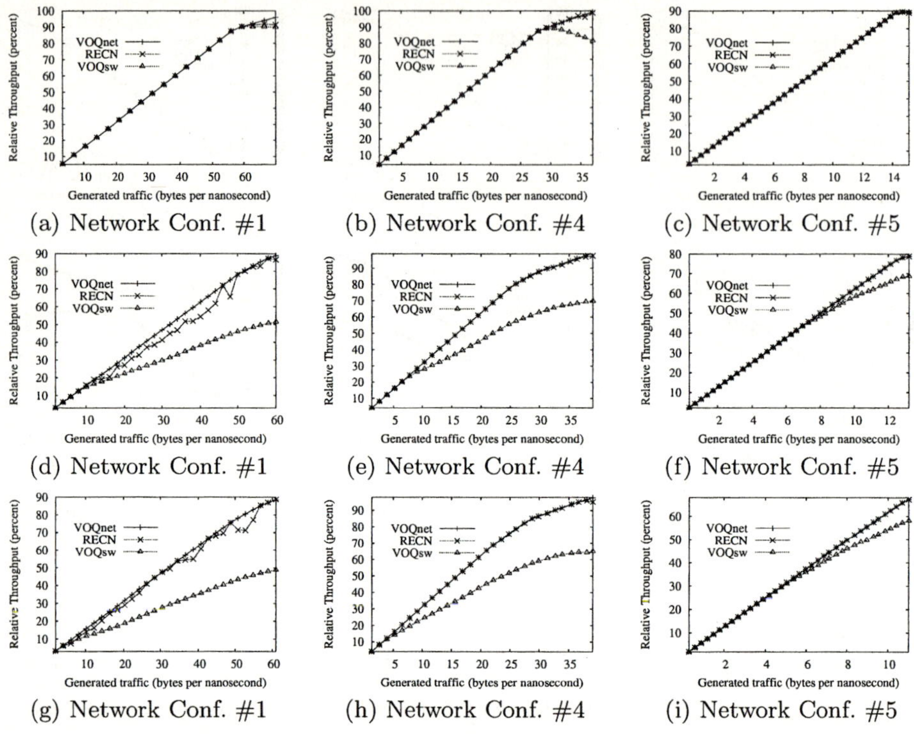

(a) Network Conf. #1   (b) Network Conf. #4   (c) Network Conf. #5

(d) Network Conf. #1   (e) Network Conf. #4   (f) Network Conf. #5

(g) Network Conf. #1   (h) Network Conf. #4   (i) Network Conf. #5

Fig. 2. Relative throughput for network configurations #1, #4 and #5, traffic case (a.b,c) #1 (uniform), (d,e,f) #2 (light hot-spot), and (g,h,i) #3 (heavy hot-spot).

#2, Figures 2.d, 2.e, and 2.f) in the network, again, the VOQsw solution suffers HOL blocking that is not solved, and therefore, it does not achieve maximum offered bandwidth. Relative network throughput is always lower that 70%.

On the other hand, RECN is able to keep the maximum performance for all the network configurations. It has to be noted that RECN achieves its goal by using a maximum of 8 SAQs. Figure 3 shows, for traffic case #3 and network configurations #1, #4 and #5, the maximum SAQ utilization at ingress and egress sides. It can be seen that the maximum number of SAQs used is below 8 for most of the traffic loads.

4   Conclusions

We have shown the importance of using a suitable congestion control mechanism for virtually eliminating the HOL blocking that appears by dimensioning in several ways interconnection networks with mesh topology. From the results presented in this paper, we can deduce that network performance is affected by HOL blocking when the network is sized in certain ways and VOQsw is used. On the contrary, the RECN mechanism allows to dimension the network in any way while keeping network performance roughly at maximum, due to the

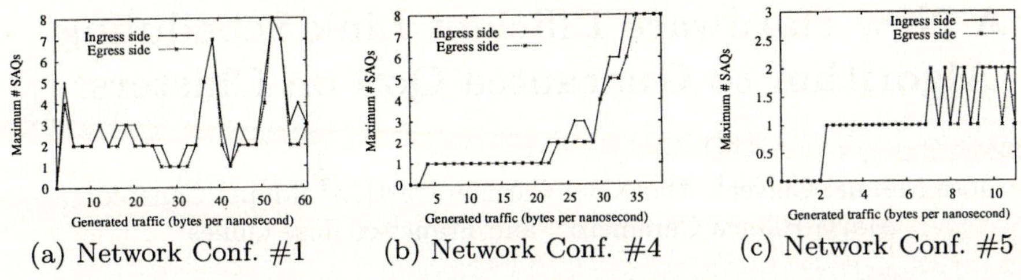

(a) Network Conf. #1   (b) Network Conf. #4   (c) Network Conf. #5

Fig. 3. Maximum number of SAQs used. Configurations #1, #4, #5, traffic case #3.

efficient handling of the HOL blocking problem. Moreover, this can be achieved in a scalable way. Therefore, RECN allows to reduce network size, cost and power consumption or to increase network size and overall bandwidth without degrading network performance.

References

1. Top 500 supercomputer list, *http://www.top500.org*.
2. N. J. Boden et al, "Myrinet - A gigabit per second local area network," *IEEE Micro*, pp. 29–36, February 1995.
3. InfiniBand^{TM} Trade Association, *http://www.infinibandta.com*.
4. W. J. Dally and C. L. Seitz, "The Torus Routing Chip," *Journal of Distributed Computing*, vol. 1, no. 3, pp. 187–196, Oct. 1986.
5. L. Shang, L. S. Peh, and N. K. Jha, "Dynamic Voltage Scaling with Links for Power Optimization of Interconnection Networks", in *Proc. Int. Symp. on High-Performance Computer Architecture*, pp. 91–102, Feb. 2003.
6. J. M. Stine and N. P. Carter, "Comparing Adaptive Routing and Dynamic Voltage Scaling for Link Power Reduction", *Computer Architecture Letters*, vol. 3, June 2004.
7. T. Anderson et al, "High-Speed Switch Scheduling for Local-Area Networks", *ACM Transactions on Computer Systems*, vol. 11, no. 4, pp. 319–352, Nov. 1993.
8. Y. Tamir and G. L. Frazier, "Dynamically-Allocated Multi-Queue Buffers for VLSI Communication Switches", *IEEE Trans. on Computers*, vol. 41, no. 6, June 1992.
9. A. Smai and L. Thorelli, "Global Reactive Congestion Control in Multicomputer Networks", in *Proc. 5th Int. Conference on High Performance Computing*, 1998.
10. W. J. Dally, P. Carvey, and L. Dennison, "The Avici Terabit Switch/Router", in *Proc. Hot Interconnects 6*, Aug. 1998.
11. J. Duato, J. Flich, and T. Nachiondo, *Cost-Effective Technique to Reduce HOL Blocking in Single-Stage and Multistage Switch Fabrics,* Euromicro Conference on Parallel, Distributed and Network-based Processing, pp. 48-53, Feb. 2004.
12. Quadrics QsNet. Available at http://doc.quadrics.com
13. J. Duato, I. Johnson, J. Flich, F. Naven, P.J. García, T. Nachiondo, "A New Scalable and Cost-Effective Congestion Management Strategy for Lossless Multistage Interconnection Networks", in *Proc. 11th Int. Symp. High-Performance Computer Architecture*, Feb. 2005.
14. "Advanced Switching for the PCI Express Architecture". White paper. Available at http://www.intel.com/technology/pciexpress/devnet/AdvancedSwitching.pdf
15. "Advanced Switching Core Architecture Specification". Available at http://www.asi-sig.org/specifications for ASI SIG.

A New Hardware Efficient Link Scheduling Algorithm to Guarantee QoS on Clusters*

José Manuel Claver[1], María del Carmen Carrión[2], Manel Canseco[1],
María Blanca Caminero[2], and Francisco José Quiles[2]

[1] Dept. of Computer Science and Engineering. E.S.T.C.E.
Univ. Jaume I, 12071 - Castellón, Spain
{claver,canseco}@icc.uji.es
[2] Dept. of Computer Science, Escuela Politécnica Superior
Univ. de Castilla-La Mancha, 02071 - Albacete, Spain
{carmen,blanca,paco}@info-ab.uclm.es

Abstract. Contemporary router/switch technology for high-performance local/system area networks (LANs/SANs) should provide the capacity to fit the high bandwidth and timing requirements demanded by current applications. The MultiMedia Router (MMR) aims at offering hardware-based QoS support within a compact interconnection component. One of the key elements in the MMR architecture is the link scheduling algorithm. This algorithm must solve conflicts among data flows that share an input physical link. Required solutions are motivated by chances for parallelization and pipelining, while providing the necessary support both to multimedia flows and to best-effort traffic. In this work, a cost-aware link scheduling based on the temperature coding of priority value associated to every head flit is presented and evaluated.

1   Introduction

Current applications include not only best-effort traffic such as ftp or e-mail but also multimedia QoS-aware applications. Numerous examples can be highlighted from web-based applications, interactive simulations, virtual meeting and collaborative design environments. Clusters are being commonly used as back-end servers for these applications, so some QoS support is needed within the underlying interconnection network.

The problem of providing architectural QoS support within switching elements in cluster and local area environments is still an open issue [1] [2] [3]. Thus, the MultiMedia Router[1] (MMR) architecture [4] arises as a solution to provide hardware-based QoS support within an interconnection component targeted for use in cluster and LAN environments. The MMR organization is based on input queues and a multiplexed crossbar internal switch.

---

* This research was partially supported by the Spanish CICYT under grants No. TIC2003-08154-C06-04 and TIC2003-08154-C06-06.

[1] The MultiMedia Router is devised as a link-layer interconnection element. The term "router" is inherited from the interconnection elements used in multicomputer and multiprocessor networks, rather than from the IP world.

J.C. Cunha and P.D. Medeiros (Eds.): Euro-Par 2005, LNCS 3648, pp. 1046–1056, 2005.

In order to achieve high link bandwidth utilization and to provide the QoS needed by the applications, the MMR requires efficient traffic scheduling algorithms. These algorithms decide which data must be transmitted at each time, so their behaviour will determine whether QoS guarantees are fulfilled or not. Most of the scheduling solutions appearing in the literature for such organization seek to maximize the link and internal switch utilization [5], and do not address QoS issues. Some recent research on scheduling algorithms tries to offer both high throughput and QoS support. However, these are almost theoretical solutions. Compact and fast hardware implementations of these algorithms are hardly feasible, which prevents their use in high-speed interconnection networks. Moreover, most scheduling solutions for input-buffered switches need to run at speeds higher than links to provide QoS guarantees and high link utilization, or lack the needed flexibility to concurrently accommodate different connection requirements. Thus, in this paper, a new link scheduling algorithm is presented, the Temperature-IABP (TIABP). The main features of this algorithm are to provide high-throughput and QoS guarantees to the different multimedia flows, according to their reservations, while being suitable for a simple low cost hardware implementation. Preliminary performance evaluation results and implementation on a Xilinx Virtex 2000E FPGA of a Temperature-IABP based link scheduling are presented. Due to the complexity of this device a high level hardware specification language, HandelC [6], is used.

The rest of the paper is organized as follows. First, Section 2 outlines the main characteristics of the Multimedia Router architecture. Then, in Sections 3 and 4 the new resource scheduling algorithm proposed for the MMR and its hardware architecture features are explained. Extensive evaluation results follow, which reveal the effectiveness of the proposed link scheduling algorithm. Finally, some conclusions are given.

2   The Multimedia Router

Figure 1 depicts the general organization of the Multimedia Router. In the following paragraphs, the basic building components will be briefly described. The interested reader is referred to [4] for a more detailed description.

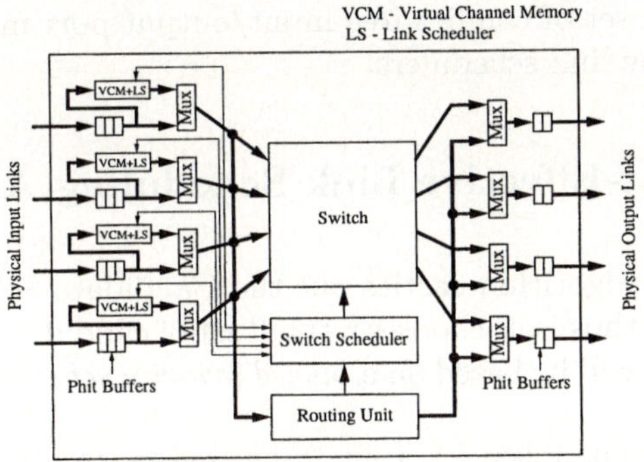

Fig. 1. MultiMedia Router organization.

a) Input Buffers: To support a large number of multimedia connections, the storage buffers at each input link are organized as a set of virtual channels. One virtual channel is provided per connection, in order to consider the QoS of each flow. This approach also avoids the head-of-line blocking.

The MMR avoids losing data due to buffer overflow by using *per connection flow control* at the link level. The selected scheme is *credit-based flow control,*as InfiniBand^{SM} does [7]. The flow control unit will be referred to as a *flit.*

b) Routing Unit: The MMR uses a *hybrid switching technique*: a connection-oriented scheme, Pipelined Circuit Switching (PCS) [8], for the multimedia flows, and Virtual Cut-Through (VCT) [9] for best-effort messages. Paths are computed according to the Exhaustive Profitable Backtracking (EPB) routing algorithm [10] for multimedia flows, while best-effort messages are routed according to a fully adaptive routing algorithm [11].

c) Internal Switch: Due to the large number of virtual channels, the MMR internal switch is a *multiplexed crossbar* with as many ports as physical channels. This crossbar organization implies several arbitration tasks. First, *the Link Scheduler module (LS)* solves conflicts at the input side selecting the virtual channel that will use the crossbar input port in the next flit cycle. Then, *the Switch Scheduler module* does a second arbitration, because several input channels might request the same output link for the same flit cycle.

The choice of link and switch scheduling algorithms are critical parameters for the MMR. Both scheduling algorithms must cooperate to guarantee that the bandwidth allocated to each connection is available during data transmission.

In the MMR, link bandwidth and switch port bandwidth are split into *flit cycles*^{2}. Flit cycles are grouped into *rounds*. The number of flit cycles in a round is an integer multiple K ($K > 1$) of the number of virtual channels per link. The allocated flit cycles will be assigned to the requesting connection every round.

The link and switch scheduling algorithms proposed for the MMR are partitioned into three basic decisions, thus being well suited to parallelization and pipelining: *Candidate Selection, Port Ordering* and *Arbitration.* The Candidate Selection phase is carried out by the link schedulers. Thus, it is performed in parallel for every input link. Its purpose is to select a set of one or more virtual channels, with flits ready for transmission, called *candidates.* The switch scheduler, also known as crossbar arbitration, tackles the other two phases. They are aimed at selecting a set of conflict-free input/output port matchings among the candidates chosen by link schedulers.

3   A New Cost-Effective Link Scheduling Algorithm

The link scheduling algorithm carries out the Candidate Selection phase of the scheduling problem, thus is, it chooses a small set of candidates per every physical input link. Selection will be based on a *biased priority scheme* associated to every

---

^{2} A flit cycle is the time taken for a flit to be transmitted through the router and
  across the physical link.

head flit. The key point in our scheme is that priorities are biased according to the ratio between the QoS a flit is receiving and the one it should receive.

We proposed the *Inter-Arrival Biased Priority (IABP)* scheme in [12]. In that case, the priority of a flit is computed as the ratio between the queuing delay, and the inter-arrival time (IAT) for the flits in the connection. The effect is that the priority grows as queuing delay grows. Moreover, priority grows faster for those flits belonging to high-bandwidth consuming connections, that is, there are more chances that they will be forwarded sooner through the switch.

In a first attempt to achieve a more practical implementation biasing function, the *Simple IABP (SIABP) algorithm* was devised [13]. The idea is to apply the same rationale introduced by the IABP algorithm, that is, to relate the bandwidth required by the connection to the experienced queuing delay, but replacing the division with some other less expensive operation. Equations 1 and 2 show how the priority of the header flit is computed and updated:

$$Priority(0) = NumCycles \tag{1}$$

$$Priority(t) = Priority(0) << n \tag{2}$$

where the symbol $<<$ represents a left shifter, $NumCycles$ is related to the bandwidth reserved for a connection and n is the position of the highest significant bit set to 1 in the register that holds the queuing delay. In this way the QoS needed (represented by the initial priority value) is also related to the QoS received by the flit (the queuing delay).

Although an important hardware area reduction is got by using this algorithm, a new link scheduling proposal is presented in this work: the *Temperature IABP*[3], *TIABP*. The purpose of the TIABP algorithm is not only to get low area cost but also to improve some critical aspects of the SIABP algorithm. Considering equations 1 and 2, which describes how is computed and updated the SIABP priority, we can be aware that inversion priority can happen. The priority of the header flit allocated in a virtual channel increases with the time being waiting for the output port. However, if the header flit remains there for a long time, its priority could decrease. In particular, when the initial priority value is a power of two, the priority updated can down to zero. In order to avoid priority inversion, extra hardware must be added to control overflow by freezing the priority of a virtual channel when the most significant bit of the register that holds the priority is equal to 1. However, this is not a solution if priorities are not power of 2. Being aware of this problem, the TIABP priority biasing function is computed as follows:

a) A counter stores the queuing delay of a flit and it is updated every router cycle, in the same way as in the IABP and SIABP algorithms.

b) The initial value for the priority is ($2^{k+1} - 1$), where k is the position of the highest significant bit set to 1 of the bandwidth required by the connection expressed as the flit cycles per scheduling round reserved to service the average bandwidth of the connection ($MSB(NumCycles)$).

---

[3] We denote this new link scheduling as Temperature IABP because it uses temperature coding for the initial priority value representing the QoS needed by a connection.

c) Next, the priority of the flit is computed as the product of the queuing delay times the bandwidth requirements. But, in order to achieve a simpler hardware design, and similar to SIABP, the product is replaced by shifting operations. More precisely, the priority value is updated by shifting to the left its current value (i.e., it is multiplied by 2) and by setting the less significant bit. We should remark that the shifter operation is done each time the queuing delay becomes greater than $1, 2, 4, \ldots, 2^n$, i.e., every time a bit in the queuing delay counter is set for the first time since it was reset.

The TIABP algorithm is summarized in equations 3 and 4:

$$Priority(0) = 2^{k+1} - 1, \ k = MSB(NumCycles) \tag{3}$$

$$Priority(t) = (Priority(0) << n) + (2^n - 1) \tag{4}$$

where n is the position of the highest significant bit set to 1 in the register that holds the queuing delay.

As in the SIABP algorithm, TIABP implementation is just reduced to a shifter and some combinational logic. Nevertheless, using TIABP priority biasing function, priority inversion is not possible because a priority value can not ever decrease. If the priority of the header flit allocated in a virtual channel increases, with the time being waiting for the output port, the maximum priority value is $2^{r+1} - 1$, where r is the priority register bitwidth. So, extra hardware to control register overflow is not necessary. Moreover, when the priority is increased on SIABP, a minimum of 2 bits change their values before overflow. By using TIABP, a maximum of 1 bit changes its value every time the priority is increased. This behavior has direct effects on power dissipation.

The priority temperature coding used on the TIABP also simplifies the sorting stage circuitry of the link scheduler which generates the *candidate vector* that is sent to the *Switch Scheduler*. This stage, called SORT on the SIABP, is a sorting bitonic network, and their basic elements are comparators. By using temperature coding, the design of this comparators is less complex and requires a reduced hardware area. We denote as TSORT (Temperature-SORT) the new implementation of the sorting bitonic network used on TIABP.

4   FPGA Hardware Architecture

In order to implement a MMR on a single chip we have studied the design and implementation of some of their more important modules on an FPGA based board. As a result of this work, a more simple and lower power dissipation circuitry has been obtained. We have focused our effort on the design of the SIABP and SORT modules, which are the core of the *Link Scheduler (LS)*. Thus, we have developed a new version of LS using TIABP and TSORT modules.

4.1   Hardware Implementation

In order to evaluate the results of TIABP design, we have used the Celoxica RC1000 PCI based FPGA board [14]. The RC1000 is a PCI bus plug-in card for PC. It has one large Xilinx FPGA (in our case a Virtex 2000E, with 2 million equivalent gates) with four banks of memory for data processing operations and two PCI Mezzanine Cards (PMC) for input/output with the outside world. The Virtex 2000E is based on slices that contain two 4-bit LUTs each one.

The FPGA Virtex has been programmed using HandelC [6] and the Celoxica DK1 environment. HandelC is a behavioral C based hardware description system developed by Celoxica that allows Co-simulation. Parallelism of process and synchronization are taken from the CSP model, in particular, from the Occam language. HandelC uses standard data types with user defined bitwidths. So, HandelC provides an efficient use of hardware resources.

The SIABP module updates the priority connection each time the queuing delay is increased to the next power of 2, following the HandelC macro

$$shl0(Priority) = ((Priority <- (width(Priority) - 1)@0b0,$$

while the TIABP module updates the priority following the HandelC macro

$$shl1(Priority) = ((Priority <- (width(Priority) - 1)@0b1,$$

where @ is the concatenation operator. Both, SIABP and TIABP modules, update $Priority$ in 1 clock cycle.

The bitonic network of SORT and TSORT modules is recursively built with basic sorter blocks which sort 2 virtual channels in function of its priority. These basic sorter blocks use a comparator, implemented as a boolean macro, called $Prio_greater$. For the SORT module this macro is expressed as

$$Prio_greater(Priority1, Priority2) = (Priority1) > (Priority2),$$

while for the Temperature-SORT (TSORT), it is expressed as

$$Prio_greater(Priority1, Priority2) = ((Priority1)|(Priority2)) == (Priority1).$$

A sorting bitonic network needs $O(log_2(ncv)^2 ncv)$ modules of $Prio_greater$ and $O(1/2log_2(ncv)^2)$ clock cycles to compute a $candidate\ vector$, where ncv is the number of virtual channels per input link.

5   Performance Evaluation

This section presents the implementation results in Xilinx Virtex 2000E FPGA and the performance evaluation of the link scheduling proposed in this paper being used inside the MMR.

We first present preliminary scaling results expressed in terms of FPGA area and maximum clock rate for the SIABP/TIABP and SORT/TSORT modules. Then, the performance evaluation of the link scheduling is done by using a discrete-event C++ simulator.

5.1   Area/Delay Results

For the purpose of this evaluation, we scaled the design of SORT/TSORT modules to handle a different number of virtual channels from 4 to 128. The bitwidth used to represent priority of each virtual channel is 16 bits in all cases. For each result, we report the FPGA area (in equivalent NAND gates) and maximum clock rate provided by the Celoxica DK1 tools.

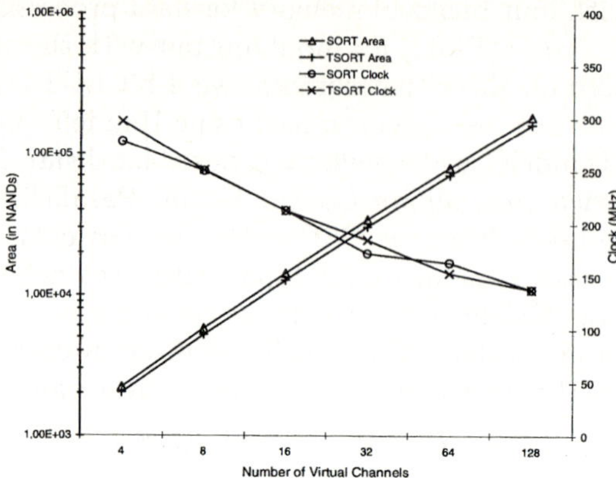

Fig. 2. Area/Clock Rate estimation for SORT and TSORT implementation with different number of Virtual Channels.

Figure 2 reports FPGA area/delay results obtained by SORT and TSORT designs for 4, 8, 16, 32, 64, and 128 virtual channels. As the number of virtual channels increase from 4 to 128, the area used by SORT and TSORT designs increases linearly. But in all cases the TSORT implementation uses between 9% and 13% less area than SORT. Respect to the maximum clock rate, both designs obtain similar results. Thus, maximum clock rate decreases logarithmically when the number of virtual channels is increased. This similar delay between both SORT and TSORT implementations is due to the fact that TSORT in completely instantiated by the DK1 tool using FPGA LUTs, but SORT is instantiated using LUTs (in all cases a number greater than in TSORT) and other FPGA specific resources that accelerate the *Prio_greater* modules used in its design.

In the case of the biasing function, we have found that there are not noticeable differences between SIABP (without support of overflow control) and TIABP. For an initial priority bitwidth of 12 bits, the SIABP module design utilizes 50 LUTs (53 LUTs if overflow control is included) while the TIABP module utilizes 51 LUTS. The maximum clock rate in both designs is 294 MHz.

5.2 Simulation Results

With the previous results in mind, we are going to evaluate the performance of a single 4×4 MMR, with full-duplex 1.24 Gbps 16 bit-wide links. This gives a router cycle of 12.9ns. The number of virtual channels per input link is 128. Flits are 1024 bit long. The MMR buffers have capacity to store one flit per virtual channel. The scheduling round size, determined by the K parameter, has been been set to $K = 16$.

Constant and Variable Bit Rate traffics (CBR and VBR traffic, respectively) have been considered in simulations. The *CBR traffic model* is composed of a mix of synthetic connections with 64 Kbps, 1.54 Mbps, and 55 Mbps average bandwidth. These are representative of several applications, such as audio, and uncompressed video and high-definition video transmission, respectively. The *VBR traffic model* is based on traces obtained from real MPEG-2 video sequences. This is a typical type of multimedia flow [15].

Traffic sources inject their flits into buffers located in the corresponding NIC attached to every input/output port. Input workload is measured as a percentage of link bandwidth. Destination ports have been randomly selected using a uniform distribution. No statistical information is gathered until some scheduling rounds have been completed in order to get data only when the system is stable. Simulations have been carried out for long enough to record significant data on applications performance. Due to space limitations, only the most significant results are presented in this paper.

First, the TIABP scheme is compared with both the IABP and SIABP algorithms. Average delay since generation for the CBR flows and the VBR traffic are presented in Figures 3(a-c). Results show little differences when load is under saturation point (less than 75% of generated load). When load is closes to saturation, the best performance is got by the SIABP algorithm. The most significant differences occurs between the IABP and the SIABP algorithms for medium CBR flows (1.54 Mbps). Regarding the TIABP algorithm, it presents results similar to that obtained with the SIABP algorithm for medium CBR flows, but performance is slightly worse for high bandwidth CBR flows (55 Mbps) and for VBR flows. Having into account that the TIABP algorithm reduces the hardware cost, this performance degradation can be tolerated only if the QoS of the traffic can be guaranteed.

Hence, in order to know whether the multimedia flows are receiving the requested QoS, we have depicted in Figures 3(d-f) the distribution of flits/frames delays, for one of the highest workload levels simulated (85% of link utilization). This distribution is computed as the percentage of flits/frames that suffered a delay lower than a set of thresholds. Threshold values are related to the QoS needs of the connections. In the case of CBR flows, their values are related to the IAT(*Inter Arrival Time*) of the connection while for VBR flows thresholds are related to the frame time (33 millisecs). For the CBR flows, Figures 3(d-e) show similar results for all the algorithms. For the VBR flows, see Figure 3(f), less frames fulfill the tightest thresholds when TIABP is used, as compared to SIABP. Nevertheless, the new proposal obtains reasonable deadlines for all the frames.

From the analysis above, it can be concluded that the router with the TIABP scheme is able to provide QoS guarantees to the multimedia flows at a lower hardware cost than than both the IABP algorithm and the SIABP algorithm.

For reference purposes, the performance obtained when using the practical TIABP algorithm is compared to that achieved when introducing a couple of classical algorithms acting as link schedulers. The algorithms chosen for this purpose are Virtual-Clock (VC) [16] and Weighted Fair Queuing (WFQ) [17]. Like TIABP does, both VC and WFQ assign priorities to flits according to the bandwidth requirements of the connection they belong to.

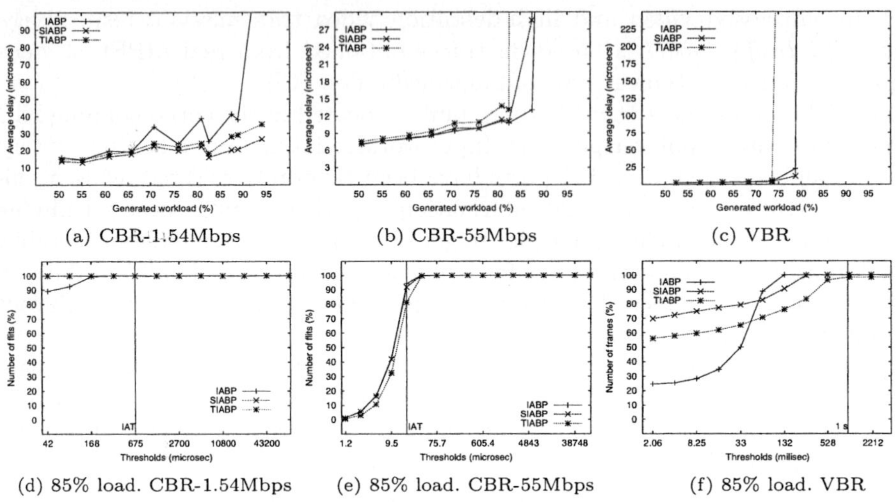

(a) CBR-1.54Mbps   (b) CBR-55Mbps   (c) VBR

(d) 85% load. CBR-1.54Mbps   (e) 85% load. CBR-55Mbps   (f) 85% load. VBR

Fig. 3. IABP vs TIABP: Delay since generation (a)-(c) and Distribution delay(d)-(f).

The plots shown in Figures 4(a) and 4(d) correspond to the average delay since generation obtained with the three link switch schedulers, for the CBR connections with medium and highest requirements. While TIABP outperforms the others algorithms for medium CBR flows, saturation is reached at lower loads than when using the VC algorithm for the most demanding CBR flows. On the other hand, considering the distribution delay of the algorithms, Figures 4(b) and 4(e), we can appreciate that little differences are found between the results obtained with VC and TIABP. Moreover, when WFQ is used a large amount of flits (30% of the generated flits, approximately) cannot fulfil even the most relaxed thresholds. Note that plots depict the distribution of flits delay for a workload of 85%.

The conclusion is that when WFQ is used, the connections with the highest bandwidth requirements cannot meet their QoS requirements, because they do not receive their share of bandwidth. On the other hand, VC and TIABP exhibit better behavior when operating in this way.

Last, Figures 4(c) and 4(f) show average jitter. TIABP is able to provide an almost constant average jitter over all the workload range, below or equal the values obtained for the other two algorithms.

6   Conclusion

The main goal pursued by the MultiMedia Router (MMR) project is to design a single-chip router able to efficiently handle multimedia flows and best-effort traffic in LAN/SAN environments. In order to achieve this goal, solutions to many difficult resource management and scheduling problems must be provided, while keeping into account that these solutions must be simple enough to permit effective single-chip implementation.

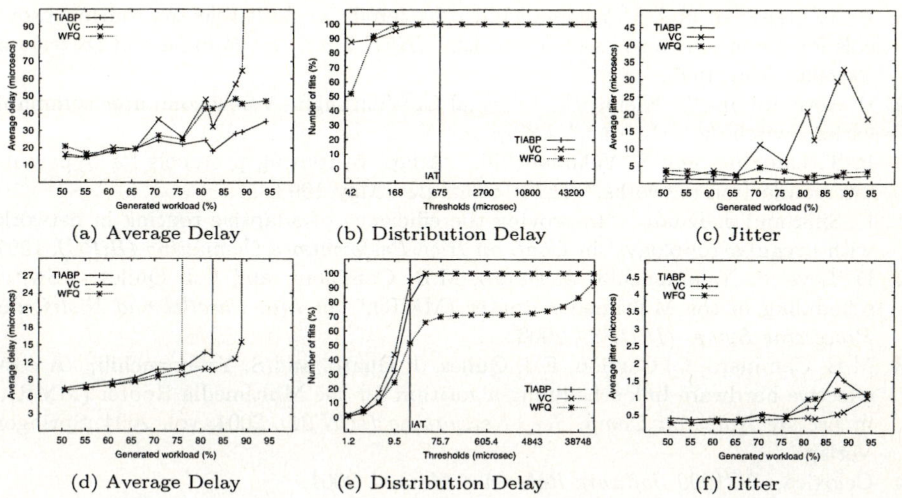

Fig. 4. TIABP versus VC and WFQ: CBR 1.54Mbps (a)-(c) and CBR 55Mbps(d)-(f).

The link scheduling algorithm is one key element on the MMR design to provide QoS guarantees to the multimedia flows. Thus, in this work the proposed Temperature-IABP scheme is analysed. We have obtained a simplified link scheduling design based on the temperature coding of the priority value associated to every head flit and used in the candidate selection phase. Thus, TSORT design uses about 10% less FPGA area than the previous SORT design. Then, the new TIABP link scheduling algorithm reduce the hardware cost while is available of provide proper QoS guarantees to multimedia flows.

References

1. Myricom, Inc, *Guide to Myrinet-2000 Switches and Switch Networks*, August 2001.
2. R. Froom, M. Flannagan, and K. Turek, *Cisco Catalyst QoS: Quality of Service in campus networks*, chapter Exploring QoS in Catalyst, Cisco Press, 2003.
3. A. Pandey and H.M. Alnuweri, "Quality of Service support over switched Ethernet," in *IEEE Pacific Rim Conf. on Communications, Computers and Signal Processing*, 1999.
4. J. Duato, S. Yalamanchili, M. B. Caminero, D. Love, and F. J. Quiles, "MMR: A high-performance multimedia router - Architecture and design trade-offs," in *Intl. Symp. on High Performance Computer Architecture (HPCA-5)*, 1999.
5. Y. Tamir and H.C. Chi, "Symmetric crossbar arbiters for VLSI communication switches," *IEEE Trans. on Parallel and Distributed Systems*, vol. 4, no. 1, 1993.
6. C. Sullivan and S. Chapell, "Handel-C for co-processing an co-design of field programmable systems on chip," in *Proc. of the JCRA'02*, 2002.
7. G. Pfister, *High Performance Mass Storage and Parallel I/O*, chapter 42: An Introduction to the InfiniBand Architecture, IEEE Press and Wiley Press, 2001.

8. P. T. Gaughan and S. Yalamanchili, "A family of fault-tolerant routing protocols for direct multiprocessor networks," *IEEE Trans. on Parallel and Distributed Systems*, May 1995.
9. P. Kermani and L. Kleinrock, "Virtual Cut-Through: A new computer communication switching technique," *Computer Networks*, vol. 3, 1979.
10. P. T. Gaughan and S. Yalamanchili, "Adaptive routing protocols for hypercube interconnection networks," *IEEE Computer*, May 1993.
11. F. Silla and J. Duato, "Improving the efficiency of adaptive routing in networks with irregular topology," in *Conf. on High Performance Computing (HiPC)*, 1997.
12. D. Love, S. Yalamanchili, J. Duato, M.B. Caminero, and F.J. Quiles, "Switch scheduling in the Multimedia Router (MMR)," in *Intl. Parallel and Distributed Processing Symp. (IPDPS)*, 2000.
13. M.B. Caminero, C. Carrión, F.J. Quiles, J. Duato, and S. Yalamanchili, "A cost-effective hardware link scheduling algorithm for the Multimedia Router (MMR)," in *Lecture Notes on Comp. Sci.: Networking (ICN'01)*. 2001, vol. 2094, Springer-Verlag.
14. Celoxica, *RC1000 Software Reference Manual*, 2001.
15. "Generic coding of moving pictures and associated audio. Rec. H.262. Draft Intl. Standard ISO/IEC 13818-2," 1994.
16. L. Zhang, "Virtual Clock: A new traffic control algorithm for packet switching networks," *ACM Trans. Comp. Sys.*, May 1991.
17. A. Demers, S. Keshav, and S. Shenker, "Analysis and simulations of a fair queuing algorithm," in *ACM SIGCOMM*, 1989.

Topic 14
Mobile and Ubiquitous Computing

Evaggelia Pitoura, Marios Dikaiakos, Valérie Issarny, and Nuno Preguica

Topic Chairs

Wireless communications along with portable computers, digital assistants and sensor devices provide a pervasive base for mobile computing. However realizing truly ubiquitous mobile computing requires innovative theories, paradigms and applications in various research areas including algorithms, networking, software architectures and data management. Topic 14 covers all such aspects. This topic attracted 33 submissions showing the increasing interest in the field. From the submitted papers, 8 were accepted as full papers (24% acceptance rate). The selected papers cover various aspects of mobile and ubiquitous computing highlighting the diversity of the field and thus making up an interesting and stimulating track.

Paper presentations are divided into three sessions. Two papers in the first session address consistency issues. In *"Efficient and Fault-Tolerant Update Commitment for Weakly Connected Replication"*, J. Barreto and P. Ferreira propose a novel epidemic weighted voting protocol for achieving eventual consistency in optimistic replication that allows multiple update candidates in an election. In *"Controlling Concurrency in Mobile Computing Environments with Broadcast-based Dissemination"*, J. M. Monteiro and A. Brayner present a new serializability-based protocol for ensuring data consistency and currency when data are disseminated to clients through wireless broadcast. The last paper in this session, *"Integrating Mobile Devices into the Grid: Design Considerations and Evaluation"* by S. Isaiadis and V. Getov, discusses implementation and performance issues for integrating mobile devices into the grid. The second session is shared with Topic 8. The two papers address aspects related to frequency utilization in wireless networking. In *"New Bounds on the Competitiveness of Randomized Online Call Control in Cellular Networks"*, I. Karagiannis, C. Kaklamanis and E. Papaioannou present new upper and lower bounds for the online version of the call control problem in wireless cellular networks. In *"A Multiple Channel Access Protocol for Ad Hoc Wireless Networks"*, K-W. Jang considers the problem of enhancing channel utilization in wireless ad hoc networks through channel exchange between neighboring nodes. In the last session, in *"Personalized Access to Semantic Web Agents Using Smart Cards"*, R. C. Erdur and G. Kardas argue for storing personal information on smart cards. The other two papers focus on secure networking. K. Kim, J. Hong and J. Lim present a *"Fast and Secure Communication Resume Protocol"* to speed up connection resume after a communication error. In *"On AAA with Extended IDK in Mobile IP Networks"*, H. Jeon, M. Y. Chung and H. Choo discuss how to attain fast and secure mobile IP networking by addressing problems of the hand-off process in a secure way.

J.C. Cunha and P.D. Medeiros (Eds.): Euro-Par 2005, LNCS 3648, p. 1057, 2005.
© Springer-Verlag Berlin Heidelberg 2005

An Efficient and Fault-Tolerant
Update Commitment Protocol
for Weakly Connected Replicas

João Barreto* and Paulo Ferreira**

INESC-ID / IST
Rua Alves Redol, 9, 1000 Lisboa, Portugal
{joao.barreto,paulo.ferreira}@inesc-id.pt

Abstract. Mobile and other loosely-coupled environments call for decentralized optimistic replication protocols that provide highly available
access to shared objects, whilst ensuring an eventual convergence towards
a strongly consistent state. In this paper we propose a novel epidemic
weighted voting protocol for achieving such goal. Epidemic weighted voting approaches eliminate the single point of failure limitation of primary
commit approaches. Our protocol introduces a significant improvement
nover other epidemic weighted voting solutions by allowing multiple,
causally related updates to be committed at a single distributed election round. We demonstrate that our proposed protocol is especially
advantageous with the weak connectivity levels that characterize mobile and other loosely-coupled networks. We support such assumptions
by presenting comparison results obtained from side-by-side execution of
reference protocols in a simulated environment.

1 Introduction

Data replication is a fundamental mechanism for most distributed systems for
performance, scalability and fault tolerance reasons. In particular, optimistic
replication protocols [1] are of extreme importance in mobile and other loosely-
coupled network environments. The nature of these environments calls for decentralized replication protocols that are able to provide highly available full access
to shared objects. Such requirement is accomplished by optimistic replication
strategies, which, in contrast to their pessimistic counterparts, enable updates
to be issued at any one replica regardless of the availability of other replicas.

As a trade-off, the issue of consistency in optimistic replication is problematic. Since replicas are allowed to be updated at any time and circumstance,
updates may conflict if issued concurrently at distinct replicas. Some optimistic
replication protocols ensure that, from such a possibly inconsistent *tentative*
state, replicas evolve towards an eventual consistent *stable* state. For this end,
a distributed consensus algorithm is executed so as to reach an agreement on a
common order in which tentative updates should be committed.

* Funded by FCT Grant SFRH/BD/13859.
** Funded by FCT Project UbiRep (POSI/CHS/47832/2002).

J.C. Cunha and P.D. Medeiros (Eds.): Euro-Par 2005, LNCS 3648, pp. 1059–1068, 2005.

There are many scenarios where users, in order to benefit from high availability, are willing to work with temporarily tentative data, provided that a commitment agreement regarding such data will eventually be reached. Consider, for instance, a laptop user that becomes disconnected from his corporate file server after leaving his office. If necessary, he may expect to be able to modify a report that is currently replicated at his laptop, even if tentatively.

Furthermore, such worker may meet other mobile team colleagues carrying their replicas and, in an ad-hoc fashion, establish a short term work group to collaboratively work on the report. A set of causally related tentative updates will result from such activity. Hopefully, if no update is concurrently issued from outside the group, such tentative work will be eventually committed by the underlying consistency protocol. Hence, the high availability provided by an optimistic replication strategy is especially interesting in such scenarios as the previous ones. However, the usefulness of one such approach strongly depends on the ability of the underlying replication protocol to efficiently achieve a commitment decision concerning the tentatively issued data. Users are typically not inclined towards working on tentative data unless they trust the protocol to rapidly achieve a strongly consistent commitment decision regarding such data.

Aiming at such central objective, this paper proposes a novel optimistic replication protocol for efficient and highly available update commitment through the use of an epidemic weighted voting protocol based on version vectors [2]. The use of a voting approach eliminates the single point of failure of primary commit approaches [3]. Hence, the unavailability of any individual replica is not prohibitive of the progress of the update commitment process. Moreover, commitment agreement is accomplished without the need for a plurality quorum of replica servers to be simultaneously accessible: voting information flows epidemically between replicas and update commitment is based solely on local information.

The solution we propose has the main contribution of introducing a significant improvement over basic epidemic weighted voting solutions by allowing multiple update candidates to participate in an election. By using version vectors, candidates consisting of one or more causally related updates may be voted and committed by running a single distributed election round. As a result, the overall number of anti-entropy sessions required to commit updates is decreased when compared to a basic weighted voting protocol. Hence, update commitment delay is minimized and so eventual strong consistency guarantees are more rapidly delivered to applications. Namely, such reduction is substantial in scenarios where frequent causally related updates are tentatively generated by applications. The examples presented above are representative of such update patterns. In worst case scenarios, our protocol behaves similarly to basic weighted voting protocols.

The paper is organized as follows. Section 2 describes related work, Section 3 introduces the protocol, evaluated in Section 4, and Section 5 concludes.

2   Related Work

The issue of optimistic data replication for mobile and loosely coupled environments has been addressed by a number of projects [1], with the common intent of offering high data availability. Most of the proposed solutions share the goal of our work by enforcing eventual convergence towards a strongly consistent stable form that is explicitly presented to applications.

Three main approaches can be distinguished. Firstly, Golding [4] proposes that each individual server commits an update when it is certain that it has been received by every replica. A main limitation is that the unavailability of any single replica stalls the entire commitment process. On the other hand, a primary commit strategy, such as the one adopted by Bayou [3], centralizes the commitment process in a single distinguished primary replica that establishes a total commit order over the updates it receives. Primary commit is able to rapidly commit updates, since it suffices for an update to be received by the primary replica to become committed, provided that no conflict is found. However, should the primary replica become unavailable, the commitment progress of updates generated by replicas other than the primary is inevitably halted.

Finally, a third approach uses voting so as to allow a plurality quorum to commit an update. In particular, Deno [5] relies on an epidemic voting protocol to support object replication in a transactional framework for loosely-connected environments. Deno requires one entire election round to be completed in order to commit each single update, if only non-commutable updates are considered. This is acceptable when applications are interested in knowing the commitment outcome of each tentatively issued update before issuing the next one. However, in the usage scenarios addressed by this paper, users and applications will often be interested in issuing multiple tentative updates before acknowledging their commitment. In such situations, the commitment delay imposed by Deno's voting protocol becomes unacceptably higher than that of primary commit.

3   Consistency Protocol

The following sections consider a model where a set of logical objects is replicated at multiple server hosts. An object replica at a given server provides local applications with access to a version of the object contents, as stored by the replica. Such accesses may read or modify the object contents. In the case of the latter, an update is issued by the server and applied to the replica.

Updates issued at a given replica are propagated to other servers in an epidemic fashion in order to eventually achieve object consistency. The local execution of an update is assumed to be recoverable, atomic and deterministic. The former means that a replica will not reach an inconsistent value if it fails before the update execution completes. It follows from the other two properties that the execution of the same ordered sequence of updates at two distinct replicas in the same initial consistent state will yield an identical final state. For simplicity and without loss of generality, we consider that each logical object is replicated at every server in the system. For the sake of generality, the set of replicas may be dynamic, and thus change with the creation or removal of new servers.

Hereafter, we assume an asynchronous system in which servers can only fail silently. Network partitions may also occur, thus restricting connectivity between servers that happen to be located in distinct partitions.

3.1   Overview

Due to the optimistic nature of the consistency protocol, an update issued at a local replica is not immediately committed at every remaining replica. Instead, such update is considered to be in a tentative form since conflicting updates may still be issued at other replicas. The consistency protocol is responsible for committing such tentative updates into a total order that will be eventually reflected at every replica.

Our protocol achieves this goal through a weighted voting approach [5]: concurrent tentative updates are regarded as rival candidates in an election. The servers replicating a given logical object act as voters whose votes determine the outcome of each election between candidate updates to the object. A candidate update wins an election by collecting a plurality of votes, in which case it is committed and its rival candidates are discarded.

Elections consider a fixed per-object currency scheme, in which each voter is associated with a given amount of currency that determines its weight during voting rounds. The global currency of a logical object, distributed among its replica servers, equals a fixed amount of 1. Currencies can be exchanged between servers and the currency held by failed servers can be recovered by running a *currency reevaluation* election, as discussed in [6].

Version Vector Candidates. In some cases, applications will be interested in generating more than one tentative update prior to its commitment decision. These may include disconnected mobile applications and ad-hoc groups of mobile applications working cooperatively in the absence of a plurality quorum. Since the commitment decision may not be taken in the short-term, these applications may wish to issue a sequence of multiple, causally ordered tentative updates.

In order to efficiently accommodate for such update scenarios, the novel solution proposed in this paper employs version vectors to identify candidate updates in a weighted voting protocol. The flexibility brought by version vectors allows a sequence of one or more updates to run for the current election as a whole. In this case, the candidate is represented by the version vector corresponding to the tentative version obtained if the entire update sequence was applied to the replica. As the next sections explain, the voting protocol relies on the causality expressiveness of version vectors to deciding if the update sequence or a prefix of it are to become committed. Consequently, candidates consisting of one or more causally related updates may be committed on a single distributed election round. In weakly connected network environments, where such update patterns are expectably dominant, a substantial reduction of the update commitment delay is therefore achievable.

Each replica r maintains the following state:

- $stable_r$, which consists of a version vector that identifies the most recent stable version that is currently known by replica r, obtained after the ordered application of all committed updates;
- $votes_r[1..N]$, which stores, for each server $k = 1, 2, .., N$, the version vector corresponding to the candidate voted for by k, as known by r; or $\perp$, if the vote of such server has not yet been known to r;

- $cur_r[1..N]$, which stores, for each server $k = 1, 2, .., N$ whose vote replica r has knowledge of, the currency associated with such vote;

Each server is able to offer two possibly distinct views over the value of a replica r to its applications and users: the stable and tentative views. The first view reflects a strongly consistent value of the replicated object that is identified by $stable_r$. On the other hand, the tentative view exposes a weakly consistent value that corresponds to the candidate version that is currently voted by the local server, $votes_r[r]$.

Issuing a tentative update on a replica r causes a new candidate to run for the current election according to the following rules:

1. If $votes_r[r] = \bot$, then $votes_r[r] \leftarrow adv_r(stable_r)^1$ and $cur_r[r] = currency_r$;
2. Otherwise, $votes_r[r] \leftarrow adv_r(votes_r[r])$;

As the next sections describe in greater detail, voting information flows in an epidemic fashion among servers and the decision to commit an update is based only on local replica information. These are important properties for operation under mobile and loosely-coupled environments. In particular, Section 3.2 addresses the storage of tentative update and their corresponding commitment upon a replica value. Section 3.3 then describes the epidemic flow of consistency information and Section 3.4 finally defines how candidates are elected.

3.2 Update Commitment

The protocol proposed hereafter is orthogonal to the issues of actual transference and storage of tentative updates. In particular, the protocol does not impose the decision of whether to transfer and store, at each individual replica, the tentative updates belonging to every candidate in the current election or, alternatively, only those concerning the replica's own candidate.

This means that, at the time a server determines that a given candidate has won the election and, thus, its updates should be committed, such updates may not be immediately available. Instead, they will be eventually collected through succeeding anti-entropy sessions with other servers. Consequently, there may occur a discrepancy between the most recent stable version identified by the consistency protocol and the actual stable value that is locally accessible. Such discrepancy is enabled by an additional element at the state of each replica r:

- c_r, which consists of an integer value representing the number of updates in the stable path that have already been committed by replica r;

The value of c_r may be lower than the number of updates that have actually been determined by the consistency protocol as belonging to the stable path. In such case, the replica's stable value does not yet reflect the most recent stable version r is aware of. As a consequence, the protocol is flexible enough to support servers with differing memory limitations.

On one hand, servers with rich memory resources may store every update associated with each candidate, hence being able to immediately gain access to

---

^{1} adv_r advances the counter corresponding to r in the supplied version vector by one.

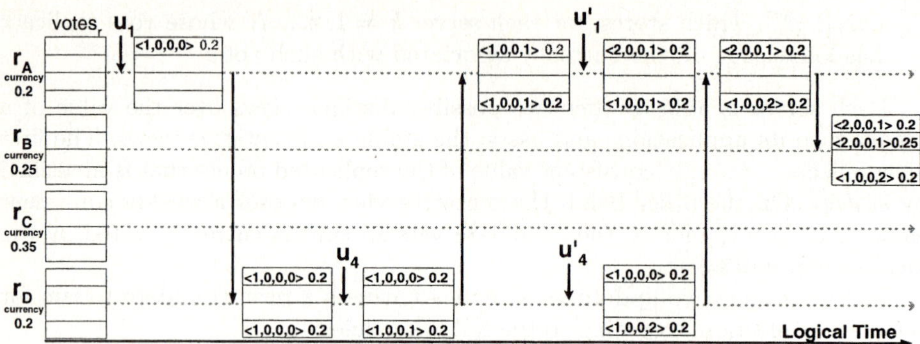

Fig. 1. Example of update generation and propagation: four replicas with unevenly distributed currencies start from a common initial stable version $stable_r = \langle 0,0,0,0 \rangle$.

the most recent known stable value as each new stable version is determined by the protocol. On the other hand, memory-constrained devices may opt to restrict themselves to storing only the updates of their own candidate and, thus, allow for occasional delays in the availability of the most recent stable value when rival candidates win an election. In either case, however, the efficiency of the protocol in taking commitment decisions is not affected. Both strategies may transparently co-exist in a system of replicas of the same logical object.

Moreover, it is assumed that a log of committed updates is maintained, including the following information:

- $gen_r[1..c_r]$, which stores, for each update committed so far in replica r, the server that generated it.

From the consistency protocol's viewpoint, the procedure for committing a sequence of updates $u_1, .., u_n$, generated by servers $i_1, ..i_n$, respectively, is therefore comprised by the following steps:

1. For each update, u_k, $gen_r[c_r + k] \leftarrow i_k$;
2. $c_r \leftarrow c_r + n$;

3.3   Anti-entropy

Voting information is propagated through the system by anti-entropy sessions established between pairs of accessible replicas. An anti-entropy session is an unidirectional pull-based interaction in which a requesting replica, A, updates its local election knowledge with information obtained from another replica, B. In case B has more up-to-date election information, it transfers such information to A. Furthermore, if A has not yet voted for a candidate that is concurrent to the one voted for by B, A accepts the latter, thus contributing to its election.

Each anti-entropy session is carried out according to the following procedure, which should be executed atomically:

1. If $stable_A < stable_B$ then
   (a) $stable_A \leftarrow stable_B$;
   (b) $\forall k$ s.t. $votes_A[k] \| stable_A$ or $votes_A[k] \leq stable_A$, then $votes_A[k] \leftarrow \bot$;

2. If $(votes_A[A] = \perp$ and $stable_A < votes_B[B])$ or $votes_A[A] < votes_B[B]$ then $votes_A[A] \leftarrow votes_B[B]$ and $cur_A[A] \leftarrow currency_A$;

3. $\forall k \neq A$ s.t. $(votes_A[k] = \perp$ and $stable_A < votes_B[k])$ or $votes_A[k] < votes_B[k]$, then $votes_A[k] \leftarrow votes_B[k]$ and $cur_A[k] \leftarrow cur_B[k]$.

4. If $c_A < c_B$ then commit update sequence issued by $gen_B[c_A+1], .., gen_B[c_B]$.

The first step ensures that, in case r_B knows about a more recent stable version, r_A will adopt it. This means that r_A will regard the elections that originated such new stable version as completed and so begin a new election from that point. Such new election is prepared by keeping only the voting information that will still be meaningful for the outcome of the election. Namely, these are the votes on candidates that causally succeed the stable version.

As a second step, r_A is persuaded to vote for the same candidate as the one voted by r_B, provided that r_A has not yet voted for a concurrent candidate. Subsequently, r_A updates its current knowledge of the current election with relevant voting information that may be held by r_B. Namely, r_A stores each vote that it is not yet aware of or whose candidate is more complete than the one it currently has knowledge of.

Finally, the set of committed updates held by B that are not yet locally available at replica A are collected and committed by the latter. An example of update generation and propagation through anti-entropy is illustrated in Fig. 1.

3.4 Election Decision

The candidates being voted in an election represent update paths that traverse through one of more versions beyond the initial point defined by the stable version, *stable*. These possibly divergent candidate update paths may share common prefix sub-paths. The following definition expresses such notion.

Definition 1: *Maximum common version.* Given two version vectors, v_1 and v_2, their maximum common version is given by a version vector, $mcv(v_1, v_2)$, s.t. $\forall k, mcv(v_1, v_2)[k] = min(v_1[k], v_2[k])$. For simplicity, we represent $mcv(v_1, v_2, ..., v_m)$ as the result of $mcv(mcv(mcv(v_1, v_2)), ...), v_m)$.

Theorem 1: Let $v_1, .., v_m \in votes_r$, be one or more candidate versions known by replica r, each connoting a tentative update path starting from the stable version, $stable_r$. Their maximum common version, $mcv(v_1, .., v_m)$, constitutes the farthest version of an update sub-path that is mutually traversed by the update paths of $v_1, .., v_m$. Complementarily, the total currency voted for such common sub-path is obtained by $voted_r(mcv(v_1, .., v_m)) = cur_r[1]+...+cur_r[N]$.

The voting protocol is responsible for progressively determining common sub-paths of candidate versions that manage to obtain a plurality of votes. This decision is based on the definition of maximum common version among the set of candidate versions voted at a given replica and on the value of uncommitted currency, $uncommitted_r = \sum cur[k] : votes_r[k] \neq \perp$, according to the following:

Definition 2: Let w be a version vector s.t. $w = mcv(w_1, .., w_m)$ where $w_1, .., w_m \in votes_r$ and $1 \leq m \leq N$. w wins an election when:

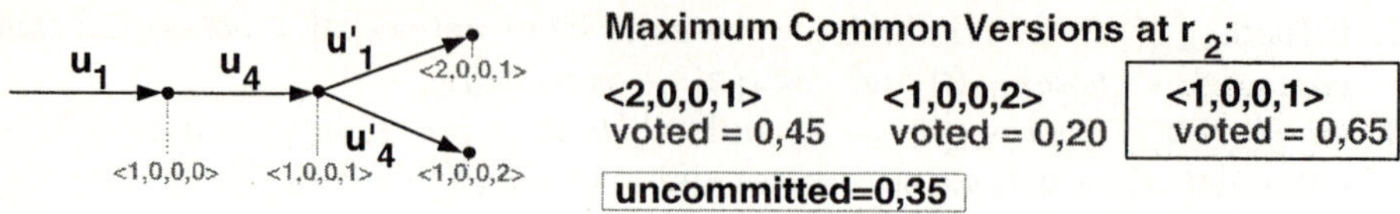

Fig. 2. Election decision for replica r_2 at the final state in Figure 1. Candidate $\langle 1, 0, 0, 1 \rangle$ has collected a plurality of votes and, thus, u_1 and u_4 will be committed in that order.

1. $voted_r(w) > 0.5$, or
2. $\forall l$ s.t. $l = mvc(l_1, .., l_k), l_1, .., l_k \in votes_r, 1 \le k \le N$ and $l \parallel w$,
   (a) $voted_r(w) > voted_r(l) + uncommited_r$, or
   (b) $voted_r(w) = voted_r(l) + uncommited_r$ and $w <_{lex} l$.

The above rules state the conditions that guarantee that a candidate has collected sufficient votes to win an election. The votes may constitute a majority, when the amount of currency voted on the winning candidate surpasses 0.5; or a simple plurality, when the voted currency is greater than the maximum potentially obtainable currency of any other rival candidate. Ties are decided by choosing the candidate whose version vector is lexically lower. If one represents each version vector as a number whose digits are the elements of the vector, such representation can be numerically compared, thus inducing a lexical order, $<_{lex}$, in the version vector space.

Determining if a candidate has won an election depends exclusively on information that is locally available at each replica. This means that, once having collected enough voting information, a given replica is able to decide, by its own, to commit a candidate version that locally fulfills the election winning conditions. Hence, update commitment is accomplished in a purely decentralized manner. An example is depicted in Fig. 2.

After finding a new winner version vector, w, a replica r atomically takes the following steps to accept the election decision and prepare for the next election:

1. $stable_r \leftarrow w$;
2. $\forall v_k \in votes_r$ s.t. $v_k \parallel w$ or $v_k \le w$, $votes_r[k] \leftarrow \perp$;
3. If the sequence of updates that comprise the update path defined between versions $stable_r$ and w is locally available, then commit it;

After accepting the election result by setting the winning version as the new stable version, the second step resets all the defeated candidates to $\perp$. Depending on the local availability of the updates that belong to the winning candidate, they may be committed into the replica's stable value; otherwise, further anti-entropy sessions will ensure that such updates are eventually collected and committed. A new election can then take place.

Theorem 2 (Correctness): After all elections have been completed at every replica and all updates belonging to the resulting stable path have been committed at every replica: $\forall r, t$, replica r has committed the same ordered sequence of updates as t.

4   Evaluation

C# implementations of the primary commit (*Primary*), basic weighted voting (*Basic WV*) and version vector weighted voting (*VVWV*) protocols were run side-by-side in a simulated environment. The simulator includes a collection of replicas of a common logical object, randomly distributed by a set of network partitions. Time is divided into logical time slices; at each time slice, each replica (1) with a given *mobility probability*, migrates to a different, randomly chosen, network partition; (2) pulls anti-entropy information from a partner, randomly selected from the set replicas present in its current partition; and, (3) generates, with a given *update probability*, one tentative update. Each replica may be active or inactive; in the case of the latter, its update probability is null. An inactive replica exchanges, with a given *activation probability*, its status with an active replica after pulling anti-entropy information from it. The differentiation between active and inactive replicas allows for non-uniform update models to be simulated, namely the hot-spot model [7], which assumes, based on empirical evidence, that updates typically occur in a small set of replicas.

The protocols were evaluated against an increasing number of partitions. Since update contention is prone to arise in a partitioned system, the update commitment delay is not a sufficiently meaningful measure for our purposes, as it does not take into account the discarded updates. Instead, a better evaluation is provided by the update commitment ratio of each protocol, i.e. the percentage of issued updates that is committed at all replicas.

The measurements were obtained with the fixed settings of 10 replicas with mobility and activation probabilities of 20% and 40%, respectively, running for 2000 time slices on each experiment; we observed that the variation of such values does not have a relevant impact on obtained results. Three update models were tested: with ten, five and just a single active replicas; a global update probability of 5%, evenly divided by the active replicas, was considered.

The commitment ratio is directly affected by the efficiency of each evaluated update commitment protocol, since if updates remain in their tentative state for longer periods, the probability of conflicts is higher; hence, lower commitment ratios reflect longer delays imposed by the update commitment process. So, as expected, update commitment ratios decrease as the connectivity among replicas is weakened by an increasing number of partitions, as shown in Figure 3.

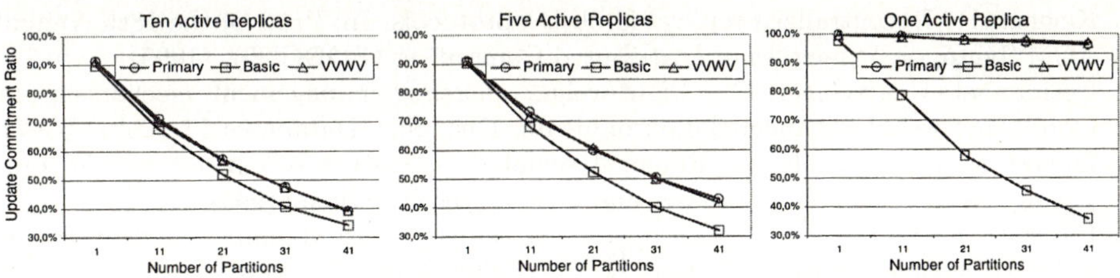

Fig. 3. Update commitment ratios versus number of partitions, for different numbers of active replicas.

However, Primary and VVWV are able to ensure higher ratios than Basic WV as partitioning grows. Situations of multiple causally related tentative updates occur more frequently as updates remain tentative for longer periods. Hence, such results are explained by the efficiency of the former protocols in the commitment of multiple causally related updates, in contrast to Basic WV. Such situations are also increased as the global update probability is distributed by a smaller number of active replicas. Accordingly, the advantage of Primary and VVWV over Basic WV is accentuated as the number of active replicas decreases. It should be noted that higher update probabilities yielded equivalent, yet magnified, conclusions. On the other hand, Primary and VVWV have similar ratios; however, VVWV has the crucial advantage of not depending on a single point of failure.

Finally, similar experiments compared the two update storage alternatives of VVWV. A maximum improvement of 0.8% was attained by storing the updates of all candidates, which suggests that the more resource-efficient alternative of storing only the updates of a replica's own candidate is acceptable.

5   Conclusions

We propose a novel epidemic weighted voting protocol, VVWV, for achieving the goal of optimistic update commitment that allows multiple causally ordered update candidates to be committed at a single election round. Simulation results show that, under weak connectivity conditions, VVWV is advantageous relatively to a basic weighted voting protocol, while attaining similar update commitment ratios to the less fault-tolerant primary commit protocol.

Additional work [8], not addressed in this paper, shows how dynamic version vector maintenance can be effectively incorporated into the proposed protocol and proves Theorems 1 and 2.

References

1. Saito, Y., Shapiro, M.: Optimistic replication. ACM Comput. Surv. **37** (2005) 42–81
2. Parker, D.S., et al: Detection of mutual inconsistency in distributed systems. Distributed systems, Vol. II: distributed data base systems (1986) 306–312
3. Petersen, K., et al: Flexible update propagation for weakly consistent replication. In: Proceedings of the 16th ACM Symp. on Operating Systems Principles. (1997)
4. Golding, R., Long, D.: Modeling replica divergence in a weak-consistency protocol for global scale dirstibuted data bases. Technical Report UCSC-CRL-93-09 (1993)
5. Keleher, P.: Decentralized replicated-object protocols. In: Proc. of the 18th Annual ACM Symp. on Principles of Distributed Computing (PODC'99). (1999)
6. Cetintemel, U., Keleher, P.: Light-weight currency management mechanisms in mobile and weakly-connected environments. Dist. Par. Databases **11** (2002) 53–71
7. D. Ratner, P.R., Popek, G.: Roam: A scalable replication system for mobile computing. In: Mobility in Databases and Distributed Systems. (1999)
8. Barreto, J., Ferreira, P.: Optimistic consistency with dynamic version vector weighted voting. Technical Report RT/008/2004, Inesc-ID (2004)

Controlling Concurrency in Mobile Computing Environments with Broadcast-Based Data Dissemination

José Maria Monteiro[1,2] and Ângelo Brayner[1]

[1] Departamento de Informática, UNIFOR
Av. Washington Soares 1321, 60811-905, Ceará, Brasil
{monteiro,brayner}@unifor.br
[2] Departamento de Informática, PUC-Rio
R. Marquês de São Vicente 225, 22453-900 - Rio de Janeiro, Brasil
monteiro@inf.puc-rio.br

Abstract. A wireless broadcast environment is defined as a mobile computing environment in which data are delivered to mobile clients by means of a broadcast-based mechanism. Of course, those applications have to see the most recent consistent database state. For that reason, in such a scenario, database servers should synchronize operations for ensuring data consistency and currency of data. However, conventional serializability-based concurrency control protocols are unsuitable for synchronizing transactions in broadcast environments. The major goal of this work is to present a new serializability-based protocol to synchronize transactions in data intensive applications. The proposed protocol saves battery power, since it ensures that mobile clients do not have to contact servers (for requiring locks, for example) to access data. Thus, mobile clients do not need to listen to the broadcast continuously; they listen to the broadcast channel to retrieve data they need. Therefore, the proposed protocol supports client disconnections. We performed simulation analysis to evaluate the performance of the new protocol. The simulation results show that the proposed protocol offers better performance than others protocols.

1 Introduction

The integration of portable computer technology with the wireless-communication technology has created a new paradigm in computer science, the so-called mobile computing. In a mobile computing environment, network nodes are no longer fixed, that means, they do not have a fixed physical location. In such an environment, mobile users using a portable computer (denoted mobile client or host) may access shared information and resources regardless of where they are located or if they are moving across different physical locations and geographical regions.

Mobile computing technology has made possible the development of new and sophisticated database applications. A particular class of such applications can be characterized by having a large number of mobile clients, a small number of servers and a relatively small database. Electronic commerce applications, such as auctions, road traffic management systems and automated industrial plants [8] are examples of database applications, which require the support of the mobile computing technology. Those applications can benefit from a broadcast mode for data dissemination (*push-*

J.C. Cunha and P.D. Medeiros (Eds.): Euro-Par 2005, LNCS 3648, pp. 1069–1079, 2005.

based approach for data dissemination). In this model, a server repetitively broadcasts data to a client population without a specific request. In turn, clients monitor the broadcast channel in order to retrieve their data items of interest. Conventionally, data are delivered to clients on demand (*pull-based* approach for data dissemination).

Broadcasting data to mobile clients instead of sending them on demand has several advantages. For instance, the database server is not overloaded with requests from a large population of mobile clients and it does not have to send individual messages to a specific client as in pull-based systems. Furthermore, data can be accessed concurrently by any number of clients without any performance degradation, since all mobile clients can simultaneously listen to the broadcast channel.

Therefore, broadcast-based data dissemination has become a widely accepted technique of disseminating data in mobile computing. Many such systems have been proposed and some commercial products for information dissemination in wireless networks already support broadcast. For example consider the AirMedia system [3], which regularly sends CNN news and information to subscribers. Such subscribers should be equipped with a receiver antenna connected to their personal computers.

A broadcast environment is defined as a mobile computing environment in which data are delivered to mobile clients by means of a broadcast-based mechanism. Applications running in a broadcast environment need to read the most recent consistent database state[1]. Therefore, the database server (database system running on a server machine) should ensure that mobile clients "see" the most recent consistent state of the database. In other words, the database server has to guarantee data consistency and currency of data. However, most of the published approaches for controlling concurrency of operations over databases in broadcast environments require that complex control structures should be sent to mobile clients. Besides having to store such structures, the clients need to be in active state for longer period of time in order to manage those structures. Approaches such as invalidation report [7] and update consistency [8] present these drawbacks. For example, in [8] an n x n matrix should be sent to all mobile clients, where n is the number of database objects.

In this paper, we propose a new concurrency control protocol for broadcast environments. The protocol, denoted *temporal serialization graph testing* (TSGT, for short), explores temporal information about database operations (read and write). The proposed protocol does not require that complex structures be sent to the mobile clients. The TSGT protocol reduces the communication traffic between server and clients and minimizes the time interval in which clients need to listen to the broadcast channel.

The rest of the paper is organized as follows. In section 2, we outline the characteristics of broadcast environments. Section 3 describes the transactional model that we will use in this work. In section 4, we describe and analyze the proposed protocol for concurrency control in broadcast environments. In section 5, the most important mechanisms for concurrency control in broadcast environments will be described and discussed. Section 6 shows the results of our simulation experiments. Section 7 concludes this work and outlines future works.

---

[1] Roughly, we can say that a consistent database state represents an acceptable view of the real world

2 Mobile Computing Environments with Broadcast-Based Data Dissemination

The main components of a broadcast-based data dissemination environment are described next. The database consists of a collection of interrelated data items. The database server (DBMS) is responsible for storing and managing data of the database. The broadcast server periodically broadcasts data items to clients. The clients, in turn, are mobile computers. Applications running on mobile clients perform read and write operations on database items which are cached by mobile clients.

The broadcast-based data dissemination differs from the traditional model for data transfer between clients and server. Traditionally, data are sent from the servers to clients on demand. In broadcast environments, the server periodically broadcasts data items to a client population without a specific request. Each broadcast period is called broadcast cycle or *bcycle*, while the content of broadcast is called *bcast*. Clients monitor the broadcast channel and retrieve data items they need.

From a transaction processing point of view, it is important to note that, when data items are broadcast to mobile clients, they are accessed by local transactions running on those clients. On the other hand, transactions running on the database server can update those data items after they were broadcast. Thus, it is likely that a mobile client reads data item instances which do not exist anymore in the database. Of course, such a phenomenon should be avoided. Furthermore, since mobile clients can be disconnected for long periods of time, transactions running on mobile clients are likely to be long-living transactions.

3 Transaction Model

A database consists of a collection of disjoint objects representing entities of the real world. The set of values of all objects stored in a database at a particular moment in time is called database state. Database states represent snapshots of the real world. They can only reflect static aspects of the world. However, a database must also reflect changes in the real world. Such changes are captured by the notion of state transition. State transitions represent "jumps" from a particular database state to another (an updated snapshot of the real world).

The real world imposes some restrictions on its entities. Additionally, databases must capture such restrictions, denoted consistency constraints. We can couple the concept of database state to consistency constraints. If the values of objects of a particular database state satisfy all the consistency constraints, the database state is said to be consistent.

Application programs containing operations on database objects are tools whereby state transitions are realized in a database. From the concurrency control perspective, not all operations of a program are relevant. Only database operations have to be considered. A transaction is an abstraction which represents a sequence of database operations resulting from the execution of an application program. Hence, transactions are modeled as finite sequences of operations on database objects. We use the notation $r_i(x)$ $(w_i(x))$ to represent a read (write) operation by a transaction T_i on object x. $OP(T_i)$ denotes the set of all operations executed by T_i. We will assume that the execution of a transaction preserves the database consistency, if this transaction runs entirely and isolation from other transactions.

We categorize transactions in a broadcast environment in two classes. One class comprises transactions executed at the mobile clients. Transactions belonging to this class are called mobile transactions. Transactions belonging to the second class are called server transactions, since they run at the database server.

Transactions are executed concurrently. The concurrent execution of a set T of transactions is realized by interleaving the operations of transactions in T. The execution of several interleaved transaction is modeled by a structure called schedule. Formally, a schedule over a set $\Im=\{T_1,T_2, ...,T_n\}$ of transactions represents an interleaved sequence of operations of transactions in $\Im$ which is an element of the shuffle product $T_1*T_2*...*T_n$. Serial executions of transactions are modeled by means of the notion of serial schedules. The precedence relation (execution order) between two operations in a schedule S is represented by $<_s$. For example, the notation $p <_s q$ indicates that operation p was executed before q in schedule S. Two operations of different transactions conflict (or are in conflict) if and only if they access the same object of the database and at least one of them is a write operation. It is important to note that not all schedules are valid; only some of them preserve database consistency. Hence, identifying whether a schedule is correct is a key point in transaction management.

Let S be a schedule over a set $\Im=\{T_1, T_2, ..., T_n\}$ of transactions. The serialization graph for S, denoted GS(S), is defined as the directed graph SG(S) = (N,E) in which each node in N corresponds to a transaction in $\Im$. The set E contains edges of the form $T_i \rightarrow T_j$, if and only if $T_i, T_j \in N$ and there are two operations $p \in OP(T_i)$, $q \in OP(T_j)$, where p conflicts with q and $p <_s q$. A schedule S is conflict serializable if and only if the serialization graph for S (SG(S)) is acyclic. A schedule S is correct if it is serial or conflict serializable.

4   Synchronizing Database Operations
in a Broadcast Environment

In this section, we will describe and analyze the concurrency control protocol we propose for synchronizing database operations (belonging to different mobile and server transactions) in a broadcast environment. The proposed protocol, called temporal serialization graph testing (TSGT), ensures that broadcast environment applications access consistent and current data.

The TSGT protocol is based on a similar strategy used by the conventional serialization graph testing protocol [5]: the dynamic monitoring and management of an always acyclic conflict graph. In contrast to the classic serialization graph testing, the TSGT exploits temporal information w.r.t. the moment in which a mobile transaction operation (read or write) is executed on a given database item.

In our approach, we have decided to distribute concurrency control functions among mobile clients and the database server. Thus, we assume that the server and the clients execute specific functionalities, in order to manage the transaction processing in a broadcast environment. In the following, we describe such functionalities.

During each broadcast cycle, the server broadcasts the data items together with a timestamp. We will assume that data item values sent in broadcast during each cycle correspond to the database state immediately before the beginning of the broadcast process. In other words, data instances sent during a broadcast correspond to them produced by all the transactions that had executed commit operations until the beginning of the broadcast cycle. Such transactions will be called "committed" transactions. Accordingly, the database server should store two versions of each data item O_i:

   (i) a version corresponding to the result yield by the last committed transaction which has updated O_i, and;

  (ii) a version corresponding to the result yield by the last non-committed transaction which has updated O_i;

The server is also responsible for building and managing the temporal serialization graph for a schedule, named *global schedule*, consisting of operations belonging to mobile and server transactions. A global schedule models the temporal execution order in which operations of mobile and fixed transactions are executed. This is possible because mobile clients send to the server timestamps for database operations they execute. In Section 4.2, we describe how those timestamps are defined.

Periodically, clients must send a package (message) to the server. Such a package consists of database objects on which a mobile transaction has executed a database operation (read or write) and the operation type. This information is sent together with the corresponding timestamp. Information of operations already informed does not need to be sent again. When a client receives a commit or an abort request of a mobile transaction T_i, it sends a message to the server consisting of request (commit or abort). After that, the client waits for an acknowledgement from the server in order to execute the commit or abort operation.

4.1 Running Example

We motivate the applicability and feasibility of our proposal by describing an application of electronic commerce. In such an application the stock of the main technology companies is available to auction in an electronic stock exchange. Now consider the following set of transactions, which read and update values of the stock: T_1: $r_1(IBM)r_1(SUN)C_1$; T_2: $w_2(IBM)C_2$; T_3: $r_3(IBM)r_3(SUN)C_3$; T_4: $w_4(SUN)C_4$; T_5: $w_5(SUN)C_5$

The transaction T_2, T_4 and T_5 are executed at the server. On other hand, the transaction T_1 is executed on the client A, while the transaction T_3 on client B. Now consider the global schedule GS presented in figure 1.

We assume that, in the execution scenario presented in figure 1, the packages containing information about the read operations of mobile transactions during the bcycle$_n$ arrive at the server before sending bcast$_{n+1}$. The serialization graph for the schedule GS is illustrated in figure 2 (a). Observe that the graph presents a cycle of the form $T_1 \rightarrow T_2 \rightarrow T_3 \rightarrow T_4 \rightarrow T_1$. Therefore, schedule SG is not conflict serializable (correct).

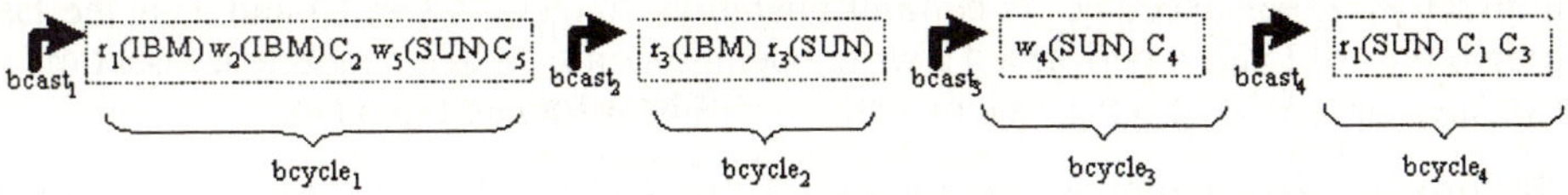

Fig. 1. Schedule GS

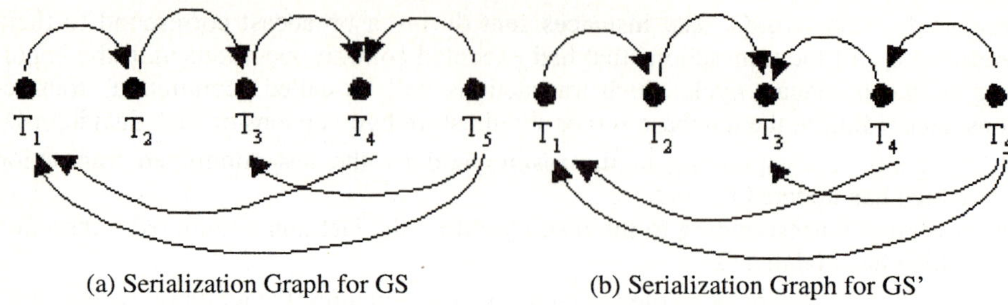

<table>
<tr><td>(a) Serialization Graph for GS</td><td>(b) Serialization Graph for GS'</td></tr>
</table>

Fig. 2. Serialization Graphs for Schedules GS and GS'

$r_1(IBM)\ w_2(IBM)\ C_2\ w_5(SUN)\ C_5$ $w_4(SUN)\ C_4$ $r_3(IBM)\ r_3(SUN)$ $r_1(SUN)\ C_1\ C_3$

Fig. 3. Schedule GS'

Now, we assume that, for some reason (problems in the communication links, for example), the package which contains the information about the read operations executed by transaction T_3 during bcycle$_2$ is late. In this case, the server will see the schedule GS' which is showed in figure 3. The serialization graph for GS' is depicted in figure 2 (b).

In figure 2 (b), the serialization graph for GS' does not present cycles. Consequently, the schedule GS' will be considered conflict serializable, that is, it is correct.

However, observing the database state we can see that it is not consistent. That means, the correct edge between T_3 and T_4 in the serialization graph should be $T_3 \rightarrow T_4$ (as depicted in figure 2 (a)) and not $T_4 \rightarrow T_3$ (figure 2 (b)). Therefore, an incorrect execution was considered correct improperly. Therefore, the conventional serialization graph is not sufficient to identify incorrect schedules in broadcast environments. To avoid the occurrence of such phenomenon, we propose the TSGT protocol, which will be described in the next section.

4.2   The Temporal Serialization Graph Testing Protocol

The TSGT protocol consists basically of monitoring and management an always acyclic graph. The graph maintained by TSGT protocol is called temporal serialization graph. This graph is constructed based on temporal information about the moment when a data item was read or updated.

Definition 1. $C(p_i(x))$ denotes the timestamp value for the operation $p_i(x)$. The value for $C(p_i(x))$ is defined as follows. If T_i is a transaction executed on the server, then $C(p_i(x))$ is the cycle number when the operation $p_i(x)$ is executed. On the other hand, if T_i it is a mobile transaction, then $C(p_i(x))$ represents the cycle number in which a transaction T_j has executed its commit operation, if $w_j(x) \in OP(T_j)$ and T_j is the last transaction which has performed a write operation on x. When a mobile transaction T_k executes an operation $p_k(x)$, this timestamp will be associated to $p_k(x)$.

Definition 2. The temporal precedence relation (temporal execution order) between two operations in a schedule S is denoted by $\prec_s^t$. For example, $p \prec_s^t q$ indicates that operation p is temporally executed before q in a schedule S. Let p and q operations in a Schedule S, we define that $p \prec_s^t q$ if and only if one of the following conditions holds:

i) $C(p) < C(q)$

ii) $C(p) = C(q)$ and $p \in OP(T_i)$, $q \in OP(T_j)$, T_i is a mobile transaction, $q <_s p$, T_j commits in a cycle, whose number is greater than $C(p)$.

iii) $C(p) = C(q)$ and $p \in OP(T_i)$, $q \in OP(T_j)$, T_i and T_j are transactions executed on the server, $p <_s q$.

iv) $C(p) = C(q)$ and $p \in OP(T_i)$, $q \in OP(T_j)$, T_i is a mobile transaction, T_j is a transaction executed on the server, $p <_s q$.

Definition 3. Let S be a schedule over a set $\Im = \{T_1, T_2, ..., T_n\}$ of transactions. We define the temporal serialization graph (TSG) for S, denoted TSG(S), as a directed graph TSG(S) = (N,E), where:

(i) $N = \Im$, that is, each node in N represents a transaction in $\Im$, and;

(ii) E represents the set of edges $T_i \rightarrow T_j$, where

- $T_i, T_j \in N$;
- there are two operations $p \in OP(T_i)$ and $q \in OP(T_j)$, which are in conflict and;
- $p \prec_s^t q$.

A schedule S is conflict serializable if and only if the temporal serialization graph for S (TSG(S)) is acyclic. A schedule S is correct if it is serial or conflict serializable.

Next, we describe how a TSGT scheduler manages the temporal serialization graph. When a scheduler starts running, the TSG is created as an empty graph. During each broadcast cycle, the server broadcasts the values of data items (last value written by committed transactions) with the respective timestamp. The timestamp for each data item can be sent in the message header or together with each data item. For each read operation, the client stores the value and the identification of the read item, together with the respective timestamp. Periodically, clients should inform to the server, which read operations they have executed. That is, a client sends periodically a package containing the item identification and the respective timestamp for each read operation. As soon as the scheduler receives the first operation of a new transaction T_i, a node representing this transaction is inserted in the TSG. For each operation $p_i(x) \in OP(T_i)$ which is received, the scheduler executes the algorithm shown in Figure 4.

In order to illustrate the correctness of the TSGT protocol, consider the example presented in Section 4.1. Observe that the graph produced by the basic TSGT protocol corresponds to the correct serialization graph for the schedule GS (see figure 1). The protocol described above ensures that the scheduler identifies that $C(w_4(SUN)) > C(r_3(SUN))$, inserting, thus, the edge $T_3 \rightarrow T_4$, and not $T_4 \rightarrow T_3$. Therefore, the TSGT protocol captures the information that the operation $r_3(SUN)$ was temporally executed before $w_4(SUN)$.

Thus far, we have analyzed global schedules with mobile transactions involving only read operations. However, the protocol proposed in this work can control concurrency in environments with mobile transactions involving write (update) operations as well, while ensuring database consistency. Next, we show how the TGST protocol can be used to control concurrency in such environments. First, we need to make the following observation. The semantic of write operations of a mobile transaction stays that an operation $w_i(x)$ is in fact executed, when it arrives at the database server.

Remark 1. Let $C(p_i(x))$ be the timestamp value for the operation $w_i(x)$, where T_i is a mobile transaction. $C(p_i(x))$ represents the cycle number when the operation $p_i(x)$ arrives at the server.

Step 1. The scheduler checks if there exists a conflicting operation $q_j(x) \in OP(T_j)$ which has been already scheduled. If there is such an operation $q_j(x)$, then the scheduler inserts an edge between T_i e T_j. In order to include such an edge correctly, two different cases should be considered:

    Case 1. T_i is a transaction executed on the server. In this case, the scheduler will execute the following temporal verification:

        If $C(q_j(x)) \leq C(p_i(x))$

            Then, the scheduler inserts an edge on the form $T_j \rightarrow T_i$.

        Else

            The scheduler inserts an edge on the form $T_i \rightarrow T_j$.

    Case 2. T_i is a mobile transaction. In this case, the scheduler will execute the following temporal verification:

        If $C(q_j(x)) < C(p_i(x))$

            The scheduler inserts an edge on the form $T_j \rightarrow T_i$

        Else

            If $C(q_j(x)) > C(p_i(x))$

                The scheduler inserts an edge on the form $T_i \rightarrow T_j$

            Else

                If T_j has already executed the *commit* operation

                    The scheduler inserts an edge of the form $T_j \rightarrow T_i$

                Else

                    The scheduler inserts an edge of the form $T_i \rightarrow T_j$

Step 2. The scheduler verifies if the new edge introduces a cycle in the temporal serialization graph. In the affirmative case, the scheduler rejects the operation $p_i(x)$, undoes the effect of the operations of T_i and removes the edge inserted. Otherwise, $p_i(x)$ is accepted and scheduled

Fig. 4. A scheduler implementing the TGST protocol

4.3   Correctness of the TGST Protocol

Next, we prove that schedules produced by the TSGT protocol are conflict serializable. That means, the TSGT protocol ensures database consistency.

Theorem 1. Let *TSGS* be the set of schedules over the set $\Im = \{T_1, T_2, \ldots, T_n\}$ of transactions produced by a TSGT protocol and CSR the set of all conflict serializable schedules over $\Im$. Then *TSGS*=CSR [6].

Sketch of Proof. It is easy to show that TSGS $\subset$ CSR. We only need to observe that every global schedule S produced by the TSGT protocol has an acyclic temporal serialization graph. By definition, TGST(S) represents the conventional serialization graph for S, with additional temporal information to capture the correct execution order of the operations in S. In other words, if the TSG for S is acyclic, the serialization graph is too. Therefore, S $\in$ CSR, consequently, *TSGS*$\subset$CSR. To prove that TSGS$\supset$CSR, we have to show that every schedule S $\in$ CSR can be produced by a TSGT protocol. We can show this by induction on the length of S that every operation p in S may not originate a cycle in the TSG and, thus, p may be executed. As already mentioned, the TSG represents the conventional serialization graph for S, with temporal information.

5 Related Work

In this section, we will describe and analyze the most important proposals for the concurrency control in broadcast environments. Initially, we will discuss the invalidation reports approach proposed in [7]. According to this approach, each *bcast* is preceded by an invalidation report. Such a report represents a list of all data items that was updated on the server during the previous broadcast cycle. Client read the invalidation report periodically. A mobile transaction T is aborted if an object x previously read by T appears in the invalidation report. This approach discards some conflict serializable schedules. Moreover, the client cannot be disconnected for long periods of time, since the client needs to read every invalidation report.

The multiversion broadcast mechanism [2] consists of keeping previous versions of data items, in order to reduce the number of aborts of mobile transactions. In this approach, the server, besides broadcasting database objects, broadcasts multiple versions for each object. Let C_0 be the broadcast cycle number during which the client transaction T performs its first read operation. To each new read operation, the transaction T tries to read the value with the largest version number C_n, such that $C_n \leq C_0$. If this version is not available the transaction is aborted. Therefore, this approach does not eliminate the necessity of aborting mobile transactions. Moreover, it generates an overhead in the execution of the read operations and to maintain the multiple versions for each database object..

Shanmugasundaram et al. [8], proposes a mechanism which uses as correctness criterion an extension of the criterion called update consistency. According to this proposal, in each *bcycle* the server broadcasts an n x n matrix, where n is the number of database objects. This matrix will be used by clients in order to ensure the consistency of read operations. For that, clients should listen to the broadcast channel during a larger period of time in order to retrieve the control matrix. It is important to note that the clients also have to store the control matrix.

As we can observe, most of the proposals described in this paper requires that control structures are transmitted by the server during each broadcast cycle and that mobile clients store and manage such structures. For this reason, we can claim that TSGT protocol is more efficient for the concurrency control in broadcast environments than the existing proposals.

6 Experimental Results

In order to evaluate the performance of the proposed protocol, we compared it with the F-MATRIX [8] and multiversion [2] protocols. We evaluated the performance of these protocols based on the following metrics:

- **Transaction Response Time:** This metric indicates the time interval between the time a transaction T is submitted by a client and the time that T ends its execution through a commit operation (including the time involved in restarts).

- **Transaction Restart:** This metric indicates the number of restarts occurred for a set of concurrent transactions. Observe that this metric indirectly measures the abort rate.

6.1 Simulation Environment

The simulation environment is based on the model used in [8]. It consists of a server, a client, and a broadcast server for transmitting both the data objects and the required control information. A mobile transaction is processed until it is committed. Only read-only transactions are executed on the client. Update transactions are executed on the server. A small database (300 data objects) helps to intensify data conflicts by creating hot-spot effect. The objects that the transactions access are determined using a random distribution function. The transaction length indicates the number of operations in a transaction. In the simulation experiments we used 8 operations with the default value for the server transaction length.

6.2 Simulation Results

Fig. 5 show the results of our simulation experiments. TSGT outperforms F-MATRIX and Multiversion in all the experiments. Furthermore, TSGT is highly scalable with respect to client transaction length and server transaction length.

Figure 5 (a) shows that our protocol presents a lower abort rate than the F-Matrix and Multiversion protocols. It shows that our protocol is scalable w.r.t the length of transactions as well.

Figure 5 (b) shows that transactions are executed in smaller time intervals than the F-Matrix and Multiversion protocols. Observe that, if we have smaller time intervals for executing transactions, we increase the throughput of the system.

Although energy usage has not been evaluated in our simulations we claim that TSGT (in comparison with F-MATRIX and Multiversion protocols) provides reduction in the use of this important and scarce resource, since it reduces the time that mobile clients need be connected (in "active" mode). This is because mobile clients do not require to listen to the broadcast channel continuously; they listen to the broadcast channel only to retrieve data they need. Moreover, with lower abort rates and smaller response times client transactions will commit more quickly, saving energy.

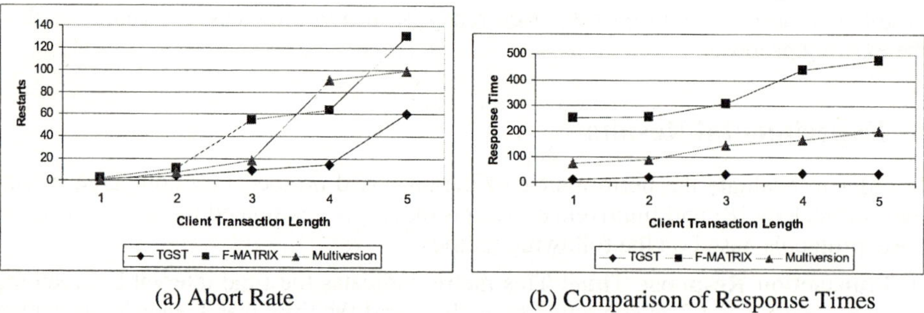

(a) Abort Rate   (b) Comparison of Response Times

Fig. 5. Simulation Results

7 Conclusions

In this paper we have proposed a new mechanism for concurrency control in broadcast environments. The proposed protocol, called temporal serialization graph testing protocol (for short, TSGT), ensures that applications in broadcast environments have access to consistent and current data. The TSGT protocol ensures that clients do not need to contact the server to perform their operations. Clients just have to listen to the broadcast channel in order to retrieve data items of interest and can be disconnected for long periods of time. Moreover, clients do not need to store nor to manage complex structures as proposed in [2], [7] and [8]. We performed simulation studies to evaluate the performance of the new protocol. The analysis of simulation results showed that the proposed protocol presents a better performance than F-Matrix [8] and Multiversion [2] protocols.

References

1. Victor C.S. Lee and Sang H. Son. On Transaction Processing with Partial Validation and Timestamp Ordering in Mobile Broadcast Environments. IEEE Transactions on Computers, Vol. 51, No. 10, 2002.
2. Evaggelia Pitoura and Panos K. Chrysanthis. Multiversion Data Broadcast. IEEE Transactions on Computers, Vol. 51, No. 10, 2002.
3. Web Page of Airmedia inc. White Paper, http:// www.airmedia.com
4. A. Brayner, T. Härder and N. Ritter. Semantic Serializability: A Correctness Criterion for Processing Transactions in Advanced Database Applications. DATA & KNOWLEDGE ENGINEERING, 31, 1999.
5. M.A. Casanova. The Concurrency Problem of Database Systems. In Lectures Notes in Computer Science, 116, 1981.
6. J.M. Monteiro. Temporal Serialization Graph Testing: An Approach to Control Concurrency in Broadcast Environments. Msc. Dissertation, Universidade Federal do Ceará, October, 2001 (in Portuguese).
7. E. Pitoura e P. Chrysanthis, Scalable Processing of Read-Only Transactions in Broadcast Push, IEEE International Conference on Distributed Computing Systems, 1999.
8. J. Shanmugasundaram, A. Nithrakashyap, R. Sivasankaran e K. Ramamritham. Efficient Concurrency Control for Broadcast Environments. Proceedings of the ACM SIGMOD Conference, 1999.

Integrating Mobile Devices into the Grid:
Design Considerations and Evaluation*

Stavros Isaiadis and Vladimir Getov

Harrow School of Computer Science
University of Westminster, London, UK
{S.Isaiadis,V.S.Getov}@westminster.ac.uk

Abstract. Mobile devices increasingly offer functionality beyond the one provided by traditional resources – processor, memory and applications. This includes, for example, integrated multimedia equipment, intelligent positioning systems, and different kinds of integrated or accessible sensors. For future generation Grids to be truly ubiquitous we must find ways to compensate for the limitations inherent in these devices and integrate them into the Grid, in order to leverage available resources and broaden the range of supplied services. The unreliability and limitations surrounding the mobile resources and services could significantly degrade the overall Grid availability and performance. In this paper, we propose the utilization of various mobile devices in the form of a single virtual wireless "cluster" that will hide the heterogeneity and dynamicity, mask the failures and quietly recover from them, provide centralized management and monitoring and allow for the federation of similar services or resources towards advanced functionality, quality of service, and enhanced performance. Finally, we present and analyze the results from the simulative performance evaluation of this design.

1 Introduction

In the last two decades we have experienced a transition towards mobility in the form of mobile devices, like cell phones initially, and hand-held computers and smartphones later. These trends are going to increase in the following years, resulting in a huge mobile computing community. An analogous transition in the field of communications and networking from the wired networks and fixed points of attachment to the flexible wireless networks has been developing rapidly. Furthermore, in addition to traditional computer resources like processors, memory, storage, and applications, new generation mobile devices provide support for integrated multimedia equipment, intelligent positioning systems, and a diverse range of sensors. A possible integration of these mobile devices into the Grid could benefit both the mobile and the Grid communities. There is no doubt about the former, since mobile users will gain access to a huge number of resources, high performance facilities, specialized hardware and software and enormous resource pools. For the latter, it is an opportunity to utilize available resources in the mobile community and increase its performance and capacity and broaden the range of services and supplied functionality. And for future generation Grids to be truly ubiquitous we must have the option of integrating mobile devices into Grid systems (in the rest of this paper we'll be using the term "mobile

---

* This research work is carried out partly under the FP6 Network of Excellence CoreGRID funded by the European Commission (Contract IST-2002-004265)

J.C. Cunha and P.D. Medeiros (Eds.): Euro-Par 2005, LNCS 3648, pp. 1080–1088, 2005.

devices" to refer to resource limited mobile devices with wireless connectivity like laptops, personal digital assistants, smart phones etc.).

Integrating mobile devices into the Grid could lead to instability and reduced overall performance, due to the inherent limitations of mobile devices and wireless communication links. Whenever a failure occurred we would have an avalanche of reactions including job rescheduling, data migration, resource re-allocation and activation of failure recovery mechanisms, and considering the increased rates of failures in such an environment, this is not acceptable. In an attempt to overcome these obstacles, we propose in this paper the clustering of all mobile devices that fall into the same subnet in order to present them as a single virtual system to the Grid. The main advantages of our approach are that we manage to hide the heterogeneity, dynamicity and complexity of mobile devices from the Grid, present a centralized point for managements and monitoring and federate similar resources or services to provide advanced operations and enhanced performance. Acting on behalf of the wireless cluster and exposing all available services to the Grid, there will be a set of dedicated proxies. We use this architecture as our vehicle towards our ultimate goal of developing and evaluating high service availability and failure resilience techniques in such hybrid Grid environments.

The rest of this paper is organized as follows. In the next section, we provide a brief overview of relevant research in this field. Section 3 gives details on the challenges we face in trying to integrate mobile devices to the Grid and the architecture we are adopting in an effort to tackle these problems. Section 4 describes the implementation details, while section 5 presents the simulative evaluation results and diagrams. Finally, section 6 concludes the paper and lists our future plans.

2 Relevant Work

A number of recent and current research projects and papers are dealing with mobility in Grids. In [2] the authors propose a clustered approach for integrating mobile devices into the Grid, without however, providing any implementation considerations or evaluation of their approach. In [10] the authors also propose a virtual cluster approach and a middleware to provide peer-to-peer operations but do not address resource virtualization or federation of similar resources or even larger scale collaboration. Also, none of the aforementioned papers address the issues of mobility and failure resilience and they do not take into consideration the widely used OGSA/OGSI (Open Grid Services Architecture / Infrastructure) or the recently adopted OGSA/WSRF (Web Services Resource Framework) specification.

Mobile OGSI.NET [11] is an implementation of an OGSI based Grid container on the .NET hosting environment on mobile devices based on Microsoft's PocketPC. Mobile OGSI.NET allows for Grid service state saving and restoring and distribution of workload among devices with the same types of services, but with the cost of having to change existing services to adhere to the specific Mobile OGSI.NET programming model. Furthermore, Mobile OGSI.NET can only be realized on PocketPCs with the .NET framework installed.

AKOGRIMO [9] is a European funded project that has recently started and is dealing with mobility issues in the Grid. The purpose of the project is to evaluate the mobile Grid introducing the notion of mobile dynamic virtual organizations through applications that highlight the challenges present in such mobile environments, like ehealth, e-learning and crisis management. AKOGRIMO favors research on ad-hoc mobile Grids and the integration of mobile IPv6 to support mobility in a Grid environment.

Finally, in [1] an agents approach is adopted to tackle device mobility, but the proposed architecture only allows mobile devices to be the consumers of services and not the providers.

The potential benefit of the integration of mobile devices into the Grid, but also the challenges that this raises, has been the theme in many other papers and research projects lately, but none of these provide any implementation methodology or propose an architecture to support this integration [3, 4, 5, 6, 7].

3 Hybrid Wired – Wireless Grid Environments

Initially the integration of mobile devices into the Grid seems highly unlikely due to inherent limitations: limited resources (CPU, secondary storage, available memory etc.); increased power consumption sensitivity; increased heterogeneity; unpredictable long periods of complete disconnectivity; unreliable, low-bandwidth and high latency communication links; and very dynamic network layout because of devices entering and leaving in a very unpredictable manner.

While directly connecting mobile devices to the Grid is very straightforward and requires no modifications to the existing infrastructure, it could significantly degrade the overall Grid performance because of the high rate of failures. Whenever a failure occurred, the Grid components would have to reschedule and reallocate resources for the active application, possibly migrating data around the Grid thus reducing the response time. Considering that in the mobile edge of the Grid, the failure rate is increased, this is not something we would like in busy, heavily loaded and complex Grid environments. In order to avoid potential performance penalties, we have decided to follow another route: group all mobile devices that fall into the same subnet and present them as one single virtual resource to the Grid. For this virtualization to be realized we need a set of dedicated proxies residing between the existing Grid infrastructure and the wireless cluster (it has to be stated here that by saying "cluster" we do not mean a cluster in its traditional form – it is more like a grouping of devices that *may*, however, borrow some ideas from the cluster domain). This method will allows us to delegate some control to the proxies, getting some load off the higher level Grid components. The actual number of proxies will be determined by early experiments, and it may be possible to distribute the proxy functionality across a number of workstations conveniently located behind the distribution network of the "cluster". Both the direct and the clustered approach are depicted in Figure 1.

By following this approach, we can encapsulate the heterogeneity and the dynamicity present in the cluster – thus imposing minimal disruption to the regular Grid operations, and leverage the resource repositories in the wireless cluster. Furthermore, we can mask all internal failures and recover from them locally without any notification of the higher-level Grid components. We deal with failures internally, re-allocating resources from within the cluster itself, locally rescheduling and migrating data if necessary.

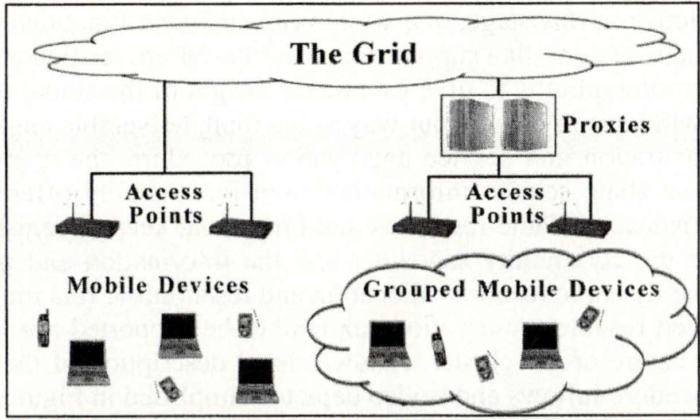

Fig. 1. Direct vs. Clustered connection to the Grid

4 Implementation Prototypes

In this architecture we are adopting a service-oriented approach currently conforming to the OGSA/OGSI specification and using the Globus Toolkit 3 (GT3) [15] implementation. In the near future we plan to migrate to the newest WSRF specification [8] and take advantage of the latest Web Services Description Language [14] (WSDL v1.2 or v2) that will provide a lot of enhancements for Grid service development.

All resources and specific functionality in the cluster are accessed through Grid service interfaces, which can fall into one of these two categories (depending on the interface type and description):

– Services that provide specific functionality, like for example access to a proprietary database, access to a sensor or a vector computer, access to other specialized hardware or software. Similar services –i.e. that offer the same functionality or implement the same interface, will be aggregated and presented through a single service interface deployed at the proxy. All requests arriving at the proxy for a specific type of service will be forwarded to an available node that supports it. This technique will give us more flexibility and will allow us to implement collective operations on all nodes offering this functionality, parallel execution whenever this is feasible and supported by the application, or even mirrored execution for higher reliability.

– Services that provide controlled access to raw resources like processor cycles, memory, secondary storage etc. Again resources are aggregated and presented through a single interface, but the aggregator service implementation is now different. Now we only publish a fraction (70-80%) of the total amount of aggregated resources, in order to expose a virtually more stable environment: there are spare resources available in the cluster to cover up any possible node failures and avoid unnecessary disturbance of the Grid by reallocating resources from within the cluster itself. Failures are now masked and dealt with locally by the "cluster" community scheduling and recovery components, thus avoiding putting more load on the higher level Grid scheduling systems, monitoring and failure recovery mechanisms.

We are implementing early prototypes of the proxy engine components, mainly the service/resource aggregation and indexing components. These early prototypes are somewhat restrictive at this stage, and we haven't included functionality that will be present at the final version –like support for mobility, failure resilience and more. The purpose of this prototyping is to give us a better insight of the actual needs and challenges that we will have to face on our way to the final deliverable engine.

After the registration and service aggregation procedure, the engine indexes the services based on static service information (like functionality offered, hosting device's characteristics, available resources and other) and keeps a registry that closely cooperates with the community scheduler and the information and monitoring services to allow for dynamic resource allocation and reservation. It is important to mention that advanced resource reservation can neither be supported nor guaranteed due to the dynamic nature of the cluster. A lower level description of the whole service aggregation procedure follows and is also depicted simplified in Figure 2.

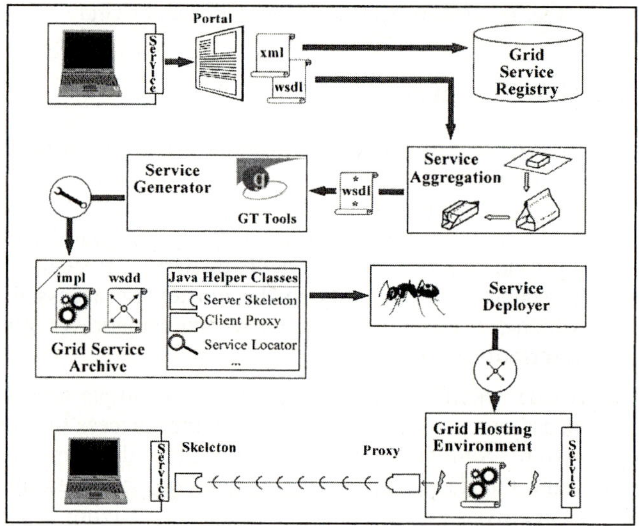

Fig. 2. The service aggregation and service generation procedure

1. The mobile node that wishes to expose a service sends the WSDL and an XML file containing service and device information to the engine by using a convenient web interface. The XML file will be used for indexing purposes by the Cluster Service Registry.
2. The WSDL file is parsed using JAXP (Java API for XML Processing) in order to extract all necessary service information, and is modified so that we can then deploy it at the proxy.
3. Both the original and the modified WSDL files will be fed to a generator that will automatically create a Java implementation for the aggregator service, the deployment descriptors and all the necessary helper Java classes like the Server-side skeleton, the client-side proxy, the Service Locator and so on.
4. Next, all these files are collected and put into a Grid Archive File, which is then deployed to the Grid hosting environment using Apache Ant tools.
5. Finally, the newly deployed service at the Proxy can be published to a UDDI registry or GT's MDS (Monitoring and Discovery Service) or any other index service.

As is easily observed from the description, both the mobile user and the cluster administrator have minimal work to do in order for the aggregation and registration to occur and for the service to be deployed and published. The mobile user merely has to fill in a small form in the web interface and submit it, while the administrator only has to deploy the service. In the near future, we plan to fully automate this procedure by making use of a dynamic resource discovery infrastructure like Jini [13].

The resulting aggregator service deployed at the proxy provides a single interface to all aggregated resources in the "cluster". This abstraction makes job submission easier for inexperienced users and simplifies the job of Grid scheduling systems, since they can now delegate finer control to the proxy engine which will take care of distributing the tasks or forwarding service requests to the mobile nodes in the "cluster".

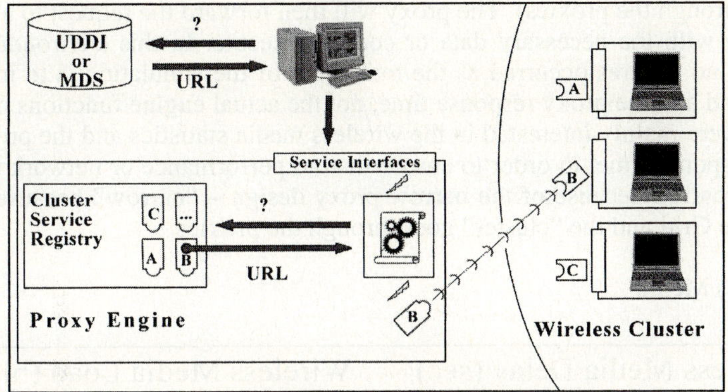

Fig. 3. Servicing requests

When a request for a service arrives at the proxy, an available mobile node that match the requirements is found by querying the Cluster Information and Monitoring Service and the Cluster Service Registry, a new service instance is created and all subsequent method calls are forwarded to that node. If no nodes can offer the requested functionality a "service unavailable" response is forwarded. Our engine also takes care of state and lifetime management and any possible notification subscriptions automatically.

5 Simulative Network Performance Evaluation

In order to ensure that this architecture will not present a bottleneck and won't cause network performance degradation, we have performed several simulations using OPNet®. We have modeled a very heavily loaded data-centric Grid environment following regular Grid behavior, i.e. light traffic initially querying service registries, followed by point-to-point data transfer, request for CPU usage, database queries and updates etc. A brief description of the simulation environment is given below, followed by some of the most important results that were generated.

- 4 Basic Service Sets each one servicing 10 mobile devices, 50% of which contribute resources to the Grid community. These sets represent the mobile resource limited edge of the Grid.

- A LAN consisting of 50 nodes all of which contribute resources to the community. The LAN acts as the main Grid infrastructure into which we want to integrate mobile devices.
- A couple of proxy servers residing in between the 4 WLANs and the LAN, acting as the interface point between the mobile "cluster" and the Grid.
- All Wi-Fi interfaces operate at a rate of 11Mb/s.
- All Ethernet interfaces operate at a rate of 10Gb/s.

During the 60 minutes of the simulation, all nodes (both wireless and wired) request services that are available either in the WLANs or the LAN. When a Grid node wants to access a service available in the cluster it sends the request to the relevant proxy aggregator interface (since the cluster is hidden and access to its services is provided through the proxies). The proxy will then forward the request to an available node along with the necessary data or code to execute. In this environment we assumed that no failures occurred as the main goal of the simulation is to measure the network load and the proxy response time, not the actual engine functions or performance. We were mainly interested in the wireless media statistics and the proxy utilization and response time, in order to ensure that no performance or network traffic bottlenecks occurred because of the narrow proxy design – "narrow" because all traffic between the Grid and the "cluster" goes through the proxies.

- Wireless Media

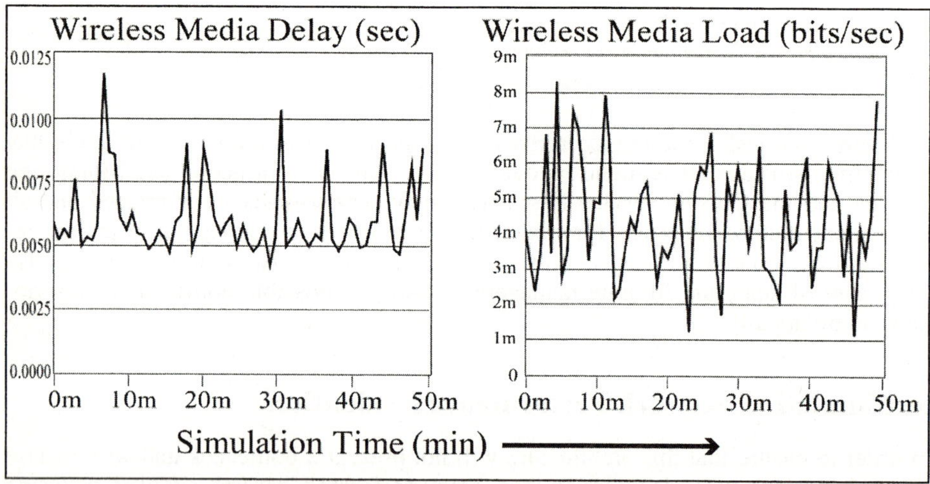

Fig. 4. Wireless Media Delay and Load

We can see from the second graph that at no point during the simulation there was any data dropped due to network traffic as the wireless media never reached its full potential (averaged at around 5Mbits/sec $\cong$ 45%) and the media access delay was kept to very low levels at around 7msecs. Our architecture, doesn't introduce any network performance bottlenecks even though the simulation environment was very heavily loaded.

- Proxy Performance

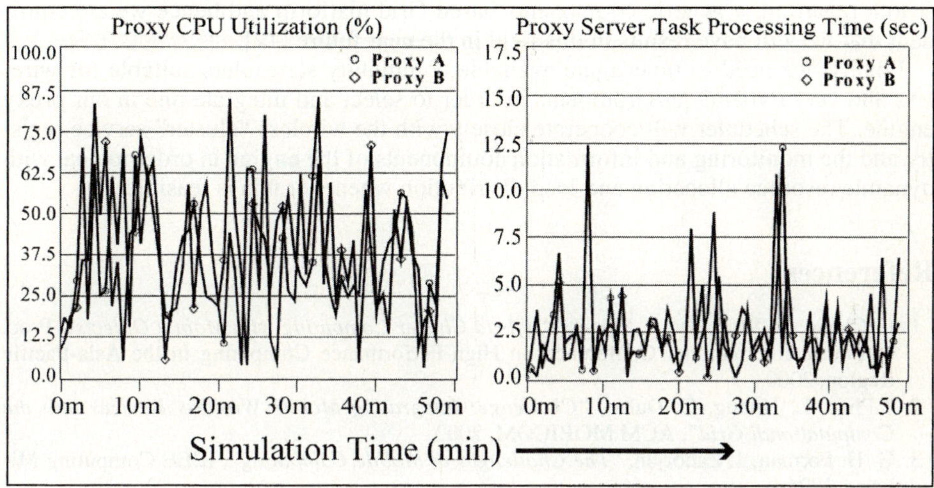

Fig. 5. Proxy CPU Utilization and Server Processing Time

The results showed that the two proxies handled very efficiently the heavy traffic we injected to the network (utilization was kept around 50% following a bursty nature) and the server performance was kept to very good levels despite the big number of service requests arriving from both the WLANs and the LAN.

6   Conclusions and Future Work

So far we have implemented the components responsible for service aggregation and indexing, and those dealing with servicing requests. Everything is achieved with minimal intervention from the mobile user and the cluster administrator. This is very important since we do not want to discourage mobile users from contributing to the Grid community by requesting lengthy and complicated procedures. Also, a cluster membership scheme – if one is deemed necessary – will be built on top of already installed Grid security infrastructure making use of available certificates so that no extra inconvenience is caused and alignment with existing authentication and authorization mechanisms is achieved.

We have tested our engine's aggregation and indexing components against a wide variety of services and it has been proved to perform efficiently. Nevertheless, there is still plenty of work to do and we have to implement failure prediction, detection and recovery components that will take us one step closer to our ultimate goal of autonomy and high service availability in this wireless environment.

The simulation reports we got were more than encouraging and showed that even for a very heavily loaded hybrid environment our proxies responded very well and there were no performance or network traffic bottlenecks introduced due to the "thin" proxy layer architecture. Near future plans include experimental evaluation of our design and addressing mobility and roaming issues between two wireless "clusters".

As we are dealing with resource limited devices, we need to develop a lightweight Grid platform suitable for such devices. The authors also participate in ongoing research regarding a generic components based Grid platform and hence we are confident that we will have results in this field in the near future [12].

Finally, we need to investigate available community schedulers suitable for wireless and very dynamic environments in order to select and integrate one in our proxy engine. The scheduler will cooperate closely with the wireless "cluster" service registry and the monitoring and information components of the engine in order to deal with dynamic resource allocation and load distribution whenever this is feasible.

References

1. L. Cheng, A. Wanchoo, I. Marsic, *"Hybrid Cluster Computing with Mobile Objects"*, Proc. of Fourth International Conference on High-Performance Computing in the Asia-Pacific Region, 2000.
2. T.Phan, L. Huang, C. Dulan, *"Challenge: Integrating Mobile Wireless Devices into the Computational Grid"*, ACM MOBICOM, 2002
3. G. H. Forman, J. Zahorjan, *"The Challenges of Mobile Computing"*, IEEE Computing Milieux, 1994
4. Chlamtac, J. Redi, *"Mobile Computing: Challenges and Potential"*, Encyclopedia of Computer Science, 4th edition, 1998
5. M. Franz, *"A Fresh Look at Low Power Mobile Computing"*, http://research.ac.upc.es/pact01/colp/paper15.pdf
6. B. Chen, C. H. Chang, *"Building Low Power Wireless Grids"*, http://www.ee.tufts.edu/~brchen/pub/LowPower_WirelessGrids_1201.pdf
7. D. Bruneo, M. Scarpa, A. Zaia, A. Puliafito, *"Communication Paradigms for Mobile Grid Users"*, IEEE/ACM International Symposium on Cluster Computing and the Grid, 2003
8. K. Czajkowski, D. Ferguson, I. Foster, J. Frey, S. Graham, T. Maguire, D.Snelling, S. Tuecke, *"From Open Grid Services Infrastructure to WS-Resource Framework: Refactoring & Evolution"*, http://www.ibm.com/developerworks/library/ws-resource/ogsi_to_wsrf_1.0.pdf, 2004
9. The *AKOGRIMO* project: http://www.akogrimo.org
10. J. Hwang, P. Aravamudham, *"Proxy-based Middleware Services for Peer-to-Peer Computing in Virtually Clustered Wireless Grid Networks"*, in Proceedings of International Conference on Computer, Communication and Control Technologies, 2003
11. D. Chu, M. Humphrey, *"Mobile OGSI.NET: Grid Computing on Mobile Devices"*, http://www.cs.virginia.edu/~humphrey/papers/MobileOGSI.pdf, 2004
12. J. Thiyagalingam, S. Isaiadis, V. Getov, "*Towards Building a Generic Grid Services Platform: A Component-Oriented Approach*", in V. Getov and T. Kielmann (Eds), "*Component Models and Systems for Grid Applications*", 39-56, Springer, 2005
13. J. Waldo, K. Arnold, *"The Jini specifications"*, Jini technology series. Addison-Wesley, Reading, MA, USA, Second edition, 2001
14. R. Chinnici, M. Gudgin, J. Moreau, S. Weerawarana, *"WebServices Description Language (WSDL) 1.2"*, World Wide Web Consortium. TUhttp://www.w3.org/TR/wsdl12/UT, 2003
15. The *Globus* Project, http://www.globus.orgUT

New Bounds on the Competitiveness
of Randomized Online Call Control
in Cellular Networks*

Ioannis Caragiannis, Christos Kaklamanis, and Evi Papaioannou

Research Academic Computer Technology Institute and
Dept. of Computer Engineering and Informatics
University of Patras, 26500 Rio, Greece
{caragian,kakl,papaioan}@ceid.upatras.gr

Abstract. We address the call control problem in wireless cellular networks that utilize Frequency Division Multiplexing (FDM) technology. In such networks, many users within the same geographical region (cell) can communicate simultaneously with other users of the network using distinct frequencies. The available frequency spectrum is limited; hence, its management should be done efficiently. The objective of the call control problem is, given a spectrum of available frequencies and users that wish to communicate in a cellular network, to maximize the number of users that communicate without signal interference. We study the online version of the problem in cellular networks using competitive analysis and present new upper and lower bounds.

1 Introduction

In this paper we study frequency spectrum management issues in wireless networks. We consider wireless networks in which base stations are used to build the required infrastructure. In such systems, the architectural approach used is the following. A geographical area in which communication takes place is divided into regions. Each region is the calling area of a base station. Base stations are connected via a high speed network. When a user A wishes to communicate with some other user B, a path must be established between the base stations of the regions where users A and B are located. Then communication is performed in three steps: (a) wireless communication between A and its base station, (b) communication between the base stations, and (c) wireless communication between B and its base station. At least one base station is involved in the communication even if both users are located in the same region or only one of the two users is part of the cellular network (and the other uses for example the PSTN). Improving the access of users to base stations is the aim of this work.

Network Model. The network topology usually adopted [8, 9] is the one shown in the left part of Figure 1. All regions are regular hexagons (cells) of the same

---

* This work was partially funded by the European Union under the IST FET Project CRESCCO.

J.C. Cunha and P.D. Medeiros (Eds.): Euro-Par 2005, LNCS 3648, pp. 1089–1099, 2005.

size. This shape results from the uniform distribution of identical base stations within the network, as well as from the fact that the calling area of a base station is a circle which, for simplicity reasons, is idealized as a regular hexagon. Due to the shape of the regions, we call these networks cellular wireless networks.

Many users of the same region can communicate simultaneously with their base station of the network via frequency division multiplexing (FDM). The base station is responsible for allocating distinct frequencies from the available spectrum to users so that signal interference is avoided. Since the spectrum of available frequencies is limited, important engineering problems related to the efficient reuse of frequencies arise. Signal interference usually manifests itself when the same frequency is assigned to users located in the same or adjacent cells. Alternatively, in this case, we may say that the cellular network has *reuse distance* 2. By generalizing this parameter, we obtain cellular networks of reuse distance k in which signal interference between users assigned the same frequency is avoided only if the users are located in cells with distance at least k.

Signal interference in cellular networks can be represented by an *interference graph* G whose vertices correspond to cells and an edge (u, v) indicates that the assignment of the same frequency to two users lying at the cells corresponding to nodes u and v will cause signal interference. The interference graph of a cellular network of reuse distance 2 is depicted in the right part of Figure 1. If the assumption of uniform distribution of identical base stations does not hold, arbitrary interference graphs can be used to model the underlying network.

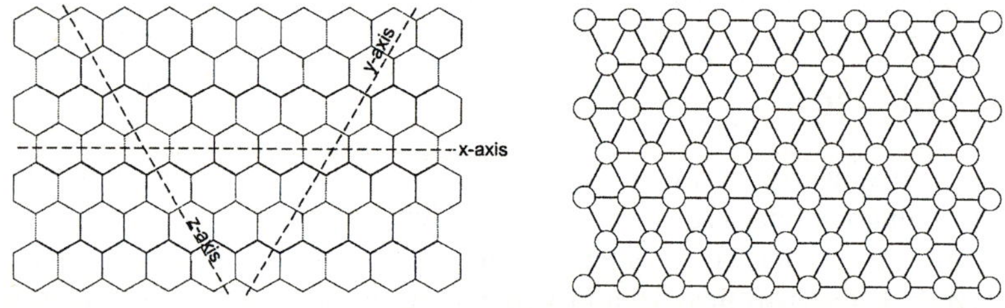

Fig. 1. A cellular network and its interference graph if the reuse distance is 2.

Problem Definition. In this paper we study the *call control* (or call admission) problem which is defined as follows: Given users that wish to communicate, the *call control* problem on a network that supports a spectrum of w available frequencies is to assign frequencies to users so that at most w frequencies are used in total, signal interference is avoided, and the number of users served is maximized.

We assume that calls corresponding to users that wish to communicate appear in the cells of the network in an online manner. When a call arrives, a call control algorithm decides either to accept the call (assigning a frequency to it), or to reject it. Once a call is accepted, it cannot be rejected (preempted). Furthermore, the frequency assigned to the call cannot be changed in the future.

We assume that all calls have infinite duration; this assumption is equivalent to considering calls of the same duration.

Competitive analysis [4] has been used for evaluating the performance of online algorithms for various problems. In our setting, given a sequence of calls, the performance of an online algorithm A is compared to the performance of the optimal algorithm OPT. Let $B(\sigma)$ be the benefit of the online algorithm A on the sequence of calls σ, i.e. the set of calls of σ accepted by A and $O(\sigma)$ the benefit of the optimal algorithm. We define the competitive ratio or competitiveness of an algorithm A as $\max_\sigma \frac{|O(\sigma)|}{\mathcal{E}[|B(\sigma)|]}$, where $\mathcal{E}[|B(\sigma)|]$ is the expectation of the number of calls accepted by A, and the maximum is taken over all possible sequences of calls. This definition applies to both deterministic and randomized algorithms. Usually, we compare the performance of deterministic algorithms against *off–line adversaries*, i.e. adversaries that have knowledge of the behavior of the deterministic algorithm in advance. In the case of randomized algorithms, we consider *oblivious adversaries* whose knowledge is limited to the probability distribution of the random choices of the randomized algorithm.

Related Work. The static version of the call control problem generalizes the famous maximum independent set problem. The online version of the problem is studied in [1–3, 5, 7, 10, 12]. [1], [2], [7] and [10] study the call control problem in the context of optical networks. Pantziou et al. [12] present upper bounds for networks with planar and arbitrary interference graphs. Usually, competitive analysis of call control focuses on networks supporting one frequency. Awerbuch et al. [1] present a simple way to transform algorithms designed for one frequency to algorithms for arbitrarily many frequencies with a small sacrifice in competitiveness (see also [7] and [13]). Lower bounds for call control in arbitrary networks are presented in [3].

The greedy algorithm is probably the simplest online algorithm. It considers frequencies as positive integers. When a call arrives, it seeks for the smallest available frequency. If such a frequency exists, the algorithm accepts the call assigning this frequency to it, otherwise, the call is rejected. As observed in [5], this algorithm has competitive ratio equal to the size of the maximum independent set in the neighborhood of any node of the interference graph (see also [12]). The competitive ratio of the greedy algorithm is a lower bound on the competitiveness of every deterministic algorithm. In particular, this gives lower bounds of 3, 4 and 5 on the competitiveness of every deterministic online call control algorithm in cellular networks of reuse distance $k = 2$, $k \in \{3, 4, 5\}$ and $k \geq 6$, respectively.

The first randomized algorithm with competitive ratio smaller than 3 in cellular networks with reuse distance 2 was presented in [5]. The main drawbacks of this algorithm are that it uses a number of random bits which is proportional to the size of the sequences of calls and, that it works in networks that support only one frequency. By extending the "classify and randomly select" paradigm [1, 2, 12], the authors in [6] present a series of simpler randomized algorithms that use a small number of random bits or comparably weak random sources, and have small competitive ratios even in the case of arbitrarily many frequencies, in cellular networks of any reuse distance k. The best competitive ratio obtained is $4 - \Omega(\frac{1}{k})$, while the randomness used is the ability to select equiprobably one out of an odd number of distinct objects. The best competitive ratio obtained for $k = 2$ is $7/3$. The best known lower bounds on the competitiveness of randomized algorithms are $13/7$ and $25/12$ for cellular networks of reuse distance $k \geq 2$ and $k \geq 5$, respectively ([5, 6]).

Our Results. In this paper, we present (Section 2) a new online call control algorithm with competitive ratio $16/7$ for cellular networks with reuse distance 2, improving the previous best known upper bound of $7/3$. Our algorithm is based on the "classify and randomly select" paradigm, uses only 4 random bits, and works in networks with arbitrarily many frequencies. Furthermore, we show new lower bounds of 2 and 2.5 on the competitiveness against oblivious adversaries of online call control algorithms in cellular networks of reuse distance $k \geq 2$ and $k \geq 6$, respectively (Section 3). Our new lower bounds improve previous ones for almost all cases of the reuse distance ($k \neq 5$).

2   The Upper Bound

In this section, we present the online algorithm CRS-D achieving a competitive ratio of $16/7$ against oblivious adversaries. The algorithm is based on the "classify and randomly select" paradigm. Such algorithms use a coloring of the cells (i.e., a coloring of the nodes of the interference graph) and a classification of the colors into not necessarily disjoint color classes. The algorithm randomly selects one out of the available color classes and executes the greedy algorithm for calls appearing in cells colored with colors from the selected color class, while it completely ignores (i.e., rejects) the calls appearing in any other cell. The following lemma gives a connection between the coloring of the interference graph and the definition of the color classes and the competitiveness of the "classify and randomly select" algorithm that uses them.

Lemma 1 ([6]). *Consider a network with interference graph $G = (V, E)$ and let χ be a coloring of the nodes of V with the colors of a set X. If there exist ν sets of colors (color classes) $s_0, s_1, ..., s_{\nu-1} \subseteq X$ and an integer $\lambda \leq \nu$ such that*

- *each color of X belongs to at least λ different color classes, and*
- *for $i = 0, 1, ..., \nu - 1$, each connected component of the subgraph of G induced by the nodes colored with colors in s_i is a clique,*

then the online call control algorithm which uses the coloring χ and the ν color classes according to the "classify and randomly select" paradigm has competitive ratio ν/λ against oblivious adversaries.

A proof was presented in [6]. The intuition behind the proof is that (1) the algorithm runs the greedy algorithm on a fraction of λ/ν of the cells, (2) the optimal solution of the subsequence defined by these calls has size at least λ/ν times the size of the original optimal solution, and (3) the greedy algorithm computes an optimal solution when applied to the subsequence.

Algorithm CRS-D uses a coloring of the cells with sixteen colors $0, \ldots, 15$ defined as follows. The cell with coordinates $(x, y, x + y)$ is colored with color $4(x \bmod 4) + y \bmod 4$. The color classes are defined as s_{4i+j} for $0 \le i, j \le 3$ as follows:

$$s_{4i+j} = \{4i + j, 4i + (j + 1) \bmod 4, 4((i + 1) \bmod 4) + j,$$
$$4((i + 1) \bmod 4) + (j + 2) \bmod 4, 4((i + 2) \bmod 4) + (j + 1) \bmod 4,$$
$$4((i + 2) \bmod 4) + (j + 2) \bmod 4, 4((i + 3) \bmod 4) + (j + 3) \bmod 4\}$$

An example of this coloring is depicted in Figure 2.

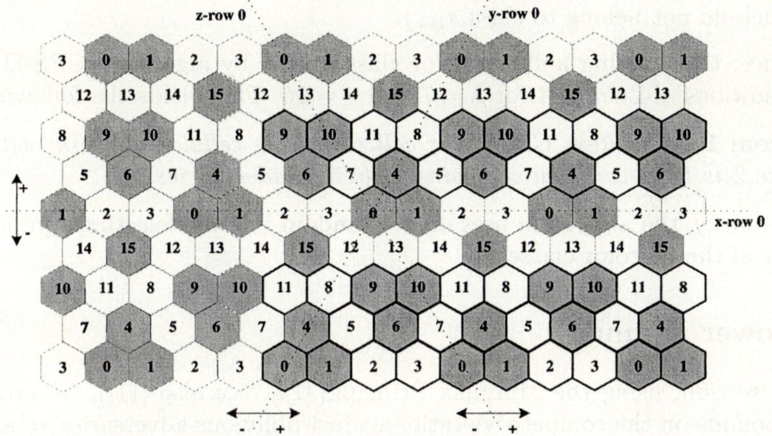

Fig. 2. The 16-coloring used by algorithm CRS-D. The grey cells are those colored with colors in the class s_0.

We now show that the coloring and the color classes used by algorithm CRS-D satisfy the conditions of Lemma 1. Each color $k = 0, 1, \ldots, 15$ belongs to 7 of the 16 color classes $s_0, s_1, \ldots, s_{15}$. For any i, j such that $0 \le i, j \le 3$, color $4i + j$ belongs to the color classes $4i + j$, $4i + (j - 1) \bmod 4$, $4((i - 1) \bmod 4) + j$, $4((i - 1) \bmod 4) + (j - 2) \bmod 4$, $4((i - 2) \bmod 4) + (j - 1) \bmod 4$, $4((i - 2) \bmod 4) + (j - 2) \bmod 4$, and $4((i - 3) \bmod 4) + (j - 3) \bmod 4$. Now, consider the cells colored with colors from the color class s_{4i+j} and the corresponding nodes of the interference graph. The connected components of the subgraph of the interference graph defined by these nodes are of the following types:

- cliques of three nodes corresponding to cells colored with colors $4i + j$, $4i + (j+1) \bmod 4$, and $4((i+1) \bmod 4) + j$, respectively. Indeed, the neighborhood of such nodes contains nodes colored with colors $4i + (j+2) \bmod 4$, $4i + (j+3) \bmod 4$, $4((i+1) \bmod 4) + (j+1) \bmod 4$, $4((i+1) \bmod 4) + (j+3) \bmod 4$, $4((i+3) \bmod 4) + j$, $4((i+3) \bmod 4) + (j+1) \bmod 4$, $4((i+3) \bmod 4) + (j+2) \bmod 4$, $4((i+2) \bmod 4) + j$, and $4((i+2) \bmod 4) + (j+3) \bmod 4$ which do not belong to class s_{4i+j}.

- cliques of three nodes corresponding to cells colored with colors $4((i+1)$ mod $4)+(j+2)$ mod 4, $4((i+2)$ mod $4)+(j+1)$ mod 4, and $4((i+2)$ mod $4)+(j+2)$ mod 4, respectively. Again, the neighborhood of such nodes contains nodes colored with colors $4i+(j+2)$ mod 4, $4i+(j+3)$ mod 4, $4((i+1)$ mod $4)+(j+1)$ mod 4, $4((i+1)$ mod $4)+(j+3)$ mod 4, $4((i+3)$ mod $4)+j$, $4((i+3)$ mod $4)+(j+1)$ mod 4, $4((i+3)$ mod $4)+(j+2)$ mod 4, $4((i+2)$ mod $4)+j$, and $4((i+2)$ mod $4)+(j+3)$ mod 4 which do not belong to class s_{4i+j}.
- isolated nodes corresponding to cells colored with color $4((i+3)$ mod $4)+(j+3)$ mod 4. The neighborhood of such a cell consists of cells colored with colors $4i+(j+2)$ mod 4, $4i+(j+3)$ mod 4, $4((i+2)$ mod $4)+j$, $4((i+2)$ mod $4)+(j+3)$ mod 4, $4((i+3)$ mod $4)+j$, and $4((i+3)$ mod $4)+(j+2)$ mod 4 which do not belong to class s_{4i+j}.

Hence, the coloring and the color classes used by algorithm CRS-D satisfy the conditions of Lemma 1 for $\lambda = 7$ and $\nu = 16$. This yields the following.

Theorem 1. *Algorithm CRS-D for call control in cellular networks with reuse distance 2 is $16/7$-competitive against oblivious adversaries.*

Obviously, the algorithm uses only 4 random bits for selecting equiprobably one out of the 16 color classes.

3   Lower Bounds

In this section, using the Minimax Principle [14] (see also [11]), we prove new lower bounds on the competitive ratio, against oblivious adversaries, of any randomized algorithm in cellular networks with reuse distance $k \geq 2$. We consider networks that support one frequency; our lower bounds can be easily extended to networks that support multiple frequencies. In our proof, we use the following lemma.

Lemma 2 (Minimax Principle [11]). *Given a probability distribution $\mathcal{P}$ over sequences of calls σ, denote by $\mathcal{E}_\mathcal{P}[B_A(\sigma)]$ and $\mathcal{E}_\mathcal{P}[B_{OPT}(\sigma)]$ the expected benefit of a deterministic algorithm A and the optimal off–line algorithm on sequences of calls generated according to $\mathcal{P}$. Define the competitiveness of A under $\mathcal{P}$, $c_A^\mathcal{P}$ to be such that*

$$c_A^\mathcal{P} = \frac{\mathcal{E}_\mathcal{P}[B_{OPT}(\sigma)]}{\mathcal{E}_\mathcal{P}[B_A(\sigma)]}.$$

Let A_R be a randomized algorithm. Then, the competitiveness of A under $\mathcal{P}$ is a lower bound on the competitive ratio of A_R against an oblivious adversary, i.e. $c_A^\mathcal{P} \leq c_{A_R}$.

So, in order to prove a lower bound for any randomized algorithm, it suffices to define an adversary which produces sequences of calls according to a probability distribution and prove that the ratio of the expected optimal benefit over the expected benefit of any deterministic algorithm (that may know the probability distribution in advance) is above some value; by Lemma 2, this value will also be a lower bound for any randomized algorithm against oblivious adversaries.

Theorem 2.
(a) No randomized online call–control algorithm can be better than 2-competitive against oblivious adversaries in cellular networks with reuse distance $k \geq 2$.
(b) No randomized online call–control algorithm can be better than 2.5-competitive against oblivious adversaries in cellular networks with reuse distance $k \geq 6$.

Proof. Due to lack of space, we prove only the first part of the theorem here. The proof of the second part which uses similar ideas in a slightly more complicated way will appear in the final version of the paper.

We present an adversary $\mathcal{ADV}$-2 which produces sequences of calls according to a probability distribution $\mathcal{P}_2$ which yields the lower bound. We show that the expected benefit of every deterministic algorithm (that may know $\mathcal{P}_2$ in advance) for such sequences of calls is at most 2, while the expected optimal benefit is at least 4. The first statement of Theorem 2 then follows by Lemma 2. First, we describe the sequences of calls produced by $\mathcal{ADV}$-2 without explicitly giving the cells where they appear; then, we show how to construct them in cellular networks of reuse distance $k \geq 2$.

We start by defining a simpler adversary $\mathcal{ADV}$-1 that works as follows: It first produces two calls in cells v_0 and v_1 which have distance at least k. Then it tosses a fair coin.

- On HEADS, it produces two calls in cells v_{00} and v_{01} which are at distance at least k from each other, at most $k - 1$ from v_0 and at least k from v_1. Then, it stops.
- On TAILS, it produces two calls in cells v_{10} and v_{11} which are at distance at least k from each other, at most $k - 1$ from v_1 and at least k from v_0. Then, it stops.

Now, consider the set of all possible deterministic algorithms $\mathcal{A}_1$ working on the sequences produced by $\mathcal{ADV}$-1. Such an algorithm $A_1 \in \mathcal{A}_1$ may follow one of the following strategies:

- It may accept both calls in cells v_0 and v_1 presented at the first step. This means that the calls presented in the second step cannot be accepted.
- It may reject both calls in cells v_0 and v_1 and then either accept one or both calls presented in the second step or reject them both.
- It may accept only one of the two calls in cells v_0 and v_1 and, if the calls produced at the second step by $\mathcal{ADV} - 1$ are at distance at least k from the accepted call, either accept one or both calls presented in the second step or reject them both.

In the first two cases, the expected benefit of the algorithm A_1 is at most 2. In the third case, the expected benefit is 1 (in the first step) plus the expected benefit in the second step. The latter is either zero with probability $1/2$ (this is the case where the cells of the calls produced by the adversary in the second step are at distance at most $k - 1$ from the cell of the call accepted by the algorithm in the first step) or at most 2 with probability $1/2$. Overall, the expected benefit of the algorithm is at most 2.

The adversary $\mathcal{ADV}$-2 works as follows: It first produces two calls in cells v_0 and v_1 which have distance at least k. Then it tosses a fair coin.

- On HEADS, it produces two calls in cells v_{00} and v_{01} which are at distance at least k from each other, at most $k-1$ from v_0 and at least k from v_1. Then, it tosses a fair coin.
  - On HEADS, it produces two calls in cells v_{000} and v_{001} which are at distance at least k from each other, at most $k-1$ from v_0 and v_{00}, and at least k from v_1 and v_{01}. Then, it stops.
  - On TAILS, it produces two calls in cells v_{010} and v_{011} which are at distance at least k from each other, at most $k-1$ from v_0 and v_{01} and at least k from v_1 and v_{00}. Then, it stops.
- On TAILS, it produces two calls in cells v_{10} and v_{11} which are at distance at least k from each other, at most $k-1$ from v_1 and at least k from v_0. Then, it tosses a fair coin.
  - On HEADS, it produces two calls in cells v_{100} and v_{101} which are at distance at least k from each other, at most $k-1$ from v_1 and v_{10}, and at least k from v_0 and v_{11}. Then, it stops.
  - On TAILS, it produces two calls in cells v_{110} and v_{111} which are at distance at least k from each other, at most $k-1$ from v_1 and v_{11}, and at least k from v_0 and v_{10}. Then, it stops.

Observe that, the subsequence of the last 4 calls produced by $\mathcal{ADV}$-2 essentially belongs to the set of sequences of calls produced by $\mathcal{ADV}$-1.

Now, consider the set of all possible deterministic algorithms $\mathcal{A}_2$ working on the sequences produced by $\mathcal{ADV}$-2. Such an algorithm $A_2 \in \mathcal{A}_2$ may follow one of the following strategies:

- It may accept both calls in cells v_0 and v_1 presented at the first step. This means that the calls presented in the next steps cannot be accepted.
- It may reject both calls in cells v_0 and v_1 and then apply a deterministic algorithm A_1 on the subsequence presented after the first step.
- It may accept only one of the two calls in cells v_0 and v_1 and, then, if the calls produced at the next steps by $\mathcal{ADV}$-2 are at distance at least k from the accepted call, apply a deterministic algorithm A_1 on the subsequence presented after the first step.

In the first case, the expected benefit of the algorithm A_2 is at most 2. In the second case, the expected benefit of A_2 is the expected benefit of A_1 on the sequence of calls presented after the first step, i.e., at most 2. In the third case, the expected benefit is 1 (in the first step) plus the expected benefit in the next steps. The benefit of the algorithm in the next steps is either zero with probability $1/2$ (this is the case where the cells of the calls produced by the adversary in the next steps are at distance at most $k-1$ from the cell of the call accepted by the algorithm in the first step) or the expected benefit of A_1 on the sequence of calls presented after the first step, i.e., at most 2 with probability $1/2$. Overall, the expected benefit of the algorithm is at most 2.

Furthermore, the expected optimal benefit on sequences produced by $\mathcal{ADV}$-2 is at least 4. Indeed, in each of the possible sequences

$$\sigma_2^{00} = \langle v_0, v_1, v_{00}, v_{01}, v_{000}, v_{001}\rangle,\ \sigma_2^{01} = \langle v_0, v_1, v_{00}, v_{01}, v_{010}, v_{011}\rangle,$$
$$\sigma_2^{10} = \langle v_0, v_1, v_{10}, v_{11}, v_{100}, v_{101}\rangle,\ \sigma_2^{11} = \langle v_0, v_1, v_{10}, v_{11}, v_{110}, v_{111}\rangle$$

generated by the $\mathcal{ADV}$-2, the calls in cells $\langle v_1, v_{01}, v_{000}, v_{001}\rangle$, $\langle v_1, v_{00}, v_{010}, v_{011}\rangle$, $\langle v_0, v_{11}, v_{100}, v_{101}\rangle$, and $\langle v_0, v_{10}, v_{110}, v_{111}\rangle$ can be accepted, respectively. Overall, the ratio of the expected optimal benefit over the expected benefit of algorithm A_2 on the sequences generated by the adversary $\mathcal{ADV}$-2 is at least 2, which (by Lemma 2) is a lower bound on the competitive ratio of any randomized algorithm for call control.

Next, we show how the adversary $\mathcal{ADV}$-2 locates the calls in cellular networks of reuse distance $k \geq 2$ completing the proof of the first part of the theorem.

The coordinates of the cells hosting possible calls produced by $\mathcal{ADV}$-2 are:

$$
\begin{aligned}
v_0 &= (0, -k, -k) & v_1 &= (0, k, k)\\
v_{00} &= (0, -2k+1, -2k+1) & v_{01} &= (0, -1, -1)\\
v_{10} &= (0, 1, 1) & v_{11} &= (0, 2k-1, 2k-1)\\
v_{000} &= (k-1, -2k+1, -k) & v_{001} &= (-k+1, -k, -2k+1)\\
v_{010} &= (k-1, -k, -1) & v_{011} &= (-k+1, -1, -k)\\
v_{100} &= (k-1, 1, k) & v_{101} &= (-k+1, k, 1)\\
v_{110} &= (k-1, k, 2k-1) & v_{111} &= (-k+1, 2k-1, k)
\end{aligned}
$$

An example of all cells hosting possible calls produced by $\mathcal{ADV}$-2 when $k = 3$ is depicted in Figure 3.

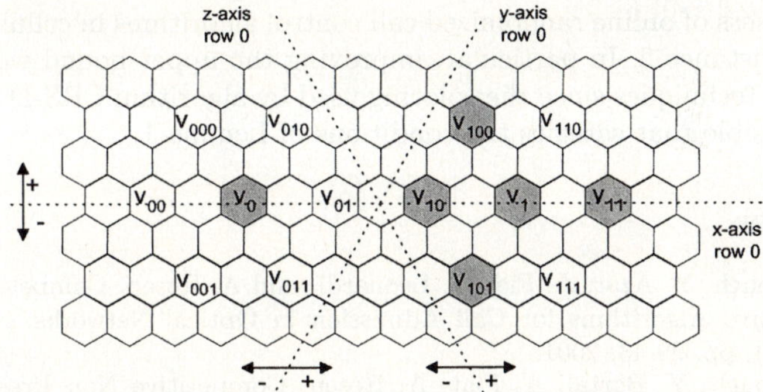

Fig. 3. The calls that may be produced by the adversary $\mathcal{ADV}$-2 in a cellular network of reuse distance 3. The grey cells host calls of the sequence σ_2^{10}.

We have to show that any of the possible sequences of calls generated according to $\mathcal{P}_2$ satisfies the constraints defined above. We have four possible sequences to examine: σ_2^{00}, σ_2^{01}, σ_2^{10} and σ_2^{11}. We show that one of them, e.g., σ_2^{10} satisfies the constraints; the proof for the other cases is similar due to symmetry. First, the cells v_0 and v_1 have distance $2k \geq k$. The cells at distance at most $k-1$ from

v_0 are those contained between the x-rows $k-1$ and $-k+1$, between the y-rows $-2k+1$ and -1, and between the z-rows $-2k+1$ and -1. The cells at distance at most $k-1$ from v_1 are those contained between the x-rows $k-1$ and $-k+1$, between the y-rows 1 and $2k-1$, and between the z-rows 1 and $2k-1$. Hence, cells $v_{10}, v_{11}, v_{100}, v_{101}$ are all at distance at least k from v_0 and at most $k-1$ from v_1. Since $k \geq 2$, the cells v_{10} and v_{11} have distance $2k-2 \geq k$. Also, the cells at distance at most $k-1$ from v_{10} are those contained between the x-rows $k-1$ and $-k+1$, between the y-rows $-k+2$ and k, and between the z-rows $-k+2$ and k. The cells at distance at most $k-1$ from v_{11} are those contained between the x-rows $k-1$ and $-k+1$, between the y-rows k and $3k-2$, and between the z-rows k and $3k-2$. Hence, cells v_{100}, v_{101} are at distance at most $k-1$ from v_{10} and at least k from v_{11}. In addition, v_{100} and v_{101} are at distance $2k-2 \geq k$ since $k \geq 2$. $\qquad\square$

4   Conclusions

In this paper, we presented a new online call control algorithm with competitive ratio 16/7 for cellular networks with reuse distance 2, improving the previous best known upper bound of 7/3. The algorithm is based on the "classify and randomly select" paradigm, uses only four random bits and works in networks with arbitrarily many frequencies. We have also presented new lower bounds of 2 and 2.5 on the competitiveness against oblivious adversaries of online call control algorithms in cellular networks of reuse distance $k \geq 2$ and $k \geq 6$, respectively. Our new lower bounds improve previous ones for almost all cases of the reuse distance $(k \neq 5)$.

An interesting open problem is to close the gap between 16/7 and 2 on the competitiveness of online randomized call control algorithms in cellular networks with reuse distance 2. In particular, improving the upper bound would require entirely new techniques since the coloring used by algorithm CRS-D seems to be the best possible that satisfies the conditions of Lemma 1.

References

1. B. Awerbuch, Y. Azar, A. Fiat, S. Leonardi, and A. Rosen. Competitive On–line Competitive Algorithms for Call Admission in Optical Networks. *Algorithmica*, Vol. 31(1), pp. 29-43, 2001.
2. B. Awerbuch, Y. Bartal, A. Fiat, A. Rosen. Competitive Non–Preemptive Call Control. In *Proc. of the 5th Annual ACM–SIAM Symposium on Discrete Algorithms (SODA '94)*, pp. 312–320, 1994.
3. Y. Bartal, A. Fiat, and S. Leonardi. Lower Bounds for On–line Graph Problems with Applications to On–line Circuit and Optical Routing. In *Proc. of the 28th Annual ACM Symposium on Theory of Computing (STOC '96)*, pp. 531-540, 1996.
4. A. Borodin and R. El-Yaniv. Online Computation and Competitive Analysis. *Cambridge University Press*, 1998.
5. I. Caragiannis, C. Kaklamanis, and E. Papaioannou. Efficient On–Line Frequency Allocation and Call Control in Cellular Networks. *Theory of Computing Systems*, Vol. 35, pp. 521-543, 2002. Preliminary versions in *SPAA '00* and *IPDPS '01*.

6. I. Caragiannis, C. Kaklamanis, and E. Papaioannou. Simple On-line Algorithms for Call Control in Cellular Networks. In *Proc. of the 1st Workshop on Approximation and Online Algorithms (WAOA '03)*, LNCS 2909, Springer, pp. 67-80, 2003.

7. T. Erlebach and K. Jansen. The Maximum Edge-Disjoint Paths Problem in Bidirected Trees. *SIAM Journal on Discrete Mathematics*, Vol. 14(3), pp. 326-355, 2001.

8. W.K. Hale. Frequency Assignment: Theory and Applications. In *Proc. of the IEEE*, 68(12), pp. 1497–1514, 1980.

9. J. Janssen, D. Krizanc, L. Narayanan, and S. Shende. Distributed On–Line Frequency Assignment in Cellular Networks. *Journal of Algorithms*, Vol. 36(2), pp. 119-151, 2000.

10. S. Leonardi, A. Marchetti–Spaccamela, A. Prescuitti, and A. Rosen. On–line Randomized Call–Control Revisited. *SIAM Journal on Computing*, Vol. 31(1), pp. 86–112, 2001.

11. R. Motwani and B. Raghavan. Randomized Algorithms. *Cambridge University Press*, 1995.

12. G. Pantziou, G. Pentaris, and P. Spirakis. Competitive Call Control in Mobile Networks. *Theory of Computing Systems*, Vol. 35(6), pp. 625-639, 2002.

13. P.-J. Wan and L. Liu. Maximal Throughput in Wavelength-Routed Optical Networks. Multichannel Optical Networks: Theory and Practice, *DIMACS Series in Discrete Mathematics and Theoretical Computer Science*, AMS, Vol. 46, pp. 15-26, 1998.

14. A. C. Yao. Probabilistic Computations: Towards a Unified Measure of Complexity. In *Proc. of the 17th Annual Symposium on Foundations of Computer Science (FOCS '77)*, pp. 222–227, 1977.

A Multiple Channel Access Protocol
for Ad Hoc Wireless Networks

Kil-Woong Jang

Dept. of Mathematical and Information Science, Korea Maritime University
1 YeongDo-Gu Dongsam-Dong, Busan, Korea
Tel: +82-51-410-4375
Fax: +82-51-404-3986
jangkw@bada.hhu.ac.kr

Abstract. We propose a new multiple channel access protocol to enhance the channel utilization and minimize the connection breakage probability in ad hoc wireless networks. In ad hoc networks with multiple channels, communication between a pair of hosts can directly be established using one of available channels. However, hosts' mobility or using channel by neighbors may cause a co-channel interference problem or the connection breakage. To solving these problems, the proposed protocol establishes the new connection using a channel exchange mechanism, which exchanges its available channel with the current channel of the neighboring host without co-channel interference. In addition, it efficiently maintains the current connection of the communicating hosts from co-channel interference caused by hosts' mobility. We evaluate the performance of the proposed protocol using simulation. Simulation results indicate that the proposed protocol may offer performance better than the conventional protocol in terms of the channel utilization and the connection breakage probability.

Keywords: ad hoc wireless networks, multiple channel access protocol, co-channel interference

1   Introduction

The demand for mobile and direct services has generated interest in ad hoc networks. Ad hoc networks have been designed that mobile host can directly communicate with each other without supporting stationary infrastructures, such as base stations or access points, in conventional wireless networks. In particular, since the stationary infrastructures are hard to establish in situations (i.e., war or natural disasters), mobile users have required the ad hoc service. In conventional wireless networks, if traffic gives overloads to the stationary infrastructures, it may cause high transmission delay and low throughput. On the other hand, mobile hosts, in ad hoc networks, can directly communicate with each other in the transmission range allowed by the transmission power. Due to the transmission range, each host acts as a router, forwarding traffic for other hosts in out of range using routing protocols [6].

Ad hoc hosts have limited channel resources and direct communication between a pair of hosts in the transmission range is established using a same

J.C. Cunha and P.D. Medeiros (Eds.): Euro-Par 2005, LNCS 3648, pp. 1100–1109, 2005.

common channel. Owing to using limited channel resources, it is important to efficiently utilize channels and reduce connection breakage in ad hoc networks. In IEEE 802.11 protocol for wireless LANs [8], RTS/CTS mechanism is proposed to avoid the occurrence of collision and increase throughput for ad hoc networks. Wireless LANs were typically defined in a single channel system. As the number of communicating hosts increases, systems with a single channel may have the disadvantage which is declination of system performance. However, IEEE Standard 802.11 already has multiple channels available for use. IEEE 802.11b physical layer has 14 channels, 5MHz apart in frequency. However, to be non-overlapping and be able to use channels, the frequency must be divided into at most 30MHz. Thus the number of available channels used for communication is 3: channel 1, 6 and 11. On the other hand, IEEE 802.11a has 12 channels, and the number of available channels is 8 for indoor use and 4 for outdoor use, respectively.

There are many related studies for multiple channels [1–5]. Dynamic Channel Assignment (DCA) [2] assigns channels in an on-demand style. This protocol maintains one control channel and other data channels. Each host has two half-duplex transceivers, and one is for control channel and the other is for data channel, respectively. The basic operation in this protocol is similar to IEEE 802.11 protocol. Using RTS/CTS packets, a pair of hosts decides which channel to communicate. Dual Busy Tone Multiple Access [4] has two common channels, which is one control channel and one data channel. To avoid hidden terminals, a control message called busy tone is transmitted on a control channel. In [5], they propose a protocol that is similar to DCA. This protocol also has one control channel and many data channels, and selects the best channel under channel condition of receiver. Anther important issue for multiple channel access protocol is the co-channel interference, which can be occurred due to host mobility or the same channel usage. Previous study [7] was proposed to use a power control mechanism to eliminate the co-channel interference. As the communicating hosts reduce the power of signal, the communication range is reduced and thus the co-channel interference can be eliminated.

In this paper, we propose the multiple channel access protocol to enhance the channel utilization and reduce the connection breakage probability. In the proposed protocol, each host stores the state of channels of communicating neighbors in the communication range and establishes the new communication link using the channel exchange mechanism, which exchanges its available channel with the occupied channel of a communicating neighbor under certain conditions. In addition, the proposed protocol is designed to minimize the co-channel interference caused by host mobility.

2   Proposed Multiple Channel Access Protocol

For simplicity, we first describe the basic operation of the proposed protocol using an example of ad hoc networks, as shown in Fig. 1. In this figure, a circle node represents a mobile host. The white-colored host is in idle state and the gray-colored one is in communication state. A dashed line between two hosts

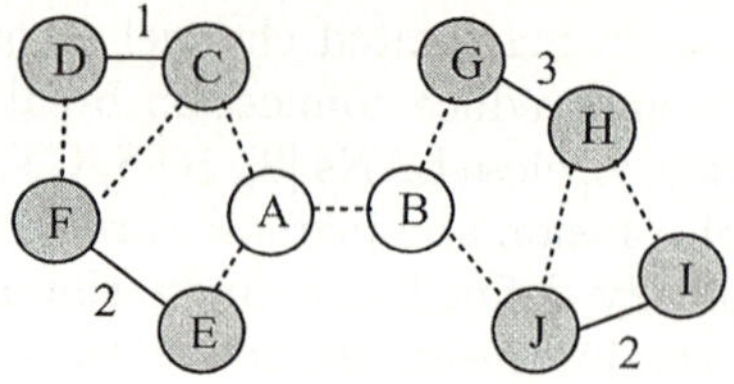

Fig. 1. An example of ad hoc networks.

indicates that each host exists in the communication range and can directly communicate with each other. For example, hosts in the communication range of host A are hosts B, C and E. A line between two gray-colored hosts represents a connecting link between a pair of hosts. A symbol in a circle node represents the address of the host, and the number on the connecting link indicates the channel ID being used by the pair of hosts.

In ad hoc networks with multiple channels, no host can use the channel that is occupied by the communicating neighbors in the communication range owing to the co-channel interference. For example, in Fig. 1, we assume that the total number of available channels in a host is three and then channel IDs are 1, 2 and 3. Hosts C and E in the communication range of host A occupy channel 1 and 2, respectively. Thus, host A is able to use only channel 3. On the other hand, host B can only use channel 1. In this situation, the conventional protocol is unable to establish communication between hosts A and B. In other words, a new connection is blocked if no common channel between a pair of hosts exists, and then both must wait until a new common channel is released.

To carry out the proposed protocol, the following operations are positively necessary. After communication is established between a pair of hosts, both inform their neighbors of a connection message. This message includes the communication state of the host, the occupied channel ID and the addresses of communicating hosts. If communication between a pair of hosts is terminated, terminated hosts inform their neighbors of a termination message. Therefore, each host can dynamically specify the state of neighbors from the received messages. To support the operations, each host maintains a table in its memory, which records the channel usage information of the neighbors, to efficiently determine an available channel. This table is called the channel information table. Fig. 2 shows the channel information table of host A in Fig. 1. For example, hosts C and D are in a communication state and use channel 1 for communication. Now, we describe the operation of the proposed protocol in the situation of Fig. 1. The proposed protocol is attempted in the following order.

1. Host A broadcasts a *connection request* message with available channel IDs and the address of host B. On receiving the message, host B compares the received channel IDs with its channel information. If there is an identical available channel, host B replies a *connection confirm* message with the identical channel ID to host A and then executes step 4; otherwise, step 2 is initiated.

Channel ID	Channel state	Address 1	Address 2
1	BUSY	C	D
2	BUSY	E	F
3	IDLE		

Fig. 2. Channel information table of host A.

2. Host B sends a *check request* message with the channel IDs received from host A to its communicating neighbors. The neighbors check whether they can continuously communicate, if they exchange its occupied channel with the received channels. If there are available channels, each neighbor replies a *check confirm* message with the available channel IDs to host B and then executes step 3; if not, it replies a *check confirm* message with a *NULL* value to host B and then executes step 3.

3. When host B receives the *check confirm* message, it replies a *connection confirm* message, including the available channel ID of host B and the available channel IDs received from neighbors, to host A and then executes step 4.

4. When host A receives the *connection confirm* message, it compares its channel information with the channel IDs received from host B. If there is an identical channel, it selects the channel for communication and sends a *connection notification* message with the selected channel ID to host B and then executes step 7; if not, we perform step 5.

5. Host A sends a *check request* message with its available channel IDs to communicating neighbors. The neighbors check whether they are continuously able to communicate, if they exchange its occupied channel with the received channels. If there are available channels, each neighbor replies a *check confirm* message with the available channel IDs to host A and then executes step 6; otherwise, it replies a *check confirm* message with a *NULL* value to host B and then executes step 6.

6. When host A receives the *check confirm* message, it compares the received channel ID with the channel IDs received from host B. If there is an identical channel, host A selects the channel for communication and sends a *connection notification* message with the selected channel ID to host B and then executes step 7; if not, we perform step 9.

7. When host B receives the *connection notification* message, it checks whether the selected channel is equal to the occupied channel of the neighbors. If so, it sends a *change request* message with one of available channels in it to the neighbor. The neighbor received the message exchanges its occupied channel with the received channel. After the channel is successfully exchanged, the neighbor replies a *change confirm* message to host B. Host B received the message sends a *channel notification* message, including the selected channel ID and addresses of hosts A and B, to the neighbors and host A.

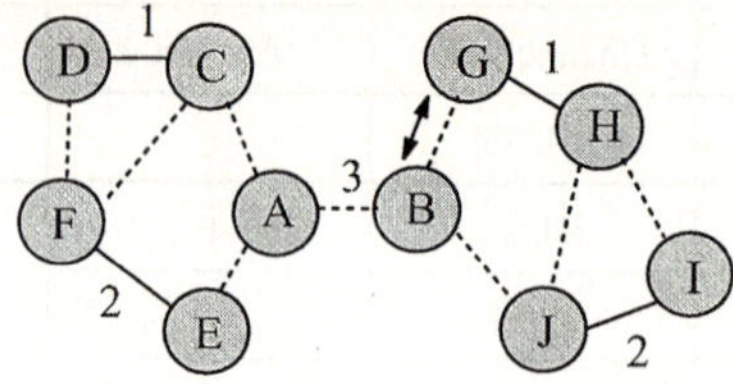

Fig. 3. An example for illustrating the channel assignment after host B exchanges its available channel with the occupied channel of host G.

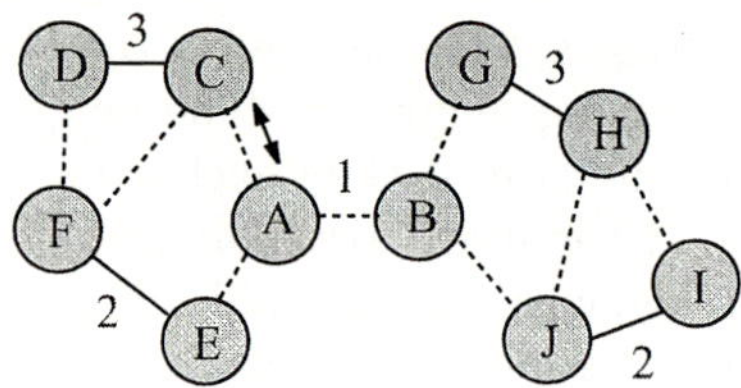

Fig. 4. An example for illustrating the channel assignment after host A exchanges its available channel with the occupied channel of host C.

8. When host A receives the *channel notification* message, it checks whether the selected channel is equal to the occupied channel of the neighbors. If so, it sends a *change request* message with one of available channels in it to the neighbor. The neighbor received the message exchanges its occupied channel with the received channel. After the channel is successfully exchanged, the neighbor replies a *change confirm* message to host A. Host A received the message sends a *channel notification* message, including the selected channel ID and addresses of hosts A and B, to the neighbors.
9. If all the above conditions do not hold, then the connection cannot be accomplished. Normally, this would result in the connection being blocked.

As carried out the operation of the proposed protocol, communication between hosts A and B is established as shown in Figs. 3 and 4. During the operation, each host checks if it can exchange its available channels with the occupied channel of neighbors. Simultaneously, it should be checked the state of channels in a host, which is communicating with the neighbor. For example, in Fig. 1, host G checks if it can exchange the occupied channel of host H with the channel received from host B. If host H is unable to exchange channels 1 and 3, host G is unable to exchange its occupied channel for the channel received from host B. Under this condition, since host B exchanges its available channel with the occupied channel of the neighboring host, Fig. 3 shows that communication between hosts A and B can be established. In Fig. 4, since host A exchanges its available channel for the occupied channel of the neighboring host, communication can be established.

In ad hoc networks, each host can directly communicate with each other and move anywhere. As a communicating host moves toward another communicating

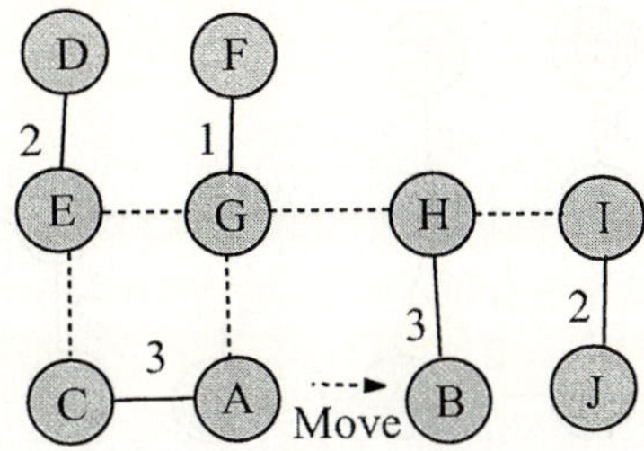

Fig. 5. An example of ad hoc networks for executing the proposed protocol due to host mobility.

host using a same channel, the co-channel interference between both is gradually arisen. For example, in Fig. 5, hosts A and C are communicating over channel 3. Hosts B and H are also using the same channel for communication. As host A moves toward host B, the co-channel interference is arisen. In this section, we present the operation of the proposed protocol to minimize the co-channel interference between a pair of hosts due to host mobility.

For simplicity, we describe the protocol operation for host mobility using an example of ad hoc networks, as shown in Fig. 5. The proposed protocol for host mobility is presented below.

1. As host A interferes with host B due to host mobility, hosts A and C check their channel information table and determine a new available channel for channel assignment. If hosts A and C have a new available channel, they change their occupied channel to the new channel. Next, hosts A and C send a *channel notification* message, including the new channel ID, the addresses of hosts A and C, to their neighbors, respectively; if not, we perform step 2.

2. Hosts A and C send a *check request* message with their occupied channel ID to their communicating neighbors except host B. The neighbors received the message check whether the co-channel interference occurs, if they exchange their occupied channel for the received channel. If no co-channel interference occurs, each neighbor replies a *check confirm* message with the received channel ID to hosts A or C; otherwise, it replies a *check confirm* message with a *NULL* value to hosts A or C.

3. When hosts A and C receive the *check confirm* message, they check if there is an available channel from the received information. If so, they send a *change request* message with the occupied channel to the neighbor. The neighbor received the message exchanges its occupied channel with the received channel. After the channel is successfully changed, the neighbor replies a *change confirm* message to hosts A or C. In addition, hosts A or C send a *channel notification* message, including the new channel ID, addresses of hosts A or C, to their neighbors, respectively; if not, step 4 is initiated.

4. Host A sends a *interference request* message to host B. On receiving the message, hosts B and H check their channel information table and determine a new available channel for channel assignment. If hosts B and H have a new

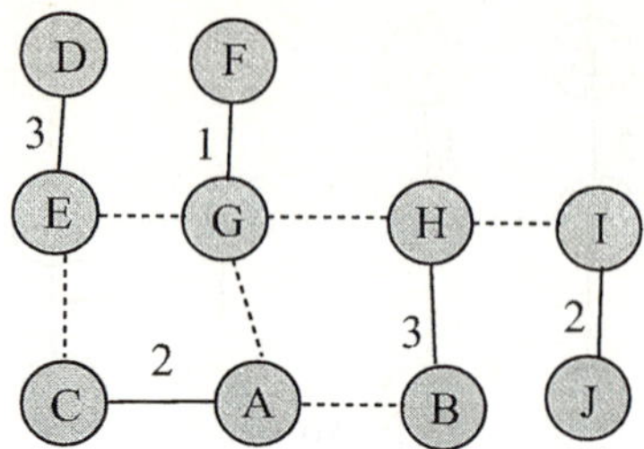

Fig. 6. An example of ad hoc networks after host A changes its occupied channel to a new available channel.

available common channel, they change their occupied channel to the new channel and then host B sends a *interference confirm* message with the new channel ID to host A. Next, hosts B and H send a *channel notification* message, including the new channel ID, the addresses of hosts B or H, to neighbors, respectively; if not, we perform step 5.

5. Hosts B and H send a *check request* message with their occupied channel ID to their communicating neighbors except host A. The neighboring hosts received the message check whether the co-channel interference occurs, if they exchange its occupied channel with the received channel. If no co-channel interference occurs, each neighbor replies a *check confirm* message with the received channel ID to hosts B or H; otherwise, it replies a *check confirm* message with a *NULL* value to hosts B or H.

6. When hosts B or H receive the *check confirm* message, they check if there is an available channel in the received information. If so, they send a *change request* message with the occupied channel to the neighbor. The neighbor received the message exchanges its occupied channel with the received channel. After the channel is successfully changed, the neighbor replies a *change confirm* message to hosts B or H. Finally, hosts B and H send a *channel notification* message, including the new channel ID, addresses of hosts B or H, to their neighbors, respectively.

As carried out the operation of the proposed protocol, we assign a new channel to hosts A and C, as shown in Fig. 6. If all the above conditions do not hold, we are unable to assign a new channel to the interfered hosts. Then, the co-channel interference between hosts A and B occurs.

3   Performance Evaluation

In this section we study the impact of the estimation process on the capacity of the proposed protocol through computer simulation. We develop a simulation model to analyze the performance of the proposed protocol under the assumption of ideal channel conditions (i.e., no hidden terminals and capture). The QoS measures that we are interested in are the connection breakage probability, P_b, and the channel utilization, U.

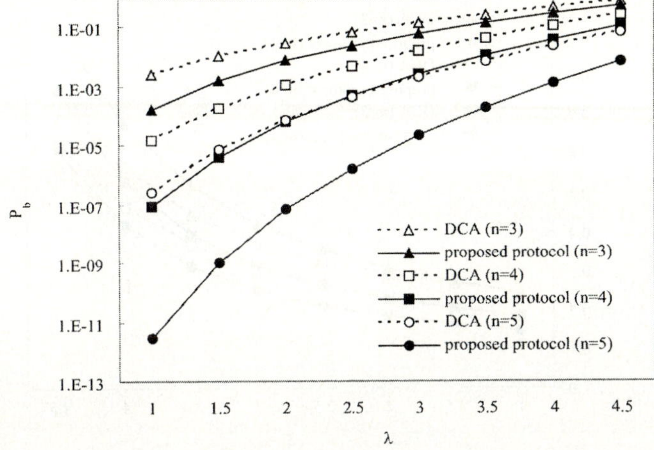

Fig. 7. Connection breakage probability for various values of n.

For simulation, we developed a discrete event driven simulator using a high level computer language, C++. The network model for simulation consists of randomly placed 100 nodes in a 1000×1000 m area. We also consider a mobile network model, where the nodes move independently of one another, which random speeds that are uniformly distributed between 0–20m/s. The mobility pattern is based on the random waypoint model [6]. Simulations are performed in wireless LAN environment. The bit rate for each channel is 11Mbps, and the transmission range of each host is approximately 100 m. We assume that each node has a constant bit rate (CBR) traffic and 512 bits control packets, respectively. We also assume that traffic at every node is generated according to Poisson processes with identical mean arrival rates λ. Each simulation was performed for duration of 60 seconds. The simulation results shown in this paper are valid for up to 95% confidence intervals.

We provide numerical results, based on simulation, to compare the performance of the proposed and DCA protocols. We obtained values for the connection breakage probability and the channel utilization under different traffic loads λ, ranging from 1 to 4.5 connections per unit time. Three difference values, 3, 4 and 5, were used for a total number of available channels, n, in a host.

Fig. 7 shows P_b for the proposed and DCA protocols under varying n. In this figure, we can observe that the proposed protocol has better performance than the DCA protocol. When n is increased to 5, the proposed protocol significantly outperforms the DCA protocol in terms of P_b. This result can be observed that the P_b curve in the proposed protocol is up to four orders of magnitude lower than in the DCA protocol. The reason the proposed protocol provides a lower P_b curve is that it can exchange its available channels with the occupied channel of neighbor and thus reduce the connection breakage probability.

Fig. 8 shows the values of U under varying λ. The proposed protocol achieves uniformly higher values of U under various loads. The difference between the pro-

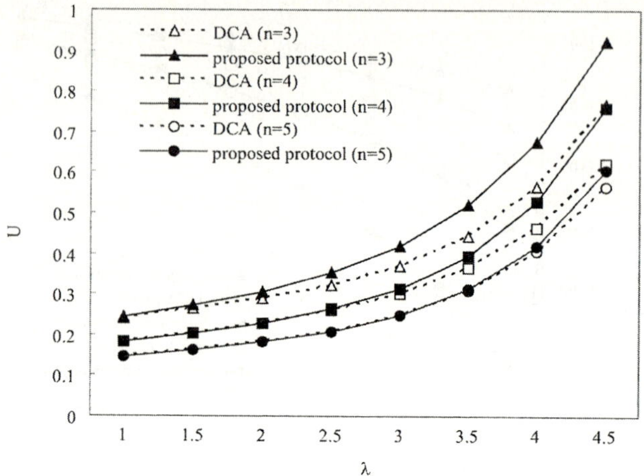

Fig. 8. Channel utilization for various values of n.

posed and DCA protocols is most apparent at high loads. Reason the proposed protocol provides a higher U value is that it exploits the channel exchange mechanism. At the same traffic loads, we can see that the protocol in $n{=}3$ has a higher values than in $n{=}5$. As λ is increased, the probability that hosts use channels in the lower number of available channels is higher than in the higher number of available channels. Therefore, in the lower number of available channels, the U value is higher than in the higher number of available channels. In addition, since the probability of the channel exchange is increased at lower channels, the proposed protocol significantly outperforms the DCA protocol. This result can be observed in Fig. 8, where the proposed protocol at high load is over 10% more than in the DCA protocol. On the other hand, at $n{=}5$, the U curve in the proposed protocol is closer than in the DCA protocol. However, the difference in the U curves will be larger as λ is increased.

4   Conclusions

We have presented a new multiple channel access protocol to improve the system performance in ad hoc wireless networks. In the proposed protocol, each host maintains the channel information and the state of the neighboring hosts. Based on this information, each host carries out a communication with other hosts without the co-channel interference. When the co-channel interference occurs due to host mobility, the proposed protocol minimizes the co-channel interference by allowing hosts to exchange their available channel with the occupied channel of their neighbor. We evaluated the performance of the proposed protocol using simulation. The simulation was focus on the connection breakage probability as well as the channel utilization and then compared with the conventional protocol. The numerical results indicated that the proposed protocol outperformed the conventional protocol over wide range of parameters.

References

1. A. Nasipuri, J. Zhuang and S. R. Das, "A multichannel CSMA MAC protocol for multihop wireless networks," Proc. WCNC'99, Sept. (1999) 1402–1406
2. S. L. Wu, C. Y. Lin, Y. C. Tseng and J. P. Sheu, "A new multi-channel MAC protocol with on-demand channel assignment for mobile Ad Hoc networks," Int'l Symposium on Parallel Architectures, Algorithms and Networks, (2000) 232–237
3. C. Y. Chang, P. C. Huang, C. T. Chang and Y. S. Chen, "Dynamic channel assignment and reassignment for exploiting channel reuse opportunities in Ad Hoc wireless networks," IEICE Trans. Commun., vol. E86-B, no. 4, April (2003) 1234–1246
4. J. Deng and Z. Haas, "Dual Busy Tone Multiple Access (DBTMA): A New Medium Access Control for Packet Radio Network," Florence, Italy, (1998)
5. N. Jain and S. Das, "A Multichannel CSMA MAC Protocol with Receiver-Based Channel Selection for Multihop Wireless Networks," in Proceedings of the 9th Int. Conf. on Computer Communications and Networks (IC3N). Oct. (2001)
6. J. Broch et al., "A Performance Comparison of Multi-Hop Wireless Ad Hoc Network Routing Protocols," in Proceedings of the 4th Int. Conference on Mobile Computing and Networking (ACM Mobicom '98), Oct. (1998) 85–97
7. Y. B. Ko, V. Shankarkumar and N. H. Vaidya, "Medium access control protocols using directional antennas in Ad Hoc network," IEEE INFOCOM 2000, (1999) 13–21
8. IEEE Standard for Wireless Medium Access Control and Physical Layer Specifications, Aug. (1999)

Personalized Access to Semantic Web Agents Using Smart Cards

Riza Cenk Erdur[1] and Geylani Kardas[2]

[1] Ege University, Department of Computer Engineering,
35100 Bornova, Izmir, Turkey
`erdur@staff.ege.edu.tr`
`http://bornova.ege.edu.tr/~erdur`
[2] Ege University, International Computer Institute,
35100 Bornova, Izmir, Turkey
`geylani@bornova.ege.edu.tr`
`http://ube.ege.edu.tr/~kardas`

Abstract. In this paper, we mainly focus on the integration of smart card based access to semantic web enabled multi-agent systems. Besides classical benefits such as smart card based authentication and authorization, integration of such a feature will make it possible for semantic web agents to take the personal knowledge stored as instances of a specific personal ontology in the smart cards into account and behave in a way that is more responsible to the individual requirements of the users. To integrate smart card based access to a semantic web agent, we need an agent plan specifically defined for that purpose. This plan will be responsible for both communicating with the smart card reader module and for semantically manipulating the personal knowledge that is transferred from the card. In the paper, we give the implementation level details for this plan. Another important aspect of the paper is that various alternatives for storing ontological knowledge on smart cards have been discussed based on some experimental results.

1 Introduction

Semantic web [2] aims to transform the World Wide Web into a knowledge representation system in which the information provided by web pages is interpreted using ontologies. This gives the opportunity for autonomous and interacting entities - semantic web agents [6] - to collect and interpret semantic content on the behalf of their users. On the other hand, smart card technology has paved the way for an individual to carry personal information in a small card with storage, data processing and security features [8].

By the marriage of agent, semantic web and smart card technologies, in addition to classical benefits such as smart card based authentication and authorization for accessing agent systems, semantic web agents can be accessed using smart cards that store users personal knowledge as instances of a specific personal ontology. By this way, agents can take the personal knowledge stored

J.C. Cunha and P.D. Medeiros (Eds.): Euro-Par 2005, LNCS 3648, pp. 1110–1119, 2005.

in the card into account in adapting their behavior to act in a way that is more responsible to the individual requirements of their users.

Here, we will describe a scenario to illustrate an example case where enabling smart card based access to semantic web agents may have potential benefits for the users: Let us think of a multi-agent system, which has been established to provide tourism related facilities in a specific geographic area. If smart card based access to this multi-agent system is provided via card terminals located in several places such as airports or various locations in city centers, then a newly arrived traveler without a pre-built travel plan can have the chance of discovering the hotel facilities that best matches against her personal knowledge stored in her smart card. In addition, after discovering the hotel facility, it will also be possible for the traveler to make an online reservation from the terminal using the credit card knowledge that is defined and stored as part of her personal knowledge. We believe that this scenario is valid, because there are always travelers around without a pre-built travel plan.

A semantic web enabled multi-agent system infrastructure that can be used in realizing the above scenario is shown in Fig. 1.

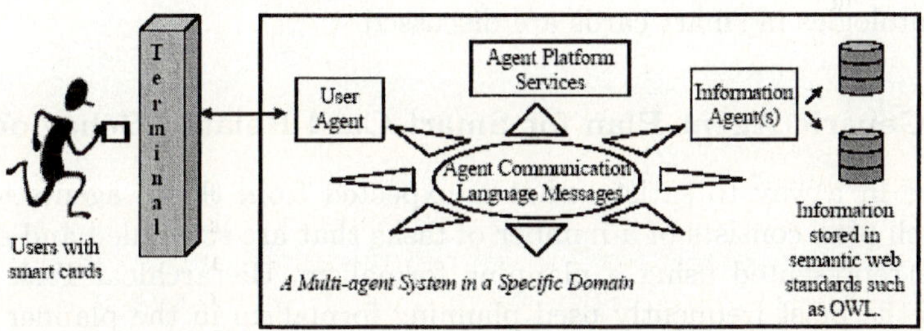

Fig. 1. The basic infrastructure for smart card based access to a multi-agent system

As shown in Fig. 1, the user agent in the multi-agent system must have the functionality to deal with the smart card based accesses. The user agent also must have the capability of understanding ontologies, since it has to understand the personal ontology instance that is transferred from the users smart card. In the multi-agent system, information is represented using semantic web standards such as OWL [12]. Hence, the information agents have the capability of manipulating semantic knowledge and answering queries over that semantic knowledge. Agent platform services, which are agent management, agent directory, and agent message transport services, are standard services that a platform, on which the multi-agent system is operating, has to provide. These services are usually provided as built-in services by the multi-agent development framework/ platform used.

In this paper, we mainly focus on the integration of smart card based access to the agents in a semantic web enabled multi-agent system. To achieve such kind of integration, we have to define a basic agent plan for that purpose. This agent plan is needed both for communicating with the smart card reader modules and for manipulating the transferred personal knowledge. Manipulation of the transferred personal knowledge includes parsing it, constructing the ontology model in memory, and preparing requests for querying the knowledge stored in information agents.

In smart card technology related literature, there are several studies covering the use of smart cards in Web applications, especially in medical healthcare systems [4]. However, there is no work discussing in detail either how smart card access support can be integrated to multi-agent systems or how ontologies are stored in smart cards, transferred from them and manipulated in agents. This paper aims to fill in this gap by discussing the implementation level details.

2   Architecture for Enabling Smart Card Based Access

In this section, first, the generic agent plan needed for smart card based access to a semantic web agent is explained. Then, details concerning the implementation of this plan are given. Finally, the personal ontology, which is used in representing the personal knowledge of each user, is given and various alternatives about storing ontologies in smart cards are discussed.

2.1   A Generic Agent Plan for Smart Card Related Behavior

To behave in a way to satisfy what is expected from them, agents formulate plans. Each plan consists of a number of tasks that are scheduled and executed. Plans are represented using a planning formalism. Hierarchical Task Network (HTN) is the most frequently used planning formalism in the planner modules [11] of agent development frameworks. Hence, the plan component defined for smart card access in this paper will be explained based on the HTN approach.

HTN structure consists of nodes that represent tasks. Since a task may be composed of subtasks, the plan structure may take the form of a tree-like structure. There are two kinds of links in a HTN representation. Reduction links describe the de-composition of high-level tasks to subtasks. Provision or outcome links represent value propagation between task nodes [11].

In an agent, smart card access related behavior can be modeled as a composite HTN task structure that consists of three subtasks, which are card reading and session opening subtask, the ontology manipulation related subtask(s) and card writing and closing the session subtask. Fig. 2 shows the HTN structure of this composite task.

As shown in Fig. 2, the smart card access task is a complex task that consists of three subtasks. The first subtask is responsible for waiting for the smart card to be inserted into card acceptance device and opening a new session when the card is inserted. After the session is opened, knowledge is read from the card and this knowledge is passed to the second subtask via the provision Knowledge_FromCard. The second subtask is a complex task that represents the behavior related with the manipulation of the semantic personal knowledge. After

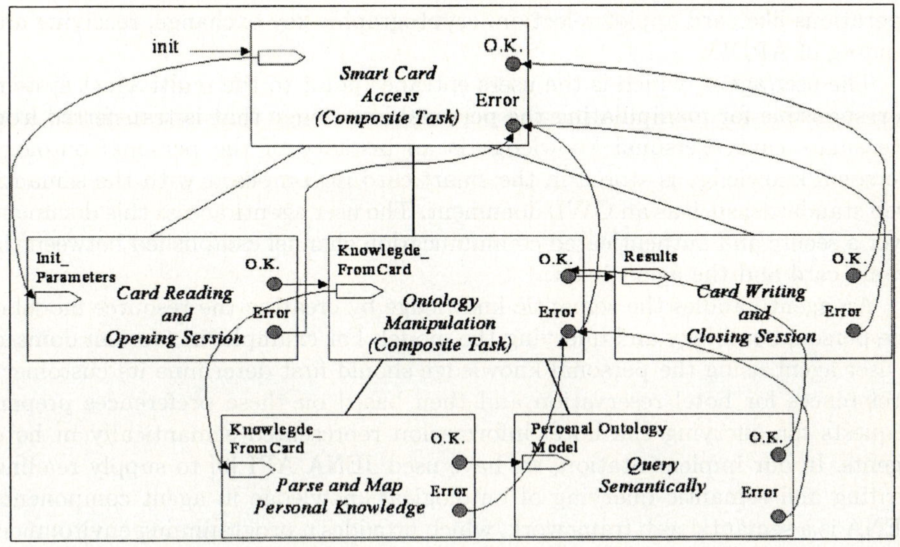

Fig. 2. HTN structure for smart card access behavior

the information agents are queried based on the concepts in the personal knowledge, incoming results are passed to the third subtask via the Results provision, which writes the necessary results back to the card and closes the session. We used a task for card writing, because we preferred the final results of a request to be written on users card so as to keep a history for the user.

2.2 Implementation Details for the Generic Plan

We need two basic components for implementing the smart card based access support plan completely. The first component is a middleware, which will enable an agent to communicate with the smart card. The second component is an API that is going to be used for the manipulation of ontological knowledge. In the following paragraphs, we discuss these two components.

In general, a smart card and an application communicate using a special protocol, which is called as Application Protocol Data Unit (APDU) and which is an application layer standard based on ISO 7816-4 [8]. In order to retrieve personal knowledge from the card, an agent needs to communicate with the smart card using APDU packages. However, we think that this task that is quite primitive and that integrating codes for pure smart card access will make an agents plan unnecessarily complicated. Hence, we have proposed a layered approach and using the OpenCard Framework (OCF) API [10] developed a middleware to handle smart card communications on behalf of the agents. OCF is an API that supports communication between the smart card host and the applications inside the card. It supports Java platform and is maintained by the OpenCard Consortium. Our middleware uses this API to control smart card events and manage operations like card applet selection, cryptographic key exchange, receiving and sending of APDUs.

The user agent, which is the users entrance point to the multi-agent system, is responsible for manipulating the personal knowledge that is transferred from the smart card. Personal knowledge is an instance of the personal ontology. Personal knowledge is stored in the smart card in compliant with the semantic web standards, such as an OWL document. The user agents access this document over a secure and authenticated communication channel established between the smart card and the agent itself.

An agent handles the semantic knowledge by creating the resource model of the personal ontology and querying the model. For example, in tourism domain, a user agent using the personal knowledge should first determine its customer's preferences for hotel reservation and then based on these preferences prepare requests for querying the hotel information represented semantically in hotel agents. In our implementation, we have used JENA API [9] to supply reading, writing and semantic querying of ontological knowledge in agent components. JENA is a semantic web framework, which provides a programming environment for RDF, RDFS and OWL and includes a rule-based inference engine. It also ensures a query language for RDF called RDQL. Our agents use RDQL to query on the ontology model to obtain the desired semantic knowledge.

During the implementation of the multi-agent system, the generic HTN structure, for which the subtask implementation details are given above, is instantiated and executed based on the plan definition and execution model of the agent development tool that is being used. For example, if JADE [1] is being used as the agent development framework/platform, then the generic HTN structure will be implemented and executed in compliant with the behavior model of JADE. How the generic plan is instantiated using an agent development framework/platform will be clear in section 3, which includes a case study explaining the instantiation of this plan in JADE environment.

2.3   Personal Ontology Example and Storing Its Instances on the Card

As mentioned before, the personal knowledge that is stored in the cards should be an instance of a specific personal ontology. Such an example personal ontology consists of two main parts: The first part is the domain independent part, where user identification, contact and payment information are kept. The other part is the domain specific part and includes knowledge represented using a specific domain ontology. For example, if the smart card is intended to be used in tourism domain for travelers who want to discover hotels and make reservations, then the knowledge belonging to this part will be represented using a specific tourism ontology defining facilities about hotels and reservations. One such ontology is introduced in [7].

Various alternatives may be considered for storing ontological knowledge in smart cards. These various alternatives are discussed below for different cases by giving the possible advantages and disadvantages:

Alternative-1: Store the personal knowledge completely in the card.

Advantages for alternative-1:

i) The knowledge stored in the card can be transferred to the agent by taking the advantage of strong security and authentication features of smart cards.

Disadvantages for alternative-1:

i) The storage capacity of smart cards is limited. However, the personal ontology is not very complex in most cases and capacity would not be a serious problem with the application of the scaling down techniques. One such technique is introduced in [3].

ii) To investigate whether response time may be a problem as the ontology file sizes get larger, we have measured the time needed to read or write ontological knowledge and have seen that as the file sizes increase, the time needed to read from or write to the card increases linearly. Measurements have been realized using Gemplus GemXpresso 211/PK model Java cards with multi applet support and with 32K ROM, 32K EEPROM and 2K RAM. JavaCard Framework 2.1 has been used during on-card software development. Gemplus GCR410 serial card read/write device has been connected to a terminal PC with 9600 baud data transmission rate. The measurement results are shown in Fig. 3. As Fig. 3 shows, larger personal knowledge file sizes may cause delays in response time. One of the most important reasons for delays is that APDU packets are maximum 255 bytes length and this requires multiple packet exchanges for large file sizes increasing the response time. In addition, the time values in Fig. 3 also include the time needed to convert the byte stream to a Java file in the agent. Please note that the ontological knowledge is stored in byte-streams in the card, which then needs a conversion into necessary Java files/objects.

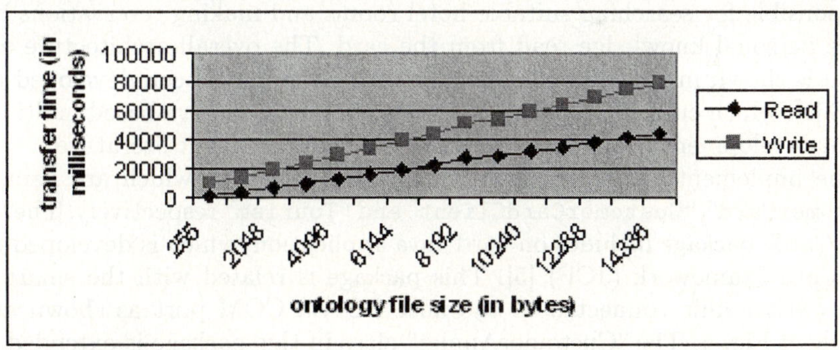

Fig. 3. Read/Write times for different ontology file sizes

Alternative-2: Store only the URL of the personal knowledge document in the smart card. In this case, the agent retrieves the URL of the ontology instance document rather than the document itself and then accesses the document over the Internet.

Advantages for alternative-2:

i) Obviously this approach decreases smart card communication time and shortens card sessions.

Disadvantages for alternative-2:

i) The security feature of smart cards cannot be used. The agent itself should manage the accessing privileges and security over the Internet. This may be a problem for the private part of personal knowledge, such as the payment information.

Conclusion: We have preferred the first alternative in our implementation although storing ontologies completely in the card may cause delays for especially large ontology files and especially for write operations. The reason for preferring the first alternative is that we think of taking the advantage of smart cards as a means of secure, portable and cheap personal knowledge storage media for accessing semantic web enabled multi-agent systems. In addition, ontology file transfer delays can be overcome using more efficient connection technologies between the card acceptance device and the host. In our experiment, the serial COM port has been used, which is another potential cause for the delays in transferring large ontology files.

3   Case Study: A Multi-agent System for Tourism Domain

As a case study, we have implemented a prototype semantic web enabled multi-agent system for realizing the infrastructure of the scenario given in the introduction section. There exist two types of agents in the developed system: hotel agents and customer agents. A hotel agent is an information type agent that represents a hotel in the system. A customer agent is an interface type agent that is responsible for searching suitable hotel rooms and making reservations based on the personal knowledge read from the card. The overall architecture of the system is shown in Fig. 4. The agents in the system have been developed using JADE [1], which currently is one of the most widely used Java based multi-agent system development framework/platform in the multi-agent literature.

The implementation consists of three Java packages, which are named as "CustomerCard", "CustomerCardClient" and "Tourism" respectively. The "CustomerCard" package includes on-card Java application, which is developed using Java Card Framework (JCF) [5]. This package is related with the smart card reader/writer unit connected to the host PC via COM port as shown at the left side of Fig. 4. The "CustomerApplet" class in this package is extended from the "javacard.framework.Applet" and it processes command APDUs sent from a customer agent. The "CustomerCardClient" package behaves like a middleware between the on-card applet and the customer agent. As it is shown in the middle part of Fig 4., this package resides on the same host with the customer agent that has smart card access and it supports functionalities such as smart card applet selection, APDU communication and hex-coded data package handling. The third package, which we named as the "Tourism" package, contains the agents in the developed multi-agent system, including the hotel agents and

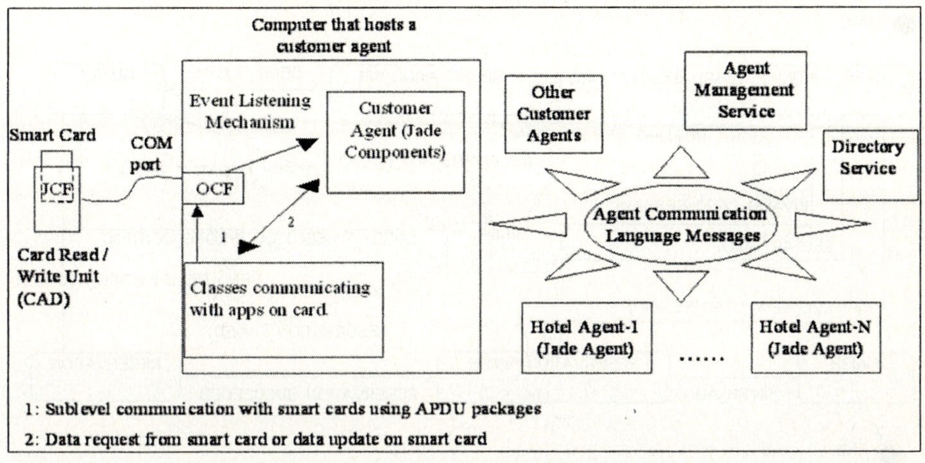

Fig. 4. A multi-agent system with a smart card access integrated customer agent

the customer agents. The agent management service maintaining life cycles of agents in a multi-agent system and the directory service providing yellow pages services to other agents are already supported by JADE platform.

The hotel agents and the customer agent are instances of the class "HotelAgent", and "CustomerAgent" respectively. These classes are derived from jade.core .Agent class and these agents behaviors are implemented using the behavior model of JADE. During initialization, a hotel agent registers itself to the platforms directory service as a Hotel Service provider. So, when a customer agent looks for a specific hotel room, it gets the hotel agents description from the directory service and can interact with this hotel agent. To be semantic web enabled, the RDQL [9] query language is supported. As mentioned before, JENA API has been used in semantic knowledge manipulations in all kinds of agents.

3.1 Instantiating the Generic Smart Card Access Plan in JADE Environment

The generic HTN structure given in section 2 can be modeled as a finite state machine (FSM) with transitions between subtasks. In JADE, there is a composite behavior subclass "jade.core.behaviours.FSMBehaviour" [1], which can be used to model behaviors as a finite state machine. For this reason, we have used JADE's FSM behavior to model the customer agents behavior as shown in Fig. 5. Each state of the FSM is implemented as a subclass of jade.core.behaviors. OneShotBehaviour.

Initially, the state is CARD_WAITING. After the card is inserted and a successful customer PIN entry is made, customer's personal knowledge is transferred from smart card and the RESERVATION_INIT becomes the new state. In that state, the customer agent searches for the descriptions of the agents providing hotel reservation services and updates its contact list based on the response from

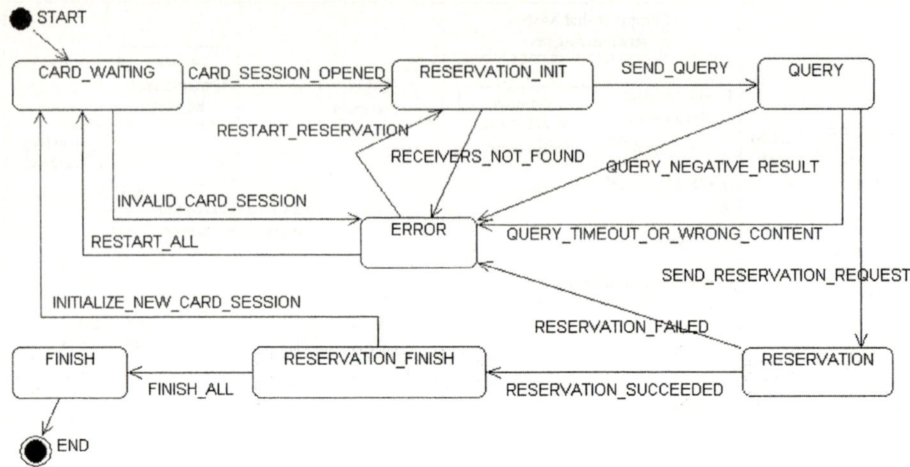

Fig. 5. Finite state machine of the customer agents behavior

the directory service so that the hotel agents that are going to be queried are
determined. Now, the state becomes QUERY. In this state, the customer agent
first constructs an ontology model for the personal ontology instance transferred
from the card. Second, based on the concepts in the domain dependent part of
users personal knowledge, the customer agent prepares requests, places them in
agent communication language messages and sends them to each hotel agent
determined before. Then, it waits for the replies from hotel agents. When a successful query result is returned from one of the hotel agents, the customer agent
begins to behave in RESERVATION state and makes a second contact with the
suitable hotel agent for reservation. Customer agent writes the reservation information to the customers smart card and finishes customers card session in
RESERVATION_FINISH state. At this point, it can again wait for a new card
session or completely terminates by calling doDelete() method in FINISH state.
The errors during the execution of the behavior are managed in ERROR state
of the behavior.

4   Conclusion

In this paper, it has been mainly discussed how to enable personalized access to
semantic web enabled multi-agent systems using smart cards that store users'
personal knowledge as instances of a specific personal ontology. For this purpose, a generic agent plan, which is responsible for communicating with the
smart card terminal via a suitable middleware and for transferring and semantically manipulating the personal knowledge, is given. This plan has been tested
with agents implemented on JADE platform. The reason for choosing JADE is
that it is currently one of the well-known and widely used agent development
frameworks/platforms for Java environments.

Another aspect that is discussed in the paper is the various alternatives for storing ontological knowledge in smart cards. As it is stated in the local conclusion part in section 3, we have preferred the approach of storing the personal knowledge completely in the card, because we wanted to take the advantage of smart cards as a means of secure, portable and cheap personal knowledge storage media for accessing semantic web enabled multi-agent systems.

References

1. Bellifemine, F., Poggi, A., and Rimassa, G.: Developing multi-agent systems with a FIPA-compliant agent framework. Software Practice and Experience, 31 (2001) 103-128.
2. Berners-Lee, T., Hendler, J. and Lassila, O.: The Semantic Web, Scientific American, 284(5), (2001), pp:34-43.
3. Bobineau, C., Bouganim, L., Pucheral, P.,Valduriez, P., PicoDMBS: Scaling Down Database Techniques for the Smartcard, VLDB (2000): 11-20
4. Chan, A.T.S.: Web-enabled smart card for ubiquitous access of patient's medical record. WWW '99: Proceeding of the eighth international conference on World Wide Web, 31 (1999) 1591-1598.
5. Chen, Z.: Java CardTM technology for smart cards architecture and programmers guide. Addison-Wesley, Massachusetts USA, (2000)
6. Dickinson, I.: The Semantic Web and Software Agents: Partners, or Just Neighbours?, AgentLink News 15, September, 3-6 (2004). http://www.agentlink.org
7. Dogac, A., Kabak, Y., Laleci, G., Sinir, S., Yildiz, A., Kirbas, S., Gurcan, Y.: Semantically enriched web services for the travel industry. SIGMOD Record 33(3): 21-27 (2004).
8. Hansmann, U., Nicklous, M.S., Schack, T. and Seliger, F.: Smart card application development using Java, Springer-Verlag, Berlin Germany, (2000)
9. JENA, A Semantic Web Framework for Java, http://jena.sourceforge.net/
10. OpenCard Consortium, OpenCard Framework 1.2 Programmer's Guide, IBM Deutschland Entwicklung GmbH, Boeblingen Germany, (1999)
11. Paolucci, M. et al., A Planning Component for RETSINA Agents, Intelligent Agents VI, LNAI 1757, N.R.Jennings and Y. Lesperance, eds., Springer Verlag, (2000).
12. Web Ontology Language (OWL), http://www.w3.org/2001/sw/WebOnt/

Fast and Secure Communication Resume Protocol for Wireless Networks

Kihong Kim[1], Jinkeun Hong[2], and Jongin Lim[1]

[1] Graduate School of Information Security, Korea University,
1, 5-Ka, Anam-dong, Sungbuk-ku, Seoul, 136-701, South Korea
`hong0612@hanmir.com, jilim@korea.ac.kr`
[2] Division of Information and Communication, Cheonan University,
115 Anse-dong, Cheonan-si, Chungnam, 330-740, South Korea
`jkhong@cheonan.ac.kr`

Abstract. There are important performance issues in secure wireless networks, such as power, bandwidth, and bit error rate (BER), that must be considered when designing a communication resume protocol. The efficiency of a secure communication resume for a fast resume of secure communication is a key point in secure connection development. In this paper, a fast secure communication resume protocol for a wireless network is presented and evaluated against the efficiency of conventional resume protocols. Our proposed resume protocol is found to achieve better performance, in terms of transmission traffic, consumed time, and BER, than conventional resume protocols.

1 Introduction

The wireless transport layer security (WTLS) provides privacy, authentication, and integrity in wireless application protocol (WAP) [1]. As the use of wireless networks becomes more widespread, the necessity of security for these networks is of increasing importance. However, in order to solve security issues in wireless networks, the efficiency of security services must be taken into account. From the point of view of wireless environmental characteristics, research on optimizing the security considerations of WTLS, such as low bandwidth, limited consumed power energy and memory processing capacity, and cryptography restrictions, has been presented [2][3][4][5]. Secure session exchange key protocol and security in wireless communications have been researched by Mohamad Badra and Ahmed Serhrouchni [6], and by Mohammad Ghulam Rahman and Hideki Imai [7]. Hea Suk Jo and Hee Yong Youn [8] examined a synchronization protocol for authentication in wireless LANs, while Min Shiang Hwang et al. [9] proposed an enhanced authentication key exchange protocol. However, in terms of efficiency, the performance considerations for secure wireless networks, such as power, bandwidth, and BER, are very important. Of particular importance for a secure connection point is the efficiency of the secure communication resume for the fast resume of secure communication. In this paper, a protocol for fast secure communication resume using IV count in wireless networks is presented and its performance is evaluated against that of conventional resume protocols.

J.C. Cunha and P.D. Medeiros (Eds.): Euro-Par 2005, LNCS 3648, pp. 1120–1129, 2005.
© Springer-Verlag Berlin Heidelberg 2005

Results shows that the proposed protocol achieves better performance in terms of transmission traffic, consumed time, and BER than conventional protocols.

The remainder of this paper is organized as follows. In the next section, detailed descriptions of the full handshaking and communication resume protocol are given. In section 3, the proposed secure communication resume protocol is illustrated. Some performance considerations are presented in section 4, and concluding remarks are provided in section 5.

2   Key Handshaking Protocol in WTLS

The WTLS protocol determines the session key handshaking mechanism for secure services and transactions in wireless networks, and consists of the following phases: the handshaking phase, the change cipher spec phase, and the record protocol phase (RP) [3][4][5]. In the handshaking phase, all the key techniques and security parameters, such as protocol version, cryptographic algorithms, and the method of authentication, are established between the client and the server. After the key handshaking phase is complete, the change cipher spec phase is initiated. The change cipher spec phase handles the changing of the cipher [6][7]. The RP phase is a layered protocol phase that accepts raw data to be transmitted from the upper layer protocols.

Secure communication necessitates the encryption of communication channels. To achieve this, a key handshaking protocol allows two or more users to share a key or an IV. A key handshaking protocol is illustrated in Fig. 1. The client sends a client hello message that includes information such as the version, acceptable cipher suites, and client random. When the server receives the client hello message, it responds with a hello message to the client and it also sends its certificate, key exchange, certificate request, and server hello done message. After receiving the server hello done message, the client responds by authenticating itself and sending its certificate. Then, the client generates the premaster secret and sends its encryption data $E_{KUS}[Premaster\ Secret]$ encrypted with the server's public key to the server. The premaster secret is used to generate a master secret that is shared between the client and the server. The client then generates the master secret using the premaster secret, client random, and server random. It also generates a sufficiently long key block using the master secret, client random, and server random [1]. The generated key block is hashed into a sequence of secure bytes, which are assigned to the message authentication code (MAC) keys, session keys, and IVs. This is represented as follows in Eq. (1).

$$MS = PRF(PS, "MS", CR + SR) \tag{1}$$
$$KB = PRF(MS, "KE", CR + SR)$$

Here, MS is the master secret, KB is the key block, PS is the premaster secret, CR is the client random, SR is the server random, PRF is the pseudo random function, and KE is the key expansion. The client sends a change cipher spec message and finished message. The server also generates MAC secrets, session keys, and IVs. Then it sends the finished message to the client. Finally, secure communication is established using session keys and IVs.

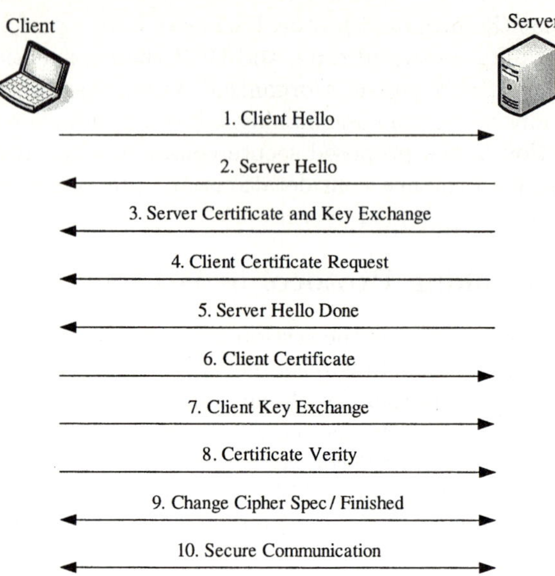

Fig. 1. Full handshaking protocol in WTLS protocol

2.1   Communication Resume Protocol Using Premaster Secret

After completion of the full handshaking protocol shown in Fig. 1, a secure communication is established. However, data frame loss occurs because of bit slips, channel loss, reflection, and diffraction in the communication channel. If a data frame is lost, the output of the decryptor will be unintelligible for the receiver and a communication resume will be required. The aim of the communication resume is to ensure that the encryptor and decryptor have the same internal state at a certain time. An internal state different from all previous sessions has to be chosen to prevent the reuse of session keys or IVs [10][11][12]. To overcome the problems caused by these data frame losses, resume protocols for secure communication have been suggested. Such protocols can be achieved by one of two methods: 1) premaster secret regeneration and retransmission, or 2) random regeneration and retransmission.

Fig. 2 shows a protocol for a secure communication resume using premaster secret regeneration and retransmission. In this protocol, a new premaster secret is generated and sent in each communication resume, and thus it results in the generation of a new master secret and new key block. Therefore, new session keys and new IVs are generated for every communication resume. However, since a new premaster secret is generated and sent in each secure communication resume, this method has disadvantages such as a large computation load, time delay, and BER.

This protocol is executed as follows. First, secure communication is performed for time Δt, and then data frame loss occurs. After the server realizes the data

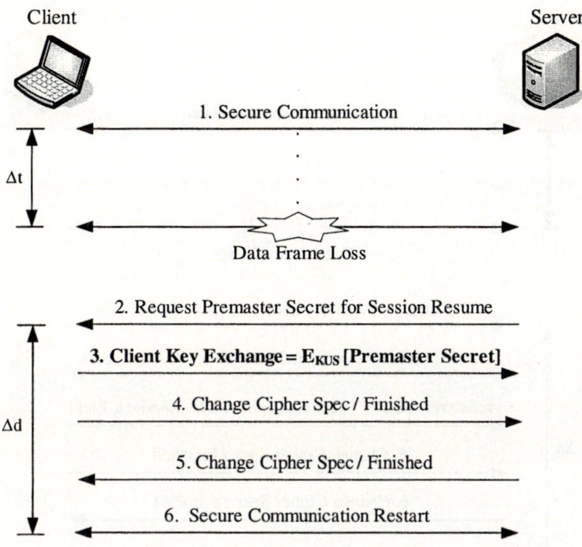

Fig. 2. Conventional protocol using premaster secret

frame loss, it requests a new premaster secret for communication resume. The client generates a new premaster secret and sends $E_{KUS}[Premaster\ Secret]$ to the server. The client then generates a new master secret using the new premaster secret and the original random cached in the initial hello message stage, and generates a new key block using the new master secret and original random. Thus, the result is the generation of new session keys and new IVs.

$$New\ MS = PRF(New\ PS,\ "MS",\ Original\ CR + Original\ SR) \quad (2)$$
$$New\ KB = PRF(New\ MS,\ "KE",\ Original\ CR + Original\ SR)$$

The client then sends the finished message to the server. The server generates a new master secret and a new key block, and then also sends the finished message to the client. After the communication resume time Δd, secure communication is reinitiated.

2.2   Communication Resume Protocol Using Random Value

On the other hand, the protocol for a secure communication resume using random regeneration and retransmission is shown in Fig. 3. In this protocol, a new random is generated and sent in each secure communication resume, which results in the generation of a new key block in each communication resume. As with premaster secret regeneration and retransmission, this protocol also suffers from time delay, and a large BER.

Secure communication is performed for time Δt, and then data frame loss occurs. After realizing the data frame loss, the server requests a new random

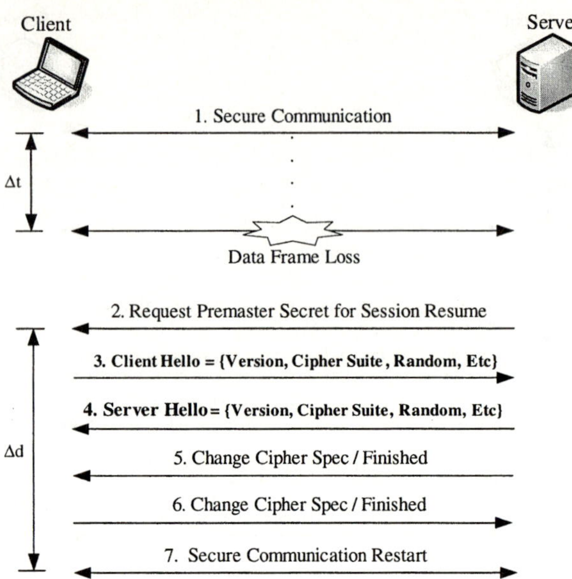

Fig. 3. Conventional protocol using random value

for communication resume. The client generates a new random and includes it in a hello message. After the server receives the hello message from the client, it sends its own hello message that includes its new random. The server also generates a new key block using the new random and cached original master secret, and then generates new session keys and new IVs. This means that a resumed communication will use the same master secret as the previous one. Note that, although the same master secret is used, new random values are exchanged in the secure communication resume. These new randoms are taken into account in the new key block generation, which means that each secure communication starts up with different key materials: new session keys and new IVs.

$$New\ KB\ =\ PRF(Original\ MS,\ "KE",\ New\ CR\ +\ New\ SR) \quad (3)$$

Finally, the server sends the finished message to the client. The server generates a new key block, and then it also sends the finished message to the client. After communication resume time Δd, secure communication is reinitiated.

3   Proposed Secure Communication Resume Protocol

3.1   Proposed Protocol Using IV Count

To overcome the problems inherent in conventional secure communication resume protocols and to reinitiate secure communication much faster than they allow, we propose a fast, efficient, and secure communication resume protocol that uses an IV count value.

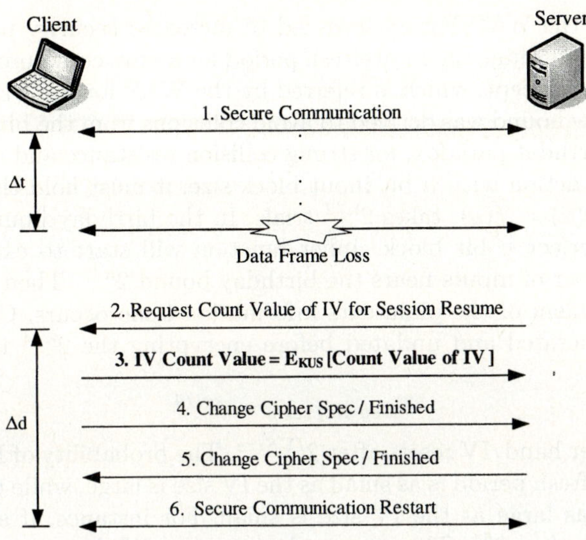

Fig. 4. Proposed protocol using IV count

Fig. 4 shows the proposed secure communication resume protocol, in which a count value of IV is sent to generate the new IVs in each secure communication. After realizing the data frame loss, the server requests a new count value of IV for communication resume. The client generates the new IV count value and sends its encrytpion data $E_{KUS}[Count\ Value\ of\ IV]$ encrypted with the server's public key to the server. It then generates new IVs using the count value. That is, the count value is used to generate new message protection materials, which means that each secure connection starts up with different IVs. Therefore, a resumed communication will use the same session keys as the previous communication. Note that, although the same session keys are used, new IVs are used in the secure communication resume. The client sends the change cipher spec and finished message to the server. The server generates new IVs using the received count value, and then sends the change cipher spec and finished message to the client. The client and server finally have the new IVs after communication resume time Δd.

$$I = I_0 + C, \ \ 1 \leq C \leq 2^{IV\ Size} - 1 \tag{4}$$

Here, I is the value of IV in each communication and I_0 represents the value of the original IV. C is a count value in each communication resume and is increased by a value of one for every communication resume.

3.2   Security Analysis

Security problems regarding attacks against the WAP WTLS were surveyed by Markku Juhani Saarinen [5], and it has been found that many of the changes that were made by the WAP Forum have led to increased security problems [1]. In this paper, to determine the key refresh period for secure communication resume, the key refresh concept, which is referred by the WAP forum, was used and the condition of low bound was derived to avoid collisions from the birthday paradox [13]. By the birthdat paradox, for strong collision resistance and a well-designed block cipher function with n bit input block size, it must hold that finding any pair $(x, y) \ni f(x) = f(y)$, takes $2^{n/2}$ trials. In the birthday bound, this means that even a perfect n bit block cipher function will start to exhibit collisions when the number of inputs nears the birthday bound $2^{n/2}$. Then, if coincidence exists, the problem of the plaintexts information issue occurs. Consequently, a new key is generated and updated before encrypting the $2^{n/4}$ input plaintext blocks.

$$T_{Key\ Refresh} = 2^{n/4} \tag{5}$$

On the other hand, IV resets after $2^{IV\,Size}$. The probability of IV reset within the $2^{n/4}$ key refresh period is as small as the IV size is large, while the probability of its reset is as large as the IV size is small. For instance, if an IV size is 8 bytes, it resets after 2^{64}. This means that 8 bytes IV do not reset within the 2^{16} key refresh period, namely, $2^{64/4}$ input plaintext blocks in 64 bits block size. Therefore, if data frame loss occurs within the key refresh period, and if then a secure communication resume is required, we have only to generate and update a new IV for every communication resume instead of a new key generation. In addition, we have only to generate and update new keys at the time of key refreshing.

4   Performance Consideration

In this paper, to prove the efficiency of the proposed protocol, we compared and analyzed the transmission message size, the consumed time, and BER for communication resume of our proposed protocol with conventional protocols.

Table 1 shows a comparison of the transmission procedure and message sizes: CLT is the client, SVR is the server, the Change Cipher Spec/Finished message is CCS/F, V is WTLS version, and SID is session ID, R is random, and SI is a security association such as key exchange suit, cipher suit, compression method, etc. In the premaster secret protocol, the transmission messages sizes is about 46 bytes. In the case of random protocol, the transmission messages are about 86 bytes in size. However, in the proposed protocol, the transmission messages are composed only of the count value of IV, CLT CCS/F message, and SVR CCS/F message and their size is about 34 bytes. This Table shows that our proposed protocol allows the establishment of secure communication in an economic way, as it has fewer transmission message flows and smaller sizes than either the premaster secret or the random protocol.

To evaluate the efficiency of secure communication resume protocol, the BER in each protocol must also be considered. The results of BER in 3G at 384Kbps according to the number of communication resume are shown in Table 2. Here, T1 are the transmission bits at each 1 iteration, TC1 are the transmission bits

Table 1. Comparison of Δd and transmission message size

Protocol for Resume	Steps	Δd	Transmission Message Size
Premaster Secret	3	Premaster Secret	20 bytes
		CLT CCS/F	13 bytes
		SVR CCS/F	13 bytes
Random	4	[V, R, SID, SI] CLT Hello	30 bytes
		[V, R, SID, SI] SVR Hello	30 bytes
		CLT CCS/F	13 bytes
		SVR CCS/F	13 bytes
Proposed	3	IV Count Value	8 bytes
		CLT CCS/F	13 bytes
		SVR CCS/F	13 bytes

Table 2. BER in 3G according to the number of communication resumes (per hour)

Protocol for Resume		♯1	♯2	♯4	♯6	♯8
Premaster Secret	T1	2.78×10^{-7}	5.56×10^{-7}	1.11×10^{-6}	1.67×10^{-6}	2.22×10^{-6}
	TC1	5.03×10^{-7}	1.11×10^{-6}	2.22×10^{-6}	3.33×10^{-6}	4.44×10^{-6}
	TC3	1.26×10^{-6}	2.78×10^{-6}	5.55×10^{-6}	8.33×10^{-6}	1.11×10^{-5}
Random	T1	4.72×10^{-7}	9.44×10^{-7}	1.89×10^{-6}	2.83×10^{-6}	3.78×10^{-6}
	TC1	9.72×10^{-7}	1.94×10^{-6}	3.89×10^{-6}	5.83×10^{-6}	7.78×10^{-6}
	TC3	2.97×10^{-6}	5.94×10^{-6}	1.19×10^{-5}	1.78×10^{-5}	2.38×10^{-5}
Proposed	T1	1.94×10^{-7}	3.89×10^{-7}	7.78×10^{-7}	1.16×10^{-6}	1.56×10^{-6}
	TC1	2.78×10^{-7}	5.56×10^{-7}	1.11×10^{-6}	1.67×10^{-6}	2.22×10^{-6}
	TC3	1.17×10^{-6}	2.33×10^{-6}	4.67×10^{-6}	7.00×10^{-6}	9.33×10^{-6}

at each 1 iteration with 50 % redundancy channel coding, and TC3 are the transmission bits at 3 iterations with 50 % redundancy channel coding. When computing the BER for 1 communication resume number per hour in the TC1 environment, the BERs in each protocol are provided: 5.03×10^{-7} in premaster secret protocol, 9.72×10^{-7} in random protocol, and 2.78×10^{-7} in the proposed protocol. This means that the proposed protocol reduces BER by over 45 % when compared with the premaster secret protocol, and by about 72 % when compared with the random protocol.

The consumed time and key refresh iteration is shown in Table 3. If a data frame loss occurs within the key refresh period, the proposed communication resume protocol is performed to reopen secure communication and a new key is generated and updated at the key refresh period. For example, at 100bps in 2G, CT is needed to 699 min and CTavg results in a key refresh of 2 iterations per day. Here, CT is consumed time for key refresh and CTavg is the average refresh iterations during one day.

Table 3. Consumed time and key refresh iteration in 2G, 3G bearer service environment (at cryptor input/output block size = 64 bits)

Consumed Time	2G at 100bps	2G at 9.6Kbps	3G at 14.4Kbps	3G at 384Kbps
CT	699 min	7.28 min	4.85 min	0.18 min
CTavg	2 iterations	198 iterations	297 iterations	7910 iterations

In the TC1 case in Table 4, the total consumed time using the proposed protocol for one day, 10.8 sec, is considerably less than the total consumed time using either the premaster secret protocol (14.7 sec) or the random protocol (27.4 sec).

Table 4. Total consumed time for communication resume for one day (at crypto input/output block size = 64 bits)

Protocol for Resume		2G at 100bps	2G at 9.6Kbps	3G at 14.4Kbps	3G at 384Kbps
Premaster Secret	T1	7.4 sec	7.5 sec	7.4 sec	7.9 sec
	TC1	14.7 sec	15.0 sec	15.1 sec	15.8 sec
	TC3	44.2 sec	45.5 sec	45.4 sec	39.6 sec
Random	T1	13.8 sec	13.9 sec	14.0 sec	13.5 sec
	TC1	27.4 sec	27.7 sec	28.1 sec	27.7 sec
	TC3	82.6 sec	85.1 sec	83.2 sec	84.6 sec
Proposed	T1	5.4 sec	5.5 sec	5.4 sec	5.5 sec
	TC1	10.8 sec	11.1 sec	10.7 sec	7.9 sec
	TC3	32.6 sec	33.7 sec	33.6 sec	33.2 sec

5   Conclusion

Most security research in wireless networks is focused on secured routing and transmitting in the network. However, because of the security issues in wireless networks, we suggest that the efficiency of security services is also an important issue. In this paper, a fast and secure communication resume protocol using IV count for wireless networks is presented and evaluated against the efficiency of conventional resume protocols. During the secure communication resume phases, we manage to reduce transferring traffic and thus also reduce the bandwidth on wireless networks. Moreover, our enhanced proposed protocol is able to reduce the consumed time or cryptographic load and the computations in order to reopen secure communication quickly.

Therefore, this proposed communication resume protocol provides a fast resume of secure communications, while reducing the transferring traffic, consumed time, and BER in a WTLS protocol environment.

References

1. WAP Forum. Wireless Transport Layer Security Spec. http://www.wapforum.org.
2. S. Jormalainen and J. Laine. Security in the WTLS.
   http://www.tml.hut.fi/Opinnot/Tik-110.501/1999/papers/wtls.htm.
3. R. Karri and P. Mishra. Optimizing the Energy Consumed by Secure Wireless Sessions-WTLS Case Study. *Mobile Networks and Applications*, No. 8, Kluwer Academic Publishers, 2003.
4. P. Mikal. WTLS : The Good and Bad of WAP Security.
   http://www.advisor.com/Articles.nsf/aid/MIKAP001, 2001.
5. M. J. Saarinen. Attacks against the WAP WTLS Protocol.
   http://www.freeprotocols.org/harm0fWap/wtls.pdf, 1999.
6. M. Badra et al.. A New Secure Session Exchange Key Protocol for Wireless Communication. *IEEE International Symposium on Personal, Indoor and Mobile Radio Communication*, 2003.
7. M. G. Rahman and H. Imai. Security in Wireless Communications. *Wireless Personal Communications*, No. 22, Kluwer Academic Publishers, 2002.
8. H. S. Jo and H. Y. Youn. A New Synchronization Protocol for Authentication in Wireless LAN Environment. *ICCSA'04*, LNCS publishers, 2002.
9. M. S. Hwang et al.. On the Security of an Enhanced Authentication Key Exchange Protocol. *AINA'04*, LNCS publishers, 2004.
10. J. Daemen, R. Govaerts, and J. Vandewalle. Resynchronization Weakness in Synchronous Stream Ciphers. *Pre-proceeding of EUROCRYPT'93*, 1993.
11. R. K. Nichols and P. C. Lekks. Wireless Security - Models, Threats, and Solutions. *McGraw-Hill Telecom*, 2002.
12. E. Amoroso. Fundamentals of Computer Security Technology. *PTR Prentice Hall*, Englewood Cliffs, New Jersey, 1993.
13. B. Schneier. *Applied Cryptography*, 2nd ed, John Wiley and Sons Inc., 1996.

On AAA Based on Brokers
and Pre-encrypted Keys in MIPv6*

Hoseong Jeon, Min Young Chung, and Hyunseung Choo

School of Information and Communication Engineering
Sungkyunkwan University
440-746, Suwon, Korea +82-31-290-7145
{liard,mychung,choo}@ece.skku.ac.kr

Abstract. For providing mobility services for users through the global
Internet, Mobile IP (MIP) has been standardized by IETF. Since conventional MIP has been investigated without the support of the security,
IETF suggests that the current servers capable of performing the authentication, authorization, and accounting (AAA) be used for secure
services. However the quality of service (QoS) may be degraded due to
inefficiency on integrating the conventional MIP and AAA. For this, we
propose a fast and secure handoff mechanism based on IDentification Key
(IDK) along with Authentication Value (AV). Also we evaluate the performance of the proposed scheme in terms of the probability of handoff
failure and average latency. The results show that our proposed mechanism yields better performance than session key exchange mechaism [11]
and ticket based one [12] while maintaining the similar level of security.

1 Introduction

Based on mobility as the essential characteristic for mobile networks, the Mobile
IP standard solution for use with the wireless Internet was developed by the
Internet Engineering Task Force (IETF) [1, 2]. However, Mobile IP does not extend properly to highly mobile users. Moreover, the term mobility implies higher
security risks than static operation in wired networks, since the traffic may at
times take unexpected network paths with unknown or unpredictable security
characteristics. Hence, there is a need to develop technologies that simultaneously enable IP security and mobility over wireless links [3].

By combining Mobile IP and AAA structure [4], the message on the Mobile
IP network can be provided with additional security through AAA protocol.
However, while an Mobile Node (MN) roams in foreign networks, a continuous
exchange of control messages is required with the AAA server in the home network [5–8]. The control message contains the confidential information to identify
the privilege of the mobile user for the service. Standard AAA handoff mechanism has inefficient authenticating procedures that limit its quality of service
(QoS). To resolve such problems, session key exchange mechanism [11] and ticket
based mechanism [12] are proposed in the literature.

---

* This paper was supported in parts by Brain Korea 21 and the MiIstry of Information
  and Communication, Korea. Corresponding author: H. Choo.

J.C. Cunha and P.D. Medeiros (Eds.): Euro-Par 2005, LNCS 3648, pp. 1130–1139, 2005.

The session key exchange mechanism, basically, reuses the previously assigned session key. In this mechanism, the handoff delay can be decreased importantly. However, it requires that the trusted third party should support key exchanges between Access Routers (ARs). For this reason, it uses only the intra-handoff within the same domain. The ticket based mechanism using an encrypted ticket that can support authentication and authorization for the MN has been proposed. It reduces the delay and the risk on MN authentication in Mobile IPv6 (MIPv6). However, it generates additional signalings and overheads of AAA server.

In order to reduce signaling delay required for performing authentication procedures we have proposed an IDK mechanism based on a pre-encrypted key [13]. However, it just uses service requests due to the mobility of MNs. For improving this shortage, an extended IDK mechanism (EIDK) has been proposed [14]. EIDK mechanism uses single AV to extend the effectiveness of IDK into the handoff process. EIDK with single AV compared with previous mechanisms is up to about 20-40% better in terms of average latency that considers hand-off latency and service latency. However, it is vulnerable to the 'replay attack.' To handle this problem, we propose EIDK mechanism with multiple AVs and evaluate its average handoff latency for the reasonable number of AVs.

The rest of the paper is organized as follows. In Section 2, an overview of the Mobile IP and AAA protocol is presented. And the session key exchange mechanism and the ticket based AAA mechanism are given. Our proposed EIDK based AAA mechanism is discussed in Section 3. After that the performance is evaluated along with previous methods in Section 4. Finally we conclude the paper in Section 5.

2   Preliminaries

The IETF AAA Working Group has worked for several years to establish a general model for authentication, authorization, and accounting. AAA in mobile environments is based on a set of clients and servers (AAA Foreign and AAA Home) located in the different domains. It operates based on the security associations (SAs) (SA_s: SA_1, SA_2, SA_3, and SA_4) as shown in Fig. 1. For the support regarding the secure communication, MN requires dynamic security associations. They are defined by sharing the session keys such as K_1, K_2, and K_3 between MN and Foreign Agent (FA), between MN and Home Agent (HA), and between HA and FA, respectively. Once the session keys have been established and propagated, the mobility devices can securely exchange data [9, 10].

Session Key Exchange Mechanism
The session key exchange mechanism is based on a variant of Diffie-Hellman key agreement protocol instead of asymmetric key cryptography [11].The protocol has two system parameters p and g. They are both public and may be used by all the users in a system. The p is a prime number and g (usually called a generator) is an integer less than p with the following property: for every number n between 1 and $p - 1$ inclusive, there is a power k of g such that $n = g^k$ mod

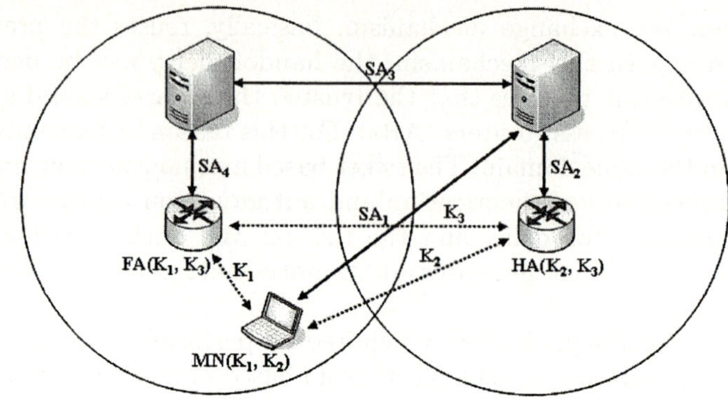

Fig. 1. AAA security association in Mobile IPv6

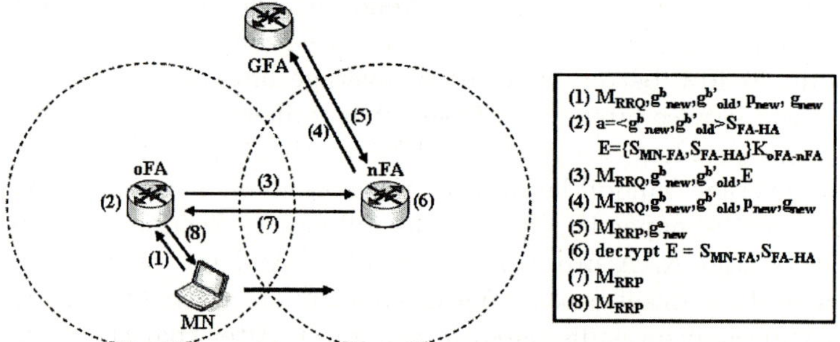

Fig. 2. Secure session key exchange procedure

p. The protocol depends on the discrete logarithm problem for its security. It assumes that it is computationally infeasible to calculate the shared secret key $k = g^a b \bmod p$ given the two public values ($g^a \bmod p$) and ($g^b \bmod p$) when the prime p is sufficiently large.

For the fast operations, this scheme reuses the previously assigned session keys, the session keys for FA(S_{MN-FA} and S_{FA-HA}). To ensure the confidentiality and integrity of the session keys, it uses the encryption and decryption under a short lived secret key, $K_{oFA-nFA}$, between oFA and nFA. The key is dynamically shared between them and can be created by only two entities.

Ticket Based AAA Mechanism

A ticket based AAA mechanism reduces the overhead on the service request by utilizing the pre-encrypted ticket without intermediate encryptions and decryptions. If the MN wants to request a service, it sends a ticket to AAAH for its authentication. The authentication of MN is performed by the Ticket Granting Service ASM (TGS ASM) in the AAA server. The result of authentication is returned to the MN, which allows the MN to request the service [12].

However, this mechanism has four additional signaling messages for the ticket issue. Fig. 3 describes exchanged additional signaling messages on initial registration. Four messages added are ticket/service request message, AAA ticket/

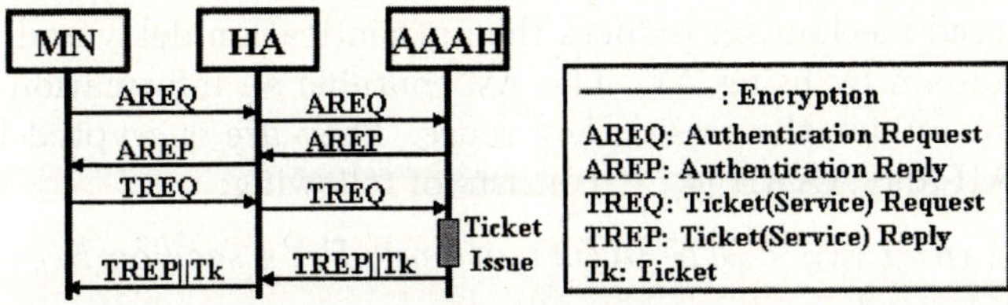

Fig. 3. Initial registration in ticket based AAA model

service request message, AAA ticket/service reply message, and ticket/service reply message. The messages between MN and HA are based on the general Mobile IP protocol, and the messages between HA and Home AAA server (AAAH) are based on the extended AAA protocol in Mobile IP networks.

EIDK Based AAA Mechanism with single AV
This mechanism reduces handoff and service signaling cost using IDK and AV [14]. However, it is vulnerable to the 'replay attack' due to a single AV its on use. It means that the malicious node traps the authentication sequence that has been transmitted by an authorized user through the network, and then has replayed the same sequence to get himself authenticated. In this case, the authorized user is attacked by the malicious one.

3   EIDK Based AAA Mechanism with Multiple AVs

This section deals with the secure enhanced EIDK mechanism using multiple AVs. Basically, this modified mechanism is identical to the previous EIDK proposed in [14], except using multiple AVs. For the proposed mechanism, we assume as follows: 1) an AAA server authenticates and authorizes subscribers, and verifies IDK. It also creates AV; 2) an AAA client is either HA or FA, which has the functionality to generate and to deliver AAA messages; 3) an AAA broker (AAAB) authenticates MN instead of AAA Home (AAAH); and 4) an MN generates IDK and delivers it.

In order to reduce the time for repeated encryptions and decryptions, an MN generates an encrypted information called IDK using authentication time (AT). This value represents the time at the initial registration of the MN. The IDK consists of the following [13]:

- Network Access Identifier (NAI) of MN
- Address of the AAA server that provides services to the MN
- Service identifier allowed for the MN
- Home network address and IP address of the MN
- IDK lifetime
- A random number (128 bits)
- The session key shared by the MN and the AAA server
- CoA of next possible area expected to be moved (optional)
- Authentication time (AT).

The proposed mechanism reduces the authentication delay and signalings at the foreign domain by using AV. The AV contains an information for MN and session keys in FA for the session key reuse. They are encrypted based on SA between AAAH and AAAB [4]. It consists of following:

$$AV = SA_{AAAH-AAAB} \{ MN\ information \parallel FA's\ session\ keys \parallel Nonce \}$$

Initial Registration to AAAH

As indicated in Fig. 4, the sequence of message exchanges for each authentication mechanism is performed for the initial registration in the home network. We assume that there is no security associate between MN and HA. This is because we do not consider the pre-shared key distribution in AAA protocol in this work.

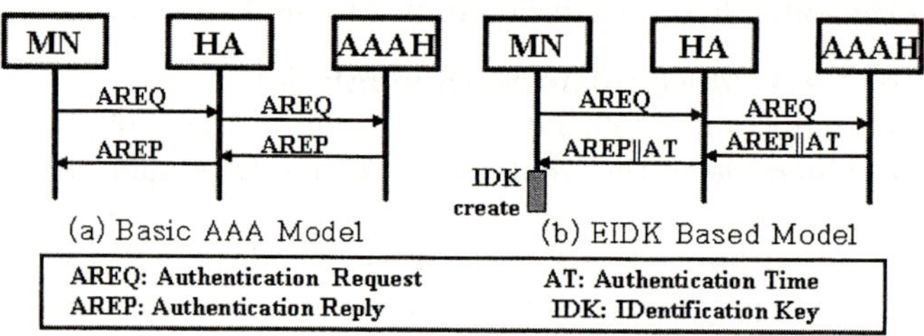

Fig. 4. Initial registration

Fig. 4(a) shows the initial registration of the basic AAA model. And both the ticket based model and the proposed EIDK based one follow the basic AAA model in the initial registration. However, as you see in Fig. 3, additional signaling for issuing a ticket is required for faster services on requests in the ticket based model.

Fig. 4(b) shows the initial registration procedure for the EIDK based mechanism. In the authentication reply phase, AT is delivered to MN together with authentication reply message (AREP). Accordingly, both the MN and AAAH server share a secret value. This one is the arrival time of the request message for the MN at AAAH. The AT would be used as a part of the encryption key value on IDK by MN and later it is used as the decryption key in AAAH. Unlike the ticket based model scheme, MN receives AT along with the authentication reply message without further additional signaling in our scheme.

Service Requests

The procedure routine of message exchanges for the service request in the home domain is in Fig. 5. The service request message (SREP) is encrypted and decrypted by the key distributed from AAAH on the authentication process in the basic AAA model. As you see in Fig. 5(a), service request message (SREQ) and SREP are encrypted and decrypted at MN, HA, and AAAH whenever they are exchanged, and these can be a significant overhead. Ticket based model in Fig. 5(b) reduces the overhead on the service request by utilizing the pre-encrypted ticket without intermediate encryptions and decryptions. This can be done by the extended AAA server structure. Also the model assumes that the

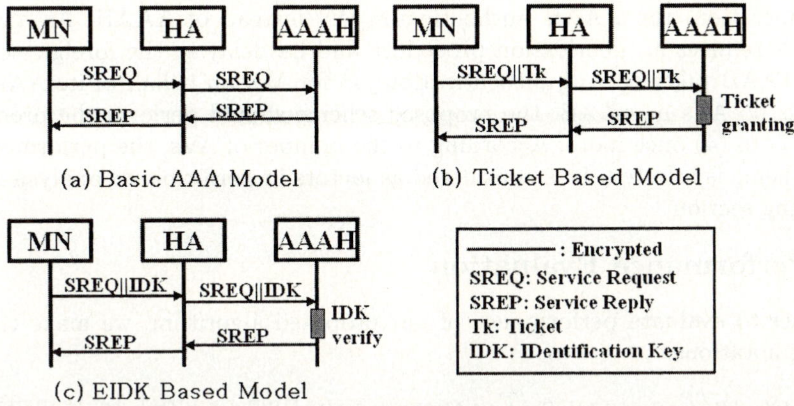

(a) Basic AAA Model   (b) Ticket Based Model

(c) EIDK Based Model

Fig. 5. Service request

time for ticket issuing and granting is not significant. However, this may not guarantee its superiority in the real world.

In Fig. 5(c), the proposed EIDK based model does not need the extended AAA server structure, but just maintains the current one. Intermediate encryptions and decryptions are not necessary on the service request in our scheme. Since we employ the pre-encrypted IDK which is created by MN beforehand. Unlike the basic AAA model, the EIDK based AAA model requires IDK creation and the time for it. But this scheme reduces the total delay since it eliminates the time for intermediate encryptions and decryptions.

Handoff Procedures Using Multiple AVs

The purpose of multiple AVs is to improve previous EIDK mechanism. We propose the usage of multiple AVs for the preventing 'replay attack.' Each AV is used only once and then it is no longer valid, so eavesdropping and replay attack are not our concern.

Fig. 6 represents the proposed handoff mechanism. It eliminates encryption and decryption delay in the authentication procedure by using pre-encrypted AV, and reduces the number of signalings due to the AAAB. When MN moves to a foreign network, AAAH creates AVs that are delivered to the AAAB. After

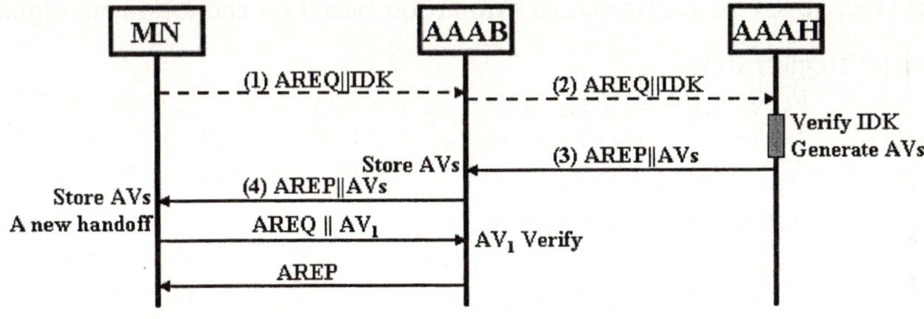

Fig. 6. Description of handoff mechanism using AVs

that operation, the AAAB authenticates MN instead of AAAH. As a result, the MN reduces authentication procedure and its delay in the foreign network since AAAB takes care of authentication job for MN on behalf of its AAAH. If there is no AVs in AAAB, the proposed scheme should perform the procedure from (1) to (4) once more. According to the number of AVs, the performance of this scheme is affected. It is considered as factors in performance analysis in the following section.

4 Performance Evaluation

In order to evaluate performance of our proposed algorithm, we make the following notations:

- $T_{MN-AR}/T_{AR-AAAH(F)}/T_{AAAH(F)-AAAB}$: time required for transfer in a message between MN and AR/AR and AAAH(F)/AAAH(F) and AAAB.
- E_{se}/E_{sd}: time required for symmetric key encryption/decryption of a message at MN/AR/AAAH/AAAF/AAAB
- BU: binding update time
- AS: authentication time in AAAH
- Tk: ticket issuance and verification time in AAAH
- IDK: IDK creation and verification time in MN/AAAH/AAAB
- AV: authentication time using AV in AAAB
- $B_{IR}/T_{IR}/E_{IR}$: time required for initial registration as basic AAA scheme/as ticket based scheme/as EIDK based scheme.
- $B_{Intra}^{H/F}/S_{Intra}^{H/F}/E_{Intra}^{H/F}$: time required for intra handoff as basic AAA scheme/ as session key exchange scheme/as EIDK based scheme in home(foreign) domain
- $B_{Inter}/S_{Inter}/E_{Inter}$: time required for inter handoff as basic AAA scheme/ as session key exchange scheme/as EIDK based scheme
- $B_{Serv}^{H/F}/T_{Serv}^{H/F}/E_{Serv}^{H/F}$: time required for service request as basic AAA scheme/ as ticket based scheme/as EIDK based scheme in home(foreign) domain

Authentication procedures can be classified into three cases: initial registration, handoff and service request. And then handoff can be also classified into another three cases by the position of the MN: intra handoff in home/foreign domain and inter handoff. Lastly, service request can be classified into two cases: service request in home/foreign domain. We calculate times required in schemes we discuss (Figs. 2–6) for performance evaluation based on the following equations:

- [Initial Registration]
$$B_{IR} = 2 \cdot T_{MN-AR} + 2 \cdot T_{AR-AAAH(F)} + 4 \cdot E_{se} + 4 \cdot E_{sd} + AS + BU$$
$$T_{IR} = 2 \cdot (2 \cdot T_{MN-AR} + 2 \cdot T_{AR-AAAH(F)} + 4 \cdot E_{se} + 4 \cdot E_{sd}) + AS + Tk + BU$$
$$E_{IR} = 2 \cdot T_{MN-AR} + 2 \cdot T_{AR-AAAH(F)} + 4 \cdot E_{se} + 4 \cdot E_{sd} + AS + BU + IDK$$

- [Intra Handoff in the home domain]
$$B_{Intra}^{H} = 2 \cdot T_{MN-AR} + 2 \cdot T_{AR-AAAH(F)} + 4 \cdot E_{se} + 4 \cdot E_{sd} + AS + BU$$
$$S_{Intra}^{H} = 4 \cdot T_{MN-AR} + 4 \cdot T_{AR-AAAH(F)} + 4 \cdot E_{se} + 4 \cdot E_{sd} + BU$$
$$E_{Intra}^{H} = 2 \cdot T_{MN-AR} + 2 \cdot T_{AR-AAAH(F)} + 2 \cdot E_{se} + 2 \cdot E_{sd} + 2 \cdot IDK + BU$$

- [Intra Handoff in the foreign domain]

$$B_{Intra}^{F} = 2 \cdot T_{MN-AR} + 4 \cdot T_{AR-AAAH(F)} + 4 \cdot T_{AAAH(F)-AAAB} + 10 \cdot E_{se} + 10 \cdot E_{sd} + AS + BU$$

$$S_{Intra}^{F} = 4 \cdot T_{MN-AR} + 4 \cdot T_{AR-AAAH(F)} + 4 \cdot T_{AAAH(F)-AAAB} + 4 \cdot E_{se} + 10 \cdot E_{sd} + AS + BU$$

$$E_{Intra}^{F} = 2 \cdot T_{MN-AR} + 2 \cdot T_{AR-AAAH(F)} + 2 \cdot T_{AAAH(F)-AAAB} + 3 \cdot E_{se} + 3 \cdot E_{sd} + 2 \cdot IDK + BU$$

- [Inter Handoff]

$$B_{Inter} = 2 \cdot T_{MN-AR} + 4 \cdot T_{AR-AAAH(F)} + 4 \cdot T_{AAAH(F)-AAAB} + 10 \cdot E_{se} + 10 \cdot E_{sd} + AS + BU$$

$$S_{Inter} = 4 \cdot T_{MN-AR} + 4 \cdot T_{AR-AAAH(F)} + 4 \cdot E_{se} + 4 \cdot E_{sd} + BU$$

$$E_{Inter} = 2 \cdot T_{MN-AR} + 2 \cdot T_{AR-AAAH(F)} + 2 \cdot T_{AAAH(F)-AAAB} + 3 \cdot E_{se} + 3 \cdot E_{sd} + 2 \cdot IDK + BU$$

- [Service request in home domain]

$$B_{Serv}^{H} = 2 \cdot T_{MN-AR} + 2 \cdot T_{AR-AAAH(F)} + 4 \cdot E_{se} + 4 \cdot E_{sd} + AS + BU$$

$$T_{Serv}^{H} = 2 \cdot T_{MN-AR} + 2 \cdot T_{AR-AAAH(F)} + 2 \cdot E_{se} + 2 \cdot E_{sd} + Tk + BU$$

$$E_{Serv}^{Home} = 2 \cdot T_{MN-AR} + 2 \cdot T_{AR-AAAH(F)} + 2 \cdot E_{se} + 2 \cdot E_{sd} + IDK + BU$$

- [Service request in foreign domain]

$$B_{Serv}^{F} = 2 \cdot T_{MN-AR} + 4 \cdot T_{AR-AAAH(F)} + 4 \cdot T_{AAAH(F)-AAAB} + 10 \cdot E_{se} + 10 \cdot E_{sd} + AS + BU$$

$$T_{Serv}^{F} = 2 \cdot T_{MN-AR} + 4 \cdot T_{AR-AAAH(F)} + 4 \cdot T_{AAAH(F)-AAAB} + 5 \cdot E_{se} + 5 \cdot E_{sd} + Tk + BU$$

$$E_{Serv}^{F} = 2 \cdot T_{MN-AR} + 4 \cdot T_{AR-AAAH(F)} + 4 \cdot T_{AAAH(F)-AAAB} + 5 \cdot E_{se} + 5 \cdot E_{sd} + IDK + BU$$

Using these equations and the system parameter in Table 1 [10, 13, 14], we compute the handoff probability and the average latency.

Table 1. System parameters

Bit rates		Processing time	
Wire links	100 *Mbps*	Routers (HA,FA)	0.5 *msec*
Wireless links	2 *Mbps*	Nodes (MN)	0.5 *msec*
Propagation time		Tk	3.0 *msec*
Wire links	500 *μsec*	IDK	3.0 *msec*
Wireless links	2 *msec*	AS	1.0 *msec*
Data size		AV	1.0 *msec*
Message size	256 *bytes*	E_{se} *and* E_{sd}	1.0 *msec*
		BU	0 *msec*

We analyze the handoff procedure to obtain the handoff failure rate for each handoff mechanism. It is influenced by few factors that are the velocity of MN and the radius of a cell. Figs. 7 and 8 show probability of handoff failure and average latency for various cell radii, respectively. From the results, secure exchange scheme shows the better performance for frequent handoff situations and ticket-based one has better result for frequent service requests. However, EIDK

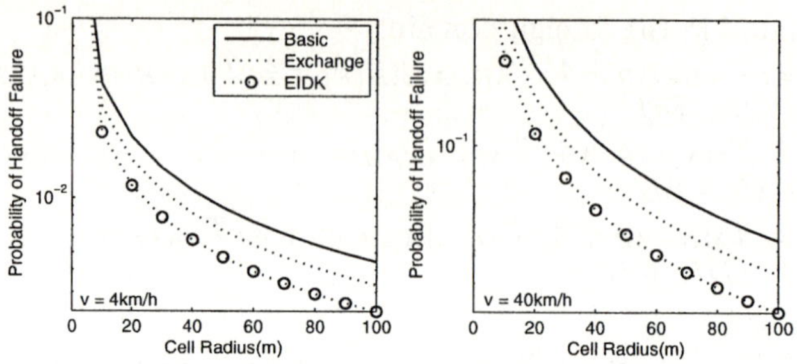

Fig. 7. The probability of handoff failure

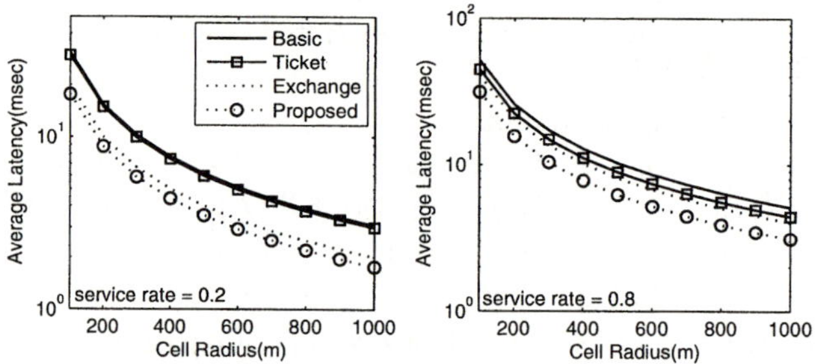

Fig. 8. Average handoff latency

based mechanism shows even better performance than previous mechanisms because it considers two factors the handoff latency and service latency at the same time.

Fig. 9 shows average latency of EIDK mechanism according to the number of AVs. It is indicated that the average latency for the modified scheme increases

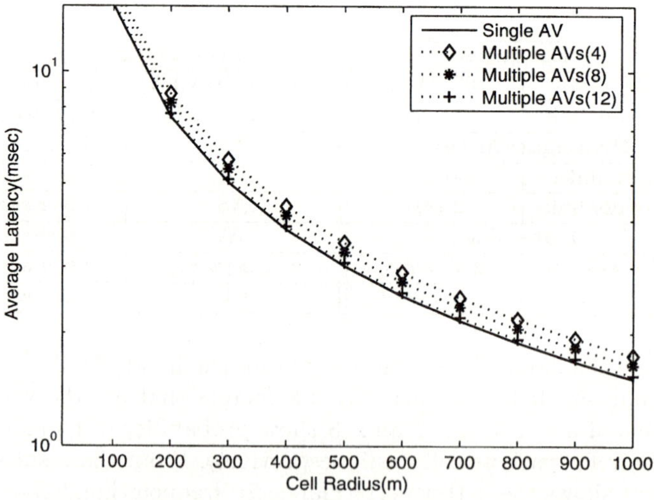

Fig. 9. Single AV versus multiple AVs

as the number of AVs is increases. For given cell radius, average latency of EIDK mechanism decreases as the number of AVs increases. However excessive number of AVs may cause additional overheads. Therefore it is important to select the appropriate number of AVs in this modified scheme.

5   Conclusion

In this paper, we proposed the EIDK based AAA mechanism with multiple AVs. This scheme prevents 'replay attack' for malicious users and reduces the latency due to handoffs and services. The performance comparison shows that the EIDK based mechanism is superior to previous schemes we disscuss in this paper in terms of latency while maintaining the same security level. Also, the performance of the proposed mechanism depends on the number of AVs employed. For further studies, researches on the optimal number of AVs are underway.

References

1. C.E. Perkins, "IP Mobility Support," IETF RFC 2002.
2. B. David, C. Perkins, and J. Arkko,"Mobility Support in IPv6," IETF draft, Internet Draft draft-ietf-mobileip-ipv6-17.txt, May 2002.
3. C. Perkins, "Mobile IP Joins Forces with AAA," IEEE Personal Communications, vol. 7, no. 4, pp. 59–61, August 2000.
4. J. Vollbrecht, P. Cahoun, S. Farrell, and L. Gommans, "AAA Authorization Framework," RFC 2904, 2000.
5. J. Vollbrecht, P. Calhoun, S. Farrell, L. Gommans, G. Gross, B. debruijn, C.de Laat, M. Holdrege, and D. Spence, "AAA Authorization Application Examples", IETF RFC 2905.
6. S. Farrell, J. Vollbrecht, P. Calhoun, and L. Gommans,"AAA Authorization Requirements," RFC 2906, August 2000.
7. S. Glass, T. Hiller, S. Jacobs, and C. Perkins, "Mobile IP Authentication, Authorization, and Accounting Requirements," RFC 2977, 2000.
8. A. Hasan, J. Jahnert, S. Zander and B. Stiller, "Authentication, Authorization, Accounting and Charging for the Mobile Internet," Mobile Summit, September 2001.
9. C. Yang, M. Hwang, J. Li, and T. Chang, "A Solution to Mobile IP Registration for AAA," Springer-Verlag Lecture Notes in Computer Science, vol. 2524, pp. 329–337, November 2002.
10. A. Hess and G. Schafer, "Performance Evaluation of AAA/Mobile IP Authentication," 2nd Polish-German Teletraffic, 2002.
11. H. Kim, D. Choi, and D. Kim, "Secure Session Key Exchange for Mobile IP Low Latency Handoffs," Springer-Verlag Lecture Notes in Computer Science, vol. 2668, pp. 230–238, January 2003.
12. J. Park, E. Bae, H. Pyeon, and K. Chae "A Ticket-based AAA Security Mechanism in Mobile IP Network," Springer-Verlag Lecture Notes in Computer Science 2003, vol. 2668, pp. 210–219, May 2003.
13. H. Jeon, H. Choo, and J. Oh, "IDentification Key Based AAA Mechanism in Mobile IP Networks," ICCSA 2004 vol. 1, pp. 765–775, May 2004.
14. H. Jeon, M. Chung, and H. Choo, "On AAA with Extended IDK in Mobile IP Networks," ICCSA 2005 vol. 3480, pp. 538–539, May 2005.

Topic 15
Peer-to-Peer and Web Computing

Anne-Marie Kermarrec, Márk Jelasity,
Antony Rowstron, and Henrique Domingos

Topic Chairs

Distributed systems have experienced a shift of scale in the past few years. This evolution has generated an interest in peer-to-peer systems and resulted in much interesting work. Peer-to-peer systems are characterized by their potential to scale due to their fully decentralized nature. They are self-organizing, adapting automatically to peer arrivals and departures, and are highly resilient to failures. They rely on a symmetric communication model where peers act both as servers and clients. As the peer-to-peer concepts and technologies become more mature, many distributed services and applications relying on this model are envisaged in the context of large-scale distributed and parallel systems. This topic examines peer-to-peer technologies, applications, and systems, and also identifies key research issues and challenges.

Twenty papers were submitted to this topic and six were accepted. These papers were organized in two sessions. The first one has three papers discussing peer-to-peer overlay construction, and the second session is devoted to applications of peer-to-peer overlays.

In *"Epidemic-style management of semantic overlays for content-based searching"*, the authors presents a proactive epidemic protocol to build semantic overlays and cluster peers having similar content. In *"Long range contacts in overlay networks"*, the authors propose a protocol to reactively add long range contacts in an overlay network based only on actual routing requests. The approach enables to tolerate non uniform distribution of the peers in the logical space. In the paper *"Combining the use of clustering and scale-free nature of user exchanges into a simple and efficient P2P system"*, the results of a depth-first search are used to create links between peers so that resulting neighbours in the overlay are peers which have successfully cooperated to find results for previous searches.

"AGNO: an adaptive group communication scheme for unstructured P2P networks" presents a group notification protocol which operates on an unstructured peer-to-peer network. In this approach, the results of previous lookups operations are also used to speed up the notification process. In *"Pastis: a highly-scalable multi-user peer to peer file system"*, the authors present the design and evaluation of a read-write peer-to-peer file sharing system based on the Pastry/Past approach. The last paper, entitled *"Semantic peer to peer overlays for publish/subscribe networks"*, presents a content-based publish-subscribe system. A peer-to-peer overlay is built so that peers get connected to neighbours having similar subscribing patterns.

J.C. Cunha and P.D. Medeiros (Eds.): Euro-Par 2005, LNCS 3648, p. 1141, 2005.
© Springer-Verlag Berlin Heidelberg 2005

Epidemic-Style Management of Semantic Overlays
for Content-Based Searching

Spyros Voulgaris and Maarten van Steen

Vrije Universiteit Amsterdam
Department of Computer Science
De Boelelaan 1081a, 1081HV Amsterdam
{spyros,steen}@cs.vu.nl

Abstract. A lot of recent research on content-based P2P searching for file-sharing applications has focused on exploiting semantic relations between peers to facilitate searching. To the best of our knowledge, all methods proposed to date suggest *reactive* ways to seize peers' semantic relations. That is, they rely on the usage of the underlying search mechanism, and infer semantic relations based on the queries placed and the corresponding replies received. In this paper we follow a different approach, proposing a *proactive* method to build a semantic overlay. Our method is based on an epidemic protocol that clusters peers with similar content. It is worth noting that this peer clustering is done in a completely implicit way, that is, without requiring the user to specify his preferences or to characterize the content of files he shares.

1   Introduction

File sharing peer-to-peer (P2P) systems have gained enormous popularity in recent years. This has stimulated significant research activity in the area of content-based searching. Sparkled by the legal adventures of Napster, and challenged to defeat the inherent limitations concerning the scalability and failure resilience of centralized systems, research has focused on *decentralized* solutions for content-based searching, which by now has resulted in a wealth of proposals for peer-to-peer networks.

In this paper, we are interested in those group of networks in which searching is based on grouping semantically related nodes. In these networks, a node first queries its semantically close peers before resorting to search methods that span the entire network. In particular, we are interested in solutions where semantic relationships between nodes are captured implicitly. This capturing is generally achieved through analysis of query results, leading to the construction of a local *semantic list* at each peer, consisting of references to other, semantically close peers.

Only very recently, an extensive study has been published on search methods in peer-to-peer networks, be they structured, unstructured, or of a hybrid form [1]. This study reveals that virtually all peer-to-peer search methods in semantic overlay networks follow an integrated approach towards the construction of the semantic lists, while at the same time accounting for changes occurring in the set of nodes. These changes involve the joining and leaving of nodes, as well as changes in a node's preferences.

The problem we are faced with is that the construction of semantic lists should result in highly clustered overlay networks. These networks excel for searching content

J.C. Cunha and P.D. Medeiros (Eds.): Euro-Par 2005, LNCS 3648, pp. 1143–1152, 2005.

when nothing changes. However, to handle dynamics requires the discovery and propagation of changes that may happen *anywhere* in the network. For this reason, overlay networks should also reflect desirable properties of random graphs and complex networks in general [2, 3]. These two conflicting demands generally lead to complexity when integrating solutions into a single protocol.

Protocols for content-based searching in peer-to-peer networks should separate these concerns. In particular, we advocate that when it comes to constructing and using semantic lists, these lists should be optimized for search only, regardless of any other desirable property of the resulting overlay. Instead, a separate protocol should be used to handle network dynamics, and provide up-to-date information that will allow proper adjustments in the semantic lists (and thus leading to adjustments in the semantic overlay network itself).

In this paper we propose such a two-layered approach for managing semantic overlay networks. The top layer contains a gossip-based protocol that strives to optimize semantic lists for searching only. The bottom layer offers a fully decentralized service for delivering, in an unbiased fashion, information on new events, similar in nature to the peer-sampling service recently described in [4]. Again, this service is implemented using a gossip-based protocol (which, by the way, is very different from those described in [4]).

Our main contribution is that we demonstrate that this two-layered approach leads to high-quality semantic overlay networks. We substantiate our claims through extensive simulations using traces collected from the eDonkey file-sharing network [5].

The paper is organized as follows. We start with presenting our protocols in the next section, followed by describing our experimental setup in Section 3. Performance evaluation is discussed in Section 4, followed by an analysis of consumed bandwidth in Section 5. We conclude with a discussion in Section 6.

2 The Protocol

2.1 Outline

In our model each peer maintains a dynamic list of semantic neighbors, called its *semantic view*, of fixed small size ℓ. A peer searches for a file by first querying its semantic neighbors. If no results are returned, the peer then resorts to the default search mechanism.

Our aim is to organize the semantic views so as to maximize the hit ratio of the first phase of the search. We will call this the *semantic hit ratio*. We anticipate that the probability of a neighbor satisfying a peer's query is proportional to the semantic proximity between the peer and its neighbor. We aim, therefore, at filling a peer's semantic view with its ℓ semantically closest peers out of the whole network.

We assume the existence of a *semantic proximity function* $S(F_P, F_Q)$, which given the file lists F_P and F_Q of peers P and Q, respectively, provides a numeric metric of the semantic proximity between the two peers. The more semantically similar the file lists of P and Q are, the higher the value of $S(F_P, F_Q)$. We are essentially seeking to pick peers $Q_1, Q_2, ..., Q_\ell$ for peer P's semantic view, such that the sum $\sum_{i=1}^{\ell} S(P, Q_i)$ is maximized.

We assume that the semantic proximity function exhibits some sort of transitivity, in the sense that if P and Q are semantically similar to each other, and so are Q and R, then some similarity between P and R is likely to hold. Note that this transitivity does not consist a hard requirement for our system. In its absence, semantically related neighbors are discovered based on random encounters. If it exists though, it is exploited to dramatically enhance efficiency.

2.2   Design Motivation

From our previous discussion, we are seeking a means to construct, for each node, a semantic view from all the current nodes in the system. There are two sides to this construction.

First, based on the assumption of transitivity in the semantic proximity function S, a peer should explore the semantically close peers that its neighbors have found. In other words, if Q is in P's semantic view, and R is in Q's view, it makes sense to check whether R is also semantically close to P. Exploiting the transitivity in S should then quickly lead to high-quality semantic views.

Second, it is important that *all* nodes are examined. The problem with following only transitivity is that we eventually will be searching only in a single semantic cluster. Similar to the special "long" links in small-world networks [6], we need to establish links to *other* semantically-related clusters. Likewise, when new nodes join the network, they should easily find an appropriate cluster to join. These issues call for a randomization when selecting nodes to inspect for adding to a semantic view.

In our design we decouple these two aspects by adopting a two-layered set of gossip protocols, as can be seen in Figure 1. The lower layer, called CYCLON [7], is responsible for maintaining a connected overlay and for periodically feeding the top-layer protocol with nodes uniform randomly selected from the network. In its turn, the top-layer protocol, called VICINITY, is in charge of focusing on discovering peers that are semantically as close as possible, and of adding these nodes to the semantic views.

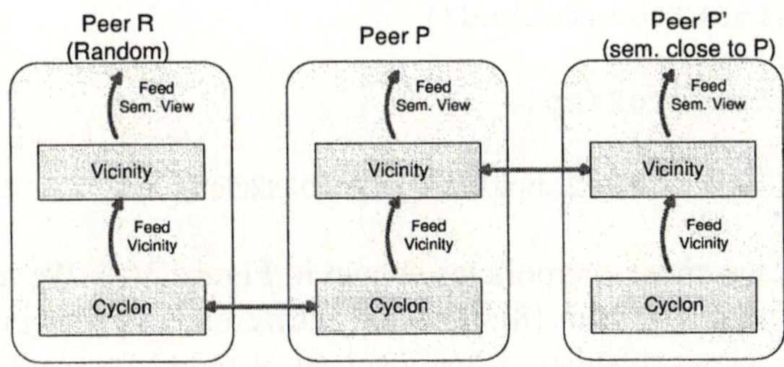

Fig. 1. The two-layered framework.

2.3   Gossiping Framework

All information exchange between peers is carried out by means of *gossip items*, or simply *items*. A gossip item created by peer P is a tuple containing the following three fields:

1. *P*'s contact information (network address and port)
2. The item's creation time
3. Application-specific data; in this case *P*'s file list

Each node maintains locally a number of items per protocol, called the protocol's *view*. This number is the same for all items, and is called the protocol's *view size* (c_v for VICINITY, and c_c for CYCLON).

Figure 2 presents a generic skeleton forming the basis for both VICINITY and CY-CLON gossiping protocols. Each node runs two threads. An *active* one, which periodically wakes up and initiates communication to another peer, and a *passive* one, which responds to the communication initiated by another peer.

The functions appearing in boldface, namely `selectPeer()`, `selectItems-ToSend()`, and `selectItemsToKeep()` form the three *hooks* of this skeleton. Different protocols can be instantiated from this skeleton by implementing specific policies for these three functions, in turn, leading to different emergent behaviors.

The number of items exchanged in each communication is predefined, and is called the protocol's *gossip length* (g_v for VICINITY, and g_c for CYCLON).

```
/*** Active thread ***/
// Runs periodically every T time units
q = selectPeer()
myItem = (myAddress, timeNow, myFileList)
buf_send = selectItemsToSend()
send buf_send to q
receive buf_recv from q
view = selectItemsToKeep()


/*** Passive thread ***/
// Runs when contacted by some peer
receive buf_recv from p
myItem = (myAddress, timeNow, myFileList)
buf_send = selectItemsToSend()
send buf_send to p
view = selectItemsToKeep()
```

Fig. 2. Epidemic protocol skeleton.

For VICINITY, we chose the policies shown in Figure 3(a). We note that the RAN-DOM protocol resembles T-Man [8]. The only difference is that in T-Man peers exchange their *whole* views, instead of just a subset of them. As we discuss below, AG-GRESSIVELY BIASED will turn out to be an excellent choice for forming semantic clusters.

Note that `selectItemsToKeep()` takes into account CYCLON's cache too in selecting the best c_v items to keep. This is the default link between the two layers.

For CYCLON, we made the choices shown in Figure 3(b). CYCLON is a protocol we previously developed, and which is extensively described and analyzed in [7].

Effectively, what `selectItemsToSend()` and `selectItemsToKeep()` establish is an *exchange* of some neighbors between the caches of the two communicating

(a)

Hook	Description
`selectPeer()`	Select peer from the item with the oldest timestamp
`selectItemsToSend()` RANDOM BIASED AGGRESSIVELY BIASED	 Randomly select g_v items Select the g_v items of nodes semantically closest to the selected peer Select the g_v items of nodes semantically closest to the selected peer from the VICINITY view *and* the CYCLON view
`selectItemsToKeep()`	Keep the c_v items of nodes that are semantically *closest*, out of items in its current view, items received, and items in the local CYCLON view. In case of multiple items from the same node, keep the one with the most recent timestamp.

(b)

Hook	Description
`selectPeer()`	Select peer from the item with the oldest timestamp
`selectItemsToSend():` active thread passive thread	 Select own item and randomly $g_c - 1$ others from the CYCLON view Randomly select g_c items from the CYCLON view
`selectItemsToKeep()`	Keep all g_c received items, replacing (if needed) the g_c ones selected to send. In case of multiple items from the same node, keep the one with the most recent timestamp.

Fig. 3. The chosen policies for (a) the VICINITY protocol and (b) the CYCLON protocol.

peers. In addition to that, the selected peer's item in the initiator's cache is always removed, but the initiator's (new) item is always placed in the selected peer's cache.

CYCLON creates an overlay with completely random, uncorrelated links between nodes, such that the in-degree (number of incoming links) is practically the same for each node. Importantly, it can achieve this property fairly quickly even when a small number of items (such as 3 or 4) is exchanged in each communication, even for large caches of several dozens of items. Therefore, it is ideal as a lightweight service that can offer a node a randomly selected peer from the current set of nodes.

3   Experimental Environment and Settings

All experiments presented here have been carried out with PeerSim [9], an open source simulator in Java for P2P protocols, developed at the University of Bologna.

To evaluate our protocol, we used real world traces from the eDonkey file sharing system [10], collected by Le Fessant et al. in November 2003 [5]. A set of 12,000 worldwide distributed peers along with the files each one shares is logged in these traces. A total number of 923,000 unique files is being collectively shared by these peers.

In order to simplify the analysis of our system's emergent behavior, we determined equal gossiping periods for both layers. More specifically, once every T time units each node initiates first a gossip exchange with respect to its bottom (CYCLON) layer, immediately followed by a gossip exchange at its top (VICINITY) layer. Note that even though nodes initiate gossiping at universally fixed intervals, they are not synchronized with each other.

Even though both protocols are asynchronous, it is convenient to introduce the notion of *cycles* in order to study their evolutionary behavior with respect to time. We define a cycle to be the time period during which *each* node has initiated gossiping exactly *once*. Since each node initiates gossiping periodically, once every T time units, a cycle is equal to T time units.

A number of parameters had to be set for these experiments, listed here.

Proximity Function S. We chose a rather simple, yet intuitive proximity function to test our protocol with. The proximity S between two nodes P and Q, with file lists F_P and F_Q respectively, is defined as the number of files that lay in both lists. More formally: $S(F_P, F_Q) = |F_P \cap F_Q|$. As stated in 2.1, the semantically closer two nodes are, the higher the value of S is. Note that our goal was to demonstrate the power of our gossiping protocol in forming a semantic network based on *a* proximity function. Even though much richer proximity functions could have been applied, it was out of the scope of this paper.

Semantic View Size ℓ. In all experiments the semantic view consisted of the 10 semantically closest peers in the VICINITY cache. As shown in [11], a semantic view size of $\ell = 10$ provides a good tradeoff between the number of nodes contacted in the semantic search phase and the expected semantic hit ratio.

Cache size. For the cache size selection, we are faced with the following tradeoff for both protocols. A large cache size provides higher chances of making better item selections, and therefore accelerate the construction of (near-)optimal semantic views. On the other hand, the larger the cache size, the longer it takes to contact all peers in it, resulting in the existence of older—and therefore more likely to be invalid—links. Of course, a larger cache also takes up more memory, although this is generally not a significant constraint nowadays.

Considering this tradeoff, and after a set of experiments that cannot be presented due to space limitations, we fixed the cache size to 100 as a basis to compare different configurations. When both Vicinity and Cyclon are used, they are allocated 50 cache entries each.

Gossip length. The gossip length, that is, the number of items gossiped per gossip exchange per protocol, is a crucial factor for the amount of bandwidth used. This becomes of greater consequence, considering that an item carries the file list of its respective node. So, even though exchanging more items per gossip exchange allows information to disseminate faster, we are inclined to keep the gossip lengths as low as possible, as long as the system's performance is reasonable.

Again, for the sake of comparison, we fixed the total gossip length to 6 items. When both Vicinity and Cyclon are used, each one is assigned a gossip length of 3.

Gossip period T. The gossip period is a parameter that does not affect the protocol's behavior. The protocol evolves as a function of the number of messages exchanged, or, consequently, of the number of cycles elapsed. The gossip period only affects how fast the protocol's evolution will take place in time. The single constraint is that the gossip period T should be adequately longer than the worse latency throughout the network, so that gossip exchanges are not favored or hindered due to latency heterogeneity. A typical gossip period for our protocol would be 1 minute, even though this does not affect the following analysis.

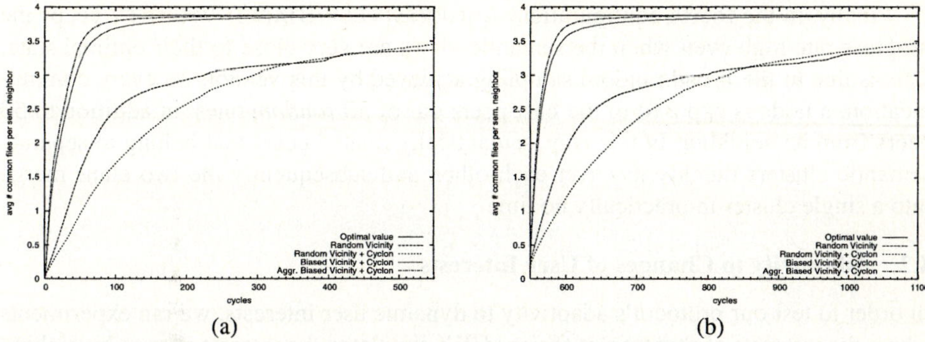

Fig. 4. (a) Convergence of sem. views' quality. (b) Evolution of semantic views' quality for a sudden change in all users' interests at cycle 550.

4   Performance Evaluation

4.1   Convergence Speed

To evaluate the convergence speed of our algorithm, we test how quickly it finds nodes having files in common. The proximity function's objective is for each node to discover the ℓ peers that have the most common files with it. Therefore, a good metric of the progress towards this goal is the average number of common files between a node and each one of its semantic neighbors. From our traces, we measured that in the optimal organization, this metric has a value of 3.88.

Figure 4(a) shows this metric as a function of the cycle for four distinct configurations. In favor of comparison fairness, the cache size and gossip length are 50 and 3, respectively, in each layer, for all configurations. The only exception is the first configuration, which has a single layer. In this case, the cache size and gossip length are 100 and 6, respectively. All experiments start with each node knowing 5 random other ones, simply to ensure initial connectivity in a single connected cluster.

In the first configuration, RANDOM VICINITY is running stand-alone. The progress of the semantic views' quality is rather steep in the first 100 cycles, but as nodes gradually concentrate on their very own neighborhood, getting to know new, possibly better peers becomes rare, and progress slows down.

In the second configuration, a two-layered approach consisting of RANDOM VICINITY and CYCLON is running. The slow start compared to stand-alone VICINITY is a reflection of the smaller VICINITY cache (3 as opposed to 6). However, the two-layered approach's advantage becomes apparent later, when CYCLON keeps feeding the RANDOM VICINITY layer with new, uniform randomly selected nodes, maintaining a higher progress rate, and outperforming stand-alone VICINITY in the long run.

In the third configuration, BIASED VICINITY demonstrates its contribution, as progress is significantly faster in the initial phase of the experiment. This is to be expected, since the items sent over in each BIASED VICINITY communication, are the ones that have been selected as the semantically closest to the recipient.

Finally, in the fourth configuration, AGGRESSIVELY BIASED VICINITY keeps the progress rate high even when the semantic views are very close to their optimal state. This is due to the broad random sampling achieved by this version. In every communication, a node is exposed to the best peers out of 50 *random* ones, in addition to 50 peers from its neighbor. In this way, semantically related peers that belong to separate semantic clusters quickly discover each other, and subsequently the two clans merge into a single cluster in practically no time.

4.2 Adaptivity to Changes of User Interests

In order to test our protocol's adaptivity to dynamic user interests, we ran experiments where the interests of some users changed. We simulated the interest change by picking a random pair of nodes and swapping their file lists in the middle of the experiment. At that point, these two nodes found themselves with semantic views unrelated to their (new) file lists, and therefore had to gradually climb their way up to their new semantic vicinity, and replace their useless links by new, useful ones.

Once again, we present the worst case —practically unrealistic— scenario, of *all* nodes changing interests at once, at cycle 550 of the experiment of figure 4(a). The evolution of the quality of the semantic views (using the metric introduced in 4.1) after the moment when all nodes change interests, is presented in figure 4(b). The faster convergence compared to figure 4(a) is due to the fact that views are already fully filled up at cycle 550, so nodes have more choices to start looking for good candidate neighbors.

Even though this scenario is very unrealistic, it demonstrates the power of our protocol in adapting to even massive scale changes. This adaptiveness is due to the priority given to newer items in `selectItemsToKeep()`, which allows a node's items with updated semantic information to replace older items of that node fast.

4.3 Effect on Semantic Hit Ratio

In order to further substantiate our claim that semantic based clustering endorses P2P searching, we conducted the following experiments. A randomly selected file was removed from *each* node, and the system was run considering proximity based on the remaining files. Then, each node did a search on that special file. We measured the semantic hit ratio to be over 36% for a semantic view of size 10.

Figure 5 presents the semantic hit ratio as a function of the cycle. Three experiments are shown, with gossip lengths for *both* layers set to 1, 3, and 5. Note that the hit ratio was autonomously computed in each cycle, without affecting the mainstream experiment's state.

5 Bandwidth Considerations

Due to the periodic behavior of gossiping, the price of having rapidly converging protocols may inhibit a high usage of network resources (i.e., bandwidth).

In each cycle, a node gossips on average twice (exactly once as an initiator, and on average once as a responder). In each gossip $2 \cdot (g_v + g_c)$ items are transferred to and from the node, resulting in a total traffic of $4 \cdot (g_v + g_c)$ items for a node per cycle. An item's size is dominated by the file list it carries. A single file is identified by its 128-bit (16-byte) MD4 hash value. Analysis of the eDonkey traces [5] revealed an average

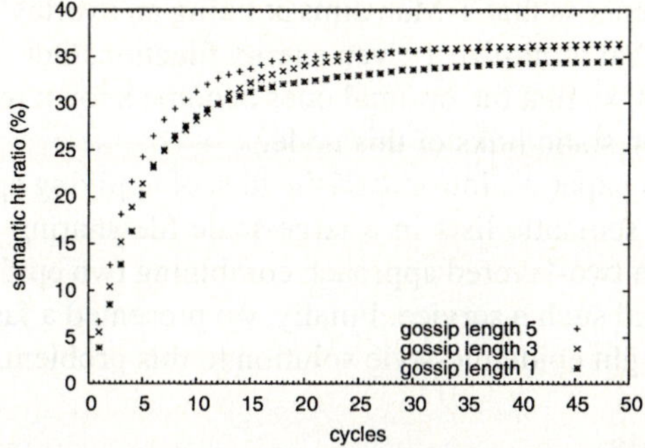

Fig. 5. Semantic Hit Ratio, for gossip lengths 1, 3, and 5 in each layer.

number of 100 files per node (more accurately, 99.35). Therefore, a node's file list takes on average 1,600 bytes. So, in each cycle, the total number of bytes transferred *to* and *from* the node is $6,400 \cdot (g_v + g_c)$.

For $g_v = g_c = 3$, the average amount of data transferred to and from a node in one cycle is 38,400 bytes, while for $g_v = g_c = 1$, it is just 12,800. Considering the gossip period T equal to 1 minute, this translates to an average bandwidth of 640 and 213 bytes per second, respectively. With $g_v = g_c = 3$ the system adapts a little faster to changes, but if bandwidth is of high concern, $g_v = g_c = 1$ can also provide very good results. Note that with a period of 1 minute, in the first 8 minutes we reach 85% of the optimal semantic hit ratio, having roughly 30% of all requests handled by the semantic neighbors.

We consider such a bandwidth consumption to be rather small, if not negligible compared to the bandwidth used for the actual file downloads. It is, in fact, a small price to pay for relieving the default search mechanism from about 35% of the search load.

6   Discussion

To the best of our knowledge, all earlier work on implicit building of semantic overlays relies on using heuristics to decide *which* of the peers that served a node recently are likely to be useful again in future queries [11–13].

However, all these techniques inhibit a weakness that challenges their applicability to the real world. They all assume a *static* network, free of node departures, which is a rather strong assumption considering the highly dynamic nature of file-sharing communities. Also, it is not clear how they perform in the presence of dynamic user preferences.

Regarding proximity-based P2P clustering, our work comes close to T-Man[8]. However, a key difference is that T-Man assumes *continuous* proximity metrics. That is, every node can point *any* other node to the right direction. This is not true in the problem we faced, i.e. in the case of completely unrelated peers. We dealt with it by harnessing CYCLON's randomness. This renders our solution more generic. Moreover, T-Man assumes a preconstructed almost random graph to start with. We make no such assumptions.

Another key difference is that T-Man aims at fixing an overlay's links to the optimal ones, that is, the ones that minimize a given energy function. Our work aims at continuously exchanging links, so that the optimal ones become known relatively soon to each node, but do not remain static links of this node.

Concluding, in this paper we introduced the idea of applying epidemics to build and dynamically maintain semantic lists in a large-scale file-sharing system. Specifically, we showed that using a two-layered approach combining two epidemic protocols is the appropriate way to build such a service. Finally, we presented a fast converging, highly adaptable, yet lightweight epidemic-style solution to this problem.

Acknowledgements

We would like to specifically thank Fabrice Le Fessant for providing us with the eDonkey2000 traces[5] he gathered in November 2003.

References

1. J. Risson and T. Moors. Survey of Research towards Robust Peer-to-Peer Networks: Search Methods. Technical Report UNSW-EE-P2P-1-1, University of New South Wales, Sydney, Australia, September 2004.
2. Reka Albert and Albert-Laszlo Barabasi. Statistical Mechanics of Complex Networks. *Reviews of Modern Physics*, 74(1):47–97, January 2001.
3. M.E.J. Newman. Random Graphs as Models of Networks. In S. Bornholdt and H. G. Schuster, editors, *Handbook of Graphs and Networks: From the Genome to the Internet*, chapter 2. John Wiley, New York, NY, 2002.
4. M. Jelasity, R. Guerraoui, A.-M. Kermarrec, and M. van Steen. The Peer Sampling Service: Experimental Evaluation of Unstructured Gossip-Based Implementations. In *Middleware 2004*, volume 3231 of *Lect. Notes Comp. Sc.*, Berlin, October 2004. ACM/IFIP/USENIX, Springer-Verlag.
5. Fabrice Le Fessant, S. Handurukande, Anne-Marie Kermarrec, and Laurent Massoulié. Clustering in peer-to-peer file sharing workloads. In *3rd International Workshop on Peer-to-Peer Systems (IPTPS)*, San Diego, USA, February 2004.
6. Duncan J. Watts. *Small Worlds, The Dynamics of Networks between Order and Randomness.* Princeton University Press, Princeton, NJ, 1999.
7. Spyros Voulgaris, Daniela Gavidia, and Maarten van Steen. Inexpensive membership management for unstructured p2p overlays. *Journal of Network and Systems Management*, To appear in 2005.
8. Márk Jelasity and Ozalp Babaoglu. T-Man: Fast gossip-based construction of large-scale overlay topologies. Technical Report UBLCS-2004-7, University of Bologna, Department of Computer Science, Bologna, Italy, May 2004.
9. Peersim. http://peersim.sourceforge.net/.
10. edonkey. http://www.edonkey2000.com.
11. S. Handurukande, A.-M. Kermarrec, F. Le Fessant, and L. Massoulié. Exploiting semantic clustering in the edonkey p2p network. In *11th ACM SIGOPS European Workshop (SIGOPS)*, Leuven, Belgium, September 2004.
12. K. Sripanidkulchai, B. Maggs, and H. Zhang. Efficient content location using interest-based locality in peer-to-peer systems. In *INFOCOM Conference*, 2003.
13. S. Voulgaris, A. Kermarrec, L. Massoulié, and M. van Steen. Exploiting semantic proximity in peer-to-peer content searching. In *10th International Workshop on Future Trends in Distributed Computing Systems (FTDCS 2004)*, Suzhu, China, November 2001.

Long Range Contacts in Overlay Networks*

Filipe Araújo and Luís Rodrigues

Universidade de Lisboa, Departamento de Informática, Faculdade de Ciências
Campo Grande, Edifício C6, 1749-016 Lisboa, Portugal
{filipius,ler}@di.fc.ul.pt

Abstract. In this paper we present and evaluate a novel mechanism,
called *Hop Level*, that creates and maintains long range contacts (LRCs)
in overlay networks. The Hop Level mechanism owns the following characteristics: *i*) lazy creation of the LRCs, *ii*) support for unbalanced
node distribution, *iii*) support for multidimensional spaces and *iv*) near-
optimal path lenght/node degree trade-off. These characteristics make
Hop Level specially suited for overlay networks that support range data
queries (as opposed to distributed hash tables that only support exact
queries) with one or more dimensions. Furthermore, and unlike previous
similar work, Hop Level can handle churn very well, because it postpones
creation of the LRCs until it is necessary. In this way, nodes that have
short lives do not overload the network with their state update requests.

1 Introduction

Distributed hash tables (DHTs) have recently emerged as an important component for distributed systems. A DHT is a dictionary that outputs values in
exchange of keys. A common aspect to the most well-known DHTs [6, 7, 9, 13,
14, 16, 17] is that they operate in the application layer as an overlay network.
To overcome the limitations inherent to DHTs, some researches have proposed
a shift to a more powerful paradigm: the distributed storage systems [4, 5, 8]
(DSSs). Unlike DHTs that only perform exact queries, DSSs allow efficient range
queries. As a consequence, when compared to a DHT, the design of a DSS is more
complex. First, in a DSS, we cannot assume that data is uniformly distributed in
space. Second, we cannot assume that entrance and departure patterns of data
items will favor balancing. On the contrary, DHTs were based on the assumption
that consistent hashing would result in a perfect balance of node identifications
and data items.

Often, in overlay networks, it is possible to distinguish between two different
types of contacts: "nearby" contacts, forming a kind of connected lattice between
nodes that have close virtual identifications, and "long range contacts" (LRCs)
between nodes that have "distant" virtual node identifications. While the former
type of contacts may be important in certain overlays, to ensure connectedness
and routing convergence, short path lengths actually depend on the latter type

---

* This work was partially supported by the LaSIGE and by the FCT project INDIQoS
  POSI/CHS/41473/2001 via POSI and FEDER funds.

J.C. Cunha and P.D. Medeiros (Eds.): Euro-Par 2005, LNCS 3648, pp. 1153–1162, 2005.

of contacts. In fact, it is the capability to "jump" over many close nodes in a single hop that makes it possible to achieve short path lengths. Therefore, in this paper, we present the Hop Level mechanism, which creates and maintains LRCs in overlay networks. The Hop Level mechanism can be used in many different overlay networks to reduce path lengths, including DHTs and DSSs. Nevertheless, it is particularly well suited to DSSs, because it can cope with unbalanced distribution of nodes and it supports single as well as multidimensional data. We believe that this is one of the most innovative aspects of Hop Level, because most overlay networks are tied to unidimensional address spaces, where nodes must be numerically or alphabetically ordered (e.g., SkipNets [6]).

Since node degree and diameter of a network cannot be arbitrarily and simultaneously reduced, the trade-off between these two metrics is often used as a fundamental efficiency measure of an overlay network. For an $O(1)$ node degree, the expected diameter can be at best $O(\log n)$, while for an $O(\log n)$ node degree, the expected diameter cannot be shorter than $O(\log n/\log\log n)$ [7], for an n-node network. Given this limitation, path lengths of Hop Level achieve a nearly optimal trade-off with node degree. Furthermore, unlike existing overlay networks that implement DHTs and DSSs, when a node using the Hop Level mechanism enters the network, it postpones the creation of the LRcs to simplify the entrance. Later, it progressively creates the LRCs as they are needed to route real messages. In fact, lazy creation of the LRCs is one of the most significant aspects of Hop Level, as this reduces control traffic with only a minor effect on routing performance. In this way, behavior of Hop Level under churn is very good.

The remainder of the paper is organized as follows: Section 2 states the problem we are solving. Section 3 overviews previous work. Our long range contact mechanism is described and evaluated, respectively in Sections 4 and 5. Section 6 concludes the paper.

2   Problem Statement

Throughout this paper we will consider that routing convergence is ensured by nearby contacts already existing in the overlay network, e.g., as in [10] or [9]. Although these are examples of two-dimensional networks (of which we tested the Delaunay triangulation of [10]), there is however no restriction to the number of dimensions of the overlay network. A crucial point here is that distribution of nodes does not need to follow any specific pattern.

Hence, we will consider the following conditions: i) nodes are organized into a multidimensional underlying overlay network having only nearby contacts; and ii) identification of nodes is arbitrary (as a result, distribution of nodes in space may be unbalanced). The goal of condition ii is to maintain locality, by preventing arbitrary conversion of node addresses from one space of identifications to another, e.g., by an hash function. There are many practical examples where this restriction holds. In a DSS, nodes may receive their identification according to the data items that they store. In [4], the overlay structure directly reflects the contents of the data, which is organized in a sequential order. In this way, it is possible to make range queries efficiently. On the contrary, hashing data to obtain some balance in a different identification space would defeat this goal. Another example where the condition *ii* holds occurs in systems where identification of a node bears some relation with its physical location, like in [3] or when using landmark ordering [13].

Furthermore, we will consider the use of a routing scheme where *i*) the preprocessing algorithm can only collect information of $O(1)$ nearby peers and $O(\log n)$ distant peers per node and *ii*) the routing algorithm will select, among the forwarding node's contacts (either short or long range), the one which is closest to destination in terms of Euclidean distances[1]. Given these conditions, our goal is to design a mechanism that creates and maintains a set of LRCs at each node such that routing convergence is guaranteed with $O(\log n)$ expected path lengths *despite* non-uniform node distribution. Moreover, each node should store $O(\log n)$ LRCs and this number must not depend on the size of the virtual identification space, but only on the nodes effectively existing in the system. Balancing the workload among the peers in the DSS is not a goal of this paper; such issue is orthogonal to our work and is already tackled in previous work, like [4].

Before presenting the Hop Level mechanism we will overview previous research in the topic of overlay networks to capture the relevant features that should be owned by efficient sets of LRCs.

3   Related Work

There is a huge body of work related with overlay networks and, in particular, with DHTs. In some DHTs it is possible to do an explicit separation between nearby and long range contacts (e.g., in DHTs based on a ring). However, there are also many other systems where this separation is only implicit or non-existent. In constrast to the previous cases, CAN [13] exhibits no LRCs but only short range contacts, thus having longer path lengths. To overcome this limitation, Xu and Zhang [17] proposed a mechanism called "expressways for CAN" that augments basic CAN with LRCs. This work and others, like [9] and [11] are very similar in spirit to the Hop Level mechanism.

All the DHTs referred before assume a balanced distribution of nodes in space. Unlike these, LAND [1] copes with unbalanced distribution of nodes, but it does not meet the conditions stated in Section 2, because it hashes identifiers of objects. SkipNet [6] was also designed from scratch to cope with the unbalanced use of identification space. In fact, SkipNet is more appropriate to support a DSS, because it supports range queries. However, the identification space of a SkipNet is unidimensional and generalization to higher-dimensional spaces does not seem trivial. Unlike SkipNet, [4, 5, 8] have explicit support for complex load balancing mechanisms without impairing efficient range queries. Of these, only Mercury [5]

---

[1] There is no loss of generality in assuming Euclidean distances, as other metrics could also be used if more appropriate to the structure of the lattice, e.g., Manhattan distance or unidimensional virtual identification distance.

supports multidimensional range queries. However, Mercury requires a different data structure (a ring of nodes) for each queriable attribute (including a copy of the data). When compared to these systems, support of multidimensional range queries is inherent to the Hop Level mechanism and does not need to be mapped to multiple unidimensional queries.

4   Hop Level LRCs Mechanism

We now describe our proposal to build LRCs in unbalanced overlays. Using our *Hop Level* mechanism, LRCs are established automatically whenever a message goes through b consecutive hops. Consider, for instance, that some node F is forwarding a message m to node N_1 originated at node S and destined to node D. If node F realizes that N_1 will be the b-th hop of m it triggers the creation of a LRC from S to N_1, denoted by $S \xrightarrow{1} N_1$. To do this F sends a control message to S. The process is repeated from N_1 onwards: if after b hops, message m reaches N_2, N_1 will create a LRC to N_2, $N_1 \xrightarrow{1} N_2$, and so on. Let us call these LRCs, *level-1 LRCs*. If the message path is very long, it may happen that a sequence of b *level-1 LRCs* occurs, for instance: $S \xrightarrow{1} N_1$, $N_1 \xrightarrow{1} N_2$, ..., $N_{b-1} \xrightarrow{1} N_b$. In this case, a new LRC from S directly to N_b should be created. This new LRC, $S \xrightarrow{2} N_b$, is one level above of the previous ones. This mechanism should be applied recursively for all levels. Hence, a LRC of level-l jumps over b^l hops.

To bound the number of LRCs per node, we limit the number of LRCs that exist in each level. This allows the number of LRCs to grow with the size of the network. The shape of this growth is evaluated in Section 5.

4.1   Algorithm

Our implementation of Hop Level algorithm requires a minimum of three variables per level l to be carried in each message m: the number of hops, $nh_m[l]$, the node that may receive a new LRC of that level, $s_m[l]$, and whether this node has space for an additional LRC, $a_m[l]$. Whenever level counter $nh_m[l-1]$ reaches the limit b, a new LRC, starting at $s_m[l]$ should be created. To conserve space we do not discuss signaling cost here, but it is possible to leave some of this temporary information at the nodes to shorten messages.

When a forwarding node uses a LRC of level-l to send a message, it must check the LRC used by the previous hop node, say level-p. If $l > p$, neither one of the LRCs that preceded this hop can be used to create new LRCs (e.g., if a level-3 LRC is being taken after a previous level-2 LRC). Now, consider that message m is going to be sent along its b-th consecutive hop of level-l to node N. In this case, forwarding node F sends a control message to the node that initiated the sequence of level-l, prompting it to create a LRC of level-$(l+1)$ to node N. Then, node F sets the number of hops of level-l to 0 and increments the number of hops of level$(l+1)$ by 1. Should this substituting hop become the b-th hop of level-$(l+1)$, the same process is repeated for level-$(l+1)$, and so on, until a level with fewer than b hops is reached.

Algorithm 1 Hop Level algorithm.

{Executed at node F when forwarding m to node N}
{Control information carried in message m:}
    {max_m — highest valid level; p_m — level of LRC used to reach F;}
    {$\forall k \in [0, max_m] : nh_m[k], s_m[k], a_m[k]$ — resp., number of hops, first node and whether there are available slots in the first node for level-k;}

1: $l \leftarrow$ level of LRC from F to N ($F \xrightarrow{l} N$)
2: **if** $p_m = \perp$ or $p_m < l$ **then**
3:    $max_m \leftarrow l + 1;\ lim \leftarrow max_m$
4: **else**
5:    $lim \leftarrow p_m$
6: **end if**
7: **for all** $k \in \{l, \ldots, lim - 1\}$ **do**
8:    $s_m[k+1] \leftarrow F;\ a_m[k+1] \leftarrow a_F[k+1];\ nh_m[k] \leftarrow 0$
9: **end for**
10: $nh_m[l] \leftarrow nh_m[l] + 1$
11: **while** $nh_m[l] \geq b$ **do**
12:    $nh_m[l] = 0$
13:    **if** $a_m[l+1] > 0$ **then**
14:      instruct $s_m[l+1]$ to create LRC $s_m[l+1] \xrightarrow{l+1} N$
15:    **end if**
16:    $l \leftarrow l + 1;$
17:    **if** $max_m == l$ **then**
18:      $max_m \leftarrow max_m + 1;\ nh_m[max_m - 1] \leftarrow 0$
19:      $s_m[max_m] \leftarrow s_m[max_m - 1];\ a_m[max_m] \leftarrow a_m[max_m - 1]$
20:    **end if**
21:    $nh_m[l] \leftarrow nh_m[l] + 1$
22: **end while**

To implement this algorithm, messages must carry the level p_m of the LRC used by the previous hop to reach F, and an indication of the highest level of the array that contains valid information, max_m. Each node F, when forwarding the message m to N, executes Algorithm 1.$a_F[k]$ is a boolean variable that indicates whether F has slots available at level k to store additional LRCs. If F is the source of the message, $F = S$, it is necessary to set previous level $p_m \leftarrow \perp$. In this case, the execution of the algorithm will initialize $max_m \leftarrow l+1$, $s_m[max] \leftarrow S$, $a_m[max] \leftarrow a_S[max]$ and $nh_m[max - 1] \leftarrow 0$.

To maintain the LRCs evenly distributed in face of membership changes, we periodically delete the least recently used LRC of some randomly selected levels. In our experiments, path lengths did exhibit low sensitivity to variations of the deletion period. Nodes should also purge hanging LRCs that point to neighbors that left. To do this, nodes can send periodic beacons to their neighbors. Alternatively, we can trade this beacon traffic by latency, by using, again, a lazy approach. In this latter solution, nodes only detect that a LRC is hanging when they try to use it. For the highest churn rates we tested in Section 5, when using a lazy apprach, 13.3% of the messages tried to follow hanging LRCs. This figure goes down to 1.2% for the lowest churn rate.

5   Evaluation

Experiment Settings. In this section we experimentally evaluate Hop Level with $b = 2$. Most experiments, including the comparison with eCAN-like mechanism (to be presented ahead) use a Delaunay triangulation as the underlying lattice [10]. However, for benchmarking purposes we have also used a mapping of a two-dimensional space into a unidimensional ring. In our experiments we evaluate the following aspects: *i*) the behavior of Hop Level, when different limits for LRCs by level are used; this includes knowing the distribution of the LRCs by the levels; *ii*) the behavior of Hop Level when compared to the eCAN-like mechanism, both in balanced and extremely unbalanced scenarios; *iii*) the behavior of Hop Level in a ring; *iv*) the cost of the bootstrap mechanism of Hop Level and, finally; *v*) the behavior of Hop Level in dynamic settings, including settings with strong membership variation, i.e., under churn.

In the tests, arbitrary pairs of nodes exchange a large number of messages in networks with sizes ranging from 100 to 50,000 nodes. To route the messages we have used the greedy routing algorithm, because it has good performance and it works both in the underlying lattice and with LRCs, without requiring any extensions. Furthermore, it agrees to the conditions of Section 2. Hence, next hop is always the neighbor (connected by a short or long range contact) closest to destination. To let Hop Level LRC scheme converge, and depending on the network size, we routed up to 1,000,000,000 different messages and only used the final 3000 paths in the evaluation of path lengths. Nevertheless, we also show that our mechanism achieves good routing performance much earlier than that. To test unbalanced distributions of nodes we used a truncated Gaussian bivariate distribution with standard deviations of 0.01 in a $[0,1] \times [0,1]$ square.

Number of LRCs per Level. The first aspect that we evaluate is the performance achieved by different configurations of the Hop Level mechanism. The goal is to determine the limit for the number of LRCs per level that ensures the most reasonable compromise between path lengths and node degrees. Figures 1(a) and 1(b) respectively show the average path lengths (in number of hops) and the average number of LRC used by each node for different network sizes and for different configurations of the Hop Level mechanism: with 1, 2, 4, 6 and 8 LRCs per level. We can see that all configurations achieve an approximately logarithmic/logarithmic trade-off (a logarithmic growth is represented by a straight line). We believe that this is quite an interesting aspect, because it minimizes the need for manual configuration of parameters. In the rest of our experiments we set the limit to 6 LRCs per level. Figure 1(c) shows how many LRCs exist on the entire network and the average length of those LRCs for each hop level. To do this evaluation, we have used a 50,000 node network with a balanced distribution of nodes, because a balanced distribution allows to reason in terms of distance. From the growth of the number of levels it is possible to determine the growth of the number of LRCs per node. Given that the distance growth from one level to the next is approximately exponential this figure points to the conclusion that the number of levels is approximately logarithmic.

Comparison with "eCAN-Like" and Hop Level in a Ring. To offer some comparative measurement, we ran our scheme against a benchmark mechanism called "eCAN-like". This benchmark results from an adaptation of the eCAN [17] logarithmic/logarithmic node degree/path length mechanism (whose applications

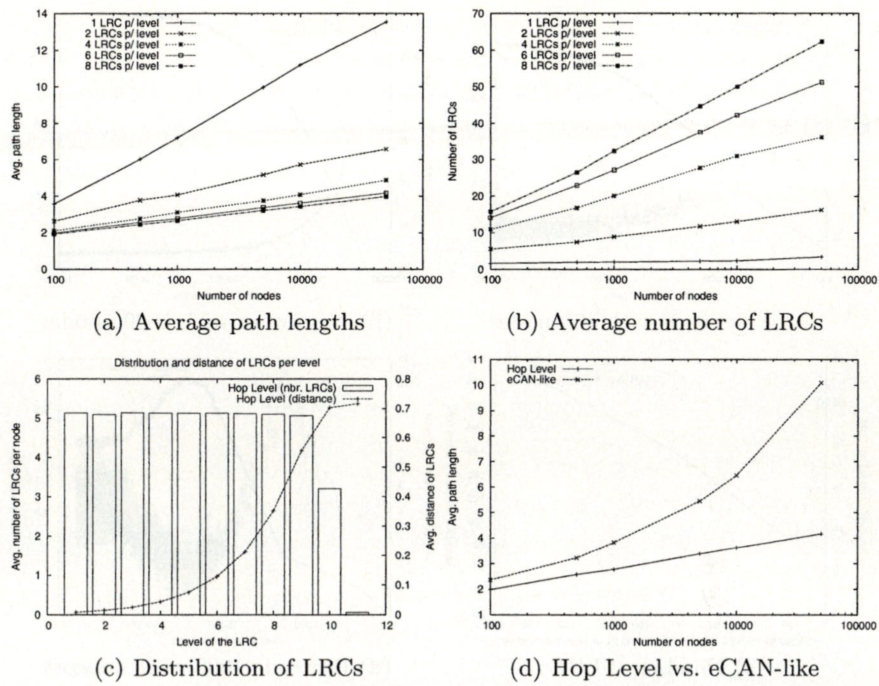

(a) Average path lengths

(b) Average number of LRCs

(c) Distribution of LRCs

(d) Hop Level vs. eCAN-like

Fig. 1. Path length and number of LRCs.

most closely resemble those of our own algorithm). Although we made some simplifications to the original eCAN, we believe that our implementation of expressways mimics the eCAN LRC mechanism with enough accuracy to allow a fair comparison. The idea in eCAN-like is to make a first level division of the entire space in four big squares. Each node keeps LRC to the two neighboring squares. Then, the four big squares are further divided in other four smaller squares. This time, nodes inside squares have a total number of four LRC (above, below, right and left). This process is repeated for as many levels as wanted. In our context, we fixed the number of levels to 8, in a total of 30 LRCs. The actual LRC will be the node responsible for the central point of each neighboring square. Comparison of Hop Level against the eCAN-like is depicted in Figure 1(d), for unbalanced networks. The number of LRCs is not depicted because it is constant in eCAN-like. Bad behavior of the eCAN-like mechanism is easily explainable: density of LRCs is no longer enough near the center and routing to nearby nodes will tend to become linear with the number of hops in the lattice, instead of logarithmic. On the contrary, node distribution has a very little impact on Hop Level.

Due to lack of space, we do not show results of mapping a two-dimensional space into a ring (the same could be done for any number of dimensions). As expected, path lengths in a ring are also logarithmic, but paths are shorter in

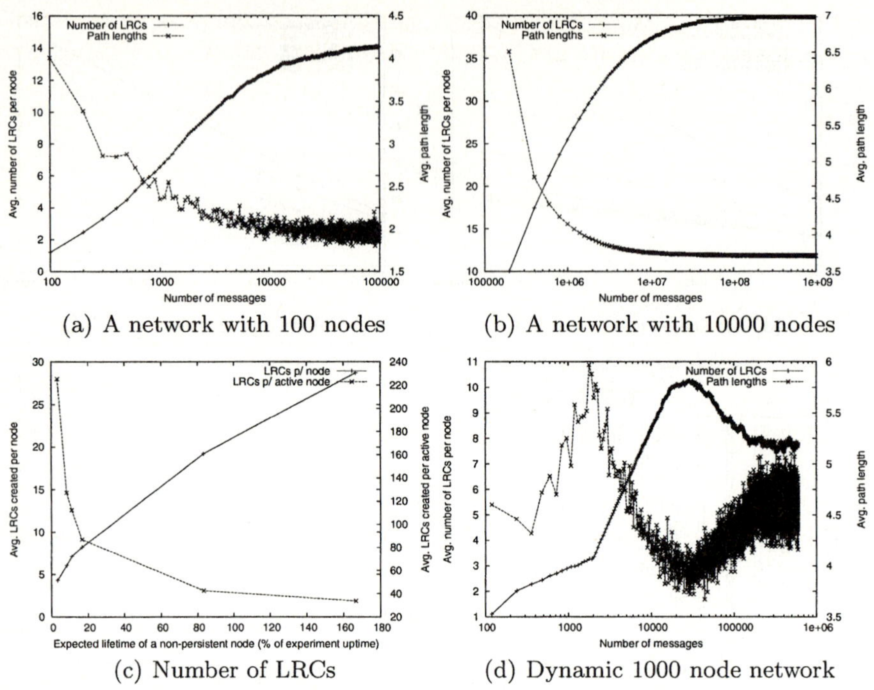

(a) A network with 100 nodes

(b) A network with 10000 nodes

(c) Number of LRCs

(d) Dynamic 1000 node network

Fig. 2. Dynamic performance.

a multidimensional space due to the higher connectivity of nodes. This result is interesting not only for benchmarking, but also because it shows that hashing nodes into a ring can have its costs in performance (not to mention a possible loss of locality information).

Network Convergence. Figures 2(a) and 2(b) depict for two network sizes the growth in the number of LRCs of the entire network and the reduction in the path lengths. In both cases, we can observe that for all network sizes under test, a short number of messages suffices to let the network reach a state similar to a steady state. For all network sizes we tested, path lengths within 3 times the optimal can be achieved before 5 messages have been generated by each node.

Dynamic Settings. In this section we will use settings similar to the ones described in Araneola [12], which are based on real measurements [2, 15]. Hence, we assume that around 7% of the nodes are permanent. The remaining 93% of the nodes are non-permanent and can enter or leave the network at any instant and repeatedly do so. When a node enters the network it becomes active, when it leaves it goes to sleep state. When network starts, non-permanent nodes are neither active nor sleeping, but in a fourth state that we can call as *out*. This means that the network starts with 7% of the permanent nodes. Then, a bootstrap process starts, bringing 50 new nodes from *out* to active or sleep states with equal probabilities at each time step[2]. After each time step, any non-permanent node that is either active or sleeping can switch from one state to the other with a given fixed probability[3] — this simulates the churn (note that nodes reenter the network in a fresh state, i.e., without any LRCs originating or pointing to it). A node can never return to the *out* state. The main parameter to vary in this experiment is the rate at which nodes enter and leave the network or, in other words, the average lifetime of non-persistent nodes. The probability of switching state after a time step is varied from 0.00005 to 0.0025. In the Hop Level mechanism, churn is associated with two types of costs: the signaling cost of changing network topology and the cost of worse routing performance.

Figure 2(c) shows the number of LRCs created in the network under churn (signaling cost). From the perspective of active non-persistent nodes, the shorter the lifetime, the fewer LRCs such a node will create. This corresponds to the line deemed "LRCs p/ node". On the other hand, the load for the network and for the persistent nodes increases with churn. This is represented in the line deemed "LRCs p/ active node", which shows the total number of LRCs created in the network, divided by the average number of active nodes. We can see that even with very small lifetimes, the growth in the number of LRCs created per active node is moderate. Churn also degrades routing performance. This is illustrated in Figure 2(d) for a non-persistent node's lifetime of 10% of experiment up time. The pattern depicted in this graphic is similar for other average lifetimes. Some time after the number of nodes stabilizes, the number of LRCs per (new entering) node starts to decay until it stabilizes to a value that depends on the churn rate.

6   Conclusions

In this paper we presented the Hop Level mechanism that manages Long Range Contacts (LRCs) in overlay networks. Experimental results showed that performance of Hop Level is nearly optimal and independent of node distribution in space. Furthermore, Hop Level resists churn very well without compromising performance in fresh networks. For these reasons, we believe that the Hop Level mechanism is applicable to a broad class of overlay networks, including multidimensional range queries in Distributed Storage Systems.

References

1. I. Abraham, D. Malkhi, and O. Dobzinski. Land: stretch $(1+\epsilon)$ locality-aware networks for dhts. In *fifteenth annual ACM-SIAM symposium on Discrete algorithms*, pages 550–559. Society for Industrial and Applied Mathematics, 2004.
2. K. Almeroth and M. Ammar. Collecting and modeling the join/leave behavior of multicast group members in the mbone. In *High Performance Distributed Computing (HPDC '96)*, pages 209–216, Syracuse, NY, USA, august 1996.

---

[2] A time step is counted after 50 messages.
[3] Hence, an exponential distribution can model joins and leaves.

3. F. Araújo and L. Rodrigues. Geopeer: A location-aware peer-to-peer system. In *The 3rd IEEE International Conference on Network Computing and Applications (NCA '04)*, pages 39–46, Cambridge, MA, USA, august 2004.

4. J. Aspnes, J. Kirsch, and A. Krishnamurthy. Load balancing and locality in range-queriable data structures. In *Twenty-Third Annual ACM SIGACT-SIGOPS Symposium on Principles of Distributed Computing (PODC 2004)*, St. Johns, Newfoundland, Canada, July 2004.

5. A. R. Bharambe, M. Agrawal, and S. Seshan. Mercury: supporting scalable multi-attribute range queries. *SIGCOMM Comput. Commun. Rev.*, 34(4):353–366, 2004.

6. N. Harvey, M. Jones, S. Saroiu, M. Theimer, and A. Wolman. Skipnet: A scalable overlay network with practical locality properties. In *Fourth USENIX Symposium on Internet Technologies and Systems (USITS '03)*, Seattle, WA., March 2003.

7. F. Kaashoek and D. Karger. Koorde: A simple degree-optimal distributed hash table, 2003.

8. D. R. Karger and M. Ruhl. Simple efficient load balancing algorithms for peer-to-peer systems. In *SPAA '04: Proceedings of the sixteenth annual ACM symposium on Parallelism in algorithms and architectures*, pages 36–43. ACM Press, 2004.

9. J. Kleinberg. The Small-World Phenomenon: An Algorithmic Perspective. In *Proceedings of the 32nd ACM Symposium on Theory of Computing*, 2000.

10. J. Liebeherr, M. Nahas, and W. Si. Application-layer multicasting with Delaunay triangulation overlays. Technical Report CS-2001-26, University of Virginia, Department of Computer Science, 5 2001.

11. G. S. Manku, M. Bawa, and P. Raghavan. Symphony: Distributed hashing in a small world. In *4th Usenix Symposium on Internet Technologies and Systems*, 2003. http://www.usenix.org/events/usits03/.

12. R. Melamed and I. Keidar. Araneola: A scalable multicast system for dynamic environments. In *The 3rd IEEE International Conference on Network Computing and Applications (NCA '04)*, pages 5–14, Cambridge, MA, USA, august 2004.

13. S. Ratnasamy, P. Francis, M. Handley, R. Karp, and S. Schenker. A scalable content-addressable network. In *Conference on applications, technologies, architectures, and protocols for computer communications*, pages 161–172. ACM Press, 2001.

14. A. Rowstron and P. Druschel. Pastry: Scalable, decentralized object location, and routing for large-scale peer-to-peer systems. *Lecture Notes in Computer Science*, 2218:329–350, 2001.

15. S. Saroiu, P. Gummadi, and S. Gribble. A measurement study of peer-to-peer file sharing systems. In *Multimedia Computing and Networking (MMCN)*, San Jose, CA, USA, january 2002.

16. I. Stoica, R. Morris, D. Karger, F. Kaashoek, and H. Balakrishnan. Chord: A scalable Peer-To-Peer lookup service for internet applications. In *ACM SIGCOMM*, San Diego, August 2001.

17. Z. Xu and Z. Zhang. Building low-maintenance expressways for p2p systems. Technical Report HPL-2002-41, HP, 2002.

Combining the Use of Clustering
and Scale-Free Nature of User Exchanges
into a Simple and Efficient P2P System[*]

Pierre Fraigniaud[1], Philippe Gauron[2], and Matthieu Latapy[3]

[1] CNRS, LRI, Univ. Paris-Sud, 91405 Orsay, France
`pierre@lri.fr`
[2] LRI, Univ. Paris-Sud, 91405 Orsay, France
`gauron@lri.fr`
[3] CNRS, LIAFA, Univ. Paris VII, 75005 Paris, France
`latapy@liafa.jussieu.fr`

Abstract. It was recently observed that the user interests in P2P systems possess *clustering* properties that can be used to reduce the amount of traffic of flooding-based search strategies. It was also observed that the user interests possess *scale-free* properties that can be used for the design of routing-based search strategies. In this paper, we show that the combination of these two properties enables the design of an efficient and simple fully decentralized search strategy. This search strategy is simple in the sense that it does not require maintaining any structured overlay network topology connecting the peers. It is efficient in the sense that simulations processed on *real-world* traces show that lookups perform in logarithmic expected number of steps.

1 Preliminaries

This paper focuses on *fully decentralized* Peer-to-Peer (P2P) systems, i.e., a (large) set of users, called *peers*, exchanging information in the absence of any central service. Such P2P systems are self-organized, and all peers play the same role. Searching for objects (files, resources, etc.) in such systems requires the use of specific algorithms: object queries are transmitted from peer to peer; if a peer p receives a query for some object that it can provide, then p simply sends the object to the demander; otherwise, p forwards the query to one or several neighboring peer(s). The way peers are connected, and the choice of the peer(s) the query is forwarded to, are essential parts of the P2P system architecture.

Two main ways of forwarding queries in fully decentralized P2P systems have been identified: either by flooding (as in, e.g., Gnutella), or by using Distributed Hash Tables and their underlying routing protocols (as in, e.g., CAN [12] or Chord [14]). Both ways present some drawbacks. In particular, the traffic induced by flooding consumes a significant portion of the bandwidth, and DHT-based

---

[*] The two first authors received supports from the INRIA project "Grand Large", and from the project PairAPair of the ACI "Grandes Masses de Données".

J.C. Cunha and P.D. Medeiros (Eds.): Euro-Par 2005, LNCS 3648, pp. 1163–1172, 2005.

protocols use ad hoc connections between peers which are generally hard to maintain and hardly support sophisticated queries. As a consequence, the design of simple search protocols insuring both quick answers and low control traffic is still an open problem. Roughly speaking, one is facing the following alternative: either connecting the peers in an unstructured manner – which is simple but requires flooding – or connecting the peers in a structured manner – which enables routing but is complex.

In this paper, our objective is to propose a method possessing both advantages: *simple* search in an *unstructured* P2P system. To achieve this, we will mainly use the statistical properties of real-world peer-to-peer queries and we use the interest graph as overlay.

1.1   The Central Idea

Peers are nodes of a *physical* underlying network, i.e., the Internet which support physical communication between peers. Fully decentralized P2P systems are based on virtual connections between peers, which form an *overlay* network on top of the physical network. Basically, a peer p_1 is connected to a peer p_2 in the overlay network if p_1 knows the physical address of p_2, and vice versa. Communications between neighboring peers in the overlay network are routed in the physical network via its communication primitives. The P2P system has no control on the way the communications are processed by the physical network but it fully controls the overlay network, and it supports the search procedure.

As argued by various researchers (see [1] and the references therein), the overlay network should better map to the physical network, so that neighboring peers in the overlay network would be close in the physical network. A communication between two neighboring peers in the overlay network would then be processed quickly by the underlying physical network. It is shown in [1] that one can dynamically maintain an overlay network such that the distance between two peers in the overlay network is no more than $1 + \epsilon$ times their distance in the physical network, for any $\epsilon > 0$. However, this approach requires using rather complex control procedures.

Alternatively, one could relate the overlay network to the interests of peers. Indeed, if two peers have interests in common, then they will probably exchange a lot, and thus they should better be close in the overlay network. In practice, peers do not exchange objects with arbitrary other peers. Instead, they tend to group themselves into communities, with lots of exchanges inside the communities, and only few exchanges between them. Several authors already noticed this fact and, based on it, have proposed some improvements for existing systems (see, e.g., [5, 8, 9, 15, 16]). In this paper, our objective is to show that these works can be pushed further while keeping our solution very simple.

1.2   Peer Interests as a Graph

One can represent the peer interests as a graph in which two peers are connected if and only if they have some interests in common. Deriving a formal definition of what is meant by "having some interests in common" is a difficult task. Various propositions have been made, based on keywords, objects in common, etc. We will here use the following simple definition: two peers are connected in the interest graph if they have exchanged an object in the past. Note that two such peers may actually have very different interests, but at least their interests are related in some way. Note also, and this is essential in our context, that, since the P2P system processes all the queries, it contains a (distributed) view of the interest graph at any time.

It has been recently shown (cf., e.g., [7–9]) that, similarly to most social networks and most real-world complex networks, the interest graph as defined above has several non-trivial statistical properties that make it very different from standard random graphs. In particular, the interest graph has:

- a low density (the average degree is very low compared to the number of nodes);
- a small average distance (it typically scales logarithmically with the size of the network);
- a clustered structure (the graph is locally dense although the (global) density is low)
- a scale-free nature (i.e., degrees are very heterogeneous, most nodes having a low degree, but some nodes having a high degree).

In our method, the overlay network will be nothing but the interest graph as described above, and the performances of our method will strongly rely on the aforementioned four properties. In fact, our method is based on both the clustering *and* the scale-free nature of the interest graph, two notions that were considered separately in several previous works, listed thereafter.

1.3   Using Scale-Free Properties

Using the scale-free nature of real-world networks for the design of efficient search strategies have been proposed in [2, 6, 13]. In these papers, the authors approximate the heterogeneous degree distributions by power laws and study the properties of some (random or deterministic) walks in random graphs with power law degree distributions (see also [10]).

In [2], at each step of the search strategy, the current node scans its neighbors and if none has the searched data, then the query is forwarded to the highest degree neighbor. A mean-field analysis, confirmed by simulations, shows that the expected number of steps required to find an object in a random power law network with n nodes and exponent α, scales sub-linearly as $n^{3(1-2/\alpha)}$ for $2 < \alpha < 3$.

In [6], the authors perform simulations on another model of power law networks, and compare a random walk search strategy with a search strategy guided by high degree nodes. They observe that the latter search strategy performs better than the former. In particular, it returns a path of polylogarithmic length between the source and the target. Nevertheless, the search strategy performs in a polynomial number of steps due to loops in the search path.

In [13], the authors propose an original approach. Every node first publishes its data at every node along a random walk of length L. The search strategy then proceeds along a random walk of same length, and every node traversed by the walk starts partially flooding the network (the search is sent through every edge with probability $< q$, where q is the percolation threshold of the network). It is then shown that this search efficiently locates the data by setting $L \sim n^{1-2/\alpha}$ for $2 < \alpha < 3$. The authors also present heuristics reducing the amount of traffic induced by this strategy.

1.4   Using Clustering Properties

Just like the heterogeneous nature of peers is captured (in part) by the degree distribution, some cultural and social factors induce a clustered structure of the interest graph. For example, if a peer p_1 is interested in an object $\mathcal{O}$ held by another peer p_2, then it will probably be interested in other objects held by p_2. Moreover, p_1 will also probably be interested in objects held by other peers interested in $\mathcal{O}$. This can be summarized by the following facts:

- peers organize themselves in communities (dense subgraphs), and
- two peers which exchanged data are likely to exchange other data in the future.

Based on these observations, [15] proposed to enhance Gnutella with an *interest-based* structure in which a link (called *shortcut*) between peers that have exchanged an object is added on top of the Gnutella network. Simulations show that the shortcuts reduce the total load of the system by a factor 3 to 7. Hence, the clustered nature of the interest graph can be used to improve search strategies. Nevertheless, this search strategy remains based on flooding the network.

Other contributions pointed out that the clustered nature of the interest graph could be used to design efficient P2P systems [5, 8, 9, 15, 16], but no protocol has actually been proposed.

1.5   Our Contribution

The previous works surveyed before gave some evidence to the fact that the scale-free nature of the interest graph, as well as its clustered structure, are two basic statistical properties that can be used for improving search strategies in P2P systems. Nevertheless, these works all used one of these two properties only. In this paper, we show that using a combination of these two properties results in even better performances.

We present the QRE (pronounced *query*) protocol, in which the overlay network is nothing but the interest graph. Despite the fact that this graph is not structured, we present a simple search procedure that is not based on flooding, and does not require any information on the global topology of the overlay network. This search procedure is simple in the sense that it does not require sophisticated publish procedures. Moreover, because of the somewhat greedy maintenance of the overlay network, joining and leaving procedures are both very simple.

To evaluate the performance of our method, we have performed intensive simulations based on *real-world traces*. These simulations demonstrates that our search procedure locates objects in a *logarithmic* expected number of steps, which outperforms all previous solutions based on only one of the two basic statistical properties of the interest graph (scale-free nature or clustered structure).

2   The QRE Protocol

This section is devoted to the description of the QRE protocol. In order to illustrate the main features of QRE, we deliberately keep it as simple as possible. Moreover, keeping QRE simple enables evaluating the direct impact of our contribution, without mixing it with other optimizations.

Connections between peers in the overlay network of QRE are driven by the queries processed in the system: a peer is connected to the peers to which it has uploaded an object, or from which it has downloaded an object. Queries are routed by a *search* procedure (described below), and are of the form $\langle @, \mathcal{O}, k \rangle$ where @ is the physical address of the source peer initiating the query, $\mathcal{O}$ is the description of an object, and $k \geq 1$ is the number of different providers of $\mathcal{O}$ the source wants to get.

We assume that each peer in the system stores the objects that it provides, as well as a (compact) description of these objects in a local lookup table. We also assume that every peer stores a local copy of the lookup table of each of its neighbors. Regular communications between a peer p and its neighbors allows the system to support this facility. Finally, we assume that every peer knows the degree (number of neighbors) of each of its neighbors.

2.1   The Search Method

For routing a query $Q = \langle @, \mathcal{O}, k \rangle$, QRE essentially executes a (distributed) depth-first search (DFS) where the priority is given to highest degree nodes. More precisely, for every peer p that is receiving a query Q, if neither p nor any of its neighbors can positively answer to Q, then p forwards Q to its highest degree neighbor among the ones which have not yet processed Q; if there is none, then p sends the query Q back to the peer from which it received Q. Figure 1 summarizes this simple search procedure. The search procedure proceeds this way until k copies of $\mathcal{O}$ have been found.

To avoid loops in the search, every peer stores the list of queries Q that it has processed so far, as well as the identity of the neighbors to which Q has already been forwarded.

Note that QRE does not use hashing. As a consequence, it can support complex queries, as wild-cards searching, regular expression searching, or interval searching. Note also that the search is exhaustive in QRE.

> (1) **if** p has $\mathcal{O}$, **then** p sends $\mathcal{O}$ to @;
> (2) **else**
>   (2.1) **if** p has a neighbor p' that stores $\mathcal{O}$
>   **then** p forwards Q to p';
>   (2.2) **else if** all the neighbors of p have already received Q
>   **then** p sends Q back to the neighbor from which it received Q;
>   (2.3) **else** p forwards Q to its neighbor of maximum degree
>   among the ones that have not yet received Q.

Fig. 1. The search procedure in QRE for the query $Q = \langle @, \mathcal{O}, 1 \rangle$.

2.2   Dynamics of the System

In QRE, any successful search induces a modification of the connections between the peers: if p_1 receives a positive answer for a query Q from another peer p_2, then a link is set between p_1 and p_2, i.e., p_1 and p_2 exchange their addresses and their lookup tables. In addition, their neighbors are informed of the changes in their degrees. This way, the system maintains an overlay network which is nothing but the interest graph as defined before.

As in most previously proposed P2P systems, we assume that any peer aiming at joining the system knows an *entry point*, i.e. a peer already in the system, whose address is publicly available. We assume that a joining peer always wants to provide or to get an object. Therefore, a joining peer is always associated to an object. The join procedure is based on such an object, say $\mathcal{O}$: the joining peer sends a query for $\mathcal{O}$ and connects, as specified before, to the peer(s) that answer(s) to this query. If no peer returns a positive answer, then the joining peer connects directly to the entry point.

In QRE, when a peer wants to leave the system, it sends a leaving message to all its neighbors, and disconnects from the system. Any peer receiving a leaving message removes the sender from its lookup table, and informs its neighbors of its new degree. Note that QRE can also handle brutal departures of peers by periodically checking the presence of neighbors.

We stress the fact that QRE does not need the use of any underlying P2P system. The joining procedure is self-contained. The overlay graph grows from the entry point, based solely on the queries, and on their answers. The first peers will typically connect directly to the entry point (because the data they look for are not in the system). However, the new peers will eventually receive positive answers, and the overlay will then grow in a non-trivial manner.

3   Performance of QRE

There is currently no satisfactory model capturing the peers behavior accurately enough to enable a formal evaluation of our method (i.e., including simultaneously: clustering properties, scale-free properties, and the fact that two neighboring peers will probably exchange objects more than once). Because of this, we performed simulations on *real-world* traces, extracted from eDonkey [3], and described in detail in [7]. The trace upon which we performed our simulations is 2h 53mn long and involves $46,202$ peers. It contains the search requests of the users, but not the connections and disconnections.

3.1   Simulation Protocol

From the trace, we extracted a (chronological) list of tuples $(p_0^i, p_1^i, p_2^i, \ldots, p_{k_i}^i)$, each associated to a query Q^i. Peer p_0^i is the source of the ith query, k_i is the number of requested providers, and the p_j^i, $j = 1, 2, \ldots, k_i$, are the providers. We have considered $342,204$ queries of that type, involving $46,202$ nodes in total.

Our simulator proceeds with each tuple, step by step, as follows. Step i considers tuple Q^i, and simulates the behavior of QRE when dealing with a request Q where $p_1^i, p_2^i, \ldots, p_{k_i}^i$ are the providers of the object requested by p_0^i. In other words, we simulated the behavior of QRE, as described in Section 2, for a query $\langle p_0^i, \mathcal{O}_i, k_i \rangle$ where $p_1^i, p_2^i, \ldots, p_{k_i}^i$ are the peers currently holding $\mathcal{O}_i$.

If p_0^i is not yet in the network at step i, then the simulator performs the join procedure for p_0^i where the entry point is chosen uniformly at random among the peers currently in the network. Then, p_0^i connects to the entry point and launches the query. Its link to the entry point is removed when p_0^i receives an answer from a peer providing $\mathcal{O}_i$ (and thus connects to it).

3.2   Simulation Results

Figure 2 displays the degree distribution of the peers (after all the requests have been processed), i.e., for $k \geq 1$, the number $\delta(k)$ of peers with degree k. This distribution is heavy tailed (there are peers with large degree). However, $\delta(k)$ does not follow a strict power law. Nevertheless, we will see that the heavy tail is sufficient for our search strategy to perform efficiently. Importantly, the maximum degree is 690, but only 0.25% of the peers have a degree larger than 300. Conversely, 2/3 of the peers have a degree ≤ 20. The average degree is $47,9$. These characteristics prove that QRE scales well with the number of nodes.

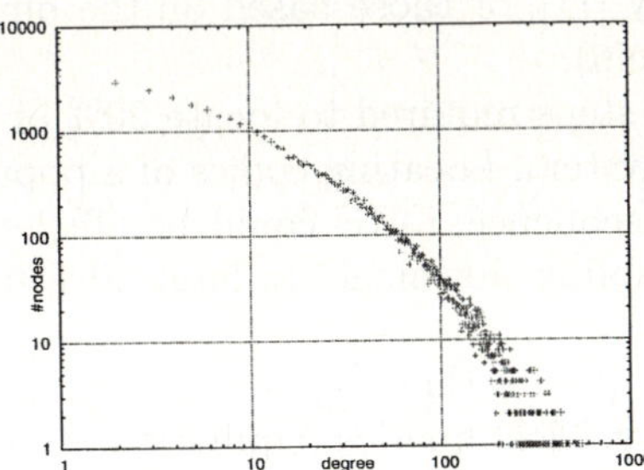

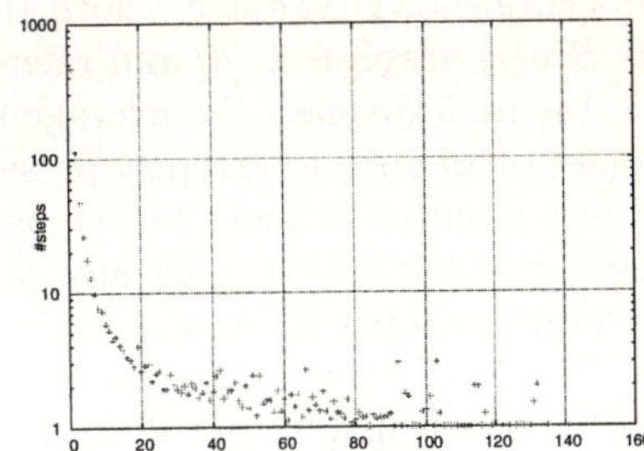

Fig. 2. Degree distribution in the overlay network of QRE.

Fig. 3. Impact of the number of providers on the search time.

Figure 3 displays the average number of steps required to locate one copy of an object, as a function of the total number of providers of this object. This number of steps decreases rapidly with the number of providers. In fact, locating an object that has at least seven copies in the network requires at most 10 steps on average, and a popular object $\mathcal{O}$ has, on average, a copy present on a node at distance at most 2 from a peer requesting $\mathcal{O}$. Importantly, $\mathcal{O}$ is not necessarily at distance at most 2 from any peer, but $\mathcal{O}$ is at distance at most 2 from any peer *interested in* $\mathcal{O}$. This demonstrates the existence of communities in the interest graph, captured and used by QRE.

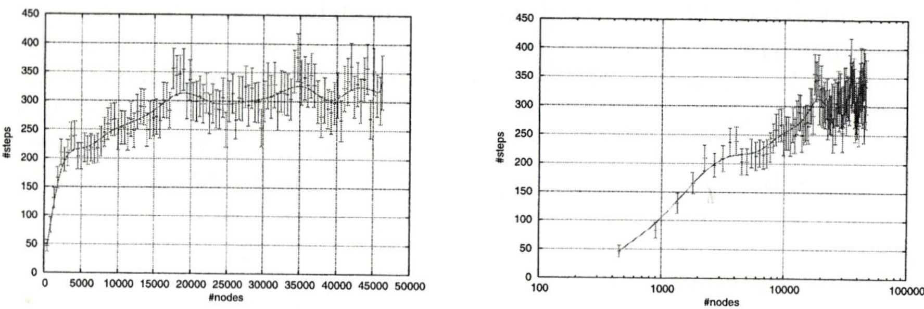

Fig. 4. Average number of steps to locate an object.

From Figure 3, rare objects are located by the search procedure of QRE after a relatively large number of steps. However, once this price has been paid by some peer, the next searches for the same object will require less and less steps while there are more and more copies of the object in the network, and while its providers are more and more connected. Figure 4 (left) displays the average number of steps $s(n)$ required to locate one copy of an object, as a function of the number n of peers in the system. Linear regression indicates that $s(n)$ scales linearly with the logarithm of the number n of peers in the system (see Figure 4 (right)). This is the most important experimental result in our contribution: *searching in QRE requires $O(\log n)$ number of steps on average*. In particular, the search procedure of QRE performs at least as well as the search procedures of DHTs like Chord [14], Viceroy [11], or those based on the binary *de Bruijn* graph (see [4] and references therein).

Figure 5 displays the average number of steps required to locate 20% of the copies of an object currently present in the system. Locating copies of a popular object requires at most ten steps. As a consequence, QRE could be efficiently used in combination with any protocol enabling downloading large files from several providers in parallel.

Finally, Figure 6 displays the number of queries that required k steps to be performed, which is well fitted by a power law. Most queries require few steps to be performed, and only very few queries require lot of steps (this corresponds to very rare data, for which it is necessary to search a large portion of the network). Typically, a TTL of 100 steps would enable most queries to be satisfied.

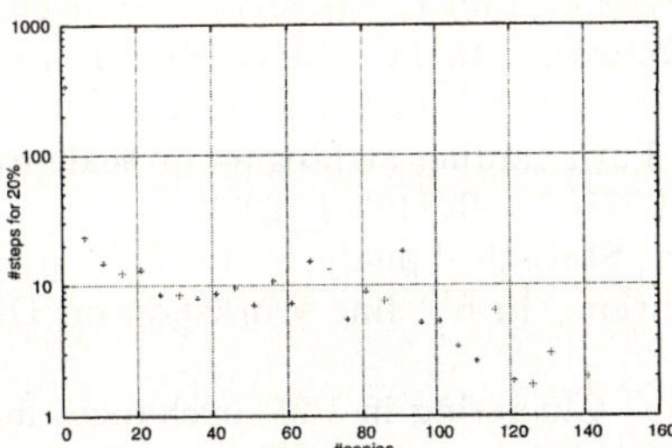

Fig. 5. Search time for 20% of the providers

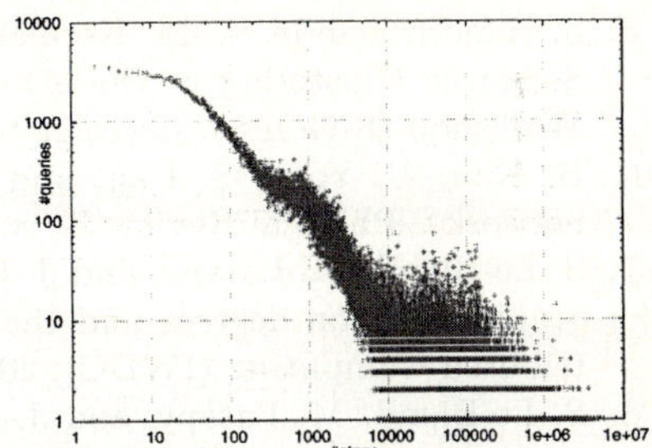

Fig. 6. Distribution of the number of hops to find an object.

4   Conclusion

In this paper, our main objective was to push further the idea of using the properties of the peer interest for the design of efficient search strategies in P2P systems. We indeed believe that these properties are among the key factors to be considered for the design of efficient and fully decentralized P2P systems. To support this belief, we demonstrate that the interest graph can be used as an overlay network supporting very simple procedures for searching, joining, and leaving the system. The main properties of the interest graph are (1) its clustered structure, and (2) the heterogeneity of the node degrees. Our search strategy uses these two properties. It locates objects in a logarithmic expected number of steps, without flooding, nor using any sophisticated routing or publish procedures.

Probably, more subtile and more efficient search strategies could be defined on the interests graph, using other properties of the graph, or combining our approach with others. In particular, it is unclear whether the DFS algorithm selecting high degree nodes first is the most appropriate search strategy for the interest graph. Further investigations are also required to measure the impact of limiting the maximum degree of the peers, as well as other issues like the the load of the high-degree nodes, the robustness of the system.

References

1. I. Abraham, D. Malkhi, and O. Dobzinski. LAND: Stretch $(1 + \epsilon)$ locality-aware networks for DHTs. In 15th ACM-SIAM Symp. on Discrete Algorithms (SODA), pages 550-559, 2004.
2. L. Adamic, R. Lukose, A. Puniyani, and B. Huberman. Search in power law networks. Physical Review E, vol. 64, pages 46135-46143, 2001.
3. eDonkey: *www.edonkey2000.com/*
4. P. Fraigniaud and P. Gauron. An overview of the content addressable network D2B. Brief announcement at the 22nd ACM Symp. on Principles of Distributed Computing (PODC), page 151, 2003.

5. S. Handurukande, A.-M. Kermarrec, F. Le-Fessant, and L. Massoulié. Exploiting Semantic Clustering in the eDonkey P2P Network. In 11th SIGOPS European Workshop (SIGOPS), pages 109-114, 2004.

6. B. Kim, C. Yoon, S. Han, and H. Jeong. Path finding strategies in scale-free networks. Physical Review E, vol. 65, pages 027103-1–027103-4, 2002.

7. S. Le-Blond, M. Latapy, and J.-L. Guillaume. Statistical analysis of a P2P query graph based on degrees and their time evolution. In 6th Int. Workshop on Distributed Computing (IWDC), 2004.

8. S. Le-Blond, M. Latapy, and J.-L. Guillaume. Clustering in P2P exchanges and consequences on performances. In 4th Int. Workshop on Peer-To-Peer Systems (IPTPS), 2005.

9. F. Le-Fessant, S. Handurukande, A.-M. Kermarrec, and L. Massoulié. Clustering in Peer-to-Peer File Sharing Workloads. In 3rd Int. Workshop on Peer-to-Peer Systems (IPTPS), 2004.

10. Q. Lv, P. Cao, E. Cohen, K. Li, S. Shenker. Search and replication in unstructured peer-to-peer networks. In 6th Int. Conf. on Supercomputing, pages 84-95, 2002.

11. D. Malkhi, M. Naor, and D. Ratajczak. Viceroy: a scalable and dynamic lookup network. In 21st Symp. on Principles of Distributed Computing (PODC), 2002.

12. S. Ratnasamy and P. Francis and M. Handley and R. Karp and S. Shenker. A scalable content-addressable network. In SIGCOMM, pages 161-172, 2001.

13. N. Sarshar, P. Boykin, and V. Roychowdhury. Percolation search in power law networks: making unstructured peer-to-peer networks scalable. In 4th International Conference on Peer-to-Peer Computing, pages 2-9, 2004.

14. I. Stoica, R. Morris, D. Karger, M. Kaashoek, and H. Balakrishnan. Chord: a scalable peer-to-peer lookup service for Internet applications. In SIGCOMM, pages 149-160, 2001.

15. K. Sripanidkulchai, B. Maggs, and H. Zhang. Efficient content location using interest-based locality in peer-to-peer systems. In INFOCOM, pages 2166-2176, 2003.

16. S. Voulgaris, A.-M. Kermarrec, L. Massoulié, and M. van Steen. Exploiting Semantic Proximity in Peer-to-peer Content Searching. In 10th IEEE Int. Workshop on Future Trends in Distributed Computing Systems (FTDCS), pages 238-243, 2004.

Pastis: A Highly-Scalable Multi-user Peer-to-Peer File System

Jean-Michel Busca[1], Fabio Picconi[2], and Pierre Sens[2]

[1] INRIA Rocquencourt
Le Chesnay, France
jean-michel.busca@inria.fr
[2] LIP6, Université Paris 6 - CNRS
Paris, France
{fabio.picconi,pierre.sens}@lip6.fr

Abstract. We introduce Pastis, a completely decentralized multi-user read-write peer-to-peer file system. In Pastis every file is described by a modifiable inode-like structure which contains the addresses of the immutable blocks in which the file contents are stored. All data are stored using the Past distributed hash table (DHT), which we have modified in order to reduce the number of network messages it generates, thus optimizing replica retrieval.

Pastis' design is simple compared to other existing systems, as it does not require complex algorithms like Byzantine-fault tolerant (BFT) replication or a central administrative authority. It is also highly scalable in terms of the number of network nodes and users sharing a given file or portion of the file system. Furthermore, Pastis takes advantage of the fault tolerance and good locality properties of its underlying storage layer, the Past DHT.

We have developed a prototype based on the FreePastry open-source implementation of the Past DHT. We have used this prototype to evaluate several characteristics of our file system design. Supporting the close-to-open consistency model, plus a variant of the read-your-writes model, our prototype shows that Pastis is between 1.4 to 1.8 times slower than NFS. In comparison, Ivy and Oceanstore are between two to three times slower than NFS.

1 Introduction

Although many peer-to-peer file systems have been proposed by different research groups during the last few years [1–5], only a handful are designed to scale to hundreds of thousands of nodes and to offer read-write access to a large community of users. Moreover, very few prototypes of these large-scale multi-writer systems exist to this date, and the available experimental data are still very limited.

This research was conducted as part of the GDS project (http://www.irisa.fr/GDS) of the ACI MD program, supported by the french Ministry of Research.

J.C. Cunha and P.D. Medeiros (Eds.): Euro-Par 2005, LNCS 3648, pp. 1173–1182, 2005.
© Springer-Verlag Berlin Heidelberg 2005

One of the reasons for this is that, as the system grows to a very large scale, allowing updates to be made anywhere anytime while maintaining consistency, ensuring security, and achieving good performances is not an easy task. Read-only systems, such as CFS [4], are much easier to design since the time interval between meta-data updates is expected to be relatively high. This allows the extensive use of caching, since cached data are either seldom invalidated or kept until they expire. Security in a read-only system is also quite simple to implement. Digitally signing a single root block with the administrator's private key and using one-way hash functions allows clients to verify the integrity and authenticity of all file system data. Finally, consistency is hardly a problem since only a single user, the administrator, can modify the file system.

Multi-writer designs must face a number of issues not found in read-only systems, such as maintaining consistency between replicas, enforcing access control, guaranteeing that update requests are authenticated and correctly processed, and dealing with conflicting updates.

The Ivy system [3], for instance, stores all file system data in a set of logs using the DHash distributed hash table. Each user has its own log to which he appends his own updates. This eliminates the need of a central serialization point, and provides high security against attacks, including attacks from users who turn out to be malicious, but also limits the number of users (more users means more logs to traverse when reading a file).

Oceanstore [2] uses a completely different approach to handling updates by introducing a centralization point called the primary tier. This set of replicas serialize updates using a Byzantine-fault tolerant (BFT) [6] algorithm. However, BFT is expensive, and primary tier nodes must be highly resilient and well-connected. Although Oceanstore has many features, the system is quite complex, and its centralized design may not be suitable for a community of cooperative users.

With the aim of finding a solution to the shortcomings of these systems we have designed Pastis, a highly-scalable, completely decentralized multi-writer peer-to-peer file system. For every file or directory Pastis keeps an inode object in which the file's metadata are stored. As in the Unix File System, inodes also contain a list of pointers to the data blocks in which the file or directory contents are stored. All blocks are stored using the Past distributed hash table, thus benefiting from the locality properties of both Past and Pastry [7].

Our system is completely decentralized. Security is achieved by signing inodes before inserting them into the Past network. Each inode is stored in a special block called User Certificate Block, or UCB. Data blocks are stored in immutable blocks, called Content-Hash Blocks, the integrity of which can easily be verified. All blocks are replicated in order to ensure fault tolerance and to reduce the impact of network latency.

Finally, we have implemented a prototype written in Java. It runs the Free-Pastry [8] open source implementation of Past and Pastry. We have modified the original FreePastry's implementation, generalizing the Past *lookup* call to efficiently support one of the consistency models provided by Pastis.

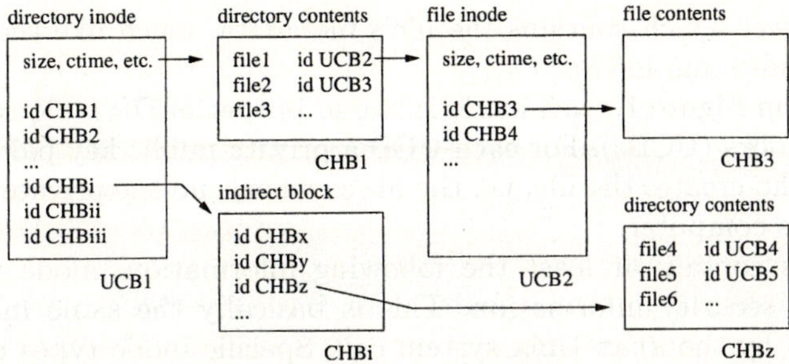

Fig. 1. File System Structure.

The remaining part of this paper is as follows: section 2 introduces Past and Pastry. Section 3 presents the design of our system in more detail. Section 4 presents our prototype, and evaluate its performance. Finally, section 5 concludes this paper. More details on Pastis can be found in [9].

2 Pastry and Past

Pastry [7] is a peer-to-peer key-based routing substrate designed to support a very large number of nodes. In a Pastry network, each node has a unique fixed-length node identifier (nodeid), whose address space can be thought of as a circle ranging from 0 to $2^{160} - 1$. Routing is performed with keys: a message is routed to the node whose nodeid is numerically closest to the specified key.

Pastry's routing algorithm is derived from the work by Plaxton [10]. This type of prefix-based routing can achieve very low hop counts, usually $O(logN)$, where N is the number of overlay nodes. However, one important feature of Pastry is that it takes into account network locality to optimize overlay routes.

Past [11] is a highly-scalable peer-to-peer storage service which provides a distributed hash table abstraction. Past uses Pastry to route messages between Past nodes, thus benefits from Pastry's properties, i.e. scalability, self-organisation, locality, etc. In addition, Past ensures high data availability through the use of replication: it implements a lazy replication protocol that maintains a constant number of replicas of each stored block, as nodes join and leave the network.

Although not detailed in this paper, we have generalized the Past *lookup* call so that it now supports one or more search predicates. This allows us to efficiently retrieve an inode replica whose version stamp is not older than a given value. We use this feature in the read-your-write consistency model, as described in 3.2.

3 Design

We begin the description of our file system by presenting how file system data are stored on the network. The data structures used in our design are similar to those of the common Unix file system (UFS). For each file the system stores an inode-like object which contains the file's metadata, much like the information found in a traditional inode.

As shown in Figure 1, each inode is stored in special DHT blocks called User Certificate Blocks (UCBs). For each UCB a private-public key pair is generated by the user who creates the file, i.e. the file's owner, and the private key is stored in the owner's computer.

All inodes contain at least the following information: inode type, file attributes, and security information. This is basically the same information as that returned by the `stat` Unix system call. Specific inode types contain additional fields which are only necessary for the corresponding file type. Regular file and directory inodes, for instance, contain a list of pointers to other blocks in which the file or directory contents are stored. Symbolic link inodes, in turn, contain only the link's destination path.

File and directory contents are stored in fixed-size DHT blocks called Content-Hash Blocks (CHBs). The DHT address of each CHB is obtained from the hash of the block's contents, and is stored within the file's inode block pointer table. As with UFS inodes, we use single, double, and triple-indirect blocks to limit the size of the inode's block pointer table.

In order to optimize directory operations, each directory inode holds a small number of directory entries in the inode itself. Therefore, clients accessing directories that contain only a few files need not retrieve any CHBs. Retrieving or inserting the UCB in which the inode is stored may be sufficient, thereby reducing operation time and increasing performance.

Our design is similar to that of CFS [4], the main difference lying in the use of modifiable blocks (UCBs) to store inodes, thus eliminating the cascade effect of CFS when updating an inode.

3.1   Updates and Conflicts

Modifying a file or directory in Pastis requires updating the UCB in which its inode is stored, but it also usually involves the insertion of new CHBs. For instance, writing to a file will usually take the following steps:

1. *Fetch the file inode (UCB) from the network*
2. *Fetch the corresponding data block(s) (CHB)*
3. *Modify the data block(s) and insert them into the DHT*
4. *Modify the inode's pointer table with the addresses of the new data block(s), and reinsert the inode into the DHT*

Note that modifying the contents of data blocks changes their DHT keys (which are obtained by hashing the block contents). Thus the need for updating the inode's pointer table.

If two or more clients update an inode concurrently, then a conflict will most probably occur. In our current design, the conflict-resolution mechanism works as follows: each inode update is uniquely identified by a version stamp, consisting of the version number of the update and the unique id of the user who issued it.

We define a total order on version stamps by first comparing version numbers, and if they are equal, by comparing user ids.

Each time a user commits an update to a file or directory, the inode's version number is incremented, and the new inode is inserted into the Past DHT. During insertion, the Past client sends the new inode to all replicas. Each replica checks that the new inode's version number is greater than the existing one before the replica is overwritten. If this check fails, the update is aborted.

3.2   Consistency

Our system currently supports two consistency models: close-to-open and a variant of the read-your-writes guarantee.

Close-to-open consistency [12] is a relaxed consistency model widely employed in distributed file systems such as AFS and NFS. In this model the *open* and *close* operations determine the moment in which files are read from and written to the network. The advantage of using close-to-open consistency is that local write operations need not be propagated to the network until the file is closed. Similarly, once a file has been opened, the local client need not check whether the file has been modified by other remote clients, an operation that would require accessing the network. In other words, the local client can cache the file's contents while it is opened, and keep this cache until the file is closed.

In our system, the close-to-open model is implemented by retrieving the latest inode from network when the file is opened and keeping a cached copy until the file is closed. Any following read requests are satisfied using the cached inode. New CHBs are also buffered locally instead of being inserted immediately into the DHT. Finally, when the file is closed all cached data are flushed to the network and removed from the local buffer.

Note that this scheme works because the immutable data blocks (CHBs) that store the contents of each different version of a given file (a new version appears each time the file is closed) are never removed from the network. If they were, then the data blocks pointed to by a cached inode could be no longer valid. Alternatively, a complex garbage collection mechanism would have to be employed to safely remove unused immutable block from the DHT.

The close-to-open consistency model may be stronger than what many applications actually need. In fact, applications which access files that are seldom shared, or that are not shared at all could benefit from a further relaxed consistency. For these applications we have implemented another consistency model, based on the *read-your-writes* session guarantee, originally introduced by Bayou [13]. Our read-your-writes model guarantees that when an application opens a file, the version of the file that it reads is not older than the version it previously wrote. Once the file is opened, file updates are performed as in the close-to-open model.

The key advantage of the read-your-writes model is that it requires fewer accesses to the DHT than the close-to-open model and thus yields lower response time. Because of possible rollback attacks and Past's lazy replication mechanism, the latter model requires all inode replicas to be retrieved to ensure that the latest

version is used. By contrast, in the read-your-writes model, it suffices to retrieve
at least one inode replica whose version stamp is not less than a given value.
Since all of the replicas are written when closing a file, all replicas normally
satisfy the search predicate, including the one that is closest to the application
opening the file. We leverage Pastry and Past's locality property and fetch this
replica in just one lookup path, using our generalized *lookup* call, thus achieving
the lowest possible latency.

Note that in practice, it is highly likely that a file open will retrieve the latest
version of the file, as in the close-to-open model. The only case the retrieved inode
is not the latest one is when the node queried during lookup is acting maliciously,
or has not been updated yet following a recent change in the set of nodes hosting
the replicas of the inode.

3.3   Security

Pastis ensures write access control and data integrity through the use of standard cryptographic techniques and ACL certificates. Pastis does not currently
provide read access control, but users may encrypt files' contents to ensure data
confidentiality if needed.

Write access control and data integrity are ensured as follows. The owner
of a file issues a write certificate for every user he allows to write to the file.
When a user modifies the file, he must properly sign the new version of the
inode and provide his write certificate along with the inode. The certificate and
the inode signature are checked by DHT nodes before they commit the update.
A user performs the same checks when reading the file in order to assert the
integrity of the file's contents. These mechanisms along with the use of replication
make Pastis tolerant to Byzantine behaviour of DHT nodes and rollback attacks,
provided at least one replica is not faulty. However, unlike Ivy, our security model
assumes that all users allowed to write to a given file trust one another regarding
update operations on that file. Because of space limitations the details of Pastis
security mechanisms will not be developed here.

4   Prototype and Evaluation

The latest version of our prototype uses FreePastry 1.3.2 and runs on any platform that supports the Java VM 5.0.

We developed a discrete event simulator, LS^3 [16], in order to conduct experiments on large-scale networks. LS^3 simulates such networks by randomly
locating nodes on a sphere and deriving network latency from the distance between source and destination nodes. The maximum network latency corresponds
to two diametrically opposed points on the sphere.

In order to evaluate the performance of our prototype we use the Andrew
Benchmark [12], which consists of five phases: (1) create directories, (2) copy files,
(3) read file attributes, (4) read file contents, and (5) run a `make` command. The
source directory we use as input to the benchmark contains two sub-directories

and 26 C source and header files, for a total size of 190 Kbytes. The benchmark executes on top of a node running a Pastis/Past/Pastry stack, which in turn communicates with other Past nodes to store and retrieve blocks. These communications can take place either through a real network, or be confined to the local Java VM when using the LS^3 simulator.

Real experiments are run on Pentiums 4 2.4 GHz with 512 Mbytes of RAM, running Linux 2.4.x. In order to simulate the latency of inter-node communications, we use a DummyNet [14] router, running FreeBSD 4.5. Experiments using the LS^3 simulator are run on a Pentium 4 1.8 GHz with 2 Gbytes of RAM.

It is important to notice that all layers from the Pastry layer upwards are unaware of whether they are executing on top a simulated or a real environment. In other words, the executed code corresponding to the DHT and Pastis layers is the same in both cases.

4.1   Network Size

This experiment evaluates Pastis' scalability with respect to the number of nodes in the network. We run the Andrew Benchmark on a simulated network of increasing size, with a constant maximum network latency of 300ms. We use the close-to-open consistency model, and Past's replication is disabled.

Figure 2 shows the total and per-phase execution time of the benchmark for network sizes ranging from 16 to 32768 nodes. We observe that the total execution time increases only by 13.5% between 16 (311 s) and 32768 nodes (353 s). This good result is mainly due to Pastry's efficient routing algorithm. This experiment confirms, however, that Pastis does not introduce any flaw in the overall design and preserves Pastry and Past's scalability over a wide range of network sizes.

4.2   Concurrent Clients

In this experiment we evaluate the performance impact of running multiple file system clients concurrently in a real environment. We run from one up to 16

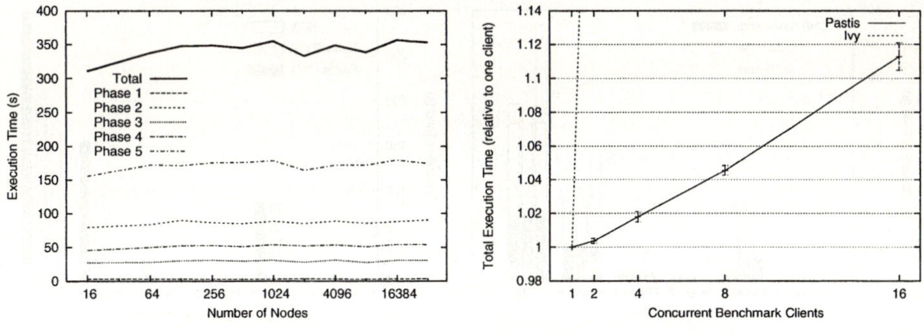

Fig. 2. Network size (left), Concurrent clients (right).

concurrent Andrew Benchmark clients, each client writing to a different directory so that no conflicts are generated. In all cases we use a Past network of 16 nodes, with a block replication factor of 4 and a 100 ms emulated inter-node delay. The consistency model is close-to-open.

Figure 2 shows that the total execution time for a single client is 311 seconds. As the number of clients increases, execution time appears to grow linearly, with only a 2% increase for 4 concurrent clients. This suggests that Pastis scales well in terms of concurrent clients. In comparison, according to [3] an equivalent test performed on the Ivy file system shows a 70% increase when going from 1 to 4 concurrent clients. This is not surprising since having multiple logs (one per participant) forces Ivy clients to traverse all the logs that have been modified, even if the records appended by the other users do not concern the files accessed by the local user. In Pastis, running 16 concurrent client produces only a 11.3% increase compared to a single client, which is very low considering that every node is running a benchmark client.

4.3   Consistency Models

This experiment compares the performance of the two consistency models that Pastis implements. We run the Andrew Benchmark on a simulated network of 32768 nodes, with a maximum network latency of 300 ms and a block replication factor of 16. We perform three test runs. The first run uses the close-to-open (CTO) consistency model, and the second uses the read-your-writes (RYW) model. The third run also uses the read-your-write model, but this time with 10% of the closest inode replicas being stale.

Figure 3 shows the total and per-phase execution time for each of these three runs. The left, middle and right bars represent the CTO, RYW and RYW with failures runs, respectively. Execution time is broken down into three categories: the lower part of each bar represents the cumulative CHB read and write time, the middle part represents the UCB write time and the upper part represents the UCB read time. We observe that in the close-to-open model, almost 40% of

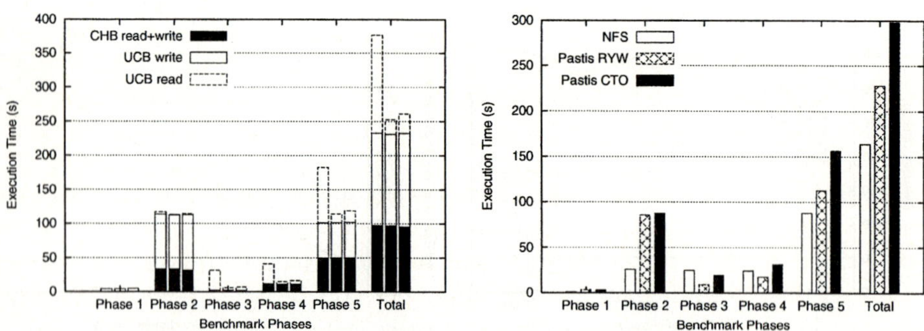

Fig. 3. Consistency models (left), NFS comparison (right).

the overall time is spent in UCB reads. This is because all of the live replicas of a given UCB must be retrieved to determine its latest version, as required by the consistency model. As expected, the read-your-writes model yield better performance than the close-to-open model by reducing UCB read time. We observe that while CHB read-write time and UCB write time remain the same as in the close-to-open model, UCB read time decreases by 85% (144 s for close-to-open, 21 s for read-your-writes), yielding a 33% increase in overall performance. Finally, the results also show that even in the presence of 10% stale UCB replicas the overall time increases by only 3% in the read-yrou-write model.

4.4   NFS Comparison

In this test we compared Pastis' performance to that of NFS v3. This allows us to make an indirect comparison to other peer-to-peer file systems for which a comparison with NFS has been performed [2, 3]. First we run a single Andrew Benchmark client on a real network of 16 machines, each running an instance of Past, with a replication factor of 4. We emulate an inter-node latency of 100 ms using the DummyNet router (a ping between any two machines yields a 200 ms round-trip time). We then run an Andrew Benchmark client on an NFS client accessing a single NFS server, and also emulate a 100 ms latency between client and server (a RPC therefore takes 200 ms).

As shown in Figure 3, total execution time is less than twice that of NFS when Pastis consistency model is set to close-to-open. With the read-your-writes model, Pastis is only 40% slower than NFS. In comparison, other peer-to-peer file systems [2, 3] are between two to three times slower than NFS.

5   Conclusion and Future Work

We have implemented a multi-user read-write peer-to-peer system with good locality and scalability properties. The use of Pastry and a modified version of Past is crucial to achieve a high level of performance, a difficult task since large-scale systems are particularly subject to network latencies.

Another equally important factor is the choice of the consistency model, as strict consistency can impair performance significantly. Therefore, a peer-to-peer file system should offer a range of different degrees of consistency, thus allowing applications to choose between various levels of consistency and performance. Pastis currently provides two relaxed consistency models and future work will involve envisaging and adding new models. Ongoing work focuses on providing support for concurrency control, through the implementation of file locks and exclusive file creation, for application requiring strict consistency.

Finally, our prototype evaluation based on simulation and real execution suggests that Pastis is only 1.4 to 1.8 times slower than NFS. However, our results are still preliminary and must be corroborated by further evaluations.

References

1. A. Adya, W. J. Bolosky, M. Castro, G. Cermak, R. Chaiken, J. R. Douceur, J. Howell, J. R. Lorch, M. Theimer, and R. P. Wattenhofer. FARSITE: Federated, Available, and Reliable Storage for an Incompletely Trusted Environment. In *5th Symposium on Operating Systems Design and Implementation (OSDI 2002)*, Boston, MA, December 2002.
2. J. Kubiatowicz, D. Bindel, Y. Chen, S. Czerwinski, P. Eaton, D. Geels, R. Gummadi, S. Rhea, H. Weatherspoon, W. Weimer, C. Wells, and B. Zhao. Oceanstore: An architecture for global-scale persistent store. In *Proc. ASPLOS'2000*, Cambridge, MA, November 2000.
3. A. Muthitacharoen, R. Morris, T. M. Gil, and B. Chen. Ivy: A Read/Write Peer-to-Peer File System. In *Proceedings of 5th Symposium on Operating Systems Design and Implementation* (OSDI 2002).
4. F. Dabek, M. F. Kaashoek, D. Karger, R. Morris, and I. Stoica. Wide-area cooperative storage with CFS. In *Proceedings of the 18th ACM Symposium on Operating Systems Principles* (SOSP '01), Chateau Lake Louise, Banff, Canada, Oct. 2001.
5. Y. Saito, C. Karamanolis, M. Karlsson, and M. Mahalingam. Taming aggressive replication in the pangaea wide-area file system. In *5th Symp. on Op. Sys. Design and Implementation (OSDI 2002)*, Boston, MA, USA, December 2002.
6. M. Castro and B. Liskov. Practical Byzantine fault tolerance. In *Proc. of USENIX Symposium on Operating Systems Design and Implementation* (OSDI 1999).
7. A. Rowstron and P. Druschel. Pastry: Scalable, distributed object location and routing orlarge-scale peer-to-peer systems. In *Proc. IFIP/ACM Middleware 2001*, Heidelberg, Germany, Nov. 2001.
8. FreePastry. `http://freepastry.rice.edu/`
9. J-M. Busca, F. Picconi, P. Sens. Pastis: a highly-scalable multi-user peer-to-peer file system. INRIA Technical Report 5288. August 2004.
10. C. G. Plaxton, R. Rajaraman, and A. W. Richa. Accessing nearby copies of replicated objects in a distributed environment. In *Proceedings of ACM SPAA. ACM*, June 1997.
11. A. Rowstron and P. Druschel. Storage management and caching in Past, a large-scale, persistent peer-to-peer storage utility. In *Proc. of the ACM Symposium on Operating System Principles (SOSP 2001)*, October 2001.
12. J. Howard, M. Kazar, S. Menees, D. Nichols, M. Satyanarayanan, R. Sidebotham, and M. West. Scale and performance in a distributed file system. In *ACM Transactions on Computer Systems*, volume 6, February 1988.
13. A. Demers, K. Petersen, M. Spreitzer, D. Terry, M. Theimer, and B.Welch. The Bayou architecture: Support for data sharing among mobile users. In *Proc. of IEEE Workshop on Mobile Computing Systems & Applications*, Dec. 1994.
14. L. Rizzo. Dummynet and Forward Error Correction. In *Proc. of the 1998 USENIX Annual Technical Conf.*, June 1998.
15. Pastis. `http://regal.lip6.fr/projects/pastis`
16. LS³. `http://regal.lip6.fr/projects/pastis/ls3`

AGNO: An Adaptive Group Communication Scheme for Unstructured P2P Networks[*]

Dimitrios Tsoumakos and Nick Roussopoulos

Department of Computer Science
University of Maryland, College Park
{dtsouma,nick}@cs.umd.edu

Abstract. We present the *Adaptive Group Notification (AGNO)* scheme for efficiently contacting large peer populations in unstructured Peer-to-Peer networks. *AGNO* defines a novel implicit approach towards group membership by monitoring demand for content as this is expressed through lookup operations. Utilizing search indices, together with a small number of soft-state shortcuts, *AGNO* achieves effective and bandwidth-efficient content dissemination, without the cost and restrictions of a membership protocol or a DHT. Our method achieves high-success content transmission at a cost at least two times smaller than proposed techniques for unstructured networks.

1 Introduction

A multicast transmission is defined as the dissemination of information to several hosts within a network. These hosts are interested in receiving the same content from an authority node (such as a web server) and naturally form a group. The lack of deployment of multicast communication in the IP layer has led to the development of various application-level multicast protocols, in which the end hosts are responsible for implementing this functionality. One-to-many communication is a very useful mechanism for a variety of network applications (e.g., [1–6]).

A number of methods have been proposed to implement multicast communication utilizing some popular P2P overlays, e.g., [3, 7–9]. Nevertheless, these approaches take advantage of the structure that DHTs (distributed hash tables) provide. In many realistic scenarios, the topology cannot be controlled and thus DHTs cannot be used (e.g., ad-hoc networks or existing large-scale unstructured overlays). Other approaches require frequent communication overhead between group members and explicit membership protocols. These schemes often prove unsuitable because of the generated traffic for large and dynamically changing group populations.

Today, many popular P2P applications operate on *unstructured* networks, where peers have a local only knowledge of a network in which nodes enter and leave frequently in an ad-hoc manner. For such systems, contacting large numbers of nodes is implemented by either broadcast-based schemes (e.g., [10, 11]), or *gossip*-based ap-

* This material is based upon work supported by, or in part by, the U.S. Army Research Laboratory and the U.S. Army Research Office under contract/grant number DAAD19-01-1-0494.

J.C. Cunha and P.D. Medeiros (Eds.): Euro-Par 2005, LNCS 3648, pp. 1183–1193, 2005.
© Springer-Verlag Berlin Heidelberg 2005

proaches, e.g., [4–6]. Both produce large numbers of messages by contacting many hosts inside the network[1].

In this paper, we present the *Adaptive Group Notification (AGNO)* method. *AGNO* combines the utilization of state accumulated during the search process together with probabilistically stored requester addresses. Our method builds its knowledge by only monitoring the independently conducted lookups and does not require any control or membership message exchange. Finally, we show that *AGNO* achieves an efficient and adaptive performance in a variety of environments and group sizes.

2   The AGNO Protocol

2.1   Our Framework and Overview of *APS*

We assume a pure P2P model, with no imposed hierarchy over the set of participating peers. All of them may equally serve and make requests for various objects using unique object IDs. Each peer retains its own collection which is locally maintained. Ignoring physical connectivity and topology from our talk, we assume that peers are aware of their one-hop neighbors in the overlay. A multicast transmission in this setting (also referred to as the *notification* or *push* phase hereafter) is initiated by an object holder (or *server* node) and its target is to reach as many group members (or *requester* nodes) as possible with the least amount of messages over the overlay. The focus of this work is to describe an efficient mechanism for such transmissions and not to define their content.

For our motivating example, we assume a distributed, unstructured P2P system, where peers share and request replicated resources. Objects are assumed to be updated often, like results of a live sports meeting, weather maps, stock quotes, real time aggregated statistics, etc. There exist some nodes (similar to the web servers or mirror sites in the Internet) that provide with fresh content, but their connectivity or availability varies, as happens with all other network nodes. Peers that are interested in retrieving the newest version of the content conduct searches for it in order to locate a fresh or closer replica. In this environment, interest in a specific object is tied to the lookups generated for it. We argue for a push-based approach, where content holders can forward notifications (or other object-specific information) towards the interested hosts. Our assumption is that peers which have recently searched or retrieved an object would also be interested in receiving such content. For example, we can assume that a user frequently asking for the temperature of an area would like to be informed about an update or another object-related notification.

It is important to note here that peers still search and retrieve objects in a distributed manner, using one of the available lookup protocols for unstructured systems. The notification itself is not necessary to be directly related to the object: A severe weather alert to be effective in the next 3 hours is not related to the current area temperature object. Similarly, a notification may state that node(s) hosting security-related software will become unavailable soon (still not related to the objects shared). A change in quote prices or breaking news inside a web page are, on the other hand, directly linked to the object's

---

[1] Due to severe space constraints, a detailed analysis of the related work appears in the accompanying technical report [12].

content. Group communication requires a considerable amount of bandwidth, therefore content providers should wisely assess the importance of various updates/notifications and choose to push those that would be the most beneficial.

The *Adaptive Probabilistic Search (APS)* [13] method is a search scheme for unstructured overlays. Each node keeps a local index consisting of one entry for each of its neighbors. These index values reflect the relative probability of a node's neighbor to be chosen as the next hop in a future request for a specific object. Searching is based on the simultaneous deployment of k walkers and probabilistic forwarding. Indices are updated along the walkers' paths according to object discovery or failure. *APS* has proved to be an adaptive, bandwidth-efficient scheme which also provides for robust behavior in dynamic environments [14]. All nodes participating in a lookup benefit from the process, while others inherit search knowledge by proximity. Index values are refined as more searches take place, enabling the network to build a useful soft-state.

2.2   AGNO Protocol Description

The main idea behind *AGNO* is to couple search knowledge with the information necessary to contact interested peers. Thus, the equivalent of group membership is demand, realized through searches and object sharing that are *independently* conducted by peers. The granularity can be as coarse or fine-grained as the application requires, but for the remainder of this paper we assume that groups are formed on a per-object basis.

After each search with the *APS* scheme, peers accumulate knowledge about the relative success of a search through their neighbors. Intuitively, after a few lookup operations have been conducted, overlay paths with high index values are the ones most frequently used to connect requesters to object holders. In *AGNO*, nodes utilize those indices in order to forward group messages towards possible group members during the push phase. Note here that, although the *APS* method is used as a means to provide with the soft state, our approach can be used with other search mechanisms, as long as they support a similar demand incentive.

APS keeps an index value for each neighbor. For *AGNO*, each peer P also maintains the index values that P's neighbors hold relative to P. If $A \rightarrow B$ denotes the index value stored at node A concerning neighbor B for a particular object, then peer P must know $X \rightarrow P$, for each neighbor X. These values can be made known to P either implicitly or explicitly during the search phase: In the first case, peer P can infer the value of $X \rightarrow P$ if it knows about the index update process used and the initial value. In the explicit approach, whenever a search is conducted and X forwards to P, it piggybacks $X \rightarrow P$. We call these new stored values the *reverse indices*, to distinguish them from the indices used by *APS* in searches. For the rest of our discussion, we assume that the explicit approach is used. Reverse indices are used by nodes during the push phase to select which neighbors will receive the notification, but they are not the only state that our method utilizes.

During searches, intermediate nodes decide with probability p_r whether to store the requester's ID or not. For a search path h hops long, the (ID, address) pair of the requester will be stored on hp_r peers on average. With this scheme, we create a number of shortcuts called *backpointers* along the search paths which point to group members. Each peer can individually decide on the maximum number of backpointers stored.

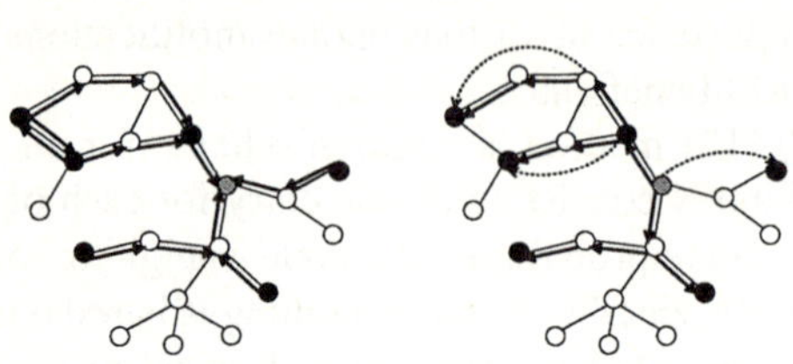

Fig. 1. Search for an object stored at the gray node and the push phase from this node towards the requesters (black nodes).

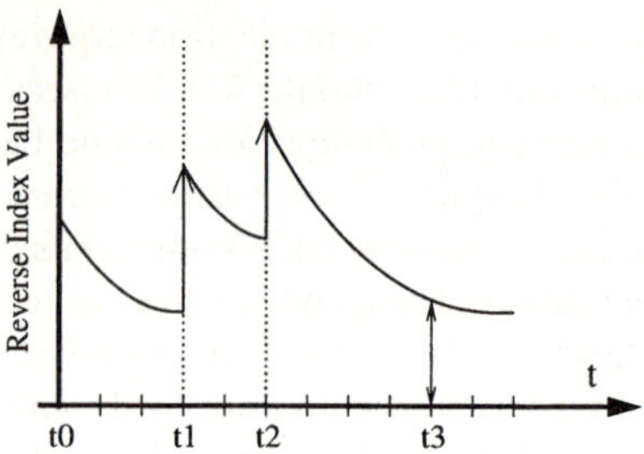

Fig. 2. Example of the reverse index value update process.

For simplicity, we assume that all nodes can store a maximum of c backpointer values. Backpointers are soft-state that gets invalidated after some amount of time.

In the push phase, a peer that receives the notification message forwards it to its neighbors consulting the respective reverse index values. Moreover, a peer forwards directly to each of its valid backpointers with probability p_n. These messages have a $TTL = 1$ and do not travel further. Notifications are discarded either when their TTL value reaches zero or if they are received by a node more than once due to a cycle. Therefore, our scheme combines a selective, modified-BFS forwarding augmented with direct messaging (backpointers) in order to contact group members. This is shown pictorially in Figure 1.

We now discuss how the aforementioned state is maintained. The backpointer values expire after a certain amount of time. Since our incentive to push a message is the demand on a per-object basis, new backpointers replace the oldest valid ones (if a node already has c valid backpointers). As searches take place inside the system, the backpointer repositories get updated, while the probabilistic fashion in which they are stored guarantees a diverse collection of (ID, address) pairs. Reverse indices get updated during searches, but this is not enough: There may be peers that have searched for an object and built large index values, but are no longer interested in receiving notifications (i.e., stop querying for that object). If searches are no longer routed through those peers, the respective reverse index values will not be updated and will remain high.

To correct this situation, we add an *aging* factor to the reverse indices, which forces their values to decrease with time. Peers need to keep track of the time that a reverse index was last updated in order to acquire its correct value before using it. When a peer receives a search message, it sets the corresponding reverse index to the piggybacked value and its last modified field to the time of receipt. We describe this process in Figure 2. The value of the index decreases exponentially, while two searches at times t_1, t_2 reset its value. A push message received at time t_3 will use the value as shown in the figure. The last modified value is also reset when a reverse index is used, since a peer computes its current value before using it.

2.3 Protocol Specifics

1) Space Requirements: The amount of space required by the peers is $O(d + c)$ per object, where d is the average node degree in the overlay and c is the maximum number

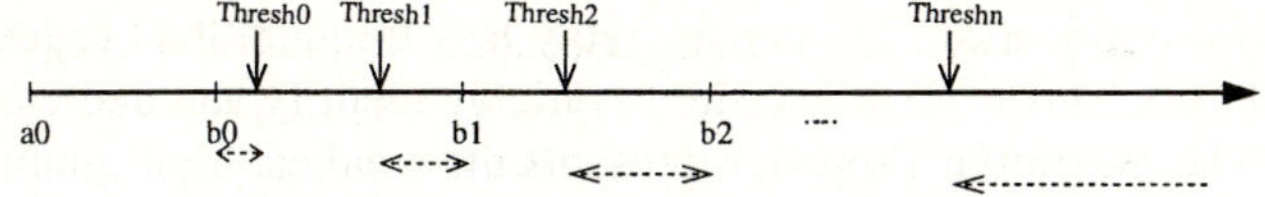

Fig. 3. Sample binning scheme with the respective threshold values for each interval.

of backpointers stored. For about 1 million objects, assuming $c = d = 4$, each peer would need approximately 48MB of memory for *AGNO*, definitely affordable by the vast majority of modern hosts.

   2) Forwarding: Nodes use a threshold parameter *Thresh* in order to choose to which neighbors the notification will be forwarded. Neither the probabilistic or the top-k value schemes are suitable, as they fail in a variety of cases. Consider for example a peer with very low values for all its neighbors. Thresholding enables peers to forward to the most "promising" (active in searches) parts of the overlay. A good first approximation is for each peer to use the average of all its neighbors' indices as *Thresh*. Nevertheless, both the average and the median values fail as well in various circumstances (e.g., when all indices have a very similar low or high value).

   3) Local Threshold Computation: Peers use a globally defined *binning* scheme to decide for the value of *Thresh*. The binning method divides the space of index values into a number of disjoint intervals. $Bin_i = ([a_i, b_i], Thresh_i)$ is characterized by its lower and upper limit values a_i, b_i ($a_0 < b_0 = a_1 < b_1 = a_2...$) and a $Thresh_i$ value. The final threshold value is $Thresh = Thresh_i$, if the average of the neighbors' reverse index values lies in $[a_i, b_i)$. Bins represent an approximation that maps reverse indices to a value representing their quality. Higher numbered bins represent higher quality indices. Figure 3 gives a graphic description of our binning scheme.

   For small i values we should pick few neighbors (therefore a high threshold relative to the bin's interval), while for large i (i.e., high quality bins), most of the neighbors need to be chosen. As a simple heuristic for selecting $Thresh_i$, their values are chosen such that $Thresh_{i-1} - b_{i-1} > Thresh_i - b_i$ and $Thresh_{i-1} < Thresh_i$, i.e., the higher the order of a bin, the smaller its threshold value is compared to the bin's upper limit.

   4) Reverse Index Aging: Peers that lose interest in an object should be left out of the push phase as quickly as possible. Our scheme uses an aging factor ξ together with the last modified time of each reverse index to reduce the influence of inactive ones. Assuming index $P \rightarrow Q$ was last modified at time t_{last}, its value at time $t \geq t_{last}$ is: $P \rightarrow Q(t) = (1 - \xi)^{t - t_{last}} P \rightarrow Q(t_{last})$, where $\xi \in [0, 1]$. For $\xi = 0.2$, a reverse index value will be at 80% of its last modified after one time unit.

   The value of ξ dictates how aggressive our aging will be. It depends on the rate at which requests (and therefore index updates) occur: The larger the rate of searches, the more aggressive the aging can be. Nevertheless, it is still application-dependent, since the rate at which notifications are issued (or even their content) largely affects the aging factor. We define λ_r, λ_n to be the average rates at which a peer or server makes requests or issues notifications respectively.

   For the remainder of this paper, we assume that peers use a value for ξ which satisfies: $(1 - \xi)^T \max_i(Thresh_i) < \min_i(Thresh_i)$ (1). In effect, we pick ξ such that any reverse index with value less or equal to $max_i(Thresh_i)$ will be reduced below the lowest threshold (and thus will not be selected) if not used for T time steps. In the vast majority of cases, notifications are considerably less frequent than requests, therefore we set $T = O(1/\lambda_r)$. This is done in order to quickly identify and decrease idle indices in the overlay. The maximum $Thresh_i$ represents the smallest high-quality index value in our binning scheme. Therefore, we choose ξ such that all reverse indices up to that level of quality are discarded after a period of time T without getting updated. Choosing larger threshold values or smaller T values results in a more aggressive aging.

5) Estimation of λ_r: In order for our scheme to work without requiring a priori knowledge of the request rate but also to be able to adapt to changes in the workload, we need an effective yet inexpensive mechanism to estimate its value and compute the new ξ before each push. This value is then piggybacked downstream and used by all receiving nodes. In order to estimate λ_r, we need the zeroth and first frequency moment (F_0 and F_1 respectively) of the request sequence arriving at a server. F_0 is the number of distinct IDs that appear in the sequence, while F_1 is the length of the sequence (number of requests). Servers can easily monitor the number of incoming requests inside a time interval. Many efficient schemes to estimate F_0 within a factor of $1 \pm \varepsilon$ have been proposed (e.g., [15, 16]). We use one of the schemes in [15], which requires only $O(1/\varepsilon^2 + \log(m))$ memory bits, where m is the number of distinct node IDs. In reality, m is in the order of the distinct peers within TTL hops from a server, since only these nodes can reach it. After each push phase, both estimates are reset and a new estimation cycle begins.

6) Backpointer Selection: Clearly, following the same number of backpointers at different peers and times is not efficient. Our method utilizes the local thresholding computation to assist in the process of selecting valid backpointers. Given that a peer's threshold bin is i at time t, the probability with which each stored backpointer will be followed is p_{ni}, given from the set $\{p_{n0}, p_{n1}, ... p_{ni}, ...\}$ (i.e., one p_n value for each bin). We choose those values such that $p_{ni} > p_{ni+1}$, since better quality bins forward to more neighbors and need not waste more bandwidth. With this scheme, *AGNO* adaptively balances the amount of forwarded messages per peer between the shortcuts and the neighbors according to the current quality of its reverse indices.

3   Simulation Results

3.1   Simulation Methodology and Compared Methods

Requesters make searches for objects using *APS* at rate λ_r (exponentially distributed interarrival times), while servers initiate push transmissions at rate λ_n. At each run, we randomly choose a node that plays the role of a server and a number of requesters, also uniformly at random. Results are averaged over several hundred runs. We present results for both *random* and *power-law* graphs, utilizing *BRITE* [17] and *Inet-3.0* [18] to create these overlays respectively.

The following metrics are used to evaluate the performance of a scheme: The *success rate*, which is the ratio of contacted group members versus the total number of group nodes and the bandwidth *stress*, which we define as the ratio of the produced messages over the minimum number of messages in order to contact all members.

AGNO Parameters: Given the value of 5 as an estimate for the TTL parameter[19], we set $p_r \geq \frac{1}{TTL}$. Given the index update policy used by *APS* as described in [13], we employ a simple 3-bin scheme. The first bin represents indices below the initial value

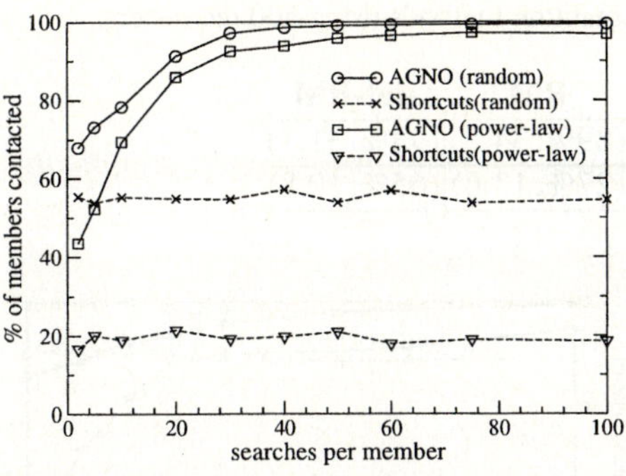

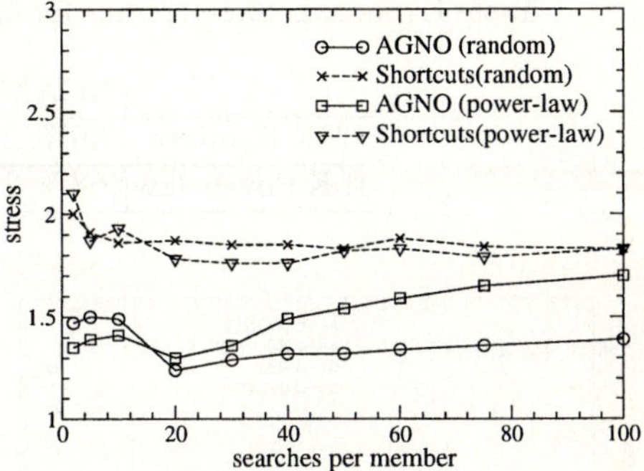

Fig. 4. Success over variable number of searches.

Fig. 5. Stress over variable number of searches.

(no successes), the second those with some hits and the last those with more successes. By default, we set $c \simeq d$ (which reserves an amount of space for backpointers roughly equal to the average node degree) and $T = 2T_r$. Thus from (1) we have:
$\xi = 1 - 0.44^{0.5\lambda_r}$. The value of λ_r (and therefore ξ) is estimated right before each server push using $\varepsilon = 0.1$. A more detailed description of the parameters chosen as well as experimental evaluations with different parameter/bin selections can be found in [12].

Compared Methods: We compare our method against 3 algorithms: The SCAMP membership protocol [5] and the two rumor-spreading schemes in [6]: *Rumor Mongering* (RM) and its deterministic version (det-RM), where peers have complete topology information. In SCAMP, joining members subscribe by contacting a random existing member. Upon receiving a subscription request, a member forwards it to all the members in its local repository. Nodes decide probabilistically whether to store or forward the subscription. For the unsubscription process, a node notifies the locally known members to replace its ID with the IDs of the members it has received messages from. Group communication is performed in the standard gossip-based manner. SCAMP is shown to converge to a local state of slightly over *log(n)* member IDs, which guarantees with high probability that all members will receive a notification. In [6], peers that have received a message less than F times, forward it to B randomly selected neighbors, but only those that the node knows have not yet received it. The deterministic version of that algorithm requires global knowledge of the overlay. Nodes forward messages to all neighbors with degree equal to 1, plus to B remaining neighbors that have the smallest degrees.

Finally, we also implement a pure shortcut forwarding scheme (*Shortcuts*), where backpointers are stored as in *AGNO*, while in the push phase a peer forwards to all valid shortcuts, using the standard *TTL* scheme. For SCAMP, we first run the membership phase, in which we favor the method by assuming joining peers know all already joined members. The parameters for those three methods are the *branching factor B*, which represents how many other peers shall be contacted per forwarding step and the *seen* value F that represents how many times a peer can receive the same message before dropping it.

Table 1. (Success, Stress) results for the remaining methods using 500 requesters.

	SCAMP	RM	det-RM
10K Random	(89%, 2.7)	(89%, 34.5)	(98%, 31.1)
10K Power-law	(68%, 2.1)	(27%, 13.6)	(65%, 10.8)

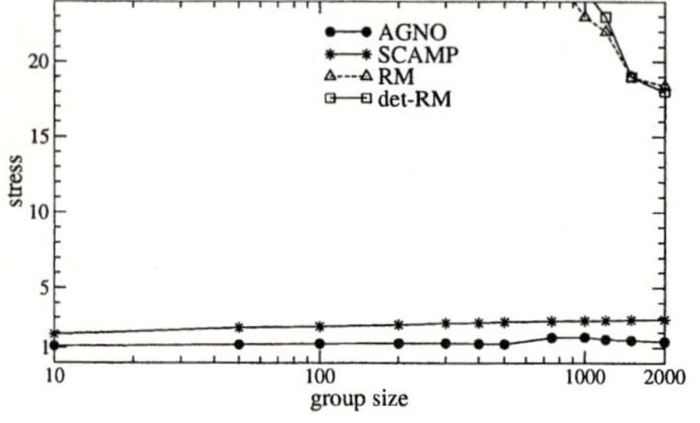

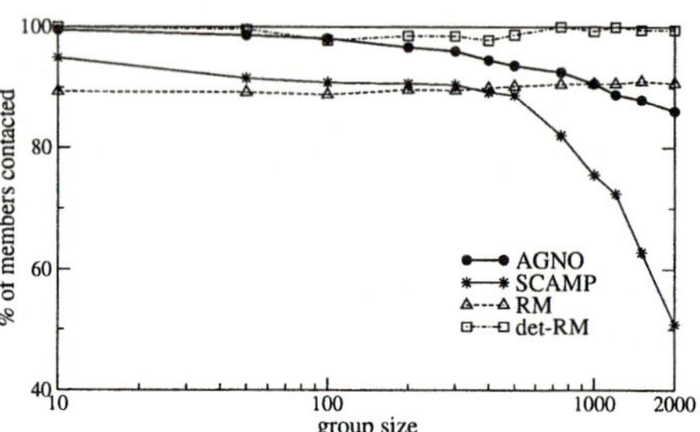

Fig. 6. Stress over variable group size.   **Fig. 7.** Success over variable group size.

3.2   Basic Performance Analysis

In this first set of experiments, using a group of 500 requesters, we vary the number of lookups each of them makes before a single push phase occurs. We report the results averaged over 10,000-Node random and power-law topologies. Figures 4 and 5 present the results for *AGNO* and *Shortcuts* which are affected by the number of searches.

We notice that the pure shortcut scheme cannot provide an efficient notification method by itself. *AGNO* quickly contacts the majority of requesters after only a few searches take place, while maintaining a low stress factor. As our scheme creates better quality indices, there exists a slight variation in the stress. This is due to the fact that after a certain number of queries, peers switch to a different (higher) bin on average.

In the power-law topologies, where about 34% of the peers have degree one, fewer paths are used compared to the random graphs. This, combined to the fact that $\xi = 0$ in these experiments, explains why the stress for *AGNO* slightly increases with more requests. The respective results for the remaining methods (not affected by searches) are shown in Table 1. *AGNO* proves very accurate (in the big majority of runs) and also the most bandwidth-efficient of the compared methods. All three rumor-spreading schemes show considerably worse numbers in the power-law topologies. *det-RM* is much more effective than *RM* in such graphs, which is in accordance to the findings of [6].

Next, we measure the scalability of our method with group sizes ranging from 10 to 2,000 peers using the random topologies. Requesters make only 10 searches on average, immediately followed by a single push phase from the server. For SCAMP, the membership protocol is run before each different group size. Figures 6 and 7 present the results.

Our method is very successful in all group sizes, deteriorating only slightly as the members increase. This happens because with more requesters, their average distance from the server increases (the number of peers reachable from a node increases exponentially with the hop distance). This makes *APS* searches (and its indices) less accurate

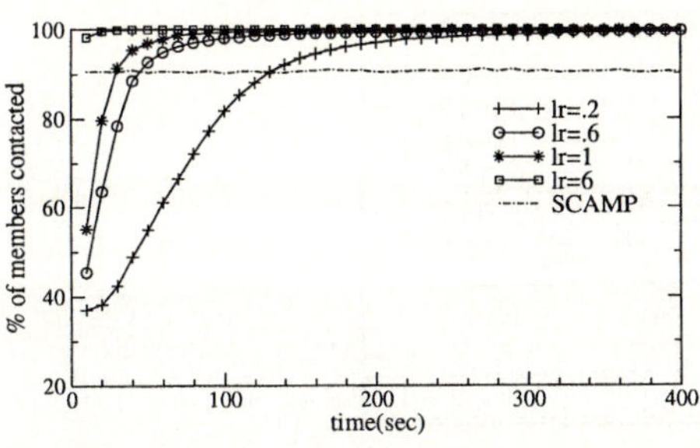

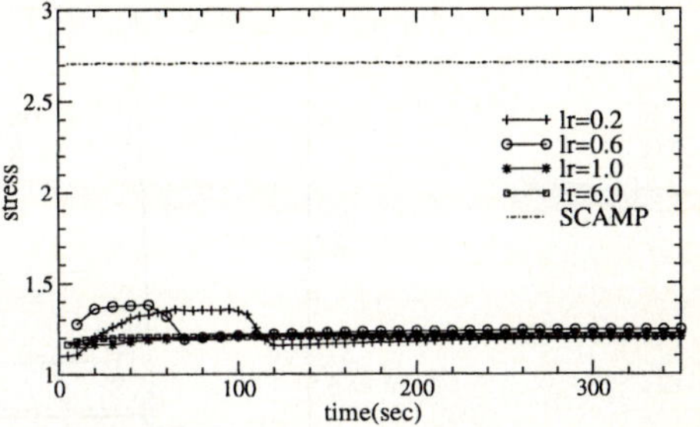

Fig. 8. Success over variable λ_r values ($T_n = 10sec$).

Fig. 9. Stress values over variable λ_r values ($T_n = 10sec$).

for some requesters. The RM schemes produce a similar number of messages regardless of the group size, while the closest competitor (SCAMP) has roughly twice the stress value of *AGNO*, without including the overhead of the membership phase. Our method manages to contact a very high percentage of the members (86-99.5%) using an almost constant message ratio over the group size.

3.3   Sensitivity to λ_r

Assuming a group size of 1,000 peers, we try to evaluate the performance of *AGNO* for different λ_r values. Figures 8 and 9 show the results. Not surprisingly, the larger the value of λ_r, the faster the increase in the success rate, since indices get accurate faster. Another observation is that, regardless of the average request rate, our method asymptotically manages to contact all interested peers and reach a very low stress level (below 1.3). For most realistic scenarios ($T_n >> T_r$), the choice of T_n does not affect *AGNO*'s performance. In the very rare cases that $T_n < T_r$, we just set $T = O(T_n)$ to achieve comparable adaptation. In all cases, our adaptive aging mechanism selects a suitable value for ξ such that the stress remains almost stable and below 1.4, half the value of the best of the remaining schemes (SCAMP). For small request rates, peers adapt using initially low and then higher quality bins (thus the slight variation in stress). The smaller the value of λ_r, the longer this adaptation takes.

3.4   Changes in Group Sizes

Figure 10 shows how our two metrics are affected by having 10%–80% of the 1,000 requesters leave the group (stop making queries) at time $t = 100sec$. We assume (worst-case scenario) that all these nodes jointly and instantly decide to leave the group. In all runs, the stress value peaks at the time of the departures, since the same number of peers are notified but fewer are now considered as members. The size of the departing sub-group directly affects the stress increase. The stress value instantly drops due to our aging mechanism, but it does not reach its previous value. This is due to the fact that a peer's indices get updated not only when it makes a request but also when any request passes through it. Therefore, while shortcuts for departing peers expire, indices leading to them may still have large values, depending on the relative positions of other requesters in the overlay. On the other hand, the success rate is hardly affected.

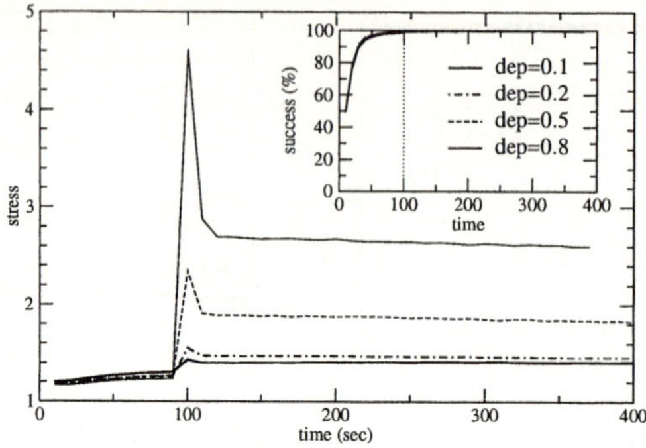

Fig. 10. Stress and success rates when a different ratio of peers depart at time t=100sec ($\lambda_r = 1, T_n = 10sec$).

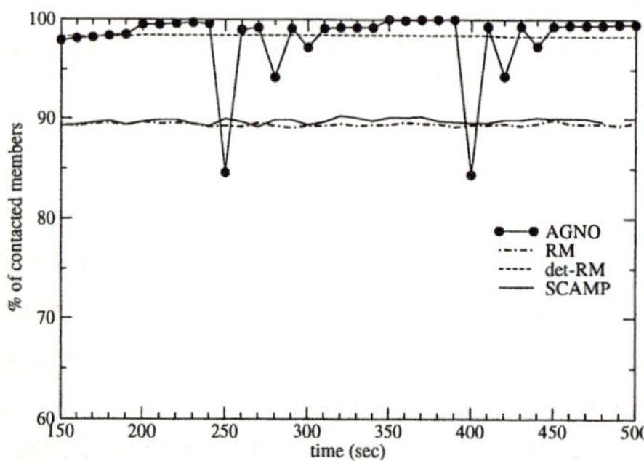

Fig. 11. Success after a series of member departures and arrivals ($\lambda_r = 0.5, T_n = 10$).

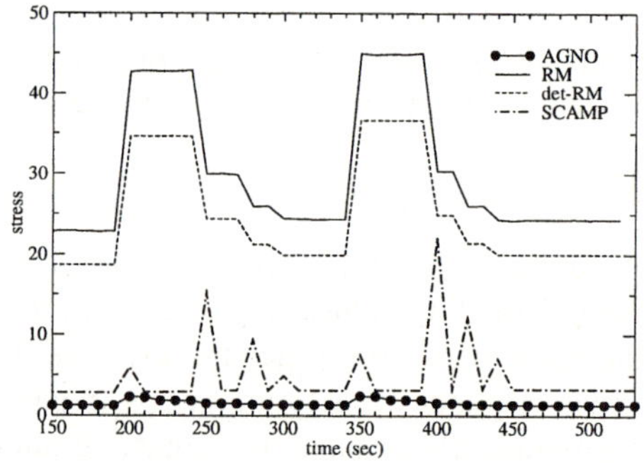

Fig. 12. Stress after a series of member departures and arrivals ($\lambda_r = 0.5, T_n = 10$).

Figures 11 and 12 display the performance of the compared methods under a combination of member joins and leaves. At times $t = \{200, 350\}sec$, 50% of the current group members decide to leave. At $t = \{250, 280, 300, 400, 420, 440\}sec$, 50% of the non-active requesters re-join the group. Members make requests at $\lambda_r = 0.5/sec$, while the group notification phase is performed every 10 secs.

The success rate shows an instant decrease at the exact time of arrival which is proportional to the number of joining peers. Nevertheless, always more than 85% of the current members are contacted, and *AGNO* has learned of their presence by the exact next transmission. In the next push phases, the method quickly reaches its previous levels. On the other hand, the value of stress is decreased after member joins and balances the small increase that occurs after member departures. SCAMP and the two rumor spreading schemes show big variations in the stress metric. For RM and det-RM, this happens because of the change in the group size (same number of messages regardless of peer membership), while for SCAMP this is due to the subscription and unsubscription processes. *AGNO* contacts the vast majority of members at a cost 1 to 10 times lower than the closest compared method (SCAMP).

4   Conclusions

In this paper we present *AGNO*, an adaptive and scalable group communication scheme for unstructured Peer-to-Peer networks. Our method integrates knowledge accumulated during searches to enable content-providers contact the large majority of interested peers with very small overhead. We described in detail our adaptive mechanisms to regulate message forwarding according to the quality of existing knowledge as well as to ensure efficient performance in all group operations. A variety of simulations showed that *AGNO* adapts quickly to variable request rates and group sizes, being at least twice as bandwidth-efficient as the compared methods.

References

1. Chu, Y., Rao, S., Seshan, S., Zhang, H.: Enabling conferencing applications on the internet using an overlay muilticast architecture. In: SIGCOMM. (2001)
2. Banerjee, S., Bhattacharjee, B., Kommareddy, C.: Scalable application layer multicast. In: SIGCOMM. (2002)
3. Zhuang, S., Zhao, B., Joseph, A., Katz, R., Kubiatowicz, J.: Bayeux: An architecture for scalable and fault-tolerant wide-area data dissemination. In: NOSSDAV. (2001)
4. Datta, A., Hauswirth, M., Aberer, K.: Updates in highly unreliable, replicated peer-to-peer systems. In: ICDCS. (2003)
5. Ganesh, A., Kermarrec, A., Massoulie, L.: SCAMP: Peer-to-peer lightweight membership service for large-scale group communication. In: Networked Group Communication. (2001)
6. Portmann, M., Seneviratne, A.: Cost-effective broadcast for fully decentralized peer-to-peer networks. Computer Communications **26** (2003)
7. Rowstron, A., Kermarrec, A., Castro, M., Druschel, P.: Scribe: The design of a large-scale event notification infrastructure. In: NGC. (2001)
8. Jannotti, J., Gifford, D., Johnson, K., Kaashoek, F., O'Toole, J.: Overcast: Reliable multicasting with an overlay network. In: OSDI. (2000)
9. Ratnasamy, S., Handley, M., Karp, R., Shenker, S.: Application-level multicast using content-addressable networks. Lecture Notes in Computer Science (2001)
10. Gnutella website: http://www.gnutella.com.
11. Kalogeraki, V., Gunopulos, D., Zeinalipour-Yazti, D.: A Local Search Mechanism for Peer-to-Peer Networks. In: CIKM. (2002)
12. Tsoumakos, D., Roussopoulos, N.: AGNO: An Adaptive Group Communication Scheme for Unstructured P2P Networks. Technical Report CS-TR-4590, University of Maryland (2004) (ext. version) http://www.cs.umd.edu/ dtsouma/objects/pbroad.pdf.
13. Tsoumakos, D., Roussopoulos, N.: Adaptive Probabilistic Search for Peer-to-Peer Networks. In: IEEE Intl Conf. on P2P Computing. (2003)
14. D.Tsoumakos, Roussopoulos, N.: A Comparison of Peer-to-Peer Search Methods. In: WebDB. (2003)
15. Bar-Yossef, Z., Jayram, T., Kumar, R., Sivakumar, D., Trevisan, L.: Counting distinct elements in a data stream. In: RANDOM. (2002)
16. Alon, N., Matias, Y., Szegedy, M.: The space complexity of approximating the frequency moments. In: STOC. (1996)
17. Medina, A., Lakhina, A., Matta, I., Byers, J.: BRITE: An Approach to Universal Topology Generation. In: MASCOTS. (2001)
18. Jin, C., Chen, Q., Jamin, S.: Inet: Internet Topology Generator. Technical Report CSE-TR443-00, Department of EECS, University of Michigan (2000)
19. Ripeanu, M., Foster, I.: Mapping the gnutella network: Macroscopic properties of large-scale peer-to-peer systems. In: IPTPS. (2002)

Semantic Peer-to-Peer Overlays
for Publish/Subscribe Networks

Raphaël Chand[1] and Pascal Felber[2]

[1] Institut EURECOM, France
`chand@eurecom.fr`
[2] University of Neuchâtel, Switzerland
`pascal.felber@unine.ch`

Abstract. Existing publish/subscribe systems suffer from several drawbacks, such as the reliance on a fixed infrastructure of reliable brokers, or the lack of expressiveness of their subscription language. Most importantly, the challenging task of routing messages based on their content remains a complex and time-consuming operation, and often provides results that are just barely better than a simple broadcast.

In this paper, we present a novel approach to publish/subscribe that was designed to specifically address these issues. The producers and consumers are organized in a peer-to-peer network that self-adapts upon peer arrival, departure, or failure. Our publish/subscribe system features an extremely simple and efficient routing process and excellent scalability to large consumer populations, both in terms of routing and peer management overhead.

1 Introduction

Motivations. The publish/subscribe paradigm is well adapted to loosely-coupled, large scale distributed systems. However, in most traditional publish/subscribe systems, the routing process is a complex and time-consuming operation. It often requires the maintenance of large routing tables on each router and the execution of complex filtering algorithms to match each incoming document against every known subscription. The use of summarization techniques (e.g., subscription aggregation [1, 2]) alleviates those issues, but at the cost of significant control message overhead or a loss of routing accuracy.

In addition, content networks usually rely on a fixed infrastructure of reliable brokers, or assume that a spanning tree of reliable brokers is known beforehand. This approach clearly limits the scalability of the system in the presence of large and dynamic consumer populations. Finally, in most existing systems, the network topology has no relationships with the subscriptions registered by the consumers. As a consequence, the process of routing an event often involves a large number of routers, some of which have no interests in the event but only act as forwarders. The routing process is then only barely more efficient than a broadcast (which benefits from a much lower processing overhead).

To address these limitations, we have designed a publish/subscribe system that follows a radically different approach to content-based networking. First,

J.C. Cunha and P.D. Medeiros (Eds.): Euro-Par 2005, LNCS 3648, pp. 1194–1204, 2005.

the routing process in our system is extremely simple and has very low resource requirements. Second, by organizing peers based on their interests, content distribution is highly efficient as compared to broadcast. Finally, instead of relying on a fixed infrastructure of reliable brokers, our system is organized as a peer-to-peer network: join and leave operations, as well as peer failures, are taken care of at the design level with efficient peers management algorithms. We present in this paper two instantiations of our system that use the same routing protocol but differ by the way peers are organized. Experimental evaluation illustrates the various trade-offs that they offer in terms of efficiency and accuracy.

We would like to emphasize that we propose a new peer-to-peer approach for publish/subscribe, which relies on a system model that differs significantly from other peer-to-peer applications like file sharing. In particular, we assume that peers are well behaved and remain online for reasonably long periods of time, in the sense that the rate of message publication is higher than the frequency of peers' arrivals or departures. Our system provides mechanisms for organizing communities of peers that wish to exchange information using the publish/subscribe paradigm, without reliance on central servers or fixed infrastructures.

Related Work. Most publish/subscribe systems use an overlay network of event brokers to implement some form of distributed content based routing, most notably IBM Gryphon [3], Siena [1], Jedi [4] and XNet [5]. As previously mentioned, these systems suffer from various limitations in terms of extensibility, scalability, and cost. To address some of these issues, a few content-based systems based on peer-to-peer (P2P) networks have been recently proposed. In [6], the authors combine the notion of rendezvous nodes and content-based multicast to implement content based routing in a peer-to-peer environment. HOMED [7] is a peer-to-peer overlay for distributed publish/subscribe systems. Peers are organized in a logical binary hypercube according to their subscriptions. Routing is achieved by propagating the event along a multicast tree embedded in the hypercube. In [8], the authors also implement publish/subscribe in a peer-to-peer environment. The system is "data-aware" in the sense that it exploits information about registered subscriptions to build hierarchical structures. However, they differ from our approach in that the system is topic-based and the routing algorithm is based on multicast. Finally, some proposals have been made to implement content based routing on top of the Chord [9] P2P network. Examples of such systems are [10] and [11]. Unlike in our approach, they consider structured P2P networks and do not take advantage of semantic communities.

2   The Routing Process

Protocol. Our system is composed of a collection of peers. Each peer has registered certain interests that specify the types of messages that it is willing to receive. Each peer is connected with a set of other peers—its neighbors—with which it exchanges messages. We initially make the natural assumption that peers publish messages that match their own interests (we can easily relax this assumption, as will be discussed later). The routing protocol in our system is entirely based on the principle that every peer forwards a message to its neighbors if and only if the message matches its own interests. The routing process starts when a peer P publishes a message m. Since P is interested in m, it forwards it to all its neighbors. Routing then proceeds trivially as shown in Algorithm 1.

Algorithm 1 Routing protocol.

1: Receive message m for the first time from neighbor n
2: **if** m matches interests **then**
3:     Forward m to all neighbors (except n)
4: **end if**

The intuition of the algorithm is to spread messages within a community that shares similar interests and to stop forwarding them once they reach the community's boundary. We emphasize the fact that the routing protocol is extremely simple and requires almost no resources from the peers. It consists of a single comparison and message forwarding operation. In addition, it requires no routing state to be maintained in the peers in the system. Each peer is only aware of its own interests and the identity of its direct neighbors, *not* their interests. The key to the protocol is the proper organization of the peers into semantic communities.

Accuracy. Clearly, the aforementioned process is not perfectly accurate and may lead to a peer receiving a message that it is not interested in—which we call a *false positive*— as well as missing a message that matches its subscriptions—a *false negative*. In other words, our system may deliver out-of-interest messages and may fail to deliver messages of interest. This is obviously due to the fact that a peer is not aware of the interests of its neighbors and forwards messages only based on its own interests. The challenge is thus to organize the peers so as to maximize routing accuracy. It should be noted that false positives are usually benign, because peers can easily filter out irrelevant messages, whereas false negatives can adversely impact application consistency.

Interest-Driven Peers Organization. Consider two neighbor peers P_1 and P_2. If P_1 and P_2 have registered close interests, it means that they are interested in similar types of messages. That is, if P_1 is interested in a message, it is likely that P_2 is also interested in it, and vice versa. It follows that neighbor peers should have close interests in order to minimize occurrences of false positives and false negatives in our system. In other words, we must organize peers based on the interests they registered: proximity in terms of neighborhood should reflect the proximity of the peers' interests.

To evaluate the proximity between two registered interests I_1 and I_2, a proximity metric must be used, that is, a function $f(I_1, I_2)$ that indicates how similar I_1 and I_2 are. Unfortunately, defining a good proximity metric is a challenging problem. It very much depends on the target application, on the language used to specify interests, and most of all on the messages being distributed in the system. The problem of interest proximity has been further discussed in [2].

3   Organizing Peers According to Containment

We now describe a hierarchical organization of the peers that yields no false negatives and only a limited amount of false positives. It uses a proximity metric based on the notion of *interest containment*, as specified in Definition 1. Note that the containment relation is transitive and defines a partial order. We define in a similar manner the relation of *interest equivalence*.

Definition 1. *Interest I_1 contains interest I_2, or $I_1 \supseteq I_2 \Leftrightarrow (\forall$ message m, m matches $I_2 \Rightarrow m$ matches I_1).*

Interest I_1 is equivalent *to interest I_2, or $I_1 \sim I_2 \Leftrightarrow (I_1 \supseteq I_2 \wedge I_2 \supseteq I_1)$. That is: $\forall$ message m, m matches $I_2 \Leftrightarrow m$ matches I_1.*

The containment-based proximity metric, which we refer to as f_c, allows us to compare interests that share containment relationships and is defined as follows. Consider the set of all registered interests $\mathcal{I} = \{I_1, I_2, \cdots, I_n\}$ that contain I. Let $\{I_i, I_j, \cdots, I_m\} \subseteq \mathcal{I}$ be the longest sequence of non-equivalent interests such that $I_i \supseteq I_j \supseteq \cdots \supseteq I_m$. Then,

$$f_c(I, I') = \begin{cases} -\infty, & \text{if } I \not\supseteq I'; \\ \infty, & \text{if } I \sim I'; \\ |\{I_i, I_j, \cdots, I_m\}|, & \text{otherwise.} \end{cases}$$

Intuitively, the objective of this metric is to favor interests that are themselves contained in many other interests, i.e., that are very specific and selective. Note that this metric is not symmetric. The containment-based proximity metric can be used with any subscription language, provided that it defines a containment relationship. Of course, it applies best to subscription languages that are likely to produce subscriptions with many containment relationships. We wish to emphasize, however, that our routing protocol can be used with any other proximity metric, as we shall see in Section 4.

Network Description. Peers are organized in a *containment hierarchy tree*, based on the proximity metric f_c defined earlier. To simplify, we assume that each peer has expressed its interests by registering exactly one subscription (if that is not the case, the peer will appear multiple times in the hierarchy). The *containment hierarchy tree* is defined as follows. A peer P that registered subscription S is connected in the tree to a parent peer P_a that registered subscription S_a if S_a is the subscription in the system closest to S according to the proximity metric f_c. Given the definition of the metric f_c, this means that S_a is the deepest subscription in the tree among those that contain S. When we have more than one peer to choose from, we select as parent the peer that has the lowest number of children in order to keep the tree as balanced as possible. Because of Definition 1, peers that have registered equivalent interests in the system are organized in specialized, balanced subtrees with limited degree that we call *equivalence trees*. From the perspective of other peers in the system, an equivalence tree is considered as a single entity represented by its root node, which is positioned in the *containment hierarchy tree* using the rules described above. Non-equivalent children of the peers in an equivalence tree are always connected to its root. To interconnect top level peers that do not share containment relationships with each others, we introduce an artificial node that we refer to as the *root node*. This node is purely virtual and is implemented by simply connecting top-level peers with each other through "sibling" links. A simple containment hierarchy tree is illustrated in Figure 1. The equivalent peers P_8, P_9 and P_{10} are organized in the *equivalence tree* rooted at P_8. Note that both P_2 and P_4 contain P_3, but P_2 has a greater depth and is hence a better parent. Similarly, P_6 is connected to P_5 rather than P_1.

Impact on the Routing Process. From Algorithm 1 and the fact that peers are organized in a containment hierarchy tree, it follows that the paths followed by a message form a content distribution tree, in which inner nodes are true positives and leaves are either false positives or leaves in the tree topology. Consequently, routing is *efficient* in terms of bandwidth usage. Besides, there are no false negatives in our system. We wish to point out that false positives can only be avoided by having each peer know about its neighbors' interests, which conflicts with our design guidelines. Finally, the construction of the containment hierarchy tree topology enables us to minimize the occurrence of false positives with uniform subscription and message workloads. Indeed, the fact that a peer P has for parent the peer of highest possible depth that contains it means that a message m has a greater chance of being discarded on the way from the root node to P. A simple example is illustrated in figure 1, where peer P_5 publishes message D. The path followed by D is highlighted by the arrows.

Maintaining the Containment Hierarchy Tree. We have implemented several peers management algorithms to maintain the containment hierarchy tree when peers dynamically join and leave the system. We now briefly discuss their basic principles and most relevant features.

The join algorithm aims at inserting a new peer P with subscription S in the tree topology. Consequently, the system is first probed to find adequate containment or equivalence relationships between S and the other registered subscriptions. This can be done by recursively propagating *join* messages in the hierarchy tree. It is important to note that a *join* message usually traverses only a fraction of the tree, very much like regular messages. As a result of the probing phase, P joins the tree by connecting to a parent that is either an equivalent peer, if any, or a peer of highest depth whose subscription contains S. Next, P proceeds to the *reorganization* phase, which might lead to moving some existing peers so as to become P's children. Indeed, when P has connected to a parent in the tree, some other peers may now be closer to P than their actual parent in the

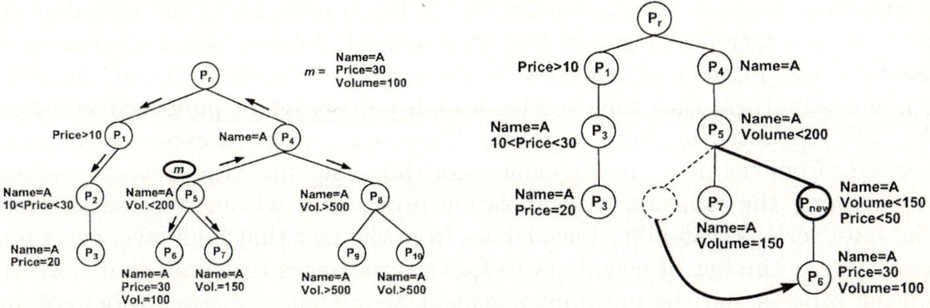

Fig. 1. A simple publish/subscribe system for stock quotes with participants organized in a *containment hierarchy tree*. The subscription registered by a peer is represented next to it.

Fig. 2. Peer P_{new} has been inserted in the network with P_5 as its closest peer. Peer P_6 is reorganized as P_{new}'s child because the latter is a better parent than P_5.

tree. The *reorganization* phase introduces significant overhead in the system, in particular because it requires additional propagations of join messages. As a consequence, we have implemented three different flavors of the join algorithm. The first variant of the algorithm *always* performs all possible reorganizations. The second variant of the algorithm *never* performs any reorganizations. Finally, the third variant of practical relevance *periodically* performs reorganizations (only a given percentage of peers are reorganized). The different variants reach different compromises between joining complexity and accuracy of the hierarchy tree.

When peer P with registered subscription S wishes to leave the system—or when it fails—each of its children has to be reconnected to another parent in the tree. If P is part of an equivalence tree, then we simply perform a *leaf promotion* (a leaf downstream P is promoted to P's position). If P is not part of an equivalence tree, then the leave algorithm consists in reconnecting P's children to their grand-parent. It follows that every peer needs to know its grand-parent (or several ancestors for increased fault-tolerance); this is achieved with trivial modifications to the join algorithm and negligible additional control traffic. Although extremely simple, this recovery technique may cause the accuracy of the containment hierarchy tree to degrade over time. This is due to the fact that P's parent may not be the closest peer in the system for P's children. In addition, P's parent may suffer from the increased number of connections that it has to manage. To address those issues or in case P's parent has also failed, P's children can look for another replacement parent by executing the join algorithm, typically starting from some ancestor, at the price of higher overhead. Note that, if we wish to maintain an optimal tree, additional peers among P's descendants might need to re-evaluate their position as well if P's departure has decreased their depth.

Scalability Issues. The instantiation of our system using the containment-based metric organizes peers in tree topologies. It follows that high-level peers receive a high number of messages. As these peers have very "broad" interests, it is not unnatural that they receive a high percentage of published messages: they are interested in those messages. They are also more exposed to control messages from the peer management algorithms, but this traffic can be reduced by confining the join and reorganization procedures within selected subtrees. The most serious scalability issue comes from the fact that high-level peers may have a large number of neighbors to forward messages to (recall that both the routing process and the peers management algorithms are straightforward and demand very little resources). To address this problem, we have performed slight modifications to our original protocol to reduce by a great deal the bandwidth utilization at the peers. Because of space requirements, we shall only briefly introduce these techniques. Informally, the principle of the improved scheme consists in connecting the children of a node with "sibling" links in a double-linked list. When a peer receives a message, it forwards it to its two neighboring siblings and one of its children, chosen uniformly at random. This approach dramatically reduces the bandwidth requirements of peers that have a large number of children, but also slows down the propagation of messages in the system. It is therefore desirable to use it only for overloaded peers.

4   Organizing Peers According to Similarity

As previously mentioned, the routing protocol used to disseminate messages does not make specific assumptions about the proximity metric used to organize the peers in semantic communities. We now present a generalization of the containment-based proximity designed to alleviate two of its limitations: (1) its poor applicability to subscription language and/or consumer workloads with little or no containment relationships, and (2) its tree topologies that may be fragile with dynamic consumer populations. This generalization is based on the general principle of interest similarity.

Similarity Metric. We first define the notion of interest similarity as follows.

Definition 2 (Interest similarity). *Consider two interests I_1 and I_2. Let $\mathcal{I}$ be the universe of all possible interests. We define the similarity between I_1 and I_2, noted $Sim(I_1, I_2)$, as a function from $\mathcal{I}^2$ in the interval $[0, 1]$ that returns the probability that a message m matching I_1 also matches I_2.*

We then define our proximity metric based on interest similarity, which we refer to as f_s:

Definition 3 (Proximity metric f_s). $f_s : \mathcal{I}^2 \mapsto [0, 1]$:

$$f_s(I_1, I_2) = \frac{Sim(I_1, I_2) + Sim(I_2, I_1)}{2}$$

Note that the proximity metric f_s is symmetric, that is, if I_1 is close to I_2 according to f_s, then I_2 is equally close to I_1. However, the similarity function is a priori *not* symmetric.

Network Description. We now briefly describe the hierarchical organization of peers based on the proximity metric f_s. A peer P with registered interest I chooses a set of n neighbors P_i, which are the n peers in the system with interests closest to I according to f_s (in case of equality, the peers with less connections are chosen). In turn, P can be chosen by some other peers as one of the n best peers according to f_s (such that I is amongst the n interests in the system closest to their subscription according to f_s).

This approach effectively organizes the peers in "interest communities," i.e., groups of peers that share similar interests. Because of the definition of the similarity function and the proximity metric f_s, this organization optimizes routing accuracy by minimizing the number of false positives and negatives exchanged by neighbor peers. To maintain good connectivity between the communities and prevent some of them from being closed (because their interests do not compare with the other communities' interests), P also chooses r neighbors at random in the system, in addition to the n peers selected with f_s. Routing proceeds as described in Section 2. Peer management algorithms are also very similar to those presented earlier, with a few differences discussed in [12]. However, the routing algorithm can be enhanced to have even better control on false positives and false negatives. For instance, we can add an indulgence factor γ that allows a peer to forward a message even if it is not interested in it. The process, which may be performed only γ times per message, is expected to reduce the false negatives ratio, notably by improving the transfer of messages between communities. Another improvement is to add a random neighbor forwarding probability ρ, which controls the probability for a peer P to actually forward a message to its r random neighbors. The base case, $\rho = 100\%$, produces fewer false negatives but more false positives; lower values of ρ have the opposite effect.

Obviously, if $n + r > 1$, the peers are organized in graphs instead of trees. We can also observe that, if we set $n = 1$, $r = 0$ and we define $Sim(I_1, I_2) = 1$ if $I_1 \supseteq I_2$ and $Sim(I_1, I_2) = 0$ otherwise, peers are organized using a containment-based metric similarly to the topology of Section 3.

The organization of peers in graphs rather than trees benefits from several advantages. It has better connectivity and is hence more resilient to failures and frequent arrivals or departures. Also, it has better flexibility and offers higher scalability since the traffic load is more evenly distributed amongst the peers. Finally, this model can be applied to any subscription languages and consumer workloads even if the subscriptions share little or no containment relationships.

5   Performance Evaluation

To test the effectiveness of our publish/subscribe system, we have conducted simulations using real-life document types and large numbers of peers. We propose an evaluation of our system when using both the containment and similarity metrics presented earlier. We are mostly interested in studying the routing process in the system. Indeed, we have seen that the cost for its extreme simplicity is that it induces a certain inaccuracy in terms of false positives and negatives but that an efficient topology enables to minimize their occurrence. In this evaluation, we quantify the accuracy of our system experimentally. In-depth evaluation of other aspects of our system is available in [12].

Peers in our system register their interests using the standard XPath language to specify complex, tree-structured subscriptions. Documents are XML documents. To evaluate our system when using the similarity metric, we have implemented a proximity metric for XML documents and XPath subscriptions. Because of space limitations, the metric is not detailed in this paper (a description can be found in [12]).

Containment Metric. We first focus on the system when peers are organized in a containment hierarchy according to the proximity metric f_c. We have seen that this topology enables to suppress all occurrences of false negatives. As a consequence, we aim at quantifying experimentally the number of false positives generated by the routing process in the system. For that purpose, we proceeded as follows. We first simulated networks of different sizes, with each version of the *join* algorithm presented in section 3, by sequentially adding peers with randomly-generated subscription (we used a reorganization rate of 10% for the *join* version with periodic reorganization). We then routed 1,000 random documents by injecting them at the root node.[1] For each document, we computed the false positives ratio as the percentage of peers in the system that received a message that did not match its interests. The results, shown in Figure 3, were obtained by taking the average of 1,000 executions.

We observe that the average false positives ratio remains small, typically less than 10% in most cases, and decreases exponentially with the size of the consumer population. This is due to the efficiency of the tree topology. By organizing peers based on their interests, documents are filtered out as soon as they reach the boundary of the community of interested consumers. The efficiency of the tree topology improves with the size of the consumer population because of the increasing number of containment relationships shared between the peers. Besides, we computed that on average, and independently of the consumer population, the percentage of uninterested peers in the system is 75%, which illustrates the benefits of our routing protocol over a broadcast. Unsurprisingly, join algorithms that reorganize the peers more frequently produce network topologies that have a lower false positives ratio. As explained in section 3, this is directly related to the number of reorganizations that are performed by each algorithm. However, the differences are very small and the benefits of the slight increase in accuracy may not justify the additional overhead of the reorganization process.

Similarity Metric. We now study the accuracy of the system when peers are organized in a graph according to the proximity metric f_s based on subscrip-

---

[1] Note that the number of false positives would not be affected when injecting the messages at another node than the root.

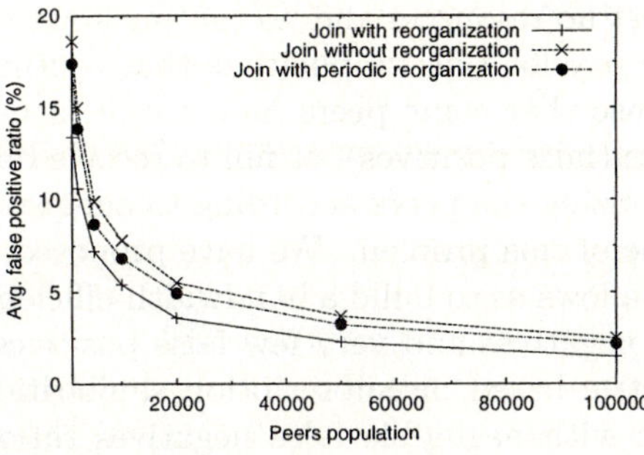

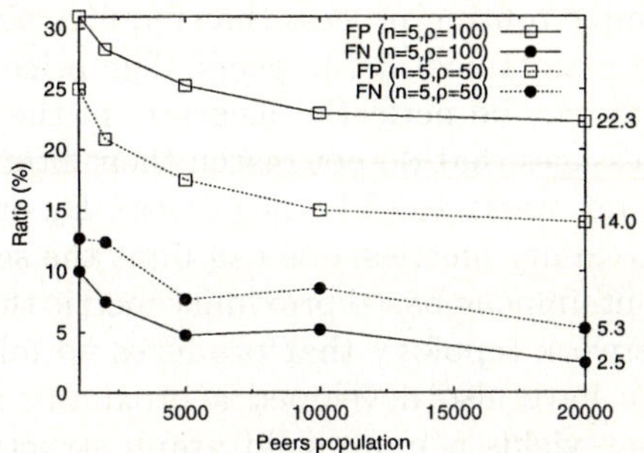

Fig. 3. False positives ratio for networks of different sizes.

Fig. 4. False positives and false negatives ratios for networks of different sizes.

tion similarities. Since this topology does not prevent the occurrence of false negatives, we are interesting in quantifying the accuracy of our system both in terms of false positives and false negatives. For that purpose, we proceeded as in the case of the metric based on containment. We first generated networks of different sizes, using a value of $n = 5$ for the number of proximity neighbors and $r = 1$ for the number of random neighbors. We then injected random documents and quantified the routing accuracy. We measured the false positives ratio as the percentage of the peers in the system that received a message that did not match their interests, and the false negatives ratio as the percentage of peers interested in a message that did not receive it. We experimented with random neighbor forwarding probabilities ρ of 100% and 50%. Results are shown in Figure 4.

We first observe that the average false negatives ratio remains small, typically less than 5%, which shows that on average, for a given document, only a small fraction of the population of interested consumers does not receive it. The false positives ratio, while significantly higher, still remains at reasonable values, typically around 25%. We also remark that, as expected, a lower value of the parameter ρ favors the false positives ratio over the false negatives ratio. For a value of $\rho = 50\%$, the false positives ratio improves significantly (14% for $20,000$ peers), at the cost of a slight increase of the false negatives ratio (5% for $20,000$ peers). Finally, all performance metrics decrease with the size of the consumer population, which shows that the routing accuracy globally improves with the consumer population. This can be explained by the fact that, in larger populations, peers are able to find better neighbors according to the proximity metric f_s and hence reduce the occurrence of false positives and false negatives.

6   Conclusion

We have designed a novel publish/subscribe system, based on the peer-to-peer paradigm, that specifically address some of the limitations of existing systems. In particular, our network does not rely on a dedicated network of content routers, nor on complex filtering and forwarding algorithms: it features an extremely simple routing process that requires almost no resources and no routing state to be maintained at the peers. The price to pay for this simplicity is that routing may not be perfectly accurate, in the sense that some peers may receive some messages that do not match their interests (false positives), or fail to receive relevant messages (false negatives). By organizing the peers according to adequate proximity metrics, one can limit the scope of this problem. We have proposed a containment-based proximity metric that allows us to build a bandwidth-efficient network topology that produces no false negatives and very few false positives. We have also developed a proximity metric based on subscription similarities that yields a more solid graph structure with negligible false negatives ratios and very few false positives. As part of our ongoing research, we are studying refinements of our proximity metrics that take into account additional factors such as physical proximity or link bandwidth, in order to minimize latency and maximize throughput.

References

1. Carzaniga, A., Rosenblum, D., Wolf, A.: Design and Evaluation of a Wide-Area Event Notification Service. ACM Transactions on Computer Systems **19** (2001)
2. Chan, C.Y., Fan, W., Felber, P., Garofalakis, M., Rastogi, R.: Tree Pattern Aggregation for Scalable XML Data Dissemination. In: Proceedings of VLDB. (2002)
3. Banavar, G., Chandra, T., Mukherjee, B., Nagarajarao, J., Strom, R., Sturman, D.: An efficient multicast protocol for content-based publish-subscribe systems. In: Proceedings of ICDCS. (1999)
4. Cugola, G., Nitto, E.D., Fugetta, A.: The JEDI event-based infrastructure and its application to the development of the opss wfms. IEEE Transactions on Software Engineering **27** (2001) 827–850
5. Chand, R., Felber, P.: A scalable protocol for content-based routing in overlay networks. In: Proceedings of NCA, Cambridge, MA (2003)
6. Perng, G., Wang, C., Reiter, M.: Providing content based services in a peer to peer environment. In: Proceedings of DEBS, Edinburgh, UK (2004)
7. Choi, Y., Park, K., Park, D.: Homed: A peer-to-peer overlay architecture for large-scale content-based publish/subscribe systems. In: Proceedings of DEBS, Edinburgh, UK (2004)
8. Baehni, S., Th, P., Guerraoui, E.: Data-aware multicast. In: Proceedings of the 5th IEEE International Conference on Dependable Systems and Networks. (2004)
9. Stoica, I., Morris, R., Karger, D., Kaashoek, F., Balakrishnan, H.: Chord: A scalable peer-to-peer lookup service for internet applications. In: Proceedings of ACM SIGCOMM. (2001) 149–160
10. Terpstra, W., Behnel, S., Fiege, L., Zeidler, A., Buchman, A.: A peer-to-peer approach to content-based publish/subscribe. In: Proceedings of DEBS, San Diego, USA (2003)
11. Triantafillou, P., Aekaterinidis, I.: Content-based publish-subscribe over structured p2p networks. In: Proceedings of DEBS, Edinburgh, UK (2004)
12. Chand, R., Felber, P.: Semantic Peer-to-Peer Overlays for Publish/Subscribe Networks. Technical report, Institut EURECOM (2005)

Topic 16
Applications of High-Performance
and Grid Computing

Ray Bair, Ed Seidel, Michel Daydé, and José Laginha Palma

Topic Chairs

The use of parallel computing and distributed information services is spreading quite rapidly, as today's difficult problems in science, engineering and industry far exceed the capabilities of the desktop PC and department file server. The availability of commodity parallel computers, ubiquitous networks, maturing Grid middleware, and portal frameworks is fostering the development and deployment of large scale simulation and data analysis solutions in many areas. This topic highlights recent progress in applications of high performance parallel and Grid computing, with an emphasis on successes, advances, and lessons learned in the development and implementation of novel scientific, engineering and industrial applications.

Today's large computational solutions often operate in complex information and computation environments where efficient data access and management can be as important as computational methods and performance, so the technical approaches in this topic span high performance parallel computing, Grid computation and data access, and the associated problem-solving environments that compose and manage advanced solutions.

This year the 23 papers submitted to this topic area showed a wide range of activity in high performance parallel and distributed computing, with the largest subset relating to genome sequence analysis. Nine papers were accepted as full papers for the conference, organized into three sessions. One session focuses on high performance genome sequence comparison. The second and third sessions present advanced approaches to scalable simulations, including some non-traditional arenas for high performance computing. Overall, they underscore the close relationship between advances in computer science, computational science, and applied mathematics in developing scalable applications for parallel and distributed systems.

J.C. Cunha and P.D. Medeiros (Eds.): Euro-Par 2005, LNCS 3648, p. 1205, 2005.
© Springer-Verlag Berlin Heidelberg 2005

Parallel Linear Space Algorithm
for Large-Scale Sequence Alignment

Eric Li, Cheng Xu, Tao Wang, Li Jin, and Yimin Zhang

Intel China Research Center, Intel Corporation, Beijing, China
`{Eric.q.li,Cheng.xu,Tao.wang,Jin.li,Yimin.zhang}@intel.com`

Abstract. Aligning long DNA sequences is a fundamental and common task in molecular biology. Though dynamic programming algorithms have been developed to solve this problem, the space and time required by these algorithms are still a challenge. In this paper we present the Parallel Linear Space Alignment (PLSA) algorithm to compute the long sequence alignment to meet this challenge. Using this algorithm, the local start points and grid cache partition the whole sequence alignment problem into several smaller independent subproblems. A novel dynamic load balancing approach then efficiently solves these subproblems in parallel, which provides more parallelism in the trace-back phase. Furthermore, PLSA helps to find k near-optimal non-intersecting alignments. Our experiments show that this proposed algorithm scales well with the increasing number of processors, and it exhibits almost linear speedup for large-scale sequences.

1 Introduction

Pair-wise sequence alignment is a fundamental operation in bioinformatics. It is useful to identify the similar and diverged regions between two sequences, e.g., DNA or protein sequences. From a biological point of view, matches may turn out to be similar functions, e.g., homology pairs and conserved regions, while mismatches may detect functional differences, e.g., Single Nucleotide Polymorphism(SNP).

Using dynamic programming, Needleman and Wunsch presented the first global alignment algorithm in 1970 [12]. Smith and Waterman improved this algorithm for local alignment to find the longest common substring [13]. As these algorithms have quadratic complexities with respect to the length of two sequences, they require excessive memory and computation time than single computation resources can afford, especially for large-scale sequences such as genome data [3, 4, 10]. E.g., aligning sequences with several megabytes length would require several Terabytes of memory.

To overcome this difficulty, Hirschberg first uses a linear space algorithm to find common substrings [6]. Myers and Miller apply Hirschberg's linear space algorithm to sequence alignment [10]. Although it reaches linear space, the algorithm is not very efficient due to significant re-computations in the backward bisecting search period. Regarding the parallel possibilities in the linear space alignment, several algorithms are also proposed to improve performance by using more processors. Martins et al. [9] calculates the whole similarity matrix by using wave front parallelism. This

J.C. Cunha and P.D. Medeiros (Eds.): Euro-Par 2005, LNCS 3648, pp. 1207–1216, 2005.

algorithm suffers from similar drawback as Myers' algorithm in that it also requires large re-computations in the backward bisecting search. FastLSA [1] improves Myers' and Martin's algorithms by incorporating some extra space called "grid cache" to save a few rows and columns of similarity matrix H, and then divides the whole problem into several subproblems. Though FastLSA has less re-computation than Martins' algorithm, the scalability performance is limited since the decomposed subproblems are independent of each other, and therefore they can only be solved sequentially. Chen and Schmidt's algorithm uses the grid cache and start points to find k near-optimal local alignments [5]. After the start and end points of an alignment are determined, they solve the alignment using Hirschberg's bisecting search algorithm. This incurs redundant re-calculations and longer execution time in the backward period.

Having studied these algorithms, we propose an efficient algorithm, i.e., "Parallel Linear Space Alignment" (PLSA) for large-scale sequence alignment. PLSA takes advantage of the grid cache and global/local start points, and it uses a dynamic load balancing approach to efficiently parallelize the trace-back phase, which provides more parallelism than conventional algorithms. Moreover, using global start points can help to find all the k near-optimal local alignments at the same time.

The rest of this paper is organized as follows. In section 2, we introduce the basic sequence alignment Smith-Waterman algorithm. In section 3 we describe our proposed PLSA algorithm and in section 4, we analyze our experimental results. Finally, in section 5 we summarize our discussion.

2 Smith-Waterman Algorithm

For two sequences S1 and S2 with length $l1$ and $l2$, the Smith-Waterman algorithm [13] computes the similarity matrix $H(i,j)$ to identify the optimal subsequences by Eq(1):

$$H(i, j) = \max \begin{cases} 0 \\ E(i, j) \\ F(i, j) \\ H(i-1, j-1) + sbt(S1_i, S2_j) \end{cases}$$

$$E(i, j) = \max \begin{cases} H(i, j-1) - \alpha \\ E(i, j-1) - \beta \end{cases}$$

$$F(i, j) = \max \begin{cases} H(i-1, j) - \alpha \\ F(i-1, j) - \beta \end{cases}$$

$$with \quad H(i,0) = E(i,0) = 0, \quad 0 \le i \le l_1$$
$$H(0, j) = F(i,0) = 0, \quad 0 \le j \le l_2$$

(1)

Where $sbt(.)$ is the substitution matrix of cost values. Affine gap costs are defined as follows: α is the cost of the first gap; β is the cost of the following gaps. $H(i, j)$ is the current optimal similarity value ending at position (i, j). E and F are the cost values from a vertical or horizontal gap, respectively. For example, the two sequences of $S1$ and $S2$ generate the similarity matrix, and the whole optimal alignment can be traced back as illustrated in Figure 1. The memory and time complexity for the Smith-Waterman algorithm is $O(l1 \times l2)$, which imposes challenging requirements both on computer memory and execution time when handling long sequences such as whole genomes.

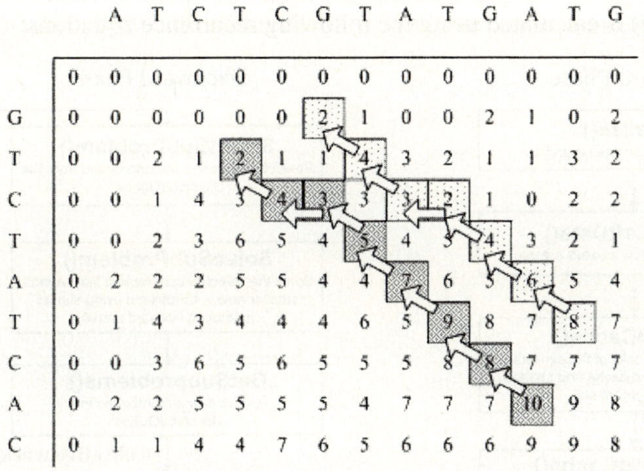

	A	T	C	T	C	G	T	A	T	G	A	T	G
	0	0	0	0	0	0	0	0	0	0	0	0	0
G	0	0	0	0	0	2	1	0	0	2	1	0	2
T	0	2	1	2	1	1	4	3	2	1	1	3	2
C	0	1	4	3	4	3	3	3	2	1	0	2	2
T	0	2	3	6	5	4	5	4	3	4	3	2	1
A	2	2	2	5	5	4	4	7	6	5	6	5	4
T	1	4	3	4	4	4	6	5	9	8	7	8	7
C	0	3	6	5	6	5	5	5	8	8	7	7	7
A	2	2	5	5	5	5	4	7	7	7	10	9	8
C	1	1	4	4	7	6	5	6	6	6	9	9	8

Similariy matrix of the alignment of sequence ATCTCGTATGATG and GTCTATCA. substitution cost of +2 if the two characters are identical and -1 otherwise. Both the first and extension gap penalty are -1. Two traceback paths show the non-intersecting optimal and near-optimal local alignments.

Fig. 1. Smith-waterman sequence alignment

3 PLSA Algorithm

Our PLSA algorithm is presented to efficiently solve the sequence alignment problem in parallel with linear space. It first uses the grid cache and global/local start points to decompose the whole problem into several smaller independent subproblems. Then a dynamic task decomposition approach is employed to solve these subproblems in parallel. The whole process consists of two parts: 1) forward calculating the whole similarity matrix by wave-front parallelism, and 2) backward solving subproblems to find k near-optimal alignment paths by using a dynamic task decomposition mechanism. Figure 2 shows the flowchart of the PLSA algorithm.

3.1 Basic Scheme in PLSA

For large-scale sequence alignment, the whole similarity matrix H is usually too large to be fully stored in the memory. In the PLSA algorithm, we use grid cache to store just a few rows and columns of H rather than storing the whole matrix. Figure 3 shows the case for the grid division $k = 3$. The entire similarity matrix is initially computed in the forward phase and only three rows and three columns are saved in the grid cache during the computation.

In order to decompose the large alignment problem into several smaller independent subproblems, PLSA uses the global/local start point to locate the ancestor point in the grid cache. The global start point of the position (i, j) denotes the starting position of the local alignment path, and the local start point of the position (i, j) is the intersection point between its left/up grid and the alignment path. Similar to Eq (1), the start point $Hst(i, j)$ is calculated using the following recurrence equations:

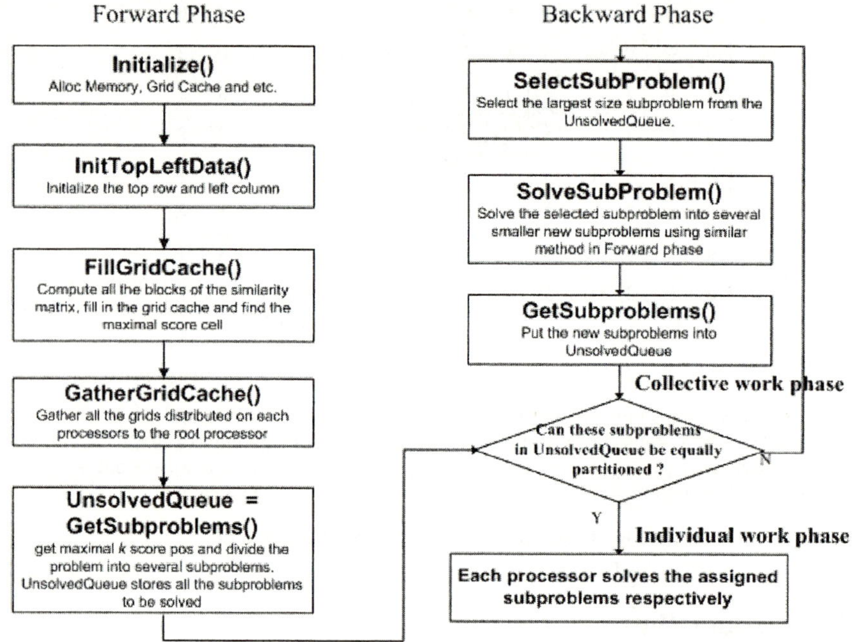

Fig. 2. Flowchart of PLSA

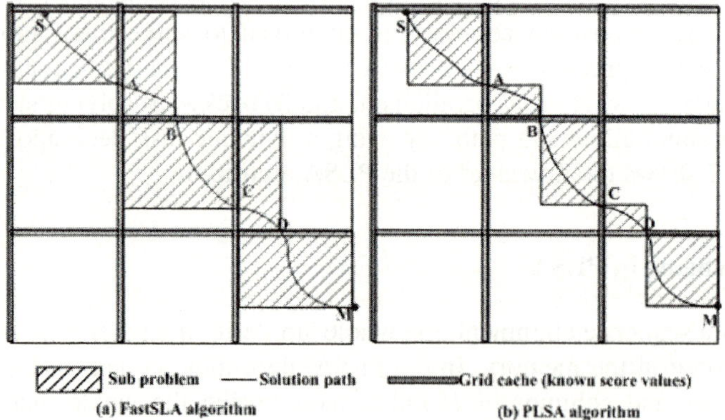

Fig. 3. Grid Cache and Start Points in PLSA

$$Est(i,j) = \begin{cases} Hst(i,j-1) & if\ E(i,j) = H(i,j-1) - \alpha \\ Est(i,j-1) & if\ E(i,j) = E(i,j-1) - \beta \end{cases}$$

$$Fst(i,j) = \begin{cases} Hst(i-1,j) & if\ F(i,j) = H(i-1,j) - \alpha \\ Fst(i-1,j) & if\ F(i,j) = F(i-1,j) - \beta \end{cases}$$

$$Hst(i,j) = \begin{cases} (i,j) & if\ H(i,j) = 0 \\ Hst(i-1,j-1) & if\ H(i,j) = H(i-1,j-1) + sbt(S1_i, S2_j) \\ Est(i,j) & if\ H(i,j) = E(i,j) \\ Fst(i,j) & if\ H(i,j) = F(i,j) \end{cases} \tag{2}$$

For example, in Figure 3(b), when obtaining the maximal score point M, we search both the last column and the last row grid to find its local start point D. Similarly, we can determine C is the local start point of D in this sub matrix. This procedure is recursively performed to find all the local start points (D, C, B, and A) and the global start point S in the optimal path. These start points form a series of independent rectangular subproblems that correspond to the subsequence alignments. Since the start and end point of these subproblems have been determined, we can get the boundary information from the status of the start point. However, as illustrated in Fig 3(a), FastSLA does not have this functionality; it has to solve these subproblems serially, extending from the bottom right to the top left boundary region. It can be observed that the dashed area in PLSA is smaller than FastLSA, which indicates that fewer re-computations are achieved in our algorithm. Theoretically, considering the calculation complexity, when sequence S_1 and S_2 are aligned using a grid cache with k rows and k columns, the worst execution time for PLSA and FastLSA are $l_1 \times l_2 \times k/(k-1)$ and $l_1 \times l_2 \times [k/(k-1)]^2$, respectively. In fact, the best execution time for FastLSA can only compete with the worst case for PLSA.

The Smith-Waterman algorithm only computes the optimal local sequence alignment result. However, the detection of near-optimal non-intersecting local alignments is particularly important and useful in practice. PLSA uses the global start point information [5] to find these near-optimal local alignments at the same time without introducing extra re-computations. In order to determine k near-optimal non-intersecting alignments, the k highest similarity scores together with different global start points are stored during the linear-space computation of the similarity matrix. For example, in Figure 5(a), two near optimal alignments can be found by tracing back from M and M' with their global start point S and S', respectively.

To improve the tradeoff between time and space, we use block as the basic matrix filling unit. A block, something like a 2D matrix, denotes a memory buffer which is available for solving the sequence alignment problem. If a problem or sub problem is small enough, it will be directly solved within a block. Otherwise it will be further decomposed into several smaller subproblems until they can be solved. After all these subproblems are solved, the sub paths are concatenated to get the final optimal alignment path.

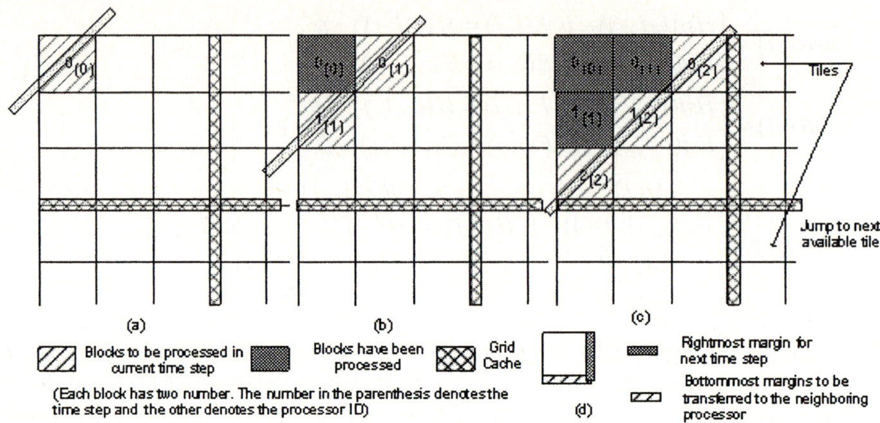

Fig. 4. Wavefront parallelism

3.2 Parallelization of the Forward Phase

In the forward phase, the computation of each block follows the dynamic programming Eq (1) to fill the block matrix. The whole similarity matrix firstly initializes the values of the left-most column and top-most row by using the *InitTopLeftData* procedure in Figure 2. After that, the top-left block can be computed immediately. Considering the dependencies among these blocks, i.e., a block depends on its adjacent left, upper-left, and upper blocks, we use wavefront parallelism to fill the whole matrix. The wavefront moves in anti-diagonals as depicted in Figure 4, and the shift direction is from the north-west to the south-east. Parallelization of the wavefront computation can be done in several different ways, according to the particular parallel architecture used. On fine-grained architectures such as shared memory system, a relatively smaller block is used, since we can exploit the parallelism with smaller granularity. On the other hand, for distributed memory system such as a PC cluster, it is more efficient to assign a relatively larger block to each processor.

In order to better exploit the data locality and minimize the communication overhead, we use a tile-based processing scheme in the wavefront parallelization. A tile is a strip of blocks consisting of a complete horizontal blocks, when a processor finishes computing a tile, it proceeds to work on the next available tile.

Along with the block computation, some temporary data are saved to the grid cache when they encounter the grid columns/rows. After the whole matrix has been filled, we can find the k maximal scores and use the local start point stored in the grid cache to decompose the whole problem into several independent subproblems. Finally, all of these subproblems are put together into the unsolved subproblem queue and are processed in the backward phase.

3.3 Parallelization of the Backward Phase

Generally there are two kinds of parallelisms in the backward phase: the inherent wavefront parallelism in each subproblem and the inter-dependencies among all these subproblems. However, both parallelisms have their deficiencies. The first one only exploits the parallelism within one subproblem, which decomposes a subproblem recursively until its descendant subproblems can be solved in the block size. This approach is too fine grained and does not scale well for more processors. The second approach, simply assigning the subproblems to all of the processors, cannot guarantee the load balance parallel performance. The limited number and different size of the subproblems may not satisfy the requirements of the equal task assignments among all the processors. Therefore, we take advantages of these two kinds of parallelism to partition these subproblems equally for all the processors in the backward phase.

In order to meet this requirement, we further separate the backward phase into two sub phases: the collective solving subproblem phase and the individual solving sub-problems phase, to dynamically assign tasks to all the processors. During the collective solving phase, we first detect whether the unsolved subproblem queue is in the "*balanced state*" or not by Eq (3), where M and N are the total processor and sub-problem number, respectively. $Size_i$ is defined as the area for each subproblem and $Size_{pj}$ is the total area of subproblems assigned to the jth processor. If the difference of the sum area of the subproblems assigned to each processor is within the *Threshold* value (default value is 10%), the subproblems can be considered to enter the "*balanced state*", indicating that the whole subproblems can be distributed equally to each processor.

$$Size_{average} = \sum_{i=0}^{N} Size_i / N$$

$$\left| \frac{Size_{pj} - Size_{average}}{Size_{average}} \right| < Threshold, 1 \leq j \leq M \tag{3}$$

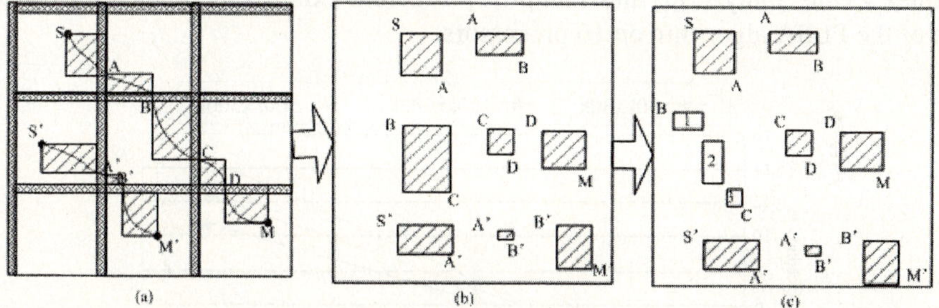

(a) Two maximal score node M and M' with their alignment paths
(b) The subproblems in the unsolved queue cannot be partitioned equally on all processors
(c) The largest subproblem (determined by B,C in (b)) is chosen and decomposed into 3 descendent subproblems (1,2,3). This process proceeds until all the subproblems can be distributed to all the processors equally

Fig. 5. Backward phase parallelism

When the unsolved subproblem queue is not in the "*balanced state*", we pop up the subproblem with the largest size from this queue and decompose it into several smaller descendant subproblems using a similar method to that used in the forward phase. After that, these descendant subproblems are pushed back into the unsolved queue. Then a new iteration begins until it reaches the "*balanced state*". Figure 5 demonstrates how the collective phase works, where the largest subproblem (B, C) is decomposed into some smaller descendent ones. This decomposition process proceeds until all the available subproblems can be equally assigned to all the processors.

After the unsolved subproblem queue is in the *"balanced state"* by recursive subproblem decompositions, PLSA enters the individual solving subproblems phase. During this period, each processor works on the assigned subproblems, in serial and independently. Finally, after all the subproblems have been solved, we can collect the sub sequence alignment results and concatenate them to the final optimal alignment paths.

4 Experimental Results

The experiments are carried out on a 16-node PC cluster interconnected with a 1Gbps Ethernet switch. Each node has a 2.8 GHz Intel Pentium 4 processor with a 1GB memory. We use the RedHat 9.0 Linux operating system and MPICH-1.2.5 [14] message passing library as the software environments. All of our implementations of the PLSA algorithm are written from scratch in C++ without reference to any free Smith-Waterman implementations. In order to evaluate the scalability of the algorithm, nucleotides ranging from 30K to 300K length are chosen from a test suite suggested by the bioinformatics group at Penn State University [11] for accurate scaling use.

4.1 Scalability Performance and Analysis

The parallel characteristics of the proposed algorithm are obtained with tools such as the Intel VTune analyzer[7] and Vampir[8]. Figure 6 exhibits the scalability performance of the PLSA algorithm on 16 processors.

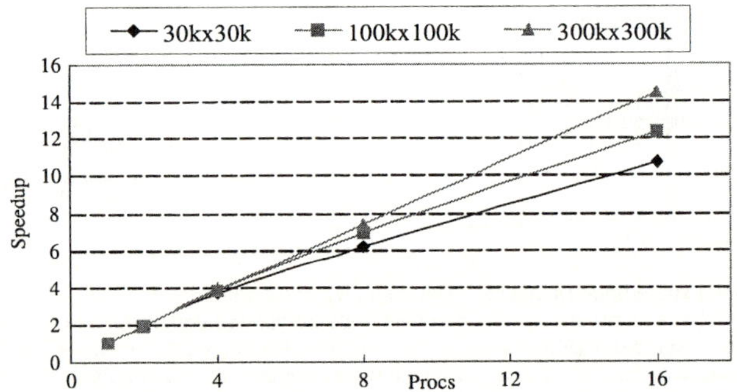

Fig. 6. Speedup performance for different size sequences (3x3 grid divisions)

For small scale sequence alignment such as a 30k*30k problem, the speedup is linear for two and four processors, but starts deteriorating when eight or more processors are used. The slowdown for more processors occurs because the granularity of the work assigned to each processor decreases. Moreover, the backward phase has to decompose the subproblems into even smaller size descendant subproblems to achieve load balance for all processors. Therefore tremendous communication overhead will occurs in this period.

The trend is similar for a 100k*100k sequence alignment. The speedup ascends almost linearly for up to eight processors but gets a little slower for 16 processors. The 12.3x speed-up number indicates that all the processors are not fully utilized. It provides much better granularity for the paralleled tasks, but not enough to satisfy 16 processors. With the increased size of the sequence alignment, we get much better speed-up curves, almost linear speed-up for 300k problems, and our statistical result shows that the parallel scheme can scale well with the problem size. More specifically, a 15.1x speedup can be obtained with 16 processors when outputting three near-optimal paths. Comparing the scalability of the forward and backward phase, Table 1 shows that the forward phase is better than the backward phase with the increase of processors. This is not surprising since the backward phase has to decompose a number of smaller subproblems which will incurs more synchronization and communication overhead.

Table 1. Execution time for 300k size sequence alignment (3x3 grids and 6000x400 block size)

300kx300k	Forward(s)	Forward-Speedup	Backward(s)	Backward Speedup	Total	Speedup
1	4318.08		1752.14		6070.56	1.00
2	2183.18	1.98	946.08	1.85	3129.98	1.94
4	1104.24	3.91	451.68	3.88	1556.35	3.90
8	584.64	7.39	238.32	7.35	822.91	7.38
16	288.34	14.98	132.48	13.23	420.82	14.43

To summarize, PLSA demonstrates very good scalability performance in the backward phase, and it achieves almost linear speed-up with large data sets. We can also expect that PLSA scales well with more processors.

4.2   The Impact of Various Parameters on Parallel Performance

There are several parameters that can be tuned to achieve the best parallel performance. The first parameter is the block size. As shown in Table 2, smaller block size has a better cache locality performance, and is well suited to shared memory parallel systems. The larger block size increases the task granularity and scales well with the typical cluster system, which has long communication latencies.

Table 2. Block size impact for 100k size sequence alignment

Block Size (*height* x *width*)	120 x 80	6000 x400
Average Time on 8 CPUs(s)	193.3	102.0
Average Time on 1 CPU (s)	621.2	743.1

The second parameter is the number of grids. With an increased number of grids, the backward phase will perform fewer re-computations, since the total area of the decomposed subproblems becomes smaller. However, there are tradeoffs between the grid number, memory size, and the additional computations involved with the grids. The last parameter is the number of near-optimal alignments k. The speedup in k

alignments ($k>1$) is usually better than the single alignment since more subproblems can be generated in the backward phase to have better load balance performance.

5   Conclusions

In this paper we propose an algorithm, PLSA, which is very efficient for large-scale sequence alignment in linear space. The basic idea is to trade space for time, where several key techniques have been put forward in our algorithm. The global/local start point and grid cache techniques can divide the whole sequence alignment problem into several smaller independent subproblems. We further exploit different parallelisms in the backward phase, and propose a novel dynamic load balancing approach to efficiently solve these subproblems in parallel. Moreover, a global start point can help to find k near-optimal non-intersecting alignments. The experiments demonstrate that PLSA exhibits good scalability performance, and it can be parameterized to tailor different parallel systems. All these advantages make the PLSA a better choice for large-scale sequence alignment.

References

1. Adrian Driga, Paul Lu, Jonathan Schaeffer, Duane Szafron, Kevin Charter and Ian Parsons, "FastLSA: A Fast, Linear-Space, Parallel and Sequential Algorithm for Sequence Alignment," in the *International Conference on Parallel Processing*, 2003.
2. Aluru, S., Futamura, N., Mehrotra, K., "Biological sequence comparison using pre•x computations," in *Proceedings 13th IEEE International Parallel Processing Symposium*, 653-659, 1999.
3. A.L. Delcher, S. Kasif, R. D. Fleischmann, J. Peterson, O. White, and S. L. Salzberg, "Alignment of whole genomes," *Nucleic Acids Research*, 27(11):2369–2376, 1999.
4. L. Delcher, A. Phillippy, J. Carlton, and S. L. Salzberg, "Fast algorithms for large-scale genome alignment and comparison," *Nucleic Acids Research*, 30(11):2478–2483, 2002.
5. Chunxi Chen and Bertil Schmidt, "Computing Large-scale alignments on a Multi-luster," in the *IEEE International Conference on Cluster Computing*, 2003.
6. D.S., Hirschberg, "A linear space algorithm for computing longest common subsequences," *Comm. ACM*, 18:341-343, 1975.
7. Intel Corp., Intel® Vtune™ Performance Analyzer, (available on-line: http://developer.intel.com/software/products/vtune/).
8. Intel Corp. and Pallas, http://www.pallas.de/pages/vampir.htm
9. Martins, W.S., del Cuvillo, J.B., Cui, W., Gao, G.R., "Whole Genome Alignment using a Multithreaded Parallel Implementation," in *Proceedings 13th Symposium on Computer Architecture and High Performance Computing*, September 10-12, 2001.
10. Myers, E., Miller, W., "Optimal alignments in linear space," *Computer Applications in the Biosciences*, 4:11-17, 1988.
11. Penn State University, *Bioinformatics Group*, http://bio.cse.psu.edu, 2001.
12. Saul B. Needleman and Christian D. Wunsch, "A General Method Applicable to the Search for Similarities in the amino acid Sequence of Two Sequences," *Journal of Molecular Biology*, 48:443-453, 1970.
13. Temple F. Smith and Michael S. Waterman, "Identification of Common Molecular Subsequences," Journal *of Molecular Biology*, 147:195-197, 1981.
14. http://www-unix.mcs.anl.gov/mpi/mpich/

Parallel Multiple Sequence Alignment
with Decentralized Cache Support

Denis Trystram[1] and Jaroslaw Zola[1,2,*]

[1] Laboratoire ID–IMAG, Grenoble, France**
zola@imag.fr
[2] Institute of Computer & Information Sciences
Czestochowa University of Technology, Poland

Abstract. In this paper we present a new method for aligning large sets of biological sequences. The method performs a sequence alignment in parallel and
uses a decentralized cache to store intermediate results. The method allows alignments to be recomputed efficiently when new sequences are added or when alignments of different precisions are requested. Our method can be used to solve important biological problems like the adaptive update of a complete evolution tree
when new sequences are added (without recomputing the whole tree).
To validate the method, some experiments were performed using up to 512 Small
Subunit Ribosomal RNA sequences, which were analyzed with different levels of
precision.

1   Introduction

Multiple sequence alignment (MSA) is one of the most commonly studied problems in
computational biology. It is a general technique utilized in biological sequences analysis
such as structure modeling, function prediction or phylogenetic analysis [1]. Unfortunately, finding an accurate multiple alignment is a hard optimization problem. Firstly,
because it is difficult to provide a formalization which would be satisfactory from the
biological viewpoint. Secondly, having a good model usually means it is algorithmi-
cally very hard to produce the best (or optimal) alignment. Indeed, the Generalized
Tree Alignment Problem (GTA), which has been shown to be the most accurate formalization of MSA, is Max–SNP–Hard [2].

Another factor making MSA a complex problem is the size of the analyzed data. Often, an input dataset contains hundreds of long sequences (e.g. longer than 1000 bp[1]).
This is especially true for biological sequence databases [3]. For example, in June
2004, the Hovergen Database contained 312987 aligned nucleic sequences classified
into 32820 families [3]. While databases like the Hovergen are very useful in molecular
phylogenetic studies, they require an enormous number of computations when updated.
Adding new sequences to an already aligned family of sequences usually requires the
entire alignment to be recomputed from the beginning.

* The work of Jaroslaw Zola has been supported by French Government.
** Laboratory ID–IMAG is funded by CNRS, INRIA, INPG and UJF.
[1] Base Pair (bp) is a basic unit used to express the sequence length. One bp corresponds to one
character.

J.C. Cunha and P.D. Medeiros (Eds.): Euro-Par 2005, LNCS 3648, pp. 1217–1226, 2005.
© Springer-Verlag Berlin Heidelberg 2005

In previous contributions [4, 5] we have reported on two heuristics aimed at solving the GTA problem. In this paper, we extend these approaches. We analyze typical application cases of our alignment procedure (called *PhylTree*) and we characterize sources of reference locality in the alignment computations. Furthermore, we propose a parallel server designed to build alignments of large sequence sets. The server utilizes a decentralized cache to store partial alignments so that they can be reused when a new alignment is requested (i.e. extended with new sequences or a more precise one).

2   A New Method for Multiple Sequence Alignment

2.1   The PhylTree Method

The *PhylTree* method is a generic multiple sequence alignment procedure which has been proposed recently [4]. As it has been presented in detail in our previous papers, we provide here only its basic concepts with no technical details.

The *PhylTree* method was designed to build a multiple alignment and the corresponding phylogenetic tree[2] simultaneously. It can be characterized as an iterative clustering method whose principle is to group closely related sequences [6]. It consists of two successive phases: first it generates a distance matrix for all input sequences (based on the all–to–all pairwise alignments). Then, it searches for the optimal partial solutions which are later combined to obtain the final phylogenetic tree and multiple alignment.

The method uses two principles: *neighbourhood* and *cluster*. A *neighbourhood* is a set of k closely related taxa[3], where k is an input parameter of the *PhylTree* procedure and is a constant integer value (typically, $k \geq 4$, not larger than 10). A *cluster* is a group of $m \leq k$ taxa which creates a part of the final phylogeny.

To determine the clusters the *PhylTree* algorithm generates a neighbourhood for every input sequence, then, it finds the best phylogenetic tree for each neighbourhood, and finally, analyzes all the found trees to extract their common subtrees. These subtrees describe a set of highly related taxa and correspond to clusters. The above process is iterative.

The accuracy of the method depends mostly on the parameter k. Increasing k will widen the search space analyzed to find the clusters. Unfortunately, while this should improve the accuracy of the final solution, it also increases the number of computations required to process a single neighbourhood. Basically, to process a neighbourhood we have to compute all possible $\frac{(2k-2)!}{2^{k-1}(k-1)!}$ trees to find the optimal one.

An important property of *PhylTree* is its genericity, which means that the method can use different alignment algorithms and different scoring functions. Moreover, it is possible to use various definitions of a neighbourhood and neighbourhoods of variable size (a variant called *QuickTree* has been proposed as a way of rapid solving large instances with decreased accuracy).

---

[2] A phylogenetic tree, or phylogeny, represents the evolutionary history of a set of species. In our consideration it is a rooted binary tree whose leaves are the input sequences.

[3] Taxon is an individual, a strain, a species or any unit of classification. In our case, a taxon is a node of the tree.

2.2 PhylTree as a Generic Scheme

As mentioned before, *PhylTree* has been designed to solve the GTA problem. This problem is a formalization of the multiple sequence alignment based on Steiner trees [2]. In general, having a set of sequences S we want to determine an optimal phylogenetic tree T and the corresponding alignment Al.

This basic use of *PhylTree* is referred to as a *single execution*. In this paper we propose an approach based on a decentralized cache support to implement efficiently *PhylTree*. It will allow many variants of the problem, which are of interest to biologists, to be handled easily.

One single execution of *PhylTree* can be formalized as follows: for a given tuple (S, F, a_F, k) find a relevant pair (T, Al), where S is the set of the sequences to align, F describes the alignment method with a scoring function, a_F is a set of scoring function arguments, and k is a tuning parameter for the precision. For example, if F denotes Sankoff's Parsimony with a linear affine gap insertion cost then a_F represents the costs of gap opening and gap continuation.

Such a formulation can be used to describe extensions of the single execution problem: It is very common for a set of sequences with already computed alignment to be extended with new elements. This is especially true for genomic sequence databases with periodic updates. Every time new sequences are added, a *single execution* is performed to obtain an alignment and its corresponding phylogenetic tree for the extended set. This example can be expressed as follows: for a given (F, a_F, k) and $\{S_0, S_1, \ldots, S_l\}$ such that $S_i \subset S_j, i < j$ find $\{(T_0, Al_0), (T_1, Al_1), \ldots (T_l, Al_l)\}$ (sometimes, a relation between sequence sets can be more general: $S_i \cap S_j \neq \emptyset, i \neq j$). Surprisingly, while this situation is very common there are no good solutions able to determine new alignments and new phylogenetic trees based on the previous results.

Another interesting extension is to build alignments with a different level of precision. A series of *single executions* is performed for the same set of input sequences, which differ only by the parameter k. That is: for a given (S, F, a_F) and $\{k_0, k_1, \ldots, k_l\}$ generate $\{(T_0, Al_0), (T_1, Al_1), \ldots (T_l, Al_l)\}$. Of course, it is possible that in some cases parameter k will be changed together with a set of input sequences. More precisely, when expanding a set of previously aligned sequences we may wish to change the precision of the new alignment.

2.3 Related Work

Both multiple sequence alignment and phylogeny are of great importance to biologists, but at the same time these problems are very computationally demanding. That is why parallel and distributed programming is often used to improve the efficiency of existing bioinformatics applications.

One of the most popular programs is the ClustalW package [6]. This tool implements basic MSA algorithms based on the phylogenetic tree approximation. In recent years a few different parallel versions of ClustalW have been proposed [7, 8], designed for both shared and distributed memory architectures. However, in most cases, the parallel approach is limited to the main ClustalW algorithm. While these approaches have been proved to be efficient for a *single execution* case, they make no assumption about possible dependencies between series of executions. Moreover, the accuracy of the solutions generated by ClustalW is usually poor for large amounts of input data. In [9] Catalyurek et al. proposed an implementation of ClustalW based on caching the alignments score. This work, however, is limited to the sequential version of the algorithm, and cache is only utilized to store the scores of the pairwise alignments.

A significant part of the research on parallel bioinformatics is concerned with maximum likelihood methods [1]. However, this class of algorithms is designed to reconstruct the evolutionary history (phylogenetic tree), and the sequence alignment is the tool required to build a proper tree. A good example of such software is the RAxML package. Recently, the authors of RAxML have reported a phylogeny inference of 10000 taxa [10]. While this result is really impressive, it cannot be directly compared with the results of the multiple alignment, since the problems are slightly different.

2.4   Alignment Reference Locality

The basic idea of the *PhylTree* method is a greedy exploration of the partial alignments search space. The exploration is performed through the analysis of the neighbourhoods. A set of multiple alignments is performed for every neighbourhood. If two neighbourhoods share two or more common elements, their analysis will require some common computations. Obviously, the alignments computed for the same elements of the first and second neighbourhood will be the same. Furthermore, the analysis of a single neighbourhood of size k requires $\frac{k!}{2 \cdot (k-2)!}$ distinct pairwise alignments to be computed. On the other hand, all possible pairwise alignments are generated in the first phase of the method. Consequently, the alignments computed in the first phase of the *PhylTree* processing can be used in the second phase, and the results computed during the second phase can be reused from one iteration to the next, or even within the same iteration. In fact, our experiments showed that in some cases only 20% of computations have to be actually performed as the others are redundant.

The properties described above relate to a *single execution* of *PhylTree*. Of course, the same features hold for the cases described in Section 2.2. If we consider a series of *single executions*, then all requests (executions) have very similar characteristics, e.g. the same evolutionary model and common (or the same) input dataset. Therefore, the intermediate results, generated by a query in the series, have the potential to be reused by other requests. For example, if a series of *single executions* is based on the same set of input sequences but different values of k are requested, then all pairwise alignments computed during the first execution can be reused in all subsequent executions. Hence, a distance matrix for the whole group of requests needs to be computed once only.

3   Parallel Server and Cache Support

The *PhylTree* method has been used to build a PhyloBuilder server. The server allows a series of *single executions*, submitted by one or several users, to be run. Our server was designed to work under the control of a batch queuing system, for example SGE or PBS. Users submit their computation (request) via some kind of interface, for example a web page, and for every request a *single execution* script is generated. This script is submitted to the dedicated scheduler queue. Next, each request is executed using our parallel server, and a persistent cache is used to store partial results. Users requests are dynamic. This means that we are not able to predict which parameters or what kind of input will be used in the request.

3.1   Parallel PhylTree

The *PhylTree* method provides good quality results at the expense of being time consuming for most real–life applications. Even if caching is applied, the number of computations to perform remains large. But, at the same time, the *PhylTree* design makes it easy to parallelize.

We have chosen a distributed master-worker architecture with arbitrary selected master node. This choice was based on the following observations: (i) The first phase of *PhylTree* is an independent task consisting of $\frac{(n^2-n)}{2}$ pairwise alignments. As a result, it is easy to parallelize. However, the parallelisation should support heterogeneous architectures. Primarily, because two different pairwise alignments may require different computation times (depending on the lengths of the input sequences). Secondly, because caching may change the actual number of computations which have to be performed by a given worker. (ii) The second phase of *PhylTree* is an iterative process. At each iteration, a set of neighbourhoods is processed, and a single iteration is completed by the distance matrix update. Thus, each iteration contains a single synchronization point. Moreover, the number of neighbourhoods per iteration is typically close to / less than the number of available workers. Hence, the assignment of neighbourhoods to workers may result in large idle times at the end of each iteration.

Summarizing, the parallel *PhylTree* method proceeds as follows: First, the master processor reads the input sequences and broadcasts them to all workers. Next, each worker receives a part of the distance matrix to compute. At this stage, we use the guided self–scheduling strategy, since it allows the load imbalance imposed by the heterogeneous environment to be minimised. Processing the distance matrix, a worker analyzes its efficiency by measuring the number of base pairs aligned per second. Thanks to this analysis, the master node can rank all the workers accordingly to their efficiency. In the second phase, only the processing of the neighbourhoods is parallelized. That is: at each iteration, the master determines the neighbourhood sets, and then, it starts to generate all possible tree topologies for each neighbourhood. Each worker receives its part of the generated topologies and looks for one with the highest alignment score. To achieve this objective, it has to compute MSA for every single topology. Because the number of created tree topologies for all neighbourhoods is usually much greater than the number of workers, again dynamic scheduling is used. However, at this stage, the host priorities computed in the first phase are utilized. Half of the total number of topologies is distributed proportionally to the workers' priorities. Then, the other part is distributed using the guided self–scheduling. When a worker completes its part of the computations it sends back the resulting best tree topology and the corresponding alignment to the master. To complete an iteration, the master computes the clusters and updates the distance matrix accordingly.

3.2   Cache of Alignments

In order to remove redundant computations, and so improve the efficiency of the *Phyl-Tree*, we have designed and implemented an *alignment cache*.

The purpose of the *alignment cache* is to store and manage all intermediate alignments so that they can be reused in future computations. An important issue here is to design a caching system is such a way that the cache management cost will not offset the performance improvement obtained from using cached results. Another concern is how to store the alignment results. Cached data should contain all the information required when it is reused, for example, a description of the underlying tree topology. This requirement is a direct consequence of the *PhylTree* genericity – some of the alignment procedures may use various help structures (e.g. alignment profile). Finally, since we are dealing with a parallel version of *PhylTree*, the cache system must be able to work in the distributed environment.

To accomplish the requirements described above, we have implemented a decentralized caching system based on the CaLi framework [11]. Our solution consists of two subsystems which differ in their management policies, and store different types of alignments. The first subsystem is dedicated to caching pairwise alignments. It is managed using the well known LRU policy and is replicated among all the workers when the first phase of *PhylTree* is completed. The second subsystem is responsible for storing only multiple alignments. It is managed using a variant of Greedy–Dual Size (GDS) policy [12] and it is distributed among the worker nodes.

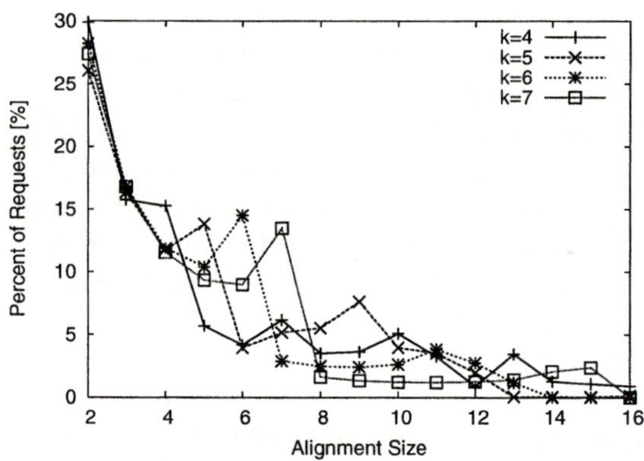

Fig. 1. Percentage of requests for an alignment of given size in a *single execution*. The same input data set was used for each execution.

The are several reasons why we distinguish between pairwise and multiple alignment. Pairwise alignments are the most frequently computed alignments in every execution (see Fig. 1). The time required to compute a pairwise alignment is much shorter than the time required to compute a multiple alignment. Finally, all pairwise alignments computed in the first phase of *PhylTree* will be reused in the second phase. On the other hand, the multiple alignment computations take place only in the second phase and we are not able to predict *a priori* which of them will be the most often requested. In addition, multiple alignment's popularity decreases with its size, but its cost (time to compute) increases. Unfortunately, because *PhylTree* may use various alignment procedures, it is not possible to describe the exact dependencies between the size, cost and popularity of alignments.

Figure 1 shows the percentage of requests for an alignment of given size in a *single execution* of various levels of precision. As it can be seen, in every execution nearly 30% of requests is related to the pairwise alignments. This can be explained by the fact that *PhylTree* creates only several sequences (clusters) per iteration. As a result, in almost every iteration some pairwise alignments are requested. Of course, this general tendency may differ slightly depending on the size and type of the input dataset.

3.3   Cache Implementation

As we have already pointed out, our caching system utilizes two distinct subsystems. While these subsystems are based on different management strategies, they use similar techniques to create and describe the entries stored in the cache.

A single alignment inserted into the cache is compressed and written in the binary file. Such an entry is identified by a unique key, generated as follows. First, the sequences to be aligned and the tree topology are digested using a hashing function, for example SHA–256. Then, the resulting fingerprint is extended with the identifier of alignment method F and its parameters a_F. Each time a worker is requested to compute an alignment, it first generates the key and then queries the proper cache (depending on the number of input sequences). If a hit occurs, the requested alignment is read and the identifiers of the component sequences are compared with the identifiers of the input sequences. This mechanism allows us to detect possible collisions of keys. If no collision is found, the alignment loaded from the cache is used. Otherwise, that is in the case of collision or cache miss, it is computed, and its result is inserted into the cache.

The pairwise alignment cache is a local storage system replicated among the workers. Because the number and the size of pairwise alignments are small compared to the total size of data generated during a *single execution*, the time of replication is not really significant compared to the total processing time. Replication guarantees that all the pairwise alignments will be served by the local cache in the second phase. The pairwise alignment cache is managed using the LRU policy. As a result, each pairwise alignment computed in the first phase remains in the cache during the second phase. Of course, we assume here that the capacity of the pairwise alignment cache is large enough to store the alignments generated by at least one *single execution*.

The management of the multiple alignment cache is a more challenging task. Because of the distribution of the computations, the same alignment may be requested or computed by different workers simultaneously. Additionally, the alignment computed by a given worker in one iteration may be requested by another one during another iteration. Therefore we have implemented the multiple alignment cache as a decentralized, content–addressable system. Every key describing the cache record is mapped to one of the workers, which becomes the *delegate worker* for a given key. When a worker requests a multiple alignment, the caching system first checks the local storage. If no proper entry is found, the request is forwarded to a *delegate worker* cache. If a miss occurs, the worker computes the requested alignment and then inserts the result into the local cache and into the *delegate worker* cache. If a remote hit occurs, the requested entry is inserted into the local storage. In this way the multiple alignment cache is partially replicated, which in turn increases the number of local hits, and the application of a good hashing function guarantees the uniform distribution of the cache entries.

The application of content–addressing has a very important advantage: If the requested alignment is not cached locally, it means that it has been not used by a given worker yet. At the same time, this alignment could have been computed by another worker and be already present in the *delegate worker* cache. Hence, in the worst case only one remote request is necessary to check if a given alignment has to be computed or not.

4   Performance Evaluation

We performed a set of experiments with actual biological data to validate our approach. We randomly created several groups of between 32 and 512 sequences, coming from SSU ribosomal RNA. The average length of the analyzed sequences was 1300 bp varying from 1200 bp to 1400 bp. These sequences were then analyzed using our parallel server.

Our experiments were performed on a small cluster of 7 SMP nodes connected by a GbitEthernet network. Every node was equipped with a dual Itanium2 CPU (one CPU used by the server and one by OS), 4GB of RAM and was running under Linux. A single node could use 32MB of storage for a pairwise alignment cache and 128MB for a multiple alignment cache. The SCSI disk storage was managed by a ReiserFS file system which allows many large files to be handled efficiently.

4.1   Experiments

In the first experiment, we generated a set of requests to compute MSA with a different level of precision for $|S| = 64$ sequences. We started our simulation with an empty cache submitting requests with parameter $k_0 = 4$. For each further request, the value of k was increased by 1. Table 1 shows the results of request processing for different values of k.

In the next experiment we examined efficiency of our approach when the set of input sequences is extended in every execution. In the first request we used a set of $|S_0| = 32$ sequences and then, we doubled the number of sequences in each request: $S_i = S_{i-1} \cup S_x$, where S_x is a group of new input sequences of the same cardinality. In every request, we have used $k = 5$ as a level of precision. Table 2 presents the time of request processing depending on the size of the input dataset.

In both experiments we utilized Sankoff's Parsimony with the gap opening cost equal to 2.0 and gap continuation cost equal to 1.0.

4.2   Discussion

As we could expect, the caching technique noticeably improved the performance of the server in both of the described experiments. Table 1 shows that the hit ratio Hr, and cost hit ratio Cr (defined as the cost of alignments found in the cache divided by the total cost of alignments requested during the execution, where the cost of alignment is the time required to compute it) increases with every execution. This can be explained by the fact that the results of previous computations are utilized, and an increase in the size of neighbourhoods results in a higher redundancy of computations. The cost saving

Table 1. Results of request processing for different values of k, where T_p is the execution time for the parallel server and T_c is the execution time for the parallel sever with cache support. Hr_p and Hr_m are respectively the hit ratio for the pairwise and multiple alignment cache. Cr_p and Cr_m describe the cost hit ratio for the pairwise and multiple alignment cache, and $Cs = 1 - \frac{T_c}{T_p}$ is the cost saving ratio for the whole cache system.

k	$T_p[s]$	$T_c[s]$	Hr_p	Hr_m	Cr_p	Cr_m	Cs	$E = \frac{T_p}{T_c}$
4	3940	2155	0.42	0.28	0.085	0.36	0.45	1.82
5	18330	8591	1.0	0.31	0.096	0.43	0.53	2.13
6	178517	74092	1.0	0.37	0.08	0.50	0.58	2.41

Table 2. Results of request processing for different sizes of the input dataset S.

$|S|$	$T_p[s]$	$T_c[s]$	Hr_p	Hr_m	Cr_p	Cr_m	Cs	$E = \frac{T_p}{T_c}$
32	6747	3430	0.87	0.28	0.099	0.39	0.49	1.96
64	18182	9123	0.82	0.29	0.079	0.41	0.49	1.99
128	55300	24796	0.74	0.34	0.066	0.48	0.55	2.23
256	156831	63631	0.63	0.34	0.067	0.53	0.59	2.46
512	438539	144090	0.43	0.45	0.048	0.63	0.67	3.04

ratio Cs is dominated by the cost hit ratio of the multiple alignment cache. This confirms our claim that it is more profitable to cache multiple alignments. The cost saving generated by the pairwise alignment cache is lesser than we expected, however it is still significant. Similar tendencies can be observed in the second experiment (Tab. 2), except that the pairwise alignment cache hit ratio decreases. This is because of a geometric increase in the input dataset size, which in turn increases the relative number of multiple sequence alignment computations. The cache hit ratio for the multiple sequence alignment cache stabilizes around 35% for the input of 256 sequences. At this point, the cache is saturated and the cache replacement policy is used. In spite of this, the cost saving ratio, as well as cache hit ratio, increases. This is possible thanks to the application of the cost–aware replacement policy, that is GDS. In fact, our trace driven simulations have shown that GDS attains the highest hit ratio and cost saving ratio in comparison with other strategies, like LRU, LFU or other size–based policies.

The presented results show that the efficiency of the server depends on the size of the input data and the required precision. More precisely, the cost saving ratio will be minor for short sequences, and analysis with small k will result in a low cache hit ratio. The same factors will influance the scalability of our system. Increasing the number of workers should allow larger problems to be solved with better precision rather than achieving better performance for small data.

In our experiments we did not compare the parallel and sequential versions of the server. This is because the sequential version is too memory consuming, e.g. for $k > 5$ alignment of a few sequences becomes an out–of–core problem. For the same reason only one server process is executed on the SMP node.

5   Conclusions

In this work we presented a new cache–based approach to solving the parallel multiple sequence alignment problem. We condacted a formal analysis and we verified by experiments that the application of decentralized caching can substantially improve the efficiency of the alignment computations in both, a *single execution* and series of *single executions*. We believe that our approach can be combined with other existing MSA software, such as, for example, ClustalW or 3D-Coffee.

In our considerations we assume that all requests share the same alignment procedure. The problem of how to use the results of previous computations when alignment with a different evolution model is requested remains open. In particular, is it possible to exclude some of the tree topologies during neighbourhood processing knowing that they were of little value when analyzed with different parameters.

Both, our parallel server and detailed results of MSA can be accessed on–line via the https://hal.icis.pcz.pl/PhyloServer web page.

References

1. Holder, M., Lewis, P.O.: Phylogeny estimation: traditional and bayesian approaches. Nature Reviews Genetics **4** (2003) 275–284
2. Jiang, T., Lawler, E.L., Wang, L.: Aligning sequences via an evolutionary tree: complexity and approximation. In: ACM Symp. on Theory of Computing. (1994) 760–769
3. Duret, L., Mouchiroud, D., Gouy, M.: HOVERGEN, a database of homologous vertebrate genes. Nucleic Acids Res. **22** (1994) 2360–2365
4. Guinand, F., Parmentier, G., Trystram, D.: Integration of multiple alignment and phylogeny reconstruction. In: Eur. Conf. on Comp. Biology, Poster Abstr. (2002)
5. Parmentier, G., Trystram, D., Zola, J.: Cache-based parallelization of multiple sequence alignment problem. In: Proc. of Euro-Par '04. (2004) 1005–1012
6. Higgins, D., Thompson, J., Gibson, T.: CLUSTALW: improving the sensitivity of progressive multiple sequence alignment through sequence weighting, position-specific gap penalties and weight matrix choice. Nucleic Acids Res. **22** (1994) 4673–4680
7. Li, K.B.: ClustalW–MPI: ClustalW analysis using distributed and parallel computing. Bioinformatics **19** (2003) 1585–1586
8. Mikhailov, D., Cofer, H., Gomperts, R.: Performance optimization of ClustalW: Parallel ClustalW, HT Clustal, and MULTICLUSTAL.
http://www.sgi.com/industries/sciences/chembio/resources/clustalw (2005)
9. Catalyurek, U., Ferreira, R., Kurc, T., Saltz, J.: Improving performance of multiple sequence alignment analysis in multi–client environments. In: Proc. of HiCOMB '02. (2002)
10. Stamatakis, A., Ludwig, T., Meier, H.: Parallel inference of a 10.000–taxon phylogeny with maximum likelihood. In: Proc. of Euro-Par '04. (2004) 997–1004
11. Zola, J.: CaLi – generic computational buffers library. http://icis.pcz.pl/~zola/CaLi (2005)
12. Balamsh, A., Krunz, M.: An overview of web caching replacement algorithms. IEEE Comm. Surv. & Tutor. **6** (2004) 44–56

Parallel Construction of Large Suffix Trees on a PC Cluster

Chunxi Chen and Bertil Schmidt

School of Computer Engineering,Nanyang Technological University, Singapore
{pg03452644,asbschmidt}@ntu.edu.sg

Abstract. The suffix tree is a key data structure for biological sequence analysis. Even though efficient algorithms for suffix tree construction exist, for long DNA sequences such as whole human chromosomes, their run-time is still very high . In this paper we introduce a new parallel algorithm for suffix tree construction. This algorithm uses a new data structure call the *common prefix suffix tree* (CPST). Our parallel implementation on a PC cluster leads to significant run-time savings.

1   Introduction

The suffix tree is a compact trie of all suffixes over a string. It is a key data structure in the field of bioinformatics, since it permits very efficient solutions to many string based problems. Examples include exact and approximate substring matching, the longest common substring problem and the maximal repetitive structures problem [8]. Consequently, many widely used large-scale bioinformatics applications have achieved amazing performance using suffix trees, such as MUMmer [5], REPuter [15], and OASIS [17].

Several linear-time algorithms for suffix tree construction have been introduced (see [8] for a summary). Among them, Ukkonen's algorithm is most widely used. The key feature of Ukkonen's algorithm is to make use of *suffix links*, which allow the incremental construction of suffix trees. Unfortunately, these algorithms are impractical for constructing large size suffix trees because of high memory overheads. For example, the suffix tree of the whole human chromosomes of length 3 Giga base pairs (GBp) using the advanced space saving optimization requires 30 to 50 gigabytes of memory [14]. Therefore, new suffix tree construction approaches are required in bioinformatics because biological sequences typically have very large size and sequence datasets are growing at an exponential rate [20].

In order to tackle the memory bottleneck problem in constructing a large size suffix tree, researchers have tried several approaches. We summarize this research work into four categories:

1) `Space saving optimizations.` This approach exploits various kinds of redundancies in suffix trees to obtain more space efficiency[14]. However, the internal structure of suffix trees doesn't permit very significant space saving optimization without any sacrifice of suffix tree virtues.

J.C. Cunha and P.D. Medeiros (Eds.): Euro-Par 2005, LNCS 3648, pp. 1227–1236, 2005.

2) `Disk-based approaches`. Disk-based approaches [7, 9, 19] hold the suffix trees in second memory. Unfortunately, suffix tree construction has poor memory locality since it requires a semi random walk over the tree as it is constructed [6]. Therefore, large-size suffix trees that will not fit in memory would take an unacceptably long time to construct and be accessed due to excessive page faulting.

3) `New data structures`. Another method is to develop alternative data structures which store less information than suffix trees and therefore have lower memory overheads. Some new data structures are *suffix array* [18], *level compressed trie* [1], *suffix binary search tree* [10], *suffix cactus* [12] and *PT-tree* [4]. This approach has the following two common shortcomings [14]. Firstly, they are specifically designed for certain applications and can not be adapted to other kinds of problems without severe performance degradation. Thus, they are not as versatile as suffix trees. Secondly, direct construction of these data structures is usually slower than suffix tree construction.

4) `Constructing suffix trees in parallel`. This approach uses the idea of processing sub-trees independently. Once all the sub-trees have been constructed it is possible to merge them together to form a complete suffix tree. We call this the *sub-tree idea*.

In this paper, we are using a PC cluster to parallelize suffix tree construction. A similar approach has been previously used in [3] and [2]. Unfortunately, [3] only gives some actual experiments on a binary alphabet, which is not relevant in practice; [2] constructs suffix trees not in a cluster, but a SMP machine with 4 CPUs and large memory. The main contributions of this paper are as follows:

1) `Presentation of a data structure with the corresponding` $O(n)$-`time construction method`. The data structure is called *common prefix suffix tree* (CPST). All suffixes in a CPST share a common prefix. A standard suffix tree can be divided into a number of CPSTs. Each CPST can be tackled independently by one node in a parallel environment. We present an algorithm that permits a linear time construction of CPSTs.

2) `Implementing the proposed method efficiently on a PC cluster`. The major difficulty of constructing a large suffix tree inside a cluster arises from the need to access the whole input sequence while constructing CPSTs. Our solution is to set aside several *data-servers* which hold the whole sequence. Processes constructing CPSTs then can access the sequence through communicating with these *data-servers*.

The rest of the paper is organized as follows. In Section 2, we provide the preliminaries of suffix trees. In Section 3, we give the description of the CPST and the algorithm for linear time construction. The parallel implementation on a PC cluster is described and evaluated in Section 4. Finally, Section 5 concludes our paper and with an outlook to further research.

2   Preliminaries

A suffix tree for a string S of length L is a rooted directed tree with exactly L leaves numbered 1 to L. For any leaf i, the concatenation of the edge labels on the path from the root to the leaf exactly spells out the suffix of S that starts from location i. Assume xs is a string over an alphabet Σ, where $x \in \Sigma$ and $s \in \Sigma^*$. In a suffix tree, for an internal node A with path-label (from the suffix tree root to the node) xs, there exists another node B with path-label s, Then the pointer from B to A is called a *suffix link*. The reason that *suffix links* are of interest is that they permit the suffix tree construction in linear time [8]. The suffix tree with corresponding *suffix links* for the string $S = accattgaagcgttaccagttat\$$ is shown in Figure 1.

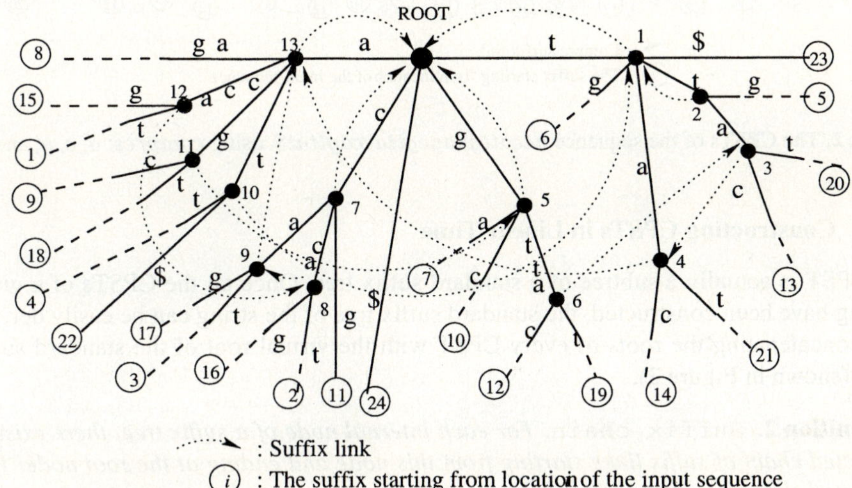

Fig. 1. The suffix tree of the sequence *accattgaagcgttaccagttat*$.

3   Constructing a Suffix Tree with CPSTs

3.1   CPST: Common Prefix Suffix Tree

A way of dividing the problem (constructing suffix trees) into smaller sub-problems is to group the suffixes of a string first and then construct suffix trees for each suffix group. All suffixes of a string can be grouped according to the prefixes of each suffix. We define a *common prefix suffix tree* (CPST) of a string S to be a compact trie of a subset of the suffixes of the string which start with a same prefix (shown in Definition 1).

Definition 1. Common prefix suffix tree (CPST)*: For a given string S and a substring "compre" of S, a common prefix suffix tree, denoted as $CPST(S, compre)$, is the compact trie of all suffixes of S which start with compre.*

Figure 2 shows four CPSTs for the string $S = accattgaagcgttaccagttat\$$ and the common prefixes a, c, g, and t, i.e. $CPST(S, a)$, $CPST(S, c)$, $CPST(S, g)$, and $CPST(S, t)$. All CPSTs have fictitious connections to a root node. Considering the case *compre* $= g$, all suffixes in S starting with the g are *gaagcgttaccagttat*$, *gcgttaccagttat*$, *gttaccagttat*$, and *gttat*$. The trie of all these suffixes are presented by he $CPST(S, g)$.

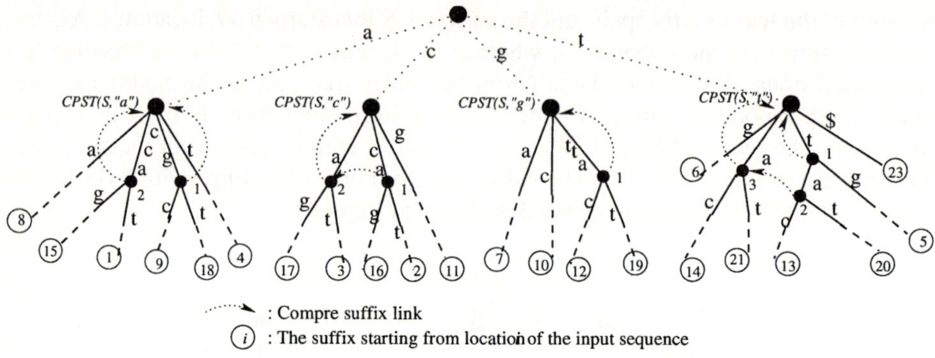

Fig. 2. The CPSTs of the sequence $accattgaagcgttaccagttat\$$. using *compres*: a, c, g, and t.

3.2   Constructing CPSTs in Linear Time

A CPST is actually a subtree of a standard suffix tree. Once all the CPSTs of a given string have been constructed, the standard suffix tree of the string can be easily derived by concatenating the roots of every CPST with the virtual root of the standard suffix tree (shown in Figure 2).

Definition 2. `Suffix chain`: *For each internal node of a suffix tree, there exists a directed chain of suffix links starting from this node and ending at the root node. This directed chain is called suffix chain.*

For example in Figure 1, The *suffix chain* for internal node 3 is: $3 \rightarrow 4 \rightarrow 13 \rightarrow root$.

Definition 3. `Compre suffix link`: *Given are two internal nodes A and B inside the same CPST. We define a compre suffix link from A to B, if A and B are part of the same suffix chain in the standard suffix tree, where A is before B and no other internal node of the CPST lies between A and B on this suffix chain. We also define a compre suffix link from B to the root of the CPST, if no other node inside the CPST is part of the suffix chain between B and the root of the standard suffix tree.*

Let's consider the *suffix chain* illustrated in Figure 3: $F \rightarrow A \rightarrow C \rightarrow D \rightarrow B \rightarrow E \rightarrow root$. Node A and B belong to $CPST_2$. We draw a *compre suffix link* from A to B, since the part of *suffix chain* between A and B ($A \rightarrow C \rightarrow D \rightarrow B$) doesn't contain any other nodes of $CPST_2$. Additionally, there is a *compre suffix link* from B to the root of $CPST_2$.

In order to simplify the description, we use $s[i, j]$ to denote the substring of S starting at location i and ending at location j. $len(compre)$ denotes the length of *compre*. $pathlabel(N)$ denotes the characters on the path from the root to the internal node N. If suffix i starts with the prefix *compre*, we say suffix i is *valid* for $CPST(S, compre)$. The edge characters from node A to B is denoted as $e(A, B)$.

Theorem 1. *Given an internal node A of CPST(S, compre) with $pathlabel(A) = S[l_1, l_2]$. Assume suffix l_3 is the next valid suffix for CPST(S, compre) and $l_3 +$*

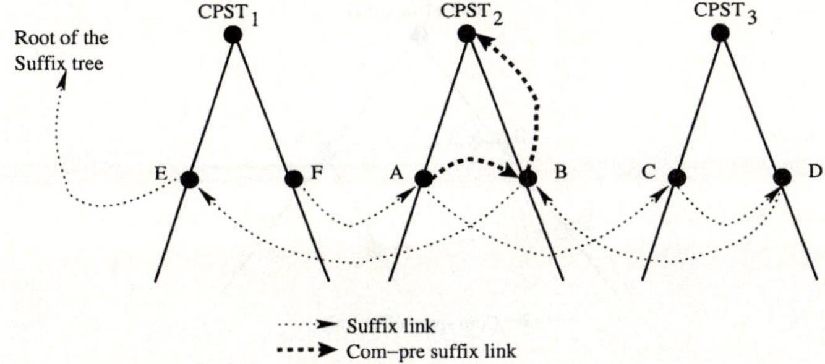

Fig. 3. An example for the *compre suffix link*.

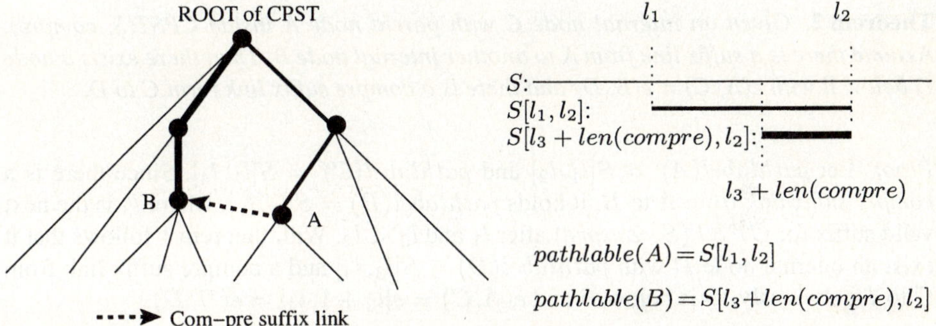

Fig. 4. The illustration for Theorem 1.

$len(compre) \leq l_2$. *Then there exists an internal node B with* $pathlabel(B) = S[l_3 + len(compre), l_2]$ *and a compre suffix link from A to B.*

Proof. The proof has two parts.

1. EXISTENCE OF B. Since A is an internal node, there exists at least two substrings of S with $pathlabel(A) = S[l_1, l_2] = S[l_4, l_5]$. Hence, there are also two substrings of S with $S[l_3 + len(compre), l_2] = S[l_6, l_5]$. Therefore, there must be an internal node B with $pathlabel(B) = S[l_3 + len(compre), l_2]$.

2. EXISTENCE OF COMPRE SUFFIX LINK FROM A TO B. We show there exists a direct chain of *suffix links* from A to B by induction over $n = l_3 + len(compre) - l_1$. The claim then follows since A and B are inside the same $CPST$.

    BASIC STEP: $n = 1$. Obviously, there is a directed *suffix link* from A to B.
    INDUCTIVE STEP: According to induction hypothesis, there is a directed chain of *suffix links* from A to a node C with $pathlabel(C) = S[l_3 + len(compre) - 1, l_2]$. Since there must also be a *suffix link* from C to B, it can be concluded that there is a directed chain of *suffix links* from A to B.

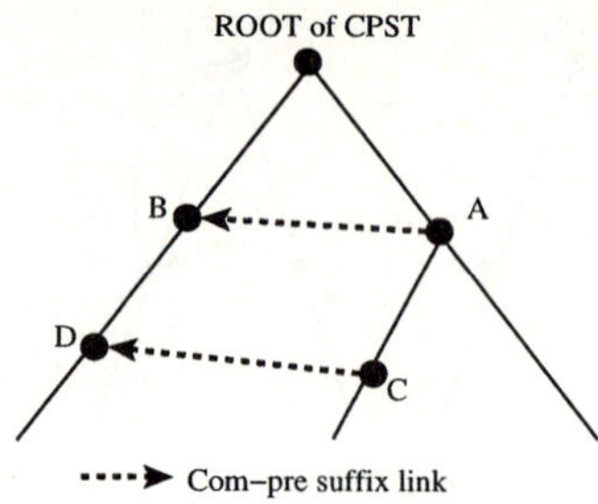

Fig. 5. The illustration for theorem 2.

Theorem 2. *Given an internal node C with parent node A inside CPST(S, compre). Assume there is a suffix link from A to another internal node B. Then there exists a node D below B with e(A, C) = e(B, D) and there is a compre suffix link form C to D.*

Proof. Let $pathlabel(A) = S[l_1, l_2]$ and $pathlabel(C) = S[l_1, l_4]$. Since there is a *compre suffix link* from A to B, it holds $pathlabel(B) = S[l_3, l_2]$, where l_3 is the next valid suffix for $CPST(S, compre)$ after l_1 and $l_3 < l_2$. With theorem 1 follows that it exist an internal node D with $pathlabel(D) = S[l_3, l_4]$ and a compre suffix link from C to D. obviously, D is below B and $e(A, C) = s[l_2 + 1, l_4] = e(B, D)$.

Our algorithm constructs a CPST through orderly inserting *valid* suffixes for the CPST. In [8], the introduction of $suffixlinks$ permits the usage of the skip/count trick which makes the Ukkonen's algorithm be in linear time. the $compre_suffix_link$ in CPSTs is the counterpart of $suffixlinks$ in standard suffix trees according to theorem 2. It can locate the next node in the CPST through using the skip/count trick instead of traversing the CPST from its root. Based on the definitions and theorems above, the algorithm of constructing $CPST(S, compre)$ can be described as follows:

CPST construction algorithm:

```
Input:    String S = α$, where α ∈ Σ*, $ ∉ Σ, and Σ is a finite alphabet.
          Common prefix compre ∈ Σ* with |compre| < |α|
Output:   CPST(S, compre)
```

$N = number_of_valid_suffixes(S, compre);$
IF $(N == 0)$ RETURN $(nil);$
FOR $i = 1$ TO N BEGIN
    $\nu(i) =$ starting position of the ith valid suffix in S;
END
$current_node = CPST_root;$
$theorem2_flag_node = current_node.father.compre_suffix_link;$
FOR $i = 1$ TO N BEGIN
    IF $((current_node == CPST_root)||(theorem2_flag_node == CPST_root))$
        $new_nodes_info = traversal(current_node, \nu(i), S);$

```
ELSE
    new_nodes_info = skip_count(theorem2_flag_node, current_node.edge_labels);
create_new_nodes(new_nodes_info)
create_new_CompreSuffixLink(new_internal_node, old_internal_node)
current_node = new_internal_node;
theorem2_flag_node = current_node.father.compre_suffix_link;
END
RETURN (CPST_root);
```

Theorem 3. *For an input sequence S and a substring compre, $CPST(S, compre)$ can be constructed in linear time.*

Proof. Our algorithm constructs a $CPST(S, compre)$ through orderly inserting *valid* suffixes for the CPST. Assume that the insertion of *valid* suffix V_i results in a new internal node A with $pathlabel(A) = S[V_i + len(compre), l_i]$. The time complexity for this assertion is $O(l_i - V_i)$. For *valid* suffixes whose starting locations are in the range $[V_i, l_i]$ (such as suffixes V_{i+1} and V_j), we can use the skip/count trick [8] to insert them according to theorem 2. The time complexity for these insertions using the skip/count strick are $O(m)$, where m is the number of suffixes whose starting locations are in the range $[V_i, l_i]$. Hence, the time complexity for inserting all *valid* suffixes whose starting locations are in the range $[V_i, l_i]$ is $O(ll_i - V_i) + O(m)$. Obviously, it is linear to the length of the range. The whole input string is composed by these ranges. Therefore, the insertions of all *valid* suffixes can be accomplished in linear time.

4   Parallel Implementation and Performance Evaluation

4.1   Input DNA Sequence

The DNA sequence used in this paper is human chromosome NC_000001.4 which is downloaded from [20]. The alphabet of actual DNA sequences consists of 16 characters, in which a, c, g, and t represent the four bases of DNA and r, y, w, s, m, k, b, d, h, and v represent undetermined base-pares. In the paper, we only consider the determined bases a, c, g, and t. For example, the human chromosome NC_000001.4 extracted by us is of length 222,827,884 bp.

4.2   Prefix Distribution in DNA Sequences

The purpose of presenting the new data structure called CPST is to divide a large-size suffix tree into a number of smaller size CPSTs first and then each CPST can be processed independently. This idea presumes that the suffix trees can be divided efficiently using CPST. However, this might not be possible for systematically biased sequences. Let's consider a worst case. For a sequence $S = aaaaaaaaaaaaaaaaaaaaaaaa$, all the suffixes of the sequence start with same prefix a. Thus, the idea of CPST is inefficient.

Fortunately, systematically biased sequences rarely occur in practice. The appearance of the 4 symbols $a, c, g, and\ g$ in actual DNA sequences is almost evenly distributed. This ensures that the number of DNA sequence suffixes starting with different possible prefixes are not severely imbalanced. Here we take human chromosome

NC_000001.4 length of 222,827,884 as an example. Table 1 shows that the number of suffixes starting with different prefixes are well balanced. This means the suffix tree can be divided efficiently into sub-problems using CPSTs.

Table 1. The number of suffixes of the human chromosome NC_000001.4 length of 222,827,884 (only consider a, c, g and t)which start with 1-letter and 2-letter prefixes.

compre	Num of suffixes	compre	Num of suffixes	compre	Num of suffixes	compre	Num of suffixes
a	64875254	c	46493994	g	46483769	t	64974866

compre	Num of suffixes	compre	Num of suffixes	compre	Num of suffixes	compre	Num of suffixes
aa	21191409	ac	11189673	ag	15878823	at	16615349
ca	16200299	cc	12132633	cg	2256627	ct	15904435
ga	13313713	gc	9838754	gg	12121539	gt	11209763
ta	14169833	tc	13332934	tg	16226780	tt	21245318

4.3 Parallelization Strategy

During the course of constructing CPSTs, the input string must reside in memory. This means every process in the parallel environment must allocate enough memory to hold the the input sequence first and then the remaindering memory can be allocated to construct CPSTs. Obviously the efficiency is low when the input sequence is large. This is the key reason some parallel implementations do not scale well.

If a cluster permits fast intra-cluster communication, it is possible that one or more nodes hold the input string while other nodes efficiently access the string by intra-cluster communication. We call this the *sharing input string idea*. Our implementation uses one or more *data-servers* which hold the whole input sequence. The processes constructing CPSTs (*constructors*) access the sequence through communication with these *data-servers*.

In order to decrease the communication between *dataservers* and *constructors*, we introduce the concept of *smallnode* and *largenode*. The communication between *dataservers* and *constructors* consists of two parts: 1) the *constructors* need to access the input string; and 2) the *constructors* need to get the edge labels of a node. The communication in Case 1 has good efficiency since the *constructors* can get a whole block of the substring a time. The highly frequent and low efficient communication comes from the Case 2, because the number of the nodes is large and the overhead of every communication is high. We classify the nodes of a CPST according to the lengths of their edge lables. The nodes whose edge-label lengths are larger than a criterion (*nodesize*) are called *largenodes*, or else called *smallnodes*. If the edge labels of the *smallnodes* are kept in their CPSTs, the access to these *smallnodes* doesn't need communication. Therefore, the communication in Case 2 will decrease.

4.4 Performance Evaluation

The cluster used in this paper consists of 10 nodes connected by a Gbit/s myrinet switch. Each node comprises two 2.6GHz CPUs and 1 Gigabytes RAM. In our experiments, the length of *compre* is set to 2 and therefore the number of CPSTs is 16. The *nodesize* is set as 10. The number of *constructors* are 4 times that of *dataservers*. Figure 6 shows

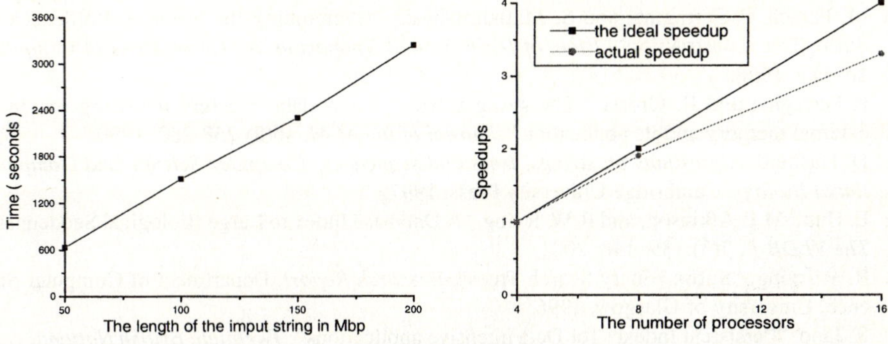

Fig. 6. The left part shows the runtimes for input strings with different lengths; The right part shows speedups using the input string length of 50M.

that the construction time of our implementation is in a linear relationship to the length of the input string. In addition, the speedup is almost linear.

5   Conclusion

The suffix tree is a key data structure for biological sequence analysis. However, construction of a suffix tree for long DNA sequences is made challenging by high memory overheads and poor memory locality. In this paper, we have introduced an efficient parallel algorithm for large-scale suffix tree construction using the CPST data structure. We have shown how a standard suffix tree can be divided into a number of CPSTs. Each CPST can then be processed independently by one cluster node. Our algorithm permits linear-time construction of CPSTs. In order to reduce space while constructing CPSTs inside a cluster, we use one or more *data-servers* which hold the whole sequence inside the cluster. *Constructors* access the input sequence through communicating with these *data-servers*. Our implementation can achieve linear space for a human chromosome DNA sequence.

References

1. A. Andersson and S. Nilsson, "Efficient Implementation of Suffix Trees", *Software-Practice and Experience*, 25(2), 129-141, 1995.
2. A.L. Brown. "Constructing Chromosome Scale Suffix Tree". *the 2nd Asia-Pacific Bioinformatics Conference*. New Zealand, 2004.
3. R. Clifford and M. Sergot. "Distributed and Paged Suffix Trees for Large Genetic Databases". *Journal of Discrete Algorithms*. Accepted.
4. L. Colussi and A. De Col, "A time and space efficient data structure for string searching on large texts", *Information Processing Letters*, 58(5), 217-222, 1996.
5. A. Delcher, A. Phillippy, J. Carlton, and S. Salzberg. "Fast Algorithms for Large-scale Genome Alignment and Comparision." *Nucleic Acids Research*, 30(11):2478-2483, 2002.

6. M. Farach, P. Ferragina, and S. Muthukrishnan. "Overcoming the Memory Bottleneck in Suffix Tree Construction". *Proc. of IEEE Annual Symposium on Foundations of Computer Science*, 1998.

7. P. Ferragina and R. Grossi. "The string B-Tree: a new data structure for string search in external memory and its application." *Journal of the ACM*, 46(2):238-280, 1999.

8. D. Gusfield. *Algorithms on strings, trees and sequences: Computer Science and Computational Biology*. Cambridge University Press, 1997.

9. E. Hunt, M.P. Atkinson, and R.W. Irving. "A Database Index to Large Biological Sequences." *The VLDB J.*, 7(3):139–148, 2001.

10. R.W. Irving, "Suffix Binary Search Trees", *Research Report*, Department of Computer Science, University of Glasgow, 1996.

11. R. Japp. "Persistent Indexes for Data intensive applications". *Twentieth British National conference on Databases*. Coventry, UK, Lecture Notes in computer Science, 2712.

12. J. Kärkkäinen, "Suffix Cactus: A Cross Between Suffix Tree and Suffix Array", *Proc. of the Annual Symposium on Combinatorial Pattern Matching (CPM'95)*, LNCS 937, 191-204, 1995.

13. J. Kärkkäinen and E. Ukkonen. "Sparse Suffix Tree". COCOON'96, LNCS1090, Hongkong, 1996.

14. S. Kurtz. "Reducing Space Requirement of Suffix Trees". *Software Practice and Experience*, 29(13):1149–1171, 1999.

15. S. Kurtz and C. Schleiermacher. "REPuter: Fast Computation of Maximal Repeats in Complete Genomes." *Bioinformatics*, 15(5):426-427, 1999.

16. U. Manber and E.W. Myers, "Sufix Arrays: A New Method for On-line String Searches", *SIAM Journal on Computing*, 22(5), 935-948, 1993.

17. C. Meek, J. M. Patel, and S. Kasetty. "OASIS: An Online and Accurate Technique for Local-alignment Searches on Biological Sequences." In *VLDB*, 2003.

18. G. Navarro, R. Baeza-Yates, and J. Tariho. "Indexing Methods for Approximate String Matching." *IEEE Data Engineering Bulletin*, 24(4):19–27, 2001.

19. S. Tata, R.A. Hankins, J.M. Patel. "Practical Sufix Tree Construction." in *proceedings of the 30th VLDB Conference*, Toronto, 2004.

20. The Growth of GenBank, NCBI, 2004. http://www.ncbi.nlm.nih.gov/genbank/

21. MPICH project: http://www-unix.mcs.anl.gov/mpi/mpich/

Parallel Edge-Based Inexact Newton Solution of Steady Incompressible 3D Navier-Stokes Equations

Renato N. Elias, Marcos A.D. Martins, and Alvaro L.G.A. Coutinho

Center for Parallel Computations and Department of Civil Engineering
Federal University of Rio de Janeiro, P. O. Box 68516
RJ 21945-970 – Rio de Janeiro, Brazil
{renato,marcos,alvaro}@nacad.ufrj.br

Abstract. The parallel edge-based solution of 3D incompressible Navier-Stokes equations is presented. The governing partial differential equations are discretized using the SUPG/PSPG stabilized finite element method [5] on unstructured grids. The resulting fully coupled nonlinear system of equations is solved by the inexact Newton-Krylov method [1]. Matrix-vector products within GMRES are computed edge-by-edge, diminishing flop counts and memory requirements. The non-linear solver parallel implementation is based in message passing interface (MPI). Performance tests on several computers, such as the SGI Altix, the Cray XD1 and a mini-wireless cluster were carried out in representative problems and results have shown that edge-based schemes require less CPU time and memory than element-based solutions.

1 Introduction

We consider the simulation of steady incompressible fluid flow governed by Navier-Stokes equations using the stabilized finite element formulation in [5]. This formulation allows that equal-order-interpolation velocity-pressure elements are employed by introducing two stabilization terms: the Streamline Upwind Petrov-Galerkin (SUPG) and the Pressure Stabilizing Petrov Galerkin stabilization (PSPG).

When discretized, the incompressible Navier-Stokes equations give rise to a fully coupled velocity-pressure system of nonlinear equations due the presence of convective terms in momentum equation. The inexact Newton method [1] associated with a proper preconditioned iterative Krylov solver, such as GMRES, presents an appropriated framework to solve nonlinear systems, offering a trade-off between accuracy and the amount of computational effort spent per iteration.

Inspired by finite volume methods, edge-based data structures have been introduced for explicit finite element computations of compressible flows on unstructured grids composed by triangles and tetrahedra. Soto et al [11] recently introduced an edge-based approach to solve incompressible flows with an uncoupled fractional step formulation. The advantages of edge-based schemes with respect to conventional element-based schemes are a major reduction in indirect addressing (*i/a*) operations and memory requirements.

When dealing with large scale problems the use of parallel solvers is an essential condition and the use of an algorithm able to run efficiently in shared, distributed or hybrid memory systems has been a motivation for many researchers to turn the solver strategy more independent of the hardware resources. Therefore, the main goal of this

J.C. Cunha and P.D. Medeiros (Eds.): Euro-Par 2005, LNCS 3648, pp. 1237–1245, 2005.
© Springer-Verlag Berlin Heidelberg 2005

work is the development of edge-based data structures for the SUPG/PSPG finite element formulation to solve steady incompressible fluid flows by a parallel inexact-Newton method.

The remainder of this work is outlined as follows: next section briefly describes the governing equations and finite element formulation. The third section presents some remarks on edge-based data structures for unstructured grids. The subsequent section treats the parallel implementation, results and final comments are summarized in the last sections.

2   Governing Equations and SUPG/PSPG Finite Element Formulation

Let $\Omega \subset \mathbb{R}^{n_{sd}}$ be the spatial domain, where n_{sd} is the number of space dimensions. Let Γ denote the boundary of Ω. We consider the following velocity-pressure formulation of the Navier-Stokes equations governing steady incompressible flows:

$$\rho(\mathbf{u} \cdot \nabla \mathbf{u} - \mathbf{f}) - \nabla \cdot \boldsymbol{\sigma} = 0, \text{ on } \Omega \tag{1}$$

$$\nabla \cdot \mathbf{u} = 0, \text{ on } \Omega \tag{2}$$

where ρ and $\mathbf{u}$ are the density and velocity, $\boldsymbol{\sigma}$ is the stress tensor.

The essential and natural boundary conditions associated with equations (1) and (2) can be imposed at different portions of the boundary Γ and represented by,

$$\mathbf{u} = \mathbf{g} \quad \text{on } \Gamma_g \quad ; \quad \mathbf{n} \cdot \boldsymbol{\sigma} = \mathbf{h} \quad \text{on } \Gamma_h \tag{3}$$

where Γ_g and Γ_h are complementary subsets of Γ.

Let us assume following [5] that we have some suitably defined finite-dimensional trial solution and test function spaces for velocity and pressure, $S_{\mathbf{u}}^h$, $V_{\mathbf{u}}^h$, S_p^h and $V_p^h = S_p^h$. The finite element formulation of equations (1) and (2) using SUPG and PSPG stabilizations for incompressible fluid flows can be written as follows: find $\mathbf{u}^h \in S_{\mathbf{u}}^h$ and $p^h \in S_p^h$ such that $\forall \mathbf{w}^h \in V_{\mathbf{u}}^h$ and $\forall q^h \in V_p^h$:

$$\int_{\Omega} \mathbf{w}^h \cdot \rho(\mathbf{u}^h \cdot \nabla \mathbf{u}^h - \mathbf{f})\, d\Omega + \int_{\Omega} \boldsymbol{\varepsilon}(\mathbf{w}^h){:}\boldsymbol{\sigma}(p^h, \mathbf{u}^h)d\Omega + \int_{\Omega} q^h \nabla \cdot \mathbf{u}^h d\Omega$$

$$+ \sum_{e=1}^{n_{el}} \int_{\Omega} \tau_{SUPG} \mathbf{u}^h \cdot \nabla \mathbf{w}^h \cdot \left[\rho(\mathbf{u}^h \cdot \nabla \mathbf{u}^h) - \nabla \cdot \boldsymbol{\sigma}(p^h, \mathbf{u}^h) - \rho \mathbf{f} \right] d\Omega \tag{4}$$

$$+ \sum_{e=1}^{n_{el}} \int_{\Omega} \frac{1}{\rho} \tau_{PSPG} \nabla q^h \cdot \left[\rho(\mathbf{u}^h \cdot \nabla \mathbf{u}^h) - \nabla \cdot \boldsymbol{\sigma}(p^h, \mathbf{u}^h) - \rho \mathbf{f} \right] d\Omega = \int_{\Gamma} \mathbf{w}^h \cdot \mathbf{h}\, d\Gamma$$

In the above equation n_{el} is the number of elements in the mesh. The first four integrals on the left hand side represent terms that appear in the Galerkin formulation of the problem (1)-(3), while the remaining integral expressions represent the additional terms which arise in the stabilized finite element formulation. Note that the stabilization terms are evaluated as the sum of element-wise integral expressions. The first summation corresponds to the SUPG (Streamline Upwind Petrov/Galerkin) term and the second to the PSPG (Pressure Stabilization Petrov/Galerkin) term. The spatial discretization of equation (4) leads to a following system of nonlinear equations,

$$\mathbf{F}(\mathbf{x}) = \mathbf{0} \qquad (5)$$

where $\mathbf{x} = (\mathbf{u}, p)$ is a vector of nodal variables comprising both nodal velocities and pressures and $\mathbf{F}(\mathbf{x})$ represents a nonlinear vector function.

Newton's method is attractive method to solve the system (5) because it converges rapidly from any sufficiently good initial guess [1]. However, one drawback of Newton's method is the need to solve a locally linear system at each stage. Computing the exact solution can be expensive if the number of unknowns is large and may not be justified when $\mathbf{x}_k$ is far from a solution. Thus, one might prefer to compute approximated linear solutions, leading to the following algorithm,

ALGORITHM IN

```
for k = 0 step 1 until convergence do
    Find some η_k ∈ [0,1) AND s_k that satisfy
    ‖F(x_k)+J(x_k) s_k‖ • η_k ‖F(x_k)‖                    (6)
    set x_{k+1} = x_k + s_k
```

for some adaptively chosen $\eta_k \in [0,1)$, where $\| \cdot \|$ is a norm of choice. The Jacobian, $\mathbf{J}(\mathbf{x}_k)$, is numerically approximated using Taylor's expansions as described by Tezduyar [6]. This formulation naturally allows the use of an iterative solver like GMRES or BiCGSTAB: one first chooses η_k and then applies the iterative solver to (6) until a s_k is determined for which the residual norm is satisfied. In this context η_k is called as a forcing term and can be specified in several different forms as described in [2].

3   Edge-Based Data Structures

Edge-based data structures operate directly in the nodal graph of the underlying unstructured grid. It was shown in [3] that for unstructured grids edge-based data structures have more advantages than element-by-element (EBE) and compressed storage row (CSR) schemes. In the edge-based strategies, global coefficients are computed and stored in single DO-LOOPS making the evaluation of the left and hand sides faster and less memory demanding. We may derive an edge-based finite element framework by noticing that the element matrices can be disassembled into their contributions as shown in [7, 8]. For the set of all elements sharing a given edge, we may add their contributions, arriving to the edge matrix, which for the problem at hand is a non-symmetric 16×16 matrix. In Table 1 we compare the storage requirements to hold the coefficients of the element and edge matrices as well as the flop count and indirect addressing (i/a) operations to compute sparse matrix-vector products in the Krylov iterative driver element-by-element or edge-by-edge, that is,

$$\mathbf{J}\mathbf{p} = \sum_{l=1}^{ne} \mathbf{J}^l \mathbf{p}^l , \qquad (7)$$

where ne is the total number of local structures (edges or elements) in the mesh and $\mathbf{p}^l$ is the restriction of $\mathbf{p}$ to the edge or element degrees-of-freedom.

Table 1. Memory to hold the matrix coefficients and computational costs for element and edge-based matrix-vector products for tetrahedral finite element meshes

	Memory	*Flop*	*i/a*
Elements	1056 *nnodes*	2112 *nnodes*	1408 *nnodes*
Edges	224 *nnodes*	448 *nnodes*	448 *nnodes*

All data in this table is referred to *nnodes*, the number of nodes in the finite element mesh. According to [9], the following estimates are valid for unstructured 3D grids, $nel \approx 5.5 \times nnodes$, $nedges \approx 7 \times nnodes$, where *nedges* is the number of edges in the mesh. We may observe that data in Table 1 favors the edge-based scheme.

4 Parallel Implementation

The parallel inexact nonlinear solver presented in the previous section was implemented based in the message passing parallelism model (MPI). The original unstructured grid was partitioned into non-overlapped sub-domains by the use of the METIS_PartMeshDual routine provided by Metis package [4]. Afterwards, the partitioned data was reordered to avoid indirect memory addressing and IF clauses inside hot loops and MPI communications. Therefore, the equation numbers shared by the partitions were relocated to the last entries of the corresponding arrays.

Most of the computational effort spent during the iterative solution of linear systems is due to evaluations of matrix-vector products or matvec for short. In our tests matvec operations achieved 92% of the total computational costs. In element-by-element (EBE) and edge-by-edge (EDE) data structures this task is message passing parallelizable by performing matvec operations at each partition level, then assembling the contribution of the interface equations calling MPI_AllReduce routine over the last array entries. Finally, it is important to note that edge (and element) matrix coefficients are computed in single DO-LOOPS also in each partition.

5 Results

This section presents two benchmark problems to analyze the parallel solver performance. The numerical procedure considers a fully coupled **u-*p*** version of the stabilized formulation using linear tetrahedron elements. The parallel solver is composed by an outer inexact-Newton loop and an inner GMRES(25) with nodal block diagonal preconditioned linear solver.

The computations were made on two SGI Altix 3700 systems (32/64 Intel Itanium-2 CPUs with 1.3/1.5 GHz and 128/256 Gb of NUMA flex memory), and a Cray XD1 system (32 AMD Opteron CPUs with 1.8 GHz). Some portability and mobile parallelism tests were performed on a mini-cluster fast-ethernet/wireless composed by 4 laptop nodes with Intel Centrino processors and Microsoft Windows platform. The same code was compiled for three different systems (Intel Fortran 8.1 on SGI Linux systems, Portland Group Fortran on Cray Linux and Compaq Visual Fortran on mini-cluster Intel/Windows). No CPU optimizations besides those provided by standard compiler flags (-O3) were made.

5.1   Three Dimensional Leaky Lid-Driven Cavity Flow

In this well known problem the fluid confined in a cubic cavity is driven by the motion of a leaky lid. Boundary conditions consist in a unit velocity specified along the entire top surface and zero velocity on the other surfaces. Table 2 shows the problem dimensions employed for all parallel performance tests, where in the label *cav-nn*, *nn* means the number of line divisions through the x, y, and z dimensions for mesh construction purposes. Fig. 1 (left) shows the streamlines for Reynolds 400 for the *cav-101* mesh. We may note the main vortex formation and the singularities at the cavity corners, typical for this problem. Fig. 1 (right) show the computed vertical and horizontal velocities at the centerline, together with the recent numerical results of Lo et al [13] and Shu et al [14]. We may observe that all results are in good agreement.

Table 2. Problem dimensions

	Elements	Edges	Nodes	Equations
cav-31	148,955	187,488	32,768	117,367
cav-51	663,255	819,468	140,608	525,556
cav-71	1,789,555	2,193,048	373,248	1,421,776
cav-101	5,151,505	6,273,918	1,061,208	4,101,106

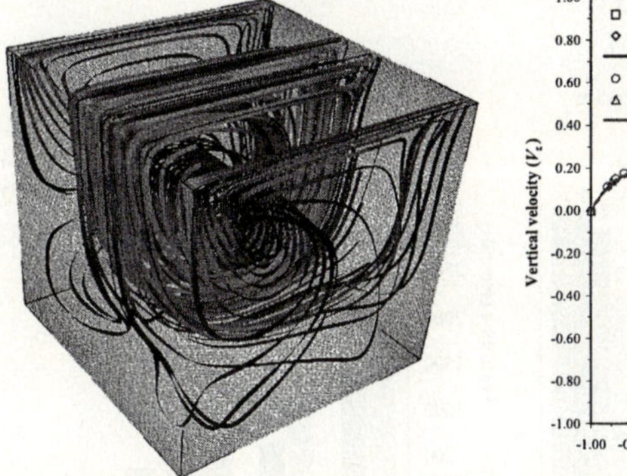
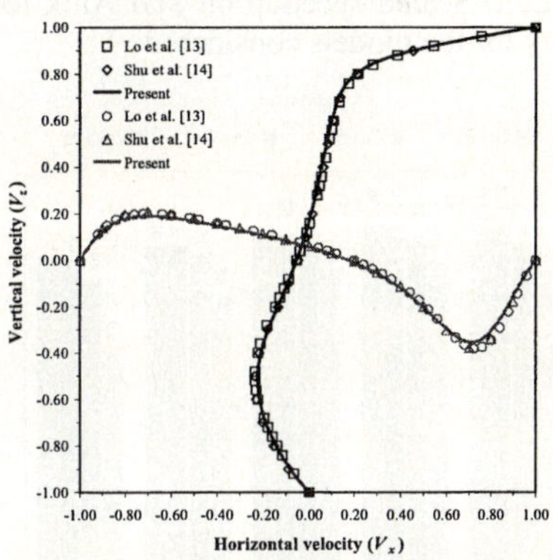

Fig. 1. (Left) Streamlines in a leaky lid-driven cubic cavity (right) Characteristic results for vertical and horizontal velocity at the centerline

In Fig. 2 (left) is shown the scaled speedup on the SGI Altix computed according to Gustafson's law [10] and defined by $S_s = n+(1-n)\,s$, where n is the number of processors and s corresponds to the normalized time spent in the serial portion of the program. The scalability reached on SGI Altix for the models listed in Table 2 is shown in Fig. 2 (right). Note that when increasing the problem size the serial fraction s tends to shrink as more processors are employed. In our tests with SGI Altix 3700 we have employed up to 32 Intel's Itanium-2 processors and according to [10] the scaled speedup should be a linear function with moderate slope $1-n$ such as the line we have measured and shown in Fig. 2 (right).

The scaled efficiency on SGI Altix for EDE data structure is presented in Fig. 3 (left). Good results may be observed for the cavity models, especially for those with larger number of degrees of freedom. In some cases efficiencies greater than 100% may be attributed to cache effects. The time spent when solving the *cav-71* problem with EBE and EDE data structures is plotted in Fig. 3 (right). We may observe that the EDE solutions were faster than the EBE in all cases. Nevertheless, the CPU time ratios between EBE and EDE solutions are around 2.5 up to 16 processors. For 32 processors this ratio decreases. This is an indication that as we refine the meshes, CPU time ratios between EBD and EDE has a tendency to remain around this value.

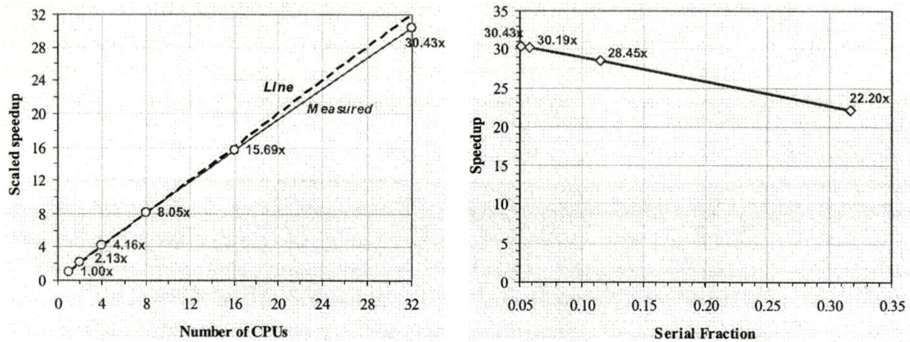

Fig. 2. (Left) Scaled speedup on SGI Altix for EDE data structure and cav101 model. (right) Scalability for the models considered

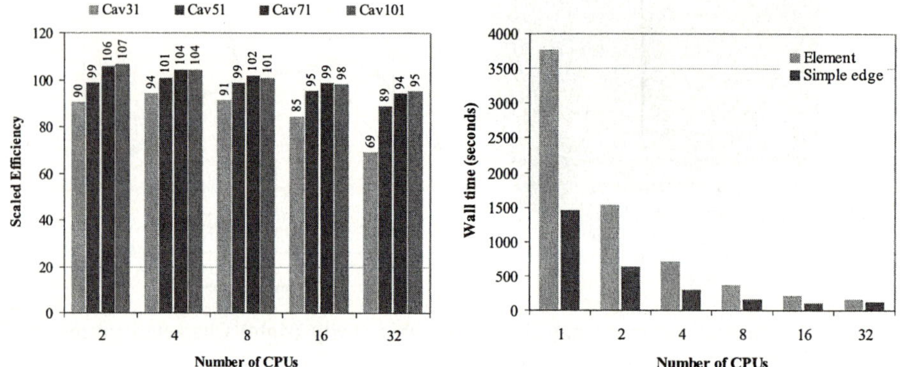

Fig. 3. (Left) Scaled efficiency for EDE data structure on SGI Altix. (right) Wall time comparisons for EBE and EDE data structures for *cav-71* mesh

The inexact nonlinear solver behavior is sketched in Fig. 4 through the decrease of relative residual (left), GMRES tolerance, and nonlinear iteration time (right). Note that at the beginning of the solution procedure the linear tolerance is large enough to allow very fast nonlinear iterations. When a sudden decay in the relative residual is detected, as may be seen in nonlinear iterations 14 to 24, the inexact nonlinear method identifies that the desired solution is imminent and the linear tolerance is tightened to entrap the final solution.

Fig. 5 shows the results of tests performed on a mini-cluster formed by 4 laptops and a wireless/fast-ethernet network (2 Intel Centrino 1.6 GHz/512Mb, 1 Intel Centrino 1.3 GHz /512Mb and 1 Intel Pentium 4 2.4 GHz/512Mb interconnected by a Linksys Wireless-B Hub, IEEE 802.11b/2.4GHz/11Mbps or Fast-Ethernet 10/100Mbps network). These tests show the versatility and portability that message passing codes can offer, making possible the solution of even large scale problems employing modest machines.

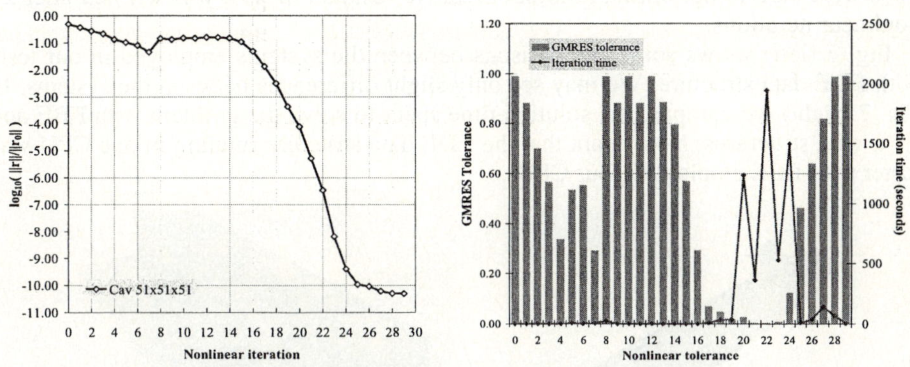

Fig. 4. (Left) Relative nonlinear residual. (right) GMRES tolerance (bars) controlled by inexact nonlinear method and time per nonlinear iteration (lines)

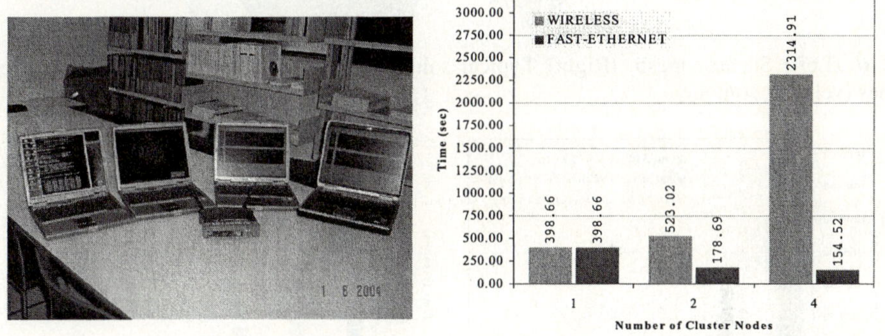

Fig. 5. (Left) Minicluster mobile wireless/fast-ethernet, (Right) Performance comparison between wireless and fast-ethernet networks

Fig. 5 (right) shows that the wireless technology employed (Wireless-B) was not able to deliver the bandwidth required to reach a desirable speedup in this irregular MPI parallel computation. However, with the increasing bandwidth in wireless technology mobile-parallel computations will be a reality in the near future. The low speedups achieved in the minicluster with fast-ethernet network were also due to the small problem size, as occurred in the case shown in Fig. 3 (left) for SGI Altix.

5.2   Flow Through a Los Angeles Class Submarine

This problem consists on a simplified three-dimensional simulation of a laminar flow around a Los Angeles class submarine. The detailed solution and discussion of this problem, involving transient and turbulent flow is given in [12]. Fig. 6 (left) shows the mesh over the submarine hull. The volume mesh comprises 504,947 tetrahedral elements, 998,420 edges and 92,564 nodes. The solution for this problem is shown in Fig. 6 (right), where the pressure contour is plotted over the submarine hull and the velocity in two longitudinal cutting planes. For this problem the linear tolerance oscillated from the maximum value of 0.99 to a minimum of 2.7×10^{-2} and the computations were carried out until a minimum relative residual of 10^{-10} was reached after 21 nonlinear iterations.

Fig. 7 (left) shows some comparisons between the systems employed in our tests with EDE data structure. We may see only slight differences between the systems. In Fig. 7 (right) we compare the solution time spent to solve the problem with EBE and EDE data structures. Note again that the EDE data structure running in one CPU was faster than EBE employing four CPUs.

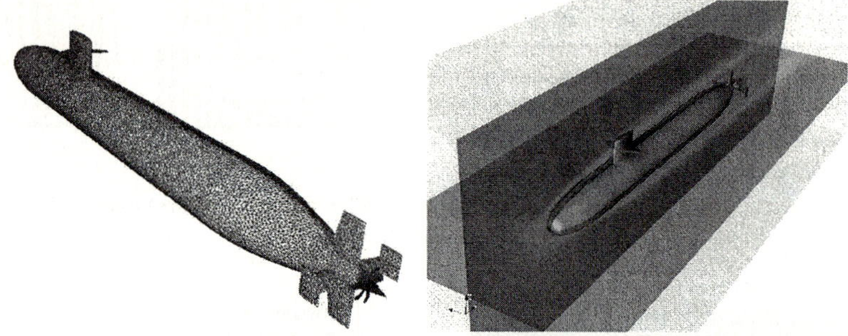

Fig. 6. (Left) Surface mesh, (Right) Typical solution - hull surface (pressure contour), cut planes (velocity contour)

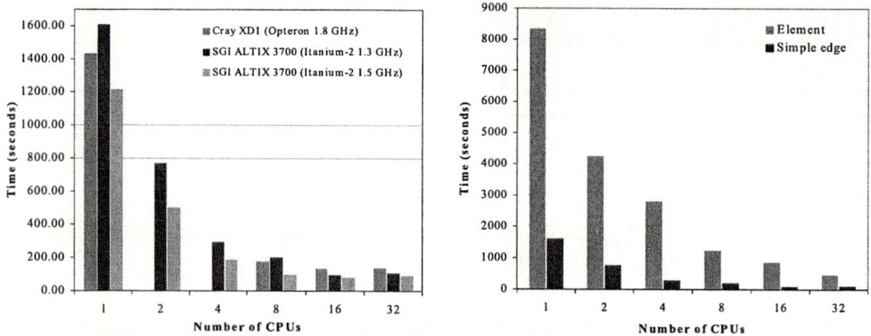

Fig. 7. (Left) Message passing performance in SGI Altix and Cray XD1 – edge-based data structure, (Right) Data structure comparisons on SGI Altix (MPI)

6 Conclusions

We have tested the performance of a parallel edge-based inexact Newton solver for fully coupled velocity-pressure nonlinear systems of equations arising from the SUPG/PSPG finite element formulation of steady incompressible flow on unstructured grids. We observed that the inexact nonlinear method employed has shown good balance between accuracy and computational effort. The edge data structure decreased the solution time even without employing any data reordering method to exploit the cache. Our tests with benchmark problems have shown good parallel performances, but interface mapping techniques could be used to reduce the amount of data communication. The code is portable across different computer platforms, ranging from a mobile wireless cluster to the SGI Altix 3700 and Cray XD1 systems without any code modifications or CPU guided optimizations.

Acknowledgements

The authors would like to thank the financial support of the Petroleum National Agency (ANP, Brazil) and the Center for Parallel Computations (NACAD) and the Laboratory of Computational Methods in Engineering (LAMCE) at the Federal University of Rio de Janeiro. The authors are very grateful for the computational resources provided by Silicon Graphics Inc. (SGI/Brazil) and Cray Inc. We are also indebted to Prof. M. Behr from RWTH Aachen University by the submarine mesh.

References

1. Dembo, R. S., Eisenstat, S. C. and Steihaug, T., Inexact Newton Methods, SIAM J. Numer. Anal. (1982) 19: 400-408.
2. Eisenstat, S. C. and Walker, H. F., Choosing the Forcing Terms in Inexact Newton Method, SIAM. J. Sci. Comput. (1996) 17–1: 16–32.
3. Ribeiro, F. L. B. and Coutinho, A. L. G. A., Comparison Between Element, Edge and Compressed Storage Schemes for Iterative Solutions in Finite Element Analyses, Int. J. Num. Meth. Engrg. 2005; 63-4:569-588.
4. Karypis G. and Kumar V., Metis 4.0: Unstructured Graph Partitioning and Sparse Matrix Ordering System. Technical report, Department of Computer Science, University of Minnesota, Minneapolis, (1998) http://www.users.cs.umn.edu/~karypis/metis.
5. Tezduyar, T. E., Stabilized Finite Element Formulations for Incompressible Flow Computations, Advances in Applied Mechanics (1991) 28: 1-44.
6. Tezduyar, T. E., Finite Elements in Fluids: Lecture Notes of the Short Course on Finite Elements in Fluids, Computational Mechanics Division – Vol. 99-77, Japan Society of Mechanical Engineers, Tokyo, Japan (1999).
7. Catabriga, L., Coutinho, A. L. G. A., Implicit SUPG solution of Euler equations using edge-based data structures. Comput. Methods in Appl. Mech. and Engrg, 2002, 32(191): 3477-3490.
8. Coutinho, A. L. G. A., Martins, M. A. D., Alves, J. L. D., Landau, L., Moraes, A., Edge-based finite element techniques for non-linear solid mechanics problems. Int. J. Num. Meth. Engrg, 2001, 50(9):2053-2068.
9. Lohner, R., Edges, stars, superedges and chains, Comput. Methods in Appl. Mech. and Engrg 1994, 111(3-4): 255-263.
10. Gustafson, J. L., Montry, G. R. and Benner, R. E., Development of Parallel Methods for a 1024-Processor Hypercube, SIAM J. on Sci. and Stat. Comp., 1988, 9(4):609-638
11. Soto, O., Löhner, R., Cebral, J. and Camelli, F., A Stabilized Edge-Based Implicit Incompressible Flow Formulation, Comput. Methods Appl. Mech. Engrg. 2004, 193:2139-2154.
12. http://manila.cats.rwth-aachen.de/developer/cases/la.0808, last visited in May 10, 2005.
13. Lo, D. C., Murugesan, K and Young, D. L., Numerical solution of three-dimensional velocity-vorticity Navier-Stokes equations by finite difference method, Int. J. Numer. Meth. Fluids 2005, 47:1469-1487.
14. Shu, C., Wang, L. and Chew Y T, Numerical computation of three-dimensional incompressible Navier-Stokes equations in primitive variable form by DQ method, Int. J. Numer. Meth. Fluids 2003; 43:345-368.

High Performance Computing for a Financial Application Using Fast Fourier Transform

Sajib Barua, Ruppa K. Thulasiram*, and Parimala Thulasiraman

Department of Computer Science, University of Manitoba
Winnipeg, MB R3T 2N2 Canada
{sajib,tulsi,thulasir}@cs.umanitoba.ca

Abstract. Fast Fourier Transform (FFT) has been used in many scientific and engineering applications. In the current study, we have applied the FFT for a novel application in finance. We have improved a recently proposed mathematical model of Fourier transform technique for pricing financial derivatives to help design and develop an effective parallel algorithm using a swapping technique that exploits data locality. We have implemented our algorithm on 20 node SunFire 6800 high performance computing system and compared the new algorithm with the traditional Cooley-Tukey algorithm We have presented the computed option values for various strike prices with a proper selection of strike-price spacing to ensure fine-grid integration for FFT computation as well as to maximize the number of strikes lying in the desired region of the asset price.

Keywords: HPC for commercial application; Option pricing; Fast Fourier transform; Mathematical modeling; Parallel algorithm; Data locality.

1 Introduction

The finance industry demands efficient algorithms and high-speed computing in solving many problems [1]. In this research we cut across two historically established, technologically evolving and most importantly traditionally different areas: computing and finance - computational finance. Specifically, this paper addresses the problem of option pricing.

Terminologies: An option is a financial contract where one of the two parties involved, known as *holder*, gets the right (but not obligation) to *buy/sell* a set of underlying financial instruments such as stocks at a preset price (known as *exercise* or *strike* price) at a preset date (known as *exercise/maturity* date) *from/to* the other party known as the *writer*. If the holder decides to exercise the option, the writer is obligated to satisfy the holder's decision. Buying/selling underlying asset through such contract is referred to as *Call/Put* option. If the option can be exercised only at the maturity date, the option contract is known as *European* option, whereas if the option can be exercised any time prior to the maturity, it is known as *American* option. Value of a call/put option (shortly call/put value

---

* Author for Correspondence: tulsi@cs.umanitoba.ca

J.C. Cunha and P.D. Medeiros (Eds.): Euro-Par 2005, LNCS 3648, pp. 1246–1253, 2005.

or call/put price) depends on the spot price of the underlying asset, strike price among other parameters such as risk-free interest rate, volatility[1] of the asset, period of the contract.

The solution for the optimal exercise policy for a financial option must typically be performed numerically, and is usually a computationally intensive problem. Pricing of options has been traditionally done using either binomial tree approach or using Monte-Carlo simulation or engineering approaches such as finite-differencing (see for example [2]). A recent addition to the numerical techniques for the option pricing problem is the use of Fast Fourier Transform (FFT) [3]. By providing an one-to-one mapping from the mathematics of Fourier space to the computational domain of the FFT, [4] explored the high performance computing for this problem.

In the current study, we develop an improved mathematical model of FFT for option pricing and a new parallel FFT algorithm. While there are many FFT algorithms available, for example, Stockham auto sort algorithm (SAS) [5] (chapter 1.7) and Bailey algorithm [6], we had to develop a new algorithm in the current study especially to satisfy the mathematics of the option pricing problem described in section 2. The structure of our new algorithm behaves similar to the SAS algorithm, however, captures the physics of option pricing closer than the SAS algorithm as explained in section 2. Due to lack of space we do not discuss the SAS or other available algorithms here. Readers are referred to [5] for an in depth look on various FFT algorithms. We leave the work on fine tuning SAS [5] (chapter 1.7) and Bailey's [6] algorithm for option pricing problem as a future study.

The rest of the paper is organized as follows. In section 2, we mention the drawback of one major related work on mathematical model of option pricing problem using Fourier transform and present an improvement to the mathematical modeling with which a finer mapping from mathematics to the FFT computational domain for option pricing is presented. In section 3, we present the new FFT algorithm, which exploits data locality to improve the performance. The results are presented in section 4, with call value results followed by the experimental results. We conclude the current study in section 5.

2 Drawback of an Existing FFT Model and Improved Model for Option Pricing

An important contribution of the current work is to alleviate the drawback in Carr-Madan (CM) model [3]. They developed a FFT model for option pricing in continuous and discrete form as follows: If $M = e^{-\alpha k}/\pi$ and $\omega = e^{-i}$ then

$$C_T(k) = M \int_0^\infty \omega^{vk} \psi_T(v) dv. \tag{1}$$

---

[1] Variation in the asset prices is generally split into two parts: (i) changes due to known factors affecting the asset price such as periodic changes - known as deterministic changes or *drift* in prices; (ii) changes due to unknown phenomena in the market place - generally known as *volatility*.

If $v_j = \eta(j-1)$ and applying trapezoidal rule for the integral on the right of equation (1), $C_T(k)$ can be written as

$$C_T(k) \approx M \sum_{j=1}^{N} \psi_T(v_j)\omega^{v_j k}\eta, k = 1, \ldots, N, \tag{2}$$

where the effective upper limit of integration is $N\eta$ and v_j corresponds to various prices with η spacing. Here $C_T(k)$ is the call option price; $\psi_T(v)$ is the Fourier transform of this call price given by $\psi_T(v) = \frac{e^{-rT}\phi_T(v-(\alpha+1)i)}{\alpha^2+\alpha-v^2+i(2\alpha+1)v}$; where α is a dampening factor and k is the logarithm of the strike price, $k = log(K))$; r is the interest rate; T is the period of the option contract. The calculation of $\psi_T(v)$ depends on the factor $\phi_T(u)$, where $u = v - (\alpha + 1)i$. We derive $\phi_T(v)$ as, $\phi_T(v) = \int_0^\lambda (\cos(vk) + i\sin(vk))q_T(s)ds$ where λ is terminal spot price and integration is taken only in the positive axis.

To calculate the call values, equation (1) has to be solved analytically. The discrete form equation (2) is not suitable to feed into the existing FFT algorithms for example,Cooley-Tukey [7], Stockham auto sort [5](chapter 1.7) and Bailey [6]. Hence, the CM model in its current form cannot be used for faster pricing. This is a major drawback of using CM model for practical purposes and for real time pricing we need to improve this mathematical model.

This leads us to state the objectives of the current work as: (1) Improving the mathematical model that will be tractable for parallel computing and for getting accurate solutions quickly; (2a) Designing an efficient parallel FFT algorithm that can map the mathematics from the improved model to the computational domain; and (2b) implementing the algorithm on distributed memory architecture to study the performance.

Improved Mathematical Model: The limits on the integral have to be selected in such a way as to generate real values for the FFT inputs. To generate the closed form expression of the integral, the integrands, especially the function $q_T(s)$, have to be selected appropriately. Without loss of generality, we use uniform distribution for $q_T(s)$. This implies occurrence of a range of terminal log prices at equal probability, which could, of course, be relaxed and a normal or other distribution could be employed. Since the volatility of the underlying asset is assumed constant (low) the variation in the drift is expected to cause a stiffness[2]

---

[2] Stiffness occurs when two processes controlling a physical phenomenon proceeds at two extremely different rates. It is common in scientific problems such as chemical reactions and high temperature physics. When a system with such physical phenomenon is manifested in mathematics such as differential or integral equations, the mathematical system is known to be *stiff*, where solution of such systems of equations would require special techniques to handle the 'stiffness'. Drift and volatility in the finance systems act as two phenomena affecting the system away from equilibrium hence may induce 'stiffness'. Our assumptions of uniform distribution for the density function to make the integration easier, in conjunction with assumed constant volatility, however, naturally avoids this issue.

in the system. However, since we have assumed uniform distribution for $q_T(s)$, variation in drift is eliminated and hence the stiffness is avoided. Therefore, use of uniform distribution would make the integration easier.

For computation purposes, the upper limit of equation (1) is assumed as a constant value and the lower limit is assumed as 0. The upper limit will be dictated based on the terminal spot price. In other words, to finish the call option in-the-money[3], the upper limit will be smaller than the terminal asset price and hence we arrive at the the modified expression for $\phi_T(v)$ presented earlier.

Without loss of generality, further modifications are required as derived below. The purpose of these modifications is to generate feasible and tractable initial input condition to the FFT algorithm from these equations. Moreover, these modifications make the implementation easier. Due to lack of space we skip the mathematical derivation and present the final improved mathematical model as

$$\psi_T(v) = \frac{A}{\{B\}\{C^2 + D^2\}} \left[\{C\Delta + D\Delta_x\} + i\{C\Delta_x - D\Delta\} \right] \tag{3}$$

where, $A = e^{-rT} q_T(s)$; $B = (\alpha + 1)^2 + v^2$; $C = \alpha^2 + \alpha - v^2$; $D = (2\alpha + 1)v$. We use this final expression for the new parallel FFT algorithm to compute the call price function. The financial input data set for our parallel FFT algorithm is the calculated data points of $\psi_T(v)$ for different values of v. We refer equation (3) as *BTT-CM Model* or *BTT-CM equation*.

We then calculate call value for different strike price values v_j where j will range from 1 to N. The lower limit of strike price is 0 and upper limit is $(N-1)\eta$ where η is the spacing in the line of integration. Smaller value of η gives fine grid integration and a smooth characteristics function of strike price and the corresponding calculated call value. If γ is the spacing in k, then the values for k can be obtained from the equation: $k_u = -p + \gamma(u-1)$, for $u = 1, \ldots, N$. Hence, the log of the ratio of strike and exercise price will range from $-p$ to p where $p = \frac{N\gamma}{2}$. Substitution of previous equation for k_u in equation (2) and replacing v_j with $(j-1)\eta$ in the equation gives $(for\ u = 1, ..., N)$

$$C_T(k_u) \approx \frac{exp(-\alpha k_u)}{\pi} \sum_{j=1}^{N} \{e^{-i\gamma\eta(j-1)(u-1)} e^{ipv_j} \psi_T(v_j)\eta\}. \tag{4}$$

Comparing equation (4) with the basic FFT equation, we note that $\gamma\eta = \frac{2\pi}{N}$. Smaller values of η will ensure fine grid for the integration. But call prices at relatively large strike spacings (γ), few strike prices will lie in the desired region near the stock price [3]. Furthermore, if we increase the values of N, we

---

[3] In-the-money call option is a situation where underlying asset price of the option is larger than the strike price; at-the-money call means asset price equals the strike price; natural extension is for out-of-the-money call, which corresponds to a situation where the asset price is smaller than the strike price. These definitions are reversed for a put option

will get more intermediate points of the calculated call prices ($C_T(k_u)$) corresponding to different strike prices (v_j). This helps the investor to capture the call price movements of an option for different strike prices in the market. In the experimental result (section(4)) of 1024 (N) numbers of calculated call values, assuming $\eta = 0.25$ with the intuition that it will ensure fine grid integration, γ is calculated as 0.02454. Similar to basic FFT equation, equation (4) can also be parallelized with an efficient parallel algorithm. In the next section we develop a data swapping technique that exploits data locality to reduce communication on a parallel computer and effectively apply our mathematical model. We implement this algorithm with the inputs derived from equation (4).

3   An Effective Parallel FFT Algorithm

Figure 1 illustrates our data swap algorithm. We assume we have N ($N = 2^m$) data elements and P ($P = 2^p$) processors where $N > P$ [8]. In our algorithm, we apply the blocked data distribution and the first ($\log N - \log P$) stages require no communication. However, in the last $\log P$ stages that require communication, we swap some data at each stage and let the data reside in the processor's local memory after swapping. Therefore, the identity of some of the data points in each processor changes at every stage of the $\log P$ stages.

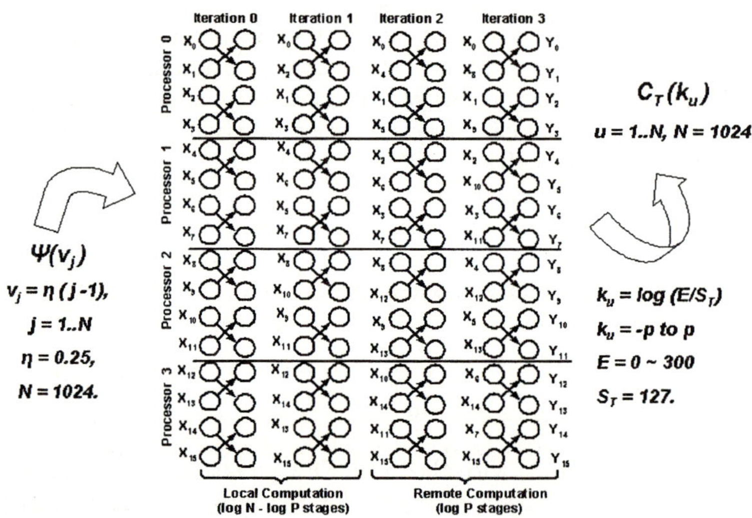

Fig. 1. Data Swap Algorithm

In figure (1), we can see that in iteration 2, processor 0 needs two input data points with index 4 and 5 and these do not reside in the local processor. Hence, we need two send operations to bring these values from processor 1. In general, for an input data point with N/P data in every processor, $N/(2P)$ communication is required. This is half of what is required in the Cooley-Tukey algorithm. That is, in the new parallel FFT algorithm, the number of communications is reduced by half. We take advantage of the the fact that communication between processors is point to point and swap the data in a similar manner. However, in this case, only $\frac{N}{2P}$ amount of data (message size) is communicated by each processor at every stage. Also note that, data swapping between processors at each location allows both the upper and lower part of the butterfly computations to be performed locally by each processor. This improvement enhances good data locality and thereby providing performance increase in the new FFT algorithm compared to the Cooley-Tukey algorithm. Analytically, the parallel runtime is given by [9] $t_c(N/P)\log N + t_s'' \log P + t_w(N/2P)\log P$, where t_s is the start up time; t_w is the per word transfer time; and t_c is the time required for the butterfly computation.

4   Results and Discussions

Option Pricing Results: Figure (1) shows how the data swap algorithm calculates the call values from the input data set generated from the BTT-CM equation. The data swap algorithm calculates N number of call values. When the call option is in-the-money, the investor would prefer to exercise the option (purchasing the option) at the strike price and immediately sell the asset in the market at the terminal spot price. Thus, the holder can profit. Figure (2) depicts the calculated in-the-money call values for different strike prices using the data swap algorithm. In the experiment of call value computation, strike price can be any value between 0 and 300. Our data swap algorithm can calculate (figure 1) call values for in-the-money, at-the-money and out-of-the-money call options. We are considering in-the-money call where the terminal spot price is always greater than the strike price. Therefore, figure 2 plots a portion of the calculated call values (in-the-money) from the output values of the data swap algorithm. The plot shows that the normalized option value is decreasing with the increase of strike price. If X, the strike price, is decreased, the call option value is expected to increase, which can be seen in figure 2. For larger values of N we can get more number of call values computed for the strike price range from 0 to 127, which makes the plot as a continuous function.

Significant Experimental Results: The experiments were conducted on a 20 node SunFire 6800 high performance computing system at the University of Manitoba running MPI. The Sunfire consists of Ultra Sparc III CPUs, with 1050 MHz clock rate and 40 gigabytes of memory and runs Solaris 8 operating system. The data generated in section 2 is used for the FFT input. Due to lack of space, we present only limited number of results.

Figure (3 a) depicts a comparison of the execution time between the swap algorithm and the Cooley-Tukey algorithm. At each iteration $\frac{N}{2P} = \frac{2^{20}}{2^5} = 2^{15}$ data points are swapped on each of the 16 processors. On a 2 processor machine, there are $\log 2^{20} - \log 2 = 19$ local computations and only 1 remote communication. However, there is a significant decrease in execution time in 16 processors. This is

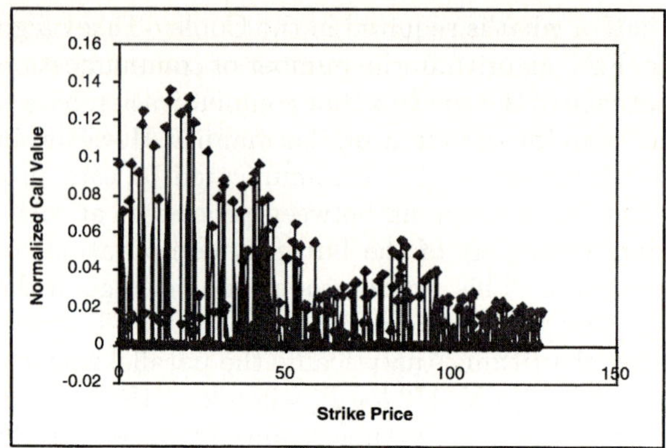

Fig. 2. Computed Call Values

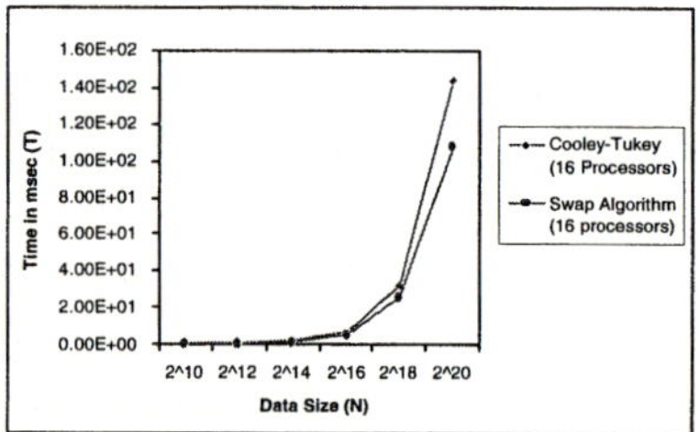

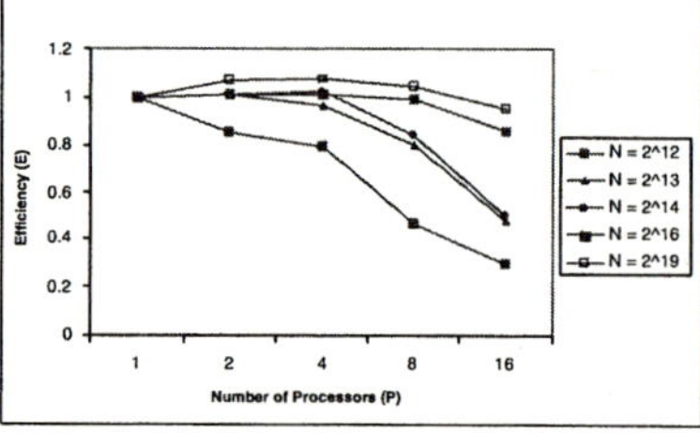

Fig. 3. a)Comparison of the execution times of swap and Cooley-Tukey algorithms, and b) Efficiency of the data swap algorithm

attributed to the fact that in MPI, the packing and unpacking of $\frac{N}{2P} = 2^{18}$ data elements for each of the 2 processors requires significant amount of time. When we compare the swap algorithm to the Cooley-Tukey algorithm in figure (3) on 16 processors, the swap algorithm performs 15% better than Cooley-Tukey algorithm on a data size of 2^{20}.

We calculated the efficiency of the swap algorithm for various processors on a fixed data size as presented in figure (3 b). The efficiency for 16 processors is close to 1. For 4, 8, and 16 processors the efficiency is 90% for data sizes 2^{14}, 2^{16}, 2^{19} respectively. Also for 8 and 16 processors the efficiency is 50% for 2^{12} and 2^{13} respectively. These results illustrate that as we increase the data size and the number of processors, the swap algorithm exhibits very good scalability.

5   Conclusions

Without loss of generality, we have improved the mathematical modeling of FFT for option pricing and we have identified appropriate values for the parameters to generate the input data set for the parallel FFT computations. We have reduced the communication latency by improving the data locality. We have presented the computed call values for various strike prices with a proper selection of strike-price spacing to ensure fine-grid integration for FFT computation as well as to maximize the number of strikes lying in the desired region of the asset price. Compared to the traditional Cooley-Tukey algorithm, the current algorithm with data swapping performs better by more than 15% for large data sizes.

Acknowledgement

The last two authors acknowledge partial financial support from Natural Sciences and Engineering Research Council (NSERC) of Canada and the University of Manitoba Research Grant Program (URGP). They also gratefully acknowledge the discussions with Prof. Sanjiv R. Das, Department of Finance, Leavey School of Business, Santa Clara University, Santa Clara, CA, USA, on the Fourier transform application for finance problems especially the option pricing problem.

References

1. E. J. Kontoghiorghes, A. Nagurnec, and B. Rustem. Parallel Computing in Economics, Finance and Decision-making. *Parallel Computing*, 26:207–209, 2000.
2. J.C. Hull. *Options, Futures and Other Derivatives*. Prentice Hall, Upper Saddle River, NJ, 5th edition, 2002.
3. P. Carr and D. B. Madan. Option Valuation using the Fast Fourier Transform. *The Journal of Computational Finance*, 2(4):61–73, 1999.
4. R. K. Thulasiram and P. Thulasiraman. Performance Evaluation of a Multithreaded Fast Fourier Transform Algorithm for Derivative Pricing. *The Journal of Supercomputing*, 26(1):43–58, Aug. 2003.
5. C.Van Loan. *Computational Frameworks for the Fast Fourier Transform*. SIAM: Frontiers in Applied Mathematics, Philadelphia, PA, 1992.
6. D. H. Bailey. FFTs in External or Hierarchical Memory Fourier. *The Journal of Supercomputing*, 4, 1990.
7. J.W. Cooley, P.A. Lewis, and P.D. Welch. *The Fast Fourier Transform and its Application to Time Series Analysis*. Wiley, New York, 1977. In statistical Methods for Digital Computers.
8. A. Grama and A. Gupta and G. Karypis and V. Kumar. *Introduction to Parallel Computing*. Addison Wesley, New York, NY, Second edition, 2003.
9. S. Barua. Fast Fourier Transform for Option Pricing: Improved Mathematical Modeling and Design of an Efficient Parallel Algorithm. Master's thesis, University of Manitoba, Winnipeg, MB, Canada, July 2004.

Parallel Simulation of the Propagation of Powdery Mildew in a Vineyard

Agnès Calonnec[1], Guillaume Latu[2], Jean-Marc Naulin[1],
Jean Roman[3], and Gaël Tessier[3]

[1] INRA Bordeaux, UMR INRA-ENITA, Santé Végétale
BP81, 33883 Villenave d'Ornon Cedex, France
[2] LSIIT UMR 7005, Université Strasbourg 1
67412 Illkirch Cedex, France
[3] INRIA Futurs and LaBRI UMR 5800, ScAlApplix project
Université Bordeaux 1 and ENSEIRB, 33405 Talence Cedex, France
http://www.labri.fr/scalapplix

Abstract. This paper describes a parallel simulator for the propagation of a parasite in a vineyard. The model considers the structure, the growth and the susceptibility of the plant which play a major role in the development of the fungus and the spread of epidemic. Two spatial scales are distinguished for the dispersal of the parasite. We use both a realistic discrete model for the local dispersal, and a stochastic model for the long-range dispersal that averages the displacement of spores. An algorithmic description of the parallel simulator is given and real life numerical experiments on IBM SP3 are provided, that use up to 128 processors.

1 Introduction

In this paper, we consider the simulation of a biological host-parasite system. The studied parasite is *powdery mildew*, a fungus of grapevine. Many epidemiological studies have been performed on this topic; however the dynamics of the spread of epidemics is not well known and powdery mildew is still the main fungus disease of grapevine in the world.

A large number of multiscale mechanisms interact in this system. A better understanding and a more effective control of epidemics will depend on our understanding of the dynamical relationships between the environment, the host and the pathogen. Knowledge obtained during experiments in vineyards will be first integrated into a model and then into a simulator. One purpose of this work is to reproduce the interactive events of the system and to synthesize them in order to understand and evaluate macroscopic emerging phenomena.

The simulation requires a large amount of computations, mainly due to the number of spores produced by the parasite and dispersed over the vineyard. An initial sequential simulator only considered only one grapevine [4]. A parallel version has been developed to model the dynamics of epidemic over a parcel. To our knowledge, this approach based on realistic simulations is rather new and has not been yet met in other research works concerning this topic.

J.C. Cunha and P.D. Medeiros (Eds.): Euro-Par 2005, LNCS 3648, pp. 1254–1264, 2005.

The biological system and its modeling will be first presented. A profiling of the sequential program will be detailled so as to identify the most time-consuming steps that will be good candidates for parallelization. Then, data distribution and parallel algorithms will be explained; results about scalability, load-balancing and performance issues of this first implementation will be given. Finally, we will conclude with the possible evolutions.

This interdisciplinary work is a collaboration between the INRIA Futurs ScAlApplix project and the LSIIT UMR 7005 for the computer science field, the INRA UMR Santé Végétale in Villenave d'Ornon for the biological investigations and the MAB UMR 5466 for the mathematical models.

2   Biological Issues and Modeling

The structure of a grapevine is strongly influenced by management practices of viticulture system. Two almost horizontal branches carry primary shoots which themselves carry secondary shoots. Apparition and growth of organs are essentially dependant on their position, on the vigour of the plant and on the temperature.

Powdery mildew [1] is a polycyclic fungus that spreads thanks to microscopic airborne spores. We break up the biological cycle in several processes: infection of leaves or clusters by spores, a latency period during which the rising colony is only growing, and a sporulation phase during which spores are released by wind.

2.1   General Modeling

The simulation covers a single season from January to the beginning of September with a time step of one day. Location and onset of primary infection are parameters of the simulation. The dynamics of epidemic is closely related to the quantitative and qualitative development of hosts: the number, the position and the age of organs. Thus the model simulates the 3D development of stocks. The computer model for a grapevine is a binary tree, in which each node represents an element of the plant. A node contains information on its spatial configuration, its biological attributes, and its possible infection state. Parameterized functions, some of them stochastic, are used to describe system growth. Host growth depends on a few magnitudes: temperature T and trophic state which is a temperature-dependant variable. Fungal colony growth depends on temperature and organ age. A vigorous grapevine can bear hardly thousand leaves whereas a weak one three to four times less. The model restricts infection to leaves and the number of colonies to one per leaf. During a day, approximatively tens of thousands of spores are possibly extracted from all the sporulating colonies of a grapevine.

As for the dispersal of spores, it is for the moment impossible for us to know the real movement of microscopic spores that can travel up to several hundreds of meters. Therefore, two scales for the dispersal of spores have been distinguished: local and long-range dispersals. The limit between these scales remains confusing as it depends on the studied pathosystem. Literature mentions the ratio 80 %/20 % between local and long-range dispersals for an optimal disease spread [9].

2.2 Local Dispersal

In the current version of the model, local dispersal was limited to the source grapevine and its two direct neighbours in the row. At this scale, the distance covered by spores is short, the dispersal occurs in the canopy, supposed to be homogeneous in a local area. Thus, we made the hypothesis that spores have linear trajectories during local dispersal.

Each day, spores are spreading from each sporulating colony. The spread is performed within a *dispersal cone*. Its axis orientation is determined by the mean wind direction of the day represented by the vector (ux, uy, uz). Its opening angle *alpha* is a simulation parameter. Algorithms and data structures have been inspired by ray-tracing methods in image synthesis [8]. A rectangular parallelepiped delimits the volume of grapevine. For efficiency, this volume was cut out with a discrete mesh size of small parallelepipeds called *voxels* [6]. Each voxel has the list of leaves contained in its volume. So computing the leaves intercepting a cone comes down to getting the voxels intercepting this cone, as shown in Fig. 1. The voxel discretization avoids to traverse the whole binary tree for the determination of all the leaves of the stock intercepting the cone. When a cone reaches one edge of the including parallelepiped, its becoming depends on the exit side: spores either fall on the ground, or they are transmitted to the contiguous grapevine, or they are dispersed over the vineyard.

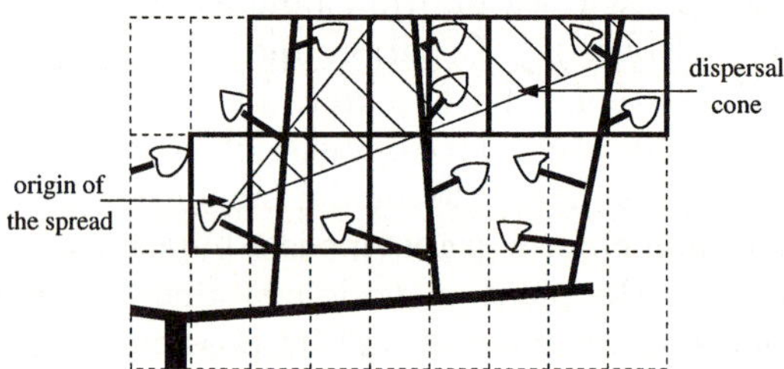

Fig. 1. Dispersal cone in a grapevine with bold voxels intercepting the cone

2.3 Long-Range Dispersal

Field data come from several campaigns of disease follow-up in vineyards. Measures of vertical and horizontal gradients of spore densities over short periods and in fixed vegetation are available, but no spore dispersal measures on a growing crop. At field scale, the previous cone-based dispersal approach is not realistic enough, because the spores have complex trajectories. A stochastic and averaged approach using distribution laws [7] has been considered.

From each grapevine, random drawings following Gaussian distributions yield a displacement; this displacement identifies the destination grapevine for each spore that has to be dispersed over the field. Then, on the destination plant, dispersal cones are used to determine the points of impact of spores, and so the infected leaves.

Set hypotheses and choices of distributions require calibrating the simulator outputs with field data. Some parameters have already been estimated, some others will be refined in the future.

3   Profiling of the Sequential Program

This section is about the time complexity and performance analysis of the sequential simulator. This simulator treats a single grapevine. Of course, there is no dispersal with neighbouring vine stocks or over the field.

During each iteration (one day) of the simulation, variables T, ux, uy, uz are fetched. Before bud break, we evaluate if there is budding or not. After budding, an iteration consists of several steps. Primary infection and management practices occur only at precise day and so, count for a little part of computation time. On the other hand, growth and local dispersal are the most costly steps, that is why they will be examined in detail. To formulate accurately their time complexities, let us call $\#nodes$ the number of nodes in the binary tree, $\#leaves$ its number of leaves and $\#voxels$ the number of voxels in the grapevine volume.

3.1   Apparition and Growth of Organs

Apparition and growth of organs are two sub-steps, each one using one recursive traversal of the binary tree. Host growth and pathogen growth are calculated simultaneously while treating infected nodes. The time complexity of the growth function is $\Theta(\#nodes)$.

3.2   Dispersal of Spores into Grapevine

Input parameters of local dispersal are a grapevine, the day, and the temperature and wind characteristics of the day. Dispersal process is based on a recursive traversal of the binary tree. For each infected node encountered, operations of the function `source_dispersal` (described in Fig. 2) are performed. Variables x, y and z are the node coordinates, and n_spores is the number of spores extracted from its colony on the current day. At function return, n_spores represents the number of spores not captured and is daily accumulated in the *vine* data structure (see Fig. 3).

The procedure `get_voxels_cone` gets the voxels intercepting the cone issued from the colony in (x, y, z), with the direction (ux, uy, uz) and the opening angle *alpha*. Its time complexity is linear with the number of voxels returned in *vox_list*. This number depends on the position of the source lesion, on the direction and opening angle of the dispersal cone and on the mesh granularity. With our set of parameters, the mesh size is 1.5 m long, 1 m large and 1.4 m high,

```
function source_dispersal(day, T, x, y, z, ux, uy, uz, n_spores)
  vox_list <- get_voxels_cone(x, y, z, ux, uy, uz, alpha)
  nodedist_list <- get_nodedist_list(vox_list, x, y, z)
  quicksort(node_list)
  for nd in nodedist_list do
    if test_node_in_cone(nd, x, y, z, ux, uy, uz, alpha)
      captured <- captured_spores(nd, x, y, z, alpha)
      n_spores <- n_spores - captured
      potential <- node_potential(nd, day, T, captured)
      if(potential >= thresthold) then
        node_infection(nd, day)
    endif
  end for
end function
```

Fig. 2. Algorithm of spore dispersal from a source point

and contains 150 voxels. In average, about one third up to half of all voxels are processed by the function. Its complexity is $O(\#voxels)$.

The function `get_nodedist_list` considers each voxel in *vox_list* and calculates for all its leaves the distance to the source lesion. Then, it merges the created list with those of other voxels. Here, computation time is proportional to the number of leaves in the returned list `nodedist_list`. Our measures with our parameters indicate 41.7 % of all leaves are intercepted in average; so we admit that the size of `nodedist_list` is proportional to $\#leaves$. The time complexity is $O(\#leaves)$.

`quicksort` sorts *nodedist_list* according to the distance of leaves to the source lesion. This is done in $O(\#leaves \cdot \log(\#leaves))$ in average.

The `for` loop first tests for each leaf in *nodedist_list*, whether this leaf is indeed intercepted by the cone, using scalar product and trigonometric functions. It calculates the number of spores captured by the leaf according to its distance to the cone axis, and its potential to be infected according to its susceptibility decreasing exponentially with its age. At end, the leaf is possibly infected. The time complexity of the `for` loop is $O(\#leaves)$.

Theoretical time complexity of local dispersal is $O(\#leaves \cdot \log(\#leaves))$. However, elementary operations in the loop are quite complex and might require a lot of time.

During our simulation on a single grapevine, almost 4000 dispersal cones were thrown, spreading two to three millions of spores, of which hardly one million left the volume of the vine. Time spent during the different operations of the `source_dispersal` function was measured. For each operation corresponding to a step of this function and for all iterations, Tab. 1 reports the total time and the time of the longest execution.

Although $\#leaves$ is most of the time bigger than $\#voxels$, the function `get_nodedist_list` requires less time than `get_voxels_cone`. Indeed, the transition from a list of voxels to its list of leaves is a cheap operation. Computing the leaves intercepting cones without using voxels would take much more time.

Table 1. Comparison of execution times of the different operations in the function `source_dispersal`

Operations	Total time (s)	Longest execution time (ms)
`get_voxels_cone`	0.156	0.086
`get_nodedist_list`	0.087	0.095
`quicksort`	0.417	0.555
`for` loop	0.259	0.340

According to theoretical complexities, the last two steps are the most costly. The `quicksort` execution time is not expected to vary, whereas the processing of leaves in the `for` loop may increase with the model refinement, especially the possible multiple infections of a same leaf.

4 Description of the Parallel Simulator

The simulation of disease spread over a vineyard does not come to only simulate the disease on each grapevine. Indeed, there exist many interactions between stocks pertaining to the parasite dispersal, which will lead to communications. We decided to develop an SPMD [2] parallel code, so we must focus on data distribution and efficient communication strategies.

4.1 Data Distribution

As illustrated in Sect. 3, costly computational steps are host and pathogen growth and local dispersal of spores. Distributing stocks over processors makes the growth trivially parallel.

Local dispersion will generate communications between adjoining stocks allocated to different processors. So as to reduce these communications, the field should be cut out in blocks of maximal size and minimal common edges.

Moreover, grapevine vigour – parameter not taken into account currently – plays a role in plant growth and so in the number of organs. A vigorous stock area will produce an higher amount of computations. This point suggests to allocate a set of uniformly distributed stocks to a processor in order to privilege a good load-balancing.

The implemented load-balancing is static and consists in a 2D block-cyclic distribution. The need of a dynamic load-balancing will be adressed later. The set of stocks allocated to a processor is called its *local_stocks*.

4.2 Parallel Algorithm

Let us first describe precisely how the dispersal with neighbouring stocks and the long-range dispersal are modeled. Each processor has two matrices, named transmission matrix *TM* and dispersal matrix *DM*. Both have the field dimensions: they have as many lines as rows in the field, and as many columns as stocks

in the rows. Each one of their elements is associated with the stock of same coordinates in the field. An element in the transmission matrix is a list of dispersal
cones transmitted to the corresponding stock by its neighbours. An element in
the dispersal matrix is the number of spores received by the corresponding stock
from all others in the vineyard.

After budding, an iteration on a processor is described in Fig. 3.

```
for vine in local_stocks do
   vine_computations(vine, day, T, ux, uy, uz, TM, DM)
end for
neighbouring_dispersal(local_stocks, day, T, TM, DM)
long_range_dispersal(local_stocks, DM)
```

Fig. 3. Algorithm of a parallel iteration after budding

`vine_computations` corresponds to the operations of Sect. 3 performed on
a single grapevine, except that dispersal cones that exit the vine volume by
lateral sides are added to the adequate cone lists of *TM*. Other exiting spores
are accumulated in the *vine* data structure for later long-range dispersal.

`neighbouring_dispersal` sends the elements of the *TM* matrix using a
`MPI_Alltoall` communication [3], and calls `source_dispersal` to propagate
cones in grapevines. Exiting spores are not transmitted a second time to neighbouring stocks, they are all accumulated for long-range dispersal.

`long_range_dispersal` considers each grapevine in *local_stocks* and accumu-
lates spores on the *DM* matrix entries by using Gaussian random drawings.
Again, a `MPI_Alltoall` communication is performed on the *DM* matrix entries,
and then spores are dispersed in grapevines thanks to the `source_dispersal`
function. Currently, the value in *DM* associated to a stock is not only a number
of spores, but n numbers corresponding to the amount of spores received by the
stock from other ones in the n uniformly spread directions.

5   Performance Analysis

The implemented parallel simulator uses MPI [3] and MPI-communications within an SMP node are performed via shared memory. The distribution is block
cyclic, and each block contains only one stock.

A platform located at CINES[1] (Montpellier, FRANCE) was used for the
numerical experiments: it is a parallel cluster with 29 nodes of 16 IBM Power 3
processors. Up to 128 processors were used for simulations.

5.1   Load-Balancing Analysis

Currently load-balancing depends totally on the quality of the initial distribution. As we have seen, grapevine vigour can generate more computations on

---

[1] Centre Informatique National de l'Enseignement Supérieur

some areas of the field. The stocks that are primary foci of the epidemic can induce load-imbalance too.

However, three different periods could be distinguished during the simulation. During about the first 80 days preceding the budding, the computation cost is very low. This period is short in time and well-balanced. A large second period corresponds to the development of the plants and to the beginning of the epidemic. Some important load-imbalance can be observed at this moment, but this period does not represent the most costly part of the simulation. Because of the rapid disease spread, the last period contains most of the computations due to the dispersal of high numbers of spores and is rather well-balanced. So, the whole simulation balancing is determined by the one of that last period.

Simulations were performed on the Power 3 platform for a 32×32 field with several configurations: 2, 4, 8 and 16 processors on a same node (SMP), 32 processors on two nodes, 64 processors on four nodes and 128 on eight nodes. Table 2 reports for the days 150 and 220, belonging to the second and third periods of the simulation respectively, the maximum (top number) and the minimum (bottom number) times of each step over all processors. It provides also the communication times that include synchronization time.

Table 2. Maximum and minimum times in milliseconds of each step over all processors for 32×32 field simulations on the Power 3 platform

	day 150				day 220			
Number of processors	2	8	32	128	2	8	32	128
Computation time								
Vine growth	270 263	60 57.6	14.9 13.6	4.03 3.18	1593 1549	396 380	98.3 85.7	24.6 19.9
Local dispersal	52.2 19.5	17.4 4.75	8.62 1.16	7.75 0.27	15120 14670	3700 3526	981 813	266 183
Neighbour dispersal	29.7 0.27	11.6 0.2	7.88 0.24	8.51 0.5	6330 6215	1619 1430	426 336	121 64.3
Global dispersal	0.013 0.009	0.005 0.004	0.004 0.003	0.004 0.003	15760 15470	4044 3749	1049 892	295 195
Final dispersal	0.15 0.15	0.052 0.051	0.014 0.012	0.009 0.005	3468 3429	848 824	211 190	56.5 43.8
Communication time								
Neighbour dispersal	59.3 0.096	24.9 0.3	14.4 2.45	30.6 19.4	588 0.11	210 1.72	200 3.27	127 24.1
Global dispersal	30 0.55	12.1 0.69	12.7 5.02	66.5 56.5	402 0.7	338 0.62	185 4.79	126 20.1

A ratio of ten to several thousands can be noticed between step times at day 150 and those at day 220. The most costly computation at day 220 is rather well balanced up to 128 processors. Vine growth is clearly the fastest computational step and represents a very small part compared to the whole dispersal process. The minimum of communication steps corresponds to the effective communication time. The maximum measures in addition the idle time of the processor that ends first the computation. Furthermore, maximum communication time is small, if not negligible, in comparison with computation time, and it is decreasing with the number of processors at day 220.

5.2   Scalability Analysis

Maximum computation, communication and total times for the complete simulations are reported in Fig. 4 and other performance measures in Fig. 5.

Processors	2	8	32	128
Computation time	2234.4	553.7	141.4	37.76
Communication and idle time	70.8	41.1	25.9	21.89
Total time	2236	563.9	153.5	55.78

Fig. 4. Maximum computation, communication and total times in seconds for a complete simulation on a 32×32 field

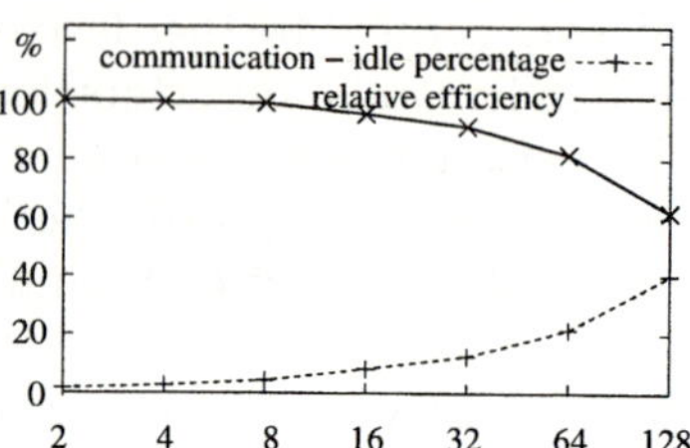

Fig. 5. Relative efficiency in comparison with 2 processors and ratio (communication and idle time) / (total time), for a complete simulation on a 32×32 field

As expected, computation time is about inversely proportional to the number of processors involved in the simulation, which is a good result.

Relative efficiency remains above 90 % up to 32 processors, it is about 80 % with 64 processors and drops to 63 % with 128 processors. Currently, there is no overlapping of communications by computations. It is only possible for the communications of the neighbour dispersal step and should be done in the future to improve performances. The ratio of communication time divided by global time increases from 2 to 128 processors because of load imbalance in computations. To enhance parallelism of the application the load-balancing should be refined.

6   Quality of the Current Biological Results

The simulator is still in the calibration phase and basic sub-models are being validated. Nevertheless, current outputs of the program are already coherent with field data. These outputs at field level consists of maps where each stock is represented by a point. The greyscale of this point depends on the severity of disease on the stock: from white to black, severity increases. Figure 6 represents three maps of a 32×32 field. Primary infection is located on four stocks in the top-left corner during days 110 and 127. The three maps correspond respectively to the days 160, 180 and 210. These results correspond qualitatively to field observations made by biologists.

7   Conclusion

These first results are encouraging: the simulator performances are quite good in terms of scalability and load-balancing.

 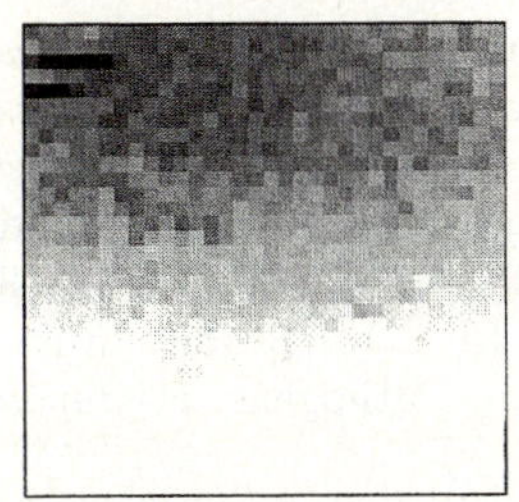 

Fig. 6. Three maps of a 32×32 field at days 160, 180 and 210

The simulator is still in the calibration process and some future adjustments could strongly modify its behaviour. Our simple modeling of the vertical wind component is being reviewed with bioclimatologists. To take into account the important variations of this component [5], several dispersal cones will be spread from each sporulating colony every day. Each cone will carry a part of the spores released by the colony during the given day. This modification will increase the computational time of local dispersal and will generate more communications between neighbouring stocks.

We also consider the extension of the local dispersal domain to all the neighbouring stocks of a source stock, including its neighbours in the previous and next rows.

Imbalance may appear when taking into account the grapevine vigour, hence it could imply to improve load-balancing strategy.

Moreover, it will be interesting to model prophylactic[2] methods.

References

1. Bulit (J.) et Lafon (R.). – Powdery mildew of the vine. *In: The powdery mildews,* éd. par Academic press, London. – DM Spencer, 1978.
2. Grana (Ananth), Gupta (Anshul), Karypis (George) et Kumar (Vipin). – *Introduction to Parallel Computing.* – Addison Wesley, 2003, second edition édition. ISBN 2-201-64865-2.
3. MPI Forum. – *Message Passing Interface MPI Forum Home Page.* Available from `http://www.mpi-forum.org/`.
4. Naulin (J.-M.), Tessier (G.), Bailey (D.), Langlais (M.) et Calonnec (A.). – A host/pathogen simulation model : Powdery mildew and vine. – february 2005. Submitted to New Phytologist.
5. Raupach (M. R.), Finnigan (J. J.) et Brunet (Y.). – Coherent Eddies and Turbulence in Vegetation Canopies: The Mixing-Layer Analogy. *Boundary-Layer Meteorology,* vol. 78, 1996, pp. 351–382.
6. Samet (Hanan). – *The Design and Analysis of Spatial Data Structures.* – University of Maryland, Addison-Wesley Publishing Company, 1989. ISBN 0-201-50255-0.

---

[2] set of measures to fight against epidemics; for example chemical treatments.

7. Shigesada (Nanaka) et Kawasaki (Kohkichi). – Invasion and the range expansion of species: effects of long-distance dispersal. *In: Proceedings of BES Annual Symposium 2001 'Dispersal'*, chap. 17, pp. 350–373. – Blackwell Science (in press), 2002.
8. Whitted (Turner). – An improved illumination model for shaded display. *Communications of the ACM*, vol. 23, n° 6, 1980, pp. 343–349. – ISSN:0001-0782.
9. Zawolek (M. W.) et Zadocks (J. C.). – Studies in Focus Development: An Optimum for the Dual Dispersal of Plant Pathogens. *Phytopathology*, vol. 82, n° 11, 1992, pp. 1288–1297.

Parallelism for Perturbation Management and Robust Plans*

Jan Ehrhoff[2], Sven Grothklags[1], and Ulf Lorenz[1]

[1] University of Paderborn
Faculty of Computer Science, Electrical Engineering and Mathematics
Fürstenallee 11, D-33102 Paderborn
[2] Lufthansa Systems Airline Services GmbH
Network Management Solutions
Am Prime Parc 9, D-65479 Raunheim

Abstract. An important insufficiency of modern industrial plans is their lack of robustness. Disruptions prevent companies from operating as planned before and induce high costs for trouble shooting. The main reason for the severe impact of disruptions stems from the fact that planners do traditionally consider the precise input to be available at planning time.
The Repair Game is a formalization of a planning task, and playing it performs disruption management and generates robust plans with the help of game tree search. Technically, at each node of a search tree, a traditional optimization problem is solved such that large parts of the computation time are blocked by sequential computations. Nevertheless, there is enough node parallelism which we can make use of, in order to bring the running times onto a real-time level, and in order to increase the solution quality per minute significantly. Thus, we are able to present a planning application at the cutting-edge of Operations Research, heavily taking advantage of parallel game tree search. We present simulation experiments which show the benefits of the repair game, as well as speedup results.

1 Introduction

An important problem in aircraft planning is to react with an instant decision, after a certain disruption hinders the company to act as planned before. This problem touches various research directions and communities. Because the problems are often computationally hard [14] the field might become an El Dorado for parallel computing and grid computing. A stochastic multi-stage fleet-assignment optimization problem is in the focus of this paper. The used solution method is based on game tree search.

Multistage Decisions Under Risk. The reason for disruptions obviously stems from the fact that planners lack information about the real behavior of the environment at

* This work has been partially supported by the European Union within the 6th Framework Program under contract 001907 (DELIS) and the German Science Foundation (DFG), SFB 614 (Selbstoptimierende Systeme des Maschinenbaus) and SPP 1126 (Algorithmik großer und komplexer Netzwerke).

J.C. Cunha and P.D. Medeiros (Eds.): Euro-Par 2005, LNCS 3648, pp. 1265–1274, 2005.

planning time. Often, data is not as fixed as assumed in the traditional planning process. Instead, we know the data approximately, we know distributions over the data. In the airline example, we know a distribution over a leg's (i.e. flight's) possible arrival times. Traditionally, plans are built which maximize profits over 'expected' or just estimated input data, but we belong to the group of people who believe that it is more realistic to optimize the expected payoff over all possible scenarios instead. This view on the world leads us to something that is often called 'multistage decisions under risk', related to linear stochastic programming [4, 16], stochastic Optimization [11], game playing [2], replanning [10] and others [18].

Current Planning Processes in Airline Industry. An airline planning process starts with the so called network design, which roughly tells the planning team which routes (so called *legs*) should be taken into account. Then, a first 'plan' is made which shows when which legs are offered to the customers. Thereafter, the process contains two layers which are of special interest for us.

Typically, airline companies have aircrafts of different types (so called *subfleets*), which differ in size and economic behavior. Given a flight schedule and a set of aircrafts, the fleet assignment problem is to determine which type of aircraft should fly each flight segment. A solution of the fleet assignment problem and the flight schedule together answers the question of how many aircrafts of which subfleet have to be at certain places at certain times.

So called time-space networks, which are special flow graphs, can be used to give a specific mathematical programming formulation for this class of problems. They were introduced by Hane et al. in [7] to solve the fleet assignment problem. On the basis of the fleet assignment, a so called rotation plan is generated. It describes which physical aircraft must be at which place in the world and at which time.

The planning is dominated by deterministic models. All uncertainties are eliminated through restrictive models. However, since some time, several large airline companies have come to the conclusion that new models and methods are necessary in order to exploit further potentials for cost reduction.

Game Tree Search. Game tree search is the core of most attempts to make computers play games. The game tree acts as an error filter and examining the tree behaves similar to an approximation procedure. At some level of branching, the complete game tree (as defined by the rules of the game) is cut, the artificial leaves of the resulting subtree are evaluated with the help of heuristics, and these values are propagated to the root [9, 15] of the game tree as if they were real ones. For 2–person zero-sum games, computing this heuristic minimax value is by far the most successful approach in computer games history, and when Shannon [19] proposed a design for a chess program in 1949 it seemed quite reasonable that deeper searches lead to better results. Indeed, the important observation over the last 40 years in the chess game and some other games is: *the game tree acts as an error filter*. Therefore, the faster and the more sophisticated the search algorithm, the better the search results! This, however, is not self-evident, as some theoretical analyzes show [1, 8, 13].

New Approach. Our approach [3] can roughly be described by looking at a (stochastic) planning task in a 'tree-wise' manner. Let a tree T be given that represents the possible scenarios as well as our possible actions in the forecast time-funnel. It consists of two different kinds of nodes, MIN nodes and AVG nodes. A node can be seen as a 'system state' at a certain point of time at which several alternative actions can be performed/scenarios can happen. Outgoing edges from MIN nodes represent our possible actions, outgoing edges from AVG nodes represent the ability of Nature to act in various ways. Every path from the root to a leaf can then be seen as a possible solution of our planning task; our actions are defined by the edges we take at MIN nodes under the condition that Natures acts as described by the edges that lead out of AVG nodes.

The leaf values are supposed to be known and represent the total costs of the 'planning path' from the root to the leaf. The value of an inner MIN node is computed by taking the minimum of the values of its successors. The value of an inner AVG node is built by computing a weighted average of the values of its successor nodes. The weights correspond to realization probabilities of the scenarios.

Let a so called *min-strategy* S be a subtree of T which contains the root of T, and which contains exactly one successor at MIN nodes, and all successors that are in T at AVG nodes. Each strategy S shall have a value $f(S)$, defined as the value of S's root. A *principle variation* $p(S)$, also called *plan*, of such a min-strategy can be determined by taking the edges of S leaving the MIN nodes and a highest weighted outgoing edge of each AVG node. The connected path that contains the root is $p(S)$. We are interested in the plan $p(S_b)$ of the best strategy S_b and in the expected costs $E(S_b)$ of S_b. The expected costs $E(p)$ of a plan p are defined as the expected costs of the best strategy S belonging to plan p, e.g. $E(p) = \min\{E(S) \mid p(S) = p\}$. Because differences between planned operations and real operations cause costs, the expected costs associated with a given plan are not the same before and after the plan is distributed to customers. A plan gets a value of its own once it is published.

This model might be directly applied in some areas, as e.g. job shop scheduling [12], not, however, in applications which are sensible to temporary plan deviations. If a job shop scheduling can be led back to the original plan, the changes will nothing cost, as the makespan will stay as it was before. That is different in airline fleet assignments. Mostly, it is possible to find back to the original plan after some while, but nevertheless, costs occur. A decisive point will be to identify each tree nodes with a pair of the system state plus the path, how the state has been reached.

1.1   Organization of This Paper

We introduce the Repair Game as a reasonable formalization of the airline planning task on the level of disruption fighting. Section 2 describes the Repair Game, its formal definition, as well as an interpretation of the definition and an example. In Section 3 we describe a prototype, which produces robust repair decisions for disrupted airline schedules, on the basis of the Repair Game. Section 4 contains details of the parallelization of the search procedure. In Section 5 we compare the results of our new approach with an optimal repair procedure (in the traditional sense). A comparison of the sequential and the parallel version of our prototype is additionally given. Section 6 concludes.

2 The Repair Game

Definitions. We define the Repair Game via its game tree. Its examination gives us a measure for the robustness of a plan and on the other hand it presents us concrete operation recommendations.

Definition 1. *(Game Tree)*
For a rooted tree $T = (V, E)$ let $L(T) \subset V$ be the set of leafs of T. In this paper, a game tree $G = (V, E, h)$ is a rooted tree (V,E), where $V = V_{MAX} \cup V_{MIN} \cup V_{AVG}$ and $h : V \to I\!N_0$.

Nodes of a game tree G represent positions of the underlying game, and edges move from one position to the next. The classes V_{MAX}, V_{MIN}, and V_{AVG} represent three players MAX, MIN, and AVG and for a node/position $v \in V_i$ the class V_i determines the player i who must perform the next move.

Definition 2. *(*Minimax Value)*
*Let $G = (V, E, h)$ be a game tree and $w_v : N(v) \to [0, 1]$ be weight functions for all $v \in V_{AVG}$, where $N(v)$ is the set of all sons of a node v. The function *minimax $: V \to I\!N_0$ is inductively defined by*

$$
\text{minimax}(v) := \begin{cases} h(v) & \text{if } v \in L(G) \\ \max\{\text{minimax}(v') \mid v' \in N(v)\} & \text{if } v \in V_{MAX} \setminus L(G) \\ \min\{*\text{minimax}(v') \mid v' \in N(v)\} & \text{if } v \in V_{MIN} \setminus L(G) \\ \sum_{v' \in N(v)}(w_v(v') \cdot *\text{minimax}(v')) & \text{if } v \in V_{AVG} \setminus L(G) \end{cases}
$$

Definition 3. *(Repair Game)*
The goal of the Repair Game $= (G,p,g,f,s)$ *is the calculation of *minimax(r) for a special game tree $G = (V, E, g + f)$ with root r and uniform depth t; $p \in L(G)$ is a special leaf, g, f and s are functions. The game tree has the following properties:*

- *Let $P = (r = v_1, v_2, \ldots, p = v_t) \in V^t$ be the unique path from r to p. P describes a traditional, original plan.*
- *V is partitioned into sets $S_1, \ldots, S_n, |V| \geq n \geq t$ by the function $s : V \to \{S_i\}_{1 \leq i \leq n}$. All nodes which belong to the same set S_i are in the same state of the system — e.g. in aircraft scheduling: which aircraft is where at which point of time —, but they differ in the histories which have led them into this state.*
- *$g : \{S_i\}_{1 \leq i \leq n} \to I\!N_0$ defines the expected future costs for nodes depending on their state; for the special leaf p holds $g(s(p)) = 0$*
- *$f : \bigcup_{1 \leq \tau \leq t}\{V\}^\tau \to I\!N_0$ defines the induced repair-costs for every possible (sub)path in (V, E) ; every sub-path P' of P has zero repair-costs, $f(P') = 0$*
- *the node evaluation function $h : V \to I\!N_0$ is defined by $h(v) = g(s(v)) + f(r \ldots v)$; note that $h(p) = 0$ holds by the definition of g and f*

2.1 Interpretation and Airline Example

A planning team of e.g. an airline company starts the game with the construction of a traditional plan for its activities. The path P represents this planned schedule, which also is the most expected path in the time-funnel, and which interestingly gets an additional value of its own, as soon as it is generated. It is small, can be communicated, and as soon as a customer or a supplier has received the plan, each change of the plan means extra costs for the change. Disruptions in airline transportation systems can now prevent airlines from executing their schedules as planned. As soon as a specific disruption occurs, the MIN-player will select a repairing sub-plan such that the repair costs plus the expected future repair costs are minimized.

As the value of a game tree leaf v depends on how 'far' the path $(r, \ldots, v)$ is away from P, it will not be possible to identify system states (where the aircrafts are at a specific time) with tree nodes. Therefore, the tree nodes V are partitioned into $S_1 \cup \ldots \cup S_n$. In S_i all those nodes are collected which belong to the same state, but have different histories. All nodes in the same state S_i have the same expected future costs. These costs are estimated by the function g. The function f evaluates for an arbitrary partial path, how far it is away from the level-corresponding partial path of P. Inner nodes of the game tree are evaluated by the *Minimax function.

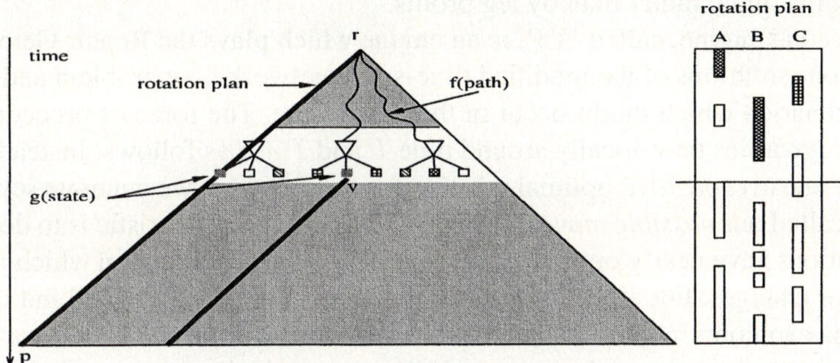

Fig. 1. The Repair Game Tree.

Figure 1 shows a rotation plan at the right. Aircrafts A, B, and C are either on ground, or in the air, which is indicated by boxes. A shadowed box means that the original plan has been changed. The time goes from the top down. The left part of the figure shows a game tree, where the leftmost path corresponds to the original plan P. When a disruption occurs, we are forced to leave the plan, but hopefully we can return to it at some node v. The fat path from node v downward is the same as in the original plan. Thus, at node v, we have costs for the path from the root r to v, denoted by $f(r \ldots v)$ and future expected costs, denoted by $g(s(v))$. If we follow the original plan from the beginning on, we will have the same expected costs at the time point represented by v, but different path costs. The only node with zero costs typically is the special leaf node p of the original plan.

3   Experimental Setup

In accordance with our industrial partner, we built a simulator in order to evaluate different models and algorithms The simulator is less detailed than e.g. SimAir [17], but we believe that it is detailed enough to model the desired part of the reality. Furthermore, it is simple enough such that the occurring problems are computationally solvable.

First of all, we discretize the time of the rotation plan into steps of $d = 15$ minutes. Every departure in the current rotation plan is a possible event point, and all events inside one d minute period are interpreted as simultaneous. When an event occurs, the leg which belongs to that event can be disrupted, i.e. it can be delayed by 30, 60, or 120 minutes, or it can be canceled with certain probabilities. Let T be the present point of time. The simulator inspects the events between T and $T + d$, and informs a repair engine about the new disruptions, waits for a corrected rotation plan, and steps d minutes forward.

The aim is to compare two different repair approaches with each other. The first one is called 'Myopic MIP' solver and it repairs a plan after a disruption with the help of a slightly modified time-space network such that the solution of the associated network flow problem is an optimal one, under the assumption that no more disruptions will ever occur. This engine represents traditional deterministic planning. It only needs a modified cost function because the repair costs are mainly determined by the changes on the original plan, rather than by leg profits.

The second engine, called 'T3', is an engine which plays the Repair Game. It compares various solutions of the modified time-space network flow problem and examines various scenarios which might occur in the near future. The forecast procedure makes use of the dynamics time-locally around time T and $T + d$ as follows: Instead of generating only one myopic MIP optimal solution for the recovery, we generate several ones. They are called our *possible moves*. A simple, certainly good heuristic is to demand that these solutions have nearly optimal costs concerning the cost function which minimizes the cost for changes. For all of these possible moves, we inspect what kind of relevant disruptions can come within the next d minutes. On all these scenarios, we repair the plan again with the help of the myopic MIP solver, which gives us value estimations for the arisen scenarios. We weight the scenarios according to their probabilities, and minimize over the expected values of the scenarios. Concerning the new part of the plan we have not optimized the costs over expected input data, but we have approximately minimized the expected costs over possible scenarios. The following algorithm is a simplified version of the algorithm shown in [2]. We refer to [3] for further details of the repair engines.

```
value *minimax(node v, value α, value β)
1  generate all successors v₁,...,vₒ of v; let b be the number of successors
2  value val := 0;
3  if b = 0 return h(v) /* (leaf eval)*/
4  for i := 1 to b
5    if v is MIN-node {
6      β := min(β, *minimax(vᵢ, α, β)); if α ≥ β return β; if i = b return β
7    } else if v is AVG-node { // let w₁,..,wₒ be the weights of v₁,..,vₒ
8      val+ = *minimax(vᵢ, α, β) · wᵢ;
9      if val + L · ∑ᵇⱼ₌ᵢ₊₁ wⱼ ≥ β return β; if val + U · ∑ᵇⱼ₌ᵢ₊₁ wⱼ ≤ α return α
10     if i = b return val
11   } else { // v is MAX-node. Analogously to v is a MIN-node }
```

4   Parallelization

The basic idea of our parallelization is to use dynamic load balancing and to decompose the search tree, to search parts of the search tree in parallel and to balance the load dynamically with the help of the work stealing concept. This works as follows: First, a special processor P_0 gets the search problem and starts performing the *minimax algorithm as if it would act sequentially. At the same time, the other processors send requests for work, the REQUEST message, to other randomly chosen processors. When a processor P_i that is already supplied with work, catches such a request, it checks, whether or not there are unexplored parts of its search tree, ready for evaluation. These unexplored parts are all rooted at the right siblings of the nodes of $P_i's$ search stack. Either, P_i sends back that it cannot serve with work with the help of the NO-WORK message, or it sends such a node (a position in the search tree etc.) to the requesting processor P_j. That is done with the help of the WORK message. Thus, P_i becomes a master itself, and P_j starts a sequential search on its own. The master/worker relationships are dynamically changed during the computation. A processor P_j being a worker of Q at a certain point of time may become the master of Q at another point of time. In general, processors are masters and workers simultaneously. If, however, a processor P_j has evaluated a node v, but a sibling of v is still under examination of another processor Q, P_j will wait until Q sends an answer message. When P_j has finished its work (possibly with the help of other processors), it sends an ANSWER message to P_i. The master-worker relationship between P_i and P_j is released, and P_j is idle again. It again starts sending requests for work into the network.

When a processor P_i finds out that it has sent a wrong local bound to one of its workers P_j, it makes a WINDOW message follow to P_j. P_j stops its search, corrects the window and starts its old search from the beginning. If the message contained a cutoff, P_j just stops its work. A processor P_i can shorten its search stack due to an external message, when e.g. a CUTOFF message comes in which belongs to a node near the root. In absence of deep searches and advanced cutting techniques, however, CUTOFF and WINDOW messages did not occur in our experiments.

In many applications, the computation at each node is fast in relation to the exchange of a message. E.g. there are chess programs which do not use more than 1000 processor cycles per search node, including move generation, moving end evaluating the node. In the application presented here, the situation is different. The sequential computations at each node takes several seconds if not even minutes, such that the latency of messages is not remarkably important. Therefore, the presented application is certainly well suited for grid computing, as well. However, it is necessary to overlap communication and computations. We decoupled the MIP solver from the rest of the application and assigned a thread to it.

In distributed systems, messages can be delayed by the system. Messages from the past might arrive, which are outdated. Therefore, for every node v a processor generates a local unique ID, which is added to every message belonging to node v. Thus, we are always able to identify misleading results and to discard invalid messages. We refer to [5] for further details.

Table 1. The daily-average profit of T3 over Myopic, two different weeks, six measure points each.

	run 1	run 2	run 3	run 4	run 5	run 6
week 1	2320	27696	32261	-9238	-15799	13150
week 2	11040	48778	11580	-1253	9144	8389

5   Results

The basis for our simulations is a major airline plan plus the data which we need to build the plan and its repairs. The plan consists of 20603 legs, operated by 144 aircrafts within 6 different subfleets. The MIP for this plan has 220460 columns, 99765 rows, and 580793 non-zeros. Partitioned into single days, the resulting partial plan for any single day consists of a corresponding smaller number of columns and non-zeros.

All experiments are performed on a 4-node-cluster of the Paderborn University. Each node consists of two Pentium IV/2.8 GHz processors on a dual processor board with 1 GB of main memory. The nodes are connected by a Myrinet high speed interconnection network. The cluster runs under the RedHat Linux operating system.

We simulated 14 days, and we divided this time period into 14 pieces such that we arrived at a test set with 14 instances. Moreover, time is divided into segments of 15 minutes, and everything happening in one 15 minute block is assumed to be simultaneous. We compare the behavior of the 'myopic MIP' and the 'T3' engines. We appropriately choose the probabilities for disruptions and control the performance of the objective function $c(TIM, ECH, CNL, revenue) = 50 \cdot TIM + 10000 \cdot ECH + 100000 \cdot CNL - revenue$, TIM being time-shifts, ECH meaning that the aircraft type of a leg had to be changed, CNL being the number of necessary cancellations and revenues being the traditional cost measure for the deterministic problem. A test run consists of about 1400 single decisions, and first tests showed benefits of more than three percent cost reductions over the myopic solver. Although a benefit of more than three percent looks already promising, statistical significance cannot be read out of a single test run. The engines make nearly 100 decisions per simulated day. However, these decisions cannot be taken as a basis for significance examinations, because the single decisions are part of a chain and not independent from each other. The results of the single days seem to form no normal distribution and, moreover, depend on the structure of the original plan. Therefore, we made further test runs and grouped those outcomes of each simulated week to one. We measure the average daily profit of a week, in absolute numbers (see Table 1). The profit, which is statistically significant, is the benefit of the T3-engine over the Myopic-MIP engine.

5.1   Speedups

We measure speedups of our program with the help the first three days of the test set which consists of the 14 single days, mentioned above. The time for simulation of the days using one processor is compared with the running time of several processors. The speedup (SPE) is the sum of the times of the sequential version divided by the

Table 2. Speedups.

# proc	simtime day 1	SPE day1	simtime day 2	SPE day2	simtime day 3	SPE day3
1	226057	1	193933	1	219039	1
2	128608	1.76	111612	1.73	126915	1.73
4	68229	3.31	59987	3.23	66281	3.30
8	46675	4.84	40564	4.78	46065	4.75

sum of the times of a parallel version [6]. Each test day consists of 96 single measure points. Table 2 shows the speedups which we could achieve. We are quite satisfied with these speedups because they bring the necessary computations on a real-time level. Of course, the question arises how the work load and the search overhead behave. As the sequential and the parallel version do indeed exactly the same, search overhead does not exist. Neither the costs for communication are relevant. The only reason that we cannot achieve the full speedup are the large sequential computing periods, caused by the commercial MIP solver.

6 Conclusion

Playing the Repair Game leads to more robust (sub-)plans in airline scheduling than traditional deterministic planning can provide. Our forecast strategy outperforms the myopic MIP solver by means of simulations. We have parallelized the search in order to drop the computation times to real time. Next, we will look for more clever and selective search heuristics, examine heuristics which give us fast new moves, and refine the simulator.

We presented an application which we think is a typical example for the benefits of cluster parallelism and grid computing. The stochastic fleet assignment problem that we presented in the frame of game tree search makes profit from its speed. The faster the application can be performed, the larger is the gained profit. Planning under uncertainty becomes more and more important in Operations Research. The resulting problems are hard to solve and can often only be approximated. We are convinced that parallel game tree search will become an important part of that area.

References

1. I. Althöfer. Root evaluation errors: How they arise and propagate. *ICCA Journal*, 11(3):55–63, 1988.
2. B.W. Ballard. The *-minimax search procedure for trees containing chance nodes. *Artificial Intelligence*, 21(3):327–350, 1983.
3. J. Ehrhoff, S. Grothklags, and U. Lorenz. Das Reparaturspiel als Formalisierung von Planung unter Zufallseinflüssen, angewendet in der Flugplanung. In *Proceedings of GOR conference: Entscheidungsunterstützende Systeme in Supply Chain Managment und Logistik*, pages 335–356. Physika-Verlag, 2005.
4. S. Engell, A. Märkert, G. Sand, and R. Schultz. Production planning in a multiproduct batch plant under uncertainty. *Preprint 495-2001, FB Mathematik, Gerhard-Mercator-Universität Duisburg*, 2001.

5. R. Feldmann, M. Mysliwietz, and B. Monien. Studying overheads in massively parallel min/max-tree evaluation. In *6th ACM Annual symposium on parallel algorithms and architectures (SPAA'94)*, pages 94–104, New York, NY, 1994. ACM.

6. P.J. Flemming and J.J. Wallace. How not to lie with statistics: the correct way to summerize benchmark results. *CACM*, 29(3):218–221, 1986.

7. C.A. Hane, C. Barnhart, E.L. Johnson, R.E. Marsten, G.L. Nemhauser, and G. Sigismondi. The fleet assignment problem: solving a large-scale integer program. *Mathematical Programming*, 70:211–232, 1995.

8. H. Kaindl and A. Scheucher. The reason for the benefits of minmax search. In *Proc. of the 11 th IJCAI*, pages 322–327, Detroit, MI, 1989.

9. D.E. Knuth and R.W. Moore. An analysis of alpha-beta pruning. *Artificial Intelligence*, 6(4):293–326, 1975.

10. S. Koenig, D. Furcy, and Colin Bauer. Heuristic search-based replanning. *In Proceedings of the International Conference on Artificial Intelligence Planning and Scheduling*, pages 294–301, 2002.

11. P. Kouvelis, R.L. Daniels, and G. Vairaktarakis. Robust scheduling of a two-machine flow shop with uncertain processing times. *IIE Transactions*, 32(5):421–432, 2000.

12. V.J. Leon, S.D. Wu, and R.h. Storer. A game-theoretic control approach for job shops in the presence of disruptions. *International Journal of Production Research*, 32(6):1451–1476, 1994.

13. D.S. Nau. Pathology on game trees revisited, and an alternative to minimaxing. *Artificial Intelligence*, 21(1-2):221–244, 1983.

14. C. H. Papadimitriou. Games against nature. *Journal of Computer and System Science*, 31:288–301, 1985.

15. A. Reinefeld. An Improvement of the Scout Tree Search Algorithm. *ICCA Journal*, 6(4):4–14, 1983.

16. W. Römisch and R. Schultz. Multistage stochastic integer programming: an introduction. *Online Optimization of Large Scale Systems*, pages 581–600, 2001.

17. J. M. Rosenberger, A. J. Schaefer, D. Goldsman, E. L. Johnson, A. J. Kleywegt, and G. L. Nemhauser. Simair: A stochastic model of airline operations. *Winter Simulation Conference Proceedings*, 2000.

18. S. Russel and P. Norvig. *Artificial Intelligence, A Modern Approach*. 2003. Prentice Hall Series in Artificial Intelligence.

19. C.E. Shannon. Programming a computer for playing chess. *Philosophical Magazine*, 41:256–275, 1950.

SPH2000: A Parallel Object-Oriented Framework for Particle Simulations with SPH*

Sven Ganzenmüller, Simon Pinkenburg, and Wolfgang Rosenstiel

Wilhelm-Schickard-Institut für Informatik
Department of Computer Engineering
University of Tübingen
Sand 13, 72076 Tübingen, Germany
{ganzenmu,pinkenbu,rosen}@informatik.uni-tuebingen.de

Abstract. A widespread method in parallel scientific computing is SPH, a grid-free method for particle simulations. Lots of libraries implementing this method evolved in the past. Since most of them are written in FORTRAN or C, there is a lack of integration of object-oriented concepts for scientific applications. These libraries are therefore hard to maintain and to extend. In this paper, we describe the design and implementation of sph2000, a parallel object-oriented framework for particle simulations written in C++. Its key features are easy configurability and good extensibility for the users to support their ongoing development of the SPH method. The use of design patterns lead to an efficient and clear design and the implementation of parallel I/O improved the performance significantly. A sample application was implemented to test the framework.

1   Introduction

Within the Collaborative Research Center (CRC) 382 physicists, mathematicians and computer scientists work together to research new aspects of astrophysics and the motion of multiphase flows, evolve them to models and run parameter studies to verify these models. Several particle codes are used to simulate these physical problems.

A well known particle simulation method is Smoothed Particle Hydrodynamics (SPH). SPH is a grid-free Lagrangian method for solving the hydrodynamic equations for compressible and viscous fluids. It was introduced in 1977 by [4] and [7] and has become a widely used numerical method for astrophysical problems. Nowadays the SPH approach is also used in fields of material sciences, for modeling multiphase flows [9] and the simulation of brittle solids [2].

Resolution and accuracy of a simulation depend on the number of used particles and interaction partners. Actual physical problems need large numbers to achieve reasonable results. Thus, high-performance computers are indispensable.

---

* This project is funded by the DFG within CRC 382: *Verfahren und Algorithmen zur Simulation physikalischer Prozesse auf Höchstleistungsrechnern* (Methods and algorithms to simulate physical processes on supercomputers).

J.C. Cunha and P.D. Medeiros (Eds.): Euro-Par 2005, LNCS 3648, pp. 1275–1284, 2005.
© Springer-Verlag Berlin Heidelberg 2005

As a result, most users in the domain of physical simulations developed self-made parallel scientific libraries. Although object-oriented programming became most recently state-of-the-art in parallel programming, almost all libraries are still implemented in FORTRAN or C, leading to a lack of integration of object-oriented concepts for parallel scientific applications.

Our group has a strong effort to develop fast parallel particle libraries, which are portable to all important parallel platforms. Therefore, we developed a parallel object-oriented SPH framework which is clearly structured, easy to configure, maintain and extendable by the advantages of object-oriented programming and design patterns. The main goals were to provide a general framework for SPH-like particle simulations and to reduce the parallel overhead and serial parts, which limit the overall speedup. To apply object-orientation to parallelization, we use TPO++, an object-oriented communication library set up on top of MPI [5]. It was developed in our working group and provides the same functionality and efficiency as MPI 1.2. Recently, we extended it by an object-oriented interface for parallel I/O. A sample application points up the usability of our approach.

This paper is organized as follows. In Section 2, we present related work. Section 3 contains the design and implementation of the framework. In Section 4 we present a sample application and in Section 5 we conclude.

2   Related Work

Cactus [1] is a large framework for parallel physical simulations with many modules, called thorns, which offers interfaces for different languages, e.g. for C++. POOMA [11] is as well a wide-spread framework for parallel physical computations. Both libraries, like most others, are laid out for grid methods. Although POOMA supports moving particles, they have to be arranged in arrays. The focus is on interactions and transformations between the particle arrays and the grid fields.

Besides other SPH libraries [6], Gadget [12] is a large SPH library which offers standard algorithms for astrophysics with self-gravitation. However, like most SPH libraries, it is not object-oriented.

The embedding of pure particle methods in object-oriented libraries is not very common. To use the given object-oriented libraries, the developer of particle methods must take along the overhead of grid methods, deal with a higher complexity and learn the concepts of the library, although not every concept is needed. Using the procedural SPH libraries means to abandon the advantages of object-oriented programming.

3   Design and Implementation

3.1   Design Goals

Our main goal was to develop a parallel object-oriented SPH-framework with extensibility, maintainability and reusability of the code. A main concern in the design was the strict decoupling of parallelization and physics. Another goal was to prove the feasibility of the object-oriented approach in the performance critical domain of particle simulations without loosing efficiency. The result is a parallel object-oriented particle simulation framework written in C++, called sph2000. Classes modeling the elements of the problem domain generate a well structured design. The use of several design patterns [3] helped to organize the classes clearly and efficient. They are introduced to structure the class library, to separate and group the application elements, as well as to define uniform interfaces. Additionally, they allow to insert extensions more simply because of decoupled elements. Table 1 shows the application elements and its corresponding design patterns.

Table 1. Application elements and corresponding design patterns.

Element	Responsibility	Design Pattern
Initialization	Configuration, Object Creation	Builder, Configuration Table
Mathematics	Time Integration	Strategy, Iterator, Index Table
Physics	Particles, Interactions, Right Hand Side (RHS)	Strategy, Compositum, Iterator, Index Table
Parallelization	Communication, Decomposition, Load Balancing	Strategy, Mediator, Proxy
Geometry	Simulation Domain, Subdomains	Strategy, Decorator

The independent elements can be extended, causing no changes in other classes. The classes within an element can be easily exchanged because of uniform interfaces. Once implemented this enables the user to form new classes by simply copying and adapting the existing ones. Thus, extensions supplement the code instead of changing it.

3.2   Configuration of Simulation Runs

Another goal was to simplify the configuration of simulations, which mainly means to configure a simulation run after the compilation at runtime by reading a parameter file. To avoid conditional compilation with preprocessor directives, as it is often seen in C libraries, the Strategy pattern, which is based on the object-oriented concept of polymorphism is used. With this pattern the program instantiates objects at runtime due to the configuration parameters.

The complete configuration with all parameters of a simulation is stored in an object of the class ParameterMap (Configuration Table pattern). Every object which needs parameters owns a reference to the ParameterMap object. Thus all objects can access the configuration uncomplicatedly. Mainly the initialization objects (Builder pattern) access the ParameterMap to realize the exchangeability of the components. Every Strategy offers an accordant parameter to determine which concrete implementation must be used in the simulation. In every simulation run, the Builders create and initialize only the needed objects.

The concept of the configuration table makes the configuration data available for all auxiliary programs which work with simulation data. Such supplementary tools adopt the existing ParameterMap class without changes.

3.3   A Quantity Index Store

The flexibility of the framework is mainly up to the used physics. Because of enhancing the framework simultaneously with the physical method, the integration of additional physical quantities and the exchangeability of different calculation methods has to be guaranteed.

An SPH particle is a sampling point of the differential equations which moves with the flow and represents a volume element of the moving fluid. It contains all physical quantities of a fluid element, their interactions between each other have to be evaluated and they have to be communicated among the subdomains.

To protect the user from adapting the Particle class for every application, we introduced the class IdStore as an index table. The Particle class contains only the administrative structure for the communication by message passing and two STL containers for the scalar and the vector variables. The initialization determines the fixed number of needed variables within a simulation and writes them into the index table. Thus the particles can adapt themselves dynamically at start time to the respective simulation.

The basis for this design are two classes, the QuantityBuilder and the IdStore. The QuantityBuilder implements references to the used physics and initializes the IdStore object. It evaluates the parameters from the ParameterMap and reserves an index in the IdStore for every physical variable (e.g. position, speed and density). The physical variables are stored inside the particle in the order of these indices. Since the particle itself has no idea about the contained variables, classes needing a particles' variable have to get the information through the IdStore, which represents the interface to the variables.

3.4   Object-Oriented View of the Right Hand Side

The QuantityBuilder class is designed according to the Builder pattern. Besides the initialization of the IdStore it establishes the QuantityList, an STL container of calculation objects for the physical quantities. Since there are no general SPH formulas for the equations of motion, many different approaches evolved in the past. To achieve a high flexibility, the calculation objects are defined as a Strategy with a Quantity base class.

The QuantityBuilder knows all possible quantities and their dependencies. By reading the physical parameters from the ParameterMap, quantities are selected and stored in the QuantityList in respect to the physical dependencies (see Fig. 1). In each time step an RhsCalculator object iterates through the QuantityList to compute the right hand side (RHS) of the differential equations.

Like most elements of the framework, the Integrator class is also implemented as a Strategy and thus configurable and extendable like the quantity classes. The several Runge-Kutta and adaptive integrators of the framework are based on an

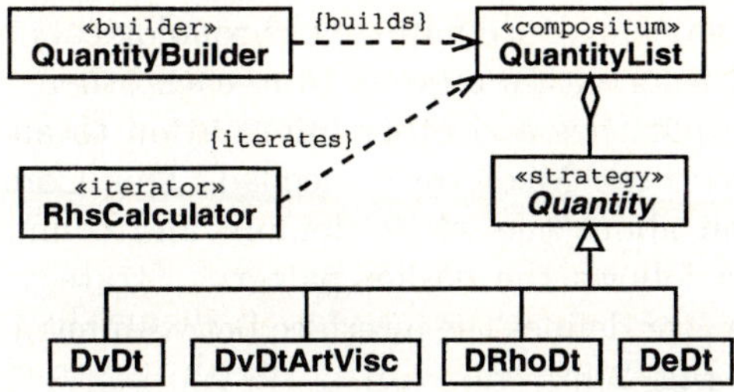

Fig. 1. Simplified class diagram of the sph2000 calculation classes.

abstract Integrator class. It defines interfaces to iterate through all particles' differential variables to integrate the right hand side and to store intermediate integration steps. The Strategy pattern is very easy to apply, since the user only has to write a configuration file, e.g. which Integrator should be utilized, and the Builders only create the needed objects of each Strategy.

3.5   Parallelization and Domain Decomposition

Geometrical Point of View. For an efficient parallelization we implemented a domain decomposition, dividing up the simulation area into several rectangular subdomains. These are equally spaced at the beginning of the simulation but dynamically change their size during runtime to keep the load balanced between all processors. The size is thereby given by the number of interaction partners, since this linear effects the calculation time.

A class SubSimulation was implemented, which defines the base methods of a subdomain like administrating the geometry, the adjacent domains and the particles within the subdomain. It is based on the Strategy pattern to be able to differentiate between subdomains with different tasks. The framework knows two specialized types of SubSimulations. The BoundarySimulation extends the Sub-Simulation by methods for reflecting and absorbing particles at the boundaries. The InletSimulation, based on the Decorator pattern, can decorate the latter with additional methods for inserting new particles to the simulation through an inlet. The ParameterMap includes the initial and behavioral values of these three types.

Domain Decomposition by Grouping Objects. In every time step each subdomain has to process the same tasks, like preparing the calculation, communicating particles to other subdomains, computing lists which contain the interaction partners, calculating the right hand side, or integrating the equations of motion. Each subdomain therefore contains classes and objects respectively for these tasks. From an object-oriented view each subdomain is a group of objects, which represent this area geometrically. The communication and application flow within a group is implemented using an *intra node* Mediator, which knows all contained objects, coordinates the chronological processing of the tasks and uncouples all objects within a group from each other.

To communicate particles and other information to and from the adjacent subdomains, an *inter node* Mediator is needed. This Communicator encapsulates the information about the whole domain and communication structure. The implementation follows the design patterns Strategy and Mediator. The class BaseCommunicator defines the interface between intra node and inter node communication (see Fig. 2).

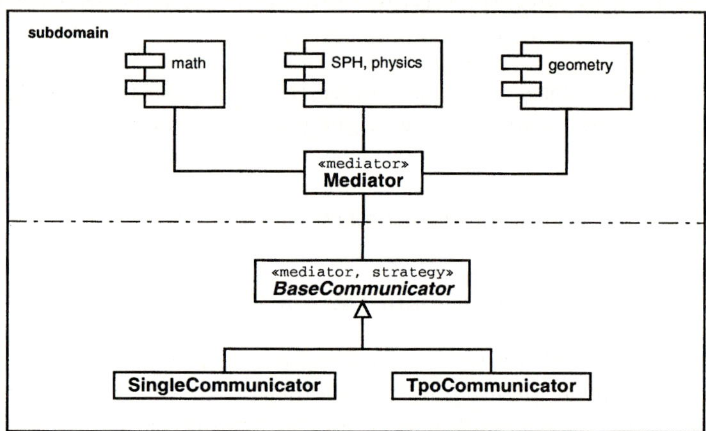

Fig. 2. Simplified diagram of a subdomain in sph2000. The upper part shows the modules with the SPH classes and the intra node mediator. The lower part shows the class diagram for the communication strategy. The Communicator classes encapsulate the whole inter node communication. The calculation objects are completely decoupled from the parallelization.

The major advantage of this concept is, that it enables the user to easily divide up the simulation domain into as many subdomains as processors are available. For communication between the processors the message passing object TpoCommunicator is generated, which uses TPO++. In case of a single processor simulation this communicator only has to be replaced by a single-node communicator. The user only has to exchange the communicators in the configuration file, leaving the code unchanged.

To coordinate all subdomains a special master subdomain is implemented, which is extended by several administration objects. These are objects for time-keeping and administrating all particles as well as particle-I/O objects for saving and restoring particle allocations.

3.6   Parallel I/O

The first results of sph2000 showed a significant lack in performance due to sequential I/O. Since current standards like MPI 2 [8] only support procedural interfaces for parallel I/O, we extended TPO++ by an object-oriented interface for parallel I/O [10].

The initial version of sph2000 implements an I/O strategy with one master process gathering the part results from all other processes and saving the whole data in ASCII format to disk. The new strategy using the parallel I/O interface was to implement collective I/O. Thereby, all processes can access the same file in parallel, which improves the performance significantly, since the communication to the master process is needless and the whole parallel I/O bandwidth can be used for transferring the data to and from disk. In addition, the library internally calculates the correct offsets within the file where each process has to place its part, avoiding any extra implementation by the user.

To provide sph2000 with parallel I/O, the particle-I/O class of the framework had to be adapted. The following listing shows the adapted method `saveDataFile` and represents the simple usage of the interface:

```
#include<tpo++.H>

void ParticleIO::saveDataFile(const ParticleContainer& particles, string name)
{
    TPO::File fh;
    int code = fh.open(TPO::CommWorld, name, TPO_MODE_CREATE);
    fh.write_all(particles.begin(), particles.end());
    fh.close();
}
```

This implementation of using a single collective call (`fh.write_all`) reduces the size of the original code by about 100 lines of code. The call is needed to save the containers of particle objects of each processor to disk simultaneously. Two iterators `begin()` and `end()` thereby define the beginning and ending of the container. This syntax also enables the user to store only a sub-set of particle objects to disk.

The performance improvement through the usage of parallel I/O within the framework depends on the application as well as on the ratio between the proportion of computation and I/O within the application. To determine the real performance gain we inserted it in our sample application, which is presented in the next section.

4   Sample Application

A first sample application was implemented to test the whole framework. It simulates the injection of diesel into an air filled chamber. Diesel engine manufactures are interested in an optimal injection of the diesel into the combustion chamber. A perfect mixing of diesel and air means an efficient use of the fuel and therefore reduces emission. For this reason the breakup of the diesel jet must be examined and understood. When injected into the cylinder of an engine, the diesel jet undergoes two stages of breakup. In the primary breakup large drops and filaments split off the compact jet. These turn into a spray of droplets during the secondary breakup. This secondary process is well known and can be modeled as a spray, but the understanding of the primary breakup is only in the initial stages. The physical effects that might influence the primary breakup are the pressure forces in the interface region of diesel and air, instabilities of the jet induced by cavitation inside the injection nozzle, surface tension and turbulence. In this area SPH simulations are not very common. There are several problems concerning the physics of multiphase flows and the requirements in terms of compute power are very high. Due to its extensibility sph2000 is very applicable for this area and enables the user to easily and fast implement new physical concepts. So far the framework provides 5 kernel functions, 6 integrators, and 20 quantities to calculate the state equations and the equations of motion for the air and diesel particles.

Fig. 3. 3D simulation of diesel injection with 2.5 million particles. The picture shows the injected diesel. First drops are already split off the jet. The injection causes a density wave traveling in front of the liquid stream.

2D and 3D simulations with up to 2.5 million particles reveal a broadening and breakup of the diesel jet leading to turbulences behind the dispartment. After a while single drops are separated from the compact jet, see Fig. 3.

The performance of the sample application was conducted on Kepler [13], a self-made clustered supercomputer based on commodity hardware. It consists of two parts: An older part with 96 dual Pentium III (650 MHz) nodes with 1 GB of memory, and a newer part with 32 dual AMD Athlon (1.667 GHz) nodes, each sharing 2 GB of memory. We measured two different simulation setups: The first running on the Pentium nodes with disabled I/O and the second on the Athlons with I/O enabled in every second time step, to compare the performance of sequential and parallel I/O. We always used only one processor per node.

Table 2 shows the performance results of the first setup. Due to memory shortage of the Pentium nodes, the measurements were made starting with 2 nodes. The parallelization scales very good up to 64 processors leading to a remarkable speedup of 44.10.

Table 2. Results of the first simulation setup with 1 million particles on Pentium III.

Processors	1	2	4	8	16	24	32	64
Time per step (in s)	-	116.9	58.8	33.8	16.4	11.3	9.7	5.3
Speedup	-	2.00	3.98	6.90	14.24	20.68	24.10	44.10

The results of the second simulation setup show the significant effect of parallel I/O on the performance (see Table 3). The I/O part could be improved by a factor of 20 when using 32 processors working on a parallel file system with 32 distributed disks. Since I/O is only a small proportion of the whole simulation, the overall gain using parallel I/O reduces to a - still remarkable - factor of 3.5 (64 processors). Note that even the sequential simulation with parallel I/O is faster than without parallel I/O, due to changing from ASCII file format to binary format and less code overhead for saving the particles.

Table 3. Results of the second simulation setup with 1 million particles on Athlon.

Processors	1	2	4	8	16	24	32	64
Execution time (in s)								
- sequential I/O	2090.2	1050.9	530.4	407.0	349.4	305.1	272.4	251.6
- parallel I/O	1223.5	617.2	312.5	191.4	130.9	96.5	79.8	71.3
Speedup								
- sequential I/O	1	1.98	3.94	5.13	5.98	6.85	7.67	8.30
- parallel I/O	1	1.98	3.91	6.39	9.34	12.67	15.33	17.15

5   Conclusion and Future Work

The application of object-oriented development methods has improved the quality of our simulation codes. The implementation is very easy to maintain and extend, e.g. to add the physics of surface tension or turbulence. The result of object-oriented techniques with design patterns is a framework, in which classes have clear and strictly separated responsibilities. Different methods can be interchanged without influencing other code. The use of our parallel I/O library and optimizations for communication reduced the sequential parts of the framework to a minimum. These lead to a well scaling parallel performance.

In the future, we focus on the development of models for simulating surface tension and turbulence. Due to an increased number of particles which is needed to simulate these effects meaningful, we are investigating solutions to decrease the amount of calculated interactions without increasing the runtime of the simulations. Since optimizing the communication is not sufficient, we furthermore try to exclude less important calculations: First, we try to separate the density wave of the injected diesel from the outer air, and second, the air particles shall be generated during runtime according to the motion of the jet. The idea is to calculate only air regions, which are affected by the diesel jet. Both leads to notedly less calculation overhead.

References

1. G. Allen, T. Goodale, E. Seidel. The Cactus Computational Collaboratory: Enabling technologies for relativistic astrophysics, and a toolkit for solving pdes by communities in science and engineering. In *7th Symposium on the Frontiers of Massively Parallel Computation-Frontiers 99*, New York, 1999. IEEE

2. W. Benz, E. Asphaug. Catastrophic Disruptions Revisited. In *Icarus*, 142: 5–20, 1999

3. Erich Gamma and Richard Helm and Ralph Johnson and John Vlissides. Design Patterns: Elements of Reusable Object-Oriented Software. Addison-Wesley, 1995

4. R. A. Gingold, J. J. Monaghan. Smoothed Particle Hydrodynamics: Theory and Application to Non-Spherical Stars. In *Monthly Notices of the Royal Astronomical Society*, 181: 375–389, 1977

5. T. Grundmann, M. Ritt, and W. Rosenstiel. TPO++: An object-oriented message-passing library in C++. In *Proceedings of the 2000 International Conference on Parallel Processing*, pages 43–50. IEEE Computer society, 2000.

6. S. Kunze, E. Schnetter, R. Speith. Development and Astrophysical Applications of a Parallel Smoothed Particle Hydrodynamics Code with MPI. In *High Performance Computing in Science and Engineering '99*, E. Krause, W. Jäger (ed.), Springer, p. 52 – 61, 2000.

7. Leon B. Lucy. A Numerical Approach to Testing the Fission Hypothesis. In *The Astronomical Journal*, 82(12): 1013–1024, 1977

8. Message Passing Interface Forum. MPI-2: Extensions to the Message-Passing Interface. *Online. URL:* *http://www.mpi-forum.org/docs/mpi-20-html/mpi2-report.html*, July 1997.

9. F. Ott, E. Schnetter. A modified SPH approach for fluids with large density differences. In *ArXiv Physics e-prints*, 3112-+, 2003

10. S. Pinkenburg and W. Rosenstiel. Parallel I/O in an Object-Oriented Message-Passing Library. In *Proceedings of the 11th European PVM/MPI Users' Group Meeting*, 2004.

11. J. V. W. Reynders, J. C. Cummings, P. J. Hinker, M. Tholburn, M. S. S. Banerjee, S. Karmesin, S. Atlas, K. Keahey, W. F. Humphrey. Pooma: A framework for scientific computing applications on parallel architectures. In *Parallel Programming using C++*, MIT Press, 1996; http://acts.nersc.gov/pooma/

12. V. Springel, N. Yoshida, S. D. M. White. GADGET: A Code for Collisionless and Gasdynamical Cosmological Simulations. In *New Astronomy*, 6(3): 51, 2001.

13. University of Tübingen. Kepler Cluster website. *Online, URL: http://kepler.sfb382-zdv.uni-tuebingen.de/kepler/start.shtml*, 2001.

Grid-BGC: A Grid-Enabled Terrestrial Carbon Cycle Modeling System

Jason Cope[1], Craig Hartsough[2], Peter Thornton[2], Henry Tufo[1,2],
Nathan Wilhelmi[2], and Matthew Woitaszek[1]

[1] University of Colorado, Boulder, CO 80309, USA
`{jason.cope,matthew.woitaszek}@colorado.edu`
[2] National Center for Atmospheric Research, Boulder, CO 80305, USA
`{craigh,thornton,tufo,wilhelmi}@ucar.edu`

Abstract. Grid-BGC is a Grid-enabled terrestrial biogeochemical cycle simulator collaboratively developed by the National Center for Atmospheric Research (NCAR) and the University of Colorado (CU) with funding from NASA. The primary objective of the project is to utilize Globus Grid technology to integrate inexpensive commodity cluster computational resources at CU with the mass storage system at NCAR while hiding the logistics of data transfer and job submission from the scientists. We describe a typical process for simulating the terrestrial carbon cycle, present our solution architecture and software design, and describe our implementation experiences with Grid technology on our systems. By design the Grid-BGC software framework is extensible in that it can utilize other grid-accessible computational resources and can be readily applied to other climate simulation problems which have similar workflows. Overall, this project demonstrates an end-to-end system which leverages Grid technologies to harness distributed resources across organizational boundaries to achieve a cost-effective solution to a compute-intensive problem.

1 Introduction

Setting up and running high-resolution simulations of terrestrial biogeochemical (BGC) processes is currently an involved process for scientists. Performing a complete simulation consists of gathering required environmental data from various storage systems onto one platform, running preprocessing software to prepare meteorological data for the target model, executing the simulation itself, and then moving the data to other systems for post-processing, visualization, and analysis. The process must then be repeated for multiple simulation tiles constituting the desired geographical region requiring mundane repetition and attention to detail. As such, the overhead to running terrestrial biogeochemical simulations is quite high, and scientists must perform many manual tasks and possess adequate data storage, computational resources, and substantial platform-specific computer expertise.

The objective of the Grid-BGC project is to create a cost effective end-to-end solution for terrestrial ecosystem modeling. Grid-BGC allows scientists to easily configure and run high-resolution terrestrial carbon cycle simulations without having to worry about the individual components of the simulation or the underlying computational and data storage systems. In order to run a simulation, the user interacts with a web-based portal to control the various stages of processing. The portal then functions

J.C. Cunha and P.D. Medeiros (Eds.): Euro-Par 2005, LNCS 3648, pp. 1285–1294, 2005.

as a grid client, submitting the simulation to a Grid-BGC tile processing grid service that gathers the required data from the storage systems and performs the simulation.

The development of Grid-BGC is a collaborative effort between the National Center for Atmospheric Research (NCAR) and the University of Colorado (CU). The Grid-BGC project uses computational and data grid technology [3] to leverage the resources available at both organizations in order to provide a cost effective and high performance solution. In particular, Grid-BGC is designed to run on commodity cluster systems such as those available at the university instead of production supercomputer systems at NCAR. Large model runs, however, produce multi-terabyte output in excess of the capacity available on the university clusters, so the system utilizes the NCAR Mass Storage System (MSS) for its storage requirements. Our software solution is also designed to provide reliable model execution tolerant of the transients present in distributed grid systems, support NCAR's operational security requirements, and be extensible enough to support running other similar scientific models.

As we engineer the Grid-BGC software, our overall goal is to develop an extensible set of grid-enabled tools that solve this problem and will be useful for subsequent similar Grid-based projects. The software infrastructure developed for Grid-BGC enables application-oriented data accessibility. Instead of requiring users to manually locate data by searching through a reference interface, entire applications can be configured to locate and download required data. In the past, these simulations would have to be performed at NCAR in order to gain access to the mass storage system. This is no longer the case, as data grid technologies allow the data to be accessed from anywhere.

The remainder of this paper is organized as follows: Section 2 describes relevant related projects in the grid community. Section 3 presents the workflow required for this terrestrial ecosystem model and forms the basis for our system requirements. Section 3 presents our solution architecture and design, and section 4 relates the current state of our prototype implementation and test grid. Section 5 describes our experiences with cluster-based grid computing. The final sections present future work and conclusions.

2 Related Work

Many other organizations are developing projects similar to the Grid-BGC execution platform. These projects, which range from holistic graphical workflow manipulation tools to client-server distributed processing systems, differ in approach and magnitude. All of these tools are service-based and allow computational platforms to expose computational resources as a commodity for the use of a community. While we are presenting our solution in respect to our targeted terrestrial climate model, our software environment is completely general and usable by applications with similar characteristics.

One example of running legacy applications in a Grid environment is the Grid Execution Management for Legacy Code Architecture (GEMCLA) project [4]. The goal of GEMCLA is to provide a framework designed to make any legacy code executable as an Globus Toolkit (GT) 3.0 [3] compliant Grid service without manually turning each application into a Grid service, access to the legacy source code, or requiring custom Java executable wrapping. GEMCLA functions as a Grid service with a front-end that interacts with the client to pass parameters and a back-end to run jobs using the Globus master managed job factory service. The interface to the legacy code is described in an XML file. GEMCLA also provides a robust graphical workflow editor, uses traditional Globus Toolkit components for job execution, and also provides workflow management and portal services.

While GEMCLA focuses on running single applications in a grid environment, other projects provide managed computing services. For example, the Distributed Infrastructure with Remote Agent Control (DIRAC) project developed by CERN coordinates computational resources for large physics simulations [9]. DIRAC is a high-throughput service oriented computational grid middleware application. In the DIRAC architecture, a user submits a series of computational jobs to a central server, much like a queue on a traditional cluster. Software running on each compute site determines its free computing resources, and then polls the central server to retrieve jobs for processing. The server runs a "Matchmaker" service to select the best jobs for the available resources. The authors assert that this pull methodology is less complex and more scalable than a traditional server-based scheduling system that must maintain the state of every compute node at all times. DIRAC implements its own data management system, including replica catalogs and a reliable data transfer service. Job execution is flexible, as each job simply installs software required for to its execution.

Similarly, NorduGrid was developed to handle large physics simulations [2]. The authors considered the use of previous grid computing tools, such as Globus and software developed by the European Data Grid, but found that they were as a whole inadequate and other components were required. NorduGrid augments the Globus Toolkit with a user interface, grid manager, replica catalog, and information dissemination service. The user interface, installed on client machines, provides the ability for users to submit job requests and obtain system information. The information service provides information on storage and computing resources on a grid using the Globus Monitoring and Discovery System (MDS). Finally, the grid manager provides an interface layer between the grid and the system software, such as a batch scheduling system. NorduGrid utilizes the Globus Replica Catalog to locate data sources and GridFTP to transfer files, but is intended for operation on cluster computer systems with locally shared file systems. Job requests are flexible and are submitted to the Grid Manager using Resource Specification Language.

Our approach to service-oriented computing is different. Instead of introducing complexity to support future arbitrary software execution, we impose *a priori* administration overhead to ensure that specific applications may be executed on resources that advertise their availability. Other solutions utilize a job description language, such as the Globus Resource Specification Language (RSL), to describe jobs in their most basic terms, such as requested architecture, requested number of nodes and processors, requested data capacity. Then, when the job is actually scheduled on a computer, the user's application and data must be transferred to that platform and executed. Many things can go wrong, ranging from compilation problems, long-distance storage system access problems, and a host of issues related to client environment configuration management.

We approach service-oriented computing from a contract perspective. In our architecture, a computational resource broadcasts that it can provide, for example, the Grid-BGC tile processing service. This broadcast availability demonstrates a commitment to provide the service with minimal details. The executable has been installed and tested, required security relationships have been established, and paths to remote storage systems have been tested. Instead of a job description language like RSL, the client submits a processing job using a generic specification format suitable for many types of executables but with additional stanzas specific to the advertised service. Our software approach provides a fault tolerant computational offloading grid service for specifically configured applications.

3 Terrestrial Ecosystem Modeling

Our software system uses two NCAR software applications, Daymet and Biome-BGC, to simulate the terrestrial ecosystem in a three step workflow (see Fig. 1). The Biome-BGC model is point-based; that is, it simulates the ecosystem at a single point on a spatial grid representing an area of planetary surface. The model itself acts on only one point at a time, but multiple points within a region are aggregated into tiles that become the unit of work for the Biome-BGC simulation. For a simulation, the area of land under analysis is broken up into manageable tiles and each tile is simulated independently.

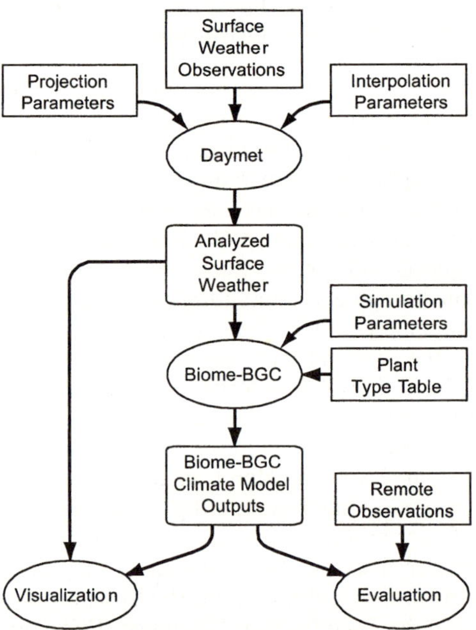

Fig. 1. Carbon cycle modeling workflow

The first stage in the workflow is preprocessing to convert raw single-site meteorological data into the spatially gridded format required by the simulator. The data ingest program Daymet [7] interpolates ground-based weather observations to produce high-resolution grids of historical surface weather data. These tile weather fields are then stored for possible later re-use. The Daymet output is then piped into the Biome-BGC model [8] in conjunction with soil and plant data. The model simulates the terrestrial carbon, water, and nitrogen cycles. The soil and plant data specification essentially defines forests and deserts, and the Daymet output describes where it rains, so the model grows trees in the forests and saguaro cacti in the deserts. The output is post-processed to display map overlays of variables of interest to climatologists such as gross primary production of carbon by photosynthesis.

The point- and tile-based nature of the Daymet and Biome-BGC models require that a scientist pay careful attention to parameter setup and spatial tile decomposition. Before Grid-BGC, the scientist running the simulation was required to manually organize the preprocessing and model execution for every tile independently. The Grid-BGC system architecture is designed to automate the process and eliminate this overhead. User interaction is constrained to a web-based portal interface, and the system automatically generates spatial tiles necessary to run the simulation over a desired area.

4 Architecture

Our software architecture is composed of several components that work together in a coordinated manner using Grid technology. GT 3.2 is used to provide grid enabled web services, authentication protocols, and data transfer facilities (see Fig. 2). These features are utilized in the four primary components of Grid-BGC: the user interface portal, the grid service, the JobManager daemon, and the DataMover file transfer utility.

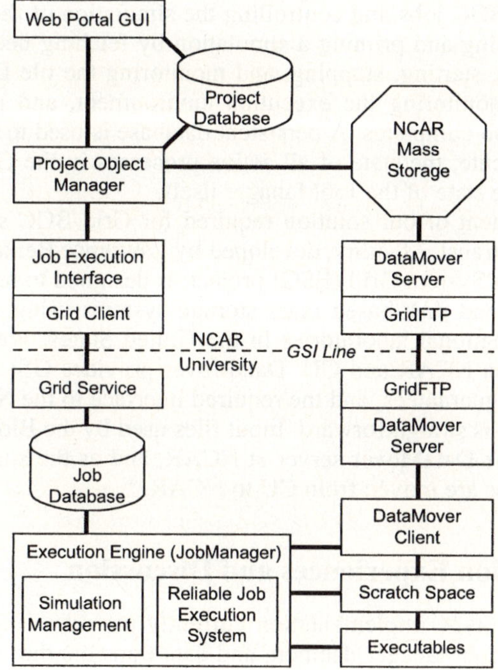

Fig. 2. System architecture

The user interface portal provides the front-end for the Grid-BGC system. The interface exposes mechanisms to define simulation parameters and support collaboration between users. A grid-enabled client is integrated into the portal, which interfaces with the remotely executing grid service. The client can interpret data received from the grid service and present it to the user and can communicate user requests to start simulations and query simulation status to the grid service. The portal essentially provides a thin web-based client that can accommodate a distributed user base with heterogeneous systems easily and efficiently.

A single grid service, the Grid-BGC Grid Service, is exposed by our execution framework. The primary responsibility of the service is to act as a link between the user interface portal and the execution environment. The service allows clients to invoke methods that start, stop, and analyze the state of Daymet and Biome-BGC jobs executing in the Grid-BGC environment. Communication between the service and the user interface portal is accomplished through the transmission of an XML based specification language. This language is not specific to Grid-BGC. Instead, the language is intended to support running any executable requiring input files, initialization files, and command line arguments. Upon receiving a message from the portal the service method parses the message, stores the parsed data into a persistent database, and then executes the appropriate action specified by the message. Once the action has been completed, the method composes a response message and sends it to the client in the portal. Typical actions include starting, stopping, and querying the state of a simulation.

The JobManager daemon is responsible for managing the high-level execution tasks of the Biome-BGC jobs and controlling the simulation of their constituent tiles. This includes preparing and priming a simulation by fetching needed input data sets from remote storage, starting, stopping, and monitoring the tile Daymet and Biome-BGC simulations, monitoring the execution environment, and performing cleanup operations when a job completes. A persistent database is used to store all the actions the server must execute, the state of all active processes in the Grid-BGC execution environment, and the state of the JobManager itself.

The final component of our solution required for Grid-BGC simulations is Data-Mover [6]. This file transfer facility, developed by Lawrence Berkeley National Laboratories for the Earth System Grid (ESG) project, is designed to replicate large sets of files between tape- and disk-based mass storage systems using GridFTP [1] and is currently in use at national laboratories in the United States. We use DataMover to transfer files between NCAR and CU. DataMover provides GSI authentication [10], reliable file transfer guarantees, and the required interface to the NCAR Mass Storage System. File staging is straightforward. Input files used by the Biome-BGC model are downloaded from the DataMover server at NCAR, and as the simulations finish, the generated output files are moved from CU to NCAR.

5 Implementation Experiences and Discussion

The Grid-BGC prototype implementation currently provides a fully functional grid service, grid client, JobManager daemon, and data transfers through the use of Data-Mover. At the present time, the prototype only executes Grid-BGC jobs, but we are working to increase the robustness of the software for application to similar projects. To date, this project has provided us with several valuable experiences while attempting to produce a working system that appeals to scientists, administrators, and software developers. In particular, we have ensured that the Grid-BGC framework fits in a managed security environment, provides reliability features to support execution on commodity cluster equipment, and is extensible so that its components are useful for future projects.

5.1 NCAR Security and Auditing Requirements

As a large government computing facility, NCAR provides computing resources to both internal projects and community users. Maintaining data security and auditing for charging purposes is required of all systems implemented at NCAR. The Grid-BGC solution is designed to meet these NCAR security and auditing requirements while functioning in a Globus Grid-based environment.

In a traditional grid environment, users authenticate with servers only using public key certificates. Because NCAR limits access to the mass storage system to users possessing a NCAR-issued "gatekeeper" account, our users must establish an account with NCAR and then use this username and password to authenticate with our portal. We internally generate a Grid-BGC certificate for all of our users. When a user logs in to the portal, their Grid-BGC certificate is used to instantiate a proxy that is uploaded into a MyProxy [5] server for later retrieval and used to contact tile processing grid services.

The authentication scheme on the cluster providing the grid service is intentionally simple to reduce administration overhead. All Grid-BGC user certificates are mapped to one UNIX user account. When a user submits a simulation request, the portal authenticates with the grid service using the user's certificate. The grid service merely stores the simulation request in a database, and the JobManager daemon then runs the simulations under the auspices of the service account. At no time does the user actually have possession of their internal Grid-BGC certificate, so they may not connect to a compute cluster directly but must use the portal interface.

In addition to the portal contacting the cluster running the grid service, it is also necessary for the JobManager daemon on the cluster to contact the NCAR mass storage system to download and upload data. Because NCAR requires complete user-based accountability, the daemon running under a service account must impersonate the user who submitted the request. To do this, we have a job request contain information about the portal's MyProxy server and the user's current stored proxy certificate. When the daemon must authenticate with the mass storage system, it first retrieves a copy of the user's proxy from the MyProxy server and uses these credentials for the data transfer.

5.2 Reliability and Fault Tolerance

Engineering fault tolerance into the Grid-BGC system is essential in our distributed grid environment. While the users of the CU cluster enjoy an uptime usually measured in months, during the course of Grid operations the end-to-end system is surprisingly prone to problems. We must cope with scheduled downtime – NCAR actually shuts down their entire facility once or twice a year for physical plant maintenance – as well as anticipated transients and genuine errors. Our software distinguishes between transients and errors so that users are not bothered with cryptic messages when a solution must be postponed due to the temporary unavailability of a required resource.

To facilitate fault tolerance, all cluster-side components including the grid service, the JobManager daemon, the data transfer system, and the models are arbitrarily restartable. The grid service is stateless and only performs atomic database transactions, so it may be restarted at any time. The remainder of the fault tolerance is built into the JobManager daemon.

Despite our best efforts, the CU cluster is still occasionally subject to node reboots with little or no warning, power failures, and students who ignore system administrator threats and circumvent the job scheduler to run code that is capable of causing kernel panics. The JobManager daemon monitors and controls the data transfer processes and simulation batch queue jobs. If, for any reason, a data transfer of a simulation job fails without completing successfully, the JobManager can restart it. If a job fails repeatedly, the system is presumed to be operating in a failure mode, and jobs are held for administrator intervention. Finally, the daemon itself maintains persistent state information in a database and all management system iterations are atomic. In the event of a system problem, the daemon may be stopped immediately. When it is restarted, the queue history is analyzed to determine if running jobs completed successfully, jobs are finalized or restarted as appropriate, and everything resumes normally. The Grid-BGC grid service provides "submit and forget" tile processing capabilities.

5.3 Expandability to Other Projects

The Grid-BGC software is designed to reliably execute the data transfers and models under its control within a completely flexible framework. Thus, the model it is running may be changed at any time. In addition to running the terrestrial ecosystem model integral to Grid-BGC, the Grid-BGC JobManager can be configured to run unrelated software applications among cooperating grid sites. We are also examining the possibility of extracting the Reliable Job Execution Service, a software component developed as part of the Grid-BGC Job Manager, and making it available by itself as part of a Grid middleware initiative.

6 Future Work and Conclusions

Work is underway to turn our Grid-BGC prototype into a fully functional system capable of running end-to-end carbon cycle simulations for the BGC user community. We believe that the entire system, including the user interface portal, will be fully operational by June 2005. At that point the system will be sufficiently developed to introduce climatologists as beta users.

After demonstrating full integrated functionality with our development cluster, we intend to expand the system to involve other clusters available via our grid. The first step is to provide the Grid-BGC tile processing service on other clusters under our control at the university and NCAR. A grid metadata publication and discovery service will be used to maintain clusters that are available to run tile simulation jobs and new simulations will be dispatched to clusters with the shortest anticipated turnaround time. The second step is to allow other collaborative institutes to instantiate their own tile processing services. In this case, users with their own clusters will be allowed to specify that their simulation jobs be run on their dedicated hardware instead of our shared resources.

One substantial component of Grid-BGC is data storage and transfer. We presently use DataMover to transmit data from one site to another. While DataMover maintains its own caching capabilities, it may be useful to analyze the operation of the Grid-BGC system in a production mode to determine if certain files should be replicated instead of transferred. This analysis will not be possible until the system is being used to run real science-based simulations instead of our test job collection.

Grid-BGC demonstrates an end-to-end system prototype leveraging Grid technologies to distribute a scientific application seamlessly across organizational boundaries. Our use of the Globus toolkit allows us to access NCAR datasets while running the computationally-intensive software on remotely administered commodity clusters. Overall, Grid-BGC provides a cost-effective, end-to-end solution for terrestrial ecosystem modeling through a straightforward and simple interface. As we have engineered the Grid-BGC execution framework to be as extensible as possible, we hope to apply our software solution for use in other similar applications.

Acknowledgements

University of Colorado computer time was provided by equipment purchased under DOE SciDAC Grant #DE-FG02-04ER63870, NSF ARI Grant #CDA-9601817, NSF sponsorship of the National Center for Atmospheric Research, and a grant from the IBM Shared University Research (SUR) program. NASA has provided funding for the Grid-BGC project through the Advanced Information Systems Technology Office (NASA AIST Grant #NAG2-1646) and the Terrestrial Ecology Program.

References

1. Allcock, B., Bester J., Bresnahan, J., Chervenak, A. L., Foster, I., Kesselman, C., Meder, S., Nefedova, V., Quesnal, D., Tuecke, S. Data Management and Transfer in High Performance Computational Grid Environments. Parallel Computing Journal, Vol. 28 (5), May 2002.
2. Eerola, P., Kónya, B., Smirnova, O., Ekelöf, T., Ellert, M., Hansen, J. R., Nielsen, J. L., Wäänänen, A., Konstantinov, A., Ould-Saada, F. The NorduGrid Architecture and Tools. Proceedings of Computing in High-Energy and Nuclear Physics (CHEP 03), La Jolla, California, March 2003.
3. Globus. The Globus Project, 2004, http://www.globus.org/A
4. Kacsuk, P., Goyeneche, A., Delaitre, T., Kiss, T., Farkas, Z., and Boczko, T. High-level Grid Application Environment to Use Legacy Codes as OGSA Grid Services. Proceedings of the 5th IEEE/ACM International Workshop on Grid Computing (GRID 2004), Pittsburgh, USA, 8 November 2004.
5. Novotny, J., Tuecke, S., Welch, V. An Online Credential Repository for the Grid: MyProxy. Proceedings of the Tenth International Symposium on High Performance Distributed Computing (HPDC-10), IEEE Press, August 2001.
6. Sim, A. J. Gu, A. Shoshani, V. Natarajan. DataMover: Robust Terabyte-Scale Multi-File Replication over Wide-Area Networks. Proceedings of the 16th International Conference on Scientific and Statistical Database Management, 403, 21 June 2004.
7. Thornton, P.E., S.W. Running, and M.A. White. Generating surfaces of daily meteorological variables over large regions of complex terrain. Journal of Hydrology, 190: 214-251, 1997.

8. Thornton, P.E., S.W. Running. An improved algorithm for estimating incident daily solar radiation from measurements of temperature, humidity, and precipitation. Agricultural and Forest Meteorology, 93: 211-228, 1999.
9. Tsaregorodtsev, A., Garonne, V., and Stokes-Rees, I. DIRAC: A Scalable Lightweight Architecture for High Throughput Computing. Proceedings of the 5th IEEE/ACM International Workshop on Grid Computing (GRID 2004), Pittsburgh, USA, 8 November 2004.
10. Welch, V., Siebenlist, F., Foster, I., Bresnahan, J., Czajkowski, K., Gawor, J., Kesselman, C., Meder, S., Pearlman, L., Tuecke, S. Security for Grid Services. The Twelfth IEEE International Symposium on High-Performance Distributed Computing, June, 2003.

Lecture Notes in Computer Science

For information about Vols. 1–3552

please contact your bookseller or Springer

Vol. 3671: S. Bressan, S. Ceri, E. Hunt, Z.G. Ives, Z. Bellahsène, M. Rys, R. Unland (Eds.), Database and XML Technologies. X, 239 pages. 2005.

Vol. 3662: C. Baral, G. Greco, N. Leone, G. Terracina (Eds.), Logic Programming and Nonmonotonic Reasoning. XIII, 454 pages. 2005. (Subseries LNAI).

Vol. 3659: J.R. Rao, B. Sunar (Eds.), Cryptographic Hardware and Embedded Systems – CHES 2005. XIV, 458 pages. 2005.

Vol. 3654: S. Jajodia, D. Wijesekera (Eds.), Data and Applications Security XIX. X, 353 pages. 2005.

Vol. 3653: M. Abadi, L.d. Alfaro (Eds.), CONCUR 2005 – Concurrency Theory. XIV, 578 pages. 2005.

Vol. 3649: W.M.P. van der Aalst, B. Benatallah, F. Casati, F. Curbera (Eds.), Business Process Management. XII, 472 pages. 2005.

Vol. 3648: J.C. Cunha, P. D. Medeiros (Eds.), Euro-Par 2005 Parallel Processing. XXXVI, 1299 pages. 2005.

Vol. 3645: D.-S. Huang, X.-P. Zhang, G.-B. Huang (Eds.), Advances in Intelligent Computing, Part II. XIII, 1010 pages. 2005.

Vol. 3644: D.-S. Huang, X.-P. Zhang, G.-B. Huang (Eds.), Advances in Intelligent Computing, Part I. XXVII, 1101 pages. 2005.

Vol. 3639: P. Godefroid (Ed.), Model Checking Software. XI, 289 pages. 2005.

Vol. 3638: A. Butz, B. Fisher, A. Krüger, P. Olivier (Eds.), Smart Graphics. XI, 269 pages. 2005.

Vol. 3636: M.J. Blesa, C. Blum, A. Roli, M. Sampels (Eds.), Hybrid Metaheuristics. XII, 155 pages. 2005.

Vol. 3634: L. Ong (Ed.), Computer Science Logic. XI, 567 pages. 2005.

Vol. 3633: C. Bauzer Medeiros, M. Egenhofer, E. Bertino (Eds.), Advances in Spatial and Temporal Databases. XIII, 433 pages. 2005.

Vol. 3632: R. Nieuwenhuis (Ed.), Automated Deduction – CADE-20. XIII, 459 pages. 2005. (Subseries LNAI).

Vol. 3629: J.L. Fiadeiro, N. Harman, M. Roggenbach, J. Rutten (Eds.), Algebra and Coalgebra in Computer Science. XI, 457 pages. 2005.

Vol. 3627: C. Jacob, M.L. Pilat, P.J. Bentley, J. Timmis (Eds.), Artificial Immune Systems. XII, 500 pages. 2005.

Vol. 3626: B. Ganter, G. Stumme, R. Wille (Eds.), Formal Concept Analysis. X, 349 pages. 2005. (Subseries LNAI).

Vol. 3625: S. Kramer, B. Pfahringer (Eds.), Inductive Logic Programming. XIII, 427 pages. 2005. (Subseries LNAI).

Vol. 3624: C. Chekuri, K. Jansen, J.D.P. Rolim, L. Trevisan (Eds.), Approximation, Randomization and Combinatorial Optimization. XI, 495 pages. 2005.

Vol. 3623: M. Liśkiewicz, R. Reischuk (Eds.), Fundamentals of Computation Theory. XV, 576 pages. 2005.

Vol. 3621: V. Shoup (Ed.), Advances in Cryptology – CRYPTO 2005. XI, 568 pages. 2005.

Vol. 3620: H. Muñoz-Avila, F. Ricci (Eds.), Case-Based Reasoning Research and Development. XV, 654 pages. 2005. (Subseries LNAI).

Vol. 3619: X. Lu, W. Zhao (Eds.), Networking and Mobile Computing. XXIV, 1299 pages. 2005.

Vol. 3615: B. Ludäscher, L. Raschid (Eds.), Data Integration in the Life Sciences. XII, 344 pages. 2005. (Subseries LNBI).

Vol. 3614: L. Wang, Y. Jin (Eds.), Fuzzy Systems and Knowledge Discovery, Part II. XLI, 1314 pages. 2005. (Subseries LNAI).

Vol. 3613: L. Wang, Y. Jin (Eds.), Fuzzy Systems and Knowledge Discovery, Part I. XLI, 1334 pages. 2005. (Subseries LNAI).

Vol. 3612: L. Wang, K. Chen, Y. S. Ong (Eds.), Advances in Natural Computation, Part III. LXI, 1326 pages. 2005.

Vol. 3611: L. Wang, K. Chen, Y. S. Ong (Eds.), Advances in Natural Computation, Part II. LXI, 1292 pages. 2005.

Vol. 3610: L. Wang, K. Chen, Y. S. Ong (Eds.), Advances in Natural Computation, Part I. LXI, 1302 pages. 2005.

Vol. 3608: F. Dehne, A. López-Ortiz, J.-R. Sack (Eds.), Algorithms and Data Structures. XIV, 446 pages. 2005.

Vol. 3607: J.-D. Zucker, L. Saitta (Eds.), Abstraction, Reformulation and Approximation. XII, 376 pages. 2005. (Subseries LNAI).

Vol. 3606: V. Malyshkin (Ed.), Parallel Computing Technologies. XII, 470 pages. 2005.

Vol. 3604: R. Martin, H. Bez, M. Sabin (Eds.), Mathematics of Surfaces XI. IX, 473 pages. 2005.

Vol. 3603: J. Hurd, T. Melham (Eds.), Theorem Proving in Higher Order Logics. IX, 409 pages. 2005.

Vol. 3602: R. Eigenmann, Z. Li, S.P. Midkiff (Eds.), Languages and Compilers for High Performance Computing. IX, 486 pages. 2005.

Vol. 3599: U. Aßmann, M. Aksit, A. Rensink (Eds.), Model Driven Architecture. X, 235 pages. 2005.

Vol. 3598: H. Murakami, H. Nakashima, H. Tokuda, M. Yasumura, Ubiquitous Computing Systems. XIII, 275 pages. 2005.

Vol. 3597: S. Shimojo, S. Ichii, T.W. Ling, K.-H. Song (Eds.), Web and Communication Technologies and Internet-Related Social Issues - HSI 2005. XIX, 368 pages. 2005.

Vol. 3596: F. Dau, M.-L. Mugnier, G. Stumme (Eds.), Conceptual Structures: Common Semantics for Sharing Knowledge. XI, 467 pages. 2005. (Subseries LNAI).

Vol. 3595: L. Wang (Ed.), Computing and Combinatorics. XVI, 995 pages. 2005.

Vol. 3594: J.C. Setubal, S. Verjovski-Almeida (Eds.), Advances in Bioinformatics and Computational Biology. XIV, 258 pages. 2005. (Subseries LNBI).

Vol. 3593: V. Mařík, R. W. Brennan, M. Pěchouček (Eds.), Holonic and Multi-Agent Systems for Manufacturing. XI, 269 pages. 2005. (Subseries LNAI).

Vol. 3592: S. Katsikas, J. Lopez, G. Pernul (Eds.), Trust, Privacy and Security in Digital Business. XII, 332 pages. 2005.

Vol. 3591: M.A. Wimmer, R. Traunmüller, Å. Grönlund, K.V. Andersen (Eds.), Electronic Government. XIII, 317 pages. 2005.

Vol. 3590: K. Bauknecht, B. Pröll, H. Werthner (Eds.), E-Commerce and Web Technologies. XIV, 380 pages. 2005.

Vol. 3589: A M. Tjoa, J. Trujillo (Eds.), Data Warehousing and Knowledge Discovery. XVI, 538 pages. 2005.

Vol. 3588: K.V. Andersen, J. Debenham, R. Wagner (Eds.), Database and Expert Systems Applications. XX, 955 pages. 2005.

Vol. 3587: P. Perner, A. Imiya (Eds.), Machine Learning and Data Mining in Pattern Recognition. XVII, 695 pages. 2005. (Subseries LNAI).

Vol. 3586: A.P. Black (Ed.), ECOOP 2005 - Object-Oriented Programming. XVII, 631 pages. 2005.

Vol. 3584: X. Li, S. Wang, Z.Y. Dong (Eds.), Advanced Data Mining and Applications. XIX, 835 pages. 2005. (Subseries LNAI).

Vol. 3583: R.W. H. Lau, Q. Li, R. Cheung, W. Liu (Eds.), Advances in Web-Based Learning – ICWL 2005. XIV, 420 pages. 2005.

Vol. 3582: J. Fitzgerald, I.J. Hayes, A. Tarlecki (Eds.), FM 2005: Formal Methods. XIV, 558 pages. 2005.

Vol. 3581: S. Miksch, J. Hunter, E. Keravnou (Eds.), Artificial Intelligence in Medicine. XVII, 547 pages. 2005. (Subseries LNAI).

Vol. 3580: L. Caires, G.F. Italiano, L. Monteiro, C. Palamidessi, M. Yung (Eds.), Automata, Languages and Programming. XXV, 1477 pages. 2005.

Vol. 3579: D. Lowe, M. Gaedke (Eds.), Web Engineering. XXII, 633 pages. 2005.

Vol. 3578: M. Gallagher, J. Hogan, F. Maire (Eds.), Intelligent Data Engineering and Automated Learning - IDEAL 2005. XVI, 599 pages. 2005.

Vol. 3577: R. Falcone, S. Barber, J. Sabater-Mir, M.P. Singh (Eds.), Trusting Agents for Trusting Electronic Societies. VIII, 235 pages. 2005. (Subseries LNAI).

Vol. 3576: K. Etessami, S.K. Rajamani (Eds.), Computer Aided Verification. XV, 564 pages. 2005.

Vol. 3575: S. Wermter, G. Palm, M. Elshaw (Eds.), Biomimetic Neural Learning for Intelligent Robots. IX, 383 pages. 2005. (Subseries LNAI).

Vol. 3574: C. Boyd, J.M. González Nieto (Eds.), Information Security and Privacy. XIII, 586 pages. 2005.

Vol. 3573: S. Etalle (Ed.), Logic Based Program Synthesis and Transformation. VIII, 279 pages. 2005.

Vol. 3572: C. De Felice, A. Restivo (Eds.), Developments in Language Theory. XI, 409 pages. 2005.

Vol. 3571: L. Godo (Ed.), Symbolic and Quantitative Approaches to Reasoning with Uncertainty. XVI, 1028 pages. 2005. (Subseries LNAI).

Vol. 3570: A. S. Patrick, M. Yung (Eds.), Financial Cryptography and Data Security. XII, 376 pages. 2005.

Vol. 3569: F. Bacchus, T. Walsh (Eds.), Theory and Applications of Satisfiability Testing. XII, 492 pages. 2005.

Vol. 3568: W.-K. Leow, M.S. Lew, T.-S. Chua, W.-Y. Ma, L. Chaisorn, E.M. Bakker (Eds.), Image and Video Retrieval. XVII, 672 pages. 2005.

Vol. 3567: M. Jackson, D. Nelson, S. Stirk (Eds.), Database: Enterprise, Skills and Innovation. XII, 185 pages. 2005.

Vol. 3566: J.-P. Banâtre, P. Fradet, J.-L. Giavitto, O. Michel (Eds.), Unconventional Programming Paradigms. XI, 367 pages. 2005.

Vol. 3565: G.E. Christensen, M. Sonka (Eds.), Information Processing in Medical Imaging. XXI, 777 pages. 2005.

Vol. 3564: N. Eisinger, J. Małuszyński (Eds.), Reasoning Web. IX, 319 pages. 2005.

Vol. 3562: J. Mira, J.R. Álvarez (Eds.), Artificial Intelligence and Knowledge Engineering Applications: A Bioinspired Approach, Part II. XXIV, 636 pages. 2005.

Vol. 3561: J. Mira, J.R. Álvarez (Eds.), Mechanisms, Symbols, and Models Underlying Cognition, Part I. XXIV, 532 pages. 2005.

Vol. 3560: V.K. Prasanna, S. Iyengar, P.G. Spirakis, M. Welsh (Eds.), Distributed Computing in Sensor Systems. XV, 423 pages. 2005.

Vol. 3559: P. Auer, R. Meir (Eds.), Learning Theory. XI, 692 pages. 2005. (Subseries LNAI).

Vol. 3558: V. Torra, Y. Narukawa, S. Miyamoto (Eds.), Modeling Decisions for Artificial Intelligence. XII, 470 pages. 2005. (Subseries LNAI).

Vol. 3557: H. Gilbert, H. Handschuh (Eds.), Fast Software Encryption. XI, 443 pages. 2005.

Vol. 3556: H. Baumeister, M. Marchesi, M. Holcombe (Eds.), Extreme Programming and Agile Processes in Software Engineering. XIV, 332 pages. 2005.

Vol. 3555: T. Vardanega, A.J. Wellings (Eds.), Reliable Software Technology – Ada-Europe 2005. XV, 273 pages. 2005.

Vol. 3554: A. Dey, B. Kokinov, D. Leake, R. Turner (Eds.), Modeling and Using Context. XIV, 572 pages. 2005. (Subseries LNAI).

Vol. 3553: T.D. Hämäläinen, A.D. Pimentel, J. Takala, S. Vassiliadis (Eds.), Embedded Computer Systems: Architectures, Modeling, and Simulation. XV, 476 pages. 2005.